W9-CQI-856

THE PEOPLE'S CHRONOLOGY

A Year-by-Year Record of Human Events from Prehistory to the Present

REVISED AND UPDATED EDITION

JAMES TRAGER

A Henry Holt Reference Book

Henry Holt and Company New York

A Henry Holt Reference Book
Henry Holt and Company, Inc.
Publishers since 1866
115 West 18th Street
New York, New York 10011

Henry Holt® is a registered trademark
of Henry Holt and Company, Inc.

Published in Canada by Fitzhenry & Whiteside Ltd.,
195 Allstate Parkway, Markham, Ontario L3R 4T8.

Library of Congress Cataloging-in-Publication Data
Trager, James.
The people's chronology: a year-by-year record of human events from
prehistory to the present/James Trager.—Rev. and updated ed.
p. cm.—(A Henry Holt reference book)
Includes index.
1. Chronology, Historical. I. Series.
D11.T83 1992 91-36734
902'.02—dc 20 CIP

ISBN 0-8050-3731-4
ISBN 0-8050-3134-0 (An Owl Book: pbk.)

Henry Holt books are available for special promotions
and premiums. For details contact:
Director, Special Markets.

First published in hardcover in 1992 by
Henry Holt Reference Books.

First Revised/Owl Book Edition—1994

Printed in the United States of America
All first editions are printed on acid-free paper. ∞

1 3 5 7 9 10 8 6 4 2

1 3 5 7 9 10 8 6 4 2
pbk.

Text Credits

Excerpts from *Rudyard Kipling's Verse,* copyright 1940, Doubleday & Co. Reprinted by permission of Doubleday & Co., Inc. Excerpt from *Salt Water Ballads* by John Masefield, copyright 1913, The Macmillan Co. Reprinted by permission of Macmillan Publishing Co., Inc. Lines from "When I Was One-and-Twenty" from "A Shropshire Lad"—Authorized Edition—excerpted from *The Collected Poems of A. E. Housman.* Copyright 1939, 1940, © 1965 by Holt, Rinehart & Winston. Copyright © 1967, 1968 by Robert E. Symons. Reprinted by permission of Holt, Rinehart & Winston, Publishers. Excerpts from *Collected Poems* by Edna St. Vincent Millay, Harper & Row. Copyright 1922, 1923, 1950, 1951 by Edna St. Vincent Millay and Norma Millay Ellis. Reprinted by permission of Mrs. Ellis. Excerpt from *Poems* by Alan Seeger, copyright 1917. Reprinted by permission of Charles Scribner's Sons. Lines from *The Collected Poems of Rupert Brooke,* copyright 1941. Reprinted by permission of Dodd, Mead & Co. Excerpt from *Chicago Poems* by Carl Sandburg, reprinted by permission of Harcourt Brace Jovanovich, Inc. Excerpts from *The Poetry of Robert Frost* edited by Edward Connery Latham. Copyright 1923, 1930, 1939, © 1969 by Holt, Rinehart & Winston. Copyright 1951, © 1958 by Robert Frost. Copyright © 1967 by Lesley Frost Ballantine. Reprinted by permission of Holt, Rinehart & Winston, Publishers. Excerpts from *The Portable Dorothy Parker,* copyright © 1973. Reprinted by permission of Viking Penguin, Inc. Excerpt from *Animal Farm* by George Orwell, copyright 1946. Reprinted by permission of Harcourt Brace Jovanovich, Inc. Excerpt from *1984* by George Orwell, copyright 1949. Reprinted by permission of Harcourt Brace Jovanovich, Inc. Lines from "The Hollow Men" excerpted from *Collected Poems 1909–1962* by T. S. Eliot, and excerpts from *Murder in the Cathedral* by T. S. Eliot reprinted by permission of Harcourt Brace Jovanovich, Inc. Excerpt from *The Poems of Dylan Thomas.* Copyright 1952 by Dylan Thomas. Reprinted by permission of New Directions. Excerpt from *Catch-22,* copyright © 1955, 1961 by Joseph Heller. Reprinted by permission of Simon & Schuster, a Division of Paramount Communications, Inc.

Picture Credits

page 6: De Forest, Julia B., *A Short History of Art.* New York: Philips & Hung, 1882; **page 7:** Kitto, John, *The Pictorial Sunday Book.* London: London Printing and Publishing Co., 19th century; **page 18:** *Münchener Bilderbogen.* Munich: Brann und Schneider, 1852–98; **page 22:** Yonge, Charlotte Mary, *A Pictorial History of the World's Great Nations.* New York: Selmer Hesse, 1882; **page 26:** from the original painting by Gustave Doré; **page 33:** Yonge, *op. cit.;* **page 38:** Brewer, Ebenezer Cobham, *Character Sketches of Romance, Fiction, and the Drama.* New York: Selmer Hesse, 1892; **page 40:** *Münchener Bilderbogen, op. cit.;* **page 44:** ibid; **page 49:** *ibid;* **page 55:** Hollander, Eugen, *Die Medizin in der klassischen Malerei.* Stuttgart: Enke, 1903; **page 72:** Myers, P.V.N., *A General History for Colleges and High Schools.* Boston: Ginn & Co., 1894; **page 88:** Ridpath, John Clark, *Cyclopedia of Universal History.* Cincinnati: Jones Bros., 1885; **page 94:** *ibid;* **page 97:** Hollander, Eugen, *op. cit.;* **page 98:** Wright, Thomas, *The History of France From the Earliest Period to the Present Time.* London: London Printing and Publishing, 1856–82; **page 100:** Stacke, Ludwig, *Deutsche Geschichte.* Leipzig: Velhagen & Lasing, 1896; **page 106:** Gilman, Arthur, *Rome: From the Earliest Times to the End of the Republic.* New York: Putnam, 1885; **page 110:** Kupetsian, A., editor, *Fifty Years of Soviet Art: Graphic Arts.* Moscow: Soviet Artist (Sovietski Khudozhnik), 1967; **page 116:** Ellis, Edward S., *Story of the Great Nations: From the Dawn of History to the Present.* New York: Niglutsch, 1901–06; **page 123:** Doyle, James E., *A Chronicle of England.* London: Longmans, 1864; **page 135:** Corvin, *Illustrirte Geschicte des Mittelalters,* 1883; **page 141:** *Century Magazine;* **page 149:** Foxe, John, *Book of Martyrs.* London: Warne & Co., 1869; **page 178:** Aus Barth, *De las Casas.* Berlin: Wahrhaftiger Bericht, 1789; **page 182:** from a painting by Holbein; **page 183:** Copernicus; **page 193:** *Century Magazine;* **page 208:** *Harper's Monthly;* **page 219:** Colange, Leo, editor, *Voyages and Travels.* Boston: E. W. Walker, no date; **page 242:** *Illustrirte Geschichte für das Volk.* Leipzig: O. Spamer, 1880–83; **page 248:** Yonge, *op. cit.;* **page 256:** Crafts, William A., *Pioneers in the Settlement of America.* Boston: Samuel Walker, 1876; **page 261:** *Harper's Monthly;* **page 267:** Lossing, Benson J., *Our Country: A History of the United States.* New York: Lossing History Co., 1905; **page 289:** Greeley, Horace, *The Great Industries of the United States.* Hartford: Burr & Hyde, 1872; **page 303:** *Harper's Monthly,* December 1875; **page 333:** Foquier, Guillaume Louis, *Les merveilles de la science.* Paris: Furne, Jouvet, 1866–69; **page 346:** Scott, David B., *A School History of the United States.* New York: Harper Bros., 1875; **page 355:** Sandhurst, Phillip T. and James Stothert, *The Masterpieces of European Art.* Philadelphia: G. Barrie, 1876; **page 364:** Scott, David B., *op. cit.;* **page 375:** Cook, Clarence, *Art and Artists of Our Times.* New York: Hess, 1888 (from the Meissonier painting); **page 423:** Lodge, Henry Cabot, *The Story of the Revolution.* New York: Charles Scribner's Sons, 1898; **page 430:** Jennings, Joseph, *Theatrical and Circus Life.* St. Louis: Sun Publishing, 1882; **page 444:** Scott, David B., *op. cit.;* **page 479:** Lossing, Benson J., *op. cit.;* **page 495:** Johnson, Robert Underwood, and Clarence Clough Buel, editors, *The Century War Book.* New York: Century, 1894; **page 491:** Guernsey, Alfred H., *Harper's Pictorial History of the Civil War.* Chicago: Star Publishing, 1866; **page 513:** Thayer, William M., *Marvels of the New West.* Norwich, Conn.: Henry Bill, 1890; **page 521:** *Century Magazine,* January 1890; **page 528:** Witkowski, G.J., comp., *Anecdotes historiques* . . . Paris: A. Malvine, 1898; **page 536:** *Harper's Weekly,* October 21, 1871; **page 548:** Currier & Ives print, 1866; **page 549:** Thomas Nast cartoon; **page 551:** *Harper's Monthly;* **page 565:** *Harper's Monthly,* 1890; **page 570:** *Harper's Weekly;* **page 573:** *Life Magazine* cartoon; **page 591:** John Tenniel cartoon from *Punch;* **page 613:** *Century Magazine,* June 1894; **page 624:** Ellis, Edward S., *op. cit.;* **page 630:** Witkowski, G.J., comp., *Anecdotes historiques* . . . Paris: A. Malvine, 1898; **page 655:** Lewis Hine photograph courtesy International Museum of Photography at George Eastman

House; **page 670:** Edward Steichen photograph courtesy The Metropolitan Museum of Art, The Alfred Stieglitz Collection, 1949; **page 676:** courtesy Ford Motor Company; **page 684:** from an etching by José Clemente Orozco; **page 689:** Culver Pictures; **page 693:** Culver Pictures; **page 695:** courtesy *Boston Globe;* **page 699:** *Life Magazine* cartoon courtesy Henry Rockwell; **page 705:** British War Museum; **page 723:** Kupetsian, A., ed., *op. cit.;* **page 725:** British War Museum; **page 745:** Kupetsian, A., ed., *op. cit.;* **page 754:** cartoon courtesy Mrs. John Held, Jr.; **page 762:** Bettmann Archive; **page 769:** *ibid;* **page 780:** *ibid;* **page 781:** courtesy *San Francisco Chronicle;* **page 782:** Bettmann Archive; **page 792:** courtesy *Variety;* **page 796:** Bettmann Archive; **page 801:** Grant Wood painting "American Gothic" courtesy The Art Institute of Chicago; **page 812:** Bettmann Archive; **page 813:** Dorothea Lange photograph "White Angel Breadline, San Francisco" from the Dorothea Lange Collection, The City of Oakland, The Oakland Museum, 1933; **page 831:** courtesy Library of Congress; **page 833:** Bettmann Archive; **page 849:** Pablo Picasso drawing "Mother With Dead Child 9 May 1937," reproduced courtesy Artists Rights Society; **page 872:** AP/Wide World Photos (official U.S. Navy photograph); **page 884:** drawing copyrighted 1944, renewed 1972, Bill Mauldin. Reproduced courtesy of Bill Mauldin; **page 887:** Robert Capra photograph from Magnum Photos, Inc.; **page 895:** David Low cartoon courtesy the *Manchester Guardian;* **page 917:** Henri Cartier-Bresson/Magnum; **page 923:** "I Have Here in My Hand—" (From Herblock's *Here and Now.* New York: Simon & Schuster, 1955; reprinted by permission); **page 923:** AP/Wide World Photos; **page 928:** UPI/Bettmann; **page 938:** UPI/Bettmann; **page 941:** UPI/Bettmann; **page 944:** Bettmann Archive; **page 959:** AP/Wide World Photos; **page 988:** reprinted with permission of *The Dallas Morning News;* **page 974:** Dickey Chapelle photograph from Nancy Palmer Photo Agency, Inc.; **page 998:** drawing by David Levine. Reprinted with permission from the *New York Review of Books.* Copyright © 1973 NYREV, Inc.; **page 1007:** Florita Botts/Nancy Palmer; **page 1009:** UPI/Bettmann; **page 1021:** courtesy *Richmond Times-Dispatch;* **page 1026:** Chie Nishio/Nancy Palmer; **page 1032:** Michael Lloyd Carlebach/Nancy Palmer; **page 1044:** AP/Wide World Photos; **page 1049:** UPI/Bettmann; **page 1055:** UPI/Bettmann; **page 1065:** AP/Wide World Photos; **page 1070:** AP/Wide World Photos; **page 1084:** AP/Wide World Photos; **page 1088:** UPI/Bettmann; **page 1095:** AP/Wide World Photos; **page 1098:** AP/Wide World Photos; **page 1106:** AP/Wide World Photos; **page 1114:** AP/Wide World Photos; **page 1120:** AP/Wide World Photos; **page 1122:** AP/Wide World Photos; **page 1123:** UPI/Bettmann; **page 1130:** AP/Wide World Photos, Ron Edmonds; **page 1131:** AP Photo/Radivoje Pavicic.

Preface

The People's Chronology is intended as a handy desk reference—accurate, precise where possible, reliable, concise, yet with more than bare-bones detail. It is a completely revised and updated edition of a book widely used in U.S., Canadian, and British newsrooms, offices, libraries, and radio and television news organizations since 1979 (and in Japan since 1985).

The People's Chronology tracks major events of international political history. It also logs landmarks in the history of human rights, science, technology, energy, transportation, medicine, religion, education, communications, literature, art, photography, theater, music, sports, architecture, agriculture, nutrition, and a dozen other fields.

One of its purposes is to make the past more accessible, to bring some sense of progress, some order of events to the milestones of human experience. Beyond that, its format permits the achievements of scientists, composers, farmers, explorers, inventors, and painters to interact contrapuntally with those of industrialists, poets, economists, athletes, novelists, and statesmen worldwide.

Presenting material chronologically has certain advantages. It serves to demonstrate that knowledge grows incrementally, that one advance leads to another, that most innovations of value come from people who have built on work done by others in years past. A chronology also helps to show interrelationships (often obscured in conventional histories) between political, economic, scientific, social, and artistic facets of life.

Entries in *The People's Chronology* are grouped by category rather than organized vertically by nation or geographical area. Except in rare instances, categories appear in a standard order. To help the reader or researcher find information quickly and easily, graphic symbols in the margins provide a handy guide to category, and entries often contain references to prior or subsequent dates. An extensive alphabetical index is also provided.

Key to Symbols

- political events
- human rights, social justice
- exploration, colonization
- economics, finance, retailing
- energy
- transportation
- technology
- science
- medicine
- religion
- education
- communications, media
- literature
- art
- photography
- theater, film
- music
- sports
- everyday life
- tobacco
- crime
- architecture
- environment
- marine resources
- agriculture
- food availability
- nutrition
- consumer protection
- food and drink
- population

Prehistory

3 million B.C. An upright-walking australopithecine ape-man appears on the earth in the late Pliocene period and has thumb-opposed hands in place of forefeet, permitting him and his female counterpart to use tools. (Fossil remains found by Carl Johanson in Ethiopia's Awash Valley, A.D. 1974; further finds in A.D. 1975.)

1.75 million B.C. Anthropoids use patterned tools (Oldowan choppers) (*see* Leakey, A.D. 1959).

1 million B.C. Australopithecine ape-man becomes extinct as the human species becomes more developed. *Homo erectus erectus* is unique among primates in having a high proportion of meat relative to plant foods in his diet, but like other primates he is omnivorous, a scavenger who competes with hyenas and other scavengers while eluding leopards (*see* 1959).

400,000 to 360,000 B.C. *Homo erectus hominid* of the Middle Pleistocene period (Peking man) may use fire to cook venison, which supplements his diet of berries, roots, nuts, acorns, legumes, and grains. By conserving his energies, he can track down swifter but less intelligent animals (but he still splits bones to get at the marrow because he does not use fire effectively to make the marrow easily available) (*see* Black, A.D. 1927).

120,000 to 75,000 B.C. Neanderthal man of the Upper Pleistocene period has large front teeth, which he may use as tools. Less than half of his surviving infants reach age 20, 9 out of 10 of these die before age 40 (*see* A.D. 1856).

75,000 B.C. Neanderthal man has become a skilled hunter, able to bring down large, hairy elephant-like mammals (*Mammonteus primigenius*), saber-toothed tigers, and other creatures that will become extinct.

Neanderthal man cares for his sick and aged but engages in cannibalism on occasion.

Neanderthal man can communicate by speech, setting him apart from other mammals.

50,000 B.C. Date palms flourish in parts of Africa and Asia, where they will become an important food source.

Neanderthal man may be on the west coast of the Western Hemisphere and may even have reached the Continent 20,000 years earlier (as determined by racemization tests that record the extent to which molecules of aspartic acid in a specimen have altered in their figuration from the form that occurs in living bone to its mirror image; such tests will be conducted in the A.D. 1970s on bones found between A.D. 1920 and A.D. 1935, but the rate of change is affected by such factors as temperature, so the tests will not be conclusive).

42,000 B.C. The continent that will be called Australia is populated by the earth's first seafaring people. Colonists arrive from the Asian mainland.

38,000 B.C. *Homo sapiens* emerges from Neanderthal man and, while physically less powerful, has a more prominent chin, a much larger brain volume, and superior intelligence. *Homo sapiens* will split into six major divisions, or stocks—Negroids, Mongoloids, Caucasoids, Australoids, Amerindians, and Polynesians, and some of these will have subdivisions (Caucasoids, for example, will include Alpine, Mediterranean, and Nordic stocks).

His control of fire, his development of new, lightweight bone and horn tools, weapons, and fishhooks, and his superior intelligence permit man to obtain food more easily and to preserve it longer. Hunters provide early tribes with meat from bison and tigers, while other tribespeople fish and collect honey, fruits, and nuts (as shown by cave paintings near Aurignac in southern France).

Increased availability of food will lead to an increase in human populations.

36,000 B.C. *Homo sapiens* reaches the northern continent of the Western Hemisphere, where Neanderthal man has probably preceded him.

33,000 B.C. *Homo sapiens* becomes the dominant species on earth, with no serious rivals to his supremacy.

28,500 B.C. The island that will be called New Guinea is populated by colonists who arrive either from Australia or from the Asian mainland (*see* 42,000 B.C).

27,000 B.C. *Homo sapiens* reaches the islands that will be called Japan and may have arrived in the islands as much as 5,000 years earlier over ice sheets or land bridges (*see* 660 B.C).

25,000 B.C. Fishermen in Europe's Dordogne Valley have developed short, baited toggles that become wedged at an angle in fishes' jaws when the line, made of plant fibers, is pulled taut.

Homo sapiens uses small pits lined with hot embers or pebbles preheated in fires to cook food that may be covered with layers of leaves or wrapped in seaweed to prevent scorching.

13,600 B.C. A Great Flood inundates much of the world following a sudden 130-foot rise in sea levels as a result of runoff from a rapid melting of a glacial ice sheet covering much of the northern continent of the Western Hemisphere (time is approximate and somewhat conjectural).

12,000 B.C. The dog is domesticated from the Asian wolf and used for tracking game. (Fossil remains found in a cave near Kirkuk in Iraq in the A.D. 1950s will be dated in the 1970s by fluorine analysis.)

11,000 B.C. Vast fields of wild grain appear in parts of the Near East as the glaciers begin to retreat.

10,500 B.C. Human habitations appear even at the southernmost parts of the Western Hemisphere, where cavemen pursue guanaco and hunt a horse species that will become extinct. (Fossil evidence found 1,200 miles south of Buenos Aires in the A.D. 1970s.)

10,000 B.C. Goats are domesticated by Near Eastern hunter-gatherer tribespeople who have earlier domesticated the dog.

Homo sapiens increases in number to roughly 3 million.

9000 B.C. The New Stone Age begins in Egypt and Mesopotamia.

8500 B.C. Goats' milk becomes a food source in the Near East, where goats have been domesticated for the past 1,500 years (as determined by carbon 14 radioactivity decay studies on fossil evidence found at Asiab, Iran) (*see* Libby, A.D. 1947).

8000 B.C. Europe's final postglacial climatic improvements begin. They will produce a movement of people to the north of the continent, where the settlers will eat fish caught in nets of hair, thongs, and twisted fiber, along with shellfish, goose, and honey.

Agriculture begins at the end of the Pleistocene era in the Near East. Women use digging sticks to plant the seeds of wild grasses.

Earth's human population soars to 5.3 million, up from 3 million in 10,000 B.C., as agriculture provides a more reliable food source. Where it has taken 5,000 acres to support each member of a hunter-forager society, the same amount of land can feed 5,000 to 6,000 people in an agricultural society.

7700 B.C. Desert predominates over fertile lands in the arc extending from the head of the Persian Gulf through the Tigris-Euphrates Basin to the eastern Mediterranean and then south to the Nile Valley. Men and animals are crowded in oases in the region that will be called the "Fertile Crescent" by U.S. archaeologist James Henry Breasted (A.D. 1865–1935).

Ewes' milk becomes a food source and supplements goats' milk and mothers' milk as lamb and mutton begin to play a large role in human diets in the Near East, where sheep are domesticated. (Sheep remains that vastly outnumber goat remains will be found at Asiab in Iran, and a large majority of the sheep remains will be from yearlings, good archaeological evidence that sheep have been domesticated.)

7300 B.C. Dogs are domesticated by tribes in the British Isles. (Evidence from carbon 14 bone studies of fossil remains found at Star Carr in Yorkshire.)

7200 B.C. Sheep are domesticated in Greece (Argissa-Magula) (*see* 7700 B.C.).

Populations in the Middle East will increase in the next 2 millennia, and more permanent camps will be established by people who have lived until now in small groups that shifted camps every 3 or 4 months. Seed collecting will become more important to the food supply.

7000 B.C. Greek seafarers sail to the Aegean island of Milos, 75 miles from the mainland, to obtain obsidian.

Glaciers recede in the northern continent of the Western Hemisphere.

Fish too large to be caught from shore are caught at sea by Greek fishermen.

Emmer wheat *(Triticum dicoccum)*, domesticated from the wild *Triticum dicoccoides*, grows in the Kurdistan area lying between what will be southeastern Turkey and northwestern Iran.

Barley (*Hordeum spolitalieum*), millet (*Panicum miliaceum*), and certain legumes, including lentils, are cultivated in Thessaly, where the Greeks may also have domesticated dogs and pigs (based on evidence found in excavations at Argissa-Magula). (Domestication of swine has been delayed by the need of pigs for shade from the sun and by the fact that they cannot be milked, cannot digest grass, leaves, or straw, and must therefore be given food that man himself can eat—acorns, nuts, cooked grain, or meat scraps.)

The Jordanian town of Jericho, 840 feet above sea level, has a population of some 2,500, attracted by the area's perennial spring. The city will soon be walled to protect it from attack.

6800 B.C. The Kurdistan village of Jarmo is founded with some 30 dwellings that cover 3 acres and house 200 people. It is one of the first permanent agricultural settlements. (Excavations in 1947 by University of Chicago team.)

6500 B.C. The wheel will be invented sometime in the next 2 centuries by Sumerians in the Tigris-Euphrates Basin and will radically change transportation, travel, warfare, and industry. Other parts of the world, including the Western Hemisphere, will never develop the wheel on their own and will not enjoy its benefits until it is introduced by foreigners. The wheel (above) will not only speed transportation but will also facilitate construction and will lead to many technological advances.

The aurochs, ancestor of domestic cattle, will be domesticated in the next 2 centuries if it has not been domesticated earlier (Obre I, Yugoslavia). The fierce beast will be the last major food animal to be tamed for use as a source of milk, meat, power, and leather.

6000 B.C. Village farmers begin to replace food-gathering tribespeople in much of Greece (*see* 7000 B.C.).

Inhabitants of the Swiss lake regions have domesticated dogs and plow oxen.

Swiss lake dwellers (above) collect wild flax *(Linum usitatissimum)* or cultivate it and use its strong fibers to make lines and nets for fishing (and for animal traps and ropes and cords for building construction and navigational purposes).

Swiss lake dwellers (above) make bread of crushed cereal grains and keep dried apples and legumes (including peas) in the houses they build on stilts. (Evidence from excavated remains of the houses and their contents.)

The first true pottery evolves, permitting new forms of cookery (although food has earlier been boiled in gourds, shells, and skin-lined pits into which hot stones were dropped).

5508 B.C. Year of Creation that will be adopted in 7th century A.D. Constantinople and used in the Eastern Orthodox Church and secularly in Russia until early in the 18th century A.D.

5500 B.C. Copper smelted from malachite (copper carbonate) by artisans in Persia produces the first metal that can be drawn, molded, and shaped, but the metal is too soft to hold an edge (*see* bronze, 3600 B.C.).

5490 B.C. Year of Creation as it will be reckoned by early Syrian Christians.

5000 B.C. Villages begin to cluster together in the Fertile Crescent, but a common need for water sometimes leads to savage warfare.

Domesticated cattle are common in the valleys of the Tigris and Euphrates rivers, and villagers often cooperate to build primitive irrigation canals and ditches.

Lands bordering the Nile River begin to dry out. The Egyptians build dikes and canals for irrigation and start to develop a civilization in North Africa.

Agricultural peoples inhabit the plains of southeastern Europe.

Corn (maize) and common beans grow under cultivation in the Western Hemisphere.

4350 B.C. Domesticated horses provide parts of Europe with a new source of power for transportation and agriculture. (Evidence from Derivka in the Ukraine) (*see* work collar, 10th century A.D.).

4004 B.C. October 23: date of Creation as it will be reckoned by Irish theologian James Ussher in A.D. 1650.

4000 B.C. Peoples of the Indus Valley raise wheat, barley, peas, sesame seeds, mangoes, and dates on irrigated fields, but the large fields of grain encourage a multiplication of insects, and the stores of dry grain bring an explosion in rodent populations. Asses, horses, buffalo, camels, and cattle are bred for meat and for use as draft animals. Bananas, lemons, limes, and oranges are cultivated as are grapes for wine, which is also made from flowers. (Evidence from excavations at Mohenjo-Daro beginning in A.D. 1922 and at Harappa 300 miles away beginning in A.D. 1945.)

The world's population reaches roughly 85 million.

3760 B.C. Year of Creation as it will be reckoned in the Hebrew calendar that will be used from the 15th century A.D. (*see* Judaism, 1700 B.C.).

3641 B.C. February 10: date of Creation as it will be reckoned by Mayan calendars in the Western Hemisphere.

3600 B.C. Bronze made by southwest Asian artisans is the first metal hard enough to hold an edge. Copper is alloyed with tin, which is even softer than copper, but the combination (5 to 20 percent tin) creates a metal with many more practical uses than copper (*see* 5500 B.C.; iron, 2500 B.C.).

3500 B.C. The Sumerian society that marks the beginning of human civilization develops in the valleys of the Tigris and Euphrates rivers, where annual floods deposit fresh layers of fertile silt. Agricultural tribespeople settle in communities and evolve an administrative system governed by priests.

Animal-drawn wheeled vehicles and oar-powered ships are developed by the Sumerians (*see* wheel, 6500 B.C.; alphabet, 2500 B.C.).

Bronze enables the Sumerians (above) to make objects that were impossible to make with softer, less fusible copper (*see* 3600 B.C.; iron, 2500 B.C.).

A written cuneiform alphabet developed by the Sumerians (above) facilitates communication (*see* 2500 B.C.).

The Sumerians (above) harness domestic animals to plows, drain marshlands, irrigate desert lands, and extend areas of permanent cultivation. By reducing slightly the number of people required to raise food, they permit a few people to become priests, artisans, scholars, and merchants.

Antiquity

3400 B.C. Egypt's 1st Dynasty (Thinite dynasty) unites northern and southern kingdoms under Menes, who has founded a city that will be called Memphis.

3000 B.C. Cotton fabric is woven in the Indus Valley (*see* 4000 B.C.; A.D. 1225).

Gilgamesh in Sumerian cuneiform is the first known written legend and tells of a great flood in which man was saved by building an ark (*see* 13,600 B.C.; Smith, A.D. 1872).

The Sahara Desert has its beginnings in North Africa, where overworking of the soil and overgrazing are in some places exhausting the land in a region that is largely green with crops and trees (*see* Lhote, A.D. 1956).

Dolphins are killed in the Euxine (Black) Sea, but in some parts of the world the mammal is considered sacred and left unmolested.

Potatoes are cultivated in the Andes Mountains of the Western Hemisphere (*see* A.D. 1530).

Sumerian foods mentioned in *Gilgamesh* (above) include caper buds, wild cucumbers, ripe figs, grapes, several edible leaves and stems, honey, meat seasoned with herbs, and bread—a kind of pancake made of barley flour mixed with sesame seed flour and onions.

The world's population reaches 100 million.

2980 B.C. Egypt's 3rd Dynasty is founded by Zoser (Tosorthros), who will rule for 30 years with help from his counselor-physician Imhotep.

Imhotep (above) will make the first efforts to find medical as well as religious methods for treating disease.

The pyramid of Zoser that Imhotep will erect at Sakkara (Step Pyramid) will be the world's first large stone structure, a tomb copied in stonework from earlier brickwork piles.

2920 B.C. The Egyptian king Snefru (or Snofru) develops copper mines in Sinai, increases sea trade by using large ships, and raises his country to new heights of prosperity. He is the last pharaoh of the Memphite 3rd Dynasty.

The two pyramids of Dahshur that will memorialize Snefru (above) will each rise more than 310 feet and will commemorate a reign that has vanquished the Nubians and Libyans and has seen the development of sea trade in cedar with Byblos.

2900 B.C. Egypt's 4th Dynasty is founded by Cheops (Khufu), who will reign for 23 years. (Dates for all early rulers are approximate and controversial.)

The Great Pyramid of Cheops at Giza will by some accounts be the work of 4,000 stonemasons and as many as 100,000 laborers working under conditions of forced servitude and given rations consisting in large part of onions and garlic. Rising to a height of 481.14 feet and covering upwards of 13 acres, the Great Pyramid will contain stones weighing as much as 5 tons, each of which will be moved into place with primitive equipment and put together with virtually no space between them. (Variance from absolute accuracy of the work is so small that the 4 sides of the base have a mean error of only 0.6 inches in length and 12 inches in angle from a perfect square.)

2850 B.C. Khafra (Khafre, or Chephren) rules as the third Egyptian king of the 4th Dynasty.

The Great Sphinx carved from rock at Giza by order of Khafra (above) is a wingless symbol of the god Harmachis in whose image the 189-foot-long monument is fashioned.

Khafra (above) erects a second pyramid at Giza.

2800 B.C. The yang and yin philosophy of nature originated by the legendary Chinese emperor Fu Hsi says that health and tranquility require perfect equilibrium—a harmonious relationship among the five elements (wood, fire, earth, metal, and water), which correspond to the five planets, the five seasons, and the five colors, sounds, senses, viscera, and tastes. The yang (male element) is always dominant, says Fu Hsi.

A third pyramid erected at Giza by Egypt's 4th Dynasty (Memphite) king Menkure is the smallest but most perfect of the pyramids at Giza. Menkure's reign marks the beginning of his dynasty's decline.

The sickle invented by Sumerian farmers of the Tigris and Euphrates valleys is a curved instrument of wood or horn fitted with flint teeth. It will remain the predominant tool for harvesting grain until it is superseded by tools with tempered metal blades.

2750 B.C. Tyre is founded by mariners on the east coast of the Mediterranean and begins its rise as a great Phoenician seapower. (The Greek historian Herodotus will say in 450 B.C. that Tyre was founded "2,300 years ago.")

2700 B.C. Principles of herbal medicine and acupuncture originated by the legendary Chinese emperor Shen Nung are based, in part, on the basic principles of yang and yin proposed a century ago by Fu Hsi. The body has 12 canals related to vital organs says Shen Nung. They circulate the two principles of yang and yin; puncturing the canals with small needles permits the escape of bad secretions or obstructions and restores the body's overall equilibrium.

2640 B.C. Silk manufacture is pioneered by the wife of the Chinese emperor Huang Ti.

2600 B.C. Oxen harnessed to plows in the Near East make it possible to plow deeper and to keep the soil productive longer.

Annual Nile floods permit the Egyptian peasant to produce enough barley and Emmer wheat to feed three with the surplus going to the builders of flood control projects, public buildings, and pyramid tombs.

The Egyptians preserve fish and poultry by sun-drying.

2595 B.C. *Nei Ching* by the legendary Chinese emperor Huang Ti is the most ancient of medical texts. Chinese medicine will contribute to the pharmacopoeia such substances as camphor, chaulmoogra, ephedrine, opium, and sodium sulfate.

2500 B.C. The Iron Age dawns in the Middle East, where artisans produce a new metal much harder than the bronze used since 3600 B.C. The men use temperatures of 1500° C, much higher than the heat needed to smelt copper, but the new metal will not come into wide use for another 1,000 years.

Egypt and Mesopotamia are well into the Bronze Age that began in 3600 B.C., but central Europe and the British Isles are only entering the Stone Age that began in 9000 B.C.

The Sumerians develop a cuneiform script alphabet of some 600 simplified signs. They have earlier developed a written language using thousands of picture-signs, or ideograms, as in the *Gilgamesh* legend of 3000 B.C., and the new alphabet is based on those ideograms (*see* 1300 B.C.; Grotefend, A.D. 1837).

2475 B.C. Maize is domesticated in primitive form in the isthmus that links the two continents of the Western Hemisphere, while potatoes and sweet potatoes are cultivated in the southern continent.

Olive trees are cultivated in Crete, which grows rich by exporting olive oil and timber.

2350 B.C. The Akkadian Empire, founded by Sumer's Sargon I, will rule Mesopotamia for the next 2 centuries. The Sumerian city-state civilization now reaches its zenith; and the empire will incorporate the advances made by the Sumerians, giving them wide currency.

2300 B.C. Rice (*Oryza sativa*) from the Indus Valley is introduced in northern China, where a civilization flourishes on a level comparable to any at Mohenjo-Daro or Harappa.

2205 B.C. The Xia (Hsia) dynasty that will rule much of China for roughly 700 years is inaugurated.

The Chinese domesticate dogs, goats, pigs, oxen, and sheep.

The Chinese demonstrate the first knowledge of milling grain.

2000 B.C. Byblos, on the Levant Coast, has grown into a port for the export of Lebanese timber to Egypt.

Phylakopi, on the Aegean island of Milos, has become a center of trade in the volcanic glass obsidian found on the island for at least 5,000 years.

Square sails on two and even three masts assist Phoenician and Cretan oarsmen.

Europe remains in the Stone Age as the Bronze Age proceeds in the Near East.

Decimal notation appears in Babylonia, which has replaced Sumer as the dominant power in the Middle East.

Farmers in the Near East raise some cattle for meat, some for milk.

The Egyptians abandon efforts to domesticate antelope, gazelle, and oryx, devoting more effort instead to hunting, fowling, fishing, and gathering wild celery, papyrus stalks, lotus roots, and other plant foods to supplement the grain and vegetables they grow on their Nile flood plains.

Watermelon is cultivated in Africa, figs in Arabia, tea and bananas in India, apples in the Indus Valley; agriculture is well established in most of the central isthmus of the Western Hemisphere.

1970 B.C. Egypt's Amenemhet I dies after a 30-year reign that has founded the 12th Dynasty. He is succeeded by his son who has served as co-regent since 1980 B.C. and who will reign alone until 1935 B.C. as Sesostris. The new ruler will complete the conquest of Nubia.

1935 B.C. Egypt's Sesostris I dies and is succeeded by his son, who has served as co-regent since 1938 B.C. and will reign alone until 1903 B.C. as Amenemhet II. The new king will increase trade with Punt.

1903 B.C. Egypt's Amenemhet II dies after a 32-year reign. He is succeeded by his son, who has served as co-regent since 1906 B.C. and will reign until 1887 B.C. as Sesostris II.

1900 B.C. Stonehenge will be erected sometime in the next 3 centuries by Bronze Age Britons, possibly as a monumental calculator to chart the movements of the sun, moon, and planets.

1887 B.C. Egypt's Sesostris II dies after a 16-year reign. He is succeeded by his son, who will reign until 1849 B.C. as Sesostris III, making Egypt a great power holding sway over 1,000 miles along the Nile.

A canal through the Nile's first cataract will be dug during the reign of Sesostris III (above).

1849 B.C. Egypt's Sesostris III dies after a 38-year reign in which he has invaded Palestine and Syria to maintain Egyptian trade routes. He is succeeded by his son, who will reign until 1801 B.C. as Amenemhet III, de-

Stonehenge, built by Indo-Europeans who invaded the British Isles, may have served as an astronomical observatory.

veloping mines in the Sinai region to keep the nation prosperous.

A vast irrigation system will be developed in the reign of Amenemhet III (above).

1801 B.C. Egypt's Amenemhet III dies and is succeeded by his son, who will reign until 1792 B.C. as Amenemhet IV.

1800 B.C. Taboos against eating pork appear among some peoples of the Near East, possibly because they are sheepherding peoples and the pig is the domesticated animal of their farmer enemies (*see* 621 B.C.; Shariah, A.D. 629; Cook, A.D. 1779).

1792 B.C. Egypt's 12th (Theban) Dynasty ends with the death of Amenemhet IV after 208 years, and the power of the Egyptian king declines.

1760 B.C. Babylon's sixth king, Hammurabi, conquers all of Mesopotamia, carries out extensive public works, and imposes an exemplary code of laws: "If a man put out the eye of another man, his eye shall be put out."

1750 B.C. The great Indus Valley cities of Mohenjo-Daro and Harappa collapse as the soil of the region becomes too saline to support extensive crop growth after centuries of crude irrigation (*see* 4000 B.C.).

1700 B.C. Babylonians employ windmills to pump water for irrigation.

An Egyptian papyrus written during the reign of Re-Ser-Ka shows that Egyptians suffer from tooth decay and ophthalmic troubles. A German Egyptologist will discover the document in A.D. 1872, and it will be called the Ebers papyrus.

Smallpox or a similar disease occurs among the Chinese.

Judaism is founded by Abraham, a prince of Ur in Mesopotamia, who moves to Canaan, replaces human sacrifice with the sacrifice of rams, and begins a religion that will attract many followers in the Middle East (*see* Jacob, 1650 B.C.).

Knossus, on the island of Crete, is destroyed either by earthquake or by troops from the rival city of Phaistos, but the Minoans (who take their name from the legendary King Minos) will rebuild the city (*see* 1600 B.C.).

Eastern Europeans cultivate rye (*Secale cereale*). It will soon become the major bread grain of the Slavs, Celts, and Teutons in northern areas where the growing season is too short for dependable wheat production (*see* ergotism, A.D. 857).

1680 B.C. Hyksos tribesmen invade Egypt from Palestine, Syria, and farther north. They wear sandals, which enable them to outfight the Egyptians on the hot sands, and introduce horses that will help them dominate the Egyptians for the next century.

Leavened (raised) bread is invented in Egypt (time approximate).

1650 B.C. The Jewish religion begun half a century ago by Abraham and carried on by his son Isaac is propagated by his grandson Jacob, whose 12 sons will come to head 12 tribes of Israel (*see* 933 B.C.).

1600 B.C. Knossus, on the island of Crete, is rebuilt within a century after its destruction in 1700 B.C. A brilliant civilization flourishes at Knossus and at Phaistos, Tylissos, Hagia, Triada, and Gornia (but *see* 1470 B.C.).

1568 B.C. The New Kingdom that will rule Egypt until 332 B.C. is inaugurated at Thebes by the Diospolite (18th Dynasty) king Amasis who begins to drive out the Hyksos who invaded Egypt in 1680 B.C. and to reunite Upper and Lower Egypt.

1545 B.C. Egypt's Amasis I dies after a 23-year reign. His son will reign until 1525 B.C. as Amenhotep I, invading Nubia and warring with the Libyans and Syrians.

1525 B.C. Egypt's Amenhotep I dies after a 20-year reign that has secured the nation's borders. His successor, who is not of royal blood, will reign until 1504 B.C. as Thutmose I, conquering Nubia.

Thutmose I (above) will restore the temple of Osiris at Abydos, will build hypostyle halls at Karnak, and will erect two pylons and two obelisks. He will have a record of his deeds preserved in rock inscriptions near the third cataract of the Nile.

1520 B.C. A volcanic eruption on the Greek island of Thera (Santorini) destroys all life on the island (*see* 1470 B.C.).

1512 B.C. Egypt's Thutmose I is deposed after a 13-year reign in which he has led successful expeditions as far as the Euphrates. His bastard son will reign until 1504 B.C. with his wife and half sister Hatshepsut as Thutmose II.

1504 B.C. Egypt's Thutmose II dies at a young age after successful military campaigns against the Nubians and Syrians. Hatshepsut rules as regent for her infant nephew Thutmose III and will assume the title of queen next year.

1500 B.C. Aryan nomads from the Eurasian steppes push into the Indian subcontinent, bringing with them flocks of sheep and herds of cattle.

Horse-drawn vehicles are used by the Chinese (*see* Sumerians, 3500 B.C.).

Silk is woven by the Chinese who also use potter's wheels.

Geometry helps the Egyptians survey boundaries of fields whose dividing lines are effaced by the annual floods of the Nile (*see* Euclid, 300 B.C.).

Water buffalo are domesticated along with several species of fowl by China's Shang dynasty, whose monarchy rules at Anyang on the Huanghe (Yellow) River.

India's Aryan invaders (above) introduce a diet heavily dependent on dairy products, using *ghee* (clarified butter) rather than whole butter, which is too perishable for India's climate.

1485 B.C. Two obelisks at Karnak are erected by the Egyptian queen Hatshepsut, who has built a magnificent temple on the west side of the Nile near Thebes and has had its walls decorated with pictorial representations of an expedition to the land of Punt.

1483 B.C. Thutmose III comes of age and begins a 33-year reign in which Egypt will reach the height of her power, extending hegemony from below the fourth cataract of the Nile in the south to the Euphrates in the east. The title "pharaoh," or "Great House," will come into use under Thutmose III.

Thutmose III builds walls around his aunt Hatshepsut's obelisks at Karnak and tries to destroy all evidence of her existence.

1470 B.C. A volcanic eruption on the Greek island of Thera that is far more violent than the eruption of 1520 B.C. deposits ashes on Crete and emits poisonous vapors that destroy the Minoan civilization of 1600 B.C. Seismic waves 100 to 160 feet high, created by the eruption, rush in to fill the void created at Thera, temporarily dropping water levels on the eastern shores of the Mediterranean.

Egyptian croplands are engulfed by seawater from seismic waves (above), the land is made uncultivatable, and famine ensues.

Mycenae is established as a new cultural center in the Greek Peloponnesus by survivors of the Minoan civilization destroyed in Crete (above).

1450 B.C. Egypt's Thutmose III dies after a splendid 33-year reign. His son, who has ruled jointly for the past year, will remain until 1424 B.C. as Amenhotep II with successful campaigns in Judea and on the Euphrates.

1424 B.C. Egypt's Amenhotep II dies after a 27-year reign and is succeeded by his son, who will reign until 1417 B.C. as Thutmose IV. The new king will marry a Mitannian princess, form alliances with Babylonia and the Mitanni, lead military expeditions into Phoenicia and Nubia, and complete the last obelisk of his grandfather Thutmose II.

1417 B.C. Egypt's Thutmose IV dies and is succeeded by his brilliant son, who will reign in luxury and peace until 1379 as Amenhotep III, the last great ruler of the New Kingdom.

1400 B.C. The Iron Age begins in Asia Minor as an economical method is found for smelting iron on an industrial scale (*see* 2500 B.C.).

The first domestic poultry is introduced into China from the Malayan Peninsula, where the jungle fowl *Gallus bankiva* has been domesticated.

1380 B.C. A canal completed by slaves of Egypt's Amenhotep III connects the Nile with the Red Sea and will remain in use for centuries (*see* 609 B.C.).

1379 B.C. Egypt's Amenhotep III dies after a 38-year reign in which Babylonia has recognized Egyptian supremacy. The pharaoh has led a successful expedition into Upper Nubia above the second cataract of the Nile, developed his capital of Thebes into a monumental city of great temples, pylons, and colossi, erected hypostyle halls at Karnak, built the Temple of Amun in Luxor, and reigned in an era of prosperity and magnificence. Amenhotep is succeeded by his son, who will reign until 1362 B.C. as Amenhotep IV (Ikhnaton), but the Hittite king Suppiluliumas will take advantage of Egypt's weakness in the next 35 years to build an empire that will extend south from Anatolia to the borders of Lebanon.

1374 B.C. Monotheism is introduced by the Egyptian king Amenhotep IV, who will be called Ikhnaton (Akhenaten or Akhnaton), meaning "Aten is satisfied." The pharaoh establishes a new cult that worships the sun god (or solar disk) Aten, and he opposes the priests of Amen, possibly due to the influence of his beautiful wife Nefertiti.

The wheel, devised by Sumerians, was slow to reach Egypt. Meanwhile, rich Egyptians rode in slave-borne palanquins.

1358 B.C. Egypt's Ikhnaton dies after a 17-year reign and is succeeded by his son-in-law, 9, who will rule until 1350 B.C. The new pharaoh Tutankhamen has accepted the sun worship faith of his wife and her father but will return to the religion of the priests of Amen and move Egypt's capital back to Thebes from the new city of Akhetaton.

1350 B.C. The Egyptian throne is seized by the soldier Harmhab, who will reorganize the country's administration and reign until 1315 B.C., founding Egypt's 19th Dynasty.

Harmhab (above) will restore worship according to the traditional tenets of the Amen priests.

1349 B.C. Egypt's late pharaoh Tutankhamen is buried at Thebes with a vast treasure of decorative art objects.

1327 B.C. Egypt's Harmhab dies and is succeeded by the aged Ramses, who will plan and begin a great hypostyle hall at Karnak.

1325 B.C. Egypt's Ramses I dies and is succeeded by his son, who has served since last year as co-regent. He will reign until 1304 B.C. as Seti I.

1304 B.C. Egypt's pharaoh Seti I dies after a reign in which he has defeated the Libyans west of the Nile Delta and made peace with the Hittites in Syria. Seti's son will reign until 1237 B.C. as Ramses II.

The pharaoh Seti (above) has completed a colonnaded hall at Karnak begun by his father Ramses I and has also built a magnificent sanctuary at Abydos dedicated to the great Egyptian gods.

1300 B.C. Alphabetic script developed in Mesopotamia is a refinement of the simplified cuneiform alphabet of 2500 B.C.

1275 B.C. A 40-year Israelite migration begins after 3 centuries of Egyptian oppression. The prophet Moses and his brother Aaron lead tribesmen and their flocks of sheep out of Egypt toward the Dead Sea in Canaan on a roundabout journey that will take them through the Sinai Peninsula, Kadesh, Aelana, and Petra.

The wandering Jews (above) will survive starvation at one point by eating "manna," possibly a kind of mushroom.

1272 B.C. Egypt's Ramses II marries a daughter of the Hittite king and arranges a permanent peace with the Hittites.

Ramses will devote the rest of his long, peaceful reign to such projects as the completion of Seti's temple at Abydos, additions to the temples at Karnak and Luxor, construction of a great mortuary temple at Thebes with colossal statues of himself, and construction of a rock-cut temple at Abu-Simbel in Nubia.

1237 B.C. Egypt's Ramses II dies after a 67-year reign in which he has used forced Israelite labor to build the treasure cities of Pithom and Ramses. His son Merneptah will reign until 1215 B.C.

1221 B.C. Egypt is invaded by Libyans, who are defeated by the pharaoh Merneptah.

1215 B.C. Egypt's Merneptah dies after a 10-year reign and is succeeded by a series of pharaohs who will rule briefly to end the 19th Dynasty founded by Harmhab in 1350 B.C. They will be followed by an interregnum that will continue until 1198 B.C.

1200 B.C. Lower Egypt's remaining Jews are expelled in the confusion following the end of the 19th Dynasty. The Jews have been active in the country's administration, arts, and trade.

The Egyptians have learned to make fine linen from flax stalks, and their high priests wear only linen, which is used also to wrap embalmed bodies (*see* flax 6000 B.C.).

1198 B.C. Egypt's 20th Dynasty entrenches itself as its second king begins a 31-year reign as Ramses III. He will rally the Egyptians against a confederation of Philistines, Sardinians, Greek Danaoi, and other sea peoples.

1193 B.C. King Priam's city of Troy at the gateway to the Hellespont in Asia Minor falls to Greek forces under Agamemnon after a 10-year siege in the Trojan War (*see* Homer, 850 B.C.; Byzantium, 658 B.C.).

1170 B.C. The first recorded strike by laboring men occurs at the Egyptian necropolis of Thebes, where acute inflation brings an organized protest by men working on a new pyramid. When the payroll is delayed, the men refuse to work.

1160 B.C. Egypt's Ramses V dies. His mummifed remains will show that the pharaoh had smallpox.

1150 B.C. Egyptian medicine splits into two basic schools. Empirico-rational medicine rests on the premise that fever, pain, or tumor is a disease rather than a symptom, but practitioners of this school charge such high fees that only the very rich can afford them. The magico-religious school of medicine relies basically on expelling demons or spirits and is popular because it is inexpensive.

Egyptian aristocrats enjoy leavened bread and drink some wine (but mostly beer) as they dine at tables and sit on chairs which they have developed, but in the bread stalls of village streets, only flat breads are commonly available.

1146 B.C. Nebuchadnezzar I begins a 23-year reign as king of Babylon.

1141 B.C. Israelite forces lose 4,000 in a battle against the Philistines and then lose another 30,000.

The Israelites' sacred Ark of the Covenant is carried off to Ashdod by the Philistines, and a plague breaks out among the Philistines, spreading with the Ark to Gathen and then to Ekron. The Philistines return the Ark to Joshua the Bethshemite in order to end the plague, but 70 Bethshemite men who peer into the Ark die of plague, which then spreads throughout Israel, killing some 50,000.

1122 B.C. The Zhou (Chou) dynasty that will rule China until 255 B.C. is founded by Wu Wang, son of Wen Wang. He overthrows the emperor Zhou Hsin, who burns himself to death.

1116 B.C. Tiglath-Pileser I begins a ruthless 38-year reign that will bring the Middle Assyrian Empire to its zenith, conquering invaders from Anatolia and elsewhere.

1100 B.C. Assyrian forces under Tiglath-Pileser I reach the Mediterranean after having conquered the Hittites. The Assyrians encounter the seafaring Phoenicians, who hunt sperm whales and conduct a farflung sea trade (*see* 2750 B.C.; 878 B.C.).

1025 B.C. The prophet Samuel anoints Saul, who will reign until 1012 B.C. as king of Hebron.

1012 B.C. The Battle of Mount Gilboa ends in defeat and death for Hebron's Saul and his eldest son Jonathan at the hands of the Philistines. Jonathan's friend David succeeds Saul and will reign until 1005 B.C. as king of Hebron.

1005 B.C. Jerusalem falls to David of Hebron, who is anointed king of Judea by the prophet Samuel and will reign until 961 B.C., breaking the power of the Philistines and defeating the Moabites, Ammonites, and Edomites.

1000 B.C. The Iron Age that began 400 years ago in the Near East moves to Europe in the Hallstatt region of what will become Austria. Iron tools and weapons begin to spread throughout Europe.

The Chinese cut down forests to create more farmland. The deforestation will lead to soil erosion, floods, and drought in millennia to come.

Land sown with grain in Egypt yields crops as bountiful as any the Egyptians will reap in the 20th century A.D.

The Chinese cut ice and store it for refrigeration.

990 B.C. Absalom, third (and favorite) son of Judea's King David, kills his half brother David's eldest son Amnon in revenge for the rape of his full sister Tamar. David banishes Absalom from Judea (date approximate).

978 B.C. Absalom regains King David's favor through the offices of David's nephew Joab, but Absalom leads a rebellion against David on the advice of David's counselor Ahithophel. Joab will suppress the rebellion and kill the fleeing Absalom and his captain Amasa (Joab's cousin), and the counselor Ahithophel will commit suicide.

961 B.C. Judea's King David dies and is succeeded by his son Solomon, who will reign until 922 B.C., making alliances with Egypt's ruling priests and with the Phoenician king Hiram of Tyre. Solomon is David's son by his second wife Bathsheba, whose first husband David murdered in order to marry her.

Solomon executes David's former army commander Joab for having killed David's son Absalom in violation of David's orders and for having killed his rival Amasa.

Solomon's fleet sails the Red Sea, trading products of Judea at Tyre and Sidon and in Africa and Arabia, where Solomon begins mining gold.

The Great Temple of Jerusalem goes up in the onetime Jebusite stronghold captured by David to house the sacred ark of Yahweh (*see* 1141 B.C.). David had proposed construction of the temple, but the prophet Nathan had thwarted him (*see* 586 B.C.).

Solomon will build a new royal palace and city wall at Jerusalem, using forced labor to erect buildings throughout his realm and introducing taxation to finance his projects.

950 B.C. The household of Judea's King Solomon includes 700 wives and 300 concubines and consumes 10 oxen on an ordinary day, along with the meat of harts, gazelles, and hartebeests.

945 B.C. Egypt's throne is usurped by the Libyan Sheshonk, who founds the 22nd (Bubastite) Dynasty that will rule Egypt for 200 years.

933 B.C. Judea's King Solomon dies and is succeeded by his son Rehoboam in Jerusalem, but 10 northern tribes secede when Rehoboam refuses their demands for relief from taxation. They establish the kingdom of Israel with Jeroboam as king.

926 B.C. Palestine is invaded by Egypt's pharaoh Sheshonk, who plunders Jerusalem and many other Judean cities.

900 B.C. The first Italian towns are established by Etruscans who have emigrated from Lydia after an 18-year famine. Lydia's King Atys, who rules the Asian country opposite the Greek islands of Chios and Samos, has commanded half his subjects to emigrate. They have journeyed to Smyrna under the leadership of Atys' son Tyrsanus, they have loaded their belongings onto ships, and they have come to the Italian Peninsula, where their towns are built mostly on hillside terraces and are enclosed with massive timbered walls (*see* 396 B.C.; Rome, 753 B.C.).

884 B.C. Assyria's Assurnasirapli II begins a 24-year reign in which he will defeat Babylonia and revive the empire.

878 B.C. The Assyrian emperor Assurnasirapli II annexes Phoenicia as he takes over the entire eastern Mediterranean coast.

859 B.C. Assyria's Shalmaneser succeeds his father Assurnasirapli to begin a 34-year reign that will end in revolution.

850 B.C. The *Iliad* and the *Odyssey* are inscribed by the blind Greek poet Homer, so the historian Herodotus will write some 4 centuries hence, but the date may be as much as a century earlier, and while references to "the deathless laughter of the blessed gods" appear in both works, they may have been written by different people employing earlier lays handed down orally before articulation in the "winged words" of Homer.

The *Iliad* is an epic poem of Ilium (Troy) and its siege by the Greeks from 1194 to 1184 B.C., a poem mixing

The Assyrians, worshiping power, extended their sway over most of southwest Asia.

gods and mortals in its history of Priam, Helen, Paris, Menelaus, Hector, Achilles, Aphrodite, Agamemnon, and Odysseus (Ulysses). "The issue is in the laps of the gods" (XVII).

The *Odyssey* is an epic poem about the wanderings of Odysseus (Ulysses) who is kept as a lover for nearly 8 years by the goddess Calypso while his wife Penelope home at Ithaca is besieged by suitors and unwanted guests and while his son Telemachus is growing to manhood. "And what he greatly thought, he nobly dared" (II); "Day-long she wove at the web but by night she would unravel what she had done" (XXIV, a reference to wife Penelope, who has vowed to accept no second husband until she has completed a winding-sheet for her aged father-in-law, a ruse she continues until she is betrayed after 3 years by one of her serving maids).

Fish cultivation is discussed in a voluminous treatise by the Chinese author Fan-Li (manuscript in British Museum, London) (*see* oyster cultivation, 110 B.C.).

814 B.C. Carthage is founded in North Africa by refugee Phoenician colonists ("Punians") (*see* 300 B.C.).

812 B.C. Assyria's Shamshiadad V, son of Shalmaneser, dies after a 12-year reign in which he has ended a revolt with Babylonian aid but has lost part of his empire. He is succeeded by his brother Adadnirari V, but the queen mother Sammuramat (Semiramis) will rule for 4 years.

801 B.C. Egypt and Greece will begin regular trade relations in the next 100 years.

Aryan religious epics, or Vedas, will lead in the next 200 years to a veneration of the cow in much of India and to a sanctification of dairy products.

800 B.C. Rice becomes an important part of Chinese diets (*see* 2300 B.C.).

776 B.C. Greece's first recorded Olympic (or Olympian) games are held where the Alpheus and Cladeus rivers converge at Olympia, although many of the great buildings within the enclosure for the games and associated religious celebrations date back as much as 500 years. Only pure Greeks may compete and only Greeks who have no police records or even any relatives with police records. The competition in the first 13 of the quadrennial Olympiads will be limited to a footrace of some 200 yards (*see* 724 B.C.).

772 B.C. Construction begins at Ephesus on the Mediterranean coast of Asia Minor on the Temple of Artemis that will be one of the wonders of the ancient world (*see* 356 B.C.).

771 B.C. China's Zhou capital at Hao on the Wei River is destroyed by barbarians from the north. The capital will be moved in the next year to Luoyang near the Huanghe (Yellow) River.

753 B.C. Rome is founded, according to legend, on a wooded Italian hilltop overlooking the Tiber. The founders are infant brothers Romulus and Remus, who are suckled by a she-wolf.

745 B.C. Assyria's Tiglathpileser III begins a 7-year reign in which he will conquer Syria, Palestine, Israel, and, finally, Babylon.

724 B.C. Greece's 14th Olympiad is held at Olympia, and a second footrace is added in which competitors run twice around the stadium to cover a distance of nearly half a mile (*see* 776 B.C., 720 B.C.).

722 B.C. Samaria, capital of Israel since 879 B.C., falls to Assyrian forces after a 3-year siege. Assyria's Shalmaneser V dies and is succeeded by his son Sargon II, who claims the victory and takes 27,290 Israelite prisoners.

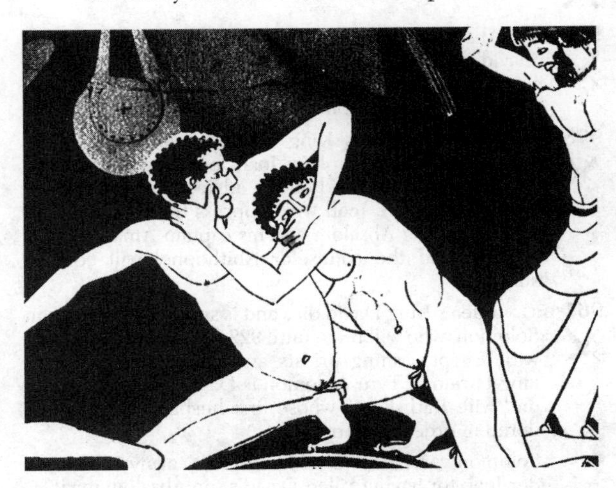

Greece's Olympic Games, inaugurated in 776 B.C., started to become war games with a boxing event in 696 B.C.

721 B.C. The kingdom of Israel, founded in 933 B.C., falls to Sargon II, who deports 27,000 people of Israel's 10 northern tribes to Central Asia; they will disappear from history (the "lost tribes of Israel").

720 B.C. Greece's 15th Olympiad is held at Olympia, and the Olympic games are extended to include a long-distance race of some 2.5 miles that requires contestants to run 12 times around the stadium.

710 B.C. Ethiopian invaders conquer Egypt.

708 B.C. Greece's 18th Olympiad is held at Olympus, and the Olympic games are enlarged by the addition of a pentathlon event in which contestants must compete in broad jumping, javelin throwing, a 200-yard dash, discus throwing, and wrestling. The javelin is a bronze-tipped elderwood spear; the discus a heavy bronze disk (*see* 720 B.C.; 696 B.C.).

705 B.C. Assyria's Sennacherib begins a 23-year reign that will see Nineveh become a city of unmatched splendor. Art and literature will flourish despite numerous wars.

7th Century B.C.

Phoenician colonists plant olive trees on the Iberian Peninsula.

China's minister of agriculture teaches the peasants crop rotation. The emperor Kuan Chung's minister also teaches them to dig drainage ditches, rents them farm equipment, and stores grain surpluses to provide free food in time of famine.

700 B.C. Aqueducts are built to carry water to the cities developing in the Near East.

Laws against animal slaughter are relaxed in India.

698 B.C. Greek colonization of the Mediterranean in the next 2 centuries will be motivated primarily by a need to find new food sources as Greece's population expands.

696 B.C. Greece's 23rd Olympiad is held at Olympia; boxing is added to the Olympic games, which are more and more intended as preparation for war.

693 B.C. Babylon is destroyed by the Assyrian king Sennacherib, but the city will be rebuilt in even greater splendor and luxury (*see* 597 B.C.).

682 B.C. Greece's 25th Olympiad is held at Olympia with the first equestrian event. A four-horse chariot race is run at the nearby Hippodrome.

660 B.C. Japan's main island of Honshu is invaded, according to legend, by Jimmu Tenno (Kami Yamato Ihare-Biko), who crosses from the island of Kyushu to establish himself as the country's first emperor (the invasion will actually occur at least 600 years hence).

658 B.C. Byzantium is founded by Greek colonists from Megara, who establish a settlement to the east (*see* 340 B.C.; Constantinople, A.D. 330).

650 B.C. Greek hillsides are bare of trees which have been cut down to provide wood for houses, for ships, and for the charcoal used by metalworkers. Loss of the trees leads in many areas to soil erosion and to a loss of fertile land (*see* Solon, 594 B.C.).

Greece's 33rd Olympiad is held at Olympus with a new event: the pancratium is a no-holds-barred contest that combines boxing and wrestling.

626 B.C. Assyria's King Ashurbanipal dies after a 43-year reign that has brought great prosperity to the country. He is the last major ruler of the Sargonid dynasty that has ruled for nearly a century, and the empire will crumble in the next 20 years.

625 B.C. Metal coins are introduced in Greece. Stamped with the likeness of an ear of wheat, the coins are a reminder that grain, usually barley, has previously served as a medium of exchange, but the new coins are lighter and easier to transport than grain and do not get moldy (*see* Croesus, 560 B.C.).

624 B.C. Corinth's tyrant Periander invites the city-state's nobility to a party and has his soldiers strip the women of their gold jewelry and of gowns adorned with golden thread. The gold will finance Periander's government for decades to come.

621 B.C. The Athenian lawgiver Draco issues a code of laws that makes nearly every offense punishable by death. Barbarously cruel or harsh punishment will forever be called "Draconian."

The Book of Deuteronomy, compiled by Israelite scribes, is among the five books of Moses containing what purports to be the dying testament of Moses to his people (*see* 1275 B.C.).

The Law of Moses in Deuteronomy (above) imposes dietary restrictions, permitting meat only from any animal "that parts the hoof and has the hoof cloven in two, and chews the cud," but proscribing meat from camels, hares, and rock badgers as well as from pigs. Also proscribed as "unclean" are fish without fins and scales, certain birds, and anything "that dies of itself." And "You shall not boil a kid in its mother's milk" (*see* Sharia, A.D. 629).

612 B.C. Nineveh falls to the Medes and the Chaldeans. The fall of the Assyrian capital will soon be followed by the disappearance of the Assyrian Empire.

609 B.C. A new canal to link the Nile with the Red Sea is begun by the Egyptian pharaoh Necho, but while the Greek historian Herodotus will write that Necho completed a channel "four days' journey in length and wide enough for two armies abreast," the canal will not be completed. More than 120,000 men will die in the effort to build it (*see* 1380 B.C.; 520 B.C.).

Necho's ships in the next 14 years will circumnavigate Africa, proceeding from east to west and taking 3 years (including a stop to plant and harvest a grain crop on the North African coast).

605 B.C. Necho is defeated at Carchemish by the Chaldean son of Nabopolassar, who begins a 43-year reign at Babylon as Nebuchadnezzar II.

6th Century B.C.

The Persian religious leader Zoroaster in this century will found a faith whose sacred literature will be the Zend-Avesta. The teachings of Zoroaster (Zarathustra) will dominate Persian religious thought for centuries.

Rome's Cloaca Maxima will be built in this century. The giant drainage system will drain the marshy area that will become the site of the Roman Forum.

Humped cattle from India become widespread in the Mediterranean countries.

Rome remains a small town but begins to experience food shortages. It will have occasional serious famines in this century.

600 B.C. Marseilles is founded by Greek colonists, who establish the settlement of Lacydon (*see* agriculture, 331 B.C.).

597 B.C. Jerusalem falls to Babylonia's Nebuchadnezzar II. The last great Chaldean king has invaded Judah, and he returns to Babylon with many Jews as prisoners (*see* 587 B.C.).

Wheeled carts drawn by onagers (small Asiatic asses) bring food into Babylon, while riverboats powered by scores of oarsmen bring copper, silver, gold, and vegetable oils from fields north of the Tigris. Camels in long caravans enter the city with dyestuffs, glassware, precious stones, and textiles.

Camel caravans (above) bring occasional plagues to Babylon; flies and mosquitoes that breed in polluted irrigation canals carry malaria, dysentery, and eye diseases.

Babylon is a magnificent city of public buildings faced with blue, yellow, and white enameled tiles that face on broad avenues crossed by canals and winding streets.

The Hanging Gardens of Babylon are one of the seven wonders of the ancient world, with exotic shrubs and flowers irrigated by water pumped from the Euphrates.

594 B.C. The Athenian statesman Solon, 42, who has regained Salamis from the Megarians, establishes a timocracy (government by the richest) and begins constitutional reforms at Athens.

Solon (above) forbids export of any Athenian agricultural produce: his well-intentioned edict will lead to more planting of olive trees. Their roots soak up deep moisture but do not hold soil together, so while olive oil and silver mines will bring riches to Athens, Solon's edict will hasten the erosion of Greek hillsides (*see* Plato, 347 B.C.).

590 B.C. The Greek poet Sappho flourishes on the island of Lesbos as priestess of a feminine love cult and celebrates the love of women for other women in poems that will survive in papyrus fragments and in quotations by later critics. The geographer Strabo will write some 600 years hence that "Sappho was something to be wondered at. Never within human memory has there been a woman to compare with her as a poet."

587 B.C. Jerusalem falls to Babylon's Nebuchadnezzar II after a 16-month siege. He carries the Jews off to exile in a "Babylonian captivity" that will continue until 538 B.C.

586 B.C. Jerusalem's Great Temple is destroyed by the forces of Nebuchadnezzar II.

573 B.C. The Phoenician city of Tyre falls to Nebuchadnezzar II after a 13-year siege. He will invade Egypt in 568 B.C.

565 B.C. Athenian forces from Megara conquer Salamis under the command of their general Peisistratus. He organizes the *diakrioi*, a new political party of small farmers, shepherds, and artisans.

Daoism (Taoism) is founded by the Chinese philosopher Lao Zi (Lao-tse), 39, in Honan province. He sets down principles of conduct in Dao De Ging (*Tao Te Ching*) ("teaching of Dao"). The liberal religion teaches that forms and ceremonies are useless; it advocates a spirit of righteousness, but it will degenerate in future centuries into a system of magic.

562 B.C. Nebuchadnezzar II of Babylon dies after a reign of 43 years. He is succeeded by his son Evil-Merodach, who will reign until 560 B.C.

561 B.C. The Athenian general Peisistratus makes himself tyrant but is driven out almost immediately by Lycurgus, who leads the city's nobility, and Megacles the Alcmaeonid, who leads the middle class.

560 B.C. Babylon's Evil-Merodach is deposed by conspirators who kill him. He has released Jehoiachin, king of Judah, who had been imprisoned for 36 years.

559 B.C. Athens restores Peisistratus, who has won the support of Megacles.

Croesus, king of Lydia, invents metal coinage to replace commodities as a medium of exchange (year approximate) (*see* 900 B.C.; 625 B.C.).

556 B.C. Athens expels Peisistratus again after he has broken with Megacles. He will spend some years amassing a fortune from his mines in Thrace and make Lygdamis tyrant of Naxos.

Babylon's last king, Nabonidus, begins a 17-year reign.

550 B.C. The king of Anshan Cambyses I dies after a long reign and is succeeded by his son Cyrus, 50. The new king will reign until 529 B.C., creating a Persian empire by uniting the Medes, Persians, and other tribes.

549 B.C. Armenia becomes a Persian satrapy after 63 years under the kings of Media. She will remain under Persian control until 317 B.C.

546 B.C. Lydia's rich King Croesus is surprised at Sardis by Persians under the command of Cyrus the Great. Croesus has ruled for 14 years, but Cyrus kills him and conquers his country.

Athens restores Peisistratus to power (he has obtained support from Thessaly and from Lygdamis of Naxos). He exiles his opponents, confiscates their lands, and uses them to benefit the poor, making the *hectemoroi* (sharecroppers) landowners and encouraging industry and trade.

Peisistratus introduces the cult of Dionysius to break down the power of the Athenian nobility through its hereditary priesthoods.

539 B.C. Persia's Cyrus the Great defeats the Babylonian king Nabonidus, who has alienated the priesthood and has built temples at the expense of the country's defenses. Cyrus enters the city of Babylon October 20 amidst wild rejoicing by the populace and either has Nabonidus killed or banishes him to Carmania.

538 B.C. The Babylonian prince Belshazzar (Bel-sharusur), son of Nabonidus, tries to expel the forces of Cyrus the Great but suffers a crushing defeat. Cyrus destroys the city of Babylon and permits its Jews to return to Jerusalem after their 49-year exile. They will rebuild the Great Temple built by Solomon in 973 B.C. (*see* 516 B.C.).

529 B.C. Persia's Cyrus the Great is killed fighting a savage tribe east of the Caspian Sea. Dead at 71 after a 21-year reign that has extended his empire from the Caucasus to the Indian Ocean and from the Indus to the Mediterranean, he is succeeded by his son, who will reign until 521 B.C. as Cambyses II.

528 B.C. Buddhism has its beginnings in India, where Siddhartha Gautama, 35, has found enlightenment after a long and severe penance at Buddh Gaya, near Benares. A prince who renounced the luxury of palace life 5 years ago, Siddhartha went into the wilderness wearing sackcloth, but he found the ascetic life futile. In the next 45 years he will travel up and down the Ganges River; he will be called the Buddha (Enlightened One) and will found monastic orders of a religion that will become dominant in China, Japan, and some other Asian countries (*see* 260 B.C.).

Vegetarianism will be an essential part of the Buddhist religion (above), although Siddhartha himself will abandon strict vegetarianism and die at age 84 after feasting on pork.

527 B.C. The Athenian tyrant Peisistratus dies and is succeeded by his sons Hippias and Hipparchus.

525 B.C. Persia's Cambyses II defeats Egypt's Psamtak III at Pelusium, adds the Nile Delta to his empire, but loses an army of 50,000 in a blistering sandstorm as he marches to conquer the Amun. He has secretly murdered his younger brother Smerdis before leaving Persia.

522 B.C. Persia's Cambyses II learns that his throne has been usurped by a "false Smerdis" (Gaumata, a Magian priest from Media) and dies en route home from Egypt.

521 B.C. Persia's "false Smerdis" is killed in battle by Darius Hystaspis, 37, son-in-law of the late Cyrus the Great. Persian noblemen make Darius king, and he will reign until 486 B.C. as Darius I.

520 B.C. A Carthaginian fleet of 60 vessels under the command of Admiral Hanno lands some 30,000 settlers at the mouth of the Rio de Oro on Africa's west coast. The colony will survive for half a century.

Phoenicia continues to grow rich on trade in grain, cloth, wine, and the purple-black dye obtained from a gland of the rare purple sea snail, or murex. Long shallow-draft Phoenician galleys powered by oarsmen slaves and large square sails may long since have circumnavigated Africa (*see* Necho, 609 B.C.). They supply Phoenicia with African gold and ivory, Spanish silver, tin from the "Tin Isles" (probably Cornwall), gold, iron, and lead from the southern shore of the Euxine (Black) Sea, copper, grain, and cypress from Cyprus, wine from southern France, slaves from everywhere.

The Persian emperor Darius digs a canal to connect the Nile with the Red Sea, continuing work begun nearly a century ago by the pharaoh Necho (*see* 609 B.C.; Suez, A.D. 1854).

Reliefs and inscriptions in Persian, Elamitic, and Babylonian chronicle the achievements of Darius (above) on a steep cliff near Behistun.

Phoenicia's island city of Tyre has a population of 25,000; Old Tyre on the mainland is even larger.

516 B.C. Jerusalem's Great Temple is rebuilt 70 years after its destruction by the Babylonian troops of Nebuchadnezzar II (*see* 165 B.C.).

509 B.C. Rome overthrows her Tarquin (Etruscan) king, becomes a republic, and begins her struggle to dominate Italy and the world. Legend will relate the uprising to the rape of Lucretia, beautiful and virtuous wife of the Roman general Lucius Tarquinius Collatinus, by Sextus Tarquinius, a son of the king Tarquinius Superbus. Lucretia tells her father and husband about the rape, makes them swear vengeance, and stabs herself to death. Her husband's cousin Lucius Junius Brutus agitates against the Tarquins, raises a people's army, and drives out the Tarquins.

5th Century B.C.

The Greek historian Herodotus will write in this century that "India is the farthest known region of the inhabited world to the East," but while natives of the Caucasus Mountains practice cannibalism, a great civilization is developing in China.

500 B.C. "In ancient times, people were few but wealthy and without strife," writes the Chinese philosopher Han Fei-Tzu. "People at present think that five sons are not too many, and each son has five sons also and before the death of the grandfather there are already 25 descendants. Therefore people are more and wealth is less; they work hard and receive little. The life of a nation depends on having enough food, not upon the number of people."

495 B.C. The Chinese philosopher Kong Fuzi (Confucius) resigns as prime minister of Lu at age 56 when the ruler gives himself up to pleasure. In the next 12 years, Confucius will wander from state to state teaching precepts dealing with morals, the family system, and statecraft, with maxims that comprise a utilitarian philosophy. A brief record of Confucian teachings will be embodied in the *Analects*, one of the Four Books of Chinese classics, and his Golden Rule will be honored (often in the breach) throughout the world: "What you do not like when done to yourself, do not do unto others."

490 B.C. The Battle of Marathon September 15 gives Athens her first great military triumph but begins a long period of conflict with Persia. A Persian army of 15,000 sent by Darius is repulsed by 11,000 Greeks under the leadership of Miltiades 25 miles northeast of Athens between Mount Pentelikon and the Gulf of Marathon.

The Greek satirist Lucian of the 2nd century A.D. will tell of a courier, Pheidippides, racing on foot to Athens (a distance of about 22 miles) with news of the victory on the plains of Marathon (above) before falling dead of exhaustion. Herodotus, in this century, will write of another courier, Philippides, who ran from Athens to Sparta (perhaps 160 miles) in less than 48 hours to seek aid against the Persians. Later historians will question both stories, but so-called "marathon" races of 25 miles and more will gain popularity in the 20th century A.D. (*see* Boston, A.D. 1897).

Persia's Darius I has 1,000 animals slaughtered each day for the royal table at his capital of Persepolis (or so the historian Xenophon will record: *see* 401 B.C.).

480 B.C. The Battle of Thermopylae August 19 ends in victory for the Persians under Xerxes, son of Darius, whose army of nearly 200,000 engulfs the force of 300 Spartans and 700 Thespians under Leonidas, who enable the main Greek force to escape but are destroyed after holding the Persians at bay. Breaking through the pass at Thermopylae from Macedonia into Greece, the Persians occupy Attica and sack Athens, whose citizens flee to Salamis and the Peloponnese.

The Battle of Salamis September 23 brings victory to the Greeks, whose Athenian general Themistocles lures the Persians into the Bay of Salamis, between Athens and the island of Salamis. More than 1,000 Persian vessels are rammed and sunk by fewer than 400 Greek ships. Xerxes is sent packing back to Persia, but he leaves behind an army under Mardonius.

479 B.C. The Battle of Plataea August 27 ends the Persian invasions of Greece as the Persian commander Mardonius is routed by the Greeks under Pausanius.

475 B.C. Iron comes into use in China nearly 1,000 years after its use became common in the Near East and half a century after its introduction into Europe (*see* 1000 B.C.).

474 B.C. The Greek poet Pindar moves to Thebes at age 44 after 2 years at the Sicilian court of Hieron at Syracuse. "Hopes are but the dreams of those who are awake," writes Pindar, who composes great lyric odes (epincia) to celebrate triumphs in the Olympian games and other athletic events.

472 B.C. Theater *The Persians (Persae)* by Athenian playwright Aeschylus, 53, who gained his first prize for drama in 484 B.C. and who has founded classical Greek tragedy by taking a relatively simple form and infusing it with heroic and unsophisticated magnificence.

470 B.C. Alfalfa is grown by the Greeks, who have been introduced to the plant by the Persians and who use it as fodder for their horses.

468 B.C. The Athenian prize for drama goes to the playwright Sophocles, 28, who defeats Aeschylus in the annual contest. Sophocles will be remembered in ages to come for his *Ajax* and other tragedies.

467 B.C. Theater *Seven Against Thebes* by Aeschylus. Performances of Greek dramas begin at sunrise.

466 B.C. Benghazi is founded on the North African coast by citizens of Cyrenaica, who will make it their capital.

458 B.C. The Roman general Lucius Quinctius Cincinnatus is summoned from his small farm by a delegation from the Senate to defend the city from attack by the approaching Aequians. Cincinnatus is named dictator of Rome, he gathers troops, he attacks and defeats the Aequians, he resigns his dictatorship, and he returns to his farm, all within 16 days (*see* Cincinnati, A.D. 1791).

458 B.C. *(cont.)* Theater Aeschylus' trilogy, the *Oresteia*, includes the plays *Agamemnon*, *The Libation Bearers* (*Choephoroi*), and *The Eumenides*; its story of the blood feud in the house of Atreus will have a powerful influence on future writers and thinkers. Clytemnestra is murdered by her son Orestes for her murder of his father Agamemnon, and although he has been urged to commit the act by the god Apollo, he is pursued by the Furies. Orestes flees to Athens, where the goddess Athena establishes the court of Areopagus, grants forgiveness to Orestes, and changes the name of the Furies to "the kindly ones." Aeschylus will die in 2 years at age 69, but he will be survived through the ages by tragedies that include *Prometheus Bound* and *The Suppliants*.

457 B.C. A 28-year Golden Age begins in Athens as the statesman Pericles makes the city preeminent in the world in architecture and the arts while preparing for the inevitable conflict with Sparta.

Athens has between 75,000 and 150,000 slaves, who represent 25 to 35 percent of the population. Some 20,000 work in the mines at Laureion producing silver which Athens trades for foodstuffs and other imports.

Pericles (above) studies with the philosopher Anaxagoras, 43, who has been teaching at Athens for the past 5 years and who introduces a dualistic explanation of the universe: all natural objects are composed of infinitesimally small particles, or atoms, containing mixtures of all qualities, and the human mind acts upon masses of these particles to produce visible objects.

450 B.C. Celts overrun the British Isles; the Indo-European people have crossed the channel separating the islands from the European mainland.

The Temple of Theseus is completed at Athens.

449 B.C. *History* by the Greek historian Herodotus, 36, gives an account of the 490-479 B.C. Graeco-Persian War along with lengthy digressions giving geographical descriptions of the eastern Mediterranean, relating anecdotes, making anthropological observations, and reporting legends. Herodotus will be called the "father of history" (date approximate).

448 B.C. Rebuilding of the Acropolis at Athens begins under the direction of architects Ictinus and Callicrates, whose work will continue over the next 15 years as Pericles rebuilds what the Persians destroyed in 480 B.C. to make Athens a brilliant city.

445 B.C. The Temple of Poseidon is completed south of Athens at Cape Sunion.

441 B.C. Theater the Athenian prize for drama is won by the playwright Euripides, 43, who made his first effort to win the prize 14 years ago but whose sensational plays have won him more notoriety than approval.

440 B.C. The Greek philosopher Heracleitus at Ephesus in Asia Minor teaches that everything is mutable, "all is flux." Principles are constantly modified through an incontrovertible law of nature that governs the universe in which worlds are alternately being created and destroyed.

Heracleitus (above) is the first to declare that dreams are not journeys into the supernatural but rather retreats into a personal world (*see* Freud, A.D. 1900).

Theater *Antigone* by Sophocles is a tragedy whose heroine defies Creon, king of Thebes, and buries a declared traitor, claiming authority higher than the king's. She is condemned to death and kills herself before Creon, who has changed his mind, can save her or save his son (who has been betrothed to Antigone and who also commits suicide).

438 B.C. The Parthenon on the Acropolis at Athens is completed by Ictinus and Callicrates and is consecrated after 9 years of construction.

435 B.C. A gold and ivory statue of Zeus, king of the gods, is completed at Elis by the Athenian sculptor Phidia. The statue will be called one of the seven wonders of the ancient world.

433 B.C. Pericles concludes a defensive alliance with Corcyra (Corfu), the strong naval power in the Ionian Sea which is the bitter enemy of Corinth. Pericles also renews alliances with the Rhegium and Leontini in the west, threatening Sparta's food supply route from Sicily. Corinth appeals to Sparta to take arms against Athens, and the appeal is backed by Megara (which has been ruined by Pericles's economic sanctions) and by Aegina (which is heavily taxed by Pericles and which has been refused home rule).

432 B.C. The Peloponnesian Wars that will occupy 20 of the next 27 years begin in Greece following a revolt in the spring by the Potidaea in Chalcidice against their Athenian masters. Pericles blockades the Potideae; they refuse to arbitrate. The naval and military demands of the blockade drain Athens; Sparta seizes the opportunity to declare war on Athens; and Corfu declares war on Corinth.

The soldier-scholar Socrates, 22, saves Alcibiades, 18, a nephew of Pericles, at Potidaea.

Spartan troops will lay waste the countryside round Athens in the Peloponnesian Wars (above), destroying not only grain fields but also olive trees, vineyards, and orchards that will not recover for decades.

431 B.C. Sparta's Archidamus II gains support by calling for the liberation of the Hellenes from Athenian despotism, and he sets out to annihilate Athens. Pericles works to make Athens-Piraeus an impregnable fortress, planning to lay waste the Megarid each spring and autumn while the Spartans are occupied with sowing and reaping their own crops.

The idea that the body has four "humors"—blood, bile, black bile, and phlegm—is propounded by the Greek physician Empedocles, whose concept will dominate medical thinking for centuries to come.

Pepper from India is fairly common in Greece, but *Piper nigrum* is used as medicine, not as a food seasoning.

Theater *Medea* by Euripides depicts the reactions of a wife discarded in favor of a younger rival. The fiery barbarian princess destroys her rival and kills her own children by the faithless Jason.

430 B.C. Athens sends a peace mission to Sparta in August but has no success. Potidaea capitulates to Athenian siege forces in the winter, but by that time the Athenian port of Piraeus is in the grip of plague.

Every natural event has a natural cause, says the Greek philosopher Leucippus.

An epidemic of blinding, paralytic deadly fever strikes Piraeus. Possibly a malignant form of scarlet fever that originated in Ethiopia, the plague has symptoms that begin with headache and progress to redness of the eyes, inflammation of the tongue and pharynx, sneezing, coughing, hoarseness, vomiting, diarrhea, and delirium. The plague does not affect the Peloponnese, but the Spartans kill everyone who falls into their hands lest they catch the disease, which rages also in the little Italian town of Rome.

429 B.C. The Athenian admiral Phormio wins naval victories at Chalcis and Naupactus at the mouth of the Corinthian Gulf, but in Athens thousands die of the plague, which kills Pericles in September, ending the Golden Age of Greece.

Plague kills at least one-third the population of Athens (and possibly two-thirds). The entire city indulges in drunkenness, gluttony, and licentiousness as the citizens lose their fear of the gods and respect for law. "As for the first," the historian Thucydides will write, "they judged it to be just the same whether they worshipped them or not, as they saw all alike perishing; and as for the latter, no one expected to live to be brought to trial for his offenses."

Spared by the plague is the physician Hippocrates the Great (as distinguished from one previous and five future Greek physicians named Hippocrates). He is the first to say that no disease is entirely miraculous or adventitious in origin and that disease is not sent as punishment by the gods. He uses dissection and vivisection of animals to study anatomy and physiology, but he often applies the results of his experiments to human bodies without further evidence. Hippocrates adds to medical terminology such words as chronic, crisis, convalescence, exacerbate, paroxysm, relapse, and resolution. Fever, he says, expresses the struggle of the body to cure itself; health results from the harmony and mutual sympathy of the humors (*see* Empedocles, 431 B.C.; Fu Hsi, 2800 B.C.). Hippocrates' cult of Aesculapius is named after a physician who may have lived about 1250 B.C.; it marks the beginning of scientific medicine.

THE HIPPOCRATIC OATH

I swear by Apollo Physician, by Aesculapius, by Health, by Heal-all, and by all the gods and goddesses, making them witnesses, that I will carry out, according to my ability and judgment, this oath and this indenture: To regard my teacher in this art as equal to my parents; to make him partner in my livelihood, and when he is in need of money to share mine with him; to consider his offspring equal to my brothers; to teach them this art, if they require to learn it, without fee or indenture; and to impart precept, oral instruction, and all the other learning, to my sons, to the sons of my teacher, and to pupils who have signed the indenture and sworn obedience to the physicians' Law, but to none other. I will use treatment to help the sick according to my ability and judgment, but I will never use it to injure or wrong them. I will not give poison to anyone though asked to do so, nor will I suggest such a plan. Similarly I will not give a pessary to a woman to cause abortion. But in purity and in holiness I will guard my life and my art. I will not use the knife on sufferers from stone, but I will give place to such as are craftsmen therein. Into whatsoever houses I enter, I will do so to help the sick, keeping myself free from all intentional wrong-doing and harm, especially from fornication with woman or man, bond or free. Whatsoever in the course of practice I see or hear (or even outside my practice in social intercourse) that ought never to be published abroad, I will not divulge, but consider such things to be holy secrets. Now if I keep this oath, and break it not, may I enjoy honor, in my life and art, among all men for all time; but if I transgress and forswear myself, may the opposite befall me.

428 B.C. The revolt of Mitylene in June opens to question the impregnability of the Athenian maritime empire. Mitylene is the chief city of Lesbos.

Theater *Hippolytus* by Euripides shows the irreconcilable conflict between sexual passion and asceticism.

427 B.C. Sparta's Archidamus II dies after a 49-year reign. The Spartan admiral Alcidas, sent to help the rebels on Lesbos, makes a hurried raid on Ionia and flees home after seeing two Athenian warships.

Mitylene surrenders to Athens in July; the Athenians punish the city with cruel severity.

Plataea surrenders in August after its garrison has come close to death from starvation. The Athenians slaughter the Plataeans in cold blood and completely destroy the city at the insistence of Thebes.

426 B.C. Corfu's democratic faction kills supporters of Sparta in a savage massacre and secures the island for Athens.

The Athenian general Demosthenes and the demagogue Cleon revitalize the city's military and naval forces despite opposition from Nicias, a rich merchant who represents the Athenian middle class. Demosthenes outlines a vigorous strategy of offense designed to make Sicily, Boeotia, and the Peloponnese itself spheres of Athenian influence.

Demosthenes proceeds in June to Acarnania with a handful of troops. He raises a large army of local levies with the hope of invading Boeotia by way of Phocis while Nicias (above) invades by way of Tanagra to threaten Thebes from the southeast, but natives in the Aetolian forests trap his army and cut it to pieces. Demosthenes barely escapes with his life; he reaches the Athenian base at Naupactus and secures it just in time to defend it against a large Spartan army from Delphi under Eyrylochus.

Victories by Demosthenes at Olpae and Idomene destroy Peloponnesian and Ambraciot influence on the Ambraciot Gulf. Demosthenes shatters Spartan prestige and returns in triumph to Athens.

425 B.C. An Athenian fleet summoned by Demosthenes bottles up the Spartan navy in Navarino Bay. Demosthenes has built and garrisoned a fort on Pylos promontory and has defended it against the attacking Spartans. A Spartan force landed on the island of Sphacteria is cut off from rescue, the Spartan naval commander surrenders under a temporary armistice, but peace talks break off when Athens refuses to surrender the Spartan ships.

Nicias resigns his generalship, his successor Cleon increases by 50 percent and more the tribute demanded from most members of the Athenian Empire, and Cleon lands reinforcements on Sphacteria (above) to overwhelm the Spartans. He brings 292 heroic Spartan defenders back to Athens and places them in dungeons to safeguard Attica from invasion. Sparta sues for peace, but Cleon refuses.

Theater *Hecuba* by Euripides; *The Acharnians* by Aristophanes, 25, a Greek playwright whose comedy is an attack on war.

424 B.C. A congress at Gela in the spring hears the statesman Hermocrates of Syracuse urge the exclusion of foreign powers. The Sicilians send home an Athenian naval force.

Pagondas of Thebes crushes an Athenian army at Delium, making skillful use of cavalrymen. The Spartan general Brasidas thwarts Athenian efforts to take Megara; he then marches rapidly through Boeotia and Thessaly to Chalcidice, where he offers liberty and protection to cities rebelling against Athens. The city of Amphipolus surrenders, and a naval force from Thasos in the north arrives under the command of Thucydides in time only to save Eion at the mouth of the Strymon (the vengeful Cleon exiles Thucydides for 20 years).

Theater *Oedipus Rex* (or *Oedipus Tyrannus*) by Sophocles (date approximate). Oedipus, king of Thebes, has left his native Corinth to escape fulfillment of a prophecy that he would kill his father and marry his mother, but Oedipus investigates the murder of his predecessor, Laius of Thebes, and discovers that he himself killed Laius, that Laius rather than the king of Corinth was his father, and that his wife Jocasta is also his mother. Jocasta kills herself, and Oedipus blinds himself (the tragedy will deeply affect future generations). *The Knights* by Aristophanes.

Athens and Sparta joined in resisting the Persians but fell to fighting each other in the Peloponnesian Wars.

423 B.C. The Truce of Laches in April is concluded by Athens to check the progress of Sparta's Brasidas who ignores the truce and proceeds to take Scione and Mende in hopes of reaching Athens and freeing the prisoners taken 2 years ago at Sphacteria. Athens sends reinforcements under Nicias, who retakes Mende.

Theater *Maidens of Trachi* by Sophocles (date approximate); *The Clouds* by Aristophanes.

422 B.C. Cleon of Athens meets Brasidas outside the city of Amphipolis; both men are killed.

Theater *The Wasps* by Aristophanes.

421 B.C. The Peace of Nicias April 11 brings a temporary end to the Peloponnesian War, but Alcibiades, nephew of the late Pericles, engineers an anti-Spartan alliance between Athens and the democracies of Argos, Mantinea, and Elis.

Theater *The Peace* by Aristophanes.

420 B.C. Corinth and Boeotia refuse to support last year's Peace of Nicias, and although Athens has released her Spartan prisoners, she has retained Pylos and Nisaea because Sparta has claimed that she is unable to turn over Amphipolus. A new Quadruple Alliance of Athens, Argus, Mantinea, and Elis, organized by Alcibiades, confronts a Spartan-Boeotian alliance in July; middle-class Athenians do not support Alcibiades, but he dominates Athenian life and politics.

Theater *The Suppliant Women* by Euripides.

419 B.C. Sparta's King Agis gathers a strong army at Philus and descends upon Argos by marching at night from the north. His Boeotian forces fail him in the clutch, but he is able to conclude a treaty with Argos.

Theater *Andromache* by Euripides (not produced at Athens). *Electra* by Sophocles, who takes his theme from *The Libation Bearers* by Aeschylus, but whose play is more melodramatic than tragic (date approximate).

418 B.C. The Battle of Mantinea in August is the greatest land battle of the Peloponnesian War and gives Sparta a stunning victory over Argos, which has broken its treaty with Sparta's Agis at the insistence of Alcibiades and his threatened Tegea. Alcibiades has not been reelected general, and Sparta breaks up the Athenian confederation with the unwitting help of Alcibiades' political enemies in Athens.

Athenian forces are turned back in Chalcidice.

Alcibiades urges the Athenians to conquer Syracuse, subdue Sicily, crush Carthage, and thus gain added forces that will enable them to finish the war against Sparta. His bold offensive plan wins the support of the Athenians.

415 B.C. Athens prepares an armada to attack Sicily, but on the eve of its departure the Athenians are given a bad omen: the Hermae busts in the streets of Athens are mysteriously mutilated May 22. Alcibiades is accused of having originated the crime and also of having profaned the Eleusinian mysteries. He demands an immediate inquiry but is ordered to set sail. When he reaches Sicily, Alcibiades is recalled to Athens to stand trial. He escapes to Sparta on the return voyage, learns that he has been condemned to death in absentia, openly joins with the Spartans, and persuades them to send Gylippus to assist Syracuse and to fortify Decelea in Attica.

Athenian forces land at Dascon in Syracuse Great Harbor in November, but their victory is of little use.

Theater *The Trojan Women* by Euripides is presented shortly after the massacre by Athenians of the male population of Melos, which has tried to remain neutral in the Peloponnesian War.

414 B.C. The Athenian fleet and army moves on Syracuse from Catana in April and begins a wall to block Syracuse's land approaches while the fleet blocks approach to the city from the sea, but the Athenian commander Lamachus is killed, the fleet is defeated, and supplies run short. Sparta's Gylippus arrives to strengthen the Syracusans, and Athens responds to appeals from Nicias by sending out 73 vessels in a second armada under the command of Demosthenes.

Theater *The Birds* by Aristophanes is presented as Athens awaits the outcome of its military expedition in Sicily.

413 B.C. Demosthenes arrives at Syracuse in July but is defeated in a nocturnal attack and sustains heavy losses. He urges Nicias to leave, Nicias delays and finally consents, but his soothsayers persuade him to remain after a lunar eclipse August 27 has aroused superstitious fears among the men. The Athenian fleet is bottled up in the harbor and is destroyed in the Battle of Syracuse in September. Demosthenes and Nicias are captured and executed in cold blood after taking to the hills with an army of foot soldiers.

Survivors of the massacre mostly die in the quarries at Syracuse.

Theater *Electra* by Euripides, who puts the legend of Orestes' revenge in modern dress and depicts the ancient matricide as a contemporary crime.

412 B.C. Alcibiades loses the confidence of the Spartans, antagonizes their king Agis, and retires to the court of the Persian satrap Tissaphernes. He advises Tissaphernes to withdraw his support from Sparta while conspiring with the oligarchic party at Athens as Sparta's allied cities break away in a series of revolts.

411 B.C. The discredited democracy of Athens is overthrown in July by the oligarchic extremists Antiphon, Peisander, and Phrynichus, who open secret and treasonable negotiations with Sparta. They are overthrown by the moderate Theramenes, who establishes the "Constitution of the Five Thousand." The Athenian navy recalls Alcibiades from Sardis, his election is confirmed by the Athenians at the persuasion of Theramenes, and a Spartan fleet in the Hellespont at Cynossema is defeated in September by an Athenian fleet.

Theater *Iphigenia in Tauris* by Euripides, whose heroine priestess of Artemis finds that the intended victims of human sacrifice are her brother Orestes and his friend. She outwits the barbarian king and organizes an escape; *Lysistrata* and *The Women at the Thesmophoria* (*Thesmophoriazusae*) by Aristophanes. The women in *Lysistrata* revolt against war by denying their sexual favors to their husbands.

410 B.C. Alcibiades crushes the Spartan navy and its supporting Persian land army in March at Cyzicus in the Sea of Marmora. Sparta makes peace overtures, but the new Athenian demagogue Cleophon lets the opportunity escape.

Democracy is restored in Athens.

409 B.C. Alcibiades recaptures Byzantium, ends the city's rebellion from Athens, clears the Bosphorus, and secures the Athenian supply route for grain from the Euxine (Black) Sea region.

Theater *Philoctetes* by Sophocles, who returns for his theme to the Trojan War of the 12th century B.C.

408 B.C. Alcibiades enters Athens in triumph June 16 after an absence of 7 years. He is appointed general with autocratic powers and leaves for Samos to rejoin his fleet. The Spartan admiral Lysander arrives at Ephesus in the fall and builds up a great fleet with help from the new Persian satrap, Cyrus.

Theater *Orestes* by Euripides, who continues the theme of his 413 B.C. tragedy *Electra*; *The Phoenician Women* by Euripides and other playwrights who have helped to complete the work. Euripides leaves Athens in dissatisfaction and travels to the court of Archelaus in Macedonia, where he will die in the winter of 407–406 B.C. at age 77, leaving incomplete his *Iphigenia at Aulis* but leaving behind also his masterpiece *The Bacchants*. The last great Greek tragedy, it will be produced at Athens by his son.

"The beauty of Helen was a pretext for the gods to send the Greeks against the Phrygians" (in the Trojan War), says Euripides in *Orestes* (above), "and to kill many men so as to purge the earth of an insolent abundance of people." Greek city-states have few resources and are not organized to support large numbers.

407 B.C. Sparta's Admiral Lysander refuses to be lured out of Ephesus to do battle with Alcibiades, who runs out of supplies and is forced to sail north and plunder some enemy coastal towns. Alcibiades leaves behind a squadron under the command of his boyhood friend Antiochus, who violates orders and taunts Lysander. The Spartan fleet responds to the challenge, sallies forth from Ephesus, and routs the Athenians at Notium, administering a defeat that gives the enemies of Alcibiades at Athens an excuse to strip him of his command.

406 B.C. Triremes, under the Spartan admiral Lysander, defeat an Ephesus-based Athenian squadron while Alcibiades is away raising funds. Disgraced, Alcibiades flees to the Hellespont and is replaced by a board of generals, one of whom is bottled up in Mitylene by Callicratidas, a Spartan admiral who has replaced Lysander. Athens raises a huge fleet and sends it off under Admiral Conon to relieve the siege of Mitylene; he is defeated, but returns with a second fleet and gains victory in the Battle of Arginusae. Sparta offers peace, which Cleophon rejects, and Sparta yields to demands by Persia's Cyrus that Lysander command a fleet in the Hellespont.

405 B.C. The Athenian fleet follows Lysander to the Hellespont where it is destroyed in September while drawn up on the beach at Aegospotami; only ten triremes escape under the command of Conon. The Spartan king Pausanius lays siege to Athens while Lysander's fleet blockades Piraeus.

Theater *The Frogs* by Aristophanes: "Alone among the gods, Death loves not gifts." (The Greeks believe the god of the underworld to be implacable and make no efforts to propitiate him.)

The Erectheum is completed in Ionian style on the Acropolis after 16 years of construction. The female figures that support its roof are called Caryatids because they are modeled after girls from the town of Caryae.

404 B.C. Cleophon is tried and executed, Athens capitulates April 25 after being starved into submission, and the First Peloponnesian War is ended. Theramenes secures terms that save the city from destruction, but her long walls are torn down to the sound of flutes, and her empire is dissolved. The oligarchy of the "Thirty Tyrants" takes power under Critias who has Theramenes forced to drink poison hemlock on charges of treason. Alcibiades is murdered in Phrygia at the behest of Sparta.

Plague sweeps Athens as hunger weakens the people's resistance.

Foreign merchants may number half the city's citizenry, and the overall population of 60,000 to 70,000 includes slaves who may number half the total.

403 B.C. The Battle of Munychia ends in victory over Sparta for Thrasybulus, who restores democratic institutions, granting amnesty to all except oligarchic extremists.

401 B.C. Persia's Cyrus the Younger leads a revolt against his brother Artaxerxes II but is defeated and killed by Artaxerxes at Cunaxa. Greek commanders in the rebel army are killed, but the essayist Xenophon, 33, takes command of 10,000 soldiers and leads them back to the Euxine (Black) Sea in a journey that he will describe in his *Anabasis*.

The Greek historian Thucydides dies at age 60, leaving behind an account of the Golden Age of Pericles and of the Peloponnesian War up to 404 B.C. He has tried to be objective and will be called "the father of scientific history."

Theater *Oedipus at Colonus* by the late Sophocles, presented at Athens, is the greatest tragedy of the playwright, who died 5 years ago at age 90. His life spanned the ascendance and decline of the Athenian Empire.

4th Century B.C.

400 B.C. London has its origins on a rise above marshy wastes at the point where the Walbrook joins the Thames River. The Celtic king Belin rebuilds an earth wall surrounding a few dozen huts and orders a small landing place to be cut into the south side of the wall, along the river front, where a wooden quay is built. The watergate cut in the wall to permit entry to the settlement from quayside will be called Belinsgate, a name that will be corrupted to "Billingsgate" (*see* A.D. 43).

399 B.C. The Greek philosopher Socrates is condemned for flouting conventional ideas and for allegedly corrupting the youth with his impiety. Imprisoned at age 70, he obediently drinks a potion made from poison hemlock (*Conium maculatum*) while some disciples, Plato not among them, look on (*see* Plato, 347 B.C.).

396 B.C. The Etruscan city of Veil in southern Etruria falls to Roman forces after a 10-year siege as the Romans begin to end the Etruscan civilization in Italy (*see* 900 B.C.; 509 B.C.).

Smallpox strikes a Carthaginian army besieging Syracuse.

393 B.C. Theater *The Ecclesiazusae* is a bawdy new comedy by Aristophanes.

388 B.C. Theater *Plutus* is a new comedy by Aristophanes.

356 B.C. The Temple of Artemis built at Ephesus beginning in 772 B.C. is burned down by a certain Herostratus, who destroys one of the seven wonders of the ancient world in a perverted bid for immortality.

354 B.C. A tomb enclosed by Ionic columns that will be called one of the seven wonders of the world is built at Halicarnassus in Caria for King Mausolus from whose name the word "mausoleum" will derive.

350 B.C. References to wheat first appear in Greek writings as wheat suitable for bread is introduced from Egypt.

347 B.C. The academy founded by the Athenian philosopher Plato will continue for 876 years. Plato's friends have purchased a suburban grove for the school dedicated to the god Academus. Philanthropists bear all costs; students pay no fees.

Plato's *Republic* will quote Socrates as saying, "Until all philosophers are kings, or the kings and princes of this world have the spirit and power of philosophy, and political greatness and wisdom meet in one, and those commoner natures who pursue either to the exclusion of the other are compelled to stand aside, cities [states] will never have rest from their evils—no, nor the human race."

Plato (above) will express certain medical beliefs along with his political philosophies. The heart is the fountainhead of the blood, says Plato, the liver mirrors the soul, the spleen cleanses the liver, and he introduces the word *anaisthesia* (*see* A.D. 1846).

Plato urges temperance and bewails the changes in the Attic landscape since his youth. Green meadows, woods, and springs have given way to bare limestone partly because the planting of olive trees has led to the ruin of the land (*see* Solon, 594 B.C.).

344 B.C. Aristotle travels from Athens to the Aegean island of Lesbos, where he will spend 2 years studying natural history, especially marine biology. Now 40, Aristotle is a follower of Plato, and his father a physician to the king of Macedon.

342 B.C. Aristotle returns to Macedon at the invitation of her king Philip and begins 7 years of teaching. His pupils will include Philip's son Alexander, now 14.

340 B.C. Philip of Macedon fails in a siege of Byzantium, whose sentries may have seen Philip's advance by the light of a crescent moon and thus have been able to save the city. The Byzantines will adopt the crescent symbol of their goddess Hecate as the symbol of Byzantium (*see* Viennese bakers, A.D. 1217).

338 B.C. Philip of Macedon defeats the Athenians and Thebans in the last struggle for Greek independence August 2 at the Battle of Chaeronea in western Boeotia (*see* Thessalonica, 315 B.C.).

336 B.C. Philip of Macedon is assassinated at Aeges during the wedding feast of his daughter. He is succeeded by his son Alexander, now 20, who will carry out Philip's planned expedition against the Persians.

335 B.C. Aristotle returns to Athens from Macedon and opens a lyceum in an elegant gymnasium dedicated to Apollo Lyceus, god of shepherds. The lyceum contains a museum of natural history, zoological gardens, and a library.

Aristotle (above) attempts to develop a deductive system as comprehensive as is possible with the scientific materials available. He concerns himself chiefly with the anatomical structures of animals, their reproduction, and their evolution, and he founds the study of comparative anatomy in an effort to categorize animal life into biological groups.

Aristotle describes various parts of the digestive canal in some detail, but his ideas of physiology are primitive in the absence of any chemical knowledge. Everything

Philip of Macedon's assassination began the reign of his son Alexander, later called Alexander the Great.

in life is subject to basic law, says Aristotle, but he believes that food is "cooked" in the intestinal tube and praises garlic for its medicinal qualities.

Aristotle (above) advises abortion for parents with too many children, and he writes in *Politics* that ". . . neglect of an effective birth control policy is a never failing source of poverty which in turn is the parent of revolution and crime." But the Greeks (and later the Romans) will encourage large families lest they have a dearth of recruits for their armies.

334 B.C. Alexander of Macedon invades Asia with an army that includes 5,000 mercenaries. The opposing Persian army includes 10,000 Greek mercenaries.

333 B.C. The Battle of Issus in October gives Alexander of Macedon a great victory over the Persians, but the emperor Darius III escapes.

332 B.C. Alexander the Great of Macedon takes over Egypt and founds a city that will be called Alexandria (*see* lighthouse, 285 B.C.).

331 B.C. The Battle of Arbela (or Gaugamela) October 1 in northern Mesopotamia gives Alexander the Great another victory over Darius III, who loses 40,000 to 90,000 men against Macedonian losses of 100 to 500. Alexander becomes master of the Persian Empire, ending the Achaemenid dynasty founded in 550 B.C.

Wheat is grown extensively in southeastern parts of the British Isles and is threshed under great barns, reports a traveler from the Greek colony at Marsilea that will later be called Marseilles (*see* Lacydon, 600 B.C.).

330 B.C. The Persian king Darius III is murdered by his satrap Bessus after a 6-year reign. Alexander the Great sacks the Persian capital of Persepolis. It takes 20,000 mules and 5,000 camels to carry off the loot.

The atomic theory developed by the Greek philosopher Democritus says that all matter is composed of tiny atomic particles, the word "atom" meaning unbreakable or indivisible (*see* Thomson, A.D. 1897). Nothing happens through chance or intention, says Democritus; everything happens through cause and of necessity. All change is merely an aggregation or separation of parts, nothing which exists can be reduced to nothing, and nothing can come out of nothing. Democritus distinguishes between vertebrate and invertebrate animals, both of which he dissects.

329 B.C. Alexander the Great conquers Samarkand (Maracanda), capital of Sogdiana, in central Asia.

327 B.C. Alexander the Great invades northern India after having gained ascendancy over all of Greece, occupied Egypt, and destroyed the power of Persia. But Alexander is persuaded to abandon his plans for invading the Ganges Valley by his army, which is tired of campaigning.

Alexander appoints Nearchus as admiral and places under his command all in the ranks with any knowledge of seafaring. Nearchus has Indian shipwrights build 800 vessels, some as large as 300 tons, and uses Indian pilots to guide his fleet through Persian Gulf waters to Babylonia.

Bananas (*Musa sapientum*) are found growing in the Indus Valley by Alexander the Great (*see* A.D. 1482).

325 B.C. The *Persae* by Timotheus of Miletus will survive as the earliest papyrus written in Greek.

The first known reference to sugar cane (*Saccharum officinarum*) appears in writings by Alexander's admiral Nearchus, who writes of Indian reeds "that produce honey, although there are no bees." The word "sugar" (adapted from the Arabic *sukhar*, which derives from the Sanskrit *sarkara*, meaning gravel or pebble) begins to appear frequently in Indian literature (*see* 300 B.C.).

When a satrap murdered the Persian emperor Darius III, Alexander the Great was moved to tears.

The poor of Athens exist mainly on beans, greens, beechnuts, turnips, wild pears, dried figs, barley paste, and occasional grasshoppers, with only sporadic welfare assistance.

324 B.C. Theater *A New Comedy* pioneered by the Athenian playwright Menander, 19, employs lighthearted humor rather than the virulent personal and political satire of the late Aristophanes and has realistic plots and characters based on the domestic life of ordinary citizens. Son of the rich Diopeithes of Cephisia, Menander will win his first dramatic prize in 316 B.C.

323 B.C. Alexander the Great dies at Babylon at age 32, and a 42-year struggle begins that will be called the Wars of the Diadochi (successors). Alexander's generals, Antigonus, Antipater, Seleucus, Ptolemy, Eumenes, and Lysimachus, contest control of the Macedonian Empire.

The Museum of Alexandria is founded by Ptolemy (above), who takes over Egypt. Like Alexander, he has studied under Aristotle, and he will staff the museum with some 100 professors paid by the state.

322 B.C. *Politics IV* by Aristotle says, "When there are too many farmers the excess will be of the better kind; when there are too many mechanics and laborers, of the worst."

321 B.C. The Battle of the Caudine Forks (Caudium) brings defeat to a Roman army trapped by the Samnites in a pass near Beneventum. The Romans are forced to pass under the yoke (a horizontal spear placed atop two upright spears) as a symbol of submission.

319 B.C. Alexander's general Antipater falls fatally ill and names as his successor the aged regent Polysperchon, whose authority is challenged by Antipater's son Cassander, 31.

317 B.C. Armenia's Persian satrap Ardvates frees his country from Seleucid control (*see* 284 B.C.)

316 B.C. Macedonia's regent Polysperchon is defeated and overthrown by Cassander, who seizes Olympias, mother of the late Alexander the Great, has her put to death, and marries Thessaloniki, half sister of Alexander, with whom he will rule until 297 B.C.

Eumenes and Antigonus, rivals to Cassander for control of Macedonia, meet in battle in Media, with Eumenes commanding a force of 36,700 foot soldiers, 6,050 cavalrymen, and 114 elephants against 22,000 foot soldiers, 900 horsemen, and 65 elephants for Antigonus. But cavalrymen sent out by Antigonus take advantage of cover provided by dust raised by the elephants and seize the baggage camp of Eumenes, whose cavalrymen desert. Antigonus offers to return the baggage camp and the wives he has captured if the enemy will desert and hand over Eumenes, who is put to death by his guard after a week's captivity.

315 B.C. The Macedonian port city of Thessalonica is founded by Cassander, whose wife Thessaloniki was named by her father Philip II of Macedon to commemorate his victory (Niki) over Thessaly in 338 B.C.

314 B.C. Antigonus promises freedom to the Greek cities in a bid to gain support against Cassander. The Aetolians enter into an alliance with Antigonus; Cassander marches against them with his allies Lysimachus, Ptolemy, and Seleucus.

312 B.C. The Battle of Gaza brings triumph to Ptolemy and Seleucus over the one-eyed Antigonus (called Antigonus Cyclops, or Monophthalmos), who is captured but is immediately released.

The Roman censor Appius Claudius Caecus begins construction of the Appian Way.

Rome gets its first pure drinking water as engineers complete an aqueduct into the city.

310 B.C. Cassander has imprisoned Roxana, widow of the late Alexander the Great, and has her put to death along with her young son Alexander IV.

308 B.C. Egypt's Ptolemy is defeated in a naval battle off Cyprus by Demetrius Poliorcetes, 29, son of Antigonus Cyclops, who was defeated with his father at Gaza in 312 B.C.

307 B.C. Rhodes is besieged by Demetrius Poliorcetes, who employs 30,000 workmen to build siege towers and engines, including the tower Helepolis that requires 3,400 men to move and a 180-foot ram that is moved on wheels by 1,000 men. The siege will fail.

305 B.C. The Seleucid Empire that will rule Babylonia and Syria until 64 B.C. is established by Seleucus, now 53, who takes the title Nicator and continues conquests that will extend the empire to the Indus River.

Egypt's governor Ptolemy makes himself king, beginning a reign as Ptolemy I Soter ("Savior") that will continue until 285 B.C.

301 B.C. The Battle of Ipsus in Phrygia ends the ambitions of Antigonus, who assumed the title of king 5 years ago, has invaded Egypt, but is slain at age 81 by the forces of Lysimachus and Seleucus, who defeat his son Demetrius.

Nomad tribes begin to occupy parts of northern China.

On Stones by the Greek philosopher Theophrastus mentions fossil substances "that are called coals, . . . found in Liguria and in Elis, on the way of Olympias, over the mountains . . . which kindle and burn like woodcoals. . .; they are used by smiths" (*see* A.D. 852).

History of Plants and *Theoretical Botany* by Theophrastus (above) mention plant diseases such as rusts and mildews and describe "caprification" of figs.

The Chinese build a vast irrigation system to reduce the flooding of Szechwan's Red Basin.

Indigenous Chinese regard the dairy products of the nomad tribes (above) as unhygienic.

The Athenian philosopher Epicurus extols luxury and indulgence in eating and drinking. Pleasure is the only good and the end of all morality, says Epicurus, but a genuine life of pleasure must be a life of prudence, honor, and justice.

3rd Century B.C.

300 B.C. Carthage in North Africa gains economic ascendancy in the Mediterranean by trading in slaves, Egyptian linen, products of the African interior that include ivory and animal skins (lion and leopard), Greek pottery and wine, iron from Elba, copper from Cyprus, silver from Spain, tin from the British Isles, incense from Arabia, and purple-black dyestuffs from Tyre.

Carthaginians are the world's greatest shipbuilders, sailing the seas in quinquiremes—ships with five banks of oars manned by well-drilled government-owned galley slaves.

Elements by the Greek mathematician Euclid is a 13-volume work that states the principles of geometry for the first time in formal style (*see* 1500 B.C.). Euclid has founded a school at Alexandria.

Carthaginian planters own fertile lands in Libya; some have as many as 20,000 slaves.

Sugar from India is introduced to the Middle East, where it is planted in areas wet enough to support its growth (*see* 325 B.C.; A.D. 1099).

297 B.C. Macedonia's Cassander dies at age 53, and Demetrius Poliocertes returns to Greece with the aim of becoming master of Macedonia and Asia.

295 B.C. Athens falls to Demetrius after a bitter siege, and its tyrant Lachares is destroyed.

The Battle of Sentinum west of Anconum ends in defeat for Samnites and Gauls at the hands of Roman legions, who lose nearly 8,000 dead but kill some 25,000 of the enemy and force peace on the Etruscans.

289 B.C. The Chinese philosopher Mencius (Meng Zu) dies at age 83 (?) after decades of trying to unify China's kingdoms (*see* 246 B.C.).

288 B.C. Demetrius is driven out of Macedonia by Lysimachus and Pyrrhus, king of Epirus, after Seleucus, Ptolemy, and Lysimachus have formed a coalition to block plans by Demetrius to invade Asia.

285 B.C. Egypt's Ptolemy Soter abdicates at age 82 after a 38-year reign that has founded a dynasty which will rule until 30 B.C. He is succeeded by his son, 24, who will rule as Ptolemy II Philadelphus until 246 B.C., first with the daughter of Lysimachus as his wife and thereafter (from 276 B.C.) with his own sister as his wife.

Demetrius Poliocertes is deserted by his troops and surrenders to Seleucus, who will keep him prisoner until his death in 283 B.C.

A 300-foot-tall lighthouse on the island of Pharos in Alexandria's harbor serves as a landmark for ships in the eastern Mediterranean. Light from its wood fire, reflected by convex mirrors at its top, can be seen for miles. Built by Sostratus of Cnidus, it is one of the seven wonders of the ancient world and will remain an important navigational aid for 1600 years.

284 B.C. Armenia's satrap Ardvates dies after a 33-year reign, having founded a dynasty that will rule until 211 B.C.

281 B.C. Seleucus Nicator defeats (and kills) Lysimachus at the Battle of Corupedium and makes himself king of Syria.

280 B.C. Seleucus Nicator tries to seize Macedonia but falls into a trap set by Ptolemy Ceraunus, who murders Seleucus and takes Macedonia for himself (*see* 307 B.C.; 279 B.C.). Seleucus Nicator is succeeded by Antiochus I Soter, who will defeat the Galatians (by terrifying them with elephants), reigning until 261 B.C. but losing Miletus, Phoenicia, and western Cilicia to Macedonia's Ptolemy II in the Damascene War (280 B.C.–279 B.C.) and First Syrian War (276 B.C.–272 B.C.).

King Pyrrhus of Epirus defeats a Roman army at Asculum and says, "Another such victory and we are ruined." A triumph that has cost the victor more than the vanquished will be called a pyrrhic victory.

The Achaean League formed by 12 towns in the northern Peloponessus will grow to include non-Achaean cities (including Corinth, in 243 B.C.). It has two generals, a federal council with proportional representation of members, and an annual assembly of all free citizens (*see* 146 B.C.)

The Colossus of Rhodes, completed by the sculptor Chares of Lindus after 12 years' work, is a bronze statue of the god Helios. Made from spoils left by Demetrius Poliocertes when he raised his siege in 305 B.C., it rises to a height of 120 feet above the harbor that has grown rich in the slave trade. Ships can pass between its legs, and it will stand as one of the seven wonders of the ancient world until an earthquake shatters it in 224 B.C.

279 B.C. Celtic tribesmen invading Macedonia kill Ptolemy Ceraunus, but fierce mountain tribesmen (Phocians, Aetolians) force them to move east.

275 B.C. The Museum of Alexandria, founded by Ptolemy II, maintains and supports scholars from all countries and encourages them in their research into all branches of known science. The museum will be the leading Greek university.

The Museum of Alexandria employs knowledge gained by the Egyptians in the practice of embalming to expand knowledge of anatomy and physiology. The museum's leading medical professor is the Greek Hippocratist Herophilus of Chalcedon, who scorns the traditional fear of dissecting human bodies and who conducts postmortem examinations that enable him to describe the alimentary canal (he gives the duodenum its name), the liver, the spleen, the circulatory system, the eye, the brain tissues, and the genital organs. Herophilus is the first to make a distinction between sensory nerves and motor nerves, and he founds the first school of anatomy.

272 B.C. Egypt's Ptolemy annexes Miletus, Phoenicia, and western Cilicia after defeating his rebellious half brother Magas and the Seleucid emperor Antiochus I Soter in the First Syrian War.

Rome's war with Pyrrhus of Epirus ends after 10 years as Tarentum is surrendered to the Romans.

270 B.C. Rome's subjugation of Italy is completed by the recapture of Rhegium from the Mamertines and the defeat of the Brutians, the Lucanians, the Calabrians, and the Samnites.

268 B.C. The Roman denarius is minted for the first time. The silver coin will become familiar throughout the Western world (*see* A.D. 81).

267 B.C. Macedonia's Antigonus II Gonatus quells a rebellion by an Athens-led Hellenic coalition of Spartans, Arcadians, and Achaeans that has tried to expel the Macedonian garrison (*see* 264 B.C.).

265 B.C. The Archimedian screw for raising water is devised by the Greek mathematician Archimedes, 22, who is studying at Alexandria. A native of Syracuse, he says he could move the earth if he had a lever long enough and a fulcrum strong enough ("Give me where to stand and I will move the earth").

Archimedes (above) has discovered ("Eureka!") the law of specific gravity while sitting in his bathtub: a body dropped into a liquid will displace an amount of liquid equal to its own weight. He considers his ingenious mechanical contrivances beneath the dignity of pure science.

264 B.C. A Punic War embroils Rome in a conflict with Carthage that will continue for 23 years (*see* 260 B.C.).

Gladiatorial combat gains huge popularity as a spectacle in Rome.

262 B.C. Athens surrenders after a long siege to the Macedonian forces of Antigonus Gonatas.

260 B.C. The Battle of Mylae off the north coast of Sicily gives Rome her first naval victory over Carthage. The Roman admiral Gaius Duilius Nepos commands quinquiremes modeled after a Carthaginian ship found stranded on the Italian coast, and he uses grappling irons and boarding bridges to revolutionize naval warfare and to defeat a larger, more maneuverable Carthaginian flotilla.

Buddhism is adopted by the third emperor of India's Mauyra dynasty, which arose following the confusion of Alexander the Great's invasions in the last century. The emperor Asoka establishes India's first hospitals and herbal gardens and places them under Buddhist control in opposition to the Hindu Brahmins (*see* 528 B.C.; China, A.D. 517).

249 B.C. The Zhou (Chou) dynasty that has ruled much of China for nearly 9 centuries ends as the last Zhou emperor is deposed (*see* 246 B.C.).

246 B.C. Modern China has her beginnings in the Qin (Chin) dynasty founded by Qin Shihuang, 28. The bastard son of a prostitute by a merchant, Shihuang will prove himself a brilliant general (*see* 221 B.C.).

Egypt's Ptolemy II dies at age 63 and is succeeded after a peaceful reign of 39 years by his son, 36, who will reign less peacefully until 221 B.C. as Ptolemy III.

Ptolemy III invades Syria, seeking vengeance for the death of his murdered sister Berenice. He throws his armies against Seleucus II.

245 B.C. Babylon and Susa fall to the Egyptian armies of Ptolemy III.

243 B.C. Ptolemy III is recalled from Syria by a revolt in Egypt and ceases his martial interests and his support of the Egyptian army.

237 B.C. A Carthaginian army under Hamilcar Barca, 33, invades the Iberian Peninsula.

228 B.C. Carthage's Gen. Hamilcar Barca falls in battle. Command of his army in the Iberian Peninsula passes to his son-in-law Hasdrubal.

223 B.C. Mesopotamia's Seleucid king Seleucus III Soter is murdered during a war with Pergamum and succeeded by his brother who will reign until 187 B.C. as Antiochus III.

222 B.C. Mediolanum (Milan) falls to Roman legions in Lombardy. The town has been occupied successively by Ligurians, Etruscans, and Celts.

221 B.C. The Carthaginian general Hasdrubal is assassinated. Command of the troops is assumed by Hannibal, 26, a son of the late Hamilcar Barca, and his brother Hasdrubal.

Egypt's Ptolemy III dies at age 61 after a 25-year reign. He is succeeded by his son, 23, who will rule with his sister-wife Arsinoe III until 203 B.C. as Ptolemy IV; court favorites will dominate the reign.

China's Qin emperor Shihuang unites the country after 25 years of fighting. He has conquered 6 warring states and his short-lived dynasty will extend the country's waterworks, build a network of roads, and raise a great defensive wall (*see* 214 B.C.).

Egyptian medical studies at Alexandria are supported by Ptolemy IV (above), who is weaker than his predecessors but devoted to the pursuit of science.

Alexandrian medical science is headed by Herophilus and his rival Erasistratus, formerly court physician to the Seleucides of Syria. Erasistratus gives heart valves the names they will henceforth carry, he establishes the connection between arteries and veins, he investigates the lymphatic ducts, he expands knowledge of the nervous system (distinguishing between motor nerves and

221 B.C. *(cont.)* sensory nerves), and he describes in detail the convolutions of the brain.

220 B.C. The Flaminian Way is completed between Rome and Rimini.

219 B.C. Antiochus III of Syria seizes the province of Coele-Syria from Egypt, initiating a Fourth Syrian War.

218 B.C. A second Punic War begins as a Carthaginian army under Hannibal attacks Rome's Hispanic allies. He besieges the town of Sagunto, whose inhabitants eat their own dead rather than surrender but are eventually forced to yield. He crosses the Alps and defeats Roman forces at the Ticino River and again at the Trebbia River.

217 B.C. The Battle of Lake Trasimene in Umbria June 24 ends in victory for Hannibal, who nearly destroys a large Roman army led by Gaius Flaminius. Carthaginians and Gauls kill some 16,000 Romans, including Flaminius, and turn the lake red with blood.

Egyptian hoplites under Ptolemy IV Philopater crush the Seleucid army at Raphia.

216 B.C. The Battle of Cannae August 2 ends in victory for Hannibal, whose 40,000-man army defeats a heavily armored and unmaneuverable Roman force of 70,000. Some 50,000 Roman and allied troops are butchered, 10,000 are taken prisoner, but Hannibal lacks the catapults and battering rams needed to besiege Rome and contents himself with laying waste the fields of Italy, forcing Rome to import grain at war-inflated prices.

Rome has been deserted by her allies and starves out Capua and loots Syracuse as an object lesson to other allies.

215 B.C. Theater *The Menaechmi* by Roman playwright Titus Maccius Plautus, 39 (dates of all Plautus' plays are conjectural).

Armored elephants helped Hannibal's Carthaginians defeat superior Roman numbers in the Second Punic War.

214 B.C. Construction begins on a great Chinese defensive wall to keep out the Mongol tribesmen who menace the territories of Qin (Chin) emperor Shihuang. Extending 2,500 miles from Mongolia to the sea, the Great Wall is made initially of earthwork but will later be of brick.

Theater *The Merchant (Mercator)* by Titus Maccius Plautus.

213 B.C. Theater *The Comedy of Asses (Asinaria)* by Titus Maccius Plautus: "Such things are easier said than done" (I, iii); "Man is a wolf to man" *(Homo homini lupus,* II, iv).

212 B.C. The Qin emperor Shihuang burns writings by dissidents and has some scholars buried alive. Shihuang, who has 3,000 concubines, saves only works on medicine, agriculture, and astrology (but other books, hidden by priests and scholars, will survive).

211 B.C. The Seleucid emperor Antiochus III removes Armenia's king Xerxes by treachery and divides the country into 2 satrapies (*see* 190 B.C.).

210 B.C. China's Qin emperor Shihuang dies at age 49 after a 36-year reign that has united the many petty states of the old Zhou dynasty, created 36 provinces with a uniform system of laws, weights, and measures, greatly expanded the empire, and kept 700,000 conscripts employed building the Great Wall and the new capital of Xian, but he has had most of the country's books collected and burned. Buried in his tomb are 8,000 life-size terra cotta figures of soldiers (each with a unique face) and horses.

207 B.C. The Battle of Metaurus in Umbria ends Hannibal's hopes of success in Italy. A Carthaginian army under Hannibal's brother Hasdrubal is defeated by the Romans under the consuls Claudius Nero and Livius Salinator. Hasdrubal is killed in the battle.

204 B.C. Roman forces under P. Cornelius Scipio (Scipio Africanus) besiege Carthage. Carthaginians immolate 100 boys of noble birth in an effort to propitiate the god Moloch to raise the Roman siege.

202 B.C. The Han dynasty that will rule China for more than 4 centuries is inaugurated as the last Qin emperor dies and one of his minor officials assumes power.

The Battle of Sama October 19 ends the Second Punic War and largely destroys the power of Carthage. Scipio Africanus defeats a combined army of Carthaginians and Numidians under the command of Hannibal and forces Carthage to capitulate.

Theater *The Casket (Cistellaria)* by Titus Maccius Plautus.

201 B.C. Carthage surrenders all her Mediterranean possessions to Rome, her Iberian territories included. The Carthaginians agree to pay Rome 200 talents per year for 50 years, make no war without Rome's permission, and destroy all but 10 Carthaginian warships (*see* 150 B.C.).

The Battle of Chios ends in the defeat of Philip V of Macedon by Rhodes and Attalus of Pergamum.

2nd Century B.C.

200 B.C. The Battle of Panium gives the Seleucid forces of Antiochus III a decisive victory over Egypt's young Ptolemy V Epiphanes in the Fifth Syrian War.

Theater *Stichus* by Titus Maccius Plautus: "An unwilling woman given to a man in marriage is not his wife but an enemy."

197 B.C. The Battle of Cynoscephalae in Thessaly gives a Roman army under T. Quinctius Flaminius victory over Philip V of Macedon. The Romans will force Philip to surrender Greece, reduce his army to 5,000 and his navy to 5 ships, promise not to declare war without Rome's permission, and pay Rome 1,000 talents in 10 years.

196 B.C. Theater *The Haunted House (Mostellaria)* by Titus Maccius Plautus: "Things which you don't hope happen more frequently than things which you do hope" (I, iii). Plautus' play *The Persian (Persa)* will appear next year.

194 B.C. Theater *The Pot of Gold (Aulularia)* and *The Rope (Rudens)* by Titus Maccius Plautus, whose play *The Weevil (Curculio)* will appear next year.

192 B.C. Syrian forces under Antiochus III invade Greece at the invitation of the Aetolians.

191 B.C. Roman forces at Thermopylae rout Antiochus III, repeating the maneuver executed by Xerxes in 480 B.C.

Cisalpine Gaul becomes a Roman province.

Theater *Pseudolus* by Titus Maccius Plautus: "If you utter insults you will also hear them" (IV); *Epidicus* by Plautus.

190 B.C. The Battle of Magnesia near Smyrna gives the Romans another victory over Antiochus III. L. Cornelius Scipio (soon to be called Scipio Asiaticus) and his brother Scipio Africanus have crossed the Hellespont to pursue Antiochus. They force the Syrian to surrender all his European and Asiatic possessions as far as the Taurus Mountains, to pay 15,000 talents over a period of 12 years, and to surrender Hannibal (who will escape).

The two Armenian satraps of Antiochus make themselves independent; their descendants will reign Armenia Major and Minor separately until 94 B.C.

189 B.C. Theater *The Two Bacchaides* by Plautus: "He whom the gods love dies young, while he is in health, has his senses and his judgment sound" (IV, vii).

188 B.C. Theater *The Captives (Captivi)* by Titus Maccius Plautus: "It is the nature of the poor to hate and envy men of property" (III); "All men love themselves" (III).

187 B.C. Theater *Three-Penny Day (Trinummus)* by Titus Maccius Plautus: "Keep what you have; the known evil is best" (I, iii); "What you lend is lost; when you ask for it back, you may find a friend made an enemy by your kindness" (IV, iii).

186 B.C. Theater *Amphitryon (Amphitruo)* by Titus Maccius Plautus: "If anything is spoken in jest, it is not fair to turn it to earnest" (III), *Truculentus* by Plautus.

185 B.C. Theater *Casina* by Titus Maccius Plautus, who will die in 184 B.C. at age 70.

Troops returning from the war with Syria's Antiochus III introduce eastern indulgence to Rome.

183 B.C. Hannibal poisons himself at the court of Bithynia's Prusia I, who was about to betray him to the Romans.

Pisa and Parma become Roman colonies.

179 B.C. Rome's Pons Aemilius, completed across the Tiber, is the world's first stone bridge.

172 B.C. A Roman army is defeated by the Macedonian king Perseus, who succeeds his father Philip V and is attacked by the Romans, beginning a war that will continue until 168 B.C.

170 B.C. The world's first paved streets are laid out in Rome. The new streets are passable in all weather and easier to keep clean, but they add to the din of traffic.

Rome's first professional cooks appear in the form of commercial bakers, but most Roman households continue to grind their own flour and bake their own bread.

168 B.C. The Battle of Pydna gives Roman forces a victory over the Macedonian king Perseus, who has succeeded Philip V. The Roman general Lucius Aemilius Paulus, 51, returns with Perseus in his triumphal procession.

Macedonians captured at Pydna (above) are sold into slavery at Rome. Females fetch as much as 50 times the price of a male.

Syria's Seleucid king Antiochus IV Epiphanes outlaws Judaism, lays waste the Great Temple at Jerusalem, and tries to hellenize the Jews by erecting idols to be worshipped by the people of Judea.

Huge amounts of booty brought home by Paulus after the Battle of Pydna (above) enrich the Roman treasury. Rome relieves her citizens of direct taxation (the *tributum*).

167 B.C. The Jewish priest Mattathias of Modin defies Syria's Antiochus IV Epiphanes, who has outlawed Judaism. Mattathias escapes into the mountains outside Lydda with his five sons and begins a revolt. He will die

167 B.C. *(cont.)* in 166 B.C., but his sons will continue the revolt, and his third son Judah will receive the surname Maccabaeus—the Hammerer.

166 B.C. Theater *The Women of Andros (Andria)* by Roman playwright Terence (Publius Terentius Afer), 24, with flute music by his fellow-slave Flaccus at the Megaleusian games in April. "Obsequiousness makes friends; truth breeds hate" (I, i).

165 B.C. Judah Maccabee (Maccabaeus) and his brothers retake Jerusalem from the Syrians.

The Maccabees (above) cleanse the Great Temple, destroy the idols erected by Antiochus IV, and restore the monotheistic religion of Judaism. It will later be said that the Maccabees found only enough oil in the temple to keep a light burning for one day, but that somehow the oil lasted for 8 days, and Jews will commemorate the event in the annual Feast of Dedication called Hanukkah.

Theater *The Mother-in-Law (Hecyra)* by Terence, who has adopted a play by Apollodorus.

163 B.C. Antiochus IV of Syria dies after a 12-year reign and is succeeded by his son, 10, who will reign briefly as Antiochus V under the regency of Lysias, who will make peace with the Jews.

Theater *The Self-Avenger (Heautontimorumenos)* by Terence: "I am a man; nothing human is foreign to me" *(Homo sum; humani nihil a me alienum puto,* I, i).

162 B.C. Syria's Antiochus V is overthrown and killed by his cousin Demetrius I Soter, who will reign until 150 B.C.

161 B.C. Theater *The Eunuch (Eunuchus)* by Terence: "I know the nature of women; / When you want to, they don't want to;/ And when you don't want to, they desire exceedingly" (IV, vii), *Phormio* by Terence: "A word to the wise is sufficient" *(Dictum sapienti sat est,* III, iii).

160 B.C. The newly appointed governor of Judea is killed in battle by the Maccabees, whose leader Judah Maccabee is killed soon after at the Battle of Elasa. Judah is survived by his older brother Simon and his youngest, Jonathan, who succeeds as leader and will make Judea a nearly independent principality by the time of his death in 143 B.C.

Theater *Brothers (Adelphoe)* by Terence: "It is better to bind your children to you by a feeling of respect, and by gentleness, than by fear" (I, i). A Carthaginian who was brought to Rome as a slave for a senator, Terence will die next year.

153 B.C. January 1 becomes the first day of the civil year in Rome. An uprising in Rome's Spanish provinces obliges Roman consuls to take office earlier than the traditional date of March 15.

150 B.C. "Carthage must be destroyed" *(Delenda est Carthago),* says the Roman censor Marcus Porcius Cato, 84, who has been urging destruction of the prosperous Punic city since 157 B.C., when he helped arbitrate a truce between Carthage and her former ally Masinna, 88, king of Numidia. Masinna is now Rome's ally, Carthage has attacked Numidia, and Cato demands that a Roman army be sent against Carthage.

The Syrian usurper Alexander Balas, who claims to be a son of the late Antiochus IV Epiphanes, defeats Demetrius I Soter in battle and kills him. The Romans support him, and he will rule Syria until 145 B.C.

149 B.C. A Roman army invades North Africa and lays siege to Carthage. The Carthaginians offer to submit but refuse to vacate their city.

Cato the Elder dies at age 85, leaving as his legacy some commentaries on agriculture. *De Agriculture* (or *De Re Rustica)* urges farmers to plant grapes and olives that draw moisture and nutrients from the subsoil rather than grain, which is more subject to drought.

146 B.C. Carthage falls to Roman legions led by Scipio Aemilianus in 6 days and nights of house-to-house fighting after a long blockade. Some 900 Roman deserters torch the Temple of Aesculapius and choose death by fire rather than execution. Hasdrubal surrenders his garrison, and his wife contemptuously throws herself and her children into the flames of the temple. The city's ashes are plowed under, its environs become the Roman province of Africa, and the Third Punic War is ended.

Corinth is sacked by legions under the Roman general Mommius on orders by the Senate to replace all democracies with oligarchies, to smash the 134-year-old Achaean League, and to place Greece under the supervision of the governor of Macedon, which becomes a Roman province.

Parthia's Mithridates I defeats Seleucid forces to conquer Babylonia and Media. He makes Ctesiphon-Seleucia his capital.

Scythian warriors (the Tochari) invade the Seleucid satrapy of Bactria.

Some 50,000 Carthaginians—men, women, and children—are sold into slavery.

145 B.C. Syria's Alexander Balas falls in battle near Antioch, his forces flee from Demetrius II and Ptolemy VI Philometor, and his son by Cleopatra Thea succeeds to the throne. The boy will rule under a regent as Antiochus VI until 142 B.C.

143 B.C. The Syrian usurper Tryphon traps Judea's Jonathan Maccabee and kills him at Bethshean. Jonathan's older brother Simon succeeds him and will drive all the Syrians out of the citadel at Jerusalem.

142 B.C. Syria's boy-king Antiochus VI dies and is succeeded by the son of Demetrius I Soter, who will reign as Demetrius II Nicator.

Judea gains independence from Syria under the leadership of Simon Maccabee. He sends an embassy to Rome and begins coinage of money.

141 B.C. Jewish forces under Simon Maccabee liberate Jerusalem while the Seleucid emperor Demetrius II Nicator is preoccupied with conquering Babylonia. Judean independence will last until 63 B.C.

The Maccabee brothers and their followers will improve agriculture in the Jerusalem region.

140 B.C. China's Han dynasty emperor Wu Di begins a 53-year reign in which he will extend the empire to the south, annex parts of Korea and Tonkin, and send his emissary Jang Qian halfway round the world to Bactria and Sogdiana to seek an east-west alliance against the Huns, or Hsiung Nu.

News of Jang Qian (above) and of Serica, "land of silk," will reach Rome, and caravans will begin to carry the first apricots and peaches (the "Chinese fruit") to Europe while Jang Qian introduces grapes, pomegranates, and walnuts to China.

135 B.C. Rome's first Servile War begins as slaves on the large Sicilian estates revolt under the Syrian Eunus, who calls himself King Antiochus and holds Henna and Tauromenium against Roman armies sent to subdue the insurrection.

134 B.C. Judea's Simon Maccabee is treacherously murdered along with 300 of his followers by his son-in-law, governor of Jericho. Simon's sons Mattathias and Judah are also killed, but he is succeeded by his surviving son John Hyrcanus, who will rule Judea until 104 B.C., extending the kingdom to include Samaria, Idumaea, and lands east of the Jordan.

133 B.C. Tiberius Sempronius Gracchus, 30, is elected Roman tribune on a platform of social reform. He proposes an agrarian law that would limit holdings of public land to 312 acres per person, with an additional 250 acres for each of two sons, but large landholders in Etruria and Campania block efforts to recover lands held in violation of the new law, and Gracchus is murdered. The great estates *(latifundia)* are not distributed among new settlers but grow at the expense of the small peasants and to some extent of Rome's urban proletariat.

132 B.C. Rome's Servile War is ended as Roman forces capture the Syrian Eunus and savagely execute him and his supporters. More than 70,000 slaves are believed to have taken part in the uprising; 20,000 are crucified.

129 B.C. Pergamum becomes the Roman province of Asia.

Rome's Scipio Aemilianus is found dead after favoring concessions to the Italian peasants who have grown increasingly bitter at their treatment by Roman landowners. Murder is suspected.

123 B.C. Gaius Sempronius Gracchus, 30, is elected Roman tribune on a platform similar to that of his late brother Tiberius. A more forceful man, Gaius puts through a far more extreme program, which includes a law obliging the government to provide grain to Rome's citizens at a price below the market average. The law protects the poor against famine and against speculators and establishes a precedent. State control of the grain supply will permit demagogues to gain popular support by distributing free grain.

110 B.C. Romans cultivate oysters in the first Western efforts to domesticate marine wildlife. Cultured oyster beds are operated in the vicinity of Baia near the town that will become Naples, where local oysterman Sergius Orata makes a fortune selling his bivalves to the luxury trade (*see* 850 B.C.; A.D. 407).

105 B.C. Two Roman armies are defeated at Arausio, on the Rhône, by the Cimbri, a Celtic or Germanic people from east of the Rhine who have moved into the Alpine regions and across the Rhône.

104 B.C. Judea's John Hyrcanus dies after a 30-year reign and is succeeded by his son, 37, who will rule briefly as Aristobulus I. He will complete the conquest of Galilee and will force the people of Hurae to embrace Judaism.

103 B.C. Judea's Aristobulus I dies at age 38 and is succeeded by his brother Alexander Jannaeus, who will extend the boundaries of the kingdom in a selfish and savage reign that will continue until 76 B.C.

A second Servile War erupts in Sicily as slaves rebel under the leadership of Tryphon and Athenion. Slaves from lands conquered by Rome's legions provide much of the power for Roman agriculture, being able to follow verbal orders even though they are less powerful and less docile than horses, whose efficiency is limited also by lack of metal horseshoes and lack of proper harnesses.

102 B.C. The Battle of Aix-en-Provence gives the Roman consul Gaius Marius, 53, a victory over the Teutons and Sciri. Marius has been reelected consul repeatedly since 107 B.C. in violation of the law of 151 B.C. The people of Provence hail his triumph, and Provençal families will name one of their sons Marius by tradition for more than 2,000 years.

101 B.C. The Battle of Campi Raudii near Vercellae gives the Roman consuls Gaius Marius and Quintus Lutatius Catulus, 51, a victory over the Cimbri. Marius becomes a national hero.

The Romans apply waterpower to milling flour and are the first people to do so.

Chinese ships reach the east coast of India for the first time with help from the navigational compass pioneered by the Chinese. They have discovered the orientating effect of magnetite, or lodestone (*see* A.D. 1086).

1st Century B.C.

Sulla, Lepidus, Crassus, Pompey, Julius Caesar, Marc Antony, and Octavian (Augustus) will rule Rome in this century.

100 B.C. Rome's plebeian tribunes Saturnius and Glaucia propose cheaper corn for the very poor plus other new social laws. Nobles in the Senate outlaw the popular leaders, and both are murdered with the complicity of Gaius Marius, the general.

99 B.C. Rome's second Servile War ends after 4 years as the consul M. Aquillius subdues an army of slaves that has put up a stubborn resistance.

96 B.C. The Seleucid king Antiochus VIII is murdered by his court favorite Heracleon after a 29-year reign in which he has been forced to divide the realm with his half brother Antiochus IX, who will reign alone until 95 B.C.

95 B.C. The Seleucid king Antiochus IX is defeated in battle and killed by the son of his late half brother who will reign briefly as Seleucus VI.

94 B.C. Armenia Minor's King Artanes is deposed by a descendant of the first king of Armenia Major, who has been held hostage for several years by Parthians but is ransomed for "70 valleys." He unites the two countries and at 45 begins a 38-year reign as Tigranes I that will make Armenia the most powerful nation in western Asia.

93 B.C. Armenia's Tigranes II seals an alliance with the Parthian king Mithridates by marrying his daughter Cleopatra. Tigranes has murdered a neighboring Armenian prince and has taken over his territory. He invades the kingdom of Cappadocia in the name of his new father-in-law, but the Roman general L. Cornelius Sulla comes to the aid of Cappadocia and forces Tigranes to retire.

92 B.C. Parthia's Mithridates II makes an alliance with Rome and prepares to invade Mesopotamia.

91 B.C. The Republic of Italia is set up by Italian insurrectionists, who establish a capital at Corfinium and begin a 3-year war against Rome.

88 B.C. Parthia's Mithridates the Great dies after a 36-year reign in the closing years of which he has conquered Mesopotamia. His son-in-law Tigranes II of Armenia invades Parthia and begins a war in which he will recover the 70 valleys paid for his ransom in 95 B.C. and overrun four Parthian vassal states, reducing the size of Parthia and extending the borders of Armenia.

The king of Pontus Mithradates VI Enpator begins the first of three wars that he will wage against Rome. Now 44, Mithradates has made himself master of Cappadocia, Paphlagonia, Bithynia, and all of the Black Sea's southern and eastern coasts.

Civil war breaks out in Rome as the legions suppress the insurrection of 3 years ago and as fresh legions are raised to fight the king of Pontus.

87 B.C. The Roman general L. Cornelius Sulla marches on Rome. He kills the demagogue P. Sulpicius Ruffus and other rebels, and he leaves for Asia as proconsul after having instituted reforms that will be short-lived.

A new Roman demagogue appears in the person of L. Cornelius Cinna, who begins a reign of terror against the reactionary Roman nobility.

86 B.C. Athens falls to Rome's General Sulla, who defeats the forces of Mithradates and his allies.

84 B.C. Sulla forces Mithradates to make peace, to evacuate all the territories he has conquered, to surrender 80 warships, and to pay an indemnity of 3,000 talents. Sulla sails for Brundisium, leaving two legions to police Rome's Asiatic territories and help L. Licinius Lucullus collect a fine of 20,000 talents from the Asiatic cities.

The Seleucid king Antiochus XII, who has seized Damascus, is killed on an expedition against the Nabataeans.

82 B.C. Sulla repels rebellious Samnites from Rome in the Battle of the Colline Gate in November. Sulla appoints himself dictator, moves to punish cities that have sided with Rome's enemies, and begins a tyranny that will continue until 79 B.C.

81 B.C. The Japanese emperor Sujin begins a great shipbuilding effort in a move to provide his people with more of the seafood on which so many depend for sustenance.

80 B.C. The Roman dictator Sulla halts public distribution of free grain (*see* 123 B.C.; 71 B.C.).

79 B.C. Sulla retires voluntarily from public life after completing substantial reforms in Rome's legal and judiciary system.

78 B.C. Sulla dies at age 60; the democratic consul M. Aemilius Lepidus immediately begins efforts to undo his work, trying to abrogate the constitution. Thwarted, Lepidus raises an army of malcontents in Etruria, but his colleague Q. Lutatius Catulus defeats him in battle outside Rome.

77 B.C. M. Aemilius Lepidus suffers total defeat at the hands of a protégé of Sulla. Gnaeus Pompeius (Pompey), 29, forces Lepidus to flee to the Hispanic provinces where he soon dies.

75 B.C. The Greek physician Asclepiades of Bithynia opposes Hippocratic medicine of 429 B.C., insisting that disease is a result of an inharmonious motion of the corpuscles that compose all bodily tissue. But Asclepiades is the first to distinguish between acute and chronic disease, and many are helped by his recommendations with regard to diet, bathing, and exercise (his patients will include Cicero, Crassus, and Marc Antony).

73 B.C. A third Servile War begins under the leadership of the Thracian slave Spartacus, a gladiator who seizes Mount Vesuvius on the Bay of Naples with the help of other gladiators and who rallies fugitive slaves to the insurrection.

72 B.C. Roman armies sent against Spartacus defeat his force of fugitive slaves.

71 B.C. The king of Pontus Mithradates VI is driven out of his country by the Roman legions of L. Licinius Lucullus and takes refuge at the court of Armenia's Tigranes II.

Spartacus is defeated by the Roman praetor M. Licinius Crassus, 41, who has enriched himself in the service of the late dictator Sulla by buying up properties of proscribed Romans. Pompey returns from the Hispanic provinces and destroys the remnants of the servile army.

70 B.C. Crassus and Pompey break with the Roman nobility and use their troops to gain the consulship. They restore the privileges of the tribunate, which were removed by Sulla.

Armenia's Tigranes II completes conquests that extend his empire from the Ararat Valley in the north to the ancient Phoenician city of Tyre on the Mediterranean coast. Calling himself "king of kings," Tigranes begins construction of a new capital to be called Tigranocerta at the headwaters of the river Tigris.

The Seleucid king Phraates III begins to restore order in Parthia but will not be able to repel the Roman legions of Lucullus and Pompey.

Crassus and Pompey resume distribution of free grain. Some 40,000 adult male citizens of Rome receive grain dispensations, and the number will rise rapidly (*see* 58 B.C.).

69 B.C. The Roman general and epicure Lucius Licinius Lucullus defeats Armenia's Tigranes II, who has seized Syria, and begins a push into the mountains of Armenia and Parthia toward Pontus.

Cherries from the Black Sea kingdom of Pontus sent back to Rome by Lucullus (above) introduce a new fruit tree to Europe.

68 B.C. Roman troops in Armenia mutiny, Lucullus is forced to retreat to the south, and the king of Pontus begins a campaign to regain his realm. Many of the Roman legions have been on campaign for 20 years.

Crete falls to the Roman legions.

67 B.C. Mediterranean pirates who have been interfering with Rome's grain imports from Egypt and North Africa are defeated by Quintus Caecilius Metellus.

The Seleucid king Antiochus XIII, who was installed last year at Antioch, is treacherously killed by the Arabian prince of Emesa.

66 B.C. Armenia's Tigranes II falls into the hands of Pompey, who has driven the king of Pontus Mithradates VI to the eastern edge of the Black Sea and who imposes a fine of 6,000 talents on Tigranes. The Armenian king will rule henceforth as a vassal of Rome; Mithradates will flee to the Crimea (*see* 63 B.C.).

Lucullus returns to Rome and begins entertaining on a lavish scale at feasts so extravagant that the word "Lucullan" will be used for millennia to denote flamboyant sumptuousness.

65 B.C. Rome's Asian and Syrian territories are reorganized by Pompey, who establishes four new Roman provinces. He leaves as client kingdoms Cappadocia, Eastern Pontus, Galatia, Judea, and Lycia.

Pompey introduces to Rome's orchards and cuisine apricots from Armenia, peaches from Persia, plums from Damascus, raspberries from Mount Ida (southeast of the old city of Troy), and quinces from Sidon.

64 B.C. Jerusalem falls to Pompey after a siege as the Romans move to subdue Judea.

The Catiline conspiracy rallies Rome's discontented debtors, veterans, ruined nobility, and others under the leadership of L. Sergius Catiline, 44. A former governor of Africa, Catiline tried 2 years ago to run for consul on a radical program but was unable to get his name presented to the Comitia Centuriata because of his alleged extortions in Africa. He runs again, but is defeated by the orator M. Tullius Cicero, 42.

63 B.C. Pompey completes his conquest of Palestine and makes it part of the Roman province of Syria.

Roman authorities arrest Lucius Sergius Catiline and his fellow conspirators, the Roman consul Cicero makes orations against Catiline in the Senate, and the conspirators are put to death as *hostes* without appeal.

Mithradates learns of a revolt by his son in Pontus and commits suicide in the Crimea (*see* 66 B.C.).

A system of shorthand notation is invented by Marcus Tullius Tiro, formerly a slave of Cicero (above).

62 B.C. Florence is founded on the Arno River in Tuscany (*see* florins, A.D. 1189).

60 B.C. A triumvirate to rule Rome is created by Gaius Julius Caesar, 42, who returns from his governorship of Rome's Hispanic provinces and forges an alliance with Pompey and Crassus. Caesar is a son-in-law of the late Cinna and an agent of Crassus.

59 B.C. Rome's new triumvirate distributes Campanian lands among Pompey's veterans. The triumvirate is solidified by the marriage of Caesar's daughter Julia to Pompey, whose eastern settlements are confirmed after years of Senatorial opposition to such confirmation, and Julius Caesar is granted Cisalpine Gaul and Illyria for a 5-year period.

58 B.C. Julius Caesar invades Gaul with the aim of enriching himself and creating a military establishment which will rival that of Pompey. He defeats the Helvetii at

58 B.C. ***(cont.)*** Bibracte (Autun) and triumphs over the German Ariovistus near Vesontio (Besançon).

The Roman demagogue Appius Claudius Pulcher distributes free grain to as many as 300,000 in a bid for the consulship.

57 B.C. Julius Caesar's legions defeat the Belgae in northwestern Gaul while the demagogue Appius Claudius Pulcher back at Rome becomes praetor.

Philosophy *De Rerum Natura* by the Roman poet-philosopher Titus Lucretius Caro, 39, is a 6-volume didactic poem of Epicurean philosophy dealing with ethics, physics, psychology, and the materialistic atoms suggested by the Greek Democritus in 330 B.C. *"Quod ali cibus est aliis fiat acre venerum,"* says Lucretius ("What is food to one man may be a fierce poison to another").

The Roman Senate gives Pompey power to supervise the city's grain supply as a grain shortage looms.

56 B.C. Julius Caesar defeats the Veneti on the southern coast of Brittany, defeats the Aquitani in southwestern Gaul, and meets with Pompey and Crassus at Luca to make plans for subduing the opposition against the triumvirate that has arisen at Rome.

Elegy *Ave atque Vale* (*Hail and Farewell*) by the Roman poet Gaius Valerius Catullus, 28, who has visited his brother's grave in Bithynia. He is famous for his epigrams, his love lyrics, his long poem *Attis*, and his epithalamium *Thetis and Peleus*.

55 B.C. Rome makes Pompey and Crassus consuls. The Senate extends Caesar's command in Gaul another 5 years, and it gives Crassus command of Syria and Pompey command of 2 Hispanic provinces. Crassus leaves for the East; Pompey remains at Rome.

54 B.C. Julius Caesar invades Britain. He fails to conquer the islands but does open them to Roman trade and influence (*see* A.D. 43).

The triumvirate that rules Rome begins to break up following the death of Caesar's daughter Julia, who is married to Pompey.

Appius Claudius Pulcher becomes consul at Rome.

Crassus plunders Jerusalem's Great Temple.

Julius Caesar (above) finds the Britons making Cheshire cheese.

52 B.C. The Gallic leader Vercingetorix, chief of the Averni, surrenders to Julius Caesar after a siege by the Roman legions.

51 B.C. Julius Caesar completes his conquest of Gaul.

49 B.C. Julius Caesar leads his forces across the Rubicon River into Italy to begin a civil war. "The die is cast *(alea iacta est),"* says Caesar.

48 B.C. Julius Caesar defeats Gnaeus Pompey June 29 at Pharsalus in southern Thessaly and becomes absolute ruler of Rome. Pompey flees to Egypt, where he is slain at Pelusium. Caesar arrives at Alexandria, learns of Pompey's murder, but remains to carry on a war in behalf of Egypt's dethroned queen Cleopatra, 20. Her brother Ptolemy XII Philopator is killed, and a younger brother succeeds as co-ruler.

Some 500 bargeloads of foodstuffs and other imports for Rome are pulled up the Tiber by oxen each month from the port of Ostia.

47 B.C. Julius Caesar moves east in June to Asia Minor. He defeats the king of Pontus Pharnaces III August 2 in a battle near Zela. Pharnaces has been an ally of the late Gnaeus Pompey, and Caesar announces his victory with the brief dispatch, "I came, I saw, I conquered *(Veni, vidi, Vici)."*

46 B.C. Julius Caesar returns to Italy, quells a mutiny of the legions in Campania, crosses to Africa, and destroys a republican army of 14 legions under Scipio at Thapsus April 6. Most republican leaders are killed; Cato commits suicide. Caesar returns triumphantly to Rome in late July with Cleopatra as his mistress, is made dictator for 10 years, and sails in November for Spain, where Pompey's sons hold out.

Julius Caesar returns in triumph to Rome with prisoners who include Vercingetorix, chief of the Averni, who is executed.

Julius Caesar grants Roman citizenship to Greek physicians, whose status has until now been that of slave or freedman.

45 B.C. Julius Caesar crushes Pompey's sons at Munda March 17 and returns to Rome in September. He adopts his great-nephew Gaius Octavius (Octavian) as his son (*see* 43 B.C.).

Caesar introduces a new "Julian" calendar of 365.25 days in which the first day of the year is January 1. He has commissioned the Greek astronomer-mathematician Sosigenes of Alexandria to reform the calendar (*see* 153 B.C.; A.D. 1582).

44 B.C. Julius Caesar is made dictator for life. He reduces the number of Romans receiving free grain from 320,000 to 150,000.

Julius Caesar is assassinated at the Senate March 15 by conspirators who include Decimus Junius Brutus and Marcus Junius Brutus, both former governors of Gaul, and Gaius Cassius Longinus, who had been pardoned by Caesar for fighting alongside Pompey at Pharsalus in 48 B.C.

Caesar's mistress Cleopatra returns to Egypt with her son Caesarion and murders her brother (and former husband) Ptolemy XIII Philopater.

Roman orator Marcus Antonius (Marc Antony), 39, persuades the Romans to expel Caesar's assassins.

43 B.C. Marc Antony marches north to dislodge Decimus Brutus from Mutina (Modena) but is defeated in two battles and forced to retire westward toward Gaul by Gaius Octavius, 19, a great-nephew of the late Julius Caesar. Octavius calls himself Gaius Julius Caesar Octavianus, forces the Senate to elect him consul, and joins with Marc Antony and Marcus Lepidus in November to form a second triumvirate.

The assassination of the dictator Julius Caesar in the Roman Senate precipitated civil war. It lasted 13 years.

Cicero is executed December 7 by agents of Marc Antony with the acquiescence of Octavian.

42 B.C. Julius Caesar is deified by the Roman triumvirate, which erects a temple to Caesar in the Forum and obliges Roman magistrates to take an oath to support the arrangements of the late Caesar.

Cassius is defeated by Marc Antony and Octavian at Philippi and commits suicide after hearing a false report that M. Junius Brutus has also been defeated. Brutus has, in fact, won a victory over Octavian, but is finally defeated 20 days later and also commits suicide.

41 B.C. Egypt's Cleopatra meets Marc Antony at Tarsus, and Antony, 42, succumbs to Cleopatra, now 28, just as Julius Caesar, at 51, succumbed 8 years ago. Antony had planned to punish Cleopatra but follows her to Egypt.

Poetry *Eclogues* by the Roman poet Virgil, 29, are pastoral poems expressing emotion at having his lands confiscated (below). Virgil will obtain restitution of his lands by personal appeal to Octavian.

Rome's triumvirate confiscates farmland in the Campania for distribution among returning legionnaires.

40 B.C. Marc Antony's wife Fulvia and his brother Lucius Antoninus make war against the faithless Antony but are defeated at Perusia (Perugsia). Fulvia dies, and Marc Antony is left free to remarry. He marries the sister of Octavian (Octavia), and Octavian takes Gaul from Lepidus, leaving him only Africa.

39 B.C. Rome's triumvirate signs the Pact of Mycaenum recognizing the Mediterranean pirate Sextus Pompey as ruler of Sicily, Sardinia, Corsica, and the Peloponnese. Pompey's fleet can interrupt Rome's grain supply, and this puts him in a position to dictate terms.

Octavian divorces his second wife and marries Livia, previously the wife of Tiberius Claudius Nero.

38 B.C. Octavian conquers Iberia and begins to gain ascendancy in the triumvirate that rules Rome. Marc Antony returns to Egypt.

Sculpture *The Laocoön* by Rhodian sculptors Agesander, Polydorus, and Athenodorus (*see* A.D. 1506).

37 B.C. Judea's Herod the Great begins a 33-year reign at age 36, 2 years after his confirmation as king by Marc Antony, Octavian, and the Roman Senate.

36 B.C. Octavian's general Marcus Vipsanius Agrippa makes the Mediterranean safe for Roman shipping by defeating the pirate Sextus Pompey with a fleet that includes ships supplied by Marc Antony. Sextus Pompey flees to Miletus, where he dies. Lepidus occupies Sicily, but his troops desert to Octavian, who ends the ambitions of Lepidus. Octavian places Lepidus in captivity at Circeii, where he will live until his death in 13 B.C.

Parthian forces defeat Marc Antony. He retreats to Armenia and openly marries his mistress Cleopatra despite his existing marriage to Octavia.

35 B.C. Octavian consolidates Roman power in the Alps and in Illyria.

32 B.C. Octavian arouses fears in Rome that Egypt's Cleopatra will dominate the empire. He publishes what he purports to be the will of Marc Antony, a will in which Antony bequeaths Rome's eastern possessions to Cleopatra. Marc Antony divorces Octavia, and her brother has the comitia annul Antony's imperium.

31 B.C. The Battle of Actium September 2 ends in a naval victory for Octavian, who becomes ruler of the entire Roman world. Cleopatra escapes to Egypt with 60 ships, followed by Antony, whose army then surrenders to Octavian.

Marble was putty in the hands of talented Greek sculptors. Their works included the Laocöon.

30 B.C. Marc Antony commits suicide after hearing a false report that Cleopatra has killed herself. She dies August 29, applying an asp to her breast after failing to seduce Octavian upon his arrival at Alexandria. Cleopatra's son Caesarion is murdered, and Egypt becomes a Roman province.

The sundial invented by the Chinese serves as a primitive clock.

Poetry *Georgics* by Virgil, completed after 7 years' work, is a didactic 4-volume work ennobling the Italian land, its trees, grapevines and olive groves, herds, flocks, and beehives.

29 B.C. Greek mariners employed by Rome's Octavian open the ancient trade routes from Egypt to India as peace returns to the Roman world.

27 B.C. The Roman Empire that will rule most of the Western world until A.D. 476 is founded January 23 by Octavian who one week earlier received the name Augustus Caesar from the Senate in gratitude for his achievements. Helped by the rich Roman merchant Mycenas, Octavian makes himself emperor at age 35 with the title Imperator Caesar Octavianus, a title he will soon change to Augustus Caesar as he begins a 41-year reign.

The number of poor Romans receiving free grain is increased by Augustus from 150,000 to 200,000.

24 B.C. The emperor Augustus acts to reduce the exorbitant price of spices in Rome. He appoints the prefect of Egypt Aeilius Gallus to lead a campaign that incorporates the south Arabian spice kingdom into the Roman Empire, but the expedition will fail.

21 B.C. Regensburg is founded on the Danube in a part of Gaul that will become Bavaria.

20 B.C. Rebuilding of Jerusalem's Great Temple begins under Herod the Great, a convert to Judaism in his youth, who has been building theaters, hippodromes, and other public buildings (*see* A.D. 64).

19 B.C. The Pont du Gard, completed by Roman engineers across the Gard River 12 miles northeast of Nîmes, is an aqueduct bridge 600 yards long with 3 tiers of arches rising 160 feet above the river.

Poetry *The Aeneid* by Virgil is a great epic about the role of Rome in world history: "I sing of arms and of the man. . ."(*Arma virumque cano* . . .). Virgil sets off for Athens to spend 3 years in the East, meets the emperor Augustus, is persuaded to return, and dies September 21 at age 50, a few days after landing at Brundisium.

18 B.C. Poetry *Amores* by the Roman poet Ovid (Publius Ovidius Naso), 25, who has been educated for a career in law but has devoted himself instead to writing love poems.

17 B.C. Ode *Carmen Saeculare* by the Roman poet-critic Horace (Quintus Horatius Flaccus), 47, is sung by a chorus of youths and maidens at a great festival of games put on by the emperor Augustus. Horace is well known for his *Odes* ["Seize the day," *carpe diem* (I, ix)], his *Satires* ["There is a certain method in his madness" (II, iii)], and his *Epistles* ["Well begun is half done" (I, ii)].

15 B.C. Rome extends her frontier to the upper Danube by annexations following an uprising by Germanic tribesmen.

Poetry *Heroides* by Ovid: "Where belief is painful, we are slow to believe" (II, ix).

The Temple of Dendur in Lower Nubia is erected by the emperor Augustus on the bank of the Nile.

14 B.C. Augsburg has its beginnings in the Gallic colony Augusta Vindelicorum founded by the Roman emperor Augustus on a 1,500-acre plateau between the rivers Lerch and Wertach, which meet in the plain below (*see* A.D. 1276).

7 B.C. The infant Jesus, who will become an important religious leader, is born at Bethlehem near Jerusalem to a Jewish carpenter's wife who professes to be a virgin at the time of her child's conception. Mary's first-born will be baptized at age 30 by her cousin John the Baptist after working as a carpenter and rabbi at Nazareth. (Year of the nativity determined by modern astronomers on the basis of a conjunction of the planets Saturn and Jupiter within the constellation Pisces, a conjunction that gives the appearance of a great new star and the fulfillment of a prophecy by Jewish astrologers at Sippar in Babylon, who have predicted the arrival of a long-awaited Messiah at some time when the two planets would meet. Some astronomers, however, will cite Chinese and Korean annals referring to a stellar flare-up, or nova, that blazed in the skies for 70 days in the spring of 5 B.C.).

The population of the world reaches roughly 250 million, up from 100 million in 3000 B.C. (*see* A.D. 1650).

4 B.C. Judea's Herod the Great dies at age 69 after a 33-year reign in which he has secured many benefits for the Jews and has begun the rebuilding of the Great Temple at Jerusalem. Herod is succeeded by his son Archelaus, who will reign until A.D. 6.

1st Century A.D.

Augustus, Tiberius, Caligula, Claudius, Nero, Vespasian, Titus, Domitian, and Trajan will be the major Roman emperors of this century.

Poetry *The Art of Love (Ars Amatoria)* by Ovid scandalizes Rome: "It is expedient that there should be gods, and since it is expedient let us believe that gods exist" (I); "Nothing is stronger than habit" (II).

De Re Rustica by Lucius Junius Moderatus Columella is a didactic poem that advises switching from grain to vines, ". . . for none in Italy can remember when seeds increased fourfold." The common yield for a bushel of seed is only 2 or 3 bushels of grain, and while an acre of land may, at best, yield 4 to 6 bushels of wheat or barley the more usual yield is 2 to 3 bushels, and most of Rome's grain comes from Egypt and North Africa *(see* A.D. 6).

"The Earth neither grows old, nor wears out, if it be dunged," writes Columella (above), who urges crop rotation that alternates grain with legumes.

Romans use blood and bones as fertilizer. They grow clover and will later grow alfalfa, but they disdain to use human excrement for fertilizer.

Rice, imported from China, is cultivated for the first time in Japan, at Kyoto.

5 Lombard tribes that have established themselves on the lower Elbe River are defeated by Roman legions.

6 China requires candidates for political office to take civil service examinations.

The number of Romans receiving free grain rises to 320,000, up from 150,000 in 44 B.C. Close to one-third of the city is on the dole.

Rome imports some 14 million bushels of grain per year to supply the city alone—an amount requiring several hundred square miles of croplands to produce. One-third comes from Egypt and the rest mostly from North African territories west of Egypt.

7 Poetry *Metamorphoses* by Ovid, whose works have been banned.

8 Augustus exiles Ovid to the Roman outpost of Tomis (Constanza) on the Black Sea, in part for his elegiac poem instructing a man in the art of winning and keeping a mistress and instructing a woman on how to win and hold a lover. Ovid destroys his new masterpiece at news of his banishment, but copies made by his friends will survive with such lines as "Love and dignity cannot share the same abode" (II).

9 The Battle of Teutoburger Wald (Forest) permanently secures the independence of the Teutonic tribes and establishes the Rhine as the boundary between Latin and German territories. Germanic tribes under Arminius annihilate three Roman legions under the command of the legate P. Quinctilius Varus, killing 20,000 legionnaires. Varus throws himself on his sword; his head is sent to Augustus.

The Hsin dynasty emperor Wong Mong grants manumission to China's slaves.

Wong Mong (above) nationalizes Chinese land, dividing the country's large estates and establishing state granaries.

12 The Chinese repeal the radical land reforms made by the emperor Wong Mong 3 years ago in response to widespread protests.

14 The Roman emperor Augustus dies at Nola August 19 at age 76 after a 41-year reign. He is succeeded despite legal obstacles to dynastic succession by his stepson Tiberius Claudius Nero, 55, who will rule until 37 as the emperor Tiberius. Son of the late emperor's widow Livia by her first marriage, he will carry on the imperial regime inaugurated by Augustus in 27 B.C.

16 Tiberius' son Drusus, 31, defeats Arminius, breaks up his Germanic kingdom, recovers the eagles of the legions lost at the Battle of Teutoburger Wald 7 years ago, and avenges the defeat of Varus.

17 The emperor Tiberius sends his nephew Germanicus to install a new king in Armenia.

Capadocia and Commagene are combined into a Roman province following the death of their king.

China enacts a tax on slaveholding; slaves do most of the menial work as they do in Rome.

Seven regional Chinese commissions are directed to establish annual high, low, and mean price levels for staples and to buy surplus goods at cost, but merchants and capitalists employed by the emperor Wong Mong as administrators will provoke revolts.

19 Germanicus is poisoned in Syria. The Syrian legate Piso, charged with the murder by enemies of the emperor Tiberius, commits suicide.

23 Drusus, son of the emperor Tiberius, is poisoned by Lucius Aelius Sejanus, the ambitious equestrian prefect of the guard, who has designs on the imperial throne and begins an 8-year domination of the emperor.

China's emperor Wong Mong is killed in a revolt after a 14-year reign in which he has attempted to curb usury and advance the welfare of the masses.

25 The eastern Han dynasty that will rule at Luoyang until A.D. 220 is founded by a collateral imperial scion, who will reign until A.D. 57 as Kwang Wudi.

Sejanus persuades the emperor Tiberius to retire from the hostile political climate of Rome and settle on the island of Caprae (Capri) in the Bay of Naples.

30 *De Res Medicos* by Roman physician Aulus Cornelius Celsus is a collection of Greek medical writings. Celsus also publishes *De Artibus*. Tracing medical history from simple remedies through Hippocratic and Alexandrian contributions, he describes in careful detail the surgical instruments of his own day after witnessing many operations and dissections. "I am of the opinion that the art of medicine ought to be rational," writes Celsus. "To open the bodies of the dead is necessary for learners" (*see* Galen, A.D. 180; "Paracelsus," A.D. 1530).

31 Lucius Aelius Sejanus is executed by order of the emperor Tiberius, who has discovered the intrigues of his former favorite (*see* A.D. 23).

A horizontal waterwheel described in Chinese writings employs a series of belts and pulleys to drive a bellows that works an iron furnace for the casting of agricultural implements.

33 Jesus of Nazareth leaves Galilee after a brief ministry and travels to Jerusalem to observe Passover at a ritual Seder meal that will be depicted by future painters as *The Last Supper*. Jesus has antagonized the priestly class by driving the money changers out of the Great Temple at Jerusalem and attacking the hypocrisy of the privileged classes. With a price on his head, he is betrayed by his disciple Judas Iscariot and seized by Roman soldiers who deliver him to the high priest and the Sanhedrin (great council and tribunal of the Jewish nation). He is condemned as a blasphemer and sent to the Roman procurator Pontius Pilate, who sends him to the ruler of Galilee Herod Antipas, who sends him back to Pilate, who lets the mob decide his fate. The mob condemns Jesus, and he is crucified, probably April 3, between two thieves on Golgotha, a knoll near the Damascus Gate.

Disciples of Jesus (above) will profess to have seen him resurrected after his burial in the tomb of Joseph of Arimathea. They will proclaim him Messiah and Savior, and they will found the Christian faith that will grow to dominate much of the world.

37 The emperor Tiberius dies March 16 at age 78 and is succeeded by Gaius Caesar, 25, youngest son of the late Germanicus Caesar and nephew of Tiberius, who is called Caligula because of the *caligae*, or soldiers' boots, he has worn. The new emperor Caligula introduces cruel oriental excesses in a monarchy that will rule Rome for more than 3 years.

40 The Greek merchant Hippalus voyages in one year from Berenice, on Egypt's Red Sea coast, to India's Madras coast and back, a journey that has previously required 2 years. Hippalus has discovered that the monsoon winds (the word derives from *mawsim*, Arabic for seasons) reverse direction twice a year, a fact the Arabs may have known for centuries. The southwest wind, favorable for voyages from Egypt to India, prevails from April to October, and the northwest wind for the return trip prevails from October to April (*see* A.D. 90).

41 The Roman emperor Caligula is murdered January 24 by a tribune of the guard after a megalomaniacal reign of savage tyranny. He is succeeded by a nephew of the late emperor Tiberius, a crippled man of 50 with a speech defect named Tiberius Claudius Drusus Nero Germanicus, who will rule until 54 as the emperor Claudius.

43 The Roman emperor Claudius leads a personal expedition to conquer Britain and Romanize her people (*see* 54 B.C.; A.D. 48).

London (Londinium) is founded by the Romans (*see* 400 B.C.; A.D. 61).

44 Judea's Herod Agrippa dies at age 54 after a 3-year reign, and Judea becomes a procuratorial province of Rome once again. Agrippa's son, 17, is studying at the court of the emperor Claudius in Rome and beginning in A.D. 48 will reign for 5 years as Herod Agrippa II.

The apostle James, who has preached the divinity of Jesus, becomes the first Christian martyr. He is executed on orders from Herod Agrippa (above) before the king's death.

Romans create the capon, gelding cocks to make them grow larger.

Vomitoriums gain popularity in Rome. The emperor Claudius and others employ slaves to tickle their throats after they have eaten their fill in order that they may return to the banquet tables and begin again. Most Romans live on bread, olives, wine, and some fish, but little meat.

46 The apostle Paul journeys to Cyprus and Galatea with the Cypriot Barnabas and with Mark, a young cousin of Barnabas.

48 Roman legions invade Wales. They will conquer the country in the next 30 years while Roman engineers construct a great network of British roads.

49 The emperor Claudius expels Jewish-Christians from Rome.

50 Cologne has its beginnings in the town of Colonia Agrippina, built on the left bank of the Rhine at the site of Oppidum Ubiorum, chief town of the Ubii. The Roman emperor Claudius fortifies the town at the request of his niece and bride Agrippina, 35, who was born in the place.

The apostle Paul travels to Greece on a journey that will take 3 years.

54 The Roman emperor Claudius dies October 13 at age 63 after eating poisonous mushrooms given him by the physician Stertinius Xenophon as part of a plot inspired by the empress Agrippina. She will seek to rule Rome through her son Nero Claudius Augustus Germanicus, 16, who will reign as the emperor Nero until A.D. 68.

Epistles to the Corinthians by the apostle Paul deal with a variety of moral and ethical questions.

55 *De Materia Medica* by Greek botanist Pedanios Dioscorides details the properties of some 600 medicinal plants and describes animal products with alleged dietetic or medicinal value. Dioscorides has served as a surgeon in Nero's army.

58 The apostle Paul arrives at Jerusalem, but authorities at Caesarea arrest him and hold him for trial before the procurator of Judea.

59 The Roman emperor Nero has his mother Agrippina put to death at the urging of his councilor Lucius Annaeus Seneca, 61.

60 Festus succeeds Felix as procurator of Judea and holds a new trial for the apostle Paul, who makes an "appeal unto Caesar" in the presence of Herod Agrippa II.

61 London is sacked by the Trinovantes of Essex and Suffolk and by the Iceni of Norfolk and Suffolk, whose queen Boudica (Boadicia) revolts upon the retirement of the Roman governor Seutonius Paulinus.

Rome's legions crush the Britons and restore the Roman authority that will continue until A.D. 407.

Roman engineers surround London with a wall 8 feet thick.

62 Roman authorities permit the apostle Paul to live in Rome but keep him under house arrest. Albinus succeeds Festus as procurator of Judea, and the Romans permit Paul to resume his travels.

63 *Epistilae Morales* by the Roman philosopher Seneca says, "All art is but imitation of nature."

64 Persecution of Christians begins at Rome.

Jerusalem's Third Temple is completed after 84 years of construction (but *see* A.D. 70).

Rome has a fire that begins the night of July 18 in some wooden booths at one end of the Circus Maximus, spreads in one direction over the Palatine and Velia hills and up to the low cliffs of the Esquiline, spreads in another direction through the Aventine, the Forus Boarium, and the Velabrum until it reaches the Tiber and the Servian Wall, destroying nearly two-thirds of the city. The emperor Nero has fretted (not "fiddled") while Rome burned and begins rebuilding to a master plan that will give the city straight, broad streets and wide squares whose cleanliness will be supervised by the *aediles*.

Vast quantities of grain are stored at Rome under the supervision of the *aediles*, who control the food supply. They introduce regulations to ensure the freshness of meat, fish, and produce sold in the city.

65 A plot to murder the Roman emperor Nero comes to light. The conspirators are executed or are forced to take their own lives.

The Gospel according to St. Mark.

66 Nero's favorite courtier Gaius Petronius is accused of treason, arrested at Cumae, and ordered to commit suicide. He leaves behind his *Satyricon*, depicting the vice and depravity of his time.

67 Roman armies under Titus Flavius Sabinus Vespasianus, 58, and his son Titus, 27, enter Galilee to put down a revolt by Jews who have massacred a body of Roman soldiers in protest against their sacrileges and extortions. All the Jews of Caesarea have been slaughtered by the town's gentile citizens. The Jews are furious, but the Roman army is overwhelming. Jewish general Joseph ben Matthias, 30, holds out in a siege of the fortress Jotapata but yields after 47 days to Vespasian and gains the favor of the Roman general.

The Christian apostle Paul is executed June 29 on the Via Ostia 3 miles from Rome. The first great Christian missionary and theologian, Paul will hold a position in the faith second only to that of Jesus.

68 The emperor Nero is sentenced to death by the Senate under pressure from the praetorian guard, which has recognized the legate Servius Sulpicius Galba, 65, as emperor. Nero commits suicide June 9 at age 30. His death ends the Julio-Claudian line of Caesars that has ruled Rome for 128 years, and he is succeeded by Galba, who will rule for less than 6 months before being challenged.

History of the Jewish People is compiled by the Jewish general Joseph ben Matthias, who has taken the Roman name Flavius Josephus *(see* A.D. 67).

69 Eight legions on the Rhine refuse allegiance to the Roman emperor Galba and salute as emperor their legate Aulus Vitellius, 54. Galba is murdered January 15 along with his newly adopted successor, Piso Licinianus. The murderer is Marcus Salvius Otho, 36, a dissolute friend of the late emperor Nero, and the Senate recognizes Otho as emperor.

Aulus Vitellius (above) sends two legions to the Po Valley. They defeat the emperor Otho April 19 in the Battle of Bedriacum near Cremona, and Otho commits suicide, leaving Vitellius to face a challenge from Titus Flavius Sabinus Vespasianus, 59, now legate of Judea. The prefect of Egypt proclaims Vespasianus emperor July 1, the legate of Syria and all the Danubian legions rally to his support, and the emperor Vitellius mobilizes forces to oppose them. Antonius Primus, commander of the Seventh Legion in Pannonia, leads other Danubian legions against Vitellius and defeats him in late October in the second Battle of Bedriacum. He sacks Cremona and forces the Senate to recognize Vespasianus as the emperor Vespasian.

The Roman emperor Vitellius dies in a street battle December 20, leaving Vespasian to begin a reign that will continue until 79.

The emperor Vespasian lays siege to Jerusalem as the Jewish Zealot leader John of Giscala continues resistance after having eliminated his rival Eleazar.

70 The emperor Vespasian returns to Rome, leaving his son Titus to continue the siege of Jerusalem. He turns his energies to repairing the ravages of civil war. He suppresses an insurrection in Gaul, restores discipline to the demoralized Roman army, renews old taxes and institutes new ones, and rebuilds the Capitol, which

70 (cont.) was burned in the fighting that raged in the city last autumn.

Jerusalem falls September 7. The Romans sack the city and destroy most of the Third Temple, which was completed only 6 years ago. The one wall left standing will become famous as the "Wailing Wall."

Titus gives some of Judea to Marcus Julius "Herod" Agrippa II but retains most as an imperial domain. Rome quarters a legion in Jerusalem under a senatorial legate whose position is higher than that of the procurator. The Romans abolish the Jewish high priesthood and Sanhedrin (Jewish national council), and they divert the 2-drachma tax paid by Jews for support of the Great Temple to a special account in the imperial treasury *(fiscus Judaicus)*.

The Jewish teacher Johanen ben Zakkai saves Judaism. A disciple of the late Babylonian Jew Hillel, who died some 60 years ago, Johanen has had himself carried out of Jerusalem during the siege and has asked Vespasian to grant him a boon. He opens a school at Jabneh with Roman permission.

Panic strikes Rome as adverse winds delay grain shipments from Egypt and North Africa, producing a bread shortage. Ships laden with wheat from North Africa sail 300 miles to Rome's port of Ostia in 3 days given good winds, and the 1,000-mile voyage from Alexandria averages 13 days. The ships often carry 1,000 tons. Shipping adds little to the price (which may double if hauled overland); wheat is a cheap commodity, but the supply depends on favorable winds.

71 The Arch of Titus erected at Rome by the emperor Vespasian celebrates the triumph of the emperor's son last year at Jerusalem.

A palatial public lavatory built by the emperor Vespasian opens in Rome, which now has an extensive system of waterworks with flush toilets and urinals.

77 The Roman governor Gnaeus Julius Agricola arrives in Britain to continue the conquest begun by the emperor Claudius 34 years ago. He finds "a fierce and savage people running through the woods."

79 The Roman emperor Vespasian dies June 23 at age 69 after a 10-year reign. He is succeeded by his son Flavius Sabinus Vespasianus Titus, now 38, who will rule as the emperor Titus until A.D. 81.

Chester is founded by Roman occupation forces in England.

Historia Naturalis by the Roman scholar Pliny the Elder (Plinius Secundus), 56, is a 37-volume encyclopedia of natural history that says of agriculture, "The farmer's eye is the best fertilizer."

Mount Vesuvius on the Bay of Naples erupts August 24 after 16 years of increasingly violent earthquakes that have seriously damaged towns in the Campania. Tons of lava, mud, and ashes bury the cities of Herculaneum and Pompeii, killing thousands. Pliny (above) witnesses the eruption from a ship in the bay, takes refuge with a friend at Stabiae, and dies of suffocation from poisonous fumes (*see* A.D. 513).

80 Anthrax sweeps the Roman Empire in epidemic form, killing thousands of humans and animals (below).

Book of Spectacles (Liber de Spectaculis) by the Roman epigrammatic poet Martial (Marcus Valerius Martialis), 40, commemorates the dedication of the Colosseum (below).

The Colosseum, dedicated at Rome by the emperor Titus, is a great Flavian amphitheater with solid masonry walls rising 160 feet above the ground and with 50,000 marble seats around the 617-foot by 513-foot oval arena built above its basements and subbasements.

Three months of celebration mark the opening of the Colosseum (above). The emperor has 500 wild beasts and many gladiators slain to amuse the populace.

Anthrax (above) strikes the cattle and horses of tribespeople on the borders of China, where an extended drought withers the grasslands. The tribespeople begin moving westward to seek new pastures.

The tribespeople moving westward from Mongolia avoid scurvy by consuming large quantities of mares' milk. The milk contains four times as much of the accessory food factor ascorbic acid as does cows' milk (no milk is notably rich in ascorbic acid but little is needed to avoid scurvy).

Some 30,000 Asian tribespeople migrate to the West with 40,000 horses and 100,000 head of cattle, joining with Iranian tribespeople and with Mongols from the Siberian forests to form a group that will be known in Europe as the Huns (*see* 140 B.C.; A.D. 451).

81 The Roman emperor Titus dies September 13 at age 40 after a 2-year reign. He is succeeded by his brother Titus Flavius Domitianus, 29, who will reign until A.D. 96 as the emperor Domitian.

The silver content of the Roman denarius will rise in the reign of Domitian to 92 percent, up from 81 percent in the reign of Vitellus.

Gladiators who died for the public's entertainment in the Roman Colosseum were generally slaves captured in battle.

The Arch of Titus, raised by the emperor Domitian at Rome, has bas-reliefs that commemorate the military triumphs of Titus and Vespasian.

83 Roman forces under Gnaeus Julius Agricola in Britain defeat the Caledonians and reach the northernmost point that they will attain in the British Isles (possibly near what will later be Aberdeen, Scotland).

84 The Roman emperor Domitian recalls Gnaeus Julius Agricola from Britain to help repel barbarian invaders near the Rhine and the Danube.

The Gospel according to St. John and the Gospel according to St. Matthew are transcribed.

90 *The Periplus of the Erythraean Sea (Sailing Round the Indian Ocean)* by a Greek sea captain gives instructions on how to use the monsoon winds to advantage (*see* Hippalus, A.D. 40).

Roman ships break the Arab monopoly in the spice trade. The ships are large enough to sail without difficulty from Egyptian Red Sea ports to India, but while spices become more plentiful, they drain Rome of her gold reserves.

Use of spices is one of the excesses that will bring about the fall of Rome, says the Christian prophet John of Ephesus in his Revelations (18:11–13). John writes metaphorically of Babylon, but he means Rome.

95 A severe form of malaria appears in the farm districts outside Rome and will continue for the next 500 years, taking out of cultivation the fertile land of the Campagna, whose market gardens supply the city with fresh produce. The fever drives small farmers into the crowded city, they bring the malaria with them, and it lowers Rome's live-birth rate while rates elsewhere in the empire are rising.

At least 10 aqueducts supply Rome with 250 million gallons of water per day, some 50 gallons per person, even after the public baths have used half the supply.

Iron plows with wheels help some of Rome's barbarian neighbors to control the depth of plowing (and to save the plowmen's energies). The barbarians use coulters to cut the soil and moldboards to turn it over. While Roman farmers practice cross-plowing, the barbarians plow deep, regular furrows that will lead to the cultivation of long strips of land rather than square blocks.

96 The Roman emperor Domitian is stabbed to death by a freedman September 18 at age 44 after a 15-year reign. The empress Domitia and officers of the court have conspired against Domitian, he is succeeded by the former Roman consul Marcus Cocceius Nerva, 60, and the new emperor Nerva recalls citizens exiled by Domitian, restoring to them what remains of their confiscated property.

97 The Roman emperor Nerva recalls the general Marcus Ulpius Trajanus, 44, from the Rhine and formally adopts him in October at ceremonies in the temple of Jupiter on the Capitol.

98 The Roman emperor Nerva dies suddenly January 25 at age 63. He is succeeded by his adopted son, who will reign until A.D. 117 as the emperor Trajan.

The silver content of the Roman denarius will rise to 93 percent under the emperor Trajan, up from 92 percent under Domitian.

2nd Century

Trajan, Hadrian, Marcus Aurelius, and Commodus will be the major Roman emperors of this century.

105 Papermaking is refined by Chinese eunuch Tsai Lun, 55, who receives official praise from the emperor for his methods of making paper from tree bark, hemp, remnant rags, and fishing nets. Crude paper has been made in China for at least 2 centuries, but bamboo and wooden slips will remain the usual materials for books and scrolls for another 2 centuries, and paper will not be made in Korea until about 600, in Japan until at least 610 (*see* 751).

106 Dacia (Romania) becomes a Roman province following the defeat of her king Decebalus by the emperor Trajan.

110 Caravans make regular departures from Luoyang with Chinese ginger, cassia (a type of cinnamon), and silk to be bartered in Central Asia for gold, silver, grape wine, glassware, pottery, asbestos cloth, coral beads, and intaglio gems from Rome.

Nonfiction *Of Illustrious Men* (*De Viris Illustribus*) by the Roman biographer Suetonius (Gaius Suetonius Tranquillis), 44.

114 Armenia is annexed to the Roman Empire by the emperor Trajan.

115 Roman legions occupy Mesopotamia up to the river Tigris.

Nonfiction *Annales* by the Roman historian Cornelius Tacitus.

116 The Roman emperor Trajan makes Assyria a province of Rome and crosses the Tigris to annex Adiabene. He proceeds to the Persian Gulf and conquers territory that becomes the province of Parthia.

117 The emperor Trajan dies August 8 at Selinus in Cilicia at age 63 while en route from Mesopotamia to Italy. His kinsman Publius Aelius Hadrianus, 41, is advised August 9 at Antioch that he has been adopted by Trajan, learns of Trajan's death August 11, and will reign until 148 as the emperor Hadrian.

Jews throughout the East rise to massacre Greeks and Romans at news of Trajan's death (above).

The silver content of the Roman denarius will fall to 87 percent in the reign of Hadrian, down from 93 percent in the reign of Trajan.

Nonfiction *Historia* by Cornelius Tacitus, who dies at age 60. His work covers the years 69 to 96, his *Annales* the preceding period from 14 to 69.

118 The Roman Forum, commissioned by the late Trajan, is completed with triumphal arches, columns, a market complex, and an enormous basilica that replace hundreds of dwellings.

Largest city in the world, Rome has a population exceeding 1 million.

120 The Delphic priest Plutarch dies in his late 70s at Chaeron, leaving behind his celebrated *Parallel Lives*—biographies of Greek and Roman legislators, orators, soldiers, and statesmen that attempt to depict character (*see* 1579).

Rome's Pantheon is completed by architects of the emperor Hadrian.

121 Nonfiction *The Lives of the Caesars (De Vita Caesarium)* by Suetonius is spiced with anecdotes gathered from private sources and public records.

122 Hadrian's Wall goes up in Britain following the arrival in the spring of the emperor Hadrian on a tour of military inspection. The 72-mile wall from the Tyne to the Solvay is built mostly of stone with at least 16 forts and will provide a defensive barrier against the Picts and other tribesmen to the north.

123 The emperor Hadrian averts a war with the Parthians by meeting in person with the king of Parthia.

125 "Bread and circuses" *(panem et circensis)* keep the Roman citizenry pacified, writes Roman lawyer-satirist Decimus Junius Juvenalis (Juvenal), 65, in his *Satires.*

Bread and circuses, including games at the great new Colosseum, kept the Roman populace in line.

Plague sweeps North Africa in the wake of a locust invasion that destroys large areas of cropland. The plague kills as many as 500,000 in Numidia and possibly 150,000 on the coast before moving to Italy, where it takes so many lives that villages and towns are abandoned.

Famine contributes to the death toll produced by the plague in North Africa and Italy (above).

128 Roman agriculture declines as imports from Egypt and North Africa depress wheat prices, making it unprofitable to farm and forcing many farmers off the land.

Roman bakeries produce dozens of different bread varieties, and the Romans distribute free bread to the poor in times of need.

130 The Temple of Olympian Zeus, begun at Athens in 530 B.C., is completed by the Roman emperor Hadrian. It is 354 feet long, 135 feet wide, and 90 feet in height, the largest temple in Greece.

132 Jerusalem's Jews rise in anger against construction of a shrine to Jupiter on the site of the Temple. Led by Simon Bar-Kokhba and the rabbi Eleazar, they gain possession of Judea, beginning a 2-year insurrection.

135 Roman legions under Julius Severus retake Jerusalem and sack the city, kill Simon Bar-Kokhba at the village of Bethar near Caesarea, and end the Jewish War of Freedom. The emperor Hadrian, who has returned to Rome, orders the site of Jerusalem plowed under and a new city, Aelia Capitolina, built on the site. Judea is renamed Syria Palestine.

A Jewish *diaspora* begins as Hadrian bars Jews from Jerusalem and has survivors of the massacre dispersed across the empire. Many pour into Mediterranean ports, only to be sold into slavery. Christians may enter the new city but only if they have not sided with the Jews in the rebellion.

136 A magnificent villa for the emperor Hadrian is completed near Tivoli (Tibur).

138 The Roman emperor Hadrian dies at Baiae July 10 at age 62 after a 21-year reign. He has adopted Titus Aurelius Fulvius Boionus Arrius Antoninus, 52, in February on condition that Antoninus adopt Marus Annius Verus, 17, nephew of his bride Faustina, and Lucius Celonius Commodus, 8, adopted 2 years ago by Hadrian himself. The new emperor Antoninus, who will reign until 161, goes to the Senate and asks in person that the senators confer divine honors on the late Hadrian, an act for which he will be called Antoninus Pius.

The silver content of the Roman denarius will fall to 75 percent in the reign of Antoninus Pius, down from 87 percent in the reign of Hadrian.

139 Rome completes a splendid mausoleum for the late emperor Hadrian and his successors (*see* Castel Sant' Angelo, 590).

140 The Antonine Wall goes up in Britain from the Firth of Forth to the Firth of Clyde. Built of turf on a stone foundation as a barrier against the Picts and Caledonians, the wall is 10 feet high, 14 to 16 feet wide, and will have 13 to 19 forts along its 37-mile length.

161 The Emperor Antoninus Pius dies of fever at Lorium in Etruria March 7 at age 73 after a 23-year reign marked by prosperity in the provinces, liberal relief to cities in distress, construction of aqueducts and baths, progress in art and science, increased importance of the Roman Senate, and construction of the wall of Antoninus from the Forth to the Clyde in Britain. Antoninus is succeeded by his adopted son Marcus Annius Verus, now 40, who will reign until 180 as Marcus Aurelius.

The silver content of the Roman denarius will fall to 68 percent under Marcus Aurelius, down from 75 percent under Antoninus Pius.

165 Roman legions returning from the East spread a plague that may be smallpox, and it moves through much of Europe and the Near East, seriously depopulating the empire.

166 Rome expropriates peasant lands and awards them to returning legionnaires, few of whom will make good farmers.

167 The first full-scale barbarian attack on Rome destroys aqueducts and irrigation conduits, but the emperor Marcus Aurelius repels the invaders.

168 Marcus Aurelius and his co-emperor Lucius Verus subdue the Marcommani, a barbarian tribe from north of the Danube that has been in northwestern Italy for 7 years.

169 The Roman co-emperor Lucius Verus dies early in the year at age 39, leaving Marcus Aurelius to reign alone.

A full Roman army moves out in the fall to repel the Marcommani, who have broken the peace concluded last year. The Romans will drive off the Marcommani in the next 3 years, and the tribe will be virtually annihilated—as much by plague (which also infects the Roman army) as by force of arms.

174 The Roman legions of Marcus Aurelius defeat the Quadi in a victory that will be commemorated by a great column at Rome.

175 Roman legions in Asia revolt under the leadership of Avidus Cassius, following rumors that the emperor Marcus Aurelius has died. Cassius, who defeated the Parthians 10 years ago, proclaims himself emperor with the encouragement of his officers, but they assassinate him and send his head to Marcus Aurelius, who persuades the Senate to pardon Cassius' family.

176 Marcus Aurelius and his son Commodus enter Rome after a campaign north of the Alps and receive a triumph for their victories over the barbarians.

177 Marcus Aurelius begins a systematic persecution of the Christians at Rome, who oppose emperor-worship and thus pose a danger to the established order. The persecution forces adherents to the new religion to practice in secret. Many take refuge in the catacombs—the underground cemetery outside the city hewn out of solid rock since Etruscan times, and the fish becomes a symbol of Christianity (the initial letters of *Iesous*

177 ***(cont.)*** *Christos Theou Uios Soter* spell *ichthus*, the Greek word for fish).

180 The emperor Marcus Aurelius dies March 17 at age 58 after a week's illness either at his camp on the Save in Lower Pannonia or at Vindobona (Vienna). He is succeeded by his son Lucius Aelius Aurelius Commodus, 18, who will rule tyrannically and recklessly until his murder in 192.

Methodus Medendo by the Greek physician Galen, 52, in Rome correlates extant knowledge of medicine into a system that will influence medical thinking for 15 centuries, but his hundreds of treatises based on empirical therapy are speculative in terms of physiology and pathology and will mislead future generations. Galen has fled the city's plague after 16 years of practice in which he has treated Roman gladiators, merchants, and emperors. He has dissected two human corpses in defiance of Roman law and dissected Barbary apes to develop a view of anatomy that is basically correct. But he mistakenly attributes gout to overindulgence (*see* Garrod, 1859), and asserts that the inner walls of the heart contain invisible pores (*see* Columbus, 1559; Harvey, 1628).

Reflections, or *Meditations*, left behind by the emperor Marcus Aurelius is a compendium of his Stoic philosophy: "Nothing can come out of nothing any more than a thing can go back to nothing"; "Observe constantly that all things take place by change, and accustom thyself to consider that the nature of the Universe loves nothing so much as to change the things which are, and to make new things like them"; "What is good for the swarm is good for the bee"; "A man's happiness is to do a man's work"; "The act of dying is also one of the acts of life."

183 The emperor Commodus escapes death at the hands of assassins who have attacked him at the instigation of his sister Lucilia and a large group of senators. He puts many distinguished Romans to death on charges of being implicated in the conspiracy and puts others to death for no reason at all.

185 Rome's capable praetorian prefect Perennis is put to death at the request of a deputation from Britain representing mutinous legionnaires.

The Roman emperor Commodus drains the treasury to put on gladiatorial spectacles and confiscates property to support his pleasures and his soldiery.

189 The Roman mob blames a grain shortage on the mercenary freedman Cleander, who succeeded Perennis as prefect 4 years ago. Cleander is sacrificed to the mob.

Plague, possibly smallpox, kills as many as 2,000 per day in Rome. Dying farmers are unable to harvest their crops, dying carters are not able to deliver what grain there is, and food shortages bring riots in the city (above).

The column of Marcus Aurelius is completed at Rome.

192 The Roman emperor Commodus is murdered December 31, ending the Antonine line that has held power since 138. The emperor's mistress Marcia, his chamberlain Eclectus, and the prefect of praetorians Laetus have found their names on the imperial execution list and have hired the wrestler Narcissus to strangle Commodus (*see* 193).

Rome's plebeians rose up against the aristocrats, but legionnaires suppressed their revolt.

193 Publius Helfius Pertinax, 67, is chosen by the Roman Senate against his will to succeed the late Commodus as emperor, but his strict and economical rule arouses quick opposition, and he is murdered March 28 by members of the Praetorian Guard who invade the imperial palace. The empire is auctioned off to the highest bidder and goes to Rome's wealthiest senator, Didius Julianus, 61, who has made a fortune in shipping and who outbids the father-in-law of the late Pertinax by offering 300 million sesterces. Legates in Britain, Syria, and Pannonia challenge Didius Julianus, and the Pannonian legate Lucius Septimius Severus, 47, offers his troops on the Danube huge bonuses if they will leave immediately for Rome. He marches them 800 miles in 40 days, enters the capital June 1 in full battle dress, has Didius Julianus put to death in the palace baths, and begins a reign that will continue until 211.

The silver content of the Roman denarius will fall to 50 percent under Septimius Severus, down from 68 percent under Marcus Aurelius.

194 The Syrian legate C. Pescennius Niger Justus is defeated by the emperor Septimius Severus and put to death near Antioch.

195 "Scourges, pestilence, famine, earthquakes, and wars are to be regarded as blessings to crowded nations since they serve to prune away the luxuriant growth of the human race," writes the Carthaginian ecclesiast Tertullian (Quintus Septimius Florens Tertullianus), 35. He was converted to Christianity 5 years ago.

196 Byzantium is sacked by the emperor Septimius Severus and reduced to the status of a village.

197 Rome's British legate D. Clodius Septimius Albinus is defeated by Septimius Severus and slain February 19 at Lugdunum (Lyons). Legionnaires sack the town, and it will never regain its prosperity.

200 The Japanese empress Jingu sends a vast fleet to invade Korea. The Koreans capitulate at sight of the huge multioared ships and offer tribute.

Huns invade Afghanistan.

Jewish Talmudic law has its beginnings in the 39 tractates of the Mishnah compiled by Palestinian scholar-patriarch Judah ha-Kadosh, 65, of Sepphoria.

3rd Century

Barbarian prisoners taken by the legions often wound up working in Roman mines, galleys, and bathhouses.

Septimus Severus, Caracalla, Heliogabalus, Decius, Valerian, Aurelian, and Diocletian will rule the Roman Empire in this century.

Some 400,000 slaves perform the menial work of Rome, with middle-class citizens often owning 8, the rich from 500 to 1,000, an emperor as many as 20,000. Free urban workers enjoy 17 to 18 hours of leisure each day, with free admission to baths, sports events, and gladiatorial contests.

Roman public roads are so good that travelers can average 100 miles per day.

Rome establishes medical licenses, awarded only to trained physicians who have passed examinations. Medical societies and civic hospitals are set up, and laws are passed to govern the behavior of medical students. They are prohibited from visiting brothels.

Rome is a city of about 1.5 million, its people housed mostly in 46,600 *insulae,* or apartment blocks, each 3 to 8 stories high, flimsily made of wood, brick, or rubble, with shutters and hangings to deaden the nightly din of iron-rimmed cartwheels in the streets.

The average Roman breakfasts on bean meal mash and unleavened breadcakes cooked on cinders and dipped in milk or honey. Midday meals, often eaten standing up in a public place, consist generally of fruit, a sweetmeat, cheese, and watered wine (the *prandium*). The evening meal, or *convivium,* may include meat, fish, broccoli, cereals, a porridge of breadcrumbs and onions fried in oil and seasoned with vinegar and chick-peas.

204 A trade recession in North Africa's Leptis Magna region is alleviated by Rome's Emperor Septimius Severus, a native of Leptis, who buys up the country's olive oil for free distribution in Rome.

211 Rome's Septimius Severus dies in Britain at Eboracum (York) February 4 at age 64 and is succeeded by his eldest son Augustus, who is called Caracalla (or Caracallus) after the long-hooded tunic he introduces from Gaul.

212 The Edict of Caracalla (*Constitutio Antoniniana*) extends Roman citizenship to all free inhabitants of the empire with the exception of a limited group that may include Egyptians.

213 The Baths of Caracalla, completed by the Roman emperor, have public baths *(Thermae),* reading rooms, auditoriums, running tracks, and public gardens that cover 20 acres. The main building alone covers 6 acres and can accommodate as many as 1,600 at one time. The poor of Rome are obliged to bathe in the Tiber, and while Rome has sewers and public health inspectors (who enforce hygiene in brothels and markets), the streets are filthy, and in small towns and villages outside the capital the streets stream with excrement.

217 The emperor Caracalla is murdered April 8 by a group of his officers as he prepares to invade Parthia. He is succeeded by the Mauretanian M. Opellius (Severus) Macrinus, 53.

218 The emperor Macrinus tries to reduce the pay of Roman troops and is defeated and slain near Antioch June 8. His successor is the Syrian Varius Avitus Bassianus, 14, a grandnephew by marriage of the late Septimius Severus, who claims to be a son of Caracalla and calls himself Heliogabalus, or Elagabalus, taking the name of the Syrian sun king.

The silver content of the Roman denarius will fall to 43 percent under Heliogabalus, down from 50 percent under Septimius Severus, as he empties the treasury with his excesses while his mother Julia Maesa runs the empire.

222 The emperor Heliogabalus is murdered March 11 by the praetorians and is succeeded by his cousin and adopted son (Gessius) Bassianus, 14, who takes the name Marcus Aurelius Severus Alexander and begins a 12-year reign that will be dominated by his mother Mamaea.

The silver content of the Roman denarius will fall to 35 percent under Severus Alexander, down from 43 percent under Heliogabalus.

Government control of Rome's trade guilds will be extended under Severus Alexander.

Gunpowder will be invented in the next half century by Chinese alchemists of the Wu dynasty, who will mix sulfur and saltpeter in the correct proportions and at the correct temperature to produce the explosive (*see* 1067).

Tea will be mentioned as a substitute for wine for the first time in Chinese writings of the next half century.

226 The Sassanian dynasty that will rule Persia until 642 is inaugurated by the rebel prince Ardashir, grandson of Sassan, who 12 years ago gained control of the region surrounding Persepolis and now defeats Artabanus at Hormuz. Artabanus is killed and the Parthian Arsacid dynasty ends after roughly 500 years.

234 Ready-made bread rather than grain is issued to the poor of Rome by decree of the emperor Severus Alexander, now 26.

235 The emperor Severus Alexander buys peace from the Alamanni, who have invaded Gaul, and is killed March 18 by his troops on the Rhine. They proclaim the Thracian Gaius Julius Verus Maximinus, 62, emperor, and he begins a 3-year reign.

238 Roman subjects in Africa revolt against the emperor Maximinus and elect as emperor their proconsul Marcus Antonius Gordianus Africanus, 80. A rich descendant of the Gracchi and the emperor Trajan, Gordianus yields to public demand that he succeed Maximinus. The Senate and most of the provinces support him, but a supporter of Maximinus besieges Gordianus for 36 days at Carthage. Gordianus commits suicide at news that his son and namesake, 46, is dead, and the Roman populace proclaims his grandson, 14, the emperor Marcus Antonius Gordianus III. The praetorians conspire to murder the emperor Maximinus in mid-June, the Praetorian Guard names young Gordianus sole emperor, and he begins a 6-year reign.

The silver content of the Roman denarius will fall to 28 percent under the emperor Gordianus III, down from 35 percent under Severus Alexander.

244 The Roman emperor Gordianus III drives a Persian army back across the Euphrates and defeats the Persians in the Battle of Resaena. Mutinous soldiers murder the emperor at the urging of the Arabian Marcus Julius Philippus. They proclaim Philippus emperor, and he makes a disgraceful peace with the Persians.

The silver content of the Roman denarius will fall to 0.5 percent under the emperor Philippus, down from 28 percent under Gordianus III.

248 The emperor Philippus holds a great exhibition of games to celebrate the 1,000th anniversary of the founding of Rome in 753 B.C.

249 The Pannonian-born Roman commander Gaius Messius Quintus Traianus Decius, 48, puts down a revolt of troops in Moesia and Pannonia. Loyal troops proclaim Decius emperor, and he kills the emperor Philippus, who has advanced to oppose him at Verona.

250 The emperor Decius institutes the first wholesale persecution of Christians in an attempt to restore the religion and institutions of ancient Rome. The persecution produces martyrs who will be revered as saints.

Arithmetica by the Greek mathematician Diophantus at Alexandria includes the first book on algebra.

251 The emperor Decius and his son die fighting the Goths on swampy ground in the Dobrudja. His treacherous general Gaius Vibius Trebonianus Gallus, 46, succeeds as emperor, makes peace with the Goths, permits them to keep their plunder, and offers them a bribe not to return.

253 Roman soldiers who have campaigned against the barbarians on the Danube elect the governor of Pannonia and Moesia as emperor. The emperor Gallus marches out to meet his rival. The new emperor Aemilianus defeats Gallus and kills him but dies himself soon after, and a supporter of Gallus who has arrived too late to save him wins the support of the legions, who elect him emperor. Publius Licinius Valerianus, 60, begins a 7-year reign as the emperor Valerian.

A 15-year plague begins in the Roman Empire.

255 *De Mortalitate* by Thascius Caecilius Cyprianus, bishop of Carthage, describes a pandemic said to have started in Ethiopia. Symptoms include diarrhea and vomiting, an ulcerated sore throat, high fever, and gangrenous hands and feet. The pandemic will pass through Egypt and extend through Europe to the northernmost reaches of the British Isles.

"The world itself now bears witness to its failing powers," writes Bishop Cyprian (above) to the Roman proconsul of Africa. "There is not so much rain in the winter for fertilizing the seeds, nor in the summer is there so much warmth for ripening them. The springtime is no longer mild, nor the autumn so rich in fruit."

256 The great pandemic of the Roman world strikes violently in Pontus on the Black Sea and causes enormous loss of life in Alexandria, encouraging thousands to embrace Christianity.

258 Bishop Thascius Caecilius Cyprianus of Carthage is beheaded (*see* 255).

260 Rome's Emperor Valerian is defeated by Persia's Shapur I at Edessa, seized treacherously at a parley, and flayed alive. His son and co-emperor Publius Licinius Egnatius Gallienus, 42, reigns alone as the empire comes under attack on all sides by Berbers, Franks, Goths, Palmyrans, Vandals, and plague.

Runaway inflation makes the Roman denarius nearly worthless, paralyzing trade. The depression ruins craftsmen, tradesmen, and small farmers, who are reduced to bartering. Large landowners grow larger by buying up cheap land.

267 Palmyra's Prince Odenaethus, a staunch ally of Rome since his rebuff by Persia's Shapur I, is assassinated along with his eldest son, evidently on orders from the

267 ***(cont.)*** emperor Gallienus. His second wife, who has borne some younger princes, takes power as Septimia Zenobia and prepares to expand her desert realm to reach from the Nile to the Black Sea.

268 The emperor Gallienus is killed by his own soldiers at Mediolanum (Milan) while besieging the pretender Aureolus, who is slain in turn by the pretender Marcus Aurelius Claudius, who will reign until 270.

269 The emperor Claudius II repels a Gothic invasion of the Balkans and is given the title Gothicus for his victory.

Palmyra's Zenobia conquers Egypt, giving her control of Rome's grain supply.

270 The Illyrian Roman emperor Claudius Gothicus dies of plague and is succeeded by his brother Quintillus, who is proclaimed Marcus Aurelius Claudius Quintillus but is deserted by his troops. He commits suicide and is succeeded by an associate of his brother, who will reign until 275 as Lucius Domitius Aurelianus.

The silver content of the denarius has fallen to 0.02 percent, down from 0.5 percent under the emperor Philippus.

271 The Alamanni are expelled from Italy by the Roman emperor Aurelianus, who has abandoned trans-Danubian Dacia, settled its Roman inhabitants in a new Dacia carved out of Moesia, begun new walls to protect Rome, and is called *restitutor orbis* (restorer of the world).

272 The Emperor Aurelianus lays siege to Palmyra and his horsemen capture Zenobia and her young son Vaballathus on the banks of the Euphrates. She is forced to march in gold chains before the emperor's chariot in his triumphal procession, but Aurelianus spares her life.

Three Christians are beheaded on the road to the Temple of Mercury that stands atop a hill that will be named Montmartre (Mountain of Martyrs) in Lutetia, later to be called Paris.

273 The emperor Aurelianus sacks Palmyra to put down a revolt.

The emperor increases Rome's daily bread ration to nearly 1.5 pounds per capita and adds pork fat to the list of foods distributed free to the populace.

274 The emperor Aurelianus recovers Gaul from insurgent forces in a battle at Chalons and returns to Rome in triumph.

A 100-foot oar-powered ship is built for Japan's Emperor Ojin. The Japanese will not use sails for another 7 centuries.

275 Rome's legions retreat from Transylvania and the Black Forest, falling back to the Rhine and Danube. The situation has become so perilous that the emperor Aurelianus pushes construction of fortifications for Rome begun 4 years ago.

Aurelianus is slain by some of his officers as he prepares to invade Persia and is succeeded by an elderly senator, who is appointed against his will and will rule briefly as Marcus Claudius Tacitus.

A plague so severe that many wonder whether mankind can survive weakens the Roman legions in Gaul and in Mesopotamia.

276 The emperor Tacitus is killed by his troops after having defeated the Goths and Alans who invade Asia Minor. He is succeeded by his brother Marcus Annius Florianus, who is soon killed and succeeded by the Illyrian Marcus Aurelius Probus, who will reign until 282.

The Persian sage Mani is executed at age 60 after 30 years of preaching his "heresy" at the court of the late Sassanian king Shapur I and on long journeys to Turkestan, India, and China. Mani has claimed that he received divine revelations and was the final prophet of God in the world; his system combines Zoroastrian dualism with Christian salvation. He has incurred the hostility of the Zoroastrian priests at Ctesiphon, but his disciples will gain wide support for Manichaeism despite opposition from Byzantine and Roman emperors.

282 The emperor Probus tries to employ his troops in such peaceful projects as clearing the canals of Egypt after having driven the Franks and Alamanni out of Gaul, suppressed pretenders in Gaul, quieted Asia Minor, and strengthened defenses on the Danube. Probus is murdered by his troops in the autumn and is succeeded by an Illyrian who has served as praetorian prefect to the late Aurelianus and who conducts a successful campaign against the Persians to begin a brief reign as M. Aurelius Carus.

283 The emperor Marcus Aurelius Carus dies and is succeeded by his son and co-emperor Marcus Aurelius Numerius Numerianus.

284 The Roman emperor Marcus Aurelius Numerius Numerianus is assassinated in late summer and succeeded by his Illyrian general Gaius Aurelius Valerius Diocletianus Jovius, 39, who is proclaimed emperor at Chalcedon August 29 and who will reign with oriental despotism at Nicomedia in Bithynia until 305 while his colleague Marcus Aurelius Valerius Maximianus Herculius controls the West from Mediolanum (Milan).

August 29: first day of the calendar used by Coptics in Egypt and Ethiopia.

4th Century

Diocletian, Constantine, Licinius, Julian, Valens, and Theodosius will be the major emperors in this century but will be challenged in Rome's outlying provinces and territories by Persians and barbarians. Most of Rome's rulers will be of peasant or barbarian origin themselves, the hereditary upper classes having become effete.

301 The emperor Diocletian at Nicomedia limits prices of goods and services in an effort to end the economic distress caused by the collapse of Roman currency, but no attempt is made to enforce his edict in the West, and in the East it soon proves impracticable.

The kingdom of Armenia makes Christianity an official state religion, the first nation to do so.

302 The public baths of the emperor Diocletian at Rome open with 3,000 rooms. They are even larger and more elaborate than the Baths of Caracalla completed in 211.

303 A general prosecution of Christians in the Roman Empire begins February 24 by edict of the emperor Diocletian, who has been persuaded to revive the old religion in a move to strengthen the empire.

305 The Roman emperor Diocletian abdicates May 1 at age 60 and retires to Salona after a reign of nearly 21 years in which the last vestiges of republican government have disappeared. Diocletian is succeeded by the Thracian Galerius Valerius Maximanus, who persuaded the emperor to persecute Christians in 303. He assumes the title Augustus and begins an 8-year reign with his Illyrian colleague Flavius Valerius Constantius.

Rich landowners dominate the Roman Empire and enjoy the title of senator, which makes them exempt from the crushing taxes imposed on the rest of the population. The Senate has lost all its power and the landowners almost never attend Senate sessions. Labor and property is evaluated in terms of a unit of wheat-producing land (*iugum*), members of municipal senates (*curiales* or *decuriones*) are charged with the responsibility of collecting taxes and paying arrears; smaller landowners are held responsible for providing recruits for the legions and with keeping wastelands under cultivation.

306 The Roman co-emperor Constantius dies July 25 at age 56 outside Eboracum (York) during a campaign against the Picts and Scots. Roman troops in Britain salute his bastard son Flavius Valerius Constantinus, 18, as emperor, but the emperor Galerius in Rome elevates Flavius Valerius Severus to the rank of co-emperor. The despot Marcus Aurelius Valerius Maxentius leads an uprising, and the praetorians proclaim him Caesar.

307 The Roman co-emperor Severus dies November 11. The emperor Galerius replaces him with the Illyrian Flavius Galerius Valerius Licinianus, who will rule until 324 as the emperor Licinius. Flavius Valerius Constantinus in Britain proclaims himself emperor but rules only in Britain and Gaul (*see* 306; 312).

308 The Roman despot Marcus Aurelius Valerius Maxentius banishes his father Maximian to Gaul.

309 Anthrax or a similar plague begins to spread across the Roman Empire. The disease will sharply reduce the empire's population in the next 5 years.

311 The Roman emperor Galerius dies in May after the despot Marcus Aurelius Valerius Maxentius has driven him out of Italy. The despot's father Maximian committed suicide in Gaul last year when authorities discovered his conspiracy against the emperor Constantine, who begins to march on Rome.

Hun (Hsiung-nu) invaders from the north pillage the Chinese city of Luoyang, slaughtering 30,000. Northern and southern dynasties will divide China from 317 to 589.

312 The Battle of Milvian Bridge (or *Saxa Rubra*) 4 miles north of Rome October 28 gives the emperor Constantine a victory over the despot Maxentius. He kills Maxentius and becomes absolute master of the western Roman Empire.

Constantine (above) will claim to have seen a vision in the sky of a luminous cross bearing the words *In hoc signo vinces* (By this sign thou shalt conquer). He will adopt the words as a motto.

313 The Roman emperor Constantine and his co-emperor Licinius accept Christianity. They return property confiscated from Christians by the Edict of Milan.

314 The Battle of Cibalae October 8 gives the Roman emperor Constantine a victory over his co-emperor Licinius, who loses all of the Balkans except for Thrace.

315 The Arch of Constantine, completed by the Roman emperor, commemorates his victory at the Milvian Bridge in 312.

320 The Gupta dynasty begins to unify northern India after 5 centuries of division.

321 The Roman emperor Constantine forbids work on the Sabbath and endorses Sunday (*see* "blue laws," 1781).

The emperor Constantine assigns convicts to grind Rome's flour in a move to hold back the rising price of food in an empire whose population has shrunk as a result of plague (*see* 309). Barbarian peoples have used waterpower for years and pressure mounts to use such power in Rome, where rulers have opposed it in the past lest it cause unemployment.

323 The Battle of Adrianople July 3 gives the emperor Constantine's son Flavius Julius Crispus a triumph over the naval forces of Licinius. The co-emperor has angered Constantine with his anti-Christian policy, and he loses again September 18 at Chrysopolis in Anatolia.

324 The emperor Constantine has his co-emperor Licinius executed. He reunites the empire, ruling singlehanded from the Clyde River to the Euphrates.

The empire will recover from the chaos of the 3rd century under Constantine. External peace, internal unity, a new coinage supported by confiscated treasure, and state revenues supported by a new, simplified taxation system will bring prosperity to the great cities of the empire. Jobs will become hereditary, even in the female line (if one marries a baker's daughter, one must become a baker).

325 The Council of Nicaea summoned by the Roman emperor Constantine is the first ecumenical council of the Church. It supports the doctrine that God and Christ are of the same substance. The priest Arius has maintained the opposite view; he will be tortured to death in 336, but Constantine and his successors will move the Church increasingly toward Arianism. Athanasianism will not become the dominant view until after 379.

330 Constantinople is dedicated May 11 as the new capital of the Roman Empire. The emperor Constantine has spent 4 years building the city on the site of ancient Byzantium (*see* 658 B.C.).

333 The Romans begin pulling troops out of Britain and abandon work on the 72-mile Hadrian's Wall begun in 122. The wall includes at least 16 forts.

335 Jerusalem's Church of the Holy Sepulchre is consecrated September 17 on the site (discovered 7 years ago) of Christ's tomb on Golgotha.

337 The emperor Constantine dies May 22 at age 49. His wife Fausta persuaded him 11 years ago to execute Flavius Julius Crispus, his son by his first wife, and he is succeeded by three sons born to Fausta.

Rome will war indecisively with Persia for the next 24 years.

340 The Roman co-emperor defeats and kills his brother Constantine II in March at Aquileia in northern Italy and unites all of the West under his rule.

341 Persecuted Christians in Mesopotamia die by the thousands.

Coptic Christianity is introduced into Ethiopia. A variant of this communion will be the state religion.

350 The emperor Constans is murdered in a coup d'état by his military commander Magnentius, who usurps the Western Empire.

351 Magnentius is defeated at Mursa by Constantius II, who pursues him into Gaul.

353 Magnentius is defeated in August and commits suicide. Constantius II becomes sole emperor.

355 Barbarian Alamanni tribesmen cross the Rhine and wreak havoc in eastern Gaul.

356 Constantius II issues a decree February 19 closing all pagan temples in the Roman Empire.

357 Constantius II visits Rome for the first time April 28; his cousin Julian defeats the Alamanni at Strasbourg August 25 and drives them back behind the Rhine.

360 Japan begins a 30-year period of great influence in Korea.

The Huns invade Europe.

Picts and Scots cross Hadrian's Wall and attack Roman forces in Britain.

Roman authorities in Britain have encouraged production of wheat, which they export to supply the legions on the Rhine.

361 The emperor Constantius II dies November 3 near Tarsus in Cilicia at age 44 as he marches to join his cousin Flavius Claudius Julianus, 30. Constantinople acknowledges Julianus as sole head of the empire, and he enters the city December 11, beginning an 18-month reign as the emperor Julian.

Constantinople enforces a strict licensing system for physicians.

The new Roman emperor Julian (above) tries to organize a pagan church and substitute it for Christianity.

363 The emperor Julian sustains a mortal wound June 26 in a battle with the Persians. The last champion of polytheism, he is succeeded by the captain of his imperial bodyguard, Flavius Iovianus, 32, who will reign for 7 months as the emperor Jovian.

364 The emperor Jovian signs a humiliating treaty with the Persian shah Shapur II, yielding the kingdom of Armenia and most Roman holdings in Persia. He is found dead in February at Dadastana en route back to Constantinople and is succeeded by the Pannonian general Valentinian, 42, who appoints his brother Valens, 36, co-emperor. Valentinian I rules from Caledonia to northwestern Africa; Valens from the Danube east to the Persian border.

371 The neo-Persian Empire reaches the height of its power under Shapur II as the Romans and Persians renew their wars. Hostilities will continue for the next 5 years.

372 The Huns begin new incursions into the West. They defeat the Alans and the Heruls, destroy the Ostrogothic Empire of Hermanric, absorb the Ostrogoths for a time, and rout the Visigoths from the Dneister River.

Buddhism comes into Korea from China.

375 The emperor Valentinian dies November 17 at age 53 in a fit of apoplexy while attending a meeting on the Danube. Extreme cruelty has marked his 11-year reign, but he has founded schools and provided physicians to serve the poor of Constantinople. Valentinian I is succeeded nominally by his son of 4, who is hailed as

Valentinian II, but the boy's half brother Flavius Gratianus, 17, assumes the real power. He will rule from Milan until 383 as the emperor Gratian.

376 Roman authorities order Visigoths who have taken refuge in the empire to disarm and settle across the Danube in Lower Moesia. The Romans fail to disarm the Visigoths, they bungle administration of the refugees, and are obliged to fight with the barbarians.

378 The emperor Gratian completely defeats the southernmost branch of the Alamanni at Argenatria, but the Battle of Adrianople August 9 ends in disaster for the Romans. Mounted Visigoths kill the emperor Valens and rout his foot soldiers in a victory that presages a revolution in the art of war. The power of cavalry will determine European military, social, and political development for the next 1,000 years.

The emperor Gratian summons his general Flavius Theodosius, 32, to replace Valens as emperor in the East. A veteran of many campaigns with his late father, who was executed at Carthage 2 years ago on charges of conspiring against Valens, the new co-emperor has fought the Picts in Britain and defeated the Sarmatians in Moesia.

379 The co-emperor Theodosius assumes office at Sirmius January 19 with power over all the eastern provinces, he comes to terms with the Visigoths, and he settles them in the Balkans as military allies *(foederati)*.

The Persian shah Shapur II dies at age 70 after a lifetime reign in which he has humbled the Romans, conquered Armenia, transferred multitudes of people from western lands to Susiana (Khuzistan), rebuilt Susa, and founded Nishapur.

383 Roman legions desert the emperor Gratian at Lutetia (Paris), he flees to Lyons, and he is delivered over to one of the generals who have risen in revolt against Rome. The emperor is assassinated August 25 at age 25 after an 8-year reign in which Christianity has become the dominant religion in the empire.

The Roman general Magnus Clemens Maximus has led the insurrection in Britain and Gaul, Gratian's younger brother Valentinian II and his co-emperor Theodosius recognize Magnus Maximus as Augustus, he receives recognition in Britain, Gaul, and the Hispanic provinces, and he begins a 5-year reign as co-emperor.

388 The co-emperor Magnus Maximus crosses into Italy, but the co-emperor Theodosius defeats him July 28 at Aquileia. He is subsequently murdered while Theodosius devotes himself to gluttony and voluptuous living. Valentinian II, now 17, continues as co-emperor.

390 An insurrection in Macedonia angers the emperor Theodosius, who has 3,000 rebels massacred at Thessalonica. Bishop Ambrose of Milan forces Theodosius to perform public penance December 25.

391 Alexandria's library, a wonder of the ancient world, is destroyed by fire. The emperor Theodosius has ordered that all non-Christian works be eliminated.

392 The co-emperor Valentinian II is murdered May 15 at Vienne in Gaul at the instigation of his Frankish general Arbogast, who sets up the grammarian and rhetorician Eugenius as emperor. The death of Valentinian at age 21 enrages the emperor Theodosius, who marches against Eugenius.

Imperial Roman legions raped and pillaged until the Germanic "barbarians" turned the tables on them.

394 Eugenius is killed in battle September 6 by the barbarian legions of the emperor Theodosius. The Frankish general Arbogast escapes into the mountains but commits suicide 2 days later.

395 The Roman Empire splits into eastern and western empires following the death at Milan January 17 of the emperor Theodosius at age 49. His son Arcadius, 17, has married Eudoxia, daughter of the Frankish leader Bauto, and will rule from Constantinople while his son Honorius, 10, rules from Milan under the dominance of his Vandal master of troops Stilicho, whose daughter Maria will be married to Honorius in 398.

The split in the Roman Empire (above) is considered temporary but will prove permanent.

An estimated 330,000 acres of farmland lie abandoned in Rome's Campania, partly as a consequence of malaria from mosquitoes bred in swampy areas, but mostly because imprudent agriculture has ruined the land.

397 The Roman master of troops Stilicho drives Alaric and his Visigoths out of Greece after a 2-year campaign.

The Scottish apostle Ninian establishes a church at Whithorn to help him in evangelizing the southern Picts. A British chieftain's son, Ninian has made a pilgrimage to Rome and been consecrated a bishop after 15 years of study.

399 *Confessions* by the North African cleric-philosopher Augustine, 44, says, "How small are grains of sand! Yet if enough are placed in a ship they sink it." Converted to Christianity by Bishop Ambrose of Milan, Augustine has served for the past 3 years as Bishop of Hippo and will continue in that position for the next 31 years (*see* 426).

5th Century

401 Visigoths penetrate the northern defenses of Italy and begin to ravage the countryside.

402 The Battle of Pollentia April 6 ends in victory for the Roman legions of Stilicho who stymie the Visigoths in their efforts to move south.

405 The emperor Honorius closes the Colosseum in an austerity move that abolishes amusements.

406 Barbarian forces led into Italy by Radagaisus meet defeat at Florence August 23 as Roman legions under Stilicho break up the invading army.

Hordes of Vandals cross the Rhine under their new king, Gunderic, who will reign until 428. Allied with the Alans and the Sciri, they follow the Moselle and the Aisne and proceed to sack Rheims, Amiens, Arras, and Tournai, and turn south into Aquitaine.

Cultivation of rye, oats, hops, and spelt (a wheat used for livestock feed) is introduced into Europe by the invading Vandals, Alans, and Sciri (above), who also introduce a heavy-wheeled plow that enables farmers to plow deeper, straighter furrows (*see* 95).

Butter, introduced by the invading Vandals, Alans, and Sciri (above), begins to replace olive oil.

407 Britain is evacuated by Roman legions who are needed closer to home. The British Isles return to Saxon rule after 360 years of Roman control.

The Romans (above) have introduced oyster cultivation into Britain (*see* 110 B.C.).

408 Visigoths under Alaric besiege Rome; the master of troops Stilicho is murdered August 22 on orders from the emperor Honorius.

The eastern emperor Arcadius, whose eunuch general Eutropius has been unable to thwart barbarian invasions, dies at age 31 after a 13-year reign. He is succeeded by his son, 7, who will reign until 450 as Theodosius II but will be dominated by his sister Pulcheria.

The Visigoth king Alaric (above) exacts a tribute from Rome that includes 3,000 pounds of pepper (5,000 pounds by some accounts). The spice is valued for alleged medicinal virtues and for disguising spoilage in meat that is past its prime.

409 Alaric's Visigoths invade Italy again, and Alaric sets up a pagan emperor whom he soon deposes.

Vandals cross the Pyrenees into the Iberian Peninsula but find food supplies short as a Roman fleet blockades imports from the North African granary.

410 Invading Huns ravage the Roman Empire and extort tribute.

Rome is sacked August 14 (or 25) after a third siege by Alaric, who dies soon afterward in southern Italy.

The Huns (above) introduce trousers, which replace togas, and stirrups (a Chinese or Korean invention), which make it easier to ride horses.

411 The self-proclaimed emperor Constantine (Flavius Claudius Constantinus), who has gained control of Britain, Gaul, and Spain in the past 4 years, is defeated near Arles by the Roman general Constantius in the service of the emperor Honorius. Taken prisoner, Constantine is put to death at Ravenna.

412 Visigoth forces move into Gaul under the leadership of Ataulf, brother of the late Alaric.

414 Ataulf, new king of the Visigoths, is married January 1 at Narbonne to Galla Placidia, sister of the Roman emperor Honorius.

The weak-minded Byzantine emperor Theodosius II yields power to his sister Pulcheria, 15, who reigns as regent.

415 Visigoths invade the Iberian Peninsula and begin to conquer territory taken earlier by the Vandals.

A mob incited by Alexandria's new bishop Cyril, 39, who has driven out the city's Jews, tears the Neoplatonic philosopher Hypatia, 45, from her chariot in March, strips her naked, scrapes her to death with oyster shells, and burns her body.

417 Galla Placidia, sister of Honorius, remarries, this time with the Roman general Constantius.

421 The emperor Honorius makes his brother-in-law Constantius co-emperor, but Constantius III dies in September.

The eastern Roman emperor Theodosius II sends his army against Persia's king Varahran, who has been persecuting Christians.

422 Theodosius II concludes peace with Persia after 2 years of war. He also agrees to pay annual tribute to the Huns in order to buy peace.

The walls of Rome's Colosseum crack during an earthquake.

423 Visigoths settled south of the Danube by Theodosius II organize a farmers' strike (*see* 379). Only payment of what amounts to a huge farm loan prevents them from occupying Rome.

425 The Huns are halted in their unopposed advance on Constantinople by a plague that decimates their hordes.

426 *The City of God* (*De Civitate Dei*) by the Christian philosopher Augustine, now 72, declares that empires like Rome are merely temporary (temporal) and that the only permanent community is the Church—visible and invisible—the city of God.

The purpose of marriage is procreation, says Augustine (above), and his view will dominate Church thinking for at least 12 centuries.

427 Korea's King Changsu moves his capital from the banks of the Yalu River to Pyongyang.

429 North Africa is invaded by 80,000 Vandals under their king, Gaiseric, who leads his forces from the Iberian Peninsula across the narrow Straits of Gibraltar (*see* 415; 430).

430 The Vandals who have invaded North Africa extend their power along the Mediterranean coast and lay siege to Hippo.

Bishop Augustine dies August 28 in the siege of Hippo (above) at age 76, leaving behind his monumental work *The City of God* and other works that will have more influence on Christianity than those of anyone else except the apostle Paul.

Britain's Bishop Patrick is sent as a missionary to Ireland. A native of the Severn Valley, he will labor for 30 years to convert the Irish to Christianity.

431 The patriarch of Constantinople Nestorius is deposed for heresy by the Council of Ephesus for preaching the doctrine that in Jesus Christ there was joined in perfect harmony a divine person (the *Logos*) and a human person but that the harmony was of action, not of a single individual. Nestorius is banished to the Libyan desert, but his followers will spread Nestorianism widely through Persia, India, Mongolia, and China (*see* 781).

The cult of the Virgin begins to spread westward from Byzantium following a decree of the Council of Ephesus (above) recognizing Mary as the Mother of God.

433 Attila becomes leader of the Huns, whom he will rule until his death in 453.

436 The Burgundian kingdom of Worms established in 406 is destroyed by Attila's Huns.

The last Roman legions leave Britain.

439 The *Codex Theodosianus* drawn up by appointees of the emperor Valentinian III is a summary of Roman law.

Carthage falls October 29 to the Vandals, who have been led since 428 by Genseric (Gaiseric). He makes Carthage his capital.

The Vandals establish a North African granary that will enable them to enforce their will on other nations by making them dependent on North Africa for food staples.

441 German Saxons establish themselves at the mouth of the Thames River.

443 The Alamanni settle in Alsace.

444 A "pestilence" that is probably bubonic plague strikes the British Isles and makes the country vulnerable to conquest (*see* 449).

The wheelbarrow is invented by the Chinese.

446 Chinese Buddhists are persecuted by the northern Wei, who have heretofore encouraged the Buddhists but whose secular government is threatened by the drain of manpower and tax money to the temples and monasteries. Monks and nuns are murdered, temples and icons destroyed, and all men under age 50 prohibited from joining any monastic order in a program that will continue until 450, helping the Confucianism of the Han Chinese to gain dominance over Buddhism.

449 Britain is conquered by Angles and Saxons who have been invited to the British Isles to fight against Pict (Caledonian) and Scottish foes of the Britons' King Vortigern, but who have wound up fighting the Britons as well. The conquest has been facilitated by an epidemic that weakened the country 5 years ago.

450 The Hawaiian Islands are discovered by Polynesian chief Hawaii-Loa, who has sailed across 2,400 miles of open water from the island of Raiatea, near Tahiti (date approximate) (*see* Cook, 1778).

Metal horseshoes come into more common use in the Near East and in Europe, increasing the efficiency of horsepower in agriculture and transportation (*see* 770).

451 The Battle of Chalons ends in defeat for Attila's Huns at the hands of a Roman force under Flavius Aetius with help from the Visigoths. The Huns have triumphed over the Alans, Heruls, Ostrogoths, and Visigoths, ravaged much of the Italian countryside, and forced people to settle on marshy islands that will become the city of Venice (*see* 687).

452 The Japanese crown prince Karu is killed by his brother Anko, who becomes emperor following the death of his father Ingyo and will reign for 44 months. Anko kills his uncle Okusaka and marries his uncle's wife Nakatarashi.

Attila's surviving Huns in Europe are reduced in numbers by plague and food shortages.

453 Attila of the Huns dies, and his followers are driven out of Italy by Roman troops with barbarian reinforcements.

455 Rome is sacked by the Vandals whose pillage is so thorough that the name "Vandal" will become a generic word for a wanton destroyer.

Chichén Itzá is founded by Mayans on the Yucatán Peninsula of the Western Hemisphere. Chichén will spread to cover 6 square miles of pyramids, temples, an observatory, ceremonial ball courts, and dwellings.

455 ***(cont.)*** Barter economy replaces organized trade as Romans and other city dwellers desert their towns for the countryside where they will be less visible targets for barbarian attack (*see* 350).

Orae Favianae (Vienna) is struck by an epidemic that spreads through the Roman provinces. The disease is probably streptococcus or a form of scarlet fever with streptococci pneumonia.

456 The Japanese emperor Anko is killed by the 10-year-old son of his late uncle Okusaka. Young Mayuwa strikes while Anko is asleep with his head on the lap of Nakatarashi, but is himself killed with other princes by Anko's brother, 38, who will reign until 479 as the emperor Yuryaku.

460 A famine that will last for several years begins in the neo-Persian Empire.

466 The Huns invade Dacia but are repelled by the eastern emperor Leo I with the help of his generals Anthemius and Anagastus.

467 The eastern emperor Leo I has his general Anthemius elected emperor of the West, and together they mount a fleet of more than 1,100 ships with an army of 100,000 to attack the pirate empire of the Vandals in North Africa.

Rome has an epidemic that takes many lives.

468 The Imperial Fleet commanded by Basiliscus, brother-in-law of the eastern emperor Leo I, is surprised by the Vandal king Genseric, and half the vessels are sunk or burned.

Leo I repels another Hun invasion of Dacia.

471 The emperor's brother-in-law Basiliscus murders the Goth Aspar, using as an excuse the failure of the Imperial Fleet against the Vandals in 468, for which he himself was responsible. The Goths use the murder as an excuse to attack the approaches to Constantinople, but the threat ends when the emperor's son-in-law Zeno has the Gothic leader Ardaburius assassinated.

472 The barbarian general Ricimer kills the emperor Anthemius and replaces him with Olybrius, Ricimer dies August 19, the Burgundian Gundobad assumes command of the western army, Olybrius dies November 2, and Rome is without a western emperor (*see* 473).

473 Gen. Gundobad nominates Glycerius as emperor, but Julius Nepos marches on Rome with backing from the eastern emperor Leo, ousts Glycerius, and makes himself emperor June 24.

474 The eastern emperor Leo I dies at age 74 after a 17-year reign. He is succeeded by his Isaurian son-in-law, 48, who will reign for all but 2 of the next 17 years as the emperor Zeno.

475 The Roman commander Orestes drives Julius Nepos out of Italy and makes his own son, Romulus Augustus, western emperor.

476 The western Roman Empire founded by Augustus in 27 B.C. ends formally August 28 at Ravenna, although the Germanic tribes have long since protected and run the empire. The emperor Augustulus (Romulus Augustus) is deposed by the Herulian (Saxon) leader Odovacar (Odoacer); because he is a mere boy, Augustulus is sent off to Naples with an annual pension of 6,000 pieces of gold.

The eastern emperor Zeno is forced to abdicate by his wife's uncle Basiliscus, who usurps the throne. Despite intrigue and corruption, the eastern (Byzantine) empire will survive for another 977 years.

477 The eastern emperor Basiliscus is deposed by the ex-emperor Zeno, who regains the throne he will hold until 491.

480 The Visigoths extend their rule from the Loire to Gibraltar and from the Bay of Biscay to the Rhine. Their seat of empire is Toulouse.

481 The king of the Salian Franks Childeric I dies at age 44 after a 24-year reign. He is succeeded by his son, 15, who will reign until 511 as Clovis I.

482 The eastern emperor Zeno issues a letter ("Henoticon") in an unsuccessful effort to resolve differences between the eastern and western churches, but the Menophysite controversy continues to divide the churches.

488 An army of Ostrogoths commanded by Theodoric invades Italy at the persuasion of the eastern emperor Zeno.

490 Theodoric and his Ostrogoths lay siege to Ravenna.

491 The eastern emperor Zeno dies at age 65 after a 17-year reign and is succeeded by a palace official who is elevated to the throne at age 61, marries Zeno's widow Ariadne, and will reign until 518 as Anastasius I.

493 The Herulian leader Odovacar surrenders Ravenna March 3 after a 3-year siege to Theodoric, who invites Odovacar to dinner and has him murdered. Theodoric will unite Italy as an Ostrogoth kingdom that will control the peninsula until 554.

495 The Wei dynasty in China moves its capital to Luoyang.

496 Clovis of the Salian Franks defeats the Alamanni near Strasbourg.

Clovis (above) is converted to Christianity and baptized by his friend Remy, bishop of Rheims.

500 Bavaria is invaded by the Marcomanni from Bohemia, whose lands are taken up by the Czechs.

Incense is introduced in Christian church services, sweetening the air of congregations of unwashed worshippers.

6th Century

Disease, war, famine, and natural disasters will take a heavy toll of the decayed Roman Empire in this century.

501 The *Susruta* medical book that will become a classic of medicine in India is compiled (date approximate).

502 Gundobad, king of Burgundy, issues a new legal code at Lyons March 29 making Romans and Burgundians subject to the same laws.

The Bulgars ravage Thrace.

Persian forces sack the town of Amida in northern Mesopotamia as they battle troops of the Eastern Empire.

China's Liang dynasty is founded by Xiao Yan, who marches on Nanjing and forces the Qi rulers, his relatives, to yield their power.

505 Rome's Colosseum suffers damage from an earthquake as it did in 422 (*see* 80, 851).

507 A Frankish army under the command of Clovis defeats the Visigoths in the Battle of the Campus Vogladensis (Vouille). Clovis and his Burgundian ally Gundobad kill Alaric II, and Clovis annexes the Visigoth kingdom of Toulouse, but the Visigoths will remain in control of the Iberian Peninsula for the next 2 centuries, even though the native population of 6 million outnumbers them 30 to 1.

508 Theodoric's Ostrogoths drive the Franks out of Provence and recover Septimania (Languedoc) from the Visigoths. Theodoric serves as regent for his infant grandson Amalaric, the Visigoth king.

Paris (Lutetia) is established as the Frankish capital by Clovis. He has triumphed over the Visigoths and wants to be close to the lands he has conquered.

510 Provence is overrun by the Italian Ostrogoths, who consolidate their gains in the region.

Roman philosopher Anicius Manlius Severinus Boethius, 30, is appointed consul by his friend Theodoric, who rules the Ostrogoths from his capital at Ravenna (*see* 522).

511 Clovis of the Salian Franks dies November 11 at age 45. His Merovingian dynasty is continued by his four sons, Theodoric, Chlodomer, Childebert, and Lothair, who divide the Frankish kingdom and rule from capitals at Metz, Orléans, Paris (Lutetia), and Soissons, respectively.

513 Mount Vesuvius erupts as it did in 79 A.D., burying Pompeii once more under lava, mud, and ashes.

516 Gundobad, king of Burgundy, dies. He is succeeded by his son Sigismund, who will reign until 524, converting his people from Arianism to Christianity.

517 Buddhism is introduced to central China by the emperor Wu D. He has ruled since 502 and is converted to the Buddhist faith (*see* 260 B.C.).

518 The Byzantine emperor Anastasius I dies July 9 at age 88 after a 27-year reign and is succeeded by his uneducated Illyrian bodyguard, 68, who will reign until 527 as Justinus I. His able nephew Justinian, 35, counsels the new emperor.

519 The eastern and western Catholic churches are reconciled, ending the schism that began in 484.

520 *Institutionis Grammaticae* by Constantinople's Latin grammarian Priscian (Priscianus Caesariensis) codifies Latin grammar in 18 volumes that will be widely used through the Middle Ages. Violating rules of grammar will be called "breaking Priscian's head."

522 Rome's consul Ancius Boethius is arrested on charges of having conspired against Theodoric the Great. An aristocrat, Boethius admits that he would like the integrity of the Roman Senate restored, but he insists that all hope for that is gone and that the letters from him addressed to Justinus at Constantinople are forgeries. He is imprisoned at Pavia (*see* 524).

524 Boethius is executed without trial after a prison term during which he has written *The Consolation of Philosophy* (below).

Burgundy's King Sigismund is killed by Chlodomer after an 8-year reign and is succeeded by Godomar.

Rome and Persia renew hostilities to begin a war that will last for 7 years.

The Consolation of Philosophy (De Consolatione Philosophiae) by Boethius (above) alternates between poetry and prose in an exposition of neo-Platonism and stoicism; "For in all adversity of fortune the worst sort of misery is to have been happy" (II, iv).

525 Ethiopian forces conquer the Yemen.

Alexandrian explorer-geographer Cosmas Indicopleustes travels up the Nile. He will venture as far to the east as Ceylon, become a monk, and write *Topographia Christiana* to vindicate the biblical account of the world.

"Easter Tables" issued by Roman theologian-mathematician Dionysius Exiguus, 25, give the birth day of Jesus incorrectly as December 25, 753 years after the founding of Rome. The error will be standardized in all Christian calendars.

526 Theodoric the Great dies of dysentery August 30 at age 72 and is succeeded as king of the Ostrogoths by his grandson Athalaric, 10, who will reign until 534 with his mother Amalasuntha as regent.

Persian forces defeat a Roman army.

A magnificent tomb is erected at Ravenna for the late Theodoric the Great (above).

An earthquake shatters Antioch and kills between 200,000 and 300,000 people.

527 Constantinople's Justinus I takes his nephew Justinian as co-emperor April 1 as an incurable wound weakens him. He dies August 1 at age 77, and Justinian (Flavius Anicius Justinianus, or Flavius Petrus Sabbatius Justinianus) will reign until 565. His wife Theodora, now 19, will have a strong influence until her death in 545.

528 The Battle of Daras ends in defeat for Persian forces by Justinian's commander Belisarius, 23, who begins an outstanding military career.

529 Justinian issues the *Codex Vitus* (Code of Civil Laws).

Ratisbon (Regensburg) is made the capital of Bavaria.

The Benedictine order of monks is established at Monte Cassino near Naples by Benedict of Nursia, 49, who founds a monastery and formulates strict rules in his *Regula Monachorum*. Benedict inaugurates monasticism in western Europe (*see* 1080; Franciscans, 1209).

The academy founded at Athens by Plato in 347 B.C. is closed by the emperor Justinian on charges of un-Christian activity. Many of the school's professors emigrate to Persia and Syria.

530 Justinian's commander Belisarius gains another victory over the Persians but is defeated at Callinicum.

532 The Nika insurrection in January destroys large areas of Constantinople as crowds shouting "Nika!" (Victory!) set fires that destroy the city. Justinian panics but is persuaded by his wife Theodora to remain. At least 30,000 rebels are put to the sword by the troops of Belisarius, and the young military commander helps Justinian begin an era of absolutism.

Justinian signs a "Perpetual Peace" with the young Persian king Chosroe to free his Byzantine armies for operations in the West.

The Franks overrun the kingdom of Burgundy.

533 Belisarius invades North Africa with a relatively small force. He defeats the Vandals and regains the region as a Byzantine province for Justinian.

534 Malta becomes a Byzantine province.

Toledo becomes the capital of the Visigoth kingdom that controls the Iberian Peninsula and will remain the capital until 711.

535 Belisarius invades Sicily and moves north to conquer the Ostrogoth kingdom of Italy.

A Christian basilica is completed at Leptis Magna in North Africa.

536 Rome falls to Belisarius December 9 as Byzantine forces recover the peninsula from the Ostrogoths.

Provence becomes part of the Frankish kingdom.

A "dry fog" covers the Mediterranean region throughout the year, ushering in the most severe winter in memory. Volcanic dust, possibly from an eruption in the East Indies, is the cause.

537 Rome resists a year-long siege by the Ostrogoth "king" Witiges, who fails to force the city's surrender.

King Arthur of the Britons is killed in the Battle of Camlan (according to legend). Later evidence will suggest that the king was not a Celtic or Welsh monarch but perhaps rather the leader of a Sarmatian tribe whose ancestors came to England as mercenaries for the Romans.

Constantinople's Church of St. Sophia is dedicated December 27 after 5 years of construction. Designed by Anthemius of Tralles and Isidore of Mieltus with a large dome, mosaics, lavish use of gold, and lacelike carving, it is the finest church in Christendom.

538 Buddhism is introduced to the Japanese court of the emperor Senka, who receives a Korean delegation that includes some Buddhists (*see* 517; 585).

539 Ravenna falls to Belisarius, who captures the Ostrogoth "king" Witiges. But Belisarius is recalled to Constantinople, Milan is starved into submission by the Ostrogoths under their new chief Totila, 300,000 Milanese are put to the sword, and the Ostrogoths begin to reconquer the peninsula.

A Byzantine-Persian war begins that will last for 23 years.

The Japanese emperor Senka dies at age 72 and is succeeded by his 30-year-old half brother Kinmei, who will reign until 571.

540 Persian forces invade Syria and take Antioch from the Byzantines.

Byzantine rule in Italy is ended by Totila the Ostrogoth.

The Monastery of Vivarium near Squillace is founded by the Roman statesman Flavius Magnus Aurelius Cassiodorus, who retires from public life to devote himself to study and writing. He directs his fellow monastics in copying and translating Greek works.

541 Justinian contracts plague (below), and although he recovers after a few months, he is obliged to abandon plans to invade Gaul and the British Isles.

The Ostrogoth king Hildebad dies and is succeeded by his nephew Totila, who will rule until 552.

The Great Plague of Justinian (bubonic plague) spreads from Egypt to Palestine and thence to Constantinople and throughout the Roman-Byzantine world, bringing agriculture to a standstill and causing widespread famine. As many as 5,000 to 10,000 die each day for a period in Constantinople, and the plague will continue with resulting famine for the next 60 to 70 years in Europe, the Near East, and Asia.

542 The Great Plague of Justinian that came into Constantinople last year by way of rats imported from Egypt and Syria fans out through Europe.

De Excidio et Conquestu Britanniae by the British monk Gildas, 26, is a history of early Britain.

543 Belisarius completes his reconquest of North Africa from the Vandals.

Disastrous earthquakes shake much of the world.

545 Justinian attempts to impose the Roman date for Easter on Constantinople in place of the Alexandrian date. The people protest by refusing to patronize butcher shops.

546 Rome falls to Totila and his Ostrogoths after a decade of rule by the Byzantine forces of Belisarius. The city's aqueducts have been cut, its aristocracy has long since fled, its population has been reduced from 500,000 to little more than 500 civilians, and these few are starved into submission by Totila.

A monastery is founded at Beneventum by Cassiodorus (*see* 540).

547 The Great Plague of Justinian reaches the British Isles, where King Ida accedes to the throne of Bernicia, the more northerly of the two Anglo-Saxon kingdoms.

Bamburgh Castle is built by Ida (above).

Ravenna's Church of St. Vitale is completed in double octagonal shape with mosaic portraits of Justinian and his wife Theodora, who has introduced long white dresses, purple cloaks, gold embroidery, tiaras, and pointed shoes into Byzantine fashion.

The Great Plague of Justinian (bubonic plague) killed hundreds of thousands in Europe, the Near East, and Asia.

548 The Byzantine empress Theodora dies of cancer June 28 at age 40, leaving Justinian to rule alone.

Topographia Christiana by the Alexandrian explorer-monk Cosmas Indicopleustes describes the importance of the spice trade (especially in cloves and sweet aloes) in Ceylon and the harvesting of pepper in the hills of India (*see* 525).

549 Petra falls to the Persians, who will hold the eastern outpost of Byzantium for 2 years.

The Church of St. Apollinare is completed outside Ravenna at Classe after 14 years of construction.

550 Toltecs overrun the Yucatán Peninsula in the Western Hemisphere and conquer the Teotihuacan civilization.

Totila the Ostrogoth captures Rome for the second time.

East Anglia and Northumbria are founded as new kingdoms in the British Isles.

Wales is converted to Christianity by David (Dewi), who will be canonized in 1120 and will be the patron saint of Wales.

De Bellis is completed by the Byzantine historian Procopius, 60, who has served as private secretary to Belisarius and who has described the Persian, Vandal, and Gothic wars.

Poetry *Hero and Leander* by the Greek poet Musaeus Grammaticus.

Mosaics installed in the new Church of St. Apollinare at Classe near Ravenna include a depiction of *The Last Supper*.

551 A Byzantine fleet defeats the Ostrogoth navy.

552 The Battle of Tagina ends in victory for a Byzantine army under the eunuch general Narses, 74, who has replaced Belisarius. Narses takes his mercenary barbarian forces overland to invade from the north and triumphs decisively over the Italian Ostrogoths, killing their king Totila.

The European silk industry is begun by the emperor Justinian, who sends missionaries to China and Ceylon to smuggle silkworms out of the Orient.

Buddhism reaches Japan in the form of sutras and images sent as gifts by the king of Paikche (part of Korea) (*see* 538). The leader of the Soga clan urges their acceptance, the emperor grants permission to build a temple to house and worship the Buddhist image, but as soon as it is enshrined an epidemic sweeps the countryside. The image is removed from the Soga's hands and cast into the Naniwa canal (*see* 585).

553 Gen. Narses annexes Rome and Naples to Byzantium.

Missionaries in the British Isles converted Anglo-Saxons to Christianity.

Justinian makes the silk industry a Byzantine imperial monopoly.

Anecdota by the Byzantine historian Procopius is full of scandalous gossip about Justinian, Belisarius, and the late empress Theodora, but later historians will challenge its veracity.

554 Narses is made prefect (exarch) of Italy and completes his reconquest of the peninsula for Justinian.

Italian lands taken from the Ostrogoths are restored to their original owners by Justinian's Pragmatic Sanction, but the landowners have become serfs and the depopulated farmlands reverted to wilderness.

558 The Frankish kingdom is reunited by Clotaire I, king of Soissons, who becomes king of all the Franks following the death of Childebert, king of Paris.

Plague takes a heavy toll throughout the Byzantine Empire.

559 An army of Huns and Slavs advances to the gates of Constantinople but is repelled by Belisarius, who comes out of retirement to drive off the barbarian invaders.

560 The Kentish king Eormenric dies and is succeeded by his son, who will reign until 616 as Ethelbert I.

The Abbey of Bangor is founded in Caernarvonshire, Wales, by Deniol.

561 The Frankish king Clotaire (Lothar) I dies after a 3-year reign, and his sons divide the kingdom once again. Sigibert will rule Austrasia until 575, Charibert will rule Paris until 567, Guntram will rule Burgundy until 592, and Chilperic will rule Soissons until 584.

563 The Irish missionary Columba (Colum), 42, founds a monastery on the island of Iona in the Hebrides and begins to convert the Picts with 12 of his disciples.

565 The Byzantine emperor Justinian dies November 14 at age 83 after a 38-year reign. His nephew succeeds to the throne and will reign until 578 as Justin II; he pays Justinian's debts and declares religious toleration.

Lombards in the north of Italy drive the Byzantines south but permit them to retain Ravenna.

568 A kingdom that will rule northern and central Italy until 774 is founded by the Lombard ruler Alboin, who lays siege to Pavia. He destroyed the Gepidae 2 years ago with help from the Avars, killed the Gepidae king Cunimund, and married his daughter Rosamund. The Gothic wars—accompanied by famine and disease—have exhausted the Italian countryside.

571 The Japanese emperor Kinmei dies at age 62 after a 32-year reign. He is succeeded by his son Bintas, 33, who will reign until 585.

572 Pavia falls to the Lombard king Alboin, who takes over almost the entire Italian Peninsula.

A new Persian-Byzantine war begins. It will continue until 591.

573 Sigibert of Austrasia (France) goes to war against his brother Chilperic of Soissons at the urging of his wife Brunhilda. Chilperic has murdered his wife Galswintha, sister of Brunhilda (both are daughters of the Visigoth king Athanagild), in order to marry his mistress Fredegund. Sigibert appeals to the Germans on the right bank of the Rhine for help, and they obligingly attack the environs of Paris and Chartres, committing atrocities of all sorts.

575 Sigibert of Austrasia pursues his brother Chilperic as far as Tournai. As the nobles of Neustria are raising Sigibert in triumph on the shield in the villa at Vitry near Arras, he is assassinated by hirelings of his brother's mistress Fredegund. Sigibert is succeeded by his young son Childebert II, with Brunhilda as regent.

The Slovenes move into Cornelia.

577 English forces from Wessex defeat the Welsh at Deorham.

578 The Byzantine emperor Justin II dies after several periods of insanity. On the advice of his wife Sophia, he has raised his general Tiberius to the rank of co-emperor. Tiberius has ruled jointly with Sophia since December 574, and he now begins a 4-year reign as Tiberius II Constantinus.

579 Persia's King Chosroes (Khosrow) I dies after a 48-year reign that has extended his realm from the Oxus to the Red Sea.

580 The Lombards drive the last of Italy's Ostrogoths across the Alps. Few in number, they will never take Rome or Naples, are bitterly opposed by the natives, but will Italianize their names.

581 The Sui dynasty that will rule China until 618 is founded at Chang-an by Yang Jian, duke of Sui and chief minister of the northern Zhou who have ruled since 557. Yang, who is of mixed Chinese and Turko-Mongol blood, kills his ruler, the last of the Zhou, along with 58 royal relatives and proclaims himself the emperor Wendi (*see* 589).

582 The Byzantine emperor Tiberius II Constantinus dies after a 4-year reign during which Thrace and Greece have been inundated by Slavs. He is succeeded by his son-in-law, 43, who will reign until 602 as Flavius Tiberius Mauricius (Maurikios).

Nonfiction *The Ostographia* by Flavius Magnus Aurelius Cassiodorus, who dies at age 92, leaving behind an account of the Ostrogoth rule in Italy.

584 Chilperic of Neustria dies and is succeeded by his son, who will make himself king of all the Franks and reign until 628 as Clotaire II.

Mercia is founded in the British Isles and becomes a new Anglo-Saxon kingdom.

585 The Visigoth king Leovigild puts down a revolt by his son Hermenegild, who has married a Catholic princess and has been converted from his father's Arian faith. Leovigild imprisons his son, has him killed, and proceeds to conquer the entire Iberian Peninsula.

The Japanese emperor Bintas dies at age 47 after a 14-year reign and is succeeded by his brother Yomei, 45, who will reign for 2 years.

The king of Paikche sends another Buddha figure to Japan along with a famous ascetic master of Buddhist meditation, a nun, a reciter of Buddhist magic spells, a temple architect, and a sculptor of Buddhist images (*see* 552). Another temple is built, the new Soga chief converts three pubescent girls and makes them nuns, a new epidemic ensues, and Moriyo Mononobe burns the Soga temple. He opposes imperial rule and the Buddhist faith that the Soga clan has adopted as a tool in its rivalry with the Mononobe family (*see* 586).

586 The Visigoth king Leovigild dies and is succeeded by Recared, who will rule until 601.

Japanese Buddhism is called a "foreign" religion by Moriyo Mononobe and Okoshi Mononobe, who say it conflicts with the native Shintoism, but the new emperor Yomei and his grandfather Iname Soga support Buddhism. The agrarian-naturist Shinto religion will adopt Buddhist imagery to embody its gods with a Buddhist counterpart to every *kami* (deity) in the Shinto iconography.

587 The Japanese emperor Yomei dies at age 47 and is succeeded by a nephew of strongman Iname Soga. The new emperor Sushun, 66, will rule until 592.

Agents of Iname Soga (above) kill the anti-Buddhist Moriyo Mononobe.

The first Japanese Buddhist monastery is founded.

The Visigoths ruled by Recared are converted to Christianity.

588 The Sassanid Persian emperor Hormizd is deposed and assassinated after suffering military reverses at the hands of the Byzantines. He is succeeded by his son, who is helped by the Byzantine emperor Maurikios to gain the throne and who begins a 39-year reign as Khusru Parviz (Chosroes II) that will see the Persian Empire reach its zenith and suffer its downfall.

Arab, Khazar, and Turkish forces invade Persia but are repelled.

The Lombards are converted to Roman Catholicism under their king and queen, Authari and Theodelinda.

589 The Chinese Empire is reunited by the Sui emperor Wendi, who defeats Chen forces at Jian-kang (later Nanjing), ending the Chen dynasty that has ruled in the south since 557.

The Persian military deposes Chosroes II, who flees to Constantinople (*see* 591).

590 Pope Gregory I (below) establishes claims to papal absolutism as he leads Italian opposition to Lombard rule.

The Lombard king Authari dies after a 6-year reign and is succeeded by Turin's Thuringian Duke Agilulf, who marries the widow of the late king Alboin's grandson, founds a Roman Catholic Lombard state, and will reign until 615.

A plague strikes Rome but subsides, allegedly after Pope Gregory (below) has received a vision of the Destroying Angel sheathing his sword atop the mausoleum of Hadrian which is renamed the Castel Sant' Angelo (*see* 139).

Gregory I becomes the 64th pope—the first monk to be elected to the papacy—and at age 50 begins a 14-year administration of rigorous discipline during which Rome's aqueducts will be repaired, her courts reformed, and her people fed with doles of grain as they were under the old imperial rule.

591 The Byzantine emperor Maurice restores Persia's Chosroes II to his throne and receives territorial concessions for his help.

Lombard forces under Agilulf extend their advances in northern Italy (*see* 598).

592 The Byzantine emperor Maurice sends troops against the Avars and Slavs who have been threatening the Balkans and Constantinople.

The Japanese emperor Sushun is murdered after 5 years on the throne by agents of his uncle Umako Soga, who is jealous of the emperor's power. Sushun is succeeded by the widow of the late emperor Bintas; now 38, she will reign for 35 years beginning next year as the empress Suiko.

593 The Japanese empress Suiko is the first to receive official recognition from China and begins a long reign during a pivotal period in which Buddhism will take firm root and Japanese culture will be sinoized. Suiko's son, Crown Prince Shotoku, 19, is made prime minister and with strongman Umako Soga will hold power for 30 years.

Japan's Shitenno-ji monastery is founded at Osaka by Crown Prince Shotoku.

Construction begins at Osaka of the Temple of Four Heavenly Kings (Shitenno-ji).

Hoko Temple is built.

594 The Japanese empress Suiko announces that she will support Buddhism.

597 Pope Gregory sends the monk Augustine with 40 other monks to convert the Jutes in the British Isles to Christianity. Augustine lands in Thanet, baptizes Ethelbert of Kent, and founds a Benedictine monastery at Canterbury.

598 The Byzantine emperor Maurice makes peace with the Lombard king Agilulf, conceding northern Italy.

The first English school is founded at Canterbury.

Food production increases in northern and western Europe as a result of agricultural technology introduced by the Slavs, who have made it possible to farm virgin lands whose heavy clay has discouraged agriculture. The Slavs employ a new, lightweight plow with a knife blade (coulter) that cuts vertically, deep into the soil, and a plowshare that cuts horizontally at grassroots level, together with a shaped board, or moldboard, that moves the cut soil or turf neatly to one side.

A population explosion begins in northern and western Europe as the new agriculture (above) increases the availability of food.

7th Century

The Franks, Merovingians, and Carolingians will successively control most of Europe in this century as royal power declines and strong feudal lords rise in power to gain the allegiance of the people and replace state governments.

A Japanese feudal nobility will rise in this century.

601 Indian physicians compile the *Vaghbata*. At least one Indian medicinal herb mentioned in the classic work—*Rauwolfia serpentina*—will be employed in Western medicine (*see* 1949).

602 The Byzantine emperor Maurice (Maurikios) is executed at Chalcedon after being forced to witness the slaughter of his five sons and all his supporters by the centurion Phocas, who has been proclaimed emperor by legions fighting the Avars on the Danube.

The archepiscopal see of Canterbury is established by Augustine of Canterbury, who is made archbishop.

603 England's bishopric of Rochester is founded, St. Andrews Church of Rochester is built, and London's first St. Paul's Church is built.

604 China's Sui emperor Wendi is assassinated by his son after a 23-year reign in which he has attacked hereditary privilege, reduced the power of the military aristocracy, and established civil service examinations. The parricide son will reign until 618 as the emperor Yangdi.

Pope Gregory the Great dies March 12 at age 64 after a 14-year papacy that has laid the foundations for claims to papal absolutism, pioneered the conversion of Britain to Roman Catholicism, and led the war against the Lombards.

The Shotoku Taishi code issued by Japan's Crown Prince Shotoku is a constitution of sorts that demands veneration of Buddha, Buddhist priests, and Buddhist laws.

605 Persia's Chosroes II resumes war with the Byzantines (*see* 591). He will soon control Armenia and Syria (*see* 608).

A Chinese Grand Canal is completed by a million laborers who link existing waterways to connect the new Chinese capital, established last year at Luoyang, to the Long River. The canal will be extended to Hangchow by 610.

606 A northern India empire is established by Harsha of Thanesar, who will reign until 647.

607 The first Japanese envoy to China's Sui Court, sent by the empress Suiko, begins a long interchange that will lead to the sinoization of Japan.

Horyuji Temple and hospital at Nara are completed by Japan's Crown Prince Shotoku.

608 Persian forces cross the Taurus mountains into Asia Minor, meeting little resistance from the Byzantines.

Hokoji Temple at Nara is completed after 20 years of construction.

609 The Pantheon at Rome is consecrated as the Church of Santa Maria Rotonda.

610 Constantinople is attacked October 5 by a fleet under the command of Heraclius, 35, son of the governor of Africa. Gaining popular support, he overthrows the Byzantine emperor Phocas and has him hanged. With his father's help, Heraclius establishes a new dynasty that will rule for 31 years.

The prophet Mohammed at Mecca begins secretly to preach a new religion, to be called Islam. Now 40, the onetime camel driver, who at 25 married the 40-year-old widow Khadija, his employer, has become a merchant. Having meditated for years on the ignorance and superstition of his fellow Arabs, he feels called upon to teach the new faith which will grow to embrace a major part of mankind in the millennium ahead (*see* 613).

611 Antioch is sacked by forces of Persia's Chosroes II (Khusru, or Khosrow, Parviz) who is creating a neo-Persian empire to rival that of more than 8 centuries ago.

612 Chinese troops cross the Yalu River into Korea and attack Pyongyang; only 2,700 return out of 300,000.

Harsha of Thanesar takes the title "Emperor of the Five Indies" (*see* 606; 635).

Arnulf, counselor to the Frankish king Clotaire II, becomes bishop of Metz, his wife enters a convent, and his son marries the daughter of Clotaire's mayor of the palace Pepin of Landen.

The Monastery of St. Gallen is founded by Gallus, a disciple of Columban.

613 Clotaire II of the Franks unites Austrasia and Burgundy. He captures the queen mother Brunhilda, 80, widow of the late Sigibert, who was his uncle, and has her dragged to death behind a wild horse.

Northumbrians under Ethelfrit defeat the Britons near Chester.

The prophet Mohammed begins to teach openly. Mecca's leaders oppose any change in the traditional tribal and religious customs, persecuting those attracted to Mohammed's teachings (*see* 610; 622).

614 Damascus is sacked by the advancing Persian armies of Khusru Parviz.

614 *(cont.)* The *Edictum Chlotacharii*, issued by Clotaire II of the Franks, defines the rights of king, nobles, and the Church.

The Italian monastery of Bobbio is founded by Columban.

615 Jerusalem is sacked by the Persians, who take the "True Cross" as part of their booty (*see* 628).

The Irish missionary Columban dies at age 72, a year after founding a monastery in the Appenines.

616 Persian forces overrun Egypt and subjugate her people.

Ethelbert of Kent dies, Kent passes to Wessex, and Northumberland becomes the leading Anglo-Saxon state of the British Isles.

618 The Tang dynasty that will rule China until 907 is founded by an official of the Sui regime, who was helped last year to seize the key city of Chang-sen. He has the emperor Yangdi murdered and establishes himself as the emperor Kao Zu, meaning High Progenitor. The Tang dynasty will bring a golden era to Chinese culture.

619 Jerusalem is sacked by Sassanid Persian forces.

620 Chosroes II captures Rhodes and restores the Persian Empire as it existed in 495 B.C. under Darius I.

621 The Chinese establish an imperial bureau for the manufacture of porcelain. Their technology will advance further under the Tang dynasty (*see* 1708).

622 Mecca's city leaders oppose Mohammed's teachings and force the prophet to flee July 16 to the town of Yathrib (it will be called Medina; his flight will be called the Hegira. Civil war begins) (*see* 613; 629).

The Byzantine emperor Heraclius lands at Issus and defeats the Persian forces of Shar-Baraz.

624 The Byzantine emperor Heraclius, who last year invaded Armenia, surprises and defeats Persian forces under Shar-Baraz.

Some 300 followers of Mohammed surprise a reinforced Meccan caravan returning from Syria and defeat more than 1,000 Meccans.

625 Avars and Persians attack Constantinople, but the emperor Heraclius repels the attacks.

626 Japanese strongman Umako Soga dies.

Edinburgh is founded by Edwin of Northumbria, who begins Christianizing his subjects.

627 The Chinese emperor Kao Zu abdicates after a 9-year reign that has inaugurated the Tang dynasty. He is succeeded by his son, who will reign until 649 as Tai Zong.

Enemies of the prophet Mohammed move on Medina from Mecca and slaughter 700 Jews.

The emperor Heraclius invades Assyria and Mesopotamia with an army aroused by religious enthusiasm. Refusing to be distracted by the Avars who are constantly attacking his Balkan territories, Heraclius gains a decisive victory December 12 in the Battle of Nineveh and saves Constantinople from the Persians.

628 The Persian king Chosroes II, who regained his throne in 591, is imprisoned after a mutiny by the military following last year's disastrous Battle of Nineveh. He is murdered April 3 by his son, who takes power as Kavadh II and makes peace with the Byzantine emperor Heraclius. Prisoners are exchanged, conquered lands are mutually restored, and the "True Cross" carried off by the Persians is returned to Jerusalem (*see* 615).

Mecca falls to the forces of Mohammed, who writes letters to all the world's rulers explaining the principles of the Muslim faith.

The Japanese empress Suiko dies at age 74 after a 35-year reign and is succeeded by a grandson of her late husband Bintas. The grandson, 35, will reign until 641 as the emperor Jomei.

Byzantine soldiers bring sugar from India to Constantinople. They found it last year upon capturing the Persian castle at Dastagerd.

629 The Frankish kingdom gets its last strong Merovingian ruler as Clotaire II dies and his son Dagobert succeeds to the throne (*see* 613). Counseled by the bishop of Metz Arnulf and by Pepin the Elder (Pippin of Landen), Dagobert will make wide dynastic alliances, but his firm rule will lead to revolt (*see* 639).

The emperor Heraclius recovers Jerusalem from the Persians, who have held it since 619, and restores it to Byzantine suzerainty.

Mohammed returns to Mecca as master with the Koran (below) in which he has established the monotheistic principles of Islam. The Koran's Arabic name, *Quran*, means "recitation," and it says, "There is no god but Allah and Mohammed is His messenger."

Muslims (Mohammedans) will recognize the authority of the *Sharia*, a complex legal system much like that of the Jewish Talmud, which will forbid gambling, public entertainment, and art that portrays the human figure.

The *Sharia* will require women to remain veiled and segregated and will order severe punishments for crimes—murderers to have their hearts cut out, thieves to have a hand severed, adulterers to be tied in a sack and stoned to death. But it will condone polygamy to a limited extent, castration to create eunuchs, torture, slavery, and rape of slaves.

The Koran (above) depicts the joys of heaven and the pains of hell in vivid imagery while warning that the last day is approaching. Written largely in the first person as if voiced by God, it requires Muslims to offer prayers 5 times a day according to an elaborate ritual, to fast during daylight hours in the lunar month of Ramadan, to give one-fortieth of their income in alms, and to make a pilgrimage to Mecca once during their lives if such a pilgrimage is not physically or financially impossible.

The *Sharia* (above) forbids eating of pork and thus follows the Mosaic law of Deuteronomy in 621 B.C., but permits camel meat. The Sharia prohibits consumption of alcohol, but Mohammed's followers circumvent his prohibition against wine by boiling it down to a concentrate and sweetening it with honey and spices; Mohammed himself drinks *nabidth,* a mildly alcoholic ferment of raisins or dates mixed with water and aged for 2 days in earthenware jugs (but not longer lest it become too strong).

630 The Chinese Tang court established 12 years ago receives its first Japanese ambassadors.

Norway has her beginnings in a colony founded in Vermeland by Olaf Tratelia, who has been expelled from his native Sweden.

632 The prophet Mohammed dies June 7 at age 63 (or 65), leaving behind an Islamic monotheism whose believers will soon conquer the Near East and North Africa. The prophet's closest followers choose Abu Bekr, 59, to succeed him as ruler of Islam and give him the title caliph. He makes Medina his seat of power.

Mohammed's youngest daughter Fatima (accent on first syllable) dies at age 26, leaving two sons—Hassan and Hussein—who will found the Fatimid dynasty that will rule Egypt and North Africa from 909 to 1171.

633 Muslim forces attack Persia, beginning the conquests that will give them power over much of the world.

Edwin of Northumbria is defeated and killed by the Mercians, who overwhelm the Roman Catholic forces in the north of England and begin a period of anarchy. The Roman Church is replaced in many places by Celtic Christianity championed by Oswald of Northumbria (Aidan from Iona).

Christian churches at Alexandria, Antioch, and Jerusalem are turned into Muslim mosques.

634 The first caliph, Abu Bekr, dies August 22 at age 61 and is succeeded by Mohammed's adviser Omar, who will reign until 644, conquering Syria, Persia, and Egypt in a "holy war."

635 Damascus falls to Muslim forces under Khalid ibn-al-Walid, the "Sword of Allah," who defeats the Byzantines at nearby Marj al-Saffar. Damascus is made the seat of the caliphate and will remain so until 750.

Gaza falls to the Muslims.

Chalukyas in India repel an invasion by Harsha.

The Byzantine emperor Heraclius makes an alliance with Kuvrat, king of the Bulgars, to break the power of the Avars.

Basra is founded at the head of the Persian Gulf where the Tigris and Euphrates rivers converge. The port will become a major trading center for commodities from Arabia, India, Persia, and Turkey.

636 The Battle of Yarmuk east of the Sea of Galilee August 15 ends in victory for Islamic forces who crush a Byzantine army and gain control of Syria.

Churches go up at Glastonbury, St. Albans, and Winchester; Christians in southern Ireland submit to Roman Catholicism.

French and German language differences appear in the Frankish Empire.

637 Persian forces are decisively beaten by the Arabs at Qadisiya (Tunis) and at Ualula. The Muslims loot Ctesiphon and go on to invade Mesopotamia.

The Muslims do not force their conquered subjects to embrace Islam, but they do require acceptance of the Koran as the doctrine of divine teaching and will oblige their subjects to learn Arabic, thus building an empire united by a common tongue.

638 Jerusalem falls to Islamic forces under the caliph Omar. Another Arab army spreads out through Mesopotamia while a third reaches into central Persia; the Persians appeal for Chinese aid.

639 The Merovingian Frankish king Dagobert I dies January 9 after a 10-year reign as king of all the Franks. His son of 6 succeeds him as Clovis II of Neustria and Burgundy.

Arab forces invade Armenia.

Arab forces invade Egypt.

640 Byzantine forces are defeated by the Arabs at Heliopolis as the Islamic conquest of Egypt continues.

A Welsh army defeats a Saxon army in the British Isles, where the Roman Church has lost power and where anarchy reigns. The Welsh victory is credited partly to the fact that each Welsh soldier has affixed a leek to his helmet so that Welshmen will not accidentally kill Welshmen, and the leek will become the national emblem of Wales.

A Chinese mission sent by the Tang emperor Tai Zong studies Indian techniques of sugar manufacturing at Behar in the Ganges Valley dating to 100 B.C. Extensive cultivation and manufacture of sugar in China will begin within a few years.

641 The Japanese emperor Jomei dies at age 48 and is succeeded by his widow, 47, who will reign until 645 as the empress Kogyoku.

642 The Battle of Nehawand (Niharvand) gives the Arabs a final victory over the Persians and ends the Sassanian dynasty that has ruled Persia for 4 centuries. The Persian king Yezdigird III is a grandson of the late Chosroes II (Khusru Parviz). He appeals to the Chinese emperor for help, but his provinces are incorporated into the Arabian caliphate, and he will be murdered in a miller's hut near Merv in 651.

Alexandria capitulates to the Arabs, who complete their conquest of Egypt. The city's patriarch Cyrus has arranged the capitulation on condition that the Muslims guarantee security of persons and property and free exercise of religion in return for a payment of tribute.

642 ***(cont.)*** The Arabs begin construction of the Amr Mosque at Cairo.

644 The caliph Omar is assassinated at Medina November 4 and succeeded by Othman, 68, who will rule until his own assassination in 656.

Japan has a terrible famine, thousands die, and a new religion springs up (*Tokoyonomushi*). Devotees worship a large worm, get drunk on sake, dance in the streets, and give away all their money.

645 Japan has a coup d'état in June. Iruka, grandson of the late Umako Soga, is killed at the imperial palace by Nakanō-ōenō-ōji, a cousin of the late Shotoku, and by Nakatomi Kumatari, who founds the Fujiwara (Kumatari) clan that will later control the Japanese throne. The empress Kogyoku is removed, the 49-year-old grandson of the emperor Bintas is made emperor to begin a 9-year reign as Kotoku, Nakanō-ōenō-ōji becomes crown prince and prime minister, and the Taika period begins as the Japanese adopt the Chinese custom of giving names to periods.

Alexandria revolts from Arab rule at the appearance of a Byzantine fleet, and the city is recaptured by Byzantine forces, but Egypt's Arab governor Abdalla ibn Sa'd mounts an assault that retakes the city, and he begins building an Arab fleet.

646 A Great Reform edict changes Japan's political order. It will lead to the establishment of a centralized government with the emperor presiding over a Chinese-style bureaucracy and ruling from a palace in a permanent capital city (*see* Nara, 710).

648 Cyprus falls to Arab invaders from North Africa. Other Arabs invade Armenia, which will be conquered by 653.

653 The Visigoth king Recessuinth at Toledo draws up the *Liber ludicorium*, a code based on Roman law that establishes equality between Goths and Hispano-Romans without regard to racial or cultural differences.

654 Arab invaders plunder Rhodes.

The Japanese emperor Kotoku dies at age 58, and the empress Kogyoku, who was removed in 645, is restored. Now 60, she begins a 7-year reign under the name Saimei.

655 An Arab fleet defeats a Byzantine fleet under the personal command of the emperor Constans II off the Lycian coast, but disaffection grows among Arab forces in Iraq and Egypt as a result of blatant nepotism on the part of Othman, who has ruled the Islamic Empire since 644.

656 The Arab caliph Othman is assassinated at Medina June 17 at age 80. He is succeeded by Mohammed's nephew and son-in-law Ali ibn Abi Talib, but the succession is disputed. The new caliph's forces defeat the rebels December 9 at the Battle of the Camel (*see* 661).

661 The caliph Ali is assassinated January 24 in Mesopotamia by a former supporter who has become a Kharajite. Ali is succeeded by Muawiya, a relative who moves his seat of government to Damascus and founds the Omayyad caliphate that will rule the Islamic Empire until 750.

Supporters of the late caliph Ali (above) and his son al-Husain, now 37, will be called Shiites (from the Arabic *shi'at Ali*, or taking the part of Ali) (*see* 680).

The Japanese empress Saimei dies at age 67 and is succeeded by a son of the late emperor Jomei. Now 35, he will reign until 671 as the emperor Tenji.

663 The Byzantine court moves from Constantinople to Italy as the emperor Constans II tries to stop the Arab conquest of Sicily and to restore Rome as the seat of empire. The Lombards resist Constans, and no Byzantine emperor will hereafter visit Rome.

664 Kabul falls to Arab forces that have invaded eastern Afghanistan.

The Synod of Whitby, which founded a monastery 7 years ago, returns the British Isles to the orbit of the Roman Church. It adopts the Roman Catholic faith, and the Northumbrian king Oswiu accepts the Roman ritual (*see* Theodore of Tarsus, 669).

668 The Byzantine emperor Constans II dies under mysterious circumstances in his bath at Syracuse July 15 at age 37 during a mutiny. The Byzantine court returns to Constantinople after an absence of 5 years in which the Arabs have made annual invasions and devastations of Anatolia. Constans is succeeded by his sons Pogonatus, Heraclius, and Tiberius, but Pogonatus (the Bearded) will reign alone beginning in 680 as Constantine IV.

669 The system introduced by the archbishop of Canterbury (below) will become the model for the secular state, creating a new concept of kingship. Rival kingdoms will come together in the British Isles in the next 30 years, beginning an evolutionary process that will create an English nationality and English national institutions destined to spread throughout much of the civilized world.

Constantinople is besieged by Arab forces that have taken Chalcedon.

Theodore of Tarsus becomes archbishop of Canterbury at age 67 and introduces a strict Roman parochial system and a centralized episcopal system (above).

670 The founding of Quairawan (Tunis) consolidates the conquest of North Africa (Ifrikquiya) by Arabs.

671 The Japanese emperor Tenji dies at age 45 after a 10-year reign in which he has given the Fujiwara fam-ily its name. He is succeeded by his son Kobun, 23, but Tenji's brother Ooama objects that Kobun's mother was the late emperor's mistress, a commoner, and insists that the emperor must be entirely of royal blood.

672 The Japanese emperor Kobun is deposed after 8 months by his uncle Ooama, he commits suicide, and Ooama makes himself emperor with support from the

Fujiwara family. He takes the name Tenmu and begins a reign that will continue until 686.

675 Childeric of the Franks is murdered while hunting; his death leads to civil war and anarchy in the Frankish kingdom.

677 A Byzantine fleet destroys the Arab fleet at Syllaeum, ending the Arab threat to Europe.

678 Arab forces lift a 5-year blockade of Constantinople and conclude a peace that will last for 30 years.

The caliph Muawiya dies at Damascus after an 18-year reign. His son Yazid succeeds as sixth caliph, but Kufans in Iraq invite al-Husain, son of the late caliph Ali, to take the throne (*see* 680).

Arab horsemen sweep across North Africa to the Atlantic, where their leader Uqbah ibn Nafi rides into the waves.

680 Bulgarian forces defeat a Byzantine army. They have occupied the territory between the Danube and the Balkan Mountains while other Bulgars control Wallachia, Moldavia, and Bessarabia (*see* 675).

The Battle of Kerbela October 10 ends in victory for the caliph Yazid over al-Husain, who has advanced from Mecca but has been deserted by the Kufans and is killed.

The martyrdom of al-Husain (above) will be celebrated annually in the month of Muharram by Shiite descendants of Mohammed's daughter Fatima (*see* 632). Shiites consider themselves to be more faithful to Islamic law than Sunnis and resent the ascendancy of the Umayyad dynasty (*see* Persia, 1512).

681 Migrating Bulgar tribes under Khan Asparukh, who have crossed the Danube from the East, subjugate the Slavs and found the kingdom of Bulgaria (*see* 811).

682 Muslim forces led by Uqbah ibn Nafi overrun the south coast of the Mediterranean and occupy Tripoli, Carthage, and Tangiers, the last Byzantine bases in Africa.

685 Justinian II assumes the Byzantine throne at age 16, following the death of his father Constantine IV and begins a 10-year reign.

Picts defeat Northumbrians May 21 in the Battle of Nechtansmere, thwarting their efforts to gain control of Scotland.

686 The Japanese emperor Tenmu dies after a 14-year reign and is succeeded by his widow (and niece), 40, who has her late husband's son executed for alleged treason in order that she may be succeeded by her own son by Tenmu, but the boy soon takes ill. His mother will reign until 697 as the empress Jito.

Sussex is converted to Christianity, ending pagan resistance to the faith.

Yakushi Temple is completed at Nara before his death by the Japanese emperor Tenmu (above).

687 The city of Venice elects its first doge and begins its rise as a major power in the Mediterranean (below).

Pepin the Younger gains a victory at Testry and unites the Frankish kingdom.

Venice (above) has been built up from fishing villages settled by fugitives from the Huns. The city occupies some 60 marshy islands near the head of the Adriatic Sea, and its citizens have grown prosperous in the fish trade, in their salt monopoly, and by virtue of their central location (*see* 1071).

688 The Wessex king Ine subdues Essex and parts of Kent.

689 Justinian II defeats the Slavs in Thrace and transfers many of them to Anatolia.

690 Wintred becomes king of Kent to begin a 35-year reign in which he will draw up a code of laws.

691 The Dome of the Rock is completed at Jerusalem by the caliph Abd al-Malik.

692 The Battle of Sevastopol in Cilicia ends in a bad defeat for Justinian II at the hands of the Arabs.

695 Byzantine army officers depose the emperor Justinian II, cut off his nose, and exile him to the Crimea (Cherson). One officer assumes the throne to begin a 3-year reign as the emperor Leontius.

The first Arab coins are minted.

697 Carthage is destroyed by the Arabs, ending Byzantine rule in North Africa forever.

The Japanese empress Jito abdicates at age 32 after an 11-year reign and is succeeded by the 14-year-old grandson of the late emperor Tenmu. He will reign until 707 as the emperor Momu.

698 The Byzantine emperor Leontius is deposed by the commander of the fleet, who begins a 7-year reign as the emperor Tiberius III Apsimar. The new emperor will gain military triumphs over the Saracens.

The island of Heligoland is discovered by the bishop of Utrecht Willibrord.

699 *Beowulf* has probably been completed by this time. The heroic epic poem of more than 3,000 lines in Old English chronicles the deeds of the Geatish (southern Swedish) hero Beowulf who freed the court of the Danish king Hrothgar from the ravages of the ogre Grendel and Grendel's mother, became king of the Geats, and ruled for 50 years until called upon to rescue the country from a fire-breathing dragon. Mortally wounded while slaying the dragon, Beowulf was cremated on a giant funeral pyre.

700 Thuringia becomes part of the Frankish Empire ruled by Childebert III.

Algiers falls to the Arabs.

Fiction *The Adventures of the Ten Princes (Dasakuitiaracharita)* by the Indian Sanskrit poet-novelist Dandin.

8th Century

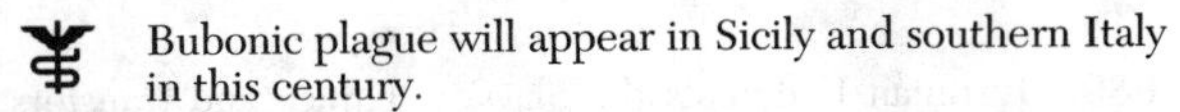

Bubonic plague will appear in Sicily and southern Italy in this century.

The Arabs in this century will lay waste the farmlands of Palestine, undoing the work of the Maccabees in the 2nd century B.C. and creating agricultural havoc that will not be repaired for 12 centuries.

Arab merchants will introduce Oriental spices into Mediterranean markets in this century.

701 The Japanese emperor Momu becomes sole proprietor of all the nation's land through a codification of political law.

Arab and Persian mariners visit the Spice Islands (the Moluccas) for the first time.

702 Justinian II Rhinotmetus regains the Byzantine throne that he lost in 695. He will rule until 711.

The Umayyad Mosque is erected at Damascus.

The circular church at Marienberg is erected near Würzburg by the duke Hetan II.

707 The Japanese emperor Momu dies at age 24 after a 10-year reign and is succeeded by his aunt, 46, who will reign until 715 as the empress Gemmei.

708 Tea drinking gains popularity among the Chinese in part because a hot drink is far safer than water that may be contaminated and may produce intestinal disease if not boiled. Tea is also valued for its alleged medicinal values (*see* Japan, 805).

709 Mont Saint-Michel has its beginnings in an oratory on Mont Tombe in the Bay of St. Michael on the coast of Normandy built by the bishop of Arrandes Aubert. Additions to the oratory will make Mont Saint-Michel an architectural masterpiece.

710 Nara (Heijo) becomes the capital of Japan, which until now has had a new capital with each new emperor or empress (*see* Kyoto, 794).

The Japanese empress Gemmei completes a palace at Nara that is a copy of the Chinese imperial palace.

Sugar is planted in Egypt (*see* 300 B.C.; A.D. 711; 1279).

711 Moors (Arabs and Berbers) from North Africa invade the Iberian Peninsula under the command of the freed slave Tariq ibn Ziyad and defeat the Visigoth king Roderick, who dies fighting in July at the Battle of Wadi Bekka near Rio Barbate. Cordova and the capital at Toledo fall to Tariq, who becomes master of half the peninsula and moves on toward Seville.

The Byzantine emperor Justinian II Rhinotmetus sallies forth from Constantinople to oppose insurgent troops who have revolted in the Crimea and marched on the capital under the leadership of a soldier named Bardanes. The troops defeat Justinian in northern Anatolia and put him to death in December, ending the house of Heraclius that has ruled since 610. They proclaim the incompetent Bardanes emperor under the name Philippicus (the empire will have two other short-lived rulers in the next 6 years).

Moors invading the Iberian Peninsula (above) introduce rice, saffron, and sugar.

712 Seville falls to the Moors, who invaded the Iberian Peninsula last year.

Samarkand falls to Arab forces led by Gen. Abu Qasim al-Thagafi, 17. The Arabs will make the city a center of Islamic culture.

The first Arab conquests in India are made by the Muslim general Mohammed ibn-Kasim, who crosses Makran (Baluchistan), invades the Indus Valley, and conquers Sind.

Nonfiction *Kojiki* by Yasumaro Ono is the first history of Japan.

713 The Chinese emperor Xuanzong (Ming Huang) ascends the throne to begin a reign that will continue until 756. His court will be a center of art and learning.

The Byzantine emperor Philippicus is deposed after a 2-year reign in which he has been defeated by the Arabs. He is succeeded by the conspirator Anastasius II, who will organize a strong army and navy and reign until 716.

715 The Japanese empress Gemmei abdicates at age 54 after an 8-year reign and is succeeded by her daughter, 35, who will reign until 724 as the empress Gensho.

716 The Byzantine emperor Anastasius is deposed in an army mutiny and is succeeded by an obscure and incapable tax official who is raised to the throne to begin a brief reign as the emperor Theodosius.

Lisbon falls to the Moors.

Chinese landscape painter Li Su Xun dies at age 65.

The new king of Neustria Chilperic II authorizes delivery of a shipment of spices from Fos to the monastery of Corbie in Normandy. Included are one pound of cinnamon, 2 pounds of cloves, and 30 pounds of pepper.

717 Constantinople is besieged by the Arabs, and the Byzantine emperor Theodosius is deposed. He is succeeded by Leo III, 37, who will reign until 741, frustrate the Arabs in their efforts to take the city, and inaugurate the Isaurian dynasty, which will control the throne until 802.

A new Arab caliph takes power in the person of Omar II (Omar ibn-al-'Aziz), who grants tax exemptions to all true believers and makes an unsuccessful effort to reorganize the finances of the empire.

Sculpture *Buddha with the Gods of the Sun and the Moon* is completed in bronze by Japanese artisans at Nara.

718 The Byzantine emperor Leo III destroys the Arab fleet in September, ending a 13-month siege of Constantinople and blocking further Arab expansion.

The Spanish kingdom of the Asturias is founded by Pelagius.

720 The Moors cross the Pyrenees and capture Narbonne.

Muslim forces invade Sardinia.

Nonfiction *Nihonshoki* by Yasumaro Ono at Nara is a chronology of Japanese history and marks the first use of the word *Nihon* (Nippon) to designate Japan.

724 The Japanese empress Gensho abdicates and is succeeded by her nephew Shomu, 23, son of Momu, who will reign until 749.

The new emperor Shomu (above) orders that houses of the Japanese nobility be roofed with green tiles as in China and have white walls with red roof poles.

725 Parsees who follow the teachings of the 6th century B.C. leader Zoroaster move to Sanjan on India's Gujurat coast and are welcomed by the local Hindu ruler. They fled Persia's Muslim invaders in the last century and have been in India since 706.

726 Greece revolts from the Byzantine rule of Leo III, who has forbidden the worship of icons (images) in a move to check superstition, miracle-mongering, and the spread of monasticism, which is draining thousands of men from active economic activity and is concentrating great wealth in the tax-exempt cloisters. A Greek fleet sets out for Constantinople with an antiemperor but is destroyed by the Byzantine imperial fleet with an incendiary mixture called Greek fire.

Pope Gregory II at Rome attacks the iconoclasm of Leo III.

730 The Alamanni are joined to the Frankish Empire as a dukedom.

Pope Gregory II excommunicates Leo III for his iconoclasm.

731 The Mayan Empire in the central Western Hemisphere begins its greatest period (*see* 1027).

Nonfiction *Historia Ecclesiastica Gentis Anglorum* by English scholar Bede, 58, of the monastery at Jarrow, marks the beginning of English literature. Bede will be known as "the venerable Bede" beginning in the next century.

732 The Battle of Tours near Poitiers October 11 ends the menace of a 90,000-man Moorish army that has invaded southern France under the Yemenite Abd ar-Rahman, who has crossed the Pyrenees, captured and burned Bordeaux, defeated an army under Eudo, duke of Aquitaine, and destroyed the basilica of St. Hilary at Poitiers. The Moors march on Tours, attracted by the riches of its famous church of St. Martin, but they are routed in battle by the Frankish leader Charles Martel (Charles the Hammer), 44, whose men kill Abd ar-Rahman. The Moors retreat to the Pyrenees, and their advance into Europe is terminated, partly by their loss to Charles Martel and partly by a revolt of the Berbers in North Africa.

The Moors continue to harry the coasts of Europe with help from the Venetians and take slaves that are sold in the markets of Venice.

Bubonic plague again strikes Constantinople as it did in 541. The plague will kill as many as 200,000 in the next 4 years.

Pope Gregory II orders the Benedictine missionary Wynfrith Boniface, archbishop of Hesse, to forbid consumption of horseflesh by his Christian converts in order that they may be seen to differ from the surrounding Vandals, who eat horsemeat as part of their pagan rites.

735 Charles Martel, mayor of Austrasia and Neustria, conquers Burgundy.

England's archbishopric of York is founded with Egbert as archbishop.

737 The Merovingian king of all the Franks Theodoric IV dies after a 16-year reign. His cousin Childeric will succeed after a 6-year interregnum.

Christians from the south invade Egypt to protect the patriarch of Alexandria.

739 Pope Gregory III asks Charles Martel to help fight the Lombards, Greeks, and Arabs.

The bishoprics of Passau, Ratisbon, and Salzburg are founded by Wynfrith Boniface.

741 The Byzantine emperor Leo III dies at age 61 after a 24-year reign that has saved the empire and has delivered eastern Europe from the threat of Arab conquest. He is succeeded by his son, 22, who will reign until 775 as Constantine V (Copronymus).

Charles Martel dies October 22 at age 53 after dividing his realms between his elder son Carloman and younger son Pepin (or Pippin), although the country has had no true king since the death of Theodoric in 737. Lands to the east, including Austrasia, Alemannia, and Thuringia, have gone to Carloman along with suzerainty over Bavaria, while Pepin has received Neustria, Burgundy, and Provence.

742 The Byzantine emperor Constantine V defeats his brother-in-law Artavasdus, who has led a 2-year insurrection in an attempt to usurp the throne, and renews his attacks on image worship.

743 The last Merovingian king of the Franks begins an 8-year reign as Childeric III succeeds to the throne left vacant by Theodoric IV in 737.

744 Swabia becomes part of the Frankish Empire of Childeric III.

Lombard rule ends in Italy as the Lombard king Liutprand dies at age 54 after a 32-year reign in which he has defeated the dukes of Spoleto and Beneventum, bringing Lombardy to the height of her power. Liutprand's policies will be continued by Aistulf, who will become king of the Lombards in 749 and reign until 756.

745 The Byzantine emperor Constantine V invades Syria, carrying the war to the Arabs.

Wynfrith Boniface, now archbishop of Mainz, receives a gift of pepper from the Roman deacon Gemmulus.

746 Cyprus is retaken from the Arabs by the Byzantine emperor Constantine V, who destroys a great Arab fleet.

Constantinople is struck by the worst plague since the 6th century Plague of Justinian (*see* 732).

747 Pepin's brother Carloman unexpectedly abdicates, becomes a monk, retires to a monastery near Rome, and leaves Pepin as sole master of the Frankish realm.

748 Todai Temple is completed at Nara after 5 years of construction.

749 The Battle of the Zab ends in defeat for the Umayyad caliph Marwin, who will be the last of his dynasty (*see* 750).

The Japanese emperor Shomu abdicates at age 48 after a 25-year reign and is succeeded by his daughter Koken, 31, who will reign until 758.

750 The Abbasid caliphate that will rule most of the Islamic empire for 350 years is inaugurated by Abu-Abbas al-Sarah, a descendant of the Prophet's uncle al-Abbas, who has most of the Umayyads massacred (*see* 754).

The kingdom of Galacia, established by the duke of Cantabria Alphonso I, conforms roughly to the Roman province of that name.

Plague follows a famine in the Iberian Peninsula, taking a heavy toll.

751 Charles Martel's son Pepin (the Short) has himself crowned at Soissons by the archbishop of Mainz Wynfrith Boniface and becomes Pepin III in a ceremony new to the Franks but one that gives the sovereign great prestige.

The first paper mill in the Muslim world is established at Samarkand. Two Chinese prisoners have revealed the technique of papermaking to their captors.

752 Japan's 55-foot Buddha statue *Rushanabutsu* is completed at Nara after 9 years of work.

754 Pepin III is crowned at St. Denis in January by Pope Stephen II; the new Frankish dynasty is proclaimed holy and its title indisputable (*see* 756). The pope has appealed for help from the Franks against the Lombards.

The Abbasid caliphate begun in 750 is firmly established by al-Mansur, 42, who succeeds his brother Abu-I-Abbas to begin a 21-year reign that will see the new caliphate recognized everywhere except in the Iberian Peninsula and Morocco. A revolt by al-Mansur's uncle Abdallah, governor of Syria, is crushed, and Abdallah is murdered on orders from al-Mansur.

A Tang census shows that 75 percent of Chinese live north of the Changjian (Yangtse) River. Chang-en has a population of 2 million, and more than 25 other cities have well over 500,000.

756 The Donation of Pepin establishes the papal states and begins the temporal power of the papacy. The Frankish king Pepin III has taken lands that legally belong to the Eastern empire, and he gives them to Pope Stephen II, thus tacitly recognizing claims of the popes to be heirs to the empire in Italy.

Abd-al-Rahman I is proclaimed emir of Cordova May 15 and makes it the capital of Moorish Spain, beginning a great period of prosperity. A Syrian prince who came to Spain last year, he is the only member of the Umayyad dynasty to escape the massacre of 750.

The Chinese emperor Xuanzong abdicates after a 44-year reign following the death of his favorite concubine, Yang Guifei, who had been his daughter-in-law. She had become the patron of his Mongol general An Lushan, who has rebelled, forcing Xuanzong and Yang Guifei to flee. The emperor's bodyguards have blamed Yang Guifei for their plight and demanded her death, and the emperor has permitted her to be strangled by his chief eunuch to appease the bodyguards. An Lushan seizes the capital and proclaims himself emperor (but *see* 757).

757 The Mongol adventurer An Lushan is assassinated at Luoyang, ending a revolt that has left millions dead with millions more fleeing as refugees to the south. Military leaders continue to hold power in much of the country.

Aethelbad of Mercia is murdered at Seckington by one of his bodyguards after a 40-year reign. He is succeeded by his kinsman Offa, who will reign until 796 (*see* 774; 779).

The Arab writer Ibn al-Mukaffa, 39, is tortured at Basra on orders from the caliph al-Mansur. His limbs are severed and he is thrown, still alive, into a burning oven.

758 The Japanese empress Koken abdicates after a 9-year reign and is succeeded by her cousin Junin, 25, who will reign until 764, but Koken and the Fujiwara family retain power.

759 Frankish forces in Gaul retake Narbonne from the Arabs, who have held it since 720, and gain control of Septimania (Languedoc).

Poetry *Manyoshu* by Japanese poets, including the emperor, noblemen, and commoners, contains some 500 poems.

761 The Japanese empress mother Koken is "cured" of a disease by the priest Dokyo, who has used prayers and potions and who becomes Koken's court favorite, arousing the jealousy of the emperor Junin.

762 The Arab seat of empire is moved to Baghdad by the Abbasid caliph al-Mansur, who starts building a new capital in the still fertile Tigris Valley (*see* 766).

764 A revolt against the Japanese empress mother Koken and her favorite Dokyo is led by Nakamaro Fujiwara, but the revolt is suppressed, the emperor Junin is forced into exile, and Koken reassumes the imperial throne. She takes the name Shotoku, and she makes Dokyo her prime minister, beginning a reign that will continue until 770.

765 European writings make the first known mention of a three-field crop-rotation system, describing a northern system in which some spring plantings supplement the traditional winter plantings of the south. The system makes a given section of land productive 2 years out of 3, instead of every other year, as one field is sown with wheat or rye at the end of the year, a second is sown in the spring, and a third is left fallow. The second field is sown with barley, broad beans, chickpeas, lentils, oats, or peas—food with more protein value.

766 Baghdad nears completion as 100,000 laborers create a circular city 1.5 miles in diameter with a palace at its center for the caliph al-Mansur (*see* 762). The city is ringed by three lines of walls, some containing bricks that weigh 200 pounds.

768 Pepin the Short dies September 24 at age 54 and is succeeded as Frankish king by his son Charles, 26, who will be called Charles the Great, or Charlemagne, in a reign that will continue until 814. Pepin's son Carloman becomes king of Austrasia.

Kasuga Shrine is erected at Nara by the Fujiwara family.

770 The Japanese empress Shotoku (Koken) dies at age 52 and is succeeded by a 62-year-old grandson of the late emperor Tenji, who will reign until 781 as the emperor Konin. The accession is engineered by Nakamaro Fujiwara, who prevents the accession of the crown prince Ochi.

Horseshoes come into common use in Europe, making horses more efficient for pulling farm implements on stony ground (*see* 450).

771 Charlemagne becomes king of all the Franks December 4 upon the death of his brother Carloman, king of Austrasia, and marries Hildegarde of Swabia.

772 Charlemagne begins a war with the Saxons that will end in 13 years with their subjugation on the Continent.

773 Charlemagne is crowned king of Lombardy after invading the country and subduing the Lombards.

774 Charlemagne becomes the first Frankish king to visit Rome, and he confirms the Donation of Pepin granted in 756 while making it clear that he is sovereign even in papal lands. Well over 6 feet tall, Charles the Great is a superb athlete. A merry man who can speak Latin and understand Greek even though he cannot learn to write, he has repudiated his 771 marriage to Hildegarde, daughter of the Swabian king Desiderius.

Charlemagne absorbs Lombardy into his Frankish Empire and establishes his rule in Venetia, Istria, Dalmatia, and Corsica.

The Mercian king Offa subdues Kent and Wessex.

775 The Byzantine emperor Constantine V Copronymus dies at age 56 after a 34-year reign in which he has suppressed monasticism and image worship, restored aqueducts, revived commerce, and repopulated Constantinople. He is succeeded by his son Leo the Khazar, 25, who will reign as Leo IV for 5 years, continuing Constantine's energetic campaigns against the Arabs and Bulgars. Byzantine forces defeat the Bulgars at Lithosaria.

The Abbasid caliph al-Mansur dies at age 63 after a 21-year reign in which he has made Baghdad the seat of a powerful empire. He is succeeded by his son, who will reign until 785 as al-Mahdi.

Tibet subdues her Himalayan neighbors and concludes a boundary agreement with the Chinese.

777 Charlemagne gains a victory over the Saxons and invades Moorish Spain, but his advance is checked at Saragossa, where he encounters a heroic defense.

778 Basque forces annihilate Charlemagne's rear guard August 15 at Roncesvalles in the Pyrenees.

Byzantine forces defeat the Arabs at Germanikeia and expel them from Anatolia.

Charlemagne's paladin Roland is killed at Roncesvalles (above); his death gives rise to the epic *Chanson de Roland (Song of Roland)*, the beginning of a great body of medieval French literature.

779 The Mercian king Offa makes himself king of all England.

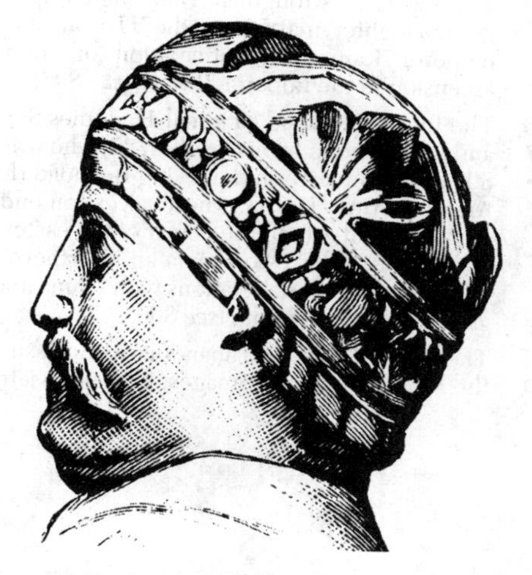

Charlemagne, the Frankish king who made himself the first of 29 Holy Roman Emperors to be crowned at Rome.

780 The Byzantine emperor Leo IV dies at age 30 after a 5-year reign in which he has been dominated by his beautiful Athenian wife Irene, now 28. He is succeeded by his son of 10, who will reign until 797 as Constantine VI, with Irene as regent until 790.

The queen mother Irene (above) restores image worship.

Charlemagne encourages the three-field system of crop rotation in his realms (*see* 765).

781 The Japanese emperor Konin dies at age 73 after an 11-year reign and is succeeded by his half-Korean son of 44, who will reign until 806 as the emperor Kanmu.

Nestorians in China build Christian monasteries. They have been proselytizing among the Chinese since 645 (*see* 431).

The bishopric of Bremen is established.

782 Arab forces advance to the Bosphorus, but agents of the Byzantine child-emperor Constantine VI and his mother Irene will buy the Arabs off.

Charlemagne executes 4,500 Saxon hostages at Verdun and issues the *Capitulatio de partibus Saxoniae,* making Saxony a Frankish province after 3 years of fighting. He imposes Christianity on the Saxons (but *see* 783).

The Mercian king Offa builds Offa's Dyke to protect his realm against Welsh attacks.

783 Saxons led by Widukind rebel against Charlemagne and massacre a Frankish army. Charlemagne kills his Saxon prisoners and launches a new invasion (*see* 782; 785).

The Byzantine general Staurakios wages a successful campaign against the Greek and Macedonian Slavs.

785 Charlemagne subjugates the Saxons once again and is reconciled with their leader, Widukind, who is baptized.

The third Abbasid caliph al-Mahdi dies after a 10-year reign and is succeeded by his son, who will reign for one year as al-Hadi.

The Japanese strongman Tanetsugu Fujiwara has his granddaughter married to the 11-year-old son of the emperor Kanmu in anticipation of young Heizei's ascension to the imperial throne (*see* 806).

786 The fourth Abbassid caliph al-Hadi dies September 24 and is succeeded by his brother, 22, who will reign until 809 as Harun al-Rashid, making Baghdad the center of Arabic culture. Harun, whose accession ends a decade of rivalry, will extend the power of the eastern caliphate over all of southwestern Asia and northern Africa. He will have diplomatic relations with China and exchange gifts with Charlemagne (*see* 809).

787 The Council of Nicaea abandons iconoclasm and orders the worship of icon images, a major victory for the monks, who will advance extensive claims to complete freedom for the Church in religious matters.

788 A Moroccan governor rebels from the caliphate that rules Islam from Baghdad to Cordova. He sets up an empire that will last for more than 1,000 years.

790 The Byzantine army mutinies against the unscrupulous queen mother Irene and the monkish party. It places Constantine VI in command of the Byzantine Empire at age 19.

Irish monks reach Iceland in hide-covered curraghs to begin settlement of that island (*see* 874).

792 Constantine VI recalls his mother Irene and makes her co-ruler of the Byzantine Empire.

793 Vikings raid the Northumbrian coast June 8, sacking the monastery on Lindisfarne and slaughtering many of the monks.

A paper mill is established at Baghdad as the Arabs spread the techniques developed by the Chinese in 105 (date approximate).

794 Japan's seat of government is transferred to Heian (Kyoto), where it will remain until 1868 (*see* 710). A golden age of Japanese culture begins that will endure for 4 centuries under the domination of the Fujiwara, Minamoto, and Taira families (*see* 1190).

Scotland has her first Norse invasion.

795 Ireland has her first Norse invasion 8 years after the first recorded raid of the Danes on England. The Danes have begun to pour into Ireland.

Charlemagne bans the export of grain from his dominions in order to avoid food shortages.

796 Charlemagne's forces reduce the Avars on the lower Danube.

King Offa of Mercia dies July 26 after a 40-year reign that has incorporated Kent, Sussex, Essex, and East Anglia into his realm. He has built a 150-mile dike to mark his border with Wales.

797 Constantine VI is seized and blinded by orders of his mother Irene, who has been placed in control of the Byzantine Empire by a new army uprising. She begins a 5-year reign as first Byzantine empress at age 45.

800 Charlemagne is crowned head of the western Roman Empire at Rome on Christmas Day but is not recognized by Irene, empress of the eastern Roman Empire.

An improved still is invented by the Arab scholar Jabir ibn Hayyan, but distillers are still able to do little more than separate liquids (like rosewater) from solids (*see* brandy, 1300).

9th Century

Scandinavian invaders in this century will overrun France, sack and burn Aachen and Cologne, and take London, Cadiz, and Pisa.

From the semibarbaric kingdoms of eastern and central Europe, and even from Europe's western edges, the Vikings will take slaves (the word comes from the same root as "Slav") and sell them to rich Muslims in the eastern Mediterranean.

Europe's peasantry is, in effect, enslaved to large landowners, who defend them in frequent wars, feed them in times of famine from stores held in reserve against such times, but generally exploit them.

Agronomic science is now in a decline, despite the growing use of the three-field system (*see* 780), and while the Scandinavians will improve agriculture, which has been unprofitable since the 4th century, Europe will experience famines from now to the 12th century.

Famines will be less severe in this century than they have been in the past and will be in the future, since almost everybody now lives on the land. Towns have practically disappeared, and since farm surpluses have no market, the farmers raise only enough for their own needs plus small quantities for barter.

Chinese aristocrats eat from translucent porcelain plates that Europe will not be able to duplicate for more than 900 years (*see* 621; 1708).

802 The Byzantine empress Irene is deposed by a conspiracy of patricians upon whom she has lavished honors and favors in a 5-year reign of prosperity. They exile her to Lesbos and oblige her to support herself by spinning (she will die there next year), replacing her with the minister of finance, who will reign until 811 as Nicephorus I.

805 Tea is introduced to Japan as a medicine. The Buddhist bonze (priest) Saicho, 38, has spent 3 years visiting Chinese Buddhist temples on orders from the emperor, and he returns with tea (*see* 708; 1191).

806 The Japanese emperor Kanmu dies at age 69 after a 24-year reign that has seen Korean culture and technology introduced into Japan. Kanmu is succeeded by his manic-depressive son Heizei, 32, who becomes hysterical at news of his father's death but will reign until 809.

Norsemen sack the monastery of Iona founded in 563.

Famine strikes Japan.

807 An Arab fleet ravages Rhodes.

808 The Byzantine emperor Nicephorus invades Arab territory.

Fez is founded in Morocco by the Abbasid king Idris. The tent colony will become the cultural center of North Africa.

809 The Abbasid caliph Harun al-Rashid dies March 24 at age 46 on an expedition to put down an uprising in Khurasan, one of many that have plagued his 23-year reign. The caliph will be glorified in Arabic legend and in the *Arabian Nights*, which will be published in English in 1888. One of his sons will reign until 813 as al-Amin, but his son Mamun is accepted as caliph in Persia and begins a revolt against al-Amin.

The Japanese emperor Heizei abdicates after a 3-year reign. His brother Saga, 23, succeeds to the imperial throne and will reign until 823.

Famine sweeps the empire of Charlemagne.

811 The Byzantine emperor Nicephorus invades Bulgaria, whose King Krum took Sofia in 809. He forces Krum to ask for terms, but the Bulgarians surprise the Byzantines and kill the emperor along with much of his army. Nicephorus has reformed Constantinople's finances in his 9-year reign, and his son Staurakios succeeds him for a few months. Staurakios is succeeded in turn by his brother-in-law, who will reign incompetently until 813 as Michael I Rhangabe.

812 Charlemagne orders that anise, coriander, fennel, flax, fenugreek, sage, and other plants be planted on German imperial farms.

813 The Byzantine army deposes the emperor Michael I Rhangabe and replaces him with an Armenian who will reign until 820 as Leo V.

The Abbasid caliph al-Amin surrenders Baghdad after his brother's general Tahir has granted peace terms, but he is treacherously murdered September 25. His brother will reign until 833 as al-Mamun (Mamun the Great), encouraging Arabian science and literature.

814 Charlemagne dies of pleurisy January 28 at age 71. The Carolingian Empire begins as Charlemagne's son, 36, becomes Holy Roman Emperor and will reign until 840 as Louis I (the Pious), a name derived from Clovis (*see* 508).

Japan's Minamoto (Genji) family has its beginnings as 32 of Emperor Saga's 50 children are made either commoners or Buddhist priests to reduce the imperial budget. The commoners will be called "Genji."

817 The Holy Roman Emperor Louis the Pious partitions the Carolingian Empire for the first time. His son Lothair receives most of Burgundy plus the German and Gallic parts of Francia, his son Louis the German receives Bavaria and the marches to the east, and his son Pepin (Pippin) receives Aquitaine and parts of Septimiania and Burgundy (*see* Treaty of Verdun, 843).

820 The Byzantine emperor Leo V is murdered in the mosque of Santa Sophia December 25 by supporters of his Phrygian general Michael, who has been sentenced to die for conspiring against Leo. The general will reign until 829 as Michael II Psellus (the Stammerer).

823 The Japanese emperor Saga abdicates at age 37 after a 14-year reign. He is succeeded by his brother, 31, who will reign until 833 as the emperor Junna.

825 Japan's Taira family has its beginnings as children of the late emperor Kenmu are made commoners to reduce the imperial budget. They are given the name Taira (*see* 1156; Minamoto family, 814).

826 Crete is conquered by Muslim pirates from Spain, who will use the island as their base until 961.

827 Saracens (Arabs) from North Africa invade Sicily, beginning a 51-year war of conquest.

Spinach is introduced into Sicily by the Saracens (above), who found the plant originally in Persia.

829 The Byzantine emperor Michael II dies. The arrogant religious fanatic Theophilus succeeds to the throne and will reign until 842.

832 The Byzantine emperor Theophilus promulgates a new edict against idolators and persecutes violators with great cruelty.

833 The Abbasid caliph al-Mamun dies after a 20-year reign and is succeeded by his brother, who will reign until 842 al-Mu'tasim, making Samarra his capital.

The Japanese emperor Junna abdicates at age 47. His nephew, 23, will reign until 850 as the emperor Ninmio.

837 Naples beats off a Saracen attack (*see* 827).

838 The Persian social and religious revolutionary Babak is cruelly executed January 4 by the Abbasid caliph al-Mu'tasim.

840 The Holy Roman Emperor Louis the Pious dies June 20 at age 62 on an island in the Rhine after suppressing a revolt by his son Louis the German. His son Lothair, 45, succeeds as emperor and tries to seize all the domains of the late Charlemagne. His son Charles, 17, succeeds as king of France and joins with Louis the German in resisting Lothair.

841 The Battle of Fontenoy ends in defeat for the Holy Roman Emperor Lothair, but while his brothers Charles and Louis triumph, Lothair will remain emperor until his death in 855.

842 The Byzantine emperor Theophilus dies in January after a 14-year reign devoted chiefly to warring against the caliphs of Baghdad, who have taken the city of Amorium, cradle of the Phrygian dynasty, slaughtered its 30,000 inhabitants, and razed it. Theophilus is succeeded by his son of 3, who will reign until 867 as Michael III (the Drunkard), his mother Theodora serving as regent in his minority and neglecting his education. He will war with the Saracens, Bulgarians, and Russians.

The Abbasid caliph al-Mu'tasim dies at Samarra; his two sons and a grandson will reign until 866.

843 The Treaty of Verdun, executed between the sons of France's late Louis the Pious, gives his son Lothair the title of emperor together with Italy, the valley of the Rhône, the valleys of the Saone and the Meuse, and the capital cities of Rome and Aix-la-Chapelle. France's Charles the Bald receives the rest of Gaul, Louis the German receives lands to the east, and the treaty breaks the unity of the Carolingian Empire.

844 The Holy Roman Emperor Lothair has his son Louis, 19, crowned king of Italy at Rome June 15 by Pope Leo IV. Louis soon marries a daughter of his uncle Louis the German.

France's Charles the Bald decrees that a traveling bishop may requisition at each halt in his journey 50 loaves of bread, 50 eggs, 10 chickens, and 5 suckling pigs.

847 Pope Leo IV builds the Leonine Wall to protect St. Peter's at Rome from Muslim invaders.

850 The Norseman Rurik makes himself ruler of Kiev. The house of Rurik will be an important Russian royal family until the end of the 16th century.

The Japanese emperor Ninmio dies at age 40 and, after a struggle over the succession, is succeeded by his son, 23, who will reign until 858 as Montoku.

Arab scientists perfect the astrolabe as a navigational aid. The compact instrument facilitates observation of the celestial bodies.

Salerno University has its beginnings.

Yiddish has its beginnings as groups of Jews settle in some of the German states and develop their own language, mixing German, Hebrew, and other languages.

Coffee is discovered, according to legend, by the Arab goatherd Kaidi in east Africa who notices that his goats become frisky after chewing the berries from certain wild bushes (*see* 1453).

851 Canterbury Cathedral is sacked by Danish forces that enter the Thames estuary, but they are defeated at Ockley by the king of the Kentishmen and West Saxons, Ethelwulf.

Pope Leo's 4-year-old Leonine Wall is damaged by a violent earthquake that further destroys the Colosseum at Rome (*see* 505).

852 The Bulgarian czar Malamir dies after a 23-year reign in which the Bulgarians have gradually expanded into upper Macedonia and Serbia. He is succeeded by Boris I, who will reign until 889 (*see* 864).

The Umayyad emir of Cordova Abd ar-Rahman II dies after a 30-year reign and is succeeded by Mohammad I, who will put down a Christian uprising and reign until 886.

Coal is mentioned for the first time in English chronicles, although the fossil fuel has been used to some extent for centuries to supplement the solar energy obtained from wood (*see* Theophrastus, 3rd century B.C.). The Saxon Chronicle of the Abbey of Peterborough states that the abbot of Ceobred has rented the land of Sempringham to one Wilfred, who is to send to the monastery each year "60 loads of wood, 12 loads of coal, 6 loads of peat, etc . . ." (*see* 1285).

853 The king of the West Franks Charles the Bald goes to war with his half brother Louis the German.

The first important Japanese painter, Kudawara Kuwanari, dies.

855 The Holy Roman Emperor Lothair dies September 29 at age 60 after dividing his lands between his three sons and is succeeded as emperor by his son of 33, who will reign until 875 as Louis II. He has been king of Italy since 844 and receives Italy along with the imperial crown while his brother Lothair II receives part of Austrasia, which is renamed Lotharingia (kingdom of Lothair), or Lorraine. A third son, Charles, receives Provence and southern Burgundy.

The king of the Kentishmen Ethelwulf makes a pilgrimage to Rome with his 6-year-old son Alfred.

856 The Kentish prince Ethelbald leads a rebellion against his father Ethelwulf.

An earthquake at Corinth kills an estimated 45,000 Greeks.

857 The first recorded major outbreak of ergotism occurs in the Rhine Valley, where thousands die after eating bread made from rye infected with the ergot fungus parasite *Claviceps* (clubheaded) *purpura* (purple). The fungus contains several alkaloid drugs including ergotamine, which is transformed in baking into an hallucinogen (*see* 943 ff.).

858 The Kentish king Ethelwulf dies after having given up the kingdom of the West Saxons to his rebellious son Ethelbald and is succeeded by his son Ethelbert, who will reign until 866.

Algeciras is sacked by Vikings, who are driven off by the Arabs.

The Japanese emperor Montoku dies at age 31 and is succeeded by his son Seiwa, 8, who will reign until 876.

860 The simplified Japanese alphabet Hiragana becomes popular among Japanese women. The phonetic alphabet will be further simplified and reduced to 51 basic characters that will be used to supplement the Konji (Chinese) alphabet, which contains thousands of characters.

861 Maurauding Norsemen sack Paris, Cologne, Aix-la-Chapelle, Worms, and Toulouse.

862 Novgorod is founded by the Scandinavian chief Rurik, who makes himself grand prince and establishes the Russian royal family that will rule until 1598.

863 The Cyrillic alphabet that will be used by Russians, Bulgarians, and other peoples is invented by the Macedonian missionary Cyril, 36, and his brother Methodius, 35, who have preached the gospel to the Khazar and who begin preaching in Moravia, where they have been sent by the Byzantine emperor Michael at the request of Moravia's ruler Rostislav.

864 The Bulgarian czar Boris I accepts Christianity.

865 Russian Norsemen sack Constantinople.

866 Byzantine strongman Caesar Bardas is murdered in April with the consent of his nephew, the emperor Michael III. The assassin is Michael's Armenian chamberlain Basil, who has divorced his own wife to marry a mistress of the emperor in order to please Michael, who rewards him by making him co-emperor.

867 The Macedonian dynasty that will rule the Byzantine Empire until 1054 is founded by the co-emperor Basil, who has Michael III murdered in September and will reign until 886 as Basil I. Raised in Macedonia, the new emperor will rebuild the Byzantine army and navy in an effort to restore the empire.

A schism with the Roman Church that will become final in 1054 is begun by the patriarch of Constantinople Photius, 47, who was excommunicated by Pope Nicholas I 4 years ago and now issues an encyclical against the pope, who is anathametized by the Council of Constantinople.

Pope Nicholas dies November 13 at age 67.

868 *The Diamond Sutra,* produced in China, is the world's first printed book (it will be found by archaeologists in the Caves of the Thousand Buddhas at Kansu).

869 The Abbasid caliph al-Mu'tazz is murdered by his mutinous troops and succeeded by a grandson of the late al-Mu'tasim (but *see* 870).

Arab invaders take Malta.

870 The Abbasid caliph al-Muqtadi abdicates under pressure from the Turks after a brief reign. The Turks choose another grandson of the late al-Mu'tasim as caliph; he will reign until 892 as al-Mu'tamid, moving his court to Baghdad.

Lorraine is partitioned under terms of the Treaty of Mersen forced upon the king of Lorraine Charles the Bald by the Holy Roman Emperor Louis II, who gains

870 ***(cont.)*** territory west of the Rhine. The realm is divided equally on the basis of revenue.

England's Ethelred I defeats the Danes December 31 in a skirmish at Englefield in Berkshire.

871 Ethelred of Wessex is defeated by Danish forces January 4 at Reading, gains a brilliant victory 4 days later at Ashdown, is defeated January 22 at Basing, triumphs again March 2 at Marton in Wiltshire, but dies in April. His brother, 22, pays tribute to the Danes but will reign until 899 and be called Alfred the Great.

874 Danish forces move into Mercia, King Burgred abdicates and makes a pilgrimage to Rome, and the Danes set up a puppet king.

Iceland is discovered by Viking Norsemen, who begin almost immediately to colonize the country (*see* 790).

875 The Holy Roman Emperor Louis II dies August 12 in Brescia at age 50 after having named his cousin Carloman, son of Louis the German, as his successor, but Charles the Bald persuades Carloman to go home, beats Louis the German to Rome, and makes himself emperor.

876 Louis the German dies at Frankfurt September 8 at age 72 while preparing for war with the Holy Roman Emperor Charles the Bald. Most competent of Charlemagne's descendants, he is succeeded by his son Carloman, who will reign as king of Bavaria until 880.

The Japanese emperor Seiwa abdicates at age 26 and is succeeded by his mentally and physically weak son Yozei, 8, who will reign until 884.

877 Exeter is seized by Danish forces while their leaders are deceptively negotiating with Alfred. He blockades the town, a relief fleet is scattered by a storm, and the Danes are forced to withdraw to Mercia.

Charles the Bald dies October 5 at age 54 while crossing the pass of the Mont Cenis en route back to Gaul. He is succeeded as king of France (but not as emperor) by his son Louis le Begue (the Stammerer), 30, king of Aquitaine, who will reign briefly as France's Louis II.

878 The English king Alfred is surprised by the Danes in January at Chippenham, where he has been celebrating Christmas. Most of his men are killed, but he escapes through woods and swamps with a small band of survivors and prepares a counteroffensive from Athelney. Troops from Somersetshire, Wiltshire, and Hampshire join Alfred in mid-May, he meets the Danes at Edington in Wiltshire, and he gains a great victory.

The Danish king Guthrum submits to Alfred (above) and accepts baptism to Christianity along with 29 of his chief officers.

879 France's Louis II (the Stammerer) dies at Compiègne April 10 at age 32 after an ineffectual reign of 18 months. He is succeeded by his teenage sons, who assume the throne jointly as Louis III and Carloman. They are crowned at Ferrière in September and divide the kingdom between them a few months later.

Alfred the Great divided parts of Mercia into shires, consolidated England, and drove off the Danes.

England's Alfred the Great clears the Danes out of Wessex and most of Mercia, but London remains in Danish hands.

881 The throne of the Holy Roman Empire, vacant since 877, is filled in February. Pope John VIII crowns the third son of the late Louis the German at Rome; the stout king of Swabia, 48, begins a 6-year reign as Charles III.

882 France's Louis III dies at St. Denis August 5 at age 19, and his brother Carloman becomes sole king.

883 The Zenj rebellion that has devastated Chaldea since 869 is quelled by al-Muwaffiq, brother of the Abbasid caliph al-Mu'tamid.

884 The Japanese emperor Yozei, who has devoted himself mainly to his horses, is forced to abdicate at age 16 after an 8-year reign and is succeeded by the 54-year-old half brother of his grandfather, who will reign until 887 as Koko.

France's Carloman dies December 12 while hunting and is succeeded as king of the West Franks by the Holy Roman Emperor Charles III (the Fat), son of the late Louis the German.

885 The Holy Roman Emperor Charles III disposes of his rival Hugh of Alsace, illegitimate son of the late Lothair II.

England's Alfred the Great retakes London from the Danes and suppresses a revolt by the East Anglian Danes.

886 Norsemen besiege Paris. The Holy Roman Emperor Charles III arrives with a large army in October but agrees to let the invaders withdraw, paying them a heavy ransom and permitting them to ravage Burgundy without interference on his part.

The Byzantine emperor Basil I dies August 29 after a 19-year reign that has been marked by rare wisdom despite its treacherous beginnings and despite Basil's lack of education. He is succeeded by a son of the late emperor Michael by Basil's widow Eudocia, who will reign until 912 as Leo VI (the Wise).

887 The Holy Roman Emperor Charles III (the Fat) is deposed in November by an assembly meeting at Tribur. There will be no emperor until 891.

The Japanese emperor Koko abdicates and soon dies at age 57. He is succeeded by his son, 20, who will reign until 897 as Uda.

888 Odo, count of Paris, is elected king of the West Franks by one faction to succeed the emperor Charles the Fat. He will be opposed by Charles the Simple, 10-year-old son of the late Louis the Stammerer, who will rule from Laon until 923 as the last Carolingian king but will have no real authority.

890 England's Alfred the Great extends the power of the king's courts and establishes a regular English militia and navy.

Fiction *Taketori Monogatari* (the story of a bamboo gatherer) by a Japanese writer.

891 The king of Italy Guy (Guido) of Spoleto is crowned Holy Roman Emperor by Pope Stephen V.

892 England is invaded in the autumn by Danish forces who arrive from the mainland in 330 ships and are accompanied by their wives and children.

The Abbasid caliph al-Mu'tamid dies after a 22-year reign in which he has lost his eastern provinces. His nephew will reign until 902 as al-Mu'tadid, restoring Egypt to the caliphate.

893 Danish forces in England are defeated by Alfred the Great's son Edward, 23, at Farnham and are forced to take refuge on Thorney Island. Another force of Danes besieges Essex, but Alfred raises the siege and the Danes receive other setbacks.

Nonfiction *The Life of Alfred the Great* by the Welsh monk Asser, bishop of Sherborne, who studies for 6 months each year in Alfred's household.

894 The Holy Roman Emperor Guy of Spoleto dies after a 3-year reign and is succeeded by the co-emperor Lambert with whom he has reigned since 892 and who will continue as emperor until 898 (but *see* Arnulf, 896).

Moravia's Svatopluk dies after a 34-year reign in which he has united Moravia, Slovakia, and Bohemia under the name Great Moravia to oppose the Germans but who has himself been overcome by the Magyars and their leader Arpád.

Danish forces in England retire to Essex after being deprived of food by Alfred the Great. They draw their ships up the Thames and the Lea and entrench themselves 20 miles above London.

895 England's Alfred the Great blocks the Lea River, captures the Danish fleet, and forces the Danes to move off to the northwest.

The Magyars are expelled from southern Russia; Arpád leads them into Hungary.

896 The Bavarian king Arnulf has himself crowned Holy Roman Emperor by Pope Formosus at Rome in February. He sets out to establish his authority in Spoleto but is seized en route by paralysis.

England's Alfred the Great ends the Danish threat to his country. The Danes abandon the struggle, and some return to the mainland, while others retire to Northumbria and East Anglia.

897 The Japanese emperor Uda abdicates at age 30 after a 10-year reign and is succeeded by his son, 12, who will reign until 930 as the emperor Daigo.

898 The Carolingian king Odo of the West Franks dies January 1, and his rival Charles III in Laon gains sole sovereignty after 5 years of civil war. He will reign until 923.

899 England's Alfred the Great dies after a 28-year reign in which he has forced invading Danes to withdraw, consolidated England round his kingdom, divided parts of Mercia into shires, compiled the best laws of earlier kings, and encouraged learning by bringing famous scholars to Wessex and by making his own translations of Latin works. Alfred is succeeded by his son Edward, now 29, who will reign until 924.

The Holy Roman Emperor Arnulf dies in December at age 49 and is succeeded as German king by his son Louis, 6, who will reign until 911 as Louis III (the Child), the last of the German Carolingian kings.

900 The Czechs assert their authority over all Bohemian tribes.

Greenland is discovered by the Norseman Gunbjorn, who is blown off course while sailing from Norway to Iceland and who comes upon the large ice-capped island (*see* 874; Eric the Red, 981).

Breslau has its beginnings in the Bohemian fortress of Wrotizlav built on the banks of the Oder River.

The first distinction between measles and smallpox is made by Rhazes (Abu-Bakr Muhammed ibn-Zakariya' al-Razi), chief physician of a busy Baghdad hospital, who gives the first description of smallpox and establishes criteria for diagnosing the disease that will be used until the 18th century (*see* inoculation, 1718).

England receives her first shipments of East Indian spices, used chiefly for their alleged medical benefits.

10th Century

The century will be marked by struggles between Muslims and Christians, Greeks and Russians, and Lombards, Bulgarians, Arabs, and Byzantines.

The work collar, or broad-breast collar, for draft animals will come into general use in this century, allowing a horse to draw from its shoulders without pressing upon its windpipe and quadrupling a horse's efficiency as compared with horsepower derived from animals harnessed with neck yokes of the kind employed on oxen.

Horses will come into wider use in those parts of Europe where the three-field system produces grain surpluses for feed, but hay-fed oxen will be more economical, if less efficient, in terms of time and labor and will remain almost the sole source of animal power in southern Europe, where most farmers will continue to use the two-field system.

Europe will suffer 20 severe famines in this century, some of them lasting for 3 to 4 years. An alimentary crisis will occur every few years for the next 3 centuries, in fact, and as populations grow and become more urbanized, the famines will strike more cruelly, although they may be fewer in number than in some earlier centuries.

901 The son of the late king of Provence (or Lower Burgundy) Boso, 21, is chosen king of the Lombards at Pavia, crowned at Rome in February by Pope Benedict IV, and will reign until 905 as the Holy Roman Emperor Louis III.

England's Edward the Elder takes the title "King of the Angles and Saxons."

904 Thessalonica is stormed July 31 by the Saracen corsair Leo of Tripoli, who plunders the town and carries off some 20,000 inhabitants as slaves.

905 The Holy Roman Emperor Louis III is surprised July 21 after having subdued the Lombards. He is blinded and sent back to Arles.

Navarre is made a kingdom as the Christian reconquest of Spain begins under Alfonso III, 57, of León and the Asturias.

906 Annam in Southeast Asia obtains independence after more than 1,000 years of Chinese domination.

The Chola king Aditya dies after a 35-year reign in which he has expanded his empire to cover all of southeast India.

907 China's Tang dynasty comes to an end after 289 years as the Khitan Mongols under Ye-lu A-pao-chi begin to conquer much of the country. Five different dynasties will assert imperial authority in the next 52 years, but none will exercise much power beyond the Yellow River Basin (*see* Song, 960).

Hungary's Magyar chief Arpád dies after having founded a dynasty that will reign until 1301 (although its first king will not be crowned as such until 997). The Magyars destroy the Moravian Empire and make raids into German and Italian territory.

Oleg, Prince of Kiev, besieges Constantinople with 2,000 ships and secures trading rights from the world's leading center of commerce and culture.

909 England's Edward the Elder defeats an army of invading Danes.

910 England's Edward the Elder gains fresh victories over the Danes and takes possession of London and Oxford on the death of his brother-in-law Ethelred of Mercia.

The kingdom of Asturias is renamed the kingdom of León with Alfonso III as its king.

The Byzantine emperor Leo VI (the Wise) is forced to pay tribute to the Magyars.

The Benedictine Abbey of Cluny is founded.

911 The Treaty of St. Clair-sur-Epte establishes the dukedom of Normandy, and Rollo the Viking (Hrolf the Ganger) becomes France's first duc d'Orléans as the Scandinavian Norsemen extend their domination over the Franks. Rollo will be baptized next year, taking the name Robert, and will acquire large parts of what will later be called Normandy.

The German king Louis the Child dies in early November at age 18, and the son of Conrad, count of Lahngau, is chosen German king November 8 at Forchheim. Lorraine transfers her allegiance to France.

912 The Byzantine emperor Leo VI dies after a 26-year reign in which he has completed the Basilian code of laws begun by his predecessor. He is succeeded by his brother, who will reign for less than a year.

913 The Byzantine emperor Alexander II dies and is succeeded by his 8-year-old nephew, son of the late Leo VI, who will reign until 959 as Constantine VII Porphyrogenitus ("born to the purple"). The government is administered by a regency composed of Constantine's mother Zoë Carbonopsina, the patriarch Nikolas, and John Eladas.

Bulgarian forces menace Constantinople. Their czar Symeon calls himself emperor of the Romans.

914 Adrianople falls to the Bulgarian czar Symeon, but Byzantine forces soon retake it.

England's Warwick Castle near Stratford-upon-Avon is fortified by a daughter of the late Alfred the Great.

915 Famine strikes the Iberian Peninsula, possibly as a result of a wheat crop failure due to rust.

917 The Bulgarian czar Symeon overruns Thrace in violation of a 913 agreement with Constantinople. The Byzantines launch a counterattack but are routed August 20 at Anchialus. Symeon gains control of the Balkans.

918 The German king Conrad I dies September 23 after a 7-year reign in which he has warred with the Danes, the Slavs, the Magyars, and the duke of Saxony, 42. The Lombard king Berengar, a grandson of Louis the Pious, was crowned Holy Roman Emperor in 915 and will continue as such until his death in 924, but Conrad has advised his nobles to make the duke of Saxony the next German king. They will elect the duke king at Fritzlar next May, and he will reign until 936 as Henry the Fowler.

919 Famine returns to the Iberian Peninsula, which is ruled in the south by the Amayyad caliph Abd ar-Rahman III and in the Christian north by princes who are establishing the beginnings of León, Castile, and Navarre.

922 France's Charles III (the Simple) is deposed by rebellious barons and replaced with Robert, brother of King Odo, who is crowned king of the Franks at Reims June 29 while Charles gathers an army to march against the usurper.

The Persian mystic Al-Hallaj (abu al-Mughith-al-Hsayn ibn Mansur), 64, is sentenced to death for heresy after a long trial and is flogged, mutilated, and beheaded March 27 at Baghdad. He has supported reform of the caliphate and been seen as a rabble-rouser.

923 The Battle of Soissons June 15 ends with the death of Robert at the hands of Charles III, but Charles is defeated and the barons elect Rudolf, duke of Burgundy, to succeed Robert. They imprison Charles, who will die at Peronne in 929.

924 England's Edward, son of Alfred the Great, dies July 17 and is succeeded by his son Athelstan, who will reign until 940, continuing his father's conquest of the Danelaw north of the Thames-Lea line from the Vikings who have been in Britain since 787.

Bulgaria's czar Symeon attacks Constantinople but is repelled at the city walls.

925 Henry the Fowler annexes Lotharingia (Lorraine) as the French fight among themselves.

926 The Bulgarian czar Symeon attacks Croat forces who have allied themselves with the Byzantines and is repulsed, meeting with his first defeat.

927 Bulgaria's Symeon dies of a heart attack May 27 after building an empire that stretches from the Ionian to the Black Sea. He is succeeded by his son Peter, who signs a peace treaty with the Byzantines in October.

Famine devastates the Byzantine Empire. Constantinople's Constantine VII and his co-emperor father-in-law Romanus Lecapenus push through stringent laws to prevent great landed magnates from buying up the small holdings of poor farmers.

928 France's Louis III (the Blind) dies at Arles in September at age 48 after a 27-year reign, of which 23 were sightless. His son Charles Constantine succeeds only to the county of Vienne.

929 Wenceslas, king of Christian Bohemia, is murdered at Prague by his pagan brother Boleslav.

Cordova's emir Abd ar-Rahman III proclaims himself caliph, establishing Spain as a Muslim power and a rival to the Fatimids of Ifriqiyah (Tunisia).

930 The Japanese emperor Daigo dies at age 45 after a 33-year reign and is succeeded by his son of 7, who will reign until 946 as the emperor Suzaku. The boy was kept indoors until age 3 by his mother, who will dominate him, but the ex-emperor Uda remains the power behind the throne and will retain power until his death next year at age 65.

932 Printing is used for the first time to reproduce Confucian classics of the 5th century B.C.

933 Henry the Fowler routs Magyar raiders March 15 at Merseburg.

936 The duke of Burgundy Rudolf dies and is succeeded as king of France by the son of the late Charles III. His mother is the sister of England's Athelstan and fled with him to England after the imprisonment of Charles in 922. Now 15, the boy is chosen king by the count of Paris, Hugh the Great, whose father Robert was killed at Soissons in 923. He is consecrated at Laon June 19 and will reign until 954 as Louis IV d'Outremer.

The German king Henry the Fowler dies at Memleben July 2 at age 60 after a 17-year reign in which he has created the Saxon army and founded Saxon town life. He is succeeded by his son, 23, who will be crowned Holy Roman Emperor in 962 and reign until 973 as Otto I.

Tatars attack Beijing and capture the town (*see* 1151).

939 Otto I gains victory in the Battle of Andernach over Eberhard of Franconia and other rebellious dukes, asserting Saxon power over all Germany after years of anarchy, internal rivalries, and foreign invasions since the breakup of the Frankish Empire.

Vietnam gains independence after nearly 1,000 years of Chinese rule (*see* Annam, 906).

Japan's Taira and Minamoto clans challenge the imperial court at Kyoto, causing chaos in the provinces.

940 England's Athelstan dies after a 16-year reign and is succeeded by his son, who will reign until 946 as Edmund.

Irish-Norse Vikings led by Olaf conquer York and force the Anglo-Saxon king Edmund to cede lands (*see* 944).

941 Igor, Prince of Kiev, crosses the Black Sea while the Byzantine fleet is in the Aegean, plunders Bithynia, and reaches the gates of Constantinople, but the Greek navy drives off the Russians and nearly annihilates their fleet with the help of "Greek fire," an incendiary liquid that burns atop water.

943 Ergotism strikes Limoges in France, killing an estimated 40,000 who have eaten bread made from diseased rye (*see* 857; 1039).

944 Danish settlers help Edmund regain the territory he ceded to Olaf in 940.

945 Igor, Prince of Kiev, is killed in battle with Drevlianian tribesmen and is succeeded by his widow, Olga, who will reign until 962. She will be the first Slavic ruler to embrace Christianity.

Baghdad falls in December to Shiite forces under Imad ibn Buwayhid, who takes the capital of the once-powerful Abbassid caliphate but keeps the caliph as a figurehead while he tries to restore peace (*see* 946).

Kyoto is invaded by several thousand farmers who demonstrate against requisition of their rice and other crops.

946 The Abbasid caliph al-Mustaqfi is blinded and deposed in January. His realm is being taken over by the three Buwayhid brothers (Imad al-Dawla, Runk al-Dawla, and Mu'izz al-Dawla) whose descendants will, in turn, lose their dominions to the Ghaznavids, Kurdish Kakwayhids, and Seljuk Turks (*see* 1055).

England's Edmund I dies after a 6-year reign and is succeeded by his brother, who will reign until 955 as Edred.

The Japanese emperor Suzaku dies at age 23 after a 16-year reign and is succeeded by his brother Murakami, 2, who will reign until 967.

950 Europe's intellectual center is Cordova in Muslim Spain. The city of 500,000 has libraries, medical schools, and a large paper trade.

954 France's Louis IV dies September 10 at age 33 and is succeeded by his son Lothair, 13, who will reign until 986, initially under the guardianship of Hugh the Great, count of Paris, and later under his maternal uncle Bruno, archbishop of Cologne.

955 The Battle of the Lechfeld August 10 ends 50 years of Magyar invasion of the West. Otto the Great of Saxony defeats the Magyars with an army recruited from all the duchies. He will go on to defeat the Wends on the Recknitz, reestablish Charlemagne's East Mark (Austria) with Bavarian colonists, and begin what Germans will call the First Reich.

England's king Edred dies after a 9-year reign and is succeeded by his nephew, who will reign until 959 as Edwig.

956 Hugh the Great dies June 17, 2 months after becoming effective master of Burgundy. He is succeeded by his son Hugh Capet, 18, who is recognized with some reluctance as duke of the Franks by his cousin Lothair, king of the Franks.

959 The first king of England is recognized as such in the person of Edgar, who ascends the throne at age 15 following the death of his elder brother Edwig in October. The king is a son of Edmund I, a great-grandson of Edward I, a great-great-grandson of Alfred the Great, and will reign until 978.

The Byzantine emperor Constantine VII dies after a reign of 47 years. His young son Romanus II will begin a 4-year reign of dissipation.

Edgar (above) recalls the monk Dunstan, 50, from exile in Flanders, makes him bishop of Worcester, and will make him bishop of London.

960 The Northern Song dynasty that will rule China until 1279 is established at Kaifeng by Zhao Kuangyin, who begins to restore China's unity. He will rule until 976 as (Song) Tai Zu to begin the dynasty that will overlap the Mongol Yuan dynasty, which will begin in 1260 (*see* 618).

The first ruler of Poland establishes himself in the person of Mieszko, who has conquered territory between the Oder and the Warthe rivers. He will be converted to Christianity in 966. He will reign until 992 but will be defeated by the margrave Gero and be forced to recognize German suzerainty.

961 Crete is reconquered from Saracen pirates by a great Byzantine armada commanded by Nicephorus Phocas, who storms Candia, expels the Muslims, and converts the people to Christianity.

Bishop Dunstan is made archbishop of Canterbury by Edgar, whose coronation will not take place until 963 but whose chief adviser will be the archbishop.

962 The Saxon Otto I is crowned Holy Roman Emperor February 2 by Pope John XII, ending Rome's feudal anarchy. Otto revives the Western Empire, beginning a period of friction with Constantinople.

963 The dissolute Byzantine emperor Romanus II dies at age 25, probably of poison administered by his wife Theophano. He is succeeded by his infant son, who will reign until 1025 as Basil II. The great general Nicephorus Phocas, now 41, will be co-emperor until 969.

967 The Japanese emperor Murakami dies at age 41 after a 21-year reign and is succeeded by his son of 17, who is insane but will nonetheless reign until 969 as the emperor Reizei.

969 The Byzantine co-emperor Nicephorus is murdered by his wife's lover, the Armenian general John Tzimiskes, 45, who will himself reign as co-emperor until 976.

Kiev's Prince Sviatoslav invades Bulgaria and captures her czar (*see* 971).

Antioch falls to Byzantine forces October 28 after a long siege, ending 300 years of Arab rule in the Syrian city.

The mad Japanese emperor Reizei is removed by the Fujiwara family after a reign of nearly 2 years and replaced by his brother, 10, who will reign until 984 as the emperor Enyu.

Cairo is founded by the Fatimids, Shiite Muslims from Ifriqiyah, who have conquered much of Egypt. They impose their will on the local Sunni.

970 A great hospital is founded at Baghdad by the vizier Abud al-Daula. Physicians are divided into the equivalent of interns and externs, a primitive nursing system is developed, and the hospital's pharmacy is stocked with drugs from all parts of the known world, including spices thought to have medicinal value.

971 The Bulgarian czar John drives out Sviatoslav, annexes much of the country, and restores trade with the Russians.

972 Kyoto's 5-story Daigo pagoda is completed after 29 years of construction.

973 The Holy Roman Emperor Otto I dies May 7 at age 60 after an 11-year reign. He is succeeded by his son, 18, who has been joint emperor since Christmas 967 and last year took as his wife the Byzantine princess Theophano, daughter of Romanus II. The new emperor will reign until late 983 as Otto II.

Cloves, ginger, pepper, and other Eastern spices are available for purchase in the marketplace at Mainz reports the Moorish physician-merchant lbrahim ibn Yaacub, who has visited that city. The spices have been brought to Mainz by Jewish traveling merchants known as "Radanites," who have kept some international trade channels open in the 3 centuries of conflict between the Christian and Islamic worlds. In addition to spices, the Radanites have traded in numerous other commodities, transporting furs, woolen cloth, Frankish swords, eunuchs, and white female slaves to the Orient, while returning to Christian Europe with musk, pearls, precious stones, aloes, and spices that include cinnamon.

Direct trade between Egypt and Italy begins.

975 The king of the English Edgar the Peaceful dies at age 31 after a 16-year reign in which he has ceded Lothian (northern Bernicia) to Scotland's Kenneth for the sake of good will and organized a fleet to resist northern pirates. He is succeeded by his son of 12, who will reign until 978 as Edward II.

William, count of Arles, takes Garde-Freinet from the Arabs.

Arabs introduce modern arithmetical notation, originally from India, into Europe making calculations much easier than with Roman numerals.

976 The Byzantine co-emperor John I Tzimisces dies January 10 at age 51 after returning from a second campaign against the Saracens. The emperor Basil II, now 20, will reign alone until 1025.

The Bulgarian Samuel sets himself up as czar to begin a 38-year reign in which he will challenge the power of the Byzantine emperor Basil II (above; *see* 996).

Modern Austria has her beginnings in a margraviate on the Danube granted by the Holy Roman Emperor Otto II to the Franconian count Leopold (Luitpold), whose family will rule the country until 1246 (*see* 1156).

Cordova's caliph Hakam II dies at age 63 after a 15-year reign in which he has ended the Fatimid dynasty in Morocco and made the University of Cordova the greatest institution of learning in the world. He is succeeded by his son Hisham, 12, who will be caliph until 1009 and then again from 1010 to 1013 but will never have any real power (*see* 978).

978 England's Edward II is assassinated March 18 at Corfe Castle in Dorsetshire at age 15 in a conspiracy engineered by his stepmother Elfthryth, who wants the crown for her son of 10. He succeeds Edward the Martyr and will reign until 1016 as Ethelred the Unready.

Aix-la-Chapelle is sacked by the Holy Roman Emperor Otto II, who is at war with the French king Lothair.

Cordova's regent al-Mansur (Mohammad ibn-abi'-Amir), 39, seizes power from the caliph Hisham II, now 14. Finance minister under Hashim's late father, Hakam II, al-Mansur will wage successful campaigns against the Christian kingdoms to the north, check separatist religious movements, and extend the power of the Umayyad caliphate in the next 24 years.

980 Kiev falls to Viking warriors called in by Vladimir of Great Novgorod, 24, who makes himself grand duke of Kiev and will reign until 1003, extending his Russian dominions (*see* 987).

English ports including Chester and Southampton come under attack in a renewal of Danish raids.

981 The Holy Roman Emperor Otto II marches into Apulia to punish the Saracens who have invaded the Italian mainland.

Norse Icelanders sail for Greenland in 25 ships that carry 700 people with cattle, horses, and other necessities for starting a colony (*see* 900). The expedition is led by Eric the Red, an exiled murderer whose family came to Iceland more than 20 years ago from Joeder. Only 14 of the 25 ships will reach their destination (*see* 982).

982 Last year's invasion of Apulia by the Holy Roman Emperor Otto II provokes the Byzantine emperor Basil II, who sends troops to support the Arabs. Otto sustains a devastating defeat in July and escapes on a Greek vessel to Rossano without revealing his identity.

Viking raiders attack Dorset, Portland, and South Wales.

Eric the Red establishes the first Viking colonies in Greenland (*see* 981).

983 The Holy Roman Emperor Otto II raises German and Italian princes for a new campaign against the Saracens, but he hears of a general rising by the Slavs east of the Elbe River and dies suddenly in his palace at Rome December 7 at age 28. He is succeeded by his son of 3, who is crowned German king Christmas Day at Aix-la-Chapelle by Germans unaware of the emperor's death and will reign until 1002 as Otto III with his mother Theophano exercising power until 991 and his grandmother Adelheid and Archbishop Willigis of Mainz taking over until 996.

984 The German boy-king Otto III is seized early in the year by the deposed duke of Bavaria, Henry the Quarrelsome, 33, who was defeated and dethroned by Otto II in 976 and forced to give up Carinthia and Verona. Henry (Heinrich der Zanker) has recovered his duchy and claims the regency as a member of the reigning

984 *(cont.)* house but is forced to hand over the boy to his mother Theophano, who arrives with the boy's grandmother Adelheid.

The Japanese emperor Enyu abdicates in favor of his son, 16, who will reign until 986 as Kazan.

985 The Danish king Harald II (Bluetooth) dies after a 35-year reign and is succeeded by his son Sweyn (Forkbeard), who will defeat the Norwegians, Swedes, and Wends, conquer England, and reign until 1014.

986 Lothair of the Franks dies March 2 at age 44 and is succeeded by his son, 19, who will reign briefly as Louis V (le Faineant), embroiling the Carolingians with Adalberon, archbishop of Reims, and Hugh Capet.

India is invaded by Subuktigan, Muslim sultan of Ghazni, a former Turkish slave who has obtained the sultanate by marriage and who founds the Ghaznevid dynasty that will rule Afghanistan and much of India.

The Japanese emperor Kazan is tricked into abdicating at age 18 and becomes a Buddhist priest one year after the death of his wife in childbirth. He is succeeded by his half brother, 6, who will reign until 1011 as Ichijo.

al-Tasrif is compiled at Cordova and will serve for centuries as a manual of surgery. The surgeon Albucasis (Abul Kasim), who served as court physician to the late caliph Hakam II, illustrates the résumé of Arabian medical knowledge in which he describes the position for lithotomy (cutting a stone out of the urinary bladder), differentiates between goiter and cancer of the thyroid gland, and instructs surgeons in the delivery of infants in abnormal positions, use of iron cautery, amputation of limbs, transverse tracheotomies, removal of goiters, tying of arteries, repair of fistulas, healing of aneurisms and arrow wounds, and other matters.

987 Louis V of the Franks dies in May, and it is alleged that his mother Emma poisoned him. His death at age 20 ends the Carolingian dynasty founded by Charlemagne in 800, and the Capetian dynasty that will rule until 1328 comes to power in the person of Hugh Capet, now 49. The archbishop of Reims declares that the Frankish monarchy is elective rather than hereditary, denies the claims of the late king's uncle Charles, duke of Lower Lorraine, and engineers the election of his friend Hugh Capet, who is crowned in July and will reign until 996.

Anatolia is overrun by the great feudal barons Bardas Phocas and Bardas Skleros, who rise against the Byzantine emperor Basil II.

Kiev's Grand Duke Vladimir acts after long consultations with the boyars and sends out envoys to study the religions of Kiev's neighbors. They report that "there is no gladness" among the Moslem Bulgarians on the Volga but "only sorrow and a great stench; their religion is not a good one." They find "no beauty" in the temples of the Germans, but they report back from Constantinople that the ritual of the Orthodox Church is so awesome that "we no longer knew whether we were in heaven or on earth, nor such beauty, and we know not how to tell it." They also return with an offer from the Byzantine emperor Basil II of his sister Anna in marriage.

988 Constantinople is threatened by the insurgents Bardas Phocas and Bardas Skleros.

Kiev's Grand Duke Vladimir has himself baptized at Cherson in the Crimea, and takes the Christian name Basil in honor of the emperor at Constantinople. He marries the emperor's sister Anna, returns in triumph to Kiev, and begins a general conversion of Russians to Eastern Orthodoxy (*see* 1054).

989 The Byzantine emperor Basil II uses 6,000 Russians to help him defeat Bardas Phocas at Abydos in Anatolia April 13, ending the threat to Constantinople. Bardas Skleros yields to Basil's superior forces.

990 Ghana forces take the Berber town of Awdaghost as the West African nation makes further gains. Ghana's fetish-worshipping king has made himself the most powerful ruler in non-Islamic Africa with a realm extending from the Sahara south to the upper reaches of the Niger and Senegal rivers.

992 Poland's first recorded ruler, Mieszko I of the house of Piast, dies after a reign of more than 30 years and is succeeded by Boleslav the Brave (Chrobry), 25, who will reign until 1025. Boleslav begins an invasion of eastern Pomerania to gain access to the Baltic.

Venice is granted extensive trade privileges in the Byzantine Empire.

993 Sweden's first Christian ruler comes to power in the person of Olaf Skutkonung, son of Eric the Conqueror, who will reign until 1024.

994 Poland's Boleslav completes his conquest of eastern Pomerania.

995 Syria is incorporated into the Byzantine Empire by Basil II, whose forces take Aleppo and Homs.

996 Otto the Great's Saxon grandson, now 16, is crowned Otto III May 21 at Rome. He has invaded the Italian Peninsula and replaced the Crescentine pope with his cousin Bruno, who is elected Pope Gregory V, but the new papacy will be short-lived.

Byzantine forces under Basil II recover Greece. Basil defeats the Bulgarian king Samuel on the Sperchelos River, bribing supporters of Samuel to defect as he proceeds to reduce Bulgarian strongholds in a campaign that will continue until 1014, causing him to be known as Basil Bulgaroktonos (Slayer of the Bulgarians).

Hugh Capet dies at Paris October 14 at age 58 and is succeeded by his son, 26, who will reign until 1031 as Robert II.

The first Christian Magyar duke Géza dies and is succeeded by his son, 20, who will rule until 1038 as Stephen I. He was married last year to the daughter of Bavaria's Duke Henry II, has chosen St. Martin of Tours as his patron saint, and is engaged in putting down a pagan uprising between the Drave and Lake Balaton.

999 Europeans and Byzantines throng monasteries and churches in the fall bringing deeds of land, valuable jewels, manuscripts, and wagonloads of possessions in hopes of being in good grace on Judgment Day. Church bells ring December 31, lords go down on their knees with peasants, and many are astonished that the world does not end.

1000 Norway's Olaf I Tryggvesson falls in battle with the kings of Denmark and Sweden after a 5-year reign in which English clergymen have helped him convert Norway, Iceland, and Greenland to Christianity. The Danes take over Norway.

Poland's Boleslav the Brave unites Bohemia and Moravia and persuades the Holy Roman Emperor Otto III to create the independent archbishopric of Gnesen.

Ceylon (Sri Lanka) is invaded by the Chola king Rajaraja the Great.

The Japanese emperor Ichijo, now 20, makes his wife Sadako (Teishi), 25, empress, but she dies after 10 months and Akiko, 12, becomes empress.

Norseman Lief Ericsson discovers the Western Hemisphere when storms drive him westward while he is attempting to return to Norway from Greenland. Landing on the coast of what later will be either Newfoundland or Nova Scotia, Ericsson makes his way back to Greenland with descriptions of Vinland (Wineland) where grapes and wild wheat grow.

Saxons settle at Bristol, England.

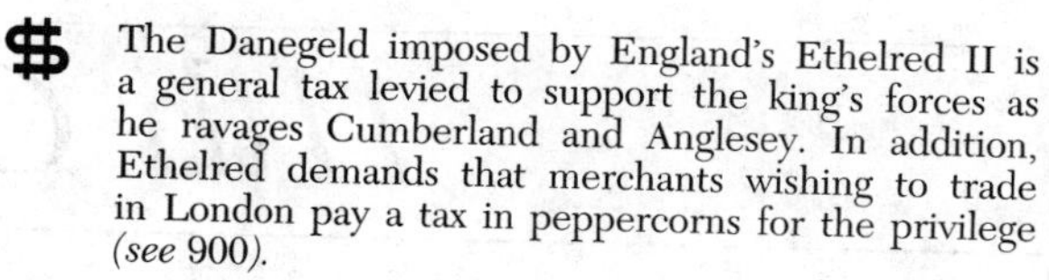

The Danegeld imposed by England's Ethelred II is a general tax levied to support the king's forces as he ravages Cumberland and Anglesey. In addition, Ethelred demands that merchants wishing to trade in London pay a tax in peppercorns for the privilege *(see* 900*)*.

The Chinese Bridge of the Ten Thousand Ages is completed at Foochow.

The Indian mathematician Sridhara recognizes the importance of the zero.

Churches go up throughout Europe, especially in France and Germany, to express thanks at the postponement of Judgment Day.

Hungary's Arpád king Stephen I founds the monastery of Gran.

The Pillow-Book of Sei Shonagon will be compiled in the next 15 years by a lady-in-waiting to the late Japanese empress Sadako (above). Daughter of the wit Mutosuke Kiyowara, Sei Shonagon, now 37, will fill her notes and comments with scathing criticisms and polished indelicacies that will make her diaries popular for centuries.

The "grapes" found by Lief Ericsson (above) are either mountain cranberries, wild currants, or gooseberries, and his wild "wheat" is Lyme grass (*Elymus arenarius, var. villosus*), a tall wild grass with a wheatlike head whose seeds are used to make flour for bread in Iceland.

11th Century

Iron plows with wheels will replace wooden plows in much of northern Europe in this century. Food will become more abundant as a result of such agricultural improvements, but France will nevertheless have 26 famines, and England will average one famine every 14 years.

1002 The Holy Roman Emperor Otto III dies of malaria January 23 at age 21 while conducting a campaign against the Romans. He is succeeded as king of the Franks and Bavarians June 7 by his cousin Henry, 28, duke of Bavaria, who will become emperor in 1014.

The Umayyad vizier al-Mansur of Cordova dies at age 63 after 21 years in power, and the caliphate begins to decline under Hisham II.

The Byzantine armies of Basil II overrun Macedonia and defeat the Bulgarians at Vidin.

Danish settlers in England are massacred on St. Brice's Day by order of Ethelred II.

1003 The Viking king Sweyn Forkbeard (Sven Tveskaeg) plunders the English coast and exacts tribute in retaliation for last year's massacre of Danish settlers by Ethelred II. The raids will continue almost annually until 1014.

Norse mariner Thorfinn Karlsefni leaves Greenland with three ships for a 3-year visit to the northern continent of the Western Hemisphere. His attempts at settlement will not be successful, nor will any other attempt for nearly 600 years (*see* Ericsson, 1000; Corte-Real, 1500).

1004 The Lombard king Ardoin is defeated by Henry of Bavaria, who has himself crowned king of Lombardy at Pavia, but Ardoin continues to make war on Henry.

1005 Scotland's Kenneth III dies after an 8-year reign and is succeeded by Malcolm II, who will rule until 1034.

1006 Burgundy's last independent king Rudolph II makes Henry of Bavaria his heir. Henry makes an alliance with France's Robert II against Baldwin of Flanders.

Muslims settle in northwest India.

Java's Mount Metrop erupts, killing the Hindu king Dharmawangs and shattering the Temple of Borbudar, largest in Southeast Asia.

China's Song emperor establishes granaries for emergency famine relief in every prefecture.

1007 England's Ethelred II pays the Danes £40,000 to gain 2 years' freedom from the attacks of the Viking Sweyn.

1008 Mahmud of Ghazni defeats Hindu forces at Peshawar as he works to extend his Afghan kingdom from the Tigris to the Ganges.

1009 Egypt's Fatimid caliph al-Hakim destroys Jerusalem's Church of the Holy Sepulcher. The act stirs demands in Europe for a Christian crusade to recover the Holy Land (*see* 1095).

1010 Poetry *The Book of the Kings* (*The Shah-nama*) by Persian poet Mansur Abu'l-Quasim Firdawsi, 75, is presented by its author to the Sultan Mahmud of Ghazni. It will survive as the great epic poem of Persia.

1011 The Japanese emperor Ichijo dies at age 31 after a 15-year reign and is succeeded by his cousin, 35, who will reign until 1016 as Sanjo but begins to lose his eyesight almost immediately.

South Wales is invaded by Ethelred II, and Canterbury is taken by the Danes in his absence.

1012 England's Ethelred II pays the Danes another £48,000 to cease their raids.

"Heretics" are persecuted for the first time in the German states.

1013 The Danes conquer England. Ethelred II takes refuge in Normandy.

Cordova's Hisham II dies after a caliphate that began in 976 but was interrupted between 1009 and 1010 as Umayyad power declined in Spain. He is succeeded by Sulaiman al-Mustain, who was installed as caliph by the Berbers from 1009 to 1010 and who will reign until 1016, but Hisham's death precipitates civil war.

1014 The German king Henry of Bavaria recognizes Benedict VIII as the rightful pope and is crowned at Rome February 14. He will reign until 1024 as the Holy Roman Emperor Henry II.

The Viking king Sweyn Forkbeard dies suddenly at Gainsborough in February and is succeeded by his son Canute, 20, who flees to Denmark as Ethelred returns from Normandy.

The Battle of Contarf April 23 ends Danish rule in Ireland, but a Dane kills the Irish king Brian Boru, 87.

The Bulgarian army is blinded on orders from Basil II, whose Byzantine armies annex the western part of Bulgaria.

1015 Olaf II restores Norwegian independence and reinstates Christianity in his realm.

Wessex yields to Canute, who has returned to England to wage war against Edmund Ironside.

Leipzig is settled by Slavs on a plain just above the junction of three small rivers.

Fiction *The Tale of Genji (Genji Monogatari)* by Japanese baroness Shikibu Murasaki, 35, whose work will be a classic of world literature. Widowed at age 21, the baroness has become a lady-in-waiting to the empress Akiko and has probably written her long novel to be read as entertainment for the empress.

1016 The blind Japanese emperor Sanjo abdicates at age 40 and is succeeded by the 8-year-old son of the late Ichijo, who will reign until 1036 as Goichijo.

England's Ethelred II dies November 30 at age 48 after a 38-year reign and is succeeded by his son Edmund Ironside, who is chosen king by Londoners, while Canute is chosen by the witan at Southampton. Edmund's brother-in-law Edric deserted him last year in anger at Edmund's marriage to the widow of a Danish earl, and while Edric now rejoins the new king, his treachery causes a rout of the English at Assundum (Ashingdon) in Essex. Canute defeats Edmund but permits him to reign in the south until his death later in the year at age 26, whereupon Canute proceeds to conquer all of England.

1017 King Canute divides England into four earldoms, and while Edric is restored to an earldom, he is found untrustworthy and is executed.

1018 Byzantine forces defeat an army of Lombards and Normans at Cannae in a great victory that assures continued domination of southern Italy by the Greeks

The Bulgarians submit to Constantinople, which regains Macedonia.

The Treaty of Bautzen ends a 15-year German-Polish war. Poland's Boleslav the Brave is given Lusatia as an imperial fief.

1019 The Prince of Kiev Jaroslav the Wise begins a 35-year reign in which he will codify Russian law and build cities, churches, and schools.

1020 Norway's Olaf II gains recognition as king from Faroe, Orkney, and Shetland islanders.

Pisa annexes Corfu.

Bemberg Cathedral is consecrated by Pope Benedict VIII.

1021 A Byzantine army sent by Basil II invades Armenia.

Epidemics of St. Vitus's dance sweep Europe. The disease is a chorea whose victims invoke the name of a Christian child martyr of the 3rd century.

1022 Byzantine forces in southern Italy are defeated by the Holy Roman Emperor Henry II.

The Synod of Pavia orders celibacy among the higher clergy.

The Japanese Buddhist Hōjō Temple is completed by the prime minister Michinaga Fujiwara, 57.

1024 The Holy Roman Emperor Henry II dies July 13 at age 51 after a 10-year reign. He is succeeded as king of the Germans and Roman emperor by his son, 34, who will reign until 1039 as Conrad II, founding the Franconian (or Salic) dynasty that will rule until 1125.

Sweden's first Christian king Olaf Skutkonung dies after a reign of 31 years, and the Goths and Swedes begin 110 years of religious wars.

1025 Poland gains independence from the Holy Roman Empire as Boleslav the Brave makes himself king. He dies within a few months at age 33, leaving a nation that is one of Europe's most powerful, extending from the Elbe to the Bug and from the Danube to the Baltic, with Russia as her vassal. Boleslav's eldest son will reign until 1034 as Mieszko II, but other sons take parts of the country, and dynastic conflicts begin that will lose most of Boleslav's territorial gains to neighboring countries.

The Byzantine emperor Basil II dies December 15 at age 69 after a 49-year reign. He is succeeded by his brother, who will reign until 1028 as Constantine VIII.

1026 King Canute fights off the kings of Norway and Sweden in their attempt to conquer Denmark. He begins a pilgrimage to Rome.

Solmization in music (do, re, mi, fa, sol, la . . .) is introduced by the Benedictine monk Guido d'Arezzo, 31.

1027 Normandy's Duke Richard the Good dies after a 31-year reign in which he has defeated English and Swedish forces and has brought Normandy to the height of her power. He is succeeded by his son, who will reign boldly but unscrupulously until 1035 as Robert I (Robert le Diable).

Japan's prime minister Michinaga Fujiwara dies at age 62.

The Mayan Empire in the Western Hemisphere suffers an epidemic that weakens her people and begins her decline (*see* 731).

1028 King Canute conquers Norway. He will rule that country along with Denmark and England until his death in 1035.

The Byzantine emperor Constantine VIII dies at age 68 and is succeeded by his daughter Zoë, 48. She marries Romanus III Argyrolpolus, 60, and makes him co-emperor.

Castile is conquered by Navarre's Sancho III (Sancho Garces el Mayor), who has also conquered León and proclaims himself "king of Spain."

Nonfiction *Eiga* is a Japanese history covering 2 centuries and 15 emperors from Uda to Horikawa, but it embellishes on true history.

1030 Norway's Olaf Haraldsson is defeated at the Battle of Stikelstad and killed by King Canute.

1030 ***(cont.)*** The Byzantine emperor Romanus III is defeated in a battle with the Muslim emirs who have attacked Syria.

Canon of Medicine by the Arab physician Avicenna (Abu Sina) follows the thinking of Aristotle and Galen but is so well written and organized that it will be a major influence on medical thinking for centuries.

1031 France's Capetian king Robert the Pious dies July 10 at age 61 after a short war in which he has been defeated by his younger sons. His son of 23 by Constance of Aquitaine has ruled with him since 1027 and will reign until 1060 as Henri I.

The caliph of Cordova Hisham III dies after a 4-year reign, ending the Umayyad caliphate that has ruled since 756.

1032 Burgundy's Rudolph III dies childless February 2. The Holy Roman Emperor Conrad II claims Rudolph's realm under terms of a 1027 treaty of succession and unites it with the empire.

Robert II, duke of Normandy, helps France's Henri I defeat his mother Constance and her younger son Robert, ending the French civil war.

1033 Poland's Mieszko II is defeated by German and Russian forces one year after the division of his realm with two relatives by order of the emperor Conrad II.

Castile regains her independence from Navarre.

1034 The Byzantine empress Zoë poisons her husband Romanus III and marries the epileptic weakling Michael IV Paphiagonian with whom she will reign until 1041.

Poland's Mieszko II dies and his death begins a 6-year period of dynastic warfare and insurrection during which peasants will rise against landlords, churches will be burned, and clergymen will be massacred.

Silesia is overrun by Bretislav the Restorer, who will rule Bohemia and Moravia until 1055 and, for a period, rule Poland as well.

Scotland's Malcolm II dies and is succeeded by his grandson Duncan, who will reign for 6 years.

1035 Castile's Sancho the Great (Sancho III of Navarre) dies at age 65 after ruling Navarre for 35 years, Castile for 8. The kingdom of Sancho Garces el Mayor is divided among his four sons, and he is succeeded in Castile by his second son, who will reign until 1065 as Ferdinand I.

King Canute dies at Shaftesbury November 12 at age 39. His four sons are unable to control England, and Norway breaks away from Denmark.

1036 The Japanese emperor Goichijo dies at age 28 after a 20-year reign and is succeeded by his brother, 27, who will reign until 1045 as the emperor Gosuzaku.

Modern musical notation is pioneered by Guido d'Arezzo, who also invents the "great scale," or gamut, the hexachord, and hexachord solmization (*see* 1026).

1037 Castile's Ferdinand I conquers the Spanish kingdom of León and makes himself king.

The *Constitutio de Feudis*, issued by the Holy Roman Emperor Conrad II, makes Italian fiefs of small landholders hereditary.

1038 Hungary's Stephen I dies at age 61 after a 40-year reign that has imposed Roman Catholicism. He is succeeded by his nephew Peter Orseolo, 27, whose father is doge of Venice (but *see* 1041).

1039 The Holy Roman Emperor Conrad II dies at Utrecht June 4 at age 49 and is buried in the cathedral he has started at Spires. Conrad's son Henry, 21, succeeds as German king.

The Welsh prince Gruffydd of Gwynedd and Powys defeats English forces.

Ergotism breaks out in parts of France (*see* 943; 1089).

1040 Scotland's young Duncan Canmore is slain by his nobles who invite the able king of Inverness Macbeth (or Malebaethe), Mormaor of Moray, to succeed Duncan. Macbeth will reign until 1057 (*see* 1054).

England's Harold Harefoot dies. His half brother Hardicanute arrives from Denmark with a large fleet and is crowned king in June to begin a 2-year reign that will antagonize his English subjects.

Lady Godiva rides naked through the streets of Coventry to persuade her husband Leofric, earl of Mercia, to remit the heavy taxes that oppress the citizens (year approximate). The earl has said he would grant his wife's request if she would ride naked through the streets. Lady Godiva (Godgifu) has issued a proclamation asking all citizens to remain indoors with their windows shut, she has only her long hair to cover her on the ride, her husband keeps his word and abolishes the taxes, but a local tailor is reputedly struck blind because he peeped.

Increased planting of oats in the three-crop system of the past 3 centuries has led to an expanded use of horses in Europe and thus to increased trade, larger towns, and more people who do not raise their own food.

1041 Hungary's Peter Orseolo is driven out by the usurper Samuel Aba, brother-in-law of his late uncle Stephen, and takes refuge with the German king Henry.

The German king Henry defeats Bohemia's Bretislav and forces him to pay homage.

Northumberland's Eardwulf is murdered by the Danish husband of his niece, who makes himself earl of Northumberland and will reign until 1055 as Siward the Strong.

The Byzantine emperor Michael IV Paphiagonian dies in the autumn and is succeeded by his nephew who will reign for 4 months as Michael V Kalaphates (the caulker).

Movable type for printing will be used in the next 8 years by the Chinese printer Pi Sheng, who will use hundreds of clay blocks bearing Chinese ideograms (*see* 1381).

1042 England's Hardicanute dies at age 23, and the son of the late Ethelred the Unready is recognized as king.

Now 42, he will reign until 1066 as Edward the Confessor, last of the Anglo-Saxon kings.

The German king Henry receives the homage of Burgundy.

The Byzantine emperor Michael V Kalaphates shuts the empress Zoë up in a cloister, but the Constantinople nobility rises against him, locks him up in a monastery, and releases Zoë. She marries the scholarly Constantine IX Monomachus, 42, with whom she will reign until 1050.

The Seljuk Turks rise against their Byzantine overlords.

1043 England's Edward the Confessor is crowned at Easter. He has been helped to gain the throne by the powerful earl of Wessex, Godwin, whose daughter Edith will be married to the king in 1045.

The German king Henry III moves against Hungary's Samuel Aba, who has attacked eastern Bavaria.

1044 Hungary's Peter Orseolo is restored by the German king Henry III, who brings Hungary under German control after gaining a great victory in July at Menfo (but *see* 1047).

Burma has its beginnings in the kingdom of Pagan founded in Southeast Asia. The Mons and the Burmese end their rivalry and join together after generations of vying for control of the region.

Korea completes a defense system that consists largely of a great wall across the northern part of the peninsula to keep out Mongol and Turkish raiders.

1045 The Japanese emperor Gosuzaku dies at age 36 after a 9-year reign and is succeeded by his son of 16, who will reign until 1069 as the emperor Goreizei.

1046 The German king is crowned Holy Roman Emperor Henry III Christmas Day at Rome by the bishop of Bemberg, who he has installed as Pope Clement II.

1047 Magnus I dies after 12 years as king of Norway and 5 as king of Denmark. He is succeeded in Norway by Harald Haardraade, 32, who will reign until 1066 as Harald III, and in Denmark by Sweyn Estrithson, grandson of Sweyn Forkbeard, who will reign until 1075 as Sweyn II but will be at war with Harald III until 1064.

The duchies of Carinthia, Bavaria, and Swabia are restored by the Holy Roman Emperor Henry III.

Hungary's Peter Orseolo is overthrown by his cousin András, who will reign until 1060 as András I.

1050 The Byzantine empress Zoë dies at age 70, and her older sister Theodora, who has shared the throne since 1042, is left to rule with Constantine IX. He has spent huge sums of money on luxuries and public buildings with a profligacy that has weakened the empire.

Oslo is founded by Norway's Harald III, who has visited Novgorod, Kiev, and Constantinople.

1052 Pisa conquers Sardinia from the Arabs, who have held the island since 720.

1053 Norseman Robert Guiscard gains victory June 23 at Civitate over papal forces raised by Leo IX, who is captured. Robert, 38, takes Benevento from the Byzantines and founds the Norman Empire that will rule southern Italy until 1194.

England's earl of Wessex dies and is succeeded by his son, 31, who becomes chief minister to his brother-in-law Edward the Confessor.

$ England abolishes the Danegeld after 33 years.

1054 Scotland's Macbeth is defeated at Dunsinane by Malcolm Canmore, son of Duncan (*see* 1040). Malcolm is aided by the Dane Siward, earl of Northumberland, who has invaded Scotland to support his kinsman (*see* 1057).

France's Henri I invades Normandy and is defeated at Mortemer.

The Eastern (Orthodox) and Western (Roman) Churches break irreparably apart as Pope Leo IX (the antipope Bruno) excommunicates Michael Cerularius and his followers. Leo declared war last year against the Normans in southern Italy, but his army of Italian and German volunteers was defeated at Astagnum near Civitella. He was captured, and he dies April 19 at age 51.

A minor star in the constellation Taurus explodes July 5 in a supernova that is visible by daylight for 23 days and remains visible at night for another 633 days, an event recorded by pictographs in China and in the Western Hemisphere. The irregular gas cloud created by the supernova will be called by astronomers the Crab nebula.

1055 Northumberland's Siward the Strong dies after having helped to rout Scotland's Macbeth and make Malcolm III king of Cumbria. He is succeeded by Tostig, a son of the late Godwin of Wessex and a brother-in-law of Edward the Confessor.

The Byzantine emperor Constantine IX dies at age 55, leaving his sister-in-law Theodora to rule alone.

Seljuk Turks under Toghril-Beg enter Baghdad to liberate the Abbasid caliphate from the Shiites. Toghril-Beg restores Sunni power and makes himself temporal master of the caliph.

1056 The Holy Roman Emperor Henry III dies October 5 at age 38 and is succeeded as German king by his son of 5, who will reign until 1106 as Henry IV, with his mother Agnes as regent until 1065.

The Byzantine empress Theodora dies at age 76 to end the Macedonian dynasty that has ruled the empire since Justinian the Great took power in 527. Theodora's successor Michael VI (Stratioticus) will be overthrown early next year by a revolt of the feudal barons of Anatolia.

1057 The Comnenus dynasty that will rule the Byzantine Empire until 1185 is founded by Isaac I Comnenus, a military leader who is proclaimed emperor by the insurgents who have overthrown Michael VI. The new emperor eliminates many sinecures and reforms Byzantine finances.

Macbeth is murdered by Malcolm Canmore, and is succeeded by his stepson Lulach (*see* 1054; 1058).

1058 Scotland's Lulach is slain by Malcolm Canmore, who makes himself king; he will reign until 1093 as Malcolm III MacDuncan.

William of Normandy defeats Godfrey of Anjou at the Battle of Varaville.

Poland's Grand Duke Casimir I dies at age 43 after having regained much of Poland's lost territory with the help of the late Henry III and restoring Christianity. Casimir is succeeded by his son of 19, who will rule until 1079 as Boleslav II (the Bold).

1059 The Byzantine emperor Isaac I Comnenus abdicates in favor of a high financial officer who begins an 8-year reign as Constantine X (Dukas). The new emperor will neglect and antagonize the army, giving more authority to the civil service, the church, and the scholars.

The Treaty of Melfi in August cements relations between the papacy and Robert Guiscard, who promises fidelity to the Church; he is recognized as duke of Apulia and of Sicily, although the latter is occupied by Saracens.

The new pope Nicholas II decrees papal election by cardinals only.

1060 France's Henri I dies August 4 at age 52 and is succeeded by his son of 8, who will reign until 1108 as Philip I.

Hungary's András I is driven from the throne by his brother and dies after a 14-year reign in which he has gained imperial recognition of Hungarian independence. The brother will reign until 1063 as Belá I.

West Africa is conquered from the kingdom of Ghana by Arabs who are supported by Berbers greedy for the salt mines of Ankar. Founded in the 4th century, the kingdom of Ghana has ruled from near the Atlantic Coast almost to Timbuktu.

Byzantine mosaic *Christ as Ruler of the World* at Daphni, Greece.

The Bayeux Tapestry recorded the Roman conquest of England, a triumph of Norsemen over Saxons.

1061 Scotland's Malcolm III MacDuncan invades Northumbria.

Bohemia's Spytihnev dies after an uneventful 6-year reign and is succeeded by Vratislav II, who will reign until 1092.

1062 The archbishop of Cologne Anno II seizes the German king Henry IV, now 11, in the coup d'état of Kaiserswerth. Henry is forced to yield control to Anno and the archbishop of Bremen Adalbert.

Marrakech is founded by sultans of the Almoravide dynasty, who start a tent settlement on the West Tensift River, northwest of the High Atlas range.

1063 Harold of Wessex subdues Wales with help from his brother Tostig of Northumbria.

1064 Castile's Ferdinand I takes Coimbra from Portuguese defenders June 9.

Hungary's new king Solomon takes Belgrade from the Byzantines.

The Seljuk Turks conquer Armenia.

"Regulative granaries" established by China's emperor Ying Zong buy up surplus grain in good harvests and release stocks in times of shortage.

1065 Castile's Ferdinand I dies after a 32-year reign in which he has made himself king of León by defeating his brother-in-law Bermudo II and begun the reconquest of Spain from the Moors. Fernando el Magno is succeeded by his son, who will reign until 1072 as Sancho II (el Fuerte).

The University of Parma has its beginnings.

Westminster Abbey is consecrated after 13 years of construction.

1066 The Battle of Hastings October 14 seals the Norman conquest of England by Norsemen under William, 39, duke of Normandy, who will be called William the Conqueror and will rule England as William I until his death in 1087. Edward the Confessor has died January 5 and been succeeded by his brother-in-law Harold Godwineson, earl of Wessex, who secured his election as king January 6 and has ruled as Harold II.

Norway's Harald III Haardraade has been invited by Tostig of Northumbria to aid in the conquest of England from Harold II. He has sailed with a large fleet but has been killed September 25 at the Battle of Stamford Bridge. William the Conqueror has landed September 28 at Pevensey, Harold II is killed in the Battle of Hastings, and William is crowned December 25.

A comet appears in the skies that will later be called "Halley's Comet" (*see* 1705).

No written account of the Battle of Hastings will appear until that of the chronicler Guillaume de Poitiers, whose *Gesta Willelmi* will be published late in the century, but the *Bayeux Tapestry* will depict events of the savage battle.

The Normans (above) will introduce many French words into the language of England from across the Channel (*see* 1100).

The French words *boeuf, mouton, veau, porc,* and *poularde* introduced by the Normans (above), will be the basis of the English words *beef, mutton, veal, pork,* and *poultry.*

1067 The Byzantine emperor Constantine X Dukas dies at age 60, and his widow Eudoxia Macrembolitissa marries a general who will reign jointly with her until 1071 as Romanus IV Diogenes.

Poland's Boleslav II takes Kiev.

A Chinese edict aimed at keeping gunpowder a state monopoly forbids export of sulfur or saltpeter (*see* 222; 1234).

The world's first leprosarium is founded by the Castilian soldier Ruy Diaz de Bivar, 27, who is known as El Cid.

Scotland's Malcolm III MacDuncan marries Margaret, 22, sister of the Saxon Edgar the Aetheling, and she brings the Black Rood to Scotland, thus beginning the country's transition from Celtic culture and Columban religious rites to the Anglicized feudal system and Roman Catholic ritual.

1068 Saxon nationalists in the north and west of England rise to challenge William the Conqueror and his Norman barons.

The Chinese Song emperor Shen Tsung begins a 17-year reign that will see radical reforms put through over bitter opposition from conservatives.

The Japanese emperor Goreizei dies at age 39 after a 23-year reign and is succeeded by his brother Gosanjo, 34, who will reign until 1072. Goreizei had married a daughter of Yorimichi Fujiwara, son of the late strongman Michinagi Fujiwara, but she has not borne a child and has thus frustrated her father's ambition to be the grandfather of an emperor.

1069 England's William I puts down a great rising of Saxons in the north who have won Danish support. The Normans confiscate Saxon lands and establish their feudal superiority in the "harrying of the north" that devastates and depopulates a strip of territory from York to Durham.

The Chinese prime minister Wang An-Shih, 48, reforms his government by cutting the budget 40 percent and raising salaries to make it possible for ordinary government officials to afford to be honest.

Wang An-Shih (above) empowers China's chief transport officer to accept taxes in cash or in kind in order to avoid excessive transport costs and to control prices.

Wang An-Shih (above) begins a radical program to reform Chinese agriculture after finding the nation's granaries stocked with emergency stocks of relief grain valued at 15 million strings of cash. He offers poor farmers loans at 2 percent interest per month in cash or grain to free them from usurers and monopolists who charge higher rates. He gives his chief transport officer power to sell from state granaries when prices are high and to buy when prices are low.

1070 The order of the Knights of St. John is founded at Jerusalem by merchants from Amalfi to care for the hospital of St. John. The Hospitalers will be militarized in 60 years along lines that will be established by the Knights Templar in 1120.

Roquefort cheese is discovered in France.

1071 Five centuries of Byzantine rule in Italy end April 16 as Bari falls to Robert Guiscard after a 3-year Norman siege.

The Battle of Manzikert (Malaz Kard) August 26 virtually ends Byzantine power in Asia Minor. The emperor Romanus IV Diogenes has led a Christian army of 60,000 to recover some fortresses from the Seljuk Turks in the spring, but he has encountered a 100,000-man army commanded by the sultan Alp Arslan. Byzantine noblemen, including Andronicus Dukas, desert Romanus, who falls into enemy hands. He gains release, but when he tries to regain his throne, his enemies put out his eyes, and he soon dies. The bureaucrats at Constantinople elevate the son of the late Constantine X to the throne, and he will reign until 1078 as Michael VII Parapinakes, with the scholar Michael Psellus as his chief adviser.

Venice completes the Cathedral of San Marco after 95 years of construction on the Piazza San Marco (the chapel will be consecrated in 1094). An annex to the Doge's Palace, it will be used for civil and religious ceremonies.

A 2-pronged fork is introduced to Venice by a Greek princess who marries the doge. Rich Venetians adopt the new fashion (*see* 1518; 1570).

1072 Palermo falls January 10 to Robert Guiscard and his brother Roger, who begin to conquer all of Sicily.

Castile's Sancho el Fuerte is assassinated at the siege of Zamora while warring with Sancho IV of Navarre. He is succeeded by his brother, who will reign until 1109 as Alfonso VI, resuming the Christian reconquest of Castile with help from El Cid (Ruy Diaz de Bivar).

The Japanese emperor Gosanjo abdicates at age 38 (he is ill and will die next year). He is succeeded by his son Shirakawa, 19, who will reign until 1086.

Parma Cathedral is completed after 16 years of construction.

1073 Robert Guiscard's Normans take Amalfi from Gisulf of Salerno, antagonizing the new pope Gregory VII.

Pope Alexander II dies after a 12-year papacy in which the archdeacon Hildebrand, now 50, has played a major role. Hildebrand is elected to succeed Alexander and will reign until 1086 as Gregory VII, attacking simony, upholding celibacy among the clergy, and fighting for papal omnipotence.

A reorganization of the English Church subordinates York to Canterbury.

1074 The Peace of Gerstungen ends conflicts between the Saxons and Henry IV. Hungary's Solomon is defeated by his cousin who succeeds him and begins a brief reign as Géza I.

Pope Gregory VII (Hildebrand) excommunicates the Norman Robert Guiscard and all married priests.

1075 Syria and Palestine are subdued by the Seljuk Turk Malik Shah.

Acoma (the Sky City) in the Western Hemisphere is founded by tribespeople who have moved 3 miles from the "Enchanted Mesa" following an earthquake that has destroyed trails to the top and left those up there to starve to death. Acoma will grow to have a population of 5,000 and be the oldest continuously inhabited town in the northern continent of the hemisphere.

The Synod of Rome called by Pope Gregory VII passes strict decrees against simony (purchase or sale of church offices; the name is derived from that of Simon Magus, a Samaritan sorcerer mentioned in the Bible). A 47-year struggle against the papacy begins as German bishops oppose Hildebrand's confirmation as pope and Henry VI ignores a papal letter of rebuke.

1076 Ghana's capital Kumbi is plundered by nomadic Almoravids who are overrunning West Africa and imposing their Islamic religion. Overgrazing will hasten the decline of the Ghana Empire.

The Synod of Worms is called by Henry IV under pressure from the German bishops, who repudiate their allegiance to Pope Gregory VII and declare him deposed. Henry demands Gregory's abdication and is supported by north Italian bishops at Piacenza, but Gregory assembles a Lenten Synod at Rome, suspends and excommunicates the rebellious German and Lombard prelates, deposes and excommunicates Henry, and creates political and ecclesiastical chaos in the German states and Lombardy. The Diet of Tribur in October orders Henry to humble himself, stand trial, and clear himself of Gregory's charges before February 22, 1077.

1077 Hungary's Géza I dies after a 3-year reign and is succeeded by the son of the late Belá I, who will reign until 1095 as Ladislas I. Ladislas, 37, will support the pope in his conflicts with Henry IV (below) and restore order and prosperity at home in the greatest reign since that of Stephen I.

Polish noblemen rebel against the autocratic rule of Boleslav II while he is in Russia reinstating a relative on the throne of Kiev. Bishop Stanislas (Szczepanski), 46, sides with the rebels.

The penance at Canossa January 21 wins absolution for Henry IV, but civil war begins in the German states. The king presents himself as a penitent at Canossa after a midwinter dash across the Alps with his wife Bertha to avoid public trial at home, but while Hildebrand grudgingly accepts Henry's promises and his solemn oaths of contrition, the German nobility elects Rudolf of Swabia as antiking with approval of the pope's legates.

The first English Cluniac monastery opens at Lewes.

1078 England's William I is defeated in battle by France's Philip I, who supports William's son Robert Curthose against Anglo-Norman pressure.

The Byzantine emperor Michael VII abdicates and is succeeded by a soldier chosen by the Asiatic troops. The new emperor will reign until 1081 as Nicephorus III Botaniates, and while some of the army mutinies, the insurrections are put down by Gen. Alexius Comnenus.

1079 Stanislas, bishop of Cracow, excommunicates Poland's Boleslav II (the Bold) after denouncing his excesses. Bishop Stanislas has joined a conspiracy of nobles against the king, who has the prelate murdered May 9 while he is saying mass in his cathedral. Pope Gregory VII excommunicates Boleslav, Polish rebels chase him out of the country, and he flees with his son to Hungary, where he will die in 2 years at age 42. He is succeeded by his lazy brother, who will reign until 1102 as Ladislas I Hermann.

Newcastle-on-Tyne is founded in England by William the Conqueror to resist the incursions of Scotland's Malcolm III MacDuncan (*see* coal, 1233).

Bishop Stanislas (above) will be canonized in 1253 and be the patron saint of Poland.

1080 Rudolf of Swabia is defeated and killed to end civil war in the German states. Henry IV has steadily gained strength, and while he is once again deposed and excommunicated by Pope Gregory, the pope himself is deposed by a synod of German and Lombard prelates who attempt to install a new pope.

Medical research progresses at the Benedictine school associated with the monastery established at Monte Cassino in 529. Arabian, Jewish, and Greco-Roman medical works are translated into Latin by Constantine the African, a physician who has studied medicine and magic at Babylon and who is now disguised as a monk. His translations of Galen and Avicenna help to emancipate medicine from the religious bonds that have held it.

Greek, Latin, Jewish, and Arabic medical teachings are employed at Salerno, some 125 miles from Monte Cassino (above), where neighboring Benedictine monasteries support a medical school whose teachers are mostly clerics but include some women physicians. Christian dogma is not strictly observed, pigs are dissected, Arabic dietetic ideas are followed, bleeding is regarded as a panacea, but the first medieval pharmacopoeia is begun (*see* 1231).

1081 The Byzantine emperor Nicephorus III abdicates under pressure from his general Alexius Comnenus, 33, who will reign until 1118 as Alexius I Comnenus. He signs a commercial treaty with Venice.

The Norman Robert Guiscard invades the Balkans, landing in Epirus and laying siege to the Byzantine city of Durazzo.

The German king Henry IV invades Italy and places Matilda, marchioness of Tuscany, under the imperial ban for supporting Pope Gregory VII. Henry accepts the Lombard crown at Pavia and forces a council to recognize the archbishop of Ravenna as Pope Clement III.

1082 The German king Henry IV lays siege to Rome and finally gains entrance after two futile attempts. He fails in his efforts at reconciliation with Pope Gregory VII, who calls on the Norman Robert Guiscard to support him, and he makes a treaty with the Romans, who agree to call a synod that will rule on the dispute between crown and papacy.

Robert Guiscard (above) defeats the Byzantine forces of Alexius I Comnenus and takes Durazzo, occupying Corfu as well.

1083 A synod meets at Rome to resolve the quarrel between Pope Gregory VII and the German king Henry, but the synod has secretly bound itself either to crown Henry emperor or select a new pope.

1084 The Synod of Rome declares Pope Gregory deposed in March and recognizes the antipope Clement, who crowns Henry March 31. The emperor attacks fortresses still in Gregory's hands but retreats back across the Alps as Norman forces under Robert Guiscard advance from the south. The Normans sack Rome, Pope Gregory is unable to remain in the ravaged city, and his supporters take him to Salerno where he will die next spring.

Defensio Henrici IV by Peter Crassus (who teaches Roman law at Ravenna) takes the position that neither the pope nor Henry's rebellious subjects have any more right to interfere with the emperor's hereditary territorial possessions than they have to take away any person's private property.

1085 Castile's Alfonso VI takes Toledo May 25; the Arab center of science falls into Christian hands.

Pope Gregory VII dies May 25 at Salerno; Henry IV extends the "Peace of God" over the entire Holy Roman Empire.

The Norman Robert Guiscard dies of fever July 15 at age 60 after regaining Corfu and Kephalonia, which his son Bohemund had lost. The duke is succeeded by his brother Roger Guiscard, now 54, who has conquered Sicily and will rule until 1101.

1086 The Oath of Salisbury makes English vassals responsible directly to the crown. Some 170 great tenants-in-chief and numerous lesser tenants have emerged to replace England's earls, and the oath is exacted from all such vassals, prohibiting them from making private wars.

A Muslim army from Africa defeats Castile's Alfonso VI October 23 at Zallaka. Yusuf ibn Tashfin, Berber leader of the Almoravids, sends cartloads of Christian heads to the chief cities of Spain and Africa's Magrib as evidence of his victory.

The Japanese emperor Shirakawa abdicates at age 33 and makes his 7-year-old son Horikawa emperor. Japan's emperor is not permitted to have private property, and Shirakawa's move begins a long era of voluntary abdications in which the ex-emperor will retain power.

The Domesday Book compiled on orders from England's William I lists more than 25,000 slaves and 110,000 villeins (serfs) among the properties and assets of English landowners.

The Domesday Book (above) lists the assets of landowners to provide a basis for taxation and administration. The royal commissioners oblige the landowners to give information under oath as to the size of every piece of land, its resources, and its ownership—past and present.

The magnetic compass is pioneered by Chinese waterworks director Shen Kua, who writes that magicians can find directions by rubbing a needle on a lodestone and hanging the magnetized needle by a thread. The needle, he says, will usually point south but sometimes will point north (*see* 1150).

1087 William the Conqueror invades the French Vexin to retaliate for raids on his territory. He sacks and burns the town of Mantes, but when he rides out to view the ruins, his horse plunges on the burning cinders. William sustains internal injuries, is carried in great pain to Rouen, and dies there September 9 at age 59 or 60, leaving his third son, William Rufus, 31, to rule England until 1100 and his eldest son, Robert Curthose, 33, to succeed him as duke of Normandy until 1106.

The Byzantine emperor Alexius I Comnenus is defeated with a large army at the Battle of Drystra by heretic Bogomils in Thrace and Bulgaria who have revolted against Constantinople.

Control of the western Mediterranean is wrested from the Arabs by Genoa and Pisa, whose forces capture Mahdiyah in Africa.

1089 Ergotism strikes a French village, whose inhabitants run through the streets in fits of madness (*see* 1039; 1581).

1091 The Treaty of Caen ends hostilities between England's William II and Normandy's Robert Curthose.

Sicily is freed of the last of the Muslims, who have held the island for 130 years. Roger Guiscard completes his conquest in February and goes on to take Malta.

1092 Bohemia's Vratislav II dies after a 31-year reign, but although he personally was made a king by Henry VI in 1086 his successor will rule until 1110 as Duke Bretislav II.

England's William II conquers Cumberland.

1093 Scotland's Malcolm III MacDuncan and his eldest son Edward are entrapped and killed November 13 at a place that will be called Malcolm's Cross. Malcolm has been laying siege to Alnwick in an invasion of England, his wife Margaret dies 4 days later, and he is succeeded by his brother Donald Bane, who will reign until 1097.

1094 Valencia falls to El Cid June 15 after a siege of 9 months by an army of 7,000 men, most of them Muslims. Violating all the conditions of surrender, El Cid has the cadi ibn Djahhaff burned alive, slaughters many of the citizens, and rules a kingdom that extends over nearly all of Valencia and Murcia.

1095 Hungary's Ladislas I conquers Croatia and Dalmatia, where he introduces Catholicism, founds the bishopric of Zagreb (Agram), persecutes the heathen, but dies suddenly July 29 at age 55 just as he is about to join the crusade (below). Ladislas is succeeded by his nephew Coloman, son of his late brother Géza, who will reign until 1116.

Pope Urban II preaches a crusade against the infidel and gains support from Peter the Hermit and others for his holy war. A learned French nobleman, the pope points out that vast areas of land are available to knights (many of whom are landless younger sons) who will join the crusade.

Pope Urban receives an appeal from the Byzantine emperor Alexius Comnenus for aid against the Seljuk Turks. He proclaims the crusade November 27 at the Synod of Clermont, and he excommunicates France's Philip I for adultery.

1096 The Byzantine emperor Alexius Comnenus provides food and escort for the Crusaders (below). He exacts an oath of fealty from the leaders in an effort to protect his title to any recovered "lost provinces" of the Greek Empire, but Count Raymond of Toulouse refuses to take any such oath.

French Jews are attacked as the crusade against the infidel empties whole villages.

The First Crusade raises more than 30,000 men and converges on Constantinople in three groups as Norman-French barons rush to take the cross. Godfrey of Bouillon and his brother Baldwin lead an army from Lorraine via Hungary, Count Raymond of Toulouse and the papal legate Adhemar of Puy lead an army from Provence via Illyria, and Bohemund of Otranto leads an army from Normandy via Durazzo, traveling both by sea and land. A leading figure is the Norman knight Tancred, 18, a nephew of Sicily's Roger Guiscard.

Crusaders from Europe sallied forth on a holy war against the Saracen "infidel," often with very worldly motives.

The death of his daughter Teishi so affects Japan's ex-emperor Shirakawa that he shaves his head and becomes a Buddhist priest, but he retains almost dictatorial control of temporal affairs.

1097 The Battle of Nicaea June 30 ends in defeat for a Muslim army at the hands of a combined force of Crusaders and Byzantine Greeks who take the Seljuk Turks' capital. The French knight Walter the Penniless (*Gautier sans avoir*) is killed after having led hordes through Europe and Asia in what will be called the Peasants' Crusade.

Scotland's Edgar is proclaimed king and begins a 10-year reign. Son of the late Malcolm III MacDuncan, Edgar has an uncle of 47, also named Edgar, who leads an expedition to Scotland but will soon join the First Crusade (above).

Half the knights of France will set off in the next 30 years either for the Levant or for Islamic lands in northern Spain, but the First Crusade has been inspired as much by population pressures (a product of the new agriculture) as by religious zeal or desire for plunder.

1098 Antioch falls after a 9-month siege by Bohemund of Otranto who has lost 5,000 of his 7,000 horses to hunger and disease. So many of his men have sickened and died so quickly that it has not been possible to bury all the corpses, and there has been a falling out between Norman and Provençal Crusaders.

Norway's Magnus III seizes the Orkneys, the Hebrides, and the Isle of Man.

France's Philip I makes his son Louis, 17, co-regent and charges him with resisting the intermittent attacks by England's William II Rufus and Normandy's William III.

The Cistercian order of monks has its beginnings in the monastery of Citeaux founded by French ecclesiastic Robert de Molesmes, 69, and English ecclesiastic Stephen Harding, 50.

1099 Jerusalem falls to the Crusaders July 15 after a siege of just over one month. The streets of the city run with blood as the Crusaders slaughter 40,000 and set fire to mosques and synagogues, and the First Crusade comes to an end.

A kingdom of Jerusalem is established under the Norman Godfrey of Bouillon, who is elected king and assumes the title Defender of the Holy Sepulcher. He defeats an Egyptian force at Askalon August 12, but disease and starvation have reduced the Crusaders to 60,000, down from an original strength of 300,000, and most of the survivors head for home.

El Cid dies in July at age 59 after being defeated by the Almoravids at Cuenca. His widow will hold Valencia against the Moors until 1102.

Arab Maronites, driven decades ago from Syria's Orontes River Valley, welcome the invading Crusaders to the fastness of Mount Lebanon, where they have found refuge. When the Church broke apart in 1054, the Maronites chose Rome over the weakening Greek Orthodox hierarchy in Constantinople.

Crusaders plant sugarcane fields in the Holy Land (*see* 300 B.C.; A.D. 1148).

1100 Jerusalem's Godfrey of Bouillon dies July 18 at age 39 after successful forays against the Seljuk Turks that have taken him as far as Damascus. He is succeeded by his older brother Baldwin, count of Flanders, who will rule until 1118 with help from Tancred, the Sicilian Norman who is now prince of Galilee.

England's William II Rufus dies August 2 at age 44 after being struck by an arrow while hunting in the New Forest. Sir Walter Tyrel is accused of having shot the arrow but flees the country to avoid a trial and then protests his innocence (Ralph of Aix is also accused). Also in the royal hunting train is William's brother, 32, who will reign until 1135 as Henry I.

Middle English begins to supersede Old English, and the dialect of the Ile-de-France begins to prevail over other French dialects.

12th Century

1101 England is invaded by Normandy's Robert Curthose, who has returned from the First Crusade and who tries to take the English throne from his younger brother Henry. He is bought off in the Treaty of Alton.

Sicily's Roger I Guiscard dies June 22 at age 70 and is succeeded by his son, 8, who beginning in 1112 will rule until 1154 as Roger II.

China's Song emperor Hui Zong ascends the throne at age 19 to begin a 24-year reign.

China's Imperial Academy of Painting will be founded by Hui Zong (above), who is himself an able painter.

Gothic architecture will appear in Europe beginning in this century (*see* 1136; 1184; 1550; Chartres, 1260).

Florence's baptistry is erected.

1102 Poland's Ladislas I Hermann resigns the royal title after a 23-year reign in an effort to secure peace by supporting the Holy Roman Emperor Henry IV. His son, 16, gains the throne after a violent struggle with another son and will reign until 1138 as Boleslav III (Wrymouth).

Hungary's Coloman I completes his conquest of Dalmatia from the Venetians.

1103 Norway's Magnus III (Magnus Barfot, or Barefoot) invades Ireland but is killed in battle.

1104 Acre surrenders to Jerusalem's Baldwin I, a Norman army is badly beaten on the River Balikh near Rakka while trying to take Harran, but Byblos falls to Raymond of Toulouse, who lays siege to Tripoli.

Bohemund of Otranto appears at Epirus with a huge army he has raised in Italy to challenge the Byzantine emperor Alexius Comnenus, who conquers the towns of Cilicia.

Iceland's Mount Hekla volcano erupts and devastates farms for 45 miles around. The volcano will have 14 eruptions in the next 850 years (*see* 1154).

1105 The Holy Roman Emperor Henry IV is captured by his son Henry, 24, who last year announced that he owed no allegiance to his excommunicated father and has gained support from the emperor's foes, especially in Saxony and Thuringia. A diet at Mainz compels the emperor to abdicate in December, but the conditions of his abdication are violated, and he is held prisoner at Ingelheim.

1106 Henry IV escapes his captors at Ingelheim, enters into negotiations at Cologne with English, French, and Danish supporters, begins to collect an army to oppose his treacherous son, but dies August 7 at age 54. His son Henry has received homage from some princes at Mainz in January, he will be crowned in 1111, and he will reign until 1125 as Henry V, last of the Salic emperors who have ruled since 1024.

The Battle of Tinchebrai September 28 ends in defeat for Normandy's Robert Curthose at the hands of England's Henry I who has crossed the Channel and who takes his brother home in chains. Robert will be imprisoned first in the 28-year-old Tower of London, later in the castles of Davizes and Cardiff, and will remain a prisoner until his death early in 1134.

1107 Scotland's Edgar dies after a 10-year reign. His brother, 29, succeeds him and will reign until 1124 as Alexander I, putting down an insurrection of northern clans.

The Sicilian Norman Tancred recovers the Cilician towns conquered by Alexius Comnenus 3 years ago.

The Japanese emperor Horikawa dies at age 28 after a 21-year reign. He is succeeded by his 4-year-old son Toba, who will reign until 1123, but Toba's grandfather Shirakawa remains the country's strongman (*see* 1118).

1108 France's Philip I dies at the end of July at age 56 after a reign of nearly 48 years. He is succeeded by his son, 27, who will reign until 1137 as Louis VI. The new king faces insurrections from feudal brigands and rebellious barons, and it will take him 24 years to root out the robber barons who depend for their livelihood on plundering travelers en route to and from Paris.

The Byzantine emperor Alexius Comnenus defeats Bohemund of Otranto at Durazzo and makes him a vassal.

1109 Poland's Boleslav III defeats the Pomeranians at the Battle of Naklo and defeats the German king Henry V at the Battle of Hundsfeld near Breslau.

France's Louis VI goes to war with England.

Castile's Alfonso VI dies after a 37-year reign and is succeeded by his daughter, who will reign until 1129 as Urraca.

1110 The dukedom of Bohemia is secured for Ladislas I following the death of Bretislav II. Ladislas is helped by the German king Henry V and will rule until 1125.

Henry V invades Italy with a large army and concludes an agreement with Pope Paschal II at Sutri. Henry renounces the right of investiture, and the pope promises to crown him emperor and to restore to the empire all lands given by kings or emperors to the German church since the time of Charlemagne.

An English miracle play presented at Dunstable is the earliest of positive record. Priests have celebrated great ecclesiastical festivals since the 9th century with presentations of the Easter story, but the new miracle plays are based on legends of the saints.

1111 Henry V presents himself at St. Peter's February 12 for his coronation, but when Pope Paschal II reads the treaty terms commanding the clergy to restore the fiefs of the crown to Henry, there is a storm of indignation. The pope thereupon refuses to crown the king, Henry refuses to renounce the right of investiture, and he leaves Rome, taking the pope with him. When Paschal is unable to obtain help, he confirm's Henry's right of investiture and crowns him emperor.

1112 The Portuguese monarchy and the Burgundian dynasty that will control it until 1385 are founded on the death of Henry of Burgundy, count of Portugal. Henry's son of 3 succeeds him, with the boy's mother Teresa as regent, and will assume authority in 1128, reigning until 1185 as Afonso Henriques.

The Holy Roman Emperor Henry V is excommunicated by the Synod of Vienna, but an uprising led by Lothair, duke of Saxony, is easily quelled.

1113 Pisa conquers the Balearic Islands (Majorca, Minorca, Ibiza, and others).

The great-grandson of Vladimir the Great becomes grand duke of Kiev at age 60 and begins a 12-year reign in which he will campaign against the Cumans on the steppes. Vladimir Monomakh will also write works depicting conditions of Russian life.

Novgorod's Church of St. Nicholas pioneers the onion-domed style of Greek Orthodox church architecture.

1116 Hungary's Coloman I dies at age 44 and is succeeded by Stephen II, who will reign until 1131.

1118 The Byzantine emperor Alexius I Comnenus dies at age 70 after having the heretic Bogomilian leader Basilius burned to death. He is succeeded after a 37-year reign by his son, 30, who will reign until 1143 as John II Comnenus.

Japan's emperor Toba, 15, takes as his wife the pretty 17-year-old mistress of his grandfather, the ex-emperor Shirakawa. Shoshi is a daughter of Kimizane Fujiwara and is pregnant with Shirakawa's son Sutoku.

Aragon's Alfonso the Battler takes Saragossa December 18 from the Almoravid king Ali ibn-Yusuf, whose father defeated Castile's Alfonso VI in 1086.

1119 Charles le Bon becomes count of Flanders at age 36. A son of Denmark's late Canute IV and a grandson of Robert I the Frisian, Charles tries to promote the welfare of the Flemish and will rule until 1127.

England's Henry I defeats his Norman foes in a skirmish at Bremule. He dissuades Pope Calixtus II from supporting the French rebels who have rallied behind William the Clito, son of his imprisoned brother Robert (Henry released Clito after the victory at Tinchebrai in 1106).

1120 England's Henry I makes peace with France's Louis VI, but his only legitimate son is drowned in the disaster of the White Ship off Harfleur.

Measurement of latitude and longitude in degrees, minutes, and seconds is pioneered by the Anglo-Saxon scientist Welcher of Malvern, who in 1092 observed the solar eclipse and tried to calculate the difference in time between England and Italy.

The order of the Knights of the Temple is founded in a house near Jerusalem by Hugh of Pajens to guide and protect pilgrims. Knights Templar will be organized among noblemen of several countries and, with their men-at-arms and chaplains, will take monastic vows of poverty, chastity, and obedience to an order that will survive until 1312 (*see* 1070).

1121 The Byzantine emperor John II Comnenus recovers southwestern Anatolia from the Seljuk Turks and then hastens to the Balkans, where the Patzinaks are continuing their incursions.

The Concordat of Worms condemns French theologian-philosopher Pierre Abelard, 42, for his teachings of the Trinity, and he is castrated by the hirelings of one Fulbert, whose niece Héloïse he has secretly married. Abelard withdraws to a monastery, pursued by his former student Héloïse, and begins a persecuted life of wandering from one monastery to another, founding the priory of the Paraclete at which Héloïse will be prioress (*see* 1140).

1122 The emperor John II Comnenus and his Byzantine hordes exterminate the Patzinak Turks in the Balkans, but John's refusal to renew exclusive trading privileges granted to Venice by the late Alexius Comnenus in 1082 precipitates a 4-year war.

The Holy Roman Emperor Henry V renounces the right of investiture with ring and crozier in the September 23 Concordat of Worms, recognizes freedom of election of the clergy, and promises to restore all church property.

1123 The Byzantine emperor John II Comnenus defeats Serbian forces in the Balkans.

The First Lateran Council forbids marriage of priests and the practice of simony.

Japan's ex-emperor Shirakawa imposes a strict Buddhist law against killing any living thing (including fish, poultry, livestock, game and game birds).

Persian poet-astronomer Omar Khayyám dies at age 96, leaving works that include the *Rubaiyat* (*see* 1859).

London's Smithfield meat market has its origin in the priory founded beside the "Smooth" field just outside the city's walls. The field will soon be the scene of St. Bartholomew's Fair, where drapers and clothiers will exchange goods and where a weekly horsemarket will be held (*see* 1639).

1124 Hungary's Stephen II is defeated in battle by the Byzantine emperor John II Comnenus, who supports claims to the throne by Belá, blinded by the late Hungarian king Coloman. John acts to keep the Hungarians from gaining control of Dalmatia, Croatia, and Serbia.

Scotland's Alexander I dies April 27 at age 46 after a 17-year reign and is succeeded by his brother David, 40, who has ruled in the south since the death of their

1124 ***(cont.)*** grandfather Edgar in 1107. The new king will reign until 1153 as David I and be called "the Scotch Justinian."

The first Scottish coins are struck.

Rochester Cathedral is completed.

1125 The Holy Roman Emperor Henry V dies at Utrecht May 23 at age 44 after leading an expedition against France's Louis VI and then against the citizens of Worms. Lothair, 55, is chosen king at Mainz August 30 with papal support and crowned at Aix-la-Chapelle September 13.

The new emperor (above) asks Frederick of Hohenstaufen to restore to the crown the estates he has inherited from the late Henry V (above). Frederick refuses and is placed under the ban.

Venetian forces pillage Rhodes, occupy Chios, and ravage Samos and Lesbos.

The Almohades conquer Morocco.

1126 Lothair III makes his son-in-law Henry the Proud (Heinrich der Stolze), 18, duke of Bavaria (*see* 1138).

The Peace of 1126 ends hostilities between John II Comnenus and the Hungarians and Venetians. The emperor secures Branicova on the Danube but is forced by Venice to renew the republic's exclusive commercial privileges.

1127 Sicily's Roger II claims the Hauteville possessions in Italy and also claims overlordship of Capua, but subjects of Apulia resist union with Sicily. They gain support from Pope Honorius II, who sets Roger II of Capua and his brother-in-law Ranulf of Alife (Avellino) against the Sicilian king, excommunicates him, and lays claim himself as feudal overlord to Norman possessions in southern Italy.

Charles the Good, count of Flanders, is murdered. He has no children, France's Louis VI tries to impose the son of Normandy's Robert Curthose as ruler, but the Flemish towns (Ghent, Bruges, Ypres) force the selection of Thierry of Alsace as the new count.

1128 Pope Honorius invests Sicily's Roger II as duke of Apulia at Benevento in August after failure of his coalition against Roger.

Portugal's Afonso Henriques assumes authority at age 19, repudiates his mother Teresa's agreement to accept Castilian domination, defeats Castile's Alfonso VII at the Battle of São Mamede, and drives his mother into exile (*see* 1112; 1139).

The Abbey of Holyrood is founded by Scotland's David I.

The order of the Knights Templar, founded in 1120, is recognized and confirmed by Pope Honorius II.

Cistercian monks from Normandy settle in England and begin an extensive program of swamp reclamation, agricultural improvement, and stock breeding (*see* 1098). The Cistercians live austerely, depend for their income entirely on the land, and will have a salutary effect in improving English and European horse and cattle breeds and raising the standards of agriculture.

1129 Sicily's Roger II gains recognition as duke at Melfi in September from the barons of Naples, Bari, Capua, Salerno, and other cities that have resisted him.

Castile's queen Urraca dies after a 17-year reign and is succeeded by her son, who will reign until 1157 as Alfonso VII.

Former Japanese emperor Shirakawa dies July 7 at age 76; his grandson Toba becomes the power behind the emperor Sutuku.

1130 Pope Honorius II dies February 13. Gregorio Papareschi is elected to succeed him as Innocent II, but Cardinal Pierleone is elected as an antipope, and Innocent flees to France. The antipope is supported by Sicily's Roger II, who is crowned at Palermo December 25 by Anacletus II, beginning a 10-year war with Pope Innocent's champion, Bernard of Clairvaux, who calls Roger a "half-heathen king" and enlists the support of France's Louis VI, England's Henry I, and the German king Lothair II.

The antipope Anacletus II (above) is more acceptable canonically than is Innocent II, but his father is a rich converted Jew; prejudice prevents him from gaining secular support.

1131 Tintern Abbey is founded in Wales by Cistercian monks in the Wye Valley. A larger structure will be erected in the 13th and 14th centuries (*see* Wordsworth, 1798).

1133 The German king Lothair II arrives at Rome in March with Pope Innocent II and a small army after a 6-month journey across the Alps. He finds St. Peter's in the hands of the antipope Anacletus II. He is crowned emperor by Innocent II at the Church of the Lateran June 4 and receives as papal fiefs the vast estates of Matilda, marchioness of Tuscany, which he secures for his daughter Gertrude and her husband Henry the Proud of Bavaria.

England's Durham Cathedral is completed after 37 years of construction.

1134 The German House of Brandenburg has its beginnings as the emperor Lothair II makes Albrecht the Bear, 34, head of the Nordmark; Albrecht founds the ruling house of Anhalt (*see* 1150).

The Pontifical University of Salamanca has its beginnings in Castile.

1135 The Holy Roman Emperor Lothair II receives homage from Denmark's Eric II, is promised tribute by Poland's Prince Boleslav III, receives Pomerania and Rugen as German fiefs, and is implored by the Byzantine emperor John II Comnenus to help oust Roger II of Sicily.

The Hohenstaufen king Conrad III submits to Lothair, is pardoned, and recovers his estates in October.

England's Henry I dies December 1 at age 67 after a 35-year reign (his brother Curthose has died earlier as a prisoner at Cardiff). He is succeeded by his nephew Stephen of Blois, 38, who asserts his claim to the throne in opposition to claims by Henry's daughter Matilda, 33, widow of the late emperor Henry V, whose successsion was accepted by England's barons in 1126 and who was married 2 years later to Geoffrey Plantagenet of Anjou.

1136 The emperor Lothair II invades southern Italy in response to last year's appeal from John II Comnenus and conquers Apulia from Sicily's Roger II.

The English princess Matilda asserts her right to the throne of her late father (*see* 1135; 1138).

Historia Calamitatus Mearum by Pierre Abelard describes his love affair with Héloïse (*see* 1121; 1140).

The French church of St. Denis, completed by the abbé Suger, 45, is a romanesque structure but includes some pointed arches and high clerestory windows that mark the beginning of Gothic architecture. Suger initiates the idea of the rose window, arguing that people can only come to understand absolute beauty, which is God, through the effect of precious and beautiful things on the senses. "The dull mind rises to truth through that which is material."

1137 France's Louis VI dies August 1 at age 56 after a 29-year reign. He is succeeded by his son of 16, who marries Eleanor, heiress to the duke of Aquitaine at Bordeaux, and will reign until 1180 as Louis VII.

The prince of North Wales Gruffydd ap Cynan dies at age 56 after having rebuilt Welsh power overturned earlier by England's late Henry I.

Gruffydd's sons Owain and Cadwaladr work to revive the power of the principality of Gwynedd behind the Snowdonian range.

The Holy Roman Emperor Lothair II dies December 4 in the Tyrolean village of Breitenwag at age 67 while retreating from Italy after a mutiny among his troops and a breach with Pope Innocent II as to who shall control Apulia.

Antioch is forced to pay homage to the Byzantine emperor John II Comnenus, who has conquered Cilician (Little) Armenia.

1138 Swabia's House of Hohenstaufen begins a 130-year domination of the German states as Conrad is chosen German king for a second time March 7 by princes meeting at Coblenz in the presence of the papal legate. Now 45, Conrad is crowned March 13 at Aix-la-Chapelle. Bavaria's Henry the Proud refuses allegiance and loses Bavaria to the Franconian margrave of Austria Leopold IV. A long struggle begins between "Ghibellines" and "Guelphs" (the Hohenstaufens take their name from their Swabian castle Staufen on their estates of Waiblingen, whose name will be corrupted by their Italian supporters into "Ghibelline," while the family name Welf of Bavaria's Henry will be corrupted into "Guelph").

Poland's Boleslav III (Wry-mouth) dies at age 62 after a 36-year reign. He has divided his realm among his five sons.

The Battle of the Standard fought in August near Northallerton ends in defeat for Scotland's David I who has invaded England in support of Matilda against Stephen, but David does take possession of Northumberland.

1139 Civil war begins in England as Matilda lands at Arundel with an army to support her claims to the throne.

Portugal gains her independence from the Moors and begins to achieve identity as a sovereign nation. Portuguese forces under Afonso Henriques defeat the Moors July 25 in the Battle of Ourique in Allentejo (*see* 1128; 1143).

1140 Bohemia's Sobeslav I dies after a 15-year reign and is succeeded by Ladislas II, who will reign until 1173.

The fortress of Weinsberg is captured from the Welf in December by Conrad III, who permits the women of the town to leave. Each is permitted to take with her as much property as she can carry on her back, and each comes out bearing on her back a husband, father, or son who thus escapes.

Pierre Abelard is condemned for his "heresies" by the Council of Sens and sets out for Rome to present his defense.

Spanish rabbi Judah ben Samuel ha-Levi (Abu'l Hasan) dies at age 55, leaving as his chief philosophical work *Sefer ha-Kusari,* which categorizes revealed religion as superior to philosophic and rational belief.

1141 England's King Stephen is surprised and captured while laying siege to Lincoln Castle (*see* 1139). Matilda reigns for 6 months as "Domina of the English," but Stephen's supporters secure his release in exchange for the earl of Gloucester, Matilda's half brother Robert, bastard son of the late Henry I (*see* 1142).

Hungary's blind king Belá II dies after a 10-year reign in which he has taken terrible revenge on his opponents: he is succeeded by Géza II, who will reign until 1161.

1142 The German king Conrad III makes peace in May at Frankfurt with the duke of Saxony, Henry the Lion (Heinrich der Lowe), 13, son of the late Henry the Proud of Bavaria. Young Henry is confirmed in his duchy, while Bavaria is given to Conrad's stepbrother Henry Jasomirgott, margrave of Austria, who marries Henry the Proud's widow Gertrude.

England's Matilda is expelled from Oxford after a long siege by Stephen, who forces her to take refuge in the western part of the country. A 5-year period of anarchy begins.

Pierre Abelard dies at age 63 while en route to Rome. His body will be given to his widow, Héloïse, who will be buried beside him in 1164 (both will be reentombed at Paris in 1817).

1143 The Treaty of Zamora arranged by the Vatican obtains Castilian recognition of Portuguese independence. Afonso Henriques, now 31, is proclaimed king by the

1143 (cont.) cortes and begins a 42-year reign that founds the Portuguese monarchy and Burgundian dynasty (*see* 1139; 1185).

The Byzantine emperor John II Comnenus (Kalojoannes) dies at age 55 after a 25-year reign. He is succeeded by his son, 23, who will reign until 1180 as Manuel I Comnenus.

Geoffrey of Anjou, son-in-law of England's late Henry I, becomes duke of Normandy upon news of the death last year of his father Foulkes le Jeune, who was king of Jerusalem from 1131 until his death at age 51.

1145 Pope Eugene III is forced into exile by Arnold of Brescia, 45, a student of the late Abelard, who has fought corruption in the clergy, was condemned along with Abelard in 1140 by the Council of Sens, and turns Rome into a republic with a government patterned on that of ancient Rome.

Almohad forces under Abd al-Mu'min defeat an Almoravid army at Tlemcen, the Almoravid ruler Tasfin ben Ali dies, and the Almohads, originally from the Atlas Mountains, gain support among other Berber groups as revolts weaken the hold of Almoravid rulers on Spain.

Cistercian French ecclesiastic Bernard of Clairvaux, 54, preaches a Second Crusade after Pope Eugene (above) overcomes his resistance to the idea.

The Danube is bridged at Ratisbon by a span begun 10 years ago.

1147 Marrakesh falls to Almohad forces in April, ending more than 80 years of Almoravid rule in North Africa and Spain (where some Muslim rulers recognize the Almohad leader Abd al-Mu'min as caliph).

A Second Crusade assembles 500,000 men under the leadership of France's Louis VII and the German Conrad III, who take separate routes but give the crusade no coherent command, achieve nothing, and lose most of their men to starvation, disease, and battle wounds.

The distraction of the Second Crusade (above), enables Sicily's Roger II to seize the Greek islands and to attack Athens, Thebes, and Corinth.

An 11-year war begins between Sicily and the Byzantine Empire.

1148 The Byzantine emperor Manuel I Comnenus purchases Venetian aid to help him resist the Norman fleets of Sicily's Roger II, who has plundered Thebes.

An Italian silk industry is started at Palermo by Roger II (above), who takes numbers of silk workers back from Greece.

Returning Crusaders bring back sugar from the Middle East. Virtually unknown in Europe, even in the greatest castles, the sweetener will soon be prized above honey (*see* 1226).

1149 Venetian mercenaries regain Corfu for the Byzantines.

Theisir (al-Taysir) by the Arab physician Avenzoar (Abu Mervan ibn Zuh) describes otitis media, paralysis of the pharnyx, and pericarditis in a work based on personal experience, independent observation, and sound judgment. Opposing purgatives, Avenzoar strongly advocates venesection (phlebotomy).

Crusaders built imposing castles in the Holy Land as they advanced against the Saracen "infidels."

1150 Sweden's Sverker is deposed after a 16-year reign that has amalgamated the Swedes and the Goths. He is succeeded by Eric IX, who will reign until 1160.

Albrecht the Bear inherits Brandenburg (*see* 1134).

Chinese sea masters and caravan leaders use magnetic compasses in crude form to guide them in their journeys (*see* 1086; 1558; Bacon, 1267).

The University of Paris has its beginnings (*see* Sorbonne, 1253).

The Black Book of Carmarthen is compiled in Wales.

Troubadour music is organized in the south of France.

Angkor Wat hails the completion of the Hindu funeral temple of Khmer king Suryavarman II. Largest and most splendid of Asia's Hindu temples, it is the centerpiece of a capital that is surrounded by a 12-mile moat.

1151 Chinese forces retake Beijing from the Tatars (*see* 936).

Ghazni is burned by the Shansabani Persian princes of Ghur (Ghor), who will drive the Yamini to the Punjab and will depose them in 1186.

Geoffrey of Anjou dies September 7 at age 38. He has been called "Plantagenet" for his habit of wearing a sprig of broom (genet) in his cap and is succeeded as count of Anjou by his son Henry, 18, to whom he gave the duchy of Normandy last year.

The first fire and plague insurance is issued in Iceland.

1152 The German Hohenstaufen king Conrad III dies at Bemberg February 15 at age 58 after a 14-year reign. He is succeeded by his red-bearded nephew Frederick, 29, duke of Swabia, who is chosen king at Frankfurt March 5 partly because he is of both Welf and Waiblingen blood and it is hoped that he will end the rivalry that has divided the German states since 1138. He is crowned at Aix-la-Chapelle March 9 and will reign until 1190 as Frederick III (Barbarossa).

The 15-year marriage of France's Louis VII and Eleanor of Aquitaine is annulled. Eleanor, 30, who has produced no male heir, marries Henry Plantagenet, now 19, count of Anjou, Maine, and Touraine and duke of Normandy, who gains by marriage domains that make him master of more than half of France (*see* 1154).

1153 Scotland's David I dies at Carlisle May 24 at age 69 and is succeeded by his grandson, 12, who will never marry but will reign until 1165 as Malcolm IV.

1154 Roger II of Sicily dies at Palermo February 26 at age 60 and is succeeded by his fourth son William, who will reign until 1156.

Damascus surrenders April 23 to the sultan Nur ad-Din of Aleppo.

England's king Stephen dies October 25 at age 54 and is succeeded by his adopted son Henry Plantagenet, who is crowned at age 21 and will reign until 1189 as Henry II, inaugurating a Plantagenet dynasty that will rule England until 1399.

The first and only English pope begins a 5-year reign as Nicholas Breakspear is elected to the papacy. Adrian IV tries to end the anarchy that has persisted in Rome.

Iceland's Mount Hekla erupts as it did 50 years ago, and word spreads through the Catholic world that Hekla's craters are the entrance to hell.

England's domestic wine industry begins to decline as cheap French wines are introduced by Eleanor of Aquitaine, wife of the new king Henry II (above).

1155 Rome's Arnold of Brescia is interdicted by the new English pope Adrian IV and forced to flee. Betrayed by Frederick Barbarossa, he is hanged at Rome.

Pope Adrian IV gives Ireland to England's Henry II.

Henry II abolishes English fiscal earldoms and restores the royal demesne.

The Carmelite order of mendicant monks is founded by the Crusader Berthold from Calabria and 10 companions who establish themselves as hermits on Mount Carmel in the kingdom of Jerusalem.

1156 Frederick Barbarossa makes the margraviate of Austria a duchy after 180 years of rule by the Babenberg family and gives it special privileges (*see* 1246). He makes Bohemia a hereditary kingdom.

William of Sicily destroys the Byzantine fleet May 28 at Brindisi and recovers Bari from Greek barons who have been encouraged by Pope Adrian IV to revolt. William and Adrian come to terms at Benevento June 18, and William is confirmed as king.

Frederick Barbarossa is married June 9 at Würzburg to Beatrix, daughter of the late count of Upper Burgundy, Renaud III, and adds Upper Burgundy (Franche Comte) to his domain.

The former Japanese emperor Toba dies at age 53 in the midst of a struggle over the succession. His son Sutoku obtains help from Tameyoshi Minamoto and Tadamasa Taira, and they rise against Sutoku's half brother Goshirakawa, who is in turn supported by Minamoto's son Yoshitomo Minamoto and Taira's nephew Kiyomori Taira. Tameyoshi Minamoto is killed by his son and by Kiyomori Taira. Minamoto's father is exiled to the island of Izo-no-Oshima, south of Edo. Sutoku fails in his attempt to take over and is exiled to Sanuko on the island of Shikoku in the Inland Sea, and Goshirakawa remains as emperor.

1157 Frederick Barbarossa invades Poland, forces the submission of Duke Boleslav IV, and in October receives the homage of Burgundian noblemen at Besançon. The Diet of Besançon permits Boleslav, duke of Bohemia, to call himself "king."

Sweden's Eric IX conquers Finland and forces baptism on the vanquished Finns.

Frederick Barbarossa's army at Rome is destroyed by typhus or some other plague.

Castile's Alfonso VII dies at age 32 in the Sierra Moreno while returning to Toledo after a campaign against the Almohades, whose power is increasing. "The Emperor" has held León through most of his 21-year reign, but León breaks away as Alfonso's son ascends the throne as Sanchez III and makes his brother king (he will reign until 1188 as Fernando II).

1158 The Japanese emperor Goshirakawa abdicates after a 3-year reign, and his son Nijo, 15, begins a 7-year reign that will be interrupted briefly next year. Goshirakawa, who retains power, gives strongman Kiyomori Taira a higher position than Yoshitomo Minamoto, antagonizing Minamoto (*see* 1159).

Castile's Sanchez III dies after a brief reign; his infant son succeeds him and will reign until 1214 as Alfonso VIII.

Frederick Barbarossa leaves in June for a second Italian expedition, Milan is taken from its insurgents by Frederick's imperial officers (podestas) in northern Italy, and a long struggle begins between Frederick and the pope.

Lübeck is founded on the Baltic by a consortium of families who deal in grain, timber, fish, and other commodities. Each family takes a large block of land, builds a house on the best corner lot, and rents the balance to newcomers, while the consortium runs bakeries and operates the town's marketplace (*see* Hanseatic League, 1241).

1159 An English army led by Henry II's chancellor Thomas à Becket, 41, invades Toulouse to assert the rights of Henry's wife, Eleanor of Aquitaine, but Louis VII drives the English off (*see* 1152; 1160).

1159 ***(cont.)*** The Japanese emperor Nijo and his father Goshirakawa are dethroned and imprisoned by Noboyori Fujiwara and Yoshitomo Minamoto, who stage a palace revolution while Kiyomori Taira is away from Kyoto visiting the Kumano shrine. Kiyomori returns at news of the coup, kills both Fujiwara and Minamoto, reinstates the emperor Nijo, and becomes the power behind the throne.

Policratus by John of Salisbury is the first medieval attempt to formulate an extended and systematic treatment of political philosophy. John denounces Henry II for exacting funds from the church to support his invasion of Toulouse (above) and presents the first explicit defense of tyrannicide.

1160 England's Henry II comes to terms with France's Louis VII after failing to enlist the aid of Frederick Barbarossa, but the peace is unstable. Desultory skirmishing continues between the French and English.

Arab forces expel the Normans from North Africa.

Kiyomori Taira is elevated to the nobility by the ex-emperor Goshirakawa.

Theater *Jeu de St. Nicholas* by French troubadour Jean Bodel 12/6 at Arras.

1162 Milan is destroyed by Frederick Barbarossa in his continuing campaign against supporters of the pope (*see* 1158; 1167).

The archbishop of Canterbury Theobald dies, and England's Henry II has his chancellor Thomas à Becket elected to the primacy. Becket immediately becomes an ardent champion of Church rights.

1163 Henry II returns from the Continent and begins almost immediately to quarrel with Thomas à Becket, who defends clerical privileges. The king proposes that a land tax, which has been used to pay part of the sheriff's salary, shall henceforth be paid into the exchequer, but Becket opposes the move and defeats the proposal in July at the Council of Woodstock.

Dissection of human bodies is discouraged by a Church dictum promulgated by the Council of Tours in connection with the practice of dismembering dead Crusaders and boiling them down for transport back home: "The Church abhors the shedding of blood."

The Spanish physician Maimonides (Moshe ben Maimon), 28, is forced to leave Cordova. He will go to Fez for 10 years and then to Cairo, where he will become physician to Saladin, sultan of Egypt. Maimonides will write works on medicine, law, mathematics, logic, and theology.

1164 England's Constitutions of Clarendon are an attempt by Henry II to delimit spiritual and temporal jurisdictions. The archbishop of Canterbury Thomas à Becket has excommunicated a tenant-in-chief who has encroached on Canterbury lands, protected a clerk charged with assaulting a royal officer, and renounced his promise to observe the constitutions. Becket flees to France in November and persuades Pope Alexander III to condemn the constitutions.

1165 The Byzantine emperor Manuel I Comnenus makes an alliance with Venice against Frederick Barbarossa, who takes an oath at the Diet of Würzburg in May to support the antipope Paschal III against Pope Alexander III.

Scotland's Malcolm IV dies December 9 at age 24 and is succeeded by his brother William the Lion, 22, who will reign until 1214.

The Japanese emperor Nijo abdicates and dies after a 7-year reign. He is succeeded by his year-old infant son, who will reign until 1168 as the emperor Rokujo.

1166 Sicily's William I dies May 7 after a 12-year reign. William the Bad is succeeded by his son of 13, who will reign until 1189 as William II, championing the papacy and joining the Lombard cities in a secret conspiracy against Frederick Barbarossa.

Ballad *The Song of Canute* by an English monk of Ely.

Saladin (Salah-al-Din Yusuf ibn-Ayyub), 28, a city official, builds the Cairo citadel.

1167 Frederick Barbarossa enters Rome by storm on a fourth expedition into Italy. He has the antipope Paschal III enthroned, he has his wife Beatrix crowned, but a sudden outbreak of pestilence destroys his army, and he returns to Germany.

Copenhagen is founded by the Danish soldier Absallon (or Axel), 39, archbishop of Lund, who has delivered Denmark from Wendish pirates and conquered much of slavonic Mecklenburg and Estonia for the late King Waldemar and for Canute VI.

Oxford University has its beginnings (*see* Balliol, 1261).

1168 The Japanese emperor Rokujo is removed at age 4 and is succeeded by his uncle, 7, who will reign until 1180 as the emperor Takakura.

1170 England's Henry II has himself crowned by Roger, archbishop of York, in violation of the rights of Thomas à Becket, who persuades Pope Alexander III to suspend Roger of York and other bishops who have supported Henry and agreed to royal control of episcopal elections and other extensions of royal power (below).

Henry II increases the crown's power by replacing England's baronial sheriffs with men of lower rank trained in the royal service, thus breaking the hold of the barons on the shrievalty. Pope Alexander III forces Henry to reconcile his quarrel with Thomas à Becket in July.

Brandenburg's Albrecht the Bear, who has founded the House of Anhalt, dies November 13 at age 70 after dividing his territories among his six sons.

Thomas à Becket returns to England in early December after more than 6 years in France. He refuses absolution to the country's bishops, and overzealous knights murder him in the cathedral at Canterbury December 29.

The "Inquest of Sheriffs" strengthens the English exchequer by its financial inquiry.

Poetry *Le Chevalier à la Charette* by French poet troubadour Chrétien de Troyes introduces the Arthurian hero Lancelot du Lac.

Japan has her first recorded instance of ritual suicide: a feudal warrior slashes his own belly to commit *seppuku*.

1171 England's Henry II lands at Waterford in response to a request for aid from Ireland's deposed king Dermot MacMurrough (Diarmaid Macmurchada), who ruled Leinster from 1126 but was banished 5 years ago by Irish chieftains and dies at age 61. Henry is greeted as "lord of Ireland" and claims the country as his own (*see* 1155).

Saladin, now vizier of Cairo, abolishes the Fatimid caliphate that has ruled since 968 and made Egypt the center of Muslim culture.

1172 England's Henry II receives homage at Tipperary's 200-foot-high Caskel Rock from Irish princes who include Donal O'Brien. Henry has gone to Ireland in part to avoid the attacks of Pope Alexander III for the assassination of Thomas à Becket.

Eleanor of Aquitaine raises Aquitaine against her faithless husband Henry II, and he is forced to reconcile his differences with the pope.

The Venetian Grand Council restricts the powers of the doges.

Poetry *Roman de Rou* by Anglo-Norman poet Wace is a chronicle of the dukes of Normandy.

1173 England's Henry II has his wife Eleanor of Aquitaine held captive while he dallies in western Herefordshire with his mistress Rosamund Clifford, daughter of Walter de Clifford. Henry's sons Henry, Richard, and Geoffrey lead a rebellion against their father, but the House of Commons gives hearty support to the king.

Hungary's Stephen III dies after a 12-year reign and is succeeded by Belá III, who has been educated at Constantinople. The new king will introduce Byzantine customs in a reign that will continue until 1196.

The Waldensian movement begins at Lyons, where local merchant Peter Waldo renounces the world, gives away all his goods and property, and devotes himself to preaching voluntary poverty (*see* 1179).

1174 Frederick Barbarossa purchases Tuscany, Sardinia, Corsica, and Spoleto from Henry the Lion's uncle, Welf VI.

England's Henry II does penance at Canterbury for the murder in 1170 of Thomas à Becket, who was canonized last year as Saint Thomas Becket.

Monks at Engelberg monastery cut woodblocks for printing elaborate capital letters in manuscripts.

The Leaning Tower of Pisa has its beginnings in a campanile (bell tower) built by architect Bonnano Pisano. The 177-foot tower will tilt until it is more than 18 feet off the perpendicular (*see* Galileo, 1592).

1175 India is invaded by the Persian sultan Mu'izz-ad-din, Mohammed of Ghor.

1176 The Battle of Legnano May 29 gives the Lombard League infantry a victory over the feudal cavalry of Frederick Barbarossa, who sustains a wound so serious that he is thought for a while to be dead.

The Egyptian sultan Saladin conquers Syria and mounts a campaign to drive the Christians from the kingdom of Jerusalem.

The University of Modena has its beginnings.

Welsh ecclesiastic Walter Map, 36, organizes legends of King Arthur and his knights of the round table at Camelot (*see* 537). Map has traveled abroad on foreign missions.

The Eisteddfod festival of Welsh music, poetry, and drama at Cardigan Castle begins an annual competition by bards that will continue at various locations for more than 800 years.

Rabbits are introduced into England as domestic livestock.

1177 Frederick Barbarossa and Pope Alexander III sign the Treaty of Venice in August to begin a 6-year period of peace between the Lombard League and the Holy Roman Emperor.

England's Henry II and France's Louis VII sign the Treaty of Ivry; Pope Alexander III has intervened to bring the two monarchs to terms.

The Khmer capital Angkor Wat falls to Champa invaders who will hold it for 4 years.

Belfast is founded in northern Ireland as John de Courcy builds a castle to command a ford near the mouth of the Lagan River.

1178 Frederick Barbarossa is crowned king of Burgundy at Arles July 30. He will repeat the ceremony in 1186.

1179 Pope Alexander III prohibits the Waldensians from preaching without permission of the bishops, but Peter Waldo replies that he must obey God, not man (*see* 1173; 1184).

Henry II of England submitted to whipping at Canterbury as penance for the murder of Thomas à Becket.

1180 France's Louis VII dies September 18 at age 59 after a 43-year reign and is succeeded by his son of 15, who was crowned last year at Reims and will rule until 1223 as Philip II Augustus.

The Byzantine emperor Manuel I Comnenus dies September 24 at age 60 after a 37-year reign of great splendor that has involved the empire in repeated conflicts with the Normans and has weakened it financially. Manuel is succeeded by his son, 12, who will reign briefly, with his mother Maria of Antioch as regent.

An uprising against Japan's ruling Taira family begins as Kiyomori Taira alienates the imperial family with his excesses following the death last year of his eldest son Shigemori at age 41. One revolt is led by an imperial prince and another by Yorimasa Minamoto, 74, who is supported by some of the monasteries but who is killed by the Taira. A general uprising is led August 17 by Yoritomo Minamoto, 33, who is helped by his father-in-law Tokimasa Hojo. The emperor Takakura abdicates at age 19 and is succeeded by his 2-year-old son Antoku, grandson of Kiyomori Taira, whose forces are opposed on the Fuji River in October by troops raised by the Minamoto family. A Minamoto detachment gets behind the Taira position, a sudden flight of waterfowl alarms the Taira in the night, they flee, and the Minamoto forces begin a series of victories.

Glass windows appear in private English houses (*see* tax, 1695).

1182 A Constantinople mob massacres the Latins who rule as agents of the regent Maria of Antioch. They kill city officials and traders and proclaim an uncle of Alexius II Comnenus co-emperor to rule as Andronicus I Comnenus together with his nephew.

Denmark's Waldemar the Great dies at age 51 after a 25-year reign. His son of 19 will reign until 1202 as Canute VI, extending Danish dominion over Pomerania and Holstein, and will call himself king of the Danes and Wends.

France banishes her Jews.

England's port of Bristol sends wooden vessels built "shipshape and Bristol fashion" out to Spain (Jerez) for sherry, Portugal (Oporto) for port wine, Iceland for stockfish (dried cod), and Bordeaux and Bayonne in Gascony for woad from the plant *Isatis tinctoria* to make blue dye for Bristol's woolens.

1183 The Byzantine emperor Alexius II Comnenus is strangled to death by agents of his uncle and co-emperor, who assumes sole power and begins a 2-year reign as Andronicus I Comnenus.

The Peace of Constance ends the conflict between the Lombard towns, the pope, and Frederick Barbarossa, but it also ends any remnant of unity in the Holy Roman Empire.

The sultan Saladin takes Aleppo.

Japan's Taira clan is driven out of Kyoto by Minamoto clansmen led by Yoshinaka, 29, a cousin of Yoritomo. The Minamoto family installs the 3-year-old half brother of the emperor Antoku as the emperor Gotoba in rivalry with the Taira family's emperor (*see* 1185).

Poetry *Senzaishu* by Japanese poet Saigyo-hosi, 75, who at age 23 abandoned his wife and family to become a priest and has spent his life wandering through Japan composing short poems of 31 syllables each, called tanka.

1184 The Great Diet of Mainz, held by Frederick Barbarossa, is a colossal medieval pageant at which his two sons are knighted in the presence of a multitude of princes and knights.

Cyprus gains her freedom from the Byzantines.

Georgians make Tamara, 25, their queen; she begins a 28-year reign that will raise Georgia to the peak of its political power.

Japan's Minamoto military leader Yoshinkaka is killed at age 30 by the Taira in the Battle of Ichinotani near Suma on the Inland Sea.

Peter Waldo is excommmunicated by Pope Lucius III (*see* 1179).

Modena Cathedral is consecrated after 85 years of construction.

Canterbury Cathedral opens in Kent after 5 years of construction to replace the cathedral in which Thomas à Becket was murdered at the end of 1170. Architect for the great Gothic structure is William of Sens, who has built the French cathedral at Sens that has helped establish the Gothic style in which pillars supported by buttresses replace walls as the main support of vaulted roofs, making possible large stained-glass windows to permit the entrance of mysteriously colored light. The cathedral will be given a nave in the late 14th century and a central tower in the 15th century.

Awe-inspiring cathedrals helped to keep religion the dominant force in medieval Europe.

1185 The Japanese emperor becomes the puppet that he will remain until 1868 as the shoguns (generals) assume power. The Minamoto family, led by Yoritomo's half brother Yoshitsune, defeats the Taira clan at the Battle of Danoura April 24 on the Inland Sea. The emperor Antoku is drowned and the imperial sword lost. The 5-year-old Gotoba reigns without the sword and with no real power, and Japan enters a centuries-long period of civil wars among the feudal lords *(daimyo)* who rule under the shoguns with support from retainer-knights (*samurai*).

Norman forces attack the Byzantine Empire. They take Durazzo, storm Thessalonica with an army and navy, and massacre the Greeks. The Greek nobility, led by Isaac Angelus, deposes the Byzantine emperor Andronicus I Comnenus, who is tortured and executed. The Byzantine general Alexius Branas defeats the Normans at Demetritsa. Isaac will reign until 1196 as Isaac II Angelus, restoring the corruption that Andronicus had begun to eliminate, and the empire will begin to disintegrate.

A great Bulgarian insurrection that will lead to the creation of an independent Bulgaria begins under the boyars Peter and John Asen, who gain support from the Cumans against the extortions of Byzantine revenue agents from Constantinople. Much of the Greek population of the Balkans will be annihilated in the revolt, and much of the region will be desolated.

The sultan Saladin seizes Mosul and begins to conquer Mesopotamia.

Portugal's Afonso Henriques dies at age 76 after a reign that has founded the Portuguese monarchy. He is succeeded by his son, 31, who will reign until 1211 as Sancho I, building roads and cities.

The Knights Templar, founded at Jerusalem in 1120, is established at London.

Japan's Kyoto is a city of 500,000 and is larger than any city in the West with the possible exceptions of Cordova and Constantinople.

1186 The Kamakura period that will dominate Japan until 1333 begins under Minamoto leader Yoritomo, 39, whose family is based in the village of Kamakura (*see* 1189).

A triple coronation occurs at Milan as Frederick Barbarossa is crowned king of Burgundy, his son Henry, 21, is crowned Caesar in a deliberate revival of the old Roman title, and Henry's bride Constance, daughter of Sicily's late Roger II, is crowned queen of the Germans.

Frederick Barbarossa takes the cross and prepares for a Third Crusade.

1187 The sultan Saladin, provoked by an attack on a caravan, cuts down a small group of Templars and Hospitalers at Tiberias in May, crushes a united Christian army July 4 in the Battle of Hittin, lays siege to Jerusalem September 20, and takes the city October 2 without sacking it.

The Punjab is conquered by Mohammed of Ghor, who rules at Ghazni as governor for his brother Ghiyas-ud-Din Mohammed.

Verona Cathedral is completed after 48 years of construction.

1188 England's Prince Richard Coeur de Lion pays homage to France's Philip II for English possessions in France and does it in the presence of his father Henry II at the November 18 Conference of Bonmolins. A struggle follows in which Henry is overpowered. He is chased from Le Mans to Angers and forced to buy peace by conceding to all demands, including the immediate recognition of Richard as his successor (*see* 1189).

The Saladin tithe imposed by Philip III to raise money for a Third Crusade is the first tax ever levied in France.

The Pont Saint Benezet spans the Rhône at Avignon for the first time. A French folk song will make it famous.

1189 England's Henry II dies July 6 at Chinon at age 54 after doing homage to France's Philip II and surrendering the territories of Gracy and Issoudon. He is succeeded by his son Richard Coeur de Lion, who will reign until 1199 as Richard I, spending only one year of the reign in England and visiting the British Isles only twice. The new king quickly ends his friendship with France's Philip by continuing his father's policy of territorial aggrandizement and refusing to honor his contract to marry Philip's sister Alais to whom he has been betrothed since age 3, but at year's end Richard and Philip exchange pledges of mutual good faith and fellowship as they prepare to join the Third Crusade (below).

Sicily's William II makes peace with the Byzantine emperor Isaac II Angelus, abandons Thessalonica and his other conquests, and dies childless in November at age 36 as he prepares to join the Third Crusade. Tancred of Lecce, bastard son of the late Roger II, will succeed William next year and reign until 1194.

Japan's shogun Yoshitsume is killed by his older brother Yoritomo, who will crush the Fujiwara clan in the north (*see* 1186; 1192).

English Jews are massacred at the coronation of Richard I (above) (*see* 1290; France, 1182).

The first silver florins are minted at Florence (*see* 1252).

German merchants negotiate a commercial treaty with Novgorod.

The Third Crusade begins in May as Frederick Barbarossa leaves Regensburg at the head of a splendid army.

The first paper mill in Christian Europe is established at Herault, in France, some 38 years after the opening of a paper mill in Muslim Spain, but parchment remains almost the only writing material in Europe.

1190 The Holy Roman Emperor Frederick Barbarossa drowns June 10 at age 67 while crossing (or bathing in) the river Calycadnus (Geuksu) near Selucia (Selefke) in Cilicia. He is succeeded by his son who was crowned Caesar 4 years ago at Milan and has been serving as regent, but while Henry will be crowned next year and reign until 1197 as Henry VI, his father's

1190 ***(cont.)*** death leaves the Third Crusade temporarily without a leader.

France's Philip II prepares to join the Third Crusade by making arrangements to rule France from the Holy Land.

England's Richard the Lion-Hearted takes desperate measures to raise and equip a force of 4,000 men-at-arms and 4,000 foot soldiers for the Third Crusade.

A massacre of some 500 Jewish men, women, and children in York Castle March 17 ends a 3-day siege by young men about to leave on the Third Crusade, urged on by people indebted to Jewish money lenders. Since last year, Jews have been attacked from Durham south to Winchester.

Nonfiction *Guide of the Perplexed* by Cairo rabbi-physician Maimonides (Moshe ben Maimon, or Abu Imran al-Kufuni), now 55, who has held that conversion was no sin so long as one remained secretly faithful to Israel.

French poet Chrétien de Troyes dies at age 60 after writing works that include *Perceval, ou Le Conte du Graal,* earliest known version of the Holy Grail legend. His works have been translated into English and German and are highly popular.

1191 England's Richard the Lion-Hearted embarks on the Third Crusade with a fleet of 100 ships but spends the winter quarreling with France's Philip II in Sicily. He leaves Messina in March, conquers Cyprus, sells it to the Knights Templar, moves on to join the siege of Acre in June, and takes a major role in reducing Acre, but he offends Leopold of Austria with his arrogance.

France's Philip II falls ill, leaves the crusade, and returns to Paris by Christmas after concluding an alliance en route with the new Holy Roman Emperor Henry VI against England's offensive Richard Coeur de Lion who meanwhile has gained a brilliant victory over the forces of Saladin at Arsuf and led the Christian host to within a few miles of Jerusalem.

Third Crusaders led by Richard the Lion-Hearted vanquished Saladin's Saracens but then came to a bad end.

The Order of the Teutonic Knights has its beginnings in a hospital founded at Acre by merchants of Lübeck and Bremen. The hospital will soon become attached to the German church of Mary the Virgin at Jerusalem (*see* 1198).

Zen Buddhism is introduced to Japan by the priest Aeisai, 50, who returns from a visit to China.

Aeisai (above) plants tea seeds, making medicinal claims for tea that will be published in 1214 (*see* 805; 1597; tea ceremony, 1484, 1591).

1192 The Third Crusade follows treacherous guides into the desert beyond Antioch, where famine, plague, and desertions reduce its numbers from 100,000 to 5,000. Richard the Lion-Hearted makes a truce with Saladin under which the Christians are permitted to keep the coastal towns they have taken and receive free access to the Holy Sepulcher at Jerusalem. Richard leaves for home October 9, traveling in flimsy disguise, and is captured December 20 at Vienna, where Leopold of Austria imprisons him in the duke's Durenstein castle on the Danube.

Yoritomo Minamoto is appointed shogun by the Japanese emperor Tsuchi and continues his efforts to crush the Fujiwara clan in the north.

1193 England's Richard the Lion-Hearted is surrendered early in the year to the Holy Roman Emperor Henry VI, who resents the support that the Plantagenets have given to the family of his German rival Henry the Lion and the recognition that Richard has given to the Norman antiking Tancred of Sicily. Henry demands a ransom of 130,000 marks for the return of the English king.

Egypt's Ayyubite sultan Saladin dies at Damascus March 4 at age 52, and his empire is divided among his relatives. Saladin has briefly united the Muslim world and stemmed the tide of Western conquest in the East.

Aztecs in the Western Hemisphere invade Chichemec territory and conquer the Chichemecs.

Licensed prostitution begins in Japan (*see* Yoshiwara, 1617).

The first known merchant guild is established at London.

Indigo from India is imported to the British Isles for dyeing textiles.

1194 Richard the Lion-Hearted returns to England in March following payment of the first installment of the king's ransom to the Holy Roman Emperor Henry VI (who will never receive the full amount he demanded). England has levied the first secular taxation to be imposed on movable property, and her resources have been strained to the utmost to raise the ransom. Richard remains for only a few weeks before returning to the Continent. He leaves the administration of England in the capable hands of the archbishop of Canterbury

Hubert Walter, who accompanied Richard on the Third Crusade, led his army back to England, levied the taxes to pay the king's ransom, and has put down a plot against Richard by the king's worthless brother John Lackland.

Richard crushes France's Philip Augustus at Freteval and regains his French fiefs by the Truce of Verneuil.

Norman rule in Italy ends after 91 years as the emperor Henry VI reduces Sicily with help from Pisa and Genoa to terminate the reign of the rich Tancred of Lecce, who has gained support from the pope and from England's Richard Coeur de Lion. Henry also reduces southern Italy and part of Tuscany. He is crowned king of Sicily, retains the Matildine lands in central Italy, organizes an imperial administration of his territories, and begins to plan a great empire with its base in Italy.

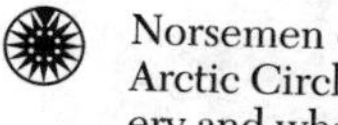

Norsemen discover Spitzbergen far to the north of the Arctic Circle. The island will later be an important fishery and whaling center (*see* 1557).

The Elder Edda is a collection of Scandinavian mythology.

China's Huanghe (Yellow) River begins flowing southward from the Shandong massif after repeated alterations of its streambed. The river will retain its new course until 1853.

1195

The Byzantine emperor Isaac II Angelus is dethroned by his brother Alexius while off on a hunting trip. Alexius is proclaimed emperor by the troops and begins a 9-year reign as Alexius III Angelus. He captures Isaac at Stagira in Macedonia, has his eyes put out, and takes him prisoner despite Isaac's redeeming him from captivity at Antioch and loading him with honors.

The Holy Roman Emperor Henry VI marries his brother Philip of Swabia to Isaac's daughter Irene and thus tries to give the Hohenstaufen a new title and a valid claim against the usurper Alexius (above).

1196

Hungary's Belá III dies after a 23-year reign in which he has helped the Byzantine emperor Isaac II Angelus against the Bulgarians. Belá is succeeded by his son Emeric, who will reign until 1204 but whose brother András, 21, will challenge his position.

The Holy Roman Emperor Henry VI persuades the diet meeting at Würzburg in April to recognize his 2-year-old son Frederick as king of the Romans, but opposition from Adolph, archbishop of Cologne, and others thwarts the will of the diet.

1197 The Holy Roman Emperor Henry VI goes to Italy to persuade Pope Celestine III to crown his son Frederick, who has been chosen king of the Romans at Frankfurt. The pope refuses. Henry takes cruel measures to put down an insurrection in the south that has been provoked by the oppression of his German officials, he prepares to embark on a crusade against the Byzantine usurper Alexius III Angelus, but he catches cold while hunting at Messina and dies September 28 at age 32.

A 15-year German civil war begins upon the sudden death of Henry VI (above), whose Waiblinger (Ghibelline) brother Philip of Swabia is supported by France's Philip II against the Welf (Guelph) Otto of Brunswick, son of Henry the Lion, who is supported by England's Richard the Lion-Hearted.

Bulgarian insurgent Peter Asen is assassinated by rival boyars and is succeeded by his younger brother Kaloyan Joannitsa), who becomes czar of the Bulgarians and who will complete the conquest of northern Bulgaria from the Byzantines.

Bohemia ends 14 years of dynastic struggle during which the country has had some 10 different rulers. Duke Ottokar I, who was deposed in 1193, is restored to power and begins a 37-year reign in which he will strengthen Bohemia by taking advantage of German civil war to make Bohemia a decisive factor in German affairs.

Château Gaillard is completed on the Seine by England's Richard the Lion-Hearted as he fights to restore Angevin power in northern France.

1198 Richard the Lion-Hearted adopts *"Dieu et mon droit"* (God and my own right) as the motto of the royal arms of England at the Battle of Gisors. He asserts that he is no vassal of France but royal by God's grace and his own, but is forced to cede Gisors to France's Philip II, along with the fortresses of Neaufles and Dangu.

The new pope Innocent III excommunicates France's Philip II for repudiating his 1193 marriage to Ingeborg, 22, sister of Denmark's Canute VI, to whom he took an almost instant dislike, but public opinion forces Philip to effect a reconciliation with the pope.

The Teutonic Knights (Order of the Knights of the Hospital of St. Mary of the Teutons in Jerusalem) is established by Germans gathered for a new crusade (*see* 1191). Headquarters of the order will remain at Acre until 1291, membership will be open only to Germans, and knighthood will be reserved for noblemen (*see* 1226; Battle of Tannenberg, 1410).

1199

Richard the Lion-Hearted lays claim to a French treasure trove, becomes embroiled in a dispute with the viscount of Limoges, harries Limousin, and lays siege to the castle of Chalus. Wounded in the shoulder by a crossbow bolt while directing an assault on the castle, Richard dies April 6 at age 32. He is succeeded by his brother John Lackland (or Softsword), who is crowned with support of the Norman barons who reject claims of primogeniture advanced by supporters of John's nephew Arthur, 12, posthumous son of the late Geoffrey, count of Brittany, who died in 1186.

The Declaration of Speyer confirms the right of German princes to elect the German king.

Japan's shogun Yoritomo Minamoto dies at age 52 after a 7-year reign that has founded the Kamakura shogunate. A council of 13 takes power under the leadership of Yoritomo's widow Masako and her father Tokimasa

1199 ***(cont.)*** Hojo, 62, who has supported Yoritomo against the Taira.

Liverpool is founded at the mouth of the Mersey River in the west of England (*see* 1207).

1200 The Peace of Le Goulet ends hostilities between England and France.

Anglesey is seized by the prince of North Wales, Llewelyn, who 6 years ago returned from exile and drove his uncle David I from his territory.

Poetry *Der Arme Heinrich* by German poet Hartmann von Aue is an epic poem.

13th Century

1201 Europe's commercial capital is Venice, whose location is midway between the East and the cities of Europe. Venice's Lido and marshes protect it from invaders, and its position gives it natural access via the Adriatic to both the eastern and western Mediterranean, with direct connections via the Po and Adige valleys to the rich north Italian cities of Brescia, Ferrara, Mantua, Milan, Padua, and Verona. The city has links via eastern Alpine passes to the cities of southern Germany and via western Alpine passes to Constance, Geneva, Lucerne, and Zurich, and while Genoa's commercial ventures receive little official encouragement, Venice's doge and senate are eager to support commercial interests.

The St. Gotthard Pass opens through the Swiss Alps between the cantons of Uri and Ticino.

1202 France's Philip II confiscates Aquitaine, Anjou, and Poitu and gives these fiefs of England's John Lackland to John's nephew Arthur of Brittany. John's troops triumph over French forces at Mirabeau, and young Arthur vanishes at Rouen, possibly murdered by order of John to block him from claiming the English throne.

Pope Innocent III gives command of a Fourth Crusade to Boniface III, count of Montferrat, who gains support from Venice's Doge Enrico Dandolo, 94. The doge has abstained from earlier crusades because the Arabs are Venice's best customers, but he now sees potential for profit and agrees to provide ships for a maritime attack on Egypt provided that he receive 85,000 marks plus half of all booty. The Crusaders gather at Venice but cannot raise the 85,000 marks. Dandolo finally agrees to transport them on condition that they first sack Zara on the Dalmatian coast.

The Crusaders sack Zara in November and are joined by Alexius, son of the deposed Byzantine emperor Isaac II Angelus. He makes large promises and persuades many of the Crusaders to follow him to Constantinople and overthrow the usurper Alexius III.

Liber Abaci by Italian traveler-mathematician Leonardo Fibonacci (Leonardo da Pisa) introduces Europe to Arabic numerals from North Africa and the zero from India, making calculation much easier than with Roman numerals.

Pope Innocent III excommunicates the Fourth Crusade following the sack of Zara (above).

The papal bull *Venerabilem* asserts the Vatican's right to pass upon the fitness of any candidate elected to rule the Holy Roman Empire and to review disputed or irregular elections (*see* 1338).

Court jesters make their first appearance in European courts.

1203 The Fourth Crusade reaches Constantinople in July, and the usurper Alexius III takes flight and dies in exile. The Crusaders retrieve the blind Isaac II Angelus from the dungeons and restore him to the Byzantine throne, but Isaac's mind has grown feeble after 8 years in prison, and his son Alexius rules as regent for the vicious Isaac.

Mohammed of Ghor completes his conquest of Upper India.

Brittany rebels against England's John of Lackland (*see* 1202).

Siena University has its beginnings.

Poetry *Parzival* by German poet Wolfram von Eschenbach, 31, is an epic romance of the Holy Grail based on *Perceval* by the late Chrétien de Troyes.

Livestock breeding and viniculture increase in parts of France and the Lowlands as cheap grain from the new granary in the Baltic Sea region makes it less profitable to grow wheat, rye, barley, and oats.

Famine ravages England and Ireland as it will repeatedly throughout this century. Recurrent crop failures will bring hardship to the British Isles, the German states, and Poland, but cheap grain from the Balkans will lower food prices in much of the Continent.

A large brewing industry develops at Hamburg and in the Lowlands as barley malt from the Baltic becomes more readily available at lower prices.

1204 The blind Byzantine emperor Isaac II Angelus is deposed again after a 6-month reign with his son Alexius IV Angelus. Isaac's general Mourzouphles usurps the throne, has the son put to death, and is proclaimed emperor with the title Alexius V Ducas. He tries to defend Constantinople against the Fourth Crusaders, but the city falls April 12 and is mercilessly sacked. Alexius is driven out of the city, arrested in Morea, and executed.

A Latin empire is established at Constantinople, and the count of Flanders Baldwin IX, 33, is elected first emperor. He begins a brief reign as Baldwin I.

Thomas Morotinople, a Venetian, is made patriarch of Constantinople, and Venice's Doge Enrico Dandolo receives "a quarter and a half" of the Eastern Empire, adding numerous islands and coastal ports to expand Venetian territory and influence.

The empire of Trebizond that will survive until 1461 is founded by Byzantine leaders who establish themselves under Alexius Comnenus on the northern coast of Anatolia at Trebizond. David Comnenus is at Sinope,

1204 ***(cont.)*** Theodore Lascaris is in Bithynia, Theodore Mancaphas is at Philadelphia, Manual Maurozomes is in the Meander Valley, while Leo Gabalas takes over the island of Rhodes.

An independent Greek empire is founded by Michael Angelos Comnenus who makes himself despot of Epirus and will reign until 1214.

Thessalonica is sacked by the count of Montferrat Boniface III, who has been chosen leader of the Fourth Crusade following the death of the count of Champagne Thibaut III. Boniface will rule Greece and Macedonia as king of Thessalonica until 1207.

Hungary's Emeric I dies after an 8-year reign and is succeeded by his infant son, who will reign briefly as Ladislas III.

Bubonic plague reduces the ranks of the Fourth Crusaders, prevents them from reaching Jerusalem, and ends the crusade.

A hospital of the Holy Spirit is founded at Rome by Pope Innocent III.

The University of Vicenza has its beginnings.

Returning Crusaders plant Damson plum trees from Damascus in France and will hasten the spread to Europe of such Arabic products as rice, sugar, lemons, and cotton.

1205 The Bulgarian king Kaloyan defeats Frankish Crusaders, the Venetian doge Dandolo, and the emperor Baldwin I near Adrianople with support from the Cumans and local Greeks. Dandolo dies at age 97, Baldwin is captured, Baldwin's able brother, 31, is named regent, and when Baldwin dies in captivity the brother begins an 11-year reign as Henry I.

Hungary's Arpád dynasty begins its most disastrous reign. The brother of the late Emeric I dethrones his infant nephew Ladislas III and will reign profligately until 1235 as András II.

France's Philip Augustus conquers Anjou from the English.

1206 Mohammed of Ghor is assassinated after a reign that has founded Muslim power in India. His former viceroy Kuth-uddin-Aibak takes over Delhi as the first independent Muslim ruler of north India. A former slave from Turkestan who was named viceroy in 1192, Aibak will be killed playing polo in 1210, but the slave dynasty he founds will rule until 1266 and be followed by five other slave dynasties, which will rule until 1526.

Uigurs on the Chinese border are overrun by Mongol forces led by their chief Temujin, 44, who is proclaimed Genghis Khan at Karakorum.

The Uigurs (above) have adopted block printing and use it to print Buddhist works in the Turkish language, using alphabet script derived from the Phoenician through Aramaic and other languages, with Sanskrit notes and Chinese page numbers. Genghis Khan (above) employs the Uigur scholar Tatatonga, who applies the Uigur script to the Mongol language.

1207 Liverpool becomes chief English port for trade with Ireland following the silting up of the river Dee, which makes the approach to Chester unnavigable.

1208 The German king Philip of Swabia is murdered June 21 at Bamberg by Otto of Wittelsbach, 26, Bavaria's count palatine, to whom he has refused the hand of his eldest daughter Beatrix, 10. Otto has been at odds with Philip since 1198, he receives overtures from many former supporters of Philip, and he is chosen king November 11 at Frankfurt. He is betrothed to Beatrix and receives large concessions from the pope, who has earlier sided with Philip.

Theodore Lascaris founds the Nicaean Empire (*see* 1204).

Genghis Khan completes his conquest of Turkestan.

A vassal of Raymond IV of Toulouse kills Peter of Castelnau January 15; Pope Innocent III excommunicates Raymond.

Students from Bologna found a school of medicine at Montpelier (*see* Arnold of Villanova, 1300).

England's John Lackland objects to the pope's choice for archbishop of Canterbury; Innocent places England under an interdict March 24, forbidding the clergy to administer sacraments.

Pope Innocent preaches an Albigensian Crusade against the heretic Catharists of Albi and a Waldensian Crusade against followers of Peter Waldo. Noblemen espousing the lower-class Waldensian movement against Church corruption have seized Church lands in the south of France.

1209 Carcassonne surrenders after a siege by the Crusader Simon de Montfort, 49, earl of Leicester and comte de Toulouse, who captures the Albigensian "heretics." But war continues between the French nobility of the north and the Provençal nobility which has protected the ascetic *"bons hommes"* and their antisacerdotal teachings.

England's John Lackland invades Scotland.

Otto of Wittelsbach invades Italy, meets with Pope Innocent III at Viterbo, refuses papal demands that he yield to the Church all territories in dispute before 1197, but agrees not to claim supremacy over Sicily and is crowned emperor at Rome October 4 to begin a reign of more than 8 years as the Holy Roman Emperor Otto IV.

The coronation of Otto IV (above) is followed by fighting between German and Roman soldiers. The pope asks the emperor to leave Roman territory, Otto refuses to leave until he receives satisfaction for losses suffered by his troops, and he then violates his treaty with the pope by taking property that Innocent has annexed to the Church and distributing disputed territories among his supporters, whom he rewards with large estates.

Innocent III excommunicates England's John Lackland in November for opposing the election of Stephen Langton as archbishop of Canterbury, an election consecrated by the pope more than 2 years ago.

The Franciscan order of monks has its origin at Assisi, where Italian friar Giovanni Francesco Bernardone, 27, obtains approval from Pope Innocent III for rules he has drawn up to administer the new order. The friar has consecrated himself to poverty and religion, he has gathered some likeminded companions at Assisi, and his order will be confirmed in 1223 by Pope Honorius III (*see* 1212; 1228).

England's Cambridge University has its beginnings (*see* 1217).

1210 The Holy Roman Emperor Otto IV is excommunicated by the pope November 18 but proceeds to complete his conquest of southern Italy. The emperor's excommunication coalesces German opposition by princes who are supported by France's Philip II Augustus.

Poetry *Tristan und Isolde* by German poet Gottfried von Strassburg, who will die with his epic incomplete.

1211 Portugal's Sancho I dies in March at age 57 after a 26-year reign and is succeeded by his stout son, 26, who will reign until 1223 as Afonso II (Afonso o Gordo).

China is invaded by Genghis Khan, who will withdraw his Mongol forces in a few years to seek opportunities for plunder in the West.

1212 German princes elect a 17-year-old grandson of the late Frederick Barbarossa to succeed Otto IV, whom they deposed late last year. Otto returns from Italy in March and makes some headway against his enemies, but his wife dies in August, and his hold on the southern duchies is thus weakened. Young Frederick arrives from Sicily in the fall, is welcomed in Swabia, and is crowned December 9 at Mainz.

The Battle of Las Navas de Tolosa July 16 breaks the power of the Almohades in Iberia. Castile's Alfonso VIII gains a great victory just south of the Sierra Moreno and drives most of the Almohades out of the peninsula.

Venice conquers Crete (Candia).

A Children's Crusade sets out for the Holy Land led by a French shepherd boy known only as Stephen and a child from Cologne named Nicholas. Slave dealers kidnap Stephen's army and sell it into Egypt, Nicholas's crusade is aborted in Italy, and some 50,000 children are lost (many are sold into slavery at Marseilles).

The Order of the Poor Clares (Franciscan nuns) is founded by Italian nun Clara of Assisi, 18, with help from Friar Giovanni Francesco Bernardone (*see* 1209).

Tofu (soybean curd) is introduced to Japan from China, where it has been eaten for more than 2,000 years.

1213 The English Parliament has its beginnings in the Council of St. Albans (*see* 1258).

The Battle of Muret September 12 ends in victory for the Crusaders of Simon de Montfort, who has used religious zeal as a pretext to invade Iberia. Aragon's Pedro II, whose 5-year-old son was betrothed 2 years ago to de Montfort's daughter, is killed at age 39 resisting de Montfort. Placed in de Montfort's hands, the young king will be put in the care of the Knights Templar at the insistence of Pope Innocent III, be brought to Saragossa in 1216, and reign until 1276 as Jaime I.

The archbishop of Canterbury Stephen Langton returns after years on the Continent and absolves John Lackland, who submits to Pope Innocent III. Both England and Ireland become papal fiefs.

The German king Frederick gains papal support by a bull he promulgates July 12 at Eger renouncing all lands claimed by the pope since the death of the emperor Henry VI in 1197. Frederick's rival Otto IV is supported by his uncle John Lackland.

1214 The Battle of Bouvines July 27 establishes France as a major European military power, brings Flanders under French domination, gains the imperial crown for Frederick II, and ruins John Lackland of England. A great anti-Capetian alliance formed by John with the emperor Otto IV, the counts of Boulogne and Flanders, and most of the Flemish, Belgian, and Lorraine feudal nobility suffers defeat near Tournai at the hands of professional cavalry fighting for Philip II Augustus in alliance with young Frederick, who will be crowned emperor in 1220.

Scotland's William the Lion dies at age 71 after a 49-year reign in which he has established a Scottish church independent of the English church and subject only to the see of Rome. He is succeeded by his son of 16, who will reign until 1249 as Alexander II.

Genghis Khan captures the Chinese town that will become Beijing (*see* 1211; 1216).

1215 The Magna Carta signed at Runnymede in mid-June limits the power of the English monarchy. Feudal barons supported by Scotland's new king Alexander II meet with England's John Lackland between Staines and Windsor and exact major concessions reaffirming traditional feudal privileges contained in the accession charter signed by Henry I a century ago. John immediately appeals to Pope Innocent III, who issues a bull annulling the charter. John imports foreign mercenaries to fight the barons, but the Magna Carta will remain the basis of English feudal justice.

German nobles crown the Hohenstaufen king Frederick II at Aix-la-Chapelle July 25. Now 20, Frederick has been king of Sicily since age 3; he was married at age 14 to the daughter of Aragon's Alfonso II (widow of Hungary's late king Emeric), and the victory of his ally Philip II Augustus of France at the Battle of Bouvines last year has strengthened his hand. His rival Otto will die in 1218, leaving him undisputed ruler of the Germans.

The Magna Carta (above) reaffirms human rights. Chapter 39 of the document states, "No freeman shall be arrested and imprisoned, or dispossessed, or outlawed, or banished, or in any way molested; nor will we set forth against him, nor send against him, unless by the lawful judgment of his peers, and by the law of the land."

The Fourth Lateran Council prohibits trial by ordeal.

The Magna Carta signed at Runnymede by John Lackland limited the power of the English monarchy.

The Dominican order of monks (*Fratres Praedicatores*, or Preaching Friars) is founded by the Spanish-born priest Dominic, 45, who was made canon of the cathedral at Osma at age 24 and 10 years later attacked the Albigenses in Languedoc.

1216 French mercenaries land in England in January to help John Lackland fight the barons, who last year forced him to sign the Magna Carta.

The French dauphin Louis Coeur de Lion, 28, lands in England in May. Some of the barons who had earlier supported John rally to Louis, and many of John's French mercenaries desert to him. Barons friendly to Louis besiege Lincoln Castle, rebel barons in Nottinghamshire take the castles of Nottingham and Newark, Louis takes Winchester in the south, but he is unable to take Windsor or Dover.

England's John Lackland comes down with dysentery in October after crushing resistance in the north. He crosses the Wash and reaches Newark Castle but dies there October 19 at age 38. The king is succeeded by his 9-year-old son, who will reign until 1272 as Henry III; the earl of Pembroke William Marshal serves as regent, and the moderate party takes control, thus ending the need for opposition to royal authority.

Norway's Eric X Cnutsson dies after an 8-year reign (*see* 1217).

Genghis Khan invades the Near East with 60,000 Mongol horsemen who destroy ancient centers of civilization, ruin irrigation works, and destroy every living thing in their path (*see* 1214; 1222).

1217 French forces leave England following defeats at Lincoln and Sandwich.

Norway's late Eric X Cnutsson is succeeded by the bastard son of the late Haakon III Sveresson, who ruled briefly from 1202 to 1204. The new king, 13, will reign until 1263 as Haakon IV Haakonsson, initially with Earl Skule as regent.

Castile's Ferdinand III is declared of age at 18 and will reign until 1252, ending the dynastic wars that have plagued the country and achieving successes against the Moors. As peace returns, there will be a renewal of Spanish agriculture, which has been disrupted.

Portugal's Afonso o Gordo defeats a Moorish army at Alcacer do Sal.

Salamanca University has its beginnings in a school founded by Alphonso VIII of León.

England's Cambridge University is founded (*see* 1209).

Vienna's bakers pay homage to Duke Leopold VI, who is about to leave on a crusade against the Moors in Egypt and the Holy Land. They offer him rolls baked in the shape of crescents (*see* 340 B.C.; 1683).

1218 The Holy Roman Emperor Otto IV dies May 19 at age 36, leaving the German king Frederick II without opposition.

The Peace of Worcester ends hostilities between England's Henry III and the Welsh.

Genghis Khan conquers Persia.

Pope Innocent III preaches a Fifth Crusade at the Fourth Lateran Council as the sultan Malik-al-Adil is succeeded by Malik-al-Kamil late in the year. The king of Jerusalem John of Brienne is displaced as leader of the crusade by the papal legate Pelagius.

Denmark adopts the Danneborg; it will survive as the oldest national flag.

Newgate Prison is completed at London. It will be a debtors' prison for more than 560 years (*see* 1782).

1219 Minamoto family control of the Japanese shogunate ends in January with the assassination of the shogun Sanetomo Minamoto while he is returning from the shrine at Kamakura. His uncle Yoshitoke Hojo, 57, has encouraged the assassins. He installs Yoritsume Fujiwara as shogun, although the real power remains in his own hands and in those of his sister Masako, and the Hojo family will rule Japan until 1333.

1220 The German king Frederick sets out for Rome in August. He promises the new pope Honorius III that he will undertake a new crusade, and the pope crowns him Holy Roman Emperor November 22, but he becomes involved in restoring order in Sicily, using cruel measures to suppress the country's anarchy.

Dresden's Boys' Choir is founded at the Kreuz-Kirche.

Geneva's St. Peter's Cathedral is completed atop the city's highest hill after 60 years of construction in Roman and Gothic styles (*see* Calvin, 1541).

1221 The Hojo family, which seized power in Japan 2 years ago, exiles the ex-emperor Gotoba and three other ex-emperors to various small islands. It installs the 4-year-old son of the exiled Juntoka as emperor in April, and replaces him in June with the 9-year-old Gohorikawa. He will reign until 1232, but he will be confined to one-quarter of the imperial palace.

Genghis Khan plunders Samarkand in his march to the West.

1222 A Golden Bull, issued by Hungary's András II under pressure from his son Belá and other feudal lords, strengthens the monarchy but limits its functions. The bull exempts the clergy and the gentry from taxation, orders periodic meetings of the diet, guarantees the nobility against arbitrary imprisonment and confiscation, and grants landowners the right to dispose of their domains as they see fit.

The Mongols make their first appearance in Europe as Genghis Khan invades Russian territories.

No lands or offices may be given to foreigners or Jews under terms of Hungary's Golden Bull (above).

England's Council of Oxford establishes April 23 as St. George's Day to honor the Christian martyr who died early in the 4th century. The patron saint of England is said to have slain a dragon (representing the Devil) to rescue the king's daughter Sabra (representing the Church).

The University of Padua has its beginnings.

1223 The Battle of the Kalka River shows Europe the power of the Mongols. Their leader Subutai defeats a Russian army supported by Cuman allies, but the Mongols do not press their victory and withdraw back into Asia (*see* 1235).

Portugal's Afonso II dies in March of leprosy at age 36 after a 12-year reign. His eldest son, barely 12, will reign until 1245 as Sancho II.

France's Philip II Augustus dies July 14 at age 57 after a reign of 43 years. His weak son of 36 will rule as Louis VIII, but only for 3 years.

1224 A new Anglo-French war begins and the French take English lands between the Loire and the Garonne.

Franciscan monks arrive in England (*see* 1209).

The True Pure Land sect (Jodo Shin) is founded by the theologian Shinran Shonin, 51, as an offshoot of his master's Pure Land sect. The Jodo Shin will be Japan's most popular religious sect, rivaled only by the Zen.

Naples University has its beginnings.

Geographical Dictionary is completed by the Greek Moslem geographer Yaqut ibn 'Abdullah (Abdullah ur-Ruml), 45, who was captured as a child in Asia Minor, sold as a slave at Baghdad to a merchant who had him educated, and later sent as the merchant's agent to the Persian Gulf and to Syria. Freed in 1199, Abdullah became a scribe and bookseller, traveled in 1213 to Syria, Egypt, Iraq, and Khurasan (northeastern Persia), spent 2 years working at libraries in Central Asia, but returned to Mosul and Aleppo to escape the Mongol invaders of Genghis Khan.

1225 English forces regain territory taken by the French last year with the exceptions of Poitou, the Limousin, and Perigord.

Cotton is manufactured in Spain. The fabric will compete with linen and wool (*see* 3000 B.C.).

Poetry *Roman de la Rose* by French poet Guillaume de Lorris whose allegorical metrical romance of courtly wooing will be completed by Jean de Meung (Jehan Clopinel).

"Sumer is icumen in" is sung as a round in England.

1226 France's Louis VIII dies at Montpensier November 8 at age 39 after conquering the South in a renewal of the 1208 to 1213 crusade against the Albigensian heretics. Louis is succeeded by his son of 12, who assumes the throne November 29 at Reims and will reign until 1270 as Louis IX.

Francis of Assisi (Giovanni Francesco Bernardone) dies October 3 at age 44, 2 years after allegedly having a vision of an angel.

The Teutonic Knights founded in 1198 are commissioned to conquer and convert Prussia.

"Kreuzlied" is sung by Middle High German lyric poet and minnesinger Walther von der Vogelweide, 56, whose religious songs follow love songs such as "Unter den Linden" and political songs ("Sprijehe") that he has sung at the courts of Philip of Swabia, Hermann of Thuringia, Otto IV, Frederick II, and others since 1198. The political songs have urged German unity and independence and opposed the more extreme claims of the popes.

England's Henry III, now 19, asks the mayor of Winchester to obtain 3 pounds of sugar, a quantity considered enormous.

1227 Genghis Khan dies at age 65, and his vast empire is divided among his three sons.

England's Henry III declares himself of age and begins a personal rule that will continue for 45 years, but he remains under the influence of Hubert de Burgh.

The landgrave of Thuringia Louis IV dies, and his widow Elizabeth, 20, daughter of Hungary's András II, is driven out along with her infant son Hermann by Henry Raspe, 25, brother of Louis, who makes himself landgrave.

Frederick II embarks September 8 from Brindisi on a new crusade to the Holy Land, but his army is stricken with fever. He lands at Otranto 3 days later, is suspected of malingering by Pope Gregory IX, is excommunicated September 29, and spends the rest of the year in violent quarrels with the pope.

The potter Toshiro returns after 4 years in China and pioneers Japanese porcelain production.

Construction begins at Beauvais of Saint-Pierre Cathedral.

Construction of Toledo Cathedral begins.

1228 Papal soldiers of Pope Gregory IX (below) invade the lands of Frederick II (below) in what the pope calls a crusade against the "non-Christian" king.

A Sixth Crusade has embarked for the Holy Land in midsummer under the leadership of the excommunicated Holy Roman Emperor Frederick II, who capitalizes on dissensions among descendants of the late Malik-al-Adil to achieve his ends through diplomacy (*see* 1229).

Francis of Assisi is canonized 2 years after his death by Pope Gregory IX. Franciscan monks will play a major role in history for the next 4 centuries.

1229 Frederick II signs a treaty February 18 with the Egyptian sultan Malik-al-Kamil, nephew of the late Saladin, who surrenders Bethlehem, Nazareth, and Jerusalem plus a corridor to the port of Acre for use by Christian pilgrims. Jerusalem's patriarchs oppose Frederick's accession, but he enters the city March 12 and crowns himself king March 18 in the Church of the Holy Sepulcher, assuming the monarchy by right of his marriage in November 1225 to the late Iolande (Isabella), daughter of Jerusalem's titular king John, count of Brienne. Frederick returns to Italy in June and easily drives out his foes.

Aragon conquers the Balearic Islands.

The Inquisition at Toulouse imposed by Albigensian Crusaders forbids laymen to read the Bible.

Turku is founded by Finns who build a cathedral and make the new town their capital.

The University of Toulouse has its beginnings.

First Legend by Franciscan monk Thomas of Celano, 29, is a biographical sketch of St. Francis of Assisi, whom Thomas joined at age 14. His *Second Legend* will appear in 1247.

1230 The Treaty of San Germano in July ends hostilities between Pope Gregory IX and Frederick II, who is absolved from excommunication after promising to respect the papal territory and to allow freedom of election to the Sicilian clergy.

Bohemia's Ottokar I dies after a 33-year reign and is succeeded by his son, who will rule until 1253 as Wenceslas I.

León's Alfonso IX dies childless September 24 after a 42-year reign. León accepts Ferdinand III of Castile as sovereign, and he unites the 2 kingdoms.

Berlin is founded on the site of former Slavic settlements.

Leprosy is introduced into England by returning Crusaders who have accompanied Frederick II to the Holy Land.

1231 The Japanese shogun orders his people not to sell their children into slavery, but poor farmers will continue for centuries to sell daughters.

A medical school is founded at Salerno by Frederick II, who decrees that the school's curriculum shall include 3 years of logic, 5 years of medicine, and one year of practice, with a diploma to be granted at the end of the 9 years. The school is directed by Nicolaus Praepositus, author of the first medieval pharmacopoeia the *Antidotorium* (*see* 1235).

Japan has a famine in the spring that will be followed by many similar famines in this century.

1232 The Nasrides dynasty that will rule Granada until 1492 comes to power in the person of Mohammed I, 29, who will rule until 1273. He stiffens Muslim resistance to Christian Spain.

Verona's Ezzelino IV da Romano succeeds to power at age 36. He will be a powerful opponent of the papacy and will lay waste northeastern Italy.

Antony of Padua is canonized by Pope Gregory IX. The Portuguese Franciscan teacher died last year at age 36.

1233 England's Henry III faces a rebellion of barons led by Richard Marshall, third earl of Pembroke, who gain support from the prince of North Wales Llewelyn the Great and demand that Henry dismiss the Poitevin foreigners with whom Peter des Roches has filled the royal administration. The king summons Pembroke to Gloucester in August, the earl fears treachery and does not go, Henry declares him a traitor, and Pembroke crosses to Ireland, where Peter des Roches instigates his supporters against the earl.

Mongol forces under Subutai capture a Chinese ammunition works at Bian Qin.

Pope Gregory IX entrusts the Inquisition to the Dominicans.

England mines coal at Newcastle for the first time. The town will become so famous for its coal that "carrying coals to Newcastle" will become a common phrase to signify superfluous effort.

The "Great Hallelujah" penitential movement begins in the north of Italy.

A Japanese royal family adopts the ancient custom of staining teeth black *(ohaguro)*. The practice begins to gain wide acceptance as a sign of beauty.

1234 The earl of Pembroke falls into enemy hands and dies in captivity. The new archbishop of Canterbury, Edmund Rich, 59, rebukes Henry III for following foreign counselors, holds him responsible for Pembroke's murder, and threatens him with excommunication. The king dismisses his Poitevin friends but replaces them with a new clique of servile and rapacious followers.

France annexes Navarre and will retain it for 2 centuries.

Pope Gregory IX, who canonized St. Francis of Assisi in 1228, canonizes the Spaniard Dominic, who began the mendicant Dominican order that rivals the Franciscans.

1235 The Mongols annex the Qin empire despite Qin efforts to resist them with explosive bombs.

Malinke tribesmen led by Sundiata Keita defeat Sosso forces at Kirina and gain freedom from Sosso's king Samanguru Kante. Sundiata Keita begins expanding his Mali kingdom, taking over parts of Ghana (*see* 1240).

Hungary's András II dies after a 30-year reign in which he has lost vast territories. His son will reign until 1270 as Belá IV, trying to recoup the losses sustained by András.

Frederick II of Hohenstaufen marries Isabella, the sister of England's Henry III.

Surgeons at the school of medicine in Salerno supported by Frederick II (above) dissect human bodies for the first time since the Ptolemic enlightenment of Alexandria in the 3rd century B.C. (*see* 1163; 1231).

Frederick II (above) employs Michael Scot to translate the writings of Aristotle from Greek into Latin.

1236 Cordova surrenders to Castile's Ferdinand III June 29 after 5 centuries in Moorish hands.

The Delhi slave dynasty sultan Altamsh dies after a 25-year reign in which he has enlarged and strengthened the Muslim Empire of northern India, has conquered the governors of Bengal and Sind, escaped destruction by the hordes of Genghis Khan (who stopped at the Indus), destroyed the capital (Ujjain) of the ancient Hindu kingdom of Vikramaditya, and built the Kutb Minar tower at Delhi. He is succeeded by his daughter Raziya (Raziy'yat-ud-din), who will rule until she is assassinated by her Hindu followers in 1240—the first Muslim woman to rule on the subcontinent.

Anesthesia is pioneered by the Dominican friar Theodoric of Lucca, who teaches at Bologna. Son of a surgeon to the Crusaders, Theodoric advocates the use of sponges soaked in a narcotic and applied to the nose in order to put patients to sleep before surgery, he favors use of the opiates mandragora and opium, and he recommends mercurial ointments for skin diseases.

Bavarian knight-poet Neidhardt von Reuenthal dies after a career in which he has founded the popular lyric poetry of the German courts. He has been nicknamed Neidhardt Fuchs (the Fox) for his poems ridiculing the rude life and manners of rich German peasants.

1237 Mongol forces use gunpowder and possibly firearms to conquer much of eastern Europe. Led by the khan Ogadai, Batu Khan, and Gen. Subutai, now 65, they devastate Poland but will fail in their efforts at conquest (*see* 1241).

A second Lombard League is shattered by Frederick II, who gains a great victory November 27 at Cortenuova. Frederick has secured the election of his 8-year-old son Conrad as king of the Germans to succeed the imprisoned Henry.

Eyeglasses and distilled alcoholic beverages will be introduced to Asia from Europe by the Mongols (above).

The Teutonic Knights merge with the Livonian Brothers and move to convert Russians from the Greek church to the Roman as Moscow falls to the Mongols (above).

1238 Novgorod elevates Aleksandr Nevski, 18, to the rank of prince. (He will be called Nevski after his victory on the Neva River in 1240.)

The Holy Roman Emperor Frederick II fails in an attempt to take Brescia from the second Lombard League. Pope Gregory IX has taken alarm at a projected marriage between Frederick's bastard son Enzio and Adelasia, heiress to Sardinia, and has made an alliance with the Lombard League.

Valencia surrenders September 28 to Aragon's Jaime I, now 30.

France's Louis IX rebuilds Foulque Castle at Angers in slate on a white stone foundation.

1239 Pope Gregory IX excommunicates the Holy Roman Emperor Frederick II for a second time March 20, calling him a rake, a heretic, and the anti-Christ.

1240 Aleksandr, prince of Novgorod, wins the name Nevski July 15 by defeating Swedish forces under Birger Jarl on the banks of the Neva River.

Mongols of the Golden Horde conquer Kiev as they sweep through southern and central Russia.

Mali forces under Sundiata Keita complete their takeover of the entire Ghana Empire, a once-mighty power whose roots are lost in legend (*see* 1235; 1255).

England and Scotland fix the border between them.

A crusade to the Holy Land that violates a prohibition by Pope Gregory IX embarks under the leadership of Richard Cornwall, 31, a brother of England's Henry III, and Simon de Montfort, 32, earl of Leicester, who has married Henry's sister Eleanor, widow of the late William Marshal, second earl of Pembroke.

The earl of Leicester (above) has expelled the Jews from Leicester.

Bolognese poet Guido Guinizelli founds the *dolce stil nuovo* school of Italian love poetry.

1241 The Battle of Liegnitz (Wahlstatt) in Silesia ends in victory for the Mongols of the Golden Horde, who cut down the feudal nobility of eastern Europe, including the German Knights Templar, but the Mongols who have invaded Poland, Hungary, and much of Russia retire to Karakorum upon the death of their khan Ughetai (Ogadai) at age 56 in December.

Pope Gregory IX dies August 22 at age 94 while the Holy Roman Emperor Frederick II is advancing against him. The pope has called a synod at Rome to depose Frederick, who ravages papal lands, nearly takes Rome, and captures a large delegation of French prelates off Genoa who are en route to the synod (he is induced to release the clergymen by France's Louis IX).

A Hanseatic League formed by Baltic trading towns will soon include Lübeck, Cologne, Breslau, and Danzig with concessions as far distant as London and Novgorod (*see* 1158; 1250).

Ships of the Hanseatic League (above) employ new navigational discoveries including the rudder and the

Mongols of the Golden Horde invaded Eastern Europe but failed to keep the territories they overran.

bowsprit. The rudder is an improvement over the oar that has been used canoe-fashion at the side of the stern, and the bowsprit permits the lower forward corner of the mainsail to be hauled beyond the bows so that a ship may sail more closely into the wind.

1242 England's Henry III invades France to support feudal lords in the south who have rebelled against Louis IX, but Aquitaine and Toulouse will soon submit to the king, and the coalition of rebellious barons will collapse.

The Holy Roman Emperor Frederick II pays a short visit to his German realms which have been threatened earlier by the Mongols. He gains some support by granting extensive privileges to certain towns.

The Mongol Golden Horde regroups at Sarai on the lower Volga under the command of Batu, a grandson of the late Genghis Khan.

The Battle on Lake Peipus April 5 ends in a victory for Aleksandr Nevski over the Teutonic Knights.

1243 A 5-year truce ends hostilities between France and England.

Famine sweeps the German states whose towns are infested like most in Europe by black rats that have come westward with the Mongols (*see* Black Death, 1340).

1244 Jerusalem is recaptured by the Muslim mercenaries of the Egyptian pasha Khwarazmi whose action will inspire a Seventh Crusade, but Jerusalem will remain in Egyptian hands until 1517 and in Muslim hands until 1918.

Rome University has its beginnings.

England holds her first Dunmow Flitch archery competition.

1245 A council at Lyons deposes Frederick II July 27, finding him guilty of sacrilege and possible heresy. War breaks out in the German states as the pope orders the Germans to elect a new king.

Portugal's Sancho II is deposed by Pope Innocent IV, who offers the crown to Sancho's brother Afonso, count of Boulogne. Sancho tries to regain the crown but will die early in 1248. The new king will reign until 1279 as Afonso III despite condemnation by the pope, who will accuse him of all manner of crimes.

A Seventh Crusade mobilizes forces under the leadership of France's Louis IX, who takes the cross against his mother's advice. He will embark in 1248 on an expedition to regain Jerusalem from the Egyptians.

1246 German electors choose the landgrave of Thuringia Henry Raspe as king May 22 to replace the deposed Frederick II. He defeats Frederick's son Conrad, 18, near August 5 at the Battle of Nidda near Frankfurt, but Conrad obtains help from the towns and from his father-in-law Otto II, duke of Bavaria, and drives Henry Raspe out of Thuringia.

The last Austrian Babenberg duke Frederick II has died after a 16-year reign to end the dynasty that began in 976. The deposed German king Frederick II seizes the vacant dukedom of Austria and Styria, using ruthless measures to suppress a formidable conspiracy of discontented Apulian barons (*see* 1278).

Franciscan monk Bartholomaeus Anglicus says leprosy is contagious and hereditary; he believes that it comes also from eating hot food, pepper, garlic, and the meat of diseased dogs. (*see* Hansen, 1874).

Poetry *Meier Helmbrecht* by German aphoristic poet Wernher der Gertenaere is the earliest German peasant romance. Wernher is one of the 12 founders of the Meistersinger guild.

1247 Parma falls to Lombard Guelphs who surprise the city while its imperial garrison is off guard. Frederick II concentrates a large army outside Parma and lays siege to the town.

The city of Buda is founded by Hungary's Belá IV to replace the city of Pest destroyed 6 years ago by the Mongols. Belá repopulates the city with foreign colonists, mostly Germans.

England's St. Mary of Bethlehem hospital for the insane has its beginnings. The word "bedlam" will be derived from its name and will be applied to situations such as those that will prevail in the new hospital.

Siena University has its beginnings south of Florence in Tuscany.

A gravestone erected in Yorkshire bears the inscription, "Hear underneath dis laihl stean/ las Robert earl of Huntingtun/ neer arcir yer az hie sa geud/ And pipl kauld im Roben Heud/ sick utlawz as he an iz men/ il england nivr si agen/ Obiit 24 kal Decembris 1247."

1248 Lombards besieged at Parma break out in February while Frederick II is off on a hunting expedition. They storm the imperial wooden town which Frederick has built and has named Vittoria in accordance with a prediction by his astrologers. The emperor's forces are scattered or destroyed, and his harem and some of his most trusted officers fall into Lombard hands along with the treasury and the imperial insignia. The Lombards hack Thaddeus of Seussa to pieces, and they place the imperial crown on the head of a hunchbacked beggar who they carry back to the city in mock triumph.

Genoese forces take Rhodes.

The Seventh Crusade embarks for Egypt in August led by France's Louis IX. He is accompanied by his three brothers and the chronicler Jean de Joinville, 24.

Seville surrenders November 23 after a 2-year siege to a Christian army under Castile's Ferdinand III. Its Muslim inhabitants flee to Granada.

The University of Piacenza has its beginnings on the Po River in northern Italy.

Spanish poet Gonzalo de Berceo dies at age 68 after writing some 13,000 verses dealing largely with the lives of the saints, miracles of the Virgin, and other devotional subjects.

1249 Scotland's Alexander II dies at Kerrera July 8 at age 51 while en route to subdue the Western Isles which are dependent on Norway. He is succeeded by his 8-year-old son, who will reign until 1285 as Alexander III.

The Hague becomes the seat of Dutch government as Count Willem II of the Netherlands builds a castle in the town.

Bolognese forces defeat and capture Sardinia's titular king Enzio, 24, a bastard son of the Holy Roman Emperor Frederick II. They will hold him in a dungeon until his death in 1272.

The Seventh Crusade led by Louis IX invades Egypt in the spring, takes Damietta without a blow, marches on Cairo, but is halted before Mansura.

Roger Bacon, 35, makes the first known European reference to gunpowder in a letter written at Oxford. The English Franciscan writes 12 years after the Mongol invasions and knows how to make the powder.

Bacon (above) fights to make science part of the curriculum at Oxford, holding that it is complementary to religion, not opposed to it.

Oxford's University College is founded.

1250 The Battle of Fariskur April 6 ends in a victory for Egyptian forces who rout the scurvy-weakened Seventh Crusaders of Louis IX and massacre them. Louis himself falls into the hands of the new Egyptian caliph Turanshah, who has arrived from Syria to claim the throne left vacant last year by the death of Malik al-Salih Najm al-din. Turanshah releases Louis after he agrees to evacuate Damietta and to pay a ransom of 800,000 gold pieces.

The new Egyptian caliph advances from Mansura to Fariskur. His conduct troubles the emirs who have raised him to the throne, and when he advances to Shajar al-durr, they overthrow him. The Mamelukes will rule Egypt until 1517 and thereafter under Ottoman suzerainty until 1805. (Originally purchased as slaves from the Georgian Caucasus, the Mamelukes have fought in the armies of the caliphs.)

The Holy Roman Emperor Frederick II dies at Fierentino December 13 at age 55 after a 38-year reign. He is succeeded as German king by his son, now 22, who narrowly escapes assassination at Regensburg and will reign until 1254 as Conrad IV.

England's Henry III confirms letters of protection to the "merchants of Germany" whose London Steelyard is an outpost of the fledgling Hanseatic League (*see* 1241; 1252).

The Crusaders have introduced Arabic numerals and the Arabic decimal system, both of Indian origin, to Europe (*see* 1202).

Cinnamon, cloves, coriander, cumin, cubebs, ginger, mace, and nutmegs carried back by returning Crusaders are now to be found in rich English and European houses but are in many cases valued more for supposed medical value than for culinary purposes.

Nonfiction *Speculum naturale, historiale, doctrinale* by Vincent of Beauvais is an early encyclopedia.

Poetry *Heike* by Japanese poet Yukinaga Shinanozenji is a popular epic account of the Taira family (*see* 825).

1251 Portuguese forces seize the Algarve in the southwest corner of the Iberian Peninsula.

Shepherds and farm workers in northern France abandon flocks and fields in a widespread insurrection. The rebels assemble at Amiens and then at Paris, they spread out to England, and the revolt spreads as far as Syria with bloody demonstrations and riots that meet with ruthless suppression.

Kyoto's Temple of Thirty-Three Spaces (Sanjusangendo), built in 1132, is rebuilt after being destroyed by fire.

1252 Castile's Ferdinand III dies at age 53 after a 35-year reign in which he has regained Cordova from the Moors and has obtained permission to establish a Christian church at Marrakah. His death aborts plans to invade Africa, and he is succeeded by his learned son of 24, who will reign until 1284 as Alfonso X.

Aleksandr Nevski becomes grand duke of Vladimir by order of the grand khan, who replaces Nevski's elder brother Andrew. The new duke works to prevent any renewal of the Mongol invasion.

India's Ahom kingdom is founded in Assam.

France's Louis IX expels the country's Jews (*see* 1182; 1269).

The Inquisition uses instruments of torture for the first time.

A Hanseatic settlement in the Flemish harbor of Bruges is established by merchants from Lübeck and Hamburg who have earlier taken joint action in Eng-

1252 ***(cont.)*** land and in the Netherlands. They obtain common rights at Bruges (*see* 1250; 1259).

Florence issues 54-grain gold "florins" in an assertion of independence (*see* 1189).

The Kamakura Daibutsu is cast by Japanese sculptor Ono Goroemon 32 miles from Edo. The 40-foot-tall 100-ton statue of the Buddhist divinity Amida has a circumference of 96 feet at its base and portrays the divinity in the traditional meditative position with hands in lap, palms up, fingers touching.

The Church of St. Francis is completed at Assisi.

Sake (rice wine) production is halted by the Japanese shogun to conserve rice.

1253 Bohemia's Wenceslas I dies after a 23-year reign in which he has encouraged wide-scale German immigration and has antagonized the nobility by seeming to favor the Germans. He is succeeded by his son the duke of Babenberg, 23, who will reign until 1278 as Ottokar II and make Bohemia one of the richest countries in Europe by opening her silver mines.

Franciscan friar Guillaume de Roubrock arrives at Karakorum on a mission from France's Louis IX and finds a silver fountain built for the Mongol prince Mangu Khan by Frankish goldsmith Guillaume Boucher. The fountain's four spouts dispense wine, mead, rice wine, and kumyss (mead and kumyss are alcoholic beverages made, respectively, from honey and mares' milk).

The Japanese shogun imposes price controls to halt inflation.

The Sorbonne founded for indigent theological students is the first college in the University of Paris. It opens under the name Community of Poor Masters and Scholars and has been started by Robert de Sorbon, 52, chaplain and confessor to Louis IX.

1254 The German king Conrad IV dies in Italy May 21 at age 26, and election of his 2-year-old son Conradin is barred by German and papal opponents of Conrad and of the late Frederick II. The death of the last Hohenstaufen king begins a 19-year Great Interregnum in the Holy Roman Empire (*see* 1273).

France's Louis IX returns after an absence of 6 years in Egypt and the Holy Land and by moral force ends the dissension that has engaged his nobles in petty civil wars since the death late in 1252 of his regent Blanche of Castile.

1255 The Mali king Sundiata Keita dies at Timbuktu after a long reign in which he has made his empire the largest and richest in sub-Saharan Africa, with copper mines at Takedda, salt mines at Taghaza, and gold mines at Bure.

England's Henry III accepts Sicily for his 10-year-old son Edmund (Crouchback), earl of Lancaster, who has been given the title King of Sicily by the new pope Alexander IV.

Poetry *Frauendienst* by German lyric poet Ulrich von Lichtenstein, 55, is an autobiographical novel in verse containing minnesongs about chivalry. Italian poet and Franciscan monk Thomas of Celano dies at age 55, leaving behind lyrics to the hymn "Dies Irae."

1256 Kublai Khan's brother Hülegü wipes out Persia's Assassins, ends the caliphate, and begins a 9-year reign, inaugurating the Ilkhan dynasty that will rule Persia until 1349 (*see* 1358).

Venice and Genoa go to war in a conflict that will last for a century.

Italian scholar-philosopher Thomas Aquinas, 31, receives papal dispensation to receive the master of theology degree at the University of Cologne which had required that a master of theology be at least 34. Aquinas is named to fill one of two chairs allotted to Dominicans at the university and begins writing scholastic disputations refuting accusations against the Dominicans and Franciscans.

The monastic order of Augustine Hermits is founded.

1257 Richard of Cornwall, 48, brother of England's Henry III, is elected Holy Roman Emperor, defeating Castile's Alfonso X, and is crowned at Aix-la-Chapelle. He establishes his authority in the Rhine Valley but soon runs out of money, and when he asks England's great council to give the pope one-third of all English revenue, the request is refused.

Economic distress troubles England.

1258 Persia's Hülegü Khan routs the last army of the eastern Abbasid caliphate January 17; his Mongols sack Baghdad February 10, massacre tens of thousands in a single week, and end the caliphate that has ruled from Baghdad since 762, making it one of the world's great centers of learning and culture. But Hülegü withdraws at news that his brother Mangu died late in 1257.

The Provisions of Oxford establish the English House of Commons in a move by Simon de Montfort and other rebellious barons to restore the Magna Carta of 1215. The Commons is a council of 15 men, most of them barons, with veto power over the king. It is to meet three times a year with a committee of 12 appointed to advise the king. Henry III, his son Edward, 19, and all other officials take an oath of loyalty to the new Provisions June 11.

A rumor that the king of Jerusalem and Sicily, Conradin, 6, is dead reaches Palermo where Conradin's regent Manfred is crowned king August 10. Manfred, 26, is the bastard son of the late Holy Roman Emperor Frederick II (Sicily's Frederick I). He refuses to abdicate when the rumor about Conradin proves false, insisting that Sicily needs a strong, native ruler. Pope Alexander IV excommunicates the handsome Manfred, but the new king rallies Ghibelline support (*see* 1260).

Mongol invaders pillage Hanoi.

A flagellant movement arises in Europe following widespread famine and disease (*see* below). Organized under masters, the flagellants wear special uniforms, live under strict discipline, and conduct public and private self-flagellation, beating themselves according to a set ritual to divert divine punishment and forestall plagues thought to be sent by heaven as chastisement for sins (*see* 1349).

England's Salisbury Cathedral is completed after 38 years of construction.

La Sainte-Chapelle is completed at Paris after 12 years of construction.

Famine and disease follow crop failures in the German and Italian states (*see* flagellant movement, above).

1259 France's Louis IX yields Perigord and the Limousin to England in exchange for renunciation of English claims to Normandy, Maine, and Poitou.

China's Song armies make the first known use of firearms that propel bullets. They repel a Mongol invasion with bullets fired from bamboo tubes.

The Hanseatic towns Lübeck, Rostock, and Wismar agree not to permit pirates or robbers to dispose of goods within their environs. Any city that violates the agreement is to be considered as guilty as the outlaw (*see* 1241; 1300).

England's Henry III grants commercial rights to merchants of Genoa.

1260 The Yuan dynasty that will rule China until 1368 is founded by a grandson of the late Genghis Khan. Kublai Khan, 44, has himself elected by his army at Shant-tu (*see* Khanbelig, 1267).

The Mamelukes, who have ruled Egypt since 1250, save the country from the Mongols at Ain Jalut, Palestine, in September and preserve the last refuge of Muslim culture. Hülegü Khan has taken Damascus and Aleppo, but the Mamelukes, led by the ex-slave Baybars, kill Hülegü's general Ket Buqa and revive the caliphate by inviting to Cairo a scion of the Abbasid house and giving him the title Mustansir l'Jllah.

Sicily's Manfred defeats a Guelph army September 4 at Monte Aperto with Ghibelline support (*see* 1258; 1265).

Venetian merchants Niccolo and Mateo Polo sail from Constantinople for the Crimean port of Sudak where a third Polo brother, Marco the Elder, has a house (*see* 1271).

England's Henry III extends to all north German cities the trading rights he granted in 1250 to the merchants of Cologne and Lübeck.

The papal bull *Clamat in suribus* issued by Alexander IV appeals for unity among the forces of Christendom against the common danger of the Mongols.

Painting *Madonna* by Italian painter-mosaicist Giovanni Cimabue (Cenni di Pepo), 20, for Florence's Trinita.

France's Chartres Cathedral is consecrated after 66 years of construction. Built largely between 1195 and 1228, the great Cathedral of Notre Dame in the Loire valley raises Gothic architecture to its greatest glory with a Vieux Clocher rising 351 feet in height (the Clocher Neuf will rise 377 feet when it is completed in the 16th century).

1261 The Latin Empire founded in 1204 ends July 25 as Constantinople falls to a Greek army under Alexius Stragopulos, who has taken advantage of the absence of the Venetian fleet to cross the Bosphorus and drive out the emperor Baldwin II, who has reigned since 1228. The Palaeologi family that will rule until 1453 takes power as Michael VIII Palaeologus, 27, begins a 21-year reign, restoring Byzantine control.

Bohemia's Ottokar the Great annexes Styria.

Balliol College is founded at Oxford, England, where a school has existed since the 9th century (*see* 1167; 1209; 1264).

1262 Pope Urban IV determines to end Hohenstaufen power in Italy and offers the kingdom of Naples and Sicily to Charles d'Anjou, 36, brother of France's Louis IX. A veteran of the Seventh Crusade, Charles has forced several communes in the Piedmont to accept his suzerainty (*see* 1265).

Crusaders on the Iberian Peninsula recover Cadiz from the Moors, who have held the city for more than 500 years.

Aleksandr Nevski persuades the Tatars to accept a smaller tribute and drop their demand that Russians render military service.

Norway's Haakon IV unites Iceland and Greenland with Norway.

1263 A Venetian war fleet off the Greek island of Spetsopoula pounces on a Genoese convoy of merchant carracks and fighting galleys in early summer and gains a victory that will bring a partitioning of naval control in the eastern Mediterranean. Venice has learned of the convoy's departure from its Monemvasia refuge for a destination either in the Cyclades, Constantinople, or the Black Sea where Genoa holds a dominant trade position.

Scotland's Alexander III defeats the Norwegian king Haakon IV to win the Hebrides. Alexander has laid formal claim to the islands on which Scandinavians have settled since the 9th century. Haakon has sailed round the west coast of Scotland to Arran; Alexander has prolonged negotiations there in anticipation of autumnal storms. Haakon finally attacks but encounters a terrible storm that severely damages his ships. The Battle of Largs October 12 is not decisive, but it leaves Haakon in a hopeless position. He turns for home but dies en route at age 59 after a 46-year reign.

The grand duke of Vladimir Aleksandr Nevski dies at age 43.

1264 England's Henry III wars with the country's barons and is captured by the earl of Leicester Simon de Montfort after an ignominious defeat at the Battle of Lewes. Leicester becomes virtual governor of England and 2 days after the battle receives the surrender of Henry's son Edward, now 25, who has contributed to his father's defeat by a rash pursuit of the Londoners and is forced to give up his earldom of Chester.

Oxford's Merton College is founded by the former chancellor of England Walter de Merton whose new

1264 (cont.) school begins Oxford's collegiate system (*see* Balliol, 1261; Queen's, 1340).

Nonfiction *Summa Contra Gentiles* by Thomas Aquinas; *De Computo Naturali* by Roger Bacon.

1265 England's Prince Edward escapes from his custodians at Whitsuntide. He gains support from lords of the Welsh march who are still at arms, and he kills the earl of Leicester Simon de Montfort August 4 at Evesham. Edward will dictate government policy for the remaining 7 years of his father's reign.

Pope Clement IV induces Charles d'Anjou to accept the Kingdom of the Two Sicilies as a papal fief, offering conditions even more favorable than those offered by Pope Urban IV in 1262. He grants Charles all the privileges of a crusade; Charles sails for Rome. He narrowly escapes capture by the fleet of the Sicilian king Manfred, but he reaches Rome safely and is crowned by the pope.

The motet musical form developed by Franco of Cologne and Pierre de la Croix will not reach mature form until the 16th century.

London's Covent Garden market has its beginnings in a fruit and vegetable stand set up on the north side of the highway between London and Westminster by monks of St. Peter's Abbey. They begin to sell surpluses from their garden that exceed the requirements of their Westminster Abbey (*see* 1552).

1266 The king of the Two Sicilies Charles d'Anjou defeats his rival Manfred in battle near Benevento February 26. Manfred is killed at age 33 after a reign of less than 8 years, and Italian Ghibellines send envoys to Bavaria to request the help of Conradin, 13, son of the late German king Conrad IV. Young Conradin crosses the Alps to Verona and issues a manifesto setting forth his claim to the Sicilian crown. His partisans take up arms in the north and south of Italy, and he receives tumultuous welcomes at Pavia and Pisa.

A new slave dynasty at Delhi succeeds the one founded in 1206. Purchased as a slave in 1233, the new sultan Balban was made chamberlain in 1242, and he married his daughter to the sultan Mahmud. The sultan dies after a 20-year reign, and Balban succeeds to the throne to begin a 21-year reign in which he will end highway robbery in the south and east and repress the Indian nobility with help from an effective army and a corps of royal newswriters.

Kyoto's Sanjusangendo Temple is completed.

English bakers are ordered to mark each loaf of bread so that if a faulty one turns up "it will be knowne in whom the faulte lies." The bakers' marks will be among the first trademarks.

1267 Pope Clement IV excommunicates Conradin in November for claiming sovereignty over Sicily, but Conradin's fleet gains a victory over that of Charles of Anjou (*see* 1266; 1268).

Beijing has its beginnings in the town of Khanbelig constructed by Kublai Khan (*see* 1260; 1271).

Cambridge, England, is chartered by Henry III.

Members of London's goldsmith and tailor guilds fight each other in fierce street battles.

Roger Bacon, the first truly modern scientist, describes the magnetic needle and reading glasses, predicts radiology, and predicts the discovery of the Western Hemisphere, the steamship, the airplane, and television (*see* 1249; 1277).

Roger Bacon (above) describes principles of a camera obscura that can project pictures (*see* Porta, 1553).

1268 Conradin receives an enthusiastic reception at Rome in July but his troops are defeated at Tagliacozzo in August by the king of the Two Sicilies Charles d'Anjou. Agents of Charles seize Conradin at Astura, he is tried as a traitor and condemned, he is beheaded at Naples with his friend Frederick of Baden, titular duke of Austria, and the house of Hohenstaufen becomes extinct. The execution of the 16-year-old Conradin has at least the tacit approval of Pope Clement IV, it shocks Europe, antagonizes England's Henry III and France's Louis IX, and begins a long-lasting alienation of Germans from the Roman Church.

Pope Clement IV dies at Viterbo November 29 and will not be replaced until 1276 except for the antipope Gregory X (Tebaldo Visconti), who will fill the vacancy from 1271 to 1276.

France's Cathedral of Notre Dame at Amiens is completed after 49 years of construction. The immense church is one of the country's finest Gothic structures.

An earthquake in Silicia, Asia Minor, kills an estimated 60,000.

1269 Bohemia's Ottokar the Great takes Carinthia and Carniola from Hungary.

France's Louis IX orders French Jews to "wear the figure of a wheel cut out of purple Woolen Cloth, sewed on the upper Part of their Garments on the Breast, and between the Shoulders" (*see* 1252; 1306).

Niccolo and Mateo Polo reach Acre, eastern outpost of Roman Christendom (*see* 1260; 1271).

The first toll roads appear in England.

1270 France's Louis IX leads an Eighth Crusade, arrives at Carthage after a 17-day voyage, but dies of plague August 25 as his army is cut down by heat and disease. Louis is succeeded after a reign of nearly 44 years by his son of 25, who will reign until 1285 as Philip III (his father will be canonized as St. Louis in 1297).

Carcassonne is walled with an outer ring of fortifications by Louis IX (above) before his departure. The city is considered impregnable and will not be attacked for centuries.

The German poet-minnesinger Tannhäuser dies at age 65.

1271 Kublai Khan assumes the dynastic title that he will hold until his death in 1294 and founds the Mongol dynasty that will rule China until 1368.

Marco Polo, 17, joins his father Niccolo and uncle Mateo as they travel once more to Acre and begin a journey to India and the Far East (*see* 1269; 1274).

1272 The king of the Romans Richard, earl of Cornwall, dies at age 62 of paralysis following the murder late last year of his eldest son Henry of Cornwall by the sons of the late Simon de Montfort.

England's Henry III dies at Westminster November 16 at age 65 after a reign of 56 years. His son Edward, now 33, hears of the death late in the year while in Sicily on an Eighth Crusade. He will do homage in late July of next year to his cousin Philip III at Paris and reign until 1307 as Edward I (*see* 1274).

Languedoc comes under French control following the death of Philip III's uncle Alphonse of Poitiers.

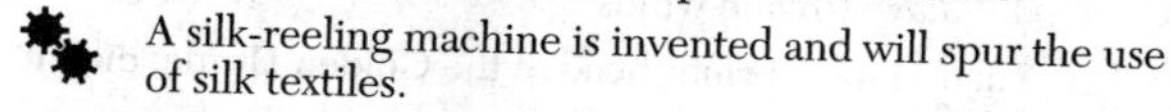

A silk-reeling machine is invented and will spur the use of silk textiles.

Caerphilly Castle is built in Wales by Gilbert de Clare, 29, earl of Gloucester and Hereford.

1273 The Great Interregnum that has left the Holy Roman Empire without a ruler since the death of the last Hohenstaufen king in 1254 ends September 29 with the election of the count of Hapsburg as German king. Having increased his estates at the expense of his uncle Hartmann of Kyburg and the bishops of Strasbourg and Basel, Rudolph, 55, has gained election largely through the efforts of his brother-in-law Frederick III of Hohenzollern, burgrave of Nuremberg, purchasing support from the duke of Saxe-Lauenburg and the duke of Upper Bavaria by betrothing two of his daughters to them. He promises to renounce imperial rights in Rome, in the papal territories, and in Sicily. He vows to lead a new crusade, and is thereupon recognized by Pope Gregory X, who crowns him October 24 at Aix-la-Chapelle; he will reign until 1291 as the emperor Rudolph I.

Nonfiction *Summa Theologica* by Thomas Aquinas, now 48, who began the work in 1267 but stops writing December 6 leaving it incomplete: "I can do no more; such things have been revealed to me that all I have written seems as straw, and I now await the end of my life" (he will die March 7 of next year) (*see* 1323).

The Persian poet Djeleddin Rumi dies. He has founded the Order of the Dancing Dervishes.

Painting Roman painter-mosaicist Pietro Cavallini, 23, completes mosaics for Rome's Church of Santa Maria Maggiore.

Beijing's Drum Tower is erected at Khanbelig (*see* 1267).

1274 England's Edward I lands at Dover August 2 and is crowned at Westminster August 18.

China's Kublai Khan sends an invasion fleet to conquer Japan, whose emperor has refused dispatches calling on him to submit. 300 large vessels plus 400 to 500 smaller vessels manned by 8,000 Korean and 20,000 Mongol troops begin landing at Hakata Bay in Kyushu October 19. Equipped with iron cannons and cannonballs, the invaders use poisoned arrows, violating traditional Japanese protocol, but a typhoon strikes November 20, sinking more than 200 Mongol ships along with 13,000 men sleeping aboard; the survivors retreat to the mainland in terror (*see* 1281).

England's Edward I (above) takes over the Kentish castle of Leeds begun by a Saxon chief and rebuilt by the Norman baron Robert de Crèvecoeur in 1119. Edward will make many improvements.

Marco Polo visits Yunnan and sees the "Tartars" eating raw beef, mutton, buffalo, poultry, and other flesh chopped and seasoned with garlic. Young Polo will enter the service of Kublai Khan next year and will continue until 1292.

1275 Amsterdam is chartered by the count of Holland Floris IV, who exempts the new town from certain taxes.

Chirurgia by William of Saliceto contains the earliest record of human dissection, a practice discouraged by the Church since 1163.

Zohar by the Jewish theologian Moses de León will endure as the fundamental work on Jewish mysticism.

Lausanne's Cathedral of Notre-Dame is consecrated with great pomp and ceremony by Pope Gregory X.

1276 The prince of North Wales Llewelyn II ap Gruffyd, grandson of the late Llewelyn the Great, refuses to pay homage to England's Edward I. The English king prepares to invade North Wales.

The German king Rudolph outlaws Bohemia's Ottokar the Great, but Ottokar submits to Rudolph, who permits him to retain Bohemia and Moravia.

Augsburg becomes a free city. Founded in 14 B.C., it is now a major textile-producing center.

Papermaking reaches Montefano in Italy (*see* 1189; 1293).

Theater *Le jeu de la feuille* by French playwright Adam de la Halle, 26.

1277 England's Edward I invades North Wales, blocks all the avenues to Snowdon, and forces the surrender of Llewelyn II ap Gruffyd for lack of supplies. Edward reduces the Welsh prince to the status of a petty chieftain and will devote the next 5 years to establishing English government in the districts ceded by Llewelyn (*see* 1282).

The Visconti family that will rule Milan for 170 years gains control in the person of the Ghibelline leader Matteo Visconti, 27, whose great uncle Ottone Visconti, 70, is Milan's archbishop (*see* 1351).

Bulgaria's Constantine Asen is killed by a peasant usurper after a 19-year reign as czar in which he has fought the Hungarians and the Byzantine emperor Michael VIII Palaeologus. The Asen dynasty that has ruled since late in the last century is extinguished, and Bulgarians will be subject to Serbian, Greek, and Mongol overlords.

Marco Polo traveled to China, served Kublai Khan, and returned to Venice with wondrous tales of the Orient.

Roger Bacon is imprisoned for heresy. He will remain incarcerated until 2 years before his death at age 80 in 1294.

1278 The Hapsburg family gains sovereignty over Austria to begin a dynasty that will continue until 1918. The Bohemian king Ottokar the Great has renewed his claims to three Swiss and Austrian duchies; the new German king Rudolph of Hapsburg defeats Ottokar August 26 at Durnkrut in Marchfeld. Ottokar is killed at age 48; his son of 7 will reign until 1305 as Wenceslas II.

Andorra in the Pyrenees becomes an autonomous republic. The French counts of Foix have been contesting possession of the mountain stronghold but agree to let it be independent.

London cracks down on coin-clippers but with unequal punishment: 278 Jews are hanged while Christians convicted of the same crime are merely fined.

The glass mirror is invented.

1279 Portugal's Afonso III dies at age 69 after a 34-year reign and is succeeded by his son of 18, who has led a rebellion against him but will reign as Diniz (the Worker) until 1325.

1280 Flemish textile workers rebel against their exploiters.

Marco Polo visits Hangchow and finds it an eastern Venice, a city of 900,000 that Marco will call "beyond dispute the greatest city which may be found in the world, where so many pleasures may be found that one fancies himself in Paradise." Having come from Europe's most sophisticated city, Marco is awed by the countless ships bringing spices from the Indies and embarking with silks for the Levant.

The merchants of Hangchow (above) use paper money, unknown in Europe.

Poetry *Oeuvres* by the French trouvère Rutebeuf is a collection of poems barbed with satire.

Kublai Khan calls in Egyptian experts to improve Chinese techniques of refining sugar (*see* 100 B.C.). The Egyptians have acquired a reputation for making exceptionally white sugar.

1281 Kublai Khan sends a second invasion fleet to conquer Japan. 42,000 Mongol troops from Korea arrive in Kyushu in May and are joined in July by 100,000 more from southern China, but the Japanese are saved by the elements as they were in 1274. A typhoon destroys most of the Mongol invasion fleet July 29, more than 2,000 of the invaders are taken prisoner, scarcely one-fifth of the Mongols and Koreans survive, and the Japanese begin calling typhoons *kamikaze* (divine winds).

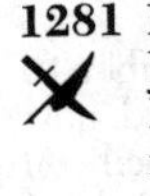

Khan Mangu Temir, head of the Golden Horde, dies at Astrakhan.

Kublai Khan employs imperial inspectors to examine the crops each year with a view to buying up surpluses for storage against possible famine.

1282 The prince of North Wales Llewelyn II ap Gruffyd leads a second rebellion against England's Edward I. He is killed in a skirmish near Builth in central Wales by Roger de Mortimer, who is himself killed as England conquers Wales (*see* 1535).

The Sicilian Vespers rebellion that begins in a church outside Palermo at the hour of vespers March 31 (Easter Tuesday) leads to a wholesale massacre of the French and triggers a war that will continue for years. A French soldier has allegedly insulted a Sicilian woman, but the real basis of the rebellion is the heavy taxation imposed by Charles I (Charles d'Anjou) to equip an expedition against Constantinople.

Sicilian noblemen support the popular uprising against French insolence and cruelty, they persuade the king of Aragon Pedro III to assert claims to the Sicilian crown (Pedro's wife Constanza is the daughter of the late Sicilian king Manfred killed in 1266), and Pedro arrives at Palermo in September.

Florence's bourgeois merchants stage an armed rebellion against the nobility and set up a reformed government with the sanction of Pope Martin IV.

Florence's upper classes are known as the *popolo grasso* (fat people), the poor are the *popolo minuto* (small, or lean people). Detached from the soil, the city's lower classes are dependent on their employers, who often force them to labor at night by torchlight.

1284 Kublai Khan leads a 500,000-man Chinese army into Vietnam. Guerrillas organized by Tran Hung Dao virtually destroy the invasion force.

Castile's Alfonso El Sabio (the Wise) dies at Seville April 4 at age 63 after a 32-year reign in which he is reputed to have said, "Had I been present at the Creation, I would have given some useful hints for the better ordering of the universe." He is succeeded by his second son Sancho whom he has tried to exclude from the succession but who will rule through violence and strife as Sancho IV (el Bravo) until his own death at age 37 in 1295.

Silver ducats are coined for the first time at Venice which prospers in trade with the East while Florence thrives on trade with the North (*see* florins, 1252).

Rats are so prevalent in Europe that a story appears about a piper who leads children into a hollow because townspeople have refused to pay him for piping their rats into the river Weser. "Rattenfänger von Hameln" may also have reference to the Children's Crusade of 1212, but much distress is caused by rats which consume grain stores, seeds, poultry, and eggs and which also bite infants and spread disease (*see* 1484; Black Death, 1340).

Ravioli is eaten at Rome where the people have been eating *lagano cum caseo* (fettucini) for years.

1285 The king of the Two Sicilies Charles I (Charles d'Anjou) dies at Foggia January 7 at age 58 as he prepares to invade Sicily again with a new fleet from Provence. His son Charles the Lame, 34, remains a prisoner of the king of Aragon but beginning in 1289 will reign as Charles II until 1309.

France's Philip III (the Bold) dies of plague October 5 while retreating from Gerona. He is succeeded after a 15-year reign by his son, 17, who will rule until 1314 as Philip IV in one of France's most momentous reigns.

The plague that kills Philip III (above) turns the French back from their march into Aragon and kills most of the army's officers and many of the men as well as the king.

Smog problems develop as Londoners burn soft coal for heat and cooking (*see* 1306).

1286 Scotland's Alexander III falls from his horse March 16 while riding in the dark to visit his new queen at Kinghorn. He dies at age 43 after a 35-year reign, two sons and a daughter have predeceased him, and his only living descendant is the infant daughter of his own late daughter by the Norwegian king Eric Magnusson. The Great Council of Scottish tenants-in-chief appoints guardians to govern in the name of Margaret of Norway, but some clans begin a rebellion (*see* 1290).

The king of Aragon Pedro the Great repels a French invasion of Catalonia but dies November 8 at age 50. He is succeeded by his weak son, who will reign until 1291 as Alfonso III and who will recognize the right of his nobility to rebel, a move that will make anarchy permanent in Aragon.

Theater *Le Jeu de Robin et Marion* by Adam de la Halle is the first French pastoral drama. The playwright presents it at Naples, where he has followed the late Charles d'Anjou, king of the Two Sicilies.

1287 Kublai Khan dispatches an army to invade Burma.

1288 The new king of Aragon Alfonso III releases Charles the Lame on condition that he retain Naples alone, that Sicily remain an Aragonese kingdom, and that he persuade his cousin Charles of Valois to renounce the kingdom of Aragon given him by Pope Martin IV.

1289 Pope Nicholas IV takes office February 22. Charles the Lame meets with him at Rieti. The new pope absolves Charles of the promises he made to the king of Aragon, crowns Charles king of the Two Sicilies, and excommunicates Alfonso III as Charles of Valois prepares to seize Aragon in alliance with Castile.

Tripoli on the coast of Asia Minor falls April 29 to Qala'un, caliph of Egypt (*see* Acre, 1291).

France's University of Montpellier has its beginnings.

Block printing is employed for the first time in Europe at Ravenna.

1290 The Ottoman Empire that will rule part of Europe and much of the Mediterranean for 6 centuries has its beginnings in a Bithynian Islamic principality founded by Osman (or Othman) al-Ghazi, 31, who 2 years ago succeeded his father Ertogrul as chief of the Seljuk Turks.

The Khalji dynasty that will rule Delhi until 1320 is founded by the Firuz shah Jalal-ud-din. A Muslim long resident among the Afghans, Jalal-ud-din overthrows the Muslim slave dynasty founded in 1266.

Cuman rebels assassinate Hungary's Ladislas IV July 10. Dead at age 28, he is succeeded by his senior kinsman, who will rule as András III until his death in 1301, the last of the Arpád dynasty founded in 907.

Sweden's Magnus Ladulos dies at age 50 after an 11-year reign. He is succeeded by his son, 10, who will be crowned Birger III in 1302 and reign until 1318.

Scotland's titular queen Margaret, the maid of Norway, reaches the Orkneys where she dies in September under mysterious circumstances at age 7. Margaret had been betrothed to Edward, 6-year-old son of England's Edward I, who intervened to secure her throne in 1286, and her death leaves Scotland without a monarch (*see* 1292).

England's Edward I exiles the country's Jews at the behest of Italians who seek to handle English banking and commerce (*see* 1259; 1306; France, 1252).

The University of Lisbon has its beginnings.

A Chinese earthquake in Chihli (Hebei, or Beijrli) province September 27 kills an estimated 100,000.

1291 The German Hapsburg king Rudolph I dies at Spires July 15 at age 73 after a reign of nearly 18 years. He has invested his son Albrecht, 41, with Austria but has been unable to obtain Albrecht's election as German king.

Switzerland has her beginnings in the League of the Three Forest Cantons formed August 1 by Uri, Schwyz, and Unterwalden. The league is for mutual defense, but the cantons do not claim independence from Austria (*see* 1315; Lucerne, 1332).

Acre falls to Mameluke forces who end Christian rule in the East as enthusiasm for the crusades wanes in

1291 *(cont.)* Europe (*see* 1271; Tripoli, 1289). The Knights of St. John of Jerusalem settle in Cyprus.

Venice moves its glass ovens to the island of Murano to remove the danger of fire. The city establishes draconian penalties for any glassmaker caught jeopardizing the Venetian monopoly in clear glass by taking production secrets abroad.

Painting Mosaics by Pietro Cavallini for the apse of Rome's Santa Maria in Trastevere.

1292 Adolf of Nassau is elected German king May 5 to succeed the late Rudolph I, but he has only nominal allegiance from many princes, including Rudolph's son Albrecht of Austria (*see* 1298).

England's Edward I resolves the Scottish succession, selecting John de Baliol, 43, to succeed the late Alexander III and Alexander's granddaughter, the maid of Norway.

John de Baliol is crowned at Scone November 30 after swearing fealty to Edward, and he pays homage to Edward at Newcastle.

Venice in this decade will develop the "great galley" that will be capable of long voyages with large cargoes. The shallow-draft single-deck merchant vessels, 120 to 150 feet in length, will have one or two masts for sails but will be powered mostly by oarsmen. Crews of 100, 200, and more will man two or three banks of oars with 25 to 30 benches of oarsmen on each side (*see* 1317).

Some 20,000 gondoliers pole their slim craft through the more than 100 canals that link Venice's 112 islands. Rich families in the lagoon city of 200,000 own as many as five gondolas, some of them up to 33 feet in length (*see* 1881).

1293 England's Edward I surrenders Gascony temporarily as a pledge following a Gascon-Norman sea battle. France's Philip IV confiscates Gascony and summons Edward to his court, and Philip's treachery precipitates a war.

A Chinese naval expedition to Java is forced to return after some initial success.

Florence excludes from its guilds anyone not actively practicing his profession. The Law of 1282 permitted nobles to participate in government only if they joined a guild, the new ordinance effectively removes the nobility from all shares in the city government, and it gives rise to two factions of Guelphs. One faction will favor repeal of the ordinance, the other (adherents will be called Ghibellines) its retention.

Paper is manufactured in the Italian province of Ancona at Fabriano (*see* 1276; 1320).

Painting Mosaics by Pietro Cavallini for Rome's Santa Cecilia in Trastevere.

A Japanese earthquake at Kamakura May 20 kills an estimated 30,000.

1294 China's Kublai Khan dies February 12 at age 78 after a 35-year reign that has established the Mongol (Yuan) dynasty, subdued Korea and Burma, and founded the city that will become Beijing. The great khan is succeeded by his grandson Timur Oljaitu, who continues the dynasty that will rule until 1368.

The Siamese kingdoms of Xieng-mai and Sukhotai are forced to pay tribute to Kublai Khan before his death (above).

England's Edward I begins a series of futile expeditions against France's Philip the Fair in an effort to regain his Gascon fortresses. He is supported by an alliance with the count of Flanders.

England's Devonshire mines in the next 7 years will provide the country's mint with an average of 500 pounds of silver per year, but most English coins will continue to be made from imported silver.

Paper money printed with both Chinese and Arabic characters is used at Tabriz in Azerbaijan (*see* 1280).

Travel in England is easier than it will be at any time for the next 500 years.

English grain exports supply the Continent with wheat from the South and barley and oats from the North. Collected in manorial barns and in market towns for carriage by wagon to the ports, the corn is carried in heavy wagons on well-maintained roads.

Famine strikes England with special severity despite the fact that more English land is tilled than ever has been before and ever will be again.

1295 England's Model Parliament sits at London, convened by a summons to bishops, abbots, earls, barons, knights, burgesses, and representatives of all the realm's chapters and parishes: *Quod omnes tangit ab omnibus approbetur*, says the summons (let that which touches all be approved by all), but most of the clergy soon quits to form its own assembly, or convocation, and the only Church people left are the great prelates who attend in a feudal rather than an ecclesiastical capacity.

Scotland's John de Baliol ignores a summons to attend England's Edward I and forms an alliance instead with France's Philip IV, thus provoking an invasion by English forces.

Castile's Sancho el Bravo dies at age 37 after an 11-year reign marked by continuous dispute and violence. He is succeeded by his son, 9, whose minority will be a period of anarchy but who will reign until 1312 as Ferdinand IV (el Emplazado, the Summoned).

Marco Polo returns to Venice after having traveled from 1275 to 1292 in the service of the late Kublai Khan. He brings home spices, oriental cooking ideas, descriptions of Brahma cattle, reports of cannibalism, and curious ideas of diet.

Painting *Madonna with St. Francis* by Giovanni Cimabue.

Theater *The Harrowing of Hell* is an English miracle play.

Construction begins at Florence on the Church of Santa Croce designed by Arnolfo di Cambio.

1296 The Battle of Curzola gives Genoa a victory over the Venetian fleet. The Genoese capture 7,000 Venetians, including Marco Polo who has commanded a galley and is thrown into prison.

Sicily elects her governor king after he has refused to yield Sicily to the pope. Now 24, he will reign until 1337 as Frederick II, but Carlos II of Naples attacks him, beginning a 6-year war.

The Marineds who will rule Morocco until 1470 capture the nation's capital from the Berber Almohades.

The senile Turkish ruler Firuz (Jalal-ud-din) at Delhi, who has ruled since 1290 to found the Khilji dynasty, is murdered by his nephew Ala-ud-din, who buys support with booty acquired in a recent surprise attack on Maharashtra. Ala-ud-din will reign until 1316, consolidating the empire.

Scotland's John de Baliol surrenders in July to a bishop representing England's Edward I, who has invaded his realm. He appears before Edward at Montrose and abdicates his throne. The English take him home in chains with his son Edward, they move the Scottish coronation stone from Scone to Westminster, and Baliol will remain in captivity until 1299.

England's clergy refuses a grant to the crown in accordance with the bull *Clericis laicos* issued by Pope Boniface VIII. Edward I withdraws protection of the clergy by the royal courts, he is supported by the public, most of the clergy evades the papal bull by making "gifts" to the Crown, recalcitrants have their lands confiscated, and the pope is forced to modify his stand.

Marco Polo (above) will dictate his *Book of Various Experiences* in the next 3 years to his fellow prisoner Rusticiano of Pisa. It will describe the Orient in which the Venetian traveler has spent one-third of his life.

Costard mongers in English streets sell costard apples at 12 pennies per hundred; they are among the earliest cultivated varieties.

1297 The Confirmation of Charters reaffirms the Magna Carta of 1215. A coalition of English barons angered by the loss of Gascony to France and of middle-class groups angered by rising taxes forces Edward I to reaffirm the great charter and to agree that the Crown may not levy a nonfeudal tax without a grant from Parliament.

England's Edward I invades northern France.

Scotland's "Hammer and Scourge of England" William Wallace, 25, ravages Northumberland, Westmoreland, and Cumberland after driving the English out of Perth, Stirling, and Lanark. Wallace routs an English army of more than 50,000 at Stirling Bridge in September.

The giant Moas giraffe bird becomes extinct in the South Pacific islands that will be called New Zealand.

1298 The bold German king Adolf of Nassau, now 48, alarms the electors with his imperialist advances in Meissen and Thuringia. They depose Adolf and kill him July 2 at the Battle of Gollheim near Worms. His rival Albrecht of Austria, 48, son of the late Hapsburg king Rudolf, takes over and will reign until 1308 as Albrecht I.

The Battle of Falkirk July 22 in Stirlingshire gives English archers a victory over Scotsmen led by William Wallace. The longbow scores its first great triumph in pitched battle, taking a heavy toll of clansmen armed with swords and spears.

The invention of the spinning wheel revolutionizes textile production.

The archbishop of Genoa Jacobus de Varagine dies at age 68 after having written a chronicle of Genoa up to the year 1296. He is more celebrated as the author of *The Golden Legend*.

1299 The Scottish patriot William Wallace begins soliciting French, Norwegian, and papal intervention in behalf of his country against England. Pope Boniface VIII persuades England's Edward I to release John de Baliol and his son Edward from captivity and let them move to France.

Rotterdam begins its rise to a position of importance. Count John grants the people of the Dutch port town the same rights as those enjoyed by the burghers of Beverwijk and Haarlem. Rotterdam expands the English trade that will help it become the leading Dutch commercial center and the largest port on the Continent (*see* 1632).

Construction begins at Florence on a palace designed by Arnolfo di Cambio (*see* 1295). The front section of the Palazzo Vecchio will be completed in 1315 (*see* 1592).

1300 The papacy at Rome reaches its zenith as Pope Boniface VIII (Gaetani) holds a great jubilee to mark the beginning of a century (which will not actually begin until 1301). Some 2 million Christians make pilgrimages to Rome, where huge donations intended for the subjection of Sicily and for a second Gaetani state in Tuscany are raked over public tables by papal "croupiers."

The Vatican's treasury has been drained as a result of resistance, led by France's Philip IV, to the 1296 papal bull *Clericis laicos*, which forbade laymen to tax the clergy and made it an excommunicatory sin for any clergyman to pay taxes and for any layman to impose them. Philip has forbidden export of precious metals.

The Hanseatic League begun in 1241 is solidified by a network of agreements among towns on the Baltic and on north German rivers. Ships of the towns import salt from western Europe, wool and tin from England, and olives, wine, and other commodities from Lisbon, Oporto, Seville, and Cadiz, to which they carry dried and salted fish, hides, tallow, and other items of trade (*see* 1344).

Florentine poet-philosopher Guido Cavalcanti dies August 29 at age 45 of malaria, leaving behind love lyrics, sonnets, and ballads. Cavalcanti has been exiled to Saranza with other leading Guelphs and Ghibellines but has returned to Florence to die.

1300 ***(cont.)*** The Baltic Sea has been rich in fish through most of this century and will remain so in the century to come. The Hanseatic fisheries have developed an efficient system for salting the herring, a fat fish, within 24 hours after catching it, and Europe holds the Hanse herring in higher esteem than fish salted at sea by Norfolk doggers or by English fishermen out of Yarmouth or Scarborough.

The first brandy is distilled at the 92-year-old Monpellier medical school by French medical professor Arnaldus de Villa Nova (Arnaud de Villeneuve, or Arnoldus Villanovanus), 65 (*see* 800).

14th Century

1301 The Arpád dynasty that has ruled Hungary since 907 ends with the death of András III after an 11-year reign. The nomadic Magyars have adopted Christianity and have accepted an absolute monarchy under the Arpád kings. András defeated the late Holy Roman Emperor Rudolph I, who claimed Hungary for his son Albrecht, but the king has been unable to control the Hungarian magnates and his death begins a civil war that will continue for 7 years.

The Ottoman sultan Osman defeats Byzantine forces at Baphaion.

1302 The Battle of the Spurs at Courtrai in Flanders July 11 follows a massacre of the French (Matin de Bruges). Count Guy de Dampierre leads Flemish burghers, alienated by Philip IV, to defeat the flower of French chivalry.

France's Estates-General is convened for the first time with representatives of the towns in their feudal capacity. They meet to show support for Philip the Fair in his struggle with Pope Boniface, whose bull *Unam sanctam* asserts papal supremacy, saying, "Men live on two levels, one spiritual, the other temporal. If the temporal power should go astray, it must be judged by the spiritual power."

The Black Guelphs (Neri) triumph in Florence and expel the White Guelphs (Bianchi), among them the poet Dante Alighieri.

De Potestate Regia et Papali by the Dominican friar John of Paris defends the authority of the king, denies the ownership of ecclesiastical property by the pope (whose control is that of an executive acting for the community and who can be held accountable for any misuse of church property), and denies the papal claim to a unique type of authority.

1303 France's Philip the Fair sends Guillaume de Nogaret to seize Pope Boniface and fetch him back to France to face trial by a general council. On the "Terrible Day at Anagni" September 8, Nogaret and Sciarra Colonna gain entrance to the papal apartment, find Boniface in bed, and take him prisoner after threatening to kill him. Nogaret and Colonna are forced to flee, but Boniface is taken to Rome and confined by the Orsini family in the Vatican, where he dies humiliated in October.

France returns Gascony to England's Edward I.

The *carte mercatoria*, or merchant's charter, granted by England's Edward I, allows foreign merchants free entry with their goods and free departure with goods they have bought or have failed to sell (with the exception of wine, which is too much in demand to be allowed to leave the country). Edward's policy, with some modifications forced by English merchants and by certain towns, will endure for nearly 2 centuries.

The universities of Avignon and Rome are founded.

1304 Persia's Mongol Ilkhan Ghazan dies after a 9-year Islamic reign. His brother Uljaitu will reign until 1316.

Franciscan missionary Oderico da Pordenone, 18, sets out to retrace Marco Polo's journey to China in the 1270s. His *Description of Eastern Regions* in 1330 will corroborate Polo's accounts and will say that Guangzhou (Canton) is "three times as large as Venice" and that Hangchow is "greater than any [other city] in the world."

Florentine painter Giotto di Bondone, 38, decorates Padua's Arena Chapel and breaks away from the flowing linear style of Byzantine painting that has prevailed for 500 years. The moneylender Enrico Scrovegni wants to atone for the sins of his father (who has been imprisoned for usury) and has commissioned Giotto to create frescoes for the chapel. Giotto portrays a genuine likeness of Scrovegni presenting a model of his chapel to three angels, and he accompanies it with scenes from the lives of the Virgin and Christ, including *Wedding Procession*, *Noli Me Tangere*, and *Lamentations over Christ*.

1305 The Scottish patriot Sir William Wallace is betrayed. Sir John Mentaith captures him near Glasgow August 5 and takes him in fetters to London, where an English court at Westminster Hall tries Wallace on charges of treason. Wallace protests that he cannot be a traitor since he has never been a subject, but the court nevertheless finds him guilty. He is hanged the same day, and his body is drawn and quartered (*see* 1299; Bannockburn, 1314).

Poland's Wenceslas II abdicates and then dies at age 34 after a 5-year reign. He is succeeded as king of Poland and Bohemia by his son of 16, who will reign until his death next year as Wenceslas III. Otto of Bavaria assumes the Hungarian throne to begin a 2-year reign as Otto I.

The French prelate Raimond Bertrand de Got, 41, is elected to succeed the late Pope Boniface VIII, who died a prisoner in the Vatican 2 years ago. He begins a 9-year reign as Clement V (but *see* 1309).

France's University of Orléans has its beginnings.

Painting *The Life of Christ, The Last Judgment, Pietà,* and other frescoes by Giotto for Padua's Church of Santa Maria dell' Arena.

Sculpture *Madonna and Child* by Giovanni Pisano.

1305 ***(cont.)*** *Opus Ruralium Commodorum* by Bolognese agriculturist Pietro Crescenzi (Petrus de Crescentiis), 75, is the first book on agriculture to appear in Europe since the 2nd century.

1306 The Premyslid dynasty that has ruled Bohemia since 1198 ends with the death of Wenceslas III at age 17 after a 1-year reign. The Holy Roman Emperor Albrecht I gives the Bohemian crown to his son Rudolf, but the Bohemians will not accept Rudolf, and an inter-regnum begins that will continue until 1310. Wenceslas is succeeded in Poland by the diminutive former king of Great Poland, who has gained the support of Pope Boniface VIII. Now 46, he will unite the principalities of Little and Great Poland, will be crowned at Cracow early in 1320, and will reign until 1333 as Ladislas IV (or I) (Lokietek).

Rhodes is purchased by the military order of the Knights of St. John of Jerusalem founded in 1113. The knights will defend the island until 1522.

France arrests her Jews, strips them of their possessions, and expels them (*see* 1269).

England whips and expels some 100,000 Jews who have remained since Edward I's expulsion order of 1290 (*see* Cromwell, 1657).

Only the tiniest fraction of the world's energy comes from sources other than manpower, animal power, and solar sources which include wood and windmills, but use of coal is increasing (below; 1615).

Painting *The Lamentation* by Giotto di Bondone for Padua's Arena Chapel.

A Londoner is tried and executed for burning coal in the city (*see* 1285).

1307 England's Edward I dies July 7 near Carlisle at age 68 while preparing to take the field against Scotland's Robert Bruce. His fourth and only surviving son assumes the throne at age 23 and will reign until 1327 as Edward II. He immediately recalls his homosexual lover Piers Gaveston from exile, abandons the campaign against Robert Bruce, and devotes himself to frivolity.

France's Philip IV seizes the property of the Order of the Knights Templar. The rich but decadent order has become the king's creditor as well as the pope's and has made itself virtually a state within the state, but Philip launches a propaganda campaign to stir the people against the knights (*see* 1308).

Constantinople turns back attacks by 6,000 Catalan mercenaries headed by German military adventurer Roger di Flor, 27, who is assassinated by order of Andronicus II and his co-emperor son Michael IX. The Catalan Grand Company turns westward and lays waste Thrace and Macedonia.

The *Commedia* that will become immortal as *The Divine Comedy (Divina Commedia)* is begun by Italian poet Dante Alighieri, 42, as a philosophical-political poem recounting an imaginary journey through Hell, Purgatory, and Paradise (*Inferno, Purgatorio, Paradiso*). Dante's work of 100 cantos begins with the "Inferno" whose first lines are, "In the middle of the road of life/ I found myself in a dark wood,/ Having strayed from the straight path. . ." and which includes the line, "Abandon hope, all ye who enter here." Immortalized in Dante's work are his Florentine friends and enemies, including the late noblewoman Beatrice Portinari de' Bardi, who died in 1290 when she was 24 and Dante 25.

1308 England's Edward II journeys to France and on January 25 marries Isabella, 15, daughter of Philip IV, while his favorite Piers Gaveston rules as regent at home. Gaveston marries the king's niece Margaret of Gloucester and receives the earldom of Cornwall.

France steps up her attack on the Knights Templar with an appeal to the Estates-General. Pope Clement V is obliged to cooperate, and torture is used to force confessions that will result in the abolishment of the order in 1312.

Albrecht I of Austria refuses demands by his nephew John, 18, for John's hereditary domains and is murdered May 1 by conspirators organized by John (the Parricide) who flees to the south (he will be arrested at Pisa in 1312 and executed the following year). Albrecht, 58, has been German king since 1298 and is succeeded by the count of Luxembourg, 39, who is elected king November 27. He will be crowned emperor in 1311 and reign until 1313 as Henry VII.

The universities of Perugia and Coimbra have their beginnings.

1309 The king of Naples Charles II dies in August at age 59 after a 20-year reign in which he has been forced to give up all claims to Sicily. He is succeeded by his cultivated son Robert, 34, who will reign until 1343, fighting the Ghibellines in behalf of the popes and supporting writers such as Petrarch.

The Babylonian Exile of the papacy that will last until 1377 begins at Avignon, an enclave in papal territory within France. Clement V moves the papal court to Avignon at the request of his friend Philip the Fair since Clement's predecessor has been exiled by the anarchy at Rome. The Avignonese papacy will embrace seven pontificates.

Venice begins construction of a Doge's Palace that will not be completed until 1483.

1310 Venice establishes a Council of Ten to rule the city.

England's barons force Edward II to appoint lords ordainers to help him rule.

1311 England's Parliament confirms reform ordinances requiring baronial consent to royal appointments, to any declaration of war, and to a departure of Edward II from his realm (*see* 1312).

The pulpit of the Pisa Cathedral is completed after 11 years of work by sculptor-architect Giovanni Pisano, now 66.

Notre Dame Cathedral at Reims is completed after 99 years of construction on the site at which Clovis was reputedly baptized in 496. The cathedral is a masterpiece of Gothic architecture.

1312 The Gascon knight Piers Gaveston, favorite of England's Edward II, is kidnapped by English barons and treacherously executed. Edward is forced to stand aside and permit 21 lords ordainers to govern the country under a series of ordinances drawn up last year.

Lyons is incorporated into France under terms of the Treaty of Vienne.

Genoese mariners rediscover the Canary Islands (*see* Portuguese conquest, 1425).

1313 The Holy Roman Emperor Henry VII of Luxembourg dies at Buonconvento near Siena August 24 at age 44. He has taken ill while en route with Venetian allies to attack Robert of Naples.

Nobles of Lower Bavaria call in Austria's Duke Frederick the Fair, but the Bavarian Duke Louis defeats him at Gammelsdorf November 9.

Nonfiction *De monarchia* by Dante Alighieri is a treatise on the need for the dominance of royal power in secular affairs.

1314 The Battle of Bannockburn June 24 assures the independence of Scotland. Thirty thousand Scotsmen under Robert Bruce VIII, 40, rout a force of 100,000 led by England's effeminate Edward II and take Stirling Castle, the last Scottish castle still in English hands.

Private wars break out in England as Thomas, earl of Lancaster, 37, takes advantage of Edward's defeat at Bannockburn (above) to wrest control of the government from Edward and his new favorite Hugh le Despenser, 52, but Lancaster is opposed by other barons who make it impossible for him to govern.

Louis of Bavaria is chosen king of the Germans October 20 at Frankfurt and is crowned November 25 at Aix-la-Chapelle to begin a reign of nearly 33 years. Louis, 27, is opposed by Austria's Duke Frederick the Handsome, 28, who was chosen by a minority of the electors and begins an 11-year civil war with Louis IV.

France's Philip IV dies November 29 at age 69 after a momentous 29-year reign. He is succeeded by his son of 25 by Jeanne of Navarre, who has been king of Navarre since his mother's death in 1305, but France's Louis X will rule for scarcely 18 months (*see* 1316).

The grand master of the French Knights Templar Jacques de Molay, 71, has been seized by order of Philip IV (above), taken before the French Inquisition at Paris, charged with heresy, found guilty, and burned at the stake in March.

St. Paul's Cathedral is completed at London (*see* 603; 1666).

1315 The Swiss infantry gains renown and begins a brilliant career by handing Leopold of Austria a thorough defeat November 15 at the Battle of Morgarten. Leopold had sought to crush the Swiss for supporting Louis of Bavaria as German king against the claims of his brother Frederick the Handsome (*see* 1314).

Italian immigrants at Lyons develop a silk industry.

Scotland gained her independence at the Battle of Bannockburn in 1314. The English conceded 4 years later.

The first public systematic dissection of a human body is supervised by Italian surgeon Mondino de Luzzi, 40, who shows his students the abdominal organs, the thorax, the brain, and the external organs. His *Anatomia* will be the first manual based on practical dissection.

English wheat prices climb to 3 shillings 3 pence per bushel as a short crop combines with export demand to inflate price levels.

Disastrous famine strikes large parts of western Europe and will continue for the next 3 years.

1316 France's Louis X dies suddenly June 5 at age 26 after an 18-month reign. His posthumous son John born November 15 briefly succeeds Louis "the Quarreler" (le Hutin) but dies November 22. Louis's brother Philip, 22, proclaims himself king and will reign until 1322 as Philip V.

Lithuania's Prince Witen dies and is succeeded by his brother (or possibly his servant) Gedymin, who inherits a domain that extends from the Baltic to Minsk but is controlled largely by the Teutonic Knights and by the Livonian Knights of the Sword. Gedymin will marry his daughter Anastasia to Muscovy's Grand Duke Simeon, he will marry another daughter to the son of Poland's Ladislas Lokietek, he will marry his son Lubart to the daughter of the prince of Halicz-Vladimir, and by thus gaining powerful allies he will assist the republic of Pskov to break away from Great Novgorod and join with Lithuania to create a vast country ruled from the capital that Gedymin will establish at Vilna in 1321.

Eight Dominicans sent to Ethiopia by the new pope John XXII check on rumors of a Christian king named Prester John (*see* 1459).

1317 France adopts the Salic Law to exclude women from succeeding to the throne (*see* 1322; 1328).

A fleet of Venetian great galleys makes the longest voyage undertaken by European trading vessels since

1317 *(cont.)* ancient times. A quarrel between Venice and France has made land travel difficult.

The Venetian galleys (above) carry sugar, spices, currants, dates, wine, paper, glass, cotton, silk, damask from Damascus, calico from Calicut, alum, dyes, draperies, books and armor. The great galleys return with hides, leather, tin, lead, iron, pewter, brass, cutlery, bowstrings, Cambrai cambric, Laon lawn, Ypres diapered cloth, Arras hangings, caps, and serges that include serge de Nîmes (denim) used for sailcloth — all obtained at Bruges, Europe's most important commercial city outside the Mediterranean.

1318 Ireland's Edward Bruce is killed in October at Dundalk 3 years after being proclaimed king. Younger brother of Scotland's Robert the Bruce, Edward has failed to conquer the country south of Ulster.

German poet Heinrich Frauenlob von Meissen dies at age 68. Founder of a meistersinger school at Mainz, von Meissen has been called *Frauenlob*, meaning praise of women because he has used the word *Frau* for woman rather than the word *Weib*.

1320 The Declaration of Arbroath asserts Scottish independence from England. The Scottish Parliament meets April 6 at Arbroath and drafts the letter to Pope John XXII reciting in eloquent terms the services which their "lord and sovereign" Robert the Bruce has rendered to Scotland (*see* 1323).

The Peace of Paris ends hostilities between France and Flanders, which retains her independence.

Poland's Ladislas IV Lokietek is finally crowned 14 years after ascending the throne.

The Muslim Tughlak dynasty that will rule western India until 1413 is founded by the elderly Turkish shah Ghiyas-ud-din Tughlak, who has gained numerous victories over the Mongols and will make Warangal and Bengal provincial states before his death in 1325. Tughlak leads a rebellion that overthrows the Khaljis who have ruled since 1290, and as Gharzi Khan he moves the capital 4 miles east from Delhi to the new city of Tughlakabad.

Paper is produced at Mainz and will lead to paper currency in Europe. First used by the Chinese in 1236, paper money was discontinued in 1311 when treasury reserves failed to keep up with the flood of paper money, depreciating the value of money through inflation and diminishing the financial (and moral) credit of Kublai Khan's Yuan dynasty.

Paper (above) will soon be made at Cologne, Nuremberg, Ratisbon, and Augsburg. Perfected by the Chinese in A.D. 105, it replaces the vellum which has given monasteries a monopoly on manuscripts and on written communication.

Poetry *Sir Gawain and the Green Knight* by the English "pearl poet" deals with the legendary King Arthur and his round table at Camelot (*see* 537; 1190; 1203; Tennyson, 1859).

Painting Frescoes by Pietro Cavallini for the Church of Santa Maria Donnaregina.

1321 Parliament exiles Hugh le Despenser at the behest of the earl of Lancaster, who continues to dominate England in opposition to Edward II.

The University of Florence has its beginnings.

Dante Alighieri dies September 14 at age 56 while attending Guido da Polenta at Ravenna. His treatise *Concerning Vernacular Eloquence (De Vulgari Eloquentia)*, left unfinished, is a pioneer philological study of literary use of the Italian vernacular.

1322 France's Philip V dies at Longchamp January 2 at age 28 after a 6-year reign in which he has ended a war with Flanders and inflicted heavy taxes on French Jews, extorting 150,000 livres from the Jews of Paris alone. Philip is succeeded by his brother, 27, who will reign until 1328 as Charles IV, last of the direct line of Capetian kings.

The accession of Charles IV (above) establishes in principle the Salic Law adopted 5 years ago that the French throne can pass only through males; it excludes England's 9-year-old prince Edward, grandson of France's late Philip IV.

England's Edward II recalls Hugh le Despenser from exile and makes war against the barons. He defeats the earl of Lancaster at Boroughbridge, 22 miles northwest of York, has him beheaded at Pontefract, restores Hugh le Despenser as his counselor, and makes him earl of Winchester.

The Battle of Muhldorf September 28 ends in victory for the German king Louis IV largely through the timely aid of Frederick IV of Hohenzollern, burgrave of Nuremberg. Frederick of Austria is taken prisoner.

Use of counterpoint in church music is forbidden by Pope John XXII.

1323 The Treaty of Northampton between Scotland and England recognizes Robert the Bruce's title to the Scottish throne and provides for the marriage of Bruce's son David to Edward II's daughter Joanna.

Pope John XXII canonizes Thomas Aquinas 49 years after the death of Aquinas.

1324 The Japanese emperor Godaigo tries to regain power from the country's feebleminded regent Tatatoki Hojo who occupies himself mainly with drinking and dogfights. Now 36, Godaigo is betrayed by an informer but denies everything (*see* 1331).

Defensor Pacis by the rector of the University of Paris Marsilius of Padua (Marsiglio dei Mainardini), 34, and John of Jandun is a juridical treatise against the temporal power of the pope. Addressed to Louis the Bavarian, the work is denounced by the Vatican, Marsilius will be forced to flee Paris in 1326, Pope John XXII will condemn and excommunicate him in 1327, but Marsilius will help Louis of Bavaria conquer Rome in 1328 and enjoy imperial protection until his death in 1343.

1325 Portugal's Diniz (the Worker) dies childless in January at age 64 after a 46-year reign that has raised the cultural level of the Portuguese court and brought prosperity through agricultural and economic development. He is succeeded by his brother Afonso, 35, who will reign until 1357 as Afonso IV.

England's Queen Isabelle travels to France and arranges the marriage of her son Edward, 12, to Philippa of Hainault, 11.

The German king Louis of Bavaria accepts Frederick of Austria as co-regent, but attempts at joint rule will not succeed.

Delhi's Ghiyas-ud-din Tughlak is murdered by his son after a vigorous 5-year rule. He is succeeded by the parricide, who will reign until 1351 as Mohammed Tughlak.

Mexico City has its beginnings in the city of Tenochtitlan founded by Aztecs in Lake Texcoco. The Aztec Empire that now begins to arise is the culmination of a history that has seen Olmec, Teotihuacan, and Toltec cultures come and go (*see* 1521).

1326 The French queen of England's Edward II invades her husband's realm, which is effectively ruled by the earl of Winchester Hugh le Despenser and his son Hugh. Vowing revenge for the execution of the earl of Lancaster in 1322, Isabelle has the support of her paramour Roger de Mortimer, 39, earl of March, who has been outlawed by Edward at the urging of the earl of Winchester, and Edward's supporters desert him. The king flees London October 2, taking refuge on the Glamorgan estates of the earl of Winchester. Isabelle's forces capture both Despensers and put them to death. Edward tries to escape by sea, Isabelle's men capture him November 16, and they imprison him at Kenilworth Castle.

Brusa capitulates to Ottoman forces April 6 after a 9-year siege has starved the ancient capital of Bithynia into submission.

The first Ottoman emir Osman I dies at age 67 and is succeeded by his eldest son, 47, who will reign until 1360 as Orkhan I, extending Ottoman sway from Angora in central Anatolia to Thrace in Europe and taking the title sultan of the Gazis (warriors of the faith).

Orkhan (above) strikes the first Ottoman coins.

The Florentine banking house of Scali fails, and its 200 creditors clamor for the 400,000 gold florins they are owed. The creditors include France's Charles IV.

Oxford University's Oriel College is founded. Cambridge University's Clare College is founded.

1327 England's Edward II is effectively deposed by his wife Isabelle and her lover Mortimer, who have the parliament of Westminster force the king's abdication and replace him with his son of 14, who will reign until 1377 as Edward III.

The deposed Edward II is imprisoned at Berkeley Castle in Gloucestershire and mistreated in hopes that he will die of disease and malnutrition, but the king, now 43, has a strong constitution, so he is put to death with cruelty September 21; it is announced that he has died of natural causes.

Munich has a great fire which destroys much of the city.

1328 The German king Louis of Bavaria is crowned Holy Roman Emperor January 17 at Rome after forcing the surrender of Pisa. The Roman nobleman Sciarra Colonna places the crown on Louis's head, and the emperor replies to attacks by Pope John XXII by proclaiming the deposition of the pope and undertaking an expedition against the pope's ally, Robert of Naples.

France's Charles IV dies February 1 at age 33, and his death without an heir ends the direct line of Capetian kings descended from Charlemagne. An assembly of French barons confirms the Salic Law that "no woman nor her son may succeed to the monarchy," and Charles the Fair is succeeded by his cousin, 35, who will reign until 1350 as Philip VI, establishing the Valois dynasty that will rule until 1589. He regards the monarchy not as an obligation but merely as a possession and leaves the country's administration to a royal bureaucracy.

Scottish independence gains formal recognition in a final treaty signed at Northampton between England's young Edward III and Robert the Bruce, who is now disfigured with leprosy. The treaty is ratified by the marriage July 12 of Bruce's young son David to Edward's sister Joanna.

1329 The king of the Scots Robert the Bruce dies of leprosy June 7 at age 54 and is succeeded by his 5-year-old son, who will reign until 1371 as David II (but *see* 1332).

England's Edward III pays homage to France's Philip VI for his French fiefs. Flanders, Guienne, and Burgundy remain outside Philip's control, but the thrones of Provence, Naples, and Hungary are occupied by rulers from the Capetian house of Anjou, the papacy at Avignon is under strong French influence, French culture is dominant in England and northern Spain, and French interests are well entrenched in the Near East.

1330 England's Edward III leads a baronial revolt against his regent Roger de Mortimer. He has Mortimer hanged and at age 17 begins a personal rule that will continue until his dotage in the late 1360s and nominally until his death in 1377.

The Hapsburgs recognize Louis IV of Bavaria as Holy Roman Emperor upon the death of Frederick of Austria.

Serbian forces defeat the Bulgars in a decisive battle, ending the power of the Bulgars.

1331 The Japanese emperor Godaigo renews his efforts to regain power from the Hojo regent, is informed upon again as in 1324, and escapes to Nara; the Hojo capture him and exile him to Oki Island, and civil war begins against the Hojo (*see* 1333).

1332 England's Edward III and Edward de Baliol invade Scotland, Edward installs Baliol as the new king of Scotland, and they oblige David Bruce to flee to France.

Denmark's Christopher II abdicates under pressure from Gerhard, count of Holstein, who parcels out crown lands, establishes German noblemen in every

1332 *(cont.)* major Danish fortress, gives German traders full reign, and begins an 8-year period of anarchy.

The Battle of Plowce September 27 gives the Teutonic Knights their first serious reverse. Forces of Poland's Ladislas IV Lokietek emerge triumphant.

Lucerne joins the three Swiss forest cantons of Uri, Schwyz, and Unterwalden in the Helvetic confederation (*see* 1291; Zurich, 1351).

Trading in bitter oranges begins at the French Mediterranean town of Nice. No other kind of orange is known in Europe (*see* 1529; 1635).

1333 Poland's Ladislas IV Lokietek dies in March at age 72 after a 13-year reign in which he has reunited the principalities of Little and Great Poland, has suppressed the magistrates of Cracow, has saved Danzig from the margraves of Brandenburg with help from the Teutonic Knights, and has given the knightly order its first major setback (1332) to lay the foundations for a strong Polish monarchy. The diminutive Ladislas (or Wladislaw) is succeeded by his son of 23, who will reign until 1370 as Casimir III.

The Battle of Halidon Hill July 19 gives England's Edward III revenge for his father's defeat at Bannockburn in 1314. A Scottish army marching to relieve Edward's siege of Berwick is dispersed by Edward's forces under Edward de Baliol and Henry Beaumont, constable of England.

The Japanese emperor Godaigo escapes from Oki Island with help from confederates and gains support from Takauji Ashikaga. The regent Takatoki Hojo—besieged at Kamakura by Gen. Yoshisada Nitti—commits suicide July 5 along with 873 soldiers and retainers. Godaigo regains power from the shogunate, takes territory from the *samurai,* levies high taxes, and starts to rebuild the imperial palaces.

The death of the Japanese regent Takatoki Hojo (above) ends the Kamakura period that began in 1185 (*see* 1336).

The Black Death begins in China as starvation weakens much of the population and makes it vulnerable to a form of bubonic plague (*see* 1340; 1343).

Famine grips China following a severe drought.

1334 The Mughal emperor Mohammed Tughlak welcomes the Moorish traveler Ibn Batuta with lavish gifts in hope that the Moor may help Mohammed conquer the world.

Poland's new king Casimir III encourages immigration of Jews needed as bankers and tax collectors and grants them extensive privileges (*see* 1348).

Pope John XXII dies December 4 at age 85 after an 18-year papacy at Avignon in which he has levied heavy taxes on Christian Europe in an effort to regain independence and prestige for the Church. He is succeeded by the Cistercian cardinal Jacques Fournier, who will serve as Pope Benedict XII until his death in 1342.

Painting *Madonna and Child with Saints and Scenes* by Florentine painter Taddeo Gaddi, 34, whose teacher Giotto di Bondone is now 68.

Florence appoints Giotto di Bondone (above) official architect, and he begins work on a campanile for the city's cathedral.

1335 Japanese troops organized by the Hojo family try to depose the emperor Godaigo in July, but the warlord Takauji Ashikaga supports Godaigo and defeats the Hojo forces in August (*see* 1333; 1336).

The Byzantine emperor Andronicus III conquers Thessaly and part of Epirus from the despot John II Orsini.

Casimir III cedes Poland's Silesian claims to Bohemia in the Treaty of Trentschin.

English grain harvests amount to only 15 bushels of wheat, barley, and oats per capita after setting aside seed for the year to come. Part of the harvest goes to sustain work animals, livestock, and the mounts of knights and barons, while nearly one-third goes into making beer and ale.

1336 The Japanese warlord (*daimyo*) Takauji Ashikaga overthrows the emperor Godaigo and his son Morinaga in January, drives them out of Kyoto, and installs a new emperor to begin the Muromachi (or Ashikaga) period that will continue until 1568 (*see* 1333). The new era will be marked by the disintegration of the shogun's effective control and the rise of independent feudal domains (*hans*) governed by local warlords (*daimyos*).

Aragon's Alfonso IV dies at age 37 after a 9-year reign. He is succeeded by his son of 17, who will reign until 1387 as Pedro IV (Pedro the Ceremonious).

France's Philip VI purchases the Dauphine, the first major imperial fief to be added to French territory.

Greek forces reconquer Lesbos from the Byzantines.

France's walled town of Carcassonne permits the Florentine banking house of Peruzzi to collect its taxes, giving the bankers a percentage of the take.

England's Edward III embargoes export of wool to frustrate the French in Flanders.

Florence's baptistery has the bronze doors at its south end decorated with reliefs of scenes from the life of John the Baptist. Sculptor-architect Andrea Pisano (Andrea da Pontadera), 46, has been working on the reliefs for 7 years.

1337 A "Hundred Years' War" between England and France begins as Philip VI contests English claims to Normandy, Maine, Anjou, and other French territories while England's Edward III denies Philip's legitimacy, assumes the title king of France, and orders Philip to yield his throne. Edward gains support from the townspeople of Flanders, who depend on English wool for their industry, and from the City of London, which is concerned about French influence in its Flemish market (*see* wool embargo, 1336).

Nicomedia (Ismid) falls to the Ottoman Turks.

English merchants contribute 20,000 sacks of wool as a gift to pay the expenses of Edward III. The merchants depend on receipts from the sale of their wool at Bruges and Ghent to pay for the casks of wine they import from Bordeaux.

Giotto di Bondone dies at Florence January 8 at age 69.

1338 France's Philip VI declares England's French territories forfeit, he besieges Guienne, and he burns Portsmouth as the Hundred Years' War grows more intense.

England's Edward III gains the support of the Holy Roman Emperor Louis IV. Louis recognizes Edward's title to the French crown in the Alliance of Coblenz and names Edward vicar of the empire.

The *Licet juris* issued by the diet of Frankfort declares that the electors are capable of choosing an emperor without papal intervention. The declaration effectively divorces the Holy Roman Empire from the papacy (*see* 1202; 1356).

Pisa University has its beginnings.

1339 *Jinno-shotoki* by the Japanese historian Kitabatake Chikafusa is a history of Japan imbued with extreme patriotic sentiments that promote nationalism as civil war continues between the southern and northern courts which will not be reunited until 1392.

Venice conquers Treviso and gains her first mainland possession.

Grenoble University has its beginnings.

1340 The Battle of Sluys June 24 gives England's Edward III victory over a French fleet after Philip VI has dismissed two squadrons of Levantine mercenary ships.

Edward gains mastery of the Channel and free access to northern France (*see* Crécy, 1346).

The Battle of Rio Salado October 30 ends forever the Moorish threat to Spain. Castile's Alfonso XI aided by Portuguese troops gains a decisive victory over a combined force of Spanish and Moroccan Moslems.

Denmark's anarchy ends with the murder of Gerhard, count of Holstein. The youngest son of Christopher II ascends the throne and will rule until his death in 1375 as Waldemar IV Atterdag (*see* 1332).

England's Parliament increases its power by requiring parliamentary sanction for nonfeudal levies and changes in levies. Parliament will appoint auditors of expenditure in the next year and will make money grants conditional on redress.

The Florentine bankers who are financing Edward III in his war with France demand and receive as collateral the person of the archbishop of Canterbury. Florence's Peruzzi banking house asks 120 percent interest and charges an additional 60 percent when payment is not prompt, taking a lien on state income and installing two merchants to supervise Edward's household accounts.

The road from the Black Sea to Cathay (China) is safe both by day and by night, reports *The Merchant's Handbook* by Pegolotti.

Travelers on the road from China (above) will return with rat-borne ticks or fleas that will bring the Black Death to Europe (*see* 1333; 1343).

Queen's College is founded at Oxford.

1341 Civil war begins in the Byzantine Empire as Andronicus III Palaeologus dies June 15 at age 45 after a 13-year reign in which he has annexed large parts of Thessaly and Epirus but has lost them to the growingly powerful Serbs. The emperor's 9-year-old son and successor is challenged by the boy's guardian John Cantacuzene, 49, who sets himself up as emperor in Thrace with support from aristocratic elements and from Greek religious zealots inspired by the mystic teachings of the Mt. Athos monasteries.

France imposes the first Gabelle (salt tax) to help defray the cost of war against England's Edward III.

The Greek zealots (above) range themselves against the rationalist Greek orthodox clergy and against Anna of Savoy, who rules as regent for her son John V (*see* 1342).

1342 Thessalonica has a popular uprising against the Byzantine emperor John V and his regent mother. The religious zealots (Hesychasts) who support John Cantacuzene in the civil war establish a nearly independent state that will continue until 1347.

Hungary's Charles I dies after a 32-year reign that has founded the Anjou line. He is succeeded by his son Louis, who will reign until 1382, making Hungary's power felt throughout the Balkans.

Louis of Bavaria acquires the Tyrol and Carinthia by marrying the "ugly duchess" Margaret of Tyrol.

Painting *Nativity of the Virgin* by Italian painter Pietro Lorenzetti, now 62, who has been helped by his younger brother Ambrogio to produce frescoes for churches at Assisi and Siena.

1343 Smyrna falls to Venetian forces which take advantage of the civil war in the Byzantine Empire to capture the valuable commercial port of Ionia.

Florence's Peruzzi banking house fails as England's Edward III repudiates his debts and the bankers are unable to collect despite measures they have taken to protect themselves from default (*see* 1340; 1344).

The Black Death that began in China 10 years ago strikes marauding Tatars who attack some Genoese merchants returning from Cathay with silks and furs. The plague, called bubonic because of its characteristic bubes, or enlarged lymph glands, is transmitted by fleas, carried by rats, and harbored perhaps in the merchants' baggage. The Tatars besiege the merchants at the Crimean trading post of Calla and before withdrawing catapult their corpses over the walls into Calla, infecting the merchants, some of whom die on the road home while others carry the plague to Constantinople, Genoa, Venice, and other ports (*see* 1340; Cyprus, 1347).

Inventorius sive Collectorium Partis Chirurgicalis Medicinae by French surgeon Guy de Chauliac, 43, will serve physicians as a manual for 3 centuries. De Chauliac has developed the treatment of fractures by slings and extension weights, he operates on hernias and cataracts,

1343 ***(cont.)*** he exorcises superficial growths, and he believes that pus from wounds serves to release the *materia peccans.*

Dialogues by English scholastic philosopher William of Ockham (or Occam) lay the foundations of modern theory of independence of church and state. Ockam, 43, has been imprisoned after defending evangelical poverty against Pope John XXII in his *Opus Nonaginta Dierum* of 1330 but has escaped and is living in Munich, where he has sided with the Holy Roman Emperor Louis IV in contesting the temporal power of the pope.

1344 The Spanish Moors lose the southern port of Algeciras to Castile's Alfonso XI.

Florentine banking prestige plummets with the failure of the Bardi banking house that comes on the heels of last year's Peruzzi collapse as England's Edward III repudiates his debts (*see* 1340). Civil war begins in Florence, the commune is restored under the sway of a businessmen's oligarchy, the smaller guilds lose power, and workers who belong to no guild are further exploited as the Florentine oligarchy seeks access to the sea, expansion in Tuscany to dominate trade routes, and support of the popes in order to retain papal banking business.

The term *Hanseatic League* is used for the first time to denote the confederation of Baltic traders now so prominent in fish export (*see* 1300; 1360).

Spanish olive oil, fruit, and fine manufactured goods are exported from the south, wool and hides from the north as Castile and Aragon increase their commerce.

1345 The Ottoman Turks make their first crossing into Europe in response to a call for support by John Cantacuzene as civil war continues in the Byzantine Empire.

Florence has her first social revolution as the rising industrial proletariat attempts to overthrow the ruling business oligarchy (*see* 1344; Strozzi, 1347).

A bridge that will be called the Ponte Vecchio is completed across the Arno at Florence to replace a 13th century bridge swept away by a flood in 1333.

Poetry *De Vita Solitaria* by Petrarch.

The Cathedral of Notre Dame on the Ile de la Cité in the Seine at Paris is completed in Gothic style after 182 years of construction.

1346 The Battle of Crécy August 26 establishes England as a great military power, reorients English social values by its joint participation of yeomanry and aristocracy, and begins the end of the era of feudal chivalry (cavalry) that has dominated warfare since the barbarian invasions of the Roman Empire. Europe's greatest army of horse soldiers is massacred near Abbeville by England's Edward III, who has landed July 12 at La Hague near Cherbourg with 1,000 ships, 4,000 knights, and 10,000 English and Welsh longbowmen equipped with the weapons acquired in the Welsh wars. France's Philip VI meets with disaster as his 12,000 men-at-arms, 6,000 mercenary Genoese crossbowmen, 20,000 *milice des communes,* and assorted feudal vassals prove easy targets for the quick-shooting English archers and spearmen who make short work of the heavily armored knights (who cannot remount once their horses go down). The French attack the English lines 16 times by midnight and are practically annihilated (Edward's heralds count 1,542 dead French knights on the battlefield while English casualties total little more than 50).

Scottish allies of Philip VI invade England but are stopped at Neville's Cross. Scotland's David II, now 22, is captured and will not be ransomed for 11 years.

Bohemia's blind John of Luxembourg is killed at age 50 while fighting in support of Philip VI at Crécy (above) and is succeeded by his son of 30, who assumes the throne as Charles I to begin a "golden age" of Bohemian history. Charles is crowned German king at Bonn November 26 and prepares to attack the Holy Roman Emperor Louis IV of Bavaria (*see* 1347).

Serbia's king Stephen Dushan, 38, proclaims himself Emperor of the Serbs, Greeks, Bulgars, and Albanians, establishes a court of Skoplye with elaborate Byzantine titles and ceremonies, and prepares to seize Constantinople and replace the Greek dynasty.

A Muslim dynasty in Kashmir that will endure for 243 years is founded by Shah Mirza, who will substitute a land tax of one-sixth for the extortionate taxes imposed by Hindu kings.

Florence's Bardi banking house fails in January, going the way of the Scali, Peruzzi, Acciauoli, and Frescobaldi houses (*see* 1344). The Bardis have had to keep loaning funds to England in order to assure continued supplies of raw wool for their textile industry.

France's Estates-General at Langue d'Oil openly refuses to continue war levies in support of Philip VI and demands reforms in the country's administration as French cloth, gold and silver table services, jewelry, and other treasure is pillaged by the invaders and shipped off to England.

1347 English forces under Edward III take the French port of Calais after a long siege and make it a military and commercial outpost that will be English for 211 years. Edward celebrates the end of the long siege by taking as servants six of the city's leading burghers, whose lives he has spared only at the request of his wife Philippa (*see* sculpture, 1895).

Florentine banker Andrea Strozzi buys up quantities of grain in the midst of a Tuscan famine and sells it at low prices to the *popolo minuto* in a bid for their allegiance, but the people see through Strozzi's scheme and refuse to follow him, following instead the cunning suggestions of the Medicis and the Capponi to attack the houses of prominent older bankers such as the Pazzi, Bardi, and Frescobaldi.

The 6-year civil war in the Byzantine Empire ends in victory for John Cantacuzene, who begins an 8-year reign as John VI Cantacuzene, nominally as co-emperor with his ward John V. Thrace and Macedonia have been ravaged by Serbs and Turks brought in to support the rival factions, and the new reign will be marked by attacks from these outsiders and also from the Genoese.

Rome's dissolute plutocracy is overthrown in a revolt led by papal courtier Cola di Rienzi (Niccolo Gabrini), 34, who heads a procession to the Capitol dressed in full armor and is given unlimited authority by the assembled multitude. Rome's nobles leave the city or go into hiding, Rienzi takes the title of tribune in late May, begins a government of stern justice in contrast to the license that has prevailed, is encouraged by the poet Petrarch, proclaims the sovereignty of the Roman people over the empire in July, is installed as tribune with great pomp in mid-August, but is obliged to impose heavy taxes in order to maintain his costly regime. Rienzi offends Pope Clement VI by proposing to set up a new Roman Empire based on the will of the people, he is ridiculed for his pretensions, the pope empowers a legate to depose Rienzi and bring him to trial, Rome's barons gather troops, but Louis of Hungary comes to Rienzi's aid, the barons are defeated November 20 outside the city gates, and Rienzi's noblest enemy Stephen Colonna is killed. Denounced by the pope as a criminal, pagan, and heretic, Rienzi devotes himself to feasts and pageants until mid-December when he panics at some disturbance, abdicates, and flees the city.

The Holy Roman Emperor Louis IV of Bavaria has died on a bear hunt outside Munich October 11 at age 60, leaving Charles of Luxembourg undisputed ruler of Germany. He will be crowned emperor in 1355 and will reign until 1378 as Charles IV.

The Black Death reaches Cyprus, whence it will spread to Florence and find thousands of victims weakened and made vulnerable by famine (*see* 1348).

Jane I, queen of both the Sicilies and countess of Provence, opens a house of prostitution at Avignon in an effort to reduce venereal disease. "The Queen commands that on every Saturday the Women in the House be singly examined by the Abbess and a Surgeon appointed by the Directors, and if any of them has contracted any Illness by their Whoring, that they be separated from the rest, and not suffered to prostitute themselves, for fear the Youth who have to do with them should catch their Distempers."

1348 The Black Death that will devastate Europe reaches Florence in April and spreads to France and England, where it arrives in July or August, although London is spared until November (*see* 1349).

French surgeon Guy de Chauliac remains at Auvergne after most physicians have fled and writes, "The visitation came in two forms. The first lasted two months, manifesting itself as an intermittent fever accompanied by spitting of blood from which people died usually in 3 days. The second type lasted the remainder of the time, manifesting itself in high fever, abscesses and carbuncles, chiefly in the axillae and groin. People died from this in 5 days. So contagious was the disease especially that with blood spitting [pneumonia] that no one could approach or even see a patient without taking the disease. The father did not visit the son nor the son the father. Charity was dead and hope abandoned."

The Black Death (bubonic plague), transmitted by rat-borne fleas, killed almost everybody in some European towns.

Guy de Chauliac (above) recommends bloodletting to those unable to flee.

Jews are blamed for spreading the Black Death by deliberately contaminating wells and by "anointing" people and houses with poison. The Jews are persecuted first at Chillon on Lake Geneva and then at Basel and Freiburg, where all known Jews are herded into wooden buildings and burned alive. At Strasbourg more than 2,000 Jewish scapegoats are hanged on a scaffold erected at the Jewish burying ground. Jews are not permitted to enter the Avignon house of prostitution established last year.

Pope Clement VI issues two bulls declaring the Jews to be innocent, but the persecution continues. Thousands of Jews flee to Poland, Russia, and other more tolerant parts of eastern Europe (*see* 1334).

The University of Prague is founded by Charles IV in Bohemia.

England has her third cold, wet summer in a row, and this one is the worst. Rain falls steadily from midsummer to Christmas, crops are poor, food is short, and the hunger makes the country vulnerable to disease (*see* 1349).

The Black Death will extinguish nearly two-thirds of the population in some parts of Europe. At Locarno on Lake Maggiore the population will fall from 4,800 to 700.

1349 The Order of the Garter founded by England's Edward III adopts the motto *Honi soit qui mal y pense* (evil to him who evil thinks).

Persia's Ilkhan Nushirwan dies after a troubled 5-year reign, ending the Mongol dynasty that began with Hülagü in 1256. The Sarbadarids will reign in Khorasan until 1381, the Muzaffarids in Fars, Kirman, and Kur-

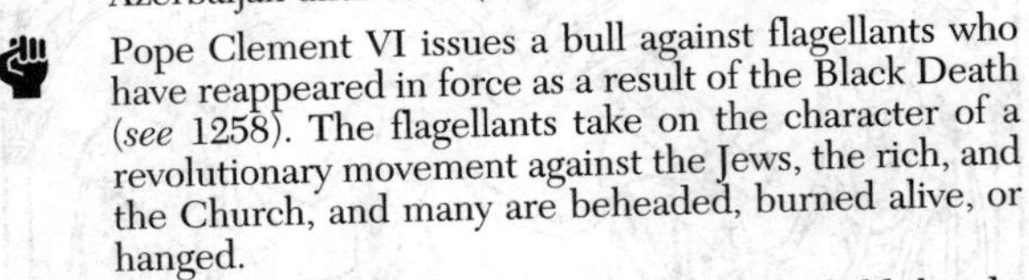

1349 ***(cont.)*** distan until 1393, and the Jalayrs in Iraq and Azerbaijan until 1411 (*see* Tamerlane, 1380).

Pope Clement VI issues a bull against flagellants who have reappeared in force as a result of the Black Death (*see* 1258). The flagellants take on the character of a revolutionary movement against the Jews, the rich, and the Church, and many are beheaded, burned alive, or hanged.

English landlords offer high wages to field hands. Reapers and mowers spared by the Black Death eat better than they ever have or ever will again.

The Black Death kills from one-third to one-half the population of England which calls a truce in hostilities with plague-stricken France (*see* 1347; 1350).

A Scottish army invades England in the autumn and is stricken with plague which the dispersing soldiers carry back to Scotland in pneumonia (person to person) form (*see* 1350).

The Black Death reaches Poland and moves on toward Russia, flourishing on poverty and malnutrition, especially in the larger cities.

1350 Castile's Alfonso XI dies of the plague March 27 at age 38 while besieging Gibraltar, the last Spanish city still in Muslim hands. Alfonso's son, 16, will reign until 1369, ruling so harshly that he will be called Pedro the Cruel.

France's Philip VI dies at Nogent-le-Roi August 12 at age 57 after a 22-year reign. He is succeeded by his son of 31, who will reign until 1364 as John II largely under the domination of evil counselors.

Rome's Cola di Rienzi emerges from hiding, reaches Prague in July, denounces the temporal power of the pope, and asks Charles IV to deliver Italy from its oppressors, but the emperor imprisons Rienzi in the fortress at Raudnitz and will turn him over to the pope next year.

The Black Death reduces population pressure on food supplies, which has been growing in England and Europe in this century, and prices drop for lack of demand. Where a good horse brought 40 shillings in England 2 years ago it now brings only 16, while a fat ox fetches only 4 shillings, a cow 1 shilling, and fat sheep sixpence, but wheat fetches 1 shilling per quarter (8 bushels), up from as little as 16 pence in good crop years, as the dearth of field hands reduces the crop and forces many landlords to turn farmland into pasturage (*see* 1349).

Salt production takes a sharp drop in northern Europe as a result of economic conditions and of the Hundred Years' War. Poor-quality salt from Brittany's Bourgneuf Bay begins to dominate the salt fish trade as good white salt becomes too costly. Great salt deposits will soon be opened in Poland, with a Genoese firm headquartered at Cracow receiving a monopoly in Polish salt production.

A wire-pulling machine invented in Europe is an early step in the development of metallurgical technology.

The Black Death reaches Scotland and Wales.

1351 The Order of the Star that will be replaced by the Order of St. Michael in 1469 is founded in France.

Florence goes to war with Milan, whose archbishop Giovanni Visconti, 61, tries to gain control of Tuscany. Florence permits her citizens to buy commutation of military service and employs mercenary captains (*condottieri*) to resist the Milanese (*see* 1353).

Zurich joins the growing Swiss Confederation.

An English Statute of Labourers fixes wages at their 1346 levels and attempts to compel able-bodied men to accept work when it is offered. The shortage of workers resulting from the Black Death has combined with prosperity produced by the Hundred Years' War to create an economic and social crisis in England.

The Statute of Labourers (above) helps keep wages within bounds as England adjusts to the fact that she no longer has a glut of labor, but while Parliament also orders victualers and other tradesmen to sell their goods at reasonable prices, the Statute of Labourers destroys the country's social unity without resolving its problems.

English landholders enclose common lands for sheep raising and begin to develop great fortunes while unrest spreads through the English yeomanry, which has emerged from its traditional passivity as it has shared in the plunder of war and in the higher wages paid to survivors of the Black Death.

1352 Cola di Rienzi is tried by three cardinals at Avignon and sentenced to death. The sentence is not carried out, Pope Clement VI dies December 6, and Rienzi is pardoned and released from prison by the new pope Innocent VI, who wants to reduce the power of the Roman barons (*see* 1350; 1354).

Glarus joins the growing Swiss Confederation.

The Black Death reaches Moscow and spreads eastward back to India and China.

Corpus Christi College is founded at Oxford.

1353 Bern joins the Swiss Confederation, which now embraces seven cantons (*see* Battle of Sempach, 1386).

Milan's Giovanni Visconti annexes Genoa which goes to war with Venice.

The Arab traveler Ibn Batuta visits Africa's Mandingo Empire.

Fiction *The Decameron* by Italian humanist Giovanni Boccaccio, 40, is a love story full of vivid description of the Black Death that has killed three out of every five Florentines. Boccaccio's 10 protagonists have fled the city to the seclusion of a villa garden on the slopes of Fiesole.

1354 England resumes the Hundred Years' War after an interruption by the Black Death. The English will ravage Languedoc in the next 2 years, they will meet little opposition, and they will reduce the prestige of the French crown.

Cola di Rienzi receives the title senator from Pope Innocent VI, the pope sends him to Rome with the papal legate Cardinal Albornoz and a few mercenaries, and he enters the city to wild acclaim in August. The Romans restore him to his position as tribune, but rioting breaks out October 8, Rienzi tries to address the ugly mob, the crowd sets the building ablaze, and Rienzi is murdered while trying to escape in disguise.

The Battle of Sapienza gives Genoa a victory over Venice, which loses its fleet. Milan's Giovanni Visconti dies at age 64 and is succeeded by his two nephews, who will mediate the war between Venice and Genoa.

Ottoman forces take the Gallipoli Peninsula and spread through Thrace.

Granada's Alhambra Palace is completed on a 35-acre plateau after 106 years of work begun in the reign of the Moor Al Ahmar. Enclosed by a fortified wall, the palace is flanked by 13 towers.

1355 Chiang-ning, (later Nanjing) falls to the monk Qu Yuanzhang, 27, who leads a revolt against Mongol domination of China (*see* Ming dynasty, 1368).

The Byzantine emperor John VI Cantacuzene is driven out of Constantinople and retires to a monastery after an 8-year reign in which he has imposed heavy taxes in a vain effort to pay his foreign mercenaries. His former ward John V Palaeologus entered the city late last year and will reign until his death in 1391, except for the years 1376 to 1379 when his rebellious son Andronicus will reign.

Milan's Matteo Visconti is assassinated by his brothers Bernabo and Galeazzo who rule, respectively, at Milan and at Pavia. Their harsh administrations will be marked by ostentatious patronage of the arts and of learning.

France's John II obtains a grant of funds to resist England's Black Prince Edward, 25, son of Edward III, who is laying waste the country.

The French king encounters opposition led by Etienne Marcel, the richest man in Paris, who persuades the Estates-General of Languedoc and Langue d'Oil to compel John to consult with the Estates before imposing new tax levies and to let a commission from the Estates supervise collection and expenditure of the tax moneys. John induces the Estates-General to adjourn, proceeds to debase the nation's coinage in the interests of his treasury, and organizes opposition to the Estates-General.

Portugal's Afonso IV has the mistress of his son Pedro murdered lest her brothers in Galicia gain control of the government. Pedro claims he has married the former Inês Pires de Castro, organizes a revolt, and lays siege to Oporto.

Painting *Madonna and Child with Angels* by Taddeo Gaddi.

1356 The Battle of Poitiers (or Maupertuis) September 19 ends in defeat for France's John II, whose army is cut to ribbons by the English under the Black Prince of Wales Edward. The Black Prince takes John as a prisoner to England along with a crowd of French aristocrats, leaving France to the regency of John's son Charles, 18, who is unable to prevent civil chaos.

The Golden Bull issued by the Holy Roman Emperor Charles IV fixes the form and place of the imperial election and coronation of electors, it establishes the duties and privileges of electors, and it transforms the empire from a monarchy into an aristocratic federation that will endure for 450 years (*see* 1338). Charles makes no secret of his view that the empire is an anachronism, but he values the emperor's right to nominate vassals to vacant fiefs, and he makes sure that the king of Bohemia is given first place among the empire's secular electors.

Korea lapses into a 36-year period of disorder. The Koryo kings who dominated the country for 235 years until 1170 have intermarried with the Mongols and have become mere satellites of the Mongol imperial family. They stage a successful revolt against the Mongols, but they have depended for their authority on Mongol prestige and are unable to suppress their vassals, who are helped by Japanese pirates to create chaos (*see* 1369).

1357 Paris merchants rebel against the dauphin and recall the Estates-General under the leadership of Etienne Marcel. The members make no effort to reduce the monarchy's traditional powers, but they order reforms that include relief for the poor and frequent meetings of the Estates on a regular basis. The Great Ordinance passed by the Estates provides for a standing committee to supervise tax levies and expenditures.

Etienne Marcel (above) makes an alliance with Charles the Bad of Navarre and thus discredits the Estates-General, which splits into many factions. The regent Charles opposes the alliance and flees Paris to form a powerful coalition against the Estates-General and their ally Charles the Bad.

Portugal's Afonso IV (the Brave) dies May 30 after a troubled 32-year reign; he made peace last year with his son, who will reign until 1367 as Pedro I.

An account of travels by a fictitious Sir John Mandeville will appear in the next 15 years written in French and probably authored by Liège physician Jehan de Bourgogne (Jehan a la Barbe), a former chamberlain to John, baron de Mowbray, in England. Borrowed from works by Giovanni de Piano Carpini, who traveled east to Karakorum in 1245, and by friar Oderic of Pordenone, whose *Travels in Eastern Regions* appeared in 1330, the John Mandeville travel accounts will describe such fanciful marvels as men whose heads grow beneath their shoulders.

1358 Etienne Marcel forces his way into the Paris palace of the dauphin February 22, and the marshals of Champagne and Normandy are murdered in the presence of young Charles, who rules in the absence of his father.

1358 ***(cont.)*** Charles the Bad of Navarre approaches Paris in league with the English and gains avowals of support from Marcel, who is assassinated July 31.

French peasants and Parisians join in a violent rising of the Jacquerie in May against oppressive taxes imposed to ransom John II and other captives taken by the English at Poitiers in 1356 and to carry on the war. The rising is ruthlessly repressed in June with the aid of the English, who pause in their war with the French to join in the wholesale slaughter of rebellious serfs: 300 peasants trapped in a monastery are burned to death.

1359 Edward III begins a final expedition to France and penetrates the walls of Paris, but the south of France has been so devastated by war that the English have trouble provisioning their forces.

France's John II virtually restores Angevin lands to England, but his son Charles rejects the preliminary peace terms made by John, who is held in luxurious captivity (*see* 1360).

Revolutionists wearing red hats storm Bruges as the artisan guilds of the Flemish city try to overturn its patrician government.

The prince of Muscovy Ivan II Krasnyi (the Red) dies after a 6-year reign. His son, 9, will reign until 1389 as Dmitri Donskoi (of the Don).

The campanile, or bell tower, is completed at Florence. It has been designed by architects who included the late Giotto di Bondone (*see* 1334).

1360 The Peace of Bretigny signed at Calais brings a brief truce to the Hundred Years' War that has exhausted both England and France. Edward III virtually renounces his claim to the French crown and Charles, regent for France's John II, promises 3 million gold crowns for his father's return and yields Calais, Guienne (southwestern France), Ponthieu, and their immediately surrounding territories to England. Leaving three sons as hostages, John returns to France and cannot raise the huge ransom. One son escapes from custody, and John is returned to England (*see* 1364).

The Ottoman sultan Orkhan dies at age 71 after a 24-year reign. He is succeeded by his eldest son, 40, who will reign until 1389 as Murad I, using an elite Janissary military corps composed of war prisoners and, later, Christians taken captive in childhood as Murad extends Ottoman power throughout Anatolia and the Balkans.

English laborers who ask wages above the legal minimum established by the 1351 Statute of Labourers are ordered to be imprisoned with bail.

The Hanseatic League grows to include 52 towns that number among them Bremen, Cologne, Danzig, Dortmund, Gronigen, Hamburg, and Hanover. The number will be enlarged to 70 or 80.

1361 Denmark's Waldemar IV begins a 2-year war against the Hanseatic League that will end with a sharp but brief curtailment of the Hansa's power (*see* 1360; 1370).

The duchy of Burgundy comes into possession of France's John II, who will give it to his son Philip in 1363.

Buda is selected as the capital of Hungary (*see* 1247).

Murad I takes Adrianople, which will be the Ottoman capital until 1453.

The Black Death strikes again in England, but less severely than in 1349, and rages also in France, Poland, and elsewhere, especially among children (*see* 1371).

The University of Pavia has its beginnings.

1362 Norse king Magnus Ericson sends an ill-fated expedition to look for would-be colonists who have failed to reach Greenland (*see* 981). The expedition may enter what will later be called Hudson's Bay (*see* 1610).

English is made the language of pleading and judgment in England's courts of law, but legal French continues to be used for documents.

The Vision of William Concerning Piers the Plowman, commonly called *The Vision of Piers Plowman*, will be written in the next 30 years, probably by English poet William Langland (or Langley), who will speak of "Lombards of Lucques that liven by lone as Jews" (*see* 1290; 1306).

The Palace of the Popes at Avignon is completed after 28 years of construction.

1363 Denmark's Waldemar IV forces the Hanseatic League to accept peace and a curtailment of its privileges. Waldemar defeated the Hansa fleets last year at Helsingborg.

Sweden's Magnus II abdicates under pressure at age 47 after a weak 44-year reign. He is succeeded by Albert of Mecklenburg, a tool of the Swedish aristocracy which will permit him to rule until 1387.

A leg of roast mutton sells in a London cookshop for as much as a farm worker earns in a day, and a whole roast pig sells for more than three times that amount to Londoners prospering by the war with France.

1364 France's John II dies April 8 at age 45 in England where he has been a prisoner since his capture at the Battle of Poitiers in 1356. His body is sent home with royal honors, and he is succeeded by his son of 27, who will reign until 1380 as Charles V.

The University of Cracow is founded by Poland's Casimir the Great.

The University of Vienna has its beginnings.

Famine strikes France following a bad harvest, and plague in epidemic form follows on the heels of hunger.

1365 France's House of Blois cedes its rights to Brittany April 12 in the Treaty of Guerande with John IV de Montfort.

1366 Amadeus of Savoy leads a crusade against the Ottoman Turks, taking Gallipoli (the Turks will recover it in 1373).

The English Parliament refuses to pay feudal dues to the pope.

Poetry *Canzoniere* by Petrarch.

Toledo's El Transito Synagogue is completed by Meier Abdeli.

Europeans commonly eat the main meal of the day at 9 o'clock in the morning.

1367 Portugal's Pedro I goes to the aid of Castile's Pedro the Cruel and is helped by the Black Prince of Wales Edward to defeat Pedro's bastard half brother Enrique, Count of Trastámara, who is forced to take refuge in France. Pedro returns to Portugal and dies after a 10-year reign. His son of 22, who will reign until 1383 as Fernão I, ratifies the existing peace with Castile and Aragon but begins to squander the wealth amassed by his father.

The Confederation of Köln (Cologne) brings together 77 German towns whose representatives organize common Hanseatic finances, form Scandinavian alliances, and prepare naval forces to counter Denmark's Waldemar IV (*see* 1363; 1370).

1368 The Ming dynasty founded by Qu Yuanzhang, now 40, will rule until 1644 and begins a resurgence of Chinese nationalism. Qu seized the city of Chiang-ning (later Nanjing) in the anarchy of 1355, set up an orderly government, and annexed the holdings of neighboring southern war lords. He drives the Mongols out of Beijing, will drive them out of Shenxi and Gansu next year and out of Sichuan in 1371, and he will rule until 1398 under the name Hung-wu.

The Ming emperor Hung-wu (above) begins restoring the Great Wall of China.

1369 Tamerlane (Timur the Lame, or Tamburlaine) makes himself master of Samarkand in Turkestan at age 33. A descendant of Genghis Khan who succeeded his father as chief of a Turkish tribe 9 years ago, the lame leader of the western house of Jagatai begins to develop an armed horde that will conquer much of the world (*see* Persia, 1380).

Korea's state of Koryo submits to the Chinese Ming forces after 13 years of rebellion, but disorder continues in the country (*see* 1392).

Castile's Pedro the Cruel antagonizes the Black Prince, who deserts him. Besieged at La Mancha (Montiel) by the Count of Trastámara, he is tempted out March 23 and knifed to death at age 34 by his bastard half brother, who will reign until 1379 as Enrique II Trastámara.

1370 Poland's Casimir III (the Great) dies in a hunting accident November 5 at age 60 after a long reign in which he has repulsed a Mongol invasion and annexed Galicia. Having no direct heir, he has promised that his crown shall pass to Louis of Anjou, son of Hungary's Charles Robert, who will reign until 1382, governing through regents.

The Hanseatic League reaches the height of its power after taking united—and successful—military action against Denmark's Waldemar IV. The Treaty of Stralsund gives the League a monopoly in the Baltic trade that will continue for 70 years, and the League gains control of Scandinavian polities (*see* 1360; 1375).

England's Exeter Cathedral is completed after 90 years of construction.

1371 Castile's Enrique II Trastámara takes Zamora February 26 and forces Portugal's Fernão I to renounce his claims to Castile in the Treaty of Alcoutin. Fernão agrees to marry Enrique's daughter Leonora.

The Battle of Chernomen (Chirmen) September 26 ends in victory for the Ottoman sultan Murad I, whose armies have swept northward to the Balkans in the past 5 years. He forces the rulers of Bulgaria, Macedonia, and the Byzantine Empire to recognize his suzerainty.

The Black Death reappears in England, but the outbreak is milder than in 1361 (*see* 1382).

1372 France's Charles V regains Poitou and Brittany from the English and regains control of the Channel after 32 years by defeating the English at La Rochelle. Charles is helped by a Castilian fleet which blocks English transport in the North.

The Channel island of Guernsey is captured by the self-styled prince of Wales Owen-ap-Thomas.

The Vatican asks an astronomer to correct the Julian calendar in use since 46 B.C. because it is too long by 11 minutes and 15 seconds each year, but the astronomer will die before he can reform the calendar (*see* Gregorian calendar, 1582).

1373 John of Gaunt, 33, duke of Lancaster, invades France, fanning out from Calais to Bordeaux.

Lisbon is burned by Castilian forces sent by Enrique II, who is at war with both Portugal and Aragon (*see* 1374).

Brandenburg is annexed to Bohemia by the Holy Roman Emperor Charles IV of Luxembourg.

1374 John of Gaunt, duke of Lancaster, returns from the war in France and becomes leader of the state, using anticlerical feeling and social unrest to pursue his ambitions to succeed his father Edward III.

London retains Oxford don John Wycliffe, 54, to conduct negotiations with Pope Gregory XI.

Castile makes peace with Aragon and Portugal at Imazan and continues to lead in the reconquest of Spain from the Muslims.

Plague precautions are instituted at Venice, where three officials are appointed to inspect vessels wishing to enter ports of the Venetian Republic and exclude any found to be infected with the Black Death (*see* waiting period, 1403).

A dancing mania sweeps the Rhenish city of Aix-la-Chapelle in July and sends hordes of men and women into a pathological frenzy of dancing in the streets. The dancing continues for hours until the crowds are too exhausted or too injured to continue; the phenomenon will never be fully explained.

1375 The Kingdom of Armenia founded by Tigranes I in 94 B.C. ends with the surrender of the new Armenian king Levon V to the governor of Aleppo, who has besieged

1375 ***(cont.)*** the Armenian capital of Sis with a force of nearly 30,000 Mamelukes. The victors slaughter many of the Armenians and convert many of the rest to Islam, they take Levon to Cairo, he will remain in royal prison until his royal Aragon and Castile in-laws ransom him in 1382, and France's Charles VI will give him a pension and a house.

The Truce of Bruges brings a pause in the Hundred Years' War between France and England.

Denmark's Waldemar IV Atterdag dies at age 55 after having recovered nearly all of Schleswig. He is succeeded by his 5-year-old grandson, who will reign until 1387 as Olaf II with his mother Margaret as regent.

The Hanseatic League wins formal recognition from the Holy Roman Emperor Charles IV of Luxembourg, who visits Lübeck (*see* 1370; 1377).

The Hanseatic League (above) establishes common weights, measures, and coinage, arranges for the settlement of disputes at home and abroad, secures new trading privileges for its member cities, protects merchants and their goods on the road and at sea, draws up a *Seebuch* to help navigators find lighthouses, harbors, and buoys from Riga to Lisbon, and opens up new lines of trade to supplement its commerce in herring, cod, salt, leather, hides, wool, grain, beer, amber, timber, pitch, tar, turpentine, iron, copper, horses, and falcon hawks.

England forbids transportation of seed oysters during the month of May. The English continue the oyster cultivation begun by the Romans in 110 B.C.

Le Viander de Taillevent by Guillaume Tirel, 49, gives a detailed account of France's developing cuisine.

1376 John Wycliffe and his Reform party win control of London and oppose ownership of property by clergymen.

England's Black Prince Edward of Wales dies at age 46 of a disease contracted while fighting in Spain. His brother John of Gaunt packs Parliament with men who will undo the "Good Parliament" 's reforms (*see* 1377).

England's "Good Parliament" appeals to Edward III to exclude foreigners from London's retail trade and asks the king to banish the Lombard bankers, enforce the 1351 Statute of Labourers, regulate fisheries, and ban export of English grain and yarn.

John Wycliffe (above) revises ordinances governing the sale of food and encounters opposition from London fishmongers and other victualers who depend on wines from France, spices from the Orient, and air-cured stockfish from Iceland (*see* Walworth, 1377).

John Wycliffe (above) expounds the doctrine of "dominion as founded in grace" by which all authority, both secular and ecclesiastical, is derived from God and is forfeited when its possessor falls into mortal sin. He attacks the worldliness of the medieval church and presents logical grounds for the refusal by England of certain tribute demanded by Rome.

1377 England's Edward III dies June 21 at age 64 shortly after a new parliament has reversed the acts of last year's "Good Parliament." Edward is succeeded by his grandson Richard, 10, son of the Black Prince. Thomas, Duke of Gloucester and John of Gaunt, Duke of Lancaster, will administer the government until Richard attains his majority, and he will reign until 1399 as Richard II.

Parliament levies a 4-shilling-poll tax that will lead to widespread rioting in 1381.

John Wycliffe's Reform party loses control of London to victualers headed by William Walworth. They raise food prices in the city.

The Hanseatic League retains its privileges in England and will keep them for nearly 2 centuries despite growing resentment against the League.

The Babylonian Exile of the papacy that began in 1306 has ended January 17 with the entry into Rome of Pope Gregory XI, who left Avignon in September of last year.

Bulls promulgated by Pope Gregory XI accuse John Wycliffe of heresy. He is summoned before the bishop of London at St. Paul's to answer the charge but escapes trial as general rioting in the streets of London ends the session of the ecclesiastical court.

Population estimates based on the English poll tax (above) suggest that the nation's population is little more than 2 million, down from at least 3.5 million and possibly 5 million before the Black Death (*see* 1546).

1378 The Holy Roman Emperor Charles IV divides his lands among his three sons and dies at Prague November 29 at age 62. He obtained the margravate of Brandenburg 5 years ago for his son Wenceslas, now 17, who turns Brandenburg over to his half brother Sigismund, 10, succeeds Charles, and will reign until 1400 as the emperor Wenceslas.

A Great Schism that will divide the Catholic Church for 39 years and bring society's highest authority into disrepute begins as Bartolommeo Prignani, 60, is elected Pope Urban VI and announces that he will reform the Church beginning with the sacred college of cardinals. Thirteen cardinals meet at Anagni and elect Robert of Geneva, 36, as Pope Clement VII, and he is supported by France's Charles V. This "antipope" establishes himself at Avignon (the Babylonian Captivity), and each pope appoints his own college of cardinals, collects tithes and revenues, and excommunicates partisans of the other (*see* 1417).

1379 Enrique II Trastámara of Castile and León dies May 30 at age 46 after a 10-year reign in which he has made vast land grants to noblemen and concessions to cities in return for support against claims by the daughters of the late Pedro the Cruel. Enrique's son of 21 inherits the crown and will reign until 1390 as Juan I in the face of claims by John of Gaunt, a son-in-law of Pedro the Cruel, who has the backing of Portugal's Fernão I.

England's Winchester College is founded by William of Wykeham, whose school will become a model for the nation's great public schools (*see* New College, 1394; Eton, 1440).

1380 Tamerlane invades Persia and his Mongol-Turkish hordes begin to overrun Khorasan, Jurjan, Mazandaran, Sijistan, Afghanistan, Fars, Azerbaijan, and Kurdistan (*see* 1369). They will invade Russia next year.

The Battle of Kulikovo September 8 ends in victory for the prince of Moscow Dmitri Donskoi, 30, who turns back the Mongols and destroys the prestige of Mongol power (although the Mongols will reach the gates of Moscow several times in the next few years).

France's Charles V dies at Vincennes September 16 at age 43 after eating poisonous mushrooms. He has ruled since the capture of his father at the Battle of Poitiers in 1356 and is succeeded by his son of 12, who will reign until 1422 as Charles VI (Charles le Bien-Aimé) despite bouts of insanity beginning in 1392 (four uncles will rule as guardians until 1388).

The *Apocalypse Tapestry* completed at Paris after 5 years of work depicts the biblical allegory. French authorities will discover the 70-panel tapestry in the 1840s, the bishop of Angers will buy it for 300 francs, he will have it restored, and it will hang at Angers Château.

The Gothic cathedral of Saint Etienne at Metz in Lorraine is completed after 130 years of work.

1381 Venice gains supremacy over Genoa after the 3-year War of Chioggia in which Genoa's admiral Luciano Doria has seized Chioggia and blockaded Venice but has been forced to surrender after being starved out by Venetians under Vittorio Pisano, who has blocked the channel.

Wat Tyler's rebellion creates anarchy in England as farm workers, artisans, and city proletarians stage an uprising against the 1351 Statute of Labourers and the 1377 poll tax. Mobs assemble in Essex, Kent, and Norfolk, sack palaces and castles at Norwich and Canterbury, take hostages, choose Wat (Walter) Tyler as their leader in June, and converge on London.

Some 30,000 rioters enter London over a drawbridge, burn John of Gaunt's Savoy Palace, the Temple used by London's lawyers, and the building at Clerkenwell of the Knights Hospitalers. Simon Sadbury, archbishop of Canterbury, is beheaded by the mob at Tower Hill, and Hanseatic traders are chased into their fortified Steelyard (*see* 1250).

Richard II, now 14, is presented with a list of demands by Wat Tyler at Mile End June 14 and replies with empty promises to demands for abolition of serfdom, the poll tax, restrictions on labor and trade, and game laws, with a ceiling of fourpence per acre on land rents and a ceiling on road tolls.

Fresh demands are made at Smithfield next day and Wat Tyler is betrayed and killed with a cutlass by William Walworth (*see* 1377), now lord mayor of London, during a conference with the young king.

Similar uprisings occur in Languedoc as France suffers economic distress and the Tuchins stage an uprising against the tax collectors.

Parliament passes the first Navigation Act as England begins her rise as a mercantilist power.

Tamerlane led his Tatar hordes from Turkestan through Persia, Mesopotamia, Afghanistan, and much of India.

French printers at Limoges use movable type (*see* 1041; Antwerp, 1417).

1382 Hungary's Louis the Great dies suddenly at Nagyszombat September 10 after a 40-year reign that also has included nearly 12 years as king of Poland. Louis has gained Dalmatia from Venice and an annual tribute of 7,000 ducats by the peace that last year ended the War of Chioggia. He is succeeded in Hungary by his daughter Maria of Anjou, whose husband Sigismund of Luxembourg will rule Hungary for 50 years beginning in 1387. He will be succeeded in Poland by his daughter Jadwiga (Hedwig), who will be married in 1386 to Jagiello, grand duke of Lithuania, who will rule Poland until 1434 as Vladislov V.

England repeals the reforms granted to Wat Tyler last year and reestablishes serfdom, but the people have lost confidence in the Crown. The balladeered idolatry of the outlaw Robin Hood expresses their bitterness (*see* 1247).

The archbishop of Canterbury William Courtenay, 40, purges Oxford of Lollardy. Courtenay was made archbishop last year following the murder of Simon Sadbury. John Wycliffe has been discredited by last year's peasant rebellion and condemned by the Church; he withdraws to Lutterworth (*see* 1409).

The archbishop of Canterbury destroys academic freedom, separating the Lollard movement from the cultured elite, but Parliament refuses to permit persecution of Wycliffe's followers.

Florence has a revolution as wool combers led by Michele di Lando seize the palace, but the labor government fails when the bourgeoisie objects to assigning Salvestro de' Medici the rents from shops on the 37-year-old Ponte Vecchio. Rival merchants and industrialists exile Michele, close their shops and factories, and

1382 ***(cont.)*** persuade neighboring landowners to cut off the city's grain and food supply; a half-century of oligarchy begins in Florence.

The Black Death sweeps Europe in a weaker epidemic than that of more than 30 years ago. It will take an especially heavy toll in Ireland in the next few years, and by the end of the century it will have killed an estimated 75 million people, leaving some areas completely depopulated (*see* 1403).

1383 Portugal's Fernão I dies without male issue October 22 at age 38 after having antagonized Juan of Castile by canceling the betrothal of his daughter Beatrix to Juan, who lays claim to the Portuguese throne. Fernão's widow Leonora reigns as regent for Beatrix, who marries Juan (*see* 1384).

The Japanese nō drama is pioneered by actor-dramatist Motokiyo Zeami, 20, who began appearing with his father Kan-ami at age 7 and will write some 240 nō plays before his death in 1443, all employing male actors who perform on a bare wooden stage and wear masks to portray women, supernatural beings, or old men. Some of Zeami's nō plays will remain in the repertoire for as long as 600 years (*see* 1418).

1384 Lisbon is besieged by forces sent by Castile's Juan, who has married the Portuguese infanta Beatrix, but the people of Portugal resist his claims to the throne (*see* 1383; 1385).

Flanders passes to Burgundy upon the death of the count of Flanders, whose rebellious followers will be pacified within a year through cruel measures of repression.

John Wycliffe dies December 31 at age 64 following a paralytic stroke, but others will carry on his work (*see* 1382; 1388).

1385 Portugal gains independence August 14 by defeating Castile in the Battle of Aljubarrota. The bastard son of the late Pedro has taken power with unanimous support from the cortes at Coimbra, disease forces the Castilians to withdraw, and João I, 28, will reign until 1433, establishing the Avis dynasty (*see* 1386).

Parliament blocks England's Richard II from setting up a personal government. Richard makes an unsuccessful expedition to Scotland.

Ottoman forces capture Sofia (*see* 1393; 1443).

Heidelberg University has its beginnings in Baden-Württemberg.

1386 The Treaty of Windsor May 9 allies England and Portugal by joining Portugal's João I in marriage with John of Gaunt's daughter Philippa.

The Battle of Sempach July 9 gives the Swiss a victory over an Austrian army, whose commander Leopold III dies fighting at age 34 as the cantons struggle to gain independence from Vienna.

1387 Denmark's Olaf II dies at age 17 after a 12-year reign and is succeeded by his mother Margaret, 34, who will unite Scandinavia under her rule. Daughter of the late Waldemar IV and widow of Norway's late Haakon VI, Margaret has served as regent of Denmark since her father's death, the Norwegians will elect her queen next year, and the disaffected Swedish magnates will help her drive out their king Albert of Mecklenburg (*see* 1389).

Poetry *Troilus and Criseyde* by English poet Geoffrey Chaucer, 47, whose patron, John of Gaunt, last year helped him secure election as one of the two knights of the shire of Kent. Chaucer's *The Book of the Duchess* was an elegy to John of Gaunt's first wife, Blanche, who died in 1369. Chaucer's wife Philippa is a sister of Katherine Swynford, who has married John of Gaunt, but his connections have not sufficed to prevent him from losing his positions as controller of the customs for wool and controller of the petty custom on wines. Chaucer lost both posts last year, but he represents Kent in Parliament and he begins work on the prologue to his great work *The Canterbury Tales* (*see* 1400).

1388 France's Charles VI begins his personal reign at age 19 following the death of the duke of Anjou which leaves Philip of Burgundy in a position of great power. Charles replaces Burgundy with his own brother Louis, duc d'Orléans, but Louis is an unpopular dandy and Burgundy gains the support of the king's wife Isabelle of Bavaria and poses as a reformer.

Venice signs a treaty with the Ottoman Turks in its first effort to assure trading privileges in the eastern Mediterranean in the face of rising Turkish power.

The University of Cologne receives a charter (*see* Thomas Aquinas, 1256).

The first complete English translation of the Bible is completed by John Purcey on the basis of work begun by the late John Wycliffe in an effort to reach the people directly with a bible they can read for themselves (date approximate).

1389 England's Richard II begins his personal rule at age 22 by dismissing from office supporters of the duke of Gloucester and concluding a truce with France's Charles VI.

Margaret of Denmark has her 7-year-old grandnephew Eric of Pomerania proclaimed her successor; he will reign until 1439 as Eric of Norway. She receives an offer of the Swedish throne, her forces defeat the Swedish king Albert of Mecklenburg at Falkoping, and they take Albert prisoner (*see* 1395).

The Battle of Kossovo June 15 ends the Serb Empire as a coalition of Serbs, Bosnians, Albanians, and Wallachians fails to stop the Ottoman sultan Murad I. Murad is assassinated September 16 by a Serbian noble posing as a deserter. Dead at 69, he is succeeded by his eldest son, 50, who will reign until 1402 as Bayazid I (the Thunderbolt).

The new Ottoman sultan (above) has the Serbian prince Lazar captured and put to death. Lazar is succeeded by his son Stephen Lazarevich, but the Serbs become Ottoman vassals.

1390 Scotland's first Stuart king Robert II dies May 13 at age 74. His legitimized son John, 50, changes his baptismal name and will reign until 1406 as Robert III.

Castile's Juan I dies October 9 at age 32 after an 11-year reign in which his claims to the Portuguese throne have been thwarted. He is succeeded by his delicate son of 11, who will reign until 1406 as Enrique III.

The Byzantine emperor John V is deposed briefly by his grandson, who assumes power as John VII but is himself soon deposed as John V is restored to the throne by his second son Manuel.

The Forme of Cury is an illuminated cookbook manuscript containing recipes of dishes prepared for England's Richard II and his barons. Possibly prepared to guide the steward who superintends the king's illiterate cooks and to help him keep track of the costly spices used, the manuscript includes recipes for macaroni adopted from the Italians.

1391 The Byzantine emperor John V Palaeologus dies at age 59 and is succeeded by his able son of 41, who will reign until 1425 as Manuel II Palaeologus.

Tamerlane defeats the khan Toqtamish of the Golden Horde (*see* 1380; 1393).

Seville has a pogrom in June that spreads throughout Andalusia as Spaniards seek scapegoats for the Black Death. Castilian sailors set fire to the Barcelona ghetto August 5 and for 4 days a mob rages out of control, killing hundreds. Many Spanish Jews will accept conversion in the next few years (*see* 1492).

The University of Ferrara has its beginnings.

1392 Japan is reunited after 56 years of civil war between northern and southern dynasties. The southern emperor abdicates in favor of the northern emperor on the understanding that the throne will alternate between the imperial family's two branches, but the northern branch will in fact never relinquish the throne.

The Yi dynasty that will rule Korea until 1910 is founded by the warlord I Songgye, who proclaims himself king after a series of coups d'état and assassinations and makes Kyonsong (Seoul) the capital.

Nearly 400 Hanse vessels embark for London from Danzig during the year with cargoes of grain, honey, salt, potash, furs, and beer.

Playing cards, designed by French court painter Jacques Gringonneur, will be employed for centuries to come in various games. The 52 cards are divided into four suits, each representing one of the four classes of French society: spades stand for pikemen or soldiery, clubs for farmers and husbandmen, diamonds evoke the diamond-shaped hats worn by artisans, and hearts represent the clergy, the word *coeur* evolving from the word *chorus* meaning the clergy (*see* Hoyle, 1742).

1393 Baghdad falls to Tamerlane, whose Tatar horsemen overrun Mesopotamia (*see* 1391; 1398).

The Ottoman sultan Bayazid I subdues Bulgaria.

A German price table lists a pound of nutmeg as being worth seven fat oxen. In Europe 1 pound of saffron is generally equal in value to a plowhorse, 1 pound of ginger will buy a sheep, and 2 pounds of mace will buy a cow.

The Holy Roman Emperor Wenceslas of Bohemia has the ecclesiastic John of Nepomuk, 54, tortured and drowned in the Moldau at Prague for refusing to reveal the confessions of the empress. John will be canonized in 1729 and will be the patron saint of Bohemia.

1394 England's Richard II sets off for Ireland to put down a rebellion.

The Holy Roman Emperor Wenceslas of Bohemia is taken prisoner by his cousin Jobst of Moravia.

New College at Oxford is founded by William of Wykeham, who 15 years ago founded Winchester College.

1395 England's Richard II forces the Irish barons to pay homage to him and grants them amnesty.

The Swedish king Albert of Mecklenburg renounces the throne and retires to Mecklenburg as Margaret of Denmark continues her conquest of his realm (*see* 1389; 1397).

Milan's tyrannical Gian Galaezzo Visconti, 44, assumes the title of duke. His 1387 marriage to Isabelle of Valois will be the basis of French territorial claims to Milan (*see* 1499).

1396 The Crusade of Nicopolis advances along the Danube, pillaging and killing under the leadership of the Hungarian king Sigismund, who is supported by both the Roman and Avignon popes and has enlisted Balkan princes, French, German, and English knights. The 20,000 Crusaders encounter an equal number of Turks 4 miles south of Nicopolis, the Turks overwhelm the Crusaders in a battle September 25 and take many of the survivors prisoner.

France's Charles VI makes a truce with England. It will last for nearly 20 years.

1397 Ottoman forces under Bayazid I lay siege to Constantinople, but the marshal of France Jean Boucicquaut, 31, defends the city. Tatars led by Tamerlane distract Bayazid from the siege (*see* 1398; Battle of Angora, 1402).

Visconti forces from Milan invade Tuscany, but Florence resists the Visconti.

Margaret of Denmark completes her conquest of Sweden; she has her grandnephew Eric crowned king of a united Scandinavia. Her dynastic Union of Kalmar will continue at least nominally until 1523, but Margaret herself will rule as the "Semiramis of the North" until her death in 1412.

Parliament demands that England's Richard II submit a financial accounting. Richard has the parliamentary leader condemned for treason (*see* 1398).

Kyoto's Golden Pavilion (Kinkaku) is completed on the Kitayama estate on the outskirts of town for the retired Ashikaga shogun Yoshimitsu, who has passed his title

1397 ***(cont.)*** on to his son in order that he may live as a monk in the Golden Pavilion.

1398 England's Richard II moves the country toward totalitarian government after executing three dissident lords for treason and packing the House of Commons with his adherents. Richard exiles his cousin Henry of Bolingbroke, 31, son of John of Gaunt, has himself voted a lifetime income by Parliament, delegates Parliament's powers to a committee friendly to his own interests, imposes heavy taxes, and pursues a reign of terror designed to make him absolute monarch (*see* 1399).

Tamerlane leads his Tatar hordes through the passes into northern India after having conquered Persia, Mesopotamia, and Afghanistan. Crossing the Indus September 24 he advances 160 miles in 2 days in early November, overtaking thousands who have fled his approach at Bhatnair. He massacres 100,000 Hindu prisoners at Delhi December 12, sacks Delhi December 17, and moves on to Meerut (*see* 1399).

Merchant Richard Whittington is made lord mayor of London, having grown rich by importing silks, damasks, velvets, and other goods.

1399 Tamerlane storms Meerut January 9 and fights his way back along the Himalaya foothills to the Indus River, which he reaches March 19 after having desolated the kingdom of Delhi.

Geoffrey Chaucer was the first great English poet, but he never completed his *Canterbury Tales*.

John of Gaunt dies at 59, Richard II confiscates his Lancaster estates, and John's son Henry of Bolingbroke returns from exile, landing at Ravenspur in July while Richard is in Ireland. When Richard returns, he is defeated and captured by Bolingbroke and deposed by Parliament, which acclaims the usurper Bolingbroke king. He will reign until 1461 as Henry IV, founding the house of Lancaster.

Richard is imprisoned in the Tower of London and signs a deed of abdication, but rebellions are launched by his supporters, notably the Welsh chieftain Owen Glendower (Owain ap Gruffydd), 40, who proclaims himself prince of Wales and begins a 16-year war.

1400 England's Richard II dies at the Tower of London in February, possibly at the hands of a murderer but more probably of illness in the cold, damp tower. Rumors spread that Richard has escaped and that the body buried without ceremony is that of another man. Supporters of Richard rise in revolt against the Lancastrian Henry IV (*see* 1399; 1402).

The Holy Roman Emperor Wenceslas of Bohemia is deposed for drunkenness and incompetence after a 22-year reign. Wenceslas refuses to accept the decision and will continue to hold the imperial crown for 10 years against the challenges of his rivals Sigismund and Jobst, who will have the support of two rival kings. But the elector palatine of the Rhine, 48, is elected German king August 21 at Rense, Pope Boniface IX will recognize him in 1403, and he will reign until 1410 as the emperor Rupert.

Nonfiction *Chronique de France, d'Angleterre, d'Ecosse et d'Espagne* by French chronicler Jean Froissart, 67, who visited England in 1360, toured Scotland in 1365, traveled to Milan in 1368 with Petrarch and Geoffrey Chaucer, and entered the Church in 1372.

Poetry *The Canterbury Tales* by Geoffrey Chaucer, who dies October 25 at age 60, leaving the work incomplete. The tales will make the poet's name immortal with such Chaucerian phrases as "bless this house," "every man for himself," "through thick and thin," "love is blind," "brown as a berry," "pretty is as pretty does," "it is no child's play," "murder will out," and "for in the stars is written the death of every man."

London's population reaches 50,000, but no other English city is as populous as Lübeck or Nuremberg, each of which has 20,000, much less as big as Cologne, which has 30,000.

15th Century

Improvements in agricultural technology will have a powerful effect on conditions of European farm labor in this century. Labor will be emancipated in the economically advanced nations of Tuscany, Lombardy, and the Low Countries, further enslaved in the less-advanced nations.

The rise of capitalism in the first half of this century will begin a 3-century advance in Europe's agricultural technology.

The Hanseatic League will lose its trade to Dutch fishing interests in this century as copepod crops fail in the Baltic, diminishing fish populations that feed on these tiny crustaceans.

Inca terraces *(andanes)* in this century will turn steep slopes in the Andes of the Western Hemisphere into arable land. Soil and topsoil will be carried up from the valleys below.

1401 Tamerlane sacks Baghdad, slaughtering thousands.

England increases the power of the Church over heresy, principally Lollardy, by the statute *de Heretico Comburendo*, first English law of its kind.

1402 The Battle of Angora (Ankara) July 28 ends in victory for Tamerlane over the Ottoman sultan Bayazid I, whose army of 120,000 is no match for Tamerlane's 800,000 warriors. Bayazid is deserted by 18,000 Tatars and captured (*see* 1403). Tamerlane restores many Turkish emirs, takes Smyrna by assault after a 2-week siege, and massacres the inhabitants.

Milan's Gian Galeazzo Visconti dies of plague September 3 while besieging Florence. His death at age 51 saves Florence but plunges Milan into anarchy as his son Giovanni Maria, 13, is proclaimed duke while his son Filippo Maria, 10, is made nominal head of Pavia (*see* 1395; 1412).

A Scottish invasion of England is stopped at Homildon Hill in Northumberland by Sir Henry Percy, 60, and his brother Sir Thomas Percy, 58, who support England's Henry IV (*see* 1403).

1403 China's Ming emperor Zheng Zu, 44, begins a 21-year reign with the dynastic title Yung Lo. His nephew disappeared in a palace fire last year, great violence marks the start of the new reign, uneasiness persists that the deposed nephew will reappear, but the Yung Lo reign will bring progress to China.

The captive Ottoman sultan Bayazid I dies in March at age 43 in Tamerlane's camp en route east. A 10-year interregnum begins as Bayazid's six sons vie for power.

England's Henry IV subdues Percy of Northumberland, who has joined the Welsh revolt of Owen Glendower that began in 1399.

The Hanseatic League gains complete control of Bergen, Norway, and enjoys a virtual monopoly in most of the commodities produced by northern Europe including fish, the salt used for curing fish, whale oil, pitch and rosin employed in shipbuilding and maintenance, and eiderdown (*see* 1406).

The doge of Venice imposes the world's first quarantine as a safeguard against the Black Death. All who wish to enter the city must wait, and the waiting time will be standardized at 40 days in 1485 (*see* 1374).

1405 Tamerlane dies suddenly at Atrar February 17 at age 68 while planning a campaign against China. The Tatar emperor has conquered Persia, Mesopotamia, Afghanistan, and much of India, but his empire quickly begins to dissolve.

Venetian forces defeat an army mounted by the Carrara family and seize Padua, Verona, Vicenza, and other domains of the Carraras and the Viscontis, whose leader Gian Galeazzo Visconti of Milan died in 1402.

Florence buys Pisa to get direct access to the sea.

French troops land in Wales to support the rebellion of Owen Glendower. The archbishop of York makes his own break with England's Henry IV.

The Chinese emperor Yung Lo orders the first Chinese sea expedition. A fleet of 63 junks carrying some 28,000 men sails for islands to the south under the command of the Muslim eunuch Ma, who calls himself Zheng He (Cheng Ho) (*see* 1407).

The University of Turin has its beginnings.

1406 Scotland's Robert III suspects complicity by the country's real ruler Robert, duke of Albany, in the mysterious death of the king's elder son David in 1402. He sends his son James, 11, to France for safety, English sailors capture James, news of the capture reaches Robert, and he dies April 4 at age 65. James I will remain in English hands until 1424.

England's French territories are attacked by Louis, duc d'Orléans, who is married to Valentina Visconti of Milan.

Castile's Enrique III (el Doliente, the Sufferer) dies Christmas Day at age 27 after a despotic reign of 16 years in which he has resolved rivalries between descendants of Pedro the Cruel and Enrique II Trastámara by marrying the granddaughter of Pedro. He is

1406 ***(cont.)*** succeeded by his infant son, who will reign until 1454 as Juan II.

London names Richard Whittington lord mayor again, and he loans money to Henry IV as he previously loaned money to Richard II (*see* 1398). Whittington will be the subject of a Mother Goose rhyme (*see Contes de ma mère l'oie,* 1697).

Hanseatic fishermen catch 96 English fishermen fishing off Bergen (*see* 1403). They bind the Englishmen hand and foot and throw them overboard to drown.

1407 France's Duc d'Orléans is assassinated November 23 at age 35 by partisans of the Burgundian John the Fearless, and civil war is precipitated between Burgundians and Armagnacs. The latter are named for the count of Armagnac, father-in-law of the new duc d'Orléans, and are a reactionary, anti-English war party supported in the south and southeast, while the Burgundians are pro-English, favor peace, support the antipope Benedict XIII, and are themselves supported by the people, the University of Paris, and the Wittelsbachs.

Zheng He returns to China with the prince of Palembang (Sumatra), who has been defeated in battle and placed in chains (*see* 1405; 1408).

The Black Death kills thousands in London (*see* 1499).

1408 Bohemian preacher Jan Hus, 35, is denounced by other clergymen who complain to the archbishop of Prague about the strong language used by Hus in reference to the sale of indulgences and other clerical abuses. Hus is stripped of his appointment as synodal preacher and forbidden to exercise priestly functions, but he is defended by the populace.

The Chinese admiral Zheng He embarks on a second great expedition from which he will return with the king of Ceylon and the Sinhalese royal family which has dared to attack He's mission (*see* 1407; 1412).

1409 Venice recovers its territories on the Dalmatian coast.

Pope Alexander V promulgates a bull ordering the surrender of all books by England's John Wycliffe (*see* 1382). Bohemia's archbishop publicly burns some 200 of Wycliffe's writings.

Bohemia's archbishop excommunicates Jan Hus along with several of his friends, who will appeal to the successor of Pope Alexander V. Hus continues to preach at the Bethlehem chapel and begins to defend in public the so-called heresies of John Wycliffe (*see* 1408; 1411).

The Council of Pisa assembles 500 prelates and delegates from all over Europe to end the schism between Rome and Avignon that has persisted since 1378. The conclave hears charges against Pope Gregory XII at Rome and the antipope Benedict XIII at Avignon. It declares both of them deposed, it elects Peter Philarges pope June 26, and he begins a brief papacy as Alexander V.

Leipzig University has its beginnings in a college founded by émigrés from Prague.

Sculpture *David* (marble) by Florentine sculptor Donatello (Donato di Noccolo di Betto Bardi), 23.

1410 The Holy Roman Emperor Rupert dies at Landskron near Oppenheim May 18 at age 58. He is succeeded by Sigismund of Luxembourg, 42, a brother of the deposed emperor Wenceslas of Bohemia. Sigismund's rival Jobst of Moravia will die next year; Sigismund will be crowned at Aix-la-Chapelle late in 1414 and will reign until 1437.

The Battle of Tannenberg July 15 gives a huge army of Poles and Lithuanians victory over the Teutonic Knights, but the victors are unable to exploit their triumph.

A translation into Latin of the *Geography* by the 2nd century astronomer-mathematician-geographer Ptolemy revives the notion that the world is round (*see* Columbus, 1474).

Pope Alexander V dies at Bologna sometime before dawn May 4, possibly the victim of poisoning. He is succeeded by the Neapolitan cardinal Baldassare Cossa, who is elected May 17 at Bologna and begins a 5-year papacy as John XXIII.

1411 The Peace of Thorn signed February 1 ends the Slavic advance but fails to give Poland access to the Baltic and costs the Teutonic Knights only the Lithuanian territory of Samogitia and an indemnity, despite their crushing defeat last year at Tannenberg.

Portugal and Castile make peace after 26 years of hostilities, and Portugal begins her rise as a great world power.

The Ottoman prince Musa enlists Serbian support to attack Suleiman June 5 at Edirne. Suleiman is defeated and killed, but Musa alienates his supporters with his radical policies. The Serbs ally themselves with Musa's brother Mohammed (*see* 1403; 1413).

A new ban is pronounced in March on Prague's Jan Hus, but he continues to preach in defense of the treatises written by John Wycliffe (*see* 1409; 1414).

A Celestine monastery is founded at Vichy by France's Charles VI.

The University of St. Andrews is founded in Scotland.

1412 Margaret of Denmark dies suddenly October 28 at age 59 aboard her ship in Flensborg harbor. Her grandnephew Eric of Pomerania continues his reign as Eric VII of Denmark and Norway, Eric XIII of Sweden, but the death of the "lady king" (as delegates from Lübeck have called her) begins a long period of dissension as Eric assumes personal power and begins an oppressive rule.

Milan's Gian Maria Visconti dies at the hands of an assassin. His brother Filippo Maria is left to rule alone and will rule the duchy until his death in 1447.

The Chinese admiral Zheng He leaves on a third expedition that will take him as far as the Hormuz Straits, gateway to the Persian Gulf (*see* 1408; 1416).

The University of Turin opens in Piedmont.

1413 England's Henry IV dies the evening of March 20 at age 45 and is succeeded by his son, 25, who will reign until 1422 as Henry V, raising England to the rank of a major European power.

Paris butchers led by skinner Simon Caboche seize Paris and try to make the government more efficient by imposing the Ordonnance Cabochienne. The Armagnacs soon regain control and revert to feudal reaction, ending all hope for reform.

The Ottoman Empire's 10-year civil war ends as Mohammed defeats and kills his brother Musa outside Constantinople. The Byzantine emperor Manuel II holds a fourth brother, Mustafa, hostage as Mohammed reunites the empire's possessions and begins an 8-year reign (*see* 1414).

Sculpture *St. Mark* by Donatello.

1414 The Ottoman sultan Mohammed I defeats a Karamanid army and restores Adrianople's power over the emirs of Anatolia.

A Lollard plot against England's Henry V comes to light, authorities seize a group of conspirators, and most are hanged.

Cologne expels its Jews.

The Council of Constance is assembled by Pope John XXIII at the insistence of Hungary's King Sigismund, the Holy Roman Emperor who wants the Church's unity restored. The purpose of the council is also to reform the Church's leadership and its members and to extirpate heresy, especially that of Jan Hus (*see* 1411), who has arrived under an imperial safe conduct that permits his free return to Bohemia whatever judgment may be passed upon him.

1415 The Battle of Agincourt October 25 ends in defeat for the French at the hands of English archers, the lowest caste in the military hierarchy. Commanded by Henry V, who has been delayed by torrential rain in his march from Harfleur to Calais, 6,000 archers, 1,000 men-at-arms, and a few thousand foot soldiers encounter a French force of 25,000 under the command of Charles d'Albret, the constable of France whose knights have not learned the lessons of Crécy and Poitiers (*see* 1346; 1356).

Henry's archers plant long pointed stakes in their midst at Agincourt (above), the French cavalry is impaled or trapped in the mud wearing heavy armor, Henry orders that prisoners be killed since he does not have enough men both to guard the prisoners and attack the foe, and the slaughter ends only when the French withdraw.

The constable of France Charles d'Albret dies in battle at Agincourt (above) along with three dukes, five counts, 90 barons, and 5,000 knights of noble birth. One thousand prisoners are taken including Duc Charles d'Orléans, 24, who will not be ransomed until 1440. English casualties number only 13 men-at-arms (including the duke of York) and 100 or so foot soldiers. France's nobility is shattered, the feudal system discredited, and Normandy lies open to reconquest by the English.

The Battle of Agincourt gave England's Henry V a great triumph over the knights of France.

Welsh chieftain Owen Glendower is defeated by the English after 16 years of rebellion and is pardoned.

News that Jan Hus has been savaged in violation of his imperial safe conduct produces a surge of Bohemian nationalism combined with demands for religious reform that will split the Holy Roman Empire (*see* 1420).

The Council of Constance unanimously condemns the writings of England's John Wycliffe May 5 and demands that Jan Hus recant in public his "heresy." Hus refuses and is burned at the stake July 6.

The Very Rich Book of Hours of the Duke of Berry is prepared for the first duc de Berry Jean de France, 75, with illustrations by Belgian painter Pol Limburg and his brothers Hermann and Jan.

Sculpture *St. John the Evangelist* by Donatello.

1416 England's Henry V begins a 3-year move across Normandy in which he will take Caen, Bayeux, Lisieux, Alençon, Falaise, and Cherbourg.

The first war between Venice and the Ottoman Empire is won by the doge Loredano, who defeats the Turks at the Dardanelles and forces the sultan to conclude peace.

A Chinese fleet under Zheng He reaches Aden (*see* 1412; 1421).

Sculpture *St. George* by Donatello.

1417 Merchants of the Hanseatic League agree not to buy wheat before it is grown, herring before it is caught, or cloth before it is woven. The League regulates city tariffs and prices to keep supplies of grain and meat cheap for townspeople even at the expense of peasants.

1417 *(cont.)* The Great Schism that has divided the Church since 1378 ends November 11. The Council of Constance, having deposed Benedict III, Gregory XII, and John XXIII, elects Ottone (Otto) Colonna, 49, who will reign until 1431 as Pope Martin V.

Printers at Antwerp use movable type (*see* Limoges, 1381; Haarlem, 1435).

1418 Rouen falls to England's Henry V December 31 after a 2-month siege in which the English have starved out a town as large as London.

The Japanese nō drama pioneered in the last century is taken up by the upper classes with support from the Ashikaga shogun Yoshimitsu. Motokiyo Zeami, now 55, has formalized the symbolic drama originally played by farmers in the 11th century (*see* 1383; Kabuki, 1603).

1419 England's Henry V conquers all of Normandy except Mt. St. Michel by July. He takes Rouen, and Burgundy's John the Fearless is murdered September 10 at Montereau after a meeting with the dauphin.The Burgundians return to the English alliance.

Portuguese explorers land in the Madeira Islands in the Atlantic off North Africa (*see* Henry the Navigator, 1421).

The Decameron completed by Giovanni Boccaccio in 1353 is published for the first time at Venice.

The Doge's Palace at Venice dating to the 9th century and rebuilt in later centuries receives a new facade that will endure for more than 580 years. The palace will be extended toward the Piazetta in 1424 and the Foscari Arch will be completed in 1471 (*see* 1520).

Sugar cane from Sicily is planted in the Madeira Islands (above) (*see* 1456; Columbus, 1493).

1420 An army of knights, mercenaries, and adventurers summoned by Pope Martin V to a crusade against heretics is defeated at Prague by a small force of Bohemian peasants and laborers led by the blind veteran Jan Zizka, 60 (*see* 1415; 1431).

England's Henry V signs a peace treaty May 21 with Philippe, the new duke of Burgundy, putting England and France under one crown. He agrees to marry Catherine of Valois, daughter of France's Charles VI, and enters Paris in triumph December 20.

Florence makes vain attempts to put a 20 percent ceiling on interest rates charged by Florentine bankers, especially on loans to the *popolo minuto*.

Painting *The Crucifixion* and *The Last Judgment* by Flemish painter Jan van Eyck, 35, and his brother Hubert, 50. They have pioneered in using oil paint on wood to achieve brilliant colors.

1421 Florence buys Livorno (Leghorn) and establishes the Consuls of the Sea.

Milan's Filippo Maria Visconti subjugates Genoa as he works to rebuild the duchy that broke up into city-states following the death of his father in 1402.

Representatives of Bohemia and Moravia meet at Caslav June 1, renounce the Emperor Sigismund, and found a government of their own.

The Ottoman sultan Mohammed I dies at age 34 after an 8-year reign in which he has consolidated the empire. He is succeeded by his son, 18, who will reign until 1451 as Murad II, extending the empire into southeastern Europe (*see* 1432).

The Chinese Ming emperor Zheng Zu (Yung Lo) moves his capital from Nanjing to Beijing as he continues to reform local governments and attempts to establish trade with islands to the south.

The emperor Zheng Zu (above) sends Zheng He on a fifth expedition abroad (*see* 1416; 1431).

The Portuguese prince Henry the Navigator, 27, assembles Europe's leading pilots, mapmakers, astronomers, scholars, and instrument makers at Sagres on the Cape St. Vincent, where they will pioneer a new science of navigation. Son of João I and grandson of John of Gaunt, Henry has his shipwrights develop a lateen-rigged caravel with three masts—a highly maneuverable vessel able to stand up to the winds of the open sea (*see* Madeira Islands, 1418; Canary Islands, 1425).

The North Sea engulfs more than 70 Dutch villages. Upwards of 100,000 die as the shallow Zuider Zee spreads over thousands of square miles.

1422 England's Henry V dies of dysentery at Vincennes August 31 at age 35, protesting that he wants to live so that he may rebuild the walls of Jerusalem. His will names his brother Humphrey, 41, duke of Gloucester, as protector of his 9-month-old son, who will reign until 1460 as Henry VI. Richard de Beauchamp, 40, earl of Warwick, is appointed the boy's preceptor.

France's Charles VI dies at Paris October 1 at age 53 after a 42-year reign. The Treaty of Troyes signed in 1420 disinherited the dauphin, now 19, and he will not be crowned until 1429. England's Henry VI is proclaimed king of France.

England resumes war with France under the leadership of the king's uncle John of Lancaster, 33, duke of Bedford, who will rule much of France in the absence of the dauphin and who will rule England in the minority of Henry VI through his brother Humphrey, duke of Gloucester.

Lisbon becomes Portugal's seat of government.

France's University of Besançon has its beginnings.

1423 Venice buys Thessalonica from Constantinople under terms of an agreement with the Byzantine emperor Manuel II to keep the Ottoman Turks from taking the city (but *see* 1430).

Scottish leaders sign a treaty at York in September undertaking to pay 60,000 marks for the "maintenance in England" of their uncrowned king James I, now 29, and agree to the marriage of James to Jane, daughter of John Beaufort, earl of Somerset.

Painting *The Adoration of the Magi* by Italian painter Gentile da Fabriano (Gentile di Niccolo di Giovanni Massi), 53, who settled last year at Florence after having worked at Venice.

Sculpture *St. Louis* by Donatello.

1424 Count Giacomuzzo Sforza drowns in the Pescara River January 4 at age 54 while engaged in a military expedition against the Spanish near Aquila. His bastard son Francesco, 23, will carry on the Sforza name adopted by his father in the field and become duke of Milan in 1450.

Scotland's James I gains his freedom at age 29 after nearly 18 years in English hands. The earl of Somerset remits 10,000 marks of ransom money as a dowry for his daughter Jane; she and James are married at Southwark February 12. James is crowned at Scone May 21, and he will reign until 1437.

England's duke of Bedford soundly defeats a combined force of French and Scottish troops August 17 at Verneuil, and his brother Gloucester invades Hainaut in October.

De Imitatione Christi (Imitation of Christ) by German ecclesiastic Thomas à Kempis, 44, says, "Man proposes, but God disposes." "And when a man is out of sight, quickly also is he out of mind." *Sic transit gloria mundi* (So passes away the glory of this world). No book except the Bible will be more widely read in Europe for more than a century.

Bronze doors designed by Florentine painter-goldsmith Lorenzo Ghiberti (Lorenzo di Clone di Ser Buonaccorso), 46, with reliefs depicting scenes from the life of Christ are installed as the north portals of the baptistery of San Giovanni at Florence. Ghiberti has worked for 21 years on the doors and is commissioned to create the east portals that he will complete in 1447.

1425 The Canary Islands less than 70 miles off the northwest coast of Africa fall to Portugal's Henry the Navigator, who wrests them from Castile. The Canaries are seven volcanic peaks that rise out of the Atlantic to snow-covered peaks as high as 12,250 feet (*see* 1496; Henry, 1421).

The Byzantine emperor Manuel II Palaeologus dies at age 75 after a 34-year reign. He has recently been forced to pay tribute to the Ottoman sultan Murad II, and he is succeeded by his son, 35, who will reign until 1448 as John VII Palaeologus.

Poetry *La Belle Dame sans Merci* by French writer-diplomat Alain Chartier, 40.

Sculpture *Habbakuk (Zuccone)* by Donatello in gilt bronze for the campanile of the Florence cathedral.

1426 The duke of Bedford returns from France, reaching London in January. He concludes an alliance with his brother Gloucester, effects a reconciliation between Gloucester and the bishop of Winchester Henry Beaufort, who is chancellor of England, and knights the infant king Henry VI.

Venice goes to war with Milan's Filippo Maria, beginning 3 years of hostilities that will end with Venice gaining control of Verona, Vicenza, Brescia, Bergamo, and Crema.

Louvain University has its beginnings.

Painting *Virgin Enthroned* by Florentine painter Masaccio (Tomasso di Giovanni di Simoe Cassai), 24, for the altarpiece of Pisa's Church of the Carmine.

1427 The duke of Bedford leaves in March to resume the war in France after promising to act in accord with the council. He has ordered his brother Gloucester to desist from further attacks on Hainaut and tries to restore prosperity to the French districts under his rule by reforming the debased coinage, removing various abuses, and granting privileges to merchants and manufacturers.

Sculpture *The Feast of Herod* by Donatello is a relief for the baptistery of Siena.

1428 The Treaty of Delft ends hostilities between England and Flanders, but English forces besiege Orléans in central France beginning in October with the reluctant consent of the duke of Bedford.

Venetian forces under the condottiere Carmagnola conquer Brescia and Bergamo.

Vietnam regains her independence from China's Ming Empire.

Japanese transport workers strike in protest against high prices as famine cripples the country. The strikers are joined by farmers who riot in the streets and wreck warehouses, temples, private houses, and sake production facilities.

The bones of England's late John Wycliffe are disinterred 44 years after his death and are burned by order of the Council of Constance.

The University of Florence begins to teach Greek and Latin literature with special emphasis on history and its bearing on human behavior and moral values.

Painting *Merode* altarpiece by Flemish painter Robert Campin, 50, the Master of Flemalle.

Florence's Church of San Lorenzo is completed by Filippo Brunelleschi after 7 years of construction.

1429 Joan of Arc (Jeanne d'Arc) becomes the heroine of France and changes the course of history. A shepherd girl of 17 from Lorraine, La Pucelle d'Orléans has heard voices and has seen visions of the Archangel Michael, Saint Catherine, and Saint Margaret, who pledged her to seek out the dauphin and deliver Orléans from the English. She persuades a royal captain to equip her with armor and have her conducted to Chinon, she finds the dauphin there, he provides her with a small army, and she liberates Orléans in early May.

The duke of Bedford sends to London for reinforcements, strengthens his hold on Paris, visits Rouen to bind the Normans closer to England, but assigns the French regency to Philip of Burgundy in compliance with the will of the Parisians.

1429 ***(cont.)*** Joan of Arc (above) persuades the dauphin that he is the legitimate son and heir of the late Charles VI despite the Treaty of Troyes that disinherited him in 1420. He is crowned at Reims July 17 and will reign until 1461 as Charles VII.

Joan of Arc has a standard embroidered for her bearing the *fleur-de-lys* and the words *Jesus Maria*.

The Order of the Golden Fleece is founded in Burgundy. It will become a Hapsburg order beginning in 1477.

Cosimo de' Medici becomes head of the Florentine banking house at age 40 upon the death of its founding father Giovanni. "I leave you with a larger business than any other in the Tuscan land," writes Giovanni, and he advises his son Cosimo the elder to "be charitable to the poor," belong to the popular political party, "speak not as though giving advice," "avoid litigation," and "be careful not to attract public attention."

The Grocers' Company is formed in London to succeed the spicers' guild of 1180. Henry VI will grant the company a charter to sell wholesale—*vendre en gros*, source of the word *grocer*—and manage the trade in spices (now used widely in medicine), drugs, and dyestuffs.

1430 Joan of Arc enters Compiègne outside Paris May 23 and is taken prisoner along with her brothers. She is delivered to Jean de Luxembourg and thence to the English, who imprison her in a tower at Rouen, intending to discredit her. Charles VII makes no effort to save Joan, nor is any interference made by the duke of Bedford (*see* 1431).

Thessalonica is taken by the Ottoman Turks under Murad II and begins nearly 500 years of Turkish rule (*see* 1423; 1913; Sofia, 1385).

Sculpture *David* (bronze) by Donatello.

Painting Mosaics for Venice's Church of San Marco by Italian painter Paolo Uccello (Paolo di Dono), 33.

1431 Joan of Arc is handed over to the former bishop of Beauvais Pierre Cauchon by the English, who vow to seize her again if she is not convicted of high treason against God. Duly condemned by an ecclesiastical court, the maid of Orléans is burned at the stake in the Old Market Square of Rouen May 30.

England's Henry VI, now 10, is crowned at Notre Dame in Paris December 16.

Bohemian peasants defeat another army of knights and adventurers at the town of Tabor (*see* 1420; 1434).

Zheng He leads a final expedition that will reach 20 states. He will exact tribute from 11, including Mecca (*see* 1421; 1433).

The French university of Poitiers has its beginnings.

1432 The Italian condottiere Carmagnola, conte di Castelnuovo, is convicted of treason by Venice's Council of Ten after leading an unsuccessful campaign against the Visconti forces of Milan whose duke he served from 1416 to 1423. The doge intercedes in his behalf, but Carmagnola is beheaded April 5 at age 41.

Constantinople withstands a siege by the Ottoman sultan Murad II, who withdraws after a stubborn defense by the Byzantine emperor John VII Palaeologus.

The Azores are discovered off the west coast of Portugal by Portuguese mariner Gonzalo Cabral.

Painting *Reform of the Carmelite Rule* (fresco) by Italian painter Fra Filippo Lippi, 26, who leaves the Florentine monastery in which he has lived since being orphaned in childhood and goes to Padua; *The Adoration of the Mystic Lamb* (24 panels) by Jan van Eyck, who completes a work begun in 1420 by his late brother Hubert, who died in 1426.

1433 Portugal's João I dies August 14 at age 76 after a 48-year reign. He is succeeded by his son Duarte (Edward), 40, who will reign until 1438.

Timbuktu falls to the Berbers.

Bohemia's Hussite Wars end after 13 years of hostilities between peasant armies and papal "crusaders." Hussites called to the Council of Basel will accept a compromise, but Bohemian nationalism has been asserted and the country has broken forever from its German ties.

Florence is defeated after a 4-year war on Lucca; Giovanni de' Medici's son Cosimo, imprisoned as a scapegoat, is sentenced to 10 years' exile (but *see* 1434).

Zheng He returns from a seventh and final expedition that has taken his fleet to 20 states. Mecca and 10 other states send tribute to the Ming emperor, who is presented with giraffes, zebras, and other curiosities. China's brief show of interest in the outside world will now end.

Africa's Cape Bojador south of the Canary Islands is rounded for the first time by Portuguese navigator Gil Eannes (or Gillianes, or Guillanes), a captain of Prince Henry who has sailed out of the port of Lagos, Portugal.

Chinese war junks commanded by the eunuch general Zheng. He subdued countries as far away as the Persian Gulf.

"De Concordantia Catholics" by the German Roman Catholic priest Nikolaus von Cusa, 32, gives the Council of Basel (above) a carefully reasoned defense of its authority but leaves open the question of whether ultimate power vests in the pope or in the council which has cited Pope Eugenius IV for failing to appear but whose deposition of the pope will have no practical effect.

Painting *Man in a Red Turban* by Jan van Eyck.

1434 Florence recalls Cosimo de' Medici from exile, he will rule the city until his death in 1464, and his family will continue to dominate the city for 30 years thereafter (*see* 1433; 1440).

Poland's Jagellon king Ladislas (Vladislav) V dies at age 84 after a 38-year reign that has seen his country become a great power. He is succeeded by his son of 10, who will reign until 1444 as Ladislas VI.

The radical Bohemian priest Prokop the Great leads his Taborites into battle at Lipany but meets defeat as civil war wracks the country.

The Swedish peasant leader Engelbrecht Engelbrechtsen marches through eastern and southern Sweden, seizing castles and driving out bailiffs in a revolt that spreads to Norway.

African slaves introduced into Portugal by a caravel returning from the southern continent are the first of millions that will be exported in the next 4 centuries (*see* 1441).

Painting *Giovanni Arnolfini and His Bride* by Jan van Eyck.

Venice's Ca' d'Oro is completed in a brilliant blend of Veneto-Byzantine and Gothic styles.

1435 The duke of Bedford dies in France September 19, leaving England without a strong hand on the Continent. France's Charles VII and Burgundy's Philippe le Bon sign a treaty at Arras September 21 by which Philippe breaks with the English, recognizes Charles as France's only king, and is himself recognized as a sovereign prince. Charles promises to punish the murderers of Philippe's father, but England refuses to make peace on terms acceptable to the French.

Naples is conquered by Aragon's Alfonso V (the Magnanimous), who ends the 21-year reign of Joanna (Giovanna) II, reunites Naples and Sicily, and makes Naples the center of a Mediterranean Aragonese empire (*see* 1442).

Sweden's diet recognizes the claims of Engelbrecht Engelbrechtsen and elects him regent for the weak King Eric, who has lost Schleswig to the dukes of Holstein.

Movable type is employed by Dutch printers at Haarlem (*see* Antwerp, 1417; Mainz, 1454).

Painting *The Descent from the Cross* by Flemish painter Roger van der Weyden, 35, on commission from the Louvain archers' guild. He is appointed city painter at Brussels; Swiss painter Konrad Witz, 35, becomes a citizen of Basel and is commissioned to paint the *Mirror of Salvation* altar (*Heilsspiegelaltar*); *The Vision of St. Eustace* by Italian painter Pisanello (Antonio Pisano), 40; *The Adoration of the Child* by Fra Filippo Lippi.

1436 France's Charles VII recovers Paris from the English as Scottish forces defeat the English near Berwick.

The Compact of Iglau ends the Hussite Wars as all parties agree to accept the Holy Roman Emperor Sigismund as king of Bohemia.

Della Pittura by Italian architect-scholar Leon Battista Alberti, 32, is the first literary formulation of the aesthetic and scientific theories embodied in Renaissance paintings.

Sculpture *Jeremiah* by Donatello at Florence.

Florence's Duomo (Cathedral of Santa Maria del Fiore) is completed after 140 years of construction with the installation of a 350-foot high dome designed either by Filippo Brunelleschi, now 59, or the 13th century sculptor Arnolfo di Cambio.

England permits the export of wheat and other corn without a state license when prices fall below certain levels (*see* 1463).

1437 Scotland's James I remains at Perth for 6 weeks despite warnings of danger and is stabbed to death February 10 at age 42 by Sir Robert Graham, a man James had imprisoned and then banished. Walter Stewart, earl of Atholl, has instigated the attack and is tortured and executed along with Graham. The king's 5-year-old son is crowned at Holyrood in March and will reign until 1442 as James II amidst civil war.

Portuguese forces suffer a crushing defeat at Tangier at the hands of the Moors. The king's youngest brother Fernando, 35, has urged the crusade. The Moors make the Portuguese promise to return Ceuta, Fernando offers himself as a hostage, but Portugal will not return Ceuta, and Fernando will die in the dungeons of Fez after 5 years of cruel punishment.

The Holy Roman Emperor Sigismund dies December 9 at age 69, and the house of Luxembourg becomes extinct; Sigismund will be succeeded as German king by his son-in-law Albrecht of Hapsburg (*see* 1438).

The University of Caen has its beginnings in France.

1438 The Inca dynasty that will rule Peru in the Western Hemisphere until 1553 is founded by Pachacutec (*see* 1525).

The Austrian duke Albrecht V of Hapsburg is elected German king to succeed his late father-in-law Sigismund (*see* 1437). Albrecht, 40, is also crowned king of Hungary and Bohemia but will not be able to obtain possession of Bohemia and will spend most of his brief reign defending Hungary against the Ottoman Turks.

Scotland and England conclude a truce that will last until 1448.

Eric VII, king of Denmark, Norway and Sweden since 1412, flees peasant rebellions, takes refuge on the

1438 ***(cont.)*** Swedish island of Gotland, and turns pirate, preying on Baltic shipping.

Portugal's Duarte I dies of plague December 9 at age 47 with his brother Fernão still unransomed. Duarte's 6-year-old son inherits the crown and will reign until 1481 as Afonso V (Afonso o Africano) with his uncle Pedro as regent until 1449.

A Pragmatic Sanction issued by France's Charles VII limits papal authority over French bishops and gives the king a voice in clerical appointments (but *see* 1461).

1439 The Hapsburg German king Albrecht II dies at Langendorf October 27 at age 42 after a reign of less than 2 years spent mainly in defending Hungary against the Ottoman Turks (he has also been king of Bohemia and Hungary and the uncrowned Holy Roman Emperor). Albrecht has led an army to the gates of Baghdad but has lost much of his force to dysentery which hastens his own end.

The death of Frederick of Tyrol, Hapsburg cousin of Albrecht II, leaves Frederick's 24-year-old ward and nephew the senior member of the Hapsburg family (*see* 1440).

Albrecht II (above) will be succeeded as king of Hungary and Bohemia by his posthumous son who will be born February 22 of next year, crowned less than 3 months later, and reign until 1457 under the guardianship of his uncle Frederick as Ladislas Posthumus.

A nephew of the Scandinavian king Eric VII assumes the throne; he will reign until 1448 as Christopher III.

The Ottoman sultan Murad II annexes Serbia and forces the Serbian despot George Brankovich, now 72, to take refuge in Hungary.

The Pragmatic Sanction of Mainz leaves the German church under imperial and princely control and postpones any reform in the church (*see* 1440).

The Council of Florence reaffirms the union between Constantinople and Rome. The Byzantine emperor John VIII Palaeologus accepts papal primacy.

Painting *Margaret van Eyck* by Jan van Eyck.

Sculpture Marble reliefs for the altar of St. Peter in the great cathedral at Florence by Florentine sculptor Luca della Robbia, 39.

1440 German electors at Frankfurt vote February 2 to make the Hapsburg duke of Styria and Carinthia successor to the late Albrecht II (*see* 1439; 1442).

The elector of Brandenburg Frederick I, who has brought the house of Hohenzollern to greatness, dies at Kadolzburg September 21 at age 48 shortly after declining an offer to accept the Bohemian throne. The king of Poland Ladislas VI has accepted the Hungarian crown in March and begun a 4-year reign in July as Ladislas VI while the posthumous son of the late Albrecht II remains under the protection of his uncle Frederick III.

The Battle of Anghiari gives the condottiere Niccolo Piccinino, 65, overlord of Bologna, a victory over the forces of Milan's Filippo Maria Visconti. Florence's Cosimo de' Medici backs the claims of Francesco Sforza to the duchy of Milan and supports a coalition of Venetians and Florentines who defeat Piccinino.

Harfleur falls to English forces under the duke of Somerset John Beaufort, 37, and his brother Edward. A peace mission to France led by Cardinal Henry Beaufort, now 67, has no more success than did his similar mission last year.

Cosimo de' Medici (above) institutes a progressive income tax to lighten the burden on the poor of Florence who support the Medici family.

Florence's Platonic Academy is founded.

England's Eton School is founded by Henry VI. The King's College of Our Lady of Eton beside Windsor will become the largest of the ancient English public schools (*see* Winchester, 1379).

Painting *Christ Appearing to His Mother* by Roger van der Weyden; *The Last Judgment* by German painter Stefan Lochner, 30; Florentine painter Andrea del Castagno, 17, is commissioned to paint a fresco on the façade of the Palazzo del Podesta. The fresco shows men hanged as traitors and Andrea will be nicknamed "Andrew of the Hanged Men" (*Andreino degl'impiccati*).

1441 African slaves are sold in the markets of Lisbon, and a trade begins that will see more than 20 million Africans transported in the next 460 years to Europe and—more especially—to the New World (*see* 1511).

The University of Bordeaux has its beginnings.

Jan van Eyck dies at Bruges, where he has developed the new art of secular portrait painting.

1442 Aragon's Alfonso V is crowned king of Naples June 12 (*see* 1435).

Austria's Hapsburg archduke Frederick III is crowned German emperor July 17 at Aix-la-Chapelle more than 2 years after being chosen at Frankfurt. Now 26, he will reign until 1493 as Frederick III (*see* 1452).

England loses all her Gascon territory except Bordeaux and Bayonne to the French.

Hungarian forces under the governor of Transylvania János Hunyadi, 55, rout the Ottoman forces of Mezid Bey, who has invaded Transylvania, and vanquish an army sent to avenge the bey's defeat.

1443 Christian forces take Sofia and Nish from the Ottoman Turks. Ladislas (Vladislav) VI, 19, of Poland and Hungary, János Hunyadi, and Serbia's George Brankovich lead the attack, but the Ottoman sultan Murad II stops the Christians in a Balkan pass at the Battle of Zlatica (Izladi).

Albania's governor George Castriota (Skanderbeg), 38, declares himself a Christian and proclaims independence from the Turks November 28 while Murad is preoccupied by the Hungarians and Serbs (above).

Painting *Madonna with Violets* by Stefan Lochner.

1444 The Truce of Adrianople (Edirne) June 12 brings temporary peace between the Christians and Ottoman Turks. George Brankovich, now 77, is restored to his Serbian despotat.

The Ottoman sultan Murad II summons the Wallachian prince Dracul across the Danube and forces him to leave as hostages his son Radu and his son Dracula, 13 (*see* 1448).

A papal representative encourages the Hungarians to break the Truce of Adrianople (above). They resume hostilities in September, advancing through Bulgaria toward the Black Sea.

The Battle of Varna November 10 ends in victory for the sultan Murad II who has been called back from retirement to head the Ottoman army. He has crossed the Bosphorus in defiance of a Venetian fleet that has blocked the Dardanelles, defeats a large force of Hungarians and Wallachians, kills Ladislas (Vladislav) VI, and takes many knights prisoner.

Portuguese explorer Nino Tristram reaches the Senegal River.

Venetian efforts to find new spice routes are spurred by Niccolo de' Conti, a traveler who has returned to Venice after 25 years in Damascus, Baghdad, India, Sumatra, Java, Indo-China, Burma, Mecca, and Egypt. Pope Eugenius IV orders de' Conti to relate the story of his travels to papal secretary Poggio Bracciolini as penance for his compulsory renunciation of Christianity.

Painting *St. Peter Altarpiece* by Konrad Witz, who uses a landscape of Lake Geneva to depict *Christ Walking on the Water*.

1445 Charles VII creates the first permanent French army: 20 companies of elite royal cavalry with 200 lances per company, 6 men per lance.

Copenhagen becomes the capital of Denmark.

Portuguese explorer Dinis Diaz rounds Africa's Cape Verde for the first time in modern history.

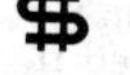

Some 25 caravels per year trade between Portugal and West Africa.

Painting *Santa Lucia* altarpiece by Italian painter Domenico Veneziano, 39, represents the Madonna and the saints in the same scale for the first time and reveals a new use of light and shadow; *The Adoration of the Magi* by Fra Angelico and Fra Filippo Lippi; *Giants* by Paolo Uccello; frescoes of Dante, Petrarch, Boccacio, and others by Andrea del Castagno for the refectory of Florence's Church of Santa Apollonia.

Sculpture *The Resurrection* by Luca della Robbia who perfected a technique of creating figures of terracotta glazed like faience in white relief against blue grounds.

1446 The *Ordenacoes Afonisans* gives Portugal her first code of laws, an amalgam of Roman and Visigothic law mixed with local custom.

Corinth falls to the Turks, who abort an effort by the Greeks to expand from the Peloponnesus (Morea) into central Greece.

A Korean alphabet devised by two research teams is proclaimed "the right language to teach the people" by the Korean king Sejong October 9. The 28-letter alphabet will be disdained by the educated elite who will retain the thousands of Chinese ideographs, the masses will be unable to use even the simple new alphabet, but a written Hangul combining ideographs and alphabet will evolve.

Painting *Edward Grymestone* and *Carthusian Monk* by Flemish painter Petrus Christus, 26; *The Patron Saints of Cologne* by Stefan Lochner.

Florentine architect Filippo Brunelleschi dies April 15 at age 69, leaving the cupola of the great Duomo incomplete (*see* 1436).

Ireland's Blarney Castle is completed by Cormac Laidhiv McCarthy, lord of Muskerry. Set in a turret below the battlements is a limestone rock that will be called the Blarney Stone and that will be reputed to confer eloquence on anyone who hangs head downward to kiss it.

1447  Milan establishes an Ambrosian republic upon the death of the last of the Visconti, Filippo Maria. Francesco Sforza is hired as military leader (*see* Sforza, 1450).

Poland's nobility chooses the grand duke of Lithuania to succeed the late Ladislas VI, his older brother. Now 20, the new king unites Lithuania with Poland and will reign until 1502 as Casimir IV, giving Poland access to the Baltic (*see* 1466).

The Turkestani shah Rukh, fourth son of the great Tamerlane, dies at Samarkand after a 43-year reign of great splendor. He is succeeded by Tamerlane's grandson Ulugh-Beg, 53, who will reign until 1449.

Painting *Presentation in the Temple* by Stefan Lochner.

1448  The Byzantine emperor John VIII Palaeologus dies at age 57 after a 23-year reign. He is succeeded by his brother, 44, who will reign until 1453 as Constantine XI Palaeologus, last of the Byzantine emperors.

The Ottoman sultan Murad II gains a victory over his Albanian governor Skanderbeg (George Castriota), Hungary's János Hunyadi marches to oppose him, treacherous Hungarian noblemen desert Hunyadi, and Murad II defeats him October 19 at the Battle of Kossovo.

The prince of Wallachia Dracula escapes his Ottoman captors and assumes the throne of his father, who was murdered last year. Fearing for his own life, Dracula vacates the throne after a few months and departs for Moldavia and Transylvania (*see* 1444; 1456).

The Union of Kalmar that has united the three Scandinavian kingdoms since 1387 begins to crumble as Christopher of Bavaria dies at age 30 after a reign of 9 years in which the Hansa towns have determined national policy. The Danish council elects Christian of

1448 ***(cont.)*** Oldenburg king but retains all real power. The Swedish nobility elects Karl Knutsson king, and he begins a 9-year reign as Charles VIII, but Christian of Denmark resists Swedish efforts to secure the Norwegian throne.

France's Charles VII takes Maine in a renewal of the Hundred Years' War with England.

Charles VII (above) puts down an insurrection of French noblemen (the Praguerie) supported by his son the dauphin. He exiles the dauphin to the Dauphine, where the son continues to intrigue against the king.

The Concordat of Vienna establishes the authority of German princes over the German church (*see* 1439). The Vatican strikes a cynical bargain with the new German king Frederick III, triumphs over the conciliar movement for reform by agreeing to divide its profits with the princes and with Frederick, and agrees to give the princes a share in episcopal tax revenues.

Painting *Adoration of the Shepherds* by Italian painter Pietro di Giovanni d'Ambrogio.

1449 The Turkestan prince Ulugh-Beg is assassinated at age 55 by a Muslim religious fanatic who ends the 2-year reign of Tamerlane's grandson.

The chief of a new Mongol federation captures the Chinese emperor Ying Zong in battle after a 13-year reign. The Mongols will hold Ying prisoner until 1450, he will not regain his throne from his brother Qing Di until 1457, but he will then rule until 1464.

The Japanese shogun Yoshimasa assumes power at age 14 to usher in a second period of Ashikaga art that will rival that of his grandfather Yoshimitsu, but Yoshimasa's 25-year shogunate will be marked by uprisings, civil war, plague, and famine.

Court favorites of Portugal's Afonso V persuade him to make war against his uncle Pedro. The regent, who has given the country able and enlightened rule since 1438, is killed with his son at the Battle of Alfarrobeira.

Ulugh-Beg (above) has been a scientist who used a curved device more than 130 feet long set on iron rails to catalog 1,018 stars in the constellations, making tables so precise that his calculations of the annual movements of Mars and Venus will differ from modern figures by only a few seconds. The Muslims have feared his learning.

1450 Francesco Sforza overthrows Milan's 3-year-old Ambrosian Republic in a February coup and makes a triumphal entry as duke March 25. Sforza and his son Galeazzo Maria will make their court a rival to that of Florence's Medicis by attracting scholars and Greek exiles (*see* 1424; 1476).

French forces triumph over the English at Formigny April 15, completing their reconquest of Normandy.

England's Henry VI banishes William de la Pole, 53, first duke of Suffolk, following the murder January 9 of former parliamentary treasurer Adam Molyneux by sailors at Portsmouth. Suffolk is widely resented for banishing Richard of York to Ireland and is accused of selling Anjou and Maine to France. Intercepted off Dover as he sails for France May 1, he is beheaded May 2. Other royal advisers are murdered.

Cade's rebellion demands English governmental reforms and restoration of power to Richard, duke of York (above), who returns from Ireland and forces his way into the Council. Kentish rebel John (Jack) Cade rallies 30,000 small Kentish and Sussex landowners in May to protest oppressive taxation and corruption in the court of Henry VI. They defeat Henry's forces June 18 at Sevenoaks, enter London July 3, and force the lord mayor and judges to pass a death sentence on Kent's sheriff and tax collector William Crowmer and on Lord Saye-and-Sele. The rebels grow violent, exact forced contributions to their cause, are denied readmission to the city, repulsed at London Bridge, and dispersed under amnesty. Cade is hunted down and killed July 12 at Heathfield.

A 14-year civil war ends in the Swiss confederacy which has been strengthened by the conflict.

Pope Nicholas V authorizes the Portuguese to "attack, subject, and reduce to perpetual slavery the Saracens, pagans, and other enemies of Christ southward from Cape Bajador and Non, including all the coast of Guinea" (*see* 1433; 1434; 1460).

England's nobility encloses more lands to raise sheep at the expense of the peasantry. The landowning class is enriched by the rapidly developing wool trade and by the continuing war with France (*see* More, 1515).

Glasgow University is founded by Bishop Turnbull under a bull of Pope Nicholas V.

The Universities of Barcelona and Trier have their beginnings.

Painting *The Flood* and *The Drunkenness of Noah* by Paolo Uccello.

Leon Battista Alberti redesigns the exterior of Rimini's Church of San Francesco.

The Porta Giova of Milan's Sforesco Castle is designed by Florentine architect Filarete (Antonio di Pietro Averlino), 50.

1451 The Ottoman sultan Murad II dies of apoplexy at Adrianopole February 2 at age 48 after a 30-year reign. His eldest son, 21, will reign until 1481 as Mohammed II, driving the Venetians and Hungarians out of Rumelia and Anatolia, reasserting Ottoman authority over rebellious Turkish emirs, and taking Constantinople (*see* 1453).

Sculpture *The Ascension* by Luca della Robbia is a glazed polychrome terracotta work above a portal of the cathedral at Florence.

Florence's Rucellai Palace is completed by Leon Battista Alberti.

1452 The German king Frederick of Hapsburg makes Austria an archduchy January 9 and is crowned Holy Roman Emperor March 19 by Pope Nicholas, the last emperor to be crowned at Rome. Now 36, Frederick has married Leonora of Portugal March 16 and will reign until 1493 as Frederick III, but his countrymen are indignant at his capitulation to the pope; their indignation will grow as they witness his apathy toward Ottoman aggression.

War begins in June between Constantinople's last Greek emperor Constantine and the new Ottoman sultan Mohammed II as the sultan completes fortifications that control the flow of supplies to Constantinople. The sultan's Castle of Europe (Rumili Hisar) opposite the older Castle of Asia (Anadoli Hisar) at the narrowest point of the Bosphorus has alarmed the Byzantine emperor Constantine.

France's Louis the dauphin strengthens ties with the Swiss cantons, making treaties also with the towns of Trier, Cologne, and others and with Saxony as part of an anti-Burgundian policy.

Scotland's James II subdues the earls of Douglas and confiscates their lands.

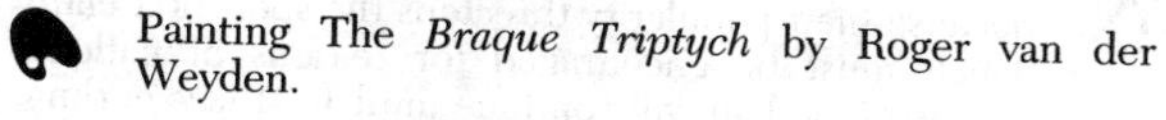

Painting The *Braque Triptych* by Roger van der Weyden.

Florence's Medici Palace is completed on the Via de Ginori by architect Michelozzi di Bartolommeo for Cosimo de' Medici whose family will occupy the palazzo for 100 years (the Riccardi family will later acquire it).

1453 Constantinople falls to the Ottoman Turks, who end the Byzantine Empire that has ruled since the fall of the Roman Empire in 476. An enormous iron chain has kept the fleet of Mohammed II out of the Golden Horn, but he has had some 70 small ships dragged overland from the Bosphorus to support the 250,000 troops that have besieged the city since April 6. Beginning April 12 he pounds the city walls with 1,200-pound cannonballs fired from a 26-foot long cannon built by the Hungarian renegade Urban. Dragged by 60 oxen and 200 men, the cannon has an internal caliber of 42 inches, it breaches the walls, the Turks force an entry at the Romanos Gate May 29, the last Byzantine emperor Constantine XI Palaeologus is killed in the fighting, the Turks sack Constantinople, and they make it the Ottoman capital.

The Hundred Years' War that has continued off and on since 1377 ends in France with the expulsion of the English from every place except Calais, which England will retain for more than a century. The earl of Shrewsbury John Talbot, 65, is the last surviving general of the late Henry V, he has come ashore in the Gironde with a force of 3,000, Bordeaux and the smaller towns have risen to his support and have driven out their French garrisons, but the French kill Shrewsbury in mid-July at Castillon with roundshot, they slaughter his army, and the Bordelais surrender October 19 when it becomes clear that they can expect no further English aid.

England's Henry VI has his first episode of insanity, inherited from his grandfather Charles VI of France. His cousin Richard of York serves as regent.

The Ottoman Turks sacked Constantinople in 1453, ending the Byzantine (eastern Roman) Empire begun in 395.

The fall of Constantinople (above) ends the Greek Empire of the East and leaves the Roman pope without any serious rival.

Mohammed II follows up his victory at Constantinople by advancing into Greece and Albania.

The White Sheep dynasty that will rule Persia until 1490 comes to power in the person of Uzun Hasan, who will extend Turkoman authority over Armenia and Kurdistan and then over Azerbaijan and Iran.

The fall of Constantinople (above) increases the need for sea routes to the Orient. Muslim rulers have imposed stiff tariffs on caravan shipments with the highest duties being levied on spices, and the sultan of Egypt exacts a duty equal to one-third the value of every cargo entering his domain.

Venice continues to import spices at higher prices and maintains her monopoly in the spice trade (*see* Lisbon, 1501).

Greek scholars fleeing Constantinople are welcomed by Cosimo de' Medici to his palazzo at Florence (*see* 1429).

Coffee is introduced to Constantinople (above) by the Ottoman Turks (*see* 850; 1475).

1454 Burgundy's Philippe le Bon takes the "vow of the pheasant" February 17, swearing to fight the Ottoman Turks.

The Peace of Lodi April 9 ends hostilities between Venice, Milan, and Florence. Pope Nicholas V has negotiated the treaty.

Venice's Doge Francesco Foscari signs a treaty April 18 with the new sultan Mohammed II on terms favorable to Venice.

Castile's Juan II dies July 21 after a 48-year reign in which he has fought Aragon and the Moors. His son will reign until 1474 as Enrique IV.

England's Henry VI recovers from a bout of insanity at year's end and dismisses the Duke of York as his protector. The duke of Somerset returns to power (*see* 1455).

1454 ***(cont.)*** Printers at Mainz use movable metal type for the first time (traditional date) (*see* 1381 ff; Gutenberg Bible, 1456).

Some 28 French musicians inside a huge pie perform at the Feast of the Pheasant for the duke of Burgundy Philippe le Bon (above). A Mother Goose rhyme about "four and twenty blackbirds baked in a pie" will commemorate the event.

1455 England's Wars of the Roses erupt in a civil war that will continue until 1471 between the houses of York and Lancaster. Excluded from the Council, the duke of York defeats royal forces at St. Albans May 22 and the earl of Dorset Edmund Beaufort is killed (*see* 1460).

Venetian navigator Alvise da Cadamosto, 23, discovers the Cape Verde Islands off the coast of Africa. He will explore the Senegal and Gambia rivers in the next 2 years in the service of Portugal's Prince Henry (*see* 1421).

Fra Angelico dies at Rome March 18 at age 55; Lorenzo Ghiberti dies at Florence December 1 at age 77.

1456 Athens falls to the Ottoman Turks, who will rule Greece and the Balkans for most of the next 4 centuries. The Hungarian prince János Hunyadi destroys the Ottoman fleet July 14, routs an Ottoman army besieging Belgrade July 21 and 22, forces Mohammed II to return to Constantinople, but dies of plague August 11 at age 69. His eldest son Laszlò will be arrested at Buda in mid-March of next year and beheaded at age 23 by order of the Hungarian governor, but the victory of János Hunyadi will keep Hungary free of Ottoman rule for 70 years.

Wallachia's Vlad Dracula, now 24, avenges his father's 1447 murder and reassumes his throne. He will reign until 1462, resisting Turkish demands for recruits (*see* 1448; 1462).

The Gutenberg Bible published at Mainz by local printer Johann Gutenberg, 56, is a Vulgate bible that marks one of the earliest examples of printing from movable type in Europe (*see* 1454). Gutenberg has taken 5 years to produce the bible, printing it in two volumes, folio, with two columns of 42 lines each per page (*see* Tyndale Bible, 1526).

Poetry *Le Petit Testament* by French poet François Villon, 25, is published in 40 stanzas. Villon stabbed a priest to death last year and is involved this year in a robbery at the College de Navarre in Paris.

Painting *Niccola da Tolentino* by Andrea del Castagno.

Florence's Church of Santa Maria Novella receives a new façade designed by Leon Battista Alberti.

Sugar from Madeira reaches Bristol, giving many Englishmen their first taste of the sweetener.

1457 Sweden's Charles VIII (Knut Knutsson) is driven out of the country in a revolt inspired by the Church, and Denmark's Christian I of Oldenburg is crowned king of Sweden while the noblemen Sten Stures, Svante Stures, and Sten the Younger hold the real power.

The king of Hungary and Bohemia Ladislas V flees to Prague to escape the storm of criticism that follows the execution of the Hungarian hero Laszlò Hunyadi March 16 (*see* 1456).

Ladislas prepares to marry Magdalena, daughter of France's Charles VII, but he dies suddenly November 23 at age 17, possibly the victim of poisoning, and will be succeeded in Hungary by the second son of the late János Hunyadi.

The University of Freiburg has its beginnings.

Donatello moves to Florence at age 71 after years of working at Rome, Naples, Padua, and Siena; painter Andrea del Castagno loses his wife to plague August 8 and dies of plague himself at Florence 11 days later at age 34.

Painting *Madonna with Saints Francis and Jerome* by Petrus Christus; *The Rout of San Romano* by Paolo Uccello.

Scotland's Parliament forbids "futeball and golfe" because their popularity threatens the sport of archery which must be encouraged for reasons of national defense. The ban will continue until 1491 (*see* curling, 1465; cricket, 1477).

1458 Matthias Corvinus wins election to the Hungarian throne at age 18 to succeed the late Ladislas. He will reign until 1490, a reign that will be among the greatest in Hungary's history, but the throne of Bohemia remains in dispute (*see* 1457; 1459).

Bohemia elects her regent George Podiebrad, 38, king March 2 (but *see* 1459).

The archduke Albrecht storms Vienna twice during the year.

Ottoman forces invade Morea; Constantinople annexes the remainder of Serbia.

Aragon's Alfonso V (the Magnanimous) dies at age 73 after a 42-year reign. He is succeeded in Aragon by his son Juan and in Naples by his unscrupulous bastard son Ferdinand (Ferrante), 35, who has a struggle to gain the succession but who enlists the support of Francesco Sforza, now 57, and Cosimo de' Medici, now 69, who are alarmed at the presence of the French at Genoa.

Delhi's Mahmud I begins a 53-year reign that will be marked by the conquests of Girnar and Champanir, the further extension of Islam in northern India, and the construction of magnificent mosques and a great palace at Sarkhej.

Work begins across the Arno River from Florence on a great palace for Luca Pitti, a rival of the Medici family.

1459 Hungary's Matthias Corvinus challenges the legitimacy of Bohemia's new king George Podiebrad, an avowed follower of Jan Hus and hence, technically, a heretic (*see* 1463).

The realm of an African king named Prester John appears on a map drawn by the Venetian monk Fra Mauro (*see* Covilhão, 1487).

Painting *The Crucifixion* by Italian painter Andrea Mantegna, 28, is completed as part of the main altarpiece triptych for Verona's Church of Saint Zeno. Mantegna becomes court painter to the Gonzaga family at Mantua, where he will remain until his death in 1506.

1460 England's deranged Henry VI is taken prisoner July 10 at the Battle of Northampton 66 miles from London by Yorkists wearing white roses who defeat the royal Lancastrians wearing red roses. Richard Plantagenet, 49, third duke of York, asserts his hereditary claim to the throne, marches on London, is assured by the lords that he will succeed to the throne upon Henry's death, but is killed at Wakefield in the West Riding, where his forces are defeated by an army raised in the north by Henry's wife Margaret of Anjou. Southern England rallies behind Richard's son Edward (*see* 1461).

Scotland's James II is killed August 3 when a cannon bursts while he is besieging Roxburgh Castle in a show of sympathy for the Lancastrian cause. He is succeeded by his son, 9, who will reign until 1488 as James III.

A Portuguese squadron returns from the west coast of Africa with a cargo of slaves.

The University of Basel has its beginnings.

Painting *The Seven Sacraments* altarpiece by Roger van der Weyden.

Venice completes its arsenal. Almost a town within a town, the heart of the republic's naval power includes a large shipyard for building the vessels that provide Venice with her wealth and power.

Mantua's Church of San Sebastian is completed by Leon Battista Alberti.

1461 Edward of York, 19, defeats an army of Lancastrians at Mortimer's Cross to avenge his father Richard's death last year but is defeated at the second Battle of St. Albans in February, loses possession of the deranged Henry VI, proclaims himself king, defeats the Lancastrians at Towton, and is crowned in June as Edward IV after Parliament calls Lancastrians usurpers. Edward will reign until 1483.

France's Charles VII dies July 22 at age 58 and is succeeded by his son Louis, 38, who made unsuccessful attempts to take the throne in 1446 and 1456, was banished to the Dauphine in 1447, and has not seen his father for 14 years. Called the Spider, the new king is crowned at Reims August 15 and will reign until 1483 as Louis XI. He begins by imposing restrictions on the French nobility, even forbidding them to hunt without his permission, and forcing the clergy to pay long-neglected feudal dues. Louis will increase the power of the Crown at the expense of local and provincial independence and urban liberty.

Trebizond falls to Ottoman forces who take over the last Greek state on the Black Sea and absorb the 257-year-old kingdom into their empire. They also seize the principality of Kastamonu.

Japan has plague and famine that bring an uprising against the Ashikaga shogun Yoshimasa.

France's new king Louis XI makes a rapprochement with the papacy by a formal revocation of his father's 1438 Pragmatic Sanction, but he sacrifices little royal power and keeps France's church under the control of the Crown.

Poetry *Le Grand Testament* by François Villon is a lyric poem of 173 stanzas containing many ballads, or *rondeaux*. French authorities will banish Villon from Paris next year after arresting him for theft and brawling and then commuting his death sentence, but he will be quoted for centuries, especially for his line, "where are the snows of yesteryear?" (*ou sont les neiges d'antan?*) from the "Ballade des Dames du Temps Jadis."

Sculpture *Judith and Holofernes* by Donatello. Domenico Veneziano dies at Florence May 15 at 55.

1462 The grand duke of Muscovy Basil II dies at age 47 after a 37-year reign marked by anarchy and civil war. Called Temny (the Blind), Basil is succeeded by his son of 22, who will reign until 1505 as Ivan III Vasilievich, the first Russian national sovereign, enlarging Muscovite territory enormously (*see* Novgorod, 1471).

Wallachia's Vlad Tepes (Dracula) slaughters 20,000 Turks along the Danube, but Vlad the Impaler is deposed and replaced by his pro-Turkish brother (*see* 1476).

The Holy Roman Emperor Frederick III is besieged in the Hofburg at Vienna by radical opponents led by his brother Albrecht.

1463 Upper and lower Austria are united by the Holy Roman Emperor Frederick III following the death of his brother Albrecht, but Bohemia's George Podiebrad and Hungary's Matthias Corvinus ravage the territories. Bohemia's Catholic nobility elects Matthias Corvinus king, but George Podiebrad defeats him in battle (*see* 1459; 1466).

Venice goes to war with Constantinople following Ottoman interference with her Levant trade. Hostilities will continue for 16 years.

English landowners gain a monopoly in the home grain market by a statute that prohibits imports of grain except when prices rise to levels at which the export of grain is prohibited. Food prices will rise in the next 2 centuries without compensating wage increases, but almost no grain will be imported (*see* 1436; 1773).

The University of Nantes has its beginnings.

1464 England and Scotland make peace, but civil war continues in England. The duke of Somerset Henry Beaufort is captured and beheaded at age 28.

Florence's Cosimo de' Medici dies August 1 at age 75 while listening to one of Plato's *Dialogues*. He is succeeded as head of the great Florentine banking house by his son Piero the Gouty.

The Poste Royale founded by France's Louis XI pioneers national postal service (*see* 1621).

Roger van der Weyden dies at Brussels June 16 at age 63.

1465 Yorkists capture England's Lancastrian king Henry VI and imprison him in the Tower of London as the Wars of the Roses continue.

1465 ***(cont.)*** The French dukes of Alençon, Burgundy, Berri, Bourbon, and Lorraine defeat Louis XI at Montl'hery and force him to sign the Treaty of Conflans that restores Normandy to the duc de Berri and restores towns on the Somme to Burgundy. Louis begins immediately to evade the treaty and to split the League of the Public Weal by means of diplomacy.

The University of Bruges has its beginnings.

Painting *Duke of Urbino* by Piero della Francesca.

England's Yorkist king Edward IV forbids the "hustling of stones" and other sports related to bowling and curling (*see* 1457; cricket, 1477).

1466 Poland gains an outlet to the Baltic October 19 under terms of the Second Peace of Thorn that ends a war against the Teutonic order. The order becomes a vassal of the Polish crown, and half the Teutonic Knights become Poles.

Pope Paul II excommunicates Bohemia's George Podiebrad and inspires a Hungarian crusade against the king (*see* 1463; 1470).

Florence's Luca Pitti fails in an effort to assassinate Piero de'Medici. He is stripped of his powers, and the Pitti Palace he began in 1458 on the left bank of the Arno remains unfinished (*see* 1549).

Florence's Medici family forms a cartel alliance with the Vatican to finance the mining of alum (aluminum potassium sulfate) deposits in the papal states (the material is used as a mordant to fix dyes in textiles). The Medicis persuade the pope to excommunicate anyone who imports alum from the infidel Turk in breach of the Vatican-Medici monopoly (*see* 1471).

Painting *Pietà* by Venetian painter Giovanni Bellini, 36 (his sister Nicolosia is married to Andrea Mantegna of Padua, and his work shows Mantegna's influence). Donatello dies at Florence December 13 at age 80.

1467 The duke of Burgundy Philippe le Bon dies June 15 at age 71 after a 48-year reign in which Burgundy has become the richest state in Europe. He is succeeded by his son Charles the Bold (Charles le Téméraire), 32, who defeats the Liègeois at St. Trond and makes a victorious entry into Liège to begin a 10-year struggle with France's Louis XI.

Turkish forces enter Herzogovina and begin to conquer the Balkan country for the Ottoman Empire.

The Japanese shogun Yoshimasa chooses his brother Yoshime as his successor and is challenged by supporters of his son Yoshihisa. A 10-year civil war begins at Kyoto, which will be nearly destroyed as both factions permit their unpaid soldiers to loot and pillage.

The Japanese Zen priest and landscape painter Sesshu, 48, goes to Beijing to study after having studied for years under Shubun, a Chinese painter who became a naturalized Japanese and brought Chinese and Japanese art closer together.

1468 The Songhai king Sonni Ali takes Timbuktu from the Tauregs (*see* 1492).

Ethiopia's Solomonic emperor Zara Yaqub dies after a 34-year reign in which he has defeated Muslim armies to protect the freedom of his country's Coptic Christians. Yaqub has brought all of the Ethiopian highlands under his rule.

Albania's prince Skanderbeg dies at age 62 after a 24-year reign. His death opens the way for reoccupation by the Ottoman Turks.

The duke of Burgundy Charles the Bold effects an alliance against France's Louis XI by marrying Margaret, sister of England's Edward IV.

Painting *Donne Triptych* by Flemish painter Hans Memling, 38, who has moved to Bruges; altarpiece for the Municipio of Massa Fermana at Ascoli by Italian painter Carlo Crivelli, 33; *Pietà* by Italian painter Cosme (Cosimo) Tura, 38, who has been court painter since 1452 to Duke Boso d'Este.

The Spanish plant rice at Pisa on the Lombardy plain. It is the first European planting of rice outside Spain (*see* 711).

1469 The Spanish crowns of Aragon and Castile join in alliance October 19 at Valladolid, where the infante Ferdinand of Aragon and León, 17, marries the infanta Isabella of Castile, 18.

Florence's Piero de' Medici dies December 3 after 5 gouty years in which he has maintained authority chiefly through the prestige of his late father Cosimo. His sons Lorenzo, 20, and Giuliano, 16, succeed to control the now mighty Florentine banking house (*see* 1478).

Fra Filippo Lippi dies at Spoleto October 9 at age 63.

1470 England's Henry VI regains his throne briefly with support from the earl of Warwick, but the duke of Clarence disapproves of the restoration, deserts Warwick, and defects to Edward IV. Edward uses artillery to help his forces defeat Warwick in a battle at Stamford. He obliges Warwick to take refuge in France, but his enemies force him to flee to Holland while Clarence takes refuge in France.

Hungary's Matthias Corvinus is proclaimed king of Bohemia and margrave of Moravia, but Bohemia's George Podiebrad gains Polish support by promising the succession to the son of Poland's Casimir IV, and he forces Matthias to come to terms.

Venice loses Euboea (Negroponte) to Constantinople. An enormous Ottoman fleet lands an army that takes the Venetian territory in the continuing war between the Turks and Venetians.

Portuguese explorers reach Africa's Gold Coast. Fernão Gomes has sent João de Santarem and Pedro de Escolar to Africa in accordance with terms of a 5-year lease on the Guinea trade granted to Gomes last year on condition that he carry explorations forward by at least 100 leagues per year.

The first French printshop is set up at Paris by three German printers (*see* 1454). It will be followed in the next few years by printshops in the Lowlands, the Swiss cantons, Castile, and Aragon (*see* Caxton, 1475).

French printer Nicolas Jensen, 50, sets up shop at Venice as a publisher and printer after having learned the art from Gutenberg at Mainz (*see* 1456). He will be the first to use purely roman letters, but most printers will continue to use Gothic letters for years to come.

England's York Minster Cathedral of Saint Peter is completed after 250 years of construction on the site of a chapel erected for the baptism of Edwin of Northumbria in the 7th century.

1471 France's Louis XI declares war on Burgundy's Charles the Bold in January and occupies towns in Picardy.

Bohemia's George Podiebrad dies at Prague March 22 at age 51 and is succeeded by the Polish prince Ladislas (Vladislav), 15, a son of Casimir IV, who will reign until 1516 as Ladislas II, dominated by nobles. The ascension of Ladislas begins a 7-year Hungarian war.

England's Edward IV lands at Ravenspur in March, defeats Warwick at Barnet April 14 (Warwick is killed), and gains victory at Tewkesbury May 6. Henry VI is murdered in the Tower of London May 21, possibly by Richard, duke of Gloucester.

The grand duke of Muscovy Ivan III forces Novgorod to renounce her ties with Lithuania and to pay tribute to Moscow. Novgorod is the center of a vast republic in northern Europe, and while Lithuania extends from the Baltic to the Black Sea, Ivan has made Muscovy a more powerful nation (*see* 1479).

Sweden's Sten Sture the Younger repels an invasion by Christian of Denmark with help from Stockholm and other towns. He returns to the reforms of Engelbrecht Engelbrechtsen (*see* 1435).

Tangier falls to Portugal, which has earlier taken Casablanca in a campaign against the North African kingdom of Fez.

Pope Paul II dies July 26 at age 54 and is succeeded by the della Rovere cardinal, who is elevated to the papacy and who will reign for 13 years as Sixtus IV.

Pope Sixtus IV cancels the Vatican alliance made with the Medici family in 1466 to monopolize the alum trade. Cosimo de' Medici's grandson Lorenzo, 22, conciliates the new pope and is appointed receiver of the papal revenues and banker to the Vatican (but *see* 1475).

San Jorge d'el Mina is founded by the Portuguese as a port to trade in gold on what will be called Africa's Gold Coast.

The University of Genoa has its beginnings.

Painting *Madonna in a Garden* by Cosme Tura.

1472 Scotland acquires the Orkney and Shetland islands from Norway.

Portuguese explorers continue to flourish. Fernando Po discovers islands off the coast of Africa that will bear his name. Lopo Goncalves crosses the equator. Ruy de Sequeira reaches latitude 2° south.

Muscovy returns to the Roman Church through the marriage of the grand duke Ivan III to Zoë (Sophia) Palaeologa, niece of Constantinople's last Greek emperor Constantine XI Palaeologus. Zoë's father Thomas is despot of the Morea, Pope Sixtus IV has arranged the marriage, later Russian rulers will use it as the basis of their claim to be the protectors of Orthodox Christianity, Ivan will establish a Byzantine autocracy in Russia, and he will take the title czar (caesar).

The University of Munich has its beginnings in Bavaria.

Painting *The Annunciation* by Florentine painter Leonardo da Vinci born out of wedlock 20 years ago at Vinci to a peasant girl and fathered by a prominent local notary who has taken the youth to Florence and apprenticed him to the master Andrea del Verrocchio.

Sculpture *Tomb of Giovanni* and *Piero de' Medici* by Florentine sculptor-painter Andrea del Verrocchio (Andrea di Michele di Francesco Clone), 37.

Mantua's Church of San Andrea is completed by Leon Battista Alberti, who dies April 25 at age 68.

The grand duke of Muscovy Ivan III (above) will use Italian architects brought in by his wife Zoë to rebuild Moscow's grand ducal palace the Kremlin.

1473 The *Dispositio Achilles* signed by the elector of Brandenburg Albrecht III, 59, legalizes the custom of primogeniture that has existed for centuries and helped motivate younger sons of Europe's nobility to join the crusades in quest of lands. Albrecht is known as Achilles, and his *Dispositio* makes the eldest surviving male in a family the heir to the power and fortune of his father.

Burgundy's Charles the Bold occupies Alsace and Lorraine.

Cyprus comes under Venetian rule (*see* 1573).

Japan's western warlord Yamana Mochitoyo dies at age 69, and his son-in-law Hosokawa Katsumoto dies at age 48 in the Onin War that has raged since 1467. Civil war will continue until 1477.

Painting *The Virgin of the Rose Garden* by German painter-engraver Martin Schongauer, 23.

Pope Sixtus IV adds the Sistine Chapel to the Vatican Palace (*see* Michelangelo, 1512).

Kyoto's *Ryoanji* (rock garden) is designed by a Zen priest.

1474 The Japanese shogun Yoshimasa abdicates at age 39 after a 25-year reign. His son Yoshihasa succeeds him, but the civil war that began in 1467 continues (*see* 1477).

1474 ***(cont.)*** The Union of Constance is formed with subsidies on France's Louis XI, and the coalition begins war on Burgundy's Charles the Bold.

Isabella succeeds to the throne of Castile and León December 13 upon the death of her half brother, who has ruled since 1454 as Enrique IV. But the succession 5 years after Isabella's marriage to Ferdinand of Aragon is challenged by Enrique's daughter, who is married to Portugal's Afonso V (*see* 1475).

Genoese seaman Christopher Columbus (Cristóbal Colón), 23, begins discussing the possibility of a westward passage to Cathay (China). Also called Cristoforo Colombo, the young navigator uses projections made by German mathematicians and Italian mapmakers at Sangres (*see* 1421) to revive the ancient Greek knowledge that the earth is round. Columbus has the advantages of the compass invented in the 12th century and of the more recently invented mariner's astrolabe by which a navigator can calculate the altitude of the sun, moon, or stars above the horizon and thus determine his distance north or south of the equator (*see* 1477; quadrant, 1731).

The Hanseatic League (Hanse) gains generous trading privileges in England by terms of the Treaty of Utrecht.

Antwerp Cathedral is completed except for its spire after 122 years of construction. The spire for the great Gothic structure will be completed in 1518.

1475 England's Edward IV invades France in support of Burgundy's Charles the Bold. Louis XI meets with him and buys him off August 29 in the Peace of Picquigny.

Portugal and Castile go to war over the succession to Enrique IV (*see* 1474), but the cortes at Segovia recognizes Isabella's right to the throne with her husband Ferdinand of Aragon.

Pope Sixtus IV replaces Florence's Medici family as papal bankers with the Pazzi family following a rapprochement with Don Ferrante of Naples as the della Rovere pope tries to consolidate the papal states.

The *Recuyell of the Historyes of Troye* published at Bruges by English printer-translator-cloth dealer William Caxton, 53, is the first book to be printed in English. Caxton has learned the printing trade at Cologne after having dealt in handwritten manuscripts, and he has spent 2 years in translating the *Recuyell* (Collection) from French (*see* 1476).

Painting *St. Justine* by Giovanni Bellini; *The Adoration of the Magi* by Florentine painter Perugino (Pietro di Cristoforo di Vannucci), 29; *Apollo and Daphne* and *The Martyrdom of St. Sebastian* by Italian painter Antonio Pollaiuolo, 44, and his brother Piero, 31. Paolo Uccello dies at Florence December 10 at age 78.

England's Winchester Cathedral is completed after 425 years of construction.

Concerning Honest Pleasure and Well-being (*De Honeste Voluptate ac Caletudine*) by Vatican librarian Platina (Bartolomeo Sacchi) is the world's first printed cookbook.

The world's first coffee house opens at Constantinople under the name Kiva Han (*see* 1453).

1476 Burgundy's Charles the Bold conquers Lorraine, makes war on the Swiss cantons that have allied themselves with France's Louis XI, but is defeated at Grandson and Morat (*see* 1474; 1477).

Wallachia returns Vlad Tepes (Dracula) to the throne January 31. Now 43, Dracula has married the sister of Hungary's Matthias Corvinus (*see* 1477).

Milan's tyrant Galeazzo Maria Sforza is assassinated December 26 at age 32 by three young noblemen on the porch of the city's cathedral. He is succeeded by his 7-year-old son Gian Galeazzo under the regency of the boy's mother Bona of Savoy (*see* 1450; 1479).

William Caxton sets up the first English press to employ movable type and prints an indulgence to raise money for a Christian fleet against the Turks (*see* 1475).

Painting *The Demidoff* altarpiece by Carlo Crivelli.

Sculpture *David* (bronze) by Andrea del Verrocchio.

1477 The Battle of Nancy January 5 ends in a triumph for Swiss pikemen who prevail over the cavalry of the duke of Burgundy Charles the Bold and demonstrate a prowess that will make them widely sought after as mercenaries. Charles, now 43, fails in his siege of Nancy, he is killed in the battle, and his body is found half-eaten by wolves.

France's Louis XI invades Burgundy, Franche-Comte, and Artois.

The Hapsburgs acquire the Netherlands August 18 through the marriage at Ghent of Maximilian, 18, son of the Holy Roman Emperor Frederick III, to Mary, 20, daughter of the late Charles the Bold (above).

Ottoman raiders reach the outskirts of Venice as the 14-year-old war with Constantinople continues.

Wallachia's Vlad Tepes (Dracula) is ambushed outside Bucharest and killed one year after regaining his throne (*see* 1476; Bram Stoker novel, 1897).

Japan's 10-year-old civil war at Kyoto ends indecisively, but a 100-year period of internal wars begins. The wars will make it impossible for the imperial family to collect rents on its lands and will impoverish the family.

Christopher Columbus visits England but is unable to obtain backing for his projected venture in quest of a new route to the Indies (*see* 1474; 1484).

Sweden's Uppsala University, Württemberg's Tübingen University, and Hesse's Mainz University all have their beginnings.

Cricket is banned in England by order of Edward IV who fixes a fine and 2 years in prison as punishment for anyone caught playing "hands in and hands out" (as the game is called) because it interferes with the compulsory practice of archery (*see* "futeball and golfe," 1457).

The diamond engagement ring tradition has its beginnings as the counsel to the court of the Holy Roman Emperor Frederick III has the young archduke Maximilian (above) use the gold and silver pressed upon him by villagers as he sets out for Burgundy to have a jeweler create a gold ring set with diamonds forming the letter *M*.

1478 Lorenzo de' Medici and his brother Giuliano are attacked while attending high mass in the cathedral at Florence April 26 in a plot engineered by the Pazzi bank and Pope Sixtus IV to remove the Medicis from power. Giuliano, now 25, is stabbed to death, but Lorenzo beats back his assailants, takes refuge in the sacristy, and wreaks cruel vengeance on the Pazzi, hanging several from the palace windows, having others hacked to pieces, dragged through the streets, and thrown into the Arno, while still other supporters of the Pazzi are condemned to death or sent into exile.

Novgorod loses her independence after a second war with the grand duke of Muscovy Ivan III (*see* 1479).

Isabella of Castile launches an Inquisition against converted Jews who secretly practice their original faith, persecuting the so-called Marranos (the word originally meant pigs). The Inquisition will be broadened to include all "heretics," including Muslims (*see* Torquemada, 1483).

Lorenzo the Magnificent (above) blandly withdraws 200,000 gold florins from the Florence city treasury to cover the default of the Medici branch at Bruges. He is excommunicated by the pope (*see* 1480).

Painting *Primavera (Spring)* by Florentine painter Sandro (Alessandro di Mariano de Filipepi) Botticelli, 34, for the villa at Costello of Lorenzo Pierfrancesco de' Medici, a cousin of Lorenzo the Magnificent (above). Botticelli has also done work for the palazzo of the great Lorenzo on the Via Larga.

Brussels becomes the center of Europe's tapestry industry following the destruction of Arras.

1479 The Treaty of Constantinople January 25 ends a 15-year war between Venice and the Ottoman Empire, whose forces have reached the outskirts of Venice. The Venetians retain Dulcigno, Antivan, and Durazzo but are forced to yield Scutari and other points on the Albanian coast, accept the loss of Negroponte (Euboea) and Lemnos, and pay a tribute of 10,000 ducats per year for the privilege of trading in the Black Sea.

Aragon's Juan II has died January 20 at age 81 and been succeeded by his son Ferdinand, who at age 26 assumes the throne as Fernando II of Aragon and—through his 1469 marriage to his cousin Isabella—as Fernando V of Castile and León, uniting the major crowns of Spain to begin a reign that will continue until 1516 (*see* Columbus, 1486).

Maximilian of Austria gains victory August 7 over France's Louis XI, stopping his incursion into Burgundian lands.

Ludovico Sforza in Milan seizes power from his young nephew Gian Galeazzo (*see* 1476).

Novgorod is annexed by the grand duke of Muscovy Ivan III, who deports her resisting aristocrats to central Russia (*see* 1478; 1494).

Ferdinand and Isabella (above) make peace with Portugal at Alcacovas September 4 after nearly 4 years of bitter fighting. Portugal acknowledges Spanish rights to the Canary Islands and is permitted a monopoly in trade and navigation on Africa's west coast.

The University of Copenhagen has its beginnings.

The Game and Playe of Chesse, first book to be printed from metal type in England, is published by William Caxton (*see* 1476), who has translated a French version by Jehan De Vigny of a Latin treatise by Dominican friar Jacobus de Cassolis.

Painting *Mystic Marriage of St. Catherine* by Hans Memling.

1480 The grand duke of Muscovy Ivan III takes advantage of disunity among the Tatars to stop their advance on Moscow and free the country of Tatar domination.

Anjou, Bar, Maine, and Provence fall to the French crown upon the extinction of the house of Anjou. René, count of Anjou, dies without an heir July 10 at age 71, and Louis XI annexes his realms.

Otranto in southern Italy falls to the Ottoman Turks August 11, but Mohammed II (the Conqueror) fails in a siege of Rhodes. The Knights of St. John of Jerusalem, who purchased the island in 1306, successfully resist the sultan from May to August. The fall of Otranto ends an Italian civil war precipitated by the Pazzi plot of 1478 that destroyed the balance of power among Florence, Naples, and Milan.

Ferdinand of Aragon helps Florence's Lorenzo de' Medici make peace with Pope Sixtus IV.

Pestilence decimates the Mayan Empire in the Western Hemisphere.

Painting *The Seven Joys of Mary* by Hans Memling for the chapel of the tanners in Notre Dame at Bruges. A rich member of the guild commissioned the work and is portrayed as the foremost of the two Nativity observers. The Venetian republic appoints Giovanni Bellini official painter at a salary.

Brussels completes its Town Hall (Hotel de Ville) in the Grand'Place after 72 years of work, with a tower designed in 1449 by Jan van Ruysbroek.

1481 The Ottoman sultan Mohammed II dies May 3 at age 49 on the eve of another campaign in Anatolia. His eldest son, 34, will reign until 1513 as Bayzid I, strengthening Turkish power in Europe through wars with Poland, Hungary, and Venice.

Milanese power changes hands as the uncle of Gian Galeazzo Sforza ousts the boy's mother from her regency and takes over as Lodovico Il Moro (*see* 1479; 1494).

Portugal's Afonso V (the African) dies at age 49 in August after a 43-year reign. He is succeeded by his son of 26, who will reign until 1495 as João II.

Portugal's new king spurs the explorations that have given his nation dominance in trade with West Africa.

1482 Mary of Burgundy, daughter of the late Charles the Bold, dies March 27 at age 27 after a hunting accident in Flanders. Her husband, Maximilian of Austria, claims power over the Lowlands as regent for their infant son Philip; Brabant and Flanders reject his claims (*see* 1485).

The Peace of Arras December 23 ends hostilities between the Hapsburgs and France's Louis XI, who has lived in isolation for the past 2 years in his spider's-nest château Plessis-les-Tours, 2 miles southwest of Tours, surrounded by astrologers and physicians. Burgundy and Picardy are absorbed into France.

The bishop of Liège is killed by "the Wild Boar of the Ardennes" Guillaume de la Marck, 36, a Belgian soldier for France's Louis XI (the Spider). La Marck will be captured in 1485 and beheaded.

Venice begins a 2-year war with Ferrara that will result in the Venetian acquisition of Rovigo, the last expansion of Venice on the mainland.

Luca della Robbia dies at Florence February 23 at age 82.

Portuguese explorers on Africa's west coast find bananas growing and adopt a version of the local name for the fruit *Musa sapientum* (*see* 327 B.C.; de Berlanga, 1516).

1483 England's Edward IV dies April 9 at age 40 after a tyrannical reign in which he has multiplied his wealth by confiscating the estates of his enemies and putting his money to work in partnership with London merchants. On June 5, a quasi-legal Parliament declares Edward's marriage invalid and his sons illegitimate. His capable brother Richard, duke of Gloucester, is proclaimed king and begins a brief reign as Richard III. Richard's nephews (Edward's sons) are found smothered in the Tower of London, and it is widely believed that they were murdered on Richard's orders.

The duke of Buckingham Henry Stafford, 29, leads a rebellion against Richard III (above), but supporters of Richard suppress the revolt, capture Buckingham, and execute him at Salisbury.

France's Louis XI (the Spider) dies August 10 at age 60 after a 22-year reign in which the nation's prosperity has revived despite oppressive taxes. The tyrannical Louis is succeeded by his son of 13, who will reign until 1498 as Charles VIII.

The Spanish Dominican monk Tomas de Torquemada, 63, takes command of the Inquisition in all Spanish possessions at the request of Ferdinand and Isabella. A nephew of Cardinal Juan de Torquemada, the first grand inquisitor will persecute "heretics" with unparalleled zeal (*see* 1478; 1487).

Pope Sixtus IV says Mass August 9 in the new Sistine Chapel (named for him).

Painting *The Magnificat* by Sandro Botticelli, who was summoned to Rome last year along with Ghirlandaio and others to help decorate the Vatican chapel of Pope Sixtus IV. Botticelli's round picture of the Madonna with singing angels will be his most copied work; *The Virgin and Child and Six Saints* by Giovanni Bellini for Venice's Church of San Job; *Coronation of the Virgin* by Piero Pollaiuolo for the church of San Agostino at San Gimignano.

1484 Pope Sixtus IV dies August 12 after a 13-year papacy marked by nepotism and political intrigue in which the pope has warred with Florence and incited Venice to attack Ferrara. The Venetians have saved Sixtus from a Neapolitan invasion, but he has turned on Venice for not halting the hostilities which he himself instigated.

Venice acquires Rovigo and reaches the height of its mainland expansion. Venetian territory will remain essentially unchanged for more than 300 years, but the city's wealth and power will soon begin to decline.

Pope Innocent VIII succeeds to the papacy and inveighs against witchcraft and sorcery. The bull *Summis desiderantes* issued December 5 initiates harsh measures against German "witches" and magicians.

Christopher Columbus asks Portugal's João II to back him in a westward voyage to the Indies, but João rejects the request (*see* 1477; 1485).

Portuguese explorer Diogo Cão (Diego Cano) discovers the mouth of Africa's Congo River. He will explore the west coast of the continent south to the 22nd parallel (*see* 1487).

So-called witches (above) are usually midwives, detested by physicians for encroaching on their obstetrical practice.

Painting *St. Anthony* by Cosime Tura; *Garden of Delights and Earthly Paradise* by Dutch painter Hieronymus Bosch (Hieronymus von Aeken), 34 (dates of all Bosch paintings are speculative).

Kyoto's Silver Pavilion is completed for Japan's eighth Ashikaga shogun Yoshimasa near the Gold Pavilion completed in 1397.

The tea ceremony has been introduced by Japan's Yoshimasa (above). Now 48, the shogun has encouraged painting and drama, his reign has otherwise been disastrous, but the tea ceremony will remain for centuries a cherished part of Japanese culture.

1485 The Battle of Bosworth August 22 ends England's Wars of the Roses that have continued since 1460. Henry, 28, earl of Richmond, has landed at Milford Haven with French support after a sojourn abroad to escape the wrath of Richard III. His Welsh allies help him defeat the king, who falls in battle; Richard's crown is found hanging on a bush and passes to the earl. He is crowned at Westminster October 30 as Henry VII, restores peace to the realm, and will reign until 1509, inaugurating a 117-year Tudor dynasty.

Bruges and Ghent capitulate to Maximilian of Austria in June and July after sieges (*see* 1482, 1488).

Hungary's Matthias Corvinus expels the Holy Roman Emperor Frederick III from Vienna; Frederick becomes an imperial mendicant.

Christopher Columbus sends his brother Bartholomew to France and England in hopes that he may interest

Charles VIII or Henry VII in outfitting an expedition to the Orient, but Bartholomew has no success (*see* 1484; 1486).

Venice standardizes its quarantine of 1403 at 40 days.

The "sweating sickness" cuts down the army of England's Henry VII (above) and postpones the king's coronation. The mysterious malady closes down Oxford University for 6 weeks, spreads quickly to London, and within a week has killed thousands.

Le Morte d'Arthur by Sir Thomas Malory is published by the English printer William Caxton (*see* 1476). Malory is probably a knight-retainer to the late earl of Warwick, his abridged compilation of French Arthurian romance has probably been completed in prison, and it includes such lines as "there syr launcelot toke the fayrest lady by the hand, and she was naked as a nedel."

Painting *Pietà Panchiatichi* by Carlo Crivelli.

1486 Maximilian of Austria is chosen king of the Romans February 16 despite opposition from his father, the emperor Frederick III. Now 33, Maximilian is crowned king of the Germans April 9 at Aix-la-Chapelle and will reign until 1519 as Maximilian I.

Hungary's Matthias Corvinus lays down a code of laws as part of a program that is making his country for a brief period the dominant state in central Europe.

African slaves in the kingdom of Gaur in India rebel and place their own leader on the throne.

Christopher Columbus submits his plan for a westward expedition to Ferdinand and Isabella May 1 and persuades them to sponsor him (*see* 1485, 1492).

Africa's kingdom of Benin begins trade with Portugal.

Painting *The Birth of Venus* by Sandro Botticelli for the villa at Castello of his patron Lorenzo Pierfrancesco de' Medici; *Annunciation* by Carlo Crivelli.

1487 A pretender to the English throne gains support from Margaret, duchess of Burgundy and sister of the late Edward IV. The imposter Lambert Simnel, 10, impersonates the earl of Warwick, 10, a son of the duke of Clarence, who is imprisoned in the Tower of London. His supporters crown him Edward VI in Dublin Cathedral, and he lands in Lancaster with a force of poorly armed Irish levies and German mercenaries. The English defeat him at Stoke June 16, Henry VII pardons him, and he will become a falconer to the king after a period as scullery boy.

Pope Innocent VIII names Tomas de Torquemada grand inquisitor, and Torquemada's Inquisition introduces measures of cruelty that will make his name infamous (*see* 1483).

The Star Chamber introduced under another name by England's Henry VII gives defendants no right to know the names of their accusers. The king moves toward royal absolutism.

Portuguese navigator Bartholomeu Dias, 37, leaves Lisbon with two caravels and a storeship. A storm drives his fleet round Africa's southernmost tip, which he names the Cape of Storms (*see* Cape of Good Hope, 1488).

Portugal's João II sends explorer Pedro de Covilhão, 37, to the Levant in search of spices and the land of a legendary Prester John mentioned by Fra Mauro in 1459. Covilhão will cross the Red and Arabian seas to India, visit Madagascar, and send home word from Cairo that if ships can round southern Africa they will find pilots who can guide them to India (*see* 1497).

Venice's Palazzo Dario is completed on the Grand Canal for Giovanni Dario, Venetian secretary to Constantinople.

1488 Scotland's nobility rebels against the lavish spending and Anglophilic leanings of James III, the clan chieftains seize the king's eldest son, and they defeat him in battle at Sauchieburn near Bannockburn. Now 36, the king is murdered June 11, possibly by a soldier disguised as a priest, and his son of 15 assumes the throne. He will reign until 1513 as James IV.

Bartholomeu Dias returns to Lisbon in December and reports to João II. The king sees hope of a sea route to India and suggests that Dias's Cape of Storms be called the Cape of Good Hope (*see* Vasco da Gama, 1497).

Andrea del Verrocchio dies at Venice October 7 at 53 leaving unfinished his painting *Baptism of Christ* and his equestrian statue of Bartolommeo Colleoni for the Campo San Zanipolo (Alessandro Leopardi will complete the piece, which will be unveiled in 1496).

Munich's Church of Our Lady (Frauenkirche) is completed off the Kaufingerstrasse.

1489 Venice buys Cyprus March 14 from Catherine Cornaro, last of the island's Lusignan dynasty, after 7 centuries of Frankish rule.

Malleus Maleficarum by the inquisitors Kramer and Sprenger is a handbook on witch-hunting that will be used to justify the burning and shackling of innocent midwives and countless mentally ill people (*see* 1484; Wier, 1563).

The first major European epidemic of typhus breaks out in Aragon, where the disease is introduced by Spanish soldiers returning from Cyprus after helping the Venetians fight the Moors.

1490 Hungary's Matthias Corvinus dies suddenly April 4 at age 50 after an illustrious 32-year reign. He is succeeded by Bohemia's Ladislas II, who will reign until 1516.

The Portuguese plant sugar cane on the island of São Tomé and bring in slaves from the kingdom of Benin and other African countries to work in the canefields (*see* Cadbury, 1901).

Portuguese explorers ascend the Congo for some 200 miles and convert the king of the Congo Empire to Christianity. They establish a post at São Salvador and begin a sphere of influence that will continue for more than a century.

The discovery of new land by Christopher Columbus opened the Western Hemisphere to European conquest.

England's Henry VII takes control of the country's wool trade out of the hands of Florentine bankers and turns it over to Englishmen. The loss in trade brings a new crisis to Florence and opens the way to foes of big business in the city, foes led by the monk Girolamo Savonarola, prior of San Marco.

Painting *The Annunciation* altarpiece by Sandro Botticelli.

1491 France's Charles VIII annexes Brittany by marrying Anne, duchess of Brittany. England's Henry VII goes to war to prevent the annexation, making peace with Scotland to release his troops for action.

The Bohemian-Hungarian king Ladislas II signs the Treaty of Pressburg acknowledging the Hapsburg right of succession.

Constantinople makes peace with Egypt's Mamelukes after a 6-year war. Egypt gains control of Cilicia in Anatolia.

Girolamo Savonarola, 39, begins denouncing the corruption at Florence and particularly that of Lorenzo de' Medici. The spiritual leader of the democratic party (the Piagnoni) preaches vehement sermons deploring the alleged licentiousness of the ruling class, the worldliness of the clergy, and the corruption of secular life (*see* 1497).

Painting *The Nativity* by Perugino, a triptych for Rome's Villa Albani; *The Last Judgment* (frescoes) for Breisach Cathedral by Martin Schongauer, who dies at Breisach February 2 at age 40.

Hyderabad City is completed on the Musi River as the capital of an autonomous Muslim kingdom by Mohammed Quli of the Turkoman Qutub Shahi dynasty. Its manmade lakes serve as reservoirs, and its new Charminar arch commemorates the end of a plague.

Spanish colonists plant sugar cane in the Canary Islands, where the cane flourishes.

1492 Christopher Columbus weighs anchor Friday, August 3, with 52 men aboard his flagship the 100-ton *Santa Maria*, 18 aboard the 50-ton *Pinta* commanded by Martin Alonso Pinzon, 52, and another 18 aboard the 40-ton *Niña* commanded by Vicente Yanez Pinzon, 32. The *Pinta* loses her rudder August 6, the fleet puts in at Tenerife for refitting, the three caravels put out to sea again September 6, and land is sighted October 12.

Financed by Castile's Isabella, who has borrowed the wherewithal from Luis de Santangel by putting up her jewels as security, Columbus has crossed the Atlantic to make the first known European landing in the Western Hemisphere since early in the 11th century. He disembarks in the Bahamas on an island he names San Salvador under the impression that he has reached the East Indies.

Columbus lands in Cuba October 28 and on December 6 lands on the island of Quisqueya, which he renames Hispaniola, but his *Santa Maria* runs aground on Christmas Day and must be abandoned.

Granada surrenders January 2 to Castile's Isabella and Aragon's Ferdinand, who take the last Muslim kingdom in Spain (*see* Inquisition, below).

Florence's Lorenzo de' Medici dies April 18 at age 43. He has helped to make the Tuscan dialect the language of Italy in place of the classic Latin and helped make his city a center of European culture.

Poland's Casimir IV dies in June at age 65 after a 45-year reign and is succeeded by his son of 33, who will reign until 1501 as John Albert, reducing the power of the Polish burghers and peasants while extending the powers of the gentry (*see* 1496).

Lithuania is invaded following the death of Casimir (above), who was grand duke of Lithuania before becoming king of Poland in 1447.

Sonni Ali dies under mysterious circumstances after a 28-year reign in which he has built Gao from a small one-city kingdom into the vast Songhai Empire. He has had the mullahs of Timbuktu murdered for defying his authority (*see* 1468; 1493).

A decree issued March 31 by Ferdinand and Isabella extends the Spanish Inquisition begun by Isabella in Castile in 1478. It orders Granada's 150,000 Jews to sell up and leave the country by July 31 "for the honor and glory of God." Thousands pretend to accept the cross (they will be called Marranos), some 60,000 pay for the right to settle in Portugal (*see* 1496), still others are welcomed by the Ottoman sultan Bajazet II.

Ferdinand decrees November 23 that all property and assets left by the Jews (above) belong to the Crown, even those now in Christian hands.

Piero della Francesca dies at his native Borgo San Sepoicro near Arezzo October 12 at age 72.

Luis de Torres and Rodrigo de Jerez make the first known reference to smoking tobacco. Sent ashore in the New World by Columbus (above), they report seeing natives who "drank smoke," and Rodrigo will later be imprisoned by the Spanish Inquisition (above) for his "devilish habit" of smoking (*see* 1518).

Sweet potatoes that came originally from the Western Hemisphere have long been grown in the mid-Pacific and for a century or two have been cultivated as far west as the islands that will be called New Zealand, where Maori tribespeople have introduced the tuberous roots.

Christopher Columbus discovers foods unknown in the Old World: maize, sweet potatoes, capsicums (peppers), allspice (*Pimenta officinalis*), plantain (*Musa paradisica*), pineapples, and turtle meat.

1493 The Treaty of Narbonne signed January 19 cedes Roussillon and Cerdagne from France to Aragon. France's Charles VIII has agreed to the treaty in hopes of obtaining Ferdinand's support for an invasion of Italy, but Ferdinand has joined with the pope, Maximilian I, Milan, and Venice to block the French plan.

The German king Maximilian I takes Artois and Franche-Comte from France under terms of the Peace of Senlis in May.

A papal bull issued by Alexander VI (Borgia) May 4 establishes a line of demarcation between Spanish discoveries and Portuguese. The Spanish are to have dominion over any lands they discover west of the line, the Portuguese over lands east of the line.

Askia Mohammed, 50, takes over the Songhai Empire built up by the late Sonni Ali, beginning a 35-year reign in which he will dominate the Mandingo Empire and extend his own territory beyond the Niger (*see* 1492, 1529).

Christopher Columbus builds a fort on the island of Hispaniola using wreckage from the *Santa Maria,* he leaves 44 men at Fort La Navidad, he sets sail for home January 4 in the *Niña,* is joined 2 days later by Martin Pinzon in the *Pinta,* but is afterwards separated from Pinzon by a storm. He reaches Lisbon March 4 after having been delayed for 6 days by the Portuguese governor of the Azores, and arrives at Palos March 15.

Columbus presents Isabella with "Indians," parrots, strange animals, and some gold, he demands and receives the reward that rightfully belongs to the sailor Rodrigo de Triana of the *Niña* who first sighted land, the queen grants Columbus enormous privileges in the territories he has claimed for the Spanish, and she sends him back as governor with 1,500 men in a fleet of 17 ships which weighs anchor September 24.

Columbus lands Sunday, November 3, on an island he calls Dominica, sights Hispaniola November 22, and sails westward to La Navidad only to find that the fort has burned down and the colony has been dispersed.

Columbus's second voyage (above) has been financed through the sale of assets formerly owned by Jews (*see* 1492).

Doctors use peppers brought back from the "Indies" by Columbus in a medicinal preparation to treat the ailing Isabella.

Painting *Madonna della Candeletta* and *Coronation of the Virgin* by Carlo Crivelli.

Sculpture *Adam and Eve* (limestone) by German sculptor Tilman Riemenschneider, 33, who last year completed an altarpiece for the parish church at Mlinnerstadt.

The horses and livestock landed by Christopher Columbus (above) at Santo Domingo (Hispaniola) are the first seen in the New World. Columbus left Palos with 34 stallions and mares, he has 20 remaining when he arrives, but his cattle weigh only 80 to 100 pounds when fully grown. (The horse originated in the Western Hemisphere and migrated to Asia before becoming extinct in its continent of origin at the close of the Ice Age.)

Sugar cane and cucumbers planted by Columbus at Santo Domingo have come from the Canary Islands. Columbus has a special interest in sugar: his first wife's mother owns Madeira canefields (*see* 1418; 1506).

Pineapples brought back to Europe by Christopher Columbus give Europeans their first taste of the fruit.

1494 The Treaty of Tordesillas divides the globe between Spain and Portugal along lines similar to those established last year by Pope Alexander VI.

The grand duke of Muscovy Ivan III gains his sobriquet Ivan the Great by driving out Novgorod's German merchants, closing the Hanseatic Kontor, and extending his realm eastward to the Urals by annexing the vast territories of Novgorod.

The Drogheda parliament summoned by the English governor of Ireland Sir Edward Paynings, 35, enacts Payning's law which provides that every act of parliament must be approved by the English privy council to be valid.

Milan's Lodovico the Moor receives the ducal crown October 22 following the death of the rightful duke Gian Galeazzo Sforza, who has probably been poisoned by his uncle Lodovico. France's Charles VIII has helped the new duke, who soon joins a league against Charles. He gives his niece Bianca in marriage to the German king Maximilian, and he receives in return imperial investiture of the duchy of Milan (*see* 1476; 1499).

Queen Isabella of Castile suspends a royal order for the sale of more than 500 Carib "Indians" into slavery. Christopher Columbus (below) has brought the Caribs home from the West Indies, but the queen suggests in a letter to Bishop Fonseca that any sale await an inquiry into the causes for the imprisonment of the docile Indians and the lawfulness of their sale. When theologians differ on the lawfulness, Isabella orders the Caribs returned to their island.

Columbus discovers the island of Jamaica May 14 and names it Santiago (St. James). He proceeds to land on islands that will be called Guadeloupe, Montserrat, Antigua, St. Martin, Puerto Rico, and the Virgin Islands on a voyage in which he goes for 33 days with almost no sleep, loses his memory, and comes close to death.

1494 ***(cont.)*** Genoese merchant Hieronomo de Santo Stephano visits Calicut on the coast of India and observes trade in ginger and pepper (*see* da Gama, 1498).

The University of Aberdeen has its beginnings in Scotland.

The first English paper mill begins operation.

Painting *Pietà* by Perugino; *Calumny* by Sandro Botticelli. Ghirlandajo dies at Florence January 11 at age 44; Hans Memling dies at Bruges August 11 at age 64.

Flemish composer Josquin des Pres, 49, accepts an invitation to be chief singer at the chapel of France's Charles VIII. In his 8 years at the papal chapel in Rome he has composed numerous masses, motets, psalms, and other sacred works.

1495 Naples surrenders in February to France's Charles VIII, who is crowned king. Alfonso of Naples has abdicated in favor of his son Ferrandino, 25, who retakes the city following the Battle of Fornovo July 6. Pope Alexander VI has organized a Holy League to drive out the French, Charles escapes to France, the Spanish general Gonzalo de Cordova helps Ferrandino, the French fleet is captured at Rapallo, and a French army capitulates at Novaro.

Portugal's João II dies in October at age 40 after a 14-year reign in which he has encouraged exploration and has overcome the feudal nobility, putting its leaders to death. João the Perfect (*O principe perfecto*) is succeeded by his brother-in-law, 26, who will reign until 1521 as Manoel I in a golden age of exploration and discovery.

A new pretender to the English throne gains financial support from the German king Maximilian, makes an unsuccessful attempt to invade Kent, but receives a welcome from Scotland's James IV, who gives him Lady Catherine Gordon in marriage. Perkin Warbeck, 21, is a Walloon who worked as a servant to a Breton silk merchant in Ireland; some people have mistaken him for the son of the duke of Clarence or of Richard III. He professes to be the second of Edward IV's sons murdered in the Tower of London in 1484 and has been acknowledged as her nephew by Margaret, dowager duchess of Burgundy. The earls of Desmond and Kildare have lent him their support, and France's Charles VIII has entertained him as Richard IV (*see* 1496; 1497).

The Diet of Worms attempts to modernize the Holy Roman Empire. The Imperial Diet proclaims Perpetual Peace, sets up an Imperial Chamber and Court of Appeal, and imposes a general tax.

Lithuania expels her Jews as does Cracow, but within 5 years Poland will be regarded as the safest place for Jews in all of Europe.

Columbus orders that every Hispaniola native over 14 pay tribute money every 3 years to the king of Spain.

Augsburg's Jakob Fugger II will earn more than 500,000 gulden in the next 15 years by selling copper and silver from mines he leases in Hungary plus 200,000 gulden per year from selling to the mints of Europe the silver he mines in the Tyrol and Carinthia. Water-driven machinery devised by his engineer associate Johann Thurzo facilitates mining the metals and refining them, enabling Fugger to supply customers at Antwerp, Danzig, Nuremberg, and Venice and in Poland, Prussia, and Russia. The Fuggers' control of mercury production in Spain, whose "quicksilver" mines are now its chief source of income, gives the family a grip on political power in Spain which it will maintain until 1634. As one of the leading European importers of Portugal's spices, the Fuggers will almost always earn a 20 percent return on their investment, and in most years their profit will exceed 50 percent (*see* 1546).

Syphilis strikes Naples in history's first recorded outbreak of the disease that will appear throughout Europe in the next 25 years, but the disease may have existed for years and been confused with leprosy. A more violent form of syphilis than the disease of later centuries, the "new" malady infects the army of France's Charles VIII (above). Frenchmen call it the Neapolitan disease, Italians the French disease (*see* Fracastoro, 1530).

Painting *Vulcan and Aeolus* by Florentine painter Piero di Cosimo, 33.

Leonardo da Vinci submits plans to control the Arno River and avoid its disastrous flooding. The Florentine artist-scientist-engineer now serves Milan's Lodovico Sforza (*see* 1679).

1496 Spanish forces complete their conquest of the Canary Islands from Portugal and from the indigenous fair-skinned Guanche who have fought off invaders for a century. Tenerife, largest of the seven volcanic islands, falls to the Spanish, who will rapidly assimilate the Guanche.

The new king of Naples Ferrandino dies September 7 at age 27, the last of the Anjou line in Italy. He is succeeded by his uncle Frederick, 44, who will reign until 1504 as king of the Two Sicilies but will be forced to yield Naples to France's Louis XII in 1501.

Poland's Statute of Piotkow restricts the burghers from buying land, deprives the peasants of freedom to move about, and gives the Polish gentry (*szlachta*) extensive privileges in what will be called the Magna Carta of Poland.

Scotland's James IV invades Northumberland to press the claims of Perkin Warbeck (*see* 1495; 1497).

Portugal's Manoel I orders the expulsion of all Jews and has many of them massacred. He takes the action to please Ferdinand of Aragon and Isabella of Castile, whose daughter Isabella he will soon marry (*see* 1492).

England's Henry VII refuses to recognize Spanish and Portuguese claims under the papal bull of 1493. He grants a patent to John Cabot, 46, to search for new lands and to rule any he may find. Born Giovanni Caboto, Cabot is a Venetian merchant who has settled in England with his sons (*see* 1497).

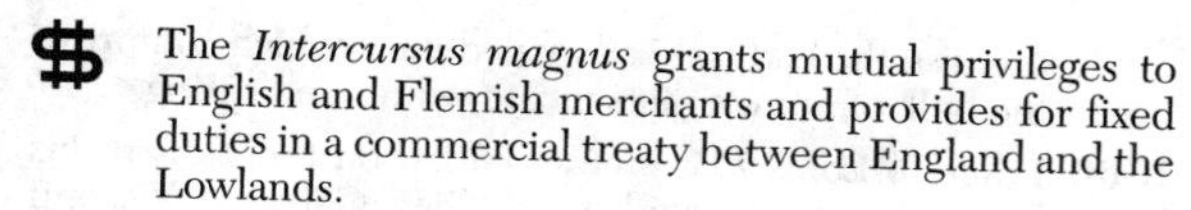

The *Intercursus magnus* grants mutual privileges to English and Flemish merchants and provides for fixed duties in a commercial treaty between England and the Lowlands.

Painting *Procession of the True Cross in St. Mark's Square* by Gentile Bellini; *The Blood of the Redeemer* by Vittore Carpaccio; *The Crucifixion* by Perugino for Florence's Church of S. Maria Maddalena dei Pazzi.

The Spanish who take Tenerife find bananas growing under intense cultivation (*see* 1482; 1516).

1497 England's Henry VII puts down an insurrection in Cornwall where the people have risen to protest taxes imposed to support English defenses against Scottish invasion forces. The Cornishmen are defeated at Blackheath.

Perkin Warbeck invades Cornwall with a small force after failing to find support in Ireland (*see* 1495). The pretender fails in a siege of Exeter, advances to Taunton, takes sanctuary at Beaulieu in Hampshire, but is forced to surrender in September and imprisoned in the Tower of London with the earl of Warwick (*see* 1499).

Denmark's John I defeats a Swedish army at Brunkeberg, enters Stockholm, revives the Scandinavian Union that ended in 1412, and begins a 4-year reign as John II.

Lucrezia Borgia, 17, duchess of Ferrara, has her father Pope Alexander VI annul the marriage he arranged for her 4 years ago to Giovanni Sforza, lord of Pesaro. He betrothes her to Alfonso of Aragon, a nephew of the king of Naples.

Persia's Rustam Shah dies after a 5-year reign that ends the dynasty of the White Sheep which has ruled since 1453 (*see* 1502).

Portuguese explorer Vasco da Gama, 28, leaves Lisbon with four ships to investigate the possibility of a sea route to India as was suggested by the voyage of Bartholomeu Dias in 1487 and by the report of Pedro de Covilhão. Da Gama rounds the Cape of Good Hope November 22 (*see* 1498).

Vasco da Gama sails along the west coast of Africa on Christmas Day and gives Natal its name.

Florentine seaman Amerigo Vespucci, 46, advances the claim that he discovered the American mainland in 1491. A resident of Seville who sails in the service of Spanish interests, he is probably an agent of Florence's Medici family (*see* 1500; Waldseemüller, 1507).

John Cabot has reached Labrador June 24 after a 35-day voyage. Accompanied by 18 men aboard the *Matthew*, he explores what later will be called Nova Scotia and Newfoundland, returning to Lisbon August 6 (*see* 1496; 1498).

The Florentine prior-dictator Savonarola celebrates the annual carnival with a "burning of the vanities" in the Piazza della Signoria. Masks related to carnival festivities, indecent books and pictures, and other items are burned, attracting crowds too large for the cathedral. The prior attacks the alleged crimes of Pope Alexander VI and indignantly spurns the offer of a cardinal's hat, but the pope is determined to silence the daring friar and issues a bull excommunicating him (*see* 1491; 1498).

Painting *The Last Supper* by Leonardo da Vinci is completed for Milan's Lodovico Sforza in the monastery refectory adjoining the Church of Santa Maria delle Grazie; *Apollo and Marsyas* by Perugino; The *Meeting of Joachim and Anne at the Golden Gate* by Florentine painter Filippino Lippi, 40, a son of the late Fra Filippo Lippi. Benozzo Gozzoli dies at Pistoia October 4 at age 77.

Sculpture *Bacchus* by Michelangelo.

John Cabot (above) notes vast codfish banks off the coast of Newfoundland. The fishing grounds have been visited in the past by Breton fishermen and will be called the Grand Banks (*see* 1504).

1498 France's Charles VIII dies April 8 at age 27 as he prepares a new expedition to invade Italy. Having no male heir, Charles is succeeded by his Valois cousin the duc d'Orléans, 36, who will reign until 1515 as Louis XII. The new king led a revolt against Charles more than a decade ago and was imprisoned from 1488 to 1491. His accession unites the duchy of Orléans with the royal domain.

The Spanish ship some 600 cannibal Caribs home to Spain to be sold into slavery (*see* 1494).

Christopher Columbus embarks June 7 on a third voyage to the New World, this time with six ships. He sights St. Vincent July 22 and Grenada August 15, discovers Trinidad, and lands at what may be the mouth of the Orinoco River on the South American mainland.

Portuguese explorer-scientist Duarte Pacheco Pareira touches the South American coast. He will write in 1505 of a vast continent extending south from 70° north latitude.

John Cabot and his son Sebastian coast Newfoundland, Nova Scotia, and points south on a second voyage with a fleet of six ships (*see* 1497).

The Spanish settle some 200 colonists in Hispaniola.

Vasco da Gama frees Europe from dependence on Venetian middlemen in the spice trade. He lands at Calicut, where Arab spice dealers fearful of losing their monopoly give him a rude reception, and he establishes a sea route between Portugal and India (*see* 1453; 1497; 1501).

Savonarola is burned at the stake for heresy May 23 in Florence's Piazza della Signoria off the Via Condotta (*see* 1497).

Painting *The Discovery of Honey* by Piero di Cosimo; *Baptism of Christ* (triptych) and *Judgment of Cambyses* by Dutch painter Gerard David; *The Apocalypse* (woodcuts) by Nuremberg painter-engraver Albrecht Dürer, 27, is the first book published by an artist from his own designs.

1499 France's Louis XII obtains a divorce from Jeanne, daughter of Louis XI, and marries Anne of Brittany, widow of the late Charles VIII, to keep the duchy of Brittany in the French crown.

Louis XII gains Venetian support for his claims to Milan, invades Italy once again, forces Lodovico Sforza to flee Milan, and accepts the city's surrender September 14.

The Swiss receive French financial support in a war with the German king Maximilian I; the southern German cities support Maximilian, but the Swiss gain a series of victories and force Maximilian to sign the Treaty of Basel September 22 granting the Swiss independence (formal independence will not come until 1648).

Venice begins a 4-year war with the Ottoman Empire. The Venetians will lose some territory and trading posts to the Turks.

The Ottoman Turks conquer Montenegro (Zeta).

Perkin Warbeck is hanged November 12 for conspiring to escape from the Tower of London with the imprisoned earl of Warwick (*see* 1497).

Granada's Moors stage a massive revolt as the Spanish Inquisitor-General Francisco Jimenez de Cisneros, 63, introduces forced conversion to Christianity on a wholesale basis.

Vasco da Gama returns to Portugal from Mozambique with pepper, nutmeg, cinnamon, and cloves after having lost 100 of his 160 men to scurvy (*see* 1498). His success encourages others to attempt the sea voyage around Africa to India (*see* Cabral, 1500).

London has another epidemic of the Black Death. It will kill thousands in the next 2 years (*see* 1407; 1603).

Nonfiction *Bellus Helveticum* by German humanist Willibald Pirkheimer, 29, is a history of the Swiss war (above) and includes his autobiography.

Poetry *The Bowge of Court* by English poet-translator John Skelton, 39, is an allegory satirizing the court of Henry VII.

Theater *La Celestina (Tragicomedia de calisto y Melibea)* by Spanish playwright Fernando de Rojas, 24.

Venice's campanile (clock tower) is completed in the Piazza San Marco after 3 years of construction. Two side wings will be added in the next 7 years, and additional floors will be added in 1755 to the campanile just west of the 400-year-old San Marco Church. Carved Moorish figures ring the campanile's great bell to provide a spectacle for the people.

Oriental spices such as those brought back by Vasco da Gama (above) are widely used to preserve meat and to disguise the bad taste of spoiled meat which comprises the bulk of human diets in late winter and spring.

1500 France's Louis XII annexes Milan on the basis of his being the great-grandson of Gian Galeazzo Visconti (*see* 1395; 1499). Lodovico Sforza is thrown into a French prison where he will die in 8 years (*see* 1512).

Persia's Turkoman dynasty of the White Sheep comes under attack from tribesmen commanded by the Safavid leader Ismail from eastern Azerbaijan. Only 14, Ismail comes out of hiding to take advantage of the confusion that has existed since the death of Rustam Shah in 1497 (*see* 1501).

Portugal's influence in Africa reaches its height.

Portuguese navigator Gaspar de Corte-Real (or Corterreal), 50, makes the first authenticated European landing on the northern continent of the Western Hemisphere since Leif Ericsson in 1000 and Thorfinn Karlsefni a few years later. Corte-Real's father João Vaz Corte-Real received an Azore Island captaincy in 1474 as a reward for having made a voyage to the "Land of the Codfish" and the younger Corte-Real explores the coast of Labrador (*see* 1501).

Portuguese explorer Pedro Alvares Cabral, 40, claims Brazil for Manoel I. Cabral has left the Cape Verde Islands with a fleet of 13 caravels bound for India, but contrary winds have driven him westward, he lands on Good Friday, and he takes possession in the name of the king Easter Monday, heads out across the South Atlantic for India, but loses four ships (including one commanded by Bartholomeu Dias) in a storm off the Cape of Good Hope. Brazil's coastline has just been explored by Alonzo de Ojeda, a Spanish navigator who may have been accompanied by the Italian Amerigo Vespucci (*see* 1497; 1501).

Spanish navigator Vicente Yanez Pinzon touches Cape St. Roque at the eastern extremity of Brazil. Pinzon captained the *Niña* on the 1492–1493 Columbus voyage.

Pedro Alvares Cabral (above) proceeds with his fleet to India in the company of Bartholomeu Dias and Duarte Pareira (*see* 1498). He will return loaded with spices to begin regular spice trade round the Cape of Good Hope.

Nonfiction *Adages* (*Adagia*) by Dutch humanist Desiderius Erasmus (Herasmus Gerardus), 34, is published at Paris. The collection of sayings from classical authors will appear in a larger edition at Venice in 1508.

Painting *Christ Crowned with Thorns* and *Ship of Fools* by Hieronymus Bosch; *Mystic Nativity* by Sandro Botticelli; *Self-portrait* by Albrecht Dürer.

16th Century

1501 Ferdinand of Aragon declares Granada a Christian kingdom but encounters resistance from the Moors.

Ferdinand (above) helps France's Louis XII conquer the kingdom of Naples from Frederick, king of the Two Sicilies. French forces enter Rome, and Pope Alexander VI declares Louis king of Naples (*see* 1504).

The German king Maximilian I recognizes French conquests in northern Italy in the Peace of Trent.

England's Henry VII declines a papal request to lead a crusade against the Ottoman Turks, who take Durazzo from Venice.

The grand duke of Muscovy Ivan the Great invades Lithuania.

Persia's Alwand of the White Sheep is defeated at the Battle of Shurur by the young Safavid leader Ismail, who takes Tabriz (*see* 1500; 1502).

Spanish settlers at Santo Domingo introduce African slaves into Hispaniola—the first importation of blacks to the New World (*see* 1503).

Gaspar de Corte-Real makes a second voyage to the northern continent of the New World, kidnaps 57 "Indians" to be sold as slaves, but drowns along with his crewmen and the slaves chained in the ship's hold as his caravel sinks in a storm on the homeward voyage. A second vessel reaches Portugal with seven Indians left alive, but Corte-Real's brother Miguel will be lost next year on a voyage in search of his lost sibling.

Amerigo Vespucci makes a second voyage to the New World, this time in the service of Portugal (*see* 1497). Vespucci's account of the Brazilian coast will express his conviction that it is not part of Asia but indeed a New World (*see* 1502).

Vasco da Gama wins control of the spice trade for Lisbon. He sets out with a fleet of 20 caravels to close the Red Sea, and he cuts off the trade route through Egypt to Alexandria, where Venetian merchants have been buying spices (*see* 1499; 1504).

The Spanish universities of Valencia and Santiago have their beginnings.

Nonfiction *Speculum Principis* by John Skelton is a moral treatise for England's Prince Henry.

Poetry *The Palice of Honour* by Scottish poet Gawin Douglas, 26, who has taken holy orders and becomes provost of Edinburgh's St. Giles Cathedral.

Painting *Life of the Virgin* by Albrecht Dürer.

Sculpture *Pietà* and *Bacchus* by Florentine artist Michelangelo (Michelagniolo Buonarroti), 26, who has created the *Pietà* on a commission from the French cardinal Jean de Villiers de la Grolaie, abbot of St. Denis, and the *Bacchus* for Roman nobleman Jacopo Galli. Michelangelo returns to his native Florence after a 5-year stay at Rome and begins work on a great statue of David.

Venetian printers use movable type to print music for the first time.

1502 Castile expels the last of the Moors, who have been in the country since 711.

A Spanish fleet seizes Taranto in March as Ferdinand of Aragon supports the claims of Louis XII to Naples.

Cesare Borgia, 27, receives French aid in putting down a revolt of his captains at Sinigaglia in December. Borgia is the son of Pope Alexander VI.

The Safavid dynasty that will rule Persia until 1736 is founded by the rebel leader Ismail, who has himself proclaimed shah (*see* 1501). He will reign until 1524.

Montezuma II ascends the throne of the Aztec Empire at Tenochtitlan at age 22 (*see* 1325; Juan de Grijalva, 1518).

Persia's new shah (above) executes Sunnis who do not accept the state Shiite brand of Islam.

Amerigo Vespucci returns in September from a voyage to the New World. An account of this voyage will be the basis of the name America (*see* Waldseemüller, 1507).

Christopher Columbus has embarked May 11 on a fourth voyage to the New World, this time with 150 men in 4 caravels that take 8 months to make the Atlantic crossing, forcing the crews to eat wormy biscuit (dried bread), sharkmeat, and ships' rats in order to survive. Columbus discovers St. Lucia, the island of Guanaja off Honduras, Honduras itself, Costa Rica, and the Isthmus of Panama (*see* 1503).

Some 2,500 new colonists arrive at Hispaniola. Ferdinand of Aragon installs Nicholas de Ovando as governor of the new colony.

The University of Wittenberg has its beginnings.

Painting *St. Jerome* by German painter-engraver Lucas Cranach, 29.

The Tempietto at St. Pietro in Montorio is completed by the Vatican architect Bramante (Donato d'Agnolo), 58, with a sculptured chapel that interrelates convex and concave elements.

1503 France's Louis XII abandons claims to Naples following the breakup of his alliance with Ferdinand of Aragon and defeat of French forces by Ferdinand's general Gonzalvo de Cordoba (*see* 1504).

The Spanish governor of Hispaniola Nicolas de Ovando receives royal authorization to relieve a labor shortage in the colony by importing African slaves (*see* 1511).

Another 1,000 to 2,000 Spanish colonists arrive at Hispaniola.

Christopher Columbus discovers Panama November 2 and finds the "beautiful port" that will be called Portobelo (*see* Darien, 1509).

Christopher Columbus observes rubber on his fourth voyage to the New World. The heavy black ball used in games played by the natives astonishes the Spaniards by bouncing as if it were alive. The Spaniards are the first Europeans to see the vegetable gum made from the latex either of the guayule shrub *Parthenium argentatus* or of *Hevea brasiliensis,* but rubber will not come into commercial use for another 3 centuries and then only to rub out pencil marks, whence it will derive its English name rubber, or india rubber (*see* 1772).

Poetry *The Thissill and the Rois* by Scottish poet William Dunbar, 43, who has composed the political allegory to honor Margaret Tudor, whose marriage to James IV he has helped to negotiate.

Painting *Crucifixion* by Lucas Cranach; *Leda and the Swan* by Leonardo da Vinci, who begins his great mural *The Battle of Anghiari* for Florence's Palazzo della Signoria (*see* battle, 1440).

England's Canterbury Cathedral is completed in Norman-Gothic style after 436 years of construction.

Portuguese caravels return from the East Indies with 1,300 tons of pepper—a quantity six times the amount that Egypt's Mameluke regime has permitted to be shipped in any one year.

1504 Ferdinand of Aragon completes his conquest of Naples January 1 with the surrender of French forces at Gaeta. France's Louis XII cedes Naples to Ferdinand in the Treaty of Lyons, the French control northern Italy from Milan, and the Spanish control Sicily and southern Italy and will hold Naples until 1707.

Albrecht of Bavaria defeats Rupert, son of the elector Palatine, in the Bavarian War. The German feudal knight Götz von Berlichingen, 24, loses his right hand at the siege of Landshut, has it replaced with an iron hand, and will be called "Götz of the Iron Hand."

The Treaty of Blois signed in September brings accord between France and the German king Maximilian I, whose 4-year-old grandson Charles is betrothed to Claude, daughter of France's Louis XII. Maximilian invests Louis with the duchy of Milan (above), and Louis promises to help Maximilian gain the imperial crown.

Castile's Isabella the Catholic dies November 24 at age 53 after a 30-year reign in which she has financed the voyages of Columbus and persecuted non-Christians with the Inquisition. She is succeeded by her daughter Juana and Juana's husband Philip the Handsome (Felipe el Hermoso), but they remain in Flanders; Isabella's widower Ferdinand of Aragon will rule Castile until 1506.

The tiny north Indian principality of Ferghana deposes her ruler Zahir-ud-din Babar, 21, for the third time from the throne he inherited at age 11. A descendant of Tamerlane, Babar escapes across the Hindu Kush and takes refuge in Kabul, where a faction of warring Muslim princes asks his help (*see* 1526).

Christopher Columbus returns from a final voyage to the Western Hemisphere, landing at Sanlucar November 7 too ill after a 9-week voyage from Hispaniola to pay his respects to the dying Isabella.

Painting *Rest on the Flight into Egypt* by Lucas Cranach; *Adam and Eve* (engraving) by Albrecht Dürer.

Breton fishermen begin making annual visits to the Grand Banks off Newfoundland which they have been visiting irregularly since before 1497.

Spices in the Lisbon market drop to 20 percent of Venetian prices, breaking Venice's monopoly in the spice trade and making her vulnerable to attack (*see* League of Cambrai, 1508).

1505 The grand duke of Muscovy Ivan the Great dies at age 65. He is succeeded by his son of 26, who will reign as Basil III Ivanovich until 1533 and incorporate the last remaining independent Russian principalities with Muscovy (*see* 1510).

Poland's Constitution of Radom makes the national diet the supreme legislative body: no laws are to be passed without its consent. The Diet of Piotrkow 12 years ago was the first *seym* to legislate for all Poland. The nobility, voting in provincial assemblies, elects the new diet; the statute *nihil novi* gives the *seym* and the senate a voice equal to that of the Crown in executive matters.

Portugal sends Francisco de Almeida to the Indies as her first governor. He takes Quiloa and Mombasa on the African coast enroute to his post, and he establishes forts at Calicut, Cananor, and Cochin on the Malabar coast (*see* 1509).

Sri Lanka (Ceylon) is discovered by the Portuguese off the southeast coast of India.

Seville University has its beginnings.

Painting *The Madonna and Child Enthroned with Saints* by Raphael; *The Virgin and Child with Four Saints* by Giovanni Bellini for Venice's church of St. Zacharias; *Portrait of Bishop Bernardo de' Rossi* by Italian painter Lorenzo Lotto, 25; *The Combat Between Love and Chastity* by Perugino.

1506 Castile's Philip I dies suddenly at Burgos September 25 at age 28, and his wife Juana loses her mind. Her father Ferdinand II of Aragon becomes regent of

Castile, marries Germaine de Foix, niece of France's Louis XII, and will rule Castile until 1516 as Ferdinand V.

Poland's Alexander I dies after a 5-year reign in which he has lost the left bank of the Dnieper to the grand duke of Muscovy Ivan the Great. He is succeeded by his brother, who will reign until 1548 as Sigismund I.

Korean rebels overthrow the cruel ruler Yonsangun and will place Chungjong on the throne next year.

A Florentine militia created by the city's vice-chancellor Niccolo Machiavelli, 37, is the first Italian national army.

Lisbon has a riot in which between 2,000 and 4,000 converted Jews are slaughtered (*see* Inquisition, 1536).

Christopher Columbus dies in obscurity at Valladolid May 21 at age 55. His bones will be removed to Santo Domingo, the island he discovered December 5, 1492, changing the course of history.

An English worker can earn enough to buy 8 bushels of wheat by working for 20 days, 8 bushels of rye by working for 12 days, 8 bushels of barley by working for 9 days (*see* 1599).

Poetry *The Dance of the Sevin Deidly Synnis* by William Dunbar.

Painting *St. Jerome in the Wilderness* by Lorenzo Lotto; *Madonna di Casa* by Raphael; *St. Catherine* altarpiece by Lucas Cranach. Andrea Mantegna dies at Mantua September 13 at age 75.

The *Laocoön* sculpture is unearthed in Rome's Esquiline Hill (*see* 38 B.C.).

Bramante begins rebuilding St. Peter's at Rome in the form of a huge Greek cross with a central dome. Bramante's successors will alter his plan.

1507 Maximilian I appoints his daughter Margaret of Austria, 26, guardian of her nephew the archduke Charles, 7, who is betrothed by treaty to the daughter of England's Henry VII. Maximilian makes Margaret regent of the Netherlands to serve until Charles is of age.

The Diet of Constance recognizes the unity of the Holy Roman Empire and founds the Imperial Chamber.

Italian adventurer Cesare Borgia is killed March 12 at age 30 while besieging the rebellious count of Lerin at his castle of Viana. Borgia has been fighting in the service of his brother-in-law the king of Navarre to whose court he fled late last year after making his escape from a Spanish prison where he had been held for 2 years.

The Portuguese capture Zafi in Morocco and begin commerce in captive Moors, Berbers, and Jews. Many are women, all are called white slaves.

Cosmographiae Introductio by German geographer Martin Waldseemüller, 37, gives Amerigo (or Americus) Vespucci credit for discovering the New World and calls it America (*see* 1501). The name will be applied at first only to the southern continent, but by the end of the century it will be generally applied to the entire Western Hemisphere.

La Prima Navigazione per l'Oceano alle Terre de' Negri della Bassa Ethiopia by Alvise Cadamosto describes his explorations.

The sweating sickness that struck London in 1485 strikes again (*see* 1518).

Painting *Mona Lisa* by Leonardo da Vinci, who has been making sketches since 1505 of Lisa di Anton, Neapolitan wife of local businessman-politician Francesco del Giocando, 46. His model leaves in the spring for Calabria on a long business trip with her husband, and the portrait is left incomplete; *Madonna with Child and Four Saints* by Lorenzo Lotto; *St. Mark Preaching in Alexandria* by Gentile Bellini, who dies February 23 at age 77 and whose work is completed by his brother Giovanni.

Florence's Palazzo Strozzi is completed after 18 years of construction.

Pope Julius II proclaims an indulgence to raise money for the rebuilding of St. Peter's.

1508 Maximilian I assumes the title of Roman Emperor Elect February 4 at Trent, and Pope Julius II confirms the fact that the German king shall hereafter automatically become Holy Roman Emperor. Maximilian has set out for Rome, the Venetians have refused to let him pass through their territories, he attacks the Venetians, but he signs a truce when he finds that the war is not popular with the southern German cities.

Pope Julius II forms the Holy League of Cambrai December 10 to recover from Venice the papal lands she has taken on the Adriatic. France's Louis XII and Aragon's Ferdinand II join the League, expecting to obtain territory themselves.

The Portuguese colonize Mozambique.

Spanish navigator Sebastian de Ocampo explores Cuba with a view to settlement.

The enigmatic smile of Leonardo's *Mona Lisa* raised questions that would defy answers down the centuries.

1508 ***(cont.)*** Puerto Rico is explored by Juan Ponce de León, 48, a Spaniard who accompanied Columbus on his second voyage to America. He plants a colony on the island (*see* 1509).

Painting *The Chess Player* and *Self-Portrait* by Dutch prodigy Lucas van Leyden (Lucas Hugensz), 14; *Madonna and Saints* by Lorenzo Lotto. Raphael enters the service of Pope Julius II.

Theater *Cassaria* by Italian playwright Lodovico Ariosto, 34.

The native population of Hispaniola in the Caribbean falls to 60,000, down from 200,000 to 300,000 in 1492 (*see* 1514).

1509 England's Henry VII dies April 22 at age 52 after a reign of nearly 24 years. He is succeeded by his athletic, well-educated son, who ascends the throne at age 17 and will reign until 1547 as Henry VIII. The new king is married June 11 to Catherine of Aragon, 23, daughter of Ferdinand II (but *see* 1533).

Pope Julius II excommunicates Venice April 27, and French forces triumph over Venetians May 14 at Agnadello.

The Battle of Diu February 2 in the Indian Ocean has brought victory to Portugal's Indian viceroy Francisco de Almeida, who destroys the Muslim fleet to establish Portuguese control over the spice trade.

Spanish forces invade North Africa. Cardinal Jimenez de Cisneros, 73, and Pedro Navarro, 49, lead a crusade against the Muslim rulers of Oran, Bougie, and Tripoli.

The Holy Roman Emperor Maximilian I joins the League of Cambrai but fails in a siege of Padua.

Portuguese navigator Ruy de Sequeira visits Malacca following the defeat of the Muslim fleet at the Battle of Diu (above).

Portuguese explorer Diego Alvaros Correa founds the first European settlement in Brazil near Porto Seguro.

Spanish explorer Alonso de Ojeda ventures into territory that will be called Colombia.

Ponce de León seizes control of Puerto Rico, making himself governor (*see* 1508; Florida, 1513; San Juan, 1521).

Spanish conquistadors found a colony at Darien on the Isthmus of Panama (*see* Balboa, 1513).

Fiction *Praise of Folly* (*Moriae Encomium*) by Erasmus is a witty satire on male idiocy. Erasmus has been teaching at Cambridge University, where he will remain until 1514.

Poetry *The Ship of Fools (The Shyp of Folys of the Worlde)* by English poet Alexander Barclay, 34, who has adapted the popular German satire *Das Narrenschiff* of 1494 by Sebastian Brant.

Engraving *The Temptation of St. Anthony* by Lucas van Leyden.

Painting *Madonna with Angels and Saints* by Gerard David; *St. Anne* altarpiece by Flemish painter Quentin Massys, 43.

1510 Russia's last free republic loses her charter January 29. Basil III Ivanovich obliges Pskov to send her assembly bell to Moscow.

Hamburg becomes a free city of the Holy Roman Empire.

Pope Julius lifts his excommunication of Venice February 10 and turns against France's Louis XII. The Swiss join his Holy League of Cambrai.

England's Parliament attaints former House of Commons speakers William Empson and Edmund Dudley who are beheaded in mid-August. Henry VIII has charged them with treason, but their real crime has been the extortionate measures they have used in administering Henry VII's arbitrary system of taxation—measures that fattened their own purses and provided them with vast estates.

Portuguese explorer Afonso de Albuquerque takes the island of Goa off India's Malabar coast. Western Europe's first toehold in India, it will remain in Portuguese hands for more than 4½ centuries.

Persia's Safavid shah Ismail defeats an Uzbek army, kills Mohammed Shaybani, and takes Herat, Bactria, and Khiva, extending his realm from the Tigris to the Oxus.

European explorers probe the east coast of North America to a point north of the Savannah River.

A horizontal water wheel designed by Leonardo da Vinci pioneers the water turbine.

Anatomy by Leonardo da Vinci contains drawings from life based on cadavers that Leonardo has somehow obtained and dissected, but he does not permit his work to be published (*see* Vesalius, 1543).

The Venetian painter Giorgione (Giorgio da Castelfranco) dies of plague at age 32, having caught the disease from "a certain lady." The artist is remembered for his landscape *Tempests*, and his death brings a rush by Italian art patrons for this and for a handful of other known Giorgione canvases including *Laura*, *The Three Philosophers*, *Trial of Moses*, *Judgment of Solomon*, *Sleeping Venus*, and *The Virgin and St. Francis and St. Liberate*.

Other paintings *Presentation in the Temple* by Venetian painter Vittore Carpaccio, 45; *The Gypsy Madonna* by Venetian painter Titian (Tiziano Vecellio), 23; *Salome* by Venetian painter Sebastiano del Piombo (Sebastiano Luciani), 25; *The Triumph of Galatea* by Raphael. Sandro Botticelli dies May 17 at age 65.

Engraving *Ecce Homo*, *The Milkmaid*, *The Return of the Prodigal Son* by Lucas van Leyden.

Theater *Everyman* is an English morality play based on the Dutch morality play *Elckerlijk* first performed in 1495.

Sunflowers from the Americas are introduced to Europe by the Spanish. In many countries they will be a major oilseed crop (*see* 1698).

1511 The papal forces of Julius II take Modena and Mirandola from the French in January, the French take Bologna May 13, Pope Julius allies himself with Venice to drive the French out of Italy, and in October he enlists Castile's Ferdinand II and England's Henry VIII in his Holy League.

Portuguese forces under Afonso de Albuquerque capture Malacca, center of the East Indian spice trade, to complete Portuguese control of Far Eastern spice sources (*see* 1509).

Spanish forces under Diego Velázquez use force to gain control of Cuba.

Spanish colonists in Cuba import African slaves as laborers because the native Carib population has died off alarmingly (*see* 1503; 1512).

Colonists in Hispaniola hear the Dominican friar Antonio de Montesinos preach a sermon against the enslavement of Indians.

Poland establishes serfdom under laws passed by the diet.

A Spanish ship bound from Darien in Panama to Santo Domingo strikes a reef and founders in the Caribbean. Survivors reach the Yucatán, where some are killed and eaten according to native ritual, and the remainder enslaved. Only two will survive (*see* Cortez, 1519).

Watches mentioned for the first time in print in the Nuremberg *Chronicles* have hour hands but no minute hands. "From day to day more ingenious discoveries are made," writes Johannes Cocleus; "for Petrus Hele, a young man, makes things which astonish the most learned mathematicians, for he makes out of a small quantity of iron horologia devised with very many wheels, and these horologia, in any position and without any weight, both indicate and strike for 40 hours, even when they are carried on the breast and in the purse" (*see* 1670).

Painting *St. Sebastian* and *Marriage of St. Catherine* by Italian painter Fra Bartolommeo (Bartolommeo di Pagolo del Fatorino, or Baccio della Porta), 36; *The Sistine Madonna* and *Julius II* by Raphael; *Procession of the Magi* by Florentine painter Andrea del Sarto (Andrea Domenico d'Agnolodi Francisco), 25; *Deposition* (triptych) by Quentin Massys.

Woodcut engraving *The Triumph of Christ* (10 blocks) by Titian (the work is nearly 9 feet long).

1512 Forces of the Holy League renewed last year by Julius II meet defeat in an Easter battle at Ravenna. French forces under Gaston de Foix have taken Brescia by storm with help from the knight Pierre Terrail, 39, seigneur de Bayard (known as *chevalier sans peur et sans reproche*), but a coalition of Swiss, papal, and imperial forces drive the French and their 5,000 German mercenaries out of Milan in May and return the Sforzas to power in the duchy (*see* 1500; 1536).

The Swiss take Locarno, Lugano, and Ossola as their reward for helping to drive the French out of Milan.

Spanish troops conquer Navarre which is annexed to Castile.

The Ottoman sultan Bayazid II is deposed by his Janissaries April 12 at age 65 and dies under suspicious circumstances. His eldest son, 47, has Bayazid's other sons strangled in November and will reign until 1520 as Selim I (the Grim), conquering Syria and Egypt and ending the Abbasid caliphate.

Poland and Muscovy begin a 10-year war over the White Russian region.

Golconda gains independence in India and will remain independent for 175 years.

Songhai's Askia Mohammed the Great conquers the Hausa states Kano, Katsina, and Zaria.

France's Louis XII imposes a tax on converted Jews from the Spanish states and Portugal.

Spanish colonists import black slaves into Hispaniola's western settlement to replace Indian slaves who have died in great numbers from disease and from being worked to death in the Spaniards' quest for gold (*see* Ovando, 1503; Las Casas, 1517).

Burgos enacts laws December 27 protecting West Indian natives from abuse and authorizing use of black slaves.

Pope Julius II convenes a council at the Lateran to counter a church council summoned at Pisa last year by France's Louis XII. The Lateran Council will remain in session until 1517, undertaking the first reforms of abuses in the Church of Rome (but *see* Luther, 1517).

Painting *The Creation of Adam* and *The Prophet Jerome* by Michelangelo for the Sistine Chapel at Rome; *Madonna and Saints* by Fra Bartolommeo.

The Newfoundland cod banks provide fish for English, French, Portuguese, and Dutch vessels which use the island as a base, drying the catch there for shipment back to Europe.

Portuguese explorers find nutmeg trees to be indigenous to the island of Banda in the Moluccas. The Portuguese will dominate the nutmeg and mace trade until 1602.

1513 French forces in Italy suffer a bad defeat June 6 at Novara (Swiss mercenaries rout the French and their Venetian allies in only an hour) and the battle at Guinegate August 17 will be called the Battle of the Spurs because of the hasty flight of the beaten French from England's Henry VIII and the Holy Roman Emperor Maximilian, who broke with Louis XII last year and has joined the Holy League. The allies force Louis to give up Milan and end his Italian invasion.

The Battle of Flodden Field September 9 just south of the Scottish border ends in victory for an English army sent by Henry VIII under the earl of Surrey. Scotland's James IV is killed at age 40 while fighting on foot, nearly all his nobles are killed, and the king's only legitimate son succeeds to the throne at age 15 months and will reign until 1542 as James V.

English forces land near Calais and take the French towns of Therouanne and Tournai.

The Scottish navy on which James IV has lavished so much money is sold to France as Scotland ends a quarter-century that has unified the country and given

1513 ***(cont.)*** it a prosperity it will not enjoy again for more than a century.

Denmark's John I, who ruled Sweden as John II from 1497 to 1501, dies after a 32-year reign. Scandinavian noblemen assemble in the Herredag at Copenhagen to confirm his son, 32, as successor. The new king will reign until 1523 as Christian II of Denmark and Norway, but the Swedes refuse to accept him (*see* 1520).

The Prince (Il Principe) by Florentine public servant Niccolo Machiavelli, now 44, is an analysis of the means by which a man may rise to power. Machiavelli has served for 14 years as vice chancellor and secretary of Florence and has observed the intrigues and machinations of the late Cesare Borgia of Romana. "Men are always wicked at bottom, unless they are made good by some compulsion," writes the cynical Machiavelli. "It is much more secure to be feared than to be loved." "Only those means of security are good, are certain, are lasting, that depend on yourself and your own vigor." "Hate is gained through good deeds as well as bad ones." "One who deceives will always find those who allow themselves to be deceived." "It is not titles that honor men, but men that honor titles." "Success or failure lies in conformity to the times." "The first method for estimating the intelligence of a ruler is to look at the men he has around him."

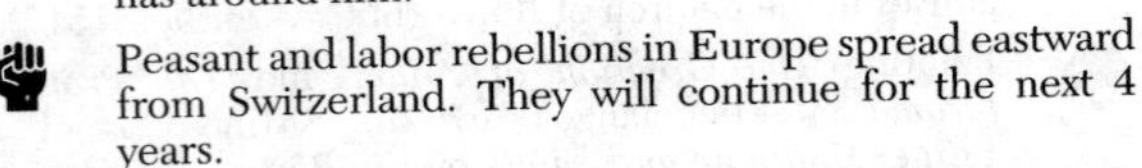

Peasant and labor rebellions in Europe spread eastward from Switzerland. They will continue for the next 4 years.

Florida is named by Puerto Rico's governor Ponce de León. He sights land on Easter Sunday, calls it Florida after Pascua Florida (the Easter season), and goes ashore.

Spanish explorer Vasco Núñez de Balboa sights the Pacific Ocean which he calls the South Sea and takes possession of El Mar de Sur September 29 in the name of Spain (*see* Magellan, 1520). Now 38, Balboa has joined an expedition from Santo Domingo as a stowaway, taken command of the 190-man force supported by 1,000 natives, crossed the 45-mile-wide thickly forested Isthmus of Darien (Panama) in 25 days to reach a peak overlooking the sea, and sent three men including Francisco Pizarro to reconnoiter (*see* 1514).

A Portuguese caravel reaches Guangzhou (Canton)—the first European ship to land in China.

Portuguese explorers discover uninhabited Pacific islands that will be called Mauritius and Réunion.

Pinturicchio dies at Siena December 11 at age 59.

Theater *Cassandra the Sibyl (Auto de la Sibila Casandra)* by Portuguese poet-playwright Raymond Gil Vicente, 43, 12/25 at the Convent of Euxobregas.

Chartres Cathedral is completed 60 miles southwest of Paris after nearly 400 years of construction. The new north tower gives the Gothic cathedral a magnificence matched only by its blue stained glass windows (*see* 1260).

Ponce de León (above) plants orange and lemon trees in Florida.

1514 England's Henry VIII concludes peace with Scotland and with France which cedes Tournai to England but will later buy it back for 600,000 crowns.

Persia is invaded by the Ottoman sultan Selim the Grim, who has slaughtered an estimated 40,000 of his heretic subjects and is determined to impose Sunnism on the Shiite Persians. His 80,000 cavalrymen rout a Persian army August 23 in the Battle of Chaldiran; Shah Ismail is wounded but escapes to Dagestan, leaving behind the favorites of his harem. Selim enters Tabriz September 15 and massacres much of its populace (*see* 1516).

Cuba is conquered from the "Indians" by the Spanish (*see* Havana, 1515).

"Not only the Christian religion but nature cries out against slavery and the slave trade" says the new pope Leo X (Giovanni de' Medici) in a bull against slavery, but the trade continues to grow (*see* 1450; 1526). The pope is a son of the late Lorenzo the Magnificent.

Some 1,500 Spanish settlers arrive at Panama (*see* 1509; 1513; Pizarro, 1524).

Henry VIII charters Trinity House at Deptford in Kent as a royal dockyard. The "guild or fraternity of the most glorious and undividable Trinity of St. Clement" is an association of mariners that will monopolize the training and licensing of pilots and masters. Given charge of directing the new naval dockyard, it will be given authority late in the century to erect beacons and other marks to guide England's coastal shipping, and it will grow in influence for centuries.

Painting *Birth of the Virgin* by Andrea del Sarto; *The Money Changer and His Wife* by Quentin Massys.

Green peas come into use in England to a limited extent, but dried peas are more commonly used and are consumed as "pease porridge"—hot, cold, even 9 days old (*see* 1555).

Hispaniola has 17 chartered Spanish towns; the island's native population falls to 14,000, down from 60,000 in 1508 (*see* 1548).

1515 France's Louis XII dies January 1 at age 52 after a 15-year reign and is succeeded by his robust son-in-law and cousin once removed. The new king, 21, will reign until 1547 as François I and begins by conquering Lombardy in northern Italy.

The Treaty of Vienna July 22 allies Maximilian I's Hapsburg family with the Jagiello family of Bohemia's Ladislas and makes Maximilian's brother Ferdinand potential heir to the Hungarian throne (*see* 1516; 1526).

French forces gain a decisive victory over the Swiss and Venetians at the Battle of Maiganano in mid-September and conclude peace September 29. The Swiss retain most of the Alpine passes and receive a French subsidy in return for French rights to enlist Swiss mercenaries. François I makes peace with Pope Leo X December 14.

Portuguese naval strategist Afonso de Albuquerque takes Hormuz at the mouth of the Persian Gulf, returns to Goa in December, receives word that he has been dismissed, and dies December 16 at age 62.

England's Henry VIII issues decrees designed to protect peasants from the results of land enclosure.

Spanish explorer Juan Diaz de Solis, 45, discovers the mouth of the Río de la Plata. Guaraní tribesmen will kill him next year when he tries to land near the mouth of the Paraná (*see* 1536).

Spanish explorer Juan de Bermudez discovers an Atlantic archipelago that will be called Bermuda.

Havana, Cuba, is founded by Spanish conquistadors.

Nonfiction *Education of a Christian Prince (Institutio Princip Christiani)* by Erasmus.

Fiction *Utopia* by English envoy to Flanders Thomas More, 38, describes an imaginary island governed entirely by reason and offers solutions to the social ills that plague England in a time when landlords are driving the peasantry off farm lands in order to develop sheep pastures for the burgeoning wool industry.

Painting *Isenheim* altarpiece by German painter Matthias Grünewald (Mathis Gothart Nithart), 39; *Portrait of Fra Teodoro* by Giovanni Bellini.

Hampton Court Palace is completed in Middlesex for England's Cardinal Wolsey. Now 40, Thomas Wolsey was made archbishop of York last year and has just been elevated to cardinal.

The Vatican appoints the painter Raphael chief architect of St. Peter's. He succeeds the late Bramante, who died last year at age 70 (*see* 1506).

1516 The Hapsburg dynasty that will rule Spain until 1700 is founded February 23 upon the death at age 63 of Ferdinand V of Castile and León (Ferdinand II of Aragon). Ferdinand's grandson, 16, a student in Flanders, succeeds to the throne, uniting Catalonia and Valencia with the kingdoms of Ferdinand, and will reign until 1556 as Carlos I.

Bold Spanish adventurers set out for the New World in tiny caravels, hoping for fabled treasure.

The Concordat of Bologna between France's François I and Pope Leo X (Giovanni de' Medici) rescinds the Pragmatic Sanction of 1438 and strengthens French royal power. It gives the French king freedom to choose bishops and abbots, and it removes the principle of the 1431–1449 Council of Basel that made the pope subordinate to an ecumenical council.

Bohemia's Ladislas II dies at age 60 after a weak reign of 45 years as king of Bohemia and 26 as king of Hungary as well. He is succeeded by his son Louis, 10, who will reign until 1526 as king of both countries.

The Battle of Marjdabik north of Aleppo August 24 gives the Ottoman sultan Selim a victory over the Mamelukes, who consider artillery a dishonorable weapon. Selim takes Aleppo with his cannon; he enters Damascus September 26 and moves on to Cairo (*see* 1517).

The Castilian regent Jiminez forbids importation of slaves into Spanish colonies, but Carlos I in Flanders (above) grants his courtiers licenses to import slaves into Spanish colonial islands.

De Rebus Oceanicus et Novo Orbe by Italian historian and royal chronicler Peter Martyr (Pietro Martire d'Anghiera), 59, is the first published account of the discovery of America in 1492.

A smallpox epidemic sweeps across Yucatán.

Poetry *Orlando Furioso* by Italian poet Ludovico Ariosto, 41, who will complete a final version in 1532.

Painting *Baldassare Castiglione* by Raphael (*see* 1528); *The Tribute Money* by Titian. Giovanni Bellini dies at Venice November 29 at age 86.

Theater *The Ship of Hell (Auto de la Barca do Inferno)* by Gil Vicente; *Magnyfycence* by John Skelton is an English morality play.

The first sugar grown in the New World to reach Europe is presented to Spain's Carlos I (above) by Hispaniola's inspector of gold mines, who gives the king six loaves.

Spanish missionary Fra Tomas de Berlanga introduces wheat, oats, and bananas into the Santo Domingo colony on Hispaniola.

The Portuguese plant maize in China.

1517 The Ottoman sultan Selim sacks Cairo January 22, the sharif of Mecca surrenders to the Turks, and the caliph Mutawakkil is sent to Constantinople as Selim secures control of the holy places of Arabia. Selim leaves Egypt under the rule of the Mameluke beys who have administered the government since 1250.

Africa's Songhai king Askia Mohammed suffers a defeat at the hands of the Hausa Confederation which gains dominance east of the Niger (*see* 1493, 1529).

London suppresses Evil May Day riots; 60 rioters are hanged on orders from Cardinal Wolsey.

The archduke Charles grants Florentine merchants a monopoly in the African slave trade.

1517 ***(cont.)*** Spanish priest Bartolomeo de Las Casas, 43, protests enslavement of Indians in the New World. Originally a Spanish planter in Hispaniola, he has been the first man to be ordained in the Western Hemisphere, has turned his efforts to serve the interests of the oppressed natives, and has voyaged back to Spain to plead the case of the Indians to Carlos I (*see* 1512; 1542).

Spanish explorer Francisco Fernandez de Cordoba observes traces of a Mayan civilization in the Yucatán. He has sailed westward from Cuba (*see* 1480).

The Fifth Lateran (18th ecumenical) Council ends 5 years of deliberations by overturning the Church's age-old prohibition against usury. The Franciscan order has demonstrated a need for the change; it has set up pawnshops for the poor and discovered that the shops are not viable unless some charge can be made for the loans extended. The growth of commerce, with its need for capital, has been making the Church's opposition to charging interest on loans quite untenable, but the Church's new position hurts Europe's Jews and Italians who have had a monopoly on moneylending.

Middle German working people and small merchants rally to the cause of Martin Luther (below) in protest against the monopolies of the Fuggers and other papal bankers who raise prices and send German gold to Rome.

Reformation of the Catholic Church begins October 31 at Wittenberg, 60 miles southwest of Berlin, on the Elbe. Augustinian monk Martin Luther, 34, nails 95 theses to the door of the Wittenberg Cathedral and challenges the excesses and abuses of the Roman Church, notably the sale of indulgences. Luther's action begins a long period of religious and civil unrest in Europe.

Poetry *The Tunnynge of Elynour Rummyng* by John Skelton is a comic poem about tavern life.

Painting *The Raising of Lazarus* by Sebastiano del Piombo; *Lo Spasimo* by Raphael; *Madonna of the Harpies* by Andrea del Sarto; *Erasmus* by Quentin Massys. Fra Bartolommeo dies at Florence October 31 at age 42.

Seville Cathedral is completed after 115 years of construction.

1518 Spanish colonists in Santo Domingo import more slave labor from Africa to perform the hard work of chopping cane in the colony's 28 sugar plantations (*see* 1501). The island's native population has dwindled as a result of disease and exploitation.

New Spain gets its name from Spanish explorer Juan de Grijalva, 29, who has been sent to follow up last year's discovery of the Yucatán by the late Francisco de Cordoba.

A third major epidemic of the sweating sickness spreads over England with more severity than that of 1507. It wipes out most of the population in some towns, many important figures succumb at Cambridge and Oxford, the disease reaches Calais, but it affects only the English there (*see* 1529).

Oxford physician-humanist Thomas Linacre, 58, founds a college of physicians with authority from Henry VIII.

Martin Luther's Reformation gains the support of Swiss clergyman Huldreich Zwingli, 34, at Zurich, who persuades the city council to forbid entrance to the Franciscan monk Bernardin Samson despite Samson's commission to sell indulgences. Zwingli becomes priest at the Great Minster of Zurich.

Painting *The Assumption* by Titian; altarpiece for Florence's Church of San Michele Visdomini by Italian painter Jacopo da Pontormo (Jacopo Carrucci), 24.

Theater *The Ship of Purgatory (Auto de la Barca do Purgatorio)* by Gil Vicente.

Tobacco is introduced to Juan de Grijalva (above) by a native chief who (according to Spanish historian Fernandez de Oviedo) "gave the general and to each of the Spaniards . . . a little hollow tube, burning at one end, made in such a manner that after being lighted they burn themselves out without causing a flame, as do the incense sticks of Valencia. And they smelled a fragrant odor. . . The Indians made signs to the Spaniards not to allow that smoke to be lost" (*see* 1492; 1531).

French silk merchant Jacques Le Saige attends a ducal banquet at Venice and notes that "these seigneurs, when they want to eat, take the meat up with a silver fork" (*see* 1071; 1570).

New Spain (above) has an estimated 11 million inhabitants while old Spain has 4.5 million (*see* 1519; 1547).

1519 The Holy Roman Emperor Maximilian I dies January 12 at age 59 in Upper Austria, and Spain's Carlos I is elected emperor as Charles V after the Fuggers and other Augsburg merchants bribe some electors. The election plunges Europe into political turmoil that will culminate in war.

Spanish adventurer Hernando Cortez, 34, sails from Cuba to conquer New Spain (*see* 1518). The Cortez expedition includes 500 Spaniards, nearly 300 Indians, and 16 horses—10 stallions, five mares, and a foal.

Cortez (above) moves up from the coast at Veracruz. Helped by an Aztec legend that the bearded white god Quetzalcoatl will return, Cortez is joined by an allied army of Totonacs. He enlists another army of Tlaxcalans, vanquishes the Cholulans, takes Montezuma II prisoner, and by year's end is ruling the country through Montezuma (*see* 1502).

Portuguese navigator Ferdinand Magellan (Fernao de Magelhaes, or Hernando de Magellanes), 39, leaves Seville August 10 with a fleet of five ships, remains at the mouth of the Guadalquivir for more than 5 weeks, and puts to sea September 20 in quest of spices from the Orient with financial support from an Antwerp banker.

The dollar has its origin in the "thaler" minted in Bohemia at Joachimsthal, where the large coin is called the Joachimsthaler.

Painting *The Raising of Lazarus* and *Christopher Columbus* by Sebastiano del Piombo. Leonardo da Vinci dies May 2 at age 67 in Amboise castle on the Loire.

Theater *The Ship of Heaven (Auto de la Barca de la Gloyia)* by Gil Vicente.

Hernando Cortés (above) hears of bearded men in Mayan towns and pays a ransom to secure the release of Geronimo de Aguilar, who was shipwrecked off the coast of Yucatán in 1511 and who describes the foods he has eaten as a slave to a Mayan chief. The foods include *cacao* (chocolate), *cacahuates* (peanuts), *camotes* (sweet potatoes), and *uahs* (tortillas). Cortés also discovers turkeys, tomatoes, vanilla, papaya, and beans which the Mayans call *avacotl*, a word the Spanish will turn into *habicuela* and the French into *haricot* (*see* 1528).

1520 Denmark's Christian II invades Sweden with a large army of French, German, and Scottish mercenaries. He has persuaded Pope Leo X to excommunicate Sten Sture the Younger and place Sweden under an interdict, he defeats Sture at Bogesund, Sture sustains a mortal wound January 19 at the Battle of Tiveden, and Christian advances without opposition on Uppsala where the Swedish Riksraad has assembled. The Swedish senators agree to accept Christian as king provided that he rule according to Swedish laws and customs and without recriminations. The king signs a convention March 31, but Sture's widow, Dame Christina Gyllenstjerna, at Stockholm has rallied the peasantry to defeat the Danish invaders at Balundsas March 19. The bloody Battle of Uppsala April 6 (Good Friday) gives Christian a narrow victory over the Swedish patriots, the Danish fleet arrives in May, Christian lays siege to Stockholm, and Christina surrenders September 7 on the promise of a general amnesty. Christian is crowned hereditary king of Sweden at Stockholm's cathedral November 4; Danish soldiers seize some of the king's guests November 7. Convicted of heresy and violence against the Church, the bishops of Skara and Stragnas are beheaded in the public square at Stockholm at midnight, November 8, and the Danes kill 80 other Swedes in the ensuing bloodbath.

Christian II (above) has his men exhume the body of Sten Sture and burn it along with that of Sture's small child. He has Sten's widow Dame Christina and other Swedish noblewomen sent to Denmark as prisoners and suppresses opposition on the pretense of defending the Church.

Victims of the massacre at Stockholm (above) have included the nobleman Erik Vasa, whose son Gustavus Eriksson, 24, hears about the massacre from a peasant while hunting near Lake Mälar. Gustavus escaped last year from the island fortress of Kalo on the east coast of Jutland where Christian II had treacherously held him hostage for 12 months. The peasant tells him that the king has put a price on his head, he rallies the yeomen of the vales, and he will begin next year to drive the Danes out of Sweden.

Last year's election of Spain's Carlos I as the Holy Roman Emperor Charles V provokes an uprising of the *communeros*, a group of cities led by Toledo's Juan Lopez de Padilla, 30. The *communeros* take exception to the king's leaving the country and using Spanish men and money for imperial purposes. They organize a Holy League (Santa Junta) at Avila in July, but radical elements soon displace the aristocratic and bourgeois leadership (*see* 1521).

The Ottoman sultan Selim dies September 21 at age 53 after an 8-year reign in which he has annexed Syria and Egypt to augment his Persian conquests. His son of 24 will reign until 1566 as Suleiman I, adding to his father's conquests and winning the soubriquet Suleiman the Magnificent.

A new Spanish army of 1,400 men arrives in New Spain under the command of Panfilo de Narvaez to challenge Hernando Cortés, but Cortés surprises Narvaez near Veracruz and captures him. He will remain a prisoner for 2 years.

Ferdinand Magellan negotiates a stormy 38-day passage through the straits at the southernmost tip of South America, he sails into the South Sea, and renames it the Pacific Ocean (*see* 1513; 1519; 1521).

England's Henry VIII and France's François I meet June 7 with 10,000 courtiers outside Calais on the Field of the Cloth of Gold. The banquets, tournaments, and spectacles that ensue for 3 weeks will leave the French treasury crippled for 10 years.

German gunsmith August Kotter invents the rifle.

Smallpox takes a heavy toll at Veracruz (above). Introduced by a black seaman on a ship carrying the troops of Narvaez, the pox will spread until it kills half the population of New Spain.

Appeal to the Christian Princes of the German Nation (An den Christlichen Adel deutscher Nation) by Martin Luther has a first printing of 4,000 copies and sells out in a week.

Painting *The Madonna with Saints Aloysius and Francis* by Titian. Raphael dies at Rome April 6 at age 36, leaving his *Transfiguration* incomplete.

Venice completes a new wing to the Doge's Palace to replace a structure destroyed by fire in 1483.

1521 The Battle of Vilialar April 23 gives the Holy Roman Emperor Charles V a victory over the insurgent *communeros* and ends the last Spanish resistance to absolutism.

Belgrade falls in August to the Ottoman sultan Suleiman after a 3-week siege. His forces make raids into Hungary (*see* Rhodes, 1522).

French support of the *communeros* (above) and French designs on Navarre precipitate a war between France and Spain that will last for 8 years. French forces take Pamplona and Fantarabia.

The colossal Central American city of Tenochtitlan (Mexico City) falls to Hernando Cortés August 13 after

1521 ***(cont.)*** an 85-day battle in which the emperor Montezuma II has been killed (*see* 1519). Cortez sees bison in Montezuma's menagerie.

Portugal's Manoel I dies in December at age 52 after a 26-year reign marked by great Portuguese explorations and discoveries. He is succeeded by his son of 19, who will reign until 1557 as João III.

San Juan, Puerto Rico, is founded by Spanish conquistadors who will pave the city's streets with stones brought as ballast in ships from Spain (*see* 1508).

Ferdinand Magellan in the Pacific claims islands that will be called the Marianas in 1668. Magellan calls them Islas de los Ladrones (Islands of Thieves) because the natives steal articles from his ships.

Ferdinand Magellan discovers the Philippine Islands March 15, tries to subdue the native chief Lapu Lapu, wades ashore on Mactan April 24 with 48 men in full armor, and is killed in a skirmish with Mactan warriors. Only three of Magellan's original five ships have made the Pacific crossing, the other two have been lost, his men have come close to starvation, but survivors of the Mactan encounter sail on in two remaining ships to the Moluccas, or Spice Islands (*see* 1522; 1564).

"Here I stand," says Martin Luther before the Diet of Worms April 18. "I can't do anything else. God help me! Amen." (*Hier stehe ich! Ich kann nicht anders. Gott helfe mir! Amen.*) Luther has come to Worms under a safe-conduct pass after the ban of the empire has been pronounced against him. Charles V's diet orders him to recant, he refuses, and the German princes back him in starting an evangelical movement that will bring turmoil to much of Europe (*see* 1517).

Frederick the Wise of Saxony takes Martin Luther to Wartburg and protects him as he translates the Bible in defiance of the Edict of Worms which prohibits all new doctrines (*see* Peasants' War, 1524).

Huldreich Zwingli at Zurich condemns the hiring of mercenaries.

Assertion of the Seven Sacraments by England's young Henry VIII is a reply to Luther. Pope Leo X gives Henry the title Defender of the Faith.

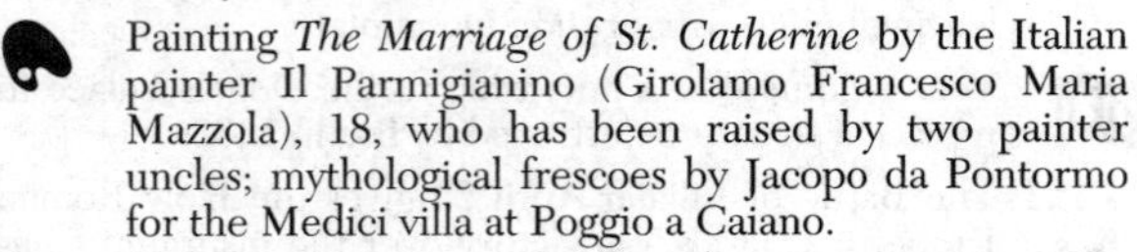

Painting *The Marriage of St. Catherine* by the Italian painter Il Parmigianino (Girolamo Francesco Maria Mazzola), 18, who has been raised by two painter uncles; mythological frescoes by Jacopo da Pontormo for the Medici villa at Poggio a Caiano.

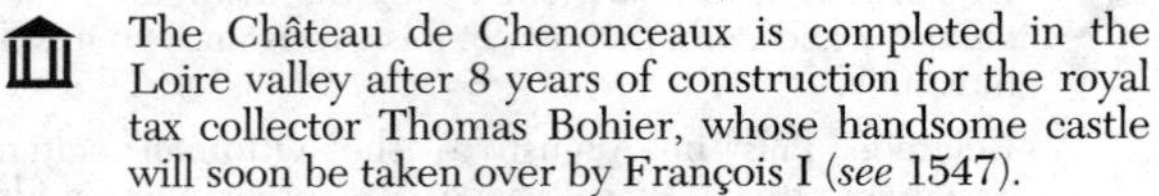

The Château de Chenonceaux is completed in the Loire valley after 8 years of construction for the royal tax collector Thomas Bohier, whose handsome castle will soon be taken over by François I (*see* 1547).

1522 The Holy Roman Emperor Charles V expels French forces from Milan with help from Florence, Mantua, and the papacy. England's Henry VIII joins the war against France.

The Knights of St. John who have held Rhodes since 1306 defend the island against the Ottoman sultan Suleiman, who triumphs December 21 after a 6-month siege (*see* Malta, 1530; Mohács, 1526).

The Ming emperor Jia Qing, who will reign until 1566, comes to power and expels the Portuguese for acts of piracy by Simao d'Andrade and others.

Hispaniola has a large-scale slave rebellion that will be followed in the next 31 years by at least 10 such uprisings in the Spanish possessions.

Ferdinand Magellan's lieutenant Juan Sebastian d'Elcano (del Cano) returns to Seville September 6 aboard the *Vittoria* with 18 surviving sailors of the Magellan expedition and with a cargo of valuable spices that more than pays for the expedition that has accomplished the first circumnavigation of the world.

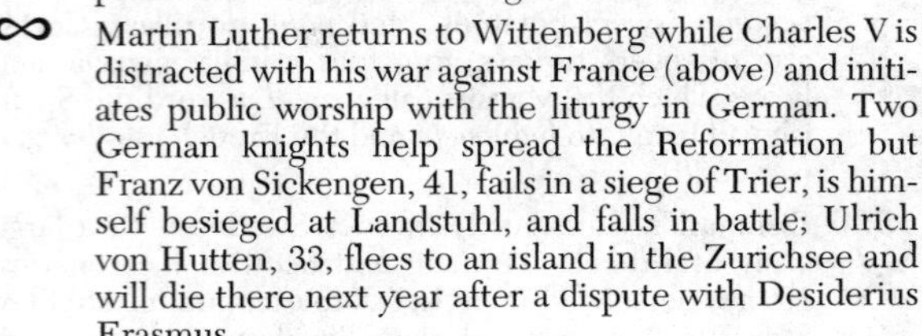

Martin Luther returns to Wittenberg while Charles V is distracted with his war against France (above) and initiates public worship with the liturgy in German. Two German knights help spread the Reformation but Franz von Sickengen, 41, fails in a siege of Trier, is himself besieged at Landstuhl, and falls in battle; Ulrich von Hutten, 33, flees to an island in the Zurichsee and will die there next year after a dispute with Desiderius Erasmus.

Huldreich Zwingli condemns celibacy and Lenten fasting, calls on the bishop of Constance to permit priests to marry (or at least wink at their marriages), and will himself marry in 1524.

Poetry *Colin Clout* and *Why Come Ye Nat to Courte?* by John Skelton are clerical satires directed against the rising power of Cardinal Wolsey.

Painting *The Resurrection* altar by Titian.

1523  England's Henry VIII tries to force a grant of funds from Parliament and provokes a rebellion that ends only when the king abandons his demand.

Denmark's Christian II is deposed by the nobility after a cruel 10-year reign. He is succeeded in January by the duke of Holstein, 52, who will reign until 1533 as Frederick I.

The house of Vasa that will rule Sweden until 1665 and make it the strongest power in the Baltic comes to the throne in the person of Gustavus Eriksson Vasa, now 27, who has led Swedish resistance to the Danes. He is crowned June 6 and will reign until 1560 as Gustavus I.

Two followers of Martin Luther—Augustine monks from Antwerp—are burnt alive at Brussels July 1.

Huldreich Zwingli at Zurich publishes his 67 *Articles* January 19, attacking transubstantiation and the authority of the pope.

Mennonite religious views have their origin at Zurich where a small community leaves the state church to pursue a form of Christianity that emphasizes the sanctity of human life and of man's word, acknowledges no authority outside the Bible and the enlightened conscience, limits baptism to true believers, and denies the Christian character of the state church and of civil authorities while recognizing a duty to obey civil laws (*see* Simons, 1536).

Pope Adrian VI dies September 14 at age 64 after 20 months as pope. The Dutch pontiff's successor, Giulio de' Medici, will reign until 1534, first in a line of Italian popes that will be unbroken until 1978.

Painting *Erasmus* by German painter-engraver Hans Holbein the Younger, 25, who has delighted the Dutch humanist with marginal drawings for his *Praise of Folly* (*see* 1509). Gerard David dies at Bruges August 13; Lucca Signorelli dies at Cortina October 16.

Turkeys from New Spain are introduced into Spain and will soon appear in England, where they will get their name from the "turkey merchants" of the eastern Mediterranean in a confusion with guinea fowl from Africa introduced by the Turks (*see* Cortez, 1519).

Conquistadors in Cuba recognize the possibilities of cultivating sugar there for the first time.

Maize grows in Crete and in the Philippines, where Magellan's men introduced the plant 2 years ago.

1524 Ferdinand of Austria makes an alliance with the two dukes of Bavaria and the bishop of southern Germany in a move taken at the instigation of the papal legate Lorenzo Campeggio to check religious changes.

The Swiss cantons of Lucerne, Uri, Schwyz, Unterwalden, and Zug join against Zurich and the Reformation movement.

French and imperial troops battle in Spain April 25. The French knight Pierre du Terrail, chevalier de Bayard, is shot in the back by a Spaniard and dies at age 51.

French forces invade Italy and retake Milan October 29 (but *see* 1525).

Persia's Shah Ismail dies May 23 at age 38 after a 22-year reign. His eldest son, now 10, will reign until 1576 as Tahmasp I.

Aden becomes a tributary of Portugal.

Vasco da Gama returns to India as Portuguese viceroy (*see* 1499; Mughal dynasty, 1526).

Spanish forces in New Spain disperse the main body of the Quiche army outside the city of Xelaju February 20. Chief Tecum Uman descends from his golden litter and kills the horse of Pedro de Alvarado, a lieutenant of Hernando Cortés, in the belief that man and horse are one. Alvarado runs the chief through with his sword, and panic spreads through the Quiche warriors.

A Peasants' Rebellion breaks out in the southern German states as Anabaptist Thomas Müntzer, 34, claims to be an apocalyptic messenger of God who brings "not peace, but the sword." Advocating social as well as religious reform, he overthrows the town government at Mühlhausen and sets up a communistic theocracy. His peasant followers demand an end to serfdom, feudal dues, and tithes, they battle Catholics, "heretical" books are burned in the marketplace at Mainz, and an orgy of pillaging and slaughter ensues (*see* 1525).

Guatemala City is founded by Pedro de Alvarado (above).

Francisco Pizarro, 54, proposes an expedition to "Piru." An officer who crossed the Isthmus of Panama with Balboa in 1513, he tells Panama's governor about a land to the south where people drink from golden vessels and have animals (llamas) that are half sheep, half camel.

Giovanni da Verrazano, 39, explores the North American coast. An Italian navigator sailing the 100-ton *Dauphine* for the French, Verrazano discovers a "beautiful" harbor in April and gives the name Angoulême to the island that will later be called Manhattan (*see* Hudson, 1609).

German mathematician Peter Bennewitz proposes a lunar observatory to produce a standard time that may help navigators determine longitude. Observation of the moon's position among the fixed stars may produce such a standard time, says Bennewitz, professor of mathematics at Ingolstadt and a friend of the emperor Charles V (*see* Frisius, 1530).

Painting *Ascension of Christ* by the Italian painter Correggio (Antonio Allegra), 30, is completed as a fresco on the cupola of the Benedictine church of San Giovanni at Parma; *The Sculptor* by Andrea del Sarto. Il Parmigianino moves to Rome and presents as his credentials to Pope Clement VII a self-portrait reflected in a convex mirror.

1525 The Battle of Pavia February 24 gives Spanish and German forces a victory over France's François I and his Swiss mercenaries. François has ordered his artillery to cease fire and has led a charge, only to have his horse shot under him by an arquebus. The battle ends the supremacy of armored knights, 6,000 Frenchmen are killed, and François, taken to Madrid a prisoner, writes to his mother (Anne of Brittany), "There is nothing left to me but honor, and my life, which is saved."

The Grand Master of the Teutonic Knights Albert von Bradenburg, 35, assumes the title Duke of Prussia April 8. Founded in 1198, the knights will break with the Church of Rome next year, beginning a long association between Lutheranism and the aristocracy controlling the vast estates of East Prussia, but Prussia remains a fief of Poland.

Peru's eleventh Inca king Huayana Capac dies at Quito, and his empire is divided between his sons Huascar and Atahualpa. Without a written language, they rule a complex, orderly society of 12 million in which each head of family is allowed enough land for his own needs and must also help till common lands that support the Inca court, the priesthood, and the engineers who build irrigation systems, stone roads, and fiber suspension bridges.

The German peasant rebellion is quelled May 14 as Philip, landgrave of Hesse, shoots down 5,000 and disperses Thomas Müntzer's army. Müntzer is beheaded May 27. Some 150,000 peasants have been killed in the uprising.

Francisco Pizarro sails from Panama November 1 in two caravels with 112 men and a few natives to explore "Piru" (above), whose natives call it Twwantinsuyu (The Four Corners of the World) (*see* 1526).

Santa Marta is founded by Spanish conquistador Rodrigo de Bastidas, 65. It is the first settlement in the territory that will become New Granada.

1525 *(cont.)* Jakob Fugger II dies December 30 at age 66 after a life that has seen the Fugger family become Europe's great financial power and Antwerp become the world's leading port, profiting from the wealth of the Indies and the New World. The merchants of Augsburg have rejected a scheme devised by Fugger to fix the price of bread at a permanently low level.

Painting *Madonna del Sacco* by Andrea del Sarto.

Theater *Don Duardas* by Gil Vicente.

The Fuggerei left by Jakob Fugger II at Augsburg consists of 106 dwellings which Jakob the Rich has built for rental at low rates to the poor of the city. He leaves it as a legacy to the people.

Chili (or chile) peppers and cayenne from the Americas are introduced by the Portuguese into India, where they will become the ingredients of the hottest curries.

1526 France's François I signs the Treaty of Madrid January 14 after being held captive for nearly a year by the emperor Charles V. François abandons Burgundy and renounces his claims to Flanders, Artois, Tournai, and Italy. Once released, he says the terms were extorted and the treaty is invalid. He forms an alliance with the Ottoman sultan Suleiman the Magnificent against Charles (*see* 1527).

The Battle of Panipat April 19 ends in victory for the Indian robber baron Zahir-ud-din Babar, now 43, whose 12,000-man army uses artillery to rout the 100,000-man force of Delhi's sultan Ibrahim shah Lodi. Having conquered the Punjab and made himself master of Lahore, Babar takes Agra and founds the Mughal Empire that will rule India until 1761 (*see* 1504; 1527).

Hungary's Louis II is killed at age 20 in the Battle of Mohács August 29 and 30, and his 20,000 poorly disciplined knights and peasants are defeated as the Turkish forces of Suleiman the Magnificent extend their European conquests from Belgrade.

Hungary's crown passes to Ferdinand, brother of Maximilian I, under terms of the 1515 Treaty of Vienna. But John Zápolya, 39, who served as regent during Louis's minority, is elected king by Hungary's nobility (*see* 1529).

Bohemia's crown goes to Louis's brother-in-law Ferdinand, brother of the emperor Charles V, who is elected king at age 23 to begin a reign that will continue until 1564.

Congolese king Mbemba Nzinga protests to Portugal's João III that his merchants are "taking every day our natives, sons of the land and sons of our noblemen and our vassals and our relatives" in exchange for goods from Europe and are selling his people as slaves to Brazilian sugar planters (*see* 1517). A convert to Christianity, the king says the kidnappers are depopulating his country.

Francisco Pizarro encounters rough seas en route to Peru and makes a landing at what will be called Port of Famine (*see* 1525). His freshwater barrels have breached, his food stores have spoiled, and 20 men die of hunger before the caravel he has sent back for fresh supplies arrives with flour and fresh meat.

Pizarro returns to Panama for reinforcements after being wounded in a skirmish with hostile natives, but on a second expedition he explores the Gulf of Guayaquil and secures some gold from natives who greet him as Viracocha. They believe him to be the fulfillment of a legend that the 14th century Inca king Viracocha Inca will return (*see* 1529).

The Tyndale Bible published in secret at Worms is an English translation of the New Testament by English linguist William Tyndale, 34, who has visited Martin Luther at Wittenberg. Tyndale has fled Cologne after the dean of Frankfurt discovered printers at work on the Bible, persuaded the senate of Cologne to forbid further printing, and warned Henry VIII and Cardinal Wolsey to watch English ports for Tyndale. The archbishop of Canterbury William Warham buys up copies of the Tyndale Bible on the Continent and burns them, but some copies are smuggled into England where bishops suppress most of them (*see* 1536).

The University of Granada has its beginnings in southern Spain.

Painting *Madonna of Burgomaster Meyer* by Hans Holbein, who visits Sir Thomas More in England with a letter of introduction from Erasmus; *The Last Judgment* by Lucas van Leyden; *The Four Apostles* by Albrecht Dürer is Dürer's last great religious painting; *The Last Supper* (fresco) by Andrea del Sarto; *Pesaro Madonna* by Titian.

The Château de Chantilly is completed in Renaissance style in Ile-de-France by the Montmorency family that will add a Petit Château in 1560.

1527 The Battle of Kanvaha March 16 gives the Mughal emperor Babar victory over Rajput forces, eliminating his chief Hindu rivals in northern India (*see* 1526; 1530).

The Treaty of Westminster signed April 30 allies England's Henry VIII and France's François I (*see* 1526). François begins new hostilities against the emperor Charles V.

The Renaissance greatness of Rome ends May 6 in a terrible sack of the city by Spanish and German mercenaries in the pay of the Holy Roman Emperor Charles V. The invaders besiege Pope Clement VII in the Castel San Angelo and take him prisoner.

England's Henry VIII appeals to Rome for permission to divorce Catherine of Aragon so that he may marry his young mistress Anne Boleyn (*see* 1533).

The Muslim Somali chief Ahmed Gran invades Ethiopia with firearms, taking a deadly toll. The negus appeals for Portuguese aid (*see* 1541).

Hernando Cortés completes his conquest of New Spain even though his power was revoked last year.

Sir Hugh Willoughby searches for a northeast passage to Cathay and the Spice Islands, but Henry VIII is more intent on building the English navy to protect his merchant trade with nearby coastal ports than with financing explorations.

Sebastian Cabot explores South America's Río de la Plata, sailing into the Paraná and Paraguay rivers.

English trade with Russia begins.

Chemotherapy and modern medical thinking are pioneeered by Basel physician Theophrastus von Hohenheim, 34, whose critics call him Theophrastus Bombastus. He rejects traditional notions of the body having four "humors," says the three prime "elements" are salt, sulfur, and mercury, and burns Greek medical books to promote his "new medicine" (*see* 1528).

Painting *The Vision of St. Jerome* by Parmigianino, who escapes to Bologna after the sack of Rome (above); *The Worship of the Golden Calf and Moses Striking Water from the Rock* by Lucas van Leyden.

Conquistadors return to Spain with avocados, papayas, tomatoes, chocolate, vanilla, and turkeys. They have found the natives of New Spain eating such foods in addition to algae, agave worms (maguey slugs), winged ants, tadpoles, water flies, larvae of various insects, white worms, and iguana (*see* 1528; tomato, 1534).

1528 French forces lay siege to Naples in the war against Charles V. Imperial troops under Philibert de Chalon, 26, prince of Orange, come close to starvation, but typhus forces an end to the siege in late August (below). After 6 weeks, the prince of Orange leads his cavalry out of the city and cuts down the retreating forces of François I and Clement VII.

Spanish conquistador Panfilo de Narvaez lands on the west coast of Florida in April with 400 prospective colonists. Most die of hunger and thirst; de Narvaez dies at sea.

Augsburg banker Bartholomaus Welser obtains rights to conquer and colonize most of northeastern South America as an hereditary fief. Welser has made a fortune trading in sugar, spices, and African slaves, and he has helped the emperor Charles V finance some Spanish expeditions to the New World (*see* 1527; 1546).

Hernando Cortés returns to Spain to confront the authorities who revoked his authority in 1526.

Giovanni Verrazzano is killed by natives on an expedition to Brazil.

A typhus epidemic sweeps the Italian states and kills an estimated 21,000 in July. The French army laying siege to Naples (above) is felled by the disease, and by late August only 4,000 of the original 25,000-man army remain.

Basel physicians force Theophrastus von Hohenheim to leave town (*see* 1527; Paracelsus, 1530).

The Reformation gains acceptance in Bern, Basel, and three other cantons, but Lucerne, Uri, Schwyz, Unterwalden, Fribourg, Solothurn, and Zug remain staunchly Catholic (*see* 1531).

Austrian evangelist Jacob Hutter founds a "community of love" whose Austrian, German, and Swiss members share everything communally. The Hutterites will be forced to seek refuge in Muscovy (*see* 1536).

The Courtier (Il Cortegiano) by Venetian diplomat Baldassare Castiglione, 50, bishop of Avila in Spain, is a dialogue on ideal courtly life and a guide to polite manners in the Spanish court.

Painting *Nicholas Kratzer* and *The Artist's Family* by Hans Holbein; *Madonna of St. Jerome (Il Giorno)* by Correggio; *The Visitation* by Jacopo da Pontormo. Albrecht Dürer dies at Nuremberg April 6 at age 56; Matthias Grünewald dies at Frankurt am Main.

Fontainebleau Palace, completed outside Paris for François I, is the work of three Italian architects.

Spanish colonists introduce wheat into New Spain (*see* Mexico, 1941).

Sweet potatoes, haricot beans, turkeys, cocoa, and vanilla are introduced to Spain by Hernando Cortés (above), who presents some of the beans to Pope Clement VII (Giulio de' Medici). Fava beans have been the only beans known to Europe until now.

1529 The Mughal emperor Babar gains victory over the Afghan chiefs of Bihar and Bengal May 6 at the Battle of Ghagra, extending his realm from Kabul to Bengal.

The Ottoman Empire reaches the height of its imperial expansion May 27 as ad-Din Barbarossa completes his conquest of Algeria. Suleiman I arrives at Buda September 3 with 250,000 troops and 300 cannon, taking the city after 6 days. John Zápolya is officially proclaimed king, and the sultan marches against Vienna, slaughtering Hungarians en route. His main army arrives September 27, Vienna's garrison of 20,000 men and 22,000 cavalry is far outmatched, but the Turks have supply problems, thousands die of cold and hunger, an infantry charge October 14 is driven back, and Suleiman's army returns to Constantinople, taking children home as slaves.

The Songhai king Askia Mohammed dies at age 86 after a 36-year reign that has raised the empire to new heights, expanding it east of the Niger (*see* 1517; 1591).

The Peace of Cambrai signed August 5 settles the conflict between France and the emperor Charles V. Louise of Savoy signs in behalf of François I, who gives up all claims in Italy, Flanders, and Artois and agrees to pay a ransom of 2 million crowns; Margaret of Austria signs in behalf of her nephew Charles V, who renounces any claims to Burgundy. Henry VIII accedes to the treaty August 27.

The emperor Charles V cedes Spanish rights in the Spice Islands to Portugal for 250,000 ducats.

Francisco Pizarro returns to Spain to claim the territory he has explored in Peru (*see* 1526). Charles V names Pizarro governor for life and captain-general of New Castile with a salary of 725,000 maravedis and nearly all the prerogatives of a viceroy, and he sends Pizarro back to Peru with 200 men and 27 horses in a fleet of three ships (*see* 1531).

South American territories of New Granada that will later be called Venezuela and Colombia are colonized

1529 *(cont.)* by the Welsers of Augsburg, whose agents explore the Andes and the valley of the Orinoco River (*see* 1528; Felderman, 1539).

London has a severe epidemic of the sweating sickness (last epidemic in 1518), and the disease spreads within 2 months to Hamburg, Lübeck, and Bremen, reaches Mecklenburg a month later (August), reaches Königsberg and Danzig in September, and then strikes Göttingen, where it rages so fiercely that corpses must be buried eight to a grave. The epidemic moves on to Marburg, where it interrupts the Council of the Reformation, to Augsburg, to Vienna, where it has some influence in raising the Turkish siege (above), and to Switzerland, but the sweating sickness somehow spares the French (*see* 1551).

Protestantism gets its name April 16 as followers of Martin Luther protest a ruling by the Diet of Speyer forbidding the teaching of Luther's ideas in Catholic states while letting Catholics teach in Lutheran states.

Henry VIII removes Cardinal Wolsey as his lord chancellor October 17 for failing to secure a papal annulment of his marriage; he replaces Wolsey with Thomas More.

Japanese monks from the Tendai monasteries on Mount Hiei sweep down on Kyoto and massacre adherents of Buddhism's Nicheren sect.

Painting *The Madonna with St. Margaret and Other Saints* by Parmigianino.

Hymns "A Mighty Fortress Is Our God!" and "Away in a Manger" by Martin Luther.

The Turks plant paprika (capsicums) from the New World at Buda (above) (*see* nutrition, 1928).

Vast fields of maize from America are grown in Turkey, whence the grain will come to England as "turkey corn."

The elector of Saxony, who supported Martin Luther, read a protest at the Diet of Speyer. Hence the word *Protestant*.

Sweet oranges from the Orient are introduced into Europe by Portuguese caravels and will forever be called portugals in the Balkans, in parts of Italy, and in the Middle East (*see* Chinese orange, 1635).

1530 Spain's Carlos I is crowned Charles V of the Holy Roman Empire and king of Italy February 23 at Bologna by Pope Clement VII, the last coronation of a German king by any pope. Charles presides over the Diet of Augsburg (below).

The Sovereign Military Hospitaler Order of Saint John founded in 1113 settles in Malta and will hereafter be called the Knights of Malta (*see* 1522; 1565).

Cardinal Wolsey dies November 29 at age 55 after a political career in which he has amassed a fortune second only to that of Henry VIII.

The Mughal emperor Babar dies at age 47. His son Muhammed Humayun will reign without distinction until 1556.

The Inca Huascar's half brother Atahualpa in Peru moves south from Quito with an army of 30,000 and destroys the town of Tumebanba (*see* 1525). Huascar's army of 10,000 falls back, leaving suspension bridges over the Apurimac River intact. Atahualpa crosses the Apurimac, his generals take Huascar prisoner, and Atahualpa rests at Cajamarca (*see* 1531).

German mathematician Gemma Frisius suggests that accurate mechanical clocks be set to the local time of a prime meridian to produce a standard time for the measurement of longitude as an aid to navigation. Frisius is professor at Louvain University and cosmographer royal to Charles V (*see* Bennewitz, 1524; Greenwich Observatory, 1676).

Die Gross Wundartznei by Paracelsus describes the doctrine of "signatures" and other specious medical ideas but does establish sound principles that bodily functions are based on chemical processes and that different diseases demand different treatments (e.g., mercury is a specific for syphilis) (below). Paracelsus is the name Theophrastus von Hohenheim has adopted to equate himself with the Roman physician Celsus of 1500 years ago.

Syphilis sive Morbus Gallicus by Italian physician Girolamo Fracastoro, 47, gives a name to the disease first observed in 1495.

The Confession of Augsburg read at the brilliant diet in that city June 25 is a detailed explanation of Lutheranism designed to reconcile the Protestants with the Catholic Church. Martin Luther's collaborator Philipp Schwarzert Melancthon, 33, has prepared the *Confession*, which fails to move the diet. It orders the abolition of all innovations.

Confessio Augustana by Philipp Melancthon (above) is published at Wittenberg. A disciple of Erasmus, Melancthon gave Lutheranism a dogmatic basis in his *Loci Communes* of 1521.

Poetry *Das Schlaraffenland* by Nuremberg cobbler-poet-dramatist Hans Sachs, 36, whose humorous anecdotes told in doggerel verse are called Schwänke. Their satirical good humor does not obscure their moralizing.

Painting *Adoration of the Shepherd* and *Assumption of the Virgin (La Notte)* by Corregio; *Man in a Red Cap* by Titian. Andrea del Sarto dies at Florence September 29 at age 44.

The potato, discovered in the Andes by Spanish conquistador Jiminez de Quesada, 30, will provide Europe with a cheap source of food and thus spur population growth. Quesada finds the natives eating only the largest of the tubers, which they call *papas*, and planting the smallest, thus steadily reducing the size of the tubers they harvest. Since the tubers are only about the size and shape of peanuts, the Spanish mistake them for a kind of truffle and call them *tartuffo*, which will persist with variations as the word for potato in parts of Europe (*see* 1539).

1531 The Schmalkaldic League organized February 6 allies the majority of Europe's Protestant princes and imperial cities against the Holy Roman Emperor Charles V, who has had his brother Ferdinand, 27, elected king of the Romans (meaning German king) at Cologne in January. The elector of Saxony leads the opposition to Ferdinand.

The Catholic cantons attack Zurich and defeat the Protestants October 11 in the Battle of Kappel. Huldreich Zwingli is killed in the fighting.

Atahualpa in the South American altiplano has his half brother Huascar put to death and becomes the Inca himself (*see* 1530; 1532).

Francisco Pizarro and his brothers Gonzalo and Hernando leave Panama for Peru with 300 men and 100 horses (*see* 1529). Reinforcements under the command of Hernando de Soto, 31, join the expeditionary force and it moves up into the Cordilleras with two artillery pieces, encountering virtually no opposition (*see* 1526; 1532).

São Vicente is founded in Brazil by Portuguese conquistador Martin Affonso de Souza. New settlers extend Portugal's Brazilian holdings.

Sugar becomes as important as gold in the Spanish and Portuguese colonial economies.

The bourse that opens at Antwerp is the forerunner of all mercantile exchanges.

German sculptor Tilman Riemenschneider dies at Würzburg July 8 at age 71.

Spain's West Indian colonists cultivate tobacco on a commercial scale (*see* 1518; 1560).

An earthquake shatters Lisbon January 26, killing 30,000.

1532 The Ottoman forces of Suleiman II invade Hungary, but Carinthia and Croatia repel the attackers.

Brittany's Duchess Anne signs the Treaty of Plessis-Mace with France's François I, who adds the duchy to his realm (final absorption of Brittany into France will come in 1547).

The Inca Atahualpa visits the Spanish camp of Francisco Pizarro, who has ascended the Andes. Pizarro seizes the Inca November 16 and holds him for ransom, paralyzing the machinery of Inca government (*see* 1533).

The religious peace of Nuremberg permits Protestants free exercise of their religion until the meeting of a new council to be summoned within a year.

Cartagena is founded by Pedro Herdia, who acts by authority of the Holy Roman Emperor Charles V (Spain's Carlos I).

Fiction *Pantagruel* by Lyons physician François Rabelais, 38, whose earthy satire is published under the pen name Alcofrybas Nasier.

Painting *Charles V* by Titian; *Madonna of St. George (Il Giorno)* by Correggio; *Martin Luther* and *Philipp Melanchthon* by Lucas Cranach the Elder.

Russia's Church of the Ascension is completed at Kolomenskoie.

The Holy Roman Emperor Charles V establishes residence at Madrid and pays for improvements to the imperial palace with taxes on Caribbean sugar.

Horses introduced into Peru by Francisco Pizarro (above) will be seen within 3 years running wild on the pampas of eastern South America. Within 70 years the horses will be running in herds of uncountable size and will be revolutionizing daily life.

Count Cesare Frangipani in Rome invents the almond pastry that will bear his name.

1533 England's Henry VIII secretly marries Anne Boleyn, 26, who has been his mistress for the past 6 years. The ceremony is performed January 25 by Thomas Cranmer, 43, who has advised Henry that his 1509 marriage to Catherine of Aragon is null and void because she was previously married to Henry's late brother Arthur, prince of Wales, even though that marriage was never consummated. Henry makes Cranmer archbishop of Canterbury March 30, and the new queen is crowned at Westminster June 1 (*see* 1536).

Catherine de' Medici of Florence marries Henri of Valois, 14, duc d'Orléans, who will become France's Henri II in 1547. Catherine is also 14, and her family has grown rich in the spice trade and by supplying alum to the textile industry.

Denmark's Frederick I dies at age 62 and is succeeded as king of Denmark and Norway by his son of 30, who will reign until 1559 as Christian III. Danish Catholics oppose Christian and hope to make his younger brother Hans king. Peasants and burghers hope to restore the imprisoned ex-king Christian II. Count Christopher of Oldenburg leads Lübeck in support of Christian II, but most of the nobility supports Christian III. He will crush the rebellion, make peace with Lübeck, and establish Lutheranism as the state religion.

The grand duke of Muscovy Basil III Ivanovich dies at age 54 and is succeeded by his 3-year-old son, who will make himself czar of Muscovy and rule until 1584 as Ivan IV (the Terrible) (*see* 1547).

1533 ***(cont.)*** Suleiman the Magnificent signs a treaty June 22 with Ferdinand of Hungary, brother of the Holy Roman Emperor, who rules part of Hungary while Suleiman's puppet John Zápolya rules the other part.

The Peruvian Empire ends August 29 with the strangulation of the last Inca, Atahualpa, who has professed himself a Christian and has received baptism. The conquistadors have accused the Inca of plotting to overthrow them, Francisco Pizarro has brought him to trial on charges of murder, sedition, and idolatry, and he has been condemned to death despite the payment of a huge ransom for his release. Pizarro's advisers (with the exception of the priest Valerde) have protested the sentence to death by fire, calling it treacherous, so the Inca has died at the hands of a strangler.

Hispaniola's slaves stage another uprising which the conquistadors suppress with great bloodshed.

The Algerian corsair Khair ad-Din, 67, evacuates Moors driven from Spain by the Inquisition. A former Greek, Khair ad-Din has commanded the pirates since the death of his brother at Spanish hands in 1518.

The ransom paid for Atahualpa (above) has included gold and silver chairs, fountains, statues, plates, and ornaments made by generations of craftsmen—literally tons of gold and silver. The Spaniards melt it into ingots worth $4 million and distribute the bullion; most goes to Pizarro and his brothers, a little to the Church, and one-fifth to the Holy Roman Emperor Charles V (*see* 1532; 1534).

Gemma Frisius announces a new method of surveying that will replace the laborious method of pacing out distances (*see* 1530).

Catherine de' Medici (above) introduces to France such vegetables as broccoli, globe artichokes, savoy cabbage, and haricot beans—*fagioli* given her by her brother Alessandro (*see* 1528; 1749).

The double boiler called by Italians the *bagno maria* after a legendary alchemist named Maria de Cleofa is introduced to the French court by Catherine de' Medici (above). The French will call it a *bain marie*.

Stuffed guinea hen is introduced by Catherine (above), and the dish will be known to the French as *pintade à la Medicis.* Catherine also introduces truffles and starts the French digging enthusiastically for truffles of their own.

Pastries such as frangipani, macaroons, and Milan cakes introduced only recently to Florence will be introduced to France by Catherine (above) (*see* frangipani, 1532).

The death of the Inca Atahualpa was the most infamous act of treachery committed by the Spanish conquistadors.

1534 Ottoman forces take Tabriz from the Persians July 13. Shah Tahmasp, now 20, has his regent executed and assumes personal power.

Tunis falls to Turkish forces led by the corsair Khair ad-Din (*see* 1533; 1535).

Ottoman forces take Baghdad December 31.

Santiago de la Vega (Spanish Town) is founded on the West Indian island of Jamaica.

French explorer Jacques Cartier, 43, sets out for the New World with two small ships and 36 men on orders from François I. Cartier sails into the Gulf of St. Lawrence, he plants the cross at Gaspé Bay, and he finds *les sauvages* growing beans.

England's Henry VIII (below) will break up the nation's Catholic monasteries, upsetting the nation's land system. A merchant class will arise, more land will be enclosed, landlords will prevent peasants from grazing their cows and sheep on common land, the need for rural labor will decline, the enclosure system will lead to increased poverty and hunger, and the breakup of the monasteries will have other effects (*see* honey shortage, 1536; Thomas More, 1515).

Francisco Pizarro returns to Spain with the royal share (one-fifth) of the huge ransom paid for the Inca Atahualpa, who was executed last year at Cajamarca.

Henry VIII breaks with the Church of Rome which has voided the annulment of his marriage to Catherine and has excommunicated him. The Reformation is established in England by the Act of Supremacy which appoints the king and his successors Protector and only Supreme Head of the Church and Clergy of England, but English dogma and liturgy will remain for a time essentially unchanged from Roman Catholic dogma and liturgy (*see* 1551).

Pope Clement VII dies September 25 after eating poisonous mushrooms. He has evaded Henry VIII's demand for nullification of the 1509 marriage to Catherine of Aragon and has refused to sanction Henry's marriage to Anne Boleyn, who last year bore Henry's daughter Elizabeth.

Pope Clement (above) is succeeded by Alessandro Farnese, 66, who assumes the papacy as Paul III.

The Society of Jesus (Jesuit order) is founded by Basque ecclesiastic Ignatius Loyola (Iñigo de Oñez y Loyola), 43, and five associates (*see* 1540; 1541).

The Dutch Anabaptist fanatic John of Leyden (Jan Beuckelszoon), 27, establishes a theocratic kingdom of Zion at Münster, saying the world will soon end but his followers will be spared. The former tailor, merchant, and innkeeper introduces communization of property and polygamy, shocks the Catholic world with his hedonistic orgies, and discredits the Reformation; Catholics charge that the Anabaptists' antinomian anarchism illustrates the evil consequences of Martin Luther's doctrine of justification by faith alone (*see* 1521; 1535).

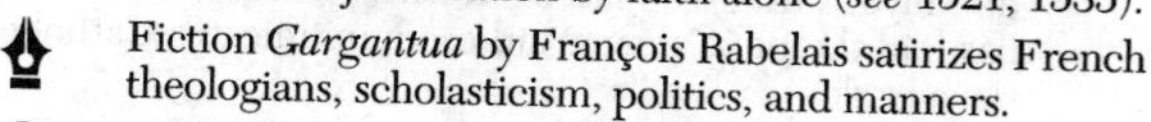

Fiction *Gargantua* by François Rabelais satirizes French theologians, scholasticism, politics, and manners.

Michelangelo moves from Florence to Rome after completing a tomb for the Medici family. Correggio dies at his native Correggio in Emilia March 5 at 45.

Chambord is completed in the Loire Valley for François I. Built over a period of 15 years by a crew of 1,800 men, the great château replaces a 50-room hunting lodge with a fantasy of steeples, turrets, bell towers, and cupolas. It will be enlarged by the king's successors, and it will eventually have 440 rooms, more than 350 chimneys, 75 staircases, and nearly 14,000 acres of gardens and game reserves surrounded by a continuous 20-mile wall.

The Farnese Palace designed by Antonio Sangallo the Younger goes up in Rome for the new pope.

English fish consumption begins to decline as the king's break with Rome (above) ends church rules against eating meat on Fridays and during Lent. England will enact new fish laws to revive her coastal towns and to encourage the fishing industry from which the English navy draws recruits.

The first written description of a tomato appears in an Italian chronicle that calls the cherry-sized yellow fruit pomo d'oro (golden apple) (*see* 1519; 1596).

1535 An Act of Union joins the principality of Wales to England (*see* Scotland, 1707).

Milan's Francesco Sforza II dies October 24 at age 43. He dies without issue, the main line of the Sforza dynasty is ended, and Milan becomes a suzerainty of the Holy Roman Emperor Charles V (*see* 1536).

Tunis falls to the forces of Charles V, who has sent a fleet commanded by the Genoese admiral Andrea Doria. The pirate leader Khair ad-Din is defeated and Tunis is sacked (*see* 1534).

The viceroyalty of New Spain is established with Mexico City (Tenochtitlan) as its capital.

Sir Thomas More is beheaded July 6 at age 57 by order of England's Henry VIII for refusing to swear an oath of supremacy as required by last year's Act of Supremacy.

Lima, Peru, is founded by Francisco Pizarro, who returns to Peru and resumes his exploitation of the Inca Empire (*see* 1534; university, 1551).

Jacques Cartier returns to North America with 110 men and sails up the St. Lawrence River to what will later be the site of Montreal (*see* 1534). Cartier's fleet is frozen solid in November at the mouth of the St. Charles River under the Rock of Quebec. "All our beverages froze in their casks," Cartier will write, "and on board our ships, below hatches as on deck, lay four fingers' breadth of ice."

Münster's Anabaptist leader John of Leyden beheads one of his four wives by his own hand in the marketplace in a fit of frenzy and justifies all his arbitrary actions on the basis of having received visions from heaven (*see* 1534). After 12 months of profligacy the besieged city falls to Francis of Waldeck June 24, and the leading Anabaptists are imprisoned (*see* 1536).

Painting *Madonna del Collo Lungo* by Parmigianino.

Scurvy breaks out among the Hurons and then among the French under Jacques Cartier (above).

1536 Geneva adopts the Reformation as her ally Bern subdues Vaud, Chablais, Lausanne, and other territories of the duke of Savoy. A long struggle begins between Bern and Savoy (*see* Calvin, 1541).

England's Henry VIII has his wife Anne Boleyn beheaded May 19 on charges of adultery. On May 20 he marries Ann's lady-in-waiting Jane Seymour, 27, whose brother is duke of Somerset.

A Catholic rebellion in England gains support from the prelate Reginald Pole, 36, son of the countess of Salisbury. The rebellion is crushed, but Pole is created a cardinal by Pope Paul III, who will dispatch him as an emissary to incite France's François I and the Holy Roman Emperor Charles V to send an expedition to depose Henry VIII.

A third war begins between Charles V and François I, who has renewed claims to Milan following the death last year of Francesco Sforza. François makes a formal alliance in March with Suleiman the Magnificent after more than a decade of discussions between French and Ottoman emissaries. Suleiman advances on Hungary, and sends fleets to ravage the coasts of Italy while French forces take Turin.

John of Leyden and some of his more prominent Anabaptist followers are tortured with exquisite cruelty and then executed in January in the marketplace of Münster (the zealot's remains will swing in a cage from a church rafter until the 20th century). The Anabaptists are butchered wholesale and will hereafter lose their identity.

The Anabaptist Jacob Hutter is arrested and burned at the stake on orders from the German king Ferdinand. Hutter has taught that his followers were God's elect and could only expect hardship and suffering (*see* 1528; 1874).

English clergyman William Tyndale of 1526 bible fame is condemned for heresy at Vilvorde Castle outside Brussels. He is strangled at the stake October 6.

Portugal installs the Inquisition and invites the Society of Jesus to send Jesuits into the country (*see* 1506).

1536 ***(cont.)*** Spanish grandee Don Pedro de Mendoza arrives early in the year at the mouth of the Río de la Plata with 1500 settlers in 11 large ships to establish the town of Nuestra Señora de Buenos Aires. But the Spaniards soon antagonize the native Guaraní population, which lays siege to the town from June to August (*see* 1537).

Hernando Cortés discovers Lower California (*see* 1528; 1540).

Spanish explorer Alvar Nuñez Cabeza de Vaca, 46, reaches Mexico City (Tenochtitlan) after having been wrecked on a Texas coastal island in 1528. Captured by the natives and held captive for 2 years, Cabeza de Vaca has escaped, has made a long overland journey through the Southwest, and sails for Spain (*see* 1542).

Jacques Cartier returns from New France with a Huron chief, the chief's two sons, three adult Hurons, two little girls, and two little boys (all of whom soon die). The Huron chief swears to François I that Saguenay is rich in cloves, nutmegs, and peppers and grows oranges and pomegranates.

Portuguese mathematician Pedro Nunes, 44, describes errors in the plain charts used by seamen. Nunes will invent the nonius for graduating instruments, and his nonius will be improved and developed into the vernier (*see* 1631).

Anabaptists (above) will be confused by many with Mennonites—followers of Friesland clergyman Menno Simons, 43, who leaves the Roman communion January 12 after several years of questioning infant baptism and getting inconsistent answers to his questions from Martin Luther and other Protestant leaders. Menno and his followers repudiate the doctrines of the Münster Anabaptists (*see* 1523; 1537).

The Practyce of Prelates by William Tyndale (above) is published in England. Tyndale lost favor with Henry VIII by criticizing the king's divorce from Catherine of Aragon.

Jacques Cartier (above) loses 25 men to scurvy in New France (*see* 1535). By mid-February, "out of the 110 men that we were, not ten were well enough to help the others, a thing pitiful to see." But although 50 Hurons in the area also die of scurvy, a Huron chief shows Cartier how to grind the bark of the common arborvitae *Thuja occidentalis,* boil the ground bark in water, drink the infusion every other day, and apply the residue as a poultice to swollen, blackened legs. Cartier digs up arborvitae saplings and transplants them in the royal garden at Fontainbleau upon his return.

England begins to suffer shortages of honey as monasteries which raised honeybees as a source of wax for votive candles are dissolved pursuant to the 1534 Act of Supremacy.

1537 The Inca Manco Capac II rebels against Pizarro and establishes a new state at Vicabamba.

French and Turkish forces lay siege to Corfu with help from the Algerian corsair Khair ad-Din, but Corfu's Venetian defenders hold fast.

Sweden's Gustavus I Ericksson ends the Hanse's Baltic monopoly after a war with Lübeck.

Henry's VIII's queen Jane Seymour dies October 24 a few days after giving birth to a son who will become Edward VI in 1547.

The papal bull *Sublimus Deus* issued June 2 by Paul III prohibits enslavement of Indians, but Charles V forces the pope to recall his briefs and imprisons Bernardino de Mayo, a friar who has been with Pizarro in Peru and has had the pope's ear. Paul excommunicates Catholic slave traders.

Slaves on the island of Hispaniola stage another revolt (*see* 1533).

Don Pedro de Mendoza leaves Buenos Aires for Spain in April but is desperately ill with syphilis and dies en route. His settlement will be taken over by the Guaraní within 4 years (*see* 1580).

Asunción is founded on the Paraguay River August 15 by Spanish explorer Juan Salazar de Espinosa.

Nova Scientia by Italian mathematician Niccolo Tartaglia, 37, discusses the motion of heavenly bodies and the shape and trajectory of projectiles. Tartaglia will discover the solution to the cubic equation in 1541, but mathematician Geronimo Cardano, now 36, will appropriate his findings.

Menno Simons preaches the Mennonite views of 1523 with modifications to some Anabaptists who have left the Münster faction (*see* 1536). Set apart to the eldership at Groningen in January, Menno repudiates the formation of a new religious sect, but he begins actively to advocate a faith that absolutely forbids oaths and the taking of life (and thus makes it impossible for a believer to be a magistrate or to serve in the army). He rejects terms such as Trinity which cannot be found in the Bible, prohibits marriage with outsiders, but insists that his followers obey the law in all things not prohibited by the Bible (*see* 1568; Amish, 1693).

The University of Lausanne has its beginnings.

The University of Strasbourg has its beginnings.

1538 The Peace of Grosswardein February 24 ends hostilities between Hungary's two kings, Ferdinand of Hapsburg and John Zápolya (*see* 1533).

The third war between France's François I and the Holy Roman Emperor Charles V ends June 18 with each side retaining possession of its conquests, but the truce of Nice is inconclusive.

A Holy League against the Ottoman Turks allies Pope Paul III, Venice, and the Holy Roman Emperor Charles V. Charles tries to buy off the corsair Khair ed-Din, who controls much of the Mediterranean. A Venetian fleet under Genoa's Doge Andrea Doria is defeated at Prevesa in September by the Ottoman Turks, who gain naval supremacy in the Mediterranean.

An Ottoman naval expedition takes over Aden and Yemen on the east coast of the Red Sea and ventures as far as the northwest coast of India.

Painting *The Venus of Urbino* and *The Allegory of Marriage* by Titian; *Christina of Denmark* by Hans Holbein.

The first definite reference to Newfoundland fishing expeditions appears in Basque records, although Basque fishermen have been visiting the Grand Banks for some decades and possibly for a century (*see* 1497; 1504).

Poule d'Inde (chicken of the Indies, meaning turkey) is served in France (*see* 1523). The French will call hen turkey *dinde*, tom turkey *dindon*.

English parish records begin to indicate births, marriages, and deaths. The records will make future population estimates more accurate.

1539 The Treaty of Toledo signed February 1 ends hostilities between Charles V and François I. Charles makes a truce at Frankfurt April 19 with the German Protestants there.

Charles puts down a rebellion in his native Ghent, whose citizens have refused to pay taxes to finance the emperor's war with France. François I has not responded to their pleas for help, and they are stripped of their privileges.

Nikolaus Felderman arrives at Bogotá as agent for the Welsers (*see* 1528). The city has been founded in the past 2 years under the name Santa Fe de Bogotá by Gonzalo Jiminez de Quesada.

Franciscan missionary Marcos de Niza explores territory that will become Arizona and New Mexico. He returns from the Zuni pueblos with glowing accounts of gold in the legendary "seven cities of Cibola."

The Bank of Naples is founded with a capital of 4,000 ducats by Neapolitans Aurelio Paparo and Leonardo di Palma to free the poor from the evils of usury by granting loans on pledges without interest or at very low rates of interest. The bank will grow to become the most powerful agricultural credit institution in the southern Italian provinces.

A fixed maximum on Spanish grain prices becomes a permanent facet of royal economic policy. Applied only sporadically until now, the *tasa del trigo* has the effect of favoring sheep raising over tillage, making Spain dependent on imports for her food, and making food prices so high that the Spanish worker can barely afford food, clothing, housing, and fuel.

England's Statute of the Six Articles makes it heresy to deny any of six positions: transubstantiation, communion in one kind for laymen, celibacy of the priesthood, inviolability of chastity vows, necessity of private masses, necessity of auricular confession.

The poet Guru Nanak dies in the Punjab at age 70 after establishing the tenets of the Sikh religion (*Sikh* is Punjabi for disciple) which is neither Hindu nor Muslim (*see* Golden Temple, 1605).

Painting *King François I* by Titian.

Mantua's Palazzo Ducale is completed by architect-painter Romano Giulio, 40, who was an assistant to the late Raphael.

Potatoes arrive in Spain with conquistadors returning from Quito (*see* 1530). Pedro de Cieza of the Pizarro expedition describes the tubers as something similar to chestnuts (*see* 1540).

Portugal's agrarian system declines as a result of dependence on slave labor introduced since 1441.

1540 England's Henry VIII marries Anne of Cleves at Greenwich January 6 less than a week after meeting her and 4 days after saying openly that she had no looks, spoke no English, and was "no better than a Flanders mare." Anne, 25, is the daughter of the German Protestant leader John, duke of Cleves, and the marriage has been arranged by the lord privy seal Baron Thomas Cromwell, 54, to give Henry an ally against the Holy Roman Emperor Charles V and France's François I, but Henry soon finds that he has no reason to fear an attack from either. The king says the marriage has not been (and cannot be) consummated; he makes Cromwell earl of Essex but has the duke of Norfolk charge him with treason June 10. Parliament sends Cromwell to the Tower of London and declares the king's marriage null and void July 9. Henry marries Norfolk's niece Catherine Howard, 25, and Cromwell is beheaded on Tower Hill July 28.

Hungary's John Zápolya dies in July at age 53. The Turks recognize his infant son Sigismund as the new king, and he will reign until 1571 as John II Zápolya, but Ferdinand of Hapsburg invades eastern Hungary and Turkish forces take over the great central plain, splitting the country in three to begin decades of conflict.

Dutch whaling captain Jon Greenlander lands in Greenland and finds the last Norse colonist lying dead outside his hut with an iron dagger in his hand (*see* 982).

Hernando Cortés, now 55, returns to Spain (*see* 1536; 1547).

Spanish explorer Hernando de Alarcon discovers the Colorado River.

Hernando de Soto lands at Tampa Bay with more than 600 men, 200 horses, and 13 hogs. Rich with gold from Peru, de Soto moves west (*see* 1541).

Spanish explorer Francisco Vazquez Coronado, 30, arrives in the American southwest with the first horses, mules, cattle, sheep, and hogs ever seen in the region. Accompanied by the missionary Marcos de Niza, he takes a pueblo July 7, thinking it is one of the Seven Cities of Cibola, and finds the natives living in poverty. His lieutenant Lopez de Cardenas discovers the Grand Canyon in December.

Waltham Abbey in Essex, the last of the great monastic houses, is seized March 23 by agents of Henry VIII, who has enriched himself and his friends with the

Henry VIII ruled England through almost 38 turbulent years, sparring with Rome and with half a dozen wives.

properties, plate, and jewels formerly owned by the Church.

United companies of barbers and surgeons are incorporated at London.

Pope Paul III recognizes the Jesuit order founded in 1534. The pope will make the Jesuits his chief agents in spreading the Counter Reformation.

Painting *Henry VIII* by Hans Holbein; *Doge Andrea Gritti* by Titian; *Adoration of the Shepherd* by Venetian painter Jacopo (or Giacomo) da Bassano (*né* Ponte), 30. Il Parmigianino dies at Casalmaggiore near Parma August 28 at age 37.

A specimen potato from South America reaches Pope Paul III via Spain. The pope gives the tuber to a Frenchman who introduces it into France as an ornamental plant (*see* 1539; 1740).

1541 Portugal sends troops to Ethiopia under Christopher da Gama, son of the late explorer Vasco da Gama. The Portuguese will expel the Somali chief Ahmed Gran, who invaded Ethiopia in 1527.

Francisco Pizarro completes his conquest of Peru but is assassinated June 26 at age 69 or 70 by supporters of Diego de Almagro, 21, whose father and namesake Pizarro defeated and executed 3 years ago. Diego is made governor of Peru at Lima June 26, but royalists will defeat him next year and put him to death.

Pizarro's lieutenant Pedro de Valdivia founds the city of Santiago in Chile February 12.

Hernando de Soto discovers the Mississippi River May 8 (*see* 1540; Marquette and Joliet, 1673).

Jacques Cartier makes a third expedition to North America. Financed by François I, he establishes a short-lived community at Quebec (*see* 1534).

John Calvin (Jean Chauvin), 32, establishes a theocratic government that will make Geneva a focal point for the defense of Protestantism throughout Europe. The French theologian was banished from Paris in 1533, his *Institution Chrétienne* was published 3 years later, he tried to establish a theocracy at Geneva with Guillaume Farel, 52, 3 years ago, a popular revolt drove him out of town, but the Swiss burghers have invited him to return.

Catholic missionary Francis Xavier, 35, sails for the Orient from Lisbon as Jesuit apostle to the Indies at the behest of Ignatius of Loyola, first superior (or general) of the Society of Jesus (*see* 1540; 1551).

Painting *The Last Judgment* by Michelangelo on the altar wall of the Sistine Chapel at Rome measures 60 feet in length, 33 feet in height, and is the largest, most comprehensive painting in the world.

1542 England's Henry VIII has his fifth wife Catherine Howard beheaded February 13 on charges of adultery. She has admitted to having had premarital intimacies with her cousin Thomas Culpepper and with Francis Dereham; she has had clandestine meetings with both since her marriage to the king, and they have been beheaded earlier.

Henry VIII makes Ireland a kingdom. The Irish summon a parliament in June, and six Gaelic chiefs approve the act that makes Henry king of Ireland.

The Battle of Solway Moss November 25 gives Henry VIII a victory over Scotland's James V. James dies December 14 at age 30 and is succeeded by his week-old daughter Mary Queen of Scots, who was born to his second wife Mary of Guise as he lay dying.

The Universal Inquisition established by Pope Paul III at Rome July 21 tries to stem the tide of the Reformation with cruel repression (*see* Jesuits, 1540). A council of Dominican cardinals conducts trials of alleged heretics and permits them no legal counsel.

Spain forbids her colonists in America to enslave Indians but does not abolish Indian slavery (*see* papal bull, 1537).

Some 150 Spanish colonists led by Alvar Cabeza de Vaca travel 600 miles inland from the coast of southern Brazil and settle at Asunción (*see* 1536; 1537).

Spanish explorer Juan Rodriguez Cabrillo probes the coast of California. His pilot Bartolome Ferrelo nearly reaches the mouth of the Columbia River (*see* Drake, 1579; Gray, 1791).

Hernando de Soto dies May 21 at age 46 after having spent the winter on the Ouachita River and is buried in the Mississippi which he discovered last year. De Soto's men descend the Mississippi under the command of Luis Moscoso de Alvarado from a point near the junction of the Arkansas River.

Spanish conquistador Gonzalo Pizarro returns to the mouth of the Amazon August 24 after an 8-month journey in which he has reached the river's headwaters and been attacked en route by Indian women with bows and arrows. He has named the river accordingly.

1543 England's Henry VIII marries Catherine Parr, 31, July 12.

The Berber pirate Kheir ad-Din Barbarossa joins with François I to bombard, besiege, and sack the imperial city of Nice.

Spanish conquistador Ruy Lopez de Villalobos is driven out of the Philippine Islands by the natives a year after discovering the islands and giving them their name. He is captured by the Portuguese.

Japan receives her first European visitors as a Chinese ship carrying two Portuguese adventurers is wrecked on an island off Kyushu. Called *namban* by the Japanese, the foreigners have arquebuses (muskets) which the local lord buys and duplicates. Firearms will henceforth be employed in Japanese warfare.

The Spanish Inquisition burns Protestants at the stake for the first time.

Juan Rodriguez Cabrillo dies in January; his pilot Bartolome Ferrelo returns to New Spain April 14 after discovering a bay that will later be called San Francisco (*see* Drake, 1579).

De Revolutionibus Orbium Coelestium by Prussian-Polish astronomer Nikolaus Copernicus (Mikolaj Kopernik), 70, defies Church doctrine that the earth is the center of the universe and establishes the theory that the earth rotates daily on its axis and, with other planets, revolves in orbit round the sun. Copernicus has not permitted the work to be published earlier, he sees the first copy on his deathbed, he dies May 24 of apoplexy and paralysis, and his idea is as revolutionary as his title (the word *revolution* will derive from his *Revolutionibus*) (*see* Tycho Brahe, 1572, 1576; Kepler, 1609; Galileo, 1613).

De Corporis Humani Fabrica by Belgian anatomist Andreas Vesalius, 29, with anatomic drawings by the Venetian painter Titian is the first accurate book on human anatomy. Vesalius has defied Church opposition to human dissection (*see* 1564).

Pope Paul III issues an *index librorum prohibitorum* forbidding Roman Catholics to read certain books (*see* 1559).

Painting *Ecce Homo* by Titian. Hans Holbein dies in England November 29 at age 45.

Wheat, barley, broad beans, chickpeas, European vegetables, and cows are introduced into New Spain by a new viceroy sent out to replace Pedro de Alvarado, who died 2 years ago at age 46. The grains, vegetables, and livestock introduced into North America in the next few decades by English explorers will have come in many cases from Spaniards to the south.

Nikolaus Copernicus said the earth revolved about the sun, defying Church teachings that insisted the opposite was true.

1544 Parliament recognizes two daughters of Henry VIII as heirs to the English throne in the event that Henry's son Edward should die without issue. Mary is Henry's daughter by Catherine of Aragon, Elizabeth his daughter by Anne Boleyn.

An English army under Edward Seymour, 38, earl of Hertford, invades Scotland in May and sacks Edinburgh; the Scots refuse to surrender.

The Battle of Ceresole south of Turin April 14 has brought victory to a French army over the imperial forces of Charles V.

English troops on the Continent join with those of Charles V to threaten Paris. The English take Boulogne September 14.

The Treaty of Crespy-en-Valois September 18 ends a 2-year war in the Netherlands between Charles V and François I, restoring unity in Europe. France loses Artois and Flanders and abandons her claims to Naples. Piedmont and Savoy are restored to their legitimate ruler, Charles renounces his claim to Burgundy, and Milan continues in the possession of Charles as Holy Roman Emperor.

Korea's Chungjong dies after a 37-year reign in which the country's great families have defeated Confucian

1544 ***(cont.)*** scholars charged by the king with curbing the families' power.

Painting *Portrait of an Old Man* by Lorenzo Lotto.

Sculpture *Nymph of Fontainebleau* by Florentine sculptor-goldsmith Benvenuto Cellini, 43.

Northern Europe suffers a honey shortage as a result of the breakup of monasteries by the Reformation (*see* England, 1536). The decline in honeybee colonies creates a growing need for cheap sugar, but sugar will remain a luxury for more than a century.

1545 Scottish forces defeat their English invaders February 25 at Ancrum Moor. The English retreat but invade once again in September.

Henry VIII's new flagship H.M.S. *Mary Rose* is warped out of Portsmouth Harbor into the Solent laden with cannon and 400 bowmen. A strong gust of wind causes her to heel, her guns break loose, and she sinks in less than a minute.

The emperor Charles V makes a truce with Suleiman the Magnificent at Adrianople in November.

The Mughal emperor Humayun takes Kandahar. His rival emperor Sher Shah, who has forced Humayun out of India and reformed the empire's administration, is killed by a cannon ball while besieging the Rajput stronghold of Kalanjar.

A Provençal baron massacres Waldensian Protestants April 20 and seizes his victims' lands.

A silver mine discovered by the Spanish at Potosí in the altiplano of New Castile will yield much of the wealth to fuel the commercial activity of Europe in the next century and prepare the way for an industrial revolution. Reports of the silver have reached conquistador Juan de Villaroel through a llama herder who said he had built a fire to keep warm while he slept and the next morning had found shining silvery threads melted out of the rocks by the heat of the fire. Accompanied by Diego and Francisco de Centeno, Juan de Villaroel has come to Potosí and found deposits that will yield an estimated $2 billion in silver and later will produce bismuth, tin, and tungsten (*see* 1611; Drake, 1573).

A typhus epidemic in Cuba kills as many as 250,000, while a similar number die in New Spain, 150,000 die in Tlascala, and 100,000 at Cholula according to the friar Geronimo de Mendieta.

The Council of Trent convened by Pope Paul III under Jesuit guidance undertakes reform of the Church. No Protestants attend, but the Tridentine Decrees issued by the Council will effect some genuine internal reforms while formulating rigid doctrines in direct opposition to Protestant teaching.

The Tridentine Decrees (above) will establish the Latin liturgy that will be used in Roman Catholic church services for more than 400 years.

Painting *Venus, Cupid, Folly, and Time* by Florentine painter Il Bronzino (Agnold di Cosimo di Mariano), 41, court painter to Cosimo de' Medici.

Fishing grows poor in the Baltic Sea while becoming good in the North Sea, a development that will have important economic and political consequences.

1546 The Peace of Ardres signed June 7 by Henry VIII and François I ends 2 years of conflict.

Pope Paul meets with Charles V June 7 and promises money and troops to help squelch the Protestant movement. Charles forms an alliance with Maurice of Saxony and on July 20 outlaws the leaders of the Schmalkaldic League, Philip of Hesse and Saxony's elector John Hesse. He conquers Saxony, which is then retaken by John Frederick while Charles is busy crushing the southern members of the League (*see* 1547).

An Ottoman army occupies Moldavia while other Ottomans capture Yemen on the Red Sea.

Portuguese forces in India rout the Gujurati army at Diu.

Scottish Lutheran reformer George Wishart is burnt to death March 1 on orders from Cardinal Beaton, archbishop of St. Andrews. The archbishop, who has persecuted Protestants, is assassinated at his castle May 29.

Parisian printer Etienne Dolet is hanged and burnt at the stake August 3. He has been denounced as a heretic and blasphemer for printing the works of Erasmus, Melancthon, and other humanists.

Mayans in New Spain stage a major uprising against the Spanish but are crushed by the conquistadors.

Charles V (above) revokes the South American charter granted in 1528 to the Welser family of Augsburg, whose banking house suffers enormous losses.

Europe's Fugger family is estimated to have a fortune of 6 million gulden (*see* 1525; 1550).

De Contagione et Contagiosis Morbis et eorum Curatione by Girolamo Fracastoro gives the first description of typhus and suggests that infections are carried from one person to another by tiny bodies (*seminaria contagionum*) capable of multiplying themselves (*see* 1530; van Leeuwenhoek, 1683; Nicolle, 1903).

The health of England's Henry VIII fails rapidly. The king has grown so grossly overweight that he must be moved up and down stairs by special machinery.

Martin Luther dies at his birthplace of Eisleben February 18 at age 63, and the emperor Charles V goes to war (above) at the urging of Pope Paul III, who wants to restore the unity of the Church.

Oxford's Christ Church is founded in a reorganization of Cardinal College.

The Proverbs of John Heywood by English epigrammatist John Heywood, 47, includes the proverb "No man ought to look a given horse in the mouth," which goes back in one form or another to St. Jerome of 400 A.D. Other proverbs cited by Heywood: "All is well that ends well"; "A penny for your thoughts"; "A man may well bring a horse to the water, but he cannot make him drink"; "Beggars shouldn't be choosers"; "Better late than never"; "Butter would not melt in her mouth";

"The fat is in the fire"; "Half a loaf is better than none"; "Haste makes waste"; "The green new broom sweepeth clean"; "It's an ill wind that blows no good"; "Look before you leap"; "Love me, love my dog"; "Many hands make light work"; "Two heads are better than one"; "When the iron is hot, strike"; "When the sun shineth, make hay"; "The tide tarrieth for no man"; "One good turn deserves another"; "One swallow maketh not a summer"; "Rome was not built in a day"; "Out of the frying pan into the fire"; "To tell tales out of school"; and "More things belong to marriage than four bare legs in a bed."

Painting *Martin Luther* by Lucas Cranach.

Theater *Orazia* by Italian playwright-poet Pietro Aretino, 54.

England's population tops 4 million, with many of her people desperately short of food after a series of bad harvests.

1547 The grand duke of Muscovy Ivan IV has himself crowned czar (caesar) January 16, the first Russian ruler formally to assume that title. Now 16, Ivan has ruled personally since age 14. He has virgins brought from all over Russia for his inspection and on February 3 selects as his wife Anastasia Zakharina-Koshkina of the ancient and noble family that will later take the name Romanov. He establishes a council of selected advisers to counter the power of the boyars in their duma.

Henry VIII dies January 28 at age 55 of syphilis and liver cirrhosis. News of the king's death is kept from his 10-year-old son by the boy's uncle Edward Seymour, earl of Hertford, until he has obtained the boy's consent to become protector of England with power to act with or without advice of counsel. Jane Seymour's son thereupon ascends the throne and will reign until 1553 as Edward VI.

François I dies March 31 at age 52 and is succeeded by his son of 28, who will reign until 1559 as Henri II. Catherine de' Medici is queen of France, but the new king is dominated by his mistress Diane de Poitiers, 48, who will have great influence over him throughout his reign.

Brittany is fully united with the French crown under Henri II (*see* 1532).

The Battle of Mühlberg April 24 ends in victory for the Holy Roman Emperor Charles V, who captures the elector of Saxony and lays siege to the elector's capital of Wittenberg. But the brief Schmalkaldic War soon ends with the Protestant German states retaining some independence.

Scottish royalist forces besieging St. Andrews castle capture the Lutheran reformer John Knox, 42, July 31. He is exiled and condemned to work on a French galley.

The Battle of Pinkie September 10 ends in victory for the earl of Hertford (above), who routs a much larger force, crushing Scottish resistance to the new boy-king. Now king in all but name, the protector (who is now Baron Seymour, duke of Somerset) defeats James Hamilton, 32, second earl of Arran, duke of Chatelherault, and regent of Mary Queen of Scots.

Although beaten at Pinkie (above), the Scots thwart a plan by Somerset to enforce a marriage treaty between young Edward and his 4-year-old half sister. They hastily arrange a marriage between Mary Queen of Scots and the 2-year-old French dauphin.

Henry Howard, 29, earl of Surrey, has been beheaded January 19 on trumped up charges of treason. He had blocked the projected marriage of his sister, the duchess of Richmond, to Thomas Seymour, 39, the lord high admiral, who now plots to displace his older brother Edward as guardian of the king.

The Mughal emperor Humayun exploits disputes among potential successors of the late Sher Shah Suri to regain his Indian territories and capture Kabul.

Parliament repeals the Statute of the Six Articles enacted in 1539 to define heresy.

England's late earl of Surrey Henry Howard (above) was a poet who translated much of Virgil's *Aeneid*, introduced into English the blank verse form of five iambic feet, and introduced from Italy the sonnet form of three quatrains and a couplet.

Sebastiano de Piombo dies at Rome June 21 at age 61.

France's new king Henri II (above) gives the Château de Chenonceaux in the Loire valley to his mistress Diane de Poitiers, who will add an Italian garden and a bridge over the Cher River (*see* 1521; 1560).

The population of New Spain falls to 6 million, down from 11 million in 1518 before the arrival of Hernando Cortés (who dies outside Seville December 2 at age 63). Economic upheaval, exploitation, and new diseases have taken a heavy toll among the Mayans and other tribespeople (*see* 1605).

1548 Poland's Sigismund I dies April 1 at age 81 after a 42-year reign that has established Catholicism in the country. He is succeeded by his only son, 28, who will reign until 1572 as Sigismund II while the Protestant Reformation spreads to Poland. The new king refuses the diet's demand that he repudiate his second wife, the beautiful Lithuanian Calvinist Barbara Radziwill, and she will die under mysterious circumstances 5 days after her coronation in early December 1550.

The Holy Roman Emperor Charles V annexes the 17 Lowland provinces to the empire's Burgundian Circle. Included are Artois, Flanders, Brabant, Limburg, Luxembourg, Gelderland, Hainault, Holland, Zeeland, Namur, Zutphen, East Friesland, West Friesland, Mechlin, Utrecht, Overyssel, and Groningen.

Ottoman forces occupy Tabriz, Persia.

The Battle of Xaquixaguane in Peru gives Pedro de la Gasca a victory over Gonzalo Pizarro, son of the late conquistador. He has Gonzalo executed.

Hispaniola has another slave uprising.

1548 *(cont.)* Painting *The Miracle of St. Mark* by Venetian painter Tintoretto (Jacopo Robusti), 30, whose father is a dyer (*tintore*); *Equestrian Portrait of Charles* V and *The Imperial Chancellor Antoine Perrenot de Granvella* by Titian; *Bevilacqua-Lazise* altarpiece by Italian painter Paolo Veronese (Paolo Caliari), 20.

Hispaniola's native population falls to 500 or less, down from 14,000 in 1514.

1549 English authorities arrest the lord high admiral Thomas Seymour in January and send him to the Tower of London on charges of having schemed to marry Edward VI to Lady Jane Grey, 12, daughter of the duke of Suffolk, and gain as his own bride Henry VIII's daughter Elizabeth, 16. Seymour is baron Seymour of Sudeley and brother of the protector, but he is convicted of treason and executed March 20. The lord protector himself is sent to the Tower October 14, having provoked aristocratic opposition.

England declares war on France August 9.

The czar of Muscovy Ivan IV summons the first Russian national assembly.

The Act of Uniformity (below) and rising prices provoke social and religious rebellions in Cornwall, Kent, and Oxfordshire (where the thirteenth Baron Grey de Wilton William Grey restores order).

Portugal's first governor of Brazil Thome de Souza founds the city of São Salvador that will later be called Bahia.

The Book of Common Prayer (first Prayer Book of Edward VI) by the archbishop of Canterbury Thomas Cranmer simplifies and condenses the Latin services of the medieval church into a single, convenient, comprehensive English volume that will serve as an authoritative guide for English priests and worshippers. Cranmer has become an influential counselor to the young king, he discussed a draft of the book last year in a conference of scholars, and Parliament has commissioned its publication. An Act of Uniformity forbids other prayer books after May 20.

The Villa d'Este at Tivoli outside Rome is completed by Piero Ligorio for the cardinal d'Este Ippolito II.

The wife of Florence's Cosimo de' Medici Eleanoro Toledo pays 9,000 florins to buy the Pitti Palace from Buonaccorso Pitti, now 79, and completes it as the Grand Ducal Palace (*see* 1466).

1550 The Treaty of Boulogne signed in March restores peace between England and France. England receives 400,000 crowns and the release of John Knox, France regains Boulogne, and English troops withdraw from Scotland.

Maurice of Saxony lays siege to Magdeburg in October.

The Japanese *daimyo* (feudal lord) who welcomed Francis Xavier to Kyushu in August 1549 makes it a capital offense to become a Christian after midsummer.

Helsinki is founded on the Gulf of Finland.

Augsburg banker Anton Fugger fails in an attempt to monopolize the tin production of Bohemia and Saxony. Fugger goes bankrupt after losing half a million gulden, and his failure precipitates bankruptcies throughout Europe, producing financial chaos at Augsburg and at Genoa, where millions of gulden are lost.

Prices in Europe and England rise as coins minted from Mexican and Peruvian gold and silver ingots devalue the old currencies and as population growth booms demand for food, clothing, and shelter (*see* 1559; Gresham, 1562).

Trigonometry tables are published by German mathematician Rhaticus (Georg Joachim von Lauchen), 36, who has studied under the late Copernicus.

A concordance of the whole English Bible is published by theologian-organist John Marbeck.

Fiction *Facetious Nights* (*Tradeci Piacevoli Notti*) by Italian writer Giovanni Francesco Straparola is the first European collection of fairy tales. A second volume will appear in 1553.

Poetry *Odes* by French poet Pierre de Ronsard, 26.

Painting *Presentation of the Virgin* by Tintoretto for Venice's Church of Santa Maria dell' Orto; *Eleanora of Toledo and Her Son* by Il Bronzino; A *Nobleman in His Study* by Lorenzo Lotto; *Deposition from the Cross* by Michelangelo.

The Lives of the Most Eminent Italian Architects, Painters, and Sculptors (Le Vite de'Piu Eccellenti Architetti, Pittori e Scultori Italiani) by Italian architect-painter Giorgio Vasari, 39, is published at Florence.

Japanese *ukiyoe* painting has its beginnings.

The Booke of Common Praier Noted by John Marbeck (above) adapts the plain chant of earlier rituals to the liturgy of England's Edward VI.

Billiards is played for the first time in Italy.

Vicenza's Villa Rotunda and Palazzo Chiericati are designed by local architect Andrea Palladio (Andrea di Pietro della Gondola), 41, who completes the city's Palazzo Thiene.

Giorgio Vasari (above) gives Gothic architecture its name and disparages it. A pupil of Michelangelo, he says that medieval cathedrals were built in a style originated by the Goths ("those Germanic races untutored in the classics"), and describes them as a "heap of spires, pinnacles, and grotesque decorations lacking in all the simple beauty of the classical orders" (*see* 1184).

Corn (maize), sweet potatoes, and peanuts will be introduced in much of China in the next half century. All will produce large yields and will spur population growth by creating abundance with declining prices.

France's population reaches 15 million. Spain's is half that; 6.5 million of Spain's population is in Castile, which has lost at least 150,000 to American emigration in the past half century.

1551 The Ottoman Turks take Tripoli after failing in an attempt to take Malta.

France's Henri II publicly disavows the Council of Trent and renews war against the emperor Charles V, seizing the bishoprics of Toul, Metz, and Verdun (*see* 1552).

Parliament enacts a law to encourage employment of England's poor.

Prosperity fueled by wars and by precious metals from America will swell the coffers of European merchants in the next 6 years.

Prudentic Tables (Tabulae Prudenticae) by German astronomer Erasmus Reinhold, 40, contains astronomical tables based on numerical values provided by the late Nikolaus Copernicus. They represent an improvement on the widely used Alfonsine Tables.

Historia Animalium by Swiss naturalist Konrad von Gesner, 35, is published in its first volume. Gesner has collected animals from the New World and the Old, and his work pioneers modern zoology.

A fifth epidemic of the sweating sickness strikes England in April, in Shrewsbury (*see* 1529). Foreigners are somehow spared, but Englishmen who flee to the Continent die there, even though Frenchmen and Lowlanders are not affected (*see* 1563; Kaye, 1552).

Forty-two articles of religion, published by the archbishop of Canterbury Thomas Cranmer, will be the basis of Anglican Protestantism (*see* Thirty-Nine Articles, 1563).

Francis Xavier leaves Japan November 21 after 2 years in which he has proselytized scarcely 150 people. He writes to the pope with advice on what trade goods should be brought to Japan, leaves behind two Jesuits and some converts, and sails for Goa. The "apostle of the Indies" will die of exhaustion in December of next year near Guangzhou (Canton) (*see* 1550; 1582).

Peru's University of San Marco has its beginnings in a school founded by Dominican priests in a Lima convent.

The National University of Mexico is founded at Mexico City in New Spain.

Painting *Prince Felipe of Spain* by Titian.

Hymn "Praise God from Whom All Blessings Flow" by French composer Louis Bourgeois.

Italian composer Giovanni Pierluigi da Palestrina, 25, is appointed *magister capellae* and *magister puerorum* at Rome's church of St. Giulia by Pope Julius III (Giovanni Maria del Monte), who succeeds the late Paul III but will be pope only until 1555.

English and Welsh alehouses are licensed for the first time.

1552 The duke of Somerset, former lord protector of England's Henry VI, is executed January 22 after a trial on trumped-up charges.

France's Henri II has signed the Treaty of Chambord with Maurice of Saxony January 15, promising to supply troops and money for the war against Charles V. Maurice takes Augsburg in May, nearly captures Charles at Innsbruck, and forces Charles's kinsman Frederick of Hapsburg to sign The Treaty of Passau August 2, permitting Protestant princes freedom of religion at least until the next diet. Charles lays siege to Metz, which is defended by François, duc de Guise, and a 4-year war begins between France and the emperor.

The czar of Muscovy Ivan IV attacks Kazan with 50 guns and an army of 150,000 after a faction in Kazan offers him the entire khanate. He lays siege August 20 to the fortress and takes it by assault October 2, using artillery to breach the walls of the Tatar capital. The Volga becomes a Russian river for the first time, and Ivan goes on to attack Astrakhan.

The Portuguese ship *São João* goes aground June 24 on the Natal coast with 610 aboard. Only 25 of the 500 who get ashore survive the long walk to the nearest Portuguese settlement at Sofola. Half a dozen major shipwrecks will land 3,000 Portuguese on Africa's east coast in the next 100 years, but scarcely 500 will be rescued (*see* 1683).

Charles V lifts his siege of Metz (above) and withdraws after losing more than 12,000 men in a single month to typhus and scurvy.

"The Sweate" by English physician John Kaye is a pamphlet describing a mysterious epidemic that may be a severe form of influenza (*see* 1551).

Tabulae Anatomicae by Italian anatomist Bartolommeo Eustachio, 28, describes what will be called the Eustachian tube in the ear and the Eustachian valve of the heart. He also describes the stapes, thoracic duct, uterus, kidney, and teeth (*see* Vesalius, 1543).

Rome's Collegium Germanicum is founded by Jesuits.

Books on geography and astronomy are destroyed in England because they are thought to be infected with magic.

Poetry *Centuries* by French physician-astrologer Nostradamus (Michel de Notredame, or Nostredame), 52, is a book of rhymed prophecies that will be interpreted centuries hence as prophesying 20th-century events; *Amours de Cassandre* by Pierre de Ronsard appears in its first volume.

Painting *Self-portrait* by Titian. He has begun sending works to the Spanish infante Felipe, 25, who confers rich rewards on the artist.

Theater *Cleopatre Captive* by French dramatic poet Etienne Jodelle, 20, Sieur de Lymodin, is the first classical French tragedy; *Ralph Roister Doister* by English cleric-playwright Nicholas Udall, 47, is the earliest English comedy that will survive (year of first performance approximate. Udall was headmaster of Eton from 1534 until his dismissal for misconduct in 1541, and he has modeled his play on a comedy by Plautus).

1552 ***(cont.)*** Palestrina wins appointment to the Sistine choir from Pope Julius, who will soon die (*see* 1551; 1555).

Scotland's Royal and Ancient Golf Club of St. Andrews has its beginnings (*see* 1754).

London's Covent Garden is granted to Sir John Russell, first earl of Bedford, whose family will retain the property until 1914. The land of the convent garden owned by the abbey of Westminster, now part of London, was confiscated along with other church properties in 1534 and will serve as London's produce and flower market beginning in 1661.

1553 England's Edward VI dies of tuberculosis at Greenwich July 6 at age 15 and is succeeded by his Catholic half sister Mary, 37, who has been raised by her mother Catherine of Aragon. A treaty of marriage is arranged between Mary and Spain's Philip, 26, son of Charles V, who is to be given the title king of England but to have no hand in government and no right to succeed Mary.

England's new queen (above) faces an insurrection led by Henry Grey, duke of Suffolk, who has married his daughter Lady Jane Grey, 16, to Lord Guildford Dudley as part of a plot to alter the succession (Dudley induced the late Edward VI to sign letters of patent, making Lady Jane Grey heir to the crown). Suffolk is supported by Sir Thomas Wyatt, 32, son and namesake of the poet-statesman who died in 1542, Lady Jane is proclaimed queen July 9, but forces loyal to Mary disperse Suffolk's troops and imprison Lady Jane July 19 (*see* 1554).

Mary enters London August 3 to begin a harsh 5-year reign. She has the duke of Northumberland arrested, tried for treason, and executed August 22.

Maurice of Saxony sustains a mortal wound July 9 at Sievershausen as his forces defeat Albert of Brandenburg-Kulmbach.

The Battle of Marciano August 2 ends in defeat for a French army that has invaded Tuscany.

English navigator Richard Chancellor reaches Moscow by way of the White Sea and Archangel. Commander of a ship in Sir Hugh Willoughby's expedition of 1527 in search of a northeast passage to Cathay, Chancellor has studied navigation under John Dee (*see* 1558; Muscovy Company, 1555).

Christianismi Restitutio by Spanish theologian Michael Servetus (Miguel Serveto), 42, is the first book to refute the Galen idea of 164 A.D. that the ventricular system is perforated. Servetus shows a familiarity with pulmonary circulation. He is arrested for his views and brought before the Inquisition at Vienna, escapes to Geneva, is imprisoned on orders from John Calvin, and burned at the stake for heresy October 27.

England's new queen has Protestant bishops arrested and restores Roman Catholic bishops.

Painting *Landscape with Christ Appearing to the Apostles at the Sea of Tiberius* by Flemish painter Pieter Brueghel, 28; *All-Saints* altar (La Gloria) by Titian.

Lucas Cranach (the Elder) dies at Weimar October 16 at age 81.

The camera obscura devised by Roger Bacon in 1267 is improved by Italian physicist Giambattista della Porta, 15, who introduces a convex lens.

Florence's Strozzi Palace is completed by the sons of the late banker Filippo Strozzi the Elder, who began construction of the palazzo in 1489 but died 2 years later at age 65.

The first written reference to the potato appears in *Chronica del Peru* by Pedro de León (Pedro Creca) which is published at Seville. The author calls the tuber a *battata* or *papa* (*see* 1540; 1563).

1554 England's Queen Mary releases the duke of Suffolk from prison in a show of clemency toward those who took arms against her last year, but she hardens her position when Suffolk again proclaims his daughter Lady Jane Grey queen and tries to rally Leicestershire to her support.

Sir Thomas Wyatt summons his Kentish followers to the Wyatt castle of Allington January 22 and raises a new insurgent army that occupies Rochester January 26. Wyatt reaches Southwark February 3, but some of his men are cut off, others desert, and Wyatt is forced to surrender.

Lady Jane Grey and her husband Guildford Dudley are executed February 12, 5 months after the execution of Dudley's father, the duke of Northumberland. Wyatt is brought to trial for treason March 15 and executed April 11.

England's Princess Elizabeth, now 20, is sent to the Tower of London in March as the Spanish cry for her execution. Edward Courtenay, 28, earl of Devonshire, has been released from the tower after 15 years of imprisonment in connection with his father's aspirations to the Crown. Foiled in his effort to marry the new queen, he has plotted to marry Elizabeth and join her on the throne. Elizabeth is released from the tower in May; Mary and the Spanish infante Felipe are married July 25. Felipe receives the kingdoms of Naples and Sicily plus the duchy of Milan from his father, Charles V.

France's Henri II invades the Netherlands.

Mehedia on the Tunisian coast falls to Algerians led by the corsair Dragut, who succeeded to the leadership of the pirates after the death of Khair ed-Din at Constantinople in 1546. The Spanish will be driven from the North African coast in the next 2 years.

French physician-astronomer-mathematician Fernelius (Jean Fernel), 57, codifies the practical and theoretical medicine of the Renaissance, rejecting astrology and magic but emphasizing the functions of the organs.

Painting *Danaë* and *Venus and Adonis* by Titian.

Cruydeboek by the Dutch botanist Dodonaeus (Rembert Dodoens), 37, makes the first mention of kohlrabi and Brussels sprouts, both varieties of cabbage (*see* Gerard, 1597).

Flemish hop growers emigrate to England and start growing hops in Kent for the English brewery trade.

Parliament reenacts the Corn Law of 1436 and enacts other statutes to encourage farming in England. The Corn Law regulates export and import of grain according to prices, other laws forbid the enlargement of farms and place restrictions on storing, buying, and selling grain, but none of the measures relieves England's food shortages or lowers food prices.

1555 The Religious Peace of Augsburg September 25 compromises differences between Catholics and Protestants, ending a period of religious wars in the German states. Each prince may choose which religion shall be followed in his realm.

An imperial army led by Cosimo de' Medici, duke of Florence, has forced a French army to end its 15-month siege of Siena and surrender in April (*see* 1557).

A long period of peace settles on Europe as Charles V turns over full imperial authority to his brother Ferdinand, 52, and sovereignty in the Netherlands to his son Philip (Felipe), now 28, in formal ceremonies October 25 in the Hall of the Golden Fleece at Brussels. Now 55, Charles is dyspeptic, gouty, and uncontrollably gluttonous. Ferdinand asks that formal abdication be delayed until 1558 but begins a reign that will continue until 1564.

The Mughal emperor Humayun defeats an Afghan claimant to the throne and reoccupies Delhi and Agra.

Japanese pirates besiege Nanjing.

Bloody Mary returns Roman Catholicism to England and persecutes Protestants. Hugh Latimer, bishop of Worcester, and Nicholas Ridley, bishop of London, have been imprisoned for 2 years on charges of heresy, and the queen has them burned at the stake October 16 at Oxford. Bishop Latimer, now 70, says to Bishop Ridley, 55, "Be of good comfort, Master Ridley, and play the man. We shall this day light such a candle by God's grace in England as I trust shall never be put out" (*see* Cranmer, 1556).

Pope Julius III has died March 23 at age 67, his successor Marcellus II has died April 30 at age 54, and *his* successor Paul IV extends the powers of the Universal Inquisition founded by Paul III in 1542. The new pope orders that Rome's Jewish quarter be surrounded by a wall, creating the ghetto of Rome. "God has imposed servitude until they should have recognized their errors," says the pope.

French Huguenots settle Rio de Janeiro Bay in Brazil under the leadership of Nicolas Durand de Villegagnon (but *see* 1567).

A translation of Peter Martyr's 1516 Italian work *De Rebus Oceanicis et Novo Orbe* by Richard Eden stimulates English interest in America.

The Muscovy Company (the Association of Merchant Adventurers) founded by Richard Chancellor and others to trade with Russia is the first of the great English trading companies (*see* 1553; 1561; East India Company, 1600).

L'Histoire de la nature des oyseaux by Pierre Belon is a pioneer study of birds.

Portuguese Jesuit missions to Ethiopia begin under the direction of Pedro Paez and will continue until 1603.

Poetry *Amours de Marie* by Pierre de Ronsard.

Painting *St. George and the Dragon* by Tintoretto.

Sculpture *Pietà* by Michelangelo, who has spent 5 years creating the work for the Duomo at Florence; *Perseus* by Benvenuto Cellini.

Palestrina is pensioned off along with two other members of the Sistine choir by the successor to Pope Julius (above). He is appointed choirmaster of the church of St. John Lateran but will give up the post in 1558 after composing a book of magnificats and lamentations—settings of the Holy Week lessons from the Lamentations of Jeremiah (*see* 1552).

Villa Giulia is completed for Pope Julius III (above), who dies before he can occupy the summer palace. It has been designed by architect Giacomo da Vignola (Giacomo Barozzi), 48, in collaboration with Giorgio Vasari and Bartolommeo Ammanati.

Rome's Palazzo Farnese is completed by Giacomo da Vignola (above) and Michelangelo.

Famine grips England; the peasantry discovers that peas taste good green as well as dried.

1556 The Holy Roman Emperor Charles V resigns his Spanish kingdoms and Sicily January 16, resigns Burgundy soon after, leaves the empire to his brother Ferdinand, and sails from Flushing September 17 to settle as a guest at a monastery in Estremadura, where he will die in 1558. His inept son Philip (Felipe) is left to rule Spain, the Netherlands, Milan, Naples, Franche-Comte, and the rich Spanish colonies. The Truce of Vaucelles signed February 5 makes peace between Philip and France's Henri II.

The Mughal emperor Humayun dies January 27 after falling from his library roof in Delhi. His son Jalal-ud-Din, 14, returns from exile and will reign until 1605 as Akbar with initial guidance from his regent Bairam Khan. Akbar defeats Hindu forces at the Battle of Panipat in the Punjab November 5 and regains the Hindustani Empire.

The czar of Muscovy Ivan the Terrible completes his conquest of Kazan and Astrakhan from the Tatars. His triumph opens the way for Russian expansion to the east and southeast.

England's archbishop of Canterbury Thomas Cranmer is degraded from his office February 14 in humiliating ceremonies conducted by papal delegates at Christ Church, Oxford, and burned at the stake March 21 at age 66 after renouncing the Church of Rome and refusing all recantations. Cranmer holds his right hand to the flame saying that the hand has offended and should therefore be burned first.

Antwerp has a financial crisis as Spain's Philip II and France's Henri II default on their war debts.

1556 *(cont.)* *Castle of Knowledge* by English mathematician Robert Recorde, 46, is a navigational guide for voyagers to Cathay. Recorde has taught at Oxford and Cambridge, served as physician to Edward VI and Mary, and is first to use the = sign to denote equality.

De Re Metallica by Bohemian physician–mine owner Georgius Agricola (Georg Bauer), 62, is the world's first textbook on mining and metallurgy and will be the only such text for centuries. It establishes the science of mineralogy.

Thomas Cranmer (above) has been succeeded as archbishop by Reginald Pole, now 56, whose criticism of Henry VIII's divorce was published abroad 20 years ago, provoked Henry to execute Pole's mother and brother in England, and encouraged Pope Paul III to create him a cardinal.

Ignatius Loyola, who founded the Society of Jesus in 1534, dies at Rome July 31 at age 65.

Theater *Der Paur im Egfeur* by Hans Sachs at Nuremberg dramatizes anecdotes and incidents of everyday life in a *Fastnachtsspiel* that will be produced each year at Shrovetide.

A book of motets by Dutch composer Roland de Lassus (or Orlando di Lasso), 26, published at Antwerp includes an adulatory motet to England's new archbishop of Canterbury Reginald Pole. Lassus will soon be invited to Munich by Albrecht IV, duke of Bavaria, and will enjoy the duke's patronage until the duke dies in 1579.

Tobacco seeds reach Europe with Franciscan monk André Thevet, who returns from Rio de Janeiro with seeds of what the Brazilian natives call *petun,* or *petum* (*see* Grijalva, 1518; Nicot, 1561).

The Château d'Anet for Diane de Poitiers is completed by French architect Philibert Delorme, 46, after 9 years of construction.

The worst earthquake in history rocks China's Shanxi province January 24, killing more than 830,000.

1557 Portugal's João III (the Pious) dies on his 55th birthday June 6 after having instituted the Inquisition. He is succeeded by his 3-year-old grandson Sebastian, who will reign until 1578, mostly under a regency.

England's Mary Tudor declares war on France June 7 in support of her husband Philip II in a conflict provoked by Pope Paul IV.

The Battle of St. Quentin August 10 ends in victory for Spanish forces under Emmanuel Philibert, duke of Savoy, over a French army led by the constable of Montmorency, 64, and drives the French from Italy.

Siena loses her independence as Florence's Cosimo de' Medici becomes ruler of the former republic.

A 14-year Livonian War begins as Muscovite forces invade Poland, the Swedes take Estonia, and the Danes acquire part of Courland in a dispute over succession to the Baltic territories ruled by the Teutonic Knights (*see* 1525; 1569).

Macão off the Chinese coast near Canton is settled by Portuguese colonists who establish regular trade with the Chinese mainland.

São Paulo is founded by the Portuguese in Brazil.

Whetstone of Witte by Robert Recorde is dedicated to the governors of the Muscovy Company which Recorde serves as technical adviser on matters of navigation (*see* 1555).

Poetry *Songs and Sonettes Written by the Ryght honorable Lorde Henry Howard Late Earle of Surrey, and Other* (*apud Richardum Tottel*) (*see* 1547); *Hundredth Good Pointes of Husbandry* by English versifier Thomas Tusser. "The stone that is rolling can gather no moss./Who often removeth is sure of loss"; "At Christmas play, and make good cheer, / For Christmas comes but once a year."

Painting *Landscape with the Parable of the Sower* by Pieter Brueghel.

Observations of whales by voyagers of the Muscovy Company lead to the opening by the English of a whale fishery at Spitzbergen (*see* 1555; 1596).

1558 England's glorious Elizabethan Age begins November 17 as Mary Tudor dies at age 42 and is succeeded by her half sister Elizabeth, now 25. Daughter of Henry VIII by the late Anne Boleyn, Elizabeth will reign until 1603 in a golden period of English arts and letters.

Calais has fallen January 20 to François, duc de Guise, after 211 years in English hands. England gives up hope of conquering France (*see* 1564).

Charles V has died September 21 at age 58 after 3 weeks of indigestion from eating an eel pie at the monastery of Yuste. His son will return to Spain from the Netherlands next year and reign until 1598 as Philip II.

English mathematician John Dee succeeds Robert Recorde as technical adviser to the Muscovy Company and invents two compasses for master pilots.

First Blast of the Trumpet Against the Monstrous Regiment of Women by Scottish clergyman John Knox, now 53, says, "The nobility both of England and Scotland are inferior to brute beasts, for they do that to women which no male among the common sort of beasts can be proved to do with their females; that is, they reverence them, and quake at their presence; they obey their commandments, and that against God." Knox fled to the Continent at Mary Tudor's accession in 1553 and has been at Geneva since 1556, meeting occasionally with John Calvin.

Painting *The Fall of Icarus* by Pieter Brueghel; *Supper at Emmaus* and *Feast in the House of the Pharisee* by Paolo Veronese.

London's population reaches 200,000, four times its 1400 level (*see* 1603).

1559 The Danish-Norwegian king Christian III dies January 1 at age 55 after a reign of nearly 24 years that has strengthened and enriched the realm. His cousin who reigned as Christian II until he was deposed in 1523 dies later in the month at age 79, still a prisoner as he has been since 1531, and his son, 24, ascends the throne. He will reign until 1588 as Frederick II.

The Treaty of Cateau-Cambresis April 3 ends the last war between the late Charles V and France. The treaty confirms Spanish possession of the Franche-Comte and the Italian states of Milan, Naples, and Sicily. Spain's Philip II marries Henri II's daughter Elizabeth; Henri's sister Margaret marries Emmanuel Philibert, comte de Savoy.

France's Henri II sustains a head wound June 30 in a tournament celebrating the Treaty of Cateau-Cambresis (above) and his daughter's marriage. The king dies in agony July 10 at age 40, fulfilling a prophecy by Nostradamus, who foretold the manner of Henri II's death 7 years ago. Henri's son, 14, who was married last year to Scotland's Mary Stuart begins an 18-month reign as François II with his uncles François, duc de Guise, and Charles, cardinal of Lorraine, as regents.

Suleiman the Magnificent helps his son Selim defeat Selim's brother Bayezid at the Battle of Konya. Bayezid and his five sons flee to Persia, where Suleiman pays to have them executed.

Mary, Queen of Scots (above) assumes the title Queen of England. England's Elizabeth I raises her court favorite Robert Dudley to the privy council.

Romans vent their anger at the late Pope Paul IV (below) by demolishing his statue, liberating the prisoners of the Inquisition, and scattering the Inquisition's records.

Some 1,500 Spanish colonists land at Pensacola, Florida, but hostile natives force them to move to Port Royal Sound in what later will become the English colony of South Carolina (see 1561; Ribault, 1562; St. Augustine, 1565; Albemarle, 1653).

Italian anatomist Realdo Columbus at Padua advances knowledge of human blood circulation. A pupil of Andreas Vesalius, he shows that the right and left ventricles of the heart are separated by an impenetrable wall, that blood is conveyed from the right side to the lungs where it is mixed with air, and that it returns in aerated form to the right side. But Columbus hews to the traditional view that the liver is the center of the venous system and is the organ that creates blood (*see* Servetus, 1553; Harvey, 1628).

Pope Paul IV founds the Order of the Golden Spur as a military body, but the Hapsburg duke of Alba defeats the papal forces. The pope dies August 18 at age 83 and is succeeded by Giovanni Medici, 60, no kin to Florence's de' Medici family, who assumes the papacy as Pius IV.

The University of Geneva has its beginnings in an academy founded by John Calvin and the French Protestant reformer Théodore de Bèze.

The *Index auctorum et librorum qui tanquam haeretici aut suspecti aut perverse ab Officio S. R. Inquisitionis reprobantur et in universa Christiana republica interdicuntur* published by Pope Paul IV before his death condemns certain authors with all their writings, prohibits certain books whose authors are known, and prohibits pernicious books published anonymously. It is the first Roman Index in the modern ecclesiastical use of the term (*see* 1543).

Painting *The Fight Between Carnival and Lent* and *Netherlandish Proverbs* by Pieter Brueghel; *The Entombment, Diana and Actaeon,* and *Diana and Callisto* by Titian.

English food prices soar to three times their 1501 levels largely because Henry VIII debased the coinage to raise quick money for his wars with Spain, France, and Scotland. The typical English wage is up only 69 percent above its 1501 level (*see* 1550).

Ice cream appears in Italy as ice and salt are discovered to make a freezing combination.

1560 The Conspiracy of Amboise organized by the French Huguenot Louis de Bourbon, comte de Condé, tries to overthrow the Catholic Guises, but the queen mother Catherine de' Medici declares herself regent and helps thwart the Huguenots. Some 1,200 are hanged at Amboise in March, and Louis is imprisoned.

French troops in Scotland try to assert the claims of France's Queen Mary Stuart, 17, against Elizabeth of England, who is considered illegitimate by Catholics. English troops besiege the French at Leith, Mary's mother Marie of Lorraine dies in June, and the Treaty of Edinburgh July 6 ends French interference in Scotland (but *see* 1561).

Sweden's Gustavus I Eriksson, now 64, abdicates June 25 after a 37-year reign that has made the country independent. His son, 27, who has been writing love letters of marriage proposals for 2 years to Elizabeth of England, will reign until 1568 as Eric XIV.

France's François II dies December 5 at age 16. His brother, 10, will reign until 1574 as Charles IX.

The Auracanian Federation destroys Spanish settlements in the interior of Chile.

Antwerp reaches the height of its prosperity. The city has a thousand foreign merchants in residence with as many as 500 ships entering its harbor each day from Danish, English, Hanse, German, Italian, Portuguese, and Spanish ports.

A smallpox epidemic decimates Portugal's Brazilian colony and increases the need for African slaves to cut sugar cane.

The Geneva Bible published by followers of John Calvin is the first bible to have both chapter and verse numbers. The same numerical divisions will be used for more than 400 years.

Painting *Children's Games* by Pieter Brueghel.

Tobacco grows in Spain and Portugal, where it is cultivated as an ornamental plant and for its alleged medicinal properties (*see* 1531; Nicot, 1561).

The Villa Foscari (Villa Malcontenta) is completed by Andrea Palladio overlooking the Brenta at Vicenza.

1560 ***(cont.)*** The Château de Chenonceaux passes into the hands of France's Queen Mother Catherine (above) who ousts her late husband's mistress and engages the architect Philibert Delorem, 35, to design a two-story gallery on the bridge erected by Diane de Poitiers (*see* 1547).

Venice gets its first coffee house. The city is a major sugar-refining center, using raw sugar imported through Lisbon, but Europe's chief sugar refiner is Antwerp (above) which also gets its raw materials from Lisbon but refines as much sugar in a fortnight as Venice does in a year.

1561 Mary, Queen of Scots returns from France August 19, landing at Leith because England has refused her passage. She becomes embroiled in argument with Calvinist John Knox, who last year drew up a Confession of Faith denying papal authority in Scotland (*see* 1558).

Spain withdraws troops from the Netherlands.

Spain abandons her 2-year-old American colony at Port Royal Sound because it is too isolated. Spain remains Europe's leading power and holds dominion over much of the world.

The Edict of Orléans suspends persecution of France's Huguenots.

England receives her first Flemish Calvinist refugees.

Anthony Jenkinson of England's Muscovy Company reaches Isfahan through Russia and opens trade with Persia (*see* 1555). Anglo-Persian trade will continue until 1581.

The Order of the Teutonic Knights in the Baltic states is secularized.

Theater *Gorboduc, or Ferrex and Porrex* by English playwrights Thomas Norton, 29, and Thomas Sackville, 25, first earl of Dorset and grandmaster of England's Freemasons. The earliest known English tragedy, it marks the first use of blank verse in English drama (all action occurs offstage).

The French ambassador to Lisbon Jean Nicot, 31, sends seeds and powdered leaves of the tobacco plant home to the queen mother Catherine de' Medici (*see* 1560; French physicians consider tobacco a panacea for many ills). The botanical name for tobacco, *Nicotiana rustica*, will be derived from Nicot's name, as will the word *nicotine* (*see* Hawkins, 1565; Thevet, 1567; Vanquelin, 1810).

Venice's Ca' Corner della Ca' Grande is completed by Florentine architect Sansovino dacopo Tatti), now 75, who has built the imposing palazzo on the Grand Canal for Jacopo Cornaro, a nephew of the queen of Corsica.

Moscow's basilica of St. Basil is completed after 27 years of construction.

1562 A massacre of Huguenots at Vassy March 1 by order of François, duc de Guise, begins a series of French civil wars. The Huguenots come largely from the nobility and the new capitalist-artisan class, with some peasant support in the southwest, while Paris and the northeast remain Catholic. The Huguenots, who retaliate by murdering priests and raping nuns, hold Lyons, Rouen, and Orléans, having been driven out elsewhere with much bloodshed. Both sides seek to control the government in the absence of a strong Crown.

A treaty signed at Hampton Court September 20 pledges Elizabeth to support France's Huguenot leader Louis de Bourbon, comte de Condé, against the Catholics. England is to occupy Le Havre pending restoration of Calais to English control (but *see* 1564).

English navigator John Hawkins, 30, hijacks a Portuguese ship carrying African slaves to Brazil, trades 300 slaves at Hispaniola for ginger, pearls, and sugar, and makes a huge profit. His enterprise marks the beginning of English participation in the slave trade (*see* 1563).

A French expedition under Jean Ribault, 42, reaches Port Royal (*see* 1559; 1561). Ribault leaves 30 men at Charlesfort, named in honor of France's 13-year-old Charles IX, but the men are without means of support and soon set out in small boats they have built. An English vessel picks them up half dead (*see* Ribault, 1564).

English financial agent Sir Thomas Gresham, 43, writes that the setbacks of the Huguenots (above) "hath made such alteracione of credit as this penn cannot write you," and he worries about what will happen to the English pound (*see* note on inflation, 1559). "Here ys soche great dowtes cast upon our Estate, as the creadyte of the Queen's Majestie and all the whole nacyon ys at a stay; and glad ys that man that maye be quit of an Englishman's bill." Gresham will note that when two coins are equal in debt-paying power but unequal in intrinsic value, the less valuable coin will have a tendency to remain in circulation while the more valuable one is hoarded (Gresham's law).

Parliament acts to aid the employment of the poor as it did in 1551, but at the same time it raises the price level at which wheat, barley, and malt may be exported, thus making food and drink more costly at home and enriching the landed gentry at the expe˜nse of the lower classes.

The University of Lille has its beginnings.

Painting *Europa and the Bull* by Titian; *Marriage at Cana* by Paolo Veronese; *The Suicide of Saul*, *The Monkeys*, *The Fall of the Rebel Angels*, and *The Triumph of Death* by Pieter Brueghel.

1563 The Peace of Amboise signed by Catherine de' Medici March 19 is an edict ending the year-long conflict between French Catholics and Huguenots. It follows by 1 month the murder of François, duc de Guise, by a Huguenot, leaving Catherine in control of Catholic forces. She grants limited toleration, but the peace will not be lasting.

Mary, Queen of Scots sends her secretary William Maitland of Lethington, 35, to England to claim the right of succession to Elizabeth, but she receives no declaration from her cousin.

China's Ming generals overcome Japanese pirates on the country's south coast.

Queen Elizabeth tolerates English dissenters but persecutes Catholics, Unitarians (who deny the Trinity), and Brownists–Puritan extremists who will form the nucleus of the Congregational Church.

Wednesdays are declared fast days in England which now has more meatless days than any other country in Europe. A London woman is pilloried for having meat in her tavern during Lent (*see* 1534).

Parliament enacts a Statute of Apprentices.

John Hawkins sells a cargo of 105 African slaves in Hispaniola (300 by some accounts) (*see* 1562).

"If any African were carried away without his free consent," says Queen Elizabeth, "it would be detestable and call down the vengeance of Heaven upon the undertaking" (but *see* 1564).

De Praestigiis Daemonum by Flemish physician Johann Wier maintains that witches are merely miserable people with distorted minds, but much of the world will for centuries remain under the influence of the 1489 handbook on witch-hunting *Malleus Maleficarum* (*see* Loudon, 1634; Salem, 1692).

A sixth (and final) epidemic of the sweating sickness devastates London, killing more than 17,000 in a population of 66,000.

The Anglican Church (Church of England, or Episcopal Church) is established by the adoption of the Thirty-Nine Articles, a modification of the 42 articles published by Thomas Cranmer in 1551. The Church is largely Protestant in dogma but has a liturgy reminiscent of the Tridentine liturgy established by the Council of Trent convened in 1545 and has a hierarchy similar to that of the Roman Catholic Church.

Painting A personification of Summer that includes an ear of maize is painted at Milan by German-born artist Guiseppe Arcimboldo, 36 (the "turkey corn" is familiar throughout much of the Mediterranean); *Landscape with the Flight into Egypt* and *The Tower of Babel* by Pieter Brueghel.

The Tuileries Palace is built at Paris for the queen mother Catherine de' Medici (*see* 1792).

Sir John Hawkins (above) brings the potato to England, but the potato from Bogotá may be a sweet potato.

France has famine which will be exacerbated by the country's religious wars (above).

Japan has rice riots at Mikawa following requisition of crops by the Tokugawa family and imposition of heavy taxes. Buddhist temples of the Ikko sect are burned in retaliation by the *daimyo* Ieyasu.

1564

The Peace of Troyes ends hostilities between France and England. Huguenots and Catholics have joined forces to drive the English out of Havre, and England renounces all claims to Calais in return for 222,000 crowns (*see* 1558).

Maize from the New World provided a cheap new way to fatten livestock for Europe's growing populations.

The Holy Roman Emperor Ferdinand I dies at Vienna July 25 at age 61 after 8 years in power. He is succeeded by his son of 36, who will reign until 1576 as Maximilian II.

England's Elizabeth takes shares in John Hawkins's second slave-running venture and loans him one of her ships as her avarice overcomes the humane antipathy to slavery she expressed last year.

Miguel Lopez de Legazpe, 54, leaves New Spain with four ships to colonize the Philippines (*see* 1521; 1565).

French Huguenot leader Gaspard de Coligny fits out a second expedition to the New World. The fleet commanded by René de Landonniére sails to Fort Carolina on the St. John's River of northern Florida.

Jean Ribault arrives at Fort Carolina (above) with seven ships, 600 men (*see* 1562).

The Inquisition forces Andreas Vesalius to make a pilgrimage to the Holy Land as a condition for the commutation of his death sentence for dissecting human bodies (*see* 1543). He will disappear on the pilgrimage.

A new disease epidemic in New Spain decimates the Aztecs.

The Council of Trent convened by Pope Paul III in 1545 is concluded by Pope Pius IV, who issues the decree *Professio Fidei Tridentine*.

John Calvin dies at Geneva May 27 at age 54. His friend Théodore de Bèze, now 44, succeeds to the leadership of French Protestantism.

Painting *The Death of the Virgin*, *The Adoration of the Kings*, and *The Slaughter of the Innocents* by Pieter Brueghel. Michelangelo dies at Rome February 18 at age 88.

The sweet potato reaches England aboard one of John Hawkins's slave ships returning from New Castile. The

1564 ***(cont.)*** tubers are planted to end England's reliance on imports for her sweet potato pies.

1565 The Battle of Talikota January 23 ends the Hindu empire of Vijayanagar. Muslim armies from Ahmadnagar, Bijapur, Bidar, and Golconda join forces to attack a Hindu army led by the warrior Ramaraja, who is captured and beheaded.

Ottoman forces lay siege to Malta in May. Some 700 Knights of St. John, led by their grand master Jean Parisot de La Valette, 71, hold out until September, when Spanish forces arrive and drive 31,000 Turks into the sea.

St. Augustine (San Agostin), Florida, is established August 28 as the first permanent European settlement in North America. Spanish conquistador Pedro Menendez de Aviles settles 600 colonists there, having attacked Fort Carolina and slaughtered all its male inhabitants (*see* 1564; 1600).

A colony at Cebu in the Philippines is established by Miguel Lopez de Legazpe, who has been sent out from New Spain by the viceroy Luis de Velasco on orders from Philip II, has conquered the islands for Philip, and heads for home with a cargo of cinnamon (*see* 1564; Manila, 1571).

The London Royal Exchange is founded by Sir Thomas Gresham (*see* 1562).

London's Royal College of Physicians receives authority to conduct dissections of human cadavers.

The first picture of a lead pencil appears in a book on fossil collecting by German-Swiss naturalist-zoologist Konrad von Gesner (who dies of plague at age 48). English shepherds in Cumberland found a large deposit of pure "wadd," or "black lead," or "plumbago" in the past decade (it will be called *graphite* beginning in 1789); Gesner's woodcut shows an ornately turned tube of wood holding a tapered piece of the "lead." Borrowdale lead, mined near Keswick, will be a Crown monopoly and the world's chief source for 2 centuries.

Painting *Haymaking, The Harvesters, The Return of the Herd, The Gloomy Day, The Hunters in the Snow,* and *Winter Landscape with a Bird Trap* by Pieter Brueghel; *The Death of Actaeon* by Titian; *The Family of Darius Before Alexander* by Paolo Veronese.

Mass "Missa Papae Marcelli" by Palestrina.

John Hawkins introduces tobacco into England from Florida and writes, "The Floridians when they travell have a kinde of herbe dried, who with a cane and an earthen cap in the end, with fire, and the dried herbs put together, doe sucke thorow the cane the smoke thereof, which smoke satisfieth their hunger, and therewith they live foure or five dayes without meat or drinke. . ." (*see* 1600).

Europe has poor harvests. Catherine de' Medici decrees that meals shall be limited to three courses.

1566 The Italian secretary to Mary, Queen of Scots is seized the evening of March 9 by the earls of Morton and Lindsay, who invade Mary's supper chamber with armed men, hack David Rizzio, 33, to death with daggers, and throw his body into the courtyard at Holyrood. The noblemen act on orders from Mary's husband Lord Darnley, who will himself be murdered next year.

A German army of 80,000 under Maximilian II prepares to fight the Turks under Suleiman the Magnificent at Komon in Hungary, but the army is stricken with typhus, and the campaign is abandoned.

Suleiman the Magnificent dies September 5 at age 70 during the siege of Szigeth in Transylvania (his death is kept secret for 7 weeks from his 150,000 troops). His indolent son of 42 inherits the greatest and best-organized empire in Europe and will reign until 1574 as Selim II, but the Ottomans now begin a long decline as Selim signs a truce with Maximilian II; both sides retain their possessions.

Calvinists ransack monasteries and churches in Antwerp and Ghent and throughout Flanders and the northern provinces as grain prices soar following a bad harvest. Margaret of Parma, regent for the Lowlands, receives a petition April 2 demanding abolition of the Inquisition. Willem of Nassau, 33, Prince of Orange; Lamoral, 44, Comte d'Egmont; and Philip de Montmorency, 48, Comte d'Hoorn lead the protest by 300 noblemen. Margaret promises to forward their petition to her natural brother Philip of Spain, but she raises an army (*see* 1567).

Philip II cracks down on Spain's Moriscos (Moors converted to Christianity), forbidding them to speak Arabic or wear traditional dress.

Painting *The Census at Bethlehem, The Sermon of Saint John the Baptist,* and *The Wedding Dance* by Pieter Brueghel.

Theater *Gammer Gurton's Nedle* by clergyman-teacher William Stevenson, perhaps in collaboration, is performed at Christ's Church, Cambridge: "I cannot eat but little meat,/My stomach is not good;/But sure I think that I can drink/With him that wears a hood./That I go bare, take ye no care, I am nothing a-cold:/I stuff my skin so full within/Of jolly good ale and old./Back and side go bare, go bare,/Both foot and hand go cold;/But belly, God send thee good ale enough,/Whether it be new or old."

1567 Lord Darnley, husband to Mary, Queen of Scots, is found murdered February 10 at Kirk o'Field, Edinburgh, and the queen is suspected of complicity. She is kidnapped April 24 by James Hepburn, 31, fourth earl of Bothwell, who marries her in a Protestant church May 15. The marriage provokes a rebellion by Scottish noblemen who desert the queen at Carberry Hill in June, force her to dismiss Bothwell, and make her sign an abdication in favor of her 13-month-old son, who is proclaimed James VI of Scotland July 24. The young king will be raised at Stirling Castle by the earl and countess of Mar.

Philip II sends 20,000 troops under the duke of Alva to the Lowlands where they capture Antwerp as the Lowlanders prepare for a struggle to gain independence from Spain (*see* 1560; 1568).

Catholic troops under the constable of Montmorency attack a Huguenot force under the Prince of Condé November 10 at St. Denis, near Paris. The constable, now 74, is killed.

French colonists in Brazil are captured and put to the sword by the Portuguese (*see* 1555).

John Hawkins joins two African kings in besieging an inland African town, he takes 470 slaves as booty, and he sells them to Spanish colonists in America (*see* 1562; 1568).

Rio de Janeiro is founded by the Portuguese who destroy the French colony under the leadership of Mem de Sa.

Caracas is founded by Spanish settlers in an area discovered by Columbus in 1498 and later called "little Venice."

Rugby School is founded in England's east Warwickshire (*see* game of rugby, 1823).

Painting *The Wedding Banquet (Peasant Wedding), The Land of Cockaigne, The Conversion of Saul,* and *Adoration of the Kings in the Snow* by Pieter Brueghel; *The Martyrdom of St. Laurence* by Titian.

"I can boast of having been the first in France who brought the seed of this plant [tobacco], who sowed it, and named the plant in question herbe Angoulmoisine," writes André Thevet in *Les Singularités de la France antartique, autrement nommée Amérique* (*see* 1556). "Since then a certain individual, who never made any voyage, has given it his name, some ten years after my return" (*see* Nicot, 1561).

Andrea Palladio completes the Rotonda (Villa Capra) outside Vicenza.

Antwerp's sugar-refining industry moves to Amsterdam following the capture of Antwerp by the duke of Alva (above).

1568 The Peace of Longjumeau signed by Catherine de' Medici with the Huguenots March 23 ends a second French religious war.

Mary, Queen of Scots escapes from captivity in May but is defeated at Langside May 13 and is placed in confinement after fleeing to England.

Leaders of the Flemish opposition to the Spanish Inquisition are beheaded as traitors at Brussels June 4, and the action precipitates a revolt of the Lowlands that will continue for 80 years. The Comte d'Egmont, the Comte d'Horn, and 18 others are executed, and there is a general confiscation of the estates of those who have failed to appear before the Council of Blood, including Willem of Orange, who has left Holland with thousands of Netherlanders (*see* Goethe, 1796).

Sweden's Eric XIV is deposed September 30 after several years of increasing insanity and is replaced by his brother, now 31, who will reign until 1592 as John III.

Sir John Hawkins leaves Veracruz in October after a third slaving expedition and is ambushed by the Spanish in West Indian waters. He loses two of his five ships, and the incident precipitates an undeclared state of war (*see* 1567; 1588).

Japan's Azuehi-Momoyama period of unification begins following the seizure of Kyoto by the Taira clan general Oda Nobunaga, 34, who begins subjugating feudal lords (*daimyo*), humbling Buddhist priests, destroying the political power of Buddhism, and establishing order in many provinces.

Mennonites fleeing Spanish persecution in the Lowlands (above) move east. The pacifist Protestant sect will settle in the German states, Switzerland, and eastern Russia (*see* 1523; 1536; Catharine, 1783).

Moriscos in Granada rebel against the restrictions placed upon them by Philip II 2 years ago.

The Solomon Islands of the Pacific are explored by Spanish mariner Albaro de Mendana de Neyra, 27, who discovered the islands last year and who will find the Marquesas just before he dies in 1595.

English explorer David Ingram travels from the Gulf of Mexico north to Canada and finds "vines which beare grapes as big as a mans thumbs."

Abridgement of the Chronicles of England is published by English printer Richard Grafton, who first records the mnemonic rhyme "Thirty days hath September,/April, June and November;/All the rest have thirty-one,/Excepting February alone,/And that has twenty-eight days clear/And twenty-nine in each leap year."

Painting *Imperial Art Dealer Jacopo della Strada* by Titian; *The Peasant Dance, The Peasant and the Bird-Nester, The Cripples, Landscape with the Magpie in the Gallows, The Misanthrope,* and *The Parable of the Blood* by Pieter Brueghel.

1569 The Battle of Jarnac March 13 gives Catholic forces under the duc d'Anjou victory over Huguenot forces. Louis de Bourbon, Prince of Condé, is murdered while crossing the river Charente.

The Union of Lublin July 1 merges Lithuania with Poland over the objections of Lithuania, strengthening resistance to attacks from Muscovy and the Crimean Tatars.

Ottoman forces drag ships overland from the Don to the Volga but fail in an attack on Astrakhan, sustain heavy losses from the Muscovite defenders, and retreat.

English earls to the north rebel against Elizabeth and sack Durham Cathedral, but the rebellion is easily suppressed.

The Mercator projection map of the world published by Flemish instrument maker Gerhardus Mercator (Gerhard Kremer), 57, represents the meridians of longitude by equally spaced parallel lines and gradually spaces out latitude lines toward the polar regions to exaggerate degrees of latitude in exactly the same proportion as degrees of longitude. Mercator has studied under the late Gemma Frisius (*see* 1530; 1533), his map enormously simplifies the job of charting a course (the course of a ship on a constant compass bearing always

1569 (cont.) appears as a straight line), and the map will contribute to the accuracy of navigation charts everywhere.

Hunger and plague kill 500 people a day at Lisbon through much of the summer.

Painting *The Storm at Sea* by Pieter Brueghel who dies at Brussels September 5 at age 44.

1570 The czar of Muscovy Ivan the Terrible enters the city of Great Novgorod January 8 and begins a 5-week reign of terror. A man of dubious character has accused the Novgorodians of being sympathetic to Poland in the ongoing Livonian War, Ivan has ravaged the approaches to the second richest city in the czardom, and he has batches of Novgorodians from all classes of society massacred each day. He has his men plunder every church, monastery, manor house, warehouse, and farm within a radius of 100 miles, destroying all cattle, leaving the structures roofless. Not until February 13 does he permit rebuilding.

The Treaty of St. Germain signed August 8 by Charles IX ends France's third religious war on terms favorable to the Huguenots.

The Ottoman sultan Selim II declares war on Venice at the persuasion of Don Joseph Nasi, a Sephardic Jew who has plans to make Cyprus a refuge for Jews. Venice has refused to cede Cyprus, Spain has come to Venice's support, an Ottoman fleet of 360 sail arrives at Limisso July 1, the Spanish and Venetian fleets are delayed, 55,000 Turks attack Nicosia, and Cyprus falls to the Turks (*see* 1571).

The Peace of Stettin December 13 ends a 7-year war between Sweden and Denmark, which recognizes Swedish independence. French and Polish diplomats have mediated to negotiate the peace, and the two Scandinavian powers will remain at peace with each other until 1611.

Five North American tribes confederate under the name Iroquois. The Mohawk brave Hiawatha and the brave Dekanawida, originally a Huron, have persuaded the Mohawk, Oneida, Onondaga, Cayuga, and Seneca to form a league with a common council, each tribe having a fixed number of chiefly delegates.

Large-scale traffic in black slaves begins between the African coast of Sierra Leone and the Brazilian bulge 1,807 miles away (*see* 1526; 1581).

Austria's Don John, bastard half brother of Spain's Philip II, puts down the 2-year-old Morisco rebellion in Granada. The Moriscos are dispersed throughout Castile and replaced in Andalusia with 50,000 settlers.

Ivan the Terrible (above) has nearly all of his closest advisers and ministers publicly executed at Moscow July 25 while he watches.

Tribesmen in the region that will be called Virginia kill Juan Bautista Seguar and other Jesuits, thus ending Spanish efforts to colonize the region (*see* 1584).

Nagasaki begins its role as Japan's major port for foreign commerce. A local lord in Kyushu opens the fishing village to foreign trade (*see* 1638).

The Royal Exchange opened at London is a prototype department store that rents stalls to retail merchants.

The Schoolmaster by the late English scholar Roger Ascham is a treatise on education: "It is costly wisdom that is bought by experience. . . Learning teacheth more in one year than experience in twenty." Ascham died 2 years ago at age 53.

Painting *Moses Striking the Rock* by Tintoretto.

I Quattro Libri dell Architettura by Andrea Palladio is published at Venice.

A North Sea tidal wave November 2 destroys sea walls from the Lowlands to Jutland, killing more than 1,000.

Slave ships returning from Brazil (above) bring maize, manioc, sweet potatoes, peanuts, and beans to supplement Africa's few subsistence crops. Manioc (cassava) will prove highly resistant to locusts and to deterioration when left in the field, and it will serve as a reserve against famine.

Cooking Secrets of Pope Pius V (*Cuoco Secrete di Papa Pio Quinto*) by Bartolomeo Scappi is published at Venice with 28 pages of copperplate illustrations that include the first picture of a fork (*see* 1518).

Only two native villages remain in Hispaniola (*see* 1548).

Maize from Brazil (above) will fuel population growth in Africa, providing a steady supply of slaves for the new trade.

1571 The Battle of Lepanto October 7 in the Gulf of Lepanto near Corinth ends in defeat for an Ottoman fleet of 240 galleys. The Maritime League of 212 Spanish, Venetian, Genoese, and Maltese ships under the command of Don John loses about 5,000 oarsmen and soldiers, but it captures 130 vessels with their stores and provisions, takes 4,000 prisoners, frees 12,000 Christian galley slaves, and kills 25,000 Turks. The merchants of Venice are eager for peace and fail to follow up on their triumph (*see* 1573). Constantinople will regain full strength within a year.

Crimean Tatars sack Moscow.

Japan's strongman Oda Nobunaga destroys the Enryakuji Buddhist monastery on Mt. Hiei, eliminating the most militarily powerful of his enemies.

Manila is founded May 19 by Miguel Lopez de Legazpe. He has subjugated the Philippine natives and moves his capital from Cebu to the new city, using it as a base for colonization (*see* 1565).

Harrow School is founded by a charter granted to John Lyon, a yeoman of Preston, by England's Queen Elizabeth. It will open its first building to scholars in 1611.

Benvenuto Cellini dies at Florence February 13 at age 70. His autobiography will be published in 1728.

1572 The Ottoman sultan Selim II rebuilds his navy following last year's Battle of Lepanto. His naval forces reach such imposing strength that Don John of Austria refuses to attack.

Dutch insurgents capture Brill as insurrection against the Spanish spreads in the north. The Spanish lay siege to Haarlem.

England's duke of Norfolk Thomas Howard III, 36, is beheaded for having conspired with the Spanish to invade England and free Mary Queen of Scots.

The Massacre of St. Bartholomew August 23 and 24 kills an estimated 50,000 Huguenots at Paris and in the provinces. Urged on by the queen mother Catherine de' Medici, Catholics disembowel the young king's adviser Gaspard, Admiral de Coligny, and throw him from his window still alive. Pope Gregory XIII and all the Catholic powers congratulate Catherine, and the pope commands that bonfires be lighted to celebrate the massacre, which he calls better than 50 Battles of Lepanto.

Huguenot leader Henri of Navarre is in Paris to marry the king's sister Margaret of Valois; he saves his life by feigning conversion to Catholicism (but later recants) (*see* 1593; Edict of Nantes, 1598).

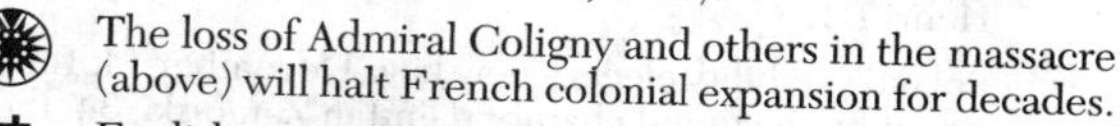

The loss of Admiral Coligny and others in the massacre (above) will halt French colonial expansion for decades.

English navigator Francis Drake, 32, leaves with two small ships carrying 73 men and boys on an expedition to capture the annual shipment of silver from Peru that mules carry across the Isthmus of Panama to be loaded aboard Spanish galleons (*see* 1573).

Danish astronomer Tycho Brahe, 26, discovers a bright new star beyond the moon in Cassiopeia and thus destroys the Aristotelian idea that no change can occur in the celestial regions. Brahe has used instruments of his own design to see the supernova that will be called Tycho's star and to accumulate data on planetary and lunar positions (*see* 1576).

The Massacre of St. Bartholomew in 1572 was the shocking nadir in the history of French religious intolerance.

French mathematician and logician Petrus Ramus (Pierre de la Ramée), 57, is killed in the massacre (above).

French surgeon Ambroise Paré, 55, wins fame by ending the brutal treatment of "infected" gunpowder wounds with boiling alder oil. Physician to Charles IX, Paré has run out of oil and has improvised a compound of egg yolks, oil of roses, and turpentine, has introduced arterial ligature to replace cauterization, will perform herniotomy without castration, treats infected burns by applying onion poultices, and will be famous for his line, "I dressed him and God healed him."

Vienna's Spanish Riding School is mentioned for the first time. The Holy Roman Emperor Maximilian II has visited a riding school in Castile, decided that his knights need a similar school, and imported Iberian horses. The riding hall's Lippizaner stallions foaled in the town of Lipizza, dark brown at birth, turn almost pure white within 5 years.

Dutch insurgents besieged at Haarlem (above) use pigeons to carry out messages.

Poetry *Os Lusiadas* by Portuguese poet Luiz Vaz de Camoes, 48, is an epic of 10 cantos, modeled on Virgil's *Aeneid* of 19 B.C., recounting in ottava rima Vasco da Gama's discovery of the sea route to the Indies.

Court painter François Clouet dies at Paris September 22 at age 62; Il Bronzino dies at Florence November 23 at age 69.

1573 Venice makes peace with the Turks March 7 at Constantinople, breaking with Spain, abandoning Cyprus, and agreeing to pay an indemnity of 300,000 ducats. Only Candia (Crete), Paros, and the Ionian Islands remain under Venetian control.

Poland elects her first king May 11, choosing Henri of Valois, brother of France's Charles IX. His mother, Catherine de' Medici, has paid heavily to secure the election. The Pacta Conventa, signed by Henri, places strict limits on royal power and formally recognizes the right of the Polish nobility to elect kings (*see* 1574).

Spanish forces under Don John of Austria take Tunis from the Ottoman Turks, who have held it since 1569. The Ottomans will regain it next year through the strategy of the Grand Vizier Sokullo, who effectively controls the empire of the drunken sultan Selim II.

The Edict of Boulogne July 8 ends a fourth war between French Catholics and Huguenots on terms favorable to the Huguenots (but *see* 1574).

China's Wan Li assumes the imperial throne at age 10 and begins a 47-year reign as Shen Zong. Ming dynasty culture will flourish in the new reign, but Manchu power will increase.

The Ashikaga shogunate that has ruled Japan since 1336 is ended as the shogun Yoshiaki takes arms against the strongman Oda Nobunaga, who took power 5 years ago. Yoshiaki is defeated, shaves his head, and becomes a Buddhist priest (*see* 1582).

1573 ***(cont.)*** Poland (above) makes all religions equal under the Constitution. A safe haven since about 1500, the country will have more than half of Europe's Jews by 1800 (*see* Russia, 1766).

Francis Drake makes the biggest haul in the history of piracy. With assistance from Indians and blacks, he captures a shipment of Spanish silver from the Potosí mines of New Castile while it is being transported across the Isthmus of Panama for shipment to Spain (*see* 1545; 1572; 1577).

A typhus epidemic strikes the area surrounding the city of Mexico (Tenochtitlan) in New Spain.

Poetry *Five hundred good points of Husbandry* by Suffolk farmer Thomas Tusser uses rhymed proverbs to guide fellow farmers.

Painting *The Battle of Lepanto* by Tintoretto for the Doge's Palace.

Theater *Amyntas (Aminta)* by Italian playwright Torquata Tasso, 29, 7/31 at a theater on Belvedere Island in the Po River.

1574 France's Charles IX dies May 30 at age 24 and is succeeded by his brother Henri of Valois, now 23, who vacates the Polish throne, returns to France, and will reign until 1589 as Henri III under the domination of his mother, Catherine de' Medici.

A new religious war begins between French Catholics and Huguenots (*see* 1573; 1576).

The departure of Henri of Valois (above) leaves Poland without a king (*see* 1575).

Leyden in the Lowlands comes under siege by Spanish forces under Luis de Requesens y Zuñiga, who has succeeded the duke of Alva, gained a victory over Protestant rebels at Mookerheide in April, and advised Philip II in June to pardon the Protestants. Willem of Orange (the Silent), who has lost two brothers in the battle, cuts the dike in several places to flood the land, and his ships sail up to Leyden's city wall October 3 to relieve the siege (*see* 1575).

The Ottoman sultan Selim II falls in his Turkish bath, cracks his skull, and dies December 12 at age 50 after a drunken 8-year reign. His eldest son, 27, has his brothers strangled in his presence and will reign until 1594 as Murad III.

Islands discovered November 22 some 365 miles off the west coast of South America will be named the Juan Fernández Islands after the 38-year-old Spanish navigator who has found them (*see* Defoe, 1719).

Poetry *La Franciade* by Pierre de Ronsard who has been court poet to Charles IX (above) since 1560 but whose national epic will remain unfinished.

Florence's Uffizi Palaces are completed after 14 years of construction by painter-architect Giorgio Vasari, who dies June 27 at age 62. The two matching palaces frame a narrow piazza and serve as offices for the grand duke of Tuscany Cosimo de' Medici.

Some 5,000 of Leyden's 15,000 people die of hunger in the 6-week siege by the Spanish (above), but burgomaster van der Werer cries defiance. He draws his sword and says, "You may eat me first; I will not surrender to the Spanish."

Ships bearing white bread and herring relieve Leyden's hunger. Together with Leyden hutsput (a stew of stale beef and root vegetables), they will be served October 3 each year after 1648 to celebrate the liberation of the Netherlands.

An estimated 152,500 Spanish settlers are now in the Americas.

1575 Protestant rebels in the Lowlands meet at Breda in February with the Spanish governor-general Requesens. The emperor Maximilian II mediates, and Requesens agrees to withdraw troops and officials from the Netherlands (but *see* 1576).

French Catholic troops under Henri, duc de Guise, defeat a Protestant army in October at the Battle of Dormans. The duc d'Alençon then allies himself with Henri of Navarre against the duke's brother, Henri III.

Poland's nobility elects a new king December 14. Influenced by the grand chamberlain Jan Zamojski, 34, they choose Transylvania's Prince Stephen Báthory, 42, who has Turkish support and will reign until 1586.

Luanda is founded by Portuguese colonists in West Africa.

The University of Leyden has its beginnings in a college founded at the Dutch town.

Poetry *Jerusalem Delivered (Gerusalemme Liberata)* by Torquato Tasso is an epic of the First Crusade.

El Greco arrives in Spain after having left his native Crete to study painting in Italy. Kyriakos Theotokopoulos, 34, who has studied under Titian, will remain in Spain, settling at Toledo.

Sculpture *Mercury* (bronze) by Flemish-born Italian sculptor Giovanni Bologna (Jean Bologna), 46, who has been attached to the Florentine court of the Medici family since 1558.

England's royal organist Thomas Tallis, 60, and his fellow organist-composer William Byrd, 33, receive a 21-year license January 22 from Elizabeth to print and sell music and music paper. They issue their joint work *Cantiones quae ab arguments Sacrae vocantur, quinque et sex partium* containing 16 motets by Tallis and 18 by Byrd, some of which will be given English translations and be sung for centuries as Anglican cathedral anthems.

The first European porcelain, created by Tuscany's Grand Duke Francesco Maria de' Medici, 34, is far inferior to the Chinese porcelain imported by Portuguese caravels. Almost all Europeans eat off earthenware plates or wooden trenchers (*see* d'Entrecolles, 1712).

1576 The Peace of Chastenoy May 6 ends France's fifth civil war since 1562, but the Paix de Monsieur concedes so much to the Huguenots that Catholics form a Holy League and make an alliance with Spain's Philip II, hoping to bring the Guises to the throne. Alarmed at the League's ambitions, Henri III proclaims himself its head and forbids exercise of Protestantism in his realm.

The Holy Roman Emperor Maximilian II dies October 12 at age 49 and is succeeded by his eldest surviving son Rudolf, 24, who will reside mostly at Prague and will reign until his death in 1612.

"The Spanish Fury" erupts in November following the death of Requesens. Left unpaid and without food, Spanish garrisons in the Lowlands mutiny and vent their rage. At Antwerp they massacre 6,000 men, women, and children, burn 800 houses, and wreak havoc on the city's commerce. Willem of Orange (the Silent) persuades the 17 Lowland provinces to unite by treaty November 8 in the Pacification of Ghent, burying religious differences and vowing resistance until the Inquisition is ended and their liberties restored.

Persia's Shah Tahmasp I is poisoned to death at age 53 after a 43-year reign. His fourth son kills off nearly all his relatives and begins a brief reign as Ismail II.

The Mughal emperor Akbar conquers Bengal, richest province in northern India.

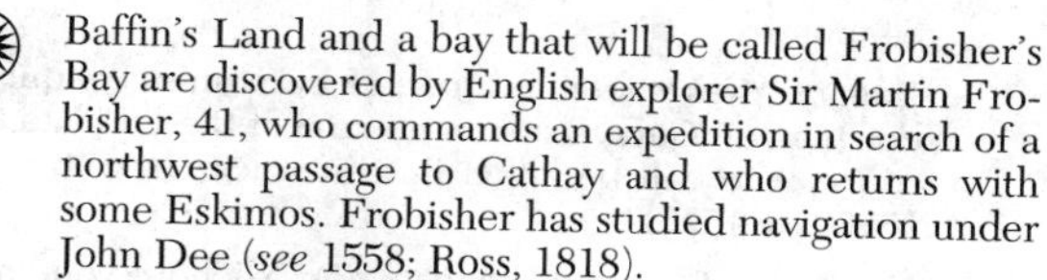

Baffin's Land and a bay that will be called Frobisher's Bay are discovered by English explorer Sir Martin Frobisher, 41, who commands an expedition in search of a northwest passage to Cathay and who returns with some Eskimos. Frobisher has studied navigation under John Dee (*see* 1558; Ross, 1818).

Tycho Brahe establishes an astronomical observatory on the island of Hven in the Sound with royal aid, but he will reject the Copernican system of 1543, holding that the five planets revolve about the sun which in turn revolves about an immobile earth (*see* Kepler, 1609; Galileo, 1613).

A typhus epidemic ravages New Spain as in 1551.

Warsaw University has its beginnings in a new Polish academy.

Six Livres de la République by French political economist Jean Bodin, 46, attempts to revive the system of Aristotle and apply it to modern politics. Bodin defines the powers of a sovereign, but his theory of sovereignty is full of discrepancies.

Titian dies at Venice August 27 in his late 80s.

Hans Sachs dies January 19 at age 81 after a career as *Meistersinger* in which he has composed innumerable *Fastnachtsspiele*—the popular plays that have emerged as drama has become more secular.

England's first playhouse opens at Shoreditch under the direction of actor-manager James Burbage, who has been one of the earl of Leicester's players (see Blackfriar's, 1596; Globe, 1599).

Japan's Azuchi Castle goes up on the shores of Lake Biwa, where the Taira strongman Oda Nobunaga builds the country's first great castle.

1577 French Catholics triumph in mid-September over Huguenot forces, but the sixth war is brief. Henri III, reluctant to let the Holy League become too powerful, grants the Huguenots generous terms.

Willem the Silent enters Brussels in triumph September 23 and becomes lieutenant to the new governor, Archduke Mathias of Hapsburg.

Francis Drake embarks December 13 with a fleet of ships and sails down the African coast en route for South America. His flagship the *Pelican* is 102 feet in length overall and carries nine "gentlemen" in addition to a crew of 80 (which includes 40 men-at-arms, a tailor, a shoemaker, an apothecary, and Drake's personal trumpeter) (*see* 1578).

The Black Assize at Oxford, England, ends with the judges, the jury, witnesses, everyone in the court except the prisoners dying of "gaol fever," a pestilence believed to have arisen out of the bowels of the earth but actually typhus carried by the prisoners, who live in filth.

Venice's Church of the Redeemer (Il Redentore) designed by Andrea Palladio goes up with funds voted by the Venetian Senate at the height of a plague.

1578 Don John of Austria receives reinforcements in the Lowlands as Spain's Philip II sends an army under his Italian cousin Alessandro Farnese, 32. Farnese attacks and defeats a patriot army at Gemblours January 31, Don John dies of fever October 1, and Farnese takes command of Spanish and Austrian forces in the Lowlands.

Portugal's Sebastian I, now 24, is killed August 4 and his army annihilated at Al Kasr al Kebir in northwest Africa where he has led a crusade in defiance of warnings by Philip II and Pope Gregory XIII. The king of Fez and the Moorish pretender are also slain in the battle, but the Portuguese people refuse to believe that Sebastian is dead. "Sebastianism" develops as a religion whose votaries believe that the king is either away on a pilgrimage or is waiting on some enchanted island for an appropriate time to return. Four pretenders will successively impersonate Sebastian and be executed, the last an Italian who speaks no Portuguese (*see* 1580).

Persia's Ismail II dies after less than 2 years. His oldest brother, half blind, has escaped his vengeance and will reign until 1587 as Mohammed Khudabanda.

Francis Drake navigates the Strait of Magellan in a 16-day passage after having lost four of his ships, he renames his flagship the *Golden Hinde,* he ravages the coasts of Chile and Peru, and he continues up the coast (*see* 1577; New Albion, 1579).

Queen Elizabeth grants a patent to Sir Humphrey Gilbert, 39, to "inhabit and possess at his choice all remote and heathen lands not in the actual possession of any Christian prince." Gilbert crosses the Atlantic in search of a northwest passage to the Indies (*see* 1583).

Chronicles of English History to 1575 by English historian-printer Raphael Holinshed continues work begun by his late employer Reginald Wolfe. Holin-

1578 *(cont.)* shed's *Chronicles* will be the source of plot material for historical dramas.

Fiction *Euphues, The Anatomy of Wit* by English author John Lyly, 24.

Poetry *Sonnets pour Hélène* by Pierre de Ronsard whose lyrical lines include *"Quand vous serez vielle, au soir à la chandelle. . ."*

China's population reaches 60 million.

1579 The Union of Arras in January joins Low Country Walloons (Catholics) with those of Hainaut and Artois, but the Union of Utrecht January 23 joins Dutch patriots to the north in opposition to the hated Spanish. This final division of the former Netherlands establishes the United Provinces and marks the birth of the Dutch Republic. The Dutch sign a military alliance with England.

Spanish troops land in Ireland's County Munster to support a rebellion against English Protestantism (*see* 1580).

Scotland enacts a law "for Punishment of the Strong and Idle Beggars, and Relief of the Poor and Impotent" (*see* 1551; 1601).

Francis Drake puts in for repairs June 17 at a point north of what will be called San Francisco and claims possession of New Albion for Elizabeth (*see* 1769; Cabrillo, 1542; Cook, 1778).

Russian pioneers move into Siberia. In the next 60 years a few thousand people will gain control of a vast region extending to the Pacific (*see* 1637; 1741).

Vindiciae contra Tyrannos is published anonymously (probably by Theodore de Bèze). The pamphlet attacks absolutism, declares that rulers must be accountable to the people, but insists that only magistrates may resist the king.

De Juri Begni apud Scotos by Scottish humanist George Buchanan, now 73, confutes absolutism with the argument that kings exist only by the will of the people. Written for the instruction of his royal pupil who will become England's James I, Buchanan's work justifies tyrannicide and states that the obligation of subjects to their king is conditioned on the performance of that king in the duties of his office.

Lives of the Noble Grecians and Romans by the Greek biographer Plutarch of the 1st to 2nd century A.D. is published in an English translation from the French of Jacques Amyot, 66, by English translator Sir Thomas North, 44, who will add additional *Lives* in 1595. North's work will influence Elizabethan poets and provide material for playwrights.

Poetry *The Shepheardes Calendar* by English poet Edmund Spenser, 27, who secures a place in the earl of Leicester's household and dedicates the work to Leicester's nephew Philip Sidney.

Painting *Trinity and Assumption* by El Greco, who has been commissioned to decorate Toledo's Church of S. Domingo el Antigum.

1580 Spain and Portugal unite under one crown following the death of Portugal's Cardinal Henri after a 2-year reign as king. Spaniards under the aging duke of Alva invade the country, they defeat the Portuguese August 25 at the Battle of Alcantara near Lisbon, and Spain's Philip II is proclaimed Philip I of Portugal (*see* 1581; 1640).

An Irish rebellion supported by the Spanish is suppressed by the English who pacify the rebels by starving them.

A seventh French civil war between Catholics and Huguenots breaks out but is ended November 26 by the Treaty of Fleix which confirms earlier treaties granted to the Huguenots (*see* 1585).

The Moroccan port of Ceuta opposite Gibraltar is occupied by the Spanish, who will hold it until 1688.

Francis Drake enters Plymouth Harbor September 26 after having made the first circumnavigation of the world by an Englishman. He has left Java March 26, has rounded the Cape of Good Hope June 15 with only three casks of water for his 57 men aboard the *Golden Hinde,* watered on the Guinea coast a month later, and has completed his round-the-world voyage in roughly 34 months.

Buenos Aires is founded June 11 by Spanish conquistador Juan de Garay, captain-general of the La Plata territory, who will be killed in a massacre by natives in 1583 (*see* 1515).

Nonfiction *Essais* by French writer Michel Eyquem de Montaigne is published in its first two volumes. Now 47, Montaigne served as a judge at Bordeaux from 1555 to 1570, he retired to Montaigne in 1570 and began work on his essays 2 years later, and while he will insist that "I do not teach, I recount," his pursuit of truth expressed in vernacular style reveals extraordinary insight into the human condition.

Fiction *Euphues and His England* by John Lyly continues his 1578 book. Dedicated to his patron the earl of Oxford, Lyly's didactic romance is intended to reform education and manners.

Cocoa gains widespread use as a beverage in Spain (*see* 1527; 1615).

1581 Willem the Silent names François of Valois, brother of France's Henri III, king of the Netherlands. François tries to regain Antwerp from the Spanish.

Poland's Stephen Báthory invades Muscovy, gains a victory over Ivan the Terrible, and advances to Pskov (*see* 1582).

Scholars attending commencement exercises at St. Mary's, Oxford, June 27 find on their benches 400 copies of *Decem Rationes*, a pamphlet attacking the Anglican Church. Authorities seize English Jesuit Edmund Campion, 41, as he preaches at Lyford, Berkshire, July 14. He is committed to the Tower of London, examined in the presence of the queen, placed on the rack three times in an effort to shake his faith in Roman Catholicism, indicted for conspiring to dethrone Elizabeth, found guilty November 20, hanged December 1, drawn, and quartered.

Spain's Philip II sends some of his black slaves to his Florida colony of St. Augustine. They are the first blacks to be landed in North America (*see* 1565; Drake, 1586).

Spain's Philip II (above) unifies control of the Oriental spice trade, eliminating the competition in sugar, spices, and slaves that existed before the Spanish takeover of Portugal last year.

Francis Drake captures the richest prize ever taken on the high seas. He returns with Spanish gold, silver, and gemstones worth as much as the Crown's total revenue for a year, the Spanish ambassador demands restitution, but Elizabeth knights Drake after a 9-month delay.

"The Newe Attractive" by London compass maker Robert Norman is a pamphlet describing Norman's discovery of the dip in the magnetic needle.

Ergotism kills thousands in the German duchy of Luxembourg and in Spain where the disease is endemic (*see* 1039; 1587).

Ballet *Ballet Comique de la Reyne* at Paris with music by the Italian composer Baltasarini who is known in Paris as Beaujoyeulx. Presented for the queen mother Catherine de' Medici, the ballet represents the first use of dance and music to convey a coherent dramatic idea (*see* first grand opera, 1600).

Andrea Palladio dies at Venice August 19 at age 71, but his books will spread his influence throughout the world.

The first potato recipes are published in a German cookbook, but few Europeans eat potatoes even in the German states.

1582 Muscovites at Pskov put up heroic resistance in February to a siege by Polish and Lithuanian troops. Ivan the Terrible makes peace with Poland and Sweden August 10 after 25 years of conflict. Pope Gregory XIII has sent the Jesuit Antonio Possevino to Muscovy in hopes of effecting a union of the Roman and Orthodox churches, which have been split since 1054, and Possevino has mediated differences between the warring powers. Ivan gives up claims to Livonia on the Baltic.

Cossack mercenaries use firearms to rout an army of Siberian Tatars, capture their khan Kuchum, and overrun his capital. The Stroganovs, a rich Muscovite merchant family, have hired the 1,500 Cossacks to safeguard their vastly profitable salt and fur empire.

Japan's Taira strongman Oda Nobunaga travels to join his general Hideyoshi, 46, who has conquered much of western Honshu from the Mo'ori family. His enemy, Gen. Akechi Mitsuhido, 56, sets fire to a monastery where Nobunaga has halted for the night. Nobunaga dies in the flames or commits *seppuku* June 21 at age 49, having destroyed the political power of the Buddhist priests. Hideyoshi kills Mitsuhido, gains support from Nobunaga's Tokugawa *daimyo* Ieyasu, 39, at Edo, and begins to eliminate the Nobunaga family in a great struggle for power (*see* 1573; 1584).

Divers Voyages Touching the Discovery of America by Oxford clergyman-geographer Richard Hakluyt, 30, is published in England (*see* 1599).

A delegation of Japanese Christian boys goes to Rome. The youths will return in April 1586 after visiting Pope Gregory XIII.

The University of Edinburgh has its beginnings in Scotland.

The University of Würzburg has its beginnings.

A new Gregorian calendar instituted by Pope Gregory abolishes the ancient Julian calendar because its error of one day in every 128 years has moved the vernal equinox to March 11. Gregory restores the vernal equinox to March 21. Devised by the late Neapolitan astronomer-physician Aloysius Lilius (Luigi Lilio Ghiraldi), the new calendar takes effect in Europe's Roman Catholic countries October 5 (which becomes October 15). Protestant countries will adhere to the Old Style Julian calendar of 45 B.C. until 1700 or later, and the Russians will retain it until 1918 (*see* Britain, 1752).

1583 Dutch forces from the seven United Provinces sympathetic to Spain occupy the mouth of the Scheldt River, blocking Protestant-held Antwerp from sea trade (*see* 1585).

Humphrey Gilbert takes possession of Newfoundland in the name of Elizabeth August 5 (*see* 1578). He tries to make the first English settlement in the New World but is lost on his return voyage. The colonists he has left behind in Newfoundland will disappear (*see* Virginia, 1584).

England's minister to Turkey William Harborne trades English woolens, tin, mercury, and amber for spices, cotton goods, silks, and dyes.

Portugal dominates the world sugar trade and sells Brazilian sugar at high prices for use by European aristocrats and capitalists (below).

The first life insurance policy of record is issued on the life of "William Gybbons, citizen and salter of London." The annual premium is £32, and when Gybbons dies within the year his beneficiaries will collect £400 (*see* Tontine, 1653).

De Plantis by Italian physician-botanist Andre Cesalpino contains the first modern classification of plants based on a comparative study of forms. Cesalpino's work will be acknowledged by Linnaeus in 1737.

Nonfiction *The Names of Christ (De los Nombres de Cristo)* by Spanish poet-theologian Fray Louis (Ponce) de León, 56, who was imprisoned by the Inquisition in 1572 but was absolved in 1576. His work is a Platonic discussion of the nature of Christ; *The Perfect Wife (La Perfecta Casada)* by Fray Louis is a handbook on the duties of a wife (his poetry will not be published until 1631).

1583 ***(cont.)*** Sculpture *Rape of the Sabine Women* by Giovanni da Bologna for the court of the Medicis at Florence. The marble work stands 13 feet 6 inches high.

1584 The czar of Muscovy Ivan the Terrible dies March 18 at age 53 after a 37-year reign and is succeeded by his somewhat feebleminded son, 27, who will reign until 1598 as Fedor Ivanovich. Ivan struck his eldest surviving son 3 years ago in a momentary fit of fury, the blow proved fatal, the czar has been inconsolable ever since, the boyars have refused to let him abdicate, but he has joined the strictest order of monks. Boris Federovich Godunov, 33, a son-in-law and court favorite of the late czar, will dominate the 14-year reign of Fedor I, trying to conciliate the boyars by regularizing the conditions of serfdom.

Willem of Orange is murdered at Delft July 10 at age 51. Spain's Philip II has for 3 years offered a large reward to anyone who would rid the world of the man he calls a traitor, the Burgundian Balthazar Gerard has shot the *stadtholder* of Holland and Zeeland, and Willem is succeeded by his son Maurice of Nassau, 17, who has been studying at the University of Leyden.

François, duc d'Alençon et Anjou, younger brother of France's Henri III, dies August 10, leaving the ruling Valois family without a successor (*see* 1585).

Bern, Geneva, and Zurich form an alliance against Savoy and the Catholic cantons. The Catholic cantons will form an alliance with Spain in 1587.

Hideyoshi gains hegemony over central Japan and on August 8 moves into Hideyoshi Castle at Osaka. Begun in September of last year, the castle has been built by 30,000 workers using stones brought to the site by 1,000 boats per day (*see* 1582; 1587).

The Virginia colony planted on Roanoke Island by English navigator-courtier Walter Raleigh, 32, is named for England's virgin queen, who knights Raleigh for his services. Raleigh has secured the renewal of the patent on colonization granted in 1578 to his late half brother Humphrey Gilbert, he has sent out an expedition in April under the command of captains Phillip Amadas and Arthur Barlow, both 34, they have sailed by way of the Canary Islands to Florida and thence up the coast in search of a suitable site for an English plantation in the New World, and have come to an inlet between Albemarle and Pamlico sounds (*see* 1585).

A pamphlet by Phillip Amadas and Arthur Barlow (above) describes the soil of the New World as "sweet smelling" and speaks of vines "bowed down with grapes, the woods abounded with game, the waters with fish" in "the goodliest land under the cope of heaven."

1585  France begins another civil war (the War of the Three Henris) as the Holy League vows to bar Henri of Navarre from inheriting the French throne (*see* 1584). Supported by the Holy League and Spain's Philip II, Henri of Guise battles Henri III of Valois and Henri of Navarre (*see* Coutras, 1587).

Antwerp surrenders August 17 to Alessandro Farnese, duke of Parma, who sacks Europe's chief commercial center, exiles its Protestants, and secures the southern Netherlands, Flanders, and Brabant for Spain.

The Treaty of Nonsuch in August allies England with the Protestant United Provinces as Elizabeth breaks with Spain. Philip II seizes all English ships in Spanish ports; Elizabeth sends Robert Dudley, 53, first earl of Leicester, with an army to aid the Lowlanders.

Sir Francis Drake sails for the West Indies with 2,300 men in a fleet of 30 ships to attack the Spanish.

Sir Walter Raleigh sends a new expedition to Virginia under the command of his cousin Sir Richard Grenville, 44, and Sir Ralph Lane, 55.

Chesapeake Bay is discovered by Sir Ralph Lane (above), who remains in the New World as governor of Raleigh's Roanoke Island colony (*see* 1584; 1586).

Davis Strait is discovered by English explorer John Davis (Davys), 35, on the first of three voyages he will make with Adrian Gilbert in search of a northwest passage through North America to the Pacific.

La disme by Dutch mathematician Simon Stevin, 37, introduces decimal fractions (*see* 1586).

France's Henri III promulgates a code of etiquette for his courtiers.

Kronborg Castle is completed at Elsinore (Helsingör) for Denmark's Frederick II on the site of an earlier castle built early in the last century by Eric of Pomerania to enforce the collection of tolls on foreign ships passing through the Oresund.

Jamaican ginger reaches Europe on a ship from the West Indies. It is the first Oriental spice to have been grown successfully in the New World.

Jesuit missionaries introduce deep-fried cookery (tempura) into Japan (*see* Xavier, 1551).

1586  Sir Francis Drake surprises the heavily fortified city of San Domingo on Hispaniola January 1 and forces its Spanish governor to pay a heavy ransom. He captures Cartagena on the Spanish Main in February, first plundering and then ransoming the city. He burns San Agostin (St. Augustine), Florida, June 7.

The Battle of Zutphen September 22 ends in victory for the earl of Leicester and the Dutch Staats-General over Spanish forces, but Leicester's nephew Sir Philip Sidney, 32, takes a bullet in the thigh and dies 26 days later at Arnhem.

A plot to assassinate England's queen is discovered by spies of Elizabeth's secretary of state Francis Walsingham (whose daughter is married to Sir Philip Sydney, above). Anthony Babbington, 25, a page to Mary, Queen of Scots, has conspired with Jesuit priest John Ballard, who is executed along with Babbington and five others. Mary is convicted October 25 of involvement in the scheme, and her life hangs in the balance (*see* 1587).

Poland's Stephen Báthory dies suddenly of apoplexy December 12 at age 53 after an 11-year reign. His death ends Polish plans to unite Poland, Muscovy, and

Transylvania into one great state, and he will be succeeded in mid-August of next year by the 21-year-old son of Sweden's John III, who will rule until 1632 as Sigismund III in a reign dominated by the Jesuits.

The Mughal emperor Akbar annexes the kingdom of Kashmir.

The Japanese emperor Goyozei makes Hideyoshi prime minister (*Dajodaijin*) at year's end and gives him the name Toyotomi (rich citizen) (*see* 1584; 1587).

Sir Francis Drake (above) moves up the coast to Roanoke Island and sets sail for England June 18, taking along Ralph Lane and other surviving settlers. Sir Richard Grenville arrives afterward and lands new settlers at Sir Walter Raleigh's colony (*see* White, 1587).

Sir Richard Grenville (above) captures a Spanish ship on his return voyage and stops to pillage in the Azores before setting to work at organizing English defenses against a Spanish invasion (*see* 1587).

English navigator Thomas Cavendish, 31, sails for Brazil with three ships, discovers Port Desire (Puerto Deseado) in Patagonia, passes through the Strait of Magellan, plunders a Spanish galleon captured in South American waters, and proceeds westward. Cavendish will arrive home by way of the Philippines with Spanish gold but with only one ship, the *Desire*, and his voyage of 2 years and 50 days will make him the third man to circumnavigate the world (*see* del Cano, 1522; Drake, 1580).

Treatises on statics and hydrostatics by Simon Stevin give mathematical proof of the law of the level, prove the law of the inclined plane, and show that two unequal weights fall through the same distance in the same time (*see* 1585; Galileo, 1592). A military and civil engineer under Maurice of Nassau, Stevin has invented a system of sluices as a means of defense. He will enunciate the theorem of the triangle of forces, and discover that downward pressure of a liquid is independent of the shape of its container.

Painting *The Legend of St. Mark* by Tintoretto, now 68, who has been working on the four panels since 1548; *The Burial of Count Orgaz* by El Greco.

The 75-ton obelisk erected in Rome's St. Peter's Square in September after 17 months of work has required 907 men and 75 horses to move 800 feet from the Circus of Nero. Pope Sixtus V (Felice Peretti) works to suppress the powerful nobility of the Papal States, to rid the territory of bandits, to encourage silk culture, to put the finances of the Papal States on a sound basis, and to beautify Rome. He will superintend construction of a Vatican Palace and Library, the Lateran Palace, the Santa Scala, and the completion of the dome of St. Peter's designed by Michelangelo before his death in 1564.

Potatoes picked up with other supplies at Cartagena are brought home by Sir Francis Drake (above) who returns also with Virginia tobacco and with starving and disheartened Roanoke colonists. The potatoes are planted on Raleigh's Irish estate near Cork.

1587 Mary, Queen of Scots is beheaded February 8 by order of her cousin Elizabeth, who has been persuaded that Mary Stuart's existence poses a continuing threat to the Protestant crown of England.

Spain's Philip II prepares an invasion fleet to take the English throne but is delayed by Sir Francis Drake, who sails into the Bay of Cadiz April 19 and burns upwards of 10,000 tons of Spanish shipping, a feat he will later call "singeing the king's beard" (*see* 1588).

Poland's nobility votes August 19 to elect the son of Sweden's John III to succeed the late John Báthory (*see* 1586). Now 21, the Swede will rule Poland until 1632 as Sigismund III in a reign dominated by the Jesuits.

The earl of Leicester returns to England in August, having failed in his mission to aid the Dutch revolt.

Henri of Navarre wins a victory over the Catholic League October 20 at Coutras, but the loss is only a temporary setback for the Catholic party (*see* 1588).

Swiss and German troops trying to link up with Henri of Navarre at Vimory and Auneau retreat under pressure from Henri of Guise.

Persia's half-blind shah Mohammed Khudabanda dies after a 9-year reign. He is succeeded by his son of 30, who will reign until 1629 as Abbas I, expanding his realm considerably (*see* Isfahan, below).

Toyotomi Hideyoshi invades Kyushu and subjugates the Shimazu family. The civil dictator (*Kampaku*) completes his mastery over Japan's feudal barons (*daimyo*).

Virginia Dare is born August 18 on Roanoke Island to the daughter of John White and is the first English child to be born in North America. White lands 150 new settlers including 17 women in the Virginia colony on orders from Sir Walter Raleigh, but the new arrivals land too late to plant crops for fall and winter, so White returns to England for supplies to see them through (*see* 1586; 1591).

Sir Francis Drake captures the Spanish treasure ship *Sao Felipe* with her rich cargo of bullion, silks, spices, Chinese porcelains, pearls, and gemstones.

Ergotism reaches endemic levels in the German states, bringing insanity and death to thousands who eat bread made from infected rye (*see* 1581, 1595).

Hideyoshi (above) discovers the strength of Christianity in Kyushu when his request for a beautiful woman to sleep with is refused. All the women have taken the cross, he is told, and he issues an edict July 25 banning Christianity and ordering all Jesuits to leave the country (but *see* 1593).

An Epistle of Comfort to the Reverend Priests by English Jesuit poet-priest Robert Southwell, 26, is a letter of encouragement to fellow Catholics including Philip Howard, 30, earl of Arundel, who became a Catholic 3 years ago through the influence of his wife but has been imprisoned for trying to escape from England; he will die in prison in 1594.

Theater *Tamburlaine the Great* by English playwright-poet Christopher Marlowe, 23, is presented in both its

Elizabethan England's most glorious military triumph was the defeat of the Spanish Armada in 1588.

parts at London by the Admiral's Men, a theatrical company sponsored by the earl of Nottingham; *The Spanish Tragedie, Containing the lamentable end of Don Horatio, and Bel-imperia: with the pittifull death of olde Hieronimo* by English playwright Thomas Kyd, 29, whose aged Hieronimo is played by actor Edward Alleyn, the same man who plays Marlowe's Tamburlaine. Both plays are swashbuckling dramas that gratify the Elizabethan taste for bloody death, murder, and suicide.

Isfahan becomes capital of Persia by order of the shah Abbas (above) who undertakes to make the city a showplace of the world. Abbas will give Isfahan so many palaces, mosques, gardens, bridges, and caravansaries that its 600,000 inhabitants will say, "Isfahan is half the world" (*Isfahan nesfe Jahan*).

Pope Sixtus V writes that Rome is swarming with vagabonds in search of food as work goes forward to make the city a showplace of baroque palaces and cathedrals (*see* 1586).

Eggplant is introduced into England. The variety is small, light brown in color, and eggshaped.

1588 An "invincible" Spanish Armada of 132 vessels sails against England under the command of Spain's "admiral of the ocean," the untrained nobleman Alonso Perez de Guzman, 38, seventh duke of Medina-Sidonia. His largest ship is a 1,300-ton vessel, but more than 30 are below 100 tons. The Royal Navy commanded by Lord High Admiral William Howard has only 34 ships, of which the largest is the 1,000-ton *Triumph.* Admiral Howard on his 800-ton flagship *Ark Royal* has the support of 163 armed merchant vessels, including the *Buonaventure,* first English vessel to round the Cape of Good Hope and sail on to India, and he has the help of Sir Francis Drake and Sir John Hawkins. The two fleets engage forces July 31, a great storm blows up in the following week, and the elements help the English defeat the armada by August 8, scoring a victory that opens the world to English trade and colonization.

The Danish-Norwegian king Frederick II has died April 4 at age 53 after a 29-year reign. The beloved Frederick is succeeded by his son of 10, who will reign until 1648 as Christian IV.

Henri of Guise has entered Paris and been acclaimed king of France to the delight of the Holy League and Spain's Philip II. A popular insurrection May 12 (the Day of the Barricades) forces Henri III to take refuge at Blois; receiving no support from the Estates-General against the Catholic League, he has Henri of Guise and his brother Louis the Cardinal murdered December 23.

English merchants found the Guinea Company to traffic in slaves from Africa's Guinea coast.

A Briefe & True Report of the New Found Land in Virginia by English mathematician Thomas Hariot, 28, is based on a visit to Sir Walter Raleigh's Roanoke colony (*see* below; 1587; 1591).

Venice's library is completed on the Piazza San Marco after more than a century of construction following a plan by the late Jacopo Sansovino.

Paolo Veronese dies at Venice April 19 at age 59.

Theater *The Tragedy of Dr. Faustus* by Christopher Marlowe at London with the Admiral's Men. "Was this the face that launched a thousand ships, / And burnt the topless towers of Ilium?" (XVIII, i, 99).

Lamentations by Giovanni Pierluigi da Palestrina is published at Rome. Now 72, Palestrina complains in his introduction to Pope Sixtus V that he has been compelled by poverty to publish the book of sacred music in small format and omit many pieces.

Potatoes are introduced into the Lowlands by Flemish botanist-physician Carolus Clusius (Charles de Lecluse), 62, who may have received the tubers from English herbalist John Gerard (*see* 1597; 1621).

Virginia fields planted Indian fashion with corn, beans, squash, melons, and sunflowers yield "at the least two hundred London bushelles" per acre, whereas in England "fourtie bushelles of wheat [per acre] . . . is thought to be much," writes Thomas Hariot (above).

1589 The Bourbon dynasty that will rule France until 1792 is founded by Henri of Navarre. The Catholic party has revolted at news that Henri of Guise and his brother have been murdered by Henri III, the king has fled to the Huguenot camp of Henri of Navarre at St. Cloud, outside Paris, and the Dominican monk Jacques Clement murders him there July 31. Henri of Navarre forces recognition of his claims to the throne and will reign until 1610 as Henri IV.

Catherine de' Medici has died at Blois January 5 at age 69; her son Henri III has reconciled himself with Henri of Navarre April 3, but nearly a century of civil and foreign war has left France close to ruin. The Battle of Arques September 21 ends in victory for Henri IV over the duc de Mayenne, a brother of the late Henri of Guise and new head of the Catholic League (*see* 1590).

Portuguese pretender Antonio of Crato obtains support from the English and marches on Lisbon, but he is defeated by the Spanish who struggle to regain their national confidence after the defeat last year of their great armada.

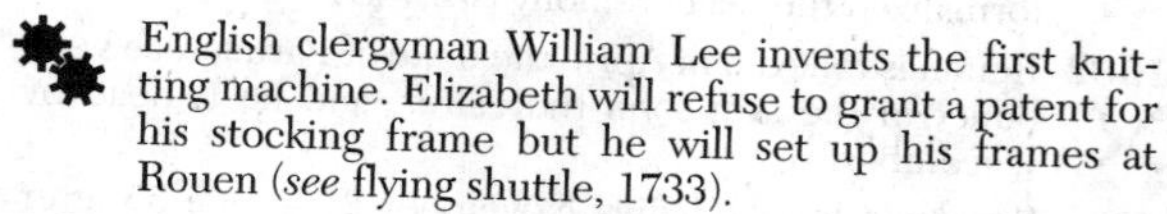

English clergyman William Lee invents the first knitting machine. Elizabeth will refuse to grant a patent for his stocking frame but he will set up his frames at Rouen (*see* flying shuttle, 1733).

The Russian Orthodox Church gains independence from Constantinople by the establishment of a separate Russian patriarchate in a move dictated by the swift decline of the Ottoman Empire through the degeneracy of its sultan Murad III and the corruption of its officials.

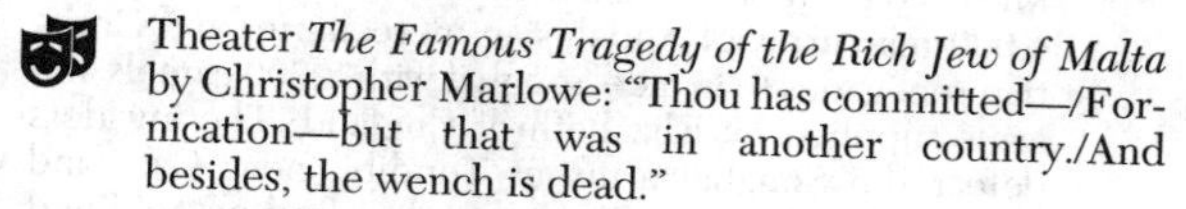

Theater *The Famous Tragedy of the Rich Jew of Malta* by Christopher Marlowe: "Thou has committed—/Fornication—but that was in another country./And besides, the wench is dead."

Kamurzell House is completed at Strasbourg.

England imports timber from western Norway because the British Isles have been so deforested by the national effort to build ships for defense against the Spanish Armada of 1588 (*see* 1605).

Destruction of forests by the Spanish in New Spain will lead to a water crisis in that country.

1590 The Battle of Ivry March 14 secures the French throne of Henri IV as the king defeats the Catholic League under the duc de Mayenne. Henri's 15,000-man royalist army besieges Paris in mid-May, the duke of Parma Alessandro Farnese arrives with 14,000 men in a few months to lift the siege, but the starvation it produces kills thousands (below).

France's Catholic party refuses to recognize Henri IV. It has proclaimed the elderly cardinal de Bourbon king in January but "Charles X" dies in May.

Persia's Shah Abbas and the Ottoman sultan Murad III end a 12-year war. Murad extends his empire to the Caucasus and the Caspian Sea by acquiring Georgia, Azerbaijan, and Shirwan. The shah has his father and brothers blinded.

The Mughal emperor Akbar conquers Orissa.

Toyotomi Hideyoshi unifies Japan. His vassal Tokugawa Ieyasu moves his capital to Edo, dominating the great eastern plain (*see* 1582; 1598).

Viaggio all'Iridia Orientali by Venetian merchant-traveler Gasparo Balbi is the first European account of India beyond the Ganges.

Baltic trade with the rest of Europe will reach its peak in the next 3 decades as measured by shipping tolls exacted in the sound (*see* Kronberg, 1585).

Venice's Rialto Bridge is completed to join the island of Rialto, financial center of the city, with the island of San Marco.

The murder of Henri III by a monk outside Paris in 1589 ended the Valois dynasty that had ruled France since 1328.

Dutch optician Zacharias Janssen invents the compound microscope (*see* van Leeuwenhoek, 1675).

The Black Death reaches Rome and other Italian cities. Ergotism remains endemic in Spain.

Poetry *The Faerie Queene* by Edmund Spenser is published in its first three books with a letter to Sir Walter Raleigh explaining its purpose and structure: "And all for love, and nothing for reward."

Painting *The Cardsharp* by Italian painter Michelangelo Merisi de Caravaggio, 17, who travels to Rome where he will paint from life, rejecting the current taste for the classics.

Theater *King Henry VI* by English playwright-poet William Shakespeare, 26, of Stratford-upon-Avon. An actor who married Anne Hathaway in 1582, Shakespeare will move to London in 1592 (dates for all Shakespearean plays are imprecise).

The cupola of St. Peter's Basilica at Rome is completed to designs by the late Michelangelo.

Palazzo Balbi is completed for Venetian merchant Nicolo Balbi after 8 years of construction on the bend of the Grand Canal.

The siege of Paris (above) brings hunger and malnutrition that kill 13,000 in the city. Food supplies are inadequate for the 30,000 inhabitants and the 8,000-man garrison, by mid-June the Spanish ambassador has proposed grinding the bones of the dead to make flour, by July 9 the poor are chasing dogs and eating grass that grows in the streets.

The archbishop of York accuses English vicar Edward Shawcross of Weaverham, Cheshire, of being an "instructor of young folkes how to comyt the syn of

1590 ***(cont.)*** adultrie or fornication and not to beget or bring forth children."

1591 French royalists besiege Rouen with help from Robert Devereux, 25, earl of Essex. The resulting starvation in Normandy's richest city resembles that seen in Paris last year.

Muscovy's epileptic czarevich Demetrius, 9, is found dead May 15 at Uglitch, his throat cut. Czar Fyodor's regent, Boris Godunov, is suspected of complicity.

North Africa's black culture is destroyed and the country made vulnerable to rival pashas. Spanish and Portuguese renegades hired by Moroccans cross the desert, use firearms to defeat a Songhai army, destroy Gao, and help the Moroccans take Timbuktu.

Portugal closes Brazil to further immigration of anyone except Portuguese, but continues to import African slaves for the plantations of its Brazilian colony (*see* 1570; 1603; 1619).

John White returns to Roanoke after having been delayed by the war with Spain and finds the colony there has vanished, possibly victims of hostile Indians. The word *Croatan,* never to be satisfactorily explained, is the only message left by the 117 lost colonists who include White's family (*see* 1587; Jamestown, 1607).

French mathematician François Viète, 51, seigneur de la Bigotière, introduces systematic use of letters in algebra to represent both coefficients and unknown quantities. Viète, who will be called the "father of algebra," has served as privy councilor to Henri IV and discovered the key to coded messages from Spain to the Netherlands, decoding them. He will apply algebra to both geometry and trigonometry (*see* Descartes, 1619).

Plague and famine strike the Italian states. Nuremberg merchant Balthaszar Paumgartner writes home to his wife Magdalene, "It is reckoned that in one year here, one-third of the folk in all Italy has died, and a highly necessary thing, too. For were it not for the pest, they must die anyway, as there would not be enough for so many to eat."

Dublin's Trinity College is founded by Queen Elizabeth.

Japan's first books to be printed from movable type are produced at Nagasaki by Portuguese Jesuit Alessandro Valegnani, a messenger from the viceroy of the Indies who arrived last year with a printing press and several printers.

Poetry *The Harmony of the Church* by English poet Michael Drayton, 28.

Theater *The Comedy of Errors* by William Shakespeare is derived from the Plautus comedy *The Menaechmi* of 215 B.C.: "There is something in the wind" (III, i); "Small cheer and great welcome makes a merry feast" (III, i).

Pamplona in northern Spain has a running of the bulls in which men run through the streets in front of the bulls being driven to the local bullring. The event will be repeated for centuries with occasional deaths and frequent injuries as men are gored or trampled.

Japanese teamaster Rikyu Sen commits ritual suicide *(seppuku)* on orders from Toyotomi Hideyoshi. Sen has formalized the tea ceremony (*see* 1484).

1592 Rouen is relieved in April, the duke of Parma receives a bullet in the arm soon thereafter, and he is dead by December.

Sweden's John III dies November 27 at age 55 after having failed to Romanize the country. He is succeeded by his son Sigismund, 26, who has been king of Poland since 1587 and marries the Austrian archduchess Anne; Sigismund will reign in Sweden until 1604, in Poland until 1632.

Morocco's Sultan Mulai Ahmed al-Mansur sends a 4,000-man army of Andalusian mercenaries and Christian renegades to invade Songhai with 8,000 camels and arms supplied by Elizabeth of England. The invaders defeat the Songhai army at Tondibi, near Gao, and again at Bamba, near Timbuktu, but find no gold and retreat, harassed by guerrillas.

The Mughal emperor Akbar conquers Sind.

Toyotomi Hideyoshi invades Korea, whose government has rejected his trade terms. His general Konishi takes Pusan Castle in May, but a Chinese ironclad dispatched by Admiral Yi-sun-sin nearly wipes out the Japanese fleet (*see* 1593).

A Russian census lists peasants under the names of landholders. The peasants will hereafter be considered the landlords' serfs (*see* 1597).

Jesuit Robert Southwell is apprehended while celebrating mass in a Catholic household. He is tortured severely and imprisoned in the Tower of London but refuses to reveal the names of fellow priests (*see* 1587; 1595).

The Malay Peninsula is rounded by English navigator Sir James Lancaster, a veteran of the 1588 battle with the Spanish Armada.

Portuguese colonists settle in Mombasa on Africa's east coast.

Spanish navigator Juan de Fuca explores North America's Pacific coast (*see* Vancouver, 1793).

University of Pisa mathematician Galileo Galilei, 28, quits under pressure and moves to the University of Padua after having defied the accepted belief that objects fall to the ground at speeds proportionate to their body weights. Galileo has purportedly climbed the spiral stairs to the top of the city's leaning tower, he has dropped cannonballs and bullets of various weights, one by one, and he has allegedly shown that except for fractional disparities caused by air resistance the objects all fall at the same rate of speed (*see* sector, 1597; Newton, 1666).

Ergotism is endemic in many of the German states, as it was 5 years ago (*see* Marbourg, 1597).

Pamphlet "A Quip for an Upstart Courtier" by Robert Greene, who dies September 3 at age 32 after having surfeited on pickled herring and Rhenish wine.

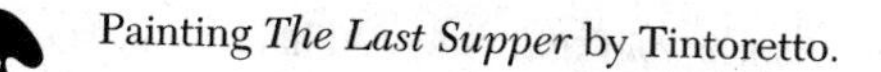

Painting *The Last Supper* by Tintoretto.

Theater *Two Gentleman of Verona* by William Shakespeare: "Who is Sylvia? what is she?/That all our swains commend her? Holy, fair, and wise is she;/The heavens such grace did lend her" (IV, ii); *Titus Andronicus* by Shakespeare: "And easy it is/Of a cut loaf to steal a shive" (II, i, meaning that adultery is easily arranged and likely to go unnoticed); *King Richard III* by Shakespeare: "Now is the winter of our discontent/Made glorious summer by this sun of York" (I, i); "A horse! A horse! My kingdom for a horse!" (V, iv); "The early village cock/Hath twice done salutation to the morn" (V, iii).

Florence's Palazza Vecchio is completed after 294 years of construction.

1593 Elizabeth reminds England's Parliament February 27 that she has the right to "assent to or dissent from anything" it may do. Statutes are pending that will impose stiff new penalties on Catholics who refuse to attend Church of England services and make it a crime to attend Catholic services.

France's parliament (Estates-General) meets in February for the first time since 1576 and holds its assembly at Paris (held by the Catholic League), rather than at Rheims as heretofore. It calls for a Catholic king.

France's Henri IV rejects Protestantism July 25 at Paris, is accepted into Catholicism, makes confession, and hears Mass. "Paris is worth a Mass," he says, and his action undermines the Catholic opposition.

Chinese troops cross the Yalu River into Korea and force the Japanese to evacuate Seoul (*see* 1592). By year's end, Toyotomi Hideyoshi has lost a third of his troop strength and retreated to the southern coast as supply ships sink, winter sets in, and Korean guerrillas harry his forces (*see* 1598).

English admiral Sir Richard Hawkins begins an abortive voyage round the world after reporting that 10,000 men have died of scurvy under his command in the Royal Navy.

English coal mining gains impetus from a shortage of firewood, which has become so costly that an Act of Parliament compels beer exporters either to fetch back their barrels or return with foreign clapboard sufficient to make the same number of barrels as was shipped (*see* 1589).

Sweden's Diet of Uppsala adopts the 1530 Confession of Augsburg and requires the king, Sigismund, to continue Lutheranism as the state religion.

Spanish Franciscans arrive in Japan to begin proselytizing in competition with the Portuguese Jesuits who arrived in 1551 (*see* 1597).

Poetry *Venus and Adonis* by William Shakespeare: "Love keeps his revels where there are but twain," "Love surfeits not, Lust like a glutton dies./Love is all truth, Lust full of forged lies." Poet-playwright Christopher Marlowe is killed May 30 at age 29 in a tavern brawl at Deptford where his companion Ingram Frizer stabs him in the eye in self-defense. Marlowe is survived by lines of verse that include "Come live with me and be my Love/And we will all the pleasures prove/That hills and valleys, dale and field,/And all the craggy mountains yield" (from *The Passionate Shepherd to His Love*).

Giuseppe Arcimboldo dies at Prague July 11 at age 66.

Theater *The Massacre at Paris* by Christopher Marlowe 1/26 at London's Rose Theatre (Lord Strange's Men); *The Troublesome Reign and Lamentable Death of Edward II* by Christopher Marlowe at London by Pembroke's Men.

Spain's Escorial Palace is completed near Madrid after 30 years of construction. Designed by architect Juan Bautista de Toledo, the palace includes a church, monastery, and mausoleum.

Sir Richard Hawkins (above) recommends orange and lemon juice as antiscorbutics (*see* Lind, 1747).

1594 France's Henri IV obtains the surrender of Paris March 22, is crowned at Chartres, and continues his campaign to win over each French province, either by negotiation or by force of arms.

Sir Richard Hawkins rounds Cape Horn, plunders the Spanish port of Valparaiso, but is defeated in Peru's San Mateo Bay and taken prisoner. He will be sent back to Spain in 1597 and not ransomed until 1602.

The Mughal emperor Akbar gains control of Baluchistan and Makran, which he annexes.

Lisbon closes her spice market to the English and Dutch, forcing creation of the Dutch East India Company to obtain spices directly from the Orient (*see* English East India Company, 1600).

The Central University of Ecuador has its beginnings in a school founded at Quito.

Fiction *The Unfortunate Traveller, or The Life of Jack Wilton* by English novelist-playwright Thomas Nashe, 27, who pioneers the adventure novel.

Poetry *The Rape of Lucrece* by William Shakespeare: "Men's faults do seldom to themselves appear."

Painting *The Musical Party* by Michelangelo da Caravaggio. Tintoretto dies at Venice May 31 at 75.

Theater *Mother Bombie* by John Lyly: "There is no fool like an old fool"; *Dido, Queen of Carthage* by the late Christopher Marlowe, whose tragedy has been completed by Thomas Nashe (above); *King John* by William Shakespeare: "Bell, book, and candle shall not drive me back/When gold and silver becks me to come on" (III, iii); "Life is as tedious as a twicetold tale/Vexing the dull ear of a drowsy man" (III, iv); "To gild refinèd gold, to paint the lily,/. . . Is wasteful and ridiculous excess" (IV, ii); *Love's Labour's Lost* by Shakespeare.

The sweet potato reaches China 30 years after being introduced by the Spanish into the Philippines.

Spain's Atlantic gateway port of Seville has a population of 90,000, up from 45,000 in 1530.

1595 The Ottoman sultan Murad III dies January 6 at age 49 after a 21-year reign of debauchery in which he has sired 102 children (20 sons and 27 daughters survive). His eldest son, 27, will reign until 1603 as Mohammed III, furthering the empire's decline. He has his 19 brothers murdered in accordance with the "law of fratricide" and has 10 of his father's concubines drowned because they are pregnant, possibly with sons. Mohammed's mother, the sultana Valide Baffo, is the power behind the throne.

The Battle of Fontaine-Française June 5 results in victory for France's Henri IV, who drives the Spanish out of Burgundy. Henri announced January 17 that he would fight Spain, which was trying to enforce the claims of a Spanish pretender to the French throne.

Irish Catholic leader Hugh O'Neill, 55, earl of Tyrone, captures Enniskellen and Monaghan castles and approaches Philip II for help against England, even though he was made earl for services to the Crown in Ireland. Elizabeth proclaims him a traitor June 30.

Spanish forces land in Cornwall and burn the English towns of Penzance and Mousehole as hostilities continue between Elizabeth and Philip II.

Transylvania's Sigismund Báthory has his opponents executed and subdues Wallachia, defeating Sinan Pasha October 28 at the Battle of Giurgevo.

The Mughal emperor Akbar annexes Kandahar.

English Jesuit poet Robert Southwell is tried for treason and found guilty. He is executed February 21 at Tyburn.

Upper Austria has a peasant revolt.

The Dutch East India Company sends its first ships to the Orient, the Dutch make their first settlements on Africa's Guinea coast, Dutch ships arrive in the East Indies, and the Dutch begin colonization.

Sir Walter Raleigh sails with four ships and 100 men to explore the Orinoco River in South America, but he returns empty-handed (*see* 1584; 1618).

Measles, mumps, and typhus (*tabardillo*) are common among the Indians of New Spain, reports Friar Mendieta (*see* 1545).

Ergotism breaks out in epidemic form at Marbourg, France, and remains endemic in many of the German states (*see* 1592; 1597).

The University of Ljubljana is founded.

The University of San Carlos is founded at Cebu City in the Philippines.

Poetry by the late Robert Southwell (above) is published posthumously and includes *St. Peter's Complaint* and short devotional lyrics among which is "The Burning Babe." "An Apologia for Poetrie" by the late Sir Philip Sidney is also published.

Painting *Venus and Adonis* by Bolognese painter Annibale Caracci, 34, who has arrived at Rome to join his younger brother Agostino.

Theater *Romeo and Juliet* by William Shakespeare whose "star-cross'd lovers" belong to the rival Montague and Capulet families of Verona: "What's in a name? That which we call a rose/By any other name would smell as sweet" (II, ii); "A plague o' both your houses!" (III, *i); Richard II* by Shakespeare: "This royal throne of kings, this sceptered isle,/This earth of majesty, this seat of Mars,/This other Eden, demi-paradise,/This fortress built by Nature for herself/Against infection and the hand of war,/This happy breed of men, this little world,/This precious stone set in the silver sea,/Which serves it in the office of a wall/Or as a moat defensive to a house/Against the envy of less happier lands—/This blessed plot, this earth, this realm, this England,/This nurse, this teeming womb of royal kings,/Feared by their breed and famous by their birth" (II, i); "For God's sake, let us sit upon the ground/And tell sad stories of the death of kings—/How some have been deposed, some slain in war,/Some haunted by the ghosts they have deposed,/Some poisoned by their wives, some sleeping killed,/All murdered. For within the hollow crown/That rounds the mortal temples of a king/Keeps Death his Court, and there the antic sits,/Scoffing his state and grinning at his pomp" (III, ii); *A Midsummer Night's Dream* by Shakespeare: "The course of true love never did run smooth" (I, i); "What fools these mortals be!" (III, ii).

England's wheat crop fails and food prices rise sharply (*see* 1596).

1596 France's Henri IV receives the surrender in January of the Catholic League's leader, the duc de Mayenne. The War of the Catholic League ends with the decrees of Folembray. France allies herself with England and the Netherlands against Spain's Philip II, whose forces capture Calais in April.

An English fleet under the earl of Essex, Lord Howard of Effingham, and Francis Vere captures Cadiz July 1 and sack the city.

Ottoman forces commanded by the new sultan Mohammed III defeat a Hungarian army in September at Erlau (Eger). The Battle of Keresztes in October starts badly for Mohammed, much of his army deserts, but the sultan is persuaded to remain; the Christians break ranks to plunder the Turkish camp and the Ottoman cavalry charges, slaughtering more than 30,000 Hungarians and Germans, capturing 100 enemy cannon, and taking other spoils.

Sir Francis Drake dies January 28 of the plague aboard his ship near the town of Nombre de Dios in the West Indies.

Dutch navigator Willem Barents lands on the islands of Newland (Spitzbergen) (*see* 1557; Hudson, 1607).

Nonfiction *The Triumphs over Death* by the late Robert Southwell is published posthumously.

Poetry *Metamorphosis of Ajax* by English poet Sir John Harington, 35, who is banished from the court for his satire: "I know that the wiser sort of men will consider, and I wish that the ignorant sort would learn, how it is not the baseness or homeliness, whether of words or matters, that makes them foul and obscene, but their base minds, filthy conceits, or lewd intents that handle them. . ." "From your confessor, lawyer and physician,/Hide not your case on no condition."

Painting *View of Toledo* by El Greco.

Theater *The Taming of the Shrew* by William Shakespeare and collaborators: "Crowns in my purse I have and goods at home/And so am come abroad to see the world" (I, ii); "Kiss me, Kate" (II, i); "Such duty as the subject owes the prince/Even such a woman oweth to her husband" (V, ii); "Thy husband is thy lord, thy life, thy keeper,/Thy head, thy sovereign, one that cares for thee,/And for thy maintenance commits his body/To painful labor both by sea and land,/To watch the night in storms, the day in cold,/While thou liest warm at home, secure and safe;/And craves no other tribute at thy hands/But love, fair looks and true obedience,/Too little payment for so great a debt" (V, ii); *The Merchant of Venice* by Shakespeare: "Hath not a Jew eyes? Hath not a Jew hands, organs, dimensions, senses, affections, passions? Fed with the same food, hurt with the same weapons, subject to the same diseases, healed by the same means, warmed and cooled by the same winter and summer as a Christian is?" (Shylock, III, i); "The quality of mercy is not strained,/It droppeth as the gentle rain from heaven/Upon the place beneath. It is twice blest;/It blesseth him that gives and him that takes" (Portia, IV, i), but many playgoers come away remembering only the insistence of Jessica's father Shylock on "a pound of flesh"; *The Blind Beggar of Alexandria* by English playwright George Chapman, 35, 2/12 at London's Rose Theatre, with the Lord Admiral's Men.

Blackfriar's Theatre opens at London where actor-theatrical manager James Burbage has converted a private house into a playhouse (*see* 1576; 1599).

A practical water closet invented by English poet Sir John Harington (above) has few buyers. The privy and chamber pot now in universal use will remain in common use for centuries (*see* Cummings, 1775).

Basque whaling captain François Sopite Zaburu devises the world's first factory ship. He builds a brick furnace on his deck and extracts whale oil from blubber which he has "tryed out" (boiled down) aboard ship, a procedure far more economical than dragging whale carcasses to shore factories.

The tomato is introduced into England as an ornamental plant (*see* 1534; 1800).

Famine brings rural unrest to Austria, but food is plentiful—if costly—at Vienna. Food shortages create misery in much of Europe, in English cities, in Asia, in the Caribbean islands, and in Peru.

1597 Spain's Philip II sends a second armada against England (*see* 1588). A storm scatters his ships once again.

Elizabeth makes her court favorite Robert Devereux, earl of Essex, earl marshal of England. He embarks for the Azores to capture the Spanish treasure fleet, but the venture will fail.

Irish rebel Hugh O'Neill, earl of Tyrone, gains support from the baron of Dungannon and is joined by Hugh Roe O'Donnell, 26, in petitioning Philip II for aid in obtaining religious and political liberty from the English (*see* 1596; 1598).

Toyotomi Hideyoshi crucifies three Jesuits, six Franciscans, and 17 Japanese converts February 5 at Nagasaki. He orders all remaining missionaries to leave the country, but when most defy his order and remain, he backs off lest he drive away Portuguese traders, especially now that direct trade with China has ceased.

A Muscovite *ukase* requires that runaway serfs be seized and returned to their masters (*see* 1592).

Akbar the Great orders that Indian peasants pay him one-third of the gross produce their fields are supposed to yield as determined by a detailed survey of the country's agricultural resources.

Willem Barents dies in the Arctic June 20, ending his effort to find a northeast passage.

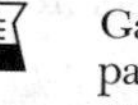

Galileo Galilei invents the sector. His draftsman's compass will, with refinements, serve for centuries as a calculating device capable of solving many algebraic problems mechanically (*see* 1592; 1610; Babbage, 1833).

Ergotism is caused by eating spurred rye, concludes the faculty of medicine at Marbourg, France (rye infected with ergot fungus appears to have spurred grain heads) (*see* 1595; 1674).

Nonfiction *Essays* by English barrister-scientist Francis Bacon, 36, who says in "Of Studies," "Some books are to be tasted, others to be swallowed, and some few to be chewed and digested: that is, some books are to be read only in parts, others to be read, but not curiously, and some few to be read wholly, and with diligence and attention"; *Herball* by English botanist John Gerard, 52, superintendent of the gardens of William Cecil, 77, first Baron Burghley and lord high treasurer of England. His *Herball* is largely a poor and thinly disguised translation of the 1554 *Cruydeboek* by Rembert Dodoens with most of its 1,800 illustrations lifted from a German work by Jacob Theodore of Bergzabern; *A Survey of London* by English historian John Stow, 72.

Theater *King Henry IV*, Part I, by William Shakespeare whose comic figure Falstaff says, "The better part of valour is discretion, in which better part I have saved my life" (V, iv).

Opera *Dafne* at carnival time in Florence's Palazzo Corsi with music by Italian composer Jacopo Peri, 36, who pioneers use of recitative in a reconstruction of musical forms used by the ancient Greeks in their tragedies; melodramatic libretto by Italian poet Ottavio Rinuccini, 35.

Iceland's Mount Hekla volcano erupts as it did in 1154 (*see* 1836).

John Gerard (above) describes potatoes, which he has received directly from Virginia, mistakenly calling them *Batata virginian sive Virginianorum, et Pappus*,

1597 ***(cont.)*** *Potatoes of Virginia.* The only potato common in England, however, is the *batata,* or sweet potato.

The blackness of Queen Elizabeth's teeth is noted by a German traveler visiting England. Paul Henter ascribes it to the queen's excessive consumption of sugar, making the first recorded association between sugar and tooth decay.

The first English mention of tea appears in a translation of Dutch navigator Jan Hugo van Lin-Schooten's *Travels.* Van Lin-Schooten calls the beverage *chaa* (*see* coffee, 1601).

1598 Muscovy's Fedor I dies January 7 at age 40 after a weak reign of more than 13 years. He is succeeded by his brother-in-law Boris Godunov, who has actually controlled the country since 1584, recovering towns taken by the Swedes, fighting off a Tatar raid on Moscow, building fortresses in the northeast and southeast to keep Tatars and Finns in order, recolonizing Siberia, supporting the middle classes at the expense of the old nobility and peasantry, and encouraging trade with English merchants by exempting them from tolls. Godunov will reign until 1605.

The Treaty of Vervins signed by Henri IV May 2 ends the war between France and Spain. Philip II is permitted to retain Flanders, Artois, and Charolais but is obliged to quit Picardy.

The Battle of Yellow Ford in Ireland August 15 ends in victory for rebel forces over the English, who lose 2,500 of a 4,000-man army.

Spain's Philip II dies at the new Escorial palace September 13 at age 71 after a 42-year reign, leaving his country in dire financial straits. He is succeeded by his indolent son of 20, who will reign until 1621 as Philip III with the duque of Lerma controlling the nation.

Japan's dictator Toyotomi Hideyoshi dies September 18 at age 62. A power struggle among his former vassals plunges Japan into turmoil (*see* Ieyasu, 1600).

The Japanese evacuate Korea after a disastrous campaign that has cost the lives of 260,000 men and has ruined the Japanese peasantry, whose crops have been commandeered to feed the troops.

The Edict of Nantes issued April 15 by France's Henri IV gives Protestant Huguenots equal political rights with Catholics, allows them to obtain some fortified towns, and opens political offices to them. The edict does not establish complete freedom of worship, but it does permit Huguenots to practice their Protestant religion in a number of French cities and towns (*see* Revocation, 1685).

Dutch admiral Wijbrand van Warwijck lands on a Pacific island, takes possession, and names it Mauritius after Maurice of Nassau.

The Marquis de la Roche tries to establish a French colony on Sable Island off Acadia (Nova Scotia). He has obtained a patent from Henri IV and lands two shiploads of vagabonds (*see* Champlain, 1603, 1604).

Pueblo territory in the American Southwest is colonized by some 400 Spanish men, women, and children who arrive in more than 80 wagons with 7,000 head of livestock. Soldiers soon follow to subdue the Pueblo.

Some 14 Dutch ships leave early in the year for India following Lisbon's closing of trade with Holland. The Dutch set out to trade directly with the East and commence a gradual conquest of Portuguese possessions (*see* 1599).

Spain's new king Philip III offers a prize of 1,000 crowns for a method of ascertaining longitude. Holland's Staats-General offers a prize of 10,000 florins for a solution to the problem (*see* 1524; 1530; Greenwich Observatory, 1676).

Fiction *Arcadia* by Spanish novelist-poet Lope de Vega, 36, who sailed with the Armada against England in 1588.

Poetry *La Dragontea* by Lope de Vega is an epic poem about Sir Francis Drake; "Address to the Nightingale" by English poet Richard Barnfield, 24, begins "As it fell upon a day/In the merry month of May . . ."; "Hero and Leander" by the late Christopher Marlowe: "Who ever loved that loved not at first sight?"

Painting *St. Martin and the Beggar* by El Greco.

Theater *The Peony Pavilion* by Chinese playwright Tang Xiansu; *Every Man in His Humour* by English playwright-poet Ben Jonson, 26, in September at London with a cast (The Lord Chamberlain's Men) that includes William Shakespeare: "The rule, Get money, still get money, boy;/No matter by what means"; "Odds me, I marvel what pleasure or felicity they have in taking this roguish tobacco! It's good for nothing but to choke a man and fill him full of smoke and embers" (III, ii); *King Henry IV,* Part II, by Shakespeare: "Out of this nettle, danger, we pluck this flower, safety" (II, iii), "Uneasy lies the head that wears a crown" (III, i).

Cuzco Cathedral is completed by Spanish architect Francisco Becerra, 53, who has been in Peru since 1573.

French tillage and pasturage are "the true mines and treasures of Peru," says Maximilien de Bethune, 38, duc de Sully. One of Henri IV's highest officials, Sully calls *le labourage et le pastourage* the two breasts of France *(les deux mamelles dont la France)* from which the nation takes her nourishment (*see* Boisguilbert, 1695).

1599 England's earl of Essex is sent by Elizabeth to subdue Ireland's rebellious earl of Tyrone but is defeated at Arklow, makes a truce with Tyrone, and leaves his post as governor general of Ireland to vindicate himself before the queen.

Sweden's Riksdag votes in November to depose the Polish king Sigismund III, who has ruled the Swedes since 1592 and supported the counter-Reformation.

Sigismund's uncle Charles is named regent for the grandson of the late Gustavus I (*see* 1604).

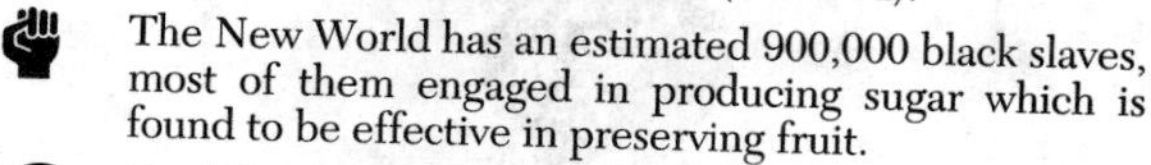

The New World has an estimated 900,000 black slaves, most of them engaged in producing sugar which is found to be effective in preserving fruit.

English geographer Richard Hakluyt publishes a recognizable map of North America (*see* 1582).

Four Dutch vessels return from India with cargoes of pepper, cloves, cinnamon, and nutmeg to establish Holland's control of the Oriental spice trade. The first Dutch trading posts are set up at Banda, Amboina, and Ternate, the Dutch raise the price of pepper from 3 shillings per pound to 6 or 8, and 80 London merchants are motivated to form their own East India Company (*see* 1598; 1600).

Prices in western Europe are generally at least six times what they were a century ago. The nobility is impoverished and in many cases is forced to sell its land to the despised middle class.

An English worker must work 48 days to buy 8 bushels of wheat, 32 days to buy 8 bushels of rye, 29 days to buy 8 bushels of barley (*see* 1506).

Essen merchant Arndt Krupp seizes the opportunity of plague in the Ruhr Valley to buy up extensive lands outside the city at giveaway prices. Krupp survives the plague to found a dynasty (*see* 1811).

The Black Death takes a heavy toll in the Hanseatic city of Essen (above), whose population of 5,000 makes it one of Europe's largest.

Plague and famine will decimate Andalusia and Castile in the next 2 years.

Theater *Much Ado about Nothing* by William Shakespeare: "Sigh no more, ladies, sigh no more,/Men were deceivers ever,/One foot in sea and one on shore,/To one thing constant never" (II, iii); *The Life of King Henry the Fifth* by Shakespeare: "Once more unto the breach, dear friends, once more,/Or close the wall up with our English dead!/In peace there's nothing so becomes a man/As modest stillness and humility./But when the blast of war blows in our ears,/Then imitate the action of the tiger,/Stiffen the sinews, summon up the blood,/Disguise fair nature with hard-favored rage." (III, i); "This day is called the Feast of Crispian./He that outlives this day and comes safe home/Will stand a-tiptoe when this day is named/And rouse him at the name of Crispin . . . . /We few, we happy few, we band of brothers . . ./And gentlemen in England now abed/Shall think themselves accursed they were not here,/And hold their manhoods cheap while any speaks/That fought with us upon Saint Crispin's Day" (IV, iii).

The Globe Theatre is opened as a summer playhouse on London's Bankside by William Shakespeare and some partners who include the brother of the late actor James Burbage, who died last year. Burbage is survived by his son Richard, 33, who, as England's most popular actor, will help establish the roles that will make Shakespeare's plays endure.

Hymn "Sleepers, Wake! The Watch-Cry Pealeth" by P. Nicolai, 43.

"It is unseasonable and unwholesome in all months that have not an R in their names to eat an oyster," writes London author Richard Buttes in *Dyet's Dry Dinner*. But while oysters spawn in such months in French and English waters, the concern of Buttes is not with any threat to their seeming inexhaustibility but rather with the difficulty of keeping oysters fresh in warm weather.

Spain grows increasingly dependent on northern and eastern Europe for grain staples. Her standard of living has declined markedly as prices have risen above those seen elsewhere in Europe, and her small landowners *(hidalgos)* are being forced off the land, which is failing into the hands of large absentee landlords. American colonial demand for Spanish cloth, wine, olive oil, and flour has slackened with the development of agriculture in Peru and New Spain, and as a result Spanish workers are increasingly idle or unproductively employed.

1600 The Battle of Sekigahara September 15 ends in a victory for the Japanese general Ieyasu over the other three regents for the 6-year-old son of the late dictator Toyotomi Hideyoshi. English navigator Will Adams reaches Kyushu and is summoned to Kyoto by Ieyasu, who retains Adams as adviser when the Englishman impresses him with his knowledge of shipbuilding and navigation. Ieyasu moves the capital from Kyoto to Edo (*see* 1603).

France's Henri IV marries his niece Marie de' Medici, 27, in October and gains control of Tuscany (*see* Opera, below).

Italian philosopher Giordano Bruno, 51, is burnt at the stake February 17 at Rome. The Inquisition has condemned him after 7 years of imprisonment for supporting Copernican astronomy and suggesting the possibility of more than one world and more than one absolute divinity (*see* 1543; Galileo, 1613).

French fur traders (below) establish a colony on the St. Lawrence River at the mouth of the Saguenay (*see* Champlain, 1603).

A French commercial partnership obtains a monopoly on fur trade in the New World.

The Honourable East India Company, the Governor and Merchants of London Trading into the East Indies is chartered December 31 to make annual voyages to the Indies via the Cape of Good Hope, challenging Dutch control of the spice trade (*see* 1599). Its initial capital is £70,000, its first governor Sir Thomas Smith (*see* 1601).

De Magnete, Magneticisique Corporibus by English physicist-physician William Gilbert, 60, is a pioneer work on electricity that introduces such terms as *electric attraction, electric force,* and *magnetic pole* (*see* Leyden jar, 1745).

A new French canal links the Rhône and the Seine Rivers (*see* 1603; 1765).

Poetry "The Nymph's Reply to the Shepherd" by Sir Walter Raleigh whose ironic pastoral poem begins: "If

1600 *(cont.)* all the world and love were young,/And truth in every shepherd's tongue . . ."

Painting *Flight into Egypt* by Annibale Caracci, who works with his younger brother Agostino on frescoes for the Farnese Gallery at Rome depicting scenes from classical mythology on great baroque ceilings.

Theater *The Shoemaker's Holiday, or The Gentle Craft* by Thomas Dekker, 1/1 at the English court, by the Lord Admiral's Men; *As You Like It* by William Shakespeare: "Sweet are the uses of adversity,/Which, like the toad, ugly and venomous,/Wears yet a precious jewel in its head" (II, i); "All the world's a stage,/And all the men and women merely players./They have their exits and their entrances,/And one man in his time plays many parts,/His acts being seven ages./At first the infant,/Mewling and puking in the nurse's arms./Then the whining schoolboy, with his satchel/And shining morning face, creeping like a snail/Unwillingly to school. And then the lover,/Sighing like a furnace, with a woeful ballad/Made to his mistress' eyebrow. Then a soldier, . . ." (II, vii); *The Merry Wives of Windsor* by Shakespeare: "We burn daylight" (II, i); "The world's mine oyster, which I with sword will open" (II, ii); *Julius Caesar* by Shakespeare: "Let me have men about me that are fat,/Sleekheaded men, and such as sleep o' nights./Yond Cassius has a lean and hungry look./He thinks too much: such men are dangerous" (I, ii); "Why, man, he doth bestride the narrow world/Like a Colossus, and we petty men/Walk under his huge legs, and peep about/To find ourselves dishonourable graves./Men at some time are masters of their fates./The fault, dear Brutus, is not in our stars,/But in ourselves, that we are underlings" (I, ii); "Upon what meat doth this our Caesar feed/That he is grown so great?" (I, ii); "Cowards die many times before their deaths;/The valiant never taste of death but once" (II, i); "Et tu, Brute?" (III, i); "Friends, Romans, countrymen, lend me your ears./I come to bury Caesar, not to praise him./The evil that men do lives after them,/The good is oft interred with their bones" (III, ii); "If you have tears, prepare to shed them now" (III, ii); "This was the most unkindest cut of all" (III, ii); "There is a tide in the affairs of men,/Which, taken at the flood, leads on to fortune;/Omitted, all the voyage of their life/Is bound in shallows and in miseries" (IV, iii).

Opera *Eurydice* 10/6 at Florence's Pitti Palace with music by Jacopo Peri: the world's first grand opera is presented on the occasion of the wedding of Marie de' Medici (above) to France's Henri IV, who has made the match to the overweight Marie to satisfy his financial obligations to her father, the grand duke of Tuscany.

Tobacco sells at London for its weight in silver shillings and is a popular extravagance among the dandies. They dip snuff and blow smoke rings as they puff on clay pipes (*see* 1565; 1604).

The Bard of Avon whose comedies and tragedies were the crowning glory of the arts in Elizabethan England.

Muscovy's Church of St. Nicholas is completed at Panilovo on the North Doina.

An estimated 350 mammal and bird species will become extinct in the next 370 years (*see* dodo, 1681).

Théâtre d'agriculture des champs by French Huguenot squire Oliver de Serres is a revolutionary treatise on agriculture. The Languedoc landowner has imported maize from Italy, hops from England, he has introduced root crops for winter fodder, and he has sown grass on land normally left fallow. While the work impresses Henri IV, who has parts of it read to him as he dines in public, only a few enthusiasts adopt the innovations of de Serres, the peasants being too conservative and the nobles too preoccupied with other matters.

Smugglers break the Arabian monopoly in coffee growing. They take "seven seeds" of unroasted coffee beans from the Arabian port of Mocha to the western Ghats of southern India (*see* 1713; 1723).

Europe's most populous nation is France with 16 million as compared to 20 million in all the German-speaking nations, principalities, and city-states, 13 million in Italy, 10 in Spain and Portugal, and 4.5 to 5 in England. China has 120 million. North America has roughly 1 million, but the New World will receive some 2.75 million slaves from Africa in the next century.

Some 54 million people will die in the world's wars over the next 345 years, but all the wars, famines, and disease epidemics will slow population growth only by some 10 years (according to a 1972 study).

17th Century

1601 Queen Elizabeth punishes the earl of Essex for his failure in Ireland, refusing to renew his patent for sweet wines. He parades through London February 7 with 300 retainers, gains no public support, returns to Essex House, surrenders to the queen's men after a brief struggle, and is put on trial. Attacked by his erstwhile friend Francis Bacon, he is condemned and beheaded February 25 at age 34.

English adventurer John Smith, 21, fights with Transylvania's Sigismund Bàthory against the Turks, but is captured and sold into slavery (*see* 1603).

England enacts a law corresponding to Scotland's 1579 Poor Law. The "Act for the relief of the poor" provides work for the able-bodied unemployed, assists those too old or too ill to work, provides for apprenticing children whose parents cannot maintain them, and specifies that support for the program must come from taxes levied in local parishes. Residence requirements prevent the needy from migrating to wealthier parishes.

The first English spice fleet sails out of Woolwich in January under the command of James Lancaster, a veteran of Francis Drake's 1588 battle with the Spanish Armada. Lancaster's 600-ton flagship *Red Dragon* is twice the size of the *Hector* or of any of the other three vessels outfitted by the East India Company (*see* 1600; 1602).

Japan's Tokugawa regent Ieyasu improves transportation by ordering that 53 stations be established between Edo and Osaka with *ryokans* (inns) at which travelers can stay overnight and obtain fresh horses (*see* paintings, 1833).

Theater *Twelfth Night, or What You Will* by William Shakespeare: "If music be the food of love, play on./Give me excess of it, that, surfeiting,/The appetite may sicken, and so die" (I, i); "Youth's a stuff will not endure" (II, iii); "O mistress mine! Where are thou roaming?/Oh, stay and hear, your true love's coming..." (II, iii); "She never told her love,/But let concealment, like a worm i' th' bud,/Feed on her damask cheek" (II, iv); "But be not afraid of greatness: some are born great, some achieve greatness, and some have greatness thrust upon 'em" (II, v); "This is very midsummer madness" (III, iv); *Blurt, Master Constable* by Thomas Middleton.

The East India Company's James Lancaster (above) doses his crew with lemon juice while at the Cape of Good Hope, then heaves to off Madagascar to take on more lemons and oranges. His 200 men are the only crew not decimated by scurvy.

The word *coffee* appears in an English account by William Parry of the Persian expedition of adventurer Anthony Sherley who 2 years ago failed in an effort to make an alliance between Elizabeth and Persia's Shah Abbas. Sherley has also failed to make military alliances for the shah against the Turks, but he has introduced coffee to London where it sells at £5 per ounce (*see* 1554; 1650).

1602 English mariner Bartholomew Gosnold explores "New England." Sponsored by William Shakespeare's patron Henry Wriothesley, 29, third earl of Southampton, he has sailed from the Azores and is the first Englishman to set foot in the region. Gosnold sails from Maine to Cape Cod which he so names after "coming to anker" in the harbor of what will be called Provincetown. He names Martha's Vineyard after his eldest child, builds a house on Cuttyhunk (which he calls Elizabeth's Island), trades with the natives, and returns with a valuable cargo of sassafras (believed to be a specific for syphilis), furs, and other commodities, leaving smallpox in his wake.

Santa Catalina Island gets its name November 25 as Basque navigator Sebastian Viscaino lands on the island off the west coast of North America on Santa Catalina's Day. Viscaino anchors a few days later some 325 miles up the coast in an "excellent harbor" which he names after the count of Monte Ray, viceroy of New Spain, who sent out the Viscaino expedition from Acapulco in April. The expedition's crews are scurvy-ridden and 16 men have died (*see* Gaspar de Portola, 1769).

James Lancaster's English East India Company fleet arrives in June at Achin in northern Sumatra. Portuguese traders have antagonized the local ruler and he is happy to meet the victors over Portugal's Spanish ally in 1588. Finding no ready market for his wrought iron and clothing stores, Lancaster engages a large Portuguese galleon, defeats her, and loots her cargo of jewels, plate, and merchandise, some of which he trades for pepper at the Dutch port of Bantam in Java.

The United East India Company chartered March 20 by the Staats-General combines various companies to eliminate cutthroat competition. It receives a 21-year monopoly and sweeping powers to wage defensive wars, make treaties, and build forts in the Indies. This new Dutch East India Company doubles and even triples European pepper prices.

Theater *All's Well That Ends Well* by William Shakespeare: "Our remedies oft in ourselves do lie,/Which we ascribe to heaven" (I, i); *Troilus and Cressida* by Shakespeare; *Hamlet, Prince of Denmark* by William Shakespeare, whose setting is the Kronborg Castle completed at Elsinore in 1585: "How weary, stale, flat and unprofitable,/Seem to me all the uses of this world!" (I, ii); "Neither a borrower nor a lender be;/For

1602 ***(cont.)*** loan oft loses both itself and friend,/And borrowing dulls the edge of husbandry" (I, iii); "There is nothing either good or bad, but thinking makes it so" (II, ii); "To be, or not to be—that is the question./Whether 'tis nobler in the mind to suffer/The slings and arrows of outrageous fortune,/Or to take arms against a sea of troubles/And by opposing, end them . . . For who would bear the whips and scorns of time,/The oppressor's wrong, the proud man's contumely,/The pangs of dispriz'd love, the law's delay,/The insolence of office, and the spurns/That patient merit of the unworthy takes,/When he himself might his quietus make/With a bare bodkin?" (III, i).

1603 Queen Elizabeth I dies March 24 at age 69 after an extraordinary 45-year reign, ending 118 years of Tudor monarchy in England. She is succeeded by Scotland's James VI, Stuart son of the late Mary Queen of Scots, who becomes James I of England and will reign until 1625 over a united kingdom.

Sir Walter Raleigh is tried for high treason on charges of complicity in the "Main Plot" to dethrone James I (above). He is sentenced to prison (*see* 1617).

John Smith kills his slave-master, escapes from captivity east of the Black Sea, and returns to England (*see* 1601; 1607).

West African ruler Idris Aloma dies after a 33-year reign in which he has rebuilt Bornu into a mighty Islamic state, the greatest power between the Niger and the Nile.

The Ottoman sultan Mohammed III dies of plague December 25 at age 37 after an 8-year reign. His son of 14 will reign until 1617 as Ahmed I.

The Tokugawa shogunate that will rule Japan until 1867 is founded at Edo by Tokugawa Ieyasu, who ushers in an era of domestic peace and prosperity (*see* 1600; 1616).

France's Henri IV names Samuel de Champlain, 36, pilot and geographer and sends him with fur trader François du Pontgrave to explore the Saguenay and St. Lawrence rivers. Champlain and his companion learn from *les sauvages* of the Great Lakes and Niagara Falls; they make an alliance with the Algonquins that will endure for 150 years (*see* 1604).

English merchant John Mildenhall reaches India via Isfahan, Kandahar, and Lahore and presents himself as self-appointed ambassador to the Great Mughal Akbar at Agra in hopes of obtaining commercial privileges for England (*see* 1608).

Japan's Tokugawa shogunate (above) will move the country from a rice economy to a money economy, and while Japan will have 125 crop failures in 265 years with some degree of famine each time, her industry, commerce, and national wealth will increase. However, a rising living standard and a large population increase will produce economic troubles.

France begins a canal project to link the Atlantic with the Mediterranean (*see* 1681).

London has an epidemic of the Black Death that kills at least 33,000 (*see* 1625).

Painting *St. Bernardino* by El Greco.

Masque *The Magnificent Entertainment Given to King James* by Thomas Dekker 3/15.

Theater *Othello, the Moor of Venice* by William Shakespeare: "Who steals my purse steals trash: 'tis something, nothing;/'Twas mine, 'tis his, and has been slave to thousands;/But he that filches from me my good name/Robs me of that which not enriches him,/And makes me poor indeed" (III, iii); "One that loved not wisely but too well" (V, ii).

Japan's Kabuki theater has its beginnings in April at Kyoto, where women led by Okuni Izumo dance at the Kitani shrine playing men's roles as well as women's (but *see* 1629).

Castel Gondolfo is completed by Carlo Maderno, 47, who has built the papal summer residence 45 kilometers southeast of Rome at Castelli Romani in Latium.

A Russian famine kills tens of thousands. Czar Boris orders distribution to the neediest of grain from the palace granaries.

Philip III sends food to starving Portuguese colonists in drought-stricken northeastern Brazil.

The English East India Company's James Lancaster returns with 278 of his original 460 men and sells a cargo of more than 1 million pounds of pepper at a good profit to the company (*see* 1601; 1602). Lancaster is knighted and becomes a proprietor of the company (*see* Bantam, 1605).

London's population reaches 210,000, up from 200,000 in 1583 (*see* 1692).

1604 Sweden's Duke John, 14, younger brother of Sigismund, formally renounces his claim to the throne. His uncle takes the title king March 6 at age 53 and will reign until 1611 as Charles IX.

Spanish forces retake Ostend from the Dutch September 20 after a 42-month siege.

The Dutch East India Company asks Dutch jurist Huigh de Groot, 21, to write a legal brief supporting the company's position in its quarrel with the Portuguese about dominion over the oceans (*see* 1609).

England's James I makes peace with Spain and directs his efforts to American colonization (*see* 1606). James is proclaimed king of Great Britain, France, and Ireland October 24.

Samuel de Champlain explores the North American Atlantic Coast from Maine to what will be called Cape Cod. He plants a colony at what will be called Nova Scotia (*see* 1603; 1609).

The Pont Neuf completed across the Seine at Paris is the city's first paved bridge and the first to be lined with houses and shops.

"The Triumphant Chariot of Antimony" introduces a new name for stibium. Probably written by a monk, the anonymous pamphlet encourages European physicians to use antimony salts as a homeopathic remedy for fever. Reputable physicians will prescribe "everlasting pellets" of antimony (swallowed, recovered from patients' stools, and used again) until the 19th century, and although it is toxic, antimony will later be found useful in treating snail fever, or schistosomiasis (*see* Bilharz, 1856).

England's James I presides over a Hampton Court Conference in January between Puritans and Anglican bishops, issues a proclamation enforcing the Act of Uniformity, banishes Jesuits and seminary priests, and commissions Sir Henry Savile, William Bedwell, Lancelot Andrews, and nearly 450 other scholars to retranslate the Old and New Testaments (the king has been disturbed by marginal notes in many bibles questioning the divine right of kings) (*see* 1611; Wycliffe, 1376).

The Granth compiled in India will serve as the sacred scripture of the Sikhs.

Poetry *The Owl* by Michael Drayton, who has been rebuffed by James I (above).

Painting *St. Ildefonso* by El Greco; *The Deposition* by Caravaggio.

Theater *Measure for Measure* by William Shakespeare: "Our doubts are traitors,/And make us lose the good we oft might win,/By fearing to attempt" (I, iv); *Westward Hoe!* by London playwrights John Webster, 24, and Thomas Dekker; *The Honest Whore,* Part I, by Thomas Dekker and Thomas Middleton is performed by the Children of St. Paul's: "The calmest husbands make the stormiest wives" (V, i).

"Counterblaste to Tobacco" by England's James I, published anonymously, makes reference to two Indians brought to London from Virginia in 1584 to demonstrate smoking: "What honor or policie can move us to imitate the barbarous and beastly manners of the wilde, godlesse, and slavish Indian especially in so vile and stinking a custome?" The king points out that tobacco was first used as an antidote to the "Pockes" (*see* Nicot, 1561), but he observes that doctors now regard smoking as a dirty habit injurious to the health and finds it on his own part "a custome lothsome to the eye, hatefull to the nose, harmefull to the braine, dangerous to the lungs and in the blacke stinking fume thereof, nearest resembling the horrible Stigian smoke of the pit that is bottomlesse" (but *see* 1612).

1605 The czar of Muscovy Boris Godunov dies April 13 at age 52. He is succeeded for a few months by his only son, but the new czar Fedor II is murdered by enemies of the Godunovs. The only legal heir of Fedor I was killed in exile at Uglich in 1591, but an imposter claiming to be that son Demetrius gains support from boyars, Cossacks, and Polish volunteers. The "false Demetrius" enters Moscow June 19 and begins a reign of 11 months.

India's Mughal emperor Akbar dies October 17 at age 62 after a 49-year reign that has established the dynasty which will rule until 1858. Possibly poisoned, Akbar is succeeded by his son, 36, who has been called Prince Selim. He will reign until 1627 as Jahangir ("Conqueror of the World") but is a drunk and will add little territory to the empire. When he marries in 1611, his empress Nur Jahan ("Light of the World") will control the realm.

Japan's Tokugawa shogun Ieyasu makes his son Hidetada, 26, co-shogun.

A "Gunpowder Plot" to blow up the Houses of Parliament in revenge against harsh penal laws enacted against Roman Catholics comes to light when English Catholic Guy Fawkes, 35, is arrested the night of November 4 while entering the gunpowder-filled cellar under Parliament. A convert to Catholicism who has served with the Spanish in Flanders, Fawkes has conspired with Robert Catesby, 32, and under torture he reveals the names of his co-conspirators who include also Thomas Percy, 45, and Thomas Winter, 33. Catesby is killed while resisting arrest, Percy mortally wounded while fleeing arrest; Winter and his brother Robert will be hanged next year along with Fawkes.

English explorer George Waymouth lands on an island off the North American coast and calls it Nanticut (*see* 1658).

French colonists in Acadia found Port Royal.

English pilot Will Adams obtains an invitation to establish a Dutch trading post in Japan. Forced to remain in Japan since his shipwreck in 1600, Adams has become foreign adviser to the Tokugawa shogun, taken the Japanese name Anshin Miura, and taken a Japanese wife (*see* 1609; 1613).

English ships of the East India Company reach Bantam and load up with pepper.

Dye wood costs English dyers six times what it cost in 1550 (*see* 1589; 1593; 1615).

Tobacco smoke infuriated England's James I and others, but tobacco exports saved the Virginia colony.

1605 *(cont.)* Pope Clement VIII dies March 5 at age 70; Camillo Borghese, 52, is elected May 16 and as Paul V begins improving the lot of Italian peasants.

Sikhs in the Punjab complete the Golden Temple at Amritsar.

Catherine de Vivonne de Savelli, 17, Marquise de Rambouillet, finds herself annoyed by the coarseness of Parisian society under Henri IV, packs her husband off to the country, and makes her home a center for refined and witty conversation—the first great Paris salon.

The world's first newspaper begins publication at Antwerp under the direction of local printer Abraham Verkoeven, a notorious drunkard.

Poetry *Poems Lyric and Pastoral* by Michael Drayton whose "Ballad of Agincourt" begins, "Fair stood the wind for France . . ."

Painting *The Crucifixion, The Adoration of the Shepherds, The Resurrection, Pentecost, St. Peter,* and *St. John the Baptist and St. John the Evangelist* by El Greco; frescoes for Rome's Farnese Palace are created by Annibale Caracci.

Theater *Volpone, or The Fox,* by Ben Jonson; *The Honest Whore,* Part II, by Thomas Dekker and Thomas Middleton.

New Spain has a native population of 1,075,000, down from 6 million in 1547 and 11 million in 1518 as a result of disease and exploitation introduced by the conquistadors.

1606 Portuguese forces in the Pacific drive off the Dutch after a 4-month siege of Malacca.

The Russian pretender is driven from the throne and is murdered May 17 by a faction of the boyars as the "time of the troubles" continues. The Muscovite boyar Basil Shuiski, who has caused the death of the "false Demetrius," makes himself czar and begins a 4-year reign as Basil IV.

The Austro-Turkish Treaty of Zsitva-Török November 11 recognizes the equality of the Austrians, relieves them of having to pay tribute for their part of Hungary, but gives Transylvania to István Bocskay, the Transylvanian prince who has led Hungarian revolutionists against the emperor Rudolf with Turkish support.

The Treaty of Vienna secured by István Bocskay (above) gives Hungarian Protestants the same religious freedom as Catholics, but Bocskay dies at age 49, possibly a victim of poisoning by Counter-Reformationists.

A Virginia charter granted by England's James I establishes the Plymouth Company and the London Company made up respectively of men from those two English cities and their environs. The two companies are authorized to establish settlements at least 100 miles apart in North America, the Plymouth Company to settle somewhere on the coast between the 38th and 45th parallels, the London Company to settle between the 34th and 41st parallels.

A Plymouth Company ship sent out August 12 under Henry Challons is captured by the Spanish. A second vessel sent out in October under Thomas Hanham and Martin Pring reaches the coast of Maine and returns with glowing accounts (*see* 1620).

Three ships of the London Company set sail December 19 for Virginia. Capt. Christopher Newport commands 144 men, including Bartholomew Gosnold and John Smith, aboard the *Godspeed*, the *Sarah Constant*, and the *Discovery* (*see* 1607).

Sailors from the Dutch ship *Duyjken (Dove)* out of Java land at Cape Keerweer, Australia, and are driven off by natives after making the first landing by Europeans on the South Pacific continent (*see* 1699).

South Pacific islands that will be called the New Hebrides in 1774 are discovered by Portuguese explorer Pedro Fernandes de Queiros.

Painting *Fray Felix Hortensio Paravicino* by El Greco.

Theater *King Lear* by William Shakespeare: "How sharper than a serpent's tooth it is/To have a thankless child" (I, iv); *Macbeth* by William Shakespeare: "Tomorrow, and tomorrow, and tomorrow,/Creeps in this petty pace from day to day,/To the last syllable of recorded time;/And all our yesterdays have lighted fools/The way to dusty death. Out, out, brief candle!/ Life's but a walking shadow, a poor player/That struts and frets his hour upon the stage/And then is heard no more: it is a tale/Told by an idiot, full of sound and fury,/Signifying nothing" (V, v); *The Whore of Babylon* by Thomas Dekker, performed by Prince Henry's Servants; A *Trick to Catch the Old One* by Thomas Middleton, performed by the Children of St. Paul's; *Eastward Ho!* by George Chapman, Ben Jonson, and John Marston.

France's Henri IV tells the king of Savoy that if he lives for another 10 years "there would not be a peasant so poor in all my realm who would not have a chicken in his pot every Sunday," but French peasants live mainly on bread and gruel (*see* 1610).

Australia (above) has an aborigine population that will be estimated to number 300,000.

1607 Gascony becomes part of France.

English vagrants outside Northampton demonstrate in June against the enclosure of common lands and other abuses by the landed gentry; several are killed in "Captain Pouch's revolt" and three are hanged as an example.

Jamestown, Virginia, is founded May 14 by Capt. Christopher Newport of the London Company who sailed into Chesapeake Bay April 26 after losing 16 men on the voyage from England. Newport has come up a river he named the James, in honor of the English king, and sails for home June 22, leaving behind colonists under Capt. John Smith.

The Plymouth Company attempts a settlement at the mouth of the Kennebec River, but the colonists will abandon George Popham's settlement after a terrible winter (*see* 1620).

French colonists abandon the settlement founded by Samuel de Champlain in 1604.

The Muscovy Company employs English navigator Henry Hudson to find a passage to China (*see* 1555). Hudson coasts the eastern shore of Greenland with 10 men and a boy until he reaches "Newland," or Spitzbergen (*see* 1596; 1608).

A third English East India Company fleet sails in March for the Indies. The fleet includes the *Red Dragon* and the *Hector,* as in 1601, and will return with cloves and other cargo that will yield a profit of 234 percent (*see* 1621).

Theater *The Knight of the Burning Pestle* by English playwrights Francis Beaumont, 23, and John Fletcher, 27, is the first of some 50 comedies and tragedies that the team will write before Beaumont's death in 1616; *Antony and Cleopatra* by William Shakespeare: "The barge she sat in, like a burnish'd throne,/Burn'd on the water; the poop was beaten gold;/Purple the sails, and so perfumed that/The winds were love-sick with them; the oars were silver,/Which to the tune of flutes kept stroke, and made/The water which they beat to follow faster,/As amorous of their strokes" (II, ii).

Opera *Orfeo* 2/24 at Mantua's Court Theater, with music by Italian composer Claudio Monteverdi, 29, who has studied with the late Giovanni Palestrina and employs a new treatment of discords to provide emotional and dramatic values.

Jamestown's colonists (above) are sick and starving by autumn, having buried at least 50. Their Capt. John Smith goes up the Chickahominy River in December to trade for corn with the Algonquins and is captured. By his own account his life is spared only by the intercession of the Powhatan chief's daughter Pocahontas, 12, who holds his head in her arms to prevent her father's warriors from clubbing Smith to death (*see* Rolfe, 1614).

1608 The elector Palatine of the Rhine Frederick IV, 34, organizes a Protestant Union (*see* 1609).

The Holy Roman Emperor Rudolf II is compelled to cede the kingdom of Hungary and the government of Austria and Moravia to his brother Matthias in June.

Paraguay is founded by Jesuits in South America (*see* Asuncíon, 1537; de Vaca, 1542).

Capt. Christopher Newport arrives at Jamestown in January with 110 new Virginia colony settlers. He finds that disease and malnutrition have reduced the original contingent to a group of 40. Jamestown's fort is destroyed by fire January 7.

Capt. John Smith surveys Chesapeake Bay and the Potomac River through much of the summer, using an open boat to look for a passage to the South Sea.

Capt. John Smith is elected president of the Jamestown Council September 10 and tries to cope with the disease and famine that have ravaged the colony since late summer.

Capt. Newport arrives at Jamestown September 29 with a second supply ship and departs in December with Smith's map of Chesapeake Bay and its rivers. He carries back a cargo of pitch, tar, iron ore, soap ashes, and clapboards in place of the worthless mica he has twice carried home under the impression it was gold.

Henry Hudson searches the Barents Sea for a Northwest Passage on a second voyage for the Muscovy Company. His failure persuades the 51-year-old company to direct its future energies to developing its profitable Spitzbergen fishery (*see* 1611).

France exports so much grain that "it robbeth all Spain of their silver and gold that is brought thither out of their Indies," says England's George Carew, 53, Baron of Clopton.

The English East India company ship *Hector* arrives at Surat after a 17-month voyage and becomes the first Company ship to reach India. William Hawkins disembarks with a letter from James I to the late Mughal emperor Akbar asking for trade (*see* 1609).

The new Mughal emperor Jahangir grants trading concessions to John Mildenhall (*see* 1603).

Theater *Timon of Athens* by William Shakespeare; *Pericles, Prince of Tyre* by Shakespeare and collaborators; *Philaster* by Beaumont and Fletcher; *The Revenger's Tragedy* by English playwright Cyril Tourneur, 33 (or by John Webster, Thomas Middleton, or John Marston).

Opera *L'Arianna* 5/28 at Mantua's Teatro della Arte, with music by Claudio Monteverdi.

1609 Spain's Philip III signs a truce with the Dutch April 9 after mediation by France's Henri IV. He recognizes the independence of the Netherlands but does not seize the opportunity to reform Spanish society, which is rapidly decaying through the extravagance of the court and nobility. Shipments of silver from the Americas have diminished, sheep herding is replacing agriculture, and the country must import large quantities of food while exporting olive oil, wine, wool, and luxury goods, much of it to America.

Bavaria's Duke Maximilian organizes a Catholic League July 10 to oppose the Protestant Evangelical Union organized last year by the Palatine elector Frederick IV.

Mare liberum by Hugo Grotius (Huigh de Groot) urges freedom of the seas to all nations. The work is premised on the assumption that the sea's major known resource—fish—exists in inexhaustible supply (*see* 1604; 1625).

A long period of hostility begins between America's Five Nations Iroquois and the French. Samuel de Champlain has precipitated the hostility by killing some Mohawks at the behest of the Hurons (*see* 1604).

A *Majestätsbrief* (letter of majesty) from the Holy Roman Emperor Rudolf II permits free exercise of religion in Bohemia to the three estates—lords, knights, and royal cities.

Expulsion of Spain's Moriscos begins as Philip III acts on the advice of Francisco Gomez de Sandoval y Rojas, 57, Duque de Lerma. Some 275,000 Moriscos (con-

1609 ***(cont.)*** verted Muslims who continue to practice their old religion in secret) will leave the country in the next 5 years, disrupting the economic life of Valencia (which will lose one-fourth of her population) and creating problems in Castile, Aragon, and Andalusia as well.

Champlain (above) has explored an area that will later be called Vermont as lieutenant to the owner of the French fur trade monopoly, and with a Huron and Algonquin war party he has discovered the lake that will later bear his name.

Henry Hudson makes a third voyage to America, this time in the employ of Dutch interests (the 7-year-old United East India Company), and explores first the Barents Sea, then the New England coast and the bay that next year will be called Delaware Bay, and finally a 200-mile tidal estuary that will be called the North (Hudson) River, ascending as far as what will one day be called Albany. Hudson's ship the *Half Moon* carries a crew of 18 or 20 (*see* Fort Nassau, 1614).

The London Company chartered in 1606 obtains a new charter, receives additional land grants, and sends out a fleet of nine ships with 800 new settlers and supplies for the Virginia colony. Among the new colonists are John Rolfe, 24, and his young wife, but their ship the *Sea Venture* is wrecked with the rest of the fleet on reefs off one of the Bermuda islands whose beauty so delights George Somers, 54, one of the ship captains, that he will return to England and form a company to colonize Bermuda (*see* 1610).

William Hawkins defies Portuguese authorities at Surat and travels overland to Agra where he confers in Turkish with the emperor Jahangir (*see* 1608; 1612).

Dutch merchants establish a trading post at Hirado in western Japan at the invitation of the Tokugawa shogun Ieyasu, who ends Portugal's monopoly in trade with Japan. The Dutch will operate the post until 1623 (*see* 1605; 1624).

The Bank of Amsterdam is founded with silver coined from South American ingots. Some 800 varieties of coins are in circulation, the new city-owned bank weighs coins to pioneer the principle of public regulation of money, and the bank soon finds that it can loan money at interest (*see* Bank of England, 1694).

Astronomia Nova . . . de Motibus Stellas Martis et Observationibus Tychonis Brahe by German astronomer Johannes Kepler, 38, establishes two of the cardinal principles of astronomy: planets travel round the sun in elliptical paths rather than in perfect circles, and they do not travel at uniform rates of speed. Kepler has headed the observatory at Prague since the death of his master Tycho Brahe in 1601 and has used data compiled by Brahe to arrive at the new astronomical laws (*see* 1619).

Aviso Relation oder Zeitung begins weekly publication January 15 at Wolfenbuttel in Lower Saxony; *Relation and Aller Furnemmen und Gedenkwürdigen Historien* begins publication at Strasbourg. They are the world's first regular newspapers (*see* 1605).

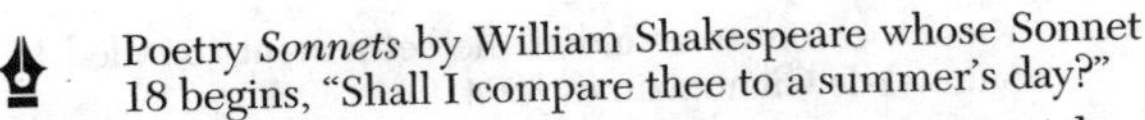

Poetry *Sonnets* by William Shakespeare whose Sonnet 18 begins, "Shall I compare thee to a summer's day?"

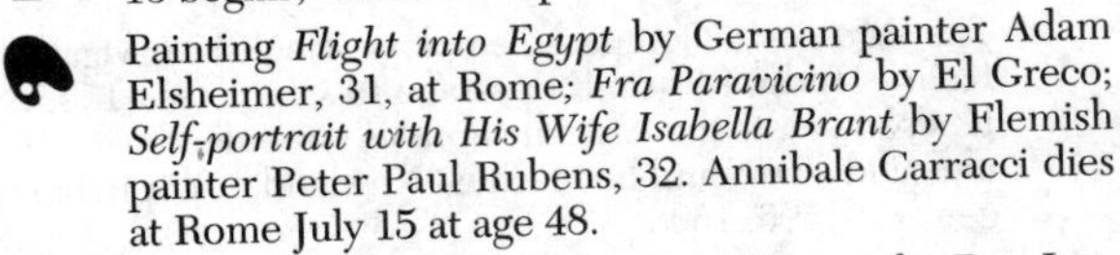

Painting *Flight into Egypt* by German painter Adam Elsheimer, 31, at Rome; *Fra Paravicino* by El Greco; *Self-portrait with His Wife Isabella Brant* by Flemish painter Peter Paul Rubens, 32. Annibale Carracci dies at Rome July 15 at age 48.

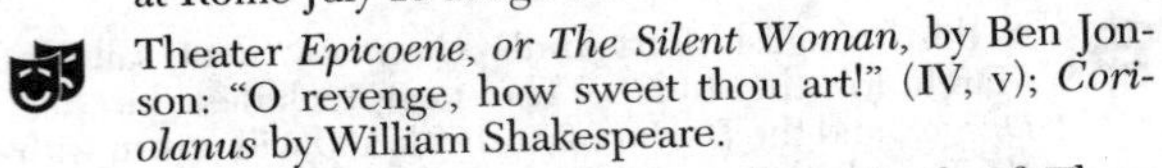

Theater *Epicoene, or The Silent Woman,* by Ben Jonson: "O revenge, how sweet thou art!" (IV, v); *Coriolanus* by William Shakespeare.

"Three Blind Mice" ("A Round or Catch of Three Voices") is published in *Deuteromelia, or The Second Part of Music's Melodic, or Melodius Musicke.*

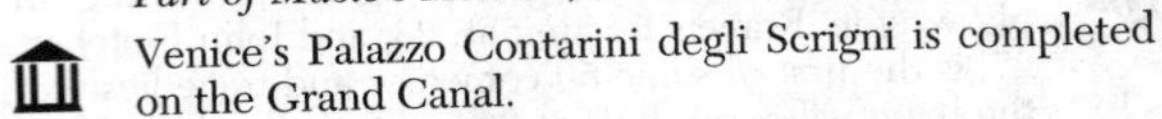

Venice's Palazzo Contarini degli Scrigni is completed on the Grand Canal.

The Virginia colony declines in population to 67 by January as food stocks run low despite the introduction of carrots, parsnips, and turnips. The colonists sustain themselves in the "starving times" until their crops ripen by gathering cattail roots, marsh marigolds, Jerusalem artichokes, and other wild plant foods. Still many die of hunger.

1610 France's Henri IV is assassinated at Paris May 14 by the fanatic François Ravaillac, 31. Dead at age 56 after a momentous reign of nearly 21 years, Henri is succeeded by his 9-year-old son, who will reign until 1643 as Louis XIII. The boy's mother Marie de' Medici will act as regent until he is of age and will exercise power for long after.

The grand duke of Muscovy Basil IV Shuiski is deposed in July by Poland's Sigismund III with help from Muscovite boyars and landlords. He is carried off to Warsaw, his throne is offered to Sigismund's son, who is proclaimed Ladislav IV Vasa, the Cossacks and peasants resist him, and Sigismund wants the throne for himself. A second "false Demetrius" has established a camp outside Moscow at Tushino and gains support, a Tatar in his retinue kills him in December, but his supporters blockade Moscow and a popular uprising

France's Henri IV brought peace between Catholics and Huguenots but met death at the hands of an assassin.

obliges the Poles inside the city to take refuge in the Kremlin (*see* 1612).

The Hausa queen Amina dies at age 34 after an 18-year reign that has expanded her realm to the mouth of the Niger.

The Virginia colonists are attacked by Powhatan chief Wahunsonacook, who may have been inspired by Spaniards to the south. A few colonists are killed.

The French queen mother Marie de' Medici removes Sully from office and installs her favorite, Concino Concini, in a position of power, ushering in an era of cruelty and oppression for the peasantry.

The duque de Lerma steps up the expulsion of Spain's Moriscos (Moors), who have contributed much to the country's culture and economy. Spain will never recover from the loss.

Henry Hudson makes another attempt to find a Northwest Passage. Backed by Sir Thomas Smith, Sir Dudley Digges, John Wolstenholme, and other English investors, Hudson crosses the Atlantic in his 55-ton ship *Discovery,* but succeeds only in entering the strait which will later bear his name and in exploring the shallow bay which will be called first Hudson's Bay and then Hudson Bay (*see* 1611; 1668).

Jamestown colonists in Virginia prepare to abandon the colony and move to Newfoundland after having buried some 500 of their men, women, and children. En route down the James River, however, they encounter the *Virginia* commanded by Thomas West, 33, the Baron De La Warr, with 150 new settlers and fresh supplies. The outward bound vessels come about and return to Jamestown to try again with De La Warr as their first governor (*see* Cole, 1611).

Survivors of last year's Bermuda shipwreck build two new ships from timbers and planks salvaged from their wrecks and arrive at Jamestown May 24. Among the arrivals are John Rolfe and his wife (*see* 1612).

Delaware Bay is named in honor of Governor De La Warr by Virginia colonist Sir Samuel Argall.

The Dutch East India Company introduces the term *share.*

Siderius Nuncius by Galileo Galilei creates a sensation with talk of the unevenness of the moon's surface as observed by Galileo through his spyglass. He has observed the three moons of Jupiter January 7, calls them the Medicean stars, dedicates his treatise to Cosimo II de' Medici, and moves to Florence (*see* 1597; 1613).

Painting *Laocoön* and *The Opening of the Fifth Seal* by El Greco; *Raising of the Cross* by Peter Paul Rubens. Michelangelo Caravaggio dies of malaria at Port 'Ercole while en route to Rome July 18 at age 36 (he fled Rome after a fatal brawl in 1606 and has worked at Naples, in Malta, and in Sicily); Adam Elsheimer dies at Rome in December at age 32.

Theater *Peribanez* by Lope de Vega, who shows 15th century peasants driven to murder their tyrannical lord in defense of their honor; *The Alchemist* by Ben Jonson,

Henry Hudson and other English navigators tried to find the fabled Northwest Passage to China. Hudson came to grief.

who refers in his prologue to "Fortune, that favors fools"; *The Roaring Girl, or Moll Cutpurse* by Thomas Dekker and Thomas Middleton at London's Fortune Theatre, with the Prince's Men; *Cymbeline* by William Shakespeare: "Hark, hark! the lark at heaven's gate sing" (II, iii); "Golden lads and girls all must/As chimney-sweepers come to dust" (IV, ii).

"Vespers" by Claudio Monteverdi are published at Mantua.

Only 150 of the 900 colonists landed in Virginia in the last 3 years have survived, the others having succumbed to starvation and disease. The colonists are primarily English but include some French, German, Irish, and Polish artisans.

1611 England's James I dissolves Parliament for the first time. His troops in Ireland force the earl of Tyrone Hugh O'Neill, now 71, to take refuge at Rome and the Plantation of Ulster is forfeited to the crown.

Denmark declares war on Sweden after more than 40 years of peace. Hostilities will continue until 1613.

Sweden's Charles IX dies October 30 at age 61 after a 7-year reign. His son of 16 has mastered Latin, Italian, and Dutch in addition to his mother tongues of Swedish and German, served as his father's co-regent for the past year, and will reign until 1632 as Gustavus II Adolphus. The new king signs a royal charter giving the Swedish council and the estates a voice in all questions of legislation and a power of veto in matters of war and peace.

English mutineers maroon Henry Hudson on the shore of James Bay (*see* 1610). The expedition has been icebound through the winter, Eskimos kill some of the mutineers, others starve to death, Hudson and eight others will never be heard of again, and English authorities will imprison the surviving mutineers upon their return.

The new governor of the Jamestown colony in Virginia introduces private enterprise. The colony's agriculture

1611 *(cont.)* has been a socialized venture until now, but Sir Thomas Cole assigns 3 acres to each man and gives him the right to keep or sell most of what he raises (*see* 1616).

Institutiones Anatomicae by Danish physician Kaspar Bartholin, 26, is published by the University of Copenhagen (*see* 1653).

The Authorized (King James) version of the Bible is published for the Church of England after 7 years of effort by English scholars.

The University of Rome is founded.

The University of Santo Tomás is founded at Manila in the Philippines.

Poetry Homer's *Iliad* in a translation by George Chapman, whose first part appeared in 1598; *The Anatomy of the World* by English poet John Donne, 38, published anonymously, is an extravagant elegy to the late daughter of his patron Sir Robert Drury of Hawsted.

Painting *Descent from the Cross* by Peter Paul Rubens.

Theater *The Atheist's Tragedy, or The Honest Man's Revenge* by Cyril Tourneur; *The Maid's Tragedy* by Beaumont and Fletcher; *A Chaste Maid in Cheapside* by Thomas Middleton at London's Swan Theatre, with Lady Elizabeth's Men; *The Winter's Tale* by William Shakespeare; *The Tempest* by Shakespeare 11/1 at Whitehall, with the King's Men: "Full fathom five thy father lies;/Of his bones are coral made;/Those are pearls that were his eyes:/Nothing of him that doth fade/But doth suffer a sea change/Into something rich and strange . . ." (Ariel's song, I, ii); "Our revels now are ended. These our actors,/As I foretold you, were all spirits and/Are melted into air, into thin air;/ . . . We are such stuff/As dreams are made on, and our little life/Is rounded with a sleep" (Prospero's farewell, IV, i, is also Shakespeare's farewell as he prepares to leave London and return to his native Stratford-upon-Avon); *Catiline, His Conspiracy* by Ben Jonson; *A King and No King* by Beaumont and Fletcher 12/26 at the court.

The Muscovy Company sends out the first English ship to be fitted out for whaling (see 1557). The 150-ton *Mary Margaret* skippered by Steven Benet kills a small whale and 500 walrus in Spitzbergen's Thomas Smyth's Bay, but the ship is lost with all hands on her return voyage (*see* 1608; 1612).

Potosí in the South American Andes reaches its population peak of 160,000 as the Spanish employ Indian labor to produce tons of silver for shipment back to Spain (*see* 1545; Cerro de Pasco mine, 1630).

1612 The Holy Roman Emperor Rudolf II dies January 20 at Prague at age 59 after a reign of more than 35 years. He is succeeded by his brother Matthias, 54, who obtains the remaining Hapsburg dominions, is crowned in June, and will reign until 1619.

Polish forces in Moscow surrender October 27 to Russian forces who have organized a militia under Prince Dmitri Pojarsky (*see* 1610; 1613).

Japanese persecution of Christians begins as a definite policy following a series of anti-Christian edicts that began in 1606. The retired shogun Ieyasu abandons his original friendly attitude toward missionaries after realizing that trade with Europe can be continued without their presence, which presents potential political dangers.

A Bermuda colony is established by a shipload of men, women, and sailors who arrive on the islands that were claimed for England 3 years ago by the late Sir George Somers, who died in 1610. The colony will have 600 settlers by 1614.

A Map of Virginia by Captain John Smith describes the physical features of the country, its climate, its plants and animals, and its inhabitants.

Two English East India Company ships defeat a Portuguese fleet of four galleons off the coast of India. The emperor Jahangir is so impressed that he grants trading rights to the English at Surat.

Theater *The White Devil* by John Webster is performed by the Queen's Men; *Cupid's Revenge* by Beaumont and Fletcher in January at the court; *The Coxcomb* by Beaumont and Fletcher with the Children of the Queen's Revels; *King Henry VIII* by William Shakespeare: "From that full meridian of my glory/I haste now to my setting: I shall fall/Like a bright exhalation in the evening,/And no man shall see me more" (III, ii); "Farewell! a long farewell, to all my greatness!/This is the state of man: today he puts forth/The tender leaves of hopes; tomorrow blossoms,/And bears his blushing honors thick upon him;/The third day comes a frost, a killing frost;/And, when he thinks, good easy man, full surely/His greatness is aripening, nips his root,/And then he falls, as I do. I have ventured,/Like little wanton boys that swim on bladders,/This many summers in a sea of glory,/But far beyond my depth: my highblown pride/At length broke under me, and now has left me,/Weary and old with service, to the mercy/Of a rude stream, that must forever hide me" (III, ii); "Men's evil manners live in brass: their virtues/We write in water" (IV, ii).

Tobacco cultivation gives Virginia colony settlers an export commodity that will provide a solid economic base for the colony. John Rolfe has obtained *Nicotiana tabacum* seed from the Caribbean islands and after 2 years in Virginia has learned from local Indians how to raise tobacco and cure the leaf that he ships to London. The James River Valley produces 1,600 pounds of leaf per acre, Jamestown becomes a boom town, the Virginia (London) Company grows prosperous, and James I is enriched by import duties that make him look more tolerantly on tobacco (*see* 1604; Rolfe, 1614).

The Muscovy Company sends out four whaling ships, takes 17 whales, and pays its shareholders a 90 percent dividend. Its whaling ventures will have only occasional success in the next dozen years, however, and the English will virtually abandon whaling, not to resume for another 150 years.

1613 The Romanov dynasty that will rule Russia until 1917 is inaugurated July 22 with the crowning in the Kremlin of Mikhail Romanov, 17, a son of the patriarch Philaret and grandnephew of Ivan IV. His election February 22 by the boyars has ended the "time of the troubles" that has persisted since the death of Boris Godunov in 1605 and he will reign, weakly, until 1645 as Czar Mikhail (Michael).

The Peace of Knared January 20 has ended a 2-year war between Sweden and Denmark. Sweden gives up Finland and allows Danish merchants into Livonia.

Bethlen Gábor (Gabriel Bethlen von Iktar), 33, overthrows Gabriel Bathory and becomes prince of Transylvania, ending the cruel Catholic reign of Bathory who is murdered at age 24. Bethlen establishes Protestantism in Transylvania and imprisons Bathory's cousin Elizabeth, who is said to have killed more than 600 maidens and bathed in their blood (she will herself die in prison next year).

The *daimyo* Date Masarnune, 48, dispatches a Japanese embassy to Spain and the pope.

Londonderry is founded in Ulster as England opens half a million acres to "plantations" of up to 3,000 acres each for Protestant settlers from England and Scotland. Irishmen who were loyal to the Crown in the recent rebellion are also eligible for such grants.

An English factory (trading post) is established at Hirado, and Will Adams attempts to establish trade between Japan and England through Capt. John Saris at Bantam (*see* 1605; 1616).

The English East India Company establishes its first factory in India (*see* 1608; 1616).

Dutch merchants establish a fur trading post at the foot of Manhattan Island (*see* 1623).

The first Japanese-built Western-style ship leaves October 28 for a 90-day voyage to Acapulco in New Spain. Built with technical advice from a Spaniard, the ship is commanded by the samurai Tsunenaga Hasekura. The 120-ton schooner carries Don Rodrigo de Vivero, Spanish governor of Luzon, who was stranded in Japan by a shipwreck.

Letters on the Solar Spots by Galileo Galilei advocates the Copernican system proposed in 1543. Galileo is admonished by the Church (*see* 1610; 1616).

Typhus strikes Württemberg and the Tyrol and appears in Magdeburg.

Plague appears at Regensburg and Leipzig and spreads through Bohemia, Austria, and eastward.

Painting *The Aurora* fresco in Rome's Raspigliosi Palace by Italian painter Guido Reni, 38, who has been influenced by the late Caravaggio.

Theater *The Masque of the Inner Temple and Gray's Inn* by Francis Beaumont 2/20 at Whitehall, on the marriage of the Prince and Princess Palatine; *The Two Noble Kinsmen* by William Shakespeare and John Fletcher. London's Globe Theatre burns down June 29 following a performance of Shakespeare's *Henry VIII*.

English merchant John Jourdain buys pepper in Sumatra and cloves at Amboina and Ceram in the Spice Islands.

Spain's population falls as a consequence of wars and emigration to overseas colonies. The country becomes largely a wool-growing nation in a period of decadence and agricultural decline (*see* 1561).

1614 French prelate Armand Jean du Plessis, 29, duc de Richelieu and bishop of Lucon, gains election to the Estates-General, representatives of the French people. Richelieu engineers the dismissal of the Estates-General, which will not reconvene until 1789 (*see* 1624; Harrington, 1656).

England's "Addled Parliament" meets, refuses to discuss finance, and dissolves.

Sweden's Gustavus II Adolphus conquers Novgorod from the Russians.

Japan's Tokugawa shogun Ieyasu bans Christian missionaries from the country (*see* 1612; 1639).

Virginia colony widower John Rolfe is married April 5 to Pocahontas, favorite daughter of the Powhatan chief Wahunsonacook. Seized last year by colonists and held for ransom, Pocahontas, 18, has become Europeanized, adopted Christianity, changed her name to Rebecca, and will never see her father again (*see* 1607; 1617).

Virginia colonists block French settlements in Maine and Nova Scotia.

Dutch traders found Albany under the name Fort Nassau on the Hudson River (*see* 1609; 1624).

Dutch navigator Adriaen Block returns to Amsterdam October 1 after exploring the New England coast, sailing up the Connecticut River, and mapping the coast of Manhattan. His name will survive in Block Island.

Mirifici Logarithmorum Canonis Descriptio by Scottish mathematician John Napier, 64, Laird of Merchiston, describes the most powerful method of arithmetical calculation. Napier has discovered logarithms that make it possible to multiply 7- and 8-digit numbers by simple addition and to raise numbers to the 15th or 16th power by simple multiplication, permitting calculations heretofore almost impossible and providing the basis of the slide rule.

Painting *Descent from the Cross* by Peter Paul Rubens; *The Last Communion of St. Jerome* by Italian painter Il Domenichino (Domenico Zampieri), 32. El Greco dies at Toledo the night of April 6 at age 72.

Theater *The Duchess of Malfi* by John Webster: "He hath put a girdle 'bout the world/And sounded all her quicksands" (III, i); "Glories, like glow-worms, afar off shine bright,/But, looked at near, have neither heat nor light" (IV, ii); *St. Joan,* Part II *(La Santa Juana)* by Spanish playwright Tirso de Molina, 30, 6/24 in the orchard of the duque de Lerma in honor of Philip III's eldest son; *Bartholomew Fair* by Ben Jonson 11/1 at London.

Venice's Church of San Giorgio Maggiore is completed after 55 years of work according to plans by the late architect Andrea Palladio.

Venice's prisons and their connecting Bridge of Sighs are completed after 41 years of work to enlarge the state prison in the Doge's Palace.

1614 ***(cont.)*** England's Levant Company brings home pepper and other spices aboard its big Indiamen from Java and Sumatra for re-export to Constantinople.

1615 France's Estates-General dissolves February 23 after 4 months in session. It has failed to gain concessions from the Crown with regard to taxation.

Dutch forces seize the Moluccas from the Portuguese.

An English fleet defeats a Portuguese armada off Bombay.

Osaka falls June 4 to the Japanese shogun Ieyasu after a 6-month siege; Hideyori, son of the late dictator Hideyoshi, commits *seppuku* and is burnt in the flames that consume the castle he has built and tried to defend.

English slaver Thomas Hunt kidnaps a Wampanoag from Patuxet (below). He sells Tisquantum (Squanto) into slavery at Malaga, Spain (*see* agriculture, 1621).

Samuel de Champlain finds Lake Huron July 28 on his seventh voyage. The lake will provide an easier inland route for fur traders.

Capt. John Smith of the Virginia colony surveys the New England coast from Maine to the cape that will be called Cape Cod. Commissioned by the Plymouth Company, Smith renames the native village of Patuxet, calling it Plymouth (Plimouth) (*see* 1606; 1620).

Rubber from South America reaches Europe but is little more than a curiosity (*see* 1503; Priestley, 1772).

England turns increasingly to cheap coal as timber grows scarce and firewood becomes costly (*see* 1658).

Fiction *El Ingenioso Hidalgo Don Quixote de la Mancha* by Spanish novelist Miguel de Saavedra Cervantes, 68, who completed the first part of the work 10 years ago. Cervantes's left hand was maimed at the Battle of Lepanto in 1571, Algerian pirates captured him in 1575 and held him for ransom until 1580. His writing has been published since 1585, and his burlesque of romantic chivalry contains phrases that will become universal: "Not to mince words" (I, preface); "Give the devil his due" (I, III, 3); "Paid him in his own coin" (I, III, 4); "A finger in every pie" (I, III, 6); "Every dog has his day" (I, III, 6); "Give up the ghost" (I, III, 7); "Venture all his eggs in one basket" (I, III, 9); "Cry my eyes out" (I, III, 11); "A bird in the hand is worth two in the bush" (I, IV, 3); "The proof of the pudding is in the eating" (I, IV, 10); "All that glitters is not gold" (II, III, 3); "Honesty is the best policy" (II, III, 33); "A word to the wise is enough" (II, IV, 30); "A blot on thy escutcheon" (II, IV, 35); "Pot calls the kettle blackarse" (II, IV, 43); "Mum's the word" (II, IV, 44); "When thou art in Rome, do as they do at Rome" (II, IV, 54); "Born with a silver spoon in his mouth" (II, IV, 73); and "There are only two families in the world . . . the Haves and the Havenots."

Poetry *The Odyssey* by Homer translated by George Chapman (*see* 1611); "Shepherd's Hunting" by English Cavalier poet George Wither, 27, who wrote the pastoral while imprisoned in Marshalsea for his libelous satire "Abuses Stript and Whipt" ("Little said is soonest mended"); *The Author's Resolution* by Wither contains the sonnet "Fidelia": "Shall I wasting in despair/Die because a woman's fair?/Or make pale my cheeks with care/'Cause another's rosy are?/Be she fairer than the day,/Or the flowery meads in May,/If she think not well of me,/What care I how fair she be?"

Painting *Scenes from the Life of St. Cecilia* by Il Domenichino for Rome's S. Luigi dei Francesi.

Recercari et Canzoni Francese with music by Italian composer Girolarno Frescobaldi, 32, is published at Rome. Frescobaldi is organist at St. Peter's.

Chocolate paste from the Spanish New World is introduced into Italy and Flanders for drinking chocolate (*see* 1580; 1659).

1616 Japan's Tokugawa shogun Ieyasu dies June 1 at age 74. His son Hidetada, now 38, will marry a daughter of the emperor Gomizno (all her siblings will be killed or aborted) and reign until 1623.

Manchu forces invade China.

The leader of a 1614 attack on Frankfurt's Jewish ghetto and some of his followers are beheaded February 28 by order of the emperor Matthias.

Persecution of Catholics intensifies in Bohemia.

Dutch East India Company mariner Willem Cornelis Schouten, 49, rounds Cape Horn for the first time and gives the cape its name (he was born at Hoorn). Navigator Jakob Le Maire, 31, accompanies Schouten and will help him explore the South Seas.

English navigator William Baffin, 32, discovers Baffin Bay on an expedition in search of a Northwest Passage. He sails to a latitude of 77°45′—farther to the north than any other explorer will venture for 236 years (*see* Frobisher, 1576).

A Description of New England by Capt. John Smith is published in London with a map of the region (*see* 1615; 1620).

Jamestown, Virginia colonists each receive 100 acres of land after having worked until now for the London Company. Each colonist will soon be given an additional 50 acres for each new settler he brings to Virginia (*see* 1611).

England's East India Company begins trading with Persia from its Indian base at Surat.

England's James I begins selling peerages to replenish the seriously depleted royal treasury.

Mitsui begins its rise toward becoming the world's largest business organization. Rebelling against his noble caste, Japanese samurai Mitsui Sokubei Takatoshi opens a sake and soy sauce establishment, and his wife Shuho finds that patrons will pawn their valuables for a drink of sake (*see* 1673).

Will Adams (Anshin Miura) embarks with Japanese vessels that will take him to Siam and Cochin China in the next 2 years (*see* 1613). Adams has built the ships to expand Japanese trade in silks and other valuable commodities from the mainland.

The Vatican orders Galileo Galilei to stop defending Copernican "heresy" and has him arrested February 26. Jesuit prelate Cardinal Bellarmine (Roberto Francesco Romolo Bellarmino), 74, has received secret reports about Galileo and declared belief in the revolutionary ideas of the late Polish astronomer Nikolaus Copernicus a violation of holy scripture (*see* 1613; 1632; Copernicus, 1543).

Dutch astronomer-mathematician Willebrord Snellius, 25, discovers the law of light refraction.

The secret Rosicrucian Society said to have been started in the 15th century by Christian Rosenkreutz (Frater Rosae Crucis) is described in the anonymous pamphlet "Chymische Hochzeit Christiana Rosenkreutz" that will be ascribed to German Protestant theologian-satirist Johann Valentin Andrea, 30.

Poetry "A Select Second Husband for Sir Thomas Overburies' Wife" by English poet John Davies, 51, of Hereford: "Beauty's but skin deep"; *Poems* by Ben Jonson include "Drink to me only with thine eyes . . ."

Painting *Banquet of the Officers of the Guild of Archers of St. George* by Dutch painter Franz Hals, 36; *The Lion Hunt* by Peter Paul Rubens.

Theater *The Scornful Lady* by Beaumont and Fletcher; Francis Beaumont dies March 6 at age 34 and his last play contains the line derived from a proverb recorded by John Heywood in 1546—"Beggars must be no choosers" (V, iii). William Shakespeare dies at Stratford-upon-Avon April 23 at age 52.

Antwerp's Notre Dame Cathedral is completed after 264 years of construction.

1617 Sweden's Gustavus II Adolphus cuts Russia off from the Baltic. He restores Novgorod March 9 by the Treaty of Stolbovo but retains Carelia and Ingria.

England's James I releases Sir Walter Raleigh from prison so that he may seek gold along the Orinoco River. The Spanish ambassador warns that Spain has settlements on the coast, but Raleigh promises to find gold for the king without encroaching on any Spanish territory. He sails with his son March 17 in an ill-equipped expeditionary fleet and reaches the mouth of the Orinoco December 31 (*see* 1603; 1618).

France's chief minister Concino Concini is arrested April 24 and executed on orders from Louis XIII, who banishes his mother Marie de' Medici to Blois. Charles d'Albert, duc de Luynes, gains power over the 16-year-old king (*see* 1618).

The Ottoman sultan Ahmed I dies November 22 at age 27 after a 14-year reign in which plague has killed 200,000 at Constantinople. His idiot brother of 16 succeeds as Mustafa I but is so imbecilic that he will be deposed in 3 months.

Japan's Yoshiwara prostitute section is established at Edo. A local vice lord has persuaded shogunate authorities to license him to operate an area of supervised brothels in exchange for his surveillance of suspicious strangers. Yoshiwara will flourish until 1958 (*see* 1789).

A smallpox epidemic in England kills Rebecca Rolfe, *née* Pocahontas, who is survived by a son.

An epidemic of what is probably smallpox sweeps the New England region and reduces the native population by some estimates from 10,000 to 1,000. A member of Thomas Hunt's 1615 slaving expedition has probably communicated the disease which leaves relatively few survivors (*see Mayflower*, 1620).

Bubonic plague becomes epidemic in India a year after being identified as such.

The Royal Mosque (Masjid-i-Shah) is completed at Isfahan for Persia's Shah Abbas.

1618 The Defenestration of Prague May 23 precipitates a Thirty Years' War that will devastate Europe. Bohemian Catholics close a Protestant church and destroy another. Protestants are further embittered by the transfer of Bohemia's administration to 10 governors, of whom seven are Catholics. Two governors are thrown from a window in the Palace of Hradcany, and although their 50-foot fall into a ditch does not kill them, their "defenestration" begins a revolt headed by Protestant leader Count Heinrich Matthias von Thurn, 51, who gave the order for the defenestration and who takes command of rebel troops to march on Austria (*see* 1619).

The duc de Richelieu negotiates a treaty between France's Queen Mother Marie de' Medici and the duc de Luynes. She returns from exile, whereupon Richelieu is exiled to Avignon for conspiring with her.

The Ottoman sultan Mustafa I is declared incompetent to rule in February (a brother of 14 will reign until 1622 as Osman II). The Turks give up Georgia and Azerbaijan by treaty with Persia's Abbas the Great.

The Virginia House of Burgesses convenes at Jamestown. It is the first legislative body in the New World.

Sir Walter Raleigh sends five small ships up the Orinoco with his son and a nephew while remaining himself at Trinidad, ill with fever. The expedition under Lawrence Keymis encounters a Spanish settlement; Raleigh's son and several Spaniards are killed in the fight that ensues. Keymis returns to Trinidad with the bad news and kills himself after being reproached by Raleigh, who returns to England, is arrested, and is executed October 29 at age 66 to fulfill the king's promise to the Spanish that any piracy on Raleigh's part would be punishable by death.

French adventurer Etienne Brulé begins to assemble furs for sale to French traders. Brulé arrived at Quebec with Champlain in 1608 and has gone into the western wilderness, where he lives with the Hurons who will kill him and eat him in 1633.

The world's first pawnshop September 28 at Brussels.

Smallpox introduced by European explorers and colonists rages through New England and spreads south to Virginia, where it kills Chief Powhatan and many of his tribesmen.

1618 ***(cont.)*** A diphtheria epidemic kills up to 8,000 at Naples.

The *London Pharmacopoeia* is published in its first edition and contains some 1,960 remedies including worms, dried vipers, foxes' lungs, oil of ants and wolves, and 1,028 simples (*see* 1677).

Painting *Old Woman Cooking* by Spanish painter Diego Rodriguez de Silva y Velázquez, 19; *Crucifixion* by Spanish painter Jusepe (José) de Ribera, 27, who has settled at Naples.

Theater *King Torrismond (Il rey Torrismondo)* by the late Torquato Tasso at Vicenza's Teatro Olympia (Tasso died in 1595 at age 51).

The population of the German states will decline by at least 7.5 million from its present level of 21 million as a result of the Thirty Years' War (above; *see* 1648).

1619 The Holy Roman Emperor Matthias dies March 20 at age 62 at Vienna after a reign of less than 7 years. He is succeeded by Bohemia's Ferdinand, 40, who has just been deposed as king of Bohemia but is elected emperor at Frankfurt August 28 and will reign until 1637 as Ferdinand II.

Protestant forces under Count von Thurn besiege Vienna in November but cold, hunger, and the Catholic forces of the Bavarian elector Maximilian soon force them to withdraw (*see* 1618; 1620).

Italian humanist Lucilio Vanini, 33, is judged guilty of magic and atheism by a Toulouse court, his tongue is cut out, and he is strangled at the stake and burnt alive February 9.

The first black slaves to arrive in the Virginia colony come ashore in August from a Dutch privateer whose booty includes Spanish plate and "twenty negars" (*see* 1649).

Enslaved Africans were brought in chains to the Americas, where their masters often worked them to death.

Most African slaves come to the West Indies and to Brazil where the sugar industry kills them faster than natural breeding can replace them. Slaves nevertheless will now begin to play a role in the North American economy.

Some 90 young women arrive at Jamestown from England to marry settlers who pay 120 pounds of tobacco each for the cost of transporting their brides.

Batavia is founded by Dutch colonists in Java, which will remain a valuable colony for 326 years.

French mathematician-philosopher René Descartes, 22, establishes the basics of modern mathematics, applying algebra to geometry and formulating analytic geometry. Serving as a soldier to the elector of Bavaria in his war against the Bohemian Protestants (above), Descartes makes the breakthrough November 19 that will provide the basis for exploring natural phenomena by mathematics, but his work will not be published until 1637.

The Harmony of the World (Harmonice Mundi) by Johannes Kepler at Linz shows that the planets move not in circles but in ellipses, traveling faster when they are close to the sun. Kepler writes out musical notations suggesting that the planets have "songs" which reach higher keys as they near the sun, his work is called heretical because only circular paths and constant speeds are considered perfect, but the celestial imperfection, Kepler explains, is to enable God to make better music.

The slaves landed at Jamestown (above) introduce to North America such African diseases as yellow fever, virulent forms of malaria, and hookworm (*see* 1647; Laveran, 1880; Ashford, 1899).

Burgundy bans the planting of potatoes on charges that the tuber produces leprosy.

The first American day of thanksgiving is celebrated November 30 by 30 Englishmen aboard the ship *Margaret* which touches land at what will be Hampton, Virginia. The new arrivals proceed up the James River and arrive December 4 at Berkeley Grant (*see* 1621).

Painting *Adoration of the Magi* by Diego Velázquez; *Diana at the Chase* by Il Domenichino; *The Rape of the Daughters of Leucippus* by Peter Paul Rubens.

Sculpture Brussels sculptor Jerome Duquesnoy casts the bronze *Mannekin Pis* to replace a stone statue dating to the mid-1400s.

Jamestown produces a large enough crop to end the threat of starvation in the Virginia colony.

1620 Spanish forces seize the vital Valtelline Pass communications link between Hapsburg Austria and the Spanish Hapsburg Italian possessions. The Spanish action against the Grisons League precipitates a 19-year conflict between Catholic and Protestant factions in Switzerland.

A Catholic League army commanded by Flemish field marshal John Tsaerelaes Count von Tilly, 61, defeats Bohemia's "Winter King" Frederick V November 8 in the Battle of the White Mountain as the Thirty Years'

War ends its third year. Bohemia loses her independence, and the lands of her native Czech nobility are confiscated wholesale by von Tilly, Maximilian of Bavaria, and the Holy Roman Emperor Ferdinand II.

The Mayflower Compact drawn up by the Pilgrims (below) establishes a form of government based on the will of the colonists rather than on that of the Crown. The Pilgrims have found that Cape Cod is outside the jurisdiction of the London Company, and they select Plymouth as the site of a settlement (*see* 1621; John Smith, 1615).

The 180-ton vessel *Mayflower* out of Southampton arrives off Cape Cod November 11 with 100 Pilgrims plus two more born at sea during the 66-day voyage. The Pilgrims are English separatists who have emigrated from Scrooby to Amsterdam and thence to Leyden but who decided 3 years ago to seek a new home in order to preserve their English identity. They have obtained a patent from the London Company to settle in America (*see* 1606).

Baltic trade will collapse in the next decade as export staples from the region shift from foodstuffs to timber, metals, and naval stores while imports of western woolens diminish. French trade with the Levant will fall by half in the next 15 years, and Dutch Levantine trade will also languish as the Thirty Years' War (above) reduces supplies of linen from Silesia and Lusatia and generally stifles commerce.

Novum Organum by Sir Francis Bacon, the first Baron Verulam, proposes an inductive method of interpreting nature as opposed to the deductive logic of Aristotle. Bacon insists on observation and experience as the sole source of knowledge (now 59, he advanced his career at age 45 by marrying a 14-year-old heiress).

Painting *The Water Carrier of Seville* by Diego Velázquez.

Pilgrims from the *Mayflower* (above) receive help from the Pemaquid Samoset and the Wampanoags Hobomah and Massasoit who have learned some English from earlier visitors and share tribal stores of maize with the new colonists to get them through the winter. But roughly half will die within 3 months of starvation, scurvy, and disease (*see* 1621).

1621 Spain's Philip III dies March 31 at age 42 after a 23-year reign. He is succeeded by his son of 15, who will reign until 1665 as Philip IV but will leave affairs of state to his prime minister Gaspar de Gusman, 34, El Conde-Duque de Olivares. The duque de Olivares breaks the truce of 1609 with Holland and resumes war, levying new taxes as he struggles to modernize the government and centralize power in the royalty.

Parliament impeaches the English lord chancellor Sir Francis Bacon on charges of having taken bribes in connection with the granting of monopoly patents that have enriched the brothers of the duke of Buckingham. Bacon confesses but protests that the presents he received from parties in chancery suits never affected his judgment. He is fined £40,000, banished from the court and from Parliament, and declared incapable of holding future office, but James I pardons him and remits his fine.

The House of Commons petitions James I against popery and any marriage of his son Charles to the Spanish infanta; James angrily rebukes the Commons for meddling in foreign affairs; and the Commons declares in a Great Protestation December 18 "That the liberties, franchises, privileges, and jurisdictions of parliament are the ancient and undoubted birthright and inheritance of the subjects of England, and that the arduous and urgent affairs concerning the king, state, and defense of the realm . . . are proper subjects and matter of council and debate in parliament" (*see* 1622).

Another 35 English colonists arrive at Plymouth.

The Dutch West Indies Company is chartered by Holland which has been making inroads into Spain's overseas empire. The new company receives a monopoly in trade with Africa and America (*see* 1628).

The English East India Company has completed 12 voyages to the Indies and has earned an average profit of 138 percent on each voyage.

An Englishman calculates that 3,000 tons of spices bought in the Indies for £91,041 will fetch £789,168 at Aleppo, the trading center at the eastern end of the Mediterranean.

The Anatomy of Melancholy by English clergyman Robert Burton, 44, is a medical treatise on the causes, symptoms, and cure of melancholy that will be widely read for its wisdom and beauty of language. Says the vicar of St. Thomas's at Oxford: "If there be a hell on earth it is to be found in a melancholy heart"; "Who cannot give good counsel? 'Tis cheap, it costs them nothing"; "Comparisons are odious"; "A good conscience is a continual feast"; "Desire hath no rest"; "Like . . . a dog in the manger, he doth only keep it because it shall do nobody else good, hurting himself and others"; "Make not a fool of thyself, to make others merry"; "Man's best possession is a loving wife"; "When thy are at Rome, they do there as thy see done"; "Every man for himself, the Devil for all"; "A loose, plain, rude writer, I call a spade a spade"; "Aristotle said . . . melancholy men of all others are most witty."

New England's Pilgrims celebrate their first Thanksgiving Day (*see* 1863; dinner, below).

Pembroke College is founded at Oxford.

Troops of Count von Tilly sack Heidelberg's University Library.

France's Estates-General establishes a national post office department at the urging of Armand Jean du Plessis, duc de Richelieu (*see* 1464).

London's first newspaper begins publication September 24: *Corante, or newes from Italy, Germany, Hungarie, Spaine, and France*.

Painting *Rape of the Sabine Women* by French painter Nicolas Poussin, 27; *Family Group* and *Rest on the*

1621 ***(cont.)*** *Flight into Egypt* by Flemish painter Anthony van Dyck, 22; *Hercules* by Guido Reni.

Sculpture *Rape of Prospero* by Italian sculptor-painter Gian (or Giovanni) Lorenzo Bernini, 22.

Theater A *New Way to Pay Old Bills* by English playwright Philip Massinger, 38, who has drawn his central situation from Thomas Middleton's play A *Trick to Catch the Old One* of 1606; *The Witch of Edmonton* by Thomas Dekker, John Ford, and William Rowley at London's Phoenix Theatre.

Potatoes are planted for the first time in the German states.

Plymouth colonists have difficulty growing wheat and barley in fields poorly cleared of stumps and rocks, but they are aided by the Wampanoag Indian kidnapped by slaver Thomas Hunt in 1615. Tisquantum (or Squanto) has made his way back from Spain, presents himself at the Pilgrim settlement on Cape Cod Bay March 16, speaks some English, and shows the Pilgrims how to catch eels and how to plant maize and beans using alewives, shad, or menhaden as fertilizer, a European idea he has learned either from French colonists in Maine and Canada or from English colonists in Newfoundland who have acquired the knowledge from the French.

Jamestown colonists build the first American grist mill to produce flour from their wheat.

New England's Pilgrim colonists entertain 92 Indian guests including Chief Massasoit at a Thanksgiving dinner-breakfast. The meal includes wild turkeys shot by the colonists and popcorn which is introduced to the Pilgrims by the chief's brother.

1622 England's James I tears out the page in the journal of the Commons bearing the Great Protestation of last December, dissolves Parliament February 8, and imprisons Sir Edward Coke, now 69, for 9 months along with John Selden, 38, earl of Southampton, and John Pym, 38.

The Battle of Wiesloch in April gives Protestant forces under Count Peter Ernst Mansfeld II, 42, a victory over Count von Tilly's Catholics. Mansfeld plunders Hesse and the Alsace, but Tilly triumphs a few weeks later over the margrave of Baden-Durlach, and he defeats Christian of Brunswick in June as the Thirty Years' War continues.

The Ottoman sultan Osman II, now 19, pretends to go on a pilgrimage to Mecca, actually hoping to raise an army that will reform Constantinople's degenerate Janissaries. They hear of his plan, march him through the streets hurling insults, strangle him May 20, and restore the imbecilic Mustafa I.

Spanish forces under the marquis of Spinola seize Bergen op Zoom from the Dutch.

France's Louis XIII lays siege to Montpelier. The duc de Rohan makes peace October 18 when Louis agrees to reaffirm the 1598 Edict of Nantes, but Louis forbids political meetings. He recalls Richelieu, bishop of Lucon, to the Royal Council and appoints him cardinal.

Persian forces take Kandahar from the Mughal Empire. With English help, they drive the Portuguese out of Hormuz on the Persian Gulf.

Execution of Christian missionaries in Japan will reach its height in the next 3 years.

The Council for New England which has succeeded the Plymouth Company grants territory between the Kennebec and Merrimack rivers to former Newfoundland governor John Mason, 36, and his rich English associate Sir Ferdinando Gorges, 56 (*see* 1629).

The Plymouth Plantation receives 67 new arrivals, and earlier colonists are forced to go on short rations.

Indian attacks March 22 destroy a number of Virginia settlements within a few hours, killing 347 colonists and destroying the first American ironworks.

Disease takes a heavy toll among Virginia colonists and among their Indian neighbors.

London's Banqueting House at Whitehall is completed by English architect Inigo Jones, 49, who has studied in Italy.

Virginia colonists who pick corn prematurely are subjected to public whipping.

Population of the Virginia colony has quadrupled to some 1,400 in the last 5 years, but is reduced this year by disease and Indian massacres (above).

1623 England's Prince of Wales travels in secret to Spain with the duke of Buckingham, who has persuaded him to seek the hand of the infanta Maria, sister of Philip IV. "Mr. Smith" and "Mr. Brown" arrive at Madrid March 7 but find the infanta and Spanish court unenthusiastic and distrustful of young Charles's promise to change English penal laws against Catholics. Charles, now 22, returns October 5 and will be betrothed instead to the sister of France's Louis XIII.

Virginia colonists found the natives less friendly than the ones who helped the Pilgrims in New England.

The Holy Roman Emperor Ferdinand II gives the Upper Palatine to the duke of Bavaria Maximilian as the Thirty Years' War continues. Papal troops occupy the Valtelline, and Count von Tilly advances to Westphalia after defeating Christian of Brunswick at Stadtlohn.

The Ottoman sultan Mustafa I is compelled to abdicate in favor of his nephew and confined in the Seraglio. The nephew, now 14, will reign until 1640 as Murad IV, his mother Kussem Sultana serving as regent until his majority.

Persia's Abbas I takes Baghdad, Mosul, and all of Mesopotamia from the Ottoman Turks. Baghdad will remain in Persian hands for 15 years.

Japan's Tokugawa shogun Hidetada abdicates at age 45 and is succeeded by his son Iemitsu, 19, who will raise the shogunate to its greatest glory in the next 28 years.

The Massacre of Amboina by Dutch East India Company agents ends English East India Company efforts to trade with the Spice Islands, Japan, or Siam. The Dutch seize 10 rival English traders and torture them before executing them. They also kill 10 Japanese and a Portuguese.

Dutch forces seize the Brazilian port of Pernambuco that will later be called Recife.

Another large group of English colonists arrives at Plymouth.

The port of Gloucester is founded.

The Dutch make New Netherlands a formally organized province, and some 30 Dutch families land on Manhattan Island (*see* 1614). Some move upriver to Fort Nassau, some to the Connecticut River Valley (*see* 1624).

English traders in Japan abandon their commercial station at Hirado (*see* 1613).

A new English patent law protects inventors.

Pilgrim fathers at Plymouth colony assign each family its own parcel of land, forsaking the communal Mayflower Compact. Given new incentive, women and children join with men to plant corn and increase production.

Italian anatomist Gasparo Aselli, 42, performs a vivisection operation on a dog which has just eaten a substantial meal and discovers "chyle" (lacteal) vessels. Aselli finds that the dog's peritoneum and intestine are covered with a mass of white threads (*see* Pecquet, 1643).

The University of Salzburg has its beginnings.

Painting *Cardinal Bentivoglio* by Anthony Van Dyck; *Baptism of Christ* by Guido Reni; Diego Velásquez gains appointment as court painter at Madrid where he will be famous for his naturalistic portraits of Philip IV, the infanta Maria, Marianna of Austria, Olivarez, court jesters, dwarfs, idiots, and beggars in addition to his religious paintings.

Sculpture *David* by Gian Lorenzo Bernini.

Theater *Love, Honor and Power (Amor, Honor y Poder)* by Spanish playwright Pedro Calderón de la Barca, 23; *The Duke of Milan* by Philip Massinger.

Mr. William Shakespeares Comedies, Histories and Tragedies According to the True Originall Copies (First Folio) is published at London.

The Paris Church of St. Marie de la Visitation is completed by François Mansart.

Brazil has 350 sugar plantations, up from five in 1550.

A Spanish edict offers rewards and special privileges to encourage large families but the offer has little effect on birth rates.

1624 France's Louis XIII makes Cardinal Richelieu his chief minister April 29. Known as "the red eminence," Richelieu begins a domination of the government that will continue until his death in 1642.

France and England sign a treaty providing for the marriage of Charles, prince of Wales, to Henrietta Maria, daughter of Henri IV and Marie de' Medici.

Dutch naval forces seize Brazil's capital Bahia May 10.

Virginia becomes a royal colony May 24 as her charter is revoked after 17 profitless years.

Eight men from the Dutch West Indies Company ship *Nieuw Nederland* land on Manhattan Island in May and move up the Hudson River estuary to Fort Nassau, later to be called Albany. The men are Walloons led by Jesse De Forest of Avesnes, Hainaut (*see* 1623; 1626).

Spain's Philip IV publishes sumptuary laws in response to popular criticism of his court's extravagance; he reduces his household staff and bans wearing of the ruff, a sartorial symbol of profligacy that will soon go out of style everywhere in Europe as luxury gives way to austerity.

Japan expels all Spanish traders and puts an end to trade with the Philippines (*see* 1609; 1641).

Italian medical professor Santorio Santorius, 63, resigns from the University of Padua, where his students have included Englishman William Harvey (*see* 1628). Santorius has invented an apparatus for measuring the pulse and a crude clinical thermometer, neither of which will be employed in medicine for centuries to come (*see* Floyer, 1707).

Poetry *Devotions upon Emergent Occasions* by John Donne, now dean of St. Paul's in London: "No man is an Iland, intire of it selfe; every man is a piece of the Continent . . . If a Clod bee washed away by the sea, Europe is the lesse . . . any man's death diminishes me, because I am involved in Mankinds; And therefore never seek to know for whom the bell tolls; it tolls for thee" (17th Meditation).

Painting *The Laughing Cavalier* by Frans Hals.

Pope Urban VIII (Mateo Barberini) threatens snuff users with excommunication as the use of tobacco from the New World gains in popularity.

1624 ***(cont.)*** The Louvre Palace at Paris is completed in part for Louis XIII by French architect Jacques Lemercier, 39. The palace will be enlarged with additions over the next 2 centuries.

1625 England's James I dies March 5 at age 58 and is succeeded by his son of 24, who ascends the throne March 27 and will reign until 1649 as Charles I. The new king is married by proxy May 1 to Henrietta Maria, 16, a sister of France's Louis XIII, and receives her at Canterbury June 13.

France has a Huguenot revolt led by Henri, duc de Rohan, 46, and his brother Benjamin de Rohan, 42, seigneur de Soubise.

Parliament rebuffs England's new king in his demands for funds to carry on the war with Spain. Charles has lent ships to France's Louis XIII for use against the Huguenots at La Rochelle and Parliament's Protestant feelings have also been aroused by the king's support of Richard Montagu, 48, the royal chaplain who has repudiated Calvinist doctrine.

The Treaty of Southampton September 8 forms an Anglo-Dutch alliance against Spain. Charles sends an expedition to Cadiz under Sir Edward Cecil, 51, viscount Wimbledon, who bungles the mission and permits Spain's treasure ships to reach port. Charles tries to raise money on the crown jewels in Holland to make himself independent of the Commons.

De Jure Belli et Pacis by Hugo Grotius is an expansion of his 1609 *Mare liberum* thesis and begins the science of international law. Grotius has been condemned to life imprisonment for leading a protest group called the Remonstrants but has escaped and emigrated to France (*see* 1702).

Breda in the Netherlands surrenders July 2 after nearly a year's siege by Spanish forces who have foiled efforts by English mercenaries to break the siege.

Fort Amsterdam is founded April 22 on the lower tip of Manhattan Island by the Dutch West India Company.

French colonists in the Caribbean settle St. Christopher (St. Kitts).

The first English colony on Barbados is established under the leadership of Sir William Courteen.

A Colonial Office is established at London.

An epidemic of the Black Death kills 41,000 at London while 22,000 die of other causes (*see* 1603; 1665).

Glauber's salt (anhydrous sodium sulfate) is discovered by German physician-chemist Johann Rudolf Glauber, 21, whose salt will be used as a laxative and for dyeing textiles.

Painting *Marie de' Medici* by Peter Paul Rubens; *Job* by Guido Reni; *The Procuress* by Dutch painter Gerrit van Honthorst, 34.

Theater *Love Tricks* by English playwright James Shirley, 28.

Moscow's Gate of Salvation in the Kremlin gets a set of bells that will peal for centuries.

Inigo Jones completes London's Covent Garden Church.

Kyoto's Katsura Rikyu (detached palace) is completed for Toshihito, a prince of the Hachijo family.

The Plymouth colony for the first time has "corn sufficient, and some to spare for others"; Governor William Bradford writes home to England and credits part of the improvement to a revised plan of communal labor with equal rations, each family being assigned land according to its size: "This . . . made all hands very industrious . . ."

Peter Paul Rubens depicts a goiter in his portrait of Marie de' Medici (above). The thyroid gland swelling is esteemed as a mark of beauty when only moderate in size (*see* Baumann, 1896).

1626 The Treaty of Monzon March 5 resolves differences between Spain, France, and the papacy over an Alpine valley used by troops to cross into and out of Italy. Spain agrees to destroy her forts in the Valtelline; France is to have free use of the passes.

The Battle of the Bridge of Dessau August 27 ends in victory for Catholic forces under the Austrian general Albrecht Eusebius Wenzel von Wallenstein, 43, and Count Tilly, who crush Danish forces and pursue the enemy through Silesia to Hungary, where Mansfeld obtains support from the Transylvanian prince Bethlen Gábor. But Bethlen soon withdraws from the war and Count Mansfeld dies, as does Christian of Brunswick.

Manhattan Island is purchased from Canarsie chiefs of the Wappinger Confederacy for fish hooks and trinkets valued at 60 guilders by Dutch colonists who call the natives *Manhattes.* Peter Minuit, Willem Verhulst, and the other Dutch leaders assign large tracts of Manhattan land to members of their company (*patroons*) on condition that they bring over stipulated numbers of settlers to the new town of Nieuw Amsterdam founded by Minuit.

The Pilgrim Fathers at Plymouth Plantation buy out their London investors November 15 for £1,800.

Madagascar is settled by its first French colonists, who seek to drive out the Hovas who have lived on the African island for 600 years.

The 60 guilders paid for Manhattan (above) is by some accounts $24 and by others $39 (roughly 0.2 cents per acre), but a 20th century economist will reckon the purchasing power of 60 guilders at several thousand dollars in modern terms (and the Canarsie recipients have sold land that belongs to the Manados).

Rome's Irish College is founded.

Fiction *La Historia de la vide del Buscon* by Spanish picaresque novelist Francisco Gomez de Quevedo y Villegas, 46, is full of grotesque atmosphere and Machiavellian characters.

Painting *The Baptism of the Moor* and *The Clemency of Titus* by Dutch painter Rembrandt van Rijn, 19; *Isaac Abrahamsz Massa* by Frans Hals; *Assumption of the*

Virgin (Antwerp altarpiece) by Peter Paul Rubens; *Drunken Silenus* by Jusepe de Ribera.

Dutch whalers establish the port of Smeerenberg in Spitzbergen to process right whales, so called by the English to distinguish them from "wrong" whales which sink when dead. The whales are prized not as food but for their oil and whalebone, used respectively for illumination and lubrication and as stays.

Sir Francis Bacon experiments with the idea of freezing chickens by stuffing them with snow. He catches pneumonia and dies April 9 at age 65.

1627 The Catholic forces of Albrecht von Wallenstein and the baron von Tilly conquer Holstein; Wallenstein subdues Schleswig and Jutland, drives the dukes of Mecklenburg out of that duchy, and forces the duke of Pomerania to submit as Denmark's Christian IV withdraws temporarily from the Thirty Years' War.

Cardinal Richelieu lays siege in late August to the Huguenot stronghold La Rochelle.

A new Bohemian constitution confirms the hereditary rule of the Hapsburgs and strengthens royal power. Most of the remaining Czech nobility emigrates; the rest have been executed or exiled.

Manchu forces invade Korea. The Manchus will make Korea a vassal state in 10 years, but Korea's court and people will remain loyal to the Ming.

The Mughal emperor Jahangir, who has ruled tyrannically from Delhi since 1605, dies in October at age 58 while returning from Kashmir. His son of 35 has effected a reconciliation following an abortive rebellion and will reign until 1658 as Shah Jahan. He has all his male collateral relatives murdered to remove any challenges to his succession and orders construction of an extravagant Peacock Throne made of emeralds, diamonds, and rubies (it will take 7 years to complete).

The Company of the Hundred Associates founded by Cardinal Richelieu is given control of New France April 25 with a monopoly on the fur trade and land from Florida to the Arctic Circle.

Barbados in the West Indies is colonized by some 80 English settlers who arrive aboard the *William and John* captained by John Powell but owned by a Dutch merchant in England. Capt. Powell captures a Portuguese ship bound for Lisbon with Brazilian sugar, he delivers her cargo to the *William and John*'s Dutch owner, the Dutchman sells the sugar for nearly £10,000, and he devotes the windfall to developing the new colony at Barbados, an island never seen by Christopher Columbus (*see* 1636).

The English East India Company will omit dividends to its stockholders in 16 of the next 50 years, but they will be the only dividendless years in the 180 years between 1602 and 1780 (*see* 1708).

Fiction *Los Sueños* by Francisco de Quevedo y Villegas; *Le Berger extravagant* by French satirist Charles Sorel, 25.

Poetry *Nymphidia* by Michael Drayton.

Painting *The Jolly Toper* and *Banquet of the Civic Guard* by Frans Hals; *The Money-Changer* by Rembrandt van Rijn; *The Mystic Marriage of St. Catherine* by Peter Paul Rubens; *Crucifixion* by Spanish painter Francisco de Zurbarán, 28.

Opera *Dafne* 4/23 at Torgau with music by German composer Heinrich Schütz, 41, libretto by German poet-critic Martin Opitz, 30, who has adapted a 1594 melodrama by Ottavio Rinuccini to write the first German opera. Schütz will move to Venice next year to study under Claudio Monteverdi.

Inigo Jones completes the Queen's Chapel in St. James's Palace, Westminster, after 4 years of work.

France's Louis XIII entrusts the planning of a great château at Versailles to Jacques Lemercier (*see* 1682; Louvre, 1624).

Beijing's Pao Ho Tien is completed. It will be one of the Three Great Halls of the Purple Forbidden City.

1628 The Duke of Buckingham George Villiers is assassinated August 23 at age 36 as he prepares to lead a final expedition for the relief of La Rochelle, center of France's Huguenot power. La Rochelle surrenders October 28 after a 14-month siege that three English fleets have not been able to lift; the Huguenots cease to be an armed political power in France.

The Austrian duke Wallenstein obtains the duchy of Mecklenburg. He assumes the title admiral of the Baltic, but suffers his first reverse when his siege of Stralsund is raised.

Dutch forces occupy Java and the Moluccas.

Salem is founded on Massachusetts Bay by some 50 colonists who arrived in September with John Endicott, 39, who with five other Englishmen has bought a patent to the territory from the Plymouth Council in England and will serve as first governor of the Massachusetts Bay colony until 1630.

The Dutch West India Company founded in 1621 declares a 50 percent dividend after its admiral Piet Heyn surrounds and captures the Spanish silver fleet August 8 off the Cuban coast.

The London Company sends ships directly to the Persian Gulf, antagonizing the Levant Company.

English engineer Edward Somerset invents the first crude steam engine (*see* Savery, 1698).

"Essay on the Motion of the Heart and Blood" (*Exercitatio de Motu Cordis et Sanguinis*) by English physician William Harvey, 50, is published at Frankfurt-am-Main. Harvey's 75-page treatise based on his Lumleian lecture of 1616 establishes that the human heart is muscular despite earlier theories to the contrary, and that its regular mechanical contractions drive the blood out into the blood vessels. Harvey acknowledges the 1559 work of Renaldo Columbus, who first used the term *circulation*, and he destroys the old idea that the liver converts food into blood, but Harvey does not offer a satisfactory new explanation for the creation of blood.

1628 ***(cont.)*** William Harvey (above) follows Galileo's principle of measuring what can be measured and shows that the heart contains 2 ounces of blood and, at 65 heartbeats per minute, that it pumps 10 pounds of blood out into the body in less than 1 minute, an amount greater by far than can be sustained by production from food consumed. But while eminent physicians and philosophers from Denmark, England, France, and the German states rally to Harvey's thesis it is roundly attacked by Scottish physician James Primrose and by the Paris Faculty of Medicine, whose dean Guy Patin calls Harvey's theory "paradoxical, useless, false, impossible, absurd, and harmful."

Bubonic plague kills half the population of Lyons. The Black Death will kill a million in the northern Italian states in the next 2 years (*see* 1630).

Painting *The Martyrdom of St. Erasmus* by Nicolas Poussin; *The Siege of Breda* by French painter-engraver-etcher Jacques Callot, 36; *Charles I and Henrietta Maria with the Liberal Arts* by Gerrit van Honthurst.

Masque *Britannia's Honour* by Thomas Dekker 10/29.

Theater *The Lover's Melancholy* by John Ford 11/24 at London's Blackfriars Theatre, with the King's Men; *The Witty Fair One* by James Shirley at London's Phoenix Theatre.

1629 Persia's Shah Abbas dies January 19 at age 72 after a 42-year reign. Two of his five sons have died, he has had two others executed and another blinded, so he is succeeded by a grandson of 13, who will reign until 1642 as Safi I. The new shah has his grandfather's counselors beheaded along with most of Persia's best generals, all the blood princes, and even some of the princesses. Kandahar's Persian governor defects to the Uzbeks, who take the city and province.

England's Charles I dissolves Parliament in March (below). It will not meet again until 1640.

The Treaty of Lübeck May 22 ends hostilities between the emperor Ferdinand II and Denmark's Christian IV, who regains his lands on condition that he abandon his allies and cease his interference in German affairs. The dukes of Mecklenburg are placed under the ban, and the Austrian Albrecht von Wallenstein (below) is confirmed as duke of Mecklenburg.

The Truce of Altmark ends hostilities between Sweden and Poland.

Transylvania's Bethlen Gábor dies November 15 at age 49 after marrying a sister-in-law of Sweden's Gustavus II Adolphus in hopes that the Swedes would help him obtain the Polish crown.

England's Charles I encounters opposition in the House of Commons where resolutions by Sir John Elliot are read while the speaker is held in his chair. The resolutions declare that anyone who advises the levy of tonnage and poundage taxes without grant of Parliament is an enemy of the kingdom and so is anyone who introduces innovations in religion or expresses opinions that disagree with those of the true church. Authorities arrest Sir John and eight other members of Parliament March 5. Elliot will be fined £2,000 next year and die in prison 2 years later.

The Edict of Restitution promulgated March 29 by the emperor Ferdinand II restores ecclesiastic estates in Europe taken since the 1552 convention of Passau and permits free exercise of religion only to adherents of the 1530 Confession of Augsburg. All other "sects" are to be broken up, troops of the Catholic League and of the Austrian duke Albrecht von Wallenstein begin enforcing the edict, and they show no mercy to "heretics."

A French edict of grace signed June 28 maintains freedom of worship for Huguenots but denies them the right of assembly and places of safety. Since the fall of La Rochelle last year, Huguenot towns in Languedoc have been suppressed.

John Mason and Sir Ferdinando Gorges obtain a new grant of land between Maine's Kennebec and Piscataqua rivers and join with others to form the Laconia Company and establish a farming community on the Piscataqua in what will become New Hampshire (*see* 1622; 1630).

The Dutch West Indies Company offers huge estates and feudal privileges to *patroons* who enlist 50 potential colonists for New Netherlands and pay their passage. The company employs 15,000 seamen and soldiers; it has more than 100 full-rigged ships, most of them fitted out to fight pirates or other merchantmen (*see* 1628).

Painting *Allegory of War and Peace* by Peter Paul Rubens; *Rinaldo and Armida* by Anthony van Dyke; *The Triumph of Bacchus* by Diego Velázquez, who gets leave from Philip IV to visit Italy.

Japan's Kabuki Theater becomes an all-male affair in October by order of the Tokugawa shogun Iemitsu, who has decided that it is immoral for women to dance in public (*see* 1603). Women's roles are performed by men as in Elizabethan England, but the Japanese will take extraordinary measures to make the men playing female roles appear as women even to other Kabuki players; *The Grateful Servant* by James Shirley, in November at London's Phoenix Theatre, with the Queen's Men.

The Barberini Pope Urban VIII appoints Giovanni Lorenzo Bernini to finish St. Peter's at Rome. Bernini will design the Scala Regia in the Vatican, complete the Barberini Palace, redecorate the Lateran and the Bridge of Sant' Angelo, and construct fountains for Rome's Piazza Navona, Piazza Trevi, and Piazza Barberini.

The joint-stock company of Massachusetts Bay organized by Anglican Puritans will bring more than 17,000 settlers to America in the next 13 years.

1630 England makes peace with France in April and with Spain in November.

Sweden's Gustavus II Adolphus lands on the Pomeranian coast in July to aid the oppressed Protestants, restore the dukes of Mecklenburg to whom he is related, and resist the extortion and cruelty of Wallenstein's army (*see* 1629; 1631).

The emperor Ferdinand II dismisses Wallenstein August 13.

Hungary's Bethlen Gábor (Gabriel Bethlen) is succeeded by George Rákóczi I, who is elected prince of Transylvania at age 39 by the Diet of Segesovar November 26. He will reign until 1648.

The Ottoman emperor Murad IV defeats a Persian army and captures Hamadan, ancient capital of Media, as he retakes conquests of the late Persian shah Abbas I. He massacres Hamadan's populace and sacks the city for 6 days, even cutting down its trees.

Boston, Dorchester, Roxbury, Watertown, Mystic, and Lynn are established as the Massachusetts Bay Colony receives 2,000 new settlers.

Puritan John Winthrop, 42, has arrived at Salem June 12 aboard the Bay Company's flagship *Arabella* and appoints himself governor.

Portsmouth is established by John Mason in what will be the New Hampshire colony (*see* 1629; 1635).

Agents of Amsterdam pearl merchant Kiliaen van Rensselaer buy a tract of land for him on the west bank of the Hudson River. They give the Indians "certain quantities of duffels, axes, knives and wampum" for a territory 24 miles in length, 48 in breadth.

Peru's Cerro de Pasco mine opens 14,000 feet high in the Andes. The mine will yield gold, silver, copper, zinc, and antimony, will become South America's largest lead producer, and will be the world's largest producer of bismuth.

Bubonic plague kills 500,000 Venetians, hastening the decline of Venice.

Repeated epidemics in the decade ahead will reduce America's Huron tribe to a third of its estimated 30,000 population.

Painting *The Triumphs of Flora* by Nicolas Poussin for his patron Cardinal Omodei; *The Vision of Blessed Alonso Rodriguez* by Francisco de Zurbarán.

Theater *Mélite, or The False Letters (Mélite, ou les fausses lettres)* by French playwright Pierre Corneille, 23, in January or February at the Berthault Tennis Court, Paris.

The Palais de Luxembourg, completed at Paris for the queen mother Marie de' Medici after 15 years' work, was designed by French architect Salmon de Brosse to resemble the Florentine Pitti Palace, where she grew up.

Lemonade is invented at Paris as sugar imported from the French West Indies drops in price.

Kikkoman soy sauce is originated at Noda, 25 miles upriver from Edo, by the Mogi and Takanashi families who also develop *miso* (bean paste) to serve the growing community of Edo (*see* 1590). They adopt the name *Kikko* (tortoise shell) *Man* (10,000) because the tortoise is thought to live for 10,000 years and many have a hexagon pattern on their shells that will be the Kikkoman trademark.

A "Great Migration" begins that will bring 16,000 people to the Massachusetts Bay Colony in the next 10 years (above).

1631 Cardinal Richelieu brings France into the war against the Hapsburgs. The Treaty of Barwalde, signed January 13, pledges French subsidies to Sweden's Gustavus II Adolphus and Bernhard, 27, duke of Saxe-Weimar.

Magdeburg falls to the forces of Count von Tilly May 20. The German cavalry general Count Gottfried Heinrich zu Pappenheim, 37, takes the city by storm and sacks it, massacring the citizenry; fires break out at scattered locations simultaneously, and the entire city burns with the exception of the cathedral as the Catholic soldiery commits terrible atrocities.

The Battle of Breitenfeld (or Leipzig) September 17 breaks the strength of revived Catholicism in central Europe. Tilly has burned Halle, Eisleben, Merseburg, and other cities, but the elector of Saxony John George has formed an alliance with Sweden's Gustavus II Adolphus, an army of Swedish Lutherans has crossed the Elbe at Wittenberg to challenge Tilly who has occupied Leipzig, and the Saxon-Swedish army of 40,000 defeats Tilly's army of equal size.

Amsterdam recalls Peter Minuit from Nieuw Amsterdam for granting undue privileges to *patroons* (landowners) and concentrating economic and political power in the hands of an elite few (*see* 1626). Minuit will enter the service of Sweden (*see* 1638).

Cautio criminalis by German Jesuit poet Friedrich von Spee, 40, is published anonymously. Von Spee has often ministered to condemned witches and he attacks the mentality behind witch-hunts and the legal use of torture to extract confessions (*see* Wier, 1563; Loudon, 1634; Salem, 1692).

The Thirty Years' War, a conflict between Catholics and Protestants, brought devastation to much of Europe.

1631 ***(cont.)*** Shipyards start up at Boston and other Massachusetts colony seaports as cheap American lumber makes an American-built ship only half as expensive as one built in England. The 30-ton sloop *Blessing of the Bay* is launched in August for Gov. John Winthrop.

The vernier scale for making accurate measurements of linear or angular magnitudes for navigation is described by French mathematician Pierre Vernier, 51, in *Construction, Usage et Propriéties du Quadrant Nouveau de Mathématiques (see* Nunes, 1536; Greenwich Observatory, 1676).

Theater *The Traitor* by James Shirley, in May at London's Phoenix Theatre; *The Humourous Courtier* by Shirley, in November at the Phoenix Theatre; *Love's Cruelty* by Shirley, in November at the Phoenix.

Potatoes are cultivated in so much of Europe, yield so much food per unit of land, and grow so well even in years when grain crops are at famine level that the tubers have encouraged the start of a population explosion in those parts of the continent where the new food is accepted and where the ravages of the Thirty Years' War (above) have not totally disrupted society.

1632 Poland's Sigismund III Vasa dies suddenly at age 65 after a reign of 44 years; Moscow immediately declares war. Sigismund's son Ladislas IV, 37, begins a 16-year reign, he engages the Muscovites in a series of bloody battles in August, and he will pursue a course opposite from that of his father.

Sweden's Gustavus II Adolphus defeats the Catholic general von Tilly at the confluence of the Lenz and the Danube as the Thirty Years' War continues. Von Tilly is mortally wounded and dies April 30 at age 72 at Ingelstodt; Munich surrenders to Gustavus II Adolphus; and the Austrian duke Albrecht von Wallenstein tries to keep him from taking Nuremberg. Both sides withdraw from the city after 18,000 have died of scurvy and typhus.

The Battle of Lutzen November 16 pits 20,000 Swedes against 18,000 Catholics. The Swedes win, but Gustavus II Adolphus is killed at age 38. The German Gottfried Heinrich, count of Pappenheim, is mortally wounded and dies the next day at Leipzig, at age 38.

Sweden's Gustavus II Adolphus is succeeded by his daughter Christina, now 6, who will ascend the throne on her eighteenth birthday December 8, 1644, after a chancellorship administered by Count Axel Gustafsson Oxenstierna, 49.

Maryland is chartered as the first of the English proprietary colonies in the New World. A grant of land from the Potomac River north to the 40th parallel is granted by Charles I to Cecilius (Cecil) Calvert, 27, whose father George, Lord Baltimore, has just died at age 52 after having rejected a grant of Newfoundland on account of its harsh climate (*see* 1633).

Nova Scotia is founded as the French colony of Acadia by settlers who will cut farms out of the region's dense forests (*see* 1710).

Galileo Galilei repeats his advocacy of the Copernican system of astronomy in *Dialogo de Massimi Sistemi del Mondo.* He is summoned to Rome (*see* 1613; 1633).

Poetry "On His Having Arrived at the Age of Twenty-Three" by English poet John Milton, 23: "How soon hath Time, the subtle thief of youth,/Stol'n on his wing my three-and-twentieth year."

Painting *De Tulp's Anatomy* by Rembrandt van Rijn, who depicts members of Amsterdam's Guild of Surgeons; *The Garden of Love* by Peter Paul Rubens; *King Charles I and Queen Henrietta* by Anthony van Dyck, who has been named court painter.

Theater *Hyde Park* by James Shirley, in April at London's Phoenix Theatre.

English Puritan William Prynne, 32, attacks the London theater in his pamphlet "Histriomastix": "It hath evermore been the notorious badge of prostituted Strumpets and the lewdest Harlots, to ramble abroad to plays, to Playhouses; whither no honest, chaste or sober Girls or Women, but only branded Whores and infamous Adulteresses, did usually resort in ancient times."

Prynne's lines (above) are construed as an aspersion on Queen Henrietta Maria, who has taken part in a performance of a play at court. Prynne will be branded, heavily fined, have his ears cut off, and serve 8 years of a life sentence beginning in 1634 (*see* theater closings, 1642).

The Palais-Royal is completed at Paris to provide a residence for Cardinal Richelieu.

Apple trees are planted by English colonists in territory that will become the New Jersey colony in 1665.

Rotterdam's population reaches 20,000. The city will become the world's largest port (*see* 1299).

1633 Russian forces lay siege to Smolensk, but Poland's new king Ladislas IV relieves the siege and obtains help from the Dnieper Cossacks.

An expedition to start the first Roman Catholic colony in North America leaves England in November as Cecil Calvert's younger brother Leonard embarks with the vessels *Ark* and *Dove* (*see* 1632; 1634).

Some 30 Dutch colonists settle in Delaware.

Dutch colonists from Nieuw Amsterdam establish a trading post on the Connecticut River at what later will be called Hartford (*see* 1636; Connecticut colony, 1635).

Galileo goes on trial at Rome April 12 and is threatened by the Inquisition with torture on the rack if he does not retract his defense of the Copernican idea that the sun is the center of the universe and the earth a movable planet. Galileo yields and is sent to his villa outside Florence, where he will be confined for the remaining 9 years of his life (*see* 1632).

René Descartes takes warning from the trial of Galileo (above). Now living in Holland, Descartes stops publishing in France (*see* 1619; 1637).

Poetry *The Temple* by the late English poet George Herbert, who has died March 1 of tuberculosis at age 39.

Painting *Henry Percy, 9th Earl of Northumberland* by Anthony van Dyck.

Theater *A Match at Midnight* by William Rowley contains the line, "He's a chip o' the old block"; *Tis Pity She's a Whore* by John Ford at London's Phoenix Theatre; *The Maidservant (La suivante)* by Pierre Corneille at the Fontaine Tennis Court, Paris; *Place Royale, or The Extravagant Lover (La place royale, ou Lamoureux extravagant)* by Pierre Corneille, in December, at the Fontaine Tennis Court.

1634

An imperial proclamation issued January 24 accuses the Catholic general von Wallenstein of involvement in a conspiracy to "rob the emperor of his crown," chief army officers are commanded to cease their obedience to Wallenstein, he is formally deposed as duke of Friedland February 18 and assassinated a week later by English and Irish officers.

The Treaty of Polianov ends a 2-year Russian-Polish war. Poland's Ladislas IV renounces his claim to the Russian crown, but Russia's Michael Romanov gives up Smolensk and surrounding territories in return for recognition of his title (*see* 1613).

The Mughal emperor Shah Jahan drives the Uzbeks out of Kandahar (*see* 1629; 1650).

Dutch forces seize Curaçao and St. Eustatius in the West Indies (*see* 1779).

French witch-hunters at Loudon crush the legs of local curate Urbain Grandier, 44, and burn him alive August 18. The vain and handsome priest of the Huguenot St. Pierre-du-March church is a notorious seducer of virgins and widows in the prosperous walled city of 20,000; he has insulted Cardinal Richelieu in a matter of church protocol, and been accused and found guilty of bewitching the convent of Ursuline nuns whose hysterical, blasphemous fits have for years attracted morbid sightseers from all over Europe. Grandier protests his innocence to the end, the nuns are exorcised, Richelieu razes the fortified castle of Loudon to prevent its use by the Huguenots, but the descendants of the judges who sent Grandier to the stake will lead tormented lives, and it will be said that Grandier put a curse on them.

French explorer Jean Nicolet, 36, makes an expedition to Lake Michigan and the Wisconsin region, the first white venture into that area. Brought to Canada by Champlain in 1618, Nicolet has lived among the Huron of the upper Ottawa River, and he now travels west in search of furs and of a Northwest Passage. Nicolet lands at what will later be Green Bay, where the Menominees have a settlement at a time when most indigenous tribes are being pushed westward by the Ottawa, Huron, and other eastern tribes. So vast is Lake Michigan that Nicolet believes he has reached Asia and is disappointed to learn otherwise.

Lord Calvert arrives in America late in March and establishes a Maryland settlement that will welcome Protestants as well as Roman Catholics (*see* 1633; 1688).

Permanent settlements are made in Connecticut by Massachusetts Puritans whose towns will later be called Windsor and Wethersfield (*see* 1635).

Charles I extends to all of England a ship-money tax formerly levied only on coastal towns; Buckinghamshire squire John Hampden, 40, defies the order, will lose his case in court, but will have popular support in his cause against the king.

Speculation in tulip bulbs reaches new heights in the Netherlands, where one collector pays 1,000 pounds of cheese, four oxen, eight pigs, 12 sheep, a bed, and a suit of clothes for a single bulb of the Viceroy tulip. Tulip bulbs have been imported from Turkey since the last century, but the tulip's attribute of variation has been discovered only recently by a professor of botany at the 59-year-old University of Leyden whose botanical garden is the first in the north (*see* 1637).

The University of Utrecht has its beginnings in a college that will become a center of Dutch learning.

Poetry *Epigrammatum Sacrorum Liber* by English poet Richard Crashaw, 21, is published anonymously.

Painting *The Surrender of Breda* by Diego Velázquez; *The Judgment of Paris* by Peter Paul Rubens.

Masque *Comus* by John Milton 9/29 at Ludlow Castle celebrates the inauguration of John Egerton, 55, first earl of Bridgewater, as Lord President of Wales (*see* literature, 1637).

Villagers at Oberammergau in the Bavarian Alps vow to enact a passion play at regular intervals if they are spared by the Black Death. The folk drama with its anti-Jewish slurs will be performed every 10 years for centuries.

The Toshogu Shrine at Nikko is completed as a mausoleum for the late Tokugawa shogun Ieyasu who died in 1616.

1635

The Peace of Prague May 30 resolves differences between the Holy Roman Emperor Ferdinand II and the elector of Saxony John George; France's Cardinal Richelieu makes an alliance with Sweden's Count Axel Gustafsson Oxenstierna. The Thirty Years' War becomes a conflict between the Franco-Swedish alliance and the Hapsburgs. France agrees to regular subsidization of Bernhard, duke of Saxe-Weimar.

Murad IV leads an Ottoman army against Persia. Erivan capitulates after a siege, Tabriz surrenders without resistance but is deliberately destroyed (the Blue Mosque is spared when the mufti observes that it was built by a Sunni, not a Shiite).

The Japanese Tokugawa shogun Iemitsu acts to prevent any feudal lord from becoming too rich and powerful. He orders that each *daimyo* must visit Edo every other year, leave his wife and children at Edo for the year he is absent, and pay all the expenses of maintaining two places of residence.

1635 *(cont.)* Dutch forces invade and occupy northern Brazil, where Dutch planters will enter the lucrative sugar industry (*see* 1623; 1654).

The Connecticut colony is created by a union of the "River Towns" Windsor and Wethersfield with English settlements at Hartford and Saybrook (*see* 1634; Hooker, 1636).

Directors of the Plymouth colony prepare to surrender their charter and draw lots for apportioning the colony's territory. Capt. James Mason, who helped found Portsmouth in 1630, has obtained a patent to the New Hampshire area from the London Company, and he receives the entire area (*see* 1680).

The General Court of Massachusetts authorizes settlement of Concord and several families move inland to obtain more pasturage on the new frontier.

Indians destroy Dutch settlements founded in Delaware 2 years ago (*see* New Sweden Company, 1638).

Yellow fever breaks out in Guadeloupe and St. Kitts in the West Indies. The French physician Duterte gives the first reliable account of the disease in western medical literature.

The world's first free medical clinic for the poor opens at Paris under the direction of physician-journalist Théophraste Rénaudot, 49.

The Académie Française is founded to establish rules of grammar and correct usage and to cleanse the French language of "impurities."

Boston Public Latin School opens. It is the first secondary school in the American colonies.

Budapest University has its beginnings.

England gets her first inland postal service as mail coach service begins between London and Edinburgh.

Painting *Self-portrait with Saskia* by Rembrandt van Rijn; *Charles I in Hunting Dress* by Sir Anthony van Dyck; *The Immaculate Conception* by Jusepe de Ribera; *Prometheus* by Italian painter Salvator Rosa, 20, who has arrived at Rome from his native Naples. Jacques Callot dies at Nancy March 24 at age 42.

Theater *Medea (Medée)* by Pierre Corneille, in January at Paris; *The Comedy of the Tuileries (La comédie des Tuileries)* by Pierre Corneille, François Le Metel de Boisrobert, 42, Guillaume Colletel, and Claude de L'Estville, in February at the Louvre Palace, Paris; *The Challenge of Charles V (El desafio de Carlos V)* by Mexican-born Spanish playwright Francisco de Rojas Zorrilla, 26, 5/28 at Madrid; *The Lady of Pleasure* by James Shirley.

A new French law restricts sale of tobacco to apothecaries on prescription from a physician.

The Mauritzhus at the Hague is completed by Dutch architect Jacob van Campen, 40.

The Chinese orange *citrus sinensis* reaches Lisbon and is even sweeter than the "Portugals" introduced to Europe by the Portuguese in 1529.

1636 The Conde-Duque de Olivares invades northwestern France in a desperate gamble to defeat the French and restore Spanish power. His soldiers burn villages in Beauvaise and Picardy, plunder crops, and force peasants to flee with food, livestock, and belongings. Basque seaports are ravaged, but the French win a narrow victory at Corbie, forcing Spain into a war of attrition that will drain her for 23 years.

Persia's Shah Safi retakes Erivan in the spring and signs a treaty with Constantinople setting western borders that will remain substantially unchanged for more than 2 centuries.

The Manchus at Mukden set up a civilian administration modeled on that of their Chinese neighbors to the south and proclaim an imperial Da Qing dynasty (*see* 1644).

Japan's shogun Iemitsu forbids his people to travel abroad (*see* 1623; 1638).

Providence is founded as a Rhode Island settlement by English clergyman Roger Williams, 33, who has been banished by the Massachusetts Great and General Court for his outspoken criticism of what he calls the "abuse of power" and who has sought a place where "persons of distressed conscience" could go. Williams selects the name in gratitude for "God's merciful providence" that the Narragansett have granted him title to the site.

Springfield, Massachusetts, is founded by colonists who follow William Pynchon west to take advantage of the abundant pasturage in the area. Springfield will become an important meat-packing center.

Hartford, Connecticut, is founded by New Towne, Massachusetts, colonists who have traveled overland with clergyman Thomas Hooker (*see* 1635).

The town of Haarlem is founded by Dutch colonists on Manhattan Island.

Another epidemic of the Black Death strikes London (*see* 1625; 1665).

Harvard College has its beginnings in a seminary founded by the Great and General Court of Massachusetts at New Towne (*see* 1638).

Painting *The Blinding of Samson* by Rembrandt van Rijn; Dutch still-life painter Jan Davidsz de Heem, 30, moves to Antwerp because "there one could have rare fruits of all kinds, large plums, peaches, cherries, oranges, lemons, grapes and others, in finer condition and state of ripeness to draw from life."

Theater *The Comic Illusion (L'illusion comique)* by Pierre Corneille in January at the Théâtre du Marais, Paris.

Requiem *Musikalische Exequien* by Heinrich Schütz is the first German requiem.

A Dutch planter introduces sugar cane from Brazil into the West Indian island of Barbados whose English settlers have been cultivating cotton, ginger, indigo, and tobacco for export while growing beans, plantains, and other food for their own consumption. Sugar will become the chief crop of Barbados and of all the Caribbean islands (*see* 1627).

1637 The Holy Roman Emperor Ferdinand II dies at Vienna February 15 at age 57. He is succeeded by his son of 28, who will reign until 1657 as Ferdinand III.

Korea becomes a vassal state of the Chinese Manchus after an invasion led by Dai Zong, but the Korean court and people remain loyal to the Ming.

Japanese peasants on Kyushu's Shimabara Peninsula rise against the Tokugawa shogun Iemitsu. Their *daimyo* (feudal lord) Arima has been baptized, many of them have embraced Christianity, but the shogun begins a siege in December to force their submission (*see* 1638).

Massachusetts colonists have their first hostile encounter with the Pequot. A force of 240 militiamen, 1,000 Narragansetts, and 70 Mohegans destroys a stockaded fort at Mystic, burning the town and slaughtering 600 inhabitants (*see* 1675).

Edinburgh has riots June 23 when clergymen at St. Giles's Cathedral attempt to read the Anglican liturgy on orders from Charles I. Many Scottish Presbyterians regard episcopacy as popery and organize to resist it (*see* 1638).

The Dutch move to assure themselves of African slaves for their sugar estates in the New World. Dutch forces take Elmina from the Portuguese and build forts along the Gold Coast.

A French expedition ascends the Senegal River for 100 miles and establishes posts.

The Swedish queen Christine charters the New Sweden Company to colonize the New World. Dutch colonists Peter Minuit and Samuel Blommaert have encouraged the queen's advisers (*see* 1638).

Russian pioneers reach the Pacific after a quick journey across all of Siberia (*see* 1579; 1689).

Dutch tulip prices collapse after years of speculation. Hundreds are ruined as the bottom falls out of the market (*see* 1634).

Discours de la Mèthode by René Descartes establishes the "Cartesian" principle of basing metaphysical demonstrations on mathematical certitude rather than on scholastic subtleties (*see* 1619). The proper guide to reason, says Descartes, is to doubt everything systematically until one arrives at clear, simple ideas that are beyond doubt. He rejects any doubt of his own existence by saying, "*Cogito, ergo sum*" ("I think, therefore I am").

Poetry *Comus* by John Milton (quarto of 35 pages); "Lycidas" by Milton, who memorializes his friend Edward King, lost with all hands when his ship bound for Ireland out of Chester struck a rock in clear weather August 10 and foundered: "Fame is the spur that the clear spirit doth raise/(That last infirmity of noble mind)," lines 70, 71; "Aglaura" by English Cavalier poet John Suckling, 28, who has invented the game of cribbage, served under Sweden's Gustavus II Adolphus, and retired 2 years ago to the large Norfolk estates he inherited from his father at age 18: "Why so pale and wan, fond lover," writes Suckling whose poem "A Session of the Poets" describes his contemporaries.

Painting *Five Children of Charles I* by Sir Anthony Van Dyck; *The Triumph of Neptune and Amphètre* by Nicolas Poussin; Japanese calligrapher and lacquerware potter Koetsu Honami dies at age 59 after a brilliant career in which he has been supported by the Tokugawa shogun Iemitsu.

Theater *No Jealousy Without Cause or The Servant Master (Donde hay agravios ho hay celos, o El amo criado)* by Francisco de Rojas Zorilla 1/29 at Madrid's El Pardo; *The Most Proper Execution for the Most Just Vengeance (El más impropio verdugo para la más justa verganza)* by Zorilla 2/2 at Madrid's Coliseo de Buen Retiro; *Le Cid* by Pierre Corneille.

1638 Japanese peasants who have occupied Hara Castle near Nagasaki for nearly 3 months yield February 28 for lack of food and musket ammunition. Most of the 37,000 have accepted Christianity and the 124,000-man siege force of the Tokugawa shogun Iemitsu annihilates most of them. Iemitsu expels Portuguese traders from Japan on suspicion of complicity in the Shimabara uprising (which has actually been supported by a Dutch vessel), he presses Japanese isolation by prohibiting construction of large ships which might carry people abroad, and he orders any farmer who does not pay his taxes to be hanged as if he were a Christian (many cannot pay because of poor crops which have produced famine).

The Ottoman sultan Murad IV retakes Baghdad from the Persians after a 40-day siege, slaughtering the city's defenders.

Scottish Presbyterians circulate a National Covenant for signature throughout the country following a threat by England's Charles I to impose the Anglican liturgy. Charles offers to withdraw the new prayer book and to permit the meeting of a free assembly and a free parliament, but Scottish opposition mounts. The assembly that meets at Glasgow in December admits only Covenanters. Presbyterian extremists defy the royal commissioner, the assembly deposes Anglican bishops, repeals all the legislation by which James VI and Charles I have established episcopacy, and its actions precipitate civil war (*see* 1639).

Dutch colonists in the Indian Ocean settle on the island of Mauritius, named after the late Prince Maurice of Nassau, and begin clubbing to death the island's indigenous bird, the dodo (*see* 1598; 1681).

Wilmington, Delaware, has its beginnings in Port Christina on the Delaware River where Peter Minuit of the New Sweden Company lands two shiploads of Swedish and Finnish colonists and builds a fort. Minuit is lost at sea soon afterward (*see* 1637; Stuyvesant, 1655).

The New Haven colony is founded on the southern New England coast by English colonists who name the colony after Newhaven in southern England and who

1638 ***(cont.)*** follow a more rigid Puritanism than that observed in the Connecticut colony (*see* 1635).

English clergyman John Harvard, 31, dies of tuberculosis September 14 after 1 year in the Massachusetts Bay Colony and leaves his library and half his estate of £800 to the "seminary" established at New Towne in 1636.

Because John Harvard (above) is a Cambridge University graduate, New Towne will be renamed Cambridge; the Great and General Court of Massachusetts next year will order that "the colledge agreed upon formerly to bee built at Cambridge shalbee called Harvard College," colonist Ann Radcliffe will contribute funds to the school, and by 1650 Harvard will have established the 4-year program that will become standard for U.S. colleges (*see* 1642).

Painting *Christ on the Cross* by Velázquez; *Et in Arcadia ego* by Nicolas Poussin; *Lords John and Bernard Stuart* by Sir Anthony van Dyck. Adriaen Brouwer dies at Antwerp in January at age 31.

Boston gets its first brick house. Brick has been used until now only for chimneys, fireplaces, and backyard baking ovens.

Honeybees will be introduced into the American colonies in the next few years, will soon escape from their domestic hives, and will establish wild colonies. Indians will call honeybees "the white man's fly," and as the bees move westward many pioneers will be led to believe that the bees are indigenous (*see* Irving, 1835).

1639 The Scots seize Edinburgh Castle and raise an army, but they make peace June 18 with England's Charles I, who has marched to meet them near Berwick. No military engagement takes place, but a new Scottish parliament assembled after the armies have disbanded is intractable in its Presbyterian opposition to English episcopacy (*see* 1640).

Swiss alpine passes are left open for use by Spanish troops under terms of a September 3 treaty.

The former Ottoman sultan Mustafa I is strangled in June. The divan convokes a special session and issues a *fatwa* establishing that hereafter the successor to a sultan shall be his oldest male relative—uncle, brother, or nephew—and not the eldest son.

Japan's Tokugawa shogun Iemitsu finds that some Christian missionaries have remained despite his 1638 expulsion order. He orders the massacre of any Portuguese who have not fled and closes Japanese ports to all foreigners except for some Chinese and Dutch (*see* 1641).

A new canal links France's Loire and Seine rivers (*see* 1603; 1681).

The Whole Book of Psalms Faithfully Translated Into English Metre (The Bay Psalm Book), published by English-American printer Stephen Day, 47, is the first book to be printed and bound in New England. Day has set up a printshop at Cambridge with encouragement from the Massachusetts Bay Colony government, and his book will be republished in frequent editions until late in the 18th century.

Paroemiologia by English-American Baptist clergyman-physician John Clarke, 30, contains the verse, "Early to bed and early to rise, makes a man healthy, wealthy and wise." Clarke helped found the Rhode Island colony last year, and his verse will be widely credited a century hence to Benjamin Franklin, as will his line, "He that will not be counseled cannot be helped."

Painting The ceiling of Rome's Barberini Palace is covered with a gigantic fresco of clouds and swirling figures by Italian painter-architect Pietro da Cortona (Pietro Berretini), 42, who has been working on the project for 6 years; *The Martyrdom of St. Bartholomew* by Jusepe de Ribera.

Philip IV completes Madrid's 353-acre Parque del Retiro.

"Smithfield" hams shipped to England from the Virginia colony are sold at London's Smithfield Market which is taken over by the city after 516 years and is reorganized as a market for live cattle.

1640 The Ottoman sultan Murad IV dies in February at age 31 after a 17-year reign in which he has restored order with a bloody reign of terror and retaken Baghdad and Ecrivan from the Persians. He has had all his brothers murdered except for Ibrahim, 25, who has feigned insanity and will reign until 1648 as Ibrahim I, dominated by his mother Kussem Sultana and story-teller Sheker-Pare.

England's Charles I appoints the earl of Strafford Thomas Wentworth, 47, his chief adviser and makes him Baron Raby, lord lieutenant of Ireland. Urging an invasion of Scotland to crush the Presbyterians and impose the English liturgy, Strafford obtains £180,000 from the Irish Parliament and prepares to lead Irish troops against the Scots (*see* 1641).

Scottish forces cross the Tweed August 20 to begin the second Bishops' War; they defeat the English at Newburn August 28 and occupy Newcastle and Durham. The Treaty of Ripon October 21 ends hostilities, the English agreeing to pay the Scots £850 per day to maintain their army.

Charles convenes the "Long Parliament" November 5; he sets free the members of Parliament he imprisoned earlier for refusing to loan him money; and he has the earl of Strafford (above) impeached and thrown into the Tower of London.

Catalonia obtains French support to begin a 19-year revolt against Spain in protest against Barcelona's taxes, quartering of troops, military recruitment, and general denial of Catalonian rights.

Portugal seizes the opportunity of the Catalonian revolt (above) to regain her independence after 60 years of Spanish rule. The Portuguese elect João da Braganza king to begin the dynasty that will reign until 1910, but Spain will not recognize Portugal's independence until 1668.

Dutch forces take Saba in the West Indies.

Gardiner's Island at the eastern end of Long Island Sound gets its name as English colonist Lion Gardiner,

41, buys Manchouake Island from the Shinnecock a year after being given the property in a land grant from Charles I. Seven miles from the nearest point of land, the island is 7 miles long and up to 3 miles wide. A military engineer, Gardiner arrived at Boston in 1635, has befriended the natives, and will ingratiate himself with the most powerful of the Long Island chiefs, Wyandanch. He will acquire 78,000 acres, permitting him to travel on his own land from Montauk to Flushing, and the island will remain in the Gardiner family for over 3 centuries.

Inflation in England reduces the value of money to one-third its value in 1540 and food prices outpace wage increases.

Copenhagen opens a stock exchange.

France enjoys power and prosperity under Cardinal Richelieu, but the mass of Frenchmen live on the edge of starvation.

A Portuguese delegation sent to reopen trade with Japan is summarily executed, leaving Japan with no means of contact with the outside world except via some Dutch traders at Hirado and a few Chinese at Nagasaki (*see* 1623; 1641).

British coal mining gains impetus from the high cost of firewood whose price has risen to nearly eight times 1540 levels while most other prices have merely tripled (*see* 1615; 1658).

The first English stagecoach lines begin as roads and coaching houses are built to accommodate coaches and travelers.

England's Puritan movement spreads to her lower classes.

Painting *The Peasant Supper* by French painter Louis Le Nain, 52; *The Inspiration of the Poet* by Nicolas Poussin; *Self-Portrait* by Rembrandt van Rijn; *The Barrel Organ Player* by Dutch painter Adriaen van Ostade, 29; *Cupid and Psyche* by Anthony van Dyck; *Three Graces* by Peter Paul Rubens, who dies at Antwerp May 30 at age 62.

Theater *The Two Households of Verona* by Francisco de Rojas Zorilla 2/4 at Madrid's Coliseo de Buen Retiro; *Horatio (Horace)* by Pierre Corneille 3/9 at the Hôtel de Bourgogne, Paris; *Keep Your Eyes Open (A bire el ojo)* by Zorilla 8/3 at Toledo; *Cinna, or The Clemency of Augustus (Cinna, ou Le clémence d'auguste)* by Corneille in December at the Hôtel de Bourgogne, Paris.

The Massachusetts Bay Colony sends 300,000 codfish to market and some colonists begin to grow rich in the cod fishery (*see* 1700).

The British West Indies have a population of 20,000, most of it employed in growing sugar cane.

1641 The Archbishop of Canterbury William Laud, 68, is sent to the Tower of London, where the earl of Strafford is executed May 12 (*see* 1640). The archbishop has sought absolutism in church and state and has used the Court of High Commission and the Star Chamber to root out Calvinism and Presbyterianism.

The "Long Parliament" abolishes the Star Chamber and the Court of High Commission in July as it shows its determination to effect a revolution against the excesses of the English constitution.

Irish peasants begin a revolt against their landlords (who are mostly English), and Catholics massacre Protestants in Ulster (*see* Cromwell, 1649).

Louis XIII's brother Jean Baptiste d'Orléans exposes a conspiracy against France's Cardinal Richelieu. He reveals that Henri Coiffier de Ruze, 21, Marquise de Cinq-Mars, has taken advantage of his position as protégé of Richelieu and king's favorite to make a secret treaty with Spain (Cinq-Mars will be executed in mid-September of next year).

Dutch forces take Malacca and begin their domination of the East Indies (*see* 1623).

The Japanese remove Dutch East India Company traders at Hirado to Deshima, and since the Dutch have no missionaries the Japanese permit them to remain on condition that company officers visit Edo once a year, turn somersaults in the street, spit on the Cross, and pay a rent in peppercorns. They may bring in three ships per year, the number will be reduced to two in 1715, and although one ship in five will be lost in the heavy Japanese seas, the surviving ships will make vast profits for the Dutch.

The Massachusetts Bay Colony lowers to 8 percent the rate of interest any lender may charge to prevent "usury amongst us contrary to the law of God" (*see* 1693).

Painting *Manoah* by Rembrandt van Rijn; *Embarkation of St. Ursula* by French painter Claude Lorrain, 42; *The Cart* by Louis Le Nain; *The Seven Sacraments* by Nicolas Poussin; *Regents of the Hospital of St. Elizabeth* by Frans Hals; *Country Fair* by Flemish painter David Teniers, 30; *Scenes from the Life of St. Januarius* by Il Domenichino, who dies at Naples April 6 at age 59;

Cardinal Richelieu ruled France with an iron hand for nearly 2 decades, but he encouraged playwrights.

1641 ***(cont.)*** *Prince Willem of Orange* by Sir Anthony van Dyck, who dies at London December 9 at age 42.

Theater *The Cardinal* by James Shirley; *Polyentes (Polyeucte)* by Pierre Corneille, in December at the Théâtre du Marais, Paris.

Russia's Michael Romanov forbids sale and use of tobacco in a decree that mentions exile as a possible punishment. Users as well as sellers are to be flogged, but the crown will violate its own law and make the "impious herb" a state monopoly, selling tobacco at high prices to produce revenue (*see* 1648).

The first sugar factory in the English New World goes up in Barbados using equipment supplied on credit by Dutch investors (*see* 1636; slavery, 1645).

1642 Persia's Shah Safi I dies at age 26 after a 13-year reign that has lost much of his empire to the Ottoman Turks. He is succeeded by his son of 10, who will reign until 1667 as Abbas II and be no more successful.

Civil war begins in England as Charles I sends his Cavaliers against the Puritan parliament at York. Supported by the gentry, the Anglican clergy, and the peasantry (especially in northern and west-central England), the king is opposed by the middle classes, the great merchants, and much of the nobility (*see* Marston Moor, 1644).

Cardinal Richelieu dies December 4 at age 57 after 18 years in power.

Some 600 Jews sail in various vessels from Holland to Brazil. They will be expelled in 12 years when the Portuguese take Pernambuco from the Dutch (*see* Nieuw Amsterdam, 1655).

Montreal is founded as Ville Marie by French colonist Paul de Chomedey, 30, sieur de Maisonneuve, on the island in the St. Lawrence River first visited by Jacques Cartier in 1535 at the point where the Ottawa and Richelieu rivers flow into the St. Lawrence.

Tasmania is discovered and named Van Diemen's Land by Dutch mariner Abel Janszoon Tasman, 39, who has been sent on an exploring expedition by Anton Van Diemen, 49, governor-general of the Dutch East Indies.

New Zealand is discovered by Tasman (above) in December, 4 months after discovering Van Diemen's Land.

French mathematical prodigy Blaise Pascal, 19, invents a machine that adds and subtracts using wheels numbered from 0 to 9 with an ingenious ratchet mechanism to carry the 1 of a number greater than 9. Pascal has made the machine to help his father compute taxes at Rouen (*see* 1692).

The first printed mention in Europe of cinchona bark, or Peruvian bark, is made by Seville physician Pedro Barba, who has relieved the Countess of Chinchon of her malaria by treating her with *polvos de la condesa.* The alkaloid drug quinine will be obtained from the bark.

German Jesuit scholar Athanasius Kircher uses a microscope to investigate the causes of disease and is the first to propound the doctrine of *cantagium animatum* (*see* Janssen, 1590). Kircher teaches mathematics and Hebrew at the College of Rome (*see* van Leeuwenhoek, 1675).

Harvard College awards its first baccalaureate degrees (*see* 1638; Eliot, 1869; William and Mary, 1693).

Nonfiction *Religio Medici* by English physician Thomas Browne, 37, appears in a pirated edition after 7 years of unauthorized circulation in manuscript (an authorized edition will be published next year): "I believe that our estranged and divided ashes shall unite again; that our separated dust, after so many pilgrimages and transformations into the parts of minerals, plants, animals, elements, shall, at the voice of God, return to their primitive shapes, and join again to make up their primary and predestinate forms" (I).

Poetry "To Althea, from Prison" by English poet Richard Lovelace, 24, who has been locked in the Gatehouse for 7 weeks after presenting to Parliament in April the Kentish petition for retention of bishops and the Prayer Book, a petition Parliament had earlier ruled seditious: "Stone walls do not a prison make,/Nor iron bars a cage;/Minds innocent and quiet take/That for an hermitage."

Painting *Night Watch* by Rembrandt van Rijn, a large canvas that will later be called *Sortie of the Banning Cock Company* (Rembrandt's wife Saskia dies, leaving him as trustee for his son Titus, the only one of their four children to survive infancy). The *Guard Room* by David Teniers the Younger. Guido Reni dies at his native Bologna August 18 at age 66.

England's great Elizabethan and Jacobean theater is ended by a September 2 Ordinance of Parliament "to appease and avert the wrath of God" (*see* Prynne, 1632).

Beriberi is observed in Java. Marked by debility, nervous disturbances, paralysis, heart weakness, swollen liver, weak and sore calf muscles, plugged hair follicles, and weight loss, the nutritional deficiency disease gets its name from the Sinhalese word *beri* for weakness.

New England has some 16,000 colonists whose transatlantic passage averaged 3 months' time.

1643 France's Louis XIII dies May 14 at age 43 after a 33-year reign dominated by his mother Marie de' Medici and Cardinal Richelieu. His son of 4 will reign until 1715 as Louis XIV, initially under the aegis of Giulio Mazarin, 41, a Sicilian who was naturalized as a Frenchman 4 years ago, was made cardinal 2 years ago, succeeded Cardinal Richelieu as prime minister, and has been retained by the queen regent Anne of Austria.

France's nobility will rise against Cardinal Mazarin (above) in the next 5 years in a final attempt to oppose the court by armed resistance and substitute rule of law for royal whim. The struggle will cause difficulties in transportation that will raise the price of flour and of bread.

The Battle of Rocroi May 19 gives a French army of 23,000 victory over a 20,000-man army of Spaniards and

Dutch, Flemish, and Italian mercenaries. Led by Louis II de Bourbon, 21, duc d'Enghien, the French gain their first great military success in decades, a triumph rivaling that of England over the Armada in 1588.

Gaspar de Guzman, count of Olivares, is driven from office and exiled from Spain.

The New England Confederation (United Colonies of New England), formed by Massachusetts, Plymouth, Connecticut, and New Haven, is the first union of English colonies in America.

Nieuw Amsterdam's Dutch governor orders a massacre of the Wappinger Indians who have sought Dutch protection from attacks by raiding Mohawks. Some 1,500 of the 15,000 Wappingers are treacherously wiped out.

English-American religious enthusiast Anne Marbury Hutchinson is killed by Indians in August at age 43 in the Long Island Sound settlement that will later be New Rochelle. Banished from the Massachusetts colony for her opposition to Puritan theocracy, Hutchinson is scorned by the Puritans, and her death is viewed as a manifestation of divine providence.

The first permanent settlement in what will be Pennsylvania is made at Tinicum Island in the Schuylkill River, where Swedish colonists build some log cabins as an extension of New Sweden (*see* 1638).

The world's first barometer is devised by Italian mathematician Evangelista Torricelli, 35, who served as amanuensis to the late Galileo when the latter lost his eyesight.

French medical student Jean Pecquet, 21, performs a dissection and discovers the ductus thoracicus that is the common trunk of the human lacteal and lymphatic systems (*see* 1623; 1653).

Typhus ravages an English Cavalier army of 20,000 and a Parliamentary army of the same size at Oxford. Charles I is forced to abandon plans to take London.

Painting *Village Fête* by David Teniers the Younger; *Three Trees* (etching) by Rembrandt van Rijn; *The Slaughtered Pig* by Adriaen van Ostade.

Theater *The Death of Pompey (La mort de Pompide)* by Pierre Corneille in January at the Théâtre du Marais, Paris; *The Liar (Le menteur)* by Corneille, in February at Paris; *Sly Gomez (Bellaco sois, Gomez)* by Tirso de Molina 4/27 at Madrid.

Opera *The Coronation of Poppy (L'Incoronozione di Poppea)* in the autumn at Venice's Teatro S. S. Giovanni e Paolo with music by Claudio Monteverdi is probably the first historical opera.

1644 The Ming dynasty that has dominated China since the 1380s ends in April with the suicide of the last Ming emperor, Chongzhen. He hangs himself as Beijing falls to the bandit and rebel leader Li Dzucheng, 39, who has overrun parts of Hopeh and Honan, conquered all of Shenxi, and proclaimed himself emperor. The Manchu regent Dagoba helps the Ming general Wu San-kuei drive Li into Hopeh (where he will be killed) and the Manchus begin the Qing (Ch'ing) dynasty that will rule China until 1912, imposing on the people the shaven head with queue (pigtail).

The Battle of Marston Moor July 2 ends in victory for Oliver Cromwell's English Roundheads over the Cavaliers of Charles I and wins the north country for the Puritan parliamentary forces that oppose the Royalist elements of gentry, peasantry, and Anglican clergy (*see* Naseby, 1645).

French forces occupy the Rhineland.

Essay "Aereopagitica" by John Milton is a defense of the freedom of the press: "Truth never comes into the world but like a bastard, to the ignominy of him that brought her forth"; "As good almost kill a man as kill a good book: who kills a man kills a reasonable creature, God's image; but he who destroys a good book, kills reason itself, kills the image of God, as it were, in the eye." (Milton has antagonized England's Presbyterians with pamphlets relating to divorce.)

Painting *Woman Taken in Adultery* by Rembrandt van Rijn; *The Kitchen of the Archduke Leopold Wilhelm* by David Teniers the Younger; *The Rest on the Flight into Egypt* and *The Three Marys at the Tomb of Christ* by Dutch painter Ferdinand Bol, 28; *Helena Van der Schalke as a Child* by Dutch painter Geraert Terborch, 27; *St. Paul the Hermit* by Jusepe de Ribera.

Rome's Church of S. Carlo alle Quattro Fontane is completed by architect Francesco Borromini (Francesco Castelli), 44, who assisted his kinsman Carlo Maderno until Maderno's death in 1629 and worked under Gian Bernini until 1633 in executing commissions from the Barberini Pope Urban VIII (who dies July 29 at age 76).

1645 The Battle of Naseby June 14 ends in a defeat of Charles I's English Cavaliers at the hands of Oliver Cromwell's Ironsides in a triumph for the English middle class and merchants who are supported by many of the country's great noblemen in the continuing civil war.

Europe's Thirty Years' War nears its end. An imperial army under Count Matthias Gallas, 61, is repulsed in January by Lennart Torstenson and Count Hans Christoph Königsmark, who block efforts by the army to relieve the hard-pressed Danes. They pursue the enemy into Germany, and nearly annihilate Gallas's army at Magdeburg. Torstenson gains a victory over the imperialists at Jankau in Bohemia in March, conquers Moravia with support from the Transylvanian prince George Rákóczi, and advances on Vienna while Turenne is defeated in Franconia but raises a French and Hessian army that defeats the Bavarians at Allershelm in August.

The Russian czar Michael Romanov dies July 12 at age 49 and is succeeded by his son, 16, who will reign until 1676 as Alexis Mikhailovich.

Ottoman forces capture Khania, in Crete, in August after a 57-day siege, beginning a 19-year war to wrest

1645 ***(cont.)*** the island from Venice. The Venetians have lost most of their commercial power.

Plague breaks out in the army of Sweden's Count Torstenson as he lays siege to Brunn and he returns to Bohemia.

Portuguese colonists in Brazil begin a popular rising against the Dutch following the return of Prince Maurice to Holland.

Some 10,000 slaves per year will be imported into the Americas in this decade, most of them for the Brazilian sugar plantations.

Barbados has 6,000 slaves (*see* sugar, 1641).

Poetry *Poems* by English poet Edmund Waller, 39, who was banished to France 2 years ago for leading a Royalist plot to seize London for Charles I; *Poems of Mr. John Milton, both English and Latin* . . . includes "Il Penseroso" and "L'Allegro": "Hence, vain deluding joys,/The brood of Folly without father bred" ("Il Penseroso"); "Where glowing embers through the room/Teach light to counterfeit a gloom,/Far from all resort of mirth,/Save for the cricket on the hearth" ("Il Penseroso"); "Hence, loathed Melancholy,/Of Cerberus and blackest Midnight born" ("L'Allegro"); "Come, and trip it as we go,/On the light fantastic toe" ("L'Allegro").

Painting *Flight into Egypt* by Spanish painter Bartolomé Esteban Murillo, 28; *The Dance in Front of the Castle* and *Tavern Scene* by David Teniers the Younger; *The Rabbi* by Rembrandt van Rijn; *King Philip IV on a Boar Hunt* by Diego Velázquez; *Wife of Candaules* by Flemish painter Jacob Jordaens, 52.

The book of *Poems* by Edmund Waller (above) contains many lyrics to songs by English composer Henry Lawes, 49, whose composer-brother William is killed fighting for the Royalist cause.

1646 England's 4-year civil war ends as Oliver Cromwell's Roundheads defeat and capture Lord Ashley March 26 at Stowe-on-the-Wold. Charles surrenders himself to the Scots May 5, but in July he rejects Parliament's Newcastle proposals that he take the Covenant and support the Protestant establishment, and that he let Parliament control the militia for 20 years. A breach between Presbyterians in Parliament and Independents in the army is clearly imminent, and Charles hopes to take advantage of differences between his opponents (*see* 1647).

Count Torstenson resigns his command on account of illness and is succeeded by Karl Gustav Wrangel, 33, count of Salmis and Sölvesborg, who joins forces with Königsmark in Westphalia, links up with Turenne at Giessen, and helps lead a Franco-Swedish army into Bavaria.

Nonfiction *Pseudoxia Epidemics (Vulgar Errors)* by Sir Thomas Browne gives careful scrutiny to a number of superstitions and popular delusions.

Poetry *Steps to the Temple* and *The Delights of the Muses* by Richard Crashaw who has gone to France during the English civil war. Crashaw has been converted to Roman Catholicism and expresses religious ecstasy in his *Steps; Poems, with the Tenth Satire of Juvenal Englished* by Welsh poet Henry Vaughan, 24, who includes "The Vanity of Human Wishes," a satire.

Painting *Count Peneranda* by Geraert Terborch, who is attending the peace conference at Münster and painting portraits of the delegates.

England has no hayfields. Grass seed has not been introduced to permit efficient raising of hay for the maintenance of livestock other than sheep.

1647 The Scots surrender England's Charles I to Parliament January 30 in return for £400,000 in back pay, and the king is brought to Holmby House in Northamptonshire as Parliament and the army become openly hostile. The army refuses to accept a Parliamentary act disbanding all soldiers not needed for garrison duty or for service in Ireland, Charles is seized at Holmby House June 4 and taken prisoner by the army, Oliver Cromwell flees Parliament to join the army at Triptow Heath June 4, two speakers of Parliament join the army in July with 14 lords and 100 members of the House of Commons, the army enters London August 6 and forces Parliament to take back Cromwell and the others, and Charles is removed to Hampton Court.

Charles I flees to the Isle of Wight November 11 and is detained by the governor of Carisbrooke Castle. The king signs a secret treaty with the Scots December 26 2 days after receiving four bills from Parliament that would give Parliament more power, bills rejected by Charles December 28 after receiving Scottish promises to restore him by force upon his agreement to abolish Episcopacy and restore Presbyterianism to Scotland.

Marshal Turenne and Sweden's Admiral Wrangel force the Truce of Ulm upon Bavaria's Elector Maximilian in the continuing Thirty Years' War, but Turenne is recalled, Wrangel goes to Bohemia, and Maximilian breaks the truce.

Peter Stuyvesant, 55, is named governor of New Netherlands. The former governor of Curaçao lost his right leg 3 years ago in a campaign against the island of St. Martin (*see* 1625; 1655; New York, 1664).

English nonconformist Sarah Wright reportedly fasts for 53 days (*see* Gandhi, 1914).

An epidemic of yellow fever sweeps the West Indies.

The Black Death strikes Spain in an epidemic that will last for 4 years, the worst since 1599.

A handbook of anatomy with a detailed account of the lacteal vessels is published at Pavia by German anatomist Johann Vesling (*see* Aselli, 1623).

The Society of Friends (Quakers) has its beginnings in the "Friends of Truth" established in Leicestershire by English clergyman George Fox, 23, who begins preaching the need for inward spiritual experience. Troubled by the deadly formalism of Puritan Christianity, Fox extends the philosophy of the Anabaptists, making conscience and self-examination supreme, and he draws recruits from the lower middle classes who heed his protest against the Presbyterian system. The

Friends will be called Quakers in 1650 by Justice Gervase Bennet either because of the Friends' vehemence in appealing to conscience, which makes them shake with emotion, or because they assert that those who do not know quaking and trembling are strangers to the experience of Moses, David, and other saints. Fox will make a missionary journey to Scotland in 1657, the first annual meeting of the Society will be held in 1669, and Fox will preach in Ireland, North America, the West Indies, and Holland until his death in 1691 despite persecution and imprisonment (*see* 1689; Penn, 1681).

1648 Europe's Thirty Years' War ends October 24 in the Peace of Westphalia whose treaties are guaranteed by France and Sweden. The long war leaves the German states destitute. Mercenary troops from Bohemia, Denmark, France, Spain, Sweden, and the German states themselves have destroyed roughly 18,000 villages, 1,500 towns, and 2,000 castles.

The Treaty of Münster recognizes the independence of the Dutch Republic of the United Provinces.

Poland's Ladislas IV has died May 20 at age 55 after a 16-year reign. His Jesuit brother, 39, will reign until 1668 as John II Casimir.

A second English civil war has begun along with an Anglo-Scottish war as Royalists battle Roundheads and Presbyterians battle Independents. A Scottish army invades England under the command of William, 32, Duke of Hamilton, and meets with defeat at the hands of Oliver Cromwell in the Battle of Preston that rages from August 17 to 20.

Parliament has renounced its allegiance to Charles I January 15 following revelations of a secret treaty signed by Charles with the Scots 20 days earlier promising to abolish Episcopacy and restore Presbyterianism. The army seizes Charles December 1, Parliament forcibly excludes 96 Presbyterian members December 6 and 7, and the remaining "Rump Parliament" of some 60 members votes December 13 that Charles be brought to trial (*see* 1649).

A French parliamentary uprising to defend the independence of magistrates is called the *Fronde* (French for sling) because stones are shot into the windows of Cardinal Mazarin in protest against the arrest of the aged magistrate Pierre Broussel. The bishop coadjutor Jean François Paul de Gondi, 35, sides with the insurgents in hopes of becoming prime minister. Broussel is released, he suggests a proclamation urging Parisians to lay down their arms, but the mob wants to get rid of Mazarin. The court takes refuge at Rueil, the Great Condé, who has just gained a great victory over the Spanish at Lens, is recalled to put down the *Fronde*, and the peace of Rueil permits the court to return to Paris (but *see* 1649).

The Janissaries at Constantinople dethrone the sultan Ibrahim August 8 following the lifting of the Ottoman siege of Candia. Ibrahim is strangled by his executioner August 18 and replaced by his eldest son, 9, who will reign until 1687 as Mohammed IV.

Dutch forces in the West Indies take St. Martin and rename it St. Maarten (*see* Stuyvesant, 1647).

Swedish Army officers back from the war receive land grants from Christina. The grants will double the land held by the nobility as of 1611 and freehold peasants will face eviction as the expanding nobility applies German customs and attitudes toward the peasantry (*see* food crisis, 1650).

A Ukrainian pogrom by Greek Orthodox peasants destroys hundreds of Jewish communities, killing all who will not accept the cross. Cossack Bogdan Chmielnicki, 55, leads the pogrom in quest of Ukrainian independence from the Polish nobility, which owns lands along the Dnieper and employs Jews to collect its taxes.

The Dutch ship *Haarlem* breaks up at Table Bay, South Africa. Ship officers Leendert Jansz and Nicholas Proot survive the wreck and are picked up 5 months later and returned to Holland where they urge authorities to establish a settlement at Table Bay for provisioning East India fleets with fresh fruit, vegetables, and other stores (*see* Cape Town, 1652).

The Treaty of Münster (above) closes the River Scheldt to navigation; Amsterdam begins to replace Antwerp as Europe's major commercial city (*see* 1567; transportation, 1815).

Ortus Mediciniae, vel Opera et Opuscula Omnia by the late Flemish physician-chemist Jan-Baptista van Helmont makes the first distinction between gases and air (*see* Boyle, 1662).

Johann Glauber invents nitric acid, which will be used chiefly in explosives (*see* Cavendish, 1766).

Digestion is the work of fermentation that converts food into living flesh, J.B. van Helmont (above) has written. He says that a different ferment causes each physiologic process and that central control is vested in the solar plexus. At a time when most physicians prescribe enormous (and often lethal) doses, van Helmont advocates only small doses of chemicals with mild therapy consisting of diet and simples (medicinal herbs and plants) (*see* 1661).

A yellow fever epidemic sweeps the Yucatán Peninsula. It is by some accounts the first definitely identifiable outbreak (*see* Duterte, 1635).

Poetry *Hesperides, or The Works Both Human and Divine of Robert Herrick, Esq.*, by English poet Robert Herrick, 57, whose Royalist sympathies cause him to be evicted from his post as vicar of Dean Prior in Devonshire: "Gather ye rosebuds while ye may,/Old Times is still a-flying:/And this same flower that smiles today,/ Tomorrow will be dying" (from "To the Virgins, to Make Much of Time").

Painting *The Peace of Münster* by Geraert Terborch (*see* 1646); *Seven Sacraments* (second series) and *Diogenes Throwing Away His Scoop* by Nicolas Poussin; *The Holy Family with St. Catherine* by Jusepe de Ribera. Louis Le Nain dies at Paris May 25 at age 60.

Russia's Czar Alexis Mikhailovich, now 19, abolishes the state monopoly in tobacco established by Michael

1648 ***(cont.)*** Romanov in 1641 and reimposes the ban on smoking (*see* 1613; 1697).

The Taj Mahal completed outside Agra in India is a red and white sandstone and marble mosque, meeting hall, and mausoleum built by the Mughal emperor Shah Jahan, now 56, for his favorite wife Mumtaz Mahal (Jewel of the Palace) who died in childbirth some 17 years ago at age 34 after bearing him 14 children.

Pilgrim colonists in the Massachusetts colony have poor crops and avoid starvation only by eating passenger pigeons, still abundant in the colony despite efforts to eradicate them (*see* Josselyn, 1672).

The end of the Thirty Years' War (above) leaves the German states with a population of less than 13.5 million and possibly a good deal less.

1649 England's Charles I blandly denies the jurisdiction of a high court but is sentenced to death and beheaded January 30 at Whitehall. His son of 18 is proclaimed Charles II at Edinburgh, in parts of Ireland, and in the Channel Islands, but England becomes a republic headed by the Lord Protector Oliver Cromwell whose Commonwealth will rule until 1660.

Oliver Cromwell (above) suppresses an Irish uprising led by the marquis of Ormonde James Butler, 39. He storms Drogheda September 12, sacks the town, massacres its garrison, and does the same to the garrison at Wexford.

France has a new *Fronde* uprising as the Great Condé turns against the court because he thinks Cardinal Mazarin does not accord him proper respect. Mazarin pretends to a reconciliation with the bishop coadjutant (*see* 1648; 1650).

A new tax on Japanese farmers allows them barely enough for their basic needs and forces them to put their wives to work weaving cloth and to send surplus children to work in the city. The Tokugawa shogun Iemitsu requisitions all rice, leaving the farmers left with little to eat but millet.

England's Charles I went to the block, ushering in 11 years of stern Puritan Commonwealth rule.

Black laborers in the Virginia colony still number only 300 (*see* 1619; 1671).

Tobacco exports bring prosperity to the Virginia colony.

Massachusetts entrepreneur John Winthrop, Jr., 43, produces more than 8 tons of iron per week at the Saugus works he has built in back of Lynn with blast furnaces and a refinery forge manned by workers obtained in England. Son of the Bay Colony's first governor, Winthrop has raised £1,000 in England to purchase his equipment.

Nonfiction *The Passions of the Soul (Les passions de l'âme)* by René Descartes.

Poetry "To Lucasta, on Going to the Wars" by Richard Lovelace, who has devoted his fortune to the Royalist cause, was wounded at Dunkirk while serving in the French army, and was imprisoned again upon his return to England: "I could not love thee, dear, so much/ Loved I not honour more."

Painting *The Four Regents of the Leper Hospital* by Ferdinand Bol; *The Vision of St. Paul* by Nicolas Poussin; *Juan de Pereja* (his slave) and *Pope Innocent X* by Diego Velázquez; *Philip IV of Spain* by Geraert Terborch; *Christ Healing the Sick* (etching) by Rembrandt van Rijn.

Opera *Giasone* 1/5 at Venice's San Cassiano, with music by Italian composer Francesco Cavalli, 46, a pupil of the late Claudio Monteverdi, who has changed his name from Caletti-Bruni in honor of his patron Federigi Cavalli, a Venetian nobleman.

The Virginia colony receives an influx of Cavalier (Royalist) refugees from England.

1650 The earl of Montrose returns from the Continent to avenge last year's execution of Charles I but loses part of his Scottish army in a shipwreck and fails to rouse the clans. Surprised and routed at Carbiesdale in Ross-shire, Montrose is betrayed by Neil McLeod of Assynt, sentenced to death by Parliament, and hanged May 21 at Edinburgh.

England's Charles II returns from the Continent, landing in Scotland June 24. He signs the National Covenant and is proclaimed king, but he is defeated September 3 at the Battle of Dunbar by Oliver Cromwell who just 1 month earlier wrote to the General Assembly of the Church of Scotland, "I beseech you, in the bowels of Christ, think it possible you may be mistaken."

Charles II is crowned at Scone despite his defeat at Dunbar and prepares to march on England (*see* 1651). Edinburgh Castle surrenders to Cromwell December 19.

France's Cardinal Mazarin has the Great Condé and his leading associates arrested January 14, Marshal Turenne is persuaded to lead an armed rebellion, the archduke Leopold sends an army from the Spanish Netherlands to aid Turenne, French peasants rise against Leopold's army, he withdraws, and Turenne's *Frondeurs* give way December 15 at the Battle of Blanc-Champ (or Rethel).

William II of Orange dies of smallpox November 6 at age 24, having failed to renew his country's war with Spain. His son and heir is born November 14.

Persia's Abbas II retakes Kandahar, but Mughal emperors will besiege the city repeatedly.

Britain has a typhus epidemic that by one account converts "the whole island into one vast hospital."

Ireland's Archbishop James Ussher, 69, calculates from biblical references that all life was created Sunday, October 23, 4004 B.C.

Nonfiction *Medulla Theologiae Morales* by German Jesuit theologian Hermann Busembaum, 50, preaches the philosophy that "the end justifies the means." Busembaum's work will be condemned for its sections on regicide, and copies will be publicly burned in 1757 by the Parlement of Toulouse.

Poetry *Silex Scintillans* by Henry Vaughan includes his poem "They are all gone into the world of light"; *The Tenth Muse Lately Sprung Up* in America by English colonist Anne Bradstreet, 38, is published at London with metaphysical Puritan poems that include "To My Dear and Loving Husband" and "Upon the Burning of Our House."

Painting *Woman With an Ostrich Fan* and *Jewish Merchant* by Rembrandt van Rijn; *Arcadian Shepherds* and *Self Portrait* by Nicolas Poussin; *View of Dordrecht* by Dutch painter Jan van Goyen, 54; *The Holy Family with the Little Bird* by Bartolomé Murillo.

Theater *Andromeda (Andromede)* by Pierre Corneille at the Théâtre Royal de Bourbon, Paris; *Nicomedes (Nicomède)* by Pierre Corneille at the Théâtre du Marais, Paris.

The minuet is introduced at the French court, where its tempo is slowed somewhat and its gaiety modified.

Sweden suffers a food crisis after the worst harvest she will have in this century. By March the bakers of Stockholm are fighting at the town gates for flour, but while the nation's sociopolitical balance is jeopardized as the clergy and burghers side with the peasants, a threatened revolution does not materialize.

Maize is eaten for the first time in Italy where "turkey corn" *(mais)* will be popular in *polenta,* in cornmeal mush, and hardened as a cake (*see* pellagra, 1749).

England's first coffee house opens at Oxford (*see* London, 1652).

London's population reaches 350,000, up from 210,000 in 1603, and the city contains 7 percent of the English population (*see* 1660).

Denmark will lose more than a fifth of its population in the next decade to disease and starvation.

Ireland will lose close to a quarter of its population in the next decade, declining from 1.3 million to less than 1 million as a result of ruinous wars, anti-Catholic penal laws, and laws that destroy the security of land-tenure and put Irishmen at the mercy of absentee landlords.

Africa occupies just over 20 percent of the earth's land surface and has roughly 20 percent of the world's population, but European slave traders in this century and the next will decimate the continent by exporting human chattels and introducing new diseases.

1651 Charles II is crowned at Scone January 1, but Oliver Cromwell takes Perth in early August and defeats royalist forces September 3 at the Battle of Worcester. Disguised as a servant to the daughter of a royalist squire, Charles escapes to France October 17 after traveling through a countryside alive with Roundheads (*see* 1650; 1660).

France's Parlement gains reluctant consent from the queen mother Anne of Austria in February to dismiss Cardinal Mazarin and release the Great Condé. Mazarin flees the country and Pope Innocent X makes his foe Jean François Paul de Gondi the Cardinal de Retz. Mazarin returns in December with 7,000 German troops to put down the rebellion led by the Great Condé, who has won support from Marshal Turenne (*see* 1652).

Leviathan, or "The Matter, Form and Power of a Commonwealth, Ecclesiastical and Civil," by English philosopher Thomas Hobbes, 63, says that people to survive must surrender their individual rights and submit to an absolute sovereign whose duty is to protect them from outside enemies much as a feudal lord protected his vassals:

"Of the Naturall Condition of Mankind, As Concerning Their Felicity, and Misery, [writes Hobbes (above),] Whatsoever therefore is consequent to a time of Warre, where every man is Enemy to every man; the same is consequent to the time, wherein men live without other security, than what their own strength, and their own invention shall furnish them withall. In such condition there is no place for Industry; because the fruit thereof is uncertain; and consequently no Culture of the Earth, no Navigation, nor use of the commodious Building; no Instruments of moving, and removing such things as require much force; no Knowledge of the face of the Earth; no account of Time; no Arts; no Letters; no Society; and which is worst of all, continuall feare, and danger of violent death; And the life of man, solitary, poore, nasty, brutish, and short."

The Japanese shogun Iemitsu dies at age 47 after a 28-year reign that has consolidated Tokugawa rule through national isolationism, oppression of the people, and suppression of Christianity. His son Ietsuna, 10, will reign until 1680, exhausting the treasury and debasing the nation's coinage.

The Navigation Act passed by Parliament October 9 forbids importation of goods into England or her colonies except by English vessels or by vessels of the countries producing the goods. The mercantilist legislation aims to help the nation's merchant marine gain supremacy over the Dutch (*see* 1660; Anglo-Dutch war, 1652).

1651 ***(cont.)*** *Exercitationes de generations animalium* by William Harvey of 1628 blood circulation theory fame is published at London. Now 73, Harvey was physician to Charles I until 1648 and he says, "All animals, even those that produce their young alive, including man himself, are evolved out of the egg," but he thinks that lower forms of life are capable of primal generation and believes the pupa to be an insect egg (*see* Redi, 1668).

Nonfiction *Jacula Prudentum* by the late George Herbert contains translations of proverbs that include "The eye is bigger than the belly"; "His bark is worse than his bite"; "Whose house is of glass, must not throw stones at another"; "God's mill grinds slow, but sure"; "For want of a nail, the shoe is lost, for want of a shoe the horse is lost, for want of a horse the rider is lost"; "One half the world knows not how the other half lives"; "He that lies with the dogs, riseth with fleas"; and "One hour's sleep before midnight is worth three after."

Painting *The Holy Family* by Nicolas Poussin; *Girl with a Broom* by Rembrandt van Rijn; *The Marriage of the Artist* by David Teniers the Younger, who moves to Brussels and becomes court painter to Archduke Leopold Wilhelm; *The Institution of the Eucharist* and *The Last Supper* by Jusepe de Ribera.

Opera *Calisto* at Venice, with a libretto about love among the gods, music by Francesco Cavalli.

"Rumbullion, alias Kill-Devill" is the "chief fudling they make in the Island" of Barbados, writes Richard Ligon. He calls rum "a hot, hellish and terrible liquor" made of "suggar canes, distilled" (*see* 1641; 1655).

1652 The Battle of Bleneau April 7 brings victory to the Great Condé over Marshal Turenne, who has switched back to the king's side (*see* 1651). Turenne forces the new *Fronde* army to break off, and both armies march to Paris to negotiate. The archduke Leopold takes more fortresses in Flanders, the duc de Lorraine marches to join Condé (his mercenaries plunder Champagne en route), Turenne intercepts Lorraine, buys him off, and hems in the *Frondeurs* July 2 in the Faubourg St. Antoine outside Paris. Anne Marie Louise d'Orléans, 25, duchesse de Montpensier, persuades Parisians to open the city gates to the *Fronde* army (which has her father's support), she turns the guns of the Bastille on Turenne's royal forces, an insurrectionist government is proclaimed, Cardinal Mazarin flees France, but the Parisian bourgeoisie quarrels with the *Fronde* and permits Louis XIV to enter the city October 21 (*see* 1653).

An Anglo-Dutch War precipitated by last year's English Navigation Act is proclaimed July 8 after the English have gained an initial victory in May off the Downs.

England's "Rump Parliament" antagonizes the army with an Act of Indemnity and Oblivion. The army charges members of Parliament with having received bribes from royalists whose estates were confiscated.

The new Japanese shogun Ietsuna survives the second of two attempted coups at Edo in the last challenge to its sovereignty that the Tokugawa family will face until the 19th century.

Cape Town, South Africa, is founded by Dutch ship's surgeon Jan van Riebeck who goes ashore at Table Bay with 70 men carrying seeds, agricultural implements, and building materials (*see* 1648).

The General Court of Massachusetts makes a 35-year contract with Boston silversmith John Hull to produce silver coins with silver to be supplied by him. Hull is appointed mint master of the colony and strikes the Pine Tree shilling, receiving 15 pence for every 20 shillings he coins.

Nonfiction *O-Dai-khi-Ran* by Japanese historian Shunsai Hayashi, 34, is a history of Japan; *The Mount Of Olives, or Solitary Devotions*, by Henry Vaughan.

Poetry *Deo Nostro Te Decet Hymnus* by the late English poet Richard Crashaw, who died 3 years ago at age 36 after being named canon of a church at Loreto. Crashaw's posthumously published work includes the hymn to St. Theresa "The Flaming Heart" and is illustrated with 12 engravings by the baroque poet.

Painting *Portrait of Hendrickje* by Rembrandt van Rijn; *A View of Delft* by Dutch painter Carel Fabritius (Carel Pietersz), 30. Georges de La Tour dies at his native Vic-sur-Seille January 30 at age 58; Jusepe de Ribera dies at Naples September 2 at age 61.

Regents of the Japanese shogun Ietsuna prohibit young boys from taking roles in the Kabuki theater. Mature men will hereafter play all roles (*see* 1629; 1673).

Beijing's Great White Dagoba is completed in the shape of a Buddhist reliquary. The structure's base, spire, crown, and gilded ball represent the five elements earth, water, fire, air, and ether.

Massachusetts colonist Joseph Russell starts an offshore whaling enterprise. Russell has founded New Bedford (*see* 1775; Nantucket, 1690).

London's first coffee house opens in St. Michael's Alley, Cornhill, under Armenian management.

The population of the Virginia colony reaches 20,000 including white settlers and black slaves.

1653 Amoy falls to the Chinese pirate-patriot Zheng Chenkong, 30 (the Portuguese call him Koxinga), who has ravaged China's coast with a fleet of 3,000 junks in a continuing fight against the Manchus who have come to power with the end of the Ming dynasty (*see* 1644; 1661).

France's Cardinal Mazarin returns to Paris unopposed, the *Fronde* is ended, and the suppression of the rebellion begins a golden age and a period of absolutism.

Oliver Cromwell is proclaimed Lord Protector of the Commonwealth of England, Scotland, and Ireland December 15. *The Instrument of Government* drawn up for the Commonwealth is a written constitution that calls for a standing army of 30,000, a cooperative council of 21 to help the Lord Protector administer the Commonwealth, and a triennial Parliament of 460 members with sole power to levy taxes and grant funds and a guarantee against dissolution for 5 months once it has been summoned.

Nieuw Amsterdam colonists build a wall across Manhattan from the North River to the East River for protection against English attacks. Wall Street will get its name from the defensive wall.

Virginia settlers enter the region north of the Roanoke River and found the Albemarle settlement in what will be called North Carolina (*see* 1663).

The tontine system of life insurance, devised at Paris by Neapolitan banker Lorenzo Tonti, is a scheme whereby a group of investors buys shares in a fund whose proceeds go to that shareholder who survives the others (*see* 1583).

Danish physician Thomas Bartholin, 37, describes the lymphatic system, enlarging on his father's 1611 *Institutiones Anatomicae* and defending William Harvey's 1628 theory of blood circulation. Swedish scientist Olof Rudbeck, 23, has discovered the system independently of Bartholin and publishes his observations.

Painting *Aristotle Contemplating the Bust of Homer* by Rembrandt van Rijn; *The Picture Gallery of Archduke Leopold Wilhelm* by David Teniers, the Younger; *The Village Wedding (The Jewish Bride)* by Dutch painter Jan Havicksz Steen, 27.

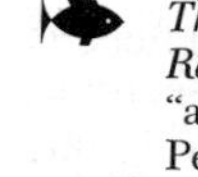

The Compleat Angler, or The Contemplative Man's Recreation by English biographer Izaak Walton, 60, is "a Discourse on Fish and Fishing, Not Unworth the Perusal of Most Anglers." A former ironmonger, Walton writes of trout, "if he is not eaten within four or five hours after he is taken [he] is worth nothing."

1654

The Treaty of Westminster April 6 ends the Anglo-Dutch War that started in 1652. The Dutch agree to recognize the English Navigation Act of 1651 and to pay an indemnity, they agree to join England in a defensive league, and the province of Holland agrees secretly to exclude members of the house of Orange from the stadtholdership in deference to Oliver Cromwell's anxieties over the marriage of Mary, daughter of England's late Charles I, to Willem II of Orange.

Sweden's Queen Christina abdicates June 6 after a 22-year reign in which she has sold or mortgaged vast amounts of crown property to support the 17 counts, 46 barons, and 428 lesser nobles that she has created. Christina leaves dressed in male attire under the name Count Dohna, she joins the Catholic Church, she settles at Rome, and she is succeeded by her cousin, 32, who will reign until 1660 as Charles X Gustavus.

The Ukrainian cossack Bogdan Chmielnicki swears allegiance to Russia's Alexis I Mikhailovich and gives up the Ukraine's aspirations to independence (*see* 1648). Russian troops seize Smolensk, beginning a 13-year war over the Ukraine between Russia and Poland. The war will bring the Russians into contact for the first time with the Turks in the Balkans.

Oliver Cromwell quarrels with Parliament September 3 and 9 days later orders the exclusion of hostile members of Parliament.

Portugal recovers the Brazilian territory taken by the Dutch in 1635.

Jesuit missionary Simon LeMoyne visits the area in upper New Netherlands that will later be the site of Syracuse (*see* Webster, 1786). He finds salt deposits that will be important to the development of the area.

The Black Death strikes eastern Europe.

Painting *Jan Stix* by Rembrandt van Rijn; *Parental Admonition* by Geraert Terborch. Rembrandt's prize pupil Carel Fabritius is killed October 12 at age 32 in an explosion at the Delft arsenal.

Sugar cane is planted by the French in Martinique which will become a major Caribbean sugar producer.

1655

Swedish forces invade Poland as Charles X Gustavus takes advantage of Poland's struggle to save her Ukrainian territories from Russia. Charles launches the first Northern War that will take away Poland's last Baltic territories (*see* 1660).

English forces under the command of Vice Admiral William Penn, 34, take Jamaica in the West Indies from the Spanish who have called the sugar-rich island San Iago. Penn's action precipitates a 3-year war with Spain.

Dutch colonists occupy New Sweden on orders from Peter Stuyvesant, governor of New Netherlands (*see* 1638).

Sephardic Jews from Pernambuco, Brazil, establish a congregation at Nieuw Amsterdam despite efforts by Peter Stuyvesant (above) to exclude them (*see* 1642).

Oliver Cromwell suppresses an uprising against the Puritan government at Salisbury and divides England into 12 military districts, each controlled by a force that is financed by a 10 percent tax on royalist estates. The Puritans order Catholic priests to leave the realm, forbid Anglican clergymen to teach or preach, censor the press, and impose rigid ("puritanical") rules.

Nonfiction *The History of Scotland 1423–1542* by the late Scottish poet-historian William Drummond, who died in 1649 at age 64.

Poetry "A Panegyric to My Lord Protector" by Edmund Waller, who has made amends to Oliver Cromwell and has returned from France.

Painting *Woman Bathing in a Stream* and *The Rabbi* by Rembrandt van Rijn.

French society uses a clean plate for each new dish, but Englishmen continue to dine off trenchers—wooden platters that give hearty eaters the name "trenchermen."

Rum from Jamaica (above) is introduced into the Royal Navy to replace beer, which goes sour after a few weeks at sea (*see* 1651; 1731).

The first known reference to the use of a sheath for contraception is made in an anonymous Parisian publication. *L'Ecole Des Filles* recommends a linen sheath to prevent passage of semen into the uterus, but while contraception will be widely employed among the upper-class French by the end of the century, intrauterine sponges will be favored over condoms (*see* Mme. de Sévigné, 1671).

1656 *The Commonwealth of Oceana* by English political theorist James Harrington, 45, takes the form of a political romance but is intended as a reply to Thomas Hobbes's *Leviathan* of 1651. The government, says Harrington, is certain to reflect a social system in which the bulk of the land is owned by the gentry rather than by the king and the Church as in ages past: "The law is but words and paper without the hands and swords of men."

England and Spain go to war. The English capture some Spanish treasure ships off Cadiz September 9 (*see* 1657; Jamaica, 1655).

Oliver Cromwell's Third Parliament convenes September 17. England has had no Parliament since January of last year.

Swedish forces invade Poland in the First Northern War. The army of Charles X Gustavus defeats the Poles at the Battle of Warsaw whereupon Russia, Denmark, and the Holy Roman Empire declare war on Sweden which is deserted by her ally Brandenburg. Poland recognizes the sovereignty of the Elector of Brandenburg over East Prussia.

Portugal's João IV dies November 6 at age 53 after a reign of nearly 16 years. He is succeeded by his son, 13, who has been paralyzed since age 3 but will reign until 1667 as Afonso VI, the second Braganza king.

Dutch forces take Colombo from the Portuguese.

The Amsterdam synagogue excommunicates rabbinical student Baruch Spinoza, 24, in July for views the elders consider heretical. Spinoza turns to grinding lenses in order to support himself (*see* 1670).

Dutch East India Company shares plummet on the Amsterdam Exchange, and many investors are ruined. Among them is painter Rembrandt van Rijn, now 50, who is declared bankrupt and whose possessions are put up for sale.

The Dutch in Ceylon make cinnamon a state monopoly.

Dutch mathematician-physicist-astronomer Christian Huygens, 27, revolutionizes clockmaking with a clock regulated by a pendulum. He applies a concept that occurred to Galileo in 1583 while watching a lamp swinging from a long chain in Pisa Cathedral (*see* Clement, 1670).

Olof Rudbeck returns to Uppsala after studying at Leyden, is appointed professor of anatomy, and builds an anatomical theater in which he performs dissections on human bodies, scorning criticism of the practice which is new to Uppsala (*see* 1653).

English physician Thomas Wharton, 42, describes the anatomy of glands. He has discovered the duct (Wharton's duct) from the submaxillary gland to the mouth.

The Hôpital Général opens at Paris. It is a combination hospital, poorhouse, and factory.

Painting *The Maids of Honor (Las Meninas)* and *The Spinners (Las Hilanderas)* by Diego Velázquez (the first depicts the family of Spain's Philip IV); *The Procuress* by Dutch painter Jan Vermeer, 23. Gerrit van Honthorst dies at Utrecht April 27 at age 65; Jan van Goyen dies at the Hague April 27 at age 60.

The Corsa al Palio held at the Italian city of Siena begins a July-August horserace event that will annually pit competitors from each of the city's *contrades* against each other.

1657 The Holy Roman Emperor Ferdinand III concludes an alliance with Poland to check the aggressions of Sweden's Charles X Gustavus, who is driven out of Poland. The emperor dies April 2 at age 48 and is succeeded by his son, 16, who will reign until 1704 as Leopold I.

Sweden and Denmark go to war as Charles X Gustavus tries to extend his holdings on the southern coast of the Baltic. The Dutch intervene to prevent the Swedes from gaining exclusive control of the Baltic fishery.

The Ottoman Turks retake Lemnos and Tenedos. A 4-year Dutch-Portuguese war begins over conflicting interests in Brazil.

Indian bandit Sivaji, a tax collector's son, raids Mughal territory in the northwest Deccan area, a small army under Afzal Khan is sent to deal with him, Sivaji calls for peace talks, and then murders Afzal Khan (*see* 1664).

An English naval force commanded by Admiral Robert Blake, 58, destroys the Spanish West Indian fleet April 20 off Santa Cruz.

Oliver Cromwell tacitly permits Jews to return to England (*see* 1306; Bevis Marks synagogue, 1701).

The Flushing Remonstrance written to Nieuw Amsterdam's Peter Stuyvesant December 27 is probably the first declaration of religious tolerance by any group of ordinary citizens in America. Townsmen of the Long Island settlement of Flushing tell Stuyvesant they will not accept his command that they "not receive any of those people called Quakers because they are supposed to be, by some, seducers of the people." Say the Flushing burghers, "We desire in this case not to judge lest we be condemned, but rather to let every man stand and fall to his own Master." But five Quakers who have arrived at Nieuw Amsterdam are shipped off to Rhode Island.

Coffee advertisements at London claim the beverage is a panacea for scurvy, gout, and other ills.

Public sale of tea begins at London as the East India Company undercuts Dutch prices and advertises tea as a panacea for apoplexy, catarrh, colic, consumption, drowsiness, epilepsy, gallstones, lethargy, migraine, paralysis, and vertigo (*see* Garraway, below).

England's Durham University has its beginnings (*see* 1833).

Painting *The Birth of Bacchus* by Nicolas Poussin; *L'Umana Fragilita* by Salvator Rosa.

Fire destroys most of Edo and its castle buildings January 18 to 19, killing more than 100,000 Japanese (*see* 1601; 1772).

The first London chocolate shop opens to sell a drink known until now only to the nobility (*see* 1656; 1659).

Tea is offered to Londoners at Thomas Garraway's coffee house in Exchange Alley between Cornhill and Lombard streets.

1658 Sweden's Charles X Gustavus makes two invasions of Denmark but fails to take Copenhagen whose garrison puts up a valiant defense.

The Battle of the Dunes June 4 gives English and French troops a victory over a Spanish relief force, and Dunkirk surrenders to the English after a siege.

Oliver Cromwell dies September 3 at age 58 after nearly 5 years as Lord Protector of the Commonwealth of England, Scotland, and Ireland. His son Richard, 31, succeeds to power and will retain the protectorship for nearly 9 months.

The Mughal emperor Shah Jahan falls ill at age 66 after a 31-year reign that has raised the power of the empire to its zenith and seen the growth of Delhi. The emperor's son Aurangzeb, 40, kills Shah Jahan's favorite son Dara Shikoh, will make himself emperor next year, and will reign until 1707, alienating Muslims and Hindus alike with his bigotry. Aurangzeb's continuous campaigns in the Deccan will reduce the country's population by close to 100,000 per year.

Salisbury, Massachusetts, colonist Thomas Macy receives sanctuary from Puritan religious intolerance on the offshore island of Nanticut (Nantucket). Algonquins on the island welcome Macy, his family, and his friend Edward Starbuck (*see* 1605; 1659).

French fur trader Pierre Esprit Radisson, 22, explores the western end of Lake Superior and trades with *les sauvages* in company with Médard Chouart, 33, Sieur de Groseillers (*see* Rupert House, 1668). Radisson came to New France 7 years ago, was captured and adopted by the Iroquois, but escaped in 1654 (*see* 1659).

English coal production at Newcastle reaches 529,032 tons, up from 32,951 in 1564, as the deforestation of England and the Continent spurs use of coal for fuel (*see* 1640; Newcomen, 1705).

Dutch naturalist Jan Swammerdam, 21, gives the first description of red blood cells.

Essays "Urne-Buriall" by Thomas Browne: "There is nothing strictly immortal, but immortality; whatever hath no beginning may be confident of no end" (V); "Hydriotaphia" by Browne; Browne's mystical treatise "The Garden of Cyprus" shows how the universe is pervaded by the "quincunx," an arrangement of five objects such as trees or buildings with one at each corner and one in the middle (*see* witches, 1664).

Painting *Courtyard of a House in Delft* by Dutch painter Pieter de Hooch, 29; *Farm with a Dead Tree* by Dutch painter Adriaen van de Velde, 27.

The London periodical *Mercurious Politicus* carries an advertisement: "That excellent and by all Physitians approved China Drink called by the Chineans Tcha, by other nations Tay, alias Tea, is sold at the Sultaness Head, a cophee-house in Sweetings Rents, by the Royal Exchange, London."

An English cookbook sanctions use of veal as food in defiance of an old Saxon tradition that had considered the killing of a veal calf a wanton act of the Norman invaders of 1066.

1659 A new English Parliament meets January 27 but soon bogs down in a dispute with the army. The new lord protector Richard Cromwell dissolves Parliament April 22 at the behest of the army, a "Rump Parliament" meets May 7 with William Lenthall as speaker and induces Cromwell to resign, the New Royalists led by George Booth, 37, join with the Cavaliers in August to restore the monarchy, the army commanded by John Lambert, 40, suppresses the insurrection in Cheshire and appoints a military committee of safety to replace the "Rump Parliament" in October, but the "Rump Parliament" is restored December 26 (*see* 1660).

The Treaty of the Pyrenees November 7 ends the ascendancy of Spain which has been exhausted by war and by domestic misgovernment that has produced the revolt in Catalonia which ends this year. Under terms of the treaty signed on the Isle of Pheasants in the Bidassoa River France's Louis XIV receives the Spanish frontier fortresses in Flanders and Artois, Spain's Philip IV cedes part of Rousillon, Contans, and Cerdagne along with some towns in Hainault and Luxembourg, he makes other concessions, and the Spanish infanta Maria Theresa, 21, is betrothed to Louis XIV with a dowry of 500,000 crowns (she renounces her claims upon her inheritance for herself or for any issue she may have by Louis).

French fur trader Pierre Radisson travels through the upper Mississippi Valley.

Thomas Macy dispatches Edward Starbuck to the mainland to recruit new settlers for Nantucket Island (*see* 1658). Starbuck returns with nine, the settlers and Macy each take partners, and together they buy 90 percent of the island—including its Indian inhabitants—from owner Thomas Mayew in the Massachusetts colony (*see* whaling, 1690).

English physician Thomas Willis, 38, gives the first description of typhoid fever.

Painting *Infanta Maria Theresa* (above) and *Infante Philip Prosper* by Diego Velázquez; *Young Girl with Flute* by Jan Vermeer; *The Letter* by Geraert Terborch.

Theater *Oedipus (Oedipe)* by Pierre Corneille 1/24 at the Hôtel de Bourgogne, Paris; *The Amorous Quarrel (Le dépit concoureux)* by the French actor-playwright Molière (Jean Baptiste Poquelin), 37, 4/16 at the Théâtre du Petit-Bourbon, Paris; *The Affected Ladies (Les précieuses ridicules)* by Molière 11/18 at the Théâtre du Petit-Bourbon: manager of a troupe known as the King's Comedians, Molière formed his own acting company at age 21 and last year gained the patronage of the court at Paris for his comedies.

Paris police raid a monastery and send 12 monks to jail for eating meat and drinking wine during Lent.

The Spanish infanta Maria Theresa (above) brings cocoa to France where it will be endorsed by the Paris

1659 ***(cont.)*** faculty of medicine and will be received with enthusiasm until it becomes surrounded with suspicion as an aphrodisiac in some circles and as a mysterious potion in others. (Mme. de Sévigné will write to her daughter that "the marquise de Coetlogon took so much chocolate, being pregnant last year, that she was brought to bed of a little boy who was as black as the devil.")

1660 Sweden's Charles X Gustavus dies the night of February 12 at age 37 and is succeeded by his son, now 4, who will take over from a corrupt regency in 1672 and will reign until 1697 as Charles XI. The Treaty of Copenhagen brings peace between Sweden and Denmark which gives up the southern part of the Scandinavian Peninsula but retains Bornholm and Trondheim. The Treaty of Oliva May 3 ends the 5-year-old Northern War, Poland's John II Casimir abandons his claims to the Swedish throne, he cedes Livonia to Sweden, and Poland loses her last Baltic territories.

The Transylvanian prince George II Rákóczi, who tried to depose Poland's John II Casimir 3 years ago, dies in May at age 38 from wounds sustained while fighting the Ottoman Turks in the Battle of Gyula.

England's civil war ends May 8 after 11 years as the son of the late Charles I is proclaimed king. Now 29, he lands at Dover May 26, arrives at Whitehall May 29 amidst universal rejoicing, and will reign until 1685 as Charles II.

Africa's Bambara kingdoms of Segu and Kaarta on the upper Niger begin their rise against the Mandingo Empire which they will replace in 1670.

Charles I (above) acts to strengthen England's Navigation Act October 1: certain "enumerated articles" from England's American colonies may be exported only to the British Isles. Included are tobacco, sugar, wool, indigo, and apples. The list will be amended to include rice, molasses, and other articles, and Virginia tobacco prices take a precipitous drop as transport bottlenecks delay shipments, producing widespread economic distress and political unrest in the colony (*see* 1663).

Charles II restored the English monarchy and reigned for nearly 25 years through war, plague, and fire.

German woodcarvers in the Black Forest town of Fürtwangen create clockworks made entirely of wood. They have invented clocks from which wooden cuckoos appear periodically to sound the hours, half-hours, and quarter-hours.

New Experiments Physics-Mechanical Touching the Spring of the Air and its Effects by Oxford chemist Robert Boyle, 33, is published (*see* 1661).

Poetry *Satires* by French poet Nicolas Boileau-Despréaux, 23, who antagonizes many bad writers by naming them.

Painting *Maidservant Pouring Milk* and *View of Delft* by Jan Vermeer; *The Morass* by Dutch painter Jacob van Ruysdael, 32; *St. Peter Denying Christ* by Rembrandt van Rijn. Diego Velázquez dies at Madrid August 6 at age 61.

Theater *The Golden Fleece (La toison d'or)* by Pierre Corneille in January at the Théâtre du Marais, Paris.

Francesco Borromini completes Rome's Church of S. Ivo della Sapienza.

Royal Navy official Samuel Pepys, 27, at London notes in his secret diary that he has drunk a "cup of tee (a China drink) of which I never had drank before."

London's population rises to roughly 450,000, up from 350,000 in 1650 (*see* 1780, 1801).

1661 Cardinal Mazarin dies at Paris March 9 at age 58; Louis XIV, now 22, assumes full power March 10.

China's Manchu regime decrees that populations within 10 miles of the coast be evacuated to points inland. The regime struggles to repel incursions by the pirate Zheng Chenkong, who is besieging the Dutch settlement at Fort Zelandia on Taiwan (Formosa) where he will establish a Chinese government next year that will resist the Manchus until Zheng's son Zheng Kuoshang surrenders in 1683.

The Manchu emperor Shun Chih dies after a 16-year reign that has launched the Qing (Ch'ing) dynasty (*see* 1662).

The Sceptical Chymist by Robert Boyle discards the Aristotelian theory that there are only four basic elements (earth, air, fire, and water) and proposes an experimental theory of the elements (*see* 1660). Boyle will be called the "father of chemistry," but he holds views that will encounter skepticism from later chemists, e.g., that plant life grows by transmutation of water, as do worms and insects since they are produced from the decay of plants (*see* Redi, 1668; Ingenhousz, 1779; Boyle's law, 1662).

Iatrochemical medical theories are consolidated as a systematic school by Prussian-born medical professor Franz de la Boe, 47, who calls himself Sylvius and teaches at Leyden. Sylvius propounds original ideas on the ductless glands, on the tactile senses, on acidosis, and he says digestion is a chemical fermentation (*see* van Helmont, 1648; Prout, 1824).

Theater *The School for Husbands (L'Ecole des maris)* by Molière 6/24 at the Palais-Royal, Paris; *The Bores (Les fâcheux)* by Molière 11/4 at the Palais Royal.

London's Covent Garden becomes a market for fruit, vegetables, and flowers (*see* 1552; theater, 1732).

1662 England's Charles II marries the Portuguese princess Catherine da Braganza, 23, who provides Charles with £300,000 in sugar, cash, and Brazilian mahogany plus the port of Tangier, the island of Bombay, and valuable trading privileges for English mariners in the New World. The May 20 marriage begins a lasting alliance between Portugal and England.

England sells Dunkirk to France for £400,000.

Holland and France form an alliance against possible attack by England.

Connecticut colonies are granted an unusually democratic charter by England's Charles II (*see* 1689).

Wampanoag Indian chief Massasoit dies and is succeeded by his son Metacum (*see* 1621; 1675).

China's Manchu emperor Shun Chih, who died last year, is succeeded by his son Hsuan Yeh, who will reign until 1722 as K'ang Hsi. Now 8, the new Qing dynasty emperor will begin his personal rule in 1667, ushering in a period of cultural achievement that will surpass the greatest achievements of earlier dynasties. Jesuit scholar-missionaries will be encouraged to bring their scientific knowledge to China.

The Royal Society for the Improvement of Natural Knowledge is chartered at London.

Boyle's Law (the volume of a gas varies inversely as the pressure varies) is enunciated by Robert Boyle, who has helped found the Royal Society (above; *see* 1661; van Helmont, 1648).

An Act of Uniformity passed by Parliament requires that England's college fellows, schoolfellows, schoolmasters, and clergymen accept everything in a newly published *Book of Common Prayer.* Those who refuse will hereafter be called "Nonconformists."

The Book of Common Prayer contains "A General Confession": "We have left undone those things which we ought to have done;/And we have done those things which we ought not to have done."

The Day of Doom, or A Poetical Description of the Great and Last Judgement, by Malden, Massachusetts, clergyman Michael Wigglesworth, 31, contains a lurid exposition of Calvinist theology that children of the Massachusetts colony will be required to memorize.

Painting *The Syndics of the Cloth Guild* by Rembrandt van Rijn.

Theater *Sertorius* by Pierre Corneille in February at the Théâtre du Marais, Paris; *The School for Wives (L'Ecole des femmes)* by Molière 12/26 at the Palais Royal, Paris.

Catherine da Braganza (above) introduces to the London court the Lisbon fashion of drinking tea; she also introduces the Chinese orange (*see* 1635).

1663 Louis XIV renews French rights to enlist Swiss mercenaries through an alliance obtained over the objection of Zurich and some of the Protestant cantons.

England's Charles II grants Carolina territory from Virginia south to Florida as a reward to eight courtiers who have aided him in his restoration. The March 24 grant includes land between 31° and 36° North latitude, and the grantees include the lord chancellor Edward Hyde, 54, first earl of Clarendon (*see* Albemarle, 1653; Charleston, 1672).

Charles II grants the Charter of Rhode Island and Providence Plantations July 8. It will remain the constitution until 1842.

A Second Navigation Act passed by Parliament July 27 forbids English colonists to trade with other European countries. European goods bound for America must be unloaded at English ports and reshipped, even though English export duties and profits to middlemen may make prices prohibitive in America (*see* 1660s; 1672).

Parliament increases the domestic price level at which grain exports may be halted in a move that serves the interest of landowners but not that of townspeople.

Jean Baptiste Colbert, 44, works to reform the finances of the world's leading nation; he will be named controller-general of French finance in 1665.

An epidemic of the Black Death kills 10,000 at Amsterdam out of the city's 200,000 people (*see* 1664).

Painting *Artist and Model* by Jan Vermeer; *Young Woman with a Water Jug* by Jan Vermeer; *Jacob and Laban* by Adriaen van de Velde.

Theater *Sophonisbe* by Pierre Corneille in January at the Hôtel de Bourgogne, Paris; *The Wild Gallant* by English playwright-poet John Dryden, 32, in February at London's Theatre Royal in Vere Street. Dryden marries the sister of Sir Robert Howard.

England's Royal Society urges that potatoes be planted to provide food in case of famine.

1664 Nieuw Amsterdam becomes New York August 27 as 300 English soldiers under Col. Mathias Nicolls take the town from the Dutch under orders from Charles II (*see* 1626). The town is renamed after the king's brother James, duke of York, who is granted the territory of New Netherland, including eastern Maine and islands to the south and west of Cape Cod, which England claims on the basis of John Cabot's explorations in the late 1490s. Hudson Valley Dutch *patroons* become English landlords.

The Duke of York (above) has granted land between the Hudson and Delaware rivers June 24 to John Berkeley, first baron Berkeley of Stratton, and Sir George Carteret, 54, formerly governor of the Isle of Jersey and now treasurer of the Royal Navy (*see* 1665).

Sivaji sacks Surat (*see* 1657, 1667).

1664 ***(cont.)*** English seamen take Africa's Cape Verde Islands from Dutch forces in Guinea although no war has been declared.

Two English women are condemned as witches on professional evidence by Norwich physician-author Thomas Browne of 1658 Urne-Buriall fame.

Traité de l'homme et de la formation du foetus by René Descartes states unequivocally that blood in the body is in a state of perpetual circulation (*see* science, 1619; Harvey, 1628).

Jan Swammerdam discovers the valves of the lymph vessels (*see* 1658; 1669; Bartholin, Rudbeck, 1653).

The Black Death kills 24,000 in old Amsterdam while the English are taking Nieuw Amsterdam (above). The plague spreads to Brussels and throughout much of Flanders, and in December it kills two Frenchmen in London's Drury Lane (*see* 1663; 1665). Men who put the dead into the deadcarts keep their pipes lit in the belief, now widespread, that tobacco smokers will be spared.

Japanese merchants establish express mail service between Edo and Osaka. The three-times-per-month service takes 6 days as compared with 30 for ordinary mail.

The Compleat Gamester by English poet Charles Cotten, 34, is more popular than his burlesque of Virgil which is also published (*see* Hoyle, 1742).

Painting *The Travellers* by Dutch landscape painter Meindert Hobbema, 25; *Young Woman Weighing Gold* by Pieter de Hooch; *The Christening Feast* by Jan Steen; *The Lacemaker* by Jan Vermeer; *Apollo and Daphne* by Nicolas Poussin, now over 70. Francisco de Zurbarán dies at Madrid August 27 at age 65.

Theater *The Indian Queen* by John Dryden and Sir Robert Howard in January at London's Theatre Royal in Bridges Street; *The Forced Marriage (Le mariage forcé)* by Molière 1/29 at the Palais-Royal, Paris; *The Comical Revenge, or Love in a Tub* by English playwright George Etherege, 29, in March at the Lincoln's Inn Fields Theatre, London; *The Royal Ladies* by John Dryden in June at London's Theatre Royal in Bridges Street; *The Thebans, or The Enemy Brothers (La Thébaïde, ou les frères ennemis)* by French playwright Jean Baptiste Racine, 25, 6/20 at the Palais-Royal, Paris, with Jean Baptiste Molière's company; *Othon* by Pierre Corneille 7/31 at the Hôtel de Bourgogne, Paris.

The French horn becomes an orchestral instrument.

Oratorio *Christmas Oratorio* by Heinrich Schütz at Dresden.

1665 English naval forces defeat a Dutch fleet off Lowestoft June 3 as a second Anglo-Dutch war begins.

Spain's Philip IV dies September 17 at age 60 after a weak reign and is succeeded by his son Don Carlos, now 4 and nearly crippled with rickets, who will reign until 1700 as Charles II, last of the Spanish Hapsburgs.

The New Haven colony accedes reluctantly to join the Connecticut colony rather than be taken over by New York (*see* 1638; 1662).

The New Jersey colony is founded by English colonists under the leadership of Philip Carteret, son of Sir George (*see* 1664). They settle at Elizabethtown and make it their capital with Carteret as governor (*see* Newark, 1666).

Compagnie Saint-Gobain is founded by royal decree to make mirrors for France's Louis XIV. It will become Europe's largest glass maker.

London has its last large outbreak of the Black Death which is introduced either by Dutch prisoners of war or in bales of merchandise from Holland that originated in the Levant. Some two-thirds of London's 460,000 inhabitants leave town to avoid contagion, but at least 68,596 die plus a few thousand at Norwich, Newcastle, Portsmouth, Southampton, and Sunderland, most of them poor people who are imprisoned in their houses (which are marked by large red crosses) and given food handed in by constables.

"Lord! how sad a sight it is to see the streets empty of people," writes Samuel Pepys in his *Diary*, "and very few upon the 'Change. Jealous of every door that one sees shut up, lest it should be the plague; and about us two shops in three, if not more, generally shut up."

English physicians institute a peer-review system.

Many physicians flee England; the court of Charles II moves first to Salisbury, then to Exeter.

A rumor that contracting syphilis will serve to ward off the more deadly plague drives the men of London to storm the city's brothels.

Réflexions ou Sentences et Maximes Morales by French writer François de La Rochefoucauld, 52, is published anonymously. The wise and witty La Rochefoucauld will be quoted for centuries: "The surest way to be deceived is to think oneself more clever than the others." "When we think we hate flattery we only hate the manner of the flatterer." "The head is always the dupe of the heart." "We would rather speak ill of ourselves than not talk of ourselves at all." "The reason why so few people are agreeable in conversation is that each is thinking more about what he intends to say than about what others are saying, and we never listen when we are eager to speak." "There are few chaste women who are not tired of their trade." "Few people know how to be old."

Painting *Juno* by Rembrandt van Rijn who has used his common-law wife Henrickje Stoffels as his model; *The Jewish Bride* by Rembrandt; *The Artist's Studio* by Jan Vermeer; *The Physician in His Study* by Adriaen van Ostade. Nicolas Poussin dies at Rome November 19 at age 71 (or 72) after a career in which he has founded French classical art.

Theater *The Indian Emperor, or The Conquest of Mexico by the Spaniards* by John Dryden in April at London's Theatre Royal in Bridges Street; *Love Is the Best Doctor (L'amour médecin)* by Molière 9/22 at the Palais-Royal, Paris; *Alexander the Great (Alexandre Le Grand)* by Jean Racine 12/4 at the Palais-Royal, Paris.

Francesco Borromini completes Rome's Church of San Andrea delle Fratte.

Bucharest's Church of the Patriarchy is completed on a three-lobed plan with an icon dating from 1463.

Potatoes are planted on a limited basis in Lorraine (*see* 1663; 1744; 1757; 1770).

England imports less than 88 tons of sugar, a figure that will grow to 10,000 tons by the end of the century as tea consumption (encouraged by cheap sugar) increases in popularity.

1666 France's Queen Mother Anne of Austria dies of breast cancer at Paris January 20 at age 64.

France and Holland declare war on England, French forces take Antigua, Montserrat, and St. Kitts in the Greater Antilles, and an English privateer takes Tobago. The Dutch sign a treaty of alliance with the elector of Brandenburg Friedrich Wilhelm and sign a quadruple alliance with Brunswick, Brandenburg, and Denmark.

Hungarian noblemen revolt against the Holy Roman Emperor Leopold I.

French explorer René Robert Cavelier, 23, sieur de La Salle, voyages to the New World to occupy a grant of land he has received on the St. Lawrence River (*see* 1669).

Newark is founded on the Passaic River in the New Jersey colony by Connecticut Puritans who have arrived under the leadership of Robert Treat.

The calculus discovered (or invented) by Cambridge University mathematics professor Isaac Newton, 23, provides rules for dealing with rates of change. The numbers in any given algebraic problem remain constant while a calculation is being made, but problems that arise in engineering and in natural phenomena such as astronomy involve constant change which have made accurate mathematical calculation impossible without calculus.

Laws of gravity established by Isaac Newton (above) state that the attraction exerted by gravity between any two bodies is directly proportional to their masses and inversely proportional to the square of the distance between them. Newton has returned to his native Woolsthorpe because the plague at Cambridge has closed Trinity College, where he is a fellow; he has observed the fall of an apple in an orchard at Woolsthorpe and calculates that at a distance of one foot the attraction between two objects is 100 times stronger than at 10 feet (*see* Galileo, 1592).

A French Academy of Sciences founded by Louis XIV at Paris seeks to rival London's 4-year-old Royal Society. Jean Baptiste Colbert has persuaded the king to begin patronizing scientists.

Destroyed in London's Great Fire (below) are thousands of old dwellings that have harbored lice-bearing rats which spread plague (*see* 1665). Some 2,000 Londoners nevertheless die of the plague, which rages also at Cologne and in the Rhine district, where it will continue for the next 5 years (*see* 1673).

Sweden's University of Lund has its beginnings.

Poetry *Instructions to a Painter* by Edmund Waller.

Frans Hals dies at Haarlem August 26 at age 86; Guercino dies at Bologna December 22 at age 75.

Theater *The Misanthrope* by Molière 6/4 at the Palais-Royal, Paris: ". . . my mind is no more shocked at seeing a man a rogue unjust, or selfish, than at seeing vultures eager for prey, mischievous apes, or fury-lashed wolves" (I, i); ". . . twenty, as everyone well knows, is not an age to play the prude" (III, v); *The Doctor in Spite of Himself (Le médecin malgré lui)* by Molière 8/6 at the Palais-Royal.

The Great Fire of London that begins early in the morning of Sunday, September 2, in Pudding Lane near London Bridge, spreads through the crowded wooden houses to the Thames wharf warehouses and continues for 4 days and nights until the flames have consumed four-fifths of the walled city plus another 63 acres of property outside the city walls. London's Guildhall, 44 of the city's rich livery company halls, the Custom House, the Royal Exchange, the great Gothic cathedral of St. Paul's, 86 other churches, and some 13,200 houses are destroyed, but new structures will spring up to make London the world's most modern metropolis (*see* Wren, 1667).

The Newton pippin will be so named to commemorate the apple that inspired Newton's law of gravity (above).

Louis XIV issues an edict encouraging large families. Drafted by Jean Baptiste Colbert, it will prove ineffectual.

1667 The Treaty of Andrussovo January 20 ends a 13-year war between Russia and Poland which cedes Kiev, Smolensk, and the eastern Ukraine to Russia.

The Treaties of Breda signed July 21 end the second Anglo-Dutch war after a Dutch fleet has broken the chain in England's Medway River, reached Chatham, and captured the flagship *Royal Charles*.

England receives New Netherlands in return for sugar-rich Surinam in South America under terms reached at Breda (above). Acadia is restored to Holland's ally France, England receives Antigua, Montserrat, and St. Kitts from France, and Charles II makes a secret treaty with Louis XIV against Spain.

French troops invade Flanders and Hainault with grenades to begin the War of Devolution. Louis XIV, now 28, takes as his mistress the marquise of Montespan Françoise Atenais Rochechouart, 26, who proceeds to bear the first of several children she will have by the king.

Persia's Shah Abbas II dies at age 35 after a weak 25-year reign. His ministers pretend that his son of 20 is blind and try to install a younger son, but a court eunuch betrays their scheme, and the dissolute elder son will reign until 1694 as Suleiman I.

1667 ***(cont.)*** The Mughal emperor Aurangzeb buys off the Maratha raider Sivaji by giving him the title rajah and allowing him to levy some taxes (*see* 1664; 1670).

Brooklyn is chartered under the name Brueckelen October 18 by the governor of New Netherlands Mathias Nicolls. The new town includes "all the lots and plantations lying and being at the Gowanus, Bedford, Walle Bocht and Ferry."

English experimental chemist-philosopher Robert Hooke, 32, shows that blood alteration in the lungs is the essential feature of respiration. He has assisted Robert Boyle in showing that air is essential to life, and he uses bellows in a dog's trachea to prove his point.

The first recorded blood transfusion is performed by Jean Baptiste Denis, who transfers blood from a lamb into the vein of a boy (*see* Lower, 1677).

Poetry *Paradise Lost* by John Milton, who has been blind since 1652 but has dictated the 10-volume work on the fall of man to his daughters. It enjoys sales of 1,300 copies in 18 months and will be enlarged to 12 volumes in 1684, the year of Milton's death; *Annus Mirabilis* by John Dryden is about the Dutch War (above) and last year's Great Fire.

Painting *Girl With a Red Hat* by Jan Vermeer.

Theater *Secret Love or The Maiden Queen* by John Dryden in March at London's Theatre Royal in Bridges Street which has escaped the Great Fire; *Sir Martin Mar-All, or The Feign'd Innocence* by John Dryden, who has adapted Molière's *L'étourdi* as translated by William Cavendish, Duke of Newcastle, in August at the Lincoln's Inn Fields Theatre, London; *Andromache (Andromaque)* by Jean Racine 11/7 at the Hôtel de Bourgogne, Paris, in a performance by the group that rivals Molière's company, whose work has not satisfied Racine.

Architect Christopher Wren, 35, is assigned the task of rebuilding London which in January is still smoldering after the Great Fire of 4 months earlier. An Oxford mathematician and astronomer as well as an architect, Wren has earlier been commissioned by Charles II to restore St. Paul's Cathedral which has been completely gutted by the Great Fire.

Rome's Piazza San Pietro (Saint Peter's Square) is completed after 11 years of work by Giovanni Lorenzo Bernini, now 68, who has designed the great square in monumental baroque style with a semicircular colonnade ringing the 83-foot Egyptian obelisk erected by Pope Sixtus V in 1586.

Rome's great baroque architect Francesco Boromini commits suicide August 3 at age 66 in a fit of depression.

The Mexico Cathedral is completed after 94 years of construction.

1668 A Triple Alliance negotiated January 23 joins England, Holland, and Sweden in resistance to France's Louis XIV in the Spanish Netherlands, but Louis will soon buy off the English and Swedes.

The Treaty of Aix-la-Chapelle May 2 ends a brief War of Devolution waged over Louis XIV's claim to the Spanish possessions in the Belgian provinces following the death of his father-in-law Philip IV of Spain in 1665. Spain regains Franche-Comte in return for Lille, Tournay, Oudenarde, and nine other fortified border towns.

Poland's John II Casimir abdicates at age 58 and retires to France as the abbé de Saint-Germain. The Lithuanian Michael Wisniowiecki, 30, succeeds to the throne and will reign until 1673 (*see* 1670).

Sault Sainte Marie is founded between Lake Superior and Lake Huron—the first permanent European settlement in the Michigan region (*see* 1701; 1855).

Rupert House opens in the Canadian northwest—the first European trading settlement in the region. England's "Mad Cavalier" Prince Rupert, 49, a cousin of Charles II, has backed Pierre Radisson and his brother-in-law Médart Chourt, sieur de Groseilliers, in a venture to trade with *les sauvages* in the area of Hudson's Bay. Their 50-ton ketch *Nonsuch* skippered by Groseilliers sails into Hudson's Bay after a voyage of 118 days and anchors at the mouth of what will be called Rupert's River (*see* 1610; Bay Company, 1670).

Spanish conquistadors in the Pacific rename the Islas de los Ladrones found by Magellan in 1521. They call them Las Marianas to honor Maria Anna of Austria, widow of Spain's Philip IV.

Italian physician-naturalist Francesco Redi, 42, disproves the notion of spontaneous generation. He shows that no maggots will develop in meat, no matter how putrified it may be, if it is covered with a thin cloth to protect it from flies that will lay eggs. But most people will continue for centuries to believe that maggots are products of spontaneous generation rather than of fly larva (*see* Needham, 1748; Pasteur, 1859).

The Black Death reaches Austria, having traveled from Flanders to Westphalia and into Normandy and Switzerland (*see* 1666; 1672).

Fiction *Fables* by French fabulist Jean de La Fontaine, 47, who will follow his six-volume work with five more between 1678 and 1679 and a twelfth in 1694.

John Dryden is named England's first Poet Laureate and renews a loan of £500 he made last year to Charles II.

Theater *She Would If She Could* by George Etherege 2/26 at the Lincoln's Inn Fields Theatre, London; *The Sullen Lovers (or, The Impertinents)* by English playwright Thomas Shadwell, 26, 5/2 at the Lincoln's Inn Fields Theatre: Shadwell has adapted the Molière play *Les fâcheux* of 1661; *George Daudin, or The Abused Husband (Georges Daudin, ou le mati confondu)* by Molière 11/19 at the Palais-Royal, Paris.

Swedish organist Dietrich Buxtehude, 31, is named organist at the Marienkirche in Lübeck, where his sacred concerto *Abendmusiken* will be presented each year during Advent.

1669 Crete falls to the Ottoman Turks who take Candia September 27 after a 21-year siege. Venice loses her last colonial possession; the Turks will rule the island until 1898.

The Mogul emperor Aurangzeb destroys Hindu temples and forbids practice of the Hindu religion. Hindus begin rebellions against the autocratic Aurangzeb.

The sieur de La Salle explores the region south of Lake Erie and Lake Ontario (*see* 1666; Louisiana, 1682).

German-American physician John Lederer leads an expedition across the Piedmont to ascend the Blue Ridge Mountains (*see* 1670).

The first French trading station in India opens.

The Hanseatic League begun in 1241 holds its last meeting.

History of the Insects by Jan Swammerdam presents a preexistence theory of genetics that the seed of every living creature was formed at the creation of the world and that each generation is contained in the generation that preceded it (*see* Maupertuis, 1745).

Robert Boyle discovers a substance that will be called phosphorus from the Latin word for morning star. Hamburg merchant-alchemist Hennig Brand distills urine and makes the same discovery, noticing a white, translucent, waxy, malodorous substance that glows in the dark.

Nonfiction "Meat out of the eater" by Massachusetts colony clergyman Michael Wigglesworth discusses the necessity of the afflictions that God visits upon His children.

Fiction *Der Abenteuerliche Simplicissimus Teutsch das ist; Beschreibung des Lebens eines Seltzamen Vagantens Genannt Melchior Sternfels von Fuchshaim* by German magistrate Hans Jakob Christoffel von Grimmelshausen, 49, is the first great German novel. Set amidst the havoc of the Thirty Years' War, the picaresque tale relates adventures of a youth who becomes successively a soldier, jester, bourgeois, robber, pilgrim, slave, and hermit.

Painting *Self-Portrait* by Rembrandt van Rijn, who dies at Amsterdam October 4 at age 63 impoverished and alone; *Girl at the Spinet* by Jan Vermeer.

Theater *Tartuffe or L'Imposteur* by Molière 2/5 at the Palais-Royal, Paris; *Tyrannic Love, or The Royal Martyr* by John Dryden in June at London's Theatre Royal; *Britannicus* by Jean Racine 12/13 at the Hôtel de Bourgogne, Paris.

The Paris Opéra has its beginnings June 28 in letters of patent granted by Louis XIV "to establish an academy to present and have sung in public operas and spectacles with music and in French verse," the Académie de Royale Musique (*see* 1671).

The first Stradivarius violin is created by Italian violinmaker Antonio Stradivari, 25, who has served an apprenticeship in his home town of Cremona in Lombardy to Nicola Amati, now 73, whose grandfather Andrea Amati designed the modern violin. The younger Amati has improved on his grandfather's design and taught not only Stradivari but also Andrea Guarnieri, 43, who also makes violins at Cremona.

Famine in Bengal kills 3 million.

London's Yeomen of the Guard at the Tower of London get the name "Beefeaters" from the grand duke of Tuscany Cosimo de' Medici. Established in 1485 to guard the tower, the warders draw large daily rations of beef, says the grand duke.

1670 Denmark's Frederick III dies at Copenhagen February 6 at age 60 after a 22-year reign in which he has used his personal popularity to make Denmark an absolute monarchy. His son, 23, will reign until 1699 as Christian V, putting the interests of the upper middle class and state officials ahead of those of the old nobility.

France's Louis XIV makes a defensive alliance with Bavaria and with England's Charles II, whose sister Henrietta, duchesse d'Orléans, obtains the signatures of the English ministers to the Treaty of Dover in May. She returns to France and dies at St. Cloud June 30 at age 26, allegedly the victim of poison administered by order of her estranged husband Philip, duc d'Orléans, brother of the Grand Monarch.

French troops occupy Lorraine.

Ukrainian Cossacks rise against their Polish overlords, but Gen. John Sobieski, 46, puts down the rebellion (*see* 1648; 1671).

The Mughal emperor Aurangzeb comes under fresh attack from the raider Sivaji, who has built up a Maratha state in the Deccan (*see* 1667). Sivaji sacks Surat again, as in 1664; Aurangzeb allows him to levy taxes in Khandesh (*see* 1674).

"The Governor and Company of Adventurers of England Trading Into Hudson's Bay" is chartered by Charles II (*see* Prince Rupert, 1668).

Two more expeditions venture into the Blue Ridge Mountains under the command of John Lederer.

French clergyman Gabriel Mouton proposes a uniform and accurate decimalized system of measurements to replace the variously defined leagues, acres, pounds, and other units now in use (*see* 1790).

London clockmaker William Clement improves the accuracy of clocks by inventing anchor-shaped gadgets (escapements) that control the escape of a clock's driving force (*see* Huygens, 1656).

Minute hands appear on watches for the first time (*see* 1511).

France's Louis XIV founds Les Invalides at Paris to house up to 7,000 disabled soldiers (below).

Tractatus Theologico-Politicus by Baruch Spinoza is published anonymously (*see* 1656). Spinoza shows that the Bible, if properly understood, gives no support to the intolerance of religious authorities and their interference in civil and political affairs. The book creates a furor, it will provoke widespread denunciations as it

1670 ***(cont.)*** goes through five editions in the next 5 years, and Spinoza, now 37 and consumptive after years of inhaling glass dust produced by his lens-grinding, moves to The Hague to gain the protection of influential friends.

Nonfiction *Pensées* by the late French philosopher-mathematician Blaise Pascal who died in August 1662 at age 39 (the book is garbled and no clearly edited version of Pascal's jaundiced thoughts will appear for 175 years): "All men naturally hate each other"; "If men knew what others say of them, there would not be four friends in the world"; "Men never do evil so completely as when they do it from religious conviction"; "Cleopatra's nose: had it been shorter, the whole aspect of the world would have been changed"; "The heart has its reasons which reason cannot know"; "Man is but a reed, the most feeble thing in nature, but he is a thinking reed. The entire universe need not arm itself to crush him. A vapor, a drop of water suffices to kill him. But if the universe were to crush him, man would still be more noble than that which killed him, because he knows that he dies, and the advantage which the universe has over him: the universe knows nothing of this"; "What a chimera, then, is man! what a novelty! What a monster, what a chaos, what a contradiction, what a prodigy! judge of all things, feeble worm of the earth, depository of truth, a sink of uncertainty and error, the glory and the shame of the universe."

English Proverbs by naturalist John Ray, 43, includes "Misery loves company," "Blood is thicker than water," "I'll trust him no further than I can fling him," "Haste makes waste, and waste makes want, and want makes strife between goodman and his wife," "He that hath many irons in the fire, some will cool," "The misery is wide that the sheets will not decide," and "The last straw breaks the camel's back."

Painting *View of Haarlem* by Jacob van Ruysdael; *The Pearl Necklace* by Jan Vermeer; *Girl and Her Duenna* by Bartolomé Murillo.

Theater *Bérénice* by Jean Racine 1/1 at the Hôtel de Bourgogne, Paris; *The Bourgeois Gentleman (Le Bourgeois Gentilhomme)* by Molière 11/23 at the Palais-Royal, Paris (below); *Titus and Berenice (Tite et Bérénice)* by Pierre Corneille 11/28 at the Palais-Royal, Paris; *The Conquest of Granada by the Spaniards* (Part I) by John Dryden, in December at London's Theatre Royal; *The Humorists* by Thomas Shadwell, in December at the Lincoln's Inn Fields Theatre, London.

Molière's *Le Bourgeois Gentilhomme* (above) includes a ballet with music by court composer Jean Baptiste Lully, 38, who has come to France from his native Florence and has changed his name from Giovanni Battista Lulli. The ballet is so popular that four performances are requested in the space of 8 days.

French court architect Jules Hardouin-Mansart, 24, designs a masterpiece of classical architecture for Les Invalides (above) and will complete the Church of the Doves for Les Invalides in 1706.

Landscape architect André Lenôtre lays out the Champs-Elysées at Paris.

"Orange Girls" in English theaters sell Spanish Seville oranges which are popular despite their sourness. One Orange Girl, Eleanor "Nell" Gwyn, 20, has become an actress—and mistress to Charles II.

Population in the Massachusetts colony reaches roughly 25,000.

1671 Welsh buccaneer Henry Morgan, 36, captures Panama City in violation of last year's Anglo-Spanish treaty. Morgan stands trial, but Charles II forgives him, knights him, will make him lieutenant-governor of Jamaica in 1674, and will charge him with the task of putting an end to piracy.

The Turks declare war on Poland.

Russia's Czar Alexis suppresses a great peasant revolt in the southeast led by the Don Cossacks, whose leader Stenka Razin is executed.

The Virginia colony's governor estimates that blacks comprise less than 5 percent of the population (*see* 1649; 1652; 1715).

A small English party penetrates the Ohio River watershed beyond the Blue Ridge Mountains.

Poetry *Paradise Regained* and *Samson Agonistes* by John Milton. He has written *Paradise Regained* at the suggestion of Quaker Thomas Ellwood, 32, who comes to his house each day to read to him in Latin. *Samson Agonistes* is a powerful drama based on Greek tragedy and comprises among other things the autobiography and epitaph of its author, who will die of gout at age 65 in 1674.

Theater *Psyché* by Pierre Corneille, Molière, and Philippe Quinault 1/17 at the Salle des Machines, Paris, with music by Jean Baptiste Lully; *Love in a Wood* by English playwright William Wycherley, 31, who acts in his coarse comedy at London's Theatre Royal in Bridges Street and gains the patronage of the duchess of Cleveland and the duke of Buckingham; *Almanzor and Almahide, or The Conquest of Granada* (Part II) by John Dryden.

The French Académie de Royale Musique opens March 3 in the Salle du Jeu de Paume de la Bouteille. Jean Baptiste Lully will take over the Paris Opéra beginning next year and run it until 1687, rebuilding the house after fires that will destroy it in 1678 and 1681 (*see* 1669; 1875).

Rice is introduced into the South Carolina colony by physician Henry Woodward, who helped found the colony in 1663. He has received some Madagascar rice from a visiting sea captain, but the grain will for years be merely a garden curiosity since nobody knows how to husk it and use it for food.

French police receive the right to search houses during Lent and give any forbidden items of food they may find to the hospitals.

Mme. de Sévigné disparages use of the condom as a means of contraception. She describes it to her daughter the Comtesse de Grignan as "an armor against enjoyment and a spider web against danger" (*see* 1655; Kennett, 1723).

1672 A French army of 100,000 crosses the Rhine without warning and invades the Dutch Republic as Louis XIV acts to punish the Dutch who have given refuge to his political critics. England's Charles II supports Louis under secret provisions in the Treaty of Dover of May 1670, the Royal Navy scores a victory at Southwold Bay May 28, and the Dutch turn for help to the prince of Orange. The staats-general revive the *stadtholderate* July 8, and they make Willem III of Orange, 21, *stadtholder*, captain-general, and admiral for life. A mob at The Hague brutally murders the grand pensionary Jan De Witt, 47, and his brother Cornelius, 49, and Willem rewards the instigators of the riot as he summons aid from the elector of Brandenburg to resist the French.

Ottoman forces invade Poland following a series of border raids by Tatars and Cossacks. A 4-year war begins over control of the Ukraine.

Russian serfs rise in revolt.

English merchants form the Royal Company to exploit the African slave trade.

The Dutch organize a system of relief for the poor, who have been provided for up to now by prosperous merchants. With Dutch trade declining and the country at war, the merchants can no longer afford to be so generous.

Charleston is founded in the Carolina colony by Puritans from the Bermudas, led by William Sayle, who arrived 2 years ago at Albemarle Point on the Ashley River and have moved to the peninsula between the Ashley and Cooper rivers. Sayle has named the town after England's Charles II (*see* 1679).

England imposes customs duties on goods carried from one of her American colonies to another.

Boston-born English merchant Elihu Yale, 23, reaches India and enters the spice trade in the employ of the East India Company (*see* education, 1718).

New York merchant Frederick Philipse begins acquiring a manorial estate of 205,000 acres in upper Yonkers (some say by altering a phrase in a contract with the Indians).

The New York to Boston Post Road is laid out to speed coach travel between the second and third largest American cities.

The Black Death flares up again in Europe, killing 60,000 at Lyons and hundreds of thousands at Naples in just 6 months.

Painting *Virgin and Child* by Bartolomé Murillo. Adriaen van de Velde dies at Amsterdam January 21 at age 41.

Theater *Bajazet* by Jean Racine 1/1 at the Hôtel de Bourgogne, Paris; *The Learned Ladies (Les femmes savantes)* by Molière 3/11 at the Palais-Royal, Paris; *Marriage à la Mode* by John Dryden, in April at the Lincoln's Inn Fields Theatre, London; *The Gentleman Dancing Master* by William Wycherley, in August at London's Dorset Garden Theatre; *Pulcheria (Pulchérie)* by Pierre Corneille 11/14 at the Théâtre du Marais, Paris; *Epsom Wells* by Thomas Shadwell in December at London's Dorset Garden Theatre.

England's Charles II has judges and barristers wear wigs in court, a custom imported from France (date approximate).

New England Rarities by English-American merchant John Josselyn says of passenger pigeons, "of late they are much diminished, the English taking them with nets" (*see* 1648; 1800).

The first Paris coffee house opens at the Saint-Germain Fair. An Armenian known only as Pascal does well at the fair, but he fails in the coffee house he opens thereafter in the Quai de L'Ecole and moves to London (*see* 1652; Vienna, 1683; Paris, 1754).

1673 Willem III of Orange saves Amsterdam and the province of Holland from France's Louis XIV by opening the sluice gates and flooding the country. He is supported by Frederick William, elector of Brandenburg, who concludes a separate peace with Louis and retains most of his possessions in Cleves.

England's duke of York marries the Catholic Maria d'Este of Modena.

Parliament passes the Test Act compelling all English officeholders to take oaths of allegiance and of supremacy, to adjure transubstantiation, and to take the sacraments of the Church of England. The Test Act will not be repealed until 1828 but will be nullified after 1689 by bills of indemnity to legalize the acts of magistrates who have not taken communion in the Church of England (*see* Papists' Disabling Act, 1678).

Polish general John Sobieski defeats a Turkish army November 10 as Poland's king Michael Wisniowiecki dies after a 4-year reign (*see* 1674).

Dutch forces retake New York and Delaware (*see* 1655; 1664; 1674).

Jesuit missionary-explorer Jacques Marquette, 36, and trader Louis Joliet, 28, ascend the Fox River from Green Bay, make a short portage to the Wisconsin River, descend the Wisconsin to the Mississippi, paddle their canoes to the mouth of the Arkansas, return to the Illinois, ascend that river, make a portage to the Chicago River, and ascend to Lake Michigan. Joliet sees the possibility of a lakes-to-gulf ship waterway (*see* 1900).

Incidence of the Black Death begins to decline in Europe and England as the brown rat replaces the medieval black rat which is more inclined to carry plague-fleas (but *see* 1675; 1679 to 1682; 1720).

The first metal dental fillings are installed by English surgeons.

Poetry "On His Blindness" (Sonnet XIX) by John Milton, who probably wrote the lines in 1655 and who begins, "When I consider how my light is

Père Marquette and Louis Joliet explored the inland waterways of the great American heartland.

spent . . ." and ends, ". . . They also serve who only stand and wait."

Painting *Madonna and Child* by Bartolomé Murillo. Salvator Rosa dies at Rome March 15 at age 57.

Theater *Mithridates (Mithridate)* by Jean Racine 1/13 at the Hôtel de Bourgogne, Paris; *The Hypochondriac (Le Malade imaginaire)* by Molière 2/10 at the Palais-Royal, Paris (Molière's own illness is not imaginary, and he dies a week later at age 51); *Amboyana, or The Cruelties of the Dutch to the English Merchants* by John Dryden in May at the Lincoln's Inn Fields Theatre, London.

Kabuki actor Sannjuro Ichikawa, 13, at Edo originates the Aragato style featuring the superman war-god. Kabuki actors will follow his style for centuries (*see* 1652).

Mitsukoshi department stores have their beginnings in the Echi-go-ya dry-goods shop founded at Edo by the Mitsui family (*see* 1616). By introducing fixed prices (*Kanane-Nashi*) and cash-down installment buying (*Gen-Gin*), Mitsukoshi (the name will be adopted in 1928) will become the largest store on the Ginza and Japan's largest department-store chain (*see* Mitsui bank, 1683).

1674 The Treaty of Westminster February 9 ends the 2-year war between England and the Dutch. It returns New York and Delaware to England (*see* 1673), freeing the English to expand their trade and grow prosperous while Europe becomes embroiled in depleting warfare. Parliament has cut off funds, forcing Charles II to cease hostilities.

French troops devastate the Palatinate. Spain and the Holy Roman Empire join with the Dutch in a coalition to frustrate the ambitions of Louis XIV.

Poland elects Gen. Sobieski king. He has intimidated other contenders by arriving with 6,000 veterans of his triumph against the Turks last year and will reign until 1696 as John III Sobieski.

A Maratha dynasty in India is founded June 6 at Raigarh, where the onetime bandit Sivaji crowns himself to begin a monarchy that will war with the Mughal emperor Aurangzeb until his death in 1707 but will ally itself at times with Aurangzeb's successors (*see* 1667; 1761).

English Puritan scholar Richard Baxter, 59, denounces slave hunters as "enemies of mankind," but he does not object to plantation slavery itself so long as the slaves are well treated.

English Quakers purchase the New Jersey colony interests of Lord John Berkeley (*see* 1665).

Oxford's Thomas Willis, now 53, establishes that the urine of diabetics is "wonderfully sweet as it were imbued with Honey or Sugar," but while he distinguishes diabetes mellitus from other forms of the disease he suggests that it is a disease of the blood. Physician to England's Charles II, Willis helped found the Royal Society in 1662 (*see* 1788; Brunner, 1683).

Dr. Willis (above) blocks the vagus nerve in a live dog and establishes the nerve's influence on the lungs and heart. He publishes works on the brain and nervous system (*see* Cannon, 1912).

Ergotism strikes French peasants at Gatinais in a severe outbreak (*see* 1597; 1722).

Poetry *L'Art poétique* by Nicolas Boileau-Despréaux is a treatise in verse expounding classical standards. The work establishes Boileau as the leading neoclassical critic of his time and his mock epic "Le Lutrin" popularizes that genre.

Theater *Iphigénia (Iphigénie)* by Jean Racine 8/18 at Versailles and late in the year at the Hôtel de Bourgogne, Paris; *Surenas (Surena)* by Pierre Corneille 10/11 at the Hôtel de Bourgogne.

Japan has a terrible famine.

1675 Marshal Turenne inflicts a heavy defeat on the Dutch at Turkheim January 5, recovers all of Alsace within a few weeks, but is killed July 27 at age 63 by almost the first shot fired in a battle at Sassbach in Baden. The French retreat across the Rhine.

Swedish allies of France's Louis XIV invade Brandenburg but are defeated June 28 by the elector of Brandenburg in the Battle of Fehrbellin; the elector retaliates by invading Pomerania (*see* 1679).

King Philip's War devastates New England as Chief Metacum rebels against a 1671 order requiring his people to pay an annual tribute of £100. Called King Philip of Potanoket by the colonists, Metacum leads the Narragansett and Wampanoag in attacks on 52 American settlements, destroying 12 or 13 of them and killing 600 of New England's finest men (*see* 1662; 1676).

Père Marquette founds a mission at the Indian town of Kaskaskia in the Illinois wilderness (*see* 1673; 1699).

Boston has 30 merchants worth £10,000 to £20,000 each and 430 large vessels at sea "so that there is little left for the merchants residing in England to import into any of the plantations."

Dutch naturalist Anton van Leeuwenhoek, 43, gives the first accurate description of red corpuscles. Van Leeuwenhoek has developed the first of more than 400 simple microscopes that he will produce (*see* 1677; Kircher, 1642).

The Black Death kills 11,000 in Malta.

Painting *The Music Lesson* and *The Concert* by Geraert Terborch. Jan Vermeer dies at Delft December 15 at age 43.

Theater *The Country Wife* by William Wycherley in January at London's Theatre Royal in Drury Lane: "Good wives and private soldiers should be kept ignorant" (I); *The Mistaken Husband* by John Dryden in September at London's Drury Lane Theatre; *Aureng-Zebe* by Dryden in November at the Drury Lane Theatre.

Construction of a new St. Paul's Cathedral begins at London June 21. Architect Christopher Wren lays the first stone of the cathedral that will replace an old Gothic church gutted in the fire of 1666. The choir of the new cathedral will open December 2, 1697, and the last stone will be set in place in 1710.

England tries to suppress the coffee houses that have become gathering places for men who neglect their families to discuss business and politics over coffee.

New England has 50,000 colonists, up from 16,000 in 1642, while its native American population has dropped to no more than 20,000.

1676

Russia's Czar Alexis Mikhailovich dies February 8 at age 47 after a reign of 31 years and is succeeded by his eldest surviving son, who assumes the throne at age 15 as Fedor II. Disfigured and half paralyzed by disease since birth, Fedor is well educated and will rule effectively until his death in 1682.

The Treaty of Zuravno October 16 ends a 4-year war between Poland's John III Sobieski and the Ottoman Empire, which acquires most of Podolia and the Polish Ukraine, thus coming into contact with Russia.

King Philip's War in New England has ended in August with the killing in battle of Chief Metacum by colonial militiamen who break up Metacum's confederacy and take his head to Plymouth, where it will be exhibited for the next 20 years. Fifty-two of New England's 90 towns have been attacked, 600 men killed, and at least 12 towns completely destroyed.

"King Philip's" widow and children are sold as slaves to the West Indies despite Increase Mather's vote that they be executed, and militiamen hunt down other survivors of the war to sell into slavery; 500 are shipped out of Plymouth alone.

Bacon's Rebellion stirs the Virginia colony, but frontiersmen fail in their struggle against the tidewater aristocracy. The colony's slave trade increases.

Dutch traders buy black slaves at 30 florins each in Angola and sell 15,000 per year in the Americas at 300 to 500 florins each.

New Jersey's western part is conveyed to English Quakers who have entered into an agreement with Philip Carteret (*see* 1665; 1702).

England's Greenwich Observatory is established to study the position of the moon among the fixed stars and to establish a standard time that will help navigators fix their longitude, but determining accurate position remains a problem (*see* Bennewitz, 1524; Frisius, 1530; Board of Longitude, 1714).

"If I have seen further, it is by standing on the shoulders of giants," writes Isaac Newton February 18. Robert Hooke has challenged some of his ideas on optics, and Newton replies by pointing out that every contributor to knowledge or learning builds on the contributions of predecessors. Newton has doubtless read Robert Burton's 1621 chrestomathy *The Anatomy of Melancholy* with its aphorism, "Pygmies placed on the shoulders of giants see more than the giants themselves," an apothegm derived from the writings of the 12th century scholastic Bernard of Chartres who said, "In comparison with the ancients we stand like dwarfs on the shoulders of giants" (a stained glass window at the Cathedral of Chartres depicts Matthew sitting astride the shoulders of Isaiah), but Newton gives the phrase new meaning, and it will be called Newton's Other Law (*see* 1666).

The University of San Carlos of Guatemala is founded at Guatemala City.

Theater *The Man of Mode or Sir Fopling Flutter* by George Etherege 3/11 at London's Dorset Garden Theatre: "Next to coming to a good understanding with a new mistress, I love a quarrel with an old one" (I, i); *The Plain Dealer* by William Wycherley 12/11 at London's Theatre Royal in Drury Lane.

A Philosophical Discourse of Earth . . . by English diarist John Evelyn, 46, is a translation from the French work by J. de la Quintinie.

1677

England and France sign a maritime agreement permitting English ships to carry Dutch cargoes without fear of French interference.

France's duc d'Orléans defeats the Dutch at Cassel.

A Dutch-Danish fleet defeats a Swedish fleet at Oland.

French forces in Africa take Dutch ports on the Senegal River and capture Gorée near Cape Verde October 30. Gorée will be a major port for the slave trade (*see* 1698).

Culpeper's Rebellion in the 14-year-old Carolina colony protests enforcement of English trade laws by the colony's proprietors. Rebellious colonists install surveyor John Culpeper as governor, but the proprietors will remove him in 1679.

1677 ***(cont.)*** Spain's territorial empire covers much of the known world, but her soil is barely cultivated, her food costly, her population dwindling, and she has become a third-rate power (*see* 1613).

Willem of Orange turns 27 on November 4 and is married that day at London to the duke of York's daughter Mary, 15, a niece of England's Charles II (*see* "Glorious Revolution," 1688).

English physician Richard Lower, 46, introduces direct blood transfusion from one animal into the veins of another. He has injected dark venous blood into the lungs and deduced that air turns blood red (*see* 1667; 1818).

The *London Pharmacopeia* of 1618 is published in a third edition which includes steel tonics, digitalis, benzoin, jalap, ipecacuanha, cinchona bark (*see* Barba, 1642), and Irish whisky *(acqua vitae Hibernoroium sive usquebaugh)*.

Painting *Musical Party in a Courtyard* by Pieter de Hooch.

Theater *Phaedra (Phèdre)* by Jean Racine 1/1 at the Hôtel de Bourgogne, Paris; *All for Love, or The World Well Lost* by John Dryden in December at London's Drury Lane Theatre.

Microscopic spermatozoa in human semen are found by Anton van Leeuwenhoek and L. Hamm at Delft (*see* 1675).

1678 The Treaty of Nimwegen signed August 10 returns to Holland the territories she has lost to France and Spain on condition that she maintain neutrality. A second treaty signed at Nimwegen September 17 returns territories that France and Spain have taken from each other, but the peace will be short-lived.

Reports of a "Popish Plot" rock England. Titus Oates, 29, is expelled from a Jesuit seminary when it is discovered that the Anabaptist preacher's son has been employed by English prelate Israel Tonge to feign conversion to Catholicism. Oates "uncovers" a plot whereby Roman Catholics are supposedly pledged to massacre Protestants, burn London, and assassinate Charles II. He swears before magistrate Sir Edmund Berry Godfrey that he is telling the truth, Sir Edmund is soon after found murdered, five Catholic lords are sent to the Tower of London, dozens of Catholics are executed, and Parliament passes the Papists' Disabling Act that excludes Roman Catholics from Parliament, an act that will stand until 1829 (*see* Test Act, 1673).

Niagara Falls is discovered by French Franciscan missionary-explorer Louis Hennepin, 38, who has been in Canada since 1675 and who is so moved by the sight that he falls upon his knees. "The universe does not afford its parallel," Hennepin will write (*see* St. Anthony's Falls, 1680).

Italian anatomist-physician Lorenzo Bellini, 35, discovers the excretory ducts of the kidneys (Bellini's ducts). He has taught medicine at Pisa since at least 1664 and publishes a book on the physiology of taste organs. He also discovers the action of nerves on muscles.

Fiction *The Pilgrim's Progress* by English preacher John Bunyan, 49, whose illustrated work inspires religious reverence while entertaining readers and achieves enormous popularity which it will continue to enjoy for 2 centuries. To Puritans kept ignorant of Homer, Shakespeare, and Cervantes the Bunyan book comes as a brilliant revelation that is enhanced by its engravings; *La Princesse de Clèves* by French novelist Marie Madeleine Pioche de La Vergne, 44, Comtesse de La Fayette.

Poetry English poet Andrew Marvell dies August 16 at age 57 after taking an overdose of an opiate for his ague. He leaves behind verses that include the lines from "To His Coy Mistress": "had we but world enough, and time,/This coyness, lady, were no crime . . ."; "At my back I always hear Time's winged chariot hurrying near."

Painting *Charles II* by German-born English painter Godfrey Kneller (*né* Gottfried Kniller), 31. Jacob Jordaens dies at Antwerp October 18 at age 85.

Theater *The Kind Keeper, or Mr. Limberham* by John Dryden in March at London's Dorset Garden Theatre.

Opera house The Hamburg Staatsoper opens January 2 with 1,675 seats. Johan Theile has composed the music for the opening opera *Adam and Eve*.

1679 The Peace of St. Germain-en-Laye forced upon the elector of Brandenburg June 29 by France's Louis XIV obliges the elector to surrender to Sweden practically all of Brandenburg's conquests in Pomerania. The elector receives almost nothing in return.

Four new treaties signed at Nijmegen, Fontainebleau, and Lund settle disputes among France, Holland, Sweden, Denmark, and the Holy Roman Empire.

The Habeas Corpus Act passed by Parliament in May obliges English judges to issue upon request a writ of habeas corpus directing a jailer to produce the body of any prisoner and to show cause for his imprisonment. A prisoner shall be indicted in the first term of his commitment, says Parliament, he shall be tried no later than the second term, and once set free by order of the court he shall not be imprisoned again for the same offense.

French explorer Daniel Greysolon, sieur Du Lhut (or Duluth) reaches the great inland sea that will be called Lake Superior and claims the region for Louis XIV (*see* La Salle, 1682).

Charleston in the Carolina colony receives a group of French Huguenots who have received permission from the English to start a silk-manufacturing industry. After the industry is suppressed, families that include the Hugers, Legares, Legendres, Manigaults, Mazycks, and Mottes will become Charleston's leading moneylenders, planters' agents, and shippers.

Scots-American Robert Livingston, 25, marries Alida Schuyler van Rensselaer, widow of the late Nicholas van Rensselaer, and extends his landholdings in the Albany, New York, area where he has made his home for the past 5 years (*see* 1630; 1686).

French physicist Edmé Mariotte, 59, announces the constant relation between the pressure and volume of an enclosed quantity of air. He has made his discovery independently of work by Robert Boyle (*see* 1662).

French Huguenot physicist Denis Papin, 32, shows that the boiling point of water depends on atmospheric pressure. He has worked since 1675 with Robert Boyle at London after working previously with Christian Huygens at Paris.

The Black Death takes at least 76,000 lives at Vienna.

Theater *Troilus and Croilus, or Truth Found Too Late* by John Dryden, in April at London's Dorset Garden Theatre.

Opera *Gli Quivoco nell'amore* in February at Rome's Teatro Capranica with music by Italian composer Alessandro Scarlatti, 20, who gains the protection of Sweden's former Queen Christina and begins a career that will produce important musical works for 44 years. Scarlatti's operas, church music, and chamber music will virtually create the language of classical music.

Sweden's Skokloster Castle is completed south of Uppsala.

Tuscany's Arno River is brought under control by Italian engineer Vincenzo Viviani, who uses a modification of a plan devised by Leonardo da Vinci in 1495. The Arno will nevertheless continue to flood periodically, inundating Florence.

1680

Sweden's Charles XI brings pressure on the estates to pass a law requiring all earldoms, baronies, and other large fiefs to revert to the Crown and legalizing wholesale confiscation of properties. The measure deals a heavy blow to aristocrats.

Japan's Tokugawa shogun Itsuna dies at age 39 after a 29-year reign. He is succeeded by his brother, 34, who will reign until 1709 as Tsunayoshi.

New Hampshire is separated by royal charter from Massachusetts, whose Bay Colony governors have bought most of Maine from the heirs of Ferdinando Gorges.

Pueblo tribesmen at Taos and Santa Fe rise against the Spaniards August 11, destroy most of the Spanish churches, and drive the 2,500 Spanish colonists from their territory.

Bohemian peasants stage a major revolt to begin an era of endemic unrest among the serfs following a shift from soil tillage to dairy farming encouraged by the Thirty Years' War that ended in 1648.

Minneapolis has its beginnings in St. Anthony's Falls, named by Father Hennepin, who has accompanied the sieur de La Salle through the Great Lakes to the upper Mississippi Valley (*see* 1678; 1819).

Europe enters a 40-year period of economic troubles that will be accompanied by wild price fluctuations, revolts, famines, and disease epidemics.

The Black Death strikes Dresden in epidemic proportions.

De Motu Animalium by the late Italian mathematician-astronomer Giovanni Alphonso Borelli expresses the view that digestion is a mechanical process with blood pressure inducing gastric secretion. Fevers, pains, and convulsions are the result of defective movements of the "nervous juices," said Borelli, who died last year at age 71 after having founded the iatrophysical school of medicine by applying mechanical principles for the first time to the study of human muscular movement (*see* hydrochloric acid, 1824; pepsin, 1835).

Ferdinand Bol dies at Amsterdam July 24 at age 64; Gian Lorenzo Bernini dies at Rome November 28 at age 81.

Theater *The Spanish Friar, or The Double Discovery* by John Dryden in March at London's Dorset Garden Theatre.

Maryland colonists complain that "their supply of provisions becoming exhausted, it was necessary for them, in order to keep from starvation, to eat the oysters taken from along their shores."

1681

The Qing (Manchu) emperor Kangxi defeats the Rebellion of the Three Feudatories in the south, kills its leader, former Ming general Wu San-kuei, and establishes Qing rule over all of mainland China.

Russia confiscates Tatar territories on the Volga, forcing the people to convert to Christianity.

Hungarian noblemen regain their constitution under terms of the treaty of Sopron with the emperor Leopold.

France annexes Strasbourg.

Pennsylvania has its beginnings in a land grant of 48,000 square miles in the New World given by Charles II to religious nonconformist William Penn, 37, whose late father has bequeathed him an immense claim of £15,000 against the king (*see* Jamaica, 1655). Penn has been the first person of means to join the Society of Friends founded by George Fox in 1647, he has served time in prison for writing and distributing pamphlets espousing the Quaker cause, and the king's generosity is motivated in part by a desire to rid England of nonconformists, but Charles honors the late admiral by prefixing "Penn" to the name "Sylvania" that William Penn gives to the new territory (*see* 1682).

England begins a period of prosperity that will be shared by the New England Confederation formed in 1643. Merchants become rich as the demand for ships and shipping increases.

The first bank checks are issued in England (*see* Bank of England, 1694; London Stock Exchange, 1698).

France's Languedoc Canal is completed to link the Atlantic with the Mediterranean. The late engineer Pierre Paul de Riquet planned the canal and spent most of his own personal fortune to pursue the project in the last 18 years but has died at age 76, 6 months too soon to see his work complete (*see* 1603).

The Black Death takes 83,000 lives at Prague.

1681 ***(cont.)*** Poetry *Absalom and Achitophel* by John Dryden, who satirizes the second duke of Buckingham George Villiers, 53. Buckingham was dismissed from his offices in 1674 for personal immorality and for promoting popery and arbitrary government.

Geraert Terborch dies at Deventer December 8 at age 64.

Christopher Wren designs Tom's Tower for Oxford's Christ Church.

Venice's Church of the Salute is completed after 50 years of interrupted activity. Ordered by a decree of the Senate in 1630 as a thanksgiving for delivering the city from the plague, the baroque church has been designed by Baldassare Longhena, now 83, who will die within a year.

The dodo becomes extinct as the last of the species dies on the Indian Ocean island of Mauritius. A flightless bird related to the pigeon but as large as a turkey, the dodo has been killed off by Europeans for food.

France's Louis XIV restricts fishing for mussels but places no restraint on dragging for oysters whose natural banks are called "inexhaustible" (*see* 1786).

1682 Russia's Fedor III dies April 27 at age 21, his half brother Peter, 9, succeeds to the throne in preference to Fedor's mentally and physically defective brother Ivan, 15, but Ivan's sister Sophia Alekseevna, 25, instigates the musketeers (*Strelitzi*) to invade the Kremlin and murder Peter's supporters. Ivan is proclaimed czar with Peter as his associate and begins a nominal 7-year reign in which Sophia will serve as regent and exercise the power.

Austrian and Polish forces begin a 17-year war to liberate Hungary from the Ottoman Turks.

A Frame of Government drawn up by William Penn for the Pennsylvania colony contains an explicit clause permitting amendments, an innovation that makes it a self-adjusting constitution.

William Penn (above) lands at New Castle on the Delaware River late in the year after a voyage from Deal in the ship *Welcome*. A third of his 100 Friends have succumbed to smallpox on the voyage which has lasted nearly 2 months.

Philadelphia—a city "of brotherly love"—is founded by William Penn.

Louisiana Territory is so named by the sieur de La Salle, who reaches the mouth of the Mississippi April 9 with a party of 50 men after descending from the Illinois River. La Salle has built Fort Prudhomme at the mouth of the Hatchie River in what later will be called Tennessee, and he claims the Mississippi Valley for Louis XIV (*see* 1669; 1763).

Norfolk, Virginia, is founded on the site of an Indian village by colonist Nicholas Wise, who will buy the land for $400 in tobacco from the government at Williamsburg.

Mennonites begin leaving the German states to settle at Philadelphia and on fertile lands that will extend from Easton through Allentown, Reading, and Lebanon to the Cumberland Valley (*see* 1537; Germantown, 1683).

Methodus Plantarum Nova by English naturalist John Ray, now 55, first demonstrates the nature of buds and divides flowering plants into dicotyledons and monocotyledons (*see* Linnaeus, 1737).

"Thoughts on the Comet of 1680" by French philosopher Pierre Bayle, 35, uses rationalism to oppose superstitions about comets.

The Black Death kills nearly half the 10,000 inhabitants of Halle and wipes out much of Magdeburg.

Fiction *The Life of an Amorous Man (Koshoku Ichidai Otoko)* by Japanese writer Saikaku Ihara, 40, who has confined himself until now to poetry.

Poetry *The Medall; a Satyre against Sedition* by John Dryden is an even more direct attack on the earl of Shaftesbury: "Successful crimes alone are justified," writes Dryden, and his barbs force Shaftesbury to flee to Holland, where he will die next year; *MacFlecknoe* by Dryden: "All human things are subject to decay."

Jacob van Ruysdael dies at Amsterdam March 14 at age 53; Bartolomé Murillo dies at Seville April 3 at age 64; Claude Lorrain dies at Rome November 21 at age 82.

Theater *Venice Preserved, or A Plot Discovered* by English playwright Thomas Otway, 30, 2/9 at London's Dorset Garden Theatre.

Versailles becomes the seat of French government as Louis XIV moves his court into a palatial château built outside Paris at a cost of $325 million and some 227 lives. Waterworks constructed at Marly-le-Roi by the royal engineer Rannequin supply water from the Seine for the Versailles gardens, and another château for the royal family goes up at Marly-le-Roi (*see* 1627; Hall of Mirrors, 1684).

The pressure cooker invented by French physicist Denis Papin employs a safety valve. Papin calls his cooker a "digester" (*see* 1940; science, 1679).

Food served at Versailles (above) is generally cold because the kitchens are so far from the dining rooms.

The Mennonites (above) will introduce Pennsylvania "Dutch" (*Deutsch*) cooking and will contribute dishes such as scrapple to the American cuisine.

1683 England's Charles II compels the City of London to surrender its charter under a writ of quo warranto, various aldermen and officers are ejected and replaced by royal nominees, municipal charters throughout England are revoked to give the Tories control over appointment of municipal officers, and some of the defeated Whigs conspire to assassinate the king. The Rye House plot is uncovered in June along with a similar conspiracy. Lord William Russell, 43, is sent to the Tower of London June 26 as are Arthur Capel, 51, first earl of Essex, and Algernon Sidney, 51. Russell is executed July 21 at Lincoln's Inn Fields after trial and conviction on the testimony of a perjured witness, Essex is discovered in his chamber July 30 with his throat slit, probably by his own hand, and Sidney is beheaded December 7.

The king's bastard son James Scott, 34, duke of Monmouth, is subpoenaed to give evidence at the trial of John Hampden, 27, who has been arrested for alleged complicity in the Rye House plot (above), partisans of the duke advance his claims to the crown, their hopes are dampened by the execution of Sidney (above), and the duke flees to Holland.

France's Jean Baptiste Colbert dies at age 64, leaving fine estates throughout the country but the French navy without a champion at court. The navy will soon begin to decline.

Spain declares war on France, and Spain's Carlos II is joined by the emperor Leopold in the League of The Hague which joins the Dutch-Swedish alliance against Louis XIV.

Portugal's dissolute Afonso VI dies at age 40 and is succeeded by his 35-year-old brother, who has served as regent since 1667 and who will reign as Pedro II until his death in 1706.

The Chinese conquer Taiwan (Formosa) and will hold the island until 1895.

The Great Treaty of Shackamaxon, signed by William Penn with the Delaware Indians, permits Penn to purchase territories that will become southeastern Pennsylvania.

Ottoman troops under the grand vizier Kara Mustafa lay siege to Vienna in July. A German-Polish army under Charles of Lorraine, 40, and John III Sobieski lifts the siege September 12 after 58 days.

The first English shipwreck on Africa's east coast strands sailors in Natal. Two more English and one Dutch wrecks in the next few years will contribute to exploration of the region, at least one man will be trampled to death by an elephant, and some will report having met the survivor of a Portuguese wreck 42 years earlier who has married an amapondo woman.

Japan's Mitsui bank is opened in Edo's Surugacho district by merchant Mitsui Hachirobei Takatoshi, youngest son of Mitsui Sokuebei (*see* 1616; Mitsukoshi, 1673). Founded with profits from selling brocades, silks, and cottons at low prices, the bank will grow into a major international trading company.

Anton von Leeuwenhoek invents an improved microscope and finds living organisms in the calculus scraped off his teeth, pioneering the germ theory of disease (*see* 1675; Fracostoro, 1546).

Swiss anatomist Johann Conrad Brunner, 30, removes a dog's pancreas and notes that the animal develops an inordinate thirst, thus pioneering knowledge of diabetes (*see* Cawley, 1788).

English physician Thomas Sydenham, 59, gives an accurate description of gout, from which he has suffered for 34 years (*see* Garrod, 1859).

The collected works of Thomas Sydenham (above), published at Amsterdam, include writings that have appeared since 1666. "Among the remedies which it has pleased Almighty God to give man to relieve his sufferings, none is so universal and so efficacious as opium," Sydenham wrote in 1680. Opiates will remain the mainstay of treating severe pain from many illnesses for more than 3 centuries, and one popular opiate will be Sydenham's drops (*see* morphine, 1803).

Iron filings steeped in Rhenish wine relieve iron-deficiency anemia (chlorosis), Sydenham has written. Returning to the 5th century B.C. principles of Hippocrates, Sydenham has built a large and lucrative practice on the premise that the cause of all diseases resides in nature and that diseases tend to cure themselves. He uses diet, discreet blood-letting, Sydenham's drops (above), cinchona bark (*see* 1642), and vegetable simples; he gives clear clinical descriptions of malaria, hysteria, scarlatina, smallpox, and St. Vitus's dance (Sydenham's chorea); and he advises horseback riding for tuberculosis, cooling measures for smallpox, fresh air for sickrooms.

Nonfiction *Dialogues des Morts* by French man of letters Bernard Le Bovier de Fontenelle, 26, is an attack on authoritarianism; *The Growth of the City of London* by Sir William Petty; *A General Description of Pennsylvania* by William Penn.

Painting *Sir Charles Cotterell* by Godfrey Kneller.

Philadelphia is laid out by colonist Thomas Holme and other members of the Society of Friends with a grid pattern employed earlier by some Spanish colonial towns.

Copenhagen's Charlottenborg Palace is completed in Dutch baroque style for Count Ulrik F. Byldenlove, illegitimate son of Frederick III. Eleven years in the building, it will be taken over by the dowager queen Charlotte Amalie in 1700 and 54 years later will become the Danish Royal Academy of Arts.

William Penn bought vast territories from the Delaware in the Great Treaty of Shackamaxon. The Delaware got little.

1683 ***(cont.)*** Christopher Wren designs London's Piccadilly Circus and St. James's Place.

Wild boars become extinct in Britain.

France's Louis XIV hears cries of hungry beggars outside the palace at Versailles and sends an order to his aged controller of finance Jean Baptiste Colbert (above): "The suffering troubles me greatly. We must do everything we can to relieve the people. I wish this to be done at once." But little in fact is done.

The Viennese have lost thousands to starvation in the 58-day siege by the Ottoman Turks (above). Survivors have sustained themselves by eating cats, donkeys, and everything else edible.

The *kipfel,* a crescent-shaped roll, is created by Viennese bakers either to celebrate the lifting of the city's siege (above) or in anticipation of a Turkish victory, but the roll that will become the French *croissant* may date to 1217 (*see* 340 B.C.; Marie Antoinette, 1770).

The first coffee house in central Europe is opened by a spy for the Viennese who has tasted coffee in his sallies among the 300,000-man Ottoman siege force and obtained some bags of coffee after the Turks retreat. His coffee house Zum Roten Kreuz (At the Red Cross) is near St. Stephen's Cathedral.

German Mennonites arrive at Philadelphia October 6 aboard the *Concord* and buy 47,000 acres 6 miles to the north. Invited by Penn and led by lawyer Franz Daniel Pastorius, 30, who arrived in August, their "Germantown" is America's first German settlement. More Germans will come to America than any other nationality group except Anglo-Scots-Irish (*see* 1586, 1682, 1710, 1882).

1684 Venice joins Austria and Poland in the Holy League against the Ottoman Turks, but Poland soon withdraws from the league sponsored by Pope Innocent XI at the insistence of France's Louis XIV.

French troops occupy Lorraine, and Louis XIV signs a 20-year truce at Regensburg with the Holy Roman Emperor Leopold. Louis retains all the territory he has obtained up to August 1, 1681, including Strasbourg, and retains Lorraine as well.

Japan's Prime Minister Hotta Masatoshi is assassinated at age 50, leaving the shogun Tsunayoshi with no able counselors. He will issue Buddhist-inspired edicts prohibiting the killing of any living creature, extend special protection and privileges to dogs, wreck the finances of Edo, and bring hardship to the Japanese.

Parliament annuls the Massachusetts Charter of 1629 following charges that colonists have usurped the rights of Mason and Gorges and their heirs in New Hampshire, evaded the Navigation Acts by sending tobacco and sugar directly to Europe, exercised power not warranted by the charter, and shown disrespect for the king's authority.

The word *American* appears for the first time in writings by Puritan minister Cotton Mather, 21, who entered Harvard College at age 12 and will become assistant next year to his father Increase Mather at Boston's North Church (*see* 1693).

The sieur de La Salle returns to America with a force of more than 400 men, but he lands in error on the Texas Gulf Coast, where he gets lost and loses most of his men to fever and desertion (*see* 1682; 1687).

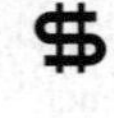

England's East India Company gains Chinese permission to build a trading station at Canton after years of having to import Chinese silks, porcelain, and tea by way of Java.

England has an outbreak of smallpox that adds to the misery of the cold (below).

The Japanese poet Saikaku composes 23,500 verses in a single day and night at the Sumiyoshi Shrine in Osaka. Unable to keep up with him, the scribes can only tally the number.

The Hall of Mirrors at Versailles is completed by the French architect Jules Hardouin-Mansart whose uncle François brought the mansard roof into use. As building superintendent to Louis XIV, Hardouin-Mansart will create the Grand Trianon, Place Vendôme, Place des Victoires, and the dome of the Hotel des Invalides (*see* 1682).

England emerges from her coldest winter in memory. The Thames has frozen over, and even the sea has frozen for 2 miles from land.

Charles II roasts oxen and feeds the poor at his own expense as bitter cold grips the country.

Tea sells on the Continent for less than 1 shilling per pound, but an import duty of 5 shillings per pound makes tea too costly for most Englishmen and encourages widespread smuggling. The English consume more smuggled tea than is brought in by orthodox routes (*see* 1784).

1685 England's Charles II dies February 6 at age 54 saying, "Let not poor Nelly starve," a reference to actress Nell Gwyn, now 34, who made her last stage appearance in 1682, has borne the king two sons, but will die in 2 years. Charles has made a profession of the Catholic faith on his deathbed, his Catholic brother James, 51, succeeds him to begin a brief reign as James II, but the new king's nephew James, duke of Monmouth, claims "legitimate and legal" right to the throne.

Monmouth (above) is the acknowledged son of Charles, he has taken the surname Scott of his wife Anne, countess of Buccleuch, his mistress Henrietta Maria Wentworth, 28, baroness Wentworth, has supplied funds, he has the support of Archibald Campbell, 56, earl of Argyll, and he lands at Lyme Regis, Dorsetshire, with 82 supporters. English troops loyal to James II capture Argyll, and he is executed June 30. James's troops easily defeat Monmouth July 6 at the Battle of Sedgemoor, last formal battle on English soil, and he is captured and beheaded.

Dahomey's King Wegbaja dies after a reign that has created the kingdom at Abomey in West Africa. His son Akaba succeeds (*see* 1698).

Titus Oates of 1678 "Popish Plot" notoriety is found guilty of perjury and sentenced to be pilloried and imprisoned for life. Oates will be pardoned in 1689 and allowed a pension of £300 per year following the accession of a new king.

The revocation of France's Edict of Nantes October 18 forbids the practice of any religion but Catholicism and forbids Huguenots to emigrate after 87 years of religious toleration (but *see* below).

A Code Noir issued at Paris establishes humane relations with respect to treatment of slaves on French colonial plantations, but planters will generally disregard the code.

French entrepreneurs establish the Guinea Company to engage in the slave trade which has become increasingly popular.

Barbados has 46,000 slaves, up from 6,000 in 1645. Blacks on the island outnumber Europeans 2 to 1.

Poetry *Of Divine Love* by Edmund Waller, now 79.

Painting *Philip, Earl of Leicester* by Godfrey Kneller. Adriaen van Ostade dies at Haarlem May 2 at age 74.

More than 50,000 French Huguenot families begin emigrating following the revocation of the Edict of Nantes (above). Half a million people will leave for England (Spitalfields), Holland, Denmark, Sweden, the Protestant German states, South Africa, and North America, many countries will attract the emigrants by offering tax exemptions and transportation subsidies, the emigrants will include so many seamen that French shipping will be hurt for generations to come, and the loss of so many craftsmen and intellectuals will leave France crippled.

1686

Austrian troops liberate Buda July 8 from the Ottoman Turks who have controlled the city since 1541, but the pashas have left it a ruin, its treasures destroyed or stolen. Moscow declares war on Constantinople.

The League of Augsburg created July 9 to oppose France's Louis XIV allies the Holy Roman Emperor Leopold I, Spain's Carlos II, Sweden's Charles XI, and the electors of Bavaria, Saxony, and the Palatine. Last year's revocation of the Edict of Nantes has aroused Protestants against France (*see* 1688).

France annexes Madagascar.

The English East India Company begins to impose its will by force after 80 years of trying to cement relations with the rulers of India. Company official Job Charnock moves his factory from the besieged town of Hooghly to an island at the mouth of the Ganges and begins a pattern of company rule that will continue until 1858 (*see* Calcutta, 1690).

The Dominion of New England is created by the consolidation of English colonies under the administration of New York governor Sir Edmund Andros, now 49. He arrives at Boston December 20 and assumes control of the government of Plymouth and Rhode Island (*see* King William's War, 1689).

England readmits Roman Catholics to the army.

Robert Livingston creates a manor out of his American landholdings which ultimately will embrace 160,000 acres (*see* 1679; Louisiana Purchase, 1803).

Nonfiction *Entrétiens sur la pluralité des mondes* by Bernard de Fontenelle popularizes Cartesian cosmology (*see* 1637); *L'Histoire des oracles* by Fontenelle attacks credulity and superstition.

Fiction *Five Women Who Chose Love (Koshoku Gonin Onna)* and *A Woman Who Devoted Her Entire Life to Lovemaking (Koshoku Ichidai Onna)* by Saikaku Ihara, a widower whose realistic fiction reflects the sentiments and manners of Japan's masses.

Theater *Successful Kagekiyo* (*Shusse Kagekiyo*) by Japanese playwright Monzaemon Chikamatsu (Sugimori Nobumori), 33, who has become associated in the puppet theater at Osaka with the chanter Takemoto Gidayu. Together they have developed an advanced style, and Chikamatsu has adapted a Japanese army play to the puppet theater.

Opera *Armide* 2/15 at Paris with music by Jean Baptiste Lully.

Cordon Bleu cookery has its origin in the Institut de Saint-Louis founded by Mme. de Maintenon for 250 daughters of the impoverished nobility, especially of titled army officers. Cookery is among the subjects taught, and the school will become known for its cooking lessons and for the *cordon bleu* (blue ribbon) which the girls wear as part of their graduation costumes.

1687

Ukrainian cossack Ivan Stepanovich Mazepa-Koledinsky, 43, visits Moscow, wins the favor of the prime minister Vasili Vasilievich Golitsyn, 44, and virtually purchases the hetmanship of the Cossacks July 25. Educated at the court of the late Polish king John II Casimir, Mazepa was caught in bed with a married Polish woman, her husband tied him naked to the back of a wild horse, Dnieperian Cossacks rescued him on the steppe, and he has risen to leadership among them (*see* 1704).

The Second Battle of Mohacs August 12 gives Charles of Lorraine a victory over the Ottoman Turks. The diet of Pressburg confers hereditary succession to the Hungarian throne upon the male line of Austria.

The Venetian general Francesco Morosini reconquers the Greek Peloponnesus from the Ottoman Turks and even captures Athens, which he shells (below).

Erlau in Hungary capitulates to Austrian forces September 14 after a century under Ottoman rule.

The Ottoman sultan Mohammed IV, now 48, is deposed by his Janissaries November 8 and thrown into prison (where he will die in 1692). His brother, 47, will reign until 1691 as Suleiman II.

The Declaration of Liberty of Conscience issued by James II April 14 grants liberty of worship to all denominations in England and Scotland.

1687 ***(cont.)*** The La Salle expedition on America's Gulf Coast is reduced to a band of 20 (*see* 1684). La Salle's men kill their leader March 19 and leave his body to the buzzards.

Philosophia naturalis principia mathematica by Isaac Newton establishes laws of gravity and universal laws of motion. Edmund Halley has inspired Newton to write *The Principia* and has it published at his own expense (*see* Halley, 1705).

A Japanese law imposed January 28 by the shogun Tsunayoshi forbids the killing of animals. Tsunayoshi's only son has died, he has become a devout Buddhist, and beginning February 27 he forbids his people to eat fish, shellfish, or birds.

The University of Bologna is founded.

Traité de l'éducation des filles by French clergyman François de Salignac de la Mothe-Fénelon, 36, shows a rare insight into psychology. Fénelon was appointed at age 27 to head a Paris institution for women converts.

Poetry *The Hind and the Panther* by John Dryden: "By education most have been misled;/So they believe, because they so were bred./The priest continues what the nurse began,/And thus the child imposes on the man"; *The Country Mouse and the City Mouse* by English poet Matthew Prior, 23, burlesques Dryden's *The Hind and the Panther* (above). Prior has written the poem in collaboration with Charles Montagu, 26.

Painting *The Chinese Convent* by Godfrey Kneller.

Paris Opéra director Jean Baptiste Lully dies March 22 at age 54 of blood poisoning after having stabbed himself in the foot with his long baton while conducting a "Te Deum" of thanksgiving for the king's recovery from an illness. Louis XIV has given Lully an unlimited budget and has at times appeared on stage himself to dance as Jupiter or Apollo with some of his top lords in operatic spectacles produced by Lully, spectacles that have included floating clouds, fireworks, volcanoes, and waterfalls.

The Temple of Athena on the Acropolis at Athens sustains severe damage as a Venetian shell scores a direct hit, exploding gunpowder stored in the temple by the Turks (above).

1688 A "Glorious Revolution" ends nearly 4 years of Roman Catholic rule in England (below), and the War of the League of Augsburg pits Protestant Europe and much of Catholic Europe against France's Louis XIV.

England's James II issues a proclamation in April ordering clergymen to read from their pulpits the king's Declaration of Indulgence of last year exempting Catholics and Dissenters from penal statutes. The birth of a son to James's Queen Mary June 10 suggests the likelihood of a Catholic succession. England's Whig leaders send an invitation to the king's son-in-law William of Orange June 30, William issues a declaration to the English people September 21, lands at Tor Bay November 5, and moves to assume the throne with his wife Mary.

James II escapes to France December 23 and begins efforts to regain the throne (*see* Battle of the Boyne, 1690).

A French army invades the Palatinate and lays waste the countryside on orders from the minister of war François Michel Le Tellier, 47, marquis de Louvois.

Ottoman forces surrender Belgrade to the Austrians August 20 after 21 days' bombardment.

Lord Baltimore loses control of the Maryland colony as a result of the Glorious Revolution in England (above; *see* 1632; 1715).

French Huguenot refugees arrive in South Africa where they will strengthen the Dutch settlement founded in 1652.

English landowners seize the opportunity of the Glorious Revolution (above) to enact a bounty on the export of grain, an act that will increase domestic prices of grain (and of food) for the next few years.

Lloyds of London has its beginnings in a society to write marine insurance formed by merchants and sea captains who gather at Edward Lloyds's coffee house near the Thames. Lloyds encourages the underwriters by providing quill pens, ink, paper, and shipping information. The term *underwriting* will derive from his patrons' practice of writing their names, one beneath the other, at the bottom of each policy, with each man writing the amount he will insure until the full amount is subscribed (*see* 1870).

Venice's Ponte de tre Archi is completed across the Cannaregio Canal.

Nonfiction *Les Caractères de Théophraste, traduits du grec, avec les caractéres et les Moeurs de ce Siécle* by French moralist-satirist Jean de La Bruyère, 43, whose portraits and aphorisms point out the arrogance and immorality of France's stupid ruling class. La Bruyère cries out against the social injustice that prevails; *Digression sur les anciens et les modernes* by Bernard de Fontenelle defends evolution in the arts.

Fiction *Oroonoko* by English novelist Aphra Behn, 48, who traveled to Surinam as a child, returned in 1658, married a London merchant named Behn, was sent to the Netherlands as an English spy, was never paid for her services, and served time in debtors' prison before starting a writing career which began with ribald stage comedies. Her novel introduces the figure of the noble savage.

Theater *The Squire of Alsatia* by Thomas Shadwell in May at London's Drury Lane Theatre. Shadwell has adopted the Terence comedy *Adelphoe* of 160 B.C., and his play gives actress Anne Bracegirdle, 25, her stage debut.

William Dampier finds breadfruit *(Artocarpus communes)* growing on the Pacific island of Guam (*see* 1697; Bligh, 1787).

1689 The War of the League of Augsburg against France's Louis XIV widens as England's new king William III forms a Grand Alliance May 12 with the Dutch and with the League that Savoy joined 2 years ago.

The War of the League of Augsburg (above) begins the involvement of the Dutch Republic in a series of land wars that will cause the Dutch to neglect their naval strength and thus facilitate English domination of the high seas.

Scotland's Convention of Estates meets in April at the summoning of the prince of Orange, declares that James VII has forfeited the Scottish Crown, offers it to William and Mary, who accept, and is converted in June into a parliament. Scottish Episcopalians form a Jacobite party to prevent the threatened abolition of episcopacy and hold out at Edinburgh Castle until forced to surrender June 13. John Graham of Claverhouse, 40, Viscount Dundee, collects 3,000 men to oppose William and Mary, the Whig general Hugh Mackay is swept back into Killiecrankie Pass July 27 and loses nearly half his men, the Bonny Dundee gains a victory, but he is killed later in the day, ending the danger of a Jacobite restoration (*see* 1707).

Russia's Regent and Czarina Sophia Alekseevna, 32, is deposed following exposure of a conspiracy to seize her half brother Peter, now 17, who is crowned czar in September and will reign until 1730 as Peter I. The new czar was forced into a marriage January 27 with the beautiful but stupid Eudoxia Lopukhina, but the marriage collapses by year's end as Peter devotes himself to sailing, shipbuilding, drilling, and mock battles while leaving administrative duties to others (*see* 1697).

The Treaty of Nerchinsk ends the first conflict between Russia and China. Accepting the advice of Jesuit negotiator Jean-François Gerbillon, the Russians withdraw from territories their pioneers have occupied in the Amur region (*see* 1637).

King William's War begins in North America as an outgrowth of the War of the League of Augsburg (above).

Bostonians revolt at news of James II's flight from England and restore charter government after imprisoning Sir Edmund Andros, but New York proclaims William and Mary the rightful monarchs of England and her colonies.

Connecticut's Charter Oak gets its name as representatives of William III try to take back the colonial charter granted by Charles II in 1662. Colonists hide the charter at Hartford in a centuries-old oak that will have a circumference of 33 feet when it is blown down in 1856.

A Toleration Act excuses England's Nonconformists from church attendance but Dissenters continue to be persecuted for such reasons as nonpayment of tithes (*see* 1662).

An Irish peasants' revolt is suppressed by William and Mary.

Edo's Yoshiwara brothels have 2,800 prostitutes, according to an official census. The "happy field" moved to the city's Asakusa district after a fire in 1656 (*see* 1617; 1789).

Indians wipe out the colony at the mouth of the Mississippi River founded by the late sieur de La Salle (*see* 1687).

France's Louis XIV has silver furniture at Versailles melted down to pay the cost of his war with the League of Augsburg (above).

A smallpox epidemic in the Massachusetts colony kills more than a thousand in 12 months.

Nonfiction *Table-Talk*, a compilation of statements recorded by the secretary of the late English jurist-antiquary John Selden, who died in 1654 at age 70: "While you are upon earth, enjoy the good things that are here"; "No man is the wiser for his learning."

Painting *The Avenue, Middelharnis* by Meindert Hobbema.

Theater *Esther* by Jean Racine 1/26 at Mme. de Maintenon's School for Young Ladies at Saint-Cyr; *Bury Fair* by Thomas Shadwell in April at London's Drury Lane Theatre; *Don Sebastian, King of Portugal* by John Dryden in October at London's Drury Lane Theatre.

Opera *Dido and Aeneas* 12/30 at the Chelsea boarding school for girls operated by English dancing master Joseph Priest, with music by English composer Henry Purcell, 30, libretto by Irish-born poet Nahum Tate, 37.

1690

The Battle of Beachy Head June 30 gives the French a triumph over an English fleet commanded by Admiral Arthur Herbert, 43, earl of Torrington, who will be acquitted by a court martial of charges that he held back.

The Battle of Fleurus gives Louis XIV's marshal François Henri de Montmorency-Bouteville, 62, duc de Luxembourg, a victory over the prince of Waldeck.

The Battle of the Boyne July 1 completes the Protestant conquest of Ireland as England's William III defeats the Catholic pretender James II and his French supporters. James flees back to France, Dublin and Waterford fall to the Protestant English, but Limerick resists, and the Irish Jacobite Patrick Sarsfield, earl of Lucan, forces William to lift his siege of Limerick.

Spain and Savoy join the League of Augsburg against France's Louis XIV.

Ottoman forces drive the Austrians out of Bulgaria, Serbia, and Transylvania, retake Belgrade, and force the Serbs to flee into southern Hungary.

Two Treatises of Civil Government by English philosopher John Locke, 57, present a theory of limited monarchy, a social contract that will greatly influence the future course of monarchical government: "The liberty of man in society is to be under no other legislative power but that established by consent in the commonwealth, nor under the domination of any will, or restraint of any law, but what that legislative shall enact according to the trust put in it."

1690 ***(cont.)*** Slaves on Jamaican sugar estates reach an estimated total of 40,000 (*see* 1789; 1820).

The city of Calcutta is founded by the East India Company's Job Charnock (*see* 1686; 1756; earthquake, 1737).

Commodity futures trading has its beginnings in Japan where merchants trade rice receipts.

A paper mill put up at Germantown in the Pennsylvania colony is the first in America. Mennonite clergyman William Rittenhouse, 46, has been in Pennsylvania for 2 years and organized a paper manufacturing company (*see* Robert, 1798).

Public Occurrences Both Foreign and Domestick begins publication September 25 at Boston, but British authorities have forbidden publication of anything without authority from the Crown and suppress printer Benjamin Harris's newspaper after one day, destroying all undistributed copies (*see* 1704).

"An Essay Concerning Human Understanding" by John Locke (above) tries to resolve the question of what human understanding is and is not capable of dealing with. Locke returned to England early last year after a self-imposed exile of nearly 6 years in Holland, he has become commissioner of appeal following the Glorious Revolution of 1688, and his essay is the fruit of 17 years' effort.

Painting *The Mill* by Meindert Hobbema. David Teniers the Younger dies at Brussels April 25 at 79.

Theater *Amphitryon, or The Two Socia's* by John Dryden in October at London's Drury Lane Theatre.

Nantucket colonists launch an offshore whaling industry. They have sent to the mainland for Cape Cod shipwright-whaler Ichabod Paddock, who has set up watch towers and instructed the islanders (*see* 1659; 1712).

Dutch mariners smuggle coffee plants out of the Arab port of Mocha. They plant some in their Java colony and send others to the botanical gardens at Amsterdam (*see* 1713).

1691 The Battle of Aughrim July 12 gives William and Mary's Dutch-born general Godert de Ginkel, 47, a victory over Ireland's earl of Lucan Patrick Sarsfield and his French allies. Gen. de Ginkel lays siege once again to Limerick, which surrenders October 3. The Treaty of Limerick which ends the Irish rebellion grants free transportation to France for all Irish officers and men who wish it (the Irish Brigade will play a prominent role in French military history) and promises religious freedom to Irish Catholics, a pledge that will be broken in 1695.

The Ottoman sultan Suleiman III dies in May at age 50 after a 3½-year reign in which he has entrusted the government to the grand vizier Zade Mustapha Kuprili, 54. His brother, 49, will reign until 1695 as Ahmed II.

The Battle of Szcelankemen August 19 gives Louis of Baden a great victory over the Ottoman Turks. Louis has continued the fight abandoned by the Austrians, who have been sidetracked by their war with France as members of the League of Augsburg. The grand vizier Zade Mustapha Kuprili (above) dies in the battle that will lead to the expulsion of the Turks from Hungary.

The Massachusetts colony receives a new charter that gives it all North American territories north to the St. Lawrence River including the Plymouth colony, Maine, and Nova Scotia. The new governor Sir William Phips, 40, is vested with power to summon and dissolve the general court, to appoint military and judicial officers, and to veto acts of the legislature.

The Massachusetts colony (above) extends religious liberty to all except Catholics.

Japanese prospectors discover a copper and silver mine that will make Osaka's Sumitomo family one of the country's dominant financial powers.

Aelbert Jacobsz Cuyp dies at his native Dordrecht in November at age 71.

Theater *Athaliah* (*Athalié*) by Jean Racine in February at Mme. de Maintenon's School for Young Ladies at Saint-Cyr, with students playing the roles.

Opera *King Arthur, or The British Worthy* in March at London's Dorset Garden Theatre, with a libretto by John Dryden, music by Henry Purcell.

German immigrant farmers in the Pennsylvania colony choose heavily wooded lands with clay loams in preference to the light, sandy uplands favored by the English. While the English girdle trees to kill them and then farm among the stumps, the Germans clear their land completely and plow deeply. Instead of planting tobacco, the Germans will stick to wheat, and instead of letting their stock roam freely they will build barns before building houses as they populate Maryland, Virginia, and other colonies (*see* 1683; 1707; 1729; 1734).

1692 The Massacre of Glencoe February 13 enrages Scotland with its treachery and will lead to years of feuding in the highlands. Ian MacDonald, chieftain of the MacDonald clan at Glencoe, has taken an oath of allegiance January 6 to England's William III, but William's agent John Dalrymple, 44, earl of Stair, has suppressed news of the oath and has conspired against MacDonald with Archibald Argyll, 41, and with John Campbell, 57, first earl of Breadalbane. A troop of soldiers whose ranks include many members of the rival Campbell clan has accepted the hospitality of the MacDonald clan at Glencoe, the Campbells have risen at a signal early in the morning, and they kill some 36 of the MacDonalds including MacIan.

The Battle of La Hogue May 29 costs France 15 ships, and Louis XIV's military advisers persuade him that great fleets are a waste of money. France will avenge her loss next year but will leave England and the Dutch to dispute supremacy of the seas.

The Battle of Steinkirk (Steenkirken) July 24 gives France's duc de Luxembourg a victory over England.

England's William and Mary deprive William Penn of his proprietorship and commission New York Governor Benjamin Fletcher as governor of Pennsylvania.

New Hampshire is made a royal colony once again.

Accusations of witchcraft by English-American clergyman Samuel Parris, 39, result in dozens of alleged witches being brought to trial at Salem in the Massachusetts colony. Nineteen will be hanged and one pressed to death in the next 2 years, many of them on the testimony of 12-year-old Anne Putnam (*see* 1664; Mather, 1693; Sewall, 1700).

Scots-American sea captain William Kidd, 38, marries New York widow Sara Oart and increases his holdings by £155 14s (*see* 1698; 1701).

Jamaica in the West Indies has a violent earthquake. Port Royal, stronghold of Spanish Main buccaneers, is shattered, and two-thirds of the pirate city falls into the sea with fortunes in booty.

A calculating machine invented by German philosopher-mathematician Gottfried Wilhelm von Leibniz, 45, multiplies by repeated addition and divides as well as doing the adding and subtracting performed by the 1642 Blaise Pascal machine. Leibniz employs a stepped drum to mechanize the calculation of trigonometric and astronomical tables (*see* Babbage, 1833).

Aesop's Fables is published at London. English journalist Sir Roger L'Estrange, 76, has translated stories said to have originated with an ugly, deformed Greek slave of the 6th century B.C. although some of the fables will be traced to earlier literature.

Hymn *Adeste Fidelis* by English clergyman John Reading. An English version beginning "Oh Come All Ye Faithful" will be published in 1841 with lyrics by another clergyman.

England's two leading provincial towns of Bristol and Norwich have some 30,000 inhabitants each. York and Exeter are the only other towns with as many as 10,000, but London has close to 550,000 and rivals Paris as a center of population.

Witch trials in the Massachusetts colony exposed the superstition and intolerance of the Puritan settlers.

England's total population approaches 6 million, of which half are farm workers in a country that is still half fen, heath, and forest.

1693 French naval forces defeat an Anglo-Dutch fleet off Cape St. Vincent May 26 to 27. The French commander Anne Hilarion de Cotentin, 50, comte de Tourville, avenges the defeat he suffered last year in the Battle of La Hogue and gains another victory June 30 at the Battle of Lagos off Portugal.

The Battle of Neerwinden July 29 gives the duc de Luxembourg another victory over the English, but William III remains in the field as the French sack Heidelberg for a second time.

Swiss Protestant cantons agree to supply mercenary troops to the Dutch after Catholic cantons have supplied mercenaries for Louis XIV to throw against the Dutch. The Catholic cantons respond by supplying mercenaries to the Spanish, a move the Protestant cantons will counter by supplying mercenaries to the English as well as to the Dutch.

Scotland's clan Macgregor mourns the death of its chief Gregor Macgregor and acknowledges his son-in-law Rob Roy (so called because of his red hair) as the new chief. Rob Roy, 22, obtains control of lands from Loch Lomond to the Braes of Balquihidder (*see* 1712).

William III initiates England's national debt by borrowing £1 million on annuities at an interest rate of 10 percent (*see* Bank of England, 1694).

The Massachusetts colony reduces maximum legal interest rates to 6 percent, down from 8 in 1641.

English naturalist John Ray tries to classify different animal species into groups largely according to their toes and teeth (*see* Linnaeus, 1737).

The Amish (or Amish Mennonite) sect has its beginnings in a schism from the Mennonite church in Switzerland led by Swiss Mennonite bishop Jacob Amman, whose followers observe strict discipline marked by opposition to change in dress or way of life (*see* 1537).

The College of William and Mary is founded by royal charter in the Virginia colony at Middle Plantation, later to be called Williamsburg (*see* 1699). The college will award its first baccalaureates in 1770 (*see* Phi Beta Kappa, 1776).

The Wonders of the Invisible World by Boston Congregationalist minister Cotton Mather analyzes with scientific detachment the work of devils among the Salem witches (*see* 1684; 1692; 1702).

Theater *The Old Batchelour* by English playwright William Congreve, 23, in March at London's Drury Lane Theatre: "Thus grief still treads upon the heels of pleasure,/Marry'd in haste, we may repent at leisure" (V, iii); *The Double Dealer* by Congreve in October at London's Drury Lane Theatre.

French composer François Couperin, 25, is named organist *du roi* and begins a notable career of producing *clavecin* works and chamber music.

1693 ***(cont.)*** Russia's Church of the Intercession of the Holy Virgin is completed at Fili near Moscow.

A November earthquake in the Italian province of Catania kills 60,000.

1694 The Royal Navy bombards Dieppe, Le Havre, and Dunkirk, but the French turn back an attack on Brest despite having been weakened by hunger and disease (below).

The Triennial Bill passed by Parliament provides for new English Parliamentary elections to be held every third year. The Place Bill prevents officers of the Crown from sitting in the House of Commons.

Persia's Shah Suleiman dies after a dissolute reign of 27 years. His son Husein, 19, succeeds to the throne and will reign until 1722.

England's Mary II dies of smallpox December 28 at age 32, leaving her husband William to rule alone.

The Bank of England chartered July 27 opens in London's Threadneedle Street to compete with small private banks that have grown out of the city's widely distrusted goldsmiths. A company of merchants headed by Scots financier William Paterson, 36, has received the charter in return for loaning the hard-pressed government £1.2 million, the government agrees to accept Bank of England notes in payment of taxes, and the new joint stock company will soon control the nation's money supply by setting the bank rate (discount rate) for commercial banks (*see* 1708; Bank of Amsterdam, 1609).

Parliament doubles the English salt tax to raise money for the continuing war with France.

Disease epidemics sweep through France's lower classes and take a heavy toll among people weakened by hunger and exhaustion.

England's press censorship ends with the expiration of the Licensing Act which is not renewed for 1695.

Painting *Hampton Court Beauties* by Sir Godfrey Kneller. Japanese *ukiyoe* painter Morinobu Hishikawa dies at age 76 after a career in which he has pioneered the art of making prints that depict the everyday life of the people.

Moravia's Frain Castle is completed with an oval Hall of the Ancestors after 6 years of construction by Austrian architect Johann Bernhard Fischer von Erlach, 38.

"All France is nothing more than a vast poorhouse, desolate and without food," writes François de Salignac de La Mothe-Fénelon, tutor to the grandson of Louis XIV.

French government economic controls prevent the free flow of food into famine districts, and speculators corner grain supplies, adding artificial scarcity to the natural famine that grips the country.

1695 Marshal Luxembourg dies January 4 at age 66 as the War of the League of Augsburg continues. François de Neufville, 51, duc de Villeroi succeeds to the command of French forces in the Low Countries but will prove far less capable.

French siege forces under the duc de Villeroi bombard Brussels in August, damaging the interior of the 150-year-old town hall and destroying the wooden guild halls.

England's William III recaptures Namur from the French in September.

The Ottoman sultan Ahmed II has died of dropsy January 27 after a 3-year reign in which he has conquered Trebizond, Albania, Euboea, upper Greece, and much of the Peloponnesus. His nephew, 32, will reign until 1703 as Mustapha II.

Russian forces under Peter the Great lay siege to Azov on the Don, but the siege fails, and the Turks inflict heavy casualties.

The Royal Bank of Scotland is founded.

The Company of Scotland trading to Africa and the Indies is created by an act of the Scottish Parliament. Bank of England founder William Paterson, who has had a falling out with his colleagues and returned to Edinburgh, has proposed a scheme to establish a settlement on the Isthmus of Darien (Panama) and "thus hold the key of the commerce of the world," but the project will end in disaster (*see* 1698; Paterson, 1694).

A nation's wealth depends not on how much money she possesses but on what she produces and exchanges, writes French economist Pierre Le Pesant, 49, Sieur de Boisguilbert, in *Le Détail de la France, la cause de la diminution de ses biens, et la facilité du remède . . .* Boisguilbert describes the ruin that the burdensome taxes of the late Jean Baptiste Colbert has brought to all classes of society and says France could regain her prosperity by abandoning war and switching from a policy of mercantilist protection to one of free enterprise based on agriculture (*see* Sully, 1598; Smith, 1776).

English botanist Nehemiah Grew, 54, isolates epsom salts (magnesium sulfate) from springs in the North Down. His four-volume *The Anatomy of Plants* in 1682 made the first observations of sex in plants.

The University of Berlin has its beginnings. It will be formally founded in 1810.

Fiction *Den Vermakelyken Avantuiler* by Dutch physician-novelist Nicolaas Heinsius, 39, is a picaresque novel modeled on a Spanish novel.

Theater *Love for Love* by William Congreve 4/30 at the Lincoln's Inn Fields Theatre, London: "You must not kiss and tell" (II).

Brussels begins rebuilding (above); by 1699 the Grand'Place will have new stone guild halls.

England imposes a window tax that will influence residential architecture until 1851.

Annapolis is laid out in the Maryland colony to serve as the colonial capital.

The duc de Montaussier invents the soup ladle, but guests at most French tables continue to dip into a common tureen with their own wooden or pewter spoons, an advance over the two-handled porringer passed round the table to be sipped from in turn.

1696 Parliament suspends the Habeas Corpus Act of 1679 following discovery of Sir George Barclay's plot to assassinate William III. Barclay is hanged along with Roman Catholic Jacobite priest Robert Charnock, 33, and the authorities also arrest Sir John Fenwick, 51, who will be hanged early next year, the last Englishman to be condemned by a bill of attainder. The Trials for Treason Act passed by Parliament requires two witnesses to prove an act of treason.

Poland's John III Sobieski dies June 17 at age 72 after a 20-year reign with the last 12 years full of disaster and humiliation (*see* 1697).

Russian forces sent by Peter I defeat Ottoman defenders July 28 to capture the fortress that commands the Sea of Azov and the Black Sea. Other Russian troops conquer Kamchatka.

Spain establishes a Florida colony at Pensacola as a defense against the French.

English journeymen hatters strike for higher wages and better working conditions.

The Navigation Act passed by Parliament April 10 forbids England's American colonists to export directly to Scotland or Ireland (*see* 1663; 1699).

Parliament reforms English coinage at the instigation of John Locke and Isaac Newton.

The first English property insurance company is founded.

England's Board of Trade and Plantations is founded.

Theater *Love's Last Shift, or The Fool in Fashion* by English playwright Colley Cibber, 24, in January at London's Royal Theatre in Drury Lane: "One had as good be out of the world as out of the fashion"; *The Relapse, or Virtue in Danger* by English playwright John Vanbrugh, 32, 11/21 at London's Drury Lane Theatre; *The Gambler (Le Joueur)* by French playwright Jean-François Regnard, 41, 12/19 at the Comédie-Française, Paris; *Woman's Wit, or The Lady in Fashion* by Cibber in December at London's Drury Lane.

Hymn "While Shepherds Watched Their Flocks by Night" by Nahum Tate and his collaborator Nicholas Brady, 37, is published in their *New Version of the Psalms of David.*

Grapefruit cultivation in America has its origin in seeds from the Polynesian pomelo tree *(Citrus grandis)* introduced into Barbados by an English sea captain named Shaddock. A sweeter and thinner mutation of the fruit, or a botanist's development of the "shaddock," will be called grapefruit (*see* Lunan, 1814).

A pioneer English statistician calculates that England's population will reach a high of 22 million in 3500 A.D. "in case the world should last so long."

1697 Sweden's Charles XI dies April 5 at age 40 after a brilliant 37-year reign. He is succeeded by his son, 14, who will reign until 1718 as Charles XII and whose deeds will eclipse those of his father.

The Russian czar Peter visits Holland, France, and England incognito, the first Russian sovereign to venture abroad. A hulking giant of 25 who towers well over 6 feet in an age when few men approach that height, Peter tastes the fruits of Western civilization and determines to westernize Russia.

Poland elects the elector of Saxony Frederick Augustus I, 27, to succeed the late John III Sobieski; he is crowned in September and will rule until 1733 as Augustus II.

The Battle of Zenta September 11 gives Eugene of Savoy a victory over the Ottoman Turks. Eugene kills 20,000 (another 10,000 drown in the river) and captures the Ottoman imperial seal, the army treasury (3,040,000 florins), all the Ottoman artillery, wagons, munitions, and provisions, thousands of camels, oxen, and horses, and 10 of the sultan's wives.

The Treaty of Ryswick September 30 ends the 11-year-old War of the League of Augsburg. France restores to Spain all conquests made since the Treaty of Nijmegen in 1679, and the French East India Company regains the Indian pepper port of Pondichery on condition that the Dutch retain commercial privileges.

France recognizes William III as king of England in the Treaty of Ryswick (above) with William's sister-in-law Anne as heiress presumptive.

Spain cedes the western third of Hispaniola (Saint-Domingue, or Haiti) to France and retains the eastern part (Santo Domingo) under terms of the Treaty of Ryswick (above) (*see* slave revolt, 1794).

Chinese forces conquer western Mongolia.

Voyage Round the World, published from the journals of William Dampier, is the first general survey of the Pacific by an Englishman in a century (*see* breadfruit, 1688; Cavendish, 1586).

France attempts to colonize West Africa.

Gold is discovered in Portugal's Brazilian colony. The extensive deposits bring thousands of prospectors from coastal towns into the Minas Gerais (General Mines) area and attract immigrants from Portugal, many of whom will die of disease, hardship, and starvation (*see* diamonds, 1729).

New York colonist Stephen (Stephanus) Van Cortlandt receives the grant of a manor north of the city.

Nonfiction "An Essay upon Projects" by English journalist Daniel Defoe, 38, who suggests such innovations as an income tax, insurance, road improvements, and an insane asylum.

Poetry *Mother Goose Tales (Contes de ma mère l'oie)* by French author Charles Perrault, 69, whose tales are in many cases based on actual events. Perrault was secretary to France's finance minister Jean Baptiste Colbert from 1664 until 1683.

Theater *The Mourning Bride* by William Congreve 2/20 at the Lincoln's Inn Fields Theatre, London, with Anne Bracegirdle, now 34, creating the role of Almeria: "Music hath charms to soothe a savage breast,/To soften rocks, or bend a knotted oak" (I, i); "Heav'n has

1697 ***(cont.)*** no rage, like love to hatred turn'd,/Nor hell a fury like a woman scorn'd" (III, viii); *The Provoked Wife* by John Vanbrugh in May at the Lincoln's Inn Fields Theatre, London.

Anthem "I Was Glad When They Said" by John Blow for the opening of Christopher Wren's Choir of St. Paul's Cathedral, London.

The Russian czar Peter (above) permits open sale and use of tobacco, imposing taxes to give the state a share in the profits from the lucrative trade (*see* Alexis, 1648).

Vienna's Palace of Prince Eugene is completed by J.B. Fischer von Erlach, who will enlarge it between 1707 and 1710.

1698 A Treaty of Partition signed by the European powers October 11 attempts to deal with the question of the Spanish succession. Spain's Carlos II is childless, and the Spanish house of Hapsburg is doomed (*see* 1700).

Discourses Concerning Government by the late English republican leader Algernon Sidney will be influential in encouraging liberal provisions in various constitutions (*see* 1683).

West Africa's rich Dahomey state is invaded by covetous Oyo cavalrymen, who will hold the country until the 1730s despite harassment from guerrillas. Dahomey has gained her wealth by supplying slave traders (*see* 1747).

Arab forces expel the Portuguese from Africa's east coast.

Parliament opens the slave trade to British merchants, who will in some cases carry on a triangular trade from New England to Africa to the Caribbean islands to New England. The merchant vessels will carry New England rum to African slavers, African slaves on "the middle passage" to the West Indies, and West Indian sugar and molasses to New England for the rum distilleries.

The first ships of the Company of Scotland trading to Africa and the Indies sail from Leith July 26 with 1,200 pioneers who include William Paterson and his wife and child (*see* 1695). They arrive November 4 at Darien having lost only 15 men en route, they call the country New Caledonia, they choose a well-defended spot with a good harbor midway between the Spanish strongholds Porto Bello and Cartagena, but lack of provisions, sickness, and anarchy will soon reduce the settlers to misery (*see* 1699).

The London Stock Exchange is founded. It is the world's first true stock exchange (*see* New York, 1792; 1825).

English engineer Thomas Savery, 48, pioneers the steam engine with a crude steam-powered "miner's friend" to pump water out of coal mines (*see* Newcomen, 1705; Watt, 1765).

Captain Kidd seizes the Armenian vessel *Auedagh Merchant,* captures other prizes, and sails for the West Indies where he will find that he has been proclaimed a pirate (*see* 1692). Commissioned in 1696 to head an expedition against pirates in the Indian Ocean, Kidd took his first prizes off Madagascar last year. He will surrender to New England authorities next year on the promise of a pardon (but *see* 1701).

Architect Jules Hardouin-Mansart lays out the Place Vendôme at Paris with a statue of Louis XIV in the center (*see* 1810).

London's Whitehall Palace burns down except for the Banqueting Hall built by Inigo Jones in 1622.

Sunflowers introduced by the Russian czar Peter on his return from the West will be the leading source of oilseeds in Russia and eastern Europe (*see* 1510).

Sparkling champagne is pioneered at the French abbey d'Hautvilliers by cellarer Dom Pierre Perignon, 60, who uses a new blend of grapes and corked bottles of strong English glass (*see* 1743).

Two-thirds of the New York colony's population remains on Long Island and near the mouth of the Hudson as the Iroquois and other hostile tribes block expansion to the north and west.

1699 The Treaty of Karlowitz January 26 ends a 17-year struggle against the Turks, who are obliged to give up all of Hungary except for the Bánat of Temesvár to Austria which also obtains Transylvania, Croatia, and Slavonia. Poland regains Podolia and the Turkish part of the Ukraine; Venice receives the Morea and most of Dalmatia.

Denmark and Russia sign a mutual defense pact.

The Treaty of Preobrazhenskoe provides for the partition of the Swedish Empire among Denmark, Russia, Poland, and Saxony.

Denmark's Christian V dies in a hunting accident August 25 at age 53. He is succeeded by his son, 28, who will reign until 1730 as Frederick IV.

William Dampier sails to the Pacific in the superannuated 290-ton ship *Roebuck* on the first Pacific expedi-

Piracy on the high seas brought infamy to Kidd, Teach, and others but was the basis of some great family fortunes.

tion to be fitted out by the Admiralty, he explores the west coast of Australia, finds it disappointing, and rounds the northern coast of New Guinea to discover a large island which he names New Britain (it will later prove to be three adjoining islands; *see* 1606; 1697; Cook, 1770).

French-Canadian explorer Pierre Lemoyne, 38, sieur d'Iberville, pioneers settlement of the North American Gulf Coast with his brother Jean Baptiste Lemoyne, 19, sieur de Bienville. They land on Dauphin island at the mouth of Mobile Bay on their way to establish French colonies for Louis XIV in the Mississippi Delta (*see* Mobile, 1702).

Cahokia is founded by French priests of the Seminary of Foreign Missions. It is the first permanent settlement in the Illinois wilderness (*see* 1675; 1717).

The Virginia colony moves its capital to Middle Plantation (Williamsburg). A fire last year drove the colonists out of Jamestown, and the 66-year-old settlement at Middle Plantation will remain the Virginia capital of Williamsburg until 1779 (*see* College of William and Mary, 1693).

Disease ravages the Darien colony of the Company of Scotland trading to Africa and the Indies, William Paterson loses his wife and child, is reduced by illness to looking "more like a skeleton than a man," and is carried weak and protesting aboard ship in June. The hapless survivors embark in three ships for home and reach England after a stormy passage in December, two shiploads of reinforcements that left England in May and four that left in August come to grief when they encounter superior Spanish forces at Darien, the Spanish oblige them to capitulate, and few of the Scotsmen will ever reach home (*see* 1698).

The Woolens Act passed by Parliament under pressure from the English wool lobby forbids any American colony to export wool, wool yarn, or wool cloth "to any place whatsoever." The act works a certain hardship on rural New Englanders, but while nearly every New England family keeps sheep and has a spinning wheel, few Americans are eager to enter the woolen manufacturing industry.

Yellow fever epidemics kill 150 at Charleston and 220 at Philadelphia.

The Massachusetts colony passes a law designed to prevent the spread of infectious diseases.

Theater *Xerxes* by Colley Cibber in February at London's Drury Lane Theatre; *The Constant Couple, or A Trip to Jubilee* by Irish playwright George Farquhar, 22, in November at the Drury Lane; *The Tragical History of King Richard III* by Cibber in December at the Drury Lane (adaptation of the 1592 Shakespeare play).

Stockholm's Drottningholm Palace is completed in French Renaissance style to provide a summer residence for the royal family on the island of Lovo in Lake Malar.

1700 Moscow and Constantinople sign a truce June 23, ending 5 years of war. The Russians retain Azov but give up their Black Sea fleet.

The Great Northern War begins in Europe as Russia, Poland, and Denmark join forces to oppose Swedish supremacy in the Baltic. Danish troops invade Schleswig, and Saxon troops invade Livonia, but Sweden's Charles XII, now 18, surprises the Danes by landing troops in Zeeland. Threatening Copenhagen, he forces Denmark to sign the Treaty of Travendal August 18 and to remove herself from the alliance against Sweden.

Charles lands 8,000 Swedish troops at Narva in Ingermanland November 30 and decisively defeats a Russian force that had been besieging that town.

Spain's Carlos II dies November 1 at age 39 after a 35-year reign. He has named as his heir Philip of Anjou, 17, grandson of France's Louis XIV; the first Bourbon king of Spain ascends the throne and will reign until 1746 as Philip V (but *see* War of the Spanish Succession, 1701).

The Selling of Joseph by Boston jurist Samuel Sewall, 48, condemns the selling of slaves. Sewell 3 years ago made public confession of error and guilt for his part in condemning to execution 19 alleged witches at Salem in 1692.

Smallpox will kill an estimated 60 million Europeans in the century ahead.

Poetry "Annie Laurie" by Scottish poet William Douglas, 28, whose song is addressed to Anne, daughter of Sir Robert Laurie of the Maxwellton family: "Maxwellton's braes are bonnie/. . . and for bonnie Annie Laurie/ I'd lay me down an' dee" (*see* song, 1838); *Hans Carvel* by Matthew Prior who writes, "What if to Spells I had Resource?/'Tis but to hinder something Worse./The End justifies the Means."

Theater *The Way of the World* by William Congreve in March at the Lincoln's Inn Fields Theatre, London: "I like her with all her faults; nay, like her for her faults" (I, iii); " 'Tis better to be left/Than never to have loved" (II); *The Ambitious Stepmother* by English playwright Nicholas Rowe, 27, in December at the Lincoln's Inn Fields Theatre; *Love Makes a Man, or The Fop's Fortune* by Colley Cibber 12/13 at London's Drury Lane Theatre.

The Gregorian calendar of 1582 is adopted by the German Protestant states by order of the Diet of Regensburg, but England and her colonies continue to use the Old Style Julian calendar as do many other countries (*see* 1752).

Boston ships 50,000 quintals of dried codfish to market, the best of it to Bilbao, Lisbon, and Oporto. The refuse goes to the West Indies for sale to slaveowners.

London's population reaches 550,000, up from 450,000 in 1660. Despite the heavy losses to plague in 1665 and the destruction by fire of much of the city in 1666, London is the largest city in Europe.

England's American colonies have an estimated population of 262,000 with 12,000 each in Boston and Philadelphia, and 5,000 in New York (*see* 1765).

18th Century

1701 The first Prussian king crowns himself at Königsberg January 18 to begin a 12-year reign as Frederick I. The Holy Roman Emperor Leopold I has given the elector of Brandenburg Frederick III, 42, sanction to assume the monarchy in return for a promise of military aid.

The Great Northern War continues with a Saxon siege of Riga. Sweden's Charles XII relieves the city June 10 and proceeds to invade Poland, beginning a 6-year string of victories over Poland and Saxony.

The War of the Spanish Succession begins in Europe as Philip of Anjou gains recognition as king of Spain, especially in Castile, and the Holy Roman Emperor Leopold I moves to take over Spain's Dutch and Italian possessions (*see* 1700). England and Holland, fearful of having the France of Louis XIV joined with Spain, form a Grand Alliance with the emperor, and Eugene, prince of Savoy, joins the alliance September 7.

The Battle of Feyiase in West Africa ends in victory for Ashanti tribesmen over their onetime overlords, the Denkyira. Osei Tutu, their leader, will build a powerful empire in the next decade, enriching his people by trading in gold and slaves (*see* 1712).

Louisiana becomes a province of France covering most of the area drained by the Mississippi, Missouri, and Ohio rivers (*see* 1682; 1699; 1763).

The Charter of Privileges gives Pennsylvania the most liberal government of any English colony in America (*see* Penn, 1682).

Detroit has its origin in the French settlement Fort Pontchartrain established July 24 on the strait between Lake Erie and Lake St. Clair by Sieur Antoine de la Mothe Cadillac, 43, who needs a fort to control the entrance to Lake Huron from Lake Erie and thus control trade with the Illinois country of Louisiana (above). The name Detroit will be based on the French word for *strait*.

London Jews build a synagogue in Bevis Marks that will stand for more than 2 centuries (*see* 1657). The congregation consists entirely of Sephardic (Spanish and Portuguese) Jews (*see* 1723).

Yale University has its beginnings in the Collegiate School established at Saybrook in the Connecticut colony (*see* Elihu Yale, 1718).

The University of Venice is founded.

Nonfiction *Hankampu* by Japanese historian Arai Hakuseki, 27, is a history of Japan's *daimyo* (feudal lords).

Theater *Sir Harry Wildair* by George Farquhar in April at London's Drury Lane Theatre; *Tamerlane* by Nicholas Rowe, in December at the Lincoln's Inn Fields Theatre; *The Funeral, or Grief à-la-mode* by English playwright-essayist Richard Steele, 29, in December at the Drury Lane.

Pirate William Kidd goes to the gallows July 6. Seized at Boston in 1699, Kidd has been sent to England for trial, but the rope breaks twice before he is dispatched (*see* 1698).

A seed-planting drill invented by Berkshire farmer Jethro Tull, 27, sows three parallel rows of seeds at once and will increase crop yields by reducing seed waste (*see* 1782; Swift, 1726).

1702 England's William III dies March 8 at age 51 after falling from his horse and suffering a chill. He is succeeded by his sister-in-law Anne, 37, who will reign until 1714, the last monarch of the House of Stuart.

The War of the Spanish Succession widens in Europe as the Grand Alliance declares war on France May 14.

Queen Anne (above) names John Churchill, 52, husband of her court favorite Sarah, as captain-general of England's land forces and raises him from earl of Marlborough to duke of Marlborough December 14 after he has forced the surrender of Kaiserswerth on the Rhine in June, Venlo on the Meuse in September, and Liège October 29.

Sir George Rooke, 52, captures part of the Spanish treasure fleet at Vigo Bay in October after failing to take Cadiz. Rooke destroys the warships of France and Spain.

Warsaw and Cracow fall to Sweden's Charles XII, who has invaded Poland in the Great Northern War that will be fought largely on Polish soil.

Dutch jurist Cornelius van Bynkershoek, 29, establishes the 3-mile territorial sea zone, ruling that a nation's territory extends 3 miles offshore (*see* Grotius, 1609). Van Bynkershoek will be made a member of the Supreme Council of Holland, Zeeland, and West Friesland next year (*see* 1793).

East and West Jersey combine to form the English colony New Jersey (*see* 1665; 1676).

The Chusingura "Forty-Seven Ronin" incident December 14 stirs Japan as retainers of the late lord of Ako, Asano Naganori, kill Kira Yoshinaka. The 47 *ronin* (unemployed samurai) have followed Confucian ethic in avenging the death at 62 of their lord, who was ordered to commit *seppuku* (ritual suicide) last year for fighting at Edo Castle but they have broken the law in killing the kinsman of the shogun and will be ordered to commit *seppuku*.

Mobile, Alabama, has its beginnings in the Fort Louis settlement founded by the Lemoyne brothers (*see* 1699). First French settlement on the Gulf Coast,

Mobile will take its name from that of the Mauvilia Indians who inhabit the region (*see* 1704).

A yellow fever epidemic kills 570 New Yorkers.

London's *Daily Courant* begins publication March 11. The city's first daily newspaper, it will have 20 competitors by the end of the century.

Fire destroys important scientific papers at Sweden's great center of learning at Uppsala.

Magnalia Christi Americana by Cotton Mather is a well-documented history of New England compiled to show that God is at work in the new land (*see* 1693; smallpox inoculation, 1721).

Japanese painter Korin Ogota, 41, unites the two imperial schools of Japanese painting, the Kano and the Yamato.

Theater *The Inconstant, or The Way to Win Him* by George Farquhar in February at London's Drury Lane Theatre; *She Wou'd and She Wou'd Not, or The King Imposter* by Colley Cibber 11/26 at the Drury Lane.

England's Queen Anne gives royal approval to horseracing and originates the sweepstakes idea of racing for cash prizes (*see* Ascot, 1711).

Salzburg's Church of the Holy Trinity is completed by J. B. Fischer von Erlach after 8 years of work.

1703 The Grand Alliance proclaims Austria's Archduke Charles, 18, king of Spain, and he prepares to invade Catalonia as the War of the Spanish Succession continues. England's Duke of Marlborough invades the Spanish Netherlands, taking Bonn, Huy, Limburg, and Guelders.

Bavarian forces invade the Tyrol but are repulsed.

Sweden's Charles XII defeats a Russian force at Pultusk April 21 and lays siege to Thorn as the Great Northern War continues.

The Ottoman sultan Mustapha II is dethroned September 3 (and soon dies of melancholia at age 41). His brother, 30, will reign until 1730 as Ahmed III.

St. Petersburg is founded May 1 by Russia's Peter I on reclaimed marshlands at the mouth of the Gulf of Finland. Peter makes the new city Russia's capital and will make it the nation's seat of power, turning Russia's focus to the West.

The Methuen Treaty between England and Portugal December 27 facilitates trade in English woolens and Portuguese wines, which come from vineyards in the Oporto area. English families own many of the vineyards, and England will admit the wine at duties one-third lower than those demanded of French wines in return for Portugal's agreement to import all her woolens from England.

Daniel Defoe is pilloried and serves a brief prison term for last year's ironic pamphlet "The Shortest Way with Dissenters," which has outraged both Whigs and Tories.

Theater *The Fair Penitent* by Nicholas Rowe, in May at the Lincoln's Inn Fields Theatre, London, introduces Rowe's rakish seducer the "gay Lothario"; *The Lying Lover, or The Ladies' Friendship* by Richard Steele 12/2 at London's Drury Lane Theatre is based on the Corneille play *Le menteur* of 1643; *Sonezakishinju* by Monzaemon Chikamatsu at Osaka's Takemoto Theater (puppet show).

The great storm that strikes England November 26 to 27 destroys the Eddystone Lighthouse and kills thousands.

A Japanese earthquake and fire December 30 destroys Edo and kills an estimated 200,000. The country will have further catastrophes soon (*see* 1707).

1704 Thorn falls to Sweden's Charles XII after an 8-month siege in which he has lost only 50 men. Charles has his ambassador at Warsaw use bribery and intimidation to secure the election July 2 of Stanislas Leszczynski, 37, to replace the elector of Saxony Augustus II as king of Poland (*see* 1705).

The Cossack hetman Mazepa helps Russia's Peter I in the Volhynian campaign as the Great Northern War continues.

Gibraltar (Jebel-al-Tarik) falls to English forces August 4 (*see* 1702); Admiral George Rooke wrests the rocky fortress from the Spanish, and the British will hold the entrance to the Mediterranean for centuries.

The Battle of Blenheim (Blindheim, or Hochstadt) August 13 gives England's Duke of Marlborough a stunning victory over the French-Bavarian-Prussian coalition. Supported by Eugene, prince of Savoy (whose realm is overrun in his absence by French forces under the duc de Vendôme), Marlborough himself leads the cavalry charge that breaks the enemy's resistance. He drives his foes into the Danube, hundreds drown, the allies sustain 4,500 casualties plus 7,500 wounded, and the English lose 670 plus 1,500 wounded. Marlborough and Eugene take 11,000 prisoners including the French general Camille de Tallard, 52, with 24 battalions of infantry and four regiments of dragoons that include the finest in the French army. The French and Bavarians lose 100 guns, and the French survivors retreat first to the Rhine, then to the Moselle.

Canadian Indians and French forces attack Deerfield in the Massachusetts colony February 29, killing 49, carrying off 112 captives, 40 of them under age 12.

The "Cassette girls" arrive at Mobile on the Gulf Coast in quest of husbands. The 25 young French women carry small trunks (*casettes*) filled with dowry gifts from Louis XIV (*see* 1702).

America's first regular newspaper begins publication at Boston. The weekly *News-Letter* published by local postmaster John Campbell, 51, consists of a single 7-by-11.5-inch sheet covered on both sides with news and rumors received from post riders, sea captains, and sailors (*see* 1690; Zenger, 1735).

Fiction *The Battle of the Books* by English satirist Jonathan Swift, 37, is a travesty on the controversy over ancient and modern learning; *A Tale of a Tub* by Swift satirizes corruption in religion and learning; *The Arabian Nights' Entertainment* translated from a 10th century

1704 *(cont.)* work is published in Europe with tales of Sinbad and his search for spices (*see* Burton, 1888).

Kabuki actor Dannjuro Ichikawa, now 44, is murdered onstage at Edo.

London's Christ Church is completed by Sir Christopher Wren, who has designed 52 London churches.

The "Cassette girls" (above) will use okra obtained from African slaves to develop a cuisine that will expand the local fare of maize products, beans, sweet potatoes, and local game and seafood (*see* 1784).

1705 The Holy Roman Emperor Leopold I dies May 5 at age 54 after a 47-year reign. He is succeeded by his son of 26, who will reign until 1711 as Josef I.

Stanislas Leszczynski is crowned king of Poland September 24 to replace the deposed Augustus II, concludes an alliance with Sweden's Charles XII, and supplies Charles with some help against Russia's Peter I in the continuing Great Northern War.

The Austrian archduke Charles lands in Catalonia and English forces help him take Barcelona October 14 in the continuing War of the Spanish Succession. Sentiment against the French has been strong in Catalonia and Valencia; both support Charles's claim to the Spanish throne.

Factum de la France by the sieur de Boisguilbert proposes a single capitation tax—10 percent of the revenues on all property to be paid to the state (*see* 1695). The farmers oppose the idea of taxes, and it finds little support from anyone (*see* George, 1879).

The Newcomen steam engine invented by English blacksmith Thomas Newcomen, 42, at Dartmouth will pave the way for an Industrial Revolution. Helped by John Calley (or Cawley), Newcomen uses a jet of cold water to condense steam entering a cylinder. He thus creates atmospheric pressure which drives a piston to produce power that will be used beginning in 1712 to pump water out of coal mines (*see* Savery, 1698; Darby, 1709; Watt, 1765).

Queen Anne confers knighthood on Isaac Newton, now 62.

Halley's Comet will receive that name on the basis of studies reported by English astronomer Edmund Halley, 49, who notes that comets observed in 1531, 1607, and 1682 followed roughly the same paths. Halley suggests that they were all the same comet and that it will reappear in 1758 (*see* 1066).

Theater *The Tender Husband, or The Accomplished Fool* by Richard Steele 4/23 at London's Drury Lane Theatre, is an adaptation of the 1667 Molière comedy *Le Sicilian; The Mistake* by John Vanbrugh 10/27 at London's Haymarket Theatre; *The Confederacy* by John Vanbrugh 10/30 at London's Haymarket Theatre; *Ulysses* by Nicholas Rowe 11/23 at London's Haymarket Theatre; *Idomeneus* (*Idomenée*) by French playwright Prosper Jolyot, 31, sieur de Crébillon, 12/29 at the Comédie-Française, Paris.

Opera *Almira* 1/8 at Hamburg, with music by German composer George Frideric Handel, 20.

First performances *St. John Passion* by George Frideric Handel.

Blenheim Palace goes up at Woodstock in Oxfordshire for the duke of Marlborough. Queen Anne has commissioned London playwright John Vanbrugh (above) to design the palace as a tribute to the victor of last year's great battle; Vanbrugh turns from drama to architecture. He gets a hand from architect Nicholas Hawksmoor, 44, who has been an aide to Christopher Wren.

Famine strikes France, causing widespread distress that will continue for years.

1706 The Battle of Ramillies May 23 gives the duke of Marlborough a victory over a French army commanded by the duc de Villeroi. Marlborough's triumph is followed by the submission of Brussels, Antwerp, Ghent, Ostend, and other major cities in the continuing War of the Spanish Succession.

English forces raise a French siege of Barcelona May 23. Portuguese forces invade Spain in June and install the Austrian archduke Charles as king at Madrid, but Philip V drives them out in October.

Savoy's Prince Eugene vanquishes a French army at Turin September 7 with help from Prussian forces under Leopold of Dessau. Lombardy submits to Eugene, Charles III is proclaimed king at Milan, and the French are driven out of Italy.

Saxony's Elector Frederick Augustus abdicates the Polish crown September 24 in the Treaty of Altranstadt, recognizing Stanislas Leszczynski as king of Poland, and breaks his alliance with Russia's Czar Peter (but *see* 1709).

Portugal's Pedro II dies December 9 at age 58 after a 23-year reign. His son of 17 will reign until 1750 as João V.

Albuquerque, New Mexico, has its origin in the town of Alburquerque founded in the northern part of New Spain. The town is named in honor of New Spain's new viceroy the duke of Alburquerque (the first *r* will be dropped in years to come).

Theater *The Recruiting Officer* by George Farquhar 4/8 at London's Drury Lane Theatre.

1707 The Mughal emperor Aurangzeb dies March 3 at age 88 while campaigning against the Mahrattas after a 49-year reign in which he has annexed Bijapur and Golconda to extend his autocratic sway over all of northern and central India up to the Himalayas, persecuting Hindus and Sikhs. The empire will quickly disintegrate as provincial governors gain virtual independence amidst wars of succession and foreign invasions.

The Battle of Almanza April 25 ends in defeat for Portuguese forces at the hands of a French army commanded by James Fitzjames, 37, duke of Berwick. Spain's Philip V has engaged the bastard son of England's late James II to head a mercenary army against the Portuguese.

The United Kingdom of Great Britain created May 1 unites England and Scotland under the Union Jack, which combines the cross of St. George and the cross of

St. Andrew. Scottish law and legal administration are to remain unchanged, but the Scottish Parliament is to be abolished, Scotland is to send 16 elective peers and 45 members of the Commons to the British Parliament at London, no further Scottish peers are to be created.

Geneva suppresses a popular uprising with help from the oligarchies of Bern and Zurich.

Philadelphia mechanics demonstrate to protest competition from Indian slaves.

A "pulse watch" invented by English provincial physician John Floyer, 58, is the first efficient clinical precision instrument for medical diagnosis. A great believer in cold baths, Floyer has designed the watch to run for exactly one minute (*see* Santorius, 1624).

Willem van de Velde the Younger dies at Greenwich, London, April 8, at age 73.

Theater *The Comical Lovers* by Colley Cibber 2/4 at London's Haymarket Theatre, a comedy made up of scenes from two of John Dryden's plays; *The Beaux' Stratagem* by George Farquhar 3/8 at London's Haymarket Theatre: "How a little love and conversation improve a woman!" (IV, ii) (paid in advance, Farquhar took ill before completing Act II but has finished the play, an extra benefit performance is given 4/29, and Farquhar dies that day at age 30); *Crispin, Rival of His Master (Crispin, rival de son maître)* by French playwright Alain-René Lesage, 34, 3/15 at the Comédie-Française, Paris; *Atreus and Thyseus (Atrée et Tysée)* by Prosper Jolyot, sieur de Crébillon at the Comédie-Française, Paris; *The Royal Convert* by Nicholas Rowe 11/25 at London's Haymarket Theatre; *The Lady's Last Stake, or The Wife's Revenge* by Colley Cibber 12/13 at the Haymarket Theatre.

Salzburg's Kollegienkirche is completed by J. B. Fischer von Erlach after 13 years of construction.

Japan's Fujiyama volcano erupts for the last time.

Fortnum & Mason's opens in Piccadilly. Started by London entrepreneurs William Fortnum, a former footman to Queen Anne, and Hugh Mason, it will make its delicacies a tradition in better English households for centuries.

An emigration begins from the Rhineland Palatinate that will bring thousands of Calvinists, Lutherans, and even some Roman Catholics to England and thence in many cases to America, where the emigrants will settle in New York's Hudson and Mohawk valleys and in Virginia (*see* 1710).

1708 Ghent and Bruges resume their allegiance to France in early July. Fearing that other cities will follow suit, the duke of Marlborough defeats the French July 11 at the Battle of Oudenarde with help from Eugene of Savoy. He lays siege to Lille for nearly 4 months, and the citadel surrenders in December after 30,000 combatants have lost their lives in the continuing War of the Spanish Succession.

Sweden's Charles XII invades the Ukraine and lays siege to the Russian fortress of Poltava, but the Russians intercept an auxiliary army carrying supplies to Charles, his own army is stricken with plague, and he is forced to yield his conquests.

Canadian Indians and French colonists massacre settlers at Haverhill in the Massachusetts colony.

The 2,400 adult whites in the 45-year-old Carolina colony are outnumbered by 2,900 black slaves and 1,100 Indian slaves (*see* 1775).

The United East India Company created by a merger of Britain's two rival East India companies is the strongest European power on the coasts of India. The company ships Indian silks, cottons, indigo (for blue dye), coffee, and saltpetre (for gunpowder) as well as China tea, and it will pay regular dividends of 8 to 10 percent (*see* 1627).

The Bank of England Act prohibits any other bank with more than six partners from issuing banknotes (*see* 1694; clearing house, 1775).

The Sot Weed Factor, or A Voyage to Maryland by English satirist Ebenezer Cook, 36, expresses disgust at the greedy colonials. Cook has gone to America to buy tobacco.

Painting *The Departing Regiment* by French painter (Jean) Antoine Watteau, 23.

Theater *Electra (Electre)* by Prosper Jolyot, sieur de Crébillon 12/14 at the Comédie-Française, Paris.

German organist-composer Johann Sebastian Bach, 23, becomes court organist at Weimar where he will be made court concertmeister in 1714. Bach traveled in the fall of 1705 from Arnstadt to Lübeck for the annual performance of the *Abendmusiken* concerto by the late Swedish organist-composer Dietrich Buxtehude, who died 2 years ago at age 70 and had great influence on young Bach.

Saxon alchemist Johann F. Böttger at Meissen, 14 miles northwest of Dresden, discovers a formula for making hard-paste porcelain of the kind imported from China—a combination of brown stone made pliable with kaolin. Ehrenfried Walther von Tschirenhausen will help Böttger open a Meissenware works at Dresden in 1710 to produce the red porcelain, using kaolin found near the Saxon town of Auenseiditz and at Elbogen in Bohemia (*see* Entrecolles, 1712).

1709 The Battle of Poltava July 8 makes Russia the dominant power in northern Europe as Peter the Great breaks the power of Sweden and forces Charles XII to take refuge in Anatolia.

The elector of Saxony Frederick Augustus I renews his claims to the Polish crown following the defeat of Sweden's Charles XII at Poltava (above). He declares the 1706 Treaty of Altranstadt void, renews his alliance with Russia and Denmark, deposes Stanislas Leszczynski, and resumes the Polish throne that he will try without success to make hereditary but will hold himself until 1733.

The Battle of Malplaquet September 11 is the bloodiest of the War of the Spanish Succession and takes 20,000 allied lives. The French retire in good order from the triumphant prince of Savoy and duke of Marlborough

1709 ***(cont.)*** (whose Tory opponents call him a butcher and use his excesses to attack the power of Britain's Whigs) (*see* 1710).

Kandahar rebels against its Persian overlords as Shah Hussein lets himself be persuaded by the mullahs to impose Shiite fundamentalism on all the Sunnis in his realm. The Ghilzai chief Mir Vais leads the Afghan revolt (*see* 1711).

Japan's Tokugawa shogun Tsunayoshi has died in January at age 62 after nearly 29 years in power. He is succeeded by his cousin, 47, who will rule until 1712 as Ienobu (below).

Portugal creates the Brazilian captaincies of São Paulo and Minas Gerais as Portuguese colonists settle in the interior of Brazil.

Britain's Caribbean colonies import 20,000 slaves per year by official estimates but many are for re-export to North and South America.

Close to 7 million Africans will come across the Atlantic in chains in this century—as many as 100,000 in a single year.

The new Japanese shogun Ienobu (above) releases nearly 9,000 prisoners, most of them victims of Tsunayoshi's laws against killing animals or eating fish or birds.

An industrial revolution begins in England with the discovery that coke, made from coal, may be substituted for charcoal, made from wood, in blast furnaces used to make pig iron and cast iron. Growth of iron smelting has been limited by the fact that it takes 200 acres of forest to supply one smelting furnace with a year's supply of charcoal, but Quaker ironmaster Abraham Darby, 31, finds that coke serves just as well for his furnaces at Coalbrookdale, Shropshire, where he makes iron boilers for the Newcomen engine, invented in 1705. Regular use of coke will not come for 50 years and will await improvements by Darby's son and namesake, but Darby's breakthrough brings an immediate surge in the demand for coal and for the Newcomen engine, whose energy will be used increasingly to permit production of coal from flooded colliery galleries.

The Black Death kills 300,000 in Prussia.

The Fahrenheit alcohol thermometer introduced by German physicist Gabriel Daniel Fahrenheit, 23, has a scale on which "normal" human body temperature is in the neighborhood of 98.6°. The freezing point at standard atmospheric pressure (sea level) is +32° Fahrenheit (F.) on Fahrenheit's scale; 212° F. is the point at which water boils (*see* Celsius, 1742).

Aphorismi de Cognoscendis et Curandis Morbis by University of Leyden medicine and botany professor Hermann Boerhaave, 41, revives the Hippocratic method of the 5th century B.C., emphasizing that medicine's principal aim is to cure the patient. The leading medical technician and clinician of his day, Boerhaave makes observations from the bedside rather than from textbooks, his slim but encyclopaedic medical text will be translated into many European languages as will his *Institutiones medicae in usus annuae exercitationis domesticos digestae* of last year, but like most others he has erroneous ideas about diseases of the solid parts as opposed to diseases of the so-called "humors."

The Tatler begins publication at London. Editor of the new triweekly journal of politics and society is playwright Richard Steele, who inserts essays under the name Isaac Bickerstaff and publishes numerous essays submitted by Joseph Addison, 37 (*see* 1711).

Meindert Hobbema dies at Amsterdam December 7 at age 71.

Theater *The Rival Fools* by Colley Cibber 1/11 at London's Drury Lane Theatre.

The *Gravicembalo col piano e forte* (pianoforte) built by the Florentine inventor Bartolommeo Cristofori, 54, is the first keyboard instrument with a successful hammer action. Cristofori is curator of a collection of musical instruments at the Medici court; his pianoforte has leather-covered poplar hammers that replace the jacks of the harpsichord and permit the player to produce gradations of tone by changing the force and manner in which the keys are struck.

Famine ravages Europe as frost kills crops, fruit trees, and domestic fowl as far south as the Mediterranean coast.

France has food riots as frost creates famine in much of Europe (*see* below).

French officials set up grain depots in an effort to make food distribution more equitable, but they continue to export grain reserves. Restrictions on internal trade strangle shipments of food stores, keep food prices high, and contribute to the starvation.

"The fear of having no bread has agitated the people to the point of fury," writes French controller-general of finances Nicolas Des Marets, 61, the seigneur de Maillebois: "They have taken up arms for the purpose of seizing grain by force; there have been riots at Rouen, at Paris, and in nearly all the provinces; they are carrying on a kind of war that never ceases except when they are occupied with the harvest."

"The winter was terrible," writes the duc de Saint-Simon Louis de Rouvroy, 34, in his journal: "The memory of man could find no parallel to it. The frost came suddenly on Twelfth Night, and lasted nearly two months, beyond all recollection. In four days the Seine and all the other rivers were frozen, and what had never been seen before, the sea froze all along the coasts, so as to bear carts, even heavily laden, upon it" (*see* England, 1684).

1710 Britain achieves her first clean-cut peaceful transfer of power. The Tory party wins a clear majority in the Commons in November and ousts the Whig government headed by the duke of Marlborough.

Britain's Tory cause has gained popular support as the result of the trial and conviction in March of political preacher Henry Sacheverell, 36, who has attacked the Whig ministry. Sacheverell had been impeached at the instigation of Sidney Godolphin, 65, lord high treasurer and an ally of Marlborough, who has remained loyal to the Jacobite cause and is dismissed by Queen Anne.

British forces occupy Acadia but leave the French Acadians in peace (*see* 1632; Nova Scotia, 1713).

Col. Peter Schuyler of the New York colony takes five Iroquois chiefs to the court of Queen Anne to impress them with Britain's power.

Pennsylvania "Dutch" gunsmiths in the next decade will develop the Pennsylvania rifle with a spiral bore for accuracy.

A British copyright law established by Queen Anne will be the basis of all future copyright laws.

Treatise concerning the Principles of Human Knowledge by Irish philosopher George Berkeley, 25, inaugurates the empiricist school of philosophy. He first propounded his system of subjective idealism (Berkeleianism) last year in an "Essay Towards a New Theory of Vision" and will popularize it in 1713 with *Three Dialogues Between Hylas and Philonus.*

Theater *Tuscaret* by Alain-René Lesage 3/15 at the Comédie-Française, Paris; *The Rival Queens,* with *The Humours of Alexander the Great* by Colley Cibber (burlesque), in June at London's Haymarket Theatre.

London's Marlborough House is completed in Westminster by Sir Christopher Wren, now 77.

Castle Howard is completed in Yorkshire for Charles Howard, 36, third earl of Carlisle. Designed by John Vanbrugh, the imposing country house has taken 9 years to build.

Venice's baroque Palazzo Ca'Pesaro designed by the late Baldassare Longhena is completed for the powerful Pesaro family.

A great German migration to America begins as Baron de Graffenreid brings 650 Swiss and German palatines to North Carolina, where they will settle at New Bern on the Neuse River. New York governor Robert Hunter brings 3,000 Germans to settle on the Hudson and produce naval stores.

1711 The Holy Roman Emperor Josef I dies of smallpox at age 32 April 17 at Vienna and is succeeded by his brother, 26, who will reign until 1740 as Charles VI. Heir to all the empire's Austrian territories, he fights to restore the empire of his Hapsburg ancestor Charles V in the continuing War of the Spanish Succession.

Hungarian followers of the patriot Francis II Rákòczi accept the peace of Szamatar May 1, the new emperor (above) agrees to redress Hungarian grievances and to respect the Hungarian constitution, and Rákòczi takes refuge in Turkey after an 8-year revolt in which his forces have threatened Vienna.

Russia's Peter the Great advances with Moldavian and Wallachian allies on the Pruth River but is surrounded there by superior Ottoman forces. Forced to sign the Treaty of Pruth July 21, Peter returns Azov to the Turks, and Sweden's Charles XII is permitted safe return to Stockholm (but *see* 1713).

Afghanistan gains independence after Persia's Shah Hussein sends a 25,000-man army to put down the Afghan uprising at Kandahar (*see* 1709). The Afghan chief Mir Vais prepares his Sunni garrison to fight to the death, beats off Persian assaults on Kandahar, the Persian's Georgian general Khusru Khan is killed, and fewer than 1,000 Persians escape.

Britain undertakes an attack on French Canada with seven of the duke of Marlborough's best regiments augmented at Boston by 1,500 colonials, but French forces abort the invasion of Quebec by sea in August, the French sink 10 ships of the fleet as it enters the St. Lawrence River, and news of the naval disaster persuades Sir Francis Nicholson to give up a projected campaign against Montreal.

Rio de Janeiro is sacked by French forces under René Duguay-Trouin, 38, as Louis XIV fights the Portuguese allies of Britain in the War of the Spanish Succession.

Queen Anne dismisses the duke of Marlborough at year's end as his enemies increase their influence on the queen. They have accused the duke of speculation, and the queen makes James Butler, 46, duke of Ormonde, the commander in chief of British forces.

Tuscorora tribesmen massacre some 200 Carolina colonists, beginning a war that will not end until forces from Virginia move south to quell the revolt.

The Landed Property Qualification Act passed by Parliament's landed excludes Britain's merchants, financiers, and industrialists from the House of Commons. The law will not be effectively enforced but will not be repealed until 1866.

German scholar G. W. Leibniz denies spontaneous generation and attempts to reconcile natural science with divine will. He expounds his conclusion that all living matter is composed not of dead atoms but of living "monads," infinite in their variety.

The Black Death kills 300,000 in Austria, 215,000 in Brandenburg.

The Spectator begins publication March 1 at London under the direction of Richard Steele and Joseph Addison whose journal *The Tatler*, begun in 1709, appeared for the last time January 2. The new nonpolitical journal publishes Steele's *Roger de Coverley Papers* and will continue for 20 months (*see* 1828).

Poetry *Essay on Criticism* by English poet Alexander Pope, 23: "A little learning is a dangerous thing;/Drink deep, or taste not the Pierian spring" (II); "Be not the first by whom the new are tried,/Nor yet the last to lay the old aside" (II); "To err is human, to forgive divine" (II); "Nay, fly to altars; there they'll talk you dead;/For fools rush in where angels fear to tread" (III, lxvi).

Theater *Rhadamisthus and Zénobia* (*Rhadamiste et Zénobie*) by Prosper Jolyot, sieur de Crébillon 1/23 at the Comédie-Française, Paris.

Opera *Rinaldo* 3/7 at London's Haymarket Theatre, with music by George Frideric Handel, libretto by Giacomo Rossi.

1711 ***(cont.)*** Queen Anne establishes the Ascot races (*see* 1702; 1807).

1712 French forces under Marshal Villars defeat British and Dutch forces commanded by the earl of Albemarle Arnold Joost van Keppel, 41, at Denain, the French recapture Douai, Le Quesnoy, and Bouchain, and the Congress of Utrecht opens to resolve the War of the Spanish Succession (*see* 1713).

Bernese forces triumph July 25 at Villmergen in a second Swiss war between Catholics and Protestants. The victory establishes the dominance of the Protestant cantons.

The Russian czarevich Alexius Petrovich infuriates his father Peter the Great by disabling his own right hand with a pistol shot. Peter has taken him on a tour of inspection in Finland, sent him to see to the building of new ships at Staraya Rusya, and ordered him to draw something of a technical nature to show his progress in mechanics and mathematics.

The Japanese Tokugawa shogun Ienobu dies at age 50 after a 3-year reign. He is succeeded by his 3-year-old son, who will reign until 1716 as Ietsugu.

The Ashanti king Osei Tutu in West Africa is killed in an ambush after a reign in which he has used war and diplomacy to create a formidable power.

North Carolina and South Carolina are created by a division of the Carolina colony founded in 1663.

A slave revolt at New York ends with six whites killed before the militia can restore order; 12 blacks are hanged July 4 (six have hanged themselves).

Pennsylvania forbids further importation of slaves.

French naturalist René Antoine Ferchault de Réaumur, 29, describes the ability of a crayfish to grow new claws (*see* steel, 1722; digestion, 1752).

Fiction *The History of John Bull* by Scottish physician John Arbuthnot, 45, satirizes the duke of Marlborough and introduces the name "John Bull" as a symbol of England.

Poetry *The Rape of the Lock* by Alexander Pope is a mock-heroic poem describing a day at Hampton Court where Queen Anne does "sometimes counsel take—and sometimes tea."

Scotland's clan Macgregor chief Rob Roy fails to repay money he has borrowed from the duke of Montrose to carry on his business as cattle dealer. The duke evicts him from his lands and declares him an outlaw, Rob Roy supports himself by depradations on the duke and his tenants, and he foils all efforts to capture him (*see* 1693; Sheriffmuir, 1715).

Vienna's Trautson Palace is completed by J. B. Fischer von Erlach after 3 years of construction.

Nantucket whaling captain Christopher Hussey harpoons a sperm whale from an open boat, probably the first such whale to be killed by man since Phoenician times. *Physeter catodon* has an oil content averaging 65 to 80 barrels—far more than other whales—and Hussey's kill initiates a new era in whaling (*see* 1690). Spermaceti from the sperm whale will be used to make millions of wax candles, its teeth are fine-grained ivory, its skin is high in glycerin, and it is the only whale that produces ambergris, valuable as a perfume fixative.

French missionary Père d'Entrecolles sends home the first accurate description of how the Chinese make porcelain from kaolin (*see* 621; 1708; 1754).

1713 Ottoman forces attack Sweden's Charles XII in his camp at Bender in Moldavia February 1 and take him prisoner; he will remain in Turkish hands for 15 months.

The first Prussian king Frederick I dies February 25 at age 55 after a 12-year reign in which he has welcomed Protestant refugees from France and elsewhere, founded the University of Halle and Berlin's Academy of Sciences, and lavished money on public buildings. His son of 24 will reign until 1740 as Friedrich Wilhelm I.

The Treaty of Utrecht April 11 ends the War of the Spanish Succession (called Queen Anne's War in North America). France's Louis XIV agrees not to unite France and Spain under one king, recognizes the Protestant succession in Britain, agrees to tear down French fortifications at Dunkirk and to fill up Dunkirk harbor, and gives up some North American territories to Britain.

Gibraltar and Minorca are ceded to Britain.

France cedes Newfoundland and Acadia to Britain, which cedes Cape Breton Island to France and changes the name of Acadia to Nova Scotia. France retains fish-drying rights on part of the Newfoundland coast and builds a fortress (Louisburg) on Cape Breton Island (*see* 1755; 1763).

France turns over her fur-trading posts on Hudson Bay to the Hudson's Bay Company, which gains control of the entire region.

The Caribbean island of St. Kitts is awarded to Britain, and the South Sea Company gains the right to send one ship per year to trade at Spain's isthmus city of Portobello (*see* 1716).

Spain cedes Sardinia and Naples to Austria and cedes Sicily to Savoy, which in 1720 will exchange Sicily for Sardinia (*see* 1735).

The South Sea Company receives *asientos* to import 4,800 African slaves per year into Spain's New World colonies for the next 30 years. Founded 2 years ago in anticipation of receiving the *asientos,* the company is essentially a British finance company, but it begins the most active period of British participation in the slave trade (*see* 1720).

The French establish a trading post at the Great Village of Natchez (*see* 1729).

Scots-American sea captain Andrew Robinson at Gloucester in the Massachusetts colony builds the world's first schooner. Its name derives from the Scottish dialect word *scoon,* meaning to skip over the surface of water. Other New Englanders quickly imitate the distinctively American vessel, and they create a fleet of schooners to fish the Grand Banks.

London physician John Woodward receives a letter from Greek physician Emanuel Timoni at Constantinople describing a method for preventing smallpox by immunization. Timoni reports that Greek physician Giacomo Pylarini at Smyrna removes some of the thick liquid from a smallpox pustule, rubs it into a small scratch made with a needle on the skin of a healthy person, and thus protects that person, who generally develops a mild case of smallpox but nothing worse (*see* 1714; Montagu, 1718).

First performances a *Te Deum* and a *Jubilate* by George Frideric Handel 7/7 at St. Paul's Cathedral, London, to celebrate the Peace of Utrecht (above). Queen Anne, who has commissioned the works and five others, grants Handel a handsome lifetime pension of £200 per year.

A privateer operated by Boston's Quincy family captures a Spanish treasure ship "worth the better part of an hundred thousand pounds sterling."

Prague's Clam-Gallas Palace is completed by J. B. Fischer von Erlach after 6 years of construction.

France's Louis XIV receives a coffee bush whose descendants will produce a vast industry in the Western Hemisphere. The 5-foot bush from the Amsterdam greenhouses will be stolen and transported to Martinique (*see* 1690; 1723).

1714 The Treaty of Rastatt March 6 ends the war between Austria and Spain, the Spanish Netherlands become the Austrian Netherlands, and Spain gives up her possessions in Italy and Luxembourg along with those in Flanders.

The House of Hanover that will rule Britain for more than two and a half centuries comes to power August 1 upon the death of the gouty Queen Anne at age 48 after a 12-year reign. Last monarch of the House of Stuart, Anne is succeeded under terms of the 1701 Act of Settlement by Hanover's Prince George Louis, 54, a great-grandson of England's James I, who speaks no English. He lands in England September 18 to assume the throne (the House of Hanover will be renamed the House of Windsor in 1917).

Sweden's Charles XII returns to Stockholm through Hungary, but the Great Northern War continues.

A Turkish-Venetian war begins as the Turks capture Corinth and Venetian stations in Candia (Crete).

London establishes a Board of Longitude and offers a £20,000 prize to anyone who can devise a means of determining a ship's longitude within 30 nautical miles at the end of a 6-weeks' voyage (*see* Greenwich, 1676; Harrison, 1728).

The Virginia colony's lieutenant governor Alexander Spotswood, 38, establishes iron furnaces on the Rapidan River and imports German ironworkers to settle in the colony (*see* 1723).

Daniel Fahrenheit devises a thermometer in which mercury replaces alcohol (*see* 1709).

The *roi de soleil* Louis XIV reigned through 72 years of often stormy French history.

London physician John Woodward publishes Emanuel Timoni's account of smallpox variolation in the Royal Society's *Philosophical Transactions*, arousing interest in Britain and America (*see* 1713; 1718).

Poetry *The Fable of the Bees: or Private Vices, Public Benefit* by Dutch-born London philosopher-satirist Bernard Mandeville, 44, whose octosyllabic verse first appeared in 1705 in *The Grumbling Hive*. Every virtue is basically some form of selfishness, says Mandeville, human nature is essentially vile, and "it is requisite that great numbers be poor" in order that society may be happy (*see* Young, 1771).

Theater *The Tragedy of Jane Shore* by Nicholas Rowe 2/2 at London's Drury Lane Theatre.

Cantata "Ich hatte viel Bekummernis" by court concertmeister Johann Sebastian Bach 6/17 at Weimar's Schlosskirche.

1715 France's Louis XIV dies September 1 at age 76 after a 72-year reign. His great-grandson, 5, will reign until 1774 as Louis XV, initially with Philip II Bourbon, 41, duc d'Orléans as regent.

French prelate André Hercule de Fleury, 62, is named tutor to the new king and will have great influence over Louis until he dies in 1743.

France's middle class, established during the Sun King's long reign, holds a solid position which will become even stronger during the next two reigns.

The death of the *roi de soleil* weakens the position of the Jacobite pretender James Francis Edward Stuart, who has challenged the House of Hanover's claim to the British throne (*see* 1707; 1714). The Scottish Jacobite John Erskine, 40, earl of Mar, raises a rebellion against Britain's George I, gaining support from Simon Fraser, 48, twelfth baron Lovat; Henry St. John Viscount Bolingbroke, 37; James Butler, 50,

1715 ***(cont.)*** duke of Ormonde; Robert Harley, 54, first earl of Oxford; and much of the English and Scottish public, but the Lowlands remain largely loyal to George I. English authorities imprison Oxford in the Tower of London and force Bolingbroke and Ormonde to flee.

The Battle of Sheriffmuir November 13 halts the earl of Mar (above) in his march on Edinburgh. Some 4,000 Royalists under John Campbell, 37, second duke of Argyll, force Mar's 12,000 Jacobites to retreat to Perth after each side has lost 500 men (the outlaw Rob Roy plunders the dead of both sides). The Royalists defeat an uprising at Preston in the north of England November 14. The change in French policy has prevented the pretender James from joining his troops, but he lands near Peterhead December 22 and tries to rally support (*see* 1716).

Kandahar's Ghilzai chief Mir Vais dies after a 4-year reign. His eldest son, 18, succeeds as Mir Mahmoud, but Mahmoud's uncle, Abdullah, seizes power with every intention of making peace with Persia. Mahmoud gathers 40 supporters, assassinates Abdullah, is proclaimed ruler, and will reign until 1725.

French forces in the Indian Ocean take Mauritius from the Dutch, who have held it since 1638.

English colonial forces in America defeat Yamassee and allied tribes in the South Carolina colony and drive them across the Spanish border into Florida.

The Maryland colony is returned to the Calvert family, which lost it in 1688 (*see* 1729).

Black slaves comprise 24 percent of the Virginia colony's population, up from less than 5 percent in 1671.

Japan becomes alarmed at the amount of copper that has been exported by Dutch and Chinese traders and reduces the amount that it will permit to be exported. The number of Dutch vessels permitted to call annually at Nagasaki will be cut to two (*see* 1641).

Methodus Incrementorum Directa et Inversa by English mathematician Brook Taylor, 30, establishes the calculus of finite differences and sets forth Taylor's theorem, a fundamental theorem concerning functions. It is the first treatise on finite differences. Taylor also publishes *Linear Perspective.*

Theater *The Tragedy of Lady Jane Gray* by Nicholas Rowe 4/20 at London's Drury Lane Theatre; *Kokusenya Gassen* by Monzaemon Chikamatsu at Osaka.

First performances *Water Music* Suite No. 1 in F major by George Frideric Handel 8/22 on the Thames at London.

French agriculture has declined in the long reign of Louis XIV (above), much of the peasantry has become impoverished, and great areas of land have gone out of cultivation (*see* Boisguilbert, 1695).

1716 The duke of Argyll's Scottish Royalists disperse Jacobite troops in January, the Old Pretender flees to France February 5, and Jacobite leaders captured at Preston last year are executed. Sir James Radcliffe, 27, third earl of Derwentwater, and William Gordon, sixth Viscount Kenmure, are both beheaded.

Simon Fraser, baron Lovat, receives a pardon and life rent on the Lovat estates as a reward for having brought in his clan on the side of George I, an action he took to obtain his cousin's estates.

Eugene of Savoy defeats Ottoman forces August 5 at Peterwardein and the Holy Roman Emperor Charles VI enters the war against the Turks.

Russia's Peter the Great receives a letter from his son Alexius Petrovich requesting permission to become a monk. Peter writes to Alexius from abroad August 26 urging him to come home and rejoin the army without delay if he wishes to remain czarevich, but Alexius flees to Vienna, places himself under the protection of his brother-in-law the emperor Charles VI, and is sent first to the Tyrolean fortress of Ahrenberg, then to the castle of San Elmo at Naples in company with his Finnish mistress Afrosina.

Japan's Tokugawa shogun Ietsugu dies at age 7 after 4 years in power and is succeeded by Yoshimune, 39, who will rule until 1745. Yoshimune will allow the Dutch at Deshima to bring Western books into Japan, help merchants establish a trading system, and build extensive irrigation projects to aid agriculture.

China outlaws Christian religious teaching.

A company of Virginia colonists crosses the Blue Ridge Mountains into the valley of the Shenandoah River under the leadership of Alexander Spotswood (*see* 1714; Iroquois treaty, 1722).

The East India Company receives exemption from Indian customs duties and gains other concessions in return for gifts and medical services.

The East India Company (above) receives permission to bring a ship of 650 tons with trading goods into Portobello for the annual Spanish-American trade fair (*see* 1713; Vernon, 1739).

"Wealth depends on commerce, and commerce depends on circulation [of money]," says Scottish promoter John Law, 45, and he persuades the French regent Philip d'Orléans to let him open the Banque Royale which pays Philip's debts and issues notes that the government accepts for taxes. Law has killed a man in a duel, is wanted for murder, and has taken refuge in France, and while the banknotes of his private bank are backed in principle by gold and silver they are in fact based largely on the gold alleged to exist in France's Louisiana territory (*see* Boisguilbert, 1695; Louisiana, 1717).

Diario di Roma begins publication. It is the first Italian newspaper.

Painting *La Leçon d'amour* by Antoine Watteau. Japanese painter Korin Ogata dies at age 58. He has gained fame for his *Tale of Ise* and screens painted with iris and red and white plum trees.

Oratorio *Juditthea* by Venetian violinist-composer Antonio Vivaldi, 41, who was ordained a priest at age 28, has taught violin since 1704 at the Conservatorio dell'

Ospedale della Pietà, and is required to supply two concerti per month to the Conservatory.

1717 The triple alliance formed January 4 by Britain, France, and the Dutch Republic forces the Old Pretender James III to leave France. James has intrigued with the Swedish king Charles XII and the Spanish prime minister Cardinal Giulio Alberoni, 53.

Cardinal Alberoni sends a secret Spanish expeditionary force to seize Sardinia and raid Sicily while Eugene of Savoy is engaged in fighting the Ottoman Turks (*see* 1718).

Eugene of Savoy takes Belgrade from the Turks.

Spain creates the viceroyalty of New Granada with its seat of power at Bogotà to reduce the viceroyalty of Peru to more manageable size (*see* 1529; 1539; 1776).

France gives Scotsman John Law a 25-year monopoly on trade and government in Louisiana on condition that he send out at least 6,000 whites and 3,000 blacks to settle the vast territory that includes the Illinois country (*see* 1716; 1720).

The Lowlands begin a long period of commercial decline after more than a century of colonization and prosperity. The 2.5 million Dutch cannot support both a major army and a powerful fleet.

Prussia makes school attendance compulsory.

Painting *Embarkation for the Isle of Cythera* by Antoine Watteau who is admitted at last to the French Academy at age 32.

Theater *Electra (Elèctre)* by Prosper Jolyot, sieur de Crébillon, with Adrienne Lecouvreur, 25, in her first appearance at the Comédie-Française.

London's St. Mary-Le-Strand is completed by John Gibbs after 3 years of construction.

John Law (above) establishes the Compagnie d'Occident to encourage emigration to Louisiana (*see* 1719).

A great emigration to the Pennsylvania colony begins among German Dunkers, Mennonites, and Moravians (*see* Schwenkenfelders, 1734).

1718 Russia's Peter the Great has his emissary Count Peter Tolstoi bring back the czarevich Alexius Petrovich, forces him to sign a "confession" February 18 implicating his friends, and has them impaled, broken on the wheel, or otherwise dispatched. The ex-czarevich Eudoxia is dragged from her monastery and publicly tried for alleged adultery, Alexius is given 25 lashes with the knout June 19 (nobody has ever survived 30), given 15 more June 24, and dies June 26 at age 28 in the guardhouse of the citadel at St. Petersburg, 2 days after being condemned by the senate for "imagining" rebellion against his father.

Spain's Philip V sends troops into Sicily in July and his seizure of the country raises fears of a new European war. The Quadruple Alliance formed August 2 by the Holy Roman Emperor, Britain, and France (Holland will join next year) determines to prevent Philip from overturning the peace of 1714 (*see* 1720).

The Treaty of Passarowitz July 21 ends a 4-year war between Venice and Constantinople. The Ottoman Turks lose the Banat of Temesvar, Northern Serbia, and Little Wallachia, but they retain the Morea while Venice retains only the Ionian Islands and the Dalmatian Coast.

Sweden's Charles XII is shot through the head December 11 as he peers over the parapet of the foremost trench 280 paces from the fortress of Fredriksten during a military expedition to Norway. Dead at age 36 after a momentous 21-year reign, Charles is succeeded by his sister Ulrika Eleanora, 30, who accepts the crown on condition that the riksdag be allowed to draft a new constitution, and who brings the Great Northern War to a close.

New Orleans is founded near the mouth of the Mississippi by the sieur de Bienville. A few Frenchmen emigrate to the Louisiana Territory (*see* 1701; 1720).

San Antonio de Valero is founded in a Texas cottonwood grove by Franciscan monks who build a chapel they call the Alamo, meaning cottonwood (*see* 1836).

William Penn dies May 30 at age 73 leaving an estate of 21 million acres.

The first English bank notes are issued (*see* Bank of England, 1694).

The first table of chemical affinities is presented to the Académie by French chemist Etienne F. Geoffroy, 46, who advances the Age of Enlightenment by showing how various chemicals react to each other.

"Innoculation Against Smallpox" by Lady Mary Wortley Montagu, 29, reports a workable method known in the East since ancient times. Wife of the English minister to Constantinople, Lady Mary describes inoculation parties she has witnessed at which a small wound is made in the arm, a few drops of smallpox pus inserted, and a walnut shell tied over the infected area, a procedure that produces a true case of smallpox but one so mild that 98 percent of those inoculated recover (*see* 1714; 1721).

A mysterious disease comparable in its effects to England's 1518 "sweating sickness" strikes Picardy in the first of a series of epidemics that will recur in France for more than a century. The "Suette des Picards" differs in its symptoms from the "sweating sickness" but is no less deadly.

Yale University renames itself as such to honor benefactor Elihu Yale, who has become an official of the East India Company serving as governor of Fort St. George at Madras and who sends a cargo of books and East India goods to the Collegiate School founded at Saybrook in the Connecticut colony 17 years ago (*see* Yale, 1672). The school sells the cargo for £560 and uses the proceeds to move to New Haven (*see* 1861).

Poetry *Alma or The Progress of the Mind* and *Solomon, or The Vanity of the World* by Matthew Prior, now 54, who played a leading role in negotiating the Peace of Utrecht 5 years ago.

1718 ***(cont.)*** Painting *Parc Fête* by Antoine Watteau; *The Duke of Norfolk* by Sir Godfrey Kneller.

Theater *The Non-juror* by Colley Cibber; *Oedipé* by French playwright Voltaire (François Marie Arouet), 23, 11/18 at the Théâtre Français, Paris. Voltaire was released from the Bastille in April after serving nearly a year for libel.

The pirate Edward Teach, known as Blackbeard, is killed in a skirmish near Ocracoke Island in the North Carolina colony after having said, "I've buried my treasure where none but Satan and myself can find it."

Carolina coastal pirate Stede Bonnet dies on the gallows after having been seized with his crew.

Potatoes arrive at Boston with some Irish Presbyterian immigrants (*see* 1719).

A great Scots-Irish emigration to America begins.

1719 The Principality of Liechtenstein created by the Holy Roman Emperor Charles VI unites Vaduz and Schellenburg into a sovereign state of 61 square miles under the Austrian count Hans Adam von Liechtenstein, who has bought the territory and its 397-year-old Vaduz Castle from another count who is heavily in debt.

The Peace of Stockholm November 20 ends hostilities between Sweden and Britain's George I, who gains Bremen and Verden for 1 million rix dollars as elector of Hanover (*see* Victoria, 1837).

Herat rebels against Persia's Shiite persecution, declares independence, and joins with Uzbeks in plundering Khorasan. A 30,000-man army sent by Shah Hussein to put down the uprising defeats a 12,000-man Uzbek force en route and joins battle with Herat's 15,000-man Afghan army. But Persian artillery accidentally hits Persian cavalrymen, the army suspects treachery and is thrown into confusion, the Afghans seize the opportunity to make a decisive charge, and the Persians flee, losing a third of their men, their general, their artillery, and their baggage. The Afghans lose 3,000 men (*see* 1722).

The Mughal emperor Mohammed Shah assumes power at age 19; he will reign until 1748 as the Great Mughal.

A British slaver anchors off Port Natal and buys 74 boys and girls for the Virginia colony's Rappahannock plantations. The Xhosa (called Bantu) are considered "better slaves for working than those of Madagascar, being stronger and blacker."

French explorer Pierre François Xavier de Charlevoix, 37, travels up the St. Lawrence River, through the Great Lakes, and down the Illinois and Mississippi rivers to New Orleans. He has been sent to find a new route westward from Acadia (*see* 1682).

John Law renames his 3-year-old Compagnie d'Occident as reports circulate that Louisiana is rich in gold and diamonds. By midyear investors who paid 500 livres for shares in Law's Compagnie des Indes are reselling them at 15,000 livres each in Paris's Rue de Quinquempoix, Parisians hear the word *millionaire* for the first time, and Law receives rights to collect all French government taxes and to issue paper currency (*see* 1720).

Italian physician Giovanni Battista Morgagni, 37, observes changes found in the bodies of disease victims, initiating the science of pathological anatomy (*see* 1760).

Fiction *The Strange Surprising Adventures of Robinson Crusoe, Mariner* by Daniel Defoe is a novel based roughly on fact. Defoe has read William Dampier's 1697 book *Voyage Round the World* and has heard accounts of Alexander Selkirk, now 43, a Scottish mariner who in October 1704 quarreled with Dampier while on a privateering expedition in the South Seas, was set ashore at his own request on Mas-a-Tierra in the Juan Fernández Isles off the coast of Chile, and remained there until picked up by another captain in February 1709 (*see* Rousseau, 1762).

Hymn "O God, Our Help in Ages Past" by English hymn writer Isaac Watts, 45, is published in *The Psalms of David*.

"Irish" potatoes are planted at Londonderry in the New Hampshire colony (*see* 1718) but will not gain wide acceptance in America for nearly a century.

1720 Spain's Philip V joins the Quadruple Alliance of Britain, France, Holland, and Austria in January. He signs the Treaty of the Hague February 17 giving up his Italian claims in return for an Austrian promise that his son Charles will succeed to Parma, Piacenza, and Tuscany. The Holy Roman Emperor Charles VI gives up his claims to Spain, and Savoy receives Sardinia from Austria in return for Sicily (*see* 1713).

Sweden's Ulrika Eleanora abdicates in favor of her husband Frederick of Hesse, 44, who will reign until 1751 as Frederick I, but a new constitution strips the new king of much of his power. Sweden's House of Nobles will largely control affairs of state through the new reign.

The British Parliament authorizes use of workhouses for confining petty offenders.

England's "South Sea Bubble" collapses in January producing widespread financial losses and a loss of confidence by investors in distant overseas enterprises. Parliament has given the South Sea Company permission to take over part of the British National Debt and to create £1 in new stock for every £1 of the National Debt it assumed. Share prices have soared to £1,050 each, but the *asiento* concession granted in 1713 has proved disappointing and so has the monopoly in British trade with Spanish America.

France names John Law controller-general with power over the entire French economy, but Law's so-called Mississippi Company has failed to attract reputable emigrants, New Orleans remains no more than a dismal collection of wooden shacks, Law's grandiose scheme turns out to be a speculative stock fraud much like the South Sea Company (above), and although the company has increased shipping between France and New Orleans and has brought some colonization of Louisiana, the collapse of the "Mississippi Bubble" ruins many French investors.

Marseilles has an epidemic of plague and more than 50,000 die in western Europe's last major epidemic of the Black Death.

A famine in Sicily weakens the population of Messina and thousands die in a typhus epidemic.

Japan's shogun Yoshimune removes the century-old ban on importing European books although he continues the prohibition against religious books. A small group of Japanese begins to study Dutch and begins through its studies to gain a knowledge of Western science, especially of medicine.

The nursery rhyme "Little Jack Horner" is published for the first time. Horner is believed to have been one Thomas Horner who was sent by the Abbott of Glastonbury to Henry VIII in 1543 with a placating gift of a pie containing deeds to valuable manors owned by the monastery (since Horner's family soon came into possession of a manor it is thought that Horner extracted one deed from the pie for himself).

Painting *Martyrdom of St. Bartholomew* by Venetian painter Giambattista (Giovanni Battista) Tiepolo, 24.

Theater *Robin, Bachelor of Love (Arlequin poli par l'amour)* by French essayist-playwright Pierre Carlet de Marivaux, 32, 10/17 at the Théâtre Italien, Paris. Marivaux has been ruined by the bankruptcy of John Law's Mississippi Company (above) but his comedy derived from the Italian *commedia dell'arte* launches him on a successful theatrical career; *The Love Suicides at Amijima (Shinju ten no Amijima)* by Monzaemon Chikamatsu at Osaka.

London's Royal Academy of Music names George Frideric Handel director and presents his oratorio *Esther.*

Belgian composer Jean Adam Joseph Faber uses the clarinet *(chalumeau)* for the first time in a serious work.

1721 The Treaty of Nystadt August 30 gives Russia a "window" on the West and makes her a European power. Peter the Great obtains territories from Sweden that include Estonia, Ingermanland, Livonia, part of Carelia, and some Baltic islands, and he is proclaimed Emperor of All the Russias.

China suppresses a revolt in Taiwan.

Britain's Chancellor of the Exchequer John Aislabie, 51, goes to the Tower of London on charges of fraud in connection with last year's "South Sea Bubble." Sir Robert Walpole, who has profited with speculations in the South Sea scheme, becomes prime minister and chancellor of the exchequer in April, reduces import and export duties to encourage trade, and averts financial panic by amalgamating South Sea Company stock with stock in the Bank of England and the East India Company. Walpole's policy of "salutary neglect" of the American colonies will slacken enforcement of the navigation laws (*see* 1663; 1672; but *see* Molasses Act, 1733).

John Law, the Scottish financier. When his "Mississippi Bubble" burst, it ruined thousands of French investors.

Philadelphia entrepreneur John Copson offers "Assurances from Losses Happening at Sea, &c.," pioneering marine insurance for American shipping.

Japan's Tokugawa shogun Yoshimune bans expensive clothing, furniture, cakes, candies, and other extravagances in an austerity decree.

A London smallpox epidemic takes a heavy toll but Lady Mary Wortley Montagu has her 5-year-old daughter inoculated in the presence of some leading physicians (*see* 1718). The child has a mild case of smallpox that immunizes her, the physicians are impressed, and George I has two of his grandchildren inoculated—but only after the procedure has been tested on 11 charity school children and on 6 inmates of Newgate Prison who volunteered in return for having their death sentences commuted.

A smallpox epidemic introduced by ships from the West Indies strikes Boston in June. Local clergyman Cotton Mather, who belongs to the Royal Society and has read the account of variolation published in 1714, writes a treatise urging inoculation. The only Boston physician who acts on Mather's advice is Zabdiel Boylston, 42, who inoculates his 13-year-old son and two black slaves June 26. All three survive mild cases of the pox, and Boylston then inoculates 247 colonists. Six die, many people accuse Boylston of spreading the pox and denounce him for interfering with nature, and he narrowly escapes being hanged by a mob. But another epidemic strikes, 5,759 Bostonians—more than half the population—come down with the pox, and 844 die—more than 14 percent, as compared with 2.42 percent of Boylston's inoculated patients. Most of the survivors are left pockmarked (*see* Jenner, 1796; Waterhouse, 1799).

1721 ***(cont.)*** Regular postal service begins between London and New England.

Fiction *Lettres Persanes* by French lawyer-philosopher Charles de Secondat, 42, the baron de La Brède et de Montesquieu, criticizes French society as seen through the eyes of two imaginary Persians traveling through the country.

Jean Antoine Watteau dies of tuberculosis at Nogent-sur-Marne July 18 at age 36.

Theater *The Refusal or The Ladies' Philosophy* by Colley Cibber 2/4 at London's Drury Lane Theatre; *The Love Suicides at Sonezaki (Sonezaki shinju)* by Monzaemon Chikamatsu, now 68, at Osaka. The play is the first of several domestic pieces based on actual incidents that Chikamatsu will write about the love affairs of Japan's growingly important middle class.

First performances *The Brandenberg Concerti* by Johann Sebastian Bach.

The Palace of La Granja built by Spain's Philip V in the Guadarama Mountains high above Segovia is a miniature Versailles modeled on the great palace of Philip's grandfather Louis XIV.

Broccoli is introduced into England some 70 years after the "Italian asparagus" became popular in France (any dish *à Parisienne* includes broccoli).

Japan's city of Edo reaches a population level of more than a million despite the earthquake and fire that killed 200,000 in 1703.

1722 Kandahar's Mir Mahmoud conquers Afghanistan, invades Persia with his Afghan army, routs a large Persian army at the Battle of Gulnabad, captures Farahabad, and takes Isfahan October 12 after a 7-month siege that has left 80,000 dead and reduced the survivors to eating human flesh. Shah Hussein abdicates in favor of his son, who escapes to Mazandaran, tries to organize resistance, and will nominally reign until 1731 as Tahmasp II, but the Safavid dynasty is virtually ended. Mir Mahmoud makes himself shah to begin a 3-year reign of terror.

Russia's Peter the Great invades Persia from the north, saying he wants to rescue the shah from Afghan tyranny. He sends a flotilla down the Volga with 22,000 men to join at Dagastan with cavalry that has marched from Astrakhan. Defeating a force of Dagastanis, he occupies Derwent. He then prepares to sweep through the Caucasus, hoping to distract the Turks with his cavalry while his navy captures the Bosphorus and the Dardanelles while the grand vizier at Constantinople is preoccupied with growing tulips (but *see* ergotism, below).

The Ottoman sultan Ahmed III takes advantage of Persia's weakness (above) to invade Shirwan, whose Sunni population has been cruelly persecuted. The mufti issues three *fatwas* ordering true believers to extirpate the heretics, and Ottoman troops enter Georgia.

China's Manchu emperor Kang Cixi dies at age 68 after a personal reign of 53 years that has added Anhui, Hunan, and Gansu to the 15 existing Chinese provinces. The emperor will be succeeded beginning next year by Shih Zhong, who will reign until 1735.

America's Iroquois tribes sign a treaty with Virginia governor Alexander Spotswood and agree not to cross the Potomac River or the Blue Ridge without the governor's permission.

Manchester, New Hampshire, has its beginnings in a settlement at Amoskeag on the Merrimack River established by Massachusetts colonists 55 miles northwest of Boston (*see* 1810).

New Orleans is made the capital of the Louisiana Territory at the persuasion of the sieur de Bienville (*see* 1718).

Easter Island is discovered Easter Sunday 2,000 miles off the coast of Chile by the Dutch explorer Roggeveen. The island has a Polynesian population of between 2,500 and 3,000 with artisans who have used easily-cut compressed volcanic ash to fashion monuments three to 36 feet tall weighing up to 50 tons. They have somehow moved the stone heads into position to stare vacantly out to sea and have developed a unique ideographic script, unparalleled in Polynesia, which will not be noted by archaeologists until 1864.

L'Art de convertier le fer forgé en acier et l'art d'adoucir le fer fondu by René A. F. de Réaumur is the first technical treatise on steelmaking. Réaumur will develop improvements in ironmaking and will work on improving steel and tinned steel.

Ergotism aborts Peter the Great's attack on the Ottoman Empire (above). He marshals his Cossacks on the Volga delta at Astrakhan, but his horses eat hay containing infected rye, his troops eat bread made from infected rye grain, hundreds of horses and men go mad, many die, and Peter has to abandon his plans.

Fiction *The Adventures and Misadventures of Moll Flanders* by Daniel Defoe.

Theater *Double Indemnity (La Double Inconstance)* by Pierre de Marivaux 5/3 at the Théâtre Italien, Paris; *The Conscious Lovers* by Richard Steele 11/7 at London's Drury Lane Theatre.

Traité de l'harmonie by French composer Jean Philippe Rameau, 39, calls attention for the first time to inverted chords but contains erroneous ideas about augmented sixths and the eleventh chord. Organist since 1715 at Clermont-Ferrand, Rameau has returned to Paris where he begins producing light dramatic pieces.

The Well-Tempered Clavier (Das wohltemperierte Klavier) by Johann Sebastian Bach is published in its first part.

Dresden's Zwinger is completed by German architect Matthaus Daniel Poppelmann, 60, who leaves its pavilion unfinished. He has spent 11 years building the Zwinger.

Durham mustard, produced at Durham, England, is the first commercial dry mustard. Mustard seeds have until now been brought to the table in their natural state, and diners have crushed them with their knife handles on the sides of their plates.

Gruyère cheese is introduced into France, which will soon be its major producer. The most famous will be made at Jura in the Franche-Comté.

Boston's population reaches nearly 12,000 with one-third of the town's almost 3,000 houses made of brick (*see* 1638).

1723 Baku capitulates to Peter the Great. Ottoman forces take Tiflis and invest Hamadan, which falls after a short siege (*see* 1722). Another Ottoman force advances on Erivan, which surrenders after a 3-month siege in which four assaults plus disease have cost the Turks 20,000 men.

Kazbin throws out its Afghan invaders, whereupon Shah Mahmoud arrives, invites the city's noblemen to dinner, has them massacred and their corpses thrown into the great square, massacres 3,000 Persian guards he has engaged, and then gives orders to shoot every Persian who has served Shah Hussein. The slaughter goes on for 15 days.

The Treaty of Charlottenburg October 10 between Britain and Prussia provides that the grandson of George I shall marry a Prussian princess and that Prussia's Prince Frederick shall marry the daughter of the Prince of Wales.

Britain banishes the English bishop Francis Atterbury, 61, for complicity in a Jacobite conspiracy.

Britain permits Jews to take oaths without using the words, "On the true faith of a Christian" (*see* Jewish Naturalization Bill, 1753).

British industrial expansion proceeds at a growth rate of 1 percent per year (but *see* Kay, 1733).

An air furnace to smelt iron near Fredericksburg in the Virginia colony uses bituminous coal which is abundant in the region. Alexander Spotswood has resigned as lieutenant governor to establish the furnace (*see* 1714; coal, 1742).

Yellow fever appears for the first time in Europe (*see* New York, 1702; Cadiz, 1741).

Sir Godfrey Kneller dies at London November 7 at age 74–77.

Opera *Ottone* 1/23 at London's Haymarket Theatre, with music by George Frideric Handel.

First performances *The St. John Passion* by Johann Sebastian Bach, who is appointed Thomascantor at Leipzig after composer Georg Philipp Telemann, 42, has refused the post.

Sir Christopher Wren dies February 25 at age 90 and is buried in his St. Paul's Cathedral while work proceeds on his Greenwich Hospital.

Austrian architect J. B. Fischer von Erlach dies April 5 at age 66 leaving his half brother Joseph Emanuel, 30, to complete Vienna's Kariskirche (Church of San Carlo Borromeo), the Hofburg, and the Imperial Library.

Boston's Christ Church is completed 300 yards from the Old North Church completed in 1676.

An American coffee industry that will eventually produce 90 percent of the world's coffee has its beginnings in a seedling planted on the Caribbean island of Martinique by French naval officer Gabriel Mathieu de Cheu, who has been helped by confederates to break into the French Jardin Royale at night and make off with the seedling (*see* 1713; 1727).

Robert Walpole reduces British duties on tea.

The condom can emancipate young English wives from fear of "big belly, and the squalling brat," writes White Kennett, son of the Bishop of Peterborough, in his satirical poem *Armour* (*see* 1655; 1671; Julius Schmid, 1883).

1724 Moscow and Constantinople sign a treaty for the dismemberment of Persia, whose territory they have invaded. Persia's Shah Mahmoud goes mad and orders a wholesale massacre at Isfahan.

Rhode Island establishes property ownership qualifications for voters.

A craft guild along the lines of European guilds is established at Philadelphia.

Black slaves outnumber whites two to one in the South Carolina colony.

A Code Noir for regulating the blacks and expelling the Jews at New Orleans is proclaimed at New Orleans by Louisiana's governor, the sieur de Bienville.

Nonfiction *The Origin of Myths (De l'Origine des fables)* by Bernard de Fontenelle explores the psychological and intellectual roots of mythology. Fontenelle refutes popular beliefs and superstitions.

Japanese *ukiyoe* painter Kiyonobu Torii dies at age 60. He has specialized in painting Kabuki actors and beautiful women.

Theater *Caesar in Aegypt* by Colley Cibber 12/9 at London's Drury Lane Theatre.

Vienna's Belvedere Palace is completed after 7 years of construction.

1725 Russia's Peter the Great dies January 28 at age 52 after a 42-year reign. His consort, 41, will reign until 1727 as Catherine I.

A 70,000-man Ottoman army takes Tabriz after a siege in which the Persians have lost 30,000 men, the Turks 20,000. Shah Mahmoud has 39 Safavid princes massacred, the first being killed by his own hand. Persia's nobility elects a nephew of the late Mir Vais to succeed the insane Mahmoud; Shah Ashraf begins a 5-year reign by killing powerful chiefs and confiscating their fortunes. Mahmoud soon dies, possibly killed by order of the new shah.

France's Louis XV, now 15, is married August 15 to the daughter of Poland's former king Stanislas Leszczynski, who was deposed in 1709. The marriage has been arranged by the inept Louis Henri, duc de Bourbon (*see* 1733).

1725 ***(cont.)*** Peter the Great (above) establishes the Russian Academy of Sciences on his deathbed. Since Russia has no scientists of her own, the academy will be staffed by foreigners for years.

The New Science (Principi di Una Scienza Nuova d'intorno alla Comune Natura delle Nazioni) by Italian philosopher Giovanni Battista Vico, 57, attempts to discover and organize laws common to the evolution of all society.

The stereotype plaster of Paris impression invented by Edinburgh jeweler William Ged can be used to cast any number of metal printing plates and will lead to faster, cheaper letterpress printing. Papier-maché versions will further reduce costs.

Opera *Giulio Cesare in Egitto* 3/2 at London's Haymarket Theatre, with music by George Frideric Handel; *Tamerlano* 11/11 at London's Haymarket Theatre, with music by Handel.

Concerti *Il Cimento dell'armonia* by Antonio Vivaldi, published in the Netherlands, contains *The Four Seasons*.

Rome's 137 "Spanish Steps" from the Piazza di Spagna up to the Church of Trinita dei Monti are completed by French architects.

France has famine. Paris workers riot when a baker raises the price of bread by 4 sous.

London has nearly 2,000 coffee houses, up from just one in 1652.

1726 Bishop André Hercule de Fleury is made a cardinal at age 73 and becomes virtual prime minister July 12 following the dismissal of Louis Henri, duc de Bourbon. Fleury will continue in power until his death at age 89 in 1743, giving France peace, economic growth, and an upsurge of religious revivalism comparable to Britain's Methodism (*see* Wesley, 1729).

Persia's Shah Ashraf defeats an Ottoman army.

Montevideo is founded by Spanish conquistadors at the mouth of the Rio de la Plata.

Travels into Several Remote Nations of the World by "English sea-captain Lemuel Gulliver" (Jonathan Swift) is a satire on cant and sham in England's courts, in her political parties, and among her statesmen. Swift receives £200 for his story of Gulliver's travels in Lilliput and Brobdingnag; it is the only payment he will ever receive for any of his writings.

London painter-engraver William Hogarth, 28, gains notice with illustrations for a new edition of *Hudibras* by the 17th-century poet Samuel Butler.

London's St. Martins in the Fields Church is completed by James Gibbs.

A character in *Gulliver's Travels* (above) "gave it for his opinion that whoever could make two ears of corn, or two blades of grass, to grow upon a spot of ground where only one grew before, would deserve better of mankind, and do more essential service to his country, than the whole race of politicians put together."

Japan's population reaches an estimated 26.5 million.

German settlers begin moving from the Pennsylvania colony into the Shenandoah Valley of Virginia.

1727 Russia's extravagant, illiterate (but shrewd) Catharine I dies May 16 at age 44 after a brief reign and is succeeded by her son Peter, 12, who will reign until 1730.

An Ottoman army of 60,000 with 70 guns advances on Isfahan. Persia's Shah Ashraf has only 20,000 men and 40 small camel-mounted field guns but kills 12,000 Turks before concluding a treaty acknowledging the sultan as caliph and being recognized himself as shah of Persia.

Tahmasp II still holds court at Farahabad, where he is joined by the former brigand Nadir Kuli, 38 (*see* 1730).

Britain's George I dies of apoplexy at 67 the night of June 10 while en route by carriage to Hanover. He is succeeded by his son, 44, who will reign until 1760 as George II in a reign that will be guided by his ministers and his wife Caroline.

The Russian-Chinese border is fixed by the Kiakhta treaty arranged by a mission to Beijing headed by Sava Vladislavich.

The Ottoman minister to Paris returns to Constantinople with a printing press. It will be instrumental in spreading enlightenment in the Muslim world.

German chemist J. H. Schulze pioneers photography; using a mixture of silver nitrate and chalk under stencilled letters, he establishes that light, not heat, causes darkening of silver salts (*see* Scheele, 1777).

Salzburg's Mirabell Palace is completed by Johann Lucas von Hildebrandt after 6 years of construction.

House rats overrun Astrakhan and appear within a few months in the German states. Originally from eastern Siberia, the rats have not been seen in Europe until now. They will be in England by next year.

Vegetable Staticks by English physiologist-parson Stephen Hales founds the science of plant physiology with its studies of the rate of plant growth (*see* 1733).

Coffee is planted for the first time in Brazil (*see* 1723).

1728 A 14-month Spanish siege of Gibraltar is raised in March (*see* 1704; 1779).

Danish explorer Vitus Bering, 47, discovers the Bering Strait between Asia and North America (*see* 1741).

Yorkshire carpenter John Harrison, 35, completes plans for a practical spring-driven marine timekeeper that will advance navigation. He obtains a £200 loan from clockmaker George Graham, 55, who has invented a mercurial pendulum and deadbeat escapement and begins work to perfect his design and create a working model that will win the prize offered in 1714 by London's Board of Longitude (*see* 1736).

English astronomer James Bradley, 35, discovers the aberration of light from fixed stars.

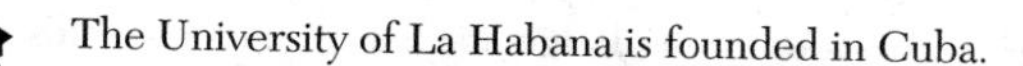

The University of La Habana is founded in Cuba.

Cyclopaedia, or an Universal Dictionary of Arts and Sciences by English encyclopaedist Ephraim Chambers is published in two volumes (*see* 1844; *Britannica,* 1771).

Fiction *Memoirs and Adventures of a Man of Quality (Mémoires et aventures d'un homme de qualité)* by French novelist Abbé Provost (Antoine François Provost d'Exiles), 31, is published in the first four of seven volumes. The novelist is a former Benedictine monk.

Poetry *The Dunciad* by Alexander Pope satirizes dullness.

Painting *The Skate* and *The Rain* by French painter Jean Baptiste Simeon Chardin, 28.

Theater *Money Makes the World Go Round (Le Triomphe de Plutus)* by Pierre de Marivaux 4/22 at the Théâtre Italien, Paris; *Love in Several Masques* by English playwright-novelist Henry Fielding, 21.

Opera *The Beggar's Opera* 2/9 at the Lincoln's Inn Fields Theatre, London, with music by Berlin-born composer John Christopher Pepusch, 51, book and lyrics by English poet John Gay, 42 (the opera begins a 2-year run) (*see* 1928); *Siroe, Re di Persia* 2/28 at London's Haymarket Theatre, with music by George Frideric Handel; *Tolomeo, Re di Egitto* 5/4 at the Haymarket, with music by G. F. Handel, who is named codirector of the King's Theatre in Covent Garden.

1729 The Treaty of Seville November 9 ends a 2-year war between Spain and an Anglo-French alliance. The alliance agrees to the Spanish succession in the Italian duchies; Britain retains Gibraltar.

The town of Baltimore first settled more than a century ago is formally established August 8. The Maryland colonial assembly purchases land near deep tidal waters that extend deep into grain and tobacco fields at the head of Chesapeake Bay, land rapidly being developed by German immigrants.

Natchez Indians in the Louisiana colony attack settlers and soldiers November 28, massacring more than 200 and taking several hundred women, children, and black slaves prisoner. The colonists had demanded that the Natchez give up their sacred burial ground.

Diamonds discovered at Tejuco, Brazil, in 1721 are confirmed as such by Portuguese-Brazilian Bernardo da Fonseca Lobo, who arrives at Lisbon with samples. A diamond rush begins to Tejuco, which will be renamed Diamintina.

A Modest Enquiry into the Nature and Necessity of a Paper-Currency is published by Philadelphia printer Benjamin Franklin, 23.

Methodism has its beginnings at Oxford University where fellow students call Charles Wesley, 22, a "methodist" because of his methodical study habits. Episcopal priest John Wesley, 26, comes into residence at Oxford in November and joins with his brother Charles and with James Hervey and George Whitefield, both 15, in a Holy Club that meets at first each Sunday evening for strict observance of sacrament and then meets every evening, reading the classics and the Greek Testament, fasting on Wednesdays and Fridays (*see* 1735).

The Pennsylvania Gazette is purchased from his former employer Samuel Keimer by Benjamin Franklin (above) and his partner Hugh Meredith. Franklin began his career as apprentice to his half brother James at Boston when James started the *New England Courant* in 1721. He took over when James was imprisoned briefly in 1722 for criticizing public officials, but he quarreled with James the following year, left for New York, moved on to Philadelphia, worked for Keimer, and in 1724 voyaged to London with the intent of buying equipment for a shop of his own. Not finding letters of credit promised by Gov. William Keith of Pennsylvania, Franklin worked for a London printshop until 1726, when he returned to Keimer's shop at Philadelphia. He went into partnership with Meredith last year and will soon buy out his partner (*see* circulating library, 1731).

First performances *The St. Matthew Passion (Grosse Passionmusik nach dem Evangelium Matthaei)* by Johann Sebastian Bach 4/25 at Leipzig.

A Modest Proposal by Jonathan Swift is an ironical tract inspired by Irish population pressures: "I have been assured by a knowing American of my acquaintance that a young healthy child well nursed is at a year old a most delicious, nourishing and wholesome food, whether stewed, roasted, baked or boiled; and I make no doubt that it will equally serve in a fricasee or a ragout: I do therefore humbly offer it to public consideration that . . . 100,000 [infants] may, at a year old, be offered in sale to the persons of quality and fortune throughout the kingdom; always advising the mother to let them suck plentifully in the last month, so as to render them plump and fat for a good table."

England's death rate begins a sharp descent, producing population pressure on food supplies, but food remains more plentiful than in Ireland.

1730 The Russian czar Peter II dies of smallpox at age 14 January 30, the very day on which he was to have married Catherine, second daughter of Alexis Dolgoruki. Peter's cousin Anna Ivanovna, 36, enters Moscow February 26, her personal friends overthrow the supreme privy council in a coup d'état March 8, and she summons her former lover Ernst Johann Biren, 39. Grandson of a groom to a former duke of Courland, Biren gained favor with Anna when she was duchess of Courland, she makes him grand-chamberlain and a count of the empire, and he adopts the arms of the French ducal house of Biron. The new czarina gives him an estate at Wenden with 50,000 crowns a year, and he will dominate her 10-year reign, antagonizing most Russians with his rapacity and treachery.

The Persian shah Ashraf, who has ruled since 1725, is defeated with his Afghans near Shiraz, some of his

1730 ***(cont.)*** followers murder him en route to Kandahar, and he is succeeded by Tahmasp II, a figurehead for the Afshar chief Nadir Kuli.

Charles Townshend, Viscount Townshend of Raynham, resigns May 15 at age 56 to devote his energies to agriculture (below), leaving Robert Walpole as sole minister in the British Cabinet.

The Ottoman grand vizier is strangled September 17 in a revolt of the Janissaries, and Ahmed III is obliged to abdicate at age 57 after a 27-year reign. The sultan's nephew, 35, ascends the throne October 18 and will reign until 1754 as Mahmud I.

Denmark's Frederick IV dies at Odense October 12 at age 59 after a 31-year reign in which he has been forced to give up some German territories. He is succeeded by his narrow-minded son of 31, who will reign until 1746 as Christian VI, following the whims of his wife Sophie Magdalene of Brandenburg-Kulmbach.

Thomas Godfrey in the Pennsylvania colony invents an improved mariner's quadrant (*see* Hadley, 1731).

Painting *The Symbolic Marriage of Venice to the Adriatic* by the Venetian painter Canaletto (Giovanni Antonio Canal), 33; *The Dancer La Camargo* by French painter Nicolas Lancret, 40.

Theater *The Game of Love and Chance* (*Le jeu de l'amour et du hasard)* by Pierre de Marivaux 1/23 at the Théâtre Italien, Paris; *The Tragedy of Tragedies, or The Life and Death of Tom Thumb the Great* by Henry Fielding, in April at London's Drury Lane Theatre; *Brutus* by Voltaire 12/11 at Paris.

"Gaudeamus Igitur" is published for the first time. The German drinking song may date to the 13th century.

The Serpentine in London's Hyde Park is created from the River Westbourne.

Scientific farming is introduced into England by Lord Townshend (above), who resigns from public life. Taking a cue from the Dutch, "Turnip Townshend" will find that he can keep livestock through the winter on his estates by feeding them turnips, thus eliminating the need to slaughter most of them each fall, making fresh meat available in all seasons, reducing the need for costly spices used to disguise the taste of spoiled meat, and permitting the development of larger cattle (*see* 1732; Bakewell, 1755, 1760; Coke, 1772).

1731 The Holy Roman Emperor Charles VI signs a second Treaty of Vienna March 16. He is 45, has no male offspring, and has decreed in a Pragmatic Sanction that if he should die without a male heir the empire shall devolve upon his daughters (the eldest, Maria Theresa, is now 14) and to their heirs by the law of primogeniture. If his daughters should die without heirs, the empire shall go to the heirs of his late brother Josef, who died in 1711. Britain recognizes this Pragmatic Sanction in the treaty, and the Dutch agree (*see* 1732).

Parma and Piacenza pass to the son of Spain's Philip V as the head of the Farnese family Antonio Farnese dies at age 52 without male issue. Philip's son Carlos has taken Elizabeth Farnese as his second wife and Spain's recognition of the Pragmatic Sanction (above) aids the succession of Charles to the Italian duchies.

Persia's Shah Tahmasp takes the field against his country's Ottoman invaders. He lays siege to Erevin but falls back to defeat a Turkish army at Arijan near Hamada, taking heavy losses. He makes peace with Constantinople, ceding large areas, but his brother-in-law Nadir Kuli denounces the peace treaty, threatens expulsion and death to all Shiites who refuse to fight, marches to Isfahan, seizes Tahmasp, locks him up at Khorasan, and sets his infant son on the throne as Abbas III while he runs the country himself (*see* 1733).

Louisiana becomes a French royal province once again upon the surrender of the charter of John Law's Compagnie d'Occident (*see* 1716; 1720; 1763).

The sieur de la Verendrye Pierre Gaultier de Varennes, 46, begins explorations of western Canada (*see* Lake of the Woods, 1732).

Welsh colonist Morgan Morgan establishes the first permanent settlement in western Virginia on Mill Creek (*see* coal, 1742).

The Ostend East India Company set up to rival Dutch and English companies is ended by the Treaty of Vienna (above).

The reflecting quadrant invented by English mathematician-mechanic John Hadley, 49, advances navigation by enabling a mariner to determine his latitude at noon or by night (*see* Godfrey, 1730). Hadley's instrument has an arc that measures one-fourth of a circle, a graduated scale is inscribed round its edge, its sliding vernier scale has minute subdivisions to facilitate measurement of the angle between the horizon and some fixed star at night, or between the horizon and the sun at noon. A Royal Navy vessel will test the quadrant next year by order of the admiralty, and its readings will be so accurate that the navy will adopt Hadley's device (*see* chronometer, 1736; sextant, 1757).

France forbids barbers to practice surgery; elsewhere they continue to double as surgeons.

A French Royal Academy of Surgery is founded under the direction of Jean-Louis Petit, who will invent the screw tourniquet and will develop a procedure for mastoidectomy.

Fiction *Histoire du Chevalier des Grieux et de Manon Lescaut* by the Abbé Provost whose work is condemned by French authorities. The story of an aristocrat who gives up everything in his passion for a demimondaine will be the basis of romantic novels, plays, and operas in the next century.

The Library Company founded July 1 by Benjamin Franklin is the first circulating library in North America. Franklin begins by having members of his Philadelphia discussion group, the Junto, pool their books for borrowing. He then obtains 50 subscribers at 40 shillings each and sends to England for 100 books.

Painting *A Harlot's Progress* by William Hogarth.

Theater *The London Merchant, or The History of George Barnwell* by English playwright George Lillo, 38, 6/22 at London's Drury Lane Theatre. Lillo's play is the first serious prose drama whose chief figures are not of the nobility.

Opera *Poro* 2/2 at London's Haymarket Theatre, with music by George Frideric Handel.

London's No. 10 Downing Street is completed in Westminster to provide a residence for the British prime minister.

Half a pint of rum in two equal tots becomes the official daily ration for all hands in the British Royal Navy (*see* 1655; Vernon, 1740).

Porcelain factories multiply in Europe as demand grows for china plates and cups.

1732 The Diet meeting at Regensburg (Ratisbon) January 11 recognizes the Pragmatic Sanction order of Charles VI (*see* 1731; 1740).

Genoa recovers Corsica from insurgent forces who revolted in 1730, but the Genoese will have to call on the French for help in controlling the island (*see* 1768).

America's Georgia colony has its beginnings in a charter granted by George II to English philanthropists who include James Edward Oglethorpe, 36, for lands between the Altamah and Savannah rivers (*see* 1733).

The sieur de la Verendrye erects a fort on Lake of the Woods (*see* 1731; 1734).

Swedish botanist Carolus Linnaeus (Carl von Linné), 25, travels 4,600 miles through northern Scandinavia studying plant life (*see* 1737).

Poor Richard's Almanack begins publication at Philadelphia. Primarily an agricultural handbook, it gives sunset and sunrise times, high and low tides, weather predictions, and optimum dates for planting and harvesting. Benjamin Franklin will publish new editions for 25 years, offering practical suggestions, recipes, advice on personal hygiene, and folksy urgings to be frugal, industrious, and orderly ("God helps those who help themselves," "Never leave till tomorrow that which you can do to-day") which are credited to a fictional "Richard Saunders." Circulation will reach 10,000; only bibles will be more widely read in the colonies.

The *Philadelphia Zeitung* published by Franklin (above) beginning in May is the first foreign-language newspaper in the British colonies.

Franklin will revolutionize colonial postal service to improve circulation of his *Almanack* (above) and *Pennsylvania Gazette* (*see* 1729). He will double and triple postal service in some areas, increase the speed of couriers, consolidate roads from Maine to Georgia into what later will be called Route 1, and produce for the Crown three times the postal revenues obtained in Ireland.

Painting *Kitchen Table with Shoulder of Mutton* by Jean Chardin.

Sculpture *Fontana di Trevi* by Italian sculptor Niccola Salvi at Rome.

Theater *Zara (Zaïre)* by Voltaire 8/13 at Paris; London's Theatre Royal in Covent Garden opens 12/7 with the 1700 Congreve comedy *The Way of the World*.

The average bullock sold at London's Smithfield cattle market weighs 550 pounds, up from 370 pounds in 1710 (*see* 1639; 1795).

Famine strikes western Japan as crops are ruined by an excess of rain and a plague of grasshoppers: 2.6 million go hungry, 12,400 die, more than 15,000 horses and work cattle starve to death in one of the worst of the 130 famines of the Tokugawa shogunate.

London's beef, pork, lamb, and mutton for the next century and more will come largely from animals raised in Ireland, Scotland, and Wales, landed by ship at Holyhead, driven across the Isle of Anglesey, made to swim half a mile across the Menai Straits to the Welsh mainland, and then driven 200 miles to Barnet, a village on the outskirts of London, to be fattened for the Smithfield market. The meat will be generally tough and costly.

1733 The War of the Polish Succession begins in Europe following the death of Poland's Augustus II February 1 at age 62. Austria and Russia demand the election of the king's only legitimate son Frederick Augustus, 36, elector of Saxony, but France's Louis XV persuades the Polish nobility to restore his father-in-law Stanislas Leszczynski to the throne (*see* 1725). A large Russian army invades Poland and obliges Stanislas to flee to Danzig, France declares war on the Holy Roman Empire October 10 and gains support from Spain and Sardinia, Russian forces lay siege to Danzig beginning in October, and Louis XV sends a French expeditionary force to relieve the Baltic city and rescue Stanislas (*see* 1734).

Philadelphia printer Benjamin Franklin worked to improve colonial transportation to help sales of his *Almanack*.

1733 ***(cont.)*** French forces occupy Lorraine (*see* 1738).

A Persian army under Nadir Kuli lays siege to Baghdad but is surrounded at Kirkuk near Shimar. The troops flee in disorder 200 miles to Hamadan, but in 3 months Nadir has rebuilt his strength and wins a victory (*see* 1739).

Smugglers evade Britain's new Molasses Act (below) and in some cases carry African slaves to Spanish colonies, trade them for sugar and molasses, and sell the cargo to New England distillers for capital with which to buy more African slaves (*see* 1698).

Savannah, Georgia, is founded by James E. Oglethorpe, who names the last of the 13 British colonies in America after George II and makes it an asylum for debtors. Oglethorpe settles at the mouth of the Savannah River with 120 men, women, and children (*see* 1732; 1742).

A Molasses Act passed by Parliament to tax British colonists imposes heavy duties on the molasses, sugar, and rum imported from non-British West Indian islands and thus effectively raises the price of the rum that Americans consume at the rate of 3 Imperial gallons (3.75 American gallons) per year for every man, woman, and child.

The Spanish Plate Fleet laden with silver and gold from South America founders in the shallow reefs of the Florida Keys and is wrecked.

The flying shuttle invented by English weaver John Kay revolutionizes the hand loom, halves labor costs, and prepares the way for further developments that will speed the industrialization of Britain's cottage industry in textiles (*see* Arkwright, 1769).

Haemostaticks by Stephen Hales reveals his findings on blood circulation. Hales has tied tubes into the arteries and veins of living animals, he has measured blood pressure and circulation rates, and he has estimated the actual velocity of the blood in the veins, the arteries, and the capillary vessels. He shows that capillaries are capable of dilating and constricting (*see* 1727; Harvey, 1628).

Poetry *Essay on Man* by Alexander Pope contains lines and couplets that will become familiar in most English households: "Hope springs eternal in the human breast;/Man never is, but always to be blest."

Painting *Southwark Fair* by William Hogarth, who sketches convicted murderess Sarah Malcolm at Newgate Prison. She is hanged 2 days later and Hogarth's engravings enjoy a large sale.

Theater *The Death of Caesar* (*La mort de César*) by Voltaire; *The Wiles of Love (L'Heureux stratagème)* by Pierre de Marivaux 6/6 at the Théâtre Italien, Paris.

First performances Mass in B minor (*Kyrie* and *Gloria*) by Johann Sebastian Bach 4/21 at Leipzig.

Opera *The Servant Mistress (La Serva Padrone)* 8/28 at the Bartholomeo Opera House in Naples, with music by Italian composer Giovanni Battista Pergolesi, 23, who establishes the comic opera form that will continue for nearly a century; *Hippolyte et Aricie* 10/1 at the Paris Opéra with music by J. P. Rameau to a libretto derived from the 1677 Racine tragedy *Phèdre; Achilles* at London's Covent Garden with a libretto by John Gay; *Rosamund* at London's Drury Lane Theatre with music by English composer Thomas Augustine Arne, 23.

1734 Danzig falls to Russian forces June 2 after a siege of nearly 8 months. Poland's Stanislas Leszczynski eludes the Russians and escapes to Prussia (*see* 1735).

Spanish forces seize Naples and Sicily while Milan falls to another Spanish force that has invaded Lombardy as the War of the Polish Succession spreads through Europe (*see* 1735).

Schwenkenfelders from Silesia emigrate to America and drop anchor near New Castle, Delaware, where they obtain the first fresh water they have had in months.

German Protestants driven out of the principality of Salzburg settle some 25 miles above the mouth of Georgia's Savannah River at Ebenezer.

A French fort on Lake Winnipeg is erected by the sieur de la Verendrye (*see* 1731; 1738).

Lettres anglaises ou philosophiques by Voltaire exalts the English constitution, making special reference to representative government.

The autumn crocus (*Crocus saliva*), whose stamens yield the spice and dyestuff saffron, will be introduced into America by the Schwenkenfelders (above) who will settle in the Pennsylvania colony.

An English seaman afflicted with scurvy is marooned on the coast of Greenland because the disease is thought to be contagious. The man cures himself by eating scurvy grass (*Cochlearia officinalis*) and is picked up by a passing ship. News of the incident will reach His Majesty's Hospital at Haslar, near Portsmouth (*see* 1747).

Considérations sur les Causes de la Grandeur des Romains et de leur Décadence by Montesquieu claims that the ancient Roman world had a population larger than that of 18th century Europe.

1735 The War of the Polish Succession ends October 5 in the Treaty of Vienna that will be ratified in 1738. The treaty legitimizes the elector of Saxony as king of Poland, and he will reign until 1763 as Frederick Augustus II. Austria cedes Naples and Sicily to the Spanish Bourbons on condition that they never be united with Spain under one crown, and the son of Spain's Philip V assumes the throne of Naples and Sicily as Charles III after giving up Parma and Piacenza to Austria (*see* 1738).

Russia gives up Peter the Great's last Persian acquisitions and joins in an alliance against the Ottoman Turks with Persia's Nadir Kuli, who wins a great victory over the Turks at Baghavand and takes Tiflis.

Trustees of the Georgia colony (below) prohibit slavery and the importation of rum.

Vincennes, Indiana, has its beginnings in a French settlement on the Wabash River.

French fur traders and lead miners found Ste. Genevieve, first permanent white settlement on the Missouri River.

Augusta, Georgia, is founded on the Savannah River by settlers in the new Georgia colony.

Scarlet fever strikes New England in a devastating epidemic (*see* 1736).

John Wesley and his brother Charles lose their father April 25 and embark in October for the Georgia colony with two other Methodists and a party of Moravians (*see* 1729). The *Simmonds* encounters storms en route; John Wesley finds his fears greater than his faith. He will provoke criticism at Savannah, but he will begin the first Methodist society (*see* 1738).

John Peter Zenger, 38, gains a landmark victory for freedom of the press. New York governor William Cosby has brought libel charges against the German-American printer for reporting that Cosby attempted to rig a 1733 election held at St. Paul's Church village green in the town of Eastchester, Philadelphia lawyer Andrew Hamilton admits that his client printed the report in his *New York Weekly Journal* as charged, but he persuades the jury that Zenger printed only the truth and wins acquittal.

Fiction *Gil Blas (L'Histoire de Gil Blas de Santillana)* by Alain René Lesage, now 67, whose four-volume pioneer picaresque novel appeared in its first volume 20 years ago.

Painting *The Pleasure Party* by Nicolas Lancret; *A Rake's Progress* by William Hogarth.

Parliament passes a Copyright Act that protects artists against pirating of their work in cheap copies.

Opera *Olimpiade* 1/8 at Rome's Teatre Tordinona, with music by Giovanni Pergolesi; *Ariodante* 1/19 at London's Covent Garden Theatre, with music by George Frideric Handel; *Alcina* 4/27 at London's Covent Garden Theatre, with music by G. F. Handel; *Les Indes galantes* 8/23 at the Paris Opéra, with music by Jean Philippe Rameau.

Pellagra is described for the first time by Don Gasper Casal, personal physician to Spain's Philip V (*see* Italy, 1749).

Botulism makes its first recorded appearance in one of the German states. An outbreak of the deadly food poisoning is traced to some sausage (*see* 1895).

English distillers produce gin at a rate of 5.4 million gallons per year, nearly a gallon for every man, woman, and child (*see* 1736).

1736 The Safavid dynasty that began its rule of Persia in 1502 ends with the death at age 6 of Shah Abbas III. The Afghan Nadir Kuli, who has ruled effectively since 1731, succeeds to the throne and will reign as Nadir Shah until 1747. A Sunni, Nadir accepts the title on condition that Persians renounce their Shiite faith, but he will have little success in converting them.

Moscow declares war on Constantinople, sends troops into Ottoman lands north of the Black Sea, but the Russians retreat to the Ukraine after sustaining heavy losses.

Qian Long ascends the Chinese imperial throne at age 25 to begin a Qing dynasty reign that will continue until 1796. He will extend Qing control throughout central Asia.

Pope Clement XII condemns Freemasonry.

Britain repeals her statutes against witchcraft.

The ship's chronometer presented to London's Board of Longitude by John Harrison is accurate to within one-tenth of a second per day (1.3 miles of longitude) and wins the prize offered by the Board in 1714 (*see* 1728). Harrison's chronometer is set to the time of 0° longitude (Greenwich time), a navigator can fix his longitudinal position by determining local time with the quadrant devised by John Hadley in 1731 and comparing it with Harrison's chronometer, the device weighs 66 pounds and is a complicated, costly, delicate affair, but it will soon be improved, reduced in size, and refined to keep it going through all weather aboard any sailing vessel (*see* 1761).

Mechanica sive motus analytice exposita by Swiss mathematician-physicist Leonard Euler, 29, is the first systematic textbook of mechanics. Euler has been at the Russian Academy of Sciences since 1727 (he arrived at St. Petersburg the day Catherine I died), has been the academy's leading mathematician since 1733 when his teacher Daniel Bernoulli returned to Switzerland, has married, but lost the sight of one eye last year (*see* 1748).

English surgeon Claudius Aymand, 56, performs the first successful appendix operation to be described in medical literature (*see* 1885).

"The Practical History of a New Epidemical Eruptive Military Fever . . . in Boston . . . 1735 and 1736" by Boston physician William Douglass gives the first clinical description of scarlet fever (*see* 1902; 1924).

Painting *The Good Samaritan* by William Hogarth.

Theater *Alzire* (*Les Américains*) by Voltaire 1/27 at the Comédie-Française, Paris.

Opera *Atalanta* 5/23 at London's Royal Theatre in Covent Garden, with music by George Frideric Handel.

The Gin Act passed by Parliament forbids public sale of gin in London. Sale continues under various aliases (*see* 1735; 1740).

Passenger pigeons sell at six for a penny in Boston (*see* 1672; 1800).

1737 Vienna declares war on Constantinople in January, coming to the aid of its Russian ally to prevent Ottoman aid to France in the renewed War of the Polish Succession.

The last of the Medici family is driven out of Tuscany following the death of Giovan (or Gian) Gastone de' Medici at age 66 leaving vacant the ducal throne. Austria takes over and will give the ducal throne next year to Franz Stefan, duke of Lorraine, who was married in February 1736 to Maria Theresa, heir apparent to the imperial throne.

1737 ***(cont.)*** *Genera Plantarum* by Carolus Linnaeus inaugurates modern botany's binomial system of taxonomy (*see* 1732). Basing his system on stamens and pistils, Linnaeus divides all vegetation into Phanerogams (seed plants) and Cryptogams (spore plants such as ferns). Phanerogams are in turn divided into Angiosperms (concealed seed plants such as flowering plants) and Gymnosperms (naked seed plants such as conifers). Angiosperms are further divided into Monocotolydons (narrow-leaved plants with parallel veins) and Dicotolydons (broad-leaved plants with net veins). Each of these is divided into Orders, Families, Genera, and Species (*see* 1738; 1789).

Theater *False Secrets* (*Les Fausses Confidences*) by Pierre de Marivaux 3/16 at the Comédie-Française, Paris.

A British Licensing Act restricts the number of London theaters and requires that all plays be subjected to the Lord Chamberlain for censorship before they may be presented (*see* 1968).

Opera *Castor and Pollux* 10/24 at the Paris Opéra, with music by Jean Philippe Rameau.

The Philadelphia police force created by Benjamin Franklin is the first city-paid constabulary. Franklin has examined the "city watch" and found it lacking. He will soon organize the first city bucket brigade (fire department).

Calcutta has an earthquake October 11 that kills an estimated 300,000.

1738 The Treaty of Vienna that ended the war of the Polish Succession in 1735 is ratified November 13. Stanislas Leszczynski receives Lorraine and Bar in compensation for renouncing the Polish throne but only on the condition that the duchies devolve upon France at his death; Spain receives Parma and Piacenza in exchange for Naples and Sicily.

The sieur de la Verendrye builds a fort on the Assiniboine River (*see* 1734; 1742).

Sur la figure de la terre by French mathematician-astronomer Pierre Louis Moreau de Maupertuis, 40, reports on an expedition to Lapland. Sent by Louis XV to measure a degree of longitude, Maupertuis confirms Isaac Newton's view that the earth is a spheroid flattened near the poles with a bulge near the equator.

Hydrodynamics by Swiss physicist-mathematician Daniel Bernoulli, 38, presents an early version of the kinetic theory of gases based on his investigations of the forces exerted by liquids (*see* d'Alembert, 1743).

Classes Plantarum by Linnaeus expands the system of taxonomy he introduced last year (*see* 1789).

Excavation begins in Naples to uncover the city of Herculaneum, buried under volcanic ash and lava by the eruption of Mount Vesuvius in 79 A.D. which killed far more people at Herculaneum than at Pompeii. The work will continue for more than 40 years.

John Wesley returns to England from the Georgia colony (his brother Charles returned in 1736 when his health failed) and starts a society that meets for the first time May 1, and both experience evangelical conversions (*see* 1735; 1739).

Painting *La Gouvernante* by Jean Baptiste Chardin.

More Frenchmen will die of hunger in the next 2 years than died in all the wars of Louis XIV, according to René Louis de Voyer, 44, the marquis d'Argenson, who will make the claim in later writings. Fragmentation of French land holdings to the point where a single fruit tree may constitute a "farm" has contributed to a decline in food production.

1739 Persia's Nadir Shah defeats a huge Mughal army near Karnal, takes Delhi March 20, sacks the city, and shatters the Mughal Empire. Afghans sweep down from the northwest frontier, Marathas come east from the great central Daccan plateau to threaten the fertile lands of Bengal, India's suahdors and nabobs establish independent states, some of these go to war with each other, and this chaos opens the way to foreign domination of the subcontinent (*see* 1746).

Ottoman forces approach Belgrade and the Holy Roman Emperor Charles VI signs the Treaty of Belgrade September 18, deserting his Russian ally and ending a 3-year war with Constantinople. Austria yields Belgrade and northern Serbia; Russia retains Azov but agrees to raze her forts and to build no fleet on the Black Sea.

The War of Jenkins's Ear between Britain and Spain begins in October as British naval squadrons receive orders to intercept Spanish galleons. English mariner Robert Jenkins picked a barroom brawl with a Spanish customs guard at Havana in 1731, he suffered a bad cut to the ear, a local surgeon amputated it, Jenkins has kept the sun-dried ear in his sea chest, and a member of Parliament has waved it in the House of Commons, demanding revenge for alleged mistreatment of British smugglers and pirates. The admiralty sends Admiral Edward Vernon to the Caribbean, and he captures Portobelo November 22 with a force of only six ships (*see* 1716; 1741).

The Cato Conspiracy at Stono, South Carolina, takes the lives of 44 blacks and 30 whites as slaves near Charleston arm themselves by robbing a store and set out for Florida, gathering recruits and murdering whites on the way. The rebellion is crushed by a hastily assembled force of whites.

French explorers Pierre and Paul Mallet reach the headwaters of the Arkansas River and sight the Rocky Mountains for the first time.

Treatise on Human Nature by Scottish philosopher David Hume, 28, challenges the prevailing monetary doctrines of mercantilism, notably the doctrine that a nation can continually increase her stock of gold and silver, and her prosperity, through surpluses in her balance of payments. Hume questions whether action by government is necessary or even helpful to the maintenance of a nation's money supply (*see* Adam Smith, 1776).

Persia's Nadir Shah seizes the Koh-i-noor diamond in the sack of Delhi (above). The 109-carat diamond has been a treasure of the Mughal emperors, the East India Company will acquire it when it conquers the Punjab, and the company will present it to Queen Victoria in 1850.

German-American glassmaker Caspar Wistar brings workers to the New Jersey colony and starts a factory.

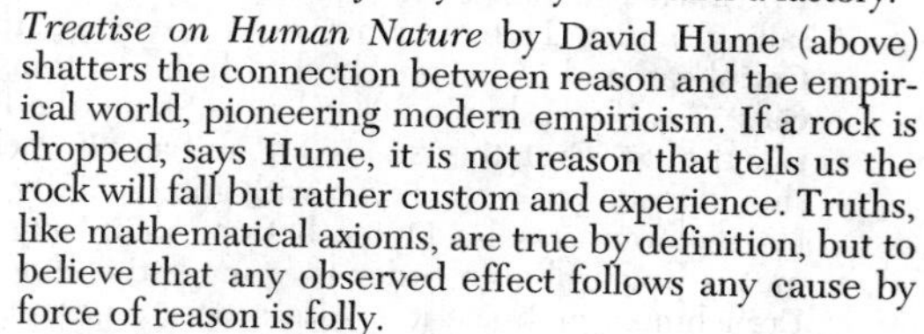
Treatise on Human Nature by David Hume (above) shatters the connection between reason and the empirical world, pioneering modern empiricism. If a rock is dropped, says Hume, it is not reason that tells us the rock will fall but rather custom and experience. Truths, like mathematical axioms, are true by definition, but to believe that any observed effect follows any cause by force of reason is folly.

John Wesley begins preaching in the fields at Bristol and buys a deserted gun factory outside London for his prayer meetings (*see* 1738; 1744).

Painting *Le Déjeuner* by French painter François Boucher, 36, who has studied at Rome and has become the most fashionable painter of his day.

Theater *The Man of the World* (*L'uomo di mondo*) by Venetian playwright Carlo Goldoni, 32, in March at Venice's Teatro San Samuele.

Opera *Les Fêtes d'Hébé*, 5/21, at the Paris Opéra with music by Jean Philippe Rameau.

English highwayman Richard "Dick" Turpin is convicted of horse-stealing at the York assizes and hanged April 7 at age 33 after a notorious career that has made him a legend in his own time.

Potato crops fail in Ireland. The effect is not calamitous since the tubers do not comprise the bulk of most people's diets as they will a century hence, but Irish cotters (tenant farmers) are becoming increasingly dependent on potatoes for food while raising cereal grains and cattle for the export market that provides them with rent money. Cotters select potatoes for high-yield varieties, thus inadvertently narrowing the genetic base of their plants, breeding potatoes with little or no resistance to the fungus disease *Phytophthora infestans* (*see* 1822).

1740

Prussia's Friedrich Wilhelm I dies May 31 at age 51 after a 27-year reign in which he has made Prussia a formidable military power with a standing army of 83,000—enormous for a country of 2.5 million. His scholarly son of 28, who will reign until 1786 as Frederick II, proceeds to occupy part of Silesia, thus precipitating a war with Austria that will continue for 15 of the next 23 years.

The Russian czarina Anna Ivanovna adopts her 8-week-old great-nephew October 5 and declares him her successor, making Count Biron regent for the infant czar Ivan VI. She dies October 17 at age 47, Field Marshal Burkhard Christoph von Münnich, 57, enters the regent's bedroom at midnight November 19 and seizes the hated Biron, and a commission is appointed to try the regent (*see* 1741).

The Holy Roman Emperor Charles VI dies October 20 at age 55, the last of the Hapsburg line that began in 1516. His daughter Maria Theresa is queen of Hungary and Bohemia, but the king of Saxony, the elector of Bavaria, and Spain's Philip V contest her right to succeed Charles. The War of the Austrian Succession will involve most of Europe's great powers in the next 8 years.

Bethlehem in the Pennsylvania colony is founded by Moravian emigrants (*see* Christmas, below).

Crucible steel, rediscovered by English clocksmith Benjamin Huntsman, 36, at Sheffield, will make Sheffield steel as famous as was Damascus steel in antiquity, described by Aristotle in 334 B.C., used for centuries to make swordblades, employed in India, but lost over the ages. Huntsman heats metal in small, enclosed earthenware cupolas which he calls crucibles, excludes air, and produces a superior steel (*see* Krupp, 1811; Sheffield plate, 1742).

Connecticut colony tinsmiths Edward and William Pattison at Berlin begin manufacturing tinware and selling it from house to house—the first colonial tin peddlers, or tinkers.

The Hydra, discovered by Swiss naturalist Abraham Trembley, 40, in Holland is a freshwater polyp that combines both animal and plant characteristics. It can regenerate lost parts.

Typhus epidemics take thousands of lives in the central German states and in Ireland.

Moravian immigrants at Bethlehem (above) introduce the celebration of Christmas with such German customs as the visit from Saint Nicholas, or Santa Claus. In Puritan New England Christmas remains a working day.

A network of post roads connects the American colonies from Charleston to Portsmouth, New Hampshire. Packet boats operate on regular schedules to link the major colonial ports (*see* Franklin, 1732).

Fiction *Pamela, or Virtue Rewarded* by English novelist Samuel Richardson, 51, who gains enormous popularity with his story in the form of correspondence of a servant who resists her master's seductions and winds up as his wife.

Theater *The Test (L'Epreuve)* by Pierre de Marivaux 11/19 at the Théâtre Italien, Paris.

Famine strikes Russia and France. French peasants are reduced to eating ferns and grassroots, but some survive by eating potatoes.

The price of a 4-pound loaf of bread rises to 20 sous at Paris, and 50 die in a prison riot protesting a cut in bread rations.

The unrestricted sale of gin and other cheap spirits reaches its peak in England. Some observers express fears that wholesale drunkenness is undermining the social fabric of the nation (*see* 1736; Hogarth, 1751).

The Royal Navy's rum ration is diluted by Admiral Edward Vernon, nicknamed "Old Grog" because he wears a grogram (grosgrain) cloak in foul weather (*see*

1740 ***(cont.)*** 1731). *Grog* will become a slang word for liquor, *groggy* for drunken dizziness.

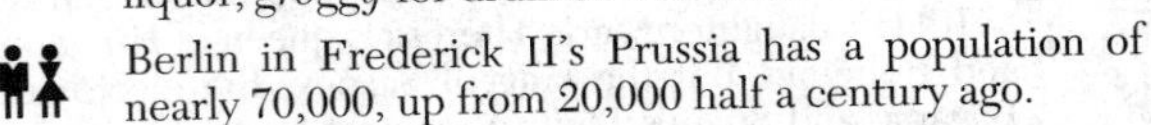

Berlin in Frederick II's Prussia has a population of nearly 70,000, up from 20,000 half a century ago.

1741 Sir Robert Walpole uses the phrase *balance of power* in a speech he delivers in the House of Commons February 13, giving expression to the principle that has long guided British foreign policy.

Prussia's Frederick II gains a victory April 10 at Mollwitz and later joins a secret alliance against Austria signed at Nymphenburg in May by France, Bavaria, and Spain who are soon joined by Saxony (*see* 1740). An allied French-Bavarian army invades Austria and Bohemia and seizes Prague with Saxon help after 30,000 Austrian defenders have died of typhus.

Frederick II (above) conquers Silesia and captures Brieg, Neisse, Glatz, and Olmutz before British diplomats mediate between Prussia and Austria.

A Russian commission finds the regent Count Biron guilty of treason April 11 and sentences him to death by quartering (*see* 1740). The sentence is commuted to banishment for life to Siberia, and the count's vast property is confiscated, including diamonds worth £600,000. The French ambassador plots to destroy the Austrian influence dominant at St. Petersburg, playing on fears of Peter the Great's daughter that she will be confined to a convent for life. He supplies Elizabeth Petrovna, 31, with money, and she drives with her supporters to the barracks of the Preobrazhensky Guards the night of December 6, arouses their sympathies with a stirring speech, and leads them to the Winter Palace for a coup d'état. She seizes the new regent Anna Leopoldovna and her children, including the infant czar Ivan VI, banishes Anna and her husband, deposes the czar and has him imprisoned, and ascends the imperial throne to begin a reign that will continue until her death early in 1762.

Admiral Vernon leads a 100-ship, 27,000-man Royal Navy armada against the Spanish fortress San Felipe at Cartagena but abandons a proposed assault on the fortress after 2 months of dodging Spanish lead and slapping mosquitoes that carry malaria and yellow fever. The British retire to Jamaica after disease and battle wounds have reduced a force of 8,000 landed ashore near Cartagena to only 3,500 effectives.

Admiral Vernon (above) decorates colonial captain Lawrence Washington of Virginia for having led three gallant charges against the walls of San Felipe. Washington returns to the Potomac River plantation he calls Green Mountain and renames it Mount Vernon (*see* 1751).

New York's journeymen bakers go out on strike.

New Yorkers charge a "Negro Conspiracy" with having started fires that break out through March and April. Roman Catholic priests are inciting slaves to burn the town on orders from Spain, they say; four whites and 18 blacks are hanged December 31, and 13 blacks are burned at the stake.

Russian pioneers in Siberia set out for North America in small boats from Kamchatka (*see* 1579). They have no navigational instruments but do have Danish navigator Vitus Bering, now 61, who will die of scurvy on the return voyage (*see* 1728).

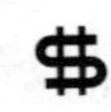

Salem sea captain Richard Derby receives instructions from the owners of his schooner *Volante* to avoid the British Navigation Act of 1660 and subsequent acts by sailing under Dutch colors on his voyage to the Caribbean and to be prepared to bribe customs officials: ". . . If you should fall so low as Statia [the Dutch island of St. Eustatius, *see* 1634], and any Frenchman should make you a good offer with good security, or by making your vessel a Dutch bottom, or by any other means practicable in order to your getting among ye Frenchmen, embrace it.... Also secure a permit so as for you to trade there next voyage, which you may undoubtedly do through your factor or by a little greasing some others."

Essai de cosmologie by P. L. M. de Maupertuis suggests a survival of the fittest concept: "Could not one say that since, in the accidental combination of Nature's productions, only those could survive which found themselves provided with certain appropriate relationships, it is no wonder that these relationships are present in all the species that actually exist? These species which we see today are only the smallest part of those which a blind destiny produced." But few people read Maupertuis (*see* Darwin, 1840).

An epidemic of exanthematous typhus kills 30,000 in France.

A typhus epidemic in Sweden comes on the heels of a poor harvest and takes a heavy toll.

A yellow fever epidemic at Cadiz kills 10,000.

"Sinners in the Hands of an Angry God" by Calvinist clergyman Jonathan Edwards, 38, of Northampton in the Massachusetts colony graphically depicts the terrors of damnation. The sermon by the Connecticut-born Edwards at Enfield is published in pamphlet form and draws opposition from Boston clergyman Charles Chauncey, 36, and other rationalists.

Painting *Autumn* by François Boucher; *The Enraged Musician* by William Hogarth.

Opera *Artaserse* 12/26 at Milan's Teatro Regio Ducal, with music by German composer Christoph Willibald Gluck, 27.

Ode "Rule, Britannia" by Thomas Arne, lyrics from the libretto of *Alfred: A Masque* by David Mallet or James Thomson presented last year.

Some 70 fishing vessels put out from Gloucester in the Massachusetts colony, 60 from Marblehead.

South Carolina colonist Elizabeth Lucas, 19, introduces indigo cultivation, beginning a dyestuffs industry.

British admiral George Anson, 44, arrives on the southwest coast of South America after a 7-month voyage from England during which he has lost three of his six ships and two-thirds of his crew to scurvy. He resumes his voyage round the world.

China's population reaches 140 million (more than 220 million by some accounts), up from 100 million in 1660. The emperor Chi'en Lung restores the *paochia* system of local registration that will make population figures more reliable from now until 1851.

1742 Bavaria's Karl Albrecht is elected Holy Roman Emperor January 24 by opponents of Austria's Maria Theresa. The new emperor is crowned February 12 at Frankfurt-am-Main and begins a 3-year reign as Charles VII, but even as he is crowned his hereditary domains are being overrun by Austrian troops in the continuing War of the Austrian Succession.

Prussia's Frederick the Great triumphs over the Austrians in May at Czaslau and Chotusitz, but his troops are stricken with violent dysentery and he withdraws from the coalition against Austria.

The Treaties of Breslau June 11 and Berlin a few weeks later end the Silesian War. Maria Theresa cedes to Prussia both upper and lower Silesia with their rich coal deposits, plus some other territory; Prussia assumes the Silesian debt of 1.7 million rix dollars held by English and Dutch creditors.

Maria Theresa makes an alliance with Britain, raises two armies, and drives her allied invaders out of Bohemia.

Britain's first prime minister Sir Robert Walpole, now 55, has resigned in February after a 21-year ministry that has created the cabinet and party systems. A series of short-lived ministries will hold power until 1783.

The Battle of Bloody Marsh gives Britain control of the Georgia colony. Troops under James E. Oglethorpe repel a Spanish attack against Fort Frederica on St. Simons Island and defeat the Spaniards.

The sieur de la Verendrye moves into the Black Hills region and adds the discovery of the Dakotas to his discoveries of Manitoba and western Minnesota.

Coal is found on the Coal River in western Virginia, the first of many deposits that will be found and exploited in the region over the next 2 centuries (*see* Morgan, 1731).

The Franklin stove invented by Philadelphia's Benjamin Franklin heats a room far more efficiently than does an open fireplace. Designed to be set inside a fireplace, Franklin's iron "Pennsylvania fireplace" represents the first application of the principle of heating by warmed air—it has a top, back, and sides with an air box inside joined to the sides but not reaching quite to the top. The fire built in front of its opening produces smoke that moves up the front of its air box, over its top, and down again behind it whence it enters the flue at floor level, thus heating the air in the air box which is released by shutters in the sides of the stove.

Sheffield silverplate, invented by Thomas Boulsover at Sheffield, England, is a bond of heavy copper sheets fused between thin sheets of silver (*see* 1740).

Cotton mills open at Birmingham and Northampton. The Lancashire millowners will import East India yarns next year to improve the quality of their textiles.

The centigrade (Celsius) scale of temperatures devised by Swedish astronomer Anders Celsius, 40, divides into 100 degrees the distance between freezing (0°) and boiling water at sea level (*see* Fahrenheit, 1709).

Fiction *The Adventures of Joseph Andrews and his Friend Abraham Adams* by Henry Fielding ridicules Samuel Richardson's 1740 novel *Pamela*.

Poetry *The Complaint—Or, Night Thoughts* by English poet Edward Young, 59, who writes, "Procrastination is the thief of time."

Theater *Mahomet the Prophet (Le fanatisme, ou Mahomet le prophète)* by Voltaire 8/9 at the Comédie-Française, Paris.

Oratorio *The Messiah* by George Frideric Handel 4/13 at the Dublin Cathedral, with Mrs. Susannah Cibber singing the contralto part in a charity concert for the benefit of "the Prisoners in several Gaols, and for the support of Mercer's Hospital in Stephen Street, and of the Charitable Infirmary on the Inn's Quay." Mrs. Cibber was disgraced 3 years ago in a sex scandal that culminated in a trial which revealed the adulterous relationship forced upon her by her husband Theophilus Cibber (son of the playwright–theatrical manager). The chancellor of the cathedral rises after her aria describing Christ's suffering and cries out, "Woman, for this all thy sins be forgiven thee," the concert raises enough money to free 142 prisoners from debtors' prison, and performances of the oratorio will become an annual Christmas tradition.

Hymn "Joy to the World" by G. F. Handel.

Keyboard practice *Goldberg Variations (Alia mit verschiedenen Veraenderungen vors Clavicimbal mit 2 manualen)* by Johann Sebastian Bach.

Short Treatise on Whist by English card-player Edmond Hoyle, 70, gives the first systematization to a game that will develop first into auction bridge and then into contract bridge (*see* 1904; Vanderbilt, 1925). Hoyle's name will survive in the phrase "according to Hoyle."

Boston's Faneuil Hall is presented to the city by local merchant Peter Faneuil, 42, whose red brick meeting house will be enlarged by architect Charles Bulfinch and be a Boston landmark for centuries.

1743 Austrian armies drive French and Bavarian troops out of Bavaria as the War of the Austrian Succession continues in Europe. An allied Pragmatic army of English, Hessian, and Hanoverian troops defeat the French June 27 in the Battle of Dettingen, the emperor Charles VII is forced to take refuge at Frankfurt, and the Dutch Republic takes sides with Britain in alliance with Austria's Maria Theresa, who makes an alliance with Saxony.

War resumes between Persia and the Ottoman Empire.

Russia has pogroms that kill thousands of Jews.

1743 ***(cont.)*** The Dutch East India Company founded in 1594 remains richer and more powerful than its slightly younger British counterpart.

Traité de dynamique by French mathematician-philosopher Jean Le Rond d'Alembert, 25, advances the theory that internal forces constitute by themselves a system of equilibrium which as a whole is statically equivalent to the system of external forces. He will apply d'Alembert's principle next year to the theory of equilibrium and the motion of fluids.

Théorie de la Figure de la Terre by French mathematician Alex Claude Clairaut, 30, reports the results of a mathematical investigation of the shape of the earth according to hydrostatic principles.

"Seasonable Thoughts on the State of Religion in New England" by Boston Congregationalist minister Charles Chauncey opposes the emotional revivalism of Jonathan Edwards at Northampton with theological liberalism (*see* 1741). The battle of pamphlets will continue until early in 1758 when Edwards will die shortly after assuming the presidency of the College of New Jersey at Princeton.

Fiction *The History of the Life of the Late Mr. Jonathan Wild the Great* by Henry Fielding.

Nicolas Lancret dies at Paris September 14 at age 53; Japanese potter Kenzan dies at age 81.

Theater *Mérope* by Voltaire 2/20 at the Comédie-Française, Paris.

Moët and Chandon has its beginnings in a champagne business founded by French entrepreneur Claude Moët, whose firm will become France's largest champagne producer (*see* 1698).

1744 Prussia's Frederick II starts a Second Silesian War by marching through Saxony with 80,000 troops. He invades Bohemia in August and takes Prague in September before the Hapsburg forces of Maria Theresa drive him back into Saxony.

France deserts Frederick II (above) and declares war on both Britain and Maria Theresa.

King George's War breaks out in the Caribbean and in North America—an offshoot of the War of the Austrian Succession between Britain and France, whose Western Hemisphere interests are now fiercely competitive. Britain has the advantage since her squadrons are permanently based in Jamaica and the Leeward Islands, with naval dockyards at Jamaica and Antigua, while the French have no dockyard facilities in the Caribbean, can rarely dispatch a ship from Europe for more than 6 months at a time, and must rely on convoys to protect their sugar shipments (French merchants pay for naval escorts).

Maria Theresa launches a pogrom to drive the Jews out of Bohemia and Moravia. Even where tolerated, Europe's Jews often live under degrading conditions with regulations designed to limit their expansion and keep their status at a low level (*see* Britain, 1753).

The gyroscope stabilizer for ships is pioneered by a Scotsman named Serson, who persuades the British admiralty to sea-test a spinning rotor that will indicate a stable reference for navigators. The rotor is supported on a pivot so the rocking or pitching of the ship will not disturb it.

London's Baltic Mercantile and Shipping Exchange has its beginnings in the Virginia and Baltic Coffee House, an outgrowth of the Virginia and Maryland Coffee House that will become the world's shipping exchange, a vast marble-floored room where shippers and their agents will match vessels and cargoes through deals ("fixtures") made in great secrecy—often "by the candle" with bids continuing until an inch of candle has burned itself out.

John Wesley holds love feasts modeled on rites of the primitive church (*see* 1739). He has held class meetings with admission by "society tickets" to exclude undesirables, journeyed on horseback through England to organize Methodist societies, and holds the first conference of Methodists. Beginning in 1747, Wesley will make 42 trips to Ireland, beginning in 1751 he will make 22 trips to Scotland, and he will use his annual income of about £1,400 from his cheap books to help the needy and unemployed, in loans to debtors and businessmen, and to open dispensaries (*see* 1784).

Islamic fundamentalism gains support in central Arabia from the sheik and emir of the ibn Saud family, who embrace the tenets of Wahhabism, named for Nejd jurist Mohammed ibn Abd al-Wahhab, who scorns the "decadent" Sunnism of the Ottomans yet differs with Persian Shiites (*see* 1773).

Tommy Thumb's Pretty Song Book is published at London with the earliest known versions of such nursery rhymes as "London Bridge is falling down"; "Who killed Cock Robin?/'I,' said the Sparrow,/With my bow and arrow, I killed Cock Robin"; "Hickory, dickory dock,/The mouse ran up the clock"; "Ladybird, ladybird,/Fly away home"; "Little Tommy Tucker/Sings for his supper"; "Oranges and lemons,/Say the bells of St. Clements./You owe me five farthings,/Say the bells of St. Martin's"; "Sing a song of sixpence/A pocketful of rye" (*see* songs, 1765).

Sotheby's art auction house has its beginnings January 7 as London bookseller Samuel Baker holds his first real auction, selling the library of a local physician for £826. Baker will retire in 1767, his nephew John Sotheby will come into the firm a few weeks later, John's nephew Samuel Sotheby will take over from John, Samuel's son Samuel Leigh Sotheby will head the firm until his death in 1861, and the house will grow to have auction rooms worldwide (*see* Christie's, 1766).

Anthem "God Save the King" is published at London with the opening words "God save our Lord the King." They will be changed next year to "God save great George our King" and will later be changed to "God save our gracious King" (see "America," 1831).

The Well-Tempered Clavier, Part II, by Johann Sebastian Bach includes new preludes, toccatas, fugues, and other works.

The first recorded cricket match pits Kent against All England.

Würzburg Residenz for the prince bishop is completed in baroque style after 25 years of construction. (Johann) Balthasar Neumann, 57, is the architect.

The 14,948-foot volcano Cotopaxi erupts in the Spanish viceroyalty of Peru.

Frederick II (above) distributes free seed potatoes to reluctant Prussian peasants, stationing armed soldiers in the fields to enforce his edict that the peasants plant potatoes or suffer their ears and noses to be cut off.

Britain's Lord Anson returns with the last ship of his original six after having lost more than four-fifths of his crew to scurvy (*see* 1741; Lind, 1747).

1745

The Holy Roman Emperor Charles VII dies at Munich January 20 at age 47 after a 3-year reign. His son Maximilian Joseph, 17, succeeds as elector of Bavaria, signs the Treaty of Fussen April 22 with Maria Theresa to end the Second Silesian War, and regains the land conquered by Austria by renouncing any claim to Austria's throne and supporting the imperial election of Maria Theresa's husband Franz Stefan. Franz Stefan is elected Holy Roman Emperor; he will reign until 1765 as Francis I.

A January alliance against Prussia has joined Austria, Saxony, Britain, and Holland. French forces under Marshal Maurice of Saxony defeat the Pragmatic army May 11 in the Battle of Fontenoy and begin conquering the Austrian Netherlands.

Frederick the Great defeats Austrian and Saxon forces June 4 at the Battle of Hohenfriedberg in Silesia, and he defeats Austrian forces September 30 at the Battle of Soor in northeastern Bohemia.

Scotland has an uprising of Highland peasants who have been driven off the land by "lairds" who want to break up the clans and use the Highlands for sheep raising. The clansmen rally behind Charles Edward "Bonny Prince Charlie" Stuart, 25, who lands in the Hebrides July 25 and raises the Jacobite standard to proclaim his father James VIII of Scotland and James III of England. The Young Pretender leads 2,000 men into Edinburgh September 11, the Jacobites win the Battle of Prestonpans September 21, and they follow Charles into England, reaching Derby December 4 (*see* Falkirk, Culloden, 1746).

The Treaty of Dresden December 25 brings peace between Prussia, Austria, and Saxony, Prussia recognizes the Pragmatic Sanction of 1731 but retains Silesia, the British army is recalled to fight "Bonny Prince Charlie" (above), but the War of the Austrian Succession that began in 1740 continues.

British and New England forces have taken the French fortress of Louisburg in Nova Scotia June 16 after a 7-week siege by William Pepperell, 49, who is promoted to the rank of colonel and next year will be created a baronet, the first American to be so honored.

Japan's Tokugawa shogun Yoshimune resigns and is succeeded after a 29-year shogunate by the far less competent Ieshige, who will reign until 1760.

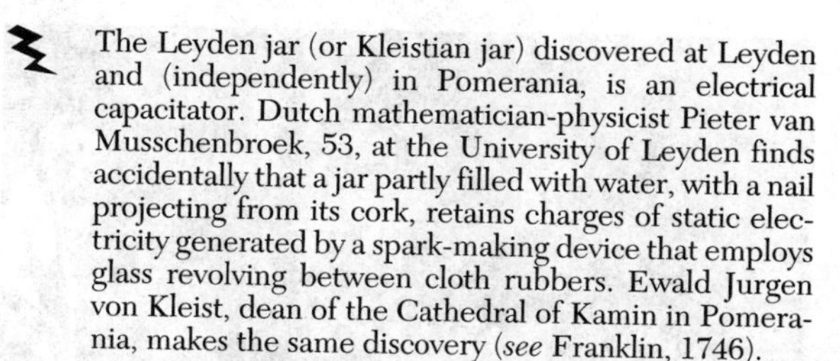

The Leyden jar (or Kleistian jar) discovered at Leyden and (independently) in Pomerania, is an electrical capacitator. Dutch mathematician-physicist Pieter van Musschenbroek, 53, at the University of Leyden finds accidentally that a jar partly filled with water, with a nail projecting from its cork, retains charges of static electricity generated by a spark-making device that employs glass revolving between cloth rubbers. Ewald Jurgen von Kleist, dean of the Cathedral of Kamin in Pomerania, makes the same discovery (*see* Franklin, 1746).

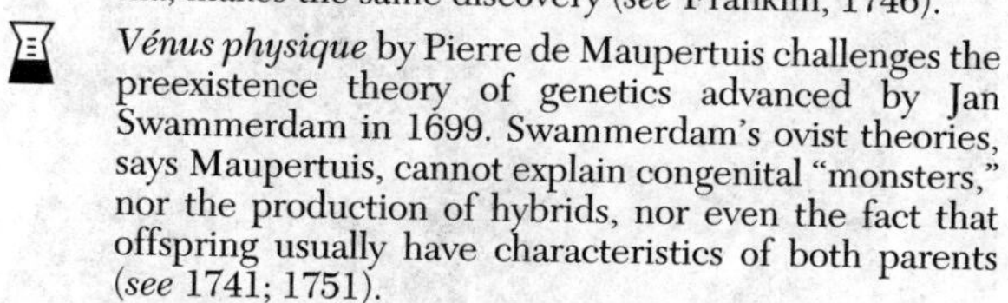

Vénus physique by Pierre de Maupertuis challenges the preexistence theory of genetics advanced by Jan Swammerdam in 1699. Swammerdam's ovist theories, says Maupertuis, cannot explain congenital "monsters," nor the production of hybrids, nor even the fact that offspring usually have characteristics of both parents (*see* 1741; 1751).

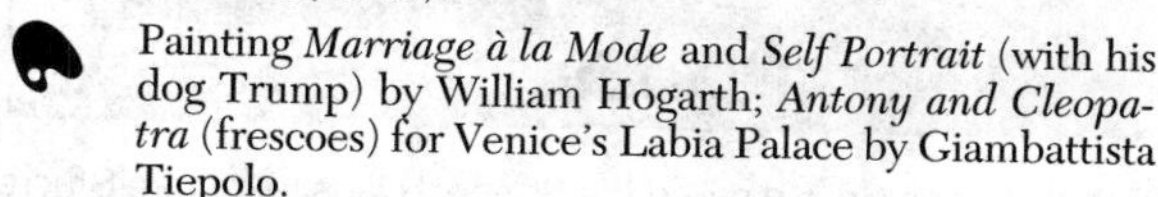

Painting *Marriage à la Mode* and *Self Portrait* (with his dog Trump) by William Hogarth; *Antony and Cleopatra* (frescoes) for Venice's Labia Palace by Giambattista Tiepolo.

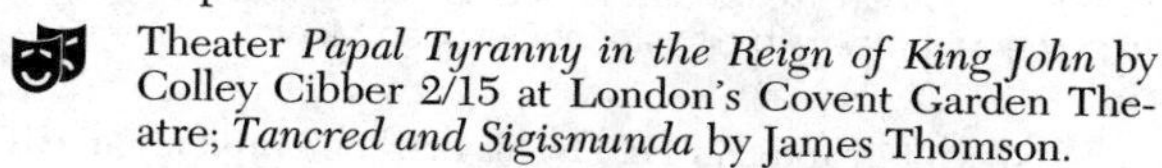

Theater *Papal Tyranny in the Reign of King John* by Colley Cibber 2/15 at London's Covent Garden Theatre; *Tancred and Sigismunda* by James Thomson.

Anthem "The Campbells Are Coming" is published in Scotland.

The quadrille becomes fashionable in France.

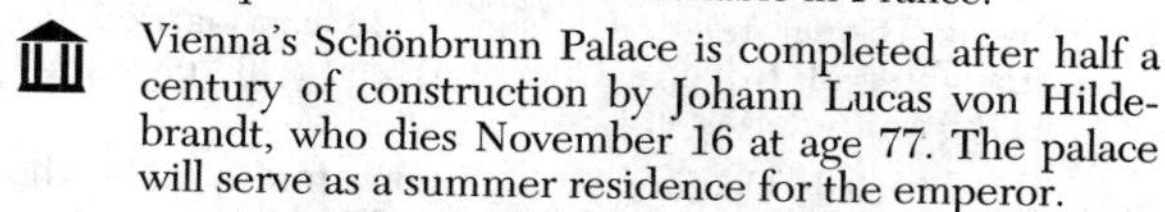

Vienna's Schönbrunn Palace is completed after half a century of construction by Johann Lucas von Hildebrandt, who dies November 16 at age 77. The palace will serve as a summer residence for the emperor.

1746

The Battle of Falkirk January 17 gives the Jacobite pretender Charles Edward Stuart and his Highlanders a victory over the British dragoons. Seven guns and 700 prisoners fall into Jacobite hands.

The Battle of Culloden Moor April 16 ends Stuart efforts to regain the British throne. The king's son William Augustus, duke of Cumberland, has stopped at Nairn to celebrate his 25th birthday, "Bonnie Prince Charlie" has marched his 7,000 starving Highlanders across the moors by night to surprise the British, the Scots have turned back exhausted, Cumberland's 9,000-man army arrives and opens fire with 10 guns to Stuart's nine, the Scots find themselves surrounded and flee, leaving 1,000 dead on the battlefield, and the dragoons hunt down and massacre those who have survived the carnage.

"Bonnie Prince Charlie" escapes to the Isle of Skye June 29 disguised as Betty Burke, maid to Flora Macdonald, 24. The young pretender gets away to France September 20, the British imprison Macdonald briefly in the Tower of London, and they forbid anyone to wear the tartan, a ban that will continue until 1782.

The Battle of Rocoux gives France's Marshal Saxe a victory that frees the Netherlands from Austria. French troops occupy Brussels February 21.

Spain's first Bourbon king Philip V dies July 9 almost completely insane at age 62 after a 46-year reign that

Bonnie Prince Charlie rallied the Scottish clans to restore Stuart to power but met with disaster at Culloden Moor.

was interrupted in 1724 by his first son's brief reign. His second son by his first wife Maria Louisa of Savoy is 33 and will reign until 1759 as Ferdinand VI.

Denmark's Christian VI dies August 6 at age 47 after a weak 16-year reign. His son of 23 has married the daughter of Britain's George II and will reign until 1766 as Frederick V.

Madras falls October 20 to French forces under the colonial governor Joseph François Dupleix in an extension of the War of the Austrian Succession.

Benjamin Franklin begins experiments with electricity at Philadelphia. Franklin will improve the Leyden jar (or Kleistian jar) invented last year by replacing its water with pulverized lead and he will invent an adaptation of the Leyden jar—a foil-coated pane of glass that will be called the Franklin pane (*see* kite, 1751; Priestley, 1767).

German physician George Erhard Hamberger, 47, gives the first description in medical literature of the duodenal ulcer that will plague modern society.

Princeton University has its beginnings in the College of New Jersey founded at Newark by Presbyterian ministers who include Aaron Burr, 30. Classes open in a building that also houses the city courthouse and jail, Burr will be president of the college from 1748 to 1757, moving it to Princeton in 1756, and it will adopt the name Princeton University in 1896 (*see* Nassau Hall, 1756).

Nonfiction *Pensées philosophiques* by French man of letters Denis Diderot, 32.

Painting *Captain John Hamilton* and *The Eliot Family* by English painter Joshua Reynolds, 23; *The Vegetable Market at San Giacomo di Rialto* by Antonio Canaletto, who moves to London where he will remain for 10 years except for two short visits home to Venice; *The Milliner* (*La Marchande de Modes*) by François Boucher.

English grain prices continue to fall as they have been doing for 30 years and will continue to do for another 10. Death rates will fall, too, as more people are able to afford better diets (*see* Burke's Act, 1773).

1747 The British admirals George Anson and Edward Hawke score smashing victories against the French in the West Indies. The Royal Navy dominates European waters and increases the peril to France's huge convoys of sugar from the Caribbean (*see* 1744).

Persia's Nadir Shah is assassinated by one of his own tribesmen June 10 at Fathabad after an 11-year reign. His death leaves Persia in anarchy (*see* 1794).

The Barkzai dynasty that will rule Afghanistan until 1929 comes to power as one of Nadir Shah's generals assumes control of the Afghan provinces. He will reign until 1773 as Ahmad Shah.

The Battle of Laufeld July 2 ends in victory for French forces under Marshal Saxe over the duke of Cumberland's Anglo-Dutch army. The French capture Bergen-op-Zoom September 16, they consolidate their occupation of Austrian Flanders, and the republic of the United Provinces is overthrown and Willem of Nassau, Prince of Orange, is made hereditary *stadtholder*.

Dahomey in West Africa accepts defeat by Oyo after half a century of effort by Oyo cavalry to take over.

Britain imposes a carriage tax.

Marine insurance companies in London charge rates as high as 11 percent on ship and cargo from New England to Madeira, 14 percent to Jamaica, and 23 percent to Santo Domingo as privateers and pirates menace shipping in the Atlantic and Caribbean, but merchantmen out of Boston and Salem carry cod from the Newfoundland banks to the West Indies and southern Europe, and they often make profits of 200 percent even after the high insurance rates.

Poetry *Ode on a Distant Prospect of Eton College* by English poet Thomas Gray, 31: "Where ignorance is bliss,/'Tis folly to be wise"; *Odes on Several Descriptive and Allegorical Subjects* by English poet William Collins, 25, includes *Ode to Evening* and *Ode Written in the Beginning of the Year 1746* ("How sleep the brave . . .").

Painting *Industry and Idleness* by William Hogarth.

Oratorio *Judas Maccabeus* by George Frideric Handel at London's Covent Garden Theatre.

Potsdam's Sans-Souci Palace is completed by Georg Wenzeslaus von Knobelsdorff.

Scottish naval surgeon James Lind, 31, pioneers the conquest of scurvy. He conducts experiments with 12 scurvy victims aboard the Royal Navy's H.M.S. *Salisbury* and finds that cider, nutmeg, seawater, vinegar, an elixir of vitriol (a sulfate), and a combination of garlic, mustard, myrrh, and balsam of Peru are all worthless as scurvy cures, but that two scurvy seamen given two oranges and a lemon each day recover in short order, an indication that citrus fruits contain an antiscorbutic element (*see* 1734; 1753).

Prussian chemist Andreas Sigismund Marggraf, 38, discovers that beets and carrots contain small amounts of sugar (*see* Achard, 1793).

1748 The Treaty of Aix-la-Chapelle signed October 18 halts the War of the Austrian Succession after 8 years, but the peace is really no more than a truce (*see* 1756). The house of Hanover retains the succession in its German states and in Britain, the Pragmatic Sanction of 1720 is sustained in Austria, Silesia is given to Prussia which has become a great power, and there is a reciprocal restoration of conquests. France regains Nova Scotia's Louisburg from Britain (*see* 1755) but gives up territory she has taken in the Lowlands and surrenders control of Madras in India (*see* 1754).

American colonists cross the Allegheny Divide and move into western lands (*see* George III, 1763).

"The only principle of life propagated among the young people is to get money," writes Scots-American Cadwallader Colden, 60, "and men are only esteemed according to what they are worth—that is, the money they are possessed of."

The Spirit of the Laws (De l'esprit des lois) by Baron de Montesquieu pioneers sociology by showing the interrelation of economical, geographical, political, religious, and social forces in history, a revolutionary concept that makes Montesquieu's most important work a best seller.

Introductio in analysin ininitorum by Leonhard Euler systematizes calculus, emphasizes the study of functions, classifies differential equations, and treats trigonometric functions and equations of curves without reference to diagrams (*see* Lagrange, 1761).

Observations upon the Generation, Composition, and Decomposition of Animal and Vegetable Substances by English naturalist John Tuberville Needham, 35, gives "proof" of spontaneous generation. Needham says he has found flasks of broth teeming with "little animals" after having boiled them and sealed them, but his experimental techniques have been faulty (*see* Redi, 1668; Maupertuis, 1751).

"Account of the Sore Throat Attended with Ulcers" by London physician John Fothergill, 36, gives the first description of diphtheria (*see* Krebs, 1883).

Fiction *The Adventures of Roderick Random* by Scottish physician Tobias George Smollett, 27, enjoys such success that Smollett will give up his London practice and devote himself to writing. "Some folks are wise and some otherwise," says Smollett in his anonymously published picaresque novel (he sailed at age 20 as a surgeon's mate to Cartagena and last year married a Jamaican heiress); *Clarissa, or the History of a Young Lady* by Samuel Richardson, whose seven-volume novel will establish his lasting reputation and that of his heroine Clarissa Harlowe.

Painting *Calais Gate* by William Hogarth.

1749 French troops move into the Ohio River Valley and claim it for Louis XV. Sent into the region by the comte de Galissonière, they win over the Indian tribes, they force British traders to leave, and they proceed to build forts (*see* 1754).

The Ohio Company chartered by Britain's George II makes its first settlements in hilly Appalachian country along the Monongahela River. The king grants half a million acres to a group of London merchants and Virginia gentlemen who include the brothers of George Washington (*see* 1752).

Halifax, Nova Scotia, is founded to rival the French port of Louisburg. Edward Cornwallis names the town after the president of the London Board of Trade George Montagu Dunk, 32, earl of Halifax.

Histoire naturelle by French naturalist George Louis Leclerc, 42, comte de Buffon, is published in the first of its three volumes. Buffon rejects any preexistence theory of genetics such as that proposed by Jan Swammerdam in 1669, but does not dispute the idea of spontaneous generation.

The University of Pennsylvania has its beginnings in the College of Philadelphia founded pursuant to a suggestion by Benjamin Franklin.

Letters on the Spirit of Patriotism: On the Idea of a Patriot King published anonymously at London advances the idea of an essential "harmony of interests" among different classes of society. Its author is former English politician Henry St. John, 71, first viscount Bolingbroke, who retired 14 years ago to his wife's French estate near Touraine.

French philosopher Jean-Jacques Rousseau, 37, wins first prize at the Academy of Dijon for his essay "Discours sur les Arts et Sciences" glorifying the savage state and calling civilized society the inevitable source of moral corruption: "The man who first had the idea of enclosing a field and saying this is mine, and found people simple enough to believe him, was the real founder of society."

Fiction *The History of Tom Jones, a Foundling* by Henry Fielding.

Painting *Mr. and Mrs. Robert Andrews* by English painter Thomas Gainsborough, 22. Allesandro Magnasco dies at Genoa March 12 at age 81.

Oxford's Radcliffe Camera is completed by James Gibbs after 12 years of construction.

London cabinetmaker Thomas Chippendale, 31, opens a factory that will be famous for its graceful, ornate furniture although many will prefer the work of his contemporary George Hepplewhite, 23, whose workshops are in the London parish of St. Giles, Cripplegate (*see* Sheraton, 1791).

1749 ***(cont.)*** New England suffers a drought so severe that farmers whose powder-dry pastures catch fire must send to Pennsylvania and even to England for hay.

Haricot beans are planted extensively for the first time in France. The flageolets planted near Soissons will soon replace fava beans in the area (*see* 1528).

Pellagra is described in Italy where polenta made of cornmeal is a dietary staple in some areas. The disease takes its name from the Italian words *pelle agra* (rough skin) and its further symptoms are diarrhea, dementia, and—ultimately—death (*see* 1735; 1907).

China's population reaches 225 million.

1750 Portugal's João V dies July 31 at age 61 after a 44-year reign and is succeeded by his son of 35, who will reign until 1777 as José Manuel. The new king appoints Sebastião José de Carvalho e Mello, marques de Pombal, as his chief minister, and Pombal strips the Inquisition of its powers.

Chippewa (Ojibway) Indians defeat Sioux tribesmen at the Battle of Kathio and gain undisputed possession of wild rice stands in the lakes of northern Minnesota.

Cumberland Gap through the Appalachians is found at an altitude of 1,665 feet by English physician Thomas Walker, agent for a Virginia land company. He names the gap for George II's third son William Augustus, duke of Cumberland.

The Iron Act passed by Parliament prohibits Britain's American colonists from manufacturing iron products while permitting them to produce bar and pig iron from their ore deposits, using the fuel which they have in abundance, and permitting them to exchange the iron for manufactured articles. The act is an expression of classic mercantilism and embodies the chief principle of the British colonial system, namely to restrict the colonies to producing raw materials which the mother country can process or resell, and to use the colonies as a market for the mother country's manufactured goods. But the Iron Act will be generally ignored, and by 1776 there will be as many iron mills in America as in England (*see* 1663; Woolens Act, 1699; Sugar Act, 1764).

Bituminous coal is mined for the first time in America at Richmond Basin in the Virginia colony (*see* Coal River, 1742).

London's Westminster Bridge opens to traffic, the first new bridge to span the Thames since London Bridge in the 10th century.

The flatboat invented by Pennsylvania colonist Jacob Yoder makes inland waterways navigable.

Anti-Senèque, System d'Epicure by French philosopher-physician Julien Offroy de La Mettrie, 41, says the only pleasures are those of the senses, life should be spent in enjoyment as advocated by the 3rd century B.C. Athenian philosopher Epicurus, and the soul dies with the body. La Mettrie has been attacked in France for his materialist teachings and lives in Berlin.

Fiction *Fanny Hill, or the Memoirs of a Woman of Pleasure* by English pornographic novelist John Cleland, 41, will be the classic of erotic literature for 2.5 centuries.

Theater *The Coffee House (La botega di caffè)* by Carlo Goldoni 5/2 at Mantua; *The Liar (Il bugiardo)* by Goldoni 5/23 at Mantua; *The Comic Theatre (Il teatro comica)* by Goldoni 10/5 at Venice's Teatro Sant' Angelo.

Munich's Residenztheater is completed for the elector of Bavaria by his court dwarf and architect François de Cuvilles, 52.

Johann Sebastian Bach completes *The Art of the Fugue* but dies July 28 at age 65.

England raises the bounty on whales to 40 shillings per ton in a move that encourages Scotsmen to enter the whaling industry.

Famine ravages France.

Massachusetts has 63 distilleries producing rum made from molasses supplied in some cases by slave traders who sell it to the Puritan distillers for the capital needed to buy African natives that can be sold to West Indian sugar planters (*see* 1733).

France's population will increase from 22 million to 27 million in the next half century, Italy's from 15.5 million to 18 million, and Spain will have a similar increase, but the total population of Europe will grow from 140 million to 188 million as births consistently exceed deaths rather than alternately falling behind deaths as in previous centuries, and as more children survive beyond age 10.

London has 11 percent of England's population while Paris has only 2 percent of France's population.

The world's population reaches 750 million.

1751 Sweden's Frederick I of Hesse-Cassel dies April 5 at age 75 after a 31-year reign and is succeeded by his wife's cousin Adolphus Frederick of Oldenberg-Holstein-Gottorp, 41, who has married a sister of Prussia's Frederick the Great and will reign until 1771 as Adolphus Frederick.

Dutch *stadtholder* Willem IV dies October 22 at age 40 after a 4-year reign and is succeeded by his 3-year-old son, who will reign until 1795 as Willem V.

British and Indian forces under Robert Clive, 26, in southeast India seize Arcot, capital of the Carnatic, August 31. A former East India Company clerk, Clive began his military career 3 years ago. He resists a 53-day siege by superior French forces, driving them off November 5 in the first serious challenge to France's hegemony in the subcontinent.

Georgetown is founded on the Maryland shore of the Potomac River. The new port is named in honor of Britain's George II (*see* 1790).

The British consol is created by a consolidation of public securities, chiefly annuities, into a single debt issue—a consolidated annuity with no maturity (*see* Rothschild, 1815).

Benjamin Franklin discovers the electrical nature of lightning by flying a kite in a thunderstorm. The kite has a wire conductor, a key at the end of its wet twine kite string, and a silk insulator which Franklin keeps dry by standing in a doorway. He sends his friends in England a paper entitled "Experiments and Observations on Electricity Made at Philadelphia" and will follow it with other papers on the subject. He works to develop a lightning rod that will prevent the fires that so often begin in thunderstorms (*see* 1752).

Système de la Nature by Pierre de Maupertuis challenges the "proof" of spontaneous generation offered by John Needham in 1748. Maupertuis also challenges orthodox ovist genetic theories. He postulates a theory of *monstres par défaut et par excés,* which anticipates the later discovery of supernumerary or missing chromosomes, and expands on his 1745 critique of ovist theories by observing that offspring reveal characteristics present in both parents, a fact well known to livestock breeders, and pointing out that ovist theories cannot explain the periodic recurrence of six-fingered hands in members of a certain Berlin family nor explain albinism in blacks.

German physicist A. Croustedt isolates nickel.

English physician Robert Whytt, 37, shows that only a segment of the spinal cord is necessary for a reflex action and draws explicit distinction between voluntary and involuntary motion. He pioneers the study of reflexes as a distinct branch of physiology.

Encyclopédie, or *Dictionnaire raisonné des sciences, des arts et des métiers* appears in its first volume in April at Paris. Editor-in-chief Denis Diderot has rejected the idea of simply translating Ephraim Chambers' 1728 *Cyclopaedia* and has enlisted the help of mathematician Jean d'Alembert, baron de Montesquieu, and others (*see* 1772).

Fiction *The Adventures of Peregrine Pickle* by Tobias Smollett; *Amelia* by Henry Fielding, who calls gin "poison" (*see* Hogarth, below).

Poetry *Elegy Written in a Country Churchyard* by Thomas Gray: "The curfew tolls the knell of parting day,/The lowing herd wind slowly o'er the lea,/The ploughman plods his weary way,/And leaves the world to darkness and to me . . ." "Full many a gem of purest ray serene/The dark unfathomed caves of ocean bear./ Full many a flower is born to blush unseen,/And waste its sweetness on the desert air."

Painting *La Toilette de Vénus* by François Boucher; *Gin Lane* (engraving) by William Hogarth, who satirizes the excesses of drinking among London's lower classes. Japanese *ukiyoe* painter Sukenobu Nishikawa dies at age 80.

At least 60 New England vessels are engaged in whaling ventures (*see* 1712; 1775).

Virginia planter George Washington, 19, visits Barbados with his brother Lawrence, who requires a warm climate to recover from an illness (*see* 1741). Young George samples such tropical fruits as shaddock (*see* 1696) but writes in his diary that "none pleases my taste as do's the pine [pineapple]."

"So vast is the Territory of North America that it will require many Ages to settle it fully," writes Benjamin Franklin (above), "and, till it is fully settled, Labour will never be cheap here, where no Man continues long a Labourer for others, but gets a Plantation of his own" (*see* 1755; Jefferson, 1782).

1752 Lahore falls to Afghanistan's Ahmed Shah Durani after a 4-month siege.

The Sudanese sultan Abu al-Qasim of Darfur falls in battle while fighting Funj forces commanded by Gen. Abu al-Kaylak.

The Logstown Treaty June 13 cedes Iroquois and Delaware lands south of the Ohio River to the Virginia colony and permits the 3-year-old Ohio Company to build a fort and to settle on lands west of the Alleghenies (*see* French, 1753).

The Philadelphia Contributionship for the Insurance of Homes from Loss by Fire is founded by Benjamin Franklin and a group of his friends for mutual protection (*see* lightning rod, below).

Public street lighting begins in Philadelphia. New York and Boston continue to light streets by placing a lamp in the window of every seventh house, but Philadelphia installs globe lamps imported from London.

Observations on the Diseases of the Army by British Army medical officer John Pringle, 45, will lead to some reforms and to rules of hygiene that will reduce the toll of typhus, which Pringle identifies with "gaol fever" and "hospital fever."

The Manchester Royal Infirmary is founded.

French naturalist René de Réaumur proves that digestion is at least partially a chemical process (*see* 1712). He places food inside tiny perforated metal cylinders and feeds these to hawks; when he recovers the cylinders and finds that the food inside is partially digested he destroys the prevailing belief that the stomach digests food simply by grinding it physically (*see* Beaumont, 1822).

Fiction *The Female Quixote, or The Adventures of Arabella* by English novelist-poet Charlotte Ramsay Lennox, 32.

Poetry *On the Prospects of Planting Arts and Learning In America* by Irish philosopher George Berkeley, now 68, who lived in the Rhode Island colony from 1728 to 1731: "Westward the course of empire takes its way . . ."; *Poems* by English poet Christopher Smart, 30.

Britain and her colonies adopt the Gregorian calendar of 1582 (the difference between the Old Style Julian calendar of 46 B.C. and the Gregorian calendar has grown to 11 days). Thursday, September 14, follows Wednesday, September 2, by decree of Parliament, the change confuses bill collectors, rumors spread that salaried employees are losing 11 days' pay and that everybody is losing 11 days of his life, Londoners cry,

1752 ***(cont.)*** "Give us back our 11 days," and there are riots to protest the calendar change.

London's Mansion House is completed by George Dance after 13 years of construction.

A French royal château at Choisy is completed by Jacques Ange Gabriel, 54, who succeeded his late father as first architect to Louis XV in 1742, completed enlargements of Fontainebleau in 1749, completed a château at Compiegne last year, and begins work on the Ecole Militaire at Paris.

Philadelphia's Liberty Bell cast by local foundrymen John Pass and John Stow for the belfry of the new Pennsylvania State House replaces a bell cast at London's Whitechapel Bell Foundry that has cracked soon after arrival. Made of a bronze alloy (77 percent copper/23 percent tin), the new 2,080-pound bell bears the same biblical inscription (from Leviticus 25:10) as the original. It cracks slightly while being tested, will crack again in July, 1835, and ring no more thereafter. Cast to celebrate next year's fiftieth anniversary of Penn's Charter, it will serve its purpose: "PROCLAIM LIBERTY THROUGHOUT ALL THE LAND UNTO ALL THE INHABITANTS THEREOF LEV. XXV VSX / BY ORDER OF THE ASSEMBLY OF THE PROVINCE OF PENNSYLVANIA FOR THE STATE HOUSE IN PHILADA."

Moscow has a fire that destroys some 18,000 houses.

The lightning rod, invented by Benjamin Franklin, is a metal conductor that will save houses from being set afire in electrical storms (*see* 1760).

George Washington inherits his brother's Mount Vernon plantation in the Virginia colony and works to improve its agricultural practices.

France's Louis XV orders that grain surpluses be stored as a reserve against famine, but government grain buying drives up bread prices. Thousands of Frenchmen are reduced to starvation, and suspicions arise that Louis is making millions of francs in profit from grain speculations (*see* 1768).

George Washington (above) builds a gristmill to produce flour from his own wheat that will be shipped in barrels from his own cooperage. He will build two more mills, his "superfine flour" will go to markets as far distant as the West Indies, and he will be the largest flour producer in the colonies, giving him a large cash income that will enable him to buy good real estate and thus hedge against inflation as colonial currencies depreciate.

The first Irish whisky distillery is founded by distiller William Jameson.

The Virginia colony has a population of nearly 250,000.

1753 Some 1,500 French troops from Canada sent by the marquis Duquesne occupy the Ohio Valley, erecting Fort Presqu'Isle and Fort Le Boeuf. The lieutenant governor of Virginia Robert Dinwiddie, 60, sends out surveyor George Washington with a demand that the French withdraw, Washington reaches Fort Le Boeuf December 12, and he is told that Dinwiddie's letter will be forwarded to Duquesne (*see* 1754).

Parliament passes a Jewish Naturalization Bill, but the measure arouses opposition. The bill will be repealed next year, and British Jews will not gain civic emancipation until 1860.

The Conestoga wagon introduced by Pennsylvania Dutch settlers in the town of Conestoga will come into wide use in the next 100 years. (Wagon drivers will smoke cheap cigars that will be called "stogies.") (*see* "The Flying Machine," 1817).

The British Museum has its beginnings January 11 as London physician Sir Hans Sloane dies at age 92, leaving the nation his library of 50,000 volumes, several thousand manuscripts, coins, curiosities, and pictures. The government buys the collection Sloane inherited in 1702 from the late naturalist William Courten and grants a royal foundation charter for a museum (*see* Royal Library, 1757).

Fiction *Ferdinand Count Fathom* by Tobias Smollett is a pioneer horror mystery story.

Poetry *Hilliad* by Christopher Smart satirizes critic John Hill.

Treatise on the Scurvy by James Lind is an account of his citrus cure (*see* 1747; 1757).

On the Numbers of Man by Robert Wallace echoes the view of Montesquieu in 1734 that the Greek-Roman world had a larger population than that of 18th-century Europe. Wallace argues against diverting capital and labor from basic commodities to manufacturing luxury goods.

1754 Britain's prime minister Henry Pelham dies March 6. His brother Thomas, duke of Newcastle, takes his place.

French troops rout a force of Virginia frontiersmen building a fort at the confluence of the Allegheny and Monongahela rivers April 17, defeating a small British colonial expeditionary force led by George Washington. The French defeat Washington again July 3 near Fort Necessity in the Ohio Valley.

The French erect Fort Duquesne at the head of the Ohio River, hoping to confine the British to the area east of the Appalachians while they build a Gallic empire in the lands to the west.

The Albany Convention June 19 assembles representatives of the American colonies in a meeting with chiefs of the Six (Iroquois) Nations in order to work out a joint plan of defense by Iroquois and British colonial forces against the advances of the French. Adopting a proposal by Benjamin Franklin, the convention issues a call July 10 for voluntary union of the 13 British colonies.

Paris recalls colonial administrator Joseph François Dupleix, 57, from India after a 12-year career as governor general of all French possessions in the subcontinent. The British are left in firm control.

The Ottoman sultan Mahmud I dismounts from his

horse and drops dead at age 60. His brother, 55, will reign until 1757 as Osman III.

Pittsburgh has its beginnings in Fort Duquesne built by the French (above).

Money from sugar, tobacco, sea-island cotton, and other commodities grown in the New World with slave labor rivals money from East India Company ventures to create a growing leisure class in England.

The first iron-rolling mill is opened by English entrepreneurs at Foreham in Hampshire.

A Society for the Encouragement of Arts and Manufactures is founded in England.

The Spanish church becomes practically independent of Rome under terms of a Concordat with the Vatican.

Columbia University has its beginnings in the King's College founded at New York (*see* 1784).

Painting *The Election* by William Hogarth; *The Judgment of Paris* by François Boucher.

Scotland's Royal and Ancient Golf Club is founded at St. Andrews (*see* 1552).

St. Petersburg's Winter Palace is completed by the Italian architect Bartolomeo Rastrelli, 54, whose late father Carlo served as sculptor to Peter the Great.

Copenhagen's Amalienborg Palace is completed by Danish architect Nikolaj Eigtoed, who dies at age 50.

Philadelphia's Christ Church is completed after 27 years of construction: the 200-foot steeple is the tallest structure in North America.

English porcelain production is pioneered by Quaker pharmacist William Cookworthy, 49, of Plymouth, who finds a deposit of kaolin in Cornwall (*see* d'Entrecolles, 1712). In 1756 Cookworthy will find a deposit of petuntse, the feldsparlike material that is mixed with kaolin to produce fine porcelain.

Paris has 56 coffeeshops, or cafés (*see* 1672).

1755 Gen. Edward Braddock lands at Hampton Roads, Virginia, February 20, with two regiments of regulars to assume command as commander in chief of British forces in America. He meets with colonial governors April 14 at Alexandria to plan a fourfold attack on French positions in Nova Scotia, at Fort Duquesne on the Monongahela River, at Crown Point on Lake George, and at Niagara.

Fort Beauséjour commanding the neck of the Acadian peninsula which links Louisburg with French Canada is besieged by British forces under Col. Robert Monckton and surrenders June 16 after a fortnight of resistance. Some 6,000 Acadians who refuse to swear allegiance to George II are sent to Georgia and South Carolina with instructions from the British governor that they are to be "disposed of in such manner as may best answer our design of preventing their reunion." The property of the 9,000 remaining Acadians is for the most part confiscated.

A British force of 1,500 from Virginia is defeated July 9 by French and Indian forces on the Monongahela River, 7 miles from Fort Duquesne. Nearly 1,000 British colonials are killed or wounded, and Gen. Braddock sustains mortal wounds.

The Battle of Lake George September 8 ends in defeat for the French, but the British under Sir William Johnson fail to reach Crown Point. Sir William erects Fort William Henry at the head of Lake George.

An expedition to Niagara under the governor of the Massachusetts colony William Shirley reaches Oswego, leaves a garrison of 700, but does not proceed.

Rangoon is founded by the Burmese king Aloung P'Houra, 44, who is fighting the French with help from the English East India Company.

"Experiments upon Magnesia, Quicklime, and other Alkaline Substances" by Scottish chemist-physician Joseph Black, 27, shows that magnesium is a distinct substance completely different from lime with which it has been confused. Black last year laid the foundations of quantitative analysis with a doctoral thesis on causticization.

Moscow State University is founded in Russia.

A Dictionary of the English Language by London lexicographer Samuel Johnson, 46, establishes the reputation of Doctor Johnson, whose work will be completed in 1773. "The chief glory of every people arises from its authors," says Doctor Johnson.

Painting *Milkmaid and Woodcutter* by Thomas Gainsborough; *A Father Explaining the Bible to His Children* by French painter Jean Baptiste Greuze, 30, who wins admission to the Paris Académie.

Theater *Miss Sara Sampson* by German playwright Gotthold Ephraim Lessing, 26, 7/10 at Frankfurt an der Oder.

Venetian adventurer Giovanni Jacopo Casanova de Seingalt, 30, is imprisoned as a spy upon his return to Venice after 14 years of traveling from one European capital to another as preacher, alchemist, gambler, and violin player. Expelled from the Seminary of St. Cyprian at age 16 for scandalous behavior, Casanova has ventured as far as Constantinople. He will effect a daring escape from prison next year and will continue his roguish life, accumulating a fortune as director of state lotteries for France's Louis XV in Paris, receiving the papal order of the Golden Spur, and engaging in duels before accepting employment as librarian to

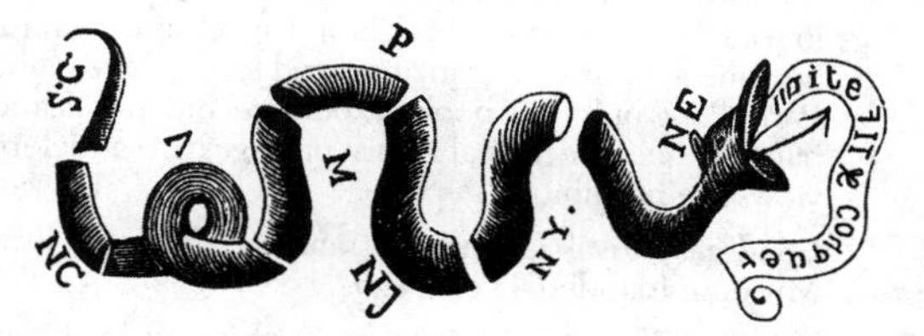

Benjamin Franklin's cartoon of May 1754 was more than 20 years premature. Colonial governments spurned his advice.

1755 *(cont.)* Bohemia's Count von Waldstein at Dux Castle, where he will write his memoirs of amours and escapades (*see* 1838).

The Golden Horn freezes over completely in one of the coldest winters on record.

An earthquake rocks northern Persia June 7 killing 40,000.

The Lisbon earthquake November 1 is the worst in Europe since the Lisbon quake of 1531. The ensuing seismic sea wave, flooding of the Tagus River, and a great fire takes 10,000 to 30,000 lives (60,000 by some accounts), and the All Souls' Day disaster shakes the confidence of all Europe (*see* Voltaire, 1759).

Portugal's chief minister Sebastião José de Carvalho e Mello, now 56, takes charge of rebuilding Lisbon (above) and seizes the opportunity to create a great square on the banks of the Tagus.

English agriculturist Robert Bakewell, 30, develops a new breed of sheep that will be called Leicester.

"Observations Concerning the Increase of Mankind" by Benjamin Franklin expresses Franklin's faith in human progress. Born the twelfth of 14 children in an age of high infant mortality, Franklin attacks the idea of "inevitable" poverty and limited growth, voicing confidence that America's colonial population will double every 20 or 25 years.

1756 A Seven Years' War begins in Europe as Prussia's Frederick the Great invades Saxony after hearing that the Saxons have formed a coalition with Austria, France, Russia, and Sweden to destroy or cripple Prussian power.

The French and Indian War brings English and French forces into conflict in a North American offshoot of the European war (above).

The "Black Hole of Calcutta" incident enrages the British. The nabob of Bengal Suraj-ud-Dowlah sacks Calcutta, forcing 146 Britons who have not fled the city into the guard room of Fort William. The chamber measures only 18 feet by 14 feet 10 inches and has but two small windows for ventilation, and when the survivors are released next morning (June 21) only 23 of the 146 remain alive (*see* 1757).

Scottish chemist Joseph Black shows that fixed air (carbon dioxide) is consistently distinguishable from normal air.

Nassau Hall is completed at Princeton for the 10-year-old College of New Jersey, which moves to Princeton from Newark.

Philadelphia's Pennsylvania State House (Independence Hall) is completed by Edmond Wolley and Andrew Hamilton (*see* Liberty Bell, 1752).

Mayonnaise, invented by the duc de Richelieu, is a mixture of egg yolks, vinegar or lemon juice, oil, and seasonings, and its name may come from that of a town in Minorca. The duke is a bon vivant who sometimes invites his guests to dine in the nude.

The Virginia colony's population reaches 250,000; more than 40 percent are slaves, up from 24 percent in 1715.

1757 Afghanistan's Ahmed Shah seizes Delhi January 28 and annexes the Punjab.

The Ottoman sultan Osman III dies at age 58 after a 3-year reign. His nephew, 40, will reign until 1773 as Mustapha III.

The Battle of Plassey June 23 establishes British sovereignty in India as Robert Clive wins a victory over a Bengalese nabob who has played him false. A British naval squadron supported by East India Company troops under Clive recovers Calcutta, avenges last year's "Black Hole" incident, and goes on to take the French settlement at Chandernagor. Clive installs a new nabob, accepts a large gift plus the quitrent of the East India Company's territory, and becomes virtual ruler of Bengal.

Prussia's Frederick the Great has invaded Bohemia and defeated an Austrian army May 6 at Prague. The Battle of Kolin June 18 ends in defeat for Frederick despite his exhortations ("Schweinhunds, would you live forever?"), and he is forced to evacuate Bohemia.

The Battle of Rossbach November 5 gives Frederick the Great a crushing victory over the French. Louis XV (or his mistress, Mme. de Pompadour) says, "Aprés moi le déluge" (After me, the flood), an expression that has long since become proverbial.

The Battle of Leuthen December 5 ends in victory for Frederick the Great over the Austrians.

Benjamin Franklin visits England and sends home advice as to how far American importers can safely go in flouting London's mercantilist acts (*see* 1750; Stamp Act, 1765).

British sea captain John Campbell creates the sextant by extending the arc of John Hadley's quadrant of 1731 to measure angles of up to 120 degrees rather than 90. Campbell's sextant has a triangular frame, with one side an arc, the arc has a scale of degrees, an index pointer pivots across the frame and the arc, and a system of reflecting mirrors brings together the two objects whose angle is to be observed by the navigator.

An essay "On the Most Efficient Means of Preserving the Health of Seamen" by James Lind establishes principles of hygiene to guard against typhus and other diseases (below).

Only a few hundred Parisians concern themselves with "literature, the arts and healthy philosophy," writes journalist-critic Baron Melchior von Grimm, who has become a naturalized citizen noted for his letters about Paris. Not only are books expensive but publications banned because of salacious passages or anticlerical views are prohibitive in price.

London's Royal Library is transferred to the British Museum founded 4 years ago.

Painting *The Artist's Daughter with a Cat* by Thomas Gainsborough; *The Fowler* and *La Paresseuse Italienne* by Jean Baptiste Greuze.

Jacques Ange Gabriel completes enlargements to the Louvre Palace at Paris.

Potato planting increases rapidly in northern Europe as the famine that accompanies the Seven Years' War gives impetus to potato culture (*see* Frederick, 1744).

James Lind describes his 1747 experiment with scurvy patients in a second edition of his 1753 *Treatise.* He reports that he divided 12 patients into pairs and experimented with six different diets, trying such things as vinegar, cider, and elixir vitriol (sulfuric acid, alcohol, and an extract of ginger and cinnamon). The best results were obtained, Lind reports, when diets were supplemented with oranges and lemons (*see* Cook, Pringle, 1775).

1758

French forces suffer reverses in the continuing Seven Years' War. The duke of Brunswick's son Ferdinand, 37, drives them back across the Rhine and defeats them June 23 in the Battle of Crefield, the French West African ports of Senegal and Gorée fall to the British, and the British score victories in America.

The French Canadian fortress of Louisburg falls July 26 to British forces under Generals Jeffrey Amherst, 41, and James Wolfe, 31, with support from Admiral Edward Boseawen as the French and Indian War continues in North America. The French have defeated a British force trying to take Fort Ticonderoga July 8, but the British take Fort Frontenac August 27 and Fort Duquesne November 25, win control of Cape Breton Island, and drive the French from other strongholds (*see* 1713; 1759).

The Battle of Zorndorff August 25 gives Frederick the Great a victory over Russian forces that have invaded Prussia, and while the Austrians defeat the Prussians October 14 at the Battle of Hochkirk, Frederick successfully resists efforts by Austria's Count Leopold von Daun, 53, to drive his forces out of Saxony and Silesia.

Tableau Economique by French surgeon-general François Quesnay, 64, propounds a system under which the products of agriculture will be distributed without government restraint among the productive classes of the community (landowners and land cultivators) and the unproductive classes (manufacturers and merchants). Agriculture is the only true source of wealth, says Quesnay, and his disciple the Marquis de Mirabeau will say that Quesnay's manifesto contributes as much to the stability of political societies as have the inventions of money and writing.

The New Jerusalem (De nova Hierosolyma) by Swedish naturalist-scientist Emanuel Swedenborg, 70, is a religious treatise. Swedenborg distinguished himself at age 30 during a siege of Frederikshall by inventing machines for carrying boats overland from Stromstadt to Iddefjord, and Queen Ulrika Eleanora elevated him to the nobility the following year. He began having visions 15 years ago, and his followers have founded the Church of the New Jerusalem, or New Church.

De l'esprit by French philosopher Claude-Adrien Helvétius, 43, extends the sensationist psychology of England's John Locke to the ethical and social fields. Helvétius produces a major scandal by propounding a utilitarianism with atheistic and materialistic implications (*see* Locke, 1690).

Painting *The Mill at Charenton* by François Boucher.

1759

French forces triumph over Ferdinand of Brunswick April 13 at Brunswick, but Ferdinand beats the French August 1 at the Battle of Minden. Russian forces defeat the Prussians July 23 at Kay, and Frederick the Great suffers a major loss August 12 as a coalition of Austrian and Russian armies gain victory at the Battle of Kunersdorf.

Austria's Count Leopold von Daun surrounds a Prussian army commanded by Friedrich August von Finck, 41, at Maxen and takes some 13,000 prisoners. Finck is court-martialed, dismissed from the army, and imprisoned in a fortress.

Spain's Ferdinand VI dies August 10 at age 45 after a 13-year reign. He has lived in a state of melancholy bordering on madness since the death last year of his Portuguese wife Barbara and is succeeded by his half brother, 43, who will reign until 1788 as Carlos III.

French Canada falls to the British September 13 following Gen. Wolfe's victory in the Battle of the Plains of Abraham outside Quebec City. Wolfe has scaled the heights on which the city stands and has come out of the darkness with an army of 5,000 men to surprise the French field marshal Louis Joseph, marquis de Montcalm de Saint-Veran, who falls mortally wounded at age 47 (Wolfe, too, is mortally wounded). British troops under Sir Jeffrey Amherst have taken Fort Ticonderoga July 27 and gone on to take Crown Point (*see* 1758).

British forces under Brig. Gen. John Forbes, 49, occupy Fort Duquesne built by the French 5 years ago, but Gen. Forbes dies along with many of his men. The British build a new, larger fortress that they call Fort Pitt after the prime minister Sir William Pitt.

Portugal expels her Jesuits (*see* France, 1762).

Voltaire reports in his novel *Candide* (below) that heretics were publicly burned after the Lisbon earthquake of 1755 because the University of Coimbra declared "that the sight of several persons being slowly burned in great ceremony is an infallible secret for preventing earthquakes."

Silhouette becomes a French word of derision meaning a figure reduced to its simplest form. The demands of war and the luxury of the court have drained the nation's treasury, Etienne de Silhouette, 50, has been named controller-general through the influence of the king's mistress Mme. de Pompadour, he has attempted reforms that have included a land tax on the estates of the nobility, a reduction of pensions, and the melting down of table silver for use as money, but his efforts arouse a storm of opposition and ridicule.

German anatomist Kaspar Friedrich Wolff, 26, observes the development of growing plants. Wolff will be a founder of modern embryology.

1759 ***(cont.)*** Tobias Smollett calls Doctor Johnson "that great Cham of literature" March 16 in a letter to John Wilkes, 32, a profligate member of Parliament who participates in the secret "Mad Monks of Medmenham" society founded by Francis Dashwood, 51, fifteenth baron le Despenser. The society conducts obscene parodies of Roman Catholic rituals at Medmenham Abbey.

Fiction *Rasselas* (*The Prince of Abyssynia: A Tale*) by Doctor Johnson (above) who wrote the two-volume novel in 8 days in January; *Candide* by Voltaire (above) whose best seller about Pangloss describes the Lisbon earthquake of 1755: the ultimate reason for things is unknown and unknowable, says Voltaire, who ridicules the optimism of the pope and his own earlier optimism: "All is for the best in the best of all possible worlds," but the predicament of man has become anything but "passable."

Painting *The 7th Earl of Lauderdale* by Joshua Reynolds; *Sigismonda* and *The Cockpit* by William Hogarth, who was named Sergeant Painter to the King in 1757.

Theater *The Lovers (Gl'innamorati)* by Carlo Goldoni, in November at Venice's Teatro San Luca.

Popular song "Heart of Oak" by English songwriter William Boyce whose "chantey" will become the unofficial anthem of the Royal Navy.

Porcelain teacups in Europe now generally have handles in a departure from Oriental design.

Irish brewer Arthur Guinness establishes a Dublin brewery that will become the world's largest. Guinness Stout will gain worldwide distribution.

1760 The Battle of Landshut June 23 in Bavaria ends with Austrian forces defeating and capturing a Prussian army, but Frederick the Great prevents a linkup of Russian and Austrian forces and defeats the Austrians August 15 at the Battle of Liegnitz.

A Russian army surprises Berlin October 9 and burns it in 3 days before retreating at news that Frederick is rushing to relieve the city.

Britain's George II dies October 25 at age 77 after a 33-year reign in which he has lost favor by favoring Hanover's interests over those of Britain. His grandson George William Frederick, 22, assumes the throne as George III of Great Britain and Ireland and Elector of Hanover to begin a disastrous 60-year reign.

The Japanese Tokugawa shogun Ieshige abdicates at age 49, drunken and ill after 15 years in power. He is succeeded by the 23-year-old son of the late shogun Yoshimune, who will rule until 1786 as the shogun Ieharu despite his mental incompetence.

On the Seats and Causes of Disease by Padua physician Giovanni Morgagni, now 78, introduces the anatomical concept that will become the chief element in medical diagnosis. Founder of pathological anatomy, Morgagni has conducted hundreds of postmortem operations, and he shows the importance of recording a patient's life history, the history of his disease, the events connected with a final illness, and the manner of death (*see* 1719).

"Some Account of the Success of Inoculation for the Smallpox in England and America" by Benjamin Franklin is published at London (*see* Lady Montagu, 1721; Jenner, 1796).

The charismatic Polish founder of Jewish hasidism Israel ben Eliezer, known as the Ba'al Shem Tov, or Besht, dies at age 60. Not a scholar, he has preached a popular, messianic people's movement outside the synagogue, reviving the idea of the *zaddik*, or superior mortal (*see* 1772).

Nonfiction *The Sermons of Mr. Yorick* by English clergyman Laurence Sterne, 46, whose book is actually a collection of his own sermons as prebendary of York. Sterne's wife Elizabeth became disturbed by his "small quiet attentions" to other women and went insane 2 years ago.

Fiction *The Life and Opinions of Tristram Shandy, Gentleman* by Laurence Sterne (above), published in its first two volumes, represents a bold and brilliant departure in the art of the novel.

Poetry *Fragments of Ancient Poetry Collected in the Highlands of Scotland* by schoolmaster James Macpherson, 24, whose alleged translations from a 3rd century Gaelic poet will be much admired for their romantic rhythm and passages of beauty but will be exposed as fraudulent by Doctor Johnson.

Painting *Mrs. Philip Thicknesse* by Thomas Gainsborough; *Giorgiana* by Joshua Reynolds. London's Royal Society of Art holds its first exhibition of contemporary art.

Theater *The Boors (I rusteghi)* by Carlo Goldoni 2/16 at Venice's Teatro San Luca.

First performances Symphonies No. 2, 3, 4, and 5 by Viennese composer Franz Joseph Haydn, 28, who started as a choir boy at St. Stephen's Cathedral.

The first roller skates are introduced at London by Belgian musical instrument maker Joseph Merlin, who rolls into a masquerade party at Carlisle House in Soho Square playing a violin. Unable to stop or turn, Merlin crashes into a large mirror valued at more than £500, smashes it to pieces, breaks his fiddle, and severely injures himself (*see* 1863).

English traveler Andrew Burnaby tours the American colonies and says of Benjamin Franklin's lightning rod, "I believe no country has more certainly proved the efficacy of the electrical rods than this. Before the discovery of them these gusts were frequently productive of melancholy consequences: but now it is rare to hear of such instances. It is observable that no house was ever struck, where they were fixed . . ." (*see* 1752).

London's Botanical Gardens open at Kew.

Robert Bakewell in Leicestershire begins experiments in the improvement of beef cattle (*see* 1755; 1769).

Large quantities of cloves and nutmegs are burned at Amsterdam to maintain high price levels.

Britain's 13 American colonies have an estimated population of 1.6 million.

1761 The Battle of Panipat in India January 14 ends in a crushing defeat of the Marathas at the hands of the Afghan chief Ahmad Shah Abdaii, who withdraws soon afterward following a mutiny of his troops. The dissension left behind plays into the hands of the British, and while the Mughal Empire that began in 1526 will continue until 1857, it is the British who will rule India from now until 1947.

French and Spanish forces invade Portugal; the Portuguese summon British aid to help repel the invaders.

France, Spain, and the Bourbon states of Italy join in a *pacte de famille* against Britain August 15, Britain's Lord Pitt resigns October 5 when George III and Parliament refuse to declare war on Spain, and Britain's Tories come to power for the first time since 1714. A new ministry headed by John Stuart, 48, third earl of Bute, will rule until 1763.

George III threatens to withdraw the subsidies that have enabled Prussia's Frederick the Great to withstand the combined forces of the coalition in the Seven Years' War that has exhausted all parties and cost hundreds of thousands of lives (*see* 1762).

Slave traders are excluded from the Society of Friends by American Quakers despite the fact that many Quakers own slaves.

Britain takes £600,000 worth of exports from Guadeloupe, most of it in sugar, while Canada yields only £14,000 (*see* 1759; 1763).

The ruin of France's merchant marine and navy frees French mariners to become privateers. They seize as much as one tenth of all British merchant shipping, but France is no longer a major commercial rival to Britain.

England's Bridgewater Canal is opened after 3 years' work to link Liverpool with Leeds.

The first test is made of John Harrison's 1736 chronometer which has been improved and which is essentially a watch 5 inches in diameter with a gridiron pendulum made of two metals that expand and contract in opposition. The chronometer is taken on a voyage to Jamaica aboard H.M.S. *Deptford* and is pronounced a success (*see* Leroy, 1765).

French mathematician Joseph-Louis Lagrange, 25, publishes a complete calculus of variations that incorporates both old and new results in a treatment both systematic and elegant. Lagrange has been a professor at an artillery school in Turin since age 19 and has corresponded with the Swiss mathematician Leonhard Euler (*see* 1748).

London physician John Hill makes the first association between tobacco and cancer, reporting six cases of "polypusses" related to excessive use of snuff in his "Cautions Against the Immoderate Use of Snuff."

Viennese physician Leopold Auenbrugger von Auenbrijgg, 39, discovers the percussion method of diagnosing lung disease (*see* Laënnec, 1819).

Fiction *Julie ou la nouvelle Héloïse* by Jean Jacques Rousseau.

Poetry *The Rosciad* by English poet Charles Churchill, 30, is published anonymously and cleverly satirizes London actors and actresses.

Painting *L'Accordée du Village* by Jean Baptiste Greuze.

Theater *The Father of the Family (Le Père de Famille)* by Denis Diderot 2/18 at the Comédie-Française, Paris; *The Rage for Country Life (Le Manie della Villeggiatura)* by Carlo Goldoni in October at Venice's Teatra San Luca.

First performances Symphony No. 7 in C minor *(Le Midi)* by Joseph Haydn, who signs a contract in May to become vice-capellmeister to the court of Hungary's Prince Miklòs Jòzsef Esterházy at Eisenstadt.

British grain prices will rise in this decade, and fewer Britons will grow their own food as public lands are enclosed and turned into pastures for sheep and more of the peasantry driven off the land.

1762 The Russian czarina Elizabeth Petrovna dies January 5 at age 52 after a 20-year reign and is succeeded by her imbecilic son of 33, who ascends the throne as Peter III.

Russia withdraws from the Seven Years' War. An admirer of Prussia's Frederick the Great, Peter III (above) makes peace, returns Pomerania to Prussia, and offers his support against Sweden, which also withdraws from the war, and against France and Austria, which soon will make peace. A military coup overthrows Peter III in July; put to death July 18, he is succeeded by his widow Sophia-Augusta of Anhalt Zerbst, 33, a convert to the Orthodox church who has changed her name to Catherine (Ekatarina Alekseevna) and will reign with "benevolent despotism" until 1796 as "the Semiramis of the North" (in Voltaire's phrase).

British rear admiral George Brydges Rodney, 43, captures Martinique, Grenada, St. Lucia, and St. Vincent in the West Indies. Another British force takes Havana despite the loss of more than half its number to fever and dysentery, and British forces in the Pacific take Manila (*see* Rodney, 1781).

France expels her Jesuits, who have been engaging in commerce and industry and have made themselves unpopular by interfering in politics (*see* Portugal, 1759; Clement XIV, 1773).

The Equitable Life Assurance Society is founded at London (*see* 1583). It will be the first company to grade premium rates according to age.

An epidemic of typhus that will continue through most of the decade sweeps across the Italian states, which are suffering famine (*see* 1764).

The Diseases of Children and Their Remedies by Swedish physician Nils von Rosenstein, 59, is the first systematic modern treatise on pediatric medical care and establishes the science of pediatrics.

1762 ***(cont.)*** Nonfiction *The Social Contract* (*Le Contrat Social*) by Jean Jacques Rousseau expresses antimonarchist views that arouse such passions that Rousseau takes refuge in Switzerland to escape prosecution. "Man is born free yet everywhere is in chains"; "The body politic, like the human body, begins to die from its birth and bears in itself the causes of its destruction."

Fiction *Emile, ou Traité de l'education* by Jean Jacques Rousseau contains the line, "I hate books, for they only teach people to talk about what they do not understand" (I), but Rousseau goes on to say, "There exists one book, which, to my taste, furnishes the happiest treatise of natural education. What then is this marvelous book? Is it Aristotle? Is it Pliny, is it Buffon? No—it is *Robinson Crusoe*" (III) (*see* 1719).

Painting Frescoes for Madrid's Royal Palace by Giambattista Tiepolo; *Garrick in "The Farmer's Return"* by German-born, Italian-trained English painter Johann Zoffany, 29; *Mares and Foals* by English painter George Stubbs, 38.

Theater *The Chioggian Brawls (La baruffe chiozzotte)* by Carlo Goldoni in January at Venice's Teatro San Luca.

Opera *Artaxerxes* 2/2 at London's Covent Garden, with music by Thomas Arne, arias that include "The Soldier Tired," "Water Parted from the Sea"; *Orfeo ed Euridice* 10/8 at Vienna's Hofburgtheater, with music by Christoph von Gluck whose expressive and dramatically forceful music will revolutionize opera, making music and drama interdependent; *Love in a Village* 12/8 at London's Covent Garden, with music by Thomas Arne, libretto by Irish playwright Isaac Bickerstaffe, 27, songs that include "There was a jolly miller once/Lived on the River Dee."

New York's first St. Patrick's Day parade steps off March 17 to celebrate a man historians will agree was not named Patrick, was not Irish, did not drive the snakes from Ireland, did not bring Christianity to Ireland, and was not born March 17. Irish-Americans will grow in number until New York has a larger Irish population than Dublin, the parade will grow to become an event of paralyzing proportions, and it will never be canceled on account of inclement weather.

English dandy Richard "Beau" Nash has died at Bath February 3 at age 87 after a career in which he has made Bath a popular watering place, has abolished the habit of wearing swords in places of public amusement, has brought dueling into disrepute, and has induced gentlemen to wear shoes and stockings rather than boots in parades and assemblies. Nash has lived on a small pension and by selling snuff boxes and trinkets since 1745 when Parliament made gambling illegal.

France's Louis XV builds Le Petit Trianon for his mistress Mme. de Pompadour (*see* 1768).

A cook to France's Duc Louis George Erasmé de Contades reinvents pâté de foie gras. Known to the ancient Egyptians, pâté is made from the enlarged livers of force-fed geese.

Frenchmen discover a kaolin deposit that will lead to increased production of porcelain (*see* 1754).

English potter Josiah Wedgwood, 32, receives an appointment from George III who names him "Potter to the Queen" (Charlotte). Wedgwood will create superior tableware (*see* 1763).

Russia's population will increase in the next 34 years from 19 million to 29 million, partly because the potato has made food cheaper and more abundant, partly through annexation of new territory.

1763 The Treaty of Paris signed February 10 ends Europe's Seven Years' War.

France cedes to Britain her territories in Canada, Cape Breton Island, Grenada in the West Indies, and Senegal in Africa.

France regains Gorée in Africa, Pondicherry and Chandernagor in India, and the sugar-rich islands of Guadeloupe and Martinique in the West Indies.

The treaty recognizes the Mississippi River as the boundary between the British colonies and the Louisiana Territory that France has ceded to Spain (*see* 1800).

Spain cedes Florida to Britain and regains all of Britain's conquests in Cuba including Havana. Spain also recovers the Philippines but loses Minorca.

Britain's prime minister Lord Bute resigns in April. A new Tory ministry headed by George Grenville, 51, brother-in-law of William Pitt, will rule as first lord of the treasury and chancellor of the exchequer until 1765.

Poland's Augustus III dies at Dresden October 5 at age 66 after a 30-year reign. He has returned to Saxony following conclusion of the Treaty of Hubertsburg in February, and he will be succeeded next year by Stanislaus Poniatowski, 31, a favorite of the Russian czarina Catherine, whose influence will gain him election to the throne. He will reign until 1795 as Stanislaus II Augustus.

Ottawa chief Pontiac leads tribesmen of the American Northwest in an uprising against Detroit and other British forts in an effort to drive the white settlers back east across the Alleghenies, but the Ottawa are deserted by their French allies in their siege of Detroit.

English philanthropist Granville Sharp, 28, finds a slave beaten and left to die in the streets of London by a Barbados lawyer. Sharp nurses the man back to health and protests when the lawyer then kidnaps the slave and ships him off to the West Indies to be sold. The case produces a public outcry (*see* 1772).

Treatise on Tolerance by Voltaire is published at Paris.

The door opened by the French and Indian War to westward movement of British colonists closes October 7 by decree of George III through George Grenville (above). Sixty percent of American colonists are British, and the king decrees that they must remain east of the sources of rivers that flow into the Atlantic, but it is a decree that will be honored in the breach. Angry frontiersmen flout western land claims made by the seaboard colonies and the "sea to sea" grants of their original charters (*see* 1748; 1769).

St. Louis has its origin in a Mississippi trading post selected by French fur trader Pierre Laclède Liguest, who names the post after France's patron saint Louis IX. He will send his stepson René Auguste Chouteau to build St. Louis next year.

Of 14,000 hogsheads of molasses brought into New England this year, only 2,500 are from British sources; smugglers account for the remainder (*see* 1764; Molasses Act, 1733).

The British Mariner's Guide by English astronomer Nevil Maskelyne, 31, contains observations that will revolutionize navigation. Maskelyne's interest in the subject was aroused 2 years ago when he voyaged to the South Atlantic island of St. Helena to observe the transit of the planet Venus, and he has used lunar distances to calculate longitude at sea (*see* Harrison, 1761; Cook, 1768; Bowditch, 1802).

A letter to the Royal Society from English clergyman Edward Stone, 61, gives the first indication that salicylic acid as contained in the bark of willow trees may be "very efficacious in curing aguish [rheumatic] and intermitting disorders [bouts of fever]." Stone has tested the remedy for 5 years (*see* Gerhardt, 1853; Bayer Aspirin, 1899).

Poetry *Song to David* by Christopher Smart: "And now the matchless deed's achiev'd,/Determined, dared, and done."

Painting *The Election of the Doge of Venice* by Venetian painter Francesco Guardi, 49, who has evolved a style of his own after years of following the photographic style of Canaletto; *The Paralytic Cared for by His Children* by Jean Baptiste Greuze; *Death of Wolfe* by English painter George Romney, 28.

La Madeleine is completed at Paris.

Touro Synagogue is completed on Touro Street at Newport in the Rhode Island colony for the town's congregation of Sephardic Jews. Its sand floor commemorates the exodus of Jews from Egypt.

Josiah Wedgwood perfects a vitrified "cream-color ware" that is not true porcelain but gives Britain a hard tableware far more durable than earthenware (*see* 1762; 1769).

Prussia's Frederick the Great estimates that the Seven Years' War produced 853,000 casualties among his troops and that it directly killed 33,000 civilians, but war-related famine in 15 of the last 23 years has reduced the population of Europe far more drastically than has war itself.

1764 The Muslim soldier Haidar Ali, 39, usurps the throne of Mysore (he will assume the title maharaja in 1766).

British forces under Sir Hector Munro defeat the nabob of Oudh at Buxar October 22 and make themselves masters of Bengal, India's richest province.

On Crimes and Punishment (*Tratto de Delittie delle Pene*) by Italian economist Cesare Bo'nesana, 26, marchese di Beccaria, condemns capital punishment, torture, and confiscation. The brief work will be translated into 22 European languages and have wide influence.

The Sugar Act passed by Parliament April 5 replaces the Molasses Act of 1733, but while it cuts in half the sixpence per gallon duty on molasses imported into British colonies from non-British islands in the West Indies, the British send customs officials to America and order colonial governors to enforce the new law and apprehend smugglers (*see* 1763).

The Currency Act passed by Parliament April 19 forbids Britain's colonies from printing paper money.

Boston lawyer James Otis, 39, denounces "taxation without representation" May 24 and urges the colonies to unite in opposition to the new British tax laws (*see* 1765).

Boston merchants organize a boycott of British luxury goods in August, initiating a policy of nonimportation.

The British pay taxes at higher rates than any Europeans and have received little tax support from their colonists. They begin to take a harder line with the colonists (*see* Townshend Act, 1767).

Pennsylvania colony mechanic James Davenport invents machinery to spin and card wool.

Naples has a violent epidemic of typhus (*see* 1762).

Russia confiscates church lands.

Brown University has its beginnings in the Rhode Island College founded at Warren. The school will move to Providence in 1770, will maintain strict nonsectarian principles despite the fact that its founders are Baptists, and will adopt the name Brown in 1804 after receiving gifts from merchant Nicholas Brown, now 35, whose 2-year-old Nicholas Brown & Co. has been joined by brothers Joseph, 31, John, 28, and Moses, 26. The Browns are the leading colonial candlemakers.

The *Connecticut Courant* begins weekly publication at Hartford October 29 (*see* 1866).

Manuel typographique by French engraver–type founder Pierre Simon Fournier, 52, is published in two volumes at Paris. Fournier has devised the first system for measuring and naming sizes of type.

Chinese novelist Ts'ao Chan (Hsijeh-ch'in) dies at age 48, leaving his novel *The Dream of the Red Chamber* incomplete. The only great novel of manners in Chinese literature, its final third will be written by novelist Kao Eh.

Poetry *The Traveler* by Irish poet Oliver Goldsmith, 35.

William Hogarth dies at London October 26 at age 66.

1764 ***(cont.)*** Sculpture *St. Bruno* by French sculptor Jean Antoine Houdon, 23, at Rome.

Theater *La Jeune Indienne* by French playwright Sebastien Roch Nicolas Chamfort, 23.

First performances Symphony No. 22 in E flat major *(The Philosopher)* by Joseph Haydn at the Esterházy court in Eisenstadt.

Madrid's Royal Palace is completed for Spain's Carlos III after 28 years of construction.

The Ruins of the Palace of the Emperor Diocletian at Spalato by Scottish architect Robert Adam, 36, establishes Adam's reputation. He designs Kenwood House in Middlesex.

London begins the practice of numbering houses.

Prussia's Frederick the Great moves to improve his country's agriculture after studying the intensive farming methods employed in the Lowlands. Frederick institutes a modified rotation system that employs clover to enrich the soil, turnips whose green tops help smother weeds, wheat, and barley (the turnips and clover both make good animal fodder; *see* Townshend, 1730). Frederick also drains the marshes along the Oder, Werthe, and Netze rivers.

1765 The Quartering Act passed by Parliament May 15 orders colonists to provide barracks and supplies for British troops (*see* Stamp Act, below).

Virginia patriot Patrick Henry, 29, protests the Stamp Act (below) in the House of Burgesses May 29: "Tarquin and Caesar each had his Brutus, Charles the First his Cromwell, and George the Third . . ." The Speaker interrupts by crying, "Treason," but Henry resumes, ". . . may profit by their example. If *this* be treason, make the most of it."

Sons of Liberty clubs formed at Boston and at other colonial towns resist the Stamp Act, and a Stamp Act Congress convenes at New York in October with delegates from nine colonies to protest taxation without representation. The delegates resolve to import no goods that require payment of duty (*see* 1766).

Britain's prime minister Lord Grenville resigns July 16. A new ministry takes office headed by Charles Watson-Wentworth, 35, second marquis of Rockingham.

The Holy Roman Emperor Francis I dies at Innsbruck August 18 at age 56 after a 20-year reign with his wife Maria Theresa. She succeeds as co-regent with their son, 24, who will reign until 1790 as Josef II.

Lisbon finally abolishes the auto-da-fé parade and ritual sentencing that for more than 2 centuries has been the occasion for violence against Jews and heretics, often with spectacular public burnings.

The Stamp Act passed by Parliament March 22 is the first measure to impose direct taxes on Britain's American colonists. Intended to raise £60,000 per year, the act requires revenue stamps on all newspapers, pamphlets, playing cards, dice, almanacs, and legal documents.

An economic depression begins at Boston that will stifle business, produce widespread unemployment, and throw many Bostonians into debtors' prisons. Workingmen and sailors average £15 to £20 per year when jobs are available, but it takes roughly £60 per year to support a family in the town, even when the wife makes all clothes except for hats, shoes, overcoats, and Sunday best.

The Mughal emperor Shah Alam II acts August 12 to give Robert Clive of the British East India Company authority to collect revenues in Bengal.

The world's first savings bank opens in Brunswick. The German bank inspires other European entrepreneurs to open thrift institutions (*see* 1799).

Scotsman James Watt, 29, invents a steam engine that produces power far more efficiently than the Newcomen engine of 1705. Mathematical instrument maker to the University of Glasgow, Watt employs a separate chamber, or condenser, to condense exhaust steam from the cylinder of his engine (*see* Boulton and Watt Foundry, 1769).

The steam-driven three-wheel gun tractor devised by French engineer Nicholas Joseph Cugnot, 40, pioneers development of the automobile. Cugnot's gun tractor can run at 2.5 miles per hour but must stop every 100 feet or so to make steam (*see* Trevithick, 1801; Benz, 1885).

The ship's chronometer invented by French horologer Pierre Leroy, 48, makes a major contribution to navigation. Leroy has discovered the isochronism of spiral springs and originated the detached escapement; his instrument embodies all the essential features of the modern ship's chronometer (*see* Harrison, 1736; 1761).

A new canal links France's Loire and Rhône rivers (*see* 1600; 1603).

Colonial American shipping interests have 28,000 tons of shipping and employ some 4,000 seamen. Exports of tobacco are nearly double in value the exports of bread and flour, with fish, rice, indigo, and wheat next in order of value. The major shippers are the Cabots and Thomas Russell of Boston, Thomas Francis Lewis of New York, and Samuel Butler of Providence.

The first American medical school opens at the College and Academy of Philadelphia. The school will become the College of Physicians and Surgeons.

Fiction *The Castle of Otranto* by English man of letters Horace Walpole, 47, fourth earl of Orford, is the first Gothic horror novel and marks the beginning of the romantic school of English fiction.

Painting *The Gladiator* by English painter Joseph Wright, 31; *Corisus Sacrificing Himself to Save Callirhoé* and *The Swing* by French painter Jean Honoré Fragonard, 33; *Mme. Pompadour* by François Boucher, who is appointed court painter at Versailles; *Young Girl Weeping Over Her Dead Bird* by Jean Baptiste Greuze.

Japanese painter Harunobu Suzuki introduces color to *ukiyoe* painting and improves *ukiyoe* printmaking technique by assigning specific functions to specialized pro-

James Watt's steam engine improved on the Newcomen engine and marked the real beginning of viable steam power.

fessionals. Suzuki specializes in painting beautiful women.

Popular songs *Mother Goose's Melodies*, published by Boston printer Thomas Fleet, includes "Baa Baa Black Sheep," "Ding Dong Bell," "Georgie Porgie," "Hickory Dickory Dock," "Hot Cross Buns," "I Saw Three Ships Come Sailing," "Jack and Jill," "Little Bo Peep," "Little Miss Muffet," "Little Tommy Tucker," "See-Saw Marjorie Daw," "Tom, Tom the Piper's Son," and other favorites sung years ago to his infant son by his late mother-in-law Elizabeth Foster Goose, who died in 1757 at age 92. She had put old French tunes to the 1697 verses of Charles Perrault.

The first true restaurant opens at Paris where a tavern-keeper named Boulanger defies the monopoly of the caterers *(traiteurs)* in the sale of cooked meat. Boulanger serves a "soup" made of sheep's meat in a white sauce, which he calls a "restorative" *(restorante)*. The *traiteurs* sue, the case goes to the French Parlement, Boulanger wins, and he gains the right to serve *restorantes* at his all-night place in the rue Bailleul.

Philadelphia's population reaches 25,000, becoming the second largest city in the British Empire after London. New York has 12,500, up from 5,000 in 1700 (*see* 1793).

1766 Denmark's Frederick V dies January 14 at age 42 after a 20-year reign. He is succeeded by his semi-idiot son of 16, who marries the 15-year-old daughter of the prince of Wales Caroline Matilda and sinks into debauchery. The new king will reign at least nominally until 1808 as Christian VII.

Lorraine reverts to France upon the death of Stanislas Leszczynski February 23 at age 89.

Parliament repeals the Stamp Act passed last year to tax the American colonists, acting partly at the persuasion of Benjamin Franklin, but the Declaratory Act passed March 18 declares that the king—by and with the consent of Parliament—has authority to make laws and to bind the British colonies in all respects (*see* Townshend Acts, 1767).

The Treaty of Oswego July 24 ends Pontiac's 3-year rebellion.

Britain's prime minister the marquis of Rockingham resigns in August. A new ministry headed by "The Great Commoner" William Pitt, earl of Chatham, will rule until December 1767.

Virginia planter-miller George Washington ships an unruly slave off to the West Indies to be exchanged for a hogshead of rum and other commodities.

Russia's Catherine II grants freedom of worship (*see* 1783).

France abolishes internal free trade in grain (*see* 1774).

English engineer James Brindley, 50, begins construction of a Grand Trunk Canal to connect the Trent and Mersey rivers. He works without drawings or written calculations, but his canal will open a water route from the Irish Sea to the North Sea.

Rutgers University has its beginnings at New Brunswick in the New Jersey colony, where members of the Dutch Reformed Church found Queen's College (*see* 1825).

Nonfiction *Confessions* by Jean Jacques Rousseau paraphrases Machiavelli with the line, "In the kingdom of the blind, the one-eyed are kings"; "It is not hard to confess our criminal acts, but our ridiculous and shameful acts," writes Rousseau.

Fiction *The Vicar of Wakefield* by poet-playwright Oliver Goldsmith (his only novel).

Painting *The Orrery* by Joseph Wright; *Queen Charlotte and the Two Eldest Princes* by Johann Zoffany.

Christie's art auction house opens at London to compete with the Sotheby firm founded in 1744. Auctioneer James Christie, 36, starts a business that will continue for centuries. His son James will carry on the business and be an expert on Etruscan and Greek vases.

Theater *The Clandestine Marriage* by English actor-playwright David Garrick, 48, and playwright George Colman, 34, 2/20 at London's Drury Lane Theatre, which Garrick has co-managed since 1747.

Austrian composer Wolfgang Amadeus Mozart, 10, returns to his native Salzburg after having toured Paris and London with his father and sister. Young Wolfgang has performed on the harpsichord and has composed violin sonatas and improvisations, beginning a career that he continues by studying counterpoint under the direction of his father Leopold, 47, who is a violinist and composer in his own right.

London merchant Richard Tattersall, who has founded the Tattersall Horse Market, orders a special pattern for his horse blankets. The check pattern designed will be called the tattersall check.

1766 ***(cont.)*** Vienna's Prater Park opens with a 1,328-acre expanse of meadowland donated to the people by the young emperor Josef II, who serves as co-regent of Austria with his mother Maria Theresa. The word *prater* comes from the Spanish *prado* for meadowland. The new park has bridle paths, a rowing lake, and other amenities, it lies on the island formed by the main stream of the Danube and its canal.

English chemist Henry Cavendish, 35, duplicates the process by which lightning in a thunderstorm produces nitrogen from the atmosphere to enrich soil. Cavendish passes electric sparks through a nitrogen-oxygen mixture, he produces nitrogen dioxide, and the NO_2 yields nitric acid when dissolved in water (*see* Glauber, 1648; Rutherford, 1772). The element nitrogen remains unknown, but the Cavendish experiment stimulates speculation on practical ways to enrich soils (*see* Wöhler, 1828; Humboldt, 1802; Haber, 1908).

Rust ruins the Italian wheat crop, food prices rise, and widespread hunger ensues.

"Let them eat cake" [if there is no bread], writes Jean Jacques Rousseau (above) in his *Confessions*. Rousseau attributes the remark to "a great princess," but it will be widely ascribed in the 1780s and 1790s to Vienna's Marie Antoinette, now 11, who will become queen of France in 1774.

1767 *Letters from a Farmer in Pennsylvania* by Philadelphia lawyer John Dickinson, 34, are published in their first installments. Dickinson drafted the resolutions and grievances of the Stamp Act Congress 2 years ago as a member of that body, and his *Letters* on the nonimportation and nonexportation agreements (below) will continue to appear through much of 1768, winning him wide popularity in the colonies.

The Mason and Dixon line between the Pennsylvania and Maryland colonies is completed after a 4-year $75,000 survey by English surveyor-astronomers Charles Mason, 37, and Jeremiah Dixon, who have been engaged to settle a century-old dispute as to ownership of lands which include the large fertile peninsula between Chesapeake Bay and Delaware Bay. Mason and Dixon's demarcation line is amazingly accurate and is marked by handsome boundary stones, with the Penn coat-of-arms on their north sides and the Calvert coat-of-arms on their south sides extending westward to the "top of the Great dividing Ridge of the Allegheny Mountains" beyond which Mason and Dixon's Indian guides have refused to proceed out of fear of the Delaware and the Shawnee.

Britain's Chatham ministry resigns in December. A new Tory ministry headed by August Henry Fitzroy, 32, third duke of Grafton, will rule until January 1770.

Burmese forces sack Siam's capital Ayuttha in August (*see* 1782).

North Carolina woodsman Daniel Boone, 33, goes through the Cumberland Gap found in 1750 and reaches "Kentucke" in defiance of King George's 1763 decree. A veteran of the 1756 French and Indian War who has learned woodcraft from the Cherokees, Boone is reputed to be able to smell salt deposits 30 miles away, and in "Kentucke" he discovers a huge brine lake, a salt lick that attracts game and makes the region a hunting ground contested by various Indian tribes (*see* 1773).

The Townshend Revenue Act passed by Parliament June 29 imposes duties on tea, glass, paint, oil, lead, and paper imported into Britain's American colonies in hopes of raising £40,000 per year. A town meeting held at Boston to protest the Townshend Act adopts a nonimportation agreement (*see* 1770).

The History and Present State of Electricity by English clergyman-chemist Joseph Priestley, 34, is published at Leeds and explains the rings, later to be called Priestley rings, that are formed by an electrical discharge on a metallic surface. Priestley proposes an explanation of the oscillatory character of the discharge from a Leyden jar (*see* 1745; 1746).

The study of "different kinds of airs" is begun by Joseph Priestley (above). Knowing that carbon dioxide, or "fixed air," is present above the open vats in which beer mash is fermenting at the Leeds brewery, Priestley holds two containers close to the surface of the fermenting mash and pours water back and forth between them; the water becomes charged with CO_2.

Fiction *The Life and Opinions of Tristram Shandy* by Laurence Sterne appears in its eighth and final volume after 7 years of earlier volumes; *l'Ingénu* by Voltaire criticizes French society from the viewpoint of an imaginary Huron *sauvage*.

Theater *Eugénie* by French playwright-watchmaker Pierre Augustin Caron de Beaumarchais, 35, 1/25 at the Comédie-Française, Paris; *The Free Thinker* (*Der Freigeist*) by Gotthold Ephraim Lessing at Frankfurt-am-Main; *Minna von Barnhelm, or The Soldier's Fortune* (*Minna von Barnhelm, oder Das Soldatenglück*) by G. E. Lessing 9/30 at Hamburg's Nationaltheater.

Opera *Apollo et Hyacinthus* 5/13 at Salzburg, with music by Wolfgang Amadeus Mozart; *Alceste* 12/16 at Vienna's Burgtheater, with music by C. W. Gluck.

A fleet of 50 American whalers makes a foray into the Antarctic, the first whaling venture into that region.

French navigator Louis Antoine de Bougainville, 38, explores Oceania in hopes of expanding French whaling operations into the Pacific.

Joseph Priestley (above) pioneers carbonated water. "Sometimes in the space of two or three minutes [I have] made a glass of exceedingly pleasant sparkling water which could hardly be distinguished from very good Pyrmont," he writes (*see* carbonic acid, 1770).

1768 Delegates from 26 Massachusetts towns meet at Faneuil Hall September 22 in response to a call by Boston selectmen following anti-British riots (below). They draw up a statement of grievances, but Royal Navy men-of-war land two infantry regiments October 1, and two more regiments are ordered sent from Halifax (*see* Boston Massacre, 1770).

The Treaty of Hard Labor signed in the South Carolina colony October 14 confirms cessions of Cherokee lands in the Virginia and Carolina colonies to the British Crown.

The Treaty of Fort Stanwix signed November 5 confirms the cession of Iroquois territories between the Ohio and Tennessee rivers to the British Crown.

Austria renounces all claims to Silesia.

France has purchased Corsica from Genoa May 15, but Louis le bienamié has lost most of his colonial empire, has taxed the people heavily to maintain his luxurious life style, and is widely hated.

The Ottoman sultan Mustapha II declares war on Russia in October, charging that Moscow's occupation of Poland violates the 1711 Treaty of Pruth (*see* 1769).

Nepal in the Himalayas is unified under King Prithwi Naryan Shah.

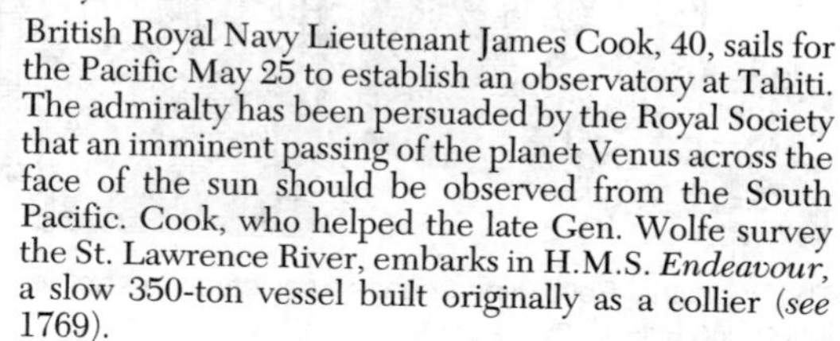

British Royal Navy Lieutenant James Cook, 40, sails for the Pacific May 25 to establish an observatory at Tahiti. The admiralty has been persuaded by the Royal Society that an imminent passing of the planet Venus across the face of the sun should be observed from the South Pacific. Cook, who helped the late Gen. Wolfe survey the St. Lawrence River, embarks in H.M.S. *Endeavour,* a slow 350-ton vessel built originally as a collier (*see* 1769).

Samuel Hearne, 23, of the Hudson's Bay Company begins a 2-year walk from Hudson Bay to the shore of the Arctic Ocean and back. He will encounter nothing that can be construed as a "Northwest Passage" from the Atlantic to the Pacific, and by scouting long-prevalent rumors, Hearne's findings will divert further explorations north to Baffin Bay (*see* Baffin, 1616).

Charlotte, North Carolina, is founded by colonists who name their settlement after George III's queen.

British customs officials at Boston seize the sloop *Liberty* June 10, the action precipitates riots, and Boston merchants adopt a nonimportation agreement August 1. Owner of the sloop is John Hancock, 31, whose late uncle has left him a fortune gained by profiteering in food supplied to the British troops and who has enhanced that fortune by smuggling wine into Boston.

Italian naturalist Lazzaro Spallanzani, 39, disproves the universally believed notion of spontaneous generation of organisms (in mutton broth) (*see* Needham, 1748; Maupertuis, 1751).

Angina pectoris gets its first accurate description. London physician William Heberdeen describes chest pains caused by insufficient oxygenation of the heart muscle (*see* nitroglycerin, 1847).

Nonfiction *A Sentimental Journey through France and Italy* by Laurence Sterne, who ridicules Tobias Smollett, calling him "Smelfungus" and, unlike Smollett, expresses delight in the French and Italian customs he has encountered between Calais and Lyons, but Sterne dies of pleurisy at London March 18 at age 54 leaving only two volumes of a projected four.

Fiction *Tales of the Rainy Moon (Ugetsu-monogatari)* by Japanese novelist Akinari Ueda, 34, who has collected stories from Japanese and Chinese tales. Born in an Osaka brothel district, Ueda was adopted at age 3 into a family of oil merchants and led a dissolute life until his marriage 3 years ago.

Painting *Garrick as Kitely* by Joshua Reynolds, London's most fashionable portrait painter, who begins a series of 15 lectures as president of the new Royal Academy of Art; *Experiment with the Air Pump* by Joseph Wright. Antonio Canaletto dies at Venice April 20 at age 70.

Theater *The Good Natured Man* by Oliver Goldsmith 1/29 at London's Royal Theatre in Covent Garden.

Opera *Lo speziale* (*The Apothecary*) in the autumn at Schloss Esterházy, with music by Joseph Haydn, libretto by Carlo Goldoni.

Italian composer-cellist Luigi Boccherini, 25, gains fame with a performance at the Concert Spirituel at Paris and receives an invitation to be composer to the court of the infante Don Luis at Madrid.

Le Petit Trianon is completed at Versailles.

London has bread riots. Government grain stores are pillaged by the mob.

The price of bread at Paris reaches 4 sous per pound and a placard appears in the city: "Under Henri IV bread was sometimes expensive because of war and France had a king; under Louis XIV it sometimes went up because of war and sometimes because of famine and France had a king; now there is no war and no famine and the cost of bread still goes up and France has no king because the king is a grain merchant."

Government regulations discourage French farmers from increasing their grain acreage, critics say, and they demand free circulation of grain (*see* 1766; Turgot, 1774).

The East India Company imports 10 million pounds of tea per year into England (*see* 1773).

1769 The Ottawa chief Pontiac is murdered April 20 by a Native American at Cahokia. Suspicions are rife that the British had him killed to prevent any repetition of the 1763–1766 rebellion.

The maharajah of Mysore, Haidar Ali, forces the British to sign a treaty of mutual assistance as famine devastates Bengal (below). The East India Company increases its demands on Bengal's remaining 20 million people to insure "a reasonable profit."

Russian troops occupy Moldavia and enter Bucharest in the continuing war with the Ottoman Turks. Austrian forces occupy parts of Poland. Prussia's Frederick the Great and the Holy Roman Emperor Josef II meet in Silesia to discuss the partition of Poland.

Virginia colonists defy George III's 1763 decree and settle on the Watauga River in what will be Tennessee.

Daniel Boone leads an expedition to the Kentucke region (*see* 1767; 1775; Harrod, 1774).

1769 ***(cont.)*** San Diego, San Francisco, and Los Angeles have their origins.

Spanish Franciscans led by Father Junipero Serra, 56, come by ship from New Spain to found the Mission of San Diego de Alcala. Gaspar de Portola, 46, governor of Baja California, leads a party north, founds a Spanish colony at San Diego, and continues on to the Monterey Peninsula, where he establishes another colony (*see* 1536; 1602).

Mallorcan friar Juan Crespi, 47, in de Portola's company (above) renames an Indian village en route Nuestra Senora la Reina de Los Angeles de Porciuncula (*see* 1781).

A Spanish scouting mission under Sgt. José Ortega discovers a bay north of Monterey and names it San Francisco Bay after St. Francis of Assisi.

Spain's Carlos III, whose empire covers more than half the known world, authorizes the founding of 21 missions in California over the next 54 years along with four presidios (armed garrison forts) at San Diego, Santa Barbara, Monterey, and San Francisco (*see* Yerba Buena, 1776).

James Cook arrives at Tahiti and sets up an observatory (*see* 1768). He proceeds westward to chart the coasts of New Zealand, which he finds fertile and well suited to European colonization (*see* 1841; Australia, 1770).

Louis Antoine de Bougainville completes the first French expedition round the world.

The East India Company opens stations in North Borneo.

Virginia's House of Burgesses issues resolutions May 16 rejecting Parliament's right to tax British colonists. The Virginia governor dissolves the House of Burgesses, but its members meet in private and agree not to import any dutiable goods.

James Watt patents his 1765 steam engine with some improvements, grants two-thirds of the profits to ironworks owner John Roebuck, who has financed his experiments, and goes into partnership with engineer Mathew Boulton, 41, to found the Boulton & Watt Foundry at Birmingham (*see* Wilkinson, 1774).

Benjamin Franklin charts the Gulf Stream, whose force is understood by most American navigators but is a mystery to many English sea captains, who make slow progress trying to sail against the current.

English inventor Richard Arkwright, 37, patents a spinning frame that can produce cotton thread hard and firm enough for the warp of woven fabric. Arkwright's invention will have a profound effect on Western society (*see* 1770; Luddites, 1811).

Dartmouth College is founded by Congregationalists at Hanover in the New Hampshire colony (*see* 1819).

Commentaries on the Laws of England by jurist Sir William Blackstone, 46, is published in its fourth and final volume. "It is better that ten guilty persons escape than that one innocent suffer," says Blackstone, and his *Commentaries* will be influential on lawmakers both in Britain and in America.

The Industrial Revolution changed society as no political revolution had ever done.

Debrett's Peerage and Baronetage by London publisher-genealogist John Debrett is published for the first time. Debrett will publish a *Peerage of England, Scotland, and Ireland* in 1802 and a *Baronetage of England* in 1808 (*see* Burke, 1826).

Painting *Offering to Love* by Jean Baptiste Greuze, who resigns from the Académie because it will not elevate him from genre painter to painter of history.

England's first Shakespeare Festival opens at Stratford-upon-Avon. David Garrick, now 52, begins an annual revival of the bard's immortal plays that will continue for more than 2 centuries.

Theater *Eloge de Molière* by Sebastien Chamfort is given its first performance.

Opera *La finta Semplice* 5/1 at Salzburg, with music by Wolfgang Amadeus Mozart.

Monticello goes up in the Virginia colony for lawyer Thomas Jefferson, 26, who has designed a house inspired by the work of the 16th century Italian architect Andrea Palladio for a site 4 miles from his father's plantation at Shadwell. A member of the House of Burgesses (above), Jefferson gives his place the Italian name "little mountain" because it provides a 20-mile view of the Blue Ridge Mountains despite its modest elevation of just 600 feet.

Father Serra (above) plants the first wine grapes, oranges, figs, and olives to grow in California at the Mission of San Diego de Alcala.

Robert Bakewell of 1755 Leicester sheep fame develops cattle with deeper, wider bodies, shorter and thicker necks, and with more flesh over their hind quarters, ribs, loins, and backs. Bakewell works on his premise that "like produces like" and often inbreeds stock selected for bulky bodies set low on short legs.

Josiah Wedgwood opens a pottery works at Etruria, near Burslem (*see* 1763).

The Great Famine of Bengal kills 10 million Indians, wiping out one-third of the population in the worst famine thus far in world history.

1770 New York's Battle of Golden Hill January 19 and 20 brings the first bloodshed between British troops and American colonists. The Sons of Liberty have put up a series of Liberty poles with pennants bearing slogans, the redcoats have torn them down, and when they tear down the fourth pole on the common west of Golden Hill two Sons of Liberty seize two soldiers. The soldiers resist, more citizens arrive with clubs, the officer in charge orders his men back to quarters, but a second skirmish occurs the following day.

The Boston Massacre March 5 leaves three dead, two mortally wounded, and six injured following a disturbance between colonists and British troops. Lawyers John Adams, 35, and Josiah Quincy, 26, defend the British captain and his men who fired on the Americans, a jury acquits them, but agitators use the incident to arouse colonial rancor against the British troops (*see* Boston Tea Party, 1773).

Britain's Grafton ministry has resigned in January. A new Tory ministry headed by Frederick Lord North, 38, as first lord of the treasury has taken office February 10 and will rule until early 1782.

The Battle of Chesme July 6 ends in defeat for an Ottoman fleet. A Russian fleet that has come to the Anatolian coast from the Baltic under the command of British officers wins a clear victory (*see* Crimea, 1771).

James Cook explores the eastern coast of New Holland (Australia), he makes a landing, and he takes possession of the island continent in the name of George III. Inhabited for more than 25,000 years, Australia has a population of more than 250,000, and the great variety of plants obtained by his naturalists leads Cook to name his landfall Botany Bay (*see* 1788).

Parliament repeals the Townshend Revenue Act of 1767 in a bill passed April 12. Prime Minister North has used his influence to have the act repealed.

English weaver-mechanic James Hargreaves patents a spinning jenny that automates part of the textile industry (*see* Arkwright, 1769; Crompton, 1779; Luddites, 1811).

Système de la nature by French philosopher Paul Henri Dietrich d'Holbach, 47, makes the first blunt denial of any divine purpose or master plan in nature.

The Black Death strikes Russia and the Balkans in epidemic form.

"If God did not exist, it would be necessary to invent him," writes Voltaire November 10.

The Massachusetts Spy begins publication at Boston. Printer Isaiah Thomas, 21, soon buys out former master Zechariah Fowle and makes his Whig newspaper a voice for the colonists against the British.

Poetry *Faust* by German poet-playwright-law student Johann Wolfgang von Goethe, 21, is completed in its first part; *The Deserted Village* by Oliver Goldsmith is an outcry against the changes in England: "Sweet Auburn! loveliest village of the plain,/ Where health and plenty cheer'd the labouring swain,/ Where smiling spring its earliest visit paid,/ And parting summer's lingering blooms delay'd"; English poet Thomas Chatterton, 17, poisons himself August 24 at Bristol in despair at his lack of recognition; his work immediately gains popularity.

Painting *Paul Revere* by Boston portrait artist John Singleton Copley, 32, who depicts a local silversmith. Giambattista Tiepelo dies at Madrid March 17 at age 74; François Boucher dies at Paris May 30 at age 66; *ukiyoe* painter Harunobu Suzuki dies at age 52.

Theater *The Two Friends, or the Merchant of Lyons* (*Les Deux Amis, ou le Negociant de Lyons*) by Pierre Augustin Caron de Beaumarchais 1/13 at the Comédie Française, Paris.

The Dutch monopoly in East Indies spice growing ends as Governor Poivre of Mauritius steals nutmeg seeds and starts new plantations on his French island.

England grows enough potatoes for public sale of the tubers for the first time.

The *kipfel* roll invented by Viennese bakers in 1683 arrives at Paris with Marie Antoinette, now 14, daughter of the Austrian empress Maria Theresa, who marries the French dauphin, 15, May 16 at Versailles. French bakers will turn the *kipfel* into the *croissant* that will become a traditional French breakfast treat.

Swedish chemist Tobern Olof Bergman, 35, discovers a way to make carbonic acid gas in commercial quantities (*see* Priestley, 1767; carbonate water, 1807).

Britain's 13 American colonies have an estimated population of 2.2 million, up from 1.6 million in 1760.

1771 Maratha forces from the Deccan Plateau drive the Afghans out of Delhi February 10, install the son of the exiled Mughal emperor Shah Alam as temporary ruler, and in April place Shah Alam himself as their puppet emperor, raising anxiety among the British who have kept Shah Alam in custody in Allahabad (*see* 1780).

Russian cossacks conquer the Crimean peninsula of the Ukraine for Catherine the Great in a triumph assisted indirectly by the British (*see* 1770). The Russian success alarms Prussia's Frederick the Great (*see* Poland, 1772; Crimean annexation, 1783).

Sweden's Adolphus Frederick dies at Stockholm February 12 at age 60 after a 20-year reign in which the riksdag has deprived him of all his powers of state. He is succeeded by his son of 25 who will regain absolute monarchical power next year by rousing fears of Russia and Prussia and will reign until 1792 as Gustavus III.

Damascus falls to the Egyptian Mameluke forces of Ali Bey.

1771 ***(cont.)*** The Battle of Alamance May 16 ends in a rout of back-country North Carolina colony farmers, known as Regulators, who have attacked the courts in a protest against taxes. British militiamen end their revolt.

Savoy abolishes serfdom as Charles Emmanuel III nears the end of a 43-year reign.

English agriculturist Arthur Young writes, "Everyone but an idiot knows that the lower classes must be kept poor or they will never be industrious . . .they must like all mankind be in poverty or they will not work" (*see* Mandeville, 1714).

Voyage autour du monde by explorer Louis Antoine de Bougainville suggests one concept of evolution (*see* 1767). Describing the Straits of Magellan, Bougainville says, "This cape, which rises to more than 150 feet above sea-level, is entirely composed of horizontal beds of petrified shellfish. I took soundings at the foot of this monument which testifies to the great changes that have happened to our globe" (*see* Maupertuis, 1741; Darwin, 1840).

Speeches made in Britain's House of Commons are ordered to be published despite protests by members of Parliament (*see* Hansard, 1774).

The Encyclopaedia Britannica, or a Dictionary of Arts and Sciences Compiled Upon a New Plan is published at Edinburgh by three men who call themselves "A Society of Gentlemen in Scotland;" chief compiler is printer-antiquary William Smellie, 26. The three-volume, 2,659-page work, whose first volume appeared in 1768, sells for £12, which is roughly what an Edinburgh artisan earns in a year, and is only casually researched and edited (*California* is described as "a large country of the West Indies" and *woman* is defined merely as "the female of man"), but the *Encyclopaedia* will be improved and expanded and go into 14 editions in the next 203 years (*see* 1784).

The French *Encyclopédie,* whose 28 volumes Denis Diderot is completing at Paris, has inaccuracies of its own despite contributory efforts to the *Dictionaire Raisonné des Sciences, des Arts et des Métier* by the comte de Buffon, the baron de Montesquieu, François Quesnay, Jean Jacques Rousseau, A. J. J. Turgot, and Voltaire. Both encyclopaedias attempt to compile in alphabetical order what 18th-century man knows but has never had readily accessible in written form (*see* 1772).

Fiction *The Expedition of Humphry Clinker* by Tobias Smollett, whose poor health has obliged him to settle in Italy and who dies September 17 at Leghorn (Livorno) at age 50.

Painting *Penn's Treaty with the Indians* by American painter Benjamin West, 32, who has been the first American to study art in Italy and who breaks away from the standard portraiture that has characterized American painting until now; *Death of Wolfe* by Benjamin West shocks London's Royal Academy by placing the hero of the 1759 Battle of Quebec and other contemporary figures in a classical composition; completed late last year, the work gives impetus to the growing movement toward realism in art and wins West an appointment as painter to George III; *The Alchymist* by Joseph Wright.

Sculpture *Diderot* by Jean Antoine Houdon, now 30, who has been making portrait busts since his return 2 years ago from 8 years in Italy.

Theater *The Natural Son, or The Proofs of Virtue* (*Le Fils Natural, ou les Epreuves de la Vertu*) by Denis Diderot 9/26 at the Comédie-Française, Paris; *The Beneficent Bear* (*Le Bourru Bienfaisant*) by Carlo Goldoni 11/4 at the Comédie-Française.

Tobias Smollett describes adulteration of British foods in his novel *Humphry Clinker* (above). His hero Matthew Bramble writes to a friend that "the bread I eat in London is a deleterious paste, mixed up with chalk, alum, and bone-ashes, insipid to the taste, and destructive to the constitution. The good people are not ignorant of this adulteration; but they prefer it to wholesome bread, because it is whiter than the meal of corn: thus they sacrifice their taste and their health, and the lives of their tender infants, to a most absurd gratification of a mis-judging eye" (*see* Accum, 1820).

1772 Poland loses one-third of her territory and half her population as she is partitioned August 5 among Russia, Austria, and Prussia. Russia acquires White Russia and all the territory to the Dvina and the Dnieper. Austria takes Red Russia, Galicia, and western Podolia with Lemberg and part of Cracow. Prussia takes Polish Prussia except for Danzig and Thorn (*see* 1793; 1831).

Sweden's Gustavus III restores absolute monarchy by an August 19 coup d'état that ends the power of the council and takes away the riksdag's power to initiate legislation. He will abolish torture, improve Sweden's poor laws, encourage trade, and proclaim religious toleration and liberty of the press, but his enlightened despotism will turn to reaction beginning in the summer of 1789.

Danish noblemen conspire against their dictator Count Johann Friedrich von Struense, 35, who has held absolute power for 10 months but has had an affair with the young queen Caroline Matilda. He is overthrown, condemned to death, tortured, and beheaded.

Warren Hastings, 40, begins a 13-year term as governor of Bengal that will be marked by reforms which will include a simplification of Indian coinage and government control of salt and opium production.

Boston patriot Samuel Adams, 50, and local physician Joseph Warren, 31, organize a Committee of Correspondence November 2; similar committees spring up throughout the colonies, relaying the anti-British polemics of Adams and others.

The Somersett case marks a turning point in British toleration of slavery. James Somersett, one of 10,000 black slaves in Britain, has escaped from his master and been apprehended. Britain's Lord Chief Justice William Murray, 67, Baron Mansfield, rules after some hesitation June 22 that "as soon as any slave sets foot in England he becomes free" (*see* 1763; 1787).

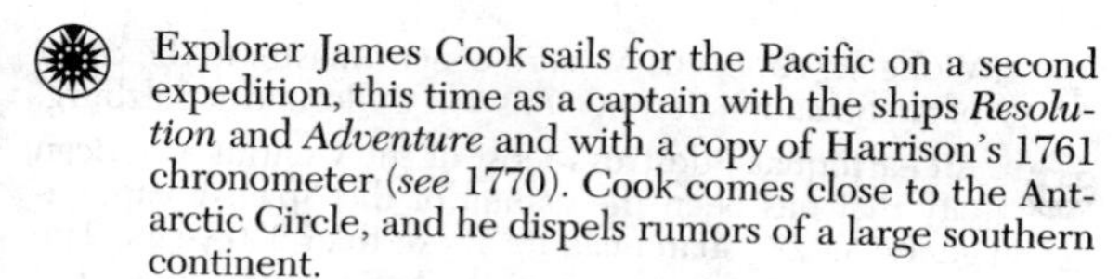

Explorer James Cook sails for the Pacific on a second expedition, this time as a captain with the ships *Resolution* and *Adventure* and with a copy of Harrison's 1761 chronometer (*see* 1770). Cook comes close to the Antarctic Circle, and he dispels rumors of a large southern continent.

Rubber gets its name from Joseph Priestley, who finds that it rubs out pencil marks (*see* 1503; macintosh, 1823; Goodyear, 1839).

Edinburgh physician Daniel Rutherford, 23, distinguishes nitrogen as a gas separate from carbon dioxide (*see* Davy, 1800; Wöhler, 1828).

Digressions Académiques by French chemist Baron Louis Bernard Guyton de Morveau, 35, sets forth original ideas about "phlogiston" and crystallization.

The brilliant *gaon* of Vilna Elijah ben Solomon Zalman, 52, issues the first *herem* against hasidic Jewry and has its books publicly burnt (*see* 1760; 1781).

Poetry *The Rising Glory of America* by College of New Jersey (Princeton) seniors Philip Morin Freneau, 20, and Hugh Henry Breckinridge whose work is published as a commencement poem; "The Progress of Dulness" by Yale graduate John Trumbull, 22, satirizes college education.

Denis Diderot's *Encyclopédie* is published in its 280th and final volume (a supplement to Bougainville's *Voyages*) after 21 years of effort. Six more supplements will appear in 1776 and 1777.

Painting *The Members of the Royal Academy* by Johann Zoffany.

Theater *Emilia Calotti* by Gotthold Ephraim Lessing 3/13 at Brunswick's Hoftheater.

First performances Symphony No. 48 in C major *(Maria Theresa)* by Joseph Haydn at Vienna; Symphony No. 21 by Wolfgang Amadeus Mozart at Salzburg; Divertimento for Strings in D major and two other divertimenti by W. A. Mozart at Salzburg.

Venice's Palazzo Grassi is completed for the Grassi family of Bologna. Designed by the late Giorgio Massari, who died 6 years ago at age 80, the trapeze-shaped palace was begun in 1748, has an inner court, and is the last of the great pieces of secular architecture in the Venetian republic that is now approaching its end.

Japan's capital of Edo is destroyed by fire as in 1657 (*see* 1923).

English agriculturist Thomas W. Coke begins a reform of animal husbandry. He will breed Southdown sheep, Devon cattle, and improved Suffolk pigs (*see* 1760; 1780).

1773 The Virginia House of Burgesses appoints a Provincial Committee of Correspondence March 12 to keep Virginia in touch with the other colonies.

Parliament passes a Regulating Act in May in an effort to bring the East India Company under government control. The company has become far more than a commercial enterprise and according to Whig leader Edmund Burke, it is "in reality a delegation of the whole power and sovereignty of this nation sent into the East."

Denmark cedes the duchy of Oldenburg to Russia.

Egypt's Ali Bey is wounded in a skirmish with Ottoman rebels and dies May 8 after 12 years of increasingly autocratic rule.

Wahhabi fundamentalists in Arabia annex Riyadh (*see* 1744; 1788).

The Ottoman sultan Mustapha III dies December 25 at age 57 after a 16-year reign. His brother, 48, will reign until 1789 as Abdul Hamid.

Yale president Ezra Stiles, 46, and Congregational theologian Samuel Hopkins, 52, urge that freed blacks be resettled in West Africa.

Captain Cook crosses the Antarctic Circle January 12, the first to do so.

Daniel Boone leads a party of settlers to the western Virginia country that will be called Kentucky. Two youths including Boone's son are killed when Indians attack the company, and the frightened pioneers scurry back east of the Appalachians (*see* 1767; Harrod, 1774; Wilderness Road, 1775).

Louisville, Kentucky, has its beginnings in a town site laid out by Virginia surveyor George Rogers Clark, 21. The town will be settled in 1779, and the Virginia Legislature will name it Louisville in 1780 to honor France's Louis XVI.

The Tea Act passed by Parliament May 10 lightens duties on tea imported into Britain to give relief to the East India Company (above) which has 7 years' supply in warehouses on the Thames and is being strained by storage charges. But the act permits tea to be shipped at full duty to the American colonies and to be sold directly to retailers, eliminating colonial middlemen and undercutting their prices.

"Two Letters on the Tea Tax" by John Dickinson are published in November.

The Boston Tea Party December 16 demonstrates against the new English tea orders (above). Led by Lendall Pitts, scion of a Boston merchant family, a group of men, including silversmith Paul Revere, 38, disguise themselves as Mohawks, board East India Company ships at Griffen's Wharf, and throw 342 chests of tea from the London firm of Davison and Newman into Boston Harbor (the tea is valued at more than £9,650). Agitator Samuel Adams has organized the action with support from John Hancock, whose smuggling of contraband tea has been made unprofitable by the new measures.

Tea is left to rot on the docks at Charleston. New York and Philadelphia send tea-laden ships back to England, but men of "sense and property" such as George Washington deplore the Boston Tea Party.

Scottish physician John Hunter, 45, begins lectures in medicine at London where he has practiced since 1763. A brilliant surgeon and investigator, Hunter performs experiments in comparative physiology, morphology,

Boston's "tea party" provoked London, but harsh measures provoked some colonists to demand independence.

and pathologic anatomy, describes phlebitis, pyemia, and shock, and saves thousands of limbs by establishing that aneurysms can be treated by a single proximal ligature (thread or wire) instead of by amputation. But like so many of his contemporaries, Hunter makes no distinction between the venereal diseases gonorrhea and syphilis.

Treatise on the Management of Pregnant Women and Lying-in Women by English surgeon Charles White devotes its first chapter to "The Causes and Symptoms of the Puerperal or Child-bed Fever" (*see* Gordon, 1795).

Baron Guyton de Morveau in France uses chlorine and hydrochloric acid gas as a disinfectant (*see* 1772; Lister, 1865).

The Holy Roman Emperor Josef II expels Jesuits from the empire, and Pope Clement XIV dissolves the Society of Jesus founded in 1534.

Dissolution of the Jesuit order (above) disrupts education in Catholic Europe.

Poems on Various Subjects, Religious and Moral, by Phillis Wheatley, Negro, Servant to Mr. John Wheatley of Boston, in New England is published at London. Born in Africa, Wheatley is about 20.

Painting *The Graces Decorating Hymen* by Sir Joshua Reynolds; *The Broken Pitcher* by Jean Baptiste Greuze.

Theater *She Stoops to Conquer* by Oliver Goldsmith 3/18 at London's Theatre Royal in Covent Garden: "This is Liberty-Hall, gentlemen. You may do just as you please here;" "Ask me no questions, and I'll tell you no fibs"; *Gotz von Berlichengen* by Johann Wolfgang von Goethe.

First performances Symphony No. 23 in D major, No. 24 in B flat major, No. 25 in G minor, No. 26 in E flat major, No. 28 in C major, and No. 29 in A major by Wolfgang Amadeus Mozart, concertmaster to the court of Archbishop Hieronymous von Colloredo at Salzburg.

An earthquake destroys most of the Central American city that has been the capital of the Spanish captain-generalcy of Guatemala for more than 230 years. The Spanish abandon the ruins at Antigua and move the seat of power to Guatemala City.

Parliament passes Burke's Act to permit the import of foreign wheat at a nominal duty when the home price reaches a certain level. The act will remain in effect long enough to establish regular imports of foreign grain, depending on the abundance of domestic harvests, and imports will for a short while exceed exports (*see* 1791).

1774 France's Louis XV dies May 10 of smallpox at age 64 after a reign of nearly 59 years. He is succeeded by his grandson, 19, who assumes the throne as Louis XVI and begins a reign with his Austrian-born queen Marie Antoinette that will last until 1792.

The Treaty of Kuchuk-Kaainardji July 16 ends a 6-year Russo-Turkish war. Moldavia and Wallachia return to Ottoman control, Russia acquires control of much of the northern Black Sea coast, and the Crimea gains independence.

Russian forces suppress a year-long Cossack revolt in September.

The Virginia House of Burgesses is dissolved May 26 by Governor Dunmore 2 days after the house has set June 1 aside as a day of fasting and prayer in sympathy with Boston (below). The Burgesses meet May 27 at Williamsburg's Raleigh Tavern in an unofficial session that adopts a resolution calling for an annual intercolonial congress. Copies are sent to other colonial legislatures.

A new Quartering Act passed by Parliament June 2 updates the 1765 act. The coercive law requires colonists to house British troops in their barns or public inns where barracks are not available.

The Quebec Act passed by Parliament June 22 extends the Province of Quebec south to the Ohio River and west to the Mississippi. It includes French-speaking settlements and ignores western land claims of Massachusetts, Connecticut, and Virginia.

The First Continental Congress assembles at Philadelphia September 5 with all colonies represented except Georgia.

The Suffolk Resolves protesting against Parliament's coercive acts are laid before the Continental Congress September 17 and approved. Drawn up by Dr. Joseph Warren at Boston, the resolves have been adopted by a convention in Suffolk County, Massachusetts.

The Continental Congress adopts a Declaration of Rights and Grievances October 14, but George Washington writes that no thinking man in all of North America desires independence.

The Association adopted October 20 by the Continental Congress is an agreement to import nothing from Great Britain after December 1 and to export nothing to Great Britain, Ireland, or the British West Indies after September 10, 1775, unless the British redress grievances against the Crown. All colonies except Georgia and New York will ratify the association within 6 months.

The Battle of Point Pleasant October 10 ends in defeat for Shawnee braves who have attacked frontiersmen at the juncture of the Ohio and Great Kanawha rivers in western Virginia. The natives yield hunting rights in Kentucky under terms of the Treaty of Camp Charlotte and agree to permit transportation on the Ohio to go unmolested.

The Connecticut and Rhode Island colonies prohibit further importation of slaves.

Philadelphia physician Benjamin Rush, 29, joins with James Pemberton to form an antislavery society.

The Fairfax Resolves signed by George Washington December 1 bar importation of slaves and threaten to stop all colonial exports to Britain.

French controller-general Ann-Robert-Jacques Turgot abolishes the Six Guilds of Paris, telling labor it must not try to restrict jobs. The Six Guilds have maintained wage levels by limiting the number of artisans, but Turgot says work is the "inalienable right of humanity," and forbids workers to "form any association or assembly among themselves under any pretext whatever."

Pioneer James Harrod makes the first English settlement west of the Alleghenies. He leads 32 other men in canoes from Pennsylvania down the Ohio River, up the Kentucky River, and into the high fertile bluegrass country plateau on which they build Fort Harrod (*see* Boonesborough, 1775).

Captain Cook in the South Pacific charts and names the New Hebrides, some 80 islands between New Caledonia and Fiji (*see* 1606; 1980).

News of last year's Boston Tea Party reaches London January via John Hancock's ship *Hayley*. Parliament passes coercive acts (below) to bring the colonists to heel.

George III gives assent March 31 to the Boston Port Bill and Boston Harbor is closed June 1 until the East India Company shall have been reimbursed for its tea and British authorities feel that trade can be resumed and duties collected (*see* 1773).

The American colonists thwart London's efforts to starve Boston into submission. Marblehead sends in codfish, Charleston sends rice, and Baltimore sends bread and rye whiskey. Col. Israel Putnam, 56, of Connecticut joins with citizens of Windham to drive a flock of 258 sheep to Boston.

The British ship *London* docks at New York April 22, and the Sons of Liberty prepare to follow the example set at Boston 4 months earlier, but while they are making themselves up as Mohawks an impatient crowd on the pier boards the vessel and heaves the tea on board into the Hudson. Colonists at York, Maine, and Annapolis, Maryland, conduct tea parties like the one at Boston.

English ironmaster John Wilkinson, 46, patents a precision cannon borer that will permit commercial development of the Watt steam engine of 1769. Wilkinson's borer permits accurate boring of cylinders (*see* steam flour mill, 1780).

Bridgewater, Massachusetts, blacksmith John Ames uses bar iron to fabricate shovel blades that will replace the hand-hewn shovels and imported English shovels now used in America (*see* 1803).

Joseph Priestley discovers oxygen, a gas he calls "dephlogisticated air." He heats red oxide of mercury in a glass tube and obtains a gas in which a candle burns more brightly than in the air. Air "spoilt" by mice is refreshed by the presence of green plants, says Priestley (*see* 1767; 1779).

Swedish chemist Karl Wilhelm Scheele, 32, discovers chlorine and manganese.

English religious mystic Ann Lee, 38, settles on a tract of land northeast of Albany in the New York colony and introduces "Shakerism" into America. A member of the Shaking Quaker sect, Lee abandoned her Manchester blacksmith husband after losing four children in quick succession. She is given to hysterics, convulsions, and hallucinations, denounces sex as a "filthy gratification," calls a consummated marriage "a covenant with death and an agreement with hell," preaches celibacy, and establishes a following (*see* 1784).

London printer Luke Hansard, 22, begins printing the House of Commons journals (*see* 1771). Official reports of parliamentary proceedings will be called Hansards for more than 2 centuries.

Letters to His Son by England's late Philip Dormer Stanhope, fourth earl of Chesterfield, are published by his widow a year after the earl's death at age 79, which came 5 years after the death at age 36 of his son Philip, born of an intimacy with a Mlle. de Bouchet. The witty and cynical letters helped the son fill diplomatic posts and a seat in Parliament obtained for him by Lord Chesterfield: "Whatever is worth doing at all, is worth doing well"; "Take care of the pence; for the pounds will take care of themselves"; "Advice is seldom welcome; and those who want it the most always like it the least"; "An injury is much sooner forgotten than an insult"; "A seeming ignorance is often a most necessary part of worldly knowledge"; "You had better refuse a favor gracefully than grant it clumsily"; "Wear your learning, like your watch, in a private pocket; and do not pull it out and strike it, merely to show that you have one. . ."; "Women are all children of a larger growth"; "Let blockheads read what blockheads wrote"; "Young men are apt to think themselves wise enough, as drunken men are apt to think themselves sober enough."

The Journal of John Woolman is published 2 years after the death of American social reformer John Woolman, who took up the Quaker ministry in 1743 at age 23 and opened a Mount Holly, New Jersey, tailor shop to support himself. Woolman has traveled throughout the colonies, spreading the Quaker doctrine from North Carolina to New Hampshire and calling for the end of slavery as an institution incompatible with religion. He kept the journal beginning at age 35 in 1756, worked to

1774 ***(cont.)*** stop the sale of liquor to the Indians, ate no sugar because it was produced by slave labor, and wore clothing made from undyed materials because textile workers were sometimes injured by fabric dyes.

"On the Rise and Progress of the Differences between Great Britain and Her American Colonies" by Benjamin Franklin whose series of articles appears in the *London Public Advertiser*.

"Observations on the Act of Parliament, Commonly Called the Boston Port Bill; with Thoughts on Civil Society and Standing Armies" by Josiah Quincy, 30, is published as a pamphlet at Boston in May.

"Considerations on the Nature and Extent of the Legislative Authority of the British Parliament" by Philadelphia lawyer James Wilson, 32, is published following Quincy's pamphlet (above).

"A Friendly Address to All Reasonable Americans" is published anonymously at New York, where it has been written by King's College president Myles Cooper, a Loyalist.

"Summary View of the Rights of Americans. . ." by Thomas Jefferson is published as a pamphlet at Williamsburg, Virginia, and at London.

"Free Thoughts on the Proceedings of the Continental Congress" by "Westchester farmer" Samuel Seabury, 45, is published as a pamphlet attacking the Congress. Loyalist Seabury is a Protestant Episcopal clergyman who was graduated from Yale in 1748.

Fiction *The Sorrows of Young Werther* (*Die Leiden des Jungen Werthers*) by Johann Wolfgang von Goethe, whose love story has been inspired by his affair with Charlotte Buff, 21.

Painting *Artiochus et Stratonice* by French painter Jacques Louis David, 26. Boston painter John Singleton Copley emigrates to Europe following a decline in his portrait commissions, partly for reasons of his Loyalist sympathies. Copley will settle permanently at London after a year's vacation in Italy.

Theater *Clavigo* by Johann Wolfgang von Goethe 8/23 at Hamburg; *Eloge de La Fontaine* by Sebastien Chamfort; *The Inflexible Captive* by English playwright Hannah More, 29.

Opera *Iphigenia in Aulis* 4/19 at the Paris Opéra, with music by C. W. Gluck.

A meager harvest in France increases the hunger that has oppressed the country since the poor harvest of last year.

Controller-general Turgot reintroduces the free trade in grain abolished in 1766, but the rivers freeze, grain cannot be moved by barge, and the price of a 4-pound loaf of bread rises in just a few months from 11 sous to 14 sous. A 4-pound loaf has sold for as much as 16 sous in earlier years, but where in the past high prices were blamed on the weather, now the government is blamed.

France has "flour wars" as rioters throw grain merchants into ponds, loot mills and bakeries, seize grain from barges, march on Versailles, and force Louis XVI to promise bread at 2 sous per pound. Some 400 rioters are arrested, and two are hanged for breaking into bakeries. No conspiracy to raise prices is uncovered, but such a conspiracy is almost universally suspected (*see* 1768; 1776).

1775 The American War of Independence begins April 19 at the Battles of Lexington and Concord. Paul Revere has ridden to Lexington April 16 to warn patriots Samuel Adams and John Hancock of Gen. Gage's intention to arrest them, and he has sent word to patriots at Concord that Gage has found out about the war matériel they have accumulated. He has ridden out with compatriot William Dawes, 30, the night of April 18 to warn Adams and Hancock that 700 redcoats have been marching from Boston since midnight; local physician Samuel Prescott, 23, has joined in spreading the alarm. Revere is captured by the British, who force Dawes to flee, and it is Prescott who rouses Samuel Hartwell of the Lincoln minutemen and other patriots.

Redcoats under Lieut. Col. Francis Smith, 51, and Marine Major John Pitcairn, 53, arrive at Lexington near dawn and encounter a force of minutemen under the command of John Parker, 45, who says to his men, "Stand your ground. Don't fire unless fired upon. But if they mean to have a war, let it begin here."

Eight minutemen fall to British musketfire at Lexington, and the rest retreat at bayonet point. The British cut down a Liberty pole at Concord and skirmish with the militia at a bridge.

British reinforcements under Brig. Gen. Sir Hugh Percy, 33, arrive at Concord with two 6-pound cannon. Although outnumbered 4,000 to 1,800, the British are well organized, Gen. Percy executes a skillful retreat, and the redcoats are back in Boston by nightfall, having lost 65 dead, 173 wounded, 26 missing (American marksmen have picked off 15 British officers).

"Give me liberty or give me death," Virginia's Patrick Henry has said March 23 in an oration. Many patriots echo his sentiments (*see* 1765).

Capt. Richard Derby at Salem assembles cannon and refuses to let British troops cross a drawbridge to seize the guns (*see* 1741). "Take them if you can," Derby shouts across the stream, "They will never be surrendered."

Britain has hired nearly 30,000 mercenaries from six small German despots, notably the landgrave of Hesse-Cassel, who has supplied 17,000 Hessians.

The Royal Navy begins an 11-month siege of Boston.

The first American victory comes May 10 as 832 Green Mountain Boys led by Ethan Allen, 37, and Benedict Arnold, 34, move on orders from Connecticut, cross Lake Champlain in the small hours of the morning, enter Fort Ticonderoga through a breach in the wall, and obtain surrender of the fort from its commandant Capt. de la Place. An attempt to take Quebec fails, and the British capture Allen at Montreal, but Seth Warner takes Crown Point on Lake Champlain May 12.

The Second Continental Congress meets at Philadelphia May 10 and 2 weeks later chooses John Hancock of Boston as its president.

The Battle of Bunker Hill June 17 ends in victory for the British, but 1,150 redcoats fall in the battle, including Major Pitcairn (above), while the colonists lose only 411, including their leader Gen. Joseph Warren, the Boston physician who was named general of militia June 14. Breed's Hill, lower than Bunker Hill and closer to the enemy, has been fortified by military engineer Col. Richard Gridley, 64, whose men have been building breastworks since midnight and who is wounded in the battle that begins shortly after dawn. "You are all marksmen. Don't fire until you see the whites of their eyes," Gen. Israel Putnam has ordered, and the militiamen have fired with deadly effect.

George Washington is appointed commander in chief of the Continental army June 17 and arrives at Cambridge, Massachusetts, July 3 to take command after a 12-day journey from Philadelphia.

An American navy is established October 13 by the Second Continental Congress, which authorizes construction of two "swift sailing vessels" (*see* 1779).

George III signs an order releasing from bondage the women and young children in British coal mines and salt mines. Many of the children are under 8, work like the women for 10 to 12 hours per day, and have been transferable with the collieries and salt works when the properties changed hands or their masters had no further use for them (*see* Scottish coal miners, 1778).

Stop discrimination against women, says English-American pamphleteer Thomas Paine, 38, in his *Pennsylvania Magazine.* Paine has failed as a corsetmaker and tax collector in England, twice failed in marriage, and arrived at Philadelphia late last year with letters of introduction from Benjamin Franklin and encouragement to try his luck in America.

The first American abolition society is founded in Pennsylvania to free the slaves, whose population below the Mason-Dixon line now exceeds 450,000. Black slaves outnumber colonists two to one in South Carolina, while in Virginia the ratio is about equal.

The Transylvania Company employs Daniel Boone to lead a party of 30 North Carolina woodsmen westward (*see* 1773). The group sets out March 10 and breaks a Wilderness Road of nearly 300 miles that will be used in the next 15 years by more than 100,000 pioneers en route to the new territories of western Tennessee and Kentucky. Boone founds Boonesborough on the Kentucky River in April (*see* Harrod, 1774).

Captain Cook returns from a second voyage to the Pacific (*see* below).

American imports from Britain decline by 95 percent as a result of efforts by the Association formed last year. English merchants feel the loss, support the colonists, and ask Parliament to repeal the so-called Intolerable Acts, but while Edmund Burke makes a speech on conciliation with America (below), Lord North tells the House of Commons that Britons are paying 50 times more tax per capita than are the American colonists.

British banks organize their first clearing house in London's Lombard Street (*see* 1708; New York, 1853).

Philadelphia merchants form the United Company of Philadelphia for Promoting American Manufactures. The colonies have more iron furnaces and forges than England and Wales combined.

News of the skirmish at Lexington, Massachusetts (above), reaches Philadelphia via post rider Israel Bissel, 23, who takes a note penned at 10 in the morning of April 19, rides 36 miles to Worcester in 2 hours, alerts Israel Putnam (above) at Brooklyn, Connecticut, during the night, and reaches Old Lyme at 1 in the morning of April 20. Bissel is ferried across the Connecticut River and reaches Saybrook by 4 in the afternoon and Guilford by 7. At noon on April 21 he arrives at Branford, and on the morning of April 22 reaches New Haven, where local apothecary Benedict Arnold (above) calls out his company of militia and marches north. Bissel reaches New York City April 23, and the intelligence from Boston causes New Yorkers to close the port, distribute arms, and burn two sloops bound for the British garrison at Boston. Bissel is ferried across the Hudson, arrives at New Brunswick at 2 in the morning of April 24, reaches Princeton by dawn, and is in Trenton by 6 and in Philadelphia soon after, having traveled in 5 days a distance the fastest stage would have taken 8 days to cover.

"Taxation No Tyranny" by Doctor Johnson is a tract against the American colonies.

Edmund Burke's parliamentary speech "On Conciliation with the Colonies" is published in London.

Nonfiction *A Journey to the Western Islands of Scotland* by Doctor Johnson, who toured Scotland in 1773 from August to November in the company of James Boswell.

Painting *Self-portrait* (pastel) by Jean Chardin, now 75; *Miss Bowles* by Sir Joshua Reynolds.

Theater *The Rivals* by Irish playwright Richard Brinsley Sheridan, 23, 1/17 at London's Royal Theatre in Covent Garden. Sheridan's Mrs. Malaprop delights audiences with "malapropisms" such as "headstrong as an allegory on the banks of the Nile" (III, iii). *The Barber of Seville, or The Useless Precaution* (*Le barbier de Séville, ou La précaution inutile*) by Pierre A. C. de Beaumarchais 2/23 at the Comédie-Française; *St. Patrick's Day, or The Scheming Lieutenant* by Sheridan 5/2 at London's Royal Theatre in Covent Garden; *The Jews (Die Juden)* by Gotthold Ephraim Lessing 9/13 at Frankfurt-am-Main (one-act comedy).

English tragic actress Sara Kemble Siddon, 20, has her debut at London's Drury Lane Theatre, beginning a 32-year career that will make her queen of the stage.

Opera *La finta Giordiniera* 1/13 at Munich, with music by Wolfgang Amadeus Mozart; *Il Re Pastore* 4/23 at Salzburg, with music by W. A. Mozart, libretto by the

1775 ***(cont.)*** Italian poet-playwright Metastasio; *The Duenna* 11/29 at London's Royal Theatre in Covent Garden, with music by English composer Thomas Linley, 42, libretto by Richard Brinsley Sheridan.

Stage musical *Maid of the Oaks* at London with music by London playwright-composer John Burgoyne, 53, and songs that include "The World Turned Upside Down."

Wolfgang Amadeus Mozart at Salzburg writes five violin concerti between April 14 and December 20, each in 1 day except for the Concerto No. 4 in D major which he works on over the course of a month and completes in October.

First performances Concerto No. 5 in A major for Violin and Orchestra *(Turkish)* by Wolfgang Amadeus Mozart 12/20 at Salzburg.

Popular song "Yankee Doodle" with lyrics to an old English tune. American colonist Edward Barnes has written the words that will be popular in the Continental army.

English inventor Alexander Cummings receives the first patent to be issued for a flush toilet, but such devices will not come into common use for more than a century (*see* Harington, 1596; Bramah, 1778).

Bath's Royal Crescent is completed by English architect John Wood the Younger, 47.

The first house rats recorded in America appear at Boston (*see* 1727).

The New Bedford, Massachusetts, whaling fleet reaches 80 vessels. More than 280 whaling ships put out from American ports, 220 of them from Massachusetts (*see* 1751; 1845).

British mercantile interests reenter the whaling industry largely abandoned by the English in 1612 as Dutch activity begins to wane.

James Cook returns to England from the Pacific, and the Royal Society awards him its gold Copley Medal for having conquered scurvy. Sir John Pringle, chief medical officer of the British army and physician to George III, hails Captain Cook's achievement in bringing 118 men through all climates for 3 years and 18 days "with the lofs of only one man by diftemper," and while Cook has succeeded with sauerkraut and may "entertaine no great opinion of [the] antiscorbutic virtue" of concentrated citrus juices, Pringle suggests that he may be mistaken (*see* 1794; Lind, 1757).

Prussia's Frederick the Great moves to block imports of green coffee which are draining his country's gold as coffee becomes nearly a match for beer as the national beverage.

1776 The Declaration of Independence signed July 4 at Philadelphia (below) follows military action in the American Revolution.

The Battle of Moore's Creek Bridge February 27 ends in defeat for Scottish Loyalists from upper North Carolina who lose 900 prisoners to Carolina patriots near Wilmington. The Americans discourage Gen. Henry Clinton, 38, from landing a British expeditionary force.

British troops evacuate Boston March 17. Gen. William Howe, 46, sails for Nova Scotia with 900 Loyalists; many will settle in New Brunswick.

Lee's Resolutions introduced into the Continental Congress at Philadelphia June 7 urge the colonies to make foreign alliances and form a confederation under a constitution to be approved by each state. The Virginia Convention has instructed its delegate Richard Henry Lee, 44, May 15 to propose independence, and his three resolutions include an announcement "That these United States are and of right ought to be free and independent states."

Patriots at Sullivan's Island off Charleston, South Carolina (Fort Moultrie), repulse a British fleet under Gen. Clinton (above) and Sir Peter Parker, 55, June 28.

Gen. Howe lands on Staten Island June 30.

Congress adopts Lee's Resolutions (above) July 2.

Continental Congress president John Hancock signs the Declaration of Independence (above), writing his name in large letters and saying, "There, I guess King George will be able to read that." The Declaration is set in type and printed in a shop at 48 High Street and signed on and after August 2 by members of Congress including Benjamin Franklin, who says, "We must all hang together, else we shall all hang separately." Thomas Paine's pamphlet "Common Sense," published at Philadelphia January 10, has persuaded many of the 56 signers.

Gen. Howe on Staten Island is joined July 12 by a British fleet under the command of his older brother Lord Richard Howe, 50, and on August 1 by Gen. Clinton (above).

The Battle of Long Island August 27 ends in defeat for 8,000 patriots under Gen. Israel Putnam at the hands of Gen. Howe's 20,000 regulars.

Gen. Howe and his secretary Sir Henry Starchey meet September 11 in a house on Staten Island with John Adams, Benjamin Franklin, and Edward Rutledge in an effort to end the war by negotiation. Howe learns for the first time that the Declaration of Independence (above) has been signed, the Americans refuse to retract the document, and negotiations break off.

Gen. Howe lands at Kips Bay September 15 and occupies New York City. He narrowly misses catching Washington, who retreats to Harlem Heights and repulses a British attack September 16 with help from his sharpshooters. Armed with Pennsylvania long rifles, they can fire accurately at 200 to 400 yards, while musket balls carry effectively only 80 to 100 yards (the British will complain that American sharpshooters are unsportsmanlike in concentrating their fire on officers).

The British capture Continental army captain Nathan Hale, 21, September 21 on his return to Manhattan from an espionage mission to gather intelligence on Long Island. Hale has set numerous fires to harry the British in New York, disguised himself as a Dutch schoolmaster to avoid arrest, and is hanged September 22 by order of Gen. Howe. His last words will be reported as "I only regret that I have but one life to lose for my country."

Polish military tactician Tadeusz Andzej Bonawentura Kosciuszko, 30, enters the Continental army as a volunteer, having been wounded by the retainers of Sosnowski of Sownowicka, the Grand Hetman, with whose youngest daughter he tried to elope. Kosciuszko has studied fortification and naval tactics at Polish government expense in Prussia, France, and Italy, and distinguishes himself in the American cause. Gen. Washington will make him his adjutant and raise him to the rank of colonel of artillery (*see* 1794).

Benedict Arnold's fleet on Lake Champlain is defeated October 11 in the Battle of the Island of Valcour, but Arnold has delayed the southward advance of Sir Guy Carleton, who takes Crown Point but soon withdraws to Canada.

The Battle of White Plains October 28 gives Gen. Howe a narrow victory over Gen. Washington.

Gen. Nathanael Greene surrenders Fort Lee to the British November 20; Washington begins a retreat across New Jersey the next day, with Gen. Charles Cornwallis, 38, in hot pursuit.

Washington crosses the Delaware Christmas night, surprises the Hessians at Trenton, and turns the tide of the war by taking more than 1,000 prisoners at the Battle of Trenton December 26.

Catherine the Great's court favorite Grigori Aleksandrovich Potemkin, 37, builds a Russian Black Sea fleet. Potemkin distinguished himself in 1769 fighting the Turks.

"We hold these Truths to be self-evident," says the Declaration of Independence (above), "that all Men are created equal, that they are endowed by their Creator with certain unalienable Rights, that among these are Life, Liberty, and the Pursuit of Happiness—That to secure these Rights, Governments are instituted among Men, deriving their just Powers from the Consent of the Governed, that whenever any Form of Government becomes destructive of these Ends, it is the Right of the People to alter or to abolish it, and to institute new Government . . ."

Thomas Jefferson's draft has been edited to delete an attack on slavery.

Delaware forbids further importation of slaves as the slave population in the colonies reaches 500,000.

Britain's House of Commons hears the first motion to outlaw slavery in Britain and her colonies. David Hartley, 44, calls slavery "contrary to the laws of God and the rights of man," but his motion fails (*see* 1772; Wilberforce, 1787, 1789).

San Francisco has its beginnings in the settlement of Yerba Buena (good herb) established by Spanish monks in California (*see* 1769; 1846).

The Continental Congress starts a national lottery to raise money for the Continental army.

Inquiry into the Nature and Causes of the Wealth of Nations by Scottish philosopher Adam Smith, 53, of Kirkcaldy, Fifeshire, proposes a system of natural liberty in trade and commerce (*see* Hume, 1739). "Consumption is the sole end and purpose of all production, and the interest of the producer ought to be attended to, only so far as it may be necessary for promoting that of the consumer," who in the long run has full control over what will and will not be produced, says Smith, who teaches at the University of Glasgow. His massive work establishes the classical school of political economy and will influence all future thinking on politics and economics, but it shows no awareness of the developing industrial revolution, and while it espouses free-market competition with limited government intervention it regards unemployment as a necessary evil to keep costs—and therefore prices—in check (*see* Ricardo, 1817).

"The discovery of America, and that of a passage to the East Indies by the Cape of Good Hope, are the two greatest and most important events recorded in the history of mankind," writes Adam Smith (above).

Bushnell's "Connecticut Turtle" pioneers the use of the submarine in warfare. Built by Yale graduate David Bushnell, 34, the pear-shaped 7-foot vessel is made of oak staves held together with pitch and iron hoops. It has ballast tanks operated by foot pumps, its conning tower has windows level with the head of the operator who uses two air tubes for intake and exhaustion of air (automatic valves close them for diving), is propelled horizontally and vertically by hand-cranked propellers and guided by a flexible rudder, carries a powder magazine with a clock timer, and goes into action the night of September 6 in New York Harbor. Bushnell's craft has an auger mounted on its top to bore a hole into the wooden hull of an enemy vessel so that it may plant its powder magazine, but many of the British vessels have copperclad bottoms to protect them against shipworms, and several attempts to plant charges prove fruitless (*see* Lake, 1897).

Karl Wilhelm Scheele discovers uric acid in a kidney stone (*see* 1774; Garrod, 1859).

Smallpox decimates the Continental army in the north. By June some 5,500 of the 10,000-man force are incapacitated, largely by the dread pox to which the British are generally immune as a result of having had mild bouts with the disease in childhood or in some cases by inoculation (*see* 1777; Franklin, 1760).

The Phi Beta Kappa Society is founded December 5 at Virginia's 83-year-old College of William and Mary in Williamsburg by five young men who have gathered at a local tavern for conviviality and to debate such subjects as "Whether French politics be more injurious than New England rum" or "Had William the Norman a right to invade England?" (Chapters of the new scholastic fraternity will be established at Harvard and Yale in 1779, Harvard men will debate the question of whether Adam had a navel, Yale men whether females have intellectual capacities equal to those of males, and election to PBK will carry great prestige in U.S. academic circles.)

1776 ***(cont.)*** "These are the times that try men's souls," writes Thomas Paine (above) in "The Crisis," first in a series of 16 pamphlets that he will publish under that title in the next 7 years. Paine's January pamphlet "Common Sense" marshalled arguments for the justice of the revolutionary cause, sold half a million copies almost overnight, and won its author a position as aide to Gen. Greene. His December pamphlet "The Crisis" has a similar electrifying effect.

The History of the Decline and Fall of the Roman Empire by English historian Edward Gibbon, 39, is published in its first volume and creates a controversy over the rise of Christianity (*see* 1781).

Painting *The Washerwoman* by Jean Honoré Fragonard.

Sculpture *Voltaire* by French sculptor Jean Baptiste Pigalle, 62.

Theater *Stella* by Wolfgang von Goethe 2/8 at Hamburg; *The Twins (Die Zwillinge)* by German playwright Friedrich Maximilian von Klinger, 24, 2/13 at Hamburg; *Sturm und Drang* (*Storm and Stress*) by Friedrich von Klinger 4/1 at Leipzig (von Klinger will move to St. Petersburg in 1780, enter the Russian army, be elevated to the nobility, and, in 1790, marry an illegitimate daughter of Catherine the Great; the title of his new play will provide the name for a German literary movement headed by Goethe and Schiller); *The Soldiers* (*Die Soldaten*) by German playwright-poet Jakob Michael Reinhold Lenz, 25, who has followed Goethe to Weimar but is forced to leave because of his bad manners and tactlessness (Lenz will suffer a mental breakdown); *Mustapha et Zeangir* by Sebastien Chamfort, whose tragedy brings him a pension from the French royal family and confirms his success in society.

The Bolshoi Theater is founded at Moscow.

Thomas Paine's fiery pamphlets "Common Sense" and "The Crisis" roused American sentiment against the British.

Vienna's Burgtheater opens to give the city a new opera house and concert hall. The Holy Roman Emperor Josef II has founded the theater and forbids curtain calls, feeling that he is more worthy of applause than his servants the actors.

First performances Serenade No. 6 in D major for two Small Orchestras *(Serenata Notturna)* by Wolfgang Amadeus Mozart in January at Salzburg; Serenade No. 7 in D major *(Haffner)* by W. A. Mozart, 7/22 at Vienna, for the marriage of Elisabeth Haffner, daughter of the late Viennese burgomaster, to F. X. Spath.

Hymn "Rock of Ages" verses by London editor Augustus Toplady are published in the February issue of *The Gospel Magazine* (*see* 1830).

England's St. Leger stakes race has its first running at Doneaster as its founder, Col. Barry St. Leger, 39, prepares to leave for America to join Gen. Burgoyne in the Hudson Valley.

Cattle ranches begin to flourish on the Argentine pampas as Spain creates the Viceroyalty of La Plata with its capital at Buenos Aires.

A good harvest in France reduces the price of bread, but the French again abolish internal free trade in grain (*see* 1774; 1787).

Philadelphia's population reaches 40,000, making it larger than Boston and New York (24,000) combined.

1777 The Battle of Princeton January 3 gives Gen. Washington a victory over three British regiments under Gen. Cornwallis. Gen. Benedict Arnold defeats the British April 27 at Ridgefield, Connecticut. The American garrison at Fort Ticonderoga abandons the fort at news that Gen. John Burgoyne is approaching, Burgoyne defeats the retreating garrison force July 7 at Hubbarton, Vermont. Gen. Philip Schuyler evacuates Fort Edward July 29, and the British take it, but Americans under Gen. Nicholas Herkimer check the British in their march down the Mohawk Valley, defeating them August 6 in the Battle of Oriskany despite the loss of Gen. Herkimer in battle.

The Battle of Bennington August 16 ends in victory for the Americans under Capt. John Stark, who receives support from Col. Seth Warner, but although the British abandon their siege of Fort Stanwix August 22 at the approach of Benedict Arnold, they defeat Gen. Washington September 22 at the Battle of Brandywine.

The Continental Congress has voted July 31 to accept the services of French captain of dragoons Marie Joseph Paul Yves Roch Gilbert du Motier, 19, marquis de Lafayette, who was orphaned 6 years ago but was left with a princely fortune. Louis XVI forbade Lafayette to pursue his plan to support the American cause, but the young man escaped from custody, picked up a little English en route to America, landed in South Carolina, hurried to Philadelphia, and has been given the rank of major general. Gen. Lafayette is wounded at Brandywine but soon secures the command of a division.

The First Battle of Saratoga (Bemis Heights) September 19 inflicts heavy losses on the British, but the redcoats hold their ground, and other British forces defeat Anthony Wayne at Paoli, Pennsylvania, September 20. Gen. Howe occupies Philadelphia September 26 and defeats Washington at Germantown October 6.

The Second Battle of Saratoga October 7 gives Gen. Horatio Gates a decisive victory over Gen. Burgoyne, who surrenders with his entire force. Benedict Arnold has inspired Gates to defeat the British, Gen. Burgoyne marches his troops to Boston, and he embarks for England, never to resume the war. The British gain control of the Delaware River in November.

John Paul Jones has embarked for France November 1 aboard his ship *Ranger* with news of the victory at Saratoga. Now 30, the Scots-American naval commander adopted his surname in his early 20s after killing a mutineer in Tobago and taking flight to avoid imprisonment while awaiting trial. He has paused to take two prizes en route to France (*see* 1779).

The Articles of Confederation adopted by the Continental Congress at York, Pennsylvania, November 15 provide that at least nine states consent to all important measures, requires unanimous consent for any changes in the articles, provides no way for the central government to coerce recalcitrant states into compliance with congressional decisions, and denies Congress any power to tax or to regulate trade. Congress submits the articles to the 13 states for ratification, makes the first of several requisitions of funds, to be paid by the states in paper money, and authorizes confiscation of Loyalists' estates.

Portugal's Brazil annexes Maranhão as the Spanish and Portuguese settle their differences over their South American colonies. The Portuguese move the capital of Brazil from Bahia to Rio de Janeiro.

Portugal's José Manuel, insane since 1774, has died February 24 at age 61 after a 27-year reign. His chief minister José de Carvalho e Mello, marques de Pombal, has broken the power of the nobility and the church, reformed the military, straightened out the nation's finances, established trade companies with monopolistic powers to encourage industry, and tried to improve education. Now 78, he is dismissed March 5 by the queen mother Marianna Victoria, whose feebleminded daughter Maria Francisca, 43, will reign until 1816 (she was married at 26 to the king's dimwitted brother, her uncle, who will die in 1777, and will go insane in 1792).

Bavaria's elector Maximilian III dies December 30 without heirs. Charles Theodore, the elector Palatine, is his legal heir (but *see* 1778).

Gen. Washington obtains congressional approval in January to inoculate the entire Continental army against smallpox, which in some cases is further spread by inoculation (*see* 1776; vaccination, 1796).

Painting *The Watering Place* by Thomas Gainsborough.

Swedish chemist K.W. Scheele undertakes studies with silver salts that advance rudimentary knowledge of photography (*see* Schulze, 1727; Wedgwood, 1802).

Theater *The School for Scandal* by Richard Brinsley Sheridan 5/1 at London's Theatre Royal in Drury Lane, which Sheridan has been managing for the past year: "Here's to the maiden of bashful fifteen;/Here's to the widow of fifty;/Here's to the flaunting, extravagant queen,/And here's to the housewife that's thrifty"; *Percy* by Hannah More 10/12 at London's Covent Garden Theatre.

First performances Concerto No. 9 for Pianoforte and Orchestra in E flat major by Wolfgang Amadeus Mozart, in January at Salzburg; Concerto for Three Pianofortes and Orchestra in F major by W. A. Mozart 10/22 at Augsburg.

The Baldwin apple is discovered by Continental army officer Laomo Baldwin, 37, who has been invalided home to Wilmington, Massachusetts.

George Washington leads his 11,000 ragged troops into Valley Forge, Pennsylvania, December 14 to spend the winter. At least 2,000 are barefoot, and food is scarce. Gen. Howe has captured his salt stocks.

1778

News of Gen. Burgoyne's defeat last fall thrills Paris, which has been secretly supplying money and supplies to the American revolutionists for 2 years. France recognizes American independence and signs a treaty with Benjamin Franklin February 6. The Continental Congress ratifies the treaty May 4 and is so assured by the French alliance that it rejects British peace offers in mid-June.

The Battle of Monmouth June 28 ends in a victory for Gen. Washington, whose army of 11,000 was reduced to 8,000 by hunger, illness, and exposure in winter quarters at Valley Forge. The troops have been drilled at Valley Forge by Friedrich Wilhelm von Steuben, 46, a Prussian general who performs valuable service.

Gen. Charles Lee, 47, reluctantly agrees to lead the American attack at the Battle of Monmouth (above) but begins a retreat in the heat of battle. Gen. Washington intervenes to save the day; Gen. Lee is court-martialed July 4, the court finds him guilty August 12 and suspends him for 12 months.

Kaskaskia falls to a troop of some 150 Virginia volunteers led by George Rogers Clark, who takes the chief British fort in the Illinois country July 4 and goes on to take Cahokia and Vincennes, but the British retake Vincennes December 17.

"Molly Pitcher" gains her sobriquet at the Battle of Monmouth (above) by carrying water to tired and wounded Continental army troops. When her husband George McCauley is overcome by the heat, Mary Ludwig "Molly Pitcher" McCauley, 23, takes over his cannon and uses it to good effect through the rest of the battle.

British Loyalists (Tories) and their Indian allies massacre settlers in Pennsylvania's Wyoming Valley July 4 and in New York's Cherry Valley November 11.

British forces take Savannah December 29.

1778 ***(cont.)*** Bengal's Gov. Warren Hastings takes Chandernagore, Pondicherry, and Mahe in the continuing Anglo-Maratha War.

The War of the Bavarian Succession begins in July as Prussia's Frederick the Great invades Bohemia. The elector Palatine has been persuaded to recognize some old Austrian claims to Lower Bavaria and part of the Upper Palatine in the January Treaty of Vienna (*see* 1777). Austrian troops have occupied Lower Bavaria, but Frederick has persuaded the elector to renounce Austrian claims and has persuaded Saxony and Mecklenburg to join him in opposing the Emperor Josef II (*see* 1779).

Scottish coal miners gain freedom from the conditions of virtual slavery under which they have worked until now (*see* 1775), but they will not have complete freedom until 1799.

Parliament acts at the urging of English prison reformer John Howard, 52, and establishes the principle of separate confinement with labor and religious instruction (but *see* 1782).

The Virginia legislature forbids further importation of slaves at the persuasion of Thomas Jefferson (*see* manumission, 1782).

The Sandwich Islands discovered by Hawaii-Loa about 450 A.D. are rediscovered in January by Captain Cook. He lands on the island of Kauai and names the islands after John Montagu, 60, fourth earl of Sandwich and first lord of the admiralty.

Captain Cook explores the Pacific coast of North America northward from Oregon, laying the basis for future British claims to the region (*see* Drake, 1579; Gray, 1791; Columbia River Valley, 1843).

Karl Wilhelm Scheele isolates molybdenum (*see* 1776).

France charters a Société Royale de Medicine and charges it with the study of epidemics.

Mesmerism attracts the gullible elite of Paris. Viennese physician Franz (or Friedrich) Anton Mesmer believes he has occult powers to effect cures with animal magnetism. He founds a Magnetic Institute with support from Louis XVI and Marie Antoinette and employs hypnotism, astrology, magic wands, and similar methods, but while he will "mesmerize" such eminent patrons as Gen. Lafayette, the medical faculties will denounce him and hound him out of France.

Fiction *Evelina, or the History of a Young Lady's Entry into the World* by English novelist Fanny (Frances) Burney, 26, is published anonymously. Burney virtually invents the social novel of domestic life, and she introduces the phrase, "Before you could say Jack Robinson."

Painting *Brook Watson and the Shark* by John Singleton Copley, who has been commissioned by the lord mayor of London to depict the grisly scene in which the lord mayor lost his legs at age 14 while swimming in Havana Harbor. Italian etcher-architect Giovanni Battista Piranese dies at Rome November 9 at age 58.

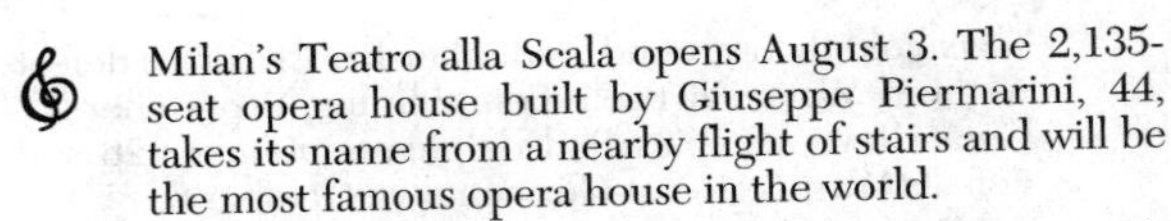

Milan's Teatro alla Scala opens August 3. The 2,135-seat opera house built by Giuseppe Piermarini, 44, takes its name from a nearby flight of stairs and will be the most famous opera house in the world.

Ballet *Les Petits Biens* 6/11 at the Paris Opéra, with music by Wolfgang Amadeus Mozart, choreography by Jean Noverre.

First performance Symphony No. 31 in D major *(Paris)* by Wolfgang Amadeus Mozart 6/18 at Paris.

A water closet patented by English engineer Joseph Bramah, 30, has a valve-and-siphon flushing system that will be the basis of all future toilet plumbing. Bramah's factory at Pimlico will train other engineers and machine tool inventors, but his flush toilet will not come into wide use for more than a century (*see* 1775).

John Montagu, earl of Sandwich (above), is said to have invented the "sandwich" (meat between two slices of bread) to eat at the gaming tables.

1779 British forces take Augusta, Georgia, January 29, they have mixed fortunes elsewhere in Georgia, Gen. William Moultrie repels them February 3 at Fort Royal, South Carolina, and they surrender Vincennes to George Rogers Clark February 25.

Spain asks Britain to cease hostilities against the American colonies and recognize their independence. The Spanish have loaned the colonists 219 bronze cannon, 200 gun carriages, 30,000 muskets, 55,000 rounds of ammunition, 12,000 bombs, 4,000 tents, and 30,000 uniforms and will give the colonists $5 million before the War of Independence is over.

Britain asks Spain to halt her aid, offering to give up Florida, Gibraltar, and codfishing rights off Newfoundland in exchange for withdrawal of the Spanish support, but Spain declares war on Britain June 21 and lays siege to Gibraltar.

The French West African port of Gorée falls to British forces, who attack the French in Senegal.

British troops take Stony Point and Verplanck, New York, May 31, but lose Stony Point July 16 to Anthony Wayne, who has attacked "Little Gibraltar" at midnight and is knocked unconscious by grapeshot. Gen. Washington orders the post evacuated, and the Americans raze its fortifications.

The largest American naval armada to be assembled during the Revolution leaves Boston July 19 under former privateer Dudley Saltonstall, 44, with orders to dislodge the British from a fort they are building at the head of Penobscot Bay. Saltonstall has 19 armed vessels with 344 guns and 24 transports with some 2,000 men, while the British have only 700 men under Brig. Gen. Francis McLean, but Saltonstall refuses to attack. His land-force commander Gen. Solomon Lovell and second-in-command Gen. Peleg Wadsworth of the Massachusetts militia put 200 men ashore but get little support from the expedition's artillery commander Paul Revere. British reinforcements arrive August 14 as seven ships from Halifax land 1,530 men and 204 guns. The Americans flee upriver toward Bangor, run their ships ashore and destroy them, and walk home to Boston through the woods. Saltonstall is cashiered for incompetence; Revere court-martialed for cowardly misconduct.

Gen. John Sullivan, Continental army, marches against the Tories and their Iroquois allies in New York's Genessee Valley. He destroys Indian villages and goes on to defeat his foes in the Battle of Newtown near Elmira.

Capt. John Paul Jones sails in August with a Continental navy squadron of five ships, rounds the British Isles, and on September 23 engages the Royal Navy vessel *Serapes* which along with the *Countess of Scarborough* is escorting Britain's Baltic trading fleet (*see* 1777). Heavily outgunned and outmanned, Jones moves in close, lashes his ship to the *Serapes,* and continues to grapple with the enemy even after the *Bonhomme Richard* begins to sink. Asked to surrender, Jones replies, "I have not yet begun to fight!" and it is the *Serapes* that surrenders after losing 100 killed, 60 wounded out of 325 in 3.5 hours of bloody battle. Jones, who has lost 150 killed and wounded out of 332, transfers the survivors to the *Serapes,* and the *Bonhomme Richard* sinks 2 days later.

The Battle of Baton Rouge September 21 ends British hopes of controlling the Mississippi Basin. The Spanish governor of Louisiana Bernardo de Galvez has moved upriver from New Orleans with a force of Spaniards, Frenchmen, Germans, Acadians, free blacks, Indians, and Americans. He tricks British forces under Lieut. Col. Dickson at Fort New Richmond into thinking he is preparing an attack from the east, he opens fire at dawn with a roundshot volley from the south, and the British surrender within a few hours.

Natchez falls to Gov. Galvez (above), and Spanish troops, landed last year in Florida, drive the British out of some Florida gulf ports (*see* 1781).

The War of the Bavarian Succession in Europe has ended May 13 with the Peace of Teschen. Austria retains only the part of Lower Bavaria bounded by the Inn, Salzach, and Danube.

A British force sent from Calcutta by Warren Hastings reaches Surat and breaks the coalition between the Marathas, the Mysore rajah Haidar Ali, and the nizam.

French forces seize St. Vincent and Grenada in the British West Indies.

British forces at Savannah come under siege by Americans under Gen. Benjamin Lincoln, 46, and the French naval commander Jean Baptiste Charles Henri Hector, 50, comte d'Estaing. They are joined October 9 by the Continental army's Polish military adviser Count Kazimierz Pulaski, who charges the British lines at the head of his cavalry. He sustains a mortal wound and dies October 11 at age 32, and the Americans abandon the siege October 20 after 34 days.

British forces evacuate Rhode Island in October.

Captain Cook is killed February 14 in a skirmish with Sandwich Island natives at Kealakekua Bay on the island of Hawaii (*see* 1778). He has introduced *taboo (kapu)* and other Polynesian words to the English language.

The capital of Virginia moves from Williamsburg to Richmond (*see* 1699; 1935).

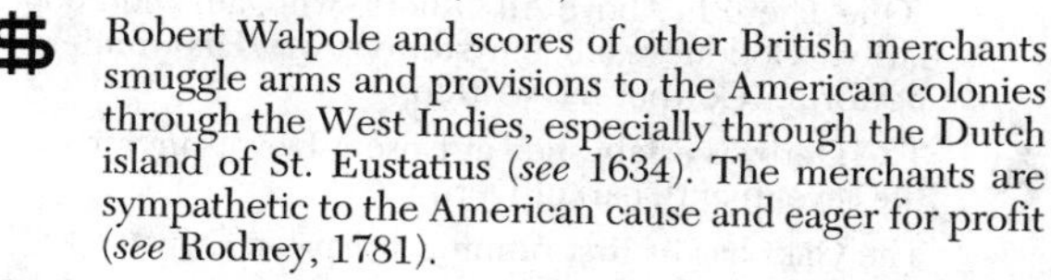

Robert Walpole and scores of other British merchants smuggle arms and provisions to the American colonies through the West Indies, especially through the Dutch island of St. Eustatius (*see* 1634). The merchants are sympathetic to the American cause and eager for profit (*see* Rodney, 1781).

English millhand Samuel Crompton, 26, devises a muslin wheel, or spinning mule, which spins yarns suitable for muslin, but he lacks the funds needed to obtain a patent for his improvement on the 1770 spinning jenny and is tricked into revealing his secret, a landmark in the Industrial Revolution (which will not be called that until 1881).

Experiments on Vegetables by Dutch plant pathologist Jan Ingenhousz, 49, establishes some of the principles of photosynthesis (which will get that name in 1898). Ingenhousz learned on a visit to England of Joseph Priestley's 1774 efforts to "improve" polluted air by introducing live plants, and he ascribes air purification to the action of sunlight on leaves to produce "dephlogisticated air" (to which French chemist Antoine Lavoisier will soon give the name "oxygen") (*see* 1784).

Poetry *The House of Night* by Philip Freneau, who has returned after 3 years as secretary to a rich planter in the West Indies to serve in the militia against the British.

Painting *George Washington at Princeton* by Philadelphia painter Charles Willson Peale, 38, who returned to the colonies in 1769 after studying at London with Benjamin West; *Blue Boy* by Thomas Gainsborough. Jean Baptiste Chardin dies at Paris December 6 at age 80.

Theater *Iphigenia in Tauris* by Johann Wolfgang von Goethe 4/6 at Weimar; *The Fatal Falsehood* by Hannah More 6/5 at London's Covent Garden Theatre; *The Critic, or a Tragedy Rehearsal* by Richard Brinsley Sheridan 10/3 at London's Royal Theatre in Drury Lane, a revision of the 1671 burlesque by the late duke of Buckingham: "Nothing is unnatural that is not physically impossible" (II, i.)

Opera *Iphigenia in Tauris* (*Iphigénie en Tauride*) 5/18 at the Académe de Musique, Paris, with music by Christoph Willibald Gluck, now 64.

First performances Concerto for Two Pianos and Orchestra in E flat major by Wolfgang Amadeus Mozart 1/15 or shortly thereafter at Salzburg, following Mozart's return after an 18-month absence in Munich, Augsburg, Mannheim, Paris, Strasbourg, and lesser towns; Symphony No. 32 in G major by W. A. Mozart; Symphony No. 33 in B flat major by W. A. Mozart, in July at Salzburg.

Olney Hymns is published with 63 verses by English poet William Cowper, 47, and 283 by the Rev. John Newton, 54, evangelical curate at Olney, Buckinghamshire, who was a slave-ship captain before he began

1779 ***(cont.)*** preaching at age 33. Hymn singing has been uncommon in England up to now but Cowper's "Light Shining Out of Darkness" and Newton's "Amazing Grace," "Glorious Things Of Thee Are Spoken" and "One There Is Above All Others" will gain wide popularity. "God moves in a mysterious way/His wonders to perform," Cowper has written.

The Derby is established at Epsom Downs in Surrey by the seventeenth earl of Derby.

The Oaks has its first running at Epsom Downs.

Richard Bagnal, an officer with Gen. John Sullivan (above), finds sweet corn along the Susquehanna River and makes the first written report of it. But although the Iroquois have been growing it for years, and Bagnal carries seeds back to Plymouth and plants them, sweet corn will be little used by Americans for another 70 years and remain largely unknown in the rest of the world for more than a century after that.

Lazzaro Spallanzani proves that semen (carrying sperm) is necessary to fertilization (*see* 1856; Spallanzani, 1768).

1780 Mobile falls to the Spanish governor Bernardo de Galvez March 14 as British forces under Sir Henry Clinton prepare to besiege Charleston. The South Carolina port city surrenders May 12 after a month, the British take some 2,500 prisoners including Gen. Benjamin Lincoln, and Clinton moves north to blockade the French fleet.

Some 6,000 French troops arrive at Newport, Rhode Island, July 10 under Jean Baptiste Donation de Vimeur, 55, comte de Rochambeau, but the British have blockaded the French fleet in Narragansett Bay; Rochambeau is loathe to abandon the fleet and will remain inactive for a year.

Gen. Greene defeats a British army June 23 at Springfield, New Jersey.

American forces under Gen. Horatio Gates suffer a disastrous defeat August 16 at Camden, South Carolina, as redcoats under Gen. Charles Cornwallis win the day.

British spy Major John André, 29, falls into American hands September 23 with papers revealing a plot by Benedict Arnold to surrender West Point to Sir Henry Clinton (above). Arnold escapes to the British ship *Vulture,* and André is hanged October 2.

The Battle of Kings Mountain October 7 marks a turning point in the war in the south. Some 900 mounted North Carolina backwoodsmen defeat a force of 900 Loyalist militiamen and kill Major Patrick Ferguson, 36, inventor of a breechloading rifle and the best marksman in the British army.

George Washington meets with defeat at Germantown in October, sustaining nearly 700 casualties; the British take some 400 prisoners, and the Americans retire to winter quarters at Valley Forge.

The Delaware River north to Philadelphia is safe to British shipping by late November.

In addition to their war with the Americans, the British go to war with the Dutch over the right to search ships at sea. The war soon ends with the Dutch losing some possessions in the East Indies and much of their power in that part of the world (*see* 1743).

A second Mysore War breaks out in India in September as Mysore's Muslim ruler Haidar Ali allies himself with the Marathas and retaliates against an attack by the British-supported nawab of Arcot. He attacks the British-held Carnatic coast while the Marathas lay siege to the East India Company headquarters city, Madras. British troops relieve the siege and the Marathas withdraw (*see* 1781).

Maria Theresa of Austria, Hungary, and Bohemia dies of smallpox November 28 at age 63, leaving her son Josef free to rule alone (he had visited Russia's Catherine the Great in April against his mother's wishes).

The Gordon "No Popery" Riots disrupt London from July 2 to 8 as Lord George Gordon, 29, heads a mob of 50,000 that marches from St. George's Fields to the houses of Parliament with petitions to repeal a 1778 act relieving Roman Catholics of certain prejudicial disadvantages. The mob destroys Catholic chapels, breaks open prisons, and attacks the Bank of England. A court acquits Gordon of treason, but he will be excommunicated in 1787 and will embrace Judaism before dying a debtor in Newgate prison in 1793.

The Holy Roman Emperor Josef II (above) abolishes serfdom in Bohemia and Hungary, completing work begun by his mother.

Nashville, Tennessee, has its beginnings in a fort that North Carolina pioneer James Robertson builds on the Cumberland River. The fort will take the name Nashville in 1784.

The American Academy of Arts and Sciences is founded at Boston.

London's first Sunday newspapers appear March 26 as the *British Gazette* and the *Sunday Monitor* begin publication.

Painting *Mary Robinson as Perdita* by Sir Joshua Reynolds; *Death of Chatham* by John Singleton Copley.

First performances Symphony No. 34 in C major by Wolfgang Amadeus Mozart, in July at Salzburg; Symphony No. 75 in D major by Joseph Haydn at Schloss Esterházy.

The first modern pianoforte is built at Paris by Sebasten Erard, 28 (*see* Cristofori, 1709).

Spanish dancer Sebastiano Carezo invents the bolero.

The Hudson River freezes over at New York, enabling British forces to cross on the ice with heavy gear that includes cannon. Blockhouses are erected on the ice to prevent Americans from crossing.

New England has a "dark day" May 19. Religious people believe the end of the world is at hand.

Britain enjoys something of an agrarian revolution as better seed, more scientific crop rotation, more efficient tools, and improved livestock increase productivity (*see* Bakewell, 1769; Coke, 1772). But the yield of European agriculture remains little more than five or six times the original seed, on average, and agricultural methods on the Continent are still largely medieval. A third or more of the land is left fallow (partly because of a manure shortage), the land is communally cultivated, and most livestock is still killed every fall (*see* 1730).

Lazzaro Spallanzani employs artificial insemination for the first time (with dogs) (*see* 1779).

Congress appeals to the states November 4 to contribute quotas of flour, hay, and pork to support the Continental armies.

James Watt designs a steam-operated flour mill (*see* 1769).

Delaware storekeeper Oliver Evans, 25, devises an automated flour mill. A cripple who has seen millers toil up long flights of stairs with heavy bags of grain to be dumped into chutes that flow into millstones, Evans invents a water-driven vertical conveyor belt that carries hoisting buckets, and he combines this with a horizontal conveyor that moves grain, meal, and flour from place to place on the ground (*see* 1787).

English sugar consumption reaches 12 pounds per year per capita, up from 4 in 1700, as Britons increase coffee and tea consumption (*see* 1872).

Of 21,000 children born in Paris, 17,000 are sent to the country to be wet-nursed, 3,000 placed in nursery homes, 700 wet-nursed at home, and only 700 wet-nursed by their own mothers, who risk the scorn of society.

Urbanization advances in western Europe. France has more than a dozen towns of 30,000 or more, but 85 percent of the population is in villages of 2,000 or less.

Italy and Spain each has nine cities of over 40,000, but no Bohemian town except Prague has more than 10,000. Warsaw has fewer than 30,000, and all the towns in Hungary put together have no more than 356,000, while Paris has nearly 700,000 and London well over that.

1781

The American Revolution ends October 19 with the surrender of Gen. Cornwallis, but the British will not evacuate Savannah until July of next year and Charleston until December and will hold New York until November 1783. The Americans win their War of Independence only after some bloody encounters (below).

Spanish forces under Don Eugenio Pourre in Michigan take Fort St. Joseph from the British in January.

British forces plunder and burn Richmond, Virginia, January 5 with help from Benedict Arnold.

The Battle of the Cowpens in North Carolina January 17 ends in victory for 800 Continental infantrymen, dragoons, and militia commanded by Brig. Gen. Daniel Morgan. They defeat 1,000 British light cavalry dragoons and infantrymen under the command of Sir Banastre Tarleton, killing 110, wounding 200, and capturing 550, while losing only 12 killed, 60 wounded.

New Jersey troops mutiny January 20, but Gen. Robert Howe soon arrives from West Point to quell the uprising. Pennsylvania troops at Morristown have broken camp with demands for back pay.

Gen. Greene reaches Virginia February 13 after a 2-week retreat from North Carolina in which he has eluded British pursuers.

The colonies adopt Articles of Confederation and Perpetual Union March 1 as delegates from Maryland add their signatures to the document after New York and Virginia have ceded their western land claims to the newly emerging republic (*see* 1777; Constitution, 1787).

Virginia's Thomas Jefferson has said in relinquishing his state's claims to western lands, "The lands . . . will remain to be occupied by Americans and whether these lands be counted in the members of this or that of the United States will be thought a matter of little moment."

The United States in Congress assembles March 2, but most Americans still call it the Continental Congress.

The Battle of Guilford Court House in North Carolina March 15 ends with the British proclaiming victory, but they are forced to leave North Carolina for Virginia. Gen. Greene turns back and retakes South Carolina and Georgia but is defeated April 25 at Hobkirk's Hill, South Carolina.

Spanish forces take West Florida from the British, who surrender Pensacola to Gov. Galvez May 9.

Admiral George Rodney sails into the harbor of the St. Eustatius capital Oranjestad, sacks the town, and seizes 250 ships of all nationalities, taking arms and provisions intended for the American rebels. Rodney's action brings denunciations in Parliament from Edmund Burke and Charles James Fox, 32, who sympathize with the American cause, and angry British merchants strip Rodney in prize courts of much of the £4 million in loot that he has seized (*see* 1762; 1779).

French and Spanish naval forces take Tobago, St. Eustatius, Demerara, St. Kitts, Nevis, and Monserrat (but *see* 1782).

The British at South Carolina's Fort Ninety-Six repulse Gen. Greene June 19.

Gen. Cornwallis repulses Gen. Lafayette July 6 at Jamestown Ford, Virginia, but is forced to retire August 1 to defensible positions at Yorktown and Gloucester Point on the York River.

A French fleet from the Caribbean arrives in Chesapeake Bay August 30 under Comte François Joseph Paul de Grasse-Tilly, 59. He badly cripples the British fleet September 5, and it is unable to stop 16,100 Allied troops under Generals Washington, Lafayette, and Rochambeau from laying siege to Yorktown beginning September 28.

Benedict Arnold helps British forces plunder and burn New London, Connecticut, September 6.

The Battle of Eutaw Springs in South Carolina Septem-

1781 ***(cont.)*** ber 8 ends with Gen. Greene retreating and the British retiring to Charleston.

A new French squadron arrives in Chesapeake Bay September 10 under Paul François Jean Nicolas de Barras, 26.

Gen. Cornwallis surrenders with 7,000 troops at Yorktown October 19. A band plays "The World Turned Upside Down."

British forces take Dutch settlements on the west coast of Sumatra.

British forces in India defeat Haidar Ali July 1 at Porto Novo in the Carnatic (*see* 1780; 1782).

Chinese imperial forces under Gao Cong suppress a Muslim revolt in Gansu Province.

Russia's Catherine the Great signs a treaty with the Holy Roman Emperor Josef II promising him the entire eastern half of the Balkans. She seeks to drive the Ottoman Turks out of Europe and establish her 2-year-old grandson Constantine as head of a new Greek empire.

An Edict of Tolerance issued by the Holy Roman Emperor Josef II October 13 ends an 8-year period in which 700 monasteries have been closed and 36,000 members of religious orders released from their vows. The edict prescribes a new organization for the remaining 1,324 monasteries with their 27,000 monks and nuns, weakens the link with Rome, introduces innovations in the form of worship, and establishes schools with church property.

The *gaon* of Vilna issues another *herem* against hasidic Jews (*see* 1772). "They must leave our communities with their wives and children," he says, and writes that "it is the duty of every believing Jew to repudiate and pursue them with all manner of affliction." He forbids his people to do business with hasidim, intermarry with them, or assist at their burials.

General Cornwallis surrendered at Yorktown, and the Americans (with French help) had won their War of Independence.

Lynch law (or Lynch's law) punishes lawlessness that accompanies the American Revolution in Virginia (above). Planter Charles Lynch, 45, presides as justice of the peace at an extralegal court, the summary punishment that follows his frequent convictions generally consists of flogging, but the term *lynching* will usually mean hanging by an unlawful mob that has taken the law into its own hands to seize a prisoner.

Los Angeles is founded in California by Spanish settlers who call the place El Pueblo de Nuestra Senora la Reina de los Angeles de Porciuncula (*see* 1769).

"Blue laws" get their name at New Haven, Connecticut, where a new town ordinance printed on blue paper prohibits work on Sunday and requires all shops to be closed on "the Lord's day."

Congress charters the Bank of North America December 31 at Philadelphia. The first incorporated bank of the new American republic, it has been founded by superintendent of finance Robert Morris, 48, who signed the Declaration of Independence in 1776 and has served the Revolution by arranging for the financing of supplies for George Washington's armies. Morris has used his own personal credit to save the government from bankruptcy.

The world's first iron bridge opens to traffic January 1 across the River Severn in Shropshire to link Benthall and Madeley Wood, a town that will be renamed Ironbridge. The 100-foot, 378-ton span designed by John Wilkinson of 1774 cannon-borer fame has been cast at Coalbrookdale by Abraham Darby III, grandson of the coke smelter and iron pioneer (*see* 1709). Built without nuts, bolts, or screws but only dovetailed joints, pegs, and keys, it has taken 3 months to erect.

Uranus, first planet to be discovered since the Babylonian era of prehistory, is identified March 31 by German-born English music teacher William Herschel, 43, who moved to England at age 19 and anglicized his name from Friedrich Wilhelm Herschel. The amateur astronomer employs a homemade telescope in the rear garden of his house at 19 New King Street, Bath.

Hasidism (above) is spreading beyond Poland and Lithuania despite orthodox efforts to stop it (which will soon end as rabbis of all stripes join to resist the Jewish enlightenment, or *haskalah*).

Nonfiction *Critique of Pure Reason (Kritik der Reinen Vernunft)* by German metaphysician and philosopher Immanuel Kant, 57, examines the limitations of human understanding and establishes the rationalism of pure experience with a scheme of transcendental philosophy; *Decline and Fall of the Roman Empire* (Volumes II and III) by Edward Gibbon, who presents Volume II on bended knee to the duke of Gloucester and is told, "Another damned thick book! Always scribble, scribble, scribble: Eh, Mr. Gibbon?" (*see* 1776).

Poetry *On the Memorable Victory of Paul Jones* by Philip Freneau.

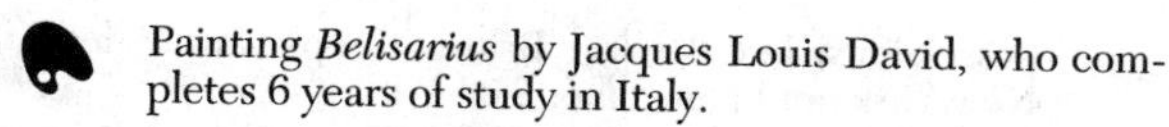

Painting *Belisarius* by Jacques Louis David, who completes 6 years of study in Italy.

Opera *Idomoneo King of Crete* (*Idomoneo, Rè di Creta*) 1/29 at Munich, with music by Wolfgang Amadeus Mozart.

First performances Concerto for Two Pianofortes and Orchestra in E flat major by Wolfgang Amadeus Mozart 11/24 at Vienna.

London's Asprey & Co. jewelers has its beginnings in a shop at Mitcham opened by metalworker William Asprey whose Huguenot ancestors arrived from France a century ago.

Some 3.5 million Americans live in the new nation, most of them on or near the Atlantic coast. Most of the country's 850,000 square miles of territory remains sparsely settled, and Europeans inhabit less than one third of the land claimed by the new republic.

1782 Admiral Rodney restores British control of the Caribbean April 12 by defeating a French fleet under de Grasse in the Battle of the Saints fought in the strait between Guadeloupe and Dominica.

Admiral Rodney (above) forestalls a union between French and Spanish forces for an assault on Jamaica, but a new British government with a mandate to make peace comes to power under the second marquis of Rockingham, Charles Watson-Wentworth, 52.

Rockingham (above) sends English M.P. Thomas Grenville, 27, to Paris to open peace talks with Benjamin Franklin (*see* Treaty of Paris, 1783).

Rockingham (above) dies July 1 at age 52. William Fitzmaurice Petty, 45, second earl of Shelburne, heads a new coalition ministry, serving as first lord of the treasury and first minister while his longtime Whig critic Charles James Fox, 33, sits in his Cabinet.

British forces evacuate Savannah in July and Charleston in December.

Spanish forces complete their conquest of Florida from the British.

Spanish forces in the viceroyalty of Peru suppress a 2-year revolt in the Audencia of Cuzco. Tupac Amaru, a descendant of Inca rulers, has made poorly armed attacks on Cuzco and has twice laid siege to La Paz, but the Spanish kill him and crush his native army (*see* San Martin, 1821).

Britain's *Royal George* sinks without warning August 29 while being repaired at Portsmouth. The flagship of Admiral Richard Kempenfeld goes down off Spithead with 800 men including the admiral.

India's Anglo-Maratha War is ended by the Treaty of Salbai after 7 years of conflict, but the 2-year-old Mysore War goes on. Mysore's Haidar Ali dies suddenly at age 60, but his son Tipu Sahib, 31, is crowned sultan and continues the war against the British, whose governor general Warren Hastings confiscates the treasure and part of the lands of the begum of Oudh, mother of the late Chait Singh, rajah of Benares.

The Chakri dynasty that will rule Siam for more than 200 years is founded April 6 by Chao P'ya Chakri, who will reign until 1809 as Rama I. He makes Bangkok his capital, will end Siam's long conflict with Burma in 1793, will reestablish royal control over local potentates, and will gain part of Cambodia by dividing that country with Annam.

Spanish forces besieging Gibraltar take Minorca from the British.

English prison reformer John Howard visits London's newly rebuilt Newgate Prison and finds inmates who have been locked up for 7 years awaiting trial. Jailers receive no formal salary, charging fees before permitting debtors, felons, and others to be released, and Howard urges that jailers be paid (*see* 1778; Newgate, below).

Austria abolishes serfdom throughout her dominions.

The Virginia legislature authorizes manumission of slaves as the "peculiar institution" begins to die out in some parts of the South (*see* 1778). Some 10,000 Virginia slaves will be freed in the next 8 years largely because they are too old, ill, or costly for their masters to maintain (but *see* 1803).

James Watt patents a double-acting rotary steam engine. He has improved on his engine of 1765 and employs the new engine to drive machinery of all kinds (*see* cotton mill, 1785).

German theologian-zoologist Johann Melchior Goeze finds a hairlike parasite in the intestine of a badger and calls it *der Haarrundwurm* (hair-round worm). It will soon be called the hookworm (*see* 1619; Froelich, 1789).

Nonfiction *Letters from an American Farmer* by French-American farmer-author J. Hector St. John de Crèvecoeur (Michel Guillaume Jean de Crèvecoeur), 47, who settled in New York's Orange County 13 years ago (below); *Notes on Virginia* by former Virginia governor Thomas Jefferson (below); *A History of the Corruption of Christianity* by English clergyman-chemist Joseph Priestley.

Fiction *Les Liaisons dangereuses* by French novelist Pierre Ambroise François Choder los de Laclos, 41; *Cecilia* by Fanny Burney.

Poetry *Poems* by William Cowper.

Painting *Fêtes for the Grand Duke Paul of Russia* by Francesco Guardi; *Colonel George K. C. Grenadier Guards* by Joshua Reynolds; *The Nightmare* by Swiss painter Henry Fuseli (Johann Heinrich Füssli), 41, who fuses Gothic elements of romanticism with classicism in his reclining figure with its supernatural presences and mood of terror. Richard Wilson dies at Lianberis, Caernarvonshire, May 12 at age 67.

Sculpture *Monument to Pope Clement XIV* by Antonio Canova.

Theater *The Robber (Die Räuber)* by German surgeon-playwright Johann Christoph Friedrich von Schiller, 22, 1/13 at Mannheim's Naturaltheater. Schiller had the play printed last year at his own expense, has left his

1782 ***(cont.)*** regiment at Stuttgart without leave to attend a performance at Mannheim, is arrested by the Duke of Württemberg, who appointed him physician to the regiment, is condemned to write nothing but medical treatises, and leaves Württemberg. Schiller will spend 9 years (the *Wanderjahre*) at Mannheim, Weimar, and Jena.

Opera *The Abduction From the Seraglio* (*Die Entführung aus dem Serail*) 7/16 at Vienna, with music by Wolfgang Amadeus Mozart; *Orlando Paladillo* in August at Vienna's Eszterházy Castle, with music by Joseph Haydn.

First performances Symphony No. 77 in B flat major by Joseph Haydn at Esterházy Castle; Serenade No. 10 in B flat major (*Gran Partita*) by Mozart 8/4 for Mozart's wedding to soprano Constanze Weber, 19, at St. Stephen's Cathedral, Vienna.

London's Newgate Prison is completed to replace an earlier prison burned down in the Gordon riots of 1780. Architect George Dance, 41, has made the new prison almost windowless (windows are not his forte) and the structure is his masterpiece.

Salem's Pierce Nichols House is completed in Federal Street for East India merchant Jerathmeel Pierce by local architect Samuel McIntire, 25.

Jethro Tull's seed-planting drill of 1701 is improved with gears for its distributing mechanism.

"The indifferent state of agriculture among us does not proceed from a want of knowledge merely," writes Thomas Jefferson in *Notes on Virginia* (above) which he has written at the request of the French government. "It is from our having such quantities of land to waste as we please. In Europe the object is to make the most of their land, labor being abundant; here it is to make the most of our labor, land being abundant."

Ice cream is served at a Philadelphia party given by the French envoy to honor the new American republic (*see* 1813).

"Here individuals of all nations are melted into a new race of men," writes St. John de Crèvecoeur (above; *see* Zangwill play, 1909).

1783 Congress proclaims victory in the American War of Independence April 19, but a mutiny of unpaid soldiers at Philadelphia has forced Congress to meet at Princeton, N.J. Some 4,435 Americans have died in battle, 6,188 have been wounded, and thousands more have died of disease and exposure. The British have lost at least as many.

Parliament has voted February 24 to abandon further prosecution of the war, Lord North has resigned March 27, and the Whig leader Charles James Fox has taken over (below).

The Treaty of Paris September 3 recognizes the independence of the 13 colonies, Britain cedes vast territories to the Americans, grants them full rights to the Newfoundland fishery, cedes Florida to Spain, and regains her West Indian possessions. France regains St. Lucia, Tobago, Senegal, Gorée, and her East Indian possessions, but the last British troops do not leave New York until November 25.

Congress meets November 26 at Annapolis, Maryland, the first U.S. peacetime capital. George Washington resigns as commander in chief of the Continental army December 23.

Sweden recognizes the new American republic; Denmark, Spain, and Russia soon follow suit, but Prussia's Frederick the Great says the United States is too large and will soon fall apart.

The Spanish siege of Gibraltar has ended February 6 after 3.5 years of hostilities in which the British defenders have used red-hot lead to repel the enemy's battering ships.

Russia's Catherine the Great annexes the Crimea that her favorite Grigori Potemkin has conquered from the Ottoman Turks, expels the Turks, and offers a large grant of land to the Mennonites, promising them religious freedom and exemption from military service (*see* 1568; 1771; 1784).

Britain's House of Lords rejects a bill to reform the colonial administration of India and criticizes the American peace settlement. Prime Minister Fox resigns December 13, and a new government takes office headed by William Pitt (the Younger), 24, who will hold power until 1801.

Some 16,000 American prisoners of war have died in British prison ships anchored in New York's East River.

English Quakers form an association "for the relief of and liberation of the negro slaves in the West Indies, and for the discouragement of the slave trade on the coast of Africa."

Maryland forbids further importation of slaves.

A postwar economic depression begins in the United States, which has foreign debts of $6.4 million to France, $1.3 million to Holland, and $174,000 to Spain.

British Orders in Council limit imports from America to naval stores and few other commodities.

Revolutionary army soldiers and officers are issued scrip certificates entitling them to tracts of land west of the Appalachians, the acreages varying according to rank and length of service. Many sell the certificates to those who want land (and to land speculators).

The phrase "not worth a Continental" is heard as inflation reduces the value of paper currency issued by the Continental Congress to finance the Revolution. Corn sells for $80 per bushel, shoes for $100 per pair, and George Washington, richest man in America, says he needs a wagonload of money to buy a wagonload of supplies.

The $5 million William Penn estate in Pennsylvania is broken up, as are other large estates held by former Loyalists (Tories) including New York's 192,000-acre Philipse manor. New York's De Lancey manor is sold to 275 individuals, a 40,000-acre North Carolina estate is divided into farms averaging 200 acres each, but most of the land goes to speculators and men who are already substantial property owners.

A crude electric cell constructed by Italian anatomist Luigi Galvani, 49, employs two different metals and the natural fluids from a dissected frog. Galvani decides that the electric current produced by the cell must derive from the frog fluids (*see* Volta, 1800).

The Montgolfier brothers give the first public demonstration of an ascension balloon June 5 at Annonay, France. Joseph Michel Montgolfier, 43, and his brother Jacques Etienne, 38, have inflated their balloon with hot air and it makes a 10-minute ascent (*see* Jeffries, 1785; Charles, 1787).

The 600,000 tons of shipping employed in Anglo-American trade are almost entirely British, but Salem merchant Elias Hasket Derby, who has learned the value of fast ships in the Revolution, has improved the art of shipbuilding and outsails the Royal Navy in his 309-ton *Grand Turk*. He will soon send her to the East Indies and China (*see* 1784).

English ironmaster Henry Cort, 43, invents a process for puddling iron that revolutionizes wrought-iron production. He will patent the process next year and will also patent the reverberatory furnace that makes his purifying process possible.

Tungsten (wolfram) is produced in metallic form by Spanish chemists Juan and Fernando d'Elhuyer, who isolate an acid from wolframite and heat it with carbon.

English physician Thomas Cawley makes the first recorded diagnosis of diabetes mellitus. He shows the presence of sugar in a patient's urine (*see* 1788; Willis, 1674).

Webster's Spelling Book by Yale graduate Noah Webster, 25, is published under the title *Grammatical Institute of the English Language.* Webster standardizes American orthography and helps make pronunciation more uniform. His *Blue-Backed Speller* will have sales of more than 65 million copies in the next 34 years and will continue to sell through much of the next century (*see* dictionary, 1806; 1828).

The Montgolfier brothers freed man from earth's gravity with their pioneer ascension balloons.

Poetry *The Village* by English poet-curate George Crabbe, 28, employs heroic couplets in a reply to Oliver Goldsmith's sentimental *The Deserted Village* of 1770.

Painting *The Duchess of Devonshire* by Thomas Gainsborough; *Lady Hamilton as a Bacchanite* by George Romney; *The Grief of Andromache* by Jacques Louis David, who gains admission to the Academy.

Theater *Nathan the Wise (Nathan der Weise)* by Gotthold Ephraim Lessing 4/14 at Berlin; *Fiesco, or The Conspiracy of Genoa (Die Verschwöriing der Fiesko zu Getza)* by Friedrich Schiller 7/20 at Bonn.

First performances Symphony No. 35 in D major *(Haffner)* by Wolfgang Amadeus Mozart 3/3 at Vienna's Burgtheater. The Holy Roman Emperor Josef II gives Mozart 25 ducats to show his appreciation; Concerto-Rondo for Pianoforte and Orchestra in D major by W. A. Mozart 3/11 at Vienna; Symphony No. 36 in C major (Linz) by W. A. Mozart 11/4 at the palace of Count Thun at Linz.

Iceland's Skaptar volcano erupts, killing 20 percent of the country's population.

Japan suffers her worst famine since 1732 following the eruption of Mount Asama (*see* 1833; riots, 1787).

1784 The Ottoman Turks accept Russian annexation of the Crimea in the Treaty of Constantinople signed January 6 (*see* 1783; Crimean War, 1854).

The U.S. Congress meeting at Annapolis ratifies the Treaty of Paris January 14, bringing the War of Independence to a formal end. Congress remains in session until August 13.

The Holy Roman Emperor Josef II abrogates the Hungarian Constitution July 4 following a revolution in Transylvania. He removes the crown of Hungary to Vienna and suppresses Hungarian feudal courts.

The India Act that becomes law in Britain August 13 establishes a new constitution for the 184-year-old East India Company. Put through by the prime minister William Pitt, it forbids company interference in native affairs, forbids any declaration of war except in the event of aggression, and makes company directors answerable to a board appointed by the Crown (*see* 1783; Sepoy Rebellion, 1857).

The British transport all Acadians who have remained in Nova Scotia and lower New Brunswick since the expulsion order of 1755 to Maine and Louisiana (*see* Cajun cookery, below; Longfellow, 1847).

Serfdom is abolished in Denmark by Andreas Bernstorm, who was dismissed as prime minister in 1780 but has been recalled to office.

Parliament further lowers British import duties on tea (*see* 1773). The lower duties end the smuggling that has accounted for so much of the nation's tea imports and

1784 *(cont.)* hurt the East India Company as the rewards become too small to justify the risks.

Britain receives her first bale of American cotton as long-staple cotton comes into widescale production in the sea islands off the Carolina and Georgia coasts (*see* Whitney, 1792).

China receives her first American ship at Canton. Financed by William Duer of New York, Robert Morris of Philadelphia, and Daniel Parker of Watertown, Mass., the 360-ton *Empress of China* out of New York carries a cargo of ginseng root, prized by Chinese physicians as a therapeutic drug and as an aphrodisiac. Major Samuel Shaw, aide-de-camp in the Revolution to Gen. Henry Knox, barters the cargo for $30,000 worth of tea and silk, the investors receive a 25 percent return on their capital, Shaw becomes first U.S. consul at Canton, and more Americans are encouraged to enter the China trade, including Salem's Elias Derby (*see* 1783; Gray, 1790).

Karl Wilhelm Scheele discovers citric acid in certain plants.

French chemist Antoine Laurent Lavoisier, 41, pioneers quantitative chemistry, introducing modern names for chemical substances while retaining the old symbols of alchemy. He relates respiration to combustion, and he demonstrates the indestructability of matter (*see* 1789; Ingenhousz, 1779).

Experiments on Air by English chemist-physicist Henry Cavendish, now 53, shows that water results from the union of hydrogen and oxygen. Cavendish has been the first to discover the true nature of hydrogen and determine its specific gravity. He has determined the specific gravity of carbon dioxide and now shows that water is produced from an explosion of two volumes of hydrogen ("inflammable air") with one of oxygen ("dephlogisticated air").

Antoine Lavoisier (above) measures the amount of "dephlogisticated air" (oxygen) consumed in respiration and combustion and the amount of carbon dioxide and heat produced; assisting is French mathematician-astronomer Pierre Simon, 35, marquis de Laplace.

Essai d'une théorie sur la structure des cristaux by French mineralogist-clergyman René Just Haüy, 41, enunciates the geometrical law of crystallization and founds the science of crystallography. Haüy has broken a crystal of calcareous spar, and the accident has led him to make experiments.

U.S. Minister to France Benjamin Franklin invents bifocal spectacles. He will describe them for the first time in a letter from Passy May 23 of next year.

Johann Wolfgang von Goethe discovers the human intermaxillary bone.

John Wesley, now 81, charters Wesleyan Methodism February 28, signing a declaration that provides for the regulation of Methodist chapels and preachers. He ordains his clerical helper September 1 and instructs the new superintendent (bishop) to ordain two colleagues despite the opposition by his brother Charles to ordination of presbyters (*see* 1744).

Shaker leader Mother Ann Lee dies at Watervliet, N.Y., September 8 at age 48, disappointing zealous followers who had believed her to be immortal. Shakerism will continue in America, setting an example with inventive, orderly methods of building, toolmaking, and furniture making, animal husbandry, cooking, and production of woodenware, yarns, textiles, and botanicals herbs (*see* 1774; round barn, 1824).

The first school for the blind opens at Paris. Valentin Haüy, 39, brother of the mineralogist (above) has invented a method for embossing characters on paper and will later open a school for the blind at St. Petersburg (*see* Braille, 1834).

New York's Columbia College opens after 6 years of wartime suspension of King's College, founded in 1754. The war has obliged King's College students such as Alexander Hamilton to educate themselves.

The Encyclopaedia Britannica first published from 1768 to 1771 appears in a new 10-volume edition (*see* 1797).

Nonfiction *Notion of a Universal History in a Cosmopolitan Sense* by Immanuel Kant; "What is Enlightenment" by Kant; *Ideas Toward a Philosophy of History (Ideen zur Philosophie der Geschichte der Menschheit)* by Johann Gottfried von Herder is published in the first of four volumes that will appear in the next 7 years.

Painting *The Oath of the Horatii* by Jacques Louis David; *Mrs. Siddons as The Tragic Muse* by Joshua Reynolds; *Don Manuel de Zuniga* by Spanish painter Francisco José de Goya y Lucientes, 38.

English portrait painter-cartoonist Thomas Rowlandson, 28, pioneers political caricature.

Theater *Cabal and Love* (*Kabale und Liebe*) by Friedrich von Schiller 4/13 at Frankfurt-am-Main. Schiller attacks the corruption of the petty German courts and their disregard of basic human values; *The Marriage of Figaro, or The Madness of a Day (La folle journée, ou Le mariage de Figaro)* by Pierre Augustin Caron de Beaumarchais 4/27 at the Comédie-Française, Paris. Beaumarchais completed the work in 1778 but has had difficulty in getting it staged because of its antiaristocratic implications.

First performances Concerto for Pianoforte and Orchestra in E flat major by Wolfgang Amadeus Mozart 2/9 at Vienna; Concerto for Pianoforte and Orchestra in B flat major by W. A. Mozart 3/17 at Salzburg; Concerto for Pianoforte and Orchestra in D major by W. A. Mozart in March at Salzburg; Concerto for Pianoforte and Orchestra in G major by W. A. Mozart 6/10 at a country house outside Vienna; Sonata for Pianoforte in A major by W. A. Mozart.

Benjamin Franklin exhorts the French to set their clocks ahead 1 hour in spring and back 1 hour in fall to take advantage of daylight. He is widely credited with having originated the lines "Early to bed and early to rise," first published in 1639, and is well known for his aphorism "Time is money" (*see* 1916; farmers, below).

The first Chinese "tree of heaven" *(Ailanthus altissima)* to grow in America is planted at Philadelphia. A Flushing nurseryman will bring the fast-growing, pollution-resistant tree to Long Island in 1820, and by 1850 it will be common in the streets of New York, admired by some, scorned by others because of the unpleasant odor given off by its male flowers in the spring.

English farmers show little interest in an iron plow developed by inventor James Small, continuing to use wooden plows (*see* Wood, 1819).

French farmers resist Benjamin Franklin's daylight-saving proposal (above), insisting that cows cannot change their habits.

U.S. Shakers (above) will develop the Poland China hog by crossing backwoods hogs with white Big China hogs to produce the breed that will be the backbone of the U.S. pork industry for generations.

The Shakers (above) will innovate the practice of retailing garden seeds in small, labeled paper packets (*see* Burpee, 1876).

Famine continues in Japan; 300,000 die of starvation, and survivors in many cases eat corpses to stay alive.

The Acadians (above) will establish Cajun cookery in Louisiana, combining their Canadian recipes with those of the native "injuns."

Abortion and infanticide become common among the poor of Japan as famine (above) discourages large families.

1785

The affair of the diamond necklace creates a sensation in France beginning August 15 when Cardinal de Rohan, 51, is arrested as he prepares to officiate at Assumption Day services for Louis XVI and Marie Antoinette. His mistress the comtesse de La Motte has led him to believe that the queen is enamored of him and has authorized him to buy a necklace of diamonds collected by the Paris jewelry firm Boehmer and Bassenge originally for Mme. DuBarry. The comtesse has presented the jewelers with notes signed by the cardinal, they have complained to the queen, she has told Boehmer that she never ordered the necklace and certainly never received it, the comtesse is also arrested, and much of France chooses not to believe the queen. The case will be tried before the Parlement next year.

The Treaty of Fontainebleau November 8 resolves Austrian-Dutch disputes. Austria receives territory in Brabant and Limburg, complete control of the Scheldt above Sanftigen, plus 10 million florins, but gives up claims to Maastricht and leaves the mouth of the Scheldt in Dutch hands.

India's governor-general Warren Hastings resigns amidst charges of imperious behavior and returns to England.

A French order in council July 17 sets strict limits on imports from Britain (*see* 1786).

Congress establishes the dollar as the official currency of the new United States, employing a decimal system devised by Thomas Jefferson, who is named minister to France to succeed Benjamin Franklin.

U.S. economic troubles worsen as English goods undercut American manufactures. States erect tariffs to keep out goods from abroad and from other states, farmers are unable to sell their tobacco and surplus food crops because foreign markets have disappeared, and New Englanders can no longer find markets for the products of their shipyards.

A U.S. land ordinance provides for the sale of public lands at auction in tracts of 640 acres at a cash price of at least $1 per acre, but few would-be western settlers can put up $640 in cash (*see* 1800).

Steam powers textile machinery for the first time. An English cotton factory at Papplewick, Nottinghamshire, installs a Boulton and Watt rotative engine (*see* 1782).

Boston-born English physician John Jeffries, 41, makes the first aerial crossing of the English Channel January 7. A Loyalist during the American Revolution who has moved to England, Jeffries ascends in a hot-air balloon from Dover with the French aeronaut François Blanchard and crosses to a forest at Guines (*see* Montgolfier brothers, 1783; Blériot, 1909).

An Account of the Foxglove by English physician William Withering, 44, introduces medical use of digitalis, obtained from dried leaves of the foxglove plant *Digitalis purpurea.*

Benjamin Franklin returns to America after 9 years in France and introduces to his countrymen the gout remedy colchicine obtained from seeds of the same autumn crocus whose stamens yield saffron.

The University of Georgia is founded at Athens.

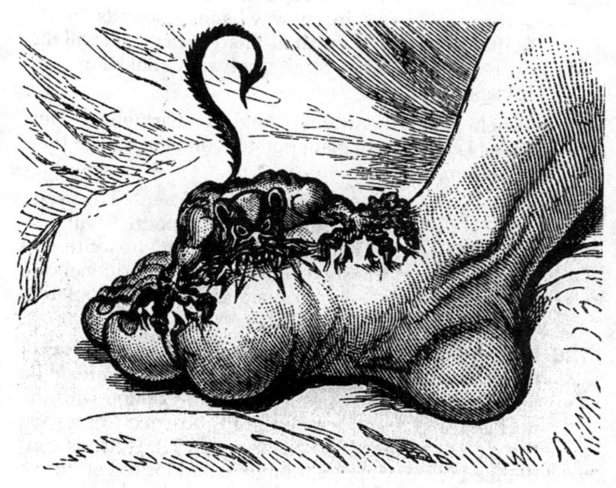

Benjamin Franklin found relief from his gout in France and returned with colchicine to ease the pain of fellow sufferers.

1785 ***(cont.)*** The University of New Brunswick is founded at Fredericton.

Nonfiction *The Journal of the Tour to the Hebrides* by Scottish lawyer James Boswell, now 46, describes a 1773 journey to his native Scotland and on to the isles of Skye and Ramasay with the late Doctor Johnson, who died last year at age 75.

Poetry *The Task, a Poem in Six Books* by William Cowper who includes "Tirocinium, or a Review of Schools" to commemorate the tyranny he suffered as a schoolboy. Cowper also includes his ballad of 1782 "The History of John Gilpin": "Variety's the spice of life,/That gives it all its flavor" (II, 606–607).

Painting *The Infant Hercules* by Sir Joshua Reynolds. Pietro Longhi dies at Venice May 8 at age 82.

Opera *The Marriage of Figaro* (*Le Nozze di Figaro*) 5/1 at Vienna's Burgtheater, with music by Wolfgang Amadeus Mozart, libretto by Lorenzo da Ponte from last year's Beaumarchais comedy.

First performances Concerto for Pianoforte and Orchestra in D minor by Wolfgang Amadeus Mozart 2/11 at Vienna's Cardisches Casino in the Mehlgrube; Concerto for Pianoforte and Orchestra in B flat major by W. A. Mozart 2/12 at Vienna; Concerto No. 21 for Pianoforte and Orchestra in C major by W. A. Mozart 3/12 at Vienna's Burgtheater; Concerto for Pianoforte and Orchestra in E flat major by W. A. Mozart 12/23 at Vienna.

Venice's Ca' Rezzonico palazzo is completed to celebrate Carlo Rezzonico's election to the papacy as Clement XIII. Begun in 1667 by Baldassare Longhena for the Priuli-Bon family, the palace had reached the first floor level by the time of Longhena's death in 1682.

Bangkok's Emerald Buddha Chapel is completed.

French explorer Jean François de Galaup, 44, comte de La Pérouse, looks for new whaling grounds in the Pacific. He embarks on an ill-fated voyage and will discover La Pérouse Strait between Sakhalin and Hokkaido.

French botanist Antoine Auguste Parmentier persuades Louis XVI to plant potatoes and encourage their cultivation.

A Dissertation on the Poor Laws by Joseph Townsend suggests that hunger, "the stronger appetite," will provide a natural restraint on "the weaker" appetite and will thus "blunt the shafts of Cupid, or . . . quench the torch of Hymen . . ." to maintain a food-population equilibrium (*see* 1791; 1798).

1786 France's diamond necklace affair ends in acquittal of Cardinal de Rohan May 31 (*see* 1785). The comte de La Motte is believed to have escaped to London with the necklace and is condemned in his absence to serve in the galleys for life. The comtesse de La Motte is condemned to be whipped, branded, and locked up in the Salpetrière. Cardinal de Rohan is exiled to the abbey of la Chaise-Dieu; Marie Antoinette is disappointed by his acquittal. It is widely believed that he was trapped by the queen, and when Mme. de La Motte escapes from the Salpetrière and takes refuge abroad, the court will be suspected of having connived in her escape.

Prussia's Frederick the Great dies at Sans-Souci in Potsdam August 17 at age 72 after a 46-year reign. He is succeeded by his inept nephew, 41, who will reign until 1797 as Friedrich Wilhelm II.

Japan's feebleminded Tokugawa shogun Ieharu dies at age 49 after a 26-year reign and is succeeded by his kinsman Ienari, 13, who will take power in 1793 after a 6-year regency and rule until 1837.

Russia's Catharine the Great issues a ukase establishing a Pale of Settlement within which Jews may live (*see* 1766). The Pale will grow to contain 25 provinces in southwestern Russia and Poland.

Syracuse, N.Y., has its beginnings in a trading post established by Yankee trader Ephraim Webster (*see* 1654).

Penang on the Malayan Peninsula is founded by the British, who will make it a port for pepper exports.

Russian sea captain Gerasim Pribylov discovers islands in the Bering Sea that will be called the Pribilofs (*see* 1741).

French manufacturers press for a measure of free trade that will give them a foreign market comparable to that of their envied British rivals. A commercial treaty is signed with London, English tariffs are lowered on French wheat, wine, and luxury goods, French tariffs are lowered on English textiles, but British imports flood the French market, undercut domestic prices, idle the looms at Troyes, and bring widespread unemployment, producing demands for renewal of tariff protection (*see* 1788).

Shays' Rebellion in Massachusetts aims to thwart further farm foreclosures in the continuing U.S. economic depression. State militia prevent seizure of the Springfield arsenal September 26 in a confrontation with 800 armed farmers led by Revolutionary War veteran Daniel Shays, 39, but the rebels succeed in having the state supreme court adjourn without returning indictments against them. Scattered fighting will continue through the winter (*see* 1787).

News of Shays' Rebellion (above) reaches a convention assembled at Annapolis to remedy the weaknesses of the 1781 Articles of Confederation (*see* 1781). Delegates will create a Congress with exclusive powers to coin money, forbidding states to levy tariffs or embargoes against each other that would restrict internal free trade (*see* Constitution, 1787).

Rhode Island farmers burn their grain, dump their milk, and leave their apples to rot in the orchards in a farm strike directed against Providence and Newport merchants who have refused to accept the paper money that has depreciated to the point of being virtually worthless. The strike has little effect, since 90 percent of Americans raise their own food, growing peas, beans, and corn in their gardens and letting their hogs forage in the woods for acorns.

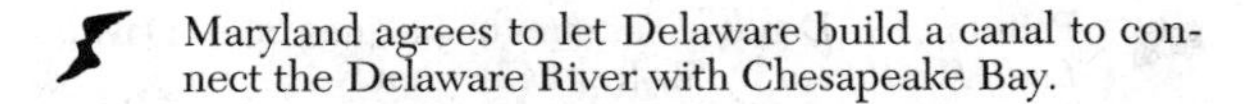

Maryland agrees to let Delaware build a canal to connect the Delaware River with Chesapeake Bay.

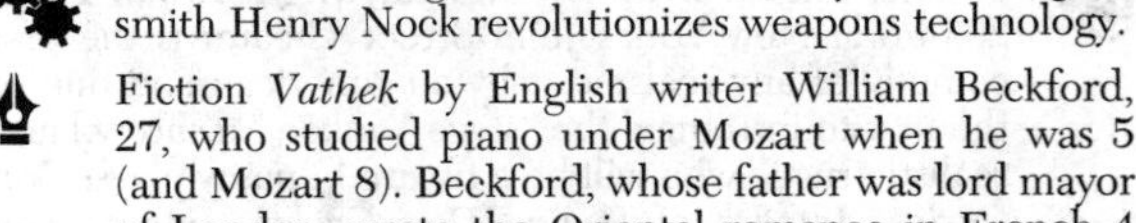

A breech-loading musket invented by London gunsmith Henry Nock revolutionizes weapons technology.

Fiction *Vathek* by English writer William Beckford, 27, who studied piano under Mozart when he was 5 (and Mozart 8). Beckford, whose father was lord mayor of London, wrote the Oriental romance in French 4 years ago.

Poetry *Poems, chiefly in the Scottish dialect* by Kilmarnock poet Robert Burns, 27, whose verse *To a Mouse* contains the line, "The best laid schemes o' mice an' men gang aft a-gley"; *To a Louse* contains the line, "Oh wad some Power the giftie gie us/ To see oursels as ithers see us!" and Burns includes also *The Cotter's Saturday Night, Halloween, Scotch Drink, The Vision,* and *Epistle to Davie.* His first edition brings in only £20 but its critical success dissuades the poet from emigrating to Jamaica, and an enlarged edition will be printed next year at Edinburgh; *The Wild Honey Suckle* by Philip Freneau.

Painting *The Duchess of Devonshire and Her Daughter* by Joshua Reynolds; *Colonel Mordaunt's Cock-Match* by Johann Zoffany, who has lived in India since 1783 and will remain until 1789.

First performances Concerto No. 24 for Pianoforte and Orchestra in C minor by Wolfgang Amadeus Mozart 4/7 at Vienna's Hoftheater; Symphony No. 38 in D major *(Prague)* by W. A. Mozart 5/1 at Vienna's Burgtheater; Concerto for Pianoforte and Orchestra in C major by W. A. Mozart in December at Vienna.

Europe's highest peak, Mont Blanc in the French Alps near Chamonix, is scaled for the first time by two French mountaineers, Jacques Balmart and Michel-Gabriel Paccard.

The first U.S. golf club is founded at Charleston's Green near Charleston, S.C., by local clergyman Henry Purcell.

Americans mix woolen yarn with linen fibers to make rough "linsey-woolsey" cloth stained with sumac or butternut dyes. They obtain the cash they need for salt, sewing needles, and land taxes by selling pot ashes (potash) produced by burning wool and leaching the ashes.

Bridgewater, Mass., inventor Ezekiel Reed patents a nail-making machine, but nails remain so costly that houses are put together in large part with wooden pegs (*see* Perkins, 1790).

London's Somerset House is completed by architect Sir William Chambers, 60, after 10 years of work.

The Pribilof Islands (above) have an estimated 5 million fur seals. Gerasim Pribylov founds a colony to harvest pelts (*see* 1867).

Paris sends the Abbé Dicquemare to report on the state of oyster beds in the gulf at the mouth of the Seine. The naturalist reports that the oysters have diminished by half "in the last forty years. . . The real causes of the deficit are the maneuvers of cupidity and the insufficiency of laws" (*see* 1681).

Scottish millwright and agricultural engineer Andrew Meikle, 67, develops the first successful threshing machine. It rubs the grain between a rotating drum and a concave metal sheet, employing a basic principle that will be used in future threshing machines (*see* 1831).

1787 A Constitutional Convention that has been meeting at Philadelphia draws up a Constitution for the new United States of America. George Washington has chaired the convention, Virginia's James Madison, 36, settles differences between proponents of a strong central government and proponents of absolute rights for the separate states, Washington resists efforts to make him king, and the Constitution establishes a balance of powers between the federal government (borrowing the idea of federalism pioneered by the Iroquois Confederation in 1570) and state governments. It establishes a bicameral legislature rather than a parliament and provides for legislators and a chief executive to be elected for limited terms rather than for life. No provision is made for political parties or a standing army.

"Federalist Papers" explaining the new Constitution appear under the name Publius in the *New York Independent Journal* beginning October 27 and will run for 7 months. James Madison (above), New York lawyer Alexander Hamilton, 32, and U.S. foreign affairs secretary John Jay, 41, have written the papers.

Delaware ratifies the Constitution by unanimous vote December 7 to become the first state of the Union.

Pennsylvania votes 46 to 23 to ratify the Constitution December 12.

New Jersey ratifies the Constitution by unanimous vote December 18.

A Northwest Ordinance adopted by Congress sitting at New York provides for a government of the Northwest Territory. The region is to gain ultimate statehood, and no slavery is to be permitted.

South Carolina cedes its western lands to the federal government, and the Ohio Company signs a contract with the Treasury Board of Congress to buy lands on the Ohio River (*see* Losantiville, 1788).

Thomas Paine travels to Europe where he will remain for the next 15 years agitating for revolution.

Russia's Catherine the Great begins a second war with the Ottoman Empire, using Turkish intrigues with the Crimean Tatars as an excuse to pursue her objective of obtaining Georgia.

France's working class is ill-paid, the typical working day is 14 to 16 hours, workers are lucky to get 250 days of work per year. The basic food of the people is bread which takes 60 percent of the average man's wages.

Rice riots at Osaka climax years of Japanese peasant unrest. Rice warehouses are broken open May 11, the riots spread throughout the city May 12, by May 18 they have spread to Edo and 30 other cities, and by May 20 nearly every large rice merchant has seen his house

1787 ***(cont.)*** destroyed as the rioters break into 8,000 establishments that include not only rice warehouses but also pawnshops, sake breweries, and shops dealing in textiles, dyes, drugs, and oils.

Britain begins clearing her prisons which are overcrowded with inmates, who before the American Revolution would have been transported to the American colonies. The prisoners are transported to Britain's new Australian penal colony at Botany Bay (*see* 1788; Cook, 1770).

English philanthropist William Wilberforce, 28, begins agitating against slavery in Britain's colonies (*see* 1789).

"Essay on the Slavery and Commerce of the Human Species" by Cambridge University student Thomas Clarkson, 27, is an expanded version of an essay originally in Latin that won Clarkson a Latin prize.

The Free African Society is founded at Philadelphia by freedman Richard Allen, 27, and other blacks who were pulled off their knees in November at a "white" Methodist church. With Absalom Jones and others, Allen establishes the African Methodist Episcopal Church while working to improve the economic and social conditions of American blacks through the Free African Society.

The Constitutional Convention (above) adopts a "three-fifths rule" as a compromise to settle differences between Northern and Southern states over the counting of slaves for purposes of representation and taxation. Slaves are to be counted as three-fifths of a free man for both purposes.

Massachusetts rebel Daniel Shays is defeated in early February and escapes to Vermont, most of the rebels are later pardoned, and Shays will be pardoned next year (*see* 1786), but hard-money interests in maritime Boston win out over inflationary-minded farmers.

English cotton goods production is 10 times what it was in 1770, and iron production has quadrupled, but cottage industry still prevails with few large power-operated mills. The Industrial Revolution is still in its early stages.

A steamboat launched on the Delaware River August 22 by U.S. inventor John Fitch, 44, is the first of several that Fitch will build in the next few years (*see* 1791; Symington, 1788).

A steamboat demonstrated on the Potomac River in December by Maryland inventor James Rumsey, 44, is driven by streams of water forced through the stern by a steam pump.

French physicist Jacques Charles, 41, launches the first hydrogen ascension balloon (*see* Montgolfier, 1783).

Charles's law, enunciated in its first steps by physicist Charles (above), states that for every degree Celsius rise in temperature, a gas expands proportionately to its volume at 0° C., an idea that he will develop into the law that will bear his name.

The University of Pittsburgh has its beginnings in a college founded at Fort Pitt (*see* 1759; 1816).

Painting *The Death of Socrates* by Jacques Louis David; *Lady Heathfield* by Sir Joshua Reynolds.

Theater *The Contrast* by U.S. playwright Royall Tyler, 29, 4/6 at New York's John Street Theater is the first comedy of any real merit by an American and one of the first to introduce the "stage Yankee," a shrewd and realistic rustic who will be a fixture in many plays; *Don Carlos, Infante of Spain (Don Carlos, Infante von Spanien)* by Friedrich von Schiller 8/29 at Hamburg's National Theater.

Opera *Don Giovanni* 10/29 at Prague's National Theater, with music by Wolfgang Amadeus Mozart, libretto by Lorenzo da Ponte, based on a literary classic about the legendary Spanish lover Don Juan dating to the Middle Ages.

First performances *Eine Kleine Nachtmusik* by Wolfgang Amadeus Mozart; Sextet for Strings and Horns in F major (A Musical Joke, *Ein Musikalischer Spass*) by W. A. Mozart. Christoph Willibald Gluck dies November 15 at age 73, the Holy Roman Emperor Josef II appoints Mozart imperial composer to succeed Gluck, but he pays him only 800 florins per year whereas Gluck received 2,000.

The first English cricket match to be played at Lord's pits Essex against Middlesex May 31. The Marylebone Cricket Club plays its first match July 30 against the Islington Cricket Club (*see* 1788; Grace, 1860).

A French government edict intended to encourage agriculture removes the requirement that grain producers take their grain to market. Producers are permitted to sell directly to consumers or even to export their grain by land or sea with no restrictions. French granaries are emptied and the country is left unprepared for a bad harvest (*see* 1788).

English sea captain William Bligh, 33, sails for Tahiti on H.M.S. *Bounty* to obtain breadfruit plants for planting in the Caribbean islands as a new source of food for British colonists. As sailing master of Captain Cook's 1772–1775 voyage, Bligh developed an enthusiasm for breadfruit and has persuaded the admiralty that *Artocarpus communis* can be a boon to the Caribbean (*see* 1789; Dampier, 1688).

An automated process for grinding grain and bolting (sifting) flour marks the beginning of automation in U.S. industry (*see* 1946). Devised by Oliver Evans (*see* 1780), it requires some capital investment but can be operated by just two men, sharply lowers the cost of milling, and will make white bread widely available in America.

1788 The Parlement at Paris presents Louis XVI with a list of grievances as the country suffers its worst economic chaos of the century. Louis recalls Jacques Necker as minister of finance, making him secretary general and virtual premier. The king calls the Estates-General to assemble in May of next year for the first time since 1614.

Sweden's Gustavus III invades Russian Finland, beginning a 2-year war that will end with Finland and Karelia still in Russian hands.

Russia's war with the Ottoman Empire gains support from Austria, which joins with Catherine the Great under terms of a 1781 alliance treaty and declares war on the Turks.

Spain's Carlos III dies December 14 at age 72 after an enlightened reign of 29 years in which he has encouraged trade and industry, suppressed lawlessness, constructed roads and canals, improved sanitation, and supported the American colonists. Greatest of the Spanish Bourbons, he is succeeded, not by his firstborn son (an epileptic imbecile), but by his slothful second son of 40, who will reign until 1808 as Carlos IV.

The United States Constitution becomes operative June 21 as New Hampshire ratifies the Constitution 57 to 47, the ninth state to ratify. Georgia, Connecticut, Massachusetts, Maryland, and South Carolina have ratified earlier in the year to become states of the Union, and Virginia and New York ratify June 25 and July 26, respectively, New York by a vote of 30 to 27.

Josef II orders Austrian Jews to adopt "proper recognizable surnames." They have been known until now largely by biblical patronyms and when some object to changing their names, the emperor's officials give them such surnames as "Fresser," "Pferd," and "Weinglas."

The first permanent Ohio Valley settlement is established at Marietta by New Englanders under the leadership of General Rufus Putnam, 50.

Cincinnati has its beginnings in a few buildings erected on the Ohio River by settlers who call the place Losantiville (*see* 1790).

Dubuque has its beginnings in an Illinois country settlement established by French-Canadian entrepreneur Julien Dubuque, 26, who negotiates an agreement with Fox tribal leaders to mine lead.

Australia's Botany Bay, discovered by Captain Cook in 1770, becomes an English penal colony as the first shipload of convicts is landed January 18. The 736 prisoners are soon augmented by new arrivals, and the colony is moved to a neighboring area that will be called Sydney after British home secretary Thomas Townshend, 55, first viscount Sydney.

New Jersey lawyer-inventor John Stevens, 39, develops a multitubular steam boiler for marine engines (*see* 1790; 1804).

Scottish engineer William Symington, 25, invents the first practical steamboat, installing a direct-action steam engine in a paddle boat. His vessel attracts no commercial attention (*see* 1787; 1791).

Traite des Affinities Chymiques, ou Attractions Electives by Swedish chemist-physicist Torbern Bergman includes affinity tables summarizing results of his studies of displacement reactions. The University of Uppsala scientist obtained nickel in its pure state for the first time 3 years ago (*see* 1770).

Mechanique Analitique by Joseph-Louis Lagrange of 1761 calculus fame is a strictly analytical treatment of mechanics. Lagrange bases statics on the principle of virtual velocities and dynamics in a work that is based in turn on d'Alembert's principle of 1743.

English physician Thomas Cawley notes abnormalities of the pancreas in the autopsy of a diabetic patient (*see* 1783). He believes diabetes to be a kidney disease and ignores his own observation, but it is the first known observation of any relationship between diabetes and the pancreas (*see* Lancereaux, 1860).

New York physicians go into hiding for 3 days in April as a mob riots in protest against grave robbers. Boys playing outside the Society of the Hospital of the City of New York on Broadway between Duane and Catherine streets put a ladder against the hospital's wall Sunday afternoon, April 13. One boy climbs up, peers in the window, and discovers medical students and physicians dissecting a cadaver for study. He runs home and tells his father, who takes some friends and visits the grave of his recently buried wife. Finding the casket open and the corpse gone, they rush to the hospital and seize four doctors (the others have fled for their lives), but the sheriff rescues the men and puts them in jail overnight to protect them. A mob of 5,000 storms the jail Monday afternoon, injuring Alexander Hamilton, John Jay, Baron von Steuben, Mayor James Duane, and Gov. George Clinton. The militia is ordered to fire into the mob, several citizens are killed or wounded, the riot continues through Tuesday, but the state will pass a law next year enabling physicians to obtain cadavers without robbing graves (*see* Burke, 1829).

Wahhabi Islamic fundamentalists in Arabia expand their influence beyond the Nejd plateau (*see* 1773; 1902).

The *Times* of London begins publication January 1. Shipping losses bankrupted coal merchant John Walter, 49, 6 years ago after 27 years in the business, he switched to publishing books, went from that into newspaper publishing, and now changes the name of the *Daily Universal Register* that he has been publishing since 1785.

Nonfiction *Classical Dictionary (Bibliotheca Classica)* by English scholar John Lempriére, 23, is a reference book in classic mythology and history; *Thoughts on the Importance of the Manners of the Great to General Society* by Hannah More; *Lettres sur le Caractère et les Ecrits de J. J. Rousseau* by French libertine Mme. de Staël (Anne Louise Germaine Necker, baronne de Staël-Holstein), 22, whose 2-year-old marriage to the Swedish ambassador has turned sour. Daughter of the finance minister Jacques Necker, Mme. de Staël is becoming notorious for her love affairs.

Poetry *The Indian Burying Ground* by Philip Freneau.

Painting *Antoine Lavoisier and his Wife* and *Love of Paris and Helena* by Jacques Louis David; *Tiger Hunt in the East Indies* by Johann Zoffany. Sir Thomas Gainsborough dies at London August 2 at age 61.

England's Marylebone Cricket Club is founded and codifies the rules of a game played since the 16th century if not longer (*see* 1787; baseball, 1839).

1788 ***(cont.)*** French wheat prices soar as drought reduces the harvest. Grain reserves are depleted as a result of last year's edict permitting grain producers to sell without restriction, but Jacques Necker (above) suspends grain exports. He requires that all grain be sold in the open market once again to allay suspicions that the endless line of heavy carts seen to be carrying grain and flour are bound for ports to be shipped abroad.

Most Frenchmen remain convinced that the king has an interest in the Malisset Company he has entrusted with victualling Paris and that the king and the aristocracy are profiting at the people's expense. Hungry peasants and townspeople seize wagons in transit even when escorted in large convoys, farmers resist bringing their grain to market lest it be commandeered, and people starve.

1789 The French Revolution that begins with the tennis court oath June 20 follows widespread rioting triggered by rumors that the nobility and the clergy (the first and second estates) have plotted to collect all the nation's grain and ship it abroad. Jacques Necker has ordered requisitioning of all grain in April to assure fair distribution and convened the Estates-General for the first time since 1614, but rumors abound that the first and second estates intend to disrupt the Estates-General and starve the people.

The Estates-General calls itself the National Assembly beginning June 17, its meetings are suspended June 20, and members repair to a neighboring tennis court where they take an oath not to split up until they have given France a constitution. They elect Count Honoré de Mirabeau, 40, to head the assembly, and some members of the nobility and clergy join the bourgeois third estate in the assembly, but the king dismisses Jacques Necker July 11, he concentrates troops near Paris, and there are rumors that he will dissolve the assembly.

Members of the third estate attack the Bastille prison at Paris July 12, it falls July 14 (only 7 prisoners are inside the fortress), and the revolutionists overthrow the regime of Louis XVI, maintaining the monarchy in name only. The National Assembly recalls Jacques Necker, names Gen. Lafayette commander of the new National Guard, and adopts *le drapeau tricolore* as the flag of France.

France's nobility begins to emigrate as peasants rise against their feudal lords.

The guillotine proposed to the National Assembly by Paris physician Joseph Ignace Guillotin, 51, is a beheading machine originally called a *louisette* after Dr. Antoine Louis (who did not invent it any more than did Guillotin). A deputy of the Estates-General who was the first to demand a doubling of third-estate representatives, Guillotin says, "My victim will feel nothing but a slight sense of refreshing coolness on the neck. We cannot make too much haste, gentlemen, to allow the nation to enjoy this advantage." Only 10 percent of guillotine victims will be of the nobility, most of the 400,000 people put to death in the revolution will be shot, burned, or drowned, and the guillotine will often require several chops to do its job.

The price of bread reaches 4.5 sous per pound at Paris in July, and in some places it is 6 sous per pound. Widespread unemployment has reduced the people's ability to avoid starvation, but the National Assembly permits duty-free grain imports to relieve the hunger.

A Paris mob riots from October 5 to 6 and a revolutionary band, mostly women, marches to Versailles. Gen. Lafayette rescues the royal family and moves it to Paris.

The Austrian Netherlands (Belgian provinces) declare their independence from Vienna. The Holy Roman Emperor Josef II has ordered peasants to pay more than 12 percent of the value of their land in taxes to the state, plus nearly 18 percent to their feudal landlords.

North Carolina ratifies the Constitution to become the 12th state of the Union.

The House of Representatives holds its first meeting April 1.

George Washington takes office April 30 at New York's Federal Hall to begin the first of two terms as first president of the United States, taking care not to set any unfortunate precedents.

The Ottoman sultan Abdul Hamid has died April 7 of poison at age 65 after a 15-year reign. His nephew of 27 will reign until 1807 as Selim III.

The Declaration of the Rights of Man adopted by the French assembly (above) declares that man has "natural and imprescribable rights. These rights are liberty, property, personal security, and resistance to oppression . . ." "No one may be accused, imprisoned, or held under arrest except in such cases and in such a way as is prescribed by law . . ." "Every man is presumed innocent until he is proved guilty. . ." "Free expression of thought and opinion is one of the most precious rights of mankind: every citizen may therefore speak, write, and publish freely. . ." "Since the ownership of property is a sacred and inviolable right, no person may be deprived of his property except with legal sanction and in the public interest, after a just indemnity has been paid . . ."

Planters in Saint Domingue on Hispaniola send delegates to the National Assembly in Paris demanding freedom to deal with the colony's affairs without interference. Saint Domingue has 480,000 slaves and is the envy of the Caribbean, but the revolutionists at Paris are not sympathetic to the slave-owning pressure group (*see* 1791).

Jamaica has 211,000 slaves, up from 40,000 a century ago. Britain's House of Commons hears from William Wilberforce that one third of the African slaves landed in the West Indies die within a few months of arrival, many of them by suicide (*see* 1787; 1791).

Japan bans streetwalkers and requires prostitutes at Edo to move into the Yoshiwara section established in 1617. Some 2,000 move into the area within a few days (mixed bathing is also prohibited, and public bath houses separate men from women).

Rochester, N.Y., is founded by Yankee miller Ebenezer Allen, who settles in a swamp on the bank of the Genesee River at the point where it empties into Lake Ontario.

A customs house opens at New York and levies duties of $145,329 its first year. The figure will grow to more than $8.3 billion in 165 years as New York becomes the major port of entry.

"Nothing is certain but death and taxes," writes Benjamin Franklin in a letter to a French acquaintance.

Derbyshire technician Samuel Slater, 21, brings England's textile technology to the United States. He has spent 6.5 years as apprentice to a partner of Richard Arkwright and is familiar also with the cotton-spinning inventions of James Hargreaves and Samuel Crompton (*see* 1769; 1770; 1779). Having posed as a common laborer to deceive English emigration inspectors, Slater arrives at Providence, R.I., and contracts with the firm Almy and Brown to construct machinery from plans he has committed to memory. He will father the American factory system (*see* 1790).

Elements of Chemistry (*Traité élémentaire de chimi*) by Antoine Lavoisier is the first modern chemical textbook. Lavoisier lists 23 elements but includes as an inorganic element a substance he calls caloric (heat) (*see* 1784; 1794).

German chemist Martin Heinrich Klaproth, 46, discovers zinc.

Genera Plantarum by French botanist Antoine Laurent de Jussieu, 41, improves on the Linnaean system of 1737 and begins the modern classification of plants.

Hookworm (*Haakenwurm*) gets its name in a report by a German zoologist named Froelich, who has discovered a hairlike parasitic worm in the intestine of a fox (*see* Goeze, 1782). Froelich notes curious hooklike structures in the worm's tail, and he assigns it to a new genus which he establishes under the name *Uticinaria* (*see* Ashford, 1899).

The University of North Carolina founded at Chapel Hill is the first U.S. state university.

Georgetown University has its beginnings in a Roman Catholic college founded by John Carroll, first archbishop of Baltimore. The Jesuit school will be prominent for its School of Foreign Service.

Nonfiction *An Introduction to the Principles of Morals and Legislation* by English barrister Jeremy Bentham, 41, expounds the basic ethical doctrine of "the greatest happiness of the greatest number" as the chief object of all conduct and legislation. Bentham returned last year from a visit to his younger brother Samuel, a naval architect in the service of Russia's Catherine the Great, and he suggests that the morality of actions is determined by utility, meaning the capacity for rendering pleasure or preventing pain. He takes his key phrase from clergyman-chemist Joseph Priestley, who used it in his 1768 "Essay on the First Principles of Government" (*see* 1774; Owen, 1809).

Poetry *Songs of Innocence* by English artist-poet William Blake, 32, includes *The Lamb* and is illustrated by Blake with water-colored etchings.

Pears's soap—oval and translucent—is introduced by London soapmaker Andrew Pears.

The first known American advertisement for tobacco appears with a picture of an Indian smoking a long clay pipe while leaning against a hogshead marked "Best Virginia." The advertisement has been placed by Peter and George Lorillard, whose Huguenot French immigrant father Pierre, then 18, opened a shop in 1760 on the High Road between New York and Boston at a point that will later be called Park Row. The senior Lorillard was killed by Hessian troops in the Revolution (*see* 1911).

Sailors aboard H.M.S. *Bounty* bound for the West Indies with breadfruit plants mutiny April 28 in protest against being deprived of water that is being lavished on the plants (*see* 1787). Having spent 23 idyllic weeks on Tahiti, the mutineers, led by master's mate Fletcher Christian, cast Captain Bligh adrift in a 22-foot open boat with 18 men near the island of Tofau, return to Tahiti for native brides, and establish a colony on the desolate 1.75-square-mile Pitcairn Island southeast of Tahiti where the sole survivor of nine mutineers, six Tahitian men, and 13 women will remain undiscovered until 1808. Bligh reaches Timor in the East Indies after a 45-day voyage across 3,600 miles of open sea in which he loses seven of his 18 men (the mutineers will burn the *Bounty* January 23 of next year) (*see* 1791).

The Panthéon is completed in Paris as a monument to Ste. Geneviève. Begun in 1764 by the late Louis XV, it was designed by the late architect Jacques Germain Soufflot, who had seen the Pantheon of Agrippa in Rome.

The first movable-frame beehive is constructed by Swiss naturalist François Huber, 39, who has discovered the aerial impregnation of the queen bee and the killing of males by worker bees. Huber is nearly blind, but his wife, his son, and a servant have helped him to study the life and habits of honeybees (*see* Langstroth, 1851).

"I am so antiquated as still to dine at four," writes English author Horace Walpole, now 72; most English people now dine at five or six o'clock.

The first bourbon whiskey is distilled by Baptist minister Elijah Craig in the bluegrass country established as Kentucky County last year by the Virginia state assembly. The territory will become Bourbon County in the state of Kentucky, and Craig's corn whiskey is so refined that it will become more popular than rum or brandy in America (*see* tax, 1791; Whiskey Rebellion, 1794).

Nine out of 10 Americans are engaged in farming and food production (*see* 1820).

1790

The Holy Roman Emperor Josef II dies February 20 at age 48, 3 weeks after withdrawing all his reforms. He is succeeded by his brother, 42, who will reign until 1792 as Leopold II.

1790 *(cont.)* France's Louis XVI accepts the constitution drafted by French revolutionists in the Festival of the Champs de Mars.

Belgians revolt against the new emperor Leopold (above), who threatens to cede the Austrian Lowlands to France if Britain supports the revolution. William Pitt yields to the pressure, he refuses to recognize Belgian independence, and the Austrian troops crush the revolt at Brussels (*see* 1830).

The British make an alliance with the nizam of Hyderabad, and a third Mysore War begins.

Benjamin Franklin dies at Philadelphia April 17 at age 84.

Rhode Island ratifies the Constitution May 29 and becomes the 13th state of the Union.

Philadelphia becomes the capital of the United States in August, but Secretary of the Treasury Alexander Hamilton has selected a new national capital site on the banks of the Potomac near the Maryland town of George Town, thus resolving a dispute between North and South. The Resident Act passed by Congress July 16 confirms Hamilton's choice, providing for a new federal city to be designed and laid out on the site (below).

Abolish slavery in the French colonies, English abolitionist Thomas Clarkson urges the revolutionists at Paris (*see* 1787; 1794).

French Jews gain a grant of civil liberties.

The Indian Nonintercourse Act passed by Congress forbids taking of lands from Indian tribes without 1791 congressional approval, but Maine, Massachusetts, and other states will continue to take Indian lands without such approval.

Virginia's slave population reaches 200,000, up from over 100,000 in 1756.

Cincinnati gets its name as the 2-year-old town of Losantiville on the Ohio River is renamed by the Northwest Territory's first governor Gen. Arthur St. Clair, who calls it Cincinnati after the Society of Cincinnati formed by officers of the Continental army at the end of the Revolution (*see* 458 B.C.).

France's revolutionary government acts to revive trade that has stagnated as a result of businessmen, capital, and gold fleeing the country. The government prints *assignats*—paper money issued against the security of public lands confiscated from noblemen who have emigrated—and the new paper money stimulates production for a while, reviving employment (but *see* 1796).

Boston sea captain Robert Gray brings his *Columbia* back to port after a 3-year voyage round the world and enriches the ship's seven stockholders, who include architect Charles Bulfinch and merchant John Derby. News that furs acquired for trinkets from Indians on the northwest American coast fetch high prices at the Chinese river marts of Canton begins a large triangular maritime trade. Boston merchants will ship cargoes of cloth, clothing, copper, and iron to the Columbia River to be bargained for furs (*see* 1792) and to the Sandwich Islands for sandalwood. The sea captains will sell the cargoes at Canton and return round the Cape of Good Hope with Chinese porcelains, teas, and textiles. In another triangular trade, the Bostonians will exchange West Indian sugar for Russian iron and linens.

The first successful U.S. cotton mill is established at the falls of the Blackstone River at what later will be called Pawtucket, R.I. Samuel Slater and ironmaster David Wilkinson set up a mill that operates satisfactorily after a correction is made in the slope of the carder teeth (*see* 1789; 1793; Whitney, 1792).

Congress establishes a patent office April 10 to protect inventors and give them an incentive to develop new machines and methods. James Stevens of 1788 steam boiler fame has petitioned the lawmakers to give him patent protection.

Newburyport, Mass. inventor Jacob Perkins, 24, invents a nail cutter and header that is effective in a single operation (*see* 1834). So costly are nails that they are carefully salvaged when a building burns down.

Some 200 English spinning mills now use the patented Arkwright frame of 1769, but France has at most only eight mills using the water frame, and the few steam engines being used in France are almost all of the crude Newcomen variety (*see* 1705).

A committee established by the 124-year-old Paris Academy of Sciences works to institute a uniform system of weights and measures at a time when a meter measures 100 centimeters at Paris, 98 at Marseilles, 102 at Lille, and 96 at Bordeaux. Charles Maurice de Talleyrand-Perigord, 36, bishop of Autun and a member of the National Assembly, has requested the committee, which includes the chemist Antoine Lavoisier (*see* 1791).

Painting *Queen Charlotte and Miss Farren* (actress Elizabeth Farren, 29) by English painter Thomas Lawrence, 21, who has received almost no formal education or artistic training except for a few lessons with Joshua Reynolds; *Lady Raeburn* and *Sir John and Lady Clerk* by "the Scottish Reynolds" Henry Raeburn, 34.

First performances Concerto for Pianoforte and Orchestra in D major *(Coronation)* by Wolfgang Amadeus Mozart 10/9 at Frankfurt-am-Main for the coronation of Leopold II (above); Concerto No. 19 for Pianoforte and Orchestra in F major by W. A. Mozart 10/15 at Frankfurt-am-Main.

President Washington appoints French-American engineer Pierre Charles L'Enfant, 35, who has served in the Continental army, to design the federal city (above) with a Capitol to be built on an 85-foot hill. Secretary of State Thomas Jefferson names as chief surveyor Andrew Ellicott, 36, of Maryland, who has helped continue the Mason and Dixon line of 1767, and Ellicott employs astronomer-surveyor Benjamin Banneker.

Pineapples are introduced into the Sandwich Islands (Hawaii; *see* 1778; first cannery, 1892).

The world's first carbonated beverage company is started at Geneva by Swiss entrepreneur Jacob

Schweppe in partnership with Jacques and Nicholas Paul (*see* Priestley, 1767; Bergman, 1770). The partnership will dissolve within a few years, and Schweppe will move to London. He will open at 11 Margaret Street, Cavendish Square, to sell through chemists (pharmacists) his Schweppe's soda water in round-ended "drunken" bottles so made in order to keep their corks damp and thus prevent the gas from escaping (*see* 1851).

The population of the United States reaches 3,929,000, 95 percent of it rural; population density is four to five people per square mile.

1791 The French National Assembly elects the comte de Mirabeau president but he dies April 2 at age 42. His death weakens support for the monarchy.

Louis XVI flees with his family to the northeast frontier to gain the protection of loyalist troops, but he is thinly disguised. Revolutionist Jean Baptiste Drouet, 28, recognizes the king June 22 at Sainte-Menehould in Lorraine, crosses the Argonne Forest, and alerts the townspeople of Verennes-en-Argonne; five National Guardsmen and a few townspeople arrest Louis June 25 and return him to Paris with Marie Antoinette and his children.

British orator Edmund Burke criticizes the French Revolution, and Thomas Paine answers him with the pamphlet "The Rights of Man." Paine returned to his native England 4 years ago to sell a pierless iron bridge he designed (*see* 1792).

Louis XVI accepts the constitution September 14, France annexes Avignon and Venaissin, and the National Assembly dissolves September 30 after voting that no member shall be eligible for election to the next assembly.

Irish revolutionists Theobald Wolfe Tone, 28, Thomas Russell, 24, James Napper Tandy, 51, and others found the Society of United Irishmen to agitate for independence from Britain (*see* 1796).

A new Polish constitution put through by patriots May 3 converts Poland's elective monarchy into an hereditary one. Frederick Augustus III of Saxony declines the Polish throne offered him, and Russia resists the new constitution (*see* 1793).

The Canada Act passed by Britain's Parliament June 10 takes effect December 26, dividing Canada at the Ottawa River into Upper Canada (which is mainly English) and Lower Canada (mainly French) with two governors and two elected assemblies (*see* Act of Union, 1840).

Vermont is admitted to the Union as the 14th state. Created from parts of New Hampshire and New York, the new state will soon be the nation's leading sheep raiser.

Blacks born of free parents in the French West Indies gain voting rights and the same privileges as all French citizens. The National Assembly at Paris (above), rejects pleas from French Caribbean colonists, but Saint Domingue's 30,000 white colonists refuse to obey the assembly's decree and prepare to secede as the colony's 480,000 blacks and 24,000 mulattoes grow restless.

Free blacks and mulattoes at Saint Domingue revolt to obtain the rights they have been granted (above), and within a few months some 2,000 whites have been killed along with 10,000 blacks and mulattoes. Sugar plantations are burned but only after 70,000 tons of sugar have been produced (*see* 1792).

A motion by William Wilberforce to prevent further importation of slaves into the British West Indies fails by a 188 to 163 vote in the House of Commons, which has been prejudiced by the slave insurrection at Saint Domingue (above) and similar revolts in Martinique and on the British island of Dominica (*see* 1823; 1833; Wilberforce, 1789).

French physician Philippe Pinel, 46, advocates more humane treatment of the insane in *Traité Medico-Philosophique sur L'Aliénation Mentale*.

The Bill of Rights becomes U.S. law December 15 as Virginia ratifies the first 10 amendments to the Constitution drawn up in 1787. The First Amendment guarantees freedom of religion, of speech, and of the press, the right of peaceable assembly, and the right to petition the government for a redress of grievances; the Second Amendment guarantees the right to keep and bear arms; the Third Amendment forbids quartering of soldiers without consent; the Fourth Amendment forbids unreasonable searches and seizures; the Fifth Amendment establishes legal rights of life, liberty, and property, saying that no person "shall be compelled in any criminal case to be a witness against himself"; the Sixth Amendment protects rights of accused persons in criminal cases; the Seventh Amendment guarantees trial by jury; the Eighth Amendment prohibits excessive bail, excessive fines, and cruel and unusual punishment; the Ninth Amendment protects rights not enumerated in the Constitution; "The powers not delegated to the United States by the Constitution, nor prohibited by it to the States, are reserved to the States respectively, or to the people," says the Tenth Amendment.

Boston shipmaster-explorer Robert Gray discovers the Columbia River, naming it after his ship *Columbia*. Gray last year completed a round-the-world voyage and embarked a month later on a second such voyage. U.S. claims to Oregon territory will be based on Gray's discoveries (*see* 1843; Cook, 1778).

Odessa is founded by Russian pioneers on the Black Sea.

The first Bank of the United States receives a 20-year charter from Congress and succeeds the Bank of North America chartered at the end of 1781. The new bank has an initial capital of $10 million provided largely by private investors who buy stock with bonds acquired under a funding plan devised by Treasury Secretary Alexander Hamilton.

A "Report on Manufactures" prepared by Secretary Hamilton (above) with help from Philadelphian Tenche Cox lists 17 viable industries in the new repub-

1791 (cont.) lic. It makes no mention of milling, although flour has long been a major U.S. export.

U.S. privateer Joseph Peabody settles at Salem, Mass., after having amassed a fortune. Peabody starts a shipping business that will soon own 83 vessels, employ 7,000 sailors, and trade with Calcutta, Canton, St. Petersburg, and dozens of other world ports.

John Fitch receives a patent on his steamboat (*see* 1787), but the wreck of his fourth vessel in a storm next year will discourage financial backers, and Fitch will move to Kentucky (*see* Stevens, 1804; Fulton, 1807).

The Leblanc process perfected by French chemist Nicolas Leblanc, 49, produces soda (sodium carbonate) for French glassmakers. The French Academy offered a prize of 12,000 francs in 1784 for a substitute for the soda imported from Spain, which is obtained from the ashes of burned seaweed. Leblanc obtains it by treating salt (sodium chloride) with sulfuric acid to produce salt cake (sodium sulfate), which he roasts with charcoal and chalk to produce a black ash from which he then dissolves out and crystallizes calcium carbonate (*see* Solvay process, 1863).

The French National Assembly sets up a General Commission of Weights and Measures to "bring to an end the astounding and scandalous diversity in our measures." The Academy of Sciences committee established last year has advised the assembly to establish a natural standard for the *metre* (*meter,* from the Greek word for measure), making it one ten-millionth part of a quadrant of the earth's circumference, and the committee has urged that the gram be standardized at 1 cubic centimeter of water at 4° Centigrade (*see* 1799; 1801).

English clergyman-chemist-mineralogist William Gregor, 30, at Manaccan in Cornwall discovers titanium in the mineral ilmenite and isolates it for the first time.

The shortage of trained French physicians that will result from the closing of the nation's universities (below) will lead to a restructuring of a late 17th-century institution to create the modern medical clinic. The clinics will advance medical knowledge by applying observations gained in actual practice to modify dusty textbook theories.

France's revolutionaries close the nation's "Gothic universities and aristocratic academies" in order to open the learned professions to competition, experiment, and the unobstructed flow of ideas.

The University of Vermont is founded at Burlington.

The Observer begins publication at London December 5 with quotations from Doctor Johnson and a motto promising that the paper will be "unbiased by prejudice, uninfluenced by party. . . and delivered with the utmost dispatch." The weekly will gain wide readership with its accounts of sensational trials.

Nonfiction *The Life of Samuel Johnson* by James Boswell, who credits the late "Great Cham of Literature" with such lines as "No nation was ever hurt by luxury; for it can reach but to a very few" (4/13/73); "Patriotism is the last refuge of the scoundrel" (4/7/75); "When a man is tired of London he is tired of life; for there is in London all that life can afford" (9/20/77).

Fiction *Justine* by French pervert Donatien Alphonse François de Sade, 51, the marquis de Sade, who has been confined for much of his life in prisons, partly for using Spanish fly made from the pulverized bodies of the blister beetle *Lytta vesicatoria* which irritates the bladder and urethra. Sexual gratification by inflicting pain on a loved one is described by the marquis in his obscene novel and will be called "sadism" (*see* "Masochism," 1874).

Poetry *Tam O'Shanter* by Robert Burns appears in the March *Edinburgh Magazine; Lines Written for a School Declamation* by English schoolboy David Everett: "You'd scarce expect one of my age/To speak in public on the stage;/And if I chance to fall below/ Demosthenes or Cicero,/Don't view me with a critic's eye,/But pass my imperfections by./Large streams from little fountains flow,/Tall oaks from little acorns grow."

Opera *Così fan tutte* (*Women Are Like That*) 1/26 at Vienna's Burgtheater, with music by Wolfgang Amadeus Mozart; *La Clemenza di Tito* 9/6 at the Prague Theater, with music by W. A. Mozart, libretto from the Pierre Corneille play *Cinna* of 1640; *The Magic Flute* (*Die Zauberflöte*) 9/30 at Vienna's Theater auf der Wieden, with music by W. A. Mozart.

First performances Concerto No. 27 for Pianoforte and Orchestra in B flat major by Wolfgang Amadeus Mozart 3/4 at Vienna. Mozart completes the motet *Ave verum Corpus,* a Quintet for Strings in E flat major, and other works, completes a Concerto for Clarinet and Orchestra in A major September 28, but dies of typhoid fever December 5 at age 35 with his Requiem still incomplete. The three great Mozart symphonies No. 31 in E flat major, No. 40 in G minor, and No. 41 in C major (*Jupiter*), all written in 1788 during a space of less than 2 months, remain unperformed (*see* Kochel listings, 1862). Symphony No. 93 in D minor and Symphony No. 96 in D major (*Miracle*) by Joseph Haydn 3/11 at London's Hanover Square Rooms, in a concert conducted by German-born London conductor Johann Peter Solomon, who has invited Haydn to England where the composer will write and conduct six new symphonies this year.

Berlin's Brandenburg Gate (Brandenburger Tor) is completed by Prussian architect Carl Gotthard Langhaus, 59, at the end of the Unter den Linden nearly a mile from the royal palace. A protégé of the king's first mistress Wilhelmine Enke, Langhaus has copied the Propylaea at Athens.

The Dublin Customs House is completed on the north bank of the Liffey by Irish architect James Gandon, 48.

Venice's island of San Giorgio gets a new campanile (bell tower) designed by Bologna architect Benedetto Buratti.

The Cabinet-Maker and Upholsterer's Drawing Book by London furniture maker Thomas Sheraton, 40, advocates a style more severe than those of Chippendale and Hepplewhite (*see* 1749).

Capt. William Bligh sails for Tahiti once again to obtain breadfruit (*see* 1789). Bligh has returned from the East Indies to England and sails now on H.M.S. *Providence* to complete the mission he began in 1787 (*see* 1792; 1793).

British landowners and farmers protest the low duty on grain imports which are depressing their prices. Parliament responds by raising the domestic price level at which imports are permitted (*see* 1773; 1797).

President Washington prepares the first U.S. crop report in response to letters from *Annals of Agriculture* editor Arthur Young, 50, in England, who has written to ask what crops are produced on American farms, what the crops are worth, and so forth. Washington has conducted a personal survey by mail and compiled the results, giving all prices in English pounds but providing a conversion ratio (the dollar is worth 7 shillings 6 pence).

France has famine, which the Legislative Assembly has little power to relieve.

Congress imposes a 9¢-per-gallon tax on whiskey to discourage frontier farmers, blacksmiths, and storekeepers from diverting grain needed for food to use as distillery mash (and from competing with rum made in New England distilleries) (*see* Craig, 1789; Whiskey Rebellion, 1794).

Oliver Evans patents an "automated mill" in which power that turns the millstones also conveys wheat (grist) to the top of the mill (*see* 1787).

Camembert cheese is invented, or reinvented, by French farmer's wife Marie Fontaine Harel at Vimoutiers in the Department of Orne.

"Increase the quantity of food, or where that is limited, prescribe bounds to population," writes Joseph Townsend in *A Journey Through Spain in the years 1786 and 1787* (*see* 1785). "In a fully peopled country, to say, that no one shall suffer want is absurd. Could you supply their wants, you would soon double their numbers, and advance your population *ad infinitum* . . . It is indeed possible to banish hunger, and to supply that want at the expense of another; but then you must determine the proportion that shall marry, because you will have no other way to limit the number of your people. No human efforts will get rid of this dilemma; nor will men ever find a method, either more natural, or better in any respect, than to leave one appetite to regulate another" (*see* 1798).

1792

The Holy Roman Emperor Leopold II dies suddenly March 1 at age 44 and is succeeded by his son, 24, who will reign until 1806 as Francis II, the last Holy Roman Emperor, and then as the first Austrian emperor Francis I until 1835.

A Swedish assassin shoots Gustavus III in the back March 16 at a midnight masquerade at the Stockholm opera house; the king dies March 29 at age 46 after a 21-year reign. His son of 13 succeeds to the throne and will reign as Gustavus IV until his forced abdication in 1809.

Austria and Prussia form an alliance against France, partly at the instigation of émigré French noblemen, and France declares war on Austria April 20.

Three French armies commanded by Lafayette, Rochambeau, and Marshal Nicolas Luckner, 70, suffer reverses. The Legislative Assembly announces July 11 that the nation is in danger.

A Parisian mob storms the Tuileries Palace August 10 at the instigation of Georges Jacques Danton, 33, after Louis XVI orders his Swiss guard to stop firing on the people. The mob massacres some 600 guardsmen, the king is confined in the Temple, and the Paris commune takes power under Danton.

The French National Assembly declares Gen. Lafayette a traitor August 19, and he flees to Liège, where the Austrians take him prisoner. Verdun falls to the Prussians August 20, but the revolutionary armies gain their first triumph a month later by defeating the Prussians at the Battle of Valmy, a fog-shrouded artillery duel.

The French National Convention meets September 21, abolishes the monarchy, decrees perpetual banishment for French émigrés, and declares September 22 the first day of the Year One for the new French Republic.

Thomas Paine enlarges upon his 1791 pamphlet "The Rights of Man" to urge the overthrow of the British monarchy, British authorities indict him for treason, he escapes to France, and the French make him an honorary citizen.

French forces under Adam Philippe, 53, comte de Custine, rout the Prussians to take Mainz and Frankfurt-am-Main. Another French army takes Brussels and the Austrian Lowlands before the Prussians retake Frankfurt.

The French National Convention issues a proclamation November 19 offering assistance to peoples of all nations who want to overthrow their governments.

Kentucky is admitted to the Union as the 15th state. It incorporates territory ceded by Virginia in 1781.

English historian Edward Gibbon writes to Lord Sheffield to approve the vote against slavery by the House of Commons, but expresses a fear: "If it [the vote] proceeded only from an impulse of humanity, I cannot be displeased. . . But in this rage against slavery, in the numerous petitions against the slave trade, was there no leaven of new democratical principles? No wild ideas of the rights and natural equality of man? It is these I fear" (*see* Paine, 1791).

Vindication of the Rights of Women by English feminist Mary Wollstonecraft, 33, at London is a blunt attack on conventions.

Denmark abandons the slave trade and becomes the first nation to do so (*see* Sweden, 1813).

Eli Whitney's cotton gin (below) will increase U.S. cotton planting, producing an increased demand for slave labor (*see* 1782; 1803).

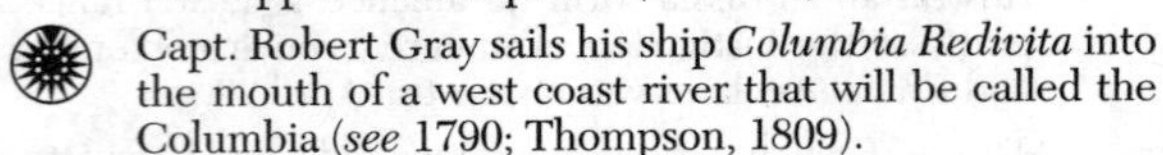

1792 ***(cont.)*** Paris sends an army to restore order in Saint Domingue, but the officers take the side of the slaves and support emancipation (*see* 1791; 1794).

Capt. Robert Gray sails his ship *Columbia Redivita* into the mouth of a west coast river that will be called the Columbia (*see* 1790; Thompson, 1809).

France's Girondist government issues more paper *assignats* as the war with Austria and Prussia creates irresistible demands for more paper currency (*see* 1790). The new *assignats* fuel the nation's inflation, diminish the people's traditional thrift, and encourage loose luxury as money declines in value and incentives to save disappear (see 1796).

The Mint of the United States established by Congress April 2 begins decimal coinage of silver, gold, and copper at Philadelphia (*see* 1873).

The New York Stock Exchange has its origin May 17 in an agreement signed by local brokers who have been buying and selling securities under an old buttonwood tree and who now formalize methods by which they will do business (*see* 1825).

U.S. money markets collapse following the failure of some land companies organized by New York speculator William Duer of 1784 *Empress of China* fame. A former assistant secretary of the treasury, Duer has sought European capital for the settlement of lands in the Ohio Valley, his scheme fails, and New York investors and speculators lose some $5 million, an amount equal to the value of all the buildings in the city (Boston and Philadelphia sustain losses of $1 million each).

Secretary of the Treasury Alexander Hamilton averts a U.S. economic depression by supporting the government's 6 percent bonds at par value.

The Lancaster Road has its beginnings at the Philadelphia State House where some 5,000 investors meet and subscribe $30 each to buy shares in the toll road that will be the first major publicly financed U.S. turnpike (*see* 1794; Cumberland Road, 1811).

Western Inland Lock Navigation Co. is incorporated through efforts by Gen. Philip Schuyler and Pittsfield, Mass., businessman Elkanah Watson, 34. The new company improves the channel of the Mohawk River that will be the basis of a great inland waterway (*see* 1815).

Eli Whitney, 26, invents a cotton "gin" that will revolutionize the economies of the United States and Britain. Just out of Yale, the mechanical genius has been visiting at Mulberry Grove on the Savannah River, Georgia plantation of Katherine Greene, widow of Gen. Nathanael Greene, who died 6 years ago at age 43. She met Whitney aboard ship while he was en route to the Carolinas for a tutoring position, she has invited him to Mulberry Grove, and he has observed that upland short-staple cotton has green seeds that are difficult to separate from the lint, quite unlike the long-staple sea-island cotton whose black seeds are easily separated and which has long been a staple of American commerce.

Eli Whitney's cotton gin made short-staple cotton profitable, increasing demand for plantation slave labor.

The simple cylinder produced by Eli Whitney (above) contains teeth made of bird cage wire bent and pointed into the shape of sawteeth, each circle of teeth moving in a slot slightly smaller than a cotton seed. When the cylinder is turned, the teeth carry the cotton through the slot, leaving the seeds behind, and a slave can clean 50 pounds of green-seed cotton per day instead of 1 pound—a great advance over the *churka* used in India since 300 B.C. to clean cotton at a rate of 5 pounds per day.

U.S. cotton production will rise from 140,000 pounds in 1791 to 35 million pounds in 1800 as the efficiency of the Whitney cotton gin leads to rapid growth of cotton planting in the South and a boom in northern and English cotton mills (*see* Lowell, 1812).

An epidemic of dysentery halts Prussia's Friedrich Wilhelm II in his march against France's revolutionary armies (above). He retreats across the Rhine with only 30,000 effectives, down from an original force of 42,000 counting Austrian allies.

An epidemic of bubonic plague in Egypt takes as many as 800,000 lives.

Smallpox strikes Boston; 8,000 volunteer for inoculation (*see* 1721; Jenner, 1796).

Poetry *Afton Water* and *Ye Flowery Banks O' Bonnie Doon* by Robert Burns appear in the *Scots Musical Museum* IV: "Flow gently, sweet Afton, among thy green braes . . ."

Sir Joshua Reynolds dies at London February 23 at age 68.

Opera *Il Matrimonio Segreto* (*The Secret Marriage*) 2/7 at Vienna's Burgtheater, with music by Italian composer Domenico Cimarosa, 43.

Venice's Teatro de la Fenice opens with 1,500 seats between the Canal della Verona and the Campo San Fantin.

First performances Symphony No. 94 in G major *(The Surprise)* by Joseph Haydn 3/23 at Vienna.

Anthem "La Marseillaise" by French army officer Rouget de Lisle, 32, who composes the song the night of April 24 at Strasbourg not as a revolutionary anthem but as a royalist patriotic song in support of the government that 4 days earlier declared war against the Holy Roman Emperor Francis II and Prussia's Friedrich Wilhelm II.

Popular song "Oh! Dear, What Can the Matter Be?" is published at London.

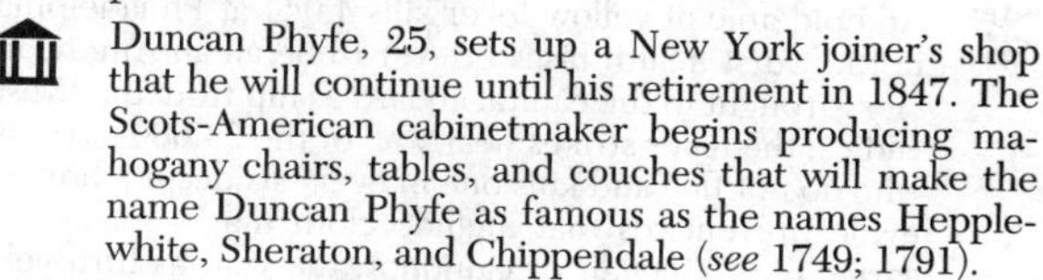

Duncan Phyfe, 25, sets up a New York joiner's shop that he will continue until his retirement in 1847. The Scots-American cabinetmaker begins producing mahogany chairs, tables, and couches that will make the name Duncan Phyfe as famous as the names Hepplewhite, Sheraton, and Chippendale (*see* 1749; 1791).

Capt. William Bligh loads Tahitian breadfruit aboard the decks of H.M.S. *Providence* for planting in the Caribbean islands (*see* 1791; 1793).

The Farmer's Almanac is founded by U.S. printer Robert B. Thomas, who establishes a format that will be continued for more than 200 years.

1793 Louis XVI goes to the guillotine January 21 in the Place de la Revolution that will later be the Place de la Concorde. The king has been tried before the convention, which has declared him guilty by a vote of 683 to 38 and voted that he be executed rather than imprisoned or banished.

The French republic declares war February 1 against Britain, Holland, and Spain and annexes the Belgian provinces. Britain, Holland, Spain, and the Holy Roman Empire join in an alliance against France with Savoy (Sardinia), which has been at war with the French since July 1792.

Louis XVI went to the guillotine in the French Revolution's bloody Reign of Terror. Marie Antoinette soon followed.

President Washington meets with department heads at his house February 25 in the first U.S. cabinet meeting. The cabinet is comprised of the secretaries of state, treasury, and war, the attorney general, and the postmaster general.

A French Committee of Public Safety organized at Paris April 6 has dictatorial powers. Leaders of the *Comité de salut public* include Georges Jacques Danton and Maximilien François Marie Isidore de Robespierre, 34.

President Washington issues a Proclamation of Neutrality in the European war (above) April 22. He has resisted pressure from Alexander Hamilton to support the British and from Thomas Jefferson to support the French.

French Girondists attack the radical Jacobin Jean Paul Marat and bring him to trial. Acquitted April 24, Marat joins with Danton and Robespierre of the *Comité de salut public* in overthrowing the power of the Girondists, 31 Girondist deputies are arrested June 2, but patriot Charlotte Corday, 25, horrified by the excesses of the Jacobin terrorists, stabs Marat to death in his bath July 13.

Prussian forces recover Mainz from the French after a 3-month siege, and the Allies take Conde and the Valenciennes. The French general Adam Philippe, comte de Custine, is accused of treason, a revolutionary tribunal finds him guilty, and he goes to the guillotine at Paris August 28 at age 53.

France begins a levy of all men capable of bearing arms August 23 as the Allies drive republican troops back on all fronts, and the British lay siege to Toulon. Fourteen new French armies succeed in taking Caen, Bordeaux, and Marseilles.

Lyons falls to the French republicans in October after a 2-month siege; the city is partially destroyed and much of the population massacred.

The Reign of Terror gathers force at Paris and elsewhere. Marie Antoinette goes to the guillotine October 16, some 15,000 are guillotined in 3 months at Nantes, and the 21 Girondist deputies arrested June 2 are guillotined October 31.

A purge of priests in the Vendée and efforts to conscript peasants into the revolutionary army incites a popular uprising. Generals Louis-Marie Turreau and François Westermann suppress the revolt with unbridled brutality, killing 300,000 to 600,000 men, women, and children. Jacobin judge Jean-Baptiste Carrier executes 13,000 at Nantes (many are drowned in the Loire in specially built boats). Gen. Westermann routs the Vendée rebel army at Savenay December 23.

Napoleon Bonaparte, 24, gains prominence for the first time as the French take Toulon from the British in December. The artillery officer from Corsica has won the favor of Robespierre's brother and is put in charge

1793 ***(cont.)*** of directing siege operations to recover Toulon from a British squadron that has come to the aid of the city, which has rebelled against the convention. His success wins Bonaparte promotion to general of brigade.

Poland has been partitioned for a second time January 23 with Russia taking most of Lithuania and the western Ukraine including Podolia, and Prussia taking Danzig, Thorn, and Great Poland. Russia has been given free entry for her troops in Poland plus the right to control Poland's foreign relations (*see* first partition, 1772; Kosciuszko, 1794).

Japan's Tokugawa shogun Ienari begins a personal reign of 45 years after a 6-year regency during which Matsudaira Sadanobu, now 34, has put through a series of reforms. The reign begins with the failure by Russian lieutenant Adam Laxman to establish friendly relations and will be marked by the collapse of military rule and the growth of extravagance and inefficiency in the Tokugawa court.

U.S. Secretary of State Thomas Jefferson suggests that the nation's sovereignty should extend offshore as far as cannonball range (*see* van Bynkershoek, 1702; Santiago Convention, 1952).

The Fugitive Slave Act voted by Congress at Philadelphia February 12 makes it illegal for anyone to help a slave escape to freedom or give a runaway slave refuge (*see* "Underground Railway," 1838).

Samuel Slater introduces child labor at the first U.S. cotton mill. He operates the mill under the name Almy, Brown and Slater and finds it advantageous to use children because they have small hands.

Scots fur trader Alexander Mackenzie, 29, reaches the Pacific after the first crossing of the North American continent by a European. Mackenzie, who has been employed by the Northwest Fur Co. since age 15, discovered the river that will be called the Mackenzie 4 years ago, has set out from Fort Chippewa, and has sustained himself on his journey by eating pemmican (dried lean meat from a large game animal pounded to shreds and mixed thoroughly half and half with melted fat, bone marrow and wild berries or cherries; the word *pemmican* comes from the Cree word for fat).

English navigator George Vancouver, 36, explores an island off the west coast of Canada that will bear his name (*see* Juan de Fuca, 1592).

Toronto has its origin in the frontier hamlet of York founded by Canadian governor John Graves Simcoe, 41. Mississauga tribesmen traded the site in 1788 for 149 barrels of provisions, blankets, and other goods. The settlement will be renamed Toronto in 1834 and will become Canada's financial capital.

The French republic imposes anti-inflation measures to avoid a catastrophic fall in the value of the *assignats* that have been issued since 1790. The government fixes wages and establishes maximum prices for many commodities, and while there are frequent violations of the maximums, the new laws permit a rationing of goods and make it possible to provision the French republican armies (but *see* 1796).

Charles Maurice de Talleyrand voyages to America to escape the Reign of Terror. He speculates in land and prospers (*see* 1790; 1796).

The Cornwallis Code imposed on India by governor-general Charles Cornwallis standardizes land assessments but fails to halt exploitation.

A cotton thread is perfected by the wife of Samuel Slater (above) (*see* 1790; Lowell, 1812).

Plymouth, Conn., clockmaker Eli Terry, 21, goes into business for himself after 7 years of apprenticeship (*see* 1803).

An epidemic of yellow fever kills 4,044 at Philadelphia in the worst health disaster ever to befall an American city. Brought to the capital aboard a ship from the West Indies, the fever strikes nearly all of the 24,000 citizens who do not flee and kills one in every six despite homemade preventives that include cloth masks soaked in garlic juice, vinegar, or camphor; the victims turn yellow, run high fevers, hiccup, and vomit blood.

France abolishes worship of God November 10. A cult of reason is founded by Commune of Paris leaders Jacques René Hébert, 38, Pierre Gaspard Chaumette, 30, and Jean Baptiste du Val-de-Grace, 38, baron de Cloots. All will be guillotined next year.

Samuel Slater (above) starts a Sunday school to teach factory children simple writing and arithmetic.

Williams College is founded at Williamstown, Mass. French is accepted for entrance in lieu of the classics.

Hamilton College is founded near Utica, N.Y., under the name Oneida Academy. The founder is Samuel Kirkland, 52, a missionary to the Iroquois who was influential in obtaining a declaration of neutrality from the Six Nations at the outbreak of the American Revolution in 1775. He names Alexander Hamilton as first trustee, and the college will take Hamilton's name when it receives a charter in 1812.

France makes education compulsory beginning at age 6.

The semaphore developed by French engineer Claude Chappe, 30, and his brother Ignace Urbain Jean, 33, is a visual telegraph system. They achieve their first success transmitting messages between Paris and Lille and see their system adopted rapidly throughout the country (*see* Henry, 1831).

Enquiry Concerning Political Justice by English philosopher William Godwin, 37, restates the political tenets of the Enlightenment and begins to develop a system of philosophical anarchy.

Advice to the Privileged Orders by Connecticut wit Joel Barlow, published at London, is a reply to Edmund Burke's tract on the French Revolution and wins Barlow French citizenship.

Poetry *Geschichte des Dreissigjahrigen Kreiges* by Friedrich von Schiller is published at Jena, where Schiller has been professor of history since 1789.

Painting *The Death of Marat* by Jacques Louis David, who has become an associate of Robespierre (above) and helps abolish the Academy. Francisco Guardi dies at Venice January 1 at age 80.

The Louvre Palace begun at Paris in 1564 and still incomplete is opened to the public as an art museum (*see* 1624).

The French republic proclaims a new Revolutionary Calendar with twelve 30-month days beginning with the Year One of the revolution (September 22, 1792). The new months are Vendémiaire, Brumaire, Frimaire, Nivôse, Pluviose, Ventose, Germinal, Floreal, Prairial, Messidor, Thermidor, and Fructidor; every tenth day is a holiday, and there are 5 *sans culottides* (intercalary) days.

Lansdown Crescent is completed at Bath to create a spectacular new English townscape.

French revolutionists turn the Tuileries gardens at Paris into a potato field. A French ordinance forbids consumption of more than 1 pound of meat per week on pain of death.

Britain suffers from lack of French food exports and establishes a Board of Agriculture to make "the principles of agriculture better known" and to increase domestic food production.

U.S. farmers export grain and other foodstuffs to Britain and the Continent as the European war booms demand (*see* 1804; 1806).

Capt. Bligh plants breadfruit in St. Vincent and Jamaica, British West Indies. He also plants seedlings of *Blighia sapida* whose fruit (ackee) will become a Jamaican food staple (*see* 1792).

Berlin chemist Franz Karl Achard, 40, reveals a process for obtaining sugar from beets (*see* 1799; Marggraf, 1747; Delessert, 1810).

1794

Robespierre crushes his rivals at Paris, has Danton and the others guillotined in late March and early April, and ends the cult of reason. He establishes himself June 8 as high priest of a new Festival of the Supreme Being; the Reign of Terror reaches its height in June and July.

The Law of 22 Prairial passed June 10 allows juries to convict without hearing evidence or argument. As many as 354 per month go to the guillotine, and opposition to Robespierre mounts.

Lord Richard Howe has defeated a French fleet off Ushant June 1, capturing nine French ships in the Channel and sinking a tenth, but the Battle of Fleurus June 26 ends in victory for the French. They force the duke of Coburg to withdraw from Belgium.

The conspiracy of 9 Thermidor (July 27) topples Robespierre from power. Moderates arrest him, his brother, and his associates Georges Couthon, 39, and Louis Antoine Leon de Saint-Just, 27. Robespierre's supporters release him but he and his companions are taken by surprise July 28 at the Hôtel de Ville and sent to the guillotine with 18 others. More than 80 of Robespierre's sympathizers go to the guillotine July 29 as opponents resist efforts to make France "a republic of virtue," but public opinion forces Robespierre's successors to end the Reign of Terror.

The Thermidoreans (moderates) close Jacobin clubs but readmit to the convention those Girondists who have escaped with their lives.

Polish revolutionists rise in March under the leadership of American Revolution veteran Tadeusz Kosciuszko, but they are no match for the Russian and Prussian forces sent to suppress the uprising (*see* partitions, 1793; 1795).

The Kajar dynasty that will rule Persia until 1925 is founded by Aga Mohammed, a brutal chieftain who takes power following the defeat and death of Lutf Ali Khan, who has reigned since 1789 and whose death ends the Zand dynasty founded in 1750. Aga Mohammed will be crowned in 1796 and reign until his assassination the following year.

The Whiskey Rebellion by U.S. frontier farmers brings the first show of force by the new U.S. government. The farmers have converted their grain into whiskey in order to transport it more efficiently to market but have resisted a cash excise tax imposed on whiskey, which is itself a medium of exchange in western Pennsylvania. Federal militiamen put down the rebellion without bloodshed.

The Battle of Fallen Timbers August 20 ends the Indian menace to American settlers in the Ohio-Kentucky region. British provocateurs have encouraged the Indians to attack the whites, but Gen. Anthony Wayne defeats the tribesmen.

Jay's Treaty signed November 19 settles outstanding disputes that remain between the United States and Britain. Resisting popular demands that the United States take France's side in her war with the British, President Washington has sent U.S. Supreme Court Chief Justice John Jay to London, the British agree to evacuate their posts in the U.S. Northwest between the Great Lakes and the Ohio River, and the treaty will spur settlement in the area.

Congress votes March 22 to forbid U.S. citizens to participate in slave trade with foreign nations.

English abolitionist Thomas Clarkson issues a pamphlet urging suppression of slavery.

The French Legislative Assembly frees slaves in all French colonies. The action comes in the midst of the Reign of Terror (above) and makes France the first nation to free her slaves (but *see* Bonaparte, 1802).

Haitian slaves in the French colony of Saint Domingue on Hispaniola rise under the leadership of Pierre Dominque Toussaint L'Ouverture, 51, Jean Jacques Dessalines, 36, and Henri Christophe, 27. They lead 500,000 blacks and mulattoes against the colony's 40,000 whites (*see* 1802).

Saint Domingue (above) has about 40,000 white Frenchmen, nearly half a million African slaves, and produces nearly two-thirds of the world's coffee, nearly

1794 *(cont.)* half its sugar, much of its cotton, indigo, and cocoa. The colony accounts for one-third of France's commerce.

A modification of Jay's Treaty (above) permits U.S. ships to carry cocoa, coffee, cotton, molasses, and sugar from the British West Indies to any part of the world.

Insurance Company of North America, chartered at Philadelphia, is the first U.S. commercial firm to offer life insurance policies, but its chief business is in fire and marine insurance.

Eli Whitney patents his 1792 cotton gin and sets up a company with Mrs. Nathanael Greene's second husband Phineas Miller to establish gins at central points for processing planters' green-seed cotton. But the patent is infringed, Whitney earns nothing from his cotton gin, and Miller loses a fortune trying vainly to fight patent infringers (*see* 1798).

The Lancaster Road opens to link the "bonnyclabber" Pennsylvania Dutch country and Lancaster with Philadelphia and the Delaware River (*see* 1792). Stockholders have subscribed $465,000 to finance it and will receive such handsome dividends—15 percent in some years—that their success will inspire similar toll road projects (*see* Cumberland Road, 1811).

The first U.S. national arsenals are established at Springfield, Mass., whose citizens were asked in 1776 to produce firearms for the Massachusetts Committee of Public Safety, and at Harper's Ferry, Virginia. The 1795 Springfield flintlock musket will be the first official U.S. weapon, and Springfield will become the small arms center of the world (*see* Springfield rifle, 1903; Harper's Ferry, 1859).

The slide-rest that will be an essential part of the modern lathe is invented in England by Joseph Bramah, or his employee Henry Maudslay, 23, or both working together. The slide-rest is a saddle which moves a cutting tool horizontally along the work being turned.

Welsh ironmaster Philip Vaughan at Carmarthen patents radial ball bearings for the axle bearings of carriages, but full development of ball bearings will await the invention of precise grinding machines that can produce accurately spherical metal balls.

The Ecole Normale is founded at Paris.

The Ecole Polytechnique is the world's first technical college. France's revolutionary government founds it to provide training for scientists with special emphasis on mathematics and applied science.

A museum for the popularization of science is started at Philadelphia by portrait painter Charles Willson Peale, a veteran of the War of Independence who is well known for his nearly 60 portraits of George Washington based on 7 he has painted from life.

The Age of Reason, Being an Investigation of True and Fabulous Theology by Thomas Paine is published in the first of its three parts. It is perceived as an attack on religion and alienates Paine's friends both in Europe and America; *Foundation of the Whole Theory of Science (Grundlage der gesammten Wissenschaftslehre)* by German philosopher Johann Gottlieb Fichte, 32, at Jena departs from Kantian transcendentalism by replacing God with an Absolute Mind *(Ur-Ich),* or Primeval Self (*see* Kant, 1781).

Poetry *Scots Wha Hae* by Robert Burns is published May 8 in the *Morning Chronicle:* "Scots, wha hae wi' Wallace bled,/Scots, wham Bruce has often led,/Welcome to your gory bed,/Or to victorie"; French poet André Marie de Chénier is guillotined July 25 at age 31; *Songs of Experience* by William Blake indicates the poet's disillusion with France and describes with stark simplicity the shattering of innocence by man and society: "Tyger! Tyger! burning bright/In the forests of the night,/ What immortal hand or eye/Could frame thy fearful symmetry?"

Painting *The Declaration of Independence* by U.S. painter John Trumbull, 38, who served as an aide to George Washington and then as adjutant to Gen. Gates, attained a colonelcy before he was 21, sailed to England in 1780 to study with Benjamin West, was imprisoned on suspicion of treason, began his large canvas in London 8 years ago, and returned to the United States in 1789 to continue work on the painting, which includes 48 portraits, two-thirds of them painted from life.

French playwright-aphorist Sebastien Chamfort dies in agony April 13 at age 53 from wounds self-inflicted late last year in a suicide attempt as he was about to be arrested for his outspoken opinions during the Reign of Terror. Chamfort's *Tableaux de la révolution française* will be published posthumously, as will his *Maximas et pensées, Caractères,* and *Anecdotes.*

First performances Symphony No. 99 in E flat major by Joseph Haydn 2/10 at London's Hanover Square Rooms; Symphony No. 100 in E major *(Military)* by Haydn at London; Symphony No. 101 in D major *(The Clock)* by Haydn 3/3 at the King's Concert Hall, a new addition to the King's Theatre, London. Haydn has come to England as in 1791 and once again writes six new symphonies for J. P. Solomon in little more than a year's time.

Popular songs "The March of the Men of Harlech" ("Gorhoffedd Gwyr Harlech") is published without words at London. The song was allegedly composed during a siege of the Welsh castle in the 15th century Wars of the Roses.

The founder of nutritional science and of modern chemistry Antoine Lavoisier goes to the guillotine May 8 at age 50 in retribution for his direction of *l'ancien regime*'s tax organization the *Fermier Général:* "La république n'a pas besoin de savants," says the vice president of the tribunal Coffinhall, but mathematician Joseph-Louis Lagrange says, "It required but a moment to sever that head, and perhaps a century will not be sufficient to produce another like it."

A British naval squadron tests the scurvy-prevention theories of James Lind, who dies July 13 at age 78 (*see* 1747; 1753). It sails for Madras with its hatches full of lemons and arrives 23 weeks later with only one seaman suffering from scurvy (*see* 1795).

Napoleon Bonaparte offers a prize of 12,000 francs to anyone who can invent a means of preserving food for long periods of time so that France's military and naval forces may be supplied on long campaigns (*see* Appert, 1795).

1795  The French capture the Dutch fleet lying frozen in the Texel January 2, Willem V takes refuge in England, and the French establish a Batavian republic as the White Terror rules Paris. The Batavian republic will continue until 1806.

The British begin seizing Dutch colonies including the Cape of Good Hope and Ceylon.

The Treaty of Basel March 5 removes Prussia from the war against France. Saxony, Hanover, and Hesse-Cassel soon follow suit.

The second Treaty of Basel June 22 gives Santo Domingo (Saint Domingue) to France and returns territory in Catalonia and Guipuizcoa to Spain.

The French Convention, threatened by a royalist revolt, calls upon the dissolute rake Paul François Jean Nicolas, 40, comte de Barras, to defend it. Barras met the Corsican artillery officer Napoleon Bonaparte during the siege of Toulon in 1793, has installed his creole mistress Josephine in Bonaparte's bed, and gives Bonaparte and other Jacobin officers command of forces defending the convention. Bonaparte drives the Paris mob from the streets with a "whiff of grapeshot" on "The Day of the Sections" October 5 (13 Vendémiaire).

The French Convention names Napoleon Bonaparte commander of the Armée d'Interieur for saving the Tuileries Palace with his artillery cannonade from the Church of St. Rochelle. The convention dissolves October 26 after voting that relatives of émigrés may not hold office.

The Directory that will govern France until 1799 takes power following the dissolution of the convention (above). Most prominent of the five directors is the comte de Barras (above).

The Treaty of San Lorenzo (Pinckney's treaty) signed by U.S. Minister to Great Britain Thomas Pinckney October 27 with Spain permits U.S. ships to store cargo at New Orleans. Spain grants free navigation of the Mississippi.

The Eleventh Amendment ratified February 7 has forbidden the federal government or foreigners to sue any U.S. state.

Poland's Stanislas II Augustus abdicates at age 63 after a 31-year reign as Russia, Prussia, and Austria partition his anarchic country for a third time October 24. Russia takes the remainder of Lithuania and the Ukraine, Prussia takes Mazovia and its city of Warsaw, Austria obtains the remainder of the Cracow region north of the Vistula and east to the Bug.

Britain's House of Lords acquits Warren Hastings of charges of "high crimes and misdemeanours" while governor-general of India from 1773 to 1785. His trial has lasted for 7 years, during which he has been made the scapegoat for the offenses of the East India Company; the expenses of the trial have ruined him financially.

Milwaukee has its beginnings in a trading post on Lake Michigan established by the North West Co. (*see* Juneau, 1818).

Theory of the Earth by Scottish geologist James Hutton, 69, appears 1 year after his *Investigations of Principles of Knowledge* and pioneers scientific geology. Hutton will be credited with originating the modern theory of the formation of the earth's crust and proposing the doctrine of uniformitarianism.

A Treatise on the Epidemic Puerperal Fever by Aberdeen surgeon Alexander Gordon suggests that the fever is contagious: "By observation, I plainly perceived the channel by which it was propagated, and I arrived at that certainty in the matter, that I could venture to foretell what women would be affected with the disease upon hearing by what midwife they were to be delivered, or by what nurse they were to be attended, during their lying in; and almost in every instance, my prediction was verified" (*see* White, 1773; Holmes, 1843).

Union University has its beginnings at Schenectady, New York.

The Institut National is founded at Paris to replace the academies closed in 1791.

Nonfiction *Tableau historique des progrés de l'esprit humain* gives a classic exposition of the idea of human progress and suggests the ultimate perfectability of mankind. Author Marie Jean Antoine Nicolas de Caritat, marquis de Condorcet, was arrested with other Girondists in the Reign of Terror, he died in prison March 28 of last year at age 50, and his work is published posthumously.

Fiction *Wilhelm Maisters Lehrjahre* by Johann Wolfgang von Goethe.

Painting *The Duchess of Alba* by Francisco de Goya; *Isabey and His Daughter* by French painter François Pascal Simon Gerard, 25, who has studied with Jacques Louis David; *Madame Seriziat and Her Daughter* by J. L. David.

First performances Symphony No. 102 in B flat major by Joseph Haydn 2/2 at London; Concerto No. 2 in B flat major for Pianoforte and Orchestra by German composer Ludwig van Beethoven, 24, 3/29 at Vienna's Burgtheater, in Beethoven's debut as composer and virtuoso (he has studied piano and violin under Haydn and, briefly, with the late W. A. Mozart); Symphony No. 103 in E major *(Drumroll)* and Symphony No. 104 in D major *(London)* by Haydn 5/12 at London.

The Paris Conservatoire de Musique is founded.

The average weight of cattle sold at London's Smithfield market is twice what it was in 1710 (*see* 1732; Bakewell, 1769; Coke, 1772).

Paris has bread riots April 1 (12 Germinal) as food prices soar, but reactionary sentiment sweeps the peasantry

1795 ***(cont.)*** which has grown rich supplying the black market and wants to maintain its position.

A poor English harvest drives up the price of bread, which now consists of 95 percent wheat flour.

English magistrates meeting at the Pelican Inn at Speenhamland, Berkshire, order a Poor Law giving bread to England's poor on a sliding scale based on the price of bread and the number of children in the family (*see* Malthus, 1798).

The Royal Navy orders lime juice rations aboard all naval vessels after the fifth or sixth week at sea following confirmation last year of James Lind's theory that citrus juice is an antiscorbutic (*see* 1794). The juice is usually combined with the rum ration (*see* 1740; 1805; 1884).

Parisian confectioner Nicolas Appert, 43, begins work on a method for preserving food in response to last year's offer by Napoleon Bonaparte of a reward for an effective process. Appert will move in the next year from Lombard Street to Ivry-sur-Seine to devote all his efforts to the project (*see* 1804).

1796 French forces under Napoleon Bonaparte invade Italy and defeat the Austrians in April at Millesimo and the Piedmontese at Modovi. On March 9 Bonaparte has married Paris society leader Josephine de Beauharnais, 33, Martinique-born widow of the late Vicomte de Beauharnais, who was blamed for the fall of Mainz to the Prussians in 1793 and went to the guillotine in July 1794.

Savoy and Nice are ceded to France, Bonaparte enters Milan May 15, he sets up the Lombard republic May 16, and he conquers all of Lombardy as far as Mantua, obtaining large sums of money and many art treasures from the pope, the king of Naples, and the dukes of Parma and Modena in return for truce agreements.

Babeuf's Conspiracy to overthrow the Directory that rules France is aborted May 10 by the arrest of socialist François Noel Babeuf, 35, whose song "Dying of Hunger, Dying of Cold" ("Mourant de faim, mourant de froid") is popular in Parisian cafés. "Nature has given to every man the right to the enjoyment of an equal share in all property," Babeuf has said on posters headed "Analyse de la doctrine de Babeuf, tribun du peuple." He has tried to incite soldiers to revolt, attracted ex-Jacobins to his cause, but will be executed next year along with an associate, and some of his followers will be exiled.

Baden, Württemberg, and Bavaria are forced to conclude truces in August with Gen. Jourdan, but he is defeated at Amberg and at Würzburg by the archduke Charles, brother of the emperor Francis II, and resigns his command. Gen. Moreau is forced to retreat through the Black Forest to the upper Rhine.

The Treaty of San Ildefonso August 19 aligns Spain with France against the British (*see* 1800).

Charles Maurice de Talleyrand has arrived at Hamburg in January after 30 months in America. He travels to Paris in September and will become foreign minister next year (*see* 1793; 1803).

Ceylon falls to the British, who gain control of all the Dutch spice islands with the exception of Java.

China's Manchu emperor Qian Long (Gao Cong) abdicates at age 85 after a 60-year reign in which he has enlarged the empire, established Chinese control over Tibet, sanctioned trade relations with the United States at Canton, invaded Burma and Nepal, and encouraged literature and the arts, especially pottery. He is succeeded by his son Jia Qing, 36, who will reign until his death in 1820.

Russia's Catherine the Great dies of apoplexy November 10 at age 67 after a 34-year reign. She is succeeded by her mentally unbalanced son of 42, who will reign until 1801 as Paul I.

A French fleet of 43 ships and 15,000 men assembles to invade Ireland in support of Irish rebels and leaves Brest December 15 under the command of Gen. Louis Lazare Hoche, 28. "Adjutant-general Smith" (T. W. Tone) has been allowed by the British to emigrate to America (*see* 1791) and accompanies the fleet. He has proceeded from America to Paris in quest of support from Napoleon and has succeeded, but a storm disperses the French ships off the coast of Kerry and no invasion is made (*see* 1798).

"It is our true policy to steer clear of permanent alliance with any portion of the foreign world," President Washington has said in his Farewell Address September 17.

John Adams of Massachusetts is elected to succeed Washington as president after a bitter contest with Thomas Jefferson, who wins 68 electoral votes to 71 for Adams. Jefferson becomes vice president.

Tennessee is admitted to the Union as the 16th state. It has been part of North Carolina.

Cleveland has its beginnings in the town of Cleaveland founded at the mouth of the Cuyahoga River on Lake Erie by Gen. Moses Cleaveland, 42, who has been sent out by the Connecticut Land Co. to survey territory the company has purchased in the Western Reserve.

The first permanent white settlement in Oklahoma Territory is established at what later will be called Salina. French explorer Jean Pierre Laclede, 28, whose father founded St. Louis in 1763, founds the settlement.

The first U.S. ship to enter California waters sails into Monterey Bay. The ship *Otter* is under the command of Yankee skipper Ebenezer Dorr.

Scottish physician-explorer Mungo Park, 24, reaches the Niger June 21, the first European to reach the West African interior.

France suffers ruinous inflation as her *assignats* of 1790 decline in value. A bushel of flour sells for the equivalent of $5, up from 40¢ in 1790; a cartload of wood for $250, up from $4; a pound of soap for $8, up from 18¢; a dozen eggs for $5, up from 24¢ (*see* 1792; Bonaparte, 1800).

A Public Land Act passed by Congress May 18 authorizes the sale of U.S. government lands in minimum lots of 640 acres each at $2 per acre with payments to be made under a credit system (*see* 1800; Land Ordinance, 1785; farmers, 1820).

English physician Edward Jenner, 47, pioneers the use of vaccination against smallpox. Testing the popular belief that a case of cowpox will avert smallpox, he takes some of the lymph from cowpox pustules on the hands of dairymaid Sarah Nelmes at Berkeley in Gloucestershire, scratches the lymph matter into the skin of 8-year-old schoolboy James Phipps, and a few weeks later inoculates the boy with matter from the smallpox pustule. Young Phipps does not develop even a mild case of smallpox, encouraging Jenner, who will continue such experiments for 2 years (*see* 1798).

The Boston Dispensary is founded to provide medical care for the poor in their homes and in clinics.

Royal Technical College is founded at Glasgow.

France restores freedom of the press.

Nonfiction *The Influence of the Passions (De L'Influence des Passions)* by Mme. de Staël, who fled to the Necker family estate at Coppet on Lake Geneva to escape the Reign of Terror, has borne two children by her lover Vicomte Louis de Narbonne and begun an affair with the courtier Benjamin Constant, 28, who has left his wife to join her, and returned to Paris to establish a salon.

Müsenalmanach is founded by Friedrich von Schiller, whose contributions to the new literary journal will include "Das Ideal und das Leben," "Die Macht des Gesanges," "Würde der Frauen," "Der Spaziergang," "Der Ring des Polykrates," "Der Handschuh," "Der Taucher," and "Das Lied von der Glocke."

Poetry *My love is like a red red rose* by Robert Burns; *Auld Lang Syne* by Robert Burns: "Should auld acquaintance be forgot,/And never brought to mind?/ Should auld acquaintance be forgot,/And auld lang syne?"; *Joan of Arc* by English poet Robert Southey, 22.

Painting *George Washington (Athenaeum Head)* by U.S. painter Gilbert Charles Stuart, 40, who studied at London during the American Revolution under the émigré master Benjamin West and returned home 3 years ago. His *Athenaeum Head* will be the most popular of all Washington portraits, and he completes also his Lansdowne portrait of Washington; *The Victor of Arcole* by French painter Antoine Jean Gros, 25, who has studied with Jacques Louis David and attracted the notice of Napoleon Bonaparte.

Theater *Egmont* by Johann Wolfgang von Goethe 3/31 at Weimar's Hoftheater.

Dublin's Four Courts are completed by James Gandon.

American Cookery by Amelia Simmons ("an American orphan") is the first cookbook to contain native American specialties. Simmons includes Indian pudding, Indian slapjack (pancakes), jonnycake, pickled watermelon rind, Jerusalem artichokes, spruce beer, and a gingerbread that is much softer than the thin European variety.

1797 The Battle of Rivoli January 14 to 15 on the river Adige gives Gen. Bonaparte his first decisive victory. He separates the enemy infantry from its artillery and routs an Austrian army east of Lake Garda.

Mantua falls to Bonaparte February 2 after a siege of more than 6 months during which time the French have beaten back four Austrian attempts to relieve the fortress. Romagna, Bologna, and Ferrara are ceded to France in the Treaty of Tolentino concluded by Pope Pius VI after Bonaparte has begun an advance on Rome.

Bonaparte crosses the Alps to challenge the archduke Charles at Vienna and is nearly cut off as the natives of Venetia and the Tyrol take arms against France.

The French proclaim a Cisalpine republic composed of Milan, Modena, Ferrara, Bologna, and Romagna July 9, and the Republic of Genoa is turned into a Ligurian republic controlled by France. A coup d'état at Paris September 4 (18 Fructidor) drives the party of reaction from power and gives the republican party control of the Council of 500.

The Treaty of Campo Formio October 17 obliges Austria to cede her Belgian provinces to France, recognize the Cisalpine republic (above), indemnify the duke of Modena with the Breisgau, and—under terms of secret articles—cede the left bank of the Rhine to France and make other concessions. Austria receives the territory of Venice as far as the Adige and obtains Dalmatia, Istria, and the city of Venice.

The XYZ affair brings France and the United States to the verge of war. France has regarded Jay's Treaty of 1794 as evidence of U.S. support for the British, President John Adams takes office March 4 and sends a conciliatory mission to Paris, three members of the French Directory known only as X, Y, and Z attempt to extort money from the Americans Charles Cotesworth Pinckney, 51, of South Carolina, John Marshall, 41, of Virginia, and Elbridge Gerry, 52, of Massachusetts. Pinckney and Marshall return home indignant, Congress and the public are outraged, but President Adams resists Federalist pressure to declare war.

Prussia's Friedrich Wilhelm II dies November 16 at age 53, leaving his country bankrupt, the monarchy discredited, and the army weakened after an 11-year reign in which Prussia has been compelled to give up her territories west of the Rhine but has gained some Polish territory. The king's grandson, 27, will reign until 1840 as Friedrich Wilhelm III but will be little better.

Persia's Aga Mohammed is assassinated after a 3-year reign. His nephew will reign until 1835 as Fath Ali Shah.

British sailors mutiny at Spithead April 15 demanding better treatment; the government meets their demands

1797 ***(cont.)*** May 17, but a more serious mutiny breaks out at the Nore and is suppressed June 30 only by force.

A poor wheat harvest forces Parliament to encourage imports, but grain prices shoot up as a result of the war with France, the breakdown in commerce, and the halt in specie payments by the Bank of England (*see* agriculture, 1804).

Britain issues her first copper pennies and £1 notes.

French chemist Louis Nicolas Vauquelin, 34, isolates chromium.

Chemist Martin Klaproth at Berlin reduces uranium oxide from pitchblende (*see* Becquerel, 1896). He has discovered and identified as separate elements cerium, titanium, and zirconium, although he does not obtain any of them in pure metallic form. A follower of France's late chemical pioneer Antoine Lavoisier, Klaproth improves and systematizes analytical chemistry and mineralogy.

Théorie des Fonctions Analytiques by Joseph L. Lagrange improves upon the Taylor series of 1715.

The Encyclopaedia Britannica is published in an 18-volume third edition (*see* 1784).

Fiction *Der blonde Eckbert* by German writer (Johann) Ludwig Tieck, 24.

Poetry *Hermann und Dorothea* by Johann Wolfgang von Goethe is an epic pastoral poem.

Painting *Millbank, Moon Light* by English painter J. M. W. (Joseph Mallard William) Turner, 22. Joseph Wright dies at Derby August 29 at age 62.

Anthem "Gott, Erhalte den Kaiser" is sung for the first time February 12 on the emperor's birthday. The "Emperor's Hymn" by Joseph Haydn with lyrics by Austrian poet Lorenz Leopold Hoschka, 48, will be Austria's national anthem.

First performances *The Emperor Quartet* by Haydn.

Opera *Medea (Medée)* 3/13 at the Théâtre Feydeau, Paris, with music by Italian composer Maria Luigi Carlo Zenobio Salvatore Cherubini, 36, who has lived at Paris since 1788. Libretto from the 1635 Pierre Corneille tragedy.

Cuban cigarmakers make "cigarettes" using paper wrappers derived from cotton to make the little cigars (*see* 1843).

Cuzco in Spain's viceroyalty of Peru and Quito in the viceroyalty of New Granada shudder in severe earthquakes that take up to 41,000 lives February 4.

The United States enters the world spice trade. Salem, Mass., sea captain Jonathan Carnes returns to port with the first large cargo of Sumatra pepper (*see* 1805; 1818).

English tea consumption reaches an annual rate of 2 pounds per capita, a figure that will increase fivefold in the next century (*see* 1840).

James Keiller starts packing orange marmalade at Dundee, Scotland—the first commercial marmalade.

1798 "Right," "Center," and "Left" political designations have their origin as France's Council of 500 meets in the palace of Louis XIV's bastard daughter Louise Françoise de Bourbon. The assembly hall in the palace's "grand apartments" is semicircular, and the representatives soon develop the habit of seating themselves with the most revolutionary on the left, the most conservative on the right.

Napoleon Bonaparte occupies Rome in February, proclaims a Roman republic, and takes Pius VI off to Valence in southern France, where the pope will die next year.

Bonaparte assembles an army of England at Boulogne, but while the English await an invasion he sails from Toulon May 19 with 35,000 men and a corps of scientists. Bonaparte takes Malta June 12, lands in Egypt July 1, and captures Alexandria July 2.

The Battle of the Pyramids outside Cairo July 21 gives the French an easy victory over the medieval Mameluke cavalry, armed only with lances. Cairo falls to Bonaparte July 22.

The Battle of the Nile August 1 destroys the French fleet in the harbor of Abukir east of Alexandria and cuts Bonaparte and his men off from their French homeland. News of Admiral Horatio Nelson's brilliant victory reaches Naples (which goes wild with joy), and swift couriers speed the news northward to the Papal States, Florence, Venice, and Vienna (*see* Haydn mass, below).

The Battle of Vinegar Hill June 21 has broken Irish resistance to British rule. Gen. Gerard Lake, 54, who savagely disarmed the north last year, defeats the rebels and enters Wexford, which the United Irishmen have held for a month. A French force sent by the Directory at Paris at the urging of T. W. Tone lands August 22, defeats several British contingents, but surrenders to Gen. Cornwallis September 15 at Ballynamuck in Connaught. Tone's brother is captured and hanged, Tone himself is taken at sea October 12 and sentenced to death, but he slits his throat with his penknife and dies November 19. James Napper Tandy is taken but released upon the intercession of Bonaparte and will be remembered as the hero in "The Wearing of the Green."

The Alien Acts approved by Congress June 25 and July 6 empower President Adams to order any alien from the country and imprison any alien in time of war. The acts force French aliens to flee the country at a time when war looms with France, Thomas Jefferson opposes the legislation and drafts Kentucky Resolutions that declare acts of Congress "void and of no force" when Congress "assumes undelegated powers," the governor of Kentucky approves Jefferson's resolutions November 16, James Madison drafts similar resolutions, and the governor of Virginia approves them December 24 (*see* Sedition Act, below).

Georgia forbids further importation of slaves (*see* 1808; South Carolina, 1803).

Swiss inventor P. L. Guinand, 50, patents a new way to make optical glass. His stirring process will be the basis of a great German optical industry.

Eli Whitney pioneers the "American system" of mass production with jigs—metal patterns that guide machine tools to make exact replicas of any part—that will doom the handicraft methods of cottage industry and have an effect on American society as profound as that of the cotton gin (*see* 1792). Whitney has made little from his cotton gin, but he devises a method for producing firearms from interchangeable parts and obtains a $134,000 U.S. Army contract to deliver 10,000 muskets in 28 months (*see* 1801).

Tableau Elementaire de l'Histoire Naturelle des animaux by French naturalist George Leopold Chrétien Frederic Dagobert, 29, baron Cuvier, founds the science of comparative anatomy (*see* 1799).

Louis Nicolas Vauquelin isolates beryllium.

Inquiry into the Cause and Effects of the Variolae Vaccinae by Edward Jenner announces his discovery of vaccination, a word that comes from the Latin *vacca* for cow and represents a much safer means of protection against smallpox than inoculation (*see* 1796). To counter widespread opposition to inoculation with an animal disease (he gives cowpox the name *Variola vaccinae),* Jenner shows that lymph from cowpox pustules on human skin is just as effective as lymph from infected cattle (*see* Waterhouse, 1799).

A typhus epidemic kills thousands of Britons as starvation makes the country vulnerable to disease (below).

Practical Education by English writer Maria Edgeworth, 32, and her father Richard, 54, is based on recorded conversations of children with their elders to illustrate a child's chain of reasoning.

Edward Jenner's vaccination helped end the horrors of smallpox after decades of chancy inoculations.

The Sedition Act passed by Congress July 14 suppresses editorial criticism of the U.S. president and his administration (*see* Alien Acts, above). Thomas Jefferson opposes the measure.

Papermaking gains impetus with the invention of a machine that makes it possible to produce paper from wood pulp in continuous rolls. The inventor is Louis Robert, a clerk employed in France's Essonne paper mills, but the machine will be generally called the Fourdrinier machine after two English brothers who will introduce it in Britain (*see* 1807).

Lithography is invented by Bavarian printer Aloys Senefelder, 25, who has developed a technique based on the incompatibility of grease and water. The Senefelder process permits ink from a grease base to be deposited on grease-treated printing areas, while dampened areas that are not to be printed reject the ink. He produces words or pictures on a flat stone covered with a greasy substance from which ink impressions are taken (*see* color, 1826).

Nonfiction *Alcuin: A Dialogue on the Rights of Women* by U.S. writer Charles Brockden Brown, 27. The first American professional writer, Brown has been influenced by the ideas of William Godwin and Mary Wollstonecraft in England.

Fiction *Wieland, or The Transformation* by Charles Brown (above) is a Gothic novel that gains its author wide popularity.

Poetry *Lyrical Ballads* by English poets Samuel Taylor Coleridge, 25, and William Wordsworth, 28, whose distinctive blank verse rebels against artificial forms. Prevented by the Napoleonic wars from marrying a young woman of Blois who has borne his child, Wordsworth lives with his sister Dorothy, 27, on a legacy of £900 received in 1795 and will spend the winter with her in Germany. *Lines Composed a Few Miles Above Tintern Abbey, on Revisiting the Banks of the Wye During a Tour, July 13, 1798* demonstrates Wordsworth's conviction that poetry should not be stylized or elaborate in language or content but should express rather the relation between man and nature. *The Rime of the Ancient Mariner* by Coleridge contains the lines, "Water, water everywhere,/And all the boards did shrink;/Water, water everywhere,/Nor any drop to drink" (II, ix). *The Battle of Blenheim* by Robert Southey whose wife Edith is the sister of Coleridge's wife Sarah: " 'And everybody praised the Duke/Who this great fight did win.'/'But what good came of it at last?'/Quoth little Peterkin./'Why, that I cannot tell,' said he/'But 'twas a famous victory' " (*see* 1704).

Painting *Love and Psyche* by Baron Gerard; *Caspar Melchor de Jovellanos* by Francisco de Goya.

Theater *Speed the Plough* by English playwright Thomas Morton, 34. "Be quiet, wull ye [says farmer Ashfield]. Always ding, dinging Dame Grundy into my ears—What will Mrs. Grundy say? What will Mrs. Grundy think?" *Wallenstein's Camp (Wallensteins Lager)* by Friedrich von Schiller 10/12 at Weimar's Hoftheater.

1798 ***(cont.)*** Opera *Leonore, ou l'amour conjugal* 2/19 at the Théâtre Feydeau, Paris, with an overture by Ludwig van Beethoven (Leonore No. 2), other music by Pierre Gaveaux, libretto by Jean Nicolas Bouilly, 34.

First performances Concerto No. 1 in C major for Pianoforte and Orchestra by Ludwig van Beethoven at Prague; The Creation Mass by Joseph Haydn 4/29 at Vienna; Mass in D minor *(Nelson Mass)* by Haydn 9/15 at the court of Prince Nicolaus II Esterházy, with Haydn playing the organ solo. He has written the work without knowing of Nelson's victory at Abukir (above) but people will soon begin calling it the Nelson Missa.

London hatter John Hetherington makes the first top hat of silk shag, or plush, thus reducing demand for American beaver pelts.

The replacement of beaver hats with silk hats in the next half century will produce a rise in U.S. and Canadian beaver populations (*see* Marsh, 1864).

Britain has a poor crop year. Wheat prices climb to £12 per cwt and there is widespread hunger (*see* typhus, above; Malthus, below).

"Essay on the Principles of Population" by English parson Thomas Robert Malthus, 32, expounds the thesis that population increases in geometrical, or exponential, progression (1, 2, 4, 8, 16, 32, etc.), while food production ("subsistence") increases only in arithmetical progression (1, 2, 3, 4, 5, 6, etc.). Resting his case on the history of the American colonies in this century, Malthus poses the inevitability of a world food crisis and attacks proposals to reform England's Poor Law (*see* 1795). "A slight acquaintance with numbers will show the immensity of the first power in comparison with the second," says Malthus, but he opposes the use of "improper arts," meaning contraception (*see* 1805; Townsend, 1785, 1791).

1799 Napoleon Bonaparte invades Syria from Egypt in February following a declaration of war by Constantinople. The French storm Jaffa and massacre 1,200 Ottoman prisoners but are unable to take Acre.

Plague breaks out among the French, and the army retreats to Egypt.

The Battle of Abukir July 25 ends in triumph for Bonaparte and his cavalry commander Joachim Murat, 32, over a Turkish force landed at Abukir with British support, but Gen. Joubert is defeated and killed August 15 at the Battle of Novi, and Bonaparte embarks for France August 24 to stiffen the sagging French armies in Europe, leaving Gen. Jean Baptiste Kléber to command his Egyptian forces.

Russian forces are driven out of Zurich September 26 by the French commander, who had been defeated in the Battle of Zurich June 4 to 7. The Russians have been reduced in number by starvation, and Moscow withdraws from the coalition October 22, complaining that Austria and the other allies have given her no support.

Bonaparte is elevated to first consul of France November 19 (18 Brumaire) in a coup d'état engineered by revolutionists who include the Abbé Sieyes (Emmanuel Joseph Sieyes), 51, and Pierre Roger Ducos, 52. The Constitution of the Year VIII receives overwhelming support from the public in a vote held December 24, the Directory that has ruled since 1795 is ended, and Bonaparte is established as dictator of France.

George Washington dies at Mount Vernon December 14 at age 67, and when the news reaches Philadelphia 4 days later his wartime cavalry general Henry "Light-Horse Harry" Lee is chosen to deliver the funeral oration: "First in war, first in peace and first in the hearts of his countrymen, he was second to none in the humble and endearing scenes of private life . . ."

Welsh industrialist Robert Owen, 28, marries the daughter of Scottish industrialist David Dale, 60, and becomes manager and co-owner of Dale's cotton mill in England's Lancashire. He determines "to make arrangements to supersede the evil conditions [of the millhands]. . . by good conditions" (*see* 1806).

Travels in the Interior of Africa by Mungo Park is published at London (*see* 1796). Park has returned from 19 months of travels that followed his escape from 4 months' imprisonment after being captured by an Arab chief. He will make a second expedition to the Niger in 1805 and will reach Bamako before perishing in the rapids during an attack by the natives in 1806.

The first British income tax bill passes Parliament January 9 and produces revenue of £6 million for the year. Introduced by the prime minister Sir William Pitt as a war measure, the law levies a standard rate of 10 percent on incomes over £200, taxes incomes of £60 to £199 at reduced rates, allows deductions for children, life insurance premiums, repairs to property, and tithes, and will raise £175 million in the next 17 years (*see* 1802).

The first British savings bank opens at Wendover, Buckinghamshire, where clergyman Joseph Smith has acted on the suggestion of Jeremy Bentham (*see* 1765; first American savings bank, 1816).

French economist Pierre Samuel du Pont de Nemours, 60, emigrates to the United States after having been imprisoned for a time as a collaborator of the late A. R. Jacques Turgot (*see* 1774). Du Pont is a disciple of the late François Quesnay (*see* E. I. Du Pont, 1802).

Pennsylvania rebel John Fries, 48, leads an armed force of German-Americans against U.S. tax assessors to protest a federal property tax levied in anticipation of a war with France. President Adams orders federal troops into the area, they arrest Fries, a court convicts him of treason and sentences him to death, but the president will pardon him next year.

The Dutch East India Company's charter expires after 198 years in which stockholders have received annual dividends averaging 18 percent.

Gas lighting is pioneered by French chemist and civil engineer Philippe Lebon, 30, who develops methods for producing inflammable gas from wood. Lebon will make important contributions to the theory of gas lighting and will foresee most 19th-century uses of illuminating gas (*see* 1801).

Scottish steam engineer William Murdock, 45, develops methods for purifying and storing gas. He has worked in Cornwall for Boulton and Watt since 1779 and carried out experiments in the distillation of coal, peat, and wood.

A letter announcing "the discovery at Rosetta of some inscriptions that may offer much interest" arrives at the Institute of Egypt founded by Napoleon Bonaparte at Paris to gather knowledge about the country he is conquering and colonizing. An Army Corps of Engineers captain named Boucard has found a large block of basalt whose polished surface has been chiseled with Greek characters, with hieroglyphs, and with characters that will later be called demotic. Boucard suspects that all three inscriptions may say the same thing and that this Rosetta stone may be the key to an understanding of the Egyptian language of antiquity and its hieroglyphic writings (*see* Champollion, 1822).

Baron Cuvier in France introduces the term *phylum* to denote a category more general than classes; he takes the word from the Greek for tribe (*see* 1798).

Prussian naturalist Friedrich Heinrich Alexander von Humboldt, 30, sails from Coruma, Spain, for Spanish America aboard the *Pizarro*. Accompanying him is French botanist Aimé Jacques Alexandre Bonpland (*né* Goujuad), 26, who will travel with von Humboldt in Cuba, Mexico, and the Andes, will be a professor of natural sciences at Buenos Aires from 1818 to 1821, and will be imprisoned by the dictator of Paraguay from 1821 to 1830 (*see* von Humboldt, 1800).

France's National Convention fixes values of meters and kilograms in the law of the 10th of December (*see* 1791; 1801).

Digitalis is related to heart disease for the first time by physician John Ferriar, who notes the effect of dried foxglove leaves on heart action and relegates to secondary importance the use of foxglove as a diuretic (*see* Withering, 1785).

English chemist Humphry Davy, 21, produces nitrous oxide ("laughing gas"), inhales 16 quarts of it, and finds that it makes him "absolutely intoxicated" and immune to pain. He suggests that the gas may have use as an anesthetic in minor surgery (*see* 1800; Wells, 1844).

Harvard medical professor Benjamin Waterhouse, 35, administers the first U.S. vaccination against smallpox (*see* 1800; Jenner, 1798).

The Royal Military Academy, Sandhurst, is founded at Camberley, England, to train officers.

Fiction *Ormond, or The Secret Witness* by Charles Brockden Brown.

Poetry *The Pleasures of Hope* by Scottish poet Thomas Campbell, 21: "Tis distance lends enchantment to the view,/And robes the mountain in its azure hue."

Theater *The Piccolominos (Die Piccolomini)* by Friedrich von Schiller 1/10 at Weimar's Hoftheater; *The Death of Wallenstein (Wallensteins Tod)* by Schiller 4/20 at Weimar's Hoftheater (the plays complete the trilogy begun last year).

First performances *The Creation (Die Schoepfung)* by Joseph Haydn 3/19 at Vienna under the direction of the composer; *Sonate Pathétique* by Ludwig van Beethoven.

The U.S. Executive Mansion at Washington, D.C., is completed after 7 years of construction by Irish-American architect James Hoban, 37, who won a $500 prize for his design in an open competition (*see* "White House," 1814).

Cultivation of beets as a root vegetable gains importance in parts of Europe, but most farmers continue to cultivate the plant only for its greens.

Prussia's Friedrich Wilhelm III receives a loaf of beet sugar from Berlin chemist Franz Karl Achard and is persuaded to give Achard some land at Cunern in Silesia and finance his work with sugar beets (*see* 1793; 1801).

1800

The Convention of El Arish concluded by Napoleon Bonaparte with the Ottoman Turks January 24 provides for French withdrawal from Egypt. Admiral George Keith Elphinstone, 54, Viscount Keith rejects terms of the treaty. Gen. Kléber defeats the Turks at Heliopolis March 20 with a force of 10,000 men that is outnumbered six to one, but Kléber is assassinated by a fanatic at Cairo June 14 at age 47.

Austrian troops have starved Genoa into surrender June 4 but come under attack from Bonaparte, who advanced through the St. Bernard Pass with 40,000 men in May and wins a narrow victory June 14 at the Battle of Marengo over Baron Michael Friedrich Benedikt von Melas, 71. Melas signs a truce giving the French all Austrian fortresses west of the Mincio and south of the Po.

Napoleon Bonaparte fought the monarchs of Europe who feared the spread of France's revolution.

1800 ***(cont.)*** Gen. Moreau takes Munich in July and goes on to defeat the archduke John December 3 in the Battle of Hohenlinden.

A Jacobin conspiracy to assassinate the dictator Bonaparte is discovered at Paris.

Osei "The whale" Bonshu assumes the Ashanti throne in West Africa and begins a reign as Asanthane. The gold-encrusted throne has been received "from heaven" by fetish prophet Okomfo Anokye to create the sacred empire of the Ashanti (*see* 1822).

Congress creates the Indiana Territory out of the western part of the Northwest Territory in the Division Act of May 7.

Spain returns the Louisiana Territory to France October 1, having received it by the Treaty of Paris in 1763. A secret agreement in the Treaty of San Ildefonso signed in 1796 has obliged her to return it; France guarantees not to transfer the territory to any power other than Spain (but *see* Louisiana Purchase, 1803).

Thomas Jefferson wins election to the U.S. presidency by winning 73 electoral votes as compared with 65 for President Adams and 64 for Charles Pinckney of South Carolina. Aaron Burr of New York also wins 73 electoral votes, the Federalists force the election into the House of Representatives in an effort to make Burr president, and 35 ballots are taken before the firm opposition of Alexander Hamilton to Burr swings the decision to Jefferson.

Washington, D.C., replaces New York as the U.S. capital, 123 federal government clerks are moved to the new city during the summer, and Congress convenes at Washington for the first time November 17.

Gabriel's Insurrection inspires Virginians to support plans for black emigration to Africa. A conspiracy organized by the slave "General Gabriel" to attack Richmond comes to light, Gov. James Monroe orders in federal militia, they suppress the insurrection, and the ringleaders are executed.

Alexander von Humboldt explores the course of the Orinoco River in South America, covering 1,725 miles of jungle by foot and canoe in 4 months and establishing the water link between the Orinoco and the Amazon (*see* 1799; nitrates, 1802).

Ottawa is founded on the Ottawa River in eastern Ontario.

The Banque de France is founded February 13 at the instigation of Napoleon Bonaparte "to counteract the displacement or dispersion of the funds serving to feed the trade of the nation, and all such influences as tended to impair the public credit and to clog the circulation of wealth in the country, influences arising from the French Revolution and expensive wars." The private bank has an initial capital reserve of 30 million francs.

Napoleon Bonaparte acts to stop France's inflation and avert national bankruptcy (*see* 1796). He raises 5 million francs from French and Italian bankers and 9 million from a national lottery, introduces a new and tighter system of income tax collections, reduces the budgets of all his ministries, and restores confidence among the bourgeoisie. French government bonds that sold at 12 francs December 23 of last year reach 44 within the year and will advance in the next 7 years to 94.40.

Congress passes legislation to reduce the minimum amount of public land that may be sold at auction under the U.S. Land Ordinance of 1785. The bill introduced by Congressman William Henry Harrison, 27, of the Northwest Territory permits sale of 320-acre tracts at $2 per acre with a down payment of one-fourth and the balance to be paid in three annual installments (*see* 1796; 1804).

The voltaic cell invented by Italian physicist Alessandro Volta, 55, pioneers the electric storage battery. It consists of several metal disks, each made of one or the other of two dissimilar metals, in alternating series, and separated by pads moistened with an electrolyte. The word *volt* will be derived from Volta's name (*see* Daniell, 1836).

English engineer Richard Trevithick, 29, builds a high-pressure steam engine that will be used to power a road vehicle (*see* 1801; Watt, 1782; Evans, 1797).

Pennsylvania engineer James Finley builds the first suspension bridge with iron chains (*see* 1825).

The Industrial Revolution spreads to Europe as Lieven Bauwens of Ghent smuggles in the mule jenny from England. Ghent will become the Manchester of the Continent (year approximate).

Infrared rays are discovered by Sir William Herschel, who discovered the planet Uranus in 1781. He uses a sensitive thermometer to show that the invisible rays beyond the red light of the sun's spectrum are essentially heat radiation.

Researches, Chemical and Philosophical, Chiefly Concerning Nitrous Oxide is published by Humphry Davy, who will soon be appointed lecturer in chemistry at the Royal Institution of Great Britain, set up this year pursuant to a proposal last year by Count Rumford (American physicist Benjamin Thompson, 47, who received his title in 1791 from the elector of Bavaria) (*see* Davy, 1799; 1807).

London's Royal College of Surgeons is founded.

A Prospect of Exterminating the Small Pox by Benjamin Waterhouse is published at Boston, but the disease will remain a leading cause of death worldwide for well over a century (*see* 1799; 1802).

The Library of Congress is established with a $5,000 appropriation to purchase some 900 books—including Adam Smith's 1776 *Wealth of Nations*—and maps that arrive from London in 11 trunks aboard a British cargo vessel, *The American* (*see* 1814).

Fiction *The Life and Memorable Actions of George Washington* by Mason Locke Weems, 43, a Philadelphia clergyman turned book peddler. Weems has fabricated stories including one about young George chopping down his father's cherry tree; *Castle Rackrent* by Maria Edgeworth is about Irish life.

Poetry *Lyrical Ballads* of 1798 by Samuel Taylor Coleridge and William Wordsworth is published in an expanded edition with a preface by Wordsworth who says, "Poetry is the breath and finer spirit of all knowledge; it is the impassioned expression which is in the countenance of all science." He includes his poems *Lucy Gray, She dwelt among the untrodden ways,* and *Strange fits of passion have I known.*

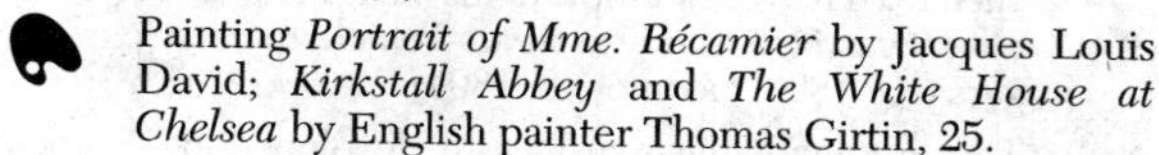

Painting *Portrait of Mme. Récamier* by Jacques Louis David; *Kirkstall Abbey* and *The White House at Chelsea* by English painter Thomas Girtin, 25.

Theater *Macbeth* by Friedrich von Schiller 5/14 at Weimar's Hoftheater, in a German version of the 1606 Shakespearean tragedy; *Maria Stuart* by Schiller 6/14 at Weimar's Hoftheater (Schiller settled at Weimar late last year to be near his friend Goethe.)

First performances Symphony No. 1 in C major by Ludwig van Beethoven 4/2 at Vienna's Hofburg Theater, in a concert that includes also Beethoven's Septet in E flat major for Strings and Winds, Air and Duet from the Creation, and Improvisations on Haydn's Emperor Hymn; Concerto in C minor for Pianoforte and Orchestra by Beethoven 4/5 at Vienna's Theater-an-der-Wien.

An octagon house is completed at Washington, D.C. (above), by Col. John Tayloe.

Passenger pigeons in America are estimated to number 5 billion and provide low-cost food in the Northwest Territory (*see* Josselyn, 1672; Wilson, 1810).

Sugar beets have a sugar content or roughly 6 percent; efforts begin to raise the figure by selective breeding (*see* Achard, 1793; Delessert, 1810).

The Northern Spy apple is originated at East Bloomfield, N.Y. The new variety is good both raw and cooked.

The New U.S. capital at Washington, D.C. (above), has 2,464 residents, 623 slaves.

The world's population reaches 870 million (960 by some estimates) with some 188 million of the total in Europe (*see* 1901).

19th Century

1801 The United Kingdom of Great Britain and Ireland is created by Parliament January 1 in an Act of Union that provides for one Parliament for both Britain and Ireland (*see* 1707; 1798; Home Rule Assn., 1870).

British prime minister William Pitt the Younger resigns February 2 after George III refuses to make concessions to Roman Catholics.

The Holy Roman Empire is practically destroyed by Napoleon Bonaparte February 9 in the Treaty of Luneville (*see* 1806).

Thomas Jefferson takes office as third president of the United States March 4, promising, "Peace, commerce, and honest friendship with all nations—entangling alliances with none" (*see* 1796).

Secretary of State John Marshall, 46, is named chief justice of the United States. Marshall's court will establish fundamental principles for interpreting the Constitution (*see* 1803; 1819; 1821; 1824).

Russia's Paul I is murdered in his bedroom at the St. Michael Palace in St. Petersburg the night of March 11 in a palace revolution that ends 4 years of insane rule. Dead at age 46, the czar is succeeded by his son, 23, who makes peace with Britain, withdraws from the Second Coalition against France, annexes Georgia in September, and will reign until 1825 as Aleksandr I Pavlovich.

Britain and France end hostilities March 14, but the peace will not last.

The Battle of Copenhagen April 2 ends in defeat for a Danish fleet by a British armada of 18 ships commanded by Sir Hyde Parker, 62, whose second-in-command Horatio Nelson pretends he does not see Parker's signal to withdraw (he holds his telescope to his blind eye), an order that would have meant disaster. The Danes lose all but three of their ships, they sign an armistice, and Nelson, who has been vice admiral since January 1, assumes command of the fleet in May following the recall of Parker. He returns home in June and is given the title viscount.

Frankfurt-am-Main moneylender Meyer Amschel Rothschild, 58, becomes financial adviser to the landgrave of Hesse-Cassel. Rothschild will serve as an agent of the British Crown in subsidizing European opposition to Napoleon Bonaparte, and his sons will establish important banking houses at London, Vienna, Naples, and Paris (*see* Wellington, 1812).

A gas-fueled thermolamp exhibited by Philippe Lebon in a Paris residence is successful both technically and aesthetically, but the French government is not forthcoming with funds to finance gas lighting on a large scale (*see* 1799; Winsor, 1804).

Richard Trevithick employs his steam engine to power a road carriage, the first steam vehicle to carry passengers (*see* 1800; locomotive, 1804).

The automatic Jacquard loom developed by French inventor Joseph Marie Jacquard, 49, employs punched cards to guide the movements of a loom that weaves figured silk fabrics and will later be used for making figured worsteds. It encounters initial resistance from weavers fearful of being displaced but will gain acceptance within the decade. France will have 11,000 Jacquard looms in operation by 1812.

Eli Whitney fails to meet his deadline for delivery of 10,000 muskets because he has taken so long to tool up a factory at the Lake Whitney fall line of the Mill River at Hamden, Conn. (*see* 1798). With only 500 muskets completed, Whitney puts on a show for President Adams and Vice President Jefferson in January, disassembling some guns, scrambling the parts, and then assembling new weapons to demonstrate the interchangeability of parts. The War Department gives Whitney an extension and advances him $134,000 so that he may fulfill his contract (*see* 1812).

Revisionists will question the story of Whitney's demonstration (above) when ancient Whitney muskets are found to have parts that have been filed and drilled here and there to make them fit, raising doubts as to their interchangeability.

English chemist Charles Hatchett, 36, discovers the metallic elements columbium and niobium.

The metric system fixed by France's National Convention late in 1799 is made compulsory by the convention in a move that most of Europe will follow but not Britain or the United States.

The University of South Carolina is founded at Columbia.

How Gertrude Teaches Her Children (Wie Gertrud ihre Kinder lehrt) by Swiss educationalist Johann Heinrich Pestalozzi, 55, shows that to educate a woman is to educate a family community.

The *New York Evening Post,* a new Federalist newspaper founded by Alexander Hamilton, begins publication November 16.

Painting *Napoleon au Grand Saint-Bernard* by Jacques Louis David; *Calais Pier* by J. M. W. Turner.

Sculpture *Perseus* by Antonio Canova.

Theater *The Maid of Orleans* (*Die Jungfrau von Orleans*) by Johann von Schiller 4/11 at Leipzig.

Ballet *The Creatures of Prometheus* (*Die Geschöpfe des Prometheus*) 3/28 at Vienna's Royal and Imperial Theater, with music by Ludwig van Beethoven, choreography by Neapolitan ballet master Salvator Vizano, 31.

Oratorio *The Seasons (Die Jahreszeiten)* by Joseph Haydn 4/24 at the Vienna palace of Prince Schwarzenberg, text from the James Thomson poem.

Italian violin prodigy Nicolo Paganini, 17, completes a concert tour in which he has dazzled audiences with his virtuosity, using a violin made by Joseph Guarnerius given him by a French merchant after Paganini had pawned his own violin to pay gambling debts. Paganini will live for the next 3 years in Tuscany with a noblewoman who has fallen for him.

"Johnny Appleseed" arrives in the Ohio Valley with seeds from Philadelphia cider presses that will make the valley as rich a source of apples as Leominster, Mass., home town of pioneer John Chapman, 26.

The world's first beet sugar factory opens in Silesia (*see* Achard, 1793, 1802).

U.S. salt prices fall to a new low of $2.50 per bushel as new sources come into production. Salt has been four times as costly as beef on the frontier but is essential to keep livestock (and people) healthy and meat from spoiling.

The first accurate census shows that China has a population of 295 million, India 131 million, the Ottoman Empire 21, Japan 15, Russia 33, France 27.4, the German states and free cities 14.1, Britain 10.4, Ireland 5.2, Spain 10.5, Egypt 2.5, and the United States 5.3 million of whom 1 million live west of the Alleghenies.

Guangzhou (Canton) is the world's largest city with 1.5 million; Nanjing, Hangchow, Kingtechchen, and the Japanese capital Edo each has 1 million.

London is by far the largest European city with a population of 864,000 as compared with 598,000 in Constantinople, 548,000 in Paris, 530,000 in Kyoto, 300,000 each in Isfahan, Lucknow, and Madras, 183,000 in Berlin, 75,000 in Manchester, 10,000 in Düsseldorf, 6,000 in Stockholm, 60,500 in New York, 1,565 in Pittsburgh.

1802

The Italian republic created January 26 from the former Cisalpine republic with Napoleon Bonaparte as president annexes Piedmont September 21 and annexes Parma and Piacenza in October.

The Treaty of Amiens March 27 brings a temporary end to hostilities among Europe's warring powers. Spain yields Trinidad to Britain but secures Minorca, the Batavian republic cedes Ceylon to Britain, the British give up all their other conquests to France and her allies, but hostilities soon begin again.

The Legion of Honor (*Légion d'Honneur*) created by Bonaparte May 19 gives him a tool comparable to the orders of knighthood so that he may avoid criticism of his government by keeping his courtiers busy with matters of etiquette and protocol.

The French make Bonaparte first consul for life August 2 and give him the right to name his successor.

Bonaparte revokes the French assembly's emancipation decree of 1794, he reintroduces slavery in the colonies, and he sends an army of 25,000 under his brother-in-law Charles Victor Emmanuel Leclerc, 30, to put down the rebellion in Haiti. Leclerc sends Toussaint L'Ouverture to France, where the black general soon dies in prison, but Leclerc himself dies of yellow fever as do 22,000 of his men (*see* 1804).

A Health and Morals of Apprentices Act voted by Parliament forbids cotton mills to hire pauper children under age 9, prohibits their working at night, and limits their work day to 12 hours (but *see* 1809).

Parliament repeals the British income tax of 1799 following the Peace of Amiens (above) and orders that all documents and records relating to the tax be destroyed in response to public outcry. Parliament will reimpose the tax next year and not remove it until 1815.

Italian scientist Gian Domenico Romagnosi advances knowledge of electricity with the observation that an electric current flowing through a wire causes a magnetic needle to line up perpendicularly to the wire (*see* Oersted, 1819).

The New American Practical Navigator by Salem, Mass., mathematican-astronomer Nathaniel Bowditch, 29, will soon be a second bible for sea captains. Bowditch has found 8,000 errors in the tables of the standard English work on navigation.

The paddle-wheel tugboat *Charlotte Dundas* built by Scotsman William Symington is the first successful steamship (*see* 1788; 1803).

E. I. du Pont de Nemours has its beginnings in a gunpowder plant built on the Brandywine River near Wilmington, Del., by Eleuthére Irénee du Pont, 31, whose father returns to France, where he will serve a dozen years hence as secretary to the provisional government (*see* 1799; 1803).

Biologie oder Philosophie de Lebenden Natur by German naturalist Gottfried Reinhold Treviranus, 26, introduces the word *biology*. The six-volume work will be completed in 1822.

Boston's Board of Health orders vaccination against smallpox (*see* 1798; Waterhouse, 1800). The board begins to improve the city's hygiene, regulate burials, and impose quarantines.

The United States Military Academy is founded at West Point, N.Y.

Fiction *Delphine* by Mme. de Staël, whose liberalism antagonizes Napoleon Bonaparte (he will exile her and her lover Benjamin Constant next year to her family estate on Lake Geneva).

Poetry *Lochiel's Warning* by Thomas Campbell: " 'Tis the sunset of life gives me mystical lore,/And coming events cast their shadows before."

Painting *Madame Récamier* by François Gérard. Thomas Girtin dies at London, November 9 at age 27;

1802 ***(cont.)*** George Romney dies at Kendal, Westmoreland, November 15 at age 67.

Sculpture *Napoleon Bonaparte* and *The Pugilists* by Antonio Canova.

British physician Thomas Wedgwood, 31, produces the world's first photograph. A son of the late potter Josiah Wedgwood, he sensitizes paper with moist silver nitrate to retain an image projected on its surface but can find no way of fixing the image, which quickly fades (*see* Schulze, 1727; Scheele, 1777). Wedgwood reports to the Royal Society that silver chloride is more sensitive than silver nitrate (*see* Niepce, 1822).

Ludwig van Beethoven's *Sonata Quasi una Fantasia* is published in March at Vienna. It will be called the *Moonlight Sonata* in midcentury by the music critic Ludwig Rellstab.

English horse racing at Goodwood is introduced by Charles Lennox, 67, duke of Richmond.

Mme. Tussaud's wax museum opens at London by Swiss wax modeler Marie Gresholtz Tussaud, 42, who was commissioned during the Reign of Terror at Paris 9 years ago to make death masks of famous guillotine victims. She will settle her collection in Baker Street in 1833 and connect it with a chamber of horrors containing relics of criminals and instruments of torture.

Arlington House is completed on a Virginia hill across the Potomac from the U.S. capital to provide a residence for Martha Washington's grandson George Washington Parke Custis, 21. The house will later be the home of Custis's son-in-law Robert E. Lee.

The nitrogen potential of guano is studied by Alexander Humboldt, who has crossed the Cordilleras to Quito, descended to Callao, and found great deposits of guano concentrated on islands off the west coast of South America (*see* 1800). The guano represents droppings from seabirds over the course of thousands of years, nitrates from guano can be used both for fertilizer and explosives, and Peru will export guano to Europe in large quantities primarily on the basis of Humboldt's reports (*see* 1809).

Soybeans are introduced into the United States via England (*see* 1920).

The world's first beet sugar factory goes into production but soon runs deep in debt (*see* 1799; 1801). A disciple of Franz Karl Achard is no more successful with a factory he sets up at Krayn, also in Silesia, but he does succeed in growing the white Silesian beet that is higher in sugar content and will be the basis of all future sugar-beet strains (*see* 1808).

1803 The Louisiana Purchase doubles the size of the United States, extending her western border to the Rocky Mountains. Desperate for funds to finance Bonaparte's military adventures, foreign minister Charles de Talleyrand recognizes that France's hold on Saint Domingue is shaky, so he ignores the terms of the 1800 Treaty of San Ildefonso with Spain and sells the 828,000-square-mile Louisiana Territory to the United States for 80 million francs ($15 million). U.S. Minister to France Robert R. Livingston negotiates the purchase, which is financed in part by the London banking house Baring Brothers.

Chief Justice John Marshall of the U.S. Supreme Court rules that any act of Congress which conflicts with the Constitution is null and void. His decision in the case of *Marbury v. Madison* February 14 establishes the Court as the ultimate interpreter of the Constitution.

Massachusetts and New York threaten to secede from the Union in protest against the Louisiana Purchase (above) which Thomas Jefferson has made without consent of the Senate.

Ohio is admitted to the Union as the 17th state.

The Hawaiian chief Kamehameha, 66, unites the eight major Sandwich Islands, employing American, English, and Welsh governors (*see* Cook, 1778; Kamehameha II, 1819).

Britain and France renew war May 16, the French complete their occupation of Hanover, and they threaten to invade England from Boulogne, where they begin to assemble an invasion fleet.

Swiss cantons regain their independence under terms of the Act of Mediation.

Irish nationalist Robert Emmet, 25, leads a rebellion against the British. He spoke with Bonaparte and Talleyrand last year in an effort to obtain French support, his small band of followers commits two murders and other acts of violence July 23 in a move to capture the viceroy, but the rioters are quickly dispersed. Emmet escapes, goes into hiding, is captured August 25, tried by a special court, and hanged September 20.

The Battle of Assaye September 23 ends in victory for British forces under Arthur Wellesley, 34, brother of India's governor-general Richard Colley, first Marquis Wellesley. He defeats the sindhia of Gwalior in the Second Mahratha War but is recalled as London becomes alarmed at his policies.

South Carolina resumes importing slaves as Eli Whitney's 1792 cotton gin makes cotton growing profitable and boosts demand for field hands.

Chicago has its beginnings in Fort Dearborn, built by U.S. federal troops on Lake Michigan and named for President Jefferson's Secretary of War Henry Dearborn (*see* 1825; steamboat, 1832).

Buffalo, N.Y., is laid out at the mouth of the Niagara River on Lake Erie by Holland Land Co. surveyor Joseph Ellicott (*see* Washington, 1790).

Harmony, Pa., is founded as a celibate religious communistic society by Harmonites led by German separatist George Rapp, 46, who emigrates to Pennsylvania's Butler County with a group of followers. The group will move in 1814 to Indiana, where it will establish another Harmony in the Wabash Valley, but in 1825 it will move back to Pennsylvania to settle in a town below Pittsburgh that will be called Economy (*see* New Harmony, 1824).

Cotton passes tobacco for the first time as the leading U.S. export crop.

Renewal of hostilities in Europe brings higher prices for American farm products and increases trade for U.S. shipping merchants such as Asa Clapp of Portland; George Cabot, Thomas Handasyd Perkins, and James Lloyd of Boston; and Cyrus Butler and Nicholas Brown of Providence.

E.I. du Pont de Nemours is founded in Delaware with $36,000 (*see* 1802; 1814).

The Ames Shovel Co. started by John Ames in 1774 is taken over by his son Oliver, who invests $1,600 to expand operations and moves the firm to Easton, Mass. Faced with little competition, Ames shovels will be the largest-selling U.S. shovels and play a major role in digging the Erie Canal (*see* 1817) and other canals, building U.S. railroads, digging the trenches and foxholes that U.S. troops will use, and digging the graves for America's dead (*see* Oakes Ames, 1862).

U.S. engineer Robert Fulton, 38, develops a small ship propelled by steam power. His *Treatise on the Improvement of Canal Navigation* appeared in 1796, and he has been in Paris since 1797 working on his submarine *Nautilus* (*see* 1807; Symington, 1788; Fitch, 1791).

Construction begins in Scotland on a 60.5-mile Caledonian Canal to connect the Atlantic with the North Sea across northern Scotland.

The Middlesex Canal opens December 21 to connect the Merrimack River with Boston Harbor. The canal has been dug with Ames shovels (above).

Connecticut clockmaker Eli Terry, 31, introduces wooden-wheeled clocks much like those produced in the Black Forest since 1660. Following the ideas of French watchmaker Frederic Japy and of Eli Whitney, Terry makes his clock parts interchangeable to permit mass production, and he will soon replace his wooden wheels with brass ones. When he encounters sales resistance to his clocks, Terry becomes the first U.S. merchant to offer merchandise on a free-trial, no-money-down basis, and he makes the discovery that farm families grow accustomed to having a clock and will sooner pay for the clock than give it up (*see* Seth Thomas, 1810).

English artillery officer Henry Shrapnel, 42, develops an explosive shell that will be called the shrapnel shell.

A table of atomic weights arranged by English chemist-physicist John Dalton, 37, at Manchester represents the first clear statement of atomic theory (*see* Mendeléev, 1870).

"Absorption of Gases by Water and Other Liquids" by John Dalton (above) states Dalton's law of partial pressures (*see* 1808).

Swedish chemist Jöns Jakob Berzelius, 24, at Uppsala, and Swedish mineralogist Wilhelm Hisinger, 37, discover cerium.

English chemist Smithson Tennant, 42, discovers osmium.

Essai de Statique Chymique by French chemist Claude Berthollet employs the system of chemical nomenclature that he devised with the late Antoine Lavoisier.

Parliament votes Edward Jenner a grant of £10,000 for his smallpox vaccination discovery of 1798. It will vote him another £20,000 in 1806.

Morphine is isolated and given its name by German pharmacist Friedrich Wilhelm Adam Saturner, 20, who uses the name of the Greek god of dreams Morpheus for his narcotic analgesic opium derivative (*see* Sydenham, 1683; codeine, 1832).

John Dalton (above) gave the first clear statement of color blindness when he was 28; the defect will be called Daltonism.

Constantinople loses 150,000 to plague.

Improvement in Education as it Respects the Industrious Classes by English Quaker schoolmaster Joseph Lancaster, 25, is based on his experience teaching a free school of 1,000 boys. Lancaster has organized a corps of elder boys as monitors to oversee and instruct and his Lancastrian system will be widely adopted in nonconformist schools.

Poetry *Minstrelsy of the Scottish Border* compiled by Scottish poet Walter Scott, 32, is published as Scott works on his own first original poem (*see* 1805); *Misfortune and Pity* by French abbé Jacques Delille, 65; "Fate makes our relatives, choice makes our friends."

Painting *The McNab* by Henry Raeburn; *Calais Pier* by J. M. W. Turner, who last year made his first trip abroad and has begun imitating the old masters: Titian, Poussin, and Claude Lorrain.

The British Museum's "Elgin Marbles" have their beginning as Thomas Bruce, 37, seventh earl of Elgin, starts shipping home from Athens some portions of the sculptured frieze of the Parthenon completed on the Acropolis in 438 B.C. Elgin will ship large parts of the frieze to London in the next 9 years.

John James Audubon, 18, arrives at Philadelphia. Son of a Haitian Creole woman who has been legally adopted by his French father and raised at Nantes, he will move to Kentucky in 1808, open a general store at Louisville, and begin painting American birds from life (*see* 1813; 1830).

Theater *The Bride of Messina (Die Braut von Messina)* by Johann von Schiller 3/19 at Weimar's Hoftheater.

First performances Symphony No. 2 in D major by Ludwig van Beethoven 4/5 at Vienna's Theater-an-der-Wien, in a concert that includes also Beethoven's only oratorio *Christ on the Mount Olives (Christus am Oelberg)* and his Concerto No. 3 in C minor for Pianoforte and Orchestra with Beethoven as soloist.

The first ice refrigerator (icebox) is patented by Maryland farmer Thomas Moore, who places one wooden box inside another, insulates the space in between with charcoal or ashes, and places a tin box container at the top of the inner box. Moore's icebox will be in common use by 1838 (*see* Perkins, 1834).

1803 ***(cont.)*** Britain makes abortion a statutory crime (*see* Pius XI, 1869).

1804 A royalist conspiracy against France's first consul Napoleon Bonaparte comes to light in February. One conspirator, Gen. Jean Victor Moreau, 41, escapes to America, but the duc d'Enghien Louis Antoine Henri de Bourbon Condé, 32, is seized in Baden, condemned by a commission acting under Bonaparte's orders without regard to law, and shot at Vincennes the night of March 20.

"It is worse than a crime; it is a blunder," says political leader Antoine Jacques Claude Joseph Boulay de la Meurthe, 42, at news of the duke's execution (above). Gen. Charles Pichegru, 43, is imprisoned and is either murdered or takes his own life April 5; Georges Cadoudal, 33, is guillotined June 5.

Napoleon is proclaimed emperor May 18 by the French senate and tribunate, a national plebiscite ratifies his elevation by a vote of 3.6 million to 2,569, and Pope Pius VII consecrates the emperor Napoleon December 2 at Paris in a ceremony imitative of Pepin III's coronation in 754 and Charlemagne's in 800.

Austria's Francis II assumes the title Emperor Francis I of Austria, which he will hold until his death in 1835. The title gives some semblance of unity to the emperor's widely extended German, Hungarian, Bohemian, Austrian, and Italian dominions which he rules with his wife (and first cousin) Maria Theresa, daughter of Ferdinand, king of Naples.

The Code Napoleon (*Code civil des Françaises*) drafted by Boulay de la Meurthe (above) and others and promulgated March 21 (30 Ventose XII) goes into force immediately throughout France, Belgium, Luxembourg, the Palatinate, those parts of Rhenish Prussia and Hesse-Darmstadt situated on the left bank of the Rhine, the territory of Geneva, Savoy, Piedmont, and the duchies of Parma and Piacenza. It is a code of civil laws common to the whole realm as promised in the constitution of 1791 and will have great influence on the legal codes of most European countries and even of countries as far distant as Japan (after 1870). It combines Roman law with some of the radical reforms brought by the French Revolution with respect to conditions affecting the individual, tenure of property, order of inheritance, mortgages, contracts, and the like, and it will make French law as much admired as French culture (but *see* below).

Spain declares war on Britain (*see* Trafalgar, 1805).

Haiti is established with a black republican government in the western part of Hispaniola following defeat of a 5,000-man army sent out by Napoleon to subdue the rebels (*see* 1794; 1802; 1822).

Former U.S. Treasury Secretary Alexander Hamilton is mortally wounded July 11 at age 49 in a duel at Weehawken, N.J. with Vice President Aaron Burr, now 48, who has heard of insults directed at him by Hamilton and demanded satisfaction (*see* 1800 election). Indicted in New York and New Jersey, Burr flees to Philadelphia and proceeds to the South where he will conspire in the next 3 years with Gen. James Wilkinson, 44, who is secretly in the pay of Spain but will soon be governor of the Louisiana Territory (see 1807).

The Twelfth Amendment to the Constitution, declared in force September 25, requires separate electoral ballots for president and vice president to prevent a recurrence of the confusion 4 years ago when Aaron Burr was nearly elected president.

Thomas Jefferson is re-elected with an overwhelming plurality of 162 electoral votes to 14 for Charles C. Pinckney.

Haiti's revolutionists (above) free all slaves and kill all whites who do not flee (many emigrate to Baltimore).

The Code Napoleon (above) regards an accused person as guilty until proven innocent. The tenet will be rejected in Britain and the United States.

The Lewis and Clark expedition to explore the Louisiana Purchase territory begins its ascent of the Missouri River May 14 seeking to determine whether the Gulf of Mexico and the Pacific Ocean are linked by a river system, and in the absence of any such water connection, to pioneer an overland route across the Rocky Mountains. Funded by a congressional appropriation of $2,500, the 35-man expedition is headed by President Jefferson's personal secretary Capt. Meriwether Lewis, 29, and William Clark, 33, a brother of George Rogers Clark of 1778 Kaskaskia and Vincennes fame.

Congress permits 160-acre tracts of U.S. public lands to be sold at auction (*see* 1841).

Gas lighting is demonstrated at London's Lyceum Theatre by a German promoter who has anglicized his

U.S. political rivalry reached a low point when Aaron Burr killed Alexander Hamilton in a duel at Weehawken, N.J.

name. Frederick Albert Winsor, 41, has seen demonstrations by the French chemist Philippe Lebon, made his own thermolamp, and brought it to England (*see* Lebon, 1801; British National Light, 1812).

The first steam locomotive to be tried on rails hauls five wagons containing 10 tons of iron and 70 men some 9.5 miles at nearly 5 miles per hour. Richard Trevithick has built the locomotive and will be the first to apply high-pressure steampower to agriculture (*see* 1801; Blenkinsop, 1811).

The S.S. *Little Juliana* launched in Hoboken, N.J., by Col. John Stevens and his two sons is a twin-screw steamship built with some parts brought from England. The vessel fails to achieve commercial success (*see* 1790; 1808; Fulton, 1807).

Smithson Tennant discovers iridium.

English chemist-physicist William Hyde Wollaston, 38, discovers palladium in platinum.

Ohio University is founded at Athens.

Painting *The Pesthouse at Jaffa* by Antoine Jean Gros.

Theater *Wilhelm Tell* by Johann von Schiller 3/17 at Weimar's Hoftheater is based on the 15th-century legend of a Swiss hero of the previous century who used his crossbow to shoot an apple placed on the head of his young son by the Austrian tyrant Gessler, bailiff of Uri (*see* Rossini opera, 1829); *Homage to the Arts (Die Huldigung der Künste)* by Schiller 11/12 at Weimar's Hoftheater. Schiller has been in poor health and will die next year at age 45.

Botanist William Bartram, now 65, declines an invitation from President Jefferson to join the Louis and Clark expedition (above), which sights frequent herds of bison, estimated to number at least 10 million on the western plains and thought by some to number five or ten times that many.

Ohio farmer George Rennic drives a corn-fed herd of cattle across the mountains to Baltimore, losing less than 100 pounds per head and making a sound profit. Many western farmers will be inspired to follow his example, but there are still food surpluses west of the Alleghenies and shortages on the seaboard.

Haitian sugar production falls to 27,000 tons, down from 70,000 in 1791, as the revolution against France (above) disrupts agriculture (*see* 1825).

Rust destroys part of England's wheat crop, bread prices rise, but Parliament imposes prohibitory duties that discourage imports of foreign grain. Customs and excise duties on food make up nearly half of Britain's total revenues (*see* Corn Law, 1815).

Nicolas Appert opens the world's first vacuum-bottling factory, or cannery, at Massey (Seine-et-Oise) near Palaiseau and Paris (*see* 1795). He has subjected his vacuum-packed foods to public tests at Brest (*see* 1810).

Capt. John Chester's schooner *Reynard* lands the first shipment of bananas to arrive at New York (*see* 1830; Baker, 1870).

1805 The emperor Napoleon crowns himself with the old iron crown of the Lombard kings May 26 in the cathedral at Milan, he appoints his stepson Eugène Beauharnais viceroy June 7, but a Third Coalition mobilizes against him; Austria, Russia, and Sweden make alliances with Britain, while Napoleon gains support from Spain, Bavaria, Württemberg, Baden, Hesse, and Nassau.

French troops mass at Boulogne to cross the Channel and invade the British Isles. Two squadrons of French ships are hastily put together under the command of Vice Admiral Pierre Charles Jean Baptiste Silvestre de Villeneuve, 42, to escort the landing barges, but the admiral and the minister of marine have no confidence in the ships, the officers, or the untrained crews. Villeneuve sails south rather than enter the Channel, and Viscount Nelson blockades him in the harbor of Cádiz.

Napoleon breaks camp at Boulogne, proceeds by forced marches into Germany, forces the surrender October 17 at Ulm of a 30,000-man Austrian army under Baron Karl Mack von Leiberich, 53, and caps the victory by taking Vienna.

The Battle of Trafalgar October 21 ends French pretensions as a sea power. Admiral Villeneuve tries to escape the British blockade of Cádiz, and the British Royal Navy commanded by Lord Nelson defeats a combined French and Spanish fleet off the Spanish coast near Gibraltar. Signal Number 16, hoisted by Nelson before the battle has said, "England expects every man to do his duty." Nelson takes 60 percent of the enemy fleet as prizes, but a French sharpshooter picks him off from his command post on the flagship *Victory*, shooting the great admiral dead during the engagement.

The Battle of Austerlitz December 2 gives Napoleon a resounding victory over combined Austrian and Russian forces. The French success dissuades Prussia from joining the Third Coalition (above).

The Treaty of Pressburg December 26 ends hostilities between France and Austria. Piedmont, Parma, and Piacenza are ceded to France; the Tyrol, free city of Augsburg, and other territory are ceded to Bavaria, which yields Würzburg to the elector of Salzburg, whose city is given to Austria along with Berchtesgaden and the estates of the Teutonic Order; Württemberg and Baden are recognized as kingdoms and given what remain of the western lands of the Hapsburgs.

Napoleon issues a proclamation at Schönbrunn in December dethroning the Bourbons in Naples.

Egypt gains independence from the Ottoman Turks, whose forces are driven out by Mohammed Ali, 36, a former tobacco merchant from Kavalla who came to Egypt in 1799 as commander of an Albanian contingent, was appointed governor by the Ottoman sultan, has intervened in political intrigues among Turkish offi-

1805 *(cont.)* cials and the Mamelukes, and is proclaimed pasha by Mameluke supporters.

A treaty signed June 4 ends hostilities between the United States and Tripoli.

Michigan is made a territory separate from the Indiana Territory, with Detroit as its capital (*see* 1701; Illinois Territory, 1809).

The District of Louisiana is made Louisiana Territory with St. Louis as its capital.

The Lewis and Clark expedition survives a bitter winter by eating wild roots which the men have been taught to find by Sacajawea, a Shoshone teenager whose French-Canadian husband serves as the expedition's guide and interpreter. The expedition reaches the Pacific in November (*see* 1804; 1806).

Lieut. Zebulon Montgomery Pike, 26, U.S. Army, ascends the Mississippi River in search of its source (*see* 1806; Marquette and Joliet, 1673).

Prussia abolishes internal customs duties to improve her domestic economy.

Philadelphia merchant Stephen Girard, 55, builds ships to enter the lucrative China and West Indian trade. Frenchman Girard has made $50,000 selling the belongings of French planters killed in the insurrection last year at Saint Domingue (*see* 1812).

The Beaufort Scale devised by British Royal Navy officer Francis Beaufort, 31, measures wind velocities and will be an aid to navigation.

Boston entrepreneur Frederic Tudor, 21, pioneers export of U.S. ice. He has declined to "waste his time" attending Harvard like his three brothers, heard that lumber shipped from New England ports in winter will sometimes still have ice on its plank-ends when unloaded weeks later at West Indian ports, and engaged men to chop ice from nearby lakes and ponds to be stored in icehouses. Undaunted by a costly initial setback, Tudor will ship thicker ice in better-insulated ships, will obtain a virtual monopoly in ice trade to Cuba, build icehouses at Caribbean and South American ports, and develop a large business (*see* 1833; Wyeth, 1825).

The first American covered bridge spans the Schuylkill River. A Philadelphia judge has persuaded Timothy Palmer to cover the three-span structure to prevent snow from blocking it (*see* Vermont, 1824).

William H. Wollaston discovers rhodium.

Haileybury College is founded to train East India Company personnel.

Poetry *The Lay of the Last Minstrel* by Walter Scott is his first original poem: "Breathes there a man, with soul so dead,/ Who never to himself hath said,/ This is my own, my native land!"

Painting *Shipwreck* by J. M. W. Turner; *Greta Bridge* by English landscape watercolorist John Sell Cotman, 23; *Doña Isabel Cobos de Procal* by Francisco Goya.

Opera *Fidelio, oder Die eheliche Liebe* 11/20 at Vienna's Theater-an-der-Wien, with music by Ludwig van Beethoven who includes a second version of his *Leonore* Overture (*see* 1798; 1814).

First performances Symphony No. 3 in E flat major *(Eroica)* by Ludwig van Beethoven 4/7 at Vienna's Theater-an-der-Wien. The composer dedicated the symphony last year to the French first consul Napoleon Bonaparte but has struck out the dedication upon learning that Bonaparte has accepted the title emperor of the French; Sonata in A major for Violin or Piano *(Kreutzer)* by Beethoven 5/17 at Bonn, in a concert given at 8 in the morning. Beethoven has dedicated the work to violinist Rudolf Kreutzer who has called it "unintelligible" and refused to play it.

Italian violinist Nicolo Paganini begins a second European tour.

The virtual abolition of scurvy from the Royal Navy has played a major role in the victory of the British at Trafalgar (above) (*see* 1794; 1884).

Tangerines from Tangier in Morocco reach Europe for the first time.

English population authority Thomas R. Malthus is named to a chair in political economy at the new Haileybury College (above) (*see* 1817).

1806 Napoleon places his older brother Joseph on the throne of Naples and his younger brother Louis on the throne of Holland, once the Batavian republic.

London announces a blockade of the European coast from Brest to the Elbe May 6 but permits ships from neutral nations to pass if they are not carrying goods to or from enemy ports.

The Holy Roman Empire that has existed since Christmas of 800 ends July 12 in the Confederation of the Rhine organized under French auspices. The confederation brings most of the German states under French domination and precipitates a war with Prussia, whose main armies are routed October 14 at the Battles of Jena and Auerstad.

Napoleon occupies Berlin October 27 and proclaims a paper blockade of Britain November 21. His Berlin Decree inaugurates the Continental system designed to deny the British food and supplies from the Continent.

British troops occupy the Cape of Good Hope.

Venezuelan Creole leader Francisco de Miranda, 50, fails in an effort to win freedom from Spain (*see* 1811).

Lancashire mill owner Robert Owen continues to pay full wages to all hands at his New Lanark mill despite its shutdown for lack of raw cotton (below). Prejudice against Owen has been strong despite his closing of shops that cheat the people and his substitution of shops of his own that sell only pure food (and pure whiskey), but his maintenance of wages wins Owen the confidence and support of his workers (*see* 1799; 1809).

The Lewis and Clark expedition returns to St. Louis September 23 after nearly 28 months of exploration. The expedition had been given up for lost, and its return is celebrated throughout the country. Only one man has been lost, and the maps, notes, and specimens brought back will be of immense value to scientists. Lewis is named governor of the Louisiana Territory and will hold the position until his mysterious death in October 1809 while traveling near Nashville, Tennessee, en route to Washington.

U.S. commerce falls to one-third its former value as a result of European trade restrictions. Thomas Jefferson introduces a bill to exclude such British goods as can be replaced by goods from other nations or can be produced domestically.

English textile mills shut down as supplies of raw cotton from the U.S. South run short (*see* Owen, above).

Coal gas is used for lamps by David Melville at Newport, R.I., but candlewax and whale oil remain the chief sources of illumination (*see* 1804).

Congress authorizes construction of a road to connect Cumberland, Md., with the Ohio River (*see* 1811).

Congress authorizes construction of the Natchez Trace to follow the 500-mile Indian trail from Nashville to Natchez on the Mississippi River.

A cotton thread strong and smooth as silk is developed at Paisley, Scotland, by Patrick Clark, who has sold silk thread for the loops (heddles) that hold the lengthwise warp threads in place on a loom. The blockade (above) has cut Britain off from supplies of French and Oriental silk, but Clark's cotton heddles are so effective that silk heddles will no longer be used in Britain. The cotton thread will prove strong enough to replace the linen thread now in common use, and Clark with his brother James will open the first factory for making cotton sewing thread.

The Compendious Dictionary of the English Language by Noah Webster is the first Webster's dictionary (*see* 1783; 1828).

Poetry *Simonidea* by English poet Walter Savage Landor, 31, contains his poem *Rose Aylmer* and gains him wide acclaim; *Original Poems for Infant Minds* by English poets Ann and Jane Taylor, 21 and 23, respectively, includes Jane's nursery rhyme "Twinkle, twinkle, little star,/ How I wonder what you are!/ Up above the world so high,/ Like a diamond in the sky."

Painting *The Battle of Aboukir* by Antoine Jean Gros. George Stubbs dies at London July 10 at 81; J. H. Fragonard dies at Paris August 22 at 74; Japanese *ukioye* painter Utamaro Kitagawa dies at 53.

First performance Concerto in D major for Violin and Orchestra by Ludwig van Beethoven 12/23 at Vienna's Theater-an-der-Wien.

Colgate-Palmolive-Peet has its beginnings in a New York tallow chandlery and soap manufactory opened at 6 Dutch Street by English-American candlemaker William Colgate, 23 (*see* 1928; Colgate University, 1890).

London architect John Nash, 54, begins work that will transform the city in the next few decades. He probably owes his new appointment as architect to the Chief Commissioner of Woods and Forests to the fact that his wife Mary Ann, 33, is the mistress of the prince of Wales, who in 5 years will become prince regent. The prince has evidently given Nash the means to afford a handsome mansion in Dover Street and a large estate on the Isle of Wight (*see* Regent's Park, 1811).

Ireland suffers a partial failure of her potato crop (*see* 1822).

The western high plains of America are "incapable of cultivation," says Zebulon Pike, who leads an expedition to the Rocky Mountains (Pike's Peak) and compares the plains to the "sandy deserts of Africa" (*see* 1805; Long, 1820).

1807 A British Order in Council January 7 prohibits neutral nation ships from trading with France and her allies, but while the Royal Navy blockades Napoleon's ports, the French are agriculturally self-sufficient and suffer less than do the British.

The Battle of Eylau February 7 to 8 pits Prussian and Russian forces against the French in a bloody slaughter that ends indecisively.

A British squadron under Admiral Sir John Thomas Duckworth, 59, appears before Constantinople and forces the Dardanelles. The Turks drive him out and he loses two ships.

British forces occupy Alexandria March 18 but are able to hold it for only 6 months before the Turks force their evacuation.

French forces take Danzig May 26 and defeat the Russians June 14 in the Battle of Friedland. Napoleon occupies Königsberg, concludes a truce after reaching the Niemen River, and meets with Russia's Aleksandr I and Prussia's Friedrich Wilhelm III on a raft in the river to conclude the Treaties of Tilsit, July 7 to 9.

The Treaties of Tilsit (above) oblige Russia to recognize the grand duchy of Warsaw created from Polish territory acquired by Prussia, recognize Jérôme Bonaparte as king of a new kingdom of Westphalia, restore the position of Danzig as a free city, and accept Napoleon's mediation in concluding peace with the Turks, while Napoleon accepts Aleksandr's mediation in concluding peace with Britain (a secret article provides for a Russo-French alliance against Britain should the British refuse to accept a peace offer). Bialystock in New East Prussia is ceded to Russia, while Prussia cedes to Napoleon all lands between the Rhine and the Elbe, makes other cessions, and recognizes the sovereignty of Napoleon's three brothers.

A British fleet bombards Copenhagen in September for joining the Continental system and carries off the Danish fleet. Denmark makes an alliance with France, Russia declares war on Britain, and French troops occupy Stralsund and Rügen.

U.S. authorities arrest former vice president Aaron Burr February 19 in Alabama and bring him to trial May 22 at Richmond, Va., in a circuit court presided over by Chief Justice John Marshall of the Supreme

1807 *(cont.)* Court. Burr has apparently schemed to establish an independent nation comprised of Mexico and parts of the Louisiana Territory, but the court acquits him for lack of evidence. He will go to Britain and France next year in an effort to gain support for his various schemes, but will return to New York in 1812, resume his law practice, and remain in obscurity until he dies in 1836.

Constantinople's Janissaries force their way into the Seraglio May 28, depose the Ottoman sultan Selim III, now 46, and replace him with the illiterate son of the late Abdul Hamid, who will reign until next year.

The British man-of-war *Leopard* fires on the U.S. frigate *Chesapeake* June 22 and removes four alleged British deserters. President Jefferson closes U.S. ports to all armed British vessels July 2 and tells France's Citizen E. Genêt that U. S. jurisdiction over territorial waters should extend to the Gulf Stream (*see* 1793).

Sierra Leone and the Gambia become a British crown colony (*see* 1843).

Portugal refuses to cooperate in Napoleon's Continental system. A French army under Andoche Junot, 36, invades the country and captures Lisbon, Junot is given the title duc d'Abrantas, and Portugal's royal Braganza family flees November 29 to Brazil (*see* 1808; 1822).

Prussian serfs are emancipated by the country's new prime minister Baron Heinrich Friedrich Karl vom und zum Stein, 60 (*see* 1810).

Britain's George III dissolves Parliament rather than grant emancipation to Catholics (*see* 1829).

Parliament passes a bill to forbid trade in slaves before it dissolves (above). The vote comes after great agitation by the humanitarian William Wilberforce (*see* 1789; antislavery society, 1823).

The Milan Decree issued by Napoleon December 17 reiterates the paper blockade against British trade with the Continent.

An Embargo Act signed by President Jefferson December 22 prohibits all ships from leaving U.S. ports for foreign ports. The act is designed to force French and British withdrawal of restrictions on U.S. trade, but its effect will be to make overseas sales of U.S. farm surpluses impossible. New England shippers protest the embargo and are joined by southern cotton and tobacco planters (*see* 1809).

The first commercially successful steamboat travels up the Hudson River August 17 and arrives in 32 hours at Albany to begin regular service between New York and Albany. Designed by Robert Fulton and backed by Robert Livingston of 1803 Louisiana Purchase fame, the *Clermont* has paddle wheels (suggested by inventor Nicholas Roosevelt, 40) powered by an English Boulton and Watt engine with a cylinder 24 inches in diameter and a 4-foot stroke. The vessel is 133 feet long, 18 feet in the beam, with a 7-foot draft and a stack 30 feet high, and is the forerunner of steamboats that soon will be carrying western grain to Gulf Coast ports via the Ohio and Mississippi (*see* 1811; Stevens, 1804, 1808).

System of Chemistry (3rd edition) by Scottish chemist Thomas Thomson, 34, gives the first detailed account of John Dalton's 1803 atomic theory.

On Some Chemical Agencies of Electricity by Humphry Davy gives sodium and potassium their names. Davy has prepared the elements by electrolysis (*see* 1800; calcium, 1808).

Phrenology arrives at Paris with German physician Franz Joseph Gall, 49, who introduced the pseudoscience in 1800. Gall claims that most emotional and intellectual functions are determined by specific areas of the brain that can be recognized by bumps in the skull.

The University of Maryland is founded outside Baltimore.

The Fourdrinier machine patented by English papermaker Henry Fourdrinier, 41, and his brother Sealy will spur the growth of newspapers (*see* Robert, 1798). Helped by civil engineer Bryan Donkin, 31, the Fourdriniers have developed an improved machine that can produce a continuous sheet of paper in any desired size from wood pulp.

Poetry *Poems in Two Volumes* by William Wordsworth brings the poet to the height of his powers although his masterpiece *The Prelude* will not appear until after his death in 1850. Included are the sonnets *London, 1802 ("Milton! thou shouldst be living at this hour . . . ")* and *Composed Upon Westminster Bridge, September 3, 1802:* "Earth has not anything to show more fair:/Dull would he be of soul who could pass by/A sight so touching in its majesty:/This city now doth, like a garment, wear/ The beauty of the morning: silent, bare. . ." Other

Robert Fulton's S.S. *Clermont* began a new era of steamboat travel, but sailing ships ruled the waves for 50 years more.

poems included by Wordsworth in his new books are *Ode to Duty, Intimations of Immortality from Recollections of Early Childhood, She Was a Phantom of Delight, I Wandered Lonely as a Cloud (Daffodils)*, all written in 1804, *My Heart Leaps Up (The Child is father of the man)*, his 1806 sonnets *The World Is Too Much with Us* and *Thought of a Briton on the Subjugation of Switzerland*, and his 1806 poem *Character of a Happy Warrior*.

Juvenile *Tales From Shakespeare* by English essayist-critic Charles Lamb, 32, and his sister Mary Ann, 43. Lamb has been taking care of his sister since she killed their invalid mother in a fit of insanity 11 years ago and has given up a projected marriage to devote himself to that responsibility, although he himself has been mentally unbalanced and was confined to an asylum from age 20 to 21.

Painting *Coronation of Napoleon and Josephine* by Jacques Louis David.

First performances Symphony No. 1 in C minor and Symphony No. 2 in C major by German composer Carl Maria von Weber, 20, at Schloss Carlsruhe in Silesia where the sickly von Weber lives with his father and aunt at the invitation of Duke Eugen of Württemberg; Symphony No. 4 in B flat major by Ludwig van Beethoven 3/5 at the Vienna palace of Prince Lobkowitz, in a program that includes Beethoven's three earlier symphonies, his *Coriolanus* Overture, and his Concerto in G major for Pianoforte and Orchestra; Mass in C major by Ludwig van Beethoven 9/13 at the palace of Nicholas Esterházy; Sonata in F minor *(Appassionata)* by Beethoven.

England's Ascot Gold Cup horse race has its first running (*see* 1711; 1872).

English pugilist Thomas Cribb, 26, wins the title from Henry Pearce and offers to defend it "against any fighter in the world." The offer will have no takers after 1811, and Cribb will retire undefeated in 1822 (*see* first world championship, 1860).

Water carbonated with carbonic gas is bottled and sold in the New York City area by Yale's first chemistry professor Benjamin Silliman, 28 (*see* Bergman, 1770; Matthews, 1833).

1808 Napoleon announces that France must guard the coast of Spain against Britain, he orders upwards of 70,000 men to cross the Spanish frontier in January, French forces occupy the fortresses of San Sebastian, Pamplona, Figuera, and Barcelona by treachery in February, and by mid-March there are nearly 100,000 French troops in Spain. Their commander in chief Joachim Murat occupies Madrid March 26, and Carlos IV abdicates in favor of his son Ferdinand, 23. Napoleon resolves to make his brother Joseph king of Spain, the Spaniards revolt May 2, Murat intimidates the Spanish council at Madrid May 13 into requesting Joseph as king, the Portuguese join in the insurrection, and they ask for British help.

Spanish forces at Bailen force a French army of 18,000 under Gen. Pierre Antoine Dupont, 43, to surrender July 23.

British forces invade Portugal August 1 under Sir Arthur Wellesley, they defeat Gen. Dupont (above) August 21, but Wellesley's superiors prevent him from following up his victory.

The defeat of Gen. Dupont at Bailen (above) forces Spain's new king Joseph Bonaparte to leave Madrid in August.

Gen. Murat succeeds Joseph Bonaparte as king of Naples, enters Naples in September as King Joachim Napoleon, and wrests Capri from the British.

Napoleon moves into Spain in the autumn with an army of 150,000, routs the Spanish in a series of engagements in November, takes Madrid December 4, and puts his brother Joseph back on the throne.

Denmark's demented Christian VII has died March 13 at age 59 after a 32-year reign. His son of 39 has ruled as regent since 1784, become an ally of Napoleon, and will reign until 1839 as Frederick VI.

The Ottoman grand vizier Mustafa Bairakdar, 33, marches on Constantinople to restore the sultan Selim III, Janissaries strangle Selim July 28 to prevent his restoration, and Bairakdar has the new sultan Mustapha IV deposed and assassinated. He places Selim's nephew, 23, on the throne, sends his troops to the Danube, and commits suicide to avoid being captured by the Janissaries. The new sultan will reign until 1839 as Mahmud II (the Reformer).

A British mission to Japan arrives August 15, but the Tokugawa shogun Ienari rejects the emissaries.

Thomas Jefferson declines a third term, he supports his Secretary of State James Madison, and Madison's Jeffersonian Republican party wins the fall election with 122 electoral votes to 47 for the Federalist Charles C. Pinckney.

Importation of slaves into the United States is banned as of January 1 by an act of Congress passed last year, but illegal imports continue (*see* 1814).

Parliament repeals an Elizabethan statute making it a capital offense to steal from the person following a campaign by London barrister Samuel Romilly, 51, who is reforming the English criminal laws and agitating against slavery.

John Jacob Astor, 45, incorporates the American Fur Co. with himself as sole stockholder. The German-American entrepreneur came to New York from his native Waldorf in 1783 at age 20, apprenticed himself to baker George Dieterich, soon quit to go into the fur trade, and has bartered firearms, "firewater" (whisky), and flannel with the Indians to obtain animal pelts, chiefly beaver for use in hats (*see* Hetherington, 1798; Pacific Fur Co., 1810).

The S.S. *Phoenix* launched by New Jersey engineer John Stevens is the first steamboat with an American-built engine (*see* 1804; Fulton, 1807). Stevens learns that his brother-in-law Robert R. Livingston and

1808 ***(cont.)*** Robert Fulton have obtained a 20-year monopoly on steamboat operations in New York State waters and prepares to have his son take the new side-wheeler to Philadelphia (*see* 1809; 1824).

A New System of Chemical Philosophy by English chemist John Dalton establishes the quantitative theory of chemistry (*see* 1863).

Humphrey Davy isolates barium, boron, calcium, and strontium (*see* 1807; Choss, 1840).

New York physician John Stearns describes amazing results in easing women's labor pains by administering a medication made according to an old midwife's instructions from diseased heads of rye (*see* ergonovine, 1935).

The *Times* of London sends journalist-diarist Henry Crabb Robinson, 33, to report on the Peninsular War in Spain. He is the world's first war correspondent.

Poetry *Marmion* by Walter Scott, who introduces the romantic hero young Lochinvar: "Oh, what a tangled web we weave/When first we practice to deceive" (VI); *Lyudmila* by Russian poet Vasily Andryevich Zhukovsky, 25, who has adapted the 1773 poem *Lenore* by the late German poet Gottfried August Burger. Zhukovsky's ballad marks the start of modern Russian poetry.

Painting *The Battle of Eylau* by Antoine Jean Gros; *Execution of the Citizens of Madrid* by Francisco Goya, now 62, who records massacres by the French, (above) that will be seen in his etchings *Disasters of the War*. Formerly court painter to Carlos IV, Goya becomes court painter to Joseph Bonaparte. Goya paintings owned by the Spanish prime minister Manuel de Godoy, 41, are inventoried and the list includes *The Naked Maja* and *The Maja Clothed*, painted nearly life-size and possibly showing the duchess of Alba who died in 1802. The word *Maja* is derived from *maya*, title of the queen of the May; it means something between wanton and wench, and the work is so scandalous that Godoy is able to keep it only by virtue of his position as prime minister and lover of the queen Maria Luisa; *La Grande Baigneuse* by French painter Jean Auguste Dominique Ingres, 28; *The Cross on the Mountains* by Dresden painter Kaspar David Friedrich, 34.

Theater *Faust* by Johann Wolfgang von Goethe is published in its first part. Goethe's drama of dissatisfaction, seduction, and infanticide is rooted in German legend, in the academic turmoil of his student days, and in the 1601 Christopher Marlowe drama (*see* Gounod opera, 1859).

First performances Symphony No. 5 in C minor and Symphony No. 6 in F major *(Pastoral)* by Ludwig van Beethoven 12/22 at Vienna's Theater-an-der-Wien, in a concert that also includes Beethoven's Concerto No. 4 in G minor for Pianoforte and Orchestra and Fantasy for Piano, Chorus, and Orchestra.

Irish Melodies is published with music by John Stevenson, lyrics by Irish poet Thomas Moore, 29. It includes "Believe Me, If All Those Endearing Young Charms" (Moore will gain fame also for "The Harp that Once Through Tara's Halls," "The Minstrel Boy," and "Oft in the Stilly Night").

French confectioner Nicolas Appert wins a prize of 12,000 francs (nearly $250,000) for his method of vacuum-packing food in jars (*see* 1804; Bonaparte, 1794). Appert's factory packs food for Napoleon's armies, and his method will remain briefly a French military secret (*see* Durand, 1810).

An article on beet sugar written by F. K. Achard in Berlin appears in the *Moniteur* and arouses French interest in domestic sugar production, as the British Orders in Council issued last year cut France off from sugar imports (*see* 1810; Achard, 1802).

1809 French forces defeat Sir John Moore and kill him in the Battle of Corunna January 16. Sir Arthur Wellesley succeeds Moore as commander of British forces in the Peninsular War (*see* 1808; 1812).

Napoleon returns from Spain and finds that the Austrians have taken advantage of his absence to push preparations for war against the diminished French forces west of the Rhine. He orders the formation of new battalions to oppose the Austrians.

Seven Swedish army officers break into the royal apartments March 13, seize the insane Gustavus IV in a coup d'état, and conduct him to the château of Gripsholm. A provisional government is proclaimed the same day under the duke of Sudermania. The king abdicates March 29 in hopes of saving the crown for his son, but the estates (dominated by the army) declare May 10 that the whole family has forfeited the throne. The duke, 60, is proclaimed king June 5 and will reign until 1818 as Charles XIII.

Austrian forces cross the Inn April 9, other Austrian regiments advance on Regensburg from Pilsen in Bohemia, while a French army advances on Regensburg under the command of Louis Nicolas Davout, 39, duc d'Auerstaedt.

Napoleon sustains a slight wound in late April while directing an assault on Regensburg but continues to direct French forces against the Austrians. Vienna falls to the French May 12, but the Battle of Aspern (Essling) May 21 to 22 brings Napoleon his first defeat.

The French retreat to the small (1 square mile) island of Lobau in the Danube, jamming it with upwards of 100,000 men plus 20,000 wounded but without medical supplies and with few provisions. Napoleon rejects his generals' advice to retreat and calls for reinforcements.

The Battle of Wagram July 5 to 6 ends in victory for the French but at a terrible cost: 23,000 French soldiers killed or wounded, 7,000 missing out of a 181,700-man army that includes 29,000 cavalrymen, while the Austrian army of 181,700 (4,600 cavalrymen) is forced to retreat after losing 19,110 killed or wounded, 6,740 missing.

Austrian statesman Clemens Wenzel Lothar Metternich-Winneburg, 36, has urged the war against Napoleon, been arrested at Paris in reprisal for Austria's action in interning two members of the French embassy in Hungary, been conducted to Vienna in June under military guard, is exchanged in July for the two French diplomats, and is present with the emperor at the Battle of Wagram (above).

Metternich (above) is made Austrian minister of state August 4, leaves soon thereafter for a peace conference at Altenburg, and is appointed minister of foreign affairs October 8 to begin a 40-year career in that post, but he has no hand in the Treaty of Schönbrunn, signed October 14, ending hostilities between France and Austria, which joins Napoleon's Continental system.

Napoleon announces to his wife Josephine that he is compelled to divorce her for reasons of state, and his lawyers find a slight technical irregularity in the marriage ceremony of December 1, 1804, which they use to have the marriage declared null and void.

Sir Arthur Wellesley has defeated the Spanish king Joseph Bonaparte July 27 to 28 at Talavera de la Reina and been made duke of Wellington for his triumph.

Napoleon annexes the Papal States. Pope Pius VII is taken prisoner July 5 and will remain in custody until 1814.

The Treaty of Amritsar in India checks the advance of a Sikh confederacy under Ranjit Singh, 29, and fixes the East India Company's northwest territorial border at the Sutlej River.

The territory of Illinois is cut from Indiana Territory and established as a separate territory (*see* 1818; Michigan Territory, 1805).

The Treaty of Fort Wayne negotiated September 30 by Gen. William Henry Harrison obtains 3 million acres of Indiana Territory Indian land on the Wabash River for the United States (*see* Tippecanoe, 1811).

Britain repeals her 1563 Statute of Apprentices in the name of "economic freedom." Parliament also repeals other measures that have protected British workers.

Robert Owen proposes to his partners that they stop employing children under age 10 in their cotton mills and build nurseries, schools, playgrounds, and lecture halls for the people of New Lanark (*see* 1806). His proposals are rejected, Owen finds new partners to buy out the old ones, but they "objected to the building of the schools, and said they were cotton-spinners and commercial men carrying on business for profit, and had nothing to do with educating children; nobody did it in the manufactories." Owen finds a London group that will go into business with him on the promise of a reasonable return on capital rather than of exorbitant profits; the group includes Jeremy Bentham (*see* 1789), Quakers William Allen and James Walker, and the future lord mayor Michael Gibbs (*see* 1814).

Hudson's Bay Company fur trader David Thompson, 37, erects a trading post that is the first building in what will be Idaho Territory. He will survey and map more than 1 million square miles of Canada between Lake Superior and the Pacific tracing lakes and rivers that will include the Columbia, Mississippi, and Saskatchewan.

The British minister to Washington assures President Madison that Britain will repeal her 1807 Orders in Council, Madison withdraws the U.S. embargo on trade with Britain imposed late in 1807, 1,200 ships sail for British ports, but London says its ambassador has exceeded his authority and trade halts once again (*see* 1811).

A pamphlet on "The High Price of Bullion a Proof of the Depreciation of Bank-Notes" by London broker David Ricardo, 37, will be influential in determining British economic policy. It establishes Ricardo as a major economic theorist (*see* 1811; 1817).

U.S. mariner Moses Rogers, 30, makes the first ocean voyage by steamboat in the *Phoenix,* launched last year by John Stevens. He steams from New York round Sandy Hook and Cape May to the Delaware River (*see* 1819).

William H. Wollaston invents the reflecting goniometer for physics research.

Acquired characteristics can be inherited, says French naturalist Jean Baptiste Pierre Antoine de Monet, 65, chevalier de Lamarck. The false ideas in his *Zoological Philosophy (Philosophic Zoölogique)* will persist for more than a century.

Danville, Ky., surgeon Ephraim McDowell, 38, performs the first recorded U.S. operation to remove an ovarian tumor. His procedure on Jane Todd Crawford at Greensburg demonstrates that surgery of the abdominal cavity need not be fatal.

Miami University has its beginnings in a college founded on the Miami River at Oxford, Ohio.

The first U.S. parochial school is founded near Baltimore by English-American widow Elizabeth Ann Seton, 36, a recent convert to Roman Catholicism who is joined by other Catholic women in starting the Sisters of Charity, a religious order that will be allied in 1850 with France's Daughters of Charity of St. Vincent de Paul. Mother of five, "Mother Seton" will be credited with several "miracles" and be canonized in 1975 as the first American saint.

Nonfiction *A History of New York from the Beginning of the World to the End of the Dutch Dynasty* by New York essayist Diedrich Knickerbocker (Washington Irving), 26, is published on St. Nicholas Day, December 6. Irving dedicates his comic history derisively to the 5-year-old New-York Historical Society and makes references to an impish, pipesmoking St. Nicholas who brings gifts down chimneys, thus beginning a legend that will travel round the world (*see* Moravians, 1740; Moore, 1823).

Fiction *Coelebs in Search of a Wife* by Hannah More; *The Elective Affinities (Die Wahlverwandtschaften)* by Johann Wolfgang von Goethe.

Painting *Malvern Hill* by English painter John Constable, 33; *Mrs. Spiers* by Henry Raeburn; *Monch am Meer* by Kaspar Friedrich.

England's Two Thousand Guineas race has its first running at the Newmarket races.

1809 ***(cont.)*** Export of Chilean nitrates to Europe begins, but the nitrates are used less for fertilizer than for explosives (*see* Humboldt, 1802).

1810 Napoleon is married March 11 by proxy at Vienna to the Austrian archduchess Maria Luisa, 18, in a match arranged by Austrian foreign minister von Metternich. The "little corporal" hopes to father an heir.

Tyrolese patriot Andreas Hofer, 41, defeats a Bavarian army in a rebellion against Bavarian rule but is later defeated by combined French and Bavarian forces, forced into hiding, betrayed, and shot February 20 at Mantua.

Prince Christian of Holstein-Augstenburg is elected heir to the Swedish throne but dies suddenly in May and replaced as crown prince by Napoleonic general Jean Baptiste Jules Bernadotte, 47, who will succeed to the throne in 1818.

France annexes Holland July 9 following the abdication and flight of Napoleon's brother Louis Bonaparte, who has been king since 1806 but refused to join Napoleon's Continental system. The emperor annexes other lands along the Channel and North Sea coasts, including Bremen, Lübeck, Lauenburg, Hamburg, and Hanover in an effort to discourage smuggling of British goods which continue nonetheless to enter European markets.

Prussia's Queen Louise dies July 19 at age 34. Napoleon has tried to destroy her reputation, but his charges have only endeared her to her people.

The British secretary of state for foreign affairs George Canning, 40, resigns September 9, saying that the War Office is hampering progress in the Peninsular War. The secretary for war and the colonies Robert Stewart, 41, marques of Londonderry and Viscount Castlereagh, thinks Canning has called him incompetent and challenges him to a duel. He wounds Canning in the thigh and disgraces him.

Spanish colonists in South America proclaim loyalty to Ferdinand VII, who remains a French prisoner at the château of Valencay. A provisional government of the Rio de la Plata is established in Ferdinand's name May 25, a similar junta is established at Caracas, and another at Santiago.

Mexican rebels led by the Creole priest Miguel Hidalgo y Costilla, 57, capture Guanajuato, Guadalajara, and Valladolid. They approach the capital with an ill-equipped force of 80,000, but a small Spanish force under Felix Maria Calleja del Rey, 60, forces them to retreat November 6 (*see* 1811).

President Madison takes advantage of an insurrection in west Florida to seize Spanish territory whose ownership has not been clearly defined in the Louisiana Purchase of 1803 (*see* 1818).

Prussia abolishes serfdom and gives ex-serfs the lands they have cultivated for their landlord masters, allowing them to sell the lands if they choose, but Prussian peasants will live for nearly a century under conditions not far removed from serfdom (*see* 1807; Russia, 1858).

U.S. Treasury Secretary Albert Gallatin, 49, estimates the total value of U.S. manufactured products at $120 million for the year. In some lines, says Gallatin, the nation is now self-sufficient.

John Jacob Astor organizes the Pacific Fur Co. (*see* 1808; Astoria, 1811).

The differential gear invented by German art publisher Rudolph Ackerman, 46, permits carriages to turn a sharp corner. His steering mechanism will be used within the century on horseless carriages.

Amoskeag Manufacturing Co. is founded on the Merrimack River in the New Hampshire town of Amoskeag that becomes Manchester, taking its name from the great English milltown. The new company will soon be operating the world's largest cotton mill.

Plymouth, Conn., clockmakers Seth Thomas, 25, and Silas Hoadley buy Eli Terry's clock business after fulfilling a 3-year contract to supply Terry with 4,000 clocks (*see* 1803). Thomas will sell his share in 1812 and go into business for himself at Plymouth Hollow (later Thomaston), organize the Seth Thomas Clock Co. in 1853, and leave it to his son upon his death early in 1859.

Patents registered by the U.S. Patent Office begin a rapid rise after 20 years of averaging 77 per year.

U.S. cotton mills grow in number to 269, up from 62 last year (*see* Slater, 1789; Lowell, 1812).

Barcelona and Cádiz lose as many as 25,000 to yellow fever.

Organon of Therapeutics (*Organon der Rationellen Heilkunde*) by German physician Christian Friedrich Samuel Hahnemann, 51, pioneers homeopathic medicine (a term that will not be used until 1826). Hahnemann announced the principle 14 years ago that if a drug produces symptoms in a healthy person then a minute quantity of that drug can cure a patient with such symptoms. He has worked with Peruvian *cinchona* and other drugs.

Nonfiction *De l'Allemagne* by Mme. de Staël appears in the first of its three volumes, introducing German romanticism to French literature.

Poetry *The Lady of the Lake* by Walter Scott: "Hail to the chief who in triumph advances"; *The Forest Minstrel* by Scottish poet James Hogg, 40.

Painting *Distribution of the Eagles* by Jacques Louis David; *The Battle of Austerlitz* by François Gerard; *Los Desastres de la Guerra* (engravings) by Francisco Goya. Johann Zoffany dies at Kew, Surrey, November 11 at age 77.

First performance Music to Goethe's play *Egmont* of 1788 by Ludwig van Beethoven 6/15 at Vienna's Burgtheater.

French chemist Louis Nicolas Vauquelin identifies the active principle in tobacco and calls it *Nicotianine* after Jean Nicot (*see* 1561) and the plant *Nicotiana rustica*.

France makes tobacco a government monopoly.

Paris inaugurates a column in the Place Vendôme August 15 to commemorate Napoleon's victory at Austerlitz in December 1805. Modeled after Trajan's column of the 2nd century A.D. at Rome, the column is girdled with a spiral band of bronze made from 1,200 enemy cannon captured at Austerlitz, covered with bas reliefs depicting military scenes, and surmounted by a statue of Napoleon as Caesar.

Work continues at Paris on the Arc de Triomphe begun 4 years ago by the sculptor Clodion (Claude Michel), now 72 (*see* 1836).

Scots-American ornithologist Alexander Wilson, 44, sights a flock of passenger pigeons 250 miles long over Kentucky and estimates it to contain 2 billion birds, nearly half of all the passenger pigeons in the country. The birds account for roughly one-third of America's avian population (*see* 1800; Audubon, 1813).

The U.S. cotton crop reaches 178,000 bales as Eli Whitney's cotton gin of 1792 encourages production.

Nicolas Appert announces his 1808 discovery of vacuum-packed food in *Le Livre de Tous Les Ménages,* which is published in June with the subtitle *L'art de Conserver, Pendant Plusieurs Annés, Toutes Les Substances Animales et Végetales.*

The first English patent for a tin-plated steel container is issued to Peter Durand.

A London cannery that will soon be shipping thousands of cases of tinned foods for use by the Royal Navy is opened under the name Donkin and Hall by Bryan Donkin, who has interests in an ironworks (*see* 1812).

French banker Benjamin Delessert, 37, sets up beet-sugar factories at Passy to supply France with sugar in the absence of imports blockaded by the Royal Navy. The factories will produce more than 4 million kilos of high-cost sugar in the next 2 years (*see* 1811; 1814).

The Munich Oktoberfest has its beginnings in a festival staged to celebrate the marriage of Bavaria's crown prince Ludwig (Louis), 24, to princess Therese of Saxe-Hildburghausen, 18, who will bear him seven children.

New York passes Philadelphia in population to become the largest U.S. city.

New Orleans has a population of 24,562, while all of Louisiana has 70,000 and no other city in the West has more than a few thousand. Pittsburgh has 4,768, Lexington 4,326, Cincinnati 2,450, Louisville, Nashville, Natchez, St. Louis about 1,000 each.

The U.S. population reaches 7,215,858 with nearly 4 million in the Northeast, 2.3 million in the South, and less than 1 million west of the Alleghenies.

France makes abortion a criminal offense (*see* 1814).

1811

The Regency Act passed by Parliament February 5 authorizes the prince of Wales, 49, to reign in place of his father George III, who has become permanently mad after losing his favorite daughter Amelia. He has had periodic bouts of insanity since as early as 1765, they have been more frequent since 1788, and he is now blind as well.

President Madison has sent Britain an ultimatum February 2 demanding that she revoke her 1807 Order in Council and cease harrassment of U.S. ships (*see* 1812).

The Ottoman viceroy Mohammed Ali invites Egypt's Mameluke leaders to a banquet at the citadel of Cairo March 1 and massacres them for having plotted against him. Only a few escape the slaughter, and Mohammed Ali will reign supreme until 1849.

British forces capture Batavia from the French in Java. English colonial officer Thomas Stafford Raffles, 30, has persuaded his superiors to take the island, accompanied the expedition, and will administer Java until 1816, introducing a new system of land tenure and removing trade barriers.

British forces on the Iberian Peninsula triumph over the French May 5 at the Battle of Fuentes de Onoro and May 16 at the Battle of Albuera where Gen. William Carr Beresford, 42, defeats the French marshal Nicolas Jean de Dieu Soult, 42.

A Venezuelan junta formed by the captain general of Caracas proclaims independence, disavowing the Spanish regency created by Napoleon, invites other Spanish colonies to follow its example, and dispatches a mission to enlist English support for the revolutionists of New Granada. Francisco de Miranda returns to Venezuela to assume leadership, a general congress proclaims Venezuelan independence July 7 and gives Miranda command of the revolutionary forces; Simon Bolívar, 28, returns with the mission sent to London, becomes one of Miranda's lieutenants, and joins the fighting that begins against Spanish forces under Juan Domingo Monteverde (*see* 1812).

Paraguay declares her independence August 14.

Spanish control ends in the Banda Oriental, where Uruguayan revolutionist José Artigas, 65, has been active.

Chilean revolutionist José Miguel Carrera, 26, overthrows a conservative junta at Santiago (*see* 1812).

Cartagena in the viceroyalty of New Granada declares its independence November 11.

Spanish troops under Felix Calleja crush the forces of the revolutionary Creole priest Miguel Hidalgo in New Spain at the Bridge of Calderon near Guadalajara (*see* 1810). Hidalgo falls into Spanish hands and is tried and executed (*see* 1812; Mexican independence, 1821).

The Battle of Tippecanoe November 7 on the Wabash River ends in complete defeat for the Shawnee who have been drawn into battle by Gen. Harrison in the absence of their chief Tecumseh (*see* 1809).

A New Orleans slave revolt is crushed January 10 after an uprising on a plantation spreads to involve more than 400 blacks. Of these, 66 are killed in the fighting or subsequently executed, their heads being displayed on the road to the plantation.

1811 ***(cont.)*** "Luddites" riot against English textile manufacturers who have replaced craftsmen with machines at Nottingham. Organized bands, taking their name from a real or imaginary "General Ludd" or "King Ludd," burn one of Richard Arkwright's factories and break into the house of James Hargreaves to smash his spinning jenny (*see* 1770; 1812; Arkwright, 1769).

Fort Ross above San Francisco Bay is founded February 2 by Russians who land at Bodega Bay and build the fort as the center of a farming colony and as a trading post for sea otter skins (*see* 1769; Sutter, 1841).

Astoria at the mouth of the Columbia River is founded by the skipper of fur trader John Jacob Astor's ship *Tonquin,* who establishes a trading post (*see* 1814; Pacific Fur Co., 1810).

Two Portuguese half-caste explorers complete the first crossing of the African continent after a 9-year journey.

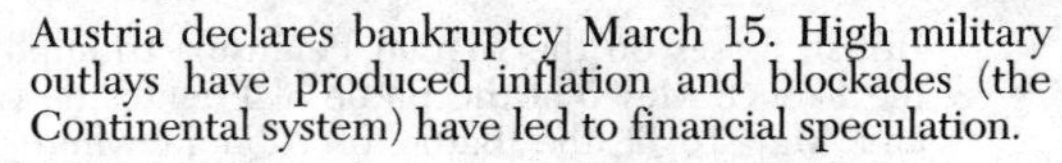

Austria declares bankruptcy March 15. High military outlays have produced inflation and blockades (the Continental system) have led to financial speculation.

Britain adopts paper money as currency May 10 to ease an economic crisis.

A bullion report issued by British government economists establishes the intrinsic-value theory of money. Banknotes may be a useful medium of exchange, but the notes must bear a definite ratio to the amount of coin and bullion in the vaults. Government cannot create money, which is rather a token of labor performed. Government can acquire money only through taxation or by borrowing, and for government to issue irredeemable paper money is to violate the sanctity of contracts, cheat creditors, increase prices, and disrupt business (*see* Smith, 1776; Ricardo, 1809; Bank Charter Act, 1844).

U.S. customs duties fall to $6 million as loss of trade with Britain and France reduces duties to half their 1806 level. The Napoleonic wars and Britain's Orders in Council have reduced trade, and Congress will impose an excise tax in 1813.

The charter of the Bank of the United States lapses (*see* 1791).

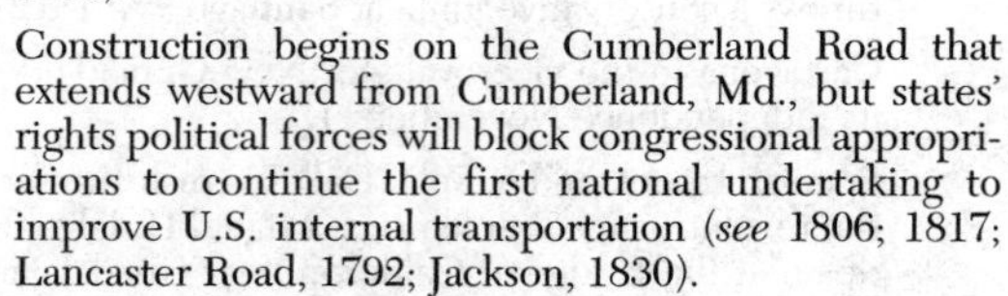

Construction begins on the Cumberland Road that extends westward from Cumberland, Md., but states' rights political forces will block congressional appropriations to continue the first national undertaking to improve U.S. internal transportation (*see* 1806; 1817; Lancaster Road, 1792; Jackson, 1830).

The S.S. *New Orleans* launched by Nicholas Roosevelt leaves Pittsburgh September 11 with Roosevelt and his wife among the passengers. First steamboat in the Mississippi Valley, the ship starts down the Ohio River for the Mississippi on a 14-day voyage to New Orleans (*see* Fulton, 1807; Shreve, 1816).

A two-cylinder steam locomotive patented by English engineer John Blenkinsop, 28, works by means of a racked rail and a toothed wheel. He will test it successfully next year and see it used in collieries (*see* Trevithick, 1804; Stephenson, 1814).

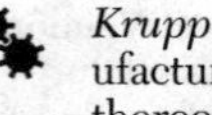

Krupp Gusstahlfabrik (Cast Steel Works) "for the manufacture of English cast steel and all articles made thereof" is founded at Essen by German industrialist Friedrich Krupp, 24, a descendant of Arndt Krupp (*see* 1599; Huntsman, 1740). The emperor Napoleon has offered a 4,000 franc reward to the first steelmaker who can match the British process, and Krupp gives up his coffee and sugar wholesaling trade to try for the reward, opening a small steelworks in a shed adjoining his house (*see* 1816).

Avogadro's hypothesis, stated by Italian chemist-physicist count Amedeo Avogadro, 35, states that equal volumes of all gases at the same temperature and pressure contain equal numbers of molecules.

French chemist-physicist Pierre Louis Dulong, 26, discovers nitrogen chloride.

French chemist Bernard Courtois, 34, isolates iodine while studying products obtained by leaching the ashes of burnt seaweed. Iodine will be used as an antiseptic and to purify drinking water.

Anatomy of the Brain by Scottish anatomist Charles Bell, 37, announces discovery of the distinct functions of sensory and motor nerves.

Fiction *Sense and Sensibility, A Novel by a Lady* is published anonymously at London. Its author is Jane Austen, 36, a spinster whose depiction of manners and mores in English country society will rank her as one of the world's great novelists.

Opera *Abu Hassan* 6/4 at Munich's Court Theater, with music by Carl Maria von Weber, libretto by Franz Carl Hiemer.

First performances Concertina for B flat Clarinet and Orchestra by Carl Maria von Weber 4/5 at Munich; Concerto for Clarinet and Orchestra No. 1 in F minor and Concerto for Violoncello and Orchestra in D major by Weber 6/13 at Munich; Concerto for Pianoforte and Orchestra No. 2 in E flat major by Weber 11/9 at Munich; Concerto No. 5 in E flat major for Pianoforte and Orchestra *(Emperor)* by Ludwig van Beethoven 11/28 at Leipzig.

John Nash designs Regent's Park and its terraces at London and will lay out Regent Street later in the decade (*see* 1806; Brighton Pavilion, 1815; Buckingham Palace, 1825).

New York adopts a Commissioners' Plan marking off future Manhattan streets and avenues in a grid pattern employed earlier in Spanish colonial towns. Above Canal Street the city remains mostly farmland.

Earthquakes beginning December 16 turn the Mississippi River to foam in places between Memphis and St. Louis, the current runs upstream for several hours, most of New Madrid, Mo., crumbles down the bluffs into the river, the sequence of shocks is felt as far away as Detroit, Baltimore, and Charleston, and the tremors continue for months.

Napoleon awards Benjamin Delessert the medal of the Legion d'Honneur for his success with beet sugar, he orders the immediate allotment of 32,000 hectares to sugar beet cultivation, and he tells the Paris Chamber of Commerce that "the English can throw into the Thames the sugar and indigo which they formerly sold on the Continent with high profit to themselves" (*see* 1810; 1813).

Iodine (above) will prove to be an essential nutrient in human diets (*see* Baumann, 1896).

The Russian ambassador to Paris Prince Aleksandr Borosovich Kurakin, 35, introduces the practice of serving meals in courses *(à la Russe)* instead of placing many dishes on the table at once (*see* 1852).

Napoleon decrees that foundling hospitals in France shall be provided with turntable devices so that parents can leave unwanted infants without being recognized or questioned. Millions of infants have been drowned, smothered, or abandoned, depriving the French army of potential recruits.

1812 Napoleon invades Russia in June as Britain and the United States go to war over the impressment of U.S. seamen (below).

The Grande Armée of 600,000 that invades Russia includes Austrians, Dutch, Germans, Italians, Poles, Prussians, and Swiss as well as Frenchmen, but 80,000 are sick with dysentery, enteric fever, and typhus after the Battle of Ostrowo in July.

British troops under the duke of Wellington (Arthur Wellesley) defeat the French July 22 at Salamanca and enter Madrid in mid-August (*see* Rothschilds, below).

The Battle of Borodino, September 7, is a bloody encounter on the Moskova River. Russia's wily field marshal Mikhail Kutusov, now 66, retreats to save his army, Napoleon enters Moscow, but most of the 300,000 inhabitants have fled, and fires set by the Russians burn much of Moscow in the next 5 days.

Napoleon begins a retreat from Moscow October 19, and the retreat turns into a rout as Marshal Kutusov defeats Marshal Ney and Marshal Davout at Smolensk in November. Crippled by hunger, cold, and lack of salt, harrassed by Cossack troops and Russian irregulars, the invading force has dwindled to no more than 100,000 by mid-December, when the survivors finally straggle across the Nieman River.

The War of 1812 begins June 18 as Washington declares war, unaware that the British Orders in Council of 1807 were withdrawn June 16.

U.S. troops invade from Detroit in July under the command of Gen. William Hull, 59, who is outmaneuvered and defeated by the British.

Fort Dearborn on Lake Michigan is evacuated by order of Gen. Hull (above). Its inhabitants are massacred August 15 by Indians while being escorted to Fort Wayne, and all are killed including 12 children.

Fort Dearborn is sacked and burned August 16 and Gen. Hull forced to surrender. He will be court-martialed, convicted of cowardice and neglect of duty, and sentenced to be shot in 1814, but the sentence will not be carried out.

Bonaparte's retreat. Russia's cold vastness and wily Cossack leaders were too much for Napoleon's Grande Armée.

The British frigate *Guerrière* that has seized so many U.S. seamen is destroyed August 19 by the 15-year-old U.S. frigate *Constitution* under the command of Isaac Hull, 39. The *Constitution* has won the name "Old Ironsides" because enemy cannonballs appear to bounce off her tumble home; her victory gives heart to the Americans.

The U.S. sloop *Wasp*, commanded by Jacob Jones, 44, captures the British frigate *Frolic* October 18, but Jones loses both his prize and his own ship when he encounters the 74-gun British vessel *Poictiers* while en route home.

The U.S. ship *United States* commanded by Stephen Decatur, 23, defeats the British frigate *Macedonian* October 25.

The British ship *Java* is defeated and destroyed December 20 by the *Constitution* (above) under the command of William Bainbridge, 38, but while the U.S. victories at sea win wide acclaim in America, the British actually enjoy more success in the sea war.

U.S. frontiersmen eager for lands that may be gained by war give James Madison his margin of victory over New York's De Witt Clinton in the November elections (below).

Louisiana is admitted to the Union as the 18th state.

Missouri Territory is established west of the Mississippi.

Eastern shipping interests support De Witt Clinton and say President Madison's neutrality policy is helping the despot Napoleon, "war hawks" support Madison, Clinton carries every northern state except Pennsylvania and Vermont, but his 89 electoral votes are topped by Madison, who receives 128 and wins re-election.

Gerrymander is introduced as a term to describe the redrawing of election district boundaries for political

1812 ***(cont.)*** purposes. When the Massachusetts legislature redraws senatorial districts to isolate Federalist strongholds and ensure Republican domination of future elections, painter Gilbert Stuart adds a head, wings, and claws to a map of the districts, saying, "That will do for a salamander." Editor Benjamin Russell of the *Centinel* replies, "Better say a Gerrymander," but Gov. Elbridge Gerry, 68, who wins election as vice president under James Madison (above) has not been responsible for the redistricting (*see* civil rights, 1960).

A March 26 earthquake in Venezuela has spared territories held by the Spanish royalists but wreaked havoc among Francisco de Miranda's revolutionists, demoralizing the liberation forces (*see* 1811). Priests calls the catastrophe divine retribution for disloyalty, Miranda is made dictator to reorganize the revolutionary elements, but he is forced to capitulate, falls into Spanish hands, and is sent as a prisoner to Cadiz, where he will die in 1816 (*see* Bolívar, 1813).

Mexican revolutionists led by José Maria Morelos take Oaxaca and move on to Acapulco (*see* 1811; 1813).

Spanish colonial troops quickly suppress revolutions in the captaincy general of Guatemala which includes territory that will become Guatemala, San Salvador, Honduras, Nicaragua, Costa Rica, and Chiapas.

Chilean revolutionist José Miguel Carrera finds his dictatorial rule opposed by patriot Bernardo O'Higgins, 34 (*see* 1811; 1817).

Spain's cortes (national assembly) promulgates a constitution May 8 providing for universal suffrage and other democratic principles (*see* 1836).

The Luddite riots that began last year in England spread to Yorkshire, Lancashire, Derbyshire, and Leicestershire as wheat prices soar and inflate the price of bread. Public opinion supports the rioters especially after soldiers shoot down a band of Luddites at the request of a threatened employer who is subsequently murdered (*see* 1813).

Gold, English guineas, Dutch gulden, and French Napoleon d'or smuggled south by the eight sons of Frankfurt's Meyer Rothschild have financed the duke of Wellington in his peninsular campaign (above) (*see* 1801). Asked by the British government to help, the Rothschilds, with no language other than Yiddish, have used fake passports, false names, disguises, and bribes to elude the French and organize what amounts to the world's first international bank clearing house (*see* Nathan Rothschild, 1815).

Philadelphia's Girard Bank is founded by merchant Stephen Girard with more than $1 million obtained in the shipping trade (*see* 1805).

New York's Citibank has its beginnings in the City Bank of New York that opens June 16 at 52 Wall Street (*see* 1865).

British National Light and Heat Co. is chartered by the German promoter Frederick Albert Winsor, who lighted one side of London's Pall Mall with gas in 1807 and has recognized the possibilities of piping gas from a central plant to consumers through a system of London gas mains. His German employers dismiss Winsor for incompetence, but the company will erect a London gasworks in 1814 and supply gas to lamps in the parish of St. Margarets, Westminster. Other companies will be formed in other cities, and the age of gas lighting will arrive in Britain (*see* Welsbach, 1885).

Scottish engineer Henry Bell, 45, pioneers European steam navigation with his three-horsepower steamboat *Comet*, which he will operate on the Clyde for the next 8 years (*see* 1816; *New Orleans,* 1811).

New Englander Francis Cabot Lowell, 37, charters a cotton fabric company in association with his brother-in-law Patrick Tracey Jackson, 33. Lowell has memorized the designs and specifications of English textile machinery (*see* 1814; Slater, 1789).

Eli Whitney receives a second government contract, this one for 15,000 muskets (*see* 1801). He will grow rich from his army supply business and attract other gunmakers to his manufacturing ideas (*see* Colt, 1836).

Clockmakers from New Haven and the Naugatuck Valley come to observe Eli Whitney's methods and his crude milling machines for chipping and planing metals. A new machine tool industry begins to flourish in New England.

The 30-month War of 1812 (above) will encourage U.S. manufacturing (it will otherwise gain nothing for either side) and U.S. agriculture will begin its long, slow yielding of supremacy to industry.

Swiss traveler-orientalist John Lewis Burckhardt, 27, rediscovers Petra in the Middle East August 22, finding ruins of a small fort built by Crusaders in the 12th century on ruins of the ancient Arab city 180 miles south of Amman. Local legend says that Moses brought forth water from a rock at Petra during the flight of the Israelites from Egypt, but an earthquake leveled the city in 350 A.D.

Récherches sur les ossemens fossiles by Baron Cuvier founds the science of paleontology (*see* 1798).

Russian chemist Gottlieb Sigismund Iorchoff suggests the first understanding of catalytic processes. He shows that starch breaks down to the simple sugar glucose when boiled with dilute sulfuric acid.

The *New England Journal of Medicine* begins publication under the name *New England Journal of Medicine and Surgery*.

Napoleon's surgeon Baron Dominique Jean Larrey performs 200 amputations in 24 hours after the Battle of Borodino (above).

Poetry *Childe Harold's Pilgrimage* by English poet George Gordon, 24, Lord Byron, who traveled through Portugal and Spain 2 years ago despite the Peninsular War. Byron journeyed on to Greece and Turkey, he swam across the Hellespont May 3, 1810, and his first two cantos deal with his travels and love affairs of 1809 and 1810.

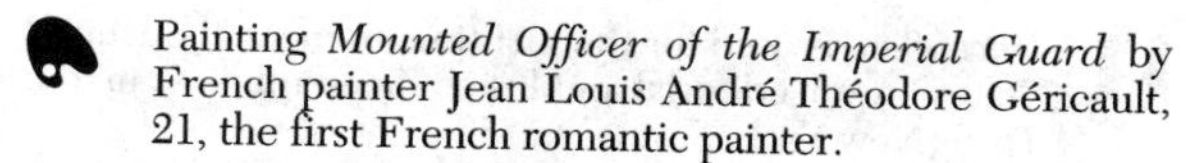

Painting *Mounted Officer of the Imperial Guard* by French painter Jean Louis André Théodore Géricault, 21, the first French romantic painter.

The camera lucida invented by William Hyde Wollaston uses mirrors to project the image of an external object on a plane surface so that its outline may be traced.

First performances Overture to *King Stephen* by Ludwig van Beethoven 2/9 at Budapest's Neues Deutsches Theater, for a new play by August Friedrich von Kotzebue.

France's Brigade de la Sureté is founded by detective François Eugène Vidocq, 37, who has been placed in charge of a plainclothes detail staffed entirely by ex-convicts. They prove far more efficient than the regular police force by infiltrating the underworld.

New York's City Hall is completed just north of the City Common at Broadway and Park Row. Designed in Federal style with French influence, its principal architect has been a Scotsman named Mangin, who has helped design the Place de la Concorde in Paris. Its north side is left undecorated, there being little but farmland in that direction.

British wheat prices soar to £30 per cwt, the highest level ever and one that will not be matched for another 160 years (*see* 1798; 1834).

An English canning factory at Bermondsey established by Bryan Donkin produces tinned foods for British naval and military forces, but such foods remain unavailable to the public, which is hard-pressed by high prices (*see* 1810; Donkin-Hall, 1814).

1813 Wars of liberation against the French begin in the wake of Napoleon's disastrous Russian expedition. Prussia's Friedrich Wilhelm III issues an appeal February 3 from Breslau calling for a volunteer corps, young men and students rally to the cause, the Treaty of Kalisch February 28 makes Russia a Prussian ally, and Friedrich Wilhelm establishes the Landwehr and the Landsturm Iron Cross March 17.

A Swedish army of 30,000 under the command of Crown Prince Bernadotte takes the field following a March 3 treaty with Britain. The British have paid 1 million rix dollars and promised not to oppose the union of Norway with Sweden.

Russian troops occupy Hamburg in late March, and the dukes of Mecklenburg withdraw from the Confederation of the Rhine. Russian and Prussian troops occupy Dresden March 27 following the withdrawal of Marshal Davout, but Napoleon raises a new army of 500,000 to replace his Russian losses.

An Allied army of 85,000 attacks Napoleon on the Elbe, but the emperor has a larger force and gains victory May 2 at the Battle of Lützen (Gross-Görschen) despite heavy losses. The Allies withdraw to Lusatia, the emperor enters Dresden, and the king of Saxony returns from Prague to form a close alliance with the French.

The Battles of Bautzen and Wurschen May 20 and 21 give Napoleon fresh victories, again at great cost in French lives (Gen. Duroc is killed in action May 22). The emperor forces the Allies back across the Spree into Silesia.

Marshal Davout takes Hamburg May 30 following the withdrawal of Russian troops, and the Armistice of Poischwitz June 4 ends hostilities until August.

Britain agrees to subsidize Prussia and Russia under terms of a June 15 treaty signed at Reichenbach.

Austria declares war on France August 12 following the breakdown of negotiations at Prague in which von Metternich has tried to mediate with France's new minister of foreign affairs Marquis Armand Augustin Louis de Caulaincourt, 41, and the Prussian envoy Baron Wilhelm von Humboldt, 46, a brother of the naturalist.

Prussian general Baron Friedrich Wilhelm von Bülow, 58, saves Berlin August 23 by defeating the French at Grossbeeren. The crown prince of Saxony looks on, refusing to aid the French general Nicolas Charles Oudinot, 46.

Prussian general Gebhard Leberecht von Blücher, 71, scores a victory August 26 at the Battle of Katzbach, defeating a French army under Jacques Etienne Joseph Alexandre Macdonald, 48. Friedrich Wilhelm III makes von Blücher prince of Wahlstatt.

The Battle of Dresden August 26 to 27 gives Napoleon his last major victory on German soil. He defeats an Allied army under the Austrian prince Karl Philipp von Schwarzenberg, 42, who served with Napoleon last year in Russia and now commands a Bohemian army.

The Battle of Dennewitz September 6 results in victory for Baron von Bülow, who prevents Marshal Ney from taking Berlin.

The Battle of Leipzig October 16 to 19 will be called the "Battle of the Nations"; it ends in defeat for the emperor Napoleon, who has lost 219,000 men to typhus and 105,000 in battle and has left Dresden in order to avoid being cut off from France by the Allied armies. His Saxon and Württemberg contingents desert him October 18, the Allies storm Leipzig October 19, capturing the king of Saxony, and Napoleon begins a retreat after losing 30,000 men.

The Austrian emperor Francis I makes Metternich a hereditary prince of the empire October 20.

The kingdom of Westphalia comes to an end following the great French defeat at Leipzig (above). Napoleon's brother Jérôme flees from Cassel, and the old rulers are restored in Cassel, Brunswick, Hanover, and Oldenburg.

A Dutch revolt November 15 expels French officials, and an Allied army under Gen. von Bülow enters Holland while Count Bernadotte invades Holstein with a Swedish army.

The duke of Wellington has advanced through northeast Spain following the recall of much of the French army to support Napoleon against the Prussians and Russians. He has triumphed over Marshal Jourdan June 21 at the Battle of Vittoria and sent Joseph Bonaparte scurrying back to Paris. Pamplona falls to British

1813 *(cont.)* and Spanish forces October 31 after a long siege, and Wellington crosses the French frontier to defeat Marshal Soult November 10.

Dresden surrenders to Allied forces November 11; Stettin, November 21; Lübeck, December 5; Zamose, Modlin, and Torgau, December 26; and Danzig, December 30 as the Allies vow to invade France if Napoleon will not make peace.

The War of 1812 continues between British and U.S. forces. U.S. naval officer James Lawrence, 32, commands the *Chesapeake* in an engagement with the British frigate *Shannon* June 1. Lawrence has been transferred to the *Chesapeake* after defeating the British brig *Peacock* February 24, but the British defeat him. Mortally wounded, he is carried below decks crying, "Don't give up the ship!"

The Battle of Lake Erie September 10 ends in victory for an improvised U.S. squadron commanded by Capt. Oliver Hazard Perry, 28, who sends a message to General Harrison of 1811 Tippecanoe fame: "We have met the enemy and they are ours: two ships, two brigs, one schooner, and one sloop."

The Battle of the Thames in Ontario October 5 reestablishes U.S. supremacy in the Northwest. Gen. Harrison defeats a British army under Henry A. Proctor whose Shawnee ally Tecumseh is killed at age 45.

Gen. Andrew Jackson defeats Creek tribesmen in the Mississippi Territory November 9 at Talledega. The Creeks attacked Fort Mims in Alabama August 30, massacring more than 500 whites, Jackson retaliates, and he will defeat the Creeks again next year.

Simon Bolívar retakes Caracas (*see* 1811). Commissioned by the Congress of New Granada to fight the Spanish in Venezuela, Bolívar becomes virtual dictator (*see* 1815).

A Mexican Congress convened at Chilpancingo September 14 declares independence from Spain November 6. It makes José Maria Morelos head of government and adopts reforms.

Leaders of England's Luddite movement are hanged or transported after a mass trial at York (*see* 1812; 1816).

Sweden abandons the slave trade.

English humanitarians organize a Society for the Prevention of Accidents in Coal Mines after a series of mine disasters (*see* Davy's safety lamp, 1815).

The British East India Company loses its monopoly in the India trade but continues to monopolize the China trade (*see* 1833).

Britain issues her first gold guineas.

English inventor William Horrocks produces the world's first power loom (*see* Bigelow, 1839).

Théorie Elémentaire de la Botanique by Swiss botanist Augustin Pyrame de Candolle, 35, advances the sciences of plant morphology, taxonomy, and physiology beyond the levels established by Linnaeus in his *Philosophia Botanica* of 1750.

Uncle Sam is used for the first time to mean the United States in an editorial published September 7 in the *Troy* (New York) *Post* (*see* 1852).

Fiction *Pride and Prejudice* by Jane Austen.

Juvenile *The Swiss Family Robinson (Der Schweizerische Robinson)* by Swiss writer-philosopher Johann Rudolf Wyss, 32, who has based the story on a tale his father Johann David, 70, has told about a family shipwrecked on a tropical island. Young Wyss 2 years ago wrote the words to the Swiss national anthem.

Poetry *The Bride of Abydos* by Lord Byron; *Queen Mab* by English poet Percy Bysshe Shelley, 21, who was expelled from Oxford 2 years ago for circulating his pamphlet "The Necessity of Atheism." In addition to his antireligious poem, Shelley issues a treatise on vegetarianism; *The Queen's Wake* by James Hogg.

Opera *Tancredi* 2/6 at Venice's Teatro de la Fenice, with music by Italian composer Gioacchino Antonio Rossini, 20; *L'Italiana in Algeri* 5/22 at Venice's San Benedetto, with music by Rossini.

The Royal Philharmonic gives its first concert March 8 at London.

First performances Symphony No. 1 in D major by Austrian composer Franz (Peter) Schubert, 16, in the autumn at Vienna's Stadtkonvikt (City Church School; private performance); Symphony No. 7 in A major by Ludwig van Beethoven 12/8 at the University of Vienna.

The waltz gains widespread popularity in Europe.

A flight of passenger pigeons seen by painter John James Audubon takes 3 days to pass overhead; Audubon describes it as a "torrent of life" (*see* 1803; Wilson, 1810; *Birds of America*, 1830).

English inventor Edward Howard devises a vacuum pan that will spur the growth of the canning industry; he will obtain a patent in 1835 (*see* Borden, 1853).

Dolley Madison serves ice cream March 4 at the inauguration party of her husband James.

Commercial salt production begins at Syracuse, N.Y., to compensate for the cutoff of salt shipments from Bermuda and Europe (*see* 1786; 1863).

France has 334 sugar plantations by year's end and has produced 35,000 tons of beet sugar (*see* 1814; Delessert, 1811).

1814 Europe ends 22 years of war as America's 3-year War of 1812 draws to a close (below).

The Battle of La Rothière February 1 is a defeat for Napoleon at the hands of the Marshal von Blücher, whose various corps Napoleon then defeats at Champaubert, Montmitrail, Château-Thierry, and Vauchamps. Napoleon beats the main Prussian army at Nangis and Montereau, and refuses an Allied offer to restore the French frontier of 1792.

The Battle of Laon March 9 to 10 begins a series of reverses for Napoleon. Bordeaux falls to the duke of Wellington March 12, Napoleon is defeated at the Battle of Arcis-sur-Aube March 20 to 21, the French are beaten March 25 at the Battle of La Fère-Champenoise, Allied troops storm Montmartre March 30, and the triumphant Allies enter Paris March 31.

Napoleon abdicates unconditionally April 11 and is awarded sovereignty of the 95-square-mile Mediterranean island of Elba with an annual income of 2 million francs to be paid by the French. His wife Marie Louise receives the duchies of Parma, Piacenza, and Guastalla and retains her imperial title as does Napoleon. He arrives at Elba May 4.

France restores her monarchy; the comte de Provence, 59, will reign until 1824 as Louis XVIII (with a 100-day interruption next year).

Charles Maurice de Talleyrand has been instrumental in restoring the Bourbon monarchy (above) and is named minister of foreign affairs to Louis XVIII.

Spain's Ferdinand VII is restored to his throne, which he will hold until 1833.

Sweden loses Finland to Russia but acquires Norway from Denmark when she agrees May 17 to a new Norwegian constitution providing for a single-chamber assembly (Storting) and denying the king his former right to dissolve the assembly or exercise absolute veto power. Norwegians try to elect the Danish prince Christian Frederick as their king but accept Sweden's Charles XII after an invasion by the crown prince Jean-Baptiste Bernadotte.

The kingdom of the Netherlands is created by a union of the Austrian Netherlands (Belgium) and Holland under terms of the June 21 Protocol of the Eight Articles concluded between the prince of Orange and the allied powers.

The Battle of Lundy's Lane near Niagara Falls in Canada July 25 ends with both British and U.S. forces claiming victory. Acclaimed as a hero is U.S. Brig. Gen. Winfield Scott, 28, who has been wounded twice in the fighting.

The Battle of Bladensburg 4 miles from Washington, D.C., August 24, ends in a rout of 7,000 U.S. militiamen by 3,000 British regulars who march into Washington, burn most of the city's public buildings (below) along with several private houses in retaliation for the burning of York (Toronto), teaching the U.S. government that green state militia cannot be relied upon for the nation's defense.

A U.S. naval force under Lieut. Thomas Macdonough, 31, defeats and captures a British squadron on Lake Champlain September 11 in the Battle of Plattsburgh, 16,000 British troops that have launched an invasion from Canada are forced to retire, and the victory assures final success for the Americans.

British ships bombard Baltimore's Fort McHenry September 14; the bombardment is witnessed by Georgetown, Md., lawyer Francis Scott Key, 34, who has been sent on a mission to obtain the exchange of an American held on a British ship (*see* national anthem below).

The Congress of Vienna convenes in September to work out the European peace settlement.

The Treaty of Ghent December 24 ends the War of 1812, but troops unaware of the peace treaty will fight the Battle of New Orleans 2 weeks after its signing (*see* 1815).

Britain and the United States agree to cooperate in suppressing the slave trade under terms of the Treaty of Ghent (above), but the trade actually expands as U.S. clipper ships built at Baltimore and Rhode Island ports outsail ponderous British men-of-war to deliver cargoes of slaves.

Holland abandons the slave trade.

The duke of Sutherland George Granville Leveson-Gower, 56, destroys the homes of Highlanders on his Scottish estates to make way for sheep. The duke is married to the countess of Sutherlandshire, and by 1822 he will have driven 8,000 to 10,000 people off her lands, which comprise two-thirds of the county.

Lancashire millowner Robert Owen joins with Quaker philanthropist William Allen and utilitarian philosopher Jeremy Bentham in a program to ameliorate living conditions of all millhands. Bentham is famous for his 1789 *Principles of Morals and Legislation*; Owen stated last year in *A New View of Society* that human character is determined entirely by environment (*see* 1809; 1824; 1828).

London banker Alexander Baring, 40, takes the position in a Parliamentary debate that the working classes have no interest at stake in the question of British wheat exports and that it is "altogether ridiculous" to argue otherwise: "Whether wheat is 130 shillings or 80 shillings, the labourer [can] only expect dry bread in the one case and dry bread in the other" (Parliament has allowed free export of wheat, and protests have come from manufacturing districts as food prices have risen) (*see* Corn Law, 1815).

The War of 1812 has inspired Americans to improve their roads, strengthen their national government, and support their toddling domestic industry.

Massachusetts becomes a cotton cloth producer to meet the pent-up demand for the cloth that came from England before the war. Francis Cabot Lowell raises $100,000 for the company he started with Patrick Jackson in 1812, uses power from the Charles River to power machines he installs in an old papermill at Waltham, employs farm girls to run the machines, and houses them six to a room while they earn their dowry money. Lowell and Jackson card and spin cotton thread and weave cotton cloth in an enterprise that is soon producing some 30 miles of cloth per day and paying dividends of 10 to 20 percent. After Lowell's death in 1817, Jackson will acquire mill sites on the Merrimack River, and two of these will become the cities of Lowell and Lawrence (*see* 1834).

E.I. du Pont de Nemours has supplied U.S. land and naval forces with 750,000 pounds of gunpowder and been especially strengthened by the War of 1812 (*see* 1803; Henry du Pont, 1861).

1814 ***(cont.)*** John Jacob Astor loses his Astoria to the British but makes large and profitable loans to the U.S. government (*see* 1811; 1817).

The world's first steam locomotive goes into service on the Killingworth colliery railway as English inventor George Stephenson, 34, applies Richard Trevithick's 1804 steam engine to railroad locomotion and replaces horses and mules for hauling coal (*see* Stockton-Darlington line, 1825).

Fire destroys most of The Library of Congress established in 1800 as British troops burn the Capitol. Only the most valuable records and papers are saved, but former president Thomas Jefferson offers his private library of some 6,500 volumes at cost, and Congress by a margin of 10 votes appropriates $23,700 to acquire Jefferson's collection as the nucleus of a new library (*see* 1851).

Fiction *Waverley* by Walter Scott, who publishes a 19-volume *Life and Works of Swift* but turns his full energies to fiction; *Mansfield Park* by Jane Austen.

Poetry *The Corsair* by Lord Byron; *The Excursion* by William Wordsworth.

England's Dulwich Gallery opens to the public, the first public art gallery.

Painting *The Wounded Cuirassier* by Théodore Géricault; *2 May 1808*, *3 May 1808*, *General Palafox on Horseback*, and *Ferdinand VII* by Francisco Goya.

Japanese *ukiyoe* painter Toyoharu Utagawa dies at age 79 after a career in which he has founded a new style by using Western perspective techniques.

Theater English actor Edmond Kean, 26, appears 1/26 at London's Drury Lane Theatre in the role of Shylock in *The Merchant of Venice,* wins great acclaim, and begins a 19-year career as England's greatest Shakespearean actor; *The Dog of Montargis, or The Forest of Bundy (Le chien de Montargis, ou La forêt de Bundy)* by French playwright René-Charles Guilbert de Pixerecourt, 41, 6/18 at the Théâtre de la Gaité, Paris. Pixerecourt will be called "the father of melodrama."

Opera *Fidelio, oder Die eheliche Liebe* 5/26 at Vienna's Kartnertor-Theater, with music (including a new overture) by Ludwig van Beethoven (*see* 1805); *The Turk in Italy (Turco in Italia)* 8/14 at Milan's Teatro alla Scala, with music by Gioacchino Rossini.

First performances Symphony No. 8 in F major by Ludwig van Beethoven 2/27 at Vienna.

Anthem "The Star Spangled Banner" by Francis Scott Key (above) is published in the *Baltimore American* a week after the bombardment of Fort McHenry. The words are soon being sung to the tune of "The Anacreontic Song" by London composer John Stafford Smith, now 64, for the 48-year-old Anacreontic Society (*see* 1931).

The Carabinieri founded by the restored king of Piedmont Victor Emmanuel I to restore law and order will grow to become an 83,000-member elite paramilitary police force with plumed cocked hats.

The White House at Washington gets its name as architect James Hoban works to rebuild the 15-year-old executive mansion and paints it white to conceal the marks of fire set by the British troops who have gutted the structure.

Hortus Jamaicensis by English botanist John Lunan uses the word *grapefruit* for the first time. The fact that the citrus fruit grows in grapelike clusters has evidently suggested the name (*see* Shaddock, 1696, 1751; Don Philippe, 1840).

England's Donkin-Hall factory introduces the first foods to be sold commercially in tins (*see* 1810; Dagett and Kensett, 1819).

French beet sugar production declines sharply as imports of cane sugar resume and undercut prices.

L'art du Cuisinier by Parisian restaurateur Beauvillier of the Grand Taverne de Londres extols such British dishes as "woiches rabettes," "plombpoutingue," and "machepotetesse."

France prohibits abortion under a new law that will remain in force for more than 162 years. Abortion is permitted only "when it is required to preserve the life of the mother when that is gravely threatened."

1815 The Battle of Waterloo (below) ends the career of Napoleon 5 months after the Battle of New Orleans (below).

The Battle of New Orleans January 8 gives Americans their chief land victory in the War of 1812. It follows by 2 weeks the Treaty of Ghent that ended the war last December, but neither side is aware of the treaty. British troops try to seize New Orleans, French pirate-smuggler Jean Laffite, 34, reveals British plans to Gen. Andrew Jackson, many of Laffite's men join in the battle on the American side, and the British retire after 700 of their troops have been killed, 1,400 wounded. U.S. losses are 8 dead, 13 wounded, bringing total battle casualties in the war to 2,260 dead, 4,505 wounded.

The emperor Napoleon hears of discontent under France's restored Bourbon monarchy, leaves Elba, and lands at Cannes with 1,500 men March 1, attracting thousands to his cause as he marches on Paris. He reaches the city March 20, and begins a new reign of 100 days as Louis XVIII flees to Ghent.

A new alliance mobilizes to oppose the Little Corporal's renewed threat to peace. Austria, Prussia, Russia, and Britain raise a million men.

The Battle of Quatre Bras June 16 pits France's Marshal Ney against the duke of Wellington, whose resistance (combined with a confusion over orders) prevents Ney from sending reinforcements to help Napoleon, who is fighting the Prussians 8 miles away at Ligny. Napoleon wins his battle but not decisively.

The Battle of Waterloo June 18 near the Belgian town of that name is fought by opposing armies of 70,000 men each amidst blinding smoke and choking fumes, with the French facing an Allied army of British, Dutch, and German troops from Brunswick, Hanover, and Nassau under the command of the duke of Wellington. The French horses balk at charging the standing squares of British infantrymen, a Prussian army under Gen. von Blücher (who was injured by his own cavalry at Ligny) arrives with vital reinforcements before dark and scatters the French.

Napoleon abdicates once more June 22. Taken as a prisoner of war to the island of St. Helena in the South Atlantic, he will live there until his death in 1821.

The Congress of Vienna confirms the kingdom of the Netherlands, makes Luxembourg a grand duchy with the Dutch king Willem I as grand duke, awards the Spice Islands to the Dutch, Ceylon and the Cape of Good Hope to Britain.

Lombardy and Venetia are awarded to Austria (*see* 1861).

Switzerland is established as an independent confederation of 22 cantons.

Legitimate dynasties are restored in Sardinia, Tuscany, Modena, and the Papal States, and the Bourbons are reinstated at Naples. The former king of Naples Joachim Murat sails for Calabria September 28 in a bid to recover the throne he held from 1808 to 1814, four of his six ships are scattered by a storm and one deserts, he lands at Pizzo with 30 men, is captured soon after, court-martialed, and executed by a firing squad October 13 at age 48.

A Germanic confederation is created to replace the Holy Roman Empire that was ended in 1806.

Sweden retains Norway, which she received last year in the Treaty of Kiel. The Norwegians will not gain independence until 1905.

The kingdom of Poland is created from the grand duchy of Warsaw, but Russia's Aleksandr I is made king of Poland. The Poles will not gain independence until 1919 (*see* 1864).

Prussia obtains part of the grand duchy of Warsaw (above) not included in the new kingdom of Poland. Prussia also obtains Danzig and other territories.

Britain retains Malta, Heligoland, and some former French and Dutch colonies.

A force of 10,000 seasoned Spanish troops released by the peace in Europe lands in Venezuela. Led by Pablo Morillo, the Spaniards defeat the revolutionary armies in New Granada, restore royal authority, and force Simon Bolívar to flee to Jamaica (*see* 1811; 1818).

U.S. Navy Commodore Stephen Decatur leaves New York May 20 with a fleet of 10 vessels bound for the Mediterranean. He signs a treaty June 30 with the Ottoman dey of Algiers, follows it up with treaties with the sovereigns of Tunis and Tripoli, and all three agree to exact no further ransoms or tributes from U.S. merchant ships and restrain their pirates from attacking U.S. vessels.

Decatur (above) proposes a toast at a banquet given to celebrate his return from the Mediterranean: "Our country! In her intercourse with foreign nations may she always be in the right; but our country, right or wrong!"

Zulus in southeast Africa remain a clan of barely 1,500 but will soon conquer their neighbors. Chief Senzangakona's son Shaka, 28, has devised a more effective *assegai* (spear), a shield that can serve as a weapon, and new military tactics. When his father dies next year, Shaka will begin a 12-year reign in which his army will increase from fewer than 400 men to more than 40,000, and he will rule with mercurial cruelty over the Zulu nation.

England's Dartmoor Prison has a massacre of inmates April 6. A slight incident irritates the warden, he has his Somerset militiamen fire on the prisoners, and hundreds of men die—most of them American seamen convicted of privateering.

Humphry Davy invents a safety lamp for miners that will prevent colliery accidents (*see* 1813). A metal gauze surrounds the flame to prevent any inflammable gas from being ignited.

France promises to end her slave trade by 1819, limiting it meanwhile to her own colonies.

London banker Nathan Meyer Rothschild, 38, receives carrier pigeon reports from Belgium advising him of Napoleon's defeat at Waterloo (above). Feigning gloom, he depresses the price of British consols by selling short, then has his agents buy them up at distress prices, and when news of Wellington's victory sends prices sky-high, Rothschild sells, reaping a great fortune on the London Exchange (*see* consols, 1751; Rothschilds, 1812).

Britain suffers economic depression as demand for military supplies abruptly ceases and as Continental markets are unable to absorb backlogged inventories of English manufactured goods. Prices fall, thousands are thrown out of work, and 400,000 demobilized troops add to the problems of unemployment.

A Corn Law enacted by Parliament helps British landlords to maintain prices by restricting grain imports. The law permits wheat to enter the country duty-free only when the average domestic price exceeds a "famine" level of 80 shillings per bushel, but unemployment reduces demand, and the price of wheat falls to 9 shillings per bushel, down from a record 14 shillings 6 pence in 1812 (*see* 1838).

The British income tax ends and will not be resumed until 1842 (*see* 1802).

The Congress of Vienna (above) relaxes the 1648 rule against navigation on the Scheldt River.

Scottish road surveyor John London McAdam, 59, employs stone broken into pieces to create a hard, smooth, water-resistant roadway that will be called Macadam paving. McAdam has spent some years in America.

New Jersey grants the first state charter for an American railroad to John Stevens and his associates (*see* 1804; T-rail, 1830).

1815 ***(cont.)*** The New York State legislature approves a plan to finance an Erie Canal with state bonds pursuant to a proposal by Governor De Witt Clinton (*see* 1817).

Jews at Berlin open what will later be called a Reform Temple, modeled on one that opened at Seesen in 1810. "Enlightened" worshipers try to bring their faith into harmony with the modern world, giving up Mosaic dietary laws, rules of dress, and other restrictions. Such Jews will remain a tiny fraction for more than half a century.

London's *Morning Chronicle* scoops the competition June 12 with news of Napoleon's defeat at Waterloo 4 days earlier (*see* Rothschild, above).

The *North American Review* begins publication at Boston.

Fiction *Guy Mannering* and *The Antiquary* by Walter Scott; *Adolphe* by French novelist-statesman Henri Benjamin Constant de Rebecque, 47.

Juvenile *Grimm's Fairy Tales (Kinder- und Hausmärchen)* by German philologists Jacob and Wilhelm Grimm, 30 and 29, who include "The History of Tom Thumb," "Little Red Riding-Hood," "Bluebeard," "Puss in Boots," "Hop o' my Thumb," "Snow White and the Seven Dwarfs," "Goldilocks and the Three Bears," "The Princess and the Pea," "Thumbelina," "The Sleeping Beauty in the Wood," and "Cinderella" (which first appeared in a centuries-old Chinese book).

Poetry *Songs (Gedichte)* by German poet (Johann) Ludwig Uhland, 21, of Tübingen whose *Songs of the Fatherland (Vaterländische Gedichte)* will appear next year. *Der Güter Kamerad* and other nationalist ballads by Uhland will become German folk songs.

Painting *Boulevard des Italiens* by English painter John Crome, 46, who visited Paris last year to see Napoleon's art collection; *Mrs. Wolf* by Thomas Lawrence, who is knighted by the prince regent. John Singleton Copley dies at London September 9 at age 77.

Theater France's Grand Guignol horror play theater has its beginnings in a puppet show theater started at Lyons by puppet maker Laurent Mourquet, 71, whose puppet hero Guignol is modeled on the popular Polichenelle (Punch) (*see* 1897).

Popular songs "Erlkönig" by Franz Schubert, who has been writing songs since age 14.

England dances the Quadrille for the first time.

John Nash completes Brighton Pavilion in a pseudo-Oriental redesign for the prince regent.

St. Petersburg's Admiralty is completed to plans by the late Adrian Dmitrievich Zakharov, who died in 1811 at age 50.

Harvard's University Hall is completed in the Yard by architect Charles Bulfinch.

The volcano Tambora on the island of Sumbawa in the East Indies erupts April 5, killing 20,000 and producing tidal waves and whirlwinds that raise enormous clouds of dust that will affect climatic conditions throughout the world (*see* agriculture, 1816).

A dinner at the Louvre Palace celebrates the restoration of the French monarchy (above). Chef Marie-Antoine (Antonin) Carême, 31, prepares food for 1,200 guests.

The population of the United States reaches 8.35 million, with 80 percent of it in New England and the Atlantic Coast states, the balance mostly in Ohio, Kentucky, and Tennessee, with scattered settlements in the Mississippi Valley.

Philadelphia is a city of 75,000 and the social capital of America, while New York has 60,000, Baltimore nearly 30,000, Boston 25,000.

1816 Brazil proclaims herself an empire January 16 with the Portuguese prince regent João as emperor.

Portugal's insane queen Maria I dies March 20 at age 81 after a 39-year reign, the last 17 years of which have been controlled by her son. He will reign until 1817 as João VI but remains in Brazil (*see* 1807; 1820).

Argentina (United Provinces of La Plata) declares herself independent of Spain July 29 in a proclamation issued by the Congress of Tucuman.

Paraguayan military leader José Rodriguez de Francia, 50, assumes supreme powers, beginning a 24-year dictatorship in which he will develop a strong army, gain absolute power over the Guaraní Indian majority, and isolate his country.

Britain returns Java to the Netherlands (*see* 1811; 1946).

A survey to establish the boundary between Canada and the United States begins under the supervision of David Thompson. It will take 10 years (*see* 1809; 1818).

Indiana is admitted to the Union as the 19th state.

President Madison's Secretary of State James Monroe, 58, wins election to the presidency, gaining 183 electoral votes as compared with 34 for Georgia states' rightist William H. Crawford, the last Federalist candidate for president.

The Luddite movement that was suppressed in 1813 revives as a dismal harvest produces economic depression throughout Britain. Well-organized efforts are made to smash machinery in riots at many industrial centers; a crowd gathers December 2 in London's Spa Fields and commits acts of violence, the rioters are vigorously prosecuted, but the movement will continue until rising prosperity ends it.

Pittsburgh is incorporated on the site of Fort Pitt erected in 1759.

Congress imposes a 25 percent duty on all imports to protect America's infant industry from foreign competition and make the nation self-sufficient, but even with the tariff wall British manufacturers will find ways to undersell the Americans. Rep. Daniel Webster, 34, (N.H.) has opposed the tariff as a representative of New England shipping interests.

The Philadelphia Saving Fund Society opens December 2 in the office of its secretary-treasurer George

Billington in Sixth Street—the first U.S. savings bank actually to accept deposits. It will receive a charter early in 1819 and grow to become the largest U.S. thrift institution.

France's Caisse des Depots & Consignations is founded. The central investment fund for savings banks must invest "in the public interest."

The Black Ball Line begins regular Baltimore clipper ship service between New York and Liverpool, but the full flowering of the clipper ship will not come for 3 decades (*see* 1832).

A French fleet of four naval vessels sets out in late June for Senegal in West Africa bearing the colony's new governor aboard the flagship *Medusa* with 240 passengers and a crew of 160. Navigated by a civilian friend of the governor who ignores the advice of seasoned naval officers, the *Medusa* founders on a sandbank off the African coast July 2. The captain refuses to jettison the ship's heavy guns or her cargo of flour, the governor orders a large raft built to accommodate all who cannot fit into the *Medusa*'s six boats, 89 of the 149 on the overcrowded raft are lost within 2 days, 12 more die the third night, but one body is saved from the sharks and eaten by the survivors. Only 15 survive after 13 days, and more die in the sands of the Sahara. The governor and his friend go free, and the captain is sent to prison for 3 years (*see* Géricault painting, 1819).

The steamboat *Washington* launched June 4 at Wheeling, Va., sets a pattern for U.S. riverboats. Designed by Capt. Henry M. Shreve, 31, the 148-foot 400-ton *Washington* has a shallow draft to clear the snags and sandbars of the Ohio and Mississippi, and her two decks rise high above the water with two chimneys on either side of her small, boxlike pilothouse, a design that inspires spectators to call her a "floating wedding cake." The ship has two high-pressure, 24-inch cylinder, 6-foot stroke engines placed horizontally on the deck instead of vertically in the hull. The engines each operate a side-wheel and are unconnected so that the pilot can reverse one engine to turn the ship in its own length. The second deck is for passengers who arrive at New Orleans on the ship's maiden voyage, October 7. By 1820, there will be more than 60 versions of Shreve's ship, whose owner will be memorialized in the name Shreveport, La.

The United Kingdom has 2.5 million tons of merchant shipping; 1,000 tons are in steam (*see* 1826).

The celeripede, invented by French physicist Joseph Nicéphore Niepce, 51, is a primitive two-wheeled bicycle propelled by the action of the feet on the ground. German inventor Karl D. Sauerbroun devises a slightly more advanced version (*see* Macmillan, 1839; Niepce, 1822).

Friedrich Krupp produces the first Krupp steel at Essen (*see* 1811). It is not cast steel, which Sheffield has resumed exporting to Europe, but consists of tiny bars of low-grade steel marketed as files for tanners (*see* 1838).

Hindu College is founded at Calcutta.

Nonfiction *Wissenschaft der Logik* by German philosopher Georg Wilhelm Friedrich Hegel, 46, whose third and final volume completes the "Hegelian" philosophy of the absolute that will dominate metaphysics during the second quarter of the century (the first volume appeared in 1812).

Fiction *Emma* by Jane Austen, who will die next year at age 41, leaving her novels *Persuasion* and *Northanger Abbey* to be published posthumously; *The Black Dwarf* and *Old Mortality* by Walter Scott.

Poetry *The Story of Rimini* by English poet James Henry Leigh Hunt, 32, who since 1808 has been editor of the *Examiner,* a witty journal owned by his brother John, and has been imprisoned with his brother for attacking the prince regent with the phrase "a fat Adonis of 50." "The two divinest things this world has got, /A lovely woman in a rural spot!" writes Hunt, who returns to the rhythms of Chaucer and Spencer to pioneer a new romantic school. *The Siege of Corinth* by Lord Byron; *Kublai Khan* by Samuel Taylor Coleridge: "In Xanadu did Kublai Khan/A stately pleasure-dome decree:/ Where Alph, the sacred river, ran/ Through caverns measureless to man/ Down to a sunless sea./ So twice five miles of fertile ground/ with walls and towers were girdled round"; English poet John Keats, 20, has his first sonnet published May 5 in Leigh Hunt's *Examiner* (above) and follows it in December with his sonnet *On First Looking into Chapman's Homer*: "Then felt I like some watcher of the skies/ When a new planet swims into the ken;/ Or like stout Cortez, when with eagle eyes/ He stared at the Pacific—and all his men/ Looked at each other with a wild surmise—/ Silent, upon a peak in Darien" (he means Balboa; *see* 1513).

Painting *Mousehold Heath* and *Porington Oak* by John Crome.

Opera *The Barber of Seville (Il Barbiere de Siviglia)* 2/5 at Rome's Teatro Argentina, with music by Gioacchino Rossini. Based on the Beaumarchais drama of 1775 and initially titled *Almaviva,* Rossini's opera creates a furor; *Othello, or the Moor of Venice (Otello, o Il Moro di Venezia)* 12/4 at the Teatro Fondo, Naples, with music by Rossini.

English dandy George Bryan "Beau" Brummell, 38, flees to Calais to escape creditors hounding him for gambling debts. A friend of the prince regent, Brummell will be named British consul at Caen from 1832 to 1840 but will die in a French insane asylum in 1840.

Providence, R.I., gets a gray stone Unitarian church with the largest belfry ever cast by Boston's Paul Revere, now 81.

Pittsburgh (above) is known even now as a "smoky city" because of all its coal-burning houses and factories (*see* 1947).

Cold weather persists through summer in much of the world's temperate zones, apparently as a result of dust

1816 ***(cont.)*** in the atmosphere following last year's volcanic eruption in the East Indies. Frost occurs from Canada to Virginia every night from June 6 to June 9, laundry laid out to dry on the grass at Plymouth, Conn., June 10 is found frozen stiff, heavy snows fall in the Northeast in June and July, and frost kills crops in what farmers will call "eighteen hundred and froze to death."

Cape Cod, Mass., cranberry production gets a boost from the discovery that cranberry vines grow more vigorously in places where the wind has blown sand over a mat of wild vines. The observation by Henry Hall at Dennis will lead to draining of swampy wastelands and development of commercial cranberry bogs.

1817 Chile wins her liberation from Spain February 12 at the Battle of Chacabuco and will proclaim independence early next year (*see* 1812). Bernardo O'Higgins has enlisted the support of Argentine military commander José de San Martin, 39, who has raised and trained a small army of Chilean and Argentinian volunteers, crossed the high Andes with O'Higgins to surprise a royalist army, and helps make O'Higgins supreme dictator of Chile.

The Rush-Bagot Treaty signed April 28 limits U.S. and British naval forces on the Great Lakes. U.S. Secretary of State Richard Rush, 37, has negotiated the treaty with British minister to Washington Sir Charles Bagot, 36.

Mississippi is admitted to the Union as the 20th state.

Ohio Indians sign a treaty ceding their remaining 4 million acres of land to the United States.

The Seminole War begins as Georgia backwoodsmen attack Indians just north of the Florida border in retaliation for depradations by tribesmen (*see* 1818).

The Obrenovic dynasty that will rule Serbia until 1842, and then again from 1858 until 1903, is founded by Milos Obrenovic, 37, who leads a second insurrection against the Ottoman Turks and gains recognition as prince of Serbia over the objections of the Karageorgevic family. The rival family's leader Kara George returns from exile in Austria but the Obrenovics murder him, beginning a long blood feud between the two families.

Jena students organize the Wartburg Festival to celebrate the Reformation and Battle of Leipzig. They burn emblems of reaction October 18 (*see* 1819).

The Coercion Acts passed by Britain's Parliament in March extend the 1798 act against seditious meetings, suspend temporarily the right of habeas corpus, renew an act punishing attempts to undermine the allegiance of soldiers and sailors, give the prince regent the same safeguards against treason as the king himself, and have the effect of stimulating activity by extremists in the radical movement (*see* 1819).

Principles of Political Economy by English economist David Ricardo contributes to Adam Smith's free trade theory of 1776 with a doctrine of "comparative costs" and a "labor theory of value." Ricardo postulates a basic antagonism between the landlords of England's "establishment" and the rising lords of English industry (*see* 1809).

John Jacob Astor gains a monopoly in the Mississippi Valley fur trade (*see* 1814; Juneau, 1818).

Ismaelis (progressive Muslims) begin contributing the *zakat*—12 percent of their income—to the Aga Khan, a direct descendant of the prophet Mohammed. He and future Aga Khans will become immensely rich.

The Cumberland Road reaches west from Cumberland on the Potomac River to Wheeling, Va., on the Ohio. The road has a 30-foot wide gravel center on a stone base (*see* 1811; 1852).

The Conestoga wagon that covers the 90 miles between Philadelphia and New York in 3 days is called "the Flying Machine" (*see* 1753; 1840).

Gov. De Witt Clinton of New York orders construction of a 363-mile Erie Canal that will connect Buffalo on Lake Erie with Troy on the Hudson River. The state legislature authorizes state funds for the canal and ground is broken July 4 (*see* 1815; 1819).

The first steam ferry between Manhattan and Staten islands goes into service. The *Nautilus* is owned by U.S. Vice President Daniel D. Tompkins.

The stern paddle steamboat *Washington* leaves Louisville March 3 for a round-trip voyage to New Orleans and completes the first such voyage. By year's end, a dozen steamboats have gone up the Mississippi and penetrated into the Ohio and other western rivers (*see* Lake Erie, 1818).

The first Waterloo Bridge is completed across the Thames at London by Scottish civil engineer John Rennie, 56, whose son John will complete his London Bridge in 1831.

Welsh inventor Richard Roberts, 28, devises a screw-cutting lathe and a machine for planing metal. He will also invent weaving improvements, advanced steam locomotives, railway cars, and steamships.

German chemist Friedrich Strohmeyer, 41, discovers cadmium.

Swedish chemist A. Arfvedson discovers lithium.

Jöns Jakob Berzelius discovers selenium.

Parkinson's disease is described by English surgeon-paleontologist James Parkinson, 62 (*see* 1967).

The University of Ghent is founded by the Dutch king Willem I.

The University of Michigan is founded at Ann Arbor.

Harvard Law School is established at Cambridge (*see* Eliot, Langdell, 1869).

The *Scotsman* begins publication as a weekly Edinburgh newspaper that will become a daily in 1855.

Blackwood's Magazine has its beginnings in the *Edinburgh Monthly Magazine* started by Scottish publisher William Blackwood, 41.

Fiction *Rob Roy* by Walter Scott.

Poetry *Poems* by John Keats, who has been helped by his friend Percy Shelley; *Beppo: A Venetian Story* by Lord Byron; *Lalla Rookh* by Thomas Moore; *Thanatopsis* by Great Barrington, Mass., lawyer-poet William Cullen Bryant, 23, whose "view of death" in the September issue of *North American Review* will appear in an expanded version in 1821.

Painting *Flatford Mill* by John Constable.

Theater *The Ancestress (Die Ahnfrau)* by Viennese playwright Franz Grillparzer, 26, 1/31 at Vienna's Theater an der Wien.

Opera *Cinderella* (*La Cenerentola*) 11/25 at Rome's Teatro Valle, with music by Gioacchino Rossini.

The British secretary for Ireland Robert Peel, 29, establishes a regular constabulary for Ireland, which is seething with discontent and on the verge of rebellion. The Irish will call the constables "Peelers" (*see* "Bobbies," 1829).

Haiti's Citadelle La Ferrière is completed for Henri I (Henri Christophe) atop a 3,100-foot peak outside Cap Haitien, where 200,000 former slaves have worked for 13 years at a cost of 2,000 lives to build the impregnable fortress that can hold 15,000 men with enough food and water for a 1-year siege.

U.S. farm prices fall as Europe's peace ends the foreign markets that have taken some of America's farm surpluses.

Hereford cattle are imported into the United States for the first time and raised in Virginia. The English breed will become the dominant cattle breed on the western plains (*see* Corning, Sotham, 1840).

"The hunter or savage state," writes President Monroe, "requires a greater extent of territory to sustain it than is compatible with progress and just claims of civilized life . . . and must yield to it."

Thomas Malthus rejects artificial contraceptive devices in a fifth edition of his 1798 *Essay*. The misery of overpopulation is necessary to "stimulate industry" and discourage "indolence," writes Malthus (*see* Place, 1822).

1818 Sweden's Charles XIII dies February 5 at age 69 after a 9-year reign and is succeeded by the crown prince Jean Baptiste Bernadotte, 55, who will reign until 1844 as Charles XIV John, founding a new Swedish dynasty.

Allied occupation troops leave France after 3 years of peacekeeping in the wake of the Napoleonic wars.

India's Rajput states, Poona, and the Holkar of Indore come under British control.

The U.S.-Canadian border is established by a convention signed October 20, making the 49th parallel the boundary from Lake of the Woods to the Rocky Mountains as David Thompson continues his survey (*see* 1816). The two countries agree to a joint occupation of the Pacific Northwest territories for a 10-year period (*see* 1828; Drake, 1579; Cook, 1778; Gray, 1791).

Illinois is admitted to the Union as the 21st state.

Congress adopts a flag with 13 alternate red and white stripes and with a blue square containing a white star for each state of the union.

The Seminole War ends in Florida. Spain cedes Florida to the United States and agrees to relinquish all claims to Pacific Coast territory north of the 42nd parallel. The United States agrees to pay Floridians $5 million for their claims against Spain.

Simon Bolívar secures control of the lower Orinoco basin from the Spanish. Bolívar returned to the region in 1816 and has gained support from José Paez, 28 (*see* 1815; 1819).

Chile proclaims independence from Spain February 12, José de San Martin defeats a royal army from Peru April 5 at Maipu, and he gains support from rebels at Buenos Aires and from former British naval commander Thomas Cochrane, now 43, who has been fined, imprisoned, and expelled from the Royal Navy on charges of having connived in a speculative fraud.

The Peul Diallo dynasty that has ruled on Africa's Niger River since the 15th century ends in a coup d'état by the Muslim usurper Marabout Cheikou Ahmadou. He will organize the movement of nomads in the region and establish patterns designed to conserve grass and water.

Scottish explorer John Ross, 41, searches for a Northwest Passage and discovers an 8-mile expanse of red-colored snow cliffs overlooking Baffin Bay. The cliffs will be called the Ross Ice Shelf (*see* 1576).

Failure of the Baltimore branch of the second Bank of the United States precipitates a U.S. financial crisis (*see* Supreme Court decision, 1819).

John Jacob Astor's American Fur Co. agent Solomon Laurent Juneau, 25, opens a trading post on Lake Michigan at a point that will later be the site of Milwaukee (*see* 1834; 1846; Astoria, 1811).

The U.S. steamboat *Walk-in-the-Water* leaves Buffalo for Detroit October 10 with 100 passengers. The 388-ton 135-foot vessel is the first steamboat on Lake Erie and rivals the 2-year-old Canadian steamboat S.S. *Frontenac* which plies only Lake Ontario.

Guy's Hospital surgeon James Blundel at London performs the first successful human blood transfusion using a syringe for the purpose. Blundel will invent two special pieces of apparatus that will permit blood to be transferred from donor to patient with a minimum of physical interference (*see* 1677; 1909).

Dalhousie University is founded at Halifax, Nova Scotia. The ninth earl of Dalhousie George Ramsay, 48, will serve as governor in chief of the Canadian colonies from 1819 to 1828.

Fiction *Frankenstein, or The Modern Prometheus* by English novelist Mary Wollstonecraft Godwin Shelley, 21, gains instant success with its story about a scientist who takes parts of corpses to manufacture a living creature. Author Shelley is the wife of poet Percy Bysshe Shelley and daughter of women's rights champion

Ludwig van Beethoven's musical genius ultimately transcended his own inability to hear what he had composed.

Mary Wollstonecraft by the philosopher William Godwin, now 62, who is supported by his son-in-law. The book will go through at least two printings per year for 40 years, will be translated into 30 languages, millions of copies will be sold, and it will make the word *frankenstein* a common noun meaning any work that controls its originator; *The Heart of Midlothian* and *The Bride of Lammermoor* by Walter Scott.

Poetry *Ozymandias* by Percy Bysshe Shelley, whose sonnet has been inspired by a huge statue erected in the desert by the ancient Egyptian pharaoh Ramses II, a work that to Shelley symbolizes the futility of all human striving to achieve immortality through earthly glory: "And on the pedestal these words appear:/My name is Ozymandias, king of kings:/Look on my works, ye Mighty, and despair!/Nothing beside remains. Round the decay/Of that colossal wreck, boundless and bare/The lone and level sands stretch far away"; *Childe Harold* by Lord Byron, who says in his fourth and final canto, "Roll on, thou deep and dark blue ocean—roll!/Ten thousand fleets sweep over thee in vain;/Man marks the earth with ruin—his control/Stops with the shore. . .; *Endymion* by John Keats contains the line, "A thing of beauty is a joy for ever;/Its loveliness increases; it will never/Pass into nothingness"; *All'Italia* and *Sopra il Monumento di Dante* by Italian poet Giacomo Leopardi, 20; *The Old Oaken Bucket* by New York printer-poet Samuel Woodworth, 34, begins, "How dear to this heart are the scenes of my childhood" (*see* song, 1822).

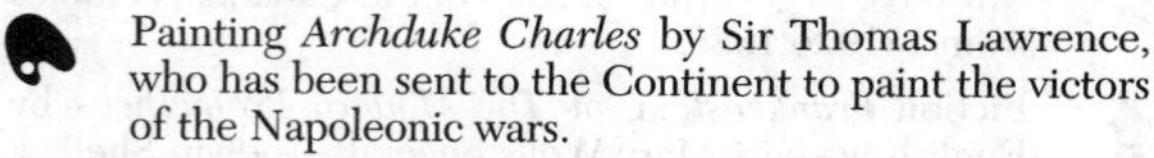

Painting *Archduke Charles* by Sir Thomas Lawrence, who has been sent to the Continent to paint the victors of the Napoleonic wars.

Theater *Sappho* by Franz Grillparzer 4/21 at Vienna's Burgtheater.

Family Shakespeare in 10 volumes by English editor Thomas Bowdler, 64, will go through four editions in 6 years and many subsequent editions "in which nothing is added to the original text; but those words and expressions are omitted which cannot with propriety be read aloud in a family." Bowdler will also edit a six-volume "bowdlerized" edition of Edward Gibbon's *Decline and Fall of the Roman Empire.*

Cantata "Jubel" ("Jubilee") by Carl Maria von Weber 9/20 at Dresden for the 50th anniversary of the accession of Saxony's Friedrich August I.

Ludwig van Beethoven becomes totally deaf after 20 years of defective hearing but continues to compose.

Hymn "Silent Night" ("Stille Nacht, Heilige Nacht") by Austrian parish priest Joseph Mohr, 26, will be set to music by composer Franz Xavier Gruber, now 31.

Brooks Brothers has its beginnings in a New York menswear shop at the corner of Catharine and Cherry streets opened by merchant Henry Sands Brooks, 46, who invests $17,000 in a stock of imported British woolens and adopts as his symbol the golden fleece symbol of a dead lamb cradled by a ribbon, the symbol used in the 15th century for a knighthood established by the duke of Burgundy Philippe le Bon. Brooks Brothers will clothe bankers, lawyers, military officers, statesmen, and upwardly mobile Americans for generations (*see* 1915).

U.S. ships make 15 voyages to Sumatra for pepper (*see* 1797; 1873).

Angostura bitters are invented at Angostura, Venezuela, by German physician Johann Gottlieb Benjamin Siegert, who served as a surgeon under von Blücher 3 years ago at Waterloo and has emigrated to South America. Siegert brews his elixir from gentian root, rum, and other ingredients as a stomach tonic to overcome debility and loss of appetite in the tropics, but it will gain wider use as a cocktail ingredient used by British colonials to make "pink gin." Siegert's son will take the formula to Trinidad and produce angostura bitters in an abandoned monastery at Port of Spain long after Angostura has become Ciudad Bolívar.

London's Donkin, Hall & Gamble "Preservatory" in Blue Anchor Road, Bermondsey, turns out satisfactory canned soup, veal, mutton and vegetable stew, carrots, and beef (both boiled and corned, meaning salted), but tinned foods will not be available in food stores until 1830.

The tin can is introduced to America by Peter Durant (*see* 1810; 1847).

Cincinnati begins packing pork in brine-filled barrels. Salt pork is a U.S. food staple, and Cincinnati will come to be called "Porkopolis."

1819 Spain cedes eastern Florida and all its possessions east of the Mississippi to the United States February 22 after months of effort by Gen. Andrew Jackson, who has occupied Pensacola and executed some Seminole chiefs along with two white sympathizers.

Alabama is admitted to the Union as the 22nd state.

Maine is separated from Massachusetts (*see* 1820).

Hawaii's Kamehameha dies May 8 at age 82 after a 24-year reign that has consolidated the Sandwich Island kingdom (*see* 1803). His son of 22 will welcome Hawaii's first missionaries next year and reign until 1824 as Kamehameha II, overthrowing the ancient Hawaiian taboo system and religion.

Simon Bolívar moves up the Orinoco River, crosses the Andes, and with help from British volunteers defeats a superior Spanish force under the command of Pablo Morillo, conde de Cartagena, August 7 at the Boyaca River. Bolívar occupies Bogotá August 10 and liberates New Granada from Spain. The Congress of Angostura, which includes representatives from New Granada, proclaims the independence of Great Colombia—comprised of New Granada, Venezuela, and Quito—and establishes Bolívar as president and military dictator (*see* 1818; 1820).

Singapore is occupied by the East India Company's Thomas Stamford Raffles, who persuades the sultan and tenggong of Johore (who has no authority to do so) to cede the fishing village to Britain. Now governor of Benkuilen in Sumatra, Raffles wants to check Dutch power in the East Indies, and he makes Singapore a British crown colony.

Reactionary German journalist August von Kotzebue, 58, is murdered March 23 at Mannheim by Jena University student Karl Ludwig Sand, 23, who calls Kotzebue a Russian spy and an enemy of liberty. Austria's Prince Metternich, alarmed, persuades Prussia's Friedrich Wilhelm to issue the repressive Carlsbad Decrees that establish an inquisition at Mainz in July to look into secret societies, impose strict censorship on all publications, and place German universities under the control of commissioners appointed by German sovereigns.

The Peterloo Massacre August 16 at St. Peter's Fields, Manchester, kills several people and injures hundreds as British soldiers are ordered to arrest a speaker urging parliamentary reform and repeal of the Corn Laws. The incident is followed by the Six Acts passed by Parliament in December, a repressive code that curtails public meetings, forbids training in the use of firearms, empowers magistrates to search citizens for arms and seize any that may be found, provides for speedy trial in "cases of misdemeanor," increases the penalties for seditious libel, and limits radical journalism by imposing the newspaper stamp duty on all periodicals containing news.

Memphis is laid out on the Mississippi River at Fort Adams. It takes its name from the Egyptian city famous for its cotton.

Minneapolis has its beginnings in Fort Snelling, built by U.S. Army engineers some 900 miles up the Mississippi from Memphis (above) near St. Anthony's Falls (*see* Hennepin, 1680; Minneapolis, 1847).

"The power to tax is the power to destroy," says John Marshall March 6. The Supreme Court rules in *McCulloch v. Maryland* that no state may tax any instrumentality of the federal government. The case specifically

England's Peterloo Massacre of 1819 typified an age of almost universal political repression.

involves taxation of the Bank of the United States and its Baltimore branch by the state of Maryland (*see* 1818).

The U.S. economic depression that began last year brings appeals from manufacturers for higher protective tariffs (*see* 1816).

Danish physicist Hans Christian Oersted, 42, advances knowledge of electromagnetic energy. He notices that a compass needle located close to a wire carrying an electric current will swing wildly but will finally settle at right angles to the wire. He suspects that the current in the wire must set up a magnetic field around it, and his observation will be the starting point for others (*see* Romagnosi, 1802; Ampère, 1820; Faraday, 1821; electromagnet, 1823).

London gas mains extend 288 miles, and some 51,000 houses in the city are equipped with gas burners. National Light and Heat Co. has become Gas Light and Coke Co., a firm that will continue for nearly 150 years.

The first stretch of the Erie Canal opens after 2 years of construction to connect Utica and Rome, N.Y. Yankee engineers have devised an endless screw linked to a roller, a cable, and a crank to pull down tall trees. They have improvised a horse-drawn crane to lift rock debris out of the cut, but most of the work is done by Irish immigrants who are paid 37.5¢ per day (plus as much as a quart of whisky per day in periodic work breaks) and die by the thousands of malaria, pneumonia, and snakebite (*see* 1825; White, 1820).

The 350-ton U.S. ship *Savannah* sails from the Georgia port May 22 and arrives June 30 at Liverpool—the first transatlantic crossing assisted by steam propulsion. Commanded by Moses Brown, who piloted the *Phoenix* in 1809, the *Savannah* has one inclined direct-acting, low-pressure steam engine to power the side-wheels

1819 ***(cont.)*** that supplement her sails, but she proceeds on steam for only 80 hours, and while she has 32 staterooms, the perils of her 90-horsepower engine have discouraged passengers, and she carries none (*see* 1826).

French scientists make it possible to determine atomic weights. Pierre Louis Dulong, now 34, joins chemist-physicist Alexis Thérèse Petit, 30, in enunciating the rule that the product of the relative atomic weight and the specific heat of any element is constant.

The stethoscope invented by French physician René Théophile Hyacinthe Laënnec, 38, is a roll of paper that avoids the indelicacy of having to place the physician's ear to the heaving bosom of a female patient. Laënnec introduces the practice of auscultation (*see* von Auenbrijgg, 1761).

Sumatra has a cholera epidemic (*see* 1820).

Unitarianism is founded by Boston Congregationalist pastor William Ellery Channing, 30, who becomes the leader of a group that denies the holiness of the Trinity and believes in only one Divine Being. Channing will organize the American Unitarian Association in 1825.

The charter of a benevolent institution is a contract and therefore inviolate, rules Chief Justice John Marshall February 2 in the case of *Dartmouth College v. Woodward.*

The University of Virginia is chartered at Charlottesville, 3 miles west of Thomas Jefferson's Monticello estate. The ex-president has been instrumental in obtaining the charter and is planning the grounds, buildings, and curriculum of the new university, whose students will begin classes March 27, 1825.

The University of Cincinnati is founded in Ohio.

Colgate University is founded as Madison College at Hamilton, N.Y. (*see* 1890).

Nonfiction *The World as Will and Idea (Die Welt als Wille und Vorstellung)* by German philosopher Arthur Schopenhauer, 30, who expounds a philosophy of pessimism that will be generally ignored for a generation but will have great influence in the latter part of the century.

Fiction *Ivanhoe* and *The Legend of Montrose* by Walter Scott.

Poetry *Ode to a Nightingale* by John Keats; *Mazeppa* by Lord Byron.

The Prado is completed at Madrid after 34 years of construction, and Ferdinand VII makes it the Spanish Royal Museum, bringing together paintings and sculpture collected by his predecessors since Carlos I.

Painting *The White Horse* by John Constable, who is elected an associate of the Royal Academy; *George IV* and *Pope Pius VII* by Sir Thomas Lawrence; *The Raft of the 'Medusa'* by Théodore Géricault (*see* 1816).

First performances *Stabat Mater* by Franz Schubert in July at Steyr in Upper Austria.

Hymn "From Greenland's Icy Mountains" by English prelate Reginald Heber, 36, who will be bishop of Calcutta from 1822 to 1826: "What though the spicy breezes/Blow soft o'er Ceylon's isle;/Though every prospect pleases,/And only man is vile . . ."

The whaling industry expands to new grounds off Japan. Many whaling ships will obtain supplies from Japanese ports.

U.S. diplomats in foreign posts receive instructions to send home any valuable new seeds and plants.

An improved plow, patented by Cayuga County, N.Y., farmer Jethro Wood, 45, is constructed in several major pieces so that a farmer who breaks one part can replace it without having to buy a whole new plow. Other plowmakers will infringe on Wood's patent, and many farmers will insist that cast-iron plows poison the soil and will refuse to give up their old wooden plows (*see* 1784; Deere, 1837).

Salmon, lobster meat, and oysters are packed in tin cans at New York by Ezra Daggett and Thomas Kensett (*see* Donkin-Hall, 1814; Kensett's patent, 1825).

Vermont inventor John Conant patents an iron cooking stove, but U.S. cooks and housewives spurn his stove, preferring the traditional practice of hearthside cooking (*see* 1850).

The world's first eating chocolate to be produced commercially is manufactured at Vevey, Switzerland by François-Louis Cailler, 23, who introduces the first chocolate to be prepared and sold in blocks made by machine. Cailler starts a company that will specialize in fondants, but his chocolate is not candy (*see* van Houten, 1828).

The population of the Hawaiian (Sandwich) Islands (above) falls to 150,000 as a consequence of diseases introduced by Europeans and Americans since the landing of Captain Cook in 1778 and the arrival of occasional visitors in the past 41 years. Some 60 Caucasians live on the island of Oahu.

1820 Britain's George III dies January 29 at age 81 and is succeeded by the prince regent, now 57, who has ruled since 1811 and will reign until 1830 as George IV. The king's wife Caroline, 51, demands recognition as queen, but the profligate George continues his efforts to obtain a divorce.

France's Duc de Berry, heir presumptive to the throne, is assassinated February 13. Charles Ferdinand de Bourbon, 41, was a nephew of Louis XVIII; his posthumous son the comte de Chambord, born September 29, becomes heir to the throne.

The Cato Street conspiracy to murder members of the British Cabinet comes to light February 23. Its ringleaders will be executed.

Maine is admitted to the Union as the 23rd state (*see* Missouri Compromise, below).

A Spanish revolution led by Col. Rafael Riego forces Ferdinand VII to restore the constitution of 1812, which he does March 7.

The king of Naples Ferdinand IV promises a constitution July 7 following a revolt instigated by secret societies that include the Carbonari.

A Portuguese revolt begins August 24 at Oporto and reaches Lisbon 5 days later as discontent grows under the regency that rules with British influence while João VI lives in Brazil. Leaders of the revolution demand a constitution.

A bill of pains and penalties against Britain's Princess Caroline, wife of George IV, has been pending in Parliament since July 5 but is dropped November 10. Inquiry into Caroline's conduct and efforts to deprive her of her titles are also dropped as the public rallies to her cause.

President Monroe is reelected with 231 electoral votes out of 232. One vote is cast for John Quincy Adams by an elector who either opposes the Virginia dynasty of presidents or holds the conviction that only George Washington should have the honor of unanimous election.

The Missouri Compromise accepted by Congress March 3 permits entry of Missouri into the Union as a slave state in exchange for Maine's entry as a free state (above) but only on condition that slavery be abolished in the rest of the Louisiana Purchase territory.

Jamaica in the West Indies has a slave population of 340,000. Some 800,000 have been imported since 1690, but adult males have far outnumbered adult females, and infant mortality among slaves on the sugar estates has been 500 per 1,000 live births.

Liberia is founded by the Washington Colonization Society for the repatriation of U.S. blacks to Africa (*see* 1847).

English Quaker minister Elizabeth Gurney marries chocolate heir Joseph Fry at age 40 and works to improve English prison conditions. She will continue her philanthropic efforts until her death in 1845.

The Spanish Inquisition begun by Castile's Isabella in 1478 is ended by Ferdinand VII (above) as a result of the revolution that will have effects in Latin America (*see* 1821).

Indianapolis is founded on the White River in Indiana. Its name combines the Greek word for city *(polis)* with the name of the state.

Edward Bransfield of the Royal Navy surveys New South Shetland in Antarctica and proceeds as far south as 64° 30'.

U.S. sealer Nathaniel Brown Palmer, 21, out of Stonington, Conn., discovers the Antarctic continent. The mountainous archipelago south of 64° 30' will be called Palmer Peninsula.

Russian prospectors discover rich deposits of platinum in the Ural Mountains.

French physicist André Marie Ampère, 45, discovers the left-hand and right-hand rules of the magnetic field surrounding a wire that carries electric current. Repeating last year's experiments by Hans Christian Oersted, Ampère reports that two current-carrying wires exercise a reciprocal action upon each other. (The unit of measurement of an electric current's intensity will be named the ampere.) (*see* Ohm, 1827).

Regent's Canal is completed in London.

Connecticut architect Ithiel Town, 36, patents a truss bridge (*see* first Vermont covered bridge, 1824).

The first modern concrete to be produced in America advances construction of the Erie Canal. Engineer Canvass White patents the hydraulic use of cement made from "hydraulic lime" found in New York's Madison, Cayuga, and Onondaga counties.

A cholera epidemic begins to kill thousands in China and the Philippine Islands (*see* 1819; 1823).

The U.S. Pharmaeopoeia established by New York physician Lyman Spalding, 45, is a government-approved list of medical drugs with their formulas, methods for preparing medicines, and requirements and tests for purity.

Quinine sulfate is discovered. The alkaloid drug is effective against malaria (*see* 1642).

Indiana University is founded at Bloomington.

Fiction *The Monastery* and *The Abbott* by Walter Scott, who is made a baronet by Britain's new king George IV; *The Sketch Book* by Geoffrey Crayon (Washington Irving) contains "The Legend of Sleepy Hollow."

Poetry *The Eve of St. Agnes*, *La Belle Dame Sans Merci*, and *Ode on a Grecian Urn* by John Keats, who writes in his *Ode*, "Beauty is truth, truth beauty—that is all/Ye know on earth, and all ye need to know"; *Prometheus Unbound* by Percy Bysshe Shelley, whose *Ode* contains the line, "If Winter comes, can Spring be far behind?"; *Meditations Poétiques* by French poet Alphonse Marie Louis de Prat de Lamartine, 30, who celebrates his deceased love Mme. Julie Charles, marries Elizabeth Birch, and enters the diplomatic service to begin 10 years in Italy during which he will publish several other books of poetry.

Paintings *Harwich Lighthouse* by John Constable; the Old Testament Book of Job is published with illustrations by William Blake. Benjamin West dies at London March 11 at age 81.

Sculpture *The Lion of Lucerne* by Danish sculptor Bertel Thorvaldsen, 52.

The *Venus de Milo* (Aphrodite of Melos) is discovered on the Aegean island of Melos by a Greek peasant named Yorgos, who dislodges a boulder and finds an underground chamber containing the armless marble figure of the 2nd century.

Theater *The Bear and the Pasha* (*L'ours et le Pacha*) by French vaudeville writer (Augustin) Eugène Scribe, 28, with Xavier Saintine 2/10 at the Théâtre des Variétés, Paris. Scribe has been writing vaudeville comedies

1820 ***(cont.)*** since 1811 and will write more than 250 plays, opera librettos, and ballets.

Bristol's Royal York Crescent is completed.

Congress enacts a land law April 24 providing for the sale of 80 acres of public land at $1.25 per acre in cash, making it possible for a man to buy a farm for $100 with no need for further payments. Thousands of farmers are in debt to the government for public lands on which they have made down payments of only $80 each, and farmers who have paid in full are angry (*see* 1804).

Nebraska Territory is "a great American desert," says U.S. Army officer Stephen Harriman Long, 36, who commands an exploring expedition into the Rocky Mountains (*see* Pike, 1806; Deere, 1837: irrigation, 1848).

Adulteration of Foods and Culinary Poisons by English chemistry professor Frederick Accum enrages the vested interests, forcing Accum to flee to Berlin to avoid public trial. A professor at the Surrey Institution, he has shown that pepper is almost invariably adulterated with mustard husks, pea flour, juniper berries, and sweepings from storeroom floors; counterfeit China tea is made from dried thorn leaves colored with poisonous verdigris; pickles are treated with copper to look green (*see* 1850).

Dutch-dominated South Africa gets its first large-scale British settlement in July. The Albany settlers, 4,000 strong, are settled by the British government at a place that will be called Grahamstown in the eastern coastal region where an autocratic British governor will rule the colonists until 1825 (*see* 1826).

The United States will average 35,000 Irish immigrants per year for the next two decades (*see* 1846; 1851).

The U.S. population reaches 9.6 million, with some 83 percent of gainfully employed Americans engaged in agriculture. Between 2.5 million and 3 million live west of the Alleghenies, up from 1 million in 1801. Pittsburgh has a population of 7,248, New York nearly 124,000.

1821

A Greek War of Independence begins following an insurrection in February against Ottoman-Greek rule in Wallachia. Greeks in the Morea (Peloponnese) slaughter the Ottoman minority; Constantinople retaliates by hanging the city's Greek patriarch and massacring its Greek population. Greek nationalist leader Alexander Ypsilanti, 29, invades Moldavia with a battalion, occupies Bucharest, and appeals to Czar Alexander for aid. The Battle of Dragasani west of Bucharest June 19 ends in victory for Ottoman forces over Ypsilanti, who escapes but is captured and imprisoned by the Austrians. Constantinople rejects a Russian ultimatum delivered July 27 demanding restoration of Christian churches and protection of the Christian religion, and on October 5 the Greeks take Tripolitsi, the main Ottoman fortress in the Morea, and massacre some 10,000 Turks. Prince von Metternich and Viscount Castlereagh have averted Russian intervention by warning Czar Aleksandr against supporting the revolution, which will continue until 1831 (*see* 1822).

Piedmont's Victor Emmanuel abdicates under pressure in March after refusing to accept a constitution, his brother Charles Felix becomes king, but in the absence of Charles Felix his regent Charles Albert of Savoy proclaims the Spanish constitution. The Battle of Novara April 8 ends in victory for combined Austrian and Sardinian forces over the Piedmontese, and Savoy (Sardinia) regains control of Piedmont. The Swiss cantons act in May to expel Italians who have taken refuge following the failure of their movements in Naples and Piedmont (*see* 1832).

Mexico declares her independence from Spain February 24, claiming freedom also for the provinces of California and Texas. Agostin de Iturbide, 38, becomes regent of Mexico pending selection of a new ruler (*see* 1822).

Simon Bolívar sends his lieutenant Antonio José de Sucre, 28, with an army to liberate Quito. Bolívar himself moves into Venezuela, joins with José Paez to defeat a royalist army at Carabob June 24, occupies Caracas 5 days later, and is named president of Venezuela August 30.

José de San Martin proclaims the independence of Peru July 22 and assumes supreme authority as protector of the new South American nation (*see* Chile, 1817). San Martin will resign in 14 months to permit Simon Bolívar to become dictator of Peru (*see* 1824).

Costa Rica, El Salvador, Guatemala, and Honduras declare independence from Spain September 14. A junta convened at Guatemala City makes the declaration (*see* Iturbide, 1822).

Panama declares independence from Spain in December and joins Colombia.

Missouri is admitted to the Union as the 24th state under terms of last year's Missouri Compromise that permits slavery in the new state (*see* Ashley, 1822).

The power of the U.S. Supreme Court is superior to that of any state court in matters involving federal rights, says Chief Justice John Marshall. He hands down the decision March 3 in *Cohens v. Virginia.*

Ohio Quaker saddlemaker Benjamin Lundy, 32, urges abolition of slavery and begins publication of his antislavery newspaper *Genius of Universal Emancipation.* He soon moves to Greenville, Tenn., and will relocate to Baltimore in 1824. A slave trader will attack and severely injure him in 1828, but Lundy will enlist the support of William Lloyd Garrison, now 16, and Garrison will serve as associate editor for 6 months beginning in September 1829 (*see* 1831).

Austin, Tex., has its origin in San Felipe de Austin founded in Mexico's Texas territory (above) by Stephen Fuller Austin, 28, whose father has just died at age 60. Young Austin carries out his father's plans by obtaining a grant of land on condition that he settle a given number of families on the land, and he founds the first permanent Anglo-American Texas settlement. Impresario Austin will bring some 8,000 colonists to the region (*see* 1836).

Minneapolis has its beginnings in the town of St. Anthony founded on the east bank of the upper Mississippi at St. Anthony's Falls (*see* Fort Snelling, 1819; Minneapolis, 1847).

English chemist-physicist Michael Faraday, 29, pioneers the electric motor with a demonstration of electromagnetic rotation. He has pondered on Oersted's 1819 discovery and conducts various experiments, including one involving a 6-inch length of copper wire suspended from a hook with its lower end dipping into a bowl of mercury. A bar magnet is fixed vertically in the center of the mercury, and Faraday finds that when he passes a continuous electric current from a battery through the hook and down through the wire to the mercury, the wire begins to rotate and continues to do so for as long as the current flows (*see* Sturgeon, 1823).

The first U.S. natural gas well is tapped at Fredonia, N.Y.

McGill University is chartered at Montreal. Scots-Canadian businessman James McGill died in 1813 at age 69, leaving £10,000 to found a college.

The University of Buenos Aires is founded.

George Washington University is founded at Washington, D.C.

Boston's English High School opens with 102 students. The first tuition-free public high school teaches no language but English, it prepares students for the professions, both "mercantile and mechanical," with a curriculum that emphasizes science, mathematics, logic, and history, and by 1827 Massachusetts will require every town of 500 families or more to support a high school from public tax revenues, making secondary education a birthright rather than a privilege (*see* normal school, 1839).

The Emma Willard School has its beginnings in the Troy Female Seminary at Troy, N.Y. Educator Emma Hart Willard, 34, will prove that young women can master subjects such as mathematics and philosophy without losing their health or charm.

News of Napoleon's death at St. Helena May 5 reaches Britain only 7 weeks later.

The *Manchester Guardian* begins publication. The weekly founded by English Liberal John Edward Taylor, 30, will appear daily beginning in 1855.

The *Saturday Evening Post* begins publication at Philadelphia. The weekly magazine founded by Samuel C. Atkinson and Charles Alexander will continue until 1969 (*see* Curtis, 1897).

Nonfiction "Confessions of an Opium Eater" by English author Thomas De Quincey, 36, appears in *London Magazine*. De Quincey began using opium in his second year at Oxford and dropped out in 1808.

Fiction *Kenilworth* and *The Pirate* by Sir Walter Scott; *The Spy* by U.S. novelist James Fenimore Cooper, 32, who has told his wife that he could write a better story than the popular English novel she was reading. His novel gains quick success.

The 1750 pornographic novel *Fanny Hill* is the subject of a Massachusetts trial following passage of a state obscenity law.

Painting *Landscape: Noon (The Hay Wain)* by John Constable. John Crome dies at Norwich April 22 at age 52.

Theater *The Golden Fleece (Das goldene Vliess)* by Franz Grillparzer 3/26 at Vienna's Burgtheater.

Opera *Der Freischutz (The Free-Shooter)* 6/18 at Berlin's Schauspielhaus, with music by Carl Maria von Weber.

First performances Konzertstück for Piano and Orchestra in C minor by Carl Maria von Weber 6/25 at Berlin; *Invitation to the Dance (Aufforderung zum Tanze)* by von Weber.

Cantata "Hinaus In's Frische Leben" by Carl Maria von Weber 11/16 at Dresden, for the birthday of the sister of Saxony's Friedrich August I.

March "Hail to the Chief" by Scottish composers James Sanderson and E. Rilley (who have adapted lines from Sir Walter Scott's *Lady of the Lake* poem of 1810) is played at the second inaugural of U.S. President James Monroe March 4.

Poker has its beginnings in a card game played by sailors at New Orleans. They have combined the ancient Persian game *As Nas* with the French game *poque*, a descendent of the Italian game *Primiera*, and a cousin of the English game *Brag*. Played with three cards from a deck of 32, the new game contains combinations such as pairs and three of a kind; the draw and the full deck of 52 cards will soon be added, but the game of stud poker will not evolve for many years nor will straights or flushes (date approximate; *see* Schenck, 1871).

New York's Fulton Fish Market opens on the East River.

Game birds shot by market hunters are mainstays of the U.S. diet. A single market hunter kills 18,000 migrating golden plover within the year.

Britain's population reaches 20.8 million, with 6.8 million of it in Ireland. France has 30.4 million, the Italian states 18, Austria 12, the German states, duchies, free cities, and principalities 26.1.

1822 A Greek assembly at Epidauros proclaims independence from the Ottoman Empire January 13 (*see* 1456; 1821), an Ottoman fleet takes the island of Chios in April, massacres much of the island's population, and sells most of the rest into slavery. Greek patriots set the Ottoman admiral's flagship afire in reprisal, a Greek flotilla under Admiral Konstantinos Kanaris, 32, destroys the Ottoman fleet June 19, but an Ottoman army of 35,000 invades Greece in July, overruns the

1822 ***(cont.)*** peninsula north of the Gulf of Corinth, and forces the new Greek government to take refuge in the islands.

Greek guerrilla leaders including Demetrius Ypsilanti, 29, Alexander's brother, gain victories over the Turks.

Viscount Castlereagh slits his throat with a penknife August 12 and dies at age 53. He is replaced as foreign minister and leader of the Commons by George Canning, now 52. Robert Peel, 34, becomes secretary for home affairs, and a liberal wing of the Tory cabinet begins to make important reforms.

Orangemen at Dublin attack the British viceroy and the city has bottle riots.

An Ashanti War begins in West Africa following an exchange of insults between an Ashanti trader and a Fanti policeman. The tribesmen go into battle blowing war horns, and the other side plays "God Save the King"; hostilities will continue for 9 years (*see* 1824).

The Battle of Pichincha outside Quito May 24 gives José de Sucre a decisive victory over a Spanish royalist army. Sucre has moved up the Andes, José de San Martin has sent forces to help him, he liberates Quito, and San Martin meets with Simon Bolívar at Guayaquil (*see* 1821; 1824).

Haiti conquers Santo Domingo from Spain and takes over the entire island of Hispaniola (*see* 1804; 1844).

Mexico crowns Agustin de Iturbide emperor July 25. Agustin I sends troops to bring other Central American nations into the new Mexican Empire (*see* 1821; 1823).

Brazil's Portuguese regent Dom Pedro, 24, proclaims independence from his father João VI, is crowned emperor December 1, and will reign until 1831 as Pedro I over the country that occupies 47 percent of South America. Pedro took refuge in Brazil with his father 15 years ago when the French invaded Portugal, has been prince regent since his father's return to Portugal last year, and has sided with the Brazilians against the reactionary policies of Lisbon (*see* 1831).

Vesey's Rebellion fails in South Carolina June 16 when authorities at Charleston arrest 10 slaves who have heeded the urgings of local freedman Denmark Vesey, 55. Vesey himself is arrested, defends himself eloquently in court, but is hanged July 2 with four other blacks. Further arrests follow, more than 30 other executions will take place, and several southern states will tighten their slave codes.

Exploration of the upper Missouri River gains impetus from Missouri Lieut. Gov. William H. Ashley, 44, who places a notice in the newspaper at St. Louis: "To enterprising young men. The subscriber wishes to engage one hundred young men to ascend the Missouri River to its source, there to be employed for one, two, or three years." Within a week, Ashley and his associate Andrew Henry, 43, have the men they want to establish a permanent fur trade, including keelboatman Mike Fink, 52, James Bridger, 18, Jedediah Strong Smith, 22, William L. Sublette, and Joseph Reddeford Walker, 23 (*see* South Pass, 1824).

The S.S. *Robert Fulton* completes the first steamboat voyage from New York to New Orleans and proceeds to Havana. Scots-American shipbuilder Henry Heckford, 47, has built the vessel.

The S.S. *Aaron Manby* that slides down the ways at Rotherhithe, England, April 30 is the world's first iron steamship. Named for the proprietor of the Horsley Ironworks in Tipton, Staffordshire, she undergoes trials on the Thames and then goes into service across the Channel, arriving at Paris June 10 with a cargo of linseed oil and iron.

George Stephenson completes the world's first iron railroad bridge for the Stockton-Darlington line (*see* 1814; 1825).

French Egyptologist Jean François Champollion, 32, deciphers the Rosetta stone found in 1799, making it possible to read the papyri and stones that will expand knowledge of ancient Egyptian civilizations. Having spent years studying the stone's hieroglyphics, Champollion completes his translations and reports that the script contains both sound-signs (phonograms) and sense-signs (ideograms) that provide the necessary clues for deciphering all Egyptian inscriptions.

Histoire Naturelle des Animaux sans vertèbres by the Chevalier de Lamarck proposes an erroneous theory of evolution that says environmental changes can produce structural changes in plants and animals (*see* 1809). Now 78, Lamarck contributes to science by introducing the classes *Annelida, Arachnids, Crustacea, Infusoria,* and *Tunicata*, but he claims that environmental changes can induce new or increased use of certain organs or parts and that these acquired characteristics can be transmitted to offspring.

U.S. Army physician William Beaumont, 37, begins pioneer observations on the action of human gastric juices. Serving on Michilimackinac (Mackinac) Island in the straits between Lakes Michigan and Huron, Beaumont has saved the life of a young French-American voyageur employed by John Jacob Astor's American Fur Co., but the shotgun blast wound in St. Martin's left side has not completely healed and the remaining fistula permits Beaumont to make his studies (*see* 1833).

Yellow fever strikes New York City; thousands flee to Greenwich Village, whose growth is spurred by the influx.

The first typesetting machine, patented in England by Connecticut inventor William Church, casts letters in lead and composes words automatically. Spacing the words to desired line measurements must still be done by hand (*see* Mergenthaler, 1884; Lanston, 1897).

The *Sunday Times* begins publication at London (*see* 1788).

Fiction *The Fortunes of Nigel* by Sir Walter Scott.

Painting *View of the Stour* by John Constable.

Sculptor Antonio Canova dies at Venice October 13 at age 64.

J. N. Niepce makes the first permanent photograph (*see* bicycle, 1816). Seeking a compound that will stabilize a photographic image on paper, the French physicist and his kinsman Claude Niepce, 17, follow Thomas Wedgwood's 1802 work and produce a heliograph, or heliotype, by using asphaltum, or bitumen of Judea, used for years in etching. The compound becomes insoluble in its usual solvents when exposed to light, they find, and in addition to giving a resist for the etching of metal, it produces transparent images on glass (*see* 1826).

Hungarian pianist Franz (Ferencz) Liszt makes his debut at Vienna December 1 at age 11 and meets composer Franz Schubert.

Popular songs "The Old Oaken Bucket" ("Araby's Daughter") by English-American composer George Kiallmark, 41, with lyrics from the 1818 Samuel Woodworth poem.

Britain's George IV has the first Royal Enclosure erected at Ascot, where horse races have been run since 1711.

New York's Greenwich Village (above) is a rural Manhattan area whose streets will never conform to the grid pattern employed elsewhere in the city (*see* 1811).

Potato crops fail in the west of Ireland as they will do again to some degree in 1831, 1835, 1836, and—most disastrously—in the mid-1840s (*see* 1739).

New York City's population reaches 124,000. A family of 14 can live comfortably on $3,000 per year.

English reformer Francis Place, 51, recommends contraception in his pamphlet "To the Married of Both Sexes of the Working People." He advises inserting a soft sponge "as large as a green walnut, or a small apple" and "tied by a bobbin or penny ribbon."

1823 French troops sent by foreign minister François René, vicomte de Chateaubriand, now 54, invade Spain to suppress her 3-year-old revolution. Chateaubriand is dismissed June 6 in a dispute that largely concerns Greece, but French forces triumph August 31 at the Battle of the Trocadero. They take Cadiz September 23, the rebels hand over Ferdinand VII, and the French restore him to power. He ignores advice to introduce a moderate constitutional regime and begins a repressive rule that will continue until his death in 1833.

U.S. forces defeat the Sac and Fox chief Black Hawk (Makataemishklakiak), 56, who has created an alliance with the Winnebago, Pottawotamie, and Kickapoo to resist white incursions into the Illinois country (but *see* 1832).

Mexico's Emperor Agustin confirms the grant of land on the Rio Brazos River made by the Spanish viceroy to the late Moses Austin and now held by his son Stephen F. Austin.

An assembly at Guatemala City July 1 declares the sovereignty of the United Provinces of Central America. Mexico recognizes its independence August 20, but the confederation of Guatemala, Honduras, Costa Rica, Nicaragua, and San Salvador will dissolve by 1840, and further attempts at union will fail.

The Monroe Doctrine enunciated by President Monroe in his annual message to Congress December 2 states a nationalistic determination to oppose any European influence in the Western Hemisphere and to remain aloof from European conflicts: "The American continents . . . are henceforth not to be considered as subjects for future colonization by any European power."

Britain abolishes the death penalty for more than 100 crimes that have been capital offenses. The mitigation of capital punishment crowns efforts by the late Samuel Romilly (*see* 1808).

An antislavery society is established at London under the leadership of William Wilberforce and Thomas Fowell Buxton, who has acquired a fortune as a brewer (*see* 1807; 1833).

The petroleum industry has its beginnings at the Caspian Sea port of Baku on Russia's Apsheron Peninsula where the "eternal fire" of natural gas holes has been known at least since the time of Alexander the Great in the 4th century B.C. Primitive drilling for oil begins at Baku, which will be yielding half the world's petroleum by 1900 (*see* 1859; 1871).

English physicist William Sturgeon, 40, devises the first electromagnet. He varnishes an iron bar to insulate it, wraps the bar with copper wire, connects the wire to the terminals of a voltaic pile, and creates a crude device that can lift a few pounds of iron (*see* Volta, 1800; Faraday, 1821; Henry, 1827).

Asiatic cholera reaches the gateway to Russia at Astrakhan in an epidemic that will affect every nation of the Western world (*see* 1820; 1829).

The Lancet begins publication October 5 at London under the direction of local physician Thomas Wakley, 28, who has started the weekly medical journal to advance knowledge and reform British medicine. Wakley will crusade against nepotism, the Royal College of Surgeons, and the adulteration of foods (*see* 1850).

Michael Faraday liquefies chlorine. The water-soluble poisonous gas will find wide use in water purification and bleaches. (It will also be used in warfare; *see* 1915.)

Trinity College is founded at Hartford, Conn.

An electric telegraph offered to the British admiralty by English inventor Francis Ronalds, 35, employs wires whose ends are attached to a battery at the sending point and to electrodes in acidified water at receiving stations, each electrode being marked with a letter or number. Ronalds has set up a complete telegraph system on his estate at Hammersmith using 8 miles of wire, but the admiralty rejects his offer, saying, "Telegraphs of any kind are wholly unnecessary and no other than the one in use [a semaphore installed in 1796] will be adopted" (*see* Chappe, 1793; Henry, 1831).

Fiction *The Pioneers* by James Fenimore Cooper begins his *Leatherstocking Tales*.

1823 *(cont.)* Poetry *Don Juan* by Lord Byron whose greatest work is based in part on the account of a 1741 shipwreck off the coast of Chile as described by his grandfather John "Foul-weather Jack" Byron, who made a round-the-world exploratory voyage from 1764 to 1766: "Man's love is of man's life a thing apart,/ 'Tis woman's whole existence." *A Visit from St. Nicholas* is published anonymously 12/23 in the *Troy* (New York) *Sentinel*. New York lexicographer Clement Clark Moore, 44, had his *Compendious Lexicon of the Hebrew Language* published in 1809, and he brings to life the figure depicted by Washington Irving in 1809: "'Twas the night before Christmas. . . Now, Dasher! now, Dancer! now, Prancer and Vixen! On, Comet, on, Cupid! on, Donder and Blitzen!"

Theater *The Truthful Liar (Le Menteur Véridique)* by Eugène Scribe 4/24 at the Théâtre du Gymnase, Paris (16 Scribe comedies are produced, up from 11 last year).

Opera *Semiramide* 2/3 at Venice's Teatro la Fenice, with a libretto based on a Voltaire tragedy of 1748, music by Gioacchino Rossini; *Clari, or the Maid of Milan* 5/8 at London's Covent Garden Theatre, with Charles Kemble, 48, music by English composer Henry Rowley Bishop, 37, libretto and lyrics by New York actor-playwright John Howard Payne, 37, arias that include "Home Sweet Home." The opera opens at New York's Park Theater 11/12, 100,000 copies of the song are sold in its first year, the London publishers net 2,000 guineas in profit, and Bishop will become the first musician to be granted knighthood (in 1842); *Euryanthe* 10/25 at Vienna's Kärntnerthor Theater, with music by Carl Maria von Weber.

First performances *Ballet-Musik aus Rosamunde* by Franz Schubert 12/20 at Vienna.

Boston cabinetmaker Jonas Chickering starts a piano company that will become the largest in America. In 1843 Chickering will develop the first practical application of the one-piece solid casting for both grand and upright pianos, and in 1850 he will develop the first overstrung bass scale, features that will become standard with all pianos (*see* 1853; Steinway, 1859).

Rugby student William Webb Ellis, 17, inaugurates a new game whose rules will be codified in 1839. Playing soccer for the 256-year-old college in East Warwickshire, Ellis sees that the clock is running out with his team behind so he scoops up the ball and runs with it in defiance of the rules. Rugby College will adopt rugby football (or rugger) officially in 1841 (*see* 1839; Association soccer, 1863; football, 1869, 1873).

The Royal Thames Yacht Club is founded.

The Macintosh raincoat has its beginnings in a waterproof fabric of rubber bonded to cloth patented by Scottish chemist Charles Macintosh, 57, who applies his research on possible uses of the coal tar distillate naphtha. It is sticky in hot weather and brittle in cold but Macintosh's cloth will make the name Macintosh a British generic for raincoat (*see* Goodyear, 1839; Aquascutum, 1851; Burberry, 1856).

New York's A.T. Stewart Department Store has its beginnings in a small dry-goods shop opened by Irish-American merchant Alexander Turney Stewart, 20, who will open the world's first true department store (*see* 1846).

John Nash completes London's Sussex Place. He also completes a new royal pavilion at Bath after 8 years of construction in a "Hindoo" Islamic style to replace a villa built for the prince regent in the late 1780s.

China's monopoly in the tea trade begins to fade as an indigenous tea bush is found growing in northern India's Upper Assam. Acting for the British government, Charles Bruce smuggles knowledgeable coolies out of China and puts them to work transplanting young tea bushes into nursery beds to begin tea plantations (*see* 1839).

Illinois opens to U.S. corn farmers following the defeat of Black Hawk (above).

Michael Faraday (above) pioneers mechanical refrigeration with the discovery that certain gases under constant pressure will condense until they cool (*see* Perkins, 1834; Gorrie, 1842).

Bourbon whiskey distilling increases in Kentucky (*see* Craig, 1789). Aging in charred oak barrels will not begin until 1860, and whiskey will not take on color until then (the color will generally derive more from added caramel than from wood char).

1824 Ashanti tribesmen defeat a Fanti detachment January 21 at Essamako, outnumbering the Fanti 10,000 to 500. The Ashanti capture the British governor Sir Charles M'Carthy, cut off his head, and will use his skull as a royal drinking cup at Coomassie, but their king Osai Tutu Kwadwo (Quamina) is killed.

A Burmese War begins February 24 as the British governor general of India Lord Amherst declares war on Burma, whose forces have captured the island of Shahpuri in violation of East India Company territorial rights. Rangoon falls to the British May 11.

Hawaii's Kamehameha II and his wife die of measles July 14 on a visit to Britain (*see* 1819; 1825).

South American liberation forces under Simon Bolívar and José de Sucre move into the Andean highlands of Charcas and defeat a Spanish force August 24 at Junin.

The Battle of Ayacucho December 9 gives José de Sucre and his 5,800 men a triumph over 9,300 Spanish royalists. The 13,000 royalists who remain in Peru are forced to withdraw (*see* 1825).

France's Louis XVIII dies September 16 at age 68 after a 10-year reign that was interrupted for 100 days in 1815. His brother, 66, will reign until 1830 as Charles X.

Greek forces at Mitylene nearly annihilate an Ottoman army in October as the Greeks mourn the loss of Lord Byron, who died at Missolonghi April 19 at age 36 while helping the Greeks in their fight for independence.

The U.S. presidential election in November ends with no candidate having a majority in the Electoral College although John Calhoun is elected vice president.

Andrew Jackson has 99 electoral votes, John Quincy Adams 81, William H. Crawford 41, Henry Clay 37 (*see* 1825).

English reformer Robert Owen promotes abolition of slavery, women's liberation, and free progressive education.

British workers gain the right to organize June 21 as Parliament repeals the Combination Acts at the urging of radical M.P. Joseph Hume, 47, who has been persuaded by reformer Francis Place, the contraception advocate (but *see* 1825).

Scottish social reformer Frances "Fanny" Wright, 29, champions women's rights and free public schools in America. She will settle in New York in 1829 and scandalize the nation by discussing contraception, a more equal distribution of wealth, an end to religion, and emancipation of the slaves with resettlement of African-Americans outside the United States.

Indiana settlers attack an Indian village and massacre two braves, three squaws, and four children. The settlers claim that "killing an Indian serves a better purpose than killing a deer," but the conscience of the white community is aroused, four of the killers are hanged, and tension grows between whites and Indians on the frontier.

The Royal Society for the Prevention of Cruelty to Animals (RSPCA) is founded at London (*see* 1866).

Jedediah Strong Smith of the Rocky Mountain Fur Co. discovers South Pass through the Rocky Mountains into the Great Basin (*see* 1822; 1826).

Robert Owen (above) purchases New Harmony, Ind., from the German Lutheran Rappites who founded it 10 years ago. He is starting communes in England, Ireland, Mexico, and the United States, all of them doomed to fail. New Harmony will be the first of many American communes.

Reflections on the Motive Power of Fire (Réflexions sur la puissance motrice du feu) by French physicist Nicolas Léonard Sadi Carnot, 28, states a principle that will be expressed as the second law of thermodynamics by William Thomson (Lord Kelvin). It is a major advance in the understanding of how heat can be used to drive engines. Carnot notes that steam engines are used to work mines, propel ships, excavate ports, forge iron, saw wood, grind grain, and spin and weave cloth, but observes that "their theory is very little understood, and attempts to improve them are still directed almost by chance." He states that "the production of motive power is then due in steam-engines not to an actual consumption of caloric, *but to its transportation from a warm body to a cold body* . . . [a principle that] is applicable to any machine set in motion by heat. According to this principle, the production of heat alone is not sufficient to give birth to the impelling power. It is necessary that there should also be cold; without it the heat would be useless" (*see* Clausius, 1850; Thomson, 1851).

In some engines, says Carnot (above), a water-cooled condenser provides the cold element, while others vent their steam into the atmosphere. For the latter the cooler element of the cycle is provided by the great volume of the atmosphere, and in the ideal "Carnot cycle" there is no heat loss through equipment, all of the heat being converted into motion. Heat used to produce motion can be regenerated by that same motion, making the cycle reversible.

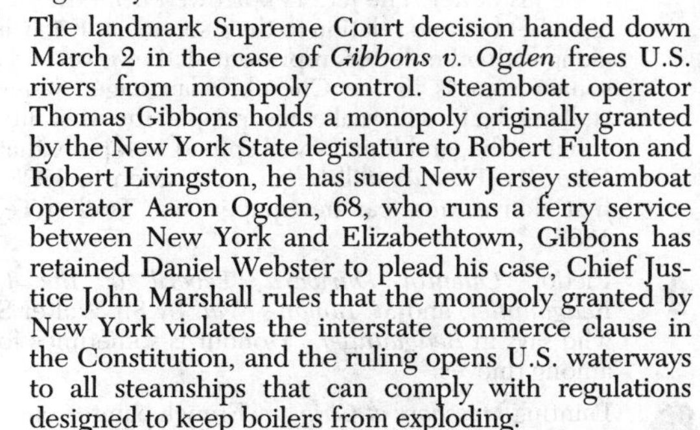

The landmark Supreme Court decision handed down March 2 in the case of *Gibbons v. Ogden* frees U.S. rivers from monopoly control. Steamboat operator Thomas Gibbons holds a monopoly originally granted by the New York State legislature to Robert Fulton and Robert Livingston, he has sued New Jersey steamboat operator Aaron Ogden, 68, who runs a ferry service between New York and Elizabethtown, Gibbons has retained Daniel Webster to plead his case, Chief Justice John Marshall rules that the monopoly granted by New York violates the interstate commerce clause in the Constitution, and the ruling opens U.S. waterways to all steamships that can comply with regulations designed to keep boilers from exploding.

The U.S. Army Corps of Engineers embarks on a career of building harbors, damming and channeling rivers, and generally developing waterways and other civil projects under terms of a bill signed by President Monroe May 24 to establish the corps that had its real beginnings in 1775 when George Washington appointed Col. Richard Gridley chief engineer to the Continental army just before the Battle of Bunker Hill. The primary mission of the corps will be one of providing support for military operations, but it will play a significant role in developing U.S. waterways and highways, often at the expense of the environment.

Vermont's first covered bridge spans the Missisquoi River at Highgate Falls (*see* Philadelphia, 1805). Scores of such bridges throughout the state will follow the arch truss bridge at Highgate Falls.

A Road Survey Act passed by Congress April 30 authorizes the Army Corps of Engineers (above) to survey possible road and canal routes.

Land grants to the Wabash and Erie Canal Co. will total 826,300 acres in the next 10 years (work on the Wabash and Erie will start in 1832). Land grants to other private canal companies will total more than 4.2 million acres in the next 42 years (*see* Miami Canal, 1827; Chesapeake and Ohio, 1828; Louisville and Portland, 1830; Cleveland to Portsmouth, 1832).

Portland cement, patented by English bricklayer Joseph Aspdin of Leeds, is impervious to water and as durable as the cement used by Roman aqueduct builders in ancient times. Aspdin has mixed chalk and clay and heated the mixture to a high temperature (*see* White, 1820; reinforced concrete, 1849).

Jöns Jakob Berzelius isolates silicon.

London physician William Prout, 39, isolates hydrochloric acid from stomach juices and establishes that it is the chief agent of human digestion (*see* pepsin, 1835).

1824 *(cont.)* Rensselaer Polytechnic Institute, founded at Troy, N.Y., is the first U.S. engineering and technical school.

A Cherokee language alphabet with 85 letters is perfected by Cherokee scholar Sequoya, 54, who has taken the name George Guess from a U.S. trader he believes to be his father. The letters borrowed from the Roman alphabet bear no relation to their sounds in English but along with other letters represent all the vowel and consonant sounds in the Cherokee language. Sequoya's "talking leaf" will make his people the first literate Indian tribe, and literacy will spread so rapidly that the *Cherokee Phoenix* will begin publication in 1828, the first Native American newspaper (*see* "Trail of Tears," 1838).

Fiction *Quentin Durward*, *Peveril of the Peak*, *Redgauntlet*, and *St. Ronan's Wall* by Sir Walter Scott who says in *Redgauntlet*, "Honour is sometimes found among thieves."

Painting *Massacre at Chios* by French painter-engraver (Ferdinand Victor) Eugène Delacroix, 26; Théodore Géricault dies at Paris January 26 at 32.

First performances Mass in D major *(Missa Solemnis)* by Ludwig van Beethoven 3/26 at St. Petersburg; Symphony No. 9 in D minor *(Choral)* by Ludwig van Beethoven 5/7 at Vienna. Now 53, Beethoven is so deaf that he must be turned round at the end to see that the audience is applauding.

Montreal's Gothic parish church of Notre Dame is completed south of the Place d'Armes to replace a structure dating to 1672.

The Rappites at New Harmony, Ind., (above) have built prefabricated houses and barns of poplar and walnut timbers fitted by mortise-and-tenon joints and sided with wood or bricks.

Shakers build the first round barn at their Hancock, N.Y., community (*see* 1784). It has a silo at its center with stalls and stanchions radiating outward to save steps in feeding the livestock. In years to come the round barn will be widely adopted by Midwestern dairy farmers.

The Rhode Island red hen is produced at Little Compton, R.I., by a sea captain who has crossbred several domestic poultry breeds with an assortment of exotic fowl brought from the Orient. The Rhode Island red will be famous for its brown eggs which New Englanders will prefer to white eggs that may not come from local hens and may therefore not be fresh.

The Royal Navy reduces its daily rum ration from half a pint to a quarter pint, and tea becomes part of the daily ration (*see* 1740; 1850).

The Glenlivet started at Glenlivet by Scottish entrepreneur George Smith, 32, is the first licensed Scotch whisky distillery. Smith will continue distilling his malt whisky until his death in 1871.

The first commercial Italian pasta factory is started at Imperia on the Italian Riviera by Paolo B. Agnese, whose family will continue the business for more than 150 years.

Cadbury's Chocolate has its beginnings in a tea and coffee shop opened by Birmingham, England, Quaker John Cadbury, 23, who has served an apprenticeship at Leeds and for bonded tea houses in London. Given a sum of money by his father and told to sink or swim, Cadbury starts his business next door to his father's draper's shop, will install Birmingham's first plate glass window, employ a Chinese to preside over his tea counter, and experiment with grinding cocoa beans, using a mortar and pestle (*see* Cadbury Brothers, 1847).

1825 The odious king of the Two Sicilies Ferdinand I dies January 4 at age 73 after a reign of suppression supported by spies and informers and by his wife Maria Carolina, daughter of the Austrian empress Maria Theresa. He is succeeded by his son of 47, who will reign until 1830 as Francesco I.

Greece's war of independence continues with a renewal of the siege of Missolonghi by Ottoman forces under Mustapha Mohammad Reshid Pasha, 23, who have invaded Greece from the north. Egyptian Ottoman forces land in the Morea and subdue the peninsula under the leadership of Mohammed Ali's son Ibrahim.

John Quincy Adams is elected U.S. president February 9 in the House of Representatives where Kentucky's Henry Clay controls the deciding block of votes. Clay chooses Adams over Andrew Jackson as the lesser of two evils and is named secretary of state.

Hawaii's Kamehameha III becomes king at age 12 upon news of his older brother's death last year in England; he will be crowned in 1833 and reign until 1854.

Uruguayans revolt against Brazil with help from Buenos Aires. Argentina hopes to annex Uruguay and begins a war with Brazil to acquire also the Banda Oriental.

Bolivia (the republic of Bolívar) is proclaimed an independent nation August 6 by a congress at Chuquisaca in upper Peru. The congress has been convened by the hero of last year's Battle of Ayacucho Antonio José de Sucre, who will be Bolivia's first president from 1826 to 1828 but will resign in the face of opposition from native Bolivians (*see* 1830).

Uruguay becomes independent of Brazil August 25, but Brazil goes to war with Argentina 3 months later over the question of Uruguay.

Portugal recognizes Brazilian independence under Dom Pedro August 29.

Russia's Aleksandr I dies in agony December 13 at age 47 after a 24-year reign. The czar, who has eaten poisonous mushrooms in the Crimea, is succeeded by his brother, 21, who will reign oppressively until 1855 as Nicholas I.

A Decembrist uprising of young Russian aristocrat army officers tired of Romanov autocracy begins December 14, the new czar hopes to avoid violence, but when the rebel troops in St. Petersburg's Senate Square refuse to listen he orders a cavalry charge to clear the square. Improperly shod, the horses of the cavalrymen slip and fall on the icy cobblestones, but the rising is ill-planned and quickly suppressed.

A French law of indemnity enacted in April compensates the country's nobility for losses of lands during the French Revolution. Compensation is at the expense of government bondholders, who are mostly members of the upper bourgeoisie.

Parliament enacts a law July 16 permitting British workers to combine in order to secure regulation of wages and hours, but the law includes a provision that prohibits use of violence or threats of violence, it effectively denies the right to strike, and it introduces summary methods of conviction.

U.S. agitation for a 10-hour day begins with a great strike by 600 Boston carpenters. Master carpenters denounce the strike, warning of the "unhappy influence" of change on apprentices "by seducing them from that course of industry" and of "the . . . many temptations and improvident practices" from which journeymen are safe so long as they work from sunrise to dusk (*see* 1831).

Creek tribesmen in Georgia repudiate a February 12 treaty at Indian Springs in which their leaders have ceded all Creek lands in Georgia to the United States and have promised to leave for the West by September 1 of next year (*see* 1826).

France makes sacrilege a capital offense.

Omaha has its beginnings in a trading post established on the Missouri River. Fur trader Pierre Cabanne opens the post in Louisiana Territory that will later be called Nebraska (*see* 1854).

Chicago (Fort Dearborn) begins to grow as the new Erie Canal (below) increases its importance. The fort has a garrison of 70 (*see* 1803; 1830).

Akron, Ohio, is founded by Gen. Simon Perkins, who served in the area during the War of 1812 and who has returned to the region south of Lake Erie to acquire vast tracts of wilderness land. Akron is Greek for high point, but the town site is only 395 feet above Lake Erie.

Scottish explorer Alexander Gordon Laing, 32, is sent to explore the Niger basin via Tripoli. He crosses the desert of Tuat and goes on to become the first European to visit Timbuktu.

The New York Stock Exchange opens as such, most of the securities traded being shares in canal, turnpike, mining, and gaslighting companies (*see* 1792). No shares in industrial corporations will appear until 1831, and even on the Boston exchange, which is closer to the nation's manufacturing center, no industrial shares will be traded until 1827. Few will appear on either exchange for another 40 years.

England's Stockton and Darlington Railway opens September 27 with the world's first steam locomotive passenger service. Planned as a tramway by promoter Edward Pease, 58, the new 27-mile rail line has been built for steam traction with the help of engineer George Stephenson, whose 15-ton locomotive *Active* pulls a tender, six freight cars, the directors' coach, six passenger coaches, and 14 wagons for workmen (*see* B&O, 1828).

The Erie Canal opens October 26 to link the Great Lakes with the Hudson River and the Atlantic. Gov. De Witt Clinton greets the first canal boat on the $8 million state-owned canal, which is 363 miles long, 40 feet wide, and 4 feet deep, with tow paths for the mules that pull barges up and down its length at 1 mile per hour. The time required to move freight from the Midwest to the Atlantic falls to 8 or 10 days, down from 20 to 30, freight rates drop immediately from $100 per ton to $5, New York City becomes the Atlantic port for the Midwest, and the canal makes boom towns of Buffalo, Rochester, Cleveland, Columbus, Detroit, Chicago, and Syracuse (*see* 1819; 1836; 1917).

The Schuylkill Canal opens to connect Philadelphia with Reading, Pa.

The world's first wire suspension bridge opens near Lyons, France, where it has been built by engineer Marc Seguin, 39 (*see* Finley, 1800; Ellet, 1842).

Rutgers College gets its name. The 59-year-old Queen's College at New Brunswick, N.J., changes its name to honor local benefactor Henry Rutgers, 80.

Amherst College is chartered at Amherst, Mass., where Congregationalist clergyman-professor Zephaniah Swift Moore began giving lessons in 1821 and continued until his death 2 years ago at age 53.

Nonfiction "Milton" by English essayist Thomas Babington Macaulay, 24, appears in the *Edinburgh Review*; *The Spirit of the Age, or Contemporary Portraits*, by English essayist William Hazlitt, 47; *The Norman Con-*

Erie Canal barges brought grain from the Midwest to the Atlantic Seaboard, spurring the growth of New York.

1825 (cont.) *quest of England (Conquête de l'Angleterre par les Normands)* by French historian Jacques Nicolas Augustin Thierry, 30.

Fiction *Tales of the Crusaders*, *The Betrothed*, and *The Talisman* by Sir Walter Scott.

Painting *The Leper's House* and *The Leaping Horse* by John Constable; *General Lafayette* by New York painter Samuel Finley Breeze Morse, 34; rural scenes by English-American artist-poet Thomas Cole, 24, who moves to a Catskill, N.Y., and founds the Hudson River school of American painting; *The Peaceable Kingdom* by Pennsylvania Quaker preacher-carriagemaker-sign-painter Edward Hicks, 45. Before his death in 1849 Hicks will have painted more than 60 variations on the theme taken from Isaiah 11:6. Henry Fuseli dies at London April 16 at age 84; Jacques Louis David dies at Brussels December 29 at age 77.

Theater *King Ottokar's Rise and Fall (König Ottokars glück und ende)* by Franz Grillparzer 2/19 at Vienna's Burgtheater.

London's Buckingham Palace is created out of Buckingham House by John Nash. The palace will be the residence of Britain's ruling family beginning in 1837.

London's Belgrave Square is laid out. Belgravia will become an exclusive residential area.

Scottish botanist David Douglas, 27, discovers a coniferous evergreen in America's Pacific Northwest that will be called the Douglas fir *(Pseudotsuga taxifolia or P. Douglasii)*.

A Canadian forest fire burns over a tract of 4 million acres in New Brunswick extending more than 100 miles, destroying the towns of Chatham, Douglas, and Newcastle, and killing 160.

British colonists in Ceylon plant coffee bushes (but *see* 1861).

Haitian sugar production falls to below 1 ton, down from 27,000 tons in 1804.

The first U.S. patent for tin-plated cans is issued to Thomas Kensett (*see* 1819; 1847).

Buffalo, N.Y., becomes the meat-packing center of the United States as the Erie Canal (above) makes it a central shipping point.

Nathaniel Jarvis Wyeth, an associate of Fredric Tudor, patents an improved ice-harvesting method (*see* 1805). Tudor will use sawdust to insulate blocks of ice at Wyeth's suggestion (*see* voyage to India, 1833).

The Physiology of Taste (Physiologie du Goût), or Meditations on Transcendental Gastronomy, by French lawyer-gastronome Jean-Anthelme Brillat-Savarin, 64, is published at Paris. "Tell me what you eat and I will tell you what you are," says Brillat-Savarin (who escaped the Reign of Terror in 1792 and taught violin at Hartford, Conn.).

The first organized group of Norwegian immigrants to the United States arrives in October from Stavenger where 52 emigrants, many of them Quakers fleeing a hostile state church, boarded the sloop *Restauration* July 4. A baby has been born aboard ship, and the 53 Norwegians who arrive at New York October 9 begin a movement that will take hundreds of thousands to New York, the Midwest, and the Northwest in the next 100 years.

1826 Portugal's João VI dies March 10 at age 56 after a 10-year reign of which half was spent in Brazil. His son Dom Pedro of Brazil succeeds to the throne as Pedro IV, draws up a charter giving Portugal a moderate British-style parliamentary government, but refuses to leave Brazil. He abdicates the Portuguese throne to his infant daughter Maria da Gloria, and she will reign until 1853, with her uncle Dom Miguel, 24, as regent (*see* 1828).

Persian troops invade Russia's transcaucasian territories, but Ivan Feodorovich Paskevich, 44, defeats the invaders at the Battle of Ganja. A 3-year war begins (*see* 1828).

Constantinople receives a Russian ultimatum April 5 demanding a return to the status quo in the Danubian principalities. The sultan Mahmud II yields on the advice of Austria and France.

Ottoman Janissaries revolt in May following a decree by Mahmud II ordering formation of a new military corps to replace them. A small corps loyal to the sultan bombards the barracks at Constantinople; the city's mob joins in the attack and massacres 6,000 to 10,000 Janissaries.

The Afghan capital Kabul falls to the ruler of Ghazni, Dost Mohammad, 33, who begins to extend his sphere of influence (*see* 1835).

The Treaty of Yandabu February 24 ends the 2-year Burmese War. Britain acquires Assam, Arakan, and Tenasserim.

British forces crush the Ashanti August 7 at Dodowa near Accra, but hostilities will continue until 1831.

Britain's Cape Colony extends her borders north to the Orange River (*see* 1820; 1834).

Bolivian independence gains recognition from Peru and Simon Bolívar organizes a government for the new mountain republic. He then departs for Colombia and the Pan American Congress at Panama in June. No U.S. representative attends (one delegate sent by Washington dies en route, the other arrives too late).

Former U.S. presidents John Adams and Thomas Jefferson die July 4, aged 90 and 83, respectively, on the fiftieth anniversary of the signing of the Declaration of Independence.

The Treaty of Washington January 24 abrogates last year's Treaty of Indian Springs with the Creek nation. The Creeks cede a smaller area to the United States, and the treaty permits them to remain on their lands until January 1 of next year.

A Pennsylvania law that makes kidnapping a felony effectively nullifies the Fugitive Slave Act of 1793 (but *see* 1842).

Batavia, N.Y., freemason William Morgan, 52, is arrested on trumped-up charges of theft and indebtedness after conspiring with a local printer to publish a book revealing the secrets of the Masonic order that has existed since ancient times. Committed to the Canandaigua jail in September, Morgan is kidnapped from the jail, spirited off to Canada, and disappears. The case brings accusations that the Masonic order has obstructed justice and murdered Morgan, charges that begin press assaults on Free-masonry.

The first overland journey to Southern California begins August 22 as Jedediah Strong Smith leaves Great Salt Lake at the head of an expedition that reaches the lower Colorado River, crosses the Mojave Desert, and arrives November 27 at San Diego where a mission was established in 1769. Smith and two partners bought out William Ashley earlier in the year and have started a trading company at Great Salt Lake (*see* 1824; Fort Vancouver, 1827).

Berlin's Unter den Linden is illuminated with gas lamps.

Electrodynamics by André Ampère expands knowledge of electricity (*see* 1820).

An astatic galvanometer devised by Italian physicist Leopoldi Nobili, 42, for measuring a small electric current minimizes the effect of the earth's magnetism.

Connecticut's 6-mile Windsor Locks Canal opens to provide safe passage round the Enfield Falls and rapids in the Connecticut River 12 miles upstream from Hartford.

The S.S. *Curaçao* crosses the Atlantic to pioneer use of steam power in oceanic transportation, but while the Dutch vessel has steam-driven paddle wheels to supplement her sails, lack of fresh water for steam boilers delays the introduction of transatlantic steamers that operate entirely on steam power (*see* 1830; 1838; *Savannah*, 1819).

Steam propulsion remains an auxiliary to sail. Of Britain's 2.3 million tons of merchant shipping only 24,000 are in steamships (*see* 1846).

U.S. silk production is advanced by Gideon B. Smith of Baltimore, who plants the first of the new quick-growing Chinese mulberry trees (*morus multicaulis*) that will spur development of the infant silk industry (*see* Cheney brothers, 1838).

Collins axes are introduced at Hartford, Conn., by local storekeepers Samuel and David Collins, who start making their own axes after some years of buying British steel to supply blacksmiths for making axe blades. The Collins brothers buy an old gristmill on the Farmington River, rig up some machinery to blow air into the forges and turn grindstones, obtain dies and forging machinery devised by Elisha King Root, and begin a business that will grow to turn out 40,000 axes per month. Standardized precision-made trademarked Collins axes will fell the trees of the American wilderness and Collins machetes (called cutlasses in the British West Indies) will clear tropical jungles for more than 165 years.

French chemist Antoine Jérome Balard, 24, discovers bromine. He will devise a process for extracting sodium sulfate directly from seawater (*see* Dow, 1889).

The lyceum movement in U.S. adult education is spurred by New Englander Josiah Holbrook, who publishes recommendations that will be adopted by associations of villagers and urban workers who have had little formal education but who seek learning. The National American Lyceum will coordinate the activities of member groups beginning in 1831, and by 1839 there will be 137 lyceums in Massachusetts alone, drawing 33,000 to lectures on the arts, sciences, and public issues; by 1860 some 3,000 lyceums will be operating in New England, New York, and the upper Mississippi Valley.

University College is founded at London.

Munich University is founded in Bavaria.

Burke's Peerage by Irish genealogist John Burke, 39, is the first dictionary of British baronets and peers in alphabetic order (*see* Debrett's, 1769). Burke will later issue a dictionary of commoners under the title *Burke's Landed Gentry*.

Fiction *Woodstock* by Sir Walter Scott whose publisher partners James Ballantyne, 53, and Archibald Constable, 51, go bankrupt in January leaving Scott with £130,000 in liabilities. Scott's wife dies, and he works feverishly despite failing health to clear his debts; *The Betrothed (I Promessi Sposi)* by Italian novelist-poet Alessandro Manzoni, 41; *The Last of the Mohicans* by James Fenimore Cooper continues the *Leatherstocking Tales* that began with *The Pioneers* in 1823.

Poetry *Pictures of Travel (Reisebilder)* by German lyric poet Heinrich Heine, 29, who last year adopted the Christian faith and changed his name from Harry. Further volumes of Heine's travel sketches will appear in the next 5 years; *Past and Present* by English lyric poet Thomas Hood, 27: "I remember, I remember/ The house where I was born. . ."; *Silva a la Agricultura de la Zona Torrida* by South American poet Andres Bello, 44, who serves at London as a diplomatic representative for Simon Bolívar.

Painting *Greece in the Ruins of Missolonghi* and *Execution of the Doge Marino Falieri* by Eugène Delacroix.

Aloys Senefelder invents a process for lithographing in color (*see* 1798). He has been inspector of maps at the royal printing office in Munich for 20 years.

Photography is advanced by Parisian printer M. Lemaître, who has received etched metal plates from J. N. Niepce with instructions to print them (*see* Niepce, 1822). Lemaître exposes the plates to iodine vapors in order to make the hardened bitumen highlights more distinct and the pictures sharper. He suggests using silvered copper plates.

Louis Jacques Mandé Daguerre, 39, approaches J. N. Niepce (above) and proposes a partnership. Daguerre paints scenery for the Paris Opéra but has dabbled with silver salts in photography (*see* 1829).

Opera *Oberon* 4/12 at London's Covent Garden Theatre, with music by Carl Maria von Weber, who dies of tuberculosis at London June 5 at age 39.

1826 *(cont.)* First performances String Quartet in B flat major by Ludwig van Beethoven 3/21 at Vienna without the Grosse Fugue that Beethoven will write for the last movement; *Marche Militaire* by Franz Schubert.

Lord & Taylor opens in New York where English-American merchant Samuel Lord, 23, borrows $1,000 from his wife's uncle John Taylor to start a shop at 47 Catherine Street. By year's end Lord has taken into partnership his wife's cousin George Washington Taylor, and by 1832 their business will have an annex in an adjoining building.

John Nash completes London's Cumberland Terrace. The new royal palace overlooks Regent's Park which Nash will complete in 5 years.

The Zoological Gardens in Regent's Park (above) is founded by the Zoological Society of London with help from Sir Thomas Raffles. It will open its "zoo" to the public for 2 days a week beginning April 27, 1828, with the first hippopotamus to be seen in Europe since the ancient Romans showed one at the Colosseum. The Society will help save some bird and animal species from extinction.

The first workable reaper joins two triangular knives to two horizontal bars at the front of a machine that is pushed through a field of ripe grain by two horses. Scotsman Patrick Bell's lower bar is fixed, while the upper bar is geared to the ground wheels to give it a reciprocal motion; revolving sails hold the grain to the knives while a canvas drum lays aside the stalks in a neat swath, but horses cannot see ahead, they resist pushing Bell's reaper, it is difficult to turn, and will achieve only moderate success (*see* McCormick, 1831).

The first commercially practicable gas stove is designed by Northampton, England, gas company executive James Sharp and is installed in the kitchen of his home. Sharp will open a factory to produce the stoves in 1836 (*see* Reform Club, 1838).

Boston's Quincy Market opens August 26 across the cobble-stoned square from Faneuil Hall. Josiah Quincy's handsome Greek revival building will soon be surrounded by pushcarts.

Herkimer, N.Y., dairyman Sylvanus Ferris and cheese buyer Robert Nesbit develop a profitable scheme. Nesbit serves as advance man, turning down cheese offered for sale by farmers, hinting at a bad market, and deprecating the cheese. Ferris comes along later and buys the cheese at low prices for resale at high prices in the New York market.

The first tea to be retailed in sealed packages under a proprietary name is introduced by English Quaker John Horniman whose sealed, lead-lined packages have been designed in part to protect his tea from adulteration (*see* Accum, 1820).

A sixth edition of the 1798 *Essay on Population* by Thomas R. Malthus expands the original pamphlet into a massive book which points out that in an industrialized society national income tends to outpace population growth. The size of the family becomes a function of choice through adequacy and prevalence of contraceptive measures. Poor laws, says Malthus, encourage large families with doles. He recommends late marriage and "moral restraint" to relieve mankind, at least briefly, of its inevitable fate (*see* 1834).

1827 Peru secedes from Simon Bolívar's Colombia January 26, charging Bolívar with tyranny.

The Battle of Itzuaingó February 20 gives the Brazilian forces of Pedro I a victory over Argentine and Uruguayan troops (*see* Uruguay, 1828).

Britain's Prime Minister Lord Liverpool suffers a stroke in February and resigns. He is succeeded by George Canning, who forms a government in April.

Turkish forces enter Athens June 5 and force the defenders on the Acropolis to capitulate.

The Treaty of London July 6 allies Britain, France, and Russia in a pledge to support the Greeks against the Turks if the Ottoman sultan will not accept an armistice. Prime Minister Canning dies August 8 at age 56, and a new Cabinet is formed under Lord Goderich.

The sultan Mahmud II rejects the allied demands for a truce August 16.

A large Egyptian fleet with transports lands at Navarino September 8, but the Egyptian and Turkish fleet is largely destroyed October 20 by British, French, and Russian squadrons in the Battle of Navarino (*see* 1828).

A Russian army defeats the Persians October 1 and takes Erivan in Armenia.

The Cantonist Decrees issued by Russia's Nicholas I conscripts all male Jews aged 12 to 25.

Ottawa has its beginnings in the Bytown settlement founded at Chaudière Falls on Canada's Ottawa River (*see* 1800; 1867).

Jedediah Strong Smith blazes the first trail from southern California north to Fort Vancouver on the Columbia River but loses most of his men to Indian attackers (*see* 1826; 1830).

German physicist Georg Simon Ohm, 38, finds that the current flowing through an electrical conductor is proportional to the voltage across it and inversely proportional to its resistance. This will be called Ohm's Law and the practical unit of electrical resistance will be called the ohm to honor Ohm's discovery of the relationship between the strength (or intensity) of an unvarying electrical current (the electromotive force) and the resistance of a circuit (*see* Ampère, 1820).

U.S. physicist Joseph Henry, 30, builds an electromagnet that can lift 14 pounds and then uses a second layer of insulated wire overlapping the first layer to build an electromagnet that can lift 28 pounds. Henry will build an electromagnet for Yale's Benjamin Silliman that can lift 2,880 pounds (*see* 1831; Sturgeon, 1823; Faraday, 1831; telegraphy, 1831).

French engineer Benoit Fourneyron, 25, devises the world's first waterwheel turbine. More powerful than a waterwheel alone, it differs from a waterwheel in that the water flows under pressure from the hub toward the rim. More than 100 of Fourneyron's turbines will be built after he demonstrates an experimental 50-horsepower unit, but lack of adequate materials and technology will stymie his efforts to develop a steam turbine (*see* Parsons, 1884).

The Baltimore & Ohio Railroad has its beginnings February 28 in a charter granted to Baltimore bankers George Brown, 40, and Philip Evan Thomas to build a 380-mile railway to the West that will compete with the 2-year-old Erie Canal, which is diverting traffic from the port of Baltimore. The new railway is to be used for cars that will be drawn by horses or propelled by sails (*see* 1828).

Cincinnati gains new commercial importance with the completion of the Miami Canal (*see* 1790). By 1830 the city will have a population of 25,000 and the "Queen City of the West" will be second only to New Orleans.

The 11-mile Terneuzen-Ghent Canal opens to give the city of Ghent shipping access to the Dutch town of Terneuzen on the Scheldt Estuary.

The screw propeller for ships is invented by Austrian engineer Joseph Ressel, 34, and—independently—by Scottish engineer Robert Wilson, 24 (*see* 1832).

A sailing ship crosses from New Orleans to Liverpool in the record time of 26 days.

London's Hammersmith Bridge opens to traffic—the world's first suspension bridge of stone and metal. It will be replaced by a new span in 1887.

New York's first public transit facility begins operations. Entrepreneur Abraham Bower runs a horse-drawn bus with seats for 12 which he calls an "accommodation," but the city's population of 200,000 depends chiefly on private carts and carriages for transportation (*see* 1832).

German chemist Friedrich Wöhler, 27, isolates metallic aluminum from clay (*see* 1855; Hall, 1886).

Bright's disease is described by London physician Richard Bright, 38, of Guy's Hospital in the first diagnosis of a kidney disease by the presence of albumin in the urine. The condition is also characterized by an elevated blood pressure and will be found to include a number of separate diseases including acute and chronic nephritis.

English astronomer Frederick William Herschel, 35, invents contact lenses (his father discovered the planet Uranus in 1781) (*see* 1971 ; photography, 1840).

Disciples of Christ is founded by Irish-American evangelist Alexander Campbell, 39, who believes in the imminent Second Coming. He practices baptism by immersion and crusades against the evils of drink.

Toronto University is founded in Ontario.

The *Evening Standard* begins publication at London (*see* Beaverbrook, 1916).

The *Freeman's Journal* published at 5 Varick Street, New York, beginning March 16 is the first U.S. black newspaper. The city's white press largely favors slavery, which is abolished in the state July 4, but the new black paper denounces slavery and urges free blacks to seek education and practice thrift. It has been started by local clergyman Samuel E. Cornish, who has founded the first U.S. black Presbyterian church, and Jamaica-born college graduate John B. Russworm, who will move to Liberia and publish the *Liberia Herald* at Monrovia (*see* Liberia, 1847).

The *Journal of Commerce* begins publication September 1 at New York. The semi-religious publication has been started by silk and textile merchant Arthur Tappan and portrait artist Samuel F. B. Morse (*see* telegraph, 1832).

The *Youth's Companion* begins publication at Boston under the direction of Nathaniel Parker Willis, 21. The magazine will continue until 1929.

Fiction *The Prairie* by James Fenimore Cooper.

Poetry *Buch der Lieder* by Heinrich Heine contains poems that will serve as lyrics for numerous composers. Heine will be famous for such lines as, "The Romans would never have found time to conquer the world if they had been obliged first to learn Latin."

Painting *Death of Sardanapalus* by Eugène Delacroix; *The Cornfield* by John Constable. Charles Willson Peale dies at Philadelphia February 22 at age 85; Thomas Rowlandson dies at London August 22 at 70.

Theater *Cromwell* by French novelist-poet-playwright Victor Marie Hugo, 25, who was granted a royal annuity 5 years ago for his *Odes et poesies diverses* and married Adèle Foucher; *Marriage for Money (Le mariage d'argent)* by Eugène Scribe 12/3 at the Théâtre Française, Paris.

First performances Overture to *A Midsummer Night's Dream* by German composer (Jakob Ludwig) Felix Mendelssohn (-Bartholdy), 18, 2/17 at Stettin. Ludwig van Beethoven dies of cirrhosis of the liver at Vienna March 26 at age 56.

New Orleans has its first Mardi Gras celebration in February. Students from Paris introduce the Shrove Tuesday event.

The "Lucifer" invented by English chemist (pharmacist) John Walker is the first friction match. Its head is made of potash, sugar, and gum arabic, and it must be drawn swiftly through a piece of folded sandpaper to ignite (*see* 1836).

Sandwich glass is introduced at Sandwich, Mass., by Deming Jarves, 37, who has improved on a crude pressing machine recently invented by Cambridge Mass., glassmaker Enoch Robinson. Jarves's Boston and Sandwich Glass Co. has a monopoly on glass compounded with red lead.

1827 ***(cont.)*** The Paris Exchange is completed in the style of Rome's Temple of Vespasian after 19 years of construction.

Britain's George IV establishes a uniform standard for an acre of land, setting the measurement for an imperial (or statute) acre at 43,560 square feet (1/640th of a square mile). The standard is to apply throughout the United Kingdom, but parts of Scotland and Ireland will continue to use somewhat different measures. Most of Europe will measure land in hectares (1 hectare = 2.47 acres).

1828 A new British ministry takes office January 25 under the duke of Wellington. He sides with reactionaries and alienates liberal Tories, who resign (*see* Corn Law, below).

The Treaty of Turkmanchai February 22 ends a 3-year Russo-Persian War that began with Russian seizure of disputed territory. Persia cedes two territories that include part of Armenia and agrees to pay a huge indemnity; the Russians obtain exclusive rights to maintain a navy on the Caspian Sea.

Russia declares war on the Ottoman Empire April 26 and sends troops across the Danube June 8, but garrisons at the south-bank fortresses Shurrila, Silistria, and Varna stop the Russians. Varna falls to the Russians October 12 and the invaders go into winter quarters.

Portugal's regent Dom Miguel stages a coup d'état in May, abolishes the constitution drafted by Pedro IV in 1826, and has himself proclaimed king July 4. His niece Maria II, now 9, is taken to England by her protectors, and a 6-year civil war begins.

Uruguay proclaims her independence August 27. The Treaty of Rio resolves differences between Brazil and Argentina after mediation by British diplomats to set up the buffer state on the Rio de la Plata.

Madagascar's Hova king Radama I dies after an 18-year reign that has encouraged the spread of British influence. He is succeeded by his queen, Ranavaloana I, who will reign for 33 years, remaining throughout hostile to both French and British influence and the efforts of missionaries.

The Zulu king Shaka who has founded the Zulu nation is assassinated September 22 at age 41 after a 12-year reign that has ended the ancient pattern of society in South Africa. Demented since his mother's death last year, Shaka has begun arbitrary executions. His brothers Dingane and Mhlangane kill him and will reign jointly (*see* 1879).

Andrew Jackson, hero of the 1815 Battle of New Orleans, is elected president in November with 171 electoral votes to 83 for Adams. Vice President John Caldwell Calhoun is reelected.

"South Carolina Exposition and Protest" drafted by Vice President Calhoun (above) has set forth a theory that states may nullify within their borders any acts of Congress which their state conventions find unconstitutional. Calhoun is motivated by the Tariff of Abominations (below) which has created popular resentment by raising the price of both raw materials and manufactured goods, and his protest is reported December 19 to the South Carolina legislature which orders it printed.

Britain and the United States extend indefinitely their agreement of 1818 on joint occupation of the Pacific Northwest Territory (*see* 1845; 1846).

English reformer Robert Owen breaks with his business partner William Allen saying, "All the world is queer save thee and me, and even thou art a little queer" (*see* 1814; 1824).

The Tariff of Abominations signed into law by President Adams May 19 has raised duties on manufactured goods. Daniel Webster, representing New England shippers and manufacturers, and Henry Clay, representing western farmers, have championed the measure. Supporters of Andrew Jackson (above) framed it to discredit President Adams and are astonished when it is passed and signed.

The Reciprocity Act passed by Congress May 24 allows for lower duties on imports from countries that reciprocate, but opponents of the Tariff of Abominations are not appeased.

Construction begins July 4 on the Baltimore & Ohio (B&O) Railroad chartered last year by the state of Maryland as the first U.S. railroad for the general transportation of freight and passengers (*see* 1815; 1825). Backed by the richest man in America Charles Carroll of Carrollton, now 90, who lays its cornerstone, the B&O has a narrow 4-foot 8.5-inch gauge that is based on the standard English track width for carriages (*see* Cooper, 1829).

Work begins July 4 on the Chesapeake & Ohio (C&O) Canal. President Adams turns the first spade of soil to start a race between the B&O and C&O across the Alleghenies (*see* 1839).

The Delaware and Hudson Canal opens between Kingston, N.Y., and Port Jervis on the Delaware River where the 59-mile canal connects with the 49-mile Lackawanna Canal to Honesdale, Pa.

Scottish inventor James Beaumont Neilson, 26, devises a blast furnace to improve the manufacture of iron.

German chemist Friedrich Wöhler produces synthetic urea (NH_2CONH_2)—the first laboratory synthesis of an organic compound. Wöhler heats ammonium cyanate and takes a pioneer step toward the production of artificial nitrogen fertilizer (*see* Cavendish, 1766; Haber, 1908).

Jöns Jakob Berzelius isolates thorium.

Estonian naturalist Karl Ernst von Baer discovers the mammalian ovum and founds the modern science of embryology.

London University is founded by men who include Scottish jurist Henry Brougham, 50, Baron Brougham of Brougham, who tells the House of Commons, "Education makes a people easy to lead, but difficult to drive; easy to govern, but impossible to enslave."

The *Spectator* begins publication in early July at London. Former *Dundee Advertiser* editor Robert S. Rintoul edits the new weekly review of politics, literature, theology, and art.

An American Dictionary of the English Language is published after 28 years of work by Noah Webster, now 70, who has studied 26 languages in order to determine the origins of English words (*see* 1806). Webster defines nearly 70,000 words, introduces Americanisms such as *revolutionary, skunk,* and *applesauce,* and gives words such as *colour* and *plough* American spellings (*color, plow*).

Fiction *The Fair Maid of Perth* and *Tales of a Grandfather* by Sir Walter Scott.

Painting *Faust* (19 lithographs) by Eugène Delacroix. Francisco Goya dies at Bordeaux April 16 at age 82; Gilbert Stuart dies at Boston July 9 at age 52; Jean Antoine Houdon dies at Paris July 15 at age 87.

Theater *A Faithful Servant of His Master (Ein truer Diener seines Atrin)* by Franz Grillparzer 2/28 at Vienna's Burgtheater.

Opera *Count Ory (Le Comte Ory)* 9/20 at the Opéra-Comique, Paris, with music by Gioacchino Rossini.

First performances Fantasy in F minor for Piano Four Hands by Franz Schubert 5/9 at Vienna; Symphony No. 6 (or No. 7) in C minor by the late Franz Schubert 12/14 at Vienna (Schubert has died of typhus November 19 at age 31, leaving his Symphony in B major unfinished).

New York's Washington Square Park is created on the site of a potter's field cemetery that has been removed to a site on 42nd Street acquired by the city in 1823 (*see* Bryant, 1844).

A Japanese earthquake at Echigo December 28 kills 30,000.

A new British corn law introduced by the duke of Wellington (above) gives consumers relief from the high food prices that have prevailed since the Corn Law of 1815. The new law imposes duties on a sliding scale based on domestic prices but prices remain too low to permit grain imports.

Dutch chocolate-maker Conrad J. Van Houten patents an inexpensive method for pressing the fat from roasted cacao beans and produces the world's first chocolate candy. Van Houten adds the fat that will be called cocoa butter to an experimental mixture of cocoa powder and sugar, and the resulting sticky substance cools into a solid form (*see* Cadbury, 1847; Nestlé and Peter, 1875).

1829 Greece gains her independence from the Ottoman Empire after 4 centuries of Ottoman rule, and Russia brings the empire to the verge of collapse. The London Protocol signed March 22 has made Greece an autonomous tributary state. Russia's German-born Count Ivan Ivanovich Dibich-Zabalanski (Hans Karl Friedrich Anton von Diebitsch), 44, has won the Battle of Kulevcha June 11 and crossed the Balkan mountains to take Adrianople. The Treaty of Adrianople is signed September 24 as disease makes the Russian army too weak to proceed.

Russia gains control of the mouth of the Danube and secures the eastern coast of the Black Sea, agreeing to pay an indemnity of 15 million ducats over the next 10 years for rights to occupy the Danubian principalities.

Romania gains autonomy from the Ottoman Empire under Russian occupation (*see* 1866).

Serbia gains autonomy with guarantees of religious liberty (*see* 1839).

The London Conference November 30 decides that Greece should have complete independence, but her frontier is moved back almost to the Gulf of Corinth. Greek patriot Count Ioannes Antonios Kapodistrias, 53, rules as president-dictator.

Venezuela secedes from Simon Bolívar's Gran Colombia and will become an independent republic next year.

Argentinian provincial leader Juan Manuel de Rosas, 34, ends his country's anarchy in April by defeating Buenos Aires provincial forces in the field and making himself governor of Buenos Aires December 8 (*see* 1835).

Tampico, Mexico, falls to Spanish forces from Cuba August 18, but the Mexicans retake the city September 11.

Mexico abolishes slavery September 15, but President Vicente Guerrero acts December 2 to exempt Mexico's Texas Territory from the antislavery decree (*see* 1830).

"Slavery is not a national evil; on the contrary, it is a national benefit," says South Carolina governor Stephen D. Miller in a message to the state legislature.

Charleston women Sarah and Angelina Grimke leave for the North where they will become Quakers and be active in the antislavery and women's rights movements.

The British Slave Trade Commission takes over the administration of the African island of Fernando Po with Spanish consent.

The Catholic Emancipation Bill pushed through Parliament by the duke of Wellington over Tory opposition gives British Catholics voting rights, the right to sit in Parliament, and the right to hold any public office except those of lord chancellor and lord lieutenant of Ireland. The right to hold office is granted only on condition that the officeholder swear an oath denying papal power to intervene in British domestic affairs, that he recognize the Protestant succession, and that he repudiate any intent to upset the established church.

At least 75,000 Americans go to prison for debt each year, says the U.S. Prison Discipline Society. More than half owe less than $20.

London's Baring Brothers finance the Planters Association in Louisiana—the first state loan in America to be

1829 ***(cont.)*** underwritten by a London firm (*see* 1803; Trans-Siberian Railway, 1891).

The *Stourbridge Lion* imported from England's Stephenson Engine Works goes into service between Carbondale and Honesdale, Pa. It is the first steam locomotive on any U.S. railroad.

The *Tom Thumb* produced by New York entrepreneur Peter Cooper, 28, is the first U.S.-built locomotive. Cooper has taken scraps of iron including pipes from musket barrels to build his steam locomotive in the shops of the Baltimore & Ohio Railroad (*see* 1828; 1830).

The Rocket built by England's George Stephenson and his son Robert wins a competition sponsored by the Liverpool and Manchester Railway. The tubular steam boiler locomotive sets a pattern for future railroad locomotive development.

The first French railway line opens to connect Lyons with St. Etienne using an English-built locomotive.

A horse-drawn omnibus goes into service at London to provide public transportation (*see* New York, 1827).

A steam-powered bus appears in Britain but draws bitter opposition from stagecoach interests (*see* Locomotives on Highways Act, 1865).

New York sailing captain Cornelius van Derbilt, 33, begins building steamboats with $30,000 he has amassed skippering for Thomas Gibbons and in his own coastal schooner ventures (*see* 1834; Gibbons, 1824).

The Welland Sea Canal opens to connect Lakes Erie and Ontario via a tortuous waterway that has 25 locks to permit ships to circumnavigate Niagara Falls (*see* 1932).

French inventor Barthélemy Thimmonier, 36, develops the world's first practical sewing machine. He will obtain a contract to produce French army uniforms, but a mob will destroy one of his new machines out of fear that French tailors will be deprived of their livelihoods (*see* Luddites, 1811; Hunt, 1832).

William Burke is hanged at Edinburgh at age 37 for having smothered victims whose bodies he has sold to physicians. The medical profession needs corpses for dissection but is unable to acquire them legally (*see* New York, 1788; anatomy act, 1832).

Cholera breaks out at Astrakhan in a more serious epidemic than one in 1823 (*see* 1830).

The Perkins Institution for the Blind is founded at Boston under the name New England Asylum for the Blind. Local benefactors include merchant Thomas Handasyd Perkins, 65, who will deed his home to the asylum in 1833; its name will be changed subsequently to the Perkins Institution and Massachusetts School for the Blind (*see* Haüy, 1784; Braille, 1834; Helen Keller, 1887).

The first Baedeker travel handbook, a guide to Coblenz, is issued by German publisher Karl Baedeker, 28, whose handbooks in German, French, and English will describe the sights, foods, and accommodations to be found in cities of Europe, North America, and the Orient which Baedeker will in many cases visit incognito.

Poetry *Al Aaraaf, Tamerlane and Minor Poems* by Boston poet Edgar Allan Poe, 20, whose *Tamerlane* appeared in a 40-page book 2 years ago but gained little attention, leaving Poe in dire financial straits. Orphaned before age 3, Poe has been raised by John Allan of Richmond, Va., with whom he has had fallings out and reconciliations and whose wife has just died. Allan helps Poe secure a discharge from the army, in which he has enlisted under the name Edgar A. Perry, and to apply for an appointment to West Point; *Casablanca* by English poet Felicia Dorothea Hemans, 36, whose narrative poem based on the 1798 Battle of the Nile begins, "The boy stood on the burning deck/ Whence all but him had fled. . ."; *Joseph Delorme* by French poet-critic Charles Augustin Sainte-Beuve, 25, whose *Poesies et Pensées* purport to be the work of an imaginary poet who has died young.

Painting *Ulysses Deriding Polyphemus* by J. M. W. Turner.

Photographer Louis Daguerre goes into partnership with J. N. Niepce and accidentally discovers the light-sensitivity of silver-iodide (*see* 1826). Daguerre finds that an iodized silver plate exposed to light in a camera will produce an image if the plate is fumed with mercury vapor (*see* Talbot, 1839).

Theater *Faust* (Part I) by Johann Wolfgang von Goethe 1/19 at Brunswick (*see* 1854).

Opera *William Tell (Guillaume Tell)* 8/3 at the Paris Opéra, with music by Gioacchino Rossini.

First performances *Passion According to St. Matthew* by Felix Mendelssohn 3/11 at Berlin; Symphony No. 1 in C minor by Mendelssohn 5/25 at London's Argyll Rooms (with the scherzo from an octet Mendelssohn wrote in 1824 replacing the minuet); Variations on "La Co Darem La Mano" for Piano with Orchestral Accompaniment by Polish composer Frédéric François Chopin, 19, 8/11 at Vienna.

The Henley Regatta has its beginnings in the Oxford and Cambridge Boat Race held June 10 over a 2-mile course 57.5 miles upriver from London at Henley-on-Thames. It will be called the Henley Regatta in 1839, the Henley Royal Regatta in 1851 (*see* 1876).

The "Siamese twins" Chang and Eng, 18, arrive at Boston on the U.S. ship *Sachem* and sail in October for England where they attract wide attention among physicians and the general public. Joined at the chest since birth, the young men have been discovered by British merchant Robert Hunter, will be exhibited by U.S. showman P. T. Barnum, now 19, will marry two North Carolina women, sire 22 children between them, maintain two separate households, and will amass a fortune of $60,000 (*see* Barnum, 1835).

London "Bobbies" introduced September 29 make the city's streets safe after dark. Named for Home Secretary Robert Peel of 1817 "Peelers" fame and headquartered at Scotland Yard, the constables are the first police force in a city that has one footpad, highwayman, or thief per 22 inhabitants.

Architecture is "frozen music," says Johann Wolfgang von Goethe, now 79, in a conversation March 23: "I have found a paper of mine in which I call architecture frozen music (*erstarrte Musik*). Really there is something in this; the tone of mind produced by architecture approaches the effect of music."

Boston's Tremont House opens in October with 170 rooms. It is the first modern hotel, and while lodgers at other hotels continue in many cases to sleep with strangers "spoon fashion," three and four to a bed, with women sometimes rooming with men, each guest or couple at the Tremont House receives a private room with a key. For $2 per day a guest receives four meals a day and is given a free cake of soap, but while the new gaslighted hotel has baths, they are all in the basement and must be entered from a separate street entrance. No hotel will have private baths until 1853.

Massachusetts Presbyterian clergyman Sylvester W. Graham, 35, attacks meats, fats, mustard, catsup, pepper, and, most especially, white bread. He calls them injurious to the health or stimulating to carnal appetites (*see* 1839).

1830 Europe has revolutions as sectional conflicts threaten U.S. unity.

"Liberty and Union, now and forever, one and inseparable!" declaims Sen. Daniel Webster (Mass.) January 27 in his "Reply to Hayne" speech against Sen. Robert Y. Hayne (S.C.) as the South threatens secession over the issues of states' rights and high tariffs. Southern leaders demand low tariffs in the interest of their slave-owning planters, and Sen. Thomas Hart Benton (Mo.) protests New England's attempts to limit the sale of western lands.

"Our federal union: It must and shall be preserved!" declaims President Jackson April 13 in a toast at a Jefferson Day dinner. The president thus signifies his determination to resist southern efforts to declare the 1828 Tariff of Abominations null and void.

"The Union, next to our liberty, most dear!" declaims Vice President Calhoun in a toast at the April 13 dinner. South Carolina's Calhoun signifies his determination to persist in his fight.

The Republic of Ecuador is created May 13 in a further breakup of Gran Colombia. The liberator Antonio José de Sucre has tried to maintain Colombian unity but is assassinated June 4 at age 35 in the forest of Berueros near Pasto, Colombia, while en route to Quito. The Great Liberator Simon Bolívar dies in Colombia December 17 at age 47 saying that America is ungovernable.

Britain's debauched George IV dies June 26 at age 67 and is succeeded by his brother William, 64, duke of Clarence, who will reign until 1837 as William IV. A general election turns out the Tories, who have run Britain almost without interruption for nearly half a century, and a Whig government takes power with Charles Grey, 66, as prime minister. Earl Grey's foreign minister is Henry John Temple, 46, viscount Palmerston.

French forces land in Algeria after a 3-year blockade, Algiers capitulates July 5, and France begins 132 years of occupation and colonization (*see* 1954).

Paris revolutionists depose France's Charles X July 29. He is succeeded after a 6-year reign by the Bourbon duc d'Orléans, 56, who is proclaimed "citizen king" by the journalist Louis Adolphe Thiers and confirmed August 7 by the Liberals sitting as a rump chamber. Thiers, 33, has drawn up a protest against the five July ordinances of July 26 that established rigid governmental control of the press, dissolved the chamber, and changed the electoral system, the radicals have raised street barricades with the intent of making France a republic with the marquis de Lafayette as president, but Thiers and other Liberals thwart them, making the duc d'Orléans king; he will reign until 1848 as Louis Philippe.

Belgian revolutionists inspired by the July Revolution at Paris demand independence from the Netherlands and win support from workers and peasants, but moderate liberal elements protest that they merely want autonomous administration with the son of the Dutch king Willem I as viceroy.

Brussels workers force Dutch troops to evacuate the city in late September, an independent Belgium is proclaimed by a provisional government October 4, the Dutch bombard Antwerp October 27, but an international conference declares dissolution of the Kingdom of the Netherlands December 20 and effectively recognizes Belgian independence (*see* 1831).

A Polish revolution against Russia begins as Polish nationalists form a union with Lithuania and declare the Romanov dynasty deposed (but *see* 1831).

The king of the Two Sicilies Francesco I dies at age 53 after a 5-year reign as reactionary as was that of his father. He has left the government to court favorites and the police while living with his mistresses surrounded by soldiers to protect him from assassination. His son of 20 will reign until 1859 as Ferdinand II, continuing Francesco's despotic rule.

The Indian Removal Act signed by President Jackson May 28 provides for the general removal of Indians to lands west of the Mississippi (*see* 1832; 1834).

President Jackson names a commissioner of Indian affairs.

The Treaty of Dancing Rabbit Creek signed September 15 cedes Choctaw lands east of the Mississippi to the United States.

Mexico passes a law forbidding further colonization of its Texas Territory by U.S. citizens and prohibiting further importation of slaves into the territory (*see* 1829; 1836).

British authorities in the Bahamas declare that slaves from the schooner *Comet* wrecked on a voyage from

1830 ***(cont.)*** Alexandria, Va., to New Orleans are free. Washington registers a protest.

Chicago has its real beginnings as a town is laid out at Fort Dearborn on Lake Michigan (*see* 1825; steamboat, 1832).

William Sublette of the Rocky Mountain Fur Co. and Jedediah Strong Smith lead the first covered wagon train from the Missouri River to the Rockies. They sell their fur-trading interests to a new company organized by Jim Bridger and others (*see* 1827; 1831).

Service begins on England's Liverpool-to-Manchester Railway (*see* 1829).

A flanged T-rail invented by Robert Livingston Stevens, 43, will be the basis of future railroad track development. Stevens is a son of steamboat pioneer John Stevens. Now president and chief engineer of the Camden and Amboy Railroad and Transportation Co. (*see* 1815), he will also invent a hookheaded spike and a metal plate to cover the joint between rails.

The first division of the Baltimore & Ohio Railroad is completed May 24 to link Baltimore with Ellicott Mills, 13 miles away (*see* 1828).

The first American-built locomotive is demonstrated August 28 by Peter Cooper who has named his steam locomotive *Tom Thumb* after the diminutive folktale hero (*see* 1829). Cooper's engine pulls a train in a race against a stagecoach, losing at the last minute when a pulley belt slips, but proving that a steam engine can pull a single car round a sharp curve.

The Boston & Maine Railroad has its beginnings June 5 in the Boston & Lowell, first steam railroad to be projected in New England (*see* 1833).

New York watchmaker Phineas Davis wins a $4,000 prize from the Baltimore & Ohio for a locomotive he has developed that can pull 15 tons at 15 miles per hour. (The B&O will not reach Wheeling on the Ohio River until 1854, but it will eventually pay off handsomely for its investors.)

The locomotive *Best Friend of Charleston* begins service December 25 on South Carolina's Charleston & Hamburg Railroad, chartered as the first road intended from the start for use by steam cars. Built by New York's West Point Foundry, the engine pulls four loaded passenger cars over 6 miles of track.

President Jackson signs an act of Congress appropriating $130,000 to survey and extend the Cumberland Road that reached Wheeling in 1817. Calling it a national road, he signs the bill 4 days after vetoing the Maysville Road Bill that would have authorized federal aid to a Kentucky turnpike from Maysville to Lexington (federal expenditure for a local project is unconstitutional, says the president).

Louisville, Ky., begins its growth as a major river port with the opening of the Louisville and Portland Canal which allows riverboats to circumvent 26-foot falls in the Ohio River.

A surface condenser is invented that will solve the problem of providing fresh water for steam boilers and make oceangoing steamships economically viable (*see* 1826; 1838).

The Fairbanks scale devised by Vermont inventor Thaddeus Fairbanks, 34, is the world's first platform scale, using a system of multiplying levers to counterpoise a heavy object on the weighing platform by a light beam with a small sliding weight and slotted weights on an extension arm.

Principles of Geometry by Russian mathematician Nikolai Ivanovich Lobachevski, 37, pioneers non-Euclidian geometry (*see* 300 B.C.).

Cholera spreads from Astrakhan throughout the interior of Russia (*see* 1829). In some areas there is widespread famine as quarantines and demoralization add to crop failures, and there are widespread riots against cholera controls as priests denounce physicians, government officials, and quarantine guards, calling them Antichrist. The pandemic kills 900,000 this year and will kill several millions in Europe before it ends (*see* 1840).

London wine merchant-optician Joseph Jackson Lister, 44, improves on the microscope invented by Anton von Leeuwenhoek in 1683. He discovers the law of the aplanaic foci, fundamental principle of the modern microscope, and will be the first to ascertain the true form of red corpuscles in mammalian blood (1834). His son will be the founder of antiseptic surgery (*see* 1865).

The Church of Jesus Christ of Latter-Day Saints is founded April 6 at Fayette, N.Y., by local farmhand Joseph Smith, Jr., 25, who has *The Book of Mormon* published in 522 pages at Palmyra, N.Y. The book claims that the Indians of the New World were originally Jews who sailed from the Near East in the 6th century B.C. and who received a visit from Jesus Christ after his resurrection. Smith says he has translated the book with miraculous help from strange hieroglyphics on some golden tablets buried near Palmyra and revealed to him by an angel named Moroni. Mormons will practice polygamy, and their marriages will make Smith and his followers unwelcome in many communities (*see* Nauvoo, 1839).

The *Boston Transcript* begins publication July 24. Henry W. Dutton and his family will run the paper for the city's Brahmins until its demise in 1940.

The *Philadelphia Inquirer* has its beginnings in the *Pennsylvania Inquirer* founded by printer Jasper Harding, whose son William will take over when he retires in 1869.

Godey's Lady's Book begins publication at Philadelphia under the name *Lady's Book*. Publisher Louis Antoine Godey, 26, introduces the first U.S. periodical for women.

Fiction *The Red and the Black (Le Rouge et le Noir)* by French novelist Stendhal (Marie Henri Beyle), 47, blends 18th-century rationalism with romantic fervor in what the author calls the emotional "crystallization" of experience.

Poetry *Poems, Chiefly Lyrical* by English poet Alfred Tennyson, 21, whose book gets a poor critical reception; *Old Ironsides* by Boston medical student Oliver Wendell Holmes, 21, appears in the *Boston Daily Advertiser* and helps save the U.S. frigate *Constitution* from being destroyed by the Navy Department which has condemned her: "Ay, tear her tattered ensign down!/ Long has it waved on high. . ."; *The Ballad of the Oysterman* by Oliver Wendell Holmes; *Mary Had a Little Lamb* by Boston editor Sarah Josepha Hale, 42, whose verse in the first issue of the magazine *Juvenile Miscellany* will be set to music from the song "Goodnight Ladies" in 1867.

Painting *Liberty on the Barricades* by Eugène Delacroix who sympathizes with the July Revolution.

Sir Thomas Lawrence dies at London January 7 at age 60.

Birds of America by John James Audubon, now 45, is published in the first of its several editions (*see* 1813). His wife's financial support has enabled Audubon to pursue his passion for painting birds from life; his work began appearing in London 3 years ago.

Theater *Hernani* by Victor Hugo 2/25 at the Comédie-Française, Paris.

Opera *I Capuleti e i Montecchi* 3/4 at Venice's Teatro de la Fenice, with music by Italian composer Vincenzo Bellini, 28, libretto from the 1595 Shakespeare play *Romeo and Juliet*.

First performances Concerto No. 2 in F minor for Pianoforte and Orchestra by Frédéric Chopin 3/17 at Warsaw; Concerto No. 1 in E minor for Pianoforte and Orchestra by Chopin 10/11 at Warsaw (the concerti will be published in the opposite order of their premieres); *Symphonie Fantastique* by French composer Hector Berlioz, 27, 12/25 at Paris. Berlioz breaks new ground in use of percussion and symphonic composition.

The Kentucky State House is completed at Frankfort by architect Gideon Shrycock.

"In whatever proportion the cultivation of potatoes prevails. . . in that same proportion the working people are wretched," writes English political journalist William Cobbett, 67, in *Rural Rides*.

The Industrial Revolution has turned England's peasantry into half-starved paupers and destroyed rural skills, says Cobbett (above).

French and Belgian bakers begin using minute quantities of highly toxic copper sulfate as well as the less toxic alum to whiten bread.

Domestic baking and brewing skills have been forgotten by the English peasantry, says William Cobbett (above). "Nowadays all is looked for at shops. To buy the thing ready made is the taste of the day: thousands who are housekeepers buy their dinners ready cooked."

Tinned foods from London's Donkin, Hall & Gamble Preservatory reach English food shops for the first time.

The continuous still patented by Irish inventor Aeneas Coffee speeds up distilling and makes for "cleaner" whisky and gin.

Congress reduces U.S. duties on coffee, tea, salt, and molasses imports.

Capt. John Pearsall's schooner *Harriet Smith* lands the first full cargo of bananas—1,500 stems—at New York (*see* Chester, 1804; Baker, 1870).

Congress makes abortion a statutory crime (*see* England, 1803; Pius XI, 1869).

The U.S. population reaches 12.9 million, including 3.5 million black slaves. The country receives 34,338 immigrants from Ireland during the year and 27,489 from England.

The population of the world reaches 1 billion, up from 750 million in 1750.

1831 Italians at Modena and Parma rise in February to demand freedom of northern Italy, and there are widespread revolts in the Papal States as the reactionary Cardinal Bartolomeo Alberto Cappellari begins a 15-year term as Pope Gregory XVI. Inspired by last year's Paris Revolution, the insurrections are put down in March with help from Austrian troops, but fresh revolts break out by year's end (*see* Mazzini, 1832).

The French Foreign Legion created March 9 to employ Louis Phillipe's Swiss and German mercenaries will serve largely in North Africa, the Middle East, and Indo-China. The Legion will attract renegades and fugitives from justice for more than 160 years.

The first king of the Belgians assumes the throne at age 40 and will reign until 1865 as Leopold I. Widower of Britain's late Princess Charlotte, the prince of Saxe-Coburg is elected to the Belgian throne June 4, a large Dutch army invades his country in August, but French forces sent by Louis Philippe force the Dutch to withdraw.

The Battle of Ostrolenka May 26 gives Russian forces under Count Dibich-Zabalanski a victory over Polish troops as internal disputes between moderates and radicals divide the Poles. Warsaw falls to the Russians September 8, the revolution that began last year collapses, most of the revolutionary leadership escapes to Paris, and the russification of Poland begins in provinces not occupied by Prussia or Austria (*see* 1772; 1918).

Greece's president Count Ioannes Antonios Kapodistrias is assassinated October 9 at age 55. The Greeks make his brother Avgoustinos, 53, provisional president (*see* monarchy, 1832).

The Brazilian emperor Pedro I abdicates April 7 at age 33 and returns to Europe. His 5-year-old son by his first wife will be crowned in 1841 and reign until 1889 as Pedro II.

Comanches on the Cimarron River in the Southwest kill Jedediah Strong Smith May 27 (*see* 1830).

A Bengali uprising against oppressive Hindu rule is suppressed, and the Muslim leader Titu Mir is killed by government troops November 19.

India's Mysore State comes under British control as the British use the excuse of domestic misgovernment to increase their power in the subcontinent.

The Bristol Riots October 29 to 31 raise fears in London that the unrest may spread to other towns and

1831 ***(cont.)*** begin a revolution. The newly appointed recorder of Bristol Sir Charles Wetherall is an outspoken critic of the Parliamentary Reform Bill that passed the Commons in September but has been rejected by the House of Lords, supporters of the measure wreck Bristol's Mansion House and set the Bishop's Palace afire, other public buildings are attacked, the city magistrates call in troops, a bloody cavalry charge restores order, four of the rioters are executed, and 22 are transported (*see* Reform Act, 1832).

Irish Catholics resort to violence in an armed protest against enforcement of tithes to support the established Episcopal Church.

The Union of Northumberland and Durham Coalminers, founded by English miner Thomas Hepburn, demands shorter hours for coal workers.

French workers at Lyons revolt. The insurrection is put down in November but only with great difficulty as secret societies proliferate among the workers.

New York has its first labor demonstrations as stone cutters riot in protest against the use of stone cut at Sing Sing prison for buildings of the new University of the City of New York (*see* NYU, below; General Trades Union, 1833).

The Supreme Court rules March 18 that an Indian tribe may not sue in federal courts since the tribes are not foreign nations *(Cherokee Nation v. Georgia)* (*see* 1832).

Chief Black Hawk agrees to withdraw his tribesmen to lands west of the Mississippi (*see* 1830; 1832).

The Liberator begins publication January 1 at Boston where local abolitionist William Lloyd Garrison, 26, advocates emancipation of the slaves who account for nearly one-third of the U.S. population.

Nat Turner's rebellion brings panic to the South as whites learn that the Virginia slave has murdered his master and all his master's family in their sleep the night of August 21. A religious zealot, Turner, 30, has convinced other slaves that he has received divine guidance, the seven companions who have helped him murder the Joseph Travis family quickly grow to a mob of more than six dozen, state militia and armed townspeople intercept the rebels 3 miles outside Jerusalem, Va., many slaves are killed including some who took no part in Turner's rebellion which has killed 60 whites in 48 hours, and although Turner escapes for 6 weeks, he is finally apprehended, tried, convicted, and hanged November 11 at Jerusalem along with 16 accomplices.

Samuel Sharp's rebellion in Jamaica beginning 2 days after Christmas kills no whites, but the slaves force sugar estate owners at gunpoint to draw up documents freeing all their workers, they burn cane fields and other property, British authorities retaliate by burning the slave shacks, hanging Sharp and hundreds of others, and flogging slaves who have joined in Sharp's refusal to work without compensation, but a fall in the price of sugar has made slavery unprofitable to many planters.

Scottish polar explorer James Clark Ross, 31, and his uncle Sir John Ross determine the position of the magnetic North Pole (*see* 1818).

Michael Faraday discovers the basic principle of the electric dynamo (*see* 1821; Sturgeon, 1823). His electromagnetic current generator consists simply of a cylindrical coil (solenoid) and a bar magnet that can be slipped into the coil, but Faraday succeeds in generating electrical current October 17 and discovers electromagnetic induction. He finds by using his galvanometer that a current is registered while the magnet is being inserted, and that the current starts again in the opposite direction when the magnet is withdrawn, but that no current is registered while the magnet is stationary (*see* galvanometer, 1826; Gramme, 1872).

Faraday finds that if the current in a wire wrapped round an iron rod is interrupted, a current will be generated in a second wire wrapped round the rod. He is excited by the discovery that electrical energy is transferred between two circuits but does not see the potential for stepping down power from a high-voltage line for use with alternating current which regularly reverses its direction (*see* 1888).

Joseph Henry discovers a method for producing induced current much like that of Faraday's (above)(*see* 1827). The unit of induction will be called a henry.

The first Baldwin locomotives are manufactured by New Jersey industrialist Matthias William Baldwin, 35, who will soon produce a steam locomotive that goes 62 miles per hour (*see* 1830).

A cowcatcher for locomotives, invented by U.S. engineer Isaac Dripps, will be used initially in 1833 on the Camden and Amboy Railroad between Bordentown and Hightstown, N.J., and when its projecting points are found to impale animals on the right of way, the prongs will be replaced by a heavy bar set at right angles to the rails.

The Mohawk and Hudson Railroad begins service August 9 as the locomotive *De Witt Clinton* built by the West Point Foundry pulls a train of cars between Albany and Schenectady in the first link of a road that will become the New York Central in 1853.

The 102-mile Morris Canal, completed between Newark, N.J., and Easton, Pa., carries anthracite coal to New York. By 1861 the canal will be handling 889,000 tons of coal (*see* Delaware-Raritan, 1834).

A feeder canal to the Ohio River and the Erie Canal spurs development of Columbus, Ohio, which will be chartered as a city in 1834.

John Jacob Astor's American Fur Co. steamboat S.S. *Yellowstone* makes the first steamboat voyage on the upper Missouri.

A new London Bridge opens across the Thames to replace the 10th-century structure now being demolished (*see* Rennie, 1817).

A patent for making malleable cast iron is issued to U.S. inventor Seth Boyden, 43, who invented his process 5 years ago. Boyden invented a process in 1819 for making patent leather and will go on to invent a process for making sheet iron, a hat-shaping machine, and improvements in railroad locomotives and stationary steam engines.

The British Association for the Advancement of Science (BAAS) is founded along lines of the earlier Gesellschaft deutscher Naturalforscher (*see* 1848).

English naturalist Charles Darwin, 22, embarks on a voyage to South America and the Galápagos Islands as ship's naturalist aboard H.M.S. *Beagle* (*see* 1840).

Chloroform is invented independently by German chemist Justus von Liebig, 28, and U.S. chemist Samuel Guthrie, 49, whose chloric ether will be widely employed as an anaesthetic (*see* 1847; "laughing gas," 1799; Long's sulfuric ether, 1842).

New York University (NYU) has its beginnings in the University of the City of New York (*see* 1896).

Wesleyan University is founded at Middletown, Conn.

The University of Alabama is founded at Tuscaloosa.

Telegraphy is pioneered by Joseph Henry (above) who sees that an electromagnet can be used to send messages over great distances by wiring the magnet to a switch, turning it on and off to attract and release a piece of iron, and thus producing a pattern of clicks. Henry exhibits his 14-inch-long device at Albany, N.Y., transmitting signals over more than a mile of wire, but does not patent the device or put it to any practical use (*see* Morse, 1832).

Nonfiction "The Autocrat of the Breakfast Table" by Oliver Wendell Holmes appears as an essay in the *New England Magazine.* Another such essay will appear next year (*see Atlantic Monthly*, 1857).

Fiction *Notre-Dame de Paris* by Victor Hugo who has become depressed over the failure of last year's Paris Revolution and the intimacies his wife Adèle has begun with the poet-critic Charles Augustin Sainte-Beuve; *The Young Duke* by English novelist Benjamin Disraeli, 27, whose father Isaac D'Israeli abandoned Judaism in 1817 and has had his children baptized in the Anglican church.

Poetry *Autumn Leaves* (*Les feuilles d'automne*) by Victor Hugo (above); *Boris Godunov* by Russian poet Aleksandr Sergeivich Pushkin, 32, whose liberal views have cost him his government post but who will regain his position in the ministry of foreign affairs next year; *Legends of New England in Prose and Verse* by *New England Weekly Review* editor John Greenleaf Whittier, 23; "To Helen" by Edgar Allan Poe: "Thy hyacinth hair, thy classic face,/ Thy Naiad airs, have brought me home/ To the glory that was Greece/ And the grandeur that was Rome."

Painting *Salisbury Cathedral from the Meadows* by John Constable; *Le 28 Juillet 1830* by Eugène Delacroix.

The Barbizon school of French painters holds its first exhibition at Paris. The painters take their name from a village near Fontainebleau, their rural genre scenes are based on direct observation of nature, and they will soon be joined by Jean François Millet, now 17, and Théodore Rousseau, now 19.

Theater *Hero and Leander (Des Meeres und der Liebe Wellen)* by Franz Grillparzer 4/5 at Vienna's Burgtheater; *The Gladiator* by U.S. playwright Robert Montgomery Bird, 25, 9/26 at New York's Park Theater, with Edwin Forrest as Spartacus.

Opera *La Sonnambula* 3/6 at Milan's Teatro Carano, with music by Vincenzo Bellini; *Robert le Diable* 11/21 at the Academie Royale de Musique, Paris, with music by German-born composer Giacomo Meyerbeer (Uakob Liebmann Beer), 40. Meyerbeer's work establishes French grand opera style with spectacular scenic effects, florid arias designed to show off the virtuosity of the singers, dramatic recitative, and romantic plot material; *Norma* 12/26 at Milan's Teatro alla Scala, with music by Vincenzo Bellini.

First performances Piano Concerto No. 1 in G minor by Felix Mendelssohn 10/17 at Munich.

Japan's Takashimaya retail store empire has its beginnings in a dry goods shop opened at Edo.

The McCormick reaper that enables one man to do the work of five is demonstrated by Virginia farmer Cyrus Hall McCormick, 22, whose father tills some 1,200 acres near Lexington with nine slaves and 18 horses. Young McCormick's crude but effective horse-drawn reaper is so devised that the horse is hitched alongside it rather than behind as in the case of the 1826 Bell reaper. His reaper's knife vibrates in a line at right angles to the direction in which the machine is moving, a divider moves ahead to separate a swath from the field and turn the grain toward the blade, and a row of mechanical fingers ahead of the blade holds the straw straight to be cut. Severed stalks are deposited neatly on a platform, and the reaper permits horsepower to replace human power (*see* 1834; Hussey, 1833).

Growing and harvesting a bushel of U.S. wheat takes 3 man-hours of work, a figure that will begin to drop through use of the McCormick reaper.

Cyrus McCormick's reaper enabled one farmer to do the work of 5, beginning a revolution in farm productivity.

1831 ***(cont.)*** Boston's S. S. Pierce Co. has its beginnings in a shop opened to sell "choice teas and foreign fruits" by local merchant Samuel Stillman Pierce who will ship wine from Madeira to New York to Buenos Aires and back to New York to satisfy an old Bostonian's taste for Madeira wines that have spent months at sea. His firm will send a dogsled team to fetch Russian isinglass (a gelatin made from sturgeon bladders) for S. S. Pierce jellies, and put up buffalo tongue, terrapin stew, and Singapore pineapple in cans.

The U.S. population reaches 13 million; Britain has 12.2 million, Ireland 7.7 million.

Robert Owen of 1824 New Harmony fame tries to popularize contraceptive measures more effective than the widely employed coitus interruptus method (*see* Edmonds, 1832; Knowlton, 1835).

1832 Piedmontese authorities learn in March of plans for a June uprising planned by the Young Italy society started last year by patriot Giuseppe Mazzini, 26, who leads an Italian unification movement. The rising is aborted (*see* 1834).

Austria's Prince von Metternich aborts a German unification movement in July after a gathering of 25,000 at the Hambach Festival in May has toasted France's Marquis de Lafayette, now 74, demanded a German republic, and threatened armed revolt. The Six Articles adopted by the German Confederation June 28 at Metternich's insistence impose on every German sovereign the duty of rejecting petitions of his estates that would impair his sovereignty. The Six Articles repudiate the right of estates to refuse supplies as a means of securing constitutional changes, and beginning in July all public meetings are forbidden, edicts against universities renewed, and suspicious political characters placed under surveillance.

Greece becomes a monarchy with the son of Bavaria's Ludwig I as king. Otto I, 17, begins a despotic reign that will continue until 1862.

The (First) Reform Act that passes the House of Lords June 4 enfranchises Britain's upper middle class, doubling the number of eligible voters to 1 million. William IV has threatened to create new peers if the Lords continued to reject the reform bill as they did last year, pocket boroughs (controlled by single individuals or families) and rotten boroughs (election districts with the same voting power as other districts but with far fewer inhabitants) are abolished, seats are redistributed to create constituencies in the new towns, and mill owners in the Midlands are enabled to agitate more effectively for removal of high tariffs on foodstuffs from abroad (*see* Ricardo, 1817; Wellington, 1828; Cobden, 1838; Second Reform Bill, 1867).

The "spoils system" in U.S. politics gets its name January 21 in a Senate speech by Sen. William Learned Marcy, 46, (N.Y.) who says that he can see "nothing wrong in the rule that to the victor belong the spoils." Politicians will continue for decades to name cronies to government jobs and dispense favors that will enrich their friends (*see* Pendleton Civil Service Reform Act, 1883).

Vice President John C. Calhoun's Fort Hill letter of August 28 to the governor of South Carolina gives a closely reasoned classic exposition of Calhoun's theory of state sovereignty (the Doctrine of Concurrent Majority).

The Democratic-Republican party that has elected every U.S. president since Thomas Jefferson renames itself the Democratic party. Its first national convention, held at Baltimore in late May, has established the two-thirds majority requirement for nomination that will be followed by all future Democratic presidential conventions, and it renominates Andrew Jackson for a second term with 219 votes against 49 for Henry Clay, 18 for others.

South Carolina legislators hold a state convention November 19 to protest the 1828 Tariff of Abominations and a new tariff law (below). The Ordinance of Nullification passed by the convention November 24 follows the principles of John C. Calhoun (above) in calling the tariff acts "null, void, and no law" in South Carolina.

President Jackson blasts South Carolina's "nullifiers" in a proclamation issued December 10. Vice President Calhoun has been replaced on the ticket by Martin Van Buren, 50, and he resigns the vice presidency December 28 to take the Senate seat of Robert W. Hayne, who has been elected governor of South Carolina.

Abolitionists at Boston form the New England Anti-Slavery Society (*see* Jim Crow, below).

The U.S. Government has exclusive authority over tribal Indians and their lands within any state, rules the Supreme Court March 3 in *Worcester v. Georgia.*

The Creek sign a treaty March 24 ceding their lands east of the Mississippi to the United States.

Seminoles in Florida cede their lands to the United States May 9 in a treaty signed by 15 chiefs who agree to move west of the Mississippi (but *see* 1835).

Chief Black Hawk in the state of Illinois leads his Sac braves back from west of the Mississippi, retakes a village, and begins a 4-month Black Hawk War that ends only when the Illinois militia massacre the warriors August 2 at Bad Axe River in Wisconsin Territory. Black Hawk surrenders after taking refuge with the Winnebagoes (*see* 1823).

Sac and Fox tribesmen agree to remain west of the Mississippi in a treaty signed September 21.

The Chicasaw cede their lands east of the Mississippi to the United States October 14.

Parliament appoints a royal commission to investigate Britain's Poor Law which costs the nation £7 million per year, or 10 shillings per capita (*see* 1834; Malthus, 1798).

A 3-year expedition to explore the Rocky Mountains begins May 1 as French-American U.S. Army captain Benjamin Louis Eulalie de Bourneville, 36, leaves Fort

Osage on the Missouri River with a wagon-train bound for the Columbia River. The expedition includes trapper-guide Joseph Reddeford Walker, now 34, of Independence, Mo.

The Tariff Act passed by Congress July 14 reduces some of the duties in the 1828 Tariff of Abominations, but retains the principle of protectionism.

The first steamboat on Lake Michigan reaches Fort Dearborn (*see* 1818; weekly service, 1834).

A canal is completed to connect Cleveland with the Ohio River at Portsmouth (*see* Cleveland, 1796).

Construction begins in Indiana on the 459-mile Wabash Canal that will be completed in 1855 to connect the Ohio River with Lake Erie at Toledo.

Sweden's Gota Canal opens to connect the North Sea with the Baltic.

French shipbuilder Pierre Louis Frédéric Sauvage, 47, patents a screw propeller for ships but others will appropriate his design (*see* Ressel, Wilson, 1827).

The *Ann McKim,* built at Baltimore for merchant Isaac McKim, is the first true clipper ship (*see* 1790, 1816). Designed for the China trade, the 143-foot 493-ton craft has yachtlike lines and three raking square-rigged masts (*see* 1845).

The Erie Railroad is incorporated April 24 by a special act of the New York State legislature under the name New York and Erie Railroad Co. (*see* 1851).

The New York and Harlem Railroad goes into service November 14, the first New York City street railway. Two horse-drawn cars travel up and down the Bowery between Prince and 14th Streets on tracks slotted deep into the pavement. Built by coachmaker John Stephenson, the horsecars that replace Abraham Bower's 5-year-old "accommodation" have a capacity of 40 passengers each, attain speeds of 12 miles per hour, run every 15 minutes, and charge a 25¢ fare (*see* 1834).

A modern sewing machine devised by New York inventor Walter Hunt, 36, has a needle with an eye in its point that pushes thread through cloth to interlock with a second thread carried by a shuttle. Hunt does not obtain a patent, and when he suggests in 1838 that his daughter Caroline, then 15, go into business making corsets with his machine, she will protest that it would put needy seamstresses out of work (*see* Thimmonier, 1829; Howe, 1843; Hunt's safety pin, 1849).

The cholera epidemic that spread through Russia in 1830 reaches Scotland. Scottish physician Thomas Latta at Leith injects saline solution to save the life of a cholera patient and pioneers a new treatment. The epidemic appears at New York in June, causes 4,000 deaths by October, and spreads south and west.

Parliament passes an anatomy act, ending the need of body snatching for medical research (*see* 1829).

English physician Thomas Hodgkin, 34, gives the first description of a disease he describes as a disorder of the "absorbent glands and spleen." Marked by sarcoma of the lymph nodes, Hodgkin's disease will be found to have a fatality rate of about 75 percent within 5 years of onset, but many victims will live 10 years and more.

French chemist Pierre Jean Robiquet, 52, isolates codeine from opium (*see* morphine, 1803; heroin, 1898).

New York University art professor Samuel F. B. Morse begins development of an electric telegraph that will speed communication (*see* 1827, 1837; Henry, 1831).

Nonfiction *Domestic Manners of the Americans* by English novelist Frances Milton Trollope, 52, who says of her subject, "I do not like them. I do not like their principles, I do not like their manners, I do not like their opinions." Trollope came to America in 1827 after divorcing the father of her son Anthony and the boy's two older brothers.

Fiction *Indiana* by French novelist George Sand (Amandine-Aurore-Lucie Dupin, Baronne Dudevant), 28, who at age 16 inherited her family's estate at Nohant, at 18 married a retired army officer to whom she has borne two children, but has lived independently at Paris for the past year, has collaborated under the pen name Jules Sand with author Jules Sandeau, 21, in writing for *Le Figaro,* and has shocked Paris by going about in trousers; *Maler Nolten* by German novelist-poet Eduard Mörike, 28.

Poetry *Faust* (second part) by Johann Wolfgang von Goethe who dies at Weimar March 22 at age 82.

Painting *Waterloo Bridge from Whitehall Stairs* by John Constable; *A Moorish Couple on Their Terrace* by Eugène Delacroix.

French caricaturist Honoré Daumier, 24, portrays Louis Philippe as Gargantua gorging himself on the earnings of the working class and spitting them back into the arms of the ruling elite; the king has him arrested and sentenced to 6 months in prison. A staffman for the periodical *La Caricature*, Daumier will later join the *Charivari* and combine caricatures of the bourgeoisie with serious painting.

Theater *The Hunchback* by Irish playwright (James) Sheridan Knowles, 48, a cousin of the late Richard Brinsley Sheridan, 4/5 at London's Royal Theatre in Covent Garden, with Knowles as the hunchback and wife Fanny (Frances Anne) Kemble, 23, as his daughter Julia. The theater managed by Fanny's father Charles Kemble, 57, is saved from bankruptcy by the successful play which opens at two New York theaters June 18.

Ballet *La Sylphide* 3/12 at the Théâtre de l'Academie Royale de Musique, Paris, with music by French composer Jean Scheitzhoeffer, choreography by Philippe Taglione (*see* 1836).

Opera *L'Elisir d'Amore* 5/12 at Milan's Teatro alla Scala, with music by Italian composer Gaetano Donizetti, 33; the libretto, based on a French comedy, introduces the comic figure Mr. Dulcamara.

First performances Overture to *The Hebrides* (*Fingal's Cave*) by Felix Mendelssohn 5/14 at London's Covent

1832 ***(cont.)*** Garden Theatre; Symphony No. 5 in D minor *(Reformation)* by Mendelssohn 11/15 at Berlin's Singakademie.

Hymn "Rock of Ages" by Utica, N.Y., composer Thomas Hastings with words from the 1776 verses by the late Augustus Toplady.

Patriotic song "America" with lyrics by Boston Baptist minister Samuel Francis Smith, 23, who has written it in half an hour to the English tune "God Save the King." Schoolchildren at Boston's Park Street Church sing "America" July 4.

Musical stage The blackface song-and-dance act *Jim Crow* wins 20 encores at the City Theater on Jefferson Street in Louisville, Ky. Minstrel show pioneer Thomas Dartmouth "Daddy" Rice, 24, has seen an elderly, deformed slave named Jim Crow perform a little jump while working in a livery stable near the theater, and he reproduces the man's hop and song: "Wheelabout, turn about,/ Do jes so;/ An' every time I wheel about/ I jump Jim Crow" (*see* Jim Crow law, 1880; Emmet's Virginia Minstrels, 1843).

Frances Trollope (above) deplores American eating habits. Suppers, she reports, are huge buffets that may include "tea, coffee, hot cake and custard, hoe cake, johnny cake, waffle cake, and dodger cake, pickled peaches, and preserved cucumbers, ham, turkey, hung beef, apple sauce and pickled oysters. . ."

The U.S. Army abolishes its daily liquor ration.

An Enquiry into the Principles of Population by English scholar Thomas Rowe Edmonds says, "Amongst the great body of the people at the present moment, sexual intercourse is the only gratification, and thus, by a most unfortunate concurrence of adverse circumstances, population goes on augmenting at a period when it ought to be restrained. . . When [the Irish] are better fed they will have other enjoyments at command than sexual intercourse, and their numbers, therefore, will not increase in the same proportion as at present."

Jim Crow began as a comic minstrel-show figure but grew to symbolize racial prejudice in America.

1833 The Compromise Tariff Act submitted by Sen. Henry Clay (S.C.) ends the threat of open conflict between the industrial North and the cotton-exporting South by defusing the nullification issue. The House of Representatives approves it February 26 by a vote of 119 to 85, the Senate approves 29 to 16 March 1, and President Jackson signs it into law.

The Convention of Kutahya May 4 brings a temporary halt in hostilities between Cairo and Constantinople. The Ottoman sultan Mahmud II agrees to let the Egyptian Pasha Mohammed Ali have sovereignty over Syria and Cilicia.

The Treaty of Unkiar-Skelessi July 8 settles disputes between the Ottoman Empire and Russia, which in a secret clause gains the right to close the Dardanelles in time of war.

Spain's Ferdinand VII dies at Madrid September 29 at age 48 after a repressive 19-year reign. His 2-year-old daughter will reign until 1868 as Isabella II, dominated first by Ferdinand's fourth wife Maria Christina of Naples (who 3 years ago persuaded Ferdinand to abolish the Salic Law in Spain by pragmatic sanction) and then by profligate courtiers. Civil war looms as Don Carlos Maria Isidro de Bourbon, 46, brother of the late Ferdinand, claims the throne. He gains support from Basques, Catalonians, the Church, and conservative elements in Aragon and Navarre (*see* 1834).

Portugal's Maria II is restored to the throne by her father Dom Pedro who has returned from Brazil and defeated her brother Dom Miguel with French and British aid. A quadruple alliance of Britain, France, Spain, and Portugal will expel Miguel from Portugal at the end of May 1834 following the death of Dom Pedro, and Maria will reign until 1853 through two insurrections.

A British gunboat claims the Falkland Islands in the South Atlantic as Crown territory. Also known as the Malvinas, the islands were claimed by Argentina in 1820 when she seceded from Spain.

Vom Krieg (*On War*) by the late Prussian general Karl von Clausewitz is a posthumous work on the science of warfare that will be a classic. Clausewitz served with the Russians against Napoleon in 1812, was made a major general and director of the Allgemeine Kriegschule in 1818, but died in Breslau's 1831 cholera epidemic at age 51.

The German foundling Kaspar Hauser, who was accepted by some as prince of Baden, dies December 17 at age 21 of stab wounds received from someone who called him to a rendezvous with promises of information regarding his parentage. Picked up by Nuremberg police 5 years ago, he was rumored to be of noble birth, possibly the prince of Baden, and has been adopted by Philip Henry, 28, earl of Stanhope.

Parliament orders abolition of slavery in the British colonies by August 1, 1834, in a bill passed August 23 after a long campaign by the humanitarian William Wilberforce who has died July 29 at age 73. Children under 6 are to be freed immediately, slaves over 6 given a period of apprenticeship that will be eliminated in 1837, slave-owners given a total of £120 million in compensation.

A factory act voted by Parliament August 29 forbids employment of children under age 9, forbids factory owners to work children between 9 and 13 for more than 48 hours per week and for more than 9 hours per day, requires that children under 13 be given at least 2 hours of schooling per day, limits working hours for children between 13 and 18 to 69 hours per week and 12 hours per day, and establishes statutory minimums of time permitted workers of whatever age to eat the food they have brought from home. Parliament passes the bill over the opposition of Tories and of many Whigs who subscribe to Adam Smith's laissez-faire principles of 1776, it sets up a system of paid inspectors, but the new law applies only to textile factories and its safeguards are inadequate.

New York trade societies join to form the General Trades' Union, trade unionism begins to supersede workingmen's political parties, but the movement will collapse in the financial panic and economic depression 4 years from now (*see* 1834; 1864).

"An Appeal in Favor of That Class of Americans Called Africans" by Boston abolitionist David Lee Child and his bride Lydia Maria (Francis) Child, 31, proposes that blacks be educated. The idea is considered outrageous in most circles, but the Appeal converts many supporters of slavery to the abolitionist cause.

Canterbury, Conn., schoolmistress Prudence Crandall is imprisoned for violating a special act of the legislature directing her not to admit black girls to her school.

The American Anti-Slavery Society is founded at Philadelphia December 4 by abolitionists who include James Mott, 45.

The Female Anti-Slavery Society is founded at Philadelphia under the leadership of Lucretia Coffin Mott, 40, wife of James (above), who finds that her husband's group bans women.

Atlanta, Ga., has its beginnings in a cabin built by pioneer Hardy Ivy on former Creek lands at the foot of the Blue Ridge Mountains. The town of Terminus will be founded at the site in 1837 (*see* 1843).

The *Zollverein* (customs union) formed at Berlin March 22 includes Prussia, Bavaria, Württemberg, Hesse-Darmstadt, but not Austria (*see* 1853).

The East India Company loses its prized monopoly in the China trade (most of it in tea) by an act of the British prime minister Charles Grey, 69, second Earl Grey.

The Compromise Tariff Act voted by Congress (above) provides for gradual reduction of U.S. tariffs until July 1, 1842, when no rate is to be higher than 20 percent.

The Philadelphia & Reading Railroad chartered April 4 includes some lines begun in 1831.

The Andover & Wilmington Railroad is chartered in Massachusetts. It will join with the 3-year-old Boston & Lowell and other roads to create the Boston & Maine with 4,077 miles of track in Massachusetts, Maine, Vermont, and New York.

The first cargo of U.S. ice for India leaves Boston to begin a voyage of 4 months and 7 days. Frederic Tudor has loaded his ship *Tuscany* with 180 tons of ice—plus apples, butter, and cheese—for Lord William Bentinck and the nabobs of the East India Company at Calcutta; half the ice is lost in transit and during unloading, but the voyage nevertheless makes a profit (*see* 1805; 1846).

English mathematician Charles Babbage, 41, proposes an "analytical engine," a large-scale digital calculator that will go far beyond a "difference engine" he proposed in 1822. Babbage will obtain some government financial support to develop his calculator but will finance the work largely with his own fortune (*see* 1842).

Experiments and Observations on the Gastric Juices and the Physiology of Digestion by William Beaumont will become a classic in clinical medicine (*see* 1822).

The Oxford Movement to restore High Church traditions of the 17th century in the Church of England begins July 14 at St. Mary's, Oxford, with an assize sermon on "national apostasy" preached by clergyman-poet John Keble, 41, who criticizes the suppression of 10 Irish bishoprics. Keble is supported by his fellow at Oxford's Oriel College John Henry Newman, 32, whose "Tracts for the Times" begin appearing and who will preach at St. Mary's for the next 8 years on the need to secure for the Church of England a firm base of doctrine and discipline lest the Church be disestablished or abandoned by High Churchmen. Tractarians will arouse such opposition that many, including Newman, will feel compelled to leave the Church of England and become Roman Catholics.

Oberlin College opens at Oberlin, Ohio. It admits qualified blacks and will admit women on an equal basis beginning in October 1838, becoming the first coeducational U.S. college.

Haverford College is founded at Philadelphia by the Society of Friends.

The University of Delaware has its beginnings in the College of Delaware founded at Newark.

Uruguay's National University of the Republic is founded at Monevideo.

England's Durham University is reorganized under that name (*see* 1657).

France's Primary Education Law enacted June 28 gives the Church extensive control over the nation's primary schools.

The *New York Sun* launched September 1 by publisher Benjamin Day, 23, is the city's first successful penny daily (the competition charges 6¢). Day's son Benjamin

1833 ***(cont.)*** will invent the Ben Day process for shading in printed illustrations.

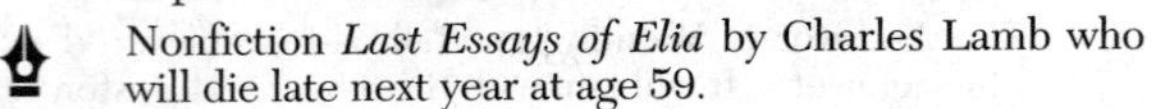

Nonfiction *Last Essays of Elia* by Charles Lamb who will die late next year at age 59.

Fiction *Lélia* by George Sand whose novel *Indiana* last year asserted the right of women to love and independence. Sand begins a love affair with playwright-poet Alfred Louis Charles de Musset, 23, whose first book of poetry appeared 3 years ago and who has recently joined the staff of *La Revue des Deux Mondes;* "Ms. Found in a Bottle" by Edgar Allan Poe appears in the October 19 *Baltimore Sunday Visitor*; "Sartor Resartus (The Tailor Retailored)" by Scottish writer Thomas Carlyle, 38, appears in *Fraser's* magazine beginning in November. The history of clothing by Carlyle's Professor Teufeldröckh is actually a speculative discussion of creeds and philosophies.

Poetry *Pauline* by English poet Robert Browning, 21, whose work is poorly received.

Painting *Fifty-three Stations of Tokaido (Tokaido Gojusansugi)* by Japanese *ukiyoe* landscape painter Ando Hiroshige, 36 (*see* 1601).

Theater *Bertand and Raton, or The Art of Conspiracy* (*Bertrand et Raton, ou l'art de conspirer*) by Eugène Scribe 11/14 at the Théâtre Françaїse, Paris.

Opera *Lucrezia Borgia* 1/26 at Milan's Teatro alla Scala, with music by Gaetano Donizetti.

First performances "Die eerst Walpurgisnacht" by Felix Mendelssohn 1/10 at Berlin's Singakademie; Symphony No.4 in A major (*Italian*) by Mendelssohn 5/13 at London's Hanover Square Rooms.

Hymn "Lead, Kindly Light" verses by Oxford University church vicar John Henry Newman (above) who has written them while traveling aboard ship from Palermo to Marseilles (*see* 1868); "God Save the Emperor" by Russian composer Alexis Feodorovich Lvov will be the Russian national hymn until 1917.

The balloon-frame house invented by Chicago carpenter Augustus Deodat Taylor will revolutionize U.S. residential construction. Taylor nails two-by-fours together to create a cage-like framework to which a roof and siding are nailed. Critics predict that prairie winds will blow the house up and away like a balloon, but it proves to be even sturdier than conventionial houses made like barns from heavy beams that require considerable time and experience to build. Within 20 years Taylor houses will be going up all over America as teams of amateurs use handy-sized precut lumber and cheap mass-produced nails to knock houses together.

A reaper patented by Cincinnati Quaker inventor Obed Hussey, 41, will rival the 1831 McCormick reaper and beat it to market. Hussey's reaper will go into production next year and be snapped up by midwestern grain farmers; McCormick's machine will not go into production until 1840 and will encounter resistance from farmers in hilly Virginia (but *see* 1847).

Farmer's Register is published for the first time in Virginia with agriculturist Edmund Ruffin, 39, as editor. Ruffin has seen farmland eroded by tobacco growers and warns planters of the "growing loss and eventual ruin of your country, and the humiliation of its people, if the long-existing system of exhausting culture is not abandoned." "Choose, and choose quickly," says Ruffin, and he urges contour plowing, crop rotation, the use of furrows for careful drainage, and the use of lime and fertilizers to rejuvenate the soil.

Avocados are introduced from Mexico into southern Florida by horticulturist Henry Perrine.

A 3-year famine begins in Japan. The suffering will be even worse than in 1783.

A Swiss millright devises an alternative to the millstones that have been used to grind grain since ancient times; he replaces the stones with rollers and uses them to reduce the grain to flour more efficiently (*see* Sulzberger, 1839).

Bottled carbonated water is sold to New York merchants by English-American entrepreneur John Matthews who opens a shop at 33 Gold Street and begins manufacturing a compact apparatus for carbonation—the first soda fountain (*see* 1891; Silliman, 1807).

The diaphragm contraceptive, invented by German physician Friedrich Adolphe Wilde, is a rubber cap that will be popularized in years to come by physician Wilhelm Mensinga (*see* Place, 1822; Sanger, 1923).

1834 A Spanish civil war begins as Don Carlos claims the throne of his niece Isabella II (*see* 1833). Supported by the Church and by Basques, Catalonians, and the conservative elements of Aragon and Navarre, the pretender faces opposition from Portugal, Britain, France, and supporters of Isabella who form a quadruple alliance. London suspends the Foreign Enlistment Act, enabling Sir George de Lacy Evans, 47, who served in the Peninsula campaign and in the War of 1812, to form a foreign legion that will support Isabella (*see* 1839).

Portugal's 6-year civil war ends May 26 with the defeat of Miguel, who leaves the country (*see* 1828).

Giuseppe Mazzini founds the Young Europe movement, expanding his Young Italy movement of 1832 by organizing Young Germany, Young Poland, and similar groups from headquarters at London, but peasants in Savoy thwart his efforts to gain control there (*see* 1848; 1849).

Sikh forces from the Punjab take Peshawar May 6. Their ruler Ranjit Singh has led the assault on the Muslim city.

South Africa has a Kaffir War as Xhosa tribesmen (called Kaffir by the Dutch, who use it in the same sense that some Americans use the word "nigger") invade eastern regions in irritation at the steady encroachment of Dutch cattlemen and farmers. The Xhosa are driven back but only with difficulty (*see* 1877; Great Trek, 1835).

Britain's Grand National Consolidated Trades Union, organized in January by Robert Owen and John Doherty, 36, is pledged to strike for an 8-hour day.

Within a few weeks the union has half a million members, but its avowed purpose of fomenting a general strike alarms the government; six workers led by George Loveless who have formed a lodge at Tolpuddle outside Dorchester are sentenced in March to be transported to New South Wales or Tasmania for 7 years. Their conviction produces demonstrations throughout the country, the prime minister Lord Melbourne refuses to receive a petition of protest bearing 250,000 signatures, a series of strikes meets with no success, Owen is intimidated, and the Grand National dissolves in October, but the Tolpuddle martyrs will be brought home in 1836 to appease the public outcry.

A new British Poor Law enacted by Parliament August 14 limits the pay of charitable doles to sick and aged paupers. It establishes workhouses where able-bodied paupers are put to work (no able-bodied man may receive help unless he enters a workhouse), and the system that has provided a dole to supplement low wages is ended.

New York's General Trades Union organizes a National Trades Union that takes in all crafts.

Workers along the Chesapeake and Ohio Canal stage a riot January 29. President Jackson orders Secretary of War Lewis Cass to send in the Army, using federal troops for the first time in a U.S. labor conflict.

A Department of Indian Affairs established by Congress June 30 sets up Indian territory west of the Mississippi. Florida Seminoles are ordered to move west October 28 in accordance with a treaty signed May 9, 1832 (*see* Osceola, 1835).

Unskilled U.S. workers demonstrate against abolitionists in fear that they will be displaced by black freedmen. Rioters break up a New York antislavery society meeting at the Chatham Street Chapel July 4 protesting the presence of some blacks in the audience, the rioting continues for more than a week, and churches and houses are destroyed.

Anti-abolitionists at Philadelphia destroy the homes of 40 blacks in October.

Some 35,000 slaves go free in South Africa August 1 as slavery is abolished throughout the British Empire amid complaints about the inadequacy of compensation to former slaveholders (*see* 1833; Great Trek, 1835).

The Spanish Inquisition instituted in the 13th century is finally abolished (*see* 1492; 1820).

South Australia is founded by followers of British colonial theorist Edward Gibbon Wakefield, 38, who obtain a charter August 2 for their South Australia Association with support from the duke of Wellington and historian George Grote, 38. The first settlers will be landed in 1836 at Kangaroo Island.

The Australian colony that will be called Victoria is settled for the first time by West Australia rancher Edward Henty and his brothers at Portland Bay. They will be joined next year by John Batman, 34, and his associates from Tasmania.

The U.S. Senate censures President Jackson March 28 for removing deposits from the Bank of the United States. The Senate approves a resolution introduced by Henry Clay late last year, Jackson enters a formal protest April 15, and the resolution will be removed early in 1837 from the Senate journal.

John Jacob Astor sells his fur interests as pelts threaten to become scarce. Beaver pelts have sold for $6 apiece in peak years, enabling trappers to make $1,000 per season, but the fur companies have charged enormous prices for supplies hauled from St. Louis to summer rendezvous points, so while the beaver has been nearly exterminated, none of the trappers have made fortunes. Astor has monopolized the upper Missouri Valley fur trade, made himself the richest man in America, and will now devote his efforts to administering his fortune, much of which he will invest in New York real estate (*see* Juneau, 1818; Astor House, 1836).

Lowell, Mass., has six corporations operating 19 mills with 4,000 looms and more than 100,000 spindles (*see* 1814; Thoreau, 1819).

The direction of an electric current induced in a circuit by moving it in a magnetic field produces an effect tending to oppose the circuit's motion, says Russian physicist Heinrich Emil Lenz, 30, who formulates a new law of electricity.

The Delaware and Raritan Canal opens between New Brunswick, N.J., and Bordentown, Pa. The 43-mile canal will be extended 21 miles to Trenton on the Delaware River, it will be the state's chief transportation corridor, mule-drawn barges will be hauling 1.2 million tons of coal and 12 million board-feet of lumber per year from Philadelphia to New York by 1860, and the canal will remain in operation until 1933 (*see* Morris Canal, 1831).

The Portage Railroad opens to connect Philadelphia with Pittsburgh via rail and canal.

The New York and Harlem Railroad of 1832 extends its horsecar route up Fourth Avenue to 84th Street (*see* 1837).

Hansom cabs are introduced in London. The patented safety cabs (Londoners call them gondolas) have been designed by architect Joseph Aloysius Hansom, 31, who last year designed the Birmingham town hall.

London's Baring Brothers acquires two ships and enters the China trade with the *Alexander Baring* and the *Falcon*. The merchant banking house competes with the East India Company (above).

Weekly steamboat service begins between Buffalo and Fort Dearborn (*see* 1832; Chicago, 1830; railroad, 1852).

Cornelius van Derbilt begins to multiply the $500,000 he has amassed in the steamboat trade (*see* 1829; 1850).

Tulane University has its beginnings at New Orleans. Local merchant Paul Tulane, 33, opened a dry-goods and clothing store 12 years ago, he will make large gifts to the school beginning in 1882, and it will take the name Tulane in 1884.

1834 ***(cont.)*** The University of Brussels has its beginnings.

The braille system of raised point writing devised by French educator Louis Braille, 25, will gain acceptance throughout the world (see Haüy, 1784). Braille has been blind since age 3, he has been teaching fellow sightless people since 1828, and his system can be used for music as well as for words.

The *New Yorker Staats-Zeitung* begins publication at New York. The German-language weekly will become a daily in 1849.

Fiction *Le Père Goriot* by French novelist Honoré de Balzac, 34; *The Last Days of Pompeii* by English novelist Edward George Bulwer-Lytton, 31.

Painting *Algerian Women at Home* by Eugène Delacroix; *The Martyrdom of Saint Symphorian* by Jean Ingres.

The Munich Glyptothek sculpture gallery is completed by Leo von Klenze, 50.

London's National Gallery is started by English architect William Wilkins, 56.

Theater *A Dream is Life* (*Der Traum ein Leben*) by Franz Grillparzer 10/4 at Vienna's Burgtheater.

First performances *Harold in Italy* Symphony by Hector Berlioz 11/23 at the Paris Conservatoire. Berlioz has based the work on the poem "Childe Harold" of 1818 by the late Lord Byron.

Hymn "Jesus, Lover of my Soul" by Simeon Butler Marsh, verses by the late Methodist leader Charles Wesley.

Popular song "Zip Coon" ("Turkey in the Straw") is published anonymously. Bob Farrel and George Washington Dixon will claim authorship.

Chantilly racetrack opens 26 miles north of Paris.

Fire destroys London's Houses of Parliament and part of the city October 16. A major reconstruction program begins (*see* Buckingham Palace, 1837; Trafalgar Square, 1843; Big Ben, 1858).

Sardines are canned for the first time in Europe (*see* first U.S. sardine cannery, 1876).

British wheat prices fall to £10 per cwt, down from £30 in 1812.

Cyrus McCormick receives a patent for his reaper of 1831 and Obed Hussey begins manufacturing his reaper (*see* 1833; 1840).

Gas refrigeration has its beginnings in a compression machine invented in England by U.S. inventor Jacob Perkins of 1790 nail-cutter fame who has lived abroad for years. Now 58, Perkins distills rubber to create a volatile liquid which is allowed to evaporate by absorbing heat from its surroundings. When the vapor is compressed it turns back to liquid, giving off heat, and by alternately compressing and expanding Perkins extracts heat from the region of expansion until he has cooled water to the point that it freezes (*see* Faraday, 1823; Gorrie, 1842; Linde, 1873).

Some 28 million acres of U.S. public lands will be offered for sale this year and next. As Americans move west of the Appalachians to take up the new lands, European immigrants will replace them.

Thomas R. Malthus dies at Haileybury, England, December 23 at age 68. Father of three (one died at age 17), Malthus has been attacked for "having the impudence to marry after preaching against the evils of a family. He has been vilified by essayist William Hazlitt, and accused of defending war, plagues, slavery, and infanticide.

1835 The last Holy Roman Emperor Francis II (Francis I as first Austrian emperor) dies March 2 at age 67 and is succeeded as emperor of Austria by his mentally-retarded son of 4 who will reign under a regency until 1848 as Ferdinand I.

Sir Robert Peel resigns as prime minister; William Lamb, Lord Melbourne, heads a new ministry beginning April 18.

The Municipal Corporations Act passed by Parliament September 9 reforms English borough government to reflect the shift of population into industrial cities and towns (*see* Reform Act, 1832).

The Barakzai dynasty that will rule Afghanistan until 1929 is founded by the khan of Kabul Dost Mohammed, now 42, who gains power over the entire country, takes the title emir, and will reign until 1839 (*see* 1826; 1838).

The governor of Buenos Aires Juan Manuel de Rosas assumes dictatorial powers, 12 provinces recognize his authority, and Rosas will rule until 1852, ruthlessly crushing all opposition.

A new Seminole War against the whites in Florida Territory begins following the arrest and imprisonment of Osceola, 31, who has thrust his knife through the 1832 treaty that ceded Seminole lands to the United States. Osecola escapes, he and his braves kill a chief who signed the 1832 treaty, killing also the U.S. Indian agent at Fort King, and begin 2 years of guerrilla activities against U.S. forces under Gen. Thomas S. Jesup while Seminole women and children remain hidden deep in the Everglades. Seminoles and their black slaves massacre a 103-man U.S. Army force under Major Francis L. Dade December 28 (*see* 1837).

Dutch (Boer) cattlemen in South Africa begin a Great Trek to the north and east of the Orange River in irritation at Britain's abolition of slavery last year.

Some 10,000 Boers will move to new lands beyond the Vaal River (the Transvaal) in the next 2 years, seriously depopulating the eastern part of the Cape Colony (*see* Orange Free State, 1854; Pretorius, 1856).

Melbourne is founded in Australia and is named for the new British Prime minister (above) (*see* Victoria, 1834).

St. Petersburg, Fla., has its beginnings in a settlement at Old Tampa Bay founded by French immigrant Odet Philippe who was once a surgeon for Napoleon.

The first passenger railroad on the Continent opens May 5 to link Brussels and Mechelen.

The St. Etienne-Lyons passenger railway opens July 9.

The first German railway links Nuremberg with Fürth and passes into private hands December 7.

The United States has 1,098 miles of railroad in operation.

Chemists synthesize the pain reliever salicylic acid but cannot produce a safe, effective anodyne (*see* Stone, 1763; Gerhardt, 1853).

London pathologist James Paget, 21, at St. Bartholomew's Hospital detects the parasite *Trichina spiralis* for the first time. The parasite will later be associated with trichinosis, a disease produced by eating raw or undercooked pork products or meat from bears, polar bears, rats, foxes, or marine animals (*see* von Zenker, 1860; Paget's disease, 1877).

Berlin naturalist Theodor Schwann extracts the enzyme pepsin from the stomach wall and proclaims it the most effective element in the digestive juices (*see* 1824). (The word "enzyme" will be coined in 1878.)

The *New York Herald* begins publication May 6 under the direction of Scots-American journalist James Gordon Bennett, 40, who has started the penny newspaper with $500, two wooden chairs, and an old dry goods box in a cellar office.

Agence France-Presse is established under the name Agence Havas. The French national news agency will take the name France-Presse in 1945.

Nonfiction *Democracy in America (La Démocracie en Amerique)* by French aristocrat Alexis Charles Henri Maurice Clerel de Tocqueville, 35, who traveled for 9 months in 1831 through eastern Canada, New England, New York, Philadelphia, Baltimore, Washington, Cincinnati, Tennessee, and New Orleans on a commission to study the U.S penitentiary system. "Nothing struck me more forcibly than the general equality of conditions. . . All classes meet continually and no haughtiness at all results from the differences in social position. Everyone shakes hands. . ."; but de Tocqueville foresees the rise of certain forces that will eventually undermine the principle of economic equality, and while the American passion for equality "tends to elevate the humble to the rank of the great. . . there exists also in the human heart a depraved taste for equality, which impels the weak to attempt to lower the powerful to their own level, and reduces men to prefer equality in slavery to inequality with freedom."

Tyranny of the majority is a possible hazard of democracy, de Tocqueville (above) warns, but he notes that law, religion, and the press provide bulwarks against democratic despotism, as he will elaborate in a second volume in 1840.

Fiction *Mirgirod* (stories) by Nikolai Gogol who will leave Russia next year to live abroad, mostly in Rome; *Arabeski* (essays and stories) by Gogol includes "Nevsky Prospekt," "The Portrait," and "Notes of a Madman"; *Mademoiselle de Maupin* by Théophile Gautier espouses the doctrine of "art for art's sake" in its preface and denounces bourgeois philistinism; *Rienzi* by E. G. E. Bulwer-Lytton is based on the 1347 Roman revolt of Cola di Rienzi; "Berenice" by Edgar Allan Poe in the March *Southern Literary Messenger*; *The Yemassee* by U.S novelist William Gilmore Simms, 29, whose story about Indians will make him widely known. Simms returns from New York to his native Charleston, S.C., where his first wife died 3 years ago and where he will soon marry the daughter of a rich planter.

Juvenile *Fairy Tales (Eventyr)* by Danish novelist Hans Christian Andersen, 30, whose novel *The Improvisators* also appears and enjoys such success that Andersen is encouraged to pursue a career as novelist and dramatist and neglect the fairy tales that will make him famous. Included in his first volume are "The Tinderbox" ("Fyrtojet"), "The Princess on the Pea" ("Princessen paa Aerten"), "Little Claus and Big Claus" ("Lille Claus og store Claus"), and "Little Ida's Flowers" ("Den lille Idas Blomster") which will be followed in time by "The Ugly Duckling," "The Fir Tree," "The Red Shoes," "The Swineherd," "The Snow Queen" ("Snedronningen"), and "The Emperor's New Clothes" ("Kejserens nye Kloeder").

Painting *Burning of the Houses of Lords and Commons* by J. M. W. Turner who witnessed last year's great fire after rushing from dinner with sketchbook in hand to find the spectacle being applauded by throngs of spectators on the Thames embankments; *The Valley Farm* by John Constable; *Homer in the Desert* by J. P. C. Corot who makes a second visit to Italy; *Hundred Views of Mount Fuji* by Japanese *ukiyoe* painter Katsushika Hokusai (Tetsuzo Nakashima), 75, whose *Ten Thousand Sketches (Manga)* will be published in its 15th and final volume next year (Hokusai's work includes *The Great Wave*). Baron Antoine Jean Gros drowns himself in the Seine at Meudon June 26 at age 64.

The Great Wave by the *ukiyoe* painter Katsushika Hokusai reflected Japan's age-old insularity.

1835 ***(cont.)*** P. T. Barnum begins a career in U.S. show business. Connecticut-born promoter Phineas Taylor Barnum, 25, buys a slave woman who is alleged to have been George Washington's nurse and to be more than 160 years old. Joice Heth will be proved to be no older than 70 when she dies next year, but Barnum will continue the well-publicized tours he has begun with a small company of carnival attractions (*see* Siamese Twins, 1829; Tom Thumb, 1842).

Opera *I Puritani* 1/25 at the Théâtre des Italiens, Paris, with music by Vincenzo Bellini; *La Juive* 2/23 at the Paris Grand Opéra, with music by French composer Jacques Halévy, 35, libretto by playwright Eugène Scribe; *Lucia di Lammermoor* 9/26 at the Teatro San Carlo, Naples, with music by Gaetano Donizetti, libretto from the 1818 Walter Scott novel *The Bride of the Lammermoor.*

First performances *Grand Polonaise Brillante* by Frédéric Chopin 4/26 at the Paris Conservatoire.

Popular songs "Long, Long Ago" by English songwriter-novelist-playwright Thomas Haynes Bayly, 38; "Kathleen Mavourneen" with lyrics by Irish author Julia Crawford, 35.

New York loses 674 buildings December 16 in a $15 million fire that rages out of control in the vicinity of Hanover and Pearl Streets (*see* Croton water, 1842).

Washington Irving describes honeybees in A *Tour of the Prairies:* "It is surprising in what countless swarms the bees have overspread the Far West within a moderate number of years. The Indians consider them the harbinger of the white man, as the buffalo is of the red man; and say that, in proportion as the bee advances, the Indian and the buffalo retire. . . I am told that the wild bee is seldom to be met with at any great distance from the frontier."

Russia's bonded peasant population reaches close to 11 million, up from fewer than 10 million in 1816. The overabundance of serfs causes problems in years when crops are short.

"Fruits of Philosophy" by Cambridge, Mass., physician Charles Knowlton, 35, is a Malthusian pamphlet advocating contraception. Knowlton is prosecuted and imprisoned for 3 months (*see* Bradlaugh, Besant, 1877).

1836 The Alamo at San Antonio falls March 6 to a 4,000-man army commanded by the president of Mexico Gen. Antonio Lopez de Santa Anna, 40, after an 11-day siege. The 188-man force that has garrisoned the Franciscan mission includes William B. Travis, 26, Bowie knife inventor James Bowie, 39, and former U.S congressman from Tennessee David "Davy" Crockett, 49. The Mexicans sustain nearly 1,600 dead plus many wounded, spare 30 U.S. noncombatants, but kill the entire garrison, shooting five prisoners.

"Remember the Alamo" is the battle cry of U.S. Army Col. Sidney Sherman, 31, as he helps former Tennessee governor Samuel "Sam" Houston, 43, defeat Santa Anna (above) April 21 at the Battle of San Jacinto and take the Mexican general prisoner (*see* Austin, 1821).

Arkansas is admitted to the Union as the 25th state June 15.

A new Republic of Texas claims all land between the Rio Grand and Neuces rivers. Sam Houston (above) is sworn in as president October 22. J. K. and A. C. Allen found the city of Houston (*see* 1845).

Vice President Martin Van Buren, 53, is elected to the presidency with support from the outgoing President Jackson. He receives more popular votes than his four opponents combined, and he wins 170 electoral votes versus the combined total of 124 for his opponents who are led by William Henry Harrison.

Congress passes a resolution May 26 stating that it has no authority over state slavery laws.

Spain has insurrections beginning August 10 in Andalusia, Aragon, Catalonia, and Madrid against the repressive regime of the regent Maria Cristina as the Carlist civil war continues. She is forced to restore the progressive constitution of 1812.

Boer farmers found Natal, Transvaal, and the Orange Free State as they continue the Great Trek that began last year in South Africa.

Adelaide is founded in South Australia on a fertile plain that sweeps up from the coast to a line of hills. The town is named for the consort of Britain's William IV (*see* 1834).

U.S. missionary Marcus Whitman, 34, takes his wife Narcissa and Eliza Spalding to the Pacific Northwest. They are the first white women to cross the Continent (*see* 1843; 1347).

English chemist John Frederick Daniell, 47, creates the first electric cell with a long working life. He improves on the 1800 cell of Alessandro Volta by producing a cell that generates a steady current for considerable periods of time (*see* Plante, 1859).

The S.S. *Beaver* tested under steam at Vancouver May 16 is the first steamboat to be seen on the Pacific Coast. Built last year in England with Boulton & Watt engines, the 101.4-foot vessel enters the Willamette River in Oregon Territory May 31 on her maiden voyage, runs down the river under steam, and enters the lower reaches of the Columbia River.

Of 2.35 million tons of U.S. merchant shipping some 60,000 tons are in steam.

The Erie Canal completed in 1825 is widened and deepened for the barge traffic that has repaid investors with toll receipts (*see* 1917).

The Long Island Rail Road begins operations April 18 (*see* 1900; 1905).

The first Canadian railway opens July 21 to link Laprairie on the St. Lawrence River with St. Johns on the Richelieu.

The hot-air balloon *Nassau* carried by 85,000 cubic feet of coal gas lifts off November 7 from London's Vauxhall Gardens with three passengers in her gondola, passes over Liège and Coblenz, and comes to rest 18 hours and 500 miles later in the German duchy of Nassau. Balloonist Charles Green has been the first to use coal gas which he employed in the *Nassau* (then called something else) to celebrate the coronation of George IV in 1821.

English chemist Edmond Davey discovers and identifies acetylene.

Galvanized iron (coated with zinc) is invented in France.

The world's largest steel manufacturing and munitions plant has its beginnings in a French ironworks at Le Creusot purchased by French industrialist Joseph Eugène Schneider, 31, and his brother Eugène. They will develop the works into a gigantic steelmaking enterprise.

The Colt six-shooter revolver patented by Hartford, Conn., inventor Samuel Colt, 22, has an effective range of only 25 to 30 yards, but improved versions will play a major role in the opening of the West. Colt establishes the Patent Arms Manufacturing Co. at Paterson, N.J.; the firm will soon fail (*see* 1846; 1871; Hays, 1844).

Evidence from the Scripture and History of the Second Coming of Christ, about the Year 1843 by U.S. religious leader William Miller, 54, will lead to the founding of the Adventist Church in 1845.

The Book of Wealth: in Which It Is Proved from the Bible that It Is the Duty of Every Man to Become Rich is published by U.S. clergyman Thomas P. Hunt.

McGuffey's First and Second *Readers* are compiled for a local publisher by Cincinnati College president William Holmes McGuffey, 36, whose *Readers* will grow to number six in the next 15 years and will be sold until 1927. They will be used to educate generations of Americans in the virtues of frugality, industry, and sobriety.

The *Toledo Blade* begins publication in Ohio.

The *Philadelphia Public Ledger* begins publication. Three journeymen printers start Philadelphia's first penny daily.

Nonfiction "Nature" by Boston essayist Ralph Waldo Emerson, 33, who says, "The inevitable mark of wisdom is to see the miraculous in the common" and advises readers to "enjoy an original relation to the universe" (*see* Thoreau, 1837).

Fiction *Sketches by 'Boz'* by London court reporter Charles (John Hulfam) Dickens, 24, whose first sketch appeared in a December 1833 periodical and who has used the pen name 'Boz' since August; *Mr. Midshipman Easy* by retired British Navy captain Frederick Marryat, 44, whose first novel was published as *The King's Own* shortly after his retirement from the navy in 1830; *Confession d'un Enfant du Siècle* by Alfred de Musset (autobiographical novel); "Maelzel's Chess-Player" by Edgar Allan Poe.

Poetry *La Ginestra* by Count Giacomo Leopardi, now 38, who will die of cholera next year at Naples.

Painting *Juliet and Her Nurse* by J. M. W. Turner, whose work is ridiculed; *Stoke-by-Nayland* by John Constable; *Diana Surprised by Actaeon* by J. P. C. Corot.

Theater *The Government Inspector* (*Revizor*) by Nikolai Gogol 4/19 at St. Petersburg's Alexandrinsky Theater.

Opera *Les Huguenots* 2/29 at the Paris Opéra, with music by Giacomo Meyerbeer; *A Life for the Czar (Ivan Susanin)* 12/9 at St. Petersburg, with music by Russian composer Mikhail Ivanovich Glinka, 32, who traveled in Italy from 1830 to 1832, met Vincenzo Bellini and Gaetano Donizetti, and fell in love with Italian opera. His opera will open the season at St. Petersburg and Moscow every year hereafter until 1917.

Ballet *La Sylphide* at Copenhagen, a new version of the 1832 ballet with music by the late German composer Christoph Willibald Gluck.

Oratorio *St. Paul* by Felix Mendelssohn 5/22 (Whitsunday) at the Düsseldorf Festival, with a chorus of 364 voices and a 172-piece orchestra.

The King's Plate and the Queen's Plate horse races at Quebec begin the oldest stakes races that will be run continuously in North America.

A phosphorus match patented by Springfield, Mass., shoemaker Alonzo D. Phillips is the prototype of matches that will be used in future to light stoves, hearthfires, and smoking tobacco (*see* 1827; 1853).

The Arc de Triomphe ordered by Napoleon in 1806 is completed at Paris in the Place de l'Etoile; 164 feet high and 148 wide, the world's largest triumphal arch has a frieze depicting the successful campaigns of France's armies between 1790 and 1814.

New York's Astor House opens on the northwest corner of Broadway and Vesey Streets. It sets new standards of hotel luxury (*see* Tremont House, 1829).

Iceland's Mount Hekla erupts as it did in 1597, creating widespread destruction (*see* 1947).

Narcissa Whitman (above) praises frontier food, writing, "I wish some of the feeble ones in the States could have a ride over the mountains; they would say like me, victuals even the plainest kind never relished so well before."

The first printed American menu is issued by New York's 5-year-old Delmonico's Restaurant and lists as one of its most expensive dishes "Hamburg steak" (*see* 1899).

Some 75 percent of gainfully employed Americans are engaged in agriculture, down from 83 percent in 1820 (*see* 1853).

1837 Britain's William IV dies June 20 at age 71 after a 7-year reign and is succeeded by his niece of 18, a granddaughter of George III, who will reign until 1901 as Queen Victoria.

Salic law forbids female succession in Hanover and that kingdom is separated from Britain. The Hanoverian crown passes to George III's eldest surviving son the

1837 ***(cont.)*** duke of Cumberland; he will reign until 1851 as Ernst August, and his son will reign as George V until Prussia annexes Hanover in 1866.

Japan's Tokugawa shogun Ienari resigns at age 64 after a 44-year rule marked by widespread disturbances produced in part by Ienari's efforts to reform government and improve education. Ienari has had 40 mistresses, half of his 54 children have died in infancy, and he is succeeded by his son Ieyoshi, 45, who will rule until 1853 as sentiment grows for restoration of imperial power and demands increase that Japan open her ports to foreign trade.

Michigan is admitted to the Union as the 26th state.

President Jackson recognizes the Republic of Texas on his last day in office March 3 after approval by Congress.

The Texas Rangers have their beginnings at Waco where Anson Darnell, Starrett Smith, and Jacob Gross establish Fort Fisher.

Congress increases Supreme Court membership March 3 from seven justices to nine.

A French-Canadian rebellion against British rule begins under the leadership of Louis Joseph Papineau, 51, and William Lyon Mackenzie, 42. Speaker of the legislative assembly of Lower Canada, Papineau is charged with high treason, escapes to the United States, and is declared a rebel after some riots and fighting in the Montreal area. Mackenzie attacks Toronto December 5, is repulsed, flees across the border December 13, seizes Navy Island in the Niagara River, and proclaims a provisional government, but he will be arrested in the United States early next year. The U.S. steamboat *Caroline* supplies the rebels but is burnt December 29 by Canadian troops who have crossed the Niagara River (*see* Durham Report, 1839).

The Seminole leader Osceola is tricked into coming out of the Florida Everglades under a flag of truce October 21 and arrested with several followers when he enters the compound at St. Augustine. The action of the commanding officer Gen. Thomas S. Jesup brings public protest, but Osceola remains a prisoner, his braves are defeated December 25 at Okeechobee Swamp by Col. Zachary Taylor, 53, who is called "Old Rough and Ready" and is promoted to brigadier general. Osceola will die early next year at Fort Moultrie near Charleston, S.C., and most of his tribespeople will be exterminated in the next few years (*see* Florida statehood, 1845).

Congress enacts a gag law to suppress debate on the slavery issue.

Alton, Ill., editor Elijah Paris Lovejoy, 34, receives a new press from the Anti-Slavery Society to replace the *Alton Observer* press that has been smashed by townspeople who favor slavery. An armed band of citizens approaches the warehouse where the new press is stored, a group of abolitionists stands guard to defend it, one of the citizens tries to set fire to the warehouse, Lovejoy dashes into the street, and he is shot dead November 7.

Boston abolitionist Wendell Phillips, 26, hears a proslavery speaker in a meeting at Faneuil Hall December 8 compare Lovejoy's murderers (above) to the patriots of the Boston Tea Party of 1773. Phillips gave up his law career 2 years ago after seeing a Boston mob drag abolitionist William Lloyd Garrison half-naked through the streets at the end of a rope, he has never spoken in public before, but he rises in fury to denounce the previous speaker, points to portraits of John Adams and John Hancock on the wall, and says, "I thought those pictured lips would have broken into voice, to rebuke the recreant American, the slanderer of the dead," and his eloquence sways the meeting.

Economic depression begins in the United States following the failure in March of the New Orleans cotton brokerage Herman Briggs & Co. Inflated land values, speculation, and wildcat banking have contributed to the crisis, New York banks suspend specie payments May 10, financial panic ensues, at least 800 banks suspend specie payment, 618 banks fail before the year is out (many have deceived bank inspectors as to the amount of gold backing their banknotes), specie disappears from circulation, employers pay workers in paper "shinplasters" of dubious value and often counterfeit, 39,000 Americans go bankrupt, $741 million is lost, and the depression reduces thousands to starvation (*see* 1839).

The depression (above) forces every New England textile mill but one to close down. Nathaniel Stevens at North Andover expands production and works 76-hour weeks, paying above average wages of $4.50 per week plus $2 for board.

Economic depression begins in Britain.

Congress authorizes the issue of U.S. Treasury notes not to exceed $10 million October 12 in a move to ease the nation's financial crisis.

Frankfurt banker August Belmont, 20, hears of the financial panic in New York (above) while en route to Havana to transact business for the younger Meyer Rothschild (*see* 1801). Belmont completes his Cuban business, sends in his resignation to Frankfurt but offers to act as agent for the Rothschilds, sails for New York, rents a small office in Wall Street, and although virtually without capital begins a business that will grow into a powerful banking house.

Completion of the Pennsylvania Canal spurs development of Pittsburgh.

The New York and Harlem Railroad reaches Harlem with its horsecars and develops the world's first steam tram (*see* 1834; 1839).

London's Euston Station opens July 20 with an enormous Doric arch to serve as the city's main-line rail terminal (*see* Waterloo Station, 1848).

Persian cuneiform inscriptions deciphered by German archaeologist Georg Friedrich Grotefend, 62, open the way to a knowledge of life among early peoples of the Near East. The inscriptions date to 3,000 B.C. (*see* Layard, 1845; Rawlinson, 1846).

French physiologist René Joachim Henri Dutrochet, 61, establishes the essential importance of chlorophyll to photosynthesis (*see* Ingenhousz, 1779).

Recherches sur la Probabilité des Jugements by French mathematician Siméon Denis Poisson, 67, establishes rules of probability based on the incidence of death from mule kicks in the French army.

A smallpox epidemic along the Missouri River kills 15,000 Indians, nearly wiping out the Arikara, Hidatsa, and Mandan.

The world's first kindergarten opens at Blankenburg, Thuringia, under the direction of German educator Friedrich Fröbel, 55 (*see* 1855; Peabody, 1860).

Third and Fourth McGuffey *Readers* are published (*see* 1836).

The University of Louisville is founded in Kentucky.

Depauw University has its beginnings in the Indiana Ashway University founded at Greencastle, Ind. The school will adopt the name Depauw in 1883.

Mount Holyoke Female Seminary, opened November 8 at South Hadley, Mass., is the first U.S. college for women. Local educator Mary Lyon, 40, proposes to give the best possible education to women of modest means and is so successful with her 80 students that she will have to turn away 400 applicants next year for lack of space.

The University of Athens is founded in Greece.

English physicist Charles Wheatstone, 35, and electrical engineer William Fothergill Cooke patent an electric telegraph (*see* 1845).

Samuel F. B. Morse gives a public demonstration of his magnetic telegraph and is granted a U.S. patent September 23 (*see* 1832). His assistant Alfred Lewis Vail, 30, devises a Morse code using dots and dashes to represent letters in place of an earlier system that assigned numbers to letters (*see* 1843; 1844).

"Post Office Reform: Its Importance and Practicality" by English schoolmaster Rowland Hill, 42, urges that Britain replace the system of charging for postage on the basis of distance and number of sheets with a uniform rate based on weight in which the sender pays in advance by buying "labels" to be affixed to each letter with "cement." Hill submits his pamphlet to the prime minister William Lamb, second Viscount Melbourne (*see* 1635; 1840).

The Pitman shorthand system devised by Englishman Isaac Pitman, 24, is the first scientific shorthand system. It is based on phonetics and employs lines, curves, and hooks with contractions and "grammalogues" for frequently occurring words (*see* Gregg, 1888).

The *New Orleans Picayune* begins publication with George Wilkins Kendall, 18, as editor. The Spanish picayune coin still in circulation is worth 6.25 cents.

The *Baltimore Sun* begins publication May 17. Founded by local printer Arunah S. Abell, it sells at one cent per copy.

Nonfiction *The French Revolution* by Thomas Carlyle whose manuscript was partly burned by the manservant of John Stuart Mill, 31, chief of relations with native states for the East India Company; *History of the Reign of Ferdinand and Isabella the Catholic* by Boston historian William Hickling Prescott, 41, who is completely blind as the result of an accident in his freshman year at Harvard when he was struck in the eye by a crust of bread thrown by another student in the commons dining room. The three-volume work establishes Prescott's reputation, and he works with assistants on a history of Cortez.

"The American Scholar," delivered as a Harvard Phi Beta Kappa address by Ralph Waldo Emerson, calls for intellectual independence from the past, from Europe, and from all obstacles to originality.

Harvard graduate Henry David Thoreau, 19, delivers a commencement address that enlarges on Emerson's 1836 essay on "Nature," saying that man should not work for 6 days and rest on the seventh but rather should work one day and leave six free for the "sublime revelations of nature."

Fiction *The Posthumous Papers of the Pickwick Club* by Charles Dickens: "Accidents will happen best regulated families"; "It's over, and can't be helped, and that's one consolation, as they always say in Turkey, ven they cuts the wrong man's head off"; "I think he's a wictim o' connubiality, as Blue Beard's chaplain said, with a tear of pity, ven he buried him"; *The Confessions of Harry Lorrequer* by Irish novelist Charles James Lever, 31; *Le Curé de Village* by Honoré de Balzac; *Twice-Told Tales* by U.S. story-writer Nathaniel Hawthorne, 33.

Poetry "Hymn, Sung at the Completion of the Concord Monument" by Ralph Waldo Emerson; "By the rude bridge that arched the flood,/Their flag to April's breeze unfurled,/Here once the embattled farmers stood,/And fired the shot heard round the world" (*see* 1775); Russian poet Aleksandr Pushkin is mortally wounded in a duel January 29 with an officer in the Horse Guards and dies February 10 at age 36; "A Song about Czar Ivan Vasilyevich, His Young Bodyguard, and the Valiant Merchant Kalashnikov" by Russian poet Mikhail Yuryevich Lermontov, 22, who is transferred to a regiment in the Caucasus after writing an anti-court poem following the death of Pushkin (above).

Baron François Gérard dies at Paris January 11 at age 66; John Constable dies at London March 31 at 60.

Theater *The Clique, or the Helping Hand (La camaraderie, ou la Courte Echelle)* by Eugène Scribe 6/19 at the Théâtre Français, Paris.

First Performance Requiem by Hector Berlioz 12/5 at the Church of Les Invalides, Paris.

The Knabe Piano is introduced by German-American piano maker William Knabe who goes into partnership with H. Gahele at Baltimore to build "pianos of quality for genteel people of means" (*see* Chickering, 1823; Steinway, 1859).

The Bayswater hippodrome opened June 3 in England is the world's first steeplechase racecourse.

1837 ***(cont.)*** France outlaws baccarat. The card game will not be legalized until 1907.

Procter & Gamble is founded at Cincinnati by English-American candle maker William Procter, 35, and his Irish-American soap-boiler brother-in-law James Gamble, 34, who peddle candles in the street and to the Ohio River trade. They gross $50,000 their first year despite competition from local soap and candle factories and from housewives who make their own soap from grease, fat, and clean wood ashes boiled in backyard kettles (*see* Ivory Soap, 1878).

Tiffany & Co. has its beginnings in a New York "Stationery and Fancy Goods Store" that opens September 18 near City Hall. Merchant Charles Lewis Tiffany, 25, has borrowed $1,000 from his father in Connecticut, he and his partner John B. Young stock Chinese bric-a-brac, pottery, and umbrellas as well as stationery; total receipts for the first 3 days are $4.98, profits the first week total 33¢, but the firm will grow to be one of the world's most prestigious jewelry retailers. The store will move up to 271 Broadway in 1847, Tiffany will start manufacturing his own jewelry in 1848 and add gold jewelry that year, open a Paris branch in 1850, adopt the firm name Tiffany & Co. in 1853, and open a London branch in 1868 (*see* Preakness Woodlawn Vase, 1873; Tiffany Diamond, Louis Comfort Tiffany, 1878).

Buildings and paved streets occupy only one-sixth of Manhattan; the rest remains planted in farms and gardens.

London completes the Thames Embankment.

London's Buckingham Palace will be the royal residence of the new queen, Victoria (above), and her successors. Designed by the late John Nash, it was completed last year in a Palladian-style reconstruction of a mansion built in 1704 for the duke of Buckingham, but the late William IV disliked the palace (*see* 1856).

A self-polishing steel plow fashioned by Vermont-born blacksmith John Deere, 32, at Grand Detour, Ill., can break the heavy sod of the Illinois and Iowa prairie. Deere chisels the teeth off a discarded circular saw blade of Sheffield steel, creates a plow with the proper moldboard curve for breaking the sod, and saves farmers from having to pull their plows out of furrows for repeated cleaning with wooden paddles. The Deere plow will permit efficient farming in vast areas that have defied earlier efforts (*see* 1839).

The first steam-powered threshing machine, patented by Winthrop, Maine, inventors John A. Pitts and Hiram Abial Pitts, separates grain from its straw and chaff with far less effort than was heretofore required (*see* Case, 1843).

The shakeout of land speculators in the economic depression (above) makes more U.S. farmland available for real farmers.

Famine strikes Japan but the Tokugawa shogun Ieyushi refuses to open storehouses at Osaka to the starving people. A riot led by the former priest Heihachiro Oshio ends with 40 percent of Osaka in ashes, the riots spread throughout Japan, but the shogun resists appeals to ease tax burdens on farmers who have little or no cash crop (*see* 1833).

Charles Dickens describes widespread hunger among Britain's urban poor in early installments of his novel *Oliver Twist.* "Please, sir, I want some more," says young Oliver, as the Industrial Revolution's dislocation of society creates problems in English agriculture.

One-third of all New Yorkers who subsist by manual labor are unemployed in the financial crisis (above) and at least 10,000 are made dependent on almshouses, which are unable to prevent many from starving to death.

Scottish chemist Andrew Ure, 59, dismisses the idea that Britain's factory children may have rickets for lack of sunlight. Gaslight is more progressive and quite as healthy, says Ure (*see* Gas Light and Coke, 1819; Steenbock, 1923).

Some 1,200 U.S. automated flour mills produce 2 million bushels of flour in the area west of the Alleghenies, with more millions being milled by millers on the eastern seaboard who have been licensed to use the 1790 Oliver Evans patent or who infringe on it.

Society in America by English novelist-economist Harriet Martineau, 35, says of corn on the cob, "The greatest drawback is the way in which it is necessary to eat it. . . It looks awkward enough: but what is to be done? Surrendering such a vegetable from considerations of grace is not to be thought of" (Martineau visited the United States from 1834 to 1836).

1838 Central America's federation breaks up after 14 years following a cholera epidemic as *mestizo* Rafael Carrera leads an Indian uprising in Guatemala. Honduras secedes from the federation (*see* 1839).

British troops invade Afghanistan and advance to Kandahar. The governor-general of India George Eden, 54, earl of Auckland, has sent in the troops out of exaggerated fears of Russian influence. They move on to Ghazni and Kabul, depose the Barakzai emir Dost Mohammed who took power in 1834, and take him to India as a prisoner (*see* 1839).

Chinese officials trying to execute an opium dealer in front of the foreign factories at Guangzhou (Canton) December 12 meet resistance from British and American opum traders, who chase them away, provoking a riot. An imperial commissioner is appointed to investigate (*see* 1839).

The Battle of Blood River December 16 in Natal gives Afrikaner forces a victory over Zulu natives. The Boers kill more than 3,000 Zulus and eliminate the main threat to their settlements.

A "People's Charter" published in London May 8 calls for universal suffrage without property qualifications. William Lovett has founded the London Workingmen's Association to mobilize Britons disappointed by the Reform Bill of 1832. The charter he has drafted contains Six Points which include payment of Members of Parliament, equal electoral areas, and other reforms, and it wins approval August 8 at a great meeting held on Newhall Hill 6 weeks after the coronation of Queen Victoria.

Britain applies the Poor Law of 1834 to Ireland and adds to the hardship of the country's famine (below). Designed to discourage paupers from seeking relief, the Poor Law makes work inside the workhouse worse than the most unpleasant sort of work to be found on the outside, and it has the effect of stimulating emigration.

The "underground railway" organized by U.S. abolitionists transports southern slaves to freedom in Canada, but slaving interests at Philadelphia work on the fears of Irish immigrants and other working people who worry that freed slaves may take their jobs. A Philadelphia mob burns down Pennsylvania Hall May 17 in an effort to thwart antislavery meetings.

The Trail of Tears takes more than 14,000 members of the Cherokee Nation from tribal lands in Georgia, Alabama, and Tennessee 800 miles westward along the Tennessee, Ohio, Mississippi, and Arkansas Rivers to Little Rock and thence to Indian territory west of the Red River. Escorted with their horses and oxen by troops under the command of Gen. Winfield Scott, the Cherokee journey by wagon and keelboat for anywhere from 93 to 139 days, an estimated 4,000—mostly infants, children, and old people—die en route of measles, whooping cough, pneumonia, pleurisy, tuberculosis, and pellagra, and the last contingents will not reach Indian territory until March 25 of next year (some have refused to budge) (*see* 1831; 1832).

St. Paul, Minn., has its beginnings in a settlement made by French-Canadian trapper Pierre Parrant who settles in the upper Mississippi Valley at Fort Snelling, formerly Fort Anthony (*see* 1819). Parrant and other squatters will be removed from the reservation in 1840, and missionary Lucian Galtier will name the neighboring squatter settlement St. Paul in 1841 (*see* Minneapolis, 1847).

Kansas City is founded in November on a hill overlooking a bend in the Missouri River by Missouri pioneers who include John McCoy, William Sublette, William Chick, and William Gillis. Their settlement is near Westport Landing, founded earlier as a post for the Indian fur trade and an outfitting station for wagons to Santa Fe and Oregon Territory.

Lancashire calico printer Richard Cobden, 34, at Manchester founds an Anti-Corn Law League to oppose British protectionism (*see* 1839).

Houston's Rice dry goods store opens in the Texas Republic. Massachusetts-born merchant William Marsh Rice, 22, will develop his enterprise into a large export-import and retail business (*see* Rice Institute, 1912).

Two British steamers arrive at New York April 23 after the first transatlantic crossings by ships powered entirely by steam. The 703-ton S.S. *Sirius,* 19 days out of London, has been used until now only for cross-channel service while the 1,440-ton S.S. *Great Western*, 15 days out of Bristol, is larger and faster. English engineer Isambard Kingdom Brunel, 32, has designed both (*see* surface condenser, 1830; Cunard, 1840).

The S.S. *Great Western* (above) has been named for the Great Western Railway which I. K. Brunel has been serving as chief engineer, and it is the first wooden steamship to cross the Atlantic. Brunel's father is at work on the Thames tunnel from Wapping to Rotherhithe and the younger Brunel has introduced broad gauge tracks and constructed the Great Western's viaducts, tunnels, and bridges, including the Royal Albert Bridge across the Tamar River into Cornwall.

The first Russian railway opens to link St. Petersburg with the czar's summer palace at Tsarskoe Selo.

Alfried Krupp, 26, visits Sheffield, England, and meets J. A. Henckel, founder of a steel mill in the Ruhr valley town of Solingen. Krupp is a son of Friedrich whose Fried. Krupp of Essen is turning out tableware on a hand mill devised by Alfried's brother Hermann (*see* 1816; 1851).

Cheney Brothers Silk Manufacturing Co. has its beginnings January 2 at South Manchester, Conn., where entrepreneurs Ralph, 31, Ward, 24, Rush, 22, and Frank Cheney, 20, join with E. H. Arnold to put up $50,000 to start the Mount Nebo Silk Mill. The factory opened March 31 is powered by a mill-race wheel and will be the first commercially successful U.S. silk mill. The monthly *Silk Growers' Manual* published by the firm beginning in July will grow to have a circulation of 10,000 as Americans rush to cultivate the fast-growing Chinese mulberry tree introduced at Baltimore in 1826, but a blight will strike the trees in 1844, forcing the industry to depend on imported raw silk.

German botanist Matthias Jakob Schleiden, 34, at Jena formulates the cell theory of physiology (*see* 1858).

German botanist Hugo von Mohl, 33, uses the word *protoplasm* for the first time to denote the substance of a cell body as differentiated from the cell nucleus (*see* Purkinje, 1840).

French physician Charles Cagniard de la Tour, 61, shows that fermentation is dependent on yeast cells (*see* Pasteur, 1857).

Crossing the Atlantic by steam was a triumph of engineering, but sailing ships often made better time.

1838 ***(cont.)*** Nonfiction *Casanova's Memoirs Ecrits par Lui Même*, twelfth and final volume. The first appeared in 1826, some 28 years after the roué's death at age 73 (*see* 1755).

Fiction *Oliver Twist, or the Parish Boy's Progress* by Charles Dickens who has serialized the work in his *Bentley's Miscellany*. Its sentimental story of Fagin, Bill Sykes, the Artful Dodger, and young Oliver is intended in part as an exposé of Poor Law evils. Illustrations are by George Cruikshank; "Ligeia" by Edgar Allan Poe, whose melodramatic short story appears in the September Baltimore American *Museum* (Poe married his cousin Virginia, now 15, 2 years ago); *The Narrative of Arthur Gordon Pym* by Edgar Allan Poe; *Münchhausen* by German novelist Karl Leberecht Immerman, 42, introduces the mendacious baron who personifies the attitudes Immerman opposes with an idealized image of patriotic peasant morality. French novelist George Sand begins a 9-year liaison with composer Frédéric Chopin 4 years after her estrangement from poet-playwright Alfred de Musset who has taken to drink after failures to achieve reconciliation (he will never recover but despair has inspired his most creative work). Chopin at 28 is, like Musset, 6 years younger than Sand, who continues to have her novels published while scandalizing Europe with her masculine attire and cigar smoking.

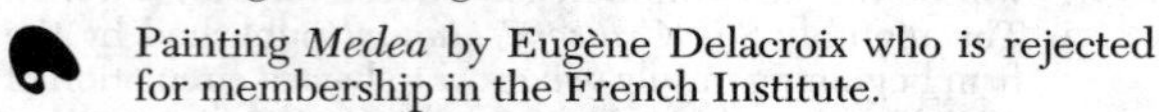

Painting *Medea* by Eugène Delacroix who is rejected for membership in the French Institute.

Theater *Thou Shalt Not Lie* (*Weh dem, der lügt*) by Franz Grillparzer 3/6 at Vienna's Burgtheater (it is whistled off the stage and then suppressed by authorities in a strict censorship that will not be relaxed until 1848); *Ruy Blas* by Victor Hugo 11/8 at the Théâtre de la Renaissance, Paris.

Opera *Benvenuto Cellini* 9/10 at the Paris Opéra, with music by Hector Berlioz.

Popular songs "Vive la Compagnie" is published at Leipzig as a French drinking song; "Annie Laurie" by Scottish composer Alicia Ann Spottiswoode, 28, Lady John Scott, lyrics from the 1700 poem by William Douglas; "Flow Gently, Sweet Afton" by Philadelphia composer James E. Spilman, lyrics from the 1792 Robert Burns poem "Afton Water."

Regent's Park opens to the London public on 410 acres of pastureland known as Marylebone Park Fields until the park was laid out in 1812.

Famine kills thousands in the north of Ireland as crops fail (*see* 1845).

Dutch chemist Gerard Johann Mulder, 36, coins the word "protein," adapting a Greek word meaning "of the first importance."

Iron in the blood is what enables the blood to absorb so much oxygen, concludes Jöns Jakob Berzelius, pioneering an understanding of hemoglobin.

London's Reform Club installs gas ovens (*see* Sharp, 1826). Coal or wood is the common cooking fuel in most of the world, but Arab nomads use camel chips, American Indians buffalo chips, and Eskimos blubber oil.

Fresh menus are printed daily at New York's 2-year-old Astor House, writes Captain Frederick Marryat (*see* 1837; Delmonico's, 1836).

1839 *Report on the Affairs of British North America* by the first earl of Durham John George Lambton, 48, proposes the union of Upper and Lower Canada and the granting of self government. Lord Durham placated the rebels 2 years ago as governor-general, the report published February 11 attempts to justify the position he took before his resignation last year, and Lord John Russell introduces a resolution in Parliament June 20 to implement Durham's recommendations (*see* 1840).

Willem I of the Netherlands recognizes Belgian independence April 19, letting the Belgians have the western part of Luxembourg, keeping the rest as a grand duchy with himself as grand duke.

Ottoman forces invade Syria in April to contest the claims of the Egyptian viceroy Mohammed Ali, 70 (*see* 1805). His general Ibrahim Pasha defeats the sultan's forces June 24 at the Battle of Nezib.

The Ottoman sultan Mahmud II, now 54, is poisoned to death July 1 at his summer residence in Scutari as his fleet surrenders without a shot at Alexandria. Greece has won independence in Mahmud's bloody 31-year reign. His son of 16, who will reign until 1861 as Abdul Mejid I, continues the war against the Egyptian viceroy.

The convention of Vergara August 31 ends Spain's Carlist War after defeats of Carlist forces in 1836 and 1837. Don Carlos emigrates to France, leaving his niece Isabella II to complete a reign that will continue until 1868.

The Serbian prince Milos Obrenovic abdicates under pressure from opponents who have conspired with the Russians to curb the autocratic powers exercised by Milos since 1817. He is succeeded by his son Milan, 20, who rules until his untimely death late in the year, and then by his son of 14 who will reign until 1842 as Michael III Obrenovic.

The British-Afghan War that began last year continues as a 5,000-man British army escorts the unpopular Shah Shuja, who reigned from 1803 to 1810, to Kabul and restores him to power in place of Dost Mohammed. Agitation rises against British influence, and Afghan patriots work to restore Dost Mohammed.

An Opium War between China and Britain begins in November after Chinese official Lin Zexu, 54, orders the destruction of 20,000 chests of illegal Indian opium stored by foreign merchants in Guangzhou (Canton) warehouses. Western physicians make wide use of opium but the East India Company has encouraged addiction to the poppy derivative in order to keep its Indian and Chinese workers subdued (*see* 1840; Hong Kong, 1842).

Denmark's Frederick VI dies December 3 at age 71 after a 31-year reign (plus 24 years as regent) in which Norway has been lost to Sweden (*see* 1814). Frederick is succeeded by his nephew, 53, who will reign until 1848 as Christian VIII.

"American Slavery as It Is" by Ohio evangelist-abolitionist Theodore Dwight Weld, 35, is a tract favoring emancipation.

Sen. Henry Clay of Kentucky delivers a conciliatory address against militant abolitionism. Consulting others before making the speech, Clay is told that it may offend ultras of both parties, but he replies, "I trust the sentiments and opinions are correct; I had rather be right than be president." (He has been trying for 15 years to gain the presidency.)

"To each according to his needs, from each according to his abilities," writes French socialist leader (Jean Joseph Charles) Louis Blanc, 28, in his essay "L'Organisation du Travail." Journalist Blanc founds *Revue du Progrès* to promote his socialist doctrines.

Britain's Anti-Corn Law League gains power by amalgamating opposition groups (*see* 1838). Cobden's league attracts workers to its cause by promising that a removal of tariffs on foreign grain will reduce food prices, give Britons more money to buy clothing, and thus create more domestic demand and more jobs in the textile industry. Promoters of the league know that lowering food prices will enable them to lower wages (*see* 1841; 1846).

Economic depression in the United States in the next few years will force Maryland and Pennsylvania to default in their interest payments to British and European bondholders.

Steam engines are introduced on the New York and Harlem Railroad (*see* 1837; van Derbilt, 1863).

The Chesapeake and Ohio Canal started in 1828 reaches 134 miles west of Georgetown but runs into financial difficulties (*see* 1850).

Italy gets her first railway as an 8-kilometer line opens between Naples and Portici (*see* 1859).

The first real bicycle is invented at Dumfries, Scotland, by local blacksmith Kirkpatrick MacMillan, 29, of Courthill who adds a pedal system and a brake to the frame and wheels designed by Sauerbroun and Niepce in 1816. MacMillan's 57-pound bike with its 32-inch iron-tired front wheel and 42-inch rear wheel makes it possible for the first time for a person to travel under his own power faster than he can run (*see* Michaux, 1861).

Charles Goodyear, 39, pioneers effective use of rubber. The former Philadelphia hardware merchant has obtained rights to a sulfur process for treating rubber, has accidentally overheated a mixture of rubber, sulfur, and white lead, and stumbles on a way to "vulcanize" rubber to make a hard and durable substance that chars but does not melt. Goodyear has sold the patent on his father Amasa Goodyear's pitchfork (the first steel-tine pitchfork) to finance his experiments with raw rubber, which has been used to some extent earlier as in Macintosh raincoats (*see* 1823). Rubber's tendency to become sticky in hot weather and lose its elasticity in cold weather has limited its commercial potential; Goodyear's vulcanization process does not solve all the problems but will extend rubber's uses (*see* 1844).

Massachusetts inventor Erastus Brigham Bigelow, 25, devises a power loom to weave two-ply ingrain carpets (*see* 1845; Horrocks, 1813).

Massachusetts inventor Isaac Babbitt, 40, patents a journal box, or housing, for the portion of a shaft or axle contained by a plain bearing. He suggests that the box be lined with an alloy of tin, antimony, and copper that will be called Babbitt's metal.

Scottish engineer James Nasmyth, 31, invents a steam hammer that will make possible larger forgings without loss of precision; he will patent the invention in 1842.

Microscopic Investigations on the Accordance in the Structure and Growth of Plants and Animals by German naturalists Theodor Schwann and Matthias Jakob Schleiden is an exegesis of the cell theory (*see* 1835; 1838).

German-Swiss chemist Christian F. Schönbein, 40, discovers and names ozone.

Swedish chemist Carl Gustav Mosander, 42, isolates lanthanum, a new metallic element.

Nauvoo, Ill., gets its name and becomes the largest city in the state as some 10,000 Mormons settle at a town on the Mississippi formerly called Commerce. The Mormons have been driven from Missouri into Illinois (*see* 1830; 1844).

New England clergyman John Humphrey Noyes, 28, establishes a bible class at Putney, Vt., where he will soon urge the establishment of a communal society (*see* Putney Community, 1846).

The first state-supported normal school (teacher's college) is founded at Lexington, Mass., through the efforts of Board of Education Secretary Horace Mann, 33 (*see* 1852).

Boston University is founded.

The University of Missouri is founded.

Virginia Military Institute is founded at Lexington.

Boston's Lowell Institute is founded by John Lowell, Jr., 70, to provide free lectures by eminent scholars.

Fiction *The Charterhouse of Parma (La Chartreuse de Parme)* by Stendhal who has written the novel of passion and political adventure in 7 weeks after 7 years of thought; *Nicholas Nickleby* by Charles Dickens who has serialized the novel for 20 months, introducing readers to the bad education administered by the exploitive Mr. Squeers of Dotheboys Hall, an institution derived from the author's childhood experience; "The Fall of the House of Usher" by Edgar Allan Poe whose melodramatic tale of an intellectual confronted with the death of a beautiful woman appears in the September *Burton's Gentleman's* magazine.

Poetry *Voices of the Night* by Harvard language professor Henry Wadsworth Longfellow, 32. His poem "A Psalm of Life" begins, "Tell me not, in mournful num-

1839 ***(cont.)*** bers,/ Life is but an empty dream!/ For the soul is dead that slumbers,/And things are not what they seem./ Life is real! Life is earnest!/ And the grave is not its goal. . ./ Let us, then, be up and doing,/ With a heart for any fate. . ."

Painting *Suzanne au Bain* and *Vénus Anadymone* by French painter Théodore Chassériau, 19; *Dignity and Impudence* by Edwin H. Landseer.

The Daguerrotype process of photographic reproduction discovered by Louis Daguerre in 1829 is described to a meeting of the French Academy of Arts and Sciences August 19 but English inventor William Henry Fox Talbott, 39, claims that he produced photographs earlier than Daguerre using a camera obscura with silver chloride and common salt or potassium iodide as fixing agents. Talbot soon afterwards reveals a Calotype process involving use of silver iodide with sodium thiosulfate used to fix the image which is a reverse image or "negative" (*see* Herschel, 1840).

Telegraph pioneer Samuel F. B. Morse makes the first Daguerrotype portraits to be produced in America. He returns from a visit to Paris with the process he has learned from Daguerre and teams up with English-American physician-scientist John William Draper, 28, who will make important contributions to photography, photochemistry, radiant energy, and the electric telegraph (*see* 1837; telegraph line, 1843).

Opera *Oberto Conte di San Bonifacio* 11/17 at Milan's Teatro alla Scala, with music by Italian composer Giuseppe Verdi, 26.

First performances *Romeo and Juliet* Symphony by Hector Berlioz 11/24 at the Paris Conservatoire.

England's Grand National Steeplechase has its first running at Aintree near Liverpool.

Rugby rules are devised by Cambridge University student Arthur Fell (*see* 1823).

Baseball rules are devised by West Point cadet Abner Doubleday, 19, of Cooperstown, N.Y. (legendary date). His dicta call for a diamond-shaped field and two teams of nine players each. Jane Austen mentioned the game in her novel *Northanger Abbey*, Oliver Wendell Holmes played it at Harvard before his graduation in 1829, and it is only distantly related to cricket (*see* 1788; 1846).

Kingscote at Newport, R.I., is completed by English-American architect Richard Upjohn, 37, for George Noble Jones of Savannah. William Henry King will acquire the house on Bellevue Avenue with money acquired in the tea trade, and it will be followed by more imposing "cottages" as Newport becomes a favorite summer resort for the rich.

The heath hen *Tympanuchus cupido cupido* once common throughout New England is now found only on Martha's Vineyard, Mass. A game bird closely related to the prairie chicken, it was once an important food source, but market hunters have reduced its numbers (*see* 1907).

Sugar production in Jamaica, British West Indies, will fall in the next two decades to between 20,000 and 25,000 tons per year, down from 70,000 tons in 1821, as a consequence of the end of slavery.

John Deere produces 10 steel plows (*see* 1837; 1842).

Congress appropriates $1,000 for the first free distribution of seeds to U.S. farmers.

Lectures on the Science of Life by Sylvester Graham urges dietary measures for good health in two volumes. Graham blames American dyspepsia and sallow complexion on fried meat, alcohol, eating too fast, and the use of "unnatural," refined wheat flour. Espousing vegetarianism, he urges readers to eat fruits, vegetables, and unbolted (unsifted) whole wheat flour in bread that is slightly stale and is to be chewed thoroughly to promote good digestion, prevent alcoholism, and diminish the sex urge.

Swiss engineer Jacob Sulzberger builds a flour mill at Budapest that employs rollers rather than millstones (*see* 1833; 1870).

Some 95 chests of Assam tea arrive at London and are sold at auction. Unlike green China tea, the leaves from India are fermented and the new black tea, less astringent than green tea, begins to gain popularity (*see* 1823; duchess of Bedford, 1840).

French chef Alexis Soyer becomes chef of London's Reform Club. Soyer fled France in the 1830 revolution and has served some of England's great houses (*see* 1847).

1840 Queen Victoria marries her first cousin Albert February 10 wearing only items of British manufacture. Her husband, 3 months her junior, is a son of the duke of Saxe-Coburg-Gotha and a nephew of the king of the Belgians Leopold I.

Prussia's Friedrich Wilhelm III dies June 7 at age 69 after a reign of more than 42 years. He is succeeded by his son of 44 who will reign until 1861 as Friedrich Wilhelm IV.

Britain declares war on China in June and British ships bombard an island at the entrance to Hanchow Bay July 5, landing troops to occupy the island of Zhoushan in the Opium War that will continue until 1842.

Maori natives in New Zealand cede sovereignty (but not land) to Britain in the Treaty of Waitangi, but settlers will take Maori land and Maoris will retaliate with attacks on their towns.

The Treaty of London July 15 joins Britain, Austria, Prussia, and Russia against the Egyptian viceroy Mohammed Ali. The Allies offer the pasha hereditary rule of Egypt and lifetime possession of Syria on condition that he give up northern Syria, Mecca, Medina, and Crete and that he return the Ottoman fleet. He rejects the offer, the British induce the boy sultan Abdul Mejid to depose the viceroy, Paris panics at news of the treaty, the minister of foreign affairs Louis Adolphe Thiers urges support of Mohammed Ali, and French newspapers echo his views.

British gunboats commanded by Sir Robert Stopford bombard Beirut September 9, and the British land troops under the command of Sir Charles James Napier, 57, a veteran of the War of 1812, the peninsular war against Napoleon, and the war of Greek independence.

Beirut falls to Gen. Napier (above) October 10, France's Louis Philippe decides not to support Mohammed Ali, Thiers resigns October 20, Acre falls to the British November 3, and Napier concludes the Convention of Alexandria November 27. Mohammed Ali agrees to return the Ottoman fleet and give up all claims to Syria in return for hereditary rule of Egypt.

Willem I of the Netherlands abdicates October 7 at age 68 after a 26-year reign in order to marry the Roman Catholic Belgian countess d'Oultremont who is unpopular with the Dutch. His son of 47 will reign until 1849 as Willem II.

The Union Act passed by Parliament July 23 unites Upper and Lower Canada, adopts various administrative reforms, but avoids the issue of government responsibility in Canada. Parliament acts in response to last year's report by Lord Durham who dies July 28 at age 48 (*see* British North America Act, 1867).

Paraguay's dictator José Gaspar de Francia dies September 20 at age 74 after a 24-year rule (*see* 1841).

Rafael Carrera makes himself dictator of Guatemala, beginning a regime that will continue until 1865 and dominate much of Central America (*see* 1838).

William Henry Harrison wins election to the U.S. presidency as Whigs sing, "Oh have you heard how Old Maine went?/ She went hell-bent for Governor Kent,/ And Tippecanoe and Tyler too." The hero of the 1811 Battle of Tippecanoe wins 234 electoral votes to Martin Van Buren's 60 (but *see* 1841).

The World's Anti-Slavery Convention opens at London, but Boston abolitionist William Garrison refuses to attend, protesting the exclusion of women (*see* 1831). The U.S. antislavery movement has split into two factions in the past year largely because of Garrison's advocacy of women's rights, including their right to participate in the antislavery movement (*see* first Women's Rights Convention, 1848).

Slaves aboard the Spanish ship *Amistad* rebel while en route from one Cuban port to another, killing the captain and all but two crew members. The slave ship arrives at a Connecticut port in August with Cinque and 52 other Africans in command. Spain demands extradition (but *see* 1841).

"Property is theft" ("la propriété, c'est le vol"), writes French socialist Pierre Joseph Proudhon, 31, in his treatise "What Is Property?" ("Quest ce-que la propriété?").

The United States has nearly 20 millionaires (*see* 1892; John Law, 1719).

"On the Production of Heat by Voltaic Electricity" by English physicist James Prescott Joule, 22, describes efforts to measure electric current by exact quantitative data (*see* Volta, 1800). Joule's paper announces the law that the rate at which heat is produced in any part of an electric circuit is measured by the product of the square of the current and the resistance of that part of the circuit. He will also formulate the first law of thermodynamics—the conservation of energy law that states that energy can be converted from one form to another but cannot be destroyed. A unit of energy will be named the joule (*see* Clausius, 1850; Thomson, 1851).

The wooden steamship *Britannia* arrives at Boston on the first voyage of the government-subsidized Royal Mail Steam Packet Co. established last year by Nova Scotia-born shipper Samuel Cunard, 43, in association with George and James Burns of Glasgow and David M'Iver of Liverpool. Cunard will introduce iron steamers in 1855 and in 1862 will replace paddle wheels with screw propellers to break the Black Ball line's dominance of the transatlantic packet trade and New York's monopoly in the trade.

More than 200 steamboats ply the Mississippi, double the number in the mid-1820s. New Orleans is the fourth largest city in America and will overtake New York in the next 10 years in volume of shipping as more than half the nation's exports move out of its port.

U.S. canals cover 3,300 miles.

U.S. operating railways cover 2,816 miles, versus 1,331 in the United Kingdom, fewer than 300 miles in France.

More than 300 U.S. railroad companies are in operation; tracks vary in gauge from 6 feet to 4 feet 8.5 inches (*see* B&O, 1828).

From New York to Boston is an overnight steamer trip or a 6-hour train journey and costs $7. From New York to Philadelphia by train and ferry takes 6.5 hours, down from 3 days in 1817, but it takes longer when the Delaware River freezes over, halting train-ferry service and forcing passengers to walk across the ice.

Zoology of the Voyage of the Beagle by Charles Darwin makes no reference to the idea that struck Darwin 2 years ago when he read the 1798 Malthus *Essay on Population*. Darwin realized that living creatures in nature must compete with each other for sustenance and that nature kills off those which cannot compete, but he will not commit the notion to paper, even in pencil, for another 2 years (*see* 1741; 1831; 1858).

Czech physiologist Johannes Evangelista Purkinje, 53, at Breslau proposes that the word "protoplasm" be applied to the formative material of young animal embryos (*see* von Mohl, 1838).

Handbuch der Physiologie der Menschen by German physiologist Johannes Peter Müller, 39, revives ideas that go back to Aristotle and includes a reaffirmation of spontaneous generation (*see* Redi, 1668). But Müller does introduce some sound new ideas about nervous energy, reflex movements, gland structures, the embryonic development of sex organs, and other matters.

1840 ***(cont.)*** A new worldwide cholera epidemic begins. It will kill millions in the next 22 years (*see* 1830; London, 1849).

The Baltimore College of Dental Surgery established in Maryland is the first regular U.S. dental school. The American Society of Dental Surgeons is organized. The *American Journal of Dental Science* begins publication (*see* calcium, below).

The world's first adhesive postage stamps go on sale in Britain May 1 for use beginning May 6 (*see* Hill, 1837). The Penny Blacks bearing the head of Queen Victoria have been printed by U.S. inventor Jacob Perkins who in 1819 developed a process and the necessary equipment for printing bank notes and has established a factory in England for printing securities. 64 million of the new stamps will be printed, and twice as many letters are posted this year as in 1839, but British postal revenues fall drastically and will not regain their 1839 level until 1875 (*see* United States, 1847; perforating machine, 1854).

Fiction *Two Years Before the Mast* by Boston novelist Richard Henry Dana, 25, who sailed on a 150-day voyage aboard the brig *Pilgrim* round Cape Horn to California in 1834, returned aboard the ship *Alert* in 1836, and has had his book published anonymously; *The Pathfinder* by James Fenimore Cooper who has received wide criticism for his 1838 novels *Homeward Bound and Home as Found* and for his 1838 nonfiction work *The American Democrat*. Cooper has won most of the libel suits brought against him and has helped establish effective rules of libel in U.S. courts; *Tales of the Grotesque and Arabesque* by Edgar Allan Poe; *A Hero of Our Time* (*Geroi Nashevo Vremeni*) by Mikhail Lermontov; *Charles O'Malley* by Charles Lever.

The Penny Black bearing Queen Victoria's profile—the world's first adhesive postage stamp.

Poetry *The Novice (Mtsyri)* by Mikhail Lermontov who is banished to the Caucasus once again after a duel with the son of the French ambassador; "Sordello" by Robert Browning.

Painting *Flight into Egypt* by Jean Baptiste Camille Corot; *Entrance of the Crusaders into Constantinople* by Eugène Delacroix; *Christ au Jardin des Oliviers* by Théodore Chassériau.

Astronomer F.W. Herschel discovers the solvent power of sodium hyposulfite (hypo) on silver salts and uses the word "negative" to describe the reverse image obtained last year by Fox Talbot. Talbot has pioneered the true photographic development process by waxing or oiling paper on the side opposite to the image, thus obtaining true copies (positives) of any negative by contact printing on another piece of sensitized paper, but his Calotypes take a long time and require specially constructed paper (*see* Daguerrotype, 1839; Archer, 1851).

Theater *The Shoemaker and the King* (*El Zapatero y el Rey*), part I, by Spanish playwright José Zorilla y Moral, 22, 3/14 at Madrid's Teatro El Principe; *Judith* by German playwright Christian Friedrich Hebbel, 27, 7/6 at Berlin's Hoftheater; *The Glass of Water, or Causes and Effects* (*Le Verre d'Eau, ou les Effets et les Causes*) by Eugène Scribe 11/17 at the Théâtre Française, Paris.

Opera *The Daughter of the Regiment* (*La Figlia del Reggimento*) 2/11 at the Opéra-Comique, Paris, is presented under the title *La Fille du Régiment* with music by Gaetano Donizetti; *The Favorite* (*La Favorita*) 12/2 at the Paris Grand Opéra, with music by Donizetti.

Belgian musical instrument maker Antoine Joseph Sax, 26, "Adolphe," invents the brass saxophone.

Popular songs "The Two Grenadiers" ("Die Grenadiere") by German composer Robert Schumann, 30, lyrics by Heinrich Heine; "Rocked in the Cradle of the Deep" by English composer Joseph P. Knight, 28, lyrics by Emma Willard who retired 2 years ago from her Emma Willard School.

The polka is introduced to the United States by Viennese ballet dancer Fanny Eissler, 30, who introduced the dance to the Paris stage 6 years ago and who begins a U.S. tour that will add to her fortune.

Oyster canning begins at Baltimore using bivalves from Chesapeake Bay.

France sends a naval vessel into Arcachon Bay to guard its oysters from draggers and rakers (*see* 1759; Coste, 1853).

Manufacture of the McCormick reaper begins with improvements added by McCormick to his original 1831 machine (*see* 1834; 1847).

A U.S. farmer requires 233 man-hours to produce 100 bushels of wheat using primitive plow and harvesting cradle, down from 300 hours in 1831.

Grapefruit trees are introduced into Florida by Don Philippe, a Spanish nobleman (*see* Lunan, 1814; sweeter grapefruit, 1900).

The first important U.S. breeding herd of quick-fattening Hereford cattle is established with stock imported from Britain by Erasmus Corning and William H. Sotham at Albany, N.Y. (*see* 1817).

Vermont and New Hampshire have some 2 million sheep, and their farmers are engaged in a profitable wool-producing industry (*see* clothing makers, 1862).

Swiss chemist Charles J. Choss demonstrates the need of calcium for proper bone development (*see* Humphry Davy, 1808; vitamin D, 1922).

Afternoon tea is introduced by Anna, the duchess of Bedford. The tea interval will become a lasting British tradition, but the English still drink more coffee than tea (*see* 1839; 1850; 1869).

Some 207,000 Irish and 76,000 English emigrants leave for the United States.

New York is a city of more than 312,000, up from fewer than 124,000 in 1820, but the U.S. population remains 90 percent rural.

1841 Chinese negotiators agree January 20 after extensive talks to pay indemnities to the British and cede the island of Hong Kong in the continuing Opium War (*see* 1840). Seamen from the survey ship *H.M.S. Sulphur* come ashore on the barren 30-square-mile island January 26 (its population of 5,000 consists mostly of fishermen and pirates), but both sides reject the January 20 draft convention. British gunboats resume attacks on Qing coastal fortifications north to Shanghai (*see* 1842).

William Henry Harrison dies of pneumonia April 4 at age 68 just 1 month after taking office as U.S. president. Vice president John Tyler, 51, moves into the White House.

The Egyptian Viceroy Mohammed Ali agrees to give up Crete as well as Syria, pay tribute to the sultan, and reduce his army to 18,000 men (*see* 1840). Now 72, he will reign until he loses his mind in 1848.

The Straits Convention signed by the five great powers July 13 guarantees Ottoman sovereignty, closes the Bosphorus and Dardanelles to all foreign warships in time of peace, and marks the return of France as an international power.

Britain's Whig prime minister Viscount Melbourne resigns August 23 after a six-term ministry. The Tories regain power in a new government headed by Sir Robert Peel.

London appoints the first British consul general to Zanzibar.

New Zealand is made a British colony (*see* Cook, 1769; Maori War, 1843).

British soldier Sir James Brooke, 38, is confirmed as rajah of Sarawak by the sultan of Borneo and will rule until his death in 1868. Sir James helped the sultan's uncle Muda Hassim suppress a rebellion by several Dyak headhunter tribes in the province 3 years ago. A veteran of the East India Company's army who sustained a serious wound in the Burmese War 16 years ago, he succeeded to a large property when his father (a civil servant) died, equipped the schooner *Royalist*, and set forth to rescue the Malay Archipelago from barbarism. He will reform the Sarawak government, will be charged in the House of Commons in 1850 with having received "head money" and otherwise abusing his power, and will return to England to defend himself; a royal commission at Singapore will rule the charges "not proven," and Brooke's nephew Charles Arthur Johnson will join him in 1852.

Afghans assassinate the puppet ruler Shah Shuja at Kabul in December and murder British envoys Sir Alexander Burnes and Sir William Macnaghten in the continuing Afghan War. British forces fall back to the Khyber Pass (*see* 1839; 1842).

Carlos Antonio López, 54, gains the presidency of Paraguay. The obese Asuncíon lawyer will rule despotically until his death in 1862 (*see* 1840; 1852).

Peru's president Agustin Camarra invades Bolivia but is defeated and killed November 20 at Ingavi. Gamarra's death at age 56 plunges Peru into a civil war that will continue until 1845.

A court at Washington, D.C., rules March 9 that Cinque and his fellow mutineers aboard the Spanish slave ship *Amistad* last year are not guilty and orders their release. Madrid protests.

Providence lawyer Thomas Wilson Dorr, 36, founds a People's Party to liberalize the Rhode Island charter of 1636. He submits a new, liberal constitution to extend suffrage in the state (*see* 1842).

Cincinnati street fights in August develop into a five-day race riot.

Irish-American trapper and Indian trader Thomas Fitzpatrick guides the first emigrant train bound for the Pacific through northwestern Montana Territory. The covered wagons with 130 settlers arrive at Walla Walla in Oregon Territory in November after a 2,000-mile journey through hostile Indian territory.

Secretary of State Daniel Webster estimates U.S. lands west of Wisconsin not yet offered for sale to have 30,000 to 50,000 settlers. Congress gives each head of family the right when the lands are opened for sale to buy as much as 160 acres at a minimum price of $1.25 per acre, provided that a house has been built on the land he holds and the land is under cultivation (*see* 1804; Homestead Act, 1862).

Dallas, Tex., has its beginnings in the independent Republic of Texas as Tennessee trader John Neely Bryan builds a house on the Trinity River to start a settlement that he will name after his friend George Mifflin Dallas, 49, who will become U.S. vice president in 1845 (*see* La Reunion, 1858).

1841 ***(cont.)*** California's Fort Ross is purchased from the Russian colonists who established it in 1811 by Swiss-American pioneer John Augustus Sutter, 38, who arrived at San Francisco 2 years ago by way of a job with the American Fur Company and a journey via Fort Vancouver, Honolulu, and Sitka. Granted lands to start a settlement by the provincial governor Juan Bautista Alvarado, Sutter has become a Mexican citizen and has thus gained title to his land grant, 50,000-acre New Helvetia, on which he has built a house and fort (*see* gold, 1848).

Britain's new Peel ministry (above) introduces the first peacetime British income tax and reduces import duties on raw materials and foodstuffs, thus encouraging grain imports which lower food prices and reduce the national debt by increasing customs revenues. The new tax law taxes incomes above £150 per year at 7d per pound sterling, and although the rate will fall as low as 2d in 1875, it will rise to 10s per pound sterling in the war a century hence.

Arc lamps for street lighting are demonstrated at Paris (*see* Brush, 1879).

The Peninsular & Oriental Steamship Navigation Co. chartered by Queen Victoria will become Britain's largest shipping firm.

A rail line opens September 19 linking Strasbourg and Basle, the first transborder railway.

Cook's Tours have their beginnings in a travel service started by English printer-temperance worker Thomas Cook, 33, who has observed that the Midlands County Railway has opened an extension between Leicester and Loughboro where a temperance conclave is to be held. He persuades the railroad to reduce its fare to Loughboro if he will guarantee 500 passengers on a route that normally carries only 50, organizes a group of 570 teetotalers, and charges them only 14 pence each for the 48-mile return trip (*see* London Great Exhibition, 1851).

The Prussian "needle gun" designed by gunsmith Johann Nikolas Dreyse, 54, is the world's first successful breech-loading military rifle. The bolt-action rifle uses a long, slender, needlelike firing pin to penetrate through a black-powder propelling charge and detonate a primer seated against the base of the bullet, and the Prussian Army will adopt it in 1848 to replace its muzzle-loading rifles (*see* Sadowa, 1866).

English mechanical engineer Joseph Whitworth, 38, proposes a uniform system of screw threads. He has introduced machine tools and methods that permit working tolerances to be reduced from the generally accepted 1/16 of an inch to a mere 1/1000 of an inch, his suggestion will eventually be accepted, and the British Standard Whitworth (BSW) system will be generally adopted.

German chemist C. J. Fritzsche shows that treating indigo with potassium hydroxide produces an oil (aniline) (*see* Perkin, 1856).

Scottish surgeon James Baird, 46, discovers hypnosis.

Brook Farm is founded at West Roxbury, Mass., by New England intellectuals who join former Unitarian minister George Ripley, 38, and his wife Sophia Dana Ripley in forming a commune and school that will pursue truth, justice, and order. Shares of stock in the enterprise are sold to members who include teachers, farmers, carpenters, printers, and shoemakers, with each man and woman receiving a dollar a day for his or her labors and with housing, food, clothing, and fuel provided at cost to all members and their families. Charles A. Dana, 21, and Nathaniel Hawthorne, now 37, are among the initial members, and by 1844 the commune will have grown to include four houses, work rooms, and dormitories with an infant school, primary school, and 6-year college preparatory course whose classes in botany, philosophy, Greek, Latin, Italian, German, music, and drawing will attract students from as far away as Havana and Manila (*see* 1846).

Fordham University is founded by New York Jesuits under the name St. John's College.

The *New York Tribune* founded by publisher Horace Greeley, 30, is a daily paper that will have great influence on U.S. opinion.

The *Brooklyn Eagle* begins publication at Brooklyn, N.Y.

The *Cincinnati Enquirer* begins publication.

The *Cleveland Plain Dealer* begins publication as the *Advertiser.*

The first advertising agency is started by Philadelphia entrepreneur Volney B. Palmer who sells newspaper space to out-of-town advertisers, charging the papers 25 percent of their space rates plus postage and stationery costs. By 1849 Palmer will have offices at New York, Boston, and Baltimore in addition to his Philadelphia office (*see* Rowell, 1869).

Punch begins publication July 17 at London. Mark Lemon edits the new humor magazine with Henry Mayhew.

Nonfiction "Heroes and Hero-Worship" by Thomas Carlyle whose moral essay contains the line, "No great man lives in vain. The history of the world is but the biography of great men"; "Self-Reliance" by Ralph Waldo Emerson whose moral essay contains the lines, "A foolish consistency is the hobgoblin of little minds, adored by little statesmen and philosophers and divines," and "Whoso would be a man, must be a nonconformist"; "Love" by Ralph Waldo Emerson: "All mankind[s] love a lover"; *Extraordinary Popular Delusions and the Madness of Crowds* by Scottish journalist Charles MacKay, 27.

Fiction *Barnaby Rudge, A Tale of the Riots of 'Eighty* by Charles Dickens (see Gordon "No Popery" riots, 1780); *The Old Curiosity Shop* by Dickens whose account of the death of Little Nell in the novel's final installment has reduced strong men to tears on both sides of the Atlantic; *The Deerslayer* by James Fenimore Cooper whose frontier figure Natty Bumppo (Leatherstocking) makes his first appearance as a young man, having been an older frontiersman in the original Leatherstocking novel *The Pioneers*, Hawkeye in *The Last of the Mohicans,* and then the Pathfinder; "The Murders in the Rue Morgue" by Edgar Allan Poe in the April *Graham's Magazine* at Philadelphia is the world's first detective story. Poe is editor of the magazine and his story "A Descent into the Maelstrom" appears in the May issue.

Poetry "Pippa Passes" by Robert Browning appears as part of the first in a series of pamphlets he will issue under the title *Bells and Pomegranates* between now and 1846: "The year's at the spring,/ And day's at the morn;/ Morning's at seven;/The hillside's dew-pearled; /The lark's on the wing;/The snail's on the thorn;/ God's in His heaven-/All's right with the world!" (I); *Ballads and Other Poems* by Henry Wadsworth Longfellow who includes his poems "Excelsior," "The Village Blacksmith," and "The Wreck of the Hesperus"; *The Demon* by Mikhail Lermontov who is killed July 27 at age 26 in a duel with a fellow officer at Pyatigorsk.

Painting *Jewish Wedding in Morocco* by Eugène Delacroix; *Père Lacordaire* by Théodore Chassériau.

Viennese mathematician Joseph Petzval, 34, produces an f/3.16 lens for portrait cameras.

Theater *London Assurance* by Irish playwright Dion Boucicault (*né* Dionysius Lardner Boursiquot), 20, 3/4 at London's Covent Garden Theater. Boucicault has written under the name Lee Morton.

Ballet *Giselle ou les Wilis* 6/28 at the Théâtre de l'Académie Royal e de Musique, Paris, with music by French composer Adolphe Charles Adam, 38, choreography by Jean Coralli, 62, libretto by Théophile Gautier. *Giselle* will be the great classic tragedy of ballet repertory.

First performances Symphony No.1 in B flat minor *(Spring)* by Robert Schumann 3/31 at Leipzig's Gewandhaus; Symphony No. 4 in D minor and Overture, Scherzo, and Finale by Schumann 12/6 at Leipzig's Gewandhaus.

Anthem "Deutschland, Deutschland Uber alles" is published with lyrics by German poet-philologist August Heinrich Hoffman, 43, a professor at Breslau University, whose words will be set to the music of Franz Joseph Haydn's *Emperors Quartet* and will become the German national anthem in 1922.

U.S. boxer Tim Hyer becomes the first recognized fisticuffs champion.

Shepheard's Hotel, Cairo, has its beginnings in the Hotel des Anglais opened by English entrepreneur Samuel Shepheard, 20, of Preston Capes, Northumberland. The hotel is housed in a building that occupies the site of a harem built in 1771 by the sheik Ali Bey and used as Bonaparte's headquarters in 1799, it will be called Shepheard's beginning in 1845 and will be the last British outpost between Gibraltar and India, a mailing address and international meeting place for the pink-gin set.

A New York State Fair at Syracuse begins the tradition of U.S. state fairs dedicated to the advancement of agriculture and the home arts.

Irish agriculture falls into decay with 663,153 out of 690,114 land holdings less than 15 acres. Tenant farmers raise grain and cattle to produce money for their rents while depending largely on potatoes for their own food. Whisky is at times cheaper than bread, and the price discrepancy helps produce widespread drunkenness (*see* potato failure, 1845).

Ireland's population reaches 10,175,000, up from 7.7 million a decade ago. The figure will be sharply reduced by famine and emigration later in this decade.

Britain's population reaches 18.5 million with 15 million of it in England and Wales.

London is a city of 2.24 million, Paris 935,000, Vienna 357,000, Berlin 300,000.

New York's population reaches 313,000 with nearly 50,000 passing through the city each year, but fewer than 140 U.S. towns have 2,500 or more.

The U.S. population reaches 17 million, with 12.5 percent of it in cities of over 8,000. The country will receive 1.7 million immigrants in this decade, up from 599,000 in the 1830s.

1842 British forces in Afghanistan fall back from Kabul January 6 under pressure from Afghans under Akbar Khan who massacre all but 121 of Lord Auckland's 16,500-man Anglo-Indian army in the Khyber Pass. Akbar's father Dost Mohammed regains the throne he held from 1835 to 1839 and will keep it until his death in 1863; a punitive force from India under Gen. Sir George Pollock, 57, reoccupies Kabul in September, but the British and their sepoy troops withdraw from the exposed position in October on orders from London.

The Treaty of Nanjing August 29 ends the Opium War that began in 1839, opening China to wholesale exploitation by the Western powers. China cedes Hong Kong to Britain, Amoy, Guangzhou (Canton), Foochow, Ningpo, and Shanghai are set apart as cities in which foreigners receive special privileges, including immunity from Chinese law, and in which foreigners may conduct trade under consular supervision. The treaty obliges China to pay the equivalent of $21 million in indemnities and forbids her to impose any tariff above 5 percent. The opium traffic continues (*see* Tai Ping Rebellion, 1850).

The young Serbian prince Michael III Obrenovic leaves the country in August under pressure from opponents who resent his heavy taxation and internal reforms. They elect Aleksandr Karageorgevic, 36, to succeed Michael and he begins a reign that will continue until 1858.

The Webster-Ashburton Treaty signed August 9 by U.S. Secretary of State Daniel Webster and Baron Ashburton Alexander Baring, now 68, finalizes the Maine-Canadian border, but the boundary of the Oregon Territory remains in dispute (*see* 1845; Oregon Trail, below).

1842 *(cont.)* A new civil war begins in South America as Peru remains locked in the war that began last year. Uruguay's exiled former president Manuel Oribe, 46, gains support from the Argentine dictator Juan Manuel de Rosas in an effort to subjugate Uruguay whose first president José Fructuoso Rivera, 52, has just completed a second term in office (*see* 1843).

The owner of a fugitive slave may recover him under the Fugitive Slave Act of 1793, the Supreme Court rules March 1 in *Prigg v. Pennsylvania*. The court overturns an 1826 Pennsylvania law that made kidnapping a slave a felony, saying an owner cannot be stopped from recovering a slave, but it says also that state authorities are under no obligation to help the slaveowner (*see* 1850).

Congressman Joshua R. Giddings of Ohio resigns his seat March 22 after being censured by the House for introducing antislavery resolutions, but his constitutents reelect Gaddings and he is back in his seat May 8.

Dorr's Rebellion in Rhode Island forces the state's conservatives to abolish the Charter of 1663 and expand suffrage (*see* Dorr, 1841).

Britain's Mines Act takes effect, prohibiting employment in mines of women, girls, and boys under age 10.

Massachusetts enacts a child labor law that limits the working hours of children under 12 to 10 per day (*see* age limit, 1848; Britain, 1802).

Massachusetts Chief Justice Lemuel Shaw, 61, upholds the legality of trade unions in *Commonwealth v. Hunt*. Reshaping the common law that has held unions to be criminal conspiracies, Shaw reasons that the purpose of the Boston Journeyman Bootmakers Society is not "to accomplish some criminal or unlawful purpose" but rather to persuade "all those engaged in the same occupation to become members," and he rules that a strike for a closed shop is quite legal and that a union is not responsible for illegal acts by individuals.

Strikes and riots disrupt industrial areas in the north of England.

The Oregon Trail mapped by U.S. Army lieutenant John Charles Frémont, 29, will take thousands of emigrants westward. Frémont eloped last year with the 16-year-old daughter of Sen. Thomas Hart Benton, 60 (Mo.). His guide is trapper Christopher "Kit" Carson, 33, who ran away from his Kentucky home at age 17 and at 20 joined an expedition to California (*see* 1845).

A wagon train of 120 people led by physician Elijah White, 36, is the first large-scale emigration to Oregon Territory. White has been appointed Indian agent for the Northwest (see Whitman, 1843).

The first U.S. wire suspension bridge opens January 2 to span the Schuylkill River at Fairmount, near Philadelphia. The $35,000 bridge is 25 feet wide, has a span of 358 feet, is supported by five wire cables on either side, and has been built by U.S. civil engineer Charles Ellet, Jr., 32, who has studied at the Ecole Polytechnique in Paris, is familiar with the Seguin bridge of 1825, and will be called the Brunel of America (*see* 1831; Roebling aqueduct, 1845).

The Paris-Versailles train pulled by two locomotives jumps the track May 8 and catches fire. Many passengers are trapped inside the wooden carriages.

A French railway act provides for government construction of roadbeds, bridges, tunnels, and stations.

A new U.S. railroad opens to connect Boston with Albany, N.Y.

The British government rejects the calculating machine on which Charles Babbage has been working since 1833 after it has advanced £17,000 to help Babbage develop the machine (he has spent some £20,000 of his own capital). Prime Minister Robert Peel has joked, "How about setting the machine to calculate the time at which it will be of use?" but an Italian engineer publishes an account of Babbage's "difference engine" in French and it is read by Augusta Ada Lovelace, 27, the only legitimate daughter of the late Lord Byron, who sees possibilities in Babbage's calculator, translates the account into English, has it published over her initials A. A. L. in Taylor's *Scientific Memoirs*, and shows the translation to Babbage. He asks why she did not write an original paper and Lady Lovelace responds with an extension of the Italian's paper that is three times longer, corrects some serious errors in Babbage's own work, and compares his machine with the Jacquard loom of 1801 which is also programmed with punched cards. "It weaves algebraic patterns just as the Jacquard loom weaves flowers and leaves," she writes, but Babbage and Lady Lovelace will lose heavily with an "infallible" betting system for horse races that employs the "difference engine" (*see* Burroughs, 1888).

Sanitary Conditions of the Labouring Population of Great Britain by English sanitary reformer Edwin Chadwick, 42, exposes the squalor of the nation's milltown slums in a report by the Poor Law commissioners. The report shows that working people have a much higher incidence of disease than do the middle and upper classes.

The Health of Towns Association, established in response to the Chadwick report (above), is a citizens group that calls for the substitution of "health for disease, cleanliness for filth, enlightened self-interest for ignorant selfishness," and demands "the simple blessings of Air, Water and Light" for everyone (*see* Engels, 1845).

Die Thierchemie by Baron Justus von Liebig applies classic methodology to studying animal tissues, suggests that animal heat is produced by combustion, and founds the science of biochemistry.

Jefferson, Ga., physician Crawford Williamson Long, 27, performs the first recorded operation under general anaesthesia. Friends familiar with Humphry Davy's 1799 paper on nitrous oxide have asked Long to stage a "laughing gas" frolic, he has suggested that sulfuric ether will work just as well, and when he finds that some of the partygoers have bruised themselves but experienced no pain, he uses ether on his patient James

Venable March 31 while removing a cyst from Venable's neck. Long will publish his findings on the anaesthetic properties of ether in surgery in 1849 (*see* 1845; Morton, 1846).

The University of Chile is founded at Santiago by poet-scholar Andres Bello, now 60.

Notre Dame University is founded at South Bend, Ind., by French Catholic missionary Edward Frederick Sorin, 28, who will serve as Notre Dame's first president from 1844 to 1865.

Villanova University is founded by Roman Catholics at Philadelphia.

The *Illustrated London News* begins publication in May.

The *Pittsburgh Post-Gazette* begins publication September 18 (*see* 1846).

Fiction *Dead Souls (Metitrye Dushi,* first part) by Nikolai Gogol whose earlier novel *Taras Bulba* appears in an enlarged, rewritten version; *Nanso Satomi Hakken* by Japanese novelist Bakin Takizawa, 75, who has taken 25 years to write the novel that is one of 300; "The Masque of the Red Death" by Edgar Allan Poe appears in the May *Graham's Magazine* which sacks Poe for drunkenness; "The Mystery of Marie Roget" by Edgar Allan Poe begins in the November *Snowden's Ladies Companion.*

Poetry *Poems* by Alfred Tennyson whose two-volume work includes "Break, Break, Break," "Godiva," "Locksley Hall," "Morte d'Arthur," and "Sir Galahad": "In the spring a young man's fancy lightly turns to thoughts of love" ("Locksley Hall"). "The old order changeth, yielding place to new,/And God fulfills Himself in many ways" ("Morte d'Arthur"). "My strength is as the strength of ten,/Because my heart is pure" ("Sir Galahad"); *Dramatic Lyrics* by Robert Browning includes "Incident at the French Camp" which ends, "'You're wounded!' 'Nay,' the soldier's pride/ Touched to the quick, he said:/ 'I'm killed, sire!' And his chief beside,/ Smiling the boy fell dead"; Browning's "The Last Duchess" begins, "That's my last Duchess painted on the wall,/ Looking as if she were still alive"; *Lays of Ancient Rome* by Thomas Babington Macaulay includes "Horatius at the Bridge"; "The Raven" by Edgar Allan Poe: "Quoth the Raven 'Nevermore' "; "The Rainy Day" by Henry Wadsworth Longfellow: "Into each life some rain must fall"; "Excelsior" by Longfellow: "The shades of night were falling fast,/As through an Alpine village passed,/A youth who wore 'mid snow and ice,/A banner with the strange device/ Excelsior!"

Theater *The Shoemaker and the King* (*El Zapatero y el Rey),* part II, by José Zorilla y Moral at Madrid's Teatro de la Cruz.

"Tom Thumb" is exhibited by P. T. Barnum who has bought Scudder's New York Museum, has discovered the midget Charles Sherwood Stratton at Bridgeport, Conn., and will use him to gain world prominence. Now 4, Stratton will stand no taller than 25 inches until he is in his teens, at maturity he will be no more than 40 inches tall with a weight of 70 pounds, and Queen Victoria will give him the title General Tom Thumb (*see* 1863).

Opera *Nabucco (Nebuchadnazzar)* 3/9 at Milan's Teatro alla Scala, with music by Giuseppe Verdi; *Linda of Chamonix (Linda di Chamounix)* 5/19 at Vienna's Kärntnerthor Theater, with music by Gaetano Donizetti; *Rienzi, the Last of the Tribunes* (*Rienzi, der Letzte der Tribunen*) 10/20 at Dresden, with music by German composer Richard Wagner, 29; *Russlan and Ludmilla* 12/9 at St. Petersburg with music by Mikhail Ivanovich Glinka.

First performances Symphony No. 3 in A minor *(Scotch)* by Felix Mendelssohn 3/3 at Leipzig's Gewandhaus.

The New York Philharmonic gives its first concert December 7 with all musicians except the cellist standing up as they play in the manner of the Leipzig Gewandhaus concerts, a practice they will continue until 1855. The cooperative will give four concerts each year until 1863 when the number will be increased to five (*see* Carnegie Hall, 1893).

Gimbels department stores have their beginnings in a Vincennes, Ind., trading post opened by Bavarian-American peddler Adam Gimbel, 26, who has acquired a 2.5-story frame house from a local dentist. "Fairness and Equality of All Patrons, Whether They Be Residents of the City, Plainsmen, Traders or Indians," Gimbel advertises (*see* Milwaukee, 1887).

Florida physician John Gorrie, 39, pioneers air conditioning (and mechanical refrigeration) with a method for lowering the temperature in his wife's sick-room at Apalachicola. Having waited in vain for ice from Maine that has been lost in the wreck of the schooner that was carrying it, Dr. Gorrie takes measures to alleviate the unbearable heat. He sets a vessel of ammonia atop a stepladder, lets it drip, and thus invents an artificial ice-making machine whose basic principle will be employed in air conditioning (and in refrigeration) (*see* Perkins, 1834; Carré, 1858; Linde, 1873; Carrier, 1902).

Paris's church of La Madeleine is completed in Greek revival style after 36 years of construction.

Hamburg is largely destroyed by a fire that rages in the German city from May 5 to 7 and in 100 hours ravages 4,219 buildings that include 2,000 dwellings. Property damage amounts to £7 million, 100 are killed, 20 percent of the city is left homeless, and contributions to help rebuild Hamburg pour in from other German states and principalities.

Croton water arrives in New York June 22 as the city gets its first good municipal reservoir system. Iron pipes lead to a new Murray Hill receiving reservoir on Fifth Avenue between 40th and 42nd streets (*see* 1892; fire, 1835).

John Deere produces 100 plows and peddles them by wagon to farmers in the area of Grand Detour, Ill. (*see* 1837; 1847).

1842 ***(cont.)*** New York gets its first shipment of milk by rail as the Erie Railroad line is completed to Goshen in Orange County. New Yorkers are unaccustomed to milk rich in butterfat and complain of the light yellow scum atop the milk.

Starch is produced from corn for the first time at Jersey City, N.J.

France has nearly 60 sugar beet factories producing 2 pounds of sugar per capita annually (*see* 1810; 1878).

Mott's apple cider and Mott's vinegar are introduced by Bouckville, N.Y., entrepreneur S. R. Mott.

Vienna has 15,000 coffeehouses (*see* 1683; 1925).

Ireland is losing some 60,000 per year to emigration (*see* 1840; 1847).

1843 India's Muslim emirs of Sind refuse to surrender their independence to the East India Company, the British commander Sir Charles Napier provokes Anglo-Sind hostilities, he attacks a 30,000-man Baluch army February 17, and his 2,800 men defeat the Baluchs in an engagement that sees generals fighting alongside private soldiers.

The Battle of Hyderabad in March destroys the army of the Sind emirs, Napier, says *Punch*, sends home the one-word dispatch "Peccari" ("I have sinned," a pun that conveys his boast, "I have Sind").

Maori natives kill New Zealand settlers in the Massacre of Wairau, antiforeign outbreaks in the North follow the massacre, and a 5-year Maori War begins.

Britain makes Natal a British colony following repulsion of the Boers who have allegedly exploited the natives and who in many cases move north across the Vaal River.

Britain separates the Gambia from Sierra Leone and makes it a separate crown colony (*see* 1807).

Manuel Oribe lays siege to Montevideo following withdrawal of French forces that have supported the Uruguayan president José Fructuoso Rivera. The siege will continue for 8 years (*see* 1842; 1852).

Hawaiian independence gains recognition November 28 in a convention between Britain and France who promise not to annex the Sandwich Islands that have been ruled since 1825 by Kamehameha III (Kauikeaouli), now 30. Kamehameha obtained U.S. recognition last year and seeks closer U.S. relations.

English social agitator Feargus O'Connor, 49, urges creation of a Chartist cooperative land association that will free working people from the tyranny of the factory and the Poor Law. Settling people on the land will take surplus labor off the market and force manufacturers to offer higher factory wages, he says (below).

English social reformer George Jacob Holyoake, 26, tells weavers at Rochdale, "Anybody can see that the little money you get is half-wasted, because you cannot spend it to advantage. The worst food comes to the poor, which their poverty makes them buy and their necessity makes them eat. Their stomachs are the waste-basket of the State. It is their lot to swallow all the adulterations on the market" (*see* cooperative society, 1844; Engels, 1845).

Former Boston schoolteacher Dorothea Lynde Dix, 41, reveals inhumane treatment of mental patients to the Massachusetts legislature. She has seen disturbed people imprisoned with criminals and left unclothed regardless of age or sex in cold, dark, unsanitary facilities, chained to the walls in some cases and often flogged. Her detailed, documented report encounters public apathy, but the legislature is moved to enlarge the Worcester insane asylum and Dix will spend the next 40 years persuading state legislators and foreign officials to build 32 new hospitals, and restaff existing facilities with intelligent, well-trained personnel.

Some 1,000 Americans who have come west on the Oregon Trail in a wagon train led by missionary Marcus Whitman settle Columbia River Valley territory that Britain claims on the basis of explorations by Francis Drake in 1579 and explorations by James Cook in 1788 (*see* 1836; 1842; Portland, 1845).

Atlanta, Ga., is named Marthasville after Martha Atlanta Thomson, daughter of Gov. Wilson Lumpkin, to replace the name Terminus used since 1837 to identify the termination point at 1,050 feet above sea level of J. E. Thomson's Western and Atlantic Railroad. The town will be named Atlanta in 1847.

The S.S. *Great Britain* launched July 19 by I. K. Brunel is the first of the large iron-hulled screw-propeller steamships that will dominate the transatlantic trade. The six-masted, single-screw, 3,270-ton vessel is 322 feet in length overall and carries a crew of 130 including 30 stewards, her dining room seats 360, she is the first propeller-driven ship (and first iron ship) to cross the Atlantic, and beginning in 1850 she will make 32 voyages in a 26-year career of carrying emigrants to Australia. She will serve as a cargo vessel from 1882 to 1883, and thereafter as a floating warehouse until she is scuttled in 1937.

The Howe Sewing Machine, invented by Boston machine shop apprentice Elias Howe, Jr., 27, uses two threads to make a stitch that is interlocked by a shuttle. Howe is not familiar with Walter Hunt's machine of 1832 (*see* 1846).

Boston physician Oliver Wendell Holmes of 1830 *Old Ironsides* fame publishes an indignant paper on the contagiousness of childbed (puerperal) fever. Dr. Holmes has quit private practice to teach medicine at Harvard (*see* 1847; Semmelweis, 1848).

Yellow fever sweeps the Mississippi Valley, killing 13,000.

The College of the Holy Cross is founded by Jesuits at Worcester, Mass.

Congress appropriates $30,000 to enable Samuel F. B. Morse to build an experimental telegraph line between Washington, D.C., and Baltimore (*see* 1837). He obtains help from Ithaca, N.Y., miller Ezra Cornell, 36, and Rochester, N.Y., banker-businessman Hiram Sibley, 36, and proceeds to construct the world's first long-distance telegraph line *(see* 1844).

The typewriter patented by Worcester, Mass., inventor Charles Thurber, 40, is a hand-printing "chirographer" with a cylinder that moves horizontally and contains a device for letter spacing (*see* Sholes, 1867).

London's *Sunday News of the World* begins publication. Circulation reaches 12,971 copies per week within a year and will grow to 6 million per week as the *News of the World* becomes the world's highest circulation paper.

The *Economist* begins publication in September under the aegis of London economist James Wilson, 38, who laments "that while wealth and capital have been rapidly increasing, while science and art have been working the most surprising miracles in aid of the human family . . . the great material interests of the higher and middle classes, and the physical condition of the labouring and industrial classes, are more and more marked by characters of uncertainty and insecurity."

La Réforme begins publication August 26 at Paris.

London museum director Henry Cole sends out the world's first Christmas cards. He has designed a three-panel card that says, "A Merry Christmas and a Happy New Year" (*see* Prang, 1875).

Nonfiction *Either-Or (Euten-Eller)* by Danish philosopher Søren Aabye Kierkegaard, 30, who repudiates the objective philosophy of the Absolute propounded by the late German philosopher G. W. F. Hegel and reaches instead a religion of acceptance and suffering based on faith, knowledge, thought, and reality that will be called "existentialism." Holding that religion is an individual matter, Kierkegaard denounces the theology and practice of the state church; *History of the Conquest of Mexico* by W. H. Prescott.

Samuel F. B. Morse persuaded congressmen to back him in a test of the telegraph that would speed communications.

Fiction *The Prose Romances of Edgar A. Poe*; "The Gold Bug" by Poe appears in the *Philadelphia Dollar Newspaper;* "The Pit and the Pendulum" by Poe appears in *The Gift;* "The Black Cat" by Poe appears in the August 19 *Philadelphia United States Saturday Post;* "Carmen" by French novelist Prosper Mérimée, 40, whose short story of a Spanish cigarette factory girl will be the basis of an 1875 opera; *A Christmas Carol* by Charles Dickens whose Ebenezer Scrooge of Scrooge & Marley shares roast goose with his clerk Bob Cratchit and the whole Cratchit family: "God bless us, every one," says Tiny Tim (*see* Christmas cards, above).

Poetry "The Song of the Shirt" by Thomas Hood appears in *Punch*, protesting against sweated labor.

Painting *The Sun of Venice Going to Sea* by J. M. W. Turner.

John James Audubon travels up the Missouri River to Fort Union at the mouth of the Yellowstone River to sketch wild animals. Now 58, Audubon completed his bird paintings in 1839. Washington Allston dies at Cambridgeport, Mass., July 9 at age 63; John Trumbull dies at New York November 10 at age 87.

Opera *Der Fliegende Hollander (The Flying Dutchman)* 1/2 at Dresden's Hoftheater, with music by Richard Wagner; *Don Pasquale* 1/4 at the Théâtre des Italiens, Paris, with music by Gaetano Donizetti; *A Midsummer Night's Dream* 10/14 at Potsdam's Neues Palais, with music by Felix Mendelssohn that includes a "Wedding March" that will be used in a British royal wedding in 1858 and will remain popular for generations; *The Bohemian Girl* 11/27 at London's Drury Lane Theatre, with music by Dublin-born composer Michael William Balfe, 37, lyrics by Alfred Dunn, arias that include "I Dreamt I Dwelt in Marble Halls," "The Heart Bow'd Down," and "Then You'll Remember Me."

First performances *Das Paradies und die Peri* by Robert Schumann 12/4 at Leipzig's Gewandhaus; Frederic Chopin continues his 5-year-old liaison with George Sand, but he has tuberculosis of the larynx and will die in the autumn of 1849 at age 39.

Popular songs "Columbia, the Gem of the Ocean" ("Columbia the Land of the Brave") by U.S. songwriter Thomas à Becket; "Stop Dat Knockin' at My Door" by U.S. songwriter A. F. Winnemore; "The Old Granite State" with lyrics by Jesse Hutchinson to the revivalist hymn "The Old Church Yard" (The Hutchinson Family singers introduce the new song at a temperance meeting in New York's Broadway Temple).

The Virginia Minstrels give the first full-scale minstrel show 2/6 at New York's Bowery Amphitheater and are the first of the Negro Minstrel troupes that will dominate U.S. musical entertainment for most of the century. Directed by Daniel Decatur Emmet, 28, the minstrels are gaudily costumed black-faced performers who sit in a semicircle of chairs, exchange jokes based on the popular white imagination of southern black

1843 ***(cont.)*** folklore, and sing songs that include Emmet's "Old Dan Tucker." The minstrel show will be the one form of American entertainment with any direct relation to American life (*see* Rice's "Jim Crow," 1832; "Dixie," 1859).

Skiing for sport has its beginnings at Tromso, Norway.

Cigarettes appear on a list of tobacco products controlled by a French government monopoly. It is the first known reference to the paper-wrapped miniature cigars that have been rolled in Cuba since the late 18th century (*see* 1856; *"Carmen"*; above).

A statue of the late Lord Nelson is hoisted atop a column erected in the center of London's new Trafalgar Square completed 2 years ago in an urban redevelopment program that has cleared out some squalid courts and cheap cookshops that had given the area the name Porridge Island.

Scottish settlers in New Zealand (above) strip millions of acres of forests to create sheep pastures. Deforestation for the country's burgeoning wool industry will lead to ruinous land erosion.

The Chartist cooperative land association urged by Feargus O'Connor (above) will, he says, encourage intensive "spade husbandry" that will double the soil's productivity.

The J. I. Case threshing machine is introduced by former Oswego County, N.Y., farmer Jerome Increase Case, 24, who last year came West with six of the best threshing machines he could buy, sold five to prairie farmers, used the sixth to thresh wheat for farmers in Wisconsin, learned from experience the deficiencies of existing machines, and used money earned while learning to develop a superior machine. Case will build a factory at Racine, Wis., and develop a sales organization that will make J. I. Case the world's largest thresher producer and a major manufacturer of farm steam engines, tractors, and other farm equipment (*see* 1837; 1908).

300 stems of Cuban red bananas are landed at New York and sold at 25¢ a "finger" wholesale by John Pearsall (*see* 1830). Now a New York commission merchant, Pearsall will go bankrupt within a few years when a shipment of 3,000 stems arrives in too ripe a state to be sold (*see* Baker, 1870).

The price of pepper imported by New Englanders from Sumatra since 1805 drops to less than 3¢ lb. (*see* 1873).

Japan's capital city of Edo has a population of 1.8 million and is second in size only to London (*see* 1801), but the nation's total population of nearly 30 million is controlled by infanticide (the Japanese call it *mabiki,* using an agricultural term that means "thinning out"). Hundreds of thousands of second- and third-born sons are killed each year, but daughters are generally spared since they can be married off, sold as servants or prostitutes, or sent off to become geishas (professional entertainers).

1844 Sweden's Charles XIV (John Bernadotte) dies at Stockholm March 8 at age 81 after a 26-year reign. He is succeeded by his son, 44, who will rule Sweden and Norway as Oskar I.

The Treaty of Tangier ends a French war in Morocco.

Santo Domingo gains her independence from Haiti and establishes the Dominican Republic on the island of Hispaniola (*see* 1822).

A band of 15 Texas Rangers led by Col. John Coffee Hays, 27, attacks a party of some 300 Comanches. The Rangers kill half the Indians and intimidate the rest with Colt revolvers that can fire six shots without reloading (*see* 1837; Colt, 1836; 1846).

U.S. Democrats deadlock in convention but finally nominate the first "dark horse" presidential candidate. James Knox Polk, 48, of Tennessee campaigns with the slogan "Reoccupation of Oregon, reannexation of Texas," Congress has delayed action on annexation, the Whig candidate Henry Clay straddles the issue, an antislavery candidate splits the Whig vote in New York, and Polk wins election with 170 electoral votes to Clay's 105.

Silesian weavers revolt against the low wages paid to hand-loom workers by employers who have switched to machine production. Authorities take harsh measures to suppress the uprising.

"Religion is the sigh of the oppressed creature, the feelings of a heartless world, just as it is the spirit of unspiritual conditions. It is the opium of the people," writes German socialist Karl Marx, 26, in his *Introduction to a Critique of the Hegelian Philosophy of the Right.* Cologne authorities suppressed Marx's newspaper *Rheinische Zeitung* last year and he has exiled himself to Paris where he meets Friedrich Engels (*see* 1845; 1847).

A prize-winning essay on Britain's "National Distress" by English author Samuel Laing, 31, reveals the effects of machinery on the country's working class: "About one-third plunged in extreme misery, and hovering on the verge of actual starvation; another third, or more, earning an income something better than that of the common agricultural labourer, but under circumstances very prejudicial to health, morality, and domestic comfort—viz. by the labour of young children, girls, and mothers of families in crowded factories; and finally, a third earning high wages, amply sufficient to support them in respectability and comfort."

The Young Men's Christian Association (YMCA) is founded at London by English dry goods clerk George Williams, 23, who has been holding meetings for prayer and bible-reading with fellow workers. The YMCA will combine interests in social welfare with those in religious welfare, will soon have branches in France and the Netherlands, and will spread round the world (*see* Boston, 1851).

Spain creates a Guardia Civil to police rural areas. "The Guardia Civil must not be fearsome, except to evildoers. Nor must it be feared, except by enemies of order," the organization's moral code will proclaim next year, but the Guardia Civil in its patent leather hats and green capes will become a symbol of governmental repression especially to Basques, gypsies, and anarchists.

A new British Bank Charter Act effectively yields control of the nation's currency to the private bankers of London's Lombard Street and permits the "Street" to share in the sovereignty of the Empire if not actually to control it. All banknotes are to be issued against securities and bullion in the bank vaults, the Bank of England may increase or decrease banknote issues at will within limits set by the act, but "the Old Lady of Threadneedle Street" may not increase the total amount of banknotes thus issued "on the Credit of such Securities, Coin, and Bullion."

Shipments of gold abroad—largely to the United States—have depleted British gold reserves, joint-stock companies have issued great quantities of paper money, the Bank Charter Act (above) separates the Bank of England's banking department from its note-issuing department, and its effect will be to eliminate by gradual degrees all notes except those of the Bank of England.

The Rochdale Society of Equitable Pioneers founded by 28 poor weavers at Rochdale, Lancashire, is the first modern cooperative society (*see* Holyoake, 1843). The society opens a store in Toad Lane, sells strictly for cash at local retail prices, and at year's end divides its profits among its members. Flour, oatmeal, butter, and sugar are its only initial wares but the store soon adds tea and groceries and will later sell coal, clothing, and furniture.

Fruitlands is founded by Concord social reformer Amos Bronson Alcott, 44, whose utopian commune in northern Massachusetts does not succeed. Alcott has recently returned from a trip to England made at the expense of his friend Ralph Waldo Emerson to observe the methods of the Alcott House School, an experimental school named in his honor. He and his fellow communards at Fruitlands neglect practical necessities in their passion for talk and social reform activities, the failure of their crops reduces them to near-starvation, and they abandon the community within a few months. (Alcott will move about New England for the next 9 years, will become superintendent of schools at Concord, Mass., in 1859, but will gain financial security only after 1869 through the literary efforts of his daughter Louisa May.)

Wells, Fargo & Co. has its beginnings in an express service between Buffalo and Detroit started by entrepreneur Henry Wells, 38, who worked briefly as a messenger for Harnden Express, quickly became a partner in another firm, and now starts Wells & Co. with William George Fargo, 26, and another partner. Wells & Co. will extend service next year to Cincinnati, Chicago, and St. Louis (*see* American Express, 1850).

Charles Goodyear obtains the basic rubber vulcanization process patent that will dominate the rubber industry (*see* 1839). He licenses the New Haven, Conn., firm L. Candee Co. to make vulcanized rubber overshoes.

Vestiges of the Natural History of Creation by Scottish author-publisher Robert Chambers, 42, anticipates some conclusions that Charles Darwin will publish in 1858 and 1859. Chambers and his brother William founded the Edinburgh publishing house W.&R. Chambers 12 years ago (below).

Boston dentist Horace Wells, 39, pioneers anaesthesiology. He has learned how to administer nitrous oxide ("laughing gas") from Gardner Q. Colton and uses it to deaden pain while extracting his own tooth (*see* Humphry Davy, 1799). He will attempt a demonstration of the gas before a Harvard Medical School class next year but his patient will not be completely anaesthetized when Wells makes the extraction and the class will not be impressed (*see* Morton's ether, 1846).

Crawford Long at Jefferson, Ga., makes the first use of ether in childbirth. He administers it to his wife during the delivery of their second child (*see* 1842; Simpson, 1847).

Mormon leader Joseph Smith and his brother Hyrum are jailed at Carthage, Ill., for wrecking the offices and press of a rival Mormon newspaper in town. A mob of 200 men drags the Smiths from their jail cell the night of June 27 and lynches them. Brigham Young, 43, is chosen August 8 to succeed Smith (*see* westward trek, 1846).

The State University of New York is founded at Albany.

Samuel F. B. Morse transmits the first telegraph message ("What hath God wrought") May 24 from the U.S. Supreme Court room in the Capitol at Washington, D.C., to his associate Alfred L. Vail at the Mount Clare Station of the B&O Railroad at Baltimore, and Vail transmits the message back (*see* 1837; 1843). Morse will establish telegraph lines with help from Ezra Cornell and Hiram Sibley (*see* 1848).

The wood pulp paper process invented by German engineer Gottlob Keller, 38, will reduce the price of newsprint and will permit the growth of cheap mass media (*see* Keen, Burgess, 1854).

Ned Buntline's Magazine begins publication at Cincinnati but soon fails. Publisher Edward Zane Carroll Judson, 21, earned a U.S. Navy midshipman's commission at age 15, his action stories have appeared in *Knickerbocker Magazine* under the pen name Ned Buntline, he will track down and capture two fugitive murderers in Kentucky next year, publish the sensational *Ned Buntline's Own* at Nashville, be arrested the following year for shooting and killing the husband of his alleged mistress, somehow survive an actual lynching, and thereafter reestablish *Ned Buntline's Own* at New York (*see* Astor Place Theater riot, 1849; Know-Nothing party, 1854).

Nonfiction *Summer on the Lake* by U.S. transcendentalist feminist Margaret Fuller, 34, who has been giving public "conversations" at Boston since 1839 to further the education of women and has become editor of the transcendentalist magazine *The Dial*. Horace Greeley invites her to become literary critic of his *New York Tribune*; *Chambers' Cyclopaedia of English Literature* by Robert Chambers (above) who has compiled the work with *Inverness Courier* editor Robert Carruthers, 45 (*see* 1868).

1844 (cont.) Fiction *The Life and Adventures of Martin Chuzzlewit* by Charles Dickens with illustrations by "Phiz"; *The Three Musketeers (Les Trois Mousquetaires)* by French novelist-playwright Alexandre Dumas, 42, whose Athos, Porthos, Aramis, and D'Artagnan cry, "All for one, one for all" (IX); *Mystères de Paris* by French novelist Eugène Sue, 39, who has changed his name from Marie Joseph to Eugène in honor of his patron Prince Eugène de Beauharnais; *Coningsby: or, The New Generation* by Benjamin Disraeli, now 40, who has been a member of Parliament since 1837 ("Almost everything that is great has been done by youth," III); "The Purloined Letter" by Edgar Allan Poe.

Poetry *Deutschland, Ein Wintermärchen,* and *Neue Gedichte* by Heinrich Heine; "Abou ben Adhem" by Leigh Hunt whose Abou wakes "from a deep dream of peace" and finds that to "love his fellow man" is even worthier than to love the Lord; *Odes to Rosa* by the late English poet Thomas Haynes Bayly who died 5 years ago at age 42: his "Isle of Beauty" contains the line, "Absence makes the heart grow fonder"; "The Day is Done" by Henry Wadsworth Longfellow ends with the lines, "And the night shall be filled with music,/ And the cares that infest the day/ Shall fold their tents, like the Arabs,/ And silently steal away"; "The Boy's Thanksgiving Day" by Boston abolitionist Lydia Child begins, "Over the river, and through the woods/ To Grandfather's house we go . . ."

Painting *Shoeing the Mare* by Edwin Landseer; *Rain, Steam, and Speed* by J. M. W. Turner.

Hartford's Wadsworth Atheneum opens in a neo-Gothic building that will house major works of art.

Theater *The Drunkard, or The Fallen Saved* by U.S. playwright William H. Smith (*né* Sedley), 37, and an anonymous "Gentleman" 2/12 at the Boston Museum. P. T. Barnum picks up the morality play about the evils of drink and presents it at Philadelphia, New York, and other cities (it will have frequent revivals; *see* 1854); *Don Juan Tenorio* by José Zorilla 3/28 at Madrid's Teatro de la Cruz.

Opera *Ernani* 3/9 at Venice's Teatro la Fenice, with music by Giuseppe Verdi; *Alessandro Stradella* 12/30 at Hamburg, with music by German composer Friedrich von Flotow, 33.

First performances Ouverture Le Carnaval Romain by Hector Berlioz 2/3 at the Salle Herz, Paris.

Popular songs "Buffalo Girls (Won't You Come Out Tonight?)" by U.S. songwriter Cool White (John Hodges); "The Blue Juniata" by U.S. songwriter Marion Dix Sullivan is about a river in Pennsylvania.

William Cullen Bryant proposes a public park for Manhattan. The poet-journalist of 1817 "Thanatopsis" fame, who is co-editor and co-owner of the *New York Post,* writes, "Commerce is devouring inch by inch the coast of the island, and if we rescue any part of it for health and recreation, it must be done now. . . All large cities have their extensive public grounds and gardens, Madrid and Mexico their Alamedas, London its Regent's Park, Paris its Champs Elysées, and Vienna its Prater" (*see* 1853; Olmsted, 1857).

The great auk *Pinguinus impennis* becomes extinct as the last one is killed by collectors on Eldey Island, 10 miles west of Iceland. A flightless bird resembling the penguin which once numbered in the millions, the auk for centuries swam 3,000 miles each year from its wintering grounds on North Carolina's outer banks to nesting sites on rocky islands off Iceland, Greenland, and Newfoundland, but while it was originally eaten by cod fishermen and used for bait, the bird has been wantonly shot and clubbed to death for flesh and for feathers to be sold in Europe.

Milk reaches Manchester by rail for the first time and shipments soon begin to London (*see* 1863). Rail transport will lower English food prices and make fresh eggs, green vegetables, fresh fish, and country-killed meat available more quickly.

1845 Mexico severs relations with the United States March 28 following U.S. Senate ratification of a treaty to annex Texas (below; *see* 1846).

Florida has joined the Union March 3 as the 27th state.

United States Magazine and Democratic Review editor John L. O'Sullivan asserts U.S. claims to Oregon Territory "by right of our manifest destiny to overspread and to possess the whole of the continent" in his July-August issue. Diplomats renew the 49th parallel as the boundary between Oregon and British territory.

U.S. title to the Oregon Territory up to the Alaskan border at 54°40' is "clear and unquestionable," says President Polk in his annual message in December, but U.S. relations with Mexico are worsening (above) and Polk does not intend war on the northwest border issue.

The first Tuesday following the first Monday in November is established by Congress as election day for electors of presidents and vice presidents.

The Republic of Texas established in 1836 is annexed to the United States over Mexican objections. Texas joins the Union as the 28th state December 29.

Peru's 3-year civil war ends in a dictatorship established by Gen. Ramon Castilla, 48, who fought under the late José de Sucre for independence in the mid-1820s. Castilla will rule until 1860.

An Anglo-Sikh War begins in India as British forces set out to conquer Kashmir and the Punjab (*see* 1849).

The Condition of the Working Class in England (Die Lage der arbeitenden Klassen in England) by German sociologist Friedrich Engels, 25, reveals exploitation of labor by capital. Engels has gone to live in England where his rich father has a cotton-spinning mill near Manchester and he has become friendly with Karl Marx in Paris (*see* 1844; 1847).

The Methodist Episcopal Church in America splits into northern and southern conferences after Georgia bishop James O. Andrews resists an order that he give up his slaves or quit his bishopric.

Portland is founded in Oregon Territory near the junction of the Columbia and Willamette Rivers. The town is named after the 213-year-old city in Maine as two New Englanders let a flip of a coin decide in favor of Portland rather than Boston.

"The Report of the Exploring Expedition to the Rocky Mountains in the Year 1842 and to Oregon and Northern California in the Years 1843–1844" by John C. Frémont is published while Frémont proceeds to California on a third congressionally-funded expedition, with Kit Carson serving once again as guide (*see* 1842; Los Angeles, 1846).

British Arctic explorer Sir John Franklin, now 59, sails in May with two Royal Navy ships on a new expedition to seek a northwest passage. Lured by unusually good weather, he ventures deep into a hitherto unknown channel where his ships will become hopelessly icebound and all hands lost.

Lawrence, Mass., is founded on the Merrimack River. The milltown for woolen production is named for the Boston firm A. (Amos) and A. (Abbott) Lawrence, the big New England textile concern founded in 1814 (*see* Lowell, 1834; Lawrence, Kansas, 1854).

British engineer William McNaughton develops a compound steam engine.

A hydroelectric machine perfected by English inventor William Armstrong, 35, produces frictional electricity by means of escaping steam.

The *Rainbow*, launched by New York naval architect John Willis Griffiths, 36, is the first of the "extreme" clipper ships that will for years be the fastest vessels afloat. Built for the China trade, the *Rainbow* is narrow in the bow, high in the stern, has aft-displaced beams, and begins a new era of improved clipper ships that will compete with the new screw-propelled iron steamships (*see* 1832; S.S. *Great Britain*, 1843; *Oriental*, 1850).

William Armstrong (above) patents an hydraulic crane (*see* Ellswick Engineering Works, 1847).

The Ames & Co. arms factory at Springfield, Mass., is acquired by Eliphalet Remington, 53, who has been making rifles at Ilion, N.Y., since 1828 and will contract for government work (*see* Oliver Ames, 1803; Remington typewriter, 1874).

E. B. Bigelow of 1830 two-ply power loom fame invents the Brussels power loom for carpet making.

French inventor Joshua Heilman, 49, patents a machine for combing cotton and wool.

Scientific American begins publication August 28 at New York in a newspaper format (*see* 1946).

British archaeologist Austen Henry Layard, 28, begins 6 years of excavations at Nimrod and Kiyunik in Iraq. His work will reveal the remains of the palaces of the Assyrian kings of Nineveh (*see* Rawlinson, 1846).

Babism is founded by Persian religious leader Ali Mohammed of Shiraz, 26, whose followers call him the Bab (the Gate). He has influenced large numbers of Persian Moslems, but Muslim leaders call his views heretical and persecute him (*see* 1850).

The Adventist Church is founded by U.S. evangelist William Miller, now, 72, whose 1836 prediction of the Second Coming in 1843 proved wrong but who has nevertheless attracted a considerable following (*see* Seventh Day Adventist Church, 1860).

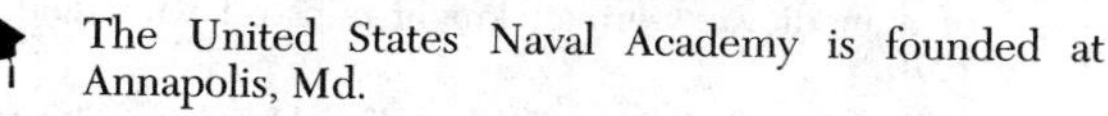

The United States Naval Academy is founded at Annapolis, Md.

Baylor University has its beginnings in the new state of Texas. State Supreme Court Justice Robert Emmet Bledsoe Baylor, 52, has helped obtain the charter for the state's first Baptist college.

Belfast's Queen's College is founded by the British government for the education of Ulstermen who do not belong to the Church of England.

The telegraphic Morse code developed by Andrew Vail in 1837 will soon come into universal use as Charles Wheatstone and W. F. Cooke in England have a falling out as to who shall receive chief credit for the improved single-key telegraph to which they are granted patent rights (*see* 1837, 1844; Electric Telegraph Co., 1846; Western Union, 1856).

The *Police Gazette* begins publication. The weekly U.S. scandal sheet carries lurid illustrations.

Nonfiction *Woman in the Nineteenth Century* by feminist Margaret Fuller whose work for the *New York Tribune* has made her the leading U.S. critic.

Fiction *The Count of Monte Cristo (Le Comte de Monte-Cristo)* by Alexandre Dumas who gains his greatest success with his adventure story about the Marseilles sailor Edmont Dantes; *The Cricket on the Hearth* by Charles Dickens who takes his title from John Milton's 1645 poem "Il Penseroso"; *Sybil: or The Two Nations* by Benjamin Disraeli whose "two nations" are the rich and the poor "between whom there is no intercourse and no sympathy; who are as ignorant of each other's habits, thoughts and feelings as if they were dwellers in different zones, or inhabitants of different planets; who are formed by a different breeding, are fed by different food . . ."; *Tales* by Edgar Allan Poe whose story "The Purloined Letter" appears in *The Gift*.

Poetry "Home Thoughts, from Abroad" by Robert Browning: "Oh, to be in England/ Now that April's there"; *The Raven and Other Poems* by Edgar Allan Poe whose title poem has appeared in the January 29 *New York Evening Mirror* where Poe is assistant editor: ". . . And his eyes have all the seeming of a demon's that is dreaming,/ And the lamplight o'er him streaming throws his shadow on the floor;/ And my soul from out that shadow that lies floating on the floor,/ Shall be lifted—nevermore!"

Painting *Catlin's North American Indian Portfolio: Hunting, Rocky Mountains and Prairies of America* by U.S. painter-author George Catlin, 49, who 20 years

1845 *(cont.)* ago went West with his wife and has gained the trust of the Indians, many of whom feared they would die if they were sketched or painted; *Fur Traders Descending the Missouri* by U.S. painter George Caleb Bingham, 34, who has turned to genre painting after 4 years as a Washington portraitist.

Opera *Tännhauser* 10/19 at Dresden, with music and libretto by Richard Wagner who has based it on legends of a medieval German knight-minstrel who died in 1270.

Ballet *Pas de Quatre* 7/12 at His Majesty's Theatre, London, with Marie Taglioni (who made her debut in 1822 at Vienna), music by Cesare Pugni, choreography by Jules Perrot.

First performances Concerto in E minor for Violin and Orchestra by Felix Mendelssohn 3/13 at Leipzig's Gewandhaus; Concerto for Pianoforte and Orchestra in A minor by Robert Schumann 12/4 at Dresden, with Clara Weik Schumann, 26, playing the work developed by her husband from a "Fantasy" he composed shortly after their marriage in 1839.

The world's first wire cable suspension aqueduct bridge opens in May to span the Allegheny River at Pittsburgh. The bridge has seven spans, each 162 feet long, and is the first built by German-American engineer John Augustus Roebling, 38, who 4 years ago founded the first U.S. factory to make wire rope (*see* Ellet, 1842; Monongahela bridge, 1846).

New Bedford, Mass., reaches the height of its whaling trade. Manned by 10,000 seamen, the New Bedford fleet brings in 158,000 barrels of sperm oil, 272,000 barrels of whale oil, and 3 million pounds of whalebone (*see* 1775).

Potato crops fail throughout Europe, Britain, and Ireland as the fungus disease *Phytophthora infestans* rots potatoes in the ground and also those in storage (*see* 1739). Irish potatoes are even less resistant than potatoes elsewhere, so up to half the crop is lost.

Famine in Ireland brought riots and death. The great hunger encouraged emigration to England and America.

Famine kills 2.5 million from Ireland to Moscow and is generally blamed on the wrath of God. The famine is especially severe in Ireland where so many peasants depend on potatoes for food while exporting their grain and meat, but British charity and British government relief do little to alleviate the suffering.

The English potato famine spurs free trade supporters Richard Cobden and John Bright at Manchester to lead a wide-scale agitation against the Corn Laws that prevent free imports of grain. Whig leader John Russell is converted to free trade (*see* 1846).

"Infanticide is practiced as extensively and as legally in England as it is on the banks of the Ganges," writes Benjamin Disraeli in his novel *Sybil* (above). Disraeli is criticizing the use of laudanum, the opium preparation commonly employed by British mothers and "nannies" to quiet their infants.

1846 The Mexican War precipitated by President Polk begins January 13. Polk has failed in an effort to buy the New Mexico Territory from Mexico, he orders Gen. Zachary Taylor to advance from the Neuces River to the Rio Grande, Taylor reaches the left bank of the Rio Grande March 28 and begins building a fort opposite Matamoras, the Mexicans order Taylor to retire beyond the Neuces April 12, Mexican troops cross the Rio Grande April 25 and kill a U.S. reconnoitering party in a cavalry skirmish.

The Battle of Palo Alto May 8 ends in victory for Gen. Taylor whose 2,000 troops defeat a Mexican force of 6,000 with help from artillery pieces near the water hole of Palo Alto at the southern tip of Texas.

The Battle at Resaca de la Palma May 9 ends in a rout of the Mexicans who fall back across the Rio Grande.

President Polk sends a war message to Congress May 11 declaring that Mexico "has invaded our territory and shed American blood upon American soil," Congress declares that a state of war exists May 13 by act of Mexico, and it votes a war appropriation of $10 million and approves enlistment of 50,000 soldiers.

California's Black Bear Revolt begins June 14 as settlers in the Sacramento Valley proclaim a republic independent of Mexico and raise a flag bearing a black bear and a star at Sonoma. U.S. commander John Drake Sloat takes possession of Monterey July 7 and claims possession of California for the United States, Sloat's subordinate John Berrien Montgomery takes San Francisco July 9, Commodore Robert F. Stockton succeeds Sloat as commander of the U.S. Pacific fleet and takes Los Angeles August 13 with help from John C. Frémont (*see* 1845), and news of the naval occupation of California reaches Washington August 31.

An Oregon Treaty signed with Britain June 15 has given territory south of the 49th parallel to the United States, overriding cries of "54°40' or Fight." Britain receives land north of the parallel on the mainland and also receives Vancouver Island (*see* 1818; 1828; 1845).

Former Mexican dictator Antonio Santa Anna, now 51, takes command of the Mexican army at Mexico City September 14, but his 10,000-man force at Monterey in northwestern Mexico sustains a defeat at the hands of 6,600 men under Gen. Taylor who occupy Monterey September 24 after a 3-day battle.

Tampico falls November 15 to American naval forces under Commodore David Conner, and Coahuilla's capital Saltillo falls to Gen. Taylor November 16.

Iowa is admitted to the Union December 28 as the 29th state with its capital at Des Moines.

The first Sino-American treaty is negotiated by the commander of the East India Squadron James Biddle, 63, who returns to command a Pacific Coast flotilla in the Mexican War.

Milwaukee is incorporated. The new city on the shore of Lake Michigan is made up of several neighboring Wisconsin Territory villages (*see* 1818).

Boston aristocrat Francis Parkman, 23, journeys westward from St. Louis on the Oregon Trail but is overcome by illness (*see* 1847).

Brigham Young leads Mormons in a trek to the Mexican Territory beyond the western limits of the United States, stopping en route to the Great Salt Lake Valley at the town of Omaha (*see* 1844; 1847).

Parliament finally repeals Britain's Corn Law June 28. Conservatives led by Benjamin Disraeli oppose Prime Minister Robert Peel's free-trade actions, denounce Peel for betraying his protectionist principles, but are unable to prevent repeal of the 1828 Corn Law which is replaced by laws that reduce (and will soon virtually eliminate) duties on grain imports, reduce duties on imported cheeses, butter, and other foods, and abolish duties on live animal imports.

Repeal of the Corn Law (above) removes the favored status that Ireland has enjoyed as a supplier to the British market. Large Irish landowners (most of them absentee landlords) switch from growing wheat to raising cattle. They throw cotters off the land so that it may be used for pasturage.

The United States moves closer to free trade July 30 with passage of the Walker Tariff Act named for President Polk's Treasury Secretary Robert S. Walker of Alabama. The act taxes luxuries at a high rate but lowers import duties and enlarges the list of items admitted duty-free.

Britain and America will both benefit in the next 15 years by the lowering and suspension of tariffs. U.S. exports will double to $306 million in the first 10 years of the Walker Tariff Act while imports will triple to $361 million, customs revenues will more than double, and U.S. consumers will enjoy cheap manufactured products while Britons enjoy low food prices.

New York's Marble Dry-Goods Palace opens on Broadway at Chambers Street where merchant A. T. Stewart will outgrow his new store and its additions in less than 15 years (*see* 1823; 1862).

The Pennsylvania Railroad is chartered April 13 for the purpose of constructing a line between Harrisburg and Pittsburgh (*see* 1851).

John A. Roebling completes a suspension bridge across the Monongahela River (*see* 1845; 1855).

Britain adopts a standard gauge for railroads in a move that will spur development of the nation's rail system (*see* Pennsylvania, 1852).

English inventor Robert William Thompson patents a pneumatic tire and equips a horse-drawn buggy with his tires (*see* Dunlop, 1888).

Of the United Kingdom's 3.2 million tons of merchant shipping some 131,000 tons are steam-driven.

Some 175 U.S. ships leave port with 65,000 tons of ice for export as Boston's Tudor Ice Co. ships large tonnages to the Far East where it is sold at high prices to buy silk which is bought cheap and sold at high prices in America (*see* 1833; 1856).

Samuel Colt receives an order for revolvers as the Mexican War (above) produces a shortage of firearms. Texas Ranger Samuel Walker has persuaded President Polk to place the order with Colt who engages E. K. Root to help him mass-produce the revolvers with interchangeable parts at a new factory he has built at Hartford, Conn. (*see* 1851; Hays, 1844; Collins brothers, 1826).

The Howe Sewing Machine of 1843 is patented September 10, but U.S. tailors and garment makers are fearful of using it lest they antagonize their workers. Howe's English agent pirates British royalties on the machine and there is wide infringement on the patent (*see* Singer, 1850, 1851).

German optical instrument maker Carl Zeiss, 30, opens a factory at Jena in Thuringia.

The Smithsonian Institution is founded by Congress at Washington, D.C., with a £100,000 bequest from the late James Smithson who died in 1829 at age 64. Illegitimate son of the late duke of Northumberland Hugh Smithson Percy who died in 1786, Smithson was a chemist-mineralogist. Calamine will be renamed smithsonite because of his paper on calamines.

English Assyriologist Henry Creswicke Rawlinson, 36, deciphers an inscription of Darius I Hystapis at Behistun and opens Assyrian history to modern understanding (*see* Grotefend, 1837; Layard, 1845; Smith, 1872).

Boston dentist William Thomas Green Morton, 27, pioneers modern anaesthesiology. Unfamiliar with C. V. Long's 1842 ether discovery, he hears a lecture by chemist Charles T. Jackson, 41, and learns that inhaling sulfuric ether will cause a loss of consciousness. He tries it on himself and on his dog, and then uses it to extract a tooth from a patient who leaks news of the painless extraction to the newspapers. Boston surgeon Henry Jacob Bigelow reads the newspaper account and persuades Morton to demonstrate his procedure at Massachusetts General Hospital where surgeon John Collins Warren uses ether October 16 to anaesthetize a

1846 *(cont.)* patient during an operation on a "congenital but superficial vascular tumor." Morton obtains a patent November 12, but actions taken by C. T. Jackson and C. W. Long will prevent him from enforcing it. Ether will be called an anaesthetic at the suggestion of Oliver Wendell Holmes, who writes to Morton the week he receives his patent, and it finds immediate use in operations on Mexican War casualties, opening a new era in surgery (*see* 1847; cocaine, 1884).

Bishop Hill is founded 100 miles west of Chicago by Swedish preacher Eric Janson who predicts an imminent apocalypse.

The Putney Community organized in Vermont by John Humphrey Noyes and his followers antagonizes its neighbors (*see* 1839). Noyes preaches promiscuity and free love in what he calls the communism of the early Christian church, he opposes monogamy, his neighbors have him arrested, and they try to break up his communal society (*see* 1848).

Brook Farm at West Roxbury, Mass., celebrates the completion of a large new central building with a dance the night of March 2, but the building catches fire and burns to the ground. The commune has attracted such eminent visitors as Amos Bronson Alcott, William Ellery Channing, Ralph Waldo Emerson, and Margaret Fuller, its weekly journal the *Harbinger* has published items contributed by Horace Greeley, James Russell Lowell, and John Greenleaf Whittier, but the new building consumed all available funds and the Brook Farm experiment will end next year.

The rotary "lightning press" patented by New York printing press manufacturer Richard Hoe, 34, can run 10,000 sheets per hour, a rate far faster than that of traditional flatbed presses. The type form of Hoe's press is attached to a central cylinder rather than to a flatbed and from 4 to 10 impression cylinders revolve about the central cylinder (*see* 1847).

The *London Daily News* begins publication January 21. Editor of the first cheap British newspaper is Charles Dickens.

The *Boston Herald* begins publication.

The *Pittsburgh Dispatch* begins publication February 18 (*see* 1842).

Electric Telegraph Co. is founded in England by W. F. Cooke and J. L. Ricardo, the latter a member of Parliament (*see* 1837; 1845). The company acquires Cooke's patents of last year for £168,000 and by September of next year two British telegraph networks will be in operation, the northern system taking in most major cities from Edinburgh to Birmingham, the southern system linking London with Dover, Gosport, and Southampton. Rates will be based on distance at first, and long-distance telegraphy will be prohibitively expensive, but beginning in March 1850 the rate will be reduced to a maximum of 10s for any distance, and competition will force the rate down to 1d or 2d for most inland telegrams by 1860.

Nonfiction *The People (Le Peuple)* by French historian Jules Michelet, 48, calls on France to unite, find her strength in the people, eliminate class divisions created by the industrial revolution, and return to the ideals of the Revolution of 1789: "I have acquired the conviction that this country is one of invincible hope. With France, nothing is finished. Always everything begins anew;" "The Biglow Papers" by Boston poet-essayist James Russell Lowell, 27, begin appearing June 1 in the *Boston Courier*. The letters in dialect by a fictional Ezekiel Biglow discuss the Mexican War (above) and the possibility of the extension of slavery.

Fiction *Typee* by U.S. novelist Herman Melville, 27, who at 19 sailed as cabin boy aboard a New York to Liverpool packet, at 20 sailed aboard the whaler *Acushnet* bound for the South Seas, and after 18 months jumped ship and found refuge with a companion on a jungle island of the Marquesas whose friendly but cannibal Taipi tribe he describes in a book that gains wide readership; *The Wandering Jew (Le Juif errant)* by Eugène Sue; "The Cask of Amontillado" by Edgar Allan Poe appears in the September *Godey's Lady's Book*.

Poetry *The Belfry of Bruges and Other Poems* by Henry Wadsworth Longfellow includes "The Arsenal at Springfield" and "The Arrow and the Song"; *Voices of Freedom* by John Greenleaf Whittier, now 38, includes "Massachusetts to Virginia."

Juvenile *A Book of Nonsense* by English poet-artist Edward Lear, 34, who made up nonsense rhymes to entertain the children of Edward Stanley, earl of Derby, while engaged in sketching the earl's menagerie at Knowsley Hall.

Painting *The Stag at Bay* by Edwin H. Landseer; *Pizarro Seizing the Inca of Peru* by English painter John Everett Millais, 17; *The Abduction of Rebecca* by Eugène Delacroix; *The Jolly Flatboatmen* by George Caleb Bingham; *Noah's Ark* by Edward Hicks.

Theater *Maria Magdalena* by Friedrich Hebbel 3/13 at Königsberg.

First performances Symphony No. 2 in C major by Robert Schumann 11/5 at Leipzig's Gewandhaus.

Oratorios *Elijah* by Felix Mendelssohn 8/26 at England's Birmingham Musical Festival; *The Damnation of Faust (La Damnation de Faust)* by Hector Berlioz 12/8 at the Opéra-Comique, Paris.

Popular song "Jim Crack Corn, or the Blue Tail Fly" is published at Baltimore.

Baseball rules are codified by New York surveyor Alexander Cartwright of the Knickerbocker Baseball Club whose members have been playing the "New York game" since 1842 in an open area near Broadway at 22nd Street. Distances between bases are set at 90 feet, each team is to have nine players and is allowed three outs per inning and three strikes per man, but Cartwright's team is defeated 23 to 1 by the New York Nine in its first game June 19 at the Elysian Fields in Hoboken, N.J. (*see* 1884; Cincinnati Red Stockings, 1867).

New York's Trinity Church is completed in Gothic revival style on Broadway opposite Wall Street by

Richard Upjohn. The spire of the new church is visible throughout the city.

New York's Grace Episcopal Church is completed on Broadway at East 10th Street by architect James Renwick, Jr., 28.

Famine sweeps Ireland as the potato crop fails again and food reserves are exhausted. British Conservatives ascribe the famine to the divine hand of Providence and say it would paralyze trade to give away food to the Irish, Britons marshal private aid programs, Americans raise $1 million and send relief ships, but the aid programs are mismanaged, the Irish lack horses and carts for carrying imported grain to inland famine areas, they lack ovens and bakers for making bread, and at least half a million die of starvation and hunger-related typhus.

A portable hand-cranked ice cream freezer is invented by Nancy Johnson in New Jersey.

Irish emigration to England, Canada, Australia, and America is spurred by the famine (above) and repeal of the Corn Law as cotters are thrown off the land and small farmers denied their favored status in the English market (*see* 1847).

1847

The Battle of Buena Vista February 23 ends in a rout of Mexican forces under Gen. Santa Anna by U.S. troops led by Gen. Zachary Taylor.

Thousands of U.S. troops led by Gen. Winfield Scott are landed on beaches south of Vera Cruz March 9 in the first large-scale amphibious operation. They gain possession of Vera Cruz March 29.

The bloody Battle of Cerro Gordo April 18 ends in another defeat for Gen. Santa Anna.

The Battle of Chapultepec September 13 ends in victory for Gen. Scott whose men have scaled a fortified hill on the outskirts of Mexico City after further U.S. victories at Contreras, Churubusco, and Molino del Rey. Gen. Santa Anna flees Mexico City which falls to Gen. Scott.

Liberia is proclaimed an independent republic July 26 under the presidency of Virginia octoroon Joseph Jenkins Roberts. The first African colony to gain independence, Liberia has been colonized since 1821 by U.S. freedmen sent across the Atlantic by the American Colonization Society which has decided that the colony should stop being dependent on American aid and which supports Roberts who has enlarged the colony and improved its economic position.

Escaped slave Frederick Douglass, 30, begins publication at Rochester, N.Y., of an abolitionist newspaper, the *North Star*. The Massachusetts Antislavery Society published Douglass' autobiography 2 years ago and he has earned enough from lecture fees in Britain, Ireland, and the United States to buy his freedom.

Parliament enacts legislation June 8 limiting working hours of women and children aged 13 to 18 to 10 per day.

Paris expels Russian anarchist Mikhail Aleksandrovich Bakunin, 33, for making a violent speech urging overthrow of absolute monarchy in Poland and Russia. Bakunin will be active in European revolutionary movements in the next 2 years, will be sentenced to death in Austria, but will be turned over to czarist authorities in 1851 and sent to eastern Siberia in 1855 (*see* 1861).

The Communist Manifesto (Manifest des Kommunismus) published late in the year says, "Let the ruling classes tremble at a Communist revolution. The proletarians have nothing to lose but their chains. They have a world to win. Workers of the world, unite!" The pamphlet is the work of Karl Marx and Friedrich Engels, who have been retained to write it by the newly formed Communist League at London. The League adopts its principles December 8.

The Manifesto (above) calls for "1. Abolition of property in land and application of all rents of land to public purposes. 2: A heavy progressive or graduated income tax. 3. Abolition of all right of inheritance. 4. Confiscation of the property of all emigrants and rebels. 5. Centralization of credit in the hands of the state, by means of a national bank with state capital and an exclusive monopoly. 6. Centralization of the means of communication and transport in hands of the state. 7. Extension of factories and instruments of production owned by the state; the bringing into cultivation of waste lands, and the improvement of the soil generally in accordance with a common plan. 8. Equal obligation of all to work. Establishment of industrial armies, especially for agriculture. 9. Combination of agriculture with manufacturing industries; gradual abolition of the distinction between town and country, by a more equable distribution of the population of the country. 10. Free educa-

Karl Marx cried out for social justice, but his economic theories were controversial to say the least.

1847 *(cont.)* tion for all children in public schools. Abolition of child factory labor in its present form." (*see* 1848).

Fur trader Charles Bent, 47, assumes his new post as governor of the New Mexico Territory, but Pueblos and Mexicans enter his house at Taos January 19 and kill him and 12 other Anglos with arrows (no women or children are touched). A punitive expedition uses howitzers and hand grenades against Pueblos holed up in a church August 19, killing 150 and wounding 300 (10 soldiers are killed in the action).

A relief expedition reaches the Donner Party in the Sierra Nevada February 19 and finds evidence of cannibalism. Trapped for 3 months by heavy snows in the worst winter ever, the wagon train has lost 12 members to starvation; 46 of the original 87 will ultimately survive.

The "Mormon battalion" under Lieut. Col. Philip St. George Cooke opens a wagon road from Santa Fe to San Diego. Cooke arrived at the San Diego mission January 29, having left Santa Fe 100 days earlier. Traveling through desert, mountains, and hostile Apache country, he and his 400 men have dug wells along the way to establish the Santa Fe Trail that thousands of California-bound émigrés will soon follow.

Nearly 15,000 Mormons led across the mountains by frontiersman Jim Bridger arrive on the shores of Great Salt Lake in Mexican territory that will soon be ceded to the United States. "This is the place," says Brigham Young July 24, and he organizes the "State of Deseret," an independent nation with himself as president (*see* 1846; polygamy, 1871).

Salt Lake City is founded by Brigham Young (above) who orders that avenues be made wide enough for a wagon and four oxen to make a U-turn.

Francis Parkman suffers a nervous breakdown in Oregon and is invalided home half-blind to Boston (*see* 1846; book, 1849).

Cayuse warriors in Oregon country kill Marcus Whitman November 27 along with his wife Narcissa and 12 other settlers who are blamed for the measles epidemic which has killed many of the tribe. Peter Skene Ogden, 53, of the Hudson's Bay Company rescues remnants of the group in December after they have been held for ransom for more than 2 weeks (*see* 1843).

Minneapolis is founded across from St. Anthony's Falls on the west bank of the upper Mississippi (*see* 1838). The town takes its name from the Falls of Minnehaha combined with the Greek word for city.

Economic depression engulfs England, provincial banks fail, and even the 153-year-old Bank of England comes under pressure.

Paris jeweler Louis François Cartier opens a small shop that will grow to have branches worldwide. Cartier and his son Alfred will open a shop in the rue Neuve des Petits-Champs, the demands of their growing clientele will force them to open a larger salon on the Boulevard des Italiens, and they will open a clock and watchmaking department in 1880 (*see* 1898).

Europe's first covered shopping arcade, the glass-enclosed Royal Galleries of St. Hubert, opens in Brussels (*see* Milan, 1867).

Steam powers a U.S. cotton mill for the first time at Salem, Mass., where the Maumkoag Steam Cotton Mill begins production.

Nitroglycerin is discovered by Italian chemist Ascanio Sobrero, 35, whose highly explosive liquid prepared from glycerol with nitric and sulfuric acid will be used chiefly in dynamite (*see* Nobel, 1866).

Ellswick Engineering Works is founded by William G. Armstrong (*see* 1845). The breech-loading gun he invents will be the prototype of all modern artillery, and his shipyard will build heavily armored Ellswick cruisers for the Royal Navy.

Butterick Patterns have their beginnings in a technique invented by U.S. tailor-shirtmaker Ebenezer Butterick, 21, for printing and cutting paper dressmaking patterns that can be used with sewing machines.

Nitroglycerin (above) will prove useful in relieving symptoms of angina pectoris (*see* 1768; amyl nitrate, 1867).

London physician John Snow, 34, introduces ether into British surgery (*see* 1846; Snow, 1853).

Ether is used as an anaesthetic in obstetrics by Scottish physician James Young Simpson, 36, who discovers the anaesthetic properties of chloroform and introduces it into obstetric practice (*see* 1831).

London has an influenza epidemic that will take 15,000 lives in the next 2 years.

Oliver Wendell Holmes becomes dean of the Harvard Medical School and his 1843 paper on childbed fever finally gains attention (*see* Semmelweis, 1848).

The American Medical Association is founded under the leadership of upstate New York doctor Nathan Smith Davis, 30, after a Philadelphia convention attended by representatives of medical societies and medical schools.

A new British Museum opens in London's Great Russell Street to replace the Montague House museum opened in 1759. The great circular reading room will be completed in 1857.

Iowa State University is founded at Iowa City.

The first U.S. adhesive postage stamps go on sale July 1 in the form of Benjamin Franklin 5¢ stamps and George Washington 10¢ stamps, but use of adhesive stamps will not be obligatory until January 1, 1856 (*see* England, 1840; Pitney-Bowes, 1920).

Arunah S. Abell's *Philadelphia Public Ledger* installs Richard Hoe's 1846 rotary press, prints 8,000 papers per hour, and becomes the first newspaper able to publish large daily editions (*see* Bullock, 1865).

The *Philadelphia Evening Bulletin* begins publication April 12.

The *Chicago Tribune* begins publication June 10 as the *Chicago Daily Tribune* (*see* Medill, 1855).

New York, with a population of 400,000, has 16 daily newspapers including the *Evening Post*, *Sun*, *Herald*, and *Tribune* (see *Times*, 1851).

Siemens & Hanske Telegraph Co. is founded by Prussian entrepreneurs Werner von Siemens and Georg Halske (*see* dynamo, 1866).

Nonfiction *History of the Conquest of Peru* by W. H. Prescott (two volumes).

Fiction *Wuthering Heights* by English novelist Emily Jane Brontë, 29, is a romance about Catherine Earnshaw and Heathcliff; *Jane Eyre* by Emily's sister Charlotte Brontë, 31, reflects the penury and unhappiness of Brontë's life as governess and school teacher in the story of an orphan girl who becomes a governess and falls in love with her sardonic employer, a married man named Rochester; *Children of the New Forest* by Captain Frederick Marryat; *Omoo* by Herman Melville who arouses controversy by revealing the hypocrisy and venality of Christian missionaries in the South Pacific. Melville moves to New York where he will write some minor novels in the next few years before moving to a farm near Pittsfield, Mass.

Poetry *Evangeline* by Henry Wadsworth Longfellow whose long poem about the expulsions of the Acadians from Nova Scotia in 1755 and 1784 begins, "This is the forest primeval,/The murmuring pines and the hemlock. . ."

Painting *Raftsmen Playing Cards* by George Caleb Bingham.

Theater *The String of Pearls or The Fiend of Fleet Street* by English playwright George Dibdin Pitt, 48, 3/8 at London, a grisly melodrama about "the demon barber" Sweeney Todd; *A Caprice* (*Un Caprice*) by Alfred de Musset 11/26 at the Comédie-Française, Paris.

Opera *Macbeth* 3/14 at Florence's Teatro della Pergola, with music by Giuseppe Verdi, libretto from the 1606 Shakespeare tragedy; *I Masnadieri* 7/22 at Her Majesty's Theatre, London, with music by Verdi; *Martha* (*oder Der Markt zu Richmond*) 11/25 at Vienna's Hofoper, with music by Friedrich von Flotow. The opera's tenor hero sings "The Last Rose of Summer," an old Irish air adapted to verse by Irish poet Thomas Moore, now 68, and incorporated by Flotow into the opera.

Popular song "Liebestraum" ("Dreams of Love") by Franz Liszt, lyrics by E. Freiligrath.

Ireland's potato crop is sound for the first time since 1844 but is small for lack of seed potatoes in the spring.

Hawaii's Parker Ranch has its beginnings in a small parcel of land at the base of Mauna Kea volcano granted January 14 by Kamehameha III to Massachusetts-born *kamaaina* (long-time resident) John Palmer Parker, 57, who jumped ship to settle in the Sandwich Islands at age 19, married a granddaughter of Kamehameha 17 years later, and has served the royal family by providing hides and meat from wild cattle put ashore originally by George Vancouver from his ship *Discovery* in 1793. Parker has developed a ranching operation that will grow to embrace 227,000 acres with more than 50,000 head of Herefords—the largest privately owned ranch in the world.

Cyrus McCormick forms a partnership with C. M. Gray and builds a three-story brick reaper factory on the north bank of the Chicago River near Lake Michigan (*see* 1834). McCormick has rejected sites at Cincinnati, Cleveland, Milwaukee, and St. Louis, deciding that Chicago may still be a swamp but is receiving great tonnages of grain via William Ogden's new Galena and Chicago Union Railroad and is clearly destined to become a grain transportation center (*see* 1848).

Obed Hussey introduces an improved version of his 1834 reaper, but he has moved his works to Baltimore and lacks the central geographical location (and capital) to compete successfully with McCormick (above).

John Deere builds a factory at Moline, Ill., to produce his self-polishing steel plows (*see* 1842; 1852).

Horticulturist Henderson Lewelling arrives in Oregon Territory after having traveled by covered wagon from Iowa. He begins an industry in the Willamette Valley by planting 700 grafted fruit trees—apples, sweet cherry, pear, plum, and quince, all less than 4 feet tall (*see* 1849).

A stamping process is developed to make tin cans cheap enough for wider sale (*see* 1830; Mason, 1858; Solomon, 1861; can opener, 1865).

Canned tomatoes are put up in small tin pails with soldered lids by the assistant steward of Lafayette College, established 21 years ago near Easton, Pa.

The first ring doughnuts are introduced by Camden, Maine, baker's apprentice Hanson Crockett Gregory, 15, who knocks the soggy center out of a fried doughnut (*see* doughnut cutter, 1872).

Cadbury Brothers moves to larger premises in Birmingham as John Cadbury takes his brother Benjamin into partnership (*see* 1824). Cadbury has been roasting and grinding cocoa since 1831, he has been preparing sugar-sweetened chocolate powder and unsweetened cocoa powder, and for the past 5 years he has been offering French eating chocolate (*see* 1866; van Houten, 1828).

Charitable Cookery, or the Poor Man's Regenerator, is published at London. Its author is Reform Club chef Alexis Soyer who establishes a soup kitchen that serves 2,000 to 3,000 starving Londoners per day as the economic depression (above) creates widespread unemployment in the city.

More than 200,000 emigrants leave Ireland, up from 60,000 in 1842, and many come to America. The poor pay a fare of between £3 and £5 ($15 to $25) per head for passage aboard small sailing vessels, few of which are inspected and many of which are not seaworthy. Passengers provide their own food, which is often inadequate when poor winds make the passage a long one.

A great migration from the Netherlands begins to the U.S. Middle West.

1847 ***(cont.)*** The New York Commissioners of Emigration begin to keep accurate records for the first time. Between now and 1860 some 2.5 million immigrants will enter the United States through the port of New York alone and more than a million of these will be Irish.

New York's first Chinese immigrants arrive July 10 aboard the seagoing junk *Kee Ying* out of Guangzhou (Canton) with 35 Cantonese whose voyage has taken 212 sailing days. Several crewmen jump ship, and their arrival marks the beginning of New York's Chinatown, which will have 1,000 residents by 1887.

1848 Denmark's Christian VIII dies January 20 at age 51 after a reign of less than 9 years. He is succeeded by his son, 39, who will rule until 1863 as Frederick VI. He is plunged into a war with Prussia over Schleswig-Holstein.

French revolutionists (below) force Louis Philippe to abdicate in February and proclaim a new republic. Prince Louis Napoleon Bonaparte, 40, is elected president December 10 (but *see* 1851).

Prince Metternich resigns under pressure at Vienna (below) and leaves for London.

The Treaty of Guadelupe Hidalgo February 2 ends the Mexican War that began in 1846. Mexico gives up claims to lands north of the Rio Grande and cedes vast territories that include California to the United States in return for $15 million and the assumption by Washington of U.S. claims against Mexico. Mexico loses 35 percent of her territory.

Wisconsin is admitted to the Union as the 30th state.

Costa Rica establishes herself as a republic.

Egypt's pasha Mohammed Ali becomes imbecilic at age 79, his adopted son Ibrahim serves as regent, but Gen. Ibrahim dies November 10 at age 59 (*see* 1841; 1854).

Persia's Kajar shah Mohammed dies at age 38 after a 13-year reign that has brought his country to the brink of revolution and bankruptcy. He is succeeded by his son Nasr-ed-Din, 17, who will reign until 1896 with help (until 1852) from his able minister Taki Khan.

Austria's emperor Ferdinand I, now 55, has escaped to Innsbruck May 17 to avoid the disorder at Vienna (below), returned in mid-August, a third insurrection by students and workers has forced him to flee once again, and he abdicates December 2 after a 13-year reign in which he has been insane much of the time and confused even in his lucid moments. Ferdinand's nephew, 18, has served under Radetzky in Italy and will reign until 1916 as Franz Josef I.

Switzerland becomes a federal union under a new constitution.

U.S. Whigs nominate Mexican War hero Zachary Taylor for the presidency in preference to the controversial party leaders Henry Clay and Daniel Webster, the Democrats nominate Lewis Cass of Michigan, Taylor receives 163 electoral votes to win the election with a 5 percent margin in the popular vote, Cass receives 127 electoral votes, losing New York because Free Soil party candidate Martin Van Buren has taken Democratic votes.

Paris students and workers seize the city in response to last year's *Communist Manifesto* and proclaim a new French Republic (above); the revolution spreads to Berlin, Budapest, Vienna, and throughout much of Europe as the *Manifesto* appears in virtually every European language.

Vienna puts down a revolution but abolishes serfdom in response to agitation.

Prague has a revolution which is suppressed by Austrian troops under field marshal Prince Alfred Candidus Ferdinand zu Windisch-Graetz, 61.

Roman revolutionists assassinate the prime minister of the papal states Count Pellegrino Rossi, 61, and force Pope Pius IX to take refuge at Gaeta.

Milan has a revolution March 17 as the Milanese hear of the revolution at Vienna 4 days earlier and rise against their Austrian overlords.

Britain's Chartist movement of 1843 revives. Feargus O'Connor, now 54, was elected to Parliament from Nottingham last year and presents a Chartist petition that is claimed to have 6 million signatures, but a procession to support Chartism that was scheduled for April 10 is called off when the government garrisons London on suspicion that a revolt will start that day. The government announces that the Chartist petition contains fewer than 2 million signatures.

Britain suspends the Habeas Corpus Act in Ireland in July as political seething follows in the wake of last year's famine and continuing high food prices despite improvement in the potato crop. Smith O'Brien, 45, leads an insurrection in Tipperary; he will be jailed, convicted, sentenced to death, but eventually pardoned.

Paris recalls army commander Louis Eugène Cavaignac, 46, from Algeria and appoints him minister of war, the government backs Cavaignac in putting down a June insurrection, and thousands of workers die at the barricades as Cavaignac quells the revolt.

Pennsylvania enacts a child labor law March 28 to restrict the age of workers (*see* Massachusetts, 1842; Owen-Keating Act, 1916).

The first Woman's Rights Convention opens at Seneca Falls, N.Y., under the leadership of Elizabeth Cady Stanton, 33, and Lucretia Coffin Mott of 1833 Anti-Slavery Society fame (*see* 1850).

Gold is discovered in California January 24 by New Jersey prospector James Marshall, 38, while working to free the wheel in the millrace for a sawmill he is building on the American River for Johann Sutter (*see* 1841). Sutter tries to keep Marshall's discovery a secret in order to avoid disruption of his farm, but the news appears August 19 in James Gordon Bennett's *New York Herald*, and by year's end some 6,000 men are working in the goldfields (*see* 1849).

John Jacob Astor dies March 28 at age 84, leaving a fortune of $20 million acquired in the fur trade and New

York real estate. Astor has been the richest man in America (*see* 1834).

Principles of Political Economy by British economist-philosopher John Stuart Mill, 42, is a classic text on economics. An employee of the East India Company, Mill follows the abstract 1817 theory of David Ricardo but applies economic doctrines to social conditions and expresses them in human terms.

"Wage, Labour and Capital" by Karl Marx is published as a pamphlet. Marx has returned to Cologne at the outbreak of revolution and founded the journal *Neue Rheinische Zeitung,* but he will be expelled from Prussia next year (*see* 1867).

The Oneida Community is founded in central New York State by sawmill owner Jonathan Burt, farmer Joseph Ackley, and Vermont preacher John Humphrey Noyes who has been pressed by his neighbors to leave Vermont and has moved with 25 adherents to the communal association founded earlier by Burt and Ackley. They have asked Noyes to join them in a community made up largely of farmers and mechanics (*see* 1846; 1851).

The 538-ton mail clipper *Deutschland* goes into transatlantic service for the Hamburg-American Packet Co. founded last year by German shipping interests.

Samuel Cunard's 9-year-old Royal Mail Steamship Line makes New York its base for transatlantic operations. His 1,422-ton wooden-hulled S.S. *Hibernia* with a service speed of 9.5 knots arrived at New York in December of last year.

Pacific Mail Steamship Co. is incorporated to employ a route across the Isthmus of Panama for California-bound gold seekers (*see* 1849).

London's Waterloo Station opens (*see* 1837; 1852).

British mathematician William Thomson, 24, proposes an absolute scale of temperatures (*see* thermodynamic theory, 1851).

English naturalists Henry Walter Bates, 23, and Alfred Russel Wallace, 25, travel to Brazil with only £100 between them, journey up the Amazon to Manaus, and go their separate ways in search of insect specimens. Wallace moves up the Rio Negro into the world's largest primeval forest and spends 40 days collecting rare butterflies and beetles, all of which will be lost when his homeward bound ship catches fire en route down the Amazon in 1852 (*see* 1858; Bates, 1863).

The American Association for the Advancement of Science (AAAS) is founded at Philadelphia on the model of the 17-year-old BAAS (*see* 1863).

U.S. anaesthesia pioneer Horace Wells is jailed in New York while under the influence of chloroform and commits suicide January 24 at age 33 in a fit of apparent despondence (*see* Simpson, 1847).

Hungarian obstetrician Ignaz Philips Semmelweis, 30, shows that childbed fever (puerperal fever) is contagious. He reduces its incidence at a Vienna hospital by requiring that attending physicians wash their hands in chlorinated water (*see* 1850; Holmes, 1843).

Parliament approves a public health act that establishes the first British sanitary regulations.

London appoints John Simon its first medical officer of health and he creates a public health service that will be a model for other nations. Simon establishes the science of epidemiology by showing how environmental conditions influence the spread and severity of disease.

The University of Mississippi is founded at Oxford.

The University of Wisconsin opens at Madison.

A telegraph line opens between New York and Chicago (*see* Morse, 1844; Western Union, 1856).

The Associated Press has its beginnings in the New York News Agency formed in May by a group of New York newspapers to avail themselves of the new telegraph connection (above). The name Associated Press will be adopted in 1851 (*see* Reuters, 1851).

New York's Astor Library is founded with a $400,000 bequest from the late John Jacob Astor (above). His son William Backhouse Astor, 56, will add another $550,000 to the library's endowment (*see* 1854).

Fiction *Vanity Fair* by English novelist William Makepeace Thackeray, 37, with the unscrupulous adventuress Becky Sharp. Thackeray's essential point is that society puts a premium on hypocrisy, and that a person without money or position must violate the ethical principles to which society pays lip service; *Dealings with the Firm of Dombey and Son, Wholesale, Retail and for Exportation* by Charles Dickens; *Mary Barton* by English novelist Elizabeth Cleghorn Stevenson Gaskell, 38, who gains immediate success with her realistic portrayal of Manchester factory life; *Agnes Grey* and *The Tenant of Wildfell Hall* by English novelist Anne Brontë, 28, youngest of the three Brontë sisters, whose second novel is the story of a young mother seeking asylum from her drunken husband; *La dame aux camélias* by French novelist-dramatist Alexandre Dumas, 24, bastard son of Alexandre Dumas père, whose doomed Marguerite Gautier and her lover Armand Duval will appear on the stage in 1852.

Poetry "The Bothie of Toberna-Vuolich" by English poet Arthur Hugh Clough, 29, who resigns as a fellow of Oxford's Oriel College to protest the Oxford movement and the establishment's Irish famine policies; "A Fable for Critics" by James Russell Lowell whose Biglow Papers continue in the *Boston Courier:* "a weed is no more than a flower in disguise"; "The Vision of Sir Launfal" by Lowell: "And what is so rare as a day in June? Then, if ever, come perfect days."

Painting *Portrait Equestre d'Ala ben Hamet, Calife de Constantine* by Théodore Chassériau; Thomas Cole dies at Catskill, N.Y., February 11 at age 47.

Theater *You Can't Be Sure of Anything* (*Il ne Faut Jurer de rien*) by Alfred de Musset 6/22 at the Comédie-Française, Paris; *The Candle-Stick* (*Le Chan-*

1848 (cont.) *delier*) by de Musset 8/10 at the Théâtre Historique, Paris; *Andrea del Sarto (André del Sarto)* by de Musset 11/21 at the Comédie-Française.

March "Radetzky March" by Johann Strauss the elder who sides with the revolutionists against the emperor. The new emperor Franz Josef (above) refuses to have Strauss waltzes played at court balls.

Popular songs "Oh! Susanna!" by Pittsburgh songwriter Stephen Collins Foster, 21, is sung February 25 by G. N. Christy of the Christy Minstrels. Foster's song will be in the repertoire of every minstrel show and gold-seekers will sing it en route to California; "Ben Bolt, or Oh! Don't You Remember" by Louisville composer Nelson Kneass, lyrics by Thomas Dunn English.

A solid gutta-percha golf ball replaces the leather-covered feather-stuffed ball used in Britain for centuries (*see* 1457). Made from a rubberlike gum produced by the sapodilla tree in Southeast Asia, the new ball travels 25 yards farther than any feather ball, professional golfer Tom Morris will begin using gutta-percha balls in 1852, but some players will claim that he takes unfair advantage and will not speak to him (*see* liquid-center ball, 1899).

Spain's Seville Fair has its beginnings in an April cattle market that will grow to become an annual 6-day *feria* of bullfighting, horsemanship, and flamenco dancing—the greatest fair in Europe.

The New York High Bridge goes up to carry an aqueduct that brings water from Croton Reservoir across the Harlem River to Manhattan (*see* 1842). A steel arch will replace five of the masonry arches in 1920.

Passenger pigeons appear to mate only occasionally in the eastern United States but remain plentiful in the new state of Wisconsin (above; *see* 1813; 1871).

The Chicago Board of Trade has its beginnings April 3 in a commodity exchange opened at 101 South Water Street by 82 local businessmen. Served by 400 vessels, 64 of them steamers, Chicago has become a major shipping point for grain and livestock from the Midwest bound for the Atlantic seaboard (*see* 1865).

Chicago's Cyrus McCormick produces 500 reapers in time for the harvest (*see* 1847; 1849).

Mormon farmers begin plowing the shores of Great Salt Lake and introduce irrigation to U.S. agriculture (*see* 1847).

"Bartlett" pears are distributed by Dorchester, Mass., merchant Enoch Bartlett, 69. He buys them from Roxbury farmer-sea captain Thomas Brewer who some years ago introduced to America the pear known in France and England as the bon chrétien.

The American Pomological Society begins to standardize the names of apple varieties, hundreds of which are grown in U.S. orchards.

Famine strikes Europe. Denmark permits grain imports free of duty to relieve the hunger.

A Treatise on the Falsification of Foods and the Chemical Means Employed to Detect Them by English analytical chemist John Mitchell is published at London (*see* Accum, 1820; Hassall and Letheby, 1851).

News of last year's Donner Party experience reaches Galveston, Tex. Local surveyor-land agent Gail Borden, 47, determines to find ways to make foods that will be long-lasting and portable (*see* 1849).

Failure of the liberal movement in the German states (above) forces hundreds of young men to flee for their lives. Many will emigrate to Wisconsin while German Jews will emigrate to New York, Boston, Cincinnati, Philadelphia, and some southern cities.

San Francisco loses three-fourths of its population in 4 months as men hurry to strike it rich in the goldfields (above).

1849 Austrian troops sent by the new emperor Franz Josef occupy Buda and Pest January 5 to suppress the revolution that began last year. The Hungarian Diet meeting at Debreczen proclaims a republic April 13 and elects Lajos Kossuth, 46, "responsible governor-president" (dictator), Kossuth issues a declaration of independence April 19, saying, "The house of Hapsburg-Lorraine, perjured in the sight of God and man, has forfeited the Hungarian throne," Franz Josef accepts an offer of help from the Russian czar Nicholas, Russian troops invade Hungary June 17 and hand the Hungarians a decisive defeat August 9 at the Battle of Temesòvár, Kossuth flees to Turkey August 11, his successor surrenders to the Russians August 13, and the victors take bloody reprisals. Nine generals are hanged, four shot.

Russia and Austria demand extradition of the Hungarian refugees, including Kossuth (above). Constantinople has imprisoned Kossuth and refuses to extradite him, the sultan appeals to Britain for aid, Lord Palmerston promises support, French and British naval forces make a show of force at Besika Bay, but a British squadron that has entered the Dardanelles November 1 to escape bad weather withdraws in response to a Russian protest.

Romans have proclaimed a republic February 9 with executive power vested in a triumvirate composed of Giuseppe Mazzini and two others with support from patriot Giuseppe Garibaldi, 42, who was forced to flee in 1834 for agitating against Austrian rule and lived in Uruguay from 1836 until last year. Marshal Oudinot leads an army of 86,000 French, Spanish, Austrian, Neapolitan, and Tuscan troops against Garibaldi's red-shirted army of 4,700, Garibaldi inflicts heavy losses on the advancing army April 30 but is wounded. He launches a costly attack on the French lines June 3, is forced to evacuate Rome in early July, Rome is returned to Pius IX, Garibaldi narrowly escapes capture in August, and flees to the United States. He will become a naturalized U.S. citizen but will return to Italy in 1854 (*see* 1857).

Sardinia's Charles Albert denounces last year's armistice with Austria under pressure from radical forces, Piedmontese forces are crushed March 23 by Austria's General Radetzky, now 83, as they were at Custozza 8 months earlier, Charles Albert abdicates, and he is suc-

California gold attracted hordes of eager prospectors. A few made fortunes—at a terrible cost to the environment.

ceeded by his son, 29, who will reign until 1861 as Victor Emmanuel II (*see* 1861).

British troops defeat Sikh forces at Chillianwalla and Gujarat and force the Sikhs to surrender at Rawalpindi. Britain annexes the Punjab by treaty with the raj (*see* 1845; 1856; 1857).

Dresden and Baden have revolutions which are quickly suppressed.

"Resistance to Civil Government" by Concord, Mass., philosopher-pencilmaker Henry David Thoreau describes the author's one-day imprisonment in 1845 for refusing to pay a poll tax to support the Mexican War which violated his antislavery views. The essay, which will be reissued under the title "On the Duty of Civil Disobedience," says every citizen has a duty to oppose bad government by acts of passive resistance such as not paying taxes, calls the state essentially a malevolent institution and a threat to the individual, and says, "that government is best which governs least" (*see* 1837; Gandhi, 1914).

Maryland slave Harriet Tubman, 29, escapes to the North and begins a career as "conductor" on the Underground Railway that started in 1838. Tubman will make 19 trips back to the South to free upward of 300 slaves including her aged parents whom she will bring North in 1857.

News of last year's gold discovery at Sutter's Mill brings a rush of 7,000 "Forty-Niners" to California, whose population will jump in the next 7 years from 15,000 to nearly 300,000 as the gold fields yield $450 million in precious metal.

Huntington & Hopkins supplies California prospectors with clothing, food, and equipment. Oneonta, N.Y., storekeeper Collis Potter Huntington, 28, has given up prospecting after one day and joined forces with Mark Hopkins, 36, who has founded the New England Trading and Mining Co. and come round Cape Horn to Sacramento with 26 men (each has put up $500 to capitalize the venture) and a year's supply of stores and equipment. Huntington & Hopkins will open an iron and hardware store at Sacramento in 1854 (*see* Central Pacific, 1861).

Canadians seek annexation to the United States as economic depression grips the country following repeal of the British Navigation Acts (below).

Britain reduces duties on food imports to nominal levels under the law passed in 1846.

U.S. commodity prices leap as a result of the California gold discoveries. Workers strike for higher wages in order to live, but wage hikes do not keep pace with rises in the cost of living.

Harrods has its beginnings in a London grocery shop at 8 Brompton Road that has been run by Philip Henry Burden. Tea wholesaler Henry Charles Harrod, 49, of Eastcheap takes over the shop that will grow to become one of the world's largest department stores (*see* 1861).

The first successful power dam across the Connecticut River is completed at Holyoke, Mass.

The first gold-seekers from the East arrive at San Francisco February 29 aboard the S.S. *California, a* 1,050-ton vessel in the service of the new Pacific Mail Steamship Co. Equipped to accommodate 60 saloon passengers and 150 in steerage, the sidewheeler left New York October 6 last year almost empty for lack of business, stopped at Rio de Janeiro, rounded Cape Horn, stopped at Valparaiso, Callao, and Paita, picked up 70 California-bound Peruvians at Callao, found 1,500 eager would-be passengers when she arrived at Panama in mid-January, left Panama February 1 with 350 passengers, and has proceeded north via Acapulco, San Blas, Mazatlan, San Diego, and Monterey.

Railroad construction begins across the Isthmus of Panama to facilitate passage to California (*see* 1855).

The Chicago and Galena Railroad reaches Chicago and locomotive number 1, the *Pioneer*, steams into town in April to begin Chicago's career as America's transportation hub. By 1869 the city will be served by 10 railroad lines.

Iron horses like the Baldwin locomotive improved inland transportation in an age of poor roads and turnpikes.

1849 ***(cont.)*** The B&O Railroad bridge completed across the Ohio River at Wheeling is the world's longest bridge. A span of more than 1,000 feet, it has been built by Charles Ellet, Jr., of 1842 Schuylkill River bridge fame who 2 years ago built a suspension bridge across the Niagara River.

Some 50,000 Forty-Niners pass through St. Joseph, Mo., a town of 3,000 whose location at the northern and western terminus of Mississippi and Missouri River steamboat transport makes it a supply center for people preparing to walk the 2,000 miles to California.

Parliament abolishes Britain's Navigation Acts June 26, ending restrictions on foreign shipping. U.S. clipper ships are permitted to bring cargoes of China tea to British ports (*see* 1850).

Budapest's Chain Bridge spans the Danube to link in the cities of Buda and Pest that will not become one city until 1873.

French physicist Armand H. L. Fizeau, 30, establishes the speed of light at approximately 186,300 miles (300,000 kilometers) per second.

A cholera epidemic at London wins support for the Health of Towns Association and its Great Sanitary Movement (*see* 1842; Snow, 1853). London clergyman Henry Moule, 48, works indefatigably to aid cholera victims; he will invent a dry-earth system of sewage disposal.

A cholera epidemic spread by gold-rush emigrants crossing the Texas Panhandle wipes out the leadership of the Comanche tribe, but the Indians continue to resist settlement of their lands (*see* 1875).

The first U.S. woman M.D. graduates at the head of her class at Geneva Medical College in Syracuse, N.Y., after having been ostracized by other students. Elizabeth Blackwell, 28, will play an important role in U.S. medicine, but while women physicians will be prominent in some European countries, they will remain an insignificant minority in U.S. medicine for more than a century (*see* New York Infirmary, 1857).

The College of the City of New York (CCNY) has its beginnings in the Free Academy that opens on Lexington Avenue at 23rd Street. The CCNY name will be used beginning in 1866. The tuition-free institution of higher learning, which will enable thousands of poor immigrants to attain high positions in life, will become City University of New York (CUNY) in 1961 (*see* 1976).

The *St. Paul Pioneer* begins publication in Minnesota Territory.

Nonfiction *The California and Oregon Trail* by Francis Parkman who has lost his eyesight but has embarked on the career of historian (*see* 1847); *Physical Geography of the Mississippi Valley* by bridge builder Charles Ellet, Jr. (above); A *Week on the Concord and Merrimack Rivers* by Henry David Thoreau; "Nemesis of Faith" by Oxford deacon James Anthony Froude, 31, who breaks with the Oxford movement, becomes a skeptic, goes to London, and marries.

Poetry *Ambarvalia* by Arthur Clough; "The Bells" and "Annabel Lee" by Edgar Allan Poe who dies October 7 at age 40 after a final spree in Baltimore; "The Ballad of the Tempest, or the Captain's Daughter," by Boston publisher James Thomas Fields of Ticknor, Reed & Fields whose poem will be published in the McGuffey *Readers* (*see* 1836, 1837): " 'We are lost!' the captain shouted,/ As he staggered down the stairs. / But his little daughter whispered,/ As she took his icy hand,/ 'Isn't God upon the ocean,/ Just the same as on the land?' "

Painting *Rienzi* by English pre-Raphaelite painter (William) Holman Hunt, 22; *Isabella* by John Everett Millais who last year joined with Dante Gabriel Rossetti and Holman Hunt to form the Pre-Raphaelite Brotherhood; *After Dinner at Ornans* by French naturalist painter (Jean Desiré) Gustave Courbet, 30. Japanese *ukiyoe* painter Katsushika Hokusai dies at Edo May 10 age 89 after an eccentric life. He has changed his pen name 33 times, giving the old name each time to a student. Edward Hicks dies at Newtown, PA, at age 69.

Theater *Herod and Marianne (Herodes und Marianne)* by Friedrich Hebbel 4/19 at Vienna's Burgtheater; *Genoveva* by Hebbel 5/13 at Prague; *The Ruby (Der Rubin)* by Hebbel 11/21 at Vienna's Burgtheater.

English actor William Charles Macready, 56, plays *Macbeth* at New York's Astor Place Opera House May 10; partisans of U.S. actor Edwin Forrest, 43, gather outside, possibly at the instigation of Forrest who was mistreated at London in 1845. The mob proceeds to wreck the theater, the police are unable to disperse the rioters, the state militia is called in, and the Astor Place riot ends with 22 dead, 36 injured. Edward Z. C. "Ned Buntline" Judson is convicted of having led the riot and is sentenced to a year in prison (*see* 1844; 1854).

Opera *The Merry Wives of Windsor (Die Lustigen Weiber von Windsor)* 3/9 at Berlin's Hofoper, with music by German composer Carl Otto Ehrenfried Nikolai, 38, libretto from the 1600 Shakespeare comedy; *Le Prophète* 4/6 at the Paris Grand Opéra, with music by Giacomo Meyerbeer; *Luisa Miller* 12/8 at the Teatro San Carlos, Naples, with music by Giuseppe Verdi, libretto from Schiller's 1784 play *Kabale und Liebe*.

First performances *Tusso: Lament and Triumph* symphonic poem No. 2 by Franz Liszt 8/28 at Weimar's Grand Ducal Playhouse.

Viennese waltz king Johann Strauss the elder dies September 25 of scarlet fever at age 45. His son and namesake, 23, broke with his father nearly 5 years ago and began conducting his own orchestra at Dommayer's Casino in Hietzing, delighted Vienna last year with his "Explosions Polka" and "Festival Quadrille," takes over his father's orchestra October 11, and pursues a career that will overshadow that of his father.

Popular songs "Frölicher Landmann" ("Happy Farmer") by Robert Schumann; "Santa Lucia" by Neapolitan songwriter Teodoro Cottrau, 22; "Nelly Was a Lady" by Stephen C. Foster; "Oh Bury Me Not on the Lone Prairie" by U.S. songwriter George N. Allen, lyrics from an 1839 poem ("The Ocean Burial") by E. H. Chapin.

Hymn "It Came Upon the Midnight Clear" by U.S. composer Richard Storrs Wills, 30, lyrics by Unitarian clergyman-poet Edward H. Sears.

A safety pin is patented by New York sewing machine inventor Walter Hunt, now 53, who sells the patent rights for $400 in order to raise money to discharge a small debt (*see* 1832; paper collar, 1854).

The bowler hat (derby) is introduced by London felt-hat makers Thomas Bowler, Ltd., of Southward Bridge Road who made the hat to fill an order placed by the 172-year old firm James Lock Co. of St. James for their customer William Coke of Holkham, Norfolk, who wants protection from low overhanging branches while out shooting. His hard shellacked derby headgear will become popular with foxhunters and businessmen.

The first modular prefabricated cast iron and glass "curtain wall" buildings are erected in New York at the corner of Washington and Murray Streets by former watchmaker and inventor James Bogardus, 49, who has designed them on commission from local merchant Edgar H. Laing. The columns and spandrels that make up the facades are simply bolted together with the bolt heads covered by cast iron rosettes and other decorative ornaments. Bogardus will obtain a patent next year to cover his revolutionary invention, and prefabricated buildings will soon go up all over Manhattan and at Philadelphia, Baltimore, St. Louis, and other cities as well (*see* department store, 1862).

Reinforced concrete containing iron bars, patented by French inventor Joseph Monier, 26, will permit construction of taller buildings, bigger dams, and other structures not heretofore possible, but no reinforced concrete building of more than two stories will be erected for 54 years (*see* Cincinnati, 1903; Otis, 1852).

Moscow's Kremlin Palace is completed after 11 years of construction.

A U.S. Department of the Interior that will eventually serve as custodian for the nation's natural resources is created March 5 by act of Congress; former Secretary of the Treasury Thomas Ewing takes office as Secretary of the Interior in President Taylor's new administration. The U.S. Patent Office established in 1790 is transferred from the State Department to Interior and the Bureau of Indian Affairs established in 1834 in the War Department is also moved to Interior, which for years will serve merely as general housekeeper for the government.

Henry David Thoreau (above) laments what dam builders are doing to the shad, a fish "formerly abundant here and taken in weirs by the Indians, who taught this method to the whites by whom they were used as food and as manure" (*see* Tisquantum, 1621). The shad are disappearing, says Thoreau, as "the dam, and afterward the canal at Billerica and the factories at Lowell, put an end to their migrations hitherward" (*see* 1834; 1858).

Thousands of U.S. farmers buy $100 McCormick reapers after being deserted by workers gone to California. McCormick has stocked warehouses throughout the upper Mississippi Valley to meet the demand, he guarantees his reapers, lets farmers buy them on an installment plan geared to harvest conditions, never sues a farmer for payment, but pays his factory workers small wages for long hours of labor (*see* 1848; 1850).

Basque shepherds from Argentina and Uruguay flock to California in quest of gold. Many will later become sheepherders on the western range.

Some 10,000 California gold-seekers will die of scurvy in the next few years. More thousands will avoid scurvy by eating winter purslane *(Montia perfoliata)*, an herb that will be called miner's lettuce.

Gail Borden invents a "meat biscuit" to provide a portable food for friends leaving in July for California (*see* 1848). He will put 6 years and $60,000 into developing and promoting his "meat biscuit" (*see* 1851).

Henderson Lewelling takes his first crop of Oregon apples to San Francisco and sells all 100 of them at $5 apiece to prospectors hungry for fresh fruit (*see* 1847; scurvy, above).

Domingo Ghirardelli arrives at San Francisco and begins selling tent stores to gold seekers. An Italian merchant who has lived in Latin America, Ghirardelli has seen cacao growing in Guatemala and will build a San Francisco chocolate factory.

1850 An "omnibus bill" introduced January 29 by Sen. Henry Clay contains compromise resolutions designed to reduce the growing polarity between North and South. California is to be admitted to the Union as a free state, territorial governments are to be set up in the rest of the territory acquired from Mexico with no congressional stipulation as to slavery in the territory, Texas is to give up her claim to part of the New Mexico Territory in return for federal assumption of the Texas state debt, slavery is to be abolished in the District of Columbia, and a stronger fugitive slave law is to be enacted.

John Calhoun attacks the Clay compromise March 4 in his "Speech on the Slavery Question," but Daniel Webster supports Clay March 7 in his speech "For the Union and Constitution."

The Clayton-Bulwer Treaty negotiated April 19 between U.S. Secretary of State John Middleton Clayton, 54, and Sir Henry Lytton Bulwer of Britain regulates the interests of both countries in Central America with special reference to a proposed canal across Nicaragua (*see* 1889; Walker, 1856).

President Taylor dies of acute gastroenteritis July 9 at age 65 after 16 months in office (he grew ill after downing large quantities of iced cherries and ice milk at a hot Fourth of July celebration in which the cornerstone was laid for the Washington Monument). His vice president Millard Fillmore, 50, moves into the White House.

California is admitted to the Union September 9 as the 31st state.

1850 ***(cont.)*** The balance of the Mexican cession is divided at the 37th parallel into the territories of New Mexico and Utah. Texas receives $10 million for relinquishing her claims to part of the New Mexico Territory (above).

China's Tai Ping Rebellion begins under the leadership of Kwangsi district schoolmaster and mystic Hong Xiuquan, 38, who believes himself the younger brother of Jesus Christ, calls himself Tin-wang (Heavenly Prince), and calls his dynasty Tai Ping (Great Peace). Helped by the strategist Yang Xiuqing, Hong leads southern peasants against the Manchu government in a civil war that will take 20 to 30 million lives in the next 14 years (*see* 1853).

The Don Pacifico affair brings Britain and Greece close to war. David Pacifico, 66, is a British subject who served as Portuguese consul general at Athens from 1837 to 1842 and has held substantial claims against the Greek government, an anti-Semitic mob burned his house at Athens in December of last year, the British foreign secretary Viscount Palmerston (Henry J. Temple), now 66, has ordered a squadron to the Piraeus to force a settlement of the Pacifico claims and others, the British embargo Greek vessels in the Piraeus and then seize some Greek ships in January, the Greeks comply with British demands April 26 under pressure, and Palmerston defends his action June 29 in a parliamentary speech that appeals to British nationalism with the phrase "civis Romanus sum."

Britain's Amalgamated Society of Engineers organizes workers using novel principles that include large contributions from members, benefits to members, and vigorous pursuit of measures to improve wages and working conditions through direct action and collective bargaining.

A French electoral law enacted May 31 requires that a voter be resident in one place for at least 3 years as attested by a tax receipt or employer's affidavit. The law is directed at radical workers who tend to be migratory.

French clubs and public meetings are forbidden June 9, Republican civil servants are dismissed at the slightest provocation, homes of Republicans are searched, and Republican newspapers are hit with fines and lawsuits (*see* 1851).

The first national women's rights convention opens at Worcester, Mass., largely through efforts by reformer Lucy Stone, 33, who was graduated 3 years ago from Oberlin College and will organize annual conventions each year for years to come (*see* 1848; Bloomer, 1851).

A new Fugitive Slave Act passed by Congress September 18 strengthens the 1793 act by substituting federal jurisdiction for state jurisdiction (*see* Clay compromise, above). A deputy U.S. marshal at New York arrests New York freedman James Hamlet as a fugitive from Baltimore in the first recorded action under the new act, but the arrest arouses so much public indignation that Hamlet is redeemed and freed. The Chicago City Council moves October 21 not to sustain the new act; a mass meeting at New York October 30 resolves that the act should be sustained (*see* Boston, 1851; Wisconsin, 1854).

Congress abolishes flogging in the U.S. Navy.

Scottish explorer Edward John Eyre, 25, arrives at Albany in Western Australia July 7 after a year-long journey across the Nullarbor Plain with his aborigine companion Wylie. Eyre left home for Australia at age 17, has discovered a lake that will be named after him, and is the first white man to cross the parched plain (he has been given up for dead) (*see* Jamaica black revolt, 1865).

Britain enters a "Golden Age" of prosperity as she embraces free trade principles that remove tariffs on foodstuffs. Self-sufficient in food until now and generally even a food exporter, Britain becomes a net food importer and depends on her manufacturing industry to provide the foreign exchange needed to feed her people. Food prices will rise far less swiftly than other prices and British wage increases will more than keep pace with price increases.

Irish entrepreneur William Russell Grace, 18, makes his way to Callao, Peru, after 2 years at sea and 2 more years with a Liverpool ship chandlery. Grace joins a Callao chandlery that he and his brother will take over within a few years (*see* 1865).

American Express Co. is formed by a merger of the 6-year-old Wells & Co., Livingston, Fargo & Co., and the firm of Butterfield, Wasson & Co. founded last year by upstate New York entrepreneur John Butterfield, 48 (*see* Wells, Fargo, 1852; Overland Mail, 1857; traveler's cheques, 1891).

New York shirtmaker Oliver Fisher Winchester, 39, sets up a New Haven, Conn., factory whose success will enable him to buy control of New Haven's Volcanic Repeating Arms Co. He will reorganize it in 1857 under the name New Haven Arms Co., and will reorganize it again in 1867 under the name Winchester Repeating Arms Co. (*see* Henry lever-action rifle, 1860).

The United States uses energy at a rate of 7,091 pounds of coal per capita, a rate many technologically advanced nations will not attain for more than 120 years, but 91 percent of U.S. energy comes from wood and the rest largely from whale oil (*see* 1859).

The first U.S. clipper ship to be seen at London arrives from Hong Kong after a 97-day voyage. The *Oriental* carries a 1,600-ton cargo of China tea and her $48,000 cargo fee nearly covers the cost of her construction. British shipbuilders are inspired to copy the *Oriental's* lines but are handicapped by English rules of taxation that consider length and beam in measuring tonnage while leaving depth untaxed. The short deep ships built at Aberdeen and on the Clyde do not approach the speed of the U.S. clipper ships, which soon abandon the China trade for the more profitable business of transporting gold seekers to California.

The S.S. *Atlantic* goes into service for Edward Knight Collins' U.S. Mail Steamship Co. in competition with

Samuel Cunard's Royal Mail Steam Packet Co. The 2,856-ton 200-passenger paddle-wheeler, 282 feet in length overall, leaves New York April 27, damages her side-wheels on ice off Sandy Hook, but breaks the Royal Mail's speed record on her return voyage, crossing in 10 days, 16 hours (*see* 1852).

Rio Grande steamboat operator Richard King, 25, and former Navy commander Mifflin Kenedy are backed by Texas merchant Charles Stillman in a steamboat venture that will modernize and monopolize commercial traffic on the Rio Grande (*see* Santa Gertrudis Creek, 1853).

Cornelius van Derbilt establishes a shipping line to California via Nicaragua and cuts the prices charged by competitors (*see* 1834, 1853; Clayton-Bulwer Treaty, above).

The Illinois Central Railway receives nearly 2.6 million acres of Illinois land as a gift from Congress to become the first U.S. railroad to obtain a land grant. The Illinois Central will be chartered next year to build lines between Cairo and East Dubuque and between Centralia and Chicago, will become the chief U.S. railroad running from North to South, and will sell much of its land to settlers at $5 to $15 per acre.

U.S. railroad trackage reaches 9,000 miles, up from little more than 3,000 in 1840. Canal mileage reaches 3,600, up from 3,000 in 1840.

The Chesapeake and Ohio Canal begun in 1828 finally reaches Cumberland, Md., which the B&O Railroad reached in 1842. The $22 million 184.5-mile canal with its 74 lift locks is obsolete, plans to continue it 180 miles westward to Pittsburgh are abandoned, but it will be used until 1924.

Land grants to U.S. railroad companies in the next 21 years will cover more territory than France, England, Wales, and Scotland combined.

German industrialist Friedrich Bayer, 25, founds a company at Elberfield that will be called Friedrich Bayer und Co. until it is renamed Farbenfabriken vorm. Friedrich Bayer und Co. in 1881 (*see* heroin, 1898; aspirin, 1899).

A trip hammer devised by Neponsit, Mass., inventor Silas Putnam will be used by the U.S. cavalry to make horseshoe nails.

The Singer Sewing Machine invented by U.S. actor-mechanic Isaac Merrit Singer, 38, will become the world's largest-selling machine of its kind (*see* Howe, 1843). A boiler explosion has destroyed Singer's patented wood-carving machine, he has watched some Boston mechanics trying to repair a primitive sewing machine, and he has been inspired to devise a better one (*see* 1851).

The second law of thermodynamics, enunciated by German mathematical physicist Rudolf Julius Emanuel Clausius, 28, states that heat cannot pass from a colder body to a warmer body but only from a warmer body to a colder body. The heat-transfer principle will be basic to all refrigeration and to many developments in chemistry and physics (*see* Carnot, 1824; Thomson, 1851).

The Bunsen burner invented by German chemist Robert Wilhelm Bunsen, 29, at Heidelberg University produces an intensely hot, almost nonluminous flame that leaves no sooty deposit on test tubes. It uses three volumes of air to every volume of coal gas and will become standard equipment in every chemical laboratory (gas stoves will work on its principle).

Only half the children born in the United States until now have reached the age of 5. The percentage will increase dramatically.

Ignaz Semmelweis enforces antiseptic practices in the obstetric ward of Budapest's St. Rochus Hospital (*see* 1848). He has returned to Budapest after failing to convince Viennese hospital authorities of the need for sterile conditions (*see* Pasteur, 1861; Lister, 1865).

The first textbook of histology (the study of organic tissues) is published at Würzburg. Its author is Swiss anatomist Rudolf Albert von Kolliker, 33, who studied under Johannes Müller (*see* 1840).

The ophthalmoscope invented by German physicist-anatomist-physiologist Hermann Ludwig Ferdinand Helmholtz, 29, is an instrument for examining the retina and the interior of the eye.

The Persian founder of Babism Ali Mohammed of Shiraz is condemned to death for heresy and executed in the public square of Tabriz July 9 at age 31. His follower Husayn Ali, 33, takes over the Bab's mission but will be exiled to Baghdad, then Constantinople, then Adrianople, and finally to Acre; he will claim to be the leader promised by the Bab, take the title Babaullah (Splendor of God), and preach the new faith Bahai (*see* 1845).

The University of Sydney is founded in Australia.

The University of Utah is founded at Salt Lake City.

The University of Rochester is founded in New York.

The United States has 254 daily newspapers, up from 138 in 1840 (*see* 1860).

The *Portland Oregonian* begins publication.

Harper's Monthly begins publication at New York (*see* *Harper's Weekly*, 1856).

Nonfiction *Representative Men* by Ralph Waldo Emerson.

Fiction *The Personal History of David Copperfield* by Charles Dickens whose "humble" Uriah Heep and Mr. Micawber are confined in the King's Bench debtors' prison as the author's father was confined in the Marshalsea prison: "Annual income twenty pounds, annual expenditure nineteen pounds six, result happiness. Annual income twenty pounds, annual expenditure twenty pounds ought and six, result misery" (illustrations by "Phiz"); *Pendennis* by William Makepeace Thackeray; *The Scarlet Letter* by Nathaniel Hawthorne is about inhumanity in Puritan New England.

1850 ***(cont.)*** Poetry *Sonnets from the Portuguese* by English poet Elizabeth Barrett Browning, 44, expresses hesitation at burdening poet Robert Browning with an invalid wife. She married Browning 4 years ago despite her infirmity, the result of an injury sustained while saddling a horse at age 15: "How do I love thee? Let me count the ways./ I love thee to the depth and breadth and height/ My soul can reach. . ."; "In Memoriam" by Alfred Tennyson who began the elegy 17 years ago upon the death of his sister's fiancé Arthur Hallam: "'Tis better to have loved and lost,/ Than never to have loved at all;" "Ring out, wild bells, to the wild sky,/ The flying cloud, the frosty light:/ The year is dying in the night;/ Ring out, wild bells, and let him die. . .;" "Ring out the old, ring in the new,/ Ring, happy bells, across the snow:/ The year is going, let him go;/ Ring out the false, ring in the true"; "The Building of the Ship" by Henry Wadsworth Longfellow: "Sail on, O Ship of State!/ Sail on, O Union, strong and great!/ Humanity with all its fears/ With all the hopes of future years,/ Is hanging breathless on thy fate!"

Painting *Morning, the Dance of the Nymphs* by Jean B. C. Corot; *The Stonebreakers* and *Burial at Ornans* by Gustave Courbet; *Christ in the House of His Parents* by John Everett Millais.

Gallery of Illustrious Americans by New York portrait photographer Mathew B. Brady, 27, gains national prominence for Brady (*see* 1864).

Opera *Genoveva* 6/25 at Leipzig, with music by Robert Schumann; *Lohengrin* 8/28 at the Weimar Theater, with music by Richard Wagner that includes a "Wedding March" (*see* royal wedding, 1858).

P. T. Barnum engages Swedish coloratura Johanna Maria "Jenny" Lind, 30, for an American tour and pays "the Swedish nightingale" $1,000 per night plus all expenses. She opens at New York's Castle Garden 9/11 in a concert that grosses $17,864.05 and begins a 2-year U.S. tour in which she will earn $130,000 of which $100,000 will go to charities.

Popular songs "De Camptown Races (Gwine to Run All Night)" by Stephen C. Foster; "Cheer, Boys, Cheer" by English composer-singer Henry Russell, 38, lyrics by Scottish journalist Charles Mackay.

Squash racquets is played for the first time by boys at England's 279-year-old Harrow School (approximate year).

Anagrams is invented by a Salem, Mass., schoolteacher whose word-building game will gain wide popularity (*see* Scrabble, 1948).

Bavarian-American entrepreneur Levi Strauss, 20, introduces "bibless overalls." He has arrived at San Francisco with a bundle of canvas fabric he hoped to sell to a tentmaker, was told by a miner that prospectors needed sturdy pants, has found a tailor and had him make up two pairs of canvas pants, wears one pair himself and gives the other to a new friend who goes around town talking about these new pants of Levi's, and sends word to his brothers in the East to "buy up all the canvas you can lay your hands on." By 1853 Strauss will be hiring all the tailors and seamstresses he can find for his California Street pants factory, he will switch from canvas to denim, and when he finds that two pieces rarely dye to the same shades of light blue, brown, or grey, he will order a deep indigo blue that assures him of a standard color for the "blue jeans" that will be worn by generations of Americans (*see* rivets, 1874).

The Pinkerton National Detective Agency opens at Chicago under the direction of Scots-American lawman Allan Pinkerton, 31, who was elected deputy sheriff of Kane County, Ill., 4 years ago after discovering a gang of counterfeiters and leading a force to capture them. Pinkerton has recently joined the Chicago police force as its first and only detective, helped to apprehend the perpetrators of a series of railway and express company robberies, and resigned to open the Pinkerton Agency which will solve a number of railroad crimes in the next decade (*see* 1861).

A 50-year period of large-scale tenement construction begins at New York (*see* tenement law, 1901).

The Brooklyn Institute imports eight pairs of English sparrows to protect Brooklyn, N.Y., shade trees from caterpillars (*see* starlings, 1890).

Fire damages San Francisco May 4, June 14, and September 17.

Salmon is taken for the last time from England's Thames River which is fast becoming heavily polluted. The fish will not reappear in the Thames for more than 120 years.

Cyrus McCormick buys out William Ogden for twice the $25,000 Ogden invested in McCormick's Chicago reaper business (*see* 1847; 1851).

The U.S. cotton crop reaches 2,136,000 bales.

California raises 15,000 bushels of wheat (*see* 1860).

German-American entrepreneur Henry Miller begins buying up California land. The immigrant butcher has arrived at San Francisco with $6 in his pocket and instead of prospecting for gold has borrowed money to start buying property along rivers, cutting off back-country homesteaders from the water, and forcing them to sell out to him at distress prices. Learning of a new swamp lands law that permits a settler to gain title to lands that are underwater and pay nothing if he agrees to drain the land, Miller loads a rowboat atop a wagon and has himself pulled across a large stretch of dry land, whereupon he files a map of the sections with a sworn statement that he has covered the land in a rowboat. By such means, and by purchasing land from speculators who have obtained rights while working as U.S. Government inspectors, Miller will one day own more than 14.5 million acres of the richest land in California and Oregon, a territory three times the size of New Jersey (*see* 1905; Haggin and Tevis, 1877).

Jersey cows are introduced into the United States where they will be an important dairy breed noted for the high butter fat content of their milk.

Open-ranging longhorn cattle herds on the western plains are estimated to number 50 million head, sharing the prairie with 20 million head of buffalo. At age 7 or 8 the cattle's horns measure from 4 to 5 feet tip to tip, and in years past the animals have been hunted for their horns, which are used for rifle racks, buttons, spoons, and forks, and for their hides which are used for rawhide ropes, whips, buckets, clothing, and—soaked and dried—even for nails (*see* Herefords, 1840; Aberdeen Angus, 1873).

Milo maize (Kaffir corn) is introduced into the United States for use as livestock forage.

The Elberta peach is imported into the United States from China by some accounts (but *see* 1870).

The Red Delicious apple is discovered as a chance seedling in Iowa. It will become the leading U.S. apple variety.

The Lancet announces appointment of an analytical and sanitary commission to study the quality of British foods (*see* 1823; Hassall and Letheby, 1851, 1855).

The Royal Navy reduces its daily rum ration from one-quarter pint to one-eighth pint to be dispensed before the midday meal (*see* 1731; 1824).

Nearly 15 percent of the world's sugar supply now comes from beets (*see* 1814; 1878).

Tea catches up with coffee in popularity among the English (*see* 1840; 1869).

U.S. wheat-flour consumption reaches 205 pounds per capita, up from 170 pounds in 1830; Americans average 184 pounds of meat, up from 178.

Millard Fillmore (above) installs the first White House cooking stove, but his cooks quit in protest, preferring to use the fireplace until an expert from the Patent Office spends a day showing them how to regulate the heat with dampers.

The population of New York City reaches 700,000 with 20 percent of it foreign-born, mostly Irish.

The U.S. South has 1.8 million black slaves, 2.1 million whites; 15.7 percent of the U.S. population is black.

The native population of the Hawaiian (Sandwich) Islands falls to 75,000, down from 150,000 in 1819 (*see* 1860). King Kamehameha III gives up rights to much of his lands in the Great Mahele and foreigners take over the lands.

The world's population reaches 1.24 billion by some estimates—more than twice its number in the 17th century.

1851 Siam's Rama III (Phra Nang Klao) dies after a 27-year reign that has seen the reopening of his country to contact with the West. He is succeeded by his half brother Phra Chom Klao Mongkut, 47, who will reign until 1868 as Rama IV. A Buddhist monk until now, the new king will build canals and roads, set up a printing press, stimulate education, reform the administration, improve the condition of slaves, issue Siam's first currency, encourage commerce with Europe and the United States, and import an English governess who will publish books about her experiences.

Victoria is proclaimed a separate Australian colony from New South Wales July 1 (*see* 1834).

France's Third Republic ends after 3 years in a December 2 coup d'état engineered by President Louis Napoleon Bonaparte and his half brother Count Morny. Army brigades occupy Paris, arresting leading deputies in the middle of the night, and the minister of war quells a popular uprising, sending his troops against the workers' street barricades. Troops fire on unarmed crowds in the Massacre of the Boulevards, and a 9-year period of repression begins (*see* Napoleon III, 1852).

A mob of Boston blacks defies last year's Fugitive Slave Act and rescues the fugitive Shadrach from jail February 15. President Fillmore calls upon Massachusetts citizens and officials 3 days later to execute the law, but abolitionists at Syracuse, N.Y., rescue another fugitive slave October 1 (*see* 1854).

Sioux chieftains cede all Sioux lands in Iowa and some in Minnesota to the federal government in a treaty signed July 23.

U.S. social reformer Amelia Jenks Bloomer, 33, urges reform of women's clothing in her magazine *The Lily*. She will be ridiculed for wearing full-cut trousers ("bloomers") under a short skirt in public—a costume designed by Elizabeth Smith Miller and introduced 3 years ago at the Women's Rights Convention at Seneca Falls, N.Y.

Men and women are treated equally and all classes of work are viewed as equally honorable in the Oneida Community which has 300 converts living in communal buildings made of timber from the community's farms (*see* 1848). Children are reared in the "children's house" operated by men and women considered best qualified (*see* Newhouse trap, 1860).

The Young Men's Christian Association (YMCA) opens its first American offices at Boston and Montreal (*see* 1844; YWCA, 1855).

Seattle is founded in Oregon Territory. The lumber town will be given its name in 1853 to honor Chief Sealth of the Duwamish and Suquamish who befriends the first settlers (*see* 1884).

An Australian gold rush follows the discovery by sheep station manager Edward Hammond Hargreaves, 35, who has found the yellow metal near Bathhurst in New South Wales. An influx of Chinese attracted by the gold strike brings demands for legislation to keep non-white immigrants out of Australia (*see* 1855; Victoria, above).

The U.S. Treasury turns out nearly 4 million $1 gold pieces, tiny coins authorized by Congress in 1849.

Congress votes March 3 to authorize minting of 3¢ silver coins to reduce the demand for large copper pennies.

The Corliss steam engine exhibited at the London Great Exhibition (below) by Providence, R.I., engineer-inventor George Henry Corliss, 34, weighs 1,700 tons and produces 2,500 horsepower from cylinders more than 3 feet in diameter operated by 30-foot gear wheels.

1851 ***(cont.)*** German physicist Franz Ernst Neumann, 53, enunciates the law of electromagnetic induction.

German physicist-manufacturer Heinrich Daniel Ruhmkorff, 48, invents the high-tension induction coil.

Visitors to the London Great Exhibition (below) have in many cases booked their travel accommodations through Thomas Cook (*see* 1841; 1866).

The Erie Railroad reaches Dunkirk on Lake Erie May 15—the first line linking New York City with the Great Lakes, providing competition for the 26-year-old Erie Canal. Begun in 1832 with state and county money, the Erie is now controlled by New York financier Daniel Drew, 53, who began his fortune as a livestock dealer who watered his stock by feeding the animals salt and letting them drink their fill to put on weight before selling them. Drew has operated steamboats on the Hudson and on Long Island Sound in competition with Commodore van Derbilt (*see* 1852).

The Pennsylvania Railroad reaches Pittsburgh (*see* 1846; 1852).

The Baltimore & Ohio reaches the Ohio River at Wheeling (*see* 1828; 1854).

The Hudson River Railroad opens to link New York City with East Albany (*see* van Derbilt, 1858).

The Missouri Pacific Railway begins laying track at St. Louis July 4 under the name Pacific Railway. The first railroad west of the Mississippi, it will grow to serve the Mississippi Valley south to Memphis and New Orleans and the Missouri Valley west to Kansas City and Pueblo, Colo., will own majority interests in other companies to serve Texas, but will never reach the Pacific.

Some 4,400 miles of railway track will be laid this year and next between the Atlantic seaboard and the Mississippi.

The *Flying Cloud*, launched by Nova Scotia-born Boston shipbuilder Donald McKay, 40, is a 229-foot clipper ship and the greatest of some 40 that will be built this year and next. 41 feet wide, 22 feet deep, and displacing 1,783 tons, she is bought by New York merchant Moses Grinnell, sails from Pier 20 on New York's East River under the command of Marblebead master Josiah Creasy, and sets a new record by reaching San Francisco in just under 90 days (*see* 1852).

The London Great Exhibition that opens May 1 is the world's first world's fair and in 141 days attracts more than 6 million admissions. It is intended to show British industrial achievement and prosperity, but foreign exhibits in many cases outshine the British.

German steel maker Alfried Krupp exhibits a 4,300-pound steel ingot cast miraculously in one piece; it dwarfs a 2,400-pound Sheffield "monster" casting and wins world acclaim for Fried. Krupp of Essen (*see* 1838).

The Colt revolver exhibited at the London Great Exhibition alarms British gun makers who fear that Colt's mass-production methods will swamp their handmade guns, but gun maker Robert Adams has patented a revolver that re-cocks itself each time the trigger is pulled, while the .36 caliber Colt is a single-action revolver and must be thumb-cocked for each shot. The British master general of ordnance conducts tests September 10, Adams circulates an account that the Colt weapons misfired 10 times while no Adams weapon misfired, the *Times* says the Colt is "very good," but no official results are published. Samuel Colt presents Prince Albert and the prince of Wales with Colt revolvers, British officers use some of the other revolvers handed out by Colt to fight the "kaffirs" who are using Sioux tactics in the Cape War by attacking the British as they reload their muskets, and Colt wins over the British when he addresses the Institution of Civil Engineers November 25 and asserts that the British will never defeat the kaffirs without Colt revolvers (*see* 1846; 1871).

Scottish industrial chemist James Young, 40, patents a method for producing paraffin by dry distillation of coal. Young will manufacture naphtha, lubricating oils, paraffin oil, and solid paraffin from Bogshead coal and, later, from Scottish shale (*see* kerosene, 1855).

William Thomson gives a complete account of thermodynamic theory that coordinates the discoveries of the past half-century. The Glasgow mathematics professor, whose 1848 absolute scale of temperature will be called the Kelvin scale after 1892 when Queen Victoria raises him to the peerage as Baron Kelvin of Largs, has developed the findings reported in 1840 by J. P. Joule, he will lay the foundations of the theory of electric oscillations, and his study of the oscillating discharge of condensers will lead to the discovery of radio waves by German physicist Heinrich Hertz in 1887.

The Cannizzaro reaction discovered by Italian chemist Stanislao Cannizzaro, 25, at the National College of Alexandria will lead to a clear definition of the distinction between molecular and atomic weights. Cannizzaro finds that alcoholic potash (potassium carbonate) will dissolve aromatic aldehydes (such as benzaldehyde) into a mixture of the corresponding acid and alcohol (benzoic acid and benzyl alcohol). He will simplify Avogadro's hypothesis of 1811 and apply it to atomic theory. He will also devise a method for deducing atomic weights of elements in volatile compounds from the molecular weights of the compounds.

Northwestern University is founded north of Chicago in an area that will be called Evanston after physician John Evans, 36, professor of obstetrics at Chicago's Rush Medical College.

The University of Minnesota is founded at Minneapolis.

Duke University has its beginnings in North Carolina's Trinity College founded at Durham (*see* 1924).

The *New York Times* begins publication September 18 with Henry J. Raymond as editor of the new daily (*see* Ochs, 1896).

The first successful cable is laid between Dover and Calais under the English Channel.

Reuters News Service is started by German entrepreneur Paul Julius Reuter (*né* Israel Beer Josaphat),

35, who last year pioneered in using carrier pigeons to convey messages, notably final stock prices, between Brussels and Aachen to plug the only gap in a telegraph system linking the commercial centers of Berlin and Paris. Reuter moves to London to take advantage of the new cable (above) and establishes a continental cable service that will be extended from stock prices to general news and will become a worldwide news agency that will compete with the Associated Press founded 3 years ago and with other news agencies (*see* INS, 1906; UP, 1907).

Fire damages the Library of Congress in the Capitol at Washington December 24, destroying two-thirds of the collection acquired from Thomas Jefferson in 1814 along with thousands of other volumes. Congress appropriates $100,000 to buy new books and create a more fireproof room for the library, which by 1865 will have more than 80,000 books, and will add the Smithsonian Institution's 40,000-volume library in 1866 (*see* 1897).

Nonfiction *Fifteen Decisive Battles of the World* by London University historian Edward Shepherd Creasy, 39; *The League of the. . . Iroquois* by Rochester, N.Y., lawyer-ethnologist Lewis Henry Morgan, 33, who will be called the "father of American anthropology."

Fiction *Uncle Tom's Cabin, or Life Among the Lowly* (first serial installments) by U.S. novelist Harriet Beecher Stowe, 40, whose sentimental tearjerker about slavery will have record sales (*see* 1852); *Moby Dick* by Herman Melville is only superficially about whaling and is possibly the greatest novel in all American literature. It will have sales of barely 50 copies in Melville's lifetime; *The House of the Seven Gables* by Nathaniel Hawthorne; *The Snow-Image and Other Twice-Told Tales* by Hawthorne.

Poetry *Romanzero* by Heinrich Heine.

Painting *Washington Crossing the Delaware* by German painter Emanuel Leutze, 35, who will emigrate to New York and Washington in 1859. John James Audubon dies at New York January 27 at age 65; J. M. W. Turner dies at Chelsea December 19 at age 76.

Sculpture *The Greek Slave* exhibited at the London Great Exhibition (above) is a marble figure completed in 1843 by U.S. sculptor Hiram Powers, now 45, who started his career in a Cincinnati waxworks, moved to Washington, D. C., in 1834 to make busts from life of U.S. statesmen, and moved in 1837 to Florence where he will live until his death in 1873. Inspired by reports that Greek prisoners were sold in slave markets by the Turks during the 1821–1829 war for Greek independence, the neoclassical nude attracts great attention, authorities set aside special hours for women who are embarrassed at viewing it in the presence of men, the work is sold in London, and Americans by the thousands pay 25¢ each to see full-sized marble copies that tour U.S. cities.

English architect Scott Archer, 38, publishes a wet collodion process for developing photographic images that will be used in photomechanical houses for nearly a century. A solution of nitrocellulose in alcohol-ether, collodion is iodized by Archer and sensitized in the darkroom by immersion in a bath of silver nitrate to form silver iodide with an excess of silver nitrate. The plate is exposed in the camera, developed by pouring on a solution of pyrogallol containing acetic acid, and fixed with a strong solution of thiosulfate of soda (cyanide of potassium will later be used instead). Archer's process, which he does not patent, will quickly replace Calotype and Daguerrotype (*see* Talbot, 1841; Daguerre, 1839; Sayce and Bolton, 1864).

Theater *Dame de Pique* by Dion Boucicault at London's Drury Lane Theatre; *Love in a Maze* by Boucicault 3/6 at the Princess' Theatre, London; *The Follies of Marianne (Les Caprices de Marianne)* by Alfred de Musset 6/14 at the Comédie-Française, Paris; *An Italian Straw Hat* (*Un Chapeau de Paille d'Italie*) by French playwright Eugène Labiche, 36, 8/14 at the Théâtre de la Montansier, Paris. The five-act comedy is one of six Labiche plays performed this year (nine were produced last year, 12 will be produced next year); *Bettine* by Alfred de Musset 10/30 at the Théâtre du Gymnase, Paris.

Opera *Rigoletto* 3/11 at Venice's Teatro la Fenice, with music by Giuseppe Verdi, libretto from a Victor Hugo drama of intrigue and treachery in the French court of François I.

First performances Symphony No. 3 in E flat major (*Rhenish*) by Robert Schumann 2/6 at Düsseldorf's Geisler Hall; Hungarian Rhapsody No. 2 by Franz Liszt who has settled at Weimar with Princess Sayn-Wittgenstein.

Popular songs "Old Folks at Home" by Stephen C. Foster.

The America's Cup won by the U.S. schooner *America* August 22 will be the most coveted trophy in world ocean racing. The U.S. schooner beats seven British schooners and eight cutters in a race round the Isle of Wight to win the Royal Yacht Squadron Cup which she carries home to the United States. It will remain there until 1983 despite repeated attempts by British, French, Australian, Canadian, and other world yachtsmen to outsail U.S. boats and gain—or regain—the America's Cup (*see* 1871).

I. M. Singer receives a patent on his sewing machine August 12. He has gone into partnership with his New York lawyer Edward Clark, 41, who will defend I. M. Singer and Co. from patent suits brought by Elias Howe (*see* 1846; 1850). Howe will eventually win a Massachusetts court decision and make a fortune from royalties that Singer will pay as the sewing machine gains worldwide distribution (*see* Hunt, 1858).

The Aquascutum raincoat challenges the Macintosh raincoat of 1823. The London firm Bax & Co. in Regent Street makes the new raincoat from a chemically treated wool fiber trademarked Aquascutum (*see* Burberry, 1856).

Boston's Jordan Marsh Co. has its beginnings in a dry goods shop opened by merchants Eben D. Jordan and Benjamin L. Marsh (*see* Boston Opera, 1909).

1851 ***(cont.)*** Lazarus Brothers opens at Columbus, Ohio, where Fred Lazarus and his brothers start a dry goods shop (*see* Federated Department Stores, 1929).

The London Great Exhibition (above) is housed in a Hyde Park pavilion constructed by English gardener Joseph Paxton, 50. The first great building not of solid masonry construction, the Crystal Palace is actually an immense 108-foot high greenhouse modeled after a conservatory designed by Paxton at Chatsworth in the late 1830s using newly developed techniques for making large sheets of glass, it has taken 2,000 men to build, has consumed one-third the nation's glass output for a whole year, and is not only the world's largest glass-walled structure but by far the largest single structure of any kind yet seen in the world, enclosing an area four times that of St. Peter's in Rome. The gas-lighted Crystal Palace will be a major influence in European railway station design for decades to come and will be copied almost literally for a New York exposition in 1853.

Cyrus McCormick exhibits his reaper at the London Great Exhibition, produces 6,000 reapers, and begins to enlarge foreign markets for his product (*see* 1849, 1854; 1879; Stanton, 1861).

A 100-acre wheat field remains the largest any one man can farm.

Ireland suffers widespread blindness as a result of the malnutrition experienced in the potato famine that began in 1846.

Articles exposing the adulteration of British foods will be published in the next 3 years by British chemist Arthur Hill Hassall and dietitian Henry Letheby who will document the whitening of bread with alum, the dilution of coffee with chicory, etc. (*see* 1850; 1855).

Gail Borden returns from the London Great Exhibition with the Great Council Gold Medal for his meat biscuit (*see* 1849). His ship encounters rough seas on the Atlantic, the two cows in the ship's hold become too seasick to be milked, an immigrant infant dies, and the hungry cries of the other infants determine Borden to find a way to produce a portable condensed milk that can keep without spoiling (*see* 1853).

The London Great Exhibition (above) forbids sale of wine, spirits, beer, and other intoxicating beverages but permits tea, coffee, chocolate, cocoa, lemonade, ices, ginger-beer, and soda water. The 61-year-old firm Messrs. Schweppe & Co. sells 177,737 dozen bottles, up from roughly 70,000 last year, and nearly 85,000 dozen of the total are sold at the Exhibition. White Rhine wine (hock) and soda is a popular British beverage, and Schweppe's soda water is a popular mixer (*see* 1858).

The first U.S. state prohibition law is voted in Maine where the mayor of Portland Neal Dow, 47, has drafted the law, submitted it to the state legislature, and campaigned for its passage. An ardent temperance advocate, Dow will see his measure followed by other states (*see* 1852).

Several tons of butter are shipped by rail from Ogdensburg, N.Y., to Boston in an ice-cooled wooden railcar insulated with sawdust.

The first wholesale ice cream business is founded by Baltimore milk dealer Jacob Fussell who receives milk in steady supply but is faced with a problem of erratic demand. Fussell sells his ice cream at less than half the price charged by others.

The Castle & Cooke food empire has its beginnings in a Honolulu mercantile house started by former missionaries Samuel Northrup Castle and Amos Starr Cooke who arrived in the Sandwich Islands aboard the *Mary Frazier* in 1837. They sever their ties with the American Board of Commissioners for Foreign Missions of Boston (*see* Dole, 1964).

China's population reaches 440 million, but Chinese census procedures have been disrupted by the Taiping rebellion that began last year.

India has a population of 205 million, Japan 33.5, Russia 65, Turkey 27, France 36, the German states and free cities 34, the Italian states 24, Britain 20.9 (with 17.9 in England and Wales), Brazil an estimated 8 million including 2.5 million slaves, the United States 23.6 million with nearly half living west of the Alleghenies.

More than 250,000 Irish emigrate and the country's population falls to 6.5 million, down from 10,175,000 in 1841. Some go to South Australia (above) which has a population of 77,000, as does Victoria, while New South Wales has 190,000.

The United States will receive 2.5 million immigrants in this decade, up from 1.7 million in the 1840s.

London is the world's largest city with a population of 2.37 million. Sochow has 2 million, Beijing 1.65, Guangzhou (Canton) 1.24; Changchow, Jiujiang (Kingtehchen), Xian (Sian), and Siangtan 1 million each, Wuhan 997,000, Constantinople 900,000, Calcutta 800,000, Hangchow 700,000, Bombay 650,000, Fuzhou (Foochow) 600,000. Paris has nearly 1.3 million and begins a rapid rise (figures include suburban environs).

1852 The Sand River Convention signed by the British January 17 recognizes the independence of the Transvaal (*see* 1843; Orange Free State, 1854).

Argentina abandons her designs on Uruguay following the defeat of the Argentine dictator Juan Manuel de Rosas February 3 in the Battle of Caseros. Insurgent Justo de Urquiza, 51, wins the battle with support from Brazilian and Uruguayan forces and Rosas flees to England.

Argentina recognizes Paraguay's independence and right of free river navigation July 17.

Burma deposes her king Pagan Min after a 6-year reign, and British forces take Rangoon as a second Burmese War begins.

New Zealand breaks apart under terms of a new constitution that provides for six provinces, each under the control of a superintendent with a provincial council elected by voters who must be property owners. A governor with a legislative council and a house of representatives exercises general administrative power, but regulation of the natives is left to the colonial office at London, and provincial governments will determine most affairs until 1875.

U.S. Democrats nominate "dark horse" Franklin Pierce of New Hampshire for president, and he wins the election by defeating Whig candidate Winfield Scott of Mexican War fame with 254 electoral votes to Scott's 42. The Whig party begins to dissolve over the slavery issue.

Louis Napoleon proclaims a second French Empire December 2 and will reign until 1870 as Napoleon III. He has exiled the Orléans family and ordered a plebiscite in November and found support for revival of the empire.

A Boston pharmaceutical firm distills "coal oil" from coal tar. It will start selling the oil under the brand name "kerosene" when it is found that the oil is not only a lubricant but will also burn in lamps (*see* Young, 1851; Gestner, 1855).

German-American inventor Christopher Dorflinger devises an improved lamp chimney.

Wells, Fargo & Co. is founded at New York by W. G. Fargo to "forward Gold Dust, Bullion, Specie, Packages, Parcels & Freight of all kinds, to and from New York and San Francisco. . . and all the principal towns of California and Oregon" (*see* 1844). Fargo will consolidate with rival stagecoach companies in 1866 and will become president of American Express in 1868 after a larger merger.

The first through train from the East reaches Chicago February 20 by way of the Michigan Southern Railway. The railroad will make Chicago more than ever the grain and meat-packing center of America.

The National Road that began as the Cumberland Road in 1811 and reached Wheeling 6 years later finally reaches Illinois.

The Pennsylvania Railroad completes trackage between Philadelphia and Pittsburgh (*see* 1851; 1854).

Pennsylvania adopts a railroad gauge different from that of New York to prevent the Erie Railroad from passing through the state into Ohio (*see* 1853).

A railroad that will become famous as the Rock Island Line runs its first train October 10. Pulled by a brightly painted locomotive named the *Rocket*, six yellow coaches travel 40 miles from Chicago to Joliet, Ill., in 2 hours to start service on the Chicago Rock Island Line, which by February 1854 will reach Rock Island in the Mississippi to give Chicago its first rail link to America's key waterway. The road will grow to have 7,560 miles of track extending north to Minneapolis-St. Paul, south to Galveston, east to Memphis, and west to Denver, Colorado Springs, and Santa Rosa (*see* Iowa City, 1855).

Studebaker Brothers is founded at South Bend, Ind., by Pennsylvania-born wagon maker Clement Studebaker, 21, with his older brother Henry. Brother John M. is making wheelbarrows for gold miners at Placersville, Calif., and will join the company in 1858 with $8,000 in earnings from the gold fields, brothers Peter I. and Jacob will join thereafter, and the company will become the world's largest wagon and carriage maker (*see* 1902).

The first Boston street railway begins operations with a single horsecar between Harvard Square, Cambridge, and Union Square, Somerville.

London's Paddington and King's Cross railway stations open (*see* 1848; 1864).

The Société Aerostatique—world's first aeronautical society—is founded at Paris (*see* Cayley, 1853).

The world's first tramp steamer is launched July 30 at Jarrow, England; the S.S. *John Bowes* goes into service as collier.

The S.S. *Pacific* goes into transatlantic service for the Collins Line (United States Mail Steamship Co.). A sister ship to the S.S. *Atlantic* that began service in 1850, the 2,856-ton *Pacific* is the first ship to cross from New York to Liverpool in less than 10 days (*see* 1854).

The *Sovereign of the Seas* launched by Donald McKay of last year's *Flying Cloud* fame confounds skeptics by showing that a 2,421-ton clipper ship can be practical (but *see* 1853).

Massachusetts adopts the first effective compulsory school-attendance law.

Tufts College is founded at Medford, Mass., by Hosea Ballou II, nephew of a prominent Universalist clergyman.

Antioch College is founded at Yellow Springs, Ohio, by Massachusetts educator Horace Mann (*see* 1839).

Mills College for Women is founded at Oakland, Calif.

"Uncle Sam" is portrayed for the first time in cartoon form by the New York comic weekly *Diogenes, Hys Lantern* (*see* 1813; 1917).

The Boston Public Library opens with funds raised by public subscription in a drive led by local financier Joshua Bates, 64, a partner in London's Baring Brothers since 1828.

Nonfiction *Thesaurus of English Words and Phrases* by English physician-scholar Peter Mark Roget, 73, whose book will go through 28 editions before its compiler dies at age 90.

Fiction *Uncle Tom's Cabin* by Harriet Beecher Stowe goes through 120 U.S. editions within a year (*see* 1851). 300,000 copies are sold in America with similar sales abroad, a popular success never before seen in publishing and rarely to be seen again; *The Blithedale Romance* by Nathaniel Hawthorne is a critique of the 1841–1847 Brook Farm experiment and analyzes the urge to seek Utopia and the possibly tragic effects of feminist emancipation; *History of Henry Esmond* by William Makepeace Thackeray; *Ange Pitou* by Alexandre Dumas père, now 50; A *Sportsman's Sketches (Zapiski okhotnika)* by Russian novelist Ivan Turgenev, 33, who is exiled to his estates for an obituary praising Nikolai Gogol, who has died February 21 at age 42.

1852 ***(cont.)*** Poetry *Emaux et Camées* by Théophile Gautier.

Painting *Pietà* by French painter Gustave Moreau, 26; *Christ Washing Peter's Feet* by English painter Ford Madox Brown, 31; *Ophelia* by John Everett Millais.

Theater *Camille, the Lady of the Camellias* (*La Dame aux Camélias*) by Alexandre Dumas *fils* 2/2 at the Théâtre du Vaudeville, Paris. Based on last year's novel, the play was published in 1848 but censors have delayed its production; *Agnes Bernauer* by Friedrich Hebbel 3/25 at Munich's Hoftheater; *Uncle Tom's Cabin* by George L. Aiken, 22, 9/27 at the Troy, N.Y., Museum. Adapted from the Harriet Beecher Stowe novel (above), the play will be produced throughout the country with enormous success in the North; *Masks and Faces* by English playwrights Charles Reade, 38, and Tom Taylor, 35, 11/20 at London's Haymarket Theatre; *The Journalists (Die Journalisten)* by German playwright Gustav Freytag, 36, 12/8 at Breslau.

Opera *Dmitri Donskoi* 4/30 at St. Petersburg, with music by Russian court pianist Anton Grigorev Rubinstein, 22, who is invited by the Grand Duchess Helena Pavlovich to be music director at Michael Palace and spend the summer with her on Kammeney Island in the Neva River (below).

First performances Overture to Byron's *Manfred* by Robert Schumann 6/13 at Weimar; Overture to *Julius Caesar* by Schumann 8/3 at Düsseldorf's Male Choral Festival; Melody in F by Anton Rubinstein (above) at St. Petersburg. The piano piece composed on Kammeny Island is played primarily with the thumbs.

Popular song "Massa's in de Cold Cold Ground" by Stephen C. Foster.

Bright tobacco is discovered by North Carolina farmers Eli and Elisha Slade near Durham and will soon be used in Bull Durham pipe tobacco. The Slade brothers heat some of their tobacco crop over hot flues and find that the clear, golden tobacco that emerges from this curing process is sweet to smoke. They apply the process to other tobaccos but find that none reacts to flue-curing in the same way (*see* Reynolds, 1875; Camels, 1913).

Chicago merchant Potter Palmer opens a shop that will develop into the city's largest department store (*see* Marshall Field, 1865).

French merchant Aristide Boucicaut joins the Bon Marché at Paris and will turn the small piece-goods shop into the world's first true department store. He institutes revolutionary retailing principles that include small markups to encourage high volume and rapid turnover to compensate for smaller gross margins, fixed and marked prices, and free entrance with no moral obligation to buy. Boucicaut will soon begin extending the right to exchange merchandise or obtain a refund, sales will increase from half a million francs this year to 5 million by 1860, and by 1870 the Bon Marché will occupy an entire city block with sales of 20 million francs in dresses, coats, underwear, shoes, millinery, and the like (*see* 1893; Louvre, 1855; Printemps, 1865).

The safety elevator invented at Yonkers, N.Y., by master mechanic Elisha Graves Otis, 41, will lead to the development of high-rise buildings. Otis sets up ratchets along each side of an elevator shaft at the Yonkers Bedstead Manufacturing Co., he attaches teeth to the sides of the cage, the rope that holds up the cage keeps the teeth clear of the ratchets so long as the rope remains under tension, and when the tension is released the teeth grip the ratchets and hold the cage securely in place. Otis has been about to join the gold rush to California but changes his mind when he receives two unsolicited orders for his "safety hoister" (*see* 1854).

Château-sur-Mer is completed at Newport, R.I., for William S. Wetmore who has made a fortune in the China trade. Richard Morris Hunt will augment the ornate Victorian mansion on Bellevue Avenue with Newport's first French ballroom in 1872 (*see* 1839; 1895).

Napoleon III (above) gives Paris the Bois de Boulogne as a public park and begins his reign with a huge program of public works under the direction of Baron Georges Eugène Haussmann, 42, who is named prefect of the Seine and starts widening streets and opening new boulevards. Named by its inhabitants in 1308 after a pilgrimage to Notre Dame de Boulogne-sur-Mer on the Manche (English Channel), the Bois has been a royal hunting ground for more than two centuries, but Baron Haussman will transform it into a 2,100-acre park modeled after London's Hyde Park.

Hydraulic mining begins in California as miners complete a 45-mile canal to bring in water at the rate of 60 million gallons per day. Gold miner Anthony Chabot devises a canvas hose with an 8-inch thick nozzle that allows gold-bearing gravel to be washed from stream banks into placer pits. This labor-saving device will devastate California hillsides as it brings tons of topsoil and gravel into bottomland for every ounce of gold extracted (*see* 1884).

Philadelphia clergyman Lorenzo Lorraine Langstroth, 42, designs the first practical movable-frame beehive. He has discovered the "bee space" of just under one-quarter inch in height and his hive will revolutionize the beekeeping industry (*see* Huber, 1789).

The Holstein cow that will be most significant to the development of the major U.S. dairy breed arrives aboard a Dutch vessel (*see* 1625).

John Deere's plow factory at Moline, Ill., produces 4,000 plows (*see* 1847; 1855).

Prohibition laws are adopted by Massachusetts, Vermont, and Louisiana (*see* 1855; Maine, 1851).

Service à la Russe arrives in England but will not be common for another 20 or 30 years (*see* 1811). Instead of dishes being placed in turn on sideboards for waiters to serve to guests, a multitude of dishes will continue to be placed before English diners simultaneously in two or three great courses and several minor courses.

Emigration from Ireland has its peak year but will continue on a large scale for years to come (*see* 1847).

The governor of California calls for land grants to encourage continued immigration of Chinese, "one of the most worthy of our newly adopted citizens." Close to 50,000 Chinese have defied China's death-penalty law against emigration to make their way to California's gold fields, and an immigration agency (called the Chinese Six Companies after the six districts of China from which most have come) has been formed to advance money to newcomers to buy picks, shovels, and other mining supplies (*see* Central Pacific, 1861).

1853 Commodore Matthew Calbraith Perry, 59, U.S. Navy, arrives in Edo Bay July 8 with 1,600 men aboard seven black ships, including the steam frigates *Mississippi* and *Susquehanna*. Dispatched in November by President Fillmore with a request for a treaty, Perry demands the treaty "as a matter of right, and not. . . as a favor;" he departs July 16, leaving word that he will return in the spring for a favorable reply.

Japan's Tokugawa shogun Icyoshi dies July 23 at age 61 after a 16-year reign. He is succeeded by his brother Iesada, 29, who is in delicate health, will live only until 1858, but will open two ports to trade next year in order to end a civil war precipitated by the demands of Commodore Perry (above).

A Russian fleet arrives at Nagasaki August 8 aiming to open trade relations and resolve boundaries in the Kuriles and Sakhalin.

British forces end the Second Burmese War, Britain annexes Pegu, and Mindon Min becomes king of Burma, beginning an enlightened reign that will continue until 1878. The new king will build Mandalay and make it the Burmese capital in 1857.

Shanghai falls to rebel forces September 7 as the Tai Ping rebellion continues (*see* 1850). Bandits in Anhuei, North Kiangsiu, and Shandong organize the peasantry and begin a 15-year Nian Rebellion, plundering villages as imperial troops withdraw to cope with the Tai Ping rebels (*see* 1860).

The French emperor Napoleon III marries the Spanish comtesse Eugènie de Montijo, 26, January 9 at the Tuileries Palace. Napoleon has failed to win the hand of a Vasa or Hohenzollern.

Portugal's Maria II (Maria de Gloria) dies November 15 at age 34 after a 27-year reign marked by insurrections in 1846 and 1851 and interrupted by a 5-year civil war during which the queen lived in England. Her son Pedro de Alcantara, 16, will assume power in 1855 after a 3-year regency by his father and reign as Pedro V until his death from cholera in 1861.

Political unrest in Europe over Turkish occupation of holy places in Palestine brings Russia, France, and Britain to the brink of war (*see* 1854).

The Gadsden Purchase treaty signed with Mexico December 30 permits the United States to annex a tract of land south of the Gila River. Mexico receives $10 million under terms of the treaty negotiated by U.S. Minister James Gadsden, 65.

Transportation of British convicts to Tasmania ends after a half-century in which some 67,000 convicts have been landed on the island that is now definitively renamed Tasmania to replace the name Van Diemen's Land (*see* 1642).

The Massachusetts Constitution Convention receives a petition to permit woman suffrage. The appeal comes from the wife of educator and social reformer Amos B. Alcott and 73 other women (*see* Lucy Stone, 1850; Elizabeth Cady Stanton, 1860).

Brunswick, Hanover, and Oldenburg join the German *Zollverein,* bringing all non-Austrian states into the customs union founded in 1833 (*see* 1867).

London bankers Samuel Montagu, 21, and his brother and brother-in-law found a firm and foreign exchange that will become the world's leading gold trader. The family name Samuel has been changed to Montagu by royal license, and the new firm will make London the clearinghouse of the international money market.

The New York Clearing House that opens October 11 at 14 Wall Street is the first U.S. bank clearing house; 38 banks use the facility, and total clearings for the first day amount to nearly $2 million.

Territory acquired in the Gadsden Treaty (above) includes the best railroad route from Texas to California. The treaty also gives the United States right of transit across the Isthmus of Tehauntepec.

Congress appropriates $150,000 March 3 for a survey of the most practicable transcontinental U.S. railroad routes to be made under the direction of the War Department (*see* Central Pacific, 1861; Union Pacific, 1862).

The New York Central, first major U.S. rail combine, is created by a consolidation of 10 short-line railroads between Albany and Buffalo, but gauge standards change at the New York, Pennsylvania, and Ohio state borders, and the city of Erie, Pa., tears up rails of identical gauge in order to force trains to stop at Erie (*see* Britain, 1846; Vanderbilt, 1867).

A railroad from Philadelphia reaches the shore of Absecon, N.J., which will become famous as Atlantic City (*see* boardwalk, 1896).

The Vienna-Trieste railway line opens through the Alps (*see* 1854; 1867).

India's first railway line opens to link Bombay with Thana 20 miles away. English engineer James John Berkley, 34, has built the line and will complete the Bombay-Calcutta-Madras-Nagpur line in 1856.

Parliament authorizes construction of a 3.75-mile London underground railway between Farringdon Street and Bishop's Road, Paddington (*see* 1863).

The Boston-built clipper ship *Northern Light* makes a record 76-day 6-hour voyage from San Francisco to Boston.

The *Great Republic* launched by Donald McKay is the world's largest sailing ship but is too big to be a commercial success. The 4,555-ton four-masted barque has patent double topsails and a 15-horsepower steam engine on deck to handle her yards and work her pumps.

1853 ***(cont.)*** I. K. Brunel begins construction of the S.S. *Great Eastern*. The passenger steamer will employ paddle wheels as well as a propeller and will be slightly larger than the *Great Republic* (above) but no more successful commercially (*see* Brunel, 1843).

The British single-screw steamship S.S. *Himalaya* goes into service, a 4,690-ton ship, 340 feet in length overall, that will remain in service for more than 70 years.

U.S. shipping magnate Cornelius van Derbilt acknowledges that he is worth $11 million and that his fortune brings him an annual return of 25 percent (*see* 1850; 1858).

The first manned heavier-than-air flying machine soars 500 yards across a valley carrying the terrified coachman of English engineer Sir George Cayley, 80, in a large glider. Cayley defined the problem of heavier-than-air flight in his 1809 paper "On Aerial Navigation" (*see* Wright brothers, 1903).

The *United States Review* predicts that "within half a century "machinery will perform all work—automata will direct them. The only tasks of the human race will be to make love, study, and be happy."

Pocket watches are produced in quantity for the first time by the American Waltham Watch Co. of Waltham, Mass. More than 30 watch factories will spring from the Waltham works but mass production will not begin for 40 years and wristwatches will not appear for 54 (*see* Ingersoll, 1892).

Samuel Colt opens a Hartford, Conn., armory with 1,400 machine tools that will revolutionize the manufacture of small arms (*see* 1846; 1871).

London physician John Snow traces a local cholera epidemic to a busy public pump in Broad Street whose water supply is contaminated by the cesspool of a tenement in which a cholera patient resides. Asked how the epidemic can be stopped, Snow replies, "Remove the pump handle" (*see* Snow, 1847).

Stephen Foster's most popular song had its debut June 15, 1853. It helped arouse sympathy for America's slaves.

A yellow fever epidemic at New Orleans kills 7,848.

Alsatian chemist Karl Frederick Gerhardt, 37, at Montpelier University produces acetyl salycilic acid but his procedure is impossibly tedious and time-consuming (*see* 1763; 1835; aspirin, 1899).

The University of Melbourne is founded in Australia.

Washington University is founded at St. Louis (*see* medicine, 1910).

The University of Florida is founded at Gainesville.

Nonfiction *The Stones of Venice* by English art and social critic John Ruskin, 34, whose three-volume study explains the development of Gothic architecture and its beauty in terms of the moral virtue of the medieval society that produced it, attacking "the pestilent art of the Renaissance" that arose, he says, from an immoral society.

Fiction *Bleak House* by Charles Dickens with illustrations by "Phiz"; *Peg Woffington* by Charles Reade (he has adapted last year's play *Masks and Faces* at the suggestion of actress Laura Seymour, who will be his housekeeper beginning next year); *Cranford* and *Ruth* by Mrs. Gaskell; *Villette* by Charlotte Brontë.

Juvenile *Tanglewood Tales for Girls and Boys* by Nathaniel Hawthorne.

Poetry *Poems* by English poet-critic Matthew Arnold, 31, include "Sohrab and Rustum," "Scholar Gypsy," and "Requiescat."

Painting *Le Tepidarium* by Théodore Chassériau.

Theater *Gold* by Charles Reade 1/10 at London's Drury Lane Theatre.

Opera *Il Trovatore* (*The Troubadour*) 1/19 at Rome's Teatro Apollo, with music by Giuseppe Verdi; *La Traviata* 3/6 at Venice's Teatro la Fenice, with a libretto based on last year's Alexandre Dumas play, *La Dame aux Camélias*, music by Giuseppe Verdi who has completed it in 4 weeks (the production in modern dress is a dismal failure).

First performances Fantasia on Hungarian Melodies for Pianoforte and Orchestra by Franz Liszt 6/1 at Pest's Hungarian National Theater, with Hans von Bülow as soloist; Symphony No. 1 in E flat major by French composer (Charles) Camille Saint-Saëns, 17, at Paris.

Popular songs "My Old Kentucky Home, Good Night" and "Old Dog Tray" by Stephen C. Foster.

Boston piano maker Jonas Chickering dies at age 55 as his burnt-out factory is being replaced by a building surpassed in size only by the U.S. Capitol at Washington. The new plant has 220,000 square feet of floor space and a huge inner court for Chickering & Sons whose Thomas E., C., Frank, and George H. Chickering carry on the 30-year-old firm.

Freehold Raceway opens at Freehold, N.J., with harness races.

The first U.S world's fair opens July 14 at New York in a Crystal Palace Exposition modeled on the 1851 London Great Exhibition.

The safety match is patented by Swedish inventor J. E. Lundstrom (*see* 1836; book matches, 1892).

The Mount Vernon Hotel that opens at Cape May, N.J., is the world's first hotel with private baths (*see* London's Savoy, 1889).

Charleston's 125-room Mills House opens at the corner of Meeting and Queen streets where it has been built at a cost of $300,000 by Otis Mills, known in South Carolina as the John Jacob Astor of Charleston.

New York's state legislature authorizes the city to purchase land for a public park, and some 624 acres are acquired from 59th Street north to 106th Street between Fifth and Eighth Avenues (*see* Bryant, 1844; Olmsted, 1851).

The world's largest tree is discovered in California and called the *Wellingtonia gigantea.*

Continued decline in French oyster production disturbs the government at Paris which sends the embryologist-ichthyologist Coste to study the remains of ancient Roman oyster culture in Lake Fusaro (*see* 175 B.C.; 1375; 1840).

Fewer than half of Americans are engaged in agriculture, down from 83 percent in 1820.

The King Ranch has its origin as Texas steamboat captain Richard King begins acquiring land on Santa Gertrudis Creek between the Rio Grande and Neuces rivers (*see* 1850). Acting on the advice of his friend Col. Robert E. Lee, 46, U.S. Army, King pays a Spanish family $300 for 15,500 acres—less than 2¢ per acre (*see* 1860).

Extreme drought parches the Southwest, lowering prices of land and cattle. Richard King (above) is able to buy longhorns at $5 per head.

Texas cattleman Samuel A. Maverick, 50, receives an anonymous note telling him to look after his strays on the Matagorda Peninsula or risk losing them. Maverick accepted 400 head of cattle in payment of a debt in 1845 and has never bothered to brand the animals on his 385,000-acre ranch, but he sends down a party of cowhands who round up about the same number that were turned loose on the peninsula 6 years ago, sells the cattle to Toutant Beauregard of New Orleans, and retires from ranching. The word "maverick" will become a generic for unbranded cattle and for political or intellectual independents who do not go along with their associates.

Concord, Mass., horticulturist Ephraim Wales Bull, 48, exhibits the first Concord grapes to the Massachusetts Horticulture Society. He has developed the slip-skin Lambrusca variety and begins selling cuttings from the parent vine at $5 each. Within 10 years Concords will be growing all across America (*see* Welch, 1869).

Potato chips are invented at Saratoga Springs, N.Y., where chef George Crum of Moon's Lake House gives a mocking response to a patron who has complained that his French fries are too thick. He shaves some potatoes paper thin and sends them out to the customers—who are delighted, order more, and encourage Crum to open a restaurant of his own across the lake. Crum's new restaurant will take no reservations and millionaires including Jay Gould and Commodore van Derbilt (above) will stand in line along with everyone else (*see* Wise, 1921; Lay's, 1939).

Keebler biscuits are introduced at Philadelphia by baker Godfrey Keebler whose Steam Cracker Bakery will become a leading U.S. producer of cookies (biscuits) and crackers.

Gail Borden succeeds in his efforts to produce condensed milk (*see* 1851). Using vacuum pails obtained from Shakers at New Lebanon, N.Y., he finds a formula for a product that has no burnt taste or discoloration and lasts for nearly 3 days without souring. Borden travels to Washington to file a patent claim (*see* 1855).

1854

The Crimean War that begins March 28 will continue until 1856. British and French fleets have entered the Black Sea January 3, Russia has broken off relations with both nations February 6, the czar has ignored a February 7 Anglo-French ultimatum to evacuate the Danube principalities by April 30, London and Paris have made an alliance with Constantinople March 12, Russian troops have crossed the Danube March 20, and Britain and France declare war March 28.

Allied troops land on the Crimean Peninsula of the Ukraine in mid-September and defeat an inferior Russian force September 30 at the Battle of the Alma River.

The Battle of Balaclava October 25 ends in victory for the allies and includes a cavalry charge of Britain's Light Brigade led by the infamous Commander James Thomas Brudenell, 57, seventh earl of Cardigan, who has made a career of purchasing military commissions, fighting duels, and womanizing. He is the first man to reach the lines and emerges unscathed, but Russian artillery cuts down 503 of his 700 men (*see* Tennyson poem, below).

The Battle of Inkerman November 5 gives the allies another major victory.

The Danube principalities have been evacuated by the Russians since August 8 and occupied with Turkish consent since August 22 by Austria, which gave Russia an ultimatum June 3 not to carry the the war across the Balkan mountains and signed a June 14 treaty with Constantinople agreeing to intervene in Bosnia, Albania, or Montenegro should any disturbance break out there.

The British withdraw from territory north of South Africa's Orange River in accordance with the Convention of Bloemfontein signed February 17 with the Boers. Boer settlers organize the Orange Free State (*see* 1852; Pretorius, 1856).

1854 ***(cont.)*** Egypt's khedive Abbas I is assassinated July 13 at age 41 by enemies in his court after a 6-year reign. His uncle, 32, will reign until 1863 as Said Pasha (*see* Suez Canal concession, below).

The Treaty of Kanagawa signed March 31 opens the Japanese ports Hakodate and Shimoda to U.S. trade, makes provision for shipwrecked sailors, and establishes friendly relations. Commodore Perry has returned in February with ten ships. Japan needs a defense force, the Tokugawa shogun Iesada decides, and he asks the emperor to order Buddhist temples to contribute their great bells for gun metal.

The Elgin Treaty signed June 5 establishes reciprocity between Canada and the United States.

The Republican party organized February 28 at Ripon, Wis., by former Whigs and disaffected Democrats opposes the extension of slavery. They take the name Republican from the party founded by Thomas Jefferson, which became known as the Democratic-Republican party but dropped the last part of its name in 1828.

The Kansas-Nebraska Act passed by Congress May 20 repeals the Missouri Compromise of 1820, opens Nebraska country to settlement on the basis of popular sovereignty, provides for the organization of Kansas and Nebraska territories, undoes the sectional truce of 1850, and effectively destroys the Whig party whose northern members join the new Republican party (above).

Commodore Perry opened Japan to the West. His Caucasian nose amused the Asians, and artists exaggerated it.

Quixiault, Lummi, Nisqualie, Puyallup, and other Puget Sound tribes that have ceded their lands in the Northwest receive treaty rights to fish for the salmon which climb the rivers of the region—rights guaranteed "in common with white men."

The Wisconsin Supreme Court releases a Wisconsin man convicted by a federal court of rescuing a runaway slave in violation of the 1850 Fugitive Slave Act; the court rules that the act is unconstitutional and therefore invalid (*see* 1842; Dred Scott decision, 1857).

A Boston mob attacks a federal courthouse May 26 in a vain attempt to rescue the fugitive slave Anthony Burns. Federal troops are called in to escort Burns to the Boston docks for return to his southern owner, and they march the prisoner through streets lined with silent but outraged citizens.

Scots-American journalist James Redpath, 21, of Horace Greeley's *New York Tribune* travels through the slave states urging slaves to run away.

U.S. Roman Catholics come under attack on numerous occasions throughout the year from members of a new political party which holds its first national convention in November at Cincinnati. Composed of "native-born Protestants," the new American party opposes immigration; New York publisher E. Z. C. "Ned Buntline" Judson calls it the Know-Nothing party (*see* Ned Buntline, 1844; 1849).

Lawrence, Kan., is founded by two parties sent out by the New England Emigrant Aid Society, a group organized by Massachusetts educator Eli Thayer, 34, to encourage emigration to Kansas. The town is named after the society's patron, Boston merchant Amos Lawrence, now 67 (*see* 1845; Brown, 1856).

Omaha is formally founded in the Nebraska Territory (*see* 1825).

The Kansas-Nebraska Act (above) signed into law by President Pierce May 30 opens to white settlement western lands that have been reserved by sacred treaty for the Indians.

The U.S. Mint opens a San Francisco branch and pays miners the official rate of $16 per ounce for gold (which can drop to as low as $8 per ounce in the open market but which can also go much higher than $16). The mint produces $4 million in gold coins in its first year, and will produce nearly $24 million by 1856.

Yale chemistry professor Benjamin Silliman the younger, 38, makes the first fractional distillation of crude petroleum. He has been asked by New Englander George Henry Bissell, 33, to analyze a sample of Pennsylvania "rock oil" which burns better than the coal oil squeezed from asphalt, cannel coal, coal tar, and "albertite" bituminous rock (*see* 1852). Pennsylvania Rock Oil of New York, founded by Bissell and a New York partner, is the first company organized to exploit the potential of Pennsylvania's petroleum deposits. Higher boiling fractions of petroleum will prove useful as lubricants, but lower boiling fractions such as gasoline will be avoided as a menace that causes lamps to explode and is difficult to eliminate without causing complaints about pollution (*see* 1859; kerosene, 1855).

The Pennsylvania Railroad headed by J. Edgar Thomson, 46, opens its line between Harrisburg and Pittsburgh (*see* 1852; 1857).

The B&O opens its line between Baltimore and Wheeling (*see* 1851).

The Chicago, Burlington & Quincy is created by a merger of four small lines. Entrepreneur James Frederick Joy, 43, who has engineered the merger (the new corporate name will be adopted February 14 of next year), has been helped in earlier effort to extend the Michigan Central to Chicago by Springfield, Ill., lawyer Abraham Lincoln, 45, and now works to extend the Burlington (*see* Quincy Bridge, 1868).

Railroads reach the Mississippi (above), but cattlemen continue to drive Texas cattle great distances overland as drought continues in the Southwest. The first Texas longhorns to reach New York arrive after a long trek that has made their meat tough and stringy.

Passage of the Kansas-Nebraska Act (above) has been pushed by Sen. Stephen Arnold Douglas, 41, (D. Ill.) who has introduced the term "popular sovereignty" to replace the less elegant term "squatter sovereignty." Douglas has been motivated in large part by a desire to facilitate construction of a transcontinental railway.

The Vienna-Trieste rail line opens over Semmering Pass with tunnels and steep gradients that make it an engineering marvel.

The side-wheeler S.S. *Arctic* sinks September 27 off Cape Race, Newfoundland, after colliding with the 250-ton French iron propeller ship S.S. *Vesta*. A sistership to the S.S. *Pacific* that went into service in 1852, the 3,000-ton S.S. *Arctic* has been the largest and most splendid of the Collins Line steamships, her casualties include 92 of her 153 officers and men, and all her women and children are lost including the wife, the only daughter, and the youngest son of E. K. Collins (*see* 1850; 1852).

Egypt's new khedive Said Pasha (above) grants a Suez Canal concession November 30 to his long-time friend Ferdinand de Lesseps, 49, a French diplomat-promoter who has dreamed of such a canal for 25 years (*see* 1859; Darius, 520 B.C.).

"An Investigation of the Laws of Thought, on Which Are Founded the Mathematical Theories of Logic and Probabilities" by self-taught English mathematician George Boole, 39, advances the first workable system substituting symbols for all the words used in formal logic. Boole's paper "The Mathematical Analysis of Logic" appeared 7 years ago and won him a professorship in mathematics at Queen's College, York, Ireland, despite his lack of any degree.

A typhus epidemic in the Russian army spreads to the British and French in the Crimea (above), it reaches Constantinople, and it is spread by merchant ships throughout Russia and Turkey.

Florence Nightingale, 34, takes 34 London nurses to Scutari. Superintendent of a London hospital for invalid women, she organizes a barracks hospital where she introduces sanitary measures that will reduce the toll of cholera, dysentery, and typhus (but disease will nevertheless take more lives in the Crimean War than will battle casualties).

Philadelphia dentist Mahlon Loomis, 28, patents a kaolin process for making false teeth.

A Vatican ruling that makes the Immaculate Conception of the Virgin an article of faith implies papal "infallibility" in all matters (*see* 1869; 1870).

Brooklyn Polytechnic Institute is founded at Brooklyn, N.Y.

University College is founded at Dublin.

Le Figaro begins publication at Paris. Local journalist Jean Hippolyte Auguste Villemessant, 42, will make the weekly paper a daily beginning in 1866 (*see* Coty, 1924).

The *Age* begins publication at Melbourne. Australian journalists John and Henry Cooke will sell out in 2 years to Ebenezer Syme for £2,000 and Syme's brother David will make The *Age* one of the world's leading dailies.

A paper mill at Roger's Ford in Chester County, Pa., produces paper from wood pulp at low cost. Philadelphia inventor Morris Longstreth Keen, 34, boils wood in water under pressure and will found the American Wood-Paper Company in 1863.

English inventor Hugh Burgess arrives at Roger's Ford (above) where he will build a paper mill to use the soda process for making paper from wood pulp which he has

Crimean War casualties brought Florence Nightingale to Russia. Most deaths came from disease, not cannon.

1854 *(cont.)* developed with inventor Charles Watt (*see* 1798; Tilghman, 1867).

New York's Astor Library opens February 1 just below Astor Place in Lafayette Street (*see* 1848). None of the library's 80,000 volumes may be taken from the building, nor can a book even be removed from the shelf unless the visitor is accompanied by a library officer, but by 1873 the library will be averaging 86 visitors per day, 5 percent of them women (*see* Dewey, 1876; New York Public Library, 1895).

Nonfiction *Walden, or Life in the Woods* by Henry David Thoreau: "The mass of men lead lives of quiet desperation" (1, "Economy") and "If a man does not keep pace with his companions, perhaps it is because he hears a different drummer. Let him step to the music which he hears, however measured or far away" (18, "Conclusion"); *Die Geschichte der Deutscheit,* volume I of a massive German language dictionary by Jacob Grimm of 1815 *Grimm's Fairy Tales* fame.

Fiction *Hard Times* by Charles Dickens; *Ten Nights in a Barroom and What I Saw There* by Philadelphia novelist Timothy Shay Arthur, 45, whose work about the evils of drink will be adapted to the stage and be the leading temperance drama for decades (*see* 1844; "Come Home Father," 1864).

Poetry "The Charge of the Light Brigade" by Alfred Tennyson glorifies the action by Lord Cardigan at the Battle of Balaclava (above): "Half a league, half a league, half a league onward,/ Into the Valley of Death rode the Six Hundred. . ."

Painting *The Reaper* by Jean François Millet; *Bonjour, Monsieur Courbet* by Gustave Courbet; *The Light of the World* by Holman Hunt; *The Falls of Tacemdama* by U.S. painter Frederic Edwin Church, 28, who has traveled to South America.

Theater *Poverty Is No Crime* (*Bednost ne porok*) by Russian playwright Aleksandr Nikolayevich Ostrovsky, 30, 1/25 at Moscow's Maly Theater; *The Courier of Lyons* by Charles Reade and Tom Taylor 6/26 at the Princess's Theatre, London; *One Must Not Live as One Likes* (*Ne tak zhvi, Kak Khochetsya*) by Ostrovsky 12/3 at Moscow's Maly Theater.

The New York Academy of Music opens at the northeast corner of 14th Street and Irving Place (*see* Metropolitan Opera House, 1883; Carnegie Hall, 1891).

German composer Robert Schumann attempts suicide February 27 in a fit of depression (he suffered a mental breakdown in 1844 but recovered). Boatmen in the Rhine pick Schumann up after he jumps from a bridge, he is committed to an asylum 5 days later, and will die there in 1856 at age 46.

First performances *Orpheus* (symphonic poem) by Franz Liszt 2/16 at Weimar (originally a new overture to the Gluck opera *Orfeo ed Euridice* of 1762); *Les Préludes* (Symphonic Poem No. 3) after Lamartine's *Méditationies Poétiques* by Franz Liszt 2/23 at Paris; *Mazeppa* by Liszt 4/16 (Easter Sunday) at Weimar's Grand Ducal Palace; *Fest-Lange* (Symphonic Poem No. 7) by Liszt 11/9 at Weimar.

Oratorio *L'Enfance du Christ* by Hector Berlioz 12/10 at the Salle Herz, Paris.

Popular song "Jeanie with the Light Brown Hair" by Stephen C. Foster.

The name of Lord Cardigan (above) will survive in the cardigan sweater; that of his commanding officer Field Marshal Fitzroy James Henry Somerset, 66, baron Raglan of Raglan, in the raglan-sleeve coat.

A disposable paper collar patented by sewing machine pioneer Walter Hunt of 1849 safety pin fame will gain great popularity, but only after Hunt's death. He takes pains to patent his new invention and secure a royalty agreement for its production (*see* 1858).

The Otis safety elevator invented 2 years ago impresses visitors to a New York industrial fair. Elisha G. Otis has thus far sold only three of his elevators, but he has himself hoisted aloft, orders the rope to be cut, and plunges melodramatically earthward as spectators gasp and scream; safety ratchets engage to halt his descent, and Otis emerges from his elevator cage saying, "All safe. All safe, ladies and gentlemen" as he sweeps the stovepipe hat from his head and takes a bow (*see* 1857).

The Supreme Court confirms Cyrus McCormick's reaper patents in the case of *Seymour v. McCormick.* Baltimore lawyer Reverdy Johnson, 60, and Washington lawyer Thaddeus Stevens, 62, defend the reaper inventor (*see* 1851; Stanton, 1861).

Henry David Thoreau (above) describes ice harvesting on Walden Pond, saying, "They told me that in a good day they could get out a thousand tons, which was the yield of about one acre. . . They told me that they had some in the ice-houses at Fresh Pond five years old which was as good as ever. . . The sweltering inhabitants of Charleston and New Orleans, or Madras and Bombay and Calcutta, drink at my well" (*see* 1856; Tudor ice ships, 1846).

1855 "The Angel of Death has been abroad throughout the land," says British orator John Bright in a February speech to the House of Commons. He urges withdrawal of British troops from Sevastopol and the Crimea. Viscount Palmerston, now 70, becomes prime minister February 5 and will retain the position until his death in 1865.

Russia's Nicholas I dies March 2 at age 58 after remarking that "Generals January and February" would be his best allies against British and French troops in the Crimea. The "Iron Czar" who has reigned since 1825 is succeeded by his son, 36, who will reign until 1881 as Aleksandr II, bringing great reforms to the nation (*see* 1861).

Savoy's premier Count Camillo Benso di Cavour, 45, has brought his country into the Crimean War on the side of England and France. A 10,000-man Piedmontese force helps win the Battle of Chermaia August 16.

"Here I am and here I stay" ("J'y suis, j'y reste"), says the French general Marie Edme Patrice Maurice de

MacMahon, 47, in the trenches before Malakoff September 8. He leads the assault on Malakoff.

Russian forces abandon Sevastopol September 11, sinking their ships and blowing up their forts to keep them from falling into enemy hands.

The Treaty of Peshawar signed March 30 creates an Anglo-Afghan alliance against Persia whose Nasr Ud-den has designs on Herat. Russian forces complete a 5-year conquest of the Syr Darya Valley that marks the beginning of a Russian advance to the Persian frontier in Central Asia (*see* 1856).

Japan's Tokugawa shogun Iesada signs treaties with Russia in February and the Netherlands in November; the foreigners call him "tycoon," mistakenly believing he is the "secular emperor."

British diplomat Harry Smith Parke, 27, negotiates the first treaty with Siam, obtaining rights to establish consuls and trade throughout the kingdom of Rama IV. Siam will sign similar treaties next year with the United States and France, and with other countries thereafter.

Ethiopian chief Ras Kassa, 37, deposes Ras Ali of Gondar, conquers the rulers of neighboring Tigre, Goijam, and Shoa, proclaims himself king of kings, and begins a 13-year reign as the emperor Theodore. Englishmen Walter Plowden and John Bell will help him put down rebellions by subordinates.

The Young Women's Christian Association (YWCA) is founded at London to improve the condition of working girls by providing good food and a decent place to sleep for young women living away from home (*see* YMCA, 1844, 1851).

Scottish missionary David Livingston, 42, discovers falls on the Zambezi River that will be called Victoria Falls. He has organized a great expedition northward from Cape Town through West Central Africa (*see* Stanley, 1869, 1871; slavers, 1871).

Personal Narrative by British explorer Richard Francis Burton, 34, describes a pilgrimage he made 2 years ago to the sacred Muslim city of Mecca disguised as a Pathan. Burton is in Africa in the company of explorer John Hanning Speke, 28 (*see* 1862).

Kerosene gets its name from Long Island, N.Y., physician Abraham Gestner of Newtown Creek who makes it from raw petroleum, promotes it as a patent medicine (in the absence of any lamp or stove that can burn it efficiently as a fuel), and coins the word "kerosene" from the Greek "keros" (wax) (*see* 1852; 1859; Young, 1851).

The Sault St. Marie ("Soo") River Ship Canal opens to link Lake Huron and Lake Superior and to make the Great Lakes a huge inland waterway navigable by large ships. Built by engineer Charles T. Harvey with capital from Chicago, New York State, and St. Johnsbury, Vt., the "Soo" will give Chicago access to the vast iron ore deposits of northern Michigan and Minnesota.

The new Cleveland Iron Mining Co. started by Samuel Livingston Mather, 38, ships the first load of ore to go through the new "Soo" Canal (above).

John A. Roebling completes a suspension bridge across Niagara Gorge after Charles Ellet has quit in a dispute over finance (*see* 1849). A 368-ton train crosses the 821-foot single-span Roebling bridge March 6 and is the first train to cross a bridge sustained by wire cables (*see* Cincinnati-Covington bridge, 1867).

Australian rail transport is thrown into confusion as New South Wales changes from a 5-foot 3-inch gauge to a narrower 4-foot 8.5-inch gauge while South Australia, West Australia, and Queensland adopt a 3-foot 6-inch gauge for reasons of economy.

The Lehigh Valley Railroad has its beginnings in a line built between Easton and Mauch Chunk by Pennsylvania coal mine operator Asa Packer, 50. The road will grow to have 1,364 miles of track from New York to Buffalo.

An accident on the Camden and Amboy Railroad near Burlington, N.J., August 29 kills 21, injures 75.

The Chicago & Rock Island Railroad works to extend its line to Iowa City, Iowa, in response to an offer of $50,000 if the road reaches Iowa City by December 3. With only minutes to spare, workers drag the locomotive the final 1,000 feet with ropes in -30° weather (*see* 1852; 1856; song, 1863).

The Paris International Exposition that runs from May to November hails French technological and economic progress as France gains military prestige in the Crimea (above).

French chemist Henri Etienne Sainte-Claire Deville, 36, pioneers commercial aluminum production with a practical method for producing the metal (*see* Wöhler, 1827; Hall, 1886).

Celluloid is patented by English chemist Alexander Parkes, 42, who has developed the first man-made plastic material from guncotton and camphor (*see* Hyatt, 1872).

Principles of Psychology by English philosopher Herbert Spencer, 35, is published at London. Spencer has developed a doctrine of evolution as applied to sociology (*see* Darwin, 1858, 1859).

Continuing epidemics of cholera and typhus in the Crimea take more lives than battle wounds or food poisoning episodes, but English physician Edmund Alexander Parkes, 35, serving as superintendent of a civil hospital in the Dardanelles, pioneers the science of modern hygiene (*see* 1854).

English physician Thomas Addison, 62, gives the first description of a kind of anemia that is inevitably fatal and will later be called *pernicious* (*see* Whipple, 1922; Minot and Murphy, 1924).

The Amana community is founded on the Iowa frontier at a site christened Amana by Prussian-American sectarian leader Christian Metz, 61, who started his Community of True Inspiration sect in 1817 and has led a communal society near Buffalo, N.Y., for the past 12 years. The community will be incorporated in 1859 as the Amana Society.

1855 ***(cont.)*** The first U.S. kindergarten opens at Watertown, Wis., where the wife of German immigrant Carl Schurz has started the school for children of other immigrants (*see* Froebel, 1837; Peabody, 1860).

Elmira Female College, founded at Elmira, N.Y., will be the first U.S. institution to grant academic degrees to women (*see* Mount Holyoke, 1837; Mills, 1852; Vassar, 1861).

Pennsylvania State University is founded.

Michigan State University is founded at East Lansing.

Congress authorizes a telegraph line to link the Mississippi River with the Pacific Coast and commissions James Eddy and Hiram Alden to construct it (*see* San Francisco, 1861).

A telegraph line opens to link London with Balaclava in the Crimea as war continues.

The *Daily Telegraph* begins publication June 29 following repeal of Britain's Stamp Act. Other British papers go from weekly publication to daily, but the *Daily Telegraph* will have double the circulation of the Times by 1860, and by 1870 will be the largest paper in the world with a circulation of more than 270,000.

The *Chicago Tribune* founded 8 years ago comes under the control of Cleveland publisher Joseph Medill, 32, who will run the paper for the next 44 years as Chicago grows from a town of 80,000 to a city of nearly 2 million (*see* McCormick, Patterson, 1910).

Leslie's Illustrated Newspaper is started at New York by English-American publisher Frank Leslie, 34, who has changed his name from Henry Carter, has gained experience as an engraver for the *Illustrated London News* and been a poster artist for P. T. Barnum (*see* 1842; Borden, 1857).

English-American inventor David Edward Hughes, 24, patents a teleprinter. He will later lay out the basic principles of the "loose contact" on which hearing aids, the telephone, and other forms of the microphone will be established (*see* 1877).

Nonfiction *Familiar Quotations* by Boston publisher John Bartlett, 35, whose work will be periodically updated and reissued for well over a century; *The Age of Fable* by Boston scholar Thomas Bulfinch, 59, a son of the architect, whose work will be popular for generations as *Bulfinch's Mythology*.

Fiction *Westward Ho!* by English clergyman-novelist Charles Kingsley, 36, rector of Eversley in Hampshire, who describes in vivid terms a South American landscape he has never seen; *North and South* by Mrs. Gaskell; *The Warden* by English novelist Anthony Trollope, 40, who has been a postal inspector in Ireland since 1841.

Poetry *Men and Women* by Robert Browning whose poem "Andrea del Sarto" contains the line, "A man's reach should exceed his grasp"; *Leaves of Grass* by former *Brooklyn Eagle* editor Walt (Walter) Whitman, 36, whose volume of 12 poems has been published at his own expense, receives mixed reviews, and sells few copies. Included are poems that will later be titled "Song of Myself," "I Sing the Body Electric," and "There Was a Child Went Forth." An expanded edition will appear next year, a still larger third edition in 1860; "Song of Hiawatha" by Henry Wadsworth Longfellow who gave up his Harvard teaching post last year: "By the shores of Gitche Gumee,/ By the shining Big-Sea-Water. . ."

Painting *Interior of the Studio: A Real Allegory Summing up Seven Years of My Life as an Artist* by Gustave Courbet whose work has been rejected by the Paris Exhibition and who has mounted an exhibition of his own in a nearby shed; *The Horse Fair* by French painter Rosa (Marie Rosalie) Bonheur, 33.

English photographer Roger Fenton captures Crimean War scenes using the slow wet collodion process of 1851 but his work cannot be published for lack of technology (*see* Ives, 1886).

Wood engravings of some of Fenton's scenes appear in the *Illustrated London News*, giving a new sense of reality to war, even though action shots are impossible (*see* Brady, 1864).

Opera *The Sicilian Vespers* (*I Vespri Siciliani*) 6/13 at the Paris Opéra, with music by Giuseppe Verdi.

First performances Concerto No. 1 for Pianoforte and Orchestra in E flat minor by Franz Liszt 2/17 at Weimar's Grand Ducal Palace, with Liszt at the piano under the direction of Hector Berlioz; "Te Deum" by Hector Berlioz 4/30 at Paris; *Prometheus* (Symphonic Poem No. 5) by Liszt 10/18 at Braunschweig; Trio in B major by German composer Johannes Brahms, 22, 11/27 at Dodsworth's Hall, New York.

Popular songs "Listen to the Mockingbird" by Philadelphia barber-composer Richard Millburn, lyrics by Alice Hawthorne (Septimus Winner, who writes under his mother's maiden name); "Star of the Evening" by James M. Sayles; "Come Where My Love Lies Dreaming" by Stephen C. Foster.

Les Grands Magazins du Louvre opens on the rue de Rivoli across from the Louvre Museum in Paris. The new dry goods store employs principles initiated by Aristide Boucicaut at the Bon Marché in 1852 and attracts visitors to the International Exposition (above).

Boston hotelman Harvey D. Parker opens the Parker House hotel in School Street (*see* 1925; meals, below).

Engineers dredge the malodorous Chicago River to create landfill on which an enlarged city will rise (*see* Union Stockyards, 1865).

Fairmount Water Works, built at Philadelphia, will be the basis of 3,845-acre Fairmount Park, largest public park within any American city (*see* Centennial Exposition, 1876).

London modernizes its sewer system following a new cholera outbreak.

The *Milwaukee Wisconsin* reports that a New York poultry dealer has received a shipment of 18,000 passenger pigeons in a single day.

Scots-American James Oliver, 32, at South Bend, Ind., invents a steel plow whose working surface is remarkably smooth without being brittle. The surface of Oliver's plow is made from a steel casting that is chilled more quickly than its back-supporting sections (*see* 1869; Deere, 1837).

Some 16 million bushels of U.S. wheat are exported to Europe, up from 6 million in 1853 (*see* 1861).

Food and Its Adulterations: Comprising the Reports of the Analytical Sanitary Commission of 'The Lancet' is published at London (*see* 1850; 1851). A. H. Hassall scandalizes England with reports that all but the most costly beer, bread, butter, coffee, pepper, and tea contains trace amounts of arsenic, copper, lead, or mercury, but Crosse & Blackwell announces that it has stopped coppering pickles and fruits and will no longer use bole armenian to color sauces.

A £25 domestic gas oven introduced by the English firm of Smith and Phillips is one of the first such stoves to be put on the market in an age of coal and wood stoves (*see* Reform Club, 1838). Cooking fuel is so costly that hot meals in most British homes are prepared only two or three times a week, but gas ovens will not come into common use for another 40 years.

A military cookstove that can prepare food for a battalion with 47 pounds of wood instead of the usual 1,760 pounds is demonstrated August 27 by Alexis Soyer who has been chef at London's Reform Club for the past 12 years.

Gail Borden obtains a patent for his condensed milk containing sugar to inhibit bacterial growth (*see* 1853; 1856). Unsweetened condensed milk will not be canned satisfactorily until 1885.

Miller's beer has its origin as Milwaukee brewer Frederick Miller buys the Best Brothers brewery which sent the first shipment of Milwaukee beer to New York in 1852. The brewery will be rebuilt in 1888 and the company reorganized under the name Frederick Miller Brewing Co. (*see* 1856).

Prohibition laws are adopted by Delaware, Indiana, Iowa, Michigan, New Hampshire, New York, and the Nebraska Territory (*see* 1852; national party, 1869).

Boston's Parker House (above) serves à la carte meals at all hours of the day instead of requiring that guests sit down at mealtimes to be served a meal dished out by the host. Its kitchen will become famous as the birthplace of the soft Parker House roll.

The New York State Immigration Commission leases Castle Garden at the foot of Manhattan to receive immigrants. 400,000 arrive in the course of the year (*see* Ellis Island, 1892).

Australia's Victoria colony acts in response to the influx of 33,000 Chinese who have helped swell the population to 333,000 from 77,000 in 1851 when gold was discovered in New South Wales. A new law restricts immigration of Chinese and provides for a poll tax of £10 on each Chinese immigrant.

1856 The Crimean War ends February 1 as Russia yields to an Austrian ultimatum and agrees to preliminary peace terms at Vienna. An Ottoman edict guarantees Christian rights (below), reassuring the European powers. The Treaty of Paris March 30 neutralizes the Black Sea and part of Bessarabia, and the powers promise to respect the independence and integrity of the Ottoman Empire.

The Victoria Cross "For Valour" established in January by royal warrant will be Britain's highest miliary decoration.

Persian forces occupy the Afghan town of Herat, precipitating a new Anglo-Persian War (*see* 1857).

Britain annexes the Indian province of Oudh west of Lucknow (*see* mutiny, 1857).

Britain makes Natal a crown colony July 12 (it has been part of the Cape Colony) (*see* 1843).

Boers in South Africa establish the South African Republic (Transvaal) with the year-old town of Pretoria as its capital and Marthinus Wessels Pretorius, 37, as president (*see* 1854; 1877; 1880).

British and French forces seize Canton, precipitating a new Anglo-Chinese War.

U.S. adventurer William Walker, 32, sacks the Nicaraguan capital Granada, gains U.S. recognition of his regime in May, and becomes president in July. Walker led an armed force into the country last year at the invitation of a native revolutionary faction and has become virtual dictator; he envisions a Central American military empire based on slave labor, agricultural development, and a cross-isthmus canal (*see* van Derbilt, 1857).

U.S. voters elect Pennsylvania Democrat James Buchanan, 65, president. He receives 174 electoral votes to 114 for Republican John C. Frémont.

The Hatt-I-Humayun edict issued February 18 by the Ottoman sultan Abdul Mejid guarantees Christian subjects life, honor, and property, ends the civil power of Christian church heads, abolishes torture, reforms prisons, guarantees full liberty of conscience, opens all civil offices to every subject of the sultan, makes Christians liable for military service but permits them to buy exemption, and permits foreigners to acquire property under certain circumstances. Austrian, British, and French ambassadors have forced the sweeping reforms on Constantinople following the Turkish defeat in the Crimean War (above).

The Kansas-Nebraska Act of 1854 was a "swindle," says U.S. senator from Massachusetts Charles Sumner, 45, May 20 during a Senate debate on the admission of Kansas to the Union. Sumner inveighs against the act's authors Stephen A. Douglas and Andrew P. Butler, 62, of South Carolina, condemning the "harlot slavery" and the "incoherent phrases" spoken in its behalf by Sen. Butler. Butler's nephew, Rep. Preston Smith Brooks, 38 (D. S.C.), physically assaults Sumner 2 days later as he sits in the Senate Chamber, beating him so badly that he will not be able to resume his duties for 3 years;

1856 ***(cont.)*** Massachusetts voters will defiantly reelect Sumner next year.

Lawrence, Kan., is sacked May 21 by pro-slavery "border ruffians" who have poured into the territory by the thousands in a move to pack the territorial legislature of "bleeding Kansas" with men who will vote to make Kansas a slave state under terms of Stephen A. Douglas's "popular sovereignty" idea.

The Kansas territorial legislature indicts Free Soil leaders for treason and pitched battles ensue between free-soilers and slavery proponents (although some of the raids, lootings, lynchings, and murders arise out of claim-jumping, not the slavery issue).

Kansas abolitionist John Brown, 56, of Osawatomie and his followers attack pro-slavery men along Pottawatomie Creek May 24. They hack five of the slavery advocates to death in revenge for the sacking of Lawrence (*see* Harper's Ferry, 1859).

South Carolina governor James H. Adams urges repeal of the 1807 law against trading in slaves.

Banque Credit Suisse is founded at Zurich.

Marshall Field, 22, moves to Chicago to begin a career that will make him a legend among merchants. The Pittsfield, Mass., store clerk turns down his employer's offer of a partnership, he takes a job clerking for the dry goods firm of Cooley, Wadsworth, works 18 hours a day at a yearly salary of $400, and by sleeping in the store and buying no clothes except overalls saves half his pay (*see* 1861).

Andrew Carnegie, 20, makes his first investment at the encouragement of his new employer and buys 10 shares of Adams Express stock at $50 per share. The Scots-American railway telegrapher has taken a position as secretary to the Pennsylvania Railroad's Pittsburgh division superintendent Thomas A. Scott, and by 1863 his $500 investment will be returning $1,500 per year in dividends (*see* 1865).

The whaling ship *E. L. B. Jennings* returns to New Bedford, Mass., with 2,500 barrels of sperm oil after a 4.5-year voyage.

The Wabash and Erie Canal opens after 24 years of construction marked by loss of life to cholera and loss of money to embezzlers. The 458-mile canal extending south to Evansville on the Ohio River is the largest ever dug in America, but the section below Terre Haute will close in 4 years and the rest in 1874 as railroads make the canal obsolete.

The first railway bridge to span the Mississippi opens April 21 between Rock Island, Ill., and Davenport, Iowa. Built of wood resting on stone piers, the bridge is tested April 22 by a train of three locomotives and eight passenger cars. The S.S. *Effie Afton* rams the 1,582-foot bridge within 2 weeks, but before the Chicago and Rock Island can sue for damages, the steamboat company files suit claiming that the bridge has blocked its right of way. The courts will call the bridge a public nuisance, but lawyer Abraham Lincoln will persuade the Supreme Court to uphold its legality.

The Illinois Central Railroad line is completed between Chicago and Cairo, Ill.

The Bessemer converter patented by English engineer Henry Bessemer, 42, de-carbonizes melted pig iron with a blast of cold air to produce low-cost steel. The air's oxygen combines with carbon in the iron and dissipates it in the form of carbon dioxide, so although the Bessemer process requires ore that is relatively free of such impurities as phosphorus, the new converter will bring down the price of steel and permit its use in many new applications (*see* Kelly, 1857).

A regenerative smelting furnace invented by German engineer Friedrich Siemens, 30, permits production of ductile steel for boiler plate. Siemens works in England at his brother Wilhelm's works; his invention will lead to development of the open-hearth process for making steel (*see* 1861).

A mauve dye produced from coal tar by English chemistry student William Henry Perkin, 18, is the world's first synthetic dye. Hoping to find a synthetic quinine that will break the Dutch monopoly in cinchona bark, Perkin winds up with a disappointing tarry black solution, but when he dips a piece of silk into the solution he finds it is a stable dye, the first ever made from anything but a root, bark, or berry (*see* Fritzsche, 1841).

W. H. Perkin (above) has been working as assistant to German chemist August Wilhelm von Hofmann, 38, who was brought to London's Royal College of Medicine by the queen's consort Prince Albert. Von Hofmann will persuade young Perkin to develop a German aniline dye industry; synthetic organic dyes will wreck the market for indigo and for the madder roots used to produce the dye alizarine (*see* 1857).

Neanderthal man fossils are found in the Neanderthal Valley of the German Rhineland. Johann C. Fuhrott discovers a human skull in a stratum of rock clearly thousands of years old; French surgeon Paul Broca, 32, world's leading authority on skull structure, maintains that the skull is from an early form of man, quite different in some ways from modern man. He disputes Berlin physician Rudolf Virchow, 35, who says the skull is from an ordinary savage with a congenital skull malformation or bone disease (*see* Cro-Magnon, 1868; Virchow, 1858).

German scientist Theodore Bilharz, 31, identifies the worm parasite that produces kidney and liver malfunctions in the deadly snail-fever disease schistosomiasis (bilharzia). He has come to Cairo to study the disease, will die himself of typhus in 1862, and while syphilis will be effectively controlled well within the next century, bilharzia will affect 70 percent of rural Egyptian men and be an important factor in keeping Egypt's life expectancy to 52 long after other countries have increased their life expectancies to 70 and more.

Auburn University is founded at Auburn, Ala.

St. Lawrence University is founded at Canton, N.Y.

Western Union is chartered as an amalgamation of small U.S. telegraph companies by Ezra Cornell and

Hiram Sibley who financed Samuel F. B. Morse in 1844. Sibley organized the New York and Mississippi Valley Printing Telegraph Co. in 1851, and he becomes president of the new Western Union whose facilities will be greatly expanded (*see* 1859).

Harper's Weekly begins publication at New York where it will continue until 1915 (*see Harper's Monthly*, 1850; Thomas Nast, 1869).

Frankfurter Zeitung begins publication at Frankfurt-am-Main.

Nouveau Dictionnaire de la Langue Française by French lexicographer Pierre Larousse, 39, contains Larousse's dictum, "Un dictionnaire sans examples est un squelette" (skeleton).

Nonfiction *The Rise of the Dutch Republic* by U.S. historian John Lathrop Motley, 42, who has served as secretary to the U.S. legation at St. Petersburg; *A Chronological History of the United States* by Boston historian Elizabeth Palmer Peabody, 52 (*see* kindergarten, 1860).

Fiction *It's Never Too Late to Mend* by Charles Reade details abuses, including torture, in English prisons; *Rudin* by Ivan Turgenev.

Poetry *The Panorama* by John Greenleaf Whittier contains the poems "Maud Muller" and "Barefoot Boy"; "A Farewell" by Charles Kingsley, is a satiric poem containing the line, "Be good, sweet maid, and let who will be clever."

Painting *La Source* by Jean Auguste Dominique Ingres, now 76.

U.S. playwrights get their first legal copyright protection under a new law that releases a flood of dramatic productions.

Moscow's 80-year-old Bolshoi Theater gets a new 2,200-seat opera house that opens in Petrosky Street to replace the house that opened in 1825 but was destroyed by fire 3 years ago.

Popular songs "Darling Nelly Gray" by Otterbein College student Benjamin Russell Hanby at Westerville, Ohio, whose song does much to arouse sympathy for America's slaves; "Gentle Annie" by Stephen C. Foster.

The Burberry raincoat, introduced by English tailor Thomas Burberry of Basingstoke, is made of water-repellent fabric rather than rubberized fabric or oilskin. It will vie with Macintosh and Aquascutum (*see* 1823; 1851; Tielocken, 1910).

Cigarettes are introduced at London clubs by Crimean War veterans who have discovered them in Russia (*see* 1843). The new smokes are generally considered to be effete and effeminate, but Crimean veteran Robert Gloag opens the first British cigarette factory (*see* Philip Morris, 1858).

I. M. Singer & Co. offers a $50 allowance on old sewing machines turned in for new Singer machines—the first trade-in allowances (*see* 1851). Singer's Edward Clark has established 14 branch stores with pretty demonstrators, he follows his trade-in offer with a pioneer installment-buying (hire-purchase) plan that allows $5 monthly rental fees to be applied toward ultimate purchase price, and Singer sales will increase by 200 percent within the year (*see* 1861).

London architects enlarge the 20-year-old Buckingham Palace to give it a new south wing with a ballroom 110 feet long.

The Agen plum that will be the basis of a large prune industry is introduced into California's Santa Clara Valley by French immigrant Pierre Pellier (*see* Chinese plum, 1870).

Xhosa prophets in South Africa encourage tribesmen to believe that legendary heroes are about to return and drive out the whites with whom they have warred for years. The Xhosa slaughter their cattle; two-thirds of them will starve to death in the next few years.

The first calf ever to be butchered in Japan is slaughtered for U.S. envoy Townsend Harris, 52, at Shimoda where he awaits recognition. A New York merchant dispatched by President Pierce to follow up Commodore Perry's opening of Japan and be the first U.S. consul general to that country, Harris has arrived at Shimoda in August. He also has a cow milked—the first cow's milk ever obtained for human consumption in Japan (*see* 1872).

Borden's condensed milk gets a cold shoulder from New York customers accustomed to watered milk doctored with chalk to make it white and molasses to make it seem creamy (*see* 1841; 1855). Borden abandons the factory he has set up with two partners at Wolcottville, Conn., and sells a half-share in his patent to one of his partners (but *see* 1857).

Boston exports more than 130,000 tons of "fine, clear" ice from Massachusetts lakes and ponds as 363 U.S. ships sail from various ports with a total of 146,000 tons of ice, up from 65,000 on 175 ships in 1846 (*see* 1880).

Mechanical ice-making is pioneered by Australian inventor James Harrison, an emigrant from Scotland whose ether compressor makes it possible to produce beer even in hot weather (*see* Perkins, 1834; Gorrie, 1851; Carré, 1858; Linde, 1873).

The German Mills American Oatmeal Factory opened at Akron, Ohio, by German-American grocer Ferdinand Schumacher, 33, employs water-powered millstones to grind 3,600 pounds of oatmeal per day. U.S. farmers grow 150 million bushels of oats per year and while most goes to feed horses, Schumacher will find a large market for his oatmeal among other German immigrants and his mills will become Akron's leading enterprise as he makes himself America's "Oatmeal King" (*see* 1875).

German botanist Nathaniel Pringsheim, 33, observes sperm entering ova, thus advancing human understanding of the reproductive process (*see* 1779; Hertwig, 1875).

1857 Afghan independence gains recognition March 4 in the Treaty of Paris forced upon Persia's Nasir Ud-Din by the British (*see* 1856).

1857 ***(cont.)*** The Sepoy Mutiny that begins May 10 at Meerut ends control of India by the East India Company. The British Army in India has introduced a new Enfield rifle adapted from designs by Samuel Colt following appearances by him before a Parliamentary Committee on Small Arms at London, but the rifle fires cartridges that are partially coated with grease and must be bitten open before being loaded. Sepoys (natives) make up 96 percent of the 300,000-man army that controls a population of 210 million. Caste Hindus among them insist that the grease is beef fat from the sacred cow, Muslims insist it is pork fat from flesh forbidden by the Koran, and local chiefs encourage scattered revolts in hopes of regaining lost privileges.

British losses in the Sepoy Mutiny (above) are small, but a massacre at Cawnpore July 15 takes the lives of 211 British women and children in one of several atrocities perpetrated by both sides.

Few in India want to restore the Mughal Empire, the rebellion is disorganized, and loyal forces from the Punjab recapture Delhi September 20, turning the tide against the rebels (*see* 1858).

The Irish Republican Brotherhood is founded (*see* Fenians, 1867).

Giuseppi Garibaldi founds the Italian National Association to work for unification of the country. Garibaldi returned 3 years ago after 5 years on Staten Island, N.Y., where he worked as a candle maker (*see* 1849; 1860).

Nicaragua's president William Walker seizes overland transportation properties between the Atlantic and Pacific belonging to Cornelius van Derbilt's Accessory Transit Co. (*see* 1856). Van Derbilt sends agents to other Central American nations and organizes an effort to oust Walker, who surrenders himself May 1 to a U.S. Navy officer in order to escape capture, is returned to the United States but goes back to Central America where he is arrested in November by Commodore Hiram Paulding. Returned once again to the United States, he will be captured by a British naval officer in 1860, turned over to Honduran authorities, court-martialed, and executed.

The Sepoy Rebellion ended the East India Company's century-old control of the Indian subcontinent.

The Mountain Meadows Massacre in Utah Territory September 11 kills 135 California-bound emigrants. The Fancher wagon-train arrived at Salt Lake City August 3 from Arkansas with 60 men, 40 women, and 50 children, most of them Methodists. They found the Mormon capital disturbed by the fact that President Buchanan has removed Brigham Young as governor and has sent a force under Gen. Albert Sydney Johnston, 54, to establish the primacy of federal rule in the territory. Some 300 to 400 Pah-Ute braves led by a few Mormons acting on orders from Brigham Young ambush the Fancher party some 320 miles south of Salt Lake City, the wagon-train's riflemen repulse the initial assault, the party holds out from Monday to Friday, Young's friend John D. Lee treacherously induces the emigrants to pile their arms into a wagon, and the Indians butcher to death all but a dozen or so children under age 8.

The Dred Scott decision announced by Supreme Court Chief Justice Roger B. Taney March 6 enrages abolitionists and encourages slaveowners. The fugitive slave Dred Scott, now 62, brought suit in 1848 to claim freedom on the ground that he resided in free territory, but the court rules that his residence in Minnesota Territory does not make him free, that a black may not bring suit in a federal court, and in an *obiter dicta* by Taney, that Congress never had the authority to ban slavery in the territories, a ruling that in effect calls the Missouri Compromise of 1820 unconstitutional.

The Impending Crisis of the South, and How to Meet It by North Carolina farmer Hinton Rowan Helper, 28, tries to show that slavery is not only economically unprofitable, but is ruinous to small farmers such as himself who do not own slaves. Helper calls slavery "the root of all the shame, poverty, ignorance, tyranny and imbecility of the South"; Southerners denounce him, many Northerners hail his courage and vision.

Britain's Matrimonial Causes Act establishes that a husband's responsibility as provider continues in perpetuity after a marriage is ended; it orders the world's first alimony payments.

Parliament eases British game laws after years of harsh penalties that made anyone caught poaching liable to transportation to Australia for 7 years. Poached game has been a dietary mainstay for many families.

A prime field hand in the South fetches $1,300 and up in the market, and critics such as Hinton Helper (above) point out that slavery ties up capital in human beings when it might better be employed in labor-saving improvements. 1,000 Southern families share an income of $50 million per year, while another $60 million is split among 660,000 families, yet even rich slaveholders have their earnings eaten up by commissions, freight charges, tariffs, imports of manufactured goods from the North, and the care and maintenance of slaves whose labor may actually be more costly than free labor in the North (*see* 1863).

Financial panic strikes New York following failure of Ohio Life Insurance and Trust Co. A severe depression ensues in which 4,932 business firms will fail after a period of speculation and overexpansion.

A new U.S. tariff act reduces import duties.

"The whale oils which hitherto have been much relied on in this country to furnish light and yearly become more scarce, may in time almost entirely fail," says the *Scientific American* June 27. The increase in U.S. literacy spurs demand for lighting (*see* 1866; Drake, 1859).

Congress abolishes shipping subsidies introduced in 1845 in response to Southern opposition.

Congress passes an Overland California Mail bill; John Butterfield of American Express organizes an Overland Mail Co. (*see* 1850). He wins the government contract to carry mail from St. Louis to San Francisco via Little Rock, El Paso, Tucson, Yuma, and Los Angeles using the route pioneered in large part by Philip Cooke's Mormon Battalion in 1847 (*see* 1858).

A nationwide celebration marks the linking by rail of New York and St. Louis.

The Pennsylvania Railroad extends its control through purchase and lease over the entire rail route between Philadelphia and Pittsburgh. It buys up the state's chief canal system to eliminate competition (*see* 1854; 1871).

France passes a law to spur railway construction, guaranteeing interest payments on railroad company bonds. By the end of next year France will have 16,207 kilometers of railway line, up from 3,627 in 1851.

Venice gains a rail link to other European cities as the Milan-Venice railway opens across a 3.5-kilometer bridge put up in 1846 to span the lagoon between San Giuliano on the mainland and the area of San Giobbea and San Lucia.

The luxury steamship S.S. *Central America* en route from California with 3 tons of gold and nearly 600 passengers and crew founders in a September hurricane 200 miles east of Charleston. A small brig arrives after 30 hours of frantic bailing by every man aboard and takes on about 170, mostly women and children. Some 420 perish, and the loss of the gold precipitates bank failures, contributing to the nation's financial panic (above).

New York to San Francisco ocean freight rates drop to $10 per ton, down from as high as $60 during Gold Rush days, as an excess of shipping intensifies competition, making the industry unprofitable.

Some clipper ships can cover more than 400 miles on a good day, a speed no steamship can match, but the day of the clipper ship is ending.

Eddyville, Ky., steel maker William Kelly patents a "pneumatic" steelmaking process he invented in 1847. His patent battle with Henry Bessemer of England will combine with a shortage of sufficiently pure iron to delay adoption of oxygen steel furnaces in the United States (*see* 1856; 1861; 1863).

The cant hook (peavey), invented on the Stillwater branch of Maine's Penobscot River, multiplies the efficiency of a logger. Joseph Peavey sees river drivers trying to break up a logjam and gets an idea; he instructs his blacksmith son Daniel to make a clasp with lips connected by a bolt, and then to attach a hook, two rings, and a pick at the end of the handle. Peavey will also invent a hoist for pulling stumps, a clapboard-making machine, a shingle-making machine, and an undershot waterwheel.

The aniline dye industry begins in England as W. H. Perkin and his father build a mauve dye works near Harrow (*see* 1856; German Chemical Society, 1868).

French chemist Louis Pasteur, 32, shows that a living organism causes the lactic fermentation that spoils milk (*see* 1859).

French physiologist Claude Bernard pioneers modern physiology by applying the biochemical methods developed by Justus von Liebig in 1842 to the living animal, combining them with physiological methods. Bernard pursues his investigations of glycogen production by the human liver (*see* 1850; 1860).

The New York Infirmary for Women and Children opens on Florence Nightingale's birthday May 12 under the direction of Elizabeth Blackwell, her younger sister Emily, and another woman doctor (*see* 1849). The loss of an eye has prevented Blackwell from achieving her ambition to practice as a surgeon, she has been unable to find a place on any other hospital staff because of her sex, and has started the new hospital that will be run entirely by women. It will move in November 1875 to 321 East 15th Street and remain there for more than a century.

U.S. cities have higher death rates than any other places in the world. Tuberculosis is the big killer, causing roughly 400 deaths per 100,000 population (the disease is not considered contagious), and while New York has the highest death rate, Philadelphia and Boston are not far behind.

The University of California is founded at Oakland, but classes will not begin until 1869 (*see* 1873).

Illinois State University is founded.

Marquette University is founded at Milwaukee.

The Universities of Bombay, Calcutta, and Madras are founded in India.

Cooper Institute is founded in New York's Astor Place by Peter Cooper of 1829 *Tom Thumb* locomotive fame. The Institute will develop into Cooper Union and will carry on extensive programs of adult education (*see* 1859).

Baltimore's Peabody Institute opens in Mount Vernon Place with a great concert hall, a lecture hall, an art collection, a reference library, and a music conservatory

1857 ***(cont.)*** established with a $1.4 million gift from financier George Peabody, 62.

The *Atlantic Monthly* begins publication at Boston under the editorship of James Russell Lowell. Oliver Wendell Holmes has suggested the magazine's name and when Lowell accepts the editorship on condition that Holmes be a contributor, Holmes revives his 1831 title "The Autocrat of the Breakfast Table" for his contributions (*see* 1858).

The *New York Tribune* sacks all but two of its foreign correspondents in an economy move (*see* financial panic, above). One of the two retained is Karl Marx.

England's *Birmingham Post* begins publication.

Fiction *Madame Bovary* by French novelist Gustave Flaubert, 36, who is prosecuted for immorality but wins acquittal; *Little Dorrit* by Charles Dickens: "Papa, potatoes, poultry, prunes, and prisms are all very good words for the lips: especially prunes and prisms"; *Barchester Towers* by Anthony Trollope who continues the Barsetshire chronicle that began with *The Warden* in 1855; "The Sad Fortunes of the Rev. Amos Barton" by English writer George Eliot (Mary Anne, or Marian, Evans), 37, who 3 years ago began living openly with philosopher-critic George Henry Lewes, 40, a married man who cannot divorce his wife because he has failed to condemn her for having had a child by another man. Eliot, whose story is published in *Blackwood's* magazine, is a puritan at heart and calls herself Mrs. Lewes, as does Lewes, but she is shunned by proper Victorian wives.

Juvenile *Tom Brown's School Days* by English jurist Thomas Hughes, 35, whose Rugby principal Dr. Thomas Arnold says, "Life isn't all beer and skittles." Hughes worked 3 years ago with London theologian Frederick Denison Maurice, 52, to found Working Men's College.

Poetry "Santa Filomena" by Henry Wadsworth Longfellow is a tribute to Crimean War nurse Florence Nightingale: "Lo! in that house of misery/ A lady with a lamp I see. . ." (*see* 1854).

Painting *Niagara Falls* by Frederic E. Church.

The first Currier & Ives prints are issued as New York lithographer Nathaniel Currier, 44, takes into partnership his bookkeeper of 5 years James Merritt Ives, 33, and begins signing all his prints with the new firm name. Currier has been issuing lithographs since 1835 that illustrate U.S. manners, personages, and notable events and has been famous since 1840 for his prints of W. K. Hewitt's lurid depiction of a fire aboard the steamship *Lexington*. Priced at 15¢ to $3, Currier & Ives prints will be sold all over the United States and will be distributed through branch offices in European cities. Currier will retire in 1880 and his son Edward West Currier will sell out in 1902 to Ives's son Chauncey whose firm will continue until 1907.

Theater *The Poor of New York* by Dion Boucicault 1/8 at Wallack's Theater, New York. Boucicault emigrated to New York in 1853 and has adapted *Les pauvres de Paris* by Edouard-Louis-Alexandre Brisebarre and Eugène Nus.

Opera *Simon Boccanegra* 3/12 at Venice's Teatro la Fenice, with music by Giuseppe Verdi to a libretto based on history concerning Genoa's first doge (who served from 1339 to 1344). The opera is a failure and will not be successful until 1881 when it will be given a new libretto by Arrigo Boito.

First performances Piano Concerto in A major and Symphonic Poem No. 1 *(Ce qu'on entend sur la montagne,* after Victor Hugo) by Franz Liszt 1/7 at Weimar's Grand Ducal Palace; Sonata in B minor by Franz Liszt 1/22 at Berlin, with pianist Hans von Bülow who marries Liszt's daughter Cosima, 19, August 18; the *Faust* Symphony in Three Characters (after Goethe) by Liszt 9/5 at Weimar's Grand Ducal Palace; Symphony to Dante's *Divine Comedy* by Liszt 11/7 at Dresden; *The Battle of the Huns (Die Hunnenschlacht)* symphonic poem by Liszt 12/29 at Weimar.

The Philadelphia Academy of Music designed by Napoleon Le Brun and C. Runge opens at the corner of Broad and Locust Streets (*see* Philadelphia Orchestra, 1900).

Hymn "We Three Kings of Orient" by English-American clergyman John Henry Hopkins, Jr., 37.

Popular song "Jingle Bells" ("One-Horse Open Sleigh") by Boston composer James Pierpont, 35.

Longchamp racetrack opens April 27 in the new Bois de Boulogne Park at Paris.

The *Harvard* constructed at Brooklyn, N.Y., is the first U.S. racing shell (*see* Henley Regatta, 1829). Built by James Mackay for the Harvard Boat Club of Cambridge, Mass., the six-oared rudderless craft is 40 feet long, 26 inches wide, weighs 50 pounds, and will be used in competition with similar white-pine racing shells that will be built for Harvard and other American colleges (*see* 1876).

The first American Chess Congress opens at New York in the fall. Top honors go to New Orleans player Paul Charles Morphy, 20, who taught himself the game in childhood by watching his father play, won two out of three games (the third was a draw) against a visiting Hungarian chess master before he was 13, and will tour Europe in the next 2 years, defeating all challengers and winning the unofficial world title (*see* Steinitz, 1886).

The world's first commercial passenger elevator is installed in the five-story New York store of E. G. Haughwort at the corner of Broadway and Broome Streets. The elevator cab is completely enclosed (*see* 1854; 1861).

Mentmore Towers goes up in Buckinghamshire for English sportsman-art collector Mayer Amschel de Rothschild, 39, who will be elected to Parliament next year. The 75-room neo-Elizabethan House designed by Joseph Paxton of 1851 Crystal Palace fame is a fantasy of pierced towers in buff sandstone and pioneers the use of large plate-glass windows.

New York appoints landscape architect Frederick Law Olmsted, 35, superintendent of the new Central Park which is under construction but whose name belies the fact that it is far to the north of the city's population center (*see* 1844). William Cullen Bryant, *Tribune* publisher Horace Greeley, botanist Asa Gray, financier August Belmont, and author Washington Irving (now 74) have endorsed Olmsted's application, but he finds the park site filled with hog farms, bone-boiling works, and squatters' shacks which make it "a pestilential spot where rank vegetation and miasmatic odors taint every breath of air." Work on the park becomes a relief project for city politicians struggling to cope with the economic depression (above) (*see* 1858).

Michigan State College of Agriculture opens to offer the first state courses in scientific and practical agriculture.

John Deere produces steel plows at the rate of 10,000 per year (*see* 1852).

The California grape and wine industries have their beginnings at Buena Vista in the Valley of the Moon near the Sonoma Mission where Count Agoston Haraszthy de Moksa, 55, plants Tokay, Zinfandel, and Shiras grape varieties from his native Hungary. The count founded Sauk City, Wis., in the 1840s, came to California in 1849, and the vines he plants are the first varietal grape vines in America (*see* 1861).

The condensed milk patented by Gail Borden in 1855 is made from skim milk devoid of all fats and of certain necessary food factors. The product will contribute to rickets in young working-class children (*see* Pekelharing, 1905).

Commercial production of his condensed milk begins at Burrville, Conn., where Borden has opened a condensing plant with financial backing from New York grocery wholesaler Jeremiah Milbank, 39, whom he has met by chance on a train. *Leslie's Illustrated Newspaper* helps Borden's sales by crusading against "swill milk" from Brooklyn cows fed on distillery mash. Samples of Borden's product are carried through the streets of New York and now meet with more success (*see* 1856; New York Condensed Milk Co., 1858).

Demand for Minneapolis flour begins in the East as a Minnesota Territory farmer ships a few barrels to New Hampshire in payment of a debt as economic depression makes specie payment difficult. The flour is made from winter wheat, which produces a whiter flour, and since Minnesota spring wheat yields a darker flour, the demand from the East will spur efforts to refine milling methods (*see* 1874).

The population of Chicago reaches 93,000, up from 4,100 in 1837 (*see* 1870).

1858

An assassination attempt on France's Napoleon III and his wife kills 10 and injures 150, but the January 14 bomb explosions leave the emperor and his Eugènie untouched. The French execute Italian revolutionist Felice Orsini, 39, and his accomplice Joseph Pieri March 13 but only after Orsini has appealed to Napoleon from prison for help in freeing Italy from Austrian rule.

Napoleon meets secretly with Count Cavour at Plombières July 10, agrees to join Piedmont in a war against Austria provided that the war can be provoked in a way that will make it appear justified in the public opinion of Europe, sends Prince Jérome on a mission to Warsaw in order to assure himself of Russian goodwill, and signs a formal treaty with Cavour December 10 (*see* 1859).

Prussia's Friedrich Wilhelm IV is declared insane a year after suffering two paralytic strokes. His brother Wilhelm, 61, is made regent October 7.

The Serbian prince Aleksindr Karageorgevic is deposed December 23 at age 52 after a weak 16-year reign. He is succeeded by Milos Obrenovic, now 79, who was deposed in 1836 and will reign until his death in 1860.

The Treaties of Tientsin signed June 26 to 29 end the Anglo-Chinese War that began in 1856. China agrees to open more ports to Britain, France, the United States, and Russia; the Russians obtain the north bank of the Amur by the Treaty of Aigun.

A U.S.-Japanese commercial treaty arranged by U.S. consul Townsend Harris is signed July 29 through the influence of Prime Minister Ii Naosuke, 43, who has won appointment as *tairo* under the shogun Iesada. Harris has pointed out the fate that China (above) has met at the hands of European imperialists and persuaded the Japanese that a treaty with the United States will be on favorable terms and will serve to protect Japan from more rapacious Western powers. The treaty covers tariffs, provides for an exchange of diplomats, and is followed by similar treaties with the Netherlands August 18, Russia August 19, Britain August 26, and France October 27.

Japan's feeble Tokugawa shogun Iesada dies at age 34 without an heir, the last Tokugawa shogun of any consequence. He has appointed Iemochi, 12, as his successor (Ii Naosuke arranged it) and Iemochi will reign until 1866.

The British Crown takes over the duties and treaty obligations of the East India Company following last year's Sepoy Rebellion and assumes responsibility for India's "protected" states (*see* 1686; 1447).

Minnesota is admitted to the Union as the 32nd state.

"Cotton is king," says Sen. James Henry Hammond, 41, (D. S.C.) in a speech March 4 taunting critics of the South (the phrase was used 3 years ago as the title of a book by David Christy): "You dare not make war upon cotton! No power on earth dares make war upon it. Cotton is king."

"A house divided against itself cannot stand," says former congressman Abraham Lincoln June 16 at Springfield, Ill., in accepting nomination as the Republican candidate for U.S. senator. "I believe this government

1858 ***(cont.)*** cannot endure permanently half slave and half free." Lincoln will lose to Democrat Stephen A. Douglas (but *see* 1860; railroads, 1854, 1856).

A fugitive slave in Ohio is rescued by Oberlin College students and one of their professors. They send the man off to safety in Canada (*see* 1854; 1859).

Russia's Aleksandr II begins emancipating the nation's serfs (*see* 1861).

Britain imposes legal trade in opium on China; by 1900, 90 million Chinese will be addicted (*see* 1842; Boxer Rebellion, 1900).

Denver is founded in Kansas Territory that will become Colorado. The town established at the junction of the South Platte River and Cherry Creek is named in honor of Kansas Territory governor James W. Denver (*see* gold strike, below).

The Texas socialist community La Reunion is abandoned by its followers, who move to nearby Dallas (*see* 1841). The communards have been followers of the late French social philosopher and reformer Charles Fourier who died in 1837 at age 65.

A placer gold strike in the eastern Colorado region of Kansas Territory 90 miles from Pike's Peak begins a new gold rush to Cherry Creek (*see* Denver, above).

American Bank Note Co. is created at New York by a merger of the seven independent engraving-printing firms that have printed U.S. paper currency since the Revolution. The company will print U.S. currency until 1879, U.S. postage stamps until 1894, and currency and stamps for scores of foreign nations thereafter.

Isaac M. Singer offers sewing machine inventor Walter Hunt of 1854 paper collar fame $50,000 in five annual payments to clear up any possible patent claims, but Hunt will die in June of next year at age 62 before the first payment falls due, having derived little benefit from his Globe stove, fountain pen, breech-loading rifle, or other inventions.

R. H. Macy Company opens October 27 with an 11-foot storefront on New York's 14th Street. Nantucket-born merchant Rowland Hussey Macy, 36, went to sea on a whaling ship at age 15, returned with a red star tattooed on his arm, opened a dry goods store at Haverhill, Mass., in 1851 selling only for cash and never deviating from a fixed price, failed after 3 years, but will use flamboyant advertising to make his New York store more successful; while his first day's sales of ribbons, trimmings, hosiery, and gloves total only $11.06, his gross sales next year will be $90,000 (*see* 1867).

The Overland Mail stage reaches St. Louis October 9 after 23 days and 4 hours on its first trip from San Francisco (*see* 1857). A westbound stage that has left at the same time reaches San Francisco October 10 after 24 days, 20 hours, 35 minutes (*see* transcontinental railroad, 1869).

The first practical sleeping car is perfected for the Chicago & Alton Railroad by Brocton, N.Y., cabinetmaker George Mortimer Pullman, 27, whose retractable upper berth doubles the sleeping car's payload (*see* 1864).

"Commodore" van Derbilt sells his New York-to-California shipping line to rivals who will operate via Panama rather than Nicaragua (*see* 1857). Van Derbilt takes the $20 million he has gained in the shipping trade and begins buying up shares in the Harlem Railroad running out of New York City and the Hudson River Railroad running north to East Albany (*see* 1863).

London's Chelsea Bridge is completed across the Thames. Tolls will be collected until 1879 and the bridge will be rebuilt in the 1930s.

The McKay machine invented by South Abingdon, Mass., shoemaker Lyman Reed Blake, 23, permits production of low-cost shoes by eliminating the heavy work of hand sewing. Gordon McKay, now 37, will promote Blake's machine and it will become famous as the McKay machine.

London's Linnaean Society hears a paper on the survival of the fittest in the struggle for existence in nature presented by English naturalists Alfred Russel Wallace and Charles Darwin (*see* 1840). Wallace has developed theories of evolution while visiting the Malayan archipelago (*see* 1859).

The Theory of the Vertebrate Skulls is published by English biologist Thomas Henry Huxley, 33.

Die Cellularpathologie by Berlin physician Rudolf Virchow calls the cell the basic element in the life process (*see* Neanderthal man, 1856). Virchow has analyzed diseased tissues from the point of view of cell formation and cell structure, and he concludes that all cells arise from cells ("omnis cellula e cellula") (*see* von Kolliker, 1850).

Uber die Konstitution und die Metamorphosen der chemischen Verbindungen und über die chemische Natur des Kohlenstoffs by German chemist Friedrich August Kekule von Stradonitz, 29, demonstrates the composition of organic molecules. He shows that they are formed by carbon atoms linked together to form long skeletal chains (*see* 1865).

Anatomy of the Human Body, Descriptive and Surgical by London physician Henry Gray, 33, is published for the first time. Gray is a Fellow of the Royal College of Surgeons and Gray's *Anatomy* will be a standard text for more than a century.

French schoolgirl Bernadette Soubirous, 14, at Lourdes has a vision February 11 of "a lady" dressed in a white robe with a blue sash, the first of 18 such visions that she will have through mid-July. The asthmatic daughter of a poor miller, Bernadette lives with her parents and six siblings in a one-room stone house. She is cured of her asthma, tells local priests that "the lady" has urged her to "tell the priests to build a chapel here" at the grotto near a bend in the Gave de Pau River, and is called a sorceress, but there are reports of miraculous cures at the grotto and a chapel will be built (*see* 1866).

E. R. Squibb & Sons has its beginnings in a chemical and pharmaceutical laboratory founded by Brooklyn,

N.Y., pharmacist Edward Robinson Squibb, 39, whose lower eyelids have been burned off by an ether explosion making it impossible for him to close his eyes completely (he covers them with court plaster to sleep).

Iowa State College is founded at Ames.

Oregon State University is founded at Corvallis.

Queen Victoria and President Buchanan exchange messages August 16 over the first transatlantic cable. New York paper merchant Cyrus West Field, 39, has promoted the project first proposed by Frederick Newton Gisborne, 39, consulted with Samuel F. B. Morse, and founded the New York, Newfoundland, and London Telegraph Co. with backing from Peter Cooper and others. Queen Victoria cables, "Glory to God in the Highest, peace on earth, good will to men," and the first commercial cable message follows (at $5 per word). London merchant John Cash whose firm J. and J. Cash makes woven name labels cables his New York representative, "go to Chicago," but its electrical insulation soon fails and by October the cable is useless (*see* 1866).

The *Territorial Enterprise* begins publication at Mormon Station in the Territory of Western Utah with a press freighted in from San Francisco. Operations will soon move to Virginia City (*see* 1859).

The pen name "Artemus Ward" appears in the 17-year-old *Cleveland Plain Dealer*. Maine-born city editor Charles Farrar Browne, 24, uses the name to sign a letter to the editor allegedly written by a shrewd showman with Yankee dialect and unconventional spelling (*see* "Josh Billings," 1865).

Street & Smith Publications, Inc., has its beginnings in the *New York Weekly* published by former *Sunday Dispatch* bookkeeper Francis S. Street and former *Sunday Dispatch* reporter Francis Shubael Smith, 38, who buy out the *New York Weekly Dispatch* for $40,000 and rename it. Begun in 1843 as the *Weekly Universe*, the fiction magazine has a small circulation and Street and Smith have less than $100 between them, but founder Amos J. Williamson agrees to wait for his money until the new owners have earned it, they boost circulation by attracting well-known writers with enormous sums, and by 1861 their *Weekly* will have a circulation of 100,000 (*see* Nick Carter, 1886).

Fiction *Dr. Thorne* by Anthony Trollope.

Poetry *The Courtship of Miles Standish* by Henry Wadsworth Longfellow; "The Deacon's Masterpiece; or, The Wonderful One Hoss Shay" by Oliver Wendell Holmes is published as a satirical poem in *Holmes' Atlantic Monthly* column "The Autocrat of the Breakfast Table": Holmes' Deacon completes his "chaise" on the day of the 1755 Lisbon earthquake and some take his "Logical Story" to be a satire on the collapse of Calvinism; "The Chambered Nautilus" by Holmes appears in his *Atlantic Monthly* column with a final stanza that begins, "Build thee more stately mansions, O my soul,/ As the swift seasons roll..."; *Legends and Lyrics* by English poet Adelaide Ann Procter, 33, whose lyric poem "The Lost Chord" begins, "Seated one day at the organ,/ I was weary and ill at ease,/ And my fingers wandered idly/ Over the noisy keys. . ." (*see* popular song, 1877).

The Defence of Guenevere and Other Poems by English poet-craftsman William Morris, 24.

Painting *Derby Day* by English painter William Powell Frith, 39 (*see* 1779); *Eight Bells* by U.S. painter Winslow Homer, 22. Ando Hiroshige dies at Edo October 12 at age 61.

Theater *Jessie Brown, or The Relief of Lucknow* by Dion Boucicault 2/22 at Wallack's Theater, New York; *Foul Play* by Charles Reade and Dion Boucicault 5/28 and London's Adelphi Theatre; *Our American Cousin* by Tom Taylor 10/18 at Laura Keene's Theater, New York.

Britain's Princess Royal is married January 25 at age 18 to Prussia's Prince Frederich Wilhelm and both the 1843 Mendelssohn "Wedding March" from *A Midsummer Night's Dream* music and the 1850 Wagner "Wedding March" from the opera *Lohengrin* are played, beginning a tradition that will endure for weddings of royalty, nobility, and commoners in much of the world.

Opera *Orpheus in the Underworld* (*Orphée aux Enfers*) 10/21 at the Bouffes Parisiens, Paris, with music by German-French composer Jacques Offenbach (Jacob Eberst), 39, who was conductor at the Comédie Française from 1849 to 1855 and has been managing his own theater for 3 years. The opera introduces "Le Cancan," a new dance.

London's Covent Garden Opera House is completed by local architect Sir Charles Barry, 63 (*see* Theater, 1732).

Popular songs "The Yellow Rose of Texas" with music by a U.S. composer identified only as "JK"; "The Old Grey Mare (Get Out of the Wilderness)" by U.S. songwriter J. Warner whose song will be adapted as an 1860 presidential campaign song by supporters of Abraham Lincoln (above).

London tobacco merchant Philip Morris opens a cigarette factory using mostly smoke-cured Latakia leaf from Turkey (*see* 1902; Gloag, 1856).

English dressmaker Charles Frederick Worth, 36, opens a Paris shop on the rue de la Paix and establishes the first house of haute couture. Instead of catering to the whims of his customers, he will create his own designs and offer them to patrons. By the 1880s the House of Worth will be the authority not only in fashion but in all questions of taste, supplying gowns to every royal court in Europe.

Boston begins filling in its Back Bay using newly developed steam shovels and rail lines to bring gravel 9 miles from Needham. By 1882 the entire area will be filled to create land on which Commonwealth Avenue, Copley Square, and much of modern Boston will stand.

1858 ***(cont.)*** Big Ben begins chiming out the hours, half-hours, and quarter-hours in the 316-foot tall clock tower of London's Westminster Palace opened by Queen Victoria in 1852. The 13.5-ton great bell that chimes the hours solo for the 23-foot clock is tuned to the note E, it is the lowest note in the chime that strikes the quarter-hours, and it is named for Sir Benjamin Hall, Baron Llanover, who is chief commissioner of works.

New York's Central Park opens to the public in the autumn although it remains 5 years short of completion (*see* 1857). F. L. Olmsted has been assisted by landscape architect Calvert Vaugh, their Park has cost $5 million, but the high cost of horsecar and elevated railway transportation will for years keep Olmsted's sylvan retreat beyond the reach of most New Yorkers (*see* 1863).

Henry David Thoreau pleads for "national preserves in which the bear and the panther, and some even of the hunter race may still exist, and not be civilized off the face of the earth—not for idle sport or food, but for inspiration mid our own true recreation" (*see* fish, 1849; Yellowstone National Park, 1872).

Grenada in the Caribbean produces its first crop of nutmeg and mace to begin the island's career as the world's leading producer of those spices. An English sea captain bound home with spices from the East Indies presented seedlings to his hosts at the port of St. George's in 1843 (*see* 1955).

French inventor Ferdinand P. A. Carré devises the first mechanical refrigerator, employing liquid ammonia in the compression machine that he will demonstrate at the 1862 London Exhibition by using it to manufacture blocks of ice (*see* Harrison, 1856; Lowe, 1865; Linde, 1873).

The Mason jar, patented by New York metalworker John Landis Mason, 26, is a glass container with a thread molded into its top and a zinc lid with a threaded ring sealer. Mason's reusable jar, made at first by Whitney Glass Works of Glassboro, N.J., will free farm families from having to rely on pickle barrels, root cellars, and smoke houses to get through the winter. Urban families, too, will use Mason jars to put up excess fruits and vegetables, especially tomatoes, berries, relish, and pickles, and the jars will soon be sealed with paraffin wax, a by-product of kerosene (*see* Drake, 1859; Ball brothers, 1887).

New York Condensed Milk Co. is established by Gail Borden (*see* 1857; 1861).

London's meat becomes fatter and juicier as the Smithfield market receives rail shipments of ready-dressed carcasses from Aberdeen 515 miles away. Beef, mutton, pork, and veal arrive in perfect condition the night after the animals are slaughtered, and the meat is far less lean than if the animals had been driven to market as in the past.

Schweppe's Tonic Water has its beginnings in "an improved liquid known as quinine tonic water" patented at London. J. Schweppe & Co. Ltd. will make and sell the product beginning in 1880 (*see* 1891; 1953).

1859 Napoleon III warns the Austrian ambassador January 1, thus leaking the secret of his alliance with Count Cavour (*see* 1858). Piedmont calls up reserves March 9, Austria mobilizes April 7, Piedmont rejects an Austrian ultimatum April 23 to demobilize within 3 days, and Cavour has the provocation he has wanted. Austria invades Piedmont April 29, but Gen. Franz Gyulai's slow advance gives France time to mobilize.

France declares war on Austria May 12 pursuant to her 1858 alliance, the Piedmontese defeat Austrian forces May 30 at Palestro after peaceful revolutions have driven out the rulers of Tuscany, Modenti, and Parma. French and Piedmontese forces cross the Ticino into Lombardy, and the Battle of Magenta June 4 ends with the withdrawal of Austrian troops.

The king of the Two Sicilies Ferdinand II has died May 22 at age 49 after a 19-year reign in which he has won the epithet "King Bomba" by having the chief cities of Sicily bombarded to enforce his reactionary control. He is succeeded by his son Francesco, 13, who will reign for less than 2 years.

The Battle of Solverino June 24 brings cruel suffering to both sides but is indecisive (*see* medicine, 1862).

The Austrian emperor Franz Josef finds himself short of funds and faced with revolution in Hungary, he meets with Napoleon III July 11 at Villafranca as Prince Metternich dies at age 86, and hostilities cease. Napoleon has become alarmed at all the bloodshed and at the revolutions that have been fomented by nationalists in the papal legations of Bologna, Ferrara, and Ravenna as well as in Tuscany, Modena, and Parma, and the Italian princes are returned to their thrones (*see* Garibaldi, 1860).

Sweden's Oskar I dies July 8 at age 60 after a 15-year reign. He is succeeded by his son of 33 who will reign until 1872 as Charles XV.

Queensland in Australia is separated from New South Wales and becomes a separate colony with its capital at Brisbane (*see* 1901).

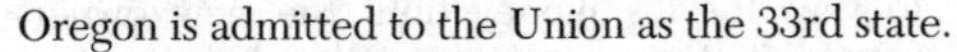

Oregon is admitted to the Union as the 33rd state.

President Buchanan orders that New Mexico Territory lands occupied by the confederated bands of Pima and Maricopa tribes near the Gila River be surveyed and set apart as a reservation under terms of a February 28 act of Congress. The reservation is not to exceed 100 square miles, and $10,000 is to be given to the tribes for tools and to cover the expenses of the land survey (the 64,000-acre reservation will be enlarged to 145,000 acres in 1869).

The Supreme Court upholds the Fugitive Slave Act of 1850 March 7 in the case of *Ableman v. Booth*, reversing the 1854 Wisconsin court decision that called the act unconstitutional (*see* Dred Scott, 1857).

Georgia prohibits the post-mortem manumission of slaves by last will and testament. The state legislature votes to permit free blacks to be sold into slavery if they have been indicted as vagrants.

John Brown's raid on Harper's Ferry, Va., October 16 stirs passions throughout the country. Abolitionist Brown of 1856 Pottawatomie Massacre fame leads 13 whites and five blacks into Harper's Ferry, they seize the town and its federal arsenal as a signal for a general slave insurrection that will establish a new state as a refuge for blacks, but federal troops under Col. Robert E. Lee, supported by U.S. marines, overpower Brown October 18, he is convicted of treason and hanged December 2 at Charlestown at age 59, but "his truth" will go marching on in the words of a popular song.

Britain's House of Commons gets its first Jewish member as Lionel Rothschild, 41, takes his seat. He has been elected and reelected since 1847 but has been unable to sit until the introduction of a new oath omitting the words "on the true faith of a Christian." His son Nathan will be the first Jewish peer in the House of Lords (*see* Nathan, 1815).

Last year's gold strike in the Colorado Rockies of Kansas Territory brings 100,000 prospectors determined to reach "Pike's Peak or bust."

The Comstock Lode discovered on Mt. Davidson in Washoe, Nev., is a vast deposit of silver and gold that will yield $145 million worth of ore in the next 11 years and $397 million by 1882. Prospectors Peter O'Reilly and Patrick McLaughlin have found pay dirt on property claimed by Canadian-American prospector Henry Tompkins Paige Comstock, 39, who has been in the area for 3 years, but a blue sand impedes mining operations until it is assayed and turns out to be three-fourths silver, one-fourth gold, and worth $4,700 per ton (*see* 1873).

Comstock (above) sells his claim for almost nothing but others make fortunes, including Missouri-born prospector George Hearst, 39, whose luck has been poor in California. He pays $450 for a half-interest in a mine believed by its owner Alvah Gould to be worthless and a few hundred dollars more for a one-sixth interest in the Ophir mine, which will prove to be the richest in the Comstock Lode, and in the Gould and Curry mine, which will be the second richest.

George Hearst (above) becomes a millionaire overnight, joins with a partner to buy the Ontario mine in Utah Territory, and will go on to acquire other partners and other mining properties, including the Homestake mine in Dakota Territory, the Anaconda copper mine in Montana Territory, and additional properties in Chile, Peru, and Mexico (*see* William Randolph Hearst, 1887).

Petroleum production begins at Titusville, Penna., giving the world a new source of energy and reducing demand for the whale oil, coal gas, and lard now used in lamps (*see* Baku, 1823). Unemployed New Haven & Hartford Railroad conductor "Colonel" Edwin Laurentine Drake, 40, has been sent to Titusville in Pennsylvania's Venange County by New York banker James Townsend, an associate of George Bissell (*see* 1854). Using salt-well drilling equipment to dig into oil-bearing strata, Drake strikes oil August 28 and his 70-foot

John Brown's raid on the federal arsenal at Harper's Ferry kindled new flames of abolitionist passion.

well is soon producing 400 gallons per day (2,000 barrels per year) to begin the first commercial exploitation of petroleum in the United States and inaugurate a new era of kerosene lamps and stoves (*see* Gesner, 1855; Rockefeller, 1860).

Electric home lighting has its first U.S. demonstration. Salem, Mass., inventor Moses Gerrish Farmer, 39, lights two incandescent lamps on his mantelpiece with platinum strip filaments powered by a wet-cell voltaic battery.

French physicist Gaston Plante, 25, invents the first practical electric storage battery (*see* Daniell, 1836; Leclanche's dry-cell battery, 1867).

Suez Canal construction begins April 25 (*see* 1854). Ferdinand de Lesseps has financed the sea-level waterway between the Mediterranean and Red seas chiefly by selling stock to French investors, but he has sold 44 percent for nearly $20 million to Egypt's khedive Mohammed Said. 30,000 Egyptian forced laborers start work at wages above prevailing rates and will be augmented by Arab, French, Italian, and Greek workers before hand labor is largely replaced beginning in 1863 by mechanical equipment brought from Europe to dig out nearly 2 million cubic centimeters of earth each month (*see* 1869).

The British clipper ship *Falcon* launches a new method of shipbuilding. Much smaller than the American clipper ships but longer lasting and more economical to operate, the *Falcon* is made of wooden planking on iron frames.

The Italian states still have only 1,759 kilometers of railway track (*see* 1839).

Brink's Express is started at Chicago by a Vermonter with a horse and wagon who in 1891 will begin moving

1859 *(cont.)* bank funds and payroll money to start a business that will grow into the leading U.S. armored truck money-transport company (*see* robbery, 1950).

On The Origin of Species by Means of Natural Selection, or the Preservation of Favored Species in the Struggle for Life by Charles Darwin creates a furor by flying in the face of fundamentalist religion. Alfred Russel Wallace has forced Darwin's hand (*see* 1858; 1871).

The spectrum analysis elaborated by German physicist Gustav Robert Kirchhoff, 35, and R. W. Bunsen of 1850 Bunsen burner fame will permit the discovery of new elements (*see* Mendeléev, 1870). They show that when light passes through a gas or any heated material only certain wavelengths are absorbed and emitted; analyzing the spectrum of emissions reveals the material's chemical composition.

Louis Pasteur disproves the chemical theory of fermentation advanced by the German chemist Baron von Liebig. Pasteur, whose daughter Jan dies of typhoid fever in September, also disproves the theory of spontaneous generation and shows that while some microorganisms exist in the air like yeast, and like yeast can transform whatever they touch, other microorganisms are anaerobic and cannot survive in the presence of air (*see* 1861).

Treatise on Gout and Rheumatic Gout by London physician Alfred Baring Garrod, 40, shows that Galen in the 2nd century was wrong in blaming gout on overindulgence. A staff doctor at University College hospital, Garrod says gout comes from an excess of uric acid in the blood and can be controlled by avoiding foods high in purine and by using drugs such as colchicine (*see* Franklin, 1785).

Cocaine is isolated from coca leaves brought home from Peru by Austrian explorer Karl von Scherzer, but physicians show little interest in the drug's potential for anaesthesia (*see* 1884).

New York's Cooper Union for the Advancement of Science and Art opens its first building at Astor Place and Fourth Avenue (*see* Cooper Institute, 1857).

Telegraph promoter Samuel F. B. Morse agrees to the formation of the North American Telegraph Association as a near-monopoly (*see* 1844; Western Union, 1856, 1866).

Rocky Mountain News begins publication, the first newspaper in what will soon be Montana Territory. U.S. surveyor William Newton Byers, 28, will be publisher and editor until 1879 (*see* Montana, 1864).

Fiction *A Tale of Two Cities* by Charles Dickens whose novel of the French Revolution begins, "It was the best of times, it was the worst of times . . ."; *Adam Bede* by George Eliot gives its author her first wide success; *The Ordeal of Richard Feverel* by English novelist George Meredith, 31; *Oblomov* by Russian novelist Ivan Aleksandrovich Goncharov, 47, a rich merchant's son who last year published *Fregat Pallada* describing a voyage to Japan as secretary to a Russian mission.

Poetry *The Idylls of the King* by Alfred Tennyson whose first volume begins a series that he will complete in 1885 to bring Arthurian legend to life with verses about Arthur, Guinevere, Lancelot, the lady of Shalott, and the towers of fabled Camelot; *The Rubáiyát of Omar Khayyám of Naishapur* by English poet-translator Edward Fitzgerald, 50, who puts into graceful English quatrains the rhymed verse of a 12th-century Persian astronomer-poet: "A Book of Verses underneath the Bough,/ A jug of Wine, a Loaf of Bread-and Thou/ Beside me, singing in the Wilderness/ Oh, Wilderness were Paradise enow!" "And much as wine has played the Infidel,/ And robbed me of my Robe of Honor—Well,/ I often wonder what the Vintners buy/ One half so precious as the stuff they sell"; "The Moving Finger writes; and having writ,/ Moves on; nor all your Piety nor Wit/ Shall lure it back to cancel half a Line/ Nor all your Tears wash out a Word of it"; "The Children's Hour" by Henry Wadsworth Longfellow begins, "Between the dark and the daylight,/ When the night is beginning to lower,/ Comes a pause in the day's occupations/ That is known as the Children's Hour"; *Mireio* by French Provençal poet Frédric Mistral, 29.

Painting *The Absinth Drinker* by French painter Edouard Manet, 27; *The Angelus* by Jean François Millet; *Le Bain Turc* by Jean Auguste Dominique Ingres, now 79; *At the Piano* by U.S. émigré painter-etcher James Abbott McNeill Whistler, 25, who settles at London. Whistler lived in Russia during his teens while his father directed construction of a railroad for Czar Nicholas I, was forced to leave West Point at age 19 for failing chemistry, and has lived abroad since age 21; *Heart of the Andes* by Frederic E. Church.

Theater *The Thunderstorm (Groza)* by Aleksandr Ostrovsky 11/16 at Moscow's Maly Theater; *The Octaroon, or Life in Louisiana* by Dion Boucicault 12/5 at New York's Winter Garden Theater. Boucicault has adapted the 1856 novel *The Quadroon* by Mayne Reid to create his melodrama.

Opera *The Masked Ball* (*Un Ballo in Maschera*) 2/17 at Rome's Teatro Apollo, with music by Giuseppe Verdi; *Faust* 3/19 at the Théâtre Lyrique, Paris, with music by French composer Charles Gounod, 41, to a libretto based on the Goethe tragedy.

First performances Serenade for Orchestra No. 1 in D major by Johannes Brahms 3/28 at Hamburg; Concerto No. 1 for Pianoforte and Orchestra by Brahms 7/22 at Hanover.

Hymns "Ave Maria" ("Hail Mary") by Charles Gounod (above) who has based it on Johann Sebastian Bach's first prelude in *The Well-Tempered Clavichord* of 1722; "Nearer My God to Thee" by Boston musician Lowell Mason, 67, lyrics by the late Sarah Adams who wrote them in 1840 or 1841.

Popular songs "Dixie" ("I Wish I Was in Dixie's Land") by Daniel Decatur Emmet of the Virginia Minstrels is performed for the first time 4/4 at New York's Mechanics Hall, 472 Broadway; "La Paloma" by Spanish songwriter Sebastian Yradier, 50.

The Steinway Piano has its beginnings in an improved grand piano developed by German-American piano maker Henry Engelhard Steinweg, 62, the first piano with a single cast-metal plate. The plate is strong enough to withstand the pull of strings which Steinweg has combined with an overstrung scale that both saves space and produces a much fuller tone than earlier pianos. Steinweg established a New York piano factory with several of his sons in 1853, he will call himself Steinway beginning in 1865, and his new grand piano will make Steinway & Sons world-famous.

The first intercollegiate baseball game July 1 ends after 26 innings with an Amherst College team defeating a Williams team 73 to 32 on the grounds of the Maplewood Female Institute at Pittsfield, Mass. (*see* 1845; Cincinnati Red Stockings, 1869).

The Cachar Polo Club founded by British colonial officers stationed in Assam, India, is the world's first polo club.

French tightrope walker Charles Blondin (Jean François Gravelet), 35, crosses Niagara Falls on a tightrope June 30.

The world's first flying trapeze circus act is performed November 12 at the Cirque Napoléon in Paris by Jules Léotard, 21, who has practiced at his father's gymnasium in Toulouse. Safety nets will not be introduced until 1871 and Léotard meanwhile will use a pile of mattresses to cushion possible falls. The song "That Daring Young Man on the Flying Trapeze" will be written about Léotard and his name will be applied to the tights worn by acrobats and dancers.

Boston's Public Garden is established on 108 acres of filled land owned by the Commonwealth of Massachusetts (*see* Swan Boats, 1877).

The New York State Legislature authorizes acquisition of land for Brooklyn's Prospect Park (*see* Central Park, 1858).

The world's first children's playgrounds open at Manchester, England, where horizontal bars and swings have been installed in Queen's Park and Philips Park.

The discovery of oil at Titusville (above) will spur mechanization of U.S. agriculture.

Australian landowner Thomas Austin imports two dozen English rabbits to provide shooting sport. In 6 years Austin will have shot 200,000 on his own estate and killed scarcely half the rabbits on his property, and since five of the fast-multiplying animals consume more grass than one sheep, the rabbits will quickly become a major problem to sheep-raisers (*see* myxomatosis, 1951).

Paraffin wax, a by-product of kerosene produced from petroleum (above), will be used to seal Mason jars used for preserving foods (*see* Mason, 1858).

Civilized America, published at London, ridicules Americans for calling a cock a rooster and using similar euphemisms.

The A&P retail food chain has its beginnings in the Great American Tea Co. store opened at 31 Vesey Street, N.Y., by local merchant George Huntington Hartford, 26, who has persuaded his employer George P. Gilman to give up his hide and leather business and go partners in buying China and Japanese tea directly from ships in New York harbor. Both originally from Maine, the two men buy whole clipper ship cargoes at one fell swoop, sell the tea at less than one-third the price charged by other merchants, identify their store with flaked gold letters on a Chinese vermilion background, and start a business that will grow into the first chain store operation (*see* 1863; A&P, 1869).

1860

Sicilian authorities abort an uprising against the Bourbon monarchy in April but Giuseppe Garibaldi organizes an army of 1,000 "Redshirts" at Genoa and sails May 5 for Marsala. He lands May 11, gathers recruits as he marches inland, defeats the Neapolitans May 15 at Calatafimi, takes Palermo June 6, crosses the Straits with British connivance August 22, takes Naples September 7, and supports Victor Emmanuel II of Piedmont as king of a united Italy. By November, he is back on the island of Caprera looking after his donkeys Pius IX, Napoleon III, Oudinot, and the Immaculate Conception (*see* 1857; 1861).

Montenegro's Danilo I is assassinated August 12 after a 9-year reign in which he has tried to reform and modernize the Balkan state. Danilo is succeeded by his nephew, 19, who will reign until 1918 as Nicholas I.

Serbia's Milos Obrenovic dies at Belgrade September 27 at age 80 after taking cruel vengeance on the enemies that deposed him in 1839. He is succeeded by his son Michael III Obrenovic who reigned briefly from 1839 to 1842 and will rule until his assassination in 1868. Michael will abolish the oligarchic constitution of 1839 and will work to consolidate the Balkan states under Serbian leadership.

Napoleon III increases the powers of the French Parlement in November in an effort to regain support lost by last year's war with Austria and Piedmont. A new commercial treaty with Britain has further weakened his support (below).

The world's first breech-loading rifled artillery is used August 12 in China as Armstrong 18-pounders bombard Sinho to force admission of foreign diplomats by Beijing. 17,000 French and British troops occupy Beijing October 12 and burn the Summer Palace to punish the Chinese for seizing Harry Parkes, 52, the diplomat who has been virtual governor of Guangzhou (Canton) and was taken prisoner despite his flag of truce.

Beijing is forced to yield to some Western demands and increase its indemnities (*see* 1842; 1853; 1864).

The U.S. Army's Fort Defiance in New Mexico Territory is attacked April 30 by 1,000 Navajo whose sheep and goats have been shot by soldiers from the fort. Led by Manuelito, the tribesmen nearly succeed in taking the fort before being driven off by musketfire (*see* 1861).

1860 ***(cont.)*** The United States becomes less united following the election of Abraham Lincoln as president in November with 40 percent of the popular vote (Lincoln has received 180 electoral votes, John C. Breckenridge 72 with 18 percent of the popular vote, Stephen A. Douglas 12 with 30 percent of the popular vote). South Carolina adopts an Ordinance of Secession December 20 to protest the election (*see* 1861).

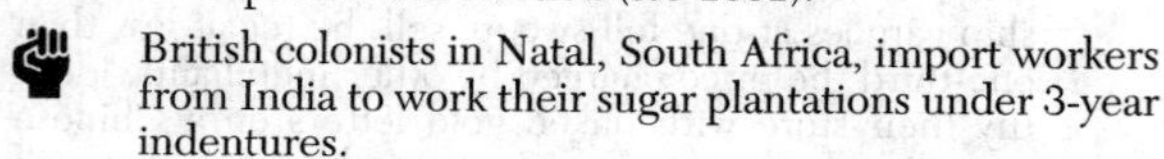

British colonists in Natal, South Africa, import workers from India to work their sugar plantations under 3-year indentures.

Elizabeth Cady Stanton urges woman suffrage in an address to a joint session of the New York State Legislature (*see* 1848; Equal Rights Association, 1866).

Vladivostok is founded by Russian pioneers on the Pacific Coast. The port is icebound 3 months of the year.

The first expedition to cross Australia from south to north leaves Melbourne with 17 men, 26 camels, and 28 horses under the leadership of Irish-Australian Robert O'Hara Burke, 40. The group will reach the Gulf of Carpentaria next year but Burke will die of starvation on the return trip, as will his lieutenant William John Wills, 26, and another man.

German adventurer Karl Klaus von der Decken, 27, explores East Africa. He will propose plans for a vast German colony in the area but natives will murder him in 1865.

U.S. cotton exports amount to $192 million out of the nation's export total of $334 million, up from $72 million in 1850 when cotton accounted for less than half the total.

An Anglo-French treaty of reciprocity signed in January has been negotiated by English free-trade advocate Richard Cobden and opens up a wider area of trade between the two countries (*see* 1845). Many Frenchmen oppose the treaty and it weakens the popularity of Napoleon III (above).

German traders sent by Woermann and Co. open a factory (trading post) on West Africa's Cameroons coast.

John Davison Rockefeller enters the oil business at age 20. A group of local businessmen has sent the junior partner in the Cleveland produce commission firm Clark & Rockefeller to investigate the potential of the petroleum found last year at Titusville, Pa., 100 miles away. He reports back that petroleum has little future, but he pools his savings with those of his partner Maurice B. Clark, they invest $4,000 in the lard oil refinery of candle maker Samuel Andrews, and Rockefeller persuades richer men to build more refineries for petroleum which he foresees as a major energy source (*see* 1863).

An internal combustion engine patented at Paris by Belgian inventor Jean Etienne Lenoir, 38, employs a carburetor that mixes liquid hydrocarbons to form a vapor. An electric spark explodes the vapor in a cylinder but Lenoir's engine works on illuminating gas and has no compression (*see* Otto, 1866).

The first Japanese-built ship to reach the United States arrives at San Francisco February 16 after a 34-day voyage. The 300-ton iron steamship S.S. *Kanrinmaru* has been accompanied by the U.S. cruiser *Powhatan* and carries a delegation of 80 which proceeds to Washington, D.C., puts up at the Willard Hotel, and drinks from the hotel finger bowls.

More than 1,000 steamboats ply the Mississippi, up from 400-odd in 1840 (*see* 1878; *Sultana*, 1865).

French railway trackage reaches 5,918 miles, up from 336 in 1842 (*see* 1870).

The Henry rifle patented by U.S. gunsmith D. Tyler Henry is the first truly practical lever-action rifle. The Winchester Repeating Arms Company will introduce an improved version, the M1866, with a .44 rim-fire cartridge.

The first Winchester repeating rifle goes into production at New Haven, Conn. (*see* 1850).

The Oneida Community grosses $100,000 from sales of the Newhouse trap named for Community member Sewell Newhouse, who invented the device that is fast becoming the standard in North America and will eventually become so for most of the world. The major enterprise of the 12-year-old Community is making traps, and men of the Community have developed dishwashers and machines for paring apples and washing vegetables to spare themselves the tedium of kitchen chores.

Diabetes is the result of a pancreas disorder, says Paris physician Etienne Lancereaux, 31 (*see* 1869).

The clinical symptoms of acute trichinosis are noted for the first time by German pathologist Friedrich Albert von Zenker, 35, but the cause-effect relationship remains unknown (*see* 1835).

The Seventh Day Adventist Church is founded at Battle Creek, Mich., by former followers of William Miller (*see* 1845; sanitorium, 1866).

Elizabeth Palmer Peabody opens the first U.S. English-speaking kindergarten at Boston, a school for preschoolers that follows the lead set by Mrs. Carl Schurz in 1855. Mrs. Peabody is a sister-in-law of Horace Mann and of Nathaniel Hawthorne.

Louisiana State University and Louisiana A&M College are founded at Baton Rouge.

The first Pony Express riders leave St. Joseph, Mo., April 3 and deliver mail to Sacramento, Cal., 10 days later. William Hepburn Russell, 48, has organized the Pony Express to carry mail at rates of $2 to $10 per ounce depending on distance, keeps 80 riders in the saddle at all times of day and night, and uses fast Indian ponies in relays to cover the 10 miles or so between each of some 190 stations (*see* 1861).

U.S. Army Signal Corps surgeon Albert J. Myer receives a $4,000 congressional appropriation to develop a wigwag system of visual signaling that he has invented.

The United States has 372 daily newspapers, up from 254 in 1850, but newspapers remain generally too expensive for the average man or woman to afford (*see* 1909).

Fiction *The Mill on the Floss* by George Eliot; *On the Eve* by Ivan Turgenev; *The Marble Faun* by Nathaniel Hawthorne; *The Woman in White*, a pioneer detective novel by English novelist William Wilkie Collins, 36, whose work is serialized in the London magazine *All the Year Round*; *Maleska: The Indian Wife of the White Hunter* by Anne S. W. Stephens is the first "dime novel." It is published at New York by Beadle & Adams whose proprietors Erastus Flavel Beadle, 39, and Robert Adams moved east from Buffalo 2 years ago, have experimented with 10¢ paper-bound song, joke, and etiquette books, but only now have a runaway best seller. Beadle & Adams will publish hundreds of dime novels and thousands of Dime Library and Half Dime Library titles including the adventures of Deadwood Dick and Nick Carter (*see* 1886).

Poetry "Tithonus" by Alfred Lord Tennyson who writes, "After many a summer dies the swan"; "Rock Me to Sleep, Mother" by New England poet Florence Percy (Elizabeth Chase Akers), 28, who writes, "Backward, turn backward, O time in your flight,/Make me a child again just for tonight."

Painting *The Guitarist* by Edouard Manet.

Theater *The Colleen Bawn, or The Brides of Garryowen* by Dion Boucicault 3/29 at Laura Keene's Theater, New York, with John Drew in a melodrama based on the 1829 novel *The Collegians* by Gerald Griffin; *Rip Van Winkle* 12/24 at New York, with Joseph Jefferson, 31, in the title role his own dramatization of the Washington Irving story.

First performances Serenade No. 2 in A major by Johannes Brahms 2/10 at Hamburg; Concerto for Cello and Orchestra in A minor by the late Robert Schumann 6/9 at Leipzig's Royal Conservatory.

Popular songs "Old Black Joe" and "The Glendy Burk" by Stephen C. Foster; "Annie Lisle" by H. S. Thompson whose melody will be the basis of the Cornell University song "Far Above Cayuga's Waters."

The first British Open golf tournament opens at Prestwick, Scotland (*see* U.S. Open, 1894).

U.S. sportsmen begin wearing knickerbockers (plus-fours).

The first world heavyweight boxing championship bout April 17 at Farnborough, Hampshire, pits U.S. champion John C. Heenan against the British champion Tom Sayers. The fight continues for 42 rounds until Sayers is unable to use his right arm, Heenan is blinded from his own blood, and the crowd breaks into the ring to stop the fight.

The longest bare-knuckle prizefight in U.S. history is held at Berwick, Me., where J. Fitzpatrick and James O'Neill battle for 4 hours, 20 minutes.

Australia's Melbourne Cup horse race has its first running in the Victoria colony.

Armstrong Cork Co. has its beginnings in a firm founded by Pennsylvania entrepreneurs Thomas Marton Armstrong and John D. Glass. It will employ machinery to cut the bark off cork trees in Spain, Portugal, and North Africa (*see* linoleum, 1907).

The word "linoleum" is coined by English inventor Frederick Walton who has devised a process for oxidizing linseed oil to produce a cheap rubber-like material for use as floor covering. Other methods will be developed for solidifying linseed oil but all linoleum production will be based essentially on Walton's invention.

Cigarette smoking increases in America. Richmond, Va., has more than 50 cigarette factories; North Carolina and Virginia together have more than 348, up from 119 in 1840 (*see* 1875).

Texas cattleman Richard King takes his old steamboat partner Miflin Kenedy into partnership and the two pool their resources to import Durham cattle that will improve their breeding stock (*see* 1853; 1868).

The growth of the railroad and steamship in the next four decades will encourage expansion of U.S. agriculture for export markets.

The U.S. wheat crop reaches 173 million bushels, more than double its 1840 level.

The U.S. corn crop reaches 839 million bushels, up from 377 million in 1840 (*see* 1870; 1885).

The U.S. cotton crop reaches 3.48 million bales, most of which is shipped to Liverpool for England's booming textile industry.

Mississippi agronomist Eugene Hilgard urges contour plowing and use of fertilizer: "Well might the Chickasaws and Choctaws question the moral right of the act by which their beautiful, park-like hunting grounds were turned over to another race. . . Under their system these lands would have lasted forever; under ours, in less than a century the state will be reduced to the . . . desolation of the once fertile Roman Campagna" (*see* Ruffin, 1833).

Parliament passes the first British Adulteration of Food Law after physician Edward Lankester quotes recent cases in which three people have died after attending a public banquet at which green blancmange was served that was colored with arsenite of copper. Lankester has shown that sulfide of arsenic is used to color Bath buns yellow (*see* 1861).

Louis Pasteur sterilizes milk by heating it to 125° Centigrade at a pressure of 1.5° atmospheres (100° C. is equal to 212°F). Pasteur will develop methods using lower heats to "pasteurize" milk (*see* 1857; 1861; cattle tuberculosis test, 1892; Bang's disease, 1896).

Godey's Lady's Book advises U.S. women to cook tomatoes for at least 3 hours.

The U.S. population reaches 31.4 million, twice its 1840 level, with 4 million of the total foreign-born. The white birth rate is 41.4 per thousand, down from 55 in 1800, but the excess of births over deaths is still the main source of population growth.

1860 ***(cont.)*** Native American population in the United States and its territories is 300,000, down from between 600,000 and 900,000 in 1607.

U.S. population density is 10.6 people per square mile, up from four to five in 1790 although 16.1 percent of the people live in cities of 2,500 or more. Only nine cities have more than 100,000 (New York has 805,651, Philadelphia 562,529, Brooklyn 266,661, Baltimore 212,418, Boston 177,812, New Orleans 168,675, Cincinnati 161,044, St. Louis 160,773, Chicago 109,260, Buffalo 81,129, Newark 71,914, Louisville 68,033, Albany 62,367, Washington 61,122, San Francisco 56,802, and Providence 50,666).

Hawaii's native population has fallen from 150,000 in 1819 and 75,000 in 1850 to less than 37,000 as a consequence of contagious diseases introduced from Europe and America (see 1959).

1861 Italy unites as a single kingdom (below) while the United States disunites.

Kansas enters the Union as a free state January 29 but Mississippi, Florida, Alabama, Georgia, Louisiana, Texas, Virginia, Tennessee, Arkansas, and North Carolina secede between January and May, following the course of South Carolina late last year.

Delegates from six seceding states meet at Montgomery, Ala., February 4 and form a provisional government, the Confederate States of America. Jefferson Davis of Mississippi, now 52, is named provisional president of the C.S.A. February 8 and confirmed by an election in October held after seizure of federal funds and property in the South.

Civil War begins April 12 as Fort Sumter on an island in Charleston harbor is bombarded by Gen. Pierre Gustave Toutant de Beauregard, 42, who has resigned as superintendent of the U.S. Military Academy at West Point to assume command of the Confederate Army.

"This country, with its institutions, belongs to the people who inhabit it," President Lincoln has said in his inaugural address March 4. "Whenever they shall grow weary of the existing government they can exercise their constitutional right of amending it, or their revolutionary right to dismember it or overthrow it."

The U.S. Army numbers 13,024 officers and men March 4. President Lincoln calls for 42,034 volunteers to serve for 3 years and by July there are some 30,000 green recruits in and about Washington under the command of Mexican War veteran Winfield Scott, but old "Fuss and Feathers" is now 75.

The Battle of Bull Run near the Manassas railway junction across the Potomac July 21 ends in a rout of Gen. Irvin McDowell's Union forces by Confederate troops under Gen. Beauregard (above), Gen. Joseph Eggleston Johnston, 54, and Brig. Gen. Thomas Jonathan Jackson, 37, who wins the nickname "Stonewall" from another general's remark ("There is Jackson standing like a stone wall"). The rebels lose 387 dead, 1,582 wounded, 13 missing; the Union 460 dead, 1,124 wounded, 1,312 missing.

Confederate forces defeat Gen. Nathaniel Lyon's Union troops August 10 at Wilson's Creek, Mo.

Union forces under Gen. Benjamin Franklin Butler, 42, capture Confederate forts on the North Carolina coast in late August.

The Battle of Ball's Bluff on the Potomac October 21 ends in defeat for the Union.

President Lincoln gives Gen. George Brinton McClellan, 34, command of all federal forces November 1. McClellan spends the winter training a 200,000-man Army of the Potomac while Union naval forces attempt to enforce an embargo on Southern ports that was ordered in April.

Detective Allan Pinkerton has gained prominence by uncovering a plot to kill president-elect Lincoln in Baltimore during Lincoln's journey to Washington for the inauguration (above). Engaged by Gen. McClellan to head a department of counterespionage, Pinkerton, disguised as a Major E. J. Allen, supplies reports that exaggerate the size of Confederate troop concentrations and discourage McClellan from taking decisive action (Pinkerton will resign next year) (*see* 1850; Molly Maguires, 1877).

The Colorado Territory is formed February 28, the Dakota Territory March 2, the Nevada Territory March 2 (*see* Comstock Lode, 1859). The Arizona Territory organized August 1 consists of all of New Mexico south of the 34th parallel (northeastern New Mexico has been included in the Colorado Territory).

The *Trent* affair November 8 precipitates an Anglo-American crisis. Arctic explorer Capt. Charles Wilkes, 63, commanding the *U.S.S. San Jacinto*, stops the British mail packet S.S. *Trent* and removes former U.S. senators John Murray Mason, 63, and John Slidell, 68, now Confederate commissioners bound for Britain and France. Secretary of State William H. Seward will avert war by releasing the two men early next year.

A united Kingdom of Italy proclaimed March 17 by the nation's parliament joins Lombardy, Piedmont, Parma, Modena, Lucca, Romagna, Tuscany, and the two Sicilies under Victor Emmanuel II of Piedmont's House of Savoy. Nice and Savoy were lost to France last year, Austria holds Venetia (it will not be united with Italy until 1866), Napoleon III holds Rome, which will not become part of Italy, and the nation's capital, until 1870. The proclamation of unification is followed within weeks by the death at age 62 of Conte Camillo Bensodi di Cavour who has been so instrumental in creating the new kingdom (*see* 1859; 1866).

Prussia's Friedrich Wilhelm IV has died January 2 at age 65 after a 2.5-year period of paralysis and insanity. He has been succeeded after a 20-year reign by his brother, 63, who has been acting as regent and will reign until 1888 as Wilhelm I.

The Ottoman sultan Abdul Mejid I dies of phthisis June 25 at age 38 after a reign marked by reforms including the repression of slavery. His brother, 31, will rule until 1876 as Abdul Aziz, obtaining enornmous wealth and squandering it recklessly.

Portugal's Pedro V dies of cholera November 11 at age 24 after a reign of less than 9 years. He is succeeded by his brother, 23, who will reign until 1889 as Luiz I.

France's Napoleon III extends the financial powers of Parlement and mounts a grandiose program of public works despite the quickly mounting expenses of his extravagant foreign policy. Political opposition to the emperor begins to grow at a rapid pace.

A French force of 5,000 relieves a blockade of Saigon February 25 and gains control of the area after defeating 20,000 Annamese regulars led by Gen. Nguyen Tri Phuong who have kept 900 French and Spanish troops pinned down. The French extend their control of Cochin China (Vietnam).

French, British, and Spanish troops land at Vera Cruz in December to force Mexico's Benito Juarez to honor his financial obligations and resume the payments he has suspended (*see* 1862).

Queen Victoria's Prince Consort Albert dies of typhoid fever December 14 at age 42.

Czar Aleksandr II completes the emancipation of Russian serfs begun in 1858. He allots land to the serfs for which they are to pay over a period of 49 years, but the land is given to village communes (*mirs*), not to individuals, and is to be redistributed every 10 or 12 years to assure equality. Since a peasant may buy only half the amount of land he cultivated as a serf and will often be unable to afford his annual "redemption" payments, he will be unable to gain economic independence much less improve the land or his methods of cultivation.

Russian anarchist Mikhail Bakunin returns to Europe after escaping imprisonment in Eastern Siberia and traveling home via Japan and the United States (*see* 1847). Bakunin will be the leading European anarchist until his death in 1876.

Chiricahua Apache chief Cochise, 49, appears at an army post in Arizona Territory to deny charges that he kidnapped a white child. Taken prisoner, he escapes and takes hostages to be exchanged for other Chiricahuas held by the U.S. Army. The exchange does not take place, the hostages are killed on both sides, and Cochise joins with his father-in-law Mangas Coloradas of the Mimbreno Apache in raids that threaten to drive the Anglos from Arizona (*see* 1862).

Congress levies the first U.S. income tax August 5 to raise funds needed for the Union Army and Navy. The law taxes incomes in excess of $800 at the rate of 3 percent (*see* 1872).

U.S. tariffs rise as Congress passes the first of three Morrill Acts which will boost tariffs to an average of 47 percent. Duties on tea, coffee, and sugar are increased as a war measure (*see* 1857; 1890).

Henry du Pont, 49, obtains enough saltpeter from England to keep his Delaware powder works producing. His firm will make 4 million pounds of gunpowder for the Union Army and sell it for nearly $1 million in the course of the war (*see* 1814; 1872).

I. M. Singer sells more sewing machines abroad than in America and has profits of nearly $200,000 on assets of little more than $1 million (*see* 1856; 1863).

Jay Cooke & Co. is founded at Philadelphia by banker Jay Cooke, 40 (*see* 1873).

U.S. banks suspend payments in gold December 30.

Britain produces 3.7 million tons of iron, France 3 million, the United States 2.8 million, the German states 200,000.

Britain produces 83.6 million tons of coal, France 6.8 million, Russia 300,000.

The Central Pacific Railroad is chartered in California to build the western section of a projected transcontinental rail link. Leland Stanford, 37, is president, Colis P. Huntington vice-president, Mark Hopkins an officer, and Charles Crocker, 39, head of construction. With manpower scarce due to the Civil War, Crocker approaches the Chinese Six Companies, which recruits workers at $35 per head from California and from China, where the Tai Ping Rebellion has disrupted the economy. Some 9,000 of the 10,000 laborers on the Central Pacific will be Chinese (*see* 1869).

U.S. entrepreneur Henry Meiggs, 50, obtains a contract to complete a Chilean Railroad between Valparaíso and Santiago. Meiggs chartered the packet ship Albany in 1849 to carry New York lumber to San Francisco and made a fortune in lumber and land speculations, a credit crisis in 1854 plunged him heavily into debt but his brother was elected city comptroller and Meiggs obtained a book of city notes signed in advance by the mayor and the outgoing comptroller, he received upward of $365,000 for the notes, and absconded September 26 aboard the bark *America;* the city did not discover its loss until October 16, Don Enrique arrived with his family in Chile, his knowledge of Spanish has won him a position superintending construction of the Santiago al Sur Railroad, he will complete the line in less than 2 years (several previous contractors have gone bankrupt), his success will bring him a profit of $1.5 million, and he will repay San Francisco (*see* 1871).

Paraguay's Ferro-Carril del Estado (state railway) begins regular operation September 21 between Asunción and Paraguarí, 72 kilometers, using British-built locomotives and rolling stock.

The velocipede developed by Parisian coach builder Pierre Michaux is the world's first true bicycle (*see* MacMillan, 1839). Michaux establishes La Compagnie Parisienne Ancienne Maison Michaux et Cie., produces two prototype models that have cranks and pedals attached directly to their front wheels, and will sell 142 cycles next year (*see* 1870).

The Gatling gun invented by U.S. engineer Richard Jordan Gatling, 43, can fire hundreds of rounds per minute. It will see service in the Civil War beginning in 1864 (*see* Maxim, 1883).

The open-hearth process for making steel developed almost simultaneously by German-born British inventor William Siemens, 38, and French engineer Pierre

1861 ***(cont.)*** Emile Martin, 37, will result in a rapid increase in steel production (*see* Siemens, 1856). Using a regenerative gas-fired furnace, the new process requires less coal than did previous methods such as the Bessemer process of 1856 (*see* Hewitt and Cooper, 1862).

Secretary of War Simon Cameron organizes a U.S. Sanitary Commission June 9. Educator Clara (Clarissa Harlowe) Barton, 40, quits her job at the U.S. Patent Office to organize facilities for recovering soldiers' lost baggage and to secure medicine and supplies for troops wounded in the Battle of Bull Run (above) (*see* American Red Cross, 1881).

Dorothea Dix, now 59, wins an appointment as superintendent of women nurses for the Union.

Bellevue Hospital Medical College is established at New York with Lewis Albert Sayre, 41, as the first U.S. professor of orthopedic surgery.

The germ theory of disease has its beginnings in a paper published by Louis Pasteur that absolutely refutes the idea of spontaneous generation (*see* 1860). *Mémoire sur les corpuscles organisés qui existent dans l'atmosphere* convincingly discredits the 1748 notion and others of its kind which are still commonly believed (*see* Lister, 1865).

MIT (The Massachusetts Institute of Technology) is founded by William Barton Rogers downriver from Harvard at Cambridge, Mass.

Vassar Female College is chartered at Poughkeepsie, N.Y., by local brewer Matthew Vassar, 69, who has made a fortune in land speculation and endows the college that will open in 1865.

Yale awards the first American degree of doctor of philosophy (Ph.D.) (*see* Bryn Mawr, 1885).

The University of Colorado is founded at Boulder.

The University of Washington is founded at Seattle.

A Western Union telegraph line opens between New York and San Francisco, one of whose hills will hereafter be called Telegraph Hill. The wire has been strung across the continent despite opposition from hostile tribes and Confederate sympathizers who have tried to prevent it, and it brings an end to the money-losing Pony Express started last year by William Hepburn Russell (*see* Holladay, 1862).

L'Osservatore Romani begins publication as the official newspaper of the Vatican.

Nonfiction *The History of England* (fifth and final volume) by Thomas Babbington Macaulay, 1st Baron Macaulay (he was raised to the peerage in 1857).

Fiction *Great Expectations* by Charles Dickens, illustrations by Marcus Stone, 21; *Silas Marner, or the Weaver of Raveloe* by George Eliot; *The Cloister and the Hearth* by Charles Reade whose romance of the Reformation is based on wide reading in literature of the period; *Framley Parsonage* by Anthony Trollops.

Poetry *The Golden Treasury of the best Songs and Lyrical Poems in the English Language* by English anthologist-poet Francis Turner Palgrave, 37, who will publish a second, enlarged edition in 1897.

Painting *St. John* by New York painter John La Farge, 26; *Orphée, Le Repos* by J. B. C. Corot.

Wood-cut engraving for Dante's *Inferno* by Gustave Doré.

Theater *Siegfried's Death (Siegfrieds Tod)* by Friedrich Hebbel 1/31 at Weimar's Hoftheater; *Kriemhild's Revenge (Kriemhildes Rache)* by Hebbel 5/18 at Weimar's Hoftheater. Irish-born dancer-courtesan Lola Montez (*née* Gilbert), former mistress to Bavaria's Louis I, dies in obscurity at Astoria, N.Y. January 17 at age 43.

First performances *Mephisto Waltz* by Franz Liszt, in January at Weimar's Grand Ducal Palace; Mass in B minor by the late Johann Sebastian Bach 4/24 at Berlin (first complete performance; *see* 1733); Quartet in G minor for piano and strings by Johannes Brahms 11/16 at Hamburg's Wormer Hall.

Hymn "Abide with Me; Fast Falls the Eventide" by English composer-organist William Henry Monk, 38, lyrics by the late Scottish Anglican clergyman Henry Francis Lyte who died in 1847 at age 54.

Popular songs "Aura Lea" by the U.S. composer George R. Poulton, lyrics by Cincinnati poet William Whiteman Fosdick, 36; "Maryland! My Maryland!" by Marylander James R. Randall whose lyric is set to the music of the German Christmas song "O Tannenbaum" (Randall has heard that Massachusetts troops were attacked while passing through Maryland and hopes his state will join the Confederacy); "The Vacant Chair, or We Shall Meet but We Shall Miss Him" by New York composer George Frederick Root, 41, lyrics by H. S. Washburn.

John Wanamaker, 23, opens a Philadelphia menswear shop with his brother-in-law Nathan Brown. Merchant Wanamaker will pioneer in selling all items at fixed prices with none of the haggling that has been customary between customer and clerk, and within 10 years he will be the largest U.S. menswear retailer (*see* 1896; Bon Marché, 1852; Strawbridge & Clothier, 1868).

Marshall Field becomes general manager of Chicago's Cooley, Wadsworth store and will become a partner next year (*see* 1856; 1865).

Charles Digby Harrod, 20, takes over his father's London grocery store in the Brompton Road and works to turn Harrods into a department store (*see* 1849). He will install a plate glass window and by 1868 will have a weekly volume of £1,000 (*see* 1883).

Elisha G. Otis patents a steam-powered elevator but dies at Yonkers, N.Y., April 8 at age 50, leaving Otis Elevator Co. to his sons (*see* 1857). The Otis elevator together with cheaper steel (above) will permit development of the high-rise building and lead to the rise of the modern city (*see* 1884).

Most Russian peasants can do no more than grow enough for their own families' needs with a small surplus in good years to sell for tax money. Most cultivate with wooden plows, harvest their crops with sickles or scythes, and thresh their grain with hand flails. One-third of the peasants have no horses, another third only one horse, and even in the richer parts of Russia the soil is drained by strip farming and a shortage of fertilizer (*see* Stolypin, 1906).

Chicago ships 50 million bushels of U.S. wheat to drought-stricken Europe, up from 31 million in 1860 despite the outbreak of the Civil War and the shortage of farm hands.

"Without McCormick's invention, I feel the North could not win and that the Union would be dismembered," says Secretary of War Edwin M. Stanton. The McCormick reaper sells for $150, up from $100 in 1849, but McCormick offers it at $30 down with the balance to be paid in 6 months if the harvest is good and over a longer term if the harvest is disappointing (*see* Singer, 1856).

The California State Legislature commissions Count Agoston Haraszthy to bring select varieties of European wine grapes to the state (*see* 1857). Haraszthy will obtain 100,000 cuttings that represent 300 varieties and will inaugurate the modern era of California wine production (*see* 1920).

A lecture to the Royal Society of Arts demonstrates that 87 percent of London's bread and 74 percent of the city's milk is still adulterated despite the Adulteration of Foods Law enacted last year (*see* 1872).

Louis Pasteur (above), confutes theories that lack of oxygen is what keeps canned food from spoiling. He also notes that since milk is alkaline rather than acid, it requires more heat for sterilization, but heating milk to 110° C. will stop the growth of "vibrios" and preserve it indefinitely, he says (*see* 1895).

Baltimore canner Isaac Solomon reduces the average processing time for canned foods from 6 hours to 30 minutes by employing Humphry Davy's discovery that adding calcium chloride raises the temperature of boiling water to 240° F. and more (*see* 1874).

Gail Borden licenses more factories to produce his condensed milk, which the Union Army is purchasing for use in field rations. Borden's son John Gail fights for the Union, his son Lee with the Texas Cavalry for the Confederacy (*see* 1858; 1866).

Van Camp's Pork and Beans helps sustain Union troops in the field. Indianapolis grocer Gilbert C. Van Camp, 37, once a tinsmith, has created a new canned food staple and secured an army contract (*see* 1882).

Philadelphia butcher Peter Arrell Brown Widener, 27, obtains a government contract to supply mutton to Union troops in and about Philadelphia. Widener will pocket a $50,000 profit which he will invest in a chain of meat stores and in street railways, building a huge fortune (*see* American Tobacco, 1890).

America's first commercial pretzel bakery opens at Lititz, Pa., with local baker Julius Sturgis in charge.

Henri Nestlé, 45, begins his rise to world prominence. The Swiss businessman has taken a financial interest in a chemical firm at Vevey operated by Christopher Guillaume Keppel, and he assumes ownership upon Keppel's retirement (*see* 1866).

Book of Household Management by English journalist Mrs. Isabella Beeton, 25, gives precise recipes for Victorian dishes in a thick 3-pound volume whose contents appeared originally in periodicals. Mrs. Beeton gives costs and cooking times as well as ingredient quantities.

Britain's population reaches 23.1 million while Ireland has 5.7 million, Italy 25, Russia 76.

America's 23 Northern states have a population of more than 22 million; the Confederacy has 10 million, over a third of them slaves.

1862 The Union gains its first major success in February as Fort Henry on the Tennessee River and Fort Donelson on the Cumberland fall to U.S. forces under Brig. Gen. Ulysses S. Grant, 39, who wins promotion to major general of volunteers.

Santa Fe falls March 4 to Confederate forces under Henry Hopkins Sibley but Southern hopes of taking over the Southwest are dashed March 27 at the Battle of Glorieta. Union forces stop Sibley in Apache Canyon at Pidgin's Ranch near Glorieta Pass.

The first naval battle between ironclad ships occurs March 8. The Confederate ironclad frigate *Merrimac* (renamed the *Virginia)* has sunk the *Cumberland* and defeated the *Congress* in Hampton Roads but is forced to withdraw March 9 after an engagement with the Union's ironclad *Monitor,* built by John Ericsson, 58, with a revolving gun turret.

The Battle of Shiloh (Pittsburgh Landing) April 6 and 7 on the Tennessee River ends after mass slaughter with both sides claiming victory. Gen. Albert Sydney Johnston, now 58, is mortally wounded the first day, Gen. Beauregard waits in vain for reinforcements from Gen. Earl Van Dorn and retreats to Corinth, but Gen. Grant loses 13,047 men to 10,694 casualties for the South.

Island Number 10 below Cairo, Ill., on the Mississippi falls to Union forces April 7. Brig. Gen. of Volunteers John Pope, 40, begins a successful effort to split the Confederacy by gaining control of the entire length of the Mississippi.

Gen. McClellan finally ends his long delay and sends Union forces up the peninsula between the James and York Rivers toward Richmond. The outnumbered Confederates under Joseph Eggleston Johnston and Robert E. Lee are obliged to pull back but much of the Union strength is diverted to the Shenandoah Valley by Gen. T.J. "Stonewall" Jackson. The Federal troops are withdrawn from the peninsula after sustaining heavy losses in the Battle of Fair Oaks May 31 and in the Seven Days' Battle from June 25 to July 1 at Mechanicsville, Gaines' Mill, White Oaks Swamp, and Malvern Hill.

The Second Battle of Bull Run (Manassas) August 30 ends in defeat for the Union as 20,000 Confederates under Stonewall Jackson and Gen. James Longstreet, 41, repulse Gen. Pope's 60,000 Federals.

1862 ***(cont.)*** The Battle of Antietam (Sharpsburg) in Maryland September 17 is indecisive but wins Gen. Joseph Hooker, 47, his nickname "Fighting Joe" (camp followers of Hooker's Massachusetts division are called "Hooker's girls," or simply "hookers"). Gen. Lee makes his first effort to carry the war into the North but is outgunned and outnumbered by 87,000 Union troops to his 41,000.

A Confederate force under Gen. Braxton Bragg, 45, threatens Cincinnati but retires after an indecisive engagement at Perryville October 8 with Union troops under Gen. Don Carlos Buell, 44. Buell is relieved of his command for not having promptly pursued Bragg, who had some success at Shiloh (above).

President Lincoln relieves Gen. McClellan of his command in November and places the Army of the Potomac under Gen. Ambrose Everett Burnside, 38. Gen. Lee hands Burnside a bad defeat December 13 at the Battle of Fredericksburg.

Gen. Grant is surprised at Holly Springs, Miss., December 20 by a Confederate force under Gen. Van Dorn which captures a huge supply of stores, takes 1,500 prisoners, and burns more than 4,000 bales of cotton.

Buell's replacement William Starke Rosecrans, 43, skirmishes with Bragg for 4 days in the Battle of Murfreesboro (Stone's River) beginning December 31.

Britain and Spain have withdrawn their troops from Mexico April 8 when it became clear that France's Napoleon III intended to establish a Catholic Latin empire in Mexico. The Battle of Puebla May 5 ends in a slaughter of crack French troops by Mexican irregulars.

Paraguay's dictator Cárlos López dies September 10 at age 75 after an 18-year rule. His megalomaniac son Francisco Solano López, 36, becomes president, makes himself absolute dictator, and begins a disastrous 8-year rule (*see* 1865).

London has decided not to recognize the Confederacy on which Britain depends for cotton (below). The British are dependent on the North for grain exports.

Greece's Otto I is deposed after a 29-year reign that has alienated the Greeks and brought an assassination attempt on his wife Amalie. Otto and his queen take refuge on a British warship and return to Bavaria (*see* 1863).

Chinese forces led by U.S. military adventurer Frederick Townsend Ward win victories over the Tai Ping rebels, but Ward is mortally wounded September 20 at age 31 (*see* 1863).

Congress abolishes slavery in the District of Columbia April 16 and in U.S. territories June 19.

"My paramount object in this struggle is to save the Union," writes President Lincoln August 22 in a letter to Horace Greeley of the *New York Tribune,* "and is not either to save or to destroy slavery. If I could save the Union without freeing any slave, I would do it; and if I could do it by freeing all the slaves, I would do it; and if I could save it by freeing some and leaving others alone, I would also do that" (but see below).

An Emancipation Proclamation issued by President Lincoln September 22 declares that "persons held as slaves" within areas "in rebellion against the United States" will be free on and after January 1, 1863. Lincoln makes emancipation a war aim.

Chiricahua and Mimbreno Apaches led by Cochise hold Apache Pass with 500 warriors against 3,000 California volunteers under Gen. James Carleton until forced out by artillery fire. The aged Mangas Coloradas is captured and killed, and Cochise leads his followers deep into the Dragoon Mountains from which he continues raids that will terrorize white settlers until 1871 (*see* 1861).

A Sioux uprising has begun in Minnesota August 17 under the leadership of Little Crow. The insurrection is suppressed, 306 tribesmen are sentenced to death, and 38 are hanged December 26 on a huge scaffold at Mankato.

The First Regiment of South Carolina volunteers organized at Boston in November is the first U.S. regiment of ex-slaves. Its commander is Col. Thomas Wentworth Storrow Higginson, 38, formerly pastor of the Free Church at Worcester, who in 1854 purchased axes and led the unsuccessful effort to liberate the fugitive slave Anthony Burns.

The Homestead Act voted by Congress May 20 declares that any U.S. citizen, or any alien intending to become a citizen, may have 160 acres of Western lands absolutely free (except for a $10 registration fee) provided he make certain improvements and live on the tract for 5 years. Enacted to redeem an 1860 Republican campaign promise after years of excessive land speculation, the law is to become effective January 1 (below).

English African explorer John Speke confirms that Lake Victoria is the source of the Nile—the world's longest river (*see* Burton, 1855).

England's Lancashire textile mills shut down as they run out of Southern lint from which they have been cut off by the Civil War.

A second London Great International Exhibition includes a display of Japanese arts and crafts that stir demand for Japanese silks, prints, porcelains, bronzes, lacquerware, and bric-a-brac (*see* 1863).

Union Army raiders led by Capt. James J. Andrews steal the Confederate locomotive The *General* April 1 and race the Western and Atlantic wood burner north in an effort to cut the rail lines and thus isolate Gen. P. G. T. Beauregard's army at Chattanooga. Confederate soldiers give chase in the locomotive The *Texas* and catch the raiders.

The Illinois-Central facilitates Union troop movements with its north-south line (*see* 1850). The company will sell off 800,000 acres of its land to settlers.

Union Pacific Railroad builders receive authority from Congress July 1 to take the timber, stone, and other material they need from public lands along the right of way they have been granted. The builders include shovel maker Oakes Ames, 58, of Oliver Ames & Sons who wins election in November to the House of Representatives (*see* 1803; 1863; Crédit Mobilier, 1867; 1872).

Congress promises up to 100 million acres of federal lands to the Union Pacific (above), the Central Pacific, and other railroads that will connect the Mississippi with the Gulf and Pacific coasts (*see* 1863).

The freighting firm Russell, Majors and Waddell has gone bankrupt financing the Pony Express that Western Union put out of business last year with its telegraph. Missouri entrepreneur Ben Holladay, 42, buys the firm at public auction in March and thus acquires the Central Overland, California, and Pikes Peak Express with its government contract for hauling overland mail between Missouri and the Pacific Coast, a contract that will soon be worth more than $1 million per year in federal subsidies as Holladay puts the stagecoach runs on schedule, builds new way stations, improves equipment, upgrades personnel and passenger accommodations, and extends his runs into the mining districts of Colorado, Idaho, and Montana Territory. The "Napoleon of the Plains" will make a fortune in the next 4 years before he sells out in anticipation of the transcontinental railroad (*see* 1863).

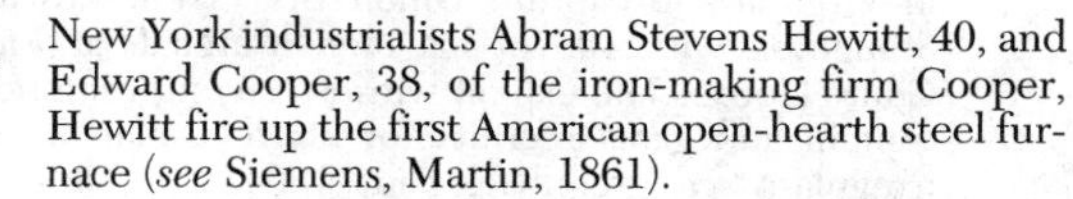

New York industrialists Abram Stevens Hewitt, 40, and Edward Cooper, 38, of the iron-making firm Cooper, Hewitt fire up the first American open-hearth steel furnace (*see* Siemens, Martin, 1861).

"Local Asphyxia and Symmetrical Gangrene of the Extremities" by French physician P. Edouard Raynaud attributes the symptoms to an arrest of the passage of blood to the affected parts as a result of a spasm of the arterioles. Reynaud's disease will prove to be one of childhood or early adulthood, affecting females more often than males.

Un Souvenir de Soverino by Swiss philanthropist Henri Dunant, 34, urges the creation of volunteer societies to help the wounded on battlefields. Dunant has written his pamphlet after witnessing terrible scenes of bloodshed and neglect at the Battle of Solverino in 1859 (*see* International Red Cross, 1863).

The Morrill Land-Grant Act provides funds to start U.S. land-grant colleges for the scientific education of farmers and mechanics (*see* below; 1890).

Fiction *Les Misérables* by Victor Hugo, now 60, whose 10-volume work depicts the injustice of French society. Hugo has been living on the island of Guernsey since his banishment by Napoleon III in 1848; *Fathers and Sons* (*Otzi i Deti*) by Ivan Turgenev introduces the term "nihilism," meaning a belief in nothing and a total rejection of established laws and institutions.

Poetry *Goblin Market and Other Poems* by English poet Christina Georgina Rossetti, 31, gives the Pre-Raphaelites their first great literary success. Rossetti's brother Dante Gabriel Rossetti, 34, is a leader of the Pre-Raphaelite Brotherhood and its rebellion against the Royal Academy of Art but his wife of 2 years succumbs to an overdose of the laudanum she has been taking to dull the pain of her tuberculosis; "Say Not the Struggle Nought Availeth" by the late English poet Arthur Hugh Clough, who died last year at age 41: his final stanza ends, "But westward, look, the land is bright."

Painting *Potato Planters* by Jean François Millet; *La Musique aux Tuileries* by Edouard Manet; *Cotopaxi* by Frederic E. Church.

Theater *East Lynne, or The Elopement* 4/21 at Boston, is an adaptation of the novel by English novelist Ellen Price Wood, 48, whose story has been running in the *Baltimore Sun*. The melodrama will be perennially revived and will be the most popular play in U.S. stage history; *Mazeppa* 6/16 at New York's New Bowery Theater, with poet-actress Adah Isaacs Menken, 27, strapped half nude to a horse in a melodramatic adaptation of the 1819 Byron poem that creates a sensation.

Opera *Beatrice and Benedict* 8/9 at Baden-Baden, with music by Hector Berlioz, libretto from the Shakespeare play *Much Ado About Nothing* of 1599; *La Forza del Destino* 11/10 at St. Petersburg's Imperial Italian Theater, with music by Giuseppe Verdi.

First performances Scenes from Goethe's *Faust* by the late Robert Schumann, in January at Cologne; *Wallenstein's Camp (Valdstynov Tabor)* by Czech pianist-composer Bedřich Smetana, 37, 1/5 at Prague, a symphonic poem based on the 1798 Schiller play.

Kochel (K.) listings for the works of the late Wolfgang Amadeus Mozart appear in *Chronologisch-thematisches Verzeichniss* by Austrian musicographer-naturalist Ludwig Ritter von Kochel, 62, whose catalogue has an appendix listing lost, doubtful, or spurious Mozart compositions. German music scholar Alfred Eisenstein will revise the work for a 1937 reissue.

"The Battle Hymn of the Republic" appears in the February *Atlantic Monthly* with lyrics by Julia Ward Howe, 43, who helps her husband Samuel Gridley Howe, 61, in publishing the Boston abolitionist journal The *Commonwealth* which is dedicated also to prison reform, improved care for the feeble-minded, and abolition of imprisonment for debt. Howe's verse is set to the tune "Glory, Glory, Hallelujah" attributed to William Steffe.

Waltzes "Perpetuum Mobile" and "Motor Waltz" by Johann Strauss in Vienna.

Popular songs "We Are Coming, Father Abraham 300,000 More" by Irish-American bandmaster Patrick Sarsfield Gilmore, 33, of the 24th Massachusetts Regiment, lyrics by Delaware Quaker abolitionist James Sloan-Gibbons, 52; "We've a Million in the Field" by Stephen Foster; "The Bonnie Blue Flag" by English-American actor Harry B. McCarthy, 27, whose verses to the melody, of "The Irish Jaunting Cart" were sung by McCarthy last year at Jackson, Miss., and have been revised by Mrs. Annie Chambers-Ketchum (the song will be the Confederacy's national anthem).

"Taps" is composed early in July by Army of the Potomac chief of staff Gen. Daniel Butterfield, 31, who writes the notes for the bugle call at Harrison's Landing

1862 ***(cont.)*** in Virginia. Butterfield has written the bugle call to be played at lights out and at funerals.

The first Monte Carlo gambling casino opens in Monaco under the direction of the former manager of the casino at Bad Hamburg (*see* 1878). Monaco sells Menton and Roquebrune to France.

An English Wholesale Society is established to supply the scores of cooperative societies that have sprung up along lines established in 1844 by the Equitable Pioneers of Rochdale, but most workers are restrained from joining such societies by the fact that they owe money to local tradespeople and must continue to live on credit.

New York department store king A. T. Stewart contributes $100,000 to the Union cause and sells uniforms at cost to the Union Army (*see* 1846). Stewart has introduced a revolutionary one-price system with each item carrying a price tag, an innovation that permits him to employ women sales clerks who can be hired more cheaply than men. Having amassed a fortune of $50 million, Stewart gives a 10 percent discount to wives and children of clergymen and schoolteachers.

The world's largest department store is erected in New York's 10th Street by A. T. Stewart (above) who has employed James Bogardus, now 62, to design the eight-story structure that is also the world's largest building with a cast-iron front (*see* 1849). It has a central court, a huge skylight, a grand stairway leading up from the spacious ground floor, and elevators to make the upper floors accessible (*see* Otis, 1861; Wanamaker, 1896).

U.S. wool production will climb from 40 million pounds to 140 million in the war years as the nation's woolen mills fill Union Army contracts and pay dividends of 10 to 40 percent.

British crops fail and hunger is widespread, especially since thousands of mill hands have been thrown out of work by a cotton famine (above).

The Morrill Land-Grant Act voted by Congress July 2 gives the states 11 million acres of federal lands to sell, nearly twice the acreage of Sen. Justin Morrill's home state of Vermont which becomes a vast sheep pasture as U.S. clothing makers shift from cotton to wool.

Nearly 470,000 settlers will apply for homesteads in the next 18 years under terms of the Homestead Act (above). Roughly one-third of these will actually receive land, but while a farmer can make a living on 80 acres in the East, land in the West is generally so arid that even 160 acres will rarely permit economically viable agriculture (*see* 1909).

Congress creates an independent U.S. Department of Agriculture (*see* 1889).

The U.S. Produce Exchange is organized at New York (*see* Chicago Board of Trade, 1865).

Union forces destroy Confederate salt works on Chesapeake Bay (*see* 1863).

Congress prohibits distillation of alcohol without a federal license but "moonshiners" continue to make whiskey.

The U.S. Navy abolishes its rum ration through the influence of Rear Admiral Andrew Hull Foote, 56, who has made his ship the first in the Navy to stop issuing rum rations. The ration will later be revived.

An Internal Revenue Act passed by Congress in July to finance the war effort taxes beer at $1 per barrel and imposes license fees on tavern owners.

Gulden's mustard is introduced by New York entrepreneur Charles Gulden whose Elizabeth Street shop has easy access to sources of mustard seed, spices, and vinegar.

Crosse & Blackwell in England introduces canned soups (*see* Campbell and Anderson, 1869).

The Homestead Act (above), war-inflated farm prices, and exemption from military service for foreigners spurs U.S. immigration.

1863 President Lincoln relieves Gen. Burnside of command January 25 and puts Gen. Hooker in charge of the Army of the Potomac.

Mosby's Rangers led by Confederate cavalry officer John Singleton Mosby, 29, operate behind Union lines in Virginia and capture Union Brig. Gen. Edwin H. Stoughton with 100 of his men March 9 at Fairfax Court House. The exploit wins Mosby a promotion to captain and gains recruits for his nine-man force of irregulars (*see* "greenback raid," 1864).

Union forces suffer defeat at Chancellorsville, Va., from May 1 to May 4 but the South loses one of its best generals. "Stonewall" Jackson is wounded by one of his own sentries May 2, his arm is amputated May 3. Robert E. Lee writes him, "You are better off than I am, for while you have lost your left, I have lost my right arm." Jackson dies of pneumonia May 10 saying, "Let us cross the river and rest in the shade."

President Lincoln replaces Gen. Hooker June 28 with Gen. George Gordon Meade, 46, who triumphs a few days later at Gettysburg. The Battle of Gettysburg in Pennsylvania July 1 to 3 marks the turning point of the war as Southern troops are routed after the heroic Pickett's Charge of July 3. Gen. George Edward Pickett, 38, leads his 4,500 muskets plus 10,000 men from Gen. A. P. Hill's command from Seminary Ridge across an open plain raked by Union musket and artillery fire from three sides and ascends the slant of Cemetery Hill. Some 6,000 Southerners fall and the Confederate advance into the North is doomed.

Vicksburg, Miss., falls to Gen. Grant July 4 after a short siege. Union forces gain control of the Mississippi, cutting Confederate supply lines.

West Virginia is admitted to the Union June 20 as the 35th state. The new state's constitution calls for gradual emancipation of slaves.

The Territory of Idaho is formed from parts of Dakota, Nebraska, Utah, and Washington.

Arizona is created as a separate territory cut from the New Mexico Territory (*see* 1912).

French forces in Mexico occupy Mexico City June 7 and a group of exiled Mexican leaders meeting under French auspices in July adopt an imperial form of government for their country. They offer the Mexican throne to Austria's Archduke Maximilian, 31, brother of Emperor Franz Josef (*see* 1862; 1864).

Conscription for the Union Army begins July 11 under legislation passed March 3 giving exemption to any man who pays $300 to hire a substitute.

Draft riots break out in Northern cities with the worst occurring at New York. Some 1,200 people are killed and thousands injured as mobs attack the Colored Orphan Asylum on Fifth Avenue between 43rd and 44th streets. The 200 orphans are evacuated before they can be harmed but blacks throughout the city are attacked and killed. Now a city of 800,000 with some 200,000 of its people Irish, New York suffers more than $2 million in property damage as rioters tear up railroad tracks, burn hotels, and create general chaos in their battles with police and federal troops, but New York State contributes more men to the Union Army than does any other state.

New York draft riots left 1,200 dead and thousands injured as the city mob vented its fears and prejudices.

Federal troops attack Battery Wagner in South Carolina at dusk July 18, sustaining 1,515 casualties including Col. Robert Gould Shaw, 25, whose 54th Massachusetts Colored Infantry gains the parapet and holds it for an hour before falling back, losing nearly half its number.

Quantrill's Raiders burn Lawrence, Kan., August 21. Led by former schoolteacher and professional gambler Capt. William C. Quantrill, 26, who helped capture Independence, Mo., last year and was mustered into the Confederate Army with his men, the 450 raiders include Jesse Woodson James, 15, his brother Frank, and Thomas Coleman "Cole" Younger, 19, who ride into Lawrence at dawn, kill 180 men, women, and children, sack the town, and burn most of its buildings, but Kansas contributes a greater percentage of its male population to the Union Army than does any other state (*see* crime, 1876).

The Battle of Chicamauga in northern Georgia September 19 and 20 ends in empty victory for Confederate troops under Gen. Bragg. While Gen. Leonidas Polk, 57, fails to support them, James Longstreet, 42, and Nathan Bedford Forrest, 42, send Union forces under Gen. Rosecrans retreating headlong to Chattanooga. Union Gen. George H. Thomas, 47, stands firm (earning the sobriquet "Rock of Chicamauga"), Bragg fails to follow up, and his costly success serves little purpose. Confederate losses total 2,132 dead, 14,674 wounded versus 1,656 Union dead, 9,756 wounded. Bragg takes thousands of prisoners.

President Lincoln dedicates a national cemetery at Gettysburg (above) November 19 and makes an eloquent address that will be memorized by generations of schoolchildren ("Four score and seven years ago . . . that government of the people, by the people, and for the people shall not perish from the earth").

The Battle of Chattanooga November 23 to 25 ends in a Confederate rout toward Chickamauga. Bragg has withdrawn from Lookout Mountain to Missionary Ridge, Grant has sent Union troops under Generals Hooker, William Tecumseh Sherman, 43, and Philip H. Sheridan, 32, to drive the rebels from the ridge. Union losses (dead and wounded): 5,824 out of 56,359 effectives; Confederate losses: 6,667 out of 64,165.

Denmark's Frederick VII dies November 15 at age 55 after a 15-year reign and is succeeded by his cousin, 45, who will reign until 1906 as Christian IX. The duke of Schleswig-Holstein-Sonderburg-Augustenburg, 34, denies Danish sovereignty and proclaims himself Frederick VIII.

The Greek Assembly chooses Britain's Prince Alfred to succeed the deposed Otto I, London rejects the election, the Greeks choose a Danish prince of 17, and he will reign until 1913 as George I.

Egypt's khedive Mohammed Said dies at age 41 after a 9-year reign and is succeeded by his cousin of 33 who will reign until 1879 as Ismail Pasha. The new khedive makes a speech about the need for economy but sets out to complete the modernization of Egypt; by year's end he is hopelessly in debt and will go further into debt as he bribes newspapers to favor his vanity and buys jewels to adorn the wives in his harem (*see* Suez Canal, 1875).

China's Tai Ping rebels lose Soochow to Manchus led by Zeng Kuofan and the British Gen. Charles G. Gordon, 30 (*see* 1862; 1864).

The Emancipation Proclamation issued last year by President Lincoln takes effect January 1, freeing nearly 4 million U.S. slaves but not those in Union-held areas. When the proclamation is instrumented universally in 1865, some 385,000 slaveowners will lose human property valued at $2 billion, but while half of all white families in South Carolina and Mississippi own slaves and

1863 ***(cont.)*** 12 percent of all slaveowners have more than 20 slaves, 75 percent of all free Southerners have no direct connection with slavery.

Kit Carson reaches Fort Defiance in Arizona Territory July 20 with federal troops that have been joined by a band of Ute tribesmen. He begins to resettle Navajo and Apache tribespeople on a reservation at Fort Sumner in New Mexico. All captives who surrender voluntarily are to be taken to the reservation to become farmers, all males who resist to be shot and their livestock and food supplies destroyed (*see* 1862; 1864).

The Nez Perce in the Northwest are forced to sign a treaty agreeing to vacate lands coveted by the whites (*see* 1877).

The Ruby Valley Treaty signed with the Shoshone, Washoe, and other tribes in the Nevada Territory gives more than 23 million acres to the tribes but most of it is desert. The white man receives rights to build railroads across the Native American lands, which account for 86 percent of the Territory (*see* 1861).

A new U.S. gold rush begins in May as prospectors discover gold at Alder Gulch in the Idaho Territory.

Britain and France sign commercial treaties with Belgium to begin a period of free trade. The River Scheldt reopens to free navigation for the first time since 1648.

Singer Manufacturing Co. is incorporated by sewing machine inventor I. M. Singer and Edward Clark who split 4,100 of the 5,000 shares between them (*see* 1861). Family incomes in America average only $500 per year and a new Singer sewing machine sells for $100 but Clark's $5 per month installment plan persuades customers to buy Singer machines which cost perhaps $40 to make including all overhead (*see* 1877).

John D. Rockefeller builds a petroleum refinery at Cleveland (*see* 1860; 1865).

President Lincoln signs a bill passed last year guaranteeing builders of the Central Pacific and Union Pacific Railroads $16,000 for every mile of track laid on the plains, $32,000 per mile for tracks laid through intermountain stretches, and $48,000 per mile for track laid through the mountains. The bill also provides for extensive land grants along the railroads' rights of way.

Ground is broken at Sacramento, Calif., February 22 for the Central Pacific (*see* 1861; 1866).

Ground is broken at Omaha in Nebraska Territory December 2 for the Union Pacific (*see* 1862; 1865).

The Atchison, Topeka and Santa Fe Railroad is organized November 24 by Kansas abolitionist Cyrus K. Holliday, 37, of Topeka with State Senator S. C. Pomeroy of Atchison and other promoters who rename a company formed earlier as the Atchison and Topeka. They lobby in the state legislature for federal land grants and county bond issues (*see* 1864).

Cornelius van Derbilt gains control of the New York and Harlem Railroad (*see* 1839; 1858; 1864).

The Milwaukee Road (Chicago, Milwaukee, St. Paul and Pacific Railroad) has its origin in the Milwaukee and St. Paul Railroad which incorporates the Milwaukee and Mississippi begun in 1851. The road will be the first to reach Minneapolis and St. Paul (in 1867), enter Chicago on its own rails in 1873, adopt the name Chicago, Milwaukee and St. Paul in 1874, reach Omaha in 1882, and by 1887 will reach Kansas City (*see* 1909).

New Zealand gets her first railroad as a state-owned line opens between Christchurch and Ferrymead.

The world's first underground railway system opens in London January 10. The London Underground carries 9.5 million passengers in its first year and will grow to cover 257 miles of track with 278 stations (*see* 1853; 1890; New York, 1904).

Twin-screw steamers appear on the high seas but Bristol, Glasgow, Liverpool, Sunderland, and other British ports will produce hundreds of iron windjammers in the next two decades while U.S. ships will continue to be made of the wood that North America's vast forests make cheap and abundant.

The Solvay process employed in a new plant at Couillet, near Charleroi, is cheaper and more effective than the 1791 Leblanc process for obtaining soda (sodium carbonate) from salt (sodium chloride). Belgian industrial chemist Ernest Solvay, 25, and his brother Alfred dissolved salt in water 2 years ago, saturated it with ammonia, and allowed it to trickle down a tower full of perforated partitions. Carbon dioxide produced by heating limestone to quicklime is blown into the bottom of the tower to produce sodium bicarbonate which is heated to make sodium carbonate, or soda, which will be widely used not only to make glass but also in paper, bleaches, water treatment, and petroleum refining.

Wyandotte Iron Works in Michigan pours the first Kelly-process steel. William Kelly's 1857 patent for the "pneumatic process" will come under the control of the only U.S. company licensed by Henry Bessemer, and the Bessemer name will come into exclusive use for the process (*see* 1856). Kelly will receive less than 5 percent of the royalties paid to Bessemer (*see* oxygen furnace, 1954).

Bethlehem Steel has its origin in the Saucon Iron Co. founded at South Bethlehem, Pa., to make rails from local iron ores. The company soon hires John Fritz of Cambria Iron who has pioneered in making Bessemer steel (*see* 1886).

Cotton textile production begins in Japan. The Japanese will be the world's largest exporters of cotton goods by 1932.

The Naturalist on the Amazon by H. W. Bates is published at London. Bates returned in 1859 with 8,000 new insect species (*see* 1848).

Congress authorizes creation of a National Academy of Sciences to advise the U.S. government in scientific matters and promote scientific research (*see* AAAS, 1848; NRC, 1916).

A scarlet fever epidemic in England takes more than 30,000 lives.

A 12-year worldwide cholera epidemic begins (*see* 1866).

The fundamental principles of the International Red Cross movement to aid wounded soldiers and other victims of war are established at Geneva's Palais de l'Athenée where 36 experts and government delegates meet in response to Henri Dunant's 1862 appeal (*see* 1864).

President Lincoln proclaims the first national Thanksgiving Day October 3 and sets aside the last Thursday of November to commemorate the feast given by the Pilgrims in 1621 for their Wampanoag benefactors. The president has acted partly in response to a plea from *Godey's Lady's Book* editor Sarah Josepha Hale, now 74, who has campaigned since 1846 for Thanksgiving Day observances and by 1852 had persuaded people to celebrate Thanksgiving on the same day in 30 of the 32 states, in U.S. consulates abroad, and on U.S. ships in foreign waters (*see* Roosevelt, 1939).

The University of Massachusetts has its beginnings in the Massachusetts Agricultural College founded at Amherst. It will be renamed in 1947.

Boston College is founded by Roman Catholics.

Le Petit Parisian begins publication under the name *Le Petite Presse;* it will be France's largest paper by 1910 with 1.5 million circulation (*see* 1944).

Advertising will not develop in France as in Britain and America; French newspapers will be obliged to support themselves by offering favorable publicity to any political group that will pay.

Fiction *Romola* by George Eliot; *Five Weeks in a Balloon (Cinq Semaines en Ballon)* by French novelist Jules Verne, 35; "The Man Without a Country" by U.S. author Edward Everett Hale, 41, whose hero Philip Nolan has been exiled for his part in the treason of Aaron Burr but has requested that he be memorialized with the epitaph, ". . .He loved his country as no other man has loved her;/but no man has deserved less at her hands"; the pen name "Mark Twain" is adopted by former Mississippi riverboat pilot Samuel Langhorne Clemens, 27, who failed as a prospector in Nevada 2 years ago and has gone to work as a Virginia City newspaper reporter using as his *nom de plume* the cry used by Mississippi riverboat leadsmen to indicate a 12-foot depth of water. Twain will move to San Francisco next year to work for the *San Francisco Call* and write short stories.

Poetry *Tales of a Wayside Inn* by Henry Wadsworth Longfellow includes "Paul Revere's Ride." Friends gathered at the Red-Horse Inn at Sudbury some 20 miles from Cambridge tell tales, the landlord going first: "Listen, my children and you shall hear/Of the midnight ride of Paul Revere,/On the eighteenth of April, in Seventy Five;/hardly a man is now alive/Who remembers that fabulous day and year. . ."; *The Water-Babies* by Charles Kingsley: "When all the world is young, lad,/And all the trees are green;/And every goose a swan, lad,/And every lass a queen;/Then hey, for boot and horse, lad,/And round the world away;/ Young blood must have its course, lad,/And every dog its day."

Painting *Le déjeuner sur l'herbe* by Edouard Manet is exhibited May 15 at the Salon des Refuses in Paris and creates a furor with its depiction of a nude picnicing with two clothed men. Other paintings: *Mlle. Victorine in the Costume of an Espada* and *Olympia* by Manet; *Man with Hoe* by Jean Millet (*see* Markham, 1899); *Icebergs* by Frederic E. Church. Eugène Delacroix dies at Paris August 13 at age 65.

James A. McNeill Whistler discovers Japanese *ukiyoe* prints at London. He will promote Japanese art in Paris and adopt a Japanese butterfly seal to sign his work (*see* 1867; 1872; Toulouse-Lautrec, 1893).

The *Winged Victory of Samothrace* turns up in excavations at the Greek city of Samothrace. The Louvre at Paris will exhibit the statue for well over a century.

First performances *Missa Solemnis* in A flat major by the late Franz Schubert 1/1 at Leipzig.

Opera *The Pearl Fishers* (*Les Pêcheurs de Perles*) 9/10 at the Théâtre-Lyrique, Paris, with music by French composer Georges (Alexander César Leopold) Bizet, 25.

Popular Songs "When Johnny Comes Marching Home" by Louis Lambert (Patrick Sarsfield Gilmore). He has adapted the song "Johnny Fill Up the Bowl" which may have been adapted from the song "Johnny, I Hardly Knew Ye"; "Clementine (Down by the River Lived a Maiden)" by U.S. songwriter H. S. Thompson; "Just Before the Battle, Mother" by George Frederick Root; "The Battle Cry of Freedom" by George Frederick Root; "Weeping, Sad and Lonely, or When This Cruel War Is Over" by Henry Tucker, lyrics by Charles Carroll Sawyer; "The Rock Island Line" by Confederate soldiers: "Now the Rock Island Line/ It is a mighty good road,/ Oh the Rock Island Line/ It is the road to ride. . ." but while the lyrics say, "She runs down to New Orleans" the Rock Island Line will in fact never go beyond Eunice, La. (*see* 1855).

A new Football Association established in England draws up definitive rules for "soccer" (*see* rugby, 1823). It will be called *association* in Britain, *football* in most other countries, and will be popular throughout Europe and South America (*see* U.S. football, 1869; World Cup, 1930).

The first major U.S. racetrack for flat racing opens at Saratoga Springs, N.Y., which will become the center of American horse racing (*see* Travers Stakes, 1864).

The first four-wheeled roller skates are patented by New York inventor James L. Plimpton whose small boxwood wheels are arranged in pairs and cushioned by rubber pads, making it possible to maintain balance and to execute intricate maneuvers (*see* 1760). Plimpton's roller skates will lead to a widespread roller skating fad later in this decade and the fad will sweep Europe in the 1870s.

General Tom Thumb is married February 10 at New York's Grace Church on Broadway and 10th Street to Lavinia Warren (Mercy Bunn) who stands 2 feet 8

1863 ***(cont.)*** inches in height (*see* 1842). Heavily promoted by showman P.T. Barnum, the wedding attracts huge crowds that jam the streets (the wedding party stands on a grand piano to receive guests at the Metropolitan Hotel).

The Capitol dome at Washington is capped December 2 to complete the great structure on which work has been continued through the war by order of President Lincoln. Architect Thomas U. Walter has unified the work of the Capitol's three previous architects William Thornton, Benjamin Latrobe, and Charles Bulfinch. The Capitol will be landscaped with terraces designed by Frederick Law Olmsted and a fresco by Constantino Brumidi will grace its rotunda in 1865.

New York's Central Park Commission reports, "If all the applications for the erection and maintenance of towers, houses, drinking fountains, telescopes, mineral water fountains, cottages, Eolian harps, gymnasiums, observatories, and weighing scales, for the sale of eatables, velocipedes, perambulators, indian work, tobacco and segars, iceboats and the use of the ice for fancy dress carnivals were granted, they would occupy a large portion of the surface of the park, establish a very extensive and very various business, and give to it the appearance of the grounds of a country fair, or of a militia training ground."

President Jefferson Davis urges Southerners to plant corn, peas, and beans. His April message gives priority to food crops over cotton and tobacco.

Widescale rustling (theft) begins on the Texas plains as cattle there are hit by the worst winter in years.

An epidemic of cattle disease in Britain over the next 4 years will boost meat prices and will cause a boom in imports of tinned meats from Australia (*see* 1866).

The British epizootic (above) strikes hardest at Dutch cows which have proved the sturdiest milkers for keeping in urban milk sheds but are the breeds most susceptible to disease. British dairymen are compelled to make wider and more efficient use of milk trains (*see* 1844), develop water coolers on farms and at milk depots, and use tinned-steel churns for transporting milk from the trains to urban markets.

Disruption of sugar plantations in the South sends U.S. sugar prices soaring and brings a vigorous increase in sugar planting in the Hawaiian Islands, dependent for years on the whaling industry which is now in a decline as a result of the growing use of kerosene for lamp oil (*see* energy, 1859).

Richmond has bread riots in April and Mobile has riots in September as the Union starves the South. Vicksburg, Miss., is starved out July 4 (above).

Bay Sugar Refining Co. is founded by German-American entrepreneur Claus Spreckels, 35, who has prospered as a San Francisco grocer and brewer since 1856 (*see* 1868).

Union forces cut the South off from its salt deposits on the Louisiana Gulf Coast and destroy all its salt works in Florida, North Carolina, and Virginia.

Salt from wells in the area of Syracuse, N.Y., and Saginaw, Mich., keep the North well supplied.

Sutlers provide Union troops with canned meat, oysters, condensed milk (*see* Borden, 1861), pork and beans (*see* Van Camp, 1861), and vegetables, including green beans.

Confederate troops eat the meat-and-vegetable stew "burgoo" created by Lexington, Ky., chef Gus Jaubert but there is widespread hunger in the South.

Granula is introduced by Dansville, N.Y., sanatorium operator James Caleb Jackson, 52, who has baked graham flour dough into oven-dried bread crumbs to create the first cold breakfast food (*see* Graham, 1837; Kellogg's Granula, 1877).

Perrier Water is introduced commercially by Source Perrier which bottles the French spring water that bubbles up from a spring near Nîmes. The water contains calcium and sparkles because of its natural carbon dioxide content.

La Villette opens at Paris. The central, hygienic slaughterhouse has been designed by Baron Georges Eugène Haussmann, prefect of the Seine since 1852.

The Great American Tea Co. founded in 1859 grows to have six stores and begins selling a line of groceries in addition to tea (*see* A&P, 1869).

The Cunard Line enters the immigrant trade with low rates for passengers aboard its new screw-propeller transatlantic ships as potato rot hits Ireland once again, restimulating the exodus to America (*see* 1846).

President Lincoln asks Congress December 8 to establish a system for encouraging immigration.

1864 President Lincoln calls February 1 for 500,000 men to serve for 3 years or the duration of the war.

Gen. Sherman succeeds Gen. Grant as commander of the Army of the Tennessee and occupies Meridian, Miss., February 14, destroying railroad tracks and supplies.

Ulysses S. Grant is commissioned lieutenant general March 9 and given command of all Union armies March 10.

Generals Grant and Meade start across the Rapidan May 3 to invade the Wilderness. Union and Confederate forces battle indecisively at the first Battle of the Wilderness near Chancellorsville, Va., May 5 and 6, and at Spotsylvania Court House May 8 to 12. "I propose to fight it out on this line if it takes all summer," says Grant May 11 in a dispatch to Secretary of War Stanton.

Gen. Lee defeats Grant June 1 to 3 at the Battle of Cold Springs Harbor, inflicting 12,000 casualties. Union forces besiege Petersburg, Va., beginning June 19.

The Battle of Kenesaw Mountain in Georgia June 27 ends in defeat for Gen. Sherman, who loses 1,999 dead and wounded. Confederate losses total 270.

Gen. Lew (Lewis) Wallace, 37, blocks an attempted raid on Washington, D.C. July 9. Wallace, who saved Cincinnati last year, loses his battle with Gen. Jubal

Anderson Early, 47, but gains time for Grant's troops to approach.

The Battle of Atlanta July 22 ends in victory for Gen. Sherman who defeats Southern troops under Gen. John Bell Hood, 33, after a 5-week siege. The Confederate Army has been augmented since June by boys of 17 and men aged 45 to 50.

The second Battle of Atlanta (Ezra Church) July 28 ends in another defeat for Gen. Hood who has sustained 10,000 casualties in the two battles with Gen. Sherman losing nearly as many.

Union sappers explode a mine beneath the Confederate fort at Petersburg, Va., July 30 but the Federals fall back with 2,864 killed and wounded, 929 missing.

"Damn the torpedoes! Four bells, Captain Drayton, go ahead!" orders Union Admiral David Glasgow Farragut, 63, at the Battle of Mobile Bay August 5. Defying Confederate torpedoes (mines) at the narrow entrance to the bay, Farragut's flagship *Hartford* proceeds into the bay and gains a victory.

Gen. Philip Sheridan defeats Gen. Jubal Early's Confederate cavalry September 19 at Winchester, Va., but 697 Union troops are killed, 3,983 wounded, 338 missing as compared with 276 Confederate dead, 1,827 wounded, 1,818 missing. Sheridan beats Early again September 22 at Fisher's Hill and October 19 at Cedar Creek as he clears the Shenandoah Valley, of Southern troops (below).

Gen. Sherman leads his army of 60,000 on a "march to the sea" beginning November 16 and proceeds to cut a mile-wide swath through Georgia.

The Battle of Franklin November 30 ends in disaster for Gen. Hood who has attempted to strike behind Sherman's lines. Hood makes a frontal assault on Gen. John McAllister Schofield, 33, south of Nashville and sustains terrible casualties: 1,750 dead including six general officers, 3,800 wounded, while Union losses total 189 dead, 1,033 wounded.

Union forces use the hand-cranked Gatling gun invented in 1861 to help defeat Gen. Hood at the Battle of Nashville in mid-December.

Sherman occupies Savannah December 22 and sends a dispatch to President Lincoln: "I beg to present you as a Christmas gift the city of Savannah."

"It is best not to swap horses while crossing the river," President Lincoln has said June 9 in an address to a delegation from the National Union League, and he has gained reelection in November with 55 percent of the popular vote, winning 212 electoral votes to 21 for the Democratic candidate Gen. George McClellan (Lincoln has run as a Union party candidate with support from "War Democrats" and Gen. Sherman's victory at Atlanta has helped him win).

Nevada has been admitted to the Union as the 36th state before the election (*see* Comstock Lode, 1859).

Montana Territory has been formed out of Idaho Territory May 26 as prospectors flock to the gold fields of Virginia City (*see* 1863).

Confederate soldiers fought stubbornly for their lost cause, but the South could not match the North's industrial resources.

Austria's Archduke Maximilian accepts the throne of Mexico April 10 and reaches Mexico City June 12. French troops help him drive Benito Juarez across the northern border into the United States (*see* 1863; 1866).

Paraguayan dictator Francisco Solano López sends Brazil an ultimatum August 31 demanding that she not intervene in Uruguay, Brazil invades Uruguay's Banda Oriental in October, and Lòpez seizes an arms-laden Brazilian steamer November 12 (*see* 1865).

The "Ever Victorious [British] Army" of Gen. Charles G. "Chinese" Gordon help Manchu forces under Tseng Kuo-fan sack Nanjing. Hung Hsiu-chuan takes poison, more than 100,000 are killed between July 19 and 21, and the Tai Ping Rebellion that began in 1850 is ended (*see* 1860; Korea, 1876; "Open Door Policy," 1899).

The Geneva Conventions signed by representatives of 26 nations pledge all parties to humanitarian rules respecting prisoners of war, wounded and sick military personnel, civilians in war zones, and Red Cross neutrality. The conventions will be expanded in 1907 and 1929 (*see* 1949).

Russian forces suppress a 17-month Polish insurrection that has spread to Lithuania and White Russia. Polish autonomy is abolished (*see* 1815; 1918).

The Russian language is made obligatory in Polish schools and proceedings are instituted against Polish Roman Catholic clergymen.

Russia reforms its judiciary, abolishing class courts and setting up new courts modeled on the French system that provides jury trials for criminal offenses and justices of the peace to deal with minor civil suits.

Russia's Zemstvo Law establishes a system of local government boards that can levy taxes for local roads,

1864 ***(cont.)*** bridges, schools, hospitals, and the like. Peasants, nobility, and townspeople are all to be represented on local *zemstvos* with no single class to have a majority.

Napoleon III acknowledges the right to strike and ends a French ban on workers' associations.

U.S. workingmen organize the Brotherhood of Locomotive Engineers, the Iron Moulders' International, and the Cigar Makers' National Union (*see* 1833; 1834; National Labor Congress, 1866).

Navajos terrorized by Kit Carson and his men are marched 300 miles to Fort Sumner in New Mexico Territory on the "Long Walk" to the Bosque Redondo concentration camp. By December the camp contains 8,354 Navajo plus 405 Mescalero Apache, but 1,200 will escape next year, and Carson will be relieved of his command in the fall of 1866 after a despotic regime in which hundreds have died of disease and starvation (*see* 1863; 1868).

Cheyennes go on the warpath and are supported by Arapahoe, Apache, Comanche, and Kiowa braves. U.S. troops under the command of John Chivington massacre many of them in November at Sand Creek in Colorado Territory.

The Confederacy suffers wild inflation as $1 billion in paper currency is circulated. Gold value of the paper money falls in the first quarter of the year to $4.60 per $100 while in the North the gold value of the greenback falls to 39¢ as of July 11.

A "greenback raid" by Mosby's Rangers October 14 seizes $168,000 in Union funds which are divided among Mosby's men to buy new equipment and uniforms (*see* 1863).

"In God We Trust" is printed on every piece of U.S. currency for the first time by order of Treasury Secretary Salmon Portland Chase, 56.

The Bank of California is founded at San Francisco by merchants and bankers who include New Yorker Darius Ogden Mills, 39, who will serve as the bank's president until 1873 (*see* Mills Hotel, 1896).

Kamehamea IV of the Sandwich Islands sells the Hawaiian island of Niihau to Mrs. Elizabeth Sinclair, an émigrée Scotswoman whose late husband acquired large holdings in New Zealand before being lost at sea. Feeling the property insufficient to keep her large family together, she has sold it, loaded her family, livestock, and movable possessions aboard her own clipper ship, and visited several places before deciding on Hawaii. The 12-mile-long, 46,000-acre island plus the 65,000-acre Great Makaweli estate on Kaui will remain in her family for more than a century after Mrs. Sinclair's death in 1890, being used to graze sheep and cattle.

A "finder" using a witch-hazel twig divining rod discovers oil along Pennsylvania's Pithole Creek; a well will be pumping 250 barrels of oil per day by early January 1865, and by the end of June four wells at the new town of Pithole will be pumping more than 2,000 barrels per day—one-third of all the oil produced in Pennsylvania. Some 3,000 teamsters will be driving wagonloads of barrels to the riverboats and the railhead (*see* pipeline, 1866).

The Chicago North Western Railway is created by a consolidation of the 16-year-old Galena & Chicago Union with the 9-year-old Chicago, St. Paul and Fond du Lac.

The Kansas state legislature helps the Atchison, Topeka and Santa Fe by accepting a federal land grant February 9 and voting to allow railroads to select additional lands within 20 miles of their lines in lieu of lands already held by settlers through preemption. The governor signs a law March 1 allowing counties to vote subsidies of up to $200,000 to railroads (*see* 1863; Atchison Associates, 1868).

Erie Railroad president Daniel Drew and New York State legisators sell New York and Hudson River Railroad stock short on the New York Stock Exchange. Cornelius Vanderbilt (who has changed his name from van Derbilt) buys up all the stock there is plus another 27,000 shares. When Drew and the Albany legislators cannot deliver the stock they have sold there is chaos. Vanderbilt sells them stock at $285 per share so they can deliver, many are driven to bankruptcy, and Vanderbilt gains control of the New York and Hudson (see 1863; 1865).

George M. Pullman and Ben Field of Chicago patent a railway sleeping car with folding upper berths (*see* 1858; 1865).

London's Charing Cross station opens.

The French Line paddle-wheeler *Washington* arrives at New York in June to begin 110 years of service between New York and the Channel ports by the Compagnie Génerale Transatlantique.

English inventor James Slater patents a precision-made drive chain that was foreseen by Leonardo da Vinci. It will be used in industrial machinery and in bicycles (*see* Renold, 1880; Starley, 1870).

Die Entwicklung der Dipteren by German biologist August Weismann, 30, refutes the notion that acquired characteristics can be transmitted to offspring (*see* Lamarck, 1809, 1822). Only variations of the germ plasma can be inherited, says Weismann.

German chemist Adolf von Baeyer, 29, synthesizes barbituric acid, the first barbiturate drug (his work will be eclipsed by that of his associate F. A. Kekule) (*see* 1858; 1865).

The Red Cross is established by the Geneva convention (above) which will be signed by every major world power. The Italian Red Cross is founded at Rome, the French Red Cross has its beginnings in a new *Société Française de Secours aux blessés militaires* which will be merged into the French Red Cross (*see* 1863; 1865; Dunant, 1862).

Swarthmore College is founded at Swarthmore, Pa., by abolitionists who include James Mott (*see* 1833).

The University of Kansas is founded at Lawrence.

The University of Denver is founded in Colorado Territory.

Illustrated weeklies such as *Harper's* and *Frank Leslie's* publish hundreds of sketches and drawings of Civil War battles, sieges, and bombardments but no magazine or newspaper has the equipment or technique needed to make halftone blocks for printing photographs (*see* 1880; Ives, 1886; Brady, below).

Fiction *Letters From the Underground* (*Zapiski zi podpoiya*) by Russian novelist Fedor Mikhailovich Dostoievski, 42, who was convicted of conspiracy in 1849, sentenced to be shot, reprieved at the last moment, sent to Siberia for 5 years, has had his review in the *Times* suppressed, and has lived abroad to escape his financial troubles; *Journey to the Center of the Earth (Voyage au Centre de la Terre)* by Jules Verne; *Renée Mauperin* by Jules Goncourt; *The Small House at Allington* by Anthony Trollope; *Can You Forgive Her?* by Trollope is the first of six Palliser novels.

Poetry *Dramatis Personae* by Robert Browning whose poem "Rabbi Ben Ezra" contains the lines, "Grow old along with me!/ The best is yet to be,/ The last of life, for which the first was made"; *In War Time* by John Greenleaf Whittier whose poem "Barbara Frietchie" contains the lines, " 'Shoot, if you must, this old gray head,/ But spare your country's flag,' she said".

Painting *The Dead Toreador* by Edouard Manet; *Souvenir de Mortefontaine* by Jean Corot; *Hommage à Delacroix* by French painter (Ignace) Henri (Joseph Theodore) Fantin-Latour, 28.

Britons B. J. Sayce, 27, and William Blanchard Bolton, 16, describe the preparation of a photographic emulsion of silver bromide in collodion (*see* Archer, 1851). But the Sayce-Bolton process leaves nitrates in the emulsion (*see* 1874; Maddox, 1871).

New York photographer Mathew B. Brady, now 41, travels through the war-torn South with a wagonful of equipment to record scenes of the conflict. Brady learned to make daguerrotypes from Samuel F. B. Morse in 1840, switched to the new wet-plate process 9 years ago, and has poured all his money into training assistants who will help him make a photographic record of the war (*see* 1850).

Popular songs "Beautiful Dreamer" by the late Stephen C. Foster who wrote the song a few days before his death in obscurity January 13 at age 37 in the charity ward of New York's Bellevue Hospital. Proceeds from Foster's songs averaged the not inconsiderable sum of $1,371.92 per year from 1849 to 1860 but Foster became ill and alcoholic and his death was hastened by poverty and drink; "All Quiet Along the Potomac Tonight" by John Hill Hewitt, now 63, lyrics by Lamar Tontaine; "Tenting on the Old Camp Ground" by Boston songwriter Walter Kittredge, 31, who wrote it one evening last year after receiving his draft notice but was rejected by the Army for reasons of health; "When the War is Over, Mary," by John Rogers Thomas, lyrics by George Cooper; "Wake Nicodemus!" by Henry Clay Work; "Come Home, Father" by H.C. Work whose new temperance song will be included in performances of *Ten Nights in a Barroom* (*see* 1854); "Der Deitcher's Dog (Where, O Where Has My Little Dog Gone?)" by Septimus Winner who has adapted a German folk song.

Opera *La Belle Hélène* 12/17 at Paris, with music by Jacques Offenbach.

The Travers Stakes has its first running at the year-old Saratoga racetrack at Saratoga Springs, N.Y.

English cricket player William Gilbert Grace, 16, plays his first county match, beginning a career that will make him famous as the greatest all-round cricketer ever. Grace will revolutionize the game as a batsman, attracting thousands with his prowess as he becomes a celebrity Empire-wide, not for his Bristol medical practice but for his performance in the test matches at Lord's (*see* 1787; the Ashes, 1882).

A cyclone destroys most of Calcutta October 1, killing an estimated 70,000.

Man and Nature by pioneer ecologist George Perkins Marsh, 63, of the U.S. foreign service warns that "the ravages committed by man subvert the relations and destroy the balance which nature had established . . .; and she avenges herself upon the intruder by letting loose her destructive energies . . . When the forest is gone, the great reservoir of moisture stored up in its vegetable mould is evaporated . . . There are parts of Asia Minor, of Northern Africa, of Greece, and even of Alpine Europe, where the operation of causes set in action by man has brought the face of the earth to a desolation almost as complete as that of the moon . . . The earth is fast becoming an unfit home for its noblest inhabitants" (*see* agriculture, 1874).

Congress protects California's Yosemite Valley, passing a bill at the urging of Frederick Law Olmsted to preserve the valley as the first U.S. national scenic reserve "for public use, resort, and recreation (*see* 1890; Yellowstone, 1872).

Chicago's Lincoln Park is designated as such. The 120-acre cemetery will have most of its graves removed and be expanded to embrace more than 1,000 acres of woodlands, bridle paths, playgrounds, golf courses, yacht basins, gardens, and museums.

Britain's first fish-and-chips shops will open in the next few years as steam trawlers are developed that can carry fish packed in ice. Taken from the North Sea, the fish will be kept chilled from the moment of catch until delivery to city retailers who will deep-fry fish fingers and serve them with deep-fried potatoes, all liberally doused with vinegar (*see* 1902).

The first U.S. salmon cannery opens on California's Sacramento River at Washington in Yolo County. Hapgood, Hume & Co. packs 2,000 cases and although half spoil because the cans have not all been properly sealed the cannery will soon be followed by others at San Francisco and at the mouth of every river north to Alaska (*see* 1866; 1878).

1864 ***(cont.)*** U.S. wheat prices climb to $4 per bushel and food prices rise as European crops fail and the North ships wheat to hungry Europe.

Virginia's Shenandoah Valley is the breadbasket of the South. Gen. Sheridan (above) has 50,000 troops and orders to "eat out Virginia clear and clean. . . so that crows flying over it for the balance of this season will have to carry their provender with them." Loss of the Shenandoah Valley as a food source forces the South to tighten its belt.

A grasshopper plague in the Great Plains shortens the U.S. wheat crop.

France's lentil industry moves to Lorraine where colder weather kills off the insect pests that have ruined lentil crops for years.

Armour Packing Co. has its beginnings in a Milwaukee pork-packing firm started in partnership with John Plankinton by local commission merchant Philip Danforth Armour, 32, who has made nearly $2 million in 90 days selling short in the New York pork market. Armour went to California in his teens, dug sluiceways for gold miners at $5 to $10 per day, saved $8,000 in 5 years to give himself the wherewithal to start his business at Milwaukee, and has traveled to New York where he found pork selling at $40 per barrel. Foreseeing a Union victory as Gen. Grant prepared to march on Richmond, Armour sold short at more than $33 per barrel, covered his sales at $18 per barrel, and has made a fortune (*see* 1868).

Heineken Beer gets its name as Dutch brewer Gerard Adrian Heineken acquires the 272-year-old De Hooiberg brewery and develops a special yeast that will give his beer a distinctive taste.

European immigrants pour into the United States to take up free land under the 1862 Homestead Act and fill farm and factory jobs left vacant by Union Army draftees and Americans gone West both to avoid the draft and claim free land.

1865 The American Civil War ends, President Lincoln is assassinated, and a 12-year Era of Reconstruction begins in the South with state legislatures run by "carpetbaggers" and "scalawags" (below).

Union forces occupy Columbia, S.C., February 17, and Charleston falls to a Union fleet February 18. Petersburg, Va., surrenders April 3 and Gen. Grant takes Richmond the same day.

Gen. Lee surrenders to Gen. Grant April 9 at Appomattox Courthouse, Va., and the war is over (although the last Confederate army will not surrender until May 26 at Shreveport, La.). Lee's 28,000 hungry men are allowed to keep their private horses and sidearms.

The Union has lost 360,222 men (110,000 of them in battle), the Confederacy 258,000 (94,000 in battle) with at least 471,427 wounded on both sides.

President Lincoln is assassinated April 14 while attending a performance at Ford's Theatre in Washington, D.C., and is succeeded by his vice president Andrew Johnson.

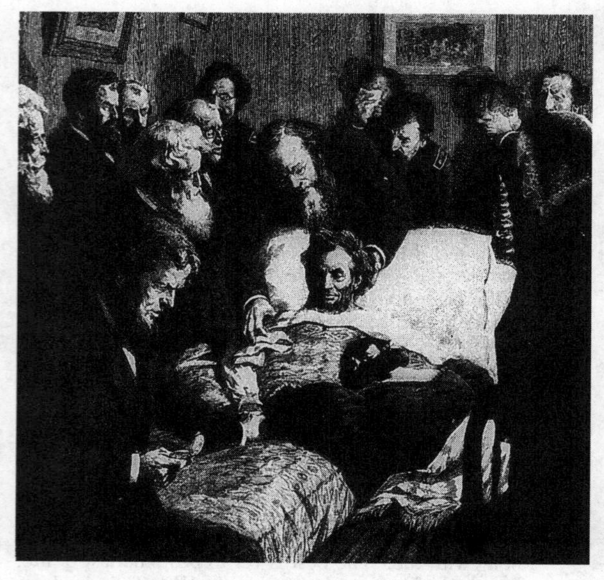

President Lincoln's tragic death at Ford's Theater came only 5 days after the Confederate surrender at Appomattox.

Actor John Wilkes Booth, 27, has climbed the stairway to Lincoln's box, he shoots the president in the head, leaps to the stage crying "Sic semper tyrannis! The South is avenged!" and escapes with a broken leg. Told of the president's death at age 66 April 15, Secretary of War Stanton says, "Now he belongs to the ages." Booth is a son of the late English-American actor Junius Brutus Booth and a brother of the eminent Shakespearean actor Edwin Thomas Booth. He is not found until April 26 when he is either killed by his captors or dies by his own hand. Several of his accomplices have tried to kill Secretary of State Seward and are hanged July 7.

The "carpetbaggers" who move into the South are so called with contempt by Southerners who say they can put all they own in the common hand luggage called carpetbags. Some become state legislators and U.S. congressmen, some are missionaries sent to help the freedmen, who are helped in some cases also by "scalawags"—Southerners who join with the ex-slaves to establish a new order in the South.

Paraguay has seized a small Argentinian river port in April and a treaty signed at Buenos Aires May 1 has created a triple alliance (Argentina, Brazil, and Uruguay) to oppose the Paraguayan dictator (*see* 1864; 1866).

Britain's prime minister Viscount Palmerston dies October 18 at age 80. He is succeeded by Lord John Russell, now, 73, who will resign next June after his reform bill fails.

The king of the Belgians Leopold I dies December 10 at age 74 after a 4-year reign that has established the Belgian monarchy. He is succeeded by his son of 30 who will reign until 1909 as Leopold II, expanding Belgian interests (and his own) in the Congo.

Boers in the Orange Free State have begun a war with the Basutos that will continue until next year.

The Colorado River Indian Reservation established by an act of Congress signed March 3 by President Lincoln will ultimately contain 264,250 acres with 225,914 in Arizona's Yuma County and 38,366 in California's San Bernardino and Riverside counties but the desolate Mojave desert reservation land has an average high of 110° to 113° F. in summer and an average low of 39° in winter.

The Ku Klux Klan organized at Pulaski, Tenn., is a secret social club of young men who hope to recapture the comradeship and excitement of the war. Their name comes from the Greek "Kuklos" meaning circle, they adopt elaborate rituals, and their curious uniform is soon discovered to terrorize superstitious blacks. A majority of Southern whites will join in the next few years as the KKK tries to return local and state government to white, Democratic party control (*see* 1915; White Camelia, 1867).

The commandant of the Confederate prison camp at Andersonville, Ga., is convicted of "murder, in violation of the laws and customs of war" and hanged November 10 in Washington's Old Capitol Prison at the foot of Capitol Hill. More than 12,000 Union prisoners of war died at Andersonville in 14 months, as many as 97 in a single day, but while 30,218 Union prisoners have died in various Confederate prison camps (as compared with 25,976 "rebels" who died in Union prison camps), the trial of Swiss-American Capt. Henry Wirz, 42, has been tainted. A priest who traveled up from Savannah to give testimony was stopped by the military judges after a few minutes when it became clear that his testimony would be favorable to the defendant. Wirz has spurned an offer of a pardon if he would testify against Jefferson Davis.

The Thirteenth Amendment to the Constitution is ratified by two-thirds of the states and beginning December 18 prohibits slavery or any other denial of liberty, "without due process of law."

Jamaican blacks revolt in the Morant Bay area but the insurrection is quickly suppressed by the governor Edward J. Eyre, now 40, who orders the execution of 450 natives, has many more flogged, and has 1,000 native homes burned. Eyre was lionized in 1850 for his courage in Australia but his latest action creates widespread indignation and he is censured and dismissed. Representative government in Jamaica has been suspended and will not be restored until 1884.

The Salvation Army has its beginnings in a London mission founded in Whitechapel by itinerant English religious revivalist William Booth, 36, who says, "A man may be down but he's never out." (*see* 1878).

Wartime inflation has reduced the value of Confederate paper money to $1.70 per $100 by January and driven the gold value of the Union greenback to 46¢. The Confederate money becomes worthless and the greenback will not regain its full value until the end of 1878 (*see* Greenback party, 1873).

New York's City Bank obtains a federal charter and becomes National City Bank (*see* First National, 1812; First National City, 1961).

W. R. Grace & Co. is founded at New York to engage in the South American trade under the leadership of William Russell Grace (*see* 1850; 1890).

Deering, Milliken is founded at Portland, Me., by local fabric salesmen William Deering, 39, and Seth M. Milliken, 29, who will soon move to New York. Deering will quit to start a farm machinery company; Milliken and his heirs will prosper by buying up financially troubled New England and Southern textile mills; being quick to adopt advanced technology will make them the largest U.S. textile manufacturers.

Andrew Carnegie enters the steel business with former blacksmith Andrew Klopman (*see* 1856; 1867).

Chicago's Marshall Field learns that local merchant Potter Palmer plans to retire from active work in the store he started in 1852 (*see* 1861). Field persuades Palmer to sell him an interest and when Palmer takes in Levi Leiter of Cooley, Wadsworth the three men organize a $600,000 firm under the name Field, Palmer, and Leiter with Field taking a $260,000 interest (*see* 1867).

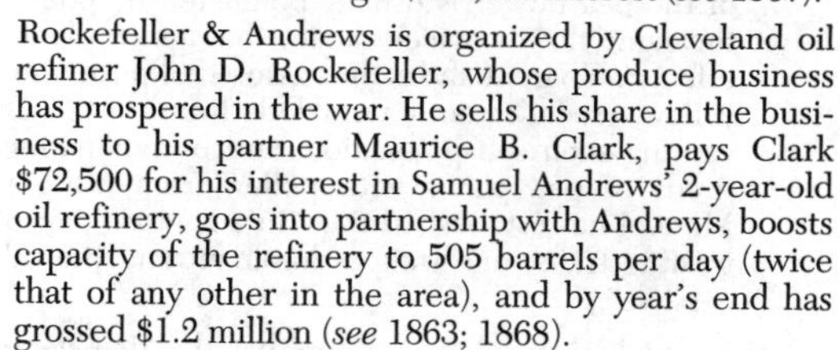

Rockefeller & Andrews is organized by Cleveland oil refiner John D. Rockefeller, whose produce business has prospered in the war. He sells his share in the business to his partner Maurice B. Clark, pays Clark $72,500 for his interest in Samuel Andrews' 2-year-old oil refinery, goes into partnership with Andrews, boosts capacity of the refinery to 505 barrels per day (twice that of any other in the area), and by year's end has grossed $1.2 million (*see* 1863; 1868).

A high vacuum mercury pump created by German chemist Herman Johann Philips Sprengel, 28, will lead to development of the electric light bulb (*see* Swan, 1878; Edison, 1879; but *see* also Langmuir, 1912).

Cornelius Vanderbilt orders his Hudson River Railroad employees not to connect with the New York Central (*see* 1864; 1867).

The Union Pacific Railroad that started at Omaha in 1863 reaches Kansas City (*see* 1867).

The Missouri-Kansas-Texas Railroad has its beginnings. The "Katy" will extend from St. Louis via 3,000 miles of track through Missouri to Kansas City, San Antonio, Houston, and Galveston.

Southern Pacific is organized to link San Francisco with San Diego (*see* 1876).

The *Pioneer* completed by George M. Pullman is the first true sleeping car with convertible berths (*see* 1864; 1867).

The 1,700-ton St. Louis-New Orleans side-wheeler steam packet *Sultana* explodes on the Mississippi April 27 killing 1,600 out of the 2,300 aboard.

A British Locomotives on Highways Act (Red Flag Act) passed by Parliament requires that steam-powered

1865 ***(cont.)*** carriages be preceded by men on foot carrying red flags (*see* 1829). Stagecoach interests have obtained passage of the act, it will not be repealed until 1896, and it will stifle further development of steam buses which have overcome many technical problems and are becoming highly successful (*see* 1896).

Pratt & Whitney is founded at Hartford, Conn., by Vermont-born machinist Francis Asbury Pratt, 38, and Amos Whitney. Pratt has initiated a system for making interchangeable parts used in Civil War rifles, he will invent a metal-planing machine in 1869, a gear cutter in 1884, and a milling machine in 1885, he will be instrumental in having a standard system of gauges adopted, and his Pratt & Whitney firm will become a major producer of engines as well as machine tools.

The Yale Lock is patented by Shelburne Falls, Mass., inventor Linus Yale, 44, whose improved cylinder lock for the doors of houses and business establishments will be widely used for well over a century.

Austrian botanist Gregor Johann Mendel, 43, elucidates natural laws of heredity in a paper read to the Brünn Society for the Study of Natural Science. Mendel entered an Augustinian order as a monk at age 21 and at 35 began studying the genetics of garden peas, using peas for his experiments because the stigma of the pea flower is usually pollinated by pollen from the same flower, which means that a new plant has in effect one parent and is thus assured of pure traits, e.g., yellow, green, smooth, wrinkled, dwarf, or tall. By preventing such self-pollination, and by pollinating hundreds of pea plants with pollen from other plants, Mendel has established that "in any given pair of contrasting traits, one trait is dominant and the other re cessive."

Mendel's law (above) states that the first generation of progeny of mixed parents will all be hybrids, but in the second generation only half will be hybrids while one quarter will reflect the true trait of one parent and a second quarter the true trait of the second parent. Mendel's recognition that various dominant or recessive characteristics depend on certain basic units, later to be called genes, will be the basis of improved varieties of plant and animal life in the century to come but established scientists will not begin to appreciate his work until the turn of the century (*see* DeVries, 1900).

F. A. Kekule von Stradonitz explains the structure of aromatic compounds, setting forth a doctrine of the linking of carbon atoms and orginates the ring (closed-chain) theory of the benzene molecule's constitution (*see* 1858).

English surgeon Joseph Lister, 38, at Glasgow discovers the value of carbolic acid as an antiseptic in treating compound fractures. He has read of Pasteur's 1861 findings in France and develops a sprayer that creates a carbolic mist, inaugurating the era of antiseptic surgery (*see* Johnson & Johnson, 1886).

The Swedish Red Cross is founded at Stockholm, the Norwegian Red Cross at Oslo.

A Sanitary and Social Chart of New York City's Fourth Ward describes overcrowding in the 30 blocks south of Chatham Square. Smallpox and typhus are found in the area where 78 people reportedly share one privy.

Cholera strikes Paris in September, the daily death toll reaches 200, and sulfur is burned to combat the "miasma" in the air that is held responsible despite John Snow's observations at London in 1853. Louis Pasteur's infant daughter Camille dies in the epidemic.

Cornell University is founded at Ithaca, N.Y., by Andrew Jackson White, 32, a state senator who last year helped codify the state's school laws and create a system of training schools for teachers. White has persuaded Ezra Cornell of Western Union to provide large sums of working capital that will make use of land granted to the state under the Morrill Act of 1862, the new university will open in 1868, and its faculty of nonresident professors will include Louis Agassiz and James Russell Lowell.

Lehigh University is founded at Bethlehem, Pa.

The University of Kentucky is founded at Lexington.

The University of Maine is founded at Orono.

Worcester Polytechnic Institute is founded at Worcester, Mass.

Reuters News Service conveys news of President Lincoln's assassination (above) to Europe a week ahead of the competition (*see* 1851). Reuters's American agent James McClean hires a tug, overtakes the mailboat bound for Britain, and throws aboard a canister containing his dispatch (*see* Atlantic Cable, 1866).

An improved rotary press devised by Philadelphia inventor William A. Bullock draws on a continuous roll of paper and cuts the sheets before they are printed. The Hoes in New York will evolve a true web press but only after U.S. paper manufacturers develop paper strong enough to form a long web that will run over many cylinders without tearing, and only after the perfection of quick-drying inks that will permit presses to print newspapers first and cut them afterward (*see* 1846; 1870).

The *Nation* begins publication at New York under the direction of Irish-American journalist E.L. (Edwin Lawrence) Godkin, 34, whose weekly journal of liberal opinion will continue for more than 125 years.

The *San Francisco Chronicle* has its beginnings in the free theater program sheet *San Francisco Dramatic Chronicle* published by local journalist Michel Harry de Young, 16, and his brother Charles, 18. They will drop the word "Dramatic" in 1868 and make the *Chronicle* a full-fledged 2¢ newspaper. Charles will be shot dead by the mayor's son in 1880, Michel by a member of the rival Spreckels family in 1884; sympathetic juries will acquit both gunmen.

The *San Francisco Examiner* begins publication (*see* Hearst, 1887).

Nonfiction *Sesame and Lilies* by John Ruskin who has put together a series of lectures he has given at

Manchester, not about art but about reading, education, woman's work, and social morals; *Josh Billings: His Sayings* by Massachusetts-born humorist Henry Wheeler Shaw, 47, who follows the deliberately bad spelling employed by Cleveland's "Artemus Ward" and takes the part of an unlettered Yankee crackerbarrel philosopher.

Fiction *Alice's Adventures in Wonderland* by English mathematician Lewis Carroll (Charles Lutwidge Dodgson), 33, who has written his fantasy for the entertainment of Alice Liddell, second daughter of classical scholar Henry George Liddell, 54, dean of Christ Church. The book is illustrated by *Punch* cartoonist John Tenniel, 45; *Our Mutual Friend* by Charles Dickens with illustrations by Marcus Stone: "The question [with Mr. Podsnap] about everything was, would it bring a blush into the cheek of the young person [Podsnap's daughter Georgiana]" (xi); *The Belton Estate* by Anthony Trollope; "The Celebrated Jumping Frog of Calaveras County" by Mark Twain.

Poetry *Drum Taps* by Walt Whitman who has added his poem "When Lilacs Last in the Dooryard Bloom'd" as a memorial to the late president. The Department of the Interior sacks Whitman from his clerkship on charges of having included explicit sexual references in his 1855 book *Leaves of Grass* into which his new book of war poems will be incorporated, the poet finds a new job in the attorney general's office, and he will remain in Washington until he suffers a paralytic stroke in 1873; *Atalanta in Calydon* by English poet Algernon Charles Swinburne, 28, whose drama in classical form with choruses includes the lines, "When the hounds of spring are on winter's traces,/ The mother of months in meadow or plain/ Fills the shadows and windy places/ With lisp of leaves and ripple of rain"; "The Hand That Rocks the Cradle Is the Hand That Rules the World" by U.S. lawyer-poet William Ross Wallace, 46, who has gained fame with his Civil War songs.

Juvenile *Hans Brinker, or The Silver Skates* by New York author Mary Elizabeth Mapes Dodge, 34.

Painting *Girl With Seagulls, Trouville* by Gustave Courbet; *Peace and Plenty* by George Innes; *Prisoners from the Front* by Winslow Homer, who produces a series of drawings for *Harper's Weekly*.

Theater *Society* by English playwright Thomas William Robertson, 36, 11/11 at the Prince of Wales' Theatre, London.

Opera *L'Africaine* (*The African*) 4/28 at the Paris Grand Opéra, with music by the late Giacomo Meyerbeer who died at Paris last year at age 72; *Tristan und Isolde* 6/10 at Munich's Royal Court Theater, with music by Richard Wagner.

Chicago's Crosby Opera House opens April 20.

Tony Pastor's Opera House opens in New York August 14 at 201 Bowery under the management of actor-clown-comedian-balladeer Antonio Pastor, 28.

Oratorio *The Legend of St Elizabeth* by Franz Liszt 8/15 at Budapest.

First performances Symphony in B minor (*Unfinished*) by the late Franz Schubert 12/17 at Vienna.

Popular songs "Ich Liebe Dich" by Norwegian composer Edvard Grieg, 22, lyrics by Hans Christian Andersen, now 60; "Marching Through Georgia" by Henry Clay Work.

A British mountain-climbing party led by artist Edward Whymper makes the first ascent of the Matterhorn in the Alps but four members fall to their deaths climbing down.

English racing's Triple Crown goes to the French-bred horse Gladiateur which wins the Two Thousand Guineas, the Derby, and the St. Leger.

The U.S. trotter Goldsmith Maid makes money at New Jersey harness tracks for her new owner Alden Goldsmith who will sell the 8-year-old mare in 1869 for $15,000. She will earn $100,000 in the following 2 years, Harry N. Smith will buy her in 1871 for $32,000 with a view to breeding her, he will race her instead, she will earn another $100,000 for Smith and go undefeated from 1871 to 1874, winning her last race at age 20; her purses will total $364,200, and she will live to age 28.

The Stetson "10-gallon" hat is created by Philadelphia hat maker John Batterson Stetson, 35, whose high-crowned "Boss of the Plains" is a modified Mexican sombrero with a 4-inch crown, a 4-inch brim that can carry 10 "galions" (ribbons), and a leather strap hatband. The $5 hat has a look of importance and is destined for fame on the Western plains (a Stetson made from better materials will sell for $10, one made from pure beaver or nutria felt for $30). Stetson worked for years in his family's hat company until he came down with tuberculosis, went West to seek a cure, joined the gold rush to Pike's Peak in 1863, and has returned home to open his own factory, which by 1900 will have a payroll of more than 3,000.

Le Printemps in Paris is started by a former Bon Marché department head named Jauzot (*see* 1852; Louvre, 1855).

The first U.S. train robbery occurs May 5. An Ohio and Mississippi passenger train bound from Cincinnati to St. Louis is derailed by objects placed on the tracks near North Bend, Ohio, and armed holdup men swarm through the cars robbing passengers of money and valuables (*see* Jesse James, 1872).

Boston City Hall is completed by architects Gridley J. Fox Bryan and Arthur Gilman.

Trading of wheat futures begins at Chicago after years of informal speculation in "to arrive" grains during the Crimean and Civil Wars, but while the Chicago Board of Trade moves into the Chamber of Commerce Building at LaSalle and Washington Streets and deals in futures it will not be used for hedging until 1879 (*see* 1848). New York and New Orleans markets continue to be "spot" markets dealing in actual grain (*see* 1872).

French-American milling engineer Edmund LeCroix revolutionizes flour milling with a small mill he builds

1865 ***(cont.)*** for Alexander Faribault at Northfield, Minn., adapting a French machine to develop a middlings purifier that improves the yield of endosperm wheat particles free of bran.

"Patent" flour is introduced using a newly patented process for separating bran from granual middlings (farina). LeCroix (above) improves on this superior quality flour with a middlings purifier that permits continuous improvement of the flour stream as it moves through the mill. He combines the bolting cloth used for centuries to sift flour for finer granulation with fan-driven air currents used for years to clean wheat.

Poor yeast is responsible for poor quality U.S. bread, says Austrian stillmaster Charles Fleischmann, 31, who visits Cincinnati to attend his sister's wedding. Salt-rising bread is preferred in the North, baking soda hot bread is preferred in the South, and what little yeast bakers use is generally "slop" yeast from potato water or from the frothy "barm" at the top of ale vats, but Fleischmann will change that (*see* 1868).

Anheuser-Busch has its beginnings at St. Louis where German-American brewer Adolphus Busch, 26, goes into business with his father-in-law Eberhard Anheuser (*see* 1873; Budweiser, 1876).

Del Monte Corp. has its beginnings in a California food brokerage firm founded by James K. Armsby whose company will become Oakland Preserving Co. in 1891, California Packing in 1916, and Del Monte in 1967—the leading U.S. packer of fruits and vegetables (*see* 1891).

Cans made of thinner steel come into general use and the rim round the top of each can will lead to the invention of the can opener that will permit cans to be opened more easily than with hammer and chisel (*see* 1870).

A compression machine built by U.S. inventor-balloonist Thaddeus Sobieski Coulincourt Lowe, 33, makes ice, pioneering artificial refrigeration. Lowe has been chief of the U.S. Army aeronautic section since 1860 (*see* water gas, 1873).

Chicago's Union Stock Yards open December 25 on a 345-acre tract of reclaimed swampland southwest of the city limits. With the Mississippi virtually shut down by the Civil War, Chicago has replaced Cincinnati, Louisville, and St. Louis as the nation's meat-packing center. The city doubled its packing capacity in a single year of the war with eight new large packing plants and many small ones to serve the nine railroads that converge on Chicago and that have spent $1.5 million to build the new Stock Yards with pens that can hold 10,000 head of cattle and 100,000 hogs at any given time (*see* Armour, 1872; Swift, 1877).

1866 Washington demands removal of French forces from Mexico February 12. Having failed to gain U.S. recognition, the emperor Maximilian sends his Belgian-born wife Carlota to seek aid from Pope Pius IX and Napoleon III but she soon finds that his cause is hopeless (*see* 1867).

Europe goes to war for 7 weeks beginning in June as French, Austrian, Hanoverian, Italian, and Prussian forces fight in three theaters with telegraphic communications playing an important part for the first time in a European war as they did in the American Civil War.

Prussian minister-president Count Otto von Bismarck-Schönhausen, now 51, has formed an alliance with the new kingdom of Italy and it is he who precipitates the conflict.

The Battle of Sadowa (Königgrätz) July 3 is a stunning defeat for the Hapsburg troops. Standing erect and firing muzzle-loaders, they are no match for the Prussian infantrymen of Count Helmuth von Moltke, 66, who have breech-loading needle guns that can be fired from prone positions (*see* 1841; chassepot muskets, below).

Austrian forces defeat Italians on both land and sea but Venetia, ceded earlier by Austria to France, is ceded to Italy July 3 and France withdraws her troops from Rome in December.

The Treaty of Prague August 23 ends the European war, terminates the Germanic Confederation of 1815, and incorporates Hanover, Electoral Hesse, Frankfurt, and Nassau into Prussia, which excludes Austria from territory north of the Main.

The first king of Romania is recognized October 24 by the Ottoman sultan Abdul Aziz and begins a reign that will continue until 1914 (although Romania will not obtain full independence until 1878). Conservatives and Liberals who want a foreign prince have kidnapped Alexander Guza and forced him to abdicate, a new constitution has been introduced in July based on the liberal but undemocratic Belgian charter of 1831, Prince Carol of Hohenzollern-Sigmaringen, 27, has been proclaimed king of Wallachia and Moldavia, and he has arrived at Bucharest after traveling incognito across Austria.

Japan's shogun Iemochi dies in August at age 20; his kinsman Keiki, 29, son of Tokugawa Nariaki of Mito, will reign briefly as Yoshinobu, the last Tokugawa shogun (*see* 1867).

Coalition forces devastate Paraguay, slaughtering much of her army, but lose 4,000 men September 22 when they attack an entrenched position at Curupaití.

The Supreme Court sets limits to the authority of martial law and to suspension of habeas corpus in time of war April 3 in the case of *Ex parte Milligan* but does not hand down its opinion until December 17: "Martial law can never exist where the courts are open in the proper and unobstructed exercise of their jurisdiction."

Congress passes a Civil Rights Act April 9 over President Johnson's veto to secure for former slaves all the rights of citizenship intended by the 13th Amendment.

The American Equal Rights Association founded May 10 at New York is an outgrowth of the Woman's Rights Society (*see* 1848; Anthony, 1869).

Reconstruction begins in the South, accompanied by racial conflicts. Efforts to introduce black suffrage into the Louisiana Constitution produce a July 30 race riot at New Orleans with some 200 casualties.

The National Labor Congress convenes August 20 at Baltimore and forms the National Labor Union. President is William H. Sylvis, 38, who early in 1861 called for a national convention of workingmen to oppose the impending civil war. Sylvis has helped reorganize the Iron-Moulders' Union and will use Washington lobbyists to promote the interests of his 600,000-member National Labor Union until his death in July 1869 (*see* Knights of Labor, 1869).

The American Society for the Prevention of Cruelty to Animals (ASPCA) is founded by New York shipbuilder's son Henry Bergh, 43, who serves as first president. Chief object of the society is to stop the abuse of horses (*see* ASPCC, 1874).

London has a Black Friday May 11 as financial panic hits the city.

Postwar economic depression begins in the United States as prices begin a rapid decline following the Civil War's inflation (*see* Black Friday, 1869).

Horses provide virtually all the power for urban transit and for agricultural production.

A crude internal combustion engine patented by Cologne engineer Nikolaus August Otto, 34, will sell by the thousands in the German states and in England in the next 10 years. Otto has read of the 1860 Lenoir engine, obtained backing from engineer-businessman Eugen Langen, 33, and shares patent rights with his brother William.

Telegraph pioneer Werner von Siemens develops the first practical dynamo-electrical machine (*see* 1847). It permits production of electricity in great quantity.

The first U.S. oil pipeline is completed to connect Pithole, Pa., with a railroad 5 miles away (*see* 1864).

Sperm oil for lighting and lubrication sells for $2.25 per gallon, up from 43¢ in 1823 (*see Scientific American*, 1857). The high price of sperm creates a growing demand for kerosene and petroleum.

The Great Tea Race from Foochow to London pits 11 clipper ships who race to minimize spoilage of the China tea in their hot holds. The skippers crowd on sail but the voyage still takes close to 3 months.

Thomas Cook initiates a system for providing hotel accommodations to implement the Cook's Tours he began in 1856 (*see* 1851; Khartoum, 1884).

The chassepot musket adopted by the French army is an improvement on the breech-loading Prussian needle gun used by the Prussians at the Battle of Sadowa (above). Gunsmith Antoine Alphonse Chassepot, 33, has developed the new musket.

The Skoda works that will supply European armaments for future wars is founded by Czech engineer Emil von Skoda, 27, who takes over a machine works at Pilsen. He will develop it into a factory famous for its cannon and other artillery.

Swedish engineer Alfred Bernhard Nobel, 33, perfects dynamite, harnessing the power of nitroglycerin discovered by Ascanio Sobrero in 1847. Nobel, who has studied mechanical engineering in the United States, mixes nitro with absorbent diatomaceous earth to create a safe blasting powder that will replace black powder (*see* 1875; 1901).

A cholera epidemic takes 120,000 lives in Prussia and 110,000 in Austria.

Cholera strikes London and an epidemic at Bristol is checked through measures instituted by physician William Budd, 55, who also stamps out an epizootic of rinderpest in British livestock (*see* 1873; Snow, 1853)

A London dispensary for women opens under the direction of local physician Elizabeth Garrett Anderson, 31, who pioneers the admission of women to the professions, including medicine.

Cholera kills some 50,000 Americans. New York, which has 2,000 fatalities, creates the first U.S. municipal board of health.

New York has had recurring epidemics of cholera, scarlet fever, smallpox, typhoid fever, typhus, and yellow fever which will grow more severe in the next 7 years in Baltimore, Boston, Memphis, New Orleans, Philadelphia, and Washington as well.

English physician Thomas Clifford Allbut, 30, invents the clinical thermometer.

Bernadette Soubirous of Lourdes enters the convent of Saint Gildard at Nevers where she will die of tuberculosis in 1879 despite administration of water from the grotto at Lourdes (*see* 1858). Reports of cures from the grotto will nevertheless continue to attract sick, blind, and infirm visitors.

The Western Health Reform Institute is founded at Battle Creek, Mich., by Seventh Day Adventist prophet Ellen G. White whose sanitorium treatment combines vegetarian diet and hydrotherapy with some accepted medical methods. The "San" will incubate the infant U.S. breakfast-food industry (*see* 1860; Kellogg, 1876; Post, 1893).

Cincinnati's Plum Street Temple is completed for the city's B'nai Yeshurun congregation whose Austrian-American rabbi Isaac Meyer Wise, 47, is pioneering a reform movement in U.S. Judaism by abandoning dietary laws and other orthodox practices such as segregating women, wearing beards, and keeping heads covered (*see* Berlin, 1815). The Moorish building with its twin minarets seats 1,400.

Beirut's American University is founded under the name Syrian Protestant College. The school will be renamed in 1920 and its graduates will include presidents, prime ministers, cabinet members, and ambassadors of several Middle Eastern nations.

Drew University is founded at Madison, N.J., with funding from financier Daniel Drew on the centennial of American Methodism.

The University of Ottawa is founded in Ontario.

1866 ***(cont.)*** A new Atlantic Cable between Britain and the United States is completed July 27 by Cyrus W. Field with backing from Peter Cooper (*see* 1858; 1869).

Western Union Telegraph absorbs two smaller telegraph companies to gain control of 75,000 miles of wire and become the first great U.S. industrial monopoly (*see* 1856; 1881).

The *Hartford Courant* begins publication on a daily basis to succeed the weekly *Connecticut Courant* that began 73 years ago.

Fiction *War and Peace (Voina i Mir)* by Russian novelist Count Leo (Lev Nikolaevich) Tolstoi, 38, is published in its first installment. Tolstoi, who will complete his long novel about the 1812 Napoleonic invasion in 3 years, lives on his country estate Yasnaya Polyana whose serfs he has liberated in accordance with the 1861 Emancipation Act; *Crime and Punishment (Prestuplertie i Nakazaniye)* by Fedor Dostoievski; *The Gambler (Igrok)* by Dostoievski, who has dictated the novel in 1 month to fulfill an obligation to a publisher (he will marry the stenographer Anna Snitkina, 20, next year and she will bear him children as he gambles away the little money they have); *Toilers of the Sea (Les travailleurs de la mer)* by Victor Hugo; *Felix Holt the Radical* by George Eliot.

Poetry *Poems and Ballads* by Algernon Charles Swinburne whose "Ballad of Burdens" carries the refrain, "Princes, and ye whom pleasure quickeneth,/ Heed well this rhyme before your pleasure tire;/ For life is sweet, but after life is death./ This is the end of every man's desire"; "Thyrsis" by Matthew Arnold; "Snowbound" by John Greenleaf Whittier; "O Captain! My Captain!" by Walt Whitman whose elegaic ode commemorates Lincoln.

Painting *The Fifer* by Edouard Manet; *Sleep* by Gustave Courbet; *A Storm in the Rocky Mountains* by Albert Bierstadt.

Opera *The Brandenburgers in Bohemia (Branibori v Cechach)* 1/5 at Prague's National Theater with music by Bedřich Smetana; *The Bartered Bride (Prodana Vevesta)* 5/30 at Prague's National Theater, with music by Smetana; *Mignon* 11/17 at the Opéra-Comique, Paris, with music by French composer Charles Louis Ambroise Thomas, 55.

Broadway musicals *The Black Crook* 9/12 at Niblo's Garden, with a Faustian plot by U.S. playwright Charles M. Barras, 40, a cast of 100 dancing girls in pink tights (clergymen denounce their costumes and indelicate postures as does James Gordon Bennett in his *New York Herald*), music adapted from various sources, 474 perfs. The first true Broadway musical, it runs 5½ hours on opening night, will see many revivals, and tour for more than 40 years.

Popular songs "Come Back to Erin" by London songwriter Claribel (Charlotte Allington Barnard); "When You and I Were Young, Maggie" by Detroit composer J. A. Butterfield, lyrics by Canadian-American journalist George Washington Johnson of the *Cleveland Plain-Dealer;* "We Parted by the River" by U.S. songwriter William Shakespeare Hays.

The New York Athletic Club is founded June 17 by John G. Babcock (who has invented the sliding seat for rowers), Henry E. Buermeyer, and William B. Curtis. They will introduce the first spiked shoes at the club's first open indoor competition November 11, 1868, at New York's Empire Skating Rink.

The Morris chair is introduced by English craftsman-poet-wallpaper designer William Morris, 32, whose reclining chair has a bar and notch arrangement that permits its back to be tilted to a 45-degree angle or folded flat to facilitate shipment.

A salmon cannery opens on the Columbia River that will be followed by canneries at the mouth of almost every river north to Alaska (*see* 1878).

The first U.S. crop report is issued by the 4-year-old Department of Agriculture. The report has been compiled by field workers who have worked to determine the nation's production of various crops.

Cattlemen discover that livestock can survive the cold of the northern Great Plains and can eat the grasses there.

Cattle from Texas are driven north for the first time on the Chisholm Trail, named after the scout Jesse Chisholm. The cattle arrive at Abilene, Kan., for shipment by rail to points east (*see* 1867; 1871; 1885).

The first San Francisco salmon cannery opens (*see* 1864).

A U.S. patent is issued for a tin can with a key opener.

British imports of tinned meat total 16,000 pounds.

More than 90 percent of Britain's tea still comes from China (*see* 1900; Great Tea Race, above).

Henri Nestlé formulates a combination of farinaceous pap and milk for infants who cannot take mother's milk and starts a firm under his own name to produce the new infant formula (*see* 1861; 1875; Liebig, 1867).

Anglo-Swiss Condensed Milk Co. is founded at Cham, Switzerland, by U.S. entrepreneurs Charles and George Page. Their supervisor John Baptist Meyenberg, now 18, will suggest the possibility of preserving evaporated milk in the same way as other canned foods, without sugar, but the Page brothers will enjoy such success with their sweetened condensed milk that they will reject Meyenberg's idea and he will emigrate to America (*see* Helvetia Milk Condensing Co., 1885; Nestlé, 1905).

Gail Borden adopts the trademark Eagle Brand to protect his condensed milk from competitors who have appropriated the name Borden (*see* 1861). Borden has made a fortune supplying his product to the Union Army; his plant at Elgin, Ill., produces at a rate of 300,000 gallons per year (*see* 1875).

Washburn, Crosby Co. erects its first flour mill on the Mississippi in Wisconsin. Former Maine timber baron

Cadwallader Washburn, 48, served as a major general in the Union Army, has settled in Wisconsin, and becomes a miller (*see* 1874).

Cadbury's Cocoa Essence is the first pure cocoa to be sold in Britain (*see* 1847). Richard Cadbury, 31, and his brother George, 27, have been running their father's Birmingham business, have come close to failing, but are saved by the new product which they advertise as "Absolutely Pure: Therefore Best." They have squeezed out excess cocoa butter to leave a pure, concentrated powder that need not be adulterated with the potato and sago flour used in all British cocoa powders until now to counteract the natural fats (*see* van Houten, 1828). The cocoa butter made available by the process enables Cadbury Brothers to increase production of chocolate candy (*see* slavery, 1901).

1867 Irish Fenians try to seize Chester in February with an attack on police barracks and kill 12 while trying to blow up Clerkenwell jail. The revolutionary movement founded among Irish-Americans by James Stephens in 1858 spread to Ireland 2 years ago but will dissolve next year as differences arise between Stephens and more radical leaders.

The British North America Act March 29 unites Ontario, Quebec, New Brunswick, and Nova Scotia in the Dominion of Canada with its capital at Ottawa, a town of 18,000. First Dominion premier is John Alexander Macdonald, 52, a Scots-Canadian who has been premier of Upper Canada and has led the federation movement.

Alaska is ceded to the United States by the Russian czar Aleksandr II in a treaty signed March 30. U.S. Secretary of State William Henry Seward has arranged the $7.2 million purchase at 1.9¢ per acre, critics ridicule "Seward's icebox," but the Senate consents April 9.

Nebraska is admitted to the Union as the 37th state.

Mexico's emperor Maximilian surrenders May 15 to Juarez forces commanded by Mariano Escobedo who have besieged Gueratero, a court martial condemns him to death, and he goes before a firing squad June 19, thus ending Napoleon III's dream of establishing an empire in Latin America.

Paraguay's President López conscripts slaves aged 12 to 60 as smallpox and cholera kill far more on both sides than do bullets in the year's single battle (*see* 1866; 1868).

Midway Islands in the Pacific are taken in the name of the United States August 28 by Capt. William Reynolds of the *U.S.S. Lackawanna*.

Japan's figurehead emperor Komei has died February 3, a movement to restore the power of the emperor gains strength under the leadership of the Mito clan in the north and the Satchō Dohi group (Satsuma, Chōshu, Tosa, and Hizen clans) in the west, the shogun Yoshinobu is virtually imprisoned in his Kyoto palace, he resigns November 9, and the emperor's son Mutsuhito, 15, takes power, ending the feudal military government that has ruled since 1185 (*see* 1868).

The Austro-Hungarian dual monarchy that will continue until 1918 is created in October by a compromise engineered by Hungarian statesman Ferencz Deak, 64, who has gained restoration of the Hungarian constitution from the Austrian emperor Franz Josef who will rule as king of Hungary until his death in 1916. The king-emperor has ruled with absolute power since 1848 and has no sympathy for the constitutional government under which he will now rule.

Prussia's Count von Bismarck organizes a North German Confederation under Prussian leadership. The socialist chairman of German workingmen's unions August Babel, 27, wins election to the Confederation's Reichstag.

Britain's Second Reform Bill passed August 15 extends suffrage, opens borough elections to all householders paying the poor rates and to all lodgers of 1 year's residence or more who pay annual rents of at least £10, opens county elections to owners of land that rents for at least £5 per year and to tenants who pay at least £12 per year, but a bill that would extend voting rights to women fails (*see* 1905).

The Knights of the White Camelia, organized in Louisiana, is similar to the 2-year-old Ku Klux Klan. The new fraternal order will spread among white supremacists throughout the former Confederate states.

Congress appoints a commission to conclude peace treaties with the Indians (*see* 1868).

Directors of the Union Pacific Railroad sell shares at par to U.S. congressmen in the Crédit Mobilier company they have formed in an effort to stave off investigation into the high personal profits they are making on

Ku Klux Klansmen and Knights of the White Camellia brought terror to Southern blacks during Reconstruction days.

1867 ***(cont.)*** the line under construction in the West (*see* 1863; 1869; 1872).

Das Kapital by Karl Marx, published in its first volume, urges an end to private ownership of public utilities, transportation facilities, and means of production.

The *Zollverein* customs union established in 1833 expands to include Baden, Bavaria, Hohenzollern, and Württemberg (*see* 1853; 1871).

South African schoolboy Erasmus Jacobs discovers a diamond as large as a child's marble on the Orange River near Hope Town, but the discovery will not be made public for some years (*see* 1871).

United Iron Mills is founded by Philadelphia scale manufacturer Henry Phipps, 28, in partnership with Andrew Carnegie (*see* 1865; Frick, 1873).

Chicago's Carson, Pirie Scott opens at 136 West Lake Street but will move to 118–120 State Street and provide serious competition for Field, Leiter and Co. (below). Samuel Carson and John Pirie have been operating a downstate Illinois chain of dry goods stores since 1854 and gone into partnership with Chicago merchants George and Robert Scott and Andrew MacLeish.

Chicago's Field, Palmer & Leiter becomes Field, Leiter & Co. as Potter Palmer sells his share in the store and Marshall Field joins with his brothers Henry and Joseph and with Levi Leiter and his brother Marshall to form the new firm which is capitalized at $1.2 million (*see* 1865; 1881).

New York's R. H. Macy Company stays open Christmas Eve until midnight and has record 1-day receipts of $6,000 (*see* 1858; 1874).

French engineer Georges Leclanche, 28, invents the first practical dry cell battery, an electrolytic cell that can provide an intermittent electric supply as required (*see* Plante, 1859). Leclanche's battery will find uses in the century to come for powering an endless array of industrial and consumer products (*see* Eveready, 1890).

Babcock & Wilcox is founded at Providence, R.I., by local engineers George Herman Babcock, 35, and Stephen Wilcox who patent a sectional industrial "safety" boiler designed to prevent any dangerous explosions. The company will become the largest U.S. producer of coalfired boilers.

Steel rail production begins in the United States, which has been using rails of iron or imported steel.

Cornelius Vanderbilt gains control of the New York Central Railroad (*see* 1865). Now 72, he issues an order January 14 that all trains on his New York & Hudson River Railroad must terminate at East Albany, 2 miles from the Albany depot, he shows protesting legislators an old law specifically forbidding the New York & Hudson River to run trains across the Hudson River, stock in the Central drops sharply, and Vanderbilt buys enough to secure control by December (*see* 1869).

Pullman Palace Car Co. is founded by George M. Pullman with Andrew Carnegie (above) who will be its major stockholder until 1873. The new company will build cars and will operate them under contract for railway companies (*see* first dining car, 1868).

The Wagner Drawing Room Car designed by New York wagon maker Webster Wagner, 50, goes into service on Commodore Vanderbilt's New York Central (above). Wagner will soon contract to use George Pullman's folding upper berth and hinged seats but when he uses the Pullman-designed equipment on Vanderbilt's Lake Shore and Michigan Southern line after 1873 Pullman will sue and the lawsuit will not yet have been settled when Wagner dies in a New York Central train collision in 1882.

Construction begins at St. Louis on the Eads Bridge that will span the Mississippi. Louisville, Ky., engineer James Buchanan Eads, 46, invented a diving bell in the 1850s, he made a fortune salvaging sunken hulls in the Mississippi, he supplied the Union with seven 175-foot ironclads in the space of 100 days in 1861, but financiers, politicians, and other engineers scoff at his proposal to sink massive stone piers through 103 feet of turbulent water to bedrock below the dense river bottom, and they ridicule his project of cantilevering three steel spans, each more than 500 feet long. Eads innovates fabrication techniques that are unique, imaginative, and effective, he will put nearly $7 million into the endeavor in the next 7 years, but 12 men will die in the huge pneumatic caisson of his east pier (*see* 1874).

A suspension bridge designed by J. A. Roebling opens to span the Ohio River between Cincinnati and Covington, Ky. (*see* 1855; Brooklyn Bridge, 1883).

The first U.S. elevated railway begins operation at New York on a single track from Battery Place to 30th Street above Greenwich Street (9th Avenue). Steam trains of the Gilbert Elevated Railway Co. provide much quicker transportation than do the horse cars below and the Gilbert line will hook up within a decade to a Third Avenue Railroad operating between City Hall Park and 42nd Street (*see* 1878; subway, 1904).

An alpine rail line with 22 tunnels is completed through the 4,500-foot high Brenner Pass between Innsbrück, Austria, and Bolzano in the Trentino-Alto Adige region of Italy (*see* Vienna-Trieste, 1854).

The Pacific Mail Steamship Company begins regular service between San Francisco and Hong Kong.

Formaldehyde is discovered by Wilhelm von Hoffmann. Occurring naturally in the environment as a gas, it will find wide uses as a preservative and germ killer and a bonding agent to strengthen paper and other materials.

The first successful gallstone operation is performed June 15 by Indianapolis physician John Stough Bobbs on McCordsville, Ind., patient Mary E. Wiggins. Bobbs will be called "the father of cholecystotomy."

Scottish physician Thomas Lauder Brunton, 23, finds amyl nitrate useful in treating angina pectoris (*see* nitroglycerin, 1847).

Two German physicians show that atropine blocks the cardial effects of vagal stimulation.

The Austrian Red Cross is founded at Vienna.

The New York State Legislature votes to establish a free public school system.

The University of Illinois is founded at Urbana.

West Virginia University is founded at Morgantown.

Howard University for Negroes is founded by white Congregationalists outside Washington, D.C. The founders, who include Gen. O. O. Howard, director of the Freedmen's Bureau, arouse ridicule with their proposal to admit students of all ages, male or female, married or single, informed or ignorant.

Atlanta University is founded in Georgia. It will become a leading institution for the higher education of blacks.

The Prussian postal monopoly, controlled since 1505 by the Thurn und Taxis family, is acquired by the government in exchange for vast tracts of land.

Milwaukee printer Christopher Latham Sholes, 48, invents the first practical "writing machine" while seeking a way to inscribe braille-like characters for use by the blind (*see* Braille, 1834). Asked to test the machine's efficiency, court reporter Charles Weller types, "Now is the time for all good men to come to the aid of the party." Sholes will call his machine a "typewriter" (*see* 1868).

U.S. inventor Benjamin Chew Tilghman, 48, devises the sulfite process for producing wood pulp for paper making (see 1870; Burgess, 1854).

Turin's *La Stampa (The Press)* has its beginnings in the *Gazette Piedmontese.*

A paperback edition of Goethe's *Faust* issued by Reclams Universal Bibliothek in Leipzig pioneers paperback book publishing (*see* Penguin, 1936).

Nonfiction *The English Constitution* by *Economist* editor Walter Bagehot, 41.

Fiction *Thérèse Raquin* by French novelist Emile Zola (Edouard Charles Antoine Zola), 27; *The Last Chronicle of Barset* by Anthony Trollope; *Under Two Flags* by English novelist Ouida (Marie Louise de la Ramée), 28, whose romance *Held in Bondage* appeared 4 years ago.

Juvenile *Ragged Dick, or Street Life in New York* is serialized in the periodical *Student and Schoolmate* published at New York under the name "Oliver Optic" by William Taylor Adams, 45. Author of the novel is Unitarian minister Horatio Alger, 35, whose more than 100 rags-to-riches novels in the next 30 years will dramatize virtues of pluck, honesty, hard work, and marrying the boss's daughter (*see* 1871).

Poetry "Dover Beach" by Matthew Arnold; "And we are here as on a darkling plain/ Swept with confused alarms of struggle and flight./ Where ignorant armies clash by night"; *The Life and Death of Jason* by William Morris.

Painting *The Execution of Maximilian* by Edouard Manet; *Rape* by French painter Paul Cézanne, 28; the Paris World's Fair introduces Japanese *ukiyoe* prints to the West; Jean Auguste Dominique Ingres dies at Paris January 14 at age 86; Théodore Rousseau dies at Barbizon December 22 at age 55.

Theater *The Death of Ivan the Terrible (Smert Ioanna Groznogo)* by Count Leo Tolstoy 1/12 at St. Petersburg's Alexandrinsky Theater; *Caste* by Thomas William Robertson 4/6 at the Prince of Wales' Theatre, London; *Dora* by Charles Reade 6/1 at London's Adelphi Theatre.

Opera *Don Carlos* 3/11 at the Paris Opéra, with music by Giuseppe Verdi, libretto from the Friedrich von Schiller play of 1787; *La Grande-Duchesse de Gerolstein* 4/12 at the Variétés Théâtre, Paris, with music by Jacques Offenbach; *Roméo et Juliette* 4/27 at the Théâtre-Lyrique, Paris, with music by Charles Gounod, libretto from the Shakespeare tragedy of 1595; *Cox and Box (or The Long-Lost Brother)* 5/11 at London's Adelphi Theatre, with music by London church organist Arthur Seymour Sullivan, 25, libretto by F. C. Burnard based on the J. Maddison Mortori farce *Box and Cox.*

First performance "The Blue Danube Waltz" ("An der Schönen Blauen Donau") by Johann Strauss 2/15 at Vienna, with lyrics from a poem by Karl Beck, 50. Sung by the Viennese Male Singing Society, the new waltz creates a sensation.

The New England Conservatory of Music is founded at Boston.

The Cincinnati Conservatory of Music is founded.

Popular songs "The Little Brown Jug" by Philadelphia songwriter R. E. Eastburn (Joseph Eastburn Winner), 32, whose older brother Septimus wrote the lyrics to "Listen to the Mocking Bird" in 1855.

Baseball's curve ball pitch is invented by Brooklyn, N.Y., pitcher William Arthur Cummings.

Washington's National baseball team tours the country, defeating the Cincinnati Red Stockings 53 to 10, the Cincinnati Buckeyes 90 to 10, the Louisville Kentuckians 82 to 21, the Indianapolis Western Club 106 to 21, and the St. Louis Union Club 113 to 26 (the teams are all amateur organizations) (*see* 1869).

Marquis of Queensberry rules for boxing are formulated by English athlete John Graham Chambers, 24, who has founded an amateur club to encourage boxing under the aegis of Scotland's Sir John Sholto Douglas, 23, 8th Marquis of Queensberry (*see* Oscar Wilde, 1895).

The Belmont Stakes has its first running and the 1.5-mile test for 3-year-old thoroughbreds is won by a horse named Ruthless. Major August Belmont II has built New York's Belmont Park racetrack and has helped to organize the Jockey Club to oversee U.S. horse racing (*see* Preakness, 1873).

Milan's Galleria Vittoria Emanuele is completed by architect Giuseppe Mengoni, 40, to connect the busy Piazza del Duomo with the neighboring Piazza della Scala (*see* Brussels, 1847). The great glass-roofed arcade (320 yards long, 16 wide, 94 feet high) in the form of a Latin cross provides a meeting place for the Milanese and is so popular that when hailstones break

1867 ***(cont.)*** every pane in the glass ceiling in 1869 the citizenry will reject proposals for an unbreakable roof and insist that the glass be replaced.

Scots-American naturalist John Muir, 29, starts out late in the year on a walk that will take him 1,000 miles through Kentucky, Tennessee, Georgia, and Florida (*see* 1868; Sierra Club, 1892).

Alaska (above) has a Pribilof seal population of 2.5 million, down from 5 million in 1786 when the islands were discovered (*see* 1911).

More than half of all U.S. working people are employed on farms.

The Grange, or the Patrons of Husbandry, is founded in the upper Mississippi Valley by former U.S. Department of Agriculture field investigator Oliver H. Kelley, 41. The organization's popular name derives from its lodges, which are designated by an archaic name for barn. Avowedly nonpolitical, the Grange's stated aims are educational and social, with emphasis on reading and discussion to increase farmers' knowledge, but the discussions will inevitably turn to freight rates, high taxes, and politics (*see* 1873).

U.S. inventor Lucien Smith files a patent application for barbed wire but no reliable machine exists to manufacture such wire in quantity (*see* 1873).

Chicago livestock dealer Joseph Geating McCoy, 29, buys 450 acres of land at $5 per acre in Abilene, Kan., builds pens and loading chutes with lumber he has brought from Hannibal, Mo., installs a pair of large Fairbanks scales, and promises Texas ranchers $40 per head for cattle the ranchers can sell at home for only $4 per head. Abilene is "a small, dead place of about one dozen log huts. . . four-fifths of which are covered with dirt for roofing," but it is the terminus of the Kansas Pacific Railway which despite its name operates only between Chicago and Abilene. The railroad has promised McCoy one-eighth of the freight charge on each car of cattle he ships East, he sends out the first shipment (20 carloads) September 5, and by year's end some 35,000 head of longhorns have passed through Abilene, a figure that will more than double next year (*see* 1871; Chisholm Trail, 1866).

Prehistoric canals in Arizona Territory are cleared by the John Swirling Company to bring water from the Salt River to the fertile lands of the valley. The town of Phoenix begins growing (*see* Roosevelt Dam, 1911).

Iowa farmer William Louden modernizes dairy farming with a rope sling and wooden monorail hay carrier. Suspended beneath the peak of a barn, Louden's carrier enables a farmer to swing a load of hay without using a pitchfork, and Louden's litter carrier hauls out manure, allowing a farmer to clean a stable or dairy barn in one-tenth the time it has taken until now. Louden's contributions will make possible large, efficient dairy herds.

French chemist Hippolyte Mège-Mouries, 50, begins development of a synthetic butter at the urging of the emperor Napoleon III who may recall his uncle's encouragement of food canning and beet sugar production (*see* 1794; 1811). Mège-Mouries works with suet, chopped cow's udder, and warm milk to begin a revolution in the butter industry (*see* margarine, 1869).

The first patent baby food is introduced by Justus von Liebig whose product will be sold commercially for decades (*see* 1842; Nestlé, 1866; Gerber, 1927).

Sugar beets are introduced into Utah Territory by Brigham Young who has machinery from Liverpool carted across the continent to Salt Lake City by ox-drawn wagon, but the beet sugar factory he builds will be abandoned within 2 years (*see* 1880).

Arm & Hammer Baking Soda has its beginnings as Brooklyn, N.Y., spice dealer James A. Church closes his factory to enter the baking soda business but retains the symbolic sign of the Roman god Vulcan that identified his Vulcan Spice Mill. Selling his product by the barrel with the help of a 7-foot 4-inch Col. Powell, Church will soon adopt the name Arm & Hammer Saleratus under which he will dominate the market.

Overpopulation is a red herring invented by the capitalists to justify poverty among the working class, says Karl Marx in *Das Kapital* (above). Marx rejects the idea of contraception and favors enhanced production and more equitable distribution of wealth as better ways to improve the lot of the working class.

1868

President Johnson dismisses Secretary of War Stanton February 18, an alleged violation of a law passed over Johnson's veto less than a year earlier forbidding removal of certain officials without the Senate's consent. Johnson's efforts to "protect" Southern whites against racial equality have antagonized Radical Republicans and the House votes February 21 to impeach the president "of high crimes and misdemeanors." A trial begins in the Senate March 30, but in the May 16 vote 7 Republicans join with Democrats to exonerate; the 35 to 19 decision fails by one vote to achieve the necessary two-thirds majority.

Military rule has continued in the South and in June Congress authorizes readmission of the seven Confederate states on condition that black suffrage be retained (below).

A Tory government headed by Benjamin Disraeli comes to power in Britain February 29. Parliament passes an Irish Reform Bill and a Scottish Reform Bill July 13, but the Liberal party scores sweeping victories in the general elections in November. Disraeli resigns December 2, and William Ewart Gladstone, 59, begins a ministry that will continue until 1874.

Japan's Meiji Restoration formally ends the Tokugawa shogunate that has held power since 1603. The emperor Matsuhito signs a Charter Oath April 6 promising to be guided in his rule by a deliberative assembly responsive to public opinion, and the last Tokugawa forces are defeated July 4 in the Battle of Ueno at Edo, which is renamed Tokyo (eastern capital) in November.

Japan's new emperor (above) says in his coronation oath, "We shall summon assemblies, and in ruling the nation we shall have regard to public opinion. . . Knowledge will be sought out among the nations of the world, and thus the well-being of the empire will be ensured." The imperial family takes over Kyuju palace, a fortress-castle in the center of Tokyo that has been used by shoguns for centuries (*see* 1869).

Spain's prime minister Ramon Maria Narvaez, duke of Valencia, dies April 23 at age 68 and a revolutionary proclamation issued at Cadiz September 18 by Admiral Juan Bautista Topete y Carballo, 47, ends absolutist rule. The press has attacked Isabella II for making her court favorite (an actor) Spain's minister of state and her royal forces are defeated September 28 at Alcolea by Marshal Juan Prim y Prats, 54.

Isabella II flees to France September 29, her enemies declare her deposed, and Marshal Prim forms a provisional government October 5 under the regency of the former governor of Cuba Francisco Serrano y Dominguez, 58. The new regime annuls reactionary laws, abolishes the Jesuit order and other religious orders, and establishes universal suffrage and a free press (*see* 1870).

The Serbian prince Michael III Obrenovic is assassinated the night of June 10 outside Belgrade at age 43. Michael has freed Serbia from Ottoman rule, he is succeeded by his cousin Milan, 13, who will gain full independence for Serbia, take the title king in 1882, and reign until 1889.

Russian forces occupy Samarkand.

The raja of Sarawak Sir James Brooke dies at age 65 and is succeeded by his nephew Sir Charles Anthony Johnson Brooke, 39, who has adopted his uncle's surname and will reign until his death in 1917 (*see* 1841; British protectorate, 1888).

The Ethiopian emperor Theodore is defeated April 10 in the Battle of Arogee by an Anglo-Indian force under Sir Robert Napier, 58. Theodore commits suicide, the British reach Magdala April 13, and they free traders, missionaries, and envoys imprisoned since 1864 by the cruel and eccentric Theodore.

Ethiopia falls into anarchy following the withdrawal of Napier who will be created first Baron Napier of Magdala by Queen Victoria (*see* 1872).

Basutoland is annexed by the British following defeat of the Basutos by the Orange Free State which protests the British action. Transvaal forces attempt to occupy Delagoa Bay but withdraw under pressure from the British (*see* 1871).

Madagascar's Hova queen Rashoherina dies after a 5-year reign and is succeeded by Ranavalona II who will reign as queen until 1883. True power remains in the hands of her husband and first minister Rainilaiarivony, a Christian who was married to Rashoherina and will retain power until 1896 by marrying Ranavalona's successsor. Paris recognizes Hova supremacy in Madagascar August 8.

Siam's Rama IV dies at age 64 after a 17-year reign and is succeeded by his son Somdeth Phra Paraminda Maha Chulalongkorn, 15, who will rule personally beginning in 1873. The new king will abolish the feudal system and slavery, improve the country's laws and education, introduce the telegraph (1883), open the first Siamese railroad, and visit European capitals (1897) in a reign that will continue until 1910.

The Republican party rejects President Johnson (above) and nominates Civil War hero Ulysses S. Grant for the presidency, selecting Schuyler Colfax of Indiana as his running mate. The Democrats nominate former New York governor Horatio Seymour but the Republicans "wave the bloody shirt" and label the Democratic party the party of treason. Grant wins 53 percent of the popular vote with 214 electoral votes to Seymour's 80.

Coalition forces occupy Asunción December 31 after 4 years of war. Nearly two-thirds of Paraguay's adult population have died or disappeared; President López remains at large (*see* 1870).

Congress makes retention of black suffrage forever a "fundamental condition" for readmission of the seven former Confederate states to the Union (above) (*see* Fifteenth Amendment, 1870).

Ratification of the 14th Amendment to the Constitution is proclaimed July 28 by Secretary of State William H. Seward: "All persons born or naturalized in the United States, and subject to the jurisdiction thereof, are citizens of the United States and of the State wherein they reside. No State shall make or enforce any law which shall abridge the privileges or immunities of citizens of the United States; nor shall any State deprive any person of life, liberty, or property, without due process of law. . ."

U.S. military authorities force Navajo chiefs to sign a treaty August 12 agreeing to live on reservations and cease opposition to whites. The treaty establishes a 3.5 million-acre reservation within the Navajo Nation's old domains (a small portion of the original Navajo holdings), and while the main reservation will grow to embrace nearly 14.5 million acres (25,000 square miles) reaching out from the Four Corners to include northeastern Arizona, northwestern New Mexico, and southwestern Utah, most of it will be desert and semidesert with only 68,000 acres of farmland.

The Navajo population has declined from 10,000 to 8,000 during 5 years of military internment. When the survivors return on foot to their homeland they find that their 200,000 sheep have dwindled to just 940 and that they must learn a new way of life if they are to endure.

Congress enacts an Eight-Hour Law for U.S. government laborers but in private industry most laborers work 10 to 12 hours per day.

Cubans begin a 10-year war with Spain, which has not adopted some promised reforms and will continue to permit slavery on the island until the 1880s.

Japan's Mitsubishi industrial empire has its beginnings as shipping magnate Yataro Iwasaki, 35, aids the Meiji Restoration (above) by transporting government troops

1868 ***(cont.)*** to fight the Tokugawa forces at Edo. The son of a ronin (unemployed samurai warrior), Iwasaki has discovered and explored two remote Japanese islands, organized a coastal shipping service, and is engaged in establishing the industrial complex that will prosper under the name Mitsubishi.

President Grant (above) has gained election with support from bankers and other creditors who want repayment of bonds in gold and who have called the Democrats disloyal for urging payment of the national debt in greenbacks.

U.S. business will use the "due process" clause in the 14th Amendment (above) to resist government intervention (but *see* 1877).

Metropolitan Life Insurance is founded at New York by a reorganization of National Travelers Insurance, started 5 years ago as National Life & Limb Insurance Co. (*see* 1879).

Rockefeller, Andrews & Flagler begins a battle to drive out competition in the chaotic U.S. petroleum industry as Rockefeller develops a huge market for his kerosene by underselling coal oil and whale oil (*see* 1865). Henry Morrison Flagler, 35, moved west from his native Canandaigua, N.Y., at age 14, he has married Mary Harkness whose uncle Stephen owns a distillery at Bellemore, Ohio, that has supplied whiskey sold by Clark & Rockefeller in Cleveland, he joined Rockefeller and Andrews last year, and he will persuade Harkness to loan the firm $70,000 that will permit acquisition of more refineries and pipelines as the firm grows larger and more efficient. With a capacity of 1,500 barrels per day when many competitors refine only one or two barrels, Rockefeller buys his wooden barrels at 96¢ each when other refiners pay $2.50 and obtains a 15¢ rebate on every barrel he ships via the Lake Shore and Michigan Southern Railroad (*see* Standard Oil, 1870).

The Westinghouse air brake devised by U.S. inventor George Westinghouse, 22, will permit development of modern rail travel, although Cornelius Vanderbilt has dismissed it as a "fool idea." Westinghouse invented a device 3 years ago for rerailing derailed cars, he will make his air brake automatic in 1872, and it will permit an engineer to set the brakes simultaneously throughout a whole train by means of a steam-driven air pump (*see* Union Switch and Signal, 1882).

An automatic railway "knuckle" coupler patented by former Confederate Army major Eli Hamilton Janney, 37, hooks upon impact and replaces the link-and-pin coupler that endangers the fingers and the lives of brakemen. Janney's coupler prevents excess sway of railcars and will become standard railway equipment in 1888.

Coil and elliptic railroad car springs invented by U.S. engineer Aaron French, 41, make rail travel more comfortable.

A bridge spans the Mississippi River at Quincy, Ill., through the efforts of Chicago, Burlington & Quincy boss James Frederick Joy (*see* 1854). He has spent $1.5 million on the project in defiance of his critics and his judgment will be confirmed when the Burlington's business in the Quincy area doubles next year (*see* Hannibal Bridge, 1869; Eads, 1874).

Atchison Associates is formed in September to build the Atchison, Topeka and Santa Fe Railroad (*see* 1864). The new group will soon run out of money and control will pass to Boston financier Thomas Nickerson. He will be president of the Santa Fe from 1874 to 1880 and will obtain backing from Boston's Kidder, Peabody and from London's Baring Brothers (*see* pooling agreement, 1880).

The U.S. Government adopts a standard system of screw threads established by Philadelphia machine-tool maker William Sellers, 44 (*see* Whitworth, 1841).

New Jersey Steel and Iron, owned by Cooper Hewitt, builds the first U.S. open hearth steel furnace at Trenton (*see* 1862).

Boston inventor William H. Remington patents a process for electroplating with nickel; he uses a solution prepared by dissolving refined nickel in nitric acid, then precipitating the nickel by the addition of carbonate of potash, washing the precipitate with water, dissolving it in a solution of salammoniac, and filtering it (*see* nickel steel, 1888).

Tungsten steel, invented by English metallurgist Robert Forester Mushet, 57, is much harder than ordinary steel (*see* 1882).

The German Chemical Society is founded by chemist Wilhelm von Hofmann, now 50, whose research in organic chemistry has advanced the aniline dye industry founded by W. H. Perkin in 1857. Hofmann has introduced violet dyes made from rosanaline and will find a way to convert an amide into an amine having one less carbon atom.

Amateur English astronomer Joseph Normann Cockyer, 32, discovers helium in the spectrum of the sun's atmosphere.

French archaeologist Edouard Armand Isidore Hippolyte Lartet, 67, goes into a cave near Perigueux and finds four adult skeletons and one fetal skeleton of Cro-Magnon man of the Upper Paleolithic period of 38,000 B.C. (see Neanderthal man, 1856; Java man, 1890).

A patent for a typewriter is issued to Christopher Sholes, Carlos G. Glidden, and Samuel W. Soule (*see* Sholes, 1867). Businessmen James Densmore and George Washington Yost encourage Sholes to construct a machine whose most commonly used keys will be widely separated on the keyboard to avoid jamming and Sholes aligns the letters in an arrangement that will permit rapid fingering:

QWERTYUIOP
ASDFGHJKL
ZXCVBNM

Densmore and Yost will buy the patent and persuade Remington Fire Arms Co. to produce the machine (*see* 1874).

The *Atlanta Constitution* is founded as a morning daily by Col. Carey W. Styles to lead the fight for reestablishment of state government by Georgians, a fight that will be won in 1871 with the routing of "scalawags" and "carpetbaggers" (*see* Grady, 1879).

The *Louisville Courier-Journal* is created by a merger of two Kentucky newspapers under the editorship of Henry Watterson who will continue to direct the paper's policy until his death in 1921.

The World Almanac is published for the first time. The *New York World* has compiled a 108-page compendium of facts (mostly political including notes on Southern Reconstruction and the extension of suffrage to blacks in the North), the *Almanac* has 12 pages of advertising, annual publication will end in 1876, but *World* publisher Joseph Pulitzer will revive the *Almanac* in 1886 and it will appear every year thereafter.

N. W. Ayer & Son is founded by Philadelphia advertising solicitor Wayland Ayer, 20, whose father has little or no part in the new advertising firm set up to represent farm journals and the religious weekly National Baptist (*see* "open contracts," 1875).

J.Walter Thompson has its beginnings as U.S. advertising pioneer James Walter Thompson, 20, persuades publishers of magazines such as *Godey's Ladies Book* to let him sell space in all the magazines as if they were one unit. Thompson persuades merchants to buy space in a group of magazines rather than in one at a time, the influx of new advertising enables the publishers to improve the quality of their printing, their paper, and their illustrations and to pay authors and artists better rates while selling their publications at 10¢ or 15¢ as advertising revenues increase their profits (*see* 1878).

Rand McNally & Co. is founded by Chicago printer William Rand and his former apprentice Irish-American printer Andrew McNally, 31, who 9 years ago assumed management of the *Chicago Tribune's* job-printing shop. Specializing initially in passenger tickets, timetables, and related print jobs, Rand McNally will publish its first map in 1872 when it issues a *Railway Guide* and the firm will grow to become the world's leading map maker.

La Prensa begins publication at Buenos Aires October 18. Argentine publisher Ottavio Paz will make his paper the largest in South America (*see* 1870).

Chambers's Encyclopaedia is published in 10 volumes (*see* 1844).

Fiction *The Moonstone* by Wilkie Collins; "The Luck of Roaring Camp" by *Overland Monthly* editor (Francis) Bret Harte, 32, in San Francisco.

Juvenile *Elsie Dinsmore* by U.S. author Martha Farquharson (Martha Farquharson Finley), 40, who will write the seven-volume *Mildred* series and the 12-volume *Pewit's Nest* series plus 26 *Elsie* novels.

Poetry *Verses on Various Occasions* by John Henry Newman who includes his 1866 poem "The Dream of Gerontius."

Painting *Zola* by Edouard Manet; *Among the Sierra Nevada Mountains, California*, by Albert Bierstadt.

Theater *Diary of a Scoundrel*, or *Enough Stupidity for Every Wise Man (Na vsyakoso mudresa devolna prostoty)* by Aleksandr Ostrovsky 11/1 at St. Petersburg's Alexandrinsky Theater.

Opera *Mefistofele* 3/5 at Milan's Teatro alla Scala, with music by Italian composer Arrigo Boito, 26; *Dalibor* 5/16 at Prague, with music by Bedřich Smetana; *Die Meistersinger von Nürnberg* 6/21 at Munich, with music by Richard Wagner; *La Perichole* 10/6 at the Théâtre des Variétés, Paris, with music by Jacques Offenbach; *Genevière de Brabart* 10/22 at New York, with music by Offenbach in the first U.S. performance of the opera that contains an aria that will become the hymn of the U.S. Marine Corps (lyrics beginning, "From the halls of Montezuma to the shores of Tripoli . . ." will be published in 1918).

First performances Symphony No. 1 in G minor *(Winter Dreams)* by Russian composer Petr Ilich Tchaikovsky, 27, 2/15 at Moscow; *A German Requiem (Ein Deutsches Requiem)* by Johannes Brahms 4/10 (Good Friday) at St. Stephans' Cathedral, Bremen; Symphony No. 1 in C minor by Austrian composer Anton Bruckner, 43, 5/9 at Linz. The work is too difficult for the orchestra but Vienna will hear a revised version in 1891; Concerto No. 2 in G minor for Piano and Orchestra by Camille Saint-Saëns 5/13 at Paris; *Tales of the Vienna Woods (Geschichten aus dem Wienerwald)* by Johann Strauss 6/9 at Vienna.

The Brahms Lullaby is published at Berlin with words from "Des Knaben Wunderhorn" (the second verse is by George Scherer and a Mrs. Natalia Macfarren writes English lyrics).

Hymns "Lead, Kindly Light" ("Lux Benigna") by English clergyman-composer John Bacchus Dykes, 45, lyrics by John Henry Newman (*see* 1833); "Yield Not to Temptation" by U.S. composer Horatio Richmond Palmer, 34; "O Little Town of Bethlehem" by Philadelphia Holy Trinity Episcopal Church organist Lewis H. Redner, 37, lyrics by church rector Philips Brooks, 33.

Popular songs "Sweet By and By" by U.S. composer Joseph P. Webster, lyrics by S. Fillmore.

The All-England Croquet Club founded at Wimbledon holds its first championship matches. Played in England for the past 12 to 15 years, the 13th century game will yield in popularity to tennis in the next 10 years (*see* 1874; 1897).

Badminton is invented at England's Badminton Hall, Gloucestershire residence of the Duke of Beaufort Henry Charles Fitzroy Somerset, 44, whose late father promoted the Badminton Hunt. The new racquet game is played with a feathered shuttlecock which is batted back and forth across a net.

The first recorded bicycle race is held at Paris over a 2-kilometer course at the Parc de St. Cloud.

1868 ***(cont.)*** Americans observe Memorial Day (Decoration Day) for the first time May 30. The holiday commemorates the Union dead of the Civil War.

Philadelphia's Strawbridge & Clothier dry goods emporium opens at 8th and Market streets not far from John Wanamaker's 7-year-old establishment. Quaker merchants Justus Clayton Strawbridge and Isaac Hollowell Clothier are both 30, sell only for cash, permit no haggling over prices, and begin a retail enterprise that will flourish.

Salt Lake City's Z.C.M.I. (Zion's Cooperative Mercantile Institution) opens under the aegis of the Church of Jesus Christ of Latter-Day Saints. Mormons are expected to trade at the new church-owned store, which will drive most "gentile" merchants out of the Utah Territory.

Earthquakes strike Ecuador and Peru in mid-August, killing 25,000.

John Muir gets his first sight of the Golden Gate at San Francisco after having journeyed to Cuba, to Panama, across the Isthmus to the Pacific, and thence north to California. Muir moves up into the Sierra Nevadas and the Yosemite Valley (*see* 1864; 1876).

U.S. wheat prices fall to 67¢ a bushel, down from a wartime high of $4 and a postwar peak of $1.50.

The Crimean War and Civil War have stimulated production of California wheat, which has a flinty character making it well suited to shipment overseas. California flour goes not only to the Rockies but also to China, Japan, Britain, and Europe.

The 300,000-acre King-Kenedy ranch in Texas is divided between Richard King and Mifflin Kenedy (*see* 1860). King prospered through the Civil War by steamboat trade in cotton carried to British ships at Brownsville and will use his fortune to expand the King Ranch on the Neuces River (*see* Kleberg, 1886).

Grape Phyloxera (plant lice) from the United States ruin European vineyards. Roots from New York State vineyards will be used to revive the European industry.

A refrigerated railcar with metal tanks along its sides is patented by Detroit inventor William Davis who dies at age 56. The tanks are filled from the top with cracked ice to transport fish, fresh meats, and fruits (*see* 1869).

A new mechanical cooler for English trains permits quick-cooled milk to be delivered by rail in new metal containers to cities such as London, Liverpool, Manchester, and Glasgow. The milk is far safer than milk from cows kept in town sheds (*see* 1863).

Chicago meat packer P. D. Armour adds a second plant as business booms. Armour, who moved from Milwaukee last year, has with his partner John Plankinton invested $160,000 to take over a slaughterhouse on Archer Avenue and set up under the name Armour & Co. (*see* 1864; 1869).

Tabasco sauce is formulated on Avery Island off Louisiana's Gulf Coast, the island that was the Confederacy's chief source of salt during the Civil War. Edmund McIlhenny has married an Avery girl, settled on the 2,500-acre island, and created a sauce from crushed, aged red peppers, vinegar, and salt to make a piquant mixture that will be popular for generations as a condiment to be served with seafood, especially shellfish.

Claus Spreckles of San Francisco patents a sugar-refining method that takes just 8 hours instead of the usual 3 weeks (*see* 1863). Spreckels has gone back to Europe to study manufacturing methods. The California Sugar Refinery he opens will win him the title "Sugar King" (*see* 1883).

The first U.S. production of compressed yeast begins at Cincinnati. Charles Fleischmann has emigrated from Hungary and joined his brother Maximilian and Cincinnati yeast maker James F. Gaff to form Gaff, Fleischmann & Co. (*see* 1865). He sells to local housewives from a basket at first, then by horse and wagon (*see* 1870).

The first regularly scheduled U.S. dining car goes into service on the Chicago-Alton Railroad. George M. Pullman's *Delmonico* is named for the New York restaurant.

Beijing appoints former U.S. minister to China Anson Burlingame, 48, to head a delegation to Europe and the United States for the purpose of making treaties. The Burlingame Treaty signed at Washington July 28 defines mutual rights of immigration and emigration (*see* Hayes, 1879).

1869 The world grows smaller with the completion of the Suez Canal and a transcontinental U.S. rail link.

Chiefs of Japan's four great clans surrender their territories to the Meiji emperor in March. He makes the Satsuma, Chōshu, Tosa, and Hizen *daimyo* governors of their former provinces in July with one-tenth of their old revenue (*see* 1868; 1871).

Hyderabad's Mahbub Ali Pasha begins a 42-year reign as Nizam of a country as large as France in central India's Deccan plateau. A benevolent Muslim despot, the Nizam will order his nobles and landlords to stop collecting rents when times are bad and will offer state grain reserves when harvests are poor so that none will starve, whether Hindu or Muslim.

The Red River Rebellion led by Louis Riel, 25, captures Fort Garry (Winnipeg) and establishes a provisional government. A Canadian of mixed Irish and Indian blood, Riel has inherited leadership from his father of Manitoba's *métis* (half-breeds) who believe the 2-year-old Dominion government has designs on their rights and land titles (the government has purchased 95 percent of the Northwest Territories from the 299-year-old Hudson's Bay Company for $1.5 million). Riel's government is short-lived but it will be revived (*see* 1885).

Boston expands by annexing Dorchester 2 years after having annexed Roxbury. Charlestown will be annexed in 1873 (*see* fire, 1872).

Greece agrees to evacuate Crete following a Turkish ultimatum (*see* 1908).

Britain abolishes debtors' prisons.

Britain's Contagious Diseases Act permits police constables to arrest female prostitutes but takes no action against their male customers (*see* 1959).

The American Woman's Suffrage Association is founded by Susan Brownell Anthony, 49, who breaks with the 3-year-old American Equal Rights Association to campaign and lecture on the need for a Constitutional amendment that will give all U.S. women the right to vote. The newspaper *Revolution* founded by Anthony last year has as its motto, "Men, their rights and nothing more; women, their rights and nothing less" (*see* 1848; 1872).

The new Wyoming Territory enacts a law giving women the right both to vote and to hold office (*see* Utah, 1870).

San Francisco has street riots July 13 against Chinese laborers (*see* Los Angeles, 1871; Chinese Exclusion Treaty, 1880, 1882).

"The only good Indians I ever saw were dead," says Civil War general Philip H. Sheridan while on a tour of the West. The Comanche chief Tochoway (Turtle Dove) has introduced himself saying he was a "good Indian."

The Noble Order of the Knights of Labor is founded by U.S. union organizers who include Uriah Smith Stephens, 48. The Garment Cutters' Association organized by Stephens in 1862 has been dissolved under pressure from Philadelphia employers, the Knights of Labor is a secret society, and its founding comes as National Labor Union president William H. Sylvis dies in July (*see* 1866; AF of L, 1886).

New York Herald publisher James Gordon Bennett, 28, who succeeded his father to the paper's editorship 2 years ago, mounts an African expedition to locate Scottish missionary David Livingstone, 56. He commissions Welsh-American newspaper correspondent Henry Morton Stanley (*né* John Rowlands), 28, to find Livingstone (*see* 1871).

Wall Street has its first "Black Friday" September 24, ruining small speculators. Financiers Jay Gould, James Fisk, and other freebooters including President Grant's brother-in-law try to corner the gold market, driving the price up to $162 per ounce by noon, and are on their way to destroying half the banks and businesses in New York when Secretary of the Treasury George Boutwell begins selling government gold, bringing the price down to $133 inside 15 minutes (*see* 1873).

T. Eaton Co., Ltd., opens December 8 at 178 Yonge Street, Toronto, with prices clearly marked and no bartering or credit allowed. Irish-Canadian merchant Timothy Eaton, 35, has had a general store at St. Mary's but his new dry goods store will grow to become the largest Canadian retail enterprise after the Hudson Bay company, with branches throughout the Dominion.

Pennsylvania oil wells produce 4.8 million barrels of crude oil (*see* 1859; Rockefeller, 1868, 1870).

The Union Pacific and Central Pacific railroads join up May 10 at Promontory Point near Ogden in Utah Territory with a ceremony that is echoed by a nationwide celebration as travel time between New York and San Francisco falls to just 8 days, down from a minimum of 3 months (and often two or three times that long). The Union Pacific has been built 1,090 miles west from Omaha with $27 million in government "loans" on 13 million acres of public lands, largely with Irish labor, the Central Pacific 680 miles east from Sacramento with similar government subsidies and land grants, largely with Chinese labor, and the road is completed 7 years short of the deadline set by Congress.

Central Pacific workers have laid a record 10 miles of track in one day. Racing to meet the Union Pacific crews, trackmen laid 35,200 60-pound rails April 28 and spiked some 24,000 6-by-8 cross ties, each 8 feet long, to cover the 10 miles.

The Virginia & Truckee Railroad founded by Bank of California officers William Ralston, William Sharon, and Darius Ogden Mills will produce $100,000 per month in dividends for its three owners beginning in 1873 when the Big Bonanza vein opens to increase production of silver from the Comstock Lode discovered in 1859. The railroad will go into receivership in 1937 and be abandoned in 1950.

Kansas City's Hannibal Bridge opens in July—the first permanent structure to span the Missouri River. James Frederick Joy of the Chicago, Burlington & Quincy has built it for the Hannibal & St. Joseph (*see* Quincy bridge, 1868).

Cornelius Vanderbilt consolidates the Hudson River and New York Central railroads to gain a monopoly in rail transport between New York and Buffalo (*see* 1867; Lake Shore and Michigan, 1873).

A cog railway up New Hampshire's 6,293-foot Mount Washington is completed by U.S. inventor Sylvester

The Iron Horse brought white ranchers and farmers to lands that had forever been Native American hunting grounds.

1869 *(cont.)* Marsh, 66, who patented a cog rail in 1867 and has devised special engines for ascending the highest peak in the northeastern United States.

The *Glory of the Seas*, launched by Boston's Donald McKay of 1851 *Flying Cloud* fame, is his last sailing ship; it will remain in service until 1923.

The clipper ship *Cutty Sark* launched in England sails for Shanghai to begin a 117-day voyage with 28 crewmen to handle the 10 miles of rigging that control her 32,000 square feet of canvas. Built for the tea trade, the ship has a figurehead wearing a short chemise, or "Cutty Sark."

The Suez Canal opens to traffic November 17, linking the Mediterranean with the Gulf of Suez at the head of the Red Sea. The canal is 103 miles long, more than 196 feet wide at its narrowest point, 38 feet deep, and brings Oriental ports 5,000 miles closer to Europe, 3,600 miles closer to America (*see* 1875; de Lesseps, 1859).

U.S. Baptist minister Jonathan Scobie at Yokohama invents the rickshaw to transport his invalid wife about town. Improved models will provide employment for Scobie's converts and the ginrickshaw will be popular in many Oriental cities.

Stanley Rule & Level Co. of New Britain, Conn., buys the patent rights and business of Leonard Bailey, who has invented the first metal plane. The Stanley-Bailey plane is cheap enough for the average craftsman to buy, has a cutter that can be adjusted easily by turning a nut, and is much easier to sharpen than the wooden planes now widely used.

Union Bag & Paper Co. has its beginnings in a patent pool formed by U.S. bagmakers to monopolize control of machines which Union Bag will lease. The company will start manufacturing its own bags in 1875 when it will make 606 million paper bags using sulfite paper from which the lignin has been removed (*see* 1899; Tilghman, 1867).

Nature begins publication November 4 at London. The 40-page "weekly illustrated journal of science," which leads off with an essay by Thomas Henry Huxley, now 44, will gain worldwide circulation and influence.

German medical student Paul Langerhans, 22, discovers tiny cells in the pancreas that produce glucagon and insulin, ductless gland secretions essential to normal human metabolism (they will be called hormones beginning in 1904 and the cells will be called the "islets of Langerhans") (see 1860; 1889).

Massachusetts establishes the first state board of health.

The Harvard Medical School rejects a demand by President Charles W. Eliot (below) that students be given written examinations. "A majority of the students cannot write well enough," says the dean, but Eliot will elevate standards at the medical school.

The Vatican Council convened December 8 by Pope Pius XI, now 78, is the first great meeting of bishops since the Council of Trent of 1545 to 1563 (*see* 1870).

Harvard College names MIT chemistry professor Charles William Eliot, 35, as president and Eliot begins a 40-year career in which he will contain all undergraduate studies within the college while developing graduate and professional schools. New York lawyer Christopher Columbus Langdell, 43, will be named dean of the Law School next year and introduce the "case study" method of teaching law. Eliot will organize a graduate school of arts and sciences in 1890 and make the Divinity School nonsectarian.

Howard College holds its first classes at Washington, D.C., where it was organized by the Freedmen's Bureau 2 years ago.

Purdue University is founded at Lafayette, Ind.

Southern Illinois University is founded at Carbondale.

The University of Nebraska is founded at Lincoln.

Japan's first public elementary school opens in May at Kyoto.

Cyrus Field completes a cable connection between France and Duxbury, Mass. His 1858 cable ceased to operate after a few weeks but his new one embodies technical improvements that will make it a great success (*see* wireless, 1901; telephone, 1927).

Harper's Weekly cartoonist Thomas Nast, 29, draws the first of many caricatures attacking New York's Tweed Ring—a group headed by State Senator William Marcy "Boss" Tweed, 46, which includes the city comptroller, city chamberlain, and Mayor Abraham Oakey Hall, 43—that has been bilking the city treasury out of millions of dollars. Nast is a German-American who has been with the *Weekly* since 1862 after 3 years with the *New York Illustrated News*. The *New York Times* will join in the attack on the Tweed Ring, and its ringleaders will be arrested late in 1871 (*see* donkey symbol for Democratic party, 1870; "Tweed" Courthouse, 1875).

The S.S. *Great Eastern* laid the Atlantic Cable that brought Europe closer to the Americas.

George P. Rowell stabilizes the U.S. advertising business by publishing the first open, accurate list of American newspapers, but agents continue to set space rates in the country's 5,411 papers (*see* Ayer, 1875; *Printer's Ink*, 1888).

Fiction *The Idiot (Idiot)* by Fedor Dostoievski; *L'homme qui rit* by Victor Hugo; *Phineas Finn* by Anthony Trollope; *Lorna Doone* by English novelist R.D. Blackmore, 44; *Innocents Abroad* by Mark Twain; "The Outcasts of Poker Flat" by Bret Harte.

Juvenile *Luck and Pluck* by Horatio Alger; *Little Women: Or Meg, Jo, Beth and Amy* by U.S. novelist Louisa May Alcott, 37, whose father Amos Bronson failed in his Fruitlands commune experiment of 1844 and who was herself forced for a time to perform menial work in order to support her family. The book will give the Alcotts financial security.

Poetry *The Ring and the Book* by Robert Browning.

Theater *The League of Youth (De unges forbund)* by Norwegian playwright Henrik (Johan) Ibsen, 41, 10/18 at Christiania's (Oslo's) Christiania Theatre.

First performances Symphony No. 4 in C minor *(Tragic)* by the late Franz Schubert (1816) 2/26 at London's Crystal Palace; Symphony No. 2 *(Antar)* by Russian composer Nikolai Andreievitch Rimski-Korsakov, 25, 3/22 at St. Petersburg; Concerto in G minor for Piano and Orchestra by Edvard Grieg 4/3 at Copenhagen; Mass in E minor by Anton Bruckner 9/30 at the still incomplete Linz Cathedral; *Liebeslieder Walzer* by Johannes Brahms 10/6 at Carlsruhe.

Vienna's Staatsoper (State Opera House) opens off the Karntner Ring Strasse with 2,263 seats.

Opera *Das Rheingold* 9/22 at Munich, with music by Richard Wagner.

Popular songs "Sweet Genevieve" by U.S. composer Henry Tucker, lyrics by George Cooper, 31; "Shoo Fly, Don't Bother Me" by U.S. composer Frank Campbell, lyrics by Billy Reeves.

The Cincinnati Red Stockings complete an undefeated season, winning 56 games and tying one. First professional baseball team, the Red Stockings will remain undefeated until they fall to the Brooklyn Athletics June 14 of next year (*see* 1867; National League, 1876).

Rutgers defeats Princeton six goals to four November 6 in the first intercollegiate football game which is actually a form of soccer with 25 men on a team (*see* 1863). Boys from the College of New Jersey at Princeton have come to New Brunswick for the game (*see* "Boston Game," 1874).

The Cardiff Giant is "discovered" at Cardiff, N.Y., where the huge stone figure of a man has been buried secretly by a group of promoters who claim the figure is a petrified man from biblical times, citing Genesis 6:4 ("There were giants in the earth in those days"). Ten feet 4.5 inches long and weighing nearly 3,000 pounds, the figure is exhibited at Syracuse, Albany, New York City, and Boston, thousands pay $1 each to view the Cardiff Giant, and the hoax takes in eminent men before it is exposed.

Washington's Pennsylvania Avenue is paved with wooden blocks for a mile between 1st Street and the Treasury Department building at 15th Street.

Garden City, Long Island, started by New York department store magnate A. T. Stewart, is a planned, community designed for families of moderate income.

Nashville's Maxwell House opens in Tennessee. Built at the start of the Civil War, the new hotel has served as barracks, prison, and hospital *(see* coffee, 1886).

Cairo's Mena House hotel opens to serve guests invited to the opening of the Suez Canal (above).

German zoology professor Ernst Heinrich Haeckel, 35, coins the word "ecology" to mean environmental balance. He is the first German advocate of Charles Darwin's organic evolution theory.

Gypsy moths *(Porthetria dispar)* are brought to Medford, Mass., by French naturalist Leopold Trouvelot who hopes to start a New England silk industry. The moths escape and their larvae (which feed on leaves) will defoliate American woodlands as the moth population explodes in the next 20 years.

U.S. geologist John Wesley Powell, 35, begins explorations of the Green and Colorado Rivers in the Utah and Colorado Territories. Powell lost an arm at Shiloh in 1862 but has led summer expeditions into the Colorado Rockies for the past 2 years between teaching terms at a college in Southern Illinois (*see* 1878).

Daily weather bulletins are inaugurated by U.S. astronomer Abbe Cleveland, 30, the first U.S. Weather Bureau meteorologist.

The first U.S. plow with a moldboard entirely of chilled steel is patented by James Oliver who has established the Oliver Chilled Plow Works (*see* 1855).

The coffee rust *Hamileia vastatrix* appears in Ceylon plantations and will spread throughout the Orient and the Pacific in the next two decades (*see* 1825). It will destroy the coffee-growing industry, and soaring coffee prices will lead to wide-scale tea cultivation (*see* 1884; tea consumption, 1898).

A British Customs Duty Act abolishes even nominal duties on food imports (*see* 1850).

Hippolyte Mége-Mouries produces margarine commercially for the first time. The product is patented in England under the name "butterine" (*see* 1867; 1871).

Boston gets its first shipment of fresh meat from Chicago by way of a refrigerated railcar developed last year by William Davis, but railroads resist losing their traffic in live animals bound for eastern markets (*see* Swift, 1877).

Armour & Co. adds beef to its line of pork products (*see* 1868). Armour will start handling lamb next year (*see* chill room, 1872).

Campbell Soup Co. has its beginnings in a cannery opened at Camden, N.J., by Philadelphia fruit whole-

1869 ***(cont.)*** saler Joseph Campbell and icebox maker Abram Anderson. They can small peas and fancy asparagus (*see* 1894).

H. J. Heinz Co. has its beginnings at Sharpsburg, Pa., where local entrepreneur Henry John Heinz, 24, goes into business with partner L. C. Noble to pack processed horseradish in clear bottles, competing with horseradish packed in green bottles to disguise the fact that it often contains turnip fillers. Heinz has employed several local women for nearly a decade to help him supply Pittsburgh grocers with the surplus from his garden (*see* 1875).

Welch's Grape juice has its beginnings at Vineland, N.J., where dentist Thomas Bramwell Welch develops a temperance substitute for the intoxicating wine used in his church's communion service. He picks 40 pounds of Concord grapes from his backyard, pasteurizes the juice in his wife's kitchen, bottles it, and begins selling "unfermented wine" to nearby churches (*see* 1896; Concord grapes, 1853).

A National Prohibition party is founded in September at Chicago (*see* WCTU, 1874).

Japan's Kirin Brewery is founded at Yokohama under the name Spring Valley Brewery by U.S. entrepreneur William Copeland.

The American Woman's Home by Catherine E. Beecher and her sister Harriet Beecher Stowe deplores the popularity of store-bought bread which accounts for perhaps 2 percent of all bread eaten in America.

The A&P gets its name as the 10-year-old Great American Tea Company is renamed the Great Atlantic and Pacific Tea Company to capitalize on the national excitement about the new transcontinental rail link (above). Proprietors George Huntington Hartford and George F. Gilman attract customers by offering premiums to lucky winners, they use cashier cages in the form of Chinese pagodas, they offer band music on Saturdays, and they employ other promotional efforts while broadening their line of grocery items to include coffee, spices, baking powder, condensed milk, and soap as well as tea (*see* 1871).

A speaker at the annual meeting of the British Medical Association condemns "beastly contrivances" for limiting the numbers of offspring (*see* U.S. "Comstock Law," 1872).

Pope Pius IX declares abortion of any kind an excommunicatory sin (*see* 1870; 1930).

1870 Spain's Isabella II, now 39, is persuaded to abdicate June 25 at Paris after a 27-year reign. Her son Alphonse, 12, succeeds in name only; the duke of Aosta, 25, son of Italy's Victor Emmanuel II, is induced to accept the crown and begins a brief reign as Amadeo I (*see* 1868; 1872).

Prussia's Wilhelm I, head of the house of Hohenzollern, has persuaded the Hohenzollern prince Leopold, 35, to withdraw his acceptance of the Spanish crown (above) following French protests.

France's Napoleon III has wired Wilhelm (who is taking a cure at Ems) demanding a letter of apology and demanding that Wilhelm prohibit Leopold from accepting any future offer of the Spanish crown.

Prussia's Count von Bismarck makes the Ems telegram (above) public with Wilhelm's permission, leaving out some details unsuitable for publication but making it clear that Napoleon has tried to humiliate the kaiser and that Wilhelm has refused Napoleon's demands.

France declares war on Prussia July 19; three German armies invade France while the French invade the Saar basin and gain a victory at Saarbrücken.

The Battle of Sedan September 1 ends in disaster for the French First Army Corps commanded by Marshal MacMahon. Napoleon himself surrenders September 2 with 80,000 men, and two German armies begin a 135-day siege of Paris September 19 as the Second Empire collapses (*see* 1871).

Italian troops enter Rome following withdrawal of French troops for the Battle of Sedan (above) and they complete the unification of the Kingdom of Italy under Victor Emmanuel II (*see* 1866).

Irish lawyer Isaac Butt, 57, forms the Home Rule Association, a coalition of Protestants and nationalists that works for repeal of the 1801 Act of Union (*see* 1879; Parnell, 1881).

Paraguay's President López is killed March 1; his 6-year war with his neighbors ends in a treaty signed June 20. Paraguay has lost 100,000 dead, 65,000 wounded, reducing her population to 28,000 men and just over 200,000 women. Brazil has lost 165,000, Argentina 20,000, Uruguay 3,000; they take 55,000 square miles of territory.

The province of Manitoba created by Canada July 15 is a 251,000-square mile section of Rupert's Land, purchased last year from the Hudson's Bay Company (*see* 1905; British Columbia, 1871).

The first black U.S. legislators take their seats at Washington, D.C., as Hiram R. Revels of Mississippi takes his seat in the Senate and J. H. Rainey of South Carolina enters the House of Representatives.

Ratification of the Fifteenth Amendment to the Constitution is proclaimed March 30 by Secretary of State Hamilton Fish. The amendment forbids denial of the right to vote "on account of race, color, or previous condition of servitude" (but *see* Mississippi, 1890; Louisiana, 1898; Alabama, 1901).

A U.S.-British convention for the suppression of the African slave trade is concluded June 3 (*see* Livingstone, 1871).

A Chinese mob at Tianjin attacks a Roman Catholic orphanage June 21 and kills 24 foreigners, including the French consul and some French and Belgian nuns accused of kidnapping children.

Women gain full suffrage in the Territory of Utah (*see* Wyoming, 1869; Anthony, 1872).

Japan's Meiji emperor issues an edict in September ordering his subjects to take last names.

Lloyds of London is incorporated 182 years after its founding but remains a society of private individuals who will increase in number to 6,000 "Names" that will include peers, members of Parliament, and Cabinet ministers. The "Names" will belong to more than 260 syndicates with each Name having unlimited liability for the insurance written by his syndicate (*see* 1887).

The New York Cotton Exchange is founded by 100 firms (*see* cotton crop, below).

New York's F. A. O. Schwarz toy shop opens on Broadway at 9th Street. German-American merchant Frederick August Otto Schwarz, 34, has worked in the Baltimore shop of his brother Henry, he will relocate to a large store in West 23rd Street, it will net nearly $100,000 per year by 1900 selling dolls, doll houses, Kiddie Kars, rocking horses, Swiss music boxes, and the like, and it will become the world's leading toy emporium.

Standard Oil Co. of Ohio is incorporated January 10 with John D. Rockefeller as president. He takes 2,667 shares in the $1 million company while 1,333 shares each are taken by his brother William, Samuel Andrews, H.M. Flagler, and Flagler's uncle Stephen Vanderburg Harkness, 52, whose loans have helped Rockefeller and his associates (*see* 1868; 1872).

Henry Clay Frick, 21, begins construction and operation of coke ovens in the Connelsville area while working for his grandfather Abraham Overholt, who dies at age 86 after 60 years of making Old Overholt Whiskey. The Pennsylvania farm hand works with associates and persuades Irish-American Pittsburgh judge-banker Thomas Mellon, 57, to loan the group money for its ventures (*see* 1873).

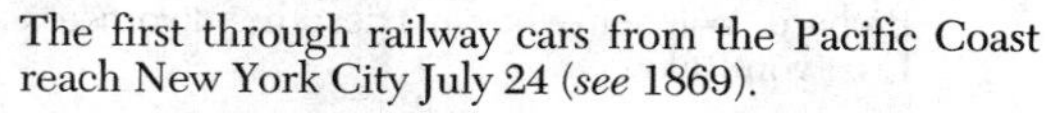

The first through railway cars from the Pacific Coast reach New York City July 24 (*see* 1869).

Construction of the Northern Pacific Railroad begins in Minnesota (*see* 1883; Villard, 1881).

London's St. Pancras Station opens with a clock from the 1851 Great Exhibition.

French railroad trackage increases to 11,000 miles, up from 5,918 in 1860.

French lawyer Leon Gambetta, 32, pioneers air travel, escaping from the besieged city of Paris by balloon October 8 to organize defenses against the Germans and lead opposition to the imperial government.

The first lightweight all-metal bicycle (and the first with wire-spoked tension wheels) is patented by Coventry machinists James Kemp Starley, 40, and William Hillman who will market their Ariel cycle in September of next year at £8 (£12 with speed-gear) (*see* velocipede, 1861; safety bicycle, 1885).

Steamships account for 16 percent of world shipping but are often slower than clipper ships.

"Go west young man," says *New York Tribune* publisher Horace Greeley who picks up the phrase first published in 1851 by John L. B. Soule in the *Terre Haute* (Indiana) *Express.* Greeley establishes the Union Colony in Colorado Territory (it will later be called Greeley). He has been inspired by his agricultural editor Nathan C. Meeker who digs the Territory's first irrigation canal (*see* 1888).

The Principles of Chemistry by Russian chemistry professor Dmitri Ivanovich Mendeléev, 36, at St. Petersburg contains the periodic table of elements that arranges the 63 known elements according to atomic weight (valency). German chemist Julius Lothar Meyer, 60, has worked independently to make important contributions to the periodic laws of Mendeléev who begins with lithium and leaves blanks for 29 elements not yet discovered but whose discovery he correctly predicts (*see* Dalton, 1803; germanium, 1886; Thomson, 1897).

DNA (deoxyribonucleic acid) is discovered by chemistry student Friederich Miescher at Tübingen but is not yet suspected of being the basic genetic material involved in conveying heritable characteristics (*see* Levene, 1909; Watson, Crick, 1953).

Archaeological excavations begin on the site of ancient Troy. German-American businessman Heinrich Schliemann, 48, acquired U.S. citizenship in California in 1850, amassed a large fortune in the next 13 years through usurious banking and Crimean War profiteering, and has devoted himself in the past 2 years to studying sites mentioned in the writings of Homer. Schliemann has married a Greek girl of 17 after studying photographs of possible candidates, Sophia Engastrontenos Schliemann helps her husband in his work, the Schliemanns will antagonize professionals with their destructive, amateur excavations, but they will prove Homer to have been accurate by finding Priam's city where others said it could not possibly be, and they will uncover at Mycenae the world's greatest golden treasure trove.

Smith Brothers Cough Drops are patented by William "Trade" and Andrew "Mark" whose bearded faces serve as a trademark. In 1872 Smith Brothers will become the first company to offer "factory-filled" packages, and production of the cough drops will jump from 5 pounds a day to 5 tons.

The British Red Cross is founded at London.

The Vatican Council votes 533 to 2 July 18 that the pope is infallible when he defines doctrines of faith or morals ex cathedra (from his throne). Such dicta are "irreformable" and require no "consent of the church." Papal nuncios have intimidated the bishops into favoring such a decree; their action produces a wave of anti-German legislation in the German states (*see* 1869; Lord Acton, 1887).

Women enter the University of Michigan for the first time since its founding at Ann Arbor in 1817. By the end of the 1870s there will be 154 U.S. coeducational colleges, up from 24 at the close of the Civil War (*see* Oberlin, 1833).

St. John's University is founded at Brooklyn, N.Y.

Syracuse University is founded at Syracuse, N.Y.

1870 *(cont.)* Ohio State University is founded at Columbus.

The University of Cincinnati is founded.

The University of Akron is founded.

Chicago's Loyola University has its beginnings in St. Ignatius College founded by Jesuits.

Texas Christian University is founded at Fort Worth.

Hampton Normal and Agricultural Institute is chartered in Virginia (*see* Booker T. Washington, 1881).

Only two Americans in 100 of 17 years and older are high school graduates, a figure that will rise to 76 percent by 1970. Sixty-seven percent of children between 5 and 17 are students, a figure that will rise to 78 percent by 1920 and to 87 percent by 1975.

Britain institutes compulsory education.

The British Post Office issues the world's first postcards October 1.

A rotary press built by Richard Hoe incorporates advances patented by William Bullock and prints both sides of a page in a single operation (*see* 1846; 1865). The *New York Tribune* will install the new press in its pressroom (*see* folding apparatus, 1875).

Production of paper from pulpwood begins in New England (*see* Burgess, 1854; Tilghman, 1867).

The donkey symbol that will identify the Democratic party in the United States for more than a century appears January 15 in *Harper's Weekly* where cartoonist Thomas Nast continues to caricature "Boss" Tweed and his cronies (*see* 1869). No Democrat will gain the presidency until 1884 but Nast will create the elephant symbol for the Republican party in 1874.

La Nación begins publication at Buenos Aires to rival the paper *La Prensa* founded last year. Descendants of founder Gen. Bartolome Mitre will control *La Naciòn* for more than a century.

The *Belfast Telegraph* begins publication in Ireland.

The *Yokohama Mainichi Shimbun* begins publication in Japan but not yet on a daily basis (*see* 1871).

Nonfiction *The History of England from the Fall of Wolsey to the Defeat of the Spanish Armada* by J.A. Froude; "Civilization" (essay) by Ralph Waldo Emerson who says, "Hitch your wagon to a star."

Fiction *The Earthly Paradise* by William Morris is a collection of classical and medieval tales with illustrations by painter Edward Coley Burne-Jones, 34, who met Morris at Oxford's Exeter College; *Lothair* by Benjamin Disraeli; *The Hunting of the Snark* by Lewis Carroll; *Twenty Thousand Leagues Under the Sea* (*Vingt mille lieues sous les mers*) by Jules Verne who delights readers with Captain Nemo and his submarine; *The Mystery of Edwin Drood* by the late Charles Dickens who has died June 9 at age 58 leaving the work unfinished (and providing in his will for his longtime mistress Ellen "Nelly" Ternan, now 31); *The Story of a Bad Boy* by Boston editor Thomas Bailey Aldrich, 33, who quit school at age 13 upon the death of his father, had a book of verse published 5 years later, and has written an autobiographical account of boyhood in Portsmouth, N.H.; *An Old Fashioned Girl* by Louisa May Alcott.

Poetry *The House of Life* by Dante Gabriel Rossetti who has been persuaded to exhume the unpublished poems he buried with his wife in 1862 and whose work establishes his reputation; "Curfew Must Not Ring Tonight" by U.S. poet Rose H. Thorpe, 20.

Painting *La perle* by J. P. C. Corot; *Un Atelier à Batignolles* by Henri Fantin-Latour.

The Boston Museum of Fine Arts is chartered in February (*see* 1909).

New York's Metropolitan Museum of Art is chartered April 13. Local real estate investor William T. Blodgett buys up three important private collections at Paris in the summer (paying $116,180 for 174 pictures), the museum will open in 1872 at 681 Fifth Avenue, and it will move the following year to a mansion on 14th Street (see 1880).

Washington's Corcoran Gallery of Art is incorporated May 14. Local banker William Corcoran, 71, made a fortune with his partner George Riggs by raising $5 million to help the government fight the Mexican War of 1846 to 1848, retired in 1854, and supported the Confederacy in the Civil War which he sat out in Europe. The gallery will open in 1874 on Pennsylvania Avenue.

Ballet *Coppélia* 5/25 at the Théâre Imperial de l'Opéra, Paris, with music by French composer Clement Philibert Léo Delibes, 34. *Coppélia* will be the comic classic of the ballet repertory as the 39-year-old *Giselle* will be its tragic classic.

Opera *Die Walküre* 6/26 at Munich, with music by Richard Wagner who marries the divorcée Cosima Liszt von Bülow.

First performances *Romeo and Juliet Overture* (*Fantasia after Shakespeare*) by Petr Ilich Tchaikovsky 3/16 at Moscow; Concerto No. 3 in E flat major for Piano and Orchestra by Camille Saint-Saëns 3/17 at the Salle Playel, Paris; *Siegfried Idyll* by Richard Wagner 12/25 (his new wife's birthday) at the Wagners' Villa Tribschen on Lake Lucerne.

The first English bid to regain the America's Cup in ocean racing won by the schooner *America* in 1851 ends in failure; the *Cambria* loses 1 to 0 to the U.S. defender *Magic* in New York Bay.

New York City gets its first luxury apartment house. Built by Rutherford Stuyvesant in East 18th Street from designs by Richard Morris Hunt, 43, the five-story walk-up building is modeled on Parisian apartment buildings, complete with concierge, but a New Yorker says, "Gentlemen will never consent to live on mere shelves under a common roof." Rents for six rooms with bath range from $1,000 to $1,500 per year, far beyond the means of New York's boarding-house and tenement-dwelling masses.

French citizens subscribe 40 million francs to build the Church of Sacré Coeur on Montmartre in Paris "in witness of repentance and as a symbol of hope" following the French defeat at Sedan (above).

Sandringham House is completed in Norfolk for Queen Victoria after 9 years of construction.

Some 4 million buffalo roam the American plains south of the Platte River by some estimates. They will be virtually wiped out in the next 4 years, leaving perhaps half a million north of the Platte (*see* 1875).

German-American lumberman Frederick Weyerhaeuser, 36, enters into a marketing agreement with other Midwestern lumber interests to form a syndicate whose headquarters Weyerhaeuser will move to St. Paul, Minn., in 1891 as he expands his timberland holdings. The Weyerhaeuser Co. will grow to control more than 2 million acres of forest land from Wisconsin and Minnesota to the Pacific Northwest, the largest holding of any U.S. company (*see* 1891).

The U.S. corn crop tops 1 million bushels for the first time (*see* 1885).

California wheat growers produce 16 million bushels and the $20 million they receive is twice the value of all the gold mined in California for the year (*see* 1868). Great wheat farms in the Livermore and San Joaquin valleys average nearly 40 bushels per acre with some producing 60 per acre.

The Chinese plum that has been domesticated in Japan is introduced into California where it will be the basis of the modern plum industry (*see* Agon plum, 1856).

The U.S. Department of Agriculture imports 300 apple varieties from Russia. Included are the Duchess, Red Astrachan, and Yellow Transparent.

The MacIntosh apple is propagated from a seedling found on his Matilda Township homestead by Ontario nurseryman Allan McIntosh. It will become the dominant variety in New England and eastern New York.

The Elberta peach is introduced according to some accounts by Marshalville, Ga., orchardman Samuel Rumph who has developed the variety and has named it after his wife (but *see* 1850).

The U.S. cotton crop is 4.03 million bales, up from 3.48 million in 1860 (*see* 1850).

Cattle drives up the Chisholm Trail from San Antonio, Tex., to Abilene, Kan., begin on a huge scale (*see* 1867). The overland treks will bring an estimated 10 million head up the trail in the next 20 years and while J. G. McCoy at Abilene does not quite fulfill his promises to Texas cattle ranchers they do receive $20 at Abilene for a steer worth $11 in Texas and that sells at Chicago for $31.50 (*see* railroad, 1871).

Parisians go hungry as German troops begin a 135-day siege September 19 (above). The people eat cats, dogs, even animals from the zoo including the beloved elephants Castor and Pollux (*see* 1871).

Italian toxicologist Francesco Selmi coins the word "ptomaine" to denote certain nitrogenous compounds easily detectible by smell and in some cases poisonous. It will be applied erroneously to food contaminants which are in many cases not detectible by smell.

European flour millers develop porcelain rollers. Flour from rolling mills is whiter than flour from grinding mills since the dark wheat germ is flattened by the rollers into a tiny flake that is sifted off with the bran, but while the whiter flour is reduced in nutrients the loss is significant only in areas where bread remains the major part of human diets.

The porcelain rollers (above) are widely adopted in Austria and Hungary where hard wheats predominate (*see* 1839; Glasgow, 1872; Minneapolis, 1879).

Cincinnati's Gaff, Fleischmann markets compressed yeast wrapped in tinfoil that permits shipment anywhere; the yeast becomes popular even with ultraconservative bakers (*see* 1868; 1876).

Canadian distiller Joseph Emm Seagram, 29, joins a 19-year-old firm of Waterloo, Ontario, millers and distillers that he will own by 1883. Joseph E. Seagram and Sons will become a leading Canadian whisky maker (*see* Bronfman, 1928).

Mott's Cider and Mott's Vinegar voyage round Cape Horn by clipper ship in 1,000-case lots (*see* 1842).

Banana traffic from the Caribbean to North American ports is pioneered by Wellfleet, Mass., fishing captain Lorenzo Dow Baker, 30, who is part owner of the 70-ton two-masted schooner *Telegraph of Wellfleet*. Baker buys 1,400 coconuts at 12 for a shilling and 608 bunches of bananas at a shilling each in Port Antonio, Jamaica, and puts in 11 days later at Jersey City, N.J., with the fruit beginning to turn yellow. An Italian merchant buys it at Baker's offering price of $2.25 per bunch while the coconuts sell at almost no profit (*see* 1871; Pearsall, 1843).

The U.S. canning industry puts up 30 million cans, up from 5 million in 1860, despite the fact that it employs scarcely 600 people (*see* 1874).

A can opener patented by U.S. inventor William W. Lyman is the first with a cutting wheel that rolls round a can's rim.

The first U.S. food trademark (#82) registered by the U.S. Patent Office (above) is a red devil. It is granted to Boston's 48-year-old William Underwood & Co. for "deviled entremets" introduced in 1867.

Kansas City's population reaches 32,260, up from 3,500 in 1865, as the year-old Hannibal Bridge across the Missouri helps attract residents.

Chicago's population reaches 300,000, up from 93,000 in 1857.

1871

A united German Empire proclaimed January 18 in the Hall of Mirrors at Versailles inaugurates the Second Reich that will continue until November of 1918 (the First Reich began in 955 by some accounts and was ended by Napoleon in 1806). Prussia's Wilhelm I is emperor and the first chancellor is Count von Bismarck, who has joined the four German kingdoms of Prussia, Bavaria, Saxony, and Württemberg with the

1871 ***(cont.)*** five grand duchies, 13 duchies and principalities, and three free cities (Bremen, Hamburg, and Lübeck) that have kept Germany fragmented.

Paris capitulates to German troops January 28, the French assembly accepts a peace treaty March 1, a definitive treaty is signed at Frankfurt May 10, and France agrees to cede Alsace and part of Lorraine, pay the Germans an indemnity of 5 billion francs, and permit an army of occupation to remain on French soil until the indemnity is paid.

Napoleon III is formally deposed March 1 by the French assembly meeting at Bordeaux because a Communist uprising (below), has taken over Paris, the assembly declares Napoleon "responsible for the ruin, invasion, and dismemberment of France," he is released from the castle of Wilhelmshohe near Cassel, where the Germans have held him prisoner, and retires to England, where he will die early in 1873 (*see* Prince Louis, 1879).

Japan abolishes feudal fiefs by imperial decree August 29, substituting prefectures.

Basutoland is absorbed by Britain's Cape Colony and Britain annexes the diamond mines of Kimberley (below; *see* 1868; Rhodes, Barnato, 1880).

The Treaty of Washington signed May 8 refers settlement of the Canadian-U.S. border to the German emperor (above), refers settlement of claims against Britain for damage by the Confederate cruiser *Alabama* in the Civil War to an international commission at Geneva, and provides for partial settlement of a dispute in the North Atlantic fishery (*see* 1877).

British Columbia joins the Dominion of Canada July 1 with the understanding that a transcontinental railroad is to be started within 2 years and completed within 10.

The Paris Commune established in March is defeated at the end of May in a "Bloody Week" that sees 20,000 to 30,000 Parisians killed at the barricades, a death toll exceeding that in the Reign of Terror of 1793 to 1794. The assembly has elected Louis Thiers, now 73, president of the Third Republic February 16, the troops he has sent to seize the Commune's cannon have refused to fire and fraternized with the crowd, Generals Lecomte and Thomas have been seized by the crowd and executed along with hostages who have included the archbishop Georges Darboy.

A *Kulturkampf* against Roman Catholics begins in Prussia (*see* Jesuits, 1872).

Parliament legalizes British labor unions.

Japan's restored Meiji government outlaws discrimination against the *burakumin*, outcasts who live in thousands of ghettos throughout the country as they have done since feudal times. Despite the edict from Tokyo, the *burakumin* will for more than a century be denied access to other jobs and to public schools and non-ghetto residential areas.

Scottish missionary-explorer David Livingstone sees hundreds of African women shot dead by Arab slavers or drowned in the Lualaba River at Nyangwe while trying to escape. He sends home an account of the incident to England where indignant efforts are made to force suppression of the slave trade by the sultan of Zanzibar (*see* 1869; 1873).

French militia squelched the Paris Commune with blood force. Tens of thousands died at the barricades.

A race riot at Los Angeles leaves more than a dozen Chinese dead and many injured (see 1877).

Mormon leader Brigham Young, now 70, is arrested at Salt Lake City in Utah Territory on charges of polygamy (he has 27 wives) (*see* 1847). The polygamy issue will delay Utah statehood until 1896.

The Indian Appropriation Act passed by Congress March 3 makes Indians wards of the federal government and discontinues the practice of according full treaty status to agreements made with tribal leaders. The act does not abrogate existing treaties but it forbids recognition of Indian tribes as nations or independent powers.

Gen. George Crook assumes command of the Army in Arizona Territory, uses other Apache as scouts, tracks down Chiricahua Apache chief Cochise, and forces him to surrender (*see* 1862; 1872).

Dodge City, Kan., has its beginnings in a sod house built on the Santa Fe Trail 5 miles west of Fort Dodge to serve buffalo hunters. Within a year the settlement will have grown into a town with a general store, three dance halls, and six saloons and will soon be shipping hundreds of thousands of buffalo hides, buffalo tongues, and buffalo hindquarters to market via the Santa Fe Railroad that will reach Dodge City next year (*see* 1868).

Birmingham, Ala., is founded and incorporated at a site surrounded by iron ore, coal, and limestone deposits that were exploited in the Civil War to produce Confederate cannonballs and rifles. The town takes its name from Birmingham, England.

Kimberley is founded in South Africa and is soon the center of a great diamond industry (*see* 1867).

New York Herald correspondent Henry M. Stanley finds David Livingstone (above) in late October at Ujiji on Lake Tanganyika. "Dr. Livingstone, I presume," says Stanley, and he joins Livingstone in exploring the northern end of the lake, proving conclusively that the Rusizi River runs into the lake and not out of it (*see* 1869).

A study undertaken to refute charges that British land ownership is excessively concentrated reveals that 710 landlords own one quarter of all the land in England and Wales and fewer than 5,000 people own three quarters of the land in the British Isles.

Japan establishes a new yen-based currency system June 29. Equitable taxation laws are instituted October 8.

Drexel, Morgan & Co. is organized with offices in Philadelphia and on New York's Wall Street, by New York banker J. P. (John Pierpont) Morgan, 34, who joins with members of Philadelphia's Drexel family to create a powerful new banking house (*see* Edison, 1882).

State supervision of grain warehouses begins in Illinois. When the Supreme Court sustains the action in 1877 it will lay the basis for all subsequent regulation of U.S. business by government.

Parliament abolishes purchase of British Army commissions (*see* Cardigan, 1854).

Germany goes on the gold standard December 4.

The first commercial Russian oil wells are sunk in the Baku suburb of Balakany on the Apsheron Peninsula that juts into the Caspian Sea (*see* 1823; 1873).

Burmah Oil Co. has its beginnings in the Rangoon Oil Co. started by British colonial interests in Burma.

Brooklyn, N.Y., oilman Henry Huttleston Rogers, 31, patents machinery for separating naphtha from crude petroleum. He went into the Pennsylvania oil fields in 1861 and has been operating a Brooklyn refinery for the past 4 years with Charles Pratt, 41. Both will join Rockefeller's Standard Oil in 1874 and Rogers will push the idea of pipeline transportation which he has originated (*see* 1864).

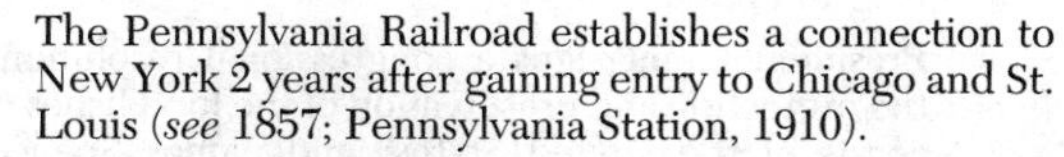

The Pennsylvania Railroad establishes a connection to New York 2 years after gaining entry to Chicago and St. Louis (*see* 1857; Pennsylvania Station, 1910).

The Texas Pacific Railroad that will end the Chisholm Trail cattle drives is chartered by Congress and is given a land grant but while the road's 1,800 miles of main fine will extend from New Orleans to El Paso via Shreveport, Texarkana, Dallas, and Fort Worth, it will never reach the Pacific (*see* 1885).

Texans drive 700,000 longhorns 700 miles from San Antonio north to the Abilene stockyards, moving at 12 miles per day through open, unsettled country that provides abundant grass and water for the livestock on the "Long Drive" up the Chisholm Trail (*see* 1870).

J. G. McCoy of Abilene has been unable to collect on his contract with the Kansas, Pacific whose officers say they never anticipated such volume, but McCoy has become mayor of Abilene and has profited in real estate and other enterprises (*see* 1870).

The New York ferryboat S.S. *Westfield* explodes July 11 blowing 104 people to pieces. Her boiler was so corroded that "a knife blade could cut through the metal" but steamboat inspection remains lax and there are no rigorous design or maintenance codes.

The first full shipload of bananas lands at Boston July 28 aboard Dow Baker's 85-ton schooner *Telegraph* 14 days out of Kingston, Jamaica. Seaverns & Co. sells the fruit on commission and the firm's young buyer Andrew Preston assures Baker that a market exists for all the bananas he can obtain (*see* 1870; 1885).

South American railroad builder Henry Meiggs wins a contract from Costa Rican president Tomas Guardia to construct a national railway from Port Limòn on the Caribbean 149 miles up to the capital of San José in the mountains. Still engaged in building railroads up and down the Peruvian coast and into the Andes (*see* 1861), Meiggs enlists the help of his nephew Minor Cooper Keith, 23, who will import 1,500 workers from New Orleans but will soon lose his 3 brothers to yellow fever (at least 4,000 men will die on the project). Keith will run out of money after building 60 miles of track, bananas he plants in the valley behind Limòn will establish a new source of funds, he will marry the daughter of ex-president José Maria Castro, and by 1883 he will be supplying bananas to companies in Costa Rica, Panama, Nicaragua, and Colombia, earning enough to complete the railroad in 1890 (*see* Meiggs, 1877; United Fruit, 1899).

The 1836 Colt revolver is redesigned to extend its effective range.

The "Mauser Model 1871" breech-loading rifle is adapted by the Prussian Army (above). Gunsmith Peter Paul Mauser, 33, will invent the Mauser magazine rifle in 1897.

Descent of Man and Selection in Relation to Sex by Charles Darwin concludes that man evolved from ape-like or monkey-like ancestors, probably 26 to 54 million years ago in Africa (*see* 1859; Scopes, 1925).

A prehistoric pterodactyl skeleton is identified for the first time by Yale paleontologist Othniel Charles Marsh, 39, the first U.S. paleontologist.

Japan reorganizes her ministry of education to promote universal, compulsory, fee-paid education.

Keio University, Japan's first private college, opens at Shiba Mita under the direction of educator Yukichi Fukuzawa, 37, who started a school at Edo in 1858, was sent by the Tokugawa shogun on a mission to America in 1860, and studied in Europe in 1861. The university will be certified as such in 1890.

Smith College is established at Northampton, Mass. The late Sophia Smith inherited a fortune from her brother Austin in 1861 and left the money at her own

1871 *(cont.)* death last year to found a college for women that will open in 1875.

The University of Arkansas is founded at Fayetteville.

Cables are laid from Vladivostok to Shanghai, Hong Kong, and Singapore via Nagasaki.

The first regular Japanese government postal service begins between Tokyo and Osaka.

Metal type and printing presses are introduced into Japan and the first Japanese daily newspaper begins publication.

Fiction *The Possessed* (or *The Demons*) (*Besy*) by Fedor Dostoievsky.

Juvenile *Tattered Tom* by Horatio Alger, who begins a new series; *Little Men* by Louisa May Alcott; *A Dog of Flanders* by Ouida.

Poetry *Obras de Gustavo Adolfo Bécquer* by the late Spanish poet who died of tuberculosis last December at age 34.

Painting *Max Schmitt in a Single Scull* by U.S. painter Thomas Eakins, 27; *Woodland Scene* by George Inness; *The Parthenon* by Frederic E. Church; *The Dream of Dante* by Dante Gabriel Rossetti.

British inventor Richard Leach Maddox, 55, makes an emulsion of silver bromide using gelatin in place of collodion (*see* Sayce and Bolton, 1864). Gelatin emulsions reduce the exposure time a photographer needs from 15 seconds to as little as 1/200th of a second and Maddox's lead will be followed by J. Kennett who will market a dry, washed emulsion which a photographer can dissolve in water and use as a coating on glass to produce his own photographic plates (*see* Bolton, 1874).

Theater *The Forest* (*Les*) by Aleksandr Ostrovsky 11/1 at St. Petersburg's Alexandrinsky Theater.

Barnum's Circus opens at Brooklyn, N.Y., and "The Greatest Show on Earth" grosses $400,000 in its first season. Showman P.T. Barnum has made a fortune promoting Jenny Lind (*see* 1850), will enlarge his show next year and make it the first circus to travel the country by rail (using 65 railcars), and by 1874 he will be playing to 20,000 people per day, charging 50¢ and taking in twice his huge $5,100-per-day operating cost (*see* Barnum & Bailey, 1881).

Opera *Indigo and the Forty Thieves* 2/10 at Vienna's Theatre-an-der-Wien, with music by Johann Strauss who includes his waltz "Ja, so singt man in der stadt wo ich geboren bin"; *Aida* 12/24 at the Cairo Opera House, with music by Giuseppe Verdi who completed it too late for the Suez Canal opening in 1869.

Stage musical *Thespis (or The Gods Grown Old)* 12/26 at London's Gaiety Theatre, with music by Arthur S. Sullivan, book and lyrics by London barrister William Schwenck Gilbert, 35. The burlesque is the first Gilbert and Sullivan collaboration.

Hymn "Onward Christian Soldiers" by Arthur S. Sullivan, lyrics by London clergyman Sabine Baring-Gould, 37.

Popular songs "Chopsticks" ("The Celebrated Chop Waltz") by English composer Arthur de Lulli (Euphemia Allen), 16.

London's Albert Hall (The Royal Albert Hall of Arts and Sciences) opens with seating for 6,036.

England's *Livonia* loses 4 to 1 in its bid to regain the America's Cup in ocean racing. The U.S. defenders *Columbia* and *Sappho* win two races each.

Poker is introduced to Queen Victoria at a royal party in Somerset by U.S. ambassador to Great Britain Robert Cumming Schenck, 62, who shows the queen how to play and at her request writes down the rules, the first written codification of the game (*see* 1821). Phrases such as "ace in the hole," "bluff," "call a bluff," "stand pat," "four-flush," and "pass the buck" will be derived from poker.

The National Rifle Association is founded by some Union Army officers to encourage marksmanship and gun safety (Civil War soldiers were often poorly trained and barely able to use their weapons). The N.R.A. will become a potent political lobby of "sportsmen."

Fire consumes the Tuileries Palace at Paris during the "Bloody Week" in May (above) and much of the Louvres Palace is gutted.

The Chicago Fire that rages from October 8 to 9 destroys 3.5 square miles of the city, killing perhaps 250. The fire has allegedly been started by a cow kicking over a kerosene lantern in Dekoven Street.

Sparks from the Chicago Fire (above) start forest fires that destroy more than a million acres of Michigan and Wisconsin timberland. They continue from October 8 to October 14 and more than 1,000 die in the logging town of Peshtigo, Wis., and in 16 surrounding communities in the worst fire tragedy in the recorded history of North America.

Passenger pigeons nesting in Wisconsin occupy 750 square miles and will continue such mass nestings until 1878 despite the October fire (above; *see* 1878).

The U.S. Arctic whaling fleet is trapped at the end of August by the earliest winter in memory. All but 7 of 39 vessels are frozen fast in the ice, New Bedford loses $1.5 million in shipping, but the surviving ships are able to carry all 1,200 men of the fleet safely back to New England ports.

President Grant signs a congressional resolution "for the protection and preservation of the food fishes of the coasts of the United States" and names Spencer F. Baird to head a new U.S. Fish Commission (*see* shad, 1873).

U.S. wheat and flour exports total 50 million bushels, while corn exports total 8 million.

Germany raises tariff barriers against food imports in a move to remain independent of France or any other country in food supply. German wheat prices will nevertheless fall 27 percent in the next 20 years and will fall 30 percent in Sweden, ruining many farmers and encouraging emigration to America.

Land planted to grain in Britain will decline by more than 25 percent in the next three decades as the railroads open up the western United States, making it unprofitable for British farmers to compete.

C. A. Pillsbury & Co. is founded by Minneapolis miller Charles Alfred Pillsbury, 29, with his brother Fred, his father George, and his uncle John Sargent Pillsbury, who has been governor of Minnesota. The founder became a partner 2 years ago in a small flour mill, increased its daily capacity from 150 barrels to 200 by improving its equipment, has turned a profit of $6,000, and uses it to start the new firm (*see* Pillsbury's Best XXXX, 1872).

Margarine production begins in the Netherlands as butter merchants Jan and Anton Jurgens open the world's first fully operative margarine factory at Oss after acquiring rights from Hippolyte Mège-Mouries for 60,000 francs per year (*see* 1869; United States, 1881; Jurgens, 1883).

Huntington Hartford of the A&P sends emergency rail shipments of tea and coffee to Chicago, most of whose grocery stores have burnt in the great October fire (above). When the city is rebuilt, Hartford will open A&P stores (*see* 1869; 1876).

The Dominion of Canada created in 1867 takes its first census. The nation has 1,082,940 French, 846,000 Irish, 706,000 English, 549,946 Scots, and 202,000 Germans (*see* Ukrainians, 1891).

The U.S. population reaches roughly 39 million while Germany, (above) has 41 million, France 36.1 (with 1.8 million in Paris), Italy 26.8, Great Britain 26, Ireland 5.4, Japan 33, and Brazil 10 including 1.5 million slaves.

Abortion is punishable by up to 5 years in prison under the criminal code enacted by the new German Empire (above).

1872 The viceroy of India Lord Mayo is murdered by a Muslim fanatic February 8 while visiting a penal colony in the Andaman Islands in the Bay of Bengal.

The Spanish pretender Don Carlos, 24, enters Navarre in May but the new king Amadeo I routs his forces at Oroquista and forces him to take refuge in the Pyrenees (*see* 1873).

Sweden's Charles IV dies at Malmö September 18 at age 46 after a 13-year reign. His brother, 43, will reign until 1907 as Oscar II.

Ethiopia's late emperor Theodore is succeeded by the ruler of Tigre who will reign until 1889 as the king of kings Johannes IV.

Japan institutes universal military service.

Radical Republicans renominate President Grant despite the corruption of his administration; liberal Republicans nominate New York publisher Horace Greeley of the *Tribune*. Greeley gains support from civil service reformer Carl Schurz, from malcontent Republicans who have not shared in the spoils of war and of politics, and from Democrats, but Grant wins reelection with 56 percent of the popular vote (and 286 electoral votes to Greeley's 66). Greeley dies November 29 at age 61.

Germany expels her Jesuits June 25 by an imperial law that is part of the *Kulturkampf* against Catholics launched May 14 by Chancellor Bismarck.

Japan permits Buddhist priests to take wives and eat meat (*see* below).

Chiricahuas in Arizona Territory give up their resistance to the whites on the promise of a reservation separate from that of the Mescalero Apache. Cochise, now 60, signs a treaty with the Indian commissioner Gen. Oliver O. Howard and retires to the new Arizona reservation where he will die in June 1874.

Congress passes a law guaranteeing equal pay for equal work in U.S. federal employment after lobbying by feminist Belva Ann Lockwood, 42.

Spiritualist Victoria Claflin Woodhull, 34, announces her candidacy for the U.S. presidency. A protégé of Commodore Vanderbilt (below), Mrs. Woodhull has made a fortune in Wall Street as has her sister the "magnetic healer" Tennessee Celeste Claflin, 26, and has demanded that women be given the right to vote. But the cause of women's rights is damaged by *Woodhull and Claflin's Weekly* which defends abortion, advocates free love, and recommends licensing and medical inspection of prostitutes.

Susan B. Anthony and other women's rights advocates are arrested at Rochester, N.Y., for trying to vote in the election November 5.

Congress abolishes the federal income tax imposed during the Civil War (*see* 1861; 1881).

The Erie Ring of Wall Street speculators James Fisk and Jay Gould collapses following the January 6 shooting of Fisk by Edward S. Stokes, a business rival and a rival for the favors of Fisk's mistress, the actress Josie Mansfield. Fisk and Gould have been looting the Erie Railroad's treasury and Fisk's death draws attention to the corruption that has flourished under the Grant administration.

The *New York Sun* begins an exposure of the Crédit Mobilier formed in 1867 by Union Pacific directors. Congress votes December 2 to appoint a committee to investigate the Crédit Mobilier.

A Geneva court of arbitration holds Britain responsible for Civil War depredations by the *Alabama* and other Confederate cruisers in a September 14 ruling that awards the U.S. government $15.5 million in damages.

A Gunpowder Trade Association organized by E. I. du Pont de Nemours controls prices of blasting and hunting powder in a market glutted with war surplus powder. Du Pont supplied the Union armies with 4 million pounds of explosives, the company is rich with profits gained in the war, it undersells competitors who do not join the association, and it forces them to yield or go out of business (*see* 1861; 1899).

John D. Rockefeller's Standard Oil Trust refines 10,000 barrels of kerosene per day and is the largest operation

1872 *(cont.)* of its kind in the world (*see* 1870). In addition to rebates received from railroads, the Trust receives drawbacks—fixed rates that the railroads pay for every barrel of oil they carry for a Standard Oil competitor—and competitors pay five times the freight rate enjoyed by Standard Oil (*see* 1877).

Belgian electrician Zenobe Theophile Gramme, 46, perfects the world's first industrial dynamo, employing a ring winding of the same type invented independently by Italian physicist Antonio Pacinotti in 1860 (*see* Faraday, 1831).

Japan gets her first railway. The 18-mile line between Tokyo and Yokohama opens October 14 with ceremonies at the Shimbashi and Yokohama terminals. Financed by the Oriental Bank of England, it has been built by 16 English engineers.

A system of automatic electric signaling for railroads patented by Irish-American engineer William Robinson, 32, will be the basis for all modern automatic railroad block signaling systems.

The New York, New Haven and Hartford Railroad Company is organized with 188 miles of track that will grow to 1,800 miles. The New York and New Haven has merged with the Hartford and New Haven to create the new company and to end a competitive struggle that has ensued for nearly 20 years (*see* Shore Line route, 1893).

Boston radiates out from its City Hall to a distance of 2.5 miles, half a mile farther than in 1850, as a consequence of horsecar transportation which by 1887 will extend the congested urban area another 1.5 miles from City Hall (*see* 1852; subway, 1897).

The Chicago & Rock Island wins a government contract for all mail west of Chicago by beating the Chicago and North Western in a race between Chicago and Council Bluffs. The Rock Island 3 years ago bought an eight-wheel locomotive built for the 1867 Paris International Exposition by the Grant Locomotive Works of Paterson, N.J., and the much-publicized steam engine with her boiler encased in German silver, with solid silver headlights, handles, whistles, and pumps, steams the distance in 27 hours with a 19-year-old engineer in the cab and flagmen at every crossing to keep livestock from blocking the rail bed.

The brigantine *Mary Celeste* clears New York harbor in mid-November for Genoa with 10 men who will never be seen again. The ship will be discovered December 4 sailing on a starboard tack with cargo and stores intact but without a soul aboard, a "ghost" ship whose mystery will never be solved.

The Farquharson Rifle patented by English gunsmith John Farquharson is a single-shot rifle with a falling-block action. George Gibbs of Bristol begins production of the rifle that will be produced in numerous variations by Webley, Bland, Westley Richards, and various European firms.

Borax ore (calcium borate) deposits discovered near Columbus, Nev., will be called colemanite after prospector William Tell Coleman, 48, who makes the find with Francis Marion Smith, 26. They will organize Pacific Coast Borax Co. and gain virtually a world monopoly in the material used to tan leather and make glass, porcelain, enamel, and soap. Smith will acquire additional deposits in California's Death Valley and haul out ore in huge wagons that will inspire the trademark "Twenty-Mule Team Borax."

Commercial production of celluloid begins under a patent obtained in 1869 by Albany, N.Y., printer John Wesley Hyatt, 35, and Rockford, Ill., inventor Isaiah Smith Hyatt (*see* Parkes, 1855). The Hyatts have organized a manufacturing company, deriving the word "celluloid" from cellulose and "oid" (meaning like), they will obtain a trademark for the word next year and sell their celluloid as a substitute for ivory, horn, amber, tortoise-shell, and the like for use in billiard balls, piano keys, men's collars, buttons, dental plates, combs, and other items (*see* Eastman, 1889).

B.F. Goodrich is founded at Akron, Ohio, by rubber maker Benjamin Franklin Goodrich, 30, with backing from local merchants. His first product is firehose to replace leather hose that cracks when frozen (*see* tires, 1899).

Pirelli and Co. has its beginnings at Milan where Italian entrepreneur Giovanni Battista Pirelli opens a shop that will grow to become a giant rubber company with large plantations in Java and factories worldwide.

U.S. inventor Luther Chicks Crowell patents a machine to make flat-bottomed paper bags. It will be improved in 1883 by a machine that will make the automatic, or "flick," bag, a self-opening bag with a flat bottom and side pleats (*see* Union Bag, 1869).

German botanist Ferdinand Julius Cohn, 44, publishes the first major work on bacteriology.

Louis Pasteur publishes a classic paper on fermentation showing that it is caused by microorganisms.

Stetigkeit und irrationale Zahlen by German mathematician Julius Wilhelm Richard Dedekind, 41, presents a theory of ideal numbers.

A cuneiform tablet deciphered by British Museum assistant George Smith, 32, bears the Gilgamesh legend of 3,000 B.C. with an Assyrian account of a great flood that conforms closely with the biblical account. Smith announces his discovery in December at a meeting of British archaeologists (*see* 1837; 1846).

The Ebers papyrus discovered at Thebes by German Egyptologist Georg Moritz Ebers, 35, is the oldest known compendium of ancient Egyptian medical writings.

"The Object and Manner of Our Lord's Return" by Pittsburgh evangelist Charles Taze Russell, 20, announces that the second coming of Christ will occur in the fall of 1874 without the awareness of mankind. Russell founds the International Bible Students' Association (*see* 1878).

Vanderbilt University is founded at Nashville, Tenn., with a grant from Commodore Vanderbilt of the New York Central.

The University of Oregon is founded at Eugene.

The University of Toledo is founded in Ohio.

Strasbourg University is founded in Alsatia.

Japan's new Meiji government issues a decree in August requiring compulsory education.

The *Boston Daily Globe* begins publication March 4 with an eight-page edition that is twice the size of any of the city's other 10 dailies. Four Boston papers have morning and evening editions, most sell for 4¢ a copy, and combined circulation is less than 170,000 in a metropolitan area of 500,000. Even the 26-year-old *Boston Herald* at 2¢ has a circulation of no more than 17,000.

Western Electric is founded April 2 to sell telegraph equipment and pursue experiments on an electric telephone. Company president Anson Stager gets sales support from Enos Melancthon Barton, 29, while cofounder Elisha Gray, 37, works on his telephone (*see* Bell, 1876).

An Australian telegraph line opens November 23 to connect Adelaide with Port Darwin and is soon extended to link Australia with Java, India, and Europe.

Fiction *Middlemarch* by George Eliot; *Under the Greenwood Tree* by English novelist-architect Thomas Hardy, 32; *Erewhon* by English satirist Samuel Butler, 37, is a utopian novel about a country devoid of machinery (the novel is issued anonymously); *Through the Looking Glass* by Lewis Carroll; *The Fiend's Delight* and *Nuggets and Dust Panned Out in California* by former *San Francisco News Letter* editor Ambrose Gwinnett Bierce, 30, who has left for a 3-year stay at London.

Painting *The Artist's Mother* (*Mrs. George Washington Whistler, or Arrangement in Grey and Black No. 3*) by James A. McNeill Whistler; *Le Foyer de la Danse* by French painter Hilaire German Edgar Degas, 38; Gustave Doré paints London scenes; George Catlin dies at Jersey City, N.J., December 23 at age 76.

Theater *Family Strife in Hapsburg* (*Ein Bruderzwist in Habsburg*) by the late Franz Grillparzer (who has died 1/21 at age 81) 9/24 at Vienna's Burgtheater; *The Woman of Arles* (*L'Arlésienne*) by Alphonse Daudet 10/1 at the Vaudeville Théâtre, Paris (*see* Bizet music, below); *The Jewess of Toledo* (*Die Jüdin von Toledo*) by the late Franz Grillparzer 11/22 at Prague.

Sarah Bernhardt (*née* Rosine Bernard), 28, begins a 10-year career with the Comédie-Française at Paris. "The Divine Sarah" will gain world renown and continue acting even after having a leg amputated in 1914.

Italian actress Eleanora Duse, 13, begins a 27-year career by appearing at Verona as Juliet in the Shakespeare tragedy of 1595.

Motion picture pioneer Eadweard Muybridge, 42, takes a sequence of photographs showing a horse running. Central Pacific Railroad president Leland Stanford has asked the English-American photographer to prove photographically that all four feet of a running horse are off the ground at the same time at some point in the animal's stride (*see* 1895).

Opera *Djamileh* 5/22 at the Opéra-Comique, Paris, with music by Georges Bizet, libretto from the Alfred de Musset poem *Namouna; La Princess Jaune* 6/12 at the Opéra-Comique, Paris, with music by Camille Saint-Saëns.

First performances *La Rouet d'Omphale* (*Omphale's Spinning Wheel*) symphonic poem by Camille Saint-Saëns 4/14 at Paris; Mass in F minor by Anton Bruckner 6/16 at Vienna; Incidental Music to *L'Arlésienne* (above) by Georges Bizet 10/1 at the Vaudeville Théâtre, Paris.

Blackjack licorice-flavored chewing gum, introduced by Staten Island, N.Y., photographer Thomas Adams, is the first chewing gum to be made from chicle and the first to be sold in stick form. Former Mexican president Antonio de Santa Anna, now 77, talked Adams into buying a ton of chicle in 1869, Adams has experimented fruitlessly with a project to make a rubber substitute from the latex of tropical trees such as the sapodilla, he has recalled that Indians used to chew chicle, has added licorice flavor to the gray gum, and goes into competition with State of Maine Pure Spruce gum magnate John B. Curtis who has been selling his gum for 20 years. Adams will sell his Tutti-Frutti gum on station platforms of the New York El beginning in 1888, employing the first vending machines for chewing gum (*see* Wrigley, 1892).

The Jesse James gang robs its first passenger train in June after hearing that the Rock Island's eastbound train from Council Bluffs, Iowa, will be carrying $75,000 in gold. The robbers part the rails in front of the oncoming train just west of Adair on the main line between Council Bluffs and Des Moines. Engineer John Rafferty is unable to stop the train in time and is killed when his locomotive is derailed, but the gold has been delayed for another train and the James gang gets away with a mere $3,000 from the express messenger plus $3,000 taken from passengers (*see* 1865; Northfield, Minn., 1876).

Congress enacts the first U.S. consumer protection law, making it a federal offense to use the mails for fraudulent purposes. Post Office Chief Special Agent P. A. Woodward proceeds under the new law to prosecute "professional cheats" who are working the most barefaced operations, mostly through the mails, without fear of punishment.

Montgomery Ward Co. is founded by Chicago mail-order pioneer Aaron Montgomery Ward, 29, whose savings were nearly consumed by last year's fire. He scrapes together $1,600 and goes into business with George R. Thorne who puts up $800. Setting up shop in a 12 by 14-foot loft over a livery stable at 825 North Clark Street, Ward and Thorne call themselves "The Original Grange Supply House" and prepare a one-sheet "catalogue" to test Ward's idea that U.S. farmers, hurt by low farm prices, will respond to bargain offerings that can be shipped by rail. They list some 50 dry

1872 ***(cont.)*** goods items, all priced at $1 or less, with savings of 40 percent. Ward will offer 10 days' grace on orders from Grange officials or countersigned with the Grange seal (*see* 1867; 1884; Sears, 1886).

Bloomingdale's has its beginnings in New York's Great East Side Store opened at 938 Third Avenue by merchant Lyman G. Bloomingdale, 31, and his brothers Joseph and Gustave (*see* 1886).

Simpson's opens in Toronto's Yonge Street where Scots-Canadian merchant Robert Simpson will have 13 clerks by 1880, rivalling T. Eaton Co. and the much older Hudson's Bay Co.

Romanesque architecture is introduced to America by Louisiana-born architect Henry Hobson Richardson, 34, who designed Boston's Brattle Street Church in 1870 and this year designs Trinity Church.

Olana on the Hudson River just south of Hudson, N.Y., is completed for painter Frederic E. Church who 4 years ago traveled to Constantinople and has commissioned architect Calvert Vaux of Central Park fame to draw up an Islamicized version of a plan made by Beaux Arts architect Richard Morris Hunt in 1867. A studio will be added in 1888 to the massive building of Persian-patterned bricks.

Boston suffers a $75 million fire that begins November 9 with an explosion in a four-story warehouse stocked with hoop skirts at the corner of Summer and Kingston Streets. The fire destroys the richest part of the city, burning 776 buildings that include granite and brick warehouses filled with merchandise and consuming 65 acres bounded by Milk and Washington Streets and Atlantic Avenue before burning itself out November 11. By 1876 the area will have been rebuilt more substantially than ever.

Yellowstone National Park is created by a March 1 act of Congress setting aside a 2-million-acre tract of wilderness on the fast-developing frontier in Wyoming Territory. Superintendent of the first national park and the first effective wildlife sanctuary is explorer Nathaniel Pitt Langford, 39, who 2 years ago made an expedition into the area.

The New York State Forest Commission halts sales of forest lands to commercial interests (*see* 1892).

U.S. Grangers form cooperative ventures to purchase farm implements, machinery, feed, seed, clothing, and other necessities. Some local Granges adopt the English Rochedale plan of 1844 and in Iowa one-third of all grain elevators and warehouses are Granger-owned or controlled (*see* 1867; 1873).

A bumper U.S. corn crop leads to the start of a stock feeder cattle industry in the Midwest.

The Burbank potato developed by Massachusetts horticulturist Luther Burbank, 23, from a chance seedling is an improved variety that will provide Burbank with funds for developing other new varieties. He will introduce not only new potato varieties but also new tomatoes, asparagus, sweet and field corn, peas, squash, apples, cherries, nectarines, peaches, quinces, ornamental flowers, and—most especially—plums, prunes, and berries (*see* 1875).

The Chicago Board of Trade moves into a building of its own at LaSalle and Washington (*see* 1865; 1930).

A strict Adulteration of Food, Drink and Drugs Act amends Britain's pure food laws of 1860, making sale of adulterated drugs punishable and making it an offense to sell a mixture containing ingredients added to increase weight or bulk without advising the consumer (chicory in coffee is a case in point) (*see* 1875).

Parliament passes a Licensing Act that sets strict limits on the number and kinds of places where alcoholic beverages may be sold and the hours such places may be open in an effort to curb excessive drunkenness. Riots occur in some places, the Liberal party legislation rallies publicans and brewers to the Conservative party, and the unpopularity of the new law will help topple the Gladstone government in the general elections of 1874.

British sugar consumption reaches 47 pounds per capita, up from 12 pounds per year in 1780.

Japan's Meiji emperor Mutsuhito starts a fad for beef eating among his more affluent countrymen but most Japanese cannot bear the smell of people who have eaten animal fats and few can afford beef (*see* 1868; priests, above; Townsend Harris, 1856).

Armour & Co. installs the world's largest chill room in a new plant at Chicago's Union Stock Yards (*see* 1865; 1869). Salt curing has been the chief way to keep perishable meats from spoiling and meat processing has been a seasonal business, but Armour employs a new method that uses natural ice to maintain operations year round (*see* refrigerated railcars, 1869).

Porcelain rollers are installed in a new flour mill built at Glasgow, Scotland (*see* 1870; Minneapolis, 1874).

Pillsbury's Best XXXX Flour is introduced by C.A. Pillsbury Co. at Minneapolis (*see* 1871; 1873).

A doughnut cutter patented by Thomaston, Me., inventor John F. Blondell has a spring-propelled rod to push out the dough of a center tube but most doughnuts continue to be made by hand (*see* 1847; Donut Corp. of America, 1920).

The Café de la Paix opens at Paris in the Grand Hotel on the Boulevard des Capucines with carved pillars and landscaped ceilings.

The Paris resturant Brasserie Lipp opens on the Boulevard Saint-Germain-de-Pres under the management of an Alsatian who names it after a famous eating place in Strasbourg.

Mennonite farmers in the Crimea send four young men on a scouting expedition to America after learning that Czar Aleksandr II plans to cancel their exemption from military service, freedom of worship, and right to have their own schools and speak their German language—rights granted by Catharine the Great in 1783. The young men are from well-to-do families in the Molotschna (Milk River) district and are led by Bernard

Warkentin, Jr., 23, a miller's son from the village of Tjerpenije. He establishes a base at Summerfield, Illinois, near East St. Louis, where some Mennonite families have settled earlier, and attends a local college for a year to improve his English. After traveling with his colleagues through Pennsylvania and west to Wyoming, and from Texas through the Dakotas into Manitoba, Warkentin will write letters home persuading many of his countrymen to resettle, and while most will go to Manitoba because of liberal laws regarding military service many will settle in Kansas (*see* agriculture, 1874).

The Ebers papyrus (above) contains a formula for a tampon medicated to prevent conception.

The "Comstock Law" enacted by Congress November 1 makes it a criminal offense to import, mail, or transport in interstate commerce "any article of medicine for the prevention of conception or for causing abortion." The law takes its name from New York moralist Anthony Comstock, 28, who heads the Society for the Suppression of Vice (*see* Mme. Restell, 1878).

1873 German occupation forces begin to evacuate France following the death January 9 of the exiled Napoleon III at age 64. Parlement's monarchist majority condemns the Thiers government, Thiers resigns May 24, he is succeeded promptly by Marshal MacMahon, now 64, who is elected almost unanimously, and the last German soldiers cross back into Germany September 16 (*see* 1877).

The first Spanish Republic is proclaimed February 12 by the radical majority in the cortes elected last year, Amadeo I abdicates the same day after a 2-year reign in which he has been opposed as a "foreigner," the foreign minister Emilio Cistelar y Ripoli, 40, becomes prime minister and tries to organize a centralized republic, but Carlist civil war continues (*see* 1872; 1874).

President Grant's outgoing vice president Schuyler Colfax has been implicated in the Crédit Mobilier scandal that came to light last year. He was not renominated, and further evidence of corruption in the Grant administration continues to emerge.

Prince Edward Island in the Gulf of St. Lawrence joins the 6-year-old Dominion of Canada July 1.

A Canadian order-in-council August 30 establishes the Northwest Mounted Police. The "mounties" are to stop trading in liquor and arms with the Indians of the Northwest Territory.

Canada's first prime minister Sir John Macdonald resigns November 7 following charges of corruption in the last election and reports of a scandal involving the projected transcontinental railway. Sir John will be reelected in 1878 and serve until his death in June of 1891.

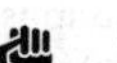

The Victoria government at Melbourne passes Australia's first factory act to protect female and juvenile mill hands and maintain safe and sanitary working conditions.

Zanzibar's public slave markets close June 5 by order of the sultan Barghash Sayyid who acts under British government pressure to prohibit the export of slaves (*see* Livingstone, 1871; Berlin Conference, 1884).

Port Moresby is founded on the Pacific island of New Guinea by English naval commander Sir Fairfax Moresby, 87, who has been admiral of the fleet since 1870 and who half a century ago suppressed the slave trade at Mauritius. Moresby raises the British flag on the island's South Coast but London refuses to support Moresby unless Australia's colonial governments agree to assume responsibility for the territory's administration (when some colonies demur, London disavows Moresby's action).

The Fourth Coinage Act passed by Congress February 12 makes gold the sole U.S. monetary standard, establishes the Bureau of the Mint, places all mint and assay office activities under control of the new Treasury Department bureau, stops coinage of 412.5-grain silver dollars, and authorizes only 420-grain trade dollars for export. The act inadvertently makes trade dollars legal tender in amounts up to $5, but a joint resolution of Congress in late July 1876 will deprive the trade dollar of its legal status (*see* Bland-Allison Act, 1878).

Demonitization of silver (above) comes amidst new silver discoveries in Nevada. Mining engineers John William Mackay, 42, James Graham Fair, 41, James C. Flood, and William T. O'Brien open the new Bonanza vein February 5 in the Panamint Mountains and it is the biggest silver strike since the Comstock Lode of 1859. Fair was named superintendent of the Ophir Mine in 1865, the other partners took over the Hale & Norcross Mine a year later and then took over the Consolidated Virginia & California Mine, Fair joined them in 1868, and the vein they open will yield $150 million for Mackay and Fair, who will establish the Bank of Nevada with Flood at San Francisco in 1878.

The Comstock Lode has yielded more than $120 million in silver, $80 million in gold (a millionaire miner named John Percival Jones has become a U.S. Senator from Nevada). The U.S. silver yield reaches $36 million for the year, up from $157,000 in 1860, and will continue to mount but gold production is declining from its 1854 peak.

"The country is fast becoming filled with gigantic corporations wielding and controlling immense aggregations of money and thereby commanding great influence and power," reports a congressional investigating committee: "It is notorious in many state legislatures that these influences are often controlling."

Germany adopts mark coinage.

Financial panic strikes Vienna in May and soon spreads to other European money centers.

Britain's "golden age" comes to an end after 23 years as Germany and the United States challenge British industrial preeminence.

European investors withdraw capital from the United States, the Wall Street banking house Jay Cooke & Co. that has been financial agent for the Northern Pacific Railway fails September 18, Black Friday on the stock

1873 *(cont.)* exchange sends prices tumbling September 19, the exchange closes for 10 days, and by year's end some 5,000 business firms have failed, millions of working Americans are obliged to depend on soup kitchens and other charities, and tens of thousands come close to starvation.

The Greenback party organized by Midwesterners claims that a shortage of money is the cause of hard times. If the government will issue greenbacks until as much money is in circulation as in the boom times of 1865 the country will again enjoy prosperity, say the Greenbackers. (They also demand control of the corporations, honesty in government, conservation of natural resources, and wide-scale reforms.)

The U.S. financial panic enables Henry Clay Frick to acquire most of the coal and coke land in the region of Connellsville, Pa., that can be operated at a profit and when Pittsburgh steel mill operators discover that Connellsville coke is the best coke for steel making, the price of coke will rise from $1 per ton to $5 (*see* 1870). Frick will have gained control of 80 percent of the Connellsville coke output, will be a millionaire by age 30, will be offered a general managership by steel magnate Andrew Carnegie, and will organize the Carnegie Co., whose basic unit will be the Homestead Works (*see* 1892).

Treatise on Electricity and Magnetism by Scottish physicist James Clerk Maxwell, 42, describes properties of the electromagnetic field and gives equations that entail the electromagnetic theory of light (*see* Hertz, 1887).

U.S. physicist Henry Augustus Rowland, 24, at Rensselaer Polytechnic Institute in Troy, N.Y., discovers the magnetic effect of electric convection. Electricity drives machinery for the first time in history, at Vienna.

Wall Street panicked when Jay Cooke & Son failed. The stock exchange closed for 10 days; 5,000 business firms failed.

Thaddeus Lowe of 1866 artificial ice fame discovers a process for manufacturing water gas that will greatly enhance use of gaslight illumination (*see* coke oven, 1897; Welsbach mantle, 1885).

Russia's Baku oil fields increase production as Alfred Nobel of 1866 dynamite fame and his brother Ludwig invest capital to build a refinery that will make Baku the world's leading petroleum producer (*see* 1871; 1901).

Cornelius Vanderbilt of the New York Central gains control of the rail line between New York and Chicago by leasing the Lake Shore and Michigan Southern (*see* 1869; 1877).

The financial panic (above) deals a heavy blow to Ben Holladay, who acquired the Russell, Majors and Waddell transcontinental freight firm in 1862, made a fortune before selling out in 1866, and has sold his steamship interests to finance a new Oregon Central Railroad. His German bondholders will force Holladay out of the company in 1876 and he will be financially ruined.

San Francisco's cable streetcar (the world's first) goes into service August 1 on Clay Street hill. English-American engineer Andrew Smith Hallidie, 37, has persuaded the city fathers to install the "endless-wire rope way" he patented in 1871 to cope with the hilly streets that defy San Francisco's eight horse-drawn lines. San Francisco will install additional cable-car lines; Los Angeles, Seattle, Denver, and Omaha will follow suit (*see* 1894; Sprague, 1888).

Rochester, N.Y., lawyer George Baldwin Selden, 27, experiments with internal combustion engines in an effort to develop a lightweight engine that will propel a road vehicle that is more efficient than the "road locomotives" now used in some farm jobs (*see* 1876; Lenoir, 1860; Otto, 1866).

The White Star liner S.S. *Atlantic* runs aground April 1 off Halifax, Nova Scotia. The 420-foot steamship founders and 502 of her 931 passengers are lost.

"Typhoid Fever: Its Nature, Mode of Spreading, and Prevention" by English physician William Budd proves the contagious nature of the disease (*see* cholera, 1866; Eberth, Koch, 1880).

Father Damien goes to the government hospital for lepers on the Hawaiian island of Molokai to care for victims of the disease that will be called Hansen's disease (*see* Hansen, 1874). Belgian missionary Joseph Damien de Veuster, 33, will contract the disease himself, will devote the rest of his life to the work, but will die in 1889.

New York's Bellevue Hospital establishes the Nightingale System of nurses' training (*see* Nightingale, 1854).

The University of California closes its Oakland campus and opens new ones at Berkeley and San Francisco (*see* 1857). The first baccalaureates will be awarded at Berkeley, which will become a leading research institution (*see* 1907).

The University of Cape Town is founded in South Africa.

Cable cars climbed San Francisco's steep hills with ease, sparing the horses. Other cities copied the idea.

The University of South Africa is founded at Pretoria.

St. Nicholas Magazine begins publication under the editorship of Mary Elizabeth Mapes Dodge of 1865 *Hans Brinker* fame who will run the new periodical for children until her death in 1905 at age 74. Dodge will attract work by such poets as Bryant, Longfellow, and Whittier.

Nonfiction *Studies in the History of the Renaissance* by Oxford classicist Walter (Horatio) Pater, 34.

Fiction *The Gilded Age* by Mark Twain and *Hartford Courant* editor Charles Dudley Warner exposes corruption in U.S. business and politics since the Civil War; *Marjorie Daw and Other People* by Thomas Bailey Aldrich; *Around the World in Eighty Days (Le tour du monde en quatre-vingt jours)* by Jules Verne *(see* Nellie Bly, 1889); *The Eustace Diamonds* by Anthony Trollope.

Poetry *Les Amours jaunes* by French poet Tristan Corbière (Edouard Joachim Corbière), 28.

Painting *Le bon Bock* by Edouard Manet; *Souvenir d'Italie* by J. P. C. Corot, now 77; *Oarsmen on the Schuylkill* by Thomas Eakins; Sir Edwin H. Landseer dies at London October 1 at age 71.

German photochemist Hermann Wilhelm Vogel, 39, discovers that adding certain dyes to photographic emulsions will increase their sensitivity.

Gelatin emulsions go on sale at London to permit any photographer "to prepare dry plates equal in sensitiveness to the best wet plates by simply pouring on the glass an emulsion and allowing it to dry," but the product introduced in July by local photographer John Burgess is not a commercial success.

Opera *Ivan the Terrible* (*The Maid of Pskov*) 1/1 at St. Petersburg's Maryinski Theater, with music by Nikolai Rimski-Korsakov; *Le Roi l'a Dit* 5/24 at the Opéra-Comique, Paris, with music by Leo Delibes.

First performances Concerto No. 1 in A minor for Violoncello and Orchestra by Camille Saint-Saëns 1/19 at the Paris Conservatoire; Symphony No. 5 by the late Franz Schubert (who wrote it in 1816) 2/1 at London's Crystal Palace; Symphony No. 2 in C minor by Anton Bruckner 10/26 at Vienna; *Variations on a Theme by Haydn* by Johannes Brahms 11/2 at Vienna; *The Tempest* (Fantasy for Orchestra after Shakespeare) by Petr Ilich Tchaikovsky 12/19 at Moscow.

Popular songs "Silver Threads Among the Gold" by English composer Hart Pease Danks, 39, lyrics by Eban E. Rexford, 25; "Home on the Range" lyrics by Kansas homesteader Bruce (Brewster) Higley appear in the December *Smith County Pioneer* under the title "Oh, Give Me a Home Where the Buffalo Roam." Kansas guitarist Daniel E. Kelly writes a tune to the lyrics, U.S. composer William Goodwin will revise it, and it will be published in 1904 under the title "Arizona Annie."

The Preakness has its first running at Baltimore's Pimlico racetrack where the 1 3/16-mile stakes race for 3-year-olds has been named for the son of Lexington that won the Dinner Party stakes on the day the track opened in 1870. The bay colt Survivor wins the first Preakness, its owner gains possession for 1 year of the Woodlawn Vase created in 1860 by New York's Tiffany and Co., the race's namesake will be shipped to England, the Duke of Hamilton will buy the horse, and he will shoot Preakness dead in a pique (*see* Belmont Stakes, 1867; Kentucky Derby, 1875).

French perfume makers at Grasse revolutionize the industry with a new process that extracts a solid essence from flower roots or from fragrant substances by placing them in contact with volatile fluids which dissolve essential ingredients and isolate them as they evaporate.

The U.S. Fish Commission carries shad across the continent and introduces it into the Pacific, but the fish will be popular on the West Coast only for its roe (*see* 1871; salmon, 1875).

DDT (dichloridiphenyl-trichlorethane) is prepared by German chemistry student Othmar Zeidler at Strasbourg who reacts chloral hydrate (the "Mickey Finn" knockout drops of the underworld) and chlorobenzene in the presence of sulfuric acid. Zeidler will describe DDT next year in the *Proceedings of the German Chemical Society (Berichte der Chemischen Gesellschaft)* but he has no idea of the significance of his discovery (*see* 1939).

Retired London silk merchant George Grant brings four Aberdeen Angus bulls from Scotland to his farm at Victoria, Kan., the first of the breed to reach the United States. Foreign investment capital is pouring into the U.S. cattle ranching industry (*see* 1881).

Barbed wire exhibited at the De Kalb, Ill., county fair by Henry Rose is studied by local farmer Joseph Farwell Glidden, 60, and his friend Jacob Haish who independently develop machines for producing coil barbed

1873 ***(cont.)*** wire by the mile and obtain patents for two separate styles of the "devil's rope" that is destined to end the open range in the West; 80.5 million pounds of barbed wire will be manufactured in the next 74 years as the steel wire becomes important not only to farmers and ranchers but also to military operations (*see* 1867; Gates, 1875).

Rep. Ignatius Donnelly, 42 (R. Minn.), complains that it costs as much to ship wheat from Minneapolis to Milwaukee as to ship the same wheat from Milwaukee to Liverpool. U.S. railroads are making deals with elevator companies, commission agents, and others to control both shipping and marketing rates and forcing farmers to sell to the nearest elevator company, agent, or railroad at rates favorable to the buyer. Grain is often downgraded as No. 2 because it is said to be wet, frozen, or weedy, and is then sold as No. 1 grade to millers, say the farmers.

A farmers' convention at Springfield, Ill., attacks monopolies, calling them "detrimental to the public prosperity, corrupt in their management, and dangerous to republican institutions."

The Northwest Farmer's Convention at Chicago urges federal regulation of transportation rates, government-built and government-owned railroads, an end to corporation subsidies and of tariff protection for industry, a revision of the credit system, and the encouragement of decentralized manufacturing, demands that bankers, industrialists, and railroad operators call "un-American."

The Grange reaches its membership peak of 750,000 but Ignatius Donnelly (above) compares a nonpolitical farmers' organization to a gun that won't shoot (*see* 1867; 1876).

Good quality Werner harvesters made by a Grange factory in Iowa sell for half the price of a McCormick harvester. When 22 plow manufacturers agree not to sell plows to Granges except at retail prices, some Granges set up factories to make their own machines and break the "machinery rings" (*see* 1872; 1875).

Louisiana's sugar crop falls to less than one-third its 1853 level as a result of the Civil War and emancipation. Sugar cane in many areas is replaced by rice which demands less labor.

The Pekin duck introduced to Long Island March 14 by Stonington, Conn., sea captain James E. Palmer begins an industry. The white birds obtained from the imperial aviaries at Beijing are a variety of mallard.

Bengal suffers a famine as rice crops fail.

Coors Beer has its beginnings in the Golden Brewery started by German-American brewer Adolph Herman Joseph Coors, 26, and Jacob Schueler at Golden, Colorado Territory.

St. Louis brewer Adolphus Busch becomes a full partner of his father-in-law and renames the Bavarian Brewery the E. Anheuser and Co.'s Brewing Association (*see* 1865; Budweiser, 1876).

Piracy and native hostility end U.S. pepper trade with Sumatra after 937 voyages (*see* 1797, 1805).

Swedish engineer Carl Linde introduces the first successful compression system using liquid ammonia as a refrigerant (*see* Perkins, 1834; Gorrie, 1842; Lowe, 1865; liquefied air, 1898).

Poultry, fish, and meat that has been frozen for 6 months is eaten at a public banquet in Australia (*see* 1856; 1925).

The thermos bottle has its beginnings in the Dewar vessel invented by Scottish chemist James Dewar, 31, who will invent the thermos flask in 1892.

The Paris Council of Hygiene rules that margarine may not be sold as butter (*see* 1869; New York, 1877).

Henri Nestlé's Infant Milk Food of 1866 is introduced in the United States (*see* 1875).

San Francisco's population reaches 188,000 and the city has a United States Mint, cobblestone streets, and tall redwood houses. Los Angeles remains a dusty town with fewer than 10,000 people.

1874

British troops under Sir Garnet Joseph Wolseley enter the Ashanti capital Coomassie February 4, ending the second Ashanti War. Wolseley, now 40, lost an eye at Sevastopol in 1855, served in the Sepoy Mutiny of 1857, and commanded an expedition to put down the Red River Rebellion in Canada 4 years ago. He sets Coomassie afire before withdrawing.

Britain annexes the Fiji Islands.

Britain's first Gladstone ministry ends February 17 as the Tories regain power. A second Disraeli ministry begins February 21 and will continue until 1880.

Japanese troops invade Taiwan in April. Tokyo excuses the invasion by citing the death of 54 Okinawan seamen who were killed in 1871 after a shipwreck. The Japanese send delegates to China to assure the Chinese that no offense is intended, but they assert claims to Okinawa and agree to recall the expedition from Taiwan in October only after China agrees to pay an indemnity (*see* 1895).

Spain's Alphonso XII, now 17, issues a manifesto from England's Sandhurst military school December 1 in response to a birthday greeting from his followers. No Alphonso has ever reigned over a united Spain but Alphonso XII proclaims himself the sole representative of the Spanish monarchy and begins a reign that will continue until his death in 1885.

A meeting of the unemployed held in New York's Tompkins Square January 13 to bring public attention to widespread poverty following last year's Wall Street collapse brings a charge by mounted police: hundreds are injured.

Massachusetts enacts the first effective 10-hour day law for women May 8 (*see* Supreme Court decision, 1908).

New York social worker Etta Angel Wheeler finds a little girl wandering naked through the city's slums after having been beaten, slashed, and turned out by her drunken foster mother. Appeals are made to the 8-

year-old ASPCA for lack of any other agency, the American Society for the Prevention of Cruelty to Animals decides that the child is an animal and deserving of shelter, the ASPCA prosecutes the foster mother for starving and abusing the 9-year-old girl, and the American Society for the Prevention of Cruelty to Children (ASPCC) is organized.

Federal troops at New Orleans put down a revolt by the White League against the black state government September 17.

Some 75 blacks are killed December 7 in race riots at Vicksburg, Miss.

Silver is found at Oro City (Leadville) in Colorado Territory where the placer gold found in 1859 has long since been exhausted. Metallurgist A.B. Wood recognizes silver-lead ores in material that gold miners have regarded merely as obstructions to their sluices. Active prospecting will begin in the spring of 1877 at Oro City, which will be renamed Leadville in 1878 (*see* 1879).

New York's R. H. Macy Co. displays its doll collection in the world's first Christmas windows, beginning a tradition that other stores will follow. Founder Macy will die at Paris in 1877 at age 55 (*see* 1867; Strauss brothers, 1888).

Sen. William Windom (R. Minn.) heads a committee that proposes a government-built, government-operated double-track freight line between the Mississippi Valley and the eastern seaboard that will prevent railroad companies from charging exorbitant freight rates (*see* 1877; farmers, 1873).

The first bridge to span the Mississippi at St. Louis is tested July 2 by seven 50-ton locomotives loaded with coal and water that chug slowly across from one bank while another seven chug across from the opposite bank, pausing at the slender junctures of the cantilevered arches on the steel arch Eads Bridge (*see* 1867). 300,000 residents and visitors crowd the levee to witness the collapse of "Eads's Folly" but the bridge holds all 700 tons. The crowd cheers, a 3-day celebration ensues with parades, band concerts, and speeches culminating in a $10,000 fireworks display, and the Eads Bridge opens to traffic July 4 with a 100-gun salvo.

Belgian-American inventor Charles Joseph Van Depoele, 28, demonstrates the practicality of electric traction (*see* trolley, 1885).

New York gets its first electric streetcar but the system invented by Stephen Dudley Field, 28, is hazardous and presents no immediate threat to the horsecar. One wheel of Field's streetcar—the first car to be run successfully with current generated by a stationary dynamo—picks up current carried by one of the rails. Conveyed to the car's motor, it flows back to the other rail via a second wheel, insulated from the first, and is returned to the dynamo (*see* 1888).

Norwegian physician Arrnauer Gerhard Henrik Hansen, 33, discovers the leprosy bacillus. The disease will hereafter properly be called Hansen's disease (*see* 1246; Father Damien, 1873).

The London School of Medicine for Women is founded by English physician Sophia Jex-Blake, 34, whose brother Thomas, 42, has just been named headmaster of Rugby. She will gain the legal right to practice medicine in Britain in 1877.

The Baltimore Eye & Ear Dispensary is founded by physicians who include local ophthalmologist Samuel Theobald, 28, who has introduced the use of boric acid for treating eye infections.

Improved surgical dressings are pioneered by East Orange, N.J., inventors Robert Wood Johnson, 29, and George J. Seabury who succeed in manufacturing an adhesive and medicated plaster with a rubber base (*see* 1885).

The U.S. public high school system wins support from the Supreme Court which rules against a citizen of Kalamazoo, Mich., who had brought suit to prevent collection of additional taxes. The court upholds the city's right to establish a high school and to levy new taxes to support the school.

The Chautauqua movement in U.S. education has its beginnings in a summer training program for Sunday School teachers started at Fair Point on Lake Chautauqua, N.Y., by Methodist bishop John Heyl Vincent and Akron, Ohio, farm machinery maker Lewis Miller: they begin an institution 10 miles from Lake Erie that will develop into a traveling tent show of lecturers that will bring culture to small-town America. President Grant's appearance at Chautauqua next year will lend prestige to the movement, more than 100,000 people will sign up for home-study correspondence courses by 1877, the Chautauqua Normal School of Languages will start in 1879, and the Chautauqua Press will list 93 titles by 1885 by which time there will be Chautauquas in more than 30 states (*see* Damrosch, 1909).

The University of Adelaide is founded in Australia.

The Remington typewriter introduced by F. Remington & Sons Fire Arms Co. begins a revolution in written communication. Philo Remington, 68, has headed his late father's company since 1868 (*see* 1845), he has acquired sole rights to the Sholes typewriter of 1868 for $12,000, but the $125 price of the Remington typewriter is more than a month's rent for many substantial business firms and Remington produces only eight machines (*see* 1876).

The *Winnipeg Free Press* begins publication in Canada.

Yomiuri Shimbun begins publication at Osaka.

Nonfiction *Encyclopaedia of Wit and Wisdom* by humorist Josh Billings who says, "It is better to know nothing than to know what ain't so."

Fiction *Far From the Madding Crowd* by Thomas Hardy; *Phineas Redux* by Anthony Trollope; "Nedda" by Sicilian short-story writer Giovanni Verga, 34, who pioneers the Italian version of naturalism that will be called *verismo; Die Messalinen Wiens* by German novelist Leopold von Sacher-Masoch, 38, from whose name the word "masochism" will be derived (*see* "sadism," 1791).

Early typewriters had capital letters only, and typists could not see what was being typed, but improvements came quickly.

Poetry *Romances sans paroles* by French poet Paul Verlaine, 30, who in 1871 left his wife after 18 months of marriage to live with the poet Arthur Rimbaud, now 20. Verlaine was encouraged and influenced by Rimbaud, but in July of last year shot Rimbaud in the wrist during a lovers' quarrel and is serving a 2-year prison sentence; *Une saison en enfer* by Arthur Rimbaud who has had his hopes dashed for a new, amoral society and for writing an unconventional new kind of poetry. Rimbaud will go into business as a North African merchant and trader, amassing a fortune before his death in 1891; *Ode* by English poet Arthur William Edgar O'Shaughnessy: "We are the musicmakers,/ And we are the dreamers of dreams. . ."

French "Impressionists" rejected by the Salon hold their first exhibition at Paris in an "independent" show of canvases which include a harbor scene entitled *Impression: Sunrise* by Claude Monet, now 40. His work prompts art critic Louis Leroy to call the painters "Impressionists," an epithet that will soon lose its pejorative overtone. The group includes Paul Cézanne, Edgar Degas, Edouard Manet, Manet's new sister-in-law Berthe Morisot, 33, Camille Pissarro, 44, Pierre Auguste Renoir, Alfred Sisley, 34, and beginning in 1879 will include American émigrée painter Mary Cassatt, now 30.

English photochemist William Blanchard Bolton, 26, shows that nitrates (by-products of the formation of silver halide and nitrate) can be washed out of photographic emulsions through a process that will be used hereafter in developing (see 1864; Maddox, 1871; Seed, 1879).

Theater *Libussa* by the late Franz Grillparzer 1/21 at Vienna's Burgtheater; *The Two Orphans* (*Les Deux orphelins*) by French playwright Adolphe D'Emery, 62, 1/29 at Paris.

Opera *Boris Godunov* 1/24 at St. Petersburg's Maryinsky Theater, with music by Russian composer Modest Petrovich Mussorgsky, 34, libretto from the Aleksandr Pushkin play of 1831; *Die Fledermaus (The Bat)* 4/5 at Vienna's Theater-an-der-Wien, with music by Johann Strauss.

First performances Overture to the *Piccolomini (Max and Thekta)* by French composer (Paul Marie Theodore) Vincent d'Indy, 22, 1/25 at Paris; *Patrie* Overture in C minor by Georges Bizet 2/15 at Paris; *Requiem* by Giuseppe Verdi 5/22 at Milan's Church of San Marco. Verdi has written the work to memorialize the late poet-novelist-patriot Alessandro Manzoni, who died May 22 of last year at age 88.

Lawn tennis is patented under the name Sphairistike by British sportsman Walter Clopton Wingfield, 41, who has codified rules for a game played indoors for at least five centuries. The new game is introduced in Bermuda and from there into the United States.

New York's Madison Square Garden opens in April under the name Barnum's Hippodrome at the north end of the city's 38-year-old Madison Square Park on Fifth Avenue. Showman P. T. Barnum has taken over a shed used until 1871 as a freight depot for the New York and Harlem Railroad, he has spent $35,000 to remodel the roofless structure, he will sell his lease in the winter to Patrick S. Gilmore, Gilmore will rename it Gilmore's Garden and will use it for flower shows, policemen's balls, America's first beauty contest, religious and temperance meetings, and the first Westminster Kennel Club Show, while Barnum will pitch his circus tent at Gilmore's Garden each spring (*see* 1879; Gilmore, 1863; Barnum, 1871; Westminster, 1877).

The first real football game May 14 at Boston is a variation of rugby (*see* 1823; Princeton vs. Rutgers, 1869). Harvard and McGill field teams of 11 men each who run with the ball as well as kicking it, Harvard wins the "Boston Game," a second match is held May 15, a third played at Montreal October 22 ends with Harvard winning 3 to 0, but scrimmage lines and "downs" will not be introduced until the 1880s (*see* Harvard-Yale game, 1875).

Levi Strauss blue jeans get copper rivets as the result of a joke about a prospector named Alkali who carries rock specimens in his pockets and has had his bibless overalls riveted by a blacksmith. The riveted denims sell at $13.50 per dozen (*see* 1850; 1937).

Texas gunman John Wesley Hardin celebrates his 22nd birthday May 26 and is confronted by Brown County Sheriff Charles Webb, who says he comes in peace. Hardin, drunk, turns toward the bar, Webb starts to draw, a bystander shouts a warning, Hardin whirls and fires before Webb's gun is out, and the sheriff falls. Hardin has killed about 20 men but never stood trial. He will be captured at Pensacola, Fla., in 1877 and charged with killing Webb. Convicted of second-degree murder, he will serve 17 years of a 25-year sentence at hard labor before winning parole (*see* 1895).

The dome of Boston's Massachusetts State House is covered in gold leaf 13 years after its first gilding.

U.S. inventor William Baldwin improves the steam radiator by screwing short lengths of one-inch pipe into a cast-iron base, but mass production of cast-iron radiators will not come for another 20 years and central heating of U.S. homes and offices not until the turn of the century.

Turkey red wheat—hard, drought-resistant, and high-yielding—is introduced into the United States by German-speaking Mennonites from Russia's Crimea (*see* 1872). The Santa Fe Railroad has brought the Mennonites to Kansas, where the road has been granted 3 million acres of land along its right of way and needs farmers who will occupy the land and produce crops that will generate freight revenue. Santa Fe official Carl R. Schmidt went to Russia last year and brought over a Mennonite delegation to see possible sites for settlement, he has obtained passage of a law in the Kansas legislature giving exemption from military service to those who oppose war on religious grounds, has offered free passage to Kansas plus free transport of furniture, set up temporary living quarters, and the first Mennonites arrive August 16 at Hillsboro in Marion County. The 163 pioneers from 34 families pay in gold to buy 8,000 acres from the Santa Fe, they found the village of Gnadenau, a second group of 600 follows, then a third group of 1,100, and by fall the Mennonites are arriving by the thousands, each family bringing its Turkey Red seed wheat obtained originally from Turkey and planted in the Crimea for years (*see* 1895).

British farm wages fall, farm workers strike in the east of England, and an agricultural depression begins that will lead to an exodus of farm workers into the growing mill towns. Agriculture has been undercut by foreign producers of grain and meat and begins a long decline.

A decade of drought begins on part of the western U.S. cattle range while other parts enjoy plenty of rain (*see* 1886).

The first shipment of Montana cattle for the East arrives at the railhead at Ogden in Utah Territory where cattleman James Forges has driven it from the Sun River range of Conrad Kohrs.

George Perkins Marsh prepares a paper on the feasibility of irrigating Western lands at the request of the U.S. Commissioner of Agriculture (*see* Marsh, 1864). Irrigation projects are possible, writes Marsh, if they are undertaken on a river-basin scale after thorough hydrological surveys "under Government supervision, from Government sources of supply" (*see* Powell, 1878).

A Minneapolis flour mill employing fluted chilled steel rollers in addition to conventional millstones is opened by C. C. Washburn who has made a fortune in Wisconsin land speculation and served as governor of Wisconsin (*see* 1866; 1879; Pillsbury, 1878; Gold Medal Flour, 1880).

New technology improves food canning—a drop press introduced by Allen Taylor and a pressure-cooking "retort" either by A. K. Shriver or Baltimore canner Isaac Solomon (*see* 1861). Live steam keeps the outside walls of the can under pressures comparable to those exerted by the heating contents of the can, thus speeding up the cooking of the contents without permitting the can to buckle or burst as it cools because of any buildup in pressure during the heating process. The retort gives canners accurate control of cooling temperatures and will lead to a large-scale expansion of the industry (*see* Howe floater, 1876).

Margarine is introduced into the United States (*see* 1881; Mège-Mouries, 1869).

The ice cream soda is invented at the semi-centennial celebration of Philadelphia's Franklin Institute. Robert Green runs out of cream after making $6 per day selling a mixture of syrup, sweet cream, and carbonated water; he substitutes vanilla ice cream and by the time the exhibition ends is averaging more than $600 per day.

The Women's Christian Temperance Union (WCTU) is founded at Cleveland where 135 women meet November 18 at the Second Presbyterian Church and dedicate themselves to ending the traffic in liquor. Frances Caroline Elizabeth Willard, 35, is named corresponding secretary (*see* National Prohibition Party, 1869; Anti-Saloon League, 1895).

The first Hutterite immigrants to America arrive at New York from Europe (*see* 1528). The Bonhomme colony at Yankton, S. Dak., is founded by a group speaking Tyrolean Hutterische dialect and will be followed by 200 colonies in the Dakotas, Minnesota, Montana, Washington, and the Canadian provinces of Alberta, Manitoba, and Saskatchewan; the Hutterites now number 20,000.

1875

China's Tonghzi emperor Mu Zung dies in January at age 19 after a 12-year reign in which the dowager empress Cixi (Ziaoqin) was co-regent until 1873. She adopts her nephew Zaitian, a cousin of Mu Zung, who formally ascends the throne February 25 at age 4 and will reign until 1908 as the Guangxu emperor.

The murder of a British legation official (Augustus Raymond Margary) by native bandits near the Sino-Burmese border February 21 has increased tensions between Beijing and London.

Gen. Tso Tsung-tang, 63, moves to suppress the Tungans of the Northern Tien Shan who have been in revolt since 1862.

A French republican constitution finalized July 16 provides for a bicameral legislature (Chamber of Deputies and Senate) with a president to be elected for a 7-year term. Marshal MacMahon will continue as president until 1879.

Bosnia and Herzogovina in the Balkans rise against the Turks until the Ottoman sultan Abdul Aziz meets insurgent demands and promises reforms.

The Suez Canal completed in 1869 comes under British control November 27 through the efforts of Prime Minister Disraeli, who obtains a loan from the

1875 ***(cont.)*** merchant banking house of Rothschild to buy up 176,752 shares in the Universal Suez Company from the profligate Egyptian khedive Ismail (*see* 1863). Ismail is deep in debt, Disraeli puts up the British Government itself as security for the £4 million loan, and he creates international consternation by purchasing control of the strategic canal (*see* 1888).

President Grant opens Oregon Territory occupied under an 1855 treaty by the Nez Perce to white settlement (*see* Chief Joseph, 1877).

Comanche chief Quanah Parker ends his resistance to settlement of the Texas prairie by white ranchers. Col. R. S. Mackenzie of the U.S. Cavalry captured and destroyed his herd of 1,400 horses last fall and the chief brings his warriors into a reservation established by the government.

Congress passes a Civil Rights Act March 1 guaranteeing blacks equal rights in public places and banning their exclusion from jury duty.

Tennessee enacts the first "Jim Crow" law, a measure prejudicial to blacks (*see* 1880; Rice, 1832; Florida, 1887).

Prudential Insurance Co. of America is founded at Newark, N.J., under the name Prudential Friendly Society by Maine-born insurance agent John Fairfield Dryden, 36, with backing from local investors. It will grow by selling sickness and accident insurance, will sell ordinary life insurance beginning in 1886, and on August 20, 1896, will advertise in *Leslie's Weekly* with a picture of the Rock of Gibraltar and the slogan, "The Prudential Has the Strength of Gibraltar" devised by J. Walter Thompson. Prudential will become a mutual company in 1915.

The New York Central lowers its tracks down New York's Fourth (Park) Avenue into a giant ditch after years of complaints about accidents. Completed after 3 years of excavation, the ditch will soon be covered over for most of its length south of 96th Street (*see* Grand Central, 1913).

London's Liverpool Street Station is completed for the Great Eastern Railway.

Swedish inventor Alfred Nobel of 1866 dynamite fame discovers blasting gelatin (*see* Baku oil fields, 1873; Nobel prizes, 1901).

The Fiji Islands lose 40,000 of their 150,000 inhabitants in a measles epidemic following the return of the Fiji king from a visit to New South Wales.

The practice of osteopathy is founded by Kirksville, Mo., physician Andrew Taylor Still, 47, who lost three of his children to spinal meningitis in 1864. He has convinced himself that all diseases are caused by abnormalities in or near bodily joints (the abnormalities will later be called osteopathic lesions or spinal joint subluxations) (*see* 1892).

Kalamazoo, Mich., inventor George F. Green patents an electric dental drill. He will assign his patent on "electro-magnetic dental tools" for sawing, filing, dressing, and polishing teeth to Samuel S. White of Philadelphia whose company will become the leading U.S. producer of dental equipment, but use of electric drills will await the development of lighter engines and less expensive batteries (*see* 1957).

Vaseline petroleum jelly is introduced by Chesebrough Manufacturing Co., founded by English-American entrepreneur Robert Augustus Chesebrough, 38, who will make kerosene and lubricating oils until 1881 but confine himself after that to producing petroleum jelly.

Science and Health with Key to the Scriptures by New England divorcée Mary Baker, 54, claims that the Bible has helped her recover from the effects of a bad fall and explains a system of faith healing (*see* Christian Science, 1879).

Wellesley College for Women, founded in 1870, begins classes in a lavishly landscaped park round Lake Waban at Wellesley, Mass., with a main building whose central hall has a four-story glass-roofed central hall. Boston lawyer Henry Durant has built the college with profits made in the Civil War.

Smith College opens at Northampton, Mass. (*see* 1871).

Brigham Young University is founded at Salt Lake City in Utah Territory.

Hebrew Union College is founded at Cincinnati by Stephen M. Wise.

Alexander Graham Bell, 28, pioneers the electric telephone that will revolutionize communication. The Scottish-American inventor came to the United States in 1871 as a teacher of speech to the deaf and conceived the idea of "electric speech" last year while visiting his parents at Brantford, Ontario. While trying to perfect a method for carrying more than two messages simultaneously over a single telegraph line, Bell hears the sound of a plucked spring along 60 feet of wire June 2 in the attic electrical workshop of Charles Williams at 109 Court Street, Boston. The spring has been plucked by Bell's young assistant Thomas A. Watson who is trying to reactivate a harmonic telegraph transmitter, one of several whose reeds or springs are each tuned to a different signal frequency; a contact screw has been screwed down so far that a circuit has been left unbroken that should have been broken only intermittently and a current is being transmitted that corresponds to a reed in Bell's room. When he hears the sound of the plucked spring he recognizes its significance and realizes that the speaking telephone can be achieved by means of a simple mechanism (*see* 1876).

New York inventor Thomas Alva Edison, 28, perfects the first duplicating process to employ a wax stencil. He has developed quadruplex telegraphy, is experimenting with paraffin paper for possible use as telegraph tape, and will receive a patent next year for "a method of preparing autographic stencils for printing." Edison will improve the process, obtain a second patent in 1880, and license Chicago lumberman Albert Blake Dick, now 19, to use his invention. Dick will construct a flat-bed duplicator suitable for office use, employing a strong stencil fabric made from a species of hazel bush that grows only in certain Japanese islands, and the first A. B. Dick Diaphragm Mimeograph will go on sale March 17, 1887 (*see* typewriter stencil, 1888).

Richard Hoe invents a high-speed newspaper folding apparatus (*see* 1870).

N. W. Ayer & Son offers advertisers "open contracts" which give them access to the true rates charged by newspapers and religious journals (*see* 1868). Rates have been set up to now by advertising agents who charged whatever the traffic would bear, but Ayer will act as the advertiser's agent rather than the newspaper's agent (*see* Rowell, 1869; Batten, 1891; ANPA resolution, 1893).

The first American-made Christmas cards appear at Boston. German-American lithographer Louis Prang, 51, who makes the cards for the English trade (*see* Cole, 1843), has been publishing $6 reproductions of famous works of art ("chromos") since 1865 but will now concentrate on introducing Christmas cards in America, developing a market that he will dominate until 1890, when cheaper imports from Germany will put him out of business (*see* Hallmark, 1910).

Painting *Boating at Argenteuil* by Claude Monet; *The Agnew Clinic* by Thomas Eakins; Jean François Millet dies at Barbizon January 20 at age 60; J. P. C. Corot dies at Paris February 22 at age 78.

Sculpture Germany's 56-foot tall statue *Hermanns-Denkmal* in the Teutoberger Forest of Lower Saxony is dedicated to celebrate the triumph of the Germanic chieftain Arminus over the Roman legions of Varus in 9 A.D.

The *Hermes* produced by the Greek sculptor Praxiteles in the 4th century B.C. is unearthed at Olympia.

The *Great Chac-Mool* reclining limestone figure from the Maya-Toltec civilization of the 10th to 12th centuries A.D. in Yucatán is discovered at the entrance to a temple at Chichén-Itzá.

German-American photographer David Bachrach, Jr., 30, and Edward Levy, 29, patent a photoengraving process that will be the basis of an industry. Bachrach will establish the portrait photography firm Bachrach, Inc. Levy and his brother Max will patent an etched glass grating or screen for making halftone engravings (*see* 1880; Ives, 1886).

Theater *Wolves and Sheep* (*Volki i ovsty*) by Aleksandr Ostrovsky 12/8 at St. Petersburg's Alexandrinsky Theater.

Opera *Carmen* 3/3 at the Opéra-Comique, Paris, with music by Georges Bizet who dies June 3 at age 36 of a throat ailment. Girls from a cigarette factory smoke on stage in Act I and the heroine sings the "Habanera;" *Trial by Jury* 3/5 at London's Royalty Theatre begins a 21-year series of comic Gilbert & Sullivan operettas with librettos by W. S. Gilbert, music by Arthur S. Sullivan.

The Paris Opéra completed by French architect Jean Louis Charles Garnier, 50, has the largest stage of its kind in the world; Paris hails the structure as a worldly reply to the 530-year-old Notre Dame Cathedral. Its huge stage occupies less than one-third of the 50 million franc building's total space of which more than half is taken up by foyers, galleries, and staircases that provide settings for tête-à-têtes.

A new Vienna Opera House opens with a performance of the 1814 Beethoven opera *Fidelio*.

First performances *Danse Macabre* (symphonic poem) by Camille Saint-Säens 1/24 at Paris; *Symphonie Espagnole* for Violin and Orchestra by French composer Edouard Lalo, 52, 2/7 at Paris; *The Moldau (Vltava)* (symphonic poem) by Bedřich Smetana 4/4 at Prague; Concerto No. 1 for Piano and Orchestra by Petr Ilich Tchaikovsky 10/25 at Boston's Music Hall with Hans von Bülow as soloist; Concerto No. 4 in C minor by Saint-Säens 10/31 at Paris; Symphony No. 3 in D major by Tchaikovsky 11/19 at Moscow.

The Boston Symphony has its genesis in the Boston Philharmonic Club founded by German-American conductor Bernhard Listemann, 34 (*see* 1881).

The first organized Canadian ice hockey match is played March 3 at Montreal's Victoria Skating Rink (*see* Stanley Cup, 1893).

The Kentucky Derby has its first running at Louisville's new Churchill Downs and a small chestnut colt named Aristides wins the 1.5-mile stakes race. Local promoter M. Lewis Clark returned from a trip to England a few years ago with plans to establish a series of races that will be comparable to the Derby Stakes at Epsom Downs, the Oaks, and other English ricing classics. The stakes race for 3-year-olds, held early each May, will be shortened to 1.25 miles and, with the 1 3/16-mile Preakness Stakes inaugurated in 1873 and the 1.5-mile Belmont Stakes begun in 1867, constitute the Triple Crown of U.S. horse racing.

The first roller skating rink opens August 2 in Belgravia, London (*see* Plimpton, 1863).

English swimmer Captain Matthew Webb, 27, crosses the English Channel from Dover to Calais August 24 in 22 hours and is the first to accomplish the feat (*see* Ederle, 1926).

The first Harvard-Yale football game is held November 12 at New Haven under "concessionary rules" that permit running with the ball and tackling (see 1874; Walter Camp, 1889).

R. J. Reynolds Tobacco has its origin in a Winston, N.C., factory set up by former tobacco peddler Richard Joshua Reynolds, 25, who has amassed $7,500 in capital working in partnership with his father. Reynolds uses $2,400 to build a 36-by 60-foot frame structure and equip it with a few crude pieces of machinery, uses the rest to buy leaf tobacco, and begins manufacturing pipe tobacco from burley leaf (Winston has become a center for bright tobacco production) (*see* Slade brothers, 1852; Prince Albert, 1907).

U.S. cigarette production reaches 50 million but will increase rapidly after 1882 when a cigarette manufacturing machine invented 3 years ago comes into com-

1875 ***(cont.)*** mercial use. The machine feeds tobacco to a ribbon of paper as it is drawn from a spool, the edges pass over a gummed wheel, and a continuous cigarette is produced that the machine then cuts into separate lengths.

The Jukes, a Study in Crime, Pauperism, Disease, and Heredity by English-born sociologist Richard Louis Dugdale, 33, purports to be an extensive criminal-genealogical study of 7 generations of a depraved U.S. family descended from a mid-18th century Dutch tavern keeper. A functionary of the Prison Association of New York, Dugdale has plowed through arrest records and interviewed prisoners, wardens, and inmates of asylums and poorhouses. He concludes that the Jukes have cost the public $1.3 million and that their immorality is hereditary (*see* jukebox, 1927; Kallikaks, 1912).

Brooklyn, N.Y., Congregational minister Henry Ward Beecher, 62, wins acquittal July 2 when a sensational 6-month adultery trial ends in a hung jury. Beecher has clearly been guilty of having sex with a parishioner's wife.

New York's onetime political boss William Marcy "Boss" Tweed is released from prison after having served a brief term for larceny and forgery (*see* 1869). Authorities rearrest Tweed in a civil action brought by the state to recover his loot, he escapes from jail in December, flees to Cuba, and proceeds to Spain disguised as a sailor; recognized from a Thomas Nast cartoon, he is apprehended and returned to New York and will die in prison in 1878 at age 55.

New York's "Tweed" Courthouse is completed north of City Hall at a cost of $13 million. The estimate had been $250,000 but "Boss" Tweed's cronies have run up huge bills including $179,729 for three tables and 40 chairs, nearly $361,000 for a month's work by a single, solitary carpenter.

Boss Tweed provoked New York cartoonist Thomas Nast to depict political corruption with savage caricatures.

A magnificent new Palmer House opens in Chicago to replace the hotel owned by merchant Potter Palmer that burned in the 1871 fire. The new hotel has a mammoth barbershop with silver dollars embedded in its floor and a lavish dining room with dishes that include broiled buffalo, antelope, bear, mountain sheep, boned quail in plumage, blackbirds, partridge, and other "ornamental dishes" (*see* 1925).

The United States Hotel opens at Saratoga Springs, N.Y., with nearly 1,000 rooms for the summer season. The new hotel calls itself the world's largest.

San Francisco's Palace Hotel opens October 2 with 755 20-by-20-foot rooms, 437 baths, five elevators, seven iron stairways, and a crystal roof over its inner court. Built on a wrought-iron foundation, it is supplied with water from four artesian wells (*see* 1906).

Bangkok's Oriental Hotel opens on the Chao Phya River.

President Grant vetoes a bill that would protect the bison from extinction (*see* 1870; 1879).

Arbor Day April 22 encourages Americans to plant trees on the open prairie. Nebraska agriculturist J. Sterling Morton, 43, has dreamed up the idea and within 12 years 40 states will have made April 22 an official Arbor Day (*see* Kilmer, 1913).

Earthquakes in Colombia and Venezuela May 16 take 16,000 lives.

Illinois promoter John Warne Gates, 20, puts on a demonstration at San Antonio to show that barbed wire is safe and effective. He turns the main plaza of the town into a giant corral that can hold long-horned cattle (*see* Glidden, Haish, 1873). Barbed wire will begin the end of the open range in America and launch the career of "Bet-you-a-million" Gates (*see* 1877; 1898).

U.S. Grangers fail in a scheme to produce farm equipment (*see* 1872). The Grange has bought up patents for cultivators, seeders, mowers, reapers, and many other machines and implements and has begun to manufacture them, but the effort falls victim to patent suits, lack of capital, defective machines, and lack of cooperation in a society of rural individualists.

Combines come into use on some U.S. wheat farms.

The first American state agricultural experiment station is established July 20 at Wesleyan University in Middletown, Conn.

Luther Burbank establishes a nursery at Santa Rosa, Calif., with money obtained from the sale of his 1872 Burbank potato. He will develop new forms of food and ornamental plant life by selection and cross— fertilization (*see* Shull, 1905).

A Chinese orchardman in Oregon develops the Bing cherry.

Apples are grown for the first time in Washington's Yakima Valley.

Navel oranges are produced at Riverside, Calif., by Jonathan and Eliza C. Tibbetts who 2 years ago obtained two trees from the USDA at Washington which in 1871 received a dozen budded seedlings from Bahia, Brazil. The seedless winter-ripening fruit are the first ever seen in the United States and will be called Washington oranges (because the first trees came from Washington) as they proliferate to dominate California groves (*see* Valencias, 1906).

The orange crate, devised by U.S. inventor E. Bean, weighs 15 pounds, holds 90 pounds (about 200 pieces), and will lead to wide-scale marketing of the fruit (*see* Orange Growers Protective Union, 1885).

Atlantic salmon are planted in inland lakes by the U.S. Fish Commission. Land-locked salmon will be an important U.S. game and food fish (*see* shad, 1873; carp, 1879).

A new British Sale of Food and Drugs Law tightens restrictions against adulteration, making any adulteration injurious to health punishable with a heavy fine and making a second offense punishable with imprisonment if the seller is proved to have guilty knowledge of the adulteration (*see* 1872).

British sugar consumption rises to 60 pounds per capita, up from 47 in 1872 (*see* 1889).

Berlin chemist Ferdinand Tiemann, 27, patents a process for making synthetic vanillin, the key flavor ingredient in natural vanilla beans.

Société Anonymé Lactée Henri Nestlé is founded by businessmen of Vevey, Switzerland, who outbid Geneva bankers to acquire the operation of Henri Nestlé which is selling hundreds of thousands of tins of baby food each year and is the town's foremost enterprise (*see* 1866). The bankers have offered a million francs, but the Vevey businessmen offer more and provide Nestlé's wife with a coach and two horses to sweeten their proposition.

The first milk chocolate for eating is invented at Vevey (above) by the Nestlé shopman in collaboration with the foreman of Daniel Peter's chocolate factory (*see* Cadbury, 1847). They hit upon the idea of mixing sweetened condensed milk with chocolate and create a product that meets with immediate commercial success (*see* Peter and Kohler, 1904; Hershey, 1893).

New York Condensed Milk Co. begins selling fluid milk in addition to its condensed milk 1 year after the death of founder Gail Borden at age 72 (*see* 1858; 1885).

H.J. Heinz's firm Heinz & Noble is forced into bankruptcy after contracting to buy some crops that have come in more abundantly than expected, but Heinz pays off the firm's creditors with help from his wife (*see* 1869; ketchup, 1876).

B&M Baked Beans, produced for use by men in the fishing fleet of Burnham & Morill Co. at Portland, Me., are the world's first canned baked beans.

A machine is invented to strip the kernels from corn cobs. It will lead to wide-scale canning of sweet corn (*see* Minnesota Valley Canning, 1903).

Ferdinand Schumacher introduces Steel Cut Oats, a new product whose flaky composition gives it a uniformly acceptable taste and consistency (*see* 1856). Schumacher uses a new machine invented largely by his employee Asmus J. Ehrrichsen whose process is the first real innovation in milling oats, employing a series of horizontal knife blades to cut the hulled oats (groats) into a meal that has little or none of the floury residues that have made oatmeal pasty, lumpy, and glutenous. Another Schumacher employee, William Heston, improves on the Ehrrichsen-Schumacher process and obtains patents which he licenses Schumacher to use while reserving the right to use the new process himself (*see* Quaker, 1877).

Hires Rootbeer has its beginnings in a recipe for an herb tea discovered by Philadelphia pharmacist Charles Elmer Hires, 24, who takes his bride on a wedding trip to a New Jersey farm. The tea is made from 16 different wild roots and berries that include juniper, pipsissewa, spikenard, sarsaparilla, wintergreen, and hops. Hires takes the recipe back to Philadelphia and begins experimenting with improved formulas (*see* 1876).

A British Trade Marks Act goes into effect; the first trademark registered is for Bass & Co. Pale Ale made at Burton-on-Trent in Staffordshire since 1777.

German embryologist Oscar Hertwig, 26, concludes that fertilization of the female egg is accomplished by a single male cell. He has worked with sea urchins (*Arbacia punctulata*) whose female produces roughly a million eggs per season (*see* Fol, 1879).

1876 Japan recognizes Korean independence from China February 26 in a treaty that opens three Korean ports to Japanese trade, permits Japan to have a resident at Seoul, and brings no protest from China.

The Royal Titles Bill passed by Parliament in April makes Queen Vicotria Empress of India. The queen elevates Prime Minister Disraeli to the peerage in August, making him earl of Beaconsfield.

Ethiopian forces defeat Egyptian troops at Gura, but the Egyptians gain a crushing victory in February as the war that began last year continues.

The Ottoman sultan Abdul Aziz is deposed May 29 at age 46 after a 15-year reign and dies 4 days later, probably by his own hand. His nephew of 35, who has been imprisoned for 15 years, reigns for 3 months as Murad V before being declared insane. Murad's brother, 33, is made sultan and will reign until 1909 as Abdul Hamid II.

A Bulgarian insurrection has begun in May and Turkish irregulars are sent in to quell the uprising.

Serbia declares war on Turkey June 30, Montenegro follows suit July 2, but the Serbs are completely defeated at Alexinatz September 1 and the Bulgarians slaughtered by the thousands to suppress their insurrection, which is ended in September.

"The Bulgarian Horrors and the Question of the East" by former British prime minister William Gladstone appears September 6; the pamphlet arouses Britons against the Turks as Russia prepares for war against the Serene Port (Constantinople) (*see* 1877).

1876 ***(cont.)*** A Turkish Constitution proclaimed December 23 by the grand vizier Midhat Pasha, 54, declares the indivisibility of the Ottoman Empire and provides for parliamentary government based on representation of all groups in society (but *see* 1877).

Mexican general Porfirio Diaz, 46, overthrows President Lerdo de Tejada and begins to restore order in a program that will stabilize the country, attract foreign capital, develop industry, build railways, promote public works, and increase commerce.

Colorado is admitted to the Union as the 38th state.

The U.S. presidential election in November ends in a dispute when neither Democrat Samuel Jones Tilden, 62, of New York nor Republican Rutherford Birchard Hayes, 54, of Ohio wins the necessary 185 electoral votes (Tilden falls one vote short): two sets of returns leave 20 votes from Florida, Louisiana, Oregon, and South Carolina in dispute and there is no constitutional provision to cover the situation (*see* 1877).

Black militiamen are massacred at Hamburg, S.C., in July, beginning a race war in that state.

Chiricahua Apaches in Arizona Territory are moved to a reservation occupied earlier by western Apaches; the Chiricahua leader Geronimo leads a band of his warriors into Mexico, beginning a 10-year reign of terror against white settlers in the Southwest (*see* Cochise, 1872). Geronimo (Jerome), 47, has been given his name by Mexicans he raided in retaliation for the 1858 killing of his family; his real name is Goyathlay (*see* 1877).

The Battle of the Little Big Horn June 25 ends in the massacre of a 264-man U.S. Seventh Cavalry force under Lieut. Col. George Armstrong Custer, 37, at the hands of Sioux chief Sitting Bull, 42, with tribesmen under chieftains Gall and Crazy Horse. The Sioux have been angered by the slaughter of buffalo in Montana Territory by the advancing whites in the Black Hills gold rush (*see* Deadwood Gulch, 1875; Homestake, below; Wounded Knee, 1890).

The Turkish Constitution (above) provides for freedom of conscience, liberty of the individual, equality of taxation, and freedom of the press and education, but the new sultan will let it lapse (*see* 1856).

A Merchant Shipping Act designed to prevent overloading of ships and use of unseaworthy vessels is passed by Parliament after a campaign by M. P. Samuel Plimsoll, 52. A retired London coal merchant, Plimsoll is known as "the Sailors' Friend" on the strength of his 1872 book *Our Seamen*, and his efforts in behalf of seamen will be commemorated in the Plimsoll line on ships' hulls (*see* gym shoes, below).

The International Association for the Exploration and Civilization of Africa is founded under the auspices of the Belgian king Leopold II.

The Reichsbank opened in January will play a major role in German economic development.

Homestake Mining Co., is founded at Lead (pronounced Leed, meaning "lode") in the Black Hills of Dakota Territory. The gold mine opened by George Hearst and Ben Ali Haggin will be the largest U.S. producer (*see* 1859; Custer, above).

The U.S. Centennial Exposition opened by President Grant at Philadelphia's Fairmount Park May 10 is called officially the "International Exhibition of Arts, Manufactures and Products of the Son and Mine"; 37 foreign nations and 26 states are represented along with innumerable private exhibitors.

All exhibits in Machinery Hall at the Centennial Exposition are powered by a great 160-horsepower engine built by Corliss Steam Engine Co. of Providence, R.I., started in 1856 by inventor George H. Corliss of 1851 London Great Exhibition fame. Harvard historian Henry Brooks Adams, 38, calls the Corliss "dynamo" the symbol of the age.

George Selden of Rochester attends the Exposition with a partner to demonstrate their patented machine for shaving and finishing barrel hoops. Selden studies the Brayton engine, developed by an Englishman in Boston, and notes that while the other engines exhibited operate on illuminating gas the Brayton engine employs crude petroleum fuel. It weighs 1,160 pounds and develops only 1.4 horsepower but Selden sees the possibility of a lighter, more efficiently designed engine (*see* 1873; 1879).

Nikolaus A. Otto invents a four-cycle gasoline engine far more advanced than his 1866 engine (*see* Benz, 1885).

Compagnie International des Wagons-Lits et des Grandes Express Européens is organized to bring sleeping cars to Europe. Engineer-founder Georges Nagelmakers has crossed the United States in Pullman cars (*see* Train Bleu, 1879).

San Francisco and Los Angeles are linked by rail September 6 as the Southern Pacific reaches Los Angeles to give that city a route to the East, but high freight rates and water shortages will retard development of Los Angeles whose population will not exceed 10,000 for another few years.

The St. Louis and San Francisco (Frisco Line) founded September 11 takes over most of the Atlantic & Pacific Railroad lines through Indian Territory.

The first carload of California fruit reaches the Mississippi Valley, but the freight rate is too steep for most growers (*see* Columbian Exposition, 1893).

Canada's Intercolonial Railroad opens to link Ontario with the Maritime Provinces.

A train wreck December 29 kills 83 at Ashtabula, Ohio, as a 13-year-old bridge on Cornelius Vanderbilt's Lake Shore & Michigan Southern gives way during a storm off Lake Erie and 10 cars of the Pacific Express pulled by two locomotives plunge into a creek 150 feet below, bursting into flames. The road's president fired a subordinate engineer in 1863 when he protested construction of the Howe truss bridge built by the road's chief engineer.

Henry A. Sherwin, 34, of Cleveland's 6-year-old Sherwin-Williams Co. pioneers prepared, ready-to-apply paint by developing a machine that grinds good-bodied pigments so finely and evenly that they will suspend in linseed oil. Sherwin-Williams will go on to make varnishes and then enamels (pigments suspended in varnish).

The stillson wrench is patented by Somerville, Mass., inventor Daniel C. Stillson who has whittled a wooden model of his pipe or screw wrench.

"On the Equilibrium of Heterogeneous Substances" by Yale professor of mathematical physics Josiah Willard Gibbs, 37, extends understanding of thermodynamics. Gibbs will follow up with another paper in 1878 as he lays the theoretical groundwork for the science of physical chemistry, but the importance of his work will not be recognized for years.

The Danish Red Cross is founded at Copenhagen.

Eli Lilly Co. is founded May 10 at Indianapolis by former Union Army officer Eli Lilly, 38, who starts with $1,400 in capital and two employees to produce reliable medications for responsible physicians in an era dominated by patent medicines.

A label for "Mrs. Lydia E. Pinkham's Vegetable Compound" is patented by Lynn, Mass., housewife Lydia Estes Pinkham, 57, whose husband Isaac lost all his money in the 1873 financial crash. Mrs. Pinkham has for more than a decade been concocting an herbal medicine for "woman's weakness" and related ills, a nostrum she will continue to brew in her own kitchen and for which she will write her own homely advertising until her death in 1883 (*see* 1898).

Seventh Day Adventist surgeon John Harvey Kellogg, 24, assumes management of the Western Health Reform Institute founded at Battle Creek, Mich., in 1866. A champion of vegetarianism, Kellogg will ask audiences, "How can you eat anything that looks out of eyes?" and he will develop vegetarian foods including a form of peanut butter and scores of dried cereals (*see* Granula, 1877).

Johns Hopkins University is founded at Baltimore with a bequest from the late Baltimore financier Johns Hopkins who died 3 years ago at age 78 (*see* hospital and medical school, 1893).

The U.S. Coast Guard Academy is founded at New London, Conn.

The University of Texas is founded at Austin.

Texas A & M (Agricultural and Mechanical College of Texas) is founded near Bryan.

"Mr. Watson, come here. I want you," says Alexander Graham Bell March 10 in the first complete sentence to be transmitted by voice over wire. Bell has improved the telephone he invented in 1875, has been granted a patent on his 29th birthday March 3, and uses the instrument at 5 Exeter Place, Boston, to speak with his assistant Thomas A. Watson. Elisha Gray of the 4-year-old Western Electric Co. will challenge the patent, the courts will uphold Bell's claim, and Western Electric will manufacture the Bell telephone (*see* 1882).

Bell demonstrates his telephone at the U.S. Centennial Exposition (above); the Brazilian emperor Dom Pedro II jumps out of his seat June 25 saying, "I hear, I hear!"

Bell's future father-in-law Gardiner Green Hubbard, 51, promotes the new telephone. A Boston lawyer whose daughter's deafness has led him to help found the Clarke Institution for the Deaf which he has headed since 1867, Hubbard borrows a telegraph line between Boston and the Cambridge Observatory, attaches a telephone at each end, and in mid-October speaks with Thomas Watson for more than 3 hours in the first sustained telephone conversation (see 1877).

Western Union president William Orton turns down an offer to acquire the Bell telephone (above) for $100,000, calling it a toy, a "scientific curiosity" that permits users to speak or listen but not both at once.

Western Union (above) retains Thomas A. Edison to improve on Bell's telephone. He builds a laboratory at Menlo Park, N.J., with $40,000 earned from a patented stock ticker, opens the world's first research laboratory, and finds that carbon black makes a perfect transmitter. Edison develops a carbon transmitter that will make the telephone commercially practicable and his invention of carbonization will soon find other applications (*see* mimeograph, 1875; Swan, 1878; Cabot, 1887).

The Remington typewriter introduced at the Philadelphia Centennial Exposition (above) does not yet have a shift key (*see* 1874; 1878). Also exhibited are an envelope-making machine and the first stamped envelopes.

Stenotypy begins to facilitate courtroom reporting and make records of legal proceedings more accurate. New York inventor John Colinergos Zachos patents a "type-

Alexander Graham Bell's telephone replaced the written word with the spoken word for instant communication.

1876 *(cont.)* writer and phonotypic notation" device with type fixed on 18 shuttle bars, two or more of which may be placed in position simultaneously, the device has a plunger common to all the bars for making impressions, and it permits printing a legible text at a high reporting speed.

McCall's magazine begins publication in April under the name *The Queen.* Scots-American entrepreneur James McCall started McCall Pattern Co. in 1870 with a small shop at 543 Broadway, New York, and uses the eight-page pattern and fashion periodical to promote his dress patterns, but it will grow to become a major magazine for women.

Corriere della Sera begins publication at Milan. It is the first reliable Italian newspaper and will be Italy's one national newspaper, but even 90 years after its founding only four Italians in 10 will read any daily paper.

"A Classification and Subject Index for Cataloguing and Arranging the Books and Pamphlets of a Library" by Amherst College librarian Melvil Dewey, 25, originates the Dewey decimal system.

Nonfiction *Robert's Rules of Order* by U.S. Army Engineer Corps officer Henry Martyn Robert, 39, who has been motivated by experience in civic and church groups to establish authoritative rules for smooth, democratic procedure in any self-governing organization. Robert will publish a revised and enlarged edition in 1915 at age 76.

Fiction *Roderick Hudson* by U.S. émigré author-critic Henry James, 35, who examines the moral and artistic disintegration of an American in Rome and begins a series of perceptive novels about European and American society; *The Prime Minister* by Anthony Trollope; *Daniel Deronda* by George Eliot; *Marthe* by French novelist Joris Karl Huysmans, 28; *The Adventures of Tom Sawyer* by Mark Twain.

Poetry *The Afternoon of a Faun (L'Après midi d'un Faune)* by French poet-schoolteacher Stéphane Mallarmé, 34.

Painting *The Spirit of '76* by Ohio painter Archibald Willard, 40, who will produce 14 versions of the canvas he has painted for the Philadelphia Centennial Exposition. The first will find a permanent home in Abbot Hall at Marblehead, Mass., but the one Willard will call his "best effort" will be a much larger canvas that will hang in the Cleveland City Hall; *Breezing Up* by Winslow Homer; *Flood at Port Marly* by Alfred Sisley; *Au Moulin de la Galette* by Pierre Auguste Renoir.

Theater *Peer Gynt* by Henrik Ibsen 2/24 at Oslo's Christiania Theater, with a "Peer Gynt Suite" by Edvard Grieg; *Truth Is Good, but Happiness Is Better (Pravda—khorosho, a schastye luchshe)* by Aleksandr Ostrovsky 11/18 at Moscow's Maly Theater.

Opera *La Giocanda* 4/8 at Milan's Teatro alla Scala, with a libretto from the Victor Hugo play *Angelo, Le Tyran de Padoue*, music by Italian composer Amilcare Ponchielli, 42, who wins acclaim for his "Dance of the Hours" ("Danze delle Ore"); *The Kiss (Hubicka)* 11/7 at Prague, with music by Bedřich Smetana; *The Golden Slippers (Vakula the Smith)* 11/24 at St. Petersburg's Maryinsky Theater, with music by Petr Ilich Tchaikovsky, libretto from a Gogol story.

The first complete performance of Richard Wagner's operatic Ring cycle opens the Festspielhaus that Wagner has built at Bayreuth. It has taken 26 years to write *Der Ring des Nibelungen*, his Bayreuth Festival begins August 13, and *Siegfried* is performed August 16, *Die Götterdämmerung* (*The Twilight of the Gods*) August 17. In the audience are the kaiser, his nephew Ludwig II of Bavaria, Friedrich Nietsche, and prominent composers.

The Goodspeed Opera House opens at East Haddam, Conn.

First performances Symphony No. 1 in C minor by Johannes Brahms 11/4 at Karlsruhe; *Marche Slav* by Petr Ilich Tchaikovsky 11/17 at Moscow.

Popular songs "I'll Take You Home Again, Kathleen" by U.S. songwriter Thomas P. Westendorf, 28; "Grandfather's Clock" by Henry Clay Work: "But it stopped, short—never to go again—when the old man died."

Hymns "What a Friend We Have in Jesus" by U.S evangelist and hymn writer Ira David Sankey, 36, who has been touring since 1870 with missionary/evangelist Dwight Lyman Moody, 39, lyrics by Horatius Bonar; "The Ninety and Nine" by I.D. Sankey, lyrics by E. C. Clephane; "Trusting Jesus, that Is All" by I.D. Sankey, lyrics by E. P. Stites.

A player piano demonstrated at St. Louis by Scots-American inventor John McTammany, 31, produces music mechanically from perforated paper rolls. McTammany, who got his idea while repairing a music box during convalescence from serious injuries sustained in the Civil War, will obtain a patent in 1881 but will lose control of his player piano manufacturing company and die a pauper after spending income from the piano to defend his patent rights against violators.

U.S. baseball's National League is founded February 2 (*see* Cincinnati, 1869; American League, 1901).

Chicago Nationals pitcher Albert Goodwill Spalding, 26, and his brother each put up $400 to start a sporting goods business that will grow to dominate the American sports scene. Spalding will be manager, then secretary, then president of the Chicago club until 1891, he will edit *Spalding's Official Baseball Guide* beginning in 1878 (and promote the myth that Abner Doubleday invented the game in 1839), the Spalding brothers will take in their banker brother-in-law William T. Brown, and he will provide capital with which the company will expand in 1885 to become a leading producer of baseball gloves by acquiring A. J. Reach, started by another ballplayer (*see* 1892).

The first Canadian bid to gain the America's Cup in ocean yacht racing ends in failure as the *Countess Of Dufferin* loses 2 to 0 to the U.S. defender *Madeleine.*

The London Rowing Club four loses to the Albany Beaverwycks at Philadelphia's Centennial Regatta. The

Times of London complains that the Albany crew's amateurs are "working mechanics" and "handicraftsmen" rather than true amateurs who row purely for recreation in leisure hours.

England's 47-year-old Henley Regatta gets its first U.S. entry as Columbia University's four-oared shell competes with British shells. The regatta will continue to be an Anglo-American event and in some years will have more entries from U.S. colleges, prep schools, and clubs than from British.

The U.S. stallion Hambletonian dies at age 27 after having sired hundreds of offspring that win make trotting horses with Hambletonian blood the dominant force in U.S. harness racing for more than a century. Hambletonian was sired by Abdallah who had been sired by Mambrino whose sire was the English horse Messenger brought to America in 1788 at age 8 (*see* Dan Patch, 1902).

Britons call rubber-soled canvas shoes "Plimsolls" after the Plimsoll line on ships (above). The gym shoes are costly and problematical.

B.V.D. underwear for men is introduced by Bradley, Voorhees, and Day of New York whose one-piece B.V.D.s will be called "Babies' Ventilated Diapers" but B.V.D. will become almost a generic term.

The Bissell Grand Rapids carpet sweeper patented by Grand Rapids, Mich., china shop proprietor M. R. Bissell is the world's first carpet sweeper (*see* vacuum cleaner, 1901). Bissell suffers from allergic headaches caused by the dusty straw in which china is packed and has devised a sweeper with a knob that adjusts its brushes to variations in floor surface and with a box to contain the dust.

Wild Bill Hickok is murdered August 2 at Deadwood in Dakota Territory at age 39. A one-time scout for the Union Army, James Butler Hickok has served as marshal at Fort Riley, Hays City, and Abilene, Kan., but he is shot from behind by Jack McCall in Deadwood's Saloon No. 10 while sitting at a poker table holding two pairs, aces and eights, that will become known as the "Dead Man's Hand."

Northfield, Minn., citizens foil an attempt on a bank September 7 by Frank and Jesse James and the Younger brothers (*see* 1872). Two of the gang are killed in a hail of bullets, four are captured including Cole Younger and his brothers James and Robert. The James brothers escape, and the Younger brothers are tried, convicted, and sent to prison. Robert Younger will die in prison in 1889, his brothers will be paroled in July, 1901, James Younger will take his own life in 1902, Cole will be pardoned in 1903 and live until 1916 (*see* James, 1882).

Newport's Walter Sherman House is completed by architect Henry Hobson Richardson.

A bentwood café chair introduced at Vienna by Gebruder Thonet will be standard furniture in world restaurants and saloons for nearly a century (*see* Le Corbusier, 1925).

New York's Central Park is completed after 17 years of work. The 840-acre park extends from 59th Street to 110th between 5th and 8th avenues.

"God's First Temples—How Shall We Preserve Our Forests" by John Muir is published in a San Francisco newspaper (*see* 1868). Muir has seen sheepmen "and their hoofed locusts" overgrazing the "gardens and meadows" of the Merced River Valley above Yosemite and has heard the lumber-mill saws "booming and moaning like bad ghosts" (*see* Yosemite Park, 1890).

Palm Beach, Fla., has its beginnings as cocoanuts from the wreck of the Spanish ship *Providencia* wash ashore. Palm trees seeded by the coconuts will proliferate in the area (*see* hotel, 1894).

A colony of passenger pigeons in Michigan reportedly covers an area 28 miles long and 3 to 4 miles wide (*see* 1871; 1878).

The first successful U.S. sardine cannery is established by New York importer Julius Wolff whose French sardine imports were interrupted by the Franco-Prussian War in 1871 (*see* 1834). Sardine canneries will soon open at Eastport, Lubec, and other points along the Maine coast.

Argentina emerges as a grain-exporting nation.

English explorer Henry A. Wickham, 30, breaks Brazil's rubber monopoly, smuggling 70,000 seeds of the rubber tree *Hevea Brasiliensis* to the Royal Gardens at Kew whence they will be transplanted in Ceylon, India, Malaya, and the East Indies on efficient plantations that will wreck Brazil's wild rubber economy within 30 years.

A treaty negotiated by Hawaiian planter Henry Alpheus Pierce Carter, 39, places Hawaiian sugar on the free list for importation into the United States (*see* politics, 1891).

The Grange departs from its apolitical stance and begins to lobby for U.S. tax law revision including an end to tax exemption for railroad properties (*see* 1873). The Grange also demands lower interest rates, better schools, cheap textbooks, and cheaper bread, coal, and clothing.

Colorado cattleman Charles Goodnight settles one of the West's first sheepherder-cattlemen wars and heads back to Texas where he discovers an enormous valley in the Palo Duro Canyon area where 10,000 bison are feeding near the Canadian River (*see* 1866). After having prospered as a Colorado Territory rancher, banker, and irrigation promoter, Goodnight was wiped out in the panic of 1873, but he has recovered with money made in cattle drives. He and his cowboys drive out the buffalo and establish the Old Home Ranch, the first in the Panhandle. Financed by Irish-American investor John G. Adair of Denver, Goodnight will develop the million-acre JA Ranch and the Goodnight-Adair partnership will net more than half a million dollars in 5 years.

1876 ***(cont.)*** Famine will kill 9.6 million in Northern China and 5 million in India in the next 3 years as drought withers wheat fields in China's Shansi province and on India's Deccan Plateau.

Canning becomes enormously more efficient with the introduction of the Howe "floater" which makes the tinsmith's soldering iron obsolete. A long line of cans, complete with tops and bottoms, floats slowly through the solder bath in the Howe machine picking up enough melted solder to make tight joints and permitting two men and two helpers to turn out more than 1,000 cans per day, 100 times the number one man and a helper could make in the first half of the century. In another 10 years the same number of men will be able to produce 1,500 cans per day (*see* 1874; 1895; 1897).

Delaware canner A. B. Richardson applies for a patent on a new can shape and a new method for canning boneless hams.

The first all-roller flour mill to be seen in America is displayed at the Philadelphia Centennial Exposition (above) but is dismantled after the fair (*see* Washburn, 1874; Pillsbury, 1878).

Philadelphia fairgoers see bread and rolls raised with compressed yeast right before their eyes at a model Viennese bakery set up by Gaff, Fleischmann and Co. which has invested all its resources in the display to spread word about Fleischmann's Yeast (*see* 1870; 1919).

Bananas fetch 10¢ each at the Philadelphia fair. The foil-wrapped novelty gives most fairgoers their first taste of the tropical fruit (*see* Baker, 1871; Boston Fruit, 1885).

Budweiser beer wins top honors in a competition at the Philadelphia fair. The new beer has been developed by Adolphus Busch whose E. Anheuser and Co. produces 15 other brands of beer, many of them pasteurized, and introduces refrigerated railcars for all its beers (*see* 1873; Michelob, 1896).

Hires Rootbeer Household Extract is promoted at the fair with an exhibit displaying packages of dried roots, barks, and herbs (*see* 1875). Charles E. Hires has adopted the name "rootbeer" at the advice of his friend Russell Conwell, 33, who says hard-drinking Pennsylvania coalminers will be more attracted to "rootbeer" than to "herb tea" (*see* 1886; Temple University, 1884).

Heinz's Tomato Ketchup is introduced by Pittsburgh's H. J. Heinz who joins with his brother and a cousin to establish F. and J. Heinz (*see* 1875). Heinz packs pickles and other foods as well as the ketchup, which is 29 percent sugar (*see* 1880).

The A&P grocery chain of Huntington Hartford opens its 67th store (*see* 1871; 1880).

Glasgow grocer Thomas Johnstone Lipton opens his first shop at age 26. Lipton sailed to America at age 15 to spend 4 years learning the merchandising methods employed in the grocery section of a New York department store and will pioneer new methods of food advertising. (He begins by having two fat pigs driven through the streets of Glasgow; painted on the pigs' scrubbed sides are the words, "I'm going to Lipton's, The Best Shop In Town for British Bacon.") (*see* 1879; tea, 1890).

The first Fred Harvey restaurant opens in the Santa Fe Railroad depot at Topeka, Kan. English-American restaurateur Frederick Henry Harvey, 41, has been in the United States since 1850 and has worked as a freight agent for J. F. Joy's Chicago, Burlington & Quincy Railroad. Joy has turned down his proposal that clean, well-run restaurants be opened for the benefit of passengers, but the Santa Fe has been receptive to the idea and Harvey's Topeka restaurant will soon be joined by others along the Santa Fe line and at other major rail depots and junctions. Harvey will also operate hotels and beginning in 1890 will start operating a fleet of dining cars. By the time of his death in early 1901 there will be 47 Fred Harvey restaurants, all serving good food on Irish linen, with Sheffield silverware, served by well-trained "Harvey Girls," plus 30 Fred Harvey dining cars and 15 Fred Harvey hotels.

1877 A U.S. electoral commission decides last year's disputed presidential election. Appointed January 29 and composed of five senators, five representatives, and five Supreme Court justices, it decides all questions along partisan lines (eight members are Republicans, seven Democrats), denying Democrat Samuel J. Tilden his marginal victory at the polls. Rutherford B. Hayes is sworn in as president March 4, 2 days after the decision.

Britain annexes Walfish (Walvis) Bay on the coast of Southwest Africa March 12, she annexes the South African Republic April 12 in violation of the 1852 Sand River Convention that recognized the independence of the Transvaal, and a new Kaffir War begins that will end with the British annexing all of Kaffraria and extending their rule toward the northeast.

Queen Victoria is proclaimed Empress of India in a ceremony (durbar) held at Delhi on Disraeli's initiative. Indian princes assemble to pay homage.

Russia declares war on Turkey April 24 as the government of Aleksandr II yields to pressure from Panslav elements. A larger war looms as the British cabinet decides July 21 to declare war on Russia if she should occupy Constantinople and if she does not make plans for immediate withdrawal. The Russians suffer reverses at Plevna, allaying fears of British intervention, but Plevna falls to the Russians December 10 and the Turks appeal for mediation (*see* 1878).

France has a May 16 crisis that threatens her survival as a republic. Premier Jules Simon is forced to resign as Marshal MacMahon becomes irritated with his weak opposition to the anticlerical position of French Leftists, but the principle of ministerial responsibility is upheld against the personal power of the Rightist president and France remains a republic (*see* 1873; Boulanger, 1889).

Japan's samurai warrior class rebels against the "evil counselors" of the Meiji emperor who have denied the samurai their pensions and have forbidden them to wear two swords. The Satsuma Rebellion that begins in January is crushed by September as a modern army of trained commoners is funded by a large issue of paper money (*see* 1869; Constitution, 1889).

The "Molly Maguires" make headlines after a decade of activity in Pennsylvania mining towns by the secret Irish-American terrorist society. They have been apprehended by Pinkerton detective James McParlan, 35, an Irish-American himself who 5 years ago gained admission to the fraternal Ancient Order of Hibernians, some of whose lodges have been under Molly Maguire control. Eleven members of the secret society are hanged June 21 and eight more will eventually be hanged, all through testimony given by McParlan who will later become head of Pinkerton's Denver branch and be involved in suppressing activities of the Western Federation of Miners in Idaho (*see* Haywood, 1907).

The New York Central, Erie, Pennsylvania, and Baltimore & Ohio call off a long-standing rate war in April and announce a 10 percent wage cut. When the B&O announces a further 10 percent cut in mid-July its workers begin a strike at Martinsburg, W. Va., that is soon joined by Pennsylvania Railroad workers. Strikers set the Pittsburgh roundhouse ablaze, the strike spreads across the nation, the railroads are shut down for a week, and the strike is not ended until federal troops are called in, firing on the workers, killing some, wounding many (both sides claim victory).

The Atchison, Topeka and Santa Fe reduces wages 5 percent August 1 and slashes conductors' salaries from a top of $120 per month to $75. The conductors strike but soon return to work at lower pay (*see* 1878).

A San Francisco mob burns down 25 Chinatown wash houses in July and anti-Chinese riots ensue for months.

Striking trainmen shut down U.S. railroads for a week until federal troops moved in and fired on the workers.

The Chinese receive little or no police protection, giving rise to the phrase "not a Chinaman's chance" (*see* 1869; Los Angeles, 1871; Exclusion Treaty, 1880).

Nez Perce tribespeople in the Northwest are ordered to leave or be removed forcibly after years of passive noncompliance with the treaty they signed in 1863. Chief Joseph, 37, says, "I would give up everything rather than have the blood of my people on my hands." But in the chief's absence his braves kill several whites, U.S. troops are sent in by Howard University founder Gen. O. O. Howard, and the troops are nearly annihilated in the Battle of White Bird Canyon. The Nez Perce are weakened after 18 subsequent engagements, Chief Joseph leads 750 tribespeople in a retreat through nearly 1,600 miles of Rocky Mountain wilderness pursued by 600 soldiers, but fresh troops under Col. Nelson Appleton Miles, 38, surround him October 5. Forced to surrender in Montana Territory less than 40 miles from the safety of the Canadian border, Chief Joseph says, "I am tired of fighting. Our chiefs are killed. . . It is cold and we have no blankets. The little children are freezing to death. My people, some of them, have run away to the hills, and we have no blankets, no food. . . I will fight no more forever."

Gen. James Carleton orders Apaches in Arizona Territory out of the Chiricahua reservation at Warm Springs in November, telling them to move to San Carlos where temperatures in summer reach as high as 140° F. in a region with no game and with no food except the meager rations issued by Indian agents. Any man, woman, or child found off the reservation is to be shot without being given a chance to surrender (*see* Geronimo, 1876, 1886).

A Swiss Emancipation Law enacted May 15 gives Jews full citizenship, ending ghetto life.

Government regulation of U.S. business gains legal support from the Supreme Court March 1 in the case of *Munn v. Illinois*. The landmark decision sustains the state law of 1871 supervising grain elevators.

Singer Manufacturing Co. cuts sewing machine prices in half as the U.S. economic depression continues. The move will increase sales enormously (*see* 1863; 1880).

Chicago's Mandel Brothers opens on State Street where Leon, Emanuel, Simon, and Solomon Mandel have been offered a lease by Marshall Field who is bent on making State Street Chicago's main shopping artery. The new department store began as Klein and Mandel in 1855.

Standard Oil of Ohio president John D. Rockefeller signs a contract with the Pennsylvania Railroad that strengthens his oil-rail monopoly (*see* 1872). He has been exploiting the Depression by buying up depressed rivals in the oil industry and gaining control of all major pipelines and of the Pennsylvania's tank cars. Secret and illegal rebates and drawbacks by the railroads give Standard Oil lower freight rates and enable the company to force competitors to merge with the Rockefeller interests (*see* 1882).

1877 ***(cont.)*** The Supreme Court upholds state power to fix freight rates for intrastate traffic and for interstate traffic originating within a state's borders March 1 in the case of *Peik v. Chicago and Northwestern Railroad Co.* But a later decision will limit state regulation to intrastate shipments, exempt the much larger volume of freight that crosses state lines, and weaken the farmers' victory in the *Peik* decision (*see* 1886; Interstate Commerce Act, 1887).

Cornelius Vanderbilt has died January 4 at age 82 leaving a fortune of $105 million to the widow of 38 and children who survive him (relatively little goes to the widow and eight daughters but his son William Henry, 55, receives $90 million and *his* four sons receive $11.5 million among them, all in stock). Vanderbilt has laid new steel rails and built new bridges to cut running time on the New York Central between New York and Chicago from 50 hours to 24 (*see* 1873; 1902).

Callao, Lima & Oroya Railroad builder Henry Meiggs dies at Lima, Peru, September 29 at age 66 after a series of paralytic strokes leaving the world's highest railroad incomplete, but his tunnel under Mount Meiggs at an altitude of 15,658 feet remains an engineering marvel (*see* 1871; Grace, 1890).

The first U.S. cantilever bridge is completed across the Kentucky River near Harrodsburg. Designed by Charles Shaler Smith for the Cincinnati Southern, the $377,500 bridge tested April 20 has three spans of 375 feet each.

Boston entrepreneur Augustus Pope converts his Hartford, Connecticut, air-pistol factory into the first U.S. bicycle factory (*see* Starley, 1870). Pope's Columbia bicycle with its high front wheel will be popular for more than a decade until safety bicycles make it obsolete, and Pope will soon be grossing $1 million per year (*see* 1896; safety bicycle, 1885).

Barbed-wire prices drop from 18¢ lb. to 8¢ as the Bessemer steel-making process of 1856 is applied to barbed-wire production. All barbed-wire patents were acquired last year by Washburn & Moen Co. of Worcester, Mass., and sales of barbed wire leap from last year's 840,000 pounds to 12.86 million pounds. A ton of barbed wire represents 2 miles of three-strand "devil's rope" and sales will climb to 26.7 million pounds next year, 50.3 million in 1879, 80.5 million in 1880, and 120 million in 1881 (*see* Gates, 1875, 1898).

German botanist Wilhelm Pfeffer, 31, discovers osmosis, finding that fluids are diffused through semipermeable membrane tissues.

Oxygen, nitrogen, and other gases not heretofore considered liquefiable will be liquefied in the next year by French physicist Louis Paul Cailletet, 45 (*see* von Linde, 1898).

German country doctor Robert Koch, 33, of Wollstein demonstrates a technique for fixing and straining bacteria. Koch, who last year obtained a pure culture of the anthrax bacillus by using the eye of an ox as a sterile medium for culturing the bacillus, will confirm Louis Pasteur's suggestion that every disease produced by microorganisms is produced by a specific bacillus (*see* tuberculosis bacillus, 1882).

English researchers A. Downes and T. P. Blunt discover the germicidal qualities of ultraviolet rays. Their findings will lead to new techniques for sterilization.

German surgeon-urologist Max Nitze, 28, constructs the cystoscope for examining the inside of the urinary bladder and introducing medication into it.

James Paget, now 63, describes the chronic bone disease *osteitis deformans* that will be called Paget's disease (*see* 1835). It will defy a century of effort to unearth its etiology and find a specific therapy.

The Japanese Red Cross is founded at Tokyo where the Meiji emperor is mobilizing forces to put down the Satsuma Rebellion (above).

Education for Italian children from age 6 to 9 becomes mandatory under terms of a new law but poor administration will make the law ineffectual.

Tokyo University is founded in Japan. The imperial university will be the nation's most prestigious.

The American Museum of Natural History opens in New York opposite Central Park at West 77th Street.

The first university press is founded at Baltimore's Johns Hopkins University where the *American Journal of Mathematics* begins publication. The new university is the first institution to subordinate student needs to research demands, inaugurating the "publish-or-perish" system that will lead to widespread publication of pedantic papers which often have little to do with real education.

The first Bell telephone is sold in May; 778 are in use by August.

The Bell Telephone Association is organized with headquarters at New York. Founder Gardiner Hubbard gives himself and his son-in-law Bell each a 30 percent interest in the Bell patent, gives Bell's assistant Thomas A. Watson a 10 percent interest, and gives a 30 percent interest to investor Thomas Sanders, who put up more than $100,000 to back Bell's early experiments.

Western Union's American Speaking-Telephone Co. competes with Bell, but the new Bell firm is deluged with applications for telephone agencies, and Hubbard starts leasing telephones at the rate of 1,000 per month.

The first telephone switchboard is installed May 17 in the Boston office of Edwin T. Holmes, proprietor of Holmes Burglar Alarm Service. Bell loans Holmes 12 telephones and the switchboard at 342 Washington Street is used for telephone service by day and a burglar alarm at night.

The first telephone exchange is organized at Lowell, Mass., by local entrepreneur Charles Jasper Glidden, 20 (*see* New Haven, 1878). He will quickly develop a system under the name New England Telephone and Telegraph and another under the name Erie Telephone and Telegraph, and will not sell out to the Bell interests until the 1900s.

German-American inventor Emile Berliner, 26, develops a loose-contact telephone transmitter superior to Bell's telephone (*see* 1878; Hughes, 1855).

A hand-cranked "phonograph or speaking machine" demonstrated by Thomas Edison November 29 records sounds on grooved metal cylinders wrapped in tinfoil. Edison has followed up on his telephone research, shouts the verses to "Mary Had a Little Lamb" into his machine, is astonished when it plays his voice back, applies for a patent in December, and is soon attracting paid audiences at "exhibitions" of the machine (*see* 1887).

The *Farm Journal* begins publication in March. Philadelphia publisher Wilmer Atkinson, 30, sends out 25,000 copies to farmers and other rural residents within a day's ride of Philadelphia.

The *Washington Post* begins publication December 6 with a four-page edition that sells at 3¢ and will have a circulation of 11,875 within a year. Publisher Stilson Hutchins, 39, calls President Hayes the "de facto president," refers to Hayes as "His Fraudulency," and uses the term "President Tilden" in his outrage over the "crime of '76." He will buy out the competing *National Union* next year and the *Daily Republican* in 1888 but will sell out in 1889 (*see* Meyer, 1933).

Britain's remaining restrictions on freedom of the press end with the court trial at London of Charles Bradlaugh, 44, and Annie Wood Besant (Ajax), 30, who have republished Charles Knowlton's Malthusian work of 1835 "Fruits of Philosophy" advocating contraception (below) (but *see* Official Secrets Act, 1911).

The Boston Library opens on Copley Square in a Romanesque building designed by Henry Hobson Richardson. Its chief benefactor has been the late Joshua Bates, a local financier who was a partner in London's Baring Brothers from 1828 until his death in 1864 at age 76 (*see* 1895).

Fiction *Anna Karenina* by Leo Tolstoy; *The American* by Henry James; *Deephaven* (stories) by South Berwick, Me., author Sarah Orne Jewett, 27.

Juvenile *Black Beauty* by English author Anna Sewell, 57, whose "Autobiography of a Horse" will be a worldwide best seller and remain popular for more than a century. Sewell gets a flat £20 for her manuscript and will receive no royalties.

Painting *Nana* by Edouard Manet; *Lever de Lune* by Charles François Daubigny; *The Cotton Pickers* by Winslow Homer; Gustave Courbet dies December 31 at age 58: imprisoned for 6 months in 1871 for destroying the Vendôme column at Paris, he fled to Switzerland in 1873 after being sentenced to pay for the column's reconstruction and dies near Vevey.

Ballet *Swan Lake* 3/4 at Moscow's Bolshoi Theater with music by Petr Ilich Tchaikovsky. The ballet is incomplete and the production not a success (*see* 1895).

The cakewalk is introduced into U.S. minstrel shows by the New York team Harrigan and Hart whose "Walking for Dat Cake" number imitates antebellum plantation "cakewalks."

Opera *The Sorcerer* 11/17 at London's Opera Comique, with Gilbert and Sullivan arias that include "My Name Is John Wellington Wells." The operetta begins a run of nearly 6 months; *Samson et Dalila* 12/2 at Weimar's Grand Ducal Palace, with music by Camille Saint-Saëns.

First performances *Francesca da Rimini* (Fantasia for Orchestra after Dante) by Petr Ilich Tchaikovsky 3/9 at Moscow; *Les Eolides (The Aeolidae)* (symphonic poem) by Belgian-French composer César (August Jean Guillaume) Franck, 54, 5/13 at Paris; Symphony No. 2 in B flat major by the late Franz Schubert (who wrote it in 1815) 10/20 at London's Crystal Palace; Symphony No. 3 in D minor by Anton Bruckner 12/16 at Vienna; Symphony No. 2 in D major by Johannes Brahms 12/30 at Vienna.

Cantata *Lallah Rookh* by English composer Frederick Clay, 39, in February at the Brighton Festival with "I'll Sing Thee Songs of Araby," lyrics by Irish poet-playwright William Gorman Wills, 49.

Hymns "Where Is My [Wand'ring] Boy Tonight?" by Robert Lowry; "Hiding in Thee" by Ira David Sankey, lyrics by William O. Cushing.

Popular songs "In The Gloaming" by English composer Annie Fortescue Harrison, 26, who has recently married the comptroller of Queen Victoria's household, Sir Arthur Hill, lyrics by Meta Orred; "The Lost Chord" by Gilbert & Sullivan's Arthur Sullivan who has been inspired by the fatal illness of his brother Frederick, lyrics by the late poet Adelaide A. Procter whose verse was published in 1858 and who died 6 years later at age 39.

The Westminster Kennel Club dog show inaugurated in May at New York's 3-year-old Gilmore's Garden attracts 1,201 entries in the "first annual N.Y. Bench Show." Dogs compete for silver trophies produced by Tiffany & Co. Fanciers have organized the club at the Westminster Hotel on Irving Place at 16th Street, and the founding members have adopted as the club's symbol the head of their pointer Sensation whose stud services they offer at $35. While no "best of show" award will be made until 1907, the first show is so popular that its scheduled 3-day run is extended for an extra day with proceeds from the fourth session going to the ASPCA.

The first Wimbledon lawn tennis championship matches are organized by the 9-year-old All England Croquet Club. Spencer Gore, 27, wins the first singles title (*see* 1874; USLTA, 1881).

The first low-rent U.S. housing "project" opens at Brooklyn, N.Y., where businessman Alfred Tredway White has financed construction of cottages for workingmen at Hicks Street and Baltic, each 11.5 feet wide with six rooms. By 1880 White will have built 44 of the cottages for $1,150 each exclusive of land and be renting them at $14 per month. He will build 226 apartments in six-story "sun-lighted tenements" round a central court with communal bathing facilities in their basements and will rent the four-room apartments for $1.93 per week.

1877 ***(cont.)*** Timber barons are "not merely stealing trees but whole forests," says Secretary of the Interior Carl Schurz, who has been named to the post by President Hayes. Schurz emigrated from his native Germany in 1848 at age 19, practiced law at Milwaukee, served as minister to Spain, fought as a Union Army brigadier general of volunteers, and has served as U.S. Senator from Missouri (*see* Muir, 1876; Forest Reserves Act, 1891).

Swan Boats designed by Robert Paget appear on the pond of Boston's 18-year-old Public Garden with official city permission and quickly become a major attraction for visitors.

The Fresh Air Fund started by Pennsylvania clergyman Willard Parsons brings needy city children for a visit to his rural community. Most have never seen a cow (*see* 1881).

The Halifax Fisheries Commission created by the Treaty of Washington in 1871 awards Britain $4.5 million for U.S. fishing rights in the North Atlantic November 23.

Congress establishes a U.S. Entomological Commission to control grasshoppers which have been devastating western farms and rangelands.

California land speculators James Ben Ali Haggin, 49, and Lloyd Tevis, 53, use political influence to have Congress pass a Desert Land Act offering land at 25¢ per acre to settlers who will agree to irrigate and cultivate the land for 3 years. Haggin and Tevis have acquired hundreds of thousands of acres by buying up old Spanish land grants, often of dubious value, and they acquire 96,000 acres in the San Joaquin Valley by hiring scores of vagabonds to enter false claims which they then transfer to themselves, ousting settlers who have not yet perfected their titles under old laws and have not heard about the new law. The Haggin-Tevis lands will become the basis of Kern County Land Co. (*see* 1890; Tenneco, 1961).

Famine kills 4 million in Bengal.

Argentina challenges Australia as a source of meat for Europe by sending its first refrigerator ship to France with a cargo of meat. The S.S. *Paraguay* employs an ammonia compression system devised by French engineer Charles Albert Abel Tellier, 49 (*see* 1880; Linde, 1873).

The first shipment of Chicago-dressed beef to reach Boston arrives in a railcar designed to prevent spoilage by Chicago meat packer Gustavus Franklin Swift, 38, who is determined to save the cost of transporting whole live animals. Swift's partner James Hathaway has sold his interest in the firm anticipating failure, but Swift has refused to be thwarted by the railroads that demand exorbitant rates to make up for lost tonnage. He has used a railroad that has not carried livestock shipments (*see* 1881).

A centrifugal cream separator invented by Swedish engineer Carl Gustaf Patrick de Laval, 32, eliminates the need of space-consuming shallow pans and the labor of skimming the cream that rises to the top. By reducing the cost of producing butter, the Laval separator will lead to a vast expansion of the Danish, Dutch, and Wisconsin butter industries.

New York State outlaws the representation of margarine as "butter," but the law will be widely ignored.

Brick cheese, invented by Wisconsin cheese maker John Jossi, is a more elastic, slightly milder version of Limburger (*see* Liederkranz, 1892).

The Quaker Mill Co. begins operations at Ravenna, Ohio, making oatmeal with a process patented by William Heston (*see* Schumacher, 1875). Heston and his partner Henry D. Seymour register as a trademark the "figure of a man in Quaker garb," but their venture suffers for lack of business acumen and will soon be acquired by distiller Warren Corning, who will use the Quaker symbol on a brand of whiskey (*see* 1884).

Granula is introduced at Battle Creek, Mich., by James Harvey Kellogg who has created the cold cereal breakfast food by mixing wheat, oatmeal, and cornmeal and baking half-inch thick biscuits which he grinds up and sells at 12¢ per pound in 1, 2, and 5-pound packages (*see* 1876; Jackson, 1863; Granola, 1881).

Distillers Company, Ltd. is created by a consolidation of six Scottish distilleries which begin a great whiskey cartel. The syndicate will be joined in 1919 by John Haig, in 1924 by Robert Burnett, and in 1925 by Buchanan-Dewars (Black & White, Dewars White Label) and John Walker (Johnny Walker Red Label and Black Label).

The taboo subject of contraception comes into the open at the Bradlaugh-Besant trial (above). The acquittal of Bradlaugh and Besant will lead to an end of the ban on disseminating contraceptive advice.

Queensland's new governor Arthur Kennedy gives approval August 20 to a drastic cut in Chinese immigration; his predecessor had refused demands for such a cut.

1878 Italy's Victor Emmanuel II dies January 9 at age 57. He has ruled the country since its unification in 1861 and is succeeded by his son of 33 who will reign until 1900 as Umberto I.

Ottoman forces capitulate at Shipka Pass and ask Russia for an armistice January 9. Russian forces take Adrianople, Disraeli sends the British fleet in response to the sultan's request, but the fleet is recalled and an armistice concluded January 31.

A London music hall ditty introduces the word "jingoism," a synonym for sabre-rattling chauvinism: "We don't want to fight, but by jingo if we do,/ We've got the men, we've got the ships, we've got the money, too." The British fleet arrives at Constantinople February 15.

The Treaty of San Stefano signed outside Constantinople March 5 ends hostilities with Russia but angers Britain, Germany, Austria, and Italy. It provides for Romanian independence (*see* 1881), Montenegrin and Serbian independence, reforms in Bosnia and Herzogovina, Bulgarian autonomy (*see* 1887), a huge Turkish indemnity, and some Russian territorial gains.

The Berlin Congress in June and July dismembers the Ottoman Empire but denies Russia most of her gains. Mediated by Bismarck and Disraeli, it divides Bulgaria in three, gives southern Bessarabia, Batum, Kars, and Ardahan to Russia but the Dobrudja to Romania, grants Austria a mandate to occupy Bosnia and Herzogovina, reduces the size of Serbia and Montenegro, and leaves Russians, Austrians, and Slavs seething.

Britain occupies Cyprus under terms of a secret June 4 agreement with the sultan to defend Turkey against further attack. France occupies Tunis.

Italian irredentists begin agitating to obtain Trieste and the southern Tyrol from Austria.

Chile and Bolivia verge on war over the control of nitrate deposits, essential to world production of explosives and fertilizer. Bolivia boosts the export tax on nitrates being mined by Chilean Nitrate Co. in Antofagasta, the company appeals to the Chilean government, and Bolivia temporarily rescinds Chilean Nitrate's contract (*see* 1879).

A trial of 193 young Russian revolutionaries ends January 23 with the acquittal of a great many of the accused but the police, using special powers of "administrative exile," pick up most of the men and women acquitted and send them off to Siberia or elsewhere. St. Petersburg's chief of police Gen. F. F. Trepov is shot and wounded in January by revolutionary Vera Zasulich, 26, as an act of protest against the flogging of a prisoner ordered by Gen. Trepov. Zasulich is clearly guilty, but a jury of her peers acquits her on the ground that her action was a legitimate form of political protest, a crowd cheers her as she leaves the courthouse, the authorities permit her to go abroad, and she will become an associate of Karl Marx. Another revolutionary stabs Gen. N. V. Mezentsev with a sword August 4 in broad daylight, and systematic terrorism will begin next year (*see* 1887).

The Atchison, Topeka and Santa Fe cuts engineers' pay 10 percent April 4, most engineers and firemen strike, the railroad is closed down for 5 days, militia is called in at Emporia, Kan., one militiaman kills an innocent bystander, and the death leads to the arrest of several strike leaders (*see* 1877).

The Bland-Allison Act passed by Congress February 28 makes the silver dollar legal tender and requires that the U.S. Treasury buy $2 million to $4 million worth of silver each month. Introduced by Rep. Richard P. Bland (Mo.) and Rep. William B. Allison (Ia.), the bill encourages prospecting for silver and will not be repealed until July 1890 (*see* Leadville, 1879).

Congress votes May 31 to reduce circulation of greenbacks; $346.7 million worth are in circulation and by December 17 they have regained their face value for the first time since 1862.

Nearly 10,500 U.S. business firms fail as the economic depression that began in 1873 continues.

The Tiffany Diamond found in South Africa's Kimberly Mine (and purchased by New York's Tiffany & Co.) is the largest and finest yellow diamond ever discovered. It weighs 287.42 carats, cutters at Paris will study it for a year before touching it, and they will turn it into a 128.51-carat stone with 90 facets.

Thomas Edison works out methods for cheap production and transmission of electrical current and succeeds in subdividing current to make it adaptable to household use. Gas company shares plummet as news of Edison's work reaches Wall Street.

Johns Hopkins professor Henry Rowland, 30, and Harvard professor John Trowbridge, 35, have helped Edison, who looks for a lamp filament that will burn for extended periods of time in a high vacuum; he experiments with molybdenum, nickel, and platinum filaments (*see* 1879).

The Edison Electric Light Co. is founded October 15 by New York investors who have raised $50,000 to support Thomas A. Edison's experiments. Edison receives half the stock in the company, incorporated later in the month with $300,000 in capital (*see* 1879).

The first carbon filament incandescent light bulb of any value is demonstrated December 18 at a meeting of the Newcastle-on-Tyne Chemical Society. English chemist Joseph Wilson Swan, 50, has produced the bulb with help from Charles H. Stearn, but it does not achieve true incandescence and Swan will not be able to perfect it or get his lamp into volume production until 1881 (*see* Edison, 1879).

The Santa Fe lays tracks through Raton Pass 15 miles south of Trinidad, Colo., into New Mexico Territory. A 2,000-foot tunnel is bored through the pass 8,000 feet high in the Sangre de Cristo Mountains.

Canadian-American entrepreneur James J. Hill, 40, and his associates buy the St. Paul and Pacific Railroad which they will reorganize and extend in 1890 to create the Great Northern Railway, a road built without government subsidy (*see* Canadian Pacific, 1881; Northern Pacific, 1901).

New York's Ninth Avenue Elevated begins running June 6 from Trinity Church to 58th Street. By next year it will be a double-track road extending 6½ miles from Morris Street up Sixth to 53rd Street, west to Ninth Avenue, and on to 155th Street; apartment houses will rise up quickly along its route (*see* 1940).

The steamboat *J. M. White* launched on the Mississippi is the grandest ever seen on the river. Her smokestacks are 80 feet high, her roof bell weighs 2,880 pounds, it takes two pilots to handle her 11-foot wheel, and the $300,000 steamboat sets a new speed record by steaming from New Orleans upriver to St. Louis in 3 days, 23 hours, and 9 minutes. Riverboat traffic at St. Louis will reach more than 1.3 million tons in 1880 but will be only half that by 1885, and will fall to 347,000 tons in 1900, 43,090 tons in 1910 as railroads displace steamboats.

New York physician William Henry Welch, 28, at Bellevue Hospital opens the first U.S. pathology laboratory (*see* Johns Hopkins, 1893).

Mississippi steamboats proceeded majestically up and down the river. Occasional races produced great excitement.

German psychologist Wilhelm Max Wundt, 46, establishes the first laboratory for experimental psychology, bringing the human mind into the realm of science (*see* Pavlov, 1879; Breuer, 1882).

A yellow fever epidemic sweeps the U.S. Gulf Coast states and Tennessee, killing an estimated 14,000. Some 4,500 die at New Orleans alone and 5,150 die up the Mississippi at Memphis (*see* Finlay, 1881).

A smallpox epidemic strikes Deadwood in the Dakota Territory. Frontierwoman Martha Jane Canary, 26, works heroically in men's clothing to nurse the ill and renders service that will help make her legendary as "Calamity Jane."

Jehovah's Witnesses has its beginnings as Pittsburgh Congregationalist minister Charles T. Russell, 26, becomes pastor of an independent church at which he will preach the doctrine that Christ's second coming came invisibly in 1874 and that the world has since then been in the "Milennial Age" that will end in 1914 and will be followed by social revolution, chaos, resurrection of the dead, and, ultimately, by the establishment of Christ's kingdom on earth (*see* 1872; 1884).

The Salvation Army adopts that name (*see* 1865; Volunteers of America, 1896; Shaw drama, 1905).

The first commercial telephone exchange opens January 28 at New Haven, Conn. The exchange has 21 subscribers and after 6 weeks it operates at night as well as during the day (*see* 1883; Glidden, 1877).

The first telephone directory, issued February 21 by the New Haven Telephone Co., lists 50 subscribers.

The Bell Co. buys Emile Berliner's loose-contact telephone transmitter. Berliner patents the use of an induction coil in a transmitter (*see* 1877; gramophone, 1887).

Remington Arms Co. improves the Remington typewriter of 1876 by adding a shift key system that employs upper- and lower-case letters on the same type bar. Wyckoff, Seamens, and Benedict buy Remington's typewriter business and found Remington Typewriter Co. (*see* Underwood, 1895).

Berlitz Schools of Languages have their beginnings at Providence, R.I., where German-American teacher Maximilian Delphinius Berlitz, 27, opens his own language school after 9 years of teaching French and German in a theological seminary. He goes out of town for a week, leaving his students in the care of a Frenchman who speaks no English; when he finds that the students have made amazing progress in his absence, he develops an easy-to-learn method using instructors who speak no English.

The *Chattanooga Times* begins publication under the direction of Cincinnati-born publisher Adolph Simon Ochs, 20 (*see* 1896).

The *Cleveland Press* has its beginnings in the *Penny Press* started by Edward Wyllis Scripps, 24, of the Detroit *Evening News*. The Detroit paper was started 5 years ago by Scripps's English-American half brother who helps Edward start the Cleveland paper with a substantial loan (*see* UP, 1907).

The *St. Louis Dispatch* begins publication December 12 and is acquired by Hungarian-American journalist Joseph Pulitzer, 31, who will merge the *Dispatch* with the *St. Louis Post* to create the *Post-Dispatch* and begin a newspaper empire. Pulitzer arrived in America in 1864, served in the First New York Cavalry in the Civil War, worked as a reporter for a St. Louis German-language newspaper, won election to the legislature in 1869, studied law, and was admitted to the bar 2 years ago (*see* 1883).

J. Walter Thompson takes over the 14-year-old advertising firm Carlton and Smith which he joined in 1868 and begins representing general magazines (*see* 1868). Commodore Thompson's first advertisement will be for Prudential Insurance Co. (*see* 1875), and in 15 years he will control the advertising in 30 publications, mostly women's magazines (*see* Curtis, 1901).

Fiction *The Return of the Native* by Thomas Hardy; *The Europeans* by Henry James.

Painting *Mme. Charpentier and Her Children* by Pierre Auguste Renoir; *Rehearsal on the Stage* by Edgar Degas; *Sierra Nevada* by Albert Bierstadt; Charles François Daubigny dies at Auvers-sur-Oise February 19 at 61.

Tiffany glass has its beginnings in a factory established by New York painter-craftsman Louis Comfort Tiffany, 30, son of the jeweler, who has studied with Venetian glass maker Andrea Boldoni (*see* 1894).

Theater *The Pillars of Society* (*Samfundets stotter*) by Henrik Ibsen 11/6 at Oslo's Mollergaten Theater; *The Girl with No Dowry* (*Bespridannitsa*) by Aleksandr Ostrovsky 11/10 at Moscow's Maly Theater.

First performances Symphony No. 4 in F minor by Petr Ilich Tchaikovsky 2/10 at Moscow; Sonata in A major for Violin and Piano by French composer Gabriel

Urbain Fauré, 33, at the Trocadero Hall built for the Paris Exposition.

Opera *H.M.S. Pinafore (or The Lass that Loved a Sailor)* 5/25 at London's Opera Comique, with the Comedy Opera Company singing Gilbert and Sullivan arias that include "We Sail the Ocean Blue," "I Am the Captain of the Pinafore," "I Am the Monarch of the Sea," "When I Was a Lad," "Things Are Seldom What They Seem," "Carefully on Tiptoe Stealing," and "He Is an Englishman;" *The Secret* (*Tagernstvi*) 9/18 at Prague, with music by Bedřich Smetana.

The Central City Opera House opens in the Colorado mining town.

The Cincinnati Music Hall opens on Elm Street with 3,600 seats for the May Festival held since 1873.

Popular songs "Carry Me Back to Old Virginny" by pioneer black songwriter James A. Bland, 24.

Herreshoff Manufacturing Co. is formed by U.S. yacht designer John Brown Herreshoff, 37, blind since age 14, and his brother Nathanael Greene, 30. Their older brother James Brown, 44, will invent a fin keel for racing yachts that Herreshoff Manufacturing will put to practical use in 1891 when it begins designing yachts for the defense of the America's Cup.

A new Monte Carlo gambling casino opens to replace the original 1862 structure. Designed by Charles Garnier of 1875 Paris Opéra fame, the new casino will be augmented in 1910 with an opera house (*see* 1913).

Frank Hadow, 23, returns to England on leave from his Ceylon tea plantation and wins in Wimbledon singles play.

Bat Masterson captures the notorious outlaw Dave Rudabaugh in Dakota Territory and is appointed U.S. marshal. William Barclay Masterson, 24, hears later in the year that his brother Edward has been gunned down while acting as marshal of Dodge City, Kan., he brings quick retribution to Edward's two killers, and he will continue as a western lawman and gambler until 1902, when he will quit to become a New York sportswriter.

Texas outlaw Sam Bass gathers a new gang at Denton 4 months after losing companions in a fight through Kansas and Missouri following a heist of $65,000 in gold and other valuables from a Union Pacific train at Big Springs, Neb. The Bass gang robs four trains near Dallas, but Bass is betrayed to the Texas Rangers, trapped while trying to rob a bank at Round Rock, and mortally wounded at age 26.

Will's Three Castles cigarettes are introduced at London (*see* 1892).

President Hayes invites the children of Washington to an Easter-egg roll on the White House lawn and begins an annual event.

Ivory Soap has its origin in The White Soap introduced by Cincinnati's 51-year-old Procter and Gamble Company whose chemists have perfected a hard, white, floating soap. Harley Procter, now 21, will be inspired to rename the soap Ivory while reading a psalm in church in 1882 and beginning in December of that year Ivory will be aggressively advertised with the slogan "99 and 44/100% pure" (*see* Crisco, 1911; Swan, 1940).

Report on the Lands of the Arid Region of the United States by John Wesley Powell points out that land west of the 96th meridian is arid, only a few areas of the Pacific Coast and the mountains receive as much as 20 inches of rain per year, an economy based on traditional patterns of farming is not possible under these circumstances, dry-land farming depends on irrigation, water rights in the West are more valuable than land titles and should be tied by law to each tract of land, only a small fraction of the land is irrigable, the Homestead Act of 1862 must be revised if it is to apply to western lands, western land policies must be geared to the region's climatic conditions, optimum irrigation will depend on large dams and canals financed and built under federal government leadership, and the best reservoir sites should be identified and reserved at the outset (*see* 1869).

The last mass nesting of passenger pigeons is seen at Petoskey, Michigan (*see* 1871; 1914).

Alaska gets its first salmon cannery (*see* 1866; 1890; 1930).

The Matador Ranch is established in the southern Texas Panhandle by Chicago promoter Henry H. Campbell who will be backed by Scottish financiers and will spread his holdings over most of the Panhandle and up into Montana Territory (*see* Prairie Cattle Co., 1881).

The first commercial milking machines are produced at Auburn, N.Y., by Albert Durant who introduces a machine invented by L. O. Colvin.

The worst famine in history kills at least 10 million Chinese and possibly twice that number as drought continues in much of Asia as it has since 1876.

Beet sugar extraction mills are demonstrated at the Paris World Exhibition. Most European countries will be encouraged by the Paris exhibit to plant sugar beets and build factories (*see* 1811; 1880).

The Hutchinson Bottle Stopper invented by Charles G. Hutchinson is made of wire with a rubber washer, it seals in the carbonation of effervescent drinks when pulled up tight, and when pushed in it permits a beverage to be poured or swallowed from the bottle, but the rubber imparts a taste to the beverage (*see* Painter, 1892).

The Silvertown Refinery owned by London sugar magnate Henry Tate, 69, introduces the first sugar cubes.

Charles A. Pillsbury at Minneapolis installs steel rollers to replace millstones and thus multiplies the output of his flour mills (*see* 1871; Washburn, 1874; A Mill, 1880).

Western Cold Storage Co. opens an ice-cooled cold store at Chicago (*see* 1887).

Salt wells go into production in New York's Wyoming County as a high tariff encourages production.

1878 ***(cont.)*** The first roasted coffee to be packed in sealed cans is packed by Boston's 15-year-old Chase and Sanborn.

Anti-vice crusader Anthony Comstock of 1872 Comstock Law fame approaches a reputed New York abortionist who has a luxurious mansion at 657 Fifth Avenue (52nd Street). Hiding his identity, he says he is desperately poor and will be ruined if his wife has another child. English-born brothel-keeper Anna A. Trow Somers Lohman, 65, alias Mme. Restell, supplies Comstock with drugs and contraceptives. He has her indicted and refuses a $40,000 offer to drop his charges, Mme. Restell posts bail and returns to her house, newspaper stories convince her that she will be convicted, and she slits her throat in her bathtub April 1.

The world's first birth control clinic is opened at Amsterdam by Dutch suffragist leader Aletta Jacobs, 29, who is the first woman physician to practice in Holland (*see* Bradlaugh, Besant, 1877; Sanger, 1914).

1879 French Republicans gain 58 seats in the January 5 senatorial elections but face a hostile majority in the senate and chamber. President MacMahon resigns January 30 with 1 year of his 7-year term still left to go, and is succeeded by Conservative Republican Jules Grévy.

The War of the Pacific begins February 14 as Chilean troops occupy Antofagasta. Bolivia proclaims a state of war, Peru refuses to guarantee neutrality, the United States is unsuccessful in its attempts at mediation, and Chilean forces proceed to occupy the entire Bolivian coast (*see* 1878; 1884).

The Irish Land League is founded to campaign for independence from Britain. Irish nationalist Michael Davitt, 33, lost his right arm in a cotton-mill accident at age 11, he was sent to Dartmoor Prison at 24 on treason-felony charges that he arranged to ship arms into Ireland, and he has recently returned from visiting his American-born mother and his siblings in the United States (*see* 1870; Parnell, 1881).

The Zulu nation founded by Shaka in 1816 ends in a blood bath as British breech-loading rifles kill some 8,000 Zulu warriors and wound more than 16,000 while 76 British officers and 1,007 men are killed in action, 17 officers and 330 men killed by disease, 99 officers and 1,286 men invalided home, 37 officers and 206 men wounded, and nearly 1,000 Natal Xhosa corpsmen killed. The British forces have been commanded by Gen. Frederic Augustus Thesiger, 51, second Baron Chelmsford, the Zulu by Cetshwayo, 51, whose father was a half brother of the late Shaka. The Zulu War ends only after an incredible stand by the British March 29 at Rorke's Drift where 140 soldiers—30 of them incapacitated—hold off 4,000 Zulus armed with spears and some rifles. Eleven Victoria Crosses (a record for any single action) will be awarded to the heroes of Rorke's Drift but later the same day the British lose 895 out of a force of 950 plus some 550 Xhosa allies in an attack by Zulu warriors at Isandhlwana.

France's Prince Imperial Louis, 23, has accompanied the British in Zululand and is killed June 2 by a Zulu assegai (spear). French Anglophobes charge that the prospective Napoleon IV was killed with the connivance of Queen Victoria.

The Zulu king Cetshwayo (above) is captured by the British August 20 and sent in ill-fitting European clothes to Cape Town. Zululand is divided into 13 separate kingdoms whose territory will eventually be absorbed in large part by Natal.

The Egyptian khedive Ismail is deposed by the sultan June 25 under pressure from the European powers after 16 years of profligacy. Ismail is succeeded by his son Tewfik, 27, who will rule until 1892.

The Afghan emir Sher Ali has died in February and been succeeded by his son Mohammed Yakub, 29, who is forced 3 months later to accept the Treaty of Gandamak, allowing the British to occupy the Khyber Pass on condition that they pay the emir an annual subsidy of £60,000. The Afghans rise against the British September 3, the British envoy Sir Louis Cavagnari and his escort are murdered, British troops advance on Kabul and take it October 12, Yakub is forced to abdicate October 19 and is succeeded by his cousin Abd-er-Rahman, 49, who will reign until 1901, playing off the British against the Russians.

The Austro-German alliance signed October 7 has been engineered by Germany's "Iron Chancellor" Bismarck after more than a year of tension between Berlin and St. Petersburg. The treaty promising mutual support in the event of an attack by Russia on either country will be renewed periodically until 1918.

A large exodus of Southern blacks to Kansas begins as restrictions against former slaves increase in states of the old Confederacy. Black migrants to Kansas early in this decade have founded the town of Nicodemus (and some have become cowboys), but their numbers have been insignificant as compared with the 6,000 freed people from Louisiana, Mississippi, and Texas who respond to the leadership of Henry Adams and Benjamin "Pap" Singleton who have selected Kansas because few blacks can afford the transportation to Liberia. While many blacks remember John Brown and regard Kansas as a modern Canaan, Southern whites resist their departure lest there be a labor shortage.

Congress gives women the right to practice law before the U.S. Supreme Court.

U.S. specie payments resume January 1 for the first time since the end of 1861.

Leadville, Colo., becomes the world's largest silver camp with more than 30 producing mines, 10 large smelters, and an output of nearly $15 million (*see* 1874; Bland-Allison Act, 1878; May Company, below).

Germany's policy of free trade ends July 13 with a new protective tariff law. German industry has been depressed since the financial crisis of 1873 with its ensuing depression, imports from foreign producers have hurt German agriculture, and protectionism will spur development of German industry, transportation, and foreign trade (*see* 1914).

An era of finance capitalism that will last for more than half a century begins in the United States.

Progress and Poverty by U.S. economist Henry George, 40, points out that while America has become richer and richer most Americans have become poorer and poorer. Since land values represent monopoly power, says George, a "single tax" should be imposed on landowners that would provide all the revenue needed by government and would free industry from taxes (*see* Boisguilbert, 1705). George has met the Irish nationalist Michael Davitt (above) on Davitt's visit to America.

Metropolitan Life Insurance Co. of New York begins offering small policies for wage-earners and pioneers mass insurance coverage in an age of high mortality. Agents collect small weekly premiums in cash from the working classes who most need insurance protection. The company will attract 554 English agents who will emigrate with their families, and within 3 years Metropolitan Life will have increased the number of its district offices from 3 to 50 (*see* 1868; Tower, 1909).

Sir Joseph Swan demonstrates a carbon-filament light bulb to 700 people at Newcastle-upon-Tyne February 5 while Thomas Edison experiments with filaments of platinum, carbonized paper, bamboo thread, and other substances (*see* 1878). The incandescent bulb that Edison demonstrates October 21 has a loop of cotton thread impregnated with lamp black and baked for hours in a carbonizing oven. This vacuum light bulb is much like the one pioneered by German chemist Herman Sprengel in 1865. After it has burned for 45 hours Edison is sure it will burn for at least 100. His bulb is announced December 21, Edison Electric Co. stock soars, and the inventor says electricity will make lighting so cheap that only the rich will be able to afford candles. But candlewax, whale oil, coal oil, coal gas, and kerosene will continue to light the world until the development of dynamos, fuses, and sockets (*see* 1880; 1882).

Cleveland and San Francisco install streetlighting systems that employ arc-lamps invented by Charles Francis Brush, 30, who has devised a method for stabilizing the electric arc between carbon electrodes (*see* 1880; Paris, 1841).

U.S. electrical wizard Elmer Ambrose Sperry, 19, invents an improved dynamo and a new type of arc lamp (*see* gyroscope, 1910).

Four California oil wells and a refinery are merged by San Francisco financial interests to create Pacific Coast Oil Co. whose production will soon go entirely to John D. Rockefeller's Standard Oil Co. (*see* 1900).

The *Orient* goes into service for the Orient Line's Australian passenger traffic and is the first ship to be lighted by electricity. Largest ship afloat except for the 26-year-old *Great Eastern,* the 445-foot iron vessel is a 11,500-ton single-screw steamship, but four square-rigged masts supplement her 3-cylinder 5,400-horsepower engine.

Electricity drives a railroad locomotive for the first time at Berlin.

Thomas Edison's brilliance and perseverance improved the telephone and produced the phonograph and lightbulb.

Le Train Bleu begins thrice-a-week service as an all-sleeper express between Calais and Rome via Nice. Georges Nagelmakers has developed the train (*see* 1876; Orient Express, 1883).

New York's Third Avenue "El" reaches 129th Street 1 year after reaching 67th Street.

George B. Selden files for a patent on a road vehicle to be powered by an internal combustion engine; he will not obtain the patent until 1895 (*see* 1876; Benz, 1885).

Scotland's Firth of Tay Bridge collapses under the weight of a train in a winter storm.

Scots-American paper-bag maker Robert Gair pioneers the low-priced cardboard carton. He has set up a factory in New York that not only produces paper bags for grocers, department stores, millers, seedmen, and other merchants but also imprints the bags with the merchants' names. When one of his workers allows a metal rule on the press to slip up so that the paper is not only printed but cut, Gair capitalizes on the blunder and designs a metal die that uses a sharp metal rule set high to cut cardboard while blunt rules are set lower to crease the cardboard for folding into cartons. Buying a secondhand press for $30, he fits his cutting and creasing rules to create a machine that can cut and crease 750 sheets per hour, each sheet providing 10 carton blanks to permit production of 7,500 cartons per hour.

National Cash Register (NCR) has its beginnings in a register patented November 4 by Dayton, Ohio, saloon-keeper James J. "Jake" Ritty whose health has been undermined by his bartenders' pilfering. Ritty took a sea voyage to Europe to recover and was inspired by a recording device on the steamship marking the revolutions of the vessel's propeller and giving its officers a complete and accurate daily record of the ship's speed. "Ritty's Incorruptible Cashier" in its first version merely registers the amount of each cash transaction on

1879 ***(cont.)*** a dial, but a second model elevates a small plate to display the amounts so both clerk and customer can see it. The inventor and his brother will develop an improved model that records each day's transactions on a paper roll that can be checked by a store owner against the amount of cash in the cashbox, but manufacturing the register will prove too big a job and Ritty will sell his business and patent rights for $1,001 (*see* Patterson, 1884).

The gonococcus bacterium *Neisseria gonorrhea* discovered by German physician Albert Ludwig Siegmund Neisser, 24, transmits the venereal infection gonorrhea.

Russian pathologist Ivan Petrovich Pavlov, 30, shows with studies on dogs that the stomach produces gastric juices even without the introduction of food. Pavlov will proceed to develop the concept of acquired, or conditioned, reflex.

Parke, Davis & Co. of Detroit introduces liquid Ergotae Purificatus and assures physicians that dosages will be of the exact strength specified, pioneering standardization of pharmaceutical drugs by chemical assay.

The Church of Christ, Scientist is chartered at Boston by Mary Baker Eddy to propagate the spiritual and metaphysical system of Christian Science (*see* 1875; newspaper, 1908).

Canadians observe their first Thanksgiving Day November 6; the day will be observed on a Monday in October beginning in 1931 (*see* U.S. Thanksgiving, 1863, 1939).

Radcliffe College has its beginnings in classes for women started at Cambridge, Mass., by Elizabeth Cary Agassiz, 51, widow of the late naturalist Louis Agassiz who died 6 years ago at age 66 (*see* 1894).

Carlisle Training and Industrial School for Indians is founded at Carlisle, Pa., by former U.S. Army officer Richard Henry Pratt, 39, who starts the first nonreservation school for Native Americans.

The multiple switchboard invented by U.S. engineer Leroy B. Firman will make the telephone a commercial success and will help increase the number of U.S. telephone subscribers from 50,000 in 1880 to 250,000 in 1890.

New York merchants urge Bell Telephone to open its exchange for calls at 5 o'clock in the morning rather than 8 and to remain open later than 6 o'clock in the evening (Fulton Fish Market dealers lead the appeal).

French interests lay the U.S. end of a new Atlantic Cable at North Eastham, Mass. (*see* Field, 1869).

The *Atlanta Constitution* founded in 1868 gets a new editor who will increase the paper's vitality. Henry Woodfin Grady, 28, has served as Atlanta correspondent for James Gordon Bennett's *New York Herald* and proclaimed the emergence of a "new South." He buys a quarter-interest in the *Constitution* with help from Atlantic Cable pioneer Cyrus Field who loans him $20,000; Grady's 10-year term as editor will be marked by public speeches aimed at turning Southern eyes to the future.

Asahi Shimbun begins publication January 25 at Osaka with 4 pages at a lower price than the city's other 4 dailies. Editors can be jailed for "dangerous thoughts" under a libel law passed in 1875 but founder Ryohei Murayama will prosper.

Fiction *The Egoist* by George Meredith; *Daisy Miller* by Henry James whose novelette will be more popular during his lifetime than anything else he will write; *The Red Room (Roda rummer)* by Swedish novelist August Strindberg, 30; *Creole Days* (stories) by New Orleans writer George Washington Cable, 35, who angers Southerners with his criticism of slavery, prison conditions, and the mistreatment of blacks; "The Tar Baby" by *Atlanta Constitution* staff writer Uncle Remus (Joel Chandler Harris), 30, who has perfected a written form of the black dialect heard on Georgia plantations.

Juvenile *Under the Window* by English book illustrator Kate Greenaway, 33, of the *Illustrated London News*. Her "toy-book" will popularize early 19th century costumes for childrenswear.

Painting *The Cup of Tea* by Mary Cassatt; Honoré Daumier dies at Valmondois February 11 at age 70 in a cottage given him by the late J. P. C. Corot; George Caleb Bingham dies of cholera at Kansas City July 7 at age 68.

The Chicago Art Institute is founded.

The Seed Dry Plate introduced by English-American photographer Miles Ainscoe Seed, 36, of St. Louis is a sensitively prepared plate that can be carried anywhere, exposed, and developed later at the photographer's leisure. The dry plate will make the wet plate obsolete and reduce the use of chemicals. Seed will incorporate his dry plate company in July 1883 and sell in 1902 to Eastman Kodak (*see* Eastman, 1880).

Theater *A Doll's House* (*Et dukkehjem*) by Henrik Ibsen 1/21 at Copenhagen's Royal Theater; *The Heart Is Not Stone* (*Serdtse ne kamen*) by Aleksandr Ostrovsky 11/21 at St. Petersburg's Alexandrinsky Theater.

Opera *Eugen Onegin* 3/29 at the Moscow Conservatory, with music by Petr Ilich Tchaikovsky, libretto from the Aleksandr Pushkin poem of 1832; *The Pirates of Penzance* (*or The Slave of Duty*) 12/31 at New York's Fifth Avenue Theater, with Gilbert and Sullivan arias that include "Model of a Modern Major-General" and "A Policeman's Lot Is Not a Happy One."

First performances Concerto in D major for Violin and Orchestra by Johannes Brahms 1/1 at Leipzig's Gewandhaus; *Slavonic Dances* by Czech composer Antonin Dvořák, 37, 5/16 at Prague. Dvořák won his first public acclaim 6 years ago with the patriotic hymn "Die Erben des weissen Berges"; *Variations on a Rococo Theme* for Cello and Orchestra by Petr Ilich Tchaikovsky 6/8 at Wiesbaden; Suite No. 1 in D major by Tchaikovsky in December at St. Petersburg.

Vaudeville *The Mulligan Guards' Ball* 1/13 at New York's Théâtre Comique with Ed Harrigan and Tony Hart who opened in a vaudeville show at Chicago's Academy of Music in mid-July of 1873 and have toured the country mocking the pretensions of men who wear military uniforms. Harrigan and Hart will dominate U.S. vaudeville until Hart dies in 1891, Dan Collyer will replace Hart, and the new team will continue for another 5 years.

Popular songs "In The Evening by the Moonlight" by James Bland who introduces the song with Callender's Original Georgia Minstrels and who also writes "Oh Dem Golden Slippers;" "Alouette" is published for the first time in *A Pocket Song Book for the Use of Students and Graduates of McGill College* at Montreal.

New York's Madison Square Garden gets its name May 30 as railroad heir William K. Vanderbilt, 30, acquires the 5-year-old Gilmore's Garden, renames it, and announces that it will be used primarily as an athletic center. The Garden will be the scene of the National Horse Show beginning in 1883 and it will be acquired in 1887 by a syndicate of horse show sponsors including financier J. P. Morgan who will buy the Garden for $400,000 in 1890 and replace it with a splendid structure of yellow brick and Pompeiian terra cotta designed by architect Stanford White (below) (*see* 1925).

Rev. John Hartley, 30, wins in Wimbledon singles play.

F. W. Woolworth Co. has its beginnings at Watertown, N.Y., where store clerk Frank Winfield Woolworth, 27, persuades his employer to install a counter at which all goods are priced at 5¢. He then induces one of the firm's principals to lend him $400 with which to open a Utica store at which all items are priced at 5¢. The store fails in 3 months, but the same partner stakes Woolworth to try a five-and-dime store at Lancaster, Pa., which opens June 21 and proves successful. Stores opened by Woolworth at Philadelphia, Harrisburg, and York, Pa., and at Newark, N.J., will have indifferent success, but when he opens five-and-dimes at Buffalo, Erie, Scranton, and some other cities Woolworth will tap the low-income market and begin to build a worldwide chain of open-shelf, self-service stores that will prosper beyond all his expectations (*see* 1900).

The May Company is started at Leadville, Colo., by German-American merchant David May, 26, who has left his Hartford City, Ind., clothing-store job with $25,000 and gone West to seek a cure for his asthma. Starting with a store of board framework covered with muslin, May sells out his stock within a few weeks as Leadville booms overnight from a town of 500 to one of 25,000 in the wake of the silver strike of 1874 (above) (*see* 1888).

McKim, Mead & White is established by New York architects Charles Follen McKim, 32, William Rutherford Mead, 33, and Stanford White, 26. The firm will become famous for such buildings as the Century, Harvard, Metropolitan, Players, and University clubs in midtown Manhattan, some Columbia University buildings, and the Boston Public Library. It will also design bridges and make restorations in the White House (*see* Newport Casino, 1881).

France's Cathedral of St. Stephen at Limoges is completed after 603 years of construction.

St. Patrick's Cathedral opens in New York after 26 years of work. Designed in Gothic Revival style by James Renwick, it occupies a full block on the east side of Fifth Avenue (50th Street to 51st).

The Boston park system is completed by Frederick Law Olmsted—an "emerald necklace" for the city.

Detroit acquires Belle Isle in the Detroit River for $200,000 and will reclaim land to enlarge the 768-acre island into a 985-acre park.

Buffalo hunters kill the last of the Southern bison herd at Buffalo Springs, Tex. (*see* Grant, 1875; North Dakota, 1883).

The U.S. Fish Commission places carp in Wisconsin lakes (*see* shad, 1873; salmon, 1875). The common carp (*Cyprinus carpio*) lives up to 25 years in the wild (up to 40 in captivity), can survive in waters above 90° F., and can even withstand freezing for short periods, powers that will help it proliferate in U.S. inland waters, growing to weights of roughly 30 pounds and providing a new food source.

Britain has her worst harvest of the century, crops fail throughout Europe, demand for U.S. wheat raises prices, and the high prices bring prosperity to U.S. farmers who increase their wheat acreage as railroads build new lines to give farmers easier access to markets.

Hedging of wheat stocks by trading in futures begins on the Chicago Board of Trade that was established in 1865 (*see* Leiter, 1897; Norris, 1903).

McCormick's reaper sells for $1,500 in the United States, up from $150 in 1861, and there are complaints that the reaper is costlier at home than in Europe (*see* 1875; Northwest Alliance, 1880).

Ireland's potato crop fails again as in 1846 and the widespread hunger produces agrarian unrest (*see* Irish Land League, above).

India has a poor crop and much of the harvest is consumed by rats which will plague many districts in the next 2 years and will always consume a significant portion of the nation's grain stores. Famine continues in China.

The first milk bottles appear at Brooklyn, N Y., where the Echo Farms Dairy delivers milk in glass bottles instead of measuring it into the pitchers of housewives and serving-maids from barrels carried in milk wagons. Some competitors will soon follow (*see* Borden, 1885) but Boston's Hood Dairy will not use bottles until 1897.

Hungarian engineers invited to America by C. C. Washburn of Washburn-Crosby Co. install steel rollers in a new Minneapolis flour mill (*see* 1866; Gold Medal, 1880).

Wheatena whole wheat cooked cereal is introduced by a small bakery owner on New York's Mulberry Street who grinds the wheat, sells it in labeled packages, and advertises it in newspapers to compete with oatmeal (*see* Quaker, 1877).

Saccharin (benzosulfamide) is discovered accidentally at Baltimore's new Johns Hopkins University by chemist

1879 ***(cont.)*** Ira Remsen, 33, and his German student Constantin Fahlberg who are investigating the reactions of a class of coal tar derivatives (toluene sulfamides). They will publish a scientific description of the new compound in February 1880 calling special attention to its sweetness. Fahlberg will file a patent claim without mention of Remsen's contribution, he will return to Germany, obtain financial backing, and organize a company to produce his sugar substitute "saccharine"—at least 300 times sweeter than sugar and a boon to diabetics (*see* Wiley, 1907; sodium cyclamate, 1937).

Thomas Lipton's British grocery store chain makes him a millionaire at age 29 (see 1876; 1890).

Swiss scientist Herman Fol observes a spermatozoan penetrating an egg, confirming Oscar Hertwig's 1875 conclusion that a single male cell performs the act of fertilization.

President Hayes vetoes an act of Congress restricting Chinese immigration, calling it a violation of the 1868 Burlingame Treaty (*see* Exclusion Treaty, 1880).

1880 Britain's Conservatives lose in the general elections March 8, Lord Beaconsfield (Disraeli), now 75, resigns April 18; William Gladstone's Liberals will hold office until 1885.

"War is hell," says Gen. William Tecumseh Sherman, now 60, in an address to a Columbus, Ohio, reunion of the G.A.R. (Grand Army of the Republic). "There is many a boy here who looks on war as all glory, but, boys, it is all hell. You can bear this warning voice to generations yet to come."

Old guard Republican "Stalwarts" try to gain a third term for President Grant, a deadlocked G.O.P. (Grand Old Party) convention selects James Abram Garfield, 48, of Ohio on the 36th ballot. He wins 214 electoral votes to 155 for the Democrats, but he beats hero Winfield S. Hancock by only 9,464 votes.

Moroccan independence gains recognition July 3 in the Madrid Convention signed by the leading European powers and the United States.

French Equatorial Africa is established as a French protectorate by French explorer Pierre Paul François Camille Brazza Savorgnani, 28, who calls himself Savorgnan de Brazza and founds Brazzaville.

France annexes Tahiti.

A Boer Republic independent of Britain's Cape Colony is proclaimed December 30 by Oom Paul Kruger who begins a short-lived revolt that will end with the establishment of the independent South African Republic under British suzerainty (*see* 1899).

The word "boycott" enters the English language through an incident in Ireland's County Mayo. Tenant farmers organized by the Irish Land League refuse to harvest crops on estates managed by Charles Cunningham Boycott, 48, a retired British Army captain (*see* Davitt, 1879). The Land League wages a campaign of economic and social ostracism against absentee landlords and their agents, but the word *boycott* will take on a wider meaning as a weapon in economic warfare (*see* Parnell, 1881).

Parliament passes Britain's first Employers' Liability Act September 13, granting compensations to workers for injuries.

A federal circuit court calls Tennessee's 1875 "Jim Crow" law unconstitutional (but *see* 1881).

The United States has more than 100 millionaires, up from fewer than 20 in 1840 (*see* 1892).

Leadville, Colo., storekeeper Horace "Hod" Tabor, 53, grubstakes two starving prospectors to $64.75 worth of provisions and ends up 10 months later owning silver mines including the Matchless that will earn him as much as $4 million per year. Tabor takes as his mistress the blonde, blue-eyed divorcée Elizabeth McCourt "Baby" Doe, 23, by buying off her previous protector. He divorces his wife and marries Baby Doe for whom he will build a Denver opera house and a mansion graced with nude statuary and 100 peacocks (*see* 1893).

De Beers Mining Corp. is founded by English diamond mine operator Cecil John Rhodes and English financier Alfred Beit, both 27. Barnato Diamond Mining Co. is founded by English speculator Barnett Isaacs "Barney" Barnato, 28, whose firm will be consolidated with De Beers in 1888 to give Cecil Rhodes a virtual monopoly in the South African diamond industry (*see* Rhodes, 1890).

Some 539,000 Singer sewing machines are sold, up from 250,000 in 1875 (*see* 1863; 1889).

Japan's Fuji Bank is founded at Tokyo.

Thomas Edison obtains a patent on his 1879 incandescent bulb (*see* Owens, 1903; Langmuir, 1912).

Electricity lights all of Wabash, Ind., in an April demonstration of the Brush arc-light system, and a mile of New York's Broadway is illuminated by Brush arc lights December 20 (*see* 1879).

Wood remains a major source of energy in the United States, but use of firewood begins to decline.

A pooling agreement signed February 2 resolves disputes among the Atchison, Topeka and Santa Fe, the Union Pacific, the Kansas Pacific, and the Denver, Rio Grande and Western. The Denver, Rio Grande obtains exclusive rights to build a line from Pueblo through the Royal Gorge to Leadville via a 3-foot narrow-gauge line, pays the Santa Fe $1.4 million for work done by the Santa Fe on the Royal Gorge, and agrees not to build a line to Santa Fe provided that the Atchison, Topeka keeps out of Denver and Leadville with its standard 56.5-inch track.

The "Chattanooga Choo-Choo" gets its name March 5 from a newspaper reporter covering the resumption of through passenger service between North and South. A Cincinnati Southern Railroad woodburner chugs out of Cincinnati and passes through Chattanooga, Tenn., en route to New Orleans.

Elevated steam trains rumble up and down New York's Second, Third, Sixth and Ninth Avenues.

A bush roller chain patented by Swiss engineer Hans Renold improves on the James Slater drive chain of 1864. Renold has acquired Slater's small textile machine chain factory at Salford and has devised a chain with an arrangement of bushes that provide a much greater load-bearing surface than did the Slater chain (*see* safety bicycle, 1885).

French Army physician Charles Louis Alphonse Laveran, 35, in Algeria traces malaria to a blood parasite (*see* Ross, 1897).

The bacillus of typhoid fever is identified simultaneously by German bacteriologist Karl Joseph Eberth, 45, and Robert Koch, who has accepted an appointment to head a Berlin University Laboratory (*see* 1877; Budd, 1873). The bacillus will be named *Eberthelia typhi* and German bacteriologist T. A. Gaffley will confirm in 1884 that it is the agent of typhoid fever (*see* Wright, 1896).

Robert Koch (above) discovers a vaccine against anthrax through the accident of an assistant (*see* 1881).

The University of Southern California is founded by educators who include Marion McKinley Bovard, 33. He becomes the university's first president.

Manchester University is created by a reorganization of England's Owens College.

England's first high schools for girls open.

Parcel Post service begins in Britain (*see* United States, 1913).

The first British telephone directory is issued January 15 by the London Telephone Co. It lists 255 names.

The first wireless telephone message is transmitted June 3 by Alexander Graham Bell on the photophone he has invented (*see* Hertz, 1887; Marconi, 1895).

The *Kansas City Evening Star* is founded by publisher William Rockhill Nelson, 39 (*see* Nelson Gallery, 1930).

Halftone photographic illustrations appear in newspapers for the first time (below).

Nonfiction *A Tramp Abroad* by Mark Twain.

Fiction *The Brothers Karamazov* (*Bratya Karamazov*) by Fedor Dostoievski who caps his career with a literary masterpiece and makes a patriotic speech at the unveiling of the Pushkin memorial at Moscow; Dostoievski will die early next year at age 59; *Nana* by Emile Zola exposes the misery of life as a Parisian prostitute; "Boule de Suif" by French short-story writer Henri René Albert Guy de Maupassant, 30, is about a patriotic prostitute in the Franco-Prussian War; it will be followed by even better stories but its author is already ill with the syphilis that will put him in a strait jacket and then kill him at age 43 just 4 years after his younger brother dies in an insane asylum of the same disease; *Niels Lyhne* by Jens Peter Jacobsen; *The Grandissimes* by George Washington Cable; *Ben Hur, A Tale Of the Christ* by former Civil War general Lew (Lewis) Wallace, now 53, who is serving as governor of the New Mexico Territory. Wallace's story of Christians in ancient Rome enjoys great popularity; *Democracy* by Henry Adams; *The Duke's Children* by Anthony Trollope.

Juvenile *The Peterkin Papers* by Boston author Lucretia Peabody Hale, 60, whose book *The Last of the Peterkin* will appear in 1886; *Kate Greenaway's Birthday Book* by Catherine Greenaway.

Poetry *Ultima Thule* by Henry Wadsworth Longfellow, now 73.

Painting *The Outer Boulevards* by Camille Pissarro; *Place Clichy* by Pierre Auguste Renoir; *Château de Medan* by Paul Cézanne; *Old Folks' Home in Amsterdam* by German painter Max Liebermann, 33.

Sculpture *Petite Danseuse de Quatorze Ans* by Edgar Degas.

New York's Metropolitan Museum of Art moves in March into a new red-brick building in Central Park at 82nd Street designed by Calvert Vaux and Jacob Wrey Mould (*see* 1870; 1888).

The first photographic reproduction in a newspaper appears March 4 in the *New York Daily Graphic*. The picture of a shanty-town is printed from a halftone produced by photographing through a fine screen with dots in the photograph representing shadows (*see* 1875; Ives, 1886; press halftones, 1897).

George Eastman, 26, perfects a process for making dry photographic plates. The Rochester, N.Y., bank clerk, who has worked at night in his mother's kitchen, goes into business producing the plates for commercial photographers (*see* Seed, 1879; roll film, 1884; Kodak, 1888).

Theater *Hazel Kirke* by U.S. playwright Steele Mackay 2/4 at New York's Madison Square Theater, 486 perfs. (a new record). Fire destroys the drop curtain 2/26 as the audience is taking its seats, but firemen extinguish the blaze and the performance goes on as scheduled.

Anthem "Oh Canada" is sung in public for the first time June 24 at Quebec City. Music is by composer Calixa Lavallèe, French lyrics by Adolphe Basile Routhier. Robert Stanley Weir will write official English lyrics in 1908 (*see* 1980).

First performances *Tragic Overture (Tragische Ouverture)* by Johannes Brahms 12/26 at Vienna.

Popular songs "Funiculi-Funicula" by Italian composer Luigi Denza, 34, lyrics by G. Turco; "Sailing (Sailing, Over the Bounding Main)" by English songwriter Geoffrey Marks (James Frederick Swift), 33.

Rev. John Hartley wins his second Wimbledon singles title.

Australian outlaw Ned Kelly, 25, is convicted of murder in a Melbourne court and hanged November 11. He has robbed two banks in northeastern Victoria and New South Wales, fought the police in homemade armor, and killed three constables.

Cologne Cathedral is completed after 634 years of construction and is the largest Gothic cathedral in northern Europe. Located near the Hohenzollern Bridge in the

1880 ***(cont.)*** oldest part of the city, the dome has twin spires 515 feet high that make it the tallest structure in the world, a distinction it will hold until 1889.

New York's 7th Regiment Armory opens April 26 on Fourth Avenue between 66th and 67th Streets with a 187-by 290-foot drill hall. Architect Charles Clinton has designed it for the regiment that protected Washington, D.C., in 1861 when it was cut off by rebel forces in Maryland.

Half of New York's population is packed into tenements on the Lower East Side, an area that accounts for a disproportionate 70 percent of the city's deaths.

Denver's Windsor Hotel, opened by Colorado cattle barons, is the most elegant hostelry between San Francisco's 5-year-old Palace and Chicago's 5-year-old Palmer House (*see* Brown Palace, 1891).

California's Del Monte Hotel on the Monterey Peninsula is opened by Central Pacific Railroad boss Charles Crocker, now 58, whose $325,000 structure accommodates 750 and will rival the finest resort hotels in the East. Guests enjoy melons, vegetables, fruits, and grapes grown in the hotel's gardens or brought down from San Francisco, and fresh flounder, pompano, rock cod, salmon, and sole from the Bay is served daily (*see* Hearst, 1887).

Tenant farmers work one-fourth of U.S. farms (which number just over 4 million).

The Northwest Alliance has its beginnings in a local founded by Chicago editor Milton George whose organization will become a leading political activist group among farmers (*see* Farmers' Alliance, 1882).

U.S. wheat production reaches 500 million bushels, up 221 percent over 1866 figures. 80 percent of the wheat is cut by machine, but wheat prices have dropped 27 percent since 1866.

U.S. export of wheat and flour combined reach 175 million bushels, up from 50 million in 1871.

U.S. corn production reaches 1.5 billion bushels, up 98 percent over 1866 figures, and corn exports total 116 million bushels, but corn prices have dropped 15 percent since 1866. While many in Europe go hungry, Midwestern farmers burn corn for fuel; prices are too low to warrant shipping the crop to market.

Cattle drives up the Chisholm Trail reach their peak (*see* 1871; railroad, 1885).

Sugar beets are raised on a commercial scale for the first time in the United States (*see* Paris World's Fair, 1878), domestic cane and beet production meets only 10 percent of U.S. sugar needs and less than 0.05 percent of this will come from beets until the end of the 1880s. Another 19 percent will come from the territories of Hawaii, Puerto Rico, and the Philippines, but sugar imported from Cuba, Germany, and Java accounts for 71 percent of U.S. sugar consumption (*see* Spreckels, 1883; Havemeyer, 1887).

Imported meat accounts for 17 percent of British meat consumption. The figure includes livestock and preserved meat, most of it salted, along with dried "Hamburg"' beef and tinned, boiled mutton and beef from Australia, eaten in quantity despite its poor quality because it is half the price of fresh meat.

The first totally successful shipment of frozen beef and mutton from Australia to England arrives in early February as the S.S. *Strathleven* docks with 400 carcasses. The meat sells in London at 5.5 shillings per pound and is soon followed by the first cargo of frozen mutton and lamb from New Zealand (*see* 1877).

U.S. ice shipments to tropical ports reach a high of 890,364 tons carried by 1,735 ships, up from 146,000 tons aboard 363 ships in 1856. On long voyages 40 percent of the cargo may melt but the remaining ice sells for as much as $56 per ton.

New England's ice crop fails due to an unseasonably warm winter, ice prices soar, and the high prices spur development of ice-making machines (*see* 1889).

The Pillsbury A mill goes up at Minneapolis and will be the largest flour mill in the world (*see* 1878).

Washburn, Crosby Company of Minneapolis prepares to market Gold Medal Flour after winning a gold, a silver, and a bronze medal for flour samples it has entered in the Miller's International Exhibition (*see* 1879; Betty Crocker, 1921).

Thomas' English Muffins are introduced in New York by English-American baker Samuel Bath Thomas who opens a retail bakery in 20th Street between 9th and 10th avenues and delivers to hotels and restaurants by pushcart.

"Philadelphia" brand cream cheese is introduced by New York distributor Reynolds who has it made for him by Empire Cheese Co. of South Edmeston, N.Y.

Heinz's White Vinegar, Apple Cider Vinegar, and Apple Butter are introduced by Pittsburgh's F. and J. Heinz (*see* 1876; 1886).

More than 95 A&P grocery stores are scattered across America from Boston to Milwaukee; the Great Atlantic & Pacific Tea Co. will not have a store on the West Coast for another 50 years (*see* 1869; 1876; 1912).

Mark Twain's new book *A Tramp Abroad* lists American dishes he has yearned for while in Europe: buckwheat cakes with maple syrup, squash, fried chicken, Sierra Nevada brook trout, Mississippi black bass, terrapin, coon, possum, soft-shelled crabs, wild turkey, canvasback duck, succotash, Boston baked beans, chitterlings, hominy, sliced tomatoes with sugar and vinegar, sweet corn, eight kinds of hot bread, fruit pies and cobblers, and porterhouse steak for breakfast.

Los Angeles has a population of 11,183, up from 5,728 in 1870, but water shortages limit growth.

A Chinese Exclusion Treaty signed at Beijing November 17 gives Washington power to "regulate, limit, or suspend" entry of Chinese laborers (*see* 1882).

Nearly 550,000 English and nearly 440,000 Irish immigrants enter the United States.

A pharmacy to dispense pessaries (vaginal contraceptive suppositories based on quinine in cocoa butter) opens at Clerkenwell, England, under the management of Walter John Rendell (*see* Place, 1882; Wilde, 1833).

1881 Russia's Aleksandr II is assassinated March 13 at St. Petersburg at age 62 after a 26-year reign. A bomb tears off his legs, rips open his belly, and mutilates his face in an act organized and carried out by Sophia Perovskaya, who heads a band of Nihilist terrorists of a populist fringe group. The czar liberator is succeeded by his six-foot four-inch son, 36, who will reign until 1894 as Aleksandr III, and the assassination precipitates a wave of repression and persecution (below).

Romania and Serbia win independence from Constantinople; the Serbs make an alliance with Austria.

President Garfield is mortally wounded July 2 by disappointed office seeker Charles J. Guiteau who shoots the president at a Washington, D.C., railway station. When Garfield dies September 19 at Elberon, N.J., at age 49, he is succeeded by his vice president Chester Alan Arthur, 50, who abandons his support of the spoils system and follows Garfield's policy of avoiding party favoritism in his appointments (*see* Marcy, 1832; Pendleton Act, 1883).

Irish Nationalist leader Charles Stewart Parnell, 35, agitates for home rule and is imprisoned on charges of obstructing operation of a new land policy. From prison Parnell directs tenant farmers to pay no rent, thus enhancing his power (*see* 1879; 1882; Boycott, 1880).

Japanese political parties are formed.

Boers in the Transvaal repulse British forces January 28 at Laing's Neck and defeat them February 27 at Majuba Hill where the governor and commander-in-chief of the Natal colony Sir George Pomeroy Colley is killed at age 45. The Treaty of Pretoria concluded April 5 gives independence to the South African Republic of the Boers but under British suzerainty.

A Century of Dishonor by U.S. author Helen Maria Fiske Hunt Jackson, 49, records government wrongs in dealing with Indian tribes. Jackson will be appointed a special commissioner next year to investigate conditions among California's Mission Indians.

A second "Jim Crow" law passed by the Tennessee legislature segregates black passengers on railroads and establishes a precedent that will be followed by similar laws throughout the South (*see* 1880; 1887).

Anti-Semitism appears in America after generations during which German Jews have mixed on equal terms with other German immigrants.

Russia's new czar Aleksandr III (above) makes Jews the scapegoats for the assassination of his father. He also persecutes Roman Catholics but vows to kill one-third of the country's Jews, drive out another third, and convert the rest, beginning a series of pogroms (massacres) that will spur emigration of millions of Jews in the next 30 years.

The U.S. federal income tax law of 1861 was unconstitutional, the Supreme Court rules (*see* 1872; 1894).

The grisly assassination of Russia's Czar Alexander II shocked the world and ushered in a long period of repression.

Meyer Guggenheim begins a mining and smelting career that will establish great fortunes. A Swiss-born Philadelphia lace and embroidery importer of 51, Guggenheim buys sight unseen a one-third interest in two lead and silver mines on the outskirts of Leadville, Colo. He travels out to inspect his holdings, finds that the A. Y. mine and the Minnie (both nearly 2 miles above sea-level) are flooded, he invests half his $50,000 capital to "unwater" them, and within a few months the A. Y. mine is producing 15 ounces of silver per ton of ore (*see* 1888).

The Wharton School of Finance and Commerce is established at the University of Pennsylvania with a gift from nickel miner Joseph Wharton, 55, whose Pennsylvania nickel mine has for years been the only U.S. producer of refined nickel. Wharton developed a process in 1875 for making pure malleable nickel (*see* Ritchie, 1885).

Marshall Field & Co. is created by a reorganization of Chicago's Field, Leiter & Co. (*see* 1867). In the next 25 years Field will make his firm the world's largest wholesale and retail dry goods store (*see* 1893; university, 1891).

Detroit's J. L. Hudson department store has its beginnings in a menswear shop opened by English-American merchant Joseph Lothian Hudson, 34, who will specialize in fire sales, acquire a site at Gratiot and Farmer Streets in 1891, and by 1911 have a store that covers an entire city block with 50 passenger elevators (*see* motor car, 1909).

Boston's William Filene Sons opens at 10 Winter Street. Prussian-American tailor-merchant William Filene, 51, and his sons Edward P., 21, and Abraham Lincoln Filene, 16, will take over a five-story building at 445–447 Washington Street for their womenswear and accessories store (*see* 1901).

A FOUL DEED.

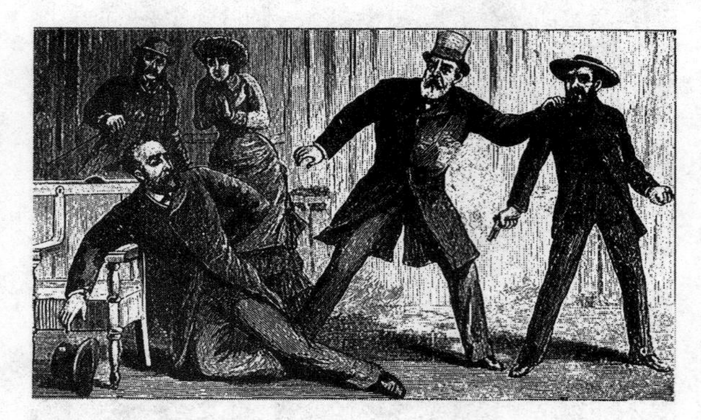

President Garfield's assassination, the second one in 15 years, followed that of the Russian czar by less than 16 weeks.

Thomas Edison's Edison Electric Light Co. creates a subsidiary (The Edison Co. for Isolated Lighting) to furnish factories and large department stores with individual power plants (*see* 1880; Edison Illuminating, 1882).

London's Savoy Theatre opens with the first electric illumination in any British public building. The theater has been built by Richard D'Oyly Carte, 37, who has been producing Gilbert and Sullivan operas since *Trial by Jury* in 1875 (below).

The Atchison, Topeka and Santa Fe links up with the Southern Pacific March 8 at Deming in New Mexico Territory, the first Santa Fe through train to California leaves Kansas City March 17, but the road still has only 868 miles of its own track and despite its name is balked by terrain from reaching Santa Fe (it will never get closer than Lamy). The 868 miles will increase to 3,600 in the next 6 years under the leadership of the Santa Fe's new president William Barstow Strong (*see* 1880; 1887).

The Southern Pacific Railway links New Orleans with San Francisco as the last spike is driven home near El Paso, Tex. Service on the Sunset route will begin in 1883 (*see* 1884).

A decade of unprecedented U.S. railroad construction begins.

Henry Villard gains control of the Northern Pacific by means of a pool he has formed to monopolize Pacific Northwest transportation. Now 46, Villard came to New York from his native Bavaria at age 18, changed his name from Ferdinand Heinrich Gustav Hilgard to avoid being returned and drafted, and worked for years as a journalist for the *New York Staats-Zeitung,* James Gordon Bennett's *Herald,* and Horace Greeley's *Tribune* (*see* 1883; 1901).

James J. Hill of the St. Paul and Pacific joins with Montreal financier George Stephen to organize a new Canadian Pacific Railway Co. after the collapse of an earlier effort. Given a $25 million pledge of government aid, a 10-year deadline, and assurance that no other railroad will be permitted in the area for 20 years, Hill and Stephen hire Chicago Milwaukee & St. Paul general manager William Cornelius Van Horne, 38, to carry out the trans-Canada project (which is long overdue in light of the Dominion government's pledge to British Columbia in 1871). Van Horne moves to Winnipeg, works his crews night and day, and by the end of summer has built nearly 500 miles of track through forest and swamp (*see* 1887).

The Norfolk and Western Railroad is created by a reorganization of roads that began with a 10-mile road in 1837. The company will augment its 479 miles of track in this decade by acquiring other roads in Maryland and Ohio and building new lines into the coal fields of Virginia and West Virginia, and within 50 years will operate 4,477 miles of track in six states with 30,000 employees.

The S. S. *Servia* goes into service for Britain's Cunard Line. It is the world's first steel ocean liner.

The motor-driven water bus (*vaporetto*) introduced at Venice will make the traditional gondola a conveyance used chiefly by tourists. By 1971 Venice will have fewer than 500 gondolas (*see* 1292).

German-American physicist Albert Abraham Michelson, 29, invents an interferometer. In 1887 he will show with Edward H. Morley that the speed of light in a vacuum is the same in all inertial reference systems (coordinate systems moving at constant velocity relative to each other). The work will lead directly to the special theory of relativity (*see* Einstein, 1905).

Scottish bacteriologist Jaime Ferran discovers a serum effective against cholera (*see* Koch, 1883).

A paper by Cuban physician Carlos Juan Finlay, 28, suggests that mosquitoes may spread yellow fever (*see* 1878; Reed, 1900).

The pneumococcus bacterium that causes pneumonia is found by U.S. Army bacteriologist-physician George Miller Sternberg, 43, who has earlier identified the malaria plasmodium and the bacilli of tuberculosis and typhoid fever.

The American Association of the Red Cross is founded by Clara Barton of Civil War fame who has done relief work in Europe during the Franco-Prussian War and has campaigned to have the United States sign a Geneva Convention (*see* 1861; 1882). Now 59, Barton will serve as president of the American National Red Cross until 1904 when lack of public support will require congressional action to keep the Red Cross from folding.

Kickapoo Indian Medicine Co. is founded by U.S. promoters John Healy and Texas Charley Bigelow who hire 300 Indians (none of them Kickapoos) from Indian agents, pay each one $30 per month plus expenses, and have them stage rodeos, rope tricks, and war dances to promote Kickapoo Indian Oil, Salve, Cough Cure, Pain Pills, Wart Killer, and the like. Healy and Bigelow will be joined in Pennsylvania by promoter Edward "Nevada Ned" Oliver (who has never been anywhere near Nevada) who will specialize in catarrh powders made of menthol, sugar, milk, and cocaine (*see* cocaine, 1884).

Tuskegee Normal and Industrial Institute is founded at Tuskegee, Ala., where local blacks have invited Booker Taliaferro Washington, 25, to start the pioneer school for blacks. Washington has been instructor to 75 Indian youths at Hampton Normal and Agricultural Institute (*see* 1870; 1895; Carver, 1896).

Spelman College has its beginnings in the Atlanta Baptist Female Seminary. The school will be renamed in 1924 to honor John D. Rockefeller's mother-in-law Lucy Henry Spelman and will be the leading U.S. college for black women.

Drake University is founded at Des Moines, Iowa.

The University of Connecticut is founded at Storrs.

The University of Pennsylvania's Wharton School is founded (above).

Western Union Telegraph is created by a consolidation of Western Union Co. with two smaller telegraph companies to form a giant monopoly (*see* 1866). Financier Jay Gould and railroad magnate William H. Vanderbilt have effected the consolidation (*see* Postal Telegraph, 1836).

"The Story of a Great Monopoly," published in the *Atlantic Monthly*, is an attack on John D. Rockefeller's Standard Oil Co. trust by *Chicago Tribune* financial editor Henry Demarest Lloyd, 34.

Harper's Monthly commissions Lloyd (above) to do a study of the Pullman Palace Car Co. but will reject his work, calling it too uncritical, and publish instead an article by Johns Hopkins economist Richard T. Ely, 27, who will praise the physical facilities of Pullman, Ill., and its absence of saloons but will call the town undemocratic, un-American, and feudalistic (*see* below).

Il Progresso Italo-Americano begins publication at New York. The paper will grow to become the largest-circulation foreign-language daily in the city.

The *London Evening News* begins publication (*see* 1894; 1896).

Andrew Carnegie donates funds for a Pittsburgh library, beginning a series of library gifts (*see* 1901).

Nonfiction *Virginibus Puerisque* (essays) by Scottish writer Robert Louis Stevenson, 30.

Fiction *The Portrait of a Lady* by Henry James; *Le Crime de Sylvestre Bonard* by French novelist Anatole France (Jacques Anatole François Thibault), 37; *The House by the Medlar Tree (I Malavoglia)* by Giovanni Verga; *Epitaph for a Small Winner (Memorias Postumas de Bas Cubas)* by Brazilian novelist Joaquim Machado De Assis, 42.

Juvenile *Heidi (Heidis Lehrund Wanderjahre, Heidi Kann Brauchen, Was es Gelernt Hat)* by Swiss author Johann Heusser Spyri, 54; *Five Little Peppers and How They Grew* by U.S. author Margaret Sidney (Harriett Mulford Stone Lathrop), 37.

Painting *Sunshine and Snow* by Claude Monet; *An Asylum for Old Men* and *Cobbler's Shop* by Max Liebermann; *Luncheon of the Boating Party* by Pierre Auguste Renoir.

Sculpture *Admiral Farragut* by Irish-American sculptor Augustus Saint-Gaudens, 33, who opened studios at New York in 1873 but will move to Cornish, N.H., in 1885. Saint-Gaudens shows a style of dramatic simplicity that has made him popular with architects such as Stanford White, who has designed a pedestal for the full-length statue of the late U.S. naval hero in New York's Madison Square.

Photographic roll film is patented October 11 by Wisconsin inventor David Henderson Houston whose "photographic apparatus" is a camera whose inner end has a receptacle containing a "roll of sensitized paper or any other suitable tissue, such as gelatine or any more durable material that may be discovered, and an empty reel, upon which the sensitized band is wound as rapidly as it has been acted upon by the light" (*see* Walker, Eastman, 1884).

Theater *Talents and Admirers* (*Talanty i poklonniki*) by Aleksandr Ostrovsky 12/20 at Moscow's Maly Theater.

Barnum & Bailey's Circus is created by a merger that joins the 10-year-old P. T. Barnum circus with that of John Anthony Bailey, 34, who will gain control in 1888 (*see* Jumbo, 1882; Ringling brothers, 1907).

Opera *The Tales of Hoffmann (Les Contes d'Hoffman)* 2/10 at the Opéra-Comique, Paris, with music by the late Jacques Offenbach who died last October at age 61: his Barbarelle aria makes the work a *succes fou*. *The Maid of Orleans (Orleanskata dieve)* 2/13 at St. Petersburg's Maryinsky Theater, with music by Petr Ilich Tchaikovsky, libretto from the Friedrich von Schiller play of 1801; *Patience (or Bunthorne's Bride)* 4/23 at London's new Savoy Theatre (above) with Gilbert and Sullivan songs that include "Prithee, Pretty Maiden"; *Libusa* 5/11 at Prague, with music by Bedřich Smetana to open the new National Theater.

Lillian Russell achieves stardom at age 19 playing D'Jemma in the comic opera *The Great Mogul* (or *The Snake Charmer*) with music by French composer Edmond Audran, 39. Born Helen Louise Leonard at Clinton, Iowa, Russell adopted her stage name last year while appearing at Tony Pastor's Opera House in New York and has toured California in the new comic opera which she will follow with roles in *Patience* (above) and other Gilbert and Sullivan operettas. Russell will be popular for more than 20 years for her clear soprano voice and well-upholstered beauty.

First performances Concerto No. 3 in B major for Violin and Orchestra by Camille Saint-Saëns 1/2 at Paris; *Academic Festival Overture (Academische Festouverture)* by Johannes Brahms 1/4 at the University of Breslau; Symphony No. 2 in C minor *(Little Russia)* by Petr Ilich Tchaikovsky 2/2 at St. Petersburg; Symphony No.

1881 ***(cont.)*** 3 in D major by the late Franz Schubert 2/19 at London's Crystal Palace; Symphony No. 4 in E flat major *(Romantic)* by Anton Bruckner 2/20 at Vienna; Concerto No. 2 in B flat major for Pianoforte and Orchestra by Johannes Brahms 11/9 at Budapest's Redountensaal; Concerto No. 2 in G major for Pianoforte and Orchestra by Tchaikovsky 11/12 at New York's Academy of Music; Concerto in D major for Violin and Orchestra by Tchaikovsky 12/4 at Vienna.

The Boston Symphony Orchestra is founded by local banker Henry Lee Higginson, 47 (*see* 1875; 1885).

Hymn "I Am Coming" by Ira David Sankey, lyrics by Helen R. Young; "Tell It Out Among the Nations" by Ira David Sankey, lyrics by Francis R. Havergal.

March "Semper Fidelis" by U.S. Marine Corps bandmaster John Philip Sousa, 27, who took the post last year and will keep it until 1892 (*see* 1889).

William Renshaw, 20, wins in singles play at Wimbledon, Richard Dudley Sears, 19, in the first national U.S. singles in August at Newport Casino, where the matches will be held until 1918. The United States Lawn Tennis Association founded May 21 by 33 U.S. clubs will control all national U.S. national tournaments.

The second Canadian bid for the America's Cup in ocean yacht racing fails. The U.S. defender *Mischief* defeats Canada's *Atalanta* 2 to 0.

Billy the Kid escapes in May while under heavy guard after 5 months in confinement on a murder conviction. He kills two deputies who try to stop him, but Sheriff Patrick F. Garrett discovers him at Fort Sumner in New Mexico Territory July 15 and shoots him dead. The New York-born western outlaw William H. Bonney, 21, has killed 21 men including a previous sheriff and within 6 weeks of his death *The True Life of Billy the Kid* is being sold in the streets of New York.

A shootout at the O.K. Corral outside Tombstone in Arizona Territory October 26 breaks up a gang headed by Ike Clanton whose brother Billy is shot along with Frank and Tom McLowry. Town marshal Virgil Earp has deputized his brothers Wyatt, 31, and Morgan; they are joined by alcoholic gambler "Doc" Holliday. Ike Clanton and Billy Claiborne escape out of town, the town suspects the Earps of murder, and Virgil Earp loses his job as marshal.

Construction of a model factory town for Pullman Palace Car employees begins on a 4,000-acre tract at Pullman, Ill., beside Lake Calumet outside Chicago (above). Architect Solon Spenser Beman, 29, has designed the first all-brick U.S. city, it will have paved streets, good water supply and sewage disposal systems, a shopping center arcade, and a hotel named after George M. Pullman's daughter. Rents will be fixed to give the Pullman company a 6 percent return on investment (*see* 1893 strike).

Newport Casino at the Rhode Island resort town is completed by McKim, Mead & White.

Ring-necked pheasants from Shanghai are introduced into Oregon's Willamette Valley. The Chinese pheasant will be an important American upland game bird.

The *New York Tribune* takes over the Fresh Air Fund for city slum children that was started 4 years ago. The newspaper appeals to the public for funds to continue the program.

Drought strikes the eastern United States. New York City runs out of water and people in many cities die of heat exhaustion.

Cattle ranges in the U.S. Southwest wither in a severe drought, but cattle-raising remains profitable.

The Prairie Cattle Co. of Edinburgh, Scotland, declares a 28 percent dividend. The company buys out Texas Panhandle rancher George W. Littlefield whose LIT squatter ranch claims no land rights but who receives more than $125,000 (*see* 1882).

The Texas state capitol burns down; Texas legislators demand the biggest capitol building in the country with a dome at least one foot higher than the one in Washington. They arrange with a group of Chicago contractors and financiers to build the new capitol, and they trade 3 million acres of Panhandle prairie to the Chicago syndicate. Chicago dry goods merchant John Villiers Farwell, 56, and his brother Charles Benjamin, 58, (a U.S. congressman) form the Capitol Freehold Land and Investment Co., wind 800 miles of barbed wire round their holdings, and the XIT Ranch they establish is the largest in the Panhandle. It will soon be driving 12,000 head per year from its 200-mile Texas range 1,200 miles up the Montana Trail to the leased XIT range in Montana Territory, the last great cattle drives from Texas.

A vaccine to prevent anthrax in sheep and hogs is found by Louis Pasteur (*see* 1861; Koch, 1883).

California imposes quarantine regulations to keep out insect pests and plant diseases.

The loganberry, introduced by Santa Cruz, Calif., judge James Harvey Logan, is a cross of the red raspberry with a California wild blackberry. The University of California will make loganberry seeds available to the public in 1883 (*see* Boysen, 1920).

The first U.S. pure food laws are passed by New York, New Jersey, Michigan, and Illinois (*see* 1904).

The first U.S.-made margarine is produced in a West 48th Street, New York, factory by Community Manufacturing Co., a subsidiary of the U.S. Dairy Co. (*see* 1874; 1886).

Jumbo Rolled Oats, introduced by Akron, Ohio, oatmeal king Ferdinand Schumacher, saves him money, gives him an advantage over competitors who have copied his steel-cut method, and permits a housewife using a double-boiler to prepare breakfast in 1 hour, less than half the time needed for steel-cut oats, and at less expense (*see* 1875; 1886).

Granola is adopted by J. H. Kellogg as a new name for his cold breakfast food. He has been sued for using the name Granula by the Dansville, N.Y., originators of that name (*see* 1877; Granose, 1895).

Chicago meat packer Gustavus F. Swift perfects a refrigerator car to take Chicago-dressed meat to eastern butchers (*see* 1877). Sides of meat hang from overhead rails inside the car, and when it reaches its destination the rails are hooked up with rails inside the customer's cold storage building, making it easy to slide the meat from railcar to cold store without loss of time or change of temperature. The efficiency of Swift's system, beginning with a disassembly line from the moment of slaughter to the butchering of carcasses into primal sections, will lower the price of meat in New York, New England, and down the Atlantic seaboard.

The first British cold store opens at London, which soon begins to receive shipments of chilled beef from America. When it is discovered that meat kept at 30° F. maintains its quality better than meat preserved with early freezing methods, U.S. meat will begin a domination of British markets that will continue for more than 25 years (*see* 1877).

Some 669,431 immigrants enter the United States, up from 91,918 in 1861, as a decade begins that will see 5.25 million immigrants arrive.

Some 500,000 Romanian Jews will emigrate to America in the next few decades and be joined by other Eastern European Jews plus 1.5 million Russian Jews, adding new diversity to the population.

The U.S. population reaches 53 million; Britain has 29.7 million, Ireland 5.1, Germany 45.2, France 37.6, Italy 28.4.

1882 Charles Stewart Parnell and his associates are released from Ireland's Kilmainham Prison May 2 after agreeing to stop boycotting landowners, cooperate with the Liberal Party, and stop inciting Irishmen to intimidate tenant farmers from cooperating with landlords (*see* 1880; 1881). Fenians murder the new chief secretary for Ireland and his permanent under-secretary May 6 in broad daylight in Dublin's Phoenix Park, and while Parnell repudiates the murder of Lord Frederick Cavendish and Thomas Burke it leads the British to suspend trial by jury and give the police unbridled power to search and arrest on suspicion. Parnell also disavows the campaign of terrorism that includes dynamiting of public buildings in England (*see* 1886).

Canada creates the District of Saskatchewan; the town of Regina is founded on the route of the Canadian Pacific Railway being constructed by Cornelius Van Horne (*see* 1881). Regina will be headquarters for the 9-year-old Northwest Mounted Police.

A Triple Alliance signed May 20 pledges Germany, Austria, and Italy to come to each other's aid should any be attacked by France within the next 5 years (*see* 1887).

Alexandria is bombarded July 11 by the British fleet under Sir Beauchamp Seymour. British troops are landed to protect the Suez Canal from nationalist forces, Sir Garnet Wolseley defeats the Egyptians September 13 at the Battle of Tel el-Kebir, British forces occupy Cairo September 15, and dual Anglo-French control of Egypt is abolished November 9.

Italy takes over Ethiopia's northern town of Assab and will make it the basis of an Eritrean colony in 1890 (*see* 1885). Ethiopia's Johannes IV makes a pact with his vanquished rival Menelek of Shoa and designates Menelek as his successor.

The International Association of the Congo is created out of the 1878 Belgian Comité d'Etudes du Haut-Congo and several companies are organized to exploit the region.

France claims a protectorate over the entire northwestern portion of Madagascar. French warships blockade the island in May, troops occupy the port of Majunga, but Madagascar's prime minister Rainilaiarivony will not capitulate.

An insurrection against the French in Indochina begins in Tonkin following Commander Henri Riviere's seizure of Hanoi April 25.

The world powers sign a Hague convention agreeing to a 3-mile limit for territorial waters.

Parliament passes the Married Women's Property Act following efforts by women's rights champion Richard Marsden Pankhurst, whose widow Emmeline, now 44, will campaign for women's suffrage after his death in 1898 (*see* 1903).

A wave of strikes for higher wages in the United States is touched off by higher prices that have resulted from last year's poor crops. New York's first Labor Day parade September 5 brings out 30,000 marchers.

Electricity illuminates parts of London beginning January 12 as power from the Edison Electric Light Co. at 57 Holburn Viaduct turns on street lights between Holborn Circus and the Old Bailey and incandescent bulbs go on in at least 30 buildings.

Electricity illuminates parts of New York beginning September 4 as Thomas Edison throws a switch in the offices of financier J. P. Morgan to light the offices and inaugurate commercial transmission of electric power from the Morgan-financed Edison Illuminating Co. power plant on Pearl Street. The company will soon supply current to all of Manhattan and it will develop into the Consolidated Edison Co., prototype of all central-station U.S. power companies (*see* 1899; Thomson, 1883).

The Electric Light Act passed by Parliament empowers local British authorities to take over privately run power stations in their areas after 21 years. By making it virtually impossible for a private electric company to recoup its investment, the new law will discourage development of power stations (no private company will generate electricity for the next 6 years and although the new law will be amended in 1888 to permit private ownership for 42 years such major cities as Manchester, Leeds, Edinburgh, and Nottingham will still have no power stations in 1890).

1882 ***(cont.)*** Swan lamps illuminate a draper's shop at Newcastle-upon-Tyne, England, making it the world's first shop to be lighted by incandescent bulbs.

The world's first electric fan is devised by the chief engineer of New York's Crocker and Curtis Electric Motor Co. The two-bladed desk fan is the work of Schuyler Skaats Wheeler, 22.

The world's first electric flatiron is patented by New York inventor Henry W. Weely.

The world's first electrically lighted Christmas tree is installed in December in the New York house of Thomas Edison's associate Edward H. Johnson.

An internal combustion engine powered by gasoline is invented by German engineer Gottlieb Daimler, 48, who has worked with Eugen Langen (*see* 1886; Otto, 1866).

The Standard Oil trust incorporated by John D. Rockefeller and his associates to circumvent state corporation laws brings 95 percent of the U.S. petroleum industry under the control of a nine-man directorate. Pennsylvania lawyer Samuel C. T. Dodd has shown Rockefeller how the idea of a trust employed in personal estate law can be applied to industry and the oil trust will soon be followed by other trusts. The richest company of any kind in the world, the Standard Oil trust controls 14,000 miles of underground pipeline and all the oil cars of the Pennsylvania Railroad (*see* 1877; 1883).

The St. Gothard tunnel that opens May 20 is the first great railroad tunnel through the Alps (*see* Orient Express, 1883).

Union Switch and Signal Co. is organized to manufacture railroad signals invented by George Westinghouse who has made a fortune from his air brake (*see* 1868; gas pipelines, 1883).

"The public be damned," says U.S. railroad magnate William H. Vanderbilt to a *Chicago Daily News* reporter who has asked, "Don't you run it for the public benefit?" when told that the fast *Chicago Limited* extra-fare mail train was being eliminated. "The public be damned," says Vanderbilt October 8 to reporter Clarence Dresser. "I am working for my stockholders. If the public want the train, why don't they pay for it?"

Electric cable cars are installed in Chicago where they travel 20 blocks along State Street in 31 minutes, averaging less than 2 miles per hour (*see* 1894; San Francisco, 1873).

English metallurgist Robert Abbott Hadfield, 24, invents manganese steel (*see* tungsten steel, 1868; nickel steel, 1888).

Robert Koch discovers the tuberculosis bacillus and establishes that the disease is communicable (*see* 1877). His findings along with those of many other bacteriologists will lead physicians to believe that certain diseases such as beriberi are caused by bacteria rather than by dietary deficiencies (*see* Takaki, 1884; diphtheria, 1883).

Viennese physician Josef Breuer, 40, discovers the value of hypnosis in treating a girl suffering from severe hysteria, pioneering psychoanalysis. Breuer induces the patient to relive certain scenes that occurred while she was nursing her sick father and succeeds thereby in relieving her permanently of her hysteria symptoms, a success he will communicate to colleague Sigmund Freud (*see* 1895).

Oscar Wilde arrives in New York in January saying, "I have nothing to declare but my genius." Irish-born essayist Oscar Fingal O'Flahertie Wills Wilde, 27, has made himself the apostle of an art-for-art's-sake cult (Gilbert and Sullivan burlesqued his affectations last year in their opera *Patience* with its character Bunthorne). Wilde will tour North America for a year lecturing on such subjects as "The English Renaissance of Art."

The University of South Dakota is founded.

Western Electric wins a contract February 6 to produce telephones for the Bell Co., which will acquire the firm (*see* 1872; 1876).

Harrison Gray Otis, 45, acquires an interest in the *Los Angeles Times*. A Union Army enlistee who rose to the rank of lieutenant colonel, Otis moved to California in 1876 and ran the *Santa Barbara Post* until 1879. The *Times* has absorbed the *Weekly Mirror;* Otis will gain full control of the Times-Mirror Co. by 1886 and will make it a powerful voice of Republican conservatism in opposition to labor unions (*see* 1910).

"When a dog bites a man that is not news, but when a man bites a dog that is news," says *New York Sun* city editor John B. Bogart.

Fiction *The Prince and the Pauper* by Mark Twain.

Juvenile *Pinocchio, la storia di un burratino* (little wooden boy) by Italian author Carlo Collodi (Carlo Lorenzini), 52. The marionette made by the woodcarver Gepetto comes to life and has a series of adventures and misadventures including one in which his nose grows longer every time he tells a lie.

Poetry *Canto novo* by Italian poet Gabriele D'Annunzio, 19.

Painting *Bar at the Folies Bergère* by Edouard Manet; *Self-Portrait* by Paul Cézanne; *Mr. and Mrs. John W. Field* by U.S. painter John Singer Sargent, 26, who was born abroad of U.S. parents, visited the United States in 1876 and obtained U.S. citizenship, but returned to Europe in 1877. Dante Gabriel Rossetti dies at Birchington April 10 at age 53 after making a desperate effort to give up his addiction to the soporific drug chloral hydrate.

Theater *La Belle Russe* by San Francisco playwright David Belasco, 28, 5/8 at Wallack's Theater, New York; *Ghosts (Gengangere)* by Henrik Ibsen 5/20 at Chicago's Arrone Turner Hall (in Norwegian); *An Unequal Match* by the late English playwright Tom Taylor 11/6 at Wallack's Theater, New York, introduces U.S. audiences to the "Jersey Lilly" Lillie Langtry (Emily Charlotte Le Breton Langtry), 30, who made her London stage debut last year. Mrs. Langtry gained her soubriquet from the painting "The Jersey Lilly" by Sir John

Everett Millais for whom she posed; *Fedora* by French playwright Victorien Sardou, 51, 12/11 at the Théâtre du Vaudeville, Paris, with Sarah Bernhardt. The melodrama provides the name for a fashionable new soft felt hat with a high roll on its side brim and a lengthwise crease in its low crown.

Jumbo the elephant appears at New York's Madison Square Garden beginning April 10 in performances of Barnum & Bailey's Circus. P. T. Barnum has imported the "largest elephant in or out of captivity" (whose name has introduced the word "jumbo" to the English language), the animal stands 11 feet tall at the shoulders and weighs 6.5 tons. Barnum claims it stands 12 feet tall at the shoulders, measures 26 feet in length if the trunk is included, and weighs 10.5 tons. Captured 20 years ago as a baby, Jumbo has been a prime attraction at London's 56-year-old Royal Zoological Gardens since 1865 when he was acquired from the Jardin des Plantes at Paris in exchange for a rhinoceros, the London zoo has sold him to Barnum for $10,000 because he has become difficult, his sale has raised a storm of protest in England, and there will be general mourning when Jumbo is killed by a freight train on the Grand Trunk Railway at St. Thomas, Ontario, in mid-September 1885.

Opera *The Snow Maiden* (*Snyegurochka*) 2/10 at St. Petersburg, with music by Nikolai Rimski-Korsakov; *Parsifal* 7/26 at the Festpielhaus in Bayreuth, Germany, with music by Richard Wagner; *The Devil's Wall* (*Centova stena*) 10/29 at Prague, with music by Bedřich Smetana; *Iolanthe* (*or The Peer and the Peri*) 11/25 at London's new Savoy Theatre, with Gilbert and Sullivan arias that include "When Britain Really Ruled the Waves" and "Faint Heart Never Won Fair Lady."

First performances Serenade in C major for String Orchestra by Petr Ilich Tchaikovsky 1/28 at Moscow; First Modern Suite for Piano by U.S. composer Edward Alexander MacDowell, 20, 7/11 at Zurich; *1812 Overture* by Tchaikovsky 8/20 at Moscow with sound effects evocative of the Battle of Borodino; *My Country (Ma Vlast)* by Bedřich Smetana (now totally deaf) 11/5 at Prague (the symphonic cycle includes The Moldau) (*see* 1875); *Song of the Fates* (*Gesang der Parzen*) by Johannes Brahms 12/10 at Basel.

William Renshaw wins in singles at Wimbledon, Richard Sears at Newport.

London's *Sporting Times* carries an advertisement August 31 with a funereal black border: "In Affectionate Remembrance of English Cricket Which Died at the Oval on 29th of August, 1882; Deeply lamented by a large circle of sorrowing friends and acquaintances. R.I.P. N.B. The body will be cremated and the ashes taken to Australia." An English team sent to Australia next year with instructions to "bring back the ashes" will be presented with a 5-inch earthenware urn filled with ashes and this most hallowed cricket souvenir will be presented to the Marylebone Cricket Club in 1928.

Jesse James dies April 3 at age 34 of a gunshot wound in the back of the head. A fugitive since the Northfield, Minn., bank robbery attempt of 1876 James has been living quietly at St. Joseph, Mo., under the name Thomas Howard. The governor has offered a large reward for the capture of James dead or alive, James has befriended his cousin and fellow outlaw Robert Ford, who has shot him to get the reward, and his murder inspires a ballad that will make future generations regard James as a Robin Hood figure.

The Hatfield-McCoy feud that has simmered for years in the southern Appalachians of West Virginia and eastern Kentucky boils over in the election day shooting of Kentucky storekeeper Ellison Hatfield, brother of William Anderson "Devil Anse" Hatfield, 43, who mortally wounded Harmon McCoy of the Union Army while fighting with Confederate forces in the Civil War. Hatfield's son Johnse has precipitated the quarrel by attempting to elope with Randall McCoy's daughter Rosanna, and when Ellison Hatfield dies of his wounds three of McCoy's sons, captured earlier by an armed posse under Anse Hatfield, are murdered in retaliation, beginning a bloodbath that will continue for years as local influence is used to obtain quick release of any arrested participants. Kentucky authorities will invade West Virginia in 1888 and seize several Hatfields, ending open warfare between the clans. "Devil Anse" Hatfield will live until late 1921.

Only 2 percent of New York homes have water connections and while tenements have some rudimentary plumbing facilities nearly every private house has a backyard privy.

The loss of Chinese labor (below) spurs development of machinery to clean and bone fish in California's salmon canneries (*see* 1866; 1895; Smith, 1903).

Drought continues on western U.S. ranch lands.

Edinburgh's Prairie Cattle Co. pays $350,000 to acquire the Quarter Circle T Ranch of Texas Panhandle rancher Thomas Bugbee who 6 years ago drove a small herd from Kansas to the Canadian River and established the ranch (*see* 1881; 1884).

The Farmers' Alliance claims 100,000 members in 8 state alliances and 200 local alliances. The Alliance is now headed by Milton George (*see* 1880; 1889).

German bacteriologist Friedrich August Johannes Löffler, 30, at Berlin's Friedrich Wilhelm Institute discovers the bacilli that produce swine fever (hog cholera), swine erysipelas, and glanders (another livestock disease) (*see* diphtheria, 1884).

Germany will assume leadership of European industry in the next two decades, but the Germans will attempt to maintain self-sufficiency in food production where the British have not. Germany will plant another 2 million acres to food crops and use high tariffs to protect her farmers from foreign competition even though such a policy means keeping domestic food prices high.

"The evidence regarding the adulteration of food indicates that they are largely of the nature of frauds upon the consumer. . . and injure both the health and morals of the people," reports a U.S. congressional committee. The practice of fraudulent substitution "has become

1882 ***(cont.)*** universal," the committee says, and it issues a list of people who have died from eating foods and using drugs adulterated with poisonous substances (*see* 1889).

Van Camp Packing Co. is incorporated and packs 6 million cans of pork and beans per year for shipment to Europe as well as to many U.S. markets (*see* 1861; 1894).

Nashville, Tenn., wholesale grocer Joel Cheek quits his firm to devote full time to developing a coffee blend (*see* Maxwell House, 1886).

U.S. immigration from Germany reaches its peak.

Congress passes the first U.S. act restricting general immigration (*see* 1875). It excludes convicts, paupers, and defectives and imposes a head tax on immigrants.

The 1880 Chinese Exclusion Act takes effect, barring entry of Chinese laborers for a period of 10 years (*see* 1884).

Crop failures in southern Japan bring widespread starvation. Recruiting agents sent by planters persuade peasants to emigrate to Hawaii's sugar fields; 100,000 Japanese workers and their families will move in the next 30 years.

1883 The Pendleton Civil Service Reform Act passed by Congress January 16 provides for competitive examinations for positions in the federal government and establishes a Civil Service Commission to end the abuses that culminated in the assassination of President Garfield 2 years ago. The act sharply reduces the number of federal appointees who get their jobs from elected officials under the "spoils system," it establishes a merit system for appointment and promotion based on competitive examinations, and the number of "classified" civil service workers employed on the basis of merit will be expanded by Congress, with such workers protected from loss of job through change in political administration.

The convention of Marsa signed June 8 with the bey of Tunis assures French control of Tunisia.

A treaty signed at Hue August 25 recognizes Tonkin, Annam, and Cochin China as French protectorates, but China regards the territory as a vassal state, rejects the treaty, and continues to resist French control. Chinese Black Flag irregulars, Vietnamese, and French forces battle outside Hanoi September 3 in a bloody confrontation.

Britain's new consul general at Cairo Sir Evelyn Baring, 42, receives his appointment with plenipotentiary diplomatic rank to advise the khedive but will effectively rule the country until his retirement in 1907 (*see* 1882).

The Mahdi Mohammed Ahmed ibn-Seyyid Abdullah of Dongola, 40, challenges Egyptian control of the Sudan. His forces ambush the British general William Hicks, 53, and kill him November 4. The Egyptians are wiped out November 5 at the Battle of El Obeid, the self-styled prophet gains control over Kordofan, the governors of Darfur and Bahr-el-Ghazal are obliged to surrender, and the Mahdi's lieutenants attack the Red Sea forts (*see* 1884).

The Treaty of Ancon October 20 ends the war between Chile, Peru, and Bolivia over the nitrate-rich Atacama Desert. Chile gains territory at Peru's expense (*see* 1879; 1884).

The League of Struggle for the Emancipation of Labor, founded in Switzerland by émigré political philosopher Georgi Valentinovich Plekhanov, 26, will make the first Marxist analysis of Russian society, economics, and politics and will issue the first directives toward the creation of a Russian labor movement (*see* 1878; Lenin, 1895).

Ediswan Co. is founded by Joseph Swan and Thomas Edison (who has beaten Swan to obtain English patent rights to the incandescent bulb; *see* 1878; 1879). Swan has developed a new lamp filament of nitrocellulose that has advantages over the carbon filament (*see* tungsten, 1913).

German Edison Co. is founded by German electrotechnician Emil Rathenau, 45, and electrical engineer Oskar Miller, 28. The new firm is the forerunner of Allgemeine Elektrizitäts-Gesellschaft (A. E. G.) which Rathenau will head.

Thomson-Houston Electric Co. is founded by Philadelphia inventor Elihu Thomson, 30, and electrical engineer Edwin James Houston, 36. Thomson has invented a transformer that steps down high-voltage alternating current (*see* 1888; General Electric, 1892).

George Westinghouse pioneers control systems for long-distance natural gas pipelines and for town gas distribution networks (*see* 1882; 1885).

John D. Rockefeller's Standard Oil Trust monopoly absorbs Tidewater Pipe. The trust takes the output of 20,000 wells and employs 100,000 people (*see* 1882; 1884).

The Brooklyn Bridge (Great East River Bridge) that opens to pedestrians May 24 links America's two largest cities. Designed by the late John Augustus Roebling, who died of a tetanus infection in 1869 at age 63 after having his leg crushed by a ferryboat while working on the bridge, the 486-meter (1,595.5-foot) span has a steel web truss to keep it from swaying in the wind, and has been completed by Roebling's son Washington Augustus, 45, who has been crippled, half-paralyzed, and made partially blind by caisson disease (the "bends") but observes the dedication ceremony from his house on Columbia Heights. A panic May 30 produces a stampede that kills 12 pedestrians.

The Northern Pacific Railroad is completed September 8 with a "last spike" ceremony at Gold Creek in Montana Territory after 13 years of work (*see* Villard, 1881; financial struggles, 1893; 1901).

U.S. railroads adopt standard time with four separate time zones: Eastern, Central, Rocky Mountain, and Pacific. Pacific Standard Time is 3 hours behind Eastern Standard Time.

Australians complete a rail link between Sydney in New South Wales and Melbourne in Victoria.

Norway completes the Oslo-Bergen railway across the Hardanger mountain range with nearly 60 miles of its track above the timber line. The new rail link ends the isolation of Bergen and brings it to within 8 hours of Norway's capital.

The Calais-Nice-Rome Express from Paris goes into service. The run to Nice takes 18 hours but will take only 14 by the end of the century.

The *Orient Express* that leaves for Constantinople October 4 from the Gare de l'Est, Paris, is Europe's first transcontinental train. The *Express d'Orient* and its route have been designed by the Compagnie Internationale des Wagon-Lits et des Grands Express Européens headed by Georges Nagelmakers and includes two sleeping cars, an elaborate dining car, a baggage wagon containing a lavish kitchen, and a mail car (*see* 1876; Train Bleu, 1879). The dining car and both sleeping cars are fitted with newly invented four-wheel bogies and superior springing that give passengers a steady ride even round the sharpest bends at speeds of close to 50 miles per hour, but since no direct rail link to Constantinople has been completed, the new train goes only to the Bulgarian border south of Bucharest where passengers are ferried across the Danube, taken by a train to Varna on the Black Sea, and there put on a ship for Constantinople (*see* 1889).

Ferdinand de Lesseps of 1869 Suez fame begins work on a Panama Canal for the Compagnie Universelle du Canal Interocéanique (*see* 1889).

The first fully automatic machine gun is invented by American-born English engineer Hiram Stevens Maxim, 43, whose Maxim/Vickers gun will be adopted in 1889 by the British army and thereafter by every other major army. Maxim's gun is an advance over the 1861 Gatling gun in that the recoil energy of each bullet is employed to eject the spent cartridge, insert the new round, and fire it. Maxim Gun Co. will be absorbed into Vickers' Sons and Maxim in 1896 (*see* Omdurman, 1898; Boer War, 1899).

An artificial silk is developed from nitrocellulose by French chemist Hilaire Bernigaud, 44, comte de Chardonnet (*see* viscose rayon, 1892).

Copper ore in the Sudbury Basin of northern Ontario is discovered by builders of the Canadian Pacific Railway (*see* 1882; 1885).

Science magazine is founded by telephone pioneer Alexander Graham Bell and his father-in-law G. G. Hubbard. The magazine will become the organ of the 35-year-old American Association for the Advancement of Science (AAAS) (*see Nature*, 1869; National Geographic Society, 1888).

A German sickness insurance law enacted in May is the first of several moves by Prince Otto von Bismarck to lead the German people away from state socialism. Workers are to bear two-thirds of the cost of the insurance program, employers one-third (*see* accident insurance law, 1884).

The Brooklyn Bridge, an engineering marvel that was soon "for sale" to unwary out-of-town visitors.

German pathologist Edwin Klebs, 49, describes the diphtheria bacillus (*see* Löffler, 1884).

Robert Koch develops a preventive inoculation against anthrax (*see* 1882).

Robert Koch identifies the comma bacillus that causes Asiatic cholera (*see* Ferran, 1881). A worldwide cholera pandemic begins that will kill millions in the next 11 years.

The University of North Dakota is founded at Grand Forks.

Telegraph service begins between the United States and Brazil.

Commercial Cable Co. is founded by Comstock Lode millionaire John W. Mackay, now 52, and *New York Herald* publisher James Gordon Bennett who challenge Jay Gould's monopoly in transatlantic cable communication. Mackay and Bennett will lay two submarine cables to Europe next year and break the Gould monopoly (*see* 1869; 1886).

Thomas Edison (above) pioneers the radio tube with a method for passing electricity from a filament to a plate of metal inside an incandescent light globe, and he patents the "Edison effect" (*see* Arnold, 1912).

Joseph Pulitzer of the *St. Louis Post-Dispatch* acquires the *New York World* from Jay Gould (above; *see* 1878). The paper has a circulation of 15,000 and is losing $40,000 per year, but Pulitzer will raise circulation to 345,000 and make the *World* profitable (*see* 1893; Hearst, 1896; Howard, 1931).

Grit begins publication at Williamsport, Pa. The weekly for rural readers will grow in the next 94 years to have a circulation of more than 5 million.

The *Ladies' Home Journal* begins publication at Philadelphia under the name *Ladies' Journal.* Maine-born

1883 ***(cont.)*** publisher Cyrus H. K. (Herman Kotzschmar) Curtis, 32, offers cash prizes to readers who submit the most names of possible subscribers (*see* patent medicine, 1892; *Post*, 1897).

Life magazine begins publication at New York. Edward Sandford Martin, 27, helped start the *Harvard Lampoon* in 1876 and his new humor magazine will continue as such for more than half a century (*see* Gibson Girl, 1890; Luce, 1936).

More than 3,000 Remington typewriters are sold, up from about 2,350 last year. By 1885 sales will reach 5,000 per year (*see* 1878; Underwood, 1895; Mark Twain, below).

Nonfiction *Life on the Mississippi* by Mark Twain is a memoir of riverboat piloting and a critique published from the world's first typewritten book manuscript; "The Forgotten Man" by Yale political and social science professor William Graham Sumner, 43, extols the "clean, quiet, virtuous, domestic citizen, who pays his debts and his taxes and is never heard of out of his little circle. . . The Forgotten Man. . . delving away in patient industry, supporting his family, paying his taxes, casting his vote, supporting the church and the school. . . He is the only one for whom there is no provision in the great scramble and the big divide. Such is the Forgotten Man: He works, he votes, generally he prays—but his chief business in life is to pay."

Nonfiction *Thus Spake Zarathustra (Also Sprach Zarathustra)* by German philosopher Friedrich Wilhelm Nietzsche, 39, is published in the first of its four parts and introduces the idea of the Superman (Ubermensch). Nietzsche's theories will be influential on German thinking 30 years hence and, more especially, a half century hence.

Fiction *Treasure Island* by Robert Louis Stevenson, an adventure novel with Long John Silver's pirate song, "Fifteen men on a dead man's chest-/Yo-ho-ho, and a bottle of rum!. . ." Stevenson has returned to Scotland after having traveled across the Atlantic in steerage and across America in an emigrant train to pursue Fanny Van de Graft Osbourne, a married woman 11 years his senior and mother of three. He met her 7 years ago at Fontainebleau, France, married her at San Francisco in 1880, and thereafter took her to a western mining town.

Poetry *The Flemish (Les Flamandes)* by Flemish poet Emile Verhaeren, 28; *Poems of Passion* by Wisconsin-born poet Ella Wheeler Wilcox, 33, whose poem "Solitude" begins, "Laugh, and the world laughs with you;/ Weep, and you weep alone"; *The Old Swimmin' Hole and 'Leven More Poems* by Indiana poet James Whitcomb Riley, 34, of the *Indianapolis Journal* who includes his poem "When the Frost is on the Punkin!"

Painting *The Swimming Hole* by Thomas Eakins; Gustave Doré dies at Paris January 23 at age 51; Edouard Manet dies at Paris April 30 at age 51.

Theater *An Enemy of the People* (*En Folkefiende*) by Henrik Ibsen 1/13 at Oslo's Christiania Theater; *The Tragedy of Man (Az ember tragédiája)* by the late Hungarian playwright Imre Madách 9/21 at Budapest's Magyar Nemzeti Szinhaz; *Lucky Per's Journey (Lycko-Pers reja)* by August Strindberg 12/22.

Buffalo Bill's Wild West Show opens at Omaha, Neb., with riding, target shooting, and showmanship in an open-air spectacle that will be America's favorite entertainment of its kind for the next 20 years. Promoter William Frederick Cody, 37, is a former buffalo hunter and cavalry scout who received his nickname "Buffalo Bill" in 1869 from E. Z. C. "Ned Buntline" Judson (*see* 1844). Cody is famous for having killed and scalped the Cheyenne leader Yellow Hand in a July 1876 duel, he will be joined in 1885 by markswoman Annie Oakley (Phoebe Mozee), "The Peerless Lady Wing-Shot," who is now 23, and for 1 year the show will include Sioux chieftain Sitting Bull (*see* Wounded Knee, 1890).

Keith & Batchelder's Dime Museum opens at 585 Washington Street, Boston, where former circus promoters Benjamin Franklin Keith, 37, and George H. Batchelder have assembled a group of sideshow freaks to support themselves during the months the circus is not on tour. Keith covers the front of the three-story building with red and blue canvas twelve-sheets advertising Siamese twins, tattooed ladies, wild men of Borneo, sword swallowers, snake charmers, Circassian beauties, and the cow that allegedly kicked over the lantern to start the Chicago fire in 1871. Keith & Batchelder employ Edward F. Albee, 26, to manage a second-floor vaudeville theater that opens at 10 in the morning with acts that include the Reed family troupe and New York song and dance comedians Joe Weber and Lew Fields (Shanfield), both 16, who go on eight times a day for $40 per week (*see* 1885).

Opera *Lakmé* 4/14 at the Opéra-Comique, Paris, with music by Leo Delibes. The Bell Song "De la fille du paria" in Act II receives special applause.

New York's Metropolitan Opera House opens October 22 with a performance of the 1859 Gounod opera *Faust* sung in Italian with Christine Nilsson, 40, of Sweden as Marguerite. The 3,700-seat Met replaces the smaller Academy of Music that opened in Irving Place in 1854 and whose Old Guard have prevented the city's new millionaires from obtaining desirable boxes (*see* 1966).

First performances *The Accursed Huntsman (Le Chasseur Maudit)* (symphonic poem) by César Franck 3/31 at Paris; "Kol Nidre" by German composer Max Bruch, 45, who has arranged a traditional Hebrew melody for cello and orchestra, 10/20; *España* orchestral rhapsody by French composer Emmanuel Chabrier, 42, 11/4 at Paris; *Husitska* (dramatic orchestral work) by Antonin Dvorák 11/18 at Prague's Bohemian Theater; Symphony No. 3 in F major by Johannes Brahms 12/2 at Vienna.

Broadway musicals *Cordelia's Aspirations* 11/5 at the Theatre-Comique, with Ned Harrigan and Tony Hart, songs by Harrigan and Hart that include "My Dad's Dinner Pail."

Hymns "When the Mists Have Rolled Away" by Ira David Sankey, verses by Annie Herbert; "Till We Meet

Again" ("God Be With You") by William Gould Tomer, verses by Jeremiah Eames Rankin.

William Renshaw wins in singles at Wimbledon, Richard Sears at Newport.

Gold Flake cigarettes are introduced at London.

Fire guts London's Harrods store December 6, but proprietor C. D. Harrod orders new stock and enjoys record Christmas sales (*see* 1861). Harrod employs 100 clerks, bookkeepers, packers, etc. and fines latecomers a penny-halfpenny for every 15 minutes they are late (*see* 1884).

Washington's Pension building opens with a grand court enclosed by four levels of galleries and with eight massive 75-foot Corinthian columns to support its roof. Designed by General Montgomery C. Meigs to provide offices for the disbursement of pensions to war veterans and widows, the building has a court that will be the scene of at least seven presidential inaugural balls beginning in 1885.

The Stoughton house at Cambridge, Mass., is completed by Henry Hobson Richardson.

Krakatoa in the Sunda Strait between Java and Sumatra erupts August 27 in the greatest volcanic explosion since the eruption of Santorini in 1470 B.C. (*see* Tamboro, 1815). Although Santorini's explosion was roughly five times greater, Krakatoa's is heard nearly 3,000 miles away some 4 hours later, the dust it throws 34 miles into the air falls 10 days later at points more than 3,000 miles away, and the explosion creates a wave that wipes out 163 Indonesian villages, taking more than 36,000 lives.

White and Cree hunters in Dakota Territory discover a herd of 10,000 buffalo on the Cannon Ball River and exterminate the animals (*see* 1875; 1893).

"Sportsmen" shooting from moving trains helped wipe out the bison that provided sustenance for the Plains Indians.

Drought strikes the northern plains states. Railroad builder James J. Hill who is pushing his Great Northern track across the plains urges diversified farming. He buys 600 purebred bulls, distributes them among the farmers in the region, and offers prizes for the best cattle produced.

Purdue University chemistry professor Harvey Washington Wiley, 39, gains appointment as chief of the Division of Chemistry in the U.S. Department of Agriculture (*see* "Poison Squad," 1902).

Claude Spreckels of San Francisco gains a monopoly on West Coast sugar refining and marketing. He has obtained a large concession of land in the Sandwich (Hawaiian) Islands to keep his big refinery supplied with cane and also uses some sugar beets (*see* 1868; 1880; politics, 1891).

An Owasco, N.Y., cannery installs the first successful pea-podder machine, replacing 600 (*see* pea viner, 1889).

Some 40,000 tons of Jurgens Dutch margarine are shipped to Britain, which imports most of her margarine and butter (*see* 1871; Unilever, 1929).

Monte Carlo's Grand Hotel engages Swiss manager César Ritz, 33, and he hires as chef Auguste Escoffier, 37, who began his career at age 13 as an apprentice at his uncle's restaurant in Nice and moved at age 19 to the popular Paris restaurant Le Petit Moulin Rouge (*see* Pèche Melba, 1894).

The most lavish party yet held in America is staged March 26 at the $2 million Gothic mansion of railroad magnate William Kissam Vanderbilt, 34, on the northwest corner of New York's Fifth Avenue at 53rd Street. Vanderbilt is chairman of the Lake Shore & Michigan Southern, and the *New York World* (above) estimates that his wife has spent $155,730 for costumes, $11,000 for flowers, $4,000 for hired carriages, $4,000 for hairdressers, and $65,270 for catering, champagne, music, and the like to make the $250,000 fancy-dress ball a success.

The Northern Pacific Railroad (above) sends agents to the British Isles to encourage immigration to the U.S. Northwest (*see* Santa Fe, 1874).

James J. Hill (above) sends agents to Scandinavia to encourage immigration to the territory that will be served by his Great Northern Railroad.

The Southern Immigration Association is founded to encourage European immigration to the U.S. South.

The United States has her peak year of immigration from Denmark, Norway, Sweden, Switzerland, the Netherlands, and China (*see* Exclusion Act, 1882).

U.S. contraceptive production is pioneered by German-American entrepreneur Julius Schmid, 18, a cripple who came to New York 2 years ago from Schemdorf. He has been making "goldbeater" capping shins for bottles of perfume and other volatile liquids, using the tissue of lamb ceca which he now uses for condoms (*see* 1723; Youngs, 1920).

1884 Egypt's khedive Mohammed Tewfik gives former Sudanese governor Charles "Chinese" Gordon executive powers and Gordon moves out to rescue Egyptian garrisons in the Sudan from the Mahdi (*see* 1883). Now 51, Gordon is refused in his requests to London for Zobeir and Turkish troops. He nevertheless manages to rescue some 2,500 women, children, and wounded men from Khartoum but is hemmed in there by the Mahdi beginning March 12 (*see* 1885).

Chancellor Bismarck cables Cape Town April 24 that Southwest Africa (Namibia) is a German colony. The German consul at Tunis proclaims a protectorate over the coast of Togoland July 5 and a protectorate over the Cameroons Coast a week later.

The War of the Pacific that began in 1879 between Chile and Peru ends April 4 with the Treaty of Valparaiso that deprives Bolivia of access to the sea (*see* 1883). Victorious Chile gained the Peruvian province of Tarapaca last year in the Treaty of Ancona and now gains nitrate-rich Bolivian territories (*see* Haber process, 1908; Grace, 1890).

China declares war on France October 26 following French bombardment of Taiwan to punish the Chinese for not agreeing to France's protectorate of Indochina and refusal to pay indemnities (*see* 1885).

The political slogan "Rum, Romanism, and Rebellion" defeats the Republican "plumed knight" presidential candidate James G. Blaine, 54, and helps elect New York governor Stephen Grover Cleveland, 47, to the presidency in the first Democratic party national victory since James Buchanan won in 1856. Allegations that Cleveland fathered an illegitimate child in his youth have produced the chant, "Ma, Ma, where's my Pa?" (Cleveland acknowledged the child and has supported it while expressing doubts in private that he was the father), and Democrats reply, "Gone to the White House, ha, ha, ha." They have won the crucial New York Catholic vote as a result of remarks made October 9 at the Fifth Avenue Hotel by local clergyman Samuel D. Burchard ("We are Republicans, and don't propose to have our party identify ourselves with the party whose antecedents have been Rum, Romanism, and Rebellion"), New York voters have taken exception, and they give Cleveland a narrow 1,149-vote edge that enables him to win with a plurality of 20,000 votes out of 10 million total, 219 electoral votes to 189 for Blaine.

Congress establishes a Bureau of Labor in the Department of the Interior as severe coal strikes occur in Pennsylvania and Ohio (*see* 1888).

A German accident insurance law passed in June extends to practically all wage-earner groups and is paid for entirely by employers. The measure continues efforts begun by Bismarck in last year's sickness insurance law to resist advocates of state socialism (*see* 1891).

Russia ends her poll tax, a relic of serfdom, but the beginning of industrialization in Russia is creating an industrial proletariat whose living and working conditions are no better than were those of rural serfs (*see* 1903; Plekhanov, 1883).

A Berlin Conference on African affairs opens November 15 with delegates from 14 nations, including the United States, who agree to work for the suppression of slavery and the slave trade (*see* 1873; 1892).

London's Toynbee Hall—the world's first settlement house—is founded to attract well-to-do young people to "settle" in the city's slums and to serve the poor (*see* New York's University Settlement, 1886).

National Cash Register Co. is founded at Dayton, Ohio, by local coal merchant John Henry Patterson, 40, who has bought a controlling interest in a company that has bought the Ritter cash register of 1879 and improved it by adding a cash drawer and a bell that rings every time the drawer is opened. Patterson is ridiculed for investing $6,500 in a company that makes anything so useless, he offers the seller $2,000 to let him out of the deal, is refused, changes the firm name, and goes to work improving the cash register. Patterson will innovate the idea of exclusive sales territories, pay large commissions to salesmen, organize a force of well-trained service men to maintain the machines he sells, and make NCR prosper (*see* 1912; Watson, 1903).

Montgomery Ward issues a 240-page catalog offering nearly 10,000 items from cutlery, harnesses, and parasols to stereoscopes, trunks, and writing paper. Still describing itself on the catalog cover as "The Original Grange Supply House," the 12-year-old mail order firm assures customers that all goods are sent "subject to examination" and that unsatisfactory items may be returned with the company paying transportation both ways (*see* 1904; Sears, 1886).

A new Harrods store opens at London in September with displays of silver and brass, trunks and saddlery, cheap patent medicines, and with pay desks (*see* 1861). C. D. Harrod more than doubles the volume of his old store, he insists on "net cash before delivery," but he will extend limited credit to approved customers such as actresses Lily Langtree and Ellen Terry and playwright Oscar Wilde beginning next year. When Harrod sells out for £120,000 in 1889, the store will be doing an annual business of nearly £500,000.

The compound steam turbine invented by English engineer Charles Algernon Parsons, 26, exerts steam power first upon a large vane wheel and then—through slots in the casing—on another vane wheel and so on until the power of the steam is spent. The turbine develops 10 horsepower at 18,000 rpm, but Parsons will add a condenser in 1891, adapt the turbine to maritime use in 1897, and be building 1,000-kilowatt turbines by 1900; his geared turbine will appear in 1910 (*see* 1897).

John D. Rockefeller's Standard Oil Trust markets more than 80 percent of the petroleum from U.S. oil wells (*see* 1882; 1892; Sherman Act, 1890).

The Central Pacific Railroad is merged into the Southern Pacific by Charles Crocker and Collis P. Huntington, who amalgamate other California railroads to create a giant competitor to the Union Pacific (*see* 1881; Santa Fe, 1887).

Seattle gets its first rail link to the East. Still a small lumber town, it begins to grow (*see* 1851; 1909).

Electric streetcars employing overhead wires appear in Germany (*see* Deft's trolley, 1885).

Gen. Charles "Chinese" Cordon (above) has been conveyed to Khartoum by travel agent Thomas Cook, now 76 (*see* 1866). John Mason Cook, 50, has been his father's partner since 1864 and extended Thomas Cook & Son service to America.

F. A. J. Löffler at Berlin isolates and cultures the diphtheria bacillus (*see* 1882; Klebs, 1883; Roux and Yersin, 1889).

Leipzig gynecologist Karl Sigismund Franz Crede, 65, discovers that a few drops of silver nitrate solution in the eyes of newborn infants will prevent blindness from gonorrheal infection (*see* Howe Act, 1890).

New York physician Edward Livingston Trudeau, 36, pioneers open-air treatment of tuberculosis in America. He developed consumption himself at age 25 and repaired to the Adirondack Mountains to regain his health. The Adirondack Cottage Sanatorium he opens at Saranac, N.Y., will later be called the Trudeau Sanatorium.

New York surgeon William Stewart Halsted, 31, injects a patient with cocaine, pioneering the practice of local anesthesia. A stickler for hygiene who sends his shirts to Paris to be laundered, Halsted becomes addicted in the course of his experiments to the drug derived from leaves of the South American Andes shrub *Erythroxylon coca* (or *E. truxillense*), and although he will recover in 2 years he will require morphine in order to function (*see* 1803; 1859; Johns Hopkins, 1893).

New York ophthalmologist Carl Killer, 27, introduces cocaine as a local anesthetic for eye surgery.

English surgeon Rickman John Godlee, 35, performs the first operation for the removal of a brain tumor November 25. Godlee is a nephew of Lord Lister.

Charles Taze Russell founds the Watch Tower Bible and Tract Society to publish his books, pamphlets, and periodicals (*see* 1878). Russell warned true Christians against political and social allegiances in his 1881 book *Food for Thinking*; he preaches that the world is on the brink of annihilation in a monstrous battle of Armageddon after which 144,000 of all who have ever lived in the history of mankind will reappear in heaven (*see* Jehovah's Witnesses, 1931).

The Fabian Society founded in January by young London intellectuals aims to reconstruct society "in accordance with the highest moral responsibilities." Recognizing that deciding what courses to pursue will require "long taking of counsel," they take the name Fabian from the 3rd century B.C. Roman statesman Quintus Fabius Maximus, called "Cunctator" (delayer) because of his cautious delaying tactics against Hannibal. George Bernard Shaw joins in May, London economist Sidney James Webb, 25, and writer Beatrix Potter, 26, will soon join, Webb will marry Potter, and the Webbs will set up a socialist salon, attracting celebrated thinkers and writers.

Temple University is founded at Philadelphia by Baptist clergyman Russell Herman Conwell, 41, a former lawyer who has made a fortune with his inspirational lecture "Acres of Diamonds," encouraging his listeners to take advantage of the opportunities in their "own backyards." Conwell will give the lecture more than 6,000 times before his death in 1925 and give all the proceeds to his university.

The Linotype typesetting machine patented by German-American mechanic Ottmar Mergenthaler, 30, will revolutionize newspaper composing rooms. The Linotype has a keyboard much like that of a typewriter (*see* Sholes, 1868). Depressing a key on the keyboard releases a matrix from a magazine, and this small rod with the die of a character on its vertical edge falls into a line-composing box where little wedges automatically adjust the spaces between the words. When a line of matrices has been set, it is carried off to be cast in metal slugs made to the width of a newspaper column, the metal slugs are assembled in a form, compositors add hand-set headlines and line illustrations, and the form is ready for the press or for casting into a stereotype plate (*see* 1725; halftones, 1886; *New York Tribune*, 1886).

Le Matin begins publication at Paris.

Svenska Dagbladet begins publication at Stockholm.

The Waterman pen invented by New York insurance man Lewis Edson Waterman, 47, is the first practical fountain pen with a capillary feed. Waterman has 200 of the pens produced by hand, but the end of the pen must be removed and ink squirted in with an eye drop-

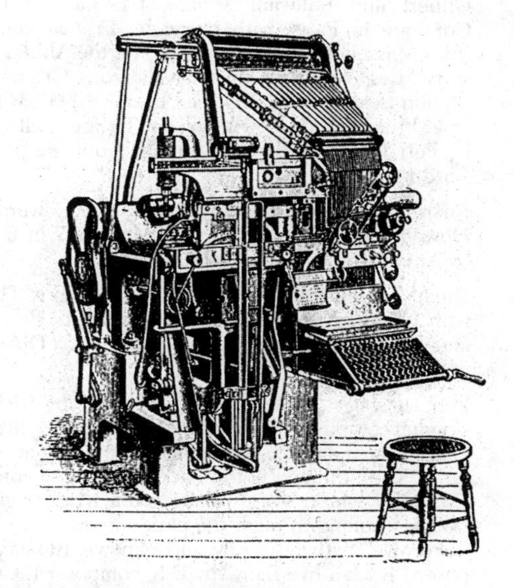

The Mergenthaler linotype machine and high-speed presses made newspapers cheap, broadening circulation.

1884 ***(cont.)*** per (the lever-fill pen will not be introduced until 1904) (*see* Parker, 1888).

Fiction *The Adventures of Huckleberry Finn* by Mark Twain; "The Lady or the Tiger?" by U.S. novelist-short story writer Frank (Francis Richard) Stockton, 50; *Against Nature (A Rebours)* by Joris Karl Huysmans; *The Death of Ivan Ilyich (Smeat'Ivana Il'icha)* by Leo Tolstoy.

Painting *Bathers at Asnières* (*Une Baignade, Asnières)* by French painter Georges Seurat, 24, who joins with Paul Gauguin, Paul Cézanne, and Dutch painter Vincent van Gogh, 31, in supporting the exhibiting society Les Vingt founded at Brussels by Belgian painter James Sydney Ensor, 24; *King Cophetua and the Beggar Maid* by Edward Coley Burne-Jones.

Sculpture *Robert Gould Shaw Memorial* by Augustus Saint-Gaudens for the Boston Common.

A photographic film roll system invented by British photography pioneer W. H. Walker uses paper coated with emulsion.

George Eastman invents a machine to coat photographic paper continuously in long rolls and will obtain a patent for the machine in 1895 (*see* 1880). Flexible roll film will permit development of cameras more convenient to use than large plate cameras (*see* Houston, 1881; Kodak, 1888).

The Playbill has its beginnings in the *New York Dramatic Chronicle,* a one-page flyer started by local printer Frank V. Strauss. It will evolve into a slick magazine with editions in every major U.S. city.

Opera *Princess Ida (or Castle Adamant)* 1/5 at London's Savoy Theatre, with book, lyrics, and music by Gilbert and Sullivan; *Manon* 1/19 at the Opéra-Comique in Paris, with music by French composer Jules Massenet, 42, libretto based on the Abbé Prevost story "Les Aventures du Chevalier des Grieux et de Manon Lescaut" of 1731 (*see* Puccini, 1893); *Mazeppa* in February at Moscow's Bolshoi Theater, with music by Petr Ilich Tchaikovsky, libretto from the poem by Lord Byron.

First performances Piano Suite No. 2 by Edward MacDowell 3/8 at New York; Symphony No. 7 in E major by Anton Bruckner 12/30 at Leipzig.

Broadway musical *Adonis* 9/24 at the Bijoux Theater, with Henry E. Dixey, music by Edward E. Rice, book and lyrics by William F. Gill and Henry Dixey, 603 perfs.

Popular songs "Love's Old Sweet Song" by Irish-born London composer James Lyman Molloy, 47, lyrics by C. Clifton Bingham, 25, beginning, "Just a song at twilight . . ."; "The Fountain in the Park" by London composer Ed Haley whose song will be given lyrics beginning, "While strolling through the park one day . . ."; "Otchi Tchorniya" with music by an unknown Russian composer; "Rock-a-bye Baby" by U.S. composer I. Canning (Effie I. Crockett), 15, who uses lyrics from Mother Goose's Melodies (*see* 1765).

William Renshaw wins in men's singles at Wimbledon, Maud Watson, 21, in women's singles; Richard Sears wins at Newport.

Overhand pitching gains acceptance for the first time in major league baseball, but a batter will be allowed until 1887 to call for a high pitch or a low pitch, and the number of balls required for a walk will not be established at four until 1889.

The "Louisville Slugger" bat is introduced by the Kentucky firm Hillerich and Bradsby. German-American woodturner J. Frederick Hillerich, 50, has made bowling balls and pins and has been asked by Louisville Eclipse player Peter "the Gladiator" Browning, 26, to make an ashwood bat that will replace one that Browning has broken. Browning has made his own bats of seasoned timber aged in his attic, but although he has averaged three hits per game with his homemade bats, he does even better with Hillerich's bat.

The first baseball playoff series is won by the National League's Providence, R.I., team, which defeats the New York Metropolitans of the American Association 3 games to 0 (*see* National League, 1876; American League, 1901; World Series, 1902).

The National Horse Show opens in October at New York's Madison Square Garden on East 27th Street with 352 animals in 105 classes including hunters, jumpers, harness horses, ponies, Arabians, police and fire horses, mules, and donkeys in an event that will be held annually the first week of November.

The first roller coaster opens at Coney Island, N.Y. It has been put up by former Elkhart, Ind., Sunday school teacher Lemarcus A. Thompson, who will soon be making improved models for amusement parks throughout America and Japan (*see* Steeplechase Park, 1897).

Gold Dust soap is introduced by Chicago's N. K. Fairbank & Co., with two black boys as the "Gold Dust

Overhand pitching and new rules helped make baseball the national pastime it would remain for more than a century.

Twins" trademark that will soon be used for Gold Dust washing powder and Gold Dust scouring cleanser as well.

New York's Dakota apartment house opens October 27 at 72nd Street and Central Park West (so called since 1882). Financed by the late Singer Sewing Machine magnate Edward Clark and designed by U.S. architect Henry Janeway Hardenbergh, 34, the $2 million nine-story building has a central court and nine hydraulic elevators to serve its marble-floored, mahogany-paneled apartments, some with 15-foot ceilings.

Chicago's Studebaker Building is completed by architect S. S. Berman whose structure will be renamed the Fine Arts building.

The first skyscraper goes up in Chicago (*see* 1885).

California outlaws the hydraulic mining that has been ruining the environment since 1852.

The soft-shell clam *Mya arenaria,* also called the longneck or steamer clam, is introduced to the U.S. Pacific Coast (*see* shad, 1873).

Ceylon's coffee output falls to 150,000 bags, down from 700,000 in 1870 when the rust disease caused by *Hamileia vastatrix* began making deep inroads (*see* 1869). The last shipment of coffee beans will leave the island in 1899.

Bordeaux mixture is invented by French botanist Pierre M. Alexis Millardet, 46, whose solution of copper sulfate, lime, and water effectively combats the fungus diseases attacking French vineyards and will prove effective against potato blight (*see* 1847; 1868).

Reaper magnate Cyrus McCormick dies March 13 at age 75, leaving a fortune of $200 million to his widow, four sons, and three daughters. His son Cyrus, Jr., 25, assumes the presidency of McCormick Harvesting Machine Co. (*see* International Harvester, 1902).

A congressional committee reports that a number of individuals, many of them "foreigners of large means," have acquired ownership of land tracts in Texas, some of them embracing more than 250,000 acres: "Certain of these foreigners are titled noblemen. Some of them have brought over from Europe, in considerable numbers, herdsmen and other employees who sustain to them a dependent relationship characteristic of the peasantry on the large landed estates of Europe." Two British syndicates hold 7.5 million acres of land (*see* 1888; Prairie Cattle Co., 1882).

The Dairylea milk producers' cooperative has its beginnings in an association of Orange County, N.Y., farmers who pool their efforts to market their milk effectively in competition with Harlem River Valley farmers, who ship milk into New York City at a rate of 50,000 quarts per day. The 40-quart milk can introduced in 1876 is now standard for shipment on night trains to the city (*see* 1842; 1907).

Chinese farm workers account for half of California's agricultural labor force, up from 10 percent in 1870. The Chinese have raised dikes at the mouths of the San Joaquin and Sacramento Rivers and are reclaiming millions of acres of rich farm lands (*see* below).

Beriberi continues to plague the Japanese navy, but the director of the Naval Medical Bureau Baron Kanehiro Takaki sends two ships on a 287-day voyage during which the men on one ship are served a diet of meat, cooked fish, and vegetables, while the men on the other ship get the usual Japanese diet of polished white rice and raw fish. Some 160 men out of 360 come down with beriberi on the second ship and 75 die, while on the first ship nobody dies, and the 16 cases of beriberi are found to occur only among men who refused to accept the prescribed diet. Takaki orders Western-style rations to be served aboard all Japanese naval vessels (*see* 1642; 1887; naval victory, 1905).

Scurvy reappears in the British navy; the Admiralty orders a daily lemon ration aboard all British warships where the old 1795 order has lapsed.

Quaker Oats becomes one of the first food commodities to be sold in packages. Put up in cardboard cannisters, the cereal will be widely and aggressively advertised throughout the United States and Britain (*see* 1877).

A second Chinese Exclusion Act passed by Congress July 5 tightens the provisions of the 1880 act (*see* 1888; California farm workers, above).

1885

Khartoum falls January 26 to the Mahdi Mohammed Ahmed whose forces massacre Gen. Charles "Chinese" Gordon and his garrison just before a British relief expedition reaches the city. The Mahdi dies June 21 and is succeeded by the khalifa Abdullah el Tasshi whose dervishes gain control of all the Sudan except for the Red Sea fortresses (*see* 1896).

Italian forces establish themselves at Massawa with British encouragement and begin to expand their holdings in the East African highlands.

The king of the Belgians Leopold II assumes the title of sovereign of the Congo Free State following French recognition of the Free State and conclusion of an agreement defining the boundary between the Free State and the French Congo. Another agreement has been concluded with Lisbon giving Portugal the Kabinda Enclave; large areas of the Congo Free State are assigned to concessionaires, but the central portion is set aside as state land and the king's private domain.

Germany annexes Tanganyika and Zanzibar.

Britain establishes protectorates in the Niger River southern region, in north Bechuanaland, and in Guinea.

British troops occupy Port Hamilton, Korea.

Qing troops attack French forces on the Vietnamese side of the Chinese border, killing Gen. François de Negrier, but the Treaty of Tianjin (Tientsin) signed June 9 recognizes France's protectorate of Tonkin in return for a promise by the French to respect China's southern frontier (*see* 1883). British diplomat Robert Hart, 50, has negotiated the treaty as China's Gen. Tso

1885 ***(cont.)*** Tsung-tang lay dying at age 73, urging modernization of his country.

Britain's second Gladstone ministry ends June 9. Robert Arthur Talbot Gascoyne-Cecil, 55, marquis of Salisbury, has headed the Tories since the death of Lord Beaconsfield in 1881 and begins a brief ministry.

Spain's Alphonso XII dies of phthisis November 24 at age 27 after a 15-year reign in which he has escaped two assassination attempts. Alphonse's mother Maria Christina becomes queen regent, but his posthumous son will be born next year, come of age in 1902, and reign until 1931 as Alphonso XIII.

Louis Riel leads another rebellion against Dominion authorities to protest the indifference of the Ottawa government toward the grievances of western Canadians (*see* 1869). Now a U.S. citizen, Riel was requested last year to give up his Montana school-teaching post and lead the protest, he has broken with the Roman Catholic Church, establishes a government of métis, and from March to May fights Dominion troops sent west on the partially completed Canadian Pacific Railway (*see* 1881). Riel is defeated at Batoche, apprehended, charged with treason, convicted by an English-speaking jury, and hanged November 16 at Regina, leaving a legacy of bitterness in Canada's western provinces.

Anti-Chinese rioting breaks out in the Washington Territory. President Cleveland issues an order November 7 calling for the troublemakers to be dispersed (*see* 1877; Seattle, 1886).

Railroad Transportation, Its History and Its Laws by Yale political science instructor Arthur Twining Hadley, 29, exposes the fallacy of the 1817 Ricardian theory of free enterprise as applied to industries having large permanent investments. Hadley pinpoints the high proportion of fixed to variable costs as a source of instability and the tendency toward combination, and is called to testify before a Senate committee drafting an interstate commerce bill (*see* 1887).

M. A. Hanna & Co. is founded by Cleveland coal, iron, and street railway magnate Marcus Alonzo Hanna, 48, who has organized the Union National Bank and acquired the *Cleveland Herald* and Cleveland Opera House (*see* McKinley, 1896).

Canadian Copper Co. is founded by Ohio businessman Samuel J. Ritchie to buy up claims in the Sudbury Basin of northern Ontario (*see* 1883). The copper ore turns out to be rich in nickel, which is used primarily for coins and nickel-plating but whose value for nickel-steel alloys will soon be recognized (*see* 1888).

Broken Hill Proprietary Co., Ltd., founded in Australia, will grow to monopolize the nation's iron and steel production. Mount Gipps sheep station manager George McCulloch, his hired hand Charles Rasp, and 12 other partners have staked claims to a tin mine found by Rasp in 1883, the mine has proved to be the world's largest silver-lead-zinc deposit, it is the basis of the new company, will gross more than £150 million before it closes in 1939, and will pay dividends of nearly £16 million.

Westinghouse Electrical & Manufacturing Co. is founded by George Westinghouse who buys up rights to the European Gaulard-Gibbs transformer and will buy patents to the Nikola Tesla induction motor and Tesla polyphase alternator, which will make it economically feasible to transmit power over long distances (*see* 1883; 1888).

U.S. electrical engineer William Stanley, 27, and George Westinghouse (above) perfect a practical transformer for large electricity supply networks. They will give the first demonstration of a practical alternating-current system in March of next year at Great Barrington, Mass., and while Thomas Edison has rejected the alternating-current system in favor of direct current, Westinghouse Co. will exploit the AC system and use it to send high-voltage current over long wires, employing transformers to step down the voltage for local distribution to houses, stores, factories, and the like (*see* 1888; Thomson, 1883).

The principle of the rotary magnetic field discovered by Italian physicist-electrical engineer Galileo Ferraris, 38, will lead to the development of polyphase motors (and of Italy's hydroelectric industry). Ferraris will devise transformers for alternating current.

Gas lighting gets a new lease on life from Austrian chemist Carl Auer von Welsbach, 27, who isolates the element praseodymium and patents a gas mantle of woven cotton mesh impregnated with thorium and cerium oxides, rare earths obtained from India's Travencore sands. The Welsbach mantle is fitted over a gas jet to increase its brilliance.

The world's first successful gasoline-driven motor vehicle reaches a speed of 9 miles per hour at Mannheim. German engineer Karl-Friedrich Benz, 41, has built the single-cylinder, chain-drive three-wheeler (*see* 1888; Mercedes, 1901; Mercedes-Benz, 1926).

The "Bicyclette Moderne" designed by French engineer G. Juzan has two wheels of equal size with a chain-driven rear wheel employing a drive chain stronger than the one on the first rear drive "Bicyclette" designed by André Guilmet in 1868 and manufactured by Meyer et Cie. The Rover Co. of Coventry, England, introduces the "safety" bicycle designed by the late J. K. Starley whose vehicle has wheels of equal size, a departure from the "ordinary" whose front wheel is much larger than its rear wheel. Sewing machine inventor James Kemp Starley died in 1881 at age 51 after having designed the bicycle that has wheels 30 inches in diameter with solid rubber tires, a chain-driven rear wheel, and—like Juzan's modern "Bicyclette"—a close resemblance to a vehicle sketched in 1493 by Leonardo da Vinci or one of his associates. The new French and English models make the bicycle suitable for general use (*see* United States, 1889; pneumatic bicycle tire, 1888).

English-American electrical engineer Leo Daft, 41, installs the world's first electric trolley line at Baltimore using double overhead wires from which the passenger cars draw electricity through a small carriage called a "troller" (*see* 1888).

The "safety bicycle" made high-wheelers obsolete—and made cycling a serious form of transportation.

Fifth Avenue Transportation Co., Ltd., is founded by New York entrepreneurs determined to forestall the introduction of trolley cars (above). The company's horsecars will remain in service until 1907.

The Congressional Limited Express goes into service on the Pennsylvania Railroad between New York and Washington, D.C.

The Missouri, Kansas, and Texas Railroad reaches the heart of the Texas cow country and ends the need to drive cattle long distances to railheads (*see* 1871).

The first successful U.S. appendectomy is performed (by some accounts) January 4 at Davenport, Iowa, by physician William West Grant who has diagnosed a perforated appendix in his patient Mary Gartside, 22. Grant is the first U.S. physician deliberately to open the abdomen and sever the appendix from the cecum (*see* McBurney, 1894).

The first anti-rabies vaccine is administered beginning July 6 to an Alsatian schoolboy of 9 who has been bitten on the hands and legs by a rabid dog. Louis Pasteur has developed the vaccine from weakened viruses that have aged on the dessicated spinal cords of rabbits which have died from rabies; the vaccine has worked with test animals but has never been tried with humans. Pasteur's assistant Pierre Paul Emile Roux, 31, on whose work the vaccine has been based, opposes its use until the reason for its effectiveness is known, but Pasteur risks criticism by the medical fraternity and permits the vaccine to be administered to young Joseph Maister, saving the boy from an agonizing death, and while the treatment requires a series of painful injections in the stomach, victims of bites from rabid animals flock from all over the world to be treated.

Moved by compassion to treat a young girl bitten 5 weeks earlier who has almost no chance of survival, Pasteur (above) is attacked when the girl dies, jealous physicians question his other successes, suggesting that perhaps Joseph Maister and the others were not really bitten by rabid animals, but Emile Roux defends Pasteur, and his critics are shamed (*see* Institut Pasteur, 1888).

Russian zoologist-bacteriologist Ilya Ilich Mechnikov, 40, discovers phagocytosis while conducting experiments with starfish larvae in Sicily where he is recovering from a suicide attempt. He finds that white blood corpuscles (leukocytes) engulf and destroy foreign microorganisms, and his work suggests means by which vaccines achieve success.

Johnson & Johnson is founded at New Brunswick, N.J., by Robert Wood Johnson with his brothers James Wood and Edward Mead Johnson to produce a full line of pharmaceutical plasters with an India rubber base (*see* 1874; 1886).

Stanford University (Leland Stanford Junior University) is founded by railroad magnate Leland Stanford, now 61, to memorialize his son Leland, Jr., who died last year at age 15 of typhoid fever at Florence after visiting with archaeologist Heinrich Schliemann (*see* 1870). The university will open outside San Francisco in 1891.

Bryn Mawr College for Women opens outside Philadelphia. Founded by former Quaker minister Joseph Taylor, the new college employs the only four women Ph.D.'s in the United States (*see* Yale, 1861) and along with Radcliffe (*see* 1894) will be the only college to prepare women for the Ph.D. for more than 50 years.

The Georgia Institute of Technology (Georgia Tech) is founded at Atlanta.

The University of Arizona is founded at Tucson.

Good Housekeeping magazine begins publication in May at Springfield, Mass.

The first-class U.S. postal rate doubles to 2¢ after a century (see 1932).

Fiction *Germinal* by Emile Zola; *Bel Ami* and *Contes et Nouvelles* by Guy de Maupassant; *Diana Of the Crossways* by George Meredith who wins his first public acclaim at age 57 for his roman à clef about the granddaughter of playwright Richard Brindsley Sheridan; *Marius the Epicurean* by Oxford essayist-critic Walter H. Pater, now 46, who expresses his ideal of the aesthetic life in his romance set in the time of ancient Rome's Marcus Aurelius; *A Mummer's Wife* by Irish novelist George Moore, 33; *King Solomon's Mines* by English novelist H. (Henry) Rider Haggard, 29, who went to Africa at age 19 as secretary to Henry Bulwer, nephew of the late novelist E. G. E. L. Bulwer-Lytton; *The Rise of Silas Lapham* by Boston novelist-editor William Dean Howells, 48, who edited the *Atlantic Monthly* from 1871 to 1881 and encouraged both Henry James and Mark Twain while writing some novels of his own.

Poetry "The Betrothed" by English journalist-poet Rudyard Kipling, 20, whose satirical poem contains the lines, "A woman is only a woman, but a good Cigar is a Smoke." Chauvinist Kipling went to India at age 15 and has worked since age 17 on the editorial staff of the *Civil & Military Gazette and Pioneer* at Lahore where

1885 ***(cont.)*** his father is curator of the Lahore Museum; *A Child's Garden of Verses* by Robert Louis Stevenson, whose poem "Bed in Summer" begins, "In winter I get up at night/ And dress by yellow candle-light./ In summer, quite the other way,/ I have to go to bed by day. In "Happy Thought" Stevenson writes, "The world is so full of a number of things,/ I'm sure we should all be as happy as kings."; "Little Orphant Annie" by James Whitcomb Riley, who writes, "An the Gobble-uns 'at gits you/ Ef you/ Don't/ Watch/ Out!"

Painting *The Potato Eaters* by Vincent van Gogh.

Amsterdam's Rijkmuseum founded by Louis Bonaparte in 1808 moves into new quarters.

Sculpture *The Puritan* by Augustus Saint-Gaudens.

Theater *The Wild Duck (Vildauden)* by Henrik Ibsen 1/11 at Oslo's Christiania Theater.

Keith & Batchelder's Dime Museum at Boston breaks up as Batchelder moves to Providence and as Keith acquires the Bijou Theater next door to the Dime Museum in Washington Street.

Opera *The Mikado* (*or The Town of Titipu*) 3/4 at London's Savoy Theatre, with book and lyrics by William S. Gilbert, music by Arthur S. Sullivan. Inspired by Commodore Perry's opening of Japan in 1853 and subsequent enthusiasm for things Japanese, the opera enjoys enormous success with arias that include "A Wand'ring Minstrel I," "Behold the Lord High Executioner," "I've Got a Little List," "Three Little Maids from School," "For He's Going to Marry Yum-Yum," "Here's a How-De-Do," "My Object All Sublime," "The Flowers That Bloom in the Spring," and "Tit-Willow"; *The Gypsy Baron (Zigeunerbaron)* 10/24 at Vienna's Theater an der Wien, with music by Johann Strauss.

First performances Suite No. 3 in G by Petr Ilich Tchaikovsky 1/24 at St. Petersburg; Prelude, Chorale, and Fugue for Piano by César Franck 1/24 at the Société Nationale, Paris; *Saugefleurie*, Legend for Orchestra, after a tale by Robert de Bonnières by Vincent d'Indy 1/25 at Paris; *Les Djinns* (symphonic poem) by César Franck 3/15 at the Société Nationale, Paris; Symphony No. 4 in E minor by Johannes Brahms 10/25 at Meiningen.

The Boston Pops is founded by the 4-year-old Boston Symphony. Audiences enjoy liquid refreshments during the May-June series of concerts (*see* Symphony Hall, 1900).

William Renshaw wins in men's singles at Wimbledon, Maud Watson in women's singles; Richard Sears wins at Newport.

The Browning single-shot rifle, introduced by the Winchester Repeating Arms Co., has been designed by Utah Territory gunsmith John M. Browning, 30, and will be enormously popular among hunters.

England's *Genesta* is defeated 2 to 0 by the U.S. defender *Puritan* in the first English effort to regain the America's Cup since 1871.

Parker Brothers is founded at Salem, Mass., by local inventor George Swinerton Parker, 18, who has devised a game he calls Banking in which the player who amasses the most wealth is the winner. Parker will be joined by his brother and will remain head of the firm for 60 years (*see* Monopoly, 1935).

Sunlight soap, introduced by Lancashire salesman William Hasketh Lever, 34, is a free-lathering yellow soap that will dominate the British market. Lever has had the soap made for him but will soon produce it himself (*see* Lifebuoy, 1897).

English scientist Francis Galton, 63, devises an identification system based on fingerprints. Galton, who founded the science of eugenics with his 1869 book *Hereditary Genius* and subsequent books, proves that fingerprints are permanent and no two people ever have the same prints.

Chicago's Home Insurance building at LaSalle and Monroe Streets is the world's first skyscraper. The 10-story marble structure completed in the fall by architect William LeBaron Jenney, 52, has a framework made partly of steel and will be enlarged by two additional stories (*see* Otis, Siemens, 1861; Tacoma building, 1888; Wainwright building, 1890).

Sagamore Hill is completed for New York widower Theodore Roosevelt, 26, who retires to a ranch in the Dakota territory. The 22-room, $16,975 house at Cove Neck, Long Island, will be electrified in 1914.

Canada's National Parks system has its beginnings in a reserve of 10 square miles set aside by Queen Victoria in the Canadian Rockies in an area where Canadian Pacific Railway surveyors discovered hot springs in 1883. A lake in the region has been named Lake Louise in honor of the queen's daughter, whose husband the marquis of Lorne was Canada's governor general from 1878 to 1883. The area set aside will become Banff National Park, and it begins a program that will respond to the promotion of Irish-Canadian trader-frontiersman John George "Kootenai" Brown. Water-

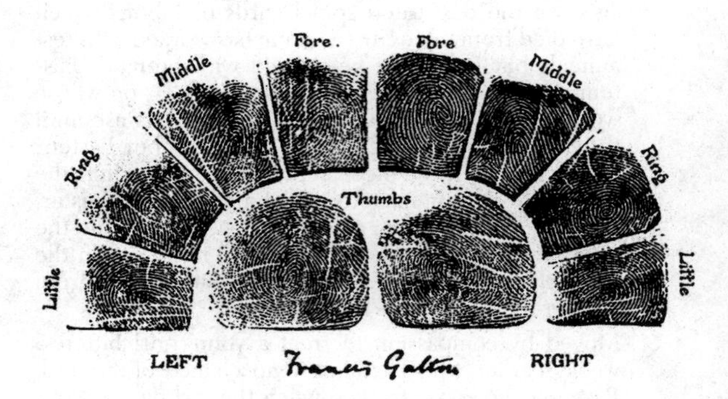

Fingerprints—more distinctive even than signatures—gave the police a new weapon against crime.

ton Lakes, Kootenay, Baner, Jasper, Yoho, Columbia Icefield, Glacier, and Mt. Revelstoke national parks will all be set aside as such in the system (*see* Waterton-Glacier, 1932).

The National Audubon Society has its beginnings in a group of U.S. bird lovers organized by *Forest and Stream* editor George Bird Grinnell, 36, to protest commercial hunting of birds and indiscriminate slaughter of U.S. wildlife in general (*see* 1905; Boone & Crockett Club, 1887).

The North American lobster catch reaches an all-time high of 130 million pounds.

The Maryland oyster catch reaches nearly 15 million bushels.

Memoirs of a Revolutionist (*Paroles d'un Revolté*) by Russian philosopher Prince Petr Alekseevich Kropotkin, 43, declares that the "Golden Age of the small farmer is over. . . He is in debt to the cattledealer, the land speculator, the usurer. Mortgages ruin whole communities, even more than taxes" (*see* 1861; 1905).

Congress prohibits barbed wire fencing of public lands in the U.S. West February 25 and President Cleveland issues a proclamation August 7 ordering removal of all unlawful enclosures (*see* 1877).

The U.S. corn crop tops 2 billion bushels for the first time in history, double the 1870 crop (*see* 1906). Most goes into hog and cattle feed, but U.S. cattle are still fattened largely on grass.

The Orange Growers Protective Union of Southern California is organized as the Santa Fe Railroad extends its service in the Los Angeles area (*see* 1881; 1887; California Fruit Growers Exchange, 1895).

Fresh milk in bottles is added to the line of the New York Condensed Milk Co. now headed by John Gail Borden (*see* 1861; 1879; evaporated milk, 1892).

Evaporated milk is produced commercially for the first time at Highland, Ill. John B. Meyenberg, now 37, has established the Helvetia Milk Condensing Co. with help from local Swiss-Americans (*see* Anglo-Swiss, 1866). When fire devastates Galveston, Tex., in November and fresh milk is hard to obtain, Helvetia Milk donates 10 cases of its product (*see* Carnation milk, 1899).

United Fruit Co. has its beginnings in the Boston Fruit Co. established by Andrew Preston and nine partners who have been persuaded to set up an independent agency to import bananas by Dow Baker (*see* 1871). Now a prosperous shipper and a partner in Standard Steam Navigation, Baker has built the *Jesse H. Freeman* schooner with auxiliary steam engines that permit him to send 10,000 stem cargoes north to Boston in 10 to 12 days from his new base in Jamaica, and the new company prospers as banana harvests increase (*see* 1899).

New York's Exchange Buffet opens September 4 across from the New York Stock Exchange at 7 New Street; it is the world's first self-service restaurant (*see* Horn & Hardart, 1902).

India has a population of some 265 million, up from 203.4 million in 1850. British physicians in this century have introduced Western medicine into India to reduce the death rate, and moralists have eliminated the slaughter of female children and other customs that once held India's population growth in check, measures that include *suttee,* the custom of burning wives along with their dead husbands on funeral pyres.

1886

The appointment of Gen. Georges Boulanger, 38, as minister of war January 4 stimulates French *revanchist* sentiment. A veteran of the 1870 siege of Metz in the Franco-Prussian War, Boulanger has been pushed into prominence by Georges Clemenceau, 44, a member of the Chamber of Deputies who served as a war correspondent with Gen. Grant's army in the U.S. Civil War (*see* 1888).

Britain's first Salisbury ministry ends January 27 after 7 months and a third Gladstone ministry begins February 12.

An Irish home rule bill introduced in Parliament April 8 by Prime Minister Gladstone provides for a separate Irish legislature but retains control of matters relating to the army and navy, trade and navigation, and the crown in the British Parliament where the Irish will no longer be represented (*see* Parnell, 1882). Conservatives attack the measure, the marquis of Huntington Joseph Chamberlain, 50, resigns from the Gladstone cabinet and leads a secession from the Liberal party, the bill is defeated in July, and the third Gladstone ministry ends July 26 when a general election gives victory to the Conservatives. A second Salisbury ministry takes power (it will continue until 1892), and Lord Salisbury's nephew Arthur Balfour, 38, is made chief secretary for Ireland (*see* 1887).

Britain annexes upper Burma following a third Anglo-Burmese war, but desultory guerrilla warfare will continue for years. China recognizes the British protectorate in Burma July 24 in return for continuation of a decennial tribute.

The Haymarket Massacre at Chicago gives the U.S. labor movement its first martyrs and marks the beginnings of May Day as a worldwide revolutionary memorial day. Chicago police fire into a crowd of strikers May 1, killing four and wounding many others. The 17-year-old Knights of Labor organization holds a peaceful rally May 4 in Haymarket Square to protest the shooting, someone throws a small bomb that knocks down 60 policemen, killing one and mortally wounding six others, the police fire into the crowd, and the workers sustain three times as many casualties as the police (*see* Altgeld, 1893).

Labor agitation for an 8-hour day and better working conditions makes this the peak year for strikes in 19th-century America (*see* 1825). Some 610,000 U.S. workers go out on strike, and monetary losses exceed $33.5 million.

A new American Federation of Labor (AF of L) is founded under the leadership of English-American cigarmaker Samuel Gompers, 35, who 5 years ago

Chicago's Haymarket Massacre grew out of a police assault on strikers. The bloodshed gave May Day new significance.

founded the Federation of Organized Trades and Labor unions at Pittsburgh; he will be AF of L president for 37 of the next 38 years (*see* 1902).

A streetcar strike ties up New York City public transit completely until motormen settle for $2 for a 12-hour day with a half hour off for lunch.

Seattle rioters drive 400 Chinese from their homes. Some are sent to San Francisco before federal troops are called in.

A municipal order discriminating against Chinese laundries violates the 14th Amendment, says the Supreme Court in *Yiek Mo v. Hopkins.*

The last major Indian war in the United States ends September 4 with the capture by U.S. troops of the Chiricahua Apache chief Geronimo after 4 years of warfare on the Mexican border (*see* 1876). Captured earlier by Gen. George Crook, Geronimo has escaped, but Gen. Henry Ware Lawton, 57, pursues him, recaptures Goyathlay, and turns him over to Gen. Nelson A. Miles, who has relieved Crook.

The U.S. settlement house movement has its beginnings in two small rooms at 146 Forsythe Street, New York, rented by ethical culture leader Stanton Coit, 29, who has returned from a visit to London's 2-year-old Toynbee Hall. An assistant to Felix Adler in the 10-year-old Society for Ethical Culture, Coit begins forming neighborhood clubs to help organize the poor to work for social improvement, and he will establish the Neighborhood Guild next year (*see* University Settlement, 1891; Hull House, 1889).

Johannesburg in the Transvaal is laid out in September and soon has a population of 100,000—half of it native workers—as the world's largest gold mines begin operations (below).

A gold rush to South Africa's Transvaal follows discovery of the yellow metal on the Witwatersrand. Diamond king Cecil Rhodes founds Consolidated Gold Fields, Ltd. (*see* De Beers, 1880; British South Africa Co., 1889).

National Carbon Co. is founded to produce carbons for electric arc streetlights and similar carbon products (*see* 1879; Ever Ready battery, 1890).

More than half of Britain's 7.36 million tons of merchant shipping is in steam, up from less than 9 percent in 1856.

The Supreme Court reverses its decision in the 1877 Peik case and forbids individual states to fix rates on shipments passing beyond their borders. The ruling handed down October 25 in *Wabash, St. Louis & Pacific Railway v. Illinois* says that only the federal government may regulate interstate railway rates, and the decision affects 75 percent of the volume on U.S. railroads (*see* Interstate Commerce Act, 1887).

The first trainload of California oranges leaves Los Angeles for the East (*see* Columbia Exposition, 1893).

New York's Metropolitan Traction Co., organized by local financier Thomas Fortune Ryan, 35, is the nation's first holding company. Ryan started a brokerage house at age 21, had his own seat on the New York Stock Exchange 2 years later, has been buying up street railway franchises, and calls his company "the great tin box." By 1900 he will have beaten his rival William C. Whitney (below) for control of the Broadway surface line and will control nearly every other line in the city (*see* Southern Railway, 1894; Royal Typewriter, 1905).

Chicago financier Charles Tyson Yerkes, 49, gains control of the city's North and West Side streetcar lines as he builds up a complex corporate empire through financial maneuvers in which he aligns himself with corrupt politicians (*see* London subway, 1890).

Flint, Mich., entrepreneur William Crapo Durant, 24, starts a buggy manufacturing company that will soon be the largest in the world. His partner is hardware merchant J. Dallas Dort who has acquired patent rights to a two-wheeled road cart (*see* Buick, 1905).

Gottlieb Daimler perfects his internal combustion engine of 1882 (*see* 1887).

Oberlin College graduate chemistry student Charles Martin Hall, 22, pioneers commercial aluminum production using methods developed by Humphry Davy in 1807 to liberate aluminum electrolytically from aluminum oxide (bauxite) (*see* Sainte-Claire Deville, 1855). Hall succeeds in February in separating the oxide from the bauxite, adds cryolite (a fluoride of aluminum), and applies electric current whose passage is resisted by the oxide, turning the electrical energy into heat and ultimately breaking down the oxide into its aluminum and oxygen components. Hall forms Pittsburgh Reduction Co. to exploit his discovery (*see* 1888).

French metallurgist Paul Louis Toussaint Heroult, 23, discovers a process almost identical to that of Hall (above) at almost exactly the same time, but litigation over patent priority will be resolved in Hall's favor in 1893. The two will eventually become close friends and develop the Hall-Heroult process used in the aluminum industry.

The first electrolytic magnesium plant opens at Hamelingen, Germany, near Bremen. The plant uses carnalite (a double salt of magnesium chloride and potassium chloride) as a raw material to make magnesium, used initially in flares.

Bethlehem Steel's John Fritz switches from commercial work to ordnance at the suggestion of Navy Secretary William C. Whitney, 45, who has made a fortune in New York City transit lines (*see* 1863; Ryan, above; U.S. Shipbuilding, 1902).

The element germanium, discovered at Freiburg in Saxony by German physicist C. Winkler in a silver thiogermanate, was unknown to Dmitri Mendeléev in 1870 when he published the periodic table of elements, but he suspected its existence and assigned it an atomic weight of 72.6.

Johnson & Johnson introduces the first ready-to-use surgical dressings (*see* 1885). Robert Wood Johnson heard an address in 1876 by England's James Lister and has developed sterile dressings wrapped in individual packages and suitable for immediate use without risk of contamination. The company has begun operations with 14 employees on the fourth floor of a former wallpaper factory (*see* 1899; Mead Johnson, 1900).

Postal Telegraph breaks Western Union's telegraph monopoly. Commercial Cable's J. W. Mackay starts the company that will be headed by his son Clarence Hungerford Mackay beginning in 1902 (see 1881; 1883; 1928).

The *New York Tribune* installs linotype machines, the first newspaper to do so (*see* Merganthaler, 1884; Fotosetter, 1949).

German inventor Paul O. Gottlieb Nipkov, 26, pioneers television with his rotating scanning device (*see* Baird, 1926).

Nonfiction *Capital (Das Kapital)* by Karl Marx is published in English (*see* 1867).

Fiction *The Mayor of Casterbridge* by Thomas Hardy; *The Bostonians* and *The Princess Casamassima* by Henry James; *Indian Summer* by William Dean Howells whose new novel makes him the acknowledged leader of the U.S. "realist" school; *An Iceland Fisherman* (*Pécheur d'Islande*) by Pierre Loti; *A Drama in Muslin* by George Moore; *The Strange Case of Dr. Jekyll and Mr. Hyde* by Robert Louis Stevenson who has written the pioneer science-fiction novel in 3 days and 3 nights and rewritten it in another 3; *Kidnapped* by Stevenson who is ill with tuberculosis and will spend the winter of 1887–1888 at the Saranac Lake, N.Y., sanatorium started 2 years ago by E. L. Trudeau; *The Romance of Two Worlds* by English novelist Marie Corelli (Mary Mackay), 31, whose works will be the favorite reading of Queen Victoria and Italy's Queen Margherita; "The Old Detective's Pupil" by New York writer John Russell Coryell, 38, whose Nick Carter story in Street & Smith's *New York Weekly* begins a series that will be revived in 1891 when S&S assigns Frederick Van Rensselaer Day to write a weekly Nick Carter Detective Library story (*see* 1858).

Juvenile *Little Lord Fauntleroy* by English-American novelist Frances Hodgson Burnett, 37, whose story of an American boy who inherits a vast estate in England is based on an actual event; *Jo's Boys* by Louisa May Alcott.

Poetry *Les Illuminations* by Arthur Rimbaud.

Painting The eighth and final show of French impressionism opens at the Maison Dorée restaurant near the Paris Opéra and includes the canvas *Sunday Afternoon on the Island of La Grande Jatte* by "neoimpressionist" Georges Seurat whose work is ridiculed by the critics. Seurat has developed a "pointillism" technique of applying paint in thousands of tiny spots rather than by brush strokes.

Other paintings *Bubbles* by Sir John Everett Millais, who was made a baronet last year; *Carnation, Lily, Lily, Rose* by John Singer Sargent; *Girl Arranging Her Hair* by Mary Cassatt; *Carnaval du Soir* by French "primitive" painter Henri Rousseau, 42.

Sculpture *The Kiss (Le Baiser)* by French sculptor Auguste Rodin, 45, whose sensuous white marble work represents the illicit Italian lovers Paolo Malatesta and Francesca da Rimini of the 13th century whose story was told by Dante Alighieri (Francesca's husband Giovanni Malatesta murdered her because she was in love with his brother Paolo).

A process for halftone engraving developed by U.S. inventor Frederick Eugene Ives, 30, uses small raised dots of varying sizes. Ives pioneered color photography 5 years ago by making the first trichromatic halftone process printing plates. He will also invent a process for gravure printing that will employ minute pits etched into a metal plate, and although rotogravure will replace his photogravure, the Ives halftone process will endure in photoengraving (*see* 1880; *Tribune*, 1897).

Theater managers B. F. Keith and Edward F. Albee found the Keith-Albee Vaudeville Circuit (*see* 1885; Orpheum Circuit, 1897).

First performances *Manfred* Symphony by Petr Ilich Tchaikovsky 2/23 at Moscow; Symphonic Variations for Piano and Orchestra by César Franck 5/1 at the Salle Playel, Paris; Symphony No. 3 in C minor by Camille Saint-Saëns 5/19 at London, with the composer conducting the London Philharmonic which has commissioned the work; "A Night on the Bald Mountain" by the late Modest Mussorgsky 10/27 at St. Petersburg; *Ophelia* by Edward MacDowell 11/4 at New York's Chickering Hall.

Popular songs "Hatikva" with lyrics by Bohemian-born English poet Naphtali Herz Imber, 30, who has come to Palestine as private secretary to a British journalist and poet. Jews throughout the world will sing Imber's lyrics to a version of Bedřich Smetana's "The Moldau" of 1879.

William Renshaw wins in men's singles at Wimbledon, Blanche Bingley, 22, in women's singles; Richard Sears wins at Newport.

1886 ***(cont.)*** The U.S. yacht *Mayflower* retains the America's Cup by defeating the English challenger *Galatea* 2 to 0.

The first world chess title goes to Bohemian chess master Wilhelm Steinitz, 50, who defeats J. H. Zukertort at London. He will hold the title until 1894.

New Yorker Steve Brodie is found in the water beneath the 3-year-old Great East River (Brooklyn) Bridge July 23 and claims to have jumped. Nobody has witnessed it, and Brodie's claim is suspect since previous jumpers have fallen to their deaths, but any suicide leap—especially one from a bridge—will in many circles be called hereafter a "brodie."

The tuxedo dinner jacket worn by tobacco heir Griswold Lorillard October 10 at the Autumn Ball of the Tuxedo Park Country Club at Tuxedo, N.Y., is a short black coat with satin lapels modeled on the English smoking jacket. It will replace the tailcoat worn until now at evening social affairs.

Johnson's Wax is introduced at Racine, Wis., by local parquet flooring peddler Samuel C. Johnson who has branched into paste wax. S. C. Johnson & Son will become the world's leading producer of floor wax.

Avon Products has its beginnings in the California Perfume Co. founded by Brooklyn, N.Y., door-to-door book salesman David H. McConnell, 28, whose firm will become the world's largest cosmetic company. McConnell has been gaining admission to his customers' parlors by offering free vials of perfume, found more response to the perfume than to *Pilgrim's Progress* or *The American Book of Home Nursing*, so abandons book selling and concentrates on selling perfume.

The William J. Burns International Detective Agency has its beginnings at Columbus, Ohio, following revelations of fraud. A citizens' "Committee of 100" that includes local tailor Michael Burns puts up money to hire a private investigator, the committee selects Michael's son William, 22, despite his father's disapproval, young Burns finds the safecracker-forger who has been spirited out of the local penitentiary to perpetrate the fraud, obtains confessions, and rounds up accomplices. He will work for the U.S. Secret Service for 14 years beginning at $3 per day in 1891, leave the Secret Service in 1905 to investigate fraud in San Francisco, start the Burns and Sheridan Detective Agency in 1909, and soon thereafter buy out his partner to operate under his own name (*see* Los Angeles Times bombing, 1910).

Sears, Roebuck has its beginnings at North Redwood, Minn., where Minneapolis & St. Louis Railroad agent Richard Warren Sears, 23, buys a consignment of $25 gold-filled "yellow watches" that a local jeweler has refused. Station agent Sears has been selling wood, coal, and lumber to local farmers and Indians, bringing in the commodities at special freight rates and shipping out the farmers' meat and berries at a profit. He uses the telegraph to offer the watches to other station agents at $14 each C.O.D. subject to examination, the other agents are able to undersell local jewelers, Sears makes $5,000 in 6 months, and he quits his job to go into business at Minneapolis (*see* 1887).

A model Bloomingdale's department store opens on New York's Third Avenue at 59th Street near a station of the Third Avenue El that opened 8 years ago (*see* 1872). Bloomingdale brothers Lyman, Joseph, and Gustave have built up a thriving enterprise, specializing in whalebone for corsets, yard goods, ladies' notions, and hoop skirts with help from the El that has contributed to an uptown movement of the city's middle class. By the turn of the century, Bloomingdale's will cover 80 percent of the block from 59th to 60th Street between Third and Lexington avenues and by 1927 will occupy the entire block (*see* Federated, 1929).

Chicago's Rookery building is completed in South LaSalle Street by Burnham and Root. Architects Daniel Hudson Burnham, 39, and John Wellborn Root, 36, started their firm in 1873, their 11-story Rookery has a graceful semiprivate square surrounded by shops and offices, and their pioneering design employs cast-iron columns, wrought-iron spandrel beams, and steel beams to support party walls and interior floors (*see* 1891).

Singapore's Raffles Hotel opens with 123 rooms to serve British colonials.

A decade of intermittent drought begins on the U.S. Great Plains after 8 years of extraordinary rainfall. Close to 60 percent of U.S. range livestock die in blizzards and from lack of grass on overgrazed lands.

North Dakota rancher Theodore Roosevelt, 28, sustains heavy losses, gives up, and returns to New York to enter politics (*see* 1887; Sagamore Hill, 1885).

Alice Gertrudis King inherits the King Ranch whose co-founder Richard King died last year at age 59. She marries King's lawyer, Robert Justus Kleberg, 31, and he assumes management of the 500,000-acre Texas ranch, which now employs 1,000 hands and grazes 100,000 head (he also assumes a $500,000 debt left by King (*see* 1868; 1906).

Manitoba farmer Angus MacKay, 45, demonstrates the Canadian prairie's ability to produce good wheat crops; he plants Red Fife wheat on a field that he left fallow last year and reaps 35 bushels of hard wheat per acre.

French chemist Henri Moissan, 34, isolates fluorine (*see* Motley, 1916).

Dutch researcher Christian Eijkman, 28, is sent to Java to study beriberi, which has become a major problem in the East Indies (*see* 1896; Takaki, 1884).

A fire destroys oatmeal king Ferdinand Schumacher's new Jumbo mill at Akron along with his Empire mill and other properties, consuming 100,000 bushels of oats plus quantities of other grains (*see* 1881). He has no insurance to cover his losses in the disastrous May fire but recovers in part by absorbing a small local competitor and entering the Consolidated Oatmeal Co. pool, whose members control half the nation's oatmeal trade.

H. J. Heinz of Pittsburgh calls on London's 179-year-old Fortnum and Mason while on holiday in England and receives orders for all the products he takes out of his Gladstone bag (*see* 1880; 1888).

Congress passes an Oleomargarine Act to tax and regulate the manufacture and sale of margarine (*see* 1881; 1890).

Maxwell House coffee gets its name. The 17-year-old hotel at Nashville, Tenn., serves its guests coffee made from the blend perfected by Joel Cheek who is persuaded by the praise of the guests to market his blend under the name Maxwell House (*see* 1882).

Coca-Cola goes on sale May 8 at Jacob's Pharmacy in Atlanta, where local pharmacist John S. Pemberton has formulated a headache and hangover remedy whose syrup ingredients include dried leaves from the South American coca shrub (*see* cocaine, 1884) and an extract of kola nuts from Africa plus fruit syrup. He has been advertising his Coca-Cola "esteemed Brain Tonic and Intellectual Beverage" since March 29. His bookkeeper Frank M. Robinson has named the product and written the name Coca-Cola in a flowing script. Pemberton has persuaded fountain man Willis E. Venable at Jacob's to sell the beverage on a trial basis, Venable adds carbonated water, and by year's end Pemberton has sold 25 gallons of syrup at $1 per gallon, but he has spent $73.96 in advertising and will sell two-thirds of his sole ownership in the product next year for $1,200 (*see* Candler, 1891).

Moxie is introduced under the name Moxie Nerve Food by Lowell, Mass., physician Augustin Thompson who was introduced by a Lieut. Moxie to the properties of gentian root, which is the beverage's chief ingredient other than sparkling water.

Dr. Pepper is introduced as "The King of Beverages, Free from Caffeine" by Waco, Tex., chemist R. S. Lazenby, who has experimented with a soft drink formula developed by a fountain man at the town's Old Corner Drug Store. The fountain man has called his drink Dr. Pepper's Phos-Ferrates.

Hires' Rootbeer is introduced in bottles, but advertising emphasizes the advantages of brewing the drink at home from Hires extract (*see* 1876).

Fauchon opens on the Place de la Madeleine in Paris as an épicerie extraordinaire.

The Statue of Liberty dedicated October 28 on Bedloe's Island in New York Harbor has been designed by French sculptor Frédéric Auguste Bartholdi, 52, and presented by the people of France. Joseph Pulitzer's *New York World* has raised $100,000 for a pedestal that Congress had refused to fund for "Liberté Eclairant le Monde," which stands more than 151 feet tall. The pedestal will be inscribed in 1903 with words written in 1883 by philanthropist Emma Lazarus, 37: "Give me your tired, your poor,/ Your huddled masses, yearning to breathe free,/ The wretched refuse of your teeming shore. / Send these, the homeless, tempest tossed, to me:/ I lift my lamp beside the golden door."

1887

Bismarck warns Europe against war January 11 in a speech advocating a much larger German army while France is agitated by nationalist sentiment and demands for revenge against her victor in the Franco-Prussian War of 1870–1871. Ententes are formed among the powers, the Triple Alliance of 1882 is renewed for another 5 years, and a secret Russian-German treaty is signed June 18 following Russia's refusal to renew the expiring 1881 Alliance of the Three Emperors.

Bulgaria's assembly elects the nation's first prince, but Ferdinand of Saxe-Coburg, 25, a grandson of France's late Louis Philippe, will not gain recognition from any of the Great Powers until 1896 (*see* 1896).

An Italian-Ethiopian War begins in January, the Ethiopians annihilate an Italian force January 25 at Dogali, but Italy strengthens her position by backing Menelek, king of Shoa, and attacks from Mahdist forces in the north divert Ethiopia's king of kings Johannes.

Britain promises to evacuate Egypt within 3 years in the May 22 Drummond-Wolff Convention with Constantinople, but only if conditions are favorable; Britain retains the right to reoccupy Egypt should the country be menaced by invasion or internal disorder. Egypt's khedive Tewfik refuses to ratify the convention.

Britain annexes Zululand to block the Transvaal government from establishing a link to the sea.

Macão off the Chinese coast is ceded to Portugal.

France's President Grévy resigns under pressure December 2 at age 80 following revelations that his son-in-law Daniel Wilson has been trafficking in medals of the Légion d'honneur. He is succeeded by Marie François Sadi-Carnot.

A "Jim Crow" law passed by the Florida legislature requires segregation of black railway passengers from whites (*see* 1881; 1891).

Lloyd's of London writes its first nonmarine insurance policy after 199 years of insuring only maritime carriers (*see* 1870). The great 120-by 340-foot Lloyd's underwriting room in Lime Street continues to be dominated by the Lutine Bell salvaged in 1859 from the frigate *H. M. S. Lutine* that was captured from the French in 1793 and sunk off the Dutch coast 6 years later with more than £1 million gold and silver. The bell is still rung before important announcements—once for bad news, twice for good. Marine insurance underwritten by Lloyd's "Names" will continue to exceed nonmarine insurance until 1949.

Tokyo Electric Light Co. (Tokyo Dento Kaisha) brings electricity to Japan in January.

The Interstate Commerce Act approved by Congress February 1 orders U.S. railroads to keep their rates fair and reasonable (*see* Hadley, 1885; ICC, 1888; Elkins Act, 1903).

The Canadian Pacific Railway reaches Vancouver May 23, the first single company transcontinental railroad in America. Built by James J. Hill and W. C. Van Horne with 2,095 miles of track, the CP joins the east and west coasts of Canada and will spur migration to Canada's

1887 ***(cont.)*** western provinces (*see* 1881; 1888; Canadian National, 1923).

The first Atchison, Topeka and Santa Fe train to reach Los Angeles via Santa Fe track arrives May 31 and a rate war begins between the Santa Fe and Collis P. Huntington's Southern Pacific. The Santa Fe has bought the Los Angeles and San Gabriel Valley Railroad to gain access to the city, passenger fares from Kansas City to Los Angeles drop as low as $1, and Los Angeles has a land boom.

The Sessions vertical end frame designed by Pullman Palace Car Co. superintendent H. H. Sessions reduces railroad-car sway and makes it safer to move from one car to another, as when going to the dining car. The company will refine the end frame to close vestibules and make a train one continuous unit.

A railroad trestle on the Toledo, Peoria & Western near Chatsworth, Ill., gives way in August and a passenger train bound for Niagara Falls with more than 600 passengers plunges into a ditch, killing more than 70, injuring nearly 300.

The first Daimler motorcar is introduced March 4 (*see* 1886; 1890).

Cabot Corp. has its beginnings in a lampblack firm founded by Boston entrepreneur Godfrey Cabot, 36, who breaks away from the family paint company to start the business that will be the basis of a vast fortune. Cabot has found that the oil and gas fields of western Pennsylvania are plagued by the sooty carbon debris of gas blowoffs and refining, he begins to build and acquire carbon black plants, will branch off into natural gas production and distribution, and will make his Cabot Corp. the major supplier of carbon black for electric lamp filaments, telephones, and other uses.

Cincinnati Milacron has its beginnings as bank clerk Frederick A. Geier joins a struggling Ohio River machine shop that he will develop into the largest U.S. producer of machine tools. Cincinnati Milling Machine will incorporate its motor drives within its machine to protect factory workers from exposed moving wheels and whirling shafts, introduce hydraulic controls to feed work to cutters and permit smoother performance, and introduce a power-controlled gear shift to eliminate the heavy manual effort needed to change the speed of tool revolutions.

The machine à calculer invented by French engineering student Leon Bolle, 18, is the first machine to automate multiplication using a direct method. Bolle's machine has a multi-tongued plate that constitutes a multiplication table and represents a marked advance over calculators that employ multiple additions for multiplication (*see* 1642; 1692; 1833; 1842).

The Comptometer introduced by the new Felt & Tarrant Manufacturing Co. of Chicago is the first multiple-column calculating machine to be operated entirely by keys and be absolutely accurate at all times. Local inventor Dorr Eugene Felt, 25, has gone into partnership with Robert Tarrant to produce the machine (*see* Burroughs, 1888).

Swedish chemist Svante August Arrhenius, 26, advances the ionic theory that electrolytes (substances that conduct electricity) split up in solution into electrically charged particles (ions) (*see* 1923).

German physicist Heinrich Rudolph Hertz, 30, shows the existence of electric or electromagnetic waves in the space round a discharging Leyden jar (*see* 1745). Investigating James Clerk Maxwell's 1873 electromagnetic theory of light, Hertz finds that the waves are propagated with the velocity of light as Maxwell had predicted.

English physician David Bruce, 32, identifies the source of "Malta fever" or "Mediterranean fever," which has caused illness and even death among British troops. He finds that milk from local goats transmits a bacterium that will later be found in cows' milk and causes undulant fever, or brucellosis (*see* Bang, 1896).

"Power tends to corrupt and absolute power corrupts absolutely," writes John Emerich Edward Dalbert-Acton, 53, April 5 to Cambridge University professor Mandell Creighton. Lord Acton is a liberal Roman Catholic and a leader of the opposition to the papal dogma of infallibility (*see* 1871).

Pratt Institute opens at Brooklyn, N.Y., to provide training for artisans and draftsmen. The school has been funded by Charles Pratt, 57, who sold his oil refinery to John D. Rockefeller in 1874 and who has been a benefactor to Amherst College and the University of Rochester.

Clark University is founded at Worcester, Mass., by educator Jonas Gilman Clark, 72.

Catholic University of America is founded at Washington, D.C.

The University of Wyoming opens at Laramie.

The Perkins Institution founded in 1829 receives a request from telephone pioneer Alexander Graham Bell to examine 6-year-old Helen Adams Keller who lost her sight and hearing at 19 months of age. Teacher Anne Mansfield Sullivan, 20, of the Perkins Institution travels to the Keller home, starts work with young Helen March 2, and quickly teaches her to feel objects and associate them with words spelled out by finger signals on the palm of her hand; Helen soon can feel raised words on cardboard and make her own sentences by arranging words in a frame (*see* 1904).

Esperanto is invented by Polish oculist-philologist Lazarus Ludwig Zemenhof, 28, who hopes his universal language will help achieve world peace and understanding (*see* Ogden, 1929; Shaw, 1950).

Heinrich Hertz's electric waves (above) will be the basis of radio communications (*see* Marconi, 1895).

San Francisco Evening Examiner publisher William Randolph Hearst, 23, scores his first big scoop April 1 by chartering a special train to cover the fire that destroys the 7-year-old Del Monte Hotel built by Charles Crocker at Monterey. Hearst has inherited the paper from his late father George Hearst of 1859 Comstock Lode fame (*see* 1895).

The *Paris Herald* begins publication October 4 with the last of its four pages devoted to advertisements directed toward English and American visitors and émigrés. The paper is a European edition of James Gordon Bennett's *New York Herald* (*see* 1967).

U.S. telephone listings reach 200,000 by December 31, with 5,767 in Boston and neighboring towns, 1,176 in Hartford, 1,393 in New Haven.

The Pratt Institute Free Library of Pratt Institute (above) is the first free library in New York State.

The Boston Public Library opens in a building designed by architects McKim, Mead & White.

Fiction *The Aspern Papers* by Henry James; *Plain Tales from the Hills* by Rudyard Kipling; *The People of Hemso (Hemsoborna)* by August Strindberg; *She* and *Allan Quatermain* by Sir H. Rider Haggard; *Thelma* by Marie Corelli; *A Study in Scarlet* by Scottish physician-novelist Arthur Conan Doyle, 28, who has written the book to supplement his small income; his otherwise undistinguished work introduces the master detective Sherlock Holmes.

Painting *Moulin de la Galette* by Vincent van Gogh; *The Flax Spinners* by Max Liebermann; *Walt Whitman* by Thomas Eakins.

Sculpture *Seated Lincoln* by Augustus Saint-Gaudens for Chicago's Lincoln Park; *Amor Caritas* by Saint-Gaudens.

Theater *Rosmersholm* by Henrik Ibsen 4/12 at Oslo's Christiania Theater; *The Father (Fadren)* by August Strindberg 11/14 at Copenhagen's Cosmo Theater; *Ivanov* by Russian playwright Anton Chekhov, 27, 11/19 at Moscow's Korsk Theater; *La Tosca* by Victorien Sardou 11/24 at the Théâtre de la Porte-Saint-Martin, Paris (*see* opera, 1900); *A Gown for His Mistress* (*Tailleur pour dames*) by French playwright Georges Feydeau, 24, 12/17 at the Théâtre de la Renaissance, Paris.

Opera *Ruddigore (or The Witch's Curse)* 1/22 at London's Savoy Theatre, with new Gilbert and Sullivan songs; *Otello* 2/5 at Milan's Teatro alia Scala, with music by Giuseppe Verdi, libretto from the Shakespeare tragedy of 1604; *The Reluctant King (Le Roi malgré lui)* 5/18 at the Opéra-Comique, Paris, with music by Emmanuel Chabrier.

First performances Suite in D for Trumpet, Flutes, and Strings by Vincent d'Indy 1/7 at Paris; Symphonie No. 1 for Orchestra and Piano (*Symphonie sur un air montagnard français*) by d'Indy 3/20 at Paris; Double Concerto in A minor for Violin, Cello, and Orchestra by Johannes Brahms 10/18 at Cologne; *Capriccio espagñol* by Nikolai Rimski-Korsakov 10/31 at St. Petersburg's Imperial Opera House; Suite No. 4 (*Mozartiana*) by Petr Ilich Tchaikovsky 11/26 at Moscow; *Pictures at an Exhibition* by the late Modest Mussorgsky 11/30 at St. Petersburg.

Thomas Edison invents the first motor-driven phonograph and opens a new laboratory at West Orange, N.J., that is 10 times the size of his Menlo Park laboratory of 1876. Edison's new phonograph plays cylindrical wax records (*see* 1877).

The gramophone patented by Emile Berliner of 1877 loose-contact telephone transmitter fame improves on Edison's phonograph (above) by substituting a disk and a horizontally moving needle for Edison's cylinder and vertically moving needle. The groove in Berliner's disk propels the arm of his gramophone automatically, eliminating the need for the separate drive mechanism that Edison's machine requires, but the pivoted tone arm of the gramophone has a fixed head and a tracking error that causes distortion in the music amplified through the large horn attached to its tone arm (*see* 1900; Victrola, 1906).

Herbert Lawford, 36, wins in men's singles at Wimbledon, Charlotte "Lottie" Dod, 15, in women's singles; Richard Sears wins in U.S. men's singles, Ellen F. Hansel in women's singles.

Tipperary beats Galway at hurling, and Limerick beats Louth at football in the first All-Ireland Championships (*see* Croke Park, 1910).

England's *Thistle* loses 2 to 0 to the U.S. defender *Volunteer* in the America's Cup ocean yacht races.

Atlanta's Piedmont Driving Club is founded by local gentlemen who restrict membership to men. The exclusive club admits members by invitation only and will exclude blacks and most Jews.

The first U.S. social register is published by New York golf promoter Louis Keller, 30, the son of a patent lawyer who owns a farm at Springfield, N.J., on which he has founded the Baltusro Golf Club. Keller has earlier helped start the scandal sheet *Town Topics.* His 100-page book sells for $1.75, contains roughly 3,600 names based largely on telephone listings which it prints in larger type than that used in the phone company directory, draws on the membership list of the Calumet Club at 29th Street and Fifth Avenue, and will be followed by social registers published in most major U.S. cities, with preference given to white, non-Jewish, non-divorced residents considered respectable by the arbiters who compile the books.

Richard Sears moves to Chicago, hires watchmaker Alvah C. Roebuck, and sells watches through clubs and by mail order, advertising in rural newspapers (*see* 1886; 1889; Sears, Roebuck, 1893).

Gimbel Bros., Milwaukee, is opened by Jacob, Isaac, Charles, Ellis, Daniel, Lewis, and Benedict Gimbel, whose father Adam Gimbel opened the first Gimbel's at Vincennes, Ind., in 1842 (*see* 1894).

The Grand Hotel on Mackinac Island, Mich., opens on a high bluff with 262 rooms and a wide 900-foot long veranda providing a view of ships passing through the Straits of Mackinac between Lakes Michigan and Huron.

Pittsburgh's Allegheny Courthouse and jail are completed by Henry Hobson Richardson.

The Boone & Crockett Club to protect American wildlife from ruthless slaughter by commercial market

1887 ***(cont.)*** hunters is organized by a group of "American hunting riflemen" who include George B. Grinnell and Theodore Roosevelt who was defeated last year in a New York mayoralty election (*see* Grinnell, 1885). Membership in the club will be limited to 100.

Congress makes Yellowstone country a refuge for buffalo and big game at the persuasion of Theodore Roosevelt (above; *see* Pelican Island, 1903).

Blizzards continue in the northern Great Plains (*see* 1886). The worst storm of the hard winter rages for 72 hours at the end of January, and millions of head of open-range cattle are killed in Montana, Kansas, Wyoming, and the Dakotas. Whole families are found frozen to death in tar-paper cabins and in dugouts, the big ranching syndicates go bankrupt, and the homesteaders that move in to start farms include former cowboys. Friction will soon begin between the farmers and the surviving ranchers.

The Hatch Act voted by Congress March 2 authorizes the establishment of agricultural experiment stations in all states having land-grant colleges.

U.S. wheat prices fall to 67¢ per bushel, lowest since 1868, as the harvest reaches new heights. Britain and other importers of American wheat enjoy lower food prices.

China's Huanghe (Yellow) River floods its banks. The resulting crop failures and famine kill 900,000.

Beriberi can be prevented by proper diet, says Baron Takaki in a report of his 1884 Japanese Navy study published in the British medical journal *Lancet* (*see* Eijkman and Grijns, 1896).

Britain's Margarine Act establishes statutory standards for margarine, but dairy magnate George Barham fails in his demand that if any coloring be allowed in margarine it should be pink, green, or, preferably, black (*see* U.S., 1950).

New York sugar refiner Henry Osborne Havemeyer, 40, founds Sugar Refineries Co. His 17 refineries account for 78 percent of U.S. refining capacity (*see* American Sugar Refining, 1891).

Log Cabin Syrup is introduced by St. Paul, Minn., grocer P. J. Towle, who blends maple syrup (45 percent) and cane sugar syrup to produce a product much lower in price than pure maple syrup.

Ball-Mason jars are introduced by Ball Brothers Glass Manufacturing Co. of Muncie, Ind. (*see* Mason, 1858). William Charles Ball, 35, and his brothers Lucius Lorenzo, Frank C., Edmund Burke, and George Alexander began making tin oil cans at Buffalo, N.Y., 10 years ago, switched to glass oil and fruit jars in 1884, and have moved to Muncie, where natural gas has been discovered and which has offered free gas and a generous land site.

Western Cold Storage Co. in Chicago installs ice-making machines, but cold stores in most places continue to be cooled by harvested ice cut from lakes and hauled ashore by teams of horses (*see* 1878; 1889).

1888

"We Germans fear God, and nothing else in the world," says Bismarck February 6 in a speech to the Reichstag at Berlin. He has allowed terms of last year's Triple Alliance renewal to leak out as a discouragement to Russian and French ambitions.

Wilhelm I dies at Berlin March 9, less than 2 weeks before his 91st birthday, after a 27-year reign of Prussia and Germany that has seen the unification of Germany under Bismarck (above). Wilhelm is succeeded by his son Friedrich Wilhelm, 57, but the new emperor dies June 15 of throat cancer and is succeeded in turn by his son of 29 who will reign until 1919 as Wilhelm II, the last German monarch.

France relieves Gen. Boulanger of his command after he has twice come to Paris without leave. He is removed from the army list on the recommendation of a council of inquiry composed of five other generals, but although wounded in an embarrassing duel with the anti-Boulangist Charles Thomas Floquet, 60, the popular "Man on Horseback" *revanchist* hero Boulanger is elected to the Chamber for the Nord (*see* 1886; 1889).

Britain establishes a protectorate over Sarawak March 17 and over North Borneo May 12, but the North Borneo Company continues to hold and administer North Borneo as it has since 1881, and Sir Charles Brooke continues to rule Sarawak as he has since 1868 (*see* 1946).

The Matebele king Lobengula, 55, accepts a British protectorate and signs a treaty October 30 giving the Cecil Rhodes interests exclusive mining rights in Matabeleland and Mashonaland.

U.S. voters elect Indiana Republican Benjamin Harrison, 55, grandson of the ninth president William Henry Harrison, although President Cleveland actually receives a 100,000-vote plurality in the popular vote. Harrison receives 233 electoral votes to Cleveland's 168, New York makes the difference as it did in 1884, and again it is the Irish vote that swings the state, this time in reaction to a statement by the British minister to Washington that Cleveland would be friendlier to Britain than Harrison, a remark the Republicans use to support their contention that Cleveland's tariff position would favor British industrial interests.

Brazil's slaves go free May 13 under terms of a law put through by a Liberal ministry under Pedro II who has ruled since 1840. The Rio Branco law of 1871 freed the children of slaves, an 1885 law freed all slaves over 60, and the new law completes emancipation without recompense to slave owners.

Congress creates a U.S. Department of Labor, restructuring the 4-year-old Bureau of Labor, but the new department will not have cabinet status until 1903.

Anti-Chinese riots break out in Seattle (*see* 1886; Chinese Exclusion Act, below).

Philadelphia Smelting and Refining is organized by Meyer Guggenheim, who last year took a venture in copper stock and did so well that he gives up the lace business he started in 1872 and goes into partnership with his four oldest sons (*see* 1881). Helped by all seven sons, Guggenheim will establish a second smelter in Mexico in 1891 and a third in 1894 (*see* ASARCO, 1899).

Andrew Carnegie gains majority ownership in the Homestead Steel Works outside Pittsburgh (*see* 1881; 1892).

An alternating-current (ac) electric motor developed by Croatian-American inventor Nikola Tesla, 31, applies a variation of the rotary magnetic field principle discovered 3 years ago by the Italian Galileo Ferraris to a practical induction motor that will largely supplant direct-current (dc) motors for most uses (*see* Stanley and Westinghouse, 1885). A former Edison Co. employee at West Orange, N.J., Tesla will make possible the production and distribution of alternating current with his induction, synchronous, and split-phase motors (he will also develop systems for polyphase transmission of power over long distances and pioneer the invention of radio), but the Tesla Electric Co. he organized last year is unsuccessful, and he will never derive much material success from his inventions (*see* 1893).

Thomson-Houston Electric of Lynn, Mass., acquires patents for an electric railway issued 5 years ago to C. J. Van Depoele and hires Van Depoele as an electrician. He receives a patent for a carbon commutator brush (*see* 1874; General Electric, 1892).

The first successful electric trolley cars go into service February 2 at Richmond, Va., where engineer-inventor Frank Julian Sprague, 31, has laid 12 miles of track between Church Hill and New Reservoir Park (*see* Deft, 1885). A former assistant to Thomas Edison, Sprague quit 4 years ago to start the Sprague Electric Railway and Motor, his trolley line has 40 four-wheeled cars that can travel at 15 miles per hour and are lighted with incandescent bulbs, and within 2 years some 200 other U.S. cities will be served by electric trolley car lines (*see* 1900).

The Interstate Commerce Commission (ICC), established under terms of last year's Interstate Commerce Act, will make some efforts to protect farmers from discriminatory freight rates (but *see* California Fruit Growers Exchange, 1895; Elkins Act, 1903).

A new Atchison, Topeka and Santa Fe line links Los Angeles to San Diego.

The Santa Fe builds a second line into Los Angeles from San Bernardino through Riverside and Orange counties.

The Santa Fe gains access to Chicago's Dearborn Station through acquisitions and new rail lines to give it routes from Chicago to points west.

The *Florida Special* leaves Jersey City in January on H. M. Flagler's Florida East Coast Railway. Flagler has contracted with George M. Pullman to build the fully vestibuled, electrically lighted train, and its 70 passengers include Pullman (*see* hotel, below).

The first Chinese railway opens between Tangshan and Tianjin (Tientsin). The 80-mile line will be extended to Shanghaikuan in 1894 and to Fengtai outside Beijing in 1896.

A railroad line between Budapest and Constantinople opens August 12 (*see* 1889).

German financiers project a Berlin-to-Baghdad railway; they obtain a concession October 6 to construct a line to Angora.

The Canadian Pacific Railway receives a British government mail subsidy to the Orient and begins to acquire ships for a Pacific mail route (*see* 1887; 1891).

A Suez Canal Convention signed at Constantinople October 29 declares the canal to be free and open to merchant ships and warships in war and peace (*see* 1875). The canal is not to be blockaded but the sultan and the khedive are to be free to take such measures as they may "find necessary for securing by their own forces the defense of Egypt and the maintenance of public order."

The first patent for a pneumatic bicycle tire is awarded October 31 to Scottish veterinary surgeon John Boyd Dunlop, 47, at Belfast, Ireland. Advised by a physician to have his sickly son ride a tricycle, Dunlop has devised the tires to cushion the boy's ride on Belfast's cobblestone streets, using rubber sheeting and strips of linen from an old dress of his wife's. Dunlop Rubber has its beginnings in a firm Dunlop establishes with entrepreneur W. H. Ducros, but he will sell the patent rights to Ducros and derive only a small profit from his invention (*see* Michelin brothers, 1895).

Benz motor carriages are advertised for the first time at Mannheim by Karl-Friedrich Benz who organizes Firma Benz & Cie. (*see* 1887; Mercedes-Benz, 1926).

The Burroughs adding machine patented by St. Louis inventor William Seward Burroughs, 31, is the first successful key-set recording and adding machine (*see* Comptometer, 1887). It is not commercially practical, but Burroughs and three partners organize the American Arithmometer Co., sell $100,000 worth of stock, develop an improved model, and while the new model will not stand up to heavy use, Burroughs will obtain further capitalization and produce a model in 1891 that will print out each separate entry plus the final result of each computation. He will be granted patents in 1893 for the first practical adding machine, American Arithmometer will begin production of the machines, move to Detroit in 1905, and become Burroughs Corp.

Nickel steel, invented in France, gives impetus to Samuel J. Ritchie's 3-year-old Canadian Copper Co. with its rich nickel ore deposits and to other nickel companies (*see* manganese steel, 1882).

C. M. Hall's Pittsburgh Reduction Co. produces the world's first commercial aluminum Thanksgiving Day (*see* 1886; Mellon, 1891).

A new law of chemistry formulated by French chemist Henri Louis Le Chatelier, 38, establishes the principle that if one of the factors in any chemical equilibrium

1888 ***(cont.)*** changes, be it pressure, temperature, or something else, the system readjusts itself to minimize the change.

The Institut Pasteur is founded at Paris with private subscriptions. The contributors have been inspired by Louis Pasteur's rabies vaccine of 1885 and include Russia's Aleksandr III, Brazil's Pedro II, and the Ottoman sultan Abdul Hamid II (*see* 1903; 1923).

Bellevue Hospital physiology professor Austin Flint, 52, at New York writes, "What has been accomplished within the past ten years as regards knowledge of the causes, prevention, and treatment of disease transcends what would have been regarded a quarter of a century ago as the wildest and most impossible speculation."

New York and New England telephone linemen work through the blizzard that strikes the Atlantic Coast in March (below) to restore service on the new line that connects Boston with New York, but the storm disrupts communications and many communities are isolated for days.

U.S. inventor John Robert Gregg introduces a new shorthand system he calls "Light Line Phonography." It will largely replace the Pitman system in America (*see* 1837).

Parker Pen Co., founded at Janesville, Wis., by local telegraphy teacher George Safford Parker, 24, will become the world's largest producer of fountain pens (*see* Waterman, 1884). Parker sells John Holland pens to his students at the Valentine School in order to supplement his meager income, the pens give trouble, Parker becomes adept at fixing them, acquires a small scroll saw, lathe, cutter, and other tools, and develops a pen of his own with a superior feed (*see* 1904).

The first typewriter stencil is introduced at London by immigrant Hungarian inventor David Gestetner who 7 years ago introduced the first wax stencil duplicating machine to be marketed commercially. Chicago's A. B. Dick Co. will introduce its first typewriter stencil in 1890 (*see* 1875).

The *Financial Times* begins publication at London.

The *National Geographic* begins publication in October at Washington, D.C., where the National Geographic Society is founded by Alexander Graham Bell's father-in-law Gardiner Greene Hubbard who founded the magazine *Science* with Bell 5 years ago. The new quarterly will begin monthly publication in 1896, publish its first color plates in 1906, and in February 1910 will adopt a yellow-and-white cover.

Printer's Ink begins publication under the direction of advertising agent George P. Rowell (*see* 1869) whose trade magazine will campaign for honest advertising.

Di Yiddishe Folkbibliothek, founded at Kiev, is the world's first Yiddish literary annual. Founder Sholem Aleichim (Sholem Rabinovitch), 29, will raise the standards of Yiddish and teach Russian Jews, who have been hounded since the assassination of Aleksandr II in 1881, to laugh at their troubles.

Nonfiction *The American Commonwealth* by Oxford professor James Bryce, 50, who sees the United States sailing in "a summer sea" and setting a course of responsible liberty that will be a model for the world. America's institutions, says Lord Bryce, are the answer to mankind's longings, "towards which, as by a law of fate, the rest of civilized mankind are forced to move," but he warns that "perhaps no form of government needs great leaders so much as a democracy," and expresses concern that "the ordinary American voter does not object to mediocrity"; *Travels in Arabia Deserta* by English explorer-poet Charles Montagu Doughty, 45, who 12 years ago learned Arabic, disguised himself as a Bedouin, and made the pilgrimage to Mecca.

Fiction *Looking Backward, 2000–1887* by U.S. socialist Edward Bellamy, 38, is a utopian novel that pictures a happy, peaceful United States of the year 2000 in which all industry has been nationalized and all wealth equitably distributed; *The Book of the Thousand Nights and a Night* by English Orientalist Richard Burton, now 67, who anticipated Doughty (above) by visiting Mecca disguised as a Pathan in 1853 and has translated in its entirety the Arabian Nights stories known in the West until now only through expurgated stories such as "Aladdin and the Lamp." Sources of the stories are largely Persian, not Arabic, and are presented within the framework of a situation involving the Persian monarch Shahriyar and his wife Shahrazad (Scheherezade). Having had faithless wives in the past, Shahriyar has been taking a new one each night and having her put to death in the morning, but Scheherezade ends this practice by keeping the monarch fascinated with stories, many of them bawdy. Burton's 16-volume work enjoys great success, but after his death in 1890 his wife Isabel will burn the manuscript containing his full translation of *The Perfumed Garden of Cheikh Nezaoui: A Manual of Arabian Erotology*.

Juvenile *The Happy Prince* by Oscar Wilde, who 4 years ago married the rich Constance Lloyd and has written the book of fairy tales for his 3-year-old son Cyril and 2-year-old son Vyvyan.

Poetry "Casey at the Bat" appears June 3 in the *San Francisco Examiner* which pays Ernest Lawrence Thayer, 24, $5 for his contribution. The comic verse about baseball will be popularized by comedian-singer William DeWolf Hopper, 30, who will recite it thousands of times to enthusiastic audiences: ". . .And somewhere men are laughing, and somewhere children shout;/ But there is no joy in Mudville—mighty Casey has struck out."

Painting *The Vision of the Sermon* (*Jacob and the Angel*) and *Still Life with Fruit à mon ami Laval* by former Paris stockbroker (Eugène Henri) Paul Gauguin, 40, who has exhibited with the impressionists, given up his job, left his wife and family, but been forced to return to Paris after running out of money on a visit last year to Martinique; *Models* and *Circus Side Show* by Georges Seurat; *Sunflowers, Arena at Arles,* and *Night*

Café by Vincent van Gogh, who has befriended Gauguin; *Place Clichy* by French painter-lithographer Henri (Marie Raymond) de Toulouse-Lautrec, 23, who has been physically deformed since youth as the result of accidents and delicate health; *The Entrance of Christ into Brussels* by James Ensor, who uses macabre skeletons and masks with strident colors to rebel against the smugness and hypocrisy of the ruling class.

Sculpture *The Thinker (Le penseur)* by Auguste Rodin.

New York's Metropolitan Museum of Art opens its first addition and begins an expansion that will continue until the museum occupies 20 acres of Central Park (*see* 1880).

The Kodak camera ("You Press the Button, We Do the Rest") introduced by George Eastman revolutionizes photography by making it possible for any amateur to take satisfactory snapshots (*see* 1884). The small, light, $25 camera comes loaded with a roll of stripping paper long enough for 100 exposures, the entire camera is sent to Rochester, N.Y., when the film has been entirely exposed, the exposed strip is developed and printed at Rochester, a new strip is inserted for $10, and the camera is returned to its owner with the finished prints. Eastman will explain the name Kodak by saying, "I knew a trade name must be short, vigorous, incapable of being misspelled to an extent that will destroy its identity, and, in order to satisfy the trademark laws, it must mean nothing" (*see* transparent negative film, 1889).

Theater *Sweet Lavender* by English playwright Arthur Wing Pinero, 33, 3/21 at Terry's Theatre, London, 684 perfs.

Opera *Le Roi d'Ys* 5/7 at the Opéra-Comique, Paris, with music by French composer Edouard Lalo who achieves his first great success at age 65; *The Yeoman of the Guard (or The Merryman and His Maid)* 10/3 at London's Savoy Theatre with Gilbert and Sullivan arias including "I Have a Song to Sing, O!"

Vienna replaces the Burgtheater of 1776 with a larger Burgtheater opera house and concert hall. The policy of permitting no curtain calls continues.

"Kaiserwalzer" by Johann Strauss restores the composer to favor with the emperor Franz Josef who stripped the waltz king of his honors 5 years ago for divorcing his second wife and marrying a young Jewish widow.

First performances *Requiem* by Gabriel Fauré in January at the Church of the Madeleine, Paris, where Fauré has been assistant organist and choirmaster since 1877; Wallenstein Trilogy by Vincent d'Indy 2/28 at Paris; *Psyche* (Symphonic Poem for Orchestra and Chorus) by César Franck 3/10 at Paris; Concerto No. 1 for Piano and Orchestra by Edward MacDowell in April at Boston; Symphony No. 5 by Petr Ilich Tchaikovsky 11/17 at St. Petersburg.

"L'Internationale" is published with music by Belgian-born woodcarver Pierre Chrétien Degeyter, 40, lyrics by Parisian transport worker Eugène Edine Pettier, 72, who wrote them during the Commune uprising of 1871: "Arise, ye prisoners of starvation;/ Arise, ye wretched of the earth. . ." It will become the Communist anthem worldwide.

Ernest Renshaw, 27 (twin brother of William) wins in men's singles at Wimbledon, Lottie Dod in women's singles; Henry Slocum, 26, wins in U.S. men's singles, Bertha Townsend, 19, in women's singles.

"There are only about four hundred people in New York Society," says social arbiter Ward McAllister, 60, in a *New York Tribune* interview (*see* social register, 1887). McAllister failed in a San Francisco law practice and traveled east to Newport, R.I., at age 22 to marry the heiress to a steamboat fortune. In 1872 he organized the Patriarchs, a group comprised of the heads of New York's oldest families on whose approval social aspirants depend.

Woodbine's cigarettes are introduced in London.

Durham, N.C., tobacco merchant Washington B. Duke, 67, produces 744 million cigarettes at Durham and at his factory on New York's Rivington Street. A Confederate army veteran, he has leased a cigarette-making machine invented by Virginian James Bonsack; W. Duke & Sons grosses $600,000 (*see* American Tobacco Trust, 1890).

Tobacco-chewer Thomas Edison refuses to hire cigarette smokers.

Jack the Ripper makes headlines. London East End streetwalkers Mary Ann Nicholls, Annie Chapman, Elizabeth Stride, and their neighbors Catherine Eddowes and Mary Kelly die at the hands of one or more killers who feed the women poisoned grapes and then disembowel them. Scotland Yard can find no solution to the mystery, and it will later be alleged that agents of Queen Victoria murdered the women to hush up a scandal involving the queen's grandson Albert, duke of Clarence.

New York's R. H. Macy & Co. takes in as partners German-American merchant Isidor Straus, 43, and his brother Nathan, 40, who will become owner of the store in 1896. Isidor came to New York at age 9, started a crockery firm with his father Lazarus at age 20, and 9 years later took over Macy's crockery and glassware department (*see* 1874; 1901; Abraham & Straus, 1893; *Titanic*, 1912).

May Company merchant David May buys a bankrupt Denver store for $31,000, hires a brass band to attract crowds, and within a week has sold out the store's old stock at bargain prices, remodeled the store, and restocked it to open the May Shoe & Clothing Co., "Goods retailed at wholesale prices" (*see* 1879; 1910).

Chicago's Tacoma building, with a steel skeleton that employs load-bearing metal throughout its structure, represents the first basic advance in building construction since the medieval Gothic arch and flying buttress. The steel skeleton method of construction for tall buildings initiated by LeBaron Jenney's Home Insur-

1888 ***(cont.)*** ance building of 1885 is firmly established by Chicago architect William Holabird, 33, whose firm Holabird & Roche will be the major developer of Jenney's idea into modern office buildings in Chicago and will revolutionize the city's skyline with the Caxton building in 1890, Pontiac building in 1891, Marquette building in 1894, Tribune building in 1901, and numerous hotels and office buildings thereafter.

The Washington Monument that will remain the world's tallest masonry structure is completed at Washington, D.C., after 40 years of on-again off-again construction. The obelisk stands 555 feet high.

The Ponce de Leon Hotel opens January 10 at St. Augustine, Fla. Standard Oil magnate H. M. Flagler has taken over the Florida East Coast Railway, is extending it south, and has built the $1.25 million hotel. Each room has electric lights and $1,000 worth of furnishings (*see* Tampa Bay Hotel, 1891; Royal Poinciana, 1894; *Florida Special,* above).

The Banff Springs Hotel opens in the Canadian Banff National Park set aside in 1885. Designed in a mixture of French château and Scottish baronial styles by New York architect Bruce Price, the five-story H-shaped wooden structure with accommodations for 280 has been built by the Canadian Pacific Railway (above) whose president William Van Horne says, "Since we cannot export the scenery we'll import the sightseers."

The Hotel del Coronado opens across the bay from San Diego, Calif., with 399 rooms around a central court in a five-story structure that has 75 baths and a main dining room whose arched ceiling is made of natural sugar pine fitted together entirely by wooden pegs.

Washington's State, War, and Navy Department building is completed across the street from the White House by architect Alfred B. Mullett who has given the building mansard roofs and dormer windows typical of structures built since the Civil War.

The New York State Capitol building nears completion on a 3.5-acre site looming hundreds of feet above downtown Albany after more than 20 years of construction by H. H. Richardson and Leopold Elditz, but large chunks of brick begin to fall from the ceiling of the Assembly chamber.

The world's first revolving door—always open, always closed—is installed in a Philadelphia office lobby by local inventor Theophilus Van Kannel, 46.

The blizzard that strikes the U.S. Northeast in March comes on the heels of the mildest winter in 17 years and follows a warm spell in which buds have opened on trees in New York's Central Park. New York's temperature drops to 10.7° F. March 12, and winds off the Atlantic build up to 48 miles per hour, bringing unpredicted snow which continues off and on into the early morning of Wednesday, March 14. The 3-day accumulation totals 20.9 inches, but snowdrifts 15 to 20 feet high bring traffic to a standstill. Washington is isolated from the world for more than a day, 200 ships are lost or grounded from Chesapeake Bay north, at least 100 seamen die in the "Great White Hurricane," pedestrians and horses freeze to death in the streets, and at least 400 die, including former U.S. Senator from New York Roscoe Conkling, 58, who catches pneumonia and dies in mid-April.

U.S. cattlemen go bankrupt and foreign investors liquidate their American holdings after a drastic decline in cattle herds following the 1886–1887 drought on the western plains (*see* 1884).

Congress enacts emergency legislation and places Major John Wesley Powell in charge of an irrigation survey to select reservoir sites, determine irrigation product areas, and carry out part of his 1878 plan (*see* 1889).

A new salt field is discovered in central Kansas by prospectors boring for natural gas.

H. J. Heinz reorganizes a food packing firm under his own name (*see* 1886; 1896).

A new Chinese Exclusion Act voted by Congress October 1 forbids Chinese workers who have left the United States to return (*see* 1882; 1884; 1894).

1889 Austria's Archduke Rudolph, 31, is found dead January 30 with his mistress Baroness Marie Vetsera, 17, at Rudolph's hunting lodge Mayerling outside Vienna. The beautiful Marie has been shot by the crown prince, 31, who has then taken his own life, leaving the Hapsburg emperor Franz Josef without an heir. His nephew Franz Ferdinand, 25, becomes heir apparent (but *see* 1914).

Serbia's Milan Obrenovic IV abdicates March 6 at age 34 and retires to Paris after a 21-year reign. The king has circulated scandalous reports about his estranged wife Natalie, extorted a divorce that is illegal in the eyes of the church, and is succeeded by their son of 13 who will reign until 1903 as Aleksandr I.

The French *revanchist* Boulanger threatens to overturn the Third Republic in a coup d'état, but a warrant is issued for his arrest and he flees the country April 1, taking refuge first at Brussels and then at London. Boulanger will commit suicide in 1891.

The first Japanese written Constitution, handed by the Meiji emperor to his prime minister, Count Kuroda, February 11, stipulates that the emperor shall exercise legislative power with the consent of the Imperial Diet, but Imperial ordinances are not valid if the Diet fails to approve them.

The Ivory Coast becomes a French protectorate January 10, and an Anglo-French agreement August 5 defines respective French and British spheres of influence on the Gold and Ivory Coasts and on the Senegal and Gambia rivers.

Ethiopia's Johannes IV is killed March 12 in the Battle of Metemma fighting the Mahdists. He is succeeded after a 17-year reign by Menelek of Shoa, whose claim to the throne is supported by the Italians against those of Johannes' son Ras Mangasha. Menelek will reign as king of kings until 1911.

The British South Africa Co. headed by Cecil Rhodes receives almost unlimited rights and powers of government in the area north of the Transvaal and west of Mozambique (*see* 1890).

A relief expedition headed by Henry M. Stanley "rescues" Equatoria's Governor Emin Pasha. Stanley is financed largely by the king of the Belgians Leopold II, king of the Congo Free State, and by Leopold's friend Sir William MacKinnon of the Imperial British East Africa Co., both desirous of securing Equatoria and giving the Congo State an outlet to the Upper Nile. The German-born Mehmed Emin Pasha, whose Egyptian province was severed from the rest of Egypt by the Mahdi revolt in 1885, arrives at Bagamoyo in December and will soon be employed by the Germans, but he will be killed by Arab slave traders west of Lake Victoria in 1892.

The first Pan-American Conference opens October 2 at Washington, D.C., to cement relations among Western Hemisphere nations. All Latin American countries except Santo Domingo send representatives, the delegates establish the Pan-American Union as an information bureau, but reject a convention calling for the promotion of peace by arbitration, and reject a plan of reciprocity.

Portugal's Luiz I dies October 19 at age 51 after a 28-year reign in which slavery has been abolished in every Portuguese colony. He is succeeded by his son of 26 who will reign until 1908 as Carlos I.

The Brazilian army deposes Pedro II November 15 after a 49-year reign and proclaims a Brazilian republic under the leadership of Gen. Manoel Deodoro da Fonseca, 62 (*see* 1891).

North Dakota, South Dakota, Montana, and Washington are admitted to the Union as the 39th, 40th, 41st, and 42nd states.

The medicine man Wawoka of the Nevada Paiutes interprets a January 1 solar eclipse as a sign from the Great Spirit and he orders a ghost dance. White men's bullets cannot penetrate the shirts worn by braves in the ceremonial dance, he promises; the War Department prohibits the dance lest it stir up violence (*see* Wounded Knee, 1890).

The Great London dock strike from August 15 to September 16 helps extend British trade unionism from the skilled classes to the less skilled.

Chicago's Hull House opens in the South Halsted Street slums under the direction of social workers Jane Addams, 29, and Ellen Gates Starr, whose settlement house will help the poor of the city (*see* Coit, 1886; University Settlement, 1891).

"The man who dies rich dies disgraced," writes steel magnate Andrew Carnegie in an article on "the gospel of wealth." Oil magnate John D. Rockefeller praises him for his philanthropies (*see* library benefactions, 1881; Carnegie Institute, 1900; Carnegie Institution, 1902; Carnegie Foundation, 1905; Carnegie Corp., 1911; Rockefeller Institute, 1901).

Oklahoma Territory lands formerly reserved for Native Americans are opened to white homesteaders by President Harrison at high noon April 22, and a race begins to stake land claims. "Sooners," who have entered the territory prematurely, claim prior rights in many areas, bitter fights break out, but by sundown there are booming tent towns at Guthrie and at Oklahoma City, formerly a small water and coaling station for the Santa Fe Railroad but suddenly a city of 10,000 (*see* 1891).

Richard Sears sells his mail-order watch business for $70,000 but soon starts another with a catalog of watches, watch chains, and other jewelry offered with the slogan "Satisfaction or Your Money Back" (*see* 1886; 1893).

Electric lights are installed at the White House in Washington, D.C., but neither President Harrison nor his wife will touch the switches. An employee turns on the lights each evening, and they remain burning until the employee returns in the morning to turn them off.

The first electric train lighting system is patented by U.S. inventor Harry Ward Leonard, 28, who worked with Thomas Edison to introduce the central station electrical system for cities (*see* 1882). Leonard will patent an electric elevator control in 1892.

I. M. Singer Co. introduces the first electric sewing machines and sells a million machines, up from 539,000 in 1880 (*see* 1903).

Union Oil has its beginnings 75 miles northwest of Los Angeles, where prospector Lyman Stewart strikes oil in Torrey Canyon.

France's Compagnie Universelle du Canal Interocéanique goes bankrupt after having spent the equivalent of $287 million in an effort to build a Panama Canal that has cost the lives of some 20,000 Frenchmen and Chinese, Irish, and West Indian laborers in 5 years. The French call the Isthmus of Panama "de Lesseps' graveyard," and thousands lose their savings in the canal company's collapse.

The Maritime Canal Co. of Nicaragua incorporated by Congress February 20 to build, manage, and operate a ship canal begins work October 22, but while supporters of the new company say Nicaragua has a healthier climate, opponents say the country has too many active volcanoes and the Nicaragua route is too long (*see* 1902).

Europe's *Orient Express* completes arrangements for travel between Paris and Constantinople without change of train (*see* 1883). The first through train leaves the Gare de l'Est June 1 and arrives at Constantinople in 67 hours, 35 minutes—more than 13 hours faster than in 1883 but 7.5 hours slower than the time that will be standard within 10 years (*see* Simplon-Orient Express, 1919).

Scotland's Firth of Forth Bridge is completed by Sir John Fowler, now 72, and Sir Benjamin Baker, 49.

Europe's first electric trolley goes into service at Northfleet, Kent, England (*see* Richmond, Va., 1888).

The Canadian Pacific opens Montreal's $2 million Windsor Station.

1889 ***(cont.)*** The "safety" bicycle patented in 1885 is introduced in the United States. Within 4 years, more than a million Americans will be riding the new bikes.

French automakers René Panhard, 48, and E. C. Levassor acquire rights to manufacture motor vehicles in France using the Daimler patents of 1882 and 1886. They will mount the Daimler engine on a chassis in 1891 (*see* pneumatic tires, 1892).

Nellie Bly leaves Hoboken, N.J., November 14 in an attempt to outdo the hero of the 1873 Jules Verne novel *Le Tour du Monde en Quatre-Vingt Jours. New York World* reporter Elizabeth "Nellie Bly" Cochrane, 22, has earlier feigned madness in a successful attempt to gain admission for 10 days to New York's insane asylum on Blackwell's Island. She has persuaded her editor to give her a bold new assignment and leaves at just after 9:40 in the morning to break the fictional record of Phileas Fogg (*see* 1890).

Bird Flight as a Basis of Aviation by German engineer Otto Lilienthal, 41, shows the advantages of curved surfaces over flat ones for winged flight. Lilienthal has studied bird flight and has built gliders to prove his thesis, but he will be killed in a glider accident in 1896 (*see* Langley, 1896).

English chemist Frederick August Abel, 62, and Scottish chemist James Dewar of 1873 Dewar's vessel fame patent cordite, a smokeless, slow-burning explosive powder made of nitroglycerin, nitrocellulose, and mineral jelly.

Canadian-American chemist Herbert Henry Dow, 23, discovers a cheap new process for producing the bromine used by the pharmaceutical and photographic industries. Dow finds that by adding certain chemicals to cold brine and passing a current of air through the solution and onto scrap iron, a moisture collects on the iron that drips into a container as ferric bromine, a solution containing a high percentage of commercial-grade bromine (*see* 1891).

Elizabeth Cochrane outdid Jules Verne's fictional Phileas Fogg by girdling the earth in little more than 72 days.

A worldwide influenza pandemic will affect 40 percent of the human race in the next 2 years (*see* 1918).

Japanese bacteriologist Shibasaburo Kitazato, 37, at Berlin isolates the bacilli of tetanus and symptomatic anthrax (*see* 1890).

Emile Roux, now 36, and his Swiss colleague Alexandre Emile John Yersin, 26, at the Pasteur Institute show that the diphtheria bacillus produces a toxin (*see* Klebs, 1883; Löffler, 1884; von Behring, 1890).

German physiologists J. von Mering and O. Minkowski remove the pancreas of a dog and observe that although the animal survives, it urinates more frequently, and the urine attracts flies and wasps. When they analyze the urine, they find the dog has a canine equivalent of diabetes, which ultimately causes it to go into a coma and die (*see* Lancereaux, 1860; Langerhans, 1869; Sharpey-Schafer, 1916).

The Mayo Clinic has its beginnings in the St. Mary's Hospital opened at Rochester, Minn., by the Sisters of St. Francis who 6 years ago ministered to the injured in the wake of a tornado that devastated the town. The hospital's medical staff consists of physician-surgeon William Worrall Mayo and his sons William James, 28, and Charles Horace, 24, whose clinic will grow to receive nearly 500,000 visitors per year (*see* Kendall, 1915).

Barnard College opens near New York's Columbia University after 3 years of effort by Annie Nathan Meyer, 24, who has charmed and cajoled Columbia trustees and alumni. The new woman's college is named for Columbia's late president Frederick August Porter Barnard who died 6 months ago at age 80 after a long career in which he favored extending the university's educational opportunities to women.

The University of Idaho is founded at Moscow.

The New York Educational Alliance organized by German-Jewish groups provides education and recreation for slum dwellers on the city's Lower East Side, most of them from eastern Europe.

The *Wall Street Journal* that begins publication July 8 is the outgrowth of a daily financial news summary distributed by Dow Jones & Company, a firm organized 7 years ago by New York economists and financial reporters Charles Henry Dow, 38, and Edward D. Jones, 33. Dow has been a member of the New York Stock Exchange since 1885 and a partner in the brokerage firm Goodbody, Glyn & Dow. The Dow Jones Industrial Average will become a measure of the stock market's performance, and the *Journal* will grow to become the second largest U.S. newspaper in terms of circulation with regional and foreign editions.

The *Memphis Commercial-Appeal* begins publication. Local cotton merchant West James Crawford, 45, has started the *Commercial* with some associates and they soon absorb the *Appeal*.

A coin-operated telephone patented by Hartford, Conn., inventor William Gray is installed in the Hartford Bank. Gray will incorporate Gray Telephone Pay Station Co. in 1891 with backing from Amos Whitney and Francis Pratt of Pratt & Whitney; it will install pay phones in stores, hotels, restaurants, and saloons, giving 65 percent of the take to the telephone company, giving 10 percent to the store, and keeping 25 percent for the Gray Co. (local calls will cost 5¢ everywhere until 1951).

Nonfiction *Time and Free Will (Essai sur les données immédiates de la conscience)* by French philosopher Henri Louis Bergson, 29.

Fiction *A Connecticut Yankee in King Arthur's Court* by Mark Twain; *The Child of Pleasure (Il Piacere)* by Gabriele D'Annunzio.

Poetry *The Ballad of East and West* by Rudyard Kipling: ". . .and never the twain shall meet"; *Hot Houses (Serres chaudes)* by Belgian poet Maurice Polydore Marie Bernard Maeterlinck, 27.

Painting *Landscape with Cypress Tree, The Laborer, L'Hôpital de St. Paul à Saint Rémy, The Starry Night*, and *Self-Portrait with Bandaged Ear* by Vincent van Gogh who is suffering from a mental disorder and has cut off his ear and presented it to a young prostitute at Arles; *Portrait of an Old Woman* by James Ensor; *Mending the Nets* by Max Liebermann; *Still Life with Japanese Print* by Paul Gauguin; *The Gulf Stream* by Winslow Homer.

The Protar f.7.5 lens designed by German physicist Paul Rudolph, 31, for Carl Zeiss of Jena is the world's first anastigmatic lens to be commercially successful (*see* Tessar, 1902).

Goodwin Film & Camera Co. is founded by New Jersey clergyman-inventor Hannibal Williston Goodwin, 67, who has been granted a patent on a "photographic pellicle produced from a solution of nitrocellulose dissolved in nitrobenzol or other non-hydrous and non-hydroscopic solvent and diluted in alcohol or other hydrous and hydroscopic diluent."

Goodwin (above) has overcome curling of film by use of a celluloid varnish which remains flat when applied to a piece of glass or metal and can be sensitized with a gelatin emulsion, he has sent George Eastman at Rochester a 17-foot length of film to be coated and sensitized, Eastman has written him in mid-February that he has succeeded in coating the film with a sensitized emulsion; 17 days later Eastman chemist Henry Reichenbock has applied for a patent on a transparent roll film produced from a solution of camphor, fusel oil, and amyl nitrate dissolved in nitrocellulose and wood alcohol (*see* Eastman, 1888, 1895; cellulose acetate film, 1924).

Theater *The Lady from the Sea (Fruen fra Haven)* by Henrik Ibsen 2/12 at Oslo's Christiania Theater; *Miss Julie (Froken Julie)* by August Strindberg 3/14 at Copenhagen's Studentersamfundet; *The People of Hemso (Hemsoborna)* by Strindberg 5/29 at Djurgardsteatern; *Before Dawn (Vor Sonnenaufgang)* by German playwright Gerhart (Johann Robert) Hauptmann, 26, 10/20 at Berlin's Lessingtheater. The play pioneers German naturalist drama and creates a sensation; *Honor (Ehre)* by German playwright Hermann Sudermann, 32, 11/27 at Berlin's Lessingtheater.

Opera *Esclarmonde* 5/18 at the Opéra-Comique, Paris, with music by Jules Massenet, who has written the title role for California soprano Sybil Sanderson; *The Gondoliers (or The King of Baratila)* 12/7 at London's Savoy Theatre, with Gilbert and Sullivan songs that include "I am a Courtier Grave and Serious," "Take a Pair of Sparkling Eyes," "There Lived a King," "O My Darling, O My Pet," "An Enterprise of Martial Kind," "I Stole the Prince," "Roses White and Roses Red," and "We're Called Gondoliers."

First performances Symphony in D minor by César Franck 2/12 at the Paris Conservatoire; Symphony No. 1 in D major by Bohemian composer Gustav Mahler, 29, 11/20 at Budapest.

English organist-electrician Robert Hope-Jones, 30, applies electricity to the organ for the first time. Chief electrician of the Lancashire and Cheshire Telephone Co. at Birkenhead, Hope-Jones moves the console to a position that would be much too far from the pipes for a pneumatic system. He will emigrate to the United States, and before his death in 1914 he will sell his patents to the Rudolph Wurlitzer Co. of North Tonawanda, N.Y. (*see* Hammond, 1934).

Japanese Musical Instrument Manufacturing Co. (Nippon Gakki Siego Kabushiki Kwaisha) is founded by organ and piano maker Torakasu Yamaha with 30,000 yen in capital. Yamaha introduced the first Japanese organ in 1885 and the Yamaha piano 2 years later.

New York's Harlem Opera House opens in 125th Street. German-American cigar maker-theatrical magnate Oscar Hammerstein, 42, has built it.

The Chicago Auditorium opens December 9 with Adelina Patti singing "Home, Sweet Home" to an audience that includes President Harrison. The new auditorium seats 3,500 and is part of a 17-story building that Louis Sullivan will complete next year.

March "Washington Post March" by John Philip Sousa.

Popular songs "Oh, Promise Me!" by U.S. composer (Henry Louis) Reginald De Koven, 40, lyrics by English drama critic Clement Scott, 48; "Down Went McGinty" by Irish-American comedian Joseph Flynn.

William Renshaw wins in men's singles at Wimbledon, Blanche Bingley Hillyard in women's singles; Henry Slocum wins in U.S. men's singles, Bertha Townsend in women's singles.

Boston pugilist John Lawrence Sullivan, 30, defeats Jake Kilrain at Richburg, Miss., in July. The 75-round fight lasts for 2 hours, 16 minutes in 106° heat; it is the last major bare-knuckle prizefight.

The first All-America football team has a backfield and linemen selected by Yale athletic director Walter Chauncey Camp, 30, from among college varsity players across the country (*see* 1875; 1891).

1889 ***(cont.)*** The Flexible Flyer sled introduced by Philadelphia farm-equipment maker Sam Leeds Allen, 48, has runners that can be flexed over their entire length to permit turns twice as tight at twice the speed possible with other sleds. Allen's daughter Elizabeth has run her non-steerable sled into a tree, inspiring Allen to invent the Flexible Flyer with double-thick runners whose concave channels give them steering bite and are mounted beneath boards of fine-grained oak varnished to a gleaming finish.

Former Texas outlaw Belle Starr (*née* Myra Belle Shirley) is shot dead February 3 in Oklahoma Territory by person or persons unknown 2 days short of her 41st birthday.

The Cleveland Street scandal titillates London. Revelations appear about a West End homosexual brothel that employs post office messenger boys to gratify the appetites of clients whispered to include the Prince of Wales, his equerry Lord Arthur Somerset, 38, son of the duke of Beaufort, and the duke of Clarence, Prince Albert Victor Christian Edward, 25 (*see* Jack the Ripper, 1888). Some of the post office boys are bribed to go abroad so that they will not give evidence, and when one boy does give evidence against the earl of Euston, the newspaper editor who reports his testimony receives a 1-year jail sentence for libel.

The Eiffel Tower designed by French engineer Alexandre Gustave Eiffel, 57, is completed at Paris for the Universal Exhibition that opens May 6. The soaring 984.25-foot tower has a wrought-iron superstructure on a reinforced concrete base, it contains more than 7,000 tons of iron, 18,038 girders and plates, 1,050,843 rivets, and has three hydraulic elevators, one of them built by Otis Co. of Yonkers, N.Y.

London's Savoy Hotel opens August 6, the first British hotel to have private baths (*see* 1853). Built by Gilbert & Sullivan impressario Richard D'Oyly Carte of 1881 Savoy Theatre fame, the new hotel is built of reinforced concrete on a steel frame, it uses timber only for floors and window frames, is Britain's first fireproof hotel, and has 70 private bathrooms while the new Hotel Victoria on Northumberland Avenue has only four baths to serve its 500 guests. Manager of the new Savoy Hotel is César Ritz (*see* 1883; 1896; Escoffier, 1894).

New York's first real skyscraper opens September 27 at 50 Broadway. Architect Bradford Lee Gilbert has climbed to the top of the 13-story structure during construction and has let down a plumb line during a hurricane to show crowds gathered to watch the building collapse that it is steady as a rock.

Otis Co. installs the world's first electric elevators in New York's Demarest building on Fifth Avenue at 33rd Street.

The Johnstown Flood May 31 kills 2,000 to 5,000 Pennsylvanians in a city of 30,000. An earthen dam 90 feet high on the Conemaugh River has given way 14 miles away in the mountains, a torrent of water roars down on the city at 50 miles per hour, and its force tosses a 48-ton locomotive 1 mile.

The Maine salmon catch reaches 150,000 pounds.

The Hudson River shad catch reaches an all-time high of 4.33 million pounds. The catch begins to decline, and although it will climb back up to 4.25 million pounds in 1942, it will then fall off drastically, and water pollution will spoil its taste.

The constitutional conventions of the new states of North Dakota and Montana (above) hear Major John Wesley Powell urge delegates to measure land values in acre-feet, the area that can be covered with 12 inches of water from irrigation or natural sources (*see* 1888). County lines should follow drainage divides, says Powell, with each river valley a political unit whose inhabitants can work cooperatively, but the politicians pay little heed.

Rust finishes off Ceylon's coffee industry (*see* 1869). Demand increases for Latin American coffee.

The first spindle-type cotton picking machine is tested by U.S. inventor Angus Campbell whose machine will not be developed and produced commercially for more than half a century (*see* Rust, 1927).

Discontented southern farmers merge their farm organizations into the Southern Alliance (*see* 1882).

Kansas and Nebraska farmers pay 18 to 24 percent interest rates on loans, with rates sometimes going as high as 40 percent. Local brokers and then local loan companies secure funds from eastern investors and take healthy cuts for themselves.

Congress votes to give the 27-year-old U.S. Department of Agriculture cabinet rank; President Harrison appoints J. M. Rusk the first secretary of agriculture.

British dairymen get their first milking machines. William Murchland at Kilmarnock, Scotland, manufactures the machines (*see* Colvin, 1878).

A pure food law is proposed in Congress but meets with ridicule (*see* 1882; 1892).

A New York margarine factory employee tells a state investigator that his work has made "his hands so sore. . . his nails came off, his hair dropped out and he had to be confined to Bellevue Hospital for general debility," but the "bogus butter" made from hog fat and bleaches is widely sold as "pure creamery butter" and is no worse than "butter" made from casein and water or from calcium, gypsum, gelatin fat, and mashed potatoes (*see* 1890).

The first large-scale English margarine plant begins production at Godley in Cheshire. Danish entrepreneur Otto Monsted competes with U.S. and Dutch firms that are producing margarine in volume (*see* 1876; 1880; 1890).

English sugar consumption rises to 76 pounds per capita, up from 60 pounds in 1875, to make Britain the heaviest sugar user in the world by far.

A U.S. ice shortage caused by an extraordinarily mild winter gives impetus to the development of ice-making plants (*see* 1880; Western Cold Storage, 1887). By year's end the country has more than 200 ice plants, up from 35 in 1879.

Aunt Jemima pancake flour, invented at St. Joseph, Mo., is the first self-rising flour for pancakes and the first ready-mix food ever to be introduced commercially. Editorial writer Chris L. Rutt of the *St. Joseph Gazette* and his friend Charles G. Underwood used their life savings last year to buy an old flour mill on Blacksnake Creek, but a falloff in westward migration by wagon train has decreased demand for mill products. Faced with disaster, Rutt and Underwood have experimented for a year and come up with a formula combining hard wheat flour, corn flour, phosphate of lime, soda, and a pinch of salt which they package in 1-pound paper sacks and sell as Self-Rising Pancake Flour, a name Rutt changes to Aunt Jemima after attending a minstrel show at which two blackfaced comedians do a New Orleans-style cakewalk to a tune called "Aunt Jemima" (*see* 1893).

The pea viner introduced to expedite pea canning takes the whole pea vine and separates peas from pods in a continuous operation (*see* 1883).

1890 Kaiser Wilhelm forces Bismarck to resign as prime minister March 18; he is caricatured for having "dropped the pilot" who united the German states and inaugurated signal reforms in German society.

The Influence of Sea Power Upon History, 1660–1783 by U.S. naval officer-historian Alfred Thayer Mahan, 50, demonstrates the decisive role of naval strength and will have enormous influence in encouraging the world powers to develop powerful navies. Mahan's objective is to have Congress strengthen the U.S. Navy, which last year launched its first battleship but depends chiefly on its White Squadron of three small cruisers and a dispatch boat. His book is eagerly read in Britain, which has long controlled the seas, Russia, Japan, and Germany, where Kaiser Wilhelm II orders a copy placed in the library of every German warship and pushes for a strong German navy.

Anglo-German disputes in East Africa are resolved July 1 in an agreement by which the Germans give up all claims to Uganda, receive the little island of Heligoland in the North Sea which Britain obtained from Denmark in 1815, and recognize the protectorate over Zanzibar established by Britain June 14.

Cecil Rhodes becomes prime minister of Africa's Cape Colony July 17 and adds the political post to his position as head of De Beers Consolidated Gold Fields and of British South African Railway.

The German East Africa Co. cedes all its territorial rights to the German government October 28.

Willem III of the Netherlands dies November 23 at age 73 after a 34-year reign. He is succeeded by his daughter, 10, whose mother Emma will rule as regent until 1898 but who will herself reign until 1948 as Wilhelmina I.

The Grand Duchy of Luxembourg splits from the Netherlands upon the death of Willem III (above).

When Kaiser Wilhelm sacked Bismarck, the man who had united Germany, he was caricatured in *Punch*.

Japan holds her first political elections July 11, but suffrage is limited to males over 25 who pay taxes of at least 15 yen per year (1.1 percent of the population), so only 453,474 drop their ballots in the small black lacquer ballot boxes (*see* 1929).

The first Japanese Diet opens November 29 as members elected in July take their seats.

Idaho is admitted to the Union as the 43rd state. Wyoming is admitted as the 44th state.

The United Mine Workers of America is organized as an affiliate of the 4-year-old American Federation of Labor (AF of L) (*see* 1897).

The Knights of Labor strike the New York Central Railroad August 8.

Mississippi institutes a poll tax, literacy tests, and other measures designed to restrict voting by blacks. Other southern states will impose similar restrictions (*see* 1898).

Sioux lands in South Dakota that were ceded to the U.S. government last year are thrown open to settlement February 10 under terms of a presidential proclamation that opens 11 million acres to homesteaders. Sioux chief Sitting Bull of 1876 Battle of the Little Big Horn fame is arrested in a skirmish with U.S. troops December 15 and shot dead by Indian police at Grand River as Sioux warriors of the Ghost Dance uprising try to rescue him (*see* 1889).

The "Battle" of Wounded Knee December 29 ends the last major Indian resistance to white settlement in America. Nearly 500 well-armed troopers of the U.S. 7th Cavalry, some with Hotchkiss artillery, massacre an estimated 300 (out of 350) Sioux men, women, and

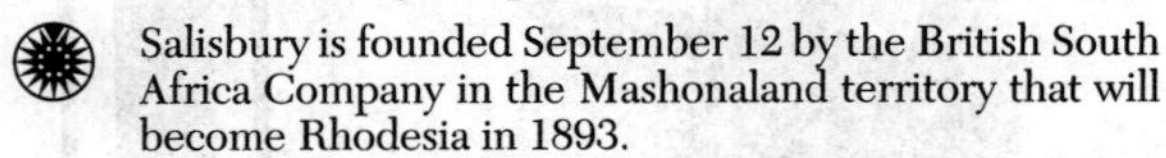

1890 ***(cont.)*** children in a South Dakota encampment. The Army takes only 35 casualties, mostly from stray bullets fired by troopers or from artillery shrapnel.

Salisbury is founded September 12 by the British South Africa Company in the Mashonaland territory that will become Rhodesia in 1893.

The Supreme Court virtually overrules its 1877 decision upholding state regulation of business March 24. The court's decision in *Chicago, Milwaukee, St. Paul Railroad v. Minnesota* denies Minnesota's right to control railroad rates and in effect reverses the court's ruling in *Munn v. Illinois*.

The Sherman Anti-Trust Act passed by Congress July 2 curtails the powers of U.S. business monopolies: "Every contract, combination in the form of trust or otherwise, or conspiracy in restraint of trade or commerce among the several States, or with foreign nations, is hereby declared to be illegal." But the new law will have little initial effect (*see* Clayton Act, 1914).

The Sherman Silver Purchase Act passed by Congress July 14 supersedes the Bland-Allison Act of 1878 but continues government support of silver prices (*see* 1893).

The McKinley Tariff Act passed by Congress October 1 increases the average U.S. import duty to its highest level yet, but while the average is roughly 50 percent, the act does provide for reciprocal tariff-lowering agreements with other countries.

Wall Street has a panic in November as the London banking house Baring Brothers fails and English investors dump U.S. securities.

The Peruvian government escapes bankruptcy with help from New York financier W. R. Grace who assumes the debt of two bond issues in return for concessions that give him virtual control of nearly all the country's developed and undeveloped resources including the rich Cerro de Pasco silver mines, all the Peruvian guano deposits, 5 million acres of oil and mineral lands, and several railroad leases (*see* Grace, 1865). Peru's treasury has been drained by the War of the Pacific that ended in 1884 and by the costs of financing the railroads built by the late Henry R. Meiggs who died in 1877. Now 58, Grace has twice been elected mayor of New York (1880 and 1884), he will organize the New York & Pacific Steamship Co. next year, will later add Grace Steamship, and in 1895 will combine his New York and Peruvian holdings into William R. Grace and Co., which will expand with agricultural, banking, chemical, mercantile, and utility investments in North and South America.

Sun Oil Co. of Ohio is founded by entrepreneur Joseph Newton Pew, 42, who in 1874 married a young Titusville, Pa., woman whose family helped develop the Pennsylvania oil fields. Pew has pioneered in pumping gas by mechanical pressure, patented a pump of his own invention to supply cities with heat and light, will acquire other companies, and will market his products under the name Sunoco.

Ever Ready batteries, the first commercial dry cell batteries, are introduced by National Carbon Co. (*see* Leclanche, 1867).

Nellie Bly boards the S.S. *Oceanic* at Yokohama January 7 and sails for San Francisco after having crossed the Atlantic, Europe, and Asia in her well-publicized attempt to girdle the earth in less than 80 days (*see* 1889). The *New York World* reporter is advised at San Francisco that the purser has left the ship's bill of health at Yokohama and that nobody may leave the ship for 2 weeks, she threatens to jump overboard and swim, she is put on a tug and taken ashore, her train across the continent detours to avoid blizzards and is almost derailed when it hits a handcar, but she pulls into Jersey City at 3:41 in the afternoon of January 25 after a journey of 72 days, 6 hours, 11 minutes, 14 seconds (*see* Mears, 1913).

Railroad-related accidents kill 10,000 Americans and seriously injure 80,000.

The United States has 125,000 miles of railroad in operation, Britain 20,073 miles, and Russia 19,000.

London's first electric underground railway goes into service, and the coke-burning locomotives that have operated since 1863 are retired. A tube ride to any point on the new underground line costs twopence. The syndicate that has built the underground system will soon be headed by Chicago traction magnate Charles T. Yerkes who will sell his Chicago interests in 1899 and move to London in 1900 (*see* 1886; Budapest, 1896).

Gottfried Daimler founds Daimler-Motoren-Gesellschaft at Bad Cannstatt in Germany (*see* 1887; 1901; British Daimler, 1896).

U.S. engineer Herman Hollerith, 30, pioneers punch-card processing by adapting techniques employed in the Jacquard loom of 1801 and the player piano of 1876 to devise a system for punching holes in sheets of paper to record U.S. census statistics. Tabulating Machine Co. will acquire patents to the Hollerith system (*see* Babbage, Lovelace, 1842; Watson, 1912).

The discovery of rich iron ore deposits in Minnesota's Mesabi region by prospector Leonidas Merritt, 46, helps U.S. steelmakers (*see* Rockefeller, 1893).

Dutch paleontologist Eugene Dubois, 32, discovers Java man fossils of a prehistoric human ancestor at Kedung Brebus, Java, while serving as a military surgeon in the East Indies. He finds the fossil evidence of *Pithecanthus erectus* in an earth stratum dating from the Upper Pleistocene era of 700,000 B.C.

Berlin bacteriologists Emil von Behring, 36, and Shibasaburo Kitazato at Robert Koch's laboratory produce the first tetanus antitoxin. They immunize animals by injecting minute amounts of toxin and find that the animals produce antibodies.

The first diphtheria antitoxin is produced by Emil von Behring (above) (*see* 1891; Roux and Yersin, 1889; Biggs, 1894).

New York State requires physicians to apply prophylactic drops to the eyes of newborn infants. The state legislature enacts the Howe Law drafted by Buffalo ophthalmologist-legislator Lucien Howe, 42, and most states will enact similar laws to combat blindness caused by gonorrheal infection (*see* Crede, 1884; Barnes, 1902).

Only 3 percent of Americans age 18 to 21 attend college. The figure will rise to 8 percent by 1930.

Colgate University takes that name after 71 years as Madison University. The late soap maker William Colgate and his sons have been large benefactors to the university at Hamilton, N.Y. (*see* 1806).

The University of Oklahoma is founded at Norman in Oklahoma Territory.

Oklahoma State University is founded at Stillwater (*see* statehood, 1907).

The *Literary Digest* begins publication with reprints of comment and opinion from other U.S. periodicals. The new weekly has been started by New York publisher Isaac Kaufmann Funk, 51, who publishes a *Standard Dictionary of the English Language* with his associate Adam Willis Wagnalls, 47 (*see* 1937; *Reader's Digest,* 1922).

Nonfiction *The Golden Bough* (first volume) by Scottish anthropologist James George Frazer, 36, who will publish 15 further volumes in the next 25 years in a monumental exploration of the cults, legends, myths, and rites of the world and their influence on the development of religion; a one-volume abridged version will appear in 1922.

Fiction *A Hazard of New Fortunes* by William Dean Howells who moved to New York from Boston 2 years ago and shows the conflict between striking workers and capitalist interests; *Hunger (Sult)* by Norwegian novelist Knut Hamsun (Knut Pedersen), 31, who returned to Norway last year after two extended visits in America during which he worked as a farmhand, Chicago streetcar conductor, and lecturer; *The Sign of the Four,* a new Sherlock Holmes detective novel, by A. Conan Doyle.

Poetry *Poems by Emily Dickinson* is published at the urging of Lavinia Dickinson, whose late sister Emily Elizabeth died at her native Amherst, Mass., 4 years ago at age 55 without ever having had any of her work appear under her own name. The book receives a cool reception from critics but has a good enough public response to warrant publication of more Emily Dickinson poems. New volumes will appear in 1891 and 1896, and by 1945 nearly all of her many hundreds of poems will have been published and her reputation as a major U.S. poet will be secure: "Because I could not stop for Death,/ He kindly stopped for me:/ The carriage held but just ourselves/ And Immortality"; *The Sightless (Les Aveugles)* by Maurice Maeterlinck; "The Lake Isle of Innisfree" by Irish poet William Butler Yeats, 25, whose first book of poetry appeared 4 years ago. His new poem begins, "I will arise and go now, and go to Innisfree,/ And a small cabin build there. . ."

Painting *Poplars* by Claude Monet; *The Cardplayers* by Paul Cézanne; *On the Threshold of Eternity, Chestnut Trees in Flower, A Field Under a Starry Sky, The Auvers Stairs, Portrait of Doctor Gachet, The Town Hall of Auvers on the Fourteenth of July,* and *Crows Over the Wheat Fields* (his last work) by Vincent van Gogh, who shoots himself July 27 at Auvers and dies there 2 days later at age 37; *Breaking Home Ties* by Irish-American painter Thomas Hovenden, 49, will be the most widely engraved painting in America.

The "Gibson Girl" created by New York illustrator Charles Dana Gibson, 22, makes her first appearance in the humor weekly *Life,* which has been buying Gibson drawings since 1886. Apprenticed briefly to sculptor Augustus Saint-Gaudens, Gibson studied at the Art Students' League. Millions will share his conception of the ideal American girl (*see* 1902).

Theater *Beau Brummel* by U.S. playwright (William) Clyde Fitch, 25, 5/17 at New York's Madison Square Garden, with English-American actor Richard Mansfield, 36, who has commissioned the play; *The Reconciliation (Das Friedensfest)* by Gerhart Hauptmann 6/1 at Berlin's Ostendtheater, four perfs.

Ballet *The Sleeping Beauty* 1/15 at St. Petersburg's Maryinsky Theater, with music by Petr Ilich Tchaikovsky, choreography by Marius Petipa.

Opera *Cavalleria Rusticana* 5/17 at Rome's Teatro Costanzi, with music by Italian composer Pietro Mascagni, 27, libretto from a play by Sicilian novelist Giovanni Verga; *Robin Hood* 6/11 at the Chicago Opera House, with music by Reginald De Koven who includes his 1889 song "Oh, Promise Me!" in the work, book and other lyrics by Harry B. Smith; *Prince Igor* 11/4 at St. Petersburg, with music by the late Aleksandr P. Borodin who died early in 1887 at age 52 after years of working on the opera for his wife, a singer. Nikolai Rimski-Korsakov and Russian composer Aleksandr Konstantinovich Glazunov, 25, have completed Borodin's score (*see* ballet, 1909); *Les Troyens (The Trojans)* 12/5 at Karlsruhe, with music by the late Hector Berlioz who died in 1869 at age 65. The opera will not be presented in its original French until a U.S. production in 1960; *Queen of Spades (Pique Dame,* or *Pikovaya Dama)* 12/19 at St. Petersburg, with music by Petr Ilich Tchaikovsky.

First performances *Lancelot and Elaine* by Edward MacDowell 1/10 at Boston.

Willoby Hamilton, 25, (Ireland) wins in men's singles at Wimbledon, Helen Bertha Grace "Lena" Rice, 24, (Ireland) in women's singles; Oliver Samuel Campbell, 19, wins in U.S. men's singles, Ellen C. Roosevelt, in women's singles.

Cy Young signs with the Cleveland team of the National League to begin an outstanding pitching career that will continue for nearly 23 years: with Cleveland through 1898, the St. Louis Cardinals through 1900, the Boston Red Sox of the American League through 1908, the Cleveland Indians of the American League through 1911, and the Boston Braves of the National League for part of the 1911 season. Denton True "Cy" Young, 23, wins both games of a

1890 ***(cont.)*** doubleheader in October and will be the first pitcher to win 500 games (*see* 1904).

The first Army-Navy football game begins a long rivalry between West Point and Annapolis; Navy wins 24 to 0.

How the Other Half Lives by *New York Evening Sun* police reporter Jacob August Riis, 51, portrays slum life and the conditions that make for crime, vice, and disease. Writes the Danish-American journalist, "When the houses were filled, the crowds overflowed into the yards. In winter [there were] tenants living in sheds built of old boards and roof tin, paying a dollar a week for herding with the rats" (*see* Steffens, 1904).

English soap maker William H. Lever builds a Sunlight soap factory and establishes the model industrial town of Port Sunlight (*see* 1885; Lifebuoy, 1897).

Rothman's is founded by a London tobacconist.

American Tobacco Co. is founded by James Buchanan "Buck" Duke, 33, who creates a colossal trust by merging his father Washington Duke's Durham, N.C., company with four other major plug tobacco firms. Helped by New York and Philadelphia financiers who include Peter A. B. Widener, Grant B. Schley, William C. Whitney, and Thomas Fortune Ryan, Duke is joined by tobacco magnate R. J. Reynolds and will pyramid a group of holdings to include American Snuff, American Cigar, American Stogie, and United Cigar Stores (*see* 1888; 1911).

New York introduces the electric chair for capital punishment. It is considered more modern and humane than hanging.

The Wainwright building completed at St. Louis by Chicago architect Louis Henry Sullivan, 33, is the first true skyscraper. Sullivan formed a partnership 9 years ago with Dankmar Adler, he finished a new addition to Chicago's Carson Pirie Scott department store at Madison and State Streets last year, his apprentices include Frank Lloyd Wright, 21, and his Wainwright building has an unbroken sweep of vertical line from base to top (tenth) story, a design that will influence architecture for years to come.

Louis Sullivan (above) completes Chicago's 17-story Auditorium building at a cost of $35 million (*see* auditorium opening, 1889).

William LeBaron Jenney completes Chicago's 16-story Manhattan building at 431 South Dearborn Street. It is the world's first tall building with metal skeleton construction throughout.

Starlings are introduced into New York by local drug manufacturer Eugene Schieffelin who has failed in efforts to introduce skylarks, song thrushes, and nightingales from his native Germany. Schieffelin releases 50 pairs of starlings, will release another 20 pairs next year, and the birds will prey upon the English sparrows that have overrun Central Park since their introduction in 1850.

Washington's Rock Creek Park is set aside after years of congressional opposition. Development is begun on an urban green belt in the nation's capital that will be 4 miles long, up to 1 mile wide, and will embrace 2,213 acres.

Yosemite National Park is created by act of Congress which also creates Sequoia National Park. Yosemite embraces 761,320 acres on California's Tuolumne River, Sequoia 386,863 acres that include Mt. Whitney in the California Sierras, highest mountain in the continental United States (14,494 feet).

Alaska has 38 salmon canneries, up from one or two in 1878, and 4,000 men sail north from San Francisco each summer to catch and can the fish, returning at the end of the 6-week season. Chinese from San Francisco's Chinatown work from 6 in the morning until 8 at night to cut, clean, and can the red salmon, chinooks, silversides, and dog salmon caught largely by Italians and Scandinavians (*see* Smith, 1903).

A second Morrill act passed by Congress August 30 supplements the 1862 law, establishing experiment stations, extension services, and agricultural research programs to aid U.S. farmers.

From 75 to 90 percent of all Kansas farms are mortgaged at interest rates averaging 9 percent. Banks have foreclosed on roughly one-third of all farm mortgages in the state in the past decade, as drought prevented farmers from producing enough to keep up interest payments on loans taken out to buy farm machinery and seed.

Kansas farmers should "raise less corn and more hell," Populist party leader Mary Elizabeth Cylens Lease, 36, tells them. She began speaking in behalf of Irish home rule in 1885 and makes 161 speeches as she stumps the state. Told by "the Kansas Python" that Kansas suffers "from two great robbers, the Santa Fe Railroad and the loan companies," the farmers vote against the Republicans and elect independent-party candidates to Congress.

Third-party platforms begin to call for federal warehouses in which farmers may store their crops and receive in return certificates redeemable in currency for 80 percent of the market price, permitting them to hold crops off the market until prices improve.

Cattlemen and sheep herders engage in open conflict as the once "inexhaustible" range of 17 western states and territories becomes fully stocked with 26 million head of cattle and 20 million head of sheep competing for grass on the prairie.

Kern County Land Co. is founded by speculators Lloyd Tevis and James Ben Ali Haggin who have battled Henry Miller and Charles Lux for Kern River water rights in California, have won 75 percent of the rights, and have seen their 400,000 acres in the lower San Joaquin Valley grow (*see* 1850; 1877). Company lands will continue to grow until they cover 2,800 square miles, an area more than twice the size of Rhode Island (*see* oil, 1936).

Milk is pasteurized by law in many U.S. communities despite opposition by some dairy interests and people who call pasteurization "unnatural" (*see* 1860; Evans, 1917).

Unsweetened condensed milk is introduced commercially by the 5-year-old Helvetia Milk Condensing Co. at Highland, Ill. (*see* Borden, 1855; Meyenberg, 1885; "Our Pet," 1894).

Margarine will be improved in the next few years by the addition of vegetable oils such as palm oil and arachis oil from peanuts (*see* 1889; Lever, 1911).

Peanut butter is invented by a St. Louis physician, who has developed the butter as a health food (*see* hydrogenation, 1901; Peter Pan, 1928; Skippy, 1932).

Cudahy Packing is founded by Irish-American meatpacker Michael Cudahy, 38, who was brought into the business 3 years ago by P. D. Armour, has gone into partnership with Armour in South Omaha, and now takes over Armour's interest in Armour-Cudahy Packing. Cudahy will introduce methods for curing meats under refrigeration and develop improved railroad cars, making it possible to cure meat all year round and transport it to distant cities.

American Biscuit & Manufacturing Co. is created by a merger of western bakeries under the leadership of Chicago lawyer Adolphus Williamson Green, 47, whose new company controls some 40 bakeries in 13 states (*see* National Biscuit, 1898).

New York Biscuit is created by a merger of some 23 bakeries in 10 states, including Newburyport's 98-year-old John Pearson & Son and Boston's 89-year-old Bent & Co. Boston's 85-year-old Kennedy Biscuit Works has become the largest U.S. bakery and joins New York Biscuit in May as Chicago lawyer W. H. Moore contributes financing and aggressive leadership in a bid to give New York Biscuit control of the entire industry.

American Biscuit (above) is headed by St. Joseph, Mo., cracker baker F. L. Sommer whose Premium Saltines are famous for their parrot trademark and their "Polly wants a cracker" slogan. A. W. Green dispatches Sommer to open a bakery at New York that will challenge New York Biscuit in its own territory, and a price war begins that will continue until 1898.

The first aluminum saucepan is produced at Cleveland by Henry W. Avery whose wife will use the pan until 1933 (*see* 1888).

Canada Dry ginger ale has its beginnings in a small Toronto plant opened by local pharmacist John J. McLaughlin to manufacture carbonated water for sale to drugstores as a mixer for fruit juices and flavored extracts. McLaughlin will soon start making his own extracts and will develop a beverage he will call McLaughlin Belfast Style Ginger Ale (*see* 1907).

Thomas Lipton enters the tea business to assure supplies of tea at low cost for his 300 grocery shops (*see* 1879). He offers "The Finest the World Can Produce" at 1d 7p lb. when the going price is roughly a shilling higher (*see* 1893).

The U.S. population reaches 62.9 million with two-thirds of it rural, down from 90 percent rural in 1840.

The population of Los Angeles reaches 50,000, up from 11,183 in 1880.

1891 Hawaii's king David Kalakahua dies January 20 at age 54 after a 16-year reign and is succeeded by his sister, 52, who will reign until 1893 as Queen Lydia Liliuokalani. The white elite which owns 80 percent of arable lands in the islands have united in the Hawaiian League to oppose Kalakahua, who has favored the interests of sugar magnate Claus Spreckels, and the sugar planters form an Annexation Club to overthrow the queen (*see* Spreckels, 1883; abdication, 1893).

Brazilians elect Gen. Deodora da Fonseca president under a new constitution, his dictatorial rule produces a naval revolt in late November, he is ousted from office, and Vice President Floriano Peixoto takes his place, but Peixoto's arbitrary rule will lead to rebellions, which will break out for decades under future presidents.

Leander Starr Jameson, M.D., 38, is made administrator of the South Africa Co.'s territories. A friend of Cecil Rhodes, Jameson is a Scottish physician who came to Kimberley in 1878 and has been employed by Rhodes to negotiate with the Matabele chief Lobengula (*see* 1893).

A Young Turk movement of exiles from the Ottoman Empire has its beginnings in a meeting at Geneva. The Young Turks subscribe to the provisions of the 1876 constitution.

French troops open fire on strikers May 1 at Fourmies as workers of the Sans Pareille factory demonstrate in the streets for an 8-hour day. Among the nine dead are two children, and about 60 are injured.

The papal encyclical *Rerum novarum* issued May 15 by Leo XIII (Gioacchino Pecei) points out that employers have important moral duties as members of the possessing class and that one of society's first duties is to improve the position of the workers.

German factory workers win the right June 1 to form committees that will negotiate with employers on conditions of employment, and factory inspection is made more efficient.

The world's first old age pension plan goes into effect in Germany. Introduced by Bismarck in the Old Age Insurance Act of June 1889, the plan compels workers to contribute if they are over age 16, are fully employed, and earn more than 2,000 marks ($500) per year. Employers must contribute equal amounts, and the pension is payable at age 70 to persons who have paid premiums for a minimum of 30 years (*see* Britain, 1908; U.S. Social Security Act, 1935).

U.S. workers strike throughout the year for higher wages and shorter hours.

New York's University Settlement is founded by philanthropists to take over the work begun 5 years ago by Stanton Coit. He has gone to England as ethical culture minister to London and will head the West London Ethical Culture Society while the new society in New York urges university students to volunteer as settlers in the city slums (*see* 1897).

1891 ***(cont.)*** A New Orleans lynch mob breaks into a city jail March 14 and kills 11 Italian immigrants who have been acquitted of murder—the worst lynching in U.S. history. Italian workers are at the bottom of the U.S. social heap and receive lower wages than any other workers, black or white.

Jim Crow laws are enacted in Alabama, Arkansas, Georgia, and Tennessee (*see* Florida, 1887; Supreme Court decision, 1896).

Chicago's Provident Hospital is the first U.S. interracial hospital. Established by black surgeon Daniel Hale Williams, 33, the new facility incorporates the first U.S. nurses training school for black women (*see* open heart surgery, 1893).

Oklahoma Territory lands ceded to the United States by the Sauk, Fox, and Potawatomie open to white settlement September 18 by a presidential proclamation that opens 900,000 acres (*see* 1889; 1892).

Union Minière du Haut Katanga is organized April 15 to mine African copper under an agreement between Leopold II of the Belgians and Cecil Rhodes aide Robert Williams, 31, who has sent an expedition north to study outcroppings observed by the late David Livingstone. The expedition has reported back that the copper lies across the border in Katanga territory owned by Leopold so Williams has gone to the king and negotiated (*see* Benguela Railroad, 1903).

A new Colorado gold rush begins as thousands rush to Cripple Creek on the slopes of Pike's Peak following a strike by prospector Robert Wommack who has found paydirt at Poverty Gulch. The Cripple Creek mines will grow in the next 20 years to become the fifth largest producer in world history, they will yield nearly $1 billion in the yellow metal, and by the end of this decade the town of Cripple Creek will have 60,000 people, 147 active mining companies, 40 grocery stores, 17 churches, 25 schools, 15 hotels, 88 doctors and dentists, nearly 100 lawyers, a stock exchange, streetcars, eight newspapers, and 139 24-hour saloons (*see* 1900).

The American Express Travelers Cheque copyrighted July 7 by the 41-year-old American Express Co. will become an important means of protecting travelers' funds from theft and loss. American Express guarantees checks countersigned by the purchaser, pays banks a commission for selling the checks, receives a 1 percent commission from the purchaser, and enjoys the use of vast sums of money interest-free. Thomas Cook & Son, which issued the first Travelers Cheques in 1874, agrees to honor the checks at its worldwide travel offices (*see* credit card, 1958).

The New York Central's *Empire State Express* travels 436 miles from New York to East Buffalo in a record-breaking 7 hours, 6 minutes.

Construction begins at Vladivostok on a Trans-Siberian Railway that has been financed in part by French financiers and by London's Baring Brothers. Russian convicts will perform most of the labor on the railway which will link Moscow with the Pacific Coast (*see* 1900).

Chicopee, Mass., bicycle designer Charles Edward Duryea, 29, and his toolmaker brother Franklin design a gasoline engine that will power a road vehicle (*see* 1892; Benz, 1885, 1888).

The S.S. *Empress of India,* S.S. *Empress of Japan,* and S.S. *Empress of China* begin service out of Vancouver as the Canadian Pacific Railway moves into shipping (*see* 1888).

The Ocean Mail Subsidy Act passed by Congress authorizes subsidization of the U.S. merchant marine (*see* 1928; 1936).

Aluminum production gets a boost from Pittsburgh banker Andrew William Mellon, 36, who invests in Pittsburgh Reduction Co. (it will soon be headed by his brother Richard) and finances the mortgage of a new plant at New Kensington, Pa., for the company that will become Aluminum Company of America (Alcoa) (*see* 1888; 1945).

The first commercial bromine to be produced electrolytically is introduced by Herbert H. Dow's Midland Chemical Co., which Dow has established at Midland, Mich., with backing from Cleveland sewing machine maker J. H. Osborn (*see* 1889). Dow's bromine will find a good market in the pharmaceutical and photographic industries, and Dow begins work that will develop a process for producing chlorine electrolytically (*see* 1897; ethyl, 1924).

German-American chemist Herman Frasch, 40, patents a process for extracting sulfur economically to permit production of cheap sulfuric acid for making superphosphate fertilizer and countless other purposes. He has noted that sulfur melts at a fairly low temperature and is not soluble in water and will use his hot-water melting process in October 1895 to extract sulfur from a huge salt dome discovered in 1869 in Louisiana.

Berlin's Koch Institute for Infectious Diseases opens with great nationalistic fanfare despite revelations that Koch's tuberculosis vaccine is ineffective.

The diphtheria antitoxin produced last year by Emil von Behring has hopelessly variable results on test animals until Koch Institute bacteriologist Paul Ehrlich, 37, shows Behring how to standardize his vaccine. Ehrlich increases the amount of antitoxin produced in the blood of a horse by steadily raising the amount of toxin injected, the diphtheria vaccine is given its first human application Christmas Eve on a child dying in a Berlin hospital, the child recovers and is discharged a week later, but while Behring will be made a baron and will grow rich on royalties from commercial sale of the vaccine, he will trick Ehrlich out of his share by persuading him that he wiil be granted an institute of his own if he renounces commercial royalties (*see* antisyphilis drug, 1909).

The University of Chicago is founded with help from merchant Marshall Field who gives 25 acres of land and will contribute $100,000 next year, but most of the funding comes from oilman John D. Rockefeller.

Drexel Institute is founded at Philadelphia with an endowment from local banker Anthony Joseph Drexel, 65, a co-owner since 1864 of the *Public Ledger.*

California Institute of Technology is founded at Pasadena.

Automatic Electric Co. is founded to exploit a dial telephone patented by Kansas City undertaker Almon B. Strowger, who has convinced himself that "central" is diverting his incoming calls to a rival embalmer (*see* AT&T, 1919).

The first full-service advertising agency opens March 15 at New York. George Batten, 36, offers "service contracts" under which his agency will handle copy, art, production, and placement, relieving clients of any need to maintain their own elaborate advertising departments. He is compensated by the media, which pay him commissions on space rates (*see* ANPA, 1893).

Fiction *Tess of the d'Urbervilles* by Thomas Hardy portrays human tragedies rooted in the conflict between the changing society of industrial England and the rural ways of an earlier age; *The Story of Gosta Berling (Gosta Berlings Saga)* by Swedish novelist Selma Lagerlöf, 32; *The Heritage of Quincas Borba (Quincas Borba)* by Joachim Machado De Assis; *New Grub Street* by English novelist George Gissing, 33; *The Light That Failed* by Rudyard Kipling, who has returned to England from India; *The Little Minister* by English novelist James M. (Matthew) Barrie, 31, who will turn to writing for the stage next year; *Peter Ibbetson* by *Punch* staff artist George Louis Palmella Busson du Maurier, 57, whose novel is autobiographical; *The Picture of Dorian Gray* by Oscar Wilde whose novel is attacked by some critics as "decadent" and "unmanly" for portraying a beautiful hedonist who retains his youthful good looks while his portrait shows the marks of time and dissipation; *Lord Arthur Savile's Crime and Other Stories* by Oscar Wilde; *News from Nowhere* (stories) by William Morris, now 57, who describes an English socialist commonwealth; *Main-Travelled Roads: Six Mississippi Valley Stories* by U.S. short story writer (Hannibal) Hamlin Garland, 31, includes "The Return of the Private"; *Tales of Soldiers and Civilians* by Ambrose Bierce is based on the author's Civil War experiences and contains the story "Occurrence at Owl Creek Bridge"; "A Scandal in Bohemia" by A. Conan Doyle in the July *Strand* magazine begins regular publication of Sherlock Holmes stories.

Poetry "The English Flag" by Rudyard Kipling: "And what should they know of England who only England know?"; *Pilgrimages* by German poet Stefan George, 23.

Painting *Toilers of the Sea* by U.S. painter Albert Pinkham Ryder, 44; *The Bath* by Mary Cassatt; *The Good Judges* and *Maskers Quarreling Over a Dead Man* by James Ensor; *The Star of Bethlehem* by Edward Coley Burne-Jones; *Hail Mary (la Ora na Maria)* by Paul Gauguin, who settles in Tahiti where he will live until 1893; Henri de Toulouse-Lautrec produces his first music hall poster; Georges Seurat dies at Paris March 20 at age 31.

Sculpture Memorial to Mrs. Henry Adams by Augustus Saint-Gaudens for Washington's Rock Creek Park.

Theater *The Duchess of Padua* by Oscar Wilde 1/26 at New York's Broadway Theater; *Hedda Gabler* by Henrik Ibsen 2/26 at Oslo's Christiania Theater; *The American* by Henry James 7/26 at London's Opera-Comique, 70 perfs. (*see* 1877 novel).

Opera *Griseldis* 5/15 at the Comédie-Française, Paris, with music by Jules Massenet.

Stage musicals *Cinder-Ellen up-too-late* 12/24 at London's Gaiety Theatre with Lottie Collins in the last successful presentation of burlesque, the light musical entertainment that has been popular for half a century but that will be replaced by musical comedy. (U.S. forms of burlesque will make the term disreputable with nudity and vulgarity.) Music and lyrics by Canadian-American songwriter-minstrel show manager Henry J. Sayers, 38, songs that include "Ta-ra-ra-boom-der-ay: did you see my wife today. No I saw her yesterday: ta-ra-ra-boom-der-ay," 236 perfs.

Carnegie Hall opens May 5 in West 57th Street, New York, with a concert conducted in part by Petr Ilich Tchaikovsky, who has come from Russia at the persuasion of Oratorio Society director Walter Damrosch. Built with a $2 million gift from steel magnate Andrew Carnegie, the new hall designed by architect William Burnett Tuthill will be New York's preeminent concert hall.

The Chicago Symphony is founded under the direction of German-American conductor Theodore Thomas, who has been bringing orchestras to Chicago since 1862 (*see* Orchestra Hall, 1904).

First performances Concerto No. 1 in F sharp minor for Piano and Orchestra by Moscow Conservatory student Sergei Vassilievich Rachmaninoff, 18, who will revise the work drastically 25 years hence; "March of the Dwarfs" by Edvard Grieg; Orchestral Suite in A minor by Edward MacDowell in September at England's Worcester Festival.

Wilfred Baddely, 19, wins in men's singles at Wimbledon, Lottie Dod in women's singles; Oliver Campbell wins in U.S. men's singles, Mabel E. Cahill in women's singles.

A new Polo Grounds is erected for the New York Giants baseball club to replace a small 6,000-seat polo stadium first used by the Giants in 1883. By 1894 the Giants will set a league attendance record of 400,000 for the year (*see* 1958; Yankee Stadium, 1923).

John Joseph McGraw, 18, joins the Baltimore Orioles to begin a 41-year career in major league baseball. McGraw will be named manager of the New York Giants in July 1902 and continue as player-manager until 1906, when he will retire with a lifetime batting average of .334, having helped the Giants win two National League pennants and a World Series victory; he will be manager of the Giants until June 1932.

Walter Camp writes the first football rule book; he has invented the scrimmage line, the 11-man team,

1891 ***(cont.)*** signals, and the quarterback position (*see* 1889; 1898).

Basketball is invented at Springfield, Mass., by Canadian-American physical education director James Naismith, 30, who is taking a course at the YMCA Training School in Springfield and who has been assigned with his classmates the project of inventing a game that will occupy students between the football and baseball seasons. Naismith sets up fruit baskets atop ladders and establishes rules that will be used in the first publicly held game in March of next year and will never be substantially changed.

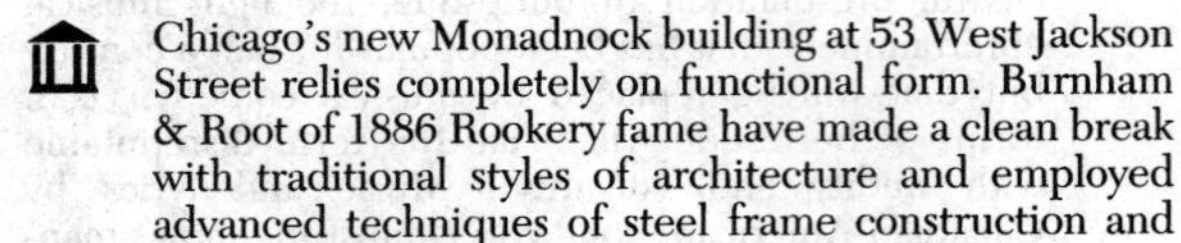

Chicago's new Monadnock building at 53 West Jackson Street relies completely on functional form. Burnham & Root of 1886 Rookery fame have made a clean break with traditional styles of architecture and employed advanced techniques of steel frame construction and utter simplicity of design.

Boston's Ames building goes up at the corner of Tremont and Washington Streets. The city's first skyscraper rises 13 stories above the street.

Ochre Court at Newport, R. I., is completed in late Gothic French château style for New York real estate magnate Ogden Goelet.

Denver's Brown Palace Hotel opens as prospectors crowd into Cripple Creek to make their fortunes in the goldfields (above). Carpenter Henry C. Brown arrived by oxcart at Cherry Creek in 1860 and by 1888 owned much of the miners' encampment that became Denver.

Florida's Tampa Bay Hotel goes up as railroad magnate Henry Bradley Plant of Atlanta's Southern Express Co. tries to make Tampa a popular winter resort in competition with the East Coast resorts of H. M. Flagler (*see* 1888). Now 71, Plant has organized the Savannah, Florida, and Western Railroad, and architect J. A. Wood has modeled the ornate $3 million hotel with its 13 minarets (and 225 rooms) on Granada's Alhambra Palace of 1354.

The Forest Reserves Act passed by Congress March 3 authorizes withdrawal of public lands for a national forest reserve as lumber king Frederick Weyerhaeuser expands his holdings (*see* 1870). Some 13 million acres will be set aside in President Harrison's administration, beginning a policy that will set aside more than 185 million acres of national forests in 40 states in the next 80 years (*see* 1905).

The New York Botanical Garden opens in the Bronx.

Tight money conditions bankrupt Kansas farmers; some 18,000 prairie schooners (covered wagons) cross the Mississippi headed back east (*see* 1889; 1890; 1894).

Russian crops fail, reducing millions to starvation; the rural peasantry raids towns in search of food.

President Harrison responds to an appeal by Cassius Marcellus Clay, 81, who was Abraham Lincoln's minister to Russia. Harrison orders U.S. flour to be shipped to the Russians (*see* 1892).

British butchers convicted of selling meat unfit for human consumption may be required on a second offense to put up signs on their shops stating their record. A Public Health Act passed by Parliament provides for the public exposure of wrongdoing.

American Sugar Refining is incorporated in New Jersey by H. O. Havemeyer whose 4-year-old Sugar Refineries Co. is dissolved by the New York courts. The new company begins taking over the entire U.S. sugar industry in one colossal trust (*see* 1892).

The "Del Monte" label that will become the leading label for U.S. canned fruits and vegetables is used for the first time. Oakland Preserving Co. executives at Oakland, Calif., have been inspired by Charles Crocker's 11-year-old Del Monte Hotel at Monterey and use its name to identify their premium quality fruit ("extras") (*see* 1865; 1895).

The first electric oven for commercial sale is introduced at St. Paul, Minn., by Carpenter Electric Heating Manufacturing Co.

Atlanta pharmacist Asa G. Candler, 40, acquires ownership of Coca-Cola for $2,300. He buys up stock in the firm but will achieve great success only when he changes Coca-Cola advertising to end claims such as "Wonderful Nerve and Brain Tonic and Remarkable Therapeutic Agent" and "Its beneficial effects upon diseases of the vocal chords are wonderful." Candler will make Coca-Cola syrup the basis of a popular 5¢ soft drink (*see* 1886; 1892).

New York City has more soda fountains than saloons. *Harper's Weekly* reports that marble chips from the construction of St. Patrick's Cathedral on Fifth Avenue are being used to make sulfuric acid for the production of 25 million gallons of soda water.

Congress votes March 3 to establish a U.S. Office of Superintendent of Immigration (*see* 1894).

Canada gets her first Ukrainian immigration influx (*see* Mennonites, 1874). Ukrainians will be Canada's fourth largest population group after the French, Irish, and English.

Overpopulation begins to cause worry in Japan.

"No human law can abolish natural and inherent rights of marriage," says Pope Leo XIII in *Rerum novarum* (above), "or limit in any way its chief and principal purpose. . .which is to increase and multiply."

1892 Egypt's khedive Mohammed Tewfik dies January 7 at age 39 after a 12-year reign in which he has been forced to let the British establish a virtual protectorate and let the Mahdi take the Sudan. Tewfik is succeeded by his headstrong son of 17 who will reign until 1914 as Abbas II and retake the Sudan.

French forces depose the king of Dahomey but encounter resistance in the form of native uprisings against imperialism.

French forces under Col. Louis Archinard defeat the Fulani on the Upper Niger and take Segu.

British explorer Sir Harry Hamilton Johnston, 34, subdues Angoni and Arab uprisings in Nyasaland.

Britain's second Salisbury ministry falls in the general election August 13 after 6 years in power; a fourth Gladstone cabinet takes office August 18.

President Harrison loses his bid for reelection as the Democrats campaign on a platform opposing the McKinley Tariff Act of 1890 and reelect Grover Cleveland with 277 electoral votes to Harrison's 145. Cleveland wins 46 percent of the popular vote, Harrison 43 percent, the Populist candidate 22 percent.

Homestead, Pa., steel workers strike the Carnegie-Phipps mill in June and are refused a union contract by managing head Henry Clay Frick who calls in Pinkerton guards to suppress the strike. Men are shot on both sides, Frick himself is shot and stabbed by Polish-American anarchist Alexander Berkman, 22, but recovers, union organizers are dismissed, and the men go back to working their 12-hour shifts November 20 after nearly 5 months of work stoppage. Andrew Carnegie's income for the year is $4 million, down only $300,000 from 1891 (*see* 1888; Carnegie Steel, 1899).

"American workmen are subjected to peril of life and limb as great as a soldier in time of war," says President Harrison.

Australia has a bitter strike that closes down ports, mines, and sheep-shearing stations and is ended only by military intervention. The 7-year-old Broken Hill Proprietary Co. and other mining companies will not make peace with the unions until 1923.

Arab slaveholders in the Belgian Congo rebel in mid-May; Belgian forces under Baron François Dhanis, 31, defeat them November 22. Forced labor is introduced December 5 in the guise of taxes to be paid in labor.

A presidential proclamation April 19 opens some 3 million acres of former Arapaho and Cheyenne lands in Oklahoma to settlement (*see* 1891; 1893).

Economic depression begins in the United States, but the country has 4,000 millionaires, up from fewer than 20 in 1840.

Shell Oil has its beginnings as English entrepreneur Marcus Samuel, 57, sends his first tanker through the Suez Canal with oil for Singapore, Bangkok, and other destinations to break the Standard Oil monopoly in the Far East. Samuel and his brother control the trading company M. Samuel & Co. begun in 1830 by their late father, Marcus's namesake, who chose the company's shell emblem because he had earlier been in the seashell business (*see* Royal Dutch-Shell, 1907).

The Ohio Supreme Court outlaws John D. Rockefeller's Standard Oil Trust under the 1890 Sherman Act but control of the trust properties are kept in Rockefeller hands by Standard Oil of New Jersey, a new company incorporated under a recent New Jersey holding-company law that permits companies chartered at Trenton to hold stock in other corporations (*see* 1884; Lake Superior Consolidated Iron Mines, 1893; Supreme Court decision, 1911).

An improved carburetor invented by Gottlieb Daimler mixes vaporized fuel with air to create a combustible or explosive gas (*see* 1890; Lenoir, 1860; Maybach, 1893).

N.V. Philips Gloeilampenfabrieken has its beginnings at Eindhoven, the Netherlands, where mechanical engineer Gerard Leonard Frederik Philips, 33, begins production of incandescent electric lamps. Gerard's youngest brother Anton Frederik, now 18, will join the firm in 1895, obtain a contract to supply the Russian court with 50,000 lightbulbs per year, and by the turn of the century Philips will be Europe's third largest lamp producer.

General Electric is created through a merger engineered by New York financier J. P. Morgan who combines Henry Villard's Edison General Electric with Charles A. Coffin's Thomson-Houston (*see* 1883; Langmuir, 1912).

The law of hysteresis discovered by German-American engineer Charles Proteus Steinmetz, 27, will improve the efficiency of electric motors, generators, and transformers. Steinmetz fled Berlin 4 years ago to escape prosecution in connection with a socialist newspaper he was editing and came to the United States a year later. A hunchbacked wizard, he changed his name from Karl August Rudolf to Charles Proteus, using the nickname Proteus given him as a student at the University of Breslau.

The Steinmetz discovery (above) permits engineers to forecast accurately how much electric power will be lost as a result of residual magnetism in electromagnets and will result in generators and motors designed to minimize such loss (*see* 1893).

A gasoline buggy produced at Springfield, Mass., by Charles and Franklin Duryea may be the first U.S. motorcar. It has a four-cycle water-cooled engine and a rubber and leather transmission (*see* 1891; Chicago-Milwaukee race, 1895).

French auto makers René Panhard and E. C. Levassor produce the first motorcar to be equipped with pneumatic tires (*see* 1889).

South Africa's first trains from Cape Town reach Johannesburg in September and begin to generate income that will be important to the Cape Colony's finances.

Chicago's first elevated railway goes into operation to begin the "Loop" that will circle the city's downtown area. The "Alley L" line on the South Side will expand to serve much of the city.

A new method for producing viscose rayon patented by English chemists Charles Frederick Cross, 37, and Edward John Bevan is safer than the Chardonnet nitrocellulose process of 1883 and cheaper than the cuprammonium process. Cross and Bevan dissolve cellulose in a mixture of carbon disulfide and sodium xanthate, and they squirt the viscous solution through fine holes to produce spinnable fibers (*see* Little, 1902; Viscose Co., 1910).

Union Carbide has its beginnings at Spray, N.C., where local entrepreneurs Thomas Leopold Willson and James T. Morehead accidentally produce calcium carbide while trying to make aluminum in an electric fur-

1892 ***(cont.)*** nace by fusing lime and coal tar. Molten slag from their operation is dumped into a nearby stream, liberating a gas, and they discover that the gas is acetylene (carbide gas) which can be used in lighting. They will establish the first commercial carbide factory at Spray in 1894 and found National Carbide Sales to market acetylene (soon found to be effective for cutting metal) (*see* 1911; Claude, ferrochrome, 1897).

Russian botanist Dmitri Iosifovich Ivanovski, 28, discovers filterable viruses, pioneering the science of virology (*see* Beljerinck, 1895).

Cholera arrives in the United States August 30 with steerage passengers from the Hamburg-Amerika line ship S.S. *Moravia*.

The American School of Osteopathy is founded at Kirksville, Mo., by Andrew T. Still (*see* 1875). His practice eschews use of drugs, it will spread quickly, and most states will eventually accord it legal recognition equivalent to that of orthodox medical practice.

A "pledge of allegiance" for U.S. schoolchildren to recite October 12 in commemoration of the discovery of America 400 years ago says simply, "I pledge allegiance to my flag, and to the Republic for which it stands, one nation, indivisible, with liberty and justice for all." Former clergyman Francis Bellamy, 36, who writes for *The Youth's Companion* has composed the pledge, generations of schoolchildren will mumble "for Richard Sands" instead of "for which it stands," and "my flag" will be changed to "the flag of the United States of America" (*see* 1954).

Telephone service between New York and Chicago begins October 18.

The *Toronto Star* begins publication.

The Ridder newspaper chain that will grow to include 19 daily newspapers, most of them in the West and Midwest, has its beginnings as German-American publisher Herman Ridder acquires the New York German-language newspaper *Staats-Zeitung und Herolt (see* Knight, 1903).

Baltimore's *Afro-American* begins publication on a semiweekly basis with John Murphy as editor.

The Addressograph invented by Sioux City, Iowa, engineer Joseph Smith Duncan prints mailing addresses automatically. Duncan will obtain a patent for his "addressing machine" in 1896, his first model employs a revolving hexagonal block of wood to which he has glued rubber type torn from rubber stamps, a new name and address advances to the printing point each time the block is turned, and the process of turning the block re-inks the type.

Fiction *The American Claimant* by Mark Twain; *The Children of the Ghetto* by English novelist-playwright Israel Zangwill, 28, whose Russian Jewish parents came to England as refugees in 1848; *The Soul of Lillith* by Marie Corelli; *The Intruder (L'Innocente)* by Gabriele D'Annunzio; *The Adventures of Sherlock Holmes* by A. Conan Doyle.

Poetry *Barrack-Room Ballads* by Rudyard Kipling includes "Gunga Din," "If," and "The Road to Mandalay"; *The Song of the Sword* by English poet William Ernest Henley, 43, includes his poem "England, My England" and his 1891 poem "Non Sum Qualis Eram Bonae Sub Regno Cynarae" with the line "I have been faithful to thee, Cynara! in my fashion." Henley lost a foot to tuberculosis at age 12, has gathered about him a group of young writers including Kipling (above) whose imperialist sentiments jibe with his own, and collaborates with Robert Louis Stevenson in three plays.

Painting *The Spirit of the Dead Watching* and *Aha cefeii (Why, you are jealous)* by Paul Gauguin; *At the Moulin Rouge* and *Au Lit, Le Baiser* by Henri de Toulouse-Lautrec; *Woman Sweeping* by French painter Jean Edouard Vuillard, 23.

Sculpture *Diana* by Augustus Saint-Gaudens for New York's Madison Square Garden.

Theater *Lady Windermere's Fan* by Oscar Wilde 2/22 at the St. James Theatre, London. Wilde calls a cynic "a man who knows the price of everything and the value of nothing"; *Walker, London* by James M. Barrie 2/25 at Toole's Theatre, London; *Monsieur Goes Hunting!* (*Monsieur chasse*) by Georges Feydeau 4/23 at the Théâtre du Palais Royal, Paris; *Champignol in Spite of Himself* (*Champignol malgré lui*) by Feydeau and a collaborator 11/5 at the Théâtre des Nouveautés, Paris; *Widowers' Houses* by George Bernard Shaw 12/9 at London's Royalty Theatre; *Charley's Aunt* by English actor-playwright Brandon Thomas, 36, 12/21 at London's Royalty Theatre, 1,466 perfs.

Opera *Pagliacci* 5/21 at Milan's Teatro dal Verme, with music by Italian composer Ruggiero Leoncavallo, 34, who delights audiences with his aria "Vesti la Giubba"; *Mlada* 10/20 at St. Petersburg's Maryinsky Theater, with music by Nikolai Rimski-Korsakov that includes "Night on Mount Triglov"; *Iolanta* 12/18 at St. Petersburg's Maryinsky Theater, with music by Petr Ilich Tchaikovsky.

Ballet *The Nutcracker Suite* (*Casse-Noisette*) 12/17 at the Maryinsky Theater, with music by Tchaikovsky.

First performances *Carneval* Overture by Antonin Dvořák 4/28 at Prague on the eve of the composer's departure for New York; Prelude in C sharp minor by Sergei Rachmaninoff 9/24 at Moscow; Symphony No. 8 in C minor by Anton Bruckner 12/18 at Vienna.

Popular songs "Daisy Bell (A Bicycle Built for Two)" by English songwriter Harry Dacre (Frank Dean) who is visiting America; "The Bowery" by English-American composer Percy Gaunt, 40, lyrics by Charles H. Hoyt.

Wilfred Baddely wins in men's singles at Wimbledon, Lottie Dod in women's singles; Oliver Campbell wins in U.S. men's singles, Mabel Cahill in women's singles.

James John "Gentleman Jim" Corbett, 26, scores a knockout in the 21st round September 7 at New Orleans to take the heavyweight title from the "Boston Strong Boy" John L. Sullivan—the "Great John L."—in the first title-match prizefight to be fought with padded gloves and according to the 25-year-old marquis of Queensberry rules.

Chicago's 16-year-old A. G. Spalding acquires Wright & Ditson, begun by cricketer-baseball player George Wright, and becomes a major factor in tennis balls and rackets. Spalding also begins making golf clubs and balls that will so improve the game that golf will gain new popularity not only in the United States but also in Britain.

Chicago salesman William Wrigley, Jr., 30, starts selling chewing gum, which is now being made by at least 12 companies (*see* Adams, 1872). Wrigley started as a soap salesman, offering baking powder as a premium to his wholesale soap customers; he then switched to selling baking powder, offering two packages of chewing gum with each can of powder, and is so encouraged by the response that he turns all his efforts to selling gum (*see* 1893).

The $1 Ingersoll pocket watch is introduced by U.S. mail-order and chain store entrepreneur Robert Hawley Ingersoll, 33. "The watch that made the dollar famous" will be a huge success, but Ingersoll's company will be insolvent by 1921 and he will sell its assets to Waterbury Clock of Connecticut in 1922 (*see* Timex, 1946).

Book matches, patented by U.S. inventor Joshua Pusey, contain poisonous white phosphorus that will be replaced by safer chemicals in 1911 (*see* Lundstrom, 1853).

Player's Navy Cut cigarettes are introduced at London.

Wills' Three Castle cigarettes are packed in the first push-up cardboard cigarette packets (*see* 1878; Imperial Tobacco, 1901).

The Dalton gang from Oklahoma Territory arrives at Coffeyville, Kan., the morning of October 5 intending to rob two banks, but the aroused townspeople meet the gang with heavy gunfire. Former train robbers Robert Dalton, 25, and his brother Grattan fall dead along with two others.

The cornerstone of New York's St. John the Divine is laid December 27 on an 11.5-acre site acquired last year on Morningside Heights by the Episcopal diocese of New York for $850,000. Planned as the world's largest cathedral, it has been designed by George Lewis Heins and Christopher Grant La Farge, who will be succeeded in 1911 by Ralph Adams Cram and Frank E. Ferguson.

Brooklyn's Grand Army Plaza is graced with a Soldiers' and Sailors' Memorial Arch dedicated to the Civil War dead. The arch serves as an entrance to Prospect Park.

Marble House is completed at Newport, R.I., by architect Richard Morris Hunt for railroad heir William K. Vanderbilt, now 42. The summer "cottage" on Bellevue Avenue has lines derived from those of the Grand Trianon and Petit Trianon at Versailles and will be augmented in 1913 by a lacquered Chinese teahouse on Cliff Walk.

The Reno Inclined Elevator patented by New York inventor Jesse W. Reno March 15 is the world's first escalator. It will be installed in the fall of 1896 at Coney Island's Old Iron Pier, but the flat-step moving staircase patented by U.S. inventor Charles A. Wheeler August 2 is the first practical escalator. It will never be built; inventor Charles D. Seeberger will buy Wheeler's patent, incorporate its basic feature in his own improved design, and have Otis Elevator produce it; the Seeberger Escalator will be used at the Paris Exposition of 1900 and at Gimbels department store at Philadelphia in 1901.

The Sierra Club is founded by John Muir and others to protect America's natural environment (*see* 1876; Muir Woods, 1908).

The New York legislature creates Adirondack Park, setting aside the nation's largest forest reserve: "The Forest Preserve shall be forever kept as wild forest lands."

New York City gets new torrents of clean drinking water as the $24 million New Croton Aqueduct is completed after 7 years of construction (*see* 1842; 1848; Catskill system, 1917).

The Populist party polls more than a million votes in the U.S. presidential election (above) as farmers register their protest against the railroads and against farm machine makers.

The first successful U.S. gasoline tractor is produced by Waterloo, Iowa, farmer John Froelich who will organize Waterloo Gasoline Traction Engine Co. early next year. John Deere Plow will acquire it (*see* 1902).

Cheap grain from America and Russia depresses French farm prices. Only one French farm holding in 15 has a horse-drawn cultivator, one in 150 has a mechanical reaper, and French farmers demand higher import duties to protect them from more efficient overseas competitors.

Canadian druggist William Saunders, 56, head of the Dominion Experimental Farms, crosses Red Fife wheat with an early ripening variety obtained from India and produces the hardy Markham wheat strain on a farm in British Columbia (*see* Mackay, 1886).

Famine cripples Russia, but by late January some 3 million barrels of U.S. flour are en route to relieve the starvation that is killing millions (*see* 1891).

Danish veterinarian Bernhard Laurits Frederik Bang, 44, develops a method for eradicating tuberculosis from dairy herds (*see* Bang's disease, 1896).

The first U.S. cattle tuberculosis test is made March 3 on a herd at Villa Nova, Pa., with tuberculin brought from Europe by the dean of the University of Pennsylvania veterinary department.

R.H. Macy partner Nathan Straus launches a campaign for pasteurized milk; he will establish milk stations in New York and other large cities.

Congress is petitioned to hold pure food law hearings (*see* 1889; Wiley, 1902).

The *Ladies' Home Journal* announces that it will accept no more patent medicine advertisements. The

1892 ***(cont.)*** *Journal* has been edited since 1889 by Dutch-American editor Edward William Bok, 29 (*see* 1883; 1904).

Asa Candler organizes the Coca-Cola Co. at Atlanta (*see* 1891; 1895).

Baltimore machine-shop foreman William Painter, 54, patents a clamp-on tin-plated steel bottle cap with an inner seal disk of natural cork and a flanged edge; he also patents a capping machine for beer and soft drink bottles. The new bottle cap will replace the unsanitary Miller Plunger patented in 1874 and the 1878 Hutchinson Bottle Stopper, both widely used. Painter will design an automatic filler and capper with a capacity of 60 to 100 bottles per minute and establish Crown Cork and Seal at Baltimore to market his bottle caps and machines (*see* Gillette, 1895).

The first Hawaiian pineapple cannery opens (*see* 1790; Dole, 1902).

American Sugar Refining controls 98 percent of the U.S. sugar industry; H. O. Havemeyer's Sugar Trust uses political influence to suppress foreign competition with tariff walls and the company saves itself millions of dollars per year in duties by having its raw sugar imports shortweighted (*see* 1891; 1895; Spreckels, 1897).

New York Condensed Milk Co. adds evaporated milk to its Borden product line (*see* 1885; Meyenberg, 1885).

Liederkranz cheese, invented by Monroe, N.Y., delicatessen man Emile Frey, 23, is a milder version of Limburger. Frey's employer gives the new cheese its name after winning approval of its taste from his choral society, which calls itself Liederkranz and has its own Liederkranz Hall in New York's East 58th Street.

Italy raises to 12 the minimum age for marriage for girls.

The *padrone* system that prevails in Italian immigration to the United States encounters resistance from Roman Catholic missionaries who open the Church of Our Lady of Pompeii in a storefront at 113 Waverly Place, New York, to help immigrants whose passage has been paid by *padrones* to whom the immigrants are bound almost as indentured servants. The clergymen will open a church at 25 Carmine Street in 1925.

New York's immigrant receiving station moves to Ellis Island in the Upper Bay of New York Harbor after the receiving station at Castle Garden established in 1855 has received nearly 7.7 million immigrants. Castle Garden will become an aquarium in 1896; 27.5-acre Ellis island will process immigrants until 1932.

The Geary Chinese Exclusion Act passed by Congress May 5 extends for another 10 years all existing Chinese exclusion laws and requires all Chinese residing in the United States to register within a year or face deportation (*see* 1902).

1893 Hawaiian annexationists overthrow Queen Liliukoalani with support from U.S. minister John Leavitt Stevens. Armed marines from the *U.S.S. Boston* are landed January 16 to "protect" U.S. interests, the queen abdicates under duress January 17 after reigning less than 2 years, and the annexationists block U.S. efforts to restore the monarchy (*see* 1894).

Laos becomes a French protectorate as France begins to develop a new colonial empire in Indochina. Col. Simon Gallieni, 44, and Louis Hubert Gonzalve Lyautey, 39, move to pacify a region as large as France herself with 16 million inhabitants whose country will be securely in French control by 1895.

France establishes French Guiana in South America and the Ivory Coast in Africa as formal colonies.

French colonial forces on the Niger River defeat Tuareg warriors.

Transvaal annexes Swaziland.

Natal gains self-government.

The king of the Matabele Lobengula leads a revolt against Cecil Rhodes's British South Africa Co., but Leander Starr Jameson cuts the Matabele down with machine-gun fire October 23, suppresses the revolt, forces Lobengula to give up his capital Bulawayo, and drives the Matabele king into exile where he will die next year (*see* 1891; 1895).

Kelly's Industrial Army marches on Washington, D.C., 1,500 strong to demand relief which is not forthcoming from Congress. Led by "General" Charles T. Kelly, the army of unemployed workers from California arrives for the most part via boxcar.

The Pullman Palace Car Co. reduces wages by one-fourth, obliging workers to labor for almost nothing while charging them full rents in company housing at Pullman, Ill., and charging inflated prices at company food stores (*see* 1881; 1894).

The American Railway Union is founded by socialist Eugene Victor Debs, 48, and others. Debs has been national secretary of the Brotherhood of Locomotive Firemen (*see* 1894).

The newly elected governor of Illinois frees three of the men convicted of conspiracy in Chicago's 1886 Haymarket riot. German-American politician John Peter Altgeld, 45, has acted at the urging of Chicago lawyer Clarence Seward Darrow, 35; his action produces a storm of criticism (*see* Haywood case, 1907).

New Zealand adopts suffrage for women, the first country to do so.

New York social worker Lillian D. Wald, 26, founds the Henry Street Settlement to help immigrants on the city's Lower East Side (*see* Chicago's Hull House, 1889).

Anti-Semitism mounts in France as Jews are blamed for the collapse in 1889 of the Panama Canal Co., whose bankruptcy has cost many French investors their savings. An international group of dubious characters controlled the company, many of its stockholders were Jewish, including Baron de Reinach and Cornelius Herz, and although Ferdinand de Lesseps and his associates are sentenced to prison in February, none of the leading figures will actually serve time (*see* Dreyfus, 1894).

Britain's Labour party is founded by socialists who include Scotsman James Keir Hardie, 37. A miner from age 10 to 22, Keir Hardie organized a labor union among his fellow miners, headed a new Scottish Labour party in 1888, and was elected last year to Parliament.

English naturalist Mary Kingsley, 30, explores West Africa. A niece of the late novelist-poet Charles Kingsley, she goes with Fang tribesmen down the Ogowe River through cannibal country, photographs wildlife, and finds beetles never before cataloged.

Norwegian Arctic explorer Fridtjof Nansen, 32, tries to drift across the polar basin in the icebound ship *Fram*. Nansen led the first expedition to cross Greenland in 1888, he will drift for 18 months in the *Fram*, and he will then travel with a companion by sled and skis across the ice to 86°14' north, the northernmost point yet reached by any explorer (*see* 1909).

"The Significance of the Frontier in American History" by University of Wisconsin historian Frederick Jackson Turner, 31, observes that the frontier has been the source of the individualism, self-reliance, inventiveness, and restless energy so characteristic of Americans but that it is now ending. He delivers his paper at a meeting of the American Historical Association held at Chicago during the Columbian Exposition that opens to show the world what progress Chicago has made since the fire of 1871.

A vast section of northern Oklahoma Territory opens to land-hungry settlers September 16. Thousands of "Boomers" jump at a signal from federal marshals and rush in by foot, on horseback, in buggies and wagons, and on bicycles to stake and claim quarter-section farms, but as in 1889 they find that "Sooners" have sneaked across the line earlier by cover of night and started building houses on the 165- by 58-mile Cherokee Strip—6 million acres purchased from the Cherokees in 1891 for $8.5 million.

Economic depression continues in America as European investors withdraw funds.

The Philadelphia and Reading Railroad goes into receivership February 20 with debts of $125 million; the Northern Pacific, Union Pacific, Erie, and Santa Fe soon follow suit. Foreclosures in this decade will affect 41,000 miles of U.S. railroad track, 15 percent of the total.

Wall Street stock prices take a sudden drop May 5, the market collapses June 27, 600 banks close their doors, more than 15,000 business firms fail, and 74 railroads go into receivership in a depression that will continue for 4 more years.

Congress repeals the 1890 Sherman Silver Purchase Act October 31, and the United States returns to the gold standard that will be retained until April 1933. Silver prices collapse, bankrupting silver barons such as Colorado's "Hod" Tabor (*see* 1880).

Lake Superior Consolidated Iron Mines is created by oil magnate John D. Rockefeller who has loaned the Merritt brothers $420,000 to develop the Mesabi iron mines of Minnesota and build a railroad to Duluth (*see* 1890). Rockefeller has called the loan on short notice, the Merritt brothers have been obliged to forfeit their properties, and Rockefeller's $29.4 million company leases the properties to Henry Clay Frick of the Carnegie-Phipps mill at Homestead, Pa. (*see* 1892; Carnegie, Oliver, 1896).

The Bon Marché in Paris has sales of 150 million gold francs, up from 20 million in 1870; it is the world's largest department store (*see* 1852).

Marshall Field and Co. occupies nearly an entire city block on Chicago's State Street and its wholesale store employs more than 3,000 on 13 acres of floor space (*see* 1881; Field Museum, above).

The name Sears, Roebuck & Co. is used for the first time as the Chicago mail-order firm racks up sales of $338,000 in baby carriages, clothing, furniture, musical instruments, sewing machines, and a wide range of other merchandise. By next year its catalog will have more than 500 pages (*see* 1889; Rosenwald, 1895).

Brooklyn's Abraham & Straus is founded as Joseph Wechsler of Wechsler & Abraham sells his interest to Macy's Isodor and Nathan Straus and their partner Charles B. Webster (*see* 1888). Abraham, 59, will bring his sons-in-law Simon F. Rothschild, 32, and Edward Charles Blum, 30, into the store to counter control by the Macy interests and by the turn of the century A&S will have entrances on four sides of the block it largely occupies (*see* Federated, 1929).

Charles Steinmetz announces a symbolic method of calculation that will bring the complex field of alternating current within the reach of the average practicing engineer. It will make the use of alternating current commercially feasible and be largely responsible for the quick introduction of apparatus using alternating current on a wide scale. General Electric (*see* 1892) has acquired Steinmetz's employer (*see* Westinghouse, Stanley, 1885; Tesla, 1888).

The Klondike oil well geysers up in a marshy field on East Toledo's Millard Avenue, beginning a rush to drill wells. The Ohio city will become a major oil refinery center.

Detroit machinist Henry Ford, 30, road-tests his first motorcar in April. An employee of the Edison Illuminating Co., Ford has been working on his "gasoline buggy" since last year (*see* 1896).

German engineer William Maybach, 46, develops a float-feed carburetor for gasoline engines. He is an associate of Gottlieb Daimler (*see* 1892).

Greece's 3.5-mile Corinth Ship Canal opens to link the Gulf of Corinth with the Gulf of Athens.

Congress makes air brakes mandatory on U.S. railroad trains (*see* Westinghouse, 1868).

The New York, New Haven and Hartford completes its Shore Line route between New York and Boston by

1893 ***(cont.)*** leasing the 500-plus miles of the Old Colony Railroad. The Old Colony controls the Boston and Providence, whose 44 miles of track are needed by the New Haven (*see* 1872; Mellen, 1903).

Chicago surgeon Daniel Hale Williams performs the world's first open-heart surgery, saving the life of a street fighter with a knife wound in an artery near his heart (*see* Williams, 1891).

The Johns Hopkins Medical School and Hospital are founded at the 17-year-old Baltimore university. Physician-in-chief is Canadian-American William Osler, 44, whose *Principles and Practices of Medicine* was published 2 years ago and who is the leading medical professor of his time. Surgery professor W. S. Halsted, pathologist W. H. Welch, and gynecological surgeon Howard Atwood Kelly, 35, will help Osler give the school a leadership unmatched in U.S. medicine (*see* Welch, 1878; Halsted, 1884; Flexner, 1910).

German physiologist Adolf Magnus-Levy, 28, devises the basal metabolism test that will be used for years to measure human metabolic rates.

A survey of Brooklyn, N.Y., schools reveals that 18 classes have 90 to 100 students each, while one classroom is jammed with 158 (*see* 1867).

The Field Museum is founded at Chicago with $1 million contributed by Marshall Field who will add a second $1 million and will leave the museum an additional $8 million in his will when he dies early in 1906.

U.S. Rural Free Delivery begins with five test postal routes in hilly West Virginia.

Rural U.S. telephone companies begin to proliferate as the Bell patents of 1876 expire.

McClure's magazine begins publication at 15¢ under the direction of Irish-American publisher S. S. (Samuel Sidney) McClure, 36, who created the first U.S. newspaper syndicate in 1884.

The American Newspaper Publishers Association (ANPA) founded in 1886 adopts a resolution agreeing to pay commission in the form of discounts to recognized independent advertising agencies and to give no discounts on space sold directly to advertisers. The resolution establishes the modern advertising agency system (*see* Batten, 1891; Curtis, 1901).

Joseph Pulitzer installs a four-color rotary press in hopes of reproducing great works of art in the Sunday supplement of his *New York World* (*see* 1883). Designed by New York's Hoe Co. of 1875 folding apparatus fame, the new press will be used to produce the first colored cartoon (see 1896).

Fiction *Maggie: A Girl of the Streets* by New York newspaper reporter Stephen Crane, 21, who has the naturalistic portrayal of slum life published at his own expense.

Poetry *Les Trophées* by Cuban-born French poet José Maria de Hérédia, 50.

Painting *The Cry* and *The Voice* by Norwegian Postimpressionist Edvard Munch, 29; *The Boating Party* by Mary Cassatt; *Rouen Cathedral* by Claude Monet; *Hina Maruru* by Paul Gauguin.

Henri de Toulouse-Lautrec produces a color lithograph poster for the Divan Japonais, a Paris café at 75 Rue des Martyrs whose waitresses wear kimonos. The Duran-Ruell Galleries uses it to advertise an exhibition of more than 300 *ukiyoe* prints by Hiroshige and Utamara. Toulouse-Lautrec produces another Japanese style-color lithograph to promote *Aristide Bruant dans son cabaret (see* 1863).

Theater *Magda (Heimat)* by Hermann Sudermann 1/7 at Berlin's Lessingtheater; *The Master Builder (Bygmester Solness)* by Henrik Ibsen 1/19 at Trondheim; *The Girl I Left Behind Me* by David Belasco and *New York Sun* critic Franklin Fyler 1/25 at New York's Empire Theater, 208 perfs.; *The Weavers (Die Weber)* by Gerhart Hauptmann 2/26 at Berlin's Neues Theater; *A Woman of No Importance* by Oscar Wilde 4/14 at London's Haymarket Theatre; *Pelléas et Mélisande* by Maurice Maeterlinck 5/17 at the Bouffe Parisienne, Paris (*see* 1898); *The Second Mrs. Tanqueray* by Arthur Wing Pinero 5/28 at the St. James's Theatre, London, with Mrs. Patrick Campbell; *The Triumph of Death (Il trionfo della morte)* by Gabriele D'Annunzio; *The Beavercoat (Der Biberpelz)* by Gerhart Hauptmann 9/21 at Berlin's Deutsches Theater; *Hannele's Trip to Heaven (Hanneles Himmelfahrt)* by Hauptmann 11/14 at Berlin's Königliches Schauspielhaus.

Opera *The Magic Opal* 1/19 at London's Lyric Theatre, with music by Spanish composer Isaac (Manuel Francisco) Albeniz, 32, who made his debut as a pianist at age 4, began running away from home at age 9, embarked for Puerto Rico at age 12 and wound up in Buenos Aires, was returned to Havana where his father had been stationed as a tax collector, persuaded his father to let him

La vie Parisienne found vivid expression in the lithographs of the painter Henri de Toulouse-Lautrec.

travel alone to New York, and played throughout the United States before returning to Spain; *Manon Lescaut* 2/1 at Turin's Teatro Reggio, with music by Italian composer Giacomo Puccini, 34, to a libretto with the same title as the Massenet opera of 1884; *Falstaff* 2/9 at Milan's Teatro alla Scala, with music by Giuseppe Verdi based on the character in the 1597–1598 Shakespeare play *Henry IV*. Verdi is 80 but his last opera shows undiminished powers of wit and composition; *Hansel und Gretel* 12/23 at Weimar, with music by German composer Engelbert Humperdinck, 39.

First performances *Hamlet and Ophelia* by Edward McDowell 1/28 at Boston; Sonata Tragica by MacDowell 3/18 at Boston's Chickering Hall; Symphony No. 6 in B minor *(Pathetique)* by Petr Ilich Tchaikovsky 10/28 at St. Petersburg 9 days before the composer dies of cholera at age 53; Symphony No. 9 *(From the New World)* by Antonin Dvořák 12/15 at New York's Carnegie Hall in a performance by the New York Philharmonic which has moved into the 2-year-old hall in West 57th Street where it will remain until 1962. Dvořák has written the symphony with snatches from black spirituals and American folk music as a greeting to his friends in Europe.

Stage musicals *A Trip to Chinatown* 8/7 at New York's Madison Square Theater, with book by Charles Hoyt, music and lyrics by Percy Gaunt, songs that include "The Bowery," "Reuben, Reuben," "Push the Clouds Away," and "After the Ball" by songwriter Chester K. Harris, 26, whose song will be the first to have sheet music sales of more than 5 million copies, 650 perfs.; *A Gaiety Girl* 10/14 at London's Prince of Wales' Theatre, with music by Sidney Jones, lyrics by Harry Greenbank, 413 perfs. (the first musical comedy to be billed as such).

Florenz Ziegfeld, 25, begins a 39-year show business career by engaging orchestras and musical attractions for the Columbian Exposition (above). A native of Chicago, he profits handsomely by exhibiting the German strong man and physical culture exponent Eugene Sandow, 26.

Popular songs "Happy Birthday to You" ("Good Morning to All") by Louisville, Ky., private kindergarten teacher Mildred Hill, 34, lyrics by her sister Patty Smith Hill, 25; "See, Saw, Margery Daw" by Arthur West; "Two Little Girls in Blue" by Charles Graham.

Hymn "When the Roll Is Called Up Yonder" by James M. Black.

The Stanley Cup ice hockey trophy has its beginnings in a silver cup presented to the winner of an amateur Canadian hockey match by the son of the earl of Denby, governor general of Canada. Frederick Arthur, Lord Stanley of Preston, has purchased the cup for £10 ($48.67), it will be replaced by a $14,000 silver bowl, and it will be the North American professional hockey trophy beginning in 1910 (*see* 1875; National Hockey League, 1917).

Joshua Pim, 24, (Ireland) wins in men's singles at Wimbledon, Lottie Dod in women's singles; Robert D. Wrenn, 19, wins in U.S. men's singles, Aline M. Terry in women's singles.

The Chicago Golf Club opens at Wheaton, Ill., where Charles B. MacDonald has laid out the first 18-hole golf course in America.

The distance from the pitcher's mound to home plate in baseball is fixed at 60 feet, 6 inches.

The U.S. racing yacht *Vigilant* defeats England's *Valkyrie* 3 races to 0 to defend the America's Cup.

Paris students witness the world's first striptease February 9 at the Bal des Quatre Arts. Gendarmes seize the artist's model who has disrobed for the art students, a court fines her 100 francs, the ruling provokes a riot in the Latin Quarter, students besiege the Prefecture of Police, and military intervention is required to restore order.

The world's first Ferris wheel goes up at the Chicago fair (above). Designed by U.S. engineer Washington Gale Ferris, 34, the Midway amusement ride is a giant $300,000 vertical power-driven steel wheel, 250 feet in diameter, with 36 passenger cars, each seating 40 people, balanced at its rim (*see* 1897).

The "Clasp Locker or Unlocker for Shoes" exhibited at the Chicago fair (above) is the world's first slide fastener. U.S. inventor Whitcomb L. Judson has patented the device and a machine to manufacture it (*see* Sundback, 1913).

Fels-Naphtha soap is introduced by Philadelphia soap maker Joseph Fels, 39, whose 17-year-old Fels & Co. has just purchased a factory in which to apply the naphtha process to making laundry soap.

Japanese entrepreneur Kokichi Mikimoto, 35, pioneers cultured pearls. He has learned from a Tokyo University professor that if a foreign object enters a pearl oyster's shell and is not expelled the oyster will use it as the core of a pearl. Five years ago he established the first pearl farm in the Shinmei inlet and when he pulls up a bamboo basket for routine inspection July 11 he finds a semispherical pearl, the world's first cultured pearl. The natural Oriental pearl business will fade in the next few decades, ruining the economies of some small countries such as Kuwait on the Persian Gulf (*see* oil 1937).

Juicy Fruit chewing gum is introduced by William Wrigley, Jr., who has been selling Lotta Gum and Vassar (*see* 1892). Wrigley's Spearmint chewing gum is introduced in the fall and will be the nation's leading brand by 1910 *(see* Wrigley Field, 1914).

Lizzie Borden makes headlines in June when she goes on trial at Fall River, Mass. Spinster Lizzie Andrew Borden, 32, is charged with having killed her stepmother and then her father on the morning of August 4 last year, the sensational trial boosts newspaper circulation figures to new heights, the jury finally rules to acquit, but street urchins chant, "Lizzie Borden took an axe/And gave her mother forty whacks./And when she saw what she had done/She gave her father forty-one."

1893 ***(cont.)*** Salt Lake City's Mormon Temple is completed on the site ordained for its construction by Brigham Young in 1847. Only "worthy Mormons" who adhere strictly to church rules may enter.

London's Piccadilly Circus is graced with a new fountain erected as a memorial to the late Lord Shaftesbury, whose puritanical attitudes are belied by the fountain's sculptured Eros.

New York's Waldorf Hotel opens March 14 on Fifth Avenue at 33rd Street where the residence of William B. Astor has been torn down to make room for the 13-story hotel designed by architect Henry Janeway Hardenbergh with 530 rooms and 350 private baths (*see* Astor House, 1836; Waldorf-Astoria, 1897).

Chicago's Congress Hotel opens on Michigan Avenue. It is an annex to the Auditorium Hotel built in connection with the Auditorium Theater Building completed by Louis Sullivan in 1890.

Quebec City's Château Frontenac Hotel opens December 18 after 19 months of construction. Modeled on a Loire Valley château by New York architect Bruce Price and built by a group of Canadians headed by Canadian Pacific Railway president William Van Horne, the 170-bedroom horseshoe-shaped hotel has an exterior of Glenboig bricks imported from Scotland. A tower will be added in 1924.

America's buffalo herd falls to 1,090 as market hunters continue to exterminate the animals (*see* 1900).

Quaker's dietetic pastry flour wins a "highest" award at the Chicago fair (above). It is an advance over self-rising flour.

Aunt Jemima pancake mix is promoted at the Chicago fair (above) by St. Joseph, Mo., miller R. T. Davis who has acquired Chris Rutt's mix of 1889 and improved it by adding rice flour, corn sugar, and powdered milk so that it can be prepared by adding only water. Davis sets up a 24-foot high flour barrel, arranges displays inside the barrel, and engages former Kentucky slave Nancy Green, 59, to demonstrate the Aunt Jemima mix at a griddle outside the barrel (*see* 1910).

C. W. (Charles William) Post, 38, develops Postum to replace coffee with a nutritious beverage of wheat, molasses, and wheat bran. Post is an ulcer patient at the Western Health Reform Institute operated since 1876 by John Harvey Kellogg (*see* 1895).

Thomas Lipton registers a new trademark for the tea he has been selling since 1890 and which is sold only in packages, never in bulk. Over the facsimile signature "Thomas J. Lipton, Tea Planter, Ceylon," Lipton prints the words "Nongenuine without this signature" (*see* 1909).

Caramel maker Milton Snavely Hershey, 36, visits the Chicago fair (above) and sees chocolate-making machinery exhibited by a Dresden firm. He has it shipped to his Lancaster, Pa., plant and begins experiments with it (*see* 1894).

1894

Britain's fourth Gladstone ministry ends March 5 after Gladstone has shattered the Liberal party with his fight for Irish home rule. The Liberals retain power with Archibald Philip Primrose, 46, earl of Rosebery, as prime minister.

Hungary's Lajos Kossuth dies at Turin March 20 at age 91; his son Ferenc, 53, returns from exile to lead the Independence party in the Parliament at Budapest.

French president Marie François Sadi Carnot dies at age 57 after being stabbed at Lyons June 24 by Italian anarchist assassin Santo Caserio. He is succeeded by Jean Casimir-Perier who will serve only 6 months.

Russia's Aleksandr III dies of nephritis at his Lavadia Palace in the Crimea November 1 at age 49 after a 13-year reign. His son, 24, will reign until 1917 as Nicholas II, the last Romanov monarch (he marries Princess Alix of Hesse-Darmstad, 22, at St. Petersburg November 26).

The Congo Treaty signed May 12 with the king of the Belgians Leopold II gives Britain a lease on a wide corridor between Lake Tanganyika and Lake Albert Edward in return for a lifetime lease to Leopold of vast territories west of the Upper Nile and north of the Congo-Nile watershed. German protests force the British to abandon their corridor and give up the possibility of a Cape Town-to-Cairo railroad, and French threats force Leopold to make an agreement August 14 giving up his claim to the northern part of his lease.

Dahomey becomes a French colony June 22 following a March 15 Franco-German agreement on the boundary between the French Congo and the Cameroons that have been a German protectorate since 1885. French imperialists have hopes of taking over the southern part of the former Egyptian Sudan and forcing the British to evacuate Egypt by threatening to divert the course of the Nile.

The Republic of Hawaii is proclaimed July 4 with Judge Sanford Ballard Dole, 50, as president. The republic gains U.S. recognition August 7, and the royalist revolt that begins December 8 is quickly suppressed (*see* 1893; 1898).

Japanese forces sink the British ship *Kowshing* carrying Chinese troops to Korea July 25, the Korean regent declares war on China July 27, China and Japan declare war on each other August 1, and the Japanese win easy victories in the ensuing months. Berlin and Washington reject a British invitation to Germany, France, Russia, and the United States to join in a united move to intervene (*see* 1895).

The Dutch East Indies rebel against the government of Queen Wilhelmina, now 14, but the revolt is suppressed. Another uprising in 1896 will be put down only with great difficulty.

Strikes cripple U.S. railroads as economic depression continues. President Cleveland takes the position that the government has authority only to keep order, thus effectively supporting the railroad operators and their strikebreakers (*see* Roosevelt, 1902).

Some 750,000 U.S. workers strike during the year for higher wages and shorter hours.

Coxey's Army arrives at Washington, D.C., April 30 after a 36-day march of unemployed workers from Massillon, Ohio, led by sandstone quarry operator Jacob Sechler Coxey, 40. The 400 marchers demand that public works be started to provide employment and that $50 million in paper money be issued, but Coxey is arrested for walking on the grass and forced to leave the Capitol grounds (*see* 1932).

Pullman Palace Car workers strike May 1 to protest wage cuts, and a general strike of western railroads begins June 26 as Eugene V. Debs orders his railway workers to boycott Pullman (*see* 1893). U.S. troops enter Chicago in July to enforce federal laws in the Pullman strike. Illinois governor John Peter Altgeld protests that President Cleveland has violated states' rights in sending in the troops, he claims that the press has exaggerated the size of Chicago's disorders, the troops are withdrawn, and the strike ends 2 weeks later. A federal grand jury indicts Debs for interfering with the mails and with interstate commerce.

Congress votes June 28 to make Labor Day a legal holiday, setting aside the Monday after the first Sunday in September to honor the contribution of labor.

A French court martial convicts Army captain Alfred Dreyfus, 35, of having passed military information to German agents. Dreyfus will later be proved innnocent, but the Dreyfus case adds to the growing anti-Semitism in France (*see* 1893; 1897).

Viennese journalist Theodore Herzl, 34, covers the Dreyfus trial (above) and hears the Paris mob cry, "Death to the Jews!" He lays the foundations of political Zionism with his book *The Jewish State: An Attempt at a Modern Solution of the Jewish Question (see* first Zionist congress, 1897).

The Wilson-Gorman Tariff Act that becomes law August 27 without President Cleveland's signature reduces U.S. tariff duties by roughly 20 percent, but lobbyists keep most of the 1890 McKinley Act's protective features intact (*see* Dingley Act, 1897).

The Wilson-Gorman Act (above) includes an income tax on incomes above $4,000 per year (*see* 1881). New York lawyer Joseph H. Choate, 62, calls the 2 percent tax "communistic, socialistic" (*see* 1895).

England's Marks & Spencer department store chain has its beginnings in the Penny Bazaar ("Don't ask the price, it's a penny") opened at Cheetham Hill, Manchester, by Polish-born merchant Michael Marks, 31, in partnership with Thomas Spencer, 42, whose employer loaned Marks £5 in 1884 when the Pole arrived at Leeds penniless, illiterate, and unable to speak English. Marks started as an itinerant peddler, opened a stall in Kirkgate Market, Leeds, has since opened stalls in six other Lancashire street markets, and has innovated the practice of arranging wares according to price. By 1900 there will be a dozen Penny Bazaar stores (three of them in London), and by 1915 there will be 140 selling haberdashery, earthenware, hardware, household goods, stationery, and toys at the fixed price of a penny per item. Marks & Spencer will open their first store under that name at Darlington in 1922, and "Marks & Sparks" will grow to become Britain's leading retail enterprise.

Harrods at London inaugurates 7 o'clock closing hours (4 o'clock Thursdays) as its new manager Richard Burlidge, 47, ends the penny-halfpenny fines for late employees that was instituted in 1883, installs display windows, and arrives himself by 7 each morning to make "Harrods Serves the World" more than a slogan for the Knightsbridge store whose cable address is "Everything." Harrods begins to acquire land for a large new store that will open in 1939.

Gimbel Brothers of Milwaukee opens a Philadelphia store that will be the city's largest retail establishment. Founding father Adam Gimbel dies at age 78, and Isaac Gimbel takes over the firm's presidency (*see* 1887; New York Gimbels, 1910).

Oil is discovered at Corsicana, Tex., as a well being bored for water suddenly begins spouting oil. In 4 years nearly every Corsicanian will have an oil well in his backyard and the field will be producing 500,000 barrels per year (*see* Spindletop, 1901).

The Southern Railway System organized by New York financier and traction magnate Thomas Fortune Ryan links the Richmond and Danville and the East Tennessee, Virginia, and Georgia systems into a company that will soon embrace the South Carolina and Georgia, Cincinnati Southern, Alabama Great Southern, New Orleans and Northeastern, Georgia Southern and Florida, and Northern Alabama to create a system that will operate 8,000 miles of road extending from the Potomac and Ohio Rivers to the Gulf and from the Atlantic to the Mississippi (*see* 1886).

The *Sunset Limited* begins service on the Southern Pacific between New Orleans and San Francisco. The once-a-week express averages 33 miles per hour to make the run in 75 hours and will run daily beginning in 1913 (*see Sunset* magazine, 1898).

Union Station opens at St. Louis. Its design has been adapted from that of the medieval French walled city of Carcasonne.

Boston's North Station opens for the Boston & Maine Railroad (*see* South Station, 1899).

The first railroad opens across the South American Andes.

Cable cars in U.S. cities haul 400 million passengers, but trolley lines will rapidly supplant the cable cars (*see* Hallidie, 1873; Richmond, 1888). The complex cable car systems are vulnerable to breakdown, they require large inputs of energy simply to move their heavy cables, but the cars will continue to operate on the hills of San Francisco.

England's 36-mile Manchester Ship Canal opens to link Manchester with the Mersey River.

London's Tower Bridge opens to span the Thames. The £1.5 million bridge has a 200-foot center span that can be raised to permit passage of vessels; its chain suspension side spans are each 270 feet long.

1894 ***(cont.)*** A chemically inert gas he calls argon from the Greek word for inactive is discovered by Scottish chemist William Ramsay, 42, who removes from the air various known gases including nitrogen, oxygen, and carbon dioxide and finds that an inert gas remains. Ramsay and his colleague Travers will discover three other inert gases in the air and will name them krypton (hidden), neon (new), and xenon (stranger) (*see* Langmuir, 1912).

Shibasaburo Kitazato and A. E. J. Yersin at Berlin discover the bacillus pestis—etiological agent of the Black Death (bubonic plague) (*see* 1665; 1900).

New York physician Hermann Michael Biggs, 35, introduces the diphtheria antitoxin developed by Emil von Behring into U.S. medical practice.

New York surgeon Charles McBurney, 49, develops a muscle-splitting incision for appendectomies; he has discovered a tender pressure point important in the diagnosis of appendicitis (it will be called McBurney's point).

Radcliffe College for Women opens at Cambridge, Mass., after 15 years of opposition by Harvard president Charles William Eliot to the classes given since 1879 by Elizabeth Cary Agassiz. The college is named for Anne Radcliffe, who in the 17th century became the first woman to make a gift to Harvard.

Austro-Hungarian-American physicist Michael Idversky Pupin, 36, improves multiplex telegraphy. He has studied under Hermann von Helmholtz.

A. C. W. Harmsworth acquires London's conservative *Evening News* and reorganizes the paper. Born in Dublin, Alfred Charles William Harmsworth, 29, began as a free-lance journalist at 17, set up a publishing house at London with his brother Harold 5 years later, and in 1888 started the periodical *Answers to Correspondents* (later simply *Answers*) *(see Daily Mail,* 1896).

Nonfiction *Wealth Against Commonwealth*, a study of John D. Rockefeller's Standard Oil Co. by Henry Demarest Lloyd who defends Eugene V. Debs (above) in legal battles arising out of the Pullman strike; *Glimpses of Unfamiliar Japan* by U.S. author (Patricio) Lafcadio (Tessima Carlos) Hearn, 44, who was born in the Ionian Islands of British-Greek parentage, has obtained a teaching position in Japan, has married a Japanese woman of a samurai family, and gives Western readers their first sympathetic view of Japanese life. Hearn will become a Japanese citizen next year, take the name Koizumi Yakumo, and teach at the Imperial University in Tokyo from 1896 to 1903.

Fiction *Pudd'nhead Wilson* by Mark Twain; *La Nonne Alfarez* by José de Hérédia; *Trilby* by George du Maurier whose malevolent hypnotist Svengali will become a generic term for anyone who tries with evil intentions to force or persuade someone else to do his bidding (the novel creates a fashion for the trilby hat); *The Prisoner of Zenda* by English novelist Anthony Hope (Anthony Hope Hopkins), 31; *The Ebb-Tide* by Robert Louis Stevenson who has based the work on a draft by his stepson Lloyd Osbourne. Stevenson suffers a stroke of apoplexy on his veranda in Samoa December 3 and dies at age 44; Samoans, who regard him as a chief, bury his body on the summit of Mount Vaea; *The Memoirs of Sherlock Holmes* by A. Conan Doyle who in December 1893 seemingly killed off his detective along with Professor Moriarty in an Alpine disaster chronicled in the Strand magazine (but *see* 1902).

Juvenile *The Jungle Book* by Rudyard Kipling.

The Venetian Painters of the Renaissance enhances the reputation Lithuanian-American art critic-historian Bernard Berenson, 29, who was graduated from Harvard 7 years ago, has studied at Oxford and the University of Berlin, and has trained himself in the galleries of Italy to recognize the works of the Italian masters by their styles and techniques. Berenson will acquire the 18th-century Florentine mansion I Tatti and be the ranking authority on classical art until his death in 1959.

The *Yellow Book* begins publication with London artist Aubrey Vincent Beardsley, 22, as art editor. Beardsley's black-and-white "art nouveau" drawings will achieve a fame far more enduring than their creator, who will die in 1898 at age 25.

Painting *Anxiety* by Edvard Munch; *Femme à sa toilette* by Edgar Degas; George Innes dies at Bridge of Allan, Scotland, August 3 at age 69.

German graphic artist Käthe Kollwitz (née Schmidt), 27, completes the first in a series of prints, *Der Weberaufstand*, inspired by last year's Gerhart Hauptmann play *The Weavers*.

Louis Comfort Tiffany trademarks favrile glass (*see* 1878). He has invented a process for staining glass by adding pigments to molten glass while it is being made, instead of applying stain or burning pigments into hardened glass, and achieves incomparable iridescences that he employs in free form "art nouveau" shapes to create brilliant lamps, chandeliers, vases, jewelry, and other objects.

Theater *How to Get Rid of Your Mistress* (*Un fil à la patte*) by Georges Feydeau 1/9 at the Théâtre du Palais Royal, Paris; *The Land of Heart's Desire* by William Butler Yeats 3/29 at London's Avenue Theatre; *Arms and the Man* by George Bernard Shaw 4/21 at London's Avenue Theatre. Shaw innovates the practice of having his plays published for general readership with elaborate prefaces that deal with political and social issues (his new play deflates romantic illusions of military "glory"); *Hotel Paradise (L'hôtel du libre échange)* by Georges Feydeau and a collaborator 12/5 at the Théâtre des Nouveautés Paris.

Billboard has its beginnings in an eight-page monthly begun by Cincinnati publishers James Hennegan and W. H. Donaldson as *Billboard Advertising.* Theater notes will be introduced beginning in late 1901, a motion picture section will be added in 1907, and *Billboard* will become a U.S. show business weekly (*see Variety*, 1905).

Boston's $1 million Keith Theater opens with Weber and Fields who make $400 per week. By 1905 they will be pulling down $4,000 per week.

Opera *Thaïs* 3/16 at the Paris Grand Opéra, with music by Jules Massenet, libretto from a novel by Anatole France; *Werther* 4/19 at New York's Metropolitan Opera House, with music by Massenet, libretto from the Goethe work of 1774; *La Navarraise* 6/20 at London's Covent Garden, with music by Massenet.

Ballet *Afternoon of a Faun (L'Aprés-midi d'un Faune)* 12/23 at Paris, with music by French composer Claude Achille Debussy, 32 (*see* Mallarmé poem, 1876).

First performances Symphony No. 5 in B flat major by Anton Bruckner 4/9 at Graz (Bruckner will die in the fall of 1896 at age 72 leaving his Symphony No. 9 in D minor incomplete); Concerto No. 2 in D minor for Piano and Orchestra by Edward MacDowell 12/14 at New York.

The Queen's Hall opens in London's Langham Place. The auditorium will be famous for its Promenade Concerts until its destruction in 1941.

Broadway musical *The Passing Show* 5/12 at the Casino Theater, with music by Viennese-American composer Ludwig Englander, 35, book and lyrics by Sydney Rosenberg. The production introduces the revue to the musical theater.

Popular songs "Humoresque" (piano composition) by Anton Dvořák; "The Sidewalks of New York" by Irish-American composer Charles B. Lawlor, 42, lyrics by native New Yorker James W. Blake, 32 ("East side, West side, all around the town. . ."); "I've Been Working on the Railroad" is published under the title "Levee Song" in *Carmina Princetoniana.* University of Texas undergraduate John Lang Sinclair will write lyrics beginning, "The eyes of Texas are upon you" for a student minstrel show that will be presented in 1903.

The Penn Relays are held for the first time in April at Philadelphia.

Joshua Pim wins in men's singles at Wimbledon, Mrs. Hillyard in women's singles; Robert Wrenn wins in U.S. men's singles, Helen R. Helwig in women's singles.

The first U.S. Open golf tournament is held at St. Andrew Golf Club in Yonkers, N.Y., at the initiative of Scots-American golfer John Reid, who helps to organize the U.S. Golf Association (USGA) (*see* British Open, 1860; liquid center golf ball, 1899).

World chess champion Wilhelm Steinitz is defeated by German chess master Emanuel Lasker, 26, who gains the world title he will hold until 1921.

Texas gunslinger John Wesley Hardin, now 42, is killed while playing poker August 19 at El Paso. He has insulted Marshal John Selman, who shoots him in the back of the head at the Acme Saloon.

A London Building Act voted by Parliament limits the height of buildings to 150 feet. A development called Queen Anne's Mansions has disturbed Queen Victoria's view, and no skyscrapers will be erected in London for nearly 60 years.

Berlin's Reichstag building is completed after 10 years of construction; the German parliament moves into its new home (*see* 1933).

The Royal Poinciana Hotel opens at Palm Beach, Fla., with 540 rooms that will be enlarged to give it 1,300. Standard Oil millionaire H. M. Flagler has extended his Florida East Coast Railway 200 miles south to Lake Worth and put up the $2 million wooden hotel to accommodate winter visitors (*see* 1888; The Breakers, 1895).

Chicago's new Reliance building designed by Daniel Burnham is "a glass tower 15 stories high."

Newport's Belcourt Castle is completed on Bellevue Avenue in the Rhode Island summer resort for Oliver H. P. Belmont. Richard Morris Hunt and John Russell Pope, 20, have designed the 52-room Louis XIII-style "cottage."

Some 6,576 New York slum dwellers are found to be living in windowless inside rooms. Landlords have installed air shafts to circumvent an 1879 law passed to ban such inside rooms, but the shafts are used in many cases as garbage chutes.

U.S. wheat sells at 49¢ per bushel, down from $1.05 in 1870. Prairie schooners headed back east have canvas covers painted with the words, "In God we trusted, in Kansas we busted."

The Carey Act passed by Congress August 18 grants 1 million acres of federally-owned desert lands to certain states on condition that they irrigate the land, reclaim it, and dispose of it in small tracts to settlers. Sen. Joseph M. Carey, 49, (R. Wyo.) has sponsored the legislation.

California's Irvine Co. ranch is incorporated by James Irvine, Jr., 27, whose late father came from Ireland to seek his fortune in the gold fields, made money as a wholesale merchant in San Francisco, joined with three other men to buy up sheep rangeland from Spanish settlers who had fallen into debt to Anglo-American merchants in Los Angeles, and bought out his partners in 1876. The younger Irvine will develop a vast ranching and farming empire, including a 100,000-acre ranch in Montana.

Ralston Purina has its beginnings in the Robinson-Danforth Commission Co. founded by St. Louis feed merchants George Robinson and William H. Danforth. They have mixed corn and oats to produce a feed that will not give horses and mules colic as corn alone may do (*see* Purina, 1897).

A study of children in London's Bethnal Green district shows that 83 percent receive no solid food besides bread at 17 out of 21 meals per week. Scurvy, rickets, and tuberculosis are widespread in such districts as they are in many British and European industrial centers.

Our Pet evaporated cream is introduced in a 5¢ miniature can by Helvetia Milk Condensing Co. whose new product will gain wide popularity (*see* 1890; Pet Milk, 1923).

1894 ***(cont.)*** U.S. ice-making plants produce 1.5 million tons of machine-made ice (*see* 1889; 1926).

German-American café owner William Gebhardt at New Braunfels, Tex., runs pepper (capsicum) bits through a small home meat grinder three times and dries the ground pepper into a powder, the first commercial chile (or chili) powder (*see* 1835).

Van Camp pork and beans are advertised in U.S. magazines for the first time. A sample can is offered for 6¢ in the first full-page food product advertisement to appear in a national publication (*see* 1882; tuna, 1914).

Joseph Campbell Preserve Co. gets a new president as Arthur Dorrance, 45, succeeds Campbell with whom he formed a new partnership in 1876 (*see* 1869; "condensed" soup, 1897).

Hershey Chocolate Company is founded as a sideline by Milton S. Hershey, whose caramel business is booming (*see* 1893; 1900).

August Escoffier creates Pèche Melba at London's 5-year-old Savoy Hotel to honor the Australian grande cantrice Madame Nellie Melba (Helen Porter Mitchell), 33, who is singing at Covent Garden. Inspired by the swan in Richard Wagner's opera *Lohengrin* of 1850, Escoffier tops a scoop of vanilla ice cream on a cooked peach half with a purée of raspberries and almond slivers.

Beijing again consents to exclusion of Chinese laborers in a Chinese Exclusion Treaty signed with the United States March 17 (*see* 1884; 1943).

Congress creates a Bureau of Immigration.

An Immigration Restriction League formed by a group of Bostonians campaigns for a literacy test that will screen out uneducated undesirables. It will direct its efforts against Asiatics, Latins, and Slavs (*see* 1917).

1895 France's President Casimir-Perier resigns in disgust January 17 and is succeeded by Félix Faure, 53.

Cuban insurgents stage an abortive revolution against Spanish rule, but the revolt is ruthlessly suppressed and the island left in turmoil (*see* 1898).

The Treaty of Shimonoseki April 17 ends a brief Sino-Japanese war that has destroyed the Chinese Army and Navy. China recognizes the independence of Korea, cedes Taiwan (Formosa), the Pescadores Islands, and the Liaotung Peninsula to Japan, and agrees to pay 200 million taels in indemnities and open four more ports to foreign commerce. Russia, Germany, and France force Japan to return the Liaotung Peninsula, but the Japanese agree only on condition that they receive another 30 million taels. France obtains territorial and commercial concessions in China's southern provinces, while Britain, France, Germany, and Russia all make huge loans to China with customs revenues as security (*see* 1897; Boxer Rebellion, 1900).

Britain's Tories regain power in the general elections and a third Salisbury ministry begins June 25. It will continue until 1902.

The name Rhodesia is given to the territory of the South Africa Co. south of the Zambezi River to honor the prime minister of the Cape Colony Cecil Rhodes. Southern Rhodesia will become a British crown colony in 1923 (*see* 1965).

Britain annexes Tongaland June 11 to block any possible access of the Transvaal to the sea via Swaziland, and Bechuanaland is attached to the Cape Colony in November.

The Jameson raid that begins December 29 is an attempt to foment rebellion against the Boer government of the Transvaal's Oom Paul Kruger, now 70. Leander Starr Jameson leads 600 men in a 140-mile dash across the Transvaal (*see* 1893; 1896).

Russian Marxist Vladimir Ilich Ulyanov, 25, travels in late April to Geneva, meets with Georgi Plekhanov, goes on to Zurich, Berlin, and Paris, returns with illegal literature in a false-bottomed trunk, organizes strikes and prints anti-government leaflets and manifestoes, is arrested in December, and will be exiled to Siberia for 3 years in 1897 after a year in prison. His older brother, Aleksandr, was executed in 1887 for plotting against Czar Aleksandr II. The younger Ulyanov will adopt the pseudonym Lenin (*see* 1900; Plekhanov, 1883).

Booker T. Washington gives a speech agreeing to withdrawal of blacks from politics in return for a guarantee of education and technical training. He makes his remarks at Atlanta's Cotton States Exposition in Piedmont Park, a fair that attracts 800,000 visitors who come to see the Liberty Bell from Philadelphia, Buffalo Bill Cody, and some 6,000 exhibits (and to hear concerts by Victor Herbert and John Philip Sousa), but Washington's speech alienates some blacks.

The Supreme Court emasculates the 1890 Sherman Anti-Trust Act January 21 in the case of *U.S. v. E. C. Knight Co.* The court upholds H. O. Havemeyer's 3-year-old Sugar Trust, dismissing an antitrust case on the ground that control of the manufacturing process affects interstate commerce only incidentally and indirectly (*see* Clayton Act, 1914).

U.S. Treasury gold reserves fall to $41,393,000 as economic depression continues, but the New York banking houses of J. P. Morgan and August Belmont join forces to loan the Treasury $65 million in gold to be paid for at a stiff price in government bonds. Morgan has made a fortune reorganizing and consolidating railroads; Belmont died late in 1890 at age 73.

The Supreme Court rules May 20 that the income tax provision in last year's Wilson-Gorman Tariff Act was unconstitutional. It hands down the 5-to-4 decision in the case of *Pollock v. Farmers' Loan and Trust Co.* (*see* 1913).

Sears, Roebuck has sales of $750,000, up from $338,000 in 1893, and A. C. Roebuck sells his interest in the firm for $2,000 to Chicago merchant Julius Rosenwald, 33, who hires his brother-in-law Aaron Nusbaum to serve as treasurer and general manager of the mail-order house (*see* 1904).

The Niagara Falls Power Co. incorporated in 1889 transmits the first commercial electric power from the Falls August 26, employing three 5,000-horsepower Westinghouse Electric generators that deliver two-phase currents at 2,200 volts, 25 cycles. Pittsburgh Reduction Co. uses the power to reduce aluminum ore (*see* Mellon, 1891).

The "diesel" engine invented by German engineer Rudolf Diesel, 37, operates on a petroleum fuel less highly refined and less costly than gasoline; it has no electrical ignition system and is simpler than a gasoline engine and more trouble-free. Diesel will work with Fried. Krupp of Essen and the Augsburg-Nuremberg machine factory to build a successful engine (*see* locomotive, 1913).

The Delagoa Bay Railway that opens July 8 gives Johannesburg and Pretoria an economic outlet to the sea independent of British control.

Germany's 53.2-mile Kiel (Nord-Ostsee) Ship Canal opens to connect the North Sea with the Rhine.

The Ohio Valley Improvement Association organized October 8 at Cincinnati works to make the Ohio a navigable channel for commerce. The U.S. Army Corps of Engineers will build 49 locks and movable dams on 980 miles of river between Pittsburgh and Cairo, Ill.

George B. Selden receives a patent for a "road locomotive" powered by a "liquid hydrocarbon engine" of the compression type (*see* 1879; 1899).

Wilshire Boulevard opens between Los Angeles and Santa Monica on the Pacific. Local gold-mine promoter, medical faddist, and socialist H. Gaylord Wilshire has built the thoroughfare.

André Michelin, 42, and his brother Edouard, 36, produce their first pneumatic tires for motorcars (*see* Dunlop, 1881; Panhard and Lavassor, 1892). Michelin & Cie. will be Europe's largest tire producer.

The first U.S. pneumatic tires are produced by the Hartford Rubber Works at Hartford, Conn., owned by bicycle maker Pope Manufacturing Co. (*see* Kelly-Springfield, 1899).

The Lanchester motorcar introduced by English engineer Frederick W. Lanchester of the Lanchester Engine Co. is the first British four-wheel gasoline-powered motorcar. It has epicyclic gearing, worm drive, and pneumatic tires.

The Wolesley motorcar introduced by the Wolesley Sheep-Shearing Machine Co. has a two-cylinder opposed air-cooled engine. Australian-born engineer Herbert Austin, 29, has designed it (*see* Austin motorcar, 1905).

An 1891 Panhard Levassor wins a Paris-Bordeaux road race; it has a two-cylinder engine with three forward speeds, no reverse, and enough speed to complete the 500-kilometer journey in 48 hours, 47 minutes.

The first U.S. automobile race takes place Thanksgiving Day on a 53.5-mile course between Chicago and Milwaukee. Herman Kohlsaat's *Chicago Times-Herald* has offered a $2,000 first prize, some 80 contestants enter, only six are able to start, average speed over the snowy roads is 5.25 miles per hour, and the winner is James Franklin Duryea, driving the only American-made gasoline-powered entry—a vehicle with a rigid front axle, water-cooled engine, and steering tiller, whose up-and-down motion controls the speed. The car has Hartford tires (above) purchased by Charles Duryea, who organizes the Duryea Motor Wagon Co. at Springfield, Mass. (*see* 1891; 1896).

Peugot Frères in Paris completes the first gasoline-powered delivery van powered in December. Its 4-horsepower Daimler motor enables it to carry a half-ton load at 9.5 miles per hour and a 650-pound load at 12 miles per hour. Two Peugot brothers converted a textile mill into a steel foundry in 1810 and produced corset frames, coffee grinders, saws, and springs (including springs for pince-nez) before turning to bicycle-making in the 1880s.

The first Benz "Omnibus" goes into service in Germany with a one-cylinder rear engine that develops between 4 and 6 horsepower and holds eight passengers.

The X-ray, or roentgen ray, discovered by Bavarian physicist Wilhelm Conrad Roentgen, 50, will revolutionize diagnostic medicine by making it possible to photograph the inner organs and bone structures of animals and humans. Professor of physics since 1888 at Würzburg's Physical Institute, Roentgen notices a phenomenon November 8 while working with the cathode-ray ultra-vacuum tube recently invented by English physicist Sir William Crookes, 63. When a current is passed through the tube, a nearby piece of paper that has been painted with barium platinocyamide appears to fluoresce brightly. The phenomenon occurs even when the tube is covered with black cardboard, and Roentgen proves that the effect is caused by an invisible ray. He announces his discovery to the Würzburg Physical-Medical Society December 28 in his paper *Eine neue Art voli Strahlen* (*see* Nobel, 1901; use in treating cancer, 1903).

Studies in Hysteria (Studien giber Hysteric) by Viennese physician Sigmund Freud, 39, is published in collaboration with Josef Breuer (*see* 1882).

Freud (above) has worked with Josef Breuer in treating hysteria with hypnosis but begins now to develop a new treatment that will be the basis of scientific psychoanalysis. "No neurosis is possible with a normal sex life," Freud will say, but he will give up sex himself at age 42, and will suffer much of his life from stomach upsets, migraine headaches, and nasal catarrh for which he will use cocaine (*see* 1900).

Dutch botanist Marinius Willem Beljerinck, 44, discovers filterable viruses. He is not familiar with Dmitri Ivanovski's 1892 discovery.

The London School of Economics is founded by Fabian socialists.

The University of Texas at Arlington is founded.

Underwood Typewriter Co. is founded by New York ribbon and carbon merchant John Thomas Underwood, 38, to develop and market a machine patented 2

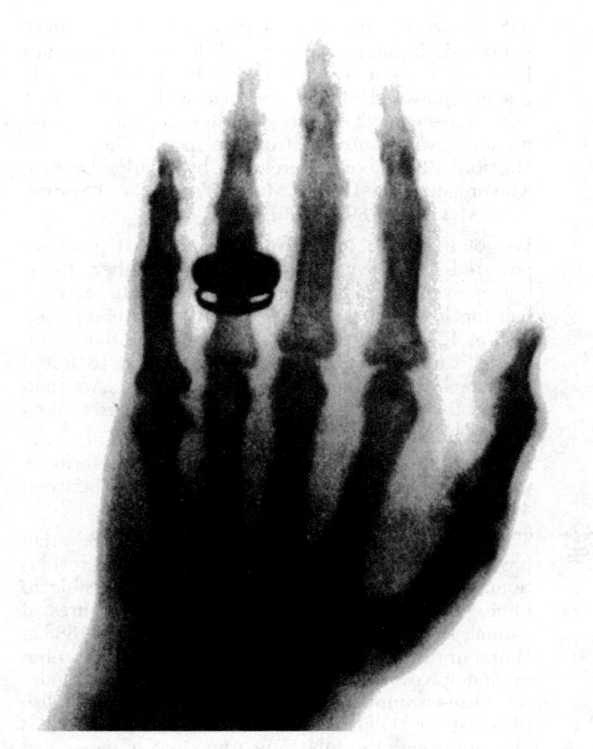

W. C. Roentgen's startling "X-ray" of a colleague's hand began a new era in diagnostic medicine.

years ago by Brooklyn inventors Franz X. and Herman L. Wagner, whose typewriter enables the typist to see what is being typed.

William Randolph Hearst acquires the *New York Morning Journal* from John R. McLean for $180,000 (*see* 1887). Hearst will call the paper's morning edition the *New York American* beginning in 1902 after acquiring the *Boston American, Chicago American,* and other papers, and will go on to publish magazines that will include *Good Housekeeping (see* 1885), *Harper's Bazaar, Town and Country,* and *Hearst's International-Cosmopolitan (see* 1905; "The Yellow Kid," 1896).

The *Denver Post* begins publication under the direction of its new proprietors Frederick Gilmer Bonfils, 35, and Harry Heye Tammen, 39, who have renamed the *Denver Evening Post.* Bonfils is an Oklahoma real estate promoter who has been running the Little Louisiana Lottery at Kansas City and has been persuaded by Denver storekeeper H. H. Tammen to buy the *Evening Post.* The two will use sensationalist techniques to multiply circulation 20-fold in the next 12 years, but their crusades in the name of justice and public good will sometimes be notable more for flamboyance than integrity.

Collier's Weekly magazine begins publication. New York publisher Peter F. Collier has changed the name of his *Once a Week* magazine and challenges the *Saturday Evening Post*.

Guglielmo Marconi, 21, pioneers wireless telegraphy. The Italian inventor has studied the Hertz discovery of 1887, set up a laboratory at his family's country house outside Bologna, and in September transmits a message to his brother who is out of sight beyond a hill. Marconi's mother is British, and he will apply in June of next year for a British patent on his wireless (*see* 1896).

The New York Public Library is created by a merger of the 41-year-old Astor Library, the Lenox Library, and the Tilden Library (*see* Carnegie, 1901).

A new Boston Public Library opens on Copley Square in a Renaissance building by Charles F. McKim of McKim, Mead & White (*see* 1877).

Fiction *Almayer's Folly* by British novelist Joseph Conrad, 37, who emigrated from the Ukraine, changed his name (from Joez Teodor Jozef Konrad Nalecz Korzeniowski), served for 4 years in the French merchant marine and then for 16 in the British merchant marine, and has been writing for just over a year, making himself a master prose stylist in his adopted language; *Jude the Obscure* by Thomas Hardy; *The Time Machine* by English journalist-science fiction writer H. G. (Herbert George) Wells, 28; *Esther Waters* by George Moore; *Celibates* (stories) by George Moore; *Bewilderments (Forvillelser)* by Swedish novelist Hjalmar Soderberg, 26; *The Promised Land (Det Forjaettede Land),* a trilogy by Danish novelist Henrik Pontoppidan; *The Red Badge of Courage* (novelette) by Stephen Crane, who will serve as a war correspondent in 1896 and 1897, move to England in 1897, and die of tuberculosis at Badenweiler, Germany, in 1900 at age 28; *The Sorrows of Satan* by Marie Corelli.

Poetry *The Tentacular Cities (Les Villes Tentaculaires)* by Emile Verhaeren; *Majors and Minors* by U.S. poet Paul Laurence Dunbar, 23, whose first poetry collection was privately printed as *Oak and Ivory* late in 1893. Dunbar is the son of former slaves and William Dean Howells warmly praises his work in *Harper's* magazine; "I never saw a Purple Cow,/ I never hope to see one;/ But I can tell you, anyhow,/ I'd rather see than be one," writes *San Francisco Lark* editor Frank Gelett Burgess, 29, whose jingle gains immediate currency.

The Venice Biennale opens. The first international show of contemporary art begins a tradition that will be continued for more than a century.

Painting *Northeaster* by Winslow Homer. Berthe Morisot dies at Paris March 2 at age 54.

Sculpture *The Burghers of Calais* by Auguste Rodin breaks with tradition by showing the burghers walking in file rather than grouped in a circle or square (*see* 1347). Commissioned by the city of Calais, the work has taken Rodin 11 years.

A Pocket Kodak introduced by Eastman Kodak gains immediate success (*see* transparent negative film, 1889; Brownie, 1900).

Theater *An Ideal Husband* by Oscar Wilde 1/13 at London's Haymarket Theatre; *The Importance Of Being Earnest* by Oscar Wilde 2/14 at London's St. James Theatre. Wilde is called a homosexual by Sir John Sholto Douglas, marquis of Queensberry, whose son is one of Wilde's intimates, Wilde brings a libel action, revelations are made about Wilde's personal morals, he is sentenced in May at the Old Bailey to 2 years' imprisonment for offenses committed, and he goes into bankruptcy soon after; *The Heart of Maryland* by David Belasco 10/22 at New York's Herald Square Theater, with Mrs. Leslie Carter, 229 perfs.

The first theater showing of motion pictures takes place March 22 at 44 rue de Rennes, Paris, where members of the Société d'Encouragement à l'Industrie Nationale see a film of workers leaving the Lumière factory at Lyons for their dinner hour. The cinematograph of inventors Louis and Auguste Lumière, 31 and 33 respectively, is a vast improvement over the kinetoscope peepshow introduced last year by Thomas Edison, whose film can be viewed by only one person at a time, and their 16-frame-per-second mechanism will be the standard for films for decades (*see* 1872; 1896).

The first commercial presentation of a film on a screen takes place May 20 at New York. An audience in a converted store at 153 Broadway views a 4-minute film of a boxing match.

A cine projector patented by U.S. inventor Charles Francis Jenkins, 28, has an intermittent motion.

A combination cine camera and projector patented May 27 by British inventor Birt Acres has an appliance for loop-forming. He has filmed the Oxford-Cambridge boat race March 30, films the Derby using film purchased from the American Celluloid Co. of Newark, N.J., but his associate Robert Paul will challenge his claim to sole proprietorship.

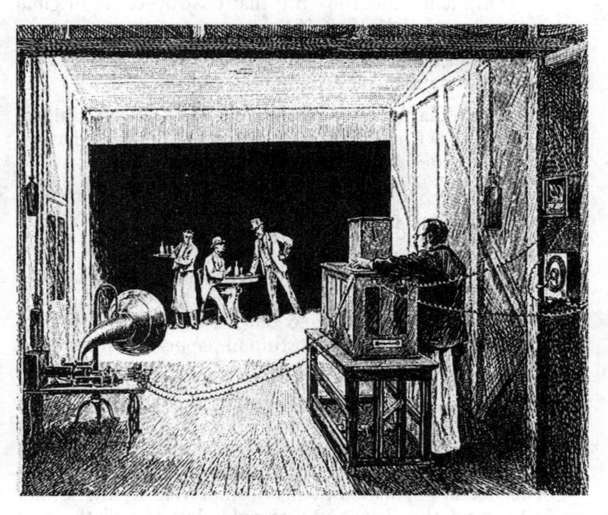

Motion pictures flickered to life with Thomas Edison's crude but prophetic kinetograph and kinetoscope.

Ballet *Swan Lake* 2/8 at St. Petersburg's Maryinsky Theater, with choreography by Marius Petipa in a complete version of the 1877 ballet with music by the late Petr Ilich Tchaikovsky.

First performances Symphony No. 2 in C minor *(The Resurrection)* by Gustav Mahler 4/14 at Berlin; "Till Eulenspiegel's Merry Pranks" (*Till Eulenspiegels Lustige Streiche*) by German composer Richard Strauss, 31, 11/5 at Cologne.

Popular songs "The Band Played On" by New York actor-composer John F. Palmer, lyrics by English-American actor Charles B. Ward, beginning, "Casey would waltz with a strawberry blonde. . ."

"America the Beautiful" by Wellesley College English professor Katharine Lee Bates, 36, will be set to the music of Samuel A. Ward's "Materna" and become an unofficial national anthem. A climb up Pike's Peak in 1893 has inspired Bates to write her poem about America's "amber waves of grain" and "purple mountain majesties above the fruited plain . . . from sea to shining sea." Revised versions will appear in 1904 and 1911.

Wilfred Baddely wins in men's singles at Wimbledon, Charlotte Cooper, 24, in women's singles; Fred H. Hovey, 26, wins in U.S. men's singles, Juliette P. Atkinson in women's singles.

The U.S. yacht *Defender* defeats *Valkyrie II* 3 to 0, thwarting the seventh English effort to regain the America's Cup.

The American Bowling Congress founded September 9 sets rules for a 10-pin bowling game devised to circumvent a congressional prohibition of 9-pin bowling. The new ABC will sponsor the first national tournament at Chicago in 1901 and the event will attract 41 five-man teams, 72 two-man teams, and 115 individual bowlers.

The Gillette razor has its beginnings in a proposal by U.S. bottle stopper salesman King Camp Gillette, 40, for a disposable razor blade. Gillette's boss William Painter has advised him to "invent something which will be used once and thrown away" so that "the customer will come back for more" (*see* 1892) and when Gillette finds his razor dull, he says to himself, "A razor is only a sharp edge, and everything back of it is just support. Why forge a great piece of steel and spend so much time and labor in hollow grinding it when the same result can be obtained by putting an edge on a piece of steel only thick enough to hold an edge?" But he is unable to find technical expertise or financial backing to develop his idea (*see* 1901).

Chicago's Marquette building is completed by Holabird & Roche with wide windows that flood the building's interior with light but create problems of partitioning office space.

Louis Sullivan completes the Prudential (Guaranty) building at Buffalo, N.Y.

New York's Washington Arch is completed at the foot of Fifth Avenue at a cost of $128,000. The 86-foot-high arch designed by architect Stanford White commemorates the first inauguration of President Washington.

1895 ***(cont.)*** Biltmore House at Asheville, N.C., completed for railroad heir George Washington Vanderbilt II, 33, is the world's largest private house. Located on an estate of 119,000 acres and built in 5 years at a cost of $4.1 million, it has 250 rooms and is surrounded by vast forests.

The Breakers is completed by Richard Morris Hunt at Newport, R.I., for railroad heir Cornelius Vanderbilt II, 52, who is president of the New York and Harlem Railroad, chairman of the boards of the New York Central & Hudson River and the Michigan Central, eldest brother of G. W. Vanderbilt II (above), and donor of an 1893 Yale dormitory that memorializes his son William Henry. The 70-room fireproof summer "cottage" on Ochre Point Road is modeled on palazzos at Genoa and Turin and has been completed in less than 3 years to replace the P. Lorillard house of the same name, which burned down in November 1892.

The Breakers completed at Palm Beach, Fla., is a vast rambling winter hotel financed by Florida East Coast Railway promoter H. M. Flagler (*see* 1903; Royal Poinciana, 1894).

Richmond's Jefferson Hotel opens with electric lights and two toilets on every floor. The electric system will cause a fire, destroying the hotel in 1901, but the Jefferson will reopen in 1910 with 300 rooms, each with private bath.

Forester Gifford Pinchot, 30, is employed to superintend the forests at Biltmore House (above) (*see* 1898).

Britain's National Trust, created with government funding to preserve country houses, parks, and gardens, will protect 150 "stately homes," 100 grand gardens and landscaped parks, 17 entire villages, more than 2,000 farms, and some nature reserves (*see* first British national park, 1950).

Massachusetts fishermen using a line trawl catch a record 211.5-pound cod measuring more than 6 feet in length.

Columbia River salmon canning reaches its peak of 634,000 cases (*see* 1866; CRPA, 1899).

The world wheat price falls to £5 per cwt. as North American producers make large initial offerings, but the price rises when it turns out that the Kansas wheat crop totals only 18 million bushels—less than half a million metric tons.

The U.S. Department of Agriculture sends agronomist Mark Alfred Carleton, 29, out to Kansas to investigate the wheat crop shortfall. Carleton reports that the only good drought-resistant wheat grown in Kansas is the Turkey Red introduced by Russian Mennonites in 1874, and his report creates a clamor for the "miracle" wheat that will soon replace spring wheat completely in Kansas (*see* 1896).

Florida ships 30 million pieces of fruit, down from a billion last year when all the state's lemon trees were killed by a freeze.

The California Fruit Growers Exchange cooperative is organized to market the produce of California growers in the face of discriminatory freight rates (*see* 1885; Sunkist, 1919).

The word "calorie" is applied to food for the first time by Wesleyan University professor Wilbur Olin Atwater, 41, who has studied in Germany and established an Office of Experiment Stations in the U.S. Department of Agriculture. An agricultural chemist, Atwater takes the kilocalorie, or large calorie (the amount of heat needed to raise the temperature of a kilogram of water 1 degree Celsius—from 15° C. to 16° C. at a pressure of 1 atmosphere—or 4 pints of water 1 degree Fahrenheit), and uses this as a measure of the energy contained in foods for a guide to economical food buying based solely on calories per dollar.

Canned foods are shown to keep from spoiling not because air has been driven out of the container but because bacteria have been killed or inhibited in their growth (*see* Appert, 1810; Pasteur, 1861). S. C. Prescott at M.I.T. and William Lyman Underwood conduct studies that will lead to safer canning methods by producing charts giving optimum processing times and temperatures.

Belgian bacteriologist Emilie Pierre Marie Van Ermengem, 44, isolates the botulism bacterium *Clostridium botulinum*. The anaerobic bacterium, whose toxin is twice as deadly as tetanus toxin and 12 times deadlier than rattlesnake venom, has been associated in Europe with sausages whose cases provide an airless environment, but botulism in the United States stems largely from nonacidic foods that have been improperly bottled or canned (*see* 1920).

Del Monte canned goods are introduced by California's Oakland Preserving Co., which also markets an ordinary line but identifies its premium grade with the name Del Monte and a label picture of Charles Crocker's hotel on Monterey Bay, which has been rebuilt since the 1887 fire that destroyed the original structure (*see* 1891; 1916).

Ohio glass maker Michael Joseph Owens, 36, patents an automatic bottle-blowing machine (*see* 1903).

"Coca-Cola is now sold in every state of the Union," boasts Asa Candler of Atlanta, but the beverage has yet to be bottled (*see* 1892; 1899).

The Anti-Saloon League of America is organized at Washington, D.C. (*see* W.C.T.U., 1874; Carry Nation, 1900).

Postum is introduced as a coffee substitute at Battle Creek, Mich., by C. W. Post, who has quit "the San" uncured and discouraged but has embraced Christian Science. Post sells his Postum in paper bags from hand carts (*see* 1893; Grape Nuts, 1897).

Granose, introduced in February, is the world's first flaked breakfast cereal. John Harvey Kellogg has run boiled wheat through his wife's machine for rolling out dough, he has baked the thin film in an oven, and he has been persuaded by his brother Will Keith, 35, not to grind the flakes into granules but to sell them as flakes. Since a 10 oz. package sells for 15¢, Kellogg and his brother sell 60¢-per-bushel wheat at $12 per bushel retail (*see* 1894; Sanitas Corn Flakes, 1898).

The first U.S. pizzeria opens in New York at 53½ Spring Street.

Emigration of Russians increases to 108,000 from scarcely 10,000 in 1882 as population pressures force Czar Nicholas II to ease restrictions against travel. Most of the emigration is to Siberia and is encouraged by the Trans-Siberian Railway whose construction began 4 years ago (*see* 1899).

1896 Boer forces in South Africa defeat L. Starr Jameson at Krugerdorp January 1, force him to surrender January 2 at Doorn Kop, and turn him over to the British for trial in England, where he is convicted but receives only a light sentence (*see* 1895). A telegram of congratulations to Oom Paul Kruger of Transvaal from the German kaiser January 3 strains Anglo-German relations, and mutual suspicions set in among the Boers and British in South Africa.

Cecil Rhodes resigns the Cape Colony premiership January 6; a committee of the Cape Assembly finds him guilty of having engineered last year's Jameson raid. The Transvaal government signs a defensive alliance with the Orange Free State in mid-March; it fortifies Pretoria and Johannesburg with munitions ordered from Europe (*see* Boer War, 1899).

British forces take Coomassie January 18 and imprison the Ashanti king in the Fourth Ashanti War.

Ethiopian warriors defeat Italian troops decisively March 1 at Adowa, forcing Rome to sue for peace. The Treaty of Addis Ababa October 6 withdraws the Italian protectorate (*see* 1895; 1898; 1934).

Matebele tribesmen in Rhodesia begin a new uprising that continues into October.

A French expedition to claim the Sudan sets out for Fashoda under the command of Jean Baptiste Marchand, 33, who has traced the Niger River to its source and explored the region from the Niger to Tengrela (*see* 1897).

France proclaims the African island of Madagascar a French colony August 6 (*see* 1897).

Anglo-Egyptian troops begin a reconquest of the Sudan under the command of Gen. Horatio Herbert Kitchener, 46, who builds a railroad as he advances. He takes Dongola September 21 (*see* 1885; 1898).

The shah of Persia Nasr-ed-Din is assassinated May 1 at age 65 after a 48-year reign that has seen the growth of Russian power in his country. He is succeeded by his incompetent son of 43 who will reign until 1907 as Muzaffar-ed-Din.

Armenian revolutionaries attack the Ottoman Bank at Constantinople August 26; Ottoman forces retaliate with a 3-day massacre of Armenians, killing at least 3,000.

Bulgaria's Prince Ferdinand gains recognition from Russia and then from other powers following the conversion of his 2-year-old son Boris to the Orthodox faith, but Bulgaria remains a principality of the Ottoman Empire (*see* 1887; 1908).

Canada gets her first French-Canadian premier as Quebec Liberal Wilfrid Laurier, 55, wins in the general elections held in June. He takes office July 1 and will hold power until late 1911.

Utah is admitted to the Union as the 45th state after Mormons agree to give up polygamous marriage.

"You shall not press down upon the brow of labor this crown of thorns, you shall not crucify mankind upon a cross of gold," says Nebraska Fundamentalist William Jennings Bryan, 36, in a speech July 8 at the Democratic National Convention. His oratory in support of free silver wins him the nomination, and the Populist party also nominates Bryan.

Ohio governor William McKinley, 53, gains the Republican nomination with support from Cleveland industrialist Mark Hanna. Employers put pressure on workers to vote against "anarchy" and "revolution" lest they jeopardize their jobs, and a rise in wheat prices before election day alleviates agrarian discontent. Bryan fails to attract urban voters, and while he wins 176 electoral votes by carrying the South and the Great Plains and Rocky Mountain states, McKinley wins 271 electoral votes and gains election with a 600,000 plurality in the popular vote.

The Supreme Court upholds racial segregation May 18, sustaining a Louisiana "Jim Crow car law" in the case of *Plessy v. Ferguson*. The Court lays down the doctrine that states may provide blacks with "separate but equal" facilities for education, transportation, and public accommodations. Justice John Harlan says, "The Constitution is color-blind" in a lone but vigorous dissenting opinion, but the majority ruling sets off a wave of new segregation measures that designate drinking fountains, public benches, rest rooms, railroad cars, hospitals, and theater sections "Colored" or "Whites Only" (*see* 1954).

Idaho women gain suffrage through an amendment to the state constitution.

Volunteers of America is founded by a grandson of the late Salvation Army founder William Booth (*see* 1878). Ballington Booth has been in charge of the Salvation Army's U.S. operations, but he has had a falling out with his father, and he organizes a rival group that resembles the Salvation Army.

Miami, Fla., is incorporated at Fort Dallas, which last year had only 3 houses. H. M. Flagler extends his Florida East Coast Railway to the new town and dredges its harbor (*see* hotel, below; Miami Beach, 1913).

A gold rush to Canada's Klondike near the Alaskan border begins following the August 17 strike by U.S. prospector George Washington Carmack, 35, who has been in Alaska since 1888 (*see* 1897).

Carnegie Steel and Henry W. Oliver of Pittsburgh buy the Mesabi Range holdings of John D. Rockefeller's

1896 ***(cont.)*** Lake Superior Consolidated Iron Mines (*see* 1893; 1901).

Yawata Steelworks, founded in March, is the first Japanese steel making firm. German engineers will complete a ¥25 million plant in Kyushu in 1901, and Japan will become a major world steel producer (*see* Fuji Steel, 1945).

Wanamaker's New York opens in East 10th Street as Philadelphia merchant John Wanamaker takes over the cast-iron A. T. Stewart retail palace of 1862. Wanamaker has long since expanded his menswear shop of 1861 into a department store (*see* 1903).

Astronomer Samuel Pierpont Langley, 61, of the Smithsonian Institution at Washington, D.C., sends a steam-powered model airplane on a 3,000-foot flight along the Potomac May 6, the first flight of a mechanically propelled flying machine. He sends an improved model on a 4,200-foot flight in November (*see* 1903).

The Haynes-Duryea motorcar produced by Duryea Motor Wagon Co. of Springfield, Mass., is the first U.S. motorcar to be offered for public sale. Total U.S. motorcar production is 25 (*see* 1895; 1897).

Henry Ford drives his tiller-steered Quadricycle through the streets of Detroit in the early hours of June 4 (*see* 1893; 1899).

British-Leyland has its beginnings in Leyland Motors, founded at the Lancashire town of Leyland (*see* 1968).

British Daimler is founded by English entrepreneur H. J. Lawson who buys a factory site at Coventry to produce a motorcar that will be favored by British royalty (*see* 1890).

Britain repeals her 1865 Red Flag Act requiring that a man on foot carrying a red flag precede all road carriages, but the suggestion that motorcars may come to rival "light railways" brings roars of laughter from the benches in Parliament.

Panhard Levassor in France introduces the first vertical four-cylinder motorcar engine; it also introduces sliding gears with a cone clutch.

The Stanley Steamer is introduced at Newton, Mass., by Francis Edgar Stanley, 47, and his twin brother Freeling O. who have prospered in making photographic dry plates but have been inspired by a DeDion *Voiture à vapeur* they saw demonstrated in Massachusetts last year (*see* 1899).

The U.S. bicycle industry has sales of $60 million, with the average bike retailing at $100. Pope Manufacturing turns out one bicycle per minute (*see* 1877; 1885; Pope motorcars, 1897).

The first underground rail service on the European continent begins at Budapest where a 2.5-mile electric subway goes into operation (*see* London, 1890; Paris, 1900; Boston, 1897).

French physicist Antoine Henri Becquerel, 44, discovers radioactivity in uranium (*see* Klaproth, 1797; Planck, 1900; Curie, 1904).

British pathologist Almroth Edward Wright, 35, originates a system of anti-typhoid inoculation (*see* Eberth, Koch, 1880).

The Canadian Red Cross is founded.

Princeton University assumes that name 250 years after being chartered as the College of New Jersey.

New York University assumes that name 65 years after being chartered as the University of the City of New York.

Marconi's Wireless Telegraph Co., Ltd., establishes the world's first permanent wireless installation in November at The Needles on the Isle of Wight, Hampshire, England. Guglielmo Marconi's British relatives have set up the firm (*see* 1895; 1901; Braun, 1897).

The *London Daily Mail* founded by *Evening News* publisher Alfred C. W. Harmsworth condenses news in "the penny newspaper for one halfpenny" and the morning paper gains wide circulation; Harmsworth continues a campaign he has begun to warn Britons of the threat from Germany (*see* 1894; 1903).

"The Yellow Kid" appears in Joseph Pulitzer's *New York World* in March. The one-panel cartoon by Richard F. Outcault will appear in strip form in Hearst's *New York Journal* beginning October 24 of next year in the *Journal's* Sunday color supplement, Pulitzer and Hearst will battle over rights to the cartoon, and the contest for readership between the two will be marked by sensationalism that will be called "yellow journalism," a term that will be applied to other sensationalist papers.

The *New York Times,* published since 1851, gets a facelift from publisher Adolph S. Ochs, now 39, who has made his *Chattanooga Times* prosper, been married since 1883 to the daughter of Cincinnati's Rabbi Isaac

Hearst and Pulitzer fought over the "Yellow Kid" cartoon and used sensationalist "yellow journalism" to gain readers.

Wise, and acquired control of the *New York Times* August 18 for $75,000, most of it borrowed. The other owners assure Ochs of a stock majority if he can make the paper profitable for 3 consecutive years, he throws out the paper's romantic fiction and tiny typefaces, improves neglected areas such as financial news, starts a weekly book review section and Sunday magazine, and adopts the slogan "All the News That's Fit to Print" (*see* 1878).

"What's the Matter with Kansas?" by *Emporia Gazette* editor William Allen White, 28, attracts national attention to White's fledgling newspaper.

Nonfiction *The Book of Masks* (*Le Livre des masques*) by French critic-novelist Rémy de Gourmont, 39, explains literary symbolism.

Fiction *Quo Vadis* by Polish novelist Henryk Sienkiewicz, 50, who will become world-famous for his story of Nero's Rome; *Country of the Pointed Firs* (stories) by Sarah Orne Jewett who portrays the loneliness of the isolated and declining coastal town of South Berwick, Me.

Juvenile *Frank Meriwell, or First Days at Fardale* by U.S. author Burt L. Standish (William Gilbert Patten), 30, who has written it at the suggestion of New York publishers Street and Smith (*see* Nick Carter, 1886). Patten will write 776 Frank Merriwell novels whose weekly sales will average 125,000 copies (*see* Stratemeyer, 1899).

Poetry *A Shropshire Lad* by University College Latin professor A. E. (Alfred Edward) Housman, 37, at London: "When I was one and twenty/I heard a wise man say,/Give crowns and pounds and guineas/But not your heart away" (XIII, 1); *Green Fire* by Scottish poet Fiona Macleod (William Sharp), 40, who has been promoting a Celtic revival (". . .But my heart is a lonely hunter that hunts on a lonely hill."); *Lyrics of Lowly Life* by Paul Laurence Dunbar whose book has a sympathetic introduction by William Dean Howells that helps make the young man fashionable; *Prosas profanas* by Nicaraguan poet Rubén Darío (Félix Rubén Garcia-Sarmiento), 29.

Painting *The Cello Player* by Thomas Eakins; *Daniel in the Lions' Den* by U.S. painter Henry Ossawa Tanner, 37, who arrived at Paris 5 years ago and will remain there; *The Vuillard Family at Lunch* by Jean Edouard Vuillard. Sir John Everett Millais dies at London August 13 at age 67.

London's National Portrait Gallery (Tate Gallery) moves from Bethnal Green into a permanent home in Westminster completed with funds contributed by sugar magnate Sir Henry Tate, now 77, who has given £80,000 and donated 57 modern paintings. The Tate will be enlarged in 1899 and again in 1908.

Theater *Michael and His Lost Angel* by English playwright Henry Arthur Jones 1/15 at London's Lyceum Theatre; *The Dupe (Le dindon)* by Georges Feydeau 2/8 at the Théâtre du Palais Royal, Paris; *Salomé* by Oscar Wilde 2/11 at the Théâtre de l'Oeuvre, Paris (Wilde wrote the play in French but cannot enjoy the furor it creates because he is confined in Reading Gaol); *Rosemary* by French-born English playwright Louis Napoleon Parker, 44, and Murray Carson 8/31 at New York's Empire Theater, with John Drew, 136 perfs.; *Secret Service* by U.S. playwright-actor William H. Gillette, 41, 10/5 at New York's Garrick Theatre, with Gillette as Captain Thorne, a secret agent for the Union Army, 176 perfs.; *The Sea Gull (Chayka)* by Anton Chekhov 10/17 at St. Petersburg's Alexandrinsky Theater (*see* 1898); *The Sunken Bell* (*Die versunkene Glocke*) by Gerhart Hauptmann 12/2 at Berlin's Deutsches Theater; *Emperor and Galilean (Kejsor og Galileer)* by Henrik Ibsen 12/5 at Leipzig's Stadttheater; *King Ubu (Ubu Roi)* by French playwright Alfred Jarry, 23 (who wrote it at age 15), 12/10 at the Théâtre de l'Oeuvre, Paris.

French comedienne Anna Held, 23, makes her U.S. debut September 21 at New York's Herald Square Theater in a lavish production of the 1884 Charles H. Hoyt play *A Parlor Match* mounted by Florenz Ziegfeld with help from Charles Dillingham, 48 perfs.

The first U.S. public showing of motion pictures April 20 at Koster and Bial's Music Hall in New York employs Thomas Edison's Vitascope, an improvement on his 1893 kinetoscope, and a projector made by Thomas Armat (*see* 1903; Lumière, 1895).

Opera *Pepita Jiminez* 1/5 at Barcelona's Liceo, with music by Isaac Albeniz; *La Bohème* 2/1 at Turin's Teatro Reggio, with music by Giacomo Puccini, libretto from "Scenes de la Vie de Bohème" by the late Henri Murger whose work appeared in the journal *Le Corsair* in the late 1840s; *The Grand Duke (or The Statutory Duel)* 3/7 at London's Savoy Theatre is the last Gilbert & Sullivan operetta (Sullivan will die in 1900, Gilbert in 1911); *Andrea Chénier* 3/23 at Milan's Teatro alla Scala, with music by Italian composer Umberto Giordano, 28; *Der Corregidor* 6/7 at Mannheim with music by German composer Hugo Wolf, 36, who is verging on insanity.

Brazil's $5 million Manaus Opera House opens 700 miles up the Amazon where it has been built for rubber barons whose business has boomed with the growth of the bicycle industry and with increased use of motorcars and has not yet been made uneconomic by East Indian rubber plantations (*see* agriculture, 1876). The ballerina Anna Pavlova will dance at Manaus and the entire company of Milan's Teatro alla Scala will cross the ocean and ascend the Amazon for annual appearances.

First performances Suite No. 2 in E minor (*Indian*) by Edward MacDowell 1/23 at New York's Metropolitan Opera House; Concerto in B minor for Violoncello and Orchestra by Antonin Dvořák 3/19 at London; Concerto No. 5 in F major for Piano and Orchestra by Camille Saint-Saëns 6/2 at Paris on the 50th anniversary of the composer's first public appearance as a pianist; "Roland (The Saracens and Lovely Alda)" by Edward MacDowell 11/5 at Boston's Tremont Theater; *Also Sprach Zarathustra* (freely after Friedrich Nietzsche) (tone poem) by Richard Strauss 11/27 at Frankfurt-am-Main.

1896 ***(cont.)*** Stage musicals *El Capitán* 4/20 at New York's Broadway Theater, with De Wolf Hopper in a comic opera with music by John Philip Sousa, 112 perfs.; *The Geisha* 4/25 at Daly's Theatre, London, with music by Sidney Jones, lyrics by Harry Greenback, 750 perfs.; *The Art of Maryland* 9/5 at New York's new Weber and Fields Music Hall, with Joe Weber and Lew Fields in a parody of David Belasco's legitimate stage hit *The Heart of Maryland* (*see* 1895).

Popular songs "Sweet Rosie O'Grady" by New York songwriter Maude Nugent, 19, who has been unable to find a publisher and introduces the song herself at Tony Pastor's Opera House; "Kentucky Babe" by German-American composer Adam Geibel, 41, lyrics by Richard Henry Buck, 26; "A Hot Time in the Old Town" by U.S. violinist-composer-minstrel Theodore August Metz, 48, lyrics by end man Joe Hayden.

Hymn "When the Saints Go Marching In" by James M. Black, lyrics by Katherine E. Purvis.

Harold Mahoney, 29 (Ireland) wins in men's singles at Wimbledon, Charlotte Cooper in women's singles; Robert Wrenn wins in U.S. men's singles, Elizabeth H. Moore in women's singles.

The Olympic Games of ancient Greece are revived through the efforts of Greek nationalists and of French educator-sportsman Pierre de Frédy, 33, baron de Coubertin, who obtained support 2 years ago for bringing back the games banned by the Romans in 194 A.D. but held on a small scale by the Greeks in 1859, 1870, 1875, and 1889 and by the French in 1892. The late Greek-Romanian Evangelios Zappas has willed his entire fortune to reestablishing the Olympic games in Greece, the Greek government has added $100,000, and Greek merchant George Averoff of Alexandria has given $390,000 to complete restoration of the Panathenaic Stadium. The first modern Olympiad opens at Athens with 484 contestants from 13 nations to begin a quadrennial event that will be broadened to include athletic events undreamed of by the Greeks of ancient times (*see* Boston Marathon, 1897).

The world's first public golf course opens in New York's Van Cortlandt Park.

Atlantic City, N.J., completes a boardwalk 41 feet wide to replace a series of four narrower boardwalks built since 1870.

Miami's Royal Palm Hotel opens to begin the city's career as a winter resort (*see* above). H. M. Flagler adds the Royal Palm to the Alcazar, Royal Poinciana, Breakers, and other hotels he has built at Palm Beach and Daytona (*see* 1903).

The Palace Hotel opens July 29 at St. Moritz, Switzerland, and gains quick popularity among European resortgoers. Attracted originally by the alleged health benefits of the town's rust-red springs, English visitors to St. Moritz have invented bobsledding and popularized tobogganing, curling, Viennese ice dancing, and Norwegian-style skiing.

New York's Hotel Greenwich opens under the name Mills Hotel No. 1 at 160 Bleecker Street. Funded by philanthropist Darius Ogden Mills of 1864 Bank of California fame, it has been designed by architect Ernest Flagg, each of its 1,500 tiny bedrooms has a window overlooking either the street or one of two grassy inner courts, and it offers lodgings at 20¢ per night to poor "gentlemen" who pay 10¢ to 25¢ per meal, on which the hotel makes a small profit.

The Olympic Games for amateur athletes, revived at Athens, grew to include events unknown to the ancient Greeks.

Cripple Creek, Colo., has a fire April 25 that begins when kerosene lanterns are knocked over during a dance-hall fight. The fire consumes more than 30 acres of buildings including the homes of new millionaires (*see* gold rush, 1891).

A Japanese earthquake and seismic wave June 15 kill an estimated 22,000.

Farmers plant Turkey Red wheat throughout much of Kansas, Nebraska, Oklahoma, and Texas where it will soon be the predominant winter wheat variety, and in Montana, Minnesota, and the Dakotas where it is planted in spring for fall harvesting, but the decline in the U.S. wheat crop last year has encouraged Australia, Canada, and Russia to enter the export market and compete actively for European sales (*see* 1895; Kubanka, 1899).

Failure of India's wheat crop raises world prices, but increased world competition (above) keeps a lid on them.

U.S. botanist George Washington Carver, 32, joins Alabama's Tuskegee Institute as director of its Department of Agriculture (*see* Booker Washington, 1881; peanuts, 1914, 1921).

L. F. Bang in Denmark discovers the bacterium of infectious abortion in cows (*see* 1892). Since it is one of several bacteria that cause undulant fever, or brucellosis, in humans, it will be called *Brucella abortus* (*see* Bruce, 1887), but the condition it produces in cattle will be called Bang's disease (*see* 1900; Evans, 1917).

Dutch physicians Christiaan Eijkman and Gerrit Grijns in Java find that chickens fed polished rice suffer from a disease resembling beriberi (*see* Takaki, 1887). They decide that the rice must contain a toxin which is neutralized by something in rice hulls (*see* 1906; Eijkman, 1886).

German biochemist E. Baumann finds iodine in the human thyroid gland. The mineral is absent from all other human body tissues (*see* Marine, 1905).

H.J. Heinz adopts "57 Varieties" as an advertising slogan (the firm produces well over that number of products). Henry J. Heinz has seen a car card advertising 21 shoe styles while riding on a New York elevated train and adapts the idea (*see* 1888; 1900).

Tootsie Rolls are introduced at New York by Austrian-American confectioner Leo Hirschfield, 29, whose penny candy is the first to be wrapped in paper. Hirschfield names the chewy, chocolaty confection after his 6-year-old daughter Clara whom he calls "Tootsie."

Cracker Jack is introduced at Chicago by German-American candy maker F. W. Rueckheim and his brother Louis whose candy-coated popcorn mixed with peanuts has been called "crackerjack!" by an enthusiastic salesman. The Rueckheim brothers have been in Chicago since shortly after the fire of 1871 and operated a popcorn stand for 20 years. They will adopt the slogan "The More You Eat, the More You Want," the 1908 song "Take Me Out to the Ball Game" will help them, they will sell Cracker Jack in boxes containing coupons redeemable for prizes beginning in 1910, pack prizes with the Cracker Jack beginning in 1912, use Jack and his dog Bingo in advertising beginning in 1916, and show them on the box beginning in 1919.

Welch's Grape juice production moves to Watkins Glen, N.Y., close to a vast grape growing region (*see* 1869). Charles Welch has been running his father's unfermented wine business since 1872 (*see* 1897).

Michelob beer is introduced by Adolphus Busch of 1876 Budweiser fame whose St. Louis brewery Anheuser-Busch is the largest in the country. Michelob will be sold only on draft until 1961 (*see* 1933; 1956).

"Little Egypt" dances nude on the table at a private dinner given at Louis Sherry's by New York playboy Barnum Seeley, grandson and heir of the late circus promoter P. T. Barnum who died 5 years ago at age 80. But while Fahrida Mahszar is widely thought to have performed the *danse du ventre*, or hootchy-kootchy, on the Midway Plaisance of Chicago's 1893 Columbian Exposition under the name "Little Egypt," she never in fact appeared there.

Fannie Farmer's Boston Cooking School Cookbook by the head of a teacher-training institution uses a precise measuring system that will make it an enduring best seller in America (but not in Europe).

1897

Crete proclaims union with Greece February 6 following months of strife between Christians and Muslims, Athens sends ships and troops February 10, the world powers announce a blockade of Crete March 18, a Greek-Ottoman war begins April 7, the Turks prevail, the Greeks withdraw, an armistice ends hostilities May 19, and a peace treaty is signed December 4 at Constantinople.

Belgian forces reach the Nile at Rejaf in February, defeat the Sudan dervishes, occupy Loda and Wadelai, reach the Bahr-el-Ghazal region in August, but are challenged in September by a great mutiny as the Batetelas revolt on the upper Congo in an insurrection that will last until 1900.

Madagascar's Queen Ranavalona is deposed by the French February 28, ending the 110-year-old Hova dynasty.

A Franco-German agreement July 23 defines the boundary between Dahomey and Togoland.

French forces have taken Busa February 23 and take Nikki November 30, bringing vigorous protests from Britain and raising the threat of war between Britain and France (*see* 1898).

The United States annexes the Hawaiian Islands under terms of a June 16 treaty that is ratified by the Hawaiian Senate September 9. Sugar planters, who proclaimed the Republic of Hawaii in 1894, have pushed for annexation, but Japan has some 25,000 nationals in the islands and files a formal protest, warning of grave consequences (*see* 1898; 1941).

Queen Victoria celebrates her Diamond Jubilee June 22.

Cubans reject an offer of self-government from Spain's Liberal premier Praxades Sagasta, insisting on complete independence (*see* 1895; 1898).

Honduras in Central America has a revolution with help from U.S. soldier of fortune Lee Christmas, who has been working as a locomotive engineer in the country. His journalist friend Richard Harding Davis, 33, celebrates him in the novel *Soldiers of Fortune* (*see* Zemurray, 1911).

German forces in China occupy Qingdao (Tsingtao) November 14 following the murder of two German missionaries in Shandong province. The Germans have selected Qingdao as their reward for having intervened in behalf of China against Japan in 1895, but their action precipitates a scramble for concessions in China by most of the great European powers (*see* 1895; 1899).

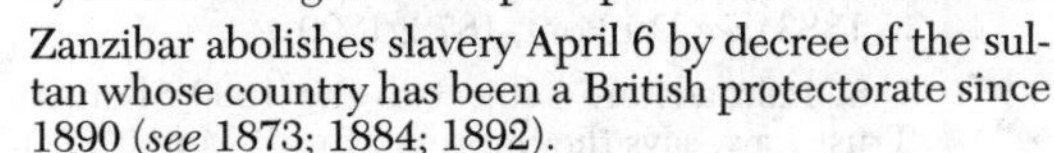

Zanzibar abolishes slavery April 6 by decree of the sultan whose country has been a British protectorate since 1890 (*see* 1873; 1884; 1892).

U.S. bituminous-coal miners leave the pits July 5, beginning a 12-week walkout. They win an 8-hour day, semimonthly pay, abolition of company stores that charge premium prices and keep them forever in debt, and biennial conferences with mine operators (*see* Virden, 1898; United Mine Workers, 1899).

New York's University Settlement moves into a $200,000 "castle" at the corner of Eldridge and Rivington streets. The settlement house started by Stanton Coit in 1886 has built public baths that will serve as models for municipal baths and leads the fight for a city subway system.

1897 *(cont.)* Alfred Dreyfus's brother Mathieu demands a new trial for Alfred after discovering that the handwriting on a key piece of evidence in the 1891 trial was not that of Alfred but of former Austrian army officer Marie Charles Ferdinand Walsin Esterházy, 50, a finding made early last year by French intelligence chief Col. Georges Picquart, 43 (*see* Zola, 1898).

The first Zionist congress opens August 31 at Basel as Theodor Herzl arouses support for his dream of a Jewish homeland in Palestine that will provide a refuge for oppressed Jews worldwide (*see* 1894). Herzl will obtain an appointment with the Ottoman sultan Abdul Hamid and although he has scant financial backing will offer to buy up the Turkish national debt in exchange for Jewish rights in "the Promised Land" of Palestine (*see* Balfour Declaration, 1917).

News of last year's Klondike gold discoveries reaches the United States in January and starts a new gold rush. The ship *Excelsior* arrives at San Francisco in July with $750,000 in Klondike gold, the *Portland* arrives at Seattle 3 days later with $800,000 in gold, and by year's end the Klondike has yielded $22 million worth of gold.

World gold production reaches nearly 11.5 million ounces, up from 5 to 6 million per year between 1860 and 1890 but still little more than half the 22 million it will reach by 1910.

Russian finance minister Sergei Yulievich Witte, 48, introduces the gold standard.

Japan adopts the gold standard March 29.

The Dingley Tariff Act passed by Congress July 24 raises U.S. living costs by increasing duties to an average of 57 percent. It hikes rates on sugar, salt, tin cans, glassware, and tobacco, as well as on iron and steel, steel rails, petroleum, lead, copper, locomotives, matches, whisky, and leather goods, but the influx of gold from the Klondike (above) helps end the 4-year economic depression and begin a decade of prosperity.

Dow Chemical Co. is founded by Herbert H. Dow at Midland, Mich. (*see* 1891; 1922).

E. I. du Pont de Nemours buys the smokeless powder patents and Squankum, N.J., powder plant of engineer-powder maker Hudson Maxim, 44, whose brother Hiram invented the first fully automatic machine gun in 1883 (*see* Du Pont, 1872; 1899).

U.S. railroads are subject to the 1890 Sherman Anti-Trust Law, says the Supreme Court March 22 in a 5 to 4 decision handed down in the *Trans-Missouri Freight* case, but the court nullifies the "long and short haul" clause of the 1887 Interstate Commerce Act November 8 in *Interstate Commerce Commission v. Alabama Midland Railway.*

German-born British banker Otto Hermann Kahn, 29, joins with his Kuhn, Loeb partner Jacob Schiff, 49, and speculator Edward Henry Harriman, 49, in reorganizing the bankrupt Union Pacific Railroad, pruning away unprofitable branch lines, and creating a simplified bond structure to replace the company's old tangle of debts (the Union Pacific will soon be able to buy control of Southern Pacific and a 40 percent interest in Northern Pacific) (*see* 1901).

Boston's Boylston Street subway line between the Public Gardens and Park Street begins service September 1—the first U.S. subway line (*see* Budapest, 1896). The line will be extended within a year to the 3-year-old North Station (*see* New York, 1904).

Trolley service begins across New York's 14-year-old Brooklyn Bridge.

A projected Cape Town-to-Cairo Railroad reaches Bulawayo in Southern Rhodesia November 4.

U.S. auto production rises to 100, up from 25 last year (*see* 1898).

Bicycle maker Albert A. Pope switches to motorcar production (*see* 1896; Pope-Tribune, 1904).

Autocar Co. has its beginnings in the Pittsburgh Motor Vehicle Co. started at Ardmore, Pa., by Louis S. Clarke, who introduces a three-wheeled vehicle powered by a one-cylinder air-cooled engine. Autocar will bring out a shaft-driven motorcar with a two-cylinder water-cooled engine in 1901 (*see* 1906).

Scots-American bicycle maker Alexander Winton, 37, at Cleveland builds the first large U.S. automobiles, organizes Winton Motor Carriage, and drives one of his cars 800 miles over dirt roads to New York. By the end of next year he will have a patent for his vehicle and will have sold nearly 30 cars.

The Royal Automobile Club is founded at London.

The S.S. *Turbinia* attains a speed of 34.5 knots, the first vessel to be propelled by a steam turbine (*see* Parsons, 1884). A jet of steam turns blades inside a rotating cylinder to drive the experimental ship.

The *Argonaut* designed by U.S. naval architect Simon Lake, 31, is the first submarine to operate successfully in open waters (*see* 1776; Holland, 1900).

French chemist-physicist Georges Claude, 27, shows that acetylene can be transported safely if dissolved in acetone, thus giving impetus to use of acetylene gas torches for cutting metal (see 1903).

The first U.S. commercial high-carbon ferrochrome for plating steel is produced by acetylene promoter James T. Morehead with help from French-American metallurgist Guillaume de Chalmot (*see* 1892).

The New Lowe Coke Oven invented by Thaddeus S. C. Lowe improves manufacture of high-grade coke for steel making (*see* 1873).

The atom, believed by the ancient Greeks to be indivisible, turns out to have a nucleus orbited by one or more electrons (*see* Democritus, 330 B.C.). English physicist Joseph John Thomson, 41, modifies the idea formulated by John Dalton in 1803 and tabularized by Dmitri Mendeléev in 1870, shows that the structure of each element is characterized not merely by a different atomic weight but also by an atomic number, and shows that this number is the number of electrons orbiting the nuclei of the element's atoms (*see* Rutherford, 1911; neutron, 1932).

The parasite that causes malaria is carried by the Anopheles mosquito, says British physician Ronald Ross, 40, in famine-stricken India. Ross has been investigating the possibility that mosquitoes may spread malaria, as was suggested in the case of yellow fever by Carlos Finlay in 1881; he shows that the Anopheles mosquito is the vector for malaria, and his discovery will lead to the draining of swamps where mosquitoes breed, to expanded use of window screens and mosquito netting, and—eventually—to widespread use of insecticides (*see* DDT, 1943; Laveran, 1881; Reed, 1900).

A cathode-ray tube (Braun tube) invented at Strassburg (Strasbourg) by German physicist Karl Ferdinand Braun, 47, pioneers development of television and other electronic communications. Braun has improved the Marconi wireless by increasing the energy of sending stations and arranging antennas to control the direction of effective radiation (*see* 1895; 1907).

The *Jewish Daily Forward* begins publication at 175 East Broadway, New York. Editor of the Socialist Yiddish newspaper is Polish-American novelist-journalist Abraham Cahan, 37, whose first novel was published last year under the title *Yekl: A Tale of the New York Ghetto* (*see* the *Day*, 1914).

The *New York Tribune* prints halftones on a power press and on newsprint for the first time, employing techniques developed by Frederick E. Ives and Stephen Horgan (*see* Ives, 1886).

William Randolph Hearst's *New York Journal* hires *Sunday World* editor Arthur Brisbane, 32, who quits the Pulitzer paper and uses sensationalist "yellow journalism" techniques to build *Jour*nal circulation from 40,000 to 325,000 in 6 weeks (*see* 1895; 1898).

Ladies' Home Journal publisher Cyrus H. K. Curtis pays $1,000 to acquire the 76-year-old *Saturday Evening Post* (*see* 1883; Lorimer, 1899).

"Yes, Virginia, there is a Santa Claus," writes *New York Sun* editor Francis Church, 58, in a September 21 editorial reply to reader Virginia O'Hanlon, age 8, who has written to inquire if Santa Claus really exists: "Not believe in Santa Claus? You might as well not believe in fairies . . . No Santa Claus! Thank God, he lives, and he lives forever" (*see* 1823).

"The Katzenjammer Kids" by German-American cartoonist Rudolph Dirks begins appearing in the *New York Journal* (above) December 12. The antics of Hans und Fritz (who torment der Captain, der Inspector, und Momma) ape those of the German cartoon characters Max und Moritz by Wilhelm Busch (*see* 1917).

A massive, ornate Library of Congress is completed at Washington, D.C. The Renaissance-style structure has been built to house the books that have piled up in a room at the Capitol since 1851.

The Monotype typesetting machine introduced by U.S. inventor Tolbert Lanston, 53, is more practical for book publishers than the Merganthaler Linotype of 1884. Lanston obtained his first patents for the machine in 1885, it incorporates a keyboard machine that perforates paper rolls in certain patterns, each pattern representing a character, the perforations go into a caster and release matrices from which the characters are cast in single metal types (but not in complete lines), the Monotype assembles the types automatically to form lines, and the lines are then made up into pages.

Fiction *Fruits of the Earth* (*Les Nourritures terrestres)* by French novelist André Gide, 28; *The Understudy* (*La Doublure)* by French novelist Raymond Roussel, 20; *The Horses of Diomede* (*Les Chevaux de Diomède*) by Rémy de Gourmont; *The Nigger of the 'Narcissus'* by Joseph Conrad; *The Spoils of Poynton* and *What Maisie Knew* by Henry James; *The Invisible Man* by H. G. Wells; *Liza of Lambeth* by English novelist William Somerset Maugham, 23; *Captains Courageous* by Rudyard Kipling; *Dracula* by Dublin-born English writer Bram Stoker, 50, who is secretary and business adviser to English actor Sir Henry Irving, 59 (the first actor to be knighted). Stoker has read up on vampire folklore and on Transylvania at the British Museum and embellished the historic Vlad Dracula who did not drink blood but did have hordes of people impaled on spears, high or low depending on their rank, and had other victims blinded, strangled, and boiled alive (*see* 1476).

Poetry *Jeanne d'Arc* by French poet Charles Péguy, 24; "Recessional" by Rudyard Kipling on the occasion of Queen Victoria's second jubilee (in the *Times* July 17): "God of our fathers, known of old/ Lord of our far-flung battle-line,/ Beneath whose awful Hand we hold/ Dominion over palm and pine/ Lord God of Hosts, be with us yet,/ Lest we forget—lest we forget!"; *Lou Pouèmo Dòu Rose* by Frédéric Mistral, now 67; *The Year of the Soul* by Stefan George.

Painting *The Sleeping Gypsy* by Henri Rousseau; *Large Interior* by Jean Edouard Vuillard; *Boulevard des Italians, Morning, Sunlight* by Camille Pissarro; *Frieze of Life* by Edvard Munch.

Theater *John Gabriel Borkman* by Henrik Ibsen 1/10 at Helsingfors; *The Little Minister* by James M. Barrie 9/27 at New York's Empire Theater, with U.S. actress Maude Adams (Maude Kiskadden), 25, who has adopted her mother's maiden name, 300 perfs.; *The Devil's Disciple* by George Bernard Shaw 10/4 at New York's Fifth Avenue Theater, with Richard Mansfield, 64 perfs.; *The Liars* by Henry Arthur Jones 10/6 at London's Criterion Theatre; *Cyrano de Bergerac* by French playwright Edmond Rostand, 29, 12/28 at the Théâtre de la Porte-Saint-Martin, Paris, 200 perfs.

The Théâtre du Grand Guignol opens in Montmartre under the direction of Paris theatrical manager Oscar Metenier, 38, whose bloodcurdling horror plays will overshadow his farces (*see* 1815).

The Orpheum Circuit is founded by German-American theatrical promoter Gustav Walter, who arrived at San Francisco in 1874, opened the Orpheum Theater in 1887 with a 22-piece Hungarian electrical orchestra, introduced Weber and Fields of New York to California 2 years later, leased Child's Opera House at Los Ange-

1897 ***(cont.)*** les last year, opened a booking office at Chicago, and now opens a third theater at Kansas City. Walter will die next year on his return from a tour of Europe in search of new talent.

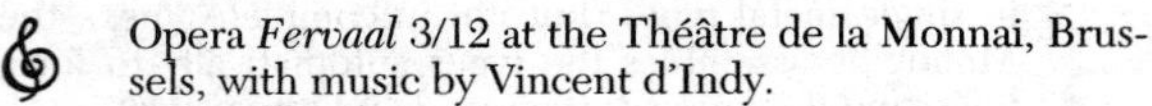

Opera *Fervaal* 3/12 at the Théâtre de la Monnai, Brussels, with music by Vincent d'Indy.

First performances *Istar* Symphonic Variations by Vincent d'Indy 1/10 at both Brussels and Amsterdam; Symphony No. 1 in D minor by Sergei Rachmaninoff 3/15 at St. Petersburg; *The Sorcerer's Apprentice (L'Apprenti sorcier, Scherzo d'après une ballade de Goethe)* by French composer Paul Dukas, 31, 5/18 at Paris; "Thanksgiving and Forefathers Day for Orchestra and Chorus" by Yale undergraduate Charles Edward Ives, 23, 11/25 at New Haven's Centre Church on the Green. The congregation is not receptive.

Broadway musicals *The Good Mr. Best* 8/3 at the Garrick Theatre, with book by John J. McNally and a demonstration in the third act of the new cinematograph (*see* 1895); *The Belle of New York* 9/18 at the Casino Theater, with Edna May, 17, as the Salvation Army girl Violet Gray, Harry Davenport, music by Gustave A. Kecker, book and lyrics by Hugh Martin, songs that include "The Anti-Cigarette Society," "You and I," 56 perfs.

Congress broadens U.S. copyright laws to give copyright owners exclusive rights to public performances of their works, but the new law will be widely flouted (*see* 1909; ASCAP, 1914; Victor Herbert, 1917).

March "The Stars and Stripes Forever" by John Philip Sousa, who quit the Marine Corps in 1892. He has been touring the world with his own band and wrote the new march last year while returning from England.

Popular songs "Take Back Your Gold" by U.S. composer Monroe H. Rosenfeld, lyrics by Louis W. Pritzkow; "On the Banks of the Wabash Far Away" by U.S. songwriter Paul Dresser (*né* Dreiser), 40; "Asleep in the Deep" by U.S. minstrel composer Henry W. Petrie, 40, lyrics by English-American writer Arthur J. Lamb, 27.

The world heavyweight boxing title changes hands March 17 in the 14th round of a match held at Carson City, Nev. New Zealand-American Robert Prometheus "Bob" Fitzsimmons, 34, knocks out the champion James J. Corbett with a "solar plexus punch" (*see* 1892; Jeffries, 1899).

The first Boston Marathon is run April 19. The Boston Athletic Association, which insists that the long-distance foot race be free of "professionalism, politics, and plugs," will extend the annual 25-mile Patriot's Day race to 26 miles, 385 yards after the 1908 Olympic Games when officials will lengthen the Olympic Course so that the British royal family may view the start at Windsor Castle.

Reginald Frank Doherty, 22, wins in men's singles at Wimbledon, Mrs. Hillyard in women's singles; Robert Wrenn wins in U.S. men's singles, Juliette Atkinson in women's singles.

The United All-England Croquet Association is founded as improved mallets, wickets, and balls revive popularity for the game (*see* 1868).

John Peter "Honus" Wagner, 23, joins the National League's Louisville team to begin a notable career in baseball. The "Flying Dutchman" will lead the league five times in stolen bases, accumulate 722 stolen bases, and have a lifetime batting average of .329.

The first Cheyenne Rodeo is held in July, beginning an annual Wyoming tradition.

Steeplechase Park opens at Coney Island, N.Y., under the management of real estate operator George Cornelius Tilyou, 35, who has invented many of the rides and fun houses himself for the large amusement area. Competing attractions will spring up in the next decade to cover 20 acres and make Coney Island the greatest amusement area in the world.

Vienna's 131-year-old Prater Park installs a 210-foot high Ferris wheel. Riders get a panoramic view of the Danube and the city from the wheel's spacious cabins (*see* 1893; St. Louis, 1904).

Lifebuoy soap is introduced by Sunlight soap producer W. H. Lever whose strong soap will be promoted as a safeguard against body odor (*see* Lux Flakes, 1906).

London's Moss Bros. of Covent Garden goes into the dresswear hire business to accommodate a customer who has gone broke as a stockbroker and obtained a professional engagement as a monologuist. Started by Moses Moss in 1860 and managed by his eldest son Alfred, the secondhand clothing shop will serve generations of Englishmen, including the nobility, who will hire from Moss Bros. not only morning dress and evening wear but also coronation robes, court dress, ball gowns, wedding dresses, ski clothes, and theatrical costumes.

Mail Pouch tobacco is introduced under the name West Virginia Mail Pouch October 15 by Wheeling, W. Va., stogie makers Aaron and Samuel Bloch who started making stogies in the back of Samuel's wholesale grocery and dry goods store in 1879. Farmers will be paid for more than half a century to let Bloch Brothers use their barns as advertising billboards for the Mail Pouch brand chewing tobacco made from flavored stogie wrapper clippings.

New York's Waldorf-Astoria Hotel is opened by John Jacob Astor IV, 30, who has had architect Henry Janeway Hardenbergh design a 17-story Astor House at 34th Street and Fifth Avenue and combined it with the 13-story Hardenbergh-designed Waldorf Hotel opened by his estranged cousin William Waldorf Astor in 1893. The 1,000-room hotel has 765 private baths and is the largest, most luxurious hotel in the world (*see* new Waldorf-Astoria, 1931).

President Cleveland sets aside 20 million acres of additional western forest reserves (*see* Harrison, 1891). Congress has passed the Forest Service Organic Act, enabling legislation for the establishment of a national forest system (it forbids government sale of National Forest timber not dead or fully mature and individually marked for felling). Some senators demand his impeachment, but Cleveland vetoes bills passed to reverse his proclamation and Congress adjourns without overriding the vetoes (*see* Pinchot, 1898).

More than 1.2 million pounds of sturgeon are landed in New York and New Jersey, but pollution begins to reduce the Hudson River's "Albany beef" industry.

Forty-nine million pounds of shad are landed on the U.S. East Coast.

The drought that began on the U.S. western plains in 1886 finally comes to an end.

Europe's wheat crop falls short of its needs, and wheat futures are bid up on the Chicago Board of Trade as speculator Joseph Leiter, 29, makes a spectacular effort to corner the market. Son of the onetime Marshall Field partner Levi Leiter, young Leiter has been given $1 million by his father to start a real estate business but instead has started to buy wheat futures at 65¢ per bushel and made a neat profit which he invests in more futures, continuing to buy until he has acquired 12 million bushels for December delivery, 9 million of them from meat packer P. D. Armour. Leiter refuses to take his profit of several million dollars in hopes of still higher profits, but Armour discovers that grain elevators at Duluth are bulging with wheat, charters 25 lake vessels and tugs that can break through the ice to Duluth, and sends them north to get wheat that will break Leiter's corner in the market (*see* 1898).

U.S. farmers reap profits of millions of dollars as wheat prices rise to $1.09 per bushel, highest since 1891, partly as a result of Joseph Leiter's buying of futures (above).

Campbell Preserve Co. chemist John T. Dorrance, 24, works to develop double-strength "condensed" soup that will give Campbell's soup dominance in the industry. A nephew of company president Arthur Dorrance, John has degrees from MIT and the University of Göttingen and persuades his uncle to hire him as researcher (*see* 1894; 1898).

The introduction of double seams and improved crimping of body and ends makes tin cans more reliable (*see* 1876; 1895; aluminum, 1960).

Grape Nuts is introduced as a health food by C. W. Post of 1893 Postum fame who includes a copy of his pamphlet *The Road to Wellsville* with each box of his ready-to-eat cold cereal of wheat and malted barley baked in the form of bread sticks. Post believes that grape-sugar (dextrose) is formed during the baking process (*see* Post Toasties, 1904).

Purina breakfast cereal is introduced by the 3-year-old St. Louis livestock food company Robinson-Danforth which names its product Purina to signify purity (*see* Ralston, 1902).

American Sugar Refining makes a deal with Claus Spreckels in San Francisco to eliminate competition (*see* 1891; 1895; 1899).

Jell-O is introduced by LeRoy, N.Y., cough medicine manufacturer Pearl B. Wait whose wife Mary gives the product its name. Wait's gelatin dessert is made from a recipe adapted from one developed by Peter Cooper in 1845. The powder is 88 percent sugar (*see* 1899).

Welch's Grape juice production moves to Westfield, N.Y. (*see* 1896). Charles Welch processes 300 tons of grapes at Westfield, a stronghold of the Women's Christian Temperance Union in the heart of the 90-mile Chautauqua and Erie grape belt along the southeastern shore of Lake Erie (*see* 1945).

Britons begin to eat lunch, dooming the classic British breakfast which still often includes kippers (smoked herring, split open and fried in butter), finnan haddie (smoked haddock, often from the Scottish fishing port of Findhorn), kedgeree (rice mixed with fish, usually salmon, and hard-boiled egg—a dish introduced from colonial India), roast beef, kidneys, bacon, sausages, porridge, snipe (Scotland), scones (Scotland), cold toast, butter, marmalade, treacle, eggs, and tea with milk.

More than 60 cases of champagne are consumed by 900 guests in Louis XV period costumes at a $9,000 ball given February 10 by Mrs. Bradley Martin, daughter of Carnegie Steel magnate Henry Phipps, who has had a huge Waldorf-Astoria suite decorated in the manner of Versailles despite the national economic depression that has persisted since 1893.

1898 The U.S. battleship *Maine* blows up in Havana harbor February 15 in an explosion that kills 258 sailors and two officers, precipitating a Spanish-American War that lasts for 112 days (*see* Sagasta, 1897). The sinking of the *Maine* follows by 6 days the publication in William Randolph Hearst's *New York Journal* of a letter stolen from the mail at Havana, a private letter from the Spanish minister to the United States calling President McKinley a spineless politician (Madrid has recalled the minister).

Commodore George Dewey, 60, receives a secret cable February 25 from Assistant Secretary of the Navy Theodore Roosevelt ordering him to proceed with his Asiatic squadron to Hong Kong and prepare for an attack on the Spanish squadron in the Philippines in the event of war.

Mobilization of U.S. Army forces begins March 9, and Congress votes unanimously to appropriate $50 million "for national defense and each and every purpose connected therewith."

A Spanish board of inquiry attributes the destruction of the *Maine* (above) to an internal explosion, but the report of the Spanish board March 22 is disputed by a U.S. naval court of inquiry which has reported March 21 that an external cause produced the explosion. The U.S. report is made public March 28, a joint note to President McKinley from ambassadors of six great powers April 6 calls for a peaceful solution to the Cuban problem, the Spanish queen Marie-Amelie acts on advice from Pope Leo XIII and orders a suspension of hostilities against Cuban rebels to "facilitate peace negotiations," but President McKinley sends a war message to Congress April 11.

1898 ***(cont.)*** A joint resolution of Congress April 19 recognizes Cuban independence, authorizes the president to demand Spanish withdrawal from the island, and disclaims any intention to annex Cuba.

Congress passes a Volunteer Army Act April 22; a Volunteer Cavalry—the "Rough Riders"—is organized by Col. Leonard Wood of the Army Medical Corps and Lieut. Col. Theodore Roosevelt (above) who resigns his Navy post in May.

The first shots of the Spanish-American War are fired April 22, Spain declares that a state of war exists April 24, Congress formally declares war April 25, Comm. Dewey receives the news in Hong Kong April 26, and he enters Manila Bay just before midnight April 30.

The Battle of Manila Bay May 1 begins at 5:40 in the morning when Comm. Dewey says to the captain of his flagship, "You may fire when you are ready, Gridley." By the time a cease-fire is ordered at 12:30 in the afternoon, all 10 ships in the Spanish squadron have been destroyed with a loss of 381 men, while eight Americans have been slightly wounded and none killed. Dewey is elevated to rear admiral May 11.

San Juan, Puerto Rico, is bombarded May 12 by the commander-in-chief of the North Atlantic Squadron William Thomas Sampson, 58, who headed the board of inquiry into the explosion that sank the *Maine*.

U.S. Marines invade Cuba's Guantanamo Bay June 10; 623 men and 24 officers go ashore.

Some 16,887 U.S. troops leave Tampa, Fla., June 14 for Santiago, Cuba, singing, "There'll Be a Hot Time in the Old Town Tonight."

Leading U.S. intellectuals meet at Faneuil Hall, Boston, June 15 and form an Anti-Imperialist League to oppose annexation of the Philippines in face of imperialist sentiment fired by the Battle of Manila (above).

Admiral Dewey's victory at Manila Bay in a brief war with Spain established the United States as a Pacific power.

A joint resolution proposing annexation of the Hawaiian Islands has been introduced in the House May 4, President McKinley signs the measure July 7, and Hawaiian sugar planters gain free access to U.S. markets (*see* 1897; 1959).

U.S. Army forces under Major Gen. of Volunteers William Rufus Shafter, 63, arrive off Santiago, Cuba, June 20 and disembark 2 days later with the Rough Riders of Col. Wood and Lieut. Colonel Roosevelt.

The Battle of Las Guasimas June 24, the first land battle of the war, ends in a Spanish defeat by U.S. forces under the command of Major Gen. of Volunteers Joseph Wheeler, 62; Leonard Wood and Theodore Roosevelt have pushed ahead of other troops with 1,000 regulars and Rough Riders.

The Battle of San Juan Hill July 1 to 2 ends in victory for U.S. forces under General Hamilton S. Hawkins, whose infantry take the hill after a charge by Col. Roosevelt and the Rough Riders who also help take Kettle Hill. Some 1,460 men and 112 officers are killed, wounded, or missing after the action.

The Battle of Santiago Bay July 3 ends with 180 Spaniards dead and 1,800 captured, while the Americans have lost only one man and sustained one wound casualty. "Don't cheer, boys, the poor devils are dying," shouts Capt. John Woodward Philip, 58, of the U.S. battleship *Texas* as the Spanish cruiser *Vizcaya* is driven ashore in flames, but only 3 percent of U.S. high-caliber shells have scored hits.

U.S. troops under Gen. Nelson A. Miles invade Puerto Rico July 25 and receive the surrender of Ponce July 28.

A peace protocol is signed with Spain August 12, and the Treaty of Paris formally ends the war December 10. Spain withdraws from Cuba and cedes Puerto Rico, Guam, and the Philippines to the United States, which pays $20 million for the Philippines (*see* 1946).

Cuba gains her independence from Spain, which loses her last dominions in the Americas (*see* 1902).

The Nishi-Rosen protocol signed April 25 by the Russian minister to Japan Baron Roman Romanovich Rosen, 51, pledges both Russia and Japan to refrain from interfering in Korea's affairs while permitting Japanese economic penetration of the country (but *see* 1904).

The Battle of Omdurman September 2 gives Gen. Kitchener a decisive victory over the khalifa of the Sudan Abdullah el Taashi. Using Maxim machine guns, the British kill 11,000 dervishes, wound 16,000, and take 4,000 prisoner, sustaining only 48 casualties. Kitchener's British army takes Khartoum from the dervishes, reaches Fashoda September 19, and finds it occupied by French forces under Major Jean-Baptiste Marchand.

France claims the left bank of the Nile, Ethiopia the right bank, London demands that the French evacuate the territory that Britain claims for Egypt by right of conquest, the French try to get Russian support but fail, and Paris orders the evacuation of Fashoda November 3.

Italian anarchist Luigi Luccheni stabs the Austrian empress Elizabeth September 10 at Geneva as she walks from her hotel to the steamer. She dies within a few hours at age 60.

New York City becomes Greater New York January 1 under terms of an 1896 law uniting Kings County (Brooklyn), Richmond County (Staten Island), Bronx County, Long Island City, Newtown (Queens County), and Manhattan to create a metropolis of just under 3.5 million inhabitants.

"J'Accuse!" headlines the Paris newspaper *L'Aurore* January 18 over an open letter by novelist Emile Zola. His attack forces a new trial of Capt. Dreyfus, who has been sent to the penal colony called Devil's Island off the coast of French Guiana *(see* 1897). It is revealed that the documents that convicted Dreyfus were forged by two other officers (a Col. Henry and Major Esterházy) and that Dreyfus was the victim of an anti-Semitic plot (*see* 1899).

Louisiana adopts a new constitution with a "grandfather clause" restricting permanent voting registration to whites and those blacks whose fathers and grandfathers were qualified to vote as of January 1, 1867, a clause that virtually disenfranchises blacks. Race riots and lynchings sweep the South (*see* Mississippi, 1890; Alabama, 1901).

Whites battle Indians in Minnesota.

Virden, Ill., coal mine operators attempt to break a strike by importing 200 nonunion black workers, an action that provokes violence: 14 miners are killed and 25 wounded in the October 12 Mt. Olive massacre that brings demands for a union (*see* 1899).

Illinois Steel of Chicago and Lorrain Steel acquire Minnesota Mining with backing from J.P. Morgan & Co., obtaining a fleet of Great Lakes ore ships and railroads in the Mesabi iron range and in the Chicago area (*see* 1896; U.S. Steel, 1901).

Republic Steel is created by a merger of Ohio and Pennsylvania firms (*see* strike, 1937).

John W. Gates becomes president of American Steel & Wire, which has a virtual monopoly in barbed wire (*see* 1875). He has established his own barbed wire company at St. Louis and negotiated a series of mergers and consolidations (*see* Tennessee Coal & Iron, 1906).

The Supreme Court sustains an 1895 Illinois inheritance tax law April 25 in *Morgan v. Illinois Trust and Savings Bank*; Congress imposes the first U.S. federal tax on legacies June 13 in a War Revenue Act that also provides for excise duties and taxes on tea, tobacco, liquor, and amusements.

Canada's Klondike yields more than $10 million worth of gold, a quantity it will sustain through 1904 (*see* 1897).

Union Carbide is founded by Chicago entrepreneurs to manufacture calcium carbide for producing acetylene gas for streetlights and home lighting, which still depends largely on coal gas and kerosene (*see* 1892; 1906).

The Supreme Court establishes the right of the courts to decide the reasonableness of U.S. railroad rates—and the doctrine of judicial review—in the case of *Smyth v. Ames* decided March 7.

Paris jeweler Alfred Cartier takes his sons Louis, Jacques, and Pierre into the business and opens a new shop at 13 rue de la Paix (*see* 1847). Louis, 23, will become Cartier's creative genius (*see*1902).

Timken Roller Bearing Axle Co. is founded by German-American carriage maker Henry Timken, 67, who opened a St. Louis carriage works in 1855, patented a special type of carriage spring in 1877, and has just patented a tapered roller bearing that will make his company the leader in its field.

The Indian motorcycle, introduced by the new Hendee Manufacturing Co. of Springfield, Mass., improves on earlier motorcycles (which have essentially been bicycles with one-cylinder gasoline engines attached). Motorcycle pioneer George Hendee will begin mass production in 1902, and his company will be renamed Indian Motorcycle in 1923 (*see* Harley-Davidson, 1903).

U.S. motorcar production reaches 1,000, up from 100 last year.

The Opel motorcar is introduced by German bicycle maker Adam Opel at Rüsselsheim (*see* 1929).

Goodyear Tire and Rubber is founded at Akron, Ohio, by Frank Augustus Sieberling, 39, who will make it the leading U.S. tire maker.

A pilot plant to produce viscose rayon yarn that can be woven and dyed opens at Kew, Surrey, England (*see* Cross, Bevan, 1892). English inventor C. H. Stearn patents a viscose filament produced by treating wood pulp with caustic soda (sodium hydroxide) (*see* Courtalds, 1905; Little, 1902).

French physical chemist Marie (Marja) Sklodowska Curie, 31, and her husband Pierre, 39, isolate radium, the first radioactive element *(see* 1904; uranium, 1896).

The Palmer School of Chiropractic is founded at Davenport, Iowa, by Canadian-American magnetic healer Daniel David Palmer, 53, who has developed a system of manually manipulating the joints, especially the spine, for which he makes extravagant claims.

Lydia Pinkham's Vegetable Compound is widely advertised as "The Greatest Medical Discovery Since the Dawn of History" (*see* 1876). The compound of black cohosh, liferoot plant, fenugreek seeds, and other herbs in a 21 percent alcohol solution promises to remedy female complaints. It will continue to be made and sold for another 76 years, but is outsold by the patent medicine Pe-Ru-Na, which has a higher alcoholic content; sold for catarrh, it has made physician-promoter Samuel Brubaker Hartman, 68, $2.5 million.

Heroin is introduced under that brand name as a cough suppressant derived from opium by the 48-year-old German chemical-pharmaceutical firm Farbenfabriken vorm. Friedrich Bayer und Co. (*see* aspirin, 1899).

1898 ***(cont.)*** Bubonic plague will kill an estimated 3 million people in China and India in the next decade (*see* 1910).

Gideons International has its beginnings at Boscobel, Wis., where traveling salesmen John H. Nicholson and Sam Hill share a room at the Central Hotel and decide to form an association of Christian businessmen (excluding those in the liquor trade) and professional men to "put the Word of God into the hands of the unconverted." The association will raise money for their work, it will place 25 bibles in a Montana hotel in 1908, and by 1976 the Gideons will be placing 16.5 million bibles per year in hotels, hospitals, prisons, schools, colleges, and other locations.

Northeastern University is founded at Boston.

A message to Garcia is delivered by U.S. Army Lieut. Andrew Rowan who is sent on a secret mission in April to obtain certain information from Cuban insurgent General Calixto Garcia. Despite a breakdown in ordinary lines of communication, Rowan delivers his message, obtains the desired information, and returns to Washington D.C. Garcia is named to represent Cuba in negotiations with the United States for Cuban independence, but dies at age 62 (*see* Elbert Hubbard, 1899).

William Randolph Hearst publishes special editions of his *New York Journal* from his private yacht anchored in Havana Harbor. He has sent cowboy painter Frederic Remington to Cuba earlier, and when Remington cabled that everything was quiet and that there would be no war, had replied, "You furnish the pictures and I'll furnish the war." *Journal* stories inflame U.S. opinion ("Remember the *Maine*!"), and the paper's circulation reaches 1.6 million May 2.

The *New York Times* drops its price from 3¢ to 1¢ and circulation triples to 75,000 within a year. By 1901 it will be over 100,000 (*see* 1896).

Sunset magazine begins publication at Los Angeles to promote business for the Southern Pacific by attracting tourists, settlers, and developers to the West. Named for the railroad's crack *Sunset Limited* between New Orleans and Los Angeles, it will become a family magazine in 1928.

The Telegraphone patented by Danish electrical engineer Valdemar Paulsen, 29, is the world's first magnetic wire recording device (*see* 1929).

Nonfiction *Mr. Dooley in Peace and in War* by *New York Evening Journal* humorist Finley Peter Dunne, 31, whose book *Mr. Dooley in the Hearts of His Countrymen* also appears, beginning a series by a fictitious Chicago bartender whose witty and skeptical remarks to his silent and gloomy companion Malachi Hennessey are published in Dunne's weekly newspaper column; *Wild Animals I Have Known* by English-Canadian author Ernest Thompson Seton, 38, launches its author on a career.

Fiction "The Turn of the Screw" (story) by Henry James; *The War of the Worlds* by H.G. Wells (*see* 1938); *David Harum, A Story of American Life* by the late U.S. banker-novelist Edward Noyes Westcott who has just died at age 52: "Them that has gits"; *Evelyn Innes* by George Moore; *The Forest Lovers* by English novelist Maurice Hewlett, 37; *The Woman and the Puppet (La Femme et le pantin)* by French novelist-poet Pierre Louÿs (Pierre Louis), 28.

Poetry *From the Dawn to the Evening Angelus* (*De l'Angelus d'aube à l'angelus du soir)* by French poet Francis Jammes, 30; "The Ballad of Reading Gaol" by Oscar Wilde, who was released from prison last year after serving the 2-year sentence imposed in 1895, has emigrated to Paris, and will die there late in November 1900 at age 44: "Yet each man kills the thing he loves,/ By each let this be heard,/ Some do it with a bitter look/ Some with a flattering word,/ The coward does it with a kiss,/ The brave man with a sword!"

Glasgow's Mackintosh School of Art is founded.

Aubrey Beardsley dies of tuberculosis at Menton, France, March 16 at age 26; Gustave Moreau dies at Paris April 18 at age 72; Sir Edward Coley Burne-Jones dies at London June 17 at 64; Pierre C. Puvis de Chavannes dies at Paris October 10 at 73.

The Graflex camera patented by U.S. inventor William F. Folmer, 36, is the world's first high-speed multiple-split focal plane camera.

Photographs taken with artificial light are produced for the first time (*see* flash bulb, 1930).

Theater *The Dream of a Spring Morning* (*Sogno d'un mattino di primavera*) by Gabriele D'Annunzio 1/1 at Rome's Teatro Valle; *Trelawney of the "Wells"* by Arthur Wing Pinero 5/27 at the St. James Theatre, London; *Drayman Henschel (Fuhrmann Henschel)* by Gerhart Hauptmann 11/5 at Berlin's Deutsches Theater; *Ghetto* by Dutch playwright Herman Heijermans, 34, 12/24 at Amsterdam's Hollandsche Schouwburg.

Russian actor-producer Konstantin Sergeevich Stanislavski (K. S. Alekseev), 35, founds the Moscow Art Theater and will head it for 40 years. He mounts a production of Chekhov's *The Seagull* in December, the first effective peformance, and it is a brilliant success where it failed dismally 2 years ago at St. Petersburg. Stanislavski's "method" will revolutionize acting with infusions of genuine emotion.

French film pioneer Léon Gaumont opens a London office in Cecil Court. English film pioneer Cecil Hepworth, 23, builds a 10-foot by 6-foot studio in his garden at Walton-on-Thames.

First performances *Pelléas and Mélisande* orchestral suite by Gabriel Fauré 6/21 at the Prince of Wales' Theatre, London, for a production of the 1893 Maurice Maeterlinck play with Mrs. Patrick Campbell.

Broadway musical *The Fortune Teller* 9/26 at Wallack's Theater, with music by Irish-American conductor-composer Victor Herbert, 30, book and lyrics by Harry B. Smith, songs that include "Gypsy Love Song," 40 perfs.

Popular songs "The Rosary" by Ethelbert Nevin, 36, lyrics by Robert Cameron Rogers, 36; "When You

Were Sweet Sixteen" by English-American songwriter James Thornton, 37.

Canadian-American yachtsman Joshua Slocum, 54, brings his 37-foot sloop *Spray* into Newport, R.I., June 27 after completing the first one-man circumnavigation of the world. He has used only a compass, sextant, and "dollar clock" as navigational instruments in his 3-year voyage.

Reginald Doherty wins in men's singles at Wimbledon, Charlotte Cooper in women's singles; Malcolm D. Whitman, 21, wins in U.S. men's singles, Juliette Atkinson in women's singles.

The touchdown in U.S. college football receives a value of 5 points, up from the 4 established in 1884, and athletic directors give the goal after touchdown a value of 1 point, down from 2 in 1884 (*see* 1912).

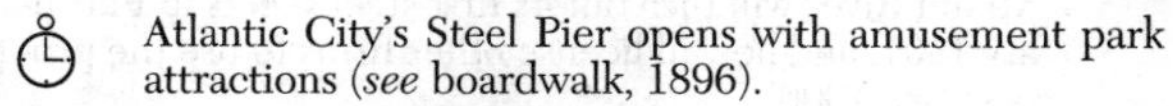

Atlantic City's Steel Pier opens with amusement park attractions (*see* boardwalk, 1896).

Louis Vuitton brands his initials on the canvas luggage he has created, hoping to discourage imitators. Once a dress packer to the empress Eugènie, the French luggage maker has gained wide acceptance for his trunks and cases, and will introduce them in the United States in 1902.

New York's Condict building designed by Chicago architect Louis Sullivan is completed at 65 Bleecker Street for Silas Alden Condict. The radical structure will later be called the Bayard building.

Brooklyn's Grand Army Plaza is graced with an 80-foot-high arch designed by architect John Duncan to commemorate Union Army forces in the Civil War. Stanford White creates four tall Doric columns to mark the entrance to Prospect Park, and sculptor Frederick MacMonnies works on a statue of Victory in a two-wheeled chariot drawn by four horses abreast to surmount Duncan's triumphal arch.

Chicago's Gage building on South Michigan Avenue is completed by Holabird and Roche.

Claridge's Hotel opens in London's Brook Street, where it has been rebuilt on a site occupied by an earlier hotel that operated under the name Claridge's for half a century.

The Paris Ritz opens in a 17th-century townhouse in the Place Vendôme. César Ritz has found that the duc de Lauzun's property was for sale, has suggested the name "Grand Marnier" to London liqueur maker Marnier La Postolle for La Postolle's orange-flavored cordial, La Postolle has given him funds to buy the townhouse, and Ritz opens the new hotel with 170 guest rooms (and with chef Auguste Escoffier of 1894 Pèche Melba fame) (*see* London Ritz, 1902).

Gifford Pinchot is appointed chief forester in the U.S. Department of Agriculture after 3 years of managing Vanderbilt forests in the Great Smokies. He had studied in France under Sir Dietrich Brandis, who founded forestry in British India, and has made long field trips (*see* 1902).

U.S. wheat futures reach $1.85 per bushel as the outbreak of the Spanish-American War (above) booms demand. Joseph Leiter has paper profits of $7 million (*see* 1897), but he holds contracts for 35 million bushels, and when wheat pours into Chicago in June, Leiter's corner on the market is broken. He not only loses his profit but winds up with a debt of $10 million which is paid by his father.

China has serious famine as the northern provinces suffer drought, while in Shandong province the Huanghe (Yellow) River overflows its banks to create disastrous floods.

More U.S. troops in the Spanish-American War (above) die from eating contaminated meat than from battle wounds. The deaths raise a public outcry for reform of the meat-packing industry (*see* 1905).

Sugar prices soar following the outbreak of the Spanish-American War (above). Sugar broker George Edward Keiser started buying stock in Cuban-American Sugar at $5 per share and continues to buy as the price soars to $300, buying so much that he winds up controlling the company.

Campbell's soups appear for the first time with red and white labels whose colors have been suggested by Cornell football uniforms (*see* 1897; 1904).

Sanitas corn flakes are introduced by Sanitas Nut Food Co., set up at Battle Creek, Mich., by J. H. Kellogg with his brother Will Keith (*see* Granose, 1895). The world's first corn flakes quickly turn rancid on grocers' shelves and have little acceptance in a market oriented toward wheat cereals (*see* 1902).

National Biscuit Co. is formed by a consolidation of New York Biscuit, American Biscuit and Manufacturing, United States Baking, and United States Biscuit (*see* 1890). Adolphus W. Green heads the new company, whose 114 bakeries comprise 90 percent of all major U.S. commercial bakeries.

Uneeda Biscuits are created by National Biscuit's A. W. Green (above) who seeks to establish a brand name that will surmount the anonymity of the cracker barrel seen in every grocery shop. Green decides to concentrate the company's efforts behind a flaky soda cracker, retains N. W. Ayer & Son as his advertising agency, and is advised by Ayer's Henry N. McKinney that a manufacturer must "insure his future sales by adopting a trade name and trade dress which will belong exclusively to him by trademark and trade right." McKinney suggests the name Uneeda Cracker, Green changes it to Uneeda Biscuit, and the name is registered December 27 with the U.S. Patent Office (*see* 1900).

Pepsi-Cola is introduced by New Bern, N.C., pharmacist Caleb Bradham who has been mixing fountain drinks since 1893 and developed a cola drink formula (*see* 1902).

Mechanical refrigeration gets a boost from Swedish inventor Carl von Linde who perfects a machine that liquefies air (*see* 1873; Freon 12, 1931).

1898 ***(cont.)*** Annual British tea consumption averages 10 pounds per capita, up from 2 pounds in 1797 (*see* 1869).

The U.S. Supreme Court rules March 28 that a child born of Chinese parents in the United States is a citizen and cannot be denied re-entrance to the United States by the Chinese Exclusion Laws of 1880 and 1892 (*United States v. Wong Kim Ark*).

1899 Two U.S. privates open fire on Filipino soldiers outside Manila on the night of February 4, beginning a 3½-year war between U.S. troops and Filipino national forces led by Emilio Aguinaldo. U.S. forces secure Luzon November 24 but many Americans oppose the "imperialist" war.

An "Open Door" policy in China proposed by U.S. Secretary of State John Milton Hay, 61, receives support from the great powers. They will agree that all the imperialist countries shall have equal commercial opportunity in spheres of special interest (*see* 1900).

France's President Faure dies suddenly February 16 at age 57, having opposed a second trial for Captain Dreyfus. Emile Loubet is elected to succeed him.

The Action Française right-wing political movement founded by Charles Maurras, 31, attracts support from the Paris newspaper *Action Française,* edited by pamphleteer Léon Daudet, 32, who rallies the defeated opponents of Alfred Dreyfus (below) with royalist and nationalist invective.

The Boer War begins in South Africa October 12 as President Kruger of the Boer republic acts to block suspected British moves toward acquiring the rich Transvaal with its gold mines (*see* 1886). Equipped with Krupp artillery, the Boers lay siege to Mafeking October 13, Kimberley October 15, and Ladysmith November 2. British forces are equipped with Vickers-Maxim weapons supplied by Basil Zaharoff, 49, a Turk who has become chairman of Vickers and who also supplies the Boers, making a fortune (*see* 1900; Maxim, 1883).

West Africa's Ashanti stage their last uprising against the British. Colonial officer Sir Fredric Hodgson has provoked the Ashanti by demanding that they surrender the Golden Stool of Friday that has been their sacred symbol of power for the past century. Hodgson wants the throne so that he may sit upon it as governor of the Gold Coast, but the tribesmen lay siege to his fortress, and it is 2 months before he and his wife can escape to the coast.

"The White Man's Burden" by Rudyard Kipling appears in *McClure's* magazine. The *Times* of London calls it "an address to the United States."

Polish-born Berlin Marxist Rosa Luxembourg, 28, attacks arguments that labor's working conditions have improved and that reforms must come from within the system. Only international revolution can help the working man, says "Bloody Rosa" in April, and she will continue to agitate until her death in 1919.

The United Mine Workers of America, organized under the leadership of Illinois coal miner John Mitchell, 29, joins anthracite coal workers with bituminous coal workers who formed a union 9 years ago and who struck the mines in 1897 (*see* Virden, Ill., massacre, 1898). U.S. mines continue to employ thousands of boys barely into their teens (*see* 1902).

Capt. Dreyfus wins a pardon September 19 after a French Army retrial forced by public opinion (*see* 1898; 1906).

Carnegie Steel is created by a consolidation of various steel properties controlled by Andrew Carnegie (*see* Homestead strike, 1892; U.S. Steel, 1901).

Armco Steel has its beginnings in the American Rolling Mill Co. founded at Middleton, Ohio, by Cincinnati entrepreneur George M. Verity, 34, who has developed a continuous wide-sheet roller mill. His process revolutionizes the manufacture of sheet steel, his mill on the Miami River will turn out its first steel sheets in February 1901, and he will license other firms to use the process (*see* 1906).

Bechtel Group has its beginnings as U.S. mule driver Warren A. Bechtel quits his job hauling train rails in Indian Territory at $2.75 per day and starts what will become a worldwide engineering concern.

U.S. copper producers merge to create the American Smelting and Refining Co. trust as growing use of electricity increases demand for copper wire.

The Guggenheims refuse to join the ASARCO copper trust (above), choosing instead to compete with it (*see* 1888). Meyer Guggenheim forms alliances with mine owners, gives them financial backing in many cases, and founds a company to seek new ore deposits (*see* 1901).

E. I. du Pont de Nemours is incorporated in Delaware (*see* 1897). Du Pont has been making dynamite since 1880 and now controls 90 percent of U.S. blasting powder production and 95 percent of U.S. gunpowder production (*see* 1902; Nobel, 1866).

International Paper Co. is created by a merger of nearly 30 U.S. and Canadian paper companies. It will purchase large timber tracts in Maine and California and build up a national marketing organization.

Union Bag & Paper is reorganized to create a $27 million trust that tries to squeeze out competitors such as International Paper (*see* 1869).

Consolidated Edison Co. is created by a merger of New York's Edison Illuminating Co. with Consolidated Gas, controlled by the Rockefeller family and William Whitney (*see* 1882; power failure, 1965).

Nippon Electric Co. (NEC) is founded with 92 employees. With 54 percent owned initially by Western Electric, it will grow to be Japan's largest producer of electrical equipment.

Renault Frères is founded in March by French auto makers Louis and Marcel Renault who produced their first *voiture* late last year (*see* 1900).

FIAT (Fabrica Italiana Automobili Torino) is founded July 1 at Turin by cavalry lieutenant Giovanni Agnelli, 33, and eight partners with a capitalization of 800,000 lire ($152,400). FIAT completes 10 3-horsepower motorcars by November and will grow to become Europe's largest auto maker and Italy's largest industrial firm (*see* tractors, 1919; "Topolino," 1936).

French racing driver Camille Jenatzy drives a La Jamais Contente at a speed of 67.79 miles per hour.

U.S. auto production reaches 2,500, up from 1,000 last year.

Henry Ford joins the new Detroit Automobile Co. as chief engineer (*see* 1893; 1901).

A group of financiers acquires the Selden road locomotive patent. They will bring a lawsuit for infringement of the patent next year and will win (*see* 1895; 1903).

A Stanley Steamer driven by F. E. Stanley climbs to the top of Mount Washington (*see* 1896).

B.F. Goodrich at Akron, Ohio, produces the first U.S. clincher tires, producing 19-ply rubber tires in sizes ranging from 28 by 2.5 inches to 36 by 3 inches (*see* 1872).

Kelly-Springfield pneumatic tires are introduced at Springfield, Ohio, by the Rubber Tire Wheel Co. (*see* Hartford Rubber Works, 1895).

American Car and Foundry (ACF) is founded at Berwick, Pa., to compete with the Pullman Palace Car Co. Founders include Charles Lang Freer, 43, who has been building railroad cars since age 17; ACF will become the world's largest maker of freight cars.

Boston's South Station opens to complement the North Station built in 1894.

Boston's last horsecar runs December 24. A trolley line replaces the horsecar as Boston extends its 2-year-old subway.

United Fruit Company is incorporated by banana exporter Minor C. Keith and the Boston Fruit Co., which has been distributing Keith's fruit since the failure of his New Orleans distributor (which cost Keith $1.5 million) (*see* 1871; 1885). Andrew Preston is president of the new firm, Keith first vice president. United Fruit controls 112 miles of railroad with 212,494 acres of land, more than 61,000 acres of it in production (*see* food, 1929).

Minor Keith (above) sees an opportunity for passenger traffic on the Fruit Company's banana boats and charters four new ships built originally for the U.S. Navy. The Great White Fleet he establishes includes the S.S. *Farragut*, S.S. *Admiral Dewey*, S.S. *Admiral Schley*, and S.S. *Admiral Sampson*, each able to carry 53 passengers and 35,000 bunches of bananas (*see* 1904).

Philadelphia engineers Frederick Winslow Taylor, 43, and Maunsel White, 43, develop the Taylor-White process for heat-treating highspeed tool steels, increasing cutting capacities of blade edges by 200 to 300 percent.

A violent hurricane sweeps Puerto Rico August 8, killing 2,100 and leaving thousands homeless. Assigned to a hospital treating native victims of the storm, U.S. Army Medical Corps assistant surgeon Bailey Kelly Ashford, 25, finds that 75 percent are anemic. When he examines their feces, he finds hookworm eggs, and when he treats patients who have not responded to arsenic, iron, or improved diet with thymol and epsom salts, he eliminates hookworm and sharply reduces the death toll from anemia (*see* Froelich, 1789; Stiles, 1902).

A cholera pandemic begins that will continue until 1923, affecting much of the world.

Aspirin (acetylsalicylic acid), perfected by German researchers Felix Hoffman and Hermann Dreser, will be marketed by prescription under the trade name Bayer Aspirin beginning in 1905 and go on to become the world's largest selling over-the-counter drug. Hoffman and Dreser have developed the powdered analgesic (painkiller) and fever reducer from coal tar; it is less irritating to gastrointestinal tracts than salicylic acid, the addition of the neutral salt calcium glutamate will make it less irritating still, and as tablets it will be consumed by the billion (*see* Gerhardt, 1853; Heroin, 1898).

Johnson & Johnson introduces zinc oxide adhesive plasters developed with help from some leading U.S. surgeons (*see* 1886). The new plasters avoid irritation to delicate skin, a blessing to patients, yet have greater strength and better sticking qualities that will make them valuable to surgeons (*see* 1916; Band-Aid, 1921).

The School and Society by University of Chicago philosopher-psychologist John Dewey, 39, pioneers progressive education by challenging traditional teaching methods based on lectures, memorization, and mechanical drill. Dewey suggests that education is a process of acculturation, an accumulation and assimilation of experience whereby a child develops into a balanced personality with wide awareness.

Saturday Evening Post publisher G. H. K. Curtis appoints his literary editor George Horace Lorimer, 32, editor-in-chief (*see* 1897). The *Post's* circulation remains below 2,000, but Lorimer will build it to more than 3 million in the next 38 years by attracting such writers as Stephen Vincent Benét, Willa Cather, Joseph Conrad, Stephen Crane, Theodore Dreiser, F. Scott Fitzgerald, John Galsworthy, O. Henry, Ring Lardner, Sinclair Lewis, Jack London, John P. Marquand, Frank Norris, Mary Roberts Rinehart, and Booth Tarkington (*see* J. W. Thompson, 1901).

Nonfiction *The Theory of the Leisure Class* by University of Chicago social scientist Thorstein Bunde Veblen, 42, says that society adopts decorum (or etiquette) and refined tastes as evidence of gentility because they can be acquired only with leisure. Corsets, white shoes, and the like are worn, and certain foods and beverages esteemed, for the same reason, as is drunkenness. Veblen introduces such concepts as "conspicuous consumption," "conspicuous waste," and "vicarious" consumption and waste to explain social behavior that has either defied explanation or gone unexamined.

Fiction *McTeague* by New York novelist Frank Norris (Benjamin Franklin Norris, Jr.), 29, is a naturalistic

Women wore corsets to show the world they didn't have to do heavy work, said Thorstein Veblen.

story of San Francisco slum life; *The Awakening* by U.S. novelist Kate Chopin (*née* Katherine O'Flaherty), 48; *The Gentleman from Indiana* by "Hoosier" novelist (Newton) Booth Tarkington, 30; *Stalky and Co.* (autobiographical stories) by Rudyard Kipling who derides the cult of compulsory games; *The Promised Land (Ziemia obiecana)* by Polish novelist Wladyslaw Reymont, 32; *Dom Casmurro* by Joaquim Machado De Assis; "A Message to Garcia" (short story) by Buffalo, N.Y., publisher Elbert Green Hubbard, 43, who 4 years ago established the Roycroft Press to publish the literary magazine *The Philistine,* whose contents he writes entirely himself. Hubbard moralizes about self-reliance and perseverance in a story based on an incident from last year's Spanish-American War; *The Amateur Cracksman* by English novelist Ernest William Hornung introduces the gentleman burglar Raffles.

Poetry *The Man With the Hoe* by California poet Edwin Markham, 46, is published in January in the *San Francisco Examiner*. Inspired by the 1863 Millet painting, it is quickly reprinted around the world as Millet's peasant becomes a symbol of exploited labor.

Juvenile The first of some 30 Rover Boys adventure novels by U.S. author Edward Stratemeyer, 36, will lead to the establishment in 1906 of the Stratemeyer Literary Syndicate with hack writers fleshing out plots and characters devised by Stratemeyer, who has been turning out books since 1894 in the Horatio Alger style. Published under various pen names, the books will include *The Motor Boys* (by "Clarence Young"), *Tom Swift* (by "Victor Appleton"), *The Bobbsey Twins* (by "Laura Lee Hope"), *The Boy Scouts* (by "Lieutenant Howard Payson"), *The Hardy Boys* (by "Franklin W. Dixon"), and *Nancy Drew* mysteries (by "Carolyn Keene").

Painting *Two Tahitian Women* by Paul Gauguin; *Paysages et intérieurs* (lithographs) by Jean Edouard Vuillard. Afred Sisley dies at Muret-sur-Loing January 29 at age 59.

Sculpture *The Puritan* by Augustus Saint-Gaudens.

Theater *Barbara Frietchie* by Clyde Fitch 10/23 at New York's Criterion Theater, with Julia Marlowe, 83 perfs.; *Uncle Vanya (Dyadya Vanya)* by Anton Chekhov 10/26 at the Moscow Art Theater; *Ashes (Cenizes)* by Spanish playwright Ramon Maria del Valle Inclan, 33, 12/7 at Madrid's Teatro de Larra; *The Tenor (Der Kammersang)* by German playwright Frank (Benjamin Franklin) Wedekind, 35, 12/10 at Berlin's Neues Theater.

Opera *Cendrillon (Cinderella)* 5/24 at the Opéra-Comique, Paris, with music by Jules Massenet.

First performances Concerto No. 2 in D minor for Piano and Orchestra by Edward MacDowell 3/5 at New York's Chickering Hall; Symphony No. 1 in E minor by Finnish composer Jean (Johann Julius Christian) Sibelius, 33, 4/26 at Helsinki; *Catalonia* Suite No. 1 for Orchestra by Isaac Albeniz 5/27 at the Théâtre Noveau, Paris; "Enigma" Variations on an Orchestral Theme by English composer Edward (William) Elgar, 42, 6/19 at London (conductor Hans Richter revises the score and adds a coda to create a version that Elgar conducts at the Worcester Festival); "Sea Pictures Songs for Chorus and Orchestra" by Elgar 10/5 at the Norwich Festival.

Stage musicals *Floradora* 11/11 at London's Lyric Theatre, with music by Leslie Stuart, lyrics by Ernest Boyd-Jones and Paul Rubens, songs that include the second-act number beginning, "Tell me, pretty maiden, are there any more at home like you?. . . ," 455 perfs.

Popular songs "My Wild Irish Rose" by New York songwriter Chauncey Olcott, 41; "O Sole Mio!" ("Oh, My Sun!") by Neapolitan composer Edoardo di Capna, 35, lyrics by Giovanni Capurro, 30; "Hello, Ma Baby" by New York composer Joseph E. Howard, lyrics by Ida Emerson that begin, "Hello, ma baby,/Hello, ma honey,/Hello, ma ragtime gal . . ."

Scott Joplin's "Original Rag" and "Maple Leaf Rag" are the first ragtime piano pieces to appear in sheet music form. Joplin, 30, was a bandleader at the 1893 Columbian Exposition at Chicago, has been taking courses in harmony and composition at the George Smith College for Negroes in Sedalia, Mo., while playing piano at local sporting houses that include the Maple Leaf Club, and has developed a ragged-type style, heavily syncopated, that publisher John Stark discovers at Sedalia and that other pianists will soon exploit.

"Lift Every Voice and Sing" by Florida songwriters John Rosamond Johnson, 26, and his brother James, 28, will be called the black national anthem. James Weldon became the first black to be elected to the Florida bar 2 years ago, the Johnson brothers will move to New York in 1901 to begin a brief career as a songwriting team, but James Weldon will soon go back to school to prepare for a larger career (*see* Fiction, 1912).

Ragtime began a new musical genre that would lead to jazz, the distinctively American music that later swept the world.

Reginald Doherty wins in men's singles at Wimbledon, Mrs. Hillyard in women's singles; Malcolm Whitman wins in U.S. men's singles, Marion Jones in women's singles.

English tea magnate Thomas Lipton has the racing yacht *Shamrock I* built for the first of five efforts he will make to regain the America's Cup, but the U.S. defender *Columbia* defeats Lipton's boat 3 to 0.

U.S. prizefighter James J. Jeffries, 24, wins the world heavyweight title from Bob Fitzsimmons June 9 with an 11th round knockout in a match held at Coney Island, N.Y. Jeffries will retire undefeated in March 1905 and be succeeded by Marvin Hart (*see* 1906).

A liquid-center gutta percha golf ball invented by Cleveland golfer Coburn Haskell with help from a B. F. Goodrich scientist replaces the solid gutta percha ball used since 1848. The "bouncing billy" is soon replaced by a ball with tightly wound rubber threads wrapped around a solid rubber core, and A. G. Spalding will acquire rights to the new ball (*see* 1892).

"I wish to preach, not the doctrine of ignoble ease, but the doctrine of the strenuous life," says Gov. Theodore Roosevelt of New York April 10 in a speech at Chicago.

Missouri becomes the "show me" state. Rep. Willard D. Vandiver, 45, (D. Mo.) addresses a naval dinner at Philadelphia and says, "I come from a state that raises corn and cotton and cockleburs and Democrats, and frothy eloquence neither convinces nor satisfies me. I am from Missouri. You have got to show me."

Washington's Post Office building is completed at Pennsylvania Avenue and 12th Street with a great glass-roofed court. The Romanesque revival granite building is topped by a clock tower.

London's Carlton House Hotel opens in Waterloo Place near the Carlton Club and Carlton House Terrace. César Ritz of the 10-year-old Savoy Hotel is manager, and Louis Escoffier of the year-old Paris Ritz is in charge of the restaurant kitchen (*see* London Ritz, 1902).

Wisconsin's last wild passenger pigeon is shot (*see* 1878; 1911).

Mount Rainier National Park, created by act of Congress, embraces 242,000 acres of dense forests and meadows radiating from the slopes of an ancient 14,408-foot volcano in Washington State and contains the greatest single-peak glacial system in the United States.

Congress passes a Refuse Act empowering the U.S. Army Corps of Engineers to prosecute polluters. The law provides for fines of up to $2,500 for oil spills and similar acts of pollution but will not be enforced.

The Columbia River Packers Association is created by seven chinook salmon canneries at the mouth of the river, one of which packs its fish under the name Bumble Bee. CRPA will acquire several sailing vessels in 1901, load them with lumber, coal, building, and canning materials, and build a cannery on the Nushagak River at Bristol Bay in northwestern Alaska that will be followed by many more CRPA Alaskan canneries (*see* 1938).

The boll weevil *Anthonomus grandis* crosses the Rio Grande from Mexico and begins to spread north and east through U.S. cotton fields. The weevil will destroy vast acreages of cotton, devastating Southern agriculture (*see* 1916; 1921).

The Russian grain harvest is 65 million tons, double the harvest of 30 years ago, as 200 million acres are planted to grain.

U.S. cerealist Mark Carleton introduces Kubanka durum wheat from southeastern Russia into North Dakota, which will be the leading U.S. producer of the wheat most suitable for macaroni and spaghetti (*see* 1895; Kharkov, 1900).

The first Colorado beet sugar refinery is built at Grand Junction; American Beet Sugar Co. is organized.

The American Sugar Refining trust has almost a 100 percent monopoly in the U.S. industry (*see* 1897; 1907).

Coca-Cola is bottled for the first time by Chattanooga, Tenn., lawyers Benjamin F. Thomas and Joseph B. Whitehead who have traveled to Atlanta and persuaded Asa Candler to let them try bottling his beverage under contract (*see* 1895). Coca-Cola Company will give seven parent bottlers contracts to establish local bottling companies and supply them with syrup, it will acquire these bottlers over the years (the last one in 1974), but most Coca-Cola continues to be dispensed by soda jerks from syrup mixed with carbonated water (*see* 1916; Britain, 1910).

Jell-O is acquired for $450 from Pearl Wait by his LeRoy, N.Y., neighbor Orator Francis Woodward who has just started a company to produce a cereal he calls Grain-O (*see* 1897; 1906).

1899 ***(cont.)*** New York Condensed Milk Co. becomes Borden's Condensed Milk Co. and will be renamed Borden Co. in 1919, and Borden, Inc. in 1968 (*see* 1892).

Carnation evaporated milk is supplied in 16-ounce cans to Klondike-bound gold seekers by Elbridge Amos Stuart who has set up a small plant at Kent, Wash., near Seattle with help from John Meyenberg of Helvetia Condensed Milk in Highland Park, Ill. (*see* Our Pet, 1894). An Indiana Quaker who became a grocer at El Paso, Tex., Stuart has come to suspect that bad milk is the cause of so many children dying of "summer complaint" and has invested his savings in a new process to manufacture canned, sterilized, evaporated milk (*see* 1906).

Wesson Oil is developed by Southern Oil Co. chemist David Wesson, 38, who introduces a new method for deodorizing cottonseed oil. His vacuum and high-temperature process will revolutionize the cooking oil industry and largely overcome the prejudice against cottonseed oil, which until now has been deodorized only by heating it with a steam coil and blowing live steam through it at atmospheric pressure.

Chopped beefsteak is called Hamburg steak in America, says a new French-German-English dictionary of foods published under the title *Blueher's Rechtschreibung* (*see* 1836; Louis Lassen, 1900).

Emigration of Russians increases to more than 223,000, up from 108,000 4 years ago, as the new Trans-Siberian Railway encourages settlement along its right-of-way.

"Conspicuous consumption" and "conspicuous waste" require such large expenditures that they are "probably the most effectual of the Malthusian prudential checks" on population growth and make for low birthrates in some classes of society, writes Thorstein Veblen in his *Theory of the Leisure Class* (above).

1900 The Boer War continues in South Africa where Frederick Sleigh Roberts (Bobs Bahadur), 67, arrives January 10 to replace Sir Redvers Henry Buller, 60, as commander-in-chief with Lord Kitchener as his chief of staff. Buller retains command of an army, but it is cavalry commander John Denton Pinkstone French, 47, who relieves Kimberley February 15 after a 4-month Boer siege. Gen. Roberts surrounds the Boer leader Piet Arnoldus Conje near Paardeberg and forces him to surrender February 27 after he has run out of food and ammunition, Gen. Buller relieves Ladysmith February 28 after nearly 4 months of siege, Bloemfontein falls to Roberts March 13, and Mafeking is finally relieved May 17 after a 215-day siege in which Col. Robert Stephenson Smyth, 43, Baron Baden-Powell of Gilwell, has resisted the Boers (*see* Boy Scouts, 1908).

Johannesburg falls to the British May 31, Gen. Buller takes Pretoria June 5, Roberts and Buller join forces at Vlakfontein July 4, Britain annexes the Orange Free State and the Transvaal (they become the Orange River Colony and the Transvaal Colony), President Kruger flees to Delagoa Bay and voyages to Europe in hopes of obtaining German support, but Kaiser Wilhelm II denies the aged Kruger an audience October 6.

Gen. Roberts hands over his command to Lord Kitchener and is created first Earl Roberts of Kandahar, Pretoria, and Waterford as the Boers resort to guerrilla warfare in their efforts to drive out the British (*see* 1901).

Britain annexes the Orange Free State in late May.

Russia annexes Manchuria May 21.

A "Boxer Rebellion" rocks China beginning June 20 as foreign legations at Beijing are besieged by members of a militia force who have murdered the German minister to Beijing with encouragement from an anti-foreigner clique at the Manchu court led by the dowager empress Cixi (Tzu Hsi), now 66, who has effectively ruled China for 39 years. An eight-nation expeditionary force lifts the siege of the legations August 14, but at least 231 foreign civilians, most of them missionaries, are killed in various parts of China, and Russian troops retaliate for mid-July Chinese bombardments across the Amur River by driving thousands of civilians to their death in the river. The Russians seize southern Manchuria in the fall (*see* 1901; Japanese, 1904).

Arthur MacArthur, U.S. military governor in the Philippines, grants an amnesty to Filipino rebels June 21 (*see* 1899).

Queen Victoria gives assent July 9 to the Commonwealth of Australia Bill, but she is old and ailing (*see* 1901).

Italy's Umberto I is assassinated July 29 at age 56 by an anarchist at Monza. Umberto is succeeded after a 22-year reign by his son of 30 who will reign until 1946 as Victor Emmanuel III.

Britain's "Khaki" election October 16 results in a victory for the Conservatives who retain power under the marquis of Salisbury, now 70.

President McKinley campaigns for re-election with "the full dinner pail" slogan symbolizing Republican prosperity and receives a plurality of nearly a million popular votes. The Democratic ticket headed by William Jennings Bryan and former vice president Adlai E. Stevenson wins 155 electoral votes versus 292 for the Republicans, but thousands of workers cast their votes for Socialist party candidate Eugene V. Debs.

Lord Kitchener (above) places some 120,000 Boer women and children in concentration camps, where 20,000 will die of disease and neglect.

V.I. Lenin returns from exile January 29 after 3 years in Siberia with his wife and mother-in-law. He sets up residence at Pskov while awaiting the release of his wife from Ufa, and on July 16 emigrates to Switzerland to begin a 5-year exile. In December, Lenin becomes an editor of the newspaper *Iskra (The Spark)* which is published in Munich for distribution in Russia and takes its name from an 1825 poem by an exile in Siberia who said, "The spark will kindle a flame!" (*see* 1895; 1903).

The International Ladies' Garment Workers Union is founded June 3 by cloakmakers who meet in a small hall on New York's Lower East Side. The union's seven locals represent 2,310 workers in New York, Newark, Philadelphia, and Baltimore. By 1904 the ILGWU will have 5,400 members in 66 locals in 27 cities (*see* 1909 strike).

Only 3.5 percent of the U.S. work force is organized. Employers are free to hire and fire at will and at whim (*see* United Mine Workers, 1899; 1902).

Storyville in New Orleans has more than two dozen ornate Basin Street "sporting palaces" in two blocks set aside 3 years ago under a plan devised by alderman Sidney Story. Poorer prostitutes operate out of "cribs" behind the "palaces."

Chicago's Everleigh Club opens at 2131–33 Dearborn Street where local entrepreneur Minna Everleigh, 23, and her sister Ada, 21, have bought and refurbished a bordello with an inheritance of $40,000 each from their late father, a Kentucky lawyer. Visitors require letters of introduction, they are entertained by string quartets and other pleasures, and the Everleigh Club will remain in business until 1911, when rival madams subsidize a crusade to force Chicago's mayor to close down the rich Everleigh sisters.

Kansas City has 147 houses of pleasure, the most elegant being that of Annie Chambers. Her $100,000 mansion is graced with a larger-than-life portrait of Annie dressed in a swirl of roses and gauze but little else.

The National Consumers League has its beginnings in the Consumers League for Fair Labor Standards founded by New York social worker Florence Kelley, 41, who last year joined the Henry Street Settlement after 8 years at Chicago's Hull House (*see* 1893; Addams, 1889).

Conscientious consumers, says Kelley, will not want to buy goods made in substandard factories, or by child labor, or finished in tenements.

The Cripple Creek gold field discovered in Colorado 9 years ago yields $20 million worth of gold per year and is second only to South Africa's Transvaal gold field discovered in 1886. It is far larger than the Klondike field discovered 4 years ago.

Moody's Manual of Railroads and Corporation Securities begins publication under the direction of New York financial analyst John Moody, 32, who has worked for 10 years in the banking house Spencer Trask & Co. Moody will establish the investor's monthly *Moody's* magazine in 1905, and his annual *Moody's Analyses of Investments* will appear beginning in 1909.

The Caisse Populaire founded at Levis across the St. Lawrence from Quebec City is the first American credit union. Parliamentary stenographer Alphonse Desjardins, 46, has long been aware of the abuses suffered by farmers, tradespeople, and laborers at the hands of loan sharks in an age when commercial banks do not make personal loans. His cooperative banking system is based on European systems set up in the past 50 years to protect working people from usurious rates. He will help start the first U.S. credit union in 1909 at Manchester, N.H., and his Quebec operation will grow enormously.

U.S. chemist Charles Skeele Palmer, 42, invents a new process for cracking petroleum to obtain gasoline. He will sell rights to the process to Standard Oil in 1916 (*see* Houdry process, 1936).

John D. Rockefeller's Standard Oil buys Pacific Coast Oil (*see* 1879; 1911).

The first British gasoline-powered motorbuses go into service in January as single-deck buses begin operating in Norfolk.

The first international championship motorcar race is held June 14 from Paris to Lyons. Five entries from Belgium, France, Germany, and the United States compete for the Gordon Bennett Cup put up by New York publisher James Gordon Bennett, who has lived abroad since 1877. The cars all finish, all run over dogs en route, and the winner is a French Panhard that has averaged 38.5 miles per hour.

Only 144 miles of U.S. roads are hard-surfaced, but by year's end there are 13,824 motorcars on the road.

Franklin, Peerless, and Stearns motorcars are introduced. The air-cooled engine Franklin is made by Herbert H. Franklin, 34, whose sales will peak at 14,000 in 1929.

The Packard motorcar introduced at Warren, Ohio, has a chain-driven, one-cylinder, 12-horsepower engine and three forward speeds. Engineer James Ward Packard, 37, and his brother William, 39, have run Packard Electric Co. since 1890.

The Auburn motorcar introduced by Auburn Automobile Co. of Auburn, Ind., is a single-cylinder runabout with solid tires and a steering tiller. Frank and Morris Eckhardt of Eckhardt Carriage Co., who have started the firm with $2,500 in capital, will produce a two-cylinder model in 1905, a four in 1909, and a six in 1912 (*see* 1924).

White steam cars and trucks are introduced by a Cleveland sewing-machine firm headed by Walter White whose New England-born father Thomas invented a sewing machine in 1859, moved to Cleveland in 1866, and has developed the company now run by his sons Windsor, Rollin, and Walter. The first gasoline-powered White motorcars will appear in 1910.

Renault Frères introduces the first glass-enclosed two-passenger motorcar (*see* 1899).

Firestone Tire & Rubber is founded August 3 at Akron, Ohio, by U.S. entrepreneur Harvey Samuel Firestone, 32, who has patented a method for attaching tires to rims. He invests $10,000 to start the new firm (*see* 1908).

The first U.S. National Automobile Show opens November 10 at New York's Madison Square Garden with 31 exhibitors. Contestants compete in starting and

1900 ***(cont.)*** braking, and exhibitors demonstrate hill-climbing ability on a specially built ramp.

The Trans-Siberian Railway opens between Moscow and Irkutsk (*see* 1891; 1904).

U.S. railroads charge a freight rate that averages 75¢ per ton-mile, down from $1.22 in 1883. The Great Northern charges only $35 for a railcar traveling 500 miles between St. Paul, Minn., and Minot, N.D.; the Chicago and North Western $40 for a railcar traveling the 750 miles between Chicago and Pierre, S.D.

Cody, Wyo., is founded by William "Buffalo Bill" Cody in order to have the Chicago, Burlington & Quincy run a spur to the large tracts of land he has acquired in the area of the Shoshone River's south fork.

The Pennsylvania Railroad acquires control of the Long Island Rail Road, which will soon be the nation's largest passenger carrier (*see* 1836; 1905).

Illinois Central engineer Jonathan Luther "Casey" Jones, 36, of Jackson, Tenn., pulls his six-coach *Cannonball Express* into Memphis on the night of April 29, learns that the engineer for the return run is ill, and volunteers to replace him. Highballing at 75 miles per hour through Mississippi to make up for lost time, Jones rounds a gentle curve at four in the morning and sees a freight train stalled on the track dead ahead. Hand upon the throttle, Jones yells at his fireman to jump, plows into the freight, is killed by a wood splinter driven through his head, and while the collision is only one of 27 rear-end collisions on U.S. railroads in the month of April, a ballad will make it famous.

Trolley cars provide transportation in every major U.S. city 12 years after the opening of the first trolley line in Richmond. Some 30,000 cars operate on 15,000 miles of track (*see* 1917).

Métro underground rail service begins at Paris July 19 (*see* London, 1891; Budapest, 1896; Boston, 1897). It will grow to become the world's third largest subway (second largest in terms of passenger traffic), carrying more than a billion passengers per year (*see* New York, 1904).

The Hay-Pauncefote Treaty February 5 includes a British renunciation of rights to build a Panama canal. Parliament rejects the treaty (*see* 1901).

The Chicago Sanitary and Ship Canal opens to link Lake Michigan with the Des Plaines River. The 33.8-mile canal connects the Great Lakes with the Mississippi and the Gulf of Mexico (*see* Joliet, 1673).

Only 25 percent of the U.S. commercial shipping fleet remains in sail, down from 56 percent in 1870. More than 60 percent of world shipping is in steamships, up from 16 percent in 1870.

The U.S. Navy purchases the first modern submarine. Invented by Irish-American engineer John Phillip Holland, 60, the *Holland* uses electric motors under water and internal combustion engines on the surface, employing water ballast to submerge (*see* Bushnell, 1776; Lake, 1897).

Retired German general Ferdinand von Zeppelin, 62, launches the first rigid airship July 2 at Friedrichshafen; he will go on to build many such lighter-than-air craft (*see* 1928).

A new quantum theory enunciated by German physicist Max Karl Ernst Ludwig Planck, 42, will have enormous impact on scientific thinking. Bodies that radiate energy do not emit the energy constantly but rather in discrete parcels which he calls quantums, says Planck (*see* Einstein, 1905).

Genetic laws revealed by Gregor Mendel in 1865 become generally known for the first time as Dutch botanist Hugo De Vries, 52, at the University of Amsterdam, German botanist Karl Erich Correns, 35, and Austrian botanist Erich Tschermak von Seysenegg, 29, working independently, discover Mendel's published work and make public his Mendelian laws (*see* Bateson, 1902).

English archaeologist Arthur John Evans, 40, unearths the palace of Knossos on the Greek island of Crete. He has been conducting excavations on the island for the past 7 years, has discovered pre-Phoenician script, and will work for 8 years to reveal the seat of an ancient "Minoan" culture.

A U.S. Public Health Commission headed by Major Walter Reed, 49, of the Army Medical Corps shows that the yellow fever virus is transmitted by the *Aëdes aegypti* mosquito. Dr. C. J. Finlay, now 47, has alerted him to the possibility (*see* 1881; Havana, 1901).

The U.S. death rate from tuberculosis is 201.9 per 100,000, from influenza and pneumonia 181.5, heart disease 123.1, infant diarrhea and enteritis 108.8, diabetes mellitus 91.7, cancer 63, diphtheria 43.3, typhoid fever and paratyphoid 35.9 (*see* 1922).

Infant mortality in the United States is 122 per 1,000 live births, in England and Wales 154, and in India 232 (*see* 1951).

Average age at death in the United States is 47.

Bubonic plague strikes Honolulu in epidemic form. A large section of the city's Chinatown is condemned and burned under fire department supervision to kill the plague-bearing rats, but the fire gets out of control and much of the city is destroyed.

The first U.S. bubonic plague epidemic begins at San Francisco. The body of a dead Chinese is discovered March 6 in the basement of Chinatown's Globe Hotel, local authorities try to hush up the cause of death, but 120 others will be stricken before the plague ends in February of 1904, and all but three will die (San Francisco will have another plague epidemic in 1907, as will Seattle; New Orleans will have one in 1914 and 1919, and Los Angeles will have one in 1924).

The Interpretation of Dreams (*Die Traumdeutung*) by Sigmund Freud is based on psychoanalytic techniques that lean heavily on dream analysis (*see* 1895; first international meeting, 1908).

The U.S. Census shows 12 million Roman Catholics, 6 million Methodists, 5 million Baptists, 1.5 million

Sigmund Freud in Vienna pioneered psychoanalysis, but his emphasis on sex antagonized opponents.

Lutherans, 1.5 million Presbyterians, 1 million Jews, 700,000 Episcopalians, 350,000 Mormons, 80,000 Christian Scientists, and 75,000 Unitarians.

A U.S. College Entrance Examination Board is founded to screen applicants to colleges. The College Board Scholastic Aptitude Tests will be graded on a scale of 200 to 800, and colleges will use SAT scores as a supplement to secondary-school records and other relevant information in judging qualifications of applicants, but the scores will never be more than approximate and will have a standard error of measurement in the area of 32 points.

Carnegie Institute of Technology is founded at Pittsburgh with a gift from steel magnate Andrew Carnegie.

England's Birmingham University is founded largely through efforts by colonial secretary Joseph Chamberlain, now 64, who will be first chancellor of the new university beginning next year.

The *Daily Express* is founded by London newspaperman Cyril Arthur Pearson, 34, who will lose his eyesight beginning in 1912 but amass a fortune that he will devote to helping blind soldiers and sailors (*see* Beaverbrook, 1916).

One U.S. home in 13 has a telephone.

Nonfiction *The Century of the Child* by Swedish feminist Ellen Karoline Sofia Key, 51, a teacher at Stockholm for 19 years and now a lecturer at the People's Institute for Workingmen.

Fiction *Sister Carrie* by U.S. novelist Theodore Dreiser, 29. Publisher Frank N. Doubleday hastily withdraws the book when his wife says it is too sordid, the small edition goes almost unnoticed, and Dreiser suffers a nervous breakdown; *Monsieur Beaucaire* by Booth Tarkington; *Lord Jim* by Joseph Conrad; *Love and Mrs. Lewisham* by H. G. Wells; *The Life and Death of Richard Yea-and-Nay* by Maurice Hewlett; *The Master Christian* by Marie Corelli; *The Flame of Life (Il fuoco)* by Gabriele D'Annunzio has been inspired by the actress Eleanora Duse, now 41, with whom D'Annunzio has had a long affair.

Juvenile *Claudine at School (Claudine à l'Ecole)* by French author Colette (Sidonie Gabriele Claudine Colette), 27, is a semi-autobiographical series of stories that will have many sequels; *The Wizard of Oz* by Chicago newspaperman Lyman Frank Baum, 44 (*see* stage musical, 1901); *Little Black Sambo* by English author Helen Bannerman, whose East Indian hero thwarts his tiger pursuers (they turn into butter).

Poetry *Almas de violeta* by Spanish poet Juan Ramon Jiminez, 19, whose impressionism is based on scenes of his native Andalusia.

Painting *Le Moulin de la Galette* by Spanish painter Pablo Ruiz y Picasso, 19, who will move from Barcelona to Paris in 1903; *La Modiste* by Henri de Toulouse-Lautrec; *Still Life with Onions* by Paul Cézanne; *Nude in the Sun* by Pierre Auguste Renoir; *The Wyndham Sisters* by John Singer Sargent; *Philosophy* (mural) by local art nouveau painter Gustav Klimt, 38, for the University of Vienna.

Noa by Paul Gauguin is an account of the painter's first years in Tahiti from 1891 to 1893. Gauguin returned to Tahiti in 1895.

The Brownie box camera introduced by Eastman Kodak sells at $1, puts photography within reach of everyone, and makes Kodak a household name (*see* Pocket Kodak, 1895). The camera's six-exposure film sells for 15¢ (*see* acetate film, 1924; Kodachrome, 1935).

Theater *Schluck und Jau* by Gerhart Hauptmann 1/3 at Berlin's Deutsches Theater; *When We Dead Awaken* (*Naar vi dode vaaguer*) by Henrik Ibsen 1/26 at Stuttgart's Hoftheater; *You Never Can Tell* by George Bernard Shaw 5/2 at London's Strand Theatre; *St. John's Fire (Johannisfeuer)* by Hermann Sudermann 10/5 at Berlin's Lessingtheater; *Mrs. Dane's Defence* by Henry Arthur Jones 10/9 at Wyndham's Theatre, London; *Michael Kramer* by Gerhart Hauptmann 12/21 at Berlin's Deutsches Theater.

"The Great Houdini" gains wide publicity by executing an escape from London's Scotland Yard, becomes a main attraction at London's Alhambra Theatre, and begins a 4-year tour of the Continent. U.S. escape artist Ehrich Weiss, 26, has adopted the name Houdini from the French magician Jean Eugene Robert-Houdin; having studied Robert-Houdin's work, his book *The Unmasking of Robert-Houdin* in 1908 will show that the Frenchman's dexterity was much exaggerated, and he will far surpass Robert-Houdin's reputation with feats such as having himself shackled in irons, locked into a roped and weighted box, dropped overboard from a boat, and emerging with a smile before baffled audiences (*see* 1926).

The Shubert brothers of Syracuse, N.Y., lease the Herald Square Theater, challenging New York's Theatrical

1900 ***(cont.)*** Syndicate headed by booking agent Charles Frohman, 40. Lithuanian-American theatrical manager Sam S. Shubert, 27, and his Syracuse-born brothers (Lee, 25, and J. J. Jacob J., 20), begin a 10-year battle with Frohman, A. L. Erlanger, and Marc Klaw by renting theaters to producers against whom the syndicate discriminates. The Shuberts will soon become producers themselves, Sam will be killed in a 1905 train wreck, the other two brothers will triumph over the syndicate (which is, in effect, a theatrical trust), and will develop the most far-flung privately-controlled theatrical organization in the world.

Opera *Louise* 1/2 at the Opéra-Comique, Paris, with music by French composer Gustave Charpentier, 39; *Tosca* 1/14 at Rome's Teatro Costanzi, with music by Giacomo Puccini. The libretto is based on an 1887 tragedy by Victorien Sardou, now 69, about the 1796 political strife in Rome between Bonapartists and monarchists; *Prométhée* 8/22 in the ancient Roman arena at Beziers, France, with music by Gabriel Fauré; *The Tale of Tsar Saltan (Skazka o Tsarye Saltanye)* 11/3 at Moscow, with music by Nikolai Rimski-Korsakov that includes his violin piece "The Flight of the Bumble-Bee."

Boston's 2,500-seat Symphony Hall designed by McKim, Mead and White of New York opens October 15 at the corner of Massachusetts and Huntington Avenues (*see* 1881).

The Philadelphia Orchestra is organized by residents who engage German conductor Fritz Scheel to direct its first concert November 16 at the city's 43-year-old Academy of Music.

First performances "Alborado de Gracioso" by French composer Maurice (Joseph) Ravel, 24, 1/6 at the Société Nationale, Paris; *The Dream of Gerontius* by Edward Elgar (text by Cardinal Newman) 10/3 at England's Birmingham Festival.

Broadway musicals *Fiddle-dee-dee* 9/16 at Weber & Fields Music Hall, with Joe Weber, Lew Fields, Lillian Russell, De Wolf Hopper, David Warfield, songs that include "Rosie, You Are My Posie (Ma Blushin' Rosie)" by John Stromberg, lyrics by Edgar Smith, 262 perfs.

Popular songs "A Bird in a Gilded Cage" by New York composer Harry Von Tilzer (Harry Gumm), 28, lyrics by Arthur J. Lamb.

The trademark "His Master's Voice" with a painting by Francis Barraud of a fox terrier named Nipper listening attentively to a horn gramophone appears on gramophone records (the first records with circular paper title-labels) issued by Eldridge Johnson's Consolidated Talking Machine Co. of Camden, N.J., which will become the Victor Co. Emile Berliner has devised the trademark (*see* 1877; Victrola, 1906).

The Olympic Games held at Paris attract 1,505 athletes from 16 nations. France is the unofficial winner, but government bureaucrats take over the games and nearly ruin the Olympics before the event is returned to the Olympic Committee headed by the Baron de Coubertin.

Christy Mathewson leaves Bucknell University to join the New York Giants, but pitches so poorly that New York sells him to the Cincinnati Reds. Right-hander Christopher Mathewson, 20, will be bought back for the 1901 season and win 20 games (*see* 1903).

Reginald Doherty wins in men's singles at Wimbledon, Mrs. Hillyard in women's singles; Malcolm Whitman wins in U.S. men's singles, Myrtle McAteer in women's singles.

The first Davis Cup tennis matches open August 8 at Brookline, Mass., and continue for 3 days. A U.S. team defeats a British team to gain possession of an $800 silver cup donated by Harvard senior Dwight Filley Davis, 21, who has commissioned the Boston jeweler Shreve, Crump & Low to design and produce the challenge cup.

The Junior League of the New York College Settlement is founded by post-debutante Mary Harriman, daughter of financier Edward Harriman, who organizes a group to aid a local settlement house. Debutantes flock to join the league, which will be followed by Junior Leagues of Baltimore, Brooklyn, Philadelphia, and (by 1906) Boston. By 1920, there will be 39 Junior Leagues engaged in civic improvement projects, and Dorothy Whitney (Mrs. Willard) Straight will organize them into the Association of Junior Leagues of America.

One U.S. home in seven has a bathtub; showers are even rarer.

F. W. Woolworth controls 59 stores, up from 28 in 1895, and has sales in excess of $5 million. Inspired perhaps by A&P storefronts, he adopts a bright carmine-red storefront with gold-leaf lettering and molding for all his stores (*see* 1879; 1909).

Earthquakes rock Ecuador and Peru in mid-August, killing thousands.

A hurricane strikes Galveston, Tex., September 8, killing between 6,000 and 8,000 people in the worst recorded natural disaster in North American history. Thousands more are injured, and property damage amounts to $17 million.

Some 2 million mustangs (wild horses) roam the U.S. prairie, but the nation has fewer than 30 head of bison, down from 1,090 in 1893. Longhorn cattle are also close to extinction (*see* 1850).

Brazil's northeastern Hump remains 40 to 50 percent forested, but the figure will drop to 5 percent in the next 70 years. Deforestation will lead to erosion and water problems.

U.S. wheat fetches 70¢ a bushel as 34 percent of the crop goes abroad, and corn brings 33¢ a bushel as 10 percent of the crop is exported.

Texas steers bring $4.25 per hundredweight, but the price index for U.S. farm products will rise by a spectacular 52 percent in the next decade as more efficient transport (above) enables Britain and Europe to import North American grain at low rates.

Mark Carleton makes another trip to Russia and returns with hard red Kharkov wheat, a winter variety that withstands winter-kill and gives high yields (*see* 1899; 1914).

Steam tractors appear on wheat fields of the U.S. Pacific Northwest, but their main use is to draw portable threshing machines, many of which are still pulled by 40 horses driven abreast.

Some 2.5 horsepower is available to each man working on a U.S. farm, up from 1.5 in 1850 (*see* 1930).

The average U.S. farm worker produces enough food and fiber for seven people, up from 4.5 in 1860.

The world has 100 million acres of irrigated cropland, up from 20 million in 1880.

The first sluice gate on the Colorado River brings irrigation water from Arizona to California's Imperial Valley (*see* Salton Sea, 1905).

Honeydew melons are introduced into the United States (year approximate).

Grapefruit becomes an important food in the United States as botanical development makes the fruit sweeter. U.S. citrus growers will account for 90 percent of world grapefruit production.

Regulations limiting bacteria in U.S. milk to 1 million per cubic centimeter prove difficult to enforce despite growing use of pasteurization (*see* 1890). Contaminated milk remains a major source of food-borne disease (*see* Evans, 1917).

Milk bottles are introduced in England but only for pasteurized milk (*see* Borden, 1885; Hood, 1897). Most British milk remains unpasteurized, and bottles will not be widely used for another two decades (*see* 1901; 1942).

Mead Johnson Co. is founded at Evansville, Ind., to produce the infant cereal Pablum that will be sold through physicians' recommendations. Founder Johnson is a brother of Johnson & Johnson's Robert Wood Johnson (*see* 1886).

Battle Creek, Mich., has 42 breakfast cereal plants (*see* Kellogg; 1898; Post, 1897).

Uneeda Biscuits have sales of more than 10 million packages per month, while all other packaged crackers sell scarcely 40,000 per month (*see* 1898; Animal Crackers, 1902).

Per capita U.S. wheat flour consumption averages 224 pounds, up from 205 pounds in 1850 (*see* 1920).

Per capita U.S. sugar consumption averages 65.2 pounds per year.

World beet-sugar production reaches 5.6 million tons, a figure that will more than quadruple in the next 64 years.

Milton S. Hershey sells his Lancaster Caramel Co. for $1 million in cash but retains his chocolate manufacturing equipment (*see* 1894). He rents a wing of his old plant from its new owners and introduces the first milk chocolate Hershey bars (*see* 1903).

Typical U.S. food prices: sugar 4¢ lb., eggs 14¢ doz., butter 24¢ to 25¢ lb. Boarding houses offer turkey dinner at 20¢ and supper or breakfast at 15¢, but a male stenographer earns $10 per week and an unskilled girl $2.50.

H. J. Heinz erects an electric sign six stories high with 1,200 lights to tell New Yorkers about the "57 Good Things for the Table" that include Heinz tomato soup, tomato ketchup, sweet pickles, India relish, and peach butter (*see* 1896; 1905).

Wesson Oil is put on the market by Southern Oil Co. (*see* 1899). David Wesson has made the pure cottonseed oil palatable and will work from 1901 to 1911 on a process for hydrogenating it (*see* 1925; Normann, 1901).

The hamburger is pioneered at New Haven, Conn., where Louis Lassen grinds 7¢/lb. lean beef, broils it, and serves it between two slices of toast (no catsup or relish) to customers at his 5-year-old three-seat Louis Lunch (but *see* 1899; 1904).

The *Guide Michelin* published at Paris is the first systematic evaluation of European restaurants. Financed by tire producers André and Edouard Michelin, the *Guide* rates restaurants using a system of three stars (worth a special journey), two stars (worth a detour), or one star, and will be updated periodically.

Kansas prohibitionist Carry Moore Nation, 54, declares that since the saloon is illegal in Kansas, any citizen has the right to destroy liquor, furniture, and fixtures in any public place selling intoxicants. She begins a campaign of hatchet-wielding through Kansas cities and towns, and although she will be arrested, fined, imprisoned, clubbed, and shot at, she will persevere (*see* Anti-Saloon League, 1895; Eighteenth Amendment and Volstead Act, 1919).

Hills Bros. in San Francisco begins packing roast ground coffee in vacuum tins to begin a new era in coffee marketing. It is the beginning of the end for the coffee roasting shops common now in every town and the coffee mill seen in almost every U.S. kitchen (*see* Chase and Sanborn, 1878).

The North American population reaches 81 million, up from 5.7 million in 1800, and begins a growth that will reach 355 million by 1978.

The U.S. population reaches 76 million with 10.3 million of the total foreign born, roughly 9 million black or of mixed blood, 237,000 native American Indian or partly Indian, 90,000 Chinese, 24,000 Japanese.

Half of all U.S. working women are farmhands or domestic servants (*see* 1960).

Some 45.8 million Americans are rural residents, 30.2 million urban.

The population of the world reaches 1.65 billion, and 16 cities of the world have populations of 1 million or more (*see* 1901; 1960).

20th Century

1901 Queen Victoria dies January 22 at age 81 after a reign of nearly 64 years in which the United Kingdom has grown from a nation of 25 million to one of 37 million. The queen's son, 59, will reign until 1910 as Edward VII.

The Commonwealth of Australia created January 1 has joined New South Wales, Queensland, South Australia, Victoria, Tasmania, the Northern Territory, and West Australia.

The British viceroy in India, George Nathaniel Curzon, 42, creates the North-West Frontier province between the Punjab and Afghanistan as he works to pacify the region.

The Boer War continues in South Africa. Lord Kitchener builds a chain of blockhouses to combat guerrilla activities and starts destroying Boer farms. Boer forces under James Hertzog and Christian de Wet invade the Cape Colony in October, coming within 50 miles of Cape Town, but British troops repel the Boers, and when Louis Botha raids Natal he has no success (*see* 1900; 1902).

U.S. Brig. Gen. Frederick Funston captures the Filipino leader Emilio Aguinaldo in March but guerrillas massacre an American garrison on the island of Samar in September.

"Speak softly and carry a big stick," says Vice President Roosevelt September 2 in a speech at the Minnesota State Fair, laying down a rule for U.S. foreign policy: "There is a homely adage which runs, 'Speak softly and carry a big stick; you will go far.' If the American nation will speak softly and yet build and keep at a pitch of the highest training a thoroughly efficient navy, the Monroe Doctrine will go far."

President McKinley is shot at point-blank range September 6 with a .32 caliber Ivor Johnson revolver fired by Polish-American anarchist Leon Czolgosz, 28, during the president's visit to the Pan-American Exposition at Buffalo, N.Y. The wounds are not properly dressed, McKinley dies of gangrene September 14 at age 58, and Teddy Roosevelt at age 42 becomes the youngest chief executive in the nation's history (Mark Hanna in Cleveland calls him "that damned cowboy").

The Peace of Beijing September 7 ends the Boxer Rebellion that began last year. China is obliged to pay indemnities to the world powers.

The Russian minister of propaganda is assassinated at age 41 February 27 in reprisal for his repression of student agitators.

British chocolate heir William Cadbury visits Trinidad and is told that cocoa workers in Portugal's African islands of São Tomé and Principe are, for all practical purposes, treated as slaves (*see* 1902).

Alabama adopts a new constitution with literacy tests and a grandfather clause designed to disenfranchise blacks (*see* Louisiana, 1898).

Booker T. Washington of Tuskegee Institute attends a White House dinner given by President Roosevelt October 16. Outraged whites take reprisals against southern blacks (*see* 1895).

Wall Street panics May 9 as brokerage houses sell off stock so they can raise funds to cover their short positions in Northern Pacific Railroad stock. Financiers Edward H. Harriman, Jacob Schiff of Kuhn, Loeb & Co., and James Stillman of New York's National City Bank vie for control of the railroad controlled by J. P. Morgan and by James J. Hill of the Canadian Pacific. Stillman's City Bank is the repository of John D. Rockefeller's Standard Oil Co. money (*see* Charles Stillman, 1850), Harriman and Schiff have acquired control of the Union Pacific, the Southern Pacific, and the SP's subsidiary the Central Pacific, they seek control of the Chicago, Burlington & Quincy to gain access to Chicago. The rivals for the Northern Pacific bid up the price of stock of the railroad to $1,000 per share, but they grow frightened when other stock prices collapse in the panic, they let short-sellers cover their positions at $150 per share, and they join forces to form Northern Securities, a holding company that controls not only

M'KINLEY SHOT

Stranger Fires Two Bullets Into His Stomach.

His Condition Is Serious But He Is Still Living.

Taken to the Hospital on the Pan American Exposition Grounds.

America's third presidential assassination in 36 years put "Rough Rider" Theodore Roosevelt in the White House.

the Northern Pacific but also the Great Northern and the Chicago, Burlington & Quincy (*see* Supreme Court ruling, 1904).

United States Steel Co. is created by J. P. Morgan (above), who underwrites a successful public offering of stock in the world's first $1 billion corporation, nets millions for himself in a few weeks of hard work, and pays $492 million to Andrew Carnegie for about $80 million in actual assets in order to eliminate the steel industry's major price cutter. Carnegie personally receives $225 million in 5 percent gold bonds and is congratulated on being "the richest man in the world" by Morgan who merges Carnegie's properties with other steel properties to create a company that controls 65 percent of U.S. steel-making capacity (*see* Bethlehem, 1905).

John D. Rockefeller's Lake Superior Consolidated Iron Mines Co., whose Mesabi range properties have been leased by Andrew Carnegie, is absorbed into United States Steel (above) to prevent Rockefeller from starting a rival company (*see* 1893).

Meyer Guggenheim and his sons gain control of the 2-year-old American Smelting and Refining Co. by buying out Leonard Lewisohn, and they merge their properties into the ASARCO copper trust (*see* 1899). Daniel Guggenheim, 45, becomes chairman of ASARCO's executive committee, four Guggenheim sons are on the board of directors, and the Guggenheims begin to extend the copper trust's operations into areas that include Chilean copper and nitrate mines, Bolivian tin mines, Alaskan gold mines, and Belgian Congo copper mines, diamond mines, and rubber plantations. A nitrate extraction process developed by E. A. Cappelen Smith will be called the Guggenheim process (*see Titanic,* 1912; Chuquicamata mine, 1910).

American Can Co. is created by a merger of 175 U.S. can makers engineered by W. H. Moore and Indiana banker Daniel Reid. The Can Trust turns out 90 percent of U.S. tin-plated steel cans.

The Spindletop gusher that comes in January 10 at Beaumont, Tex., gives John D. Rockefeller's Standard Oil Trust its first major competition. One-armed lumberman Patillo Higgins, now 36, located the Gulf Coast oil field in 1892 and has leased some 600 acres to Slavic-American Anthony F. Luchich (Lucas), who has been drilling since July 1899 into a salt dome on the field abandoned as unproductive by Standard Oil prospectors. Backed by Pittsburgh financiers John H. Galey and Col. J. M. Guffey, Lucas has drilled some 700 feet into the Big Hill and struck oil that spouts 110,000 barrels per day, flowing wild for 9 days before it can be brought under control.

The Beaumont Field (above) contains more oil than the rest of the United States combined, Spindletop establishes Texas as the major petroleum-producing state, Higgins has a 10 percent interest in the Lucas lease and becomes a millionaire overnight, and the Gulf Oil Co. has its beginnings as Galey and Guffey (above) get backing from Pittsburgh banker Andrew W. Mellon and his brother Richard, who take 40 percent of the new J. M. Guffey Petroleum Co. (*see* aluminum, 1891; Texas Co., 1902).

Persia sells a 60-year concession to explore for oil in four-fifths of the country to New Zealander William Knox D'Arcy who has made a fortune in Australian gold mining. D'Arcy pays $20,000 for the concession, he will sell it in 1908 to Burmah Oil Co., which will be backed by the British government in forming the Anglo-Persian Co., which will begin Middle Eastern petroleum production (*see* 1908).

More than half the world's oil output is from Russia's Baku oil fields which have been developed by Ludwig Nobel, brother of dynamite inventor Alfred Nobel, and by Rothschild interests (*see* 1871). Nobel has devised the world's first oil tankers and tank cars and has installed Europe's first pipeline, but the world's major supplier of petroleum is the United States, which will produce as much as two-thirds of the world's export oil for 20 years (*see* 1946).

U.S. electrical engineer Peter Cooper Hewitt, 40, invents mercury-vapor electric lamp (his father Abram Stevens Hewitt produced the first American-made open-hearth steel in 1870).

The Second Hay-Pauncefote Treaty November 18 gives the United States sole rights to construct, maintain, and control a trans-isthmian canal (*see* 1900). Washington assumes the obligation to assure passage to ships of all nations (*see* 1902).

The Mercedes motorcar introduced by German auto maker Gottlieb Daimler is named for the 11-year-old daughter of Emil Jellinek, Austrian consul at Nice. He has offered to buy and distribute 36 Daimlers on condition that the cars be named for Mercedes, thus undertaking to sell nearly an entire year's production of the motorcars (*see* 1890; Mercedes-Benz, 1926).

Detroit auto maker Ransom E. Olds, 36, moves his assembly plant to his hometown of Lansing, Mich. Copper and lumber baron Samuel L. Smith has financed the Olds Motor Works, and Olds markets 600 curved-dash "Oldsmobile" runabouts, a number he will increase to 5,000 by 1904 (*see* Reo, 1904).

Detroit Automobile Co. goes bankrupt after selling only four or five cars in 2 years. Chief engineer Henry Ford is hired as experimental engineer by the men who buy Detroit Automobile's assets, and when a Ford-designed car wins a major race, some former Detroit Automobile stockholders form the Henry Ford Co., giving Ford one-sixth of the stock in the new company, which he will soon quit (*see* 1903).

The Apperson motorcar is introduced.

The Pierce "motorette" is introduced by Buffalo, N.Y., auto maker George N. Pierce, 55 (*see* 1904).

John North Willys enters the motorcar business by selling two of the new Pierce "motorettes" (above). Willys, 27, is an Elmira, N.Y., sporting goods retailer who saw his first automobile (a Winton) on a trip to Cleveland last year and has become a Pierce dealer (*see* 1902).

1901 ***(cont.)*** New York City streetcars and elevators are converted to electric power, but horsecars continue to move up and down Fifth Avenue (*see* 1885; 1907).

The Uganda Railway completed December 26 links Mombasa with Lake Victoria.

A hydrogenation process invented by English chemist William Normann saturates unsaturated and polyunsaturated fatty acids to keep them from turning rancid. The process will find wide application in soapmaking and the production of edible fats and foods containing oils (*see* below).

Nobel Prizes are awarded for the first time from a fund (initially $9.2 million) established by Alfred B. Nobel of 1866 dynamite fame, who has added to his fortune by investments in Russia's Baku oil fields (above). "Inherited wealth is a misfortune which merely serves to dull a man's faculties," said Nobel in 1895, and he has willed that his fortune be invested in safe securities "the interest accruing from which shall be annually awarded in prizes."

The first Nobel Prize in physics goes to W. C. Roentgen for his X-ray discovery.

The first Nobel Prize for medicine goes to Emil von Behring for his diphtheria antitoxin.

The Rockefeller Institute for Medical Research is founded by John D. Rockefeller. Unlike European laboratories which are built around individuals such as the Pasteur Institute founded in 1888, the Rockefeller Institute offers facilities to groups of collaborating investigators and establishes a new pattern that others will follow (*see* McCormick, 1902; Phipps, 1903; Rockefeller Sanitary Commission, 1909).

Life expectancy at birth for U.S. white males is 48.23 years, for white females, 51.08 years.

Adrenaline (epinephrine) is isolated by Japanese-American chemist Jokichi Takamine, 47, and Armour & Co. scientists working with Johns Hopkins Medical School chemist-pharmacologist John Jacob Abel, 44. Secreted by the medullary portion of the adrenal glands, levo-methylaminoethanolcatchetol ($C_9H_{13}ON_3$) is the first ductless gland secretion to be isolated by man. It will be used in medicine as a heart stimulant, to constrict the blood vessels, and to relax the bronchi in asthma; it will be synthesized, but it will be obtained more economically from the glands of cattle, even though it takes 12,000 head of cattle to produce 1 pound of the hormone that raises blood pressure (*see* Bayliss, Starling, 1902; Dakin, 1904).

Mosquito controls virtually rid Havana of yellow fever. General Leonard Wood, now 41, and Major William C. Gorgas, 47, of the U.S. Public Health Service have put in the controls (*see* 1900; Panama, 1904).

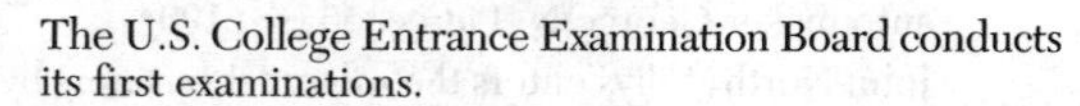

The U.S. College Entrance Examination Board conducts its first examinations.

The U.S. Army War College founded at Washington, D.C., by Secretary of War Elihu Root will move in 1951 to Carlisle, Pa.

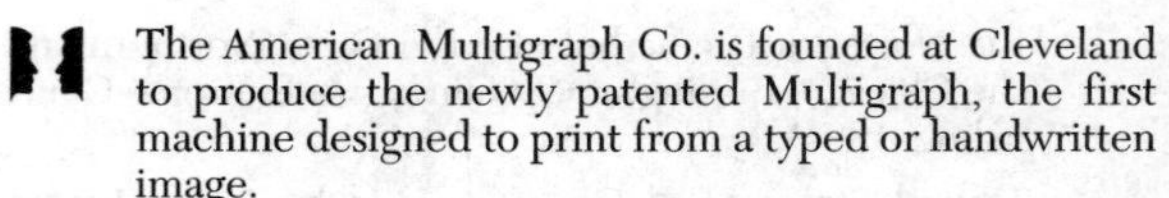

The American Multigraph Co. is founded at Cleveland to produce the newly patented Multigraph, the first machine designed to print from a typed or handwritten image.

J. Walter Thompson Co. receives a letter from Curtis Publishing in Philadelphia which applies the ANPA rules of 1893 to its *Ladies' Home Journal*, *Saturday Evening Post*, and *Country Gentleman* magazines. The 10 percent agency commission for placing advertisements in Curtis magazines will be followed by other magazines and will rise to 15 percent (*see* 1878).

Guglielmo Marconi receives the first transatlantic wireless message December 12 in Newfoundland (*see* 1896). An English telegrapher at Poldhu, Cornwall, has tapped out the letter "S," and Marconi picks it up with a kite antenna. He will build a station at Glace Bay, Nova Scotia, next year, and he will send the first readable message across the Atlantic to begin regular transatlantic wireless service (*see* 1903).

Andrew Carnegie gives the New York Public Library $5.2 million to open its first branches (*see* 1895). Within 55 years, the library will have 75 branches staffed by 2,000 plus three mobile units, and its operating costs will be $8 million per year (*see* 1911).

Nonfiction *Up from Slavery* by Booker T. Washington (above).

Fiction *Buddenbrooks* by German novelist Thomas Mann, 26, who has rejected his publisher's order to cut the manuscript in half; *Martin Birck's Youth (Martin Bircks Ungdom)* by Hjalmar Soderberg; *Kim* by Rudyard Kipling; *Erewhon Revisited* by Samuel Butler; *Sister Teresa* by George Moore; *Graustark* by Lafayette, Ind., *Daily Courier* city editor George Barr McCutcheon, 35, whose novel about a fictional kingdom achieves great popular success; *The Octopus* by Frank Norris dramatizes the struggle of California wheat growers against the Southern Pacific Railroad.

Juvenile *Mrs. Wiggs of the Cabbage Patch* by Alice Hegan Race.

Painting *The Gold in Their Bodies* by Paul Gauguin; *Girls on the Bridge* by Edvard Munch; *Medicine* (mural) by Gustav Klimt; *Cat Boats, Newport* by U.S. painter Childe Hassam, 41; *Femme Retroussant Sa Chemise* by Henri de Toulouse-Lautrec, who dies in his mother's arms September 9 at the Château de Malrome at age 36.

Sculpture *Mediterranean* by French sculptor Aristide (Joseph Bonaventure) Maillol, 39.

Theater *The Climbers* by Clyde Fitch 1/21 at New York's Bijou Theater, with Amelia Bingham, 163 perfs.; *The Three Sisters (Tri Sestry)* by Anton Chekhov 1/31 at the Moscow Art Theater; *Lover's Lane* by Fitch 2/6 at New York's Manhattan Theater; *The Wedding (Wesele)* by Polish playwright-painter Stanislaw Wyspianski, 32, 3/16 at Cracow's Teatr Slawackiego; *The Marquis of Keith (Der Marquis von Keith)* by Frank Wedekind 10/11 at Berlin's Residenztheater; *If I Were King* by Irish playwright-novelist Justin Huntly McCarthy, 40,

10/14 at New York's Garden Theater, with Edward H. Sothern, 56 perfs.; *The Last of the Dandies* by Fitch 10/24 at His Majesty's Theatre, London; *The Way of the World* by Fitch 11/4 at Hammerstein's Victoria Theater, New York, with Elsie de Wolfe, 35 perfs.; *The Conflagration (Der rote Hahn)* by Gerhart Hauptmann 11/27 at Berlin's Deutsches Theater; *The Girl and the Judge* by Fitch 12/4 at New York's Lyceum Theater, with Anne Russell, 125 perfs.

Opera *Fire Famine* (*Feuersnot*) 11/21 at Dresden, with music by Richard Strauss.

First performances Symphony No. 6 in A major by the late Anton Bruckner 3/14 at Stuttgart (first complete performance); *Menuet Antique* by Maurice Ravel 4/13 at Paris; *Cockaigne Overture (In London Town)* by Edward Elgar 6/20 at the Queen's Hall, London; Concerto No. 2 in C minor for Piano and Orchestra by Sergei Rachmaninoff 10/14 at Moscow; Pomp and Circumstance March in D major by Edward Elgar 10/19 at Liverpool; Symphony No. 4 in G major by Gustav Mahler 11/28 at Munich.

Broadway musicals *Captain Jinks of the Horse Marines* 2/4 at the Garrick Theatre, with H. Reeves Smith, Ethel Barrymore, book by Clyde Fitch, music by T. Maclaglen, lyrics by female impersonator William H. Lingard, and a song that begins "I'm Captain Jinks of the Horse Marines,/ I feed my horse on pork and beans,/ And often live beyond my means,/ I'm a captain in the army," 168 perfs.; *The Governor's Son* 2/25 at the Savoy Theatre, with music, book, and lyrics by George Michael Cohan, 23, who began touring in vaudeville with his parents, began writing songs for the act in his teens, gained his first popular success at age 17 with "Hot Tamale Alley," and followed it 3 years later with "I Guess I'll Have to Telegraph My Baby," 32 perfs.

Popular songs "I Love You Truly" by U.S. songwriter Carrie Jacobs-Bond, 39; "Mighty Lak' a Rose" by the late Ethelbert Nevin, who has just died at age 39, lyrics by *Atlanta Constitution* poet-journalist Frank L. Stanton, 44; "Boola Boola" by Yale undergraduate Allan M. Hirsch.

Arthur Wentworth Gore, 33, wins in men's singles at Wimbledon, Charlotte Cooper Sterry in women's singles; William A. Larned, 28, wins in U.S. men's singles, Elizabeth Moore in women's singles.

Baseball's American League is organized by teams whose annual pennant winner will compete beginning in 1903 with the top team of the 25-year-old National League in World Series championships.

The U.S. America's Cup defender *Columbia* with rigging by Nat Herreshoff defeats Sir Thomas Lipton's *Shamrock II* 3 to 0.

The Arlberg Ski Club is founded at St. Christoph.

The first practical electric vacuum cleaner is invented by British bridge builder Hubert Booth. His Vacuum Cleaner Co. Ltd. sends vans round to houses and uses the Booth machine to suck dust out of houses via tubes (*see* Spangler, Hoover, 1907; Electrolux, 1921).

King C. Gillette raises $5,000 to start a safety razor company and sets up a factory above a Boston fish store (*see* 1895). MIT graduate William E. Nickerson refines Gillette's idea for a safety razor and develops processes for hardening and sharpening sheet steel (*see* 1903).

Imperial Tobacco Company of Great Britain and Ireland is created by Sir William Henry Wills, 71, first Baron Winterstroke, whose grandfather and father-in-law started a tobacco and snuff business at Bristol in the 18th century. Baron Winterstroke's cousin Sir George Alfred Wills, 47, helps him effect a merger of British manufacturers and establish a company that will produce pipe tobacco and cigarettes.

New York's R. H. Macy moves into a new building on Herald Square between 34th and 35th Streets that replaces the old 14th Street Macy's and will grow to become the world's largest department store building (*see* 1888; 1912).

Boston's William Filene, Sons & Co. becomes William Filene's Sons following the death of founder William Filene at age 71. Filene is succeeded by his son Edward, who moves the store to 453–463 Washington Street, trebles its floor space, and soon releases the former store at 445–447 Washington Street as an annex for babies' and children's wear (*see* 1881; Automatic Bargain Basement, 1909).

The Nordstroms retail chain has its beginnings in a Seattle shoe shop opened by Swedish immigrant John W. Nordstrom, 30, and a partner. Nordstrom, who arrived at New York in 1887 with $5, has made $13,000 mining gold in Alaska and the Klondike.

The Elms at Newport, R.I., is completed for Philadelphia coal magnate Edward Julius Berwind, 53—the world's largest individual owner of coal mines. Architect Horace Trumbauer has modeled the huge Bellevue Avenue "cottage" on the 18th-century Château d'Agnes at Asnières near Paris and has surrounded it with a high wall that encloses extensive gardens.

New York's Carnegie mansion for steel magnate Andrew Carnegie (above) is completed on Fifth Avenue at 92nd Street by Babb, Cook and Willard. The six-story 64-room neo-Georgian house has wood panels carved by Scottish and Indian craftsmen, a utilities basement fitted like a steamship engine room, a conservatory, and a large garden opposite Central Park.

New York's Benjamin N. Duke mansion on Fifth Avenue at 78th Street is completed for a founder of the American Tobacco Company trust.

A new tenement house law is passed in New York, whose 83,000 "old law" masonry and wood tenements house 70 percent of the city's population. Strung end-to-end like railroad cars, the old six-story tenements have only tiny air shafts to provide light and air between their 90-foot ends, they crowd 10 families into 25-by-100-foot lots, and their minimal communal plumbing is indoors only because outhouses would consume land used for building.

1901 ***(cont.)*** New York's Harlem begins its rise following the start of construction of a Lenox Avenue subway line that triggers a real estate boom. The uptown Manhattan area will be overbuilt with apartment houses, many buildings will be unoccupied, and the razing of structures for Macy's department store (above), for Pennsylvania Station, and for large hotels, office buildings, and lofts in the Tenderloin area west of Herald Square is forcing blacks to seek new homes. Black real estate operator Philip A. Payton will seize the opportunity, guarantee premium rates to landlords, and make good housing available for the first time to black New Yorkers. Harlem will become America's largest black community—a model community until overcrowding in the 1920s turns it into a ghetto.

Johns Manville Co. is created by Milwaukee's 21-year-old Manville Covering Company which buys out New York's 43-year-old Johns Manufacturing Co. Johns Manville will import asbestos (45 percent silica, 45 percent magnesia, 10 percent water) from Canadian mines and be the world's largest insulation company.

The first Statler Hotel opens for Buffalo's Pan-American Exposition. Local restaurateur Ellsworth Milton Statler, 37, has erected a temporary 2,100-room structure that will be replaced by a more solid Statler Hotel in 1904 (*see* St. Louis, 1904).

One fourth of U.S. agricultural produce is exported, according to testimony given at hearings of the U.S. Industrial Commission.

The Industrial Commission (above) hears a government witness testify that a steam sheller can shell a bushel of corn in 1.5 minutes versus 100 minutes for the same job done by hand and that a wheat combine can do in 4 minutes what it would take a man 160 minutes to reap, bind, and thresh by hand.

Obesity and heart disease are observed for the first time to have a strong correlation.

The hydrogenation process invented by William Normann (above) turns polyunsaturated fats into saturated fats that will be linked to heart disease when it is found that the human liver can synthesize serum cholesterol from saturated fats (*see* 1913).

Beriberi kills thousands in the Philippines following introduction of polished white rice by U.S. occupation authorities (*see* 1898; Grijns, 1906).

Britain establishes statutory standards for milk to protect consumers, but pasteurization is not required. British milk remains a source of diseases that include undulant fever and bone tuberculosis (*see* Evans, 1917).

The hydrogenation process invented by William Normann (above) extends the shelf life of foods containing fats.

The first soluble "instant" coffee is invented by Japanese-American chemist Satori Kato of Chicago, who sells the product at the Pan American Exposition at Buffalo (*see* G. Washington, 1909).

The Settlement Cookbook by Milwaukee settlement house worker Lizzie Black (Mrs. Simon Kander) is published with funds raised by volunteer women through advertisements after the settlement house directors have refused a request for $18 to print a book that would save students in a class for immigrants from having to copy recipes off the blackboard. Using the slogan "The way to a man's heart," the book will earn enough money in 8 years to pay for a new settlement house building and be a perennial best seller.

London's population reaches 6.6 million, while New York has 3.44, Paris 2.7, Berlin 1.9, Chicago 1.7, Vienna 1.7, Wuhan 1.5, Tokyo 1.45, St. Petersburg 1.3, Philadelphia 1.3, Constantinople 1.2, Moscow 1.1, Xian (Sian) 1, Calcutta 950,000, Guangzhou (Canton) 900,000, Los Angeles 103,000, Houston 45,000, Dallas 43,000.

Europe's population reaches above 400 million, up from 188 million in 1800, with 56.4 million in Germany, 39.1 million in France, 34 in Austria, 33.2 in Italy. China has an estimated 373 million, India 284, Japan 44, Russia 117, Great Britain and Ireland 41.4, the United States more than 76.

Some 9 million immigrants will enter the United States in this decade.

Girls in Western cultures menstruate for the first time at about age 14, down from age 17 in the late 18th century. The average age of menarche will decline to 13 in this century.

1902 The Anglo-Japanese Alliance signed January 20 ends the "splendid isolation" of Britain and recognizes Japan's interests in Korea. If either party should become involved in war with a third party, its ally is to remain neutral, but if the war should expand to involve any other power or powers, then the ally is obligated to enter the conflict (*see* 1904).

Spain declares the posthumous son of the late Alphonso XII of age on his sixteenth birthday May 16. Alphonso XIII begins a reign that will continue until 1931 despite several attempts on his life.

Cuba gains independence from Spain May 20 and establishes a republic. U.S. troops are withdrawn.

The Treaty of Vereeniging signed May 31 ends the Boer War. The Boers accept British sovereignty in South Africa, the British promise £3 million for rebuilding Boer farms.

President Roosevelt officially ends the "great insurrection" in the Philippines July 4 and commends U.S. troops for upholding America's "lawful sovereignty."

Britain's prime minister Lord Salisbury retires July 11 at age 72 (he will die next summer) and is succeeded by his nephew Arthur James Balfour, 53, who will head the government until 1905.

Saudi Arabia has her beginnings as Bedouin warrior Abdul-Aziz ibn Saud, 20, emir of the Wahabi, comes out of exile in Kuwait and seizes Riyadh, a mud fort that serves as capital of the Nejd, which is still controlled by a rival clan. Abdul-Aziz sets himself up as leader of an Arab nationalist movement (*see* 1924; railway, 1908).

Venezuela refuses to meet her debt obligations, British and German warships seize the Venezuelan Navy December 9, and Italian warships join in a blockade of Venezuelan ports December 19. The blockade will continue until February of next year, when Venezuelan dictator Cipriano Castro will agree to arbitration by a Hague Tribunal commission.

Italian railway workers strike through January and February in a demand for recognition of their union. A general strike threatens as Turin gas company employees walk out, but all railway workers who belong to the army reserve are called up by the government, and a settlement is mediated in June.

More than 30,000 Russian students demonstrate in February to protest government efforts to curb student organizations. Socialist revolutionaries murder the head of the secret police April 15. Thousands die in riots, and Czar Nicholas II offers talks July 2 to quell the disturbances (*see* 1903).

United Mine Workers leader John Mitchell leads his 147,000 anthracite coal workers out of the pits May 12 to begin a 5-month strike that cripples the United States. Mine operators and railroad presidents have rejected an invitation from Mitchell to attend a conference, they continue to oppose unionization, and George F. Baker of Philadelphia and Reading Coal & Iron says July 17, "The rights and interests of the laboring man will be protected and cared for—not by the labor agitators, but by the Christian men to who God in His infinite wisdom has given the control of the property interests of this country. . ."

By September, the price of anthracite in New York has climbed from $5 per ton to $14, and the poor who buy by the bucket or bushel pay a penny per pound, $20 per ton. Schools close to conserve fuel, people buy oil, coke, and gas stoves, mobs in western towns seize coal cars, and by October the price of coal in New York is $30 per ton.

President Roosevelt brings about a settlement of the coal strike that one writer hails as "the greatest single event affecting the relations of capital and labor in the history of America." But while the strikers gain pay raises, they do not win recognition of their union as bargaining agent for their rights.

Smuggler-Union mine manager Arthur Collins is assassinated November 19 in his home at Telluride, Colo., following settlement of a violent strike against the mine by the Western Federation of Miners. Four hundred national guardsmen will go into Telluride next year to suppress the WFM.

AF of L membership reaches 1 million (*see* Gompers, 1886; Department of Labor, 1913).

William Cadbury sees an advertisement for the sale of a São Tomé cocoa plantation. Workers are listed as assets at so much per head—a strong suggestion of slavery (*see* 1901; 1903).

Portugal declares national bankruptcy.

Imperialism: A Study by English economist John Atkinson Hobson, 44, is a comprehensive critique of economic imperialism that will influence German theorists Rosa Luxembourg and Rudolf Hilferding. Through their writings it will shape the thinking of V.I. Lenin.

E.I. du Pont's Eugene du Pont dies January 28 at age 68. His sons Thomas Coleman, 39, Pierre Samuel, 32, Irenee, 26, and Lammot, 22, pay $22 million to acquire the firm founded by their great-grandfather in 1802 (*see* 1899; 1906).

President Roosevelt works through his attorney general Philander C. Knox, 49, to institute antitrust proceedings against various U.S. corporations. Roosevelt departs from the policies of the late President McKinley which he had pledged himself to continue.

International Nickel Co. is created by a merger of Canadian Copper and Orford Copper effected by J. P. Morgan at the behest of Charles M. Schwab and other steel magnates. Orford, at Bayonne, N.J., has found a way to refine nickel from ore it buys from Canadian Copper (*see* 1885) and sells it in Europe, competing with the Rothschild-controlled firm Le Nickel, which obtains its ore from New Caledonia. Despite such competition, INCO will meet 90 percent of world nickel needs.

United States Steel Co. has two-thirds of U.S. steelmaking capacity. Only public opinion and a sense of noblesse oblige restrain its near-monopoly (*see* 1901; Bethlehem Steel, 1905).

History of the Standard Oil Company by Ida Minerva Tarbell, 44, appears in *McClure's* magazine installments, revealing that John D. Rockefeller controls 90 percent of U.S. oil-refining capacity and has an annual income of $45 million.

The Texas Company is founded to battle Rockefeller interests. Former Standard Oil executive Joseph Cullinan receives backing from financier Arnold Schlaet and enlists the support of Texas governor James Hogg and John W. Gates (*see* 1898). Now 47, Gates has agreed to stop his "bear" raids on the stock market (after losing a fortune) and helps Cullinan buy up oil fields near Gulf Oil's Spindletop Hill in East Texas (*see* 1901) and build a marketing organization that will cover the United States and much of the world (*see* 1903; CalTex, 1933).

France's Panama Canal Co. offers to sell its interests to the United States and reduces its asking price January 4 from $109 million to $40 million (*see* de Lesseps, 1883). The Isthmian Canal Commission recommends adoption of the Panama route in preference to a Nicaraguan route in a supplementary report to President Roosevelt January 18. The Isthmian Canal Act passed by Congress June 28 authorizes the president to acquire the Canal Company's concession, arrange terms with the Colombian government, and proceed with construction. If it is not possible to obtain title and necessary control, the canal is to be built through Nicaragua (*see* 1903).

United States Shipbuilding Co. is organized by U.S. Steel executive Charles Michael Schwab, 40, who

1902 *(cont.)* merges Bethlehem Steel into the new company as a source of plates for ships (*see* 1886; 1905).

The *Twentieth Century Limited* goes into service June 15 to begin a 65-year career on the route between New York and Chicago. The New York Central's new luxury express has two buffet, smoking, and library cars, two observation cars, and 12 drawing-room and stateroom cars. It reduces running time on the 980-mile "water-level route" to 20 hours—down from 24 in 1877—but the *Limited*'s 49-mile-per-hour average is well below the 61.3 miles per hour averaged on the 184-mile route of the *Paris-Calais Express.*

The *Broadway Limited* goes into service June 16 for the Pennsylvania Railroad to begin a career of more than 75 years on the route between New York and Chicago. Staffed with barbers, maids, valets, and secretaries to rival the Central's *Twentieth Century Limited* (above), the new luxury express makes the 904-mile run in 20 hours.

Egypt's *Cairo-Luxor Express* goes into service with sleeping cars.

The rail link between Salisbury, Rhodesia, and Bulawayo is completed October 6.

The Isotta-Fraschini motorcar introduced by Fabbrica Automobili Isotta-Fraschini of Milan is a powerful, costly machine with a four-cylinder engine that develops 29 horsepower. Cesare Isotta and Vincenzo Fraschini started out as agents for Renault Frères, have assembled the Renault Voiturette, and will open a large factory in 1905, where production will continue until 1949.

Henry Ford's 70-horsepower "999" racing car driven by Bernard Eli "Barney" Oldfield, 24, sets an unofficial speed record on Detroit's Grosse Point track, covering 5 miles in 5 minutes, 20 seconds (*see* 1901; Ford Motor Company, 1903).

The American Automobile Association (AAA), founded March 4 by nine automobile clubs meeting at Chicago, will provide members with emergency road service, help them plan tours, and work in behalf of new highway construction.

The first Studebaker motorcar, introduced at South Bend, Ind., is an electric car. Studebaker Bros. has produced more than 750,000 wagons, buggies, and carriages since 1852.

The Marmon motorcar, introduced by Indianapolis auto maker Howard C. Marmon, 26, has an air-cooled overhead valve V-twin engine and a revolutionary lubrication system that uses a drilled crankshaft to keep its engine bearings lubricated with oil fed under pressure by a gear pump. Marmon's father Daniel is a leading producer of milling machinery.

The Overland motorcar is introduced at Indianapolis (*see* Willys, 1907).

John North Willys sells four Pierce motorcars and obtains the Elmira, N.Y., agency for the Rambler car made at Kenosha, Wis. (*see* 1901; 1903).

A new Locomobile is introduced with a front-mounted, vertical, four-cylinder, water-cooled engine and is the first U.S. motorcar to use heat-treated steel alloys.

Rayon is patented by U.S. chemist A. D. (Arthur Dehon) Little, 39, who has produced the new cellulose fiber with a new process (*see* Cross, Bevan, 1892; Courtauld, 1905; Viscose Co., 1910; Celanese, 1918; nylon, 1935).

Mendel's *Principles of Heredity—A Defence* by English biologist William Bateson, 41, supports the work by Hugo De Vries and others published 2 years ago. Bateson has explored the fauna of salt lakes in western Central Asia and in northern Europe and will introduce the term *genetics* (*see* 1926).

Elementary Principles in Statistical Mechanics by Yale physicist Josiah Willard Gibbs, 63, helps establish the basic theory for physical chemistry. Gibbs's work On *the Equilibrium of Heterogeneous Substances* appeared in the late 1870s.

The Carnegie Institution of Washington, established with a $10 million gift from Andrew Carnegie, will devote its efforts to scientific research (*see* agriculture, 1905; medicine, 1910).

English physiologists William Maddock Bayliss, 42, and Ernest Henry Starling, 36, discover the hormone secretin manufactured by glands on the wall of the small intestine (they will introduce the word *hormone* in 1904). Working at London's University College, they find that secretin acts on the liver to increase the flow of pancreatic juice when the acid contents of the stomach enters the duodenum.

Hookworm disease is endemic in the U.S. South, reports Charles Wardell Stiles, 35, chief of the Division of Zoology of the U.S. Public Health Service, in an address to the Sanitary Conference of American Republics meeting at Washington, D.C.

John D. Rockefeller establishes the General Education Board that will help reform U.S. medical education and work to conquer hookworm disease in the South (*see* 1901; 1909).

The John McCormick Institute for Infectious Disease is established by reaper heir Harold Fowler McCormick and his wife Edith in memory of a son who has died in infancy of scarlet fever. An antitoxin for scarlet fever will be developed at the institute.

Philadelphia chemist Albert C. Barnes, 31, and his German-American colleague Herman Hille, 31, announce discovery of a noncaustic antiseptic that can be used to prevent eye infections such as gonorrheal infection in newborn infants (*see* Crede, 1884). Barnes and Hille quit their jobs 3 days later and form a partnership. The "silver vitellin" they say they have developed is really a silver-gelatin colloid that they will market as Argyrol; it will find almost universal use in protecting newborn infants from blindness.

The Canadian Medical Act establishes a central national registration for physicians and uniform standards for all licensed practitioners. Passage of the act has been pushed by physician Thomas George Roddich, 56, a

Conservative member of the Dominion House of Commons.

Disease has cost the British 5 times as many men as enemy action in the Boer War (above).

Cecil Rhodes dies in South Africa March 26 at age 49 leaving a bequest to endow 3-year Rhodes Scholarships at Oxford University. Awarded each year to 60 young men from the British colonies, 100 from the United States, and 15 from Germany, the scholarships pay £250 per year; the trustees will increase the amount and, beginning in 1976, will award some scholarships to women.

"Buster Brown" debuts May 4 in the *New York Herald*. Richard F. Outcault's comic-strip adventures of the middle-class boy and his dog Tige will be far more successful than his "Yellow Kid" strip (*see* 1896; "Mutt and Jeff," 1908).

Nonfiction *Rebellion in the Backlands (Os Sertos)* by Brazilian engineer-journalist Eyclydes da Cunha, 36, who covered an 1896 uprising led by a religious fanatic in Brazil's *sertão* (backlands). His sympathetic treatise on history, sociology, geology, and geography amounts to a national epic; *Varieties of Religious Experience* by Harvard philosopher-psychologist William James, now 60, whose Gifford Lectures at the University of Edinburgh constitute a classic reconciliation of science and religion.

Fiction *The Wings of the Dove* by Henry James, younger brother of the Harvard professor (above); *Anna of the Five Towns* and *The Grand Babylon Hotel* by English novelist (Enoch) Arnold Bennett, 35; *Mrs. Craddock* by W. Somerset Maugham; "Youth and Two Other Stories" by Joseph Conrad; *The Immoralist* by André Gide; *The Supermale* (*Le Surmale*) by Alfred Jarry; *The Virginian* by U.S. novelist Owen Wister, 42, whose account of a greenhorn's misadventures in Wyoming will be famous for its line, "When you call me that, smile!"; "To Build a Fire" by California writer John Griffith "Jack" London, 26, who at 17 shipped out as a common seaman aboard a sailing ship to Japan and at 21 joined the Klondike gold rush; *The Pothunters* by English humorist P. G. (Pelham Grenville) Wodehouse, 21, who will quit his bank job next year to conduct a column in the *London Globe* and will become famous for his characters Bertie Wooster, Jeeves, Mr. Multiner, and Psmith; *The Hound of the Baskervilles* by A. Conan Doyle who has been forced by public demand to announce that his Baker Street detective Sherlock Holmes escaped death when he fell into the mountain crevasse with the arch-criminal Professor Moriarty in December of 1893 (the new book has been serialized in the *Strand* magazine).

The *Times Literary Supplement (TLS)* begins publication January 17 as part of the *Times* of London. The journal of reviews, poems, and commentary will continue for more than 75 years.

Juvenile *Just So Stories for Little Children* by Rudyard Kipling; *The Tale of Peter Rabbit* by English artist-writer-naturalist Beatrix Potter, 36, who was the first person in England to establish the fact that lichens represent a merging of algae and fungi but whose efforts in botany have been frustrated by the male scientific establishment (*see* Reinke, 1894). The story of Peter, Mopsy, Flopsy, and Cottontail will be followed by 22 more books that Potter will write and illustrate in the next 11 years.

Poetry *Salt Water Ballads* by English poet John Masefield, 24, who ran away to sea at age 13. His "Sea Fever" contains the lines, "I must down to the sea again, to the lonely sea and the sky,/And all I ask is a tall ship and a star to steer her by . . ."; *Songs of Childhood* by English poet Walter (John) de la Mare, 29; *Captain Craig* by U.S. poet Edwin Arlington Robinson, 32.

Painting *Waterloo Bridge* by Claude Monet; *Horseman on the Beach* by Paul Gauguin, who moved last year to the Marquesa Islands; *Ladies Acheson*, *Misses Hunter*, *The Duchess of Portland*, and *Lord Ribblesdale* by John Singer Sargent. *Beethoven Frieze* by Gustav Klimt. Albert Bierstadt dies at New York February 18 at age 72.

Illustrator Charles Dana Gibson signs an agreement October 23 accepting an offer from *Collier's Weekly* to draw only for *Life* and *Collier's* for the next 4 years and do 100 double-page drawings for *Collier's* for $100,000 (*see* 1890).

Crayola brand crayons are introduced by the Easton, Pa., firm Binney & Smith. Founder's son Edwin Binney has developed the crayons by adding oil and pigments to the black paraffin and stearic acid marking devices sold by the firm, and his mother has suggested the name Crayola.

Sculpture *Comin' Through the Rye* (*Off the Range*) by Frederick Remington (bronze).

Carl Zeiss of Jena introduces the Tessar f4.5 antistigmatic lens designed by Paul Rudolf (*see* 1889).

The "Gibson girl" epitomized feminine pulchritude throughout the early decades of America's twentieth century.

1902 ***(cont.)*** Theater *The Joy of Living (Es lebe das Leben)* by Hermann Sudermann 2/1 at Berlin's Deutsches Theater; *Cathleen ni Houlihan* by William Butler Yeats and Lady Gregory (Isabella Augusta Persse Gregory, 50) 4/2 at St. Teresa's Hall, Dublin; *The Lower Depths (Na dne)* by Maksim Gorki 10/25 at the Moscow Art Theater (Gorki was arrested last year after publication of his seditious prose-poem "Burevestnik"); *The Admirable Crichton* by James M. Barrie 11/4 at the Duke of York's Theatre, London; *Old Heidelberg* by German playwright Wilhelm Meyer Forster, 40, 11/22 at Berlin; *Henry of Arnë (Der arnë Heinrich)* by Gerhart Hauptmann 11/29 at Vienna's Hofburgtheater; *The Darling of the Gods* by David Belasco and John Luther Long 12/3 at New York's Belasco Theater, 182 perfs.; *Earth Spirit (Der Erdgeist)* by Frank Wedekind 12/17 at Berlin's Keines Theater; *The Girl with the Green Eyes* by Clyde Fitch 12/25 at New York's Savoy Theatre, with Lucile Watson, 108 perfs.

Opera *Le Jongleur de Notre Dame* 2/18 at Monte Carlo with music by Jules Massenet; *Pélleas et Mélissande* 4/30 at the Opéra-Comique, Paris, with music by Claude Debussy, libretto by Maurice Maeterlinck, Scots-American soprano Mary Garden, 28, sings the female lead; *Adriana Lecouvrer* 11/6 at Milan's Teatro Lirico, music by Italian composer Francesco Cilea, 36; *The Girls of Vienna (Wiener Frauen)* 11/25 at Vienna's Theater-an-der-Wien, music by Franz Lehar.

Enrico Caruso makes his first phonograph recordings March 18 at Milan's Hotel di Milano where U. S. recording engineer Fred Gaisberg, 30, of the Emile Berliner London branch and his brother Will have converted a room into a recording studio using six large crates of equipment. Now 29, the tenor is appearing at the Teatro alla Scala, he has agreed to perform 10 songs and arias for £100, the Gaisbergs have wired London asking authority to spend that much, the London office has wired a flat refusal, the Gaisbergs proceed nevertheless, the disks go on sale in London record shops in May, and they will be reissued periodically for more than 75 years.

First performances Symphony No. 2 in C minor by Russian composer Aleksandr Nikolaievitch Skriabin, 30, 1/25 at St. Petersburg; *L'aprés-midi d'un faune* (Symphonic Poem) by Claude Debussy *(see* 1894); Symphony No. 2 by Jean Sibelius 3/8 at Helsinki; *Pavane pour une Infante defunte* by Maurice Ravel 4/5 at the Societé Nationale, Paris; Symphony No. 3 in D minor by Gustav Mahler 6/9 at Krefeld.

Broadway musicals *A Chinese Honeymoon* 6/22 at the Casino Theater, with music by Hungarian-American composer Jean Schwartz, 23, lyrics by William Jerome, songs that include "Mr. Dooley," 376 perfs.; *Twirly Whirly* 9/11 at the Weber and Fields Music Hall, with Lillian Russell, music by John Stromberg, lyrics by Robert B. Smith, songs that include "Come Down Ma Evenin' Star," 244 perfs.

Popular songs "In the Good Old Summertime" by Welsh-American composer-minstrel George Evans, 42, lyrics by Ren Shields, 34; "Bill Bailey, Won't You Please Come Home" by Hughie Cannon; "In the Sweet Bye and Bye" by Harry Von Tilzer, lyrics by Vincent P. Bryan; "On a Sunday Afternoon" by Harry von Tilzer, lyrics by Andrew B. Sterling; "Oh, Didn't He Ramble" by Will Handy (Bob Cole and J. Rosamond Johnson); "Under the Bamboo Tree" by Bob Cole; "The Land of Hope and Glory" by Edward Elgar, lyrics by Eton master Arthur Christopher Benson, 40.

The Rose Bowl football game has its beginnings in a game played January 1 at Pasadena, Calif., as part of the Tournament of Roses held since 1890. Michigan defeats Stanford 49 to 0 as halfback William M. "Willie" Heston leads a backfield coached by Fielding "Hurry Up" Yost, whose team scores 644 points in the fall against 12 for its opponents (*see* 1916).

Hugh Lawrence Doherty, 25, wins in men's singles at Wimbledon, Muriel Evelyn Robb, 24, in women's singles; Bill Larned wins in U.S. men's singles, Marion Jones in women's singles.

Dan Patch breaks the harness racing mark set by his sire Joe Patchen. The 6-year-old mahogany stallion runs a mile in 2:00 3/4, a full second below his sire's fastest speed, he goes on to equal Star Pointer's record of 1:59.25, and Minneapolis feed merchant Marion Willis Savage of the International Stock Food Co. acquires the trotter for $60,000.

The Teddy Bear introduced by Russian-American candy story operator Morris Michtom and his wife of Brooklyn, N.Y., has movable arms, legs, and head. The Michtoms have seen a cartoon by Clifford Berryman in the November 18 *Washington Evening Star* showing President "Teddy" Roosevelt refusing to shoot a mother bear while hunting in Mississippi. They obtain the president's permission to use his nickname, and their brown plush toy Teddy Bear creates a sensation (*see* 1903).

The brassiere, invented by French fashion designer Charles R. Debevoise, will have little popularity until the introduction of elastic (*see* 1914).

New York Philip Morris Corp. is founded with the help of local tobacco importer Gustav Eckmeyer, who has been exclusive agent for English Ovals, Marlboro (for women) and other cigarettes produced by Philip Morris of London (*see* 1858; 1933).

Paris jeweler Louis Cartier creates 27 diamond tiaras for the coronation of Britain's Edward VII. Cartier opens a London branch in New Bond Street with his brother Jacques in charge (*see* 1898; 1917).

Tiffany & Co. founder Charles Lewis Tiffany dies at Yonkers, N.Y., February 18 at age 90, leaving an estate of $35 million.

The Dayton Company is founded by Minneapolis merchant and banker George D. Dayton. His department store will become the city's largest.

J. C. Penney Co. has its beginnings in the Golden Rule store opened at Kemmerer, Wyo., April 14 by merchant James Cash Penney, 26, and his wife. Penney went to work at age 19 for J. M. Hale & Brother at Hamilton, Mo., for $2.27 per month, he moved to Colorado 5 years ago at the advice of his physician, and got a job at Longmont near Denver working for merchant T. M. Callahan, and the firm sent him to work in its store at Evanston, Wyo. Callahan has given him an opportunity to open a store on his own and has sold him a one-third interest for $2,000. The first day's receipts amount to $466.59 after remaining open from dawn to midnight, by the end of the first year it has done $29,000 worth of business, and Penney will move to a better location in 1904 (*see* 1908).

New York's "Flatiron" building is completed southwest of Madison Square by Chicago architect Daniel Burnham, whose steel-frame Fuller building soars 20 stories high and brings people from far and near to ascend to its top for the panoramic view. The building's shape inspires its nickname.

Hungary's Houses of Parliament are completed after 9 years of construction on the Rudolph Quay of the Danube at Budapest. Designed in an imposing Gothic style by Aimery Steindl, the building has a dome 325 feet high.

The London Ritz Hotel built by owners of the 3-year-old Carlton House opens in Piccadilly on the site of Walsingham House (*see* Paris Ritz, 1898). Hotel accountant William Harris organizes a company to build other Ritz caravansaries that will soon rise at Evian, Mentone, Lucerne, Salsomaggiore, Rome, Naples, Madrid, Barcelona, Buenos Aires, Montreal, Boston, and Atlantic City, N.J. César Ritz will die in an insane asylum in 1912.

New York's Algonquin Hotel opens at 59 West 44th Street with 143 suites and 192 rooms to begin a career as gathering place for actors, painters, musicians, and literate. Desk clerk Frank Case, 32, has persuaded owner Albert Foster to name it the Algonquin rather than the Puritan, he will himself acquire ownership for $717,000 in 1927, and he will operate the hotel until his death in 1946.

New Hampshire's Mount Washington Hotel opens at Bretton Woods with a long veranda from which guests may gaze at the mountains of the Presidential Range (*see* world monetary conference, 1944).

Old Faithful Inn at Wyoming's Yellowstone National Park is completed by architect Robert C. Reanor. Wings will be added in 1913 and 1927.

Rosecliff is completed at Newport, R.I., by New York architect Stanford White for Mrs. Hermann Oelrichs, daughter of the late San Francisco mining and banking magnate James Fair. Last of the great Newport "cottages," Rosecliff is a 40-room French château with a Court of Love designed by sculptor Augustus Saint-Gaudens after one at Versailles.

Whitehall is completed at Palm Beach by Carrère and Hastings for Florida East Coast Railway magnate Henry M. Flagler and his third bride. Flagler has told the architects to "build me the finest home you can think of," they have put up a 73-room, $2.5 million Spanish-inspired temple with massive Doric columns and gigantic urns, the 6-acre waterfront home has been finished in 8 months, and the *New York Herald* calls it "more wonderful than any palace in Europe, grander and more magnificent than any other private dwelling in the world."

U.S. engineer Willis Haviland Carrier, 26, designs a humidity control process to accompany a new air-cooling system for a Brooklyn, N.Y. printing plant and pioneers modern air conditioning (*see* 1911; Gorrie, 1842).

"Forest and water problems are perhaps the most vital internal questions of the United States," says President Roosevelt in his first State of the Union message. The president asks for a federal reclamation program and a sound plan to use forest reserves (he has been influenced by Gifford Pinchot and F. H. Newell, an aide to the late Major Powell).

The National Reclamation Act (Newlands Act) passed by Congress June 17 authorizes the federal government to build great irrigation dams throughout the West. The act is based on ideas championed by Major John Wesley Powell (who dies at Haven, Me., September 23 at age 78), but small farmers in years to come will charge that the legislation's chief beneficiaries are large farmers (below).

Crater Lake National Park is created by act of Congress on 160,290 acres of Oregon wilderness.

The Boundary Waters Canoe Area created in northern Minnesota embraces roughly a million acres of virgin pine forest.

Mont Pelée on the French Caribbean island of Martinique erupts May 8, inundating the capital and commercial center St. Pierre with molten lava and ashes that destroy the city's harbor and kill between 30,000 and 50,000. The 4,430-foot volcano has given premonitory signals, but the people of St. Pierre have been preoccupied with an imminent local political contest, they have ignored the warnings, 10 percent of the island is devastated as further eruptions occur May 20 and August 30, the *Roddam* is the only ship in the harbor that escapes destruction, and Martinique's capital shifts to Fort de France.

European countries sign a convention to eliminate the capture of migrating birds with lime-smeared sticks and iron traps, but the pact does not prohibit use of nets (*see* 1960).

The National Reclamation Act (above) encourages family farms by limiting to 160 acres the size of individual holdings entitled to federal water and by specifying that such water is to go only to owners who are bona fide residents. A 1926 amendment will let larger landowners have water provided that they sign contracts agreeing to sell acreage exceeding the 160-acre limit within 10 years at a government-approved price that does not include the value added by the federal water.

International Harvester is founded by Cyrus McCormick, Jr., who persuades J. P. Morgan to underwrite a trust that merges the four top U.S. harvesting machine

1902 ***(cont.)*** makers. Charles Deering of William Deering & Co. is chairman of the board, McCormick president, and the company controls 85 percent of all U.S. reaper production (*see* 1884).

Hart-Parr Co. of Charles City, Iowa, builds the first gasoline tractors, some of them 11-ton monsters so difficult to start that farmers leave them running all night.

Hookworm disease is a major cause of malnutrition in the U.S. South, says Charles W. Stiles (above). Southerners are not inherently lazy, he says, but the disease is sapping their energy (*see* Rockefeller Sanitary Commission, 1909).

A survey in the English city of Leeds shows that in the poorest sections, half the children are marked by rickets and 60 percent have bad teeth.

A survey of English milk consumption shows that in middle-class families the average person drinks 6 pints per week, while in the lower middle class the average is 3.8 pints, among artisans 1.8 pints, and among laborers 0.8 pints. Britain is Europe's biggest meat eater and smallest milk drinker.

Fish-and-chips shops (below) will make a significant contribution to raising protein levels of urban English diets.

A "Poison Squad" of young volunteers tests the safety of U.S. foods for Harvey W. Wiley of the Department of Agriculture (*see* 1883; 1907).

Congress limits substitution of margarine for butter. Britain establishes statutory limits for butter.

Britain imports £50 million worth of meat and £4.1 million worth of fish. The average Englishman consumes more than 56 pounds of cheap imported meat per year as U.S. pork, Argentine beef, and New Zealand lamb flood the market.

Corn Products Co. is created by a merger of United States Glucose with National Starch (*see* 1900). The new starch trust controls 84 percent of U.S. starch output (*see* Corn Products Refining Co., 1906).

Barnum's Animal Crackers are introduced by the National Biscuit Co., which controls 70 percent of U.S. cracker and cookie output (*see* 1898). The new animal-shaped crackers appear just before Christmas in a box topped with a white string so that it may be hung from Christmas trees, and it joins the line of Nabisco products that includes Uneeda Biscuits, Premium Saltines, Social Tea Biscuits, Ginger Snaps, Nabisco Sugar Wafers, Jinjer Wafers, Zuzus, Lemon Snaps, Vanilla Wafers, Saltinas, Crown Pilot Crackers, and other brands including Fig Newtons, many in In-er-Seal cartons (*see* Oreo, 1912).

Ralston Purina gets its name as officials of the 8-year-old Robinson-Danforth Commission Co. at St. Louis make an arrangement with officials of the Ralston Health Club, a worldwide organization established by Washington, D.C., university professor Albert Webster Edgerly, 50, who is sometimes called "Dr. Ralston" and whose book *Complete Life Building* propounds simple rules and facts about common foods and promotes good nutrition, urging readers to avoid heavy use of preservatives and artificial sweeteners. Ralston Purina promotes its 3-year-old Purina breakfast food and introduces Ralston cereals that will compete with Quaker Oats, Post, and Kellogg (*see* 1932; Purina, 1899).

Kellogg's makes its Sanitas corn flakes lighter and crisper and gives them a malt flavoring to help them compete with their many imitators (*see* 1898; 1906).

Force Wheat Flakes are introduced in Britain in June, giving Britons their first taste of a dried breakfast food. The cereal is made by the Force Food Co. of Canada, which advertises Force in the United States with widely read verses about "Sunny Jim."

"Prices That Stagger Humanity," headlines Joseph Pulitzer's *New York World* in a campaign against the "beef trust" (24¢ a pound for sirloin steak, 18¢ for lamb chops, pork chops, or ham).

Hawaiian Pineapple Co., Ltd., is founded by James Drummond Dole, 25, whose father is a first cousin of Hawaii's governor Sanford Dole. Young Dole visited the islands during his years at Harvard, he gave up his original plan of starting a coffee plantation, and last year he staked out a 12,000-acre homestead 23 miles outside Honolulu at Wahiawa. He has set out 75,000 pineapple plants, he has financed the new company by selling $20,000 worth of stock to friends in Boston and San Francisco, and he will pack 1,893 cases of fruit next year (*see* 1909).

Pepsi-Cola Co. is founded in North Carolina (see 1898). Caleb Bradham gives up his pharmacy to devote full time to Pepsi-Cola (*see* 1903).

Horn & Hardart Baking Co.'s Automat opens in Philadelphia's Chestnut Street. The first automatic restaurant is so popular that Philadelphia will soon have scores of Automats, and the first New York Automat will open in 1912 at 1551 Broadway (*see* Exchange Buffet, 1885). Horn & Hardart has paid a German importer $30,000 for the mechanism that permits patrons to drop nickels into slots to open glass doors and obtain food from compartments that are refilled by employees behind the scenes.

Fish-and-chips shops open to serve London's working class. Fast deep-sea trawlers extend British fishing operations to Iceland and the White Sea, they pack the fish in ice, they ship it by rail the day it is landed, and the London shops serve it with fried potatoes and vinegar (*see* 1864).

Immigration to the United States sets new records. Most of the arrivals are from Italy, Austro-Hungary, and Russia.

Congress revises the Chinese Exclusion Act of 1882 to prohibit immigration of Orientals from U.S. island territories such as Hawaii and the Philippines and makes the exclusion permanent (*see* 1943).

1903 U.S. gunboat diplomacy expedites construction of a Panama Canal. The Hay-Herran Treaty signed January 22 by Secretary of State John Hay and the foreign minister of Colombia provides for a 6-mile strip across the Isthmus of Panama to be leased to the United States for $10 million plus annual payments of $250,000. Washington had favored a route across Nicaragua, but French engineer-promoter Philippe Jean Bunau-Varilla, 42, has made a deal with New York lawyer William Cromwell of Sullivan and Cromwell, approached Cleveland industrialist Mark Hanna, and persuaded him to favor the route across Panama. Cromwell has made a $60,000 contribution to the Republican party. The Senate consents to the treaty March 17, and the treaty gives $40 million to stockholders in the French canal company, many of whom are now U.S. speculators, while stipulating that Colombia is to give up all rights to sue for any portion of the $40 million and give up all police powers in the contemplated canal zone.

Colombia's Senate votes unanimously August 12 to reject the Hay-Herran Treaty (above); President Roosevelt calls the Colombians "Dagos," "cat-rabbits," "homicidal corruptionists," and "contemptible little creatures." Bunau-Varilla (above) has his wife stitch up a flag for a new Panamanian republic, meets in room 1162 of New York's Waldorf-Astoria Hotel with a physician who works for the cross-isthmus railroad represented by William Cromwell (above), and provides him with a secret code, a declaration of independence, the flag, the draft of a Panamanian constitution, and transportation back to Panama.

The U.S. cruiser *Nashville* arrives in Panamanian waters November 3, Panama declares her independence, the railroad refuses to transport Colombian troops sent to put down the provincial revolt, Secretary Hay recognizes Panama's independence November 6, Bunau-Varilla (above) is named Panamanian ambassador to the United States November 6, and on November 18 he signs a treaty on the same general terms as those rejected by Colombia in the Hay-Herran Treaty—granting "rights in perpetuity . . . as if [the United States] were the sovereign" but not granting sovereignty.

The "almighty dollar" often dictated foreign as well as domestic policy.

New York financier J. P. Morgan receives $40 million for transfer to French canal company stockholders whose names lawyer William Cromwell refuses to divulge to a Senate committee. Cromwell receives $800,000 for his legal services.

Britain, France, and Italy sign a treaty February 13 agreeing to lift the blockade of Venezuelan ports (*see* 1902).

Serbian conspirators assassinate Aleksandr I Obrenovic June 10. They also murder his wife Draga and some 20 members of the court of the king, 27, who has reigned since 1889 and ruled personally since 1893. The Serbian assembly votes June 15 to elect Prince Peter Karageorgevic, 59, to succeed Aleksandr, he will reign until 1921 as Peter I, but the new king is a puppet of the conspirators who assassinated Aleksandr and they restore the constitution of 1889.

A Macedonian insurrection against Constantinople ends after roving bands of Serbs, Greeks, and Bulgarians have ravaged the country. A Russo-Austrian reform program organizes a gendarmerie with Muslim and Christian constables assigned according to the makeup of local populations, the reformers appoint foreign officers, and they reorganize the Macedonian financial system.

Failure of the Russians to evacuate Manchuria under last year's Russo-Chinese agreement brings Japanese notes which the Russians contemptuously ignore (*see* 1904).

British forces complete the conquest of northern Nigeria.

Boston authorities indict local politician James Michael Curley, 29, and his brother Thomas on charges of having taken civil service examinations for job applicants (*see* 1904).

"Everybody is talking about Tammany men growing rich on 'graft,' " says New York political boss George Washington Plunkitt, 61, "but nobody thinks of drawing the distinction between honest graft and dishonest graft. There's an honest graft, and I'm an example of how it works. I would sum up the whole thing by saying, 'I seen my opportunities and I took 'em.' " A long-time opponent of civil service reforms, Plunkitt will leave an estate of $500,000 to $1.5 million when he dies late in 1924.

A manifesto issued by Czar Nicholas II March 12 concedes reforms, including religious freedom, but resentment against the czar mounts as famine takes a heavy toll (below) and Russian industrial wages fall beginning in October while food prices rise (*see* 1905).

Bolsheviks (extremists) led by V. I. Lenin split off from Mensheviks (moderates) at the London Congress of the Social Democratic party November 17 (*see* 1900; 1905).

U.S. anthracite coal miners win shorter hours and a 10-percent wage hike from President Roosevelt's

1903 *(cont.)* Anthracite Coal Commission, but the commission refuses to recognize the United Mine Workers as bargaining agent (*see* 1902; Colorado, 1913, 1914).

Congress votes February 14 to create a Department of Commerce and Labor, its secretary to be a member of the president's cabinet (*see* 1888; Department of Labor, 1913).

The Supreme Court upholds a clause in the Alabama Constitution denying blacks the right to vote. The court rules April 27 in the case of *Giles v. Harris*.

Britain's WSPU (Women's Social and Political Union) is founded October 3 by Emmeline Goulden Pankhurst, 45, and like-minded women who meet at Pankhurst's house and begin to work through the Independent Labour party to achieve their goal of voting rights for women (*see* 1882; 1905).

Agents of the king of the Belgians Leopold II commit atrocities against natives in the Congo, reports British consul Roger David Casement, 39. He has conducted a government investigation of the rubber trade in lands of the Upper Congo owned personally by the king (*see* Casement, 1916).

William Cadbury visits Lisbon to investigate the question of alleged slavery in the Portuguese African cocoa islands São Tomé and Príncipe (*see* 1902). Portuguese authorities tell Cadbury his suspicions are unfounded and invite him to see for himself (*see* 1905).

$ India's Tata iron and steel empire has its beginnings in Orissa where Dorabji Jamsetji Tata, 44, and his kinsman Shapurji Saklatvala, 29, discover a hill of almost solid iron ore. Son of a cotton mill magnate, Tata starts an iron and steel company that together with his father's cotton mills will be the basis of India's modern industrial development.

National Cash Register's J. H. Patterson gives his executive Thomas John Watson, 29, a budget of $1 million to start a company that will pose as a rival to NCR but will actually take control of the U.S. used cash-register business (*see* 1884). Watson's Cash Register and Second Hand Exchange opens on New York's 14th Street, undersells competitors, and drives them out of business or forces them to sell out. Watson will set up similar operations in Philadelphia and Chicago (*see* THINK, 1908).

Singer Manufacturing sells 1.35 million sewing machines, up from 1 million in 1889 (*see* 1913).

Utah Copper Co. is organized to mine ore at Bingham, Utah. The growing use of electricity and telephones has created a shortage of copper wire, and the Bingham mine provides cheap production from low-grade ore.

American Brass Co. is created by a merger of U.S. copper companies that include the Guggenheim family's American Smelting and Refining, United Copper (controlled by Montana mine operator Frederick Augustus Heinze, 34), and Amalgamated Copper (controlled by Standard Oil's H. H. Rogers and Anaconda's Marcus Daly).

A new Wanamaker Department Store at New York replaces the A. T. Stewart store of 1862 which Wanamaker has operated since 1896. Designed by Chicago's Daniel H. Burnham and Co., the enormous 15-story department store fills the block between Broadway and Fourth Avenue from 9th to 10th Streets and has no inner courtyards.

Texaco (the Texas Company) brings in its first oil well in January as prospectors make a major strike at Sour Lake, Tex. (*see* 1902; SoCal's Bahrain strike, 1932).

Oil gushes out of a new well on Osage Nation land in Oklahoma Indian Territory. The town of Bartlesville springs up near Jake Bartle's trading post (*see* 1905; Phillips Petroleum, 1917).

A new bottle-blowing machine permits volume production of electric light bulbs, whose high cost has discouraged widespread use of electric lighting. Michael J. Owens has improved his 1895 machine to create a completely automatic mechanism containing more than 9,000 parts. With the new machine, two men can produce 2,500 bottles per hour and as many light bulbs (*see* Langmuir, 1912).

The Elkins Act passed by Congress February 19 strengthens the Interstate Commerce Act of 1887. The new law forbids railroads to deviate from published rate schedules, and holds railroad officials personally liable in cases of rebating (*see* Hepburn Act, 1906).

The New York, New Haven, and Hartford begins a campaign to take over trolley and steamship lines and monopolize New England transportation in violation of the Interstate Commerce Act of 1887, the Sherman Act of 1890, and the laws of Massachusetts (*see* 1893). New Haven director J. P. Morgan installs Charles Sanger Mellen as president and helps him expand the road's interests (*see* Boston & Maine, 1909).

Construction begins in March on a 1,200-mile railroad from Africa's west coast through the center of Angola into the copper-rich Katanga district. The Benguela Railroad begun by Robert Williams on a concession from Portugal's queen Marie-Amélie will reach 300 miles inland by 1913 but will not reach the copper mines until 1931.

The Trans-Siberian Railway is completed with the exception of a 100-mile stretch along the mountainous shores of Lake Baikal. Built in 12 years, the new rail line will bring hundreds of thousands of settlers into the black-soil wheat-growing areas of central Asia (*see* 1904).

The Chadwick motorcar introduced by U.S. engineer Lee Sherman Chadwick is a four-cylinder, 24-horsepower vehicle that goes 60 miles per hour and sells for $4,000 (*see* 1906).

Ransom E. Olds drives a racing car over a measured mile at Daytona Beach, Fla., in 1 minute, 6 seconds.

Alexander Winton drives a car at 68 miles per hour at Daytona Beach (*see* 1897). Winton has enlarged his factory to cover 13 acres and employs 1,200 men.

Ford Motor Company is incorporated June 16 with $28,000 raised by 12 stockholders who include John and Horace Dodge. Henry Ford receives 225 shares in exchange for his design and 17 patents on its mechanism, production begins in a converted wagon factory on Detroit's Mack Avenue, and the $750 Model A Ford introduced by the company has a two-cylinder, 8-horsepower chain-drive engine mounted under its seat. The half-ton vehicle has a 72-inch wheelbase, is 99 inches in length overall, its steering wheel is on the right, and 658 will be sold by next March (*see* Oldfield, 1902; Model T, 1908).

Ford Motor refuses to join a new Association of Licensed Automobile Manufacturers formed to purchase rights to use the 1895 Selden patent at a royalty of 1.25 percent of the retail price of each automobile sold. Selden will soon be receiving a royalty on nearly every U.S.-made motorcar (*see* 1899; 1911).

John North Willys sells 20 cars and takes on the Detroit car made by Detroit Auto Vehicle Co. (*see* 1902; 1907).

Massachusetts issues the first true automobile license plates September 1; all states will soon follow its example.

Two-thirds of all U.S. motorcars sell at prices below $1,375, but most are too small, too light and unreliable to present serious competition to the horse and buggy, much less the railroad.

A Packard leaves San Francisco for New York May 23 and arrives after 52 days in the first successful transcontinental trip by a motorcar under its own power, but the poor quality of roads discourages sales of motorcars in the United States (*see* New York to Seattle race, 1909).

The Harley-Davidson motorcycle is introduced by Milwaukee draftsman William Harley, pattern maker Arthur Davidson, mechanic Walter Davidson, toolmaker William Davidson, and a German draftsman familiar with European motorcycles and with the DeDion gasoline engine. Some 50 Harley-Davidson bikes will be produced by 1906, and by 1917 production will reach 18,000 per year as "the hog" becomes America's top motorbike.

New York's Williamsburg Bridge opens to traffic December 19, providing a second link between Brooklyn and New York to supplement the 20-year-old Brooklyn Bridge. The new $24.1 million 488-meter span is the first major suspension bridge with steel towers instead of masonry (*see* Queensboro, Manhattan bridges, 1909).

The Wright brothers make the first sustained manned flights in a controlled gasoline-powered aircraft. Dayton, Ohio, bicycle mechanics Wilbur Wright, 36, and Orville Wright, 32, have built their *Flyer I* with a chain-drive 12-horsepower motorcycle engine whose cast aluminum engine block gives it a high strength-to-weight ratio. Near Kill Devil Hill at Kitty Hawk, N.C., December 17, they achieve (on their fourth effort) a 59-second flight of 852 feet at a 15-foot altitude, but while a number of newsmen witness the event only three U.S. newspapers report it (*see* 1905).

German aviation pioneer Karl Jatho, 30, claims he made a successful flight August 5, more than 4 months before the Wright brothers (above). Jatho built a gasoline-powered biplane in 1899 and will establish an aircraft works at Hanover in 1913.

The Springfield Rifle developed at the 108-year-old U.S. arsenal at Springfield, Mass. will become general issue in the U.S. Army and some foreign armies. The M1903 gets its name from the marking on its receiver ring (*see* 1926).

The Fouce torch invented by French inventor Edmond Fouce is the first safe, practical oxyacetylene torch for welding metals (*see* Claude, 1897). It will increase demand for acetylene and for oxygen (*see* 1907; Union Carbide, 1892, 1911).

Typhus is transmitted only through the bite of the body louse, says French physician Charles Jean Henri Nicolle, 37. He reports on studies of the ancient plague that he has made as director of an outpost in Tunis, North Africa, of the Pasteur Institute.

The electrocardiograph pioneered by Dutch physiologist Willem Einthoven, 43, at Leyden will expand knowledge of the heart's functioning and be used routinely to examine patients with potential or actual heart disease. Einthoven invents a string galvanometer that will be developed into a more sophisticated instrument.

German surgeon Georg Clems Perthes, 34, makes the first observations that X-rays can inhibit carcinomas and other cancerous growths (*see* Roentgen, 1895).

Steel magnate Henry Phipps establishes the Phipps Institute for the Study, Treatment, and Prevention of Tuberculosis at Philadelphia (*see* 1905).

John D. Rockefeller of Standard Oil contributes $7 million to fight tuberculosis.

"Typhoid Mary" gets her name as New York has an outbreak of typhoid fever with 1,300 cases reported. The epidemic is traced to one Mary Mallon, a carrier of the disease (but not a victim) who takes jobs that involve handling food, often using assumed names. Typhoid Mary refuses to stop, will be placed under detention in 1915, and will remain confined until her death in 1938.

The University of Puerto Rico begins classes at Rio Piedras.

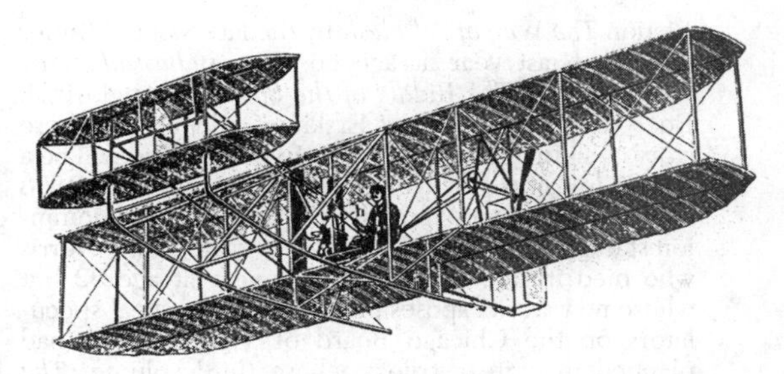

Wilbur and Orville Wright made the first sustained manned flights in a controlled gasoline-powered aircraft.

1903 ***(cont.)*** Guglielmo Marconi sends a wireless greeting January 19 from President Roosevelt to Britain's Edward VII (*see* 1901). Marconi has built four 250-foot wooden towers anchored in cement at South Wellfleet, Mass., he uses electrical waves generated by a 3-foot spark gap rotor, an operator working a huge key transmits the message at the rate of 17 words per minute, and while Marconi has arranged for a station at Glace Bay, Nova Scotia, to relay the message, the signal is so powerful that the station 3,000 miles away at Poldhu in Cornwall is able to pick it up direct (*see* 1907).

The Pacific Cable links San Francisco with Honolulu January 1 and Manila July 4. President Roosevelt sends a message from San Francisco to Manila; it is relayed round the world and returned to the president in 12 minutes.

The Knight newspaper chain that will merge with the Ridder chain in 1974 has its beginnings at Akron, Ohio, where local lawyer Charles L. Knight buys a part interest in the *Akron Beacon Journal.* By 1974 there will be 16 Knight newspapers up and down the U.S. East Coast (*see* Ridder, 1892).

The *London Daily Mirror* begins publication. *Daily Mail* publisher Alfred C. W. Harmsworth and his brother Harold hire a woman editor, address the tabloid to women, but will make it a general-interest "picture newspaper" next year to pick up lagging sales. It will have a circulation of more than a million by 1909 and spawn two competitors. A.C. Harmsworth will be made a baronet (Viscount Northcliffe) in 1905, his brother Viscount Rothermere.

Pulitzer prizes have their beginning in an agreement signed by *New York World* publisher Joseph Pulitzer April 10 to endow a school of journalism at Columbia University. Pulitzer has made a fortune since acquiring the *World* in 1883 and specifies that $500,000 of his $2 million gift shall be allotted for "prizes or scholarships for the encouragement of public service, public morale, American literature, and the advancement of education."

The Prix Goncourt is awarded for the first time from a fund established in 1895 by the late French writer Edmond de Goncourt who died the following year. His Académie Goncourt will honor the best French novel published each year, and the prix will be France's most prestigious literary prize.

Fiction *The Way of All Flesh* by the late Samuel Butler who died last year at age 66; *The Ambassadors* by Henry James; *The Riddle of the Sands* by Anglo-Irish Boer War veteran Robert Erskine Childers, 33, whose story of an imaginary German raid on England will be a classic for small boat enthusiasts (beginning in 1908 Childers will devote himself to agitating for full dominion status for Ireland); *The Pit* by the late Frank Norris who died in late October of last year at age 32 but whose new work exposes price rigging by wheat speculators on the Chicago Board of Trade. Norris had planned to write a trilogy whose third volume *(The Wolf)* would have been an account of U.S. wheat relieving a European famine, but the author was sent by the *San Francisco Chronicle* to report a trip from Cape Town to Cairo in 1899, enlisted in the British Army at the outbreak of the Boer War, was captured by the Boers, contracted fever, and died at San Francisco leaving his "epic of wheat" incomplete; *The Call of the Wild* by Jack London; *The People of the Abyss* by Jack London who has lived in the London slums.

Juvenile *Rebecca of Sunnybrook Farm* by New York author Kate Douglas Wiggin, 47.

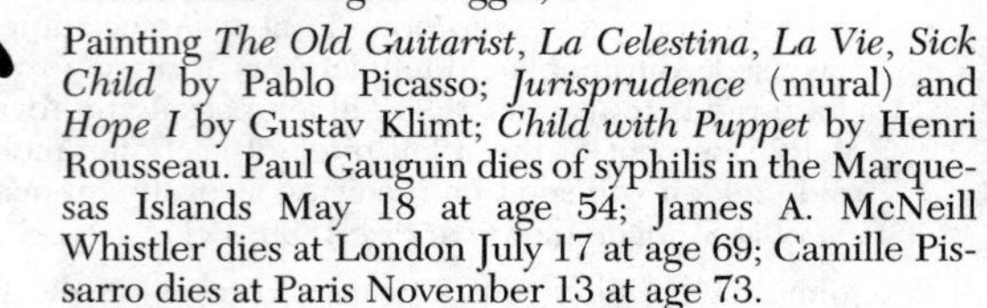

Painting *The Old Guitarist*, *La Celestina*, *La Vie*, *Sick Child* by Pablo Picasso; *Jurisprudence* (mural) and *Hope I* by Gustav Klimt; *Child with Puppet* by Henri Rousseau. Paul Gauguin dies of syphilis in the Marquesas Islands May 18 at age 54; James A. McNeill Whistler dies at London July 17 at age 69; Camille Pissarro dies at Paris November 13 at age 73.

Britain creates the National Art Collections Fund to prevent works of art from leaving the country.

Boston's Fenway Court is completed for department store heiress Isabella Stewart Gardner to house her art collection.

Steuben Glass Works is founded by British-American designer Frederick Carder, 35, who will head the firm until 1932 and will not retire until he is 96. Carder's lustrous Aurene technique will produce glass rivaling in its iridescent color the Favrile designs of Louis Comfort Tiffany, now 55 (*see* 1878).

Theater *The Hour-Glass* by W.B. Yeats and Lady Gregory 3/14 at Dublin's Moleworth Hall; *The King's Threshold* by Yeats and Lady Gregory 3/14 at Dublin's Moleworth Hall; *In the Shadow of the Glen* by Irish playwright John Millington Synge, 32, 10/8 at Moleworth Hall; *Rose Bernd* by Gerhart Hauptmann 10/31 at Berlin's Deutsches Theater; *The County Chairman* by George Ade 11/24 at Wallack's Theater, New York, 222 perfs.; *Glad of It* by Clyde Fitch 12/28 at New York's Savoy Theater, with John Barrymore, 21, 32 perfs. A nephew of actor John Drew, Barrymore is the younger brother of Lionel, 25, and Ethel, 24, who have been on the stage since ages 6 and 14, respectively. "The Great Profile" will become a matinee idol and will be famous for his Hamlet beginning in 1924.

Fire engulfs Chicago's Iroquois Theater December 30. Matinee patrons rush for the exits and 602 die in the panic despite heroic efforts by comedian Eddie Foy, 47, to calm the crowd.

Film Edwin S. Porter's *The Great Train Robbery* is the first motion picture to tell a complete story (*see* 1896). Produced by Edison Studios and photographed at the Paterson, N.J., freight yards of the Delaware & Lackawanna Railroad, the 12-minute film establishes a pattern of suspense drama that future moviemakers will follow.

Opera *L'Etranger* (*The Stranger*) 1/7 at the Théâtre de la Monnaie, Brussels, with music by Vincent d'Indy; *Tiefland* (*The Lowlands*) 11/15 at Prague's Neues Deutsches Theater, with music by Eugen d'Albert.

Enrico Caruso makes his New York debut November 21 singing *Rigoletto* at the 20-year-old Metropolitan Opera House. The Italian tenor has made a worldwide reputation since his debut at Milan's Teatro alla Scala in 1899, will be the Met's leading tenor for years, and will have a repertoire of more than 40 operatic roles (*see* 1902; Victrola, 1906).

Cantata "Hiawatha" by English composer Samuel Coleridge-Taylor, 28, 11/16 at Washington, D.C.

First performances *Transfigured Night (Verklärte Nacht)* sextet by Austrian composer Arnold Schoenberg, 28, at Vienna. The work is based on the Richard Dehmel poem *Weih und die Welt*.

Broadway musicals *The Wizard of Oz* 1/21 at the Majestic Theater, with Fred Stone and Dave Montgomery in a musical extravaganza based on the L. Frank Baum novel of 1900, 293 perfs.; *Whoop-Dee-Doo* 9/24 at the Weber & Fields Music Hall, with Weber and Fields, music by W. T. Francis, book and lyrics by Edgar Smith, 151 perfs.; *Babette* 11/16 at the Broadway Theater, with Vienna-born Metropolitan Opera soprano Fritzi Scheff, 21, music by Victor Herbert, book and lyrics by Harry B. Smith, 59 perfs.; *Babes in Toyland* 10/13 at the Majestic Theater, with music by Victor Herbert, songs that include the "March of the Toys" and the title song (lyrics by Glen MacDonough), 192 perfs.

Popular songs "Dear Old Girl" by Theodore F. Morse, lyrics by Richard and Henry Buck; "Bedelia" by Jean Schwartz, lyrics by William Jerome; "Ida, Sweet as Apple Cider" by U.S. minstrel Eddie Leonard; "(You're the Flower of My Heart) Sweet Adeline" by U.S. composer Henry W. Armstrong, 24, lyrics by Richard H. Gerard (R. G. Husch), 27, whose words, inspired by the farewell tour of Italian diva Adelina Patti, will be sung by generations of barbershop quartets (the song will be used as a campaign theme for Boston mayoral candidate John W. "Honey Fitz" Fitzgerald); "Waltzing Matilda" by Australian Marie Cowan who has adapted the 1818 Scottish song "Thou Bonny Wood of Craigie Lee" by James Barr, lyrics by Andrew Barton "Banjo" Peterson, 39, who has adapted an Australian bush ballad. ("Matilda" is Aussie slang for a knapsack; a "swagman" is a worker.)

New York's "Tin Pan Alley" gets its name from songwriter-publisher Harry von Tilzer. He receives a visit from songwriter-journalist Monroe H. Rosenfeld, who has earned little from his 1897 song "Take Back Your Gold" and other hits. Von Tilzer's piano has a peculiarly muffled tone and he explains that other tenants of the building at Broadway and 28th Street have demanded that piano players make less noise, he has used newspapers to reduce volume. Rosenfeld says, "It sounds like a tin pan," von Tilzer says, "Yes, I guess this is tin pan alley," and Rosenfeld repeats the phrase in his *New York Herald* music columns.

Waltz "Gold and Silver Waltz" by Franz Lehar who wrote it for an elaborate ball given last year by the Princess Metternich.

H. L. Doherty wins in men's singles at Wimbledon, Mrs. Dorothea Lambert Douglas Chambers, 30, in women's singles; Doherty wins in U.S. singles, the first non-American champion, Elizabeth Moore wins in women's singles.

The Tour de France bicycle race, organized as a publicity stunt by sports journal editor Henri Desgranges of *L'Auto*, covers 1,515 miles. It will grow to 2,500 miles and cyclists will cover it in 26 days at speeds averaging 22–23 miles per hour.

New York Giants pitcher Christy Mathewson wins 30 games with a 267-strikeout record that will stand for 50 years (*see* 1900). He will be nicknamed "Big Six" after a famous fire engine and will have a lifetime record of 373 wins, 188 losses, and 2,511 strikeouts.

The first World Series baseball championship pits the American League's Boston team against the National League's Pittsburgh team. Boston wins 5 games to 3.

The U.S. yacht *Reliance* rigged by Nat Herreshoff and built by a syndicate headed by J. P. Morgan defends the America's Cup by defeating Sir Thomas Lipton's *Shamrock III* 3 to 0.

The Teddy Bear introduced last year by Morris Michtom encounters a challenge from the German firm Steiff Co. founded in 1880 at Giengen in Swabia by crippled seamstress Margarete Steiff. Her nephew Richard claims to have designed a plush bear last year with jointed limbs and a movable head, he shows it at the Leipzig Fair with little success, but a U.S. buyer sees the bear on the last day of the fair and orders 3,000. The wholesale firm Butler Brothers takes Michtom's entire output of Teddy Bears (it will call itself Ideal Novelty and Toy beginning in 1938, will later shorten it to Ideal Toy, and will be the world's largest manufacturer of dolls).

King Gillette goes $12,000 into debt but escapes bankruptcy. Boston investor John Joyce advances funds to permit marketing of the safety razor; Gillette sells 51 razors and 168 blades (*see* 1901; 1904).

England's Westminster Cathedral is completed in Byzantine style with striped brick and stone. Architect John Francis Bentley died last year at age 63.

Cincinnati's Ingalls building is the world's first skyscraper with a reinforced concrete framework (*see* Monier, 1849). First reinforced concrete building of more than two stories, the 16-story $400,000 structure at Fourth and Vine Streets revolutionizes the construction industry. It is named for Big Four Railroad president Melville E. Ingalls, but the Cincinnati Transit Co. will acquire it in 1959 and will rename it the Transit building.

The New York Stock Exchange building is completed at 8 Broad Street between Wall Street and Exchange Plaza.

Kykuit is completed for John D. Rockefeller on the 4,180-acre estate he acquired 10 years ago in the Pocantico Hills north of New York. The 14-room Victorian stone house commands a view of the Hudson and

1903 ***(cont.)*** Saw Mill Rivers; its Dutch name means "lookout" (*see* 1908).

Grand Canyon's Angel Lodge and El Tovar are opened by the Santa Fe Railroad which has built a branch from its main line at Williams to bring tourists to the canyon's rim. The new hotels are operated by the sons and son-in-law of the late Fred Harvey who died early in 1901 (*see* 1876). Some 17 hotels, 47 restaurants, and 30 dining cars are under Fred Harvey management as are food services on San Francisco Bay ferries (*see* national park, 1919).

Fire destroys Florida's 8-year-old Breakers at Palm Beach. H. M. Flagler orders construction of a new hotel that will replace the vast wooden structure and will become a winter resort hotel favored by millionaires (*see* 1925).

Florida acquires title to the Everglades swamp, the largest area ever conveyed in a single patent issued by the U.S. Land Department. Drainage operations will begin in 1906 (*see* national park, 1947).

President Roosevelt designates Florida's Pelican Island a National Wildlife Refuge, a seabird sanctuary where the fowl will be safe from the plume hunters who shoot birds to obtain feathers for milliners. Roosevelt thus begins a fish and wildlife reserve system that will set aside 30 million acres of U.S. lands in the next 70 years.

Wind Cave National Park is established in South Dakota's Black Hills by act of Congress. The 28,000-acre park has herds of bison, elk, deer, and pronghorn antelope with limestone caverns and prairie dog towns.

A machine devised by U.S. inventor A. K. Smith cuts off a salmon's head and tail, splits the fish open, cleans it, and drops it into hot water all in one continuous operation. The Smith machine adjusts itself automatically to the size of the fish and will be called the "Iron Chink" because it replaces Chinese hand labor in West Coast canneries (*see* Chinese Exclusion Act, 1882; 1902).

Japan proved herself a great naval power, overcoming beriberi in the ranks and sinking a Russian fleet.

A San Pedro, Calif., packer puts white albacore tunafish into cans and launches a major industry (*see* Van Camp, 1914).

Russia's harvest fails again as in 1891. Since millions live at the edge of starvation even in the best of years, the crop failure produces famine that kills millions.

The Philippine Islands are self-sufficient in rice. They will be net importers of rice from 1904 to 1968.

Sanka Coffee is introduced by German coffee importer Ludwig Roselius who has received a shipload of beans that were soaked with seawater by a storm and who has turned the beans over to researchers. They have perfected a process to remove caffeine from coffee beans without affecting the delicate flavor of the beans and Roselius has named the product Sanka, using a contraction of the French *sans caffeine* (*see* 1923).

William Normann receives a patent for his 1901 process of hydrogenating fats and oils.

Milton Hershey breaks ground at Derry Church, 13 miles east of Harrisburg, Pa., for a chocolate factory whose products will dominate the chocolate candy and beverage industry by taking milk chocolate out of the luxury class. Hershey lays out Chocolate Avenue and Cocoa Avenue, builds new houses for his workers to rent or buy, and constructs a street railway to connect his cornfield at Derry Church (soon to be called Hershey) with five neighboring towns in the area which he has selected for milk production, water supply, and rail links to U.S. ports and consumer markets (*see* 1900; 1911).

Caleb Bradham trademarks the name Pepsi-Cola and sells 7,968 gallons of his syrup as compared with 881,423 for Coca-Cola syrup (*see* 1902; 1907).

Members of the New York Riding Club assemble at Louis Sherry's 5-year-old Fifth Avenue restaurant for a dinner given by millionaire horseman C. K. G. Billings. The floor of the banquet room is sodded, and the guests sit on their horses, eating off small tables attached to their saddles while sipping champagne from tubes connected to their saddlebags.

1904

Japanese naval forces commanded by Admiral Heichiro Togo, 57, attack Port Arthur (Lushun) in southern Manchuria February 8, bottling up a Russian squadron and launching the first war in which armored battleships, self-propelled torpedoes, land mines, quick-firing artillery, and modern machine guns will be used. Concerned at Russia's failure to withdraw from Manchuria and her continuing penetration of Korea, the Japanese follow their sneak attack with a declaration of war February 10, defeat the Russians at the Yalu River May 1, occupy Dairen May 30, besiege Port Arthur and occupy Seoul, defeat a Russian force at Liaoyang in China in October, and force the Russians to pull back to Mukden (*see* 1905).

Russia's able but ruthless minister of the interior Vyacheslav Plehve is assassinated July 28. A Zemstvo congress meets at St. Petersburg in November, demanding that civil liberties be granted and a representative assembly convened (*see* 1905).

Herero and Hottentot tribesmen in German South-West Africa rise against German colonial forces in an insurrection that will continue until early 1908. The revolt will be suppressed only after methodical campaigns by 20,000 German troops.

The new republic of Panama adopts a constitution February 13 and elects Manuel Amador Guerrero president (*see* 1903). Colombia's president José Marroquin is succeeded by Rafael Reyes, 54, who represented his country at Washington last year in the negotiations relating to Panama; he assumes dictatorial rule.

President Roosevelt wins reelection with 336 electoral votes to 140 for the Democrats' Judge Alton B. Parker of New York, who has been nominated in preference to William Jennings Bryan.

James Michael Curley and his brother are convicted in November of taking civil service examinations for job applicants in Boston, but on the day they begin serving time they are elected to the offices of alderman and state representative (*see* 1903). New York's Tammany Hall supports the Curley brothers and the public excuses their offense on the ground that "they did it for their friends."

An Easter Sunday pogrom at Kishenev in Bessarabia kills 45 Jews and destroys 600 houses; on orders from Vyacheslav Plehve (above), the police turn a blind eye at the mob running wild in the streets. Plehve does not allow any political assemblies, requires written police permission for even a small social gathering, and does not permit students to walk together in the streets of Moscow or St. Petersburg (*see* revolution, 1905).

U.S. reformers organize the National Child Labor Committee to promote protective legislation (*see* Owen-Keating Act, 1916).

Fall River, Mass., textile workers strike in August and will not return until January (*see* 1905).

$

Montgomery Ward distributes its first free catalogs after having sold catalogs for years at 15¢ each. The company mails out more than 3 million of the 4-pound books (*see* 1884).

Sears, Roebuck distributes more than a million copies of its spring catalog (*see* 1895). Sears has set up its own printing plant but still uses woodcuts for more than half its illustrations and may be keeping the art of wood engraving alive (*see* 1907).

The Supreme Court rules in a 5 to 4 decision that the Northern Securities Company of 1901 violates the Sherman Act of 1890. The court decision orders dissolution of the railroad trust.

The first tunnel under the Hudson River is completed March 8 but will not open officially until February 25, 1908. The Hudson & Manhattan Railroad Co. headed by William Gibbs McAdoo, 41, has built two single-track tubes, each more than a mile long, to connect Jersey City with Manhattan.

The first New York City subway line of any importance opens to the public October 27. The Interborough Rapid Transit (IRT) line runs from Brooklyn Bridge north under Lafayette Street, Fourth Avenue, and Park Avenue to 42nd Street, west to Broadway, and north to 145th Street, completing the run in 26 minutes with a train of five copper-sheathed wooden cars, each seating 56, with red-painted roofs. Tiles bearing visual symbols identify unnumbered stations to aid the city's polyglot population. Tickets cost 5¢ each, 111,000 people ride the first day, 319,000 the second, and 350,000 the third as New York inaugurates a system that will grow to cover more than 842 miles—the largest rapid transit system in the world.

Child labor operated factories and mines without government interference despite humanitarian protests.

The Trans-Siberian Railroad opens to link Moscow with Vladivostok some 3,200 miles away (*see* 1903). It is the longest line of track in the world.

A new rail line reaches the lower Rio Grande Valley of Texas and opens up the region for citrus fruits and vegetables on lands irrigated with river water.

The New York excursion boat S.S. *General Slocum* catches fire in the Hudson June 15 while carrying 1,400 German-Americans from the Lower East Side to Locust Grove on Long Island Sound for a Wednesday picnic sponsored by the German Lutheran Church of Sixth Street. More than 1,030 die, the tragedy shatters the German community in the Lower East Side, a court finds the paddle-wheeler's captain guilty of criminal negligence and sentences him to 10 years at hard labor in Sing Sing, and more Germans move uptown to settle in Yorkville.

The British passenger liner S.S. *Baltic* goes into service to begin a 29-year career. The largest passenger ship yet built, the 23,884-ton vessel is 726 feet in length overall.

The S.S. *Kaiser Wilhelm II* built 2 years ago for the North German Lloyd line sets a new transatlantic record in June by steaming from Sandy Hook to Eddystone, off Plymouth, in 5 days, 12 hours.

The Tropical Fruit Steamship Company, Ltd., is founded by the United Fruit Company which has con-

1904 ***(cont.)*** tracted with shipyards to build three new refrigerator ships (*see* 1899). The company will operate its new *S.S. San José, S.S. Limon,* and *S.S. Esparta* under British colors (*see* radio equipment, below).

Cadillac Motor Car Co. is created by a reorganization of Detroit Automobile under the aegis of Henry Martyn Leland, 61 (*see* 1901; General Motors, 1909; self-starter, 1911).

The Pierce Arrow and Great Arrow begin the George N. Pierce Co. on a career as the leading U.S. producer of luxury motorcars (*see* 1901). The company produces 50 Great Arrows and sells them at $3,000 to $5,000 each (*see* Glidden Tour, 1905).

A. A. Pope opens a plant at Toledo, Ohio (*see* 1897). His new Pope-Tribune costs $650 (*see* Willys, 1907).

The Maxwell motorcar introduced by J. D. Maxwell gains quick popularity. The 1914 model will have the first adjustable front seat (*see* Chrysler, 1923).

The Reo introduced by R. E. Olds has a steering wheel instead of a tiller (*see* 1901). Olds has left Oldsmobile in a dispute with Frederick Smith on the size and price of motorcars to be produced by the company.

Marie Curie discovers two new radioactive elements—radium and polonium—in uranium ore (pitchblende) and receives her doctorate at the Sorbonne (*see* 1898; Becquerel, 1896; McMillan, 1940).

German physicists Julius Elster, 50, and Hans Friedrich Geitel, 49 devise the first practical photoelectric cell (*see* radio tube, below).

U.S. consul Edward H. Thompson, 44, discovers ruins of Chichén Itzá in Mexico's Yucatán. Sponsored by Harvard's Peabody Museum (to which he will ship gold objects, making illicit use of his diplomatic pouches), archaeologist Thompson finds pyramidal Mayan temples, an astronomical observatory, and a ceremonial ball court dating from the Middle Ages (*see* 455). He concentrates his search on an underground waterhole, or cenote, which he explores with a dredge and diving gear to bring up tools, vessels, statuettes, jewelry, and ornaments.

William C. Gorgas is sent to Panama to eliminate the yellow fever that is discouraging construction of a cross-isthmus canal (*see* 1901; 1906).

English chemist Henry Drysdale Dakin, 24, and German chemist F. Stolz isolate adrenaline (epinephrine) (*see* Takamine, 1901).

The National Tuberculosis Association is established in the United States to raise money for research that will fight consumption.

The Chinese Red Cross is established at Beijing and Shanghai.

Sales of Argyrol and Ovoferrin total more than $93,000, netting Albert Barnes and his colleague Herman Hille more than $44,000 in profits. They open offices at London and at Sydney, Australia (*see* 1902).

Helen Keller is graduated magna cum laude from Radcliffe College at age 23 and begins to write about blindness, a subject taboo in women's magazines because so many cases are related to venereal disease (*see* Argyrol, 1902; Keller, 1887). Keller has learned to speak at Boston's Horace Mann School for the Deaf by feeling the position of the tongue and lips of others, making sounds, and imitating the lip and tongue motions. She has learned to lip-read by placing her fingers on the lips and throat of the speaker while the words spoken were spelled out on the palm of her hand, her book *The Story of My Life* appeared in 1902, she will begin lecturing in 1913 to raise money for the American Foundation for the Blind, her teacher Anne Sullivan will remain with her until she dies in 1936, and Keller will continue to work for the blind and deaf until her own death in 1968.

Bethune-Cookman College has its beginnings in a cabin rented at Daytona, Fla., by Mary McLeod Bethune, 29. She opens her Daytona Normal and Industrial Institute for Negro Girls with five students, makes potato pies for workmen building the nearby Clarendon Hotel, earns $5 for a down payment on some land, and solicits funds to build Faith Hall (*see* 1923).

The wireless radio distress signal letters CQD adopted January 1 mean "stop sending and listen" (*see* SOS, 1906).

United Fruit Company (above) installs commercial radio equipment on its ships and is the first to do so. The firm depends on radio communications to determine when and where bananas are available for loading (*see* 1910).

The diode thermionic valve invented by English electrical engineer John Ambrose Fleming, 55, is the first electron radio tube (*see* De Forest, 1906).

Parker Pen's George S. Parker obtains a patent May 3 for a lever mechanism that makes it easier to fill his pen's rubber sac (*see* 1888). Parker has gained success with his "Lucky Curve" hard rubber feed bar which minimizes leakage since it provides a continuous capillary passage between nib and barrel wall (the feed bar is curved at the end, it extends up into the ink chamber to touch the wall of the barrel, and when the pen is placed in the pocket, nib up, the ink all drains into the pen's reservoir). All fountain pens including Parker's have barrels of black hard rubber (the only available ink-resistant material that can be easily heated and formed), the upper end of the pen barrel is sealed to maintain a vacuum equal to the weight of the column of ink, careful adjustment of the feed-bar groove's cross-sectional area allows a slow flow of air bubbles that reduce the vacuum and release more ink, and tiny slits cut in the bottom of the ink channel aid capillarily (*see* "Big Red," 1921).

Nonfiction *The Shame of the Cities* by *McClure's* magazine managing editor (Joseph) Lincoln Steffens, 38, whose exposé of squalor in America's urban centers creates a sensation; *Mont Saint Michel and Chartres* by Henry Adams is privately printed; his study of medieval life, art, and philosophy will be published in 1913.

Fiction *The Late Mattia Pascal* (*Il fu Mattia Pascal*) by

Italian novelist Luigi Pirandello, 37, whose beautiful wife became an incurable schizophrenic last year after hearing that his father's sulfur mine had flooded, ruining her husband's family and her own; *Peter Camenzind* by German novelist-poet Hermann Hesse, 27, who has quit a theological seminary in rebellion against his Protestant missionary father and worked as a mechanic and bookseller; *Nostromo* by Joseph Conrad; *The Golden Bowl* by Henry James; *Hadrian the Seventh* by English novelist Frederick William Rolfe, 44 (who calls himself Baron Corvo and fictionalizes the life of the English pontiff Adrian IV of 1154 to 1159); *Green Mansions* by English naturalist-writer W. H. Hudson, 63, whose Rima the bird-girl gains him his first large success; *Reginald* by English short-story writer Saki (H. H. [Hector Hugh] Munro), 35, who takes his pen name from the female cup-bearer in the last stanza of Edward Fitzgerald's translation of *The Rubáiyát of Omar Khayyám* published in 1859; *Lucky Peter (Lykke Per)* by Henrik Pontoppidan; *Esau and Jacob (Esau e Jaco)* by Joachim Machado De Assis; *Cabbages and Kings* by New York short story writer O. Henry (William Sydney Porter), 42, who was charged 8 years ago with embezzling funds from an Austin, Tex., bank in which he worked from 1891 to 1894. Porter fled to Honduras to escape prosecution, returned when he heard his wife was mortally ill, was convicted after her death, and served 3 years of a 5-year prison term; *The Sea Wolf* by Jack London who spends 6 months covering the Russo-Japanese war (above); *Freckles* by Indiana novelist Gene Stratton Porter, 41.

Poetry "Alcione" by Gabriele D'Annunzio.

The Salon d'Automne at Paris exhibits paintings by Paul Cézanne, Odilon Redon, and others but is dominated by the canvases *Charity* and *Hope* by the late Pierre Cecile Puvis de Chavannes who died in 1898.

Other paintings *Mont Sainte Victoire* by Paul Cézanne; *Thames* by Claude Monet who begins to concentrate on the *Nymphaes* (water lilies) which will occupy him until his death in 1926; *Harvest Day* by German impressionist Emil Nolde (Emil Hansen), 37, who will soon turn to religious paintings; *The Wedding* by Henri Rousseau. Henri Fantin-Latour dies at Bure, Orne, August 25 at age 68.

Sculpture *The Hand of God* (bronze) by Auguste Rodin; *General Sherman Memorial* by Augustus Saint-Gaudens for New York's Central Park.

The *Christ of the Andes* is dedicated at Uspallato Pass on the Chilean-Argentine border to honor the peaceful settlement of disputes between the two countries since declaring their independence from Spain in 1817 and 1816. The 14-ton, 26-foot high monument is the work of Argentine sculptor Mateo Alonzo who has molded the figure from the bronze of old Argentine cannons.

Theater *The Cherry Orchard* (*Vishnyovyy Sad*) by Anton Chekhov 1/17 at the Moscow Art Theater (Chekhov dies July 2 at Badenweiler in the Black Forest at age 44); *Riders to the Sea* by John Millington Synge 1/25 at Dublin's Moleworth Hall; *The Daughter of Jorio* (*La figlia di Iorio*) by Gabriele D'Annunzio 3/2 at Milan's Teatro Lirico, with Eleanora Duse; *Pandora's Box (Die Büchse der Pandora)* by Frank Wedekind at Nuremberg's Intimes Theater; *Candida* by George Bernard Shaw 4/20 at London's Royal Court Theatre, with Kate Rorke in the title role; *The College Widow* by George Ade 9/20 at New York's Garden Theater, 278 perfs.; *The Music Master* by Charles Klein 9/26 at New York's Belasco Theater, with David Warfield, 288 perfs., *How He Lied to Her Husband* by George Bernard Shaw 9/26 at New York's Berkeley Lyceum, with Arnold Daley; *John Bull's Other Island* by George Bernard Shaw 11/1 at London's Royal Court Theatre, with Granville Barker in a satire about Shaw's native Ireland; *Leah Kleschna* by C. M. S. McClellan 12/12 at New York's Manhattan Theater, with Mrs. Fiske, Charles Cartwright, George Arliss, 631 perfs.; *On Baile's Strand* by W. B. Yeats and Lady Gregory 12/27 at Dublin's new Abbey Theatre off O'Connell Street; *Peter Pan, or The Boy Who Would Not Grow Up* by James M. Barrie 12/27 at the Duke of York's Theatre, London, with Nina Boucicault.

Opera *Jenufa (Jeji Pastorkyna Jenufa, Her Foster-daughter)* 1/21 at Brno in Austrian Moravia, with music by Czech composer Leos Janacek, 49; *Madama Butterfly* 2/17 at Milan's Teatra alla Scala, with music by Giacomo Puccini who has seen John Luther Long's play *Madame Butterfly* at London. Staged originally by David Belasco in New York, the play about the opening of Japan is based on the short story "Madame Butterfly" which appeared in the January 1900 issue of *Century* magazine (the new opera is a fiasco in its initial production).

First performances *Kossuth* Symphony by Hungarian composer Béla Bartók (Nagyszentmiklòs), 23, 1/13 at Budapest; Symphony No. 2 in B flat major by Vincent d'Indy 2/28 at Paris; Quartet in F by Maurice Ravel 3/5 at the Société Nationale, Paris; *In the South (Alassio)* Overture by Sir Edward Elgar (knighted this year) 3/18 at London's Covent Garden Theatre; Symphony No. 5 in C sharp minor by Gustav Mahler 10/18 at Cologne.

Chicago's $1 million Orchestra Hall is completed by Daniel Burnham in French Renaissance style at 216 South Michigan Avenue.

Russian ballet dancer-choreographer Michel Mikhaylovich Fokine, 24, proposes sweeping changes to the management of the Imperial Ballet at St. Petersburg. Fokine made his debut on his 18th birthday, the management rejects all his suggestions, but he will see them accepted. "Dancing should be interpretive," says Fokine. "It should not degenerate into mere gymnastics. The dance should explain the spirit of the actors in the spectacle. . . The ballet must no longer be made up of 'numbers,' 'entries,' and so on. It must show artistic unity of conception. The action of the ballet must never be interrupted to allow the danseuse to respond to the applause of the public" (*see* 1907).

Broadway musicals *Piff! Paff! Pouff!* 4/2 at the Casino Theater, with Eddie Foy, music by Jean Schwartz, lyrics by William Jerome, 264 perfs. (plus a week at the Majestic Theater beginning 12/26); *Higgledy-Piggledy*

1904 ***(cont.)*** 10/12 at the Weber Music Hall, with Anna Held, Marie Dressler, Fred M. Weber, music by Maurice Levy, lyrics by Edgar Smith, 185 perfs.; *Little Johnny Jones* 11/17 at the Liberty Theater, with book, music, and lyrics by George M. Cohan, songs that include "Give My Regards to Broadway" and "Yankee Doodle Boy," 52 perfs.

Popular song "Frankie and Johnny" by songwriter Hughie Cannon whose song is published under the title "He Done Me Wrong, or Death of Bill Bailey."

H. L. Doherty wins in men's singles at Wimbledon, Mrs. Chambers in women's singles; Holcombe Ward, 25, wins in U.S. men's singles, May Godfray Sutton, 16, (Br) in women's singles (the first non-American woman to win).

The Olympic Games held at St. Louis attract 1,505 contestants from seven countries. The games are part of the Louisiana Purchase Exposition that belatedly celebrates the centennial of the 1803 purchase of the Louisiana Territory from France, and French athletes take most of the medals.

Cy Young pitches the first major league "perfect" game May 5 for the Boston Red Sox, facing 27 batters in nine innings and not letting one of them reach first base (*see* 1890). After Young's death in 1955 the Cy Young Award will be established for the best pitcher in the major leagues each year, and the award will later be given to the best in each league.

No World Series is played because John McGraw of the New York Giants refuses to have his team face the Boston Red Sox.

Harvard builds the first cement football stadium; it holds 40,000 and a colonnade will be added in 1910 to provide more seats as college football grows to become a major spectator sport (*see* Yale Bowl, 1914).

Auction bridge is invented as a variation on the card game Whist played since the early 16th century if not longer. The highest bidder names the trump suit (*see* contract bridge, 1925).

An International Exposition opens a year late at St. Louis to commemorate the centennial of the 1803 Louisiana Purchase. The Ferris wheel from the 1893 Chicago Midway is moved to St. Louis.

The Gillette razor is patented November 15, sales soar to 90,844 (up from just 51 last year), blade sales reach 123,648 (up from 168), and the figures will soon be multiplied a thousandfold. King C. Gillette pays $62,500 to buy out an investor who acquired 500 shares for $250 in 1901.

The Hotel Jefferson opens in St. Louis in time to receive visitors to the city's International Exposition.

Ellsworth M. Statler of 1901 Buffalo Exposition fame opens a 2,257-room Inside Inn for St. Louis fairgoers and has a profit of $300,000 by fall (*see* 1905).

Philadelphia's Bellevue Stratford Hotel opens in September at South Broad Street and Walnut to replace an earlier structure. The 400-room hotel will be nearly doubled in size and given a roof garden in 1913 as it becomes the city's social center.

Buffalo's Martin House is completed by Frank Lloyd Wright.

The *New York Times* moves into a new 25-story Times Tower at Broadway and 42nd Street December 31 and Longacre Square becomes Times Square. The midnight fireworks display that mark the move will become, in modified form, a New Year's Eve tradition, continuing even after the *Times* moves to larger quarters in West 43rd Street and the Tower changes hands.

Fire destroys 75 city blocks in the heart of downtown Baltimore February 7 and 8. More than 1,300 buildings are gutted.

Much of Toronto is destroyed by fire April 19.

A Japanese exhibit at the New York Botanical Garden introduces a blight that begins to wipe out the American chestnut tree *Castanea dentata* whose timber has been a valuable building material and whose nuts have been an important food source.

An *entente cordiale* settles Anglo-French disputes in the Newfoundland fishery.

Most English farmers have turned from raising grain and meat animals which cannot compete with imports to dairy farming, market gardening, poultry raising, and fruit growing. They are unable to meet the demand for butter, cheese, and lard, much of which is imported—along with Dutch margarine-from Denmark and North America. Most English fruit goes into jams whose manufacture is being improved.

Britain's Interdepartmental Committee on Physical Deterioration publishes a report demonstrating the alarming extent of poverty and ill health in British slums. One out of four infants dies in its first 12 months in some areas, the committee reports, and it deplores the rapid decline in breast-feeding, due in part to the employment of married women in industry, but due mostly to the chronic ill-health that makes women incapable of producing milk. Of the children surveyed 33 percent are actually hungry (*see* School Meals Act, 1906).

Xerophthalmia is reported among Japanese infants whose diets are devoid of fats (*see* Denmark, 1917).

The *Ladies' Home Journal* launches an exposé of the U.S. patent-medicine business (*see* 1892; 1898; Adams, 1906).

Pure Food Law advocates take space at the St. Louis Exposition to dramatize the fact that U.S. foods are being colored with potentially harmful dyes.

The hamburger gains popularity at the St. Louis Exposition (above) where the chopped beef specialty is fried and sold by German immigrants who live in South St. Louis (*see* Lassen, 1900).

The ice cream cone is introduced at the St. Louis fair by Syrian immigrant pastry maker Ernest A. Hamwi who sells wafer-like Zalabia pastry at a fairground concession, serving them with sugar and other sweets. When a neighboring ice cream stand runs out of dishes, Hamwi rolls some of his wafers into cornucopias, lets them cool, and sells them to the ice cream concessionaire. But an ice cream cone mold patent has been issued earlier in the year to Italian immigrant Italo Marchiony who claims he has been making ice cream cones since 1896; other claimants challenge Hamwi's right to call himself the ice cream cone originator.

Iced tea is created at the St. Louis fair by English tea concessionaire Richard Blechynden when sweltering fairgoers pass him by, but as in the case of the ice cream cone (above), evidence will be produced of prior invention.

Green tea and Formosan continue to outsell black tea five to one in the United States.

Tea bags are pioneered by New York tea and coffee shop merchant Thomas Sullivan who sends samples of his various tea blends to customers in small hand-sewn muslin bags. Finding that they can brew tea simply by pouring boiling water over a tea bag in a cup, the customers place hundreds of orders for Sullivan's tea bags, which will soon be packed by a specially developed machine.

Campbell's Pork and Beans is introduced by the Joseph Campbell Preserve Co. (*see* 1900).

French's Cream Salad Mustard, developed by George F. French and introduced by R. T. French Co. of Rochester, is milder than other mustards; it will become the world's largest selling prepared mustard, outselling all others combined in the U.S. market.

Postum Co. introduces Elijah's Manna, the name arouses the wrath of clergymen, Britain denies C.W. Post a trademark for his new corn flakes, and he quickly renames them Post Toasties (*see* 1897). Postum is the world's largest single user of molasses, it netted $1.3 million last year, and C. W. Post's personal fortune is estimated at $10 million (*see* 1906).

1905 Russian forces at Port Arthur surrender to Japanese infantry January 2 (*see* 1904) as St. Petersburg verges on revolution (below). The Japanese have lost 57,780 men in assaults on the fortified heights, the Russians 28,200. Five Japanese armies defeat the Russians at Mukden in early March.

The Battle of Tsushima Straits May 27 between Kyushu and Korea ends in victory for Admiral Togo, who arrays his ships 7,000 yards in front of a fleet Russian vessels sent round the Cape of Good Hope to relieve Port Arthur. Bringing his guns to bear on one after the other, he fires into the eight Russian battleships, 12 cruisers, and nine destroyers, most of which explode, capsize, or stop inside 45 minutes.

President Roosevelt mediates the dispute between Russia and Japan and a treaty of peace is signed September 5 at Portsmouth, N.H., after a month of deliberations. Both nations agree to evacuate Manchuria, Russia cedes the southern half of Sakhalin Island, recognizes Japan's paramount interest in Korea, and transfers to Japan the lease of the Liaotung Peninsula and the railroad to Ch'angchun, but rioters at Tokyo protest the failure of Japanese diplomats to obtain an indemnity.

Norway obtains her independence from Sweden after 91 years of union. A popular plebiscite August 13 ratifies a decision of the Storting, the Swedish Riksdag acquiesces September 24, a treaty of separation is signed October 26, Denmark's Prince Charles, 33, is elected king of Norway and will reign until 1957 as Haakon VII.

Alberta and Saskatchewan are established as new Canadian provinces September 1 (*see* 1869).

Britain's Balfour ministry ends December 4 and the Liberals take over with a cabinet headed by Sir Henry Campbell-Bannerman, 69.

A Russian revolution begins as news of the loss of Port Arthur (above) and of "Bloody Sunday" incites the nation. A young priest (Father George Gapon) has led a peaceful workers' demonstration in front of the Winter Palace at St. Petersburg January 9, guards have machine-gunned the demonstrators on orders from the czar, revolutionary terrorists at Moscow murder the Grand Duke Serge, uncle of the czar, February 4, and peasants seize their landlords' land, crops, and livestock.

Russian sailors aboard the armored cruiser *Potemkin* mutiny in July at Odessa, a general strike is called in October, and on October 17 Nicholas II is obliged to grant a constitution, establish a parliament (Duma), and grant civil liberties to placate the people.

V. I. Lenin returns from exile and hails the workers' first Soviet (council) at St. Petersburg, but the czar withdraws his concessions one by one, the revolt begun December 9 under the leadership of the Moscow Soviet is bloodily repressed by Christmas, and the Duma will be suspended by June 1906 (*see* 1917; Lenin, 1903: *Pravda*, 1912. Dates are Julian calendar dates—13 days behind those in the Gregorian calendar of 1582 which Russia will not use until January 31, 1918).

New pogroms begin in Russia (*see* 1904). Terrorists of the anti-Semitic "Black Hundreds" will kill an estimated 50,000 Jews by 1909.

Last year's strike by Fall River, Mass., textile workers ends successfully January 8 after 5.5 months.

The Industrial Workers of the World (IWW) joins U.S. working men in "one big union for all the workers" and attacks the AF of L for accepting the capitalist system. Most of the "Wobblies" are members of the radical Western Federation of Miners (*see* Haywood, 1907; Lawrence, 1912).

A New York law limiting hours of work in the baking industry to 60 per week is unconstitutional, the Supreme Court rules April 17 in the case of *Lochner v. New York*. The owner of a Utica bakery has been convicted of violating the law, but the court rules that Lochner's right of contract is guaranteed under the 14th Amendment and "the limitation of the hours of

1905 ***(cont.)*** labor does not come within the police power." A dissenting opinion by Justice Oliver Wendell Holmes says the word " 'liberty' is perverted" when used to prevent the state from limiting hours of work as "a proper measure on the score of health," and "a constitution is not intended to embody a particular economic theory, whether of paternalism . . . or of laissez-faire."

The Niagara Movement to abolish all racial distinctions in the United States joins 29 black intellectuals from 14 states who meet at Fort Erie, N.Y., under the leadership of Atlanta University professor William Edward Burghardt DuBois, 37 (*see* NAACP, 1909).

English Quaker Joseph Burtt spends 6 months at cocoa plantations on the African islands of São Tomé and Príncipe at the suggestion of cocoa magnate William Cadbury and observes that nearly half of the newly arrived laborers at one plantation died within a year. Working conditions are tantamount to slavery, Burtt reports (*see* 1903; 1909).

English suffragist Emmeline Pankhurst begins propagandizing her cause with sensationalist methods that will include arson, bombing, hunger strikes, and window smashing (*see* 1903; 1906).

"Sensible and responsible women do not want to vote," writes former president Grover Cleveland in the April *Ladies' Home Journal.* "The relative positions to be assumed by men and women in the working out of our civilization were assigned long ago by a higher intelligence than ours."

Bethlehem Steel Co. is founded by Charles M. Schwab who determines to build a great competitor to United States Steel (*see* 1902). Bethlehem begins as the parent company of Schwab's United States Shipbuilding Co. (*see* 1907).

Bolivian store clerk Simon Ituri Patiño, 43, at Cochabamba is sacked by his German employer for extending $250 in credit to a poor Portuguese miner and forced to pay the miner's bill. The miner moves away, leaving his mine to Patiño, who recruits coca-drugged local Indians to work the mine's outcroppings; the property turns out to be a fabulous deposit of tin ore that will be the basis of a $200 million fortune for Simon Patiño and make Bolivia a rival of Malaya, Burma, and the Dutch East Indies as a world tin source. Patiño's Consolidated Tin Smelters, Ltd., will be the world's largest company of its kind.

The Cullinan Diamond found in South Africa weighs 3,100 carats—the largest gem-quality diamond ever discovered. It will be cut into stones for the Crown jewels and the British royal family's collection.

Wildcat oil prospectors in Oklahoma Territory make a big strike in the Glenn Pool near a village on the Arkansas River called by its residents Tulsey Town. It will soon be Tulsa, "Oil Capital of the World."

U.S. auto production reaches 25,000, up from 2,500 in 1899 (*see* 1907).

The Society of Automotive Engineers (SAE) is founded to standardize U.S. automotive parts. Andrew L. Riker of Locomobile is president, Henry Ford first vice president, John Wilkinson of Franklin third vice president (*see* 1911).

A Napier driven by Arthur MacDonald sets a speed record of 104.651 miles per hour at Daytona Beach.

A Stanley Steamer sets a speed record of 127 miles per hour, but its quiet, non-polluting external combustion engine requires a 30-minute warm-up and must take on fresh water every 20 miles (*see* 1899).

The floundering Buick Motorcar Co. started by Flint, Mich., plumbing supply merchant David Dunbar Buick, 50, is taken over by local millionaire William C. Durant whose 19-year-old Durant-Dorn Carriage Co. has made him the richest man in town (*see* 1906).

A French Panhard averaging 52.2 miles per hour wins the first Vanderbilt Cup race October 8 at Hicksville, Long Island, on a 10-lap course over a 30-mile circuit. Also-rans: five Mercedes machines, two other Panhards, two Fiats, two Pope-Toledos, and a Renault, Packard, Clement-Bayard, Simplex, De Dietrich, and Royal Tourist.

Britain holds her first Tourist Trophy motorcar race.

The Austin motorcar is introduced by a new company started by former Wolesley designer Herbert Austin (*see* 1895; British Motors, 1952).

Renault Frères introduces the first automobile taxicabs, in Paris.

London's Inner Circle rail lines are electrified.

The Long Island Rail Road, controlled since 1900 by the Pennsylvania, installs a low-voltage third rail system, becoming the first U.S. road completely to abandon steam locomotion.

"The Railroads on Trial" by Ray Stannard Baker, 35, appears in *McClure's* magazine. The series will be instrumental in the fight for railroad regulation (*see* Hepburn Act, 1906).

The Wright brothers improve their flying machine of 1903 to the point where they can fly a full circle of 24.5 miles in 38 minutes, a feat they demonstrate at Dayton, Ohio. Their machine will be patented in May of next year, and they will give a series of exhibitions in the United States and Europe to popularize flying (*see* Wilbur, 1908; Orville, 1909).

Commercial rayon production begins in July at an English factory, built on the outskirts of Coventry by Samuel Courtauld, 29, who has bought English rights to the Stearn patent of 1898 for £25,000. French and German companies have bought foreign rights and begin production in competition with Courtauld (*see* Little, 1902; Viscose Co., 1910).

Swiss theoretical physicist Albert Einstein, 26, at Bern publishes a paper on the special theory of relativity that revises traditionally held Newtonian views of space and time. Einstein introduces the equation $E = mc^2$ in "The Electrodynamics of Moving Bodies" ("*Zur Elektrodynamik bewegter Körper*") and predicts that light (or radio) waves moving from one planet to another will be bent off their path by a massive body like the sun, slowing them down for a fraction of a second (*see* 1916).

The first major epidemic of poliomyelitis since the disease was first identified in 1840 breaks out in Sweden (*see* 1916).

German surgeon Heinrich Friedrich Wilhelm Braun, 43, introduces procaine (novocaine) into clinical use. The local anaesthetic will find wide use in dentistry.

A U.S. Public Health Service anti-mosquito campaign ends a yellow fever epidemic that has killed at least 1,000 at New Orleans, the last U.S. epidemic of the disease (*see* Havana, 1901).

The Phipps Tuberculosis Dispensary is established by steel magnate Henry Phipps at Baltimore's Johns Hopkins Hospital (*see* 1893; 1903).

Compulsory vaccination laws are upheld by the Supreme Court which rules in the case of *Jacobson v. Massachusetts* that enacting and enforcing such laws is within the police power of any state.

Vicks VapoRub is introduced under the name Vick's Magic Croup Salve by the new Vick Chemical Co. founded at Greensboro, N.C., by former drug wholesaler Lunsford Richardson, 52. The mentholated salve is compounded of camphor, menthol, spirits of turpentine, oil of eucalyptus, cedar leaf oil, myristica oil, nutmeg, and thymol mixed in a petroleum base.

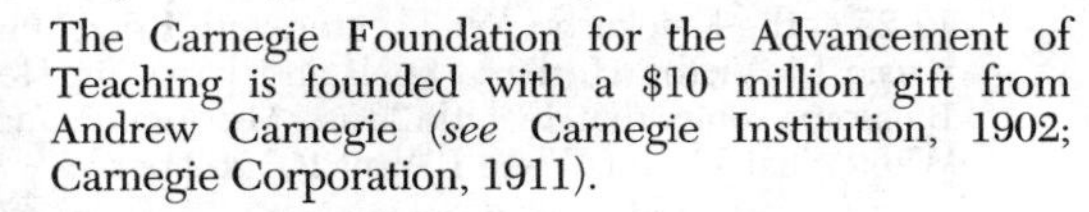

The Carnegie Foundation for the Advancement of Teaching is founded with a $10 million gift from Andrew Carnegie (*see* Carnegie Institution, 1902; Carnegie Corporation, 1911).

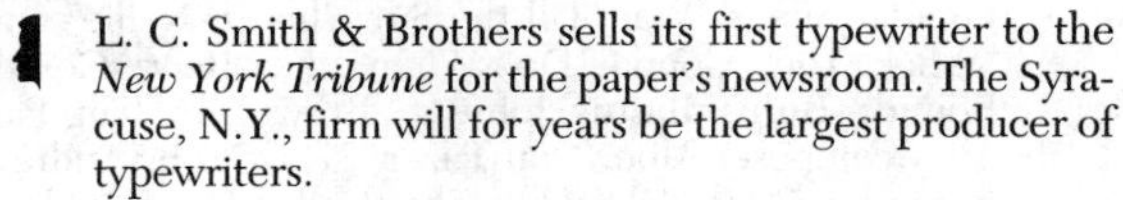

L. C. Smith & Brothers sells its first typewriter to the *New York Tribune* for the paper's newsroom. The Syracuse, N.Y., firm will for years be the largest producer of typewriters.

Royal Typewriter Co. is founded by New York financier Thomas Fortune Ryan who puts up $220,000 to back inventors Edward B. Hess and Lewis C. Meyers (*see* Southern Railway, 1894). The Royal typewriter has innovations that include a friction-free ball-bearing one-track rail to support the weight of the carriage as it moves back and forth, a new paper feed, a shield to keep erasure crumbs from falling into the nest of type bars, a lighter and faster type bar action, and complete visibility of words as they are typed.

The *Chicago Defender* begins publication May 5. Publisher Robert S. Abbott, 35, has the first issues printed on credit and sells just 300 copies of the 2¢ sheet, but the *Defender* is the first important black newspaper and within 12 years it will be so widely read in the South that critics will credit (and blame) it for inspiring a migration of blacks to the North (it will become a daily beginning in 1956).

William Randolph Hearst acquires *Cosmopolitan* magazine for $400,000.

Nonfiction *The Life of Reason* (volume I of five) by Spanish-American philosopher George Santayana (Jorge Ruis Santayana y Burrais), 42, who writes, "Those who cannot remember the past are condemned to repeat it."

Fiction *The House of Mirth* by New York novelist Edith Wharton, 43, who gains her first popularity with a fictional analysis of the stratified society she knows so well as the wife of a rich banker and of that society's reaction to change; *The Tower of London* (*Rondon-to*) and *I Am a Cat (Waga hai wa neko de aru)* by Japanese novelist Soseki Natsume, 38, who was graduated in English at Tokyo University in 1893, studied at London from 1900 to 1902, and in his second novel presents a witty, satirical view of the world as seen through the eyes of his cat; *Where Angels Fear to Tread* by English novelist E. M. (Edward Morgan) Forster, 26, of the Bloomsbury Group; *Kipps* by H. G. Wells; *The Lake* by George Moore; *The Four Just Men* by English novelist Edgar Wallace, 30, who gains fame with a thriller based on an antiestablishment concept; *Professor Unrat* by German novelist Heinrich Mann, 34, who will be overshadowed by his older brother Thomas, but whose novel will be made into the 1929 film *The Blue Angel; The Jungle* by U.S. novelist Upton Beall Sinclair, 27, who has been given a $500 advance by the Socialist periodical *The Appeal to Reason.* He has lived for 7 weeks among the stockyard workers of Chicago, his Lithuanian hero Jurgis is lured by steamship company posters to emigrate and takes a job in the stockyards of "Packington," where he encounters a variety of capitalistic evils (*see* meat packing, below).

Poetry *The Book of Hours (Stundenbuch)* by Austrian poet Rainer Maria Rilke, 30; *Präludien* by German poet Ernst Stadler, 22; *Cantos of Life and Hope (Canto de vida y esperanza)* by Rubén Darío.

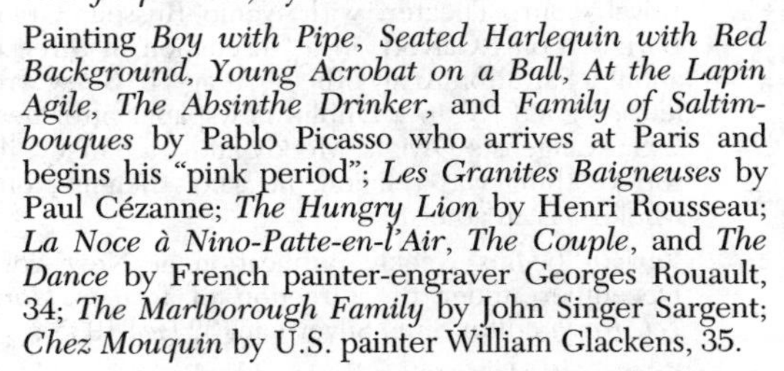

Painting *Boy with Pipe, Seated Harlequin with Red Background, Young Acrobat on a Ball, At the Lapin Agile, The Absinthe Drinker,* and *Family of Saltimbouques* by Pablo Picasso who arrives at Paris and begins his "pink period"; *Les Granites Baigneuses* by Paul Cézanne; *The Hungry Lion* by Henri Rousseau; *La Noce à Nino-Patte-en-l'Air, The Couple*, and *The Dance* by French painter-engraver Georges Rouault, 34; *The Marlborough Family* by John Singer Sargent; *Chez Mouquin* by U.S. painter William Glackens, 35.

Les Fauves create a sensation at the Salon d'Automne in Paris with paintings that free color to speak with new and unprecedented intensity. Artists who include Henri Matisse, 35, Maurice de Vlaminck, 29, and André Derain, 25, pioneer the magnification of Impressionism, critic Louis Vauxcelles calls them "wild beasts" ("les fauves"), and their works include Matisse's *The Open Window*, *Collioure*, *Girl Reading*, and *La Joie de Vivre*, Vlaminck's *The Pond at Saint-Cucufa,* and Derain's *The Mountains, Collioure.*

New York's 291 Gallery at 291 Fifth Avenue is opened by local photographer Alfred Stieglitz, 41, with a group of other photographers Stieglitz cans the "photo-secessionists" (they include Luxembourg-born photographer Edward Jean Steichen, 26, who opened a studio at 291 Fifth in 1902 after 2 years of study in Europe

1905 ***(cont.)*** with sculptor Auguste Rodin). The new gallery will show only photographs until 1908 but will then begin to give Americans their first look at work by Henri Matisse (above), Pablo Picasso, and the late Henri de Toulouse-Lautrec, among others.

Sculpture *The Bather* by German sculptor Wilhelm Lehmbruck, 24.

Theater *The Well of the Saints* by John Millington Synge 2/4 at Dublin's Abbey Theatre; *Hidada (Hidada, oder Sein und Haben)* by Frank Wedekind 2/18 at Munich's Schauspielhaus, under the title *Karl Hetmann, the Dwarf-Giant (Karl Hetmann, der Zwergriese); Elga* by Gerhart Hauptmann 3/4 at Berlin's Lessingtheater; *Alice-Sit-by-the-Fire* by James M. Barrie 4/5 at the Duke of York's Theatre, London; *Man and Superman* (minus Act III) by George Bernard Shaw 5/23 at London's Royal Court Theatre, with Granville Barker, Edmund Gwenn; *The Dance of Death (Dodsdansen forsta delen and anla delen)* by August Strindberg in September at Cologne's Residenztheater; *No Mother to Guide Her* by U.S. actress-melodramatist Lillian Mortimer 9/5 at Detroit's Whitney Theater; *The Squaw Man* by Edward Milton Royle 10/23 at Wallack's Theater, New York, with William S. Hart, 722 perfs.; *Mrs. Warren's Profession* by Bernard Shaw 10/30 at New York's Garrick Theater, with Mary Shaw, Arnold Daly (who are prosecuted for performing in an "immoral" play), 14 perfs. (London's Lord Chamberlain has refused to license the play about a woman whose income is derived from houses of prostitution. London will not see a production until 1926); *The Girl of the Golden West* by David Belasco 11/14 at New York's Belasco Theater, with Blanche Bates, 224 perfs.; *Major Barbara* by Bernard Shaw 11/28 at London's Royal Court Theater, with Annie Russell, Granville Barker, Louis Calvert, in a "discussion in three acts" about a Salvation Army officer. At age 18 Shaw wrote a letter signed "S" to a Dublin newspaper after hearing U.S. evangelists Dwight Moody and Ira Sankey; if this sort of thing was religion, he said, then he, on the whole, was an atheist.

Variety begins weekly publication at New York in December under the direction of former *Morning Telegraph* editor Sime Silverman, 32 (*see* 1935).

Film Cecil Hepworth's *Rescued by Rover.*

Opera *Salomé* 12/9 at Dresden's Hofoper, with music by Richard Strauss, libretto based on the 1896 play by the late Oscar Wilde. Soprano Marie Wittich had originally refused to sing the title role's sensuous song to the imprisoned John the Baptist praising his body, his hair, his mouth, saying, "I won't do it, I'm a decent woman." After the opera's first performance at New York's Metropolitan Opera House January 22, 1907, J. P. Morgan will forbid further performances and *Salomé* will not be revived at the Met until 1934; *The Merry Widow (Die lustige Witwe)* 12/28 at Vienna's Theater-an-der-Wien, with music by Franz Lehar, libretto by Victor Leon and Leo Stein.

First performances *Pélleas and Mélissande* (symphonic poem) by Arnold Schoenberg 1/26 at Vienna; Introduction and Allegro in G minor for Strings by Edward Elgar 3/8 at the Queen's Hall, London; *The Divine Poem* Symphony No. 3 in C minor by Aleksandr Skriabin 5/29 at the Théâtre du Châtelet, Paris. (Skriabin has deserted his wife and four children in Russia to live openly with his pretty mistress Tatyana Schloezer in Switzerland. She believes in his messianic visions of transforming the world through music and will bear him three children); *La Mer* (tone poem) by Claude Debussy 10/15 at Paris; Symphony No. 5 in C sharp minor by Gustav Mahler 10/18 at Paris.

Isadora Duncan, 27, opens a dancing school for children at Berlin. The U.S. dancer, who gives birth to a child out of wedlock, has developed a spontaneous style that tries to symbolize music, poetry, and elements of nature.

March "Die Parade der Zinnsoldaten" ("Parade of the Wooden Soldiers") by German composer Leon Jessel, 34.

"Clair de Lune" by Claude Debussy is published as part of a suite for piano.

New York's Juilliard School of Music has its beginnings in the Institute of Musical Art founded by Frank Damrosch and James Loeb (*see* 1920).

Broadway musicals *Fontana* 1/14 at the Lyric Theater, with Douglas Fairbanks, 21, in a minor role, music by Raymond Hubbell, book by S. S. Shubert and Robert B. Smith, lyrics by Smith, 298 perfs.; *Mlle. Modiste* 12/25 at the Knickerbocker Theater, with Fritzi Scheff, music by Victor Herbert, book and lyrics by Henry Blossom, songs that include "Kiss Me Again" and "I Want What I Want When I Want It," 262 perfs.

Popular Songs "Wait Till the Sun Shines, Nellie" and "What You Gonna Do When the Rent Comes Round?–Rufus Rastus Johnson Brown" by Tin Pan Alley composer Albert von Tilzer, 27, lyrics by Andrew B. Sterling; "My Gal Sal (or, They Called Her Frivolous Sal)" by Paul Dresser; "Daddy's Little Girl" by Theodore F. Morse, lyrics by Edward Madden; "In My Merry Oldsmobile" by Gus Edwards, 26, lyrics by Vincent Bryan, 22; "I Don't Care" by Harry O. Sutton, lyrics by Jean Lewis, that will be popularized by vaudeville star Eva Tanguy; "In the Shade of the Old Apple Tree" by Egbert Van Alstyne, lyrics by Henry Williams (both are song pluggers for the publishing house Remick); "Will You Love Me in December as You Do in May?" by Ernest R. Ball, lyrics by New York law student James J. Walker, 24.

H. L. Doherty wins in men's singles at Wimbledon, May Sutton in women's singles; Beals Coleman Wright, 25, wins in U.S. men's singles, Elizabeth Moore in women's singles.

Ty Cobb signs with the Detroit Tigers to begin an outstanding baseball career. Sportswriter Grantland Rice has nicknamed outfielder Tyrus Raymond Cobb, 18, "the Georgia Peach" for his prowess with the Augusta team of the South Atlantic League; he will play 22 seasons for Detroit and two more for the Philadelphia Athletics before retiring in 1928 with a lifetime batting average of .367, a record 4,191 hits, a record 2,244 runs scored, and a record 3,033 games played. Cobb will bat above .300 for a record 23 games and steal a record 829 bases.

The New York Giants win the World Series by defeating the Philadelphia Athletics 4 games to 1.

U.S. "boy wonder" billiard player Willie Hoppe, 18, wins his first world championship after 10 years of barnstorming the country with pool and billiard exhibitions. William Frederick Hoppe won the Young Masters 18.2 balkline championship at 16 and will win the world's 18.1 title next year at Paris and the world's 18.2 title in 1907. By the time he retires in October 1952, he will have won 51 world billiard titles.

Palmolive Soap is introduced by Milwaukee's 42-year-old B. W. Johnson Soap Co. whose Caleb Johnson has bought a new soap milling machine he saw at last year's St. Louis Exposition. The toilet soap is made from a formula that includes some hydrogenated palm and olive oils, using the process invented by William Normann in 1901.

The first Spiegel mail-order catalog offers furniture at terms as low as 75¢ down and 50¢ per month. Started just before the Civil War, the Chicago furniture store will give up its retail store and broaden its catalog offerings to include apparel and general merchandise.

New York's Tiffany & Co. sells a pearl necklace for $1 million—the largest single sale in the store's 68-year history and one that will not be surpassed for at least 86 years (*see* 1906).

A new Statler Hotel opens at Buffalo, N.Y., to replace the temporary Outside Inn put up by E. M. Statler for the Pan American Exposition of 1901. Statler advertises "a room and a bath for a dollar and a half," and he will develop a chain of Statler Hotels in other cities.

Cleveland's Shaker Heights suburb has its beginnings in a 1,400-acre tract of land owned by the Shaker church east of the city. Real estate developers Oris Paxton Van Sweringen, 26, and his brother Mantis, 24, have secured parcels of land in the tract, they borrow heavily to acquire a total of 4,000 acres, they will develop it into a model suburban community with winding roadways, round artificial lakes, and with restrictive covenants to control architecture, but their investment will not pay off until they persuade the Cleveland Street Railways Co. to extend a line to Shaker Heights by offering to pay 5 years' interest on the cost of laying tracks (*see* Nickel Plate Railroad, 1916).

President Roosevelt creates a Bureau of Forestry in the U.S. Department of Agriculture February 1 with Gifford Pinchot as chief forester (*see* 1902). Congress passes a bill transferring U.S. forest reserves from the Department of the Interior to the USDA, and the reserves are soon designated National Forests as Pinchot works to make the Forest Service a defender of the public lands against private interests (*see* "conservation," 1907).

The Wichita Mountains Refuge for big game is established by President Roosevelt in Oklahoma Territory.

The National Audubon Society is founded (*see* Grinnell, 1885).

Water from the Colorado River has been diverted to California by sluice gates built since 1900 and is creating the Salton Sea (*see* Laguna Dam, 1909).

U.S. botanist George Harrison Shull, 31, journeys to Santa Rosa, Calif., on a grant from the Carnegie Institution established 3 years ago and studies plant hybridization methods pioneered by Luther Burbank in 1875 (*see* 1921).

Just 100 men own more than 17 million acres of California's vast Sacramento Valley while in the arid San Joaquin Valley many individuals own tracts of 100,000 acres and more.

Roughly 12 acres of farmland are cultivated for every American. The number will fall rapidly as the population increases and as land is made more productive.

Vitamin research is pioneered at the University of Utrecht where Dutch nutritionist C. A. Pekelharing finds that mice die on a seemingly ample diet but survive when their diet includes a few drops of milk. Pekelharing concludes that "unrecognized substances" must exist in food (*see* 1906; Funk, 1911).

Iodine compounds prove useful in treating goiter. Cleveland physician David Marine, 25, has moved West from Baltimore and been struck by the fact that every dog in town—and many of the people—have goiters. He begins a campaign for the iodization of table salt (*see* Baumann, 1896; Kendall, 1915).

The Massachusetts legislature rejects a bill that would require patent medicine bottles to carry labels showing their ingredients. The $100 million-a-year patent medicine industry spends some $40 million a year in advertising, and newspapers lobby against legislation that would hinder sales of "cancer cures" and other nostrums, many of which contain cocaine, morphine, and up to 40 percent alcohol. The Proprietary Association of America has devised advertising contracts that will be canceled if legislation hostile to patent medicines is enacted. Journalist Mark Sullivan, 31, exposes the newspapers' conflict of interest in *Collier's* November 4 (*see* Pure Food and Drug Act, 1906).

Upton Sinclair exposes U.S. meat-packing conditions in *The Jungle* (above). The 308-page best seller has eight pages devoted to such matters as casual meat inspection, lamb and mutton that is really goat meat, deviled ham that is really red-dyed minced tripe, sausage that contains rats killed by poisoned bread, and lard that sometimes contains the remains of employees who have fallen into the boiling vats. Many readers turn vegetarian, sales of meat products fall off, and Congress is aroused (*see* Meat Inspection Act, 1906).

Articles on the Beef Trust by Charles E. Russell appear in *Everybody's* magazine.

The Supreme Court orders dissolution of the Beef Trust January 20 in *Swift v. United States*, but the court order will merely change the form of the Trust.

1905 ***(cont.)*** Some 41 million cases of canned foods are packed in the United States. The figure will rise in the next 25 years to 200 million (*see* 1935).

Heinz Baked Beans are test marketed in the north of England by Pittsburgh's H. J. Heinz Co. Advertising advises British working class wives that baked beans make a nourishing meal for men returning from work (Heinz beans will be canned at Harlesden beginning in 1928, and Britons will consume baked beans at twice the U.S. rate per capita).

Nestlé merges with its major competitor Anglo-Swiss Condensed Milk Co. (*see* 1875; 1918).

Louis Sherry's 7-year-old Fifth Avenue, New York, restaurant is the scene of a ball given by Equitable Life Assurance Society heir James Hazen Hyde who denies reports that the ball cost $100,000 (he says it cost only $20,000).

The Asiatic Exclusion League founded by U.S. racists seeks to ban Japanese immigration (*see* 1908).

China boycotts U.S. goods to protest U.S. laws barring entry of educated Chinese along with Chinese workers (*see* 1902).

1906 Denmark's Christian IX dies January 29 at age 87 after a 42-year reign. His son of 62 will reign until 1912 as Frederick VIII.

The British battleship H.M.S. *Dreadnought* launched February 10 has ten 12-inch guns—the first battleship whose guns are all so large. Berlin decides in May to increase tonnages of German battleships, add six cruisers to the fleet, and widen the Kiel Canal to permit passage of larger ships.

The British foreign secretary Sir Edward Grey, 44, assumes a "moral obligation" to support France in the event of a German attack, but the Cabinet will not learn of his pledge to the French until 1911.

London forces the Ottoman Turks to cede the Sinai Peninsula to Egypt.

The sultan of Morocco accepts French and Spanish control of reforms in his country following the Algeciras Conference at which France and Spain have agreed to reaffirm Morocco's independence and integrity. The country will nevertheless be the scene of international crises for years to come.

A Tripartite Pact July 4 declares the independence of Ethiopia but divides the country into British, French, and Italian spheres of influence.

The All-India Muslim League is founded by the Aga Khan, 29 (Aga sultan Sir Mahomed Shah).

Japanese authorities abort an anarchist plot to assassinate the emperor.

A world diplomatic conference revises the Geneva Convention of 1864.

Czar Nicholas II grants universal suffrage May 24 but rejects the Duma's suggestion that he grant amnesty to political prisoners. The Duma is dissolved July 21 and martial law declared.

Brownsville, Tex., has a race riot following an August 13 incident in which the town of 7,000 has allegedly been shot up by a few blacks from a crack all-black infantry battalion stationed on the outskirts of Brownsville 19 days earlier. One resident has reportedly been killed and a woman raped, the townspeople make reprisals, the truth of the situation will never be established, but President Roosevelt orders November 6 that three companies of black troops be given dishonorable discharges.

Johannesburg lawyer Mohandas K. Gandhi, 37, speaks at a mass meeting in the Empire Theater September 11 and launches a campaign of nonviolent resistance (*satyagraha*) to protest discrimination against Indians. Gandhi has been in racially segregated South Africa since 1893 (*see* 1913).

A *London Daily Mail* writer coins the term "suffragettes" for women such as Emmeline Pankhurst and her daughters Christabel, 26, and Sylvia, 24, who campaign for women's suffrage (*see* 1905). London barrister Frederick Pethick-Lawrence, 35, and his wife Emmeline, 39, support the Pankhursts (*see* 1913).

Capt. Dreyfus is restored to rank July 12 following vindication of his case and given the Légion d'Honneur (*see* 1898). Now 47, Dreyfus will live until 1935.

Mother Earth is founded by Lithuanian-American anarchist Emma Goldman, 37, and Alexander Berkman who has just been released from prison after having served time for an attempt on the life of Henry Clay Frick in the Homestead Strike of 1892. The anarchist journal will be suppressed in 1917 (*see* 1919).

The American Jewish Committee is founded by U.S. Jews, mostly of German descent, to protect the civil and religious rights of Jews and to fight prejudice.

San Francisco's board of education orders segregation of Oriental children in city schools 6 months after the April earthquake and fire that devastate the city (below). The Japanese ambassador registers an emphatic protest October 25, and the order is rescinded (*see* Hayashi, 1907).

Gary, Ind., has its beginnings as U.S. Steel Corp. breaks ground for a new mill town on Lake Michigan. The town will be named to honor Big Steel's chairman Elbert Henry Gary, 60.

Congress appropriates $2.5 million for relief following the San Francisco earthquake (below). New York bankers loan hundreds of millions to rebuild the city (*see* 1907).

San Francisco banker A. P. (Amadeo Peter) Giannini, 35, manages to rescue the gold and securities in his vault. Giannini built his stepfather's wholesale produce business into the largest in the city, sold out his share in 1901, founded his Bank of Italy in 1904, has been making loans to small farmers and businessmen, has defied banking orthodoxy by actively soliciting customers, and is the only banker in town with enough capital to make loans for rebuilding. Giannini's reliance on gold reserves will enable him to survive next year's financial panic.

J. P. Morgan and steel magnate John W. Gates purchase Tennessee Coal & Iron Co. (*see* 1898; 1907).

E. I. du Pont de Nemours has bought up or otherwise absorbed the other members of the 34-year-old Gunpowder Trade Association (Powder Trust) and has a near-monopoly in the U.S. powder industry. It produces 100 percent of the privately-made smokeless powder and from 60 to 70 percent of five other kinds of explosives (*see* 1902; 1912).

New York's Tiffany & Co. moves into a new building at Fifth Avenue and 37th Street where it will remain until it moves to 57th Street in 1940.

Fuller Brush Co. has its beginnings in a firm founded at Hartford, Conn., by Nova Scotia-born merchant Alfred Carl Fuller, 21. He quits his job as salesman for another brush company, develops a machine for making twisted-wire brushes, produces them at night, sells them door-to-door by day, and builds a line of brushes with specific uses (*see* 1910).

A 3-day Senate speech on railroad rates in April by the junior senator from Wisconsin Robert M. LaFollette, 51, elicits the comment from Pennsylvania Railroad president Alexander Johnston Cassatt, 67, "I have for several years believed that the national government, through the Interstate Commerce Commission, ought to be in a position to fix railroad rates."

The Hepburn Act passed by Congress June 29 extends jurisdiction of the Interstate Commerce Commission established in 1887 and gives the ICC powers to fix railroad rates (*see* Elkins Act, 1903).

The War Department begins excavation of the Panama Canal (*see* 1903; 1914; Gorgas, 1904).

The first Grand Prix motorcar race is organized by the Automobile Club de France which announced after last year's Gordon Bennett Cup race that it would take no further part in races which required that every part of a competing vehicle be made in the country represented by that vehicle. The 2-day 1,248-kilometer race pits 23 French cars against six Italian and three German cars; the winner is a Renault with Michelin detachable rims that can be removed by unscrewing eight nuts.

Rolls-Royce, Ltd., is incorporated March 16 by English balloonist and motoring enthusiast Charles Stewart Rolls, 29, son of Lord Llangattock, and auto maker Frederick Henry Royce, 43, who has produced an almost silent-running motorcar with a six-cylinder engine that produces from 40 to 50 horsepower. C. S. Rolls and Co. has sold F. H. Royce and Co. motorcars on an exclusive basis for nearly 2 years, the new firm will move from Manchester to Derby in 1908 and will discontinue all previous models to concentrate on the Silver Ghost, a machine that has been driven "14,371 miles nonstop," has proved that a gasoline-powered car can run as smoothly as a steam-driven car, and will remain in production for 19 years while Rolls-Royce, Ltd., develops a reputation for making the world's most luxurious motorcars.

Lagonda Motor Co., Ltd., is founded at Staines, Middlesex, by Ohio-born English entrepreneur Wilbur Gunn, who has won the London-Edinburgh motorcycle trial and will make a fortune supplying motorcars to the Russian nobility.

The Autocar Co. of Ardmore, Pa., improves its 1897 model by adding acetylene headlights and kerosene side lamps, but like most cars the new model still has a steering rudder not a steering wheel.

The Chadwick Six built by L. S. Chadwick runs more smoothly than any four-cylinder motorcar and will have a major influence on U.S. auto makers (*see* 1903). Chadwick will quit making four-cylinder cars, incorporate at Pottstown, Pa., in March of next year, road-test the first production model Chadwick Six in April, and sell it at $5,500. He will quit in 1911 but Chadwick motorcars will remain in production until 1915.

W. C. Durant's Buick Co. produces 2,295 motorcars, up from no more than 28 in 1904 (*see* 1905; General Motors, 1908).

The *North American Review* reports that more Americans have been killed by motorcars in 5 months than died in the Spanish-American War.

Princeton University president (Thomas) Woodrow Wilson, 49, says of the motorcar, "Nothing has spread socialistic feeling in this country more than the use of the automobile. To the countryman, they are a picture of the arrogance of wealth, with all its independence and carelessness" (but *see* 1909).

The Mack truck is introduced by former Brooklyn, N.Y., wagon builders John, William, and Augustus Mack who have moved to Allentown, Pa., and produced a 10-ton vehicle of considerable power that begins a reputation for "built like a Mack truck" bulldog stamina.

American Rolling Mills produces the first silicon steel for electrical use to be made in America. George Verity has adopted a process invented by Robert Hadfield of 1882 manganese steel fame and promoted by George Westinghouse (*see* Verity, 1899).

The third law of thermodynamics, formulated by German chemist Walter Herman Nernst, 42, states that specific heats at the absolute zero of temperature (minus 273.12° C.) become zero themselves. Nernst's discovery will make it possible to calculate the actual values of constants that characterize chemical reactions (*see* Haber, 1908).

Psychology of Dementia Praecox by Swiss psychologist Carl Gustav Jung, 31, breaks new ground. Jung meets Sigmund Freud for the first time and has doubts as to the sexual aspects of Freud's theory of psychoanalysis (*see* 1895; 1908).

The Wasserman test developed by German physician August Wasserman, 40, and (independently) by Hungarian physician Lazlo Detre is a specific blood test for syphilis (*see* Ehrlich, 1909).

The word *allergy* coined by Austrian pediatrician Clemens von Parquet, 32, begins to replace the word *anaphylaxis* introduced a few years ago.

1906 ***(cont.)*** U.S. pathologist Howard Taylor Ricketts, 35, discovers that Rocky Mountain spotted fever is transmitted by the bite of a cattle tick; the toxin spread by ticks will be called *Rickettsia.*

"The men with the muckrakes are often indispensable to the well-being of society," says President Roosevelt April 14 in an address to the Gridiron Club at Washington, D.C., "but only if they know when to stop raking the muck and to look upward . . . to the crown of worthy endeavor." Roosevelt has in mind John Bunyan's reference in *Pilgrim's Progress* (1684) to "a man who could look no way but downwards, with a muckrake in his hand," and has in mind also such magazines as *McClure's* (1902, 1905) and books such as *The Pit* (1903) and *The Jungle* (1905).

International News Service (INS) is founded by William Randolph Hearst whose wire service will compete with the AP founded in 1848 (*see* Brisbane, 1897; United Press, 1907).

U.S. inventor Lee De Forest, 33, develops a three-electrode vacuum tube amplifier that will be the basis of an electronics revolution. His Audion will permit the development of radio and he will broadcast the voice of operatic tenor Enrico Caruso in 1910 *(see* 1916; Tesla, 1888; Fleming, 1904).

The International Radio Telegraph Convention at Berlin adopts the distress call SOS to replace the CQD (Stop Sending and Listen) call adopted 2 years ago by Marconi International Communications and agrees to drop the CQD call completely after 1912.

The first radio broadcast of voice and music booms out of Brant Rock, Mass., Christmas Eve and is picked up by ships within a radius of several hundred miles. Canadian-American engineer Reginald Aubrey Fessenden, 40, patented a high-frequency alternator in 1901 that generates a continuous wave instead of the intermittent damped-spark impulses now being used for transmission of wireless signals; he organized the National Electric Signalling Corp. in 1902 to sponsor his experiments, he has engaged Swedish-American electrical engineer Ernst Frederik Werner Alexanderson, 28, to build a large 80-kilohertz alternator for the station he has set up at Brant Rock, his equipment features a receiver circuit of his own invention that blends an incoming radiofrequency signal with another, slightly different, signal produced locally to yield a beat frequency in the audible range, and this heterodyne circuit will be generally adopted in radio broadcasting *(see* Armstrong's superheterodyne, 1917; RCA, 1919).

New York's Pierpont Morgan Library is completed by McKim, Mead & White adjacent to J. P. Morgan's brownstone in East 36th Street to house the banker's prodigious collection of rare books, manuscripts, documents, and incunabula (books printed before 1501). A north wing will enlarge the Renaissance palazzo in 1962 and it will be further enlarged in 1976.

Nonfiction *The Education of Henry Adams* by Henry Brooks Adams, now 68, whose classic autobiographical work, privately printed, will be published in 1918 to acquaint the world with his dynamic theory of history; *The Devil's Dictionary* by Ambrose Bierce, now 64, who is the Hearst correspondent at Washington, D.C.: "Admiration: our polite recognition of another's resemblance of ourselves"; "Bore: a person who talks when you wish him to listen"; "Cynic: a blackguard whose faulty vision sees things as they are, not as they ought to be"; "Destiny: a tyrant's authority for crime and a fool's excuse for failure"; "Epitaph: an inscription on a tomb, showing that virtues acquired by death have a strong retroactive effect"; "Edible: good to eat, and wholesome to digest, as a worm to a toad, a toad to a snake, a snake to a pig, a pig to a man, and a man to a worm"; "Once: enough"; "Politics: the conduct of public affairs for private advantage."

Fiction *The City of the Yellow Devil (Gorod Zholtogo D'yavola)* by Maksim Gorki who was imprisoned last year for his revolutionary activity but has had his sentence commuted to exile at the intervention of Western writers. Gorki has come to America to collect funds for the Bolshevik cause, he has been ostracized since the revelation that his traveling companion is not his wife, and his novel is an attack on the American way of life; *The Man of Property* by English novelist John Galsworthy, 39, who introduces readers to Soames Forsyte and begins *The Forsyte Saga* that he will continue until 1922; *Young Torless (Die Verwirrungen des Zöglings Torless)* by Austrian novelist Robert Musil, 25; *Puck of Pooks Hill* by Rudyard Kipling who will win next year's Nobel prize for literature; *Broken Commandment (Hakai)* by Japanese novelist Shimazaki Toson (Shimazaki Haruki), 34, whose autobiographical work breaks his father's commandment not to reveal that he and his family are members of Japan's outcast *burakumin* group (*see* 1871); *With Grass for Pillow* (*Kusamakura*) by Soseki Natsume; *Little Boy (Bocchan)* by Soseki Natsume who makes use of his experience as a schoolteacher in the 1890s; *The Four Million* (stories) by O. Henry who gains renown for "The Furnished Room" and "The Gift of the Magi"; *White Fang* by Jack London.

Poetry *Poems* by English poet Siegfried Sassoon, 20; "The Highwayman" by English poet Alfred Noyes, 26: "The wind was a torrent of darkness among the gusty trees,/The moon was a ghostly galleon tossed upon cloudy seas,/The road was a ribbon of moonlight over the purple moor,/And the highwayman came riding, up to the old inn-door"; *Portage de midi* by French poet Paul Claudel, 38.

Painting *Portrait of Gertrude Stein, Two Nudes, Self-Portrait with a Palette, The Organ Grinder, The Flower Vendors,* and *Boy Leading a Horse* by Pablo Picasso; *Le Bonheur de Vivre* by Henri Matisse; *The Red Trees* by Maurice Vlaminck; *Westminster Bridge, The Houses of Parliament,* and *Port of London* by André Derain; *Street Decked Out with Flags at Le Havre* by Raoul Dufy; *At the Theater* and *Tragedian* by Georges Rouault; *Self-Portrait in Weimar* by Edvard Munch; *Young Men by the Sea* by German painter Max Beckmann, 22; *Coltaltepetl* by Mexican painter Diego

Rivera, 20; *Large Bathers* by Paul Cézanne who dies of diabetes at Aix-en-Provence October 22 at age 67.

Sculpture *Chained Action* by Aristide Maillol.

Theater *And Pippa Passes (Und Pippa tanzt!)* by Gerhart Hauptmann 1/19 at Berlin's Lessingtheater; *The Marquis of Bradomin (El marqués de Bradomin)* by Ramon Maria del Valle Inclan 1/25 at Madrid's Teatro de la Primers; *Captain Brassbound's Conversion* by George Bernard Shaw 3/20 at London's Royal Court Theatre, with Ellen Terry, Frederick Kerr, Edmund Gwenn; *Caesar and Cleopatra* by Shaw 3/31 at Berlin (in German), 10/31 at New York's New Amsterdam Theater, with Forbes Robertson, Gertrude Elliott, 49 perfs.; *The Silver Box* by John Galsworthy 9/25 at London's Royal Court Theatre; *The Great Divide* by U.S. playwright William Vaughan Moody, 37, 10/3 at New York's Princess Theater with Henry Miller, 238 perfs.; *The New York Idea* by U.S. playwright Langdon Elwyn Mitchell (John Philip Varley), 44, 11/18 at New York's Lyric Theater, with Mrs. Fiske, George Arliss; *The Doctor's Dilemma* by Bernard Shaw 11/20 at London's Royal Court Theatre, with William Farren, Jr., Granville Barker; *Brewster's Millions* by U.S. playwrights Winchell Smith, 35, and Byron Ongley (who have adapted the 1902 George Barr McCutcheon novel) 12/31 at New York's New Amsterdam Theater, with a cast that includes "George Spelvin," a name Smith has invented to designate a player who doubles in another role.

First performances *Jour d'été à la montagna* by Vincent d'Indy 2/18 at Paris; Symphony No. 6 by Gustav Mahler 5/27 at Essen.

Opera *Quattro Rusteghi* (*The School for Fathers*) 3/19 at Munich, with music by Ermanno Wolf-Ferrari; *The Wreckers* (*Standrecht*) 11/11 at Leipzig, with music by English composer Ethel Smyth, 48.

The Victrola is introduced by the Victor Talking Machine Co. whose Eldridge Johnson has enclosed the gramophone horn in a cabinet (*see* "His Master's Voice," 1900). Johnson signed up operatic tenor Enrico Caruso in 1901 and makes recordings by having Caruso sing directly into a horn while solo instrumentalists play into other horns (*see* 1922).

Broadway musicals *Forty-Five Minutes from Broadway* 1/1 at the New Amsterdam Theater, with songs by George M. Cohan that include "Mary's a Grand Old Name" and the title song, 90 perfs.; *George Washington, Jr.* 2/12 at the Knickerbocker Theater, with George M. Cohan, songs by Cohan that include "You're a Grand Old Flag," 81 perfs.; *His Honor the Mayor* 5/28 at the New York Theater, with songs that include "Waltz Me Around Again, Willie (Around, Around, Around)," by Ren Shields, lyrics by Will D. Cobb, 30, 104 perfs.; *The Red Mill* 9/24 at the Knickerbocker Theater, with music by Victor Herbert, book and lyrics by Henry Blossom, songs that include "Every Day Is Ladies' Day with Me," 274 perfs.

Popular songs "China Town, My China Town" by Jean Schwartz, lyrics by William Jerome; "School Days" by Will D. Cobb, lyrics by Gus Edwards; "Anchors Aweigh" (march and two-step) by U.S. Naval Academy musical director Charles A. Zimmerman, 45, and midshipman Alfred H. Miles, 23.

The National Collegiate Athletic Association (NCAA) founded January 12 works to make football safer after a 1905 season that saw 18 American boys killed playing the game and 154 seriously injured. President Roosevelt has called representatives of Harvard, Yale, and Princeton to the White House to find ways to stop the growing roughness of football, a game still played with little protective gear, but in 1909, 33 boys will die playing football and 246 will sustain major injuries as the sport grows in popularity (*see* 1939).

The forward pass is legalized in football (*see* 1913).

Canadian prizefighter Tommy Burns wins the world heavyweight championship February 23 by defeating Marvin Hart in a 20-round match at Los Angeles.

H. L. Doherty wins in men's singles at Wimbledon, Mrs. Chambers in women's singles; William J. Clothier, 24, wins in U.S. men's singles, Helen Homans in women's singles.

The first Newport-Bermuda yacht race is held over a 635-mile course to begin an ocean sailing competition that will continue for more than 85 years.

"Tinker-to-Evers-to-Chance" becomes a sportswriter catchphrase to denote efficient teamwork. Chicago Cubs shortstop Joe Tinker, 26, second baseman Johnny Evers, 27, and first baseman Frank Chance, 29, give the club an airtight infield that enables the Cubs to win 116 games, lose only 36, and take the National League pennant.

The Chicago White Sox win the World Series by defeating the Cubs (above) 4 games to 2.

The world's first ski course is laid out at Zürs by Vorarlberg skier Viktor Sohm whose pupils include Hannes Schneider, 16.

The permanent wave introduced by Swiss-born London hairdresser Charles Nestler takes 8 to 12 hours and costs $1,000. Only 18 women avail themselves of Nestler's service, but by the time the permanent wave is introduced in the United States in 1915, it will be much cheaper and will take far less time (*see* Buzzacchini, 1926; Nestle Colorinse, 1929).

Lux Flakes are introduced by Lever Brothers for chiffons and other fine fabrics (*see* Lifebuoy, 1897; Rinso, 1918; Lux Toilet Form, 1925).

New York architect Stanford White, now 52, is shot dead June 25 at the roof garden restaurant atop Madison Square Garden, which he designed in 1889. His murderer is Pittsburgh millionaire Harry K. Thaw whose wife Evelyn Nesbit Thaw was a chorus girl and White's mistress before her marriage. Thaw will win acquittal on an insanity plea.

The murder of Cortland, N.Y., factory girl Grace Brown, 19, makes world headlines in July. A capsized rowboat found in Big Moose Lake in the Adirondacks

1906 ***(cont.)*** the morning of July 13 leads to a search, a boy of 13 spots a young woman's body in 8 feet of water early in the afternoon, police arrest Chester Gillett, 22, at a nearby resort July 14 and charge him with murder. Raised by missionary parents at Kansas City, young Gillett took a job last year as foreman at his uncle's shirtwaist factory in Cortland, where the employees included Grace Brown, daughter of a local farmer. Pregnant by Gillett, Brown wrote him pitiful letters begging him to marry her and finally threatening to go to his family with her story. Gillett has ambitions to marry a society girl he has met, he wrote to Brown asking her to travel with him to the Adirondacks, where he said he would marry her. He signed the register at the Glenmore Hotel "Carl Graham and wife of Albany" upon his arrival by train the morning of July 12 and took Brown out on the lake that afternoon in a hired rowboat. Gillett will be executed next year at Auburn prison (*see* novel *An American Tragedy*, 1925).

The San Francisco earthquake and fire (below) destroy all the city's Nob Hill mansions with the exception of the 20-year-old $1 million house of silver king James C. Flood.

San Francisco's 31-year-old Palace Hotel uses up the water it has stored in vain attempts to save neighboring structures. Fire consumes the hotel but construction begins on a new Palace Hotel and on other new hotels, houses, and office buildings.

The San Francisco earthquake April 18 is the worst ever to hit an American city. The violent tremor of the San Andreas fault jolts the city at 5:13 A.M., it lasts less than a minute, but its strength will later be calculated as registering 8.3 on the Richter scale (*see* 1935). It cracks water and gas mains, and the ensuing 3-day fire razes two-thirds of the city, kills an estimated 2,500, leaves 250,000 homeless, and destroys property estimated at more than $400 million in value.

Mesa Verde National Park is created by act of Congress. The 52,074-acre park in southern Colorado contains prehistoric cliff dwellings.

The Call-Chronicle-Examiner

SAN FRANCISCO, THURSDAY, APRIL 19, 1906

EARTHQUAKE AND FIRE: SAN FRANCISCO IN RUINS

San Francisco shook and burned, thousands died, and property losses exceeded $400 million.

A typhoon hits Hong Kong September 19 killing some 50,000.

Agrarian reforms to end the communal (mir) system of landholding established in Russia in 1861 are introduced by the country's new prime minister Pyotr Arkadevich Stolypin. Peasants are permitted to withdraw from a commune, to receive their shares of the land, and to own and operate the land privately (*see* Lenin, 1922; Stalin, 1928).

The U.S. corn crop exceeds 3 billion bushels, up from 2 billion in 1885 (*see* 1963).

The King Ranch of Texas grows by acquisition to cover nearly a million acres with 75,000 head of cattle and nearly 10,000 horses (*see* 1886; 1940).

Florida's citrus groves are frozen by an ice storm.

California orange growers begin to set out Valencia orange seedlings to raise fruit that will mature in the summer and will make the state a year-round source of oranges, but most California oranges continue to be the Washington navels introduced in 1875.

Cambridge University biochemists Frederick Gowland Hopkins, now 45, and H. G. Willcock find that mice fed "zein" from corn plus the amino acid tryptophan (which Hopkins has isolated with S. W. Cole) will live twice as long as mice fed "zein" alone.

A British School Meals Act passed by the Liberal government under pressure from the emerging Labour party provides free meals for children. Many have been unable to profit from their education because they have come to school without breakfast, and the new measure permits local authorities to add up to a halfpenny per pound to taxes in order to fund a school meals program.

Gerrit Grijns suggests a new explanation for beriberi (*see* 1896). It may be caused not by a toxin in rice hulls, he says, but rather by the absence of some essential nutrient in polished rice (*see* Suzuki et. al., 1912).

A Pure Food and Drug bill introduced by Sen. Weldon B. Heyburn (D. Idaho) encounters Republican opposition. Sen. Nelson W. Aldrich (R. R.I.) asks, "Is there anything in the existing condition that makes it the duty of Congress to put the liberty of the United States in jeopardy? . . . Are we going to take up the question as to what a man shall eat and what a man shall drink, and put him under severe penalties if he is eating or drinking something different from what the chemists of the Agricultural Department think desirable?"

But the Heyburn Bill (above) regulates producers and sellers, not consumers, and its prohibitions are only against selling diseased meat, decomposed foods, or dangerously adulterated food, and it requires only that labels give truthful descriptions of contents (*see* Wiley, 1902; Sinclair, 1905).

The Senate approves the Heyburn Bill (above) February 21 by a 63 to 4 vote following pressure by the 59-year-old American Medical Association on Sen. Aldrich, the House approves it 240 to 17 June 23, and President Roosevelt signs it into law (*see* 1911).

A meat inspection amendment to the Agricultural Appropriation Bill, introduced by Sen. Albert J. Beveridge (R. Ind.), passes without dissent, but the measure does not provide for federal funding of meat inspection.

The Neill-Reynolds report made public by President Roosevelt June 24 inspires Rep. James W. Wadsworth (R. N.Y.) to introduce a meat inspection amendment that provides for federal funding of meat inspection. Settlement-house workers Charles P. Neill and James Bronson Reynolds prepared the report; the Wadsworth measure passes both houses of Congress and is signed into law.

The Great American Fraud by New York journalist Samuel Hopkins Adams, 35, exposes the fraudulent claims and hazardous ingredients of U.S. patent medicines. The book has been serialized in *Collier's*.

Battle Creek Toasted Corn Flake Co. is incorporated February 19 by W. K. (Will Keith) Kellogg, 46, and St. Louis insurance man Charles D. Bolin who puts up $35,000 to start the company whose stock is owned largely by Kellogg's brother J. H. even though the younger Kellogg has his signature printed on each package. W. K. runs the company and devotes two-thirds of his budget to advertising (*see* 1902).

C.W. Post perfects his Post Toasties corn flakes (*see* 1904; 1914).

Corn Products Refining Co. is created by Standard Oil director E. T. Bedford whose "gluten trust" controls 90 percent of U.S. corn refining capacity. The company's Karo syrup will soon be a household name.

Planters Nut and Chocolate Co. is founded at Wilkes-Barre, Pa., by Italian-American entrepreneur Amedeo Obici, 29, who has expanded a peanut stand into a store and restaurant. In partnership with his future brother-in-law Mario Peruzzi, he starts a firm that by 1912 will be using so many raw peanuts that Obici will open his own shelling plant and buy direct from farmers to avoid being squeezed by middlemen (*see* Mr. Peanut, 1916).

E. A. Stuart's Carnation Milk adopts the slogan "The milk from contented cows" (*see* 1899).

The "hot dog" gets its name from a cartoon by Chicago cartoonist Thomas Aloysius "Tad" Dorgan, 29, who shows a dachshund inside a frankfurter bun.

Sales of Jell-O reach nearly $1 million. Francis Woodward, who once offered to sell the brand for $35, now stops making Grain-O to devote all his efforts to Jell-O (*see* 1899; Postum Co., 1925).

The U.S. population reaches 85 million, Britain has 38.9 million, Ireland 4.3, France 39.2, Germany 62, Russia over 120. China's population approaches 400 million.

India's population nears 300 million, up from little more than 200 million in 1850. Mohandas Gandhi in South Africa (above) takes a vow of sexual abstinence (*brahmacharya*) and will oppose other means to limit population in India. (In 1885, at 16, Gandhi had sex with his wife of 14 while his father died in another room of the house.) British physicians in the last century introduced Western medicine to reduce India's death rate, and British moralists eliminated the slaughter of female children and other customs that held the population in check.

1907 A Franco-Japanese treaty signed June 10 guarantees "open door" access of both France and Japan to China and recognizes the special interests of Japan in Fukien and in parts of Manchuria and Mongolia.

Korea's emperor Kojong (I T'ae-wang) abdicates under pressure July 19 at age 55 after 43 years in power. The Japanese who occupy his country set up a figurehead and Korea becomes a Japanese protectorate under terms of a treaty signed July 25 (*see* 1904; 1909).

The Persian shah Muzaffar ud-Din dies at age 54 after a weak reign of 11 years. He is succeeded by his son, 35, who will reign until 1909 as Mohammed Ali. His reactionary prime minister Atabegi-Azamis is assassinated in August and a liberal ministry headed by Nasir ul-Mulk takes over.

An Anglo-Russian entente concluded August 31 resolves differences between Britain and Russia in Persia and elsewhere in Asia. The entente follows by 2 days a Russian note recognizing British preeminence in the Persian Gulf.

The new Persian shah attempts a coup d'état December 15 and imprisons the new liberal prime minister, but popular uprisings throughout the country force him to yield (*see* 1908).

Sweden's Oskar II dies December 8 at age 78 after a 35-year reign in which he also ruled Norway until 1905. His popular son, 49, will reign until 1950 as Gustavus V.

Britain grants dominion status to New Zealand.

Oklahoma is admitted to the Union as the 46th state.

Austria institutes universal direct suffrage.

Norway grants suffrage for women June 14.

Western Federation of Miners leader "Big Bill" Haywood goes on trial at Boise, Idaho, on charges of having helped murder the former governor of Idaho Frank R. Steunenberg in 1905. Now 38 and a miner since age 15, William Dudley Haywood helped found the IWW in 1905. Local lawyer William E. Borah, 42, heads the prosecution (and later wins election to the U.S. Senate), Haywood's lawyer Clarence S. Darrow of Chicago shows that Haywood's accuser has perjured himself, Haywood wins acquittal, and the WFM quits the IWW in protest against the violent direct-action tactics favored by the "Wobblies" (*see* Darrow, 1893, 1910).

A U.S. economic crisis looms as a result of drains on the money supply by the Russo-Japanese War of 1905, the demands imposed by the rebuilding of San Francisco following last year's earthquake and fire, several large railroad expansion programs, and the fact that a late season has tied up farmers' cash. New York Stock Exchange prices suddenly collapse March 13, a July 4th address by Princeton University president Woodrow

J. P. Morgan dominated U.S. finance for years, creating trusts, averting panics, making himself immensely rich.

Wilson urges an attack on the illegal manipulations of financiers rather than on corporations, and President Roosevelt attacks "malefactors of great wealth" in a speech August 20.

Financier J. P. Morgan acts singlehandedly to avert financial panic following a rush by depositors October 23 to withdraw their money from New York's Knickerbocker Trust, a bank run that is followed by others at Providence, Pittsburgh, Butte, and at New York's Trust Co. of America.

J. P. Morgan (above) obtains a pledge of $10 million in John D. Rockefeller money from National City Bank president James Stillman, obtains $10 million in specie from the Bank of England which arrives via the new ship *Lusitania* (below), gets support from First National Bank president George Baker and from financiers Edward H. Harriman and Thomas Fortune Ryan, promises New York mayor George McClellan $30 million at 6 percent interest to keep the city from having to default on some short-term bonds, and locks up leading New York trust company presidents overnight in his new library in East 36th Street until five o'clock in the morning of November 4.

President Roosevelt permits United States Steel to acquire the Tennessee Coal, Iron and Railroad Co. for $35.3 million in U.S. Steel bonds despite questions as to the legality of the acquisition under the 1890 Sherman Anti-Trust Act. The president's action saves the Wall Street brokerage firm of Moore & Schley from collapse, faith is restored, and the stock market recovers.

Bullock's opens March 4 at Los Angeles. Canadian-American merchant John Gillespie Bullock, 36, has backing from his former employer Arthur Letts, 45, an English-American whose dry goods store at Broadway and 4th Street will become The Broadway Store; Bullock will erect a 10-story building on Broadway in 1912 (*see* Federated, 1944).

Neiman-Marcus opens September 10 at Dallas, and its initial stock is practically sold out in 4 weeks. The two-story fashion emporium at the corner of Elm and Murphy Streets has been started by former Atlanta advertising agency president A. L. Neiman, his wife Carrie, and his brother-in-law Herbert Marcus who have been offered the choice of selling out for $25,000 in cash or receiving stock in the Coca-Cola Company plus the Missouri franchise for Coca-Cola syrup. They have opted for the cash, start the store with $30,000 in capital ($22,000 of it from Neiman), invest $12,000 in carpeting and fixtures for the premises they lease at $750 per month, and Carrie Neiman stocks it with $17,000 worth of tailored suits for women, evening gowns, furs, coats, dresses, and millinery for the women of Dallas, a city of 86,000 with 222 saloons (*see* 1913).

Sears, Roebuck distributes more than 3 million copies of its fall catalog (*see* 1904; 1927).

Standard Oil of Indiana is indicted for having received secret rebates on shipments of crude oil over the Chicago and Alton Railroad. Federal district court judge Kenesaw Mountain Landis, 41, imposes a fine of more than $29 million on 1,462 separate counts, but the decision against the subsidiary of John D. Rockefeller's Standard Oil of New Jersey will be reversed on appeal.

Gulf Oil Co. is founded by a reorganization of Guffey Oil (*see* Spindletop, 1901).

Royal Dutch-Shell is created by a merger of Henri Deterding's Royal Dutch Oil with the 10-year-old Shell Transport and Trading Co. of English oilman Marcus Samuel, now 72 (*see* 1892; Anglo-Persian, 1914).

U.S. motorcar production reaches 43,000, up from 25,000 in 1905 (*see* 1908).

Willys-Overland Motors is established by auto dealer John North Willys who buys Overland Automobile of Indianapolis and moves it to the old Pope-Toledo plant at Toledo, Ohio. Willys has opened an auto agency in New York and cannot obtain enough cars to fill his orders (*see* 1903; 1910).

A "valveless" engine perfected by Wisconsin agriculturist Charles Knight has a windowed sleeve between piston and cylinder. He engineers it to cover and uncover inlet and outlet ports, eliminating the valve gear that makes engines noisy with "play" between cams, tappets, and the valves needed to cope with expansion caused by heating (*see* Willys-Knight, 1915).

The Brush motorcar is introduced at $500. A Cadillac is advertised at $800, a Ford Model K at $2,800, but horses sell for $150 to $300.

New York City horsecars give way to motorbuses, but most urban transit is by electric trolley car, "El," and subway (*see* 1904; 1932; Fifth Avenue Transportation Co., 1885).

Taximeter cabs imported from Paris appear in New York in May to give the United States its first successors to hansom cabs.

The Bendix Co. is founded by U.S. inventor Vincent Bendix, 25, who left his Moline, Ill., home at age 16 to study mechanics at New York. The Bendix starter drive that Bendix will perfect in the next 5 years will lead to the development of a self-starter for motorcars (*see* brake, 1912; Kettering, 1911).

Trials by Britain's Royal Automobile Club begin a shift from steam wagons to trucks and tractors powered by internal combustion engines.

The Lancia motorcar has its beginnings at Turin where Italian engineer Vincenzo Lancia and his partner Claudio Fogolin produce a 24-horsepower Alfa which is soon followed by a six-cylinder Delta, a Beta, and a Gamma. They will move into a larger factory and produce the first commercially successful Lancia (the Theta) which will develop 70 horsepower at 2,200 rpm.

The first long-distance motorcar rally begins June 10 as five cars leave Beijing for Paris. The newspaper *Le Matin* has promoted the event, and the winner is a 40-horsepower Italia driven by Prince Scipione Borghese who arrives at Paris August 10 (*see* 1908).

The S.S. *Lusitania* launched by Britain's Cunard Line makes her maiden voyage in September and follows it soon after with a voyage in which she carries specie from the Bank of England to relieve the U.S. financial crisis (above). The 31,550-ton *Lusitania* is 790 feet in length overall with four screws, can carry 2,000 passengers and a crew of 600, and is by far the largest liner afloat (but *see* 1915).

The S.S. *Mauretania* launched by Cunard Line will remain in service until 1935. Slightly smaller than her sister ship *Lusitania* (above), the four-screw *liner* burns 1,000 tons of coal per day and requires a "black squad" of 324 firemen and trimmers to feed her furnaces (*see* 1910).

Britain and Ireland have 23,100 miles of operating railway, Canada 22,400, Austria-Hungary 25,800, France 29,700, Germany 36,000, Russia 44,600, India 29,800, the United States 237,000.

Union Station at Washington, D.C., is completed by Chicago architect Daniel Burnham of Burnham & Root. The new terminal is modeled on the Baths of Diocletian and the Arch of Constantine at Rome.

SKF (*Svenska Kullager Fabriken*, or Swedish Ball Bearing Factory) is founded at Göteborg by Swedish engineer Sven Gustav Wingquist, 30, who has perfected almost frictionless ball bearings using a steel alloy that contains chrome and manganese.

Bethlehem Steel's Saucon Mills open at Bethlehem, Pa., to roll wide-flanged girders and beams that are lighter (and therefore cheaper) than conventional riveted girders but just as strong. Bethlehem has acquired patents from inventor Henry Grey and will license other steel mills to manufacture Grey beams on a royalty basis (*see* 1905; 1912).

The first U.S. company to produce oxygen for oxyacetylene torches is founded by a group of manufacturers who make carbon electrodes used to power electric furnaces that produce alloying metals (*see* Fouce, 1903; Union Carbide, 1911).

The American Cyanamid Co. founded July 22 by U.S. entrepreneur Frank Sherman Washburn, 46, builds a plant on the Canadian side of Niagara Falls to produce calcium cyanamid for nitrogen fertilizer using the European Frank-Caro process for fixation of atmospheric nitrogen. German chemists Adolf Frank, now 73, and Nidodem Caro, now 36, developed the process in the late 1890s, it requires vast amounts of power, Washburn has obtained backing from tobacco magnate James B. Duke, he uses local limestone, and by the end of 1909 Cyanamid will have an annual production capacity of 5,000 tons (*see* Haber process, 1908).

Yale zoologist Ross Granville Harrison, 37, perfects a method of culturing animal tissues in a liquid medium, making it possible to study tissues without the variables and mechanical difficulties of *in vivo* examinations. The method will permit Harrison to make discoveries about embryonic development and facilitate research in embryology.

Radio pioneer Lee De Forest invents electrical high-frequency "radio" surgery (*see* 1906).

Bubonic plague kills 1.3 million in India (*see* 1910).

The University of California opens a Riverside campus (*see* 1873; 1908).

The University of Saskatchewan is founded at Saskatoon.

The University of Hawaii is founded at Honolulu.

Wireless telegraphy service begins October 18 between the United States and Ireland (*see* Marconi, 1901; United Fruit, 1910; Sarnoff, 1912).

Russian physicist Boris Rosing links the Braun tube of 1897 to "electric vision" (*see* Baird, 1926).

The United Press is founded to compete with the Associated Press founded in 1848 and the Hearst International News Service (INS) started last year. The new wire service is created by a merger of Publishers' Press, owned since 1904 by newspaper publisher Edward Wyllis Scripps, now 53, and the Scripps-McRae Press Association in which Scripps has been a partner since 1897. Roy Wilson Howard, 24, heads UP's New York office and will be UP president in 5 years (*see* 1925; *Cleveland Press*, 1878; UPI, 1958).

Rube Goldberg starts work with the *New York Evening Journal* to begin a career as cartoonist. San Francisco graduate engineer Reuben Lucius Goldberg, 24, will have his cartoons syndicated beginning in 1915, he will often show elaborate machines he has devised to perform simple tasks, and his ludicrous inventions will appear for nearly 60 years.

Nonfiction *Pragmatism* by William James who retires from Harvard at age 65.

1907 *(cont.)* Fiction *The Secret Agent* by Joseph Conrad whose "agent" is no individual but rather the anarchy that works unseen within any society; *Mother* by Maksim Gorky; *The Longest Journey* by E. M. Forster; *The Iron Heel* by Jack London.

Juvenile *The Wonderful Adventures of Nils (Nils Holgerssons underbars resa)* by Selma Lagerlof.

Poetry *The Seventh Ring* by Stefan George; "Barrel-Organ" by Alfred Noyes: "Go down to Kew in lilac-time, in lilac-time, in lilac-time;/Go down to Kew in lilac-time (it isn't far from London)"; *Songs of a Sourdough* by English banker-poetaster Robert William Service, 33, whose bank transferred him to the Yukon late in 1904 and who will leave the North for good in 1912. His poems include "The Cremation of Sam McGee" and "The Shooting of Dan McGrew": "A bunch of the boys were whooping it up in the Malamute saloon;/The kid that handles the music-box was hitting a jag-time tune;/Back of the bar, in a solo game, sat Dangerous Dan McGrew,/And watching his luck was his light-o'-love, the lady that's known as Lou."

An exhibition of cubist paintings at Paris begins a radical departure from expressionist and impressionist painting. Georges Braque, 25, Fernand Léger, 26, and others lead the movement to reduce nature to its geometric elements.

Other paintings *Les Demoiselles d'Avignon* by Pablo Picasso explodes and rearranges the human form to pioneer a revolutionary style; *Young Sailor II* and *Luxe, Calme et Volupté* by Henri Matisse; *Blackfriars Bridge* and *The Bathers* by André Derain; *Amor and Psyche* by Claude Monet; *Rooftops of Montmagny* by French painter Maurice Utrillo, 23; *The Snake Charmer* by Henri Rousseau; *Cavalry Charge on the Southern Plains* by Frederic Remington; *Forty-Two Kids* by New York painter George Wesley Bellows, 25, who gains his first success with a picture of urchins tumbling, swimming, diving, and playing round a New York dock and follows it with *Stag at Sharkey's* inspired by Tom Sharkey's boxing arena at 127 Columbus Avenue.

Sculptor Augustus Saint-Gaudens dies at Cornish, N.H., August 3 at age 59.

The North American Indian by U.S. photographer Edward S. Curtis, 39, is published in its first volume. J. P. Morgan has pledged up to $75,000 to support his project and Curtis continues to photograph Indians who include Chief Joseph of the Nez Perce. His work will appear in 300 20-volume sets, printed at a cost of nearly $1.5 million, that will sell for $3,000 to $3,500 each, depending on binding.

Bell & Howell Co. is founded by Chicago movie projectionist Donald H. Bell, 39, and camera repairman Albert S. Howell, 28, with $5,000 in capital. The firm will pioneer in improving equipment for motion picture photography and projection (Bell will sell out to Howell in 1921) (*see* 1909; 1911; Kodak, 1929).

Theater *Don Juan in Hell* (Act III of *Man and Superman)* by George Bernard Shaw 1/4 at London's Royal Court Theatre, with Ben Webster, Mary Barton; *The Truth* by Clyde Fitch 1/7 at New York's Criterion Theater, 34 perfs.; *The Playboy of the Western World* by John M. Synge 1/26 at Dublin's Abbey Theatre (the play produces riots); *The Maidens of the Mount (Die Jungfrau vorn Bischofsberg)* by Gerhart Hauptmann 2/2 at Berlin's Lessingtheater; *The Philanderer* by Bernard Shaw 2/25 at London's Royal Court Theatre; *The Life of Man (Zhizn cheloveka)* by Russian playwright Leonid Nikolayevich Andreyev, 35, 2/22 at St. Petersburg's V.F. Kommissarzhevsky Theater; *A Flea in Her Ear (La puce à l'oreille)* by Georges Feydeau 3/2 at the Théâtre des Nouveautés, Paris; *The Rising of the Moon* by Lady Gregory 3/9 at Dublin's Abbey Theatre; *The Witching Hour* by U.S. playwright Augustus Thomas, 50, 4/18 at Hackett's Theater, New York, 212 perfs.; *The Man of Destiny* by Bernard Shaw 6/4 at London's Royal Court Theatre, with Dion Boucicault as young Napoleon Bonaparte; *A Grand Army Man* by David Belasco and others 10/16 at New York's new Stuyvesant Theater on 44th Street, with David Warfield, William Elliott, Antoinette Perry (the theater will be renamed the Belasco in September 1910).

The Barnum & Bailey Circus created in 1881 becomes the Ringling Bros. Barnum & Bailey Circus. Albert C. Ringling, 55, and his brothers Otto, 49, Alfred T., 45, Charles, 44, and John, 41, have purchased The Greatest Show on Earth and will continue touring it round the country (*see* "Weary Willie," 1921).

Samuel Lionel "Roxy" Rothafel, 25, turns a dance hall in back of his father-in-law's Forest City, Pa., saloon into a motion picture theater with a secondhand screen and projector. The Minnesota-born showman charges 5¢ admission and will soon attract the attention of vaudeville magnate B. F. Keith.

Broadway musicals *The Honeymooners* 6/13 at the Aerial Garden, with George M. Cohan and his songs, 167 perfs.; *The Follies of 1907* 7/18 at the Jardin de Paris on the roof of the New York Theater, with 50 "Anna Held Girls" in an extravaganza staged by Florenz Ziegfeld, now 38, who married comedienne Anna Held 10 years ago. She writes lyrics to Vincent Scotto's song "It's Delightful to Be Married (The Parisian Model)," show girls appear in a swimming pool in a simulated motion picture production, and opulent new editions will appear each year until 1931 (with the exceptions of 1926, 1928, and 1929) featuring chorus lines composed of beautiful *Follies* girls chosen with an eye to slenderness of figure as Ziegfeld creates a new ideal to replace the ample figure now in vogue, 70 perfs.; *The Talk of the Town* 12/3 at the Knickerbocker Theater, with Victor Moore, music and lyrics by George M. Cohan, songs that include "When a Fellow's on the Level with a Girl That's on the Square," 157 perfs.

Opera *The Legend of the Invisible City of Kitesch and of the Maiden Fevrona* 2/7 at St. Petersburg's Maryinsky Theater, with music by Nikolai Rimski-Korsakov; *Ariane et Barbe Bleue* 5/10 at the Opéra-Comique, Paris, with music by Paul Dukas; *The Merry Widow* 6/8 at Daly's Theatre, London, with music by Franz Lehar, 778 perfs. (*see* 1905).

First performances Introduction and Allegro for Harp and String Orchestra, Flute, and Clarinet by Maurice Ravel 2/22 at Paris; Symphony No. 3 in C major by Jean Sibelius 9/25 at Helsinki.

Ballet *The Dying Swan* (*Le Cygne*) 12/22 at St. Petersburg's Maryinsky Theater, with Anna Pavlova dancing to music by Camille Saint-Saëns for *Le Carnival des Animaux* adapted by Michel Fokine.

Popular songs "Marie of Sunny Italy" by Russian-American singing waiter-songwriter Irving Berlin, 19, who has changed his name from Israel Baline; "Dark Eyes" ("Sérénade Espagnole") by Neil Moret; "(The) Glow Worm" by German songwriter Paul Lincke, English lyrics by Lila Cayley Robinson; "On the Road to Mandalay" by composer Oley Speaks, 33, lyrics from the 1890 poem by Rudyard Kipling; "The Caissons Go Rolling Along" by Edmund L. Gruber, who has written it for a reunion in the Philippine Islands of the U.S. Fifth Artillery that helped defeat the Spanish in 1898 (it will not be published until February 1918).

The world's first permanent ski school opens at St. Anton in the Austrian Arlberg region of the Alps under the direction of Hannes Schneider, who will develop one system for teaching the sport to tourists and another for teaching selected local skiers who will spread the Schneider techniques throughout Europe (*see* 1906).

Sir Norman Everard Brookes, 28, (Australia) wins in men's singles at Wimbledon (the first non-Briton to do so), May Sutton in women's singles; Bill Larned wins in U.S. men's singles, Evelyn Sears in women's singles.

U.S. West Coast pitcher Walter Perry Johnson, 19, signs with the Washington Senators of the American League to begin a 21-year career with the Senators. Johnson will win the nickname "Big Train" for his fast ball, pitch 802 games, win 414, and record 3,497 strikeouts in 5,923 innings (his 1913 record of pitching 56 consecutive scoreless innings will stand for 55 years).

The Chicago Cubs win the World Series with Tinker, Evers, and Chance in the infield helping to defeat the Detroit Tigers 4 games to 0 after one game has ended in a tie.

Australian long-distance swimmer Annette Kellerman, 22, is arrested for indecent exposure at Boston's Revere Beach where she has appeared in a skirtless one-piece bathing suit. Even infants will be required to wear complete bathing costumes on U.S. beaches for more than 25 years.

Uncle Sam's 3 Coin Register Bank is introduced by the Durable Toy and Novelty Co.; 6 inches high, 4 wide, 5 deep, and made of heavy-gauge steel, it takes nickels, dimes, and quarters, registering the input with figures displayed in a little window.

France's l'Oréal perfume and beauty product empire is started by chemist Eugène Schueller whose firm will become the world's largest supplier of hair products, employing 1,000 biologists.

Persil, introduced by Henkel & Cie. of Dusseldorf, is the world's first household detergent but is not suitable for heavy-duty laundry use (*see* Tide, 1946).

The Thor washing machine, introduced by Hurley Machine Co. of Chicago, is the first complete, self-contained electric washer.

The Maytag Pastime washer introduced by Parsons Band Cutter and Self-Feeder Co. of Newton, Iowa, is a sideline to the farm equipment produced by the company which Frederick L. Maytag, 50, has headed since 1893. The washer has a corrugated wooden tub with a hand-operated dolly inside, Maytag will add a pulley mechanism in 1909 to permit operation of the machine from an outside power source, he will introduce an electric washer in 1911, but Maytag's entry into the washer field is primarily to solve the problem of seasonal slumps in the farm implement business (*see* 1922).

Armstrong Linoleum is introduced by the Armstrong Cork Co. founded in 1860.

The Hoover Vacuum Cleaner has its beginnings in an electric vacuum cleaner invented by J. Murray Spangler who has improved on the cleaner patented by H. C. Booth in 1901. U.S. industrialist W. H. Hoover, now 58, will manufacture Spangler's machine; his Hoover Suction Sweeper Co. will become the Hoover Co. in 1922.

Alfred Dunhill, Ltd., opens in London's Duke Street near Piccadilly Circus. Former harness maker Dunhill has founded a tobacco shop that will become world famous.

Prince Albert Tobacco is introduced by R. J. Reynolds. It will soon be the leading U.S. pipe tobacco (*see* Camels, 1913).

The Washington Cathedral that will be completed in 1976 goes up in the nation's capital. London architect George F. Bodley and his Bostonian pupil Henry Vaughan have designed the great Gothic structure, architect Philip Hubert Frohman will take over in 10 years and lengthen the nave from 246 feet to 534 (*see* 1964).

The New York Customs House on the south side of Bowling Green is completed by Cass Gilbert.

Berlin's Hotel Adlon opens on Pariser Platz.

Rome's Excelsior Hotel is opened by Swiss hotelman Baron Pfyffer d'Altishofen.

New York's Plaza Hotel opens October 1 on the Grand Army Plaza just south of Central Park on a site formerly occupied by the Alfred Gwynne Vanderbilt mansion. Henry Janeway Hardenbergh has designed the $12 million 1,000-room hotel in the French Renaissance Beaux-Arts style he used for the Waldorf of 1893, Astor House of 1897, Willard of 1901, and Astor of 1904.

San Francisco's Fairmont Hotel has opened April 17 on the Nob Hill site formerly occupied by the mansion of the late James G. "Bonanza Jim" Fair. Nearly completed before last year's earthquake and fire, the Fairmont will be augmented in 1961 by a 29-story tower containing 252 additional rooms.

1907 ***(cont.)*** A new St. Francis Hotel goes up on San Francisco's Union Square to replace the hotel opened in 1904 but destroyed by last year's fire.

San Francisco's mayor urges damming of the Tuolumne River in Yosemite Park's Hetch Hetchy valley to give the city a reliable source of water in the wake of last year's disastrous fire. John Muir rallies opposition to the proposal, Gifford Pinchot supports it and seizes upon the word "conservation" to describe his philosophy (*see* 1905, 1913).

U.S. senators from Idaho, Montana, Oregon, Washington, and Wyoming force President Roosevelt to sign an appropriations bill containing a rider that repeals the Forest Reserves Act of 1891. The president signs a proclamation 10 days later creating 16 million acres of new forests in the five states, forests that cannot be cut for timber, and before the Roosevelt administration expires early in 1909 it will have set aside 132 million acres in forest reserves.

The first canned tunafish is packed at San Pedro, Calif., by A. P. Halfhill (*see* Van Camp, 1914).

Pellagra is observed for the first time in Mississippi, where cornmeal is a dietary staple (*see* 1749; 1913).

Saccharin and other benzoic acid derivatives should be banned from use in food, says Department of Agriculture chemist Harvey W. Wiley. Rep. James S. Sherman (R. N.Y.) says saccharin saves his canning firm thousands of dollars per year, Wiley interrupts him at a White House conference to say, "Yes, Mr. President, and everyone who eats these products is deceived, believing he is eating sugar, and moreover his health is threatened by this drug!" President Roosevelt replies angrily, "Anybody who says saccharin is injurious is an idiot! Dr. Rixey gives it to me every day." Wiley has developed important refining techniques for the sugar industry, Ira Remsen heads the scientific board named by the president to review the data on saccharin, and the board's final report in 1910 will conclude that a continuing daily consumption of 300 milligrams of saccharin presents no hazard (*see* 1879; sodium cyclamate, 1937).

The American Sugar Refining trust is found to have defrauded the government out of import duties (*see* 1899). Several company officials are convicted, and more than $4 million is recovered (*see* Domino Sugar, 1911).

The Coca-Cola Company buys out its Atlanta advertising agency (*see* 1899; Neiman-Marcus, above; Britain, 1909; distinctive bottle, 1916).

Pepsi-Cola sales increase to 104,000 gallons, up from 7,968 gallons in 1903, as Caleb Bradham establishes a network of 40 bottling plants (*see* 1920).

Canada Dry Pale Dry Ginger Ale is registered as a trademark by John J. McLaughlin who has eliminated the dark brown color of his Belfast Style Ginger Ale and has also eliminated the sharpness found in other ginger ales (*see* 1890). McLaughlin calls his product "the Champagne of Ginger Ales" and demand for the beverage will soon oblige him to bottle it at Montreal and Edmonton (*see* 1922).

An immigration act passed by Congress February 20 excludes undesirables, raises the head tax on arrivals to $4, and creates a commission to investigate.

An executive order by President Roosevelt in mid-March excludes all Japanese laborers coming from Canada, Hawaii, or Mexico from entry to the continental United States. The Japanese diplomat Tadasu Hayashi, 57, has sent Washington a note in mid-February (*see* "gentlemen's agreement," 1908).

Nearly 1.29 million immigrants enter the United States, a new record that will not be surpassed.

The British Empire, occupying 20 percent of the world's land surface, has a population of 400 million.

1908 Portugal's licentious Carlos I is assassinated February 1 along with the crown prince on a Lisbon street after a reign of nearly 19 years. Dead at 44, the king is succeeded by his younger son, 18, who will reign until 1910 as Manoel II. He ends the dictatorship of his father's prime minister João Franco that began in 1906.

Egypt's Mustapha Kamel dies February 10 at age 34 and his death deals the reviving nationalist movement a heavy blow. The khedive moves to counteract nationalism November 10 by appointing as premier the Christian Copt Butros Ghali; Egyptian Muslims stage violent demonstrations.

Britain's prime minister Sir Henry Campbell-Bannerman, recuperating from a heart attack suffered last November, resigns April 5 and dies April 22. He is succeeded by Herbert Henry Asquith, 55, earl of Oxford and Asquith, whose Liberal ministry will continue until 1916.

Persia's shah Mohammed Ali succeeds in a coup d'état June 23 with help from the Cossack Brigade and secret support from the Russian legation (*see* 1907). He shuts down the national assembly and imposes martial law at Teheran; many liberal leaders are killed (*see* energy, below).

The Battle of Marrakesh August 23 ends in defeat for the Moroccan sultan Abd-al-Aziz IV who has met the ransom demands of the brigand Raisuli to avert foreign wars, but whose actions have provoked his older brother Mulay Hafid to revolt. Insurgents proclaimed Mulay Hafid sultan at Marrakesh in May of last year, Fez has proclaimed him sultan in January, Berlin announces September 3 that Germany will recognize the new sultan, the French seize German deserters from the French Foreign Legion, Franco-German relations grow tense, but Mulay Hafid agrees to respect French and Spanish interests as accepted by his brother after the Algeciras Conference in 1906, and he will reign until 1912 as Abd-al-Hafiz.

Bulgaria declares independence October 5 as Prince Ferdinand assumes the title "czar" (he will reign until 1918). Austria annexes Bosnia and Herzogovina October 6, producing consternation in Russia, Serbia, Montenegro, and Turkey. Berlin supports Vienna, London and Paris support objections by Russia, Turkey, and the Balkan nations, Crete aggravates the crisis October 7 by proclaiming union with Greece, Turkey boycotts Austrian goods (*see* 1909).

Belgium's Parliament annexes the Congo State of the aged Leopold II October 18 and begins to remove the abuses that have existed since 1885.

President Roosevelt adheres to the tradition against a third term. Republicans nominate Roosevelt's secretary of war William Howard Taft, 50, of Cincinnati who has served as governor of the Philippines, quelled a potential rebellion in Cuba, and organized construction of the Panama Canal. He easily defeats his Democratic rival William Jennings Bryan (who runs for the third and last time), winning 321 electoral votes to 162 for the "Great Commoner."

President Roosevelt visits Panama November 15, the first sitting president to travel abroad.

China's dowager empress Cixi dies of dysentery November 15 at age 73 after 52 years as the power behind the Qing throne.

The Supreme Court upholds railroad official William Adair who has fired an employee for belonging to a union. A law that prohibits discrimination against union labor in interstate commerce violates the Fifth Amendment, the court rules January 27 in *Adair v. United States.*

The Danbury Hatters' case *(Loewe v. Lawlor)* brings a Supreme Court ruling February 3 that the Sherman Act of 1890 applies to combinations of labor as well as to management combinations. The court rules that a nationwide secondary boycott against D. E. Loewe & Co. by the United Hatters of North America in support of a striking Danbury, Conn., local is a conspiracy in restraint of trade.

The Supreme Court sustains Oregon's 10-hour day law for women in industry February 4 in *Muller v. State of Oregon.*

The Federal Bureau of Investigation (FBI) established as a division of the Department of Justice will be used in many cases against labor organizers (*see* J. Edgar Hoover, 1919).

Parliament enacts legislation granting noncontributory old-age pensions of 5d per week to needy people over age 70 (*see* Germany, 1891; British taxes, 1909; U.S. Social Security Act, 1935).

Emmeline Pankhurst and her daughter Christabel draw prison sentences October 24 after a sensational trial in which two cabinet members have testified for the defense.

The Supreme Court hands down prison sentences December 3 to AF of L officers Samuel Gompers, John Mitchell, and Frank Morrison for violating an injunction against a boycott of Buck's Stove and Range Co.

New York journalist Willing English Walling reports a race riot at Springfield, Mass., for the liberal weekly *The Independent*. He describes the plight of U.S. blacks, asking "What large and powerful body of citizens is ready to come to their aid?" Walling gets a response from New York social worker Mary White Ovington (*see* NAACP, 1909).

U.S. banks close as the economic depression continues. Westinghouse Electric Co. goes bankrupt.

Thomas J. Watson makes an easel presentation to National Cash Register salesmen and writes the word "THINK" at the head of every sheet of paper (*see* 1903). NCR president J. H. Patterson sees the presentation and orders that "THINK" signs be made up for every NCR office (*see* 1910).

Namibian railway worker Zacharias Lewala finds a small diamond in the desert and turns it over to his supervisor, August Stauch, who obtains a prospecting license from the German colonial government. Consolidated Diamond Mines of South-West Africa, a De Beers subsidiary, will monopolize the new industry (*see* De Beers, 1880).

J. C. Penney buys out T. M. Callahan's interest in two western stores and begins a chain that will have 22 stores by 1911 with headquarters at Salt Lake City (*see* 1902). By the time Penney moves to New York in 1913 he will have 48 stores; by 1916 he will have 127 (*see* 1927).

Petroleum production begins in the Middle East May 26 as drillers employed by William Knox D'Arcy strike oil at Masjid-i-Salaman (Mosque of Solomon) and begin tapping what will prove in the 1930s to be the world's largest reservoir of oil. British cabinet member Winston Churchill, now 33, persuades London to buy up D'Arcy's 1901 concession from the shah of Persia (above) and establish Anglo-Persian Company, which begins commercial exploitation (*see* Gulbenkian, 1914; Anglo-Iranian, 1935; British Petroleum, 1954).

Hughes Tool Co. is founded by Houston entrepreneur Howard Robard Hughes whose steel-toothed rock-drilling bits will enjoy a monopoly in the petroleum industry. Hughes runs the first successful rock bit in an oil well at Goose Creek, Tex., and revolutionizes oil-drilling technology that has been based until now on a pulverizing technique. The device he patents is the first to utilize rolling cone cutters, and by the time his son and namesake inherits the business in 1924 Hughes Tool will be making profits of $1 million per year.

Continental Oil has its beginnings in Oklahoma where Pittsburgh lawyer-oilman Ernest Whitworth Marland, 34, strikes oil on Ponca Indian lands. He used geological surveys to open West Virginia's Congo oil field in 1904, sank 54 wells without a dry hole and made a fortune, lost his money last year in the depression, has come to Oklahoma, noticed a geologic outcropping in a Ponca cemetery, and been helped by rancher George Miller to obtain drilling rights in the cemetery and on surrounding leased lands. Marland will build a refinery at Ponca City, found Marland Oil Co. in 1920, and increase his holdings to control an estimated 10 percent of world oil (*see* 1929).

Henry Ford's Model T flivver came in any color you wanted, so long as it was black. Almost anybody could afford one.

The Model T Ford introduced August 12 will soon outsell all other motorcars. Ford's $850.50 "flivver" has a wooden body on a steel frame that makes it "stronger than a horse and easier to maintain." It comes only in black (*see* 1903; 1911).

U.S. auto production reaches 63,500 with at least 24 companies producing motorcars.

Harvey S. Firestone sells Ford 2,000 sets of tires for the Model T (above) and begins a lasting relationship (*see* 1900). Firestone will acquire a 2,000-acre Liberian rubber plantation in 1924 and 2 years later will lease a million acres for 99 years.

AC Spark Plug Co. is founded by Buick Motor Car president W. C. Durant with French-American motorcycle specialist Albert Champion.

Champion Spark Plug Co. has its beginnings in a Boston garage where Frank D. Stranahan, 26, and his brother Robert Allen, 22, just out of Harvard, begin manufacturing spark plugs, magnetos, coils, and other electrical equipment for motorcars. They will move to Toledo in 1910 and will go into business with Albert Champion (above) under the name Champion. Robert Stranahan will secure an order from Henry Ford, and Champion will be Ford's sole supplier of spark plugs until 1961.

W. C. Durant's Buick Motor Car Co. introduces a White Streak model and builds 8,820 motorcars.

General Motors is created September 16 by W. C. Durant (above), who brings other auto makers together into a holding company (*see* 1906). His bankers tell him that Henry Ford's company (above) is not worth the $8 million in cash that Ford demands, so Ford does not join (*see* Cadillac, Oakland, 1909; Storrow, 1910).

AC Spark Plug Co. (above) will become a General Motors subsidiary. Together with Champion (above) it will dominate the business.

Automotive pioneer Charles Stewart Mott, 33, sells a 49 percent interest in his Flint, Mich., auto factory to General Motors (above) and will sell the remainder in 1913. A scion of the Mott cider and vinegar family, Mott went to work at age 24 for an uncle at Utica, N.Y., who made bicycle wheels and wheels and axles for motorcars, he has taken over the enterprise and moved it to Flint, and he will be a GM director from 1913 until his death in 1975.

A high-wheeled International Harvester auto buggy is introduced at $750 while other simply constructed cars sell at prices between $750 and $900—more than most Americans earn in a year.

The Hupmobile is introduced by Detroit's Hupp Motor Car Co. at the Detroit Automobile Show.

Less than 2 percent of all U.S. farm families own motorcars, but there are 200,000 cars on the road, up from 8,000 in 1900 (*see* 1923).

A New York-to-Paris motorcar race sponsored by the *New York Times* and *Le Matin* begins February 12 at Times Square, 250,000 spectators line 8 miles of snowy New York roads to watch the contestants set out for Paris via Detroit, Cleveland, Chicago, Salt Lake City, San Francisco, Tokyo, Vladivostok, Irkutsk, Novosorirs, Moscow, Berlin, Bonn, and Brussels, a distance of 13,341 miles. U.S. contestant George Schuster, 35, driving a Thomas Flyer made by the 9-year-old E. R. Thomas Motor Co. of Buffalo, N.Y., thwarts a scheme by a French contestant to buy up all available gasoline at Vladivostok; he wins the race, arriving at Paris in 169 days.

The gyrocompass invented by German engineer Hermann Anschutz-Kampfe, 36, indicates true north by sensing the rotation of the earth and then pointing the rotor axis toward the North Pole even when a ship is rolling in high seas.

French aviation pioneer Henri Farman, 34, flies a Voison biplane from Bony to Reims in the first city-to-city flight. Farman will develop the Farman biplane with his brother Maurice, 31, and will build an aircraft factory at Boulogne Billancourt (*see* 1916).

Wilbur Wright completes a flying machine for the War Department, it crashes September 17, killing Lieut. Thomas A. Selfridge, who has flown as a passenger on the test flight, but Wright will repair the plane, it will pass U.S. Army tests in June of next year, and the Wright brothers will obtain the first government contract by producing a plane that can carry two men, fly for 60 minutes, and reach a speed of 40 miles per hour.

Wilbur Wright (above) wins the Michelin Cup in France by completing a 77-mile flight December 21 in 2 hours, 20 minutes (*see* 1905; 1909).

Arab engineers complete the Hejaz Railway from Damascus to Mecca and Medina (833 miles) with a single narrow-gauge track after 8 years of construction (*see* T. E. Lawrence, 1917).

New York's 4-year-old IRT Broadway subway line is extended to Kingsbridge in the Bronx.

Two subway tunnels open to traffic at New York. The McAdoo Tunnel goes under the Hudson River to Hoboken (*see* 1904), another tunnel connects Bowling Green in Lower Manhattan with Brooklyn's Joralemon Street across the East River.

The King Edward VII Bridge opens to span the River Tyne at Newcastle.

The Haber process for synthesizing ammonia invented by German chemist Fritz Haber, 40, and his colleague W. H. Nernst will free the world from its dependence on Chilean nitrates for making explosives and nitrogen fertilizers. Using far less energy and at much lower cost than the Frank-Caro process used earlier (*see* American Cyanamid, 1907), the Haber process combines nitrogen and hydrogen directly, using as a catalyst iron (plus some aluminum, potassium, and calcium) and employing high temperatures. Since ammonia is one part nitrogen to three parts hydrogen; it can easily be reduced to nitric acid for munitions or to sulfate of ammonia or sodium nitrate for fertilizers.

German industrial chemist Karl Bosch, 34, will adapt the Haber process (above) and Badische-Anilin-und-Soda Fabrik will employ it to produce sulfate of ammonia and sodium nitrate but mostly to make nitric acid (*see* war, 1914).

Swiss-born French chemist Jacques Edwin Brandenberger, 35, patents cellophane, a transparent wrapping material (*see* 1912).

The Geiger counter developed by German physicist Hans Geiger, 26, and New Zealand-born British physicist Ernest Rutherford, 37, at Manchester University detects radioactive radiations. A high-voltage wire runs down the center of a cylinder in a near vacuum, alpha particles passing through the gas in the cylinder cause it to ionize into charged particles, and a pulse of electrical current for each alpha particle can be observed on a dial. Geiger will improve the counter in the 1920s with help from W. Muller to distinguish between alpha particles and beta and gamma rays by reduced voltage and to produce clicks through a loudspeaker (*see* Rutherford, 1911).

The first international meeting of psychiatrists opens at Salzburg, Austria, with participants who include Sigmund Freud and Carl Jung (*see* 1906). Also present are Eugen Bleuler, 51, Alfred Adler, 38, Abraham Arden Brill, 34, and Ernest Jones, 30. Psychoanalysis will begin to gain repute and support in reputable international medical circles next year when Freud and Jung are invited to speak at Clark University in Worcester, Mass., but resistance to Freudian and Jungian techniques and philosophies will remain strong.

Buerger's disease gets its name as Viennese-American physician-surgeon Leo Buerger, 29, describes a vascular disease (thromboangiitis obliterans) that affects chiefly the peripheral arteries and sometimes the veins, especially in young men, and will be confused with atherosclerosis.

Ex-Lax Co. is founded by Hungarian-American pharmacist Max Kiss, 25, who came to New York penniless 10 years ago, learned English, and heard from a physician aboard ship while traveling home for a visit to his family about the newly developed Bayer laxative phenopthalein. Kiss has developed a chocolate-flavored phenopthalein formula that he will promote in movie theaters with a film he will make using neighborhood children as actors.

The University of California opens a Davis campus (*see* 1907; 1912).

The University of Alberta is founded at Edmonton.

The University of British Columbia is founded at Vancouver.

The University of the Philippines is founded at Manila.

The first professional school of journalism opens at the University of Missouri.

The Christian Science Monitor begins publication November 25 at Boston (*see* Eddy, 1879).

"Mutt and Jeff" in William Randolph Hearst's *San Francisco Examiner* March 29 is the first comic strip to appear daily with the same cartoon figures. Cartoonist Harry Conway "Bud" Fisher, 23, will continue the strip until his death in 1954.

Nonfiction *Autobiography of a Super-Tramp* by English vagabond-poet William Henry Davies, 37, with an introduction by George Bernard Shaw.

Fiction *Penguin Island* (*L'ile des pingouins*) by Anatole France, who satirizes short story writer Guy de Maupassant; *La Rétraite Sentimentals* by Colette who last year divorced Henry Gauthier-Villars; *The Outcast* (*L'esclusa*) by Luigi Pirandello; *Old Wives' Tale* by Arnold Bennett who has quit as assistant editor of *Woman*; *A Room with a View* by E. M. Forster; *The Man Who Was Thursday* by English novelist-poet-essayist G. K. (Gilbert Keith) Chesterton, 34; *An Ocean Tramp* by English marine engineer-novelist William McFee, 27, who will continue writing at sea until 1922; *Spring (Haru)* by Shimazaki Toson; *Ayres Memorial (Memorial de Aires)* by Joaquim Machado De Assis; *Holy Orders* by Marie Corelli.

Juvenile *The Wind in the Willows* by Scottish writer Kenneth Grahame, 49, whose stories will be dramatized by A. A. Milne; *Anne of Green Gables* by Canadian novelist Lucy Maud Montgomery, 35, who begins a series of Anne books about the Canadian wilderness.

Poetry *New Poems (Neue Gedichte)* by Rainer Maria Rilke.

Painting Cubism appears at the Paris Salon in an exhibition of paintings in "little cubes" by Pablo Picasso and Georges Braque.

Other paintings *Murneau* by Russian-born painter Wassily Kandinsky, 41, *The Doge's Palace, The Contarini Palace, The Palazzo Dario, San Giorgio Maggiore,* and *The Grand Canal, Venice* by Claude Monet; *Nude Against the Light* and *After Dinner* by Pierre Bonnard; *The Kiss* by Gustave Klimt; *Nu Rouge* by Russian painter Marc Chagall, 21; *Père Juniet's Cart* and *Zeppelins* by Henri Rousseau; *Harmony in Red* by Henri Matisse who gains international fame by exhibiting his work at New York's Steiglitz-Photo-Secession 291 Gallery; *North River* by George Bellows.

The Ash Can school of U.S. realist art has its beginnings in an exhibition of work by The Eight—a group that

1908 ***(cont.)*** includes cartoonist George Luks who has worked on "The Yellow Kid," William Glackens, John French Sloan, 37, and Everett Shinn, 32. Influenced by Thomas Eakins, they have revolted against academicism with encouragement from painter Robert Henri, 43.

Sculpture *The Kiss* by Romanian-French sculptor Constantin Brancusi, 32, who has evolved a "primeval" style of stone and wood carving.

Theater *The Ghost Sonata (Spoksonaten)* by August Strindberg 1/21 at Stockholm's Intimate Theater; *Getting Married* by George Bernard Shaw 5/12 at London's Haymarket Theatre; *The Tragedy of Nan* by John Masefield 5/24 at London's New Royalty Theatre; *The Man from Home* by Booth Tarkington and Harry Leon Wilson 8/17 at New York's Astor Theater, with William Hodge, 496 perfs.; *What Every Woman Knows* by James M. Barrie 9/3 at the Duke of York's Theatre, London; *The Blue Bird (L'Oiseau bleu)* by Maurice Maeterlinck 9/30 at the Moscow Art Theater; *Days of Our Life (Dui nashey zhizni)* by Leonid Andreyev 11/6 at St. Petersburg's Novy Theater.

Stage musicals *The Three Twins* 6/15 at New York's Herald Square Theater, with music by Karl Hoschna, book and lyrics by U.S. playwright Otto Abels Harbach, 35, songs that include "Cuddle Up a Little Closer," 288 perfs.; *The (Ziegfeld) Follies* 6/15 at New York's Jardin de Paris, with comedienne Nora Bayes (Dora Goldberg Norworth), 28, book and lyrics mostly by Harry B. Smith, music by Maurice Levi and others, songs that include "Shine On, Harvest Moon" by Nora Bayes, lyrics by her husband Jack, 120 perfs.; *The Chocolate Soldier (Der tapfere soldat)* 11/14 at Vienna's Theater-an-der-Wien, with music by Oscar Strauss, libretto from the 1894 Bernard Shaw play *Arms and the Man*.

First performances *Brigg Fair* by Frederick Delius 1/18 at London; Symphony No. 2 in E minor by Sergei Rachmaninoff 1/26 at Moscow; *Romanian Rhapsody* No. 1 by Romanian composer Georges Enesco, 27, 2/7 at Paris; *Rapsodie Espagnole by* Maurice Ravel 3/15 at Paris; Symphony No. 7 in D major by Gustav Mahler 9/19 at Prague; *Poem of Ecstasy* by Aleksandr Skriabin late in the year at St. Petersburg; Symphony No. 1 in A flat major by Sir Edward Elgar 12/3 at Manchester; *In a Summer Garden* by Frederick Delius 12/11 at London.

The Brooklyn Academy of Music opens November 4 on Lafayette Avenue between Ashland Place and St. Felix Street, replacing a structure that opened in 1861.

New York's Metropolitan Opera House hires Giulio Gatti-Casazza, 40, from Milan's Teatro alla Scala. He will mastermind the Met until 1935, raising it to new heights of artistic and financial success.

Popular songs "Take Me Out to the Ball Game" by Albert von Tilzer, lyrics by Jack Norworth (above) that popularize the confection Cracker Jack introduced in 1896; "Sunbonnet Sue" by Gus Edwards, lyrics by Will D. Cobb; "It's a Long Way to Tipperary" by English songwriters Harry Williams and Jack Judge.

Columbia Phonograph Co. introduces the first two-sided disks.

Arthur Gore wins in men's singles at Wimbledon, Penelope Dorothea Harvey Boothby, 27, in women's singles; Bill Larned wins in U.S. men's singles, Maud Bargar Wallach, 36, in women's singles.

The Olympic Games at London attract 2,666 contestants from 22 countries.

The Chicago Cubs win the World Series by defeating the Detroit Tigers 4 games to 1 with help from Tinker and Evers and Chance.

Texas prizefighter Jack Johnson, 30, wins the world heavyweight title December 25 by knocking out Tommy Burns in the 14th round of a championship bout at Sydney, Australia. Johnson is the first black titleholder (*see* 1910).

The Boy Scouts of Britain is founded under the leadership of Boer War hero Robert Stephenson Smyth, Baron Baden-Powell of Gilwell, who held Mafeking through a 215-day siege in 1900 (*see* Boy Scouts of America, 1910; Girl Guides, 1910).

Mother's Day is observed for the first time at Philadelphia. Suffragist-temperance worker Anna May Jarvis, 44, attended a memorial service last year at the Methodist Church in Grafton, W. Va., for her mother Anna Reeves Jarvis who died at Philadelphia May 9, 1905, conceived the idea of an annual worldwide tribute to mothers, and will agitate for a national U.S. Mother's Day (*see* 1913).

Palm Beach brand cloth has its origin in a lightweight fabric of long-staple cotton and mohair from Angora goats created by U.S. chemist William S. Nutter who sells the patent to Goodall Worsted Co. of Sanford, Me. The fabric will revolutionize summer menswear beginning in the 1930s.

A Pocantico Hills, N.Y., mansion of fieldstone is completed in Georgian style for Standard Oil magnate John D. Rockefeller, Jr., on the vast 4,180-acre estate acquired by his father in 1893 (*see* Kykuit, 1903).

Minnesota's National Farmers Bank building is completed by Louis Sullivan at Owatonna.

Boston's neighboring city of Chelsea has a fire April 12 (Palm Sunday) that destroys one-third of the town before burning itself out in late afternoon.

Muir Woods National Monument is created by presidential proclamation. The 500-acre forest of California redwoods and sequoias is on land donated to the federal government by William Kent, U.S. congressman from Marin County (*see* Hetch Hetchy, 1913).

Grand Canyon National Monument is created by President Roosevelt who acts under provisions of the 1906 Antiquities Act to protect Arizona's spectacular canyon from private land speculators (*see* 1919).

President Roosevelt calls a White House Conference on Conservation to publicize the cause.

Less than 60 heath hens remain on Martha's Vineyard. They constitute the world's last colony of the birds once plentiful in New England (*see* 1839; 1929).

A mysterious fireball explodes the morning of June 30 over Tunguska in Siberia, it creates shock waves felt miles away, its thermal currents set great tracts of tundra woodlands afire, and the mushroom cloud and "black rain" that follow it inflict a scabby disease on reindeer herds. Irkutsk, Batavia, Moscow, St. Petersburg, Jena, and Washington, D. C., record seismic shocks. Russian scientists will not visit the sparsely populated area until 1927, they will find no meteorite fragments, and some will later speculate that the fireball was a crippled alien space vehicle powered by atomic energy.

An earthquake rocks Sicily December 28, killing some 75,000 in and about Messina in the worst quake ever recorded in Europe.

Nearly 90 percent of the horsepower used on English and Welsh farms comes from horses (*see* 1939).

J. I. Case Co. turns to selling gasoline tractors after 66 years of producing farm equipment that has made it the leading U.S. maker of farm steam engines (*see* Froelich, 1892; Edison, 1895).

Half of all Americans live on farms or in towns of less than 2,500, and the country has 6 million farms.

Tokyo University chemist Kikunae Ikeda isolates from seaweed the flavor enhancer monosodium glutamate (MSG) that gives a meaty flavor to vegetable diets. He calls the white salt "aginimoto," meaning "the essence of taste." The Ajinimoto company will become the world's leading MSG producer, and the name *Ajinimoto* will become a generic for MSG in Japan (*see* 1934).

A gentlemen's agreement concluded February 18 binds Japan to issue no further passports to workers for immigration directly to the United States; the Japanese acquiesce to last year's order by President Roosevelt barring Japanese immigration (*see* 1924).

1909 Constantinople recognizes Austrian annexation of Bosnia and Herzogovina January 12, Vienna pays the Turks a £2.2 million indemnity, the Russians cancel a £20-million Turkish indemnity in return for Constantinople's recognition of Bulgarian independence, and internal strife disrupts the Ottoman Empire.

The Ottoman grand vizir Kiamil Pasha, 76, is deposed February 13 and replaced by Hussein Hilmi Pasha, 50. The First Army Corps composed chiefly of Albanians revolts at Constantinople April 13 and forces Hilmi Pasha to resign, a 25,000-man army of liberation arrives from Macedonia April 24, a 5-hour battle ensues, and leaders of the April 13 revolt are executed.

The Ottoman sultan Abdul Hamid II is deposed April 26 at age 66 after a 33-year reign by unanimous vote of the Turkish parliament. His helpless brother of 64 will reign until 1918 as Mohammed V.

Bulgarian independence gains German, Austrian, and Italian recognition April 27.

Persia's shah Mohammed Ali Shah is deposed July 16 by the Bakhtaiari tribal chief Ali Kuh Khan, who took Teheran 4 days earlier. A Russian force has invaded northern Persia, raised the siege of Tabriz March 26, and occupied the city for the shah with savage brutality, arousing the Bakhtaiari. They replace the shah with his son Ahmad, 12, who will reign until 1925 as a puppet of radical elements (*see* 1921).

Britain, France, Russia, and Italy withdraw their forces from Crete in July and the island becomes part of Greece.

Leopold II of the Belgians dies December 17 at age 74 after a reign of nearly 41 years in which he has exploited the Congo and amassed great wealth at the expense of the Africans. His nephew, 34, will reign until 1934 as Albert I.

U.S. Marines oust Nicaragua's president José Zeiaya December 16. He is succeeded December 21 by Dr. José Madriz.

Japanese forces begin a 36-year occupation of Korea which is in a state of insurrection following the abdication of her emperor in 1907. Prince Hirobumi Itō, now 73, resigns in June after confessing failure to reform Korea's administration; a Korean nationalist assassinates him at Harbin October 26 (*see* 1910).

The Valor of Ignorance by U.S. military expert Homer Lea, 32, makes predictions about future world history. Japan will be an aggressive power, says Lea, and will make war on the United States beginning with an attack on Hawaii (*see* 1941).

Nearly two decades of Hawaiian plantation disturbances begin with a strike by exploited Japanese workers. It is the first major Hawaiian strike (*see* 1882).

Labour in Portuguese West Africa by William Cadbury draws attention to conditions of slavery in São Tomé and Príncipe (*see* 1905). Cadbury has visited both places and persuaded two other Quaker cocoa and chocolate firms (Fry and Rowntree) to join in a boycott of cocoa from the Portuguese African islands, but while working conditions in São Tomé improve, the system of cocoa slavery remains.

The NAACP (National Association for the Advancement of Colored People) is organized at New York following a January meeting in the apartment of W. E. Walling with social worker Mary W. Ovington and immigrant leader Henry Moscowitz who begin "a revival of the Abolitionist spirit." The new organization is headed by W. E. B. DuBois and is supported by social worker Jane Addams of Chicago, educator John Dewey, journalist Lincoln Steffens, Rabbi Stephen Wise, 35, and 49 others, including six blacks (*see* 1908; 1910).

Muslim fanatics murder 30,000 Armenians in April.

A 3-month strike of some 20,000 U.S. garment workers begins November 22. The workers belong to the Ladies' Waist Makers' Union Local 25 of the 9-year-old International Ladies' Garment Workers Union (ILGWU) and will win most of their demands (but *see* Triangle fire, 1911).

1909 ***(cont.)*** "Reached North Pole April 21, 1909. . .," says a wire received September 1 by the International Bureau for Polar Research at Copenhagen from Brooklyn, N.Y., surgeon Frederick A. Cook, 44, whose steamer *Hans Egede* has put in at Lerwick in the Shetland Islands. "I have the Pole April 9, 1909. . .," says a wire received September 6 by the *New York Times* from U.S. Navy engineer Robert E. Peary, 52. Neither will ever prove that he reached latitude 90° North.

Britain institutes new tax measures to finance her social security programs (*see* 1908). Chancellor of the Exchequer David Lloyd George's budget imposes a supertax at the rate of 6d per pound on incomes exceeding £5,000 per year and levies steep estate taxes (*see* 1915).

A gold strike by prospector Benny Hollinger will lead to Canada's becoming the world's third largest producer after South Africa and Russia. Holtinger has pursued reports of "white rocks" in the Porcupine Lake area of eastern Ontario. When Canadian gold production peaks in 1941, the country will have 146 gold mines including the Hollinger and Dome mines and will be producing 5.3 million ounces per year.

The Payne-Aldrich Tariff Act signed into law by President Taft August 5 abolishes the import duty on hides to help the U.S. shoe industry but maintains high duties on iron and steel and raises duties on silk, cotton goods, and many minor items (*see* 1897; Underwood-Simmons Act, 1913).

The Sixteenth Amendment to the U.S. Constitution providing for an unapportioned income tax is submitted to the 46 states for ratification (*see* 1913).

Selfridge's opens in London's Oxford Street and is Britain's first large department store. U.S. merchant Gordon Selfridge, 52, has had a successful career at Chicago's Marshall Field & Company, his new store will rival Harrods and Marks & Spencer, and it will become as famous as Marshall Field's.

The first British F. W. Woolworth store opens on Liverpool's Church Street (*see* 1900). By 1958 there will be 1,000 British Woolworth stores (*see* 1912).

Filene's Automatic Bargain Basement at Boston is the first of its kind (*see* 1901; 1912).

New York's 91-year-old Brooks Brothers closes its two downtown stores and reopens at Broadway and 22nd Street near the city's most fashionable residential district. A Brooks Brothers branch opens at Newport, R.I. (*see* 1858; 1915).

French engineer Louis Blériot, 36, makes the first crossing of the English Channel in a heavier-than-air machine July 25. He has built a monoplane and flies it from Calais to a field near Dover in 37 minutes (*see* Alcock and Brown, 1919).

Orville Wright demonstrates the success of the Wright brothers' airplane and wins assurance of its acceptance by the U.S. Army in July (*see* 1908). The brothers will establish the American Wright Co. to manufacture aircraft (*see* Curtiss-Wright, 1929).

The first international air races are held at Reims in France to compete for a cup offered by *New York Herald* publisher James Gordon Bennett. U.S. inventor-aviator Glenn Hammond Curtiss, 31, wins with an airplane and motor of his own design (*see* 1910).

U.S. engineer Glenn L. Martin, 23, opens an aircraft factory. He will receive his first government contract in 1913, found the Glenn L. Martin Co. at Cleveland in 1917, and move the company to Baltimore in 1929 (*see* Douglas, 1920).

The New York, New Haven, and Hartford buys a majority interest in the Boston and Maine, but Boston lawyer Louis Dembitz Brandeis, 50, fights the aggrandizement of New Haven president C. S. Mellen, whose poor administration is wrecking his railroad's finances (*see* 1903). Both the New Haven and the B&M will omit dividends to their stockholders in 1913, Mellen will have to resign, and in 1914 the New Haven will be forced to divest itself of its trolley-line and steamship interests.

The Chicago, Milwaukee, and St. Paul reaches Seattle, becoming the seventh line to link the Mississippi to the Pacific Coast (*see* 1863).

Alfa-Romeo has its origin in the Anonima Lombarda Fabbrica Automobila (ALFA) at Milan. Entrepreneurs have split with Alexandre Darracq, take over Darracq's Milan branch, will move to Rome in 1920 and will operate under the name Alfa-Romeo.

Hudson Motor Car Co. is founded by R. D. Chapin and Howard Earle Coffin, 35, with backing from Detroit department store magnate J. L. Hudson (*see* 1881). Coffin is a veteran of the Olds Motor Works and has designed the Chalmers motorcar for Chalmers-Detroit Motor Co. (*see* 1916).

Marmon motorcars switch to water-cooled engines and the first Marmon to compete at the new Indianapolis Speedway performs well enough to bring the car some publicity (*see* 1902; 1916; Indianapolis 500, 1911).

General Motors acquires Cadillac from Henry M. Leland for $4.5 million (*see* 1904). Cadillacs have interchangeable parts, unusual in the fledgling automotive industry (*see* 1916; self-starter, 1911; Lincoln, 1922).

General Motors acquires Oakland Motor Car, which has a factory at Pontiac, Mich., producing motorcars designed by A. P. Brush (*see* 1924; Pontiac, 1926).

The first transcontinental U.S. motorcar race pits two Model T Fords against an Acme, an Itala, a Shawmut, and a Stearns (which fails to start). Five cars leave New York June 1, and a Ford wins the race, arriving at the Alaska-Yukon Pacific Exposition in Seattle June 22.

U.S. automobile production reaches 127,731, up from 63,500 last year.

One Iowa farmer in 34 has a motorcar, while in New York City only one family in 190 has one, reports *Collier's* magazine. The magazine calls the motorcar the greatest social force in America—greater even than rural free mail delivery, the telephone, or university extension services (*see* Wilson, 1906).

Western Auto Supply Co. is founded by Kansas City bookkeeper George Pepperdine, 23, who opens a mail-order house to supply parts for Model T Fords which are sold minus tires, fenders, tops, windshields, and lights—items that can cost a buyer as much as the car itself. By 1915 Pepperdine's company will be grossing $229,000 per year and will have a second plant at Denver to produce parts for the Model T (*see* 1927).

New York's Queensboro Bridge opens March 30 to carry traffic across the East River between Manhattan and Queens. Built primarily to carry trolley cars, the $17 million span is the first important double-deck bridge.

New York's Manhattan Bridge opens December 31 to carry traffic between Manhattan and Brooklyn (the third such link). The $31 million span is the first important double-deck suspension bridge.

Bakelite, developed by Belgian-American chemist Leo Hendrik Baekeland, 46, is the world's first polymer. Baekeland's synthetic shellac plastic material is made from formaldehyde and phenol, Bakelite products will be used initially for electrical insulation, and the chemist starts a company to market a molding powder used for shaping Bakelite products (*see* Union Carbide and Carbon, 1939).

Synthetic rubber is produced by German chemist Karl Hoffman of Farbenfabriken Bayer from butadiene, a gas derived from butane (*see* 1925).

A theory of the gene formulated by Columbia University zoologist Thomas Hunt Morgan, 43, breaks new ground in the study of heredity.

The nucleic acids RNA and DNA discovered by Russian-American chemist Phoebus Theodore Levene, 40, will be the basis of major genetic discoveries (*see* 1870; Watson and Crick, 1953).

An arsenic compound formulated by Paul Ehrlich to fight syphilis is the first antibacterial therapeutic drug and pioneers chemotherapy in medicine (*see* diphtheria, 1891). The German bacteriologist's compound Number 592 is effective in destroying the trypanosomes that cause syphilis in mice and produces no side effects or aftereffects, his purer, more soluble variant Number 606 will prove effective next year in treating human victims of syphilis, and the drug will be marketed under the name Salvarsan. Ehrlich's arsphenamine will reduce the incidence of syphilis in England and France by 50 percent in the next 5 years, but while his work will lead to the discovery of new antibacterial agents, many will attack his syphilis cure on the ground that it encourages sin (*see* Fleming, 1928; Domagk, 1935).

Austrian pathologist Karl Landsteiner, 41, looks into the reasons why donors' blood sometimes causes clotting in recipients' blood; he establishes the existence of different blood types (*see* 1818). Further research will reveal that there are four essential blood types: O, A, B, and AB; knowledge of compatibility will make blood transfusions safe (*see* Rh factor, 1940).

Landsteiner (above) isolates the poliomyleitis virus (*see* 1905, 1916).

Congress bans the import of opium for anything but medical purposes. Opium derivatives are widely used in the anodynes codeine and morphine as they were in the earlier anodyne laudanum, and a tincture of opium is effective in treating some intestinal disorders (*see* heroin, 1898; Harrison Act, 1916).

John D. Rockefeller gives $530 million for worldwide medical research. The Standard Oil Company head is the world's first billionaire (*see* Rockefeller Institute, 1901).

The Rockefeller Sanitary Commission that will become the Rockefeller Foundation in 1913 begins a campaign to eradicate hookworm disease in the South (*see* 1619; Stiles, 1902).

Walter Reed Army Medical Center opens at Washington, D.C.

Kansas bans public drinking vessels, the first state to take such action.

The United States has 2,600 daily newspapers to serve her 90 million people (*see* 1958).

New York's weekly *Amsterdam News* begins publication in December. The paper's circulation will peak at 100,000, the nation's largest nonreligious black weekly.

Condé Nast takes over *Vogue* magazine, started late in 1892 as a fashion and society weekly. Now 35, the former *Collier's* advertising manager has organized a company to make and sell dress patterns under an arrangement with Cyrus Curtis's *Ladies' Home Journal*; *Vogue* has a circulation of only 22,500 but Nast will make it a monthly and in 1914 will hire Edna Woolman Chase as editor; she will build circulation to more than 130,000, launching British and French editions (*see* 1915).

The United States Copyright Law passed by Congress March 4 takes effect July 1, protecting U.S. authors, publishers (and composers) under terms that will remain unchanged for 68 years. The law gives copyright owners exclusive rights "to print, reprint, publish, copy, and vend the copyrighted work." The courts will develop an ill-defined doctrine of "fair use" by which to decide cases involving charges of copyright infringement, taking into account the nature of the copyrighted work, the amount of material copied, and the effect of the use on the copyright owner's potential market (*see* below, music).

A futurist manifesto published by Italian poet-publicist Emilio Filippo Tommaso Marinetti, 33, advocates rejection of the past, including abandonment of syntax and grammatical rules.

Fiction *Melanctha* by émigrée U.S. novelist Gertrude Stein, 35; *Strait Is the Gate* (*La Porte étroite*) by André Gide; *The Peasants (Chlopi)* by Wladyslaw Reymont; *The River Sumida (Sumida-gawa)* by Japanese novelist Kafu Nagai, 30, who has returned to Tokyo after 5 years in America and France to find the city horribly modernized; "The Tatooer" (story) by Japanese writer Jamichiro Tanizaki, 23; *Actions and Reactions* (stories) by Rudyard Kipling; *Tono-Bungay* by H. G. Wells who

1909 *(cont.)* shows the excesses of patent medicine exploitation; *Martin Eden* by Jack London; *A Girl of the Limberlost* by Gene Stratton Porter; *The Circular Staircase* by U.S. mystery novelist Mary Roberts Rinehart, 32.

Poetry *Handful of Songs (Gitangali)* by Bengali poet Rabindranath Tagore, 48, whose work will be promoted in the West by William Butler Yeats.

Painting *Eiffel Tower* by French painter Robert Delaunay, 24, who introduces brilliant color to cubism after having experimented with post-impressionism and fauvism (he will paint the Eiffel Tower at least 30 times); *The Dance* (panel) and *Woman in Green with a Carnation* by Henri Matisse; *Standing Nude* by Pierre Bonnard; *The Equatorial Jungle* by Henri Rousseau; *Violoncellist* by Amedeo Modigliani; *El Picador* by Diego Rivera; *Both Members of This Club* (prizefighters) by George Bellows. Alfred Steiglitz introduces U.S. water-colorist John Cheri Marin, 38, to New York with a show at his Photo-Secession 291 Gallery. Frederic Remington dies near Ridgefield, Conn., December 26 at age 48.

Munich's Modern Gallery opens under the direction of local art dealer Justin Thannhauser, 17, who turns his father's 5-year-old gallery into a focal point for work by "The Bridge" ("Die Brücke") expressionists Edvard Munch, Wassily Kandinsky, Swiss painter Paul Klee, 29, German expressionist Ernst Ludwig Kirchner, 29, and others.

The Boston Museum of Fine Arts is completed (see 1870).

Theater *The Easiest Way* by U.S. playwright Eugene Walter, 35, 1/19 at New York's Stuyvesant (Belasco) Theater, with Joseph Kilgour, 157 perfs.; *Strife* by John Galsworthy 2/21 at the Duke of York's Theatre, London; *Griselda* by Gerhart Hauptmann 3/6 at Vienna's Hofburgtheater; *Earl Birger of Bjalbo (Bjalb-jarle-ti)* by August Strindberg 3/26 at Stockholm's Swedish Theater; *The Fortune Hunter* by Winchell Smith 9/4 at New York's Gaiety Theater, with John Barrymore, 345 perfs.; *The Melting Pot* by Israel Zangwill 9/17 at New York's Artef Theater after a year on the road, with Walker Whiteside, 136 perfs. The play introduces the "melting pot" phrase to describe America's amalgam of nationalities and races (*see* de Crèvecoeur, 1782); *Anathema (Anatema)* by Leonid Andreyev 10/2 at the Moscow Art Theater; *The Tinker's Wedding* by the late J. M. Synge 11/11 at His Majesty's Theatre, London. Synge died of Hodgkin's disease 3/24 at age 37; *Liliom* by Hungarian playwright Ferenc Molnár, 31, 12/7 at Budapest's Vigszinhaz; *The City* by Clyde Fitch 12/21 at New York's Lyric Theater, with Walter Hampden, 190 perfs.

Film William Ranon's *Hiawatha.*

Bell & Howell Company's Albert S. Howell eliminates the "flicker" from motion pictures with a standard camera that permits precise control of film movement (*see* 1907; 1911).

Opera *Elektra* 1/25 at Dresden's Konigliches Operhaus, with music by Richard Strauss, libretto by Hugo von Hofmannsthal; *Le Coq d'Or (The Golden Cock)* 9/24 at Moscow's Zimin Theater, with music by the late Nikolai Rimsky-Korsakov.

The Boston Opera House opens November 8 with ovations to merchant Eban Jordan of Jordan-Marsh whose benefactions have made the house possible.

Ballet *Prince Igor* 5/18 at the Théâtre du Châtelet, Paris, with music by the late Nikolai Rimsky-Korsakov from the opera of 1890, choreography by Russian impressario Sergei Pavlovich Diaghilev, 37, who founds the Ballet Russe and employs dancer Waslaw Nijinsky, 19, an incredible performer who made his debut 2 years ago with Diaghilev's Imperial Ballet at St. Petersburg. Michel Fokine takes over the Imperial Ballet.

The London Symphony gives its first concert June 9.

First performances *Die Todtinsel (The Island of the Dead)* (Symphonic Poem, after a picture by A. Bocklin) by Sergei Rachmaninoff 5/1 at Moscow; Concerto No. 3 in D minor for Piano and Orchestra by Rachmaninoff 11/28 at New York.

The new United States Copyright Law (above) secures exclusive rights to composers and/or publishers to print, publish, copy, vend, arrange, record by means of gramophone or any other mechanical device, and perform publicly for profit original musical compositions, and affords protection against infringement for a period of 28 years and a renewal period of the same length (*see* 1897; ASCAP, 1914).

The Chautauqua movement founded in 1874 gains a musical aspect as Walter Damrosch, 47, brings his New York Symphony Orchestra to Chautauqua, N.Y. The Chautauqua will become an important training center for U.S. musical talent.

Broadway musicals *The (Ziegfeld) Follies* 6/14 at the Jardin de Paris, with Nora Bayes, Jack Norworth, Eva Tanguy, Russian-American actress-singer Sophie Tucker (Sonia Kalish), 25, music by Maurice Levi and others, book and lyrics chiefly by Harry B. Smith, songs that include "By the Light of the Silvery Moon" by Gus Edwards, lyrics by Edward Madden, 64 perfs.; *Old Dutch* 11/22 at the Herald Square Theater, with Lew Fields, English-American dancer Vernon Blythe Castle, 22, and Helen Hayes (Brown), 9, in her first New York appearance (she has been a professional actress since age 5), book by Edgar Smith, music by Victor Herbert, lyrics by George V. Hobart, 88 perfs.

Popular songs "Meet Me Tonight in Dreamland" by Leo Friedman, 40, lyrics by Beth Slater Whitson, 30; "I Wonder Who's Kissing Her Now" by Joseph E. Howard and Harold Orlob, 24, lyrics by William M. Hough, 27, and Frank R. Adams, 26; "Put on Your Old Grey Bonnet" by Percy Wenrich, 24, lyrics by Stanley Murphy; "Casey Jones" by Eddie Newton, lyrics by T. Lawrence Seibert (*see* 1900); "On Wisconsin!" (march) by W. T. Purdy.

Arthur Gore wins in men's singles at Wimbledon, Dorothea Penelope Boothby, 27, in women's singles; Bill Larned wins in U.S. men's singles, Hazel V. Hotchkiss, 22, in women's singles.

"Tris" Speaker signs with the Boston Red Sox to begin a 19-year American League career. Texas-born center fielder Tristram E. Speaker, 21, will have a lifetime batting average of .344.

The Pittsburgh Pirates win the World Series by defeating the Detroit Tigers 4 games to 3.

A field goal in football receives a value of 3 points, down from 4 in 1904, 5 in 1883.

The celluloid Kewpie Doll with a head that comes to a point is patented by New York author-illustrator Rose Cecil O'Neill, 35, whose creation will be the basis of a mold that will be made in 1913 by Pratt Institute art student Joseph L. Kallus. The doll will earn $1.5 million for O'Neill, who wears a toga at her Greenwich Village salons.

The first International Conference on City Planning meets in late May at Washington, D.C. It has been organized largely at the initiative of the secretary of New York's Committee on Congestion of Population Benjamin C. Marsh who has helped prepare a book on city planning which begins, "A city without a plan is like a ship without a rudder."

"Make no little plans," says Chicago architect Daniel Burnham in his "Plan for Chicago of 1909." Little plans "have no magic to stir men's blood," argues Burnham, but while his Chicago plan will spur that city and others to adopt sweeping master plans to guide their growth, some utopian master plans will be too inflexible to allow for the unpredictable social, economic, and political forces that interact to shape cities.

Chicago's Frederick G. Robie house completed by Frank Lloyd Wright at 5757 South Woodlawn Boulevard is the world's first house to be built on a slab foundation, to incorporate garages in its structure, to have roof overhangs calculated on an astronomical basis for maximum lighting in winter and maximum shade in summer, and to employ indirect lighting and rheostats (dimmers) for lighting control.

New York's 42-story Metropolitan Life Insurance Tower is completed by Napoleon Le Brun & Sons with a 40th-story observation floor on Madison Square at 23rd Street. The tower will be the world's tallest building until 1913.

Congress enacts legislation to prevent any private builder in Washington, D.C., from putting up a structure more than 130 feet high, but the 11-story Cairo Hotel erected in 1894 is allowed to stand.

A U.S. National Bison Refuge is created near Moise, Mont. (*see* 1900; Glacier Park, 1910).

U.S. lumber production reaches its peak.

Soil is indestructible, says a report issued by the U.S. Bureau of Soils which has made the first National Soil Survey (*see* 1933; 1934).

A river and harbors bill enacted by Congress empowers the U.S. Army Corps of Engineers to construct locks and dams on U.S. waterways.

Smelt are planted in the Great Lakes where they will become an important food fish species.

Laguna Dam is completed on the Colorado River north of Yuma, Ariz., to irrigate more desert land.

Colorado is the most irrigated state in the nation with more than 3 million acres under irrigation.

Acreage allotted under the Homestead Act of 1862 is doubled by Congress following the failure of tens of thousands of homesteaders in arid regions of the West for lack of enough land. But it takes at least four sections of land (2,560 acres) to support a family raising livestock in the West, so the new act is inadequate (*see* 1916; Powell, 1878).

The first Kibbutz is started at the Jordan Valley village of Degania Aleph in Palestine which is part of the Ottoman Empire.

Hawaiian Pineapple's James Dole summons his competitors to a meeting as a glut of pineapple production depresses prices (*see* 1902). Dole packs 242,822 cases and the growers agree to undertake a program of advertising to win acceptance for canned pineapple in the big eastern U.S. markets, the first advertising campaign for a commodity by any growers' association (*see* 1911; Sunkist, 1919).

Strawberries are frozen for market in the Pacific Northwest (*see* Birdseye, 1914).

U.S. ice cream sales reach 30 million gallons, up from 5 million in 1899. Philadelphia has 49 ice cream manufacturing plants and 52 ice cream "saloons."

Coca-Cola is exported to Britain for the first time (*see* 1899; 1907; distinctive bottle, 1916).

Thomas Lipton begins blending and packaging his tea at New York. His U.S. business will be incorporated in 1915, and 3 years after his death in 1931 his picture will begin appearing on the red-and-yellow packages that identify Lipton products (*see* 1893; 1914).

G. Washington soluble coffee powder is introduced by Brooklyn, N.Y., kerosene lamp maker George Constant Louis Washington, 38. Born in Belgium, Washington settled in Guatemala 2 years ago after making a small fortune in kerosene lamps, and when he noticed a fine powder on the spout of a silver coffee carafe he began experiments that led to the development of the powder for making instant coffee (*see* Kato, 1901; Nescafé, 1938).

1910

Egypt's British-supported Coptic premier Butros Ghali is assassinated by a nationalist fanatic February 20 as Islamic agitation increases.

A coalition of rebellious U.S. congressmen led by Republican George W. Norris, 48, of Nebraska curtails the powers of Speaker Joseph Gurney Cannon, 73, and excludes him from the House Rules Committee March 19. The congressmen establish a system of seniority that will control committee chairmanships for decades.

Britain's Edward VII dies May 6 at age 68 after a 9-year reign of peace and prosperity. He is succeeded by his second son, 44, who will reign until 1936 as George V.

The Republic of South Africa, independent of Britain, is established May 31 under terms of the South Africa Act approved by Parliament in September of last year.

1910 ***(cont.)*** The new Union of South Africa has dominion status, it unites the Cape Colony, Orange River Colony, Natal, and Transvaal, its legislative seat is at Cape Town, its seat of government is at Pretoria, and its prime minister is Boer statesman Louis Botha, 47, who will continue in the post until his death in 1919.

France renames the French Congo French Equatorial Africa and redivides it into the colonies of Gabon, Middle Congo, and Ubanghi-Shari.

Montenegro proclaims herself an independent Balkan kingdom August 28, and prince who obtained recognition of his country's independence in the Treaty of Berlin in 1878 receives the title king by a vote of the national legislature. Now 69, he will reign until 1919 as Nicholas I.

The Portuguese monarchy founded in 1143 by Afonso Henriquez ends October 4 in a revolution at Lisbon after a 2-year reign by Manoel II. He flees to England (where he will live as a country gentleman until his death in 1932) and a republic is proclaimed with a provisional government headed by scholar-writer Teofilo Braga, 67.

Japan formally annexes Korea by treaty August 22 and calls it Chōsen.

China abolishes slavery March 10.

A Mexican social revolution begins as Francisco Indalecio Madero, 37, leads opposition to President Porfirio Diaz who has controlled the country since 1876 and allowed white landowners to take over the lands of its 6 million Indians and 8 million *mestizos*. The exploited peons start breaking up the large landholdings and distributing farmland among the *campesinos* (*see* 1911).

German peasants still work 18-hour days and are treated little better than serfs.

Mexican revolutionists divvied up the country's large land holdings among the long-disenfranchised *campesinos*.

France's prime minister Aristide Briand, 48, averts a general strike and forces an end to a railroad strike begun by the Conféderation Genérale du Travail (CGT). He calls all railroad workers "to the colors" and the CGT orders the men back to work.

The average U.S. workingman earns less than $15 per week, working hours range from 54 to 60 hours, and there is wide irregularity of employment.

Chicago clothing workers who include Sidney Hillman, 23, revolt against low wages and piecework as they begin agitating for a stronger union in the men's clothing industry. U.S. clothing workers participated in more strikes during the last quarter of the 19th century than did workers in any other industry, and their activity accelerates (*see* Amalgamated Clothing Workers, 1914).

The International Ladies' Garment Workers Union (ILGWU) wins a 9-week strike for New York cloak makers (*see* 1909; Dubinsky, 1932).

The Mann White Slave Traffic Act passed by Congress June 25 discourages interstate transportation of women for immoral purposes. Newspaper stories about prizefighter Jack Johnson have inspired passage of the law (*see* 1915).

The "great white hope" James J. Jeffries comes out of retirement to challenge Johnson but loses July 4 (below). Race riots ensue at Boston, Cincinnati, Houston, New York, and Norfolk, three blacks are killed at Uvalda, Ga., as white bigots vent their rage at the continuing supremacy of the first black prizefight champion.

The National Association for the Advancement of Colored People (NAACP) is founded at New York (*see* Dubois, 1909).

Eight out of 10 U.S. blacks still live in the 11 states of the Old Confederacy, but a "great migration" begins that will bring more than 2 million blacks to the North (*see* 1917; census, 1940).

A bomb exploded October 1 at the *Los Angeles Times* kills 20 men. Authorities arrest James McNamara, 28, and his brother John, 27, on charges of having placed the bomb to silence opposition to organized labor by Harrison Gray Otis, now 73 (*see* 1882), Chicago lawyer Clarence Darrow defends the McNamara brothers, but they will confess their guilt next year following exposure of the facts by detective William J. Burns who in the past 5 years has uncovered political corruption in San Francisco, kickback schemes in the railroad industry, land frauds in the Northwest, and other crimes including bank embezzlements. Burns will become head of the FBI for 4 years beginning in 1921 (*see* 1886; 1908).

Women in Washington State gain the right to vote in a constitutional amendment adopted November 8 (*see* California, 1911).

Chile's enormously productive Chuquicamata copper mine is acquired by the U.S. copper trust ASARCO, controlled since 1901 by the Guggenheim family (*see* 1923).

National Cash Register has sales of $100,000 (*see* THINK, 1908). The registers have been improved by the addition of a small electric motor, invented by Dayton, Ohio, electrical engineer Charles Franklin Kettering, 34, that eliminates manual operation (*see* 1912; self-starter, 1911).

The first "Morris Plan" bank opens at Norfolk, Va., and pioneers in granting personal bank loans at a time when people who want to borrow money must generally ask family or friends or go to loan sharks, pawn shops, or eleemosynary institutions that make compassionate loans to relieve distress. The Fidelity Loan and Trust Co. founded by local lawyer Arthur J. Morris, 29, makes one-year loans to locally employed citizens of good character, requires two cosigners, obliges borrowers to repay in monthly installments, and deducts 6 percent in advance to give the bank an actual interest return of 11.6 percent. By 1920 there will be Morris Plan banks in 37 states (*see* National City Bank, 1928).

Hitachi, Ltd. is founded in Hitachi City, 80 miles northeast of Tokyo, by engineer Namihei Odaira, 36, who begins a motor repair shop to serve a nearby copper mine. Upset that all the mine's equipment is imported, Odaira makes three 5-horsepower electric motors that will soon be part of a whole line of electrically powered industrial machines.

May Department Stores is incorporated to replace the partnership of founder David May and his three brothers-in-law. The corporation adds the M. O'Neill store at Akron, Ohio, to its Denver and St. Louis stores (*see* 1879; Cleveland, 1914).

New York's Gimbel Brothers Department Store opens on Greeley Square between 32nd and 33rd Streets (*see* 1894; Saks Fifth Avenue, 1924).

The Fuller Brush Co. is incorporated in a reorganization of the 4-year-old firm. By 1920 it will have sales of nearly $12 million per year as A. C. Fuller builds a corps of independent door-to-door salesmen.

New York's Pennsylvania Station opens to Long Island Rail Road commuter traffic September 8 and to long-distance trains November 27. The Pennsylvania Railroad's late president Alexander Johnston Cassatt, who died in 1906 at age 67, had proposed erecting a hotel on the terminal's air rights but was persuaded that the railroad had an obligation to give New York a monumental gateway. Covering two square blocks between Seventh and Eighth Avenues from 31st Street north to 33rd, the $112 million granite and travertine terminal has been modeled by McKim, Mead, and White on the warm room of Rome's ancient Baths of Caracalla with 84 doric columns each 35 feet high and 150-foot ceilings in its vast waiting room (*see* 1966).

A trans-Andean railroad linking Argentina with Chile is completed following plans of U.S. entrepreneur William Wheelwright who began the road before his death at age 75 in 1873.

President Taft intercedes on behalf of U.S. financiers to let them join a consortium of French, British, and German financiers in underwriting China's Hukuang Railroad.

Sperry Gyroscope Co. is established at Brooklyn, N.Y., by Elmer A. Sperry to manufacture a gyroscopic compass and other instruments invented by Sperry to stabilize ships in rough waters (*see* 1879; patent, 1913).

The Cunard liner S. S. *Mauretania* sails from Cobb to New York in 4 days, 20 hours, 41 minutes to set a new transatlantic speed record (*see* 1907).

The *London Daily Mail* offers a £10,000 prize for the winner of an air race from London to Manchester. The prize goes to French aviator Louis Paulhan, 27, who this year reaches a height of 4,149 feet in a plane he flies at Los Angeles.

Rolls-Royce's C. S. Rolls flies from Dover to Calais and back without stopping but is killed later in the year, becoming the first English aviation victim (*see* 1906; Blériot, 1909).

Glenn Curtiss flies from Albany to New York in 150 minutes to break the long-distance speed record and win a $10,000 prize put up by Joseph Pulitzer's *New York World* (*see* 1909; 1919).

"Barney" Oldfield drives a mile in 27.5 seconds at Daytona Beach, Fla., setting a 131.724-mile-per-hour speed record that will stand for years.

The United States has 1,000 miles of concrete road, up from 144 in 1900 (*see* surfaced roads, 1921).

Steel begins to replace wood in U.S. automobile bodies (*see* Ford Model T, 1908).

Safety glass is patented by French poet-chemist Edouard Benedictus, who has accidentally knocked over a test tube lined with a film left by evaporation of a nitrocellulose mixture. Benedictus has observed that the cracked glass has not shattered (*see* Triplex Safety Glass, 1926).

General Motors directors oust W. C. Durant and replace him with Boston financier James J. Storrow of Lee, Higginson (*see* 1912; Chevrolet, 1911).

John North Willys produces 18,200 motorcars, up from 4,000 last year and 465 the year before (*see* 1917; 1915).

A Sears, Roebuck Model L sells for $370, a Reo runabout for $500, a Maxwell for $600, a Hupmobile for $750.

American Viscose Co. is founded at Marcus Hook, Pa., by Courtaulds, Ltd. of Britain (*see* 1905). It will start making rayon from spruce pulp next year, will be the first successful U.S. producer, and will control U.S. rayon production for years, protected by patents and tariff laws (*see* Little, 1902; Celanese, 1918; Du Pont, 1920).

Minnesota Mining and Manufacturing Co. is founded at St. Paul, Minn., in a reorganization of a firm set up 3 years ago. The "corundum" has turned out to be a low-grade anorthosite unsuitable for heavy-duty abrasives, and the company has quit mining to turn its efforts to producing a poor-quality sandpaper (*see* Scotch brand tape, 1925).

1910 *(cont.)* British-made steel is one-third more costly than German or U.S.-made steel; Britain's steel mills have failed to install coke ovens or employ other technological advances.

Medical Education in the United States and Canada shows that three-fourths of North American medical schools are inadequate and that only the Johns Hopkins school founded at Baltimore in 1893 is a match for the great medical schools of Europe. U.S. physician Abraham Flexner, 44, has used a $14,000 grant from the 8-year-old Carnegie Institution to inspect 155 medical schools, and the Flexner Report spurs a $600 million reform program in medical education.

Washington University president Robert Somers Brookings, 60, at St. Louis is inspired by the Flexner Report (above) and embarks on a program that will make his university's medical school second to none within 3 years. Other U.S. universities are similarly motivated to elevate medical school standards.

"Every day, in every way, I'm growing better and better," says French pharmacist Emile Coué, 28. He has studied hypnotism, suggests the slogan for autosuggestive healing, and will develop a system of psychotherapy called Couéism.

Chicago physician James Bryan Herrick, 49, examines a West Indian student of 20 and makes the first diagnosis of sickle cell anemia (it will be given that name in 1922). The hereditary disease of the blood protein hemoglobin will be found to affect up to 50 percent of people in some African tribes and from 0.25 to 3 percent of U.S. blacks, although as many as 10 percent may carry the sickle cell trait which in itself has no ill effects and will be found common among non-blacks wherever malaria is prevalent (children with the trait will prove to have a high degree of resistance to some kinds of malaria). The actual anemia in which the blood cells take on a crescent-shaped appearance occurs only in the offspring of parents who both have the trait, it is often extremely painful, it kills most of its victims before age 21, and few live beyond age 40.

Tularemia afflicts ground squirrels of Tulare County, Calif. Physician George Walter McCoy, 34, and his colleague Charles Willard Chapin recognize the disease and will name the responsible organism *Bacterium tularense.* An epizootic of wild rabbits and other animals that is communicable to humans, tularemia is the first distinctly American disease and will remain a threat to people preparing wild rabbits for cooking without taking proper precautions.

A Paris fashion for imitation sable and sealskin encourages amateur Chinese hunters to trap Manchurian marmots, many of which are infected with bubonic plague. An epidemic of the plague transmitted by unhealthy marmots will kill 60,000 in Manchuria and China in the next 2 years, and in the next 9 years will kill 1.5 million in China and India.

Southern Methodist University is founded at Dallas.

Kent State University is founded at Kent, Ohio. Bowling Green State University is founded at Bowling Green, Ohio.

The Mann-Elkins Act passed by Congress June 18 amends the Interstate Commerce Act of 1887 to regulate telephone, telegraph, and cable companies. Such companies are now subject to ICC regulations (*see* Federal Communications Act, 1934).

American Telephone and Telegraph chief Theodore N. Vail has himself elected president of Western Union and abolishes the 40¢ to 50¢ charge for placing telegraph messages by telephone. Vail has acquired a controlling interest in Western Union from the Jay Gould estate for $30 million in AT&T stock, and while the courts will force AT&T to sell its Western Union stock in 1914, free placement of telegraph messages by telephone will continue even after Vail resigns (*see* 1913).

Tropical Radio Telegraph Co. is founded by United Fruit Company to provide uninterrupted radio contact between the United States and Central America (*see* 1904). United Fruit has been handling 77 percent of world banana exports, it is shipping fruit to Europe, and it will make Tropical Radio Telegraph a subsidiary in 1913 (*see* 1929; ITT, 1920).

Pathé Gazette, shown in Britain and the United States, is a pioneer film newsreel. French cinematographer Charles Pathé, 47, and his brother Emil have become Paris agents for the Edison phonograph, visited London to acquire film-making equipment invented by British instrument maker Robert Paul, obtained financial support, and will set up production units in Britain, the United States, Italy, Germany, Russia, and Japan.

The *Pittsburgh Courier* begins publication March 10.

Chicago Tribune editor Robert W. Patterson dies and is succeeded by Robert Rutherford McCormick, 30, and his cousin Joseph Medill Patterson, 31, who will be coeditors until 1925 (*see* 1855). A nephew of the reaper inventor, McCormick will, in the next 45 years, build circulation from 200,000 to 892,000 (1.4 million on Sundays), make the *Tribune* number one in advertising revenue, and establish a radio station with the call letters WGN (for "World's Greatest Newspaper") (*see* 1917; 1919; Tribune Tower, 1925).

Women's Wear Daily begins publication at New York July 13 under the direction of journalist Edmund Fairchild, 44, whose trade paper for the garment industry will be the basis of a publishing empire.

The *Miami Herald* begins publication December 1 to serve a city whose population is still under 6,000.

Hallmark, Inc., has its beginnings in a wholesale card jobbing company started at Kansas City by Nebraskan Joyce Clyde Hall, 18, and his brother who will soon start dealing in greeting cards (*see* Pring, 1875). They will buy their own printing plant in 1916 and become the world's largest maker of greeting cards.

Fiction *The Village* by Russian novelist Ivan Bunin, 40; *Clayhanger* by Arnold Bennett begins a trilogy about the ugly life in the "five towns" that are the center of England's pottery industry; *Howards End* by E. M. Forster; *Prester John* by Scottish novelist John Buchan, 35; *The History of Mr. Polly* by H. G. Wells; *La Vagabonde* by Colette, who marries Henry de Jouvenal; *The Notebook of Malte Laurids Brigge (Die Aufzeichnungen des Malte Laurids Brigge)* by Rainer Maria Rilke.

Juvenile *The Secret Garden* by Frances Hodgson Burnett; *Rewards and Fairies* by Rudyard Kipling.

Poetry A *Handful of Sand (Ichiaku no suna)* by Japanese poet Takuboku Ishikawa, 25, whose tanka in the traditional 31-syllable form have a content that is not traditional (Ishikawa will die of tuberculosis in 1912); *Five Great Odes (Cinq Granites Odes)* by Paul Claudel; *The Town Down the River* by Edwin Arlington Robinson includes his poem "Miniver Cheevy" (who "loved the days of old" but was "born too late . . . called it fate,/ And kept on drinking").

Painting *The Enigma of an Autumn Afternoon* and *The Enigma of the Oracle* by Italian painter Giorgio de Chirico, 22; *Nudes in the Forest (Nue dans la forêt)* by Fernand Léger; *Blue Nude*, *The Dance*, and *Music* by Henri Matisse, who shows the influence of a Munich exhibition of Near Eastern art; *Yadwiga's Dream*, *Exotic Landscape*, and *Horse Attacked by Jaguar* by Henri Rousseau, who dies at Paris September 2 at age 66. Holman Hunt dies at London September 7 at age 83; Winslow Homer dies at Prout's Neck, Me., September 29 at age 74.

Wassily Kandinsky produces the world's first nonrepresentational paintings.

Theater *Deirdre of the Sorrows* by the late J. M. Synge 1/13 at Dublin's Abbey Theatre; *Chantecler* by Edmond Rostand 2/7 at the Théâtre de la Porte-Saint Martin, Paris, with Lucien Guitry in the title role; *Justice* by John Galsworthy 2/21 at the Duke of York's Theatre, London; *Misalliance* by George Bernard Shaw 2/23 at the Duke of York's Theatre, London; *Old Friends* and *The Twelve-Pound Look* by James M. Barrie 3/1 at the Duke of York's Theatre, London; *The Dragon's Head (La farsa infantil de la cabeza del dragon)* by Ramon Valle Inclan 3/5 at Madrid's Teatro de la Convents; *Get-Rich-Quick Wallingford* by George M. Cohan 9/19 at New York's Gaiety Theater, with Hale Hamilton as J. Rufus Wallingford in a comedy based on a novel by George Randolph Chet, 424 perfs.; *The Guardsman (A Tester)* by Ferenc Molnár 11/19 at Budapest's Vigszinhaz.

Films Joseph A. Golden's *The New Magdalene* with Pearl White; D. W. Griffith's *Ramona* with Mary Pickford (Gladys Smith), 17.

Brooklyn Eagle cartoonist John Randolph Bray, 31, pioneers animated motion picture cartoons, using a "cel" system he has invented and that will be used by all future animators. Each cartoon frame is a photograph of several layers of celluloid transparencies, the only layers that change from frame to frame are those that involve movements of figures, backgrounds (and some figures) remain constant, and the technique avoids the distracting moves that existed when each frame was drawn entirely by hand (it also reduces production costs enormously). Bray's cartoon "The Dachshund and the Sausage" is acquired by Pathé (above), Bray develops a "Colonel Heeza Liar" cartoon based roughly on Theodore Roosevelt, he will employ animators Max and David Fleischer, who will create "Popeye" cartoons (*see* 1929), Paul Terry, who will produce "Terry Toons," and Walter Lantz, who will create "Bugs Bunny" (*see* 1937; Disney, 1928).

Ballet *Scheherezade* 6/4 at the Théâtre National de l'Opéra, Paris, with Waslaw Nijinsky of Serge Diaghilev's Ballet Russe dancing the role of the Favorite Slave, music by the late Nikolai Rimski-Korsakov, choreography by Michel Fokine; *The Firebird* 6/25 at the Paris Opéra, with Tamara Karsavina, 25, dancing the role of Ivan Czarevich, music by Russian composer Igor Federovich Stravinsky, 28, choreography by Michel Fokine.

Opera *Macbeth* 11/30 at the Opéra-Comique, Paris, with music by Swiss composer Ernest Bloch, 30. *The Girl of the Golden West (La Fanciulla del West)* 12/10 at New York's Metropolitan Opera with Enrico Caruso as Dick Johnson, music by Giacomo Puccini.

First performances Three Piano Pieces by Arnold Schoenberg 1/14 at Vienna; *Mother Goose (Ma Mère l'Oye)* by Maurice Ravel 4/20 at Paris. The four-handed piano piece will have its first orchestrated performance early in 1912; Fantasia on a Theme by Thomas Tallis for Double String Orchestra by English composer Ralph Vaughan Williams, 37, 9/6 at Gloucester Cathedral; Symphony No. 8 in E flat major by Gustav Mahler 9/12 at Munich's Exposition Concert Hall, with 146 orchestral players, two mixed choruses of 250 voices each, a children's choir of 350, and seven vocal soloists.

The London Palladium opens December 26 with 2,500 seats. It will be London's most popular theater for vaudeville and revues beginning with one-night musical comedy turns.

Broadway musicals *Tillie's Nightmare* 5/5 at the Herald Square Theater, with Marie Dressler, music by A. Baldwin Sloane, lyrics by Edgar Smith, songs that include "Heaven Will Protect the Working Girl" based on the 1898 song "She Was Bred in Old Kentucky," 77 perfs.; *The (Ziegfeld) Follies* 6/20 at the Jardin de Paris, with singer Fannie Brice (Fannie Borach), 18, who won a Brooklyn talent contest 5 years ago singing "When You Know You're Not Forgotten by the Girl You Can't Forget," left school to start a theatrical career, and has been hired at $75 per week by Florenz Ziegfeld, music by Gus Edwards and others, book and lyrics by Harry B. Smith, 88 perfs.; *Madame Sherry* 8/30 at the New Amsterdam Theater, with songs that include "Every Little Movement Has a Meaning All Its Own" by Karl Hoschna and Otto Harbach, "Put Your Arms Around Me, Honey" by Albert von Tilzer, lyrics by Junie McCree, 45, 231 perfs.; *Naughty Marietta* 11/7 at the New York Theater, with music by Victor Herbert, book and

1910 ***(cont.)*** lyrics by Rida Johnson Young, songs that include "Tramp! Tramp! Tramp!," "I'm Falling in Love with Someone," "Ah, Sweet Mystery of Life," 136 perfs.

Popular songs "Come, Josephine, in My Flying Machine" by German-American composer Fred Fisher, 35, lyrics by Canadian-American Alfred Bryan, 39; "Mother Machree" by Chauncey Olcott and Ernest R. Ball, lyrics by Rida Johnson Young; "Down by the Old Mill Stream" by Tell Taylor, 34; "A Perfect Day" by Carrie Jacobs-Bond; "Let Me Call You Sweetheart" by Leo Friedman, lyrics by Beth Slater Whitson; "Some of These Days" by Shelton Brooks; " 'Opie'-The University of Maine Stein Song" by Norwegian-American composer E. A. Fenstad, 40, lyrics by 1906 University of Maine graduate Lincoln Colcord, 27 (*see* Rudy Vallée, 1929).

World heavyweight champion Jack Johnson handily defeats former titleholder James J. Jeffries July 4 at Reno, Nev. (above).

Anthony Frederick Wilding, 26, (New Zealand) wins in men's singles at Wimbledon, Mrs. Chambers in women's singles; Bill Larned wins in U.S. men's singles, Hazel Hotchkiss in women's singles.

Spanish matador Juan Belmonte kills his first bull July 24 at age 18 in the new ring at El Arahal.

The All-Ireland Championship finals played since 1887 move into Dublin's Croke Park where they will be played each year (except in 1947 when New York's Polo Grounds will host the games in an effort to revive interest in Gaelic football among Irish-Americans).

Connie Mack's Philadelphia Athletics win the World Series by defeating the Chicago Cubs 4 games to 1.

The Boy Scouts of America is founded by U.S. painter-illustrator Daniel Carter "Uncle Dan" Beard, 60, whose book *Boy Pioneers and Sons of Daniel Boone* was published last year. Beard has been inspired by the 2-year-old British organization founded by Lord Baden-Powell.

The Camp Fire Girls of America is founded by Luther Halsey Gulick, 45, who helped James Naismith invent the game of basketball in 1891 at Springfield, Mass. Now director of physical education for New York City public schools and a social engineer for the Russell Sage Foundation, Gulick gets help from his wife.

The Girl Guides is founded by Lord Baden-Powell (above) with his sister Agnes, 52 (*see* Girl Scouts of America, 1912).

Father's Day is observed for the first time June 19 at Spokane, Wash., where the local YMCA and the Spokane Ministerial Association have persuaded the city fathers to set aside a Sunday to "honor thy father." The idea has come from local housewife Mrs. John Bruce Dodd, 28, who has been inspired by the selflessness and responsibility of her father William Smart, a Civil War veteran who raised his daughter and her five brothers after the early death of his wife.

The Elizabeth Arden beauty-salon chain has its beginnings in a New York beauty treatment parlor started by Canadian-American beauty shop secretary Florence Nightingale Graham, 25, who first goes into business with Elizabeth Hubbard, has a falling out with her partner, borrows $6,000 from a cousin, and opens a Fifth Avenue shop under the name Elizabeth Arden inspired by the 1864 Tennyson poem *Enoch Arden*. Graham repays the loan within 4 months, will move farther uptown and open a Washington, D.C., branch in 1915; by 1938 there will be 29 Elizabeth Arden salons.

The Tielocken coat introduced by Burberry's will be called the trenchcoat beginning in 1914 (*see* 1856). Tied and locked closed with a strap and buckle, it will be given buttons, epaulettes, and rings for hanging grenades.

U.S. cigarette sales reach 8.6 billion with 62 percent of sales controlled by the American Tobacco Trust created in 1890. The tobacco companies spend $18.1 million to advertise their brands (*see* 1911; 1913).

Barcelona's Casa Mila is completed by Spanish architect Antonio Gaudi, 58, after 5 years of construction.

Madrid's Ritz Hotel opens October 23 with 200 bedrooms and salons, 100 baths.

Boston's Charles River Dam is completed to maintain the water level of the Charles River and remove the flats exposed at low tide.

Wyoming's Shoshone Dam is completed by the Bureau of Reclamation. The arch of rubble on the North Platte River rises 328 feet high.

Glacier National Park in Montana is created by act of Congress setting aside more than a million acres of lakes, peaks, glaciers, and Rocky Mountain flora and fauna (*see* 1932).

Florida orange shipments finally regain their 1894 levels, but Florida's northern groves have been abandoned and population in the Orlando area has suffered a huge decline.

A blueberry developed by New Jersey botanist Frederick Covine is plump, almost seedless, and will revolutionize the industry, creating the basis of a multimillion-dollar industry.

The world's first glass-filled milk car goes into service on the Boston & Maine Railroad for Boston's Whiting Milk Co.

Seventy percent of U.S. bread is baked at home, down from 80 percent in 1890 (*see* 1924).

Aunt Jemima pancake flour is sold throughout the United States as pancakes become a year-round staple served at many meals rather than just at winter breakfasts (*see* 1893).

The U.S. population reaches 92 million with 13.5 million of it foreign-born. Just over half live in cities and towns of 2,500 or more, up from 21 percent in 1860 (in Germany, 34.5 percent of the people live in cities of 20,000 or more, up from 18.4 percent in 1885).

The U.S. Immigration Commission winds up nearly 4 years of study with a 41-volume report that recommends restricting immigration, especially of unskilled labor (*see* 1917).

1911 Mexico's president Porfirio Diaz is overthrown May 25 (*see* 1910). Revolutionist Francisco Madero, who opposed the reelection of Diaz last year but was forced to flee to the United States, has led a military campaign against Diaz and established a new capital May 11 at Ciudad Juarez; he makes himself president November 6 and begins an administration that will continue for 15 months (*see* 1913).

A military coup establishes a new Honduras "banana republic" favorable to the interests of Bessarabian-American planter Samuel Zemurray, 34, who for the past 6 years has been shipping bananas from lands he has bought along the country's Cuyamel River. He loans former Honduras president Gen. Manuel Bonilla enough money to buy the yacht *Hornet,* sends him out to the yacht in his own private launch at Biloxi, Miss., with a case of rifles, a machine gun, and ammunition, and provides him with the services of soldiers of fortune Guy "Machine Gun" Molony and Lee Christmas (*see* 1897). When Bonilla and his cohorts oust the old Honduras government, Zemurray gains valuable concessions (*see* United Fruit, 1929).

Britain's House of Lords gives up its veto power August 10 under the Parliament Act passed under pressure from Prime Minister Asquith, who threatens to create enough peers to carry the bill.

Britain's House of Commons for the first time votes salaries for its members. Each M.P. is to receive £400 per year under terms of the measure adopted August 10, 6 weeks after the coronation of George V.

Russian Premier Pyotr Arkadevich Stolypin is assassinated September 1 in the presence of Nicholas II at the Kiev Opera House. An informer for the czar's secret police fires a revolver at point-blank range.

The Black Hand Serbian secret society founded at Belgrade in May works to reunite Serbs living within the Austrian and Ottoman Empires with their kinspeople in Serbia. "Unity or Death" ("Unedinjenje ili Smrt") is the society's slogan, but while the Black Hand will influence Serbian policies in the Balkan wars, its leader Col. Apis Dimitrievic will clash with the Serbian government and be executed with two colleagues in June 1917.

Italy declares war on the Ottoman Turks September 9, lands a force at Tripoli October 5, and occupies other coastal towns. Italian planes bomb an oasis on the Tripoli coast November 1 (the first offensive use of aircraft), and Cairo declares martial law November 2 to quell unrest in Egypt. Rome announces annexation of Libya, Tripolitania, and Cyrenaica November 5. Constantinople refuses to recognize the action (*see* 1912).

A revolution begins in China that will end the 267-year-old Qing dynasty of the Manchus, propel China into the 20th century, and begin the decline of such customs as having men wear humiliating pigtails and women's feet painfully deformed by binding them.

Chinese revolutionary leader Sun Yat-sen, 45, returns from 16 years of exile in Hawaii, England, and the United States. The first graduate of Hong Kong's new College of Medicine, Sun is elected president of the United Provinces of China December 29 by a revolutionary provisional assembly at Nanjing.

The literary magazine *Seito* (*Bluestocking*) that begins publication in February at Tokyo marks the start of a women's liberation movement launched by feminist Raicho (Haru) Hiratsuka, 25. Her father helped draft the Constitution of 1889 and the civil code that is so heavily weighted against women. *Seito* will be suppressed after its February 1916 issue and the status of Japanese women will remain low.

New York's Triangle Shirtwaist Factory at Washington Place and Greene Street has a fire March 25 and 146 people are killed, most of them sweatshop seamstresses who are unable to escape. The tragedy brings new demands for better working conditions (*see* IWW Lawrence strike, 1912).

Portuguese women get the vote April 30.

California women gain suffrage by constitutional amendment (*see* New York, 1917).

Britain has a nationwide strike of transport workers in August as Labour party leader Keir Hardie exhorts workers. With famine looming, riots erupt at Liverpool August 8 and troops open fire, killing two. Some 50,000 troops are rushed to London to restore order.

A British Trade Disputes act guarantees the right to strike and to picket peacefully.

A limited British National Insurance program legislated by Parliament in December provides unemployment insurance for 2.25 million workers in trades such as building and engineering that are especially subject to layoffs. Funded by contributions from workers, employers, and the state, the program provides benefits for a maximum of 15 weeks.

Norwegian explorer Roald Amundsen, 39, arrives at the South Pole December 14 with four fellow Norwegians

Amundsen of Norway reached the South Pole with sledge dogs. Scott and his British companions died trying.

1911 ***(cont.)*** and 17 ravenous huskies. The men survive on seal blubber, they appease the sledge dogs by killing one and feeding it to the others, they are the first to reach the Pole, but five British explorers led by Robert Falcon Scott, now 43, will be less lucky. They will reach the Pole January 18 of next year, but with each man pushing a sledge whose load averages 190 pounds the daily ration of 4,800 calories will be inadequate and the men will starve to death.

U.S. Postal Savings service opens January 3 at 48 second-class post offices under terms of legislation signed last June by President Taft. Deposits reach $11 million in 11 months, and the money is distributed among 2,710 national and state banks.

The Purchasing Power of Money by Yale political economy professor Irving Fisher, 44, advances the thesis that prices rise in proportion to the supply of money and the velocity with which money circulates. Fisher pioneers in "indexing" the economy with price indexes, cost-of-living indexes, etc. (*see* Bureau of Labor Consumer Price Index, 1913).

Theory of Economic Development by Moravian economist Joseph Alois Schumpeter, 28, will influence other economists' thinking especially after it is translated into English in 1934. A professor at the University of Graz, Schumpeter will be appointed to a Harvard chair in 1932.

The Supreme Court breaks up John D. Rockefeller's Standard Oil Company trust May 15, ruling in the case of *Standard Oil Co. of N.J. v. United States,* but the court rules only against "unreasonable" restraints of trade where a company has "purpose or intent" to exercise monopoly power in violation of the Sherman Act of 1890 (*see* tobacco trust decision, below). The trust is reorganized into five separate corporations plus some smaller ones—Standard Oil of New Jersey will later be called Esso and then Exxon, Standard of California (SoCal), Standard of Indiana, Standard of Ohio (Sohio), and Standard of New York (later Socony-Vacuum, then Mobil). Also Atlantic Refining (Atlantic Richfield beginning in 1967), Vacuum Oil, Prairie Oil & Gas, Buckeye Pipe Line, and Anglo-American Oil (*see* 1902).

An electric self-starter for motorcar and truck engines invented by C. F. Kettering improves automobile safety. Cadillac boss Henry M. Leland gives Kettering's Dayton Engineering Laboratories (Delco) a contract to supply 4,000 self-starters after losing a good friend who was killed trying to crank a woman's balky engine (*see* Leland, 1909; Ford, 1919; General Motors, 1918).

Electric Auto-Lite Co. is founded at Toledo, Ohio, by Clement O. Miniger to make ignition sets (starting motors, distributors, coils, generators) and spark plugs. By 1929 Electric Auto-Lite will be a major supplier to Ford, Essex, Hudson, Hupmobile, Jordan, Nash, Packard, Peerless, and Pierce and will rival Delco (above) (*see* 1930).

The first SAE handbook on standardization is published by the 6-year-old Society of Automotive Engineers. Beginning with spark plugs and carburetor flanges, the SAE will standardize screw threads, bolts, nuts, and all other automotive components (*see* lubricating oil, 1926).

Ford Motor Company wins a court decision that the Ford engine is fundamentally different from the G. B. Selden engine on which patents are about to expire anyway. Ford has refused to pay royalties to Selden's 8-year-old association which has received $5.8 million from other auto makers (Selden himself has received only $200,000, the rest having gone to lawyers, management executives, and the like).

Chevrolet Motor Co. is founded by former General Motors head W. C. Durant (*see* 1910). He teams up with Swiss-American racing car driver Louis Chevrolet, 32, who has been in the United States since 1901, and they produce the first Chevrolet motorcars in a New York plant at 57th Street and 11th Avenue (see 1915).

The first Indianapolis 500 (-mile) motorcar race is held May 30. A Marmon Wasp averages 75 miles per hour to win (*see* 1909).

Finishing eleventh in the Indianapolis 500 (above) is the first Stutz motorcar which averages 68 miles per hour despite tire troubles that have forced it to make persistent pit stops. Built by Indianapolis engineer Harry C. Stutz, 35, it earns the slogan "the car that made good in a day" (*see* 1913).

A transcontinental U.S. motorcar race organized by Philadelphia auto distributor John Guy Monihan begins June 26 at Atlantic City, N.J. Finalists arrive at Los Angeles August 10 after covering 4,731 miles.

A 25-horsepower Turcat-Mery wins the first Monte Carlo Rally.

An eight-valve four-cylinder Bugatti wins its class in the Grand Prix du Mans. The first all-Bugatti car, it has been designed by Italian auto maker Ettore Bugatti, 30, who has lived in France since his teens, built his first motorcar at age 17, and will dominate auto racing for more than 20 years.

Tolls on New York's Williamsburg Bridge of 1903 are removed at midnight July 19.

The last horse-drawn bus of the London General Omnibus Company goes out of service. The city had 1,142 licensed horse-drawn buses and the same number of motor buses at the end of October of last year, but while at least one horse-drawn bus will continue running (over Waterloo Bridge) until 1916, the horse is now rapidly being replaced by motorbuses and motorcars and trucks in major world cities.

A six-passenger Blériot *Berline* built for Parisian Henri Deutsch de la Neurthe is the first airplane with an enclosed passenger cabin.

Nellie Bly's 1889–1890 round-the-world record of 72 days falls to André Jaeger-Schmidt, who circles the earth from Paris to Paris in 39 days, 19 hours, traveling by ship and rail (*see* Mears, 1913).

The British White Star passenger liner S.S. *Olympic* arrives at New York June 21 to begin service that will continue until 1935. With an overall length of 892 feet, the 45,300-ton triple-screw ship is by far the largest liner yet built; she carries 2,500 passengers and is the first to have a swimming pool.

The Santa Fe *Deluxe* goes into service December 2 between Chicago and Los Angeles. Passengers pay $25 extra fare each on the weekly 63-hour passenger express to enjoy services that include a barbershop, a library, a stenographer, ladies' maids, daily market reports, bathing facilities, a club car, a Fred Harvey dining car, four Pullman sleeping cars whose compartments and drawing rooms have brass beds, and telephones at terminals en route (*see* 1926).

Union Carbide Co. acquires the oxygen company founded 4 years ago to supply users of oxyacetylene torches (*see* 1892; 1917).

Southern pine trees, used now only for turpentine and naval stores, prove useful as pulp sources for kraft paper, a strong brown packaging paper made by a sulfate process that uses an alkaline instead of an acid (as in the sulfite process) to remove the noncellulose lignin from pulp (*Kraft* is German for strength). The discovery will revolutionize the paper industry and bring new prosperity to parts of the South.

A nuclear model of the atom proposed by Ernest Rutherford at the University of Manchester is comprised largely of a positively charged nucleus surrounded by electrons (*see* Thomson, 1897; Geiger counter, 1908; Bohr, 1913).

U.S. explorer Hiram Bingham, 35, discovers the Inca city of Macchu-Pichu at an altitude of 8,200 feet in the Peruvian Andes. Jungle growth covers the long-deserted city that escaped notice by the Spanish conquistadors. Bingham's missionary grandfather developed a written language out of Hawaiian; his missionary father did the same for the language of the Gilbert Islands.

The Carnegie Corporation of New York is created with a $125 million gift from Andrew Carnegie to encourage education (*see* 1905; medicine, 1910).

The first direct telephone link between New York and Denver opens May 8.

The Official Secrets Act passed by Parliament August 22 makes it a criminal offense to publish any official government information without permission. No British publisher will defy the law until 1977.

Olivetti Co. is founded at Ivrea west of Milan in the Piedmont by electrical engineer Camillo Olivetti, 43, who has designed and built the first Italian typewriter. Few Italian firms have accepted steel-nibbed pens, much less typewriters, but by year's end Olivetti has received an order for 100 machines from the Italian Navy and by 1933 his plant will be producing 24,000 machines per year. The company will sell its typewriters in 22 foreign countries, but most will be sold through Olivetti retail shops in Italy.

The New York Public Library main branch opens on Fifth Avenue between 40th and 42nd streets, replacing a reservoir (*see* 1895). Lions sculpted by E. C. Potter guard the entrance to the white marble palace designed by Thomas Hastings of Carrère and Hastings.

Fiction *Death in Venice (Der Tod in Venedig)* by Thomas Mann; *Ethan Frome* by Edith Wharton; *Under Western Eyes* by Joseph Conrad; *Zuleika Dobson* by English critic-essayist-caricaturist Max Beerbohm, 39, whose satire about an adventuress at Oxford will be his only novel; *Life Everlasting* by Marie Corelli; *The Innocence of Father Brown* by G. K. Chesterton, whose detective hero is modeled on his friend Father O'Connor, who will receive Chesterton into the Church in 1922; *The Phantom of the Opera* by French mystery writer Gaston Leroux, 43; *Death of a Nobody (Mort de quelqu'un)* by French novelist Jules Romains (Louis Farigoule), 26; *Sanders of the River* by Edgar Wallace; *Mr. Perrin and Mr. Traill* by New Zealand-born English novelist Hugh Walpole, 27; *The Wild Geese (Gan)* by Japanese surgeon-novelist Ogai Mori, 49, whose 1909 autobiographical work *Vita Sexualis (Ita Sekusuarisu)* attacked the naturalist contention that man has no control over his sexual instinct; *Jennie Gerhardt* by Theodore Dreiser; *Mother* by U.S. novelist Kathleen Thompson Norris, 31, whose novels will be serialized in women's magazines for more than 30 years.

Juvenile *Mother Carey's Chickens* by Kate Douglas Wiggin; *Jim Davis* by John Masefield.

Poetry *The Everlasting Mercy* by John Masefield; "World's End" ("Weltende") by German poet Jacob van Hoddis (Hans Davidsohn), 24, is the world's first expressionist poem; "The Female of the Species" by Rudyard Kipling: "But the she-bear thus accosted rends the peasant tooth and nail./ For the female of the species is more deadly than the male."

Painting *The Red Studio* and *The Blue Window* by Henri Matisse; *Gabrielle with a Rose* by Pierre Auguste Renoir; *Man with a Guitar* by Georges Braque; *Accordionist* by Pablo Picasso; *Church at Chatillon* by Maurice Utrillo; *I and the Village* by Marc Chagall, who arrived at Paris last year.

Theater *The Rats (Die Ratten)* by Gerhart Hauptmann 1/13 at Berlin's Lessingtheater; *The Scarecrow* by U.S. playwright Percy Mackaye, 35, 1/17 at New York's Garrick Theater, 23 perfs.; *Mrs. Bumpsted-Leigh* by U.S. playwright Harry James Smith, 31, 4/3 at New York's Lyceum Theater, with Mrs. Minnie Maddern Fiske, 64 perfs.; *Fanny's First Play* by George Bernard Shaw 4/19 at London's Little Theatre, with Christine Silve, Harcourt Williams, Reginald Owen, 622 perfs.; *The Martyrdom of Saint Sebastian* by Gabriele D'Annunzio 5/22 at the Théâtre du Châtelet, Paris, with music by Claude Debussy. Ballet dancer Ida Rubinstein plays the lead role but D'Annunzio's works were placed on the index of forbidden books by the Vatican in early May and the archbishop of Paris warns Roman Catholics not to attend performances on threat of excommunication; *Disraeli* by Louis Napoleon Parker 9/18 at Wallack's Theater, New York, with George Arliss, 280 perfs.; *The Return of Peter Grimm* by David

1911 ***(cont.)*** Belasco 10/17 at New York's Belasco Theater, with David Warfield, 231 perfs.; *The Garden of Allah* by English novelist Robert Smythe Hichens, 47, and Mary Hudson 10/21 at New York's Century Theater, is based on Hichens's 1905 novel, 241 perfs.; *Everyman (Jedermann)* by Hugo von Hofmannsthal 12/1 at Berlin's Zirkus Schumann; *Oaha, the Satire of Satire (Oaha, die Satire der Satire)* by Frank Wedekind 12/23 at Munich's Lustspielhaus; *Kismet* by English-American playwright Edward Knoblock, 37, 12/25 at New York's Knickerbocker Theater, with Otis Skinner, 184 perfs.

Films D. W. Griffith's *Fighting Blood* and *The Lonedale Operator,* both with Blanche Sweet; William Humphreys's *A Tale of Two Cities* with Maurice Costelli, Norma Talmadge.

Bell & Howell's Albert S. Howell develops a continuous printer that makes copies of motion pictures automatically and economically to permit mass distribution (*see* 1909).

Keystone Co. is founded by Canadian-American motion picture pioneer Mack Sennett, 27, who changed his name from Michael Sinnott when he came to New York in 1904 to begin a stage career in burlesque and the circus. Sennett has been working with D. W. Griffith at Biograph Studios in New York (*see* Chaplin, 1913).

Opera *Der Rosenkavalier* (*The Knight of the Rose*) 1/26 at Dresden's Hofoper, with music by Richard Strauss, libretto by Hugo von Hofmannsthal; *Natoma* 2/8 at New York's Metropolitan Opera House, with Mary Garden, John McCormack, music by Victor Herbert; *L'Heure Espagnole* 5/19 at the Opéra-Comique, Paris, with music by Maurice Ravel; *The Jewels of the Madonna (I Gioielli della Madonna, or Der Schmuck der Madonna)* 12/23 at Berlin's Kurfuersten Oper, with music by Ermanno Wolf-Ferrari.

Ballet *Le Spectre de la Rose* at Monte Carlo's Théâtre de Monte Carlo, with music by the early 19th-century composer Carl Maria von Weber, choreography by Michel Fokine; *Petrouchka* 6/13 at the Théâtre du Châtelet, Paris, with Waslaw Nijinsky dancing the title role opposite Tamara Karsavina, music by Igor Stravinsky, who departs from his earlier romanticism to introduce bitonality and employ syncopation with unconventional meters.

First performances *Prometheus, the Poem of Fire* by Aleksandr Skriabin 3/15 at Moscow's Bolshoi Theater, in a pioneer multimedia effort to blend color and music; Symphony No. 4 by Jean Sibelius 4/3 at Helsinki; Valses Nobles et Sentimentales for Piano by Maurice Ravel 5/9 at the Salle Gaveau, Paris; Symphony No. 2 in E flat major by Sir Edward Elgar 5/24 at a London music festival.

Broadway musicals *The Pink Lady* 3/13 at the New Amsterdam Theater with Hazel Dawn, music by Belgian-American composer Ivan Caryll, 312 perfs. (the hit makes pink the year's fashion color); *La Belle Parée* 3/20 at Lee Shubert's Winter Garden Theater, with Russian-American blackface minstrel singer Al Jolson (*né* Asa Yoelson), 24, vaudeville entertainer George White (*né* Weitz), 21, music by New York composer Jerome David Kern, 26, and others, lyrics by Edward Madden and others, 104 perfs.; *The Ziegfeld Follies* 6/26 at the Jardin de Paris with the Dolly Sisters (Viennese dancers Jennie [Jan Szieka Deutsch] and Rosie [Roszka Deutsch], both 18), Leon Errol, 80 perfs.; *Betsy* 12/7 at the Casino Theater, with music by Leslie J. Stuart.

Popular songs "Alexander's Ragtime Band" by Irving Berlin popularizes the ragtime music that Scott Joplin pioneered in 1899: "Everybody's Doin' It" by Irving Berlin popularizes the Turkey Trot dance invented by dancer Vernon Castle and his 18-year-old bride Irene Foote Castle; "I Want a Girl Just Like the Girl that Married Dear Old Dad)" by Harry von Tilzer, lyrics by Will Dillon, 34; "Memphis Blues" by W. C. (William Christopher) Handy, 38, who adapts a campaign song he wrote in 1909 for a Memphis mayoralty candidate. Handy has assimilated black spirituals, work songs, and folk ballads with a musical form called "jass" and has employed a distinctive blues style with a melancholy tone achieved by use of flatted thirds and sevenths; "All Alone" by Albert von Tilzer, lyrics by Will Dillon; "When I Was Twenty-One and You Were Sweet Sixteen" by Egbert Van Alstyne, lyrics by Henry H. Williams; "Goodnight Ladies" by Egbert Van Alstyne, lyrics by Henry Williams; "Oh, You Beautiful Doll" by Nathaniel Davis Ayer, 24, lyrics by Seymour Brown, 26; "My Melancholy Baby" by Erne Burnett, 27, lyrics by Maybelle E. Watson that will be replaced next year with new lyrics by George A. Norton, 33; "Down the Field" (Yale football march) by Stanleigh P. Freedman, lyrics by C. W. O'Connor; "Little Grey Home in the West" by English composer Hermann Lohr, 40, lyrics by Miss D. Eardley Wilmot, 28; "Roamin' in the Gloamin'" by Scottish songwriter and music hall entertainer Harry Lauder (MacLennan), 41; "A Wee Doch-an-Doris" by Harry Lauder, lyrics by Gerald Grafton.

Tony Wilding wins in men's singles at Wimbledon, Mrs. Chambers in women's singles; Bill Larned wins in U.S. men's singles, Hazel Hotchkiss in women's singles.

Connie Mack's Philadelphia Athletics win the World Series by defeating John McGraw's New York Giants 4 games to 2.

The Supreme Court breaks up James B. Duke's American Tobacco Co. Trust of 1890 May 29, ruling in the case of *United States v. American Tobacco Co.* Emerging from the trust are American Tobacco, P. Lorillard (*see* 1789), Liggett & Myers, R. J. Reynolds, and British-American Tobacco Co. (*see* Brown & Williamson, 1927). The American Snuff Co., which has been part of the Tobacco Trust, breaks into three equal parts—American Snuff, G. W. Helme, and United States Tobacco, whose major brands include Copenhagen Snuff and Bruton.

German architect Walter Adolf Gropius, 28, designs a steel skeleton building whose only walls are glass "curtain walls" for the Fagus factory at Alfeld. Gropius will create a stir with his model factory at the 1914 Werkbund Exhibition at Cologne (*see* Bauhaus, 1918).

Irving Berlin wrote ragtime and other songs that kept the world singing through good times and bad.

Rome's Victor Emmanuel II monument designed by Giuseppe Sucani is completed after 28 years of construction. The ornate structure symbolizes the achievement of Italian unity in 1861.

Frank Lloyd Wright completes Taliesin East at Spring Green, Wis. (*see* 1909; Taliesin West, 1938).

A paper presented to the American Society of Mechanical Engineers by Willis H. Carrier will be the basis of modern air-conditioning (*see* 1902). "Rational Psychometric Formulae" is based on 10 years of work that have initiated a scientific approach to air-conditioning (*see* Carrier Corp., 1915).

China's Yangzi (Yangtze) River floods its banks in early September, killing an estimated 100,000.

Roosevelt Dam on the Salt River in Arizona Territory is completed by the Bureau of Reclamation as part of the first large-scale irrigation project. The 284-foot high rubble masonry arch has a crest length of 1,125 feet, creates a reservoir with a capacity of 1.64 million acre-feet, and will make the Phoenix area a rich source of fruits and vegetables.

Forest Physiography by U.S. geographer Isaiah Bowman, 33, is the first detailed study of the physical geography of the United States.

Congress passes the Appalachian Forest Reserve Act to appropriate funds for purchasing forest lands to control the sources of important streams in the White Mountains and southern Appalachians (*see* Appalachian Trail, 1921).

The Pribilof Island seal population that had fallen to 2.5 million when Alaska was acquired by the United States in 1867 falls to 150,000 following years of operations by the Alaska Commercial Co. The United States, Canada, Japan, and Russia sign a Fur Seal Treaty agreeing to save the seals from extinction.

The Columbia River salmon catch reaches a record 49 million pounds that will never be reached again (*see* 1866; 1938).

Albert I of the Belgians grants Lever Brothers' head W. H. Lever, Lord Leverhulme, a Belgian Congo concession to develop a plantation of 1.88 million acres on condition that Lever pay minimum wages and establish schools and hospitals. Leverhulme has bought 200,000 acres in the Solomon Islands of the Pacific and planted 17,000 acres to coconut palms; he founds Huiléries du Congo Belge and plants oil palms to create oils for his soaps and margarines (*see* Lux flakes, 1906; Rinso, 1918).

U.S. entrepreneur Haroldson Lafayette Hunt, 22, invests a small inheritance in Arkansas delta cotton land after having worked his way through the West and Southwest since leaving home at age 16. Hunt will increase his holdings in Arkansas and Louisiana to some 15,000 acres in the next few years (*see* oil, 1920).

U.S. inventor Benjamin Holt devises an improved combine that harvests, threshes, and cleans wheat (*see* 1904; 1905; Industrial Commission, 1901).

Famine reduces 30 million Russians to starvation. Nearly one quarter of the peasantry is affected, but 13.7 million tons of Russian grain, mostly wheat, are shipped abroad (*see* 1947).

Polish biochemist Casimir Funk, 27, at London's Lister Institute introduces the word *vitamines*. If the enzymes discovered by the late Wilhelm Kuhne in 1878 are to work properly they sometimes require coenzymes, Funk discovers, and he calls these coenzymes "vitamines" in the mistaken belief that they all contain nitrogen and are all essential to life (*see* 1912; Drummond, 1920).

The 1906 Pure Food and Drug Law provision against advertising that is "false or misleading in any particular" does not apply to far-fetched declarations of therapeutic or curative effects, rules the Supreme Court (*see* Sherley Amendment, 1912).

Crisco, introduced by Cincinnati's Procter & Gamble, is the first solid hydrogenated vegetable shortening (*see* Normann, 1901; Spry, 1936).

Domino brand sugar is introduced by the American Sugar Refining Co. (*see* 1907).

Battle Creek, Mich., plants produce corn flakes under 108 brand names but Kellogg's and Post Toasties lead the pack (*see* 1906). W. K. Kellogg buys out the last of his brother's holdings in the Battle Creek Toasted Corn Flake Co., going $330,000 into debt in order to gain virtually complete ownership of the company that will become Kellogg Corp. (*see* 40% Bran Flakes, 1915).

Mass production of canned pineapple is made possible by Henry Ginaca's invention of a machine that can remove the shells from 75 to 96 pineapples per minute, coring each one and removing its ends.

1911 ***(cont.)*** James Dole's Hawaiian Pineapple Co. packs more than 309,000 cases, up from less than 234,000 last year (*see* 1909; 1922).

M. S. Hershey, Inc., has sales of $5 million (*see* 1903; 1918).

New York's Ellis Island has a record one-day influx of 11,745 immigrants April 17 (*see* 1892). Two steamship companies have more than 5,000 ticket agents in Galicia alone, some Germans buy tickets straight through from Bremen to Bismarck, N.D., and the U.S. population is at least 25 percent foreign-born in nearly every part of the country outside the South.

The U.S. population reaches 94 million while Britain has 40.8 million (36 million in England and Wales), Ireland 4.3, France 39.6, Italy 34.6, Japan 52, Russia 167, India 315, China 425.

1912 China becomes a republic January 1 with Sun Yat-sen as provisional president. Sun founds the Guomindang (Nationalist party); Gen. Yuan Shih-kai, who controls the army and police, has his queue cut to symbolize the end of the Qing, or Manchu, dynasty (*see* 1911; 21 demands, 1915).

French statesman Raymond Poincaré, 51, forms a ministry January 14 that will wrestle with problems of Morocco, Italy's Tripolitanian war (below), and the Balkan wars (below).

Denmark's Frederick VIII dies June 14 at age 69 after a 6-year reign. His son, 41, will reign until 1947 as Christian X.

New Mexico and Arizona are admitted to the Union as the 47th and 48th states.

U.S. Marines land in Honduras in February, in Cuba 4 months later, and in Nicaragua in August to protect American interests. They will not be withdrawn from Nicaragua until 1933.

Japan's Meiji emperor Mutsuhito dies July 30 at age 60 after a 45-year reign that has restored imperial power (*see* 1867; 1868). Gen. Maresuke Nogi, principal of a school for the imperial family, commits *seppuku* with his wife by immolation. Mutsuhito is succeeded by his son Yoshihito, 33, who will reign as the Taisho emperor Taisho until 1926, a period that will see Japan emerge as a world power of the first rank.

The Treaty of Lausanne October 18 ends the 12-month war between Italy and the Ottoman Turks. Italian forces have occupied Rhodes and the other Dodecanese Islands in May, peace talks began in July, and Constantinople, under pressure in the Balkans (below) has finally agreed to give up Tripoli, Italy to exit the Dodecanese Islands as soon as the Turks exit Tripoli.

Montenegro has declared war on Turkey October 8, the Turks have declined October 12 to undertake reforms in Macedonia demanded by the great powers, Constantinople has declared war on Bulgaria and Serbia October 17, the Greeks join the conflict, and the Turks suffer major reverses.

Bulgarian forces triumph October 22 at Kirk Kilisse in Thrace, the Serbs gain a victory October 24 to 26 at Kumanovo, and Bulgarian troops under Mikhail Savov, 55, win the Battle of Lule Burgas October 28 to November 3, bringing them to the last lines of defense before Constantinople. Moscow warns that it will use its fleet to resist a Bulgarian occupation of Constantinople, and the Bulgarians fail in an attack on the Turkish lines.

The Serbs overrun northern Albania and reach the Adriatic November 10, they gain victory at Monastir November 15 to 18, but Austria declares her opposition to Serbia having access to the Adriatic and announces her support for an independent Albania. Austria and Russia begin to mobilize but an armistice December 3 ends most of the conflict.

Former U.S. president Theodore Roosevelt bolts the Republican party to run for reelection as a Progressive (his party is called the "Bull Moose" party because Roosevelt has said he felt "fit as a bull moose"); he is wounded in Wisconsin October 14 by a would-be assassin but wins 88 electoral votes and 28 percent of the popular vote. President Taft wins only 8 electoral votes and 23 percent of the popular vote, Socialist Eugene Debs polls more than 900,000 votes, and New Jersey governor Woodrow Wilson receives 435 electoral votes with 42 percent of the popular vote to give Democrats control of the White House for the first time since Grover Cleveland left office in 1897.

Lawrence, Mass., textile workers walk out January 12 in a strike led by IWW agitators (*see* 1905). The strike succeeds, giving prominence to the International Workers of the World among textile workers throughout the East (*see* 1913).

Congress extends the 8-hour day to all federal employees; in private industry most workers labor 10 to 12 hours per day 6 days per week.

A minimum wage law for women and children legislated in Massachusetts is the first state law of its kind.

Italians gain almost universal suffrage under legislation adopted June 29. The measure also provides that salaries be paid members of parliament.

E. I. du Pont de Nemours divests itself of some explosives factories by court order. Stockholders receive securities in two new companies, Atlas Powder and Hercules Powder, but the courts permit Du Pont to retain its 100 percent monopoly in military powder (*see* 1906; 1919).

Some 30 National Cash Register officers including J. H. Patterson and Thomas J. Watson are indicted for criminal conspiracy in restraint of trade (*see* 1910). Patterson fires Watson, who will become general manager of the Computer-Tabulating-Recording Co. (C-T-R), a holding company with a subsidiary that employs Herman Hollerith and has acquired the Hollerith punched card patents of 1890 (*see* IBM, 1924).

Bethlehem Steel's Charles M. Schwab journeys to France and buys Chile's Tofo Iron Mines from the Schneider interests. The Chilean mines contain 50 million tons of ore with iron content 10 percent better than Lake Superior ores (*see* 1907; 1913).

R. H. Macy's Nathan Straus is shaken by the loss of his brother Isidor on the *Titanic* (below) and sells his interest in the store to Isidor's sons Jesse Isidor Straus, 40, Percy Seldon Straus, 36, and Herbert N. Straus, 30 (*see* 1901; Thanksgiving Day parade, 1924).

L. L. Bean, Inc., of Freeport, Me., is founded by local merchant Leon Leonwood Bean, 40, who has invented the Maine Hunting Shoe and who goes into partnership with his brother Gus to open a small clothing store. Ninety of the first 100 hunting shoe orders are returned because their leather tops have separated from their rubber bottoms, but Bean solves the problem with a combination of stitching and glue, makes good on all the returns, and begins a mail-order sporting goods empire whose sales will top $1 million per year by 1937, $3.5 million by the time Bean dies in 1967 with sales of chamois shirts and down vests supplementing those of the Maine Hunting Shoe.

A new Filene's with a 7-foot doorman opens in Boston September 3 at the corner of Washington and Summer Streets in a building designed by Daniel Burnham (*see* Automatic Bargain Basement, 1909). Edward Filene will hold free tea dances beginning in 1919, and although his Automatic Bargain Basement is losing money and will continue to lose for 7 more years, it will then become profitable and endure as a Boston institution (*see* Federated, 1929).

F. W. Woolworth Co. is incorporated by Frank Woolworth who absorbs the five-and-dime chains established by four of his erstwhile partners (*see* 1909; 1919; Woolworth Building, 1913).

Electric light bulbs last longer thanks to General Electric research chemist Irving Langmuir, 31, who discovers that filling incandescent bulbs with inert gases will greatly increase the illuminating life of tungsten filaments developed by his colleague W. D. Coolidge (*see* 1913). Confuting conventional wisdom that the vacuum in the light bulb is what permits its filament to burn so long, Langmuir shows that while bulbs with poor vacuums are no worse than those with the best, adding nitrogen gas (which does not react with the tungsten filament) will avoid evaporation of the filament and prolong the life of the bulb. Langmuir will substitute argon for nitrogen (*see* 1894).

Oklahoma's Wheeler Well No. 1 comes in and begins making a fortune for wildcatter Thomas Baker Slick, 28, who has leased more than 100,000 acres of the state's oil lands. The discovery well that starts gushing in March is in the Cushing field.

The U.S. Navy establishes petroleum reserves at Elk Hills and Buena Vista Hills, Calif, as it converts its ships from coal to diesel (*see* Teapot Dome, 1914).

Turkish Petroleum Co. is founded to exploit reserves discovered in Mesopotamia.

The S.S. *Titanic* of the White Star Line scrapes an iceberg in the North Atlantic on her maiden voyage, sustains a 300-foot slash, and sinks in 2½ hours on the night of April 15. The three-screw passenger liner of 46,328 tons, 882 feet in length overall, is the world's largest passenger liner, she has been called "unsinkable," but only 711 of the 2,224 aboard survive, and the 1,513 lost include such prominent millionaires as John Jacob Astor IV, 47; R. H. Macy's Isidor Straus, 67; copper heir Benjamin Guggenheim, 47, and traction heir Harry Elkins Widener, 27 (*see* Sarnoff, below).

German-American engineer Grover Loening, 24, designs and builds the world's first amphibious aircraft. He was graduated 2 years ago from Columbia University with the first U.S. master's degree in aeronautics and his "aeroboat" brings him to the attention of Orville Wright, who will hire him next year as his assistant and manager of Wright Aircraft's Dayton, Ohio, factory.

Dutch aircraft designer Anthony Herman Gerard Fokker, 22, introduces the Fokker aeroplane, opens a factory at Johannesthal, Germany, and will build another next year at Schwerin (*see* 1916; 1922).

English aeronaut Thomas Octave Murdoch Sopwith, 24, founds Sopwith Aviation at Kingston-on-Thames (*see* Hawker, 1921).

Bendix Brake Co. is founded by Vincent Bendix, who has begun production of his Bendix starter drive. His new firm will be the first mass producer of four-wheel brakes for motorcars (*see* 1907; Bendix Aviation, 1929).

Buick Division chief Charles William Nash, 48, becomes president of General Motors and brings in American Locomotive Works manager Walter P. Chrysler, 37, to head Buick (*see* Durant, 1910; Nash Motors, 1916; Chrysler, 1923).

Seven companies produce half of all U.S. motorcars (*see* 1923).

More than 22 percent of all U.S. motorcars are Fords, which come out of Ford factories at the rate of 26,000 per month (*see* 1911; assembly line, 1913).

The Nissan automobile has its beginnings in the Kaishinsha Motor Car Works founded at Tokyo. The company will produce its first vehicle in 1917, it will merge in 1926 with Jitsuyojidosha Seizo of Osaka, will

The Boston Daily Globe.

TITANIC SINKS, 1500 DIE

Carpathia Picks Up 675 Out of 2200---Races for New York--Survivors Mostly Women and Children.

POLICE ORDER DORR'S ARREST

Lynn Chief Accuses Him of the Murder of George E. Marsh.

Suspect Said to Have Left Boston Thursday Night--Auto Found Here.

Giant Steamer Goes Down Before Help Arrives.

Virginian or Parisian May Have Some Survivors

White Star Officials Admit "Horrible Loss of Life".

Greatest Sea Tragedy in History Off Newfoundland Coast.

Grazing an iceberg proved fatal to the "unsinkable" *Titanic*—and to 1,513 of her 2,224 passengers.

1912 ***(cont.)*** make only trucks until 1930, and will enter the U.S. market in 1960 under the name Datsun.

Brazil's Madeira-Mamoro Railway opens after 6 years of construction that has cost 6,000 lives and the equivalent of 3 tons of gold. Rubber barons have financed the 255-mile road to circumvent 19 major waterfalls, the road permits resumption of Bolivian rubber shipments, but malaria, yellow fever, beriberi, snake bites, wild animals, and curare-tipped arrows from hostile natives have taken a heavy toll, and the Amazon Basin rubber boom will soon collapse as East Indian and African plantations undercut the price of wild rubber.

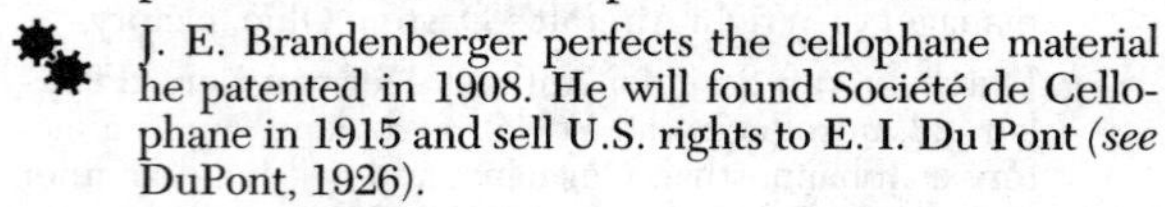

J. E. Brandenberger perfects the cellophane material he patented in 1908. He will found Société de Cellophane in 1915 and sell U.S. rights to E. I. Du Pont (*see* DuPont, 1926).

The Piltdown man hoax deceives world paleontologists. An English amateur claims to have discovered the "missing link" between man and ape. The ruse will not be exposed until 1953.

The first diagnosis of a heart attack in a living patient appears in the December 7 *Journal of the American Medical Association (JAMA)*. Chicago physician James B. Herrick points out that what may seem to be acute indigestion, food poisoning, angina pectoris, or something else may in fact be due to a blood clot (thrombosis) in the coronary artery which is what generally destroys a segment of the heart muscle to produce a myocardial infarction, considered until now merely a curiosity seen on autopsy as an inevitable consequence of aging. Herrick's patient (a banker of 55) has survived only 52 hours, but Herrick shows that heart attacks need not be fatal. Further work will show that clots generally occur in coronary arteries damaged by arteriosclerosis (or atherosclerosis), but heart disease remains a relatively minor cause of death among Americans as compared with tuberculosis and pneumonia (*see* Herrick, 1910; 1921).

The Pituitary Body and Its Disorders by Boston surgeon Harvey Cushing, 43, advances knowledge of the pituitary gland and its relation to diabetes.

Rice University is founded at Houston under the name Rice Institute with a bequest from the late Texas merchant William Marsh Rice, who was chloroformed to death by his valet in New York 12 years ago at age 84 (*see* 1838). Boston architect Ralph Adams Cram has designed the campus buildings.

The University of California opens a San Diego campus (*see* 1908; 1919).

The Montessori Method by Italian educator Maria Montessori, 42, describes her success at teaching slum children between the ages of 3 and 6 how to read. First woman graduate of Rome University medical school, Montessori begins a movement of Montessori schools for teachers.

The Kallikak Family by U.S. psychologist Henry Herbert Goddard, 46, relates feeblemindedness to crime and causes a sensation. Goddard, who invented the word "moron" in 1910, last year tested the school population of a New Jersey town and revised the Binet-Simon I.Q. test for U.S. use (*see* 1916).

The wireless message "S.S. *Titanic* ran into iceberg. Sinking fast" is picked up accidentally by Russian-American wireless operator David Sarnoff, 20, who is manning a station set up by John Wanamaker in his New York store window ostensibly to keep in touch with the Philadelphia Wanamaker's but actually as a publicity stunt. Sarnoff relays the message from the *Titanic* (above) to another steamer, which reports that the liner has sunk but that some survivors have been picked up. President Taft orders other stations to remain silent and Sarnoff remains at his post for 72 hours, taking the names of survivors and making his own name familiar to millions of newspaper readers (*see* 1917).

An SOS in Morse code—three dots, three dashes, three dots—is adopted as a universal distress signal by an International Radio-Telegraph Conference.

Irving Langmuir's work on vacuums (above) will be applied to the development of an improved radio tube whose patents will be major assets of Radio Corp. of America (*see* 1919).

Bell Laboratories physicist H. D. Arnold produces the first effective high-vacuum tube for amplifying electric currents (*see* Edison, 1883; Langmuir, above; transistor, 1948).

Parliament nationalizes Britain's telephone system. The action follows that of most European countries and brings demands for the nationalization of American Telephone and Telegraph (*see* 1910; Burleson, 1913).

AT&T's Western Union buys U.S. rights to a multiplex device that permits up to four messages to be sent at once over the same circuit. The multiplex takes advantage of the difference between the speed of mechanical impulses and the speed of electrical impulses.

Pravda begins publication to voice the ideas of Russia's underground Communist party (*see* 1905) with Vyacheslav Molotov (Vyacheslav Mikhailovich Skryabin), 22, as editor. V. I. Lenin uses the Russian word for truth and while *Pravda* will publish only one view of the truth, the newspaper's circulation will grow to become the world's largest.

The S. I. Newhouse publishing empire has it beginnings at Bayonne, N.J., where a local lawyer has acquired controlling interest in the foundering *Bayonne Times* as payment for a legal fee and sends his office boy Samuel I. Newhouse, 17, to take charge, offering him half the paper's profits in lieu of salary. By the time he is 21 Newhouse will have earned a law degree at night school and be earning $30,000 per year from the newspaper. He will persuade his employer to join him in buying the *Staten Island Advance* in 1922, and by 1955 the Newhouse chain will include the *Long Island Press, Newark Star-Ledger, Syracuse Post-Standard, Herald-Journal,* and *Herald-American, St. Louis Globe-Democrat, Portland Oregonian*, and *Portland (Oregon) Journal* (*see* 1967).

Fiction *The Autobiography of an Ex-Colored Man* by U.S. poet-essayist James Weldon Johnson of 1899 "Lift Every Voice and Sing" fame (he will not acknowledge authorship until 1927); *Jean-Christophe* by French novelist-playwright Romain Rolland, 46; *The Sea and the Jungle* by *London Morning Leader* staff writer H. M. (Henry Major) Tomlinson, 39; *Twixt Land and Sea* (stories) by Joseph Conrad includes "The Secret Sharer" and "The Inn of the Two Witches"; *The Unbearable Bassington* by Saki (H. H. Munro); *The Financier* by Theodore Dreiser; *Riders of the Purple Sage* by former New York dentist Zane Grey, 37.

Poetry *Duino Elegies* by Rainer Maria Rilke who enters a period of existential self-revelation and will not complete the *Elegies* until 1922; *Ripostes* by U.S. émigré poet Ezra Loomis Pound, 27, who went abroad to live 4 years ago and published his first book at Venice a year later under the title *A lume spento*. Pound pioneers an anti-poetical imagism which he will abandon in 2 years for a vorticism that will relate to painting and sculpture (*see* 1925); *A Dome of Many-Coloured Glass* by Boston poet Amy Lowell, 36, sister of Harvard president Abbott Lawrence Lowell, 56. Barely 5 feet tall and weighing 250 pounds, the poet defies convention by smoking cigars, cursing in public, and taking another woman into her life and home.

Painting *The Violin* by Pablo Picasso; *Homage à Picasso* by Spanish cubist Juan Gris (José Victoriano Gonzalez), 25; *Woman in Blue* by Fernand Léger who has a one-man show at the Galerie Kahnweiler; *The Cattle Dealer* by Marc Chagall; *The Life of Christ* and *St. Mary of Egypt Among Sinners* by Emil Nolde; *Women Harvesting* by Kasimir Malevich; *September Morn* (*Matinée de Septembre*) by French painter Paul Emile Chabas, 43 (*see* 1913); *McSorley's Bar* and *Sunday, Women Drying Their Hair* by John French Sloan; *Woolworth Building* by water-colorist John C. Marin. Georges Braque invents collage.

Theater *The Marchioness Rosalind (La Marquess Rosalinda)* by Ramon del Valle Inclan in January at Madrid's Teatro de la Princess; *Death and Damnation (Tod und Teufel)* by Frank Wedekind 4/29 at Berlin's Künstlerhaus Werkstatt der Werdenden; *Epic Voices (Voces de gesta)* by Ramon del Valle Inclan 5/26 at Madrid's Teatro de la Princess; *Gabriel Schilling's Flight* (*Gabriel Schillings Flucht*) by Gerhart Hauptmann 6/14 at Bad Lauchstadt's Goethe Theater; *The Yellow Jacket* by U.S. playwrights George C. Hazelton and J. Harry Beurimo 11/4 at New York's Fulton Theater, 80 perfs.; *The Wolf (A farkas)* by Ferenc Molnár 11/9 at Budapest's Magyar Szinhaz; *The Tidings Brought to Mary (L'Annonce faite à Marie)* by Paul Claudel 12/20 at the Théâtre de l'oeuvre, Paris; *Peg O' My Heart* by J. Hartley Manners 12/20 at New York's Cort Theater, with the playwright's bride Laurette Taylor, 605 perfs.

The Minsky brothers take over New York's Winter Garden Theater for bawdy burlesque productions.

Films *Queen Elizabeth* with Sarah Bernhardt, shown July 12 at New York's Lyceum Theater, is the first feature-length motion picture seen in America. U.S. rights to the 40-minute French film have been acquired by Hungarian-American nickelodeon-chain operator Adolph Zukor, 39, a former furrier who has persuaded theatrical producer Charles Frohman to join him in investing $35,000 in the venture. Zukor and Frohman earn $200,000 showing the film on a reserved-seat basis in theaters across the country, and they form Famous Players Co. to produce films of their own (*see* Lasky, 1913).

Other films D. W. Griffith's *Her First Biscuit* with Mary Pickford and *The Musketeers of Pig Alley* with Lillian Gish.

Universal Pictures Corp. is created by a merger of U.S. film producers who include German-American cinema pioneer Carl Laemmle, 45. He will have sole control of Universal from 1920 to 1936 and be the first to promote the personalities of his film performers as "movie stars," hiring Mary Pickford, now 19, from the Biograph Studios of D. W. Griffith. (*see* United Artists, 1919).

Ballet *Daphnis et Chloe* 6/8 at the Théâtre du Châtelet, Paris, with Waslaw Nijinsky in the role of Daphnis, Tamara Karsavina as Chloe, music by Maurice Ravel, choreography by Michel Fokine.

First performances Symphony No. 9 in D flat major by the late Gustav Mahler 6/26 at Vienna; *Shepherd's Key* by Australian-English composer Percy Aldridge Grainger, 30, 8/19 at Queen's Hall, London; *Five Pieces for Orchestra (Fünf Orchesterstücke)* by Arnold Schoenberg 9/3 at a London Promenade Concert; *Two Pictures (Két Kép) for Orchestra* by Béla Bartòk 10/5 at Budapest; *Pierrot Lunaire: Three Times Seven Melodies for Reciter, Piano, Piccolo, Bass Clarinet, Viola, and Cello* by Arnold Schoenberg 10/16 at Berlin's Choralionsaal.

Broadway musicals *The Isle o' Dreams* 1/27 at the Grand Opera House, with Chauncey Olcott, music by Ernest R. Ball, lyrics by Olcott and George Graf, Jr., 26 (who has never been to Ireland), songs that include "When Irish Eyes Are Smiling," 32 perfs.; *The Passing Show* 7/22 at the Winter Garden Theater, with Willie and Eugene Howard, music by Louis A. Hirsch, lyrics by Harold Atteridge, 136 perfs.; *The Ziegfeld Follies* 10/21 at the Moulin Rouge, with Leon Errol, Vera Maxwell, 77 perfs.; *The Firefly* 12/12 at the Lyric Theater, with music by Prague-born U.S. composer Rudolf Friml, 32, lyrics by Otto Harbach, songs that include "Giannina Mia," 120 perfs.

Hymn "(He Walks With Me) In the Garden" by U.S. songwriter C. Austin Niles.

Popular songs "Waiting for the Robert E. Lee" by Lewis F. Muir, lyrics by L. Wolfe Gilbert; "When the Midnight Choo Choo Leaves for Alabam" by Irving Berlin; "Moonlight Bay" by Percy Wenrich, lyrics by Edward Madden; "That Old Gal of Mine" by Egbert Van Alstyne, lyrics by Earle C. Jones; "The Sweetheart of Sigma Chi" by Albion College sophomores F. Dudleigh Vernor, 20, and Byron D. Stokes, 26, who wrote the song last year; "Bulldog" and "Bingo Eli Yale" by Yale sophomore Cole Porter, 20.

1912 ***(cont.)*** Tony Wilding wins in men's singles at Wimbledon, Mrs. D.R. Larombe (*née* Ethel Warneford Thomson), 33, in women's singles; Maurice Evans McLoughlin, 22, wins in U.S. men's singles, Mary K. Browne, 15, in women's singles.

English golfer Harry Vardon, 42, wins the British PGA 12 years after winning the U.S. Open. He has invented the overlapping "Vardon grip" and will be among the top five players in the British Open 16 times in 21 years.

The Boston Red Sox win the World Series by defeating the New York Giants 4 games to 3.

The Olympic Games at Stockholm attract 4,742 contestants from 27 countries.

Oklahoma Indian James Francis "Jim" Thorpe, 24, wins both the pentathlon and decathlon at the fifth Olympiad (above), scoring 8,412 points out of a possible 10,000 in the decathlon and winning four firsts in the pentathlon. Thorpe returns to score 25 touchdowns and 198 points for the Carlisle Indian School at Carlisle, Pa., and is named halfback for the second year on Walter Camp's All-America team. Of Sac and Fox ancestry, Thorpe will lose his Olympic gold medals when he admits to having played semi-professional baseball during his summer vacation last year, thus losing his amateur status, and his name will be stricken from the Olympic record books at the insistence of the Amateur Athletic Union (AAU), but he will be rated the greatest athlete and football player of the first half of the twentieth century.

The fourth down is added to U.S. football and a touchdown is given a value of six points, up from the five-point value established in 1898. The football field is standardized at 360 feet by 160 including end zones 10 feet deep, a pass completed to the end zone is scored as a touchdown, and the football is reduced in girth from 27 inches at the middle to between 22 ½ and 23 inches (*see* 1929).

The first electric iron is introduced by Rowenta, a German manufacturer.

The Girl Scouts of America has its beginnings March 12 at Savannah, Ga., where Juliette Gordon Low, 52, starts the first troop of Girl Guides in America (*see* England, 1910). The Girl Guides will be renamed Girl Scouts next year and headquarters will be established at New York.

Hadassah is founded by Baltimore-born Henrietta Szold, 52, who will make the sisterhood of U.S. Jewish women a Zionist organization beginning in 1916 and head the group until 1926. Szold has been disappointed in love by Talmudic scholar Louis Ginzberg, 39, whose first books she translated but who 4 years ago met Adele Katzenstein in Berlin and married her.

New York's 998 Fifth Avenue luxury apartment house, designed in Italian Renaissance style by W. S. Richardson of McKim, Mead and White with one apartment per floor, opens opposite the Metropolitan Museum of Art at 81st Street, the avenue's first apartment building.

The Beverly Hills Hotel opens on a hilltop outside Los Angeles where it will be a favorite haunt of motion picture people.

The Manila Hotel opens formally July 4. A dance for 500 waltzers inaugurates the five-story 265-room hotel on Manila Bay.

France's Normandy and Golf hotels open with a gambling casino at Deauville.

Boston's Copley-Plaza Hotel opens in Copley Square on a site formerly occupied by the Museum of Fine Arts.

Montreal's Ritz Hotel opens New Year's Eve with 267 rooms.

Alaska's 6,715-foot Katmai volcano erupts and buries Kodiak Island 100 miles away under 3 feet of ashes while darkening skies over most of Alaska.

Japanese cherry trees given to the United States as a goodwill gesture from the royal family are destroyed when found to be infected with insects and disease (*see* 1904 chestnut blight). A second shipment of trees is planted at Washington, D.C., where they will blossom each spring to make the Tidal Basin a perennial tourist attraction.

Wild boars are introduced into the United States from Germany by an American representing a group of English investors. George Gordon Moore encloses a game refuge in the North Carolina Appalachians, builds a large log lodge with indoor plumbing, and stocks the refuge with 12 bison, 14 elk, 6 Colorado mule deer, 34 black bears, and 14 young wild boars, most of which will escape to mate with free-roaming "mountain-rooter" sows to begin a feral swine population that will grow to number more than 1,200.

A new Homestead Act reduces from 5 years to 3 years the residence requirement of U.S. homesteaders (*see* 1909; 1916).

Japanese biochemists J. Suzuki, T. Shimamura, and S. Ohdake extract an anti-beriberi compound from rice hulls (*see* 1906; 1933).

U.S. biochemists Elmer Verner McCollum, 33, and Marguerite Davis discover in butter and egg yolks the fat-soluble nutrient that will later be called vitamin A. They establish that it was a lack of this nutrient that caused C. A. Pekelharing's Dutch mice to die prematurely in 1905 when given no milk. Yale biochemists Thomas B. Osborne, 53, and Lafayette B. Mendel, 40, make a similar discovery.

Die Vitamine by Casimir Funk suggests that beriberi, rickets, pellagra, and sprue may all be caused by "vitamine" deficiencies (*see* 1911).

The Sherley Amendment to the U.S. Pure Food and Drug Law of 1906 prohibits farfetched declarations of therapeutic or curative effects (*see* 1911; Wheeler-Lea Act, 1938).

The Associated Advertising Clubs of America adopts a Truth in Advertising code.

A U.S. postal regulation requires all advertising in the media to be labeled "advt."

Hellmann's Blue Ribbon Mayonnaise is introduced by German-American New York delicatessen owner Richard Hellmann, 35, who by 1927 will have plants at Chicago, San Francisco, Atlanta, Dallas, and Tampa.

Oreo Biscuits are introduced by National Biscuit Company whose two chocolate-flavored wafers with a cream filling compete with the Hydrox "biscuit bon bons" introduced in 1910. Nabisco will change the name in 1958 to Oreo Cream Sandwich.

The A&P begins an expansion program under John Hartford, a son of the founder (*see* 1880). From its present base of nearly 500 stores, the chain will open a new A&P store every 3 days for the next 3 years as it stops providing charge accounts and free delivery and bases its growth on one-man "economy" stores that operate on a cash-and-carry basis (*see* 1929).

The first self-service grocery stores open independently in California. The Alpha Beta Food Market opens at Pomona, Ward's Groceteria opens at Ocean Park, and they are soon followed by Bay Cities Mercantile's Humpty Dumpty Stores (*see* Piggly-Wiggly, 1916).

1913 France's Premier Poincaré takes office as president January 19, beginning a 7-year term in which his country will go through a terrible war.

A peace conference at London has broken down January 6 over Turkish refusal to yield Adrianople, the Aegaean islands, and Crete, but Constantinople is finally induced January 22 to give up Adrianople. A coup d'état at Constantinople January 23 topples Kiamil Pasha and gives power to extreme nationalists headed by Enver Bey. War resumes February 3 with Greece, Bulgaria, and Serbia.

Greece's George I is assassinated at Salonica March 18 at age 67 after a reign of nearly 50 years. The former Danish prince is succeeded as king of the Hellenes by his son of 44 who will reign until 1917 as Constantine I (*see* 1920).

Allied coalition troops take Adrianople after a 155-day siege; Scutari falls to Montenegrin forces after a 6-month siege.

The Treaty of London May 30 resolves the Balkan War, but a 32-day Second Balkan War begins June 29 when a Bulgarian commander orders an attack on Serbo-Greek positions. Bucharest and Constantinople declare war and Bulgaria is quickly defeated. The Treaty of Bucharest August 10 ends hostilities, Bulgaria and Turkey settle their frontier dispute September 21 (the Turks keep Adrianople) but the Balkans continue to fester (*see* 1914).

Mexico City is seized February 9 by a nephew of ex-president Diaz, President Madero flees, military leader Victoriano Huerta, 58, proclaims himself president February 18, Madero is imprisoned and shot dead February 22 at age 39 while allegedly trying to escape. Washington refuses to recognize Huerta's regime. Insurgent general Francisco "Pancho" Villa, 36, takes Juarez November 15 and vows to conquer the country (*see* 1914).

The Seventeenth Amendment ratified April 8 provides for direct election of U.S. senators, beginning the end of the "Millionaire's Club" that has dominated the Senate.

A U.S. Department of Labor is created by the new Wilson administration in response to demands by the AF of L which now has 2 million members.

The IWW takes Paterson, N.J., silk workers out on 5-month strike to protest installation of improved machinery (*see* 1912; Luddites, 1811).

The United Mine Workers strike Colorado Fuel and Iron to protest policies of the company controlled by Standard Oil's John D. Rockefeller; two mines are set afire, 27 strikers are killed (*see* Ivy Lee, below; Ludlow massacre, 1914).

U.S. suffragettes march 5,000-strong down Washington's Pennsylvania Avenue March 3 following Alice Paul, 27, who founds the National Woman's Party to spearhead the movement for woman suffrage. Crowds of angry, jeering men slap the demonstrators, spit at them, and poke them with lighted cigars, a brawl stops the march before it can reach the White House, 40 people are hospitalized, and it takes a cavalry troop from Ft. Myer to restore order.

English suffragist Emmeline Pankhurst draws a 3-year jail sentence April 3 for arson (she has incited her supporters to place explosives in the house of the Chancellor of the Exchequer David Lloyd George) but will serve only one year and devote part of it to a hunger strike (*see* 1906). Parliament passes the Prisoners' Temporary Discharge for Ill Health Act in April to thwart hunger strikes; militant suffragists call it the Cat and Mouse Act (*see* 1918).

Suffragist Emily Davison, who has been imprisoned several times and force-fed, is killed June 4 when she

"The Temptation of Saint Pankhurst." *Life* magazine spoofed the woman suffrage movement's leader Emmeline Pankhurst.

1913 ***(cont.)*** runs in front of the king's horse at the Derby. Sylvia Pankhurst is sentenced July 8 to 3 months in prison.

Social reformer Marcus Moziah Garvey, 26, founds the Universal Negro Improvement Association in Jamaica, British West Indies (*see* 1916).

Atlanta pencil factory worker Mary Phagan, 14, is murdered April 26. Plant superintendent Leo Moses Frank, 28, is indicted July 28, convicted on little evidence and sentenced to death August 26 (*see* 1915).

The Anti-Defamation League to fight anti-Semitism in America is founded by the 70-year-old B'Nai B'rith (Sons of the Covenant).

Mohandas Gandhi in South Africa leads 2,500 Indians into the Transvaal in defiance of a law, they are violently arrested, Gandhi refuses to pay a fine, he is jailed, his supporters demonstrate November 25, and Natal police fire into the crowd, killing two, injuring 20 (*see* 1906; 1914).

$ The Sixteenth Amendment to the Constitution proclaimed in force February 25 by Secretary of State Philander C. Knox empowers Congress to levy graduated income taxes on incomes above $3,000 per year. The income tax will bring drastic changes to the lifestyles of a few thousand rich Americans, but the very rich will find loopholes for avoiding taxes.

A sensational report published February 28 by the House Committee on Banking and Currency exposes the "money trust" that controls U.S. financial power. U.S. congressman from Louisiana Arsene P. Pujo heads the committee that has called witnesses including financier J. P. Morgan (who dies at Rome March 31 at age 75 leaving a vast estate).

Economic Interpretation of the Constitution by Columbia University history professor Charles A. Beard, 39, points out that America's founding fathers were all men of property when they drafted the Constitution in 1787. Beard suggests certain consequent inequities.

The Underwood-Simmons Tariff Act passed by Congress May 8 lowers import duties by an average of 30 percent, the first real break in tariff protection since the Civil War. Agricultural implements, raw wool, iron ore, pig iron, and steel rails are admitted duty-free under the new law, and tariffs on raw materials and foodstuffs are substantially reduced. The new measure hurts many U.S. manufacturers; pressure mounts for restoration of tariff protection (*see* 1922).

The United States has 40 percent of world industrial production, up from 20 percent in 1860.

The average British worker still earns less than £1 per week ($5 in U.S. currency) while American workers average more than $2 per day (*see* Ford, 1914).

The Federal Reserve System created by a measure signed into law by President Wilson December 23 will reform U.S. banking and currency. The Glass-Owen Currency Act, drafted to prevent panics such as the one in 1907, establishes 12 Federal Reserve banks in America's 12 major cities and requires member banks to maintain cash reserves proportionate to their deposits with the Fed—which loans money to the banks at low rates of interest relative to the rates the banks charge customers. The Fed's Board of Governors determines the amount of money in circulation at any given time; it provides elasticity to the supply of currency and can act to control inflation (*see* Banking Act, 1935).

The Rockefeller Foundation chartered by Standard Oil's John D. Rockefeller, Sr., will be a major force in improving world health, world agriculture, and education and will work toward world peace (*see* 1914). Rockefeller has acted on the advice of his confidant Frederick T. Gates, formerly head of the American Baptist Education Society (*see* Ivy Lee, below).

Bethlehem Steel's Charles M. Schwab acquires Fore River Shipbuilding and makes Eugene Grace, 37, president of Bethlehem. Grace will develop the company into the world's second largest steel maker (*see* 1912; Sparrows Point, 1916).

The U.S. Bureau of Labor Statistics computes its first monthly Consumer Price Index to help determine the fairness of wage levels (*see* Fisher, 1911).

Singer Manufacturing Co. sells 2.5 million sewing machines, 675,000 of them in Russia.

Neiman-Marcus in Dallas burns in the spring. Filene's Basement in Boston (*see* 1909) sells the stock while Neiman-Marcus builds a new store (*see* 1907).

Germany produces 34 percent of the world's electrical equipment, the United States 19 percent.

General Electric researcher William David Coolidge, 39, receives a patent on "tungsten and method for making the same for use as filaments of incandescent electric lamps" 5 years after producing the first ductile tungsten by using high temperatures to draw the metal into fine filaments. Long-burning tungsten filament bulbs will replace the Swan-Edison carbon filament bulbs in use until now (*see* 1879; Owens, 1903; Langmuir, 1912).

The Keokum Dam on the Mississippi River is completed to produce hydroelectric power.

An International Geological Congress at Toronto estimates world coal reserves, taking 1 foot as the minimum workable thickness of seams down to depths of 4,000 feet and 2 feet as the minimum for seams at lower depths. North America is judged to have more than 5 trillion metric tons of coal, Asia 1.3 trillion, Europe 784 billion, South America 33 billion, Africa 58 million, Oceania 170 million.

World petroleum production reaches 407.5 million barrels, up from 5.7 million in 1870. Most comes from wells in the United States and in Russia's Caspian Sea and Caucasus Mountain oil fields (*see* Persia, 1908, 1914; Venezuela, 1922).

New York's Grand Central Terminal opens February 1 at 42nd Street and Park Avenue, replacing a 41-year-old New York Central and New Haven train shed with the world's largest railway station. Trains under Park Avenue have been electrified since December 1906. Far bigger than Penn Station, Grand Central has 31 tracks on its upper level, 17 on its lower (which opened for commuter service in October 1912).

Henry M. Flagler's Florida East Coast Railway reaches Key West, but Flagler dies at West Palm Beach May 20 at age 83. Three thousand men have worked for 7 years to complete the rail link to the Keys, yellow fever has killed hundreds, and hurricanes will destroy the line in 1926 and 1935.

The first diesel-electric locomotives go into service in Sweden. Diesel-electrics will generally replace steam locomotives, but inventor Rudolf Diesel jumps or falls overboard from the Antwerp-Harwich mail steamer the night of September 30 and is lost at age 55 (*see* 1895; first U.S. diesels, 1924).

New York Evening Sun reporter John Henry Mears 35, circles the world in a record 35 days, 21 hours, 31 seconds (*see* Jaeger-Schmidt, 1911). He leaves New York on the Cunard Line's S. S. *Mauretania*, arrives at Fishguard, Wales, proceeds by rail to London, crosses the Channel by ship, proceeds by rail to Paris and on to Berlin, St. Petersburg, and thence to Manchuria via the Trans-Siberian Railway, crosses by ship to Japan, proceeds by ship to Vancouver, flies to Seattle, and completes the 21,066-mile journey by rail via Chicago (*see* Wiley Post, 1931).

The German passenger liner S. S. *Imperator* that will become the S. S. *Berengaria* in 1919 goes into service on the North Atlantic. The 52,022-ton ship has four screws, is 919 feet in length overall, and will continue in service for 25 years (*see* 1914).

Elmer Sperry receives a patent for his stabilizer gyroscope (*see* 1910; rail defect detection, 1923).

Russian aeronautical engineer Igor Ivan Sikorsky, 24, builds and flies the world's first multimotored aircraft. He built a prototype helicopter 4 years ago (*see* 1939; autogyro, 1923).

Lord Northcliffe of the *London Daily Mail* offers a £10,000 prize for the first nonstop transatlantic crossing by aeroplane. No pilot has topped the distance record of 628.15 miles set last year, and many regard the offer as an empty publicity stunt (*see* 1919).

Sales of U.S. luxury motorcars priced from $2,500 to $7,500 reach 18,500 with Packard leading the field (2,300 cars), followed by Pierce Arrow (2,000). White, Franklin, Locomobile, and Peerless provide healthy competition.

Cadillac is the leading make in the $1,500 to $2,500 range and will enter the luxury market next year with a V8 model.

Duesenberg Motor Co. is founded at St. Paul, Minn., by Fred S. and August S. Duesenberg. The German-American bicycle makers (and bike racers) opened a garage at Des Moines, Iowa, in 1903, produced their first motorcar in 1904 under the name Mason, and have sold their Iowa company to Maytag Washing Machine Co. which is selling their original car as a Maytag (*see* 1907; 1919).

Stutz Motor Car Co. is founded by Harry C. Stutz, who merges his Stutz Auto Parts with Ideal Motor Car (*see* 1911; Stutz Bearcat, 1914).

Some new Ford models sell for as little as $550 F.O.B. Detroit; the price will soon drop to $490 (*see* 1925; assembly line, below).

The Dodge brothers John and Horace break with Ford and become independent while continuing to own stock in Ford which builds a big new plant at Highland Park, Mich. (*see* 1914; 1919).

The assembly line introduced at Ford Motor Company October 7 reduces the time required to assemble a motorcar from 12.5 hours to 1.5 hours. Devised with help from engineer Clarence W. Avery, Ford's line reverses the disassembling line used by Chicago and Cincinnati meat packers, factories producing a wide variety of products will convert to assembly line production, the line will revolutionize much of industry, but increase worker boredom and permit industry to hire unskilled and semiskilled workers at wages lower than those of skilled workers.

A commercial ammonia plant employing the Haber process of 1908 begins production at Ludwigshafen, Germany. The ammonia is used primarily for making explosives.

A new model of the atom devised by Danish physicist Niels Bohr, 28, violates classical electromagnetic theory but successfully accounts for the spectrum of hydrogen. Bohr applies Max Planck's quantum theory of 1900 to Ernest Rutherford's nuclear atom of 1911 (*see* 1939; Schrödinger, 1926).

G. E.'s W. D. Coolidge (above) invents the modern X-ray tube (*see* Roentgen, 1895).

The American Cancer Society is founded at a time when nine out of ten cancer patients die of the disease. The percentage of mortality will fall sharply but the incidence of cancer will increase (*see* cigarettes, below).

Hungarian-American physician Bela Schick, 36, at New York's Mount Sinai Hospital develops the Schick test for determining susceptibility to diphtheria.

French Protestant missionary-physician Albert Schweitzer, 38, founds Lambarene Hospital in French Equatorial Africa; he will be famous for his "reverence for life."

U.S. Parcel Post service begins January 1. The U.S. Postal Service has inaugurated the service after more than two decades of agitation by Populist politicians and farmers who hope to ship produce to market via Parcel Post. American Express, Adams Express, United States Express, and Wells Fargo have opposed the service, but by year's end packages are being mailed at the rate of 300 million per year, mail-order firms such as Montgomery Ward and Sears, Roebuck receive up to five times as many orders as last year, sales of rural merchants are depressed.

1913 *(cont.)* U.S. Postmaster General Albert Burleson proposes nationalization of telephone and telegraph communications, noting that the Constitution provided for federal control of the postal system and would have provided for federal control of other means of communication had they existed *(see* Britain, 1912). AT&T president Theodore N. Vail tells stockholders, "We are opposed to government ownership because we know that no government-owned system in the world is giving as cheap and efficient service as the American public is getting from all its telephone companies," but while U.S. long-distance rates are indeed lower than in Europe, local calls generally cost more.

AT&T president Vail (above) has his vice president N. C. Kingsbury meet with Attorney General George Wickersham, the company divests itself of its Western Union holdings to avoid antitrust action, losing $7.5 million by the divestiture; it halts plans to take over some midwestern telephone companies but retains its valuable Western Electric acquisition (*see* 1910; dial telephone, 1919).

New York public relations pioneer Ivy Ledbetter Lee, 36, persuades John D. Rockefeller, Jr., to travel to Colorado and speak personally to the miners, an act that improves relations between the Rockefeller family and the outraged strikers (above). An advocate of frank and open dealing with the public, Lee will help make the Rockefellers famous for their philanthropies—partly by having John D., Sr., hand out dimes to schoolchildren.

"Toonerville Folks" introduces newspaper comic strip readers to the Powerful Katrinka, the Terrible Tempered Mr. Bang, Mickey "Himself" McGuire, Aunt Eppie Hogg, and the Toonerville Trolley that Meets All the Trains. The new strip by cartoonist Fontaine Fox, 29, will be retitled the "Toonerville Trolley" and will continue until 1955.

Harper's Bazar is purchased by William Randolph Hearst, who will change its name to *Harper's Bazaar* in 1929. The magazine for women has been published since 1867 by Harper Bros.

The first U.S. crossword puzzle appears December 21 in the weekend supplement of the *New York World*. English-American journalist Arthur Wynne has seen similar puzzles in 19th-century English periodicals for children and in the *London Graphic and* has arranged squares in a diamond pattern with 31 clues which are for the most part simple word definitions: "What bargain hunters enjoy," five letters; "A boy," three letters; "An animal of prey," four letters (sales, lad, lion).

Nonfiction *The Tragic Sense of Life (Del sentimento tragico de la vida)* by Basque philosopher-novelist-poet-dramatist Miguel de Unamuno, 49, whose early realistic novel *Paz en la guerra* appeared in 1897. Man can exist, says Unamuno, only by resisting the absurdity and injustice of the fact that death brings oblivion, and he urges readers to base their "faith" in "doubt," which alone gives life meaning.

Fiction *Swann's Way* (*Du Côte de chez Swann*) by French novelist Marcel Proust, 42, whose memories of childhood have been revived by tasting shell-shaped madeleine cakes dipped in tea and who will follow his psychological novel with six more, all under the title *Remembrance of Things Past* (*A la Recherche du temps perdu*); *The Wanderer* (*Le Grand Meaulnes*) by French novelist Alain-Fournier (Henri Alban Fournier), 26; *The Boys in the Back Room (Les Copains)* by Jules Romains; *The Happy-Go-Lucky Morgans* by Welsh poet-novelist Edward Thomas, 35; *Sons and Lovers* by English novelist D. H. (David Herbert) Lawrence, 38; *O Pioneers* by former *McClure's* magazine managing editor Willa Sibert Cather, 39, who grew up on the Nebraska frontier; *Virginia* by Virginia novelist Ellen Anderson Gholson Glasgow, 39; *Polyanna* by U.S. novelist Eleanor Porter, 45; *Dr. Fu Manchu* by English mystery novelist Sax Rohmer (Arthur Sarsfield Wade), 30.

Poetry *General William Booth Enters Heaven* by U. S. poet (Nicholas) Vachel Lindsay, 33; *The Tempers* by Rutherford, N.J., poet-physician William Carlos Williams, 40; *Dauber* by John Masefield; *Stone* (*Kamen*) by Russian poet Osip Mandelstam, 22; *Eve* by Charles Péguy.

The Armory Show that opens February 17 at New York's huge 69th Regiment Armory on Lexington Avenue at 25th Street gives Americans their first look at cubism. Represented are the late Paul Cézanne, Henri Paul Gauguin, Vincent Van Gogh, and other impressionists, but what shocks most of the 100,000 visitors are works by French-American sculptor Gaston Lachaise, 31, U.S. cubist Charles Sheeler, 30, U.S. realist Edward Hopper, 31, U.S. water-colorist John Cheri Marin, and the canvas *Nude Descending Staircase* by French-American dadaist Marcel Duchamp, 26. The show moves to the Chicago Art Institute, where students burn post-impressionist Henri Matisse in effigy, and then to Boston. Roughly 250,000 see the show, and while most are upset the show ushers in a new era of acceptance for unromantic art expression.

Other paintings *Henry James* by John Singer Sargent; *L'Ennui* by Walter Sickert; *The Violin* by Pablo Picasso; *Aridne* by Giorgio de Chirico; *Coney Island, Battle of the Lights* by Italian-American painter Joseph Stella, 36.

Anthony Comstock of the New York Society for the Suppression of Vice sees a copy of last year's Paul Chabas painting *September Morn* May 13 in the window of a New York art dealer in West 46th Street (*see* Comstock Law, 1872). Now 69, Comstock demands that the picture be removed from the window because it shows "too little morning and too much maid," Chicago alderman "Bathhouse John" Coughlin vows that the picture will not be displayed publicly in Chicago, but oilman Calouste Gulbenkian will acquire the oil and it will wind up at New York's Metropolitan Museum of Art.

Sculpture *The Little Mermaid* (bronze) by Danish sculptor Edvard Eriksen, 37.

Theater *Vladimir Mayakovsky—A Tragedy* by Russian playwright-poet Vladimir Vladimirovich Mayakovsky, 20, who performs in his own play at Moscow; *Romance* by Edward Sheldon 2/10 at Maxine Elliott's Theater, New York, with Doris Keane, 160 perfs; *Jane Clegg* by Irish playwright St. John Ervine, 29, 6/19 at London, with Sybil Thorndike; *Androcles and the Lion* by George Bernard Shaw 9/1 at the St. James's Theatre, London, with O. P. Heggie, Ben Webster, Leon Quartermaine, 52 perfs.; *The Seven Keys to Baldpate* by George M. Cohan, who has adapted a novel by Earl Der Biggers, 9/22 at New York's Astor Theater, with Wallace Eddinger, 320 perfs.

New York's Palace Theater opens March 24 at Broadway and 47th Street. Other vaudeville houses have a top price of 50¢, the Palace charges $2, and B. F. Keith of Boston and his partner Edward F. Albee in the Keith Circuit's United Booking Office present a bill that includes a wire act, a Spanish violinist, a one-act play by George Ade, and comedian Ed Wynn, 26. The Palace has no success, however, until May 5 when Sarah Bernhardt opens for a 2-week engagement that is extended for another week and a half. Pantomime juggler W. C. Fields (William Claude Dukenfield), 33, who joins the act in May began his career at age 14 and played a command performance before England's Edward VII in 1901.

Actors' Equity Association is founded May 26 at New York's Pabst Grand Circle Hotel with Francis Wilson as president, Henry Miller as vice president. Actors' Equity will set up contracts under which actors are employed, will maintain a benefit system for its membership, but will not obtain recognition as the trade union of the acting profession until its members strike in August 1919.

Films D. W. Griffith's *Judith of Bethulia* with Blanche Sweet, Henry B. Walthall, Mae Marsh in the first American-made four-reel film; Cecil B. DeMille's *The Squaw Man* with Dustin Farnum in the first full-length feature to be filmed at Hollywood, Calif.; *The Count of Monte Cristo* with James O'Neill whose son Eugene has been released from a tuberculosis sanatorium and will enroll next year in George Pierce Baker's 47 Workshop at Harvard; *The Perils of Pauline* produced by a Hearst-controlled company whose cliff-hanger films are designed to bring audiences back each week; Colin Campbell's *The Spoilers* with William Farnum; Mack Sennett's *Barney Oldfield's Race for Life* with Sennett, Mabel Normand.

Charlie Chaplin signs a $150 per week contract with movie maker Mack Sennett who has discovered English actor-dancer Charles Spencer Chaplin, 24, in New York.

The Jesse L. Lasky Feature Play Co. is founded by vaudeville producer Jesse Louis Lasky, 32, his Polish-American brother-in-law Samuel Goldwyn, 30, and playwright Cecil Blount DeMille, 31, who has selected Hollywood for making *The Squaw Man* (above) because California has an abundance of sunshine. Goldwyn (*né* Schmuel Gelbfisz) is a Polish-American glove maker whose business has been ruined by the new Underwood-Simmons Tariff Act (above) which has lowered duties on imported gloves. The Lasky firm will merge with Adolph Zukor's Famous Players to create Famous Players-Lasky and will become Paramount Pictures in 1932.

Opera *Pénélope* 3/4 at Monte Carlo, with music by Gabriel Fauré (who is now too deaf to hear it); *La Vida Breve (The Short Life)* 4/1 at Nice, with music by Spanish composer Manuel de Falla, 36; *L'Amore dei tre re* (*The Love of Three Kings*) 4/10 at Milan's Teatro alla Scala, with music by Italian composer Italo Montemezzi, 37.

Ballet *The Rites of Spring* (*Le Sacre du Printemps*) 6/13 at the Théâtre du Châtelet, Paris, with Serge Diaghilev's Ballet Russe, choreography by Waslaw Nijinsky, music by Igor Stravinsky. Outraged audiences stage violent demonstrations at Stravinsky's wild dissonances and tumultuous rhythms which begin a new epoch in music.

First performances *Gurre-Lieder* for Vocal Soloists, Mixed Chorus, and Orchestra by Arnold Schoenberg 2/23 at Vienna with a text based on an 1868 poem by the late Danish poet Jens Peter Jakobsen; *Deux Images* (symphonic diptych) by Béla Bartók 2/26 at Budapest; *Kammersimphonie* by Arnold Schoenberg 3/31 at Vienna's Musikvereinsaal; *Six Pieces for Orchestra* by Schoenberg's student Anton von Webern, 29, 3/31 at Vienna's Musikvereinsaal; *Falstaff* Symphonic Study in C minor (with two interludes in A minor) by Sir Edward Elgar in October at England's Leeds Festival.

Broadway musicals *The Sunshine Girl* 2/3 at the Knickerbocker Theater, with Vernon and Irene Castle doing the Turkey Trot, music by John L. Golden, lyrics by Joseph Cawthorne, 160 perfs.; *The Passing Show* 6/10 at the Winter Garden Theater, with Charlotte Greenwood, a runway to bring the show's scantily clad chorus girls close to the audience, music chiefly by Viennese-American composer, Sigmund Romberg, 25, lyrics chiefly by Harold Atteridge, 116 perfs.; *The Ziegfeld Follies* 6/16 at the New Amsterdam Theater, with Ann Pennington, Fanny Brice, Leon Errol, 96 perfs.; *Sweethearts* 9/8 at the New Amsterdam Theater, with music by Victor Herbert, lyrics by Robert B. Smith, 136 perfs.; *High Jinks* 12/10 at the Lyric Theater, with music by Rudolf Friml, book and lyrics by Otto Hauerbach (later Harbach) and Leo Ditrichstein, songs that include "Something Seems Tingle-angeling," 213 perfs.

Popular songs "Peg O' My Heart" by Fred Fisher, lyrics by Alfred Bryan; "You Made Me Love You (I Didn't Want to Do It)" by Italian-American composer James V. Monaco, 28, lyrics by Joe McCarthy, 28; "Ballin' the Jack" by Chris Smith, lyrics by James Henry Burris; "If I Had My Way" by James Kendis, lyrics by Lou Klein; "He'll Have to Get Under—Get Out and Get Under" by Grant Clarke and Byron Gay, lyrics by Byron Gay that make fun of the growingly popular motorcar; "Now Is the Hour (Maori Farewell Song)" by New Zealand songwriters Maewa Kaihan, Clement Scott, Dorothy Stewart.

1913 ***(cont.)*** Hymn "The Old Rugged Cross" by U.S. clergyman George Bennard.

Anthony Wilding wins in men's singles at Wimbledon, Mrs. Chambers in women's singles; Maurice McLoughlin wins in men's singles at Newport, Mary Browne in women's singles.

Connie Mack's Philadelphia Athletics win the World Series, defeating the New York Giants 4 games to 1.

Knute Rockne revolutionizes football by making brilliant use of the forward pass to defeat an Army eleven and change the course of the game. Norwegian-American Knute Kenneth Rockne, 25, is captain of the Notre Dame varsity, he receives 17 out of 21 passes thrown by Charles "Gus" Dorais, he gains 243 yards, and Notre Dame beats Army 35 to 13.

The first U.S. deck shuffleboard game is played on a hotel court at Daytona, Fla. The game will be popular with elderly retirees (*see* Miami Beach, below).

The Casino of Monte Carlo has an historic run of one color. Black comes up 26 times in succession at a roulette wheel August 18, theoretically returning 1.35 billion francs on a 20 franc stake, but gamblers bet so heavily on red that the house wins several million francs for La Société Anonyme des Bains de Mer et Cercle des Etrangers à Monte Carlo which always averages about 125 percent return on invested capital each year.

Coco Chanel pioneers sportswear for women at a new boutique in Deauville that features berets and open-necked shirts in an age when women of fashion adorn themselves with feathers and huge hats. Gabrielle Chanel, 30, gained her nickname at age 20 while entertaining the 10th Light Horse Regiment in the small garrison town of Moulins, became a Paris milliner while mistress to Etienne Balsan, and is now mistress to English businessman Arthur Capel; her lovers will include England's second duke of Westminster, who owns 17 Rolls-Royces and has mammoth greenhouses in which pears and peaches ripen all year round, and she will make sweaters the fashion of the rich as Chanel knitwear gains world fame (*see* Chanel No. 5, 1921).

Pennsylvania makes Mother's Day a state holiday (*see* Jarvis, 1908; Wilson, 1914).

Brillo Manufacturing Corp. is founded by New York lawyer Milton B. Loeb, 26, whose client (a manufacturing jeweler) has developed a way to clean aluminum cooking utensils using steel wool with a special reddish soap that gives the vessels a brilliant shine.

Swedish-American inventor Gideon Sundback, 33, develops the first dependable slide-fastener and efficient machines to manufacture it commercially. He attaches matching metal locks to a flexible backing, each tooth being a tiny hook that engages with an eye under an adjoining hook on an opposite tape. He will patent improvements on his slide fastener in 1917 and assign the patents to the Hookless Fastener Co. of Meadville, Pa., which will manufacture the Talon slide fastener (*see* Judson, 1893; "zipper," 1926).

Camel cigarettes are introduced by R. J. Reynolds which made no cigarettes before joining the Tobacco Trust in 1890 (*see* 1911). Camels are made largely from a flue-cured bright tobacco much like the tobacco discovered in 1852 by Eli and Elisha Slade, but they also contain a sweetened burley from Kentucky combined with some Turkish tobaccos that make them the first of the modern blended cigarettes. Their package design has been inspired by "Old Joe," a dromedary in the Barnum & Bailey circus menagerie (*see* 1917).

Chesterfield cigarettes are introduced by Liggett & Myers which is well known for its Fatima, Picayune, and Piedmont brands (*see* 1926). Tobacco companies spend $32.4 million to advertise their various brands in the United States, up from $18.1 million in 1910 (*see* Lucky Strike, 1916).

New York's 60-story Woolworth building opens in April near City Hall. The 792-foot "Cathedral of Commerce" designed by Cass Gilbert will be the world's tallest habitable structure until 1931.

New York's Equitable building goes up on lower Broadway. The world's largest office building has been financed by a syndicate that includes Thomas Coleman of E. I. du Pont de Nemours who has put up $30 million (*see* 1915).

New York's General Post Office building opens on Eighth Avenue between 31st and 33rd Streets across from the 3-year-old Pennsylvania Station. Designed by McKim, Mead and White with two blocks of Corinthian columns, the building's front is inscribed with lines written by the Greek historian Herodotus about the 5th-century B.C. couriers of Xerxes: "Neither snow, nor rain, nor heat, nor gloom of night stays these couriers from the swift completion of their appointed rounds."

New York's Grand Central Terminal (above) has a mammoth concourse—470 feet long, 160 wide, 125 high—designed by Warren and Wetmore with a Beaux Arts façade.

Carrère and Hastings completes a marble Fifth Avenue mansion for steel magnate Henry Clay Frick between 70th and 71st streets.

Cincinnati's Union Central Insurance building is completed on the site formerly occupied by the Chamber of Commerce building at Fourth and Vine Streets. The $2.5 million 34-story skyscraper is second in height only to the Woolworth building (above).

Miami Beach, Fla., begins its career as a winter resort. Avocado grower John S. Collins, 76, has begun a bridge across Biscayne Bay to link the sandbar to the mainland. Carl Graham Fisher, 41, who has just made $5 million by selling the Prest-O-Lite acetylene automobile lamp company he acquired for almost nothing, loans Collins $50,000 to complete the bridge, and in February he goes into business with Collins, selling lots at auction while building concrete bulkheads along the shore. Fisher and Collins will build 27 miles of bulkheads, they will add 2,800 acres to Miami Beach's present 1,600 acres in the next 10 years by dredging, pumping, and filling in land over mangrove roots. By 1923 they will have built five hotels to augment the Royal Palm put up by H. M. Flagler in 1896 (*see* 1924).

Ohio's Great Miami River floods its banks at Columbus March 25 after a 5-day rainfall that has dropped 10 inches of water. The flood kills 351, injures hundreds more, and creates property damage amounting to $100 million.

Congress passes a bill to protect migratory game and insectivorous birds.

"Trees" by *New York Times Book Review* editor and poet Alfred Joyce Kilmer, 27, appears in the August issue of *Poetry* magazine and wins worldwide acclaim: "I think that I shall never see, a poem lovely as a tree. . ." (*see* song, 1922).

Forest Lawn Cemetery at Tropico (Glendale), Calif., is taken over by Missouri-born metallurgist-promoter Hubert Eaton, 30, who embraces the new concept of selling burial plots "before need" and will expand the tiny cemetery into a vast enterprise with branches in Hollywood Hills, Cypress, and Covina Hills. Eaton will make Forest Lawn a showplace of more than 1,200 acres of towering trees, sweeping lawns marked with bronze memorial tablets rather than with tombstones, splashing fountains, flowers, birds, bronze and marble statues, and small churches modeled on famous ones in Europe. He will forbid artificial flowers.

The Los Angeles Owens River Aqueduct opens November 5, bringing in at least 260 million gallons of water per day from the Sierras via 234 miles of gravity-powered pipeline and ditch. City manager William Mulholland has planned the $25 million cement-and-steel project that permits irrigation of the San Fernando Valley and makes L.A. a boom town (*see* farmers, 1924; Parker Dam, 1939).

The U.S. Senate votes December 16 to fund construction of California's Hetch Hetchy Dam despite public opposition (*see* 1907). John Muir, now 75, says it will destroy his beloved "Tuolumne Yosemite."

Rabbits fed large amounts of cholesterol and animal fats are shown to develop hardening of the arteries (arteriosclerosis) by Russian pathologist Nikolai Anichkov (*see* Keys, 1953).

Pellagra kills 1,192 in Mississippi. The U.S. Public Health Service sends Hungarian-American physician Joseph Goldberger, 39, to study the disease which is known to affect at least 100,000 in the Deep South (*see* 1907; 1915).

Congress amends the Pure Food and Drug Law of 1906 to make it more effective as was done last year (*see* 1938).

Quaker's Puffed Rice and Quaker's Puffed Wheat are introduced by Quaker Oats Co. (*see* 1893; Aunt Jemima, 1926).

Peppermint Life Savers are introduced by Cleveland, Ohio, chocolate manufacturer Clarence Crane as a summer item to sell when chocolate sales decline. Crane uses a pharmaceutical company's pill-making machine to create the uniquely-shaped peppermint with the hole in the middle which he advertises "for that stormy breath," and when streetcar advertising salesman Edward J. Noble, 31, tries to persuade Crane to use car cards for advertising Crane offers to sell the product for $5,000. Noble finds a partner, buys Crane's trademark and a small stock of peppermints for $2,900, and starts producing hand-wrapped rolls of Life Savers in a New York loft (*see* 1920).

1914

A World War begins in Europe July 28 one month after the assassination of the heir to the Austrian throne in Bosnia. Riding in a 1912 Graf und Stift motorcar at Sarajevo, the archduke Franz Ferdinand, 51, and his wife are killed by tubercular high school student Gavrilo Prinzip, who has been hired by Serbian terrorists to kill the nephew of the emperor Franz Josef.

Austrian militarists have been spoiling to enter the wars that have embroiled the Balkans since 1912, and Vienna uses the incident at Sarajevo in June as an excuse to declare war on Serbia.

The war quickly widens as Germany declares war on Russia August 1 and on France August 3. German troops invade neutral Belgium August 4 and Britain declares war on Germany.

French passions against the "Bosch" have eclipsed the scandal of March 16, when Henrietta Caillaux, 39, wife of the minister of finance Joseph Caillaux, 51, walked into the office of Gaston Calmette, 55, editor of *Le Figaro*, and shot him dead. Calmette has been attacking Caillaux as a secret German agent and threatening to publish love letters between him and his wife.

"The lamps are going out all over Europe; we shall not see them lit again in our lifetime," says British secretary of state for foreign affairs Sir Edward Grey.

German forces slip past Belgian fortifications at Liège in a night attack and the Belgians fall back to Brussels.

Montenegro declares war on Austria August 5, Serbia on Germany August 6, Austria on Russia August 6,

"Your Government Needs YOU!" Alfred Leslie's poster helped Britain recruit men for the muddy trenches of France.

1914 ***(cont.)*** Montenegro on Germany August 8. Britain and France declare war on Austria August 12.

French defensive strategy has ignored the possibility of a German invasion through Belgium and counted on a Russian advance in the East. French forces invade Lorraine, hoping to regain territory lost in the Franco-Prussian War of 1870 to 1871 but are driven out with heavy losses.

Belgian forces fail back on Antwerp and destroy the bridges over the Meuse as the Germans enter Brussels August 20.

The Battle of the Frontiers August 14 to August 25 costs 250,000 French lives; news of the staggering loss is censored and will not appear until after the war. The French government moves to Bordeaux, making it hard for civilian ministers to control army generals.

Japan declares war on Germany August 23 and on Austria August 25. The Japanese produce munitions for Russia and the other Allied powers, they begin to land forces in Shantung for an attack on the German position at Tsingtao, and a British detachment joins them (*see* 21 demands, 1915).

German colonial forces in Togoland surrender August 26 to an Anglo-French force. Britain and France divide the German African colony between them.

The Battle of Tannenberg August 26 to 30 ends in crushing defeat for a large Russian army that has invaded east Prussia under the command of Gen. Aleksandr Vasilievich Samsonov, 55, to take pressure off the French on the western front. Another Russian army under Gen. Paul Rennenkampf, 60, fails to support Samsonov, German forces under Gen. Paul von Hindenburg, 66, and Gen. Erich Ludendorff, 49, surround the Russians and take 100,000 prisoners with help from Gen. Hermann von François, 48, executing a plan developed by Col. Max Hoffman, 45. Gen. Samsonov shoots himself, Gen. Rennenkampf will be executed in 1918.

The Russians rename their 211-year-old capital Petrograd because St. Petersburg sounds German (*see* Leningrad, 1924).

The first German air raid strikes Paris August 30 as German aircraft drop small explosives on the city.

The Battle of the Marne September 5 to 12 ends the German advance. French and British troops force the Germans to withdraw west of Verdun but are unable to dislodge them from north of the Aisne.

The Battle of the Masurian Lakes September 6 to 15 brings further defeat to the Russians, this time at the hands of Gen. August von Mackensen, 65. Taken prisoner are 125,000 Russians and the Germans advance to the lower Niemen River.

The German submarine U-9 sinks three British cruisers in September to retaliate for the sinking of three German cruisers at Helgoland bight August 28.

Ghent falls to the Germans October 11 as they try to reach the Channel ports in a rush to the sea, but while Bruges is taken October 14 and Ostend October 15, the Belgians flood the district of the Yser and the German push fails.

The Battle of Ypres October 30 to November 24 pits German troops against French *poilus* and British Tommies in trench warfare that will consume huge numbers of soldiers on both sides in the next 4 years as the conflict becomes a war of position in which the front line will not shift more than 10 miles.

A Turkish fleet that includes two German cruisers bombards Odessa, Sevastopol, and Theodosia on the Black Sea October 29. Russia declares war on Turkey November 2, Britain and France do likewise November 5, and Britain proclaims the annexation of Cyprus which the British have occupied since 1878.

German vice admiral Graf Maximilian von Spee, 63, surprises a British squadron under Rear Admiral Sir Christopher George Francis Maurice Craddock, 62, off the Chilean coast November 1. The British are defeated off Coronel, Craddock loses the *Monmouth* and goes down with his flagship the *Good Hope*; the *Gloucester* escapes.

The Battle of the Falkland Islands December 8 ends in victory for a British squadron under Rear Admiral Sir Frederick Charles Doveton Sturdee, 65, who surprises Graf Spee as he prepares to attack the Falklands while en route home from his triumph over Admiral Craddock (above). Spee goes down with his flagship the *Scharnhorst,* his two sons and 1,800 other men are also lost, the *Gneisenau, Leipzig,* and *Nürnberg* are sunk, only the *Dresden* escapes.

Russian forces meanwhile have forced von Mackensen to fall back from Warsaw and the Austrians to retreat to Cracow. The Battle of Cracow from November 16 to December 2 brings heavy casualties to both sides. New divisions diverted from the western front bolster German resistance in early December after Russian troops threaten to surround the Germans following the November battles of Lodz and Lowicz. Lodz falls to the Germans December 6, but Austrian forces fail to break through the Russian lines before Cracow and the Russians will remain within 30 miles of the city through the winter.

Serbian troops drive out the Austrians in early December after winning the Battle of Kolubara, and on December 15 retake Belgrade, which has been in Austrian hands since December 2.

U.S. forces occupy Vera Cruz April 21 following hostile acts by Mexicans. President Wilson sends a fleet to Tampico, President Huerta resigns July 15 under pressure from Vesustiano Carranza, 55, who opposes U.S. interests, and civil war erupts between Carranza and his lieutenant Pancho Villa (*see* 1913; 1916).

Northern and southern Nigeria are united.

South Africa's prime minister Louis Botha puts down a pro-German Boer revolt. He takes command of troops that enter German South-West Africa.

Britain proclaims a protectorate over Egypt December 18. The khedive Abbas Hilmi who has reigned since 1892 is deposed December 19 and succeeded by Hussein Kamil, 60, who will be khedive until 1917.

Swedish-American IWW leader Joe Hill, 34, is arrested January 13 on charges of having been one of two masked men who gunned down Salt Lake City grocer John Morrison and his son the night of January 10. Hill had sought medical attention 90 minutes after the murder, telling a physician on the other side of town that he had been shot in an argument over another man's wife, but while Morrison was wounded in a similar attack long before Hill came to work in the Utah copper mines, and his killing has apparently been connected with the earlier attack ("We've got you now," one of the masked men had said), and while no motive can be shown for Hill to have committed the crime when he is brought to trial in June, he is nevertheless convicted. Born Joel Hagglund, Joe Hill called himself Joseph Hilstrom when he came to the United States in 1903, he has written essays, letters, and songs for the Wobbly newspapers *Industrial Worker* and *Solidarity*, his songs have included "The Preacher and the Slave" which contained the phrase "pie in the sky," and he has had a major influence in furthering the IWW cause (*see* 1915).

The Ludlow massacre April 20 climaxes a struggle by Colorado coal miners struggling for recognition of their United Mine Workers union (*see* 1913). A battle with state militia near Trinidad ends with 21 dead including two women and 11 children caught in tents that have been set ablaze, angry strikers take possession of the Colorado coal fields, and they do not yield until federal troops move in June 1.

The Amalgamated Clothing Workers of America is formed by a dissident majority within the manufacturer-oriented United Garment Workers Union (*see* 1910; 1918).

Suffragettes march on the Capitol at Washington June 28 to demand voting rights for U.S. women. The march is staged within hours of the assassination at Sarajevo (above) (*see* Jeanette Rankin, 1916).

Mohandas Gandhi returns to India at age 45 after 21 years of practicing law in South Africa where he organized a campaign of "passive resistance" to protest his mistreatment by whites for his defense of Asian immigrants. Gandhi has read Henry David Thoreau's 1849 essay "On the Duty of Civil Disobedience" and attracts wide attention in India by conducting a fast—the first of 14 that he will stage as political demonstrations and that will inaugurate the idea of the political fast (*see* 1913; 1930; Sarah Wright, 1647; Irish demonstrators, 1920).

Threats of labor troubles in early January have led Henry Ford to offer workers a minimum wage of $5 per day—more than twice the average U.S. wage and more than the average English worker earns in a week.

Montreal, Toronto, and Madrid stock exchanges close July 28, half a dozen other European bourses close July 29, the London exchange closes July 31, and the New York Stock Exchange immediately follows suit (it does not reopen until December 12).

Merrill Lynch has its beginnings in a New York brokerage firm started by former semipro baseball player Charles E. Merrill, 28, who will team up next year with Johns Hopkins graduate Edmund C. Lynch, 29 (*see* 1941).

A Federal Trade Commission established by Congress September 26 to police U.S. industry is designed only to prevent unfair competition (*see* Raladam case, 1931; Wheeler-Lea Act, 1938).

The Clayton Anti-Trust Act voted by Congress October 15 toughens the federal government's power against combinations in restraint of trade as outlawed by the Sherman Act of 1890. The new law named for Rep. Henry De Lamar Clayton of Alabama, exempts labor unions (*see* Celler-Kefauver Amendment, 1950).

"The maxim of the British people is 'business as usual,' " says First Lord of the Admiralty Winston Churchill November 9 in a speech at London.

Membership in British cooperative societies reaches 3.8 million (*see* Rochdale, 1844).

Japanese prime minister Count Gombei Yamamoto, 62, and Japanese navy minister Viscount Makoto Saito, 56, are held "morally responsible" for having permitted high-ranking navy personnel to accept bribes for placing large orders with Germany's Siemens-Schuckertwerke AG and then with England's Vickers' Sons and Maxim for communications equipment. Several scapegoats go to jail but Yamamoto and Saito are merely retired to the naval reserve. Yamamoto will become prime minister again in 1923, Saito will become prime minister in 1932.

A new eight-story May Company department store goes up on Cleveland's Public Square as the chain adds the Forest City to its Denver, St. Louis, and Akron locations. The company's Famous-Barr store in St. Louis is the city's leading retailer, up from seventh position when it was acquired in 1892 (*see* 1910; Los Angeles, 1923).

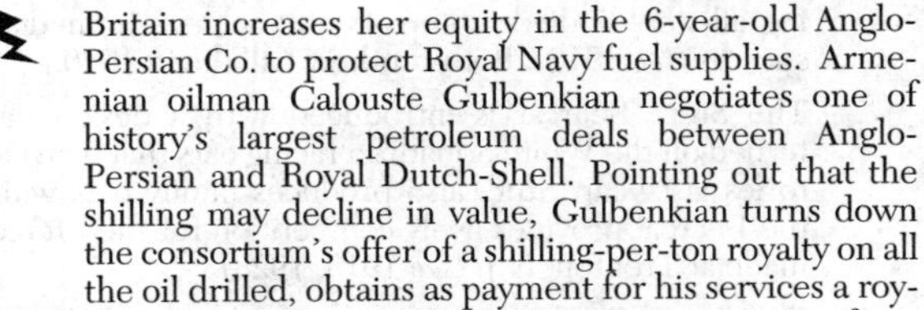

Britain increases her equity in the 6-year-old Anglo-Persian Co. to protect Royal Navy fuel supplies. Armenian oilman Calouste Gulbenkian negotiates one of history's largest petroleum deals between Anglo-Persian and Royal Dutch-Shell. Pointing out that the shilling may decline in value, Gulbenkian turns down the consortium's offer of a shilling-per-ton royalty on all the oil drilled, obtains as payment for his services a royalty to be paid in actual oil—2.5 percent to come from Anglo-Persian, 2.5 percent from Royal Dutch-Shell, and will receive 5 percent of all the oil produced in Middle East for the rest of his 86 years (*see* Iraq, 1927; Anglo-Iranian, 1935).

The U.S. Navy establishes a new oil reserve at Teapot Dome, Wyo. (*see* 1912; 1921).

The Panama Canal opens to traffic August 3 just as Germany declares war on France (above). Built essentially on French plans at a total cost of 30,000 lives and some $367 million in U.S. money (on top of the $287 million lost by the French in the 1880s), the canal uses a system of locks to carry ships 50.7 miles between deep water in the Atlantic and deep water in the Pacific (*see* 1903).

1914 ***(cont.)*** The Houston Ship Canal opens to give Houston an outlet to Galveston Bay on the Gulf. The 50-mile canal will make the Texas city a deep-water port and a major shipping point for U.S. grain.

The Cape Cod Ship Canal opens to link Buzzards Bay with Cape Cod Bay, Mass. The 17.4-mile canal enables coastal shipping to avoid the voyage round the Cape.

The White Star Line passenger vessel S.S. *Britannic* launched February 26 is a sister ship of the S.S. *Titanic* that went down in 1912 (*see* 1916).

The Hamburg-American Line's S.S. *Vaterland* that will become the S.S. *Leviathan* in 1919 arrives at New York from Cherbourg May 21 on her maiden voyage to begin a 24-year career on the North Atlantic. The 56,282-ton ship is 950 feet in length overall, has four screws, and can accommodate 4,000 passengers, 1,134 in crew.

The S.S. *Bismarck* that will become the S.S. *Majestic* in 1919 begins a 26-year career on the North Atlantic. The 56,621-ton German passenger liner is 954 feet in length overall and has four screws.

The Canadian Pacific's S.S. *Empress of Ireland* collides with another ship May 29 in the Gulf of St. Lawrence and sinks with a loss of 1,023 lives.

The Cunard Line's S.S. *Aquitania* arrives at New York June 15 on her maiden voyage. The 47,000-ton British passenger liner is 901 feet in length overall, accommodates 3,250 passengers, 1,000 in crew, and is double-skinned with an average of 15 feet between her inner and outer shells.

U.S. auto production reaches 543,679 with Model T Fords accounting for more than 300,000 of the total. The country has fewer than 100,000 trucks, most of them delivery vans (*see* 1921).

A large Dodge Brothers factory goes up at Hamtramck, Mich., where Horace and John Dodge pioneer in making all-steel-bodied motorcars and achieve immediate success (*see* 1913). Both brothers will die in 1920.

The Stutz Bearcat is introduced with a design patterned on the White Squadron racing cars that won victories last year. Stutz also produces family cars while the Bearcat provides lively competition for the Mercer made at Trenton, N.J. (*see* 1913; 1928)

Greyhound Bus has its beginnings in the Mesabi Transportation Co. founded at Hibbing, Minn., by Swedish-American diamond-drill operator Carl Eric Wickman, 30, who had opened a Hupmobile and Goodyear Tire agency but was unable to sell the Hupmobile. He runs the car on a regular schedule across the range to nearby Alice charging iron miners 15¢ one way, 25¢ roundtrip (local taxis charge upwards of $1.50 each way). Wickman takes in a partner, is soon building his own bodies and mounting them on truck chassis, and by 1918 will have 18 buses operating in northern Minnesota with annual earnings of $40,000 (*see* 1922).

Cleveland rigs up the world's first red-green traffic lights August 5 at Euclid Avenue and East 105th Street.

Gulf Oil distributes the first U.S. automobile maps (*see* Gulf, 1901). The 10,000 maps show roads and highways in Allegheny County, Pa., other oil companies soon follow Gulf's lead, and maps distributed free by service stations will encourage automobile travel (and gasoline consumption).

The hookworm control program started in the South by the Rockefeller Foundation 5 years ago reaches its peak of activity, but in many areas barefoot hookworm victims are reinfected soon after treatment.

New York City has only 38 more public schools than in 1899 despite an increase of more than 300,000 in enrollment. The overcrowded schools turn away 60,000 to 75,000 children each year for lack of space.

The teletype machine introduced by German-American inventor Edward E. Kleinschmidt, 38, speeds communications.

The *Sunday Post* begins publication at Glasgow and gains a readership that will grow until it reaches close to 80 percent of everyone in Scotland over age 15.

The *New Republic* begins publication at New York. The political weekly will compete with the *Nation* founded in 1865.

The *Day* begins publication at New York to compete with the *Forward* founded in 1897. The new Yiddish newspaper incorporates a section in English.

U.S. newspapers, magazines, advertisers, and ad agencies set up the Audit Bureau of Circulation to produce accurate data.

Nonfiction *A Preface to Politics* by New York writer Walter Lippmann, 25.

Fiction *Dubliners* (stories) by émigré Irish author James Joyce, 32, who has had trouble getting published in Ireland and who ekes out a living as an English teacher at Trieste; *A Portrait of the Artist as a Young Man* by James Joyce; French novelist Henri Alain Fournier is killed early in the war; *The Vatican Swindle (Les Caves du Vatican)* by André Gide; *Recaptured (L'Entrave)* by Colette; *Mist (Niebla)* by Miguel de Unamuno; *The Titan* by Theodore Dreiser; *Penrod* by Booth Tarkington; *A Springtime Case* (*Otsuyagorishi*) by Junichiro Tanizaki; *Kokoro* by Soseki Natsume; *Tarzan of the Apes* by U.S. novelist Edgar Rice Burroughs, 39, whose story about the son of an English peer raised by apes will have 26 sequels (Burroughs will also write science fiction novels).

Poetry *The Congo and Other Poems* by Vachel Lindsay; *North of Boston* by San Francisco-born poet Robert Lee Frost, 40, whose poem "The Death of the Hired Man" contains the line, "Home is the place where, when you have to go there, they have to take you in," his "Mending Wall" contains the line, "Good fences make good neighbors"; *Canti Orfico* by Italian poet Dino Campana, 29; *Starting Point (Der Aufbrach)* by Ernst Stadler, who wins the Iron Cross but is killed at Ypres at age 31; French poet Charles Péguy dies September 5 leading his troops at the Battle of the Marne.

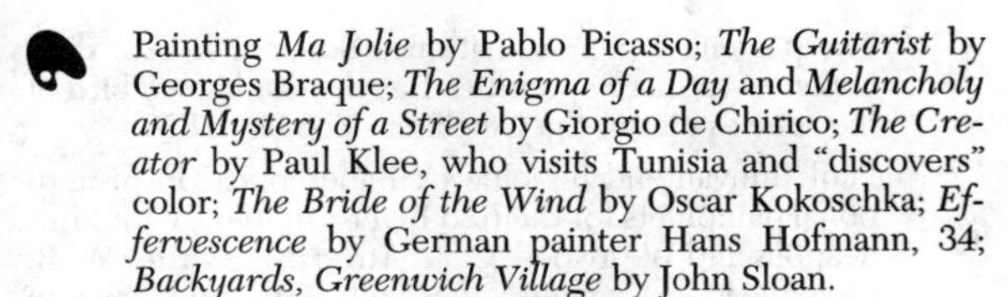

Painting *Ma Jolie* by Pablo Picasso; *The Guitarist* by Georges Braque; *The Enigma of a Day* and *Melancholy and Mystery of a Street* by Giorgio de Chirico; *The Creator* by Paul Klee, who visits Tunisia and "discovers" color; *The Bride of the Wind* by Oscar Kokoschka; *Effervescence* by German painter Hans Hofmann, 34; *Backyards, Greenwich Village* by John Sloan.

London's Old Vic Shakespeare Company opens at a theater that was given its name in 1833 for the princess who became Queen Victoria 4 years later.

Theater *The Bow of Odysseus (Der Bogen des Odysseus)* by Gerhart Hauptmann 1/17 at Berlin's Deutsches Künstlertheater; *The Exchange (L'échange)* by Paul Claudel 1/22 at the Théâtre du Vieux Colombes, Paris; *Too Many Cooks* by U.S. playwright Frank Craven, 33, 2/24 at New York's 39th Street Theater, with Craven, 223 perfs.; *Pygmalion* by George Bernard Shaw 4/11 at His Majesty's Theatre, London, with Sir Herbert Beerbohm Tree, 61, as Professor Higgins, Mrs. Patrick Campbell, now 47, as the Covent Garden flower girl Eliza Doolittle whose shocker "Not bloody likely" in Act III helps give Shaw another commercial success, 118 perfs; *The Hostage (L'otage)* by Paul Claudel 6/5 at the Théâtre de l'oeuvre, Paris; *On Trial* by U.S. playwright Elmer Rice (*né* Reizenstein), 21, 8/19 at New York's Candler Theater, 365 perfs.

Films Mack Sennett's *Tillie's Punctured Romance* with Charles Chaplin, Marie Dressler, Mabel Normand, Chester Conklin, Mack Swain, Charles Bennett, and the Keystone Kops in the first U.S. feature-length comedy; Mack Sennett's *Curses! They Remarked;* Charles Chaplin's *Between Showers*; D. W. Griffith's *The Mother and the Law* attacks factory owners who pose as public benefactors; D. W. Griffith's *The Battle of Elderberry Gulch*.

Opera *Francesco da Rimini* 2/19 at Turin's Teatro Regio, with music by Italian composer Riccardo Zandonai, 30; *Marouf, Savetier de Caire* 5/15 at the 0péra-Comique, Paris, with music by Henri Rabaud.

First performances "On Hearing the First Cuckoo in Spring" by Frederick Delius 1/20 at London. Delius is losing his eyesight to syphilis; *London* Symphony by Ralph Vaughan Williams 3/27 at London.

Broadway musicals *Sari* 1/3 at the Liberty Theater, with music by Emmanuel Kelman, lyrics by C. C. S. Cushing and Eric Heath, 151 perfs.; *Shameen Dhu* 2/2 at the Grand Opera House, with Chauncey Olcott, Rida Johnson Young, songs that include "Too-Ra-Loo-Ral, That's an Irish Lullaby" by Michigan-born songwriter James Royce Shannon, 32 perfs.; *The Ziegfeld Follies* 6/1 at the New Amsterdam Theater, with Ed Wynn, Ann Pennington, Leon Errol, music by Dave Stamper, Raymond Hubbell, and others, lyrics by Gene Buck and others, 112 perfs; *The Passing Show* 6/10 at the Winter Garden Theater, with Marilyn Miller making her debut at age 14, barelegged chorus girls, music chiefly by Sigmund Romberg, lyrics chiefly by Harold Atteridge, 133 perfs.; *The Girl From Utah* 8/24 at the Knickerbocker Theater with songs that include "They Didn't Believe Me" by Jerome Kern, lyrics by English writer Michael E. Rourke, 47, 120 perfs.; *Chin-Chin* 10/20 at the Globe Theater, with music by Henry Williams, lyrics by Joe Judge, songs that include "It's a Long Way to Tipperary," 295 perfs.; *The Only Girl* 11/2 at the 39th Street Theater, with music by Victor Herbert, lyrics by Henry Blossom, songs that include "You're the Only Girl for Me," 240 perfs.; *Watch Your Step* 12/8 at the New Amsterdam Theater, with Vernon and Irene Castle doing the Castle Walk, Fanny Brice, music and lyrics by Irving Berlin, songs that include "Play a Simple Melody," 175 perfs.

Popular songs "Keep the Home Fires Burning" by London composer David Ivor Davies, 21, lyrics by American-born poetess Lena Guklbert Ford, who will die in a zeppelin raid on London in 1918; "There's a Long, Long Trail" by Zo (Alonzo) Elliott, 23, who wrote the music while attending college in the United States, lyrics by W. Stoddard King, 25; "Colonel Bogey March" by London composer Kenneth J. Alford (F. J. Ricketts); "St. Louis Blues" by W. C. Handy; "12th Street Rag" by Euday L. Bowman, 27; "The Missouri Waltz" by John Valentine Eppel, 43; "When You Wore a Tulip and I Wore a Big Red Rose" by Percy Wenrich, lyrics by Jack Mahoney, 32; "A Little Bit of Heaven (And They Called It Ireland)" by Ernest R. Ball, lyrics by J. Keirn Brennan; "By the Beautiful Sea" by U.S. composer Harry Carroll, 22, lyrics by Harold R. Atteridge.

The American Society of Composers, Authors, and Publishers (ASCAP) is founded in New York to protect the interests of music writers, lyricists, and publishers (*see* 1909 copyright law). ASCAP will defend its members against illegal public performances for profit of copyrighted musical compositions, will protect them against other forms of infringement, and will collect license fees for authorized performances (*see* Victor Herbert, 1917; BMI, 1939).

The Millrose Games are held for the first time January 28 at New York's Madison Square Garden. Named for the country estate of department store executive Lewis Rodman Wanamaker, 50, the indoor track meet includes as a major event the Wanamaker Mile foot race.

Sir Norman Brookes wins in men's singles at Wimbledon, Mrs. Chambers in women's singles; Richard Norris Williams II, 22, wins in men's singles at Newport, Mary Browne in women's singles.

Upstate New York golfer Walter Hagen wins the U.S. Open at age 21 to begin a remarkable career.

The Boston Braves win the World Series by defeating Connie Mack's Philadelphia Athletics 4 games to 0.

Barcelona's El Sport bullring opens with 19,582 seats, the second largest in Spain. Matador Joselito (José Miguel Isidro del Sagrado Corazon de Jesus Gòmez y Gallito), 19, kills six giant bulls in the arena at Madrid July 3, making 25 quites, placing 18 banderillos, and making 242 passes with cape and muleta.

1914 ***(cont.)*** The Yale Bowl is completed at New Haven, Conn. The $650,000, 61,000-seat football stadium will be enlarged to seat 74,786 as college football becomes a major sporting attraction.

The first national Mother's Day is proclaimed by President Wilson whose wife Ellen will die August 6. The second Sunday in May will become the biggest business day of the year for U.S. restaurants and flower shops (*see* 1913; "M-O-T-H-E-R," 1915).

Grossinger's opens at Ferndale, N.Y., to provide New York sweatshop workers with a vacation boarding house that will grow to become a 1,500-acre Catskill Mountain resort. Austro-Hungarian immigrants Malke and Selik Grossinger begin an enterprise that will one day serve nearly 1,500 guests per week.

Helena Rubinstein challenges Elizabeth Arden for leadership in the fledgling U.S. cosmetics industry (*see* 1910). Now 44, Rubinstein left her native Poland to seek a husband in Australia at age 19, made $100,000 in 3 years selling a skin cream she formulated from ingredients that included almonds and tree bark, has opened a salon in London, and is the reigning beauty adviser to French and British society. The 4-foot-11-inch cosmetician will introduce medicated face creams and waterproof mascara, pioneer in sending saleswomen out on road tours to demonstrate proper makeup application, and have a $60 million business by the time of her death at age 94 in 1965.

The elastic brassiere that will supplant the corset now in common use is patented in November by Mary Phelps Jacob who as a New York debutante devised the prototype bra with her French maid before a dance, using two pocket handkerchiefs, some pink ribbon, and thread. A descendant of steamboat pioneer Robert Fulton, Jacob was asked by friends to make bras for them, a stranger asked for a sample and enclosed a dollar, she has been encouraged to engage a designer to make drawings, will make a few hundred samples of her Backless Brassiere with the help of her maid, will find them hard to sell, but will sell her patent to the corset maker Warner Brothers Co. of Bridgeport, Conn., which will acquire for $15,000 a patent that will later be estimated to be worth $15 million.

Doublemint chewing gum is introduced by William Wrigley, Jr. (*see* 1893).

Germany's 107th Regiment of Leipzig leaves its trenches December 14 and follows a band playing Christmas carols. Singing lustily and distributing gifts to the enemy, the Germans play football (soccer) with the enemy in the afternoon; hostilities resume December 15.

New York's St. Thomas Episcopal Church at Fifth Avenue and 53rd Street is completed by architect Bertram Goodhue of Cram, Goodhue & Ferguson.

New York's Municipal building is completed by McKim, Mead and White near the Manhattan end of the Brooklyn Bridge in Chambers Street at Centre Street. The 25-story office complex houses city administrative offices.

The passenger pigeon that once dominated U.S. skies becomes extinct September 1 as the last known bird of the species dies in the Cincinnati Zoo (*see* 1878).

Paul Bunyan and his blue ox Babe appear in promotional pamphlets for the Red River Lumber Co. of Minneapolis and Westwood, Calif. Advertising writer W. B. Laughhead has created the "folk hero" unknown to lumberjacks.

Fish freezing is pioneered by New Yorker Clarence "Bob" Birdseye, 28, who trades furs in Labrador. He notices that fish caught through the ice freeze stiff the instant they are exposed to the air and taste almost fresh when defrosted and cooked weeks later (*see* 1917).

Van Camp Seafood is founded by Indianapolis packer Frank Van Camp whose father Gilbert began packing pork and beans in 1861. His son Gilbert persuades him to sell the family's Midwestern operation and enter the fast-growing California tunafish packing business (*see* 1903). Father and son buy California Tunny Packing of San Pedro, contract with operators of 35 to 40 small Japanese-owned albacore boats to buy all the albacore the boats can catch, accept deliveries 7 days a week, and by October have developed a new method of purse seining to increase catches. Within a year the Van Camps will be building a new fleet of 15 modern 40-foot albacore boats, will turn the new boats over to the fishermen at no cost, and by finding new tuna banks off the Mexican Coast, rushing the catch back to the San Pedro cannery by tender, and lowering the price of canned tuna will pioneer in making tunafish an American staple rather than a costly delicacy (*see* 1922).

British farmers produce less than one-fourth the nation's grain needs. Nearly 4 million acres of arable British lands have passed out of cultivation as Canada and India have emerged as major suppliers.

German scientists develop mercury fungicide seed dressings that will prevent losses to plant diseases. In years to come, the mercury will take severe tolls of wildlife and some human life.

U.S. ranchers herd cattle with Model T Fords.

U.S. farm hands begin annual migrations north from Texas, traveling in Model T Fords and harvesting crops as they ripen up to the Canadian border.

The Smith-Lever Act passed by Congress May 8 provides for agricultural extension services by USDA county agents working through the land-grant colleges established under the 1890 Morrill Act. An outgrowth of work done by Seaman Knapp and Theodore Roosevelt, the act authorizes federal appropriations to support state and local funding of farm and home demonstration agents.

A national 4-H Club is founded 14 years after the first such club was organized in Macoupin County, Ill. Sponsored by the U.S. Department of Agriculture, the 4-H Clubs direct their activities toward youths in rural areas, trying to improve "head, heart, hands, and health," and will help end prejudice against "book farming" along scientific lines.

George Washington Carver reveals the results of experiments that show the value of peanuts and sweet potatoes in replenishing soil fertility (*see* Carver, 1896). Southern planters have in many instances been ruined by the boll weevil and begin turning to peanut and sweet potato culture, especially when Carver shows the many peanut by-products he has produced in his laboratory—not only flour, molasses, and vinegar but also cheese, milk, and coffee substitutes, synthetic rubber, plastics, insulating board, linoleum, soap, ink, dyes, wood stains, metal polish, and shaving cream (*see* cotton production, 1921).

The relief ship S.S. *Massapequa* reaches Belgium in November with a cargo of food sent by the Rockefeller Foundation to aid the starving Belgians.

Berlin places German food supplies and allocations under government control December 26, and London thereupon declares all foodstuffs on the high seas to be contraband.

Some 200,000 British schoolchildren receive free meals under the School Meals Act of 1906 (*see* Milk in Schools Scheme, 1931, 1934).

National Dairy Corp. has its beginnings as Chicago businessman Thomas H. McInnerney, 47, acquires ownership of Hydrox, a wholesaler of artificial ice, ginger ale, and ice cream. Formerly general manager of New York's Siegel Cooper department store, McInnerney will increase Hydrox sales in the next 9 years from $1 million to $4 million (*see* 1923).

Some margarine is now made entirely of vegetable oil combinations, but Americans consume 5 pounds of butter for every pound of margarine (*see* 1890; Lever, 1911).

Postum Cereal founder C. W. Post commits suicide May 9 at his Santa Barbara, Calif., home in a fit of depression at age 59. Marjorie Merriweather Post, 27, his only child, inherits the company (*see* Post Toasties, 1906; incorporation, 1922; Jell-O, 1925).

Large-scale pasta production begins in the United States which has imported almost all of its macaroni and spaghetti from Naples but which has been cut off from Italian sources by the outbreak of the European war (above) (*see* Carleton, 1899).

Thomas Lipton expands his retail grocery empire to include some 500 shops (*see* 1909). Lipton has acquired tea, cocoa, and coffee plantations, meat-packing plants, and factories, and has at least 20 British retail grocery rivals.

The Woman Rebel by U.S. feminist Margaret Higgins Sanger, 31, introduces the term "birth control." Sanger exiles herself to England to escape federal prosecution for publishing and mailing *Family Limitation*, a brochure dealing with contraception (*see* Brooklyn clinic, 1916).

1915 The Great War in Europe intensifies. Casualty lists mount for both sides on the eastern and western fronts and a German U-boat (submarine) blockade of Britain begins February 18.

British naval forces attack the Dardanelles beginning February 19 to prevent the Germans who have seized control of Turkey from blocking supplies to Russia via the Bosphorus and the Black Sea. Russia has requested the British action to relieve pressure on her defenses in the oil-rich Caucasus, but four British warships trying to force the Narrows are sunk by mines March 18, and British admiral John de Robeck withdraws, thus losing an opportunity to seize the Dardanelles at a time when the Turks have no supplies and are practically defenseless (*see* Gallipoli, below).

Turkey says Armenians side with Russia and begins April 24 to deport them, putting to death all who resist. Some 1.75 million Armenians will be deported, 600,000 will starve to death in the Mesopotamian desert, one-third will survive.

British troops land on Turkey's Gallipoli Peninsula beginning April 25, but Turkish artillery pins down British, French, and Austrian units and the Turkish commander Mustafa Kemal, 34, says to his troops, "I do not order you to attack. I order you to die."

The first sea lord of the admiralty Sir John Arbuthnot Fisher, 74, had proposed the Gallipoli campaign (above) but has turned against the idea and resigned ("Damn the Dardanelles. They will be our grave.") When the British abandon the campaign in December, blame for its costly failure falls on the first lord of the admiralty Winston Churchill, now 41.

The Germans use chlorine gas April 22 at the Second Battle of Ypres, the first use of poison gas by any warring power. Greenish-yellow clouds of gas choke French colonial troops who flee in panic, but Canadians plug the 4.5-mile gap left by the routed Africans.

Torpedoes from the German submarine U-20 hit the Cunard Line passenger ship S.S. *Lusitania* at 2:10 P.M.

Poison gas that seared the lungs and blistered the skin brought new horrors to trench warfare in Europe.

1915 ***(cont.)*** May 7 off the coast of Ireland, and the huge vessel sinks in 18 minutes killing 1,198 who include 128 U.S. citizens, among them railroad magnate Alfred Gwynne Vanderbilt, 38, and New York theatrical magnate Charles Frohman, 75. It will turn out that the 8-year-old *Lusitania* carried 173 tons of rifle ammunition, shrapnel casings, fuses, and contraband food from the United States but had no escort and remained on course despite recent U-boat sightings in the area.

"There is such a thing as a man being too proud to fight," says President Wilson May 10. "There is such a thing as a nation being so right that it does not need to convince others by force that it is right." But the sinking of the *Lusitania* (above) with so many women and children fires anti-German sentiment. "We have come to a parting of the ways," says presidential aide Col. Edward M. House, 57, a formal protest is sent to Berlin May 13, pacifist Secretary of State William Jennings Bryan resigns June 8.

Italy declares war on Austria May 23, turning against her former ally in hopes of obtaining territory along the Adriatic and in the Alps. The Italians will launch four offensives during the war but will accomplish nothing but the loss of 250,000 troops on a 60-mile front along the Isonzo River (*see* Caporetto, 1917).

France gives Gen. Henri Philippe Benoni Omer Joseph Pétain, 59, command of an army June 15.

German authorities in occupied Brussels arrest British Red Cross nurse Edith Louisa Cavell, 50, August 4 on charges of having assisted Allied military prisoners to escape. Tried and convicted, she is executed by a firing squad October 12 along with a Belgian who has provided guides; British propagandists use her death to inflame public sentiment against the savage Boche and ignore the fact that the French have shot a woman for a similar offense.

Warsaw falls to the Germans August 7, and by September the Russians have lost all of Poland, Lithuania, and Courland, along with nearly 1 million men.

Gen. Pétain (above) launches a great attack in Champagne September 25 and follows it with a report containing principles of a new tactical doctrine.

Italy has declared war on Turkey August 21, Britain declares war on Bulgaria October 13 and is joined by Montenegro, Bulgaria declares war on Serbia October 14, France declares war on Bulgaria October 16, Russia and Italy follow suit October 19.

The 21 demands presented by Tokyo to Beijing January 18 have included the demands that Japan take over German rights in Shandong, that Japanese leases in southern Manchuria be extended to 99 years with commercial freedom for the Japanese in Manchuria, that China give Japan a half-interest in the Han-yehping Co. which operates iron and steel mills at Han-yang, iron mines at Ta-yeh, and a colliery at P'ingshan, and that China declare her determination not to lease or cede any part of her coast to any power. The Chinese have given modified acceptance to these four demands but have set aside others calling for railway concessions in the Yangzi River Valley, which is within Britain's sphere of influence, and for Japanese advisers in Chinese political, financial, and military affairs.

Henry Ford charters the ship S.S. *Oscar II*, calls it the *Peace Ship*, and sails for Europe December 4 with a party of advisers in a vain but much-publicized attempt to "get the boys out of the trenches by Christmas."

The Thompson submachine gun (Tommy gun) is introduced by U.S. Brig. Gen. John Taliaferro Thompson, who organizes the Auto-Ordnance Co. to build the lightweight, semiautomatic infantry shoulder rifle.

The La Follette Seaman's Act passed by Congress March 4 abolishes prison sentences for U.S. merchant marine sailors who desert ship and contains other provisions that improve the lot of merchant seamen. The act crowns a long campaign by International Seamen's Union chief Andrew Furuseth, 61, who works ceaselessly in behalf of his membership, living a monastic life and accepting a salary equal only to the union scale wage of an able-bodied seaman.

U.S. Pullman car porters (called "George" because they are George Pullman's "boys") earn a minimum wage of $27.50 per month. Wages will rise to $48.50 per month by 1919 and increase in that year to $60 per month for nearly 400 hours of work or 11,000 miles, whichever is logged first, the Pullman Palace Car Co. will pay overtime at the rate of 60¢ per 100 miles or 25¢ per hour, but porters will still have to buy their own meals, uniforms, and equipment out of wages and tips (*see* Randolph, 1925).

A Georgia lynch mob seizes Leo Frank from a state prison August 16 and hangs him at Marietta (*see* 1913).

A firing squad executes IWW organizer Joe Hill November 19 (*see* 1914). He has been in prison for 22 months, Utah officials have rejected appeals for a new trial (the Swedish government, AF of L president Samuel Gompers, and President Wilson have all appealed), and on the eve of his execution Hill has wired "Big Bill" Haywood, "Don't waste any time in mourning. Organize." The message will help make Joe Hill a legend among working men (*see* Haywood, 1907).

D. W. Griffith's full-length motion picture *The Birth of a Nation* (below) heightens U.S. racial tensions.

Prizefighter Jess Willard takes away Jack Johnson's heavyweight title (below). He is widely called the "great white hope" because Johnson has inflamed prejudice by his marriage to a white woman and his lavish life style. Convicted in 1912 of violating the Mann Act of 1910, Johnson skipped bond and has been living abroad with his wife; the fight is held at Havana because Johnson is a fugitive from justice.

A new Ku Klux Klan inaugurated Thanksgiving night on Stone Mountain near Atlanta will incite bigotry. William Joseph Simmons, who revives the name from the KKK of the 1860s, has been a preacher, traveling salesman, and promoter of fraternal organizations; he obtains a legal charter from the state of Georgia, calls his KKK a "high-class, mystic, social, patriotic, benevolent association" dedicated to "white supremacy," the protection of Southern womanhood, and "Americanism," and will promote his "invisible empire" of Kleagles and Grand Dragons through advertising developed by Edward Young Clark. It will attract a membership of nearly 100,000 throughout the country within 6 years, concentrating political power to elect sympathetic U.S. senators, congressmen, and state officials, and employ terrorist tactics against blacks, Jews, and Roman Catholics, most especially in the South and Midwest (*see* 1925).

The Wealth and Income of the People of the United States by economist Wilford Isbell King points out the increasing concentration of income in the hands of the few, an indication of the need for a graduated income tax (*see* 1913). In 1910, King observed, the richest 1.6 percent of U.S. families received 19 percent of the national income, up from 10.8 percent in 1890. While the richest 2 percent of the population received 20.4 percent of the national income and averaged $3,386 per capita in income, the poorest 65 percent received 38.6 percent and averaged $197.

British income taxes rise to an unprecedented 15 percent as the Great War drains the nation's financial resources (*see* 1909; 1918).

The state of Delaware begins revising its corporation laws to adopt "suggestions" forwarded by New York lawyers. New Jersey has had the lion's share of corporate charters and has been called "the Mother of Trusts," but President Wilson has persuaded New Jersey legislators to outlaw a series of abusive practices, the New York lawyers recognize that corporate franchise taxes will be valuable to a state as tiny as Delaware, and that state's legislators liberalize their laws to attract corporations which will soon make Delaware the leading charterer of the largest industrial corporations.

Metropolitan Life Insurance becomes a mutual company with profits going to policyholders instead of stockholders (*see* 1879).

New York's Brooks Brothers moves from Broadway and 22nd Street into a large new Madison Avenue store at 44th Street (*see* 1909).

The worst train wreck in British history kills 227 and injures 246 as a Scottish express plows into an earlier minor wreck May 22 at Gretna Junction north of Carlisle at Quintinshill. A Scottish regiment of 500 men is reduced to 72.

The Pennsylvania Railroad electrifies 20 miles of main line track between Philadelphia and Paoli (which becomes a switching point for steam engines to and from the West).

The Tunkhamnock Viaduct completed across a valley in Pennsylvania cuts 39 miles off the route of the Delaware, Lackawanna & Western Railroad. The 2,375-foot concrete bridge has cost $12 million.

The Fruehauf tractor trailer has its beginnings at Detroit where local blacksmith-wagon maker August Fruehauf, 47, and his associate Otto Neumann create the first trailer for the Model T Ford of local lumber dealer Frederick M. Sibley, Jr., who wants to transport a boat to his new summer place on an upstate Michigan lake. Fruehauf Trailer Co. will be founded early in 1918 and become the world's largest maker of tractor trailers pulled by sawed-off trucks (tractors) to compete with railroads.

New York's Fifth Avenue Coach Co. starts designing and assembling its own buses to replace the French DeDions and English Daimlers imported since 1907. The new 34-seat double-decker coaches are powered by 40-horsepower 4-cylinder Knight sleeve-valve engines (below) (*see* 1936).

The one millionth Ford motorcar rolls off the assembly line and Ford produces more than 500,000 Model T cars, up from 300,000 last year (*see* 1913; 1927).

Willys-Overland produces 91,780 motorcars, up from 18,200 in 1910, and is the world's second largest auto maker. Willys has absorbed several other firms and its sales are helped by the patented Willys-Knight sleeve-valve engine that employs two simple sleeves in each cylinder, moving silently up and down on a film of oil without springs, valves, or camshafts (*see* 1907).

Chevrolet Motor Co. is incorporated in Delaware by W. C. Durant after having sold 16,000 motorcars since 1911. Durant offers General Motors stockholders five shares of Chevrolet stock for every share of GM, whose stock is being acquired in the open market by Pierre S. du Pont; GM stock climbs from 82 January 2 to a high for the year of 558 (*see* 1916).

The excursion boat *Eastland* suddenly turns over July 24 in the Chicago River while carrying 2,000 picnickers bound for a Western Electric Co. outing; 812 drown.

A typhus epidemic in Serbia kills 150,000.

Mayo Foundation researcher Edward Calvin Kendall, 29, isolates the thyroid hormone thyroxine. Deficiencies produce goiters which C. H. Mayo has made a career of removing surgically (*see* 1889; Marine, 1905; Akron, 1916).

U.S. inventor Morgan Parker, 23, patents a disposable scalpel.

Emory University is founded at Atlanta under the name Methodist College with support from Coca-Cola's Candler family. The university is an outgrowth of Emory College, founded as a manual labor school founded at Oxford in 1836.

Long-distance telephone service between New York and San Francisco begins January 25. Alexander Graham Bell, now 68, repeats the words of 1876 ("Mr. Watson, come here . . .") to Thomas Watson in San Francisco, the call takes 23 minutes to go through, and it costs $20.70.

"Adventures of Teddy Tail—Diary of the Mouse in Your House" by English cartoonist Charles Folkard appears in the *London Daily Mail* April 5, the mouse ties

1915 ***(cont.)*** a knot in its tail April 9 to rescue a beetle that has fallen down a hole, and this first British newspaper comic strip will continue for more than 40 years.

Direct wireless communication between the United States and Japan begins July 27.

Japanese engineer Tokuji Hayakawa invents the first mechanical pencil and markets it under the brand name Sharp. Proceeds will enable the inventor to start the Hayakawa Electric Company that will market a wide range of products under the Sharp name.

Condé Nast takes over *House and Garden* magazine which has only 10,000 readers and little advertising *(see Vogue,* 1909).

Fiction *The Metamorphosis* (*Der Verwandlung*) by Czech novelist Franz Kafka, 32; *The Good Soldier* by former *English Review* editor Ford Madox Ford (*né* J. L. Ford Hermann Hueffer), 41; *Of Human Bondage* by W. Somerset Maugham; *The "Genius"* by Theodore Dreiser; *Wood and Stone* by English novelist John Cowper Powys, 43; *Pauline* by English novelist Compton Mackenzie, 32; *The Rainbow* by D. H. Lawrence whose book is suppressed in November 2 months after publication; *Rashomon* by Japanese novelist Ryunosuke Akutagawa, 23, who gives a cynical, psychological twist to an old folktale; "The Gentleman from San Francisco" ("Gospodin iz San-Francisco") by Ivan Bunin satirizes Western bourgeois civilization; *The Thirty-Nine Steps* by John Buchan; *The Sea Hawk* by English novelist Rafael Sabatini, 39; *The Star Rover* by Jack London; *Gino Bianchi* by Italian poet Piero Jahier, 31 (prose satire).

Poetry *Chicago Poems* by *Chicago Daily News* writer Carl Sandburg, 37, who calls his city "Hog Butcher for the World/ Tool Maker, Stacker of Wheat,/ Player with Railroads and the Nation's Freight Handler;/ Stormy, husky, brawling,/ City of the Big Shoulders"; *A Spoon River Anthology* by Chicago lawyer-poet Edgar Lee Masters, 46, whose somber free-verse monologues are written as if spoken from the graveyard by former residents of a fictitious Illinois town. English poet Charles Sorley dies in battle at age 19; English poet Rupert Brooke of the Royal Navy dies April 23 of blood poisoning at age 27 after having been weakened by sunstroke. Buried on the Aegean island of Skyros, Brooke has written, "If I should die, think only this of me:/ That there's some corner of a foreign field/ That is for ever England."

Painting *Harlequin* by Pablo Picasso; *The Bride Stripped Bare by Bachelors* by Marcel Duchamp is the world's first Dada-style painting; *Birthday* by Marc Chagall; *Self-Portrait as a Soldier* by Ernst Ludwig Kirchner.

Sculpture *Woman Combing Hair* by Russian sculptor Aleksandr Archipenko, 28.

Theater *The Unchastened Woman* by Louis Kaufman Anspacher 10/9 at New York's 39th Street Theater, 193 perfs.; *He Who Gets Slapped* (*Tot, kfopoluchayet poshicheedony)* by Leonid Andreyev 10/27 at Moscow's Dramatic Theater; *John Ferguson* by St. John Ervine 11/30 at Dublin's Abbey Theatre.

Films D. W. Griffith's *The Birth of a Nation* with Lillian Gish, Mae Marsh, Henry B. Walthall has such cinematic innovations as the close-up, pan (panoramic shot), flashback, and use of a moving camera in addition to its element of racism (above). Also: Frank Powell's *A Fool There Was* with Theda Bara (Theodosia Goodman) who is billed as "the vamp" ("Kiss me, my fool").

Opera *Madame Sans Gêne* 1/25 at New York's Metropolitan Opera House, with Geraldine Farrar as Catherine Huebscher, music by Umberto Giordano, libretto from a play by Victorien Sardou and E. Moreau.

Ballet *El Amor Brujo* 4/15 at Madrid's Teatro Lara with music by Manuel de Falla.

First performance Symphony No. 5 in E flat major by Jean Sibelius 12/8 at Helsinki.

Broadway musicals *The Passing Show* 5/29 at the Winter Garden Theater, with Marilyn Miller, baritone John Charles Thomas in his Broadway debut, music chiefly by Leo Edwards and W. F. Peter, lyrics by Harold Atteridge, 145 perfs.; *The Ziegfeld Follies* 6/21 at the New Amsterdam Theater, with W. C. Fields, Ed Wynn, George White, Ann Pennington, music by Louis A. Hirsch, Dave Stamper, and others, songs that include "Hello, Frisco" by Hirsch and Gene Buck (*see* telephone, above), 104 perfs.; *The Blue Paradise* 8/5 at the Casino Theater, with Vivienne Segal, music by Edmund Eysler, book by Edgar Smith based on a Viennese opera, lyrics by M. E. Rourke, 356 perfs.; *The Princess Pat* 9/29 at the Cort Theater, with Al Shean, music by Victor Herbert, book and lyrics by Henry Blossom, 158 perfs.; *Very Good Eddie* 12/23 at the Princess Theater, with music by Jerome Kern, lyrics by Schuyler Greene, 341 perfs.; *Stop! Look! Listen!* 12/25 at the Globe Theater, with music and lyrics by Irving Berlin, book by Harry B. Smith, 105 perfs.

Popular songs "Pack up Your Troubles in Your Old Kit Bag" by Welsh-born London composer Felix Powell and his brother George Asaf (George Henry Powell), 35; "I Didn't Raise My Boy to Be a Soldier" by Al Piantadosi, who has plagiarized a melody by Harry Haas), lyrics by Alfred Bryan; "Paper Doll" by Johnny Black; "There's a Broken Heart for Every Light on Broadway" by Fred Fisher, lyrics by Howard Johnson; "M-O-T-H-E-R, a Word that Means the World to Me" by Theodore Morse, lyrics by Howard Johnson; "Nola" by New York Tin Pan Alley pianist-composer Felix Arndt, 26.

Jess Willard, 32, wins the world heavyweight title April 15 at Havana with a 26th-round knockout over Jack Johnson (above). Willard will hold the title until 1919.

Wimbledon tennis is canceled for the duration; William M. Johnston 20, wins in U.S. men's singles, Norwegian-American Molla Bjurstedt, 23, in women's singles.

The New York Yankees and the American League franchise purchased for $18,000 in 1903 by Frank Farrell and William Deveny is purchased January 11 for $460,000 by local brewer Jacob Ruppert, 73, and engineer Tillinghast l'Hommidieu Houston (*see* Babe Ruth, 1920).

Rogers Hornsby signs with the St. Louis Cardinals and makes his major league debut late in the season at age 19. The Texas-born athlete will lead the National League in batting, home runs, and runs batted in (the triple crown) in 1922 and 1925, will set a league record batting average of .424 in 1924, and by 1937 will have a lifetime batting average of .358, surpassed only by Ty Cobb's .367.

The Boston Red Sox win the World Series by defeating the Philadelphia Phillies 4 games to 1.

New York's Equitable building is completed at 120 Broadway between Pine and Cedar Streets (*see* 1913). The new office building contains 1.2 million square feet of floor space on a plot of just under one acre, has nearly 30 times as much floor space as was contained in its site, exploits its site as no building has ever done before, and raises demands for a zoning law that will prohibit such buildings (*see* 1916).

Carrier Corp. is founded under the name Carrier Engineering by air-conditioning pioneer Willis H. Carrier and six other young engineers who pool $32,600 to start the company (*see* 1911; Empire Theater, 1917).

An earthquake in the Avezzano area east of Rome January 13 leaves 29,800 dead.

Rocky Mountain National Park is created by act of Congress on 262,000 acres of Colorado wilderness land that includes 107 named peaks above 10,000 feet in elevation.

The U.S. wheat crop totals 1 billion bushels for the first time as American farmers respond to high wartime wheat prices by plowing and planting millions of acres never before planted (*see* 1921; 1934).

The Hessian fly introduced in 1776 causes $100 million worth of damage to the U.S. wheat crop which nevertheless breaks all records.

The Cortland apple is created in upstate New York by crossing a Ben Davis with a McIntosh.

Cuban sugar production is owned 40 percent by U.S., 20 percent by other foreign interests.

Germany's War Grain Association confiscates all stocks of wheat, corn, and flour at fixed prices, suspends private transactions in grain, fixes a bread ration, and orders municipalities to lay up stores of preserved meat as Britain blockades German ports.

A U.S. vessel with a cargo of wheat for Britain is sunk by a German cruiser.

The German warship *Kronprinz Wilhelm* puts into Newport News, Va., with 50 sailors suffering from beriberi (*see* Suzuki, 1912; Williams, 1933).

Pellagra is shown to be the product of Southern diets based largely on corn and not a disease spread by any bacteria (*see* 1913). Joseph Goldberger of the U.S. Public Health Service has conducted tests using inmates of Mississippi's Rankin Prison who have volunteered to go on corn product diets in return for sentence reductions (*see* Elvehejm, 1936).

The cancer-causing properties of coal tar are demonstrated by Japanese chemists Katsusaburo Yamagiwa and Koichi Ichikawa who paint rabbits' ears with coal tar in the first experiments that produce cancer in animals (*see* Kennaway, 1933).

France outlaws sale of absinthe. The aromatic liqueur made from wormwood (*Artemisia absinthium*) and other herbs produces blindness and death among habitual users.

Kraft processed cheese is introduced by Chicago's 5-year-old J. L. Kraft and Bros. which has developed a process that arrests the bacterial curing of cheese without subjecting it to such high temperatures as to cause oil separation (*see* 1909). Processed cheese packed initially in tins will hereafter comprise a growing part of Kraft's cheese business (*see* Kraft-Phenix, 1928; National Dairy Corp., 1928).

Kellogg's 40% Bran Flakes are introduced by the Battle Creek Toasted Corn Flake Co. (*see* 1911; 1919).

Franco-American is acquired by the Joseph Campbell Co. which will market the canned soups, canned spaghetti, and other products of the New Jersey firm (*see* 1922).

Corning Glass researchers Eugene C. Sullivan and William C. Taylor develop Pyrex glass whose heat-resistant and shock-resistant borosilicate glass will be used in oven dishes and coffeemakers.

1916 The Great War in Europe takes a heavy toll, the United States remains neutral while chasing Mexican bandit Pancho Villa, the Irish rise against the British in a great Easter rebellion, and an Arab revolt against the Turks begins in the Hejaz.

Britain withdraws her forces from the Gallipoli by early January without further losses but the Battle of Verdun on the western front from February 21 to July 11 takes 350,000 French lives and nearly as many German lives. Krupp of Essen supplies the Germans with 3,000 new cannons per month while Allied artillery fire Vickers shells produced under a fuse patent licensed by Krupp in 1904.

Berlin has notified Washington that German U-boats will treat armed merchantmen as cruisers, the "extended" U-boat campaign has begun March 1, three Americans have perished in the sinking of the unarmed French steamer S.S. *Sussex* March 24 in the English Channel, and Washington has warned Berlin that the United States will sever diplomatic relations unless the Germans abandon "submarine warfare against passenger and freight-carrying vessels." Berlin replies in early May that merchant vessels "shall not be sunk without warning and without saving human lives unless they attempt to escape or offer resistance."

The British hospital ship S.S. *Britannic* sinks in 50 minutes November 21 after hitting a mine en route to pick up 3,500 wounded from the Aegean island of Lemnos

1916 ***(cont.)*** for transfer to Naples. The ship has 1,136 aboard and 30 are killed, most of them by the *Britannic's* own propellers.

Pancho Villa has raided Columbus, N. Mex., March 9, killing 17 Americans, and a U.S. punitive expedition has moved into Mexico March 15 under the command of Gen. John J. Pershing, 55 (*see* 1914). Pershing has orders to "capture Villa dead or alive" but is unable to catch him and will withdraw in early February of next year.

U.S. Marines land in Santo Domingo in May to restore order. U.S. occupation will continue until 1924.

Ireland's Easter rebellion beginning April 24 lasts a week but has little popular support. Former British consular official Roger Casement has had no success in raising a brigade of Irish war prisoners in Germany, a U-boat has landed him April 20 to support the Irish Republican Brotherhood led by Patrick Henry Pearse, 37, but German aid fails to materialize (*see* Casement, 1903). While 150,000 Irish volunteers fight for the king in Flanders, some 2,000 rebels rise at Dublin, police arrest the rebel leaders. People hiss them but they become martyrs when convicted of treason and hanged August 3 (*see* 1919).

The Battle of Jutland (Skagerrak) May 31 to June 1 ends with heavy losses on both sides, but the German fleet escapes the larger British fleet.

A U.S. National Defense Act passed by Congress June 3 provides for an increase in the standing army by five annual stages to 175,000 men with a National Guard of 450,000 and an officers' reserve corps. The Wilson administration emphasizes "preparedness" as sentiment against neutrality increases.

Britain's refusal to permit U.S. imports of German knitting needles needed in U.S. mills has drawn a sharp protest from Washington in May, and Washington protests again when the *London Official Gazette* blacklists some 30 U.S. firms under the British Trading with the Enemy Act of July 18.

French pilots in Spads and Nieuports gain control of the air from the Germans who have held mastery for a year flying Fokkers equipped with A. H. G. Fokker's device that permits pilots to fire through their revolving propellers (*see* 1912). The British introduce new De Havillands and Farman Experimentals in July, and the German ace Max Immelman, 26, is shot down July 18. (He has invented the air combat maneuver that will be known as the Immelman turn.)

French aircraft designer Marcel-Ferdinand Bloch, 24, has introduced the first variable-pitch propeller to give French pilots an edge over the Germans (*see* 1947).

The Escadrille Americaine has gone into combat May 15 with seven U.S. volunteer pilots flying Nieuports in support of the French. Kiffin Rockewell downs a German plane May 18 for the Escadrille's first victory. The German ambassador at Washington complains in November that the U.S. flyers violate American neutrality, Paris orders that the group be called simply Escadrille 124, its members protest and call themselves the Lafayette Escadrille. The group will never number more than 30 men at a time, but 45 men will serve in it.

The Battle of the Somme from July to mid-November is the bloodiest battle in history and follows the largest artillery barrage in history (1,437 British guns rain 1.5 million shells on the enemy along an 18-mile front in the course of 7 days). July 1 is the bloodiest day in British history with 57,470 British casualties, including 19,240 dead, 35,493 wounded, 2,152 missing, 585 taken prisoner. The Germans sustain 8,000 casualties July 1 and the artillery is heard as far away as Hempstead Heath, England. The 140-day offensive involves 3 million men along a front of some 20 miles, the Allied armies lose 794,000 men, the Central Powers lose 538,888, the Allies drive the Germans back no more than 7 miles at any point, and the Germans will regain most of the lost ground in 1918.

The British attack July 1 (above) begins 2 minutes after five gigantic mines dug under the German lines blow up at 7:28 in the morning, Boer War veteran Gen. Henry Seymour Rawlinson has ordered the daylight attack to accommodate French artillery observers despite the known advantages of attacking at first dawn, 66,000 British troops come out of the trenches and advance on the enemy in a ceremonial step of one yard per second, Tyneside Scotsmen march to bagpipes while the 8th East Surreys come out kicking footballs, 14,000 fall in the first 10 minutes by one account, Gen. Rawlinson has not coordinated his artillery barrage with the needs of his infantry, and one-third of all the British shells fired are duds.

Commander-in-chief of British forces Gen. Douglas Haig is found to have said at a War Council April 14 of last year, "The machine gun is a much overrated weapon; two per battalion is more than sufficient." Like other top officers, Haig does not visit the front lines, saying he considers it his duty not to lest the sight of wounded men affect his judgment.

The Black Tom explosion July 30 blows up munitions loading docks at Jersey City, N.J., killing 7 men, injuring 35, and destroying $40 million worth of property. German saboteurs are generally considered responsible.

More of Europe is drawn into the Great War. Germany and Austria have declared war on Portugal in March, Romania declares war on Austria August 27, Italy on Germany August 28, Germany, Turkey, and Bulgaria on Romania.

The first tanks to be used in warfare go into action September 15 in the Battle of the Somme (above). British writer and Boer War veteran Ernest Dunlap Swinton, 48, has invented the machines.

"I believe that the business of neutrality is over," says President Wilson in late October to the Cincinnati Chamber of Commerce. "The nature of modern war leaves no state untouched."

President Wilson wins reelection on a platform that includes the slogan "He kept us out of war," but he believes he has lost until late returns from California give him 23 more electoral votes than his Republican opponent Justice Charles Evans Hughes of the Supreme Court. Wilson's 4,000-vote edge in California gives him 277 electoral votes to 254 for Hughes, who receives 46 percent of the popular vote to Wilson's 49 percent.

German Zeppelins follow up last year's raids on England with 41 more such raids. The worst comes October 13, and on November 28 the Germans make their first airplane raid on London. The Germans have introduced Albatross and Halberstadt planes and have begun flying in formation.

The secret Sykes-Picot Treaty signed by Britain and France has called for a division of the Middle East between the two powers.

An Arab revolt against the Ottoman Turks begins June 5 with an attack on the garrison at Medina, which has surrendered June 10. The grand sherif of Mecca Husein Ibn-Ali, 60, had supported the Turks but has switched sides at the persuasion of his third son Faisal, 31, and of English archaeologist-soldier T. E. (Thomas Edward) Lawrence, 27. Husein is proclaimed king of the Arabs October 29 and founds the Hashemite dynasty that will rule the newly independent Hejaz until 1924.

Polish general Jozef Pilsudski, 49, obtains recognition of an independent Poland from the Central Powers November 16. Pilsudski 2 years ago organized an independent Polish army of 10,000, acting in secret. He has fought with Austria against Russia but has resigned his command because of German and Austrian interference in Polish affairs (*see* 1918).

The Austro-Hungarian emperor Franz Josef dies November 21 at age 86 after a 68-year reign. He is succeeded by his grandson of 29 who became heir to the throne in 1914 when the archduke Franz Ferdinand was assassinated at Sarajevo; he will reign until 1918 as Karl I.

Britain's Asquith government resigns December 4 and a war cabinet takes over. The new prime minister is Welshman David Lloyd George, 63, who succeeded Lord Kitchener as secretary of state for war in July when Kitchener went down with H.M.S. *Hampshire* off the Orkneys, torpedoed while en route to Russia for secret talks.

The odious Russian faith-healer Grigori Efimovich Rasputin, 45, dies December 31 at Petrograd at the hands of a group of noblemen bent on ridding Russia of the monk's corrupting influence on Nicholas II and the czarina Aleksandra. He had ingratiated himself with the court by promising a cure for the hemophilia that afflicts the czarevich (*see* 1918; Bolshevik revolution, 1917).

U.S. iron and steel workers return to work January 13 at East Youngstown, Ohio, after receiving a 10 percent wage hike, but there are 2,000 strikes by U.S workers in the first 7 months of the year alone.

A San Francisco Preparedness Parade July 22 is disrupted by a bomb explosion that kills 9 and wounds 40. Labor leader Thomas J. Mooney, 34, is accused along with Warren K. Billings, 22, of having planted the bomb, both protest their innocence, Mooney is convicted and condemned to death, Billings is given life imprisonment. Mooney's sentence will be commuted late in 1918, he will be released early in 1939, Billings will be released late in 1939 and pardoned in 1961, but acrimony over the affair will continue for the next 25 years.

The Owen-Keating Act passed by Congress September 1 forbids shipment in interstate commerce of goods on which children under 14 have worked or on which children from 14 to 16 have worked more than 8 hours per day (but *see* Supreme Court, 1918).

The Adamson Bill signed by President Wilson September 3 averts a strike by railroad brotherhoods by providing for an 8-hour day on interstate railroads with time and a half for overtime.

A Workmen's Compensation Act passed by Congress September 7 brings 500,000 federal employees under a program to protect them from disability losses.

The first U.S. congresswoman is elected by Montana voters. She is Jeanette Rankin, 36.

Marcus Garvey moves to New York and establishes U.S. headquarters for the Universal Negro Improvement Association (*see* 1913; 1920).

The Russian port of Murmansk is founded on the Barents Sea 175 miles north of the Arctic Circle to receive Allied war matériel. Warm currents from the Gulf Stream keep the port open year round even when some ports to the south are frozen.

V. I. Lenin puts out a pamphlet under the title "Imperialism, the Highest Stage of Capitalism." He makes the ideas expressed by John Hobson in his 1902 book *Imperialism* the official view of the Communist party (*see* 1917; *Pravda*, 1907).

The Supreme Court upholds the constitutionality of the U.S. federal income tax (Sixteenth Amendment) January 24 in the case of *Brushaber v. Union Pacific Railroad.*

Hetty Green dies July 3 at age 80 leaving an estate of more than $100 million that has made her the richest woman in America. Henrietta Howland Robinson Green inherited $10 million at age 29 from her father and a maternal aunt, kept her finances separate from those of her late husband Edward H. Green, whom she married in 1867, and multiplied her fortune with investments in stocks, government bonds, mortgages, and Chicago real estate while living penuriously; her son lost a leg because she would not hire a physician to treat him, and the eccentric "witch of Wall Street" has occupied a small Hoboken, N.J., apartment since 1895.

Bethlehem Steel's Charles M. Schwab pays $49 million to acquire the Pennsylvania Steel Co. formerly controlled by the Pennsylvania Railroad. The company has a plant at Steelton, ore mines in Cuba, more in Pennsylvania, and—most important—a tidewater steel mill on Chesapeake Bay at Sparrows Point, Md., where

1916 ***(cont.)*** Bethlehem will create a vast shipyard as it continues to prosper on government shipbuilding contracts.

U.S. petroleum companies raise gasoline prices 7¢ per gallon above 1914 levels as the European war and mounting domestic demand create shortages. Engineers predict that world petroleum reserves will be exhausted within 30 years (but *see* Lake Maracaibo, 1922; Iraq, 1927; east Texas, 1930; Bahrain, 1932; Saudi Arabia, 1933; Kuwait, 1938; Alaska, 1968; North Sea, 1969).

U.S. entrepreneur Harry Ford Sinclair, 40, founds Sinclair Oil and Refining. He has put together some small Midwestern refineries plus a pipeline company, floats a $16 million bond issue on Wall Street, and by 1923 will have added Freeport and Tampico Oil and Pennsylvania's Union Petroleum (*see* Teapot Dome, 1921).

A Federal Highway act voted by Congress July 11 authorizes a 5-year program of federal aid to the states for construction of post roads on a 50–50 basis.

Henry Ford buys a site on Detroit's River Rouge for a gigantic new motorcar plant. Ford's Model T gets 20 miles per gallon of gasoline.

W. C. Durant regains control of General Motors whose president Charles W. Nash resigns (below). GM is reorganized as a Delaware corporation.

Cadillac Motor Car is set up as a division of General Motors (above) (*see* 1909).

General Motors will absorb the manufacturing facilities of W. C. Durant's Chevrolet Co. (*see* 1915). It will also absorb the Hyatt Roller Bearing Co., whose chief Alfred P. Sloan, 43, is a protégé of Cadillac's Henry M. Leland; Delco, whose C. F. Kettering is a friend of Leland (*see* 1911); and New Departure Manufacturing, a producer of roller bearings (*see* Fisher Body, 1918).

Nash Motors is founded by former GM chief Charles W. Nash (above), who purchases the Thomas B. Jeffery Co. of Kenosha, Wis., which started making bicycles in 1879 and introduced the Rambler motorcar in 1902.

The Marmon 34 priced at $2,700 and up is introduced with a "scientific lightweight" engine of aluminum. Designed by Howard Marmon with his Hungarian-American engineer Fred Moskovics and Alanson P. Brush, its only cast-iron engine components are its cylinder sleeves and one-piece "firing head." Body, fenders, hood, transmission case, differential housing, clutch cone wheel, and radiator shell are all of aluminum, but the car has problems and is unable to compete with the new Cadillac V8.

The Hudson Super-Six is introduced by the 7-year-old Hudson Motor Car Co. which sells 26,000 vehicles and becomes the leading U.S. maker of high-priced motorcars (*see* closed sedan, 1922).

The first mechanically operated windshield wipers are introduced in the United States. Electric windshield wipers will not be produced until 1923.

U.S. railroad trackage reaches its peak of 254,000 miles, up from 164,000 in 1890. The total will decline, but only marginally.

A Naval Appropriations Act passed by Congress August 29 authorizes $313 million for a 3-year naval construction program. The Battle of Jutland (above) has shown the value of dreadnoughts and light cruisers.

The German freight submarine *Deutschland* that arrived at Baltimore July 9 with a cargo of dyestuffs arrives in November at New London, Conn., with a cargo of chemicals, gems, and securities.

The Bryan-Chamorro Treaty concluded August 5 gives the United States exclusive rights to build a canal through Nicaragua, thus preventing anyone else from building a rival to the 2-year-old Panama Canal. Nicaragua receives $3 million and is thus enabled to pay off its debts to U.S. bankers, but the *New York Times* criticizes the pact, saying that President Wilson has made the "dollar diplomacy" of former president Taft and his secretary of state Philander C. Knox "more nearly resemble ten-cent diplomacy."

A general theory of relativity announced by Albert Einstein revolutionizes the science of physics (*see* 1905). Now at the University of Berlin and a director of the Kaiser Wilhelm Institute, Einstein has evolved the theory from his work in the geometrization of physics and the integration of gravitational, accelerational, and magnetic phenomena which he will try to unite into a unified field theory represented by a single set of equations (*see* 1919; 1929).

The U.S. National Academy of Sciences founded in 1863 starts a National Research Council (NRC).

English physiologist Edward A. Sharpey-Schafer, 60, introduces the word "insuline" for the hormone produced by the islets of Langerhans in the human pancreas (*see* 1869; Banting, 1922).

A medical student accidentally discovers the anticoagulant powers of the drug heparin, obtained from animal lungs. Swedish biochemist J. Erik Jorpes, 21, will identify the structure of heparin in the early 1930s and help make the potent blood thinner a standard therapy for blood clots in the leg caused by thrombophlebitis, for some heart attacks, and for certain lung conditions, as well as for preventing formation of blood clots after major surgery.

GE's William D. Coolidge revolutionizes X-ray technology. He patents a hot-cathode X-ray that will replace the cold aluminum-cathode tube and be the prototype of all future tubes (*see* 1895).

A U.S. poliomyelitis epidemic strikes 28,767 in midsummer and fall. Some 6,000 die, 2,000 of them in New York, and thousands more are crippled (*see* 1905; Franklin Roosevelt, 1921).

"Peace, it's wonderful," says New York's "Father Divine" and he organizes the Peace Mission movement. Local evangelist George Baker, 42, and his followers will establish a communal settlement at Sayville, Long Island, that they will soon have to abandon, but Father

Divine will garner a substantial following in the next 45 years preaching a renunciation of personal property, complete racial equality, and a strict moral code in the more than 170 Peace Mission settlements, or "heavens," that he will open. Tobacco, cosmetics, liquor, motion pictures, and sex will be totally banned.

The Measurement of Intelligence by Stanford University psychologist Lewis Madison Terman, 39, introduces the term *I.Q.* (intelligence quotient) and presents the first test for measuring intelligence that will be widely used. The Stanford-Binet test is a revision of the Binet-Simon scale (*see* 1912).

Russell Sage College for Women is founded at Troy, N.Y., by philanthropist Margaret Olivia Slocum Sage, 88, whose late husband died 10 years ago at age 90 leaving her $70 million amassed as an associate of Jay Gould in his railroad security manipulations. Sage's second wife, she established the Russell Sage Foundation for improving social and living conditions in the United States with a $10 million gift in 1907 and has given liberally to Cornell, Princeton, Rensselaer Polytechnic Institute, and the Emma Willard School.

W. M. A. Beaverbrook begins a vast publishing empire by gaining control of London's 16-year-old *Daily Express.* Canadian cement magnate William Maxwell Aitken Beaverbrook, 37, has been representing the Ottawa government at the western front, by 1918 he will be British minister of information and will control the *Sunday Express,* and by 1923 will control the *Evening Standard.*

El Universal begins publication at Mexico City. Local journalist Felix F. Palavaicini, 35, sets new standards by publishing news rather than editorial opinion on his daily front page (*see* 1917).

The first radio news is broadcast by Lee De Forest who has established a radio station (*see* 1906; 1922; Conrad, 1920).

Nonfiction *A Heap o' Livin'* by English-American journalist Edgar Albert Guest, 35, of the *Detroit Free Press* who says, "It takes a heap o' livin' to make a house a home."

Fiction *The Golden Arrow* by English novelist Mary (Gladys Meredith) Webb, 35; *Grass on the Wayside (Michigusa)* by Satsume Soseki; *Gharer baire* by Rabindranath Tagore, whose novel in the Bengali vulgate creates a controversy by moving away from the traditional Bengali literary language in which Tagore has written numerous poems, plays, songs, and short stories that helped him win a 1913 Nobel prize; *Casuals of the Sea* by William McFee; *Greenmantle* by John Buchan who is serving on the British headquarters staff in France; *Kingdom of the Dead (De Dodes Rige)* by Henrik Pontoppidan; *You Know Me Al* (stories) by *Chicago Tribune* sportswriter Ring (Ringgold Wilmer) Lardner, 31.

Poetry *The Man Against the Sky* by Edwin Arlington Robinson; *The Jig of Forslinby* by U.S. poet Conrad Potter Aiken, 27, who calls his work a "poetic symphony"; *Six Poems* by Edward Thomas; *Sea Garden* by U.S. poet H. D. (Hilda Doolittle Aldington), 30, who married the English imagist poet Richard Aldington in 1912 but will divorce him after the war and return to America; *Twentieth Century Harlequinade and Other Poems* by English poet Edith Sitwell, 29, and her brother Osbert, 23; *The Buried Harbor (Il Porto Sepolto)* by Italian poet Giuseppe Ungaretti, 28. U.S. poet Alan Seeger, 28, is killed in action July 4 at Befloy-en-Santerre. Seeger enlisted in the French Foreign Legion at the outbreak of the war and leaves behind lines that include, "I have a rendezvous with Death/ At some disputed barricade,/ When Spring comes back with rustling shade/ And apple-blossoms fill the air."

Painting *The Three Sisters* by Henri Matisse; *The Disquieting Muse* by Giorgio de Chirico; *Naturaleza muerta, El rastro* by Diego Rivera. Thomas Eakins dies at Philadelphia June 25 at age 71; Odilon Redon dies at Paris July 6 at age 76.

The Dada artistic and literary movement launched at Zurich will lead to surrealism next year. German pacifist writer Hugo Ball, 30, Alsatian painter-sculptor-writer-poet Hans (or Jean) Arp, 29, Romanian born French poet Tristan Tzara (Sami Rosenstock), 20, and their adherents believe that Western culture has betrayed itself by its easy acceptance of the World War. They protest against all bourgeois notions of meaning and order with chaotic experiments in form and language (*see* Duchamp, 1915).

The *Saturday Evening Post* buys its first Norman Rockwell illustration from *Boys' Life* illustrator Norman Rockwell, 22, who dropped out of Mamaroneck, N.Y., High School 6 years ago to work for the monthly magazine of the Boy Scouts of America. Rockwell has been art director of *Boys' Life* since 1913, but he will now draw mostly for the *Post,* which will buy an average of 10 Norman Rockwell covers per year until it ceases weekly publication in 1969.

Theater *Moses* by English poet-artist Isaac Rosenberg, 26; *A Kiss for Cinderella* by James M. Barrie 3/16 at Wyndham's Theatre, London; *A Night at an Inn* by Lord Dunsany 4/23 at New York's Neighborhood Playhouse; *The Mask and the Face (La maschera e il volto)* by Italian playwright Luigi Chiarelli, 35, 5/31 at Rome's Teatro Argentina; *Good Gracious Annabelle* by U.S. playwright Clare Kummer 10/31 at New York's Republic Theater, with Lola Fisher, Roland Young, Walter Hampden, 111 perfs.; *Bound East for Cardiff,* a one-act play by U.S. playwright Eugene Gladstone O'Neill, 27, 11/16 at New York's Provincetown Playhouse; *The White-Haired Boy* by Irish playwright Esme Stuart Lennox Robinson, 30, 12/13 at Dublin's Abbey Theatre.

Films D. W. Griffith's *Intolerance* with Lillian Gish, Robert Harron, Mae Marsh, Constance Talmadge. Also: Herbert Brenon's *War Brides* with Alla Nazimova.

Opera *Ariadne auf Naxos* 10/4 at Vienna, with music by Richard Strauss, libretto by Hugo von Hofmannsthal.

First performances *Scythian Suite* by Russian composer Serge Sergeevich Prokofiev, 24, 1/29 at Petrograd's Imperial Maryinsky Theater; *Nights in the Gardens of Spain* by Manuel de Falla 4/9 at Madrid's Teatro Real.

1916 ***(cont.)*** Stage musicals *Robinson Crusoe, Jr.* 2/17 at New York's Winter Garden Theater, with Al Jolson, music by Sigmund Romberg and James Hanley, book and lyrics by Harold Atteridge and Edgar Smith, 139 perfs.; *The Ziegfeld Follies* 6/12 at New York's New Amsterdam Theater, with W. C. Fields, Fanny Brice, Ina Claire, Ann Pennington, music by Louis Hirsch, Jerome Kern, and Dave Stamper, 112 perfs.; *The Passing Show* 6/22 at New York's Winter Garden Theater, with Ed Wynn, music chiefly by Sigmund Romberg, lyrics chiefly by Harold Atteridge, songs that include "Pretty Baby" by Tony Jackson and Egbert van Alstyne, lyrics by German-American songwriter Gus Kahn, 30, 140 perf.; *Chu Chin Chow, A Musical Tale of the East* 8/31 at His Majesty's Theatre, London, with a book by Oscar Asche, 45, who has created an extravaganza out of the pantomime *The Forty Thieves,* music by Frederick Norton, 2,238 perfs. (the musical will continue through the war and beyond); *The Century Girl* 11/16 at New York's Century Theater, with Hazel Dawn, Marie Dressler, Elsie Janis, Leon Errol, music by Irving Berlin and Victor Herbert, lyrics by Berlin and Henry Blossom, songs that include Berlin's "You Belong to Me," 200 perfs.

Popular songs "Roses of Picardy" by London composer Haydn Wood, 34, lyrics by lawyer-publisher Fred E. Weatherly, 68; "I Ain't Got Nobody" by Spencer Williams, 27, lyrics by Roger Graham, 31; "La Cucaracha" ("The Cockroach") is published at Mexico City and at San Francisco; "Ireland Must Be Heaven for My Mother Came from There" by Fred Fisher, lyrics by Joe McCarthy and Howard E. Johnson.

The first Rose Bowl football game January 1 pits Washington State against Brown in a huge new stadium that seats 101,385. The game is part of the Tournament of Roses (*see* 1902; Orange Bowl, Sugar Bowl, 1935).

Richard Williams wins in U.S. men's singles, Molla Burjsted in women's singles.

The first U.S. PGA (Professional Golfers Association) tournament opens at the Siwanoy Country Club, Bronxville, N.Y.

The Boston Red Sox win the World Series by defeating the Brooklyn Dodgers 4 games to 1.

The Converse basketball shoe, with canvas upper and rubber sole, is introduced by an 8-year-old Massachusetts company. Sneakers (plimsolls in Britain) have been available since the late 1860s but brand names have been almost non-existent.

U.S. Keds with canvas uppers, rubber soles, are introduced by United States Rubber Co.

Lucky Strike cigarettes are introduced by American Tobacco Co. (*see* 1911). The new brand will soon outsell the company's Sweet Caporal and Pall Mall brands and will challenge the Camels brand launched 3 years ago by R. J. Reynolds (*see* 1918; 1925).

U.S. cigarette production reaches 53.1 billion, up from 35.3 billion in 1912.

The Harrison Drug Act signed into law by President Wilson May 2 requires all persons licensed to sell narcotic drugs to file an inventory of their stocks with the Internal Revenue Service. The Supreme Court rules June 5 that users and sellers of opium are liable to prosecution.

New York City enacts the first U.S. zoning law, partly as a response to the Equitable building completed last year (*see* set-back law, 1923).

Tokyo's Imperial Hotel, completed by Frank Lloyd Wright near the Imperial Palace with Mayan architectural features, is ornate and sprawling but soundly engineered (*see* 1923 earthquake).

Miami's $16 million James Deering mansion Vizcaya is completed for the farm equipment heir. Roofing tile for the coraline stone Florida house once covered an entire Cuban village (the Cubans have been given new roofs), a 300-year-old travertine marble fountain from the town square of the Italian village Bassano di Surti near Rome has been placed in Vizcaya's Fountain Garden (the Italians have been given a modern water system), the estate has 10 acres of formal gardens including the Fountain Garden modeled after Rome's Villa Albani, and the house has 45 telephones, oversize brine-cooled refrigerators, marble floors, and two elevators.

Double-shell enameled bathtubs go into mass production in the United States to replace the cast-iron tubs with roll rims and claw feet that have been standard for decades.

The U.S. National Park Service in the Department of the Interior is created by act of Congress August 20.

California's Lassen Volcanic National Park is created by act of Congress. The 107,000-acre park embraces a still-active volcano.

Hawaii Volcanoes National Park is created under the name Hawaii National Park. The 220,345-acre park on the island of Hawaii will be renamed in 1961.

Acadia National Park has its beginnings. John D. Rockefeller, Jr., donates 5,000 acres of his family's Mount Desert Island in Maine to the nation; President Wilson proclaims it a national monument (*see* 1919).

Florida's Gulf Coast has a "red tide" caused by a proliferation of the dinoflagellate plankton *Gymnodinium breve*. The red tide kills millions of fish whose nervous systems are immobilized by the expelled waste of *G. breve* (*see* 1932).

British fuel shortages motivate Parliament to pass a "summer time" act and most European governments do the same, advancing clocks one hour to make the most of available light. British farmers protest that they must milk their cows in the dark and then wait idly until the sun has evaporated the dew before they can harvest hay (*see* Franklin, 1784; Congress, 1918).

A Federal Farm Loan Act passed by Congress July 17 provides for Farm Loan Banks throughout the United States and creates a Federal Farm Loan Board. The legislation makes it easier for U.S. farmers to obtain credit with which to buy land, acquire farm machinery, and generally improve their productivity (*see* Agricultural Credits Act, 1923).

A Stockraising Homestead Act doubles the amount of land allowed under the Homestead Act of 1909, but even 640 acres—a full square mile—is too small in many and areas (*see* 1862). Failed homesteaders will soon sell their land to stockmen bent on acquiring large ranches (*see* Taylor Grazing Act, 1934).

U.S. wheat farmers have another bumper crop but the average corn farmer produces little more than 25 bushels per acre.

Funk Brothers of Bloomington, Ill., ships the first hybrid seed corn to a Jacobsburg, Ohio, farmer. He pays $15 per bushel (*see* East and Shull, 1921; Golden Bantam, 1933).

The boll weevil reaches the Atlantic coast (*see* 1899; 1921).

The Japanese beetle *Popilla japonica* whose grubs have arrived in the roots of imported nursery stock appears for the first time in America at Riverton, N.J. The beetle will proliferate and damage millions of dollars worth of fruits and vegetables.

Germany establishes a War Food Office as Britain's naval blockade forces strict rationing of food.

A German potato blight contributes to starvation that kills 700,000 and weakens morale in the army.

France sets a maximum wheat price of 33 francs per quintal (220 pounds) to the farmer and sets controls on butter, cheese, and oil cakes. Parisians line up in milk queues.

A British Departmental Committee on Food Prices is established after a sharp rise in food prices by June has brought complaints of profiteering. Two-thirds of Britain's sugar came from Austria-Hungary before the war and sugar prices have risen steeply.

British bread prices begin to move up from 9½ d per 4-pound loaf, an Order in Council empowers the Board of Trade to regulate food supplies and fix prices, but the president of the Board of Trade tells the House of Commons in October that there is no need to establish a Ministry of Food or to appoint a Food Controller: "We want to avoid any rationing of our people in food."

Port of London Authority chairman Hudson Ewbanke Kearley, 60, viscount Davenport, is appointed Food Controller in December but the government is not prepared to ration food. Lord Davenport appeals to the public to make voluntary sacrifices (*see* 1917).

Vitamin B is isolated by E. V. McCollum who believes it to be just one coenzyme (*see* 1912; Drummond, 1920; Smith, Hendrick, 1928).

The first large-scale study of iodine's effect on human goiter is conducted by U.S. physicians David Marine and E. C. Kendall (*see* 1905; Mayo Clinic, 1915). Girls in some Akron, Ohio, schools are given tablets containing 0.2 gram of sodium iodide, and the incidence of goiter in susceptible teenage girls given the tablets is found to drop markedly (*see* table salt, 1921).

Colorado Springs dentist Frederick S. Motley finds his patients' teeth are discolored but have few cavities. He begins to trace this to the fact that the city's drinking water contains fluoride salts in a concentration of 2 parts per million (*see* 1945).

A mechanical home refrigerator is marketed for the first time in the United States, but its $900 price discourages buyers, who can buy a good motorcar for the same money (*see* 1925; Frigidaire, 1919; GE, 1927).

U.S. factories produce quantities of filled milk made with vegetable oil instead of butterfat to the consternation of dairymen (*see* 1923).

California Packing is incorporated at San Francisco with a $16 million Wall Street underwriting promoted by George N. Armsby, son of Jacob (*see* 1865; Del Monte, 1891). Now the largest U.S. fruit and vegetable canner, Calpak operates 61 canneries including some in Washington, Oregon, and Idaho, and it owns a 70 percent interest in the Alaska Packers Association whose Star Fleet continues as the last commercial sailing fleet to fly the American flag (*see* 1895; 1917; Del Monte, 1967).

Planters Nut and Chocolate introduces "Mr. Peanut" (*see* 1906). It has conducted a contest among Suffolk (Va.) high school students to find an appropriate trademark, the winning sketch has been an anthropomorphized peanut, and a commercial artist has added a monacle and a crooked leg to the initial sketch (which won its designer a $5 prize).

Coca-Cola adopts the distinctive bottle shape that will identify it for years—a slimmer model of the waist-bulging bottle of 1913 made at Terre Haute, Ind. The Supreme Court rules that if Coca-Cola is to claim a distinctive name it must contain some derivatives of cola beans and coca leaves but does not rule against the use of these substances in denatured form (*see* 1910; Woodruff, 1919).

The Piggly-Wiggly opened at Memphis, Tenn., by food merchant Clarence Saunders begins the first supermarket chain (*see* California, 1912; San Francisco, 1923; King Kullen, 1930).

Nathan's Famous frankfurters have their beginning in a Coney Island, N.Y., hot dog stand at the corner of Stillwell and Surf Avenues opened by Polish-American merchant Nathan Handwerker, 25, who sells his franks at 5¢ each—half the price charged by Feltman's German Gardens on Surf Avenue where Handwerker has worked weekends as a counterman while making $4.50 per week as a delivery boy in the Lower East Side. Handwerker has invested his life savings of $300 in the hot dog stand, works 18 to 20 hours per day with his 19-year-old bride Ida, who laces the franks with her secret spice recipe, and prospers (*see* 1917).

The first birth control clinic outside Holland opens at 46 Amboy Street, Brooklyn. Margaret Sanger distributes circulars printed in English, Italian, and Yiddish to announce the opening (*see* 1914). Police raid the clinic, Sanger is jailed for 30 days, founds the New York Birth

1916 ***(cont.)*** Control League after her release, and begins publication of the *Birth Control Review* (*see* 1921).

1917 The German high command resumes unrestricted U-boat attacks despite the probability that such attacks will bring the United States into the European war. Bolstering German confidence is the strength of the Hindenburg line on the western front and the collapse of Russian opposition on the eastern front as revolution begins in Russia (below).

British intelligence intercepts a wireless message January 17 from the German foreign secretary Arthur von Zimmermann to the German ambassador Count Johann von Bernstorff at Washington. Decoded, the Zimmermann note says, "We intend to begin unrestricted submarine warfare. We shall endeavor to keep the United States neutral. In the event of this not succeeding we make Mexico a proposal of alliance on the following basis: Make war together, make peace together, generous financial support, and an understanding on our part that Mexico is to recover the lost territory in Texas, New Mexico, and Arizona."

Berlin notifies Washington January 31 that unrestricted submarine warfare will begin the next day, the United States severs relations with Germany February 3, Latin American nations including Brazil and Peru follow suit, China severs relations with Germany March 14.

Eleven U.S. senators conduct a filibuster to block arming of U.S. merchant ships running the gauntlet of German U-boats. President Wilson denounces the senators March 4 saying, "A little group of willful men, reflecting no opinion but their own, have rendered the great government of the United States helpless and contemptible."

Russian troops mutiny March 10 following 2 days of strikes and riots at Petrograd; Czar Nicholas II en route home by private train March 15 has the train pulled into a siding and abdicates in favor of his brother Michael; Michael abdicates March 16 in favor of a provisional government headed by Prince Georgi Evgenievich Lvov, 55; and the Romanov dynasty founded in 1613 comes to an abrupt end.

V. I. Lenin and other Bolshevik leaders arrive at Petrograd in April. The German high command has sent them by sealed railroad carriage from Switzerland across Germany in a calculated move to undermine the pro-Ally provisional government (above) (*see* revolution, below; Lenin, 1905; 1916).

The United States declares war on Germany April 6, 4 days after President Wilson has sent a war message to the Senate. Gen. "Blackjack" Pershing has been recalled from his pursuit of Pancho Villa and named to head an American Expeditionary Force as the French and British sustain enormous losses on the western front.

Canadian troops take Vimy Ridge in the Battle of Arras that lasts from April 9 to May 4, but while Gen. Edmund Allenby advances 4 miles, his British army is unable to effect a breakthrough.

French troops mutiny after sustaining cruel losses in April and hearing of mutiny among Russian troops on the eastern front. French authorities relieve Gen. Robert Georges Nivelle, 61, of his command but exonerate him of blame for the failure of his plans, they replace him with Gen. Pétain who tries to restore order as the mutiny sweeps through 16 corps, the government executes 23 socialist and pacifist agitators, but the brunt of the Allied war effort falls on the British Tommy as the French *poilu* loses heart for continuing the war.

More than half the 875,000 tons of Allied shipping lost in April is British, but Prime Minister Lloyd George prevails on the Admiralty to try convoying merchant vessels. Convoys begin May 10.

An embargo proclamation issued by President Wilson July 9 places exports of U.S. foodstuffs, fuel, iron, steel, and war matériel under government control. Wilson sends U.S. destroyers to help blockade German ports and convoy merchant ships bound for Britain.

The Jones Act passed by Congress March 2 makes Puerto Rico a U.S. territory and Puerto Ricans U.S. citizens. The law makes voting compulsory and applies the Selective Service Act to Puerto Rico at the request of the San Juan government; Puerto Rico drafts 18,000 men into the U.S. Army.

The Danish West Indies which include the Virgin Islands of St. Thomas, St. Croix, and St. John with 26,000 inhabitants become U.S. territory March 31 upon Senate ratification of a treaty giving Denmark $25 million for the 132 square miles of land involved.

"Lafayette, we are here," says Col. Charles E. Stanton July 4 at the tomb of Lafayette in Paris but U.S. troops do not go into action until October 27.

A new Russian provisional government headed by Minister of War Aleksandr Feodorovich Kerenski, 36, replaces the Lvov government (above) July 7 as Russian troops pull back on all fronts (*see* revolution, below).

The Battle of Passchendaele from July 31 to November 10 costs the British 400,000 men.

British forces occupy Baghdad March 11. Aqaba in Arabia falls to Arab forces led by Col. T. E. Lawrence who attacks Turkish garrisons and breaks the Turkish communications link by disrupting the Hejaz Railway. German colonial troops gain a victory in mid-October at Mahiwa and invade Portuguese East Africa.

French premier Alexandre Ribot resigns September 9 and a new cabinet takes over headed by the minister of war Paul Painlevé.

French dancer Mata Hari (Gertrud Margarete Zelle), 41, is convicted of having spied for the Germans and executed October 15 at St. Lazare.

German U-boats continue their depredations despite convoys, sinking 8 million tons by October 10, but the German submarine campaign has lost force, permitting an increase in U.S. troop shipments.

Caporetto in the Julian Alps near Italy's Austrian border is bombarded in late October by German and Austrian artillery. Troops move in beginning October 24 in heavy fog in an effort to reach the Tagliamento River, but while nearly 300,000 Italians are taken prisoner in the 2-month Caporetto campaign and more than 300,000 desert, the Germans and Austrians make little real progress.

The Balfour Declaration issued November 2 by Foreign Secretary Arthur J. Balfour, now 69, says the British government favors "the establishment in Palestine of a national home for the Jewish people and will use their best endeavours to facilitate the achievement of that object, it being clearly understood that nothing shall be done which may prejudice the civil and religious rights of existing non-Jewish communities in Palestine." British troops have invaded Palestine and take it from the Ottoman Turks who have held it since 1516. Jerusalem falls to the British December 9 (*see* Herzl, 1897; Passfield Paper, 1930).

The Battle of Passchendaele ends November 10 with the British completely demoralized.

French premier Paul Painlevé resigns November 16; Georges Clemenceau, now 76, will serve until 1920 as premier and minister of war, revitalizing France's war effort and cementing relations with Britain.

The Battle of Cambrai that begins November 20 gives the British some initial success; 300 British tanks launch a surprise raid in the first great tank attack of the war, but the Germans counterattack November 30 and recover most of the territory they have lost.

A Bolshevik revolution begins at Petrograd the night of November 6 (October 24 by the Julian calendar still used in Russia). On orders from V. I. Lenin, Commissar Aleksandr B. Belyshev, 24, fires a blank shot from the foredeck gun of the cruiser *Aurora* anchored in the Neva River. Housewives waiting in endless bread-shop queues have been demonstrating in protest, they are joined by soldiers from the Petrograd garrison, sailors from Kronstadt, and the factory workers' Red Guards, who seize government offices and storm the Winter Palace of the Romanovs.

Aleksandr Kerenski (above) has been forced to accept Bolshevik support to resist the efforts of Gen. Lavr Georgievich Kornilov, 64, to make himself dictator, the Bolsheviks have secured a majority in the Petrograd Soviet under the leadership of revolutionist Leon Trotsky (Lev Davydovich Bronstein), 40, who was expelled from France last year and has returned from exile in the United States and England to support Lenin.

The Kerenski government falls, Kerenski himself goes into hiding (he will later take sanctuary abroad), and a new government headed by Lenin takes office November 7 under the name Council of People's Commissars. Trotsky is commissar for foreign affairs and a Georgian who calls himself Josef Stalin (Iosif Vissarionovich Dzhugashvili), 39, is commissar for national minorities.

Russian peasants seize landlords' fields as the Bolsheviks initiate plans to make Moscow the nation's capital.

The Extraordinary Commission to Combat Counter-Revolution founded by revolutionist Feliks Dzerzhinsky, 40, December 20 is a secret police organization designed to protect the Council of People's Commissars (above) and crush opposition. The Cheka will become the GPU, the NKVD, and, later, the KGB.

Race riots at East St. Louis, Ill., July 2 leave 39 dead, hundreds injured.

New York blacks led by W. E. B. Du Bois and James Weldon Johnson of the NAACP walk in silence 15,000 strong down Fifth Avenue to protest the violence at East St. Louis (above) (*see* 1910; 1948).

The Congressional Union for Woman Suffrage pickets the White House to urge presidential support of the woman suffrage amendment. President Wilson endorses equal suffrage October 25 when he speaks at the White House before a group from the New York State Woman Suffrage party, and 20,000 women march in a New York suffrage parade October 27.

New York adopts a constitutional amendment November 6 granting equal voting rights to women (*see* 1918).

Charlie Chaplin, Mary Pickford, Douglas Fairbanks, and other stars speak at bond rallies that will help sell $18.7 billion in Liberty Bonds this year and next despite their low 3.5 percent interest rate (railroad bonds yield 4.79 percent).

The Anglo-American Corp. of South Africa is founded by German-born South African mining magnate Ernest Oppenheimer, 37, whose firm will grow to control 95 percent of the world's diamond supply (*see* Rhodes, Barnato, 1880).

Union Carbide and Carbon is created in November by a merger of Union Carbide, Linde, Prest-O-Lite, and National Carbon, famous for its Eveready flashlights and batteries. The new company finds itself swamped immediately with government orders for activated carbon for gas masks, helium for dirigibles, ferrozirconium for armor-plating, and other war-related products (*see* Union Carbide, 1911; petrochemicals, 1920; Prestone, 1926).

U.S. tax revenues from income taxes for the first time pass revenues from customs duties (*see* 1913). By 1920

Russia's Bolshevik revolution replaced czarist tyranny with a tyrannical "dictatorship of the proletariat."

1917 ***(cont.)*** income tax revenues will be 10 times those from customs duties.

The United States has more than 40,000 millionaires, up from 4,000 in 1892. Millionaire railroad equipment salesman and trencherman James Buchanan "Diamond Jim" Brady dies April 13 at Atlantic City, N.J., at age 60, 3 months after the death at Denver of "Buffalo Bill" Cody (who died bankrupt at age 70).

Paris jeweler Pierre Cartier, who opened the first New York Cartier salon in 1908 on a Fifth Avenue mezzanine, gives financier Morton F. Plant a two-strand Oriental pearl necklace in exchange for Plant's Renaissance-style mansion at Fifth Avenue and 52nd Street *(see* 1902). Pierre's brother Louis, now 42, designed the first wristwatch 10 years ago for French aviator Santos Dumont; he will introduce the Tank watch next year as a tribute to the men of the American Tank Corps.

Phillips Petroleum is founded at Bartlesville, Okla., as the war doubles oil prices. Local banker Frank Phillips, 43, and his brother Lee Eldas Phillips, 40, have made money in local wells (*see* 1903, 1954; North Sea, 1969).

Oil prospectors discover the Gerber-Covington oil pool east of Enid in Oklahoma's Cherokee Strip.

Use of electric signs in the United States is restricted November 16 to conserve energy.

U.S. trolley car ridership reaches 11 billion, with 80,000 electric streetcars plying 45,000 miles of track, up from 30,000 trolley cars on 15,000 miles of track in 1900. Connecting lines make it possible to travel from New York to Boston by trolley and even for more than 1,000 miles from eastern Wisconsin to central New York State, paying a nickel to ride to the end of each line (*see* 1939).

New York's Erie Canal is replaced after 92 years by a modern barge canal 12 feet deep, 75 to 200 feet wide, 524 miles long. Steam power has made the old canal obsolete; and the new barge canal makes use of an additional 382 miles of canalized lakes and rivers between Albany and lakes Ontario and Erie *(see* 1825; 1836).

New York City's Hell Gate Bridge (the New York connecting railroad bridge) opens April 1 over the East River to give Pennsylvania Railroad trains access to New England.

Kansas City's Union Station is completed. It is the nation's second largest railway depot.

A "Stop-Look-Listen" warning sign for railroad crossings designed by a Seattle, Wash., safety lecturer for the Puget Sound Power Co. begins to replace the "Watch for Engines" sign in common use.

The worst train wreck in world history occurs December 12 at Modane, France, where a passenger train jumps the tracks killing 543, injuring hundreds more. Three U.S. train wrecks in Pennsylvania, Oklahoma, and Kentucky kill 20, 23, and 46, respectively.

Rail equipment shortages delay movement of U.S. troops and freight to ports; President Wilson takes over the railroads December 17 and places his son-in-law William G. McAdoo, secretary of the Treasury, in charge.

U.S. auto production reaches 1,745,792, up from 543,679 in 1914. Model T Fords account for 42.4 percent of the total. Apperson, Biddle, Buick, Cadillac, Chalmers, Chevrolet, Cole, Crow-Elkhart, Daniels, Detroit Electric, Doble Steam, Essex, Fageol, Franklin, Haynes, Hudson, Hupmobile, Jordan, Kissel, Lenox, Maxwell, McFarlan, Mitchell, National, Oakland, Ohio Electric, Paige-Detroit, Pathfinder, Peerless, Pierce-Arrow, Premier, Pullman, Reo, Saxon, Scripps-Booth, Willys-Knight, Willys-Overland, and others vie for customers.

BMW (Bayerische Motoren Werke) is created at Munich by a reorganization of Bayerische Flugzeugwerke. The company will introduce the world's first six-cylinder high-altitude aircraft engine and a 550-horsepower nine-cylinder one, but will be best known for its high-speed motorcycles before turning to sports-car production under the name Bavarian Motor Works.

Chance Vought Co. is founded by U.S. aeronautical engineer and designer Chance Milton Vought, 27 *(see* United Aircraft, 1929).

The Electron by University of Chicago physicist Robert Andrews Millikan, 49, describes the atomic particle which Millikan has been the first to isolate. He has succeeded in measuring the electron's charge (*see* Rutherford, 1911).

Typhus sweeps Russia, compounding problems of defeat and revolution (above). The epidemic will kill up to 3 million in the next 4 years (*see* 1918; Armand Hammer, 1921).

A superheterodyne circuit developed by U.S. Army Signal Corps major Edwin Howard Armstrong, 26, will become the basic design for all amplitude modulation (AM) radios. It greatly increases the selectivity and sensitivity of radio receivers over a wide band of frequencies *(see* 1906; FM, 1933).

David Sarnoff urges marketing of a simple "radio music box." The American Marconi Co. says his plan will make the radio "a 'household utility' in the same sense as the piano or phonograph" (*see* 1912; 1920).

U.S. journalist Dorothy Day, 19, quits her job on the IWW newspaper *The Call* and joins *The Masses*, a U.S. Communist party newspaper that is soon suppressed (*see* 1933).

Excelsior begins publication at Mexico City. The daily will become Mexico's leading newspaper.

"The Gumps" appears February 12 in the *Chicago Tribune* whose publisher Joseph M. Patterson has dreamed up the comic strip and named it *(see* 1910). Cartoonist Sidney Smith, 30, has created the characters based on real people, and in mid-March 1922 he will be signed to the first $1 million cartoon contract with a guarantee of $100,000 per year and a new Rolls-Royce *(see* 1955).

The "Katzenjammer Kids" launched in 1897 is renamed "The Captain and the Kids" as German names become unpopular. The comic strip moved 5 years ago from Hearst's *New York Journal* to Pulitzer's *New York World.*

The *World Book* encyclopedia, published in its first edition, will grow to 22 volumes in the next 60 years and outsell any other encyclopedia.

Nonfiction A *Son of the Middle Border* (reminiscences) by Hamlin Garland; *Parnassus on Wheels* by U.S. essayist Christopher Darlington Morley, 27, is the odyssey of a horse-drawn library.

Fiction *The Book of the Martyrs (Vie des Martyrs)* with war stories by French novelist Denis Thévenin (Georges Duhamel), 35; *South Wind* by Scottish novelist-scientist Norman Douglas, 48, who has settled on the Italian island of Capri which he idealizes as "Nepenthe"; *Nocturne* by English novelist Frank Swinnerton, 33; *The Rise Of David Levinsky* by *Jewish Daily Forward* editor Abraham Cahan.

Juvenile *Understood Betsy* by Vermont novelist Dorothy Canfield Fisher, 38.

Poetry *Prufrock and Other Observations* by émigré U.S. poet T. S. (Thomas Stearns) Eliot, 28, who expresses his disenchantment in modern life with "The Love Song of J. Alfred Prufrock"; *The Chinese Nightingale and Other Poems* by Vachel Lindsay; *Merlin* by Edwin Arlington Robinson; *Renascence and Other Poems* by Maine-born poet Edna St. Vincent Millay, 25, whose title poem appeared 5 years ago and won her a sponsor who has put her through Vassar; "La Jeune Parque" by French poet Paul Valéry, 45, who dedicates his first great poem to André Gide and follows it with "Aurore"; *Diary of a Newlywed Poet* (*Diario de un Poeta Recién Casatio*) by Juan Ramon Jiminez. English poet T. E. Hulme is killed in action after 3 years with the artillery in France.

"I Want You," says Uncle Sam in a recruitment poster painted by New York illustrator James Montgomery Flagg, 40, who has used his own face as a model for the stern-visaged Uncle Sam (*see* 1813; 1852). U.S. Army recruitment officers distribute 4 million copies.

The term "surrealism" coined by French man of letters Guillaume Apollinaire (Wilhelm de Kostrowitzky), 37, denotes anti-establishment art (*see* Dada, 1916; Breton, 1924).

Dutch painter Piet Mondrian (Pieter Cornelis Mondriaan), 45, founds the art review *De Stihl* with Theo van Doesburg. Mondrian has evolved a purely abstract style using only horizontal and vertical lines, the three primary colors, and shades of black, white, and gray.

Painting *John D. Rockefeller* by John Singer Sargent; *Piano Lesson* by Henri Matisse; *Nude at the Fireplace* by Pierre Bonnard; *Crouching Female Nude* by Amedeo Modigliani; *Forward* by Marc Chagall; *Ladies in Crinoline* by Wassily Kandinsky. Albert Pinkham Ryder dies at Elmhurst, N.Y., March 28 at age 70; Edgar Degas dies at Paris September 27 at age 83 (he has been blind since 1909).

Sculpture *Man with Mandolin* by Jacques Lipschitz. Auguste Rodin dies at Mendon November 17 at age 77.

Theater *The Old Lady Shows Her Medals* and *Seven Women* by James M. Barrie 4/7 at London's New Theatre; *Right You Are—If You Think You Are [Cosi e (se vi pare)]* by Luigi Pirandello 6/18 at Milan's Teatro Olimpio; *The Silken Ladder (La scala di seta)* by Luigi Chiarelli 6/28 at Rome's Teatro Argentina; *Dear Brutus* by Barrie 10/17 at Wyndham's Theatre, London; *In the Zone* by Eugene O'Neill 10/31 at New York's Provincetown Playhouse (one-act play); *The Long Voyage Home* by O'Neill 11/2 at the Provincetown Playhouse (one-act play); *Why Marry?* by U.S. playwright Jesse Lynch Williams, 46, 12/25 at New York's Astor Theater, with Estelle Winwood.

The Empire Theater at Montgomery, Ala., installs the first theater air-conditioning (*see* Carrier, 1911). Within 40 years every major U.S. theater will be air-conditioned.

Films D. W. Griffith's *Hearts of the World* with Lillian and Dorothy Gish (financed by London to arouse U.S. sympathies against the Germans but released after U.S. entry into the war); Clarence Badger's *Teddy at the Throttle* with Wallace Beery, Gloria Swanson; Charles Chaplin's *The Cure* with Chaplin.

Ballet *Les Femmes de Bonne Humeur* 4/12 at Rome's Teatro Constanza, with Leonide Massine, music by the 18th-century composer Domenico Scarlatti, book by French playwright Jean Cocteau, 27, choreography by Massine; *Parade* 5/18 at the Théâtre du Châtelet, Paris, with music by Erik Satie, choreography by Leonide Massine, scenery by Pablo Picasso (the first cubist ballet).

First performances *Trois Poèmes Juifs* by Ernest Bloch 3/23 at Boston's Symphony Hall; *Israel* Symphony by

"I Want You." James Montgomery Flagg's poster, based on the 1914 British poster, attracted U.S. recruits in 1917.

1917 ***(cont.)*** Bloch 5/3 at New York's Carnegie Hall; *Schelomo (Solomon)* Hebrew Rhapsody for Violoncello and Orchestra by Bloch 5/13 at New York's Carnegie Hall.

Austrian pianist Paul Wittgenstein, 30, loses his right arm at the Russian front, thus apparently ending a promising career that was interrupted by the war. But in the postwar years Wittgenstein will commission and perform works for the left hand by composers such as Richard Strauss, Maurice Ravel, Sergei Prokofiev, Paul Hindemith, and Benjamin Britten.

Stage musicals *Have a Heart* 1/11 at New York's Liberty Theater, with Louise Dresser, Thurston Hall, book and lyrics by Guy Bolton and P. G. Wodehouse, music by Jerome Kern, 76 perfs.; *Love o' Mike* 1/15 at New York's Shubert Theater, with Peggy Wood, Clifton Webb, book by Thomas Sydney, music by Jerome Kern, lyrics by Harold B. Smith, 192 perfs.; *The Maid of the Mountain* 2/10 at Daly's Theatre, London, with music by Harold Fraser-Simson and James W. Tate, book by Frederick Lonsdale, lyrics by Harry Graham, F. Clifford Harris, (Arthur) Valentine, songs that include "Love Will Find a Way," 1,352 perfs.; *Oh, Boy!* 2/20 at New York's Princess Theater, with Marion Davies, Edna May Oliver, music by Jerome Kern, book and lyrics by Guy Bolton and P. G. Wodehouse, songs that include "Till the Clouds Roll By," 463 perfs.; *The Passing Show* 4/26 at New York's Winter Garden Theater, with De Wolf Hopper, music chiefly by Sigmund Romberg, lyrics chiefly by Harold Atteridge, songs that include "Goodbye Broadway, Hello France" by Billy Baskette, lyrics by C. Francis Reisner and Benny Davis, 196 perfs.; *The Ziegfeld Follies* 6/12 at New York's New Amsterdam Theater, with blackface comedian Eddie Cantor (*né* Isidore Itzkowitz), 25, W. C. Fields, Will Rogers, Fanny Brice, music by Raymond Hubbell and Dave Stamper with a patriotic finale by Victor Herbert, 111 perfs.; *Maytime* 8/17 at New York's Shubert Theater, with Peggy Wood, Charles Purcell, music by Sigmund Romberg, book and lyrics by Rida Johnson Young, songs that include "Will You Remember (Sweetheart)," 492 perfs.; *Leave It to Jane* 9/28 at New York's Longacre Theater, with music by Jerome Kern, book and lyrics by Guy Bolton based on George Ade's 1904 play *The College Widow*, 167 perfs.; *Over the Top* (revue) 11/28 at New York's 44th Street Roof Theater, with comedian Joe Laurie, dancer Fred Astaire (*né* Austerlitz), 18, and his sister Adele, music by Sigmund Romberg and others, 78 perfs.; *Going Up* 12/25 at New York's Liberty Theater, with Frank Craven, Donald Meek, music by Louis A. Hirsch, book and lyrics by Otto Harbach based on the 1910 play *The Aviator*, 351 perfs.

Popular songs "Over There" by George M. Cohan, who writes it for the American Expeditionary Force embarking for the war in Europe; "You're in the Army Now" by Isham Jones, lyrics by Tell Taylor and Ole Olssen; "Hail, Hail the Gang's All Here" by U.S. composer Theodore Morse who has adapted a melody by the late Arthur S. Sullivan from Act II of the 1879 Gilbert & Sullivan opera *The Pirates of Penzance*, lyrics by D. A. Esrom (Morse); "Back Home in Indiana" by James F. Hanley, lyrics by Ballard MacDonald; "Smiles" by Lee S. Roberts, lyrics by O. Will Callahan; "For Me and My Gal" by George W. Meyer, 33, lyrics by Edgar Leslie, 32, and E. Ray Goetz, 31; "They Go Wild, Simply Wild, Over Me" by Fred Fisher, lyrics by Joe McCarthy; "Oh Johnny, Oh Johnny, Oh" by Abe Olman, 29, lyrics by Edward Rose, 42; "The Bells of St. Mary's" by Australian-born London composer A. Emmet Adams, 27, lyrics by actor-playwright Douglas Furber, 31.

"The Darktown Strutters' Ball" by Shelton Brooks is the first jazz record; "Tiger Rag" is published as a one-step with music by D. J. LaRocca of the Original Dixieland Jazz Band.

New Orleans jazz pioneer Joseph "King" Oliver, 32, moves to Chicago and Ferdinand J. La M. "Jelly Roll" Morton, 31, to California following the U.S. Navy's shutdown of Storyville sporting houses. A cornetist in Kid Orly's band, Oliver is replaced by his student Louis Daniel Armstrong, 17, who will join Oliver's Creole Jazz Band in Chicago in 1922.

Composer Victor Herbert is upheld by the Supreme Court in a suit against Shanley's Café in New York for using his songs without permission. The decision by Justice Holmes supports ASCAP, which Herbert helped start 3 years ago and which issues licenses to hotels, restaurants, dance halls, cabarets, motion picture theaters, and other establishments to play music controlled by ASCAP members.

Robert Murray, 23, wins in men's singles at the Patriotic Tournament held in place of regular USLTA championships; Molla Bjurstedt wins in women's singles.

The Chicago White Sox win the World Series by defeating the New York Giants 4 games to 2.

The National Hockey League (NHL) is organized November 22 at Montreal with teams from Montreal (the Canadians and the Wanderers), Quebec, Ottawa, and Toronto. The first U.S. team to join will be the Boston Bruins in 1924, followed by the Pittsburgh Pirates (later Penguins) in 1925 and the New York Rangers, Chicago Black Hawks, and Detroit Cougars (later Redwings) in 1926, but the U.S. teams will be comprised almost entirely of Canadian-born players.

The first NHL hockey team to use artificial ice is the Toronto Arenas, who will become the St. Patricks in 1919 and the Maple Leafs in 1926. Hockey has been played until now only on natural ice, warm spells bringing cancellations.

U.S. cigarette production reaches 35.3 billion, up from below 18 million in 1915. Camels have 30 to 40 percent of the U.S. market 4 years after their introduction.

Gen. Pershing cables Washington, "Tobacco is as indispensable as the daily ration; we must have thousands of tons of it without delay," but V. I. Lenin (above) finds smoking intolerable and forbids smoking on the train that takes him to Petrograd.

Invaders of the czar's Winter Palace at Petrograd (above) find themselves in a building of 1,050 rooms with 117 staircases, 1,786 windows, and a 290-square foot map of Russia in emeralds, rubies, and semiprecious stones. Awed by the opulence, the soldiers and sailors leave the palace unmolested.

A new Catskill water system aqueduct opens to provide New York City with 250 million gallons daily of the purest, best-tasting water of any major American metropolis (*see* 1892).

Alaska's Mount McKinley National Park is established by act of Congress. The park embraces nearly 2 million acres which include large glaciers, spectacular wildlife (caribou, Dall sheep, grizzly bears, moose, wolves), and the highest mountain (20,300 feet) in North America. Called Denali (the Great One) by local tribesmen, the mountain was given its name in 1896 by William A. Dickey, then just out of Princeton and prospecting for gold on the Sustina River, after hearing of William McKinley's nomination for the presidency.

Halifax, Nova Scotia, is destroyed December 6 by an explosion that levels two square miles, kills 1,654, and blinds, maims, or disfigures 1,028. The Norwegian relief ship *Imo* loaded with supplies for war-torn Europe has plowed into the French munitions ship *Mont Blanc* loaded with 4,000 tons of TNT, 2,300 tons of picric acid, 61 tons of other explosives, and a deck cargo of highly flammable benzene which has been ignited and has touched off the explosives. A tidal wave created by the explosion washes the city's remains out to sea.

Drought begins on the Western plains.

Britain's House of Commons is stunned by news that 2 million tons of shipping have been lost to German U-boats and that the country has only 3 to 4 weeks' supply of food in stock (above).

Britain's war bread beginning in February is made from flour milled at an extraction rate of up to 81 percent with a compulsory admixture of 5 percent barley, oat, or rye flour to stretch available supplies. Lord Devenport asks heads of families to limit their family bread consumption to 4 pounds per person each week, meat to 2.5 pounds, sugar to three-quarters of a pound.

British retail food prices rise to 94 percent above July 1914 levels by the end of March. A 4-pound loaf of bread fetches a shilling and a Royal Commission reports that rising prices and faulty food distribution are creating industrial unrest.

For a week in April 200,000 Berlin factory workers strike to protest a reduction in bread rations.

Well-to-do Britons miss sugar, butter, and white bread, but price increases after July are at only one-fourth the rate earlier and government subsidization of bread keeps food relatively cheap as compared with other commodities. Mostly because the war has ended unemployment, British working class diets actually improve.

U.S. mining engineer Herbert C. Hoover, 43, is named Food Administrator having served as chairman of the Commission for Relief in Belgium. Hoover encourages U.S. farmers with the slogan "Food Can Win the War."

The U.S. Food Administration's Grain Corp. buys, stores, transports, and sells wheat and fixes its price at $2.20 per bushel under terms of the Lever Act passed August 10 to establish government control over food and fuel.

British military conscription finds that only three out of nine Britons of military age are fit and healthy. Poor nutrition is blamed.

Britain's Medical Council reports that at least half the children in industrial towns have rickets (*see* 1902).

Xerophthalmia is observed in Danish children whose diets are lacking in butterfats. Denmark has been exporting her butter to the warring powers even at the expense of her domestic needs, and dietary deficiency is endangering the children's eyesight (*see* Japan, 1904; McCollum, 1912).

U.S. bacteriologist Alice Evans, 36, begins work that will show the ability of the bacterium which causes contagious abortion (Bang's disease) in cattle to be passed to human beings, notably via raw milk, and produce undulant fever or brucellosis (*see* Bruce, 1887; Bang, 1896). The dairy industry and medical profession will oppose Evans, but compulsory pasteurization of U.S. milk in the late 1920s will be achieved largely through her efforts in the next 9 years (*see* dairy practices, 1921).

Clarence Birdseye returns to his native New York to pursue the commercial exploitation of his food-freezing discoveries (*see* 1914). Birdseye has spent 3 years in Labrador and he has learned to freeze cabbages in barrels of seawater (*see* 1923).

The first national advertising for Del Monte brand canned fruits and vegetables appears April 21 in the *Saturday Evening Post* (*see* California Packing Corp., 1916). Calpak acquires the Hawaiian Preserving Corp. which owns pineapple fields and a cannery on Oahu, will acquire canneries in the next decade in Wisconsin, Minnesota, Illinois, and Florida, will start growing and packing pineapples in the Philippines, and will enter the canned sardine and tuna industries (*see* Del Monte, 1967).

Nathan Handwerker counters rumors spread by his Coney Island rivals who say that 5¢ hot dogs cannot be of the best quality (*see* 1916). He hires college students to stand at his counters wearing white jackets with stethoscopes hanging out of their pockets and word spreads that doctors from Coney Island Hospital are eating Nathan's hot dogs.

A U.S. immigration bill enacted January 29 over a second veto by President Wilson requires that immigrants pass a literacy test in any language. Inspired by the Immigration Restriction League founded in 1894, it becomes law February 5, excluding immigrants from most of Asia and the Pacific Islands, but to relieve the distress of Russian Jews it exempts ref-

1917 ***(cont.)*** ugees from religious persecution (*see* 1921). The U.S. population passes 100 million.

1918 The Fourteen Points for a just and generous peace outlined by President Wilson January 8 in a message to Congress are intended to counter the Russian Bolsheviks, who have released secret agreements revealing Allied plans to carve up the German Empire. Wilson calls for "open covenants openly arrived at" and for self-determination of government by Europe's peoples, asks for the creation of a League of Nations to preserve the peace, but has failed to obtain advance Allied agreement to his proposals.

The Treaty of Brest-Litovsk signed March 3 ends Russian participation in the "capitalist-imperialist" war. Russia's new Bolshevik regime abandons all claims to Poland, Lithuania, the Ukraine, the Baltic provinces, and Finland.

Finnish soldier-statesman Baron Carl Gustaf Emil von Mannerheim, 51, recaptures Helsinki from the Bolsheviks who had seized it. Mannerheim will obtain Russian recognition of Finland's sovereignty in 1919 and will head an independent republic (*see* 1939).

German and Austrian expeditionary forces go in to clear the Bolsheviks out of the Ukraine, and 7,000 U.S. troops occupy the Russian Pacific port of Vladivostok.

German forces have launched a major offensive on the western front from March 21 to April 5, gaining ground against French and British armies depleted of able fighting men and almost devoid of reserves, but the tide of war has turned with the entry of U.S. troops.

The Lafayette Escadrille of 1916 becomes the U.S. Pursuit Squadron February 18. The 94th Pursuit Squadron founded in March under the command of French-American pilot Raoul Lufbury, 32, includes U.S. pilot Eddie (Edward Vernon) Rickenbacker, 27.

German air ace Baron Manfred von Richthofen, 26, leads the "Flying Circus" that downs dozens of Allied aircraft but is shot down himself April 21. The Red Baron is credited with 80 kills and the Allies bury him with full military honors.

Raoul Lufbury (above) is killed in action May 7 and French pilot René Fonck, 24, of "Les Cigones" Groupe de Combat No. 12 shoots down six German fighter planes in 45 seconds May 9. Fonck repeats the feat September 26 and by war's end has shot down 127 enemy planes.

The U.S. Second and Third Divisions stop a German offensive in early June with support from some disorganized French units. U.S. Marines capture Belleau Wood north of Château-Thierry June 6 and hold it through 19 days of repeated German attempts to dislodge them.

Eddie Rickenbacker engages seven German planes in a dogfight September 25; he will win the Congressional Medal of Honor for his performance in the encounter, and by war's end his combined balloon and aircraft kills will total 26.

U.S. ace Frank Luke, 22, takes off in his Spad September 29 in defiance of his commanding officer who has grounded him. Luke has shot down 16 enemy planes, 10 German Fokkers have gone up expressly to seek him out, he downs two of the Fokkers before an antiaircraft shell fragment hits him in the shoulder, he goes down behind enemy lines, empties his pistol at approaching soldiers, and is mortally wounded 16 days after he first went into combat.

Russia's royal Romanov family is shot to death July 16 at Yekaterinburg by order of the Bolsheviks 4 days after others of the Russian nobility have been assassinated elsewhere. Nicholas II, the czarina Aleksandra, their daughters Olga, Tatiana, Marie, and Anastasia, their son Alexis, Prince Dolgorolkoff, their physician, a nurse, and a lady in waiting are all put to death within a few weeks of July 16, but rumors will persist that one or more of the daughters has somehow been spared.

The British attack with 450 tanks at the Battle of Amiens in August and the Germans are forced to fall back to the Hindenburg line after British and French offensives in September.

Gen. Pershing makes his first major offensive in mid-September at Saint-Mihiel and forces the Germans to give up salients they have held since 1914. The Americans take 15,000 prisoners and the Allies push north and east in the battles of the Argonne Forest and Ypres from late September to mid-October.

Bulgaria's Ferdinand I abdicates October 4 at age 57 after a 32-year reign following defeat of Bulgarian forces in September at Dobropole in Macedonia. Ferdinand leaves Sofia and is succeeded by his son, 22, who will reign until 1943 as Boris I.

Mutiny breaks out in the German fleet at Kiel October 28 and spreads quickly to Hamburg, Bremen, Lübeck, and all of northwestern Germany. Revolution breaks out November 7 at Munich.

Germany's Wilhelm II abdicates November 8 and hostilities on the western front end November 11 in an armistice signed by Germany and Allies at Compiègne outside Paris.

The Great War has killed 1.8 million Germans, 1.7 million Russians, 1.4 million French, 1.2 million Austrians and Hungarians, between 750,000 and 950,000 British, 460,000 Italians, 325,000 Turks, and 115,000 Americans. Some 20 million have been blinded, maimed, mutilated, crippled, permanently shell-shocked, or otherwise disabled.

A Congress of Oppressed Nationalities, most of them ruled by the Hapsburgs, has been held at Rome in April, and Italy has recognized the unity and independence of the new Yugoslav nation organized in 1917. The Kingdom of the Serbs, Croats, and Slovenes is proclaimed December 4 under the regency of Serbia's Prince Aleksandr Garageorgevic, 30, who will become Yugoslavia's Aleksandr I in 1921.

The Czechoslovak National Council at Paris has organized a provisional government October 14 with Thomas Garrigue Masaryk, 68, as president and Eduard Beneš, 34, as foreign minister. The council has issued a declaration of independence October 28.

Hungary has a revolution that ends the Hapsburg monarchy October 31; a republic is proclaimed November 16.

The Austrian emperor Karl I who ascended the throne in 1916 at the death of Franz Josef is forced to renounce participation in the government of Austria November 11 and of Hungary November 13.

A Polish republic is proclaimed November 3 at Warsaw. Poland quickly comes under the control of Gen. Josef Pilsudski who was imprisoned by the Germans in 1916 when his troops refused to join the Central Powers against the Allies but who has been released after the collapse of the Central Powers.

Hailed as a U.S. war hero is Tennessee doughboy Alvin Cullum York, 30, whose draft board denied his petition for exemption as a conscientious objector. In the battle of the Argonne Forest October 8, Private York of the 82nd Infantry Division led an attack on a German machine gun nest that killed 25 of the enemy and he then almost singlehandedly captured 132 prisoners and 35 machine guns. Promoted to sergeant November 1, York will be awarded the Congressional Medal of Honor and French Croix de Guerre.

The Owen-Keating Child Labor Law of 1916 is an unconstitutional encroachment on states' rights, the Supreme Court rules June 3 in the case of *Hammer v. Dagenhart.* Justice Holmes dissents *(see* Fair Labor Standards Act, 1938).

The Amalgamated Clothing Workers Union begins striking against open shops, sweat shops, and piecework pay *(see* 1914). The union will stage at least 534 strikes in the next 6 years and claim victory in 333 *(see* 1937).

A Chicago jury finds IWW leaders guilty of conspiring against the prosecution of the war. Heavy fines and prison sentences handed down to some 100 leaders by Judge Kenesaw Mountain Landis will be sustained by the U.S. Circuit Court of Appeals.

British women over age 30 gain the right to vote under terms of the Fourth Franchise Bill which also grants suffrage to all men over age 21. Emmeline Pankhurst has favorably influenced masculine opinion by persuading women to do war work and has helped obtain passage of the bill *(see* 1913; 1928).

A resolution providing for a U.S. Woman Suffrage Amendment passes the House of Representatives in January, but the Senate rejects the resolution for the third time October 1 *(see* 1919).

Inflation begins in Germany. The mark will decline in value by 99 percent in terms of gold over the course of the next few years *(see* 1922).

British income taxes rise to 30 percent, up from 15 percent in 1915, as the national debt climbs to an unprecedented £8 thousand million (£8 billion).

U.S. Fuel Administrator Harry A. Garfield orders all plants east of the Mississippi to shut down January 16 if they are not producing essential war materials. The purpose of the order is to conserve coal needed for troop transports, warships, and merchant vessels, the order remains in effect for 5 days, and it is followed by nine Monday closings.

The Fuel Administration orders four lightless nights per week beginning July 24 to conserve fuel for the coming winter. The order remains in effect until November 22.

Matsushita Electric Co. has its beginnings as Japanese entrepreneur Konosuke Matsushita, 27, invests $50 to start a firm that will manufacture electrical adapter plugs *(see* rice saucepan, 1946).

A head-on collision between two trains on the Nashville, Chattanooga, and St. Louis Railway in Tennessee June 22 kills 99 and injures 171, an all-time high for U.S. railroad mismanagement.

New York's worst subway accident kills 97 and injures 100 November 2 as a train jumps the track at 30 miles per hour approaching the new Prospect Park station of the Brooklyn Rapid Transit Service. Manned by supervisors in the absence of striking motormen who belong to the Brotherhood of Locomotive Engineers, the train is traveling at five times the speed limit, many of its cars are wooden, and it is wrecked at the Malbone Street Tunnel of the Brighton Beach line that runs along Fulton Street and Franklin Avenue to Coney Island.

General Motors acquires a controlling interest in Detroit's Fisher Body Co., which will become a GM division *(see* 1916).

Celanese Corp. of America is founded at Cumberland, Md., by Swiss-American chemist Camille Edward Dreyfus, 40. It will become the largest producer of acetate rayon and a major factor in viscose rayon and other synthetic fibers *(see* American Viscose, 1910; nylon, 1935).

The worst pandemic to afflict mankind since the Black Death of the mid-14th century sweeps through Europe, America, and the Orient killing 21.64 million—more than 1 percent of the world's population—while the European War ends after having killed some 10 million *(see* below). The "Spanish" influenza (which actually began in China) affects 80 percent of the people in Spain, the first U.S. cases appear August 27 at Boston where two sailors report to sick bay at Commonwealth Pier, by August 31 the Navy Receiving Ship has 106 flu cases, by mid-September the flu has spread all over the East Coast, nearly 25 percent of Americans fall ill, some 500,000 die including 19,000 at New York, schools are closed, parades and Liberty Loan rallies banned, hospitals jammed, coffin supplies exhausted at Baltimore and Washington.

Emergency tent hospitals go up throughout America as the Spanish influenza epidemic taxes regular hospital facilities. In many countries trolley conductors, store clerks, and others wear masks to avoid contracting the flu.

Some 46 percent of fracture cases in the U.S. Army result in permanent disability, chiefly by amputation, and 12 percent of fracture cases are fatal *(see* 1937).

1918 *(cont.)* Typhus takes a heavy toll in Galicia, the Ukraine, and the Black Sea region (*see* 1917). The International Red Cross helps fight the disease.

The alkaloid drug ergotamine is extracted from the ergot fungus *Claviceps purpura (see* 857). Useful in small doses as a muscle and blood vessel contractant, the drug will be used to treat migraine attacks and induce abortion (*see* ergonovine, 1935).

The International Church of the Four-Square Gospel is founded at Los Angeles by Canadian-American evangelist Aimee Semple McPherson, 28, a divorcée who has arrived penniless but offers hope and salvation to Southern and Midwestern migrants newly arrived in Southern California. McPherson will build a large following that will provide funds to build the huge Angelus Temple from which her sermons will be broadcast over a radio station she will purchase with funds contributed by the faithful. Patriotic-religious music played by a 50-piece band will precede the sermons, and the McPherson movement, based largely on faith healing, adult baptism, and Fundamentalist spectacle, will attract thousands *(see* 1923).

The first U.S. airmail stamps are issued as regular service begins between Washington, D.C., and New York (the 24¢ stamps show a biplane in flight, but some of the stamps are accidentally printed with the plane inverted and the faulty stamps become rare collectors' items).

The first official U.S. mail flight never makes it to New York. The pilot discovers on his first takeoff attempt that someone has forgotten to fill the fuel tank, and after he does take off he becomes lost over Maryland and has to land in a cow pasture (*see* 1921).

Henry Ford's Stout Air Line has been the first airline to carry mail but Ford's refusal to open his Detroit airport on Sundays hampers operations.

Airmail service got off the ground with difficulties that included a printing error in the first airmail stamps.

David Low joins the *London Star* at age 27 to pursue a career as political cartoonist. The New Zealand-born Australian will transfer to the *Evening Standard* in 1927 and become famous for his Colonel Blimp satire of the British ruling class.

The *Boston Post* reaches a circulation of 540,000 and claims the widest readership in America. The *Post* sells for 1¢ while the *Globe* and the *Herald* have joined most other U.S. dailies in going to 2¢.

"Believe It or Not!" is published for the first time by *New York Globe* sports cartoonist Robert LeRoy Ripley, 24, who sketches figures of men who have set records for such unlikely events as running backward and broad jumping on ice. Encouraged by reader response to pursue his quest for oddities, Ripley will move to the *New York Post* in 1923, syndication of his cartoons will begin soon after, and "Believe It or Not!" will eventually be carried by 326 newspapers in 38 countries.

Nonfiction *Eminent Victorians* by English writer (Giles) Lytton Strachey, 58, introduces a new genre of biography by injecting imagined conversations and taking other liberties while debunking the Victorian eminences of Florence Nightingale and such.

Fiction *The Four Horsemen of the Apocalypse* by Spanish novelist Vicente Blasco-Ibañez, 51, breaks all world records for book sales; *Man of Straw (Der Untertan)* by Heinrich Mann; *Tarr* by English poet-novelist-painter (Percy) Wyndham Lewis, 33; *My Antonia* by Willa Cather; *The Magnificent Ambersons* by Booth Tarkington.

Poetry *Poems* by the late English poet Gerard Manley Hopkins who died in 1889 at age 45. His rich rhythmic system will inspire a new generation of poets; *The Dark Messengers (Los Heraldos Negros)* by Peruvian poet César Vallejo, 26; *Cornhuskers* by Carl Sandburg whose poem "Prairie" says, "I speak of new cities and new people./I tell you the past is a bucket of ashes. . ." Sandburg's poem "Grass" says, "Pile the bodies high at Austerlitz and Waterloo./Shovel them under and let me work—/I am the grass; I cover all;" English poet-playwright-painter Isaac Rosenberg is killed in action April 7 at age 27; poet Wilfred Owen is awarded the Military Cross for gallantry and is killed in action November 4 at age 25; Guillaume Apollinaire, now 38, dies November 9 of influenza and a head wound received in March 1916. He leaves behind erotic poems relating battle action and sexual desire: "Two shell-bursts/A pink explosion/ Like two bared breasts/ Snooking their tips/ HE KNEW LOVE/ What an epitaph."

Painting *Gartenplan* by Paul Klee; *Engine Rooms* by Fernand Léger; *Act* and *Portrait of Mme. Zboroski* by Amedeo Modigliani; *Scottish Girl* by Juan Gris; *Friends* and *Saxonian Landscape* by Oskar Kokoschka; *Bathing Man* by Edvard Munch; *Odalisques* by Henri Matisse; *Portrait of Igor Stravinsky* by Robert Delaunay; *Still Life: Dead Game* by André Derain; *Still Life with Coffee* by Spanish painter Joan Miró, 25; *Landscape of Piquey* by Diego Rivera. Gustav Klimt dies at Vienna November 2 at age 55.

Theater *Deburan* by French playwright-actor Sacha Guitry, 32, 2/9 at Paris, with Guitry himself (matinees only since Paris is under bombardment by German cannon); *Sea Battle (Seeschlacht)* by German physician-playwright Richard Goering, 30, 3/3 at Berlin's Deutsches Theater; *Lightnin'* by Winchell Smith and Frank Bacon 8/26 at New York's Gaiety Theater, with Bacon as Lightnin' Bill Jones, 1,291 perfs.; *Mystery Bouffe (Misteriya Buff)* by Vladimir Mayakofsky 11/7 at Petrograd's Malevich Communal Theater; *The Moon of the Caribees* by Eugene O'Neill 12/20 at New York's Playwrights Theater.

Warner Brothers Pictures is incorporated in California by Polish-American cinematists Harry Warner, 36, and Albert, 34, with their Canadian-born brother Jack, 26, and a fourth brother Sam. Harry and Albert grossed a handsome return last year filming Ambassador James W. Gerard's book *My Four Years in Germany*, produced at New York's Vitagraph studios, but will have little success in Hollywood until 1926.

Louis B. Mayer Pictures is organized at Los Angeles by Russian-American movie theater operator Louis Burt Mayer, 33, who in 1915 bought New England rights to D. W. Griffith's *Birth of a Nation* for an unprecedented $25,000 and has decided to make films himself *(see* M-G-M, 1924).

Films Charles Chaplin's *Shoulder Arms* with Chaplin; J. Gordon Edwards' *Salome* and *When a Woman Sins*, both with Theda Bara; Edwards's *A Pair of Silk Stockings* with Constance Talmadge; F. Richard Jones's *Mickey* with Mabel Normand.

Opera *Where the Lark Sings (Wo die Lerche singt)* 2/1 at Budapest's Königstheater, with music by Franz Lehar; *Duke Bluebeard's Castle (A Kék szakállu Herceg Vára)* 5/24 at Budapest, with music by Béla Bartòk; *Il Trittico* 12/14 at New York's Metropolitan Opera House, with music by Giacomo Puccini who has written a trio of short operas—*Il Tabarro*, *Suor Angelica*, and *Gianni Schicchi*.

First performances Classical Symphony by Serge Prokofiev 4/21 at Petrograd; *The Legend of the World at Play (Le Dit des Jeux du Monde)* by Swiss composer Arthur Honegger, 26, 12/2 at Paris.

Broadway musicals *Oh, Lady! Lady!* 2/1 at the Princess Theater, with music by Jerome Kern, book and lyrics by Guy Bolton and P. G. Wodehouse, 219 perfs.; *Sinbad* 2/14 at the Winter Garden Theater, with Al Jolson, book by Harold Atteridge, music by Sigmund Romberg and others, songs that include "My Mammy" by U.S. composer Walter Donaldson, 25, lyrics by Joe Young, 20, and Sam M. Lewis, 33, "Rock-a-by Your Baby with a Dixie Melody" by Jean Schwartz, lyrics by Joe Young and Sam M. Lewis, 164 perfs.; *Oh, Look!* 3/7 at the Vanderbilt Theater, with songs that include "I'm Always Chasing Rainbows" by Harry Carroll, who has adapted Chopin's Fantaisie Impromptu in C sharp minor, lyrics by Joseph McCarthy, 68 perfs.; *The Passing Show* 7/25 at the Winter Garden Theater, with Marilyn Miller, Fred and Adele Astaire, Eugene and Willie Howard, Charles Ruggles, Frank Fay, music chiefly by Sigmund Romberg and Jean Schwartz, lyrics chiefly by Harold Atteridge, songs that include "Smiles" by J. Will Callahan and Lee G. Robert, and with June Caprice singing "I'm Forever Blowing Bubbles" by John William Kellette, lyrics by Jean Kenbrovin (James Kendis, James Brockman, Nat Vincent), 124 perfs.; *The Ziegfeld Follies* 6/18 at the New Amsterdam Theater with Eddie Cantor, Will Rogers, Marilyn Miller, W. C. Fields, music by Louis A. Hirsch, 151 perfs.; *Yip Yip Yaphank* 8/19 at the Century Theater with a cast of 350 recruits from Camp Upton at Yaphank, Long Island, music and lyrics by Irving Berlin, songs that include "Oh, How I Hate to Get Up in the Morning," "Mandy," "Soldier Boy," 32 perfs.; *Sometime* 10/4 at the Shubert Theater, with vaudeville trooper Mae West, 26, singing songs that include "Any Kind of Man" by Rudolf Friml, 283 perfs.

Popular songs "Till We Meet Again" by Richard A. Whiting, 27, and Canadian-American Richard B. Egan; "K-K-K-Katy" by Canadian-American songwriter Geoffrey O'Hara, 36; "After You're Gone" by Henry Creamer, 39, lyrics by Turner Layton, 24; "Somebody Stole My Gal" by Leo Wood; "Beautiful Ohio" by Robert A. King, lyrics by Ballard MacDonald.

The Original Dixieland Jazz Band tours Europe at year's end, spreading appreciation for the American musical idiom.

Exterminator wins the Kentucky Derby after starting as a late entry 30 to 1 longshot. The chestnut gelding goes on to win a record 34 stakes races.

Robert Murray wins in U.S. men's singles, Molla Burjsted in women's singles.

The Boston Red Sox win the World Series by defeating the Chicago Cubs 4 games to 2.

The Raggedy Ann Doll introduced by a New York firm is based on a doll produced to promote sales of the first book of Raggedy Ann stories. *Indianapolis Star* political cartoonist John Gruelle, 37, found the handmade rag doll in his attic last year and gave it to his tubercular daughter Marcella, they named it Raggedy Ann through a combination of neighbor James Whitcomb Riley's "Little Orphan Annie" of 1885 and Riley's poem "The Raggedy Man," the doll inspired Gruelle to make up stories which he told to entertain Marcella, she died holding the doll in March of 1916 after being vaccinated by a contaminated needle, Gruelle's book will have 25 sequels, and the Raggedy Ann doll will grow to become a $20 million-per-year business.

Rinso, introduced by Lever Brothers, is the world's first granulated laundry soap *(see* Lux Flakes, 1906; Lux Toilet Form, 1925; detergent, 1946).

Kotex is introduced under the name "Celucotton" by Kimberly & Clark Co. of Neenah, Wis., whose German-American chemist Ernst Mahler, 31, has developed a wood-cellulose substitute for cotton to fill the desperate need for dressings and bandages in European field hospitals. When word reaches Wisconsin that Red Cross nurses are using Celucotton for sanitary napkins,

1918 ***(cont.)*** the company begins development of the first commercial sanitary napkin; it will be sold as Kotex beginning in 1921 (*see* Kleenex, 1924; Tampax, 1936).

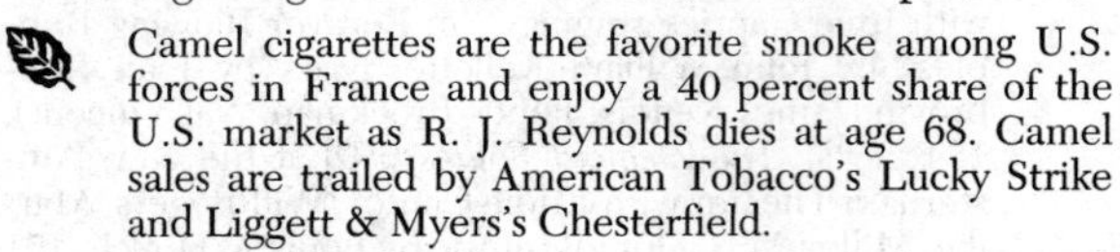

Camel cigarettes are the favorite smoke among U.S. forces in France and enjoy a 40 percent share of the U.S. market as R. J. Reynolds dies at age 68. Camel sales are trailed by American Tobacco's Lucky Strike and Liggett & Myers's Chesterfield.

The Staatliches Bauhaus founded at Weimar by German Architect Walter Gropius of 1911 "curtain wall" fame combines two art schools in a revolutionary center that interrelates art, science, technology, and humanism. The Bauhaus will move to a new Gropius-designed building at Dessau in 1925 but will be closed in 1933 (*see* 1937).

The Dutch Staats-General appropriates money to drain the Zuider Zee for the creation of fertile dry land. The Dutch will build dams, dykes, sluiceways, watergates, canals, and locks to "impolder" more than 800,000 acres that will increase the territory of the Netherlands by 7 percent (*see* 1932).

The North American lobster catch falls to a low point of 33 millon pounds, down from its all-time high of 130 million in 1885.

Prime Minister David Lloyd George asks British women to help bring in the harvest.

Congress enacts a "summer time" daylight saving law that follows the example of Britain and Europe. The new law angers U.S. farmers as it has angered the British (*see* 1916) and Congress will repeal the law next year, overriding President Wilson's veto.

More horses and mules are on U.S. farms than ever before, and enormous acreages are planted to oats for the animals, whose numbers will now decline.

International Harvester's share of the U.S. harvesting machine industry falls to 65 percent, down from 85 percent in 1902.

U.S. Corn Belt acreage sells for two and three times 1915 prices.

British food rationing begins with sugar January 1 and is extended in February to include meat, butter, and margarine. By summer it includes all important foodstuffs except bread and potatoes. Fresh meat is rationed by price and consumers are required to buy from a particular butcher to avoid deception. Ration books are issued in July and the bacon ration is raised from 8 ounces per week to 16 ounces. Other rationed commodities include 8 ounces of sugar, 5 ounces of butter and/or margarine, 4 ounces of jam, 2 ounces of tea (weekly).

Britain's milling extraction rate is raised to 92 percent, up from 81 percent last year, with soy and potato flour mixed in to stretch Britain's short wheat supplies. The bread is dark and unattractive.

France rations bread and restricts consumption of butter, cream, and soft cheese. Manufacture and sale of all confectionery is banned.

U.S. sugar rationing begins July 1 with each citizen allowed 8 ounces per week. U.S. Food Administrator Herbert Hoover asks for voluntary observance of wheatless Mondays and Wednesdays, meatless Tuesdays, porkless Thursdays and Saturdays, and the use of dark "Victory bread."

All food regulations are suspended in the United States in late December but remain in effect in Britain and Europe.

English researcher Edward Mellanby shows that cod-liver oil contains an agent that cures rickets, giving recognition to a treatment used by North Sea mothers for generations (*see* McCollum, 1921).

Yale biochemists Lafayette Benedict Mendel and B. Cohen show that guinea pigs cannot develop vitamin C (ascorbic acid) and fall prey to scurvy even more easily than do humans (*see* Drummond, 1920; Szent-Györgyi, 1928).

The U.S. Food Administration orders severe limitations on use of sugar in less essential food products, including soft drinks. As shortages of sugar continue, the Administration prepares to declare the entire soft drink industry nonessential and to order the industry closed down for the duration, but the threat fails to materialize and sugar prices begin to soar (*see* 1920; American Sugar Refining, 1919).

Hershey Chocolate Co. is donated to the Milton Hershey School for orphan boys established by Hershey in 1909. The school will retain 100 percent ownership until 1927, when some stock will be sold to the public to raise funds.

Wartime demand has multiplied Nestlé sales like Hershey's (*see* 1905). Nestlé has 100 milk condenseries in the United States plus scores in other countries.

U.S. inventor Charles Strite patents the first automatic pop-up toaster.

Americans call sauerkraut "liberty cabbage;" German toast becomes "French toast."

1919 The Versailles Peace Conference opens January 18 outside Paris with delegates from 27 victorious nations and one week later adopts a unanimous resolution to create a League of Nations whose members will protect each other against aggression and will devote itself to such matters as disarmament, labor legislation, and world health (*see* Wilson's Fourteen Points, 1918).

A new German republic is established with its constituent assembly at Weimar, a Thuringian city far to the southwest of Berlin where the traditions are humanistic rather than militaristic. The national assembly meets February 6 to draw up a constitution which is adopted July 31 but the country is embroiled in Communist uprisings mixed with conspiracies to reestablish the monarchy.

The Third International founded at Moscow March 2 is an organization dedicated to propagating communist doctrine with the avowed purpose of producing worldwide revolution. This Comintern will unite Communist groups throughout the world.

"I have been over into the future and it works," says author-journalist Lincoln Steffens to U.S. financier Bernard Baruch after a visit to Russia, but that country is embroiled in civil war between the Bolshevik Red Army and a White Russian Army led by Gen. Anton Denikin with Allied support.

Gen. Denikin (above) routs Red Army forces in the Caucasus February 3 to 9 but the Red Army captures Kiev February 3, invades Estonia, enters the Crimea April 8 one day after the Allies have evacuated Odessa, and takes Ufa June 9. Anglo-White Russian forces defeat the Red Army August 10 in North Dvina but the Red Army takes Omsk November 15 and Kharkov December 13.

British troops occupy Archangel September 27 but withdraw from Murmansk October 12.

Sinn Fein party members of Parliament have proclaimed an independent Irish Republic January 21 and organized a parliament of their own. The Dail Eireann includes Sinn Fein president Eamon de Valera, 37, but the British suppress it September 12 and war begins November 26 between the Sinn Fein and British regulars (*see* 1916; 1920).

Romania has annexed Transylvania January 11, a Bolshevist coup overthrows the Hungarian government March 21, monarchists regain control August 1, Romanian forces invade Hungary, take Budapest August 4, force Hungary's Communist premier Béla Kun, 34, to take refuge at Vienna, plunder Hungary, but are induced to withdraw November 14 (*see* 1920).

Montenegro deposes her king Nicholas I April 20 and votes for union with Yugoslavia.

Greek troops land at Smyrna May 15 with Allied support and Italian troops land in Anatolia. Turkish war hero Mustapha Kemal organizes resistance to further dismemberment of the Ottoman Empire beginning May 19 but the new sultan Mohammed VI dismisses him July 8 and outlaws him July 11. Kemal declares himself independent of the sultan August 5 at the Turkish Nationalist Congress (*see* 1915; 1923).

The Treaty of Versailles signed June 28 obliges Germany to accept sole responsibility for causing the Great War. Germany returns to France the Alsace-Lorraine conquests of 1871, cedes other territories to Belgium and Poland, cedes her colonies to the Allies to be administered as mandates under the League of Nations, and agrees to pay large reparations (below).

The Treaty of Saint-Germain signed September 10 obliges Austria to recognize the independence of Czechoslovakia, Hungary, Poland, and Yugoslavia. Austria agrees to pay heavy reparations and cede the territories of Galicia, Istria, the Trentino, South Tyrol, and Trieste.

The Treaty of Neuilly signed November 27 obliges Bulgaria to recognize Yugoslavian independence, give up her Aegean seaboard, and pay $445 million in reparations.

France acquires mandate control of Syria from Turkey and of Togo and Cameroon in Africa from Germany.

The U.S. Senate rejects the Versailles Treaty (above) and rejects U.S. membership in the League of Nations. President Wilson has suffered a stroke October 2, his left side is paralyzed, and he is powerless to fight isolationists led by Republican Senator Henry Cabot Lodge of Massachusetts who have particularly attacked Article X of the treaty which insures the permanence of the territorial boundaries agreed upon at Versailles. The president has called this "the heart of the covenant" but critics assail it and the Senate vote November 19 fails to come up with the two-thirds majority needed for ratification.

Berlin authorities arrest socialist agitators Karl Liebknecht, 48, and Rosa Luxemburg, now 49, after an insurrection inspired by their Spartacus party. They are murdered January 15 while being transferred from military headquarters to prison.

U.S. authorities release anarchists Emma Goldman and Alexander Berkman from prison but deport them with more than 200 others to Soviet Russia which Goldman will describe in her 1923 book *My Disillusionment with Russia* (*see* 1906).

Supreme Court Justice Oliver Wendell Holmes formulates a "clear and present danger" test for defining conditions under which the constitutional right of freedom of the speech may be abridged. "When a nation is at war," says Holmes March 3 in the espionage case *Schenck v. United States*, "many things that might be said in times of peace are such a hindrance to its effort that their utterance will not be endured so long as men fight."

The Supreme Court upholds conviction of Russian propagandists opposing U.S. intervention in Russia. Justices Holmes and Brandeis dissent from the ruling handed down November 10 in *Abrams v. United States*. Holmes says, "Only the emergency that makes it immediately dangerous to leave the correction of evil counsels to time warrants making any exception to the sweeping command, 'Congress shall make no law. . . abridging the freedom of speech,'" but federal authorities deport 249 Russians December 21 aboard the transport ship S.S. *Buford*.

The House of Representatives unseats Wisconsin socialist congressman Victor Berger November 10, Berger's district reelects him in December, but the House will declare his seat vacant next January.

The International Labor Office (ILO) is created by the League of Nations.

Labor unrest rocks the United States. Four million workers strike or are locked out.

Boston police strike September 9 in protest against pay scales that range from 25¢ per hour for 83-hour weeks down to 21¢ per hour for 98-hour weeks despite wartime inflation. Only 427 of the city's 1,544-man force remain on duty to protest the orgy of lawlessness that ensues, gangs of men, often drunk, roam the streets robbing, looting, raping, and beating up other citizens, Gov. (John) Calvin Coolidge goes to bed at ten

1919 ***(cont.)*** and does not learn about the rioting until next morning, he sends state militia into Boston September 12, the 1,117 strikers are dismissed, AF of L leader Samuel Gompers asks that they be reinstated, Coolidge wires Gompers September 14, "There is no right to strike against the public safety by anybody, anywhere, at any time."

Steel workers at Gary, Ind., strike September 22 to force United States Steel to recognize their union. The walkout ends in 110 days without success; one-third of the workers labor 7 days a week.

AF of L leader in the steel strike (above) is former IWW organizer William Zebulon Foster, 38, who will soon emerge as the leading figure in the American Communist movement.

United Mine Workers Union leader John Llewellyn Lewis, 39, begins a strike of bituminous coal miners November 1 but Attorney General Alexander Palmer obtains an injunction under the 1917 Lever Act whose wartime powers remain in effect. The strike ends November 9.

Serving as special assistant to A. M. Palmer (above) is lawyer John Edgar Hoover, 24, who in 2 years will become assistant director of the Federal Bureau of Investigation (FBI) (*see* 1908). Hoover handles deportation cases involving alleged Communist revolutionaries (above).

The Amritsar Riots for Indian self-government produce a massacre April 10. The local British commander calls out his troops and without giving the demonstrators adequate warning has the men fire on an unarmed mob, killing 379, wounding 1,200. Gen. Sir Reginald Dwyer is forced to resign after a commission of inquiry has censured his action.

The *Dearborn Independent* is published for the first time by auto maker Henry Ford and reveals a tone of anti-Semitism.

A Nineteenth Amendment to the Constitution granting women suffrage is adopted by a joint resolution of Congress June 20 and sent to the states for ratification (*see* 1920).

Race riots erupt in 26 U.S. cities throughout the year. Washington, D.C., has riots July 19 as white soldiers and sailors attack black ghetto sections. Chicago has riots beginning July 27 as a disturbance at a Lake Michigan beach spreads to the city's Black Belt leaving 15 whites and 23 blacks dead, hundreds seriously injured.

A "Method of Reaching Extreme Altitudes" by U.S. physicist Robert Hutchings Goddard, 36, predicts the development of rockets that will break free of Earth's gravitational pull and reach the moon. Goddard has devised the first two-stage rocket and received a grant from the Smithsonian Institution 2 years ago (*see* 1926; *New York Times*, 1921).

English economist John Maynard Keynes, 36, foresees that high reparations imposed on Germany will upset the world economy. Leader of the British Treasury team at Versailles, Keynes resigns to write in opposition to the reparations plan (*see* 1936).

The Treaty of Versailles (above) obliges Germany to pay large reparations that include not only billions of dollars, francs, pounds, lire, etc., but also merchant ships and fishing ships plus large quantities of coal to be delivered in the next 10 years to France, Belgium, and Italy.

Germany loses 72 percent of her iron ore reserves as Lorraine is given back to France under terms of the Versailles Treaty. The second largest iron producer in 1913, Germany will be seventh largest as the Germans concentrate mining and steel making in the Ruhr Basin.

Inflation begins in France which has suffered enormous property damage as well as human destruction in the Great War. The buying power of the franc in terms of gold will decline by 78 percent in the next 8 years (*see* 1926; 1928).

The war has cost the United States nearly $22 billion from April 1, 1917, through April of this year and nearly $9 billion has been loaned to the Allied powers. Americans have paid high taxes to fund the war effort and contributed $400 million in cash and materials to the Red Cross War Council.

The cost of living in New York City is 79 percent higher than it was in 1914, says the Bureau of Labor Statistics (*see* Consumer Price Index, 1913).

A Wall Street boom in "war baby" stocks sends prices to new highs. Baldwin Locomotive is up 360 percent over 1917 levels, Bethlehem Steel up 1,400 percent, General Motors up 940 percent.

E. I. du Pont de Nemours has $49 million in wartime profits even after paying out dividends of $141 million to stockholders. The company's explosives have fired 40 percent of all Allied shells in the war, it has met at least half the domestic U.S. requirements for dynamite and black blasting powder, and it begins to acquire other chemical companies and to enlarge its holdings in General Motors, which it will soon control (*see* 1912; GM, 1908, 1957; Sloan, 1929).

F. W. Woolworth dies April 8 at age 67 leaving a fortune of $67 million. There are 1,000 Woolworth stores in the United States plus many in Britain, and the chain has annual sales of $106 million (*see* 1912).

Andrew Carnegie dies August 11 at age 83 in his Berkshire Hills mansion at Lenox, Mass., after having given away some $350 million in philanthropic contributions that inspired other U.S. millionaires to be philanthropic (*see* 1889).

Oregon imposes the first state gasoline tax February 25.

The NC-4 flying boat designed by U.S. aeronautical engineer Jerome Clarke Hunsaker, 33, for Curtiss Aeroplane & Motor makes the first transatlantic crossing by a heavier-than-air machine, leaving Newfoundland May 16 and arriving at Lisbon May 27 with a five-man crew under U.S. Navy, Lieutenant Commander Albert C. Read (Curtiss has produced more than 4,000 JN-4

"Jenny" biplanes for training army and navy pilots in the war) (*see* 1911).

The Sopwith biplane *Atlantic* takes off from Newfoundland May 18 powered by a 350-horsepower Rolls-Royce engine and with 400 gallons of fuel. Pilot Harry G. Hawker, 27, and navigator K. Mackenzie Grieve fly 1,000 miles toward Ireland in quest of the £10,000 prize offered by Lord Northcliffe in 1913 for the first nonstop transatlantic flight, but engine trouble forces them to ditch, a Danish tramp steamer picks them up, London gives them a hero's welcome, and the *Daily Mail* awards a £5,000 consolation prize (*see* 1921).

A Vickers Vimy bomber powered by two Rolls-Royce Eagle engines achieves the first nonstop transatlantic flight June 14 and wins the *Daily Mail*'s £10,000 prize. Royal Flying Corps pilot John William Alcock, 27, was captured by the Turks after bombing Constantinople during the war, he makes the 1,960-mile flight from Newfoundland to Ireland in 16 hours, 12 minutes accompanied by navigator Arthur Whitten-Brown, but they land ingloriously in a bog and Alcock crashes to his death later in the year (*see* Lindbergh, 1927).

German aircraft engineer Hugo Junkers, 60, opens an aircraft factory at Dessau. Junkers will design the first all-metal airplane to fly successfully and will establish one of the first regular mail and passenger airlines in Europe.

KLM (Koninkhje Luchtvaart Maatschappig voor Nederland an Kolonien) is founded by Dutch banking and business interests. The national airline will begin scheduled flights between Amsterdam and London May 17 of next year.

Europe's *Orient Express* becomes the *Simplon-Orient Express*. Suspended in 1914, the fabled 36-year-old luxury train avoids passing over any German or Austrian soil by using the 12-mile Simplon Tunnel that opened through the Alps in 1906 between Brig, Switzerland, and Domodossola, Italy (*see* 1929).

The 5-year-old German passenger liners S.S. *Vaterland* and S.S. *Bismarck* are handed over to the British as war reparations and renamed the S.S. *Leviathan* and S.S. *Majestic*.

Scots-American aviation engineer Malcolm Lockheed (*né* Loughhead) sets up the Lockheed Hydraulic Brake Co. at Detroit. He will have little success until 1923, when Walter Chrysler buys Lockheed brakes for the first Chrysler motorcars.

Henry Ford assumes full control of the now enormous Ford Motor Company. Ford has lost a lawsuit brought by stockholders who include the Dodge brothers but he takes all his capital plus some borrowed funds and buys up all the outstanding shares in the company, paying more than $105 million, and returning roughly $12.5 million for every $5,000 invested in 1908 (*see* 1920).

Electric starters become optional on the Model T Ford but most Model Ts are still started by handcranking the engine (*see* Kettering, 1911).

Henry Ford builds a motorcar factory in his family's ancestral home town of Cork in Ireland.

The Bentley motorcar is introduced by former London taxi fleet manager W. O. (Walter Owen) Bentley, 31, who created the B.R. 2 "Merlin" rotary engine that powered the Sopwith Camel during the war. Bentley's 3-liter racing car is the forerunner of machines that will win at Le Mans and that will lead to luxury sedans (*see* Rolls-Royce, 1931).

The Citroën motorcar is introduced by French munitions-maker André Gustave Citroën, 41, who turns his factory over to producing motorcars that can be sold at low prices. Citroën will go bankrupt in 1934 and lose control of his company but the Citroën will remain a major French automobile (*see* 1955).

The Duesenberg brothers set up shop at Indianapolis to make motorcars after selling John North Willys the war plant at Elizabeth, N.J., they set up to build aviation engines and a 160-horsepower tractor engine (*see* 1913; 1922; Willys, 1915; Durant, 1921).

General Motors Acceptance Corp. (GMAC) is founded by GM, now 28.7 percent owned by du Pont interests (above). The GM subsidiary will become the world's largest automobile financing company (see 1939).

Albert Einstein's 1905 theory of relativity is confirmed May 29 by English astronomer Arthur Eddington who photographs a solar eclipse on the island of Príncipe off West Africa. Einstein has insisted that his equations be verified by empirical observation and has devised specific tests; e.g., a ray of light just grazing the surface of the sun must be bent by 1.745 arcs of light—twice the gravitational defection estimated by Newtonian theory (*see* 1929).

The Eastman School of Medicine and Dentistry is founded at the University of Rochester by photography pioneer George Eastman whose company supplies the new motion picture industry with its raw film stock (*see* School of Music, below).

UCLA (University of California at Los Angeles) opens (*see* 1912; 1944).

Dial telephones are introduced November 8 at Norfolk, Va., by American Telephone & Telegraph Company which has earlier rejected dial phones but which has been threatened by a telephone operators' strike.

U.S. Navy officials advise General Electric president Owen D. Young, 45, that GE's high-frequency Alexanderson alternator is vital to long-distance wireless communications and must remain in U.S. hands. British Marconi has offered $5 million for rights to the alternator but the Navy urges GE to start its own radio company.

Radio Corp. of America (RCA) is founded by Owen D. Young (above) who loans Ernst Alexanderson to RCA which will employ him as chief engineer for 5 years (*see* 1906). RCA will acquire the Victor Co. and become a radio-phonograph colossus but anti-trust court actions will separate RCA from GE (*see* Victrola, 1906; NBC, 1926).

Haaretz (The Land) begins publication at Tel Aviv under the direction of Palestinian journalists who include Gershon Agronsky.

1919 *(cont.)* The *New York Daily News* begins publication June 26 under the direction of *Chicago Tribune* veteran Joseph Medill Patterson, a Socialist who tells his editors, "Tell it to Sweeney—the Stuyvesants will take care of themselves." Patterson will make the *Daily News* the first successful U.S. tabloid and the most widely read U.S. newspaper of any kind (*see* 1910; 1921).

"Gasoline Alley" by *Chicago Tribune* staff cartoonist Frank King, 36, capitalizes on America's growing fascination with the motorcar. Popularity of the new comic strip will zoom when its main character Walt adopts foundling Skeezix in 1921, and it will be the first strip in which characters grow and age.

True Story magazine is started by New York physical culture enthusiast Bernarr Macfadden, 51, who has been publishing *Physical Culture* magazine since 1898. *True Story* will reach a peak circulation of more than a million as it titillates readers with suggestive morality stories while Macfadden goes on to publish a host of movie, romance, and detective story magazines plus 10 daily newspapers including the *New York Evening Graphic* (*see* Fawcett, 1922).

Captain Billy's Whiz Bang is launched at Minneapolis by former World War army captain Wilford H. Fawcett, who quickly turns his mimeographed off-color joke sheet into a 25¢ pocket-size monthly which his sons haul to local dealers on coaster wagons (*see True Confessions*, 1922).

Nonfiction *Ten Days That Shook the World* by U.S. war correspondent John Reed, 32, who witnessed the 1917 October revolution in Russia. Reed will organize and lead a U.S. Communist Labor party, edit its journal, The *Voice of Labor*, be indicted for sedition, escape to Russia, and die of typhus in October of next year at age 32; *Economic Consequences of the Peace* by John Maynard Keynes (above) who says the Germans will be unable to pay the reparations demanded at Versailles; *Introduction to Mathematical Philosophy* by British philosopher-mathematician Bertrand Arthur William Russell, 47; *Epistle to the Romans (Der Römerbrief)* by Swiss Protestant Reformed theologian Karl Barth, 33, who champions dialectic theology; *The Waning of the Middle Ages* by Dutch historian Johan Huizinga, 47, whose work will appear in English translation in 1924; *The American Language* by *Baltimore Evening Sun* staff writer H. L. (Henry Louis) Mencken, 39, who co-edits *The Smart Set* with drama critic George Jean Nathan. Mencken's scholarly analysis of the differences between English English and American English will be supplemented in 1945 and 1948; *Prejudices* (essays) by H. L. Mencken who will publish five additional books under the same title in the next 8 years.

Fiction *La Symphonie pastorale* by André Gide; *A New Life (Shinsei)* by Shimazaki Toson; *Within a Budding Grove (A l'ombre des jeunes filles en fleurs)* by Marcel Proust; *The Moon and Sixpence* by W. Somerset Maugham; *Jurgen* by Virginia author James Branch Cabell, 40, whose medieval romance is suppressed on charges of obscenity but will win wide acclaim through the 1920s; *Winesburg, Ohio* by Chicago writer Sherwood Anderson, 43, whose vignettes refute sentimental notions of untroubled rural life; *Java Head* by U.S. novelist Joseph Hergesheimer, 39, who makes vivid the glory days of Salem, Mass., in the flush of its China trade prosperity; *My Man Jeeves* by P. G. Wodehouse; *Lad, a Dog* by New Jersey author Albert Payson Terhune, 47.

Poetry *War Poems* by Siegfried Sassoon who won the Military Cross in 1917, threw it into the sea, made a public announcement of his refusal to serve further, but returned to France last year and was wounded a second time; *Marlborough* by the late Charles Sorley; *Boy (Ragazzo)* by Piero Jahier; *Altazor* by Chilean poet Vicente Huidobro, 26; *Poems* by U.S. poet Michael Strange (Blanche Oelrichs), who has divorced Philadelphia millionaire Leonard Thomas and will marry John Barrymore next year.

Painting *The Murder* by Edvard Munch; *La Marchesa Casata and Gypsy Woman with Baby* by Amedeo Modigliani, who is dying of tuberculosis, alcoholism, and drug addiction; *Pierrot* and *Harlequin* by Pablo Picasso; *Follow the Arrow* by Fernand Léger; *Dream Birds* by Paul Klee; *The Return of the Prodigal Son* by Giorgio de Chirico; *Dreamy Improvisation* and *Arabian Cemetery* by Wassily Kandinsky; *Nympheas* by Claude Monet; *Anywhere Out of the World* by Marc Chagall. Pierre Renoir dies at Caques-sur-Mer December 17 at age 78.

Sculpture *Bird in Space* (bronze) by Constantin Brancusi (*see* 1926).

Theater *The Last Days of Mankind (Die letzten Tage der Menschheit)* by Austrian playwright-poet Karl Kraus; *King Nicholas, or Such Is Life (König Nicolo, oder so ist das Leben)* by Frank Wedekind 1/15 at Leipzig's Schauspielhaus; *Up in Mabel's Room* by U.S. playwrights Wilson Collison, 26, and Otto Harbach 1/15 at New York's Eltinge Theater, 229 perfs.; *Augustus Does His Bit* by George Bernard Shaw 3/12 at New York's Guild Theater, with Norma Trevor, 111 perfs.; *Elius Erweckung* by Wedekind 3/16 at Hamburg's Kammerspiele; *Tête d'Or* by Paul Claudel 3/30 at the Théâtre du Gymnase, Paris; *The Fall and Rise of Susan Lenox* by George V. Hobart 6/9 at New York's 44th Street Theater, with Alma Tell; *Transfiguration (Die Wandlung)* by German playwright Ernst Toller, 26, 8/30 at Berlin's Tribune Theater. Toller is sentenced to 5 years in prison for his political activities; *Hercules (Herakles)* by Wedekind 9/1 at Munich's Prinzregenztheater; *Adam and Eva* by Guy Bolton and George Middleton 9/13 at New York's Longacre Theater, 312 perfs.; *The Gold Diggers* by Avery Hopwood 9/30 at New York's Lyceum Theater, with Ina Claire, Ruth Terry, 282 perfs.; *Clarence* by Booth Tarkington 9/20 at New York's Hudson Theater, with Alfred Lunt, 306 perfs.; *Mr. Pim Passes By* by English playwright A. A. (Alan Alexander) Milne, 37, 12/1 at Manchester's Gaiety Theatre. *Punch* contributor Milne wrote two comedies while fighting in France; *Aria Da Capo* by Edna St. Vincent Millay 12/15 at New York's Provincetown Playhouse.

Films Robert Wiene's *The Cabinet of Dr. Caligari* with Werner Krauss, Conrad Veidt. Also: Lambert Hillyer's *Branding Broadway* with William S. Hart; D. W. Griffith's *Broken Blossoms* with Lillian Gish, Richard Barthelmess; Griffith's *Fall of Babylon* with Constance Talmadge, Elmer Layton; Griffith's *Scarlet Days* with Richard Barthelmess, Carol Dempster; Cecil B. DeMille's *Male and Female* with Gloria Swanson, Bebe Daniels, Thomas Meighan; Charles Chaplin's *Sunnyside*; Victor Fleming's *Till the Clouds Roll By* with Douglas Fairbanks, Kathleen Clifford.

United Artists is founded at Hollywood by D. W. Griffith, Charles Chaplin, romantic star Mary Pickford, and matinee idol Douglas Fairbanks who have set up the firm to increase their share of the profits from the films they make. Now 36, Fairbanks will marry Pickford next year and both will continue making films selected, financed, and distributed (but not produced) by United Artists.

Paramount-Famous Players-Lasky Corp. opens a 6-acre Astoria, N.Y., film studio with a stage 225 feet by 126 by 60. Paramount will sell the studio to the U.S. Army in 1942.

Ballet *The Three-Cornered Hat* 7/22 at London's Alhambra Theatre, with Leonide Massine of the Ballet Russe as the Miller, Tamara Karsavina as the Miller's wife, music by Manuel de Falla, choreography by Massine.

Opera *Die Frau ohne Schatten* 10/10 at Vienna's Opernhaus, with music by Richard Strauss, libretto by Hugo von Hoffmansthal. The St. Louis Municipal Opera opens its first season with a performance of the light opera *Robin Hood*.

The Eastman School of Music is founded at the University of Rochester with a gift from photography pioneer George Eastman (*see* medicine above).

Scottish music hall favorite Harry Lauder is knighted at age 48 by George V for his recruitment efforts in the Great War. Sir Harry lost his only son in the war.

Broadway musicals *The Velvet Lady* 2/3 at the New Amsterdam Theater, with Jed Prouty, Eddie Dowling, music by Victor Herbert, lyrics by Henry Blossom, 136 perfs.; *The Whirl of Society* 3/5 at the Winter Garden Theater, with Al Jolson, George White, Blossom Seeley (who runs up and down the aisles via a runway across the pit which also gives audiences a closer look at the chorus girls), music by Louis A. Hirsch, lyrics by Harold Atteridge, 136 perfs.; (George White's) *Scandals* 6/2 at the Liberty Theater, with White, "dimple-kneed" Ann Pennington, music by Richard Whiting, book and lyrics by White, Arthur Jackson, 128 perfs.; *The Ziegfeld Follies* 6/16 at the New Amsterdam Theater, with Marilyn Miller, Eddie Cantor, Eddie Dowling, Van and Schenck, music and lyrics by Irving Berlin, Dave Stamper, Gene Buck, songs that include "A Pretty Girl Is Like a Melody" by Berlin whose song will be the *Follies* theme, 171 perfs.; *The Ziegfeld Midnight Frolic* 10/2 at the New Amsterdam Roof, with Fanny Brice, Ted Lewis, Will Rogers, W. C. Fields, Chic Sale, music and lyrics chiefly by Dave Stamper and Gene Buck, songs that include "Rose of Washington Square" by James F. Hanley, lyrics by Ballard MacDonald; *Apple Blossoms* 10/7 at the Globe Theater, with Fred and Adele Astaire, music by Austrian-American violinist Fritz Kreisler, 44, and U.S. composer Frederick Jacobi, 28, book and lyrics by William Le Baron, songs that include "I'm in Love," "Who Can Tell," 256 perfs.; *The Passing Show* 10/23 at the Winter Garden Theater, with Charles Winninger, James Barton, Reginald Denny, music chiefly by Sigmund Romberg, lyrics chiefly by Harold Atteridge, 280 perfs.; *Buddies* 10/26 at the Selwyn Theater, with Peggy Wood, Roland Young, music and lyrics by B. C. Hilliam, 259 perfs.; *Irene* 11/18 at the Vanderbilt Theater, with Edith Day, music by Harry Tierney, lyrics by Joseph McCarthy, songs that include "In My Sweet Little Alice Blue Gown," 670 perfs.

Popular songs "Swanee" by New York Tin Pan Alley composer George Gershwin, 21, lyrics by Irving Caesar, 24, inspired by the 1851 Stephen Foster song "Old Folks at Home." Al Jolson will record the song early next year; "Dardanella" by Felix Bernard, 22, and Johnny Black, lyrics by Fred Fisher; "How 'Ya Gonna Keep 'Em Down on the Farm? (After They've Seen Pa-ree)" by Walter Donaldson, lyrics by Sam M. Lewis and Joe Young; "Oh, What a Pal Was Mary" by Pete Wendling, lyrics by Edgar Leslie and Bert Kalmar; "Cielito Lindo" ("Beautiful Heaven") by Mexican composer Quirino Mendoza y Cortez; "The World Is Waiting for the Sunrise" by English composer Ernest Seitz, lyrics by Eugene Lockhart; "Let the Rest of the World Go By" by Ernest R. Ball, lyrics by J. Keirn Brennan.

New York's Roseland Ballroom opens December 31 at 1658 Broadway (51st Street) where it will remain (one flight up) until 1956 when it will move to 239 West 52nd Street. Proprietor Louis J. Brecker, 21, opened a Philadelphia Roseland 3 years ago while attending the Wharton School but has moved to New York to escape Philadelphia's blue laws.

Gerald Leighton Patterson, 23, (Australia) wins in men's singles at Wimbledon, Suzanne Lenglen, 20, (Fr) in women's singles; William Johnston wins in U.S. men's singles, Hazel Hotchkiss Wightman in women's singles.

Jack Dempsey wins the world heavyweight boxing championship July 4 by a third round knockout over champion Jess Willard at Toledo, Ohio. William Harrison "Jack" Dempsey, 24, of Mannassa, Colo. (the "Manassa Mauler") will hold the crown until 1926 (*see* 1921).

U.S. racing has its first Triple Crown winner as the chestnut colt Sir Barton wins the Kentucky Derby, Preakness, and Belmont Stakes in a feat that will not be repeated until 1930.

The Cincinnati Reds win the World Series, defeating the Chicago White Sox 5 games to 3 but heavy wagering on the underdog Reds raises suspicions (*see* 1920).

The Green Bay Packers is founded as a professional football team by halfback-coach Curly Lambeau who

1919 ***(cont.)*** will remain head coach of the Wisconsin team until 1949 (*see* NFL, 1920).

The pogo stick patented by U.S. inventor George B. Hansburg, 32, is a bouncing metal stick that thousands will ride in the 1920s and will be the basis of a dance number in *The Ziegfeld Follies.*

Hilton Hotel Corp. has its beginnings at Crisco, Tex., where entrepreneur Conrad Nicholson Hilton, 31, uses his life savings of $5,000 to finance the purchase of the small Mobley Hotel. Hilton was a partner in his father's mercantile interests at San Antonio, N. Mex., before the war, he will soon acquire several other small Texas hotels, and he will build a new hotel at Dallas in 1925 (*see* 1945).

Grand Canyon National Park is established by act of Congress. The 673,575-acre park in Arizona includes the most spectacular part of the Colorado River's great 217-mile canyon (*see* hotels, 1903).

Zion National Park is established by act of Congress. The 147,035-acre park in Utah is rich in colorful canyon and mesa scenery.

Lafayette National Park is established by act of Congress on Maine's 41,634-acre Mount Desert Island (*see* Rockefeller, 1916). First national park east of the Mississippi, it will be renamed Acadia National Park in 1929.

Famine in eastern Europe (below) boosts demand for U.S. wheat. Prices soar to $3.50 per bushel and farmers are encouraged to plant wheat on land never before tilled (*see* 1921; 1934).

Di Giorgio Farms is organized by Italian-American fruit merchant Giuseppe Di Giorgio, 45, who started the Baltimore Fruit Exchange in 1904 and has begun buying acreage in California's San Joaquin Valley.

Belgian food prices drop by 50 percent, British by 25 percent but prices remain high elsewhere in Europe. German ports are blockaded until July 12.

French beef, mutton, pork, and veal prices have increased nearly sixfold since 1914, reports *Le Petit Journal*, and egg, cheese, and butter prices have increased fourfold. France removes restrictions on the sale of beans, peas, rice, eggs, and condensed milk, and then on bread, but while the wartime ban on sugar imports is lifted the import duty on sugar is raised to almost prohibitive levels.

Eastern Europe has famine as a result of poor crops and the shortage of manpower. Vienna cuts bread rations to 4 ounces per week in late December.

U.S. food prices remain far above 1914 levels. Milk is 15¢ qt., up from 9¢; sirloin steak 61¢ lb., up from 32¢, fresh eggs 62¢ doz., up from 34¢.

A team of scientists from Britain's year-old Accessory Food Factors Committee studies malnutrition in famine-stricken Vienna (above). Harriette Chick and her colleagues find scurvy common among infants, note a serious increase in rickets, and prove the success of good nutrition in curing rickets.

Edward Mellanby finds that rickets is not an infection as has been widely believed but rather the result of a dietary vitamin deficiency (*see* 1918; Drummond, McCollum, 1920; McCollum, 1922).

The U.S. Pure Food and Drug Law of 1906 is amended once again to eliminate loopholes (*see* 1912; 1913; but *see* Kallet, Schlink, 1932).

U.S. margarine production reaches a level of 1 pound to every 2.4 pounds of butter, up from 1 to every 5 in 1914 (*see* 1881; 1921).

An eighteenth amendment to the Constitution prohibiting sale of alcoholic beverages anywhere in the United States is proclaimed in January to go into effect January 16, 1920; a War Prohibition Act passed last year goes into effect at the end of June to continue until demobilization.

The Prohibition Enforcement Act (Volstead Act) passed over President Wilson's veto October 28 defines as "intoxicating" any beverage containing 0.5 percent alcohol or more (*see* 1920).

Thirsty Americans flock to Havana where Sloppy Joe's is called the national American saloon.

Compania Ron Bacardi, S.A., is incorporated in Cuba by Facundo and José Bacardi and Henri Schueg, sons and son-in-law of Catalan Facundo Bacardi who started the distillery. They have run the rum company as a partnership for some years following the death of the senior Bacardi.

American Sugar Refining enters production by acquiring Central Cunagua, a Cuban sugar plantation of 100,000 acres with a huge mill (central) a railroad, and a town of 1,000 (*see* 1911; 1920).

Coca-Cola's Asa G. Candler turns 68 and sells out for $25 million to a group headed by Ernest Woodruff (*see* 1916; 1923).

"Sunkist" is burned into the skin of a California orange with a heated fly swatter by California Fruit Growers Exchange executive Don Francisco, 27, who urges that citrus fruit marketed by the Exchange be stamped with the name. It will be the first trademark to identify a fresh fruit commodity (*see* 1895; Chiquita Banana, 1944).

The Fleischmann Co. launches a national advertising campaign to urge housewives to buy bakery bread instead of baking at home. The company now sells most of its yeast to commercial bakers (*see* 1876; Standard Brands, 1929).

Kellogg's All-Bran is introduced by the Battle Creek Toasted Corn Flakes Co. (*see* 1915; 1922).

Frigidaire is selected as the name of a new refrigerator produced by W. C. Durant's General Motors (above) which has paid $56,000 to acquire Guardian Refrigerator Co., a one-man firm that was close to bankruptcy. The name has come out of a contest sponsored by GM, which will make Frigidaire almost a generic term for refrigerator.

U.S. ice cream sales reach 150 million gallons, up from 30 million in 1909.

Massive relocations of ethnic minorities begin in Europe. The Treaty of Neuilly (above) sanctions the exchange of 46,000 Bulgarian Greeks for 120,000 Greek Bulgarians (*see* Lausanne, 1923).

1920 Russia's civil war continues as the Bolsheviks struggle to consolidate control of the country in the face of opposition from the Poles and Letts who capture Dvinsk January 5. The White Russian admiral Aleksandr Kolchak is defeated at Krasnoyarsk and executed by the Red Army February 7 at age 45, one day before the Bolsheviks take Odessa.

Novorossiisk on the Black Sea falls to the Red Army March 28 and the White Russian Army of Anton Denikin collapses. Polish troops under Marshal Jozef Pilsudski overrun the Ukraine in late April, take Kiev May 7, but are driven out June 11 despite support from the Ukrainians.

The Red Army takes Pinsk July 27 and crosses into Poland but Polish forces supported by French general Maxime Weygand defeated the Bolsheviks August 14 to 16 and force them to give up their Polish conquests.

Russia's Polish War ends October 12 in a peace treaty signed at Tartu, freeing the Red Army to push back the White Russian forces of Baron Petr Nikolaevich Wrangel, 42, who has taken over much of southern Russia. Wrangel is forced back to the Crimea November 1, loses Sevastopol November 14, and has to evacuate his troops to Constantinople, ending the Russian counterrevolution.

The Baltic states Estonia, Latvia, and Lithuania created during the Bolshevik Revolution of 1917 gain recognition as independent nations by the Bolshevik government after unsuccessful efforts to make them part of the new U.S.S.R. (Union of Soviet Socialist Republics) (*see* 1940).

Berlin is seized March 13 in a right-wing Putsch. American-born journalist Wolfgang Kapp, 51, receives support from irregular troops under Gen. von Luttwitz, 60, in a move to restore the monarchy. Back from fighting in the Baltic provinces to destroy "Bolshevik republicanism," the disbanded troops are led by the Erhardt Brigade wearing swastikas on their helmets. The legitimate government escapes to the provinces, Kapp is made chancellor and orders a general strike, he gets support from Gen. Ludendorff but fails to gain foreign recognition, the army remains generally uncommitted, the general strike hampers Kapp, the Security Police oppose him, he soon finds he has no authority, and he flees the city March 17 (*see* Hitler, 1923).

French voters have chosen Paul Deschanel in the January 17 presidential election, rejecting Georges Clemenceau because they feel the Treaty of Versailles too lenient; former prime minister Joseph Caillaux draws a 3-year prison sentence April 23 for dealing with the enemy, President Deschanel resigns for reasons of health September 15, and his prime minister Alexandre Millerand assumes the presidency (*see* 1924).

Hungary loses nearly three-quarters of her territory and two-thirds of her population in the Treaty of Trianon signed June 4 by Admiral Nikolaus Horthy (Miklòs de Nagybánya Horthy), 52, who will serve as regent until 1944. Western Hungary is ceded to Austria, Slovakia to Czechoslovakia, Croatia-Slavonia to Yugoslavia, the Bánat of Temesvár in part to Yugoslavia and in part to Romania, which also receives Transylvania and part of the Hungarian plain.

Greek forces advance against Turkish nationalist forces June 22 with British encouragement, defeat the Turks June 24 at Alashehr, take Bursa July 9, and accept the surrender of Adrianople July 25.

Britain accepts mandate over Iraq in Mesopotamia May 5 from the supreme allied council, which excludes the small Persian Gulf shiekdom of Kuwait controlled since 1756 by the al-Sabbah family. A great Arab insurrection against the British begins in July, some garrisons are besieged for weeks, Sir Percy Cox is named high commissioner October 1, and the uprising is suppressed in December.

The Treaty of Sèvres signed by the feeble Ottoman sultanate August 10 obliges Constantinople to renounce all claims to non-Turkish territory. It recognizes the independent kingdom of Hejaz, provides for a Kurdish homeland, makes Syria a French mandate and Mesopotamia and Palestine British mandates, gives Rhodes and the Dodecanese islands to Italy, eastern Thrace, Smyrna, the other Aegean islands, Imbros, and Tendedos to Greece, but Turkish nationalists find the terms unacceptable (*see* Treaty of Lausanne, 1923).

Paris proclaims the creation of Lebanon September 1 with its capital at Beirut.

Greece's King Alexander dies of blood poisoning October 25 at age 27 after a 3-year reign (he has been bitten by a pet monkey). His father Constantine, who abdicated in 1917, resumes the reign he began in 1913 and continues hostilities against Turkey.

British reinforcements have arrived in Ireland May 15 to support His Majesty's forces against attacks by Sinn Fein political militants who continue resistance to British regulars and to the new "Black and Tans"—Royal Irish Constabulary recruits whose khaki tunics and trousers and dark green caps are almost black. Nicknamed after a familiar breed of Irish hound, the Black and Tans have been helping the British suppress Irish nationalists since March and take reprisals for nationalist acts of terrorism.

The Government of Ireland Act passed by Parliament December 23 gives Northern Ireland and Southern Ireland the right to elect separate parliaments of their own with each to retain representatives in the British Parliament at London (*see* 1919; 1921).

Two agitators for Irish independence (above) die in an English prison after fasting for 75 days in a protest demonstration (*see* Gandhi, 1914). Cork Lord Mayor Terence James MacSwiney, 41, collapses on the fifteenth day, physicians are unable to save him, and his death in October produces widespread rioting.

Mexico's president Venustiano Carranza is killed May 21 by an assassin in the employ of the insurgent gen-

1920 ***(cont.)*** erals Adolfo de la Huerta, Alvaro Obregon, and Plutarco Elias Calles. Obregon, 39, has taken Mexico City and is elected president August 31 following the surrender of Pancho Villa, who is given a handsome estate on which he will live until his assassination in 1923.

"America's present need is not heroics but healing; not nostrums but normalcy," says presidential hopeful Warren Gamaliel Harding, 54, of Ohio in a speech at Boston May 10. The handsome, genial, but colorless senator is nominated by the Republicans and opposed by Ohio's Democratic governor James M. Cox, 50, who wins only 34 percent of the popular vote and receives only 127 electoral votes to Harding's 404.

The League of Nations meets for the first time November 15 in its new headquarters at Geneva but its membership includes neither the U.S.S.R. nor the United States. The Senate has finally rejected U.S. membership March 19 in a victory for opponents led by Henry Cabot Lodge of Massachusetts (*see* 1919).

Five men kill a factory guard and paymaster April 15 at South Braintree, Mass., and escape in a stolen motorcar with a steel box containing a payroll of $15,776.51. Factory worker Nicola Sacco, 20, and fish peddler Bartolomeo Vanzetti, 32, are picked up May 5 in a car that contains propaganda leaflets attacking the U.S. government and all other governments, they are arrested for the April 15 murder and robbery, and they are prosecuted as "Reds" by U.S. Attorney General Mitchell Palmer (*see* 1927).

President Wilson rules May 5 that the Communist Labor Party of America is outside the scope of U.S. deportation laws.

The American Civil Liberties Union is founded by social reformers who include Methodist minister Harry F. Ward, 47, Clarence Darrow, Upton Sinclair, Jane Addams, Helen Keller, Roger Williams, 36, former War Labor Policies Board chairman Felix Frankfurter, 37, and socialist Norman Mattoon Thomas, 35. FBI agents soon infiltrate the ACLU (*see* 1919).

Norman Thomas (above) is a former clergyman who will staunchly oppose both communism and fascism but advocate such radical ideas as low-cost public housing, a 5-day work week, minimum wage laws, and the abolition of child labor (Congress will ultimately legislate them all), and he will be a quadrennial Socialist party candidate for president from 1928 through 1948.

World economies struggle to revive in the wake of the Great War. Demobilized soldiers find few job openings.

Britain orders unemployment insurance for all workers except domestic servants and farm workers (*see* allowances for dependent relatives, 1921).

The first Universal Negro Improvement Association international convention opens at Liberty Hall in New York's Harlem under the leadership of Marcus Garvey who has founded UNIA branches in nearly every U.S. city with a sizeable black population (*see* 1916). Some 25,000 delegates from 25 nations attend and Garvey begins to exalt African beauty and promote a "back to Africa" movement with a plan for resettlement in Liberia (*see* 1847; 1924).

Woman suffrage is proclaimed in effect August 26 following Tennessee's ratification of the nineteenth amendment; women voters help elect Harding (above).

The League of Women Voters, founded by suffragist Carrie Chapman Catt, 61, will give impartial, in-depth information on candidates, platforms, and ballot issues.

A bomb explosion September 16 sears the J. P. Morgan bank building, kills 30, injures 200, and causes $2 million in property damage.

British miners strike in October, demanding a 2-shilling-per-hour raise. The lack of coal idles dockers and other workers.

The Amalgamated Clothing Workers of America begins a 6-month strike in December against sweat shops and nonunion shops. Now 100,000 strong, the union workers find themselves locked out by New York, Boston, and Baltimore clothing manufacturers who charge the union leadership with "Sovietism" and demand a return to piecework pay. The manufacturers will lose an estimated $10 million before the strike ends in June of next year, the workers will accept a 15 percent wage cut while promising to increase productivity by 15 percent, but clothing makers will agree to a union shop and a continuation of the 44-hour work week (*see* 1918; 1937).

U.S. wells supply nearly two-thirds of the world's petroleum (*see* Venezuela, 1922).

IF YOU DON'T
COME IN
SUNDAY
DON'T COME
IN MONDAY.
THE
MANAGEMENT

U.S. sweatshop operators resisted efforts by union men to organize garment workers in the big-city loft factories.

Hunt Oil has its beginnings as the collapse of farm prices impels Arkansas-Louisiana planter H. L. Hunt to start dealing in oil leases in the new Arkansas oil fields (*see* agriculture, 1911). Hunt will build up capital, will begin drilling his own wells, and within 4 years will have made his first million dollars (*see* 1936).

London designates Croydon Airport in March as the city's terminal for air traffic to and from the Continent. Regular KLM Royal Dutch Airline flights between Rotterdam and London begin May 17 (*see* 1919).

QANTAS (Queensland and Northern Territories Air Service) begins service in Australia in November.

Regular air service begins between Cuba and Key West, Fla. (*see* Trippe, 1923).

Douglas Aircraft is founded by former Glenn L. Martin aircraft designer Donald Douglas, 28, who has quit to start his own firm to produce the large, safe, relatively slow commercial plane that he has designed but which Martin has refused to produce (*see* 1924).

U.S. motorcar production increases with Ford Model T cars accounting for 54.57 percent of all cars sold.

Toyo Kogyo Co., founded at Akigun in Hiroshima Prefecture, Japan, will produce Mazda trucks and motorcars (*see* Wankel rotary engine, 1957).

Suzuki Motor Co., founded at Hamanagunin, Shizuoka Prefecture, will become a leading producer of light four-wheeled vehicles and motorcycles.

E. I. du Pont de Nemours acquires U.S. rights to the Chardonnet viscose rayon process of 1883 from the French company Comptoir des Textiles Artificiels and sets up the DuPont Fibersilk Co. at Buffalo, N.Y. (*see* 1919; cellophane, 1924).

Union Carbide and Carbon establishes a chemical company that will pioneer development of petrochemicals. It acquires another company to expand its metallurgical capabilities in the field of corrosion- and heat-resistant metal alloys.

The world's first radio broadcasting station goes on the air November 2 to give results of the Harding-Cox election (above). Westinghouse engineer Frank Conrad, 46, has set up KDKA at East Pittsburgh but only about 5,000 Americans have radio receivers, mostly "catswhisker" crystal sets (*see* radio advertising, 1922).

David Sarnoff proposes a plan for making $75 radio music boxes; at least a million could be sold in 3 years, he predicts (*see* 1917; NBC, 1926).

International Telephone and Telegraph (ITT) is founded by Puerto Rican sugar broker Sosthenes Behn, 38, and his brother Hernand to run telephone and telegraph operations in Cuba and Puerto Rico (*see* Tropical Radio Telegraph, 1910). Born in the Virgin Islands of Danish-French parents, the Behn brothers have acquired a small San Juan telephone business to satisfy a bad debt (*see* 1925).

"Winnie Winkle the Breadwinner" by cartoonist Martin Michael Branner, 31, begins in September in the 15-month-old *New York Daily News.*

Congress enacts legislation enabling the Post Office to accept first class mail not stamped with adhesive stamps (*see* 1847).

Pitney-Bowes Postage Meter Co. is founded to produce a meter acceptable to the U.S. Post Office. Arthur H. Pitney, 49, saw his firm's postage stamps being stolen by office workers and developed an early postage meter in 1901; Walter Bowes, 38, is a former Adressograph salesman who has been selling a high-speed postmarking and stamp-canceling machine to the Post Office (which has rejected a postage meter Bowes proposed because it did not completely protect government stamp revenues). Bowes teams up with Pitney, whose self-locking meter does protect stamp revenues, they take over the Universal Stamping Machine plant at Stamford, Conn., it produces meters that rent for $10 per month, and the first metered letter is posted December 10.

Nonfiction *Outline of History* by H. G. Wells who intends it to replace "narrow nationalist history by a general review of the human record." "Human history becomes more and more a race between education and catastrophe," he says; *The Storm of Steel (In Stahlgewittem)* by German author Ernst Jünger, 25.

Fiction *This Side of Paradise* by former New York advertising copywriter F. (Francis) Scott (Key) Fitzgerald, 23; *The Age of Innocence* by Edith Wharton; *Main Street* by U.S. novelist Sinclair Lewis, 35, who satirizes small-town America; *Chéri* by Colette; *Women in Love* and *The Lost Girl* by D. H. Lawrence.

Poetry *A Few Figs from Thistles* by Edna St. Vincent Millay: "My candle burns at both ends;/It will not last the night;/But, ah, my foes, and, oh, my friends—/It gives a lovely light."

New York's Société Anonyme is founded by Marcel Duchamp, Katherine S. Dreier, and Philadelphia-born dadaist photographer-painter-sculptor Man Ray (*né* Emmanuel Radinski), 30, who will open the city's first modern art museum (*see* Phillips Collection, 1921; MOMA, 1929).

Painting *Portrait of Josie West* by Missouri painter Thomas Hart Benton, 31; *Church* by U.S. cubist Lyonel (Charles Adrian) Feininger, 49; *Here Everything is Still Floating* by German dadaist Max Ernst, 29; *Guitar, Book, and Newspaper* by cubist Juan Gris; *The Tug Boat* by Fernand Léger; *L'Odalisque* by Henri Matisse; *Brooklyn Bridge* by Joseph Stella; *Reclining Nude* by Amedeo Modigliani who dies in Paris January 25 at age 35.

Theater *Beyond the Horizon* by Eugene O'Neill 2/2 at New York's Morosco Theater, with Richard Bennett in O'Neill's first full-length play, 111 perfs.; *The Outlaw (Loupeznik)* by Czech journalist-playwright Karel Capék, 29, 3/2 at Prague's National Theater; *The White Savior (Der Weisse Heiland)* by Gerhart Hauptmann 3/28 at Berlin's Grosses Schauspielhaus; *The Skin Game* by John Galsworthy 4/21 at the St. Martin's Theatre, London; *Mary Rose* by James M. Barrie 4/22 at London's Haymarket Theatre; *I'll Leave It to You* by

1920 *(cont.)* English actor-playwright Noël (Pierce) Coward, 20, 7/21 at London's New Theatre, with Coward, Moya Nugent, 37 perfs.; *The Bat* by Mary Roberts Rinehart and Avery Hopwood 8/23 at New York's Morosco Theater, with Effie Ellsier in a dramatization of Rinehart's 1908 novel *The Circular Staircase,* 867 perfs.; *East of Suez* by W. Somerset Maugham 9/2 at His Majesty's Theatre, London; *The First Year* by Frank Craven 10/20 at New York's Little Theater, with Craven, 760 perfs.; *The Emperor Jones* by Eugene O'Neill 11/1 at New York's Provincetown Playhouse, with Charles S. Gilpin as the Pullman porter who becomes dictator of a tropical island and disintegrates morally (the play moves uptown 12/27), 204 perfs.; *Heartbreak House* by George Bernard Shaw 11/10 at New York's Garrick Theater, with Helen Westley, Dudley Digges, 125 perfs.; *Masses and Men (Massen Menschen)* by Ernst Toller 11/15 at Nuremberg's Stadttheater; *The Makropolous Secret (Vec Makropulos)* by Karel Capék 11/21 at Prague's National Theater.

Films D. W. Griffith's *The Idol Dancer* with Richard Barthelmess, Clarine Seymour; Griffith's *The Love Flower* with Barthelmess, Carol Dempster; Fred Niblo's *The Mark of Zorro* with Douglas Fairbanks; George Melford's *The Round Up* with Roscoe "Fatty" Arbuckle; Buster Keaton's *The Saphead* with Keaton.

Ballet *Le Boeuf sur le toit* (*The Nothing-Doing Bar*) 2/21 at the Théâtre des Champs-Elysées, Paris, with famous circus clowns and acrobats, music by Darius Milhaud, scenario by Jean Cocteau; *Pulcinella* 5/15 at the Théâtre National de l'Opéra, Paris, with Leonide Massine, Tamara Karsavina, music by Igor Stravinsky based on scores by Giambattista Pergolesi, choreography by Massine, scenery and costumes by Pablo Picasso.

First performances *Chant de Nigamon* (Symphonic Poem) by Arthur Honegger 1/3 at Paris; *Le Tombeau de Couperin* Suite for Orchestra by Maurice Ravel 2/28 at Paris; *La Valse,* a choreographic poem by Maurice Ravel 12/12 at Paris.

New York's Juilliard Foundation is established with a legacy left by the late French-American textile merchant Augustus D. Juilliard who in 1874 started a New York firm that earned him a fortune. Juilliard died last year at age 83 and stipulated in his will that income from the legacy be used to further music in America. The Juilliard Graduate School will be founded in 1924, the foundation trustees 2 years later will take over the Institute of Musical Art founded in 1905, and in 1946 the two schools will be combined under the name Juilliard School of Music.

Broadway musicals (George White's) *Scandals* 6/7 at the Globe Theater with Ann Pennington, comic Lou Holtz, music by George Gershwin, lyrics by Arthur Jackson, 134 perfs.; *The Ziegfeld Follies* 6/22 at the New Amsterdam Theater with Fanny Brice, W. C. Fields, Moran and Mack, Van and Schenck, music and lyrics by Irving Berlin, Victor Herbert, Joseph McCarthy, Harry Tierney, Gene Buck, and others, 123 perfs.; *Sally* 12/21 at the New Amsterdam Theater with Marilyn Miller, Leon Errol, music by Jerome Kern, lyrics by B. G. (George Gard) DeSylva, 25, songs that include "Look for the Silver Lining," 570 perfs.; *The Passing Show* 12/29 at the Winter Garden Theater, with Eugene and Willie Howard, Marie Dressler, Janet Adair, music by Jean Schwartz, lyrics by Harold Atteridge, songs that include "In Little Old New York," 200 perfs.

Popular songs "Avalon" by Al Jolson and Vincent Rose who have lifted the melody from the aria "E Lucevan Le Stelle" in Giacomo Puccini's 1900 opera *Tosca*; "I'll Be with You in Apple Blossom Time" by Albert von Tilzer, lyrics by Neville Fleeson; "The Japanese Sandman" by Richard Nutting, lyrics by Raymond Egan; "Margie" by U.S. songwriter Con Conrad (Conrad K. Dober), 28, lyrics by J. Russel Robinson, 28, and Benny Davis, 27; "Whispering" by U.S. composer John Schonberger, 28, lyrics by Richard Coburn (Frank D. de Long), 34; "When My Baby Smiles at Me" by U.S. composers Bill Munro and Harry von Tilzer, lyrics by Andrew B. Sterling and comedian Ted Lewis (Theodore Leopold Friedman), 28, who introduced the song 2 years ago at Rector's Café in New York: "Mon Homme" by French composer Maurice Yvain, 29, lyrics by Jacques-Charles.

Babe Ruth signs with the New York Yankees January 3 to begin a 14-year career as "Sultan of Swat" for New York. Yankee owner Jacob Ruppert has acquired Boston Red Sox pitcher George Herman Ruth, 24, for $125,000 plus promises of a large loan, his new left-hander sets a slugging average record of .847 in his first season with New York, hitting 54 home runs, batting .376, scoring 158 runs, and making the Yankees the first team in any sport to draw more than a million spectators, nearly double the team's 1919 gate (*see* 1923).

The Cleveland Indians win the World Series by defeating the Brooklyn Dodgers 5 games to 2 in the last Series that will require 5 victories.

The "Black Sox" scandal threatens to undermine the prestige and popularity of America's national pastime. Eight members of last year's Chicago White Sox baseball team are indicted in September for fraud in connection with last year's 5-to-3 World Series loss to Cincinnati (*see* 1921).

Spanish bullfighter José Gomez y Ortega, 25, is fatally gored May 16 at Talavera de la Reina to end the career of "Joselita" (or "Gallita"), who has killed more than 1,565 bulls in the arena.

Man o' War does not run in the Kentucky Derby but wins both the Preakness and the Belmont Stakes to close out a racing career in which he has been the odds-on favorite in every race he entered. The big chestnut colt is retired to stud at year's end, having won 20 of his 21 races.

William Tatem "Bill" Tilden, Jr., 27, (U.S.) wins in men's singles at Wimbledon (the first American to do so), Suzanne Lenglen in women's singles; Tilden wins in U.S. men's singles, Molla Burjstedt Mallory in women's singles.

British golf champion Joyce Wethered, 19, wins the English Ladies championship to begin a decade of triumph in which she will be virtually unbeatable.

The Olympic Games at Antwerp attract 2,741 athletes from 26 nations. U.S. athletes excel.

Sir Thomas Lipton's *Shamrock* IV is defeated 3 to 2 by the U.S. defender *Resolute* in its challenge for the America's Cup.

The National Football League (NFL) is organized as the American Professional Football Association by men who include Joe Carr who founded the Columbus Panhandles in 1904, and 1918 University of Illinois star end George Stanley Halas, 25 (*see* Green Bay Packers, 1919). Halas will move the team to Chicago in 1922, rename it the Bears, make it commercially viable in 1925 by hiring "Red" Grange, and revolutionize football in the 1930s by perfecting the T formation that will make the game a contest of speed and deception with increased use of the forward pass.

Antoine de Paris creates the mannish shingle bob haircut that will replace the conventional bob introduced by dancer Irene Castle in 1914. Polish-born hairdresser Antek Cierplikowski, 35, calls himself Monsieur Antoine.

The first Miss America beauty queen is crowned at Atlantic City, N.J., to begin a lasting tradition.

The "Ponzi scheme" bilks thousands of Bostonians out of their savings. Italian-American "financial wizard" Charles Ponzi, 38, offers a 50 percent return on investment in 45 days, 100 percent in 90 days, and by late July his Securities and Exchange Co. at 27 School Street is taking in hundreds of thousands of dollars per day. Ponzi says he has found that a 1¢ reply coupon issued in Spain as a convenience by international postal agreement is exchangeable at any U.S. post office for a 6¢ stamp and that he has agents buying up millions of coupons throughout Europe and is making a 5¢ profit on each. The idea is ridiculous, says Boston financier Clarence Walker Barron, 65, who publishes a financial daily that will later become *Barron's Weekly*.

Ponzi (above) replies to critics that he has invented his postal coupon story to keep Wall Street speculators from divining his real secret, but by late afternoon of July 26 the Suffolk County district attorney announces that he has agreed to accept no more funds until his books can be audited.

Ponzi continues to reimburse investors at 50 to 100 percent interest, his supporters claim he is being attacked by "unscrupulous bankers," but it soon develops that he has been convicted of swindling in Montreal and that the "wizard" with the mansion in Belmont and the chauffeur-driven Locomobile has merely been paying out part of what he has been taking in. At least $8 million of the $15 million he has accepted will never be accounted for, his noteholders will receive 12¢ on the dollar, six banks will fail as a result of the Ponzi scheme, and Ponzi will serve 3.5 years in Plymouth County jail, escaping a further 7- to 9-year term by jumping bail of $14,000 in 1925.

The New York State Legislature enacts a real estate tax abatement law that will stimulate construction of more than $1 billion worth of housing in New York City alone. Gov. Alfred E. Smith, 46, has warned that the state has a serious housing shortage (as does Britain, most of Europe, and much of America).

Barely 20 percent of America's virgin forest lands remain uncut.

Overfishing of the Sacramento River forces the closing of San Francisco's last salmon cannery (*see* 1866).

Russia suffers a disastrous drought. V. I. Lenin appeals to the U.S. Communist party for a report on American agricultural methods (*see* famine, 1921).

The American Farm Bureau Federation is founded. It will become the driving force behind mobilizing U.S. farmers' political efforts.

The U.S. soybean harvest reaches 1 million bushels, a figure that will increase more than 600 times in the next half-century (*see* 1924).

The boysenberry, developed by U.S. breeder Rudolph Boysen, is a cross of blackberries, raspberries, and loganberries. Boysen does not capitalize on the new variety (*see* Knott, 1933).

Prohibition (below) will force California vineyard owners to diversify their production, to market as table grapes much of the fruit that has gone into wines and brandies, and to improve their methods of producing raisins, which will soon be marketed under the Sunmaid label.

"Vitamines" become vitamins through efforts by British biochemist J. C. Drummond who observes that not all coenzymes are amines (*see* Funk, 1911). Drummond labels the fat-soluble vitamin "A," the water-soluble vitamin "B," and the antiscorbutic vitamin "C," and with help from O. Rosenheim finds that the human liver can produce vitamin A from the provitamin carotene widely available in fruits and vegetables.

E. V. McCollum discovers a substance in cod-liver oil that can cure rickets and xerophthalmia. He has been at Johns Hopkins since 1917 (*see* 1917; Mellanby, 1918). Xerophthalmia is an abnormal dryness of the eye membranes and cornea that can lead to blindness (*see* 1921; vitamin D, 1922).

Botulism from commercially canned food strikes 36 Americans, 23 die, and the U.S. canning industry is motivated by failing sales to impose new production safety standards (*see* 1895).

National Prohibition of sales of alcoholic beverages in the United States goes into effect January 16. A mock funeral for "John Barleycorn" is held January 15 at Norfolk, Va., by evangelist William Ashley "Billy" Sunday, 57, who has agitated for Prohibition but whose popularity now begins to fade.

Prohibition booms sales of coffee, soft drinks, and ice cream sodas, but consumption of alcoholic beverages will continue through illegal sales and homemade "bathtub gin."

1920 ***(cont.)*** The world sugar price drops from 30¢ per pound in August to 8¢ in December.

M. S. Hershey loses $2.5 million in the collapse of world sugar prices and other large sugar consumers also take heavy losses.

Pepsi-Cola's Caleb Bradham has bought sugar at 22¢ per pound, he loses $150,000, and Pepsi-Cola heads toward bankruptcy (*see* 1907; Guth, 1933).

Chero-Cola sales top $4 million, up more than 100 percent over 1918 sales, but the company ends the year with debts of over $1 million that will not be entirely paid off until 1926 (*see* 1918; 1928).

Howdy is introduced by St. Louis soft drink maker Charles L. Rigg with backing from local coal merchant Edmund Ridgway. The new drink will have some success until citrus growers push through laws in some legislatures requiring that orange-flavored drinks contain real orange pulp and orange juice (*see* 7-Up, 1929).

American Sugar Refining sells the sugar from its newly acquired Central Cunagua at an average price of 23.5¢ for the year. The company nets $5.4 million, reinvests it in 200,000 additional acres of sugar fields, buys two small islands with a combined area of 350 square miles, and builds a new *central* (sugar mill).

A new Life Savers plant opens at Port Chester, N.Y., and Edward J. Noble's Mint Products, Inc., is renamed Life Savers, Inc. (*see* 1913). Noble has increased sales by packaging his candy in resealable foil to preserve their flavor, placing his nickel Life Savers next to the cash register at cigar stores and restaurants, and having the cashiers include nickels in every customer's change (*see* American Broadcasting, 1943).

The Good Humor, created by Youngstown, Ohio, confectioner Harry Burt, is a chocolate-covered ice cream bar on a stick. Burt has read about the "I-Scream-Bar" patented last year by Christian Nelson and will obtain his own patent in 1924 (*see* 1926).

Baby Ruth is introduced by Chicago's Curtiss Candy Co., founded 4 years ago in a back room over a North Halstead Street plumbing shop by candy maker Otto Young Schnering, now 28. Schnering has used his mother's maiden name for the company and names his 5¢ chocolate-covered candy roll of fudge, caramel, and peanuts after the daughter of the late President Cleveland (*see* 1923).

U.S. flour consumption drops to 179 pounds per capita, down from 224 pounds in 1900, as Americans eat more meat, poultry, fish, and vegetables while reducing consumption of baked goods and pasta.

Donut Corp. of America is founded by New York baker Adolph Levitt, 37, who has seen Red Cross workers dispensing doughnuts during the war and who sees an opportunity for a company to develop an automatic doughnut-making machine whose buyers will provide a market for a standardized mix.

U.S. food prices will fall 72 percent in the next 2 years as farm prices plummet.

The U.S. population reaches 105.7 million. Urban residents (54 million) for the first time exceed rural residents (51.5 million) but one in every three Americans still lives on a farm, a proportion that will drop in the next 50 years to one in 22.

The world population reaches 1.86 billion.

Soviet Russia legalizes abortion by a decree of the Lenin government at Moscow, but while Russian physicians are not permitted to refuse an abortion where a woman is no more than 2.5 months pregnant they are ordered to discourage patients from having abortions, especially in first pregnancies (*see* 1936).

France makes abortion illegal to compensate for the population loss experienced in the Great War. Severe fines and prison sentences are ordered for anyone who administers or receives an abortion, but the law will be widely flouted. Within 50 years the number of illegal abortions will climb to an estimated 500,000 per year with bungled abortions causing an estimated 500 deaths per year.

Youngs Rubber Co. is founded by U.S. entrepreneur Merle Leland Youngs, 33, to make Trojan brand condoms that will compete with Julius Schmid's Shiek and Ramses brands (*see* 1883). Trojans will become the largest-selling condoms sold through U.S. drugstores (*see* 1923).

1921 Persia has a bloodless coup February 20 as former Cossack trooper Reza Khan Pahlevi, 44, expels all Russian officers from the country and starts a new regime. V. I. Lenin gives up special rights in Persia, frees her of obligations to Russia, and gives Persia joint command of the Caspian Sea.

A Russo-Turkish treaty signed March 16 recognizes Turkish possession of Kars and Ardaha in return for Turkey's retrocession of Batum. The treaty follows by 3 days an Italian agreement to evacuate Anatolia in return for economic concessions, but hostilities continue between Greece and Turkey. The Turks prevent the Greeks from reaching Angora only by desperate efforts.

An Iraqi plebiscite conducted by the British high commissioner Sir Percy Cox shows 96 percent approval of installing Syria's Faisal I as king. Faisal arrives at Basra June 23, is placed on the throne of Iraq 2 months later, and will reign until his death in 1933, establishing the Hashimite dynasty that will continue until 1958.

The Battle of Anual in Morocco July 21 brings defeat to a Spanish army at the hands of Rifs under the command of Abd-el-Krim, 36. An anarchist has assassinated the Spanish prime minister Eduardo Dato Iradier March 8, Gen. Fernandez Silvestre commits suicide when the Rifs kill 12,000 of his 20,000 men, and news of the disaster at Anual produces a political crisis at Madrid (*see* 1923).

Portugal has a revolution following the assassination in October of a founder of the 11-year-old republic.

Yugoslavia's Peter I has died August 16 at age 77 after less than 3 years as monarch of the new nation. He is succeeded by his son of 33 who will reign until 1934 as Aleksandr I (*see* 1929).

Japan's prime minister Takashi Hara is assassinated by a fanatic November 4 at age 67.

Billy Mitchell sinks a former German battleship July 21 to prove his contention that a strategic air force makes large navies obsolete. Brig. Gen. William Mitchell, 41, led a mass bombing attack of nearly 1,500 planes in 1918, his Martin M-2 bombers are twin-engined biplanes whose 418-horsepower Liberty engines give them a top speed of 98 miles per hour, and the 22,800-ton *Ostfriesland* is the first capital warship to be sunk by bombs from U.S. planes.

The U.S.S. *Jupiter* converted from an 11,050-ton collier is the first U.S. aircraft carrier. She has a stem-to-stern flight deck 534 feet (163 meters) in length and will be commissioned for fleet service in September of next year as the U.S.S. *Langley*.

The Washington Conference on Limitation of Naval Armaments that opens November 12 hears U.S. Secretary of State Charles Evans Hughes say, "Competition in armament must stop." Hughes announces that the United States is prepared to scrap immediately 15 old battleships and to cancel construction of nine new battleships and six new cruisers, some of them 85 to 90 percent complete. Naval power is generally believed to have given the Allies the margin of victory in the Great War and the announcement that Washington is prepared to scrap two-thirds of the U.S. battleship fleet stuns the conferees, many of whom believe that America would be committing naval suicide (*see* 1922).

Southern Ireland gains Dominion status December 6 in a treaty signed with Britain. The Catholic Irish Free State (Eire) embraces 26 of Ireland's 32 counties but the six counties in Protestant Northern Ireland remain part of the United Kingdom and conflict between the two will continue (*see* 1920; IRA, 1922).

Russian sailors mutiny at Kronstadt beginning February 23 as the nation's economy collapses (below). Bolshevik authorities put down the mutiny by March 17 but only with considerable bloodshed.

The Holmogor concentration camp, opened by Bolshevik authorities at Archangel, is the world's first such camp since the Boer War of 1899–1902. Moscow will establish many camps and at least 10 million prisoners will die in them in the next 32 years.

British coal miners walk out of the pits March 31.

U.S. workers strike as employers cut wages by 10 to 25 percent. United States Steel Corporation cuts wages three times during the year and the Railroad Labor Board cuts wages 12 percent July 1.

The AF of L elects Samuel Gompers president for the 40th time at age 71.

The Supreme Court calls an Arizona picketing law unconstitutional. The decision comes December 19 in the case of *Truax v. Corrigan*.

A New Economic Policy (NEP) sponsored by V. I. Lenin is announced by Russia's Communist Party as industrial and agricultural production fall off and shortages become acute in food, fuel, and transportation.

The NEP (above) abolishes food levies to placate Russia's peasants and a limited grain tax is substituted. Some freedom of trade is restored to enable the peasants to dispose of their surpluses and some private commercial enterprises are permitted in the cities (*see* Armand Hammer, below).

Secretary of State Hughes rejects a Russian plea for resumption of trade relations March 25. There will be no such trade so long as communism prevails, Hughes declares.

The United States has nearly 20,000 business failures and by September nearly 3.5 million Americans are out of work. President Harding calls an Unemployment Conference under the chairmanship of Commerce Secretary Herbert Hoover.

British unemployment reaches 2.5 million in July but then begins a gradual drop to 1.2 million where it will remain until 1930. Seven million Britons take wage cuts; the unemployed receive allowances for dependent relatives (*see* 1920).

The British Medical Association estimates that a family of five needs 22s 6½d per week for food to maintain proper health, but as unemployment reaches its peak in July the dole is 29s 3d per week and rents in even the worst slum tenements average 6s per week.

The U.S. General Accounting Office created by act of Congress is the first federally funded arm of Congress since the Library of Congress was established in 1800. The G.A.O. will be a congressional watchdog on federal spending.

V. I. Lenin rallied Russians against evil Western "imperialists," but he welcomed Western capital investment.

1921 ***(cont.)*** The New York Curb Exchange moves into a new building of its own on Trinity Place following decades in which members have transacted business by signaling from curbstone street positions to clerks located in windows of surrounding buildings (*see* American Stock Exchange, 1953).

Corning Glass Works improves light-bulb production efficiency with a new ribbon machine that turns out between 21,000 and 31,000 bulbs per hour (*see* 1903; Langmuir, 1912).

Congress grants a depletion allowance for petroleum-producing companies to encourage exploration of potential new reserves as America increases her consumption of gasoline. The tax incentive will stand at 27.5 percent until it is cut to 22 percent in 1969.

The Teapot Dome scandal that will help tarnish the Harding administration has its beginnings as Navy Secretary Edwin Denby transfers control of naval oil reserves to the Department of the Interior whose secretary Albert Fall secretly leases Teapot Dome to private oil operators Harry Sinclair and Edward Doheny (*see* 1914). Sinclair and Doheny will lose their leases after a congressional investigation in 1923 but they will be acquitted in 1928 of charges that they bribed Secretary Fall.

U.S. chemist Thomas Midgley, 32, discovers the antiknock properties of tetraethyl lead as a gasoline additive; General Motors helps him establish the Ethyl Corp. (*see* 1923).

Berlin's Avus Autobahn opens to traffic September 10. Originally planned in 1909 and nearly complete when the Great War stopped construction, the Autobahn is the world's first highway designed exclusively for motor traffic and with controlled access, it runs 6.25 miles from the Grunewald to the suburb of Wannsee, a 26-foot grass mall separates its two 26-foot wide carriageways, and it is spanned by 10 ferroconcrete bridges.

A Federal Highway Act passed by Congress November 9 begins to coordinate state highways and to standardize U.S. road-building practice (*see* 1944; Route 66, 1932).

The United States has 387,000 miles of surfaced road by year's end, up from 190,476 in 1909 (*see* 1930).

More than a million trucks travel on U.S. roads and highways, up from 100,000 in 1909.

Generators become standard equipment on Model T Fords which account for 61.34 percent of all U.S. motorcars sold.

Durant Motors, Inc., is founded by W. C. Durant who has again been forced out of General Motors but has raised $7 million to start a new venture (*see* 1916). Durant acquires the Bridgeport, Conn., plant of the bankrupt Locomobile Co. to build a Durant luxury car that will supplement the $850 Durant Four and the medium-priced Durant Six made at Muncie, Ind., and he pays $5.25 million to acquire the Willys plant at Elizabethport, N.J., which is the world's most modern auto factory (*see* Duesenberg, 1919). Willys-Overland has gone bankrupt and its receivers sell Durant the Willys design for a new medium-priced Flint motorcar that Durant intends to market (*see* Star, 1922; Willys-Overland, 1936).

GM's share of the U.S. motorcar market reaches 12 percent; the company begins a rapid expansion under the leadership of Alfred Sloan.

A *New York Times* editorial explains that Robert H. Goddard's rockets cannot possibly work because there is nothing in outer space for the rockets' exhausts to push against (*see* 1919; 1926).

English pilot Harry G. Hawker dies July 12 at age 29 in a crash of the Nieuport Goshawk he is testing for T. O. M. Sopwith's Hawker Engineering Co., Ltd., which Sopwith renamed in Hawker's honor last year (*see* 1912; 1919).

The first transcontinental U.S. day-night flight encourages Congress to support airmail service. Mail pilot Jack Knight lands at Omaha February 22 after a 248-mile flight from North Platte, Neb., learns that no pilot is available for the next leg, flies on through heavy snow to Iowa City and refuels, and proceeds to Chicago, covering 672 miles in 10 hours; two other pilots take the plane through to New York, setting a transcontinental flight record of 33 hours, 20 minutes (*see* Doolittle, 1922).

Heart disease becomes the leading cause of death in America after 10 years of jockeying for the lead with tuberculosis. Coronary disease accounts for 14 percent of U.S. deaths and the figure will increase to 39 percent in the next 50 years.

The Council of the American Medical Association refuses to endorse an A.M.A. resolution proposed in 1917 that would oppose use of alcohol as a beverage and would discourage its use as a therapeutic agent. More than 15,000 physicians and 57,000 druggists and drug manufacturers applied for licenses to prescribe and sell liquor in the first 6 months after passage of the Volstead Act in 1919 and by 1928 physicians will be making an estimated $40 million per year by writing prescriptions for whisky.

Poliomyelitis strikes former U.S. Secretary of the Navy Franklin Delano Roosevelt, 39, while he vacations at Campobello Island off Nova Scotia. Popularly called infantile paralysis, the disease is widespread in North America; it will leave Roosevelt crippled for life (*see* 1916; Georgia Warm Springs Foundation, 1927).

Columbia University medical school graduate Armand Hammer, 22, goes to Russia to help the Lenin government cope with its postwar disease plagues and collect $150,000 owed to his father's company for drugs shipped during the Allied blockade of Russian ports. Dr. Hammer, whose Russian-American father is in Sing Sing prison on an abortion conviction, travels to the Urals, barters a million tons of U.S. wheat for a fortune in furs, caviar, and precious stones, and meets late in August with V.I. Lenin who persuades him to take a concession to operate an asbestos mine in the Urals. The mine will not be successful but Hammer and his brother Victor, 21, will open an export-import business at Moscow that will serve as agents for Ford Motor, Parker Pen, U.S. Rubber, Underwood Typewriter, and nearly three dozen other foreign companies.

Psychodiagnostics by Swiss psychiatrist Hermann Rorschach, 36, introduces the Rorschach Test, based on subjects' reactions to inkblots, to probe the unconscious.

Band-Aid brand adhesive bandages are introduced by Johnson & Johnson of New Brunswick, N.J., which will diversify its product line in this decade to offer items such as Johnson's Baby Cream.

The Oxford Group is founded at Oxford University by German-American evangelist Frank Nathan Daniel Buchman, 43, who claims to have had a vision of the Cross in the English lake district in 1908. Buchman has served as a missionary for the YMCA in Japan, Korea, and India, he works to "change" lives through his First Century Christian Fellowship, and he will organize groups in the Scandinavian countries, Canada, the United States, South America, and Latin America (*see* Moral Rearmament, 1939).

Chicago's Field Museum of Natural History founded in 1893 opens in a new building overlooking Grant Park.

Two-way weekly airmail service between Cairo and Baghdad begins in June. The Royal Air Force reduces the time it takes a letter from London to reach Baghdad to between 5 and 10 days, down from 28 days before the airplane.

The British Broadcasting Company (BBC) has its beginnings (*see* 1922).

Pittsburgh's radio station KDKA is founded (*see* 1920; 1922).

Cincinnati's WLW radio station is started by local auto parts producer Powel Crosley, 35, who moves his ham station 8CR transmitter from his house to his factory and has it licensed under the call letters WLW. Crosley will organize Crosley Radio Corp. in 1923 (*see* first night baseball game, 1935).

The *Sunday News* is launched by New York's 2-year-old *Daily News*. By 1925 it will be outselling all other Sunday papers with help from its comic strips.

"Tillie the Toiler" by cartoonist Russ Westover challenges the 2-year-old comic strip "Winnie Winkle."

"Smitty" by cartoonist Walter Bendt appears for the first time November 29 in the *Chicago Tribune*.

The Parker Duofold fountain pen ("Big Red") is introduced by the Parker Pen Co. of Janesville, Wis., at $7 but is considered by many to be too large for comfortable writing.

Nonfiction *Tractatus Logico-Philosophicus* by Viennese philosopher Ludwig Wittgenstein, 31, of Cambridge University. A brother of the pianist, he wrote the work while on active service with the Austrian Army in 1918; *The Story of Mankind* by Dutch-American historian Henrik Willem Van Loon, 39.

Fiction *Chrome Yellow* by English novelist-critic Aldous Huxley, 27; *Héloïse and Abelard* by George Moore; *Memoirs of a Midget* by Walter De La Mare; *Scaramouche* by Rafael Sabatini; *Alice Adams* by Booth Tarkington; *Three Soldiers* by U.S. novelist John Dos Passos, 25; *Mavis of Green Hill* by U.S. novelist Faith Baldwin, 28, whose books will be serialized in magazines for women and be second in popularity only to those of Kathleen Norris (*see* 1911); *The Mysterious Affair at Styles* by English mystery writer Agatha (Mary Clarissa Miller) Christie, 30, who introduces the detective Hercule Poirot, late of the Belgian Sûreté.

Poetry *Il Canzoniere* by Italian poet Umberto Sàba (Umberto Poli), 38; *Second April* by Edna St. Vincent Millay; *English Poems* by Portuguese poet Fernando Pessoa, 33, who was brought up in South Africa; *The Pillar of Flame (Ognenny stolp)* by Russian poet Nikolai Gumilyov; *Anno Domini* by Russian poet Anna Akhmatova (Anna Gorenko), 32, who has defied her father's objections to her career.

The Phillips Collection opens at Washington, D.C., the first U.S. modern art museum. Collector Duncan Phillips, 35, will be director until his death in 1966.

Painting *Gray Day* by German painter George Grosz, 28, is a savage portrayal of German society; *The Elephant of the Célèbes* by Max Ernst; *Odalisque in Red Trousers* by Henri Matisse; *Still Life With Guitar* by Georges Braque; *Three Women* by Fernand Léger; *Three Musicians* by Pablo Picasso; *Composition with Red, Yellow and Blue* by Piet Mondrian; *The Fish* by Paul Klee; *The Kiss* by Edvard Munch; *Bull Durham* by U.S. cubist Stuart Davis, 26.

Theater *Diff'rent* by Eugene O'Neill 1/4 at New York's Provincetown Playhouse, 100 perfs.; *The Green Goddess* by Scottish playwright William Archer, 65, 1/18 at New York's Booth Theater with George Arliss, 440 perfs.; *R.U.R. (Rossum's Universal Robots)* by Karel Capék 1/25 at Prague's National Theater; *A Punch for Judy* by U.S. playwright Philip James (Quinn) Barry, 24, 4/18 at New York's Morosco Theater; *Six Characters in Search of an Author (Sei Personaggi in Cerca d'autore)* by Luigi Pirandello 5/10 at Rome's Teatro Valle; *If* by Lord Dunsany 5/30 at London's Ambassadors Theatre; *Dulcy* by U.S. playwrights George Simon Kaufman, 31, and Marc (Marcus Cook) Connelly, 30, 8/13 at New York's Frayzee Theater, 246 perfs.; *Getting Gertie's Garter* by Wilson Collison and Avery Hopwood 8/21 at New York's Republic Theater, with Hazel Dawn, 120 perfs.; *A Bill of Divorcement* by English playwright Clemence Dane (Winifred Ashton), 33, 10/10 at New York's George M. Cohan Theater, with Katherine Cornell, Charles Waldron, 173 perfs.; *Anna Christie* by Eugene O'Neill 11/2 at New York's Vanderbilt Theater, with Pauline Lord as Anna Christopherson in the story of a prostitute's fight for redemption, 177 perfs.

"Weary Willy" makes his bow at Ringling Bros. Barnum & Bailey's circus. Clown Emmet Kelly, 22, will play the role of sad-faced character for 50 years.

1921 ***(cont.)*** Films Fritz Lang's *Destiny* and *Spies*; D. W. Griffith's *Dream Street* with Carol Dempster, Ralph Graves; Griffith's *Way Down East* with Lillian Gish, Lowell Sherman, Richard Barthelmess; Fred Niblo's *The Four Horsemen of the Apocalypse* and *The Sheik*, both with Italian-American vaudeville dancer Rudolph Valentino (*né* Rodolpho d'Antonguolla), 26; Charlie Chaplin's *The Kid* with Chaplin and Jackie Coogan; Alfred E. Green and Jack Pickford's *Little Lord Fauntleroy* with Mary Pickford; Sam Wood's *Peck's Bad Boy* with Jackie Coogan; James Cruze's *Leap Year* with Fatty Arbuckle, whose arrest after the death of a young woman at a Labor Day drinking party in San Francisco's St. Francis Hotel will ruin Arbuckle's career and suggests wide use of drugs in the film capital.

New York musicians organize Local 802 of the American Federation of Musicians. The powerful local will negotiate contracts with the Metropolitan Opera, the League of New York Theaters, and recording companies.

Isadora Duncan opens a Moscow school for the dance at the invitation of the Soviet government (*see* 1905). Now 43, Duncan meets and marries Russian poet Sergei Aleksandrovich Esenin, 26, who 2 years ago founded the imagist group and is known as the "poet laureate of the Revolution." He and Duncan, neither speaking the other's language, will tour the United States; many will call them Bolshevist spies (*see* 1927).

First performances *Pastorale d'Eté* by Arthur Honegger 2/17 at Paris; *Le Roi David* Symphonic Psalm for Chorus and Orchestra by Arthur Honegger 6/11 at the Jorot Theater in Mezieres outside Lausanne; *The Lark Ascending* Romance for Violin and Orchestra by Ralph Vaughan Williams 6/14 at London; Concerto No. 3 for Piano and Orchestra by Sergei Rachmaninoff 12/16 at Chicago.

Ballet *The Buffoon (Le chout)* 5/17 at the Théâtre Gaité-Lyrique, Paris, with music by Sergei Prokofiev; *The Sleeping Princess* 11/2 at London's Alhambra Theatre, with Bronislava Nijinska, music by Igor Stravinsky, choreography by Marius Petipa.

Opera *Mörder, Hoffnung der Frauen, and Das Nusch-Nuschi* 6/4 at Stuttgart, with music by German composer Paul Hindemith, 25; *Kát'a Kabanová* 10/23 at Brno, with music by Leos Janacek, libretto from an 1859 Ostrovsky play; *The Love for Three Oranges* 12/30 at Chicago, with music by Sergei Prokofiev.

Stage musicals *Shuffle Along* 5/23 at New York's 63rd Street Music Hall, with teenager Josephine Baker in an all-black revue, songs that include "Love Will Find a Way" and "I'm Just Wild About Harry" by Eubie (James Hubert) Blake, 38, and Noble Sissle, 32, 504 perfs.; *Two Little Girls in Blue* 5/31 at New York's George M. Cohan Theater, with music by Vincent Youmans, 23, and P. Lannin, lyrics by Arthur Francis (Ira Gershwin), songs that include "Oh Me! Oh My!" 134 perfs.; *The Ziegfeld Follies* 6/21 at New York's Globe Theater, with Fanny Brice singing the torch song "My Man" published at Paris last year under the title "Mon Homme" and made poignant by the fact that Brice is married to gambler Nick Arnstein (English lyrics by Alfred Willametz); "Second Hand Rose" by James F. Hanley, lyrics by Grant Clark, is another *Follies* hit, 119 perfs.; *The Co-Optimists* 6/27 at London's Royalty Theatre, with music and lyrics by Melville Gideon, songs that include "When the Sun Goes Down," "Tampa Bay," 500 perfs.; (George White's) *Scandals* 7/11 at New York's Liberty Theater, with White, Ann Pennington, Lou Holtz, music by George Gershwin, lyrics by Arthur Jackson, 97 perfs.; *Tangerine* 8/9 at New York's Casino Theater, with book by Philip Bartholomae and Guy Bolton, music by Carlo Sanders, lyrics by Howard Johnson, 337 perfs.; *The Greenwich Village Follies* 8/31 at New York's Shubert Theater, with Ted Lewis, songs that include "Three O'Clock in the Morning" by Spanish composer Julian Robledo, 34, lyrics by New York lyricist Dolly Morse (Dorothy Terrio), 31, 167 perfs.; *Blossom Time* 9/21 at New York's Ambassador Theater, with book and lyrics by Dorothy Donelly who has adapted the German operetta *Das Dreimädelhaus* based on the life of composer Franz Schubert, music by Sigmund Romberg, 592 perfs.; *The Music Box Revue* 9/22 at New York's Music Box Theater, with Irving Berlin, songs by Berlin that include "Say It with Music," 440 perfs.; *Bombo* 10/6 at New York's Jolson Theater, with Al Jolson singing numbers by Sigmund Romberg, with book and lyrics by B. G. DeSylva. Jolson soon interpolates "My Mammy," "Toot, Toot, Tootsie" by Dan Russo, lyrics by Gus Kahn and Ernie Erdman, "April Showers" by Louis Silvers, 32, lyrics by Jack Yellen, "California, Here I Come" by Joseph Meyer, lyrics by Jolson and B. G. DeSylva, 219 perfs.; *The Perfect Fool* 11/7 at New York's George M. Cohan Theater, with Ed Wynn, book, music, and lyrics by Wynn; *Kiki* 11/29 at New York's Belasco Theater, with Lenore Ulric, songs that include "Some Day I'll Find You" by Zoel Parenteau, lyrics by Schuyler Green, 600 perfs.

Popular songs "Ain't We Got Fun" by Richard A. Whiting, lyrics by Gus Kahn and Raymond Egan; "There'll Be Some Changes Made" by U.S. composer Benton Overstreet, lyrics by Billy Higgins; "I'm Nobody's Baby" by Milton Ager and Lester Santly, lyrics by Benny Davis; "All by Myself" by Irving Berlin; "Ma! (He's Making Eyes at Me)" by Con Conrad, lyrics by Sidney Clare (for *The Midnight Rounders* with Eddie Cantor); "Careless Love" by W.C. Handy, Spencer Williams, Martha Koenig; "The Sheik of Araby" by Ted Snyder, Harry B. Smith, Francis Wheeler; "Fire Dance" ("Danse Rituelle du Feu") by Manuel de Falla, lyrics by Martinez Serra, 41.

Spanish bullfighter Juan Belmonte retires at age 29 after matador Manuel Granero is killed in the ring at Madrid in May, but the millionaire founder of modern bullfighting will return several times before finally devoting himself exclusively to breeding bulls for the *corrida.*

Boxing has its first $1 million gate July 2 at Boyle's Thirty Acres in Jersey City where heavyweight champion Jack Dempsey defeats French challenger Georges Carpentier by a fourth round knockout. The handsome Carpentier won the world light-heavyweight title by defeating Battling Levinsky in a Jersey City bout held October 20 of last year and Tex Rickard of Madison Square Garden has promoted the bout with Dempsey. It begins at just after 3 o'clock in the afternoon, draws more than 90,000 spectators who pay $1.7 million for tickets, and earns $300,000 for Dempsey, $215,000 for Carpentier.

Bill Tilden and Suzanne Lenglen win the singles titles at Wimbledon, Tilden and Mrs. Mallory the U.S. titles at the West Side Tennis Club courts in Forest Hills, N.Y.

Chicago judge Kenesaw Mountain Landis, now 53, has been named first commissioner of professional baseball in January. Eight members of the 1919 White Sox team go to trial in June on charges of accepting bribes from gamblers to throw the Series (*see* 1920), all eight are acquitted, but Judge Landis, who presided over the grand jury that indicted them, bans them all from organized baseball for life. Included is Joseph Jefferson "Shoeless Joe" Jackson, 33, who played flawless ball in the 1919 Series and hit the only home run.

The New York Giants win the World Series by defeating the New York Yankees 5 games to 3.

Cuban chess master José Raul Capablanca y Granperra, 33, wins the title he will hold until 1927. He defeats Emanuel Lasker, who has been world champion for 27 years.

Chanel No. 5, introduced May 5 by "Coco" Chanel, has none of the "feminine" floral scent found in other fragrances and will become the world's leading perfume (*see* 1913).

Drano is introduced by Cincinnati's P. W. Drackett & Sons, whose drain cleaner consists of crystals containing sodium hydroxide (caustic lye). The company was founded in 1910 by Philip Wilbur Drackett to distribute chemicals (*see* 1933).

Electrolux vacuum cleaners are introduced by Swedish electric lamp salesman Axel Wenner-Gren, 40, who has founded the Electrolux Co. to produce the machines that will be the world's top-performing vacuum cleaners. Wenner-Gren's company will also be a major factor in refrigerators.

The Van Heusen collar is introduced by Phillips Jones Corp. which has acquired rights to the starchless but stiff collar invented by John M. van Heusen, 52. The collar is made from multiple ply, interwoven fabrics and van Heusen has patented collars, cuffs, neckbands, and other articles made in whole or in part from such fabrics.

The Arrow shirt is introduced by Cluett, Peabody Co. of Troy, N.Y. as demand increases for collar-attached shirts. Research director Sanford Lockwood Cluett, 47, develops a "Sanforizing" process to limit shrinkage.

U.S. cigarette consumption reaches 43 billion, up from 10 billion in 1910 despite the fact that cigarettes are illegal in 14 states. Anti-cigarette bills are pending in 28 other states, college girls are expelled for smoking, but tobacco companies promote the addiction to nicotine.

Watts Tower goes up at Los Angeles where Italian-American tile-setter Simon Rodia, 41, uses only hand tools, shells, glass, and other found objects to build a structure that will awe visitors for decades.

Hot Springs National Park is created by act of Congress. Set aside as a government reservation in 1832, the 3,535-acre park in Arkansas contains 47 mineral hot springs.

U.S. forester Benton MacKaye, 42, proposes an Appalachian Trail for hikers. Extending nearly 2,000 miles from Georgia's Springer Mountain to Maine's Mt. Katahdin, the footpath will in large part follow the Great Indian Warpath that existed in the 18th century from Creek Territory in Alabama into Pennsylvania. MacKaye is first president of the Wilderness Society and will unify the Trail's diverse parts.

U.S. Communist Harold M. Ware persuades Lincoln Steffens to donate more than $70,000 in lecture fees to buy tractors and other farm equipment for Russia. Ware has made a study of U.S. farming methods in response to last year's request by V.I. Lenin and notes that Russia has fewer than 1,500 tractors, many of them unrepairable wrecks (*see* 1922).

U.S. farmers overproduce on acreage planted during the war and farm prices in July fall to levels 85 percent below 1919 highs. Cotton is 11¢ lb., down from 42¢; corn 42¢ per bushel, down from $2; wheat $1, down from $3.50.

The Harding administration tries to relieve the agricultural depression with an emergency tariff act imposing high duties on imports of wheat, corn, meat, sugar, wool, and other farm products.

Boll weevils halve cotton production in Georgia and South Carolina (*see* 1916; 1924).

Botanist George Washington Carver of the Tuskegee Institute testifies before the House Ways and Means Committee on the value of peanuts which are planted increasingly by Georgia and South Carolina farmers (*see* Carver, 1914).

Harvard biologist Edward Murray East, 42, of the Connecticut Agricultural Experiment Station, and Princeton botanist George Harrison Shull perfect a hybrid corn variety that will yield far more per acre than conventional corn (*see* 1905; 1930).

Famine kills 3 million Russians. Frozen corpses are piled 20 feet high in the streets waiting for the ground to thaw so that burial pits can be dug. U.S. Secretary of Commerce Herbert C. Hoover organizes the American Relief Administration which raises more than $60 million in response to an appeal by Russian writer Maksim Gorki and saves many from starvation.

A Russian Famine Relief Act passed by Congress December 22 authorizes the expenditure of $20 million for the purchase of seed, grain, and preserved milk to be sent to Russia.

Vitamin pioneer E. V. McCollum finds that cod-liver oil is a preventive against rickets; he observes that rats fed large amounts of calcium develop bone deformities

1921 ***(cont.)*** much like those seen in rickets but that the deformities are prevented if a little cod-liver oil is added to the rats' rations (*see* 1918; 1922).

U.S. table-salt makers introduce salt iodized with potassium iodide but iodination is not mandatory (*see* Akron study, 1916).

Several states legislate good dairy practices to improve the safety of U.S. milk which often reaches consumers with a bacteria count of 500,000 or more per cubic centimeter (*see* 1900). Contaminated milk transmits undulant fever, brucellosis, bacillary dysentery, infectious hepatitis' typhoid fever, tuberculosis, and other diseases.

Wholesale butter prices in the New York market fall to 29¢ lb., down from 76¢ during the war. Imports of Danish butter have increased and Americans are using more margarine.

U.S. butter is "scored" on a point system that gives 45 points for top flavor, 25 points on the basis of composition, 15 on color, 10 on salt content, 5 on the wooden butter tub used for packaging. Butter that scores 92 out of a possible 100 is called extra grade or "New York extras," but out of 622 Minnesota creameries only 200 make butter good enough to command the premium price paid for 92-score "New York extras."

Land O' Lakes Butter, Inc., has its beginnings in the Minnesota Cooperative Creameries Association formed by Meeker County, Minn., creameries. The federation will burgeon into a Minneapolis-based empire of midwestern dairy farmers.

Bel Paese semisoft cheese is introduced by Italian cheese maker Egidio Galbani.

Cambridge, Mass., chemist Bradley Dewey, 34, discovers the first water-based latex sealing compound. The sealant will be used inside the crimped rims by which tops and bottoms are attached to cans and Dewey's firm Dewey & Almy will be a major producer of the sealant and of the machines used to apply it (*see* synthetic rubber, 1939).

The name "Betty Crocker" is developed by Washburn, Crosby Co. of Minneapolis. The firm runs a contest to promote its Gold Medal Flour and letters from contestants are answered by the fictitious food authority Betty Crocker (*see* 1880; 1936).

The first automatic doughnut-making machine is introduced (*see* Donut Corp. of America, 1920).

Molasses prices drop to 2¢ per gallon, down from 20¢ last year.

World cocoa prices collapse in the absence of any futures market mechanism. Speculative "hoarders" have used bank credits to buy up cocoa beans produced from October through January in anticipation of the usual price advances but when the banks suddenly begin calling the loans the cocoa is thrown on the market, prices break nearly 20¢ per pound, and cocoa merchants sustain losses as do importers and outside speculators, many of whom are ruined (*see* Cocoa Exchange, 1925).

The Mounds bar introduced by Peter Paul is a coconut and bittersweet chocolate bar made from a formula devised by former chemist George Shamlian.

Wise Potato Chips are introduced by Berwick, Pa., grocer Earl V. Wise who finds himself overstocked with old potatoes. He peels them, slices them with an old-fashioned cabbage cutter, follows his mother's recipe for making potato chips, packages the chips in brown paper bags, and sells them in his grocery store (*see* George Crum, 1853; peeling machines, 1925).

Lindy's Restaurant opens at 1626 Broadway in New York under the management of German-American restaurateur Leo "Lindy" Lindeman, 33, whose cheesecake will be favored by newspapermen, politicians, and theater people for half a century.

Sardi's Restaurant opens in West 44th Street, New York, under the direction of Italian-American restaurateur Vincent Sardi, 35, whose place will be an unofficial theatrical club for more than half a century. Sardi will move to 234 West 44th Street in 1927 when his first building is razed to make way for construction of the St. James Theater.

Some 900,000 immigrants enter the United States in the fiscal year ending June 30.

The Dillingham Bill (Emergency Quota Act) enacted by Congress May 19 establishes a quota system to restrict immigration. Entry is permitted only to 3 percent of the people of any nationality who lived in the United States in 1910 (*see* 1917; 1924).

Margaret Sanger founds the American Birth Control League (*see* 1916; 1923).

The Mothers' Clinic for Constructive Birth Control opens March 17 at London. Founder of the first British birth-control clinic is paleobotanist Marie Carmichael Stopes, 41, whose book *Radiant Motherhood* appeared last year.

Britain's population reaches 42.7 million, Germany's 30 million, France's 39.2, Italy's 38.7, Brazil's 30, Japan's 78, Russia's 136.

1922 New treaties, new violence, political and economic convulsions, and the emergence of new nations and regimes follow in the wake of the Great War—an independent Egyptian monarchy, an Italian fascist dictatorship, and an independent Irish Free State.

Japan agrees February 4 to return Shandong province to China, where rival war lords engage in civil war.

The Washington Conference ends February 6 after nearly 3 months with a naval armaments treaty that provides for a 10-year period during which no new ships of more than 10,000 tons with guns larger than 8 inches in width are to be built by Britain, France, Italy, Japan, or the United States. Britain and the United States are then to be permitted totals of 525,000 tons each, Japan 315,000 tons, France and Italy 175,000 tons each. The Conference restricts submarine warfare and use of poison gas.

The Permanent Court of International Justice opens February 15 at The Hague.

Germany recognizes the Soviet Union as a great power in the Treaty of Rapallo signed April 16 and resumes relations with the Lenin government.

Germany cedes Upper Silesia to Poland May 15.

Germany's foreign minister Walter Rathenau is murdered June 24 at age 55 by nationalist reactionaries. Rathenau's father helped found the German Edison Co. in 1883 (*see* Kapp, 1920; Hitler, 1923).

The Kingdom of Egypt is proclaimed March 15; the sultan Ahmed Fuad assumes the title of king 2 weeks after the termination by Britain of her protectorate over Egypt. The Sudan remains under joint Anglo-Egyptian sovereignty and Fuad I begins a reign that will continue until 1936.

The Irish Republican Army (IRA) is formally constituted in March "to safeguard the honor and independence of the Irish Republic." Responsible Irish political leaders will disavow the IRA, but the militant arm of the Sinn Fein political party that has stood for a free, undivided Ireland since the Easter Rising of 1916 will continue to employ terrorist tactics in a civil war within the Irish Free State (Eire) and in Ulster.

President de Valera has resigned January 9 and organizes a Republican Society, rejecting the dominion status granted last year by London. He begins an insurrection against his erstwhile colleagues Arthur Griffith and Michael Collins who have formed a new government, and the terrorism used earlier against the British is now used also against Irishmen as well.

Sinn Fein terrorists murder British field marshal Sir Henry Hughes Wilson, 58, the president of the Dail Eireann Arthur Griffith dies suddenly August 12, and the prime minister Michael Collins is mortally wounded August 22 at age 30 while repelling an IRA assault. William T. Cosgrove is elected September 9 to succeed Griffith, the Dail adopts a Constitution for the Irish Free State October 24, the IRA leader Erskine Childers of 1903 *Riddle of the Sands* fame is court-martialed and executed by a firing squad November 24, the Irish Free State is officially proclaimed December 6, the parliament of Northern Ireland votes December 7 to remain outside the Free State, and the last British troops leave the Free State December 17.

Greece's Constantine I abdicates under pressure for a second time in late September (he will die of a brain hemorrhage at age 54 early next year at Palermo.) His son of 32 will reign briefly as George II, regain the throne in 1935, and reign until 1947.

Britain's Lloyd George cabinet resigns October 19 after Conservatives, meeting at the Carlton Club, vote to quit the coalition government. Canadian-born Scotsman Andrew Bonar Law, 64, heads a new Conservative government.

The Turkish sultanate at Constantinople ends November 1. Mustafa Kemal's forces have taken Smyrna from the Greeks in September (but have seen the city largely destroyed by fire) and Kemal has accepted neutralization of the Dardanelles in exchange for the return of Adrianople and Eastern Thrace. The sultan Mohammed VI flees and his cousin Abdul Mejid is proclaimed caliph (*see* 1923).

A fascist dictatorship in Italy begins in late November as Victor Emmanuel III summons Benito Mussolini, now 39, to form a ministry and grants him dictatorial powers so that he may restore order and bring about reforms. The journalist-politician has campaigned in his *Popolo d'Italia* against Communism, organized his Fascio di Combattimento in Milan into a political party, won support from Italian business interests fearful of Communism, and led his black-shirted Fascisti in a "March on Rome" October 28, but his dictatorial powers are to expire at the end of next year (*see* 1923).

The Amoskeag textile mill in Manchester, N.H., announces February 2 that it is cutting wages 20 percent and increasing weekly hours from 48 to 52. French-Canadian financier Frederic C. Dumaine, 56, controls the mill and finds that demand for gingham has shrunk, Southern mills are better equipped and more efficient. Amoskeag workers begin a 9-month strike.

U.S. coal miners strike for nearly 6 months to protest wage cuts. The massive strike by the United Mine Workers AF of L cripples industry and begins a period of chronic depression in the coal mining industry, whose operators will employ cutthroat competition to remain viable.

A 13 percent wage cut announced by the Railroad Labor Board May 28 affects 400,000 U.S. railway workers; they strike from July 1 through the summer.

Coal miners at Herrin, Ill., riot in protest against the use of strike breakers June 22 to 23, 26 men are killed, but both coal and railroad workers return to their jobs, defeated, in September.

Moscow signs commercial treaties and trade agreements with Italy and Sweden.

A modified Franco-German reparations agreement permits Germany to pay with raw materials such as coal but Britain places the burden of war debts on the United States. The Balfour note states that Britain would expect to recover from her European debtors only the amount which the United States expects from Britain.

The Reparations Commission adopts a Belgian proposal August 31 permitting Germany to pay reparations in installments on Treasury bills.

Germany's stock market collapses in August; the mark falls in value from 162 to the dollar, down to more than 7,000 to the dollar (*see* 1923).

The Fordney-McCumber Tariff Act passed by Congress September 19 returns tariffs to the levels of the 1909 Payne-Aldrich Act. The new law gives the president power to raise or lower duties by 50 percent in order to equalize production costs but presidents will use that authority in 32 of 37 cases to further increase duties in response to appeals for protection by U.S. business and labor.

The Federal Reserve Board created in 1913 sets up a bank-wire system to eliminate physical transfer of

1922 ***(cont.)*** securities from one city to another and thus avoid theft, loss, or destruction of negotiable Treasury certificates. A Federal Reserve Bank taking in a certificate for delivery to another Federal Reserve Bank retires the certificate and instructs its sister bank by teletype to issue a new one.

A U.S. business revival led by the automobile industry begins 7 years of prosperity.

An oil well near the shores of Lake Maracaibo, Venezuela, gushes for 9 days, spills nearly a million barrels into the lake before it is capped. A company controlled by Royal Dutch-Shell has brought in the discovery well (*see* 1914). Gulf Oil and Standard Oil of Indiana hold adjacent leases.

Henry Ford makes more than $264,000 per day; the Associated Press declares him a billionaire.

Ford Motor Company president Edsel Ford pays $12 million to acquire the Lincoln Motor Co. started last year by Cadillac founder Henry M. Leland (*see* Lincoln-Mercury, 1939).

The Star, introduced by Durant Motors, Inc., is a $348 motorcar built to compete with Ford's Model T, but Ford reduces the price of the Model T to stifle Durant's threat of competition (*see* 1921).

General Motors gives control of its Chevrolet division to Danish-American engineer William Knudsen, 43, who has left Ford. Using mass assembly methods and Du Pont paints in a variety of colors, Knudsen will soon overtake Ford's black Model T (*see* Model A, 1927).

Hudson introduces the first closed sedan at a price only slightly above that of a touring car (*see* 1918; 1932).

The Model A Deusenberg introduced by Deusenberg Automobile and Motor Co. of Indianapolis is the first U.S. production motorcar with hydraulic brakes, the first with an overhead camshaft, and the first U.S. straight eight. Ninety-two of the luxury cars are sold, a number that will rise to 140 in 1923 (*see* 1919).

Minnesota bus pioneer Carl Wickman sells his interests in Mesabi Transportation Co. for something over $60,000 and begins to buy out small bus lines between Duluth and Minneapolis (*see* 1914; 1926).

The first all-metal U.S. airplane is built by engineer William Bushnell Stout, 42, who will sell his commercial aircraft company to Ford Motor Company in 1925, start a passenger airline in 1926, and sell it to United Aircraft and Transport in 1929.

Dutch aircraft designer A. H. G. Fokker emigrates to America where he will establish the Fokker Aircraft Corp. of America at Hasbrouck Heights, N.J. (*see* 1912; 1916; Juan Trippe, 1927).

U.S. Army Air Corps Lieutenant James Harold Doolittle, 25, makes the first coast-to-coast flight in a single day in September, flying a DH4b 2,163 miles from Pablo Beach, Fla., to San Diego in 21 hours, 28 minutes' flying time.

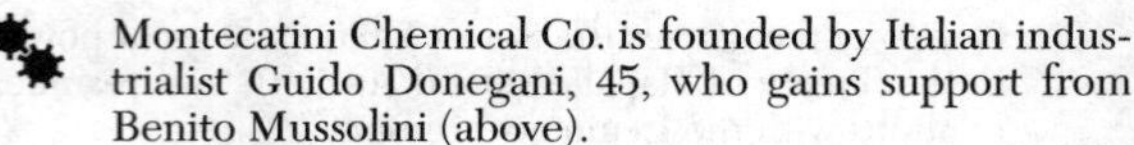

Montecatini Chemical Co. is founded by Italian industrialist Guido Donegani, 45, who gains support from Benito Mussolini (above).

Dow Chemical chemists find a way to make phenol production more efficient. The Hale-Britton process developed by William J. Hale (son-in-law of founder Herbert H. Dow) and Edgar C. Britton, 31, will permit cheap production of orthophenylphenol and paraphenylphenol that Dow will market as insecticides, germicides, and fungicides (*see* 1897, 1924).

An enzyme discovered by Scottish bacteriologist Alexander Fleming, 41, breaks down bacterial cell walls. Fleming calls his lysozyme "the dissolving enzyme" (*see* penicillin, 1928).

English archaeologist Charles Leonard Woolley, 42, discovers Ur on the Euphrates River in Iraq, finds Sumerian temple ruins dating to 2600 B.C., and gives historical reality to the ancient Mesopotamian civilization of Sumer of which there has been only legendary knowledge.

English Egyptologist Howard Carter, 49, and his patron George Edward Stanhope Molyneux, 56, earl of Carnarvon discover the tomb of Egypt's King Tut November 26 at Luxor in the Valley of the Kings. Of the 27 pharaohs buried near Thebes, only the minor 18th Dynasty king Tutankhamen (1358 B.C. to 1350 B.C.) has been spared having his tomb looted.

The U.S. death rate from diphtheria falls to 14.6 per 100,000, down from 43.3 in 1900; from influenza and pneumonia to 88.3, down from 181.5 (and much higher in the 1918–1919 epidemic); from infant diarrhea and enteritis to 32.5, down from 108.8; from tuberculosis to 97, down from 201.9; from typhoid and paratyphoid fever to 7.5, down from 35.9; but from cancer the death rate is 86.8 per 100,000, up from 63 in 1900; from heart disease 154.7, up from 123.1; and from diabetes mellitus 91.7, up from 18.4.

Insulin gives diabetics a new lease on life. Canadian medical researchers Frederick Grant Banting, 31, and Charles Herbert Best, 23, isolate the hormone (Banting calls it "isletin") from canine pancreatic juices, use it to save the life of Leonard Thompson, 14, who is dying in Toronto General Hospital, and will license Eli Lilly to make the first commercial insulin—the first treatment for diabetes other than diet restrictions (*see* 1924; 1937; Sharpey-Schafer, 1916).

U.S. electrical engineer Vannevar Bush, 32, helps start a company to produce the S-tube, a gaseous rectifier developed by inventor C. G. Smith that greatly improves the system of supplying electricity to radios (*see* analog computer, 1930).

New York station WEAF (later WNBC) airs the first paid radio commercials, setting a pattern of private control of U.S. public airwaves: "What have you done with my child?" radio pioneer Lee De Forest will ask. "You have sent him out on the street in rags of ragtime to collect money from all and sundry. You have made of him a laughingstock of intelligence, surely a stench in the nostrils of the gods of the ionosphere" *(see* 1906; NBC, 1926).

The BBC (British Broadcasting Corp.) is founded under the leadership of English engineer John Charles Reith, 33, a six-foot-six misanthrope who will run BBC for the next 16 years and make it one of Britain's most revered institutions, supported by the public with license fees.

The *New York Daily Mirror* begins publication as a tabloid competitor of the 3-year-old *Daily News*. Publisher is William Randolph Hearst, who buys the *Washington Herald, Oakland Post-Enquirer, Los Angeles Herald, Syracuse Telegram,* and *Rochester Journal (see* 1963).

The *Reader's Digest* appears in February with articles "of lasting interest" condensed from books and from other magazines into a pocket-size monthly. Former St. Paul, Minn., book salesman De Witt Wallace, 32, and his bride Lila Acheson Wallace, 32, have 1,500 subscribers whose numbers will grow to more than 200,000 by 1929 as the *Digest* plants some of its articles in other magazines and then condenses them. The magazine will accept no advertising until 1955 and by the 1970s will be publishing 29 million copies per month in 13 languages *(see* 1933).

True Confessions magazine begins publication to compete with Bernarr MacFadden's 3-year-old *True Story*. Publisher is Wilford H. Fawcett, whose 3-year-old *Captain Billy's Whiz Bang* will reach a circulation of 425,000 per month next year *(see* 1928).

Better Homes and Gardens magazine begins publication at Des Moines, Iowa, under the direction of Edwin Thomas Meredith, now 46, who started *Successful Farming* in 1902 and served as President Wilson's Secretary of Agriculture.

Krocodil begins publication at Moscow. The illustrated supplement to the newspaper *Worker (Rabochy)* will develop into a political humor magazine.

Hearst circulation director Moses Louis Annenberg, 44, and two colleagues pay $500,000 to acquire *The Racing Form,* a daily tip-sheet established a few years ago by Chicago newspaperman Frank Buenell. Annenberg will buy the competing *Morning Telegraph*, quit Hearst in 1926, go into partnership with Chicago gambler Monte Tennes to control bookie joints with leased wires to racetracks, and force his *Racing Form* partners to sell their interests to him for more than $2 million *(see* 1936).

Charles Atlas wins the "World's Most Perfectly Developed Man" contest sponsored by *Physical Culture* magazine publisher Bernarr MacFadden. Italian-American Angelo Siciliano, 28, is a former 97-pound weakling who has built himself up with "dynamic tension" exercises which he claims to have developed after watching a lion at the zoo, he will win MacFadden's contest again next year, will open a Manhattan gymnasium in 1926, and by 1927 his Charles Atlas, Ltd., will be taking in $1,000 per day from students who subscribe to his mail-order physical culture course.

Nonfiction *Decline of the West (Der Untergangdes Abendlandes. Umrisse einer Morphologic der Weltgeschichte)* by German writer Oswald Spengler, 42, who predicts the eclipse of Western civilization; *Public Opinion* by Walter Lippmann; *Etiquette—The Blue Book of Social Usage* by New York divorcée Emily (Price) Post, 48, who has found the manners of Americans far inferior to those of their social counterparts in Europe.

Fiction *Ulysses* by James Joyce. All reputable publishers have refused the stream of consciousness account of a day in the lives of Dubliner Leopold Bloom and his wife Molly, but Paris bookseller Sylvia Beach has published it under the imprint of her book shop's name Shakespeare & Co. Irish critic Æ (George William Russell) of the *Irish Homestead* calls it "the greatest fiction of the twentieth century"; *Siddhartha* by Hermann Hesse, who visited India in 1911, has undergone psychoanalysis, and has developed a Jungian mixture of depth psychology and Indian mysticism to probe the fantasies of adolescents; *The Bridal Canopy (Haknasath Kallah)* by Hebrew novelist Samuel Joseph Agnon, 44, who changed his name from Czaczkes when he moved from his native Galicia in 1909; *The Dream (Le Songe)* by French novelist Henry de Montherlant, 26; A *Kiss for the Leper (Le Baiser au lépreux)* by French novelist François Mauriac, 37; *The Enormous Room* by U.S. poet E. E. (Edward Estlin) Cummings, 27, who drove an ambulance in France and served afterward in the U.S. Army (he will sign his poetry "e. e. cummings"); *The Beautiful and Damned* by F. Scott Fitzgerald; *Tales of the Jazz Age* (stories) by F. Scott Fitzgerald with illustrations by John Held, Jr., 32; *Babbitt* by Sinclair Lewis, who examines the conformist, complacent, conservative American businessman and introduces a new pejorative; *Captain Blood* by Rafael Sabatini; *Jigsaw* by London debutante Barbara Cartland, 21, whose first novel enjoys huge success and launches her on a long career.

Juvenile *The Voyages of Dr. Doolittle* by English civil engineer Hugh Lofting, 36, who was wounded in 1917 and has turned to writing and illustrating. His book about a physician who can talk to the animals begins a series.

Poetry *The Wasteland* by T. S. Eliot indicts the 20th century in a style perfected with the help of Ezra Pound. The eloquent work links numerous allusions explained in lengthy notes; *Anabasis (Anabase)* by French poet-diplomat Saint-John Perse (Marie René Auguste Alexis Saint-Léger Léger), 34, whose epic poem about the Greek Xenophon of 401 B.C. has landscapes reflective of the Gobi desert through which the poet traveled while stationed in China; *Tristia* by Osip Mandelstam, who has developed an elliptical and metaphorical language; *Clouds (Moln)* by Swedish poet Karin Boye, 22; *Trilce* by César Vallejo.

The Fugitive magazine appears in April. The poetry monthly will continue until December 1925 with contributions by Vanderbilt University poets and

The Jazz Age portrayed by cartoonist John Held, Jr., who sometimes teamed up with F. Scott Fitzgerald.

writers who include John Crowe Ransom, 33, (John Orley) Allen Tate, 22, and Robert Penn Warren, now 16.

Painting *Twittering Machine* by Paul Klee; *Before the Bell* by Max Beckmann; *The Farm* by Joan Miró; *Still Life with Guitar, Harlequin,* and *Mother and Child* by Pablo Picasso; *Sunset* and *Maine Islands* by John Marin.

Theater *The Cat and the Canary* by U.S. playwright John Willard 2/7 at New York's National Theater with Florence Eldridge, Henry Hull; *To the Ladies* by Marc Connelly and George S. Kaufman 2/20 at New York's Liberty Theater, 128 perfs.; *Back to Methusaleh* by George Bernard Shaw 2/27 at New York's Garrick Theater (Parts I and II) with Dennis King, Margaret Wycherley, (Parts III and IV) 3/6, (Part V) 3/13, 72 perfs. total; *Loyalties* by John Galsworthy 3/8 at St. Martin's Theatre, London; *The Hairy Ape* by Eugene O'Neill 3/9 at New York's Provincetown Playhouse (and then uptown at the Plymouth Theater) with Louis Wolheim, 127 perfs.; *Abie's Irish Rose* by U.S. playwright Anne Nichols, 30, 5/23 at New York's Fulton Theater, a play about a mixed marriage, 2,532 perfs. (a new record); *The Machine Wreckers (Die Maschinensturmer)* by Ernst Toller 6/30 at Berlin's Grosses Schauspielhaus; *The Fool* by Channing Pollock 10/23 at New York's Times Square Theater with James Kirkwood, Lowell Sherman, 773 perfs.; *The World We Live In (The Insect Comedy)* by Karel and Josef Capék 10/31 at New York's Al Jolson Theater, with Mary Blair as a butterfly, Vinton Freedley as a male cricket, 112 perfs.; *Rain* by U.S. playwrights John Colton and Clemence Randolph 11/7 at New York's Maxine Elliott Theater, with Jeanne Eagels as Somerset Maugham's Sadie Thompson, 321 perfs.; *The '49ers* by Marc Connelly and George S. Kaufman 11/13 at New York's Cort Theater, with Howard Lindsay, Ronald Young, Ruth Gilmore, 16 perfs.; *The Texas Nightingale* by U.S. playwright Zoë Akins, 36, 11/20 at New York's Empire Theater, 31 perfs.

Life magazine drama critic Robert Charles Benchley, 33, presents *The Treasurer's Report* at an amateur revue; his comedy monologue that launches Benchley on a new career as humorist.

"Texas" Guinan begins her career as Prohibition era New York nightclub hostess. Mary Louise Cecilia Guinan, 38, goes to work as mistress of ceremonies at the Café des Beaux Arts and is soon hired away by Larry Fay of El Fey from whose club she will move to the Rendezvous, 300 Club, Argonaut, Century, Salon Royal, Club Intime, and to several Texas Guinan Clubs that will serve bootleg Scotch at $25 a fifth, bootleg champagne at $25 a bottle, plain water at $2 a pitcher. The clubs will charge from $5 to $25 cover to the "butter-and-egg men"; Texas Guinan ("give the little girl a great big hand") will welcome customers from her seat atop a piano with the cry, "Hello, sucker!"

Films Fred Niblo's *Blood and Sand* with Rudolph Valentino; Fritz Lang's *Dr. Mabuse: Der Spieler* with Rudolf Klein-Rogge; Frederick W. Murnau's *Nosferatu* with Max Schreck; *Nanook of the North* by British explorer-writer-documentary director Robert J. Flaherty, 28; D. W. Griffith's *Orphans of the Storm* with Dorothy and Lillian Gish; Rex Ingram's *The Prisoner of Zenda* with Lewis Stone, Alice Terry; Allan Dwan's *Robin Hood* with Douglas Fairbanks.

Hollywood Bowl opens. The 17,000-seat outdoor California amphitheater will present concerts, opera, and ballet.

Austria's Salzburg Mozart Festival has its first season to begin a lasting August tradition in the city's Festspielhaus (Festival Hall), Mozarteum, Landestheater, Mirabell Castle, and Marionette Theater.

First performances Pastoral Symphony by Ralph Vaughan Williams 1/26 at London.

Broadway musicals *The Ziegfeld Follies* 6/5 at the New Amsterdam Theater, with Will Rogers, Ed Gallagher, Al Shean, Olsen and Johnson, music by Dave Stamper, Louis A. Hirsch, Victor Herbert, and others, book and lyrics by Ring Lardner, Gene Buck, and others, songs that include "Mister Gallagher and Mister Shean," 541 perfs.; *George White's Scandals* 8/22 at the Globe Theater, with W. C. Fields, music by George Gershwin, lyrics by E. Ray Goetz, B. G. De Sylva, and W. C. Fields, Paul Whiteman and His Orchestra, songs that include "I'll Build a Stairway to Paradise" with lyrics by B. G. De Sylva and Arthur Francis (Ira Gershwin), 88 perfs.; *The Gingham Girl* 8/28 at the Earl Carroll Theater, with music by Albert von Tilzer, lyrics by Neville Fleeson, songs that include "As Long as I Have You," 322 perfs.; *The Passing Show* 9/20 at the Winter Garden Theater, with Eugene and Willie Howard, Janet Adair, Fred Allen, music largely by Alfred Goodman, lyrics mainly by Harold Atteridge, songs that include "Carolina in the Morning" by Walter Donaldson, lyrics by Gus Kahn, 95 perfs.; *Little Nellie Kelly* by George M. Cohan 11/13 at the Liberty Theater, with Elizabeth Hines, 276 perfs.

Popular songs "Way Down Yonder in New Orleans" by Henry Creamer and Turner Layton; "Chicago" by Fred Fisher; "My Buddy" by Walter Donaldson, lyrics by Gus Kahn; "Blue" by Lou Handman, lyrics by Grant Clarke and Edgar Leslie; "I'll See You in My Dreams" and "On the Alamo" by Isham Jones, lyrics by Gus Kahn; "Trees" by Oscar Rasbach, lyrics by Joyce Kilmer (*see* 1913 poem); "Goin' Home" by Anton Dvořák, lyrics by William Arms Fisher (*see* 1893 symphony).

A Western Electric Company research team led by J. P. (Joseph Pease) Maxfield, 34, invents a phonograph record graver that permits recording in acoustically correct studios rather than by singing or playing directly into horns. It chisels vibrations into wax at the rate of 30 to 5,500 wiggles per second (*see* Victrola, 1906). Electric impulses derived from sound waves as in the telephone vibrate the graver with augmented power to improve the fidelity of phonograph records (*see* vinylite records, 1946, 1948).

G. L. Patterson wins in men's singles at Wimbledon, Suzanne Lenglen in women's singles; Bill Tilden wins in U.S. men's singles, Mrs. Mallory in women's singles.

Harrison, N.Y., caddy Gene Sarazen (*né* Saracini), 21, wins the U.S. Open to begin a notable career.

The New York Giants win the World Series by defeating the New York Yankees 4 games to 0 after one game has ended in a tie.

Mah-Jongg is introduced in America and a nationwide craze begins for the ancient Chinese game. The 144-tile sets will outsell radios within a year.

The Maytag Gyrofoam washing machine introduced by the Newton, Iowa, firm outperforms all other washing machines yet takes up only 425 square inches of floor space (*see* 1907). Now 65, F. L. Maytag has built his own aluminum foundry to cast the tubs for his new machine.

The first Thom McAn shoe store opens October 14 on New York's Third Avenue near 14th Street with men's shoes at $3.99 per pair. Melville Shoe Co. vice president and founder's son (John) Ward Melville, 35, served during the Great War under J. Franklin McElwain, head of the Quartermaster Corps shoe and leather division. He and McElwain have developed the idea of a mass-produced shoe to be sold through a chain of low-priced stores with Melville Shoe and J. J. McElwain Co. of Nashua, N.H., sharing in the profits.

The Lincoln Memorial dedicated May 30 at Washington, D.C., contains a 19-foot seated figure of the sixteenth president carved out of Georgia marble by sculptor Daniel Chester French, now 72. The $2,940,000 monument beside the Potomac has taken 7 years to build.

Kansas City's Country Club Plaza Shopping Center designed by developer Jesse Clyde Nichols, 42, is the world's first large decentralized shopping center. Nichols wrote a Harvard graduate thesis on the economics of land development, he has visited English "garden city" developments, he has collected architectural ideas in Spain, and he pursues construction of the 6,000-acre Country Club district that will occupy 10 percent of downtown Kansas City, Mo.

"Ribbon windows"—rows of glass evenly divided by vertical concrete slabs—are innovated in a plan for an office building by German architect Ludwig Mies Van Der Rohe, 35, who will employ the design in many European and American buildings.

Horse-drawn fire apparatus makes its final appearance in New York December 22 as equipment from Brooklyn's Engine Company 205 at 160 Pierrepont Street races to put out a fire.

Australia moves to protect her koala bears after fur trappers have killed 8 million in less than 4 years and have nearly wiped the marsupial out of existence.

Van Camp Sea Food merges with White Star Canning and two other nearly bankrupt San Pedro, Calif., tuna packers after being hurt by the 1921 economic deflation and collapse of wartime markets (*see* 1914). The new company increases production and sales but lacks the equipment needed to make extended cruises and overcome the circumstances that make tuna fishing a seasonal business limited to albacore (*see* 1924).

V. I. Lenin permits small private farms to help the Soviet Union produce more food (*see* Stalin, 1928).

Harold M. Ware arrives in Russia with 21 tractors, other farm machinery, seeds, supplies, and food (*see* 1921). Ware is assigned to lands in the Urals where he and his volunteer associates seed most of 4,000 acres in winter wheat and encourage peasants to give up their holdings and join a collective (*see* 1925).

U.S. farmers remain in deep depression with a few exceptions such as California citrus growers (below).

The Capper Volstead Act passed by Congress February 18 exempts agricultural marketing cooperatives from antitrust law restrictions. The co-ops are subject to supervision by the Secretary of Agriculture who is to prevent them from raising prices "unduly."

A Grain Futures Act passed by Congress September 21 curbs speculation thought to have contributed to the collapse of grain prices in 1920. The new act, which supersedes an August 1921 act that has been declared unconstitutional by the Supreme Court, limits price fluctuations within any given period on U.S. grain exchanges.

The first U.S. soybean refinery is built at Decatur, Ill., by corn processor A. E. Staley, founded in 1898. Staley removes at least 96 percent of the oil and sells the residue, or cake, to the feed industry for use in commercial feeds, or to farmers to mix with other ingredients as a protein supplement for livestock (*see* 1923).

California becomes a year-round source of oranges as Valencia orange production catches up with navel orange production (*see* 1875; 1906; Sunkist, 1919).

James Dole's Hawaiian Pineapple Co. buys the 90,000-acre Hawaiian island of Lanai, 60 miles from the company's Honolulu cannery, to give Dole 20,000 acres of land that can be planted in pineapples (*see* 1911). Fourteen other Hawaiian packers now can pineapple but

1922 ***(cont.)*** Dole's company is by far the largest, producing 1.5 million cases out of the 4.8 million industry total, and Hawaiian Pineapple shareholders receive a 23 percent dividend (*see* Castle & Cooke, 1964).

E. V. McCollum isolates vitamin D and uses it in successful treatment of rickets. The newly found vitamin is found to play some role in raising the amount of calcium and phosphate deposited in bones but the mechanism remains a mystery (*see* 1923).

Experimental dogs produce hemoglobin quickly and avoid anemia when fed liver in tests made by University of Rochester pathologist George Hoyt Whipple, 44 (*see* Minot and Murphy, 1924).

Pep breakfast food is introduced by the Battle Creek Toasted Corn Flake Co. which changes its name to Kellogg Co. (*see* 1919; 1928).

Postum Cereal Co. is incorporated (*see* Post, 1914; Jell-O, 1925).

Campbell Soup Co. is established by a renaming of Joseph Campbell Co. (*see* 1932; Franco-American, 1915).

Canada Dry Ginger Ale opens its first U.S. bottling plant on New York's 38th Street to avoid the long costly haul from Canada and the high tariff on imports (*see* 1907; Fordney-McCumber Act, above).

1923 French troops occupy Germany's rich Ruhr Basin beginning January 11. The Germans have defaulted on coal deliveries promised at Versailles in 1919 and German inflation soars out of control (below).

Adolf Hitler, 34, stages a "Beer Hall Putsch" at Munich November 8 as the mark falls to below 1 trillion to the dollar. Hitler's National Socialist German Workers' (Nazi) political party founded early in 1919 by Munich locksmith Anton Drexler is not socialist and does not represent the workers but capitalizes on Germany's social unrest. Hitler takes over Munich's vast Bürgerbraükeller with help from Gen. Erich Ludendorff and seizes the city government; the authorities promptly oust the Nazis, a court next year will sentence Hitler to 5 years in prison, and he will serve 9 months before being paroled (*see* 1925; 1933).

Benito Mussolini secures his fascist dictatorship by dissolving Italy's nonfascist parties July 10 and forcing a law through the parliament November 14. Any party that wins the largest number of votes in an election shall have two-thirds of the seats in the parliament even though it may have gained no more than 25 percent of the popular vote (*see* 1922; 1924).

A Bulgarian coup d'état June 9 overthrows Prime Minister Aleksandr Stamboliski whose dictatorial policies in behalf of the peasants have antagonized the army, the Macedonians, and others. Stamboliski is shot dead June 14 while allegedly trying to escape.

The Union of Soviet Socialist Republics (Russia, the Ukraine, White Russia, and Transcaucasia) established on paper December 30, 1922 becomes a reality July 6.

Jordan (Transjordania) becomes an autonomous state May 26. Occupying 80 percent of Palestinian territory, the British protectorate is headed by Emir Abdullah ibn Husein, 40, whose father is king of Hejaz. Britain will recognize Jordanian independence early in 1928 but retain military control and some financial control.

The Treaty of Lausanne signed July 24 returns eastern Thrace, Imbros, and Tenedos to Turkey, while Greece gets the other Aegean islands, Italy the Dodecanese Islands, Britain retains Cyprus, the Straits are demilitarized, Turkey pays no reparations, and no provision is made for the Kurds as in the repudiated 1920 Treaty of Sèvres (*see* population, below).

Allied forces evacuate Constantinople August 23, Angora is made capital of the Turkish state October 14, and the Turkish Republic is formally proclaimed October 29 with Gen. Mustafa Kemal as president (*see* 1915; 1922). He has destroyed the political power wielded for centuries by Muslim leaders of the old Ottoman Empire, defeated the Greek army bent on occupation, and will make Turkey a modern nation—abolishing the veil for women, ordering the Turks to dress in Western clothes, using Roman letters instead of Arabic, and introducing the 1582 Gregorian calendar (*see* 1924; population, below).

The Spanish garrison at Barcelona mutinies September 12 and the Marquis de Estella Gen. Miguel Primo de Rivera, 53, issues a manifesto that day suspending the constitution and proclaiming a directorate comprised of army and navy officers. Gen. Primo has the approval of Alfonso XIII for his bloodless coup, he takes Barcelona, proclaims martial law throughout Spain, dissolves the cortes, suspends trial by jury, imposes rigid press censorship, and jails or exiles his liberal opponents (*see* 1921; 1930).

Britain's prime minister Andrew Bonar Law has resigned May 20 after 209 days in office (he dies October 30 at age 65). His chancellor of the exchequer Stanley Baldwin, 55, has headed a new Conservative cabinet since May 22, but the general elections in November give the Labour party its first great victory (*see* 1924).

V.I. Lenin establishes the first Soviet forced-labor camp in the Solovetsky Islands northwest of Archangel (*see* Holmogor concentration camp, 1921). Slave labor in the next 30 years will build nine new Russian cities, 12 railway lines, six heavy industry centers, three large hydroelectric stations, two highways, and three ship canals (*see* Stalin ship canal, 1933).

A minimum wage law for women in the District of Columbia passed by Congress in 1918 is unconstitutional, the Supreme Court rules April 9 in the case of *Adkins v. Children's Hospital* (*see* 1937).

The National Woman's party founded by Alice Paul in 1913 meets at Seneca Falls, N.Y., and endorses an Equal Rights Amendment drafted by Paul who calls it the Lucretia Mott amendment.

United States Steel reduces its 12-hour day to 8 hours August 2 following the lead set by American Rolling Mill in 1916. Big Steel will hire an additional 17,000 workers in the next year, raise wages, and still increase its profits.

The German mark falls to 7,260 to the U.S. dollar January 2 and Berlin has food riots when 14 municipal markets close because of a strike against higher prices charged at the stalls and booths in the markets.

The mark falls to 160,000 to the dollar by July and unemployment combines with inflation to create social unrest (*see* Hitler, above); 1.5 million are unemployed, 4.5 million employed only part time, yet prices continue to rise and by July 30 the mark has depreciated to 1 million to the dollar.

The mark falls to 13 million to the dollar in September, to 130 billion to the dollar by November 1, and to 4.2 trillion to the dollar by the end of November. Prices rise so fast that workers are paid daily—and then several times a day. Middle-class savers and pensioners are wiped out, formerly affluent Germans dispose of their possessions in order to eat, German peasants refuse to part with their eggs, milk, butter, or potatoes except in exchange for articles of tangible value and they fill their houses with pianos, sewing machines, Persian rugs, even Rembrandts (*see* 1924).

Federal income taxes paid by thousands of leading Americans appear in newspapers pursuant to terms of a tax disclosure law passed by Congress through efforts by Sen. George W. Norris (Rep. Neb.). John D. Rockefeller, now 83, reportedly paid only $124,266, but his son John D., Jr., has paid $7.4 million under the relatively low prevailing rates. Henry Ford reportedly paid nearly $2.47 million.

U.S. corporate profits will increase by 62 percent in the next 6 years and dividends to stockholders by 65 percent. Income of U.S. workers will increase by 11 percent.

Anaconda Co. acquires Chile's Chuquicamata copper mine for $70 million from the Guggenheim family which retires from the copper business (*see* 1910; first Guggenheim foundation, 1924).

A Siberian gold rush begins as Soviet prospector Voldimar P. Bertin encounters independent Yakut

Germany's postwar inflation reduced the value of the mark until it took more than 4 trillion marks to buy $1.

prospector Mikhail P. Tarbukin working a rich stake he has found near Aldan and persuades him June 19 to join forces with the state prospecting party. Tarbukin will receive an Order of Lenin and an Order of the Red Banner, Soviet troops will move into Aldan early in 1925 to restore order after a series of robberies and murders, and the Aldan fields will produce hundreds of tons of gold per year for more than half a century.

Russia recovers rapidly from her 1921–1922 famine and approaches prewar levels of agricultural and industrial production.

U.S. cotton prices fall to 11¢ lb., down from 44¢ last year.

Textile executive J. Spencer Love, 27, of Gastonia, N.C., sells his mill at auction for $200,000, retains his outworn machines, moves them to Burlington whose Chamber of Commerce has agreed to underwrite a $250,000 stock offering and sell the stock to local investors, produces a coarse cotton dress fabric that promptly goes out of fashion, but will make his company the largest U.S. rayon producer. Burlington will become the world's largest diversified textile producer.

The May Company acquires the 44-year-old Hamburger & Sons store in downtown Los Angeles for $4.2 million and adds a Los Angeles May Company to its Denver, St. Louis, Akron, and Cleveland stores (*see* 1914; Baltimore, 1927).

U.S. oil magnate Harry F. Sinclair wins the Kentucky Derby with his horse Zev and sails for Russia with an offer to buy the Baku oil fields (*see* 1901). Laughing off a congressional investigation of his role in the Teapot Dome scandal that broke 2 years ago, Sinclair and his retinue occupy two decks of the S.S. *Homeric* and by some accounts his attempt to buy the Baku fields comes close to succeeding (*see* 1916).

Ethyl Corp. introduces tetraethyl lead as a fuel additive (*see* Midgley, 1921). The lead is mixed three parts to two with ethylene dibromine, a derivative of bromine, to eliminate engine "knock" and yet leave no lead deposit in an engine (*see* Dow, 1924).

Aeroflot begins operations July 15 with a six-passenger ANT-2 flight from Moscow to Nizhny Novgorod. The plane takes 3.5 hours to cover the 200-mile distance, but Aeroflot will grow to become the world's largest, making parts of Russia accessible within hours where it once took days and even weeks to reach them overland.

Sabena (Société Anonyme Belge d'Exploitation de la Navigation Aérienne), founded at Brussels, will be the Belgian national airline.

Pan American World Airways has its beginnings in a New York City plane taxi service started by local bond salesman Juan Terry Trippe, 24, who quits his job and joins his friend John Hambleton in buying nine flying boats the U.S. Navy was about to scrap (*see* 1925).

German aircraft designer Wilhelm Messerschmitt, 27, establishes an aircraft manufacturing firm under his own name while continuing to work as an engineer for

1923 *(cont.)* Bayerische Flugzeugwerke. Messerschmitt designed his first plane during the war at age 18.

The autogyro, invented by Spanish aeronaut Juan de la Cierva, 27, has a large horizontal free-running rotor plus a conventional propeller for forward power but no conventional wings *(see* Sikorsky helicopter, 1913; 1939).

Canadian National Railways is created by a takeover of bankrupt roads and a merger of the roads with the government-owned Grand Trunk and Intercolonial roads to form a government-owned rail network larger than any other in the Western Hemisphere. Canadian National has 20,573 miles of track—7,000 miles more than Canadian Pacific.

Elmer Sperry invents a device for detecting and measuring defects in railroad rails. He will perfect the device in 1928 and the first Sperry detector cars will go into service in November of that year.

U.S. auto production reaches 3,780,358, up from 543,679 in 1914; 51.85 percent are Fords.

More than 13 million cars are on U.S. roads and 108 U.S. companies are engaged in adding to the total, but 10 auto makers account for 90 percent of sales.

Major U.S. auto makers inaugurate annual model style changes that make older models stylistically obsolete in a move that will force smaller companies out of the market and prevent new ones from entering. Forty-three U.S. companies will be making motorcars by 1926, there will be only 10 by 1935, and no new domestic manufacturer will crack the market successfully after this year *(see* Kaiser, 1946). Planned obsolescence will be a major part of automotive marketing.

Walter P. Chrysler becomes president of a reorganized Maxwell Motor Co. and begins to develop a line of innovative new motorcars *(see* Buick, 1912). Now 48, Chrysler quit his position as president of Buick division 3 years ago in a dispute with General Motors president W. C. Durant and has helped reorganize Willys-Overland *(see* 1924).

The Hertz Drive-Ur-Self System is founded at Chicago by Yellow Cab Co. president John D. Hertz, who buys a company that started in 1918 with 12 used cars operating out of a lot in South Michigan Avenue. Hertz will be the world's largest auto rental concern.

Dutch-American physicist Peter Joseph Wilhelm Debye, 39, works out a mathematical explanation for electrolysis. Metallic salts disassociate in solution into ions (charged atoms), Debye explains, and he shows why the disassociation does not always appear to be complete *(see* Arrhenius, 1887).

Tetanus toxoid, developed by French bacteriologist Gaston Ramon, 37, of the Pasteur Institute, will lead to wide-scale immunization against the infection.

Scots-Canadian physiologist John James Rickard McLeod, 47, who interpreted last year's Banting-Best insulin discovery shares the Nobel award with Frederick G. Banting.

Aimee Semple McPherson dedicates Angelus Temple at Los Angeles January 1 with a large rotating illuminated cross visible for 50 miles (*see* 1918). McPherson uses special effects to produce thunder, lightning, and wind that illustrate her "foursquare gospel" and help fill her 5,000-seat temple.

Bethune-Cookman College is founded at Daytona, Fla., by a merger of Mary McLeod Bethune's 19-year-old Daytona Normal and Industrial Institute for Negro Girls with the Cookman Institute for Men at Jacksonville. Bethune will head the college until 1942, making its slogan "Enter to Learn, Depart to Serve."

Zenith Radio is founded by Chicago auto-finance entrepreneur Eugene F. McDonald, Jr., 33, who obtains exclusive rights to market products of the Chicago Radio Laboratory founded by former U.S. Navy radio electricians Karl E. Hassel, 27, and R. H. G. Matthews. They constructed a longwave radio receiver for the *Chicago Tribune* in 1919 and have developed the trademark Z-Nith from the call letters of their amateur radio station 9ZN. The *Tribune* was able to pick up news dispatches from the Versailles peace conference and thus gain a 12- to 24-hour lead over papers using the jammed Atlantic Cable, Major Edwin H. Armstrong has licensed Chicago Radio to produce sets using his patents, McDonald has raised $330,000 to start Zenith, and he also starts the National Association of Broadcasters with himself as president *(see* Armstrong, 1917; FM, 1940).

London Radio Times begins publication. Circulation will reach 9 million by 1950.

Published March 3 at New York is Volume 1, No. 1 of newsweekly *Time,* a venture that will mushroom into a vast publishing empire. Based on rewrites of wire service stories with little additional reporting, the magazine is put out by Henry Robinson Luce, 24, and his Yale classmate Briton Hadden who resigned last year from their jobs as reporters for the *Baltimore News.* Hadden will die in 1929 after having established a distinctive Timestyle by inverting sentences and inventing such words as "socialite," "GOPolitician," "cinemaddict," and "tycoon" (meaning a business magnate) (*see Fortune*, 1930; *Newsweek*, 1933).

The Gannett Co. founded by upstate New York publisher Frank Ernest Gannett, 47, combines four regional newspapers including the *Rochester Times-Union* and *Rochester Democrat and Chronicle* (*see USA Today*, 1982).

"Skippy," by comic-strip artist Percy Crosby, 32, features the adventures of a winsome young boy.

"Moon Mullins," by comic-strip artist Frank Willard, 32, features the adventures of a roughneck ne'er-do-well, his kid brother Kayo, Uncle Willie and Aunt Mamie, and Lord and Lady Plushbottom.

A. C. Nielsen Co. is founded by Chicago electrical engineer-market researcher Arthur Charles Nielsen, 26, who will develop indexes to provide information on distribution in various industries including food, drugs, and pharmaceuticals. Nielsen ratings for the broadcast media will be used by advertisers and their agencies in selecting shows for sponsorship and placement of commercials.

Nonfiction *I and Thou (Ich und Du)* by Austrian philosopher Martin Buber, 45; *The Theme of Our Time (El Tema de Nuestro Tiempo)* by Spanish essayist José Ortega y Gasset, 40; *The Prophet* by Syrian-American mystic Kahlil Gibran, 40, whose work about man's relation to his fellow man will sell 5 million copies in the next 50 years; *The Dance of Life* by English scientist (Henry) Havelock Ellis, 64, who says, "Freud regards dreaming as fiction that helps us to sleep; thinking we may regard as fiction that helps us to live; Man lives by imagination"; *My Life and Loves* (first volume) by U.S. journalist and womanizer Frank Harris.

Fiction *Confessions of Zeno (La conscienza di Zeno)* by Italian novelist Italo Svevo (Ettore Schmitz), 62, who studied English at Trieste under James Joyce; *Genitrix* by François Mauriac; *The Good Soldier Sveiyk (Osudy dobreho vojaka Svidka za svet ove valky)* by the late Czech novelist Jaroslav Hasek who has died at age 40 leaving his bawdy attack on bourgeois values incomplete; *The Marsden Case* by Ford Madox Ford; *Riceyman Steps* by Arnold Bennett; *Antic Hay* by Aldous Huxley.

Poetry "Stopping by Woods on a Snowy Evening" by Robert Frost: "The woods are lovely, dark and deep./ But I have promises to keep,/ And miles to go before I sleep,/ And miles to go before I sleep"; *Harmonium* by Connecticut poet-life insurance company lawyer Wallace Stevens, 44; *The Harp-Weaver and Other Poems* by Edna St. Vincent Millay who marries and moves to Austerlitz, N.Y., in the Berkshires: "Euclid alone has looked on Beauty bare" (Sonnet 22, II, 11–12); *Sonnets to Orpheus* by Rainer Maria Rilke; *Red Candle* by Chinese poet Wen I-tuo, 24, who came to America last year but will return to China in 1925; *Collected Poems* by Vachel Lindsay, which includes "In Praise of Johnny Appleseed"; *Spring and All* by William Carlos Williams; *Come Hither*, a "collection of rhymes and poems for the young of all ages" by Walter de la Mare.

Painting *The Lovers* (neoclassical), *Lady with the Veil* (neoclassical), *Melancholy* (expressionist), and *Women* (surrealist) by Pablo Picasso; *Circles in the Circle* by Vassily Kandinsky; *Ivry Town Hall* by Maurice Utrillo; *Village in Northern France* by Maurice de Vlaminck; *On the Banks of the River Marne* by Raoul Dufy; *Love Idyll* by Marc Chagall; *The Trapeze* by Max Beckmann.

Murals *Entering the Mine*, *Miners Being Searched*, *Freeing of the Peon*, *The First of May*, *The Revolution Will Bear Fruit!* and others by Diego Rivera at Mexico City; *Revolutionary Trinity* by José Clemente Orozco; *The Workman Sacrificed* by David Afaro Siqueiros; Diego Rivera begins work on 239 frescos for the Secretariat of Public Education.

Theater *The Young Idea* by Noël Coward 2/1 at London's Savoy Theatre, with Coward, Herbert Marshall, Ann Trevor, 60 perfs.; *Icebound* by U.S. playwright Owen Davis, 49, 2/10 at New York's Sam Harris Theater, with Boots Wooster, 171 perfs.; *You and I* by Philip Barry 2/19 at New York's Belmont Theater with Lucile Watson, 140 perfs.; *The Adding Machine* by Elmer Rice 3/9 at New York's Garrick Theater, with Dudley Digges as Mr. Zero, Edward G. Robinson as Shrdlu, Helen Westley, 72 perfs.; *The Incorruptible Man (Der Unbestechliche)* by Hugo von Hofmannsthal 3/16 at Vienna's Raimund Theater; *The Shadow of a Gunman* by Irish playwright Sean O'Casey (*né* Cassidy), 43, 4/9 at Dublin's Abbey Theatre, a play about the Easter Rising of 1916; *Poppy* by U.S. playwright-lyricist Dorothy Donelly 9/3 at New York's Apollo Theater, with former *Ziegfeld Follies* juggler W. C. Fields who has been helped by showman Philip Goodman to develop the fraudulent character Eustace McGargle that Fields will portray for 20 years, 346 perfs.; *Outward Bound* by English playwright Sutton Vane, 34, 9/17 at London's Everyman Theatre; *Hinkemann (Der deutsche Hinkemann)* by Ernst Toller 9/19 at Leipzig's Altes Theater; *White Cargo* by U.S. playwright Leon Gordon 11/5 at New York's Greenwich Village Theater, with Richard Stevenson, Annette Margules as Tondeleyo, 678 perfs.; *Saint Joan* by George Bernard Shaw 12/28 at New York's Garrick Theater, with Winifred Lenihan, Morris Carnovsky, 214 perfs.

Films Buster Keaton's *Our Hospitality* with Keaton, Natalie Talmadge. Also: John Griffith Wray's *Anna Christie* with Blanche Sweet; James Cruze's *The Covered Wagon* with J. Warren Kerrigan; Wallace Worsley's *The Hunchback of Notre Dame* with Lon Chaney; Cecil B. DeMille's *The Ten Commandments* with Theodore Roberts, Estelle Taylor, Richard Dix, Nita Naldi; Sam Taylor and Tim Whelan's *Safety Last* with Harold Lloyd; Rex Ingram's *Where the Pavement Ends* with Ramon Novarro, Alice Terry; Fred Newmeyer and Sam Taylor's *Why Worry?* with Harold Lloyd; Charlie Chaplin's *A Woman of Paris* with Edna Purviance, Adolphe Menjou.

First performances Symphony No. 6 by Jean Sibelius 2/19 at Helsinki; Symphony No. 1 in E minor *(Nordic)* by U.S. composer Howard Hanson, 26, 5/30 at Rome; *The Black Maskers* suite by U.S. composer Roger Huntington Sessions, 26, 6/23 at Smith College, Northampton, Mass. Sessions has written the incidental music for a performance of the play by Leonid Andreyev; *Dance Suite (Tancsuit)* for Orchestra by Béla Bartòk 11/19 at Budapest.

Ballet *Façade* 6/12 at London's Aeolian Hall, with Edith Sitwell reading her poetry, music by English composer William Walton, 21 (*see* 1931); *Les Noces (The Wedding)* 6/14 at the Théâtre Gaiété-Lyrique, Paris, with Felicia Dubrovska, music and lyrics by Igor Stravinsky, choreography by Bronislava Nijinska; *La Création du Monde* 10/25 at the Théâtre des Champs Elysées, Paris, with Jean Borlin, 30, music by Darius Milhaud, choreography by Borlin, scenery, costumes, and curtain by Fernand Léger.

Stage musicals André Charlot's revue *Rats* 2/21 at London's Vaudeville Theatre, with Gertrude Lawrence

1923 ***(cont.)*** (Gertrud Alexandra Dagmar Lawrence Klasen), 24, music by Philip Braham, book and lyrics by Ronald Jeans; *The Passing Show* 6/14 at New York's Winter Garden Theater, with George Jessel, music by Sigmund Romberg and Jean Schwartz, book and lyrics by Harold Atteridge, 118 perfs.; *George White's Scandals* 6/18 at New York's Globe Theater, with music by George Gershwin, lyrics by E. Ray Goetz, B. G. DeSylva, Ballard MacDonald, 168 perfs.; *Helen of Troy, N. Y.* 6/19 at New York's Selwyn Theater, with Helen Ford, music and lyrics by Bert Kalmar and Harry Ruby, book by George S. Kaufman and Marc Connelly, 191 perfs.; *Artists and Models* 8/20 at New York's Shubert Theater, with Frank Fay, seminude showgirls, music by Jean Schwartz, book and lyrics by Harold Atteridge, 312 perfs.; André Charlot's revue *London Calling* 9/4 at the Duke of York's Theatre, London, with Noël Coward, Gertrude Lawrence, music and lyrics mostly by Coward, songs that include "You Were Meant for Me" by Eubie Blake, lyrics by Noble Sissle; *The Greenwich Village Follies* 9/20 at New York's Winter Garden Theater, with music by Louis Hirsch and Con Conrad, lyrics by Irving Caesar and John Murray Anderson, 140 perfs.; *The Music Box Revue* 9/22 at New York's Music Box Theater, with Grace Moore, music and lyrics by Irving Berlin, 273 perfs.; *The Ziegfeld Follies* 10/20 at New York's New Amsterdam Theater, with Fanny Brice, Paul Whiteman and his Orchestra, music by Dave Stamper, Rudolf Friml, Victor Herbert, and others, 233 perfs.; *Runnin' Wild* (black revue) 10/29 at New York's Colonial Theater, with a title song by A. Harrington Gibbs, lyrics by Joe Grey and Leo Wood, "Charleston" by Cecil Mack (Richard C. McPherson), 40, and Jimmy Johnson, 29, whose song launches a national dance craze; *Kid Boots* 12/31 at New York's Earl Carroll Theater, with Eddie Cantor, music by Harry Tierney, lyrics by Joseph McCarthy, songs that include "Polly Put the Kettle On," 479 perfs.; *The Song and Dance Man* 12/31 at New York's George M. Cohan Theater, with Cohan, Robert Cummings, music and lyrics by Cohan, 96 perfs.

Popular songs "Yes, We Have No Bananas" by Frank Silver and Irving Cohn; "Nobody's Sweetheart" by Gus Kahn, Ernie Erdman, Billie Meyers, Elmer Schwebel; "Who's Sorry Now?" by Ted Snyder, lyrics by Bert Kalmar, Harry Ruby; "I Cried for You" by Gus Arnheim and Abe Lyman, lyrics by Arthur Freed; "Nobody Knows You When You're Down and Out" by Jimmy Cox; "Barney Google" and "You Gotta See Mamma Ev'ry Night, Or You Can't See Mamma At All" by Con Conran and New York songwriter Billy Rose (*né* William Samuel Rosenberg), 23; "Mexicali Rose" by Jack B. Tenny, lyrics by Helen Stone; "It Ain't Gonna Rain No Mo'" by Wendell Woods Hall, 27, who has adapted an old folk song; "Down Hearted Blues" by U.S. blues singer Alberta Hunter, 24, lyrics by Lovie Austin (Philadelphia blues singer Bessie Smith, 23, records the song, which has sales of 2 million copies).

William M. Johnston, 28, wins in men's singles at Wimbledon, Suzanne Lenglen in women's singles; Bill Tilden wins in U.S. men's singles, Helen N. Wills, 17, in women's singles.

The Wightman Cup donated by Hazel Hotchkiss Wightman for the winner of a U.S.-British women's tennis tournament will do for women's tennis what the Davis Cup is doing for men's.

New York's Yankee Stadium opens April 19, draws a sell-out crowd of more than 60,000, and turns away thousands for lack of seats. Col. Jacob Ruppert has built the $2.5 million stadium with help from Tillinghast I'Hommedieu Houston (who will soon sell his share in the club to brewer Ruppert), Babe Ruth hits a three-run homer in the third inning of the inaugural game, and the Yankees beat the Boston Red Sox 4 to 1.

The New York Yankees win their first World Series by defeating the New York Giants 4 games to 2.

The Dempsey-Firpo fight at the New York Polo Grounds September 14 sees Argentine fighter Luis Angel Firpo, 29, knock Jack Dempsey out of the ring and into the laps of ringside sportswriters, but Dempsey knocks the "Wild Bull of the Pampas" down nine times, wins in two rounds, and retains his title.

The Schick magazine razor is patented April 24 by Col. Jacob Schick, 45, who during the war invented a machine that enabled a worker to fill 25 gas masks per minute where it had earlier taken 35 minutes to fill one mask.

The Schick dry shaver patented November 6 by Colonel Schick (above) is the world's first practical electric shaver (*see* 1930).

Maidenform brassieres are introduced by Russian-American entrepreneur Ida Rosenthal, 36, and her English-American partner Enid Bissett who last year opened a dress shop on New York's West 57th Street and gave away sample brassieres with a little uplift because they did not like the fit of their dresses on flat-chested "flappers" (*see* 1914).

President Coolidge lights the first White House Christmas tree to begin a lasting tradition.

U.S. cigarette production reaches 66.7 billion, up from 52 billion in 1921, as overseas sales increase.

New York City adopts a new setback law that limits the height and configuration of buildings as luxury apartment houses and hotels go up on Park Avenue *(see* Grand Central, 1913; zoning laws, 1916; 1961).

Japan's Great Kanto earthquake and fire September 1 destroy Tokyo and Yokohama. Six prefectures are affected, 100,000 killed, 752,000 injured; 83,000 houses are completely destroyed, 380,000 damaged, but the Imperial Hotel by Frank Lloyd Wright completed at Tokyo in 1916 survives intact. Tokyo will rebuild on the traditional pattern of Edo with houses numbered according to the years they were built.

An Agricultural Credits Act provides for 12 Federal Intermediate Banks to help U.S. agricultural and livestock interests which remain in deep depression *(see* 1921; 1922).

U.S. wheat farmers try to persuade each other to plant less but overproduction continues in the absence of any effective farm organization.

Grasshoppers plague Montana. Forming a cloud 300 miles long, 100 miles wide, and half a mile high, the locusts devour every green blade, leaf, and stalk, leaving holes in the ground where green plants grew.

Soybean cultivation increases in eastern states with encouragement from higher freight rates that make it costly to feed cows with cottonseed from the Cotton Belt or with bran from Minneapolis flour mills (*see* Staley, 1922; McMillen, 1934).

The Macoun apple is created in upstate New York by crossing a McIntosh with a Jersey Black.

Irradiating foods with ultraviolet light can make them rich sources of vitamin D, says University of Wisconsin biochemist Harry Steenbock, 37, and he files for a patent on his discovery. Researchers including Alfred Hess at Columbia University have found that ultraviolet light from the sun and from mercury vapor lamps can cure rickets (*see* McCollum, 1922), Steenbock has followed up on their studies, and he has found that stimulating the provitamins in foods can enable the human liver to convert them into vitamin D (*see* 1927; Rosenheim and Webster, 1926).

Congress calls filled milk a menace to health. The Filled Milk Act passed after pressure from dairy interests forbids use of nondairy ingredients (including low-fat vegetable oil to replace butterfat) in anything that is called (or looks like) a milk product (*see* 1916). The courts will overturn the law in 1973.

Pet Milk Co. is created by a reorganization of the Helvetia Milk Condensing Co. which moved to St. Louis in 1921 (*see* "Our Pet," 1894).

National Dairy Corp. is organized at New York by Thomas McInnerny who says the dairy industry is bigger than the automobile or steel industry but needs some organization to control the quality and service of its many small, local companies (*see* 1914). McInnerny's giant holding company consolidates 168 companies and will acquire many others (*see* 1928).

Birdseye Seafoods opens on New York's White Street. Clarence Birdseye has received patents on his quick-freezing process but lacks the means to gain public acceptance and is forced bankrupt (*see* 1917; 1925).

U.S. sugar consumption reaches 106.39 pounds per capita, up from 65.2 pounds in 1900. Some sugar goes into illegal "moonshine" whiskey.

Sanka Coffee is introduced in the United States (*see* Roselius, 1903).

The Milky Way candy bar developed in the Midway district between Minneapolis and St. Paul by confectioner Frank C. Mars, 39, is a mixture of milk chocolate, corn syrup, sugar, milk, hydrogenated vegetable oil, cocoa, butter, salt, malt, egg whites, etc. Inspired perhaps by his own astronomical name, Mars names his creation after the distant star galaxy and will see Milky Way sales leap in one year from $72,800 to $792,000 (*see* 1930).

The Butterfinger candy bar introduced by Chicago's Curtiss Candy Co. is a chocolate-covered honeycomb peanut butter confection. Otto Schnering has promoted Baby Ruth by chartering an airplane and parachuting candy bars down over Pittsburgh, causing a massive traffic jam as people rushed into the streets to scramble for the free candy. He will have Butterfingers dropped along with Baby Ruths on cities in 40 states and the stunt will help make Curtiss brands national favorites (*see* 1920).

A supermarket of sorts opens in San Francisco where a large steel-frame building opens on the site of a former baseball field and circus ground with 68,000 square feet of selling space and room to park 4,350 automobiles (shoppers are offered free parking for one hour). The Crystal Palace sells food, drugs, cigarettes, cigars, and jewelry and has a barber shop, beauty parlor, and dry cleaner (*see* Piggly Wiggly, 1916; King Kullen, 1930).

The Treaty of Lausanne (above) requires a massive transfer of 190,000 Greeks from Turkey to Greece and of 388,000 Muslims from Greece to Turkey (*see* Neuilly, 1919).

The American Birth Control League opens the Sanger Research Bureau with a birth control clinic at New York. Sanger persuades ABCL president James F. Cooper and his associate Herbert Simonds to start Holland-Rantos, first U.S. company to manufacture rubber diaphragm contraceptives for women (*see* Wilde, 1833; Sanger, 1921). The diaphragm provides women with their first means of avoiding pregnancy independent of a man's use of a condom (but *see* 1929).

1924 V. I. Lenin dies of sclerosis January 21 at age 53, Petrograd is renamed Leningrad, and a triumvirate takes power as Josef Stalin begins a power struggle with Leon Trotsky (*see* 1918). Ruling with Stalin are Lev Borisovich Kamenev (*né* Rosenfeld), 40, and Grigori Evseevich Zinoviev (*né* Hirsch Apfelbaum), 40 (*see* 1926; Zinoviev letter, below).

Russia gives up the czar's "ill-gotten gains" at the expense of China, returns her Boxer Rebellion indemnity of 1900 for use in Chinese education, and sends advisers to China's Guomindang national congress at Guangzhou (Canton). Sun Yat-sen admits Communists to the Guomindang (*see* 1911; Chiang Kai-shek, 1925).

Britain's first Labour government takes office January 22 under James Ramsay MacDonald, 57, who opposed British participation in the war. He recognizes the U.S.S.R. February 1, Britain signs a commercial treaty in which the Soviet Union gives British goods most-favored nation treatment, but the United States refuses recognition unless Moscow acknowledges its foreign debts and restores alien property, a position stated last year by Secretary of State Charles Evans Hughes.

The Ottoman dynasty founded in 1290 by Osman the Conquerer ends March 3 as Turkey's president Mustafa Kemal abolishes the caliphate and banishes all members of the house of Osman (*see* 1923; 1930).

1924 ***(cont.)*** Benito Mussolini's Italian fascists use pressure tactics to gain control of the electoral machinery and poll 65 percent of the vote in elections held early in April.

Italian Socialist deputy Giacomo Matteotti, 39, is murdered June 10 by Mussolini's Fascisti whose illegal acts of violence have been detailed in Matteotti's book *The Fascisti Exposed* (*see* 1923). Most of the nonfascist third of the Italian chamber secedes June 15, vowing not to return until the government has been cleared of complicity in the Matteotti murder, demands are made that the fascist militia be disbanded, Mussolini disavows any connection with the murder, dismisses all those implicated, but imposes strict press censorship July 1 and forbids opposition meetings August 3. Prominent Fascist party members will be tried for the murder in 1926 but none will receive more than a light sentence.

France's President Millerand resigns June 11 under pressure from Radical Socialist party leader Edouard Herriot, who says Millerand has tilted to the right. Gaston Doumerge becomes president June 13, Herriot prime minister June 14.

Persia's premier Reza Khan Pahlevi establishes government control throughout the country after having subdued the Bakhtiari chiefs of the southwest and the Sheik Khazal of Mohammerah who has been supported by the British and Anglo-Persian Oil Co. (*see* 1921; 1925).

The Wahabi sultan of Nejd Abdul-Aziz ibn Saud conquers the Hejaz, forces the Hashimite king Husein ibn Ali, 68, to abdicate in favor of his eldest son Ali ibn Husein, 46, and enters Mecca October 20 (*see* 1902; Saudi Arabia, 1926).

President Coolidge wins reelection on a platform of "Coolidge Prosperity." Unable to decide between President Wilson's son-in-law William Gibbs McAdoo of California and New York's Catholic governor Alfred E. Smith, the Democrats have nominated New York corporation lawyer John W. Davis on the 103rd ballot at the convention in New York's Madison Square Garden, but Davis wins only 136 electoral votes and 29 percent of the popular vote against 382 electoral votes and 54 percent of the popular vote for Coolidge. (Progressive candidate Robert M. La Follette has Socialist support and wins 13 electoral votes, 17 percent of the popular vote.)

Benito Mussolini glorified himself with propaganda techniques that masked the brutality of his Fascist rule in Italy.

Britain's Labour government falls November 4 after the general election October 29 has given the Conservatives a great victory, partly through the release October 25 of the so-called Zinoviev letter, possibly a forgery, allegedly written by Comintern chief Grigori E. Zinoviev (above) and calling on British Communists to "have cells in all units of the troops," especially those based in big cities or near munitions plants. Stanley Baldwin heads a new government that denounces British treaties with Russia November 21. He will remain prime minister until 1929.

Albania proclaims herself a republic December 24.

Liberia rejects Marcus Garvey's plan for resettlement of U.S. blacks, fearing that his motive is to foment revolution (*see* 1920). Garvey will be convicted next year of fraudulent dealings in the now-bankrupt Black Star Steamship Co. he has founded, President Coolidge will commute his 5-year sentence, but Garvey will be deported back to Jamaica in 1927 (*see* Rastafarians, 1930).

Germany's ruinous inflation ends as Berlin issues a new Reichsmark, imposes strict new taxes, and makes a sharp cut in the availability of credit for business expansion. The Reichsmark is backed 30 percent by gold and its value is set at 1 billion old marks, which cease to be legal tender and are withdrawn from circulation under terms of a plan devised by a commission headed by Chicago banker Charles Gates Dawes, 59.

The Dawes Plan, which goes into effect September 1, reorganizes the Reichsbank under Allied supervision, sets World War reparations to be paid by Germany, and provides for an Allied loan to Germany of 800 million gold marks, $110 million of it to come from the United States.

International Business Machines Corp. (IBM) is organized at New York by former National Cash Register executive Thomas J. Watson who has changed the name of C-T-R to IBM (*see* 1912; Mark I computer, 1944).

Barney's opens on New York's Seventh Avenue at 17th Street. Merchant Barney Pressman will make his discount store the nation's largest menswear retailer.

Saks Fifth Avenue opens just south of St. Patrick's Cathedral with window displays featuring $1,000 raccoon coats, chauffeurs' livery, and a $3,000 pigskin trunk. New York merchant Bernard F. Gimbel, 39, and Horace Saks, 42, merged their stores at Greeley Square last year (*see* Gimbels, 1910), their store is the first large specialty shop north of 42nd Street, and it sells out its stock of silver pocket flasks the first day.

New York's R. H. Macy Co. opens a new building west of its existing store on Herald Square. Two further additions in 1928 and 1931 will make Macy's the world's largest department store under one roof, with more than 2 million square feet of floor space (*see* 1901; 1912).

The first Macy's Thanksgiving Day parade moves 2 miles from Central Park West down Broadway to Herald Square, beginning an annual promotion event designed to boost Christmas sales.

Ethyl Corp.'s Thomas Midgley, Jr., and C. F. Kettering of General Motors make a deal with Dow Chemical to obtain 100,000 pounds of ethylene dibromide per month at 58¢ lb. Ethyl combines the chemical with tetraethyl lead to make a gasoline additive that eliminates engine knock (*see* 1923).

Germany's I. G. Farben chemical cartel starts a synthetic gasoline development program. Chief executive Karl Bosch, 50, has been advised by his experts that rising gasoline prices will make gasoline derived from coal competitive with that derived from petroleum. Germany has no domestic sources of petroleum and lacks foreign exchange but does have abundant coal reserves and faces an energy crisis. Bosch is impressed by forecasts that world petroleum reserves will be exhausted within a few decades (*see* 1916); an industrial chemist who has adapted the Haber process of 1908 to commercial ammonia production, he predicts that synthetic gasoline will soon undersell gasoline produced from petroleum.

The first round-the-world flights are completed in August as two plywood, spruce, and linen canvas *World Cruisers* arrive in California after a 30,000-mile, 5-month journey (15 flying days). A third U.S. pilot crashed early on in Alaska (but survived) and a fourth had engine trouble; British, French, and Portuguese pilots have all come to grief. Donald Douglas, who designed and built the planes and powered them with 12-cylinder Liberty engines, wins a U.S. Army order for 50 XO observation planes (*see* 1920; DC-1, 1933).

The first U.S. diesel electric locomotive goes into service for the Central Railroad of New Jersey (*see* Sweden, 1913). The 300-horsepower Ingersoll-Rand engine uses GE electrical components and is installed in a locomotive built by American Locomotive. It will remain in service until 1957 (*see* 1948).

The first Chrysler motorcar, introduced by Maxwell Motor Co., has four-wheel hydraulic brakes, a high compression engine, and other engineering advances (*see* 1923, 1925).

Chicago executive E. L. (Erret Lobban) Cord, 30, joins Auburn Automobile, gives its unsold inventory of 700 cars some cosmetic touch-ups, nets $500,000, and breathes new life into the company which is now owned by Chicago financiers including William Wrigley, Jr., but producing only six cars per day. Cord will double sales next year, introduce a new model, outperform and undersell the competition, and become president of Auburn in 1926 (*see* Duesenberg, 1926; 8–115, 1928).

Oakland automobiles made by General Motors become the first cars finished with DuPont Duco paints, which cut days off the time required to paint a car (*see* 1909; Du Pont, 1915; Pontiac, 1926).

Ford produces nearly 2 million Model T motorcars for the second year in a row and drops the price of a new touring car to a low of $290, making a durable automobile available to Americans even of modest means. More than half the cars in the world are Model T Fords (*see* 1926; Model A, 1927).

The 497-meter Bear Mountain Bridge opens at Peekskill, N.Y., to carry motorcar traffic across the Hudson River (*see* George Washington Bridge, 1931).

New York's Bronx-Whitestone, Cross-Bay, Henry Hudson, Throgs Neck, Triborough, and Verrazano bridges will be built in the next 44 years by Robert Moses, 35, who is named president of the Long Island State Park Commission by Governor Alfred E. Smith. Moses will also build the Bruckner, Clearview, Brooklyn-Queens, Cross-Bronx, Gowanus, Long Island, Major Deegan, Prospect, Nassau, Sheridan, Staten Island, Throgs Neck, Van Wyck, and Whitestone expressways, the Harlem River Drive and West Side Highway, and the Belt, Cross Island, Cross-County, Grand Central, Interborough, Northern State, Meadowbrook, Laurelton, Sunken Meadow, Mosholu, and Hutchinson River parkways, but he will build with no master plan and his projects will favor private automobile travel at the expense of mass transit.

U.S. physician George Frederick Dick, 43, and his wife Gladys, 42, isolate the streptococcus that incites scarlet fever. They will devise the Dick skin test for susceptibility to the disease.

Argentine physiologist Bernardo Alberto Houssay, 37, demonstrates that the pituitary gland is involved in human sugar breakdown, not just the pancreas (*see* 1937; Banting, Best, 1922).

Duke University gets its name and begins the rise that will make it a great seat of learning. American Tobacco Co. president James Buchanan "Buck" Duke, who has made earlier gifts to the 86-year-old Trinity College at Durham, N.C., announces that he will donate $40 million if the school will change its name, the offer is widely criticized but Duke adds another $7 million October 1 and dies 9 days later at age 67; Trinity College becomes Duke University.

The Brookings Institution is founded under the name Robert Brookings Graduate School of Economics and Government with a gift from St. Louis philanthropist-educator Robert Somers Brookings, 74, who has made a fortune in woodenware.

U.S. radio set ownership reaches 3 million but most listeners use crystal sets with earphones to pick up signals from the growing number of radio stations.

The *New York Herald-Tribune* is created by a merger of the Gordon Bennett-Horace Greeley newspapers. *Tribune* editor Ogden Mills Reid, 42, heads the new daily that will give the *New York Times* healthy competition until 1966.

1924 *(cont.)* *Le Figaro* in Paris is acquired after 70 years of publication by Corsican-born French perfume maker François Coty (*né* Francesco Giuseppe Spoturno), 50, and will make the paper a vehicle for his pro-fascist views.

"Little Orphan Annie" appears October 5 in the *New York Daily News. Chicago Tribune* staff cartoonist Harold Lincoln Gray, 30, has helped Sidney Smith draw "The Gumps" and has developed a blank-eyed, 12-year-old character together with her dog Sandy, her guardian Oliver "Daddy" Warbucks, and his manservant Punjab. The comic strip will campaign against communism, blind liberalism, and other threats to free enterprise and rugged individualism, continuing beyond Gray's death in 1968.

Nonfiction *Manifesto of Surrealism (Manifeste de surréalisme)* by French poet-essayist André Breton, 28, who has been a member of the dadaist movement and whose October Manifesto inaugurates a new movement (*see* Apollinaire, 1917). Breton becomes editor of *La Revolution Surréaliste; With Lawrence in Arabia* by U.S. journalist Lowell Jackson Thomas, 32, brings prominence to English archaeologist-soldier T. E. Lawrence who has refused a knighthood and the Victoria Cross for his World War exploits in the Arabian desert (*see* 1916; 1926).

Fiction *The Magic Mountain (Der Zauberberg)* by Thomas Mann who uses the setting of a sanatorium to explore the sickness of contemporary European civilization; A *Passage to India* by E. M. Forster who has produced no new novel since 1910; *Precious Bane* by Mary Webb; *The Green Hat* by Armenian-English novelist Michael Arlen (*né* Dikran Kuyumijian), 30; *Some Do Not* by Ford Madox Ford; *The Fleshly and the Meditative Man (L'homme de chair et l'homme reflet)* by French (Breton) novelist-poet Max Jacob, 48; *Billy Budd* by Herman Melville, who died in 1891; *The Green Bay Tree* by U.S. novelist Louis Bromfield, 27; "Salt," "The Letter," and other stories by Russian author Isaak Emmanuilovich Babel, 30, in the magazine *Left; So Big* by U.S. novelist Edna Ferber, 37.

Juvenile *Bambi* by Austrian critic-magazine writer Felix Salter, 57, whose anthropomorphic tale of a fawn will be translated into English by U.S. magazine editor Whittaker Chambers; *When We Were Very Young* by A. A. Milne who has written the verses to amuse his 4-year-old son Christopher Robin, illustrations by Ernest H. Shepard.

Poetry *Veinte poemas de amor y una cancion des esperada* by Chilean poet Pablo Neruda (Neftali Ricardo Reyes), 20, who won recognition last year with his *Crepusculazio* and will go abroad in 1927 as Chilean consul first at Rangoon, then in Java, then at Barcelona; *Tamar and Other Poems* by California poet Robinson Jeffers, 37, who has built a high rock tower at Carmel Bay and lived in seclusion with his wife since the outbreak of war in 1914; *The Man Who Died Twice* (novel in verse) by Edwin Arlington Robinson; *Chills and Fever* by Vanderbilt University teacher-poet John Crowe Ransom, now 40; *The Ship Sails On* by Nordahl Grieg.

Painting *Sugar Bowl* by Georges Braque; *Daughter Ida at the Window* and *The Birthday* by Marc Chagall; *Catalan Landscape* by Joan Miró; *Venice* by Oskar Kokoschka; *The Dempsey-Firpo Fight* by George Bellows (*see* sports, 1923).

Eastman Kodak replaces its highly flammable celluloid film with a photographic film made of cellulose acetate (*see* Reichenbach, Goodwin, 1889; Kodachrome, 1935).

Theater *Hell Bent fer Heaven* by U.S. playwright Hatcher Hughes, 43, 1/4 at New York's Klaw Theater, 122 perfs.; *The Show-Off* by U.S. playwright George Kelly, 37, 2/4 at New York's Playhouse Theater, with Louis John Bartels, Lee Tracy, 575 perfs.; *Beggar on Horseback* by George S. Kaufman and Marc Connelly 2/12 at New York's Broadhurst Theater, with Osgood Perkins, Spring Byington, 164 perfs.; *Juno and the Paycock* by Sean O'Casey 3/3 at Dublin's Abbey Theatre. O'Casey has been supporting himself with pick and shovel on a road gang; *The Farmer's Wife* by English novelist-playwright Eden Phillpotts, 61, 3/11 at London's Court Theatre, with Cedric Hardwicke, Evelyn Hope, Melville Cooper, Maud Gill, 1,329 perfs; *Welded* by Eugene O'Neill 3/17 at New York's 39th Street Theater, 24 perfs.; *All God's Chillun Got Wings* by O'Neill 5/15 at New York's Provincetown Playhouse, with former Rutgers football star Paul Bustill Robeson, 26, playing a black man married to a white woman (Mary Blair). The Ku Klux Klan threatens reprisals, the Salvation Army and the Society for the Suppression of Vice warn that "such a play might easily lead to racial riots or disorder" but there are no disturbances; *Each in His Own Way (Ciaseuno a suo modo)* by Luigi Pirandello at Milan's Teatro dei Filodramatici; *What Price Glory?* by New York playwrights Laurence Stallings, 28, and Maxwell Anderson, 35, 9/3 at New York's Plymouth Theater, with Louis Wolheim as Captain Flagg, Brian Donlevy as Corporal Gowdy, 299 perfs.; *Desire Under the Elms* by O'Neill 11/11 at New York's Greenwich Village Theater, with Walter Huston, 208 perfs. (the play moves uptown after 2 months); *They Knew What They Wanted* by U.S. playwright Sidney Coe Howard, 33, 11/24 at New York's Garrick Theater, with Pauline Lord, Glenn Anders, Richard Bennett, 414 perfs.; *The Man With a Load of Mischief* by English playwright Ashley Dukes, 39, 12/7 at London; *The Vortex* by Noël Coward 12/16 at London's Royalty Theatre, with Coward, Mary Robson, 224 perfs.; *The Youngest* by Philip Barry 12/22 at New York's Gaiety Theater, with Henry Hull, 104 perfs.

Films Erich von Stroheim's *Greed* with Gibson Gowland as McTeague, ZaSu Pitts, Jean Hersholt; Buster Keaton's *Sherlock, Jr.* with Keaton. Also: D. W. Griffith's *America*; Griffith's *Isn't Life Wonderful* with Carol Dempster; Fernand Léger's *Ballet Mécanique*; Henry Beaumont's *Beau Brummel* with John Barrymore, Mary Astor; Jean Renoir's *La Fille de l'Eau*; Mauritz Stiller's *Gosta Berling's Saga* with Swedish actress Greta Garbo (*née* Gustaffson), 19; Victor Seastrom's *He Who Gets Slapped* with Lon Chaney,

Norma Shearer, John Gilbert; John Ford's *The Iron Horse* with George O'Brien, Madge Bellamy; Hal Roach's *Jubilo Jr.* with Our Gang (Mary Korman, Mickey Davids, Joe Cobb, Farina, Jackie Condon); Fritz Lang's *Kriemheld's Revenge*; F.W. Murneau's *The Last Laugh* with Emil Jannings; James Cruze's *Merton of the Movies* with Glenn Hunter; Sidney Olcott's *Monsieur Beaucaire* with Rudolph Valentino; Buster Keaton and Donald Crisp's *The Navigator* with Keaton; Herbert Brenon's *Peter Pan* with Betty Bronson; Henry King's *Romola* with Lillian and Dorothy Gish, William Powell, Ronald Colman; Frank Lloyd's *The Sea Hawk* with Morton Sills; Serge Mikhailovich's *Strike;* Raoul Walsh's *The Thief of Baghdad* with Douglas Fairbanks; Paul Leni's *Waxworks* with Conrad Veidt as Haroun-al-Raschid, Werner Kraus as Jack the Ripper.

Metro-Goldwyn-Mayer (M-G-M) is founded by vaudeville theater magnate Marcus Loew, 54, who has absorbed Louis B. Mayer Pictures Corp. and buys Goldwyn Pictures Corp. from which Samuel Goldwyn has resigned (*see* 1917). Goldwyn will control Samuel Goldwyn, Inc. from 1926 to 1941 and continue to produce films through the 1950s, winning fame as a producer and notoriety for such expressions as "Include me out" and "In two words, impossible."

Columbia Pictures is founded by Harry Cohn, 33, who 6 years ago joined Universal Pictures as secretary to Carl Laemmle, learned the rudiments of picture making, and 2 years later formed CBS Sales Co., which he has reorganized.

Music Corp. of America (MCA) is founded by Chicago physician Jules C. Stein, 28, who has just published a definitive work on telescopic spectacles after doing postgraduate work at Vienna. To help pay his way through medical school Stein has played violin and saxophone in his own dance band, he has booked dance bands into hotels, nightclubs, and summer resorts on a commission basis, and he starts a company that will innovate the one-night stand at a time when most bookings have been for the season. Beginning with Guy Lombardo and His Royal Canadians, Stein will sign up most of the big bands before moving on to sign variety acts, radio talent, and movie stars to exclusive MCA contracts.

Opera *Ertwartung* (*Expectation*) 6/6 at Prague's Neues Deutsches Theater, an expressionistic one-character mimodrama with music by Arnold Schoenberg; *Hugh the Drover, or Love in the Stocks* 7/4 at His Majesty's Theatre, London, with music by Ralph Vaughan Williams; *Die glückliche Hand* 10/14 at Vienna's Volksoper, with music by Schoenberg.

Ballet *Les Biches* 1/6 at the Théâtre de Monte Carlo, with Polish dancer Leon Woizikovsky (*né* Wojcikowski), 24, and other members of Sergei Diaghilev's Ballet Russe, music by French composer Francis Poulenc, 24, choreography by Bronislava Nijinska; *Les Fâcheux* 1/19 at the Théâtre de Monte Carlo, with Sergei Diaghilev's Ballet Russe, music by French composer Georges Auric, 24; *Schlagobers* 5/9 at the Vienna State Opera with Tilly Losch, 19, music by Richard Strauss; *Mercure* ("poses plastiques") 6/14 at the Théâtre de la Cigale, Paris, with music by Eric Satie, sets and curtains by Pablo Picasso that overshadow Leonide Massine's choreography; *Le Train Bleu* 6/20 at the Théâtre des Champs-Elysées, Paris, with Bronislava Nijinska and other members of the Ballet Russe, music by Darius Milhaud, choreography by Nijinska, libretto by Jean Cocteau, curtain by Pablo Picasso, costumes by Coco Chanel.

George Balanchine quits the Soviet State Dancers while on tour in Paris, joins Sergei Diaghilev's Ballet Russe, changes his name from Georgi Meilitonovitch, and at age 20 begins a career that will make him the world's leading choreographer (*see* 1932).

Emigré Russian conductor and bass fiddle virtuoso Serge Koussevitsky, 50, begins a 25-year career as director of the Boston Symphony (*see* Tanglewood, 1934).

Philadelphia's Curtis Institute of Music is founded by publishing heiress Mary Louise Curtis Bok.

First performances *Rhapsody in Blue* by George Gershwin 2/12 at New York's Aeolian Hall with Paul Whiteman's Palais Royal Orchestra accompanying the pianist-composer who completed the piano score January 7 after 3 weeks of work. Composer Ferde Grofe, 31, has orchestrated the *Rhapsody* for piano and jazz ensemble; Symphony No. 7 in C major by Jean Sibelius 3/24 at Stockholm; Tzigane, Concert Rhapsody for Violin and Orchestra by Maurice Ravel 4/26 at London, with Hungarian virtuoso Yelly d'Arany as soloist; *Pacifica 231* by Arthur Honegger 5/8 at Paris; Concerto for Piano, Wind Instruments, and Double-Basses by Igor Stravinsky 5/22 at Paris; *The Pines of Rome (Pini di Roma)* (Symphonic Poem) by Ottorino Respighi 12/14 at Rome.

Broadway musicals *André Charlot's Revue of 1924* 1/9 at the Times Square Theater, with Jack Buchanan, Beatrice Lillie, Douglas Furber, music by Philip Braham, lyrics by Furber, songs that include "Limehouse Blues," 138 perfs.; *The Ziegfeld Follies* 6/24 at the New Amsterdam Theater, with Will Rogers, Vivienne Segal (W. C. Fields and Ray Dooley will join next spring), music by Raymond Hubbell, Dave Stamper, Harry Tierney, Victor Herbert, and others, 520 perfs.; *George White's Scandals* 6/30 at the Apollo Theater, with music by George Gershwin, lyrics by B. G. DeSylva, 192 perfs.; *Rosemarie* 9/2 at the Imperial Theater, with Mary Enis, Dennis King, music by Rudolf Friml, lyrics by Otto Harbach and Oscar Hammerstein II, 29, songs that include "Indian Love Call" and the title song, 557 perfs.; *The Passing Show* 9/3 at the Winter Garden Theater, with James Barton, Harry McNaughton, music by Sigmund Romberg and Jean Schwartz, lyrics by Harold Atteridge, 104 perfs.; *Greenwich Village Follies* 9/16 at the Shubert Theater, with the Dolly Sisters, Moran and Mack, music and lyrics by Cole Porter, songs that include "I'm in Love Again," "Babes in the Woods," 127 perfs.; *Dixie to Broadway* 11/22 at the Broadhurst Theater, with Florence Mills in an all-black revue, music by George Meyer and Arthur Johnston,

1924 ***(cont.)*** lyrics by Grant Clarke and Roy Turk, 77 perfs.; *Lady Be Good* 12/1 at the Liberty Theater, with Fred Astaire and his sister Adele dancing to music by George Gershwin, lyrics by Ira Gershwin, songs that include "Fascinating Rhythm," "The Man I Love," "Somebody Loves Me," 184 perfs.; *The Student Prince of Heidelberg* 12/2 at the Jolson Theater, with music by Sigmund Romberg, lyrics by Dorothy Donelly, songs that include "Deep in My Heart, Dear," "Welcome to Heidelberg," "Student Life," 608 perfs.

Bix Beiderbecke begins a 6-year career that will make the name Bix immortal in the annals of jazz. Iowa-born cornetist Leon Bix Beiderbecke, 20, records "Fidgety Feet" February 18 at the Gennett Studios in Richmond, Ind.

Popular songs "It Had to Be You" by Isham Jones, lyrics by Gus Kahn; "Tea for Two" by Vincent Youmans, lyrics by Irving Caesar and Clifford Grey; "How Come You Do Me Like You Do?" by Gene Austin and Roy Bergere; "Does the Spearmint Lose Its Flavor on the Bedpost Over Night?" by Ernest Breuer, lyrics by Billy Rose and Marty Blevin; "What'll I Do" by Irving Berlin; "When My Sugar Walks Down the Street" by Gene Austin, Jimmy McHugh, Irving Mills; "Everybody Loves My Baby (But My Baby Don't Love Nobody But Me)" by Jack Palmer and Spencer Williams; "Hard Hearted Hannah, The Vamp of Savannah" by Jack Yellen, Bob Bifelow, and Charles Bates; "There's Yes Yes in Your Eyes" by Joseph H. Saintly, lyrics by Cliff Friend; "Amapola (My Pretty Little Poppy)" by James M. LaCalle, lyrics by Albert Gamse; "When Day Is Done" ("Madonna") by Viennese composer Robert Katscher (B. G. DeSylva will write English lyrics).

Jean Borotra, 25, (Fr) wins in men's singles at Wimbledon, Kathleen "Kitty" McKane, 27, (Br) in women's singles (beating Helen Wills; Lenglen defaults); Bill Tilden wins in men's singles at Forest Hills, Wills in women's singles.

Forest Hills, Long Island, completes a tennis stadium for the West Side Tennis Club which held USLTA championship matches from 1915 to 1920 at Forest Hills. The matches that began in 1881 have been held for the past 3 years at Philadelphia's Germantown Cricket Club but will be held until 1980 at Forest Hills.

Wembley Stadium opens for the British Empire Exhibition at Middlesex, England, with seating capacity for nearly 100,000.

The first winter Olympics open at Chamonix; the games at Paris attract 3,385 contestants from 45 nations. Finnish track star Paavo Nurmi, 27, wins the 1,500-meter and 5,000-meter runs, but U.S. athletes win the most medals overall.

The Washington Senators win the World Series by defeating the New York Giants 4 games to 3.

University of Illinois halfback Harold "Red" Grange, 21, receives the opening kickoff from undefeated Michigan State, runs it back 95 yards for a touchdown, scores three more touchdowns in the next 12 minutes, and a fifth later in the game. Sportswriter Grantland Rice will nickname him the "Galloping Ghost" and select Grange for his All-America team 3 years in a row.

Notre Dame University has an undefeated season thanks to a backfield that Grantland Rice (above) calls the "Four Horsemen": Don Miller (162 pounds), Elmer Layden (162), Jim Crowley (164), and Harry Stuhldreher.

Kleenex, introduced under the name Celluwipes by Kimberly & Clark, is the first disposable handkerchief. Ernst Mahler of 1918 Kotex fame has developed the product from celucotton, the name will be changed to Kleenex 'Kerchiefs, and it will be shortened subsequently to the one word Kleenex.

Chicago students Nathan Leopold and Richard Loeb, both 19, confess May 31 that they have murdered their cousin Robert "Bobby" Franks, 14, "in the interests of science." Both are sons of rich families (Loeb's father is a Sears, Roebuck vice president). Lawyer Clarence Darrow, now 67, saves them from the gallows with his eloquence and they are sentenced to life imprisonment September 10; Loeb will be killed in a prison fight (*see* Darrow, 1911, 1925).

Federal agents break up Florida's notorious Ashley gang after 14 years of bank robberies, hijackings, and (more recently) rum running on the Atlantic Coast.

Chicago florist-bootlegger Dion O'Banion is shot dead November 10 by a gunman who walks into his flower shop with a companion and guns him down as the companion shakes O'Banion's hand. The ensuing funeral sets new records for gaudy display as Hymie Weiss (Earl Wajchiechowski) takes over the O'Banion mob to challenge Johnny Torrio and Al Capone for control of the liquor traffic.

The Federal Reserve Bank of New York building at 35 Liberty Street is completed in the style of a massive Florentine palazzo of Indiana limestone, Ohio sandstone, and ironwork (*see* Fort Knox, 1935).

Chicago's Wrigley Building is completed in May with a 32-story tower on Michigan Avenue for the William Wrigley Jr. Co. Designed by Graham, Anderson, Probst, and White, it has 442,000 square feet of office space and is the first large building north of the Chicago River.

Lima's Gran Hotel Bolivar opens with 350 rooms to begin half a century as Peru's central meeting place for businessmen, social leaders, and politicians.

Palm Beach, Fla., mansions designed by architect Addison Mizner, 52, go up for U.S. millionaires. Mizner builds Villa de Sarmiento for A. J. Drexel Biddle, Casa Bendita for John S. Philips, Concha Marina for George and Isabel Dodge, La Guerida for Rodman Wanamaker II, and El Sclano for Harold S. Vanderbilt, but his greatest project is El Mirasol, an enormous Moorish palace with 25 acres of lawn and a 20-car garage for J. P. Morgan partner E. T. (Edward Townsend) Stotesbury of Philadelphia (*see* Boca Raton, 1925).

Congress passes an Oil Pollution Act to update the Refuse Act of 1899 but prosecution of violators will be lax (*see* 1966).

Robert Moses (above) will create Jones Beach and other New York area parks and beaches including Belmont Lake, Captree, Hecksher, Hither Hills, Hempstead Lake, Valley Stream, and Wildwood.

Miami Beach promoter Carl Fisher hires publicity man Steve Hanagan who will deluge newspapers with pictures of Florida "bathing beauties" to convey the idea that the Atlantic at Miami Beach in January is warmer than the Pacific at Los Angeles in August.

Van Camp Sea Food shuts down and goes into receivership after having its credit with American Can Co. canceled for not paying long overdue bills (it had diversified to pack summer citrus fruits and produce). A heavy run of sardines encourages Frank Van Camp to borrow money and go back into business; he makes enough profit to pay off much of the firm's debts, and will regain control next year (*see* 1922; 1926).

The first effective chemical pesticides are introduced (*see* DDT, 1939).

Nearly two-thirds of the commercial fertilizer sold in the United States goes to Southern farmers who use it to push cotton to earlier maturity in order to frustrate the timetable of the boll weevil (*see* 1921).

Owens Valley, Calif., farmers armed with shotguns sit atop the Los Angeles water gates and blow up the city's aqueduct 17 times to prevent drainage of water from their lands (*see* 1913). State militia drive off the farmers but the conflict will continue for decades as the Owens Valley dries up.

Boston physicians George Richards Minot, 39, and William Parry Murphy, 32, show that eating liver prevents and cures pernicious anemia (*see* Addison, 1855). Having studied C. H. Whipple's 1922 discoveries, they show that liver feedings have at least limited value in treating the disease in humans (*see* Cohn, 1926; Lilly, 1928; Castle, 1929).

Continental Baking Co. is incorporated with headquarters at Chicago to manage a sprawl of nearly 100 plants that produce bread and cake under dozens of different labels. Continental will become the leading U.S. bakery firm (*see* Wonder Bread, Hostess Cakes, 1927).

Thirty percent of U.S. bread is baked at home, down from 70 percent in 1910.

Wheaties is introduced by Washburn, Crosby Co., which will soon advertise the wheat-flake cereal (4.7 percent sugar) as "the breakfast of champions" (*see* 1921; General Mills, 1929).

Distillers Corp, Ltd., is founded by Canadian distiller-bootlegger Samuel Bronfman, 33, and his brother Allan, 28, who borrow the name of Britain's 47-year-old Distillers Co., Ltd., and build their first distillery at Ville La Salle outside Montreal (*see* Distillers Corp.-Seagram, 1928).

Americans consume on average 17.8 pounds of butter per year, 6.8 pounds of ice cream, 4.5 pounds of cheese, more than 350 pounds of fluid milk (nearly one pint per day).

Harry Burt patents the Good Humor (*see* 1920; Brimer, 1926).

The Johnson-Reed Immigration Act passed by Congress May 26 limits the annual quota from any country to 2 percent of U.S. residents of that nationality in 1890 (*see* 1948; Dillingham Bill, 1921). The act totally excludes Japanese despite a warning by the Japanese ambassador of grave consequences should the United States abandon the Gentlemen's Agreement.

The Johnson-Reed Act (above) permits close relatives of U.S. citizens to enter as non-quota immigrants and places no restrictions on immigration from Canada or Latin America. Immigration from Canada, Newfoundland, Mexico, and South America has its peak year.

1925 China's republican leader Sun Yat-Sen dies of cancer at Beijing March 12 at age 57. Sun's Guomindang army controls Guangzhou (Canton) and surrounding areas, students at Shanghai demonstrate against "unequal treaties" May 30 until British forces open fire to disperse the demonstrators, Guangzhou has similar demonstrations June 23, China boycotts British goods and shipping, the Guomindang appoints Gen. Chiang Kai-shek, 39, commander in chief in September, Chiang makes a Soviet general his unofficial chief of staff, and by year's end he has defeated Sun's opponent Chen Chiung-ming, 50, to bring Gwangdong and Gwangxi provinces under Guomindang control (*see* 1924; 1926).

U.S. forces have landed in Shanghai January 15 to protect American nationals in the violence occasioned by factual disputes. They remain until March 12.

Japan has recognized the U.S.S.R. January 20 and agreed to return the northern part of Sakhalin Island.

Mein Kampf (My Battle) by Adolf Hitler is published in its first part (*see* beer hall putsch, 1923). Hitler has dictated the book to his aide (Walter Richard) Rudolf Hess, 31, while in prison and will complete it in 1927, saying, "The great masses of the people. . . will more easily fall victims to a great lie than to a small one" (*see* 1933).

The first president of the German Reich Friedrich Ebert dies February 28 at age 53 after 6 years of fighting left- and right-wing extremists.

Soviet authorities catch and shoot British spy Sidney Reilly (*né* Sigmund Georgevich Rosenblum), who once persuaded the British mission to Moscow to finance a scheme to destroy the Bolsheviks by parading Lenin and Trotsky without their trousers through the streets of Moscow.

Turkmenistan and Uzbekistan become Soviet Socialist republics.

Britain's Conservative government rejects a Geneva protocol for the peaceful settlements of international disputes. Czechoslovakia's Eduard Benes and Greece's Nicolaos Sokrates Polites have drafted the protocol, but it has met with opposition from the Dominion governments such as that of Canada's Raoul Dandurand, who has said, "We live in a fireproof house, far from inflammable materials."

1925 ***(cont.)*** A protocol signed by the world powers June 17 bars use of poison gas in war. An arms traffic convention signed the same day governs international trade in arms and munitions.

The Locarno Conference of October 5 to 16 results in the Locarno Treaty signed December 1 to guarantee Franco-German and Belgo-German frontiers by mutual agreement, by arbitration treaties between World War antagonists, and by mutual assistance treaties that give Europe some sense of security.

France recognizes that Germany has not bound herself against aggression to the south and east. She begins construction along the Franco-German border of the heavily fortified Maginot Line named for French politician André Maginot, 48.

Norway annexes Spitzbergen.

Britain declares Cyprus a crown colony (*see* 1914; 1960).

Syria is created January 1 by a union of Damascus and Aleppo with French general Maurice Sarrail as high commissioner. A People's Party organized February 9 demands unity of Syrian states and independence. A great insurrection of the Druse against French rule in Lebanon begins July 18 following the arrest of Druse notables who had been invited to a conference at Damascus. The French show favoritism toward Christians, the Druses charge, and they soon control the countryside. Damascus rises October 14 at the sight of rebel corpses, the French withdraw from the city, French cannon bombard Damascus October 18 and 19, tank and air attacks follow, and Henri de Jouvenal is appointed high commissioner November 6.

Persia's majlis gives Reza Khan Pahlevi dictatorial powers in February, it declares the absent shah deposed October 31, the assembly proclaims Reza Khan shah December 13, and Reza Shah Pahlevi begins a reign that will continue until he abdicates in 1941 (*see* 1924).

The Ku Klux Klan founded in 1915 stages a parade August 8 at Washington, D.C., with 40,000 marchers in white hoods.

The Brotherhood of Sleeping Car Porters is organized August 25 at Harlem by A. (Asa) Philip Randolph, 36, who publishes *The Messenger*, a New York monthly devoted to black politics and culture. Monthly wages for porters have risen in the past year to $67.50 per month, up from a minimum of $27.50 in 1915; Randolph will work to raise them further.

The Autobiography of Mother Jones by Irish-American union organizer Mary Harris Jones, 95, says, "I am always in favor of obeying the law, but if the high-class burglar breaks the law and defies it, then I say we must have a law that will defend the nation and our people." Jones has grown up in the U.S. labor movement and been imprisoned on occasion for violent activities in behalf of West Virginia and Colorado coal miners, her slogan has been "Pray for the dead and fight like hell for the living," and to illustrate the abuses of child labor she tells of meeting a boy of 10 and asking him why he was not in school. "I ain't lost no leg," says the boy, and Jones tells her readers that "lads went to school when they were incapacitated by accidents" in the mines and mills.

"The business of America is business" says President Coolidge January 17 in an address to the Society of American Newspaper Editors.

The Corrupt Practices Act passed by Congress makes it "unlawful for any national bank, or any corporation. . . to make a contribution or expenditure in connection with any election to any political office" but the new law permits an individual donor to give up to $5,000 to a political campaign committee and places no limit on the number of committees that may receive such contributions (*see* 1972).

President Coolidge tells Congress he opposes cancellation of French and British war debts. The foreign press has called Uncle Sam "Uncle Shylock" for refusing even to scale down the debts but Coolidge says privately, "They hired the money, didn't they?"

Britain's chancellor of the exchequer Winston Churchill returns to the gold standard at the prewar gold and dollar value of the pound—123.27 grains of gold, 11/12 fine, or 4.87 dollars to the pound. Parliament and the people respond with enthusiasm but the move makes British coal, steel, machinery, textiles, ships, cargo rates, and other goods and services 10 percent above world prices, and the result is unemployment and wage cuts (*see* strike, 1926).

Gimbel Brothers acquires Philadelphia's Kaufmann & Bauer store which will be renamed Gimbel Brothers in 1928 (*see* Saks Fifth Avenue, 1924; Pittsburgh Gimbels, 1940).

The Kelly Air Mail Act passed by Congress authorizes the U.S. Post Office Department to sign contracts with private companies for carrying the mail at rates ranging up to $3 per pound, rates that amount to government subsidies for airlines.

Henry Ford inaugurates commercial air service between Detroit and Chicago April 13.

Colonial Air Transport starts carrying mail between New York and Boston. The company has been created by a merger of Boston's Colonial Airways with Eastern Air Transport, a line organized by Juan Trippe and John Hambleton with backing from Cornelius Vanderbilt Whitney and William H. Vanderbilt (*see* 1923; Cuba, 1927).

Chrysler Corp. is created by a reorganization of Maxwell Motor (*see* 1924; 1926).

Chrysler, Ford, and General Motors produce at least 80 percent of U.S. motorcars and will increase their share of the market.

Henry Ford has 10,000 U.S. dealers, up from 3,500 in 1912. Each agency is required to stock at least one of each of the 5,000 parts in a Model T (43 percent of the parts retail for 15 cents each or less).

Most Ford dealers receive their cars knocked down since seven knocked-down cars can be shipped in the railroad boxcar space required for only two assembled cars and a good mechanic can assemble the Model T in half a day. A new Ford roadster sells for $260 F.O.B. Detroit.

General Motors acquires Britain's Vauxhall Motors.

The Crescent Limited goes into service between New Orleans and New York via the Southern Railway, West Point Route, Louisville & Nashville, and Pennsylvania railroads. The luxury train charges a $5 premium over the regular fare.

Synthetic rubber is pioneered by Belgian-American clergyman-chemist Julius Arthur Nieuwland, 47, at Notre Dame University. Nieuwland passes acetylene gas through a solution of ammonium chloride and gas and his casual reference to the resulting product will inspire Du Pont chemists to develop a product that Du Pont will introduce in 1931 under the name DuPrene *(see* Semon, 1926).

Malignant diphtheria strikes Nome, Alaska. Local physician Curtis Welch of the U.S. Public Health Service fears an epidemic that will jeopardize 11,000 lives in the area. He wires for antitoxin, a dog team sets out from Shaktolik January 27, fresh teams relay the antitoxin as temperatures drop below -50°F, "musher" Leonard Seppala crosses Norton Sound on pack ice heaving in the ground swell of the Bering Sea as a blizzard sweeps in with winds of 80 miles per hour, and veteran musher Gunnar Kasson arrives at Nome February 2 at 5:36 in the morning. The teams have covered 655 miles in 5.5 days (the previous record was 9 days), and the antitoxin arrives in time to prevent an epidemic *(see* Behring, Ehrlich, 1891).

Soviet Russia has her last major outbreak of cholera (*see* 1855).

Cooley's anemia, described by Detroit physician Thomas Benton Cooley, 54, is an incurable, hereditary blood disorder that impairs the synthesis of hemoglobin, kills its victims at an average of 20, and has been common for thousands of years in the Mediterranean, the Middle East, the Philippines, and parts of India and China.

The Menninger Clinic opens in a farmhouse at Topeka, Kan., where local country doctor Charles F. Menninger, 63, starts a group practice for the mentally ill with his sons Karl, 32, and William, 26. Operating on C. F.'s premise that no patient is untreatable, they combine a family atmosphere with physical exercise and a team of multi-discipline doctors for each patient—a "total-environment" approach inspired by a visit C. F. made in 1908 to the Mayo Clinic at Rochester, Minn.—and begin a revolution in the treatment of mental illness.

The Scopes "Monkey Trial" makes headlines in July as Dayton, Tenn., schoolteacher John T. Scopes, 25, goes on trial for violating a March 13 law against teaching evolution in the state's public schools. Backed by the American Civil Liberties Union, Scopes has tested the law by acquainting his classes with the 1859 teachings of Charles Darwin. Defended by Chicago attorneys Clarence Darrow and Dudley Field Malone, he is prosecuted by former secretary of state William Jennings Bryan, found guilty, and fined $100. Bryan dies of apoplexy July 26.

The John Simon Guggenheim Foundation is established by mining magnate Simon Guggenheim, 57, and his wife in memory of their son John who died 3 years ago at age 17. Guggenheim Fellowships endowed by the foundation will enable thousands of Americans to pursue studies that will result in works which will win at least seven Nobel prizes and dozens of Pulitzer prizes.

The Brooklyn Museum is completed by McKim, Mead and White.

Bennington College for Women is founded at Bennington, Vt.

The University of Miami is founded at Coral Gables, Fla., with support from local real estate promoter George Edgar Merrick (below).

Hebrew University is founded at Jerusalem by émigré U.S. Zionist Judah Leon Magnes, 48.

Freedom of speech and of the press "are protected by the First Amendment from abridgment by Congress [and] are among the fundamental personal rights and 'liberties' protected by the due process clause of the Fourteenth Amendment from impairment by the states," rules the Supreme Court June 8, but the Court upholds the conviction of a man under the New York Criminal Anarchy Act of 1902. Benjamin Gitlow's pamphlet "The Left Wing Manifesto" called for "mass strikes," "expropriation of the bourgeoisie," and the establishment of a "dictatorship of the proletariat," the Court rules in *Gitlow v. New York* that this "is the language of direct incitement," but Justice Holmes observes that "every idea is an incitement" and Justice Brandeis joins in the dissent.

Tennessee's "Monkey Trial" pitted Scripture against Darwin. Teacher John Scopes (center) had dared to defy the law.

1925 ***(cont.)*** The Scripps-Howard newspaper chain is created as New York newspaperman Roy Howard moves up to co-directorship of the Scripps-McRae chain. Howard broke the news of the 1918 World War armistice 4 days early by acting on a tip and will bring new dynamism to the newspapers in the chain.

The *New Yorker* begins publication in February with backing from Fleischmann's Yeast heir Raoul Fleischmann, 39. Edited by journalist Harold Wallace Ross, 33, the 32-page, 15¢ weekly magazine of satire, fiction, social commentary, and criticism has a cover drawing by Rea Irvin of a Regency dandy (who will be named Eustace Tilley) contemplating a butterfly through a magnifying glass. The magazine quickly attracts the talents of cartoonists Peter Arno (Curtis Arnoux Peters), 21, and Helen Hokinson, 26, and advertising writer E. B. (Elwyn Brooks) White, 26, who will write most of the "Talk of the Town" items that begin each issue. Humorist James Grover Thurber, 30, will become a *New Yorker* regular after receiving rejection slips for his first 20 submissions.

Cosmopolitan magazine is created by a merger of *Hearst's International Magazine* with an earlier *Cosmopolitan* under the editorship of Raymond Land, 47, who will publish works by Louis Bromfield, Edna Ferber, Fannie Hurst, Rupert Hughes, Sinclair Lewis, Somerset Maugham, Dorothy Parker, O. O. McIntyre, Booth Tarkington, and P. G. Wodehouse (*see* 1905).

Collier's editor William Ludlow Chenery sends three staff writers on a nationwide tour to report on Prohibition. They find a breakdown in law enforcement of all kinds and *Collier's* becomes the first major magazine to call for a repeal of the Eighteenth Amendment that has been in effect since January 1920. The magazine loses 3,000 readers but gains 400,000 new ones (*see* 1919).

Tokyo Shibaura, Japan's first radio station, begins broadcasting March 22. NHK (Nippon Hoso Kyokai) will be founded next year (*see* television, 1953).

American Telephone and Telegraph's Western Electric Co. splits off its international holdings in anticipation of antitrust action. ITT's Sosthenes Behn and his brother Hernand get backing from the J. P. Morgan banking house to buy International Western Electric for $30 million and become owners of a network of overseas manufacturing companies that will compete with Siemens, the Swedish firm Ericson, but not with AT&T, which will use ITT as its export agents under a mutual agreement not to compete with ITT abroad while ITT refrains from competing with AT&T in the United States (*see* 1920; Geneen, 1959).

AT&T establishes Bell Laboratories to consolidate research and development for the Bell System's telephone companies (*see* transistor, 1948).

Nonfiction *Mein Kampf* (above); *The Tragedy of Waste* by New York economist Stuart Chase, 37, who deplores not only the waste of U.S. natural resources but of manpower and of the human spirit that is being sacrificed to total mechanization in industry and agriculture; *In the American Grain* by poet William Carlos Williams.

Fiction *The Trial (Der Prozess)* by the late Franz Kafka captures the nightmare of injustice that pervades German society under the Weimar Republic. Kafka died last year of tuberculosis at age 41 in an Austrian sanatorium; his friend Max Brod, 41, has edited his incomplete manuscript; *Childhood (Detstvo Luvers)* by Soviet novelist Boris Pasternak, 35, who has revised the first chapter of a novel whose remaining manuscript pages he has lost; *The Desert of Love* (*Le Désert de l'amour*) by François Mauriac; *No More Parades* by Ford Madox Ford; *Pastors and Masters* by English novelist Ivy Compton-Burnett, 33; *Mrs. Dalloway* by English novelist Virginia Woolf, 43, who experiments with stream-of-consciousness technique; *Serena Blandish, or The Difficulty of Getting Married* (anonymous) by English novelist Enid Bagnold, 36; *Sorrow in Sunlight* by English novelist (Arthur Annesley) Ronald Firbank, 39, whose novel will be titled *Prancing Nigger* in the United States; *Pauline* by French novelist-poet Pierre Jean Jouve, 38; *Jew Süss* by German novelist-playwright Lion Feuchtwanger, 41; *An American Tragedy* by Theodore Dreiser is based on the 1906 Gillett-Brown murder case; *Barren Ground* by Ellen Glasgow who gains her first critical success at age 51; *The Great Gatsby* by F. Scott Fitzgerald; *Arrowsmith* by Sinclair Lewis; *The Professor's House* by Willa Cather; *Manhattan Transfer* by John Dos Passos; *The Making of Americans* by Gertrude Stein; *Gentlemen Prefer Blondes* by California novelist-scriptwriter Anita Loos, 32, who started writing scenarios for D. W. Griffith at age 15; her gold-digging Lorelei Lee will be famous for generations; *The House Without a Key* by U.S. playwright-novelist Earl Der Biggers, 41, whose Charlie Chan detective mystery is the first of many; *The Fellowship of the Frog* by Edgar Wallace.

Juvenile *The Childhood and the Enchanters (L'Enfant et les Sortilèges)* by Colette.

Poetry *The Cantos (I)* by Ezra Pound; "The Hollow Men" by T. S. Eliot whose poem has reference to the straw figures hung in London streets each year to celebrate Guy Fawkes Day commemorating discovery of the 1605 Gunpowder Plot: "This is the way the world ends, not with a bang but a whimper"; *Cuttlefish Bones* (*Ossi di seppia*) by antifascist Italian poet Eugenio Montale, 29; *Color* by Harlem poet Countee Cullen, 22.

Painting *Three Dancers* by Pablo Picasso; *The Apprentice* by Georges Rouault; *The Drinking Green Pig* by Marc Chagall; *Tower* by Lyonel Feininger. George Bellows dies at New York January 8 at age 42; John Singer Sargent dies in England April 15 at age 69.

Proletariat (woodcut series) by Käthe Kollwitz.

The Leica introduced by E. Leitz G.m.b.H. of Wetzlar, Germany, is a revolutionary miniature 35 millimeter camera invented by Oskar Bernack. It takes a 36-frame film roll and has a lens that can be closed down to take pictures with great depth of field or opened for dim lighting conditions, fast and slow shutter speeds that permit action pictures, and interchangeable lenses that permit close-ups and telephotography (*see* "candid camera," 1930; Nikon, 1948).

Theater *Is Zat So?* by James Gleason and Richard Taber 1/5 at New York's 39th Street Theater, with Gleason, Eleanor Parker, Robert Armstrong in a prize-fight comedy, 618 perfs.; *Processional* by U.S. playwright John Howard Lawson, 29, 1/12 at New York's Garrick Theater, with George Abbott, June Walker, Ben Grauer, 96 perfs.; *Fallen Angels* by Noël Coward 4/21 at London's Globe Theatre, with Tallulah Bankhead, Edna Best, 158 perfs.; *Hay Fever* by Coward 6/8 at London's Ambassadors Theatre, with Marie Tempest, Ann Trevor, 337 perfs.; *Veland* by Gerhart Hauptmann 9/19 at Hamburg's Deutsches Schauspielhaus; *The Butter and Egg Man* by George S. Kaufman 9/23 at New York's Longacre Theater, 243 perfs.; *Craig's Wife* by George Kelly 10/12 at New York's Morosco Theater, 289 perfs.; *Lucky Sam McGarver* by Sidney Howard 10/21 at New York's Playhouse Theater, with John Cromwell, Clare Eames, 29 perfs.; *In a Garden* by Philip Barry 11/16 at New York's Plymouth Theater, with Laurette Taylor, Louis Calhern, 73 perfs.; *Easy Virtue* by Coward 12/7 at New York's Empire Theater, with Jane Cowl, Joyce Carey, 147 perfs.

Films King Vidor's *The Big Parade* with John Gilbert, Renée Adorée; Harold Lloyd's *The Freshman* with Lloyd; Charlie Chaplin's *The Gold Rush* with Chaplin, Mack Swain; Sergei Eisenstein's (*The Armored Cruiser) Potemkin*. Also: Fred Niblo's *Ben Hur* with Ramon Novarro, Francis X. Bushman, May McAvoy; George Fitzmaurice's *The Dark Angel* with Vilma Banky, Ronald Colman; Fred Niblo's *Don Q* and *Son of Zorro*, both with Douglas Fairbanks; Clarence Brown's *The Eagle* with Rudolph Valentino, Vilma Banky, Louise Dresser; Sam Taylor and Fred Newmeyer's *Go West* with Buster Keaton; Herbert Brenon's *A Kiss for Cinderella* with Betty Bronson, Esther Ralston; Harry Hoyt's *The Lost World* with Wallace Beery, Lewis Stone; Erich von Stroheim's *The Merry Widow* with Mae Murray, John Gilbert; Robert Flaherty's documentary *Moana* set in the Samoas; D.W. Griffith's *Poppy* with W.C. Fields; Josef von Sternberg's *Salvation Hunters* with vagrants from the Long Beach waterfront; Rupert Julian's *Three Faces East* with Julian, Jetta Goudal; E.A. Dupont's *Variety* with Emil Jannings; Malcolm St. Clair's *Woman of the World* with Pola Negri, Charles Emmet Mack.

Opera *Wozzeck* 1/14 at Berlin's State Opera is the first full-length atonal opera. Austrian composer Alban Berg, 40, has based his music on the George Buchner tragedy of 1836; *L'Enfant et les Sortilèges* 3/21 at Monte Carlo, with music by Maurice Ravel.

Ballet *Les Matelots* (*The Sailors*) 6/17 at the Théâtre Gaiété-Lyrique, Paris, with music by Georges Auric, choreography by Leonide Massine.

First performances Symphony for Organ and Orchestra by U.S. composer Aaron Copland, 24, 1/11 at New York, with Nadia Boulanger; Concerto for Violin and Orchestra in D minor (*Concerto Accademico*) by Ralph Vaughan Williams 11/6 at London's Aeolian Hall.

Stage musicals *Big Boy* 1/7 at New York's Winter Garden Theater, with Al Jolson, songs that include "It All Depends on You" by Ray Henderson, lyrics by B. G. DeSylva and Lew Brown (Russian-American writer Lewis Bronstein, 31), "If You Knew Susie Like I Know Susie" by Joseph Meyer, lyrics by B. G. DeSylva, 188 perfs.; *No, No Nanette* 3/11 at London's Phoenix Theatre, with Binnie Hale (Beatrice Mary Hale-Munro), 25, music by Vincent Youmans, lyrics by Irving Caesar and Otto Harbach, songs that include "Tea for Two," "I Want to Be Happy"; *On with the Dance* (revue) 4/30 at London's Pavilion Theatre, with Nigel Bruce, Hermione Baddeley, music by Noël Coward and Philip Braham, book and lyrics by Coward, 229 perfs.; *The Garrick Gaieties* 6/8 at New York's Garrick Theater, with Sterling Holloway, Romney Brent, music by local composer Richard Rodgers, 23, lyrics by Lorenz Hart, 27, songs that include "Sentimental Me," 14 perfs. (plus 43 beginning 5/10/26); *Artists and Models* 6/24 at New York's Winter Garden Theater, with music by J. Fred Coots, Alfred Goodman, Maurice Rubens, lyrics by Clifford Grey, 411 perfs.; *Dearest Enemy* 9/18 at New York's Knickerbocker Theater, with Helen Ford, music by Richard Rodgers, lyrics by Lorenz Hart, book by Herbert Fields based on the Revolutionary War legend about Mrs. Robert Murray delaying General Howe, songs that include "Here in My Arms," 286 perfs.; *The Vagabond King* 9/21 at New York's Casino Theater, with Dennis King, music by Rudolf Friml, lyrics by Bryan Hooker, songs that include "Only a Rose," 511 perfs.; *Sunny* 9/22 at New York's New Amsterdam Theater, with Marilyn Miller, Jack Donahue, Clifton Webb, music by Jerome Kern, lyrics by Otto Harbach and Oscar Hammerstein II, songs that include "Who," 517 perfs.; *The Charlot Revue* 11/10 at New York's Selwyn

Charlie Chaplin's tragicomedic antics silently captured the world's hearts in Hollywood's early days.

1925 *(cont.)* Theater, with Beatrice Lillie, Gertrude Lawrence, Jack Buchanan, songs that include "Poor Little Rich Girl" by Noël Coward, "A Cup of Coffee, a Sandwich, and You" by Joseph Meyer, lyrics by Billy Rose and Al Dubin, 138 perfs.; *The Cocoanuts* 12/8 at New York's Lyric Theater, with the Four Marx Brothers, Margaret Dumont, book by George S. Kaufman, music and lyrics by Irving Berlin, 218 perfs.; vaudeville monologist Art Fisher has given the Marx Brothers their nicknames and has helped develop their characters: Groucho (Julius Henry), 30, is a moustached wit, Harpo (Arthur), 32, an idiotic kleptomaniacal mute harpist, Chico (Leonard), 34, a pianist and confidence man who serves as Harpo's interpreter, Zeppo (Herbert) is straight man; *Tip-Toes* 12/28 at New York's Liberty Theater, with Queenie Smith, Jeanette MacDonald, Robert Halliday, music by George Gershwin, lyrics by Ira Gershwin, book by Guy Bolton and Fred Thompson, songs that include "Sweet and Low-Down," 194 perfs.

"The Charleston" is introduced to Paris by "Bricktop," a red-headed American who arrived penniless from her native Harlem last year and has become hostess at a Place Pigalle nightclub. Ada Beatrice Queen Victoria Louisa Virginia Smith du Conge, 30, begins a half century as nightclub hostess.

Popular songs "Yes, Sir, That's My Baby!" by Walter Donaldson and Gus Kahn; "I Love My Baby (My Baby Loves Me)" by Harry Warren, 32, lyrics by Austrian-American writer Bud Green, 28; "Sleepy Time Gal" by Ange Lorenzo and Richard A. Whiting, lyrics by Joseph R. Alden and Raymond B. Egan; "Sweet Georgia Brown" by Ben Bernie, Maceo Pinkard, and Kenneth Casey; "Alabamy Bound" by Ray Henderson, lyrics by B. G. DeSylva, Bud Green; "Dinah" by Harry Akst, 31, lyrics by Sam M. Lewis and Joe Young; "Sometimes I'm Happy" by Vincent Youmans and Irving Caesar; "Always" and "Remember" by Irving Berlin; "Five Feet Two, Eyes of Blue" and "I'm Sitting On Top of the World" by Ray Henderson, lyrics by Sam M. Lewis and Joe Young; "Don't Bring Lulu" by Henderson, lyrics by Billy Rose and Lew Brown; "My Yiddishe Momme" by Jack Yellen and Lew Pollak, lyrics by Yellen (for Sophie Tucker); "I Dreamed I Saw Joe Hill Last Night" by Earl Robinson, lyrics by Alfred Hayes (*see* 1915); "Valentine" by French composer Henri Christine, 58, lyrics by Albert Willemetz; "Jealousy" ("Jalousie, a 'Tango Tzigane' [Gypsy Tango]") by Danish violinist-composer Jacob Gade, 46; "Show Me the Way to Go Home" by London songwriters Reg Connelly, 27, and Irving King (Jimmy Campbell), 22.

Grand Ole Opry goes on the air November 28 as *WSM Barn Dance* over Nashville, Tenn., radio station WSM owned by National Life and Accident Insurance Co. ("We Shield Millions"). When the Saturday-night show adopts the name *Grand Ole Opry* late in 1927 it will be featuring Uncle Dave Macon, now 55, as "the Dixie Dewdrop" and he will be supported or followed by such country-music stars as the Fruit Jar Drinkers, the Gully Jumpers, Dr. Humphrey Bates and the Possum Hunters, Uncle Jimmy Thompson, Fiddling Arthur Smith, and Roy Acuff, now 22, who will run for governor of Tennessee in 1948.

The Wood Memorial introduced at New York in April is a test for 3-year-old thoroughbreds prior to the Kentucky Derby in May.

Jean René LaCoste, 19, (Fr) wins in men's singles at Wimbledon, Suzanne Lenglen in women's singles; Bill Tilden wins in men's singles at Forest Hills, Helen Wills in women's singles.

Former Columbia baseball star Henry Louis "Lou" Gehrig, 22, joins the New York Yankees as a first baseman and begins a 14-year career of 2,130 consecutive games in which he will have a batting average of .341 (*see* 1939).

"Lefty" Grove is acquired for a record $100,600 by Connie Mack for the Philadelphia Athletics. Pitcher Robert Moses Grove, 25, begins a 17-year career in which he will strike out 2,266 batters and record a lifetime earned run average of 3.06.

The Pittsburgh Pirates win the World Series by defeating the Washington Senators 4 games to 3.

Contract bridge begins to replace auction bridge (*see* 1904). Railroad heir and yachtsman Harold S. Vanderbilt, 41, invents the variation while on a Caribbean cruise; it will eclipse auction bridge and whist beginning in 1930 when Romanian-American expert Eli Culbertson defeats Lieut. Col. W. T. M. Butler in a challenge match at London's Almack's Club that will bring the game wide publicity (*see* Goren, 1936).

Philadelphia's Soldiers Field opens with 165,000 seats. The new stadium will be the scene of Army-Navy football games and other events.

The New York Giants professional football team is founded by Timothy J. Mara.

New York's Madison Square Garden is demolished after a 35-year career in which it has housed prizefights, six-day bicycle races, rodeos (beginning in 1923), horse shows, and last year's Democratic National Convention. The new $5.6 million Garden designed by Tex Rickard's architect Thomas W. Lamb opens November 28 on Eighth Avenue between 49th and 50th Streets; will be used for circuses, hockey games, horse shows, basketball games, ice shows, political rallies, and track meets until February 1968, but it is designed principally for boxing and wrestling matches and has poor sightlines for other events.

Floyd Collins makes headlines in late January after being trapped inside a Kentucky cave he had been exploring with companions. Search parties try to reach him before it is too late but Collins is found dead February 16 under a boulder that has pinned him fast for 18 days.

Scotch Tape gets its name from a Detroit auto-plant car painter. Minnesota Mining and Manufacturing Co. has developed a masking tape to facilitate two-tone paint jobs, but a 3M employee has eliminated the adhesive from the center of the tape to save money. The new tape does not stick, an angry painter tells a

3M salesman to "take this tape back to those Scotch bosses of yours and have them put adhesive all over the tape," 3M replaces the missing adhesive and will make Scotch the brand name for a full line of adhesive tapes.

Lever Brothers introduces Lux toilet soap under the name Lux Toilet Form. The white milled soap will challenge Palmolive and Procter & Gamble's Cashmere Bouquet (*see* 1905; Lux Flakes, 1906; Rinso, 1918; Unilever, 1929).

U.S. cigarette production reaches 82.2 billion, up from 66.7 billion in 1923.

Old Gold cigarettes are introduced by P. Lorillard. The firm is well known for its Murad cigarettes and their slogan "Be nonchalant—light a Murad," and Old Golds will gain popularity with the slogan "Not a cough in a carload" (*see* 1911; Kent, 1952).

"Blow some my way," says a woman to a man lighting a cigarette in advertisements by Liggett & Meyers for its 12-year-old Chesterfield brand. The advertisement breaks a taboo by suggesting that women smoke.

Lucky Strike cigarettes are promoted by George Washington Hill, 41, who succeeds his father Percival as head of American Tobacco (*see* 1918). The younger Hill will introduce the slogan "Reach for a Lucky instead of a sweet" (*see* 1931).

"Pretty Boy" Floyd begins a 9-year career in which he will rob more than 30 Midwestern banks and kill ten people. Charles Arthur Floyd, 18, robs a local post office of $350 in pennies at St. Louis.

Al Capone takes over as boss of Chicago bootlegging from racketeer Johnny Torrio who retires after sustaining gunshot wounds. Italian-American gangster Alphonse Capone, 26, of Cicero, Ill., has been a lieutenant of Torrio and last year had Torrio's chief rival Dion O'Bannion gunned down in his flower shop. Capone soon controls not only bootlegging but also gambling, prostitution, and the Chicago dance-hall business while his agents on the Canadian border and in the Caribbean help him grow rich by smuggling huge quantities of whiskey and rum into the United States (*see* 1927).

Izzy and Moe are dismissed as Prohibition agents after making 4,392 arrests in just over 5 years with a 95 percent record of convictions. Isadore Einstein and Moe Smith are both in their mid-forties, weigh more than 225 pounds each although neither is taller than 5 feet 7; they have antagonized their superiors by getting too much personal publicity for their ingenious and successful, if undignified, stratagems for apprehending Prohibition law violators.

A Paris Exposition Internationale des Arts Décoratifs displays a two-story apartment designed by Swissborn architect Charles Edouard Jeanneret, 38, who calls himself Le Corbusier and calls his apartment Le Pavillon de l'Esprit. Its spacious rooms are furnished with a minimum of furniture which includes bentwood chairs introduced by Gebrüder Thonet of Vienna in 1876 but never used as modern furniture until now. Le Corbusier also employs as decorative objects bistro wine glasses, laboratory flasks, seashells, industrial equipment, and ethnic rugs and hangings.

The term "Art Deco" will be derived half a century hence from the Paris Exposition (above) but not from the pioneering ideas of Le Corbusier.

Potsdam's Einstein Tower is completed by German architect Eric Mendelsohn, 38.

Chicago's Tribune Tower is completed in May on Michigan Avenue at Tribune Place for "The World's Greatest Newspaper." New York architects John Mead Howells and Raymond M. Hood have designed a 36-story Gothic skyscraper 58 feet taller than the 32-story Wrigley building completed last year.

Washington's Mayflower Hotel opens February 18 with 1,000 rooms. The inaugural ball for President Coolidge is held at the Mayflower in March.

Paris's Bristol Hotel opens.

Boston's Parker House opens on Tremont Street to replace the hotel opened in 1855.

Chicago's Palmer House opens at State and Monroe Streets to replace the hotel built in 1875.

Memphis's Peabody Hotel opens with 617 rooms to replace the Peabody that had been the city's social center for 56 years.

The Breakers at Palm Beach, Fla., is replaced after 22 years by a $6 million hotel with 600 guest rooms and suites, two conference wings, two 18-hole golf courses, 12 tennis courts, indoor and outdoor swimming pools and a private beach, a fountain modeled after one in the Boboli Gardens of Florence, a lobby inspired by Genoa's Palazzo Larega, and a general design inspired by Rome's Villa Medici.

Florida's Boca Raton Hotel is completed 25 miles south of Palm Beach by Addison Mizner with an artificial lake, islands, and electric-powered gondolas. The $1.25-million 100-room hotel will open February 6 of next year as the Ritz-Carlton but will be bankrupt by year's end.

The Miami Biltmore at Coral Gables, Fla., is completed by John McEntee Bowman of 1914 New York Biltmore fame and Coral Gables promoter George Edgar Merrick whose late father Solomon Greasley Merrick was a Congregationalist minister who moved to Florida from Cape Cod in 1898, acquired 160 acres of land near the Everglades, and built a homestead he called Coral Gables. The younger Merrick acquired an adjacent tract in 1922, he called it Coral Gables, and set out to build the first Miami subdivision as a resort for respectable older people who could not afford Carl Fisher's Miami Beach (*see* 1913). The new $7 million Miami Biltmore has a 250-foot outdoor swimming pool and a 300-foot replica of Spain's Tower of Giralda at Seville.

The bubble of inflated Florida land values breaks as investors discover the lots they have bought are in many cases underwater. The hordes of Florida real estate promoters turn out to include a man named Charpon

1925 ***(cont.)*** who is actually Charles Ponzi of 1920 "Ponzi scheme" notoriety.

The first motel opens December 12 at San Luis Obispo, Calif. James Vail's Motel Inn with accommodations for 160 guests is located on one of the busiest U.S. motor routes.

William Randolph Hearst opens his San Simeon castle La Cuesta Encantada overlooking the Pacific on 240,000 acres of ranch land. He has spent an estimated $50 million to build the Hispano-Moorish Casa Grande and its adjoining buildings (the place requires a staff of 50).

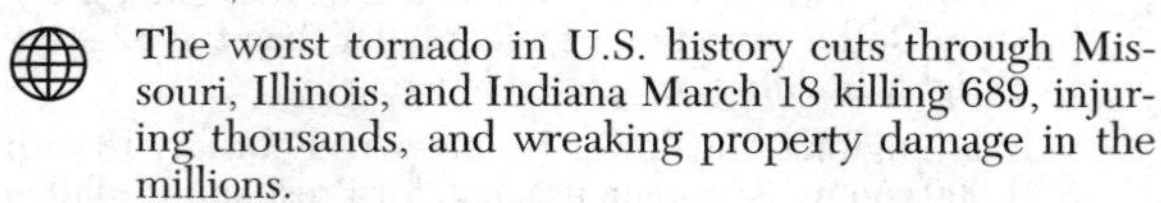

The worst tornado in U.S. history cuts through Missouri, Illinois, and Indiana March 18 killing 689, injuring thousands, and wreaking property damage in the millions.

Harold Ware returns to Russia with 26 volunteers and $150,000 worth of U.S. farm equipment, seeds, and supplies (*see* 1921; 1922). Ware demonstrates mechanized farming on 15,000 acres of land in the northern Caucasus and he is selected by Moscow to oversee a program of organizing huge mechanized state farms for the improvement of agricultural efficiency throughout Russia, a post he will hold until 1932 (*see* Stalin, 1928).

U.S. physiologists L. S. Fredericia and E. Holm find that rats do not see well in dim light if their diets are lacking in vitamin A (*see* McCollum, 1920; Wald, 1938).

U.S. refrigerator sales reach 75,000, up from 10,000 in 1920 (*see* 1929; GE, 1927).

Clarence Birdseye and Charles Seabrook develop a deep-freezing process for cooked foods (*see* 1923). Tightly packed in a small carton, the food is encased in a metal mold and several such molds are placed in a long metal tube that is immersed in a low-temperature brine solution to produce a quick-contact process that Birdseye patents (*see* 1926).

Jell-O is merged into the Postum Co. (*see* 1906; 1926).

Cocoa merchants, importers, and brokers establish the New York Cocoa Exchange to avoid repetition of the disastrous price break of 1921; they adopt rules modeled on those of the Cotton Exchange and Chicago Board of Trade.

Automatic potato-peeling machines are introduced that will permit wide-scale production of potato chips.

Vienna has 1,250 coffeehouses, down from 15,000 in 1842.

Birth control wins the endorsements of the New York Obstetrical Society, New York Academy of Medicine, and American Medical Association after persuasion by New York obstetrician Robert L. Dickinson who obtains a $10,000 Rockefeller Foundation grant for research in contraception.

1926 Josef Stalin establishes himself as virtual dictator of the Soviet Union, beginning a 27-year rule that will de-emphasize world revolution but bring new repression to Soviet citizens and terror to Russia's neighbors. The Politburo expels Leon Trotsky and Grigori Zinoviev in October.

Italy's Benito Mussolini assumes total power October 7, making the Fascist party the party of the state and brooking no opposition.

Abdul-Aziz ibn Saud, now 43, has proclaimed himself king of Hejaz January 8 and renamed it Saudi Arabia (*see* 1924; 1927).

The French fleet bombards Damascus May 8 in an effort to suppress the great insurrection of the Druses that began last year. Paris proclaims Lebanon a republic May 23.

U.S. troops land in Nicaragua beginning May 2 to preserve order and protect U.S. interests in the face of a revolt against the new president Emiliano Chamorro, who resigns under pressure in the fall.

Chiang Kai-shek succeeds the late Sun Yat-sen as leader of China's revolutionary party and begins unification under the Guomindang, which holds power only in the south. Chiang takes Wuchang in October and establishes that city as his seat of power (*see* 1925; 1927).

Japan's Taisho emperor Yoshihito dies December 25 at age 47 after a 14-year reign. His son of 25 has been acting as regent during the 5-year illness of Yoshihito and will reign until 1989 as the Showa emperor Hirohito. The influence of Japan's army and navy will increase enormously in the next 15 years (*see* Mukden, 1931).

A British General Strike cripples the nation from May 3 to May 12 as members of the Trade Union Congress rally to the slogan, "Not a penny off the pay; not a minute on the day." The strike begins with a lockout May 1 by private coal mine operators whose workers refuse to accept a pay cut averaging 13 percent and refuse to work an extra hour each day. Chancellor of the Exchequer Winston Churchill's return to the gold standard last year has squeezed export industries such as coal (which must compete with German and Polish producers and with oil) and forced mine operators to cut wages and increase hours. Railwaymen, printing trade workers, building trade workers, truckdrivers, dock workers, iron and steel workers, chemical industry employees, and some power company workers walk out in sympathy with the miners in accordance with strategy devised by Transport Workers chief Ernest Bevin, 42.

The Royal Navy trains its guns on British strikers who try to prevent ships from off-loading food and other relief materials in support of the great strike (above), economist John Maynard Keynes has called Churchill's action "featherbrained" and "silly," but while 3 million workers respond to Ernest Bevin's strike call, volunteers who include most of the students of Oxford and Cambridge maintain essential services. The sympathy strike is called off May 12, but coal miners, who average $14.60 per week, continue to resist any wage cut or lengthening of hours until November 19 when they capitulate unconditionally. The mine workers accept the terms of the mine operators plus drastic layoffs and will not strike again until 1972.

The Railway Labor Act (Watson-Parker Act) passed by Congress sets up a five-man board of mediation appointed by the president to settle railway labor disputes (*see* 1934).

Ford Motor Company plants introduce an 8-hour day and a 5-day work week beginning September 5.

The rocket launched March 16 by physicist Robert H. Goddard is the first liquid-fuel rocket; it demonstrates the practicality of rockets and convinces Goddard that rockets will one day land men on the moon (*see* 1921). Goddard sends his device on a 2.5-second flight from a field on his Aunt Effie's farm near Auburn, Mass., it travels 184 feet at a speed of only 60 miles per hour and reaches a height of only 41 feet, but Goddard writes in his diary, "It looked almost magical as it rose, without any appreciably greater noise or flame." He continues his research, and beginning in 1930 will get financial support from copper heir Harry Guggenheim.

France stabilizes the franc at 20 percent of its prewar value, thus virtually wiping out the savings of millions of Frenchmen. The devaluation will have the effect of making the French Europe's greatest gold hoarders (*see* 1928).

A merger of German industrial giants creates the mammoth United Steel Works (Vereinigte Stahlwerke) cartel, rivaled only by the great Krupp works at Essen. Major mining enterprises join with the Rhine-Elbe Union steel company that Albert Vogler, 49, has built up with help from the late Hugo Stinnes, Nazi industrialist Emil Kirdorf, 79, and Nazi steel maker Fritz Thyssen, 53.

U. S. Navy explorer Richard Evelyn Byrd, 36, and pilot Floyd Bennett, 35, take off May 9 from Spitzbergen in a trimotor Fokker monoplane, fly 700 miles to the North Pole, circle the pole 13 times, and return in 15.5 hours.

Scheduled U.S. airline service begins April 6. A Varney Air Lines two-seat Laird Swallow biplane piloted by Leon D. Cuddeback, 28, flies 244 miles on a contract mail route from Pasco, Wash., to Boise, Idaho, and proceeds to Elko, Nev., with 200 pounds of mail.

Charles Augustus Lindbergh, 24, takes off from St. Louis April 15 on the first regularly scheduled mail flight between St. Louis and Chicago. Lindbergh is chief pilot for Robertson Aircraft, whose owners Frank and William Robertson have three DH-4 biplanes (*see* 1927; American Airways, 1930).

Trans World Airlines has its beginnings in the Western Air Express Co. (*see* 1930).

Northwest Airlines has its beginnings in the Northwest Airways Co. that begins service between Chicago and St. Paul.

The Air Commerce Act passed by Congress encourages the growth of commercial aviation by awarding mail contracts.

Philadelphia's 533-meter Benjamin Franklin Bridge opens to traffic.

The Florida Keys link of the Florida East Coast Railway completed in 1912 is destroyed September 19 by a hurricane which further deflates the boom in Florida real estate (*see* 1925; 1935).

Chicago's Union Station opens to serve four large railroads.

The Chief departs from Chicago's Dearborn Station November 14 to begin daily Santa Fe service between Chicago and Los Angeles with seven cars including four sleeping cars, a diner, a club car, and an observation lounge car. The 63-hour run will be reduced to 58 hours in 1929 and 56 hours in 1930 (*see* 1911; Super Chief 1936).

Greyhound Corp. is incorporated to compete with intercity passenger rail service (*see* 1922). General Motors will be the largest Greyhound stockholder until 1948 (*see* 1929).

The Delta Queen is launched at Cincinnati and the paddle-wheel passenger steamer goes into service between the Queen City and New Orleans. Fare for the 7-day trip on the Ohio and Mississippi rivers is $35.

The Pontiac motorcar introduced by General Motors is a renamed Oakland (*see* 1924).

The Chrysler Imperial is a luxury model that will compete with Cadillac, Lincoln, Packard, and Pierce-Arrow.

The Model T Ford sells for $350 new with a self-starter but is losing ground to GM's Chevrolet (*see* 1922; Model A, 1927).

E. L. Cord's Auburn Automobile Co. acquires Duesenberg Automobile and Motor Co. (*see* 1924; Model J, 1929).

The Society of Automotive Engineers establishes S.A.E. viscosity numbers to standardize engine lubricating oils, assigning heavy oil the number 50, light oil the number 10 (*see* 1911).

Prestone, introduced by Union Carbide and Carbon, is the first ethylene glycol antifreeze for motor vehicle radiators; it retails at $5 per gallon (*see* 1920).

The Triplex Safety Glass Co. of North America is founded by U.S. entrepreneur Amory L. Haskell who has obtained U.S. rights to the 1910 patent of Edouard Benedictus. Haskell begins production on one floor of the Lipton Tea factory at Hoboken, N.J., and receives assistance from Henry Ford to build his own factory, but the initial price of safety glass is $8.80 per square foot and it costs a Cadillac owner $200 to replace all the glass in his car with safety glass (*see* 1927; Sloan, 1929).

Safety-glass windshields are installed as standard equipment on high-priced Stutz motorcar models.

Waltham, Mass., inventor Francis Wright Davis, 38, patents a power- steering unit and installs it in a 1921 Pierce-Arrow Runabout, but commercial production of cars with power steering will not begin until 1951.

U.S. auto production reaches 4 million, up nearly eight-fold from 1914 (*see* 1927).

The Mercedes-Benz supercharged SS sportscar is introduced by the new German motorcar giant Daimler-Benz created by a merger of Daimler

1926 ***(cont.)*** Motoren-Gesellschaft with Firma Benz & Cie. (*see* 1886; Jellinek, 1901; diesel, 1936).

The Maserati Neptune and Trident racing cars are introduced by Italian auto makers Alfieri, Ettore, and Ernesto Maserati who have built cars for other racing companies since 1914.

B. F. Goodrich chemist Waldo Lonsbury Semon, 28, pioneers synthetic rubber, using catalysts in an effort to extract the chlorine from the polymer polyvinyl chloride. He polymerizes PVC into a white powder, plasticizes the PVC powder with agents such as tricreylphosphate, and produces a workable synthetic that can be rolled and treated like rubber. The product is odorless, weatherproof, age- and acid-resistant, and will be introduced commercially in 1933 under the name Koroseal (*see* Nieuwland, 1925; butadiene, 1939).

Sintered carbide, introduced by Fried. Krupp of Essen is a nonferrous metal alloy on a tungsten carbide base that enables machine tools to cut steel at 150 meters per minute instead of 8 meters.

Canadian-American inventor John C. Garand, 38, patents the semi-automatic .30 M1 rifle that will be adopted by the U.S. Army in 1936.

An improved waterproof cellophane developed by E. I. du Pont chemists William Hale Church and Karl Edwin Prindle will revolutionize packaging (*see* Brandenberger, 1912). Du Pont has been making cellophane at Buffalo, N.Y., since early 1924, selling it initially for $2.65 lb.

English physicist Paul Adrien Maurice Dirac, 24, advances a formal theory that will hereafter govern the study of submicroscopic phenomena. A co-founder of the modern theory of quantum mechanics, Dirac will extend his theory in the next 4 years to embrace ideas of relativity (*see* Einstein, 1916; 1929; Heisenberg, 1927).

A wave model of the atom constructed by Austrian physicist Erwin Schrödinger, 39, makes a major contribution to modern quantum theory (*see* Planck, 1900). Schrödinger has seized upon a radical theory advanced by the French physicist Louis-Victor de Broghe that the electron, like light, should exhibit a dual nature, behaving both as particle and as wave, and in the Schrödinger model electrons wash round the nucleus (*see* Bohr, 1913; Einstein, 1929).

The Theory of the Gene by Columbia University zoologist Thomas Hunt Morgan, now 60, proves a theory of hereditary transmission that will be the basis for all future genetic research. Morgan's book *The Physical Bases of Heredity* appeared in 1919 and he has conducted experiments with fruit flies to pinpoint the location of genes in the chromosomes of the cell nucleus (*see* Bateson, 1902; Watson, Crick, 1953).

U.S. biologist Herman Joseph Muller, 35, finds that X-rays can produce mutations. His work will speed up the process of mutation for gene studies and makes him a leading advocate for limiting exposure to X-rays and for sperm banks to conserve healthy genes.

U.S. biochemist James Batchellor Sumner, 38, crystallizes an enzyme and proves that enzymes are proteins.

Ergotism from infected rye breaks out in the Soviet Union; in some places half the population is affected (*see* 1862; 1951; LSD, 1943).

Father Coughlin makes his first radio broadcast October 17 over Detroit's station WJR to begin a career of nearly 20 years. Detroit priest Charles Edward Coughlin, 34, will broadcast sermons marked by racial bigotry and right-wing sentiments (*see* 1934).

Reading University is founded in England.

Long Island University (LIU) is founded at Brooklyn, N.Y.

Sarah Lawrence College for Women is founded at Bronxville, N.Y., by local real estate developer William V. Lawrence who names the college for his wife.

The first contract airmail flight takes off February 15 from Dearborn, Mich. The all-metal Ford Pullman monoplane lands at Cleveland.

Scottish inventor John L. Baird, 28, gives the first successful demonstration of television, but his mechanical system is based on the van Nipkov rotating disk of 1886 and has serious limitations (*see* 1927).

The first motion picture with sound is demonstrated (*see* below; 1927; de Forest, 1923).

The National Broadcasting Company (NBC) is founded November 11 by David Sarnoff whose nine-station network soon has 31 affiliates (*see* 1922; 1927).

Amazing Stories magazine is founded by Luxembourg-born U.S. inventor and science-fiction enthusiast Hugo Gernsback, 42, who came to America at age 20 to promote an improved dry cell battery he had invented.

The Los Angeles Central Library is completed by New York architect Bertram Goodhue in a mixture of Spanish colonial and Beaux Arts styles.

Nonfiction *History of England* by Cambridge University historian G. M. (George Macaulay) Trevelyan, 50; *Dictionary of Modern English Usage* by English lexicographer H. W. (Henry Watson) Fowler, 68; *The Story of Philosophy* by New York educator-author Will (William James) Durant, 40; *Seven Pillars of Wisdom* by T. E. Lawrence (*see* Lowell Thomas, 1924). His 280,000-word account of desert fighting appears in a costly limited edition (a 130,000-word abridged version will appear next year under the more appropriate title *Revolt in the Desert*, and the full book will be issued for general circulation after Lawrence's death in a motorcycle accident in 1935); *Microbe Hunters* by U.S. bacteriologist Paul de Kruif, 36, who helped Sinclair Lewis with his 1925 novel *Arrowsmith* and now gives a popular account of the discoveries by Louis Pasteur, Robert Koch, Paul Ehrlich, and others.

Fiction *The Counterfeiters (Les Faux Monnayeurs)* by André Gide; *A Man Could Stand Up* by Ford Madox Ford whose title refers to the end of trench warfare in 1918; *Debits and Credits* (stories) by Rudyard Kipling (whose son was killed in 1915) includes "The Garden" and "The Wish House"; *The Plumed Serpent* by D. H. Lawrence; *The Castle (Das Schloss)* by the late Franz Kafka; *The Izu Dancer* (*Izu no odoriko*) by Japanese novelist Yasunari Kawabata, 27; *The Bullfighters (Les Bestiares)* by Henry de Montherlant; *Blindness* by English novelist Henry Green (Henry Vincent Yorke), 21; *The Sun Also Rises* by U.S. novelist Ernest Miller Hemingway, 27, who quotes Gertrude Stein in an epigraph that says, "You are all a lost generation," a line Stein heard her garageman use in scolding a young mechanic who did not make proper repairs on her Model T Ford; *The Cabala* by U.S. novelist Thornton Niven Wilder, 29; *Soldier's Pay* by Mississippi novelist William Cuthbert Faulkner, 28, who added the "u" to his name when he published a volume of poetry 2 years ago; *Under the Sun of Satan (Sous le soleil de Satan)* by French novelist Georges Bernanos, 38; *Before the Bombardment* by English novelist Osbert Sitwell, 34; *Stamboul Train* by English novelist Graham Greene, 21; *The Murder of Roger Ackroyd* by Agatha Christie; *Clouds of Witness* by English writer Dorothy Leigh Sayers, 33; *Show Boat* by Edna Ferber; *Smoky* by U.S. cowboy novelist Will James, 34.

Poetry *White Buildings* by U.S. poet Hart Crane, 27 (son of Life Savers creator Clarence Crane); *A Drunk Man Looks at the Thistle* by Scottish poet Hugh MacDiarmid (C. M. Grieve), 34; *Capitale de la douleur* by French poet Paul Eluard (Eugène Grindel), 30; *The Close Couplet* by U.S. poet Laura Riding, 25; *The Weary Blues* by U. S. poet Langston Hughes, 24, who has had help from Vachel Lindsay; *Enough Rope* by U.S. poet-author Dorothy Rothschild Parker, 33, who is well-known for her 1920 advertising line "Brevity is the soul of lingerie" and will be better known for "Men seldom make passes/At girls who wear glasses," for putting down an actress with the line, "She ran the whole gamut of emotions from A to B," and for the verse, "Guns aren't lawful; Nooses give;/Gas smells awful;/ You might as well live."

Juvenile *Winnie-the-Pooh* by A. A. Milne who delights readers with Pooh-bear, Tigger, Piglet, Eeyore, Kanga and baby Roo, Owl, and other companions of Christopher Robin; *Fairy Gold* by Compton Mackenzie; *The Little Engine that Could* by U.S. author Watty Piper (Mabel C. Bragg), illustrations by George and Doris Haumon.

Painting *Lover's Bouquet* by Marc Chagall; *Several Circles* by Wassily Kandinsky; *Still Life with Musical Instruments* by Max Beckmann; *The Red House* by Edvard Munch; *Terrace in Richmond* by Oskar Kokoschka; *The Last Jockey* by Belgian surrealist René Magritte, 27. Mary Cassatt dies at her country house outside Paris June 14 at age 81 (she has been blind for years); Claude Monet dies at Giverny December 5 at age 86.

Sculpture *Draped Reclining Figure* by Henry Moore. New York's Collector of Customs classifies Constantin Brancusi's 1919 *Bird in Space* "a manufacture of metal" subject to 40 percent duty, photographer Edward Steichen who has imported the *Bird* claims it is a work of art and thus exempt from duty under existing laws, a judge rules that while the *Bird* may not imitate nature it is a work of art according to the "so-called new school of art" which espouses abstraction.

Theater *The Great God Brown* by Eugene O'Neill 1/23 at New York's Greenwich Village Theater, with William Harrigan, 171 perfs.; *The Plough and the Stars* by Sean O'Casey 2/8 at Dublin's Abbey Theatre, 133 perfs.; *The Queen Was in the Parlour* by Nöel Coward 8/24 at Martin's Theatre, London; *Broadway* by U.S. playwrights Philip Dunning and George Abbott 9/16 at New York's Broadhurst Theater, with Lee Tracy, 332 perfs.; *And So to Bed* by English playwright J. B. Fagan 9/26 at the Queen's Theatre, London, is about the 17th-century diarist Samuel Pepys; *White Wings* by Philip Barry 10/15 at New York's Booth Theater, 31 perfs.; *Caponsacchi* by U.S. playwright Arthur Frederick Goodrich, 48, 10/26 at Hampden's Theater, New York, with Goodrich's brother-in-law Walter Hampden as the canon Giuseppe Gaposacchi, from Robert Browning's dramatic monologues *The Ring and the Book* of 1868 and 1869, 269 perfs.; *Dorothea Augermann* by Gerhart Hauptmann 11/20 at Vienna's Theater in der Josefstadt, Munich's Kammerspiele, and 15 other German theaters; *This Was a Man* by Nöel Coward 11/23 at New York's Klaw Theater, 31 perfs.; *The Constant Wife* by W. Somerset Maugham 11/29 at New York's Maxine Elliott Theater, 295 perfs.; *The Silver Cord* by Sidney Howard 12/20 at New York's John Golden Theater, with comedienne Sarah Hope Crews, Margalo Gilmore, Earle Larimore in a play about mother-son love, 112 perfs.; *In Abraham's Bosom* by U.S. playwright Paul Eliot Green, 32, 12/30 at New York's Provincetown Playhouse, with Jules Bledsoe, 123 perfs. (many after it moves uptown).

Films Alan Crosland's *Don Juan* with John Barrymore opens August 6 at New York's Manhattan Opera House and is accompanied by sound electrically recorded on disks in the Warner Brothers Vitaphone process developed by Western Electric engineers. Film czar Will H. Hays, 47, who has headed the Motion Picture Producers and Distributors of America (MPPDA) since 1922, appears on screen to predict that Vitaphone will revolutionize the industry, the film gets an enthusiastic response in October when shown at Grauman's Egyptian Theater in Hollywood, but few exhibitors are willing to install the costly equipment needed, and better established film studios will have no part of it (*see* 1927).

Other films Sam Taylor's *For Heaven's Sake* with Harold Lloyd; Fritz Lang's *Metropolis* with Brigitte Helm, Alfred Abel, Rudolf Klein-Rogge. Also: Buster Keaton's *The Battling Butler* with Keaton; Herbert Brenon's *Beau Geste* with Ronald Colman; King Vidor's *A Bohème* with Lillian Gish, John Gilbert, Renée Adorée; Lewis Seiler's *The Great K & A Train Robbery* with Tom Mix; Vsevlod Pudovkin's *Mother* with Vera Baranovskaia; Serge Eisenstein's *Oktober*; Victor Seastrom's *The Scarlet Letter* with Lillian Gish, Lars Hanson, Henry B. Walthall; John Ford's *Three Bad Men* with George O'Brien; Raoul Walsh's *What Price Glory?* with Victor McLaglen, Edmund Lowe, Dolores Del Rio; Hal Roach's *Putting Pants on Philip* with English-born comic Stan (Arthur Stanley Jeffer-

1926 ***(cont.)*** son) Laurel, 36, and Georgia-born comic Oliver Hardy, 34; George Fitmaurice's *Son of the Sheik* with Rudolph Valentino, who dies of peritonitis in New York August 23 at age 31 after surgery for an inflamed appendix and two perforated gastric ulcers. Press agents hired by Joseph Schenck of United Artists have the matinee idol's body placed in an ornate coffin at Frank E. Campbell's Broadway funeral parlor; 100,000 mourners line up for 11 blocks to view the remains.

Harry Houdini makes headlines August 6 by remaining underwater for 91 minutes in an airtight case containing only enough air to sustain a man for 5 or 6 minutes. The 52-year-old escape artist has practiced breath control and has remained absolutely still in order to minimize his oxygen consumption, but the Great Houdini suffers a subsequent stomach injury and dies of peritonitis October 31.

Opera *Judith* (drama with music) 2/13 at Monte Carlo, with music by Arthur Honegger; *Turandot* 4/25 at Milan's Teatro alla Scala, with music by the late Giacomo Puccini who died 2 years ago at age 66.

Ballet *The Miraculous Mandarin* 11/27 at Cologne, with music by Béla Bartòk.

First performances Symphony No. 7 in C major by Jean Sibelius 4/3 at Philadelphia's Orchestra Hall; Symphony No. 1 by Soviet composer Dmitri Shostakovich, 19, 5/12 at Moscow (the student's diploma piece); *Ballet Mécanique* by U.S. composer George Antheil, 26, 6/19 at Paris (the work is scored for instruments that include a player piano and an airplane engine); *Portsmouth Point* Overture by William Walton 6/22 at Zurich's Tonhalle; *Concerto for Harpsichord, Flute, Oboe, Clarinet, Violin, and Cello* by Manuel de Falla 11/5 at Barcelona with harpsichordist Wanda Landowska.

Broadway musicals *The Girl Friend* 3/17 at the Vanderbilt Theater, with music by Richard Rodgers, lyrics by Lorenz Hart, songs that include "The Blue Room," "The Simple Life," "Why Do I," 301 perfs.; *The Garrick Gaieties* 5/10 at the Garrick Theater, with Sterling Holloway, Rodney Brent, music by Richard Rodgers, lyrics by Lorenz Hart, songs that include "Mountain Greenery," 174 perfs.; *George White's Scandals* 6/14 at the Apollo Theater, with Ann Pennington introducing the Black Bottom dance step that will rival the Charleston, songs that include "The Birth of the Blues" by Ray Henderson, lyrics by B. G. DeSylva and Lew Brown, 424 perfs.; *Countess Maritza* 9/18 at the Shubert Theater, with Yvonne D'Arle, music by Viennese composer Emmerich Kalman, book and lyrics by Harry B. Smith who has adapted a Viennese operetta, 318 perfs.; *Honeymoon Lane* 9/20 at the Knickerbocker Theater, with Eddie Dowling, Pauline Mason, Kate Smith, music by James F. Hanley, lyrics by Dowling, songs that include "The Little White House (At the End of Honeymoon Lane)," 317 perfs.; *Oh, Kay* 11/8 at the Imperial Theater, with Gertrude Lawrence, music by George Gershwin, songs that include "Do, Do, Do" and "Someone to Watch over Me" with lyrics by Ira Gershwin, "Heaven on Earth" and the title song with lyrics by Ira Gershwin and Howard Dietz, 256 perfs.; *The Desert Song* 11/30 at the Casino Theater, with Vivienne Segal, Robert Halliday, music by Sigmund Romberg, lyrics by Oscar Hammerstein II and Otto Harbach, songs that include "Blue Heaven" and "One Alone," 471 perfs.; *Peggy-Ann* 12/27 at the Vanderbilt Theater, with Helen Ford, music by Richard Rodgers, lyrics by Lorenz Hart, songs that include "Where's that Rainbow?" 333 perfs.; *Betsy* 12/28 at the New Amsterdam Theater, with Al Shean, Belle Baker, music by Richard Rodgers, lyrics by Lorenz Hart, songs that include "Blue Skies" by Irving Berlin (who eloped in January with New York society girl Ellin Mackay, 22, daughter of Postal Telegraph president Clarence Mackay), 39 perfs.

Josephine Baker opens her own Paris nightclub at age 20 after having risen to fame in *La Revue Nigre* and starring at the *Folies Bergère* in a G-string ornamented with bananas. The U.S. émigrée darling of European café society will begin a professional singing career in 1930, be naturalized as a French citizen in 1937, and continue performing until shortly before her death in 1974.

Popular songs "Muskrat Ramble" by New Orleans jazz cornetist Edward "Kid" Ory (lyrics will be added in 1937); "Baby Face" by Benny Davis and Harry Akst; "(What Can I Say Dear) After I've Said I'm Sorry?" by Walter Donaldson and bandleader Abe Layman; "Charmaine" by Hungarian-American composer Erno Rapee, 35, lyrics by Lew Pollak, 31; "'Gimme' a Little Kiss, Will 'Ya' Huh?" by Roy Turk, Jack Smith, and Maceo Pinkard; "Bye Bye Blackbird" by Ray Henderson, lyrics by Mort Dixon; "If I Could Be with You One Hour Tonight" by Henry Creamer and Jimmy Johnson; "In a Little Spanish Town" by Mabel Wayne, lyrics by Sam M. Lewis and Joe Young; "When the Red, Red Robin Comes Bob, Bob, Bobbin' Along" by Harry Woods.

Jean Borotra wins in men's singles at Wimbledon, Kitty McKane Godfree in women's singles; René LaCoste wins in men's singles at Forest Hills, Mrs. Mallory in women's singles.

U.S. golfer Bobby Jones loses the U.S. Golf Association Amateur championship to George Von Elm but wins the U.S. Open.

Miniature golf is invented by Tennessee entrepreneur Frieda Carter, part owner of the Fairyland Inn resort, who will patent her "Tom Thumb Golf" in 1929. By 1930 there will be 25,000 to 50,000 miniature golf courses.

Gertrude Ederle, 19, becomes the first woman to swim the English Channel. The New York Olympic champion arrives at Dover August 6 after 14.5 hours in the water, has been forced by heavy seas to swim 35 miles to cover the 21 miles from Cape Gris-Nez near Calais, still beats the world record by nearly 2 hours, and suffers permanent hearing loss (*see* Chadwick, 1951).

German swimmer H. Vierkotter breaks Gertrude Ederle's Channel record (above) by swimming the Channel in 12 hours, 40 minutes.

Gene Tunney wins the world heavyweight boxing championship held by Jack Dempsey since 1919. New York prizefighter James Joseph Tunney, 28, gains a 10-round decision September 23 at Philadelphia, will beat Dempsey again next year, and will retire undefeated in 1928 (*see* 1927).

English cricket ace Sir John Berry Hobbs, 44, scores 16 centuries in first-class cricket.

Cushioned cork-center baseballs are introduced.

The St. Louis Cardinals win the World Series by defeating the New York Yankees 4 games to 3.

Italian hairdresser Antonio Buzzacchino invents a new permanent waving method that will make the "permanent" widely fashionable (*see* 1906).

Slide fasteners get the name "zippers" after a promotional luncheon at which English novelist Gilbert Frankau, 42, has said, "Zip! It's open! Zip! It's closed!" (*see* Sundback, 1913). Elsa Schiaperelli will use zippers in her 1930 line and when the general patents expire the following year the zipper will come into wide use in men's trousers, jeans, windbreakers, and sweaters and in women's dresses and other apparel.

Chicago bootlegger Al Capone's Hawthorne Hotel headquarters are sprayed with machine gun fire in broad daylight September 20 by gunmen firing from eight touring cars that parade single file through the streets of the suburban area but no one is killed and the cars disappear into the traffic.

Illegal liquor traffic is estimated to be a $3.6 billion business and has spawned a gigantic underworld of criminal activity since 1919. Widespread defiance of the Prohibition laws is encouraging citizens to flout other laws and the "Noble Experiment" is clearly a failure (*see* Capone, 1927; Wickersham, 1931).

Municipal zoning ordinances authorized by state governments are upheld by the Supreme Court in the case of *Euclid v. Ambler Realty Company* on the ground that "states are the legal repository of police power." The Department of Commerce publishes a "Standard State Zoning Enabling Act" to serve as a model for state legislatures (*see* 1916; 1961).

A Limited Dividend Housing Companies Law passed by the New York State legislature permits condemnation of land for housing sites, abatement of local taxes, and other measures to encourage housing construction. Limits on rents and profits are required, and income limitations are set for tenants.

Otis Elevator is challenged by Westinghouse, which acquires the patents and engineering skills of several sizeable companies that include Otis's chief competitors. Westinghouse elevators will vie with Otis elevators in America's proliferating skyscrapers.

The U.S. Forest Service identifies 55 million acres of U.S. wilderness area. The largest area found in the survey covers 7 million acres (*see* 1961).

California's tuna packing industry collapses as albacore disappear from offshore waters. San Pedro packs only 51,223 cases, down from 358,940 last year, and the San Diego pack drops to less than 14,600, down from 95,000.

Van Camp Sea Food sales manager Roy P. Harper, 33, decides to promote yellow-fin tuna which can be packed all year round, and by eliminating all flakes, packing the yellow-fin in solid pieces, and calling it "Fancy" tuna he creates a demand for fish heretofore scorned because its flesh is not the white meat associated with albacore. Harper has introduced the brand names Chicken of the Sea and White Star (*see* 1924, 1962).

Trofim Denisovich Lysenko, 28, gains notice for the first time in the Soviet Union. The agronomist puts ideology ahead of science and will have enormous influence on Soviet farm policies (*see* 1935).

Harvey Firestone opens Liberia to rubber cultivation. He has promoted rubber-tree plantations in the Philippines and in South America since 1900, become Ford Motor Company's major tire supplier, and will plant 6,000 acres to rubber trees in the next decade.

Harvard physician Edwin Joseph Cohn, 33, develops an oral liver extract to treat pernicious anemia (*see* 1930; Minot and Murphy, 1924; Castle, 1929; blood plasma, 1940).

The B vitamin proves to be more than one vitamin (*see* McCollum, 1916). Joseph Goldberger shows that rats can be cured of pellagra on a diet from which the heat-labile part of vitamin B has been removed (*see* 1915; Smith and Hendrick, 1928).

British biochemists O. Rosenheim and T. A. Webster show that sunlight converts the sterol ergesterol into vitamin D in animals (*see* Steenbock, 1923). The findings explain why children who do not get enough sunlight are vulnerable to rickets (*see* 1930).

Clarence Birdseye develops a belt freezer for his General Seafoods Co., a small freezing plant on Fort Wharf at Gloucester, Mass. (*see* 1925). Postum Cereal boss Marjorie Merriweather Post puts in at Gloucester to have her yacht provisioned, her chef obtains a frozen goose from General Seafoods, she eats goose and seeks out Birdseye, who is selling quantities of frozen fish but is on the verge of bankruptcy, his freezing process impresses her, but her stockbroker husband E. F. Hutton and board of directors oppose paying $2 million to buy Birdseye's business (*see* General Foods, 1929).

Hormel Flavor-Sealed Ham is the first U.S. canned ham. George A. Hormel & Co. has been slaughtering upwards of a million hogs per year since 1924, produces its canned ham by a process patented by German inventor Paul Jorn, and enjoys immediate success (*see* Spam, 1937).

Machine-made ice production in the United States reaches 56 million tons, up from 1.5 million in 1894 (*see* 1889). Much of it is used to chill illegal beer, highballs, and cocktails (above).

The Good Humor Corp. is formed by Cleveland businessmen following the death of Harry Burt (*see* 1924). Tennessee entrepreneur Thomas J. Brimer, 26, and

1926 ***(cont.)*** two brothers buy franchises and patent rights (*see* 1929).

1927 Generalissimo Chiang Kai-shek takes Hangchow in February with combined units of the Guomindang and the Gungchantang (Communists), Shanghai and Ghangzhou (Canton) fall a few weeks later, and Nanjing is looted and burned March 24, ending the warlord era of Chinese history. Chiang negotiates with bankers and industrialists at Shanghai, where he once ran a brokerage business; promised $3 to $10 million if he will break with Moscow, he reverses his earlier political philosophy, overthrows the leftist government at Hangchow, and establishes a rightist National Revolutionary Government at Nanjing with Communists and left-wing elements excluded.

An "autumn harvest uprising" led by Communist Mao Zedong is crushed September 19 and Chiang expels Russians from Shanghai December 15 following an attempted coup in Guangzhou. Despite questions as to the legality of his divorce from the mother of his son, Chiang has married Wellesley-educated Christianized Song Mei-ling, 26, December 1 and allied himself with one of China's richest, most powerful families (*see* 1926; Beijing, 1928).

London breaks relations with Moscow May 24 following accusations of Bolshevik espionage and subversion throughout the British Empire. The Russians respond June 9 by executing 20 alleged British spies.

Romania's Ferdinand I of Hohenzellern dies of cancer July 21 at age 61 after a 13-year reign. He is succeeded by his nephew Mihai, 5, who will reign until 1930 as Michael I under the regency of his uncle Prince Nicholas, whose brother Carol renounced his right of succession in December 1925 but proclaims himself king (*see* 1930).

Saudi Arabian independence gains British recognition May 20 in the Treaty of Jedda (*see* 1926; oil, 1933).

Chiang Kai-shek fought rival warlords to unify China but betrayed the revolutionary ideals of Sun Yat-sen.

Syria's 2-year Druse insurrection ends in June after a major French military campaign. The Druse leaders flee to Transjordan.

"I do not choose to run," says President Coolidge in an August 2 statement to the press that takes him out of contention for the 1928 presidential nomination.

Josef Stalin expels Leon Trotsky from the Central Committee of the Communist party in November.

The Trade Disputes Act passed by Parliament June 23 prohibits sympathy strikes.

Radical Viennese workers march on the Palace of Justice July 15 and set it afire, enraged because the killers of workers shot down in Burgenland have been acquitted.

Sacco and Vanzetti die in the electric chair at Dedham Prison August 23 despite worldwide efforts to have Massachusetts authorities drop charges against the two for lack of evidence (*see* 1920). A note passed to Nicola Sacco in 1925 from convicted murderer Celestino F. Madeiros said, "I hereby confess to being in the South Braintree shoe company crime and Sacco and Vanzetti was not in said crime" but the district attorney refused to investigate. The defense has filed eight motions for a new trial since 1921 but the state's Supreme Judicial Court has upheld Justice Webster Thayer in early April, ruling on law, not on evidence. "I know the sentence will be between two classes, the oppressed class and the rich class," Sacco has said. Bartolomeo Vanzetti has appealed to Gov. Alvan T. Fuller for a pardon and a Dartmouth professor has filed an affidavit alleging that Judge Thayer said to him, "Did you see what I did with those anarchistic bastards the other day?" Gov. Fuller (a millionaire Cadillac dealer) has appointed a review committee at the request of 61 law professors, the committee comprised of Harvard president A. Lawrence Lowell, MIT president Samuel W. Stratton, and former probate judge Robert A. Grant has upheld the jury verdict July 27, public demonstrations in behalf of Sacco and Vanzetti have no effect, but efforts to gain retroactive pardons will continue until 1977, when a new Massachusetts governor will grant the posthumous pardons.

J. C. Penney opens his five-hundredth store and goes public (*see* 1908). By the time the stock is listed on the New York Exchange in 1929 the company will have 1,495 stores (*see* 1971).

Sears, Roebuck distributes 15 million of its general catalogs, 23 million semiannual and other catalogs. Virtually every rural American home has a Sears or Montgomery Ward mail-order catalog (*see* 1907).

The May Company acquires a major Baltimore store for $2.3 million, renames it the May Company, and adds a Baltimore outlet to its other stores (*see* 1923).

Schlumberger, Ltd. has its beginnings September 5 in the first measurement of an oil well using an electrical-resistance log.

Petroleum prospectors in northern Iraq strike oil October 14 near Kirkuk. Baba Gurgur No. 1 gushes oil 140 feet in the air at a rate of 80,000 barrels per day, it takes the drillers 10 days to bring the well under control and prevent the oil from reaching the Eternal Fire (Baba Gurgur) a mile and a half away (it has been burning for thousands of years at a place where natural gas seeps out of the earth). The well has been financed by Iraq Petroleum Co., Ltd., a new cartel created by French, British, U.S., and Iraqi interests with help from Calouste Gulbenkian, who receives a 5 percent interest in the oil field (*see* 1914).

Charles A. Lindbergh lands his single-engine monoplane *Spirit of St. Louis* at Le Bourget Airfield, Paris, May 21 at 10:24 in the evening after completing the first nonstop solo transatlantic flight (*see* 1926; Alcock, Whitten-Brown, 1919). Backed by St. Louis businessmen, Lindbergh paid $10,580 for the plane built by San Diego aircraft designer T. Claude Ryan, 29, with a 220-horsepower, nine-cylinder Wright Aeronautical Whirlwind engine, he has spent eight weeks at the Ryan factory supervising every detail of construction and made his first test flight April 28, a crude periscope enables him to see what lies ahead since his forward vision is blocked by the gasoline tank and engine, he has declined a radio in order to save weight for 90 more gallons of gasoline, he took off in the rain from Roosevelt Field, Long Island, at 7:55 in the morning of May 20 so heavily laden with 451 gallons of gasoline that he barely cleared some telephone wires, has navigated by dead reckoning to cover 3,600 miles (1,000 miles of it through snow and sleet) in 33 hours, 29 minutes, wins a $25,000 prize offered by Raymond Orteig in 1919, and becomes a world hero. Hailed as "The Lone Eagle," Lindbergh rejects motion picture, vaudeville, and commercial offers totalling $5 million.

French flying ace Charles Nugesser, 35, vanishes in May while trying to fly the Atlantic. He shot down 45 enemy planes during the war.

San Francisco Examiner

VOL. CXXVI. NO. 142. CC SUNDAY, MAY 22, 1927—ONE HUNDRED AND THIRTY-EIGHT PAGES DAILY 5 CENTS, SUNDAY 10 CENTS

LINDBERGH REACHES PARIS IN 33½ HOURS; RIOTOUS WELCOME GIVEN WEARY AIRMAN

"Lucky"— Smilin' Through!

ADMIRERS WRECK FENCE, ROUT POLICE TO GREET SEA HERO

By WILLIAM HILLMAN,

LE BOURGET FIELD, PARIS, May 21.—(By Special Cable Dispatch.)—Swooping down from the skies in a magnificent landing, cheered to the echo by thousands upon thousands of welcoming Americans and French residents, Capt. Charles Lindbergh, the intrepid "Flyin' Fool" from St. Louis, Missouri landed here tonight at 10:22 Paris time, completing his marvelous non-stop flight across the Atlantic Ocean from New

Flying the Atlantic solo on one engine made "the Lone Eagle" a world hero and gave aviation a dramatic boost.

U.S. pilot Clarence Duncan Chamberlin, 34, takes off from New York, June 4 and flies 3,911 miles nonstop in 42.5 hours to Saxony, Germany, carrying junk dealer Charles A. Levine who has financed the flight.

Commander Richard E. Byrd makes the first radio-equipped transatlantic flight June 29 in a trimotored Fokker monoplane piloted by Lieut. Bernt Balchen and Bert Acosta with radioman George O. Noville completing the crew.

U.S. Army lieutenants Lester J. Maitland and Albert F. Hegenberger make the first successful flight from San Francisco to Honolulu June 28.

The Norden bombsight invented by Dutch-American inventor Carl Lucas Norden, 47, and his engineer associate Theodore H. Barth will remain a closely guarded military secret until it is patented in 1947 (by which time it will be obsolete). As perfected in 1931 by Norden and Frederick I. Entwhistle of the U.S. Navy, the 90-pound Mark XV bombsight will have 2,000 parts.

Juan Trippe founds Pan American Airways and obtains exclusive rights from Cuban president Gerardo Machado y Morales to land at Havana; he begins mail service between Key West, Fla., and Havana with Fokker F-7 single-engine monoplanes (*see* 1925; 1929).

U.S. mail planes fly nearly 6 million miles and carry 37,000 passengers (*see* 1929).

Tokyo's Chikatetsu subway opens. The system will grow to have eight lines with 102.9 miles of track.

U.S. railroads begin to introduce centralized traffic control (CTC) that permits automatic control of two-way traffic on single tracks and makes single-track operations nearly as efficient as double-track.

The 46-mile Chesapeake and Delaware Ship Canal opens to link the bay with the river.

Ford introduces the Model A to succeed the Model T that has been the U.S. standard for nearly 20 years. U.S. auto production falls to 3,093,428, down nearly 900,000 from 1926 on account of Ford's stoppage for retooling to produce a car that will compete with Chevrolet. Henry Ford and his son Edsel, now 34, drive the 15 millionth Ford out of the Ford plant.

More than 20 million cars are on the U.S. road, up from 13,824 in 1900.

The Volvo introduced by Göteborg industrialist Assar Gabrielsson and engineer Gustaf Larsson is the first Swedish motorcar. They have built a factory to challenge the dominance of imported cars in the Swedish market but their four-cylinder engine touring car has a top speed of only 37 miles per hour.

A 1,000 horsepower "Mystery Sunbeam" driven May 29 at Daytona Beach, Fla., by Major Henry O'N. de H. Segrave averages 203.79 miles per hour.

The Buffalo Peace Bridge opens August 4 to link Canada and the United States.

The Holland Tunnel opens November 12 to connect Canal Street, Manhattan, with Jersey City, giving motor vehicles a road link under the Hudson River, the first

1927 ***(cont.)*** alternative to New York-New Jersey ferry boats. Clifton Milburn Holland, the tunnel's chief engineer, died in 1924 just 2 days before diggers from east and west met below the Hudson (*see* Lincoln Tunnel, 1940).

Massachusetts enacts the first compulsory state automobile insurance law.

E. I. du Pont and Pittsburgh Plate Glass create Duplate Corp. to make safety glass using Du Pont pyrolin and Pittsburgh plate (*see* 1883). PPG will buy out Du Pont's interest in Duplate in 1930 (*see* 1929).

Henry Ford orders safety glass windshields for Model A Fords and Lincolns. Triplex Safety Glass contracts to supply half of Ford's needs and licenses Ford to produce the rest himself (*see* 1926).

The end of the Model T (above) deals a heavy blow to the 18-year-old Western Auto Supply Co. whose $11.5 million in annual sales derive largely from Model T parts. The company has 31 retail stores and will become a vast retail chain enterprise dealing not only in automobile parts but also in hardware, cheap watches, flashlights, electrical appliances, and drugstore items (*see* Pepperdine College, 1937).

The S.S. *Ile de France* arrives at New York on her maiden voyage. The 43,000 ton French Line "Boulevard of the Atlantic" has a plane-launching catapult on her afterdeck to speed mail delivery.

An uncertainty principle announced by German physicist Werner Heisenberg, 26, melds physics and philosophy. He states that certain pairs of variables describing motion-velocity and position, or energy and time cannot be measured simultaneously with absolute accuracy because the measuring process itself interferes with the quantity to be measured, so while quantum mechanics provides valuable information it is useful only within limits of tolerance since no events can be described with zero tolerance. Heisenberg has been working with Max Born, 45, at Göttingen University (*see* Planck, 1900; Dirac and Shrödinger, 1926).

Fossil remains of Pithecanthropus pekinsis (Peking man) dating to between 400,000 and 300,000 B.C. are found at Choukoutien, near Beijing, by Canadian anatomist Davidson Black, 42 (*see* 1890; 1959).

A stone spearhead found embedded between the ribs of an Ice Age bison skeleton near Folsom, N. Mex., is the first of many such finds that will indicate the presence of glacial man in 8,000 B.C.

Peruvian aerial survey pilot Toribio Mexta Xesspe flying over the barren plains of southern Peru sees long lines in the shape of birds, animals, and reptiles. Visible only from the air, the mysterious drawings, enormous in size, were made by the pre-Inca Nazca civilization that flourished for a millennium before being absorbed by the Inca about 700 A.D.

The Iron Lung invented by Harvard professor Philip Drinker, 40, has an airtight chamber that employs alternating pulsations of high and low pressure to force

The "talkies" gave Hollywood its voice and made motion pictures a powerful influence throughout the world.

air in and out of a patient's lungs. Drinker uses two discarded household vacuum cleaners and other cast-off machinery to create the Drinker Respirator that will be manufactured in Boston and used for the first time in mid-October of next year to treat a small girl at Boston Children's Hospital who is suffering from respiratory failure due to poliomyelitis.

Franklin D. Roosevelt founds the Georgia Warm Springs Foundation treatment center for fellow victims of poliomyelitis (*see* 1921). The center will be operated in years to come by the National Foundation March of Dimes (*see* 1938).

Transatlantic telephone service begins January 7 between London and New York: 3 minutes of conversation cost $75, or £15.

Utah engineer Philo Taylor Farnsworth, 21, invents a television image dissector tube. He has been inspired by accounts of work done in the Soviet Union by Boris Bosing to transmit moving pictures by electricity (*see* Baird, 1926).

Television gets its first U.S. demonstration April 7 in the auditorium of New York's Bell Telephone Laboratories by AT&T president Walter S. Gifford who lets a large group of viewers see Commerce Secretary Herbert C. Hoover in his office at Washington while hearing his voice over telephone wires (*see* BBC, 1936).

Development of television is thwarted by the fact that it takes a frequency band of 4 million cycles, versus only 400 for an ordinary radio band, to transmit the 250,000 elements needed for a clear picture.

David Sarnoff's year-old National Broadcasting Co. has so many radio stations that it splits up into a Blue Network and a Red Network (*see* American Broadcasting, 1943; CBS, 1928).

The Free Library of Philadelphia building is completed in French Renaissance style on Vine Street between 19th and 20th Streets.

Nonfiction *Main Currents in American Thought: An Interpretation of Literature From the Beginning to 1920* by University of Washington professor Vernon Louis Parrington, 56, who sees the development of American thought as based on a concept of democratic idealism; *The Story of a Wonder Man* (autobiography) by Ring Lardner.

Fiction *Der Steppenwolf* by Hermann Hesse; *Flight Without End (Die Flucht ohne Ende)* by Austrian novelist Joseph Roth, 33, who shows the collapsed Austro-Hungarian Empire as seen through the eyes of a returned officer; *Amerika* by the late Franz Kafka; *The Bridge of San Luis Rey* by Thornton Wilder; *Jalna* by Canadian novelist Mazo de la Roche, 42; *Mr. Weston's Good Wine* by English novelist T. F. Powys, 52, brother of J. C. Powys; *Celibate Loves* (stories) by George Moore; *The Left Bank* (stories) by Welsh-Creole author Jean Rhys, 33; *Death Comes for the Archbishop* by Willa Cather is based on the 19th-century French-American clergyman Jean Baptiste Lamy who built the first cathedral in the Southwest at Santa Fe; *Elmer Gantry* by Sinclair Lewis; *The Treasure of the Sierra Madre (Der Schatz der Sierra Madre)* by German novelist B. Traven (Ret Marut); *archy and mehitabel* by *New York Tribune* humorist Donald Robert Perry "Don" Marquis, 44, whose cockroach archy is the reincarnation of a poet and whose alley cat mehitabel has rowdy misadventures; *The "Canary" Murder Case* by U.S. critic-novelist S. S. Van Dine (Willard Huntington Wright), 39, whose detective Philo Vance will win a wide following; *The Unpleasantness at the Bellona Club* by Dorothy Sayers.

Poetry "American Names" by U.S. poet Stephen Vincent Benét, 29; *The Women at Point Sur* by Robinson Jeffers; *Tristram* by Edwin Arlington Robinson is a verse novel; "Launcelot" by E.A. Robinson; *Leitenant Schmidt* by Boris Pasternak is a verse epic.

Painting *Seated Woman* by Pablo Picasso; *Glass and Fruit* by Georges Braque; *Figure with Ornamental Background* by Henri Matisse; *Man with Newspaper* by René Magritte; *The Great Forest* by Max Ernst; *Manhattan Bridge* by Edward Hopper; *Radiator Building, New York* by U.S. painter Georgia O'Keeffe, 39, who married photographer Alfred Stieglitz of the 291 Gallery 3 years ago; *Lonesome Road* by Thomas Hart Benton. Juan Gris dies at Boulogne-sur-Seine outside Paris May 11 at age 40.

Theater *Saturday's Children* by Maxwell Anderson 1/26 at New York's Booth Theater, with Ruth Gordon, Roger Pryor, 310 perfs.; *The Marquise* by Noël Coward 2/16 at London's Criterion Theatre, 129 perfs.; *The Butterfly's Evil Spell* (*El maleficio de la mariposa*) by Spanish poet-playwright Federico García Lorca, 27, 3/22 at Madrid's Teatro Esclava; *The Second Man* by U.S. playwright S. N. (Samuel Nathan) Behrman, 34, 4/11 at New York's Guild Theater, with Alfred Lunt, Lynn Fontanne, Margalo Gilmore, Earle Larimore, 178 perfs.; *Porgy* by U.S. novelist-playwright DuBose Heyward, 42, and his wife Dorothy 10/27 at New York's Guild Theater, 367 perfs. (*see* Gershwin opera, 1935); *The Oil Islands (Die Petroleuminseln)* by Lion Feuchtwanger 10/31 at Hamburg's Deutsches Schauspielhaus; *The Road to Rome* by U.S. playwright Robert Emmet Sherwood, 31, 11/31 at The Playhouse, New York, with Jane Cowl, 440 perfs.; *Paris Bound* by Philip Barry 12/27 at New York's Music Box Theater, with Hope Williams, 234 perfs.; *The Royal Family* by George S. Kaufman and Moss Hart, 23, 12/28 at New York's Selwyn Theater is based on the Barrymores, 345 perfs.

Grauman's Chinese Theater opens in Hollywood.

Films Alan Crosland's *The Jazz Singer* is the first full-length talking picture to achieve success. Executive producer Darryl Francis Zanuck, 25, has given Al Jolson his film debut, the film opens at New York October 6, it contains only brief sequences of dialogue and singing, but the sound-on-disk Warner Brothers Vitaphone system introduces a new era of "the talkies" that will end the careers of some movie stars (*see* 1926; 1929).

Other films (all silent) Buster Keaton's *The General* with Keaton; Ted Wilde and J.A. Howe's *The Kid Brother* with Harold Lloyd; Abel Gance's *Napoléon* with Albert Dieudonné; Ernst Lubitsch's *The Student Prince in Old Heidelberg* with Ramon Novarro, Norma Shearer; F.W. Murneau's *Sunrise* with Janet Gaynor, George O'Brien; Erich von Stroheim's *Wedding March* with von Stroheim, Fay Wray, ZaSu Pitts. Also: Alan Crosland's *The Beloved Rogue* with John Barrymore, Conrad Veidt; Walter Ruttman and Karl Freund's *Berlin: The Symphony of a Great City*; Fred Niblo's *Camille* with Norma Talmadge, Gilbert Roland; Paul Leni's *The Cat and the Canary* with Creighton Hale, Laura LaPlante; Ernst Schoedsack and Merian C. Cooper's *Chang*, shot on location in Siam; René Clair's *Le Chapeau de Paille d'Italie* with Albert Prejean, Olga Tschechova; James W. Horne's *College* with Buster Keaton; Sam Wood's *The Fair Co-Ed* with Marion Davies; Clarence Brown's *Flesh and the Devil* with Greta Garbo; Frank Borzage's *Seventh Heaven* with Janet Gaynor, Charles Farrell; Serge Eisenstein's *The Ten Days That Shook the World*; Victor Fleming's *The Way of All Flesh* with Emil Jannings; William K. Howard's *White Gold* with Jetta Goudal, George Bancroft; William A. Wellman's *Wings* with Clara Bow, Charles "Buddy" Rogers, Richard Arlen, Gary Cooper.

The Academy of Motion Picture Arts and Sciences is founded May 11 by Louis B. Mayer of M-G-M (*see* 1924). Annual awards of the Academy will be called "Oscars" by movie columnist Sidney Skolsky (now a press agent of 22), first president of the Academy is Douglas Fairbanks, first winners of the gold statuette (which is 92.5 percent tin) will be William Wellman for *Wings*, Emil Jannings for best actor, Janet Gaynor, 21, for best actress.

Opera *The King's Henchman* 2/17 at New York's Metropolitan Opera House, with music by U.S. composer Deems Taylor, 41, libretto by Edna St. Vincent Millay who contributes proceeds of her pamphlet "Justice Denied in Massachusetts" to the defense of Sacco and Vanzetti (above), makes a personal appeal to the

1927 ***(cont.)*** governor that he spare the men, and is arrested in the deathwatch outside the Boston Court House the night of their execution; *Mahagonny* 7/17 at Germany's Baden Baden festival, with music by German composer Kurt Weill, 27, libretto by German playwright Bertolt Brecht, 29: the one-act singspiel (song-play) will be expanded into *The Rise and Fall of the Town of Mahagonny* which will open 3/9/30 at Leipzig; *Le Pauvre Matelot* 12/16 at the Opéra-Comique, Paris, with music by Darius Milhaud.

First performances Concerto No. 4 in G minor for Piano and Orchestra by Sergei Rachmaninoff 3/18 at Philadelphia's Academy of Music, with Leopold Stokowski conducting the Philadelphia Orchestra; *Arcana* by French-American composer Edgar Varese, 34, 4/8 at Philadelphia's Academy of Music; Symphony in E minor by Roger Sessions 4/22 at Boston's Symphony Hall.

Broadway musicals *Rio Rita* 2/2 at the new Ziegfeld Theater on Sixth Avenue at 54th Street, with music and lyrics by Harry Tierney and Joseph McCarthy, songs that include the title song, 494 perfs. Now 58, Florenz Ziegfeld has hired *Follies* set designer Joseph Urban, a Viennese sculptor-painter-architect, to design the modern theater that will stand for 40 years; *Hit the Deck* 4/25 at the Belasco Theater, with music by Vincent Youmans, lyrics by Leo Robin and Clifford Grey, songs that include "Join the Navy," "Hallelujah," 352 perfs.; *The Ziegfeld Follies* 8/16 at the Ziegfeld Theater, with Eddie Cantor, Ruth Etting, music and lyrics entirely by Irving Berlin, 167 perfs.; *Good News* 9/6 at the 46th Street Theater, with music by Ray Henderson, lyrics by B. G. DeSylva and Lew Brown, songs that include "The Best Things in Life Are Free," "Lucky in Love," "Flaming Youth," 551 perfs.; *My Maryland* 9/12 at the Jolson Theater, with a book based on the 1899 Clyde Fitch play *Barbara Frietchie,* music by Sigmund Romberg, 312 perfs.; *A Connecticut Yankee* 11/3 at the Vanderbilt Theater, with William Gaxton, Constance Carpenter, music by Richard Rodgers, lyrics by Lorenz Hart, songs that include "My Heart Stood Still," "Thou Swell," 418 perfs.; *Funny Face* 11/22 at the Alvin Theater, with Victor Moore, William Kent, music by George Gershwin, lyrics by Ira Gershwin, songs that include "The Babbitt and the Bromide," 244 perfs.; *Delmar's Revels* 11/28 at the Shubert Theater, with Frank Fay, Bert Lahr, Patsy Kelly, music by Jimmy McHugh, lyrics by Dorothy Fields, 112 perfs.; *Show Boat* 12/27 at the Ziegfeld Theater, with Helen Morgan as Julie LaVerne, book by Edna Ferber, music by Jerome Kern, lyrics by Oscar Hammerstein II, songs that include "Bill," "Can't Help Lovin' that Man," "Ol' Man River," "Only Make Believe," "Life on the Wicked Stage," "Why Do I Love You," 527 perfs.

The Varsity Drag is introduced to U.S. dance floors.

Radio music *The A&P Gypsies* debut in January over NBC's Blue Network. The violinists will give way to Harry Horlick's Orchestra which will continue for 10 years; *Cities Service Concerts* debut 2/18 and will continue each Friday evening into the mid-1950s (60 minutes until 1940, 30 minutes thereafter).

Popular songs "Black and Tan Fantasy" by Edward Kennedy "Duke" Ellington, 28, (who begins a 5-year engagement at New York's Cotton Club) lyrics by "Bubber" Miller; "Me and My Shadow" by Al Jolson and pianist Dave Dreyer, 33, lyrics by Billy Rose; "Chloe" by Neil Moret, lyrics by Gus Kahn; "Diane" by Irving Rapee, lyrics by Lew Pollak to exploit the film *Seventh Heaven;* "Girl of My Dreams" by Sunny Clapp; "Ain't She Sweet" by Milton Ager, lyrics by Jack Yellen; "I'm Looking Over a Four Leaf Clover" by Harry Woods, lyrics by Mort Dixon; "Let a Smile Be Your Umbrella" by composer Sammy Fain, 25, lyrics by Irving Kahal, Francis Wheeler; "My Blue Heaven" by Walter Donaldson, lyrics by George Whiting, 43; " 'S Wonderful" and "Strike Up the Band" by George Gershwin, lyrics by Ira Gershwin; "Mississippi Mud" by Harry Burris of the Rhythm Boys (Burris, Al Rinker, and "crooner" Harry Lillis "Bing" Crosby, 26) who sing with Paul Whiteman and his Orchestra, lyrics by James Cavanaugh; "The Song Is Ended But the Melody Lingers On" by Irving Berlin.

"He's Got the Whole World in His Hands" is published for the first time in the *Journal of American Folklore.*

The first all-electric jukeboxes are introduced by the Automatic Musical Instrument Co. of Grand Rapids, Mich., and Seeburg Co. of Chicago. Large-scale mass production will begin in 1934 and by 1939 there will be 350,000 jukeboxes in U.S. bars, restaurants, and other establishments.

Capehart Automatic Phonograph Co. is founded by Indianapolis entrepreneur Homer Earl Capehart, 30, who has acquired the patent rights of a Cleveland inventor to manufacture a coin-operated phonograph with an automatic device that not only selects a record and places it on the turntable but then also turns the record over and plays its flip side. Capehart will acquire rights to a device invented by Columbia. Phonograph engineer Ralph Erbe in 1929 will introduce an improved record changer, shifting from jukeboxes to home consoles for the luxury market.

Dancer Isadora Duncan is strangled to death at Nice September 14 at age 49 when her long scarf is entangled in a rear wheel of the sports car being demonstrated to her by an automobile salesman with whom she has become enamored (*see* 1921).

The first Golden Gloves boxing tournament opens March 11 at New York's Knights of Columbus center and at Brooklyn's Knights of St. Anthony's center, the finals are held March 28 at Tex Rickard's new Madison Square Garden, witnessed by a record crowd of 21,954, with an estimated 10,000 turned away for lack of space. Proposed by *New York Daily News* sportswriter Paul Gallico in a February 14 back page headline and feature story, the event will continue for more than 50 years, attracting thousands of contenders who will include Emile Griffith, Gus Lesnevich, Floyd Patterson, Ray Robinson, and José Torres.

Henri Cochet, 25, (Fr) wins in men's singles at Wimbledon, Helen Wills in women's singles; René LaCoste wins in men's singles at Forest Hills, Wills in women's singles. LaCoste beats Bill Tilden at Philadelphia's Germantown Cricket Club to end domination of Davis Cup play by English-speaking countries.

Golfer Tommy Armour wins the U.S. Open at age 30. The Scotsman goes on to win the Canadian Open.

The first Ryder Cup golf match ends in victory for a team of U.S. professionals led by Walter Hagen who defeat a British team at Worcester, Mass., to gain possession of the £750 silver cup put up by English seed magnate Samuel Ryder, 54. A British team won an unofficial match held last year at Wentworth, Surrey, the U.S. team includes Tommy Armour (above), British teams will win in 1929 and 1933, but U.S. teams will dominate the biennial matches.

Gene Tunney retains his world heavyweight boxing crown by surviving Jack Dempsey's seventh round knockout punch at Chicago's Soldier Field September 22. A new rule requires that a fighter go to a neutral corner after he has knocked down his opponent, but Dempsey ignores the referee's order and stands over the fallen Tunney. Paul Gallico of the *Daily News* (above) will write that Tunney "was out. No question about it. His mind was gone," but 15 seconds elapse before the referee reaches the count of nine, Tunney rises from the "long count" and goes on to win the decision.

Babe Ruth hits his sixtieth home run of the season September 30 off a pitch by Washington's Tom Zachary to set a record that will stand for 30 years.

The New York Yankees win the World Series by defeating the Pittsburgh Pirates 4 games to 0.

World chess champion José Raoul Capablanca loses to Russian-born French master Aleksandr Aleksandrovich Alekhine, 35, who will hold the title until 1935, will regain it in 1937, and will hold it until his death in 1946 (*see* 1921).

Frances Heenan "Peaches" Browning, 16, sues millionaire New York real estate operator Edward W. "Daddy" Browning, 52, for divorce in a White Plains, N.Y., courthouse after less than a year of marriage. The January trial produces testimony that titillates newspaper readers.

Super Suds introduced by Colgate and Company is a laundry and dishwashing soap product composed of quick-melting hollow beads rather than flakes or powder (*see* Rinso, 1918; Tide, 1946).

The Cyclone roller coaster opens June 26 at Coney Island, N.Y., with a 100-second ride that takes screaming passengers up and down nine hills and over connecting tilted curves for 25 cents each.

Brown & Williamson Tobacco is acquired by British-American Tobacco which will increase production and distribution of Sir Walter Raleigh and Kool cigarettes.

The Snyder-Gray murder trial makes world headlines. Queens Village, Long Island, housewife Ruth Snyder and her corset-salesman lover Henry Judd Gray kill New York magazine art editor Albert Snyder, with a sashweight March 20 in a suburban sex triangle. A jury convicts the pair, and they will die in the electric chair at Sing Sing early next year.

Al Capone has an income for the year of $105 million, the highest gross income ever received by a private U.S. citizen (*see* 1925; 1926). Most of the Chicago gangster's money derives from bootleg liquor operations and he takes in $35 million more than Henry Ford will make in his best year (*see* 1929; tax evasion, 1931).

The Supreme Court rules that illegal income is taxable, thus giving the federal government a powerful new weapon against the underworld.

The prefabricated Dymaxion House built by visionary U.S. architect R. Buckminster Fuller, 32, has rooms hung from a central mast with outer walls of continuous glass.

Radburn, N.J., is designed by U.S. architects Clarence S. Stein, 46, and Henry Wright whose aim is not to make the most economical use of land but the most economical use of people with details planned to protect residents from air pollution, the abrasive effects of noise, needless tensions, fears, and alienation. Completion of the first modern "new town" will be delayed for economic reasons but Stein will go on to design Chatham Village outside Pittsburgh, Sunnyside Gardens in Queens (N.Y), and Kitimat in British Columbia.

The American Insurance Union Citadel at Columbus, Ohio, opens with 47 floors at Broad and Front streets.

New York's Graybar building opens on Lexington Avenue adjoining Grand Central Station. The 30-story building named for the Graybar Electric Co. has more than 1 million square feet of rentable floor space—more than any other office building in the world.

Boston's Ritz Hotel opens May 18 at the corner of Arlington and Newberry streets overlooking the Public Garden.

Boston's Statler Hotel opens with 1,150 rooms on 14 floors. Hotel pioneer Ellsworth M. Statler will die in April of next year at age 65.

Washington's Hay-Adams Hotel opens on Lafayette Square. Armenian-American architect Mihrail Mesrobian, 38, has designed the hotel to replace the Romanesque Hay-Adams houses built in the 1880s by, the late Henry Hobson Richardson for author Henry Adams and former secretary of state John Hay.

Chicago's Stevens Hotel opens on Michigan Avenue. The 3,000-room Stevens is the largest hotel in the world (*see* Hilton, 1945).

Honolulu's Royal Hawaiian Hotel opens at Waikiki Beach. Alexander & Baldwin has financed the new hotel.

Mar-a-Lago is completed at Palm Beach, Fla., for Postum Cereal head Marjorie Merriweather Post and her husband E. F. Hutton. Shaped like a crescent with a 75-foot tower and surrounded by 17 acres with guest houses, staff quarters, cutting gardens, and a nine-hole

1927 ***(cont.)*** golf course, the 115-room mansion has taken 4 years to complete, three ships from Genoa have brought the Dorian stone for its outer walls, the ancient red roofing tiles have come from Cuba along with black and white marble inlay for the floors, Post has picked up 38,000 antique Spanish tiles for indoor wall decoration and has helped Palm Beach survive an economic depression by hiring all available local craftsmen to work on the $7 million project.

April flood waters in the lower Mississippi Valley cover 4 million acres and cause $300 million in property loss (*see* Flood Control Act, 1928).

More than 100 people are swept to their deaths by a Vermont flood in early November after 11 inches of rainfall in 2 days. The Winooski, Lamoille, and White rivers rise to destroy most of Vermont's covered bridges, the flood waters wash away houses and mills, and they cover 7,000 acres with rock gravel.

The mechanical cotton picker perfected by Texas inventor John Daniel Rust, 35, and his 27-year-old brother Mack will have a profound social impact on the South when marketing of the machine begins in 1949 (*see* Campbell, 1889). The Rust cotton picker inserts a long spinning spindle with teeth into the cotton boll, winds up the cotton, picks it out, and is kept wet to facilitate removal of the cotton from the teeth. It picks a bale of cotton in one day, and it will spur migration of blacks to northern cities as it reduces the need for field hands (*see* 1949).

The Supreme Court denies a Federal Trade Commission request for further dissolution of the International Harvester Trust whose McCormick and Deering harvesting machines have been separated by the court order of 1920.

Harry Steenbock patents his 1923 discovery of vitamin D irradiation and assigns the patent to the Wisconsin Alumni Research Foundation, rejecting all commercial offers (*see* Borden's vitamin D fortified milk, 1933; but *see* also 1946).

Your Money's Worth by U.S. economist Stuart Chase and engineer F. J. (Frederick John) Schlink, 35, says, "We are all Alices in a Wonderland of conflicting claims, bright promises, fancy packages, soaring words, and almost impenetrable ignorance." Schlink has spent several years with the National Bureau of Standards which tests products before the federal government will buy them and he wonders why private citizens do not have the advantage of such testing (*see* Consumers' Research, 1929).

Wonder Bread is introduced in a balloon-decorated wrapper by Continental Baking Co. (*see* 1924; sliced bread, 1930).

Hostess Cakes are introduced by Continental Baking (above; *see* Twinkies, 1930).

Lender's Bagel Bakery is founded at West Haven, Conn. It will become the largest U.S. bagel baker.

American Sugar Refining's share of the U.S. sugar market falls to 25 percent, down from nearly 100 percent in 1899, but its 16-year-old Domino brand remains the top-selling table sugar. National Sugar Refining with its 14-year-old Jack Frost brand controls 22 percent of the market and while 13 other U.S. companies refine imported raw sugar none has more than a 7 percent share.

Borden introduces homogenized milk; other U.S. milk is still sold with cream at the top that must be mixed before the milk is poured.

Gerber Baby Foods has its beginnings at Fremont, Mich., where local food processor Daniel F. Gerber, 28, is told by doctors to feed his sick daughter Sally strained peas. Gerber finds that strained baby foods are commercially available but are expensive, sold only in a few parts of the country, and available only at pharmacies only by doctors' prescription (*see* 1928).

General Electric introduces a refrigerator with a "monitor top" containing an hermetically sealed compressor. The 14-cubic-foot refrigerator sells for $525, few can afford it, but it will make GE the industry leader by 1930 (*see* 1929).

Marriott's Hot Shoppe opens at Washington, D.C., at the corner of 14th Street and Park Road Northwest. Utah-born entrepreneur John Willard Marriott, 27, has invested his savings of $1,000 plus $1,500 in borrowed capital to obtain an exclusive franchise to sell A&W Root Beer in Washington, Baltimore, and Richmond using syrup obtained from two Westerners named Allen and Wright, his business has fallen off in the fall, he has hired a barbecue cook and has borrowed some recipes from the chef of the nearby Mexican Embassy, and he puts his bride Alice behind the stove to make hot tamales and chile con carne. The nine-seat root beer stand will grow into the Marriott Corp., a worldwide empire of Hot Shoppes, Big Boy Coffee Shops, Roy Rogers Family Restaurants, Farrell's Ice Cream Parlours, and numerous business, hospital, and other institutional food service operations plus hotels.

German physiologists Bernard Zondek and S. Ascheim discover a sex hormone that will lead to the first early test for pregnancy.

Birth control pioneer Margaret Sanger organizes the first World Population Conference (*see* 1921; International Planned Parenthood, 1948).

1928 Beijing surrenders to Chiang Kai-shek, who last year renamed the city Peiping (northern peace). Chiang is elected president of China (*see* 1931).

The Kellogg-Briand Pact (Pact of Paris) is signed August 27 by 63 world powers whose representatives renounce war. Devised by U.S. Secretary of State Frank B. Kellogg, 72, and French foreign minister Aristide Briand, 66, the pact is implemented in September by the League of Nations.

The Albanian Republic proclaimed in 1925 becomes a kingdom under former prime minister Ahmed Bey Zogu, 33, who has changed his name to Scanderbeg II and is crowned as Zog I to begin an 18-year reign of which the last seven will be spent in exile.

Mexico's president Alvaro Obregon is assassinated July 17 at age 48. Former president Plutarco Elias Calles, 51, dominates the country, he will form the Partido Revolucionario Institucional (PRI) next year, and it will control Mexico for at least 65 years (*see* Cardenas, 1934).

U.S. voters elect Herbert Hoover president with 444 electoral votes to 87 for his Democratic opponent Alfred E. Smith, governor of New York. Smith loses five states of the "Solid South" and obtains only 41 percent of the popular vote to Hoover's 58 percent (*see* below).

Bolivia and Paraguay go to war December 6 over the Chaco Territory. Paraguay appeals to the League of Nations; the Pan-American Conference that held its sixth meeting early in the year at Havana offers to mediate, but skirmishes will continue until April 4, 1930, when the two countries will reach a temporary truce agreement.

Italy abolishes women suffrage May 12 under a new law that restricts the franchise to men 21 and over who pay syndicate rates or taxes of 100 lire. The law reduces the electorate from nearly 10 million to only 3 million and requires that voters approve or reject in toto the 400 candidates submitted by the Fascist grand council.

British women gain the vote on the same terms as men under provisions of a July 2 act of Parliament (*see* 1918).

The first Soviet Five-Year Plan launched by Josef Stalin in October will bring liquidation or exile to Siberia for millions of *kulaks* (rich peasants) who resist collectivization of agriculture (below).

The National Conference of Christians and Jews is founded to fight U.S. bigotry following the defeat of Catholic presidential candidate Alfred E. Smith (above). No Roman Catholic will win election to the presidency until 1960.

France devalues the franc June 24 from 19.3¢ U.S. to 3.92¢. The disguised repudiation of the national debt wreaks havoc on France's rentier class (*see* 1926).

"We in America today are nearer the final triumph over poverty than ever before in the history of any land," says Republican presidential candidate Herbert Hoover (above). He hails "the American system of rugged individualism" October 22 in a speech at New York.

New York's National City Bank goes into competition with the Morris Plan started in 1910, offering loans to employed borrowers on terms similar to those offered by Morris Plan banks.

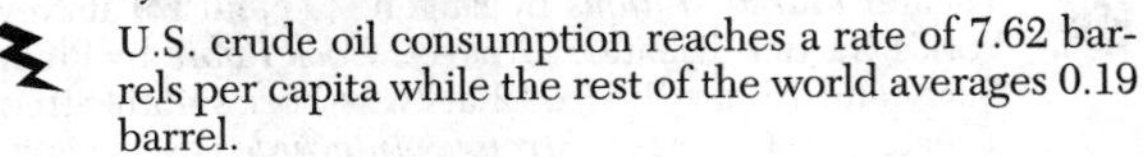

U.S. crude oil consumption reaches a rate of 7.62 barrels per capita while the rest of the world averages 0.19 barrel.

The Boulder Dam Project Act passed by Congress December 21 commits the federal government to participate in the production of hydroelectric power (*see* Kaiser, 1931).

The Jones-White Merchant Marine Act passed by Congress May 22 gives private U.S. shipping interests federal subsidies to help them compete with foreign lines (*see* 1891; 1936).

Amelia Earhart, 30, flies the Atlantic June 17 as a passenger with two men who pilot a multi-engined Fokker. She will make her own solo flight in 1932 (*see* 1937).

Newark Airport opens in September. The 68-acre, $1.75 million field will grow to cover 2,300 acres (*see* La Guardia, 1939).

The German dirigible *Graf Zeppelin* arrives October 15 at Lakehurst, N.J., after covering 6,630 miles in 121 hours on her first commercial flight. The voyage from Friedrichshafen inaugurates transatlantic service by lighter-than-air craft (*see* 1900; 1937).

Lockheed Aircraft designer John Northrop quits the small Burbank, Calif., company to start his own firm. Gerard Vultee, 27, who succeeds Northrop as chief engineer, will redesign the Lockheed Vega for speed and design the Sirius for Charles Lindbergh and his wife Anne (*see* 1927; Lockheed, 1932; American Airlines, 1930).

New York's Goethals Bridge and Outerbridge Crossing open to connect Staten Island with New Jersey.

Plymouth motorcars are introduced by Chrysler Motors to compete with Ford and Chevrolet. "Look at all three," says Walter Chrysler in advertisements featuring Plymouths with high compression engines and hydraulic brakes.

The DeSoto, introduced by Chrysler, is a medium-priced car that will remain in production until 1961.

A new 112-horsepower Chrysler Imperial "80" is advertised as "America's most powerful automobile."

Stutz advertises a 113-horsepower model.

Auburn comes out with an 8-cylinder, 115-horsepower model advertised with a picture of 115 stampeding horses. Its boat-tailed speedster travels at 108.6 miles per hour at Daytona, Fla., in March and later in the year averages 84.7 miles per hour for 25 hours at Atlantic City, N.J. (*see* 1924; 1930).

Florida's Tamiami Trail opens in April to link Miami with Fort Myers on the Gulf Coast. The 143-mile highway through the Everglades has taken 11 years to build and cost more than $48,000 per mile.

The Pioneer Yelloway bus that reaches New York from California is the first transcontinental bus. Wesley E. Travis founded Pioneer Yelloway in 1919 over his father's old stagecoach routes. His larger competitor Pickwick Stages operates as far east as Kansas City (*see* Greyhound, 1929).

American Negro: A Study in Racial Crossings by Northwestern University anthropology professor Melville Jean Herskovits, 33, advances the thesis that U.S. blacks constitute a homogeneous and culturally definable population group.

Coming of Age in Samoa by American Museum of Natural History anthropologist Margaret Mead, 26, is based on studies made while living with the natives.

1928 ***(cont.)*** Mead's book on the development of social behavior among adolescents on the Pacific island stresses the impermanence of human values.

Penicillin proves to have antibacterial properties that will launch an "antibiotic" revolution in medicine. Scottish bacteriologist Alexander Fleming at St. Mary's Hospital, London, who discovered "the dissolving enzyme" in 1922, notices that no bacteria have grown in the vicinity of some *Penicillium notatum* fungus which has accidentally fallen into a preparation of bacteria he was about to throw away (*see* 1929; Florey, 1940).

The Basis of Sensation by English physiologists Edgar Douglas Adrian, 39, and Charles Scott Sherrington, 67, presents work that will enable physicians to understand disorders of the nervous system. Adrian has worked out the mechanism by which nerves carry messages to and from the brain.

Eli Lilly introduces Liver Extract No. 43 to treat pernicious anemia (*see* Cohn, 1926; Castle, 1929).

The world's first combined radio, cable, and telegraph service company is created by a merger of Clarence Mackay's Commercial Cable-Postal Telegraph with International Telephone and Telegraph (ITT). Mackay, whose father broke Western Union's telegraph monopoly in 1886, inherited the J. W. Mackay communications empire in 1902, completed a transpacific cable in 1903, established wire communications with southern Europe via the Azores in 1907, put a cable into service between New York and Cuba in 1907 and from Miami to Cuba in 1920, and in 1923 completed a wire link to northern Europe via Ireland.

Columbia Broadcasting System (CBS) is founded by Congress Cigar Co. advertising manager William S. Paley, 27, who has been receiving $50,000 per year from his father's firm and last year committed the company to an advertising contract of $50 per week with Philadelphia's 225-watt radio station WCAU while his father was away on vacation. Young Paley was criticized for making the contract but has seen sales of La Palina cigars soar in response to radio advertising; he sells some of his stock in Congress Cigar to raise upwards of $275,000, buys into financially ailing United Independent Broadcasters (which controls Columbia Phonograph), is elected president of the 22-station network September 26, and keeps CBS solvent by selling a 49 percent interest to Adolph Zukor's Paramount-Publix motion picture firm (broadcasting 16 hours per day over long-line telephone wires costs $1 million per year). He will move CBS to New York next year and make it a rival to David Sarnoff's NBC (*see* ABC, 1943).

General Electric station WGY, Schenectady, N.Y., broadcasts the first regularly scheduled television programs beginning May 11 (*see* 1927; BBC, 1936).

The *Atlanta World* begins publication as a 5¢ black daily.

The Oxford English Dictionary (O.E.D.) appears in 12 volumes after 44 years of work, most of it by Scottish editor-in-chief James A. H. Murray who died in 1915 at age 78. A two-volume Compact Edition of the 15,487-page work will be marketed with magnifying glass in 1971 and it will be further supplemented in 1972 and thereafter to provide authoritative information on English word origins, usages, and pronunciations.

Fiction *Lady Chatterley's Lover* by D. H. Lawrence, who is dying of tuberculosis at Florence; his explicit account of the sex relations between the wife of a crippled English peer and their lusty gamekeeper Mellors is privately printed because it is denied publication in England (*see* 1959); *The Well of Loneliness* by English novelist (Marguerite) Radclyffe Hall, 44, whose novel about a lesbian attachment between a young girl and an older woman encounters censorship problems; *Orlando* by Virginia Woolf; *The Last Post* by Ford Madox Ford; *The Children* by Edith Wharton; *All the Conspirators* by English novelist Christopher Isherwood, 24; *Nadja* by surrealist André Breton; *Southern Mail (Courrier-Sud)* by French aviator-novelist Antoine de Saint-Exupéry, 28; *The Cockpit* by U.S. novelist James Gould Cozzens, 25; *Mr. Buttsworthy on Rampole Island* by H. G. Wells; *Envy (Zavist)* by Soviet novelist Yuri Oleska, 29; *Point Counter Point* by Aldous Huxley; *A Modern Comedy* by John Galsworthy; *The Childermass* by Wyndham Lewis; *Some Prefer Nettles (Tade kuu muhi)* by Junichiro Tanizaki; *The Eclipse* by Chinese novelist Shen Yen-ping (Mao Tun), 32; *Time Regained (Le Temps Retrouvé)* by Marcel Proust is the last of his *Remembrance of Things Past* novels; *Decline and Fall* by English novelist Evelyn Waugh, 25; *The Mystery of the Blue Train* by Agatha Christie.

Juvenile *The House at Pooh Corner* by A. A. Milne; *Millions of Cats* by U.S. author-illustrator Wanda Gág, 35.

Poetry *Gypsy Ballads (Romancers guano)* by Federico García Lorca whose imaginary gypsies are based only in part on the real gypsies of Andalusia; *Nine Experiments* by English poet Stephen Spender, 19; *John Brown's Body* by Stephen Vincent Benét; *Mr. Pope and Other Poems* by U.S. poet (John Orley) Allen Tate, 28; *The Dead Water (Ssu-shui)* by Chinese poet Wen-I-To, 29; *The Cantos* (II) by Ezra Pound.

Painting *Seated Odalisque* by Henri Matisse; *Still Life with Jug* by Georges Braque; *Wedding* by Marc Chagall; *Black Lilies* by Max Beckmann; *The Titanic Days* by René Magritte; *Girl on Sofa* by Edvard Munch; *Nightwave* by Georgia O'Keeffe; *Conversation, Country Dance, Louisiana Rice Fields,* and *Crapshooters* by Thomas Hart Benton; *Sixth Avenue and Third Street* by John Sloan.

The Philadelphia Museum of Art opens.

Theater *Marco Millions* by Eugene O'Neill 1/9 at New York's Garrick Theater, 92 perfs.; *Cock Robin* by Philip Barry and Elmer Rice 1/12 at New York's 48th Street Theater, 100 perfs.; *Strange Interlude* by Eugene O'Neill 1/30 at New York's John Golden Theater, with Lynn Fontanne, Glenn Anders, Earle Larimore, and Tom Powers in a Freudian study of women with Elizabethan monologistic asides (the curtain rises at 5:30, descends at 7 for an 80-minute dinner interval, rises again at 8:20, and does not fall until after 11), 426 perfs.; *The Captain from Kopenick (Der Hauptmann* von Kopenick) by Karl Zuckmayer 3/5 at Berlin's Deutsches Theater; *Lazarus Laughed* by O'Neill 4/9 at the Pasadena Playhouse in California; *Skidding* by U.S. playwright Aurania Rouverol 5/21 at New York's Bijou Theater, with Marguerite Churchill, 448 perfs.; *The Front Page* by former *Chicago Daily News* staff writer Ben Hecht, 34, and former *Hearst's International Magazine* staff writer Charles MacArthur, 33, 8/14 at New York's Times Square Theater, with Lee Tracy, Osgood Perkins, Dorothy Stickney, 276 perfs. (MacArthur marries actress Helen Hayes 8/17); *Machinal* by Sophie Treadwell 9/7 at New York's Plymouth Theater, with Jean Adair, Clark Gable, Zita Johnson, is based on last year's Snyder-Gray murder case, 91 perfs.; *Elmer the Great* by George M. Cohan and Ring Lardner 9/24 at New York's Lyceum Theater, with Walter Huston, 40 perfs.; *Squaring the Circle (Kvadratura Kruga)* by Soviet playwright Valentin Katayev, 31, 9/28 at the Moscow Art Theater, more than 800 perfs.; *Topaze* by French playwright Marcel Pagnol, 33, 10/9 at the Théâtre des Variétes, Paris; *The Far-Off Hills* by Esme Stuart Lennox Robinson 10/22 at Dublin's Abbey Theatre; *The Humiliation of the Father (Lepdrehumilig)* by Paul Claudel 11/26 at Dresden's Schauspielhaus; *Holiday* by Philip Barry 11/26 at New York's Plymouth Theater, with Hope Williams, 229 perfs.; *Wings Over Europe* by English playwrights Robert M. B. Nichols, 31, and Maurice Browne 12/12 at New York's Martin Beck Theater, with Alexander Kirkwood as physicist Francis Lightfoot, who has discovered the secret of an atomic bomb, 90 perfs.

Films Walt Disney's *Steamboat Willie* introduces Mickey Mouse in the first animated cartoon with a sound track. U.S. animator Walter Elias Disney, 26, has produced the film under license from J. R. Bray who patented the "cel" system in 1910, his Mickey Mouse will be the basis of a vast entertainment and promotion empire, he will start a *Silly Symphony* series of black and white cartoon short subjects featuring Donald Duck, Pluto, and other cartoon characters, and they will rival Mickey Mouse in worldwide popularity *(see* 1931; 1938).

Other films King Vidor's *The Crowd* with Eleanor Boardman (silent); Josef von Sternberg's *Docks of New York* with George Bancroft and *The Last Command* with Emil Jannings; Carl Theodor Dreyer's *The Passion of Joan of Arc* with Maria Falconetti; Victor Seastrom's *The Wind* with Lillian Gish, Lars Hanson. Also: Luis Buñuel's *Un Chien Andalou* with Pierre Batcheff, painter Salvador Dali, and Buñuel; Charles Chaplin's *The Circus* with Chaplin; Juri Taritch's *Czar Ivan the Terrible*; Vsevlod Pudovkin's *The End of St. Petersburg*; F. W. Murnau's *The Four Devils* with Janet Gaynor; F. Richard Jones' *The Gaucho* with Douglas Fairbanks; Joe May's *Homecoming* with Lars Hanson; George Fitzmaurice's *Lilac Time* with Gary Cooper, Colleen Moore; Fred Niblo's *The Mysterious Lady* with Greta Garbo; Henry Beaumont's *Our Dancing Daughters* with Joan Crawford, John Mack Brown; Ernst Lubitsch's *The Patriot* with Emil Jannings; Frank Borzage's *Street Angel* with Janet Gaynor; King Vidor's *Show People* with Marion Davies, William Haines; Ted Wilde's *Speedy* with Harold Lloyd; Charles F. Reisner's *Steamboat Bill, Jr.* with Buster Keaton; W. S. Van Dyke's *White Shadows in the South Seas* with Raquel Torres, Monte Blue.

Radio comedy, variety *Amos 'n' Andy* 3/19 over Chicago's WMAQ, with white actors Freeman Fisher Gosden, 28, and Charles J. Correll, 38, impersonating the black co-owners of the "Fresh-Air Taxicab Company of America" along with nearly all the other roles including Kingfish, Lightnin', and Madame Queen. The daily 15-minute comedy show will air nationally on NBC beginning next year and will soon reach 40 million listeners, two-thirds of the available audience, as movie theaters interrupt their programs to let audiences hear the show that will air for a half hour weekly beginning in 1943 (*see* 1965); *The National Farm and Home Hour* debuts to attract rural family audiences.

Opera *The Three-Penny Opera* (*Die Dreigroschenoper*) 8/31 at Berlin's Theater am Schiffbauerdamm, with Lotte Lenya as Jenny, music by her husband Kurt Weill, libretto by Bertolt Brecht, who has transposed the *Beggar's Opera* of 1728 into the idiom of Germany's Weimar Republic.

Stage musicals *Rosalie* 1/10 at New York's New Amsterdam Theater, with Marilyn Miller, Frank Morgan, Jack Donahue, music by George Gershwin and Sigmund Romberg, lyrics by Ira Gershwin and P. G. Wodehouse, book by Guy Bolton and William Anthony McGuire, 335 perfs.; *The Three Musketeers* 3/13 at New York's Lyric Theater, with Dennis King as D'Artagnan, Vivienne Segal, Reginald Owen as Cardinal Richelieu, Clarence Derwent as Louis XIII, music by Rudolph Friml, songs that include "March of the Musketeers," 318 perfs.; *This Year of Grace* 3/22 at the London Pavilion, with Jessie Matthews, Tilly Losch, Melville Cooper, book, music, and lyrics by Noël Coward, songs that include "A Room with a View," "World Weary," 316 perfs.; *Present Arms* 4/26 at New York's Mansfield Theater, with Charles King, Busby Berkeley, music by Richard Rodgers, lyrics by Lorenz Hart, songs that include "You Took Advantage of Me," 155 perfs.; *Blackbirds of 1928* 5/9 at New York's Liberty Theater, with Bill "Bojangles" Robinson, songs that include "Digga Digga Do" and "I Can't Give You Anything but Love" by Jimmy McHugh, lyrics by Dorothy Fields, 518 perfs.; *George White's Scandals* 7/2 at New York's Apollo Theater, with Harry Richman, Ann Pennington, Willie and Eugene Howard, music by Ray Henderson, lyrics by B. G. DeSylva and Lew Brown, 240 perfs.; *Good Boy* 9/15 at New York's Hammerstein Theater, with Eddie Buzzell, songs that include "I Wanna Be Loved by You" by Herbert Stothard, Bert Kalmar, and Harry Ruby, 253 perfs.; *The New Moon* 9/10 at New York's Imperial Theater, with Robert Halliday, Evelyn Herbert, music by Sigmund Romberg, lyrics by Oscar Hammerstein II and others, songs that include "One Kiss," "Lover, Come Back to Me," "Softly, as in a Morning Sunrise," "Wanting You," "Stout-Hearted Men," 509 perfs.; *Paris* 10/8 at New York's Music Box Theater, with Irene Bordoni, music and lyrics by Cole Porter

1928 ***(cont.)*** and E. Ray Goetz, songs that include "Let's Fall in Love" with lyrics that begin, "Let's do it . . ." 195 perfs.; *Hold Everything* 10/10 at New York's Broadhurst Theater, with Bert Lahr, Jack Whiting, Victor Moore, Ona Munson, music by Ray Henderson, lyrics by B. G. DeSylva and Lew Brown, songs that include "You're the Cream in My Coffee," 413 perfs.; *Animal Crackers* 10/23 at New York's 44th Street Theater, with the Four Marx Brothers, Margaret Dumont, book by George S. Kaufman and Morrie Ryskind, music and lyrics by Bert Katmar and Harry Ruby, 191 perfs.; *Treasure Girl* 11/8 at New York's Alvin Theater, with Gertrude Lawrence, Clifton Webb, music by George Gershwin, lyrics by Ira Gershwin, songs that include "I've Got a Crush on You," "I've Got a Feeling I'm Falling," "Oh, So Nice," 68 perfs., *Whoopee* 12/4 at New York's New Amsterdam Theater, with Eddie Cantor, music and lyrics by Walter Donaldson and Gus Kahn, songs that include "Love Me or Leave Me," "Makin' Whoopee," 379 perfs.

Hollywood musicals Victor Fleming's *The Awakening* with Vilma Banky as Marie, Walter Byron, Louis Wolheim, and the song "Marie" by Irving Berlin.

First performances Symphonic Piece by Harvard music professor Walter Hamor Piston, 34, 3/6 at Boston's Symphony Hall. Piston was graduated summa cum laude from Harvard in 1924 and has studied at Paris with Paul Dukas and Nadia Boulanger; Variations for Orchestra by Arnold Schoenberg 12/2 at Berlin; *America: An Epic Rhapsody in Three Parts for Orchestra* by Ernest Bloch, text from poems by Walt Whitman 12/20 and 12/21 at New York, Boston, Philadelphia, Cincinnati, Chicago, and San Francisco; *An American in Paris* by George Gershwin 12/13 at New York's Carnegie Hall with Gershwin at the piano.

Rudy Vallée forms his own band and opens at New York's Heigh-Ho Club with a megaphone to amplify his voice. The first "crooner," Hubert Prior Vallée, 27, Yale '27, is a self-taught saxophonist who worked his way through a year at the University of Maine and 3 years at Yale with interruptions for musical engagements, sings the Maine "Stein Song," and makes his theme song "My Time Is Your Time."

Lawrence Welk, 24, starts a small band and launches a half-century career with broadcasts to rural South Dakota audiences from a Yankton radio station. The accordionist will call his music "Champagne Music" beginning in 1938 and write his own theme song "Bubbles in the Wine."

The Mills Brothers cut their first record to begin a career that will continue for more than half a century. The Piqua, Ohio, singers include Herbert, 18, Harry, 16, Donald, 15, and John, Jr.

Popular songs "Puttin' On the Ritz" by Irving Berlin; "The Breeze and I" by Cuban composer Ernesto Lecuona, lyrics by Al Stillman; "Sweet Sue—Just You" by Victor Young, 28, lyrics by Will J. Harris, 28; "Sweet Lorraine" by Rudy Vallée's piano player Cliff Burwell, 28, lyrics by Mitchell Parish, "I'll Get By" by Fred Ahlert, 36, lyrics by Roy Turk, 36; "She's Funny That Way" by Neil Moret, lyrics by Billy Rose and E. Y. Harburg.

René LaCoste wins in men's singles at Wimbledon, Helen Wills in women's singles; Henri Cochet wins in U.S. men's singles, Wills in women's.

The Olympic Games at Amsterdam attract 3,905 contestants from 46 countries. U.S. athletes win the most medals.

U.S. Olympic swimming champion Johnny Weissmuller, 25, retires after having set 67 world records and having won three Olympic gold medals for the U.S. swimming team. Weissmuller will have a Hollywood screen test in 1930 and make 19 films in 18 years portraying Tarzan, the jungle man created by novelist Edgar Rice Burroughs in 1914.

Spanish picadors and their horses get some protection from a law requiring that horses wear a mattress-like armor strapped over one side and under the belly and that picadors wear steel armor over their right legs.

Portugal outlaws the killing of bulls in the ring but permits bullfighting to continue without killing.

Jimmy Foxx joins Connie Mack's Philadelphia Athletics to begin a career in which he will break Babe Ruth's home run average (*see* 1920). The Athletics will trade James Emory Foxx, 20, to the Boston Red Sox in 1935, he will move to the Chicago Cubs in 1942, and will play his final season for the Philadelphia Phillies in 1945.

The New York Yankees win the World Series by defeating the St. Louis Cardinals 4 games to 0.

The Boston Garden opens. The new arena will host basketball, hockey, track, and other events.

Colgate-Palmolive-Peet is created by a merger of the 122-year-old Colgate Co. with the 2-year-old Palmolive-Peet Co. The new firm has combined sales of $100 million and is in a position to challenge the leadership of Cincinnati's 91-year-old Procter & Gamble, whose sales have grown to $156 million.

Welcome Wagon International, Inc., has its beginnings at Memphis, Tenn., where advertising agency head Thomas W. Briggs introduces new residents to his clients by hiring agents to make personal calls.

Bubble gum is perfected by Frank H. Fleer Co. cost accountant Walter Diener, 23, who has been experimenting for more than a year. Fleer's Dubble Bubble is test-marketed December 26 at Philadelphia.

New York's 42-story Pierre Hotel is completed by Schultze & Weaver on Fifth Avenue at 61st Street.

The Beverly Wilshire Hotel is completed on Rodeo Drive at Beverly Hills, Calif.

London's Grosvenor House hotel opens in Park Line with 467 rooms on eight floors.

Detroit's 47-story Greater Penobscot building is completed at Griswold and First streets.

The Breuer chair designed by Hungarian-born architect Marcel Lajos Breuer, 26, is introduced by Vienna's Gebruder Thonet (*see* bentwood chair, 1876).

California's 3-year-old St. Francis Dam near Saugus gives way March 12, killing 450.

A Flood Control Act passed by Congress May 15 provides an estimated $325 million to be spent over the course of 10 years for levee work on the lower Mississippi (*see* 1927).

A Florida hurricane in September causes Lake Okeechobee to overflow, killing 1,836.

Bryce Canyon National Park is created by act of Congress. The park in southern Utah embraces 36,010 acres of amphitheaters, pinnacles, spires, and walls carved by natural erosion.

Josef Stalin orders collectivization of Soviet agriculture in the first Five-Year Plan (above). Peasants burn crops, slaughter livestock, and hide grain from state collectors (but *see* 1931).

Soviet scientists lead the world in plant breeding and genetic livestock development, largely as a result of biological research programs developed by Nikolay Ivanovich Vavilov to benefit socialist agriculture (but *see* 1940).

Vitamin C (ascorbic acid) is isolated from capsicums (peppers) at the Cambridge, England, laboratory of Frederick Gowland Hopkins by Hungarian-born biochemist Albert Szent-Györgyi von Nagyrapolt, 35, who finds the capsicums nearly four times as rich in the anti-scurvy vitamin as lemons (*see* 1918; Lind, 1747; Hopkins, 1906; King, 1932).

U.S. biochemists M. I. Smith and E. G. Hendrick show that vitamin B is at least two vitamins and name the heat resistant vitamin G, or B_2 (*see* 1926; McCollum, 1916; Warburg, 1932).

At least 1,565 Americans die from drinking bad liquor, hundreds are blinded, many are killed in bootlegger wars. Federal agents and Coast Guardsmen are making 75,000 arrests per year and enforcement of Prohibition laws costs U.S. taxpayers millions of dollars, but the laws are openly flouted.

Distillers Corp.-Seagram is founded in March by Samuel and Alan Bronfman who last year acquired Joseph E. Seagram & Sons, founded in 1870 (*see* 1924). The Bronfmans are shipping thousands of gallons of liquor that winds up in the United States despite Prohibition laws and they will make Seagram the world's largest producer of alcoholic beverages (*see* 1933).

U.S. companies use glyceride emulsifiers for the first time in baked goods, shortenings, margarines, and ice creams, and other foods.

Peter Pan Peanut Butter is introduced by Swift Packing Co.'s E. K. Pond division and is the first hydrogenated, homogenized peanut butter and the first to have national advertising (*see* 1890; Normann, 1901). E. K. Pond has been licensed to use a process invented in 1923 by Alameda, Calif., food processor J. L. Rosefield to stabilize peanut butter by replacing a significant part of its natural peanut oil with hydrogenated peanut oil (*see* Skippy, 1932).

National Biscuit Co. introduces peanut butter cracker sandwich packets under the name NAB, selling the 5¢ packets through newsstands, drugstores, candy counters, and other such retail outlets (*see* Ritz, 1933).

Kellogg's introduces Rice Krispies.

Daniel Gerber improves methods for straining peas for baby foods and finds by a market survey that a large market exists for such foods if they can be sold cheaply through grocery stores (*see* 1927). Gerber advertises in *Child's Life* magazine and offers six cans for a dollar (less than half the price of baby foods sold at pharmacies) to customers who will send in coupons filled out with the names and addresses of their grocers (*see* 1929).

J. L. Kraft & Co. buys out its leading competitor, Phenix Cheese, which has been making processed cheese under license from Kraft, and creates the largest U.S. cheese company under the name Kraft-Phenix (*see* 1900; 1915; National Dairy Corp., 1929).

"I say it's spinach and I say the hell with it," reads E. B. White's caption to Carl Rose's *New Yorker* magazine cartoon December 8 showing a child refusing to eat broccoli. The vegetable has only recently been introduced from Italy by an enterprising grower in northern California's Santa Clara Valley.

1929

Wall Street's Dow Jones Industrial Average reaches 381 in September, up from 88 in 1924, but breaks in October following a drop in U.S. iron and steel production and a rise in British interest rates to 6.5 percent that has pulled European capital out of the U.S. money market. A record 16.4 million shares trade Tuesday, October 29; the Dow plummets 30.57 points, and liquidation continues despite assurances by leading economists that no business depression is imminent. Speculators who have bought on margin are forced to sell and $30 billion disappears—a sum almost equal to what the 1914–18 war cost America.

Some 513 Americans have incomes above $1 million for the year and a dozen top $5 million. The rate of normal income tax on the first $7,500 of net income is 1.5 percent with exemptions of $1,500 for a single person and $3,500 for a married person or head of household. At $12,000 per year the tax rate climbs to 5 percent plus a surtax that is graduated up to 20 percent on $100,000, but while a married man with an income of $100,000 may pay as much as $16,000 in tax he will probably find legal ways to avoid paying that much.

Seventy-one percent of U.S. families have incomes below $2,500 which is generally considered the minimum necessary for a decent standard of living. The average weekly wage is $28 (*see* 1932).

British unemployment tops 12.2 percent with more miners and workmen idle than in the General Strike of 1926. Most working-class families in the Welsh coal

STAGE | BROADWAY | SCREEN

VARIETY

PRICE 25¢

VOL. XCVII. No. 3 — NEW YORK, WEDNESDAY, OCTOBER 30, 1929 — 88 PAGES

WALL ST. LAYS AN EGG

Going Dumb Is Deadly to Hostess In Her Serious Dance Hall Profesh

DROP IN STOCKS ROPES SHOWMEN

Kidding Kissers in Talkers Burns Up Fans of Screen's Best Lovers

Hunk on Winchell

Many Weep and Call Off Christmas Orders — Legit Shows Hit

MERGERS HALTED

Talker Crashes Olympus

The crash of New York Stock exchange values wiped out many investors and began a decade of economic depression.

fields exist on a dole of about $7 per week and corner pubs close down for lack of trade.

Edsel Ford announces an increase in the minimum daily wage at Ford Motor Company plants December 2. It rises from $6 to $7, but wages in virtually all industries will soon decline (*see* 1933).

Josef Stalin expels Leon Trotsky from the Soviet Union in January, 14 months after having him thrown out of the Communist Party on charges of engaging in antiparty activities. Stalin removes Trotsky's threat to his dictatorship (*see* 1924; 1940).

Nikolai Bukharin is expelled from the Communist Party November 17 after having headed the Third International since 1926. Other members of the rightist opposition are expelled, leaving Stalin to rule as undisputed dictator.

Tadzhikistan becomes a Soviet socialist republic.

Yugoslavia's Aleksandr I proclaims a dictatorship January 5 and dissolves the Croat and other parties January 21. The Kingdom of the Serbs, Croats, and Slovenes becomes Yugoslavia officially October 3 in a move to end historic divisions of the realm (*see* 1921; 1934).

Britain's Labour Party wins the general election May 30 and Ramsay MacDonald forms a second cabinet June 5 that will hold power for more than 2 years. Diplomatic relations with Moscow resume October 1.

The Lateran Treaties ratified by the Italian parliament June 7 restore temporal power to the pope over the 108.7-acre Vatican City in Rome, the Italian government agrees to pay an indemnity of 750 million lire in cash and 1 billion lire in government bonds, and the pope leaves the Vatican July 25 after years of virtual imprisonment.

Japan's last parliamentary election for more than 16 years is held with big red lacquer ballot boxes. Twenty percent of the population is eligible to vote as compared with 1.1 percent in 1891, but suffrage remains limited to men of 25 and older. Japanese militarists will soon take over the government and put an end to free elections.

A dispute over Jewish use of Jerusalem's Wailing Wall leads in August to the first large-scale Arab attacks on Jews, many of whom are killed (*see* 1930).

A Mississippi lynch mob of 2,000 burns an accused black rapist alive, a coroner's jury returns a verdict of death "due to unknown causes," and Mississippi governor Theodore G. Bilbo says the state has "neither the time nor the money" to go into the matter. But total U.S. lynchings for the year number 10, down from 23 in 1926, 97 in 1909.

Federated Department Stores is created by a loose confederation of three stores put together by Columbus, Ohio, retail merchant Fred Lazarus, Jr., 44, of Lazarus Brothers. Brooklyn's Abraham & Straus and Boston's Filene's join with Lazarus Brothers in the chain that will be joined next year by Bloomingdale's (New York) and thereafter by Shillito's (Cincinnati), Rikes (Dayton), the Boston Store (Milwaukee), Goldsmith's (Memphis), Foley's (Houston, 1951), Burdine's (Miami, 1956), Sanger-Harris (Dallas), Levy's (Tucson), Bullock's (Los Angeles), I. Magnin (San Francisco), and others to make Federated the largest U.S. department-store chain.

Continental Oil Corp. (CONOCO) is created by J. P. Morgan & Co. The New York banking house took over Marland Oil 2 years ago from polo player E. W. Marland and merges it with Continental Oil of Maine, which it has also taken over (*see* 1908).

U.S. auto production tops 5 million with minor makes accounting for 25 percent of sales. The 5 million figure will not be reached again for 20 years (*see* 1932).

General Motors buys the German Opel Motorcar Co. founded in 1898.

GM's Chevrolet Division introduces a six-cylinder model and advertises that the new Chevy costs no more than a four-cylinder car.

General Motors president Alfred P. Sloan opposes a suggestion by Lammot du Pont, 49, that Chevrolets be equipped with safety glass (*see* Duplate, 1927). GM's Cadillacs and La Salles have recently been equipped with Duplate glass but Sloan points out that Packards have not been equipped with safety glass and their sales have not suffered. "I do not think that from the stockholder's viewpoint the move on Cadillac's part has been justified," writes Sloan.

The first motorcar with front-wheel drive is introduced by E. L. Cord's Auburn Automobile Company (*see* 1928).

The Model J Duesenberg introduced by E. L. Cord's Duesenberg, Inc., is a "real Duzy" (*see* 1926). The costly 265-horsepower luxury car can go 112 to 116 miles per hour and will be built until 1936 for affluent Americans and Europeans.

Ford introduces the first station wagon, equipping a Model A with a boxy wooden body that provides extra space for cargo and passengers.

The first mobile home trailer, devised by aviation pioneer Glenn H. Curtiss, is displayed in New York showrooms by Hudson Motor Car (*see* 1922).

Greyhound Corp. buys out Pioneer Yelloway for $6.4 million (*see* 1926; 1928; 1934).

Scuderia Ferrari is founded at Modena December 1 by former Alfa Romeo racing driver Enzo Ferrari, 31, with a few partners to sell Alfas in three provinces (*Scuderia* means stable, or racing team) (*see* 1940).

Curtiss-Wright Corp. is created by a merger of American Wright Co. with Curtiss Aircraft (*see* 1909; 1910; General Motors, 1933).

Grumman Aircraft has its beginnings in a Baldwin, Long Island, aircraft repair shop opened by Leroy Grumman and Leon "Jake" Swirlbul. William T. Schwendler, 24, is chief engineer.

United Aircraft & Transport is created by U.S. aeronautical engineer and designer Chance Vought, now 39, who merges his 12-year-old Chance Vought aircraft manufacturing firm with Pratt & Whitney Aircraft and Boeing Airplane.

Vincent Bendix founds Bendix Aviation; he will merge his various aviation, motorcar, and radio equipment-making firms into the new corporation (*see* 1912; Bendix Trophy, 1930).

U.S. commercial airlines fly 30 million miles, up from 6 million in 1927, and carry 180,000 passengers, up from 37,000.

Delta Air Lines begins passenger service June 17 under the name Delta Air Service with three six-passenger Travelaire monoplanes powered by 300-horsepower Wright "Whirlwind" engines flying at 90 miles per hour between Dallas and Jackson, Miss., via Shreveport and Monroe, La. Delta was organized late last year under the leadership of former agricultural extension service county agent C. E. Woolman, 39, who pioneered in using airplanes to dust cotton crops with arsenate of lead and calcium arsenate in order to protect them from boll weevil damage.

Pan American Airways starts daily flights between Miami and San Juan, Miami and Nassau, and San Juan and Havana (*see* 1927; 1935).

Pan American consultant Charles A. Lindbergh opens a route through Central America to the Panama Canal Zone. Pan Am acquires Cia Mexicana de Aviacion, wins a mail contract to Mexico City, and by year's end Pan Am has routes totaling 12,000 miles, up from 251 at the end of last year (*see* 1930).

London's Heathrow Airport has its beginnings in Richard Fairey's Great West Aerodrome used mainly for experimental flights (*see* 1946; Croydon, 1920; Gatwick, 1936).

Lufthansa is organized as the German national airline, giving Berlin's 6-year-old Tempelhof Airport new importance. By 1936 it will be Europe's busiest air-travel center.

The *Graf Zeppelin* that was launched last year completes the first round-the-world flight of any kind. The dirigible carries nine commercial passengers 19,000 miles in 21 days, 7 hours.

The S.S. *Bremen* goes into service July 10 and sets a new transatlantic speed record for passenger liners. The North German-Lloyd line vessel crosses from Cherbourg to New York in 4 days, 17 hours, 42 minutes, cutting 3 hours off the record set by the S.S. *Mauretania* in 1910.

Albert Einstein announces January 30 that he has found a key to the formulation of a unified gravitational field theory—a group of equations applicable not only to gravitation but also to electromagnetics and subatomic phenomena (*see* 1916). His six pages of equations, however, are unprovable, incomprehensible, ignore quantum mechanics, and are incorrect (*see* 1939).

The first clinical application of crude penicillin is made January 9 at St. Mary's Hospital, London, by Alexander Fleming who treats an assistant suffering from an infected antrum by washing out the man's sinus with diluted penicillin broth, successfully destroying most of the staphylococci (*see* 1928; 1931).

The first Blue Cross nonprofit tax-exempt health insurance association is organized at Dallas, Tex., where local schoolteachers whose unpaid bills have been a burden to Baylor University Hospital make an arrangement with the hospital. Each teacher is guaranteed up to 21 days' free use of a semiprivate room and other hospital services on condition that small monthly fees be paid in advance on a regular basis. The program will quickly spread to all hospitals in the area with a central agency to collect the fees and the Blue Cross trademark of the American Hospital Association (the national association of voluntary hospitals) will be used by agencies throughout the country that meet AHA standards. By 1935 Blue Cross will have half a million subscribers and a Blue Shield program set up by medical societies and local doctors' guilds will provide surgical insurance (*see* 1940).

Harvard physician Samuel Albert Levine, 38, notes that 60 out of 145 heart attack patients have been hypertensive—the first link between hypertension (high blood pressure) and fatal heart disease.

Austrian-American psychiatrist Manfred J. Sakel, 29, uses overdoses of insulin to produce shock and finds it effective in many cases of schizophrenia (*see* Banting, Best, 1922; Thorazine, 1954).

"The so-called method of coeducation is false in theory and harmful to Christian training," proclaims Pope Pius XI December 31 in his encyclical *Divini illius magistri*.

Basic English by Cambridge University philologist C. K. (Charles Kay) Ogden, 40, proposes universal adoption of a basic international vocabulary of 850 English words he has developed after working with I. (Ivor) A.

1929 ***(cont.)*** Richards, now 36, on their 1923 book *The Meaning of Meaning.* Ogden points out that 30 percent of the world's population has English as its mother tongue or as the government language (he has India in mind) and says his easily-learned Basic English could help promote world peace.

An airmail letter crosses the United States in 31 hours at a cost of 25¢ postage for 3 ounces.

The United States has 20 million telephones, up from 10 million in 1918 and twice as many as all the rest of the world combined. Most are wooden boxes hung on walls with cranks to ring Central, but many tall tube-like phones are in use with hooks to hold their receivers, and French phones are beginning to appear with mouthpiece and receiver in one piece.

The Blattnerphone designed by German film producer Louis Blattner is the world's first tape recorder. Based on patents obtained by German sound engineer Kurt Stille, the device employs steel tape rather than the wire used in the Poulsen recorder of 1898, it is the first successful magnetic recorder with electronic amplification, and Blattner uses it for adding synchronized sound to the films he makes at the Blattner Color and Sound Studios at Ellstree, England. The BBC will acquire the first commercially produced Blattnerphone in 1931 *(see* plastic tape, 1935).

U.S. radio sales total nearly $950 million as Americans pay $118 and up for elaborate Atwater Kent and Stromberg-Carlson console sets with seven tubes and built-in superpower magnetic speakers to hear *Amos 'n' Andy*, Rudy Vallée, and Graham MacNamee's NBC sports broadcasts *(see The Goldbergs,* below).

The Motorola auto radio invented by University of Illinois engineering school graduate Paul Vincent Galvin, 34, is the first commercially successful radio for automobiles. Galvin started a business in a Chicago garage last year with $565 in capital, he will develop it into Motorola Co., but while his first radio plays in a moving car it is twice the size of a tackle box, its bulky speaker is stuffed under the floorboards, and its audio qualities leave much to be desired.

Business Week magazine begins publication September 7 at New York. The new McGraw-Hill publication will have lost $1.5 million by the end of 1935 but will turn the corner to become the leading magazine of its kind. In many years it will be the leading magazine of any kind in terms of advertising pages.

"Herblock" begins a career as political cartoonist that will continue for half a century. Herbert Lawrence Block, 19, draws editorial page cartoons for the *Chicago Daily News*, will move to Cleveland in 1933 to work for the Newspaper Enterprise Association (NEA), and in 1946 will join the *Washington Post* which will syndicate Herblock cartoons to at least 200 U.S. newspapers.

"Popeye" appears for the first time in the 10-year-old "Thimble Theatre" drawn by New York cartoonist Elzie Crisler Segar, 34, whose one-eyed, spinach-loving sailor with corncob pipe and outsize sense of chivalry has a girlfriend, Olive Oyl. His hamburger-loving pal J. Wellington Wimpy will soon join the strip *(see* spinach growers, 1937).

Nonfiction *Middletowne—A Study in Contemporary American Culture* by U.S. sociologists Robert Staughton Lynd, 37, and his wife Helen Merrell Lynd, 32, who break new ground in American sociology by applying methods and approaches used in studying primitive peoples to the Midwestern city of Muncie, Ind., and allying sociology with anthropology; *Goodbye to All That* (autobiography) by English critic-novelist Robert Graves, 34; *Is Sex Necessary?* by *New Yorker* magazine writers E. B. White and James Thurber; *The Adventurous Heart (Das abenteuerliche Herz),* essays by Ernst Jünger whose ideas have been taken over in part by the Nazis.

Fiction *The Time of Indifference (Gli indifferenti)* by Italian novelist Alberto Moravia (Alberto Pincherle), 22, whose portrayal of petty middle-class corruption in Rome is interpreted by many as a criticism of Italian society under Benito Mussolini. The government has persuaded the literary establishment to avoid references to real problems and issues and the Moravia novel creates a sensation. *The Holy Terrors (Les Enfants Terribles)* by Jean Cocteau; *All Quiet on the Western Front (Im Westen nichts Neues)* by German novelist Erich Maria Remarque, 30, whose novel sells 2.5 million copies in 25 languages in 18 months; *Look Homeward, Angel* by NYU English instructor Thomas Clayton Wolfe, 29 (Scribner's editor Maxwell Evarts Perkins, 45, who has also been F. Scott Fitzgerald's mentor, organizes and edits Wolfe's inchoate autobiographical typescript about the Gant [Wolfe] family of Asheville, N.C.); *Sartoris* by William Faulkner who depicts the fictional town of Jefferson in the fictional Yoknapatawpha County of Mississippi and begins an account of antebellum society's fall and the rise of the unscrupulous Snopes clan; *The Sound and the Fury* by Faulkner; *A Farewell to Arms* by Ernest Hemingway; *The Last September* by Irish-born English novelist Elizabeth Bowen, 30; *Living* by Henry Green; *Sido* by Colette; *Berlin-Alexanderplatz* by German novelist-psychiatrist Alfred Doblin, 51; *Wolf Solent* by John Cowper Powys; *The Innocent Voyage* by English novelist Richard Hughes, 29, whose novel will be retitled *A High Wind in Jamaica* next year; *The Near and the Far* by English novelist L. H. (Leo Hamilton) Myers, 48; *The Fortunes of Richard Mahoney* by Australian novelist Henry Handel Richardson (Ethel Florence Roberta Richardson), 59; *I Thought of Daisy* by U.S. novelist-critic Edmund Wilson, 34; *The Magnificent Obsession* by U.S. novelist Lloyd Douglas, 52; *Dodsworth* by Sinclair Lewis; *Dawn Ginsbergh's Revenge* (magazine pieces) by New York humorist S. J. (Sidney Joseph) Perelman, 25; *Claudia* by U.S. novelist Rose Franken, 31; *Red Harvest* and *The Dain Curse* by former Pinkerton detective (Samuel) Dashiell Hammett, 35; *The Roman Hat Mystery* by New York advertising and publicity writers Frederick Dannay and Manfred Bennington Lee, both 29, who write under the name Ellery Queen.

Poetry *Blind Fireworks* by Irish poet Louis MacNeice, 22, expresses social protest; *Poets, Farewell* by Edmund Wilson.

Painting *Woman in Armchair* by Pablo Picasso; *Fool in a Trance* by Paul Klee; *Love Idyll* by Marc Chagall; *Black Square* by Kazimir Malevich; *Composition with Yellow and Blue* by Piet Mondrian; *Sailing Boats* by Lyonel Feininger; *Black Flower* and *Blue Larkspur* by Georgia O'Keeffe; *Cotton Pickers (Georgia)* by Thomas Hart Benton; *John B. Turner—Pioneer* and *Woman with Plants* (a portrait of his mother) by Iowa genre painter Grant Develson Wood, 37. Robert Henri dies at New York July 12 at age 64.

New York's Museum of Modern Art opens November 8 on the twelfth floor of the Hecksher building at 730 Fifth Avenue with an exhibition of works by the late French impressionists Paul Cézanne, Paul Gauguin, Georges Seurat, and Vincent Van Gogh. Director of the museum is Alfred H. Barr, Jr., 27, who will be its guiding spirit for some 38 years (*see* 1939).

Eastman Kodak introduces 16-millimeter film with motion picture cameras and projectors for home use (*see* Kodachrome, 1935).

Theater *Street Scene* by Elmer Rice 1/10 at The Playhouse in New York, with Mary Servoss, Leo Bulgakov, Erin O'Brien Moore, Beulah Bondi, 601 perfs.; *Journey's End* by London insurance man-playwright Robert C. Sherriff, 33, 1/21 at London's Savoy Theatre, 594 perfs. (Sherriff served as a lieutenant at Vimy Ridge, his play fills audiences with the horror of wartime trench life, and it is staged in Paris, Berlin, and New York); *Dynamo* by Eugene O'Neill 2/11 at New York's Martin Beck Theater, with Dudley Digges, Glenn Anders, Helen Wylie, Claudette Colbert, 50 perfs.; *The Bedbug (Klop)* by Vladimir Mayakofsky 2/13 at Moscow's Meyerbold Theater; *Marius* by Marcel Pagnol 3/9 in Paris, with Raimu as Cesar, Pierre Fresnay as Marius; *Strictly Dishonorable* by U.S. playwright Preston Sturges, 31, 9/18 at New York's Alvin Theater with Muriel Kirkland who falls ill in December and is replaced by Antoinette Perry's 17-year-old daughter Marguerite, 557 perfs.; *The Silver Tassle* by Sean O'Casey 10/11 at London's Apollo Theatre; *Berkeley Square* by John L. Baldridge 11/4 at New York's Lyceum Theater, with Leslie Howard, Margalo Gilmore, 229 perfs.; *Amphitryon '38* by French playwright Jean Giraudoux, 47, 11/8 at the Comédie des Champs-Elysées, Paris.

Radio drama *The Goldbergs* 11/20 with Gertrude Berg, 30, in the role of Molly Goldberg in a series initially entitled *The Rise of the Goldbergs* (*see* 1949).

Films G. W. Pabst's *Pandora's Box* and *Diary of a Lost Girl*, both with Louise Brooks. Also: Josef von Sternberg's *The Blue Angel* with Marlene Dietrich, Emil Jannings; Alfred E. Green's *Disraeli* with George Arliss, Joan Bennett; King Vidor's *Hallelujah* with Daniel L. Haynes, Nina Mae McKinney; Lionel Barrymore's *His Glorious Night* with John Gilbert, whose silent-screen lover image loses something when audiences hear his tenor actor's voice; Alan Dwan's *The Iron Mask* with Douglas Fairbanks; Erich von Stroheim's *Queen Kelly* with Gloria Swanson.

Hollywood musicals Roy Del Ruth's *The Desert Song* is the first "all-talking and singing operetta"; Henry Beaumont's *The Broadway Melody* ("100% All Talking + All Singing + All Dancing!") with songs by Irving Thalberg protégés Nacio Herb Brown, 33, and Arthur Freed (*né* Grossman), 35, that include "You Were Meant for Me" and "The Wedding of the Painted Doll" (a color sequence helps the film gross $4 million for M-G-M at a time when movie tickets average 35¢); Charles Riesner's *Hollywood Revue of 1929* with Conrad Nagel, Jack Benny, Rudy Vallée, Joan Crawford, Norma Shearer, John Gilbert, John Barrymore, Marie Dressler, Marion Davies, Buster Keaton, songs by Nacio Herb Brown and Arthur Freed that include "Singin' in the Rain," some sequences in Technicolor (the musical numbers are recorded first and then synchronized to the action on the screen rather than recorded directly on the sound track as in *The Broadway Melody*, above); King Vidor's *Hallelujah* with an all-black cast (exhibitors in the South refuse to show the film), Irving Berlin songs that include "Swanee Shuffle"; Ernst Lubitsch's *The Love Parade* with Maurice Chevalier and Jeanette MacDonald singing "Dream Lover" and "The March of the Grenadiers" by Victor Schertzinger; Rouben Mamoulian's *Applause* with Helen Morgan; Roy Del Ruth's *The Gold Diggers of Broadway* with songs that include "Tip Toe Through the Tulips" and "Painting the Clouds with Sunshine" by Joe Burke, 45, lyrics by Al Dubin, 38; David Butler's *Sunny Side Up* with Janet Gaynor, Charles Farrell, music by Ray Henderson, lyrics by B. G. DeSylva and Lew Brown, songs that include "I'm a Dreamer" and the title song.

Stage musicals *Follow Thru* 1/9 at New York's 46th Street Theater, with Jack Haley, Irene Delroy, music by Ray Henderson, lyrics by B. G. DeSylva and Lew Brown, songs that include "Button up Your Overcoat," 403 perfs.; *Spring Is Here* 3/11 at New York's Alvin Theater, with Glenn Hunter, music by Richard Rodgers, lyrics by Lorenz Hart, songs that include "With a Song in My Heart," 104 perfs.; *Sing for Your Supper* 4/24 at New York's Adelphi Theater, with Earl Robinson, music by Lee Warner and Ned Lieber, lyrics by Robert Sous, songs that include "Ballad for Americans," 60 perfs.; *The Little Show* 4/30 at New York's Music Box Theater, with Fred Allen, Portland Hoffa, Libby Holman, Bettina Hall, Peggy Conklin, Clifton Webb, Romney Brent, music by Arthur Schwartz and others, lyrics by Howard Dietz, songs that include "I Guess I'll Have to Change My Plan," "Moanin' Low" (music by Ralph Rainger), "Caught in the Rain" (music by Henry Sullivan), "Can't We Be Friends" (music by Kay Swift), 321 perfs.; *Hot Chocolates* 6/20 at New York's Hudson Theater, with Louis Armstrong in an all-black revue, songs that include "Ain't Misbehavin' " by Thomas "Fats" Waller, 25, and Henry Brooks, lyrics by Andy Razaf, 219 perfs.; *Show Girl* 7/2 at New York's Ziegfeld Theater, with Ruby Keeler, Jimmy Durante, Duke Ellington and his Orchestra, music by George

1929 ***(cont.)*** Gershwin that includes the ballet "An American in Paris," lyrics by Ira Gershwin and Gus Kahn, songs that include "Liza" sung by Al Jolson in the audience to his wife Ruby Keeler onstage, 111 perfs.; *Bitter Sweet* 7/18 at His Majesty's Theatre, London, with Peggy Wood, book, music, and lyrics by Noël Coward, songs that include "I'll See You Again," "Zigeuner," 697 perfs.; *Sweet Adeline* 9/3 at New York's Hammerstein Theater, with Helen Morgan as Addie Schmidt, music by Jerome Kern, book and lyrics by Oscar Hammerstein II, songs that include "Why Was I Born," "Don't Ever Leave Me," "Here Am I," 234 perfs.; *George White's Scandals* 9/23 at New York's Apollo Theater, with Willie Howard, songs by Irving Caesar, George White, and Cliff Friend, 161 perfs.; *June Moon* 10/9 at New York's Broadhurst Theater, with book by George S. Kaufman and Ring Lardner, music and lyrics by Lardner, 273 perfs.; *Fifty Million Frenchmen* 11/27 at New York's Lyric Theater, with William Gaxton, music and lyrics by Cole Porter, songs that include "You Do Something to Me," 254 perfs.; *Wake Up and Dream* 12/30 at New York's Selwyn Theater, with Jack Buchanan, music and lyrics by Cole Porter, songs that include "I'm a Gigolo," "What Is this Thing Called Love?" 136 perfs.

The Chicago Civic Opera building completed at 20 West Wacker Drive has a 3,800-seat auditorium to challenge the Chicago Auditorium that opened late in 1889. The building also has a penthouse suite for utility magnate Samuel Insull.

Opera *Sir John in Love* 3/21 at London's Royal College of Music, with music by Ralph Vaughan Williams that includes "Fantasia on Greensleeves"; *The Gambler* 4/29 at the Théâtre Royal de la Monnaie, Brussels, with music by Serge Prokofiev, libretto from the Dostoievski novel of 1866; *Happy End* 8/31 at Berlin's Theater am Schiffbauerdamm, with music by Kurt Weill, book and lyrics by Bertolt Brecht.

Ballet *The Prodigal Son* 5/21 at the Sarah Bernhardt Theater, Paris, with the Ballet Russe, music by Serge Prokofiev.

First performances Symphony No. 3 by Serge Prokofiev 5/17 at Paris; *Amazonas* (symphonic poem) by Heitor Villa-Lobos 5/30 at Paris; Concerto for Viola and Orchestra by William Walton 10/3 at London, with Paul Hindemith as soloist; Concerto for Viola and Orchestra by Darius Milhaud 12/18 at Amsterdam; Symphony for Chamber Orchestra by Anton von Webern 12/18 at New York's Town Hall.

Guy Lombardo and his Royal Canadians open at New York's Roosevelt Hotel where the group directed by Canadian-American bandleader Guy Albert Lombardo, 27, will play dance music each winter for decades. A December 31 radio broadcast begins a national New Year's Eve tradition.

Popular songs "Stardust" by Hoagy (Hoagland) Carmichael, 29, lyrics by Mitchell Parish; "St. James Infirmary" by Joe Primrose; "Honeysuckle Rose" by "Fats" Waller, lyrics by Andy Razaf; "Am I Blue?" by Harry Akst, lyrics by Grant Clarke (for Ethel Waters to sing in the film *On With the Show*); "Louise" by Jack Whiting, lyrics by Leo Robin (for Maurice Chevalier to sing in the film *Innocents of Paris*); "Falling In Love Again" by German songwriter Friedrich Hollaender, 33, (for Marlene Dietrich to sing in the film *The Blue Angel*); "Mean to Me" Fred E. Ahlert, lyrics by Roy Turk; "I'm Just a Vagabond Lover" by Rudy Vallée and Leon Zimmerman; "More Than You Know" by Vincent Youmans, lyrics by Billy Rose and Edward Elison; "Pagan Love Song" by Nacio Herb Brown, lyrics by Arthur Freed (for the film *The Pagan*); "Siboney," "Say Si Si," by Ernesto Lecuona; "Just a Gigolo" by Italian composer Leonello Casucci, English lyrics by Irving Caesar; "Wedding Bells Are Breaking Up That Old Gang of Mine" by Sammy Fain, lyrics by Irving Kahal; "Happy Days Are Here Again!" by Milton Ager and Jack Yellen (for the film *Chasing Rainbows*).

The Orthophonic phonograph developed by Western Electric Company engineer H. C. Harrison is an improved electric gramophone that will replace wind-up mechanical record players.

English racetrack bookmakers at Newmarket rage at the installation of a totalizator that gives odds 40 percent longer than those offered by the bookies. Invented by Australian engineer J. Cruickshank, the tote introduces parimutuel betting and will be installed at U.S. tracks beginning in 1931 by British Automatic Totalisator, Ltd.

Jean Cochet wins in men's singles at Wimbledon, Helen Wills in women's singles; Bill Tilden wins in men's singles at Forest Hills, Wills in women's singles.

Australian cricketer Donald G. Bradman achieves a world record by scoring 452 not out.

Connie Mack's Philadelphia Athletics win the World Series by defeating the Chicago Cubs 4 games to 1.

"His Master's Voice" stopped winding down as electricity made the wind-up gramophone obsolete.

The football used in U.S. intercollegiate championship play is reduced in girth to between 22 and 22½ inches.

The Japanese Mah-jongg Association is founded as the Chinese game gains popularity.

Williams Electronics Co. has its beginnings in a pinball machine company founded by U.S. entrepreneur Harry Williams, 24, who will invent the "tilt" mechanism that makes the machine go dead temporarily when a player nudges it too hard. Williams will also invent the kickout hole and will make his machines "talk" with bells and gongs.

The yo-yo is introduced to America by U.S. entrepreneur Donald F. Duncan, who has acquired rights to the string-and-spool toy (based on a weapon used by 16th-century Filipino hunters) from Filipino immigrant Pedro Flores and produces it at Columbus, Ind.

The first crease-resistant cotton fabric is introduced by Tootal's of St. Helens, England.

Unilever, the first multinational firm in the consumer products industry, is created by a merger of Lever Brothers with the Dutch margarine maker Anton Jurgens founded in 1883.

Nestle Colorinse, introduced in 10 shades, is the first home-applied hair coloring; one shade is a blue-gray that will be popular among older women (*see* permanent wave, 1906; Clairol, 1931).

Gang warfare in Chicago reaches a peak of brutality the morning of February 14 in the "St. Valentine's Day massacre." Seven members of the George "Bugs" Moran gang are rubbed out at 10:30 in a North Clark Street garage and bootleg liquor depot as mobsters vie for control of the lucrative illicit liquor trade (police suspect that members of the Al Capone gang have done the killing) (*see* 1927; 1931).

Chicago has 498 reported murders, New York 401, Detroit 228, Philadelphia 182, Cleveland 134, Birmingham 122, Atlanta and Memphis 115 each, New Orleans 111.

American Radiator & Standard Sanitary Corp. is created by a merger of New York's American Radiator and Pittsburgh's Standard Sanitary to join the world's two leading companies in their respective fields. The merger has been engineered by American Radiator's Clarence Mott Woolley, 64, who has spent 43 years promoting cast-iron radiators in Europe and the United States and who acquires C. F. Church, a maker of toilet seats.

Prague's Cathedral of St. Vitus is completed after 585 years of construction.

The Tugendhat house at Brno in Czechoslovakia is completed by Mies van der Rohe, now 43, and his mistress Lilly Reich, 44.

Xanadu is completed 80 miles east of Havana, Cuba, for Irenée du Pont, 52, with high wood-beamed ceilings, seven bedrooms, and a nine-hole golf course on a 450-acre estate near Varadero Beach.

Atlantic City, N.J., opens an $8 million Convention Hall on the Boardwalk; its arena can seat 41,000.

The Barcelona chair designed by Mies van der Rohe and Lilly Reich (above) makes its debut at the German pavilion of the Barcelona Fair. The chair has a chrome steel X frame with leather-covered foam rubber cushions.

Africa's Serengeti National Park has its beginnings in a 900 square mile lion sanctuary set aside by the colonial government after complaints by professional hunters that people are going into the bush with Model T Fords to slaughter lions (*see* 1940).

One heath hen is left in America, down from 2,000 in 1916, and the male of the species seen on Martha's Vineyard has no mate. It will be extinct after 1931.

The Agricultural Marketing Act passed by Congress June 15 encourages farmers' cooperatives and provides for an advisory Federal Farm Board with $500 million in revolving funds to buy up surpluses in order to maintain prices. The funds are inadequate, and since farmers cannot be persuaded to produce less, farm prices continue to drop.

Half of all U.S. farm families produce less than $1,000 worth of food, fiber, or tobacco per year; 750,000 farm families produce less than $400 worth.

Signs of drought begin to appear in the U.S. Southwest and upper Great Plains (*see* 1930).

Soviet biologist Nikolai Vavilov endorses the government's push toward collectivized farms as a shortcut to scientific agriculture (*see* 1928; 1930).

Danish biochemist Carl Peter Henrik Dam, 34, finds he can produce severe internal bleeding in chickens if he puts them on a fat-free diet (*see* vitamin K, 1935).

Harvard physician William B. Castle finds that pernicious anemia can be prevented only if the gastric juices contain an "intrinsic" factor necessary for the absorption of an "extrinsic" factor in foods (*see* Minot and Murphy, 1924; Cohn, 1926; vitamin B_{12}, 1948).

A survey in Baltimore shows that 30 percent of children have rickets. A similar study in London's East End shows that 90 percent of children are rachitic (*see* Steenbock, 1923; 1927).

Consumers' Research is founded at Washington, N.J., by F. J. Schlink who sets up a testing organization that will do for consumers what the Bureau of Standards does for government purchasing agents (*see* 1927). Schlink begins publication of the magazine *Consumer Bulletin* (*see* 1932).

General Mills is created by a merger of the 63-year-old Minneapolis milling firm Washburn, Crosby with 26 other U.S. milling companies. General Mills is the world's largest miller.

General Foods is created by a merger of the 34-year-old Postum Co. with Clarence Birdseye's firm. Postum has acquired Jell-O, Minute Tapioca, Swans Down cake flour (Igleheart Bros.), Hellmann's Mayonnaise, Log Cabin Syrup, Walter Baker Chocolate, Franklin Baker Coconut, Calumet Baking Powder, Maxwell House Coffee (Cheek-Neal), and rights to Sanka Coffee.

Standard Brands is created by a merger that combines Chase and Sanborn, Fleischmann, and Royal

1929 ***(cont.)*** Baking Powder. Planter's will be added in 1961.

National Dairy Corp. acquires Kraft-Phenix Cheese, which sells well over 200 million pounds of product—close to 40 percent of all the cheese consumed in the United States (*see* 1923; 1928).

U.S. electric refrigerator sales top 800,000, up from 75,000 in 1925, as the average price of a refrigerator falls to $292, down from $600 in 1920. The average price will fall to $169 by 1939 and the new refrigerators will use less electricity (*see* GE, 1927).

United Fruit Company acquires Samuel Zemurray's Cuaymel Fruit after having acquired 22 other competitors since 1899 (*see* Honduras military coup, 1911). Now 52, Zemurray receives 300,000 shares of United Fruit stock with a value in November of $31.5 million and is the largest stockholder (*see* 1932).

Coca-Cola has gross sales of $39 million and adopts the slogan "The Pause That Refreshes."

Seven-Up is introduced under the name Lithiated Lemon by St. Louis bottler Charles Grigg of 1920 Howdy fame who markets the highly carbonated lemon-lime soft drink in 7-ounce bottles and sells 10,500 cases (*see* 1933).

Good Humor franchiser Tom Brimer has eight of his trucks blown up after refusing to pay protection money to Chicago racketeers but the trucks are insured (*see* 1926). The resulting publicity is so good for sales that Brimer is able to pay stockholders a 25 percent dividend; one stockholder acquires a 75 percent interest in the company for $500,000.

Daniel Gerber begins selling strained baby foods through grocery stores (*see* 1928). Using leads supplied by mail-order customers, Gerber salesmen drive cars whose horns play "Rock-a-Bye Baby," they sell 590,000 cans in one year, and other food processors are inspired to enter the baby food market.

A&P stores grow to number 15,709; the chain has sales of more than $1 billion (*see* 1912; 1937).

German chemist Adolph Butenandt, 26, and U.S. biochemist Edward Adelbert Doisy, 36, isolate the sex hormone "estrone" (*see* androgen, 1931).

New York police raid "Sanger's Clinic" but Sanger continues to campaign for birth control (*see* 1923; International Planned Parenthood, 1948).

1930 A London Naval Conference convened January 21 ends 3 months later with a treaty signed by Britain, the United States, France, Italy, and Japan who agree to limit submarine tonnage and gun-caliber and scrap certain warships. Japanese militarist factions attack the treaty (*see* below; 1931).

Spanish dictator Miguel Primo de Rivera resigns January 28 for reasons of health and dies March 16 at age 59 after having ruled the country since September 1923. Students agitate for a republic, denouncing the monarchy of Alfonso XIII (*see* 1931).

The Ethiopian empress Zanditu dies April 2 after a troubled reign of 14 years. Ras Tafari, 39, who took the name Haile Selassie when he was proclaimed king (negus) in June 1928, is crowned king of kings at Addis Ababa November 2 and will reign until 1974 as Haile Selassie I (*see* 1935).

Italy's Benito Mussolini comes out May 24 in favor of revising the Versailles Treaty.

Romania's boy king Michael is removed June 8 after a 3-year reign and succeeded by his father, now 37, who returned from exile June 6 and will reign until 1940 as Carol II. The new king has electrified the country by arriving from Paris by airplane and is soon joined by his mistress Magda Lupeseu, 26, who will have great influence.

The last Allied troops leave the Rhineland June 30, 5 years before the date set by the Treaty of Versailles.

Germany's National Socialist (Nazi) party wins 6,409,000 votes in the mid-September Reichstag elections, up from 800,000 in 1928, and are second only to the Socialists, who win 8,400,000. The Nazis gain 107 seats, up from 12 in the old Reichstag, but Adolf Hitler is barred from taking his seat because he is an Austrian citizen. Nazi deputies show up in uniform October 13, violating the rules and creating an uproar.

Turkey's President Mustafa Kemal renames Constantinople Istanbul and Angora Ankara (*see* 1924; 1935).

Turkish and Russian forces launch an offensive against Kurdish rebels August 12.

Peru's president Augusto Leguia resigns August 25 and flees the country after 11 years in office. Col. Luis Sanchez Cerro has led a military revolt and will be elected president next year (*see* 1933).

Argentina has a military coup September 5. President Hipollito Irigoyen is ousted by Gen. José Francisco Uriburu, 62, who returns the big landowners and business interests to power after 14 years of social reform.

Brazil has a revolution in October led by the governor of Rio Grande do Sul Getulio Dornelles Vargas, now 47, who accepts the presidency October 26, forces Washington Luis Pereira de Souza to resign 4 days later, and dissolves the congress November 1. Vargas will be dictator until 1945.

The Passfield Paper on Palestine presented October 20 suggests a halt in Jewish immigration to Palestine so long as unemployment persists among the Arabs. Sidney James Webb, 71, first baron Passfield, is the British secretary for colonies (*see* Balfour Declaration, 1917; White Paper, 1939).

Chinese Communists have joined forces in July to attack Hankow; Nationalist troops launch an encirclement campaign November 5 in parts of Hunan, Hubei, and Jianzi provinces.

Japan's prime minister Yuko Hamaguchi is shot by a right-wing militant November 14 and will die in 6 months. He has supported acceptance of the London Naval Conference program (above) (*see* 1931).

The last Allied troops leave the Saar December 12 (*see* 1935).

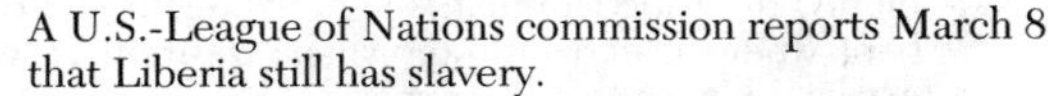

A U.S.-League of Nations commission reports March 8 that Liberia still has slavery.

A civil disobedience campaign against the British in India begins March 12. The All-India Trade Congress has empowered Mahatma Gandhi to begin the demonstrations (*see* 1914). Called Mahatma (meaning great-souled, or sage) for the past decade, Gandhi leads a 165-mile march to the Gujurat Coast of the Arabian Sea and produces salt by evaporation of seawater in violation of the law as a gesture of defiance against the British monopoly in salt production.

France enacts a workmen's insurance law April 30.

South Africa's white women receive the vote May 19 but blacks of both sexes remain disenfranchised.

Wall Street prices break again in May and June after an early spring rally that has seen leading stocks regain between one-third and one-half their losses of last year. As more investors realize the economic realities of the business depression, stock prices begin a long decline that will carry them to new depths (*see* 1932).

The Smoot-Hawley Tariff Bill signed into law by President Hoover June 17 raises tariffs to their highest levels in history—higher even than under the Payne-Aldrich Act of 1909—and embraces a Most Favored Nations policy first introduced in 1923. Certain nations are granted large tariff concessions under an arrangement that will continue for more than 60 years. Congress has approved the measure in a special session called by President Hoover despite a petition signed by 1,028 economists.

Other countries raise tariffs in response to the Smoot-Hawley Tariff Act (above).

A general world economic depression sets in as world trade declines, production drops, and unemployment increases.

U.S unemployment passes 4 million and national income falls from $81 billion to less than $68 billion.

Congress has voted a $230 million Public Buildings Act March 31 and a $300 million appropriation for state road-building projects April 4 to create jobs.

President Hoover says 4.5 million Americans are unemployed. He appoints a Committee for Unemployment Relief in October and requests $100 million to $150 million in appropriations for new public works construction December 2 in his message to Congress. Congress passes a $116 million public works bill December 20.

More than 1,300 banks close during the year. New York's Bank of the United States with 60 branches and 400,000 depositors—more than any other U. S. bank—closes December 11.

Russian-American economist Simon Kuznets, 29, begins economic studies that will culminate in the formulation of a Gross National Product (GNP) index of national wealth. Kuznets begins teaching economic statistics at the University of Pennsylvania; his studies will embrace national earnings, income levels, and productivity.

U.S. gasoline consumption rises to nearly 16 billion gallons, up from 2.5 billion in 1919.

World oil prices collapse after wildcatter Columbus M. Joiner, 71, brings in a gusher October 3 in Rusk County, eastern Texas, to open a huge new field. Joiner will sell out to H. L. Hunt for $40,000 cash, $45,000 in short-term notes, and a guarantee of $1.2 million from future profits, his field will produce at least 3.6 billion barrels of oil, and it will make Hunt the richest man in America (*see* 1920; 1936).

United Airlines is created by a merger of Boeing Transport with National Air Transport. United uses Ford trimotor planes to cut flying time from New York to San Francisco to 28 hours.

TWA (Transcontinental and Western Air) is created by a merger and receives a government mail contract (*see* 1926; Hughes, 1939).

American Airlines has its beginnings in American Airways founded by Auburn motorcar boss Errett Cord who merges Robertson Aircraft with other small firms (*see* Lindbergh, 1926). Needing 20 single-engine, 12-place planes for a Midwest shuttle service, Cord starts a company to produce the planes, he places his brother-in-law Donald Smith in charge, and Smith hires Gerard Vultee from Lockheed and sets him up in a small hangar at Grand Central Airport, Los Angeles (*see* 1934; Vultee, 1928).

The first airline stewardess begins work in mid-May for United Airlines (above). Boeing agent Steve Stimpson at San Francisco has suggested that commercial aircraft carry "young women as couriers," United has hired Ellen Church and told the registered nurse and student pilot to hire seven other nurses—all aged 25, all single, all with pleasant personalities, none taller than five feet four or heavier than 115 pounds—and they serve cold meals and beverages, pass out candy and chewing gum, and comfort airsick passengers. Women cabin attendants will serve to help allay public fears of flying.

Three transcontinental air routes are in service by year's end but passenger flights are irregular (*see* 1936).

Pan Am begins flying to South America (*see* 1929; 1935).

Wiley Post, 31, wins the first Bendix Trophy Air Race flying the Lockheed Vega *Winnie Mae* owned by his Oklahoma oilman employer Floyd C. Hall. Bendix has offered a prize for the winner of the Los Angeles to Chicago race that will become a transcontinental race next year (*see* Bendix, 1929; Post, 1931).

British aviatrix Amy Johnson, 27, arrives in Australia May 24 after making the first solo flight by a woman from London. She follows the 19.5-day flight with a record six-day solo flight from England to India.

The British dirigible R-101 burns October 5 northwest of Paris at Beauvais on her maiden voyage to Australia. The disaster takes 54 lives.

1930 *(cont.)* Aristotle Socrates Onassis, 29, buys six freighters for $20,000 each as the worldwide economic depression reduces demand for cargo vessels. Built at a cost of $2 million each, the ships are the nucleus of a large fleet for the shrewd Greek tobacco and shipping millionaire, who will buy his first tanker in 1935.

General Motors acquires two locomotive manufacturing companies that will give GM a virtual monopoly in the market (*see* 1961).

An electric passenger train travels on Delaware and Lackawanna tracks between Hoboken and Montclair, N.J., in a test run by Thomas Edison, now 84, who will die next year.

Atlanta's Union Station opens April 18 at Pryor and Wall streets.

The first South American subway line opens October 18 at Buenos Aires; 3,000 men have built the Lacroze line of the Ferrocarril Terminal Central de Buenos Aires in 21 months (*see* Mexico City, 1970).

The Oakland-Alameida Tunnel opens to traffic under San Francisco Bay.

The Detroit-Windsor Tunnel opens to traffic under the Detroit River November 3 to supplement the Ambassador Bridge that opened late last year. Like the Oakland-Alameida Tunnel (above), it has been built of prefabricated tubes floated over a trench and sunk like sections of pipeline.

The Longview Bridge opens to span the Columbia River in Washington State with 1,034 feet of cantilevered steel.

The Mid-Hudson Bridge opens at Poughkeepsie, N.Y., to span the Hudson. Builder of the suspension bridge is Polish-American engineer Ralph Modjeski, 69, who has gained a reputation with the McKinley Bridge at St. Louis, the Broadway Bridge at Portland, Ore., and others, including some at Chicago (*see* 1935).

U.S. motorcar sales drop off sharply but the country has an average of one passenger car for every 5.5 persons, up from one for every 13 in 1920. Ford sells one million Model A's, down from 2 million last year. Chevrolet sales drop only 5 percent.

Japan has 50,000 motorcars as compared with 23 million in the United States, 18,000 in China, 125,000 in India, 4,822 in Syria.

The United States has 694,000 miles of paved road, up from 387,000 in 1921, plus 2.31 million miles of dirt road (*see* 1940).

A 16-cylinder Cadillac is introduced by General Motors, whose founder is sold out by his bankers as Wall Street prices collapse again. W. C. Durant liquidated his fortune in common stocks early last year but plunged back into the market when prices plummeted, his Durant Motors will be liquidated in 1933, and Durant will be personally bankrupt by 1936.

Studebaker introduces free-wheeling.

A "differential analyzer" devised by Vannevar Bush and some colleagues is the first analog computer (*see* Babbage, 1833, 1842; Bush, 1922; Mark I, 1944).

Plexiglass, invented by McGill University research student William Chalmers, is a thermoplastic polymer of methyl methacrylate that is light in weight and can be bent when heated into any shape desired. It will be marketed in Britain as Perspex.

Astronomers at the Lowell Observatory in Flagstaff, Ariz., discover Pluto February 18 and give the ninth planet its name.

U.S. astronomer Annie Jump Cannon, 66, completes cataloging and classifying some 400,000 astronomical objects. The "Census Taker of the Sky" started her work in 1897.

Chicago's Adler Planetarium opens with a projector manufactured by Carl Zeiss of Jena. First planetarium in the Western Hemisphere, it has been built with a gift from local merchant Max Adler.

The Institute for Advanced Study at Princeton University is established with an initial endowment of $5 million from department store magnate Louis Bamberger, now 75, and his sister Mrs. Felix Fuld. Abraham Flexner of 1910 Flexner Report fame has urged them to charter a new type of institution dedicated to "the usefulness of useless knowledge"; it will attract such minds as Albert Einstein, Thorstein Veblen, J. Robert Oppenheimer, Wolfgang Pauli, and John von Neumann.

The Human Mind by Menninger Clinic psychiatrist Karl Menninger explains that psychiatry is a legitimate and relatively uncomplicated source of help for mentally disturbed persons, it popularizes a previously arcane subject, and although written for medical students finds a wide market (*see* 1925).

Tincture of merthiolate, introduced by Eli Lilly, quickly gains widespread popularity for painting cuts and scratches although it had no success until given a little alcohol to make it sting and some vegetable dye to make it show on the skin.

Rastafarians in Jamaica, British West Indies, hail the new Ethiopian emperor Haile Selassie (above) as the living God, the fulfillment of a prophesy by Marcus M. Garvey who is said to have declared, "Look to Africa, where a black king shall be crowned, for the day of deliverance is near" (*see* Garvey, 1924). Members of the new sect will withdraw from Jamaican society, call white religion a rejection of black culture, insist that blacks must leave "Babylon" (the Western world) and return to Africa, and contribute to Jamaican culture (notably to the island's reggae music), but Rasta extremists will traffic in *ganja* (marijuana) and engage in acts of violence.

Some 4.3 percent of Americans are illiterate, down from 20 percent in 1870. The figure will fan to 2.4 percent in the next 30 years but nearly 5 million Americans will remain illiterate.

U.S. radio set sales increase to 13.5 million, up from 75,000 in 1921. U.S. advertisers spend $60 million on radio commercials (*see* 1922; FCC, 1934).

Lowell Thomas begins a nightly radio network news program September 29. The program will run on both NBC and CBS for a year, NBC will carry it from 1931 to 1946, CBS from 1946 until May 14, 1976 (*see Lawrence of Arabia* book, 1924).

The *Daily Worker* begins publication at London. The Communist party organ will continue until 1941.

Fortune magazine begins publication in February. The new Henry Luce business monthly has 182 pages in its first issue and sells at newsstands for $1 ($10 per year by subscription) (*see* 1923; 1936).

William Randolph Hearst owns 33 U.S. newspapers whose circulation totals 11 million.

"Blondie" debuts September 8. Chicago cartoonist Murat Bernard "Chic" Young, 29, has been drawing "Dumb Dora." His jazz-age flapper heroine is married to playboy Dagwood Bumstead, their antics will make "Blondie" the most widely syndicated of all cartoon strips, more than 1,600 U.S. newspapers plus some foreign papers will carry the strip, and the huge sandwiches created by Bumstead on evening forays to the refrigerator will become widely known as "dagwoods."

Nonfiction "A Room of One's Own" (essay) by Virginia Woolf champions the cause of independence for women; *The Revolt of the Masses* (*La rebelion de las masas*) by José Ortega y Gasset; *Bring 'Em Back Alive* by U.S. zoo and circus supplier Frank Buck, 46, who has been making expeditions since 1911 to capture wild birds and animals.

Fiction "Machine" ("Kikai") by Japanese short-story writer Riichi Yokomitsu, 32, for whom the 1923 Tokyo earthquake is a symbol of the machine culture and modern utilitarianism; *Mitsou* by Colette; *Flowering Judas* (stories) by U.S. author Katherine Anne Porter, 40; *Brief Candles* by Aldous Huxley; *Vile Bodies* by Evelyn Waugh; *Thy Servant a Dog* (stories) by Rudyard Kipling; *As I Lay Dying* by William Faulkner; *The Apes of God* by Wyndham Lewis; *Job (Hiob)* by Joseph Roth; *East Wind, West Wind* by U.S. novelist Pearl Comfort Sydenstricker Buck, 38, who was raised in China by her Presbyterian missionary parents and wrote the book aboard ship en route to America; *Laughing Boy* by U.S. novelist Oliver Hazard Perry La Farge, 28, who writes about the Navajo; *Cimarron* by Edna Ferber; *Destry Rides Again* by U.S. pulp writer Max Brand (Frederick Schiller Faust), 38, who averages roughly two books per month; *Murder at the Vicarage* by Agatha Christie whose detective Jane Marple will appear in 14 other mystery novels; *It Walks by Night* by U.S.-English detective novelist John Dickson Carr, 25, who will sometimes write as Carter Dickson or Carr Dickson; *The Maltese Falcon* by Dashiell Hammett.

Poetry "Ash Wednesday" by T. S. Eliot is a religious poem of self-abnegation; *20 Poems* by Stephen Spender; *The Cantos* (III) by Ezra Pound; *The Bridge* by Hart Crane who has taken the 47-year-old Brooklyn Bridge as a symbol of man's creative power. The book's many poems explore the essence of the American destiny, but its mixed reception contributes to the poet's growing insecurity and he will jump or fall from a ship bound for the United States from Mexico and drown April 27, 1932; "Spring Comes to Murray Hill" by New York poet Ogden Nash, 28, appears in the *New Yorker* magazine May 3; Nash will soon join the magazine's staff and be celebrated for such verses as "Candy is dandy/ But liquor is quicker," which will often be misattributed to Dorothy Parker.

Painting *American Gothic* by Grant Wood portrays his sister and his dentist as rural farm folk and helps launch the American native regionalist style; *Early Sunday Morning* by Edward Hopper; *Fall of Cuernevaca* and *Cortez and his Mercenaries* by Diego Rivera for Mexico City's Palacio de Cortez; *Cold Buffet* by George Grosz; *Self-Portrait with Saxophone* by Max Beckmann.

Sculpture *Tiare* by Henri Matisse.

Kansas City's Nelson Gallery is founded with a $13 million trust fund set up by the late *Kansas City Star* editor-publisher Col. William Rockhill Nelson who died in 1915 at age 74 (*see* 1880). The art gallery will be housed in the Nelson mansion surrounded by 20 acres of lawn and be noted for its Oriental collection, but the benefactor's will stipulates that no work may be purchased until the artist has been dead for at least 30 years.

A *London Graphic* writer coins the phrase "candid camera" to describe the intimate and often irreverent Leica portraits of statesmen by German photographer

Artist Grant Wood portrayed vanishing rural archetypes in his genre painting *American Gothic*.

1930 *(cont.)* Erich Solomon who in 1932 will take a rare (and illegal) picture of the U.S. Supreme Court in session (*see* Leica, 1925).

Flashbulbs, patented September 23 by German inventor Johannes Ostermeir, have been introduced 7 weeks earlier by General Electric. A small filament in the "flash lamp" heats up when connected to a source of electricity, the heat ignites foil inside the bulb, and the foil flashes to produce a fireless, smokeless, odorless, noiseless (but blinding) light.

Theater *Children of Darkness* by U.S. playwright Edwin Justus Mayer, 33, 1/7 at New York's Biltmore Theater; *Tonight We Improvise (Questa sera Si recite a soggetto)* by Luigi Pirandello 1/25 at Königsberg's Neues Schauspielhaus, 4/14 at Turin's Teatro di Turino; *The Bathhouse (Banya)* by Vladimir Mayakovsky 1/30 at Leningrad's People's House, 3/16 at Moscow's Meyerhold Theater; *As You Desire Me (Come tu mi vuoi)* by Luigi Pirandello 2/18 at Milan's Teatro de Filodrammatici; *The Green Pastures by* Marc Connelly 2/21 at New York's Mansfield Theater, is an adaptation of a 1928 collection of tales by Roark Bradford, 34, depicting heaven, the angels, and the Lord as envisioned by a black country preacher for a Louisiana congregation, 640 perfs.; *Hotel Universe* by Philip Barry 4/14 at New York's Martin Beck Theater, with Ruth Ford, Glenn Anders, Earle Larimore, Morris Carnovsky, 81 perfs.; *Moscow Is Burning (Moskva golid)* by Vladimir Mayakovsky 4/21 at Moscow's Circus I (Gostsirk I); *The Barrets of Wimpole Street* by Dutch-English playwright Rudolph Besier, 52, at England's Malvern Festival (the play will open next year at New York's Empire Theater with Katherine Cornell and have 370 performances); *Private Lives* by Noël Coward 9/24 at London's Phoenix Theatre, with Coward, Gertrude Lawrence, Laurence Kerr Olivier, 23, Coward's song "Someday I'll Find You," 101 perfs.; *Once in a Lifetime* by George S. Kaufman and Moss Hart 9/24 at New York's Music Box Theater, with Kaufman and Spring Byington, 401 perfs.; *The Greeks Had a Word for It* by Zoë Akins 9/25 at New York's Sam H. Harris Theater, with Dorothy Hull, Varree Teasdale, Muriel Kirkland, 253 perfs.; *The Sailors of Cattaro (Die Matrosen von Cattaro)* by German playwright Friedrich Wolf, 42, 11/8 at Berlin's Lobe Theater; *Grand Hotel* by W. A. Drake who has adapted the Vicki Baum novel 11/13 at New York's National Theater, with Henry Hull, Sam Jaffee, 459 perfs.; *Alison's House* by U.S. novelist-playwright Susan Glaspell, 48, 12/1 at New York's Civic Repertory Theater, 41 perfs.; *The Shoemaker's Prodigal Wife* (*La Zapatera prodigiosa*) by Federico García Lorca 12/24 at Madrid's Teatro Español.

Radio drama *The Lone Ranger* 1/20 over Detroit's WXYZ begins with the overture from the 1829 Rossini opera *Guillaume Tell* and an announcer saying, "A fiery horse with the speed of light, a cloud of dust and a hearty 'Hi-yo Silver'—the Lone Ranger rides again." The serial has been developed from *Curly Edwards and the Cowboys* by James E. Jewell, 23, who will soon give his masked hero an Indian friend—Tonto—who will call the Lone Ranger "Kemo Sabe" (trusting brave); *Death Valley Days* 7/30 on NBC Blue Network stations. New York advertising writer Ruth Cornwall Woodman has researched background for the show by visiting Panamint City and introduces "The Old Ranger" to give the story authenticity; the series will move to CBS in 1941 and continue until 1945.

U.S. moviegoing increases as all studios rush to make "talkies" and a new Vitascope widens theater screens.

Films Luis Buñuel's *L'Age d'Or* with Gaston Modot, Max Ernst; Lewis Milestone's *All Quiet on the Western Front* with Lew Ayres. Also: D. W. Griffith's *Abraham Lincoln* with Walter Huston, Una Merkel; Clarence Brown's *Anna Christie* with Greta Garbo ("Garbo Talks," advertisements cry); George Hill's *The Big House* with Wallace Beery; Raoul Walsh's *The Big Trail* with John Wayne (*né* Marion Michael Morrison), 23; Howard Hawks's *The Dawn Patrol* with Richard Barthelmess, Douglas Fairbanks, Jr.; Howard Hughes's *Hell's Angels* with Ben Lyon, Jean Harlow; Henry King's *Lightnin'* with Will Rogers, Louise Dresser, Joel McCrea; Mervyn LeRoy's *Little Caesar* with Edward G. Robinson; George Hill's *Min and Bill* with Marie Dressler, Wallace Beery; Josef von Sternberg's *Morocco* with Marlene Dietrich, Gary Cooper, Adolphe Menjou; Alfred Hitchcock's *Murder* with Herbert Marshall, Norah Baring; Robert Milton's *Outward Bound* with Leslie Howard; Richard Wallace's *The Right to Love* with Ruth Chatterton is the first picture made with a new process that eliminates popping, cracking, grating, and other surface noises; George Cukor and Cyril Gardner's *The Royal Family of Broadway* with Fredric March, Ina Claire; John Cromwell's *Tom Sawyer* with Jackie Coogan, Mitzi Green.

Stage musicals *Strike Up the Band* 1/14 at New York's Times Square Theater, with Bobby Clark, Red Nichols's Band (which includes Benny Goodman, Gene Glenn Miller, Jimmy Dorsey, and Jack Teagarden), music by George Gershwin, lyrics by Ira Gershwin, book by George S. Kaufman, songs that include "I've Got a Crush on You," and the title song, 191 perfs.; *Nine-Fifteen Revue* 2/11 at New York's George M. Cohan Theater, with Ruth Etting, music by local composer Harold Arlen (*né* Hyman Arluck), 24, lyrics by Ted Koehler, songs that include "Get Happy," 7 perfs.; *Lew Leslie's International Revue* 2/15 at New York's Majestic Theater, with Gertrude Lawrence, Harry Richman, music by Jimmy McHugh, lyrics by Dorothy Fields, songs that include "Exactly Like You," "On the Sunny Side of the Street," 96 perfs.; *Simple Simon* 2/18 at New York's Ziegfeld Theater, with Ed Wynn, Ruth Etting, music by Richard Rodgers, lyrics by Lorenz Hart, songs that include "Ten Cents a Dance," "I Still Believe in You," 135 perfs.; *Fine and Dandy* 9/23 at New York's Erlanger Theater, with Joe Cook, Eleanor Powell, Dave Chasen, music by Kay Swift, lyrics by Paul James (Swift's husband James Warburg), book by Donald Ogden Stewart, songs that include "Can This Be Love?" 255 perfs.; *Girl Crazy* 10/14 at New York's Alvin Theater, with Ethel Merman, music by George Gershwin, lyrics by Walter Donaldson and Ira Gershwin, songs that include "I Got Rhythm," "Embraceable You," "Little White Lies," "But Not for Me," "Bidin' My Time," 272 perfs.; *Three's a Crowd* 10/15 at New York's Selwyn Theater, with Libby Holman, Fred Allen, Clifton Webb, songs that include "Body and Soul" by Johnny Green, lyrics by Robert Sourt, 25, and Edward Heyman, 23, "Something to Remember You By" and "The Moment I Saw You" by Arthur Schwartz, lyrics by Howard Dietz, 272 perfs.; *The Garrick Gaieties* 10/16 at New York's Guild Theater, with Sterling Holloway, Rosalind Russell, Imogene Coca, songs that include "I'm Only Human After All" by Russian-American composer Vernon Duke (Vladimir Dukelsky), 27, lyrics by E. Y. Harburg and Ira Gershwin, "Out of Breath and Scared to Death of You" by Johnny Mercer and Everett Miller, 158 perfs.; *Lew Leslie's Blackbirds of 1930* 10/22 at New York's Royale Theater, with Ethel Waters, Cecil Mack's Choir, songs that include "Memories of You" by Eubie Blake, lyrics by Andy Razaf, 57 perfs.; *Sweet and Low* 11/17 at New York's 46th Street Theater, with George Jessel, Fanny Brice, songs that include "Outside Looking In" by Harry Archer and Edward Ellison, "Cheerful Little Earful" by Harry Warren, Ira Gershwin, and Billy Rose, 184 perfs.; *Smiles* 11/18 at New York's Ziegfeld Theater, with Marilyn Miller, Bob Hope, Eddie Foy, Jr., Fred and Adele Astaire, songs that include "Time on My Hands" by Vincent Youmans, lyrics by Harold Adamson and Polish-American writer Mack Gordon, 26, "You're Driving Me Crazy" by Walter Donaldson, 63 perfs.; *Ever Green* 12/3 at London's Adelphi Theatre, with Jessie Matthews, music by Richard Rodgers, lyrics by Lorenz Hart, songs that include "Dancing on the Ceiling," 254 perfs.; *The New Yorkers* 12/8 at New York's Broadway Theater, with Hope Williams, Ann Pennington, Jimmy Durante, Lew Clayton, Eddie Jackson, music and lyrics by Cole Porter, songs that include "Love for Sale," 162 perfs.

Nazi thug Horst Wessel dies in a Berlin hospital February 23 at age 22 of blood poisoning. Painter Sol Epstein, tailor Peter Stoll, and barber Hans Ziegler will be convicted of having shot Wessel the night of January 14 in a street brawl over a "Lucie of Anderplatz," the Nazis will make a martyr of Wessel, and his "Horst-Wessel-Lied," set to music plagiarized from a Hamburg waterfront ballad and containing such lines as "When Jewish blood drips off your trusted pen-knife/We march ahead with twice as steady step," will be the Nazi anthem.

Opera *The Nose* 1/13 at Leningrad, with music by Dmitri Shostakovich (who includes an orchestral sneeze and other novel effects), libretto from a story by Nikolai Gogol; *Von heute auf morgen (From Today until Tomorrow)* 2/1 at Frankfurt's Municipal Theater, with music by Arnold Schoenberg; *Neues vom Tage (The News of the Day)* 6/8 at Kroll's Theater, Berlin, with music by Paul Hindemith.

Ballet *The Golden Age (Zolotoy Vyek)* 10/26 at Leningrad, with music by Dmitri Shostakovich.

First performances Suite for Orchestra by Walter Piston 3/28 at Boston's Symphony Hall; Symphony No. 2 *(Romantic)* by Howard Hanson 11/28 at Boston's Symphony Hall.

The BBC forms its own symphony orchestra under Adrian Boult.

Popular songs "Georgia on My Mind" by Hoagy Carmichael, lyrics by Stuart Gorrell; "When It's Sleepy Time Down South" by Leon Rene, Otis Rene, and Clarence Muse; "My Baby Just Cares for Me" by Gus Kahn and Walter Donaldson; "Dream a Little Dream of Me" by Fabian Andrée and Wilbur Schwandt, English lyrics by Gus Kahn; "Three Little Words" by Harry Ruby, lyrics by Bert Kalmar (for the film *Amos 'n'Andy*); "Walkin' My Baby Back Home" by Roy Turk, Fred E. Ahlert, Harry Richman; "Little White Lies" by Walter Donaldson; "It Happened in Monterey" by Mabel Wayne, lyrics by William Rose (for the film *King of Jazz* with Paul Whiteman); "Sing You Sinners" by Franke Harting, lyrics by Sam Coslow; "Them There Eyes" by Maceo Pinkard, William Tracey, and Doris Tauber; "I'm Confessing That I Love You" by Doc Dougherty and Ellis Reynolds; "You Brought a New Kind of Love to Me" by Sammy Fain, lyrics by Irving Kahal and Pierre Norman (for the film *The Big Pond*); "Beyond the Blue Horizon" by Richard A. Whiting and W. Frank Harting, lyrics by Leo Robin (for the film *Monte Carlo*).

Gallant Fox wins racing's Triple Crown by taking the Kentucky Derby, the Preakness, and the Belmont Stakes.

U.S. golfer Robert Tyre "Bobby" Jones, 28, wins golf's "Grand Slam," taking the British and U.S. Open and Amateur tournaments.

German prizefighter Max Schmeling, 24, is fouled by world heavyweight champion Jack Sharkey, 27, in the fourth round of a title bout at New York June 12. Schmeling wins the title but will lose it back to Sharkey in 1932.

Bill Tilden, now 37, wins in men's singles at Wimbledon, Helen Wills Moody in women's singles; John H. Doeg, 21, wins in men's singles at Forest Hills, Betty Nuthall, 19, (Br) in women's singles (the first non-American U.S. champion).

The Philadelphia Athletics win the World Series by defeating the St. Louis Cardinals 4 games to 2.

Uruguay wins the first World Cup football (soccer) competition, organized by the Federation Internationale des Associations Football headed by Jules Rimet, 57, of France. (The cup is officially the Jules Rimet Trophy.) Playing against Argentina at Montevideo, the home team wins 4 to 2.

Sir Thomas Lipton loses his final bid to gain the America's Cup as his *Shamrock V* is defeated 4 to 0 by the U.S. defender *Enterprise*. Lipton will die next year at age 81.

Schick Dry Shaver, Inc., is founded by Col. Jacob Schick who will begin production next year at Stamford, Conn., and sell 3,000 of his electric shavers at $25 each (*see* 1923). Sales will reach 10,381 in 1932, Schick will die in July of 1937 at age 59, and nearly 1.85 million

1930 ***(cont.)*** of his shavers will have been sold by the end of that year.

U.S. tobacco companies produce 123 billion cigarettes, up from less than 9 billion in 1910. Hollywood helps make smoking seem sophisticated as film directors have actors light up to fill awkward interludes with "business."

Judge Crater disappears in early August. New York Tammany lawyer Joseph Force Crater, 41, has been appointed in April by Governor Franklin D. Roosevelt to fill an unexpired term on the New York Supreme Court, interrupts his summer vacation to return to New York, withdraws all the cash in his bank account (just over $5,000), sells $16,000 worth of stock, dines with friends, steps into a taxi outside the restaurant, and is never heard of again.

Le Corbusier completes Maison Savoye at Poissy-sur-Seine, France.

Cleveland's 52-story Terminal Tower building opens January 26. The north wing of the $110 million building put up by the Van Sweringen brothers is a hotel built several years ago.

Atlanta's $1 million City Hall is completed in March by Ten Eyck Brown.

The Chicago Board of Trade building completed by Holabird and Roche at 141 West Jackson Boulevard towers 45 stories above the city and will be the city's tallest structure for 40 years.

Chicago's Merchandise Mart is completed for Marshall Field and Co. which will sell it to financier Joseph P. Kennedy in 1945. The world's largest commercial building, the $32 million Mart has 4.2 million square feet of floor space and rises 18 stories with a 25-story tower.

Chicago's Palmolive building is completed by Holabird and Roche. A Lindbergh Beacon flashing 500 miles out over Lake Michigan tops the $6 million, 37-story structure.

The *New York Daily News* building opens June 25 with 718,000 square feet of floor space. Rising 467 feet high above East 42nd Street, the skyscraper has been designed by John M. Howells, Raymond M. Hood, and J. André Fornithran.

Lefrak City raises the first of 20 18-story apartment house towers that the Lefrak Corporation will erect on 32 acres of the old Astor estate in New York's Corona, Queens, section. The apartment complex will have 20,000 middle-class residents; occupancy is now all white but will be 25 percent black by 1972 and 70 percent by 1976.

Congress establishes Carlsbad Caverns National Park on 46,753 acres of underground chambers in southern New Mexico.

Dutch elm disease kills trees in Cleveland and Cincinnati in its first attacks on the American elm *ulmus americanus*. A fungus first observed in Holland in 1919, the blight has been introduced in elm burl logs shipped from Europe to be made into veneer for furniture. Within 40 years it will have killed some 13 million trees and be destroying 400,000 elms each year in 30 states and Canada, wilting or shriveling leaves in June or July, yellowing them prematurely, killing young trees sometimes in one year and older ones in 3 or 4 as fungus spores are spread by underground root grafts and by two kinds of bark beetles. The loss of America's foremost shade tree will have a marked effect on the environment of many American cities and towns.

The *Star of Alaska* makes her final trip north from San Francisco to the Alaskan salmon canneries, ending the era of the great Star square-riggers (*see* 1893; California Packing, 1916).

Twenty-five percent of Americans live on or from the farm, many of them on subsistence farms that operate outside the money economy. The total number of U.S. farms is roughly 6.3 million (*see* 1931).

Major U.S. crops produce barely $8 per acre and the average annual farm income i$ $400 per family.

Tenant farmers work four out of ten U.S. farms as compared with nine out of ten British farms, one-third of farms in western Europe, 78 percent of Australian farms, and 40 percent of Argentine farms.

Fifty-five percent of agricultural workers on Soviet farms are employed on collective farms (*see* Stalin's Five-Year Plan, 1928).

The mortgage debt of U.S. farms is $9.2 billion, up from $3.2 billion in 1910. Farmers have borrowed heavily to buy the machines they need to remain competitive.

Twenty percent of the North American work force is employed in agriculture as compared with 7 percent in Britain, 25 percent in France, 40 percent in Japan.

Farmers in the Corn Belt begin to drop their resistance to hybrid corn as test plantings demonstrate dramatically how the new corn can increase yields (*see* 1921; Bantam, 1933).

Colorado-Utah rancher James Monaghan, 38, opens northwest Colorado cattle ranges to sheep grazing. He started working as a cowhand in summer roundups at age 16 and now heads the Rio Branco Wool Growers' Association.

Poultry farmers on the Delmarva (Delaware-Maryland-Virginia) Peninsula develop the broiler chicken that will replace the spring chicken that has kept chicken a seasonal food and limited consumption of chicken, which remains costlier than red meat.

Unprecedented drought parches the U.S. South and Midwest. Congress votes a $45 million Drought Relief Act December 20.

Vitamin D is isolated in its crystalline form calciferol and is soon being used to fortify butter, margarine, and other foods (*see* Borden's milk, 1933).

Edwin J. Cohn at Harvard Medical School develops a concentrate 100 times as potent as liver for treating pernicious anemia (*see* 1926; 1929; 1948).

The McNary-Mapes Amendment to the 1906 Food and Drug Act requires labeling of substandard canned goods. The National Canners Association has urged the move as a safety measure and to eliminate competition from shoddy goods.

Birds Eye Frosted Foods go on sale for the first time March 6 at Springfield, Mass. General Foods introduces frozen peas, spinach, raspberries, cherries, loganberries, fish, and various meats, but the packets, kept in ice-cream cabinets, are not readily visible, and sell at relatively high prices (35¢ for a package of peas) (*see* 1929; 1931).

Dry ice (solid carbon dioxide) is introduced commercially in the United States for purposes such as keeping ice cream cold.

Sliced bread is introduced under the Wonder Bread label by Continental Baking (*see* 1927). Battle Creek, Mich., inventor Otto Frederick Rohwedder perfected the commercial bread slicer in January 1928 after 15 years of work. Consumers are suspicious at first (sliced bread does grow stale faster) but soon accept the product enthusiastically.

The Toastmaster automatic toaster is introduced by McGraw-Electric of Elgin, Ill.

Hostess Twinkies are introduced by Continental Baking (above). Continental bakery manager James A. Dewar, 33, at Chicago has been turning out little sponge cakes for shoppers to use as the basis of strawberry shortcakes; when the strawberry season ends he hits on the idea of filling the cakes with sugary cream to keep his cake line in production year round. A St. Louis sign advertising "Twinkle Toes Shoes" inspires him to call the cakes Twinkies.

The International Apple Shippers Association offers its fruit on credit to jobless men who will peddle apples on street-corners and will help the association dispose of its vast surplus. By November some 6,000 men are selling apples on New York sidewalks and more thousands are selling apples in other cities, but by the spring of next year the apple sellers will be called a nuisance and City Hall will order them off the streets of New York.

Snickers candy bars are introduced by Mars, Inc., which obtains national distribution (*see* 1923; 1932).

The first true supermarket opens in August at Jamaica, Long Island, where former Kroger store manager Michael S. Cullen, 46, opens the King Kullen Market in an abandoned garage and meets with instant success (*see* 1923; Big Bear, 1932).

Jack & Charlie's 21 Club opens in January at 21 West 52nd Street, New York. Restaurateur John Karl "Jack" Kriendler, 31, and his accountant cousin Charles "Charlie" Berns, 28, have been operating speakeasy saloons since 1923 and in effect been subsidized by the Rockefellers, who have purchased their lease at 42 West 49th Street to clear the block for construction of Rockefeller Center (*see* 1931). Kriendler and Berns store their illicit liquor behind secret walls and devise an arrangement that permits the bartender to push a button at the first sign of a raid by federal Prohibition enforcement agents, tilting the shelves of the bar to send all bottles down a chute to smash in the cellar and thus destroy evidence the agents need for a conviction. They will prosper at the new location and acquire the building at 19 West 52nd Street in 1935.

Benito Mussolini's Italy makes abortion a crime "against the integrity and health of the race" but illegal abortions continue at a rate of more than half a million per year.

A human egg cell is seen for the first time through a microscope (*see* spermatozoa, 1677). B. Kraus and D. Ogino in Japan discover that human ovaries generally release maternal eggs 12 to 16 days before menstruation but about one woman in four will be found to be wildly erratic in her menstrual cycle and all women will be found to be erratic for a few months after the birth of a child.

A U.S. court decision extending trademark protection to contraceptive devices encourages production of better quality condoms. Economic depression makes large families increasingly burdensome to breadwinners.

The encyclical *Casti connubii* issued December 31 by Pope Pius XI says, "Any use whatsoever of matrimony exercised in such a way that the [sex] act is deliberately frustrated in its natural power to generate life is an offense against the law of God and of nature, and those who indulge in such are branded with the guilt of a grave sin." But the "rhythm" system of birth control favored by the Church is an inadequate form of contraception for reasons suggested above and increasing numbers of Roman Catholics are employing artificial methods of birth control including abortion.

The world's population reaches close to 2 billion with much of it in the grip of economic depression.

Emigration from the United States for the first time in history exceeds immigration.

1931

Spain's Alfonso XIII leaves the country April 14 after a 45-year reign. Forces favoring a republic have won big in municipal elections, the king is declared guilty of high treason November 12 and forbidden to return, the royal property is confiscated, a new constitution is adopted December 9, and Spain creates a republic that will continue until 1939 (*see* 1936).

France elects Paul Domer president in June to succeed Gaston Doumergue (but *see* 1932).

Britain's Labour government resigns August 24 in a disagreement over remedies for the nation's financial crisis (below), but Prime Minister Ramsay MacDonald heads a new coalition cabinet that will retain power until mid-1935.

Chinese rebels under Gen. Chen Jitang split with Gen. Chiang Kai-shek and take control of Guangzhou (Canton) April 30.

British authorities in China arrest Vietnamese Communist leader Ho Chi Minh (*né* Nguyen That Than), 41, June 17.

1931 ***(cont.)*** Japanese militarists use the Mukden Incident September 19 as an excuse to occupy Manchuria. They cite an alleged railway explosion as provocation, seize Kirin September 21, and within 5 months will have taken Harbin and the three eastern provinces (the action is in large measure a reprisal for China's boycott of Japan's cotton textiles).

Chiang Kai-shek comes under pressure from Communist forces led by Mao Zedong, now 37. Preoccupation with the Communists and Yangzi floods (below) prevent him from mounting any military effort against the Japanese (*see* 1934).

Scottsboro, Ala., makes headlines as a shouting mob surrounds the county courthouse where nine black youths are being tried on charges of having raped two white women aboard a freight car March 25 after the blacks allegedly threw some white hoboes off the train. A white mob pulled the blacks from the train when it arrived at Paint Rock, Ala., the defendants Harwood Patterson, Olen Montgomery, Clarence Norris, Ozie Powell, Willie Robertson, Charlie Weems, Eugene Williams, Andy Wright, and Roy Wright have been transferred to Scottsboro to prevent their being represented by counsel, they are defended only by a reluctant lawyer who has been assigned to the case by the presiding judge and given no preparation, the jury discounts testimony by a physician, and the 3-day trial ends April 9 with eight of the defendants sentenced to death and the ninth to life imprisonment. The physician has testified that he examined Victoria Price, 21, and Ruby Bates, 17, of Huntsville shortly after the alleged rape and that while he found dried semen he found no live spermatozoa and no blood.

The Supreme Court will reverse the Scottsboro convictions (above) in October of next year in a landmark ruling that defendants in capital cases in state courts must have adequate legal representation.

The "Scottsboro Boys" (above) will have a new trial in Alabama in 1933, it will end in conviction, and the Supreme Court will again reverse the convictions with a landmark ruling that blacks may not be systematically excluded from grand and trial juries. New York civil rights lawyers Shad Polier, 25, and Samuel S. Leibowitz, 38, will take up the cause, a third trial with one black on the jury will end in conviction, but some indictments will be dropped, the sentences will be commuted to life imprisonment, and the defendants will serve a total of 130 years behind bars (one will not be paroled until 1951).

Harlan County, Ky., coal miners and deputized guards have a gunfight May 4 at Evarts following 4 years of strikes and labor-management battles. The Harlan County Coal Operators Association began spending heavily in 1927 to terrorize miners and their families with strong-arm tactics in a move to combat efforts by the United Mine Workers to organize, the conflict has come to a head, the fight ends with three guards and one miner left dead, untold numbers of others are carried dead or dying into the hills.

Financial panic and economic depression engulf most of the world. Vienna's Kreditanstalt goes bankrupt May 11, possibly as a result of French opposition to an Austro-Hungarian credit union, and the panic spreads to Germany where the Darmstadter und National Bank closes.

The Bank of England advances money to Austria but Britain's own financial position is shaky (below).

Canada raises tariffs June 18, estimating that the new customs duties will cut off two-thirds of goods imported from the United States.

President Hoover proposes a one-year moratorium on payments of war debts and reparations June 20.

Germany's Danatbank goes into bankruptcy July 13, precipitating a general closing of German banks that continues until August 5.

Britain receives a French-American loan August 1 but London and Glasgow have riots September 10 to protest government economy measures, naval units mutiny September 15 to protest pay cuts, the pound sterling is devalued September 20 from $4.86 to $3.49, and Britain is forced to abandon the gold standard once again September 21.

Japan abandons the gold standard December 11.

U.S. motorcar sales collapse and Detroit lays off another 100,000 workers, reducing employment in auto plants to 250,000, down from 475,000 in 1929. Two out of three Detroit workers (and eight out of ten Detroit blacks) are totally or partially unemployed, and while the situation will improve unemployment will remain high in Detroit until 1942 (*see* 1933).

The "Swope Plan" for economic recovery outlined by General Electric president Gerard Swope, 58, says, in effect, leave the problem to business and let trade associations develop national economic plans to revive the economy.

In the U.S. 2,294 banks fail, up from 1,352 last year.

The Nevada State Legislature enacts a 6-week residency law for divorce-seekers in a move to generate revenue for the Depression-struck state which has almost no tax base since most of its lands are federally owned. Nevada grants 5,260 divorces, up from 2,500 in 1928, and 4,745 of the decrees are handed down at Reno, which becomes the boom-town divorce capital of the world (*see* gambling, below).

U.S. unemployment tops 8 million. Hunger marchers petition the White House December 7 for a guarantee of employment at a minimum wage but are turned away (*see* "Bonus Marchers," 1932).

President Hoover recommends an emergency Reconstruction Finance Corp. and a public works administration in his annual message to Congress December 8 (*see* 1932).

Columbia University physicist Harold Clayton Urey, 38, and two colleagues pioneer production of atomic energy. They discover heavy water (deuterium oxide) whose molecules consist of an atom of oxygen and two atoms of deuterium (heavy hydrogen, a rare isotope of hydrogen), a discovery that will lead to the separation of other isotopes, including the separation of the fissionable uranium 235 from the more common U238 (*see* Cockroft, Chadwick, 1932).

U.S. engineer Henry J. Kaiser, 49, organizes the Six Companies to construct Boulder Dam on the Colorado River. Having built miles of highway from California up the West Coast into British Columbia and in Cuba plus many cement plants, Kaiser coordinates the capabilities of six contractors (*see* 1936).

Swissair has its beginnings in an airline started with private capital that will continue to control at least 75 percent of the national airline.

Seversky Aircraft is founded by Russian-American aeronautical engineer Alexander Procofieff de Seversky, 37, who lost a leg in the World War. His company will manufacture pursuit planes (*see* Republic, 1939).

New York's Floyd Bennet Field opens at the south end of Brooklyn's Jamaica Bay. The first city-owned airport takes its name from the late aviation pioneer who died in 1928 at age 37 (*see* 1926; La Guardia, 1939).

Wiley Post circles the earth in the *Winnie Mae* (*see* 1930). Flying with Harold Gatty as navigator, Post flies across the Atlantic to the British Isles, across Europe and the U.S.S.R., and back to the United States in 8 days, 15 hours, and 51 minutes, a feat he will repeat in 1933 in 7 days, 18 hours, and 49 minutes flying solo to become the first pilot to circle the earth alone (*see* 1935; Hughes, 1938).

Omaha's Union Station opens. The magnificent new terminal serves seven railroads.

New York's George Washington Bridge opens October 24 to span the Hudson River between Manhattan and New Jersey. The 1,644-foot bridge is the world's largest suspension bridge and its 90,000 tons of deadweight are suspended from four 3-foot cables made by John A. Roebling's Sons of Trenton, N.J. (*see* Brooklyn Bridge, 1883; Golden Gate, 1937).

The Bayonne (Kill Van Kull) Bridge opens to link Staten Island and New Jersey; the 3,492-foot span is the world's longest steel arch bridge.

Auburn motorcar sales soar to 28,130 and profits equal those of 1929 after a depressing 1930 sales year. E. L. Cord signs up 1,000 new dealers as his car climbs from 23rd place in retail sales to 13th on the strength of the new Auburn 8–98, whose 286.6-cubic-inch Lycoming engine is far more powerful than any other in its price class (the costliest sells for $1,395). The new Auburn is the first rear-drive motorcar with a frame braced by an X cross member and the first moderately-priced car with L.G.S. Free Wheeling (but *see* 1937).

The four-cylinder Ford Model A has a 103.5-inch wheelbase, develops 40 horsepower at 2,200 rpm, and sells at between $430 and $595.

The four-cylinder Plymouth has a 110-inch wheelbase, develops 56 horsepower at 2,800 rpm, and sells at between $535 and $645.

The six-cylinder Chevrolet has a 109-inch wheelbase, develops 50 horsepower at 2,600 rpm, and sells at $475 to $635 and up.

Rolls-Royce acquires Bentley Motors from W. O. Bentley (*see* 1919). It will market a Rolls-Royce under the Bentley name using a different front grill.

Penicillin relieves two cases of gonococcal ophthalmitis in children at England's Royal Infirmary in Sheffield. The children have contracted gonorrhea from their mothers at birth (*see* Argyrol, 1902; Fleming, 1928; 1929). Penicillin is also used at the Royal Infirmary to relieve a colliery manager suffering from severe pneumococcal infection of the eye but no further efforts will be made until the end of the decade to pursue the antibiotic's chemotherapeutic potentials (*see* Florey, 1940).

Alka-Seltzer, introduced by Miles Laboratories of Elkhart, Ind., is an antacid and analgesic tablet made from sodium bicarbonate, monocalcium phosphate, acetyl-salicylic acid (aspirin), and citric acid that fizzes when dropped in water. It gains quick acceptance for headaches, hangovers, and upset stomachs even though its aspirin content may cause dyspepsia.

Jehovah's Witnesses adopt that name under the leadership of Joseph Franklin "Judge" Rutherford, 62, who has headed the International Bible Students' Association since the death of Charles Taze Russell in 1916. Rutherford proselytizes with the slogan "Millions now living will never die" (*see* 1884).

The Black Muslims have their beginnings at Detroit where local Baptist teacher Elijah Poole, 34, becomes an assistant to Wali Farad who has founded the Nation of Islam. Poole changes his name to Elijah Muhammad, he will establish Muhammad's Temple of Islam No. 2 at Chicago in 1934, Farad will thereupon mysteriously disappear, Muhammad will assume command of the growing sect as the Messenger of Allah, and by 1962 there will be at least 49 Temples of Islam with an estimated 250,000 followers (*see* Malcolm X, 1952).

The Supreme Court rules June 1 that a 1925 Minnesota "gag law" that banned publication of a "malicious, scandalous and defamatory newspaper, magazine, or other periodical" is unconstitutional. *Near v. Minnesota* is the first case in which provisions of a state law are held to restrict personal freedom of speech and press without due process, and a decision that incorporates provisions of the First Amendment into the Fourteenth.

"Dick Tracy" has its beginnings in the *Chicago Tribune's* new comic strip "Plainclothes Tracy" introduced by former Hearst syndicate cartoonist Chester Gould, 30, whose jutting-jawed detective is never without his snap-brim hat. The strip will be popular for more than 45 years, reaching nearly 100 million readers each day in 600 U.S. newspapers and hundreds of foreign papers.

"Mickey Mouse" by Walt Disney makes his first appearance as a comic strip figure (*see* 1928).

"Felix the Cat" by Otto Mesmer makes his first appearance as a comic strip drawn by Pat Sullivan. The cat

1931 ***(cont.)*** has appeared in animated cartoons since 1919.

Nonfiction *Axel's Castle* (critical pieces) by Edmund Wilson.

Fiction *Night Flight (Vol de nuit)* by Antoine de Saint-Exupéry; *L'Homme approximate* by Tristan Tzara; *African Confidence (Confidence Africaine)* by French novelist Roger Martin du Gard, 30; *Afternoon Men* by English novelist Anthony Powell, 26; *Prince Jali* by L. H. Myers; *The Waves* by Virginia Woolf; *Sanctuary* by William Faulkner; *Tropic of Cancer* by U.S. émigré novelist Henry Miller, 39, whose earthy novel will not be permitted in the United States for decades; *Boy* by Irish novelist James Hanley, 30, whose book is suppressed; *The Dream Life of Balso Snell* by U.S. novelist Nathanael West (Nathan Wallenstein Weinstein), 27, who has developed a style derived from dadaism and surrealism in the arts; *After Leaving Mr. Mackenzie* (stories) by Jean Rhys; *The Good Earth* by Pearl Buck; *Grand Hotel* by Austrian-American novelist Vicki Baum, 43 (*see* play, 1930). *Hatter's Castle* by Scottish physician-novelist A. J. (Archibald Joseph) Cronin, 35, who has given up his lucrative London West End practice for reasons of health; *S. S. San Pedro* by James Gould Cozzens; *The Glass Key* by Dashiell Hammett.

Poetry *Strophe* by Greek poet George Seferis, 31; *The Loosening* by English poet Ronald Bottrall, 25.

Painting *The Persistence of Memory* by Spanish surrealist Salvador Dali, 27, whose limp watches and clocks draped over barren landscapes gain wide attention and ridicule; *The Breakfast Room* by Pierre Bonnard; *The Trick-Riders* by Marc Chagall; *Still Life with Studio Window* by Max Beckmann; *Market Church* by Lyonel Feininger; *The Ghost Vanishes* by Paul Klee who begins teaching at Düsseldorf; *The Dance* by Henri Matisse who has been commissioned to paint the murals by Argyrol king Albert Barnes; *Route 6, Eastham* by Edward Hopper; *Man, Woman, and Child* by Joan Miró; *Zapatistas* by José Clemente Orozco; *Portrait of Frida and Diego* by Mexican painter Frida Kahlo, 21, who married muralist Diego Rivera in 1929.

The Whitney Museum of American Art is founded by railroad heiress-sculptor Gertrude Vanderbilt Whitney, 54, whose husband Harry Payne Whitney died last year at age 58 (*see* 1966).

Sculpture *Mlle. Pognany* by Constantin Brancusi.

Rio de Janeiro's *Christ the Redeemer is* dedicated atop Corcovado (Hunchback Mountain). The 125-foot tall concrete work by French sculptor Paul Maximilian Landowski, 56, has outstretched arms that span 92 feet and weighs 1,145 tons.

The Photronic Photoelectric Cell invented by Weston Electrical Instrument engineer William Nelson Goodwin, Jr., is the first exposure meter for photographers. It contains a dial-calculating device for translating brightness values into camera aperture settings, needs no batteries since it changes light energy directly into electrical energy, and will be introduced commercially in February of next year.

Theater *Tomorrow and Tomorrow* by Philip Barry 1/13 at New York's Henry Miller Theater, with Zita Johnson, Osgood Perkins, Herbert Marshall, 206 perfs.; *Green Grow the Lilacs* by U.S. playwright Lynn Riggs, 30, 1/26 at New York's Guild Theater, with Helen Westley, Lee Strasberg, June Walker, Franchot Tone, 64 perfs.; *Fanny* by Marcel Pagnol; *Farce of the Chaste Queen (Farsa y licencia de la rena castize)* by Ramon de Valle Inclan 6/3 at Madrid's Teatro Munoz Seca; *The House of Connelly* by Paul Green 10/5 at New York's Martin Beck Theater, with Stella Adler, Franchot Tone, Clifford Odets, Rose McClendon, 81 perfs.; *Cavalcade* by Noël Coward 10/13 at London's Drury Lane Theatre, 405 perfs.; *The Left Bank* by Elmer Rice 10/5 at New York's Little Theater, 242 perfs.; *Mourning Becomes Electra* by Eugene O'Neill 10/26 at New York's Guild Theater, with Alla Nazimova and Alice Brady in a play based on the 5th-century B.C. Greek tragedies, 150 perfs.; *Judith* by Jean Gênet 11/4 at the Comédie des Champs-Elysées, Paris; *Counsellor-at-Law* by Elmer Rice 11/6 at New York's Plymouth Theater. with Paul Muni, 292 perfs.; *Springtime for Henry* by English playwright Benn W. Levy, 31, 12/9 at New York's Bijou Theater, with Leslie Banks, Nigel Bruce, 199 perfs.

Films René Clair's *A Nous la Liberté* with Raymond Cordy, Henri Marchand; Charles Chaplin's *City Lights* with Chaplin; Fritz Lang's *M* with Peter Lorre. Also: Jean Renoir's *Bondel Saved from Drowning* and *La Chienne*; Josef von Sternberg's *An American Tragedy* with Sylvia Sidney, Phillips Holmes, Frances Dee, and *Dishonored* with Marlene Dietrich; Tod Browning's *Dracula* with Bela Lugosi; James Walsh's *Frankenstein* with Boris Karloff; Frank Borzage's *History Is Made at Night* with Charles Boyer, Jean Arthur; Karl Froelich's *Mädchen in Uniform* with Hertha Thiele, Dorothea Wieck; René Clair's *Le Million* with Annabella; Norman Z. McLeod's *Monkey Business* with the Marx Brothers; William Wellman's *Public Enemy* with James Cagney, Jean Harlow, Mae Clark (who gets half a grapefruit in her face); Normal Taurog's *Skippy* with Jackie Cooper; King Vidor's *Street Scene* with Sylvia Sidney, William Collier, Jr.; Archie Mayo's *Svengali* with John Barrymore; F.W. Murneau's documentary *Tabu* with Reri; Sergei Eisenstein's *Thunder Over Mexico*.

U.S. movie theaters show double features to boost business. Many unemployed executives spend their afternoons at the movies.

Radio *Little Orphan Annie* 4/6 over NBC Blue Network stations. Sponsored by the makers of Ovaltine and based on the comic strip that began in 1924, the show will continue until 1943 ("Who's that little chatterbox?/ The one with pretty auburn locks?/Who can it be?/It's Little Orphan Annie"); *The Easy Aces,* with comedian Goodman Ace, 32, and his wife Jane Epstein Ace whose *Easy Aces* will continue until 1945 and whose *Mr. Ace and Jane* will continue until 1955.

Congress votes March 3 to designate "The Star Spangled Banner" the U.S. national anthem (*see* 1814).

Kate Smith makes her radio debut May 1 singing "When the Moon Comes over the Mountain" on CBS stations. Kathryn Elizabeth Smith, 22, has played comic fat girl roles in Broadway shows, she has finally won a singing role at New York's Palace Theater, she begins a radio career that will continue for nearly half a century, she will sing Irving Berlin's "God Bless America" in a 1938 Armistice Day broadcast, and will acquire exclusive air rights to the song Berlin wrote originally for his 1918 show *Yip-Yip Yaphank* but laid aside.

Cantata *Belshazzar's Feast* for Baritone, Chorus, and Orchestra by William Walton 10/10 at Leeds.

First performances Concert Music for String Orchestra and Brass Instruments by Paul Hindemith 4/3 at Boston's Symphony Hall in a concert celebrating the Boston Symphony's 50th anniversary; "Amphion Suite" by Arthur Honegger 6/23 at the Paris Opéra; "Grand Canyon Suite" by Ferde Grofe 11/22 at Chicago's Studebaker Hall in a concert by Paul Whiteman and his Orchestra.

Ballet *Façade* 4/26 at London's Cambridge Theatre, with Antony Tudor, Alicia Makarova, Frederick Ashton, music by William Walton, choreography by Ashton (*see* 1923); *The Lady of Shalott* 11/12 at London's Mercury Theatre, with Frederick Ashton, music by Jean Sibelius, choreography by Ashton.

London's Royal Ballet has its beginnings in the Sadler's Wells Ballet founded by Nanette de Valois whose small opera ballet ensemble will grow to become Britain's national ballet company.

Broadway musicals *America's Sweetheart* 2/10 at the Broadhurst Theater, with music by Richard Rodgers, lyrics by Lorenz Hart, songs that include "I've Got Five Dollars," 135 perfs.; *Billy Rose's Crazy Quilt* 5/19 at the 44th Street Theater, with Rose's wife Fanny Brice, music by Harry Warren, lyrics by Rose and Mort Dixon, songs that include "I Found a Million Dollar Baby—in a Five and Ten Cent Store," 79 perfs.; *The Little Show* 6/1 at the Music Box Theater, with Beatrice Lillie, Ernest Truex, music and lyrics by Noël Coward, songs that include "Mad Dogs and Englishmen," written by Coward last year while motoring between Hanoi and Saigon, 136 perfs.; *The Band Wagon* 6/3 at the New Amsterdam Theater, with Fred and Adele Astaire in their last appearance together on the first revolving stage to be used in a musical, music by Arthur Schwartz, lyrics by Howard Dietz, songs that include "Dancing in the Dark," "I Love Louisa," 260 perfs.; *The Ziegfeld Follies* 7/1 at the Ziegfeld Theater, with Helen Morgan, Ruth Etting, Harry Richman, music by Walter Donaldson, Dave Stamper, and others, lyrics by E. Y. Harburg and others, 165 perfs.; *Earl Carroll's Vanities* 7/27 at the new 3,000-seat Earl Carroll Theater on 7th Avenue at 50th Street, with William Demarest, Lillian Roth, naked chorus girls, music mostly by Burton Lane plus Ravel's "Bolero," lyrics by Harold Adams, songs that include "Goodnight, Sweetheart" by English songwriters Ray Noble, James Campbell, and Reg Connelly, 278 perfs.; *George White's Scandals* 9/14 at the Apollo Theater, with Willie Howard, Rudy Vallée, Ray Bolger, Ethel Merman, music and lyrics by Ray Henderson and Lew Brown, songs that include "Life Is just a Bowl of Cherries," 202 perfs.; *Everybody's Welcome* 10/13 at the Shubert Theater, with Tommy and Jimmy Dorsey, Ann Pennington, Harriet Lake (Georgia Sothern), songs that include "As Time Goes By" by Herman Hupfield (see film *Casablanca*, 1942), 139 perfs.; *The Cat and the Fiddle* 10/15 at the Globe Theater, with Bettina Hall, Georges Metaxa, Eddie Foy, Jr., music by Jerome Kern, lyrics by Otto Harbach, songs that include "She Didn't Say 'Yes,'" "The Night Was Made for Love," 395 perfs.; *The Laugh Parade* 11/2 at the Imperial Theater, with Ed Wynn, music by Harry Warren, lyrics by Mort Dixon and Joe Young, songs that include "You're My Everything," 231 perfs.; *Of Thee I Sing* 12/26 at the Music Box Theater, with Victor Moore as Alexander Throttlebottom, William Gaxton as John P. Wintergreen, music by George Gershwin, lyrics by Ira Gershwin, songs that include "Love Is Sweeping the Country," "Wintergreen for President," "Who Cares?" and the title song, 441 perfs.

Popular songs "Mood Indigo" by Duke Ellington, lyrics by Albany Bigard, Irving Mills; "(Potatoes Are Cheaper—Tomatoes are Cheaper) Now's the Time to Fall in Love" by Al Sherman, lyrics by Al Lewis; "Heartaches" by Al Hoffman, lyrics by John Klenner; "Marta (Rambling Rose of the Wildwood)" by Moises Simons, lyrics by L. Wolfe Gilbert; "I Don't Know Why I Love You Like I Do" by Fred E. Ahlert, lyrics by Roy Turk; "Where the Blue of the Night Meets the Gold of the Day" by Ahlert and Turk (crooner Bing Crosby records the song and will make it his theme song); "All of Me" by U.S. songwriters Seymour Simon and Gerald Marks; "I Love a Parade" by Harold Arlen, lyrics by Ted Koehler; "When I Take My Sugar to Tea" by Sammy Fain, lyrics by Irving Kahal; "Sweet and Lovely" by Gus Arnheim, Harry Tobias, and Jules Lemare; "Take Me in Your Arms" by Fred Markush, lyrics by Mitchell Parish; "I Surrender, Dear" by Harry Burris of the Rhythm Boys, lyrics by Gordon Clifford; "Love Letters in the Sand" by J. Fred Coots, lyrics by Nick and Charles Kenny; "Out of Nowhere" by Edward Heyman and John Green; "(I'll Be Glad When You're Dead) You Rascal You" by Sam Theard; "That Silver Haired Daddy of Mine" by Texas-born Oklahoma radio "singing cowboy" Orvon "Gene" Autry, 24, who will sell more than 5 million copies of the song; "Got a Date with an Angel" by English composers Jack Waller and Joseph Tunbridge, lyrics by Clifford Grey and Jimmie Miller; "Sally" by English songwriters Harry Leon (Sugarman), Will E. Haines, and Leo Towers (for the film *Sally in Our Alley* with Gracie Fields); "Lady of Spain" by English composer Tolchard Evans, 30, lyrics by Stanley Demerell, 51, and Bob Hargreaves.

Knute Rockne dies in an airplane crash in Kansas March 31 at age 43 (*see* 1913). He took over as head coach at Notre Dame in 1918 and his teams have won 105 games and tied 5 out of 122.

Sidney Burr Beardsley Wood, 19, (U.S.) wins in men's singles at Wimbledon, Cilly Aussem, 22, (Ger) in women's singles; H. Ellsworth Vines, Jr., 19, in men's

1931 ***(cont.)*** singles at Forest Hills, Helen Wills Moody in women's singles.

Francis Ouimet wins the U.S. Golf Association Amateur.

The St. Louis Cardinals win the World Series by defeating the Philadelphia Athletics 4 games to 3.

The Nevada State Legislature legalizes gambling. California entrepreneur Raymond Smith opens Harolds Club in Reno's Virginia Street, naming the gambling casino after one of his sons; he advertises Harolds Club with billboards along western highways, he begins a thriving enterprise.

Starr Faithfull, 25, makes lurid headlines June 8 when her body—clad only in a silk dress—washes up on the beach at Long Beach, N.Y. The beautiful woman has been dead since June 5, suicide notes are produced some weeks later, but the cause of her death will remain a mystery.

Clairol hair dye is introduced by U.S. chemists' broker Lawrence Gelb, 33, who has acquired the formula in Europe (*see* Nestle Colorinse, 1929). Gelb builds a research laboratory and plant at Stamford, Conn., and tries to create a U.S. market for hair dyes (*see* Miss Clairol, 1950).

Lucky Strike cigarettes outsell Camels for the first time as American Tobacco challenges the lead of R. J. Reynolds (*see* Hill, 1925). The lead will alternate between the two brands for the next 20 years as cigarette advertising helps build U.S. radio and radio encourages cigarette use.

Prohibition is not working, reports President Hoover's Wickersham Committee headed by former Attorney General George Woodward Wickersham, 73.

U.S. underworld boss Salvatore Maranzano is assassinated September 10 in a murder engineered by Italian-American Charles "Lucky" Luciano, 34, who restructures the U.S. Mafia into a federation of "families" (*see* 1936).

Chicago mobster Al Capone is found guilty of evading $231,000 in income taxes and is sentenced by a Chicago federal court October 24 to 11 years' imprisonment and a $50,000 fine (*see* 1939).

New Delhi opens in India. The planned capital has been designed by English architects Sir Edwin Landseer Lutyens, 62, and Sir Herbert Baker, 69, who have taken some inspiration from Christopher Wren and Pierre L'Enfant.

New York's Chrysler Building opens at the corner of Lexington Avenue and 42nd Street and is for a few months the tallest building in the world. The 77-story, 1,048-foot structure of steel and concrete is capped with a stainless steel crown made by Fried. Krupp of Essen.

New York's Empire State Building opens April 30 on the site formerly occupied by the Waldorf-Astoria Hotel of 1897 at Fifth Avenue and 34th Street. Lamb & Harmon have designed the 102-story skyscraper that will be the world's tallest for more than 40 years but it has trouble finding tenants.

New York's Waldorf-Astoria Hotel opens October 1 on a full-block site leased from the New York Central Railroad from 49th to 50th Street between Park and Lexington avenues. The 2,000-room hotel by Schultze & Weaver has twin 47-story towers and will lose money for 12 years.

London's Dorchester Hotel opens in Park Lane.

Rockefeller Center construction begins on a three-block midtown Manhattan site leased by John D. Rockefeller in 1928 with the intention of building a new Metropolitan Opera House. The 23-year lease given by Columbia University will be renewed periodically and the university will receive rent on the $125 million office building development until 1985 (when the Rockefeller family will buy the land for $400 million).

Philadelphia's Convention Hall is completed at 34th Street and Vintage Avenue with seats for 13,000 and room on stage for another 1,000. U.S. architect Philip Cortelyou Johnson, 25, has designed the Renaissance-style hall.

The U.S. Public Health Service issues a study early in the year reporting that "the contamination of the atmosphere by smoke from the chimneys of private houses, office buildings, industrial plants, steam engines, tugs and steamships has become a serious matter in several of the larger cities. . ."; but while smoke cuts off an estimated one-fifth of the natural light of New York City, the chief concern in most places is that factories are idle and not belching forth the smoke that came with prosperity.

China's Yangzi River bursts a dam during a typhoon August 3, flooding 40,000 square miles, killing hundreds, and causing widespread famine that will kill more (*see* Chiang Kai-shek, above).

Some 75 percent of Soviet farms remain in private hands despite the Five-Year Plan begun by Josef Stalin in 1928 to encourage collective farms and state farms. Liquidation of *kulaks* is intensified.

Brazil establishes a National Coffee Department. Collapse of the world coffee market has brought economic disaster and has helped precipitate a revolt in the southern provinces. The Coffee Department will supervise the destruction of large quantities of Brazil's chief export item in order to maintain good prices in the world market (*see* Nescafé, 1938).

The U.S. wheat crop breaks all records, driving down prices. Many Kansas counties declare a moratorium on taxes to help farmers survive, but elsewhere farmers are forced off the land as banks foreclose on mortgages (*see* Farm Mortgage Financing Act, 1934).

The United States has 7.2 million operating farms, up from some 6.3 million last year, with more than a billion acres of farmland in production and with the world's highest rate of cultivated food crop acreage per capita. By 1935 the number of farms will be down to 6.84 million and by 1969 3.45 million.

A British Advisory Committee on Nutrition is organized to address the problem of poor nutrition among the poor. It devises new marketing schemes for farmers to put food within reach of more Britons and a Milk in Schools Scheme subsidizes milk so that children may have it at cheap prices (*see* 1906; 1934).

Swiss chemist Paul Karrer, 42, isolates vitamin A (*see* 1920; 1925; 1937).

The Supreme Court upholds a company that has been sued by the FTC for false advertising (*FTC v. Raladam*). The company has advertised its thyroid-extract product, Marmola, as "a scientific remedy for obesity," and while the Court agrees with the American Medical Association May 25 that the ads are dangerously misleading it rules that such advertising is not unfair to competition (*see* Clayton Act, 1914; Whealer-Lea Act, 1938).

Daniel Gerber's Fremont Packing Co. adds salt to its baby foods (*see* 1929). Infants' taste buds are not sufficiently developed to discern any difference but sales resistance has come from mothers who taste their babies' food (*see* 1941; MSG, 1951).

U.S. mechanical refrigerator production tops one million units, up from 5,000 in 1921. By 1937 the industry be producing refrigerators at the rate of nearly 3 million per year (*see* 1929).

Frigidaire division of General Motors adopts Freon 12 (dichlorodifluoromethane) refrigerant gas, invented by Thomas Midgley of Ethyl Corp. and C. F. Kettering of GM and produced by a Du Pont-GM subsidiary. Most other refrigerator makers follow suit, replacing ammonia and other more dangerous gases.

Birds Eye Frosted Foods go on sale across the United States as General Foods expands distribution, but only a few retail grocers have freezer cases for displaying frozen foods (*see* 1930; 1940).

Bisquick, introduced by General Mills, is a prepared mix for biscuits containing sesame seed oil to keep its flour and shortening from turning rancid; the product competes with Jiffy brand biscuit mix introduced last year by Chelsea Milling.

Ballard & Ballard of Louisville, Ky., introduces Ballard Biscuits made from refrigerated dough and packaged under pressure in cylindrical containers.

Adolph Butenandt isolates the male sex hormone androgen (*see* 1929).

U.S. birth rates and immigration to the United States will decline in this decade for economic reasons.

The U.S. population reaches 124 million. Britain has 46 million, Germany 65, France 42, Italy 41, the Dutch East Indies 61, Japan 65, Soviet Russia 148, India 338, China perhaps 440.

1932

"The United States cannot admit the legality nor does it intend to recognize" the legitimacy of any arrangement with Japan which impairs Chinese sovereignty and threatens the Open Door Policy, Secretary of State Henry L. Stimson announces January 7, but the Stimson Doctrine has no practical effect (*see* 1931). China's boycott of Japanese goods continues and Japan retaliates January 29. Troops landed from warships attack Chapei, the Chinese district of Shanghai, and planes bomb the district, killing thousands in the first terror bombing of civilians. Japanese publicists suggest that any U.S. attempt to interfere with Japan's "destiny" in Asia would be cause for war.

The assassination of Japan's prime minister Ki Tauyoshi Inukai, 75, by a reactionary May 15 effectively ends party government. The former governor-general of Chōsen (Korea), Viscount Makoto Saito, 73, succeeds Inukai.

Latin-America's first Communist revolt, in January, follows the military overthrow of El Salvador's President Arturo Araujo, who was freely elected last year on a platform to reform the nation's feudal system. He has upset oligarchs and the army that have so long dominated El Salvador. His vice president, Gen. Maximiliano Hernandez Martinez, suppresses the rebellion, leaving 10,000 to 30,000 dead in what critics will call the *matanzas*, or butchery.

The British Union of Fascists is founded by former Labour party M.P. Sir Oswald Mosley, 36, who last year quit the party in protest against its defeatist attitude toward unemployment and started a new party based on principles more socialistic than fascist. Mosley will become a supporter of Italy's Benito Mussolini and Germany's Adolf Hitler and will demand expulsion of Britain's Jews (*see* 1936; 1940).

Eamon de Valera is elected president of Ireland in March and suspends Irish land annuity payments (*see* below).

France's President Paul Doumer is assassinated by a Russian émigré May 6 and is succeeded by Albert Lebrun. May elections give leftist parties a majority; Edouard Heriot begins a second ministry but resigns in December over refusal by the Chamber to support his government's proposal to pay installments on France's war debt to the United States.

Germany's President Hindenburg asks Franz von Papen to form a government May 31, von Papen does so but excludes Nazis, voters in the general election July 31 make the National Socialist (Nazi) party the biggest in the Reichstag but with no overall majority, Adolf Hitler announces August 13 that he will not serve as vice chancellor under von Papen, Nazi leader Herman Goering, 39, is elected president of the Reichstag August 30, von Papen resigns as prime minister November 17 (*see* 1933).

Siam's absolute government ends June 24 as European-educated radicals capture Rama VII (Prajadhipok) and hold him captive briefly until he agrees to a constitution and the establishment of a senate (*see* 1935).

A Swedish socialist government comes to power September 24 with Per Albin Hansson, 47, as prime minister. He will continue as prime minister until his death in 1946, and the Social Democratic party will retain power until 1976.

1932 ***(cont.)*** "I pledge you, I pledge myself, to a new deal for the American people," says New York governor Franklin D. Roosevelt as he accepts the Democratic party nomination for president at Chicago. He wins election by a landslide, gaining 472 electoral votes and 57 percent of the popular vote versus 59 electoral votes and 40 percent of the popular vote for President Hoover, who carries only six states as economic depression worsens.

New York City's playboy mayor James John "Jimmy" Walker, 51, has resigned September 1 during an investigation of corruption by a state legislative commission headed by Judge Samuel Seabury, 59.

Mahatma Gandhi begins a "fast unto death" to protest the British government's treatment of India's lowest caste "untouchables" whom Gandhi calls *harijans*—"God's children." Gandhi's campaign of civil disobedience has brought rioting and has landed him in prison, but he persists in his demands for social reform, he urges a new boycott of British goods, and after 6 days of fasting obtains a pact that improves the status of the "untouchables" (*see* 1930; 1947).

Dearborn, Mich., police fire into a crowd of 3,000 men, women, and children demonstrating outside the Ford Motor Company plant March 7. Four are killed, 100 wounded, and the wounded are handcuffed to their hospital beds on charges of rioting.

The Norris-La Guardia Act passed by Congress March 23 prohibits the use of injunctions in labor disputes except under defined conditions and outlaws "yellow-dog contracts" that make workers promise not to join any labor union. Sponsored by George William Norris (R. Neb.), now 70, and Fiorello Henry La Guardia (R. N.Y.), 39, the bill helps establish labor's right to strike, picket, and conduct boycotts.

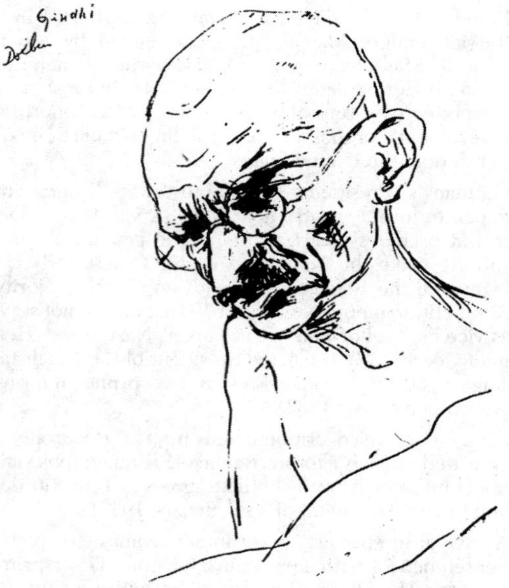

Mahatma Gandhi in India defied the British raj with "passive resistance" that would culminate in independence.

David Dubinsky (*né* Dobnievski) becomes president of the International Ladies' Garment Workers Union (ILGWU) at age 40. The Polish-American organizer launches a membership drive that will triple the union's rolls in 3 years.

"Bonus Marchers" descend on Washington beginning in May as some 25,000 poverty-stricken World War veterans demonstrate to obtain "bonuses" authorized by the Adjustment Compensation Act of 1924 but not due until 1945. Hoping to get roughly $500 each, the veterans camp out with wives and children in the city's parks, dumps, empty stores, and warehouses. *Baltimore Sun* reporter Drew Pearson, 34, sees "no hope on their faces."

Army tanks, gas grenades, cavalry, and infantry armed with machine guns and bayonetted rifles disperse the demonstrators (above) from their main camp on Anacostia Flats and elsewhere in an operation commanded by U.S. Army Chief of Staff Gen. Douglas MacArthur, 52, Major Dwight David Eisenhower, 41, Major George Smith Patton, Jr., 45, and other officers. The July 28 violence produces 100 casualties, but President Hoover maintains that the Bonus Marchers are "communists and persons with criminal records" rather than veterans.

Josef Stalin cracks down on *kulaks* in the Ukraine and Caucasus who resist collectivization. He sends in troops to requisition all foodstuffs and prevents trains carrying food from reaching the areas (*see* below).

A tariff war between Britain and Ireland begins in July (*see* 1922). The loss of the country's chief export market brings a collapse of its cattle industry and worsens its economic depression (*see* 1935).

Britain abandons free trade for the first time since 1849. She imposes a 10 percent tariff on most imported goods but agrees at the Imperial Economic Conference at Ottawa to exempt Canada, Australia, New Zealand, and other Commonwealth nations, which in turn will provide markets for Britain's otherwise uncompetitive textiles, steel, motorcars, and telecommunications equipment, discouraging innovation in many industries (*see* agriculture, below).

Germany has 5.6 million unemployed, Britain 2.8 million.

The average U.S. weekly wage falls to $17, down from $28 in 1929. "Breadlines" form in many cities.

Some 1,616 U.S. banks fail, nearly 20,000 business firms go bankrupt, there are 21,000 suicides, and expenditures for food and tobacco fall $10 billion below 1929 levels.

U.S. industrial production drops to one-third its 1929 total, and the U.S. Gross National Product (GNP) sinks to $41 billion, just over half its 1929 level.

Wall Street's Dow Jones Industrial Average plummets to 41.22 by July 7, down from its high of 381.17 September 3, 1929. It is the nadir.

Congress has authorized a Reconstruction Finance Corp. January 22 to help finance industry and agriculture in accordance with President Hoover's request of last year.

An Emergency Relief and Reconstruction Act passed by Congress July 21 gives the RFC (above) power to lend $1.8 billion to the states for relief and self-liquidating public works projects.

A Home Loan Act passed by Congress July 22 establishes 12 federal home loan banks that will lend money to mortgage loan institutions. The measure is designed to rescue the banks that are being forced to close. The controller of the currency orders a moratorium on first-mortgage foreclosures August 26.

Prominent U.S. intellectuals endorse communism, saying that only the Communist party has proposed a real solution to the nation's problems. Endorsers include Sherwood Anderson, Erskine Caldwell, John Dos Passos, Theodore Dreiser, Waldo Frank, Granville Hicks, Sidney Hook, Matthew Josephson, and Lincoln Steffens.

Chicago utilities magnate Samuel Insull, now 73, runs into financial difficulties, and three of his largest companies go into receivership. Once Thomas Edison's private secretary, Insull is indicted on charges related to his activities as president of Chicago Edison, Commonwealth Edison, Peoples Gas Light and Coke, and other companies, but he will avoid arrest for 2 years and be acquitted after trials in 1934 and 1935.

The AF of L reverses its long-standing position against unemployment insurance and urges that work be spread through a 30-hour week with some "economic planning" by the federal government.

U.S. unemployment reaches between 15 and 17 million by year's end, 34 million Americans have no income of any kind, and Americans who do work average little more than $16 per week.

Splitting the atom for the first time in history presents the possibility of a vast new energy source. English physicist John Douglas Cockroft, 35, and his associate Ernest Walton use a voltage multiplier they developed in the 1920s to accelerate charged subatomic particles to extremely high velocities. They use this "atomic gun" to bombard lithium with protons, and the alpha particles (helium nuclei) they produce show that the protons have reacted with the lithium nuclei to produce helium (*see* Thomson, 1897; Joliot-Curie, 1934).

The neutron, discovered by English physicist James Chadwick, 41, is a particle similar in mass to the proton but without an electric charge (*see* electron, 1897). Since it is neutral, the neutron easily penetrates atoms and permits efficient splitting of atomic nuclei for developing atomic reactors.

University of California physicist Ernest Orlando Lawrence, 31, builds the world's first practical cyclotron—an accelerator that propels atomic particles in spiral paths by use of a constant magnetic field.

Out of work and out of funds: the grim realities of the Great Depression made radical proposals sound appealing.

Oklahoma City lawyer Robert Samuel Kerr, 35, and his brother-in-law James L. Anderson drill oil wells within the city limits. They have acquired control of a contract drilling firm, borrowed money to post $200,000 liability bonds in a gamble that large oil companies have refused to take, and will make a $2 million profit on the venture (*see* Kerr-McGee, 1937).

Standard Oil of California (SoCal) prospectors in Bahrain strike oil in early June with their first well in the British Persian Gulf protectorate. Production will reach 20,000 barrels per day by 1936, and SoCal will take in the Texas Company as an equal partner to avail itself of Texaco's marketing facilities in the Far East (*see* 1903; 1933).

U.S. motorcar sales fall to just over one million, down from more than 5 million in 1929.

Sales of Ford passenger cars to farmers fall to 55,000, down from 650,000 in 1929.

Ford halts production of its Model A, introduced in 1927, as it tools up to introduce the first low-priced V8. Ford loses millions of dollars and lays off workers to reduce payroll costs from $145 million (1929) to $32 million (1933).

The Hudson Terraplane is introduced with streamlined styling (*see* 1922; AMC, 1954).

Some 25 percent of U.S. auto glass is safety glass. Many states enact legislation requiring it in windshields (*see* Sloan, 1929).

General Motors forms a subsidiary to acquire electric streetcar companies, convert them to GM motorbus operation, and resell them to local entrepreneurs who will agree to buy only GM buses as replacement vehicles (*see* New York, 1936; conspiracy conviction, 1949).

1932 ***(cont.)*** U.S. Route 66 opens to link Chicago and Los Angeles with a 2,200-mile continuous highway that will be called the "Main Street of America." Soon lined with motor courts, Burma-Shave signs, two-pump service stations, and curio shops, Route 66 carries truckers and motorists west via St. Louis, Joplin, Oklahoma City, Amarillo, Gallup, Flagstaff, Winona, Kingman, Barstow, and San Bernardino.

Oklahoma City News editor Carl C. Magee files the first patent application for a parking meter in December. He will be granted rights in 1936, 174 Dual Park-O-Meters will be installed in Oklahoma City in July 1935, 300 more will be ordered when their success is demonstrated (police officers will have to explain that no jackpots can be expected by those who deposit coins), and meters of improved design will appear in streets of major world cities in the next few decades.

Washington's Arlington Bridge across the Potomac is completed by McKim, Mead and White of New York.

London's Lambeth Bridge of 1862 is replaced by a new bridge across the Thames.

Sydney's Harbour Bridge opens March 18, connecting the city with its suburban areas. It is the world's longest arch bridge.

Bridgestone Tire Co. is founded by Japanese entrepreneur Shojiro Ishibashi (the name means "stone bridge") who began by making rubber work shoes.

Europe's Simplon-Orient Express becomes the Arlberg-Orient Express as it inaugurates a new route closer to the original route of 1883. The six-mile Arlberg Tunnel avoids the long detour into Italy, taking the luxury train from Switzerland eastward to Austria and thence to Istanbul *(see* 1919; 1952).

New York's Independent subway system opens September 10, extending service into areas not served by the BMT or IRT (*see* 1933).

The Welland Sea Canal of 1829 opens to connect Lake Erie and Lake Huron after having been rebuilt to accommodate vessels up to 600 feet in length with a 22-foot draft.

Aircraft designer Lloyd Carl Stearman, 33, becomes president of a revived Lockheed Co. after it is acquired in bankruptcy for $40,000 by new investors (*see* Northrop, Vultee, 1928). Stearman worked in partnership with Clyde Cessna and Walter Beech in 1924 to produce the Travel Air, a pioneer civilian production plane, and after being nearly wiped out in 1929 merged his company at Wichita with United Aircraft, which produces his training planes.

The Benzedrine Inhaler introduced by Smith Kline & French is a nasal decongestant whose active ingredient is amphetamine. It will be used in treating hyperkinetic children, prescribed for obesity, and abused as "speed."

The first-class U.S. postal rate goes to 3¢ in July. It was 3¢ from 1917 to 1928 but then returned to 2¢ (*see* 1885; 1958).

The *Family Circle* begins publication at Newark, N.J., early in September with 24 pages. The first magazine to be distributed exclusively through grocery stores, the weekly is given free to shoppers at two eastern chains and will have a circulation of 1.44 million by 1939 (*see* 1946; *Woman's Day,* 1937).

BBC takes over the responsibility of developing television from the Baird Co. *(see* 1926; 1927; 1936).

The *Walter Winchell Show* begins on U.S. radio December 4 featuring *New York Daily Mirror* columnist Walter Winchell, 35: "Good evening, Mr. and Mrs. America and all the ships at sea. . ."

Nonfiction *Death in the Afternoon* by Ernest Hemingway indulges the novelist's obsession with bullfighting.

Fiction *Laughter in the Dark* by Russian novelist Vladimir Nabokov, 33; *Brave New World* by Aldous Huxley, who projects a world in the "Year of Our Ford" when people will go not to the movies but to the "feelies," where men will be attended by "pneumatic" girls (a word borrowed from T. S. Eliot's poem "Whispers of Immortality") and human reproduction will be controlled by the state; *A Lesson in Love (La Naissance du jour)* by Colette; *Salavin* by George Duhamel; *Radetzkymarsch* by Joseph Roth; *The Knot of Vipers (Le Noeud de Viperes)* by François Mauriac; *Journey to the End of the Night (Voyage au bout de la nuit)* by French physician-novelist Louis-Ferdinand Céline (Henri-Louis Destouches), 38; *Little Man, What Now? (Kleiner Mann—was nun?)* by German novelist Hans Fallada (Rudolf Ditzen), 30; *The Memorial* by Christopher Isherwood; *Snooty Baronet* by Wyndham Lewis; *Black Mischief* by Evelyn Waugh; *Poor Toni* by Scottish novelist-poet-critic Edwin Muir, 45; *Limits and Renewals* (stories) by Rudyard Kipling; *Light in August* by William Faulkner; *Tobacco Road* by U.S. novelist Erskine Caldwell, 28, who depicts the depravity of poor whites in Georgia's sharecropper society; *The Case of the Velvet Claws* by California lawyer Erle Stanley Gardner, 43, who started writing occasional detective stories in 1921, has been averaging more than a million words of pulp magazine writing each year since 1928, and will follow his Perry Mason detective novel with dozens in the same vein; *Death Under Sail* by English physicist-novelist C. P. (Charles Percy) Snow, 27; *The Crime of Inspector Maigret* by Franco-Belgian novelist Georges Simenon, 29.

Juvenile *Little House in the Big Woods* by Missouri novelist Laura Ingalls Wilder, 65, who begins a series of eight volumes that will appear in the next 11 years to recount her girlhood in the Midwest of the late 19th century; *L'Histoire de Babar (The Story of Babar, the Little Elephant)* by French artist-writer Jean de Brunhoff, 33, whose book will sell 50,000 copies in French before being translated into English. De Brunhoff will die of tuberculosis in 1937 but his book and its sequel, *Travels of Babar,* will be followed by further Babar adventures produced by his son.

Poetry "Anna Livia Plurabelle" by James Joyce, whose daughter Lucia is troubled by mental disease that will be diagnosed as schizophrenia; *Poems* by Stephen Spender; *Late Arrival* (*Sent pa jorden*) by Swedish poet Gunnar Ekelöf, 25.

Painting *Girl Before a Mirror* by Pablo Picasso; *Christ Mocked by Soldiers* by Georges Rouault; *Sacré Coeur* by Maurice Utrillo; *Professor Sauerbruch* by Max Liebermann, now 85; *Sacco and Vanzetti* by Lithuanian-American painter Ben (Benjamin) Shahn, 33, who produces 23 gouaches inspired by the execution in 1927; *The Bowery* by New York painter Reginald Marsh, 34; *Daughters of the American Revolution* by Grant Wood; *Butterfly Chaser* by Thomas Hart Benton.

Sculpture *Head of a Woman* by Pablo Picasso. Flushing, Queens, sculptor Joseph Cornell, 28, exhibits his first boxes containing found objects.

Motorized and hand-cranked "stabiles" by U.S. sculptor-painter Alexander Calder, 34, create a stir in Paris. "Le Cirque Calder" exhibited in 1926 consisted of small wire sculptures of animals and acrobats; dadaist-surrealist Jean Arp has suggested the word "stabile." Calder follows his stabiles with hanging "mobiles" that move with air currents (Marcel Duchamp has suggested the name).

Polaroid film, invented by Harvard College dropout Edwin Herbert Land, 23, is the world's first synthetic light-polarizing film. Land opens the Land-Wheelwright Laboratories at Boston with former Harvard physics instructor George Wheelwright and will produce polarizing filters for cameras beginning in 1935 (*see* sunglasses, 1936; camera, 1947).

Theater *The Animal Kingdom* by Philip Barry 1/12 at New York's Broadhurst Theater with William Gargan, Leslie Howard, Ilka Chase, 183 perfs.; *There's Always Juliet* by English-American playwright John Van Druten, 31, 2/15 at New York's Empire Theater, with Edna Best, Herbert Marshall, 108 perfs.; *Too True to Be Good* by George Bernard Shaw 4/4 at New York's Guild Theater, with Beatrice Lillie, Hope Williams, Leo G. Carroll, Claude Rains, 57 perfs.; *Another Language* by Rose Franken 4/25 at New York's Booth Theater, with Margaret Hamilton, Margaret Wycherly, Dorothy Stickney, 344 perfs.; *The Ermine* (*L'ermine*) by French playwright Jean Anouilh, 21, 4/26 at the Théâtre de l'oeuvre, Paris; *Dinner at Eight* by George S. Kaufman and Edna Ferber 10/22 at New York's Music Box Theater, with Constance Collier, 232 perfs.; *Autumn Crocus* by U.S. playwright C. L. Anthony 11/19 at New York's Morosco Theater, with Patricia Collinge, Francis Lederer, 210 perfs.; *Biography* by S. N. Behrman 12/12 at New York's Guild Theater, with Earle Larimore, Ina Claire, 283 perfs.

Washington's Folger Library opens. The great Shakespearean collection has been funded by the late Standard Oil Co. chairman Henry Clay Folger who died 2 years ago at age 73.

Radio comedy *The Jack Benny Show* 5/2 on NBC stations with violinist-comedian Benny (*né* Benjamin Kubelsky), 38, who has toured in vaudeville (originally as Ben K. Benny), appeared last year in *Earl Carroll's Vanities*, and begins a radio show that will continue for 23 years with the help of Benny's wife Mary Livingston, Don Wilson, Dennis Day, Eddie "Rochester" Anderson, and scriptwriters who will work endless variations on the themes of Benny's stinginess, his Maxwell car, his violin playing, and his age. Perpetually 39, Benny will later move to CBS, will go on television in 1955 and continue for another 10 years; *The Fred Allen Show* 10/23 on CBS stations with former vaudeville juggler Allen (*né* John Florence Sullivan), 38. His cast will include his wife Portland Hoffa, he will have a half-hour show beginning 6/28/42 and will continue until 6/29/49, when guests will include Jack Benny (above).

Radio serials *One Man's Family* 4/28 on West Coast NBC stations (to 6/5/1950); *Buck Rogers in the Twenty-Fifth Century* 11/7 on CBS stations (to March 1947).

Films Edmund Goulding's *Grand Hotel* with Greta Garbo, Joan Crawford, John and Lionel Barrymore, Lewis Stone, Wallace Beery, Jean Hersholt; Mervyn LeRoy's *I Am a Fugitive From a Chain Gang* with Paul Muni; Merian C. Cooper and Ernest Schoedsack's *King Kong* with Fay Wray, Bruce Cabot; Ernst Lubitsch's *Trouble in Paradise* with Miriam Hopkins, Kay Francis, Herbert Marshall. Also: George Cukor's *A Bill of Divorcement* with John Barrymore, Katharine Hepburn (Bryn Mawr '28); Josef von Sternberg's *Blonde Venus* with Marlene Dietrich, Herbert Marshall, English-born actor Cary Grant (*né* Archibald Alexander Leach), 28; Clarence Brown's *Emma* with Marie Dressler, Jean Hersholt; Marc Allegret's *Fanny* with Raimu, Pierre Fresnay, Charpin; Norman Z. McLeod's *Horse Feathers* with the Marx Brothers; Edward Cline's *Million Dollar Legs* with W. C. Fields, Jack Oakie; Tay Garnett's *One-Way Passage* with Kay Francis, William Powell; Howard Hawks's *Scarface* with Paul Muni, Ann Dvorak.

Stage musicals *Face the Music* 2/17 at New York's New Amsterdam Theater, with Mary Boland, music and lyrics by Irving Berlin, songs that include "Let's Have Another Cup of Coffee," 165 perfs.; *Words and Music* 9/16 at London's Adelphi Theatre, with Joyce Barbour, Ivy St. Helier, John Mills, music and lyrics by Noël Coward, songs that include "Mad About the Boy," "Journey's End," "Let's Say Goodbye," 164 perfs.; *Earl Carroll's Vanities* 9/27 at New York's Broadway Theater, with Milton Berle, Helen Broderick, music by Harold Arlen and Richard Myers, lyrics by Ted Koehler and Edward Heyman, songs that include "I Gotta Right to Sing the Blues," 87 perfs.; *Americana* 10/5 at New York's Shubert Theater, with music by Jay Gorney, Harold Arlen, Richard Myers, Herman Hupfield, lyrics by E. Y. Harburg, songs that include "Brother, Can You Spare a Dime" (music by Gorney), 77 perfs.; *Music in the Air* 11/8 at New York's Alvin Theater, with Al Shean, Walter Slezak, music by Jerome Kern, lyrics by Oscar Hammerstein II, songs that include "I've Told Every Little Star," 334 perfs.; *Take a Chance* 11/26 at New York's Apollo Theater, with Jack Haley, Ethel Merman, Sid Silvers, music by Nacio Herb Brown, Vincent Youmans, and Richard Whiting, lyrics by B. G. DeSylva, songs that include "You're an Old Smoothie," 243 perfs.; *Gay Divorce* 11/29 at New York's Ethel Barrymore Theater, with Fred Astaire, Claire Luce, music and lyrics by Cole Porter, songs that include "Night and Day," 248 perfs.; *Walk a Little Faster* 12/7

1932 *(cont.)* at New York's St. James Theater, with Beatrice Lillie, Bobby Clark, music by Vernon Duke, lyrics by E. Y. Harburg, songs that include "April in Paris," "That's Life," 119 perfs.

Ballet *Cotillon* 4/12 at Monte Carlo's Théâtre de Monte Carlo, with the new Ballets Russe de Monte Carlo organized by George Balanchine, music by the late Emmanuel Chabrier; *Jeux d'Enfants (Children's Games)* 4/14 at Monte Carlo, with Irina Baronova, Tamara Tommanova, David Lichine, music by the late Georges Bizet, choreography by Leonide Massine, scenery and costumes by Joan Miró.

Hollywood musicals Frank Tuttle's *The Big Broadcast* with Kate Smith, George Burns and Gracie Allen, the Mills Brothers, the Boswell Sisters, Cab Calloway, and Bing Crosby, who begins regular radio broadcasts; Rouben Mamoulian's *Love Me Tonight* with Maurice Chevalier, Jeanette MacDonald, music by Richard Rodgers, lyrics by Lorenz Hart, songs that include "Lover," "Mimi," "Isn't It Romantic."

New York's Radio City Music Hall opens December 27 in Rockefeller Center with 6,200 seats. A mighty Wurlitzer organ and a 100-piece orchestra accompany vaudeville acts presented by showman Samuel "Roxy" Rothafel, but the world's largest indoor theater fails to attract crowds and will not gain success until it renames its high-kicking "Roxyette" chorus girls "Rockettes" and shows motion pictures.

First performances Four Orchestral Songs by Arnold Schoenberg 2/21 at Frankfurt-am-Main; Suite for Flute and Piano by Walter Piston 4/30 at Trask's Yaddo outside Saratoga Springs, N.Y.; *From the Gayety and Sadness of the American Scene* by U.S. composer Roy Harris, 34, 12/29 at Los Angeles.

Popular songs "Say It Isn't So" and "How Deep Is the Ocean" by Irving Berlin; "I Don't Stand a Ghost of a Chance with You" by Victor Young, lyrics by Bing Crosby and Ned Washington; "I'm Getting Sentimental Over You" by George Bassman, lyrics by Ned Washington; "(I'd Love to Spend) One Hour With You" by Richard Whiting, lyrics by Leo Robin; "Don't Blame Me" by Jimmy McHugh, lyrics by Dorothy Fields; "Shuffle Off to Buffalo" by Al Dubin and Harry Warren; "It Don't Mean a Thing if It Ain't Got that Swing" by Duke Ellington, lyrics by Irving Mills; "Willow Weep for Me" by Ann Ronell; "Maria Elena" by Mexican songwriter Lorenzo Barcelata.

Ellsworth Vines wins in men's singles at Wimbledon, Helen Wills Moody in women's singles. Vines wins in men's singles at Forest Hills, Helen Hull Jacobs, 23, in women's singles.

The Olympic Games at Los Angeles attract 2,403 contestants from 39 countries. California athlete Mildred "Babe" Didrikson, 18, wins the javelin throw and sets a new record of 11.7 seconds for the 80-meter hurdles.

Jack Sharkey regains the world heavyweight boxing title June 21 by a 15-round decision over Max Schmeling at the new Garden Bowl in Long Island City, New York. "We wuz robbed!" cries Schmeling's manager Joe Jacobs.

Pitcher J. H. "Dizzy" Dean plays his first full season with the St. Louis Cardinals to begin a brief but dazzling career with the Gas House Gang. Dean, 22, will lead the National League in strikeouts for the next 3 years.

Ducky Medwick joins the Cardinals (above) to begin an 8-year career with the team. Joseph Michael Medwick, 20, will be traded to the Brooklyn Dodgers in 1940, then to the New York Giants, then to the Boston Braves, then back to the Dodgers before retiring in 1948 with a lifetime batting average of .348

The New York Yankees win the World Series by defeating the Chicago Cubs 4 to 0. With the third game tied 4 to 4 October 1 and two strikes against him, Babe Ruth leers at his hecklers, points to the flagpole at the right of the scoreboard in center field, and hits the next pitch out of Wrigley Field.

The Washington Redskins professional football team has its beginnings in the Boston Braves team organized by George P. Marshall.

Revlon is founded by New York cosmetics salesman Charles Revson, 26, with his brother Joseph, 28, and chemist Charles Lachman, 35. When employer Elka Cosmetics rejects his ultimatum that he be made national distributor, young Revson rents a loft in the New York garment district, borrows money at 2 percent per month, and with Lachman's help develops a superior opaque nail enamel which he promotes with exotic names such as Tropic Sky rather than with the descriptive identifications dark red, medium red, pink, etc., that have been traditional. Focusing on beauty salons, Revlon uses intimidation to obtain distribution. Volume for the first 10 months is only $4,055.09 but Revson will start selling through drugstores in 1937, employing salesmen who "accidentally" destroy displays set up by the competition; by 1941 it will have a virtual monopoly on beauty salon sales.

The Zippo lighter designed by Blaisdell Oil Co. owner George Grant Blaisdell, 37, is introduced by Zippo Manufacturing of Bradford, Pa.

The kidnapping of Charles A. Lindbergh, Jr., March 1 makes world headlines. The aviation pioneer pays a $50,000 ransom for his child's safe return, but the body of the 20-month old infant is found 2 months later on Lindbergh's New Jersey estate while 5,000 federal agents comb the country (*see* 1934).

The suicide of Swedish "match king" Ivar Kreuger, 51, at Paris March 12 reveals a gigantic stock swindle that rocks the financial world and bankrupts the Boston banking house Lee, Higginson & Co.

Philadelphia's Savings Fund Society building at 12 South 12th Street, completed by George Howe and William Lescaze, is the first major international-style skyscraper to be built in the United States. The 36-story structure has 32 office floors which are visually separate from a vertical circulation spine, floor and spine floating on a curved-corner base.

Nebraska's State Capitol building is completed at Lincoln to designs by the late Bertram Grosvenor Goodhue, who died in 1924 at age 55.

A dike closed May 28 after 9 years of work reclaims the Zuider Zee (Ijsselmeer) to create new Dutch farmland (*see* 1918).

"Every European nation has a definite land policy and has had one for generations," Franklin Roosevelt tells the Democratic convention (above). "We have none. Having none, we face a future of soil erosion and timber famine . . ." (*see* Civilian Conservation Corps, 1933; Soil Erosion Service, 1933).

Waterton-Glacier International Peace Park, created by the United States and Canada, embraces 22-year-old Glacier National Park.

Britain imposes tariffs and quantitative restrictions on many farm imports while subsidizing British farmers to help them survive the Depression. Imperial preferences are introduced to favor imports from colonial and Commonwealth countries, with special preferences given to dairy products, meat, and wheat from Australia, Canada, and New Zealand at the expense of Denmark and Argentina.

U.S. farm prices fall to 40 percent of 1929 levels. Wheat drops to below 25¢ a bushel, oats 10¢, sugar 3¢ lb., cotton and wool 5¢ lb.

Iowa Farmers' Union militants start a 30-day strike August 9 to protest low farm prices. "Stay at home! Sell nothing!" say union members, who smash the windshields and headlights of farm trucks they catch going to market and block highways with chains and logs to enforce their strike.

Famine in the Ukraine and in the Caucasus begins to kill millions (*see* above). It is a deliberate, man-made famine, not caused by nature as Josef Stalin moves to starve *kulaks* who resist collectivization, seizing their grain and livestock, keeping out relief supplies.

More Americans are hungry or ill-fed than ever before in the nation's history. Nearly a million go back to the land.

Congress votes March 7 to authorize giving needy Americans 40 million bushels of wheat held by the Federal Farm Bureau with the distribution to be handled by the Red Cross. Another 45 million bushels of wheat are added July 5, plus 250 million pounds of cotton.

Nobel laureate Otto H. Warburg, 49, and his colleagues at Germany's Kaiser Wilhelm Institute for Biology find a yellow coenzyme that catalyzes the transfer of hydrogen atoms (*see* vitamin B_2—riboflavin, 1935).

University of Pittsburgh biochemist Charles Glen King, 36 isolates vitamin C (ascorbic acid) (*see* 1933; Szent-Györgyi, 1928).

100,000,000 Guinea Pigs by Consumer's Research engineer Arthur Kallet, 29, and F. J. Schlink relates horror stories dramatizing the dangers in some foods, drugs, and cosmetics sold despite the Pure Food and Drug Act of 1906 and its amendments (*see* Schlink, 1929). The book becomes a best seller, and Kallet breaks with Schlink to start Consumers Union, a rival organization, at Mount Vernon, N.Y.

Florida's Gulf Coast has its worst red tide since 1916. Millions of fish are killed (*see* 1946).

United Fruit profits drop, and the company's stock falls in value to $10.25 per share, down from $105 in November of 1929. Samuel Zemurray sees his holdings decline in value to $4.5 million, down from $31.5 million, persuades United Fruit's other directors to make him general manager, regains control of the company's Latin American properties, and the stock climbs back to $26 within a few weeks on the strength of Zemurray's prestige alone (*see* 1929; Chiquita banana, 1944).

Ralston Purina begins a sharp upward course under the direction of Donald Danforth, 33, who takes over from his father (*see* 1902). Danforth will move Ralston Purina into new consumer food areas in the next 24 years and increase sales from $19 million to more than $400 million.

Frito corn chips are introduced by San Antonio, Tex., candy maker C. Elmer Doolin who goes into a local café for a sandwich, is served a side dish of corn chips created by Mexican cartoonist-café owner Gustave Olguin, acquires the recipe for $100, and goes into business with his mother and his brother Earl using a converted potato ricer to cut *tortilla* dough into strips. Brother Earl develops a machine to force the dough out under pressure, and the Doolins license local companies to make Fritos using machinery purchased at cost with Doolin-trained personnel, taking a royalty on each package sold (*see* Frito-Lay, 1961).

Best Foods acquires Hellmann's mayonnaise from General Foods (*see* 1903; 1927; Corn Products Refining Co., 1958).

Skippy peanut butter is introduced by Rosefield Packing Co. of California, whose J. L. Rosefield cancels his exclusive licensing agreement with Peter Pan peanut butter (*see* 1928). Rosefield markets his own brand named after the comic strip started the same year he invented his process (1923) and he also introduces the first smooth-base peanut butter containing chunks of roasted peanuts. Rosefield will patent a new type of cold-processed hydrogenated peanut oil in 1950, and by 1954 Rosefield Packing will have nearly 25 percent of the U.S. peanut butter market.

3 Musketeers candy bars are introduced at a nickel by Mars, Inc. (*see* 1930; M&Ms, 1940).

Big Bear Super Market opens at Elizabeth, N. J., and is the first large cut-rate self-service grocery store (*see* King Kullen, 1930). Local entrepreneurs Robert M. Otis and Roy O. Dawson have put up a total of $1,000 to take over an empty car factory on the outskirts of town, their Big Bear the Price Crusher store has 50,000 square feet of pine tables displaying meats, fruit, vegetables, packaged foods, radios, automobile accessories, and paints, and after one year they have made a profit of $166,000 selling Quaker Oats oatmeal at 3¢ a box, pork chops at 10¢ lb. Traditional grocers persuade local newspapers to refuse Big Bear advertising and

1932 ***(cont.)*** push through a state law against selling at or below cost, but Big Bear is quickly imitated by Great Tiger, Bull Market, Great Leopard, and others as the supermarket revolution gathers force.

1933 Adolf Hitler begins 12 years as dictator of the German Reich, Franklin D. Roosevelt begins 12 years as president of the United States (he is inaugurated March 4 but the "Lame Duck" (Twentieth) Amendment ratified January 23 ends terms of all elected federal officers at noon January 20), Antonio de Oliveira Salazar begins 37 years as dictator of Portugal.

Adolf Hitler comes to power as chancellor January 30 on a rising tide of German nationalism and economic unrest. Fire destroys the Reichstag at Berlin February 27, Hitler's Nazis immediately accuse the Communists of having started it and fabricate a case against Dutch Communist Marinus van der Lubbe, 23, who is tried and will be guillotined early next year, but Nazi leader Hermann Goering is accused at the trial of having set the fire himself (his accuser is Georgi Dimitrov, who will later be premier of Bulgaria; *see* 1946).

The National Socialist (Nazi) party wins 44 percent of the votes in the Reichstag elections March 5, the Nationalist party wins only 8 percent despite support from industrialists who include Fritz Thyssen and Alfried Krupp, Adolf Hitler proclaims the Third Reich March 15, the Nationalists throw their support to Hitler by helping to pass an Enabling Act March 23, and the Nazi regime is given dictatorial powers.

Portuguese prime minister and finance minister Antonio de Oliveira Salazar, 44, drafts a new constitution that will make Europe's most backward country a "unitary and corporative republic"—a fascistic dictatorship based on political repression that dictator Salazar will head for more than 37 years, keeping his country backward.

Adolf Hitler promised Germans a new order based on "racial purity," repression, and preparation for military aggression.

Austrian chancellor Engelbert Dollfuss, 40, proclaims a dictatorship March 7 and soon after dissolves the Bundesversammlung as anti-government agitation increases following the Nazi success in Germany. Dolfuss prohibits parades and assemblies as of March 8, but Austrian Nazis stage a giant demonstration and riot March 29; when the government forbids wearing of uniforms by members of any political party, Hitler retaliates by imposing a tax of 1,000 marks on any German who visits Austria, thus ruining Austria's tourist business (*see* 1934).

Peru's President Sánchez Cerro is assassinated April 30 and succeeded by Oscar Benevides, who will serve until 1939.

Iraq's King Faisal dies at Berne September 8 and is succeeded by his son Ghazi who will reign until 1939.

Japan announces that she will withdraw from the League of Nations in 1935. The League suffers its first serious setback. Berlin announces October 14 that Germany, too, will withdraw.

Afghanistan's king Nadir Shah is assassinated at Kabul November 8. His son Mohammed Zahir Shah succeeds.

Washington and Moscow establish relations November 16 for the first time since the 1917 revolution.

Chicago mayor Anton Joseph Cermak, 59, has been mortally wounded February 15 at Miami by an assassin thought to have been aiming at President-elect Roosevelt but possibly just a hired gangland hit man.

New York's Tammany Hall is unseated by Fiorello La Guardia of 1932 Norris-La Guardia Act fame who wins election as mayor on a reform-fusion ticket and begins 12 years at City Hall.

The Nazis (above) open the first German concentration camp March 20 at Dachau near Munich. The facility is for Jews, gypsies, and political prisoners (*see* Buchenwald, 1937).

A National Labor Board established by President Roosevelt August 5 under the NRA (below) works to enforce the right of collective bargaining under the chairmanship of U.S. Sen. Robert F. Wagner, 56, of New York (*see* NLRB, 1934; Wagner Act, 1935).

Lynchings spread across the South. Forty-two blacks are killed by lynch mobs.

Detroit defaults on its $400 million debt February 14 when bank closings prevent completion of a $20 million emergency loan. The city has been caught between reduced tax revenues and the need for greater welfare support arising from the collapse of the automobile industry, and the debt will not be fully repaid until 1963.

President Roosevelt enters the White House with more than 15 million Americans out of work and says, "The only thing we have to fear is fear itself." Blacks, unskilled workers, and workers in heavy industry are especially affected by the widespread unemployment, but even Americans with jobs have had their wages and hours reduced. Wage-earner incomes are 40 percent below 1929 levels.

President Roosevelt proclaims a nationwide bank holiday March 5 and forbids gold exports.

The Emergency Banking Act passed by Congress March 6 gives the president control over banking transactions and foreign exchange, forbids hoarding or export of gold, and authorizes banks to open as soon as examiners determine them solvent. Banks begin to reopen March 13; about 75 percent are open by March 16.

An April 5 presidential order requires that all private gold holdings be surrendered to Federal Reserve banks in exchange for other coin or currency.

The United States abandons the gold standard April 19 by presidential proclamation but Roosevelt rejects a currency stabilization plan proposed by the gold standard countries meeting in July at a World Monetary and Economic Conference at London. In October he authorizes the Reconstruction Finance Corp. to buy newly mined gold at $31.36 per ounce, 27¢ above the world market.

The U.S. national income falls to $40.2 billion, down from $87.8 billion in 1929; the national wealth falls to $330 billion, down from $439 billion.

A Federal Emergency Relief Act passed by Congress May 12 establishes a $500 million fund that can be distributed in grants to the states. Former social worker Harry Lloyd Hopkins, 43, is appointed federal administrator of emergency relief and persuades the president to establish a Civil Works Administration (CWA) November 9 to provide emergency jobs for 4 million unemployed Americans through the winter (*see* 1934; WPA, 1935).

Congress establishes the Home Owners Loan Corp. June 13 as mortgage foreclosures average 1,000 per day. HOLC will grant emergency loans of more than $4 billion in the next 4 years to avert foreclosures (*see* 1931).

The Glass-Steagall Act signed into law by President Roosevelt June 16 forbids banks to deal in stocks and bonds (J.P. Morgan splits off Morgan Stanley to comply) and insures bank deposits. Sen. Carter Glass (D. Va.), 76, and Rep. Henry Bascom Steagall (D. Ala.), 60, have proposed the law, bankers denounce it.

A National Industrial Recovery Act (NIRA) passed by Congress June 16 provides for "codes of fair competition" in industries and for collective bargaining with labor, whose unions have dropped in membership from 3.5 million to less than 3 million. Industry agrees to shorten working hours and in some cases limit production and fix prices (*see* court decision, 1935).

The first NRA Blue Eagle signs of cooperation with the National Recovery Administration appear in store and factory windows August 1: "We Do Our Part. "

Typical annual U.S. earnings: congressman $8,663, lawyer $4,218, physician $3,382, college teacher $3,111, engineer $2,250, public school teacher $1,227 (5,000 Chicago schoolteachers storm the banks for back pay April 24 after being paid for 10 months in scrip), construction worker $907, sleep-in domestic servant $260 ($21.66 per month), hired farm hand $216.

A Stetson hat sells for $5, a gas stove for $23.95 (*see* truck, motorcar, and food prices below).

Dun & Bradstreet is created by a merger of New York's R. G. Dun & Co. with Cincinnati's Bradstreet Co. Dun was founded in 1859 by mercantile authority Robert Graham Dun and has published *Dun's Review* since 1893; Bradstreet was founded by John M. Bradstreet in 1849 under the name Bradstreet's Improved Mercantile Agency; Dun & Bradstreet will provide financial data and credit ratings of U.S. business firms and business executives, many of them now in dire straits.

Saudi Arabia's Abdul-Aziz ibn-Saud gives Standard Oil of California a 60-year exclusive concession to explore for oil on a 320,000-square-mile tract of desert (*see* 1932). He receives a loan of $170,327 from SoCal, which has recently begun pumping oil on the Persian Gulf island of Bahrain and will soon find that its concession overlies the world's richest petroleum reserve. To help exploit the Saudi Arabian reserve, SoCal makes a 50–50 arrangement with the Texas Co., forming Arabian American Oil Co. (Aramco) to produce the petroleum and CalTex to market it through Texaco outlets in Europe, Africa, and Asia (Aramco will later take in other partners, including the Saudi Arabian government, which will acquire full ownership in 1976).

Gasoline sells for 18¢ a gallon in America.

The Tennessee Valley Authority Act passed by Congress May 18 creates a Tennessee Valley Authority (TVA) to maintain and operate a power plant at Muscle Shoals, Ala.

The Dnieper River Dam is completed in the Soviet Ukraine. Hugh Lincoln Cooper, 68, who built the Muscle Shoals Dam in Alabama, has directed construction of the 810,000-horsepower hydroelectric installation which will save 3 million tons of coal per year.

The Baltic-White Sea Stalin Ship Canal opens in the Soviet Union to link Povenets on Lake Onega with Belmorsk on the White Sea. Slave labor has built the 140-mile canal in 2 years at a cost of some 250,000 lives, but its shallow draft and primitive wooden locks will soon make it obsolete (*see* 1923; Moscow-Volga Canal, 1937).

The Illinois Waterway linking the Great Lakes with the Gulf of Mexico via the Mississippi River opens officially at Chicago June 22.

The S. S. *Europa* of the North German-Lloyd line crosses from Cherbourg to New York in 4 days, 16 hours, 48 minutes, breaking the transatlantic speed record set by her sister ship the S. S. *Bremen* in 1929.

The Italian line's S. S. *Rex* sets a new transatlantic speed record by crossing from Cherbourg to New York in 4 days, 13 hours, 58 minutes.

1933 ***(cont.)*** United Fruit Co.'s Great White Fleet grows to number 95 vessels (*see* 1899; Zemurray, 1932).

Air France is founded August 30 by a merger of France's Aeropostale with another line.

The DC-1 introduced by Douglas Aircraft can carry 12 passengers at 150 miles per hour. TWA has suggested construction of the Douglas Commercial aircraft and orders 25 (*see* 1924; 1930; DC-3, 1936).

General Motors gains control of North American Aviation, a holding company that controls Curtiss-Wright, Sperry Gyroscope, Ford Instrument, Berliner-Joyce Aircraft, and Eastern Air Transport and is a major stockholder in most other important U.S. aviation companies (*see* 1934).

General Motors assumes leadership in U.S. motorcar sales, with Chrysler second, Ford third.

Marmon Automobile Co. goes into receivership in May (*see* 1916).

A new Pontiac coupe sells for $585, a new Chevrolet half-ton pickup truck for $650, a 1929 Ford for $58.

New York's new Independent subway line begins the F train from 179th Street in Jamaica to Manhattan and thence to Coney Island's Stillwell Avenue. The trip takes 85 minutes.

The A train goes into service on the BMT. The new express will be extended from 207th Street in Manhattan to Far Rockaway or Lefferts Boulevard—a journey of 100 minutes for 5¢ (*see* song, 1941).

The D train goes into service on the BMT from 205th Street in the Bronx to Coney Island's Stillwell Avenue—an 85-minute journey for 5¢.

Dow Chemical starts building a plant at Long Beach, Calif, to produce iodine from oil field brine. Dow will break the British-Chilean nitrate monopoly in iodine and bring the price of iodine down from $4.50 per pound to 81¢.

British biochemist Ernest Kennaway isolates the first pure chemical carcinogen (cancer-causing chemical) and shows that a polycyclic aromatic hydrocarbon from subtotal combustion will cause cancer in test animals (*see* 1915). Such hydrocarbons are found in air pollution, auto exhausts, and cigarette smoke.

The influenza virus is isolated for the first time following a London epidemic (*see* 1918 pandemic).

French researchers develop a neurotropic vaccine against yellow fever using a strain of attenuated yellow fever virus (*see* Walter Reed, 1900).

"Fireside chats" start on radio March 12. "My friends . . .," President Roosevelt begins, and urges listeners to have faith in the banks, trying to calm Depression fears and win support for New Deal measures being undertaken to restore the nation's economic health.

Frequency modulation (FM) provides static-free radio reception. Edwin H. Armstrong of 1917 superheterodyne circuit fame proved in 1915 that radio waves and static have the same electrical characteristics. Having insisted that any attempt to eliminate static without some radically new principle would be fruitless, and that hundreds of patents for static eliminating devices are worthless, he perfects FM (*see* Zenith, 1940).

The U.S. Army Corps of Engineers develops the first "walkie-talkie" portable two-radio sets with help from Paul V. Galvin of Motorola (*see* 1929).

IBM enters the typewriter business by acquiring a firm that has been trying for 10 years to perfect an electric office typewriter (*see* 1924; IBM Selectric, 1961).

Newsweek magazine has its beginnings February 17 in a weekly news magazine published under the name *News-Week* by English-American journalist Thomas John Cardel Martyn who has created the magazine to rival Henry Luce's 10-year-old *Time* (*see* 1937; 1961).

The *Washington Post* changes hands June 1 as California-born Wall Street copper and chemical millionaire Eugene Meyer, 57, outbids Evalyn Walsh McLean, wife of owner Edward B. (Ned) McLean, who wears the Hope diamond but cannot match Meyer's bid of $825,000 for the bankrupt paper (*see* 1877). Meyer was the chief force behind President Hoover's creation of the Reconstruction Finance Corp. last year, serves as an RFC director and will battle the New Deal, but his daughter Katharine will marry New Dealer Philip L. Graham, now 17, in 1940 and the *Post* will become a progressive force in U.S. journalism (*see Newsweek*, 1961).

The *Catholic Worker* begins publication with the aim of uniting workers and intellectuals in joint efforts to improve farming, education, and social conditions. Founder Dorothy Day gets support from French-American editor Peter Maurin, who has developed a program of social reconstruction he calls "the green revolution." The new monthly will be a voice for pacifism and social justice and will have a circulation of 150,000 within 3 years.

The *Reader's Digest* publishes its first original signed articles, departing from its 11-year-old policy of digesting from other publications, and begins to plant feature stories for subsequent publication in *Digest* versions (*see* 1922; 1955).

Esquire magazine begins publication in October. Editor-publisher Arnold Gingrich, 29, is a former copywriter who has been editing *Apparel Arts* at Chicago since 1931. Intended initially as a men's fashion quarterly distributed through retail stores, *Esquire* publishes a story by Ernest Hemingway in its first issue, features risqué cartoons and drawings of scantily clad women, quickly sells out on newsstands despite its high 50¢ cover price, begins monthly publication, and in its first 3 years will sell 10 million copies with help from good writing by Albert Camus, Clarence Darrow, John Dos Passos, William Faulkner, Scott Fitzgerald, Aldous Huxley, H. L. Mencken, Dorothy Parker, Damon Runyon, William Saroyan, Georges Simenon, John Steinbeck, and other leading writers.

The American Newspaper Guild is founded at Washington, D.C., in December to protect city room employees. *New York World-Telegram* reporter Heywood C. Broun, 45, is elected president.

Nonfiction *The Autobiography of Alice B. Toklas* by Gertrude Stein whose book is actually her own autobiography rather than that of the woman who has been her companion and secretary since 1907. "Rose is a rose is a rose," writes Stein who is famous for being more concerned with the sound and rhythm of words than with intelligibility; *Down and Out in Paris and London* by English writer George Orwell (Eric Blair), 30, whose health was broken by the climate while serving in the Burma Imperial Police and who has experienced poverty as a struggling writer.

Ulysses by James Joyce is acceptable for publication in the United States, rules Justice John M. Woolsey of the U.S. District Court at New York. The "dirty" words in the 1922 book are appropriate in context and not gratuitous, he decides.

Fiction *Man's Fate* (*La Condition Humaine*) by French novelist André Malraux, 32, deals with the undercover struggle against imperialism in Indochina; *The Tales of Jacob (Die Geschichten Jaacobs)* by Thomas Mann; *The Bark Tree (Le Chiendent)* by French novelist Raymond Queneau, 30; *La Chatte* by Colette; *Fontamara* by Italian novelist Ignazio Silone (Secondo Tranquili), 33, whose anti-fascist work exposes the plight of Italy's southern peasants; *The Forty Days of Musa Dagh (Die vierzig Tage des Musa Dagh)* by Austrian poet-novelist Franz Werfel, 43; *All Night at Mr. Stanyhurst's* by English novelist Hugh Edwards, 35; *Opening Day* by English novelist David Gascoyne, 17; *A Glastonbury Romance* by John Cowper Powys; *A Nest of Simple Folk* by Irish novelist Sean O'Faolain, 33; *From a View to Death* by Anthony Powell; *Lost Horizon* by English novelist James Hilton, 33, who fascinates readers with a fictional Shangri-la in the Himalayas of Tibet; *Miss Lonelyhearts* by Nathanael West whose advice-to-the-lovelorn newspaper columnist becomes involved with some of his correspondents; *God's Little Acre* by Erskine Caldwell; *The Last Adam* by James Gould Cozzens; *Murder Must Advertise* and *Hangman's Holiday* by Dorothy Sayers.

Poetry *Residence on Earth* (*Residencia en la tierra*) by Pablo Neruda, who has published the work in individual pieces between 1925 and 1931 during a period of spiritual nihilism.

Painting *The Dance* by Henri Matisse.

Germany suppresses all modernistic painting in favor of superficial realism.

Murals *Man at the Crossroads* by Diego Rivera for New York's Radio City Music Hall. The mural is destroyed because it contains a portrait of V. I. Lenin, but Rivera uses the proceeds of the work to pay for the expenses of frescoes that he executes gratis for New York's New Workers School. *Quetzelcoatl and the Old Order, The Legend of the Races, Stillborn Education,* and *Latin America* by José Clemente Orozco for Dartmouth College.

Sculpture *The Palace at Four A.M* by Swiss sculptor-painter Alberto Giacometti, 31.

Theater *Design for Living* by Noël Coward 1/24 at New York's Ethel Barrymore Theater, with Coward, Alfred Lunt, Lynn Fontanne, 135 perfs.; *Alien Corn* by Sidney Howard 2/20 at New York's Belasco Theater, with Katherine Cornell, 98 perfs.; *The Enchanted (Intermezzo)* by Jean Genet 2/27 at the Comédie des Champs-Elysées, Paris; *Both Your Houses* by Maxwell Anderson 3/6 at New York's Royale Theater, with Morris Carnovsky, Walter C. Kelly, Mary Phillips, Jerome Cowan, Sheppard Strudwick in a polemic against political corruption, 120 perfs.; *Men in White* by U.S. playwright Sidney Kingsley, 27, 9/26 at New York's Broadhurst Theater, with Morris Carnovsky, Luther Adler, Elia Kazan, Clifford Odets, 367 perfs.; *Ah, Wilderness* by Eugene O'Neill (his only comedy) 10/2 at New York's Guild Theater, with George M. Cohan as Nat Miller, William Post, Jr., Elisha Cook, Jr., Gene Lockhart, Philip Moeller, Ruth Gilbert, 289 perfs.; *Tovarich* by French playwright Jacques Duval Jacques Boularan), 42, 10/13 at the Théâtre de Paris, 800 perfs.; *Mulatto* by Langston Hughes 10/24 at New York's Vanderbilt Theater, with Rose McClendon, 270 perfs.; *Tobacco Road* by U.S. playwright Jack Kirkland, 31, who has adapted last year's Erskine Caldwell novel 12/4 at New York's Masque Theater, with Henry Hull as Jeeter Lester, 3,182 perfs.; *Twentieth Century* by Ben Hecht and Charles MacArthur 12/29 at New York's Broadhurst Theater, with Eugenia Leontovich as Lilly Garland, Moffat Johnston as Oscar Jaffe, 152 perfs.

Radio *Oxydol's Own Ma Perkins* continues the soap opera serial genre with daily 15-minute shows sponsored by Procter & Gamble. Actress Virginia Payne, 23, plays the title role she will continue for 27 years; *The Romance of Helen Trent* 7/24 is a soap opera by Frank Hummert and Anne Ashenhurst, whose show is carried on CBS network stations beginning 10/3; *Jack Armstrong the All-American Boy* 7/31 on Chicago's KBBM with St. John Terrell, who is soon replaced by Jim Ameche. Creator of the show is Robert Hardy Andrews, sponsor is the 9-year-old General Mills "Breakfast of Champions" Wheaties, and the theme song begins, "Wave the flag for Hudson High, boys/ Show them how we stand,/ Ever shall our team be champions,/ Known throughout the land" (to 1951).

Sally Rand attracts thousands to the Chicago World's Fair that opens May 27 to celebrate a Century of Progress. The fan dancer, 29, gets star billing at the "Streets of Paris" concession on the Midway, does a slow dance to Debussy's "Clair de Lune" wearing only ostrich plumes, and is credited with making the fair a success as jobless Americans flock to Chicago in search of fun.

Films Frank Lloyd's *Cavalcade* with Diana Wynyard, Clive Brook, Ursula Jeans; William Wyler's *Counsellor-at-Law* with John Barrymore, Bebe Daniels; George Cukor's *Dinner at Eight* with John Barrymore, Jean Harlow; Leo McCarey's *Duck Soup* with the Marx Brothers; Frank Capra's *Lady for a Day* with May Robson, Warren William; George Cukor's *Little Women*

1933 ***(cont.)*** with Katharine Hepburn, John Bennett, Frances Dee, Jean Parker, Paul Lukas; Alexander Korda's *The Private Life of Henry VIII* with Charles Laughton; Rouben Mamoulian's *Queen Christina* with Greta Garbo, John Gilbert; Lowell Sherman's *She Done Him Wrong* with Mae West, now 41, as Diamond Lil ("Come up and see me sometime"), Cary Grant; Jean Vigo's *Zero for Conduct* (*Zero de conduit*) with Jean Daste, Robert le Flon. Also: Marcel Pagnol's *César* with Raimu, Pierre Fresnay; Ernst Lubitsch's *Design for Living* with Gary Cooper, Fredric March, Miriam Hopkins; Stuart Walker's *The Eagle and the Hawk* with Fredric March, Cary Grant, Jack Oakie, Carole Lombard; Wesley Ruggles's *I'm No Angel* with Mae West, Cary Grant; A. Edward Sutherland's *International House* with W. C. Fields, Peggy Hopkins Joyce, Stuart Erwin, George Burns, Gracie Allen; Frank Borzage's *Man's Castle* with Spencer Tracy, Loretta Young; W. S. Van Dyke's *Penthouse* with Warner Baxter, Myrna Loy; William A. Seiter's *Sons of the Desert* with Laurel and Hardy; John Cromwell's *Sweepings* with Lionel Barrymore, William Gargan; Fritz Lang's *The Testament of Dr. Mabuse* with Rudolf Klein-Rogge; Francis Martin's *Tillie and Gus* with W. C. Fields, Alison Skipworth, Baby LeRoy.

Twentieth Century Pictures is organized by Hollywood film producers who include Darryl Zanuck of Warner Brothers. He will merge the new studio with Fox Studios in 1935 to create Twentieth Century Fox.

Hollywood musicals Lloyd Bacon's *42nd Street* with Warner Baxter, Bebe Daniels, Dick Powell, Ruby Keeler, George Brent, music and lyrics by Al Dubin and Harry Warren, songs that include "Shuffle Off to Buffalo" and "You're Getting to Be a Habit with Me"; Mervyn LeRoy's *Gold Diggers of 1933* with Ginger Rogers, Dick Powell, Joan Blondell, choreography by Busby Berkeley, music and lyrics by Al Dubin and Harry Warren, songs that include "We're in the Money," "Shadow Waltz"; Raoul Walsh's *Going Hollywood* with Bing Crosby, Marion Davies, music by Nacio Herb Brown, lyrics by Arthur Freed, songs that include "Temptation"; Lloyd Bacon's *Footlight Parade* with James Cagney, Joan Blondell, Ruby Keeler, Dick Powell, choreography by Busby Berkeley, music by Sammy Fain, lyrics by Irving Kahal, songs that include "By a Waterfall," "Honeymoon Hotel"; Thornton Freeland's *Flying Down to Rio* with Dolores Del Rio, Gene Raymond, Fred Astaire and Ginger Rogers (a new dance team), music by Vincent Youmans, lyrics by Gus Kahn and Edward Ellison, 31, songs that include "Carioca," "Orchids in the Moonlight," and the title song; Walt Disney's *The Three Little Pigs* (animated) with music and lyrics by Frank E. Churchill, songs that include "Who's Afraid of the Big Bad Wolf."

Broadway musicals *Strike Me Pink* 3/4 at the Majestic Theater, with Lupe Velez, Jimmy Durante, Hope Williams, music by Ray Henderson, lyrics by B. G. DeSylva and Lew Brown, 105 perfs.; *As Thousands Cheer* 9/30 at the Music Box Theater, with Marilyn Miller, Clifton Webb, Ethel Waters, book by Irving Berlin and Moss Hart, music and lyrics by Berlin, Edward Heyman, and Richard Myers, songs that include "Easter Parade," 400 perfs.; *Let 'Em Eat Cake* by George S. Kaufman and Morrie Ryskind 10/21 at the Imperial Theater, with William Gaxton as John P. Wintergreen, Victor Moore as Alexander Throttlebottom, music by George Gershwin, lyrics by Ira Gershwin, 90 perfs.; *Roberta* (initially *Gowns by Roberta)* 11/18 at the New Ambassadors Theater, with Ray Middleton, George Murphy, Bob Hope, Fay Templeton, Tamara Geva, Sydney Greenstreet, music by Jerome Kern, lyrics by Otto Harbach, songs that include "Smoke Gets in Your Eyes," "The Touch of Your Hand," ("Lovely to Look At" will be added for a 1935 film version), 295 perfs.

Opera *Arabella* 7/1 at Dresden's Staatsoper, with music by Richard Strauss, libretto by Hugo von Hofmannsthal.

Ballet *Les Presages* (*Destiny*) 4/13 at Monte Carlo's Théâtre de Monte Carlo, with music by Petr Ilich Tchaikovsky, choreography by Leonide Massine.

First performances Concerto for Piano and Orchestra No. 2 by Béla Bartók 1/23 in a Frankfort radio broadcast; Concerto in C major for Two Pianos and Orchestra by Ralph Vaughan Williams 2/1 in a BBC broadcast; *School for Scandal* Overture by Samuel Barber 8/30 at Philadelphia's Robin Hood Dell; Concerto for Piano, Trumpet, and String Orchestra by Dmitri Shostakovich 10/15 at Leningrad; *Charterhouse Suite* by Williams 10/21 at the Queen's Hail, London.

Popular songs "Basin Street Blues" by Spencer Williams whose work was published in small orchestra parts 4 years ago; "Only a Paper Moon" by Harold Arlen, lyrics by E. Y. Harburg, Billy Rose; "Lazybones" by Hoagy Carmichael, lyrics by vocalist Johnny Mercer, 23; "Love Is the Sweetest Thing" by Ray Noble; "Stormy Weather—Keeps Rainin' All the Time" by Harold Arlen, lyrics by Ted Koehler; "It's a Sin to Tell a Lie" by U.S. songwriter Billy Mayhew; "Dolores" by Paramount songwriter Frank Loesser, 23, lyrics by Louis Alter (for the film *Las Vegas Nights);* "Sophisticated Lady" by Duke Ellington, lyrics by Irving Mills, Mitchell Parish; "Did You Ever See a Dream Walking?" by Harry Revel and Mack Gordon (for the film *Sitting Pretty*); "Everything I Have Is Yours" by Burton Lane, lyrics by Harold Adamson (for the film *Dancing Lady*); "Let's Fall in Love" by Harold Arlen, lyrics by Ted Koehler (title song for film); "I Wanna Be Loved by You" by Johnny Green, lyrics by Billy Rose, Edward Heyman; "In a Shanty in Old Shanty Town" by Joe Young, John Siros, Little Jack Little; "I Cover the Waterfront" by Johnny Green, lyrics by Richard Heyman; "It Isn't Fair" by bandleader Richard Himber, Frank Warshauer, Sylvester Sprigato, lyrics by Himber; "Minnie the Moocher" by bandleader Cabell "Cab" Calloway, 26, who plays at Owney Madden's Cotton Club in Harlem, lyrics by Irving Mills (when Calloway forgets the lyrics he "scat" sings "Hi-de-hi-hi" and gets an enthusiastic audience response); "The Old Spinning Wheel" by Billy Hill; "I Like Mountain Music" by Frank Weldon, lyrics by James Cavanaugh.

Italian prizefighter Primo Carnera, 26, of Argentina wins the world heavyweight title June 29. The 260-pound fighter knocks out Jack Sharkey in the sixth round of a championship bout at Long Island City.

John Herbert Crawford, 25, wins in men's singles at Wimbledon, Helen Wills Moody in women's singles; Frederick J. "Fred" Perry, 24, (Br) wins in men's singles at Forest Hills, Helen Jacobs in women's singles.

The American League wins baseball's first All-Star Game July 6 at Chicago's Comiskey Field, defeating the National League 4 to 2.

The New York Giants win the World Series by defeating the Washington Senators 4 games to 1.

The Philadelphia Eagles football team is founded by Bert Bell; the Pittsburgh Steelers team is founded by Art Rooney.

Helen Jacobs (above) is the first woman to wear shorts in tournament play. Others still wear skirts knee-length or longer.

The Dy-Dee-Doll that sucks water from a bottle and wets its diaper is introduced by New York's Effanbee Doll Co., which has acquired patent rights from the doll's Brooklyn inventor Marie Wittman. More than 25,000 are sold in its first year.

Dreft is introduced by Procter & Gamble for dishwashing in hard-water areas west of the Appalachians. The hymolal-salt detergent is costlier than soap but will prove a popular addition to Ivory soap, Oxydol, Camay, and Crisco (*see* Persil, 1907; Tide, 1946).

Windex for cleaning windows is introduced by Drackett Co. of 1921 Drano fame.

Philip Morris cigarettes are introduced in the United States (*see* 1858; 1902).

Gangsters thought to include Pretty Boy Floyd open fire with machine guns June 17 in front of Kansas City's 16-year-old Union Station and kill hoodlum Frank Nash together with four law enforcement officers who were escorting Nash.

Sheraton Corp. of America has its beginnings at Cambridge, Mass., where Boston entrepreneur Ernest Henderson, 36, and his Harvard classmate Robert Lowell Moore take over the Continental Hotel that opened the day the stock market crashed in 1929. By 1937 Henderson and Moore will have bought up stock for as little as 50¢ a share in Standard Investing Corp., used the firm's capital to finance acquisitions that include a small Boston residential hotel called the Sheraton despite its Hepplewhite decor, and used the name Sheraton to identify all their properties.

New York's 70-story RCA Building (GE Building beginning in 1990) opens in May at 30 Rockefeller Plaza as Rockefeller Center construction proceeds under the direction of Reinhard and Hoffmeister, Corbet Harrison and MacMurray, and Hood and Fouilhoux who will design all the buildings put up in this decade. The small British building opens in May, La Maison Française in September, both on Fifth Avenue.

The Tennessee Valley Authority (above) will control floods in the Tennessee Valley region.

The Civilian Conservation Corps (CCC) created by Congress March 31 under the Unemployment Relief Act (Reforestation) will enlist more than 3 million young men in the next 8 years. The CCC will plant more than 2 billion trees, build small dams, aid in wildlife restoration, and tackle erosion problems (*see* 1935).

Oregon's Tillamook Burn in August wipes out 12 billion board feet of virgin timber. Friction from a falling tree has touched off the fire that advances along an 18-mile front, littering beaches for 30 miles with charred fragments and raining still smoldering debris on ships 100 miles at sea.

A Soil Erosion Service started by the U.S. Department of Agriculture helps farmers learn tilling methods that will minimize erosion. The seriousness of the problem gains recognition when a great dust storm November 11 to 13 sweeps South Dakota topsoil as far east as Albany, N.Y. (*see* 1909; 1935).

The Tennessee Valley Authority (above) will improve agriculture in the Tennessee Valley region.

U.S. farm prices drop 63 percent below 1929 levels as compared with 15 percent for industrial prices since industrialists can control production more effectively than can farmers. Cotton sells in February at 5.5¢ per lb., down from a 1909–1914 average of 12.5¢, and wheat sells for 32.3¢ per bushel, down from an average of 88.4¢.

An Agricultural Adjustment Act (AAA) voted by Congress May 12 authorizes establishment of an Agricultural Adjustment Administration (AAA) within the Department of Agriculture.

An Emergency Farm Bill becomes law in June, and the AAA (above) sets out to restore farm income by reviving 1909–1914 average prices for wheat, dairy products, cotton, tobacco, corn, rice, sugar, hogs, and peanuts.

Secretary of Agriculture Henry Agard Wallace, 44, orders limitation procedures to compensate for the fact that the AAA (above) was established after most crops were planted and pigs were farrowed. He has some 400,000 farrowed pigs destroyed and 330,000 acres of cotton plowed under.

Congress creates the Commodity Credit Corp. in the U.S. Department of Agriculture to buy surpluses and handle any subsidies payable to farmers for reducing their crop acreage (*see* 1930; 1935).

A Farm Credit Act passed by Congress June 16 consolidates rural credit agencies under a Farm Credit Administration.

Midwestern farmers strike in October to force up prices by withholding products from market. They complain that the AAA has no real power (*see* 1932).

The world wheat price falls to an historic low of £4 per cwt, where it will remain through 1934.

At least a million destitute U.S. farm families receive direct government relief. While farm prices begin to

1933 ***(cont.)*** inch up, a farmer must pay the equivalent of 9 bushels of wheat for a pair of work shoes, up from 2 bushels in 1909.

Roughly 25 percent of U.S. workers are engaged in agriculture, down from nearly 50 percent in 1853 (*see* 1950).

Golden Cross Bantam corn is introduced (*see* East, Shull, 1921). The first commercial hybrid grain to be planted on a large scale, it permits farmers to increase their yields, which now average 22.8 bushels of corn per acre of land.

Anaheim, Calif., grower Walter Knott, 43, rediscovers the boysenberry developed in 1920 and makes it the basis of Knott's Berry Farms. He harvests 5 tons of the long, reddish-black berries per acre.

U.S. biochemist Robert R. Williams, 47, isolates an anti-beriberi vitamin substance from rice husks. He was born in India and began his professional career in Manila, where beriberi is a major health problem. It takes a ton of rice husks to produce less than 0.2 ounce of the substance (*see* Suzuki, 1912; vitamin B_1—thiamine, 1936).

Austrian chemist Richard Kuhn, 33, at Berlin's Kaiser Wilhelm Institute isolates a heat-stabile B vitamin (*see* 1932; vitamin B_2—riboflavin, 1935).

Swiss chemist Tadens Reichenstein and his colleagues synthesize vitamin C (ascorbic acid) (*see* King, 1932).

Borden Co. introduces the first vitamin-D fortified milk (*see* 1930). Children are the chief victims of vitamin D deficiency and the chief consumers of milk.

Ritz crackers, introduced by National Biscuit Co., will be the world's largest selling crackers within 3 years. Crisper and less fluffy than soda crackers, they are butter crackers made with more shortening, no yeast, spread with a thin coating of coconut oil and sprinkled with salt after baking. Nabisco bakes 5 million of the crackers; by 1936 production will be at the rate of 29 million per day.

Campbell's chicken noodle and cream of mushroom soups are introduced (*see* 1922; can production, 1936).

Stokely-Van Camp is created by a merger of Stokely Bros. and Van Camp (*see* Frank Van Camp, 1914). Stokely has grown by acquiring other canning companies, Van Camp has added to its pork and beans and catsup line by canning New Orleans-style kidney beans, hominy, Vienna sausage, and other items.

The Ford Motor Company's Industrialized American Barn at the Chicago fair (above) demonstrates margarine made from soybeans. Ford chemical engineer Robert Boyer, 34, has developed soy plastics for horn buttons, gearshift handles, and control knobs and is working on ways to substitute soybean forms for conventional foods (*see* 1937).

More than 11 percent of U.S. grocery store sales are made through 18,000 A&P stores (*see* 1929; 1937).

Typical U.S. food prices: butter 28¢ lb., margarine 13¢ lb., eggs 29¢ doz., oranges 27¢ cents doz., milk 10¢ qt., bread 5¢ per 20-oz. loaf, sirloin steak 29¢ lb., round steak 26¢, rib roast 22¢, ham 31¢, bacon 22¢, leg of lamb 22¢, chicken 22¢,, pork chops 20¢, cheese 24¢, coffee 26¢, rice 6¢, potatoes 2¢, sugar 6¢ (but *see* earnings, above).

Sugar falls to 3¢ lb. in New York, and Puerto Rico has widespread hunger as wages are reduced (*see* 1920).

The prohibition against sale of alcoholic beverages in the United States that began early in 1920 ends December 5 as Utah becomes the thirty-sixth state to ratify the Twenty-first Amendment repealing the Eighteenth Amendment after an estimated 1.4 billion gallons of hard liquor have been sold illegally. Twenty-two states and hundreds of counties remain dry by statute or local option, some states permit sale only in stores controlled by a state monopoly.

Repeal finds roughly half the whisky aging in U.S. rack warehouses owned by National Distillers (Old Grand Dad, Old Taylor, Old Crow, Old Overholt, Mount Vernon).

Schenley Distillers is founded at New York by entrepreneur Lewis S. Rosenstiel, 42, who takes the name from the Schenley, Pa., location of Joseph S. Finch Co., a rye distillery that forms part of the new corporation that will be the leading U.S. distiller until 1937 and lead again from 1944 to 1947. Schenley owns 25 percent of the whisky in U.S. rack warehouses.

Most of the remaining whisky in U.S. rack warehouses is owned by Frankfort Distilleries (Four Roses) and Glenmore (Kentucky Tavern) which will both be acquired by Distillers Corp.-Seagram (*see* 1928). Seagram will build a large distillery at Lawrenceburg, Ind., to produce 5 Crown, 7 Crown, Kessler, and other whisky brands that will dominate the market (*see* 1934).

Prohibition's "Texas" Guinan has died November 5 at age 49, leaving an estate of only $28,173 although she is known to have banked well over $1 million in the mid-1920s (*see* 1922).

Congress has legalized sale of 3.2 beer April 7 and the first team of Clydesdale horses to promote Budweiser beer has appeared a day later. Anheuser-Busch has survived Prohibition by producing commercial yeast and sarsaparilla.

7-Up is promoted as a mixer for alcoholic beverages by Charles Grigg of St. Louis who renames his Lithiated Lemon and sells 681,000 cases of the soft drink (*see* 1929).

Repeal (above) dampens the boom in most soft drinks and in ice cream sodas.

Coca-Cola sales tumble to 20 million gallons of syrup (2.5 billion drinks), down from a 1930 peak of 27.7 million gallons (3.5 billion drinks).

A new automatic fountain mixer introduced by Coca-Cola Co. at the Chicago Fair (above) mixes the exact amounts of syrup and carbonated water, eliminating the chance of error by a counterman.

Pepsi-Cola is acquired by Loft candy store chief Edward Guth who begins to bottle the drink in 12-ounce bottles (*see* 1920; Mack, 1934).

The E. and J. Gallo winery is founded at Modesto, Calif., by Ernest Gallo, 24, and his brother Julio, 23, whose father committed suicide 6 weeks earlier after killing their mother. The Gallo brothers, who have learned about winemaking from a book in the public library, sink their savings of $5,900 into an operation that will become the largest U.S. wine maker.

Grape production in the Chautauqua-Erie grape belt of New York State declines precipitously as the repeal of Prohibition (above) reduces consumer demand for grapes to be used for home wine making. The New York growers use little or no fertilizer or insect spray, average little more than 1 ton per acre, receive only $20 per ton, and have lost the table grape market to California growers.

New York entrepreneur Jacob M. Kaplan, 40, acquires a small winery at Brocton, N.Y., and establishes a floor price of $50 per ton for grapes, guaranteeing the price before the grapes are picked. Kaplan, who has prospered in West Indian sugar and rum ventures during Prohibition, hires chemists and mechanical engineers to produce a uniform beverage efficiently and economically and encourages growers to improve their vineyards, concentrating on Concord grapes *(see* 1945).

Adolf Hitler (above) encourages production of more "Aryans" by offering cash incentives to Germans who will marry and cash awards for each new infant born. The policy will increase German birth rates in the next few years (*see* Lebensborn, 1935).

1934 France verges on civil war following the suicide at Chamonix January 3 of Russian-born promoter Serge Stavisky who has been accused of issuing fraudulent bonds against the security of the municipal pawnshop at Bayonne, evidently engaged in other dubious speculations, and been protected from legal action by corrupt ministers and deputies. A high official in the public prosecutor's office at Paris is found murdered, allegations are made that he was killed to protect some well-known government leaders, both communists and right-wing fascist and royalist groups say the scandal demonstrates the corruption and inefficiency of the democratic government, Paris has serious riots February 6, 7, and 9, a general strike ensues, and the republic is saved only by the establishment of a coalition government of officials untouched by the Stavisky affair.

The king of the Belgians Albert I dies February 17 in a mountain-climbing accident at age 58 after a 25-year reign; he is succeeded by his son of 32 who will reign until 1950 as Leopold III.

Austria's chancellor Engelbert Dollfuss issues a decree in February dissolving all political parties except for his own Fatherland Front. He antagonizes Vienna's working class by ordering police raids on socialist headquarters and a bombardment of the socialist housing unit, the Karl Marx Hof *(see* below).

Bulgarian fascists stage a coup May 19, seizing power with help from King Boris.

A Nazi blood purge June 30 kills at least 77 party members alleged to have plotted against Adolf Hitler in Germany. Raids by Nazi storm troopers eliminate the party's radical wing; victims include the Sturm Abteilungen group of Ernst Röhm and supporters of Gregor Stresser, both of whom are killed.

Austrian Nazis seize the Vienna radio station July 25, force the staff to broadcast a report that Chancellor Dollfuss (above) has resigned, enter the chancellory, and shoot Dollfuss dead. Italy and Yugoslavia concentrate troops on the frontier; Berlin disavows any connection with the coup attempt. The government orders the roundup of all Nazis.

Power and Earth (Macht und Erde) by German geographer Karl Haushoker, 67, implies that a dynamic Germany has the natural right to grasp all of Eurasia and dominate the oceanic countries. Based in part on British political geographer Halfor John Mackinder's 1904 paper "The Geographical Pivot of History," Haushoker's theories of geopolitics will help shape Adolf Hitler's demands for *lebensraum* (living space) (*see* Saar Basin, 1935).

Germany's President von Hindenburg dies August 2 at age 87, and a plebiscite August 19 gives Adolf Hitler 88 percent of the votes needed to assume the presidency. Hitler retains the title *der führer*.

The Soviet Union joins the League of Nations September 18.

Yugoslavia's Aleksandr I arrives at Marseilles October 9 and is assassinated at age 45 along with French Foreign Minister Louis Barthou. The assassin is a Macedonian terrorist working with Croat revolutionists headquartered in Hungary, the League of Nations averts war between Yugoslavia and Hungary, and Aleksandr is succeeded by his son of 11 who will reign until 1945 as Peter II, with Aleksandr's cousin Prince Paul as regent.

Turkey's Mustafa Kemal issues orders November 25 that all Turks must assume surnames by January 1. His own, he says, will be Atatürk, meaning "Father of the Turks."

India's All-India Congress Socialist party is founded.

Chinese Communist forces under Mao Zedong leave Kiangsi in October with Nationalist Guomindang armies in pursuit. In the Long March that will last until October of next year, Mao's forces will be on the move for 268 days out of 368, they will travel 6,000 miles to Yunnan, crossing 18 mountain ranges and six major rivers, but 68,000 of Mao's 90,000 men will be lost in continual rear-guard actions against Guomindang troops *(see* 1931; 1937).

Nicaraguan National Guard Gen. Antonio Somoza, 38, invites guerrilla leader Gen. Augusto Cesar Sandino to a meeting in February. Sandino, who fought the U.S. occupation force from 1927 until its withdrawal in 1933, is killed in cold blood (*see* 1936).

Mexico amends her constitution to extend the term of the presidency to 6 years, the minister of war and marine Gen. Lazaro Cardenas, 39, is elected president

1934 *(cont.)* July 2, he promises to "revive the revolutionary activity of the masses," and although he has been chosen by strongman Plutarco Calles, he regards Calles as too conservative and will force him into exile next year.

Josef Stalin's close collaborator Sergeo Mironovich Kirov, 46, is assassinated December 1 at Leningrad. The assassination reveals the strength and desperation of the opposition to Stalin within the Communist party; many of the party's most prominent older leaders will be convicted in spectacular trials that follow the incident at Leningrad (*see* 1937).

Italian and Ethiopian troops clash at Ualual December 5 on the frontier between Italian Somaliland and Ethiopia (*see* 1896). Benito Mussolini will use the incident as an excuse to invade Ethiopia next year.

Heinrich Himmler, 33, is appointed head of German concentration camps July 13.

The United States joins the International Labor Organization (ILO) started in 1919.

A National Labor Relations Board (NLRB) created June 19 by a joint resolution of Congress replaces the National Labor Board established under the NRA last year.

The Dill-Crozier Act, signed into law June 21, creates a National Railroad Adjustment Board to guarantee railway workers the right to organize and to bargain collectively through representatives of their own choosing (*see* 1926).

A summertime wave of strikes takes U.S. workers off the job and includes the nation's first general strike, called at San Francisco July 16 to show sympathy for a strike of 12,000 International Longshoremen's Association stevedores led by Australian-American organizer Harry Bridges, 32.

President Roosevelt tells Congress January 4 that costs of the national economic recovery program will reach $10.5 billion by June 20, 1935.

The price of gold progressively raised since late October 1933 from its long-maintained level of $20.67 per ounce is stabilized by the U.S. Treasury in January at $35 per ounce, a price that will hold for 40 years.

Josef Stalin imposed ruthless totalitarian rule that stifled opposition to his control of Soviet Russia.

The Export-Import Bank of Washington created February 12 with $11 million in capital helps finance and facilitate exports and imports of commodities; $10 million has come from the Reconstruction Finance Corp. (RFC) organized in 1932.

The Costigan-Jones Act, signed into law May 9, establishes annual U.S. sugar import quotas based on anticipated consumption. The act allows prices far above world levels to help friendly nations that sell to the United States.

The Johnson Debt Default Act passed by Congress April 13 forbids additional loans to any country in default of debt payments to the United States.

Secretary of State Cordell Hull persuades Congress that the high tariffs of the 1930 Smoot-Hawley Act have contributed to world depression and should be replaced by mutual tariff relaxation. The Reciprocal Trade Agreement Act passed by Congress June 12 gives the president power to negotiate trade pacts without advice or consent of the Senate, and the first such pact is signed August 24 with Cuba. Reciprocal tariff-cutting agreements with 18 foreign nations in the next 4 years will bring an increase in world trade.

The Swiss Parliament enacts a Bank Secrecy Law to protect the accounts of Jews in Nazi Germany. Numbered accounts have four or five-digit numbers that depositors write out in script, the number in script becomes the depositor's signature, only two or three senior bank officials know the name of the depositor, and while Swiss law says that banks may not furnish information about such accounts even to Swiss tax authorities (tax evasion is not a criminal offense in Switzerland) and while numbered accounts will be used to safeguard the fortunes of U.S. racketeers, Communist espionage agents, and Latin American and Oriental dictators, the Swiss banks will insist on background knowledge of depositors and cooperate with police or authorized investigators in cases of criminal fraud and forgery.

A Securities and Exchange Commission (SEC) created by Congress June 6 limits bank credit for speculators and polices the securities industry. Wall Street speculator Joseph Patrick Kennedy, 45, is named in July to head the new commission despite opposition from New Dealers and from leading newspapers. (Kennedy is the father of nine and will hold the post for 431 days before being named to head the U.S. Maritime Commission.)

Security Analysis by English-American financier Benjamin Graham, 40, pioneers modern security analysis; it will remain the standard text for more than 55 years and have sales of more than 100,000 copies. Starting out at $12 per week with a job writing stock and bond prices on a Wall Street brokerage house blackboard, Graham went into business with Jerome Newman in 1926 to start an investment fund, he was a millionaire before age 35, and his book emphasizes the importance of a stock's book value—the physical assets of the company issuing the stock.

The Liberty League formed in August to oppose New Deal economic measures is a bipartisan group that includes the du Ponts, Alfred Sloan and William Knudsen of General Motors, J. Howard Pew of Sun Oil, Sewell Avery of Montgomery Ward, and politicians such as Democrats Alfred E. Smith and John Jacob Raskob.

A Civil Works Emergency Relief Act passed by Congress February 15 appropriates another $950 million for the continuation of civil works programs and direct relief aid.

Relief Administrator Harry L. Hopkins reports April 13 that 4.7 million families are on relief.

Demagogic welfare schemes proliferate. Sen. Huey Pierce Long, 40 (D. La.), presents his Share Our Wealth Program, an "every man a king" wealth redistribution scheme.

Massachusetts Democratic party boss James Michael Curley is elected governor at age 59; the large, costly public works programs he undertakes in the next 2 years will be compared to those of Huey Long.

Long Beach, Calif., physician Francis E. Townsend urges an Old Age Revolving Pension Plan to be financed by a 2 percent transaction tax. The Townsend Plan is to assure $200 per month for every unemployed American over age 60.

Upton Sinclair campaigns as Democratic candidate for governor of California and announces an End Poverty in California (EPIC) Plan.

Royal Oak, Mich., radio priest Father Charles Coughlin organizes a National Union for Social Justice based on radical inflation (*see* 1926).

Italian physicist Enrico Fermi, 33, bombards uranium with neutrons and produces transuranium elements (*see* 1938; Becquerel, 1896; Chadwick, 1932; Segre, 1937).

The Warsaw Convention signed by representatives of the United States and many European countries provides for a uniform liability code and uniform documentation on tickets and cargo for international carriers. It limits liability for loss of life or injury to $8,300 except where willful misconduct can be proved (the convention is designed to keep large damage claims from putting fledgling airlines out of business).

American Airlines is created by a reorganization of the 4-year-old American Airways Co. E. L. Cord loses control of the company.

Continental Air Lines has its beginnings in the Southwest division of Varney Air Transport founded by former Varney president and former United Airlines director Louis H. Mueller, 38, who flies mail between El Paso and Pueblo, Calif. Continental will become a major domestic passenger carrier.

Gerard Vultee designs a V-11 attack bomber that will be purchased by Brazil, China, Turkey, and the U.S.S.R., but he and his young wife will be killed in 1938 when their Stinson crashes into an Arizona peak in a snow squall (*see* 1930).

North American Aviation begins production of a military training plane designed by J. H. "Dutch" Kindelberger, 39, who was chief draftsman for Glenn L. Martin and chief engineer for Douglas Aircraft (*see* General Motors, 1933; Mustang, 1940).

The Chrysler Airflow is introduced by Chrysler Motors which also introduces overdrive. One model of the streamlined Airflow has the world's first curved one-piece windshield, and the car contains many innovative advances, but only 11,292 will be sold. By 1937 Chrysler will abandon the radical design as a costly miscalculation of public acceptance.

Greyhound cuts fares between New York and Chicago to as little as $8 in a rate war with competing bus lines (regular fare is $16). The company operates 1,800 33-passenger nickel-plated buses in 43 states, buys the buses from General Motors for $10,000 on a cost-plus contract, pays its drivers $175 per month plus a mileage rate, and grosses $30 million (*see* 1929).

The *City of Salina* goes into service between Kansas City and Salina, Kan., to begin a new era of streamlined passenger trains on U.S. railroads. The new Union Pacific train has air-conditioned Pullman cars of aluminum alloy.

A new Burlington Zephyr train goes into service between Chicago and Denver using diesel engines that begin the end of steam power on U.S. railroads (*see* 1924). The streamlined passenger train hits a top speed of 112.5 miles per hour and averages 77.6 miles per hour on a 1,017-mile nonstop run between the two cities May 26 (*see* 1935).

The Chinese ship *Weitung* burns January 21 on the Yangtze River; 216 lives are lost.

The S.S. *Morro Castle* catches fire off Asbury Park, N.J., September 8; 125 lives are lost.

The Cunard Line that began in 1839 as the Royal Mail Steam Packet Co. becomes the Cunard-White Star Line by merging with the White Star Line, which it acquired in 1927 (*see Queen Mary*, 1936).

French physicist Irene Joliot-Curie, 38, and her husband Frederic, 34, bombard aluminum with alpha particles (helium nuclei) from a naturally radioactive source and turn the aluminum into a radioactive form of phosphorus, but Mme. Joliot-Curie's mother Marie Curie dies July 4 of pernicious anemia at age 66 (*see* 1904; Cockcroft, Walton, 1932).

Patterns of Culture by Columbia University social anthropologist Ruth Benedict, 47, breaks new ground. Benedict studied anthropology for the first time at age 32, enrolled at Columbia 2 years later to study under Franz Boas, and will replace Boas as head of the department upon his retirement in 1936 (*see* 1928; 1946).

The Dionne quintuplets born May 28 at Callander, Ontario, are the world's first five infants on record to be

1934 ***(cont.)*** born at one delivery and survive. Elzire Dionne, 24, who already has six children, is delivered of five girls by her physician Allen R. Dafoe, 51; the infants Emilie, Yvonne, Cecile, Marie, and Annette average 2 pounds 11 ounces each.

The Worldwide Church founded by U.S. evangelist Herbert W. Armstrong, 37, adjures its members to observe the sabbath on Saturday, celebrate Passover but not Christmas, and follow kosher dietary laws. Armstrong and his son Garner Ted will use radio to attract a wide following.

German navy Signals Research chief Rudolf Kuhnold conducts the first practical radar tests March 20 at Kiel Harbor. He has developed a 700-watt transmitter working on a frequency of 600 megacycles plus a receiver and disk reflectors, succeeds in receiving echoes from signals bounced off the battleship *Hesse* anchored 600 yards away, and in October picks up echoes from a ship 7 miles away in a demonstration before high-ranking naval officers at Pelzerhaken near Lübeck. Signals are also received accidentally from a seaplane flying through the beam (the first detection of aircraft by radar) and the Reichstag appropriates 70,000 Reichsmarks ($57,500) to develop Kuhnold's idea *(see* Watson-Watt, 1935).

A Federal Communications Commission (FCC) created by Congress June 10 supervises the U.S. telephone, telegraph, and radio industries. Radio stations receive licenses, licensees and license-applicants must show that they serve, or intend to serve, the "public interest, convenience, and necessity," but licenses will, in fact, be granted and renewed almost automatically over the next 25 years.

France bans advertising on government radio stations.

"Terry and the Pirates" by comic-strip cartoonist Milton Arthur Caniff, 27, establishes an adult adventure story line with an Oriental setting and reflects such current events as the Japanese invasion of China. The *Chicago Tribune-New York Daily News* syndicate strip will employ explicit sexiness to attract a wide following.

"Li'l Abner" by comic-strip cartoonist Al Capp (Alfred Gerald Caplin), 35, is set in the fictitious hamlet of Dogpatch, Ky. Capp has been ghost-drawing "Joe Palooka" for Ham Fisher; his new strip will soon be syndicated to hundreds of papers.

"Flash Gordon" by King Features comic-strip cartoonist Alex Raymond is a world-of-tomorrow adventure strip that competes with the popular 25th-century strip "Buck Rogers."

Nonfiction *Leaves and Stones (Blätter und Steine,* essays) by Ernst Jünger.

Fiction *Seven Gothic Tales* by Danish novelist Isak Dinesen (Karen Dinesen, 49, Baroness Blixen) who married her cousin in 1914, lived with him until 1921 on the British East African coffee plantation her parents purchased for them, and has written her book in English; *I, Claudius* and *Claudius the God and His Wife Messaline* by Robert Graves; *And Quiet Flows the Don* (*Tikhiy Don*) by Soviet novelist Mikhail Aleksandrovich Sholokhov, 29; *Rickshaw Boy (Lo-lo hsiang-tzu)* by Chinese novelist Lao She (Shu Ch'ing-ch'un), 36, who has returned to China after studying in England from 1924 to 1930 (an English translation will appear in 1945 with a happy ending added by the translator); *Burmese Days* by George Orwell; *Weymouth Sands* by John Cowper Powys; *The Bachelors (Les Celibataires)* by Henry de Montherlant; *Tender Is the Night* by F. Scott Fitzgerald, whose central character Dick Diver is modeled on former émigré Gerald Murphy, 46, a dilettante painter of some talent who last year returned to New York to take over management of his late father's Mark Cross luggage and haberdashery shop; *Appointment in Samarra* by *New Yorker* magazine writer John Henry O'Hara, 29; A *Cool Million* by Nathanael West; *The Young Manhood of Studs Lonigan* by Chicago novelist James Thomas Farrell, 30, whose *Young Lonigan* appeared in 1932 and whose *Judgment Day* will appear next year to complete the depiction of a deteriorating Chicago and a brash, aimless blue-collar Chicagoan; *Summer in Williamsburg* by Brooklyn, N.Y., novelist Daniel Fuchs, 25; *Call It Sleep* by New York novelist Henry Roth, 28; *I Can Get It for You Wholesale* by New York novelist Jerome Weidman, 22; *Such Is My Beloved* by Canadian novelist Morley Callaghan, 31; *Seven Poor Men of Sydney* and *The Salzburg Tales* by Australian novelist Christina Stead, 32; *The Materassi Sisters (Sorelle Materassi)* by Italian novelist Aldo Palezzeschi (Aldo Giurlani), 49; *A Handful of Dust* by Evelyn Waugh; *Goodbye, Mr. Chips* by James Hilton; *Long Remember* by U.S. novelist MacKinlay Kantor, 30; *Daring Young Man* (stories) by California-born short story writer William Saroyan, 26, whose lead story is "The Daring Young Man on the Flying Trapeze;" *Lust for Life* by California-born novelist Irving Stone, 32, is a fictionalized life of Vincent van Gogh; *The Postman Always Rings Twice* by U.S. detective novelist James Mallahan Cain, 42; *Fer-de-lance* by U.S. detective novelist Rex (Todhunter) Stout, 48, whose overweight detective Nero Wolfe will make him a fortune; *Murder on the Orient Express* by Agatha Christie, whose new Hercule Poirot novel is published in the United States as *Murder on the Calais Coach; The Nine Tailors* by Dorothy Sayers; *The Thin Man* by Dashiell Hammett.

Juvenile *Mary Poppins* by Australian-born English author P. (Pamela) Travers, 28, whose wonder-working English governess will appear in a series of books.

Poetry *The Cantos* (IV) by Ezra Pound; *Variations on a Time Theme* by Edwin Muir; *The Master's Hammer* (*Le Marteau sans maître*) by French poet René Char, 21.

Painting *The Bullfight* by Pablo Picasso; *Homage to Mack Sennett* by René Magritte; *William Tell* by Salvador Dali; *Lord, Heal the Child, Homestead, Ploughing It Under,* and *Going Home* by Thomas Hart Benton.

Muralists Diego Rivera, José Clemente Orozco, and David Afaro Siqueiros return to Mexico following the election of Gen. Cardenas to the presidency (above).

Fuji Film, founded in January to make movie film, will grow to dominate Japan's still-film market.

Theater *Thunderstorm* by Chinese playwright Tsao Yu (Wan Chia-pao), 24, who adopts a style similar to that of Greek tragedy and awes Beijing audiences with the disruptive powers abroad in the world, but his play is perceived as an attack on the tyranny of the traditional family system; *Days Without End* by Eugene O'Neill 1/18 at Henry Miller's Theater, New York, with Earle Larimore, Stanley Ridges, Ilka Chase, 57 perfs.; *Yellowjacket* by Sidney Howard and Paul de Kruif 3/6 at New York's Martin Beck Theater, with James Stewart, Myron McCormick is about the conquest of yellow fever, 79 perfs.; *Merrily We Roll Along* by George S. Kaufman and Moss Hart 9/9 at New York's Music Box Theater, with Kenneth McKenna, Jessie Royce Landis, Mary Philips, 155 perfs.; *The Distaff Side* by John Van Druten 9/25 at New York's Booth Theater, with Mildred Natwick, Sybil Thorndike, Estelle Winwood, 177 perfs.; *The Children's Hour* by U.S. playwright Lillian Hellman, 29, 11/20 at Maxine Elliott's Theater, New York, with Eugenia Rawls hints at sexual abnormalities, 691 perfs.; *Rain from Heaven* by S. N. Behrman 11/24 at New York's Golden Theater, with Jane Cowl protests Nazi treatment of German Jews, 99 perfs.; *Professor Mamlock* by Friedrich Wolf 12/8 at Zurich's Schauspielhaus; *Accent on Youth* by U.S. playwright Samson Raphaelson, 38, 12/25 at New York's Plymouth Theater, with Constance Cummings, 229 perfs; *Yerma* by Federico García Lorca 12/29 at Madrid's Teatro Español.

British National Films is founded at Ellstree, England, by English miller J. Arthur Rank, 46, who will soon gain control of the Odeon Circuit with its 142 sumptuous theaters plus the Gaumont theatre chain founded in the 1890s. Rank will have studios at Denham, Pinewood Studios, and Shepherd's.

The Hays Office created by the U.S. film industry's MPPDA (Motion Picture Producers and Distributors of America, Inc.), hires former Postmaster General Will H. Hays, now 55, to administer a production code established by the industry. It will enforce such prohibitions as: no exposure of female breasts, no suggestion of cohabitation or seduction, and no unconventional kissing.

Shirley Temple makes her first full-length film at age six and follows Hamilton McFadden's *Stand Up and Cheer* (in which she steals the show by singing "Baby Take a Bow") with Alexander Hall's *Little Miss Marker* and David Butler's *Bright Eyes* (in which she sings "The Good Ship Lollipop"). All dimples and curly hair, Temple will go on to star in half a dozen other Hollywood films in the next 4 years.

Other films Jean Vigo's *L'Atalante* with Jean Dasté; Frank Capra's *It Happened One Night* with Clark Gable, Claudette Colbert; Norman Z. McLeod's *It's a Gift* with W. C. Fields; John Ford's *The Lost Patrol* with Victor McLaglen, Boris Karloff; Howard Hawks's *Twentieth Century* with John Barrymore, Carole Lombard; Robert Flaherty's documentary *Man of Aran*. Also: Gus Meins and Charles R. Rogers's *Babes in Toyland* with Laurel and Hardy; Rowland V. Lee's *The Count of Monte Cristo* with Robert Donat, Elissa Landi; Mitchell Leisen's *Death Takes a Holiday* with Fredric March; Frank Borzage's *Little Man, What Now?* with Margaret Sullavan, Douglass Montgomery; Richard Wallace's *The Little Minister* with Shirley Temple (above); Cedric Gibbons and Jack Conway's *Tarzan and His Mate* with Johnny Weissmuller, Maureen O'Sullivan; W. S. Van Dyke's *The Thin Man* with William Powell, Myrna Loy; Victor Fleming's *Treasure Island* with Wallace Beery, Jackie Cooper; Jack Conway's *Viva Villa!* with Wallace Beery, Leo Carillo; Gregory LaCava's *What Every Woman Knows* with Helen Hayes, Brian Aherne.

Opera *Lady Macbeth of the Mzensk District* 1/22 at Leningrad's Maly Opera House, with music by Dmitri Shostakovich. *Pravda* denounces the music, calling it "a deliberate and ugly flood of confusing sound . . . a pandemonium of creaking, shrieking, and clashes," another critic calls it "un-Soviet, unwholesome, cheap, eccentric, tuneless, and leftist"; *Merry Mount* 2/10 at New York's Metropolitan Opera House, with music by Howard Hanson; *Four Saints in Three Acts* 2/20 at New York's 44th Street Theater, with music by U.S. composer Virgil Garnett Thomson, 37, libretto by Gertrude Stein whose opera is not about saints, is not presented in three acts, but adds to the fame of Gertrude Stein with such bewildering lines as "Pigeons in the grass, alas."

The Glyndebourne Festival Opera has its first season 54 miles south of London on the 640-acre estate of Audrey and John Christie. The London Philharmonic provides music for the June to August performances in the Christies' 800-seat opera house.

The Berkshire Music Festival has its first season on the 210-acre Tappan family estate outside Lenox, Mass., whose grounds can accommodate up to 14,000 concertgoers. The Boston Symphony will take over the Festival in 1936 under the direction of Serge Koussevitzky, a disastrous rainstorm will halt a concert August 12, 1937, and the B.S. will erect a 6,000-seat "Shed" with fine acoustics in 1938.

First performances *Sacred Music (Avodath Hakodesh), a Sabbath Morning Service* for Baritone Cantor, Chorus, and Orchestra by Ernest Bloch 1/12 at Turin (radio broadcast); *Symphony—1933* by Roy Harris 1/26 at Boston's Symphony Hall; *The Enchanted Deer (Cantata Profane)* by Béla Bartòk 5/25 at London in a performance by the BBC Wireless Chorus and BBC Symphony; Concerto for String Quartet and Orchestra after the Concerto Grosso Op. 6 No. 7 by G. F. Handel by Arnold Schoenberg 9/26 at Prague (Schoenberg has been in the United States since October of last year); *Rhapsody on a Theme of Paganini* for Piano and Orchestra by Sergei Rachmaninoff 11/7 at Baltimore in a concert by the Philadelphia Orchestra.

Stage musicals *The New Ziegfeld Follies* 1/4 at New York's Winter Garden Theater, with Fanny Brice, Jane Froman, Vilma and Buddy Ebsen, Eugene and Willie Howard in a production staged by Ziegfeld's widow Bil-

1934 ***(cont.)*** lie Burke, music by Vernon Duke and others, lyrics by E. Y. Harburg and others, songs that include "I Like the Looks of You" by Billy Hill, 182 perfs.; *Conversation Piece* 2/16 at His Majesty's Theatre, London, with Noël Coward, Louis Hayward, George Sanders, book, music, and lyrics by Coward, songs that include "I'll Follow My Secret Heart," 177 perfs.; *Life Begins at 8:40* 8/27 at New York's Winter Garden Theater, with Bert Lahr, Ray Bolger, Brian Donlevy, music by Harold Arlen, lyrics by Ira Gershwin and E. Y. Harburg, songs that include "You're a Builder Upper," "Let's Take a Walk Around the Block," 237 perfs.; *Anything Goes* 11/21 at New York's Alvin Theater, with William Gaxton, Ethel Merman, Victor Moore, book by Guy Bolton, P. G. Wodehouse, Howard Lindsay, and Russel Crouse, music and lyrics by Cole Porter, songs that include "The Gypsy in Me," "I Get a Kick Out of You," "You're the Top," "Blow, Gabriel, Blow," "All Through the Night," and the title song, 420 perfs.; *Thumbs Up* 12/27 at New York's St. James Theater, with Bobby Clark, Ray Dooley, Sheila Barrett, songs that include "Autumn in New York" by Vernon Duke, "Zing! Went the Strings of My Heart" by James Hanley, 156 perfs.

Hollywood musicals Elliott Nugent's *She Loves Me Not* with Bing Crosby, Miriam Hopkins, Kitty Carlisle, music by Ralph Rainger, lyrics by Leo Robin, songs that include "Love in Bloom."

Popular songs "The Beer Barrel Polka" *("Skoda Lasky")* by Czech songwriters Jaromir Vejvoda, Wiadimir A. Timm, and Vasek Zeman (English lyrics by Lew Brown will appear in 1939); "What a Difference a Day Makes" *("Cuando vuelva a tu lado")* by Spanish composer Marcia Greves, English lyrics by Stanley Adams; "Miss Otis Regrets" by Noël Coward, who has written it for Bricktop (*see* 1925); "June in January" by Ralph Rainger, lyrics by Leo Robin; "Blue Moon" by Richard Rodgers, lyrics by Lorenz Hart; "Stars Fell on Alabama" by Frank Perkins, lyrics by Mitchell Parish; "Solitude" by Duke Ellington, lyrics by Eddie De Lange and Irving Mills; "I Only Have Eyes for You" by Harry Warren, lyrics by Al Dubin; The Very Thought of You" by Ray Noble; "Love Thy Neighbor" by Harry Revel, lyrics by Mack Gordon; "Deep Purple" by Peter DeRose, lyrics by Mitchell Parish; "Little Man You've Had a Busy Day" by Mabel Wayne, lyrics by Maurice Sigler and Al Hoffman; "All I Do Is Dream of You" by Nacio Herb Brown and Arthur Freed; "The Object of My Affection" by U.S. songwriters Pinky Tomlin, 25, Coy Poe, and Jimmie (James W.) Grier, 36; "Hands Across the Table" by French composer Jean Delettre, English lyrics by Mitchell Parish; "Alla en el Rancho Grande" by Mexican composer Sylvano R. Ramos, lyrics by Bartley Costello; "Tumbling Tumbleweeds" by Bob Nolan; "Carry Me Back to the Lone Prairie" by Carson Robinson; "You and the Night and the Music" by Arthur Schwartz, lyrics by Howard Dietz (for the film *Revenge With Music*); "With My Eyes Wide Open I'm Dreaming" by Mack Gordon and Harry Revel (for the film *Shoot the Works);* "You Oughta Be in Pictures" by U.S. composer Dana Suesse, 22, lyrics by Edward Heyman; "On the Good Ship Lollipop" by Richard Whiting, lyrics by Sidney Clare (for the film *Bright Eyes,* above); "Isle of Capri" by Will Grosz, lyrics by Jimmy Kennedy; "Winter Wonderland" by Felix Bernard, lyrics by Dick Smith.

Le Jazz Hot by French critic Hugues Panassié, 22, is the first book of jazz criticism.

The Hammond organ patented by Chicago clock maker Laurens Hammond, 39, is the world's first pipeless organ. The 275-pound instrument has a two-manual console with pedal clavier and power cabinet but no reeds, pipes, or vibrating parts; it costs less than 1¢ per hour to operate and will lead to a whole generation of electrically amplified instruments.

Harlem's Apollo Theater is opened by entrepreneurs Leo Brecher and Frank Schiffman who have bought the failing burlesque theater Hurtig and Seaman's, changed its name, and reversed its policy of barring black patrons. Schiffman books blues singer Bessie Smith into the Apollo and makes it the leading U.S. showcase for black performers.

U.S. prizefighter Max Baer, 25, knocks out Primo Carnera June 13 in the eleventh round of a title bout at Long Island City and wins the world heavyweight championship he will hold for exactly one year.

Fred Perry wins in men's singles at Wimbledon, Dorothy Round, 24, in women's singles; Perry wins in men's singles at Forest Hills, Helen Jacobs in women's singles.

Pitcher Leroy Robert "Satchel" Paige, 28, breaks Dizzy Dean's 30-game winning streak as he strikes out 17 men and allows no runs in a Hollywood All Stars game while Dean strikes out 15 and lets one run score. Paige starts 29 games in 29 days, and his team wins 104 out of 105 games as he continues to pitch for the Bismarck, S. Dak., team for which he pitched 31 games last year and won 27. Paige will pitch in the Negro National League for the Crawford Giants of Pittsburgh, Homestead Grays of Baltimore, and Kansas City Monarchs, clinching the Negro World Series title for the Monarchs in 1942 and pitching 64 scoreless innings for the team in 1946 *(see* 1948).

The St. Louis Cardinals win the World Series, defeating the Detroit Tigers 4 games to 3 after winning the pennant on the last day of the season.

The U.S. ocean yacht *Rainbow* defeats England's *Endeavour* 4 to 2 to retain the America's Cup.

The Masters golf tournament for professionals at Georgia's Augusta National Golf Club will be one of golf's annual classics, along with the U.S. Open, the British Open, and the Professional Golfers Association (PGA) contests.

Italy defeats Czechoslovakia 2 to 1 in overtime to win the World Cup football (soccer) competition at Rome's Stadio Torino.

U.S. men's underwear sales slump after moviegoers see Clark Gable remove his shirt in the film *It Happened*

One Night (above) to reveal that he wears no undershirt. Sales of men's caps have declined as a result of the association of men's caps with hoodlums in gangster movies.

The Washeteria that opens April 18 at Fort Worth, Tex., is the first launderette. Proprietor J. F. Cantrell has installed four washing machines and charges by the hour.

Clyde Barrow, 24, and Bonnie Parker die in a hail of bullets May 23 on a road 50 mies east of Shreveport, La., after a 2-year career in which they have casually killed 12 people in Texas, Oklahoma, Missouri, and Iowa. Their most successful robbery netted them no more than $3,500 and the father of one of their gang members has told the police where to watch for them. Texas Ranger Frank Hamer, sheriff's deputy Ted C. Hinton, and four other sheriff's deputies have set up an ambush, Bonnie and Clyde drive into the trap at 85 miles per hour, the lawmen riddle them with 50 bullets, and they die holding a machine gun and a sawed-off shotgun, respectively.

John Dillinger, 32, leaves the Biograph Theater in Chicago July 22 not knowing that a woman friend has betrayed him. FBI agents shoot him dead after a brief career in which he has robbed banks in Indiana, Ohio, Illinois, and Wisconsin.

New York police arrest German-American furrier Bruno Richard Hauptmann September 20 for possession of ransom money paid to recover Charles A. Lindbergh, Jr., in 1932. Hauptmann says he received the gold certificates from a former partner in the fur business and denies any connection with the Lindbergh kidnapping (*see* 1936).

The U.S. Army turns over Alcatraz Island in San Francisco Bay to the Bureau of Prisons which will convert the "Rock" into a federal prison. The army has used the island since 1854, it is 1,650 feet long and 450 feet wide, the army built a three-story cell block for military prisoners in 1909, and before the Bureau of Prisons abandons it in 1963, Alcatraz (Spanish for pelican) will house the most notorious U.S. criminals, including Al Capone, George (Machine-gun) Kelly, and Robert Stroud.

San Francisco's Coit Tower is completed by Arthur Brown, Jr.

A Field Guide to the Birds by Boston painter-ornithologist Roger Tory Peterson, 26, takes up where John James Audubon left off (*see* 1843).

Great Smoky Mountains National Park is created by act of Congress for full development. Established for protection and administration in 1930, the park along the North Carolina-Tennessee border embraces 515,226 acres in the highest mountain range east of the Mississippi.

Dust storms in May blow some 300 million tons of Kansas, Texas, Colorado, and Oklahoma topsoil into the Atlantic. At least 50 million acres lose all their topsoil, another 50 million are almost ruined, and 200 million are seriously damaged (*see* Soil Conservation Act, 1935).

The Taylor Grazing Act passed by Congress June 28 sets up a program to control grazing and prevent erosion of western grasslands. The new law effectively closes the public domain to homesteading (*see* Stockraising Homestead Act, 1916).

Western ranchers begin two decades of wholesale slaughter of wild horses to clear the disintegrating ranges. The horse meat fetches 5¢ to 6¢ per pound (it is sold for human food, dog food, and chicken feed), the government pays a bounty to encourage the horse hunters, and by 1952 the wild horse population will be 33,000 down from 2 million in 1900.

A Farm Mortgage Financing Act passed by Congress January 31 creates a Federal Farm Mortgage Corporation to help farmers whose mortgages are being foreclosed (*see* 1931).

A Crop Loan Act passed February 23 authorizes loans to farmers to tide them over until harvest time.

The Frazier-Lemke Farm Bankruptcy Act passed June 28 allows mortgage foreclosures to be postponed for 5 years (but *see* 1935).

The Taylor Grazing Act (above) establishes grazing districts under the control of the Department of the Interior.

The western dust storms (above) are an aftermath of imprudent plowing during the Great War, when farmers planted virgin lands in wheat to cash in on high grain prices.

"Okies" and "Arkies" from the Dustbowl begin a trek to California that will take 350,000 farmers west within the next 5 years.

Dust storms ruined 100 million acres of U.S. croplands, forcing "Okies" and "Arkies" to move west to California.

1934 ***(cont.)*** Mexico's agrarian revolution advances under President Cardenas (above), who will resume distribution of the land to the pueblos and work also to build the power of organized labor.

Nazi Germany starts the *Erzengungsschlacht* program to expand domestic food production. By 1937 the country will be producing 90 percent of the food it consumes.

Drought reduces the U.S. corn crop by nearly a billion bushels, and the average yield per acre falls to 15.7 bushels, down from 22.8 last year (wheat yields average 11.8 bushels per acre).

U.S. soybean acreage increases to one million (*see* 1923; 1944). Fort Wayne, Ind., sugar beet processor Dale W. McMillen sees the potential of soybeans as a livestock feed, buys German processing machinery, and founds Central Soya (*see* Staley, 1922).

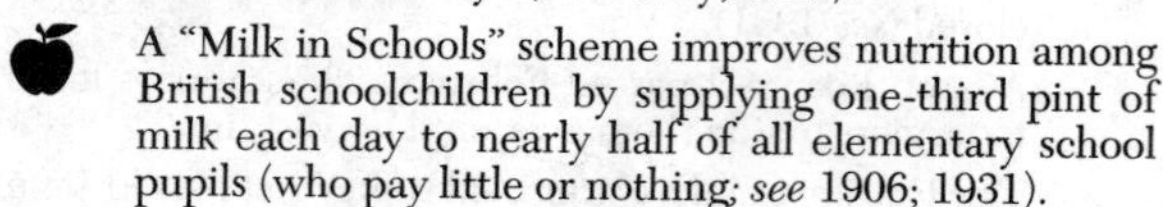

A "Milk in Schools" scheme improves nutrition among British schoolchildren by supplying one-third pint of milk each day to nearly half of all elementary school pupils (who pay little or nothing; *see* 1906; 1931).

U.S. food-buying patterns begin shifting to larger consumption of red meats (especially beef and pork), fruits, green vegetables, and dairy products as industrial earnings start to improve.

MSG (monosodium glutamate) is produced commercially for the first time in the United States, which has depended until now on the flavor enhancer Ajinimoto imported from Japan (*see* 1908; 1947).

A new chilling process for meat cargoes improves on the process used in 1880 on the S.S. *Strathleven.*

Pepsi-Cola is acquired by Walter Mack, who will promote Pepsi's 12-ounce bottle to challenge Coca-Cola (*see* 1933; 1939).

Seagram's 7 Crown, introduced by Distillers Corp.-Seagram, will become the top-selling U.S. whiskey (*see* 1933; 1947).

The Los Angeles Farmers Market opens at 6333 West Third Street with stalls that rent for 50¢ a day on 30-day leases. Started by 17 farmers who have had trouble selling their produce, the market will grow to cover 3 acres of buildings with 18 acres of parking space and with 160 individually owned stalls.

1935 A plebiscite conducted by the League of Nations January 13 shows that voters in the Saar Basin prefer reunion with the German Reich 9 to 1 over union with France or continuation of rule by the League, which has administered the region since 1919. The League returns the Saar to Germany March 1.

Adolf Hitler denounces Versailles Treaty clauses providing for German disarmament, but Berlin signs an agreement with London promising not to expand the German navy to a size larger than 35 percent of the Royal Navy.

Hitler creates the Luftwaffe to give Germany a military air capability. Reich minister for air forces is Reichstag president Hermann Goering, who served in the air force in the Great War.

The U.S. Army Air Corps has its beginnings in the General Headquarters Air Force established under the command of Brig. Gen. Frank Maxwell Andrews, 51. It is the first independent U.S. strategic air force.

Persia becomes Iran by order of Reza Shah Pahlevi, who has controlled the country since 1925.

Siam's Rama VII (Prajadhipok) abdicates March 2 after a 10-year reign in which the Siamese have abolished absolute royal power. He is succeeded by his nephew of 10 who will reign until 1946 as Rama VIII (Ananda Mahidol).

France's Flandin ministry is overthrown in May following demands that it be given near-dictatorial powers to save the collapsing franc. Pierre Laval, 51, forms a new cabinet. Socialist groups merge November 3, form a Popular Front with communists and radical socialists to counteract agitation by reactionaries, the government orders political leagues dissolved December 28 (*see* 1936).

The Soviet Comintern founded in 1919 responds in June to the growth of fascism abroad. It approves Communist participation in "Popular" front governments with other leftists or moderates (*see* 1943).

Britain's third Baldwin ministry begins June 7 as Stanley Baldwin, now 67, replaces Ramsay MacDonald as head of the coalition cabinet. An Anglo-German naval pact is signed June 18 (*see* above).

Italian troops invade Ethiopia October 3. France has ceded part of French Somaliland to Italy and sold her shares in the Ethiopian Railway, but Benito Mussolini has rejected further concessions offered by France and Britain in mid-August. His troops seize the provincial capital Makale November 8, and the League of Nations imposes economic sanctions on Italy November 18 (*see* 1934; 1936).

Czechoslovakia's first president Thomas Masaryk, now 85, resigns December 14 after 17 years in power. He is succeeded by his minister of foreign affairs Eduard Beneš (*see* 1918; 1938).

Bolivia and Paraguay end a 3-year war June 12 under pressure from the United States and five Latin neighbors. The League of Nations and Pan-American Union were unable to avert hostilities over the disputed Chaco area and a definitive peace treaty will not be signed for 3 years.

Venezuela's President Juan Vicente Gomez dies December 18 at age 78, ending a 26-year dictatorship that has seen his country become a major oil producer. The death of the strongman produces disorders. Gen. Eleazar Lopez Contreras becomes provisional president and restores order.

Josef Stalin decrees that Soviet children above age 12 are subject to the same punitive laws that apply to adults—8 years in a labor camp for stealing corn or potatoes, for example, or 5 years for stealing cucumbers.

Nazi SS (*Schutzstaffel*) leader Heinrich Himmler starts a *Lebensborn* (life source) state breeding program to produce an "Aryan super race." He encourages young women of "pure blood" to volunteer their services as mates of SS officers to contribute blond, blue-eyed babies who will grow into thin-lipped narrow-nosed "Nordic beings" to carry on the "thousand-year Reich."

The Nuremberg Laws enacted by the Nazi Party congress meeting at Nuremberg September 15 deprive Jews of German citizenship, forbid intermarriage with Jews, and make intercourse between "Aryans" and Jews punishable by death to prevent "racial pollution."

The basic definition of a Jew published November 14 defines the categories of mixed offspring, or *mischlinge*. The first degree includes anyone with two Jewish grandparents, the second degree anyone with one Jewish grandparent.

Belfast has anti-Catholic riots in July. Northern Ireland expels Catholic families; Catholics in the Irish Free State retaliate.

The National Labor Relations Act (Wagner-Connery Act) passed by Congress July 5 creates a National Labor Relations Board and reasserts the right of collective bargaining. The right was contained in section of the NRA code of 1933 that the Supreme Court has struck down (*see* below).

The Committee for Industrial Organization (CIO) is founded November 9 by dissidents within the AF of L who advocate industrial unionism but have been outvoted by craft union advocates at the AF of L's November convention at Washington, D.C. The dissidents elect United Mine Workers president John L. Lewis chairman and Scots-American UMW vice president Philip Murray, 39, chairman of the Steel Workers Organizing Committee (*see* 1936).

The United Automobile Workers (UAW) holds its first convention at Detroit and receives a charter from the AF of L (*see* 1936).

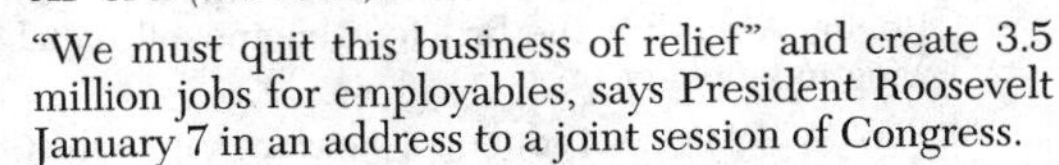

"We must quit this business of relief" and create 3.5 million jobs for employables, says President Roosevelt January 7 in an address to a joint session of Congress.

An Emergency Relief Appropriation Act passed by Congress April 8 authorizes nearly $5 billion to provide "work relief and to increase employment by providing useful projects."

A Works Progress Administration (WPA) created by executive order May 6 is headed by Harry L. Hopkins.

A National Youth Administration (NYA) is created June 26 as a division of the WPA (above) to provide jobs for 4.2 million young people seeking work.

The National Industrial Recovery Act (NRA) of 1933 is unconstitutional, the Supreme Court rules May 27 in the case of *Schechter Poultry Corp. v. United States.*

The Railway Pension Act of 1934 is unconstitutional, the Supreme Court rules June 27 in the case of *Railroad Retirement Board v. Alton Railway Co.*, but Congress passes a Railway Retirement Act to replace the 1934 law. Railroads and their employees are each to contribute 3.5 percent of the first $300 earned each month to build a retirement fund.

Franklin D. Roosevelt's New Deal measures helped Americans survive the Depression and averted a revolution.

The Social Security Act signed into law August 14 provides a system of old-age annuities and unemployment insurance benefits. Drafted by Eastman Kodak treasurer Marion Bayard Folsom, 41, the act provides for state aid to be matched by federal aid of up to $15 per month for needy retirees over age 65, with employers and employees taxed equally to support the program. The tax is to begin in 1937 at 1 percent and rise by steps to 3 percent in 1949; qualified employees are to be able to retire at age 65 beginning January 1, 1942, and receive payments of $10 to $15 per month for the rest of their lives; widows and orphans are to receive benefits beginning in 1940.

The Social Security Act (above) provides for a fund to help states pay allowances to workers who have lost their jobs. Beginning with 1936 payrolls, employers are to set aside for this purpose a 1 percent tax in the first year, 2 percent the second, and 3 percent the year after, with employees contributing half the amount set aside.

President Roosevelt reports Internal Revenue Service figures for the year in a monopoly message: one-tenth of 1 percent of U.S. corporations own 52 percent of all corporate assets reported and earn 50 percent of all corporate income. Less than 5 percent own 87 percent of all corporate assets, and less than 4 percent earn 87 percent of all net profits reported by all U.S. corporations.

A Revenue Act passed by Congress August 30 seeks to achieve some diffusion of wealth. The law provides for inheritance and gift taxes; critics protest that the law means double taxation for recipients of stock dividends and clippers of bond coupons.

1935 ***(cont.)*** The Banking Act signed into law August 23 reorganizes the Federal Reserve System created in 1913 and increases its authority. It establishes an open market committee to buy and sell government securities held by the Federal Reserve Banks and thus control the nation's money supply. The banking reform also regulates checking accounts and requires that banks contribute to the support of a Federal Deposit Insurance Corp. that protects depositors from default or theft.

Fort Knox is established in Kentucky to serve as a repository for U.S. gold bullion. The Army guards the gold and the government assures foreign bankers that they can redeem every $35 in U.S. paper currency for an ounce of gold from Treasury Department stores at Fort Knox (which do not approach the stores held for the accounts of various nations in the vaults of the Federal Reserve Bank in New York).

Britain and Ireland conclude a cattle and coal agreement that enables the Irish to ship some of their meat surplus, but the disastrous Anglo-Irish tariff war will not end until February of next year when Dublin agrees to pay land annuities (*see* 1932; 1938).

Josef Stalin urges productivity increases in November and encourages individual initiative with extra pay for notable efficiency, giving official recognition to the Stakhanovite movement based on the achievement of coal miner Aleksei Grigorievich Stakhanov, 29. Working at a coal face in the Donets Basin the night of August 30, Stakhanov and his crew reportedly mined 102 tons of coal in a single shift of 5 hours 45 minutes—14 times the normal rate. Virtually illiterate, the stocky miner has used other members of the crew to shore up the shaft with timbers while he hewed the coal, departing from the previous time-consuming process of alternating between cutting coal and erecting timbers. But most workers live by the motto, "As long as the bosses pretend they are paying us a decent wage we will pretend that we are working"; productivity will remain low.

The Rural Electrification Administration established by President Roosevelt's executive order May 11 will underwrite rural electric cooperatives and provide loans for transmission lines. Only 10 percent of 30 million U.S. rural residents have electrical service, but with help from the REA, 90 percent of U.S. farms will have electricity by 1950.

An estimated 6.5 million windmills have been produced in the United States in the past half century. Most have been used to pump water or run sawmills, but some "wind generators" have produced small amounts of electricity.

Britain's Anglo-Persian Co. of 1914 becomes the Anglo-Iranian Co. as Persia becomes Iran (above) (*see* British Petroleum, 1954).

The French Line passenger ship S.S. *Normandie* goes into service on the North Atlantic, arriving at New York June 3 after crossing from Southampton in a record 4 days, 11 hours, 42 minutes. The 79,280-ton luxury liner with four screws is 1,029 feet in length overall and has an 80-foot swimming pool, 23 elevators, and a dining room modeled after the Hall of Mirrors at Versailles (*see* 1942).

Greek entrepreneur Stavros S. Niarchos, 26, begins a shipping fleet that will rival the maritime empire of Aristotle Onassis (*see* 1930).

The Lower Zambezi railroad bridge opens January 14 in Africa and will be the world's longest railroad bridge until December (*see* below).

A railroad bridge designed by McKim, Mead and White opens to span the Cape Cod Canal of 1914 at Buzzards Bay, Mass.

Newark's Pennsylvania Station opens March 23. Lawrence Grant White of McKim, Mead and White has designed it.

A hurricane September 2 completely wipes out the Florida East Coast Railroad line between Florida City and Key West that was repaired after the hurricane of 1926. The new storm sweeps across the keys and kills 400 (*see* overseas highway, 1938).

The Huey P. Long Bridge completed December 10 at Metairie, La., is 4.35 miles long—the world's longest railroad bridge. It was designed by Ralph Modjeski (*see* 1930).

General Motors begins a program of dieselizing U.S. railroad locomotives (*see* 1930; 1941).

Twin stainless steel Zephyrs go into service on the Burlington Route. The trains have a cruising speed of 90 miles per hour and cut 3.5 hours off the 882-mile round trip between Chicago and the twin cities of Minneapolis and St. Paul (*see* 1934).

Moscow's Metropol subway opens in May with 5.5 miles of track that will grow to become a 105-mile system with more than 90 stations. The new subway has stations modeled after the great U.S. railroad terminals with magnificent murals, and chandeliers.

The B-17 bomber demonstrated for the U.S. Army by Boeing Aircraft is the first four-engine, all-metal, low-wing monoplane.

Pilot Wiley Post sets out on a pleasure trip to the Orient with his humorist friend Will Rogers, now 55. Their plane crashes near Point Barrow, Alaska, August 15, killing both men (*see* 1931).

The first official transpacific air mail flight leaves San Francisco November 22, and the China Clipper flying boat of Juan Trippe's Pan American Airways arrives November 29 at Manila after flying 8,210 miles with stops at Honolulu, Midway, Wake, and Guam (*see* 1930). The new San Francisco-Manila route gives Pan Am a total of 40,000 miles versus 24,000 for Air France, 23,600 for Lufthansa, 21,000 for British Imperial, 11,700 for KLM, 10,500 for Soviet Russia's Aeroflot.

The Toyota motorcar is introduced in prototype by Japan's Toyoda Automatic Loom Works whose Sakichi Toyoda received £100,000 from a Lancashire firm in 1929 for rights to produce his advanced loom in England. Toyoda has used the money to develop a motorcar design and has put his son Kiichiro in charge of the venture. An engineer sent by Kiichiro to the Packard works at Detroit as a tourist has acquired enough information to set up an assembly line.

English racing driver Sir Malcolm Campbell, 50, drives his *Bluebird* at 276 miles per hour over sand at Daytona Beach, Fla.

The Motor Carrier Act passed by Congress August 9 places U.S. interstate bus and truck lines under control of the Interstate Commerce Commission.

Polyethylene, developed by the Alkali Division of Britain's Imperial Chemical Industries, is the first true "plastic" ever made from the polymerization of ethylene. Company chemists 2 years ago found small specks of a white solid material when they opened their retort after an attempt to force a copolymerization between liquid ethylene and an aldehyde, using extremely high temperatures to link small molecules into long chains; their polyethylene will find wide use in packaging (*see* 1953).

Nylon, developed by E. I. du Pont chemist Wallace Hume Carothers, 39, is a synthetic polymide that will replace silk, rayon, and jute in many applications. Carothers has combined adipic acid and hexamethylenediamine to form long filaments of what he calls "polymer 66." Drawing the filaments out to a certain length aligns the polymer chains and pulls them to their full extent, making the filaments strong and durable but giving them many of the characteristics found in silk and wool (*see* 1937).

Sulfa drug chemotherapy introduced by German biochemist Gerhard Domagk, 40, launches a new era in medicine that will revolutionize treatment of infectious diseases including certain forms of pneumonia and reduce the hazards of peritonitis in abdominal surgery. Director since 1927 of I. G. Farbenindustrie at Elberfeld, Domagk and his colleagues have injected 1,000 white mice with fatal doses of streptococci and then treated them with prontosil, an azo dye patented in 1931 by Farbenfabriken Bayer, which has turned out to have antibacterial properties (all the white mice recovered, and the same results have been obtained with rabbits). Prontosil will be developed into sulfanilamide, which will be followed by other sulfonamides (*see* Ehrlich, 1909; Dubos, 1939; penicillin, 1928, 1940; streptomycin, 1943).

Rockefeller Institute biochemist Wendell Meredith Stanley, 31, demonstrates the proteinaceous nature of viruses, showing that they are not submicroscopic organisms as is commonly believed.

The alkaloid ergonovine, extracted from the ergot fungus *Claviceps purpurea*, proves an effective tool in obstetrics (*see* Stearns, 1808; ergotamine, 1918). Selective breeding will raise the alkaloid content of the fungus from 0.02 percent to 0.5 percent, and the product will be marketed under the name Ergothetrine in Britain (*see* ergotism, 857).

British inventor A. Edwin Stevens produces the first wearable electronic hearing aid and founds a company he calls Amplivox to produce the 2-pound device (*see* Zenith, 1943).

Scottish physicist Robert Alexander Watson-Watt, 43, of Britain's Government Radio Research Station begins setting up radar (radio detecting and ranging) warning systems on the British Coast. He has pioneered radar, but his installations require enormous antenna towers (the longer the wavelength transmitted, the larger the antenna needed), and he uses wavelengths of roughly 50 centimeters (*see* Hertz, 1887; Kuhnold, 1934; MIT, 1940).

The Magnetophon produced by AEG in Berlin is the first tape recorder to use plastic tape. Its tape speed is 30 inches per second, its performance is inferior to that of the 1920 Blattnerphone, but its operating cost is lower (*see* 1940).

The American Institute of Public Opinion (Gallup poll) founded by Iowa statistician George H. Gallup, 34, gauges reader reaction to newspaper features. *Des Moines Register and Tribune* publisher Gardner Cowles, 32, has hired Gallup and launches him on his career as pollster (*see* 1936).

Gardner Cowles (above) acquires the *Minneapolis Star* as he builds an empire of publishing and broadcasting properties (*see Look*, 1937).

This Week is launched by Crowell-Collier, whose Sunday newspaper supplement will be carried by scores of metropolitan papers (*see Collier's*, 1919).

Nonfiction *My Country and My People* by Chinese scholar Lin Yutang, 40.

Fiction *Gyakko* (*Dokenohana*) by Japanese novelist Osamu Dazai (Shoji Tsushima), 26; *Before the Dawn* (*Yoakemae*) by Shimasaki Toson; *Wheel of Fortune* (*Le ambizoni sbagliate*) by Alberto Moravia whose second novel is ignored by reviewers under pressure from Fascist authorities; *It Can't Happen Here* by Sinclair Lewis projects a fascist takeover of America, warning against complacency about the totalitarian governments that are tightening their grip on Japan and much of Europe; *The House in Paris* by Elizabeth Bowen; *Holy Ireland* by Irish novelist Norah Hoult, 37; *A Clergyman's Daughter* by George Orwell; *Mr. Norris Changes Trains* by Christopher Isherwood; *Rajah Amahr* by L. H. Myers; *The Stars Look Down* by A. J. Cronin; *A Crime* (*Un Crime*) by Georges Bernanos; *The Last Puritan* by philosopher George Santayana; *Tortilla Flat* by California novelist John Ernst Steinbeck, 33, is about the Spanish-speaking "paisanos" of Monterey; *Of Time and the River* and *From Death to Morning* (stories) by Thomas Wolfe.

Juvenile *National Velvet* by Enid Bagnold.

Poetry *Monologue* (*Alleenspraak*) by Afrikaans poet N. P. van Wyk Louw, 29; *Poems* by English poet William Empson, 29.

Painting *Minotauromachy* by Pablo Picasso. Max Liebermann dies at Berlin February 8 at age 87; Kazimir Malevich dies at Leningrad May 15 at age 57; Childe Hassam dies at East Hampton, N.Y., August 27 at age 75.

1935 ***(cont.)*** The WPA (above) makes jobs for artists in a program to decorate post offices and other federal buildings.

Eastman Kodak introduces Kodachrome for 16-millimeter movie cameras. Eastman has acquired production rights from U.S. concert violinist Leopold Godowsky, Jr., 35, and his pianist co-inventor Leopold Damrosch Mannes, 35, who have devised a three-color dye-coupling process with help from a team of Eastman scientists. Godowsky and Mannes met as students in 1916 and have studied physics and chemistry at university while continuing their music careers; Godovsky married George Gershwin's sister Frances in 1931 and has been working with Mannes at Rochester since July 1931 *(see* 1936).

Theater *Waiting for Lefty* by U.S. playwright Clifford Odets, 28, 1/5 at New York's Civic Repertory Theater on 14th Street, 168 perfs.; *The Old Maid* by Zöe Akins 1/7 at New York's Empire Theater, with Judith Anderson, Helen Mencken, 305 perfs.; *Point Valaine* by Noël Coward 1/16 at New York's Ethel Barrymore Theater, with Alfred Lunt, Lynn Fontanne, Broderick Crawford, 55 perfs.; *Awake and Sing* by Odets 2/19 at New York's Belasco Theater, with Morris Carnovsky, Stella Adler, John Garfield, 209 perfs.; *Till the Day I Die* and *Waiting for Lefty* by Odets 3/26 at New York's Longacre Theater with a top price of $1.50 per seat in a production by the Group Theater founded 4 years ago by Odets and director Lee Strasberg, 33, 136 perfs.; *There Was a Prisoner (Y avait un presonnier)* by Jean Anouilh 3/21 at the Théâtre des Ambassadeurs, Paris; *Night Must Fall* by Welsh-born English playwright Emlyn Williams, 29, 5/31 at London's Duchess Theatre; *Winterset* by Maxwell Anderson 9/25 at New York's Martin Beck Theater, with Burgess Meredith, Richard Bennett, Margo is based on the Sacco-Vanzetti case, 195 perfs.; *Dead End* by Sidney Kingsley 10/28 at New York's Belasco Theater, with Joseph Dowling, Sheila Trent, 268 perfs.; *Murder in the Cathedral* by T. S. Eliot 11/1 at London's Mercury Theatre is based on the death of the archbishop of Canterbury at the hands of Henry II's knights in 1170: "The last temptation is the greatest treason:/ To do the right deed for the wrong reason" (I), 180 perfs.; *Tiger at the Gates (La Guerre de Troie n'aura pas lieu)* by Jean Giraudoux 11/21 at the Théâtre de l'Athenée, Paris; *Boy Meets Girl* by U.S. playwrights Bella (Cohen) and Samuel Spewack, both 36, 11/27 at New York's Cort Theater, with Jerome Cowan, Garson Kanin, Everett Sloane, 669 perfs.

Radio comedy and drama *Fibber McGee and Molly* 4/16 on NBC Blue Network stations with Jim and Marian Driscoll Jordan (to 1952); *Gang Busters* 7/20 on NBC stations. Developed by Phillips H. Lord, the show will move to CBS early next year and continue for more than 20 years.

Films Clarence Brown's *Anna Karenina* with Greta Garbo, Fredric March; James Whale's *The Bride of Frankenstein* with Elsa Lanchester, Boris Karloff; George Cukor's *David Copperfield* with Freddie Bartholomew, W. C. Fields, Lionel Barrymore; John Ford's *The Informer* with Victor McLaglen; Henry Hathaway's *Lives of a Bengal Lancer* with Gary Cooper, Franchot Tone, C. Aubrey Smith; Frank Lloyd's *Mutiny on the Bounty* with Charles Laughton, Clark Gable, Franchot Tone; Sam Wood's *A Night at the Opera* with the Marx Brothers; Leo McCarey's *Ruggles of Red Gap* with Charles Laughton, Mary Boland, Charles Ruggles; Jack Conway's *A Tale of Two Cities* with Ronald Colman; Alfred Hitchcock's *The 39 Steps* with Robert Donat, Madeleine Carroll; Leni Riefenstahl's propaganda documentary *Triumph of the Will* extolling last year's Nuremberg rallies. Also: William Keighley's *"G"-Men* with James Cagney; William Wyler's *The Good Fairy* with Margaret Sullavan, Herbert Marshall; Clyde Bruckman's *The Man on the Flying Trapeze* with W.C. Fields; Richard Boleslawski's *Les Misérables* with Fredric March, Charles Laughton; Harold Young's *The Scarlet Pimpernel* with Leslie Howard, Merle Oberon; William Dieterle's *The Story of Louis Pasteur* with Paul Muni.

"STICKS NIX HICK PIX" headlines *Variety* July 17. The 30-year-old show business newspaper reports that rural audiences reject motion pictures with bucolic stories and characters.

Hollywood musicals Mark Sandrich's *Top Hat* with Fred Astaire and Ginger Rogers, Irving Berlin songs that include "Cheek to Cheek," "Isn't It a Lovely Day to Be Caught in the Rain," "Top Hat, White Tie, and Tails"; A. Edward Sutherland's *Mississippi* with Bing Crosby, W. C. Fields, Joan Bennett, music by Richard Rodgers, lyrics by Lorenz Hart, songs that include "It's Easy to Remember but So Hard to Forget"; Busby Berkeley's *Gold Diggers of 1935* with Dick Powell, music by Harry Warren, lyrics by Al Dubin, songs that include "Lullaby of Broadway"; Roy Del Ruth's *Folies Bergère* with Maurice Chevalier, Ann Sothern, Merle Oberon.

Stage musicals *Spread It Abroad* at London with Cyril Ritchard, Madge Evans, songs that include "These Foolish Things Remind Me of You" by Jack Strachey and Harry Link, lyrics by Holt Marvell; *At Home Abroad* 9/19 at New York's Winter Garden Theater, with Beatrice Lillie, Eleanor Powell, Ethel Waters, Eddie Foy, Jr., music by Arthur Schwartz, lyrics by Howard Dietz, songs that include "Hottentot Potentate," 198 perfs.; *Jubilee* 10/12 at New York's Imperial Theater, with Melville Cooper, Mary Boland, Montgomery Clift, music and lyrics by Cole Porter, songs that include "Begin the Beguine,""Just One of Those Things," 169 perfs.; *Jumbo* 11/16 at the New York Hippodrome, with Jimmy Durante, a live elephant, music by Richard Rodgers, lyrics by Lorenz Hart, songs that include "Little Girl Blue," "The Most Beautiful Girl in the World," 233 perfs.; *George White's Scandals* 12/24 at New York's New Amsterdam Theater, with Bert Lahr, Willie and Edgar Howard, Rudy Vallée, music by Ray Henderson, lyrics by Jack Yellen, 110 perfs.

Opera *Porgy and Bess* 10/10 at New York's Alvin Theater, with Todd Duncan, J. Rosamond Johnson, music by George Gershwin, lyrics by Ira Gershwin and

DuBose Heyward whose 1927 stage play *Porgy* has provided the libretto, arias that include "I Got Plenty o' Nuttin," "Summertime," "Bess, You Is My Woman Now," "I Loves You, Porgy," "It Ain't Necessarily So," 124 perfs.

Your Hit Parade debuts on radio in April. Sponsored by American Tobacco, the program of song hits promotes Lucky Strikes.

Benny Goodman opens at the Palomar Ballroom in Los Angeles August 21 and is hailed as the King of Swing. Clarinetist-bandleader Benjamin David Goodman, 26, uses arrangements by Fletcher Henderson to introduce a new "big band" jazz style, he starts a *Let's Dance* radio program of "swing" music, and he begins a long career that will be helped by such sidemen as Lionel Hampton, Gene Krupa, and Teddy Wilson (*see* 1938).

Jazz music of black or Jewish origin is banned from German radio beginning in October.

Popular songs "The Music Goes Round and Round" by Edward Farley and Michael Riley, lyrics by "Red" Hodgson; "(Lookie, Lookie, Lookie) Here Comes Cookie" by Mack Gordon and Harry Revel (for the film *Love in Bloom*); "About a Quarter to Nine" and "She's a Latin from Manhattan" by Al Dubin, lyrics by Harry Warren (for the film *Go Into Your Dance*); "I Won't Dance" by Jerome Kern, lyrics by Otto Harbach and Oscar Hammerstein II; "Stairway to the Stars" by violinist Matt Malneck and Frank Signorelli, lyrics by Mitchell Parish; "Goody Goody" by Matt Malneck and Johnny Mercer; "When I Grow Too Old to Dream" by Sigmund Romberg, lyrics by Oscar Hammerstein II; "I'm in the Mood for Love" by Jimmy McHugh and Dorothy Fields; "A Little Bit Independent" by Joseph A. Boyle, lyrics by Edgar Leslie; "Moon Over Miami" by Joe Burke, lyrics by Edgar Leslie; "In a Sentimental Mood" by Duke Ellington; "I'm Gonna Sit Right Down and Write Myself a Letter" by Fred E. Ahlert, lyrics by Joe Young; "Red Sails in the Sunset" by Hugh Williams (Will Grosz), lyrics by Jimmy Kennedy; "Roll Along Prairie Moon" by Ted Fiorito, Harry McPherson, and Albert von Tilzer.

The first Orange Bowl game is played January 1 at Miami. Bucknell defeats Miami 26 to 0.

The first Sugar Bowl game is played January 1 at New Orleans. Tulane beats Temple 20 to 14.

College football's Heisman Memorial Trophy is awarded for the first time by New York's Downtown Athletic Club, whose first athletic director is former University of Pennsylvania football coach John Heisman, 58. Heisman has started the Touchdown Club of New York to honor all college players who have scored on the gridiron, the first award winner chosen by sportswriters and sportscasters is University of Chicago halfback John J. "Jay" Berwanger, 21, and the trophy will be awarded annually after Heisman's death next year to honor not only the recipient but also the man who introduced the center snap, the vocal "hike" as a signal for starting play, the hidden-ball play, the scoreboard listing downs and yardage, etc., and who was a leader in the fight to legalize the forward pass and to reduce games into quarters instead of halves.

James J. Braddock, 29, wins the world heavyweight title from Max Baer June 13 at Long Island City, N.Y. Braddock gains a 15-round decision over Baer.

Fred Perry wins in men's singles at Wimbledon, Helen Wills Moody in women's singles; Wilmer Lawson Allison, 30, wins in men's singles at Forest Hills, Helen Jacobs in women's singles.

Hot Springs, Va., golf caddy Samuel Jackson "Sammy" Snead, 22, becomes a professional golfer to begin an outstanding career.

Eddie Arcaro wins a race on a gelding named No More at Chicago's Washington Park to begin a 26-year career that will win stakes races of well over $23 million. George Edward Arcaro, 19, will be the top money winner 6 years, will twice bring home Triple Crown winners, and will generally pocket a flat 10 percent of the winnings.

The first major league night baseball game is played May 14 at Cincinnati's Crosley Field. A button pressed by President Roosevelt at Washington turns on 363 lights of 1,000 kilowatts each mounted on eight giant towers to illuminate the field and begin a new era in "America's favorite pastime."

The Detroit Tigers win the World Series by defeating the Chicago Cubs 4 games to 2.

Monopoly is introduced by Parker Brothers of Salem, Mass., which rejected the board game in 1933 because it took too long to play, involved concepts such as mortgages and interest that players would find confusing, and had rules that were too complex. Unemployed engineer Charles B. Darrow, 36, of Germantown, Pa., has adapted the Landlord's Game devised at the turn of the century by one Lizzie J. Magie to popularize the ideas of Henry George (*see* 1879), who believed that capitalism could work only if no one were permitted to profit from the ownership of land. Quakers at a school in Atlantic City, N.J., have given local place names to the board properties, but Darrow gets the patent (it will make him a millionaire).

Max Factor of Hollywood opens a super-colossal salon in the film capital. Now 58, the Russian-born cosmetics magnate came to America at age 27, became a Hollywood makeup man in his late thirties, and has a factory that employs 250 people producing cosmetics and wigs.

Federal agents kill Fred and "Ma" Barker of the notorious Karpis-Barker gang January 16 outside Ocklawaha, Fla.

Louisiana governor Huey Long, now 42, is shot to death September 8 by physician Carl Austin Weiss who has determined to end the dictatorial ambitions of the Kingfish, who dies in the arms of Nazi sympathizer Gerald L. K. Smith. Underworld boss Frank Costello has installed slot machines by the hundreds in New Orleans with Long's approval.

Newark, N.J., mobster Dutch Schulz (Arthur Flegenheimer), 33, is mortally wounded October 23 by gun-

1935 ***(cont.)*** men from a newly organized New York crime syndicate who walk into his saloon headquarters and shoot him down along with his three companions. A major figure in liquor bootlegging during Prohibition, Flegenheimer has twice in the last 2 years been tried and acquitted for evading income taxes on the proceeds of his various rackets.

Allied Stores Corp. is created by a reorganization of Hahn Department Stores whose stock plummeted from $57 per share in 1929 to less than $1 in 1932. Headed since 1933 by B. Earl Puckett, 37, Allied's profits have increased from their 1933 low of $25,000 and will be up to $20 million by 1948 as Puckett decentralizes management responsibility.

The League of Nations moves into a gleaming white $6 million palace completed for the League at Geneva.

The 41-story International Building at 630 Fifth Avenue opens in New York's Rockefeller Center in May.

The U.S. Supreme Court moves from its chamber in the Capitol building to a dazzling white Vermont marble building of its own designed by the late New York architect Cass Gilbert who died last year at 75.

Williamsburg, Va., regains its colonial splendor after 8 years of reconstruction funded by a $94.9 million gift from John D. Rockefeller, Jr. Boston architect William G. Perry, 52, has supervised the work of restoring 40 buildings on a 178-acre site (*see* 1699).

The first U.S. public housing project opens December 3 on New York's Lower East Side, where a row of tenements owned by Vincent Astor at 3rd Street and Avenue A has been demolished. Gov. Herbert Lehman, Eleanor Roosevelt, and Mayor La Guardia attend the opening of the eight-building complex built by the City Housing Authority that was founded in February of last year with a promised grant of $25 million. In the next 40 years the CHA will build 228 projects with 167,000 apartments to house 560,000 persons; other cities will build similar projects.

The Richter scale devised by U.S. seismologist Charles Richter, 35, measures the intensity of earthquakes by recording ground motion in seismographs. Each increase of one number means a ten-fold increase in magnitude (the San Francisco earthquake of 1906 is judged to have been a quake of 8.3 on the scale—10 times greater than a quake of 7.3).

Shenandoah National Park in Virginia's Blue Ridge Mountains is created by act of Congress. John D. Rockefeller, Jr., has used matching funds to encourage North Carolina, Tennessee, and Virginia to preserve the Blue Ridge; the new 193,593-acre park provides vistas of the Piedmont and the Shenandoah Valley.

The Civilian Conservation Corps (CCC) created in 1933 reaches its peak enrollment of 500,000.

Dust storms in western states stop highway traffic, close schools, and turn day into night.

A Soil Conservation Act is signed by President Roosevelt who names Hugh Hammond Bennett, 54, to head the new Soil Conservation Service (*see* Soil Erosion Service, 1933). Bennett estimates that soil erosion is costing the United States $400 million in diminished productivity alone each year; he works to make Americans soil-conscious.

U.S. fish cutters launch a boom with the discovery that the small white fillet of ocean perch from the Atlantic tastes much like freshwater perch.

A record 42-pound lobster is caught off Virginia.

U.S. farm prices begin to rise as the Commodity Credit Corp. purchases surplus farm commodities for distribution among the needy (*see* 1933; AAA, 1938).

The United States has 6.81 million operating farms, down from 7.2 million in 1931.

A Resettlement Administration created by executive order April 30 works to move people from poor lands to better lands under the direction of administrator Rexford Guy Tugwell (*see* 1933).

The Frazier-Lemke Farm Bankruptcy Act of 1934 is unconstitutional, the Supreme Court rules May 27 in *Louisville Joint Stock Land Bank v. Radford.*

One-third of U.S. farmers receive U.S. Treasury allotment checks for not growing food and other crops or are committed to receive such checks under terms of the AAA law of 1933, but drought holds down production more than New Deal planting restrictions do (*see* Supreme Court decision, 1936).

Canada's Parliament establishes a Wheat Board with headquarters at Winnipeg, Manitoba, to handle barley, oat, and wheat exports and to tell farmers how much they can plant each year with a guarantee that will sell their crops.

Soviet agronomist T. D. Lysenko calls his more scientific critics "Trotskyite bandits" and enemies of the state. "Bravo, comrade Lysenko, bravo," says Josef Stalin, and Lysenkoism becomes Soviet agricultural gospel (*see* 1926; Vavilov, 1940).

German researchers synthesize vitamin B_2 (*see* 1932; 1933). The vitamin will be called riboflavin.

University of California, Berkeley, biochemists Herbert McLean Evans, 53, and Oliver and Gladys Emerson isolate vitamin E (alpha-tocopherol) from wheat germ oil.

Vitamin K—essential for blood "Koagulation"—is isolated by C. P. H. Dam and by Edward A. Doisy at St. Louis University (*see* 1937; Dam, Doisy, 1929).

Eat, Drink and Be Wary by consumer advocate F. J. Schlink relates horror stories related to food contamination (*see* 1932; 1933; Sinclair, 1905).

U.S. food packers ship 240 million cases of canned goods, up from 160 million in 1931.

Adolph's Meat Tenderizer, devised by San Francisco chef Adolphe Alfred Rempp, 24, is a powdered product that is easier to use than conventional liquid tenderizers.

Krueger Beer of Newton, N.J., introduces the first canned beer.

The U.S. government sues Canadian distillers for $60 million in taxes and duties evaded during Prohibition. The amount will be lowered to $3 million, and Distillers Corp. Seagram will pay half of it (*see* 1933; 1943).

Alcoholics Anonymous is founded at New York June 10 by ex-alcoholic Bill Wilson, 40, and his former drinking companion Dr. Robert H. Smith, who takes his last drink and works with Wilson to share with other alcoholics the experience of shaking the disease. John D. Rockefeller, Jr., rejects appeals to fund AA, saying money will spoil its spirit (the organization will spread through church groups). The anonymity of "Bill W." will not be broken until his death in 1971; English novelist Aldous Huxley will call him "the greatest social architect of our time" (*see* Synanon, 1958).

Pan Am Clipper flights (above) provide the first hot meals to be served in the air.

Ireland makes it a felony to sell, import, or advertise any form of birth control device or method.

Japan's population reaches 69.2 million, up from 34.8 million in 1872. Emigration to Manchuria eases population pressures somewhat, and the birth rate falls to 14.4 per thousand, down from its peak of 15.3 in 1930, but overpopulation causes mounting anxiety.

1936 Britain's George V dies January 20 at age 70 after a reign of nearly 26 years and is succeeded by his son David, 41, who assumes the throne as Edward VIII. The new king is in love with the American divorcée Wallis Warfield Simpson, 39, he abdicates December 10 so that he may marry her, and he is succeeded by his brother Albert, 41, who will reign until 1952 as George VI.

Young Japanese army officers mutiny at Tokyo February 26 and assassinate former premier Saito, finance minister Korekiyo Takahashi, and others in an abortive attempt to set up a military dictatorship. A military court hands down 17 death sentences July 7.

Adolf Hitler takes advantage of the Ethiopian crisis (below) to denounce the 1925 Locarno Pacts and to reoccupy the Rhineland March 7.

A Rome-Berlin Axis is formed October 25; Japan signs an Anti-Comintern Pact with Germany November 25.

Armenia, Azerbaijan, Georgia, Kazakhstan become Soviet Socialist republics.

Egypt's Ahmed Fuad dies April 28 at age 68 after a reign of more than 18 years, first as sultan, then as king; his dissolute son Farouk, 16, will reign until 1952. Britain agrees August 27 to withdraw from all of Egypt except for the Suez Canal zone.

Italian forces take Addis Ababa May 5, and Rome proclaims the annexation of Ethiopia May 9. The African nation is joined with Eritrea and Italian Somaliland to create Italian East Africa (*see* 1941).

Italian bombs and mustard gas have killed barefoot Ethiopian warriors by the hundreds of thousands while Benito Mussolini's son-in-law Conte Galeazzo Ciano expressed rapture at the beauty of bombs "opening like red blossoms" on Ethiopia's highlands. "I am here today to claim the justice that is due to my people," says the Ethiopian emperor Haile Selassie June 30 in an address at Geneva to the League of Nations. "God and history will remember your judgment," says the emperor, and as he steps down he murmurs, "It is us today. It will be you tomorrow."

France's Laval government has fallen January 22 amidst suspicions that it supported reactionaries. A stopgap cabinet headed by Albert Sarraut takes over, the May 3 parliamentary elections gives the Popular Front a majority, and the first Popular Front ministry takes office June 5 with Socialist Party leader Léon Blum as premier. He promptly ends the strikes that have crippled the nation.

A Spanish civil war begins July 18 as army chiefs at Mililla in Spanish Morocco start a revolt against the weak government at Madrid. Most of the army and air force supports insurgent generals Francisco Franco, 44, and Emilio Mola, 49, and the revolt spreads rapidly to the garrison towns of Cadiz, Seville, Saragossa, and Burgos in Spain. Commanding large Moorish contingents, Franco and Mola form a junta of National Defense July 30 at Burgos. German and Italian "volunteers" soon join them, and while Moscow supplies the government loyalists with advisers and equipment, the nonintervention policies of Britain, France, and the United States serve the interests of Franco and Mola.

The Spanish insurgents (above) obtain arms from Germany and Italy, they lay siege to Madrid November 6, and the republican government immediately begins executing rightist political prisoners and military officers lest they be freed by the Fascists.

"The fifth column" will take Madrid, says Gen. Mola (above) when asked which of his four columns will capture the city. He refers to sympathizers within Madrid, but resistance stiffens when Communist orator La Pasionaria (Dolores lbarruri, 41), exhorts housewives to take lunch to their husbands in the trenches and pour boiling oil on any fascists who enter the city. "They shall not pass" (No pasaran), she has said July 18.

An International Brigade of anti-Fascists rallies to the Spanish Loyalist cause, 82 Americans embark for France as "tourists" on the S.S. *Normandie* Christmas Day to join the fight. By February of next year several thousand Americans will be in the Lincoln Brigade, and by April 1939 some 3,100 Americans will have fought and half will have died.

Nicaragua has a coup d'état June 2 as the National Guard, led by Gen. Antonio Somoza, deposes President Juan Sacasa (*see* 1934). Somoza will make himself president next year, he will rule for 10 years, and his family will continue his dictatorship until July 1979, looting the country of its wealth.

President Roosevelt wins reelection to a second term with 61 percent of the popular vote, which reaches an unprecedented 45 million. Gov. Alfred M. Landon, 59, of Kansas denounces New Deal encroachments on American business and American institutions, but Democrats denounce "economic royalists" indifferent

1936 *(cont.)* to the needs of the people, point to the economic progress made since 1933, and win 523 electoral votes, capturing every state but Maine and Vermont. Landon wins just 37 percent of the popular vote, eight electoral votes *(see* Gallup poll, below).

Heinrich Himmler takes over Nazi Germany's Gestapo and combines it with the regular police force. Started by Hitler in 1933, the secret police *(Geheime Staatspolizei,* or Gestapo) will compile dossiers on virtually every German to suppress political opposition.

A Great Purge begins in the Soviet Union. The *Yezhovschina* will take an estimated 8 to 10 million lives in the next 2 years as Josef Stalin liquidates his political enemies (*see* 1934; 1937).

United Rubber Workers of America employees receive pink layoff slips at three in the morning of February 14 and refuse to leave Goodyear Tire and Rubber Plant No. 2, pioneering the sit-down strike.

Philip Murray of the CIO organizes the United Steel Workers of America *(see* 1933; Homestead Strike, 1892; Republic Steel, 1937).

The AF of L suspends CIO unions *(see* 1935; 1955).

The Supreme Court rules June 1 that a New York minimum wage law for women passed in 1933 is unconstitutional. The Court hands down the decision in the case of *Morehead v. New York ex. rel. Tipaldo (see* Fair Labor Standards Act, 1938).

French sit-down strikes involving 300,000 workers lead to social reforms: a 40-hour week June 12, reorganization of the Banque de France June 30, nationalization of the munitions industry July 17, compulsory arbitration of labor disputes, paid vacations, etc.

Sir Oswald Mosley leads an anti-Jewish march down London's Mile End Road October 12 (*see* 1932, 1940).

Sit-down strikers occupy the General Motors Chevrolet body plant at Flint, Mich., with CIO backing December 31 after five members of the United Automobile Workers Union have been laid off at GM's Fisher Body plant in Atlanta for wearing union buttons *(see* 1937).

The General Theory of Employment, Interest and Money by John Maynard Keynes advances the idea that massive economic depressions such as the one now engulfing the world are unnecessary. Depressions can be avoided, says Keynes, by fairly simple governmental action such as investing in public works, encouraging capital goods production, and stimulating consumption to help restore the economic equilibrium. Once this is achieved, the capitalist system will resume its normal long-term growth rate with the Gross National Product increasing by 2.5 percent each year to double national income every 30 years or so, a growth that is much faster than can be achieved under any other economic system *(see* British general strike, 1926).

Thirty-eight percent of U.S. families (11.7 million families) have incomes of less than $1,000 per year. The Bureau of Labor Statistics places the "poverty line" at $1,330.

French prices skyrocket as labor reforms (above) increase production costs and international tensions drive the government to increase spending on rearmament.

The Robinson-Patman Act passed by Congress June 20 to supplement the Clayton Anti-Trust Act of 1914 forbids U.S. manufacturers to practice price discrimination including use of advertising allowances to favored customers such as chain stores.

France, Switzerland, and the Netherlands abandon the gold standard September 27.

The French franc is devalued October 2, the Italian lira October 4.

Germany embarks on a Four-Year Plan October 19, with Hermann Goering as economic minister.

Edsel Ford establishes the Ford Foundation for Human Advancement *(see* 1927; 1943).

Hunt Oil Co., founded by H. L. Hunt, will soon be the largest independent U.S. producer of oil and natural gas *(see* Joiner, 1930). Hunt, who has expanded his operations into Oklahoma and Louisiana and then into the great East Texas field, will move his headquarters to Dallas next year and develop properties that will yield him an income of roughly $1 million per week.

A California discovery well brought in at 8,000 feet by Shell Oil drillers using new seismograph techniques will lead to major oil production on land owned by Kern County Land Co. *(see* 1890). Seventy-five percent of the company's revenues will soon be coming from oil well royalties, but it will continue to be a major producer of cattle, potatoes, and other farm products *(see* Tenneco, 1968).

Sun Oil and Socony-Vaccum introduce the Houdry process for cracking oil to produce gasoline. They are joint owners of the catalytic process invented by French-American engineer Eugene F. Houdry, 44 (the *fils* in the French steel-making firm Houdry & Fils), who has used bauxite as a catalyst to produce upwards of 80 percent gasoline from the lowest grade crude oil and refinery residuum without producing a gallon of fuel oil, on which profit is nil or minimal. Thermal cracking of crude oil by the Palmer method of 1900 or the Bosch method of 1924 has required pressures of up to 3,000 pounds per square inch and temperatures of up to 1,200° F.; the Houdry process requires pressures of only 20 to 40 psi and temperatures of only 900° F.

Boulder Dam (Hoover Dam) is completed on the Colorado River after 21 months of construction on the Arizona-Nevada border. Concrete for the $175 million project has been set with cooling tubes to hasten a process that would otherwise have taken a century, and its giant hydroelectric generators produce cheap power for southern California, Arizona, and Nevada. The dam is 726 feet high, 1,244 feet wide, and will be renamed in 1947 to honor the president who authorized its construction *(see* Kaiser, 1931; Bonneville, 1937).

The Supreme Court upholds the TVA's right to dispose of surplus hydroelectric power February 17 in the case of *Ashwander v. Tennessee Valley Authority.* The ruling is a victory for the New Deal.

The Supreme Court rules May 18 that last year's Bituminous Coal Conservation Act (Guffey Act), designed to replace the NRA code in the bituminous coal industry, was unconstitutional (*Carter v. Carter Coal Co. et al*).

The Merchant Marine Act passed by Congress June 29 provides for federal subsidies to U.S. ship owners to make up the difference between what foreign shipping lines pay to man and maintain their ships and the higher operating costs U.S. shipping lines must pay (*see* Jones-White Act, 1928).

National Bulk Carriers is founded at New York by U.S. shipbuilder Daniel Keith Ludwig, 39, who has pioneered in developing a technique for side-launching newly built ships and a welding process to replace riveting in shipyards (*see* 1941).

39,000 U.S. maritime workers tie up all West Coast ports beginning October 30 in a strike that spreads to eastern and Gulf Coast ports. The strike will last for nearly 3 months before seamen vote to accept tentative agreements.

The S.S. *Queen Mary* goes into service on the North Atlantic for the Cunard-White Star Line. The 80,774-ton passenger liner has four screws, measures 1,019.5 feet overall, and was built at Clydebank by John Brown & Co., Ltd. which built the *Lusitania, Aquitainia, Empress of Britain,* and H.M.S. *Hood.*

London-Paris night ferry service begins in October to provide the cheapest luxury route between the two capitals. Passengers leaving Victoria Station at nine o'clock in the evening (ten o'clock in summer) have their sleeping cars loaded aboard Channel steamers at Dover and offloaded at Dunkirk, permitting morning arrival at the Gare du Nord in Paris.

The *Super Chief* leaves Chicago's Dearborn Station May 12 with nine Pullman cars pulled by diesel locomotives. The new weekly luxury train put into service by the Santa Fe reduces traveling time between Chicago and Los Angeles to just over 39 hours, much faster than *The Chief* that has been in service since 1926. Santa Fe officials contract to buy nine new stainless steel air-conditioned cars, each accommodating 104 passengers and a crew of 12.

The DC-3 introduced by Douglas Aircraft is a powerful two-engine, 21-passenger aircraft built at the urging of C. R. Smith of American Airlines. The Model T of commercial aviation will prove that commercial aircraft can be profitable if flown fully loaded and will make Douglas the leader in commercial aircraft. Douglas will sell more than 800 within 2 years and produce more than 10,000 military versions (Britons will call them Dakotas, the U.S. Navy R-4Ds, the Air Transport Command C-47s) (*see* 1933; 1938).

The American Airlines Mercury flight to Los Angeles leaves Newark Airport in late October carrying 12 passengers who occupy sleeper berths plus two motion picture celebrities who pay a premium over the standard $150 fare to occupy the private Sky Room compartment of the Douglas Sleeper Transport version of the new DC-3 (above). The flight departs at 5:10 in the afternoon for Memphis, Dallas, Phoenix, and Glendale, Calif., where it lands at 9:10 the following morning, just 20 minutes after estimated time of arrival.

London's Gatwick Airport opens with direct rail connections to the city and with covered access to aircraft (*see* Heathrow, 1929).

Soviet aviator Valeri Pavlovich Chkalov, 32, flies nonstop from Moscow to Nikolaevsk on the Amur River within a year of receiving the Order of Lenin for his exploits as a test pilot (*see* 1937).

New York City's electric streetcar system is converted in part to one employing GM buses. General Motors has obtained control of the city's surface transit lines through stock purchases (*see* 1932; 1949; Pacific Coast Lines, 1938).

New York's Fifth Avenue Coach Co. replaces its open-air, double-decker omnibuses with closed double-deckers that are called "Queen Marys" and will remain in operation until 1953 (some open-air double-deckers will remain in service until late in 1948). The fare on Fifth Avenue is 10¢; subways and other bus lines continue to charge 5¢ (*see* 1907; higher fares, 1948).

New York's Triborough Bridge opens July 11 to link Manhattan, Queens, and the Bronx. The 1,380-foot suspension bridge spans the East River and charges a 25¢ toll for passenger cars.

New York's Henry Hudson Bridge opens December 12 to link Manhattan's Upper West Side with the Riverdale section of the Bronx and the Westchester parkways beyond. The 800-foot steel arch bridge across the Harlem River has a 10¢ toll.

The Cord 810 introduced by Auburn-Duesenberg-Cord of Auburn, Ind., is a sleek modern motorcar with advanced features that include disappearing headlights (*see* 1937).

Willys-Overland is reorganized and renews production of low-priced motorcars (*see* 1915; Durant, 1921; Jeep, 1940).

The Ford V8 is unveiled in November (*see* 1932).

The Fiat Topolino (Little Mouse) is a popular-priced "people's car" that will not be replaced on assembly lines until 1948. Fiat's CR-32 biplane is Francisco Franco's chief fighter plane in the Spanish Civil War (above).

Mercedes-Benz introduces the world's first production passenger car to operate on diesel fuel (*see* 1926; Diesel, 1895).

Sulfa drugs discovered in Nazi Germany come into use in the United States after further development by U.S. drug firms (*see* 1935; 1937).

Dilantin (diphenylhydantoin), developed by Harvard researchers H. Houston Merritt, 35, and Tracy J. Putnam, is the first anti-convulsive treatment for epilepsy since phenobarbitol. It will also be used to treat abnormal heart beats.

BBC sets up the world's first electronic television system (*see* 1939; Baird, 1926).

1936 ***(cont.)*** Canadian Broadcasting Corp. is founded.

The *Philadelphia Inquirer* is acquired for a reported $15 million, including $4 million in cash, by publisher Moses L. Annenberg, whose *Racing Form*, *Morning Telegraph*, and racing wire activities are netting him $1 million per year (*see* 1922; 1939).

The Gallup poll begun last year gains prominence by forecasting accurately the outcome of the November presidential elections (above) and discredits the *Literary Digest* which had predicted a victory for Gov. Landon on the basis of mailings to telephone company subscribers and automobile owners, most of whom are Republicans. The 46-year-old *Digest* will fold next year; Gallup poll predictions will become a permanent part of U.S. political life.

LIFE magazine appears November 23, beginning weekly publication that will continue through 1972. The new picture magazine gains enormous success from the start for *Time-Fortune* publisher Henry Luce, who has acquired the name from the publishers of a now-defunct humor magazine begun in 1883.

Penguin Books, Ltd., begins a paperback revolution in book publishing. Bodley Head managing director Allen Lane, 34, quits to launch the new English firm with £100 in capital, issuing paperback editions of good literary works at 6p per copy (they will find their chief retail outlet initially in F. W. Woolworth Co. stores) (*see* Pocket Books, 1939).

Nonfiction *Mathematics for the Millions* by English zoology professor Lancelot Hogben, 40; *Inside Europe* by *Chicago Daily News* correspondent John Gunther, 35; *How to Win Friends and Influence People* by U.S. public-speaking teacher Dale Carnegie, 48; *The Flowering of New England* by U.S. scholar Van Wyck Brooks, 50, is a study of U.S. literary history.

Fiction *U.S.A.* by John Dos Passos whose novel *The Forty-Second Parallel* completes a trilogy about postwar America that began with *The Big Money* in 1930 and *1919* in 1932; *Absalom! Absalom!* by William Faulkner; *In Dubious Battle* by John Steinbeck; *Nightwood* by U.S. novelist Djuna Barnes, 44; *Eyeless in Gaza* by Aldous Huxley who calls chastity "the most unnatural of the sexual perversions" (XXVII); *Black Spring* by Henry Miller; *Keep the Aspidistra Flying* by George Orwell; *The Weather in the Streets* by English novelist Rosamond Lehman, 33; *Strange Glory* by L. H. Myers; *Check to Your King* and *Passport to Hell* by New Zealand novelist-poet Robin Hyde (Iris Wilkinson), 30; *Death on the Installment Plan* (*Mort à credit*) by L. F. Celine; *Mr. Visser's Descent into Hell (Meer Visser's hellevaarb)* by Dutch novelist Simon Vestdijk; *The Gift* by Vladimir Nabokov; *The Brothers Ashkenazi (Di brider Ashkenazi)* by Polish novelist Israel Joshua Singer, 43; *Gone with the Wind* by Atlanta novelist Margaret Mitchell, 36. Macmillan editor Harold Latham, 49, has changed the title from *Tomorrow Is Another Day*, the heroine's name from Pansy to Scarlett O'Hara; *Drums Along the Mohawk* by U.S. novelist Walter D. Edmonds, 33; *Double Indemnity* by James M. Cain.

Juvenile *The Story of Ferdinand* by New York writer Munro Leaf, 31, with illustrations by Robert Lawson is the story of a Spanish bull who rejects the macho tradition and refuses to fight in the ring.

Poetry "The People, Yes" by Carl Sandburg; *Look Stranger!* by English poet W. H. (Wystan Hugh) Auden, 29; *Twenty-Five Poems* by Welsh poet Dylan Thomas, 22; *Reality and Desire (La realidad y el deseo)* by Spanish poet Luis Cernuda 32. Falangists seize poet-playwright Federico García Lorca August 19 and kill him as the civil war devastates Spain.

Painting *The Black Flag* by René Magritte; *Composition in Red and Blue* and *Composition in Yellow and Black* by Piet Mondrian; *La Ville Entière* by Max Ernst; *The Old King* by Georges Rouault; *George C. Tilyou's Steeplechase Park* and *End of the 14th Street Crosstown Line* by Reginald Marsh; *Mujer tehuana* and *El vendeder de coles* by Diego Rivera.

Sculpture *Fur-lined Teacup* (*Lunchin fur*) by Swiss sculptor Meret Oppenheim, 23.

Eastman Kodak introduces Kodachrome in 35-millimeter cartridges and a paper-backed rollfilm (*see* 1935). The 18-exposure 35-millimeter cartridge goes on sale in August for $3.50, processing included, and makes color photography as easy for an amateur as black-and-white (*see* Kodacolor, 1944).

Asahi Optical Co. is founded by Japanese entrepreneur Saburo Matsumoto, whose late uncle started a company in 1919 to make eyeglass lenses. Matsumoto pursues his uncle's dream of marketing an entirely Japanese-made camera but will soon be making optical equipment and binoculars for the military (*see* 1952).

Theater *Sunrise* by Wan Chia-pao at Beijing; *The Astonished Heart* and *Red Peppers* by Noël Coward 1/9 at London's Phoenix Theatre; *The Horns of Don Friolera (Las Cuernos de don Friolera)* by Ramon Maria de Valle Inclan 2/14 at Madrid's Teatro de la Zarzuela; *End of Summer* by S. N. Behrman 2/17 at New York's Guild Theater, with Ina Claire, Osgood Perkins, Mildred Natwick, Van Heflin, Sheppard Strudwick, 121 perfs.; *Idiot's Delight* by Robert Sherwood 3/29 at New York's Shubert Theater, with Alfred Lunt and Lynn Fontanne in an antiwar drama, 300 perfs.; *The Women* by U.S. playwright Clare Boothe (Luce), 33, 7/9 at New York's Ethel Barrymore Theater, with Ilka Chase, Jane Seymour, Margalo Gilmore, Arlene Francis, Audrey Christie, Doris Day, Marjorie Main, Ruth Hammond, 657 perfs.; *Reflected Glory* by George Kelly 9/21 at New York's Morosco Theater, with Tallulah Bankhead, 127 perfs.; *Stage Door* by George S. Kaufman and Edna Ferber 10/22 at New York's Music Box Theater, with Margaret Sullavan, Tom Ewell, 169 perfs.; *French Without Tears* by English playwright Terence Rattigan, 25, 11/6 at London's Criterion Theatre, 1,039 perfs.; *Tonight at 8:30* by Noël Coward 11/24 at New York's National Theater, 118 perfs.; *You Can't Take It With You* by George S. Kaufman and Moss Hart 12/14 at New York's Booth Theater with Frank Wilcox, Josephine Hull, Ruth Attaway, George Tobias, 837 perfs.; *Brother Rat* by V.M.I. graduates John Monks,

Jr., and Fred F. Finklehoff 12/16 at New York's Biltmore Theater, with Eddie Albert, Frank Albertson, Ezra Stone, José Ferrer, 577 perfs.

Films William Wyler's *Dodsworth* with John Huston, Paul Lucas, Maria Ouspenskaya; Frank Capra's *Mr. Deeds Goes to Town* with Gary Cooper, Jean Arthur; Charles Chaplin's *Modern Times* with Chaplin; William Wyler's *These Three* with Miriam Hopkins, Merle Oberon, Joel McCrae. Also: Mervyn LeRoy's *Anthony Adverse* with Fredric March, Olivia de Havilland; George Cukor's *Camille* with Greta Garbo, Robert Taylor, Lionel Barrymore; Fritz Lang's *Fury* with Sylvia Sidney, Spencer Tracy; Robert Z. Leonard's *The Great Ziegfeld* with William Powell, Myrna Loy, Frank Morgan; Henry King's *Lloyds of London* with Tyrone Power, Freddie Bartholomew, Madeleine Carroll; Gregory La Cava's *My Man Godfrey* with Carole Lombard, William Powell; Leni Riefenstahl's documentary *Olympia*; Archie Mayo's *The Petrified Forest* with Leslie Howard, Humphrey Bogart; Pare Lorentz's documentary *The Plow That Broke the Plains*; John Ford's *The Prisoner of Shark Island* with Warner Baxter; Richard Boleslawski's *Theodora Goes Wild* with Irene Dunne, Melvyn Douglas.

Hollywood musicals Mark Sandrich's *Follow the Fleet* with Fred Astaire, Ginger Rogers, Randolph Scott, Harriet Hilliard, Betty Grable, songs by Irving Berlin that include "Let's Face the Music," "Let Yourself Go," "We Saw the Sea"; W. S. Van Dyke's *San Francisco* with Clark Cable, Jeanette MacDonald, Spencer Tracy, music by Bronislaw Kaper and Walter Jurmann, lyrics by Gus Kahn, songs that include the title song; George Stevens's *Swing Time* with Ginger Rogers, Fred Astaire, music by Jerome Kern, lyrics by Dorothy Fields, songs that include "A Fine Romance" and "The Way You Look Tonight"; Henry Koster's *Three Smart Girls* with Deanna Durban; Roy Del Ruth's *Born to Dance* with James Stewart, tap dancer Eleanor Parker, songs by Cole Porter that include "Easy to Love" and "I've Got You Under My Skin"; Roy Del Ruth's *Broadway Melody of 1936* with Eleanor Parker, Jack Benny, Robert Taylor, songs by Nacio Herb Brown and Arthur Freed that include "I Gotta Feelin' You're Foolin'," "You Are My Lucky Star"; James Whale's *Show Boat* with Paul Robeson, Irene Dunne, Helen Morgan, songs by Jerome Kern and Oscar Hammerstein II from the 1927 Broadway musical.

Broadway musicals *The Ziegfeld Follies* 1/30 at the Winter Garden Theater, with Bob Hope, Eve Arden, Josephine Baker, Judy Canova, Fanny Brice, music by Vernon Duke, George Gershwin and others, lyrics by David Freedman, Ira Gershwin, and others, songs that include "I Can't Get Started" by Vernon Duke, lyrics by Ira Gershwin, 227 perfs. (after an interruption due to Fanny Brice's illness); *On Your Toes* 4/11 at the Imperial Theater with Ray Bolger, Tamara Geva, and George Church dancing in the ballet *Slaughter on Tenth Avenue,* music by Richard Rodgers, lyrics by Oscar Hammerstein II, 315 perfs.; *Red, Hot and Blue* 10/29 at the Alvin Theater with Ethel Merman, Jimmy Durante, Grace and Paul Hartman, Bob Hope, book by Howard Lindsay and Russel Crouse, music and lyrics by Cole Porter, songs that include "De-Lovely," "Down in the Depths on the 90th Floor," 183 perfs.; *The Show Is On* 12/25 at the Ethel Barrymore Theater with the same cast as *The Women* (above), songs that include "By Strauss" by George Gershwin, lyrics by Ira Gershwin, "Little Old Lady" by Hoagy Carmichael, lyrics by Stanley Adams, 202 perfs.

Ballet *L'Epreuve d'Amour* (*The Proof of Love*) 4/4 at the Théâtre de Monte Carlo, with Vera Nemchinova as Chung-Yang, André Eglevsky as the lover, music by Mozart, choreography by Michel Fokine; *Don Juan* 6/25 at London's Alhambra Theatre, with Anatole Vilzak, André Eglevsky, music by Gluck, choreography by Michel Fokine.

Opera *The Poisoned Kiss (or The Empress and the Necromancer)* 5/12 at Cambridge (and later at the Sadler's Wells Theatre, London), with music by Ralph Vaughan Williams.

First performances *Peter and the Wolf March* by Sergei Prokofiev 5/2 at a Moscow children's concert; Prokofiev returned from exile last year and will remain in the U.S.S.R. until his death in 1953; Symphony No. 3 in A minor by Serge Rachmaninoff 11/6 at Philadelphia's Academy of Music; *Prelude Arioso and Fughetta in the Name of Bach* by Arthur Honegger 12/5 at Paris.

Negro Folk Songs as Sung by Lead Belly is published with songs that include "Good Night, Irene" by U.S. chain-gang veteran Huddie "Lead Belly" Ledbetter, now 48. Song collector Alan Lomax, 20, and his father John Avery Lomax met Lead Belly 2 years ago while recording American ballads, they begin touring the country with Lead Belly and introduce him to audiences on radio, on college campuses, in concert halls, and at night clubs where he plays a 12-string guitar.

"Roll On, Columbia" by U.S. folk singer Woody Guthrie is based on the music of "Goodnight, Irene" (above) by Huddie Ledbetter and John Avery Lomax. Now 24, Woodrow Wilson Guthrie quit school in his native Oklahoma at age 15 to play the harmonica in Houston barbershops and pool halls, returned home to study guitar, and has appeared on radio singing his song "Those Oklahoma Hills" and other original works. Alan Lomax has suggested to the Department of the Interior that it hire Guthrie to propagandize in the Northwest against private power companies that use Hollywood stars to attract crowds to public meetings where speakers urge the people to vote against public power projects. Guthrie has hitchhiked to Portland, the government provides him with a car and chauffeur to follow the course of the river, and he writes 26 songs in 26 days (*see* 1938).

Popular songs "Pennies from Heaven" by Hollywood composer Arthur Johnston, lyrics by Johnny Burke (title song for film); "Until The Real Thing Comes Along" by Mann Holiner, Alberta Nichols, Sammy Cahn, Saul Chaplin, and L. E. Freeman; "Sing, Sing, Sing" by Louis Prima; "Stomping at the Savoy" by Benny Goodman, Edgar Sampson, Chick Webb, lyrics by Andy Razaf; "Ramblings on My Mind" and "Walking

1936 ***(cont.)*** Blues" by U.S. blues writer Robert Johnson, 25; "The Night Is Young and You're So Beautiful" by Dana Suesse, lyrics by Billy Rose and Irving Kahal; "Moonlight and Shadows" by Frederick Hollander, lyrics by Leo Robin (for the Dorothy Lamour musical *The Jungle Princess*); "Cool Water" by Bob Nolan; "I'm an Old Cowhand from the Rio Grande" by Johnny Mercer (for the film *Rhythm on the Range*); "Poinciana" by Nat Simon, lyrics (Spanish) by Manuel Llisco, (English) Buddy Denis.

Fred Perry wins in men's singles at Wimbledon, Helen Jacobs in women's singles; Perry wins in U.S. men's singles, Alice Marble, 22, in women's singles.

The Olympic Games at Berlin attract 4,069 contestants from 51 countries. U.S. track star Jesse Owens, 22, wins four gold medals, setting new Olympic and world records for the 200-meter sprint and the running broad jump (26 feet, 5 5/16 inches). He runs anchor on the 400-meter relay team that breaks the world record, and ties the Olympic record for the 100-meter sprint. Embarrassed at the defeat by a black of Germany's "master race" "Aryan" athletes, Chancellor Hitler leaves the stadium and lets someone else present Owens with his medals.

Bob Feller signs with the Cleveland Indians, strikes out seven St. Louis Cardinal batters including Leo Durocher, 29, in an exhibition game, makes his major league debut in August, and within a few weeks has tied the major league record of 17 strikeouts in one game. Robert W. A. Feller, 17, will pitch for the Indians for 20 years.

Joe DiMaggio signs with the New York Yankees and begins a 13-year career in which his batting average will average .325, peak at .381 in 1938, and never fall below .300. San Francisco-born Joseph Paul DiMaggio, Jr., 21, will play until 1952, with a 3-year hiatus from 1943 to 1945.

The New York Yankees win the World Series by defeating the New York Giants 4 games to 2.

Sun Valley Lodge ski resort opens near Ketchum, Idaho, where Union Pacific chairman W. Averell Harriman has built it to compete with the Canadian Pacific and its Banff–Lake Louise resorts (*see* 1885). Harriman, 46, takes up skiing to promote the sport and Sun Valley.

Winning Bridge Made Easy by Philadelphia lawyer Charles Henry Goren, 35, makes a radical departure from the dominant "honor-trick" method devised by Ely Culbertson for evaluating a hand (*see* Vanderbilt, 1925). Goren gives up the law to devote himself to refining his bidding system.

Butlinlands holiday resort empire has its beginnings at Skegness on England's east coast where entrepreneur William "Billy" Butlin, 35, opens a holiday camp with accommodations for 1,000 and enjoys such success that he will have to double the camp's size within a year and open another 2 hours from London at Clacton. Butlin will invite every member of Parliament who voted for the new Holiday with Pay Bill to opening ceremonies at Clacton and within 40 years will be providing low-cost family vacations for more than a million Britons each year at camps of 5,000 to 12,000 each.

Polaroid lens sunglasses are introduced by Land-Wheelwright Laboratories, which will rename itself Polaroid Corp. next year (*see* 1932; camera, 1947).

Bass Wee'juns, introduced at $12 a pair by G. H. Bass & Co. of Wilton, Me., begin a unisex fashion for slip-on moccasin "loafers."

Jesse Owens (above) has won at least one of his four gold medals wearing lightweight shoes designed by German sports-shoe maker Adolf ("Adi") Dassler, 36, whose Adidas Sportschufabriken of Herzogenaurach, near Nuremberg, will become the world's largest maker of athletic shoes (*see* Nike, 1972).

Tampax, Inc., is founded at New Brunswick, N.J., to produce a cotton tampon with string attached patented by Denver physician Earl Haas. Women since ancient times have used absorbent rags during their menstrual periods but no commercial tampon has been available until now (*see* Kotex, 1918).

Charles "Lucky" Luciano draws a 30- to 50-year prison sentence after a New York jury convicts him of compulsory prostitution (*see* 1931). Authorities will release Luciano in 1946 and deport him to Italy.

Bruno Richard Hauptmann dies in the electric chair at the New Jersey State Prison in Trenton April 3. Hauptmann protests his innocence of having kidnapped and killed Charles A. Lindbergh, Jr., in 1932; the evidence against him is far from conclusive, but anti-German sentiment has helped to seal his fate.

Japanese authorities arrest Tokyo geisha Sada Abe, 31, May 21 on charges of having stabbed her unfaithful lover Kichizo Ishida to death in his sleep, castrating him, and carrying his penis about in her *obi* (sash) for 3 days while eluding police.

Ice chokes the Ohio River for months.

President Roosevelt signs an Omnibus Flood Control Act into law.

The Agricultural Adjustment Act of 1933 is unconstitutional, the Supreme Court rules January 6 in *United States v. Butler,* but the New Deal uses soil conservation as the basis for a new effort to limit planting.

The Soil Conservation and Domestic Allotment Act passed by Congress February 29 pays farmers to plant alfalfa, clover, and lespedeza rather than soil-depleting crops such as corn, cotton, tobacco, and wheat.

Drought reduces U.S. crop harvests and brings new hardship to farmers and consumers.

Boulder Dam on the Colorado River (above) creates a lake 115 miles long (Lake Mead) whose waters irrigate 250,000 acres.

T. D. Lysenko's sensationalist agricultural nostrums gain full sway over Soviet agriculture (*see* 1935; 1940).

A 4-year program begins in Nazi Germany to produce synthetic (ersatz) replacements for raw materials such as fats and livestock fodder.

Robert R. Williams synthesizes vitamin B_1 and gives it the name thiamine (*see* 1933; 1937).

University of Wisconsin biochemist Conrad Arnold Elvehjem, 35, isolates vitamin B_3 (nicotinic acid, or niacin) from liver extract. It has an instant positive effect on a dog sick with a canine version of pellagra (*see* 1915; 1938).

Campbell Soup begins manufacturing its own cans; it has been buying them from Continental Can but will now become the world's third largest producer of food cans simply by supplying its own needs (*see* 1933).

Spry is introduced by Lever Brothers to compete with the hydrogenated vegetable shortening Crisco introduced by Procter & Gamble in 1911.

General Mills uses the name "Betty Crocker" as a signature for responses to consumer inquiries. The fictitious authority is portrayed as a gray-haired homemaker, an image that will see numerous revisions as Betty Crocker becomes a major brand name for various General Mills products (*see* 1929; Kix, 1937).

The Waring Blender, designed by U.S. bandleader Fred M. Waring, 36, begins mechanizing kitchen chores.

Howard Johnson restaurants have their beginnings at Orleans on Cape Cod where Wollaston, Mass., restaurateur Johnson, 39, has persuaded young Reginald Sprague to open an eating place that Johnson will supply with ice cream, fried clams, and frankfurters (*see* 1937).

The Joy of Cooking by St. Louis housewife Irma S. Rombauer gives recipes in the most minute detail, telling the cook exactly what to look for.

Moscow revokes the 1920 decree legalizing abortion; a new law restricts abortions to cases in which pregnancy endangers the life of the woman or in which the child is likely to inherit a specified disease (*see* 1955).

A U.S. Circuit Court of Appeals judge rules that the purpose of the 1872 Comstock law, liberalized in 1929, "was not to prevent the importation, sale, or carriage by mail of things which might intelligently be employed by conscientious and competent physicians for the purpose of saving life or promoting the well-being of their patients." The ruling is made in a case brought by birth-control advocate Hannah Stone (*see* Connecticut, 1965).

The U.S. population reaches 127 million. Britain has 47 million, France 44, Germany 70, the Soviet Union 173, Japan 89, China 422, India 360, Brazil 35.

1937 Moscow "show trials" begin January 23 as Josef Stalin purges the Communist party and Soviet army of alleged Trotskyites. Grigori L. Pyatakov, 47, is executed January 31, Karl Bernardovich Radek, 52, condemned to serve 10 years in prison, Marshal Mikhail Nikolaevich Tukhachevski, 44, convicted of treason by a military tribunal and executed June 12 along with seven other generals; hundreds of others are liquidated or sent to Siberia.

Gen. Mola dies in a plane crash as Spain's civil war continues, with more than 10,000 Germans and from 50,000 to 75,000 Italians supporting Gen. Franco. He commands several hundred thousand insurgents against the Republican Army, which is supported by an International Brigade of Russians, Britons, other Europeans, and Americans (*see* 1936).

Malaga falls to Gen. Franco February 8 as the insurgents advance with Italian aid, but the road from Madrid to Valencia remains intact, and Loyalists defeat Italian troops March 18 at Brihuega, capturing large stores of equipment.

The defenseless Basque town of Guernica is annihilated on the afternoon of April 26 by German Junker and Heinkel bombers that drop explosives and thousands of aluminum incendiary projectiles for more than 3 hours while Heinkel fighters strafe civilians who have fled into the fields (*see* Picasso mural, below).

Four German warships bombard Almeria May 31 in reprisal for a Loyalist air attack on the *Deutschland*, Bilbao falls to the insurgents June 18 after weeks of heavy fighting and aerial bombing, Basque resistance collapses, Gijon falls October 21 as Franco breaks resistance in the Asturias. The Spanish government moves from Valencia to Barcelona October 28, Franco announces a naval blockade of the entire Spanish coast November 28, but a Loyalist counter-offensive begins December 5, and Teruel falls to the Loyalists December 19.

The British prime minister Stanley Baldwin retires May 28 at age 69 and is succeeded by his chancellor of the exchequer (Arthur) Neville Chamberlain, 58, who attempts to appease Adolf Hitler.

Iraq's dictator Gen. Bake Sidqi is assassinated by a Kurd August 11 following conclusion of a nonaggression pact signed in July by Turkish, Iraqi, Iranian, and Afghan diplomats.

Japanese forces invade China July 7 as the new prime minister Prince Fumimaro Konoye, 46, embarks on an undeclared war that will continue until 1945. Beijing falls to the Japanese July 28, Tientsin July 29, Shanghai November 8, Nanjing December 13, and Hanchow December 24. Some 200,000 civilians are executed at Nanking, and this massacre, combined with merciless bombing of the Chinese cities, rouses world opinion against Japan.

The *Panay* incident of December 12 produces tension between Japan and Britain and the United States. Japanese bombers attack British and U.S. ships near Nanking but Washington accepts Japan's explanations and events in Europe distract the world powers from the aggression in China.

Italy has joined the German-Japanese anti-Comintern pact November 6 and withdrawn from the League of Nations December 11. Mussolini began the year by concluding a "gentlemen's agreement" with Britain in which each has agreed to maintain the independence and integrity of Spain and respect each other's interests and rights in the Mediterranean, and an Italian-Yugoslav treaty signed March 25 has guaranteed existing frontiers and maintenance of the status quo in the Adriatic.

"Put your trust in Japanese military," this poster advised Chinese peasants. "Return to your homes."

Brazil's president Getulio Vargas proclaims a new constitution November 10 after 7 years in office, closes the Congress, and begins nearly 15 years of dictatorship which he says is not fascist but which will have all the earmarks of fascism.

Albanian Muslims rebel in mid-May against a government decree that forbids the veiling of women and against the dictatorial rule of Zog I.

Germany's Buchenwald concentration camp opens July 16 on a plateau overlooking Weimar. The first inmates are mostly political prisoners of every religious belief, but most of the 238,980 inmates that will ultimately be sent to Buchenwald will be Jews, and 56,545 will die in the camp's gas chambers (*see* Dachau, 1933).

Germany evicts Jews from trade and industry, orders them to wear yellow badges displaying the six-pointed "star of David," and bars them from all parks, places of entertainment, health resorts, and public institutions (*see* Nuremberg laws, 1935; Kristalnacht, 1938).

Romania forbids Jews to own land and bars them from the professions at year's end under legislation put through by the newly installed prime minister Octavian Goga, 56, who takes office despite the fact that his National Christian party gained only 10 percent of the votes in the election.

General Motors recognizes the United Automobile Workers (UAW) as sole bargaining agent for workers in all GM plants February 11. GM takes the action to end the union's 44-day sit-down strike at Flint, but 4,470 other strikes idle plants nationwide, and most are sit-down strikes.

Henry Ford says, "We'll never recognize the United Auto Workers Union or any other union"; he employs a "service department" of 600 goons armed with guns and blackjacks to prevent unionizing of Ford workers (*see* 1932).

United States Steel permits unionization of its workers March 2 to avoid a strike (*see* 1936).

The United Steel Workers union meets with resistance from "little steel" firms, which include Bethlehem with 82,000 workers, Republic with 53,000, Youngstown Sheet and Tube with 27,000, and National, American Rolling Mills, and Inland with a combined total of some 38,000.

The Supreme Court upholds the principle of a minimum wage for women March 29; its ruling in the case of *West Coast Hotel v. Parrish* reverses some earlier decisions (*see* 1923).

The Supreme Court upholds the National Labor Relations Act of 1934 in a series of 5 to 4 decisions beginning April 12.

Republic Steel workers strike and picket, singing "Solidarity Forever" and "I Dreamt I Saw Joe Hill Last Night" (*see* 1915). Republic's $130,000-a-year boss Tom Girdler says he would rather go back to hosing potatoes than give in to union organizers. On May 30 he has Chicago police attack the demonstrators. Four are killed, three others mortally wounded, and 84 injured as in the Memorial Day massacre as police fire on unarmed strikers and brutally assault wives and children.

The CIO claims a membership of 3.5 million by September with more than 500,000 in the United Steel Workers Union as other incidents of police repression accompanied by further deaths and injuries follow the violence at Republic Steel (above).

"I am the law," says Jersey City, N.J., mayor Frank Hague, 61, when questioned in a legislative investigation as to his right to forbid picketing and distribution of labor circulars on the city's streets. He has been mayor for 20 years and will remain mayor for another 10.

Amalgamated Clothing Workers president Sidney Hillman obtains a settlement with the Clothing Manufacturers Association and establishes a bargaining pattern that will be used throughout the industry (*see* 1918). Hillman has organized all but a handful of men's clothing workers (*see* Office of Production Management, 1941).

"I see one-third of a nation ill-housed, ill-clad, ill-nourished," President Roosevelt has said in his second inaugural address January 20. The inauguration date has been moved from March 4 under terms of the Twentieth Amendment.

John D. Rockefeller dies May 23 at age 97. Andrew W. Mellon dies August 26 at age 82 (*see* art, 1941).

The Supreme Court upholds the 1935 Social Security Act May 24.

The U.S. economic recovery that has progressed for 4 years falters beginning in midyear. Business activity suffers a sharp drop; Wall Street's Dow Jones Industrial Average falls from its post-1929 high of 194.40.

Congress passes the Miller-Tydings Act, allowing manufacturers to fix resale prices of brand-name merchandise in states where legislatures have authorized price-fixing contracts. Designed to prevent predatory pricing and ruinous price wars, the new fair trade law has been devised as an exception to the Sherman Anti-Trust Law of 1890 and permits a manufacturer to determine the minimum price at which his products may be sold at retail outlets.

The Folklore of Capitalism by Washington lawyer Thurman Arnold, 46, shows how antitrust laws have actually promoted the growth of industrial monopolies in the United States "by deflecting the attack on them into purely moral and ceremonial channels." Arnold will be named U.S. assistant attorney general next year and in the next 5 years will file more than 200 suits alleging conspiracies in restraint of trade.

Kerr-McGee Oil is founded by Oklahoma City's Robert S. Kerr and former Phillips Petroleum geologist Dean Anderson McGee, 34, in a reorganization of Kerlyn Oil (*see* 1932).

Standard Oil of New Jersey drills the first offshore Louisiana oil wells.

Bonneville Dam is dedicated on the Columbia River in Oregon. Henry Kaiser has masterminded the new hydroelectric installation (*see* 1931; 1941).

Germany retires her 9-year-old *Graf Zeppelin* after 144 ocean crossings that have carried more than 13,000 passengers. The new *Hindenburg* carries 50 passengers in private cabins and 47 in crew, she moves noiselessly at 78 miles per hour, but she is filled with hydrogen gas and explodes and burns on arrival at Lakehurst, N.J., May 6, killing 36, 13 of them passengers, and ending the brief era of transatlantic travel by rigid airship (*see* Pan Am, 1939).

Amelia Earhart disappears July 2 on a Pacific flight from New Guinea to Howland Island (*see* 1928). She was married in 1931 to New York publisher George Putnam.

Soviet aviator V. P. Chkalov flies nonstop over the North Pole from Moscow to Vancouver, covering 5,400 miles that include 3,100 over ice fields (*see* 1936).

Soviet Russia opens the 80-mile Moscow-Volga Ship Canal to give Moscow access to the Volga (*see* Stalin Ship Canal, 1933).

Hong Kong's C. Y. Tung shipping line is founded by Chinese shipowner Chao-Yung Tung, 25, whose father-in-law has a monopoly on much of the Chinese coastal shipping trade and was himself made vice president last year of the Tientsin Ship Owners Association. Tung has moved to Hong Kong at the outbreak of hostilities with Japan (above) and will build a fleet that will rival those of the Greek ship owners Aristotle Onassis and Stavros Niarchos.

The Golden Gate Bridge opens May 27 across San Francisco Bay to link San Francisco with Marin County. Like the Longview cantilever bridge of 1930 and George Washington Bridge of 1931, the new 4,200-foot Golden Gate Bridge has been designed by U.S. bridge engineer Joseph Baermann Strauss, now 67, and is the world's longest suspension bridge (*see* Verrazano Bridge, 1964).

New York's West Side Highway opens as an elevated six-lane motorcar and truck route along the Hudson River from the Battery to 72nd Street (where it becomes the Henry Hudson Parkway).

The Lincoln Tunnel between New York and Weehawken, N.J., opens to traffic December 22. A second tube will open in December 1940.

Isuzu Motors, founded in April at Tokyo, will become Japan's third largest truck maker.

General Motors introduces an automatic transmission for automobiles under the name Hydramatic Drive as optional equipment for 1938 Oldsmobiles. Similar transmissions have been used on London buses for 12 years and will be employed increasingly on U.S. passenger cars, first as optional equipment, then as standard.

Packard sells a record 109,518 cars, most of them medium-priced "120" models.

Stutz files for bankruptcy. The firm has produced fewer than 700 cars since 1930 and none since 1935.

Pierce-Arrow ceases production. The company will be sold at auction in May of next year.

Auburn, Cord, and Duesenberg motorcars pass into history as Auburn's E. L. Cord returns from his Surrey estate and sells his motorcar holdings. Cord moved to England to escape kidnap threats against his children.

A magnetic resonance method for observing the spectra of atoms and molecules in the radio-frequency range is invented by Austrian-American physicist Isidor Isaac Rabi, 39, at Columbia University. Rabi's invention will make it possible to deduce the mechanical and magnetic properties of atomic nuclei (*see* medicine, 1977).

Italian physicists Emilio Gino Segre, 32, and Enrico Fermi produce the first laboratory-made element (*see* 1934). They bombard molybdenum with deuterons and neutrons to produce an element with the atomic number 43 that probably does not exist in nature but will be called technetium (*see* nuclear fission, 1938; McMillan, 1940).

U.S. astronomer Grote Reber, 25, builds the world's first radio telescope. He will be the world's only radio astronomer until 1945, mapping high-frequency sources.

A closed-plaster method of treating compound fractures, introduced by Barcelona physician Jose Trutta Raspall, uses principles developed by Lincoln, Nebraska, physician H. Winnett Orr to save fracture victims in the Spanish Civil War (above) and reduce the need for amputation.

Diabetics are treated successfully for the first time with zinc protamine insulin, which reduces the need for diet therapy (*see* 1924; 1942).

The Neurotic Personality of Our Time by German-American psychoanalyst Karen Danielsen Horney, 52, attacks Freudian anti-feminism.

1937 *(cont.)* U.S. children die after treatment with an elixir of the antibacterial drug sulfanilamide containing the solvent diethylene glycol (*see* 1936; Food, Drug and Cosmetics Act, 1938).

Queens College is founded at Flushing, N.Y.

St. John's College is founded at Annapolis, Md., where former University of Virginia professor Stringfellow Barr, 40, and University of Chicago administrator Scott Buchanan, 42, take over the King William's School founded in 1696 and begin a Great Books program that revives the principles of a classical education.

Pepperdine College is founded at Los Angeles with a gift from Western Auto Supply magnate George Pepperdine, now 51, who has amassed a fortune of $10 million (*see* 1909; 1927). Pepperdine will lose his money and testify in 1950 that his personal assets are no more than $1.

The emergency three-digit telephone number 999 comes into use in Britain to summon police, firefighting, or ambulance aid. Britain's example will be followed by countries in Europe, the Far East, and South America (*see* New York, 1968).

Xerography, pioneered by New York pre-law student Chester Floyd Carlson, 31, is a dry-copying process that will revolutionize duplication of papers in offices, schools, and libraries. Photostats are costly, carbon copies often blurred, and few can be made at one time, but Carlson has observed the demand for multiple copies of patent specifications and other documents while working in the patent department of a New York electronics firm, and he sees possibilities in a process based on principles of photoconductivity and electrostatics. Taking a sulfur-coated zinc plate, he gives it an electrostatic charge by rubbing it with a handkerchief in the dark, places over it a transparent celluloid ruler, and then exposes the zinc plate to light for a few seconds, neutralizing the charge except where the markings of the ruler have blocked the light. Dusting lycopodium powder over the plate and blowing away the excess, Carlson is left with a perfect image of the ruler (*see* 1938).

Newsweek magazine begins publication at New York to compete with Henry Luce's *Time.* Real estate heir Vincent Astor, 42, and railroad heir W. Averell Harriman, 45, merge their news weekly *Today* with T. J. C. Martyn's 4-year-old *News-Week,* and bring in as editor former assistant secretary of state Raymond Moley, 51, installing McGraw-Hill president Malcolm Muir as publisher (*see* 1961).

Look magazine begins publication at New York to compete with Henry Luce's LIFE. Des Moines *Register* publisher Gardner Cowles launches the biweekly picture magazine that will continue until 1971.

Woman's Day appears in October as the A&P launches a 3¢ monthly women's service magazine for distribution in A&P stores. The food chain will sell the magazine to Fawcett in 1958.

Nonfiction *Spanish Testament* by Hungarian-born French journalist Arthur Koestler, 32, who fell into fascist hands at the fall of Malaga while covering the Spanish Civil War for the *London News Chronicle,* was condemned to death as a spy and tortured, and gained release after British authorities brought pressure (an abridged version will appear in 1942 under the title *Dialogue with Death); Four Hundred Million Customers* by U.S. businessman Carl Crow, 54, is about the potential in trade with China.

Fiction *The Road to Wigan Pier* by George Orwell who examines the conditions of the unemployed in the north of England; *The Snow Country* (*Yukiguni*) by Yasunari Kawabata; *Coming from the Fair* by Norah Hoult; *Mouchette (Nouvelle Histoire de Mouchette)* by Georges Bernanos; *Mad Love* (*L'amour fou*) by André Breton; *The Revenge for Love* by Wyndham Lewis; *The Citadel* by A. J. Cronin; *The Young Desire It* by Australian novelist Seaforth Kenneth Mackenzie (Kenneth Mackenzie), 24; *To Have and Have Not* by Ernest Hemingway; *Of Mice and Men* by John Steinbeck; *Their Eyes Were Watching God* by Harlem novelist Zora Neale Hurston, 46; *Low Company* by Daniel Fuchs; *The Late George Apley* by U.S. novelist John Phillips Marquand, 43, who is best known for his Mr. Moto detective stories in the *Saturday Evening Post*; *Serenade* by James M. Cain; *The Education of H*y*m*a*n K*a*p*l*a*n* by U.S. humorist-sociologist Leonard Q. Ross (Leo Calvin Rosten), 29, who came to America from Poland as an infant and finds poignant humor in the efforts of New York immigrants to gain education in the city's free adult education classes; *Remembering Laughter* by U.S. novelist Wallace Stegner, 28; *Beat to Quarters* by English novelist C. S. (Cecil Scott) Forester, 38, who begins a Captain Horatio Hornblower trilogy that will be completed with *Ship of the Line* next year and *Captain Horatio Hornblower* in 1939; *The Hobbit* by Oxford philologist J. R. R. (John Ronald Reuel) Tolkien, 45; *Dark Frontier* and *Uncommon Danger* by London advertising agency director-detective novelist Eric Ambler, 28, whose second book is published in the United States as *Background to Danger; Busman's Honeymoon* by Dorothy Sayers.

Juvenile *And to Think That I Saw It on Mulberry Street* by U.S. writer-illustrator Dr. Seuss (Theodor Seuss Geisel), 33.

Poetry *Collected Poems* by T. S. Eliot includes "Chamber Music," "Pomes Penyeach," and "Ecce Puer"; *The Cantos* (V) by Ezra Pound; *The Man with the Blue Guitar* by Wallace Stevens; *The Semi-Circle (Die Halve Kring)* by N. P. van Wyk Louw.

Painting *Guernica* by Pablo Picasso who has been commissioned to produce a mural for the Spanish Republic's pavilion at the Paris World's Fair. After 6 months of doing nothing, Picasso has worked in a 6-week burst of outraged energy to complete a cubist canvas nearly 26 feet long filled with bedlam and terror that expresses the painter's horror at the brutality of modern warfare; *Revolution of the Viaducts* by Paul Klee; *Still Life with*

Old Shoe by Joan Miró; *Woman with a Mandolin* by Georges Braque; *The Pleasure Principle* and *Not to Be Reproduced* by René Magritte; *Echo of a Scream* by David Alfara Siqueiros. H. O. Tanner dies at Paris May 25 at age 77.

London's Tate Gallery opens a Duveen Gallery endowed by local art dealer Joseph Duveen, Baron Duveen of Milbank, now 68.

An exhibition of "degenerate art" opens at Munich.

Agfacolor film in 35-millimeter cartridges is introduced by A. G. Fur Analin (Agfa) and by Ansco (*see* Kodachrome, 1935; 1936).

The fastest film made is Eastman Kodak's black-and-white Super-X with a Weston rating of 32 (*see* Weston exposure meter, 1931; ASA ratings, 1947).

Popular Photography magazine begins publication at Chicago in May.

Theater *High Tor* by Maxwell Anderson 1/9 at New York's Martin Beck Theater, with Burgess Meredith, Peggy Ashcroft, 171 perfs.; *The Masque of Kings* by Maxwell Anderson 2/8 at New York's Shubert Theater with Dudley Digges, Henry Hull, Margo, 89 perfs.; *Yes, My Darling Daughter* by U.S. playwright Mark Reed 2/9 at The Playhouse, New York, with Peggy Conklin, Lucile Watson, 405 perfs.; *Traveler Without Luggage (Le voyageur sans baggage)* by Jean Anouilh 2/16 at the Théâtre de Mathurius, Paris; *"Having Wonderful Time"* by Austrian-American playwright Arthur Kober, 36, 2/20 at New York's Lyceum Theater, with Katherine Locke, Jules Garfield, Cornel Wilde, 132 perfs.; *The Ascent of F6* by Christopher Isherwood and W. H. Auden 2/26 at London's Mercury Theatre; *Electra (Electre)* by Jean Genet 5/13 at the Théâtre de l'Athenée, Paris; *Room Service* by John Murray and Allen Boretz 5/19 at New York's Cort Theater, with Sam Levine, Eddie Albert, Betty Field, 500 perfs.; *Golden Boy* by Clifford Odets 11/23 at New York's Belasco Theater, with Jules Garfield, Lee J. Cobb, Karl Malden, Elia Kazan in a play about prizefighting, 250 perfs.; *Of Mice and Men* by John Steinbeck 11/23 at New York's Music Box Theater, with Art Lund as Lennie, 207 perfs. (director George S. Kaufman has polished Steinbeck's efforts to present his story on the stage while Steinbeck gathers material for his novel *The Grapes of Wrath*).

Pablo Picasso's *Mother with Dead Child* expressed rage at the 1937 Fascist bombing of Spanish civilians at Guernica.

Radio *The Guiding Light* by Irma Phillips, 27, 1/25 over NBC features the Rev. John Rutledge (to CBS radio in 1947 until 1956, on TV in 1952); *Our Gal Sunday* by Chicago advertising agency writers Frank and Anne (Ashenhurst) Hummert 3/29: "the story of an orphan girl named Sunday from the little town of Silver Creek, Colorado, who in young womanhood married England's richest, most handsome lord, Lord Henry Brinthorpe—the story that asks the question, Can this girl from a mining town in the West find happiness as the wife of a wealthy and titled Englishman?" (to 1959); *The Charlie McCarthy Show* 5/9 over NBC, with ventriloquist Edgar Bergen, 34, and W. C. Fields, who appears until December (to 1948, on CBS 1949 to 1954).

Films Victor Fleming's *Captains Courageous* with Spencer Tracy, Freddie Bartholomew; John Ford's *The Life of Emile Zola* with Paul Muni; Sidney Franklin's *The Good Earth* with Luise Rainier, Paul Muni; Jean Renoir's *Grand Illusion* with Erich von Stroheim, Jean Gabin; Marcel Pagnol's *Harvest* with Gabriel Gabrio, Orane Demazis, Fernandel; Frank Capra's *Lost Horizon* with Ronald Colman, Sam Jaffe, Thomas Mitchell; Julien Duvivier's *Pepe Le Moko* with Jean Gabin; Gregory La Cava's *Stage Door* with Katharine Hepburn, Adolphe Menjou, Lucille Ball, Ginger Rogers; Mervyn LeRoy's *They Won't Forget* with Claude Rains, Lana Turner. Also: Leo McCarey's *The Awful Truth* with Irene Dunne, Cary Grant; Sam Wood's *A Day at the Races* with the Marx Brothers; John Ford's *The Hurricane* with Dorothy Lamour, Jon Hall, Raymond Massey; Leo McCarey's *Make Way for Tomorrow* with Victor Moore, Beulah Bondi; John Cromwell's *The Prisoner of Zenda* with Ronald Colman, Madeleine Carroll, Douglas Fairbanks, Jr.; Pare Lorentz's documentary *The River*; William Wellman's *A Star Is Born* with Fredric March, Janet Gaynor; Norman Z. McLeod's *Topper* with Constance Bennett, Cary Grant, Roland Young.

George B. Seitz's A *Family Affair* with Mickey Rooney as Andy Hardy, Lionel Barrymore as Judge James Hardy is the first of a series that will continue off and on for 21 years, with Lewis Stone as Judge Hardy, Fay Holden as Mrs. Hardy, and girlfriends who will include Ann Rutherford, Judy Garland, Lana Turner, Esther Williams, Kathryn Grayson, and Donna Reed.

The first Bugs Bunny cartoon is released by Warner Brothers. *Porky's Hare Hunt* features the voice of Mel Blanc, 29, who creates the voices of Bugs Bunny, Porky Pig, and others who will include Daffy Duck, Woody Woodpecker, and Speedy Gonzales. It takes 125 people to make one 6½-minute cartoon, but audiences are delighted with Bugs Bunny's "What's up, Doc?" and "That's all, folks" (*see* 1910).

1937 *(cont.)* The Thalberg Memorial Award inaugurated by the 10-year-old Academy of Motion Picture Arts and Sciences honors the late M-G-M producer Irving Grant Thalberg, who died of pneumonia in September of last year at age 37 after a brilliant career that began as secretary to Universal Pictures president Carl Laemmle.

Hollywood musicals Mark Sandrich's *Shall We Dance* with Ginger Rogers, Fred Astaire, music by George Gershwin (who dies July 11 of a brain tumor at age 38), lyrics by Ira Gershwin, songs that include "They Can't Take That Away from Me," "They All Laughed," "Slap That Bass," "Let's Call the Whole Thing Off," and the title song; Henry Koster's *One Hundred Men and a Girl* with Deanna Durbin, Leopold Stokowski, Adolphe Menjou, score by Charles Previn.

Stage musicals *Babes in Arms* 4/14 at New York's Shubert Theater, with Ray Heatherton, Alfred Drake, music by Richard Rodgers, lyrics by Lorenz Hart, songs that include "My Funny Valentine," "The Lady Is a Tramp," "Johnny One Note," "I Wish I Were in Love Again," "Where or When," 289 perfs.; *I'd Rather Be Right* 11/2 at New York's Alvin Theater, with George M. Cohan as President Roosevelt, book by George S. Kaufman, music by Richard Rodgers, lyrics by Lorenz Hart, songs that include "Have You Met Miss Jones?" 290 perfs.; *Pins and Needles* 11/27 at New York's Labor Stage Theater, with music and lyrics by Harold Rome, songs that include "Nobody Makes a Pass at Me," 1,108 perfs. (the International Ladies' Garment Workers Union [ILGWU] sponsors the production and no cast member receives more than $55 per week); *Me and My Gal* 12/16 at London's Victoria Palace Theatre, with Lupino Lane as the Cockney from Lambeth who turns out to be seventeenth baron and eighth viscount of Hareford, book and lyrics by English playwright L. Arthur Rose, 50, and Douglas Furber, music by Nöel Gay, songs that include "The Lambeth Walk," 1,646 perfs.; *Between the Devil* 12/22 at New York's Imperial Theater, with Jack Buchanan, Evelyn Laye, music by Arthur Schwartz, lyrics by Howard Dietz, songs that include "By Myself," 93 perfs.

Ballet *Les Patineurs* (*The Skaters*) 2/16 at the Sadler's Wells Theatre, London, with Margot Fonteyn, music by Giacomo Meyerbeer, choreography by Frederick Ashton; *Francesca da Rimini* 7/15 at London's Royal Opera House in Covent Garden, with Lubov Tchernicheva, music by Tchaikovsky, choreography by David Lichine, 36.

Opera *Lulu* 6/2 at Zurich's Municipal Theater, with music by the late Alban Berg, who died of an infected bee sting late in 1935 at age 50.

The National Broadcasting Co. starts the NBC Symphony with Arturo Toscanini as conductor. Now 70, Toscanini has been replaced as conductor of the New York Philharmonic after 8 years but will conduct the NBC Symphony until his retirement in 1954.

Avery Fisher, 31, invests $354 to start a hi-fi business that will make him a millionaire. He rents a small loft on New York's West 21st Street, buys RCA Photophone amplifiers that are being used in movie theaters, adds Western Electric speakers, an FM tuner, and a turntable, creates some of the world's first component systems, and by the early 1960s will account for 50 percent of U.S. hi-fi component sales. Along with Herman Hosner Scott, Paul Klipsch, Frank McIntosh, and others, Fisher breaks new ground in raising the quality of musical sound reproduction (*see* 1969).

First performances *Voice in the Wilderness (La Voix dans le desert)* Symphonic Poem for Orchestra and 'Cello Obligato by Ernest Bloch 1/21 at Los Angeles; Concertino by Walter Piston 6/20 in a CBS radio broadcast from New York; *Variations for String Orchestra on a Theme by Frank Bridge* by English composer Benjamin Britten, 23, in August at Salzburg. Britten has been a student of Bridge since age 12 and has had his own work performed since 1934; Concerto for Violin and Orchestra in D minor by the 19th century German composer Robert Schumann (who wrote it in 1853) 11/26 at Berlin's Deutsches Opernhaus; Symphony No. 5 by Dmitri Shostakovich 11/21 at Leningrad (critics hail it as a model of Soviet music and Shostakovich regains his ideological good standing); A *Boy Was Born* by Benjamin Britten 12/21 in a BBC broadcast.

Popular songs "Nice Work If You Can Get It" and "A Foggy Day in London Town" by George Gershwin, lyrics by Ira Gershwin (for the film *Damsel in Distress*); "In the Still of the Night" and "Rosalie" by Cole Porter (for the film *Rosalie*); "I've Got My Love to Keep Me Warm" by Irving Berlin (for the film *On the Avenue*); "That Old Feeling" by Sammy Fain, lyrics by Lew Brown; "Once in a While" by Michael Edwards, lyrics by Bud Green; "The Nearness of You" by Hoagy Carmichael, lyrics by Ned Washington; "Somebody Else Is Taking My Place" by Russ Morgan, lyrics by Dick Howard and Bob Ellsworth; "The Joint Is Jumpin'" by Fats Waller, lyrics by Andy Razaf; "(Up) The Lazy River" by Hoagy Carmichael and Sidney Arodin; "Bei Mir Bist Du Schön (Means I Love You)" by Sholom Secunda from a 1933 Yiddish musical, English lyrics by Sammy Cahn and Saul Chapin; "Me and the Devil Blues" and "Hell Hound on My Trail" by Robert Johnson; "The Dipsy Doodle" by bandleader Larry Clinton; "Moon of Manakoora" by Alfred Newman, lyrics by Frank Loesser (for the film *The Hurricane*); "Sweet Leilani" by Harry Owen and "Blue Hawaii" by Leo Robin, lyrics by Ralph Rainger (for the film *Waikiki Wedding*); "Too Marvelous for Words" by Richard Whiting, lyrics by Johnny Mercer (for the film *Ready, Willing and Able*); "Good Morning" by Sam Coslow (for the film *Mountain Music*); "Harbour Lights" by Jimmy Kennedy and Hugh Williams (Will Grosz).

War Admiral wins the Kentucky Derby and goes on to win the Triple Crown despite the challenge of Seabiscuit. The Bay Dancer son of Man o' War breaks the record for a mile and a half as he wins the Belmont Stakes (*see* 1938).

Joe Louis gains the world heavyweight title June 22 by knocking out James J. Braddock in the eighth round of a title bout at Chicago. Joseph Louis Barrow, 23, is the youngest fighter ever to win the championship and will hold it for exactly 12 years—longer than any other man; but while he will have grossed an estimated $4.23 million by the time he retires undefeated in 1949, the Brown Bomber will never be a millionaire.

John Donald "Don" Budge, 22, (U.S.) defeats Baron Gottfried von Cramm, 27, (Ger), to win in men's singles at Wimbledon, Dorothy Round wins in women's singles; Budge wins in men's singles at Forest Hills, Anita Lizana in women's singles.

The U.S. ocean yacht *Ranger* built and skippered by Harold S. Vanderbilt successfully defends the America's Cup against England's *Endeavour II* skippered by Sir Thomas Sopworth in the last America's Cup competition between J class boats (*see* 1958).

The New York Yankees win the World Series by defeating the New York Giants 4 games to 1.

Nylon is patented by E. I. du Pont's W. H. Carothers, who assigns the patent to Du Pont (*see* 1935). The first completely man-made fiber will have wide uses not only in clothing but also as a substitute for canvas in sailboat sails, sisal in ships' hawsers, hog bristles in brushes, etc. (*see* stockings, 1940; Terylene-Dacron, 1941).

Levi Strauss modifies its blue jeans, covering hip-pocket rivets with thread following complaints by schoolteachers that the rivets scratch desk seats (*see* 1874; 1964).

The Babee-Tenda infant chair invented by George B. Hansburg of 1919 pogo stick fame will not tip over. Now 50, Hansburg has devised the chair for his first granddaughter Norma.

The Marijuana Traffic Act signed into law by President Roosevelt August 2 outlaws possession and sale of *cannabis sativa*.

Frank Lloyd Wright completes "Falling Water" at Bear Run, Pa. The dramatic country house is for department store magnate Edgar Kaufmann.

The U.S. Housing Authority (USHA) created by Congress September 1 in the Wagner-Steagall Act provides financial assistance to the states in an effort to remedy the nation's housing shortage (and to provide work in the construction trades).

New York's Harlem River Houses open at 151st Street and the Harlem River Drive, where seven fourto five-story red-brick buildings have been built as the city's first federally financed and federally constructed public housing. Graced with trees and spacious plazas, the projects have been erected following 1935 Harlem riots demanding decent low-cost housing; rents are low, and the carefully screened tenants are provided with a nursery, a health clinic, and social rooms.

German architect Walter Gropius is called to Harvard, where he will continue Bauhaus methods as professor of architecture (*see* 1911). In self-imposed exile from Nazi Germany since 1934, Gropius has been working with London architect Maxwell Fry to design the Village College residence in Cambridgeshire. He joins Hungarian-American designer-photographer Laszlo Moholy-Nagy, 42, in organizing the New Bauhaus at Chicago that will be reorganized in 1939 as the Institute of Design and that will have a major impact on architectural design.

The Palais de Chaillot is completed at Paris on a site formerly occupied by Trocadero Castle.

Congress sets aside Cape Hatteras National Seashore as the first national seashore.

Drought ends in the United States, but stem rust attacks the wheat crop as it did in 1935 (*see* 1941).

U.S. spinach growers erect a statue to the comic-strip sailor Popeye, who is credited with having boosted consumption of the vegetable (*see* 1929).

U.S. biochemists A. N. Holmes and R. E. Corbet isolate vitamin A crystals from fish liver oils (*see* Karrer, 1931).

Vitamin K is produced in crystalline form (*see* 1935).

Synthetic thiamine (vitamin B_1) is manufactured in the United States and Switzerland at $450 per pound, but further research will soon reduce the price to a few cents per pound (*see* 1936; 1938).

Pepperidge Farm bread is introduced by Connecticut entrepreneur Margaret Fogarty Rudkin, 40, who sets up an oven in her husband's former polo pony stable on the family's 120-acre Pepperidge Farm and makes whole-wheat bread which she sells first to neighbors and then through a New York City fancy food retailer (*see* 1940).

Spam is introduced by George A. Hormel & Co., whose pork-shoulder-and-ham product will become the world's largest selling canned meat.

Ford Motor Company scientists trying to develop a synthetic wool fiber produce soy protein "analogs" that will be used as substitutes for bacon and other animal protein foods. They spin a textile filament from soybean protein and create a vegetable protein that can be flavored to taste much like any animal protein food (*see* 1949; Boyer, 1933).

The artificial sweetener sodium cyclamate discovered by University of Illinois chemist Michael Sveda is 30 times sweeter than sugar with none of the bitter aftertaste of the saccharin discovered in 1879 (*see* 1950).

Britain gets her first frozen foods as Wisbech Produce Canners Ltd. introduces frozen asparagus in May at 2d 3p per pack, strawberries in June at 1d 2p per 8-ounce pack, garden peas in July at 9p per 6 oz. pack, and sliced green beans in August at 1d 2p per 6 oz. pack. Wisbech's S. W. Smedley has developed his own freezing process after studying American techniques on a visit to the United States, but Britain has only 3,000 home refrigerators as compared with more than 2 million in the United States.

Home freezers become commercially important for the first time in the United States as frozen food sales increase, but relatively few Americans have anything more advanced than an icebox. Icemen continue regular deliveries.

1937 ***(cont.)*** The A&P begins opening supermarkets, as do other major U.S. food chains. Three or four smaller stores are closed down for every A&P supermarket opened (*see* 1929; King Kullen, 1930; Big Bear, 1932).

The supermarket shopping cart introduced at Oklahoma City June 4 begins a revolution in food buying. Sylvan N. Goldman, 38, who owns Standard Food Markets and Humpty Dumpty Stores, has created the cart to enable customers to buy more than can fit in the wicker baskets they carry; he has taken some folding chairs, put them on wheels, raised the seats to accommodate a lower shopping basket, placed a second basket on the seat, and used the chair back as a handle. Four U.S. companies will develop the shopping cart into a computer-designed chromed-steel cart that can be nested in a small area.

Howard Johnson restaurants become a franchised operation as restaurateur Johnson hits upon the idea of franchising after having found it impossible to obtain bank financing for restaurants (*see* 1936). Johnson locates property which he leases to a Cambridge, Mass., widow who invests first $5,000 and then another $25,000, he constructs a building, hires employees, charges a small fee for the franchise but takes no part of the profits, retains exclusive rights to supply food including 28 flavors of ice cream, menus, table mats, and other items, and starts a chain of franchised orange-roofed restaurants whose design, and operation he will rigidly control (*see* 1960).

Contraception receives virtually unqualified endorsement from an American Medical Association committee on birth control (*see* 1940).

The first state contraceptive clinic opens March 15 at Raleigh, N.C. The State Board of Health introduces a program for indigent married women in its regular maternity and child health service.

Islam's Grand Mufti issues a fatwa permitting Muslims to take any measure to avoid conception to which both man and woman agree.

1938 Adolf Hitler vows to "protect" the 10 million Germans living outside the Reich, annexes Austria March 14, and engineers an April 10 plebiscite which shows that 99.75 percent of Austrians desire union with the Third Reich.

France's Chautemps government falls March 10; Léon Blum tries to form a new Popular Front cabinet but resigns under pressure April 10; Edouard Daladier, now 52, heads a new Radical Socialist cabinet that is farther to the right.

The Austrian *Anschluss* (above) draws protests from Britain and France, but they appease *der führer* at Munich September 29 by permitting him to take the Sudetenland, a 16,000-square-mile territory that covers nearly a third of Czechoslovakia and contains a third of her inhabitants.

"I believe it is peace for our time . . . peace with honor," says Prime Minister Chamberlain on his return from Munich (above; *see* 1939).

Adolf Hitler's annexation of Austria brought Europe closer to war. Appeasement at Munich delayed the inevitable.

"I am convinced that it is wiser to permit Germany eastward expansion than to throw England and France, unprepared, into a war at this time," writes Charles A. Lindbergh September 23 to U.S. Ambassador to Britain Joseph P. Kennedy. Lindbergh, who has been living abroad to escape publicity since the kidnapping and murder of his infant son in 1932, has made surveys of British, German, and Soviet airpower; Britain could not possibly win a war in Europe even with U.S. aid, he argues.

London has announced postponement of any Palestine partition January 4, Jewish terrorist Solomon ben Yosef is executed June 29, Arab markets are bombed in Jerusalem, Haifa, and Jaffa either by Jewish or Arab terrorists, 20 Jews are massacred October 2 at Tiberias, Arab extremists seize Bethlehem and the old section of Jerusalem, which are retaken by British troops October 10 and October 18, respectively. A new British commission reports November 9 that all partition proposals are impractical, and by year's end there are 25,000 to 30,000 British troops in Palestine (*see* 1930; 1939).

Japanese forces in China follow up their 1937 successes by taking Qingdao (Tsingtao) January 10 after the Chinese have destroyed some Japanese factories in the area. Japanese troops advance along the Hankow Railway and through Shansi Province, reach the Huanghe (Yellow) River March 6, but suffer several reverses as Communist guerrillas retain control of the countryside.

Amoy falls to the Japanese May 10, Japanese and Russian forces clash on the Chinese-Siberian border from July 11 to August 10, Japanese troops land at Bias Bay near Hong Kong October 12, Guangzhou (Canton) falls October 21, Hankow October 25. Capturing Guangzhou permits the Japanese to cut the Guangzhou-Hankow Railway that supplies Chinese forces in the interior with war matériel from abroad.

Spanish insurgents sever Loyalist territory in Castile from Barcelona and Catalonia, the opposing forces remain deadlocked along the Ebro River through most of the summer, a great insurgent drive begins in Catatonia December 23, and the Loyalists are forced back toward Barcelona. Italy has withdrawn some troops following an Anglo-Italian pact signed April 16, but a force of 40,000 remains to support Generalissimo Francisco Franco.

Turkey's President Kemal Atatürk dies November 10 at age 57 after a 15-year regime in which he has established a modern republic. Unanimously elected to succeed him in a vote by the national assembly is Ismet Inonu, 54, who has been responsible for many of the reforms accomplished since 1923 and will be president until 1950.

Benito Mussolini demands France's colonies of Corsica and Tunisia in December.

Polish forces have occupied the Teschen area of Czechoslovakia October 2 after having sent a note September 29 demanding cession of the territory seized by the Czechs in the Polish-Russian war of 1920. Poland champions Hungarian claims in Slovakia and Ruthenia.

Britain and Ireland resume friendly relations after concluding a 3-year agreement to remove tariff barriers (*see* 1935). Britain turns over coast defense installations to Eire, and Dublin agrees to pay £10 million to satisfy land-annuity claims.

The Declaration of Lima adopted December 24 by representatives of 21 American nations meeting at a Pan-American conference in Peru pledges the Western Hemisphere neighbors to consult in the event that the "peace, security, or territorial integrity" of any state is threatened; they reaffirm absolute sovereignty in the face of fears that Europe's fascist powers may attempt takeovers in the Americas.

The Nazis deprives Austria Jews of their civil rights and means of livelihood (above); they plunder Jewish shops and homes.

Italy enacts anti-Jewish legislation.

The worst pogrom in German history begins November 9 following the assassination of Paris embassy official Ernst Edouard vom Rath November 7 by German-born Polish Jew Herschel Grynzpan, 17, who has heard of the mistreatment of thousands of Polish Jews (including members of his own family) after their deportation from Germany. The Nazis smash Jewish shop windows ("crystal") in the *Kristalnacht* riots, shops, homes, and synagogues are looted, demolished, and burned, and 20,000 to 30,000 Jews are carried off to concentration camps (*see* 1939; Dachau, 1933; Buchenwald, 1937).

The Dies Committee (to Investigate Un-American Activities) begins studying Nazi activities in the United States but soon turns to investigating communist activities. Rep. Martin Dies, 36 (D. Texas) has turned against the New Deal which he originally supported, and while most of his committee's charges will be based on hearsay, circumstantial evidence, or slander, Dies will claim that it is more effective than the FBI in exposing communist subversion (*see* 1947).

The Fair Labor Standards Act (wage and hour law) passed by Congress June 15 limits working hours of some 12.5 million U.S. workers in the first national effort to place a floor under wages and a ceiling on hours. Working hours for the first year after the new law takes effect are limited to 44 per week with the limit to be reduced to 42 for the second year and 40 for every year thereafter. Longer work weeks are permitted only if overtime work is paid for at one and one-half times the regular rate. Minimum wage is to be 25¢ per hour for the first year, 30¢ for the next 6 years (*see* Minimum Wage Act, 1949).

The new minimum wage law (above) wipes out Puerto Rico's 40,000-worker needlework industry where 25¢ has been the hourly rate for *skilled* workers.

The Congress of Industrial Organizations (CIO) formally organized in November succeeds the Committee for Industrial Organization formed in 1935. John L. Lewis is elected CIO president November 18 (*see* 1946).

The Supreme Court orders equal accommodations for Missouri law students regardless of race December 12 in *Missouri ex. rel. Gaines v. Canada.*

Germany's trade balance shows a deficit of 432 million marks. The country has been bankrupt since 1931 by ordinary capitalist standards, and Reichsbank president Horace Greeley Hjalmar Schacht, 61, warns that the country's enormous armament program must be curtailed lest the catastrophic inflation of 15 years ago recur.

President Roosevelt asks Congress for help in stimulating the U.S. economy as business remains in recession with 5.8 million Americans still unemployed. Congress reduces corporation profits taxes May 27 and passes an Emergency Relief Appropriations Act June 21.

Wall Street's Dow Jones Industrial Average falls to 98.95, down from a 1937 high of 194.40, but it rebounds to 158.41 by year's end.

An oil strike in southeastern Kuwait February 23 begins to revolutionize the emirate's economy (Mikimoto's cultured pearls have ruined its pearl fishery, once its leading industry; *see* 1893). Kuwait Oil Co., jointly owned by Anglo-Iranian and Gulf, will develop the giant petroleum reserve in the British protectorate.

Mexico nationalizes her petroleum industry March 18, revoking licenses granted to British and U.S. oil companies to operate in Mexico. Oil is a natural resource that belongs to all the Mexican people, the Cardenas government says; it expropriates properties valued at $450 million and proposes oil barter agreements with Germany, Italy, and other nations to exchange oil for manufactured goods imported up to now largely from Britain and the United States.

The first nuclear fission of uranium is produced December 18 by German chemist Otto Hahn, 59, who has found that the nucleus of certain uranium atoms

1938 ***(cont.)*** can be split into two approximately equal halves, releasing not only energy but also neutrons that can, in turn, split further uranium atoms (*see* Fermi, 1934). Assisting Hahn are his colleague Fritz Strassman and Austro-Swedish physicist Lise Meitner, 60, whose nephew Otto Frisch, 34, will help her work out the implications of Hahn's observations (*see* Bohr, Einstein, 1939).

A Civil Aeronautics Authority (CAA) created by act of Congress June 23 moves to regulate the growing U.S. aviation industry. Named to head the new authority is former Life Savers magnate Edward J. Noble, now 56.

Howard Hughes sets a new round-the-world speed record July 15, flying a twin-engine Lockheed plane from California to California in 3 days, 19 hours, 14 minutes, 28 seconds. Now 32, Hughes has $3 million per year in Hughes Tool Co. profits to finance his flying exploits and Hollywood film-making (*see* 1908; Post, 1931; Odum, 1947; TWA, 1939).

"Wrong-Way" Corrigan makes headlines July 19. Douglas Gorce Corrigan, 31, who has flown nonstop from Los Angeles to New York July 10, took off in his 1929 $900 Curtiss Robin July 16, presumably on a return flight. When he lands in Dublin after a 28-hour, 13-minute flight he insists that he intended to fly west but had compass trouble. Authorities say his flight and landing were illegal, but Corrigan is lionized.

Eastern Airlines is created out of North American Aviation's Eastern Air Transport by World War flying ace E. V. "Eddie" Rickenbacker who buys into North American with backing from Standard Oil heir Laurence Rockefeller, 28 (*see* North American, 1933). Now 48, Rickenbacker has worked in the auto industry and for several aircraft and airline companies; he will make Eastern a major carrier, and he will obtain routes up and down the East Coast and to Mexico and the Caribbean.

Northrop Aircraft is founded by John Knudsen Northrop, now 42, whose company was acquired last year by Douglas Aircraft (*see* 1928).

Douglas Aircraft has sales of $28.4 million as its DC-3 gains popularity. It solicits orders for a new four-engine DC-4, but Boeing goes into production with a four-engine 307 that challenges Douglas for leadership in commercial aircraft (*see* B-17, 1935; TWA, 1940).

McDonnell Aircraft is founded by former Glenn L. Martin project engineer James Smith McDonnell, Jr., 39, who will make McDonnell a leading producer of military aircraft (*see* McDonnell-Douglas, 1967).

El Capitán goes into service February 22 between Chicago and Los Angeles for the Santa Fe. The new once-a-week all-coach express has two chair cars, a baggage-dormitory coach, lunch-counter/tavern car, and chair observation car, carries 192 passengers, and operates on the same schedule as the *Super Chief* (*see* 1936).

The San Diegan goes into service March 27 for the Santa Fe. The new express goes between Los Angeles and San Diego in 2.5 hours and makes two round-trip runs per day.

General Motors joins with Standard Oil of California to organize Pacific Coast Lines, a firm that will convert West Coast electric street railways into motorbus lines (*see* 1932; 1939).

The Florida Overseas Highway brings Key West to within 5 hours' driving distance of Miami. The new road employs spans built by Henry Flagler for his Florida East Coast Railway of 1912, which was repaired after the hurricane of 1926 but completely wiped out by the hurricane of 1935.

The Volkswagen ("people's car"), with an air-cooled rear engine, is assembled by hand in Nazi Germany and the cornerstone for a Volkswagen factory is dedicated May 26 at Wolfsburg on the Mittelland Canal 40 miles east of Hanover in Lower Saxony. Austrian automotive engineer Ferdinand Porsche, 63, has designed the low-cost "beetle" on commission from Adolf Hitler; it will not go into mass production for more than 10 years, but more than 18 million of the "beetles" will eventually be sold, exceeding the Model T Ford's record (*see* 1949).

Fiberglas, perfected by Owens-Illinois and Corning Glass Works, can be spun into yarn and woven into fabrics or used as insulating material. The material is made of fine glass filaments.

Teflon (Fluon), discovered accidentally by Du Pont chemist Roy Joseph Plunkett, 28, is an excellent electrical insulation material, stable over a wide range of temperatures and resistant to most corrosive agents. Found while working on refrigerants, the polytetrafluoroethylene plastic will have many industrial uses; it will be marketed under the name Fluon by Britain's Imperial Chemical Industries (*see* cooking utensils, below).

New York enacts the first U.S. state law requiring medical tests for marriage license applicants April 12.

The March of Dimes to finance research into poliomyelitis is founded under the leadership of President Roosevelt's former law partner Basil O'Connor, 46.

John L. Baird gives the first demonstration of high-definition color television February 4 at London's Dominion Theatre, Tottenham Court Road. He transmits color films and shows them on a 9- by 12-foot screen via a 120-line-per-inch system (*see* 1926). Within 2 weeks he transmits live action in color from the Baird Studios at Crystal Palace, but his refusal to consider electronic transmission in place of mechanical transmission blocks commercial development (*see* Goldmark, 1940).

The first true Xerox image appears October 22 at Astoria, Queens. The electrophotographic image is imprinted on wax paper which has been pressed against an electrostatically charged 2- by 3-inch sulfur-coated zinc plate that has been dusted with lycopodium powder. Chester Carlson, who has been helped by a German refugee physicist, attends New York Law School night classes, will be admitted to the bar in 1940, and will receive his first patent that year for the process he will call "xerography," using the Greek word *xeros* for dry, but he will fail in his initial attempts to get financial backing (*see* 1937; 1946).

The ballpoint pen patented by Hungarian chemist George Biro, 41, and his brother Ladislao Biro, 39, a Budapest proofreader, has a vein-like tube that fits inside its barrel and moistens the ball at its tip by capillary attraction, but their pen will achieve its potential only after Austrian-American chemist Franz Seech in California develops a viscous fluid with a dye that forms a film on any surface when exposed to air (*see* 1945). The Biro brothers will emigrate in 1943 to Argentina, where they will get financial backing from British financier Henry Martin; he will set up a factory in England with Frederick Miles to produce pens that will not leak at high altitudes for use by the Royal Air Force. The French company Bic will acquire the Martin-Miles firm and develop a cheaper throwaway pen.

Picture Post is launched by British publisher Edward Hulton, who models the photo weekly on Henry Luce's LIFE (*see* 1936).

"Superman" is introduced in comic books by Cleveland cartoonists Jerry Siegel and Joseph Shuster, both 24, who developed their superhuman Clark Kent newspaperman hero while in high school and have finally sold it to Detective Comics, Inc., which publishes the first "Superman" episode in the June issue of *Action Comics,* paying the two young men $10 per page to give them an income of $15 each per week. "Superman" will appear in newspapers beginning in the early 1940s and be the basis of radio and television serials and endless merchandise spinoffs, but Detective Comics has acquired all rights, and "Superman's" originators will derive little financial reward until the syndicators agree late in 1975 to provide them with pensions.

The War of the Worlds broadcast October 30 over CBS stations gives a dramatic demonstration of the power of radio. Orson Welles's Mercury Theater of the Air presents a radio version of the 1898 H. G. Wells novel, and its "news" reports of Martian landings in New Jersey are so realistic that near-panics occur in many areas despite periodic announcements that the program is merely a dramatization.

Nonfiction *It Is Later Than You Think* by Russian-American critic Max Lerner, 36; *Homage to Catalonia* by George Orwell antagonizes British leftists by showing how Stalinists have suppressed Trotskyist and anarchist elements in Spain's independent left. Orwell joined the Republican side after going to Spain as a journalist late in 1936, a left-wing publisher has rejected his ms., it will sell only 600 copies in his lifetime, and will not be published in the United States until after his death in 1950; *With Malice Toward Some* by U.S. writer Margaret Halsey, 28, who spent a year in Devon, England.

Fiction *Nausea (La Nausée)* by French novelist Jean-Paul Sartre, 33; *But the World Must Be Young (Men ung ma verden ennu vaere)* by Nordahl Grieg is a novelistic critique of Stalinism; *The Death of the Heart* by Elizabeth Bowen; *Brighton Rock* by Graham Greene is his first explicitly "Catholic" novel; *Ferdyduke* by Polish novelist Witold Gombrowicz, 34; *Scoop* by Evelyn Waugh is a spoof on British foreign correspondents that thinly veils actual fact; *Out of Africa* by Isak Dinesen; *The Forest of a Thousand Demons (ogboju ode iinn igbo irummale)* by Nigerian novelist (and Yoruba chief) Daniel O. Fagunwa, 28; *Tropisms (Tropismes)* by Russian-born French novelist Nathalie Sarraute, 36; *Out of the Silent Planet* by English novelist C. S. (Clive Staples) Lewis, 40; *Chosen People* by Seaforth Mackenzie; *The Code of the Woosters* by P. G. Wodehouse; *The Fathers* by Allen Tate; *Young Man with a Horn* by U.S. novelist Dorothy Dodds Baker, 31, has been inspired by the music, if not the life, of the late Bix Beiderbecke who died of lobar pneumonia at age 28 in August 1931 (*see* 1924); *Epitaph for a Spy* and *Cause for Alarm* by Eric Ambler; *Dynasty of Death* by U.S. novelist Taylor Caldwell, 38.

Juvenile *The Man of Bronze* by former New York telegrapher Kenneth Robeson (Lester Dent), 32, whose "Doc Savage" adventure novel will be followed in the next 7 years by another 164 such novels as Dent turns out a new 60,000-word "Doc Savage" adventure almost every month; *The 500 Hats of Bartholomew Cubbins* by Dr. Seuss.

Poetry *The Odyssey (I Odysseia)* by Greek (Cretan) poet Nikos Kazantzakis, 55, whose 33,333-line "modern sequel" has been influenced by the French philosopher Henri Bergson, now 79; *Dreams Begin Realities* by U.S. poet Delmore Schwartz, 25.

Painting *Italian Women* by Georges Rouault; *Cradling Wheat* by Thomas Hart Benton. William Glackens dies at Westport, Conn., May 22 at age 68; Ernst Ludwig Kirchner dies by his own hand at Davos, Switzerland, June 15 at age 58 (the Nazis last year confiscated more than 600 of his works).

The Cloisters opens in New York's Tryon Park. A gift from the Rockefeller family to the Metropolitan Museum of Art, the medieval European nunnery is filled with art treasures that include a unicorn tapestry.

Theater *The Restless Heart (La Sauvage)* by Jean Anouilh 1/10 at the Théâtre de Mathurius, Paris; *Bachelor Born* by U.S. playwright Ian Hay 1/25 at New York's Morosco Theater, with Peggy Simpson, Helen Trenholme, 400 perfs.; *Shadow and Substance* by Paul Vincent Carroll 1/26 at New York's Golden Theater, with Cedric Hardwicke, Sara Allgood, Julie Haydon, 274 perfs.; *On Borrowed Time* by Paul Osborn 2/3 at New York's Longacre Theater, with Dudley Digges, Dorothy Stickney, Dickie Van Patten, 321 perfs.; *Our Town* by Thornton Wilder 2/4 at New York's Henry Miller Theater, with Martha Scott, Frank Craven, Philip Coolidge, 336 perfs.; *What a Life* by Clifford Goldsmith 4/13 at New York's Biltmore Theater, with Ezra Stone as Henry Aldrich, Eddie Bracken, Betty Field, 538 perfs.; *Purgatory* by William Butler Yeats, now 73, 8/10 at Dublin's Abbey Theatre; *Thieves' Carnival (Le bal des valeurs)* by Jean Anouilh 9/17 at the Théâtre des Arts, Paris; *The Corn Is Green* by Emlyn Williams 9/20 at London's Duchess Theatre; *Kiss the*

1938 ***(cont.)*** *Boys Goodbye* by Clare Boothe 9/28 at Henry Miller's Theater, New York, with Millard Mitchell, Helen Claire, Benay Venuta, 286 perfs.; *The Fabulous Invalid* by George S. Kaufman and Moss Hart 10/8 at New York's Broadhurst Theater (the "invalid" is the legitimate theater), 165 perfs.; *Abe Lincoln in Illinois* by Robert Sherwood 10/15 at New York's Plymouth Theater, with Raymond Massey in the title role, 472 perfs.; *Rocket to the Moon* by Clifford Odets 11/24 at New York's Belasco Theater, with Morris Carnovsky, 131 perfs.; *Here Come the Clowns* by Philip Barry 12/7 at New York's Booth Theater, with Eddie Dowling, Madge Evans, Russell Collins, 88 perfs.

Radio *Young Widder Brown* 7/26 over NBC stations. Created by Frank and Anne Hummert, the soap opera will continue daily until June 1956.

Films Serge Eisenstein's *Aleksandr Nevsky* with Nikolai Cherkassov; Howard Hawks's *Bringing Up Baby* with Cary Grant, Katharine Hepburn; Alfred Hitchcock's *The Lady Vanishes* with Michael Redgrave, Margaret Lockwood, Paul Lukas, Dame May Witty; Anthony Asquith and Leslie Howard's *Pygmalion* with Howard, Wendy Hiller; Michael Curtiz's *The Adventures of Robin Hood* with Errol Flynn. Also: Marcel Pagnol's *The Baker's Wife* with Raimu, Ginette Leclerc; Edmund Goulding's *The Dawn Patrol* with Errol Flynn, Basil Rathbone, David Niven; George Cukor's *Holiday* with Katharine Hepburn, Cary Grant; Henry King's *In Old Chicago* with Tyrone Power, Alice Faye; William Wyler's *Jezebel* with Bette Davis, Henry Fonda, George Brent; Lloyd Bacon's *A Slight Case of Murder* with Edward G. Robinson; Frank Borzage's *Three Comrades* with Robert Taylor, Margaret Sullavan, Franchot Tone; Norman Taurog's *The Adventures of Tom Sawyer* with Tommy Kelly, Jackie Moran; Frank Capra's *You Can't Take It With You* with Jean Arthur, James Stewart, Lionel Barrymore.

Hollywood musicals Walt Disney's *Snow White and the Seven Dwarfs* is the first full-length animated cartoon feature, Snow White sings "Some Day My Prince Will Come" and "One Song," dwarfs Doc, Grumpy, Happy, Sleepy, Dopey, Sneezy, and Bashful sing "Heigh-Ho" and "Whistle While You Work," music by Frank Churchill, lyrics by Larry Mose; Ray Enright's *Hard to Get* with Dick Powell, Olivia de Havilland, music by Harry Warren, lyrics by Johnny Mercer, songs that include "You Must Have Been a Beautiful Baby"; Wesley Ruggles's *Sing You Sinners* with Bing Crosby, Fred MacMurray, Donald O'Connor, songs that include the Franke Harling-Sam Coslow title song, "Small Fry" and "Two Sleepy People," by Hoagy Carmichael, lyrics by Frank Loesser.

Stage musicals *Operette* 3/16 at His Majesty's Theatre, London, with Peggy Wood, music and lyrics by Noël Coward, songs that include "The Stately Homes of England," 133 perfs.; *I Married an Angel* 5/11 at New York's Shubert Theater, with Dennis King, Vera Zorina, Vivienne Segal, music by Richard Rodgers, lyrics by Lorenz Hart, songs that include "Spring Is Here" and the title song, 338 perfs.; *Helzapoppin* 9/22 at New York's 46th Street Theater, with Ole Olsen and Chic Johnson who delight audiences with their slapstick and sight gags, music by Sammy Fain, lyrics by Irving Kahal and Charles Tobias, songs that include "I'll Be Seeing You," 1,404 perfs.; *Knickerbocker Holiday* 10/19 at New York's Ethel Barrymore Theater, with Walter Huston, music by Kurt Weill, book and lyrics by Maxwell Anderson, songs that include "September Song," 168 perfs.; *Leave It to Me* 11/9 at New York's Imperial Theater, with ingénue Mary Martin, 24, doing a simulated striptease to Cole Porter's song "My Heart Belongs to Daddy" (other songs include "Most Gentlemen Don't Like Love,") 307 perfs. (Porter was injured last year in a fall from a horse and will be crippled for the rest of his life); *The Boys from Syracuse* 11/23 at New York's Alvin Theater, with Jimmy Savo, Eddie Albert is about the Shubert brothers, music by Richard Rodgers, lyrics by Lorenz Hart, songs that include "Falling in Love with Love," "Sing for Your Supper," "This Can't Be Love," 235 perfs.

Opera *Mathis der Maler (Matthias the Painter)* 5/28 at Zurich's Stadtheater, with music and libretto by Paul Hindemith based on the life of the German painter Matthias Grünewald who died in 1528. The opera had been scheduled to open at Berlin in 1934 but was banned by the Nazis.

Ballet *St. Francis* 7/21 at London's Drury Lane Theatre, with music by Paul Hindemith, choreography by Leonide Massine; *Billy the Kid* 10/9 at the Chicago Civic Opera House, with music by Aaron Copland, choreography by Eugene Loring.

First performances Symphony No. 3 by Howard Hanson 3/26 in an NBC Orchestra radio concert; *The Incredible Flutist* (ballet music) by Walter Piston 3/30 at Boston's Symphony Hall; Symphony No. 1 by Piston 4/8 at Symphony Hall; Quartet in G minor for Piano and Strings by Ernest Bloch 5/5 at Los Angeles; Concerto No. 1 in D major for Piano and Orchestra by Benjamin Britten in August at Queen's Hall, London.

Benny Goodman and His Orchestra give the first Carnegie Hall jazz concert January 17 with guest performers who include Count Basie and members of the Basie and Duke Ellington orchestras *(see* 1935). Pianist Jess Stacy plays "Sing Sing Sing."

Glenn Miller begins touring with a big band of his own after years of playing trombone and arranging music for Tommy and Jimmy Dorsey and for Ray Noble. Now 39, Miller will achieve enormous success next year with his recordings of "In the Mood," "Sunrise Serenade," and "Moonlight Serenade" (which will become his theme song) *(see* 1942).

Woody Guthrie identifies himself with the labor movement and travels the country singing his songs "Hard Traveling," "Blowing Down This Dusty Road," "Union Maid," and "So Long (It's Been Good to Know You)." Guthrie will support the cause of organized labor with his "Talking Union" album and with personal appearances (*see* 1936).

Popular songs "I Can Dream, Can't I?" by Sammy Fain, lyrics by Irving Kahal (for the short-lived Broadway musical *Right This Way*); "That Old Feeling" by Fain, lyrics by Kahal (for the film *Vogues of 1938*); "Love Walked In" and "Our Love Is Here to Stay" by the late George Gershwin, lyrics by Ira Gershwin (for the film *The Goldwyn Follies*); "Jeepers Creepers" by Harry Warren, lyrics by Johnny Mercer (for the film *Going Places*); "I Get Along Without You Very Well (Except Sometimes)" by Hoagy Carmichael, lyrics by Jane Brown Thompson; "You Go to My Head" by J. Fred Coots, lyrics by Haven Gillespie; "I Let a Song Go Out of My Heart" by Duke Ellington, lyrics by Irving Miller, Henry Nemo, John Redmond; "One O'Clock Jump" by Count Basie; "Camel Hop" by Mary Lou Williams; "Sent for You Yesterday (and Here You Come Today)" by Eddie Durham, Count Basie, and vocalist Jimmy Rushing; "F. D. R. Jones" by Harold Rome; "Music, Maestro, Please" by Broadway composer Allie Wrubel, lyrics by Herb Magidson; "The Flat Foot Floogie" by Slim (*né* Bulee) Gaillard, 23, Slam Stewart, Bud Green (who have been forced to change the word "floozie" to "floogie"); "A-Tisket, A-Tasket" by Ella Fitzgerald and Van Alexander; "Cherokee" by Ray Noble (whose song will be the theme for bandleader-tenor saxophonist Charlie Barnet); "I Hadn't Anyone Till You" by Ray Noble; "Thanks for the Memory" by Ralph Rainger, lyrics by Leo Robin (title song for a film starring comedian Bob Hope, who will make it his theme song).

The samba and the conga are introduced to U.S. dance floors.

Don Budge wins in men's singles at Wimbledon, Forest Hills, France, and Australia, the first "grand slam" and one not to be duplicated for 24 years; Helen Wills Moody wins in women's singles at Wimbledon, Alice Marble at Forest Hills.

The New York Yankees win the World Series by defeating the Chicago Cubs 4 games to 0.

Italy wins the third World Cup football (soccer) finals, defeating Hungary 4 to 2 at Colombes, France; no further World Cup games will be held until 1950.

War Admiral starts as a 1–4 favorite over Seabiscuit in a November match race at Pimlico before a crowd of 40,000; Seabiscuit wins (*see* 1937).

Fiberglas (above) will largely replace wood in pleasure boat hulls, surfboards, skis, etc.

Gropius House is completed by architect Walter Gropius, whose new residence at Lincoln, Mass., gives new direction to residential architecture.

Frank Lloyd Wright completes Taliesin West at Paradise Valley outside Phoenix, Ariz. His new home will also be a teaching center (*see* 1911).

The Hardoy sling chair is introduced by New York's Knoll Associates, which also distributes the 1929 Barcelona chair designed by Mies van der Rohe.

Olympic National Park is created by act of Congress to protect 896,660 acres of Pacific Northwest rain forest, glaciers, and meadows that support a herd of rare Roosevelt elk.

Congress passes a Flood Control Act authorizing public works on U.S. rivers and harbors.

A tropical hurricane strikes Long Island and much of New England without warning September 21, and in a few hours wreaks more havoc than the Chicago fire of 1871 or the San Francisco earthquake and fire of 1906. The storm takes 680 lives, destroys 2 billion trees, causes $400 million in property damage, and inflicts great and lasting environmental damage, but 60 colonies of beavers in New Jersey's Palisades Park busily maintain their dams through the blow, minimizing flooding of 42,000 acres of land and highways.

An International Convention for the Regulation of Whaling institutes voluntary conservation quotas, but the killing of whales by Japanese, Norwegian, and Russian ships goes on virtually unabated.

Efforts begin to help salmon and steelhead trout ascend the Columbia River via fish "ladders" as the river's fishing industry is threatened by power dams, such as the year-old Bonneville dam, that have cut the fish off from spawning grounds.

Albacore tuna are found in large schools off the Oregon Coast, and salmon fishermen who have been supplying the Columbia River Packers Association since 1899 turn to tuna fishing (*see* Van Camp, 1926). The Association adopts the brand name Bumble Bee for its canned tuna after having used it for years on its canned salmon and builds the first tuna cannery in the Northwest, setting it up alongside its salmon cannery at Astoria, Ore.

A revised Agricultural Adjustment Act eases restrictions on planting most U.S. crops, even providing for lime and mineral fertilizer supplies to farmer to increase yields. The new AAA begins direct crop-subsidy "parity" payments to farmers based on 1910–1914 farm prices.

The Commodity Credit Corp. established by the new AAA (above) supports U.S. farm prices by buying up surpluses for an "Ever-Normal Granary" designed to protect the nation against drought and plant diseases, distribute surpluses among the needy, and pay export subsidies that will encourage foreign sales by "equalizing" U.S. farm prices with lower world prices (*see* 1940 wheat exports; food stamps, 1939).

Arizona's 286-foot high Bartlett Dam is completed.

Nicotinic acid (niacin) is found to prevent pellagra (*see* 1915; Elvehejm, 1936).

Enrichment of bread with the B_1 vitamin thiamine is proposed at a scientific meeting in Toronto.

Studies of the chemistry of vision in poor light begun by Harvard biologist George Wald, 31, will show the importance of vitamin A in avoiding nyctalopia, or night blindness (*see* 1925).

The United States has her last reported case of "the milksick" as dairies virtually wipe out the often fatal dis-

1938 ***(cont.)*** ease by pooling their milk from a variety of herds. Also called the slows or the trembles, the disease is transmitted in the milk of cows that have eaten white snakeroot (*Eupatorium rugosum*).

A new U.S. Food, Drug and Cosmetic Act, signed into law June 27, updates the Pure Food and Drug Act of 1906 (*see* 1912; 1913). It establishes standards of identity for most food products, requiring that basic ingredients of about 400 such products be listed in the Code of Federal Regulations, but only "optional" ingredients such as salt, sugar, and spices be listed on labels. The new law is stricter than the old one but critics say it is still not sufficiently protective of consumers' health.

The Wheeler-Lea Act gives the Federal Trade Commission jurisdiction over advertising that may be false or misleading even if it does not represent unfair competition (*see* 1931). The FTC receives special powers to regulate advertising of foods, drugs, cosmetics, and therapeutic devices.

A can of meat put up in 1824 is discovered and its contents fed to test rats with no observable ill effects.

Dewey and Almy in Boston develop the Cryovac deep-freezing method of food preservation (*see* Birdseye, 1925).

Mott's apple juice is introduced by the Duffy-Mott Co.

Nescafé is introduced in Switzerland by Nestlé, which has been asked by the Brazilian government to help find a solution to Brazil's coffee surpluses. Nestlé has spent 8 years in research to develop the instant coffee product (*see* G. Washington, 1909; Instant Maxwell House, 1942).

Teflon (above) has a low coefficient of friction that will make it popular as a coating for nonstick frying pans and other cooking utensils.

1939 World War II begins one week after the August 23 signing of a mutual nonaggression pact between Nazi Germany and Soviet Russia. Adolf Hitler has occupied Bohemia and Moravia and annexed Memel; his foreign minister Joachim von Ribbentrop, 46, signs the pact with Josef Stalin's new commissar of foreign affairs V. M. Molotov, and German troops and aircraft attack Poland September 1.

Manchurian and Mongolian forces have been fighting since May in a conflict between Soviet Russia and Japan. Tokyo scraps the anti-Comintern pact of 1936 at news of the Molotov-von Ribbentrop pact, protesting the German action.

Britain and France declare war on Germany September 3 as a U-boat sinks the British ship *Athenia* off the Irish Coast. H.M.S. *Courageous* is sunk September 19.

Soviet troops invade Poland from the east September 17, Warsaw surrenders to the Germans September 27, and Poland is partitioned September 28 between Germany and the U.S.S.R.

The Spanish Civil War has ended March 28 with the fall of Madrid to Francisco Franco. Germany and Italy have withdrawn their forces by late June, and Spain will be neutral in the new European war after having lost upwards of 410,000 in battle or by execution (more than 200,000 have died of starvation, disease, or malnutrition, and some estimates put the total dead at well over one million).

Iraq's king Ghazi is killed in a Baghdad auto accident April 4 and the British consul is stoned to death in the ensuing riots (rumor says the British arranged the accident). Ghazi is succeeded by his 3-year-old son who will reign until 1958 as Faisal II.

Romania's premier Armand Calineseu is assassinated September 21 by members of the pro-fascist Iron Guard (*see* 1940).

A British expeditionary force of 158,000 is in France by late September, but former president Herbert Hoover spearheads a U.S. nonintervention movement with support from Theodore Roosevelt, Jr., Sen. Harry F. Byrd (D. Va.), Sen. William Borah (R. Idaho), Sen. Burton K. Wheeler (R. Mont.), Henry Ford, and Charles A. Lindbergh.

Lindbergh makes his first anti-intervention speech on U.S. radio September 17, arguing that Stalin is as much to be feared as Hitler (*see* America First, 1941).

"I cannot forecast to you the action of Russia. It is a riddle wrapped in a mystery inside an enigma," says Winston Churchill October 1 in a radio broadcast. Churchill has been first lord of the admiralty since September 3 (*see* 1915; 1926; 1940).

Soviet troops invade Finland November 30 (*see* 1940). Herbert Hoover organizes a drive for Finnish relief December 5, and Congress grants Finland $10 million in credit for agricultural supplies.

The Hatch Act passed by Congress August 2 bars U.S. federal employees from taking an active role in political campaigns. The purpose of the measure is to keep party politics out of government agencies and prevent a return of the spoils system (*see* 1883).

The Battle of the River Plate December 13 off the South American coast ends with the British cruiser H.M.S. *Exeter* sustaining heavy damage from the guns of the German pocket battleship *Graf Spee* which is driven into Montevideo Harbor by the Exeter's sister cruisers H.M.S. *Ajax* and H.M.N.Z.S. *Achilles*. The *Graf Spee* is scuttled in the harbor December 17 by order of Adolf Hitler to keep her from falling into British hands.

German foreign minister von Ribbentrop sends a circular to diplomatic and consular offices January 25 under the title "The Jewish Question, a Factor in Our Foreign Policy for 1938": "It is not by chance that 1938, the year of our destiny, saw the realization of our plan for Greater Germany as well as a major step towards the solution of the Jewish problem. . . . The spread of Jewish influence and its corruption of our political, economic, and cultural life has perhaps done more to undermine the German people's will to prevail than all the hostility shown us by the Allied powers since the Great War. This disease in the body of our people had first to be eradicated before the Great German Reich could assemble its forces in 1938 to overcome the will of the world."

The S.S. *St. Louis* of the Hamburg-Amerika Line leaves Hamburg May 13 with 937 Jewish refugees from Nazi oppression and is the last major shipload to leave before the war begins. The passenger list has been approved by Joseph Goebbels, all passengers hold seemingly valid Cuban visas, but they are refused admission at Havana. The United States accepts 25,957 German immigrants per year under the Immigration Act of 1924 and refuses admittance to the passengers of the *St. Louis,* which heads back for Germany; Britain, France, Belgium, and Holland agree at the last moment to admit the refugees, most of whom will die in the next 6 years.

Brazil agrees June 24 to permit entry of 3,000 German Jewish refugees.

An international conference convened at Evian by President Roosevelt tries to organize facilities for Jewish emigration; delegates from 30 nations hold discussions, but nothing is accomplished.

The United Jewish Appeal (UJA) is founded at New York to raise funds for Jewish relief.

A British White Paper issued by London May 17 in effect repudiates the Balfour Declaration of 1917 (*see* 1938). Britain, which rules Palestine under a League of Nations mandate, limits admissions of Jews to 50,000 for the next 5 years. While the White Paper does authorize admission of 25,000 Jewish refugees, it envisions the establishment of an independent nation that will be predominantly Arab with Jewish immigration restricted (*see* 1941).

The Jewish population of Europe is 9.5 million but will decline sharply in the next 6 years. Few will escape the Holocaust that begins now for Czechoslovakian and Polish Jews who suffer at the hands of the Nazis as the German and Austrian Jews have suffered for the past year and more.

Marian Anderson tries to rent Constitution Hall at Washington, D.C., for a concert and is refused because of her race by the Daughters of the American Revolution, who own the hall. Now 42, the black contralto has been acclaimed by European critics as the world's greatest, Eleanor Roosevelt and other DAR members resign to show support, and Anderson draws an audience of 75,000 at the Lincoln Memorial Easter Sunday.

Britain's upper 10 percent holds 88 percent of the nation's wealth, down from 92 percent in 1912, and commands 34.6 percent of the nation's after-tax personal income (*see* 1960).

France is hardly any richer than she was in 1914 and is ill-prepared to fight a war.

Seventeen percent of the U.S. work force remains largely unemployed, but while the actual number of unemployed men and women has fallen from 15 million in 1933 down to 9.5 million, even Americans with jobs have relatively low average incomes.

More than 4 million Americans declare incomes above $2,000 for the year, 200,000 declare more than $10,000, 42,500 more than $25,000. Only 3 percent have enough income to pay any tax at all, and 670,000 taxpayers account for 90 percent of all income taxes collected.

Your Income Tax by New York accountant J. K. (Jacob Kay) Lasser, 42, helps Americans understand and prepare their tax returns.

U.S. Steel reports a net income of $41 million on sales of $857 million after a 1938 deficit (*see* 1964). The average U.S. Steel employee works just over 25 hours per week at a wage of just under 90¢ per hour, and his annual wage of about $1,600 is $100 more than the average earned by General Motors employees. Both companies employ roughly 220,000 people.

Thanksgiving Day is celebrated November 23—the fourth Thursday in the month rather than the last. Federated Department Stores chief Fred Lazarus, Jr., has persuaded President Roosevelt that a longer Christmas shopping season will help the economy, the president has issued a proclamation, and within a few years most states will pass laws making November's fourth Thursday Thanksgiving Day (*see* 1863; Federated, 1929).

The cyclotron of John Ray Dunning, 31, splits an atom for the first time in America January 25 in Room 128 of Columbia University's Pupin Physics Laboratory, suggesting the possibility of self-sustaining nuclear fission. When the cyclotron is shut down, stable cobalt-59 within its disks beget unstable cobalt-60, which emit gamma rays. Enrico Fermi, Hungarian-American physicist Leo Szilard, 41, Walter N. Zinn, 32, C. B. Pegram, and Dunning repeat the experiment March 3 with the same result.

The nuclear research at Columbia (above) confirms European findings that the absorption of a neutron by a uranium nucleus sometimes causes the nucleus to split into approximately equal parts with the release of enormous amounts of energy. Their Danish colleague Niels Bohr discusses the 1938 findings by O. R. Frisch and L. Meitner with Albert Einstein at Princeton, N.J., and discusses the work at Columbia as well (*see* Bohr, 1913; Einstein, 1929; U-235, 1940).

Albert Einstein writes to President Roosevelt, "Some recent work by E. Fermi and L. Szilard which has been communicated to me in manuscript leads me to expect that the element uranium may be turned into a new and important source of energy in the near future. Certain aspects of the situation which has arisen seem to call for watchfulness and, if necessary, quick action on the part of the Administration. . . . In the course of the last four months it has been made almost certain . . . that it may become possible to set up a nuclear chain reaction in a large mass of uranium, by which vast amounts of power and large quantities of radium-like elements would be generated. . . . This new phenomenon would lead also to the construction of bombs."

President Roosevelt appoints an Advisory Committee on Uranium in the autumn after being advised that

1939 ***(cont.)*** German physicists have split the uranium nucleus by assaulting it with neutrons (*see* first controlled chain reaction, 1942).

General Electric introduces fluorescent lighting, which is far more energy-efficient than incandescent. GE has developed it with help from University of Chicago physicist Arthur Holly Compton, 47.

Toshiba (Tokyo Shibaura Electric Co.) is created by a merger of the 64-year-old Shibaura Engineering works and the 49-year-old Tokyo Electric Co. Toshiba's Heavy Electric Equipment Department completes five 100 MVA water wheel generators, the largest of their kind in the world.

The first commercial transatlantic passenger air service begins June 28 as 22 passengers and 12 crew members take off from Port Washington, N.Y., for Marseilles via the Azores and Lisbon aboard the Pan American Airways *Yankee Clipper*, a Boeing aircraft powered by four 1,550-horsepower Wright Cyclone engines (*see* 1935). Pan Am has been providing air service to the Caribbean, South America, and the Pacific but Anglo-American disputes over airport landing rights have delayed the start of transatlantic service. The plane has separate passenger cabins, a dining salon, ladies' dressing room, recreation lounge, sleeping berths, and a bridal suite, the flight takes 26.5 hours, and the one-way fare is $375.

British Airways has its beginnings in British Overseas Airways (BOAC), created by a reorganization of Imperial Airways under the leadership of former BBC director J. C. W. Reith, who has been chairman of Imperial since last year and will be BOAC chairman until next.

The first turbojet aircraft is tested August 24 at Rostock-Marienehe and demonstrated in October for top Luftwaffe officials. Hans J. P. von Ohain has designed the Heinkel He-178 with centrifugal flow engine.

British aircraft designers work on a jet plane that uses a turbojet engine designed in 1930 by Frank Whittle, 32, whose Gloster-Whittle E.28/39 will be test flown for the first time in mid-May 1941.

Howard Hughes buys control of Transcontinental and Western Airlines (TWA) from the Wall Street banking house Lehman Brothers (*see* 1931; 1938). Hughes will develop TWA into Trans World Airways and control it until 1966, by which time he will have expanded his patrimony into a fortune of $1.5 billion and made TWA a transatlantic competitor of Pan Am (above).

New York's La Guardia Airport opens in the fall under the name North Beach Airport (*see* Newark, 1928; Idlewild, 1948).

Republic Aviation is created by a reorganization of Seversky Aircraft (*see* 1931; 1942).

The first American-made helicopter is flown by Igor Sikorsky, now 50, who has been in the United States since 1919 and has made the craft for United Aircraft at Bridgeport, Conn. (*see* 1913; 1929).

The Trans-Iranian Railway, completed in January after nearly 12 years of construction, links the Caspian Sea with the Persian Gulf. It has been built entirely with Iranian capital.

The Pacemaker goes into service on the New York Central for the Chicago run with a fare of just over $30 round trip in fancy coaches.

The Trail Blazer goes into service on the Pennsylvania Railroad to compete with the New York Central on the Chicago run.

U.S. interurban streetcar trackage falls to 2,700 miles as buses replace trolleys (*see* 1917; General Motors, 1932; criminal conspiracy conviction, 1949).

New York's Bronx-Whitestone Bridge opens to connect the Bronx with Queens; 2,300 feet in length, it facilitates access to the new La Guardia Airport (above) and to the World's Fair that opens in April.

Connecticut's Merritt Parkway opens June 22 with 38 miles of landscaped road winding through Fairfield County to link New York's Hutchinson River Parkway with Milford.

The Lincoln Mercury is introduced by Ford Motor Company's Edsel Ford (*see* Continental, 1941).

Fewer than 60 percent of U.S. families own automobiles. The figure will rise to 80 percent by 1964.

The U.S. Department of Justice indicts General Motors, Ford, and Chrysler for attempting to monopolize automobile financing by allegedly coercing dealers to use GMAC, Ford, or Chrysler financing facilities (*see* GMAC, 1919). The indictments against Ford and Chrysler are dropped in exchange for promises that they will stop coercing their dealers if GM is convicted, but while that conviction will be handed down in 1941, GM will not be required to give up its GMAC subsidiary, nor will Ford or Chrysler have to divest itself of its financing subsidiary. By the mid-1950s, GMAC will be the world's largest auto-financing company, averaging 18.7 percent per year in net profits.

U.S. chemist Bradley Dewey opens a pilot plant for making synthetic rubber (*see* food canning, 1921). He has conducted research on rubber-like elastomers and will complete one of the first U.S. synthetic rubber plants in 1942 (*see* Semon, 1926; Goodrich tire, 1940).

Union Carbide and Carbon resumes synthetic rubber research, acquiring Bakelite Corp. to pursue studies of butadiene as a source of synthetic rubber (*see* Bakelite, 1909; Dow, Goodyear, 1940).

Nylon is introduced commercially by E. I. du Pont (*see* Carothers, 1937; first nylon stockings, 1940).

Hewlett-Packard is founded by California engineers William Redington Hewlett, 26, and David Packard, 27, whose firm will become the leading U.S. producer of electronic instruments.

U.S. chemist Linus Carl Pauling, 38, develops a resonance theory of chemical valence by which he has been able to construct models for a number of anomalous molecules, notably the benzene molecule, which have not been explainable by conventional chemical terms. Pauling has applied quantum mechanics to the problem of how atoms and molecules enter into chemical combination with each other, he produces a full account of chemical bonding, and his paper "The Nature of the Chemical Bond and the Structure of Molecules and Crystals" will win him a 1954 Nobel Prize (*see* Heisenberg, 1925; Dirac and Schrödinger, 1926).

A rabies epidemic begins in Poland and spreads among foxes, bats, and other mammals, including dogs and cows. By 1950 the disease will be in Germany; it will advance at the rate of about 25 miles per year, and the first case of a rabid dog in more than 50 years in Britain will be reported in 1969.

French-American bacteriologist René Jules Dubos, 38, at the Rockefeller Institute for Medical Research, isolates two agents from swamp soil that are effective against a broad spectrum of gram-positive bacteria but are too toxic for internal use by humans. In isolating tyrocidine and gramicidine, Dubos establishes procedures that his former teacher Selman A. Waksman will adopt (*see* streptomycin, 1943; penicillin, 1940).

The Oxford Group's Moral Rearmament movement gains a wide following in Britain (*see* 1921).

NBC televises opening ceremonies of the New York World's Fair at Flushing Meadows April 30. One thousand viewers see the telecast, which is picked up by from 100 to 200 experimental receivers set up in the metropolitan area (*see* 1945; BBC, 1936).

FM radio receivers go on sale for the first time (*see* Armstrong, 1933; Zenith, 1940).

Some 27.5 million U.S. families have radios, up from 10 million in 1929, and 45 million sets are in use in the United States.

"This is London," says CBS correspondent Edward R. Murrow, 31, who ends his broadcasts with the tagline "Goodnight and good luck." Morrow's will become the best known voice on U.S. radio in the next 7 years.

"Batman" is launched by DC Comics artist Bob Kane, 18, whose comic-book hero, accompanied by a youthful Robin, will soon be syndicated to newspapers.

Pocket Books, Inc., Americanizes the paperback revolution in publishing begun by Britain's Penguin Books in 1936. The new company puts out 25¢ reprints of major and minor literary classics using enormous and fast new rubber-plate rotary presses that permit printing of paperbacks in large, economical quantities at a unit cost of pennies. Most authors, agents, and hardcover publishers agree to permit paperback reprints.

Nonfiction *Language in Action* by Canadian-American semanticist S. I. (Samuel Ichiye) Hayakawa, 33, whose lucid popularization will be retitled *Language in Thought and Action*.

Fiction *Finnegans Wake* by James Joyce, who will die of a perforated ulcer in 1941; *On the Marble Cliffs* (*Auf den Marmorklippen*) by Ernst Jünger, whose allegorical novel is a protest against Germany's Nazis; *Mister Johnson by* Anglo-Irish novelist Joyce Cary, 51; *Alberte* (trilogy) by Norwegian novelist Cora Sandel (Jara Fabricius), 59; *The Grapes of Wrath* by John Steinbeck, whose novel is based on observations made while traveling with migrant farm families from the Oklahoma Dust Bowl and produces a reaction almost comparable to that of *Uncle Tom's Cabin* in 1852; *The Day of the Locust* by Nathanael West, who portrays the moral degeneration of Hollywood where he has worked as a screenwriter since 1936 (West will be killed in an automobile accident near El Centro, Calif., in late December of next year at age 37); *The Web and the Rock* by the late Thomas Wolfe, who died in mid-September of last year at age 37 (Harper Bros. editor Edward Aswell will work his surviving mss. into two further novels); *Night Rider* by U.S. novelist-poet Robert Penn Warren, now 34; *Wind, Sand and Stars* by Antoine de Saint Exupéry; *Goodbye to Berlin* by Christopher Isherwood; *How Green Was My Valley* by British novelist Richard Llewellyn (Richard David Vivian Llewellyn Lloyd),32; *Mrs. Miniver* by English novelist-poet Jan Struther (Joyce Mastone Anstruther Graham), 38; *Black Narcissus* by English novelist Rumer Godden, 31; *Hangover Square* by English novelist Patrick Hamilton, 34; *The Nazarene* by Polish-born U.S. Yiddish novelist Sholem Asch, 59; *Party Going* by Henry Green; *Love and Death* by English novelist Llewelyn Powys, who dies of tuberculosis at age 55; *The Girls (Les Jeunes Filles)* by Henry de Montherlant; *Tropic of Capricorn* by Henry Miller; *Pale Horse, Pale Rider* (stories) by Katherine Anne Porter; "The Secret Life of Walter Mitty" by James Thurber in the March 18 *New Yorker*; *Wickford Point* by John P. Marquand; *Mask of Dimitrios* by Eric Ambler (*A Coffin for Dimitrios* in the United States); *The Big Sleep* by English-American detective novelist Raymond Thornton Chandler, 50, introduces the private eye Philip Marlowe.

Poetry *Autumn Journal* by Louis MacNeice; *Occasions (le occasioni)* by Eugenio Montale; *Human Poems (Poemas humanas)* by César Vallejo; *Return to My Native Land (Cahier d'un retour au pays natal)* by Martinique poet Aimé Césaire, 26; *A Heart for the Gods of Mexico* by Conrad Aiken.

Juvenile *Madeline* by Austrian-American novelist Ludwig Bemelmans, 41, who illustrates his own work (which will outlive his new novel *Hotel Splendide); Mike Mulligan and His Steam Shovel* by U.S. author Virginia Lee Burton.

New York's Museum of Modern Art (MOMA) moves into a handsome new building at 11 West 53rd Street (*see* 1929).

Grandma Moses gains overnight fame after engineer-art collector Louis Caldor sees work by the primitivist Anna Mary Robertson, 79, displayed in a drugstore window at Hoosick Falls, N.Y. Caldor drives to Robertson's farm, buys all 15 of her paintings, and exhibits three of them at the new Museum of Modern Art (above) in a show entitled "Contemporary Unknown Painters."

Other paintings *Seated Man* by Dutch-American painter Willem de Kooning, 35; *Ubermut* by Paul Klee; *Poison* and *Objective Stimulation* by René Magritte;

1939 ***(cont.)*** *Persephone*, *Susannah and the Elders*, *Threshing Wheat*, and *Weighing Cotton* by Thomas Hart Benton; *Handball* by Ben Shahn; *Retrato de Pita Amor* by Diego Rivera.

Theater *The White Steed* by Paul Vincent Carroll 1/10 at New York's Cort Theater, with Barry Fitzgerald, Jessica Tandy, George Coulouris, 136 perfs.; *The American Way* by George S. Kaufman and Moss Hart 1/21 at New York's Center Theater in Rockefeller Center, with Fredric March, Florence Eldredge, 164 perfs.; *The Little Foxes* by Lillian Hellman 2/15 at New York's National Theater, with Tallulah Bankhead, Carl Benton Reid, Dan Duryea, Patricia Collinge, 191 perfs.; *Family Portrait* by Lenore Coffee and William Joyce Cowan 3/8 at New York's Morosco Theater, with Judith Anderson, Philip Truex, 111 perfs.; *The Philadelphia Story* by Philip Barry 3/28 at New York's Shubert Theater, with Katharine Hepburn, Lenore Lonergan, Shirley Booth, Van Heflin, Joseph Cotten, 96 perfs.; *My Heart's in the Highlands* by William Saroyan 4/13 at New York's Guild Theater, 43 perfs.; *No Time for Comedy* by S. N. Behrman 4/17 at New York's Ethel Barrymore Theater, with Katherine Cornell, Laurence Olivier, Margalo Gilmore, 185 perfs.; *After the Dance* by Terence Rattigan 6/21 at the St. James's Theatre, London; *Ondine* by Jean Genet 8/3 at the Théâtre de l'Athenée, Paris; *Skylark* by Samson Raphaelson 10/11 at New York's Belasco Theater, with Gertrude Lawrence, 250 perfs.; *The Man Who Came to Dinner* by George S. Kaufman and Moss Hart 10/25 at New York's Music Box Theater, with Monty Wooley as Sheridan Whiteside and with Cole Porter's song "What Am I to Do," 739 perfs.; *The Time of Your Life* by William Saroyan 10/25 at New York's Booth Theater, with Eddie Dowling, Julie Haydon, Gene Kelly, Celeste Holm, William Bendix, Reginald Beane, 185 perfs.; *Margin for Error* by Clare Boothe 11/3 at New York's Plymouth Theater, with Otto Preminger, 264 perfs.; *Life with Father* by Howard Lindsay and Russel Crouse 11/8 at New York's Empire Theater, with Lindsay and Dorothy Stickney in a comedy based on the book by Clarence Day, 3,244 perfs.; *Key Largo* by Maxwell Anderson 11/27 at New York's Ethel Barrymore Theater, with José Ferrer, Paul Muni, Uta Hagen, 105 perfs.; *Mornings at Seven* by Paul Osborn 11/30 at New York's Longacre Theater, with Dorothy Gish, Russell Collins, Enid Marley, 44 perfs.

Films Victor Fleming's *Gone with the Wind* with Vivien Leigh as Scarlett O'Hara, Clark Gable as Rhett Butler, Leslie Howard as Ashley Wilkes, Olivia de Havilland as Melanie Hamilton. Produced by David O. Selznick with a screenplay by Sidney Howard based on the 1936 novel by Margaret Mitchell, GWTW has its world premiere December 15 at Atlanta, runs 222 minutes, and must be interrupted with an intermission.

Other films George Marshall's *Destry Rides Again* with James Stewart, Marlene Dietrich; Zoltan Korda's *Four Feathers* with Ralph Richardson, C. Aubrey Smith, June Duprez, Clive Baxter; George Stevens's *Gunga Din* with Cary Grant, Victor McLaglen, Douglas Fairbanks, Jr., Joan Fontaine, Sam Jaffe; Frank Capra's *Mr. Smith Goes to Washington* with Gary Cooper, Jean Arthur; Lewis Milestone's *Of Mice and Men* with Burgess Meredith, Lon Chaney, Jr.; Jean Renoir's *The Rules of the Game* (*La Règle du Jeu*) with Marcel Dalio, Nora Gregor, Mila Parely, Renoir; John Ford's *Stagecoach* with John Wayne, Claire Trevor; Carol Reed's *The Stars Look Down* with Michael Redgrave, Margaret Lockwood; William Wyler's *Wuthering Heights* with Laurence Olivier, Merle Oberon. Also: Edmund Goulding's *Dark Victory* and *The Old Maid*, both with Bette Davis; John Ford's *Drums Along the Mohawk* with Henry Fonda, Claudette Colbert; Sam Wood's *Goodbye, Mr. Chips* with Robert Donat, Greer Garson; Garson Kanin's *The Great Man Votes* with John Barrymore; Sidney Lanfield's *The Hound of the Baskervilles* with Basil Rathbone, Nigel Bruce; William Dieterle's *The Hunchback of Notre Dame* with Charles Laughton; Marcel Carne's *Le Jour Se Lève* (*Daybreak*) with Jean Gabin, Jules Berry, Arlette; Leo McCarey's *Love Affair* with Irene Dunne, Charles Boyer; Mitchell Leisen's *Midnight* with Claudette Colbert, Don Ameche, John Barrymore; Ernst Lubitsch's *Ninotchka* with Greta Garbo, Melvyn Douglas; Howard Hawks's *Only Angels Have Wings* with Cary Grant, Jean Arthur, Richard Barthelmess; Michael Curtiz's *The Private Lives of Elizabeth and Essex* with Bette Davis, Errol Flynn; Edmund Goulding's *We Are Not Alone* with Paul Muni; George Marshall's *You Can't Cheat an Honest Man* with W. C. Fields, Edgar Bergen.

Hollywood musicals Victor Fleming's *The Wizard of* Oz with Judy Garland (Frances Gumm, 17), Ray Bolger, Bert Lahr, Jack Haley, Frank Morgan, Margaret Hamilton, music by Harold Arlen, lyrics by E. Y. Harburg, songs that include "Somewhere Over the Rainbow," "Follow the Yellow Brick Road," "We're Off to See the Wizard," "If I Only Had a Brain;" Walt Disney's animated *Saludos Amigos* with songs that include "Brazil" by Brazilian composer Ary Barroso (who has adapted "Arguela do Brasil"), English lyrics by Bob Russell.

Radio music *The Dinah Shore Show* 8/6 on NBC Blue Network stations with singer Dinah Shore, 18.

Broadcast Music, Inc. (BMI) is founded October 14 by U.S. radio networks to build "an alternate source of music suitable for broadcasting" in competition with the ASCAP monopoly founded in 1914 *(see* 1917). ASCAP has boosted its license fees and the networks balk at paying the higher fees.

Broadway musicals *The Streets of Paris* 6/19 at the Broadhurst Theater, with Brazilian Carmen Miranda singing "South American Way," music by Jimmy McHugh and others, lyrics by Al Dubin and others, 274 perfs.; *George White's Scandals* 8/28 at the Alvin Theater, with Willie and Eugene Howard, Ben Blue, Ann Miller, Ella Logan, the Three Stooges in the 13th and final edition of the *Scandals*, music by Sammy Fain, lyrics by Jack Yellen, songs that include "Are You Having Any Fun," 120 perfs.; *Too Many Girls* 10/18 at the Imperial Theater, with Desi Arnaz, Eddie Bracken, Van Johnson, Richard Kollmar, Marcy Wescott, music by Richard Rodgers, lyrics by Lorenz Hart, songs that include "I Didn't Know What Time It Was," 249 perfs.; *Very Warm for May* 11/17 at the Alvin Theater, with Grace McDonald, Jack Whiting, Eve Arden, June Allyson, Vera Ellen, music by Jerome Kern, lyrics by Oscar Hammerstein II, songs that include "All the Things You Are," 59 perfs.; *Du Barry Was a Lady* 12/6 at the 46th Street Theater, with Bert Lahr, Ethel Merman, Betty Grable, music and lyrics by Cole Porter, songs that include "Friendship," "Do I Love You?," 408 perfs.

First performances Sonata No. 1 for Piano and Orchestra *(Concord)* by Charles Ives (who wrote it between 1911 and 1915) 1/20 at New York's Town Hall; Symphony No. 3 by Roy Harris 2/24 at Boston's Symphony Hall; Concerto No. 2 for Violin and Orchestra by Béla Bartòk 4/23 at Amsterdam (Bartòk flees the Germans in the fall); Symphony No. 6 by Dmitri Shostakovich 12/3 at Moscow.

Oratorio *Joan of Arc at the Stake* (*Jeanne d'Arc au bûcher*) 5/6 at the Théâtre Municipal, Orleans, with music by Arthur Honegger, libretto by Paul Claudel.

Frank Sinatra joins bandleader Harry James at age 23 to sing with a new band that James is assembling. The New Jersey roadhouse singer will leave within a year to join the Tommy Dorsey band, will break his contract with Dorsey in 1942 and take an 8-week engagement at New York's Paramount Theater to begin a career as idol of teenage "bobby-soxers."

Popular songs "There'll Always Be an England" by English songwriters Ross Parker and Hughie Clark; "I'll Never Smile Again" by U.S. songwriter Ruth Lowe, a former pianist in an all-girl band who has written the song in memory of her late husband; "Heaven Can Wait" by Jimmy Van Heusen, lyrics by Eddie De Lange; "And the Angels Sing" by trumpet player Ziggy Elman (Harry Finkelman), who has adapted a Jewish folk tune, lyrics by Johnny Mercer; "Ciribiribin" by composer A. Pestalozza, who has adapted an Italian folk tune, lyrics by Jack Lawrence; "Undecided" by U.S. bandleader Charlie Shavers, lyrics by Sid Robin; "Moonlight Serenade" by Glenn Miller, lyrics by Mitchell Parish *(see* 1938); "In the Mood" by Joe Garland, lyrics by Andy Razaf, "All or Nothing at All" by Arthur Altman, lyrics by Jack Lawrence; "The Lady's in Love with You" by Burton Lane, lyrics by Frank Loesser; "Scatterbrain" by Kahn Keene, Carl Bean, Frankie Masters, lyrics by Johnny Burke; "South of the Border (Down Mexico Way)" by Jimmy Kennedy and Michael Carr; "Three Little Fishes" by Saxie Dowell; "We'll Meet Again" by London songwriter Hugh Charles.

Robert L. "Bobby" Riggs, 21 (U.S.), wins in men's singles at Wimbledon and Forest Hills, Alice Marble takes the women's titles.

The Boston Red Sox bring up Ted Williams as outfielder and hitter. Theodore Samuel Williams, 20, started with the San Diego Padres at age 16, was acquired by the Red Sox for their farm club 2 years later, and will play 19 seasons for Boston. His 1941 batting average of .406 will be the highest since the 1924 Rogers Hornsby record of .424 and his lifetime batting average for 2,292 games will be .344.

Yankee "Iron Horse" Lou Gehrig is stricken with a rare and fatal form of paralysis (amyotropic lateral sclerosis), takes himself out of the lineup April 30, and bids a tearful farewell to Yankee fans July 4, retiring with a lifetime batting average of .340. Appointed parole commissioner by Mayor La Guardia, he will serve until his death in June 1941.

Lundy Lumber beats Lycoming Dairy 23 to 8 at Williamsport, Pa., June 6 in the first Little League baseball game. Local lumberyard clerk Carl Stotz, 29, has shrunk a regulation field to boys' size and formed the league, which will hold its first World Series in 1949.

The New York Yankees win the World Series by defeating the Cincinnati Reds 4 games to 0.

A Baseball Hall of Fame is established at Cooperstown, N.Y., to celebrate the "centennial" of the national pastime.

Football helmets are made mandatory in U.S. college football competition (*see* NCAA, 1906).

Cup-sizing for brassieres is introduced by Warner Brothers of Bridgeport, Conn., whose designer Leona Gross Lax, 48, has developed the concept (*see* 1914).

Pall Mall cigarettes in a new 85-millimeter length are the first U.S. "king-size" cigarettes. American Tobacco's George Washington Hill lengthens the 2-year-old brand to make it 15 millimeters longer than other cigarettes and claims that the longer length "travels the smoke, and makes it mild."

Tabakmissbrauch und Lungencarcinom by German physician F. H. Muller relates smoking to lung cancer (*see* Ochsner, 1945).

A federal grand jury at Chicago indicts *Philadelphia Inquirer* publisher Moses L. Annenberg on charges of having evaded $1.26 million in income taxes between 1932 and 1936 *(see* 1936). Including interest and penalties, Annenberg owes the government more than $5.5 million in the largest case of tax evasion in history, will pay $9.5 million in taxes, penalties, and future interest, and will serve time in federal prison at Lewisburg, Pa., until shortly before his death in July 1942 (*see Seventeen* magazine, 1944).

Al Capone is released from Alcatraz, his mind destroyed by syphilis *(see* 1931). The gangster retires to his estate at Miami Beach and will vegetate there until his death in 1947.

The Johnson Wax Research Tower designed by Frank Lloyd Wright is completed at Racine, Wis.

U.S. motels and tourist courts number 13,500, while the country has 14,000 year-round hotels each with 25 guest rooms or more. The number of motels will increase to more than 41,000 in the next 20 years as America embraces the automobile culture and the number of year-round hotels will fall to just over 10,000 (*see* Holiday Inns, 1952).

Earthquakes kill thousands in Chile.

1939 ***(cont.)*** Parker Dam is completed across the Colorado River near Parker, Ariz., as part of the $220 million Colorado River Aqueduct System which taps billion gallons of water per day from the river, lifts it nearly one-quarter of a mile high, and transports it across 330 miles of desert and mountain to the 3,900-square mile Metropolitan Water District of Los Angeles (*see* 1913). Located 155 miles below the 3-year-old Boulder Dam, the new $13 million project has been built up from 234 feet below the river bed.

Fort Peck Dam is completed with WPA funds to control the Missouri River in Montana. The earth-filled dam rises 250 feet.

DDT is introduced in Switzerland and applied almost immediately and with great success against the Colorado potato beetle which is threatening the Swiss potato crop. Swiss chemist Paul Muller, 40, of the Geigy Co. has developed the persistent, low-cost hydrocarbon pesticide (*see* 1943; Zeidler, 1873; typhus, 1944; Osborne, 1948; Carson, 1962).

Japanese beetles menace crops in much of the United States (*see* 1916).

Nearly 25 percent of Americans are still on the land, but the average farm family has a cash income of only $1,000 per year.

Germany has stockpiles of 8.5 million tons of grain as World War II begins (above).

Russia is obligated under the August 23 nonaggression treaty (above) to supply Germany with 1 million tons of wheat per year and expedite delivery of Manchurian soybeans.

The U.S. Department of Agriculture introduces the first food stamp program in May to feed the needy of Rochester, N.Y. The program will continue until 1943 (*see* 1964).

Britain imposes rationing of meat, bacon, cheese, fats, sugar, and preserves in fixed quantities per capita, allocating supplies equally and keeping prices at levels people can afford. Britain is the largest buyer of food in the world market, absorbing 40 percent of all food sold in international trade.

British bread and margarine are enriched and fortified with vitamins and minerals. Britons increase consumption of milk, potatoes, cheese, and green vegetables.

General Foods introduces the first precooked frozen foods under the Birds Eye label, marketing a chicken fricasee and criss-cross steak (*see* 1931). Competitors come out within the year with frozen creamed chicken, beef stew, and roast turkey.

A 5-minute Cream of Wheat is introduced to compete with quick-cooking Quaker Oats, which reduced its cooking time in 1922 from 15 minutes to 5.

Lay's potato chips are introduced by Atlanta's H. W. Lay Co. founded by Herman Warden Lay, 30, who has taken over Barrett Food Products for which he was a route salesman in 1932 (*see* Frito-Lay, 1961).

Pepsi-Cola challenges Coca-Cola with a radio jingle written for $2,500 by Bradley Kent and Austen Herbert Croom to the old English hunting song "D'ye Ken John Peel": "Pepsi-Cola hits the spot/Twelve full ounces, that's a lot/Twice as much for a nickel, too/Pepsi-Cola is the drink for you" (*see* 1934; 1949).

The pressure cooker, introduced at the World's Fair (above) by National Presto Industries, is a saucepan-like pot with a locking swivel lid. It does in minutes what used to take hours (*see* 1682).

The population of Palestine (above) reaches 1.4 million, including 900,000 Muslims, 400,000 Jews, 100,000 Christians (see 1948).

1940 The fall of France, Belgium, the Netherlands, Luxembourg, Denmark, Norway, and Romania to the Germans leaves Britain dependent on U.S. aid to resist the Nazi attempt at world domination (*see* below).

Soviet Russia takes 16,173 square miles of Finland and annexes the Baltic nations Estonia, Latvia, and Lithuania with German concurrence (*see* below).

France's Premier Daladier resigns March 20, finance minister Paul Reynaud, 51, forms a new cabinet and tries to rally French defenses.

German troops seize Denmark and Norway in April a few weeks after peace is concluded between Finland and the U.S.S.R. A Nazi government is set up at Oslo under Norwegian Nazi Vidkun Quisling, 52, but native resistance elements work to frustrate the Quisling regime.

The *Blitzkrieg* (lightning war) that begins May 10 ends the "phony war" and rains destruction on the Lowlands from the air as German mechanized divisons sweep without warning into Belgium and Holland. They cross the Meuse at Sedan May 12, Rotterdam is nearly obliterated by bombing raids May 14, Wilhelmina and her court escape to London, the Dutch armies surrender May 14, and the Germans race into Belgium and northern France toward the Channel ports, cutting off British and Belgian forces from Gen. Maxime Weygand, now 72, who has taken command of the collapsing French army.

Winston Churchill succeeds Neville Chamberlain as Britain's prime minister after Chamberlain's resignation May 7 following bitter attacks in the House of Commons. "I have nothing to offer but blood, toil, tears, and sweat," Churchill tells the Commons May 13, but he makes clear the British objective: "Victory: victory at all costs, victory in spite of all terror, victory however long and hard the road may be: for without victory there is no survival."

Amiens and Arras fall to the Germans May 21, Belgium capitulates May 28, and evacuation of British and French troops from Dunkirk begins May 29.

"We shall not flag or fail," says Prime Minister Churchill to the Commons June 4: "We shall fight in France, we shall fight on the seas and oceans, we shall fight with growing confidence and growing strength in the air, we shall defend our island, whatever the cost may be, we shall fight on the beaches, we shall fight on the landing grounds, we shall fight in the fields and in the streets, we shall fight in the hills; we shall never surrender."

Dunkirk is evacuated by June 4 after 5 days of frenzied efforts to take some 200,000 British troops and 140,000 French from the beaches before they can be captured by the advancing Germans. Ships and small boats of all sorts are employed to ferry the men across the Channel to England, but 30,000 are killed or taken prisoner.

Italy declares war on France and Britain June 10. "The hand that held the dagger has struck it into the back of her neighbor," says President Roosevelt.

German troops enter Paris June 14, a French appeal for U.S. aid is declined June 15 as the French fortress at Verdun falls to the Germans, Premier Reynaud resigns June 16, Marshal Pétain, now 83, succeeds as head of state and asks for an armistice a day later.

France signs an armistice with Hitler June 22 at Compiègne and signs an armistice with Mussolini June 24 as the French capital moves to Vichy.

Soviet troops move into Estonia, Latvia, and Lithuania beginning June 17 following an ultimatum that charged the Baltic countries with hostile activities. They are incorporated into the U.S.S.R. July 21 and will remain Soviet fiefs until 1991.

Soviet troops invade Romania June 27 after Carol II has refused demands that he cede Bessarabia and Bukovina to the U.S.S.R. Berlin rejects Carol's appeals for aid (*see* below).

French fleets at Oran in North Africa are sunk July 3 by the Royal Navy and France's Vichy government breaks off relations with Britain 2 days later. Gen. Charles André Joseph Marie de Gaulle, 49, has escaped to Britain and forms a Free French government in exile.

Night bombing of German targets by the Royal Air Force begins July 9 as Hitler's Luftwaffe intensifies air attacks on Britain. Ninety German bombers are shot down over Britain between July 15 and 21, and 180 German planes are shot down August 15 as the Battle of Britain reaches its peak, with Spitfires attacking Stukas (*Sturzkampflugzeng*) equipped with wind-whistles to terrify the populace.

"Never in the field of human conflict was so much owed by so many to so few," says Churchill August 20 in a speech to the Commons praising the Royal Air Force (RAF) whose Spitfire pilots have stymied the Luftwaffe, but an all-night German air raid on London 3 days later begins the Blitz.

British Fascist Sir Oswald Mosley has been taken into custody in May and locked up in Brixton Prison after saying, "I know I can save this country and that no one else can" (*see* 1936). The former Lady Cynthia Curzon has long since divorced Mosley, and when he married Diana Mitford at Berlin 4 years ago his only guests were Hitler and Goebbels.

Russian revolutionary Leon Trotsky has escaped a raid on his Mexican villa in May when Soviet agents armed with machine guns invaded the premises, but NKVD agent Ramon Mercader, 26 (alias Frank Jackson, alias Jacques Mornard), posing as a friend, stabs Trotsky with an ice pick August 21, killing him at age 61. Mercader will serve 20 years in prison.

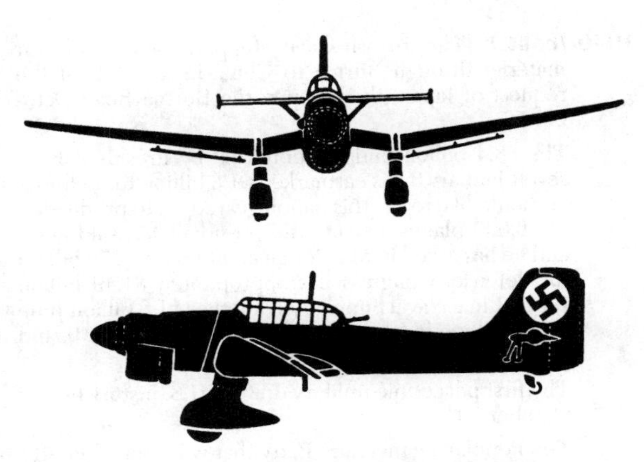

Stuka dive bombers of the German *Luftwaffe* rained death and destruction on European cities in World War II.

Hitler launches an all-out air attack on Britain September 15 as he prepares an invasion fleet. Londoners rush to the underground when air-raid sirens blow, but hundreds die each night from aerial bombs.

Berlin has proclaimed a blockade of Britain August 17.

Washington hands over 50 U.S. destroyers to London September 3 in exchange for rights to build air and naval bases on British territory in Newfoundland and in the Caribbean, but U-boats sink 160,000 tons of British shipping in September and on October 2 sink the S.S. *Empress of Britain,* carrying children to Canada.

Soviet authorities in Romania have arrested Ion Antonescu in July for opposing territorial concessions to the U.S.S.R. (above) but have released him. Now 50, he is appointed premier September 5 and forces the abdication of Carol II, who flees the country September 6 after a 10-year reign. Carol's son Michael, now 19, will reign until the end of 1947 (*see* below).

The Axis created September 27 at Berlin joins Germany, Italy, and Japan in a 10-year military and economic alliance. Hitler and Mussolini meet a week later in the Brenner Pass, German forces seize Romania's Ploesti oil fields October 7, and Italy demands strategic points in Greece October 28.

Britain responds to a Greek appeal for aid by postponing a Middle East offensive and sending troops in early November to occupy Suda Bay, Crete.

Japan's Prince Fumumaro Konoye has become premier in mid-July with a mandate to reorganize the government along totalitarian lines, Britain withdraws garrisons August 9 from Shanghai and northern China, and Japanese forces begin occupying French Indo-China September 26 after receiving permission from the Vichy government to use several Indo-Chinese ports and three airfields.

1940 *(cont.)* The British have stopped passage of war matériel through Burma to China in mid-July at the request of Japan; they reopen the Burma Road October 18.

The $8.4 billion budget submitted by President Roosevelt January 3 has earmarked $1.8 billion for defense. "I should like to see this nation geared . . . to production of 50,000 planes a year," the president has said later, and he has asked in May for an additional $1.275 billion for defense. Congress has appropriated $1.49 billion June 11 in a Naval Supply Act, another $1.8 billion June 26, $4 billion for the Army and Navy September 9, and another $1.49 billion for defense October 8.

The first peacetime military draft in U.S. history begins October 29.

The Popular Democratic Party that will control Puerto Rico until 1977 comes to power with the election of party founder Luis Muñoz Marin, 42, who started the PDP in 1938. President Roosevelt next year will appoint former Columbia University economics professor Rexford Guy Tugwell, now 48, governor of Puerto Rico and he will work with Muñoz Marin to improve the island's economy.

President Roosevelt wins reelection to a precedent-shattering third term with 54 percent of the popular vote. His Republican opponent Wendell L. Willkie, 48, of New York is a former utilities company president who has attacked the Tennessee Valley Authority but whose generally liberal, internationalist views have won him the nomination against isolationists Robert A. Taft of Ohio, Arthur Vandenberg of Michigan, and Thomas E. Dewey of New York. Willkie wins 44 percent of the popular vote but only 82 electoral votes to FDR's 449.

The British battleship H.M.S. *Jervis Bay* is sunk in the Atlantic November 4, but a British attack on Taranto a week later cripples the Italian fleet.

The Luftwaffe pulverizes Coventry in England's industrial Midlands November 10. British intelligence has alerted Prime Minister Churchill to the fact that the raid on Coventry was imminent, but Churchill has decided not to evacuate the city lest the Germans discover that their secret code has been broken with help from a device acquired last year in the fall of Poland.

Hungary joins the Axis November 20, Romania November 23. Rioting spreads across Romania beginning November 27 after Ion Antonescu's Iron Guard executes 64 officials of the Crown.

November air raids kill more than 4,550 Britons, but the anticipated German invasion of Britain does not materialize.

The British Eighth Army commanded by Gen. Archibald Percival Wavell, 57, opens a North African offensive December 9 with an attack on Sidi Barrani, which has been in Italian hands since mid-September. The Italians are driven across the Libyan border December 15, and Mussolini's forces in Italian Somaliland are driven out of El Wak a day later.

The Nazis extend persecution of Jews to Poland, Romania, the Netherlands, and other occupied territories. SS troops surround a densely populated Jewish area in Czestochowa in January, herd thousands of half-naked Polish men and women into a large square, beat them bloody, and keep them standing for hours in the frosty night air while young girls are taken into the synagogue, forced to undress, raped, and tortured.

The Katyn Forest Massacre in March kills 4,143 Polish officers captured by the Red Army when it entered Poland 6 months ago. Imprisoned 15 miles west of Smolensk, the Poles are gunned down and buried in mass graves after having their hands bound behind their backs with a rope looped round their necks.

The Alien Registration Act (Smith Act) passed by Congress June 28 requires that aliens be fingerprinted and makes it unlawful to advocate overthrow of the U.S. government or belong to any group advocating such overthrow.

The first Social Security checks go out January 30 and total $75,844, a figure that will rise into the billions as more pensioners become eligible for benefits. The first check goes to Vermont widow Ida May Fuller, 35, who receives $22.54 and will receive more than $20,000 before she dies in 1975 at age 100 (*see* 1935).

U.S. unemployment remains above 8 million, with 14.6 percent of the work force idle.

A Revenue Act passed by Congress June 25 to raise $994.3 million increases U.S. income taxes. A second Revenue Act passed October 8 provides for excess profits taxes on corporation earnings.

An Export Control Act passed by Congress July 2 gives the president power to halt or curtail export of materials vital to U.S. defense. Export of aviation gasoline outside the Western Hemisphere is embargoed July 31, and export of scrap iron and steel to Japan embargoed in October after large quantities of the metals have been shipped to the Axis partner, whose ambassador in Washington calls the action "an unfriendly act."

The U.S. Gross National Product (GNP) is $99 billion, down from $103 billion in 1929. Government spending accounts for 18 percent of the total, up from 10 percent in 1929 (*see* 1950; Kuznets, 1930).

Tiffany & Co. of New York moves uptown into a large new store at Fifth Avenue and 57th Street after 34 years at 37th Street. The new store is the first fully air-conditioned store of any kind.

A Columbia University research team isolates the rare isotope uranium-235 from its more abundant chemical twin uranium-238 by means of a gaseous diffusion process developed by J. R. Dunning and his colleagues (*see* 1939). Dunning shows that the isotope is the form of uranium that readily undergoes fission into two atoms of nearly equal size, thus releasing prodigious amounts of energy (the gaseous diffusion process will be the major source of uranium-235 used for fueling atomic reactors).

University of California, Berkeley, physicist Edwin Mattison McMillan, 32, discovers the first transuranium element. It is a radioactive element heavier than uranium, he calls it neptunium, and he will soon discover another, still heavier, transuranium element that he will call plutonium (*see* Seaborg, 1941).

A continuous coal-digging machine developed by Consolidation Coal president Carson Smith and engineer Harold Farnes Silver, 39, will revolutionize coal mining. Six banks of cutter chains moving at 500 feet per minute will enable the Joy machine to dig out a series of vertical slices 18 inches deep and to bore a tunnel up to 18 feet wide. Joseph Joy, now 56, will purchase the rights in 1947; his Joy Manufacturing Co. will dominate coal-mining equipment production.

Britain completes the world's largest passenger liner and puts her to use as a troop transport. Powered by steam turbines that develop 168,000 horsepower and give her a normal sea speed of 28.5 knots (32.8 miles per hour), the 83,673-ton ship, 1,031 feet in length overall, will go into commercial service for the Cunard Line after the war as the S.S. *Queen Elizabeth*.

The MIG-1 fighter plane introduced by Moscow's Aircraft Factory No. 1 is named for Soviet mathematician and aircraft designer Mikhail I. Gurevich, 48. Soviet pilots will fly a refined MIG-3 beginning next year.

The P-51 Mustang fighter plane designed and produced in 127 days by North American Aviation's "Dutch" Kindelberger and John Leland Atwood, 35, is powered by the same 1,000-horsepower Rolls-Royce engine that powers the Spitfire, which is winning the Battle of Britain (above) (*see* 1934).

The first commercial flight using pressurized cabins takes off July 8 as a Transcontinental & Western Air Boeing 307-B Stratoliner goes into service between La Guardia Airport and Burbank, Calif., with a stop at Kansas City. The plane carries 33 passengers by day and 25 at night (24 of the seats are in compartments convertible into 16 sleeping berths), and flying time is 14 hours going west, 11 hours, 55 minutes going east.

The first Ferrari racing car competes without success in the final prewar Mille Miglia race in April (*see* 1929). Ferrari calls his cars simply 815 because his contract with Alfa-Romeo bars him from using the name Ferrari on any car. Scuderia Ferrari will produce oleodynamic grinding machines until 1945.

The Jeep, designed by Karl K. Pabst, consulting engineer to Bantam Car Co. of Butler, Pa., is a lightweight, four-wheel-drive, general-purpose (GP) field vehicle powered by a four-cylinder Continental engine. The U.S. Army places orders for 70 pre-production models, Ford and Willys-Overland both submit prototype models for testing by the army in November, the Willys MB design will be accepted as standard in the summer of 1941, both companies will get contracts, and they will turn out nearly 649,000 Jeeps in the next 5 years, with Willys completing a new Jeep every 80 seconds at the peak of its production (*see* 1945).

The United States has 1.34 million miles of surfaced road, up from 694,000 in 1930, plus 1.65 million miles of dirt road (*see* 1950).

The Pennsylvania Turnpike that opens October 1 is the first tunneled U.S. superhighway. Of its 7 tunnels, 6 were drilled in 1883 by William H. Vanderbilt's engineers for a rail line that was never used. Begun as a WPA project, the 4-lane, 160-mile, $70 million toll road (toll: $1.50) links Carlisle with Irvin, 11 miles east of Pittsburgh, and speeds traffic through the Alleghenies in 2.5 hours, bypassing towns. The remaining 40 miles to Pittsburgh takes another hour or more.

The Arroyo Seco Parkway dedicated in December is the first Los Angeles freeway. In 30 years, some 60 percent of downtown Los Angeles will be taken up with highways and parking lots as the freeways transform the city to make L.A. a sprawling collection of suburbs dependent on the automobile and doomed to a smoggy environment.

New York's Queens Midtown Tunnel opens to link Manhattan and Queens.

New York's Sixth Avenue "El" of 1878 comes down; the scrap metal is sold to Japan before enactment of the Export Control Act (above).

Some 26,630 trolley cars serve U.S. transportation needs, down from 80,000 in 1917. The number will drop to 1,068 by 1975.

A new Chevrolet coupe sells for $659 F.O.B. Flint, a Pontiac station wagon for $1,015 F.O.B. Pontiac, a Studebaker Champion for $660 F.O.B. South Bend, a Nash sedan for $795 F.O.B. Kenosha, and a Packard for between $867 and $6,300 F.O.B. Detroit (lowest prices advertised).

B. F. Goodrich exhibits the first commercial synthetic rubber tires. Its Ameripol tires are made of butadiene synthesized from soap, gas, petroleum, and air.

Goodyear and Dow Chemical form Goodyear Dow Corp. in a joint venture to produce synthetic rubber from styrene and butadiene. The only U.S. producer of synthetic rubber has been Du Pont, which makes 2,468 long tons of neoprene this year employing research done by the late J. A. Nieuwland (*see* 1925; 1942).

President Roosevelt acts to build U.S. natural crude rubber reserves in the event of a cutoff of supplies from the Far East. The Rubber Reserve Co. created by the president in June is under the Reconstruction Finance Corp. (*see* 1942).

German-American biologist Max Delbruck, 34, Italian-American biologist Salvador Edward Luria, 28, and U.S. biologist Alfred Day Hershey, 31, pioneer molecular biology, making basic discoveries in bacterial and viral reproduction and mutation involving nucleic acid in cells (*see* Watson, Crick, 1953).

The carbon-14 isotope discovered by Canadian-American biochemist Martin David Kamen, 27, will be a basic tool in biochemical and archaeological research (*see* Libby, 1947).

The world's first electron microscope, demonstrated April 1 at the RCA Laboratories at Camden, N.J., can magnify up to 100,000 diameters. RCA engineers

1940 ***(cont.)*** supervised by Russian-American researcher Vladimir Kosma Zworykin, 50, have developed the microscope, which is 10 feet high, weighs 700 pounds, and will permit scientific studies impossible with conventional microscopes.

The Lascaux caves discovered by French schoolboy Jacques Marsal near Périgueux have wall drawings that show how man lived at least 16,000 years ago.

Race, Language, and Culture by Franz Boas, now 82, crowns a long career in anthropology (*see* 1934).

"Penicillin as a Chemotherapeutic Agent" in the August 24 issue of *The Lancet* reports studies by Australian-born British pathologist Howard W. Florey, 42, at the Sir William Dunn School of Pathology. Florey and his German refugee colleague Ernest Chain have developed Alexander Fleming's penicillin of 1928 into an agent that will become the world's major weapon against infections and chronic diseases (*see* Dubos, 1939; world supply, 1942).

Lederle Laboratories chemist Richard Owen Roblin, 33, develops the sulfa drug sulfadiazine (*see* sulfanilamide, 1935).

U.S. Blue Cross health insurance programs have 6 million subscribers, up from 500,000 in 1935, but Blue Shield surgical insurance covers only 260,000 (*see* 1950).

The Rh factor in blood (named for the rhesus monkeys used in research) is discovered by blood type pioneer Karl Landsteiner, who is now with the Rockefeller Institute for Medical Research at New York (*see* 1909). Landsteiner's colleague Alexander S. Wiener has participated in the discovery.

Edwin J. Cohn at Harvard separates the albumin, globulin, and fibrin fractions of blood plasma (*see* pernicious anemia, 1930). The American Red Cross and U.S. Navy will use albumin in the next 6 years to treat shock, globulin to treat various forms of infection, and fibrin to stop hemorrhages while serum gamma globulin will be used in years to come for mass immunization against measles, poliomyelitis, and other epidemic diseases.

New York surgeon Charles Richard Drew, 36, opens the first blood bank. Segregation rules prevent him from donating his own blood.

The letter V begins appearing on walls in German-occupied Belgium. Two Belgians working for the BBC in London have instigated the campaign, knowing that for Flemish-speaking Belgians the V stands for *vrijheid* (freedom), and for French-speaking Belgians the V stands for *victoire* (victory). The letter is soon scrawled on walls all over Europe, even where the words for freedom and victory do not begin with V.

MIT establishes a radiation laboratory in November to pursue experiments in radar (*see* Watson-Watt, 1935). The British invention is helping to win the Battle of Britain against the Luftwaffe (above).

German technicians H. J. von Braunmuhl and W. Weber improve magnetic plastic tape recording by applying a high-frequency bias to the oxide-coated tape of the 1935 AEG Magnetophone (*see* Sony,1950).

Resistance to German occupation authorities remained strong in Europe through the bleakest years of the war.

Zenith Radio's Eugene McDonald starts an FM radio station that will survive to become the world's oldest. Zenith will produce the lion's share of U.S. FM radio sets (*see* 1923; Armstrong, 1933; hearing aid, 1943).

Hungarian-American engineer Peter Carl Goldmark, 34, of CBS pioneers color television, but his system requires special receivers. It will give way in the 1950s to an RCA system whose signals will be compatible with conventional black and white TV signals (*see* 1939; long-playing records, 1948).

Newsday begins publication at Long Island City, N.Y., where copper heir Harry F. Guggenheim and his third wife Alicia Patterson start a daily paper.

Nonfiction *To the Finland Station* by Edmund Wilson.

Fiction *The Man Who Loved Children* by Christina Stead; *The Thibaults* (in eight volumes) by Roger Martin du Gard; *The Heart Is a Lonely Hunter* by U.S. novelist Carson Smith McCullers, 23; *The Tartar Steppe* (*Il deserta de Tartari*) by Italian novelist Dino Buzzati, 34; *Kallocain* by Karin Boye; *The Seed (Kimen)* by Norwegian novelist Tarjei Vesaas, 43; *The Ox-Bow Incident* by U.S. novelist Walter van Tilburg Clark, 31; *For Whom the Bell Tolls* by Ernest Hemingway; *You Can't Go Home Again* by the late Thomas Wolfe; *My Name is Aram* (stories) by William Saroyan; *The Power and the Glory* by Graham Greene; *The Pool of Vishnu* by L. H. Myers; *On a Darkling Plain* by Harvard English instructor Wallace Earle Stegner; *Native Son* by U.S. novelist Richard Wright, 31, who moved to New York in 1937 to edit the *Daily Worker; Journey into Fear* by Eric Ambler; *Ten Little Niggers* by Agatha Christie, whose new detective novel will be published in America first as *And Then There Were None*, later as *Ten Little Indians.*

Poetry *The Man Coming Toward You* by U.S. poet Oscar Williams, 40; *Spain, Take Thou This Cup from Me (España, aparta de me este caliz)* by César Vallejo; *Cantos* by Ezra Pound.

Painting *The Romanian Blouse* by Henri Matisse; *Sky Blue* by Wassily Kandinsky; *Circus Caravan* by Max Beckmann; *The Red Mask* by Mexican painter Ruffino Tamayo, 40 ; *Stump in Red Hills* by Georgia O'Keeffe; *T.B. Harlem* by "bohemian" New York painter Alice Neel, 40; *Garden in Sochi* by Armenian-American painter Arshile Gorky (*né* Vosdanig Adoian), 35; *Death and Fire* by Paul Klee who dies at Muralto-Locarno, Switzerland, June 29 at age 60. Jean Edouard Vuillard has died at La Baule, France, June 21 at age 71.

Theater *The Male Animal* by James Thurber and playwright-actor Elliott Nugent, 40, 1/9 at New York's Cort Theater, with Nugent, Gene Tierney, 243 perfs.: *My Dear Children* by playwrights Catherine Turney and Jerry Horwin 1/31 at New York's Belasco Theater, with John Barrymore, 117 perfs.; *The Fifth Column* by Ernest Hemingway and Benjamin Glazer 3/26 at New York's Alvin Theater, with Lenore Ulric, Lee J. Cobb, Franchot Tone, 87 perfs.; *There Shall Be No Night* by Robert Sherwood 4/29 at New York's Alvin Theater, with Alfred Lunt, Lynn Fontanne, Sydney Greenstreet, Richard Whorf, Montgomery Clift, 115 perfs.; *Johnny Belinda* by U.S. playwright Elmer Harris, 62, 9/18 at New York's Belasco Theater, with Helen Craig, Horace McNally, 321 perfs.; *George Washington Slept Here* by George S. Kaufman and Moss Hart 10/18 at New York's Lyceum Theater, with Ernest Truex, Jean Dixon, 173 perfs.; *Old Acquaintance* by John Van Druten 12/23 at New York's Morosco Theater, with Jane Cowl, Kent Smith, 170 perfs.; *My Sister Eileen* by Joseph A. Fields and Jerome Chodorov 12/26 at New York's Biltmore Theater, with Shirley Booth, Jo Ann Sayers, Morris Carnovsky in a play based on Ruth McKinney's book, 865 perfs.

Films John Cromwell's *Abe Lincoln in Illinois* with Raymond Massey, Gene Lockhart, Ruth Gordon; Eddie Cline's *The Bank Dick* with W.C. Fields; Alfred Hitchcock's *Foreign Correspondent* with Joel McCrae, Laraine Day; John Ford's *The Grapes of Wrath* with Henry Fonda, Jane Darwell; Howard Hawks's *His Girl Friday* with Cary Grant, Rosalind Russell; George Cukor's *The Philadelphia Story* with Katharine Hepburn, Cary Grant, James Stewart; Robert Z. Leonard's *Pride and Prejudice* with Laurence Olivier, Greer Garson, Edna May Oliver; Michael Curtiz's *The Sea Hawk* with Errol Flynn; Ludwig Berger, Tim Whelan, and Michael Powell's *The Thief of Baghdad* with Sabu, Rex Ingram, June Duprez. Also: William Dieterle's *Dr. Ehrlich's Magic Bullet* with Edward G. Robinson, Ruth Gordon, Otto Kruger; Sam Wood's *Kitty Foyle* with Ginger Rogers, Dennis Morgan; William Wyler's *The Letter* with Bette Davis, Herbert Marshall; John Ford's *The Long Voyage Home* with John Wayne, Thomas Mitchell; Rouben Mamoulian's *The Mark of Zorro* with Tyrone Power, Linda Darnell; Frank Borzage's *The Mortal Storm* with Margaret Sullavan, James Stewart; King Vidor's *Northwest Passage* with Spencer Tracy, Robert Young; Sam Wood's *Our Town* with William Holden, Martha Scott, Frank Craven; Raoul Walsh's *They Drive by Night* with George Raft, Humphrey Bogart, Ann Sheridan, Ida Lupino; Mervyn LeRoy's *Waterloo Bridge* with Vivien Leigh, Robert Taylor; William Wyler's *The Westerner* with Gary Cooper, Walter Brennan.

Radio *The Road to Happiness* debuts 1/22 on CBS stations; *Young Dr. Malone* 4/29 on CBS stations. *Truth or Consequences* has begun 3/23 over four CBS stations; originated, produced, and directed by Ralph L. Edwards, 26, it is a spoof on giveaway quiz shows.

Hollywood musicals Walt Disney's *Fantasia* with animated visualizations of musical classics that include "The Sorcerer's Apprentice" featuring Mickey Mouse and "A Night on Bald Mountain"; Ben Sharpsteen and Hamilton Luske's *Pinocchio* with Disney animation, music by Leigh Harline, lyrics by Ned Washington, songs that include "When You Wish Upon a Star."

Broadway musicals *Two for the Show* 2/8 at the Booth Theater, with Eve Arden, Betty Hutton, Keenan Wynn, Alfred Drake, songs that include "How High the Moon" by Morgan Lewis, lyrics by Nancy Hamilton, 124 perfs.; *Higher and Higher* 4/4 at the Shubert Theater, with Jack Haley, Marta Eggert, music by Richard Rodgers, lyrics by Lorenz Hart, songs that include "It Never Entered My Mind," 84 perfs.; *Louisiana Purchase* 5/28 at the Imperial Theater, with William Gaxton, Victor Moore, Vera Zorina, Irene Bordoni, music and lyrics by Irving Berlin, songs that include "It's a Lovely Day Tomorrow," 444 perfs.; *Cabin in the Sky* 10/25 at the Martin Beck Theater, with Ethel Waters, Todd Duncan, Dooley Wilson, Rex Ingram, J. Rosamond Johnson, music by Vernon Duke, book and lyrics by John Latouche and Ted Felter, songs that include "Taking a Chance on Love," 156 perfs.; *Panama Hattie* 10/30 at the 46th Street Theater, with Ethel Merman, Arthur Treacher, Betty Hutton, music and lyrics by Cole Porter, songs that include "Let's Be Buddies," 501 perfs.; *Pal Joey* 12/25 at the Ethel Barrymore Theater, with Gene Kelly, Vivienne Segal, June Havoc, music by Richard Rodgers, lyrics by Lorenz Hart, book based on the new collection of stories by John O'Hara, songs that include "Bewitched," "I Could Write a Book," "Zip," 344 perfs.

First performances *Les Illuminations* song cycle by Benjamin Britten 1/30 at London, based on the 1886 poems by Arthur Rimbaud; Concerto for Violin and Orchestra by Paul Hindemith early in the year at Amsterdam (Hindemith moves to the United States); Concerto in D minor for Violin and Orchestra by Benjamin Britten 3/30 at New York's Carnegie Hall; Divertimento for String Orchestra by Béla Bartòk 6/11 at Basel; *Fantasia de Movementos Mixtos* for Violin and Orchestra by Heitor Villa-Lobos 11/1 at the Colon Theater, Buenos Aires; Concerto for Violin and Orchestra by Arnold Schoenberg 12/6 at Philadelphia; Kammer-

1940 ***(cont.)*** simphonie No. 2 by Schoenberg 12/15 at New York.

Opera *Izaht* 4/16 at Rio de Janeiro's Municipal Theater with music by Heitor Villa-Lobos.

The Metropolitan Opera of the Air sponsored by the Texas Company (later Texaco Inc.) begins Saturday afternoon broadcasts December 7 with Ezio Pinza, Licia Albanese, and others singing Mozart's *Nozze di Figaro*.

Popular songs "The Last Time I Saw Paris" by Jerome Kern, lyrics by Oscar Hammerstein II; "Bless 'em All" by English songwriters Jimmie Hughes and Frank Lake; "Strange Fruit" by U.S. songwriter Lewis Allan (the subject is lynching); "Tuxedo Junction" by U.S. bandleader Erskine Hawkins, William Johnson, and Julian Dash, lyrics by Buddy Feyne; "Do I Worry?" by Stanley Cowan and Bobby North (for the film *Pardon My Sarong*); "Imagination" by Jimmy Van Heusen, lyrics by Johnny Burke; "You're the One (for Me)," by Jimmy McHugh, lyrics by Johnny Mercer; "Because of You" by Arthur Hammerstein and Dudley Wilkinson; "When the Swallows Come Back to Capistrano" by Leon René; "Beat Me, Daddy, Eight to the Bar" by Don Raye, Hughie Prince, and Eleanore Sheehy; "How Did He (She) Look" by U.S. composer Abner Silver, lyrics by Gladys Shelley; "I Hear a Rhapsody" by George Frogos and Jack Baker; "Blueberry Hill" by Al Leurs, Larry Stock, and Vincent Rose; "Wabash Cannonball" by A. P. Carter; "San Antonio Rose" by Bob Wills; "Back in the Saddle Again" by Ray Whitley and cowboy singer Gene Autry; "You Are My Sunshine" by Louisiana singer James Houston (Jimmie) Davis, 37, and Charles Mitchell. Davis will use the song to win the Louisiana governorship in 1944 and 1960.

Muzak is acquired by former New York advertising man William Benton, 40, who pays $135,000 to take over a small company that pipes "background music" into restaurants and bars on leased telephone wires under a system originated in World War I by Brig. Gen. George O. Squire and put into operation in 1925 by a public utility holding company. Benton will expand Muzak by piping it into factories, retail stores, offices, even elevators and will sell the company at a $4.2 million profit.

The Olympic Games Committee cancels the games scheduled for Tokyo and Helsinki on account of the war.

Wimbledon tennis matches are canceled for the duration. W. Donald McNeil, 22, wins in men's singles at Forest Hills, Alice Marble in women's singles.

The Cincinnati Reds win their first World Series since the infamous 1919 "Black Sox" Series by defeating the Detroit Tigers 4 games to 3.

The first nylon stockings go on sale in the United States May 15. Competing producers have bought their yarn from E. I. DuPont, whose nylon production will go almost entirely into parachutes beginning next year (*see* 1937; Britain, 1946).

Cotton fabrics hold 80 percent of the U.S. textile market at the mills, down from 85 percent in 1930. Manmade fabrics, most of them cellulose fabrics such as rayon and acetate, have increased their market share to 10 percent (*see* 1950).

Swan soap introduced by Lever Brothers lathers better than the Ivory soap introduced by Procter and Gamble in 1878. P&G has stolen Lever's secret and improved Ivory, an out-of-court settlement in 1944 will oblige P&G to pay Lever some $10 million for its industrial espionage, but Lever's new floating soap will never provide serious competition to Ivory.

Isle Royale National Park is established by act of Congress on the largest island in Lake Michigan. The 539,341-acre park has an outstanding moose herd.

Serengeti National Park is created in Tanganyika by the British, who have added land to the 900-square-mile lion sanctuary set aside in 1929, included the Ngorongoro Crater, and extended the list of protected species to include giraffes, buffalo, rhinoceroses, and most carnivores. Serengeti is the first national park in East Africa (*see* 1961).

Some 3 million tons of U.S. wheat are exported under the export subsidy program started in 1938. The subsidy averages 27.4¢ per bushel to equalize the higher U.S. price with the world price of wheat.

Argentina ships wheat and other foodstuffs to Europe. She will become the richest nation in South America in the next 5 years.

Berlin orders wholesale slaughter of Danish livestock to boost German morale with extra "victory rations" and to save on fodder, gambling that the war will be short. Cattle numbers are reduced by 10 percent, hogs by 30 percent, poultry by 60 percent, and by next year Danish output of animal products will fall to 40 percent below 1939 levels.

Soviet geneticist Nikolai Vavilov is arrested and imprisoned in an underground cell for his opposition to T. D. Lysenko, who will dominate Soviet agriculture until 1964 with his unscientific ideas (*see* 1936; 1942).

Santa Gertrudis cattle, developed over the last 18 years by King Ranch boss Robert J. Kleberg, Jr., 44, are recognized as a new breed by the U.S. Department of Agriculture. The cattle have the cherry red coloring and beef quality of the English shorthorn first imported in 1834. The genetic stock of the new breed is five-eighths shorthorn, but the animals have the small hump, the heavy forequarters, and the hardiness of the Indian Brahma breed introduced in 1854. Steers reach a weight of 2,300 pounds and give up to 71.9 percent meat.

Kleberg's late father countered King Ranch drought problems by digging deep artesian wells. He developed trench silos, brought in the first British breeds to run open on the western range, persuaded the Missouri Pacific to run a rail line through the ranch, established the town of Kingsville, and cultivated citrus, palm, and olive trees (*see* 1886). The younger Kleberg has made an arrangement with Humble Oil to receive royalties on oil and gas pumped from King Ranch lands, which now cover nearly a million acres, an area roughly the size of Rhode Island. The ranch is the largest single beef-cattle producer in America.

Shasta Dam is completed on the Sacramento River in northern California to help irrigate Sacramento Valley farms.

U.S. farmers have withdrawn from cultivation nearly one-third of the farmland tilled in 1930—some 160 million acres—under programs instituted by the Soil Conservation Administration (see 1935).

The U.S. soybean crop reaches 78 million bushels, up from 5 million in 1924 (*see* 1945).

The United States has more than 6 million farms, but while 100,000 farms have 1,000 acres or more, some 2.2 million have fewer than 50 acres (*see* 1959).

Black markets flourish in occupied and unoccupied France. Bread and other food staples are rationed, and consumers pay premium prices to obtain more than their allotted rations.

The U.S. Food and Drug Administration (FDA), established in 1906, is transferred from the Department of Agriculture to the Federal Security Agency, which will become the Department of Health, Education, and Welfare in 1953 and the Department of Health and Human Services in 1980.

U.S. red meat consumption reaches 142 pounds per capita on an average annual basis. The figure will increase to 184 pounds in the next 30 years.

The United States imports 70 percent of the world coffee crop, up from 50 percent in 1934.

Controlled atmosphere (CA) storage is used for the first time on McIntosh apples whose ripening is slowed by storage in gas-tight rooms from which the oxygen has been removed and replaced with natural gas. The process has been developed by Cornell University professor Robert M. Smock.

Nearly 15,000 U.S. retail stores are equipped to sell frozen foods, up from 516 in 1933.

Demand for Pepperidge Farm bread obliges Margaret Rudkin to rent some buildings at Norwalk, Conn., and to expand her baking facilities (*see* 1937; 1961).

Owens-Illinois Glass introduces Duraglas for deposit bottles. The new bottle glass can stand up to repeated use by beer and soft drink bottlers.

McDonald hamburger stands have their beginning in a drive-in opened near Pasadena, Calif., by movie theater co-owners Richard and Maurice McDonald of Glendora (*see* 1948).

M&M candies have their beginnings in a candy-coated chocolate produced for the U.S. military by Forrest E. Mars and Bruce E. Murrie (*see* Mars, Inc., 1923; 1930; 1932).

Seven out of 10 U.S. blacks still live in the 11 states of the old Confederacy, down from 8 out of 10 in 1910 (*see* 1960).

Birth control receives public endorsement from Eleanor Roosevelt.

The world's population reaches 2.3 billion, with roughly 500 million in China. Eighty percent of the world still lives in agricultural villages, says French geographer Max Sorre.

1941 World War II explodes into a global conflict as German troops invade Soviet Russia and Japanese forces attack Pearl Harbor (below).

President Roosevelt recommends a Lend-Lease program to aid the Allies in a January 6 congressional message that defies widespread isolationist sentiment.

The $17.5 billion budget submitted to Congress by President Roosevelt January 8 includes nearly $11 billion for national defense.

Congress authorizes $7 billion in Lend-Lease aid March 27 and authorizes another $6 billion October 28 and an additional $10 billion December 15 for Lend-Lease and for the U.S. armed forces.

Yugoslavia has a revolution March 27; German troops invade April 6, Croats support them, and hundreds of thousands of Serbs are killed, mostly by Croatian irregulars.

German troops invade Greece April 6.

The America First Committee founded in April unites isolationists under the leadership of *Chicago Tribune* publisher Robert R. McCormick and Sears, Roebuck chairman Robert E. Wood, who are joined by Charles A. Lindbergh, Senators Borah, Byrd, Wheeler, and others.

Hitler aide Rudolf Hess flies to Scotland May 10 on an undisclosed mission. The British intern him (*see* 1946).

German paratroops invade Crete May 20. The British withdraw 12 days later.

The 41,700-ton German battleship *Bismarck* sails out of Gdynia, Poland, May 18, sinks the British dreadnought *Hood* May 24 with great loss of life, but is herself sunk May 27 in the North Atlantic by British air and naval units. Only 100 of the *Bismarck's* 1,300-man crew survive.

British forces arrive in Iraq May 31 and prevent Axis sympathizers from taking over the government.

British and Free French troops invade Syria and Lebanon in early June to prevent a German takeover of those countries. Lebanese independence is proclaimed November 26.

Jewish terrorists in Palestine use violence to fight for independence from Britain. The Stern Gang founded by Polish-born underground leader Abraham Stern assassinates officials and bombs military installations and oil refineries, and although the British will kill Stern next year, his followers will continue the struggle (*see* 1939; 1945).

German troops invade Soviet Russia June 22, advance rapidly against the Soviet army, whose leaders have been to a large extent liquidated in Josef Stalin's political purges of the late 1930s, and lay siege to Leningrad. British intelligence predicts a Soviet collapse in 10 days, U.S. intelligence says 3 months. German soldiers

1941 ***(cont.)*** control an area twice the size of France by October, but they carry no winter clothing, and by December the *Wehrmacht* is fighting 360 Soviet divisions when it had anticipated fighting fewer than 200.

The Atlantic Charter comes out of secret meetings held off the coast of Newfoundland from August 9 to 12 by President Roosevelt and Prime Minister Churchill. The charter contains eight articles of agreement on war aims.

British and Soviet troops invade Iran in late August. Reza Shah Pahlevi abdicates September 16 after a 16-year reign and is succeeded by his son Mohammed Reza Pahlevi, 21, who is more inclined to cooperate with the Allies.

President Roosevelt issues an order September 11 that German or Italian vessels sighted in U.S. waters are to be attacked immediately.

The U.S. destroyer *Kearny* is torpedoed by a German U-boat off Iceland October 17.

The U.S. destroyer *Reuben James* is sunk by a German U-boat October 31; 100 lives are lost.

Pearl Harbor on the Hawaiian island of Oahu comes under attack Sunday morning, December 7 (December 8, Tokyo time) from 360 carrier-based Japanese planes led by Mitsuo Fuchida, 38, the Japanese lose only 29 planes, and the attack corps cripples the U.S. Pacific fleet, sinking the battleships U.S.S. *Arizona, Oklahoma, California, Nevada,* and *West Virginia,* damaging three other battleships, inflicting major damage on three cruisers and three destroyers, destroying 200 U.S. planes, and killing 2,344 men. The American public hears only that the U.S.S. *Arizona* has been sunk and the *Oklahoma* capsized.

The attack on Pearl Harbor at 7:50 A.M. Honolulu time (1:20 P.M. in Washington) has come without declaration of war. It is only at 9 P.M. Washington time that the Japanese foreign minister advises the U.S. Embassy at Tokyo that a state of war exists between the United States and Japan.

Pearl Harbor drew America into the war, exploding isolationist sentiment in no uncertain way.

December 7 is "a date which will live in infamy," says President Roosevelt in an address to Congress December 8. The Senate votes 82 to 0 for a declaration of war on Japan, the House votes 388 to 1 for war (Rep. Jeanette Rankin votes nay as she did in 1917), and President Roosevelt signs the declaration at 4:10 P.M.

Japan declares war on Britain as well as on the United States. Japanese forces land in Malaya and Siam December 8.

H.M.S. *Prince of Wales* and H.M.S. *Repulse* are sunk by the Japanese off the Malayan coast December 9.

Japanese forces land on Luzon in the Philippines December 10; they take Guam in the Marianas December 11.

Germany declares war on the United States December 11, thus making it possible for President Roosevelt to end U.S. neutrality in the European war. Italy echoes the German declaration, and Congress declares war on Germany and Italy.

Romania declares war on the United States December 12 and Bulgaria follows suit December 13.

Admiral Chester W. Nimitz, 56, is given command of the Pacific Fleet December 17, replacing Admiral H. E. Kimmel, who has been relieved of his command as has Lieut. Gen. Walter C. Short. Both are held responsible for being caught unprepared by the surprise attack on the U.S. naval base at Pearl Harbor (above).

Some 80,000 to 100,000 Japanese troops are landed at the Gulf of Lingayen December 22 for a major invasion of the Philippines.

Wake Island falls to the Japanese December 23. Hong Kong is taken by the Japanese December 25.

Ethiopia regains her independence in December. British troops help oust Italian occupation forces (*see* 1936).

President Roosevelt's January 6 message to Congress calls for a world with the Four Freedoms protected—freedom of speech and expression, freedom of religion, freedom from want, freedom from fear.

Adolf Hitler has given the signal for the "Final Solution" to Europe's "Jewish problem." Adolf Eichmann proposes "killing with showers of carbon monoxide while bathing" but the cyanide gas Zyklon B is found more effective.

German SS units machine-gun some 3,000 Jewish men, women, and children to death in the suburbs of Minsk and Mogilev while German military authorities stand by.

Babi Yar 30 miles outside Kiev is the scene of a September 29 massacre as Nazi invasion forces machine-gun to death between 50,000 and 96,000 Ukrainians of whom at least 60 percent are Jews.

Estonian, Galician, Latvian, Lithuanian, Polish, and Russian Jews flee at the approach of advancing German troops. Jews caught by the Germans are drafted into labor gangs, driven into ghettos, forced into military brothels, massacred with machine guns, or shipped in freight cars to detention camps where many are found dead on arrival. Jewish shops, factories, department stores, libraries, synagogues, and cemeteries are plundered and destroyed.

President Roosevelt appoints a Fair Employment Practice Committee (FEPC) June 25. A. Philip Randolph of the Brotherhood of Sleeping Car Porters has threatened a march on Washington by 50,000 blacks to protest unfair employment practices in war industry and the government (*see* Randolph, 1925, 1948).

The Office of Production Management (OPM) created by the president January 7 is headed by William S. Knudsen of General Motors with Sidney Hillman of the Amalgamated Clothing Workers Union as associate director general.

The Office of Price Administration and Civilian Supply (OPA) created by the president April 11 has New Deal economist Leon Henderson, 45, as its head, but wartime inflation increases the general price level by 10 percent for the year.

The U.S. Treasury issues war bonds to help reduce the amount of money in circulation. Priced to yield 2.9 percent return on investment, the Series E savings bonds come in denominations of $25 to $10,000.

U.S. tax rates rise sharply in nearly all categories September 20 in an effort to raise more than $3.5 billion for war-related expenditures.

Merrill Lynch Pierce Fenner & Beane is created by a merger (*see* 1914). Charles Merrill, now 58, transferred his brokerage business and much of his staff to Pierce & Co. to enter investment banking in 1930, merged with Pierce and E. A. Cassatt last year, and now merges with Fenner & Beane. Winthrop Hiram Smith, now 48, will join the firm in 1958 and it will become Merrill Lynch Pierce Fenner & Smith, the world's largest investment banking house.

The transuranium element plutonium (atomic number 94) isolated at Berkeley by physicists Glenn Theodore Seaborg, 29, and Edwin M. McMillan has a Pu-239 isotope that shows promise of having a higher energy yield from nuclear fission than uranium (*see* 1945).

Grand Coulee Dam begins generating power on the Columbia River in eastern Washington. Contractors organized by Henry Kaiser have built the world's largest hydroelectric installation.

Henry Kaiser (above) establishes a chain of seven Pacific Coast shipyards north of Richmond, Calif., at the outbreak of the Pacific war in December. Kaiser shipyards will use prefabrication and assembly line methods to set new speed records in shipbuilding (*see* 1942).

Welding Shipyards, Inc., is founded at Norfolk, Va., by Daniel K. Ludwig (*see* 1936).

President Roosevelt establishes an Office of Defense Transportation December 18.

U.S. motortruck production reaches 4.85 million, up from 3.5 million in 1931 (*see* 1951).

U.S. auto production reaches 3.3 million.

The Lincoln Continental, introduced by Ford Motor Company's Edsel Ford, is the sleekest U.S. car yet made (*see* 1939).

There are 32.6 million cars on the U.S. road, up from 20 million in 1927 (*see* 1950).

The first U.S. diesel freight locomotives go into service for the Atchison, Topeka and Santa Fe (*see* General Motors, 1935). Built by General Motors' Electromotive Division, the new 5,400-horsepower diesels eliminate water problems in desert country and reduce hotbox problems on downgrades with a dynamic braking system. Running time between Chicago and California drops from 6 days to 4, with only five brief stops en route, and the sound of the steam engine whistle begins to fade from the American scene (*see* 1957).

English chemist John Rex Whinfield, 40, invents Terylene (Dacron in the United States), a new polyester fiber made from terephthalic acid and ethylene glycol. Imperial Chemical Industries will market the fiber after the war under the name Terylene, E. I. du Pont will market it as Dacron (*see* nylon, 1939; Orion, 1948).

The Office of Scientific Research and Development created by President Roosevelt June 28 is headed by Vannevar Bush (*see* analog computer, 1930).

London's *Daily Worker* is suppressed January 21.

"The Sad Sack" is originated by former Walt Disney cartoonist George Baker, 25, whose cartoon strip will appear in *Yank* beginning next year (*see* 1942).

Parade magazine, founded by Marshall Field III, begins publication in May for newsstand distribution. The *Washington Post* takes the picture magazine in August for distribution with its Sunday edition and other papers acquire distribution rights.

Germany abandons Gothic type May 31 in favor of Roman type.

The Brooklyn Public Library's central branch opens in April on Grand Army Plaza. A second floor will open in 1955 and there will be further expansions in 1972 and 1989.

Nonfiction *The Mind of the South* by U.S. journalist W. (Wilbur) F. Cash who hangs himself in a Mexico City hotel room July 2 at age 40; *Let Us Now Praise Famous Men* by *Time* magazine film critic James Agee, 31, and *Fortune* magazine photographer Walker Evans, 37, who have studied the plight of Alabama sharecroppers and confront Americans with rural poverty in the South; *Black Lamb and Gray Falcon* by English woman of letters Rebecca West (Cecily Andrews), 49, is a two-volume diary of her 1937 trip to Yugoslavia; *Escape From Freedom* by German-American psychoanalyst and philosopher Erich Fromm, 41, examines the meaning of freedom and authority.

Fiction *The Fancy Dress Party* (*La mascherata*) by Alberto Moravia enrages Benito Mussolini who personally censors the novel when he realizes that Moravia's South American dictator represents Il Duce; *Conversa-*

1941 *(cont.)* *tion in Sicily (Conversazione in Sicilia)* by Italian novelist Elio Vittorini, 37; *Darkness at Noon* by Arthur Koestler, who completed the novel in a Vichy concentration camp; *Scum of the Earth* by Koestler, who shows the misery inside concentration camps; *A Woman of the Pharisees (La Pharisienne)* by François Mauriac; *The Pasquier Chronicles (Chronique des Pasquier)* by Georges Duhamel; *The Blind Owl (Buf-i-Kur)* by Iranian novelist Sadig Hidayat, 37; *Herself Surprised* by Joyce Cary; *A Curtain of Green* (stories) by Mississippi writer Eudora Welty, 32; *The Seventh Cross (Das Siebente Kreuz)* by émigrée German novelist Anna Seghers (Netty Radvanyi), 41; *Reflections in a Golden Eye* by Carson McCullers; *The Last Tycoon* by the late F. Scott Fitzgerald, who died of a heart attack at age 44 in December of last year, leaving incomplete his novel based roughly on the late MGM producer Irving Thalberg; *What Makes Sammy Run?* by U.S. novelist Budd Schulberg, 27, whose father is a leading Hollywood producer; *Mildred Pierce* by James M. Cain; *The Keys of the Kingdom* by A. J. Cronin; *H. M. Pulham, Esq.* by John P. Marquand; *Saratoga Trunk* by Edna Ferber; *The Song of Bernadette* by Franz Werfel.

Poetry *55 Poems* by U.S. poet Louis Zukofsky, 37; *Poems de la France malheureuse* by Jules Supervielle; *Raka* by N. P. van Wyk Louw; *La Crève-Coeur* by French novelist-poet Louis Aragon, now 44, who has served with distinction against the Germans.

Juvenile *Curious George* by German-American writer-illustrator H. A. (Hans Augusta) Rey, 43, and his wife Margaret, whose monkey George will appear in six sequels (the Reys were living in Paris in June of last year, bicycled south before the *Wehrmacht* arrived, reached Lisbon, sailed to Rio, and arrived at New York in October of last year).

Mount Rushmore National Monument attracts visitors to South Dakota following the death of native American sculptor John Gutzon de la Mothe Borglum who suffered a fatal heart attack March 6 at age 73. Borglum has developed new methods of stoneworking to carve 60-foot high heads of George Washington, Thomas Jefferson, Abraham Lincoln, and Theodore Roosevelt out of the living rock. His son Lincoln works to improve the collar and lapels on Washington's coat, Jefferson's collar, Lincoln's head, and Roosevelt's face but will leave the massive work incomplete.

The National Gallery of Art opens at Washington, D.C., where a $15 million bequest from the late Pittsburg financier Andrew W. Mellon has financed construction of the rose-white Tennessee marble building. When he died in 1937, Mellon left an art collection acquired at the persuasion of the London art dealer Baron Duveen of Milbank, now 72. The Gallery will house other major collections as well.

Painting *Nighthawks* by Edward Hopper; *New York Under Gaslight* by Stuart Davis; *Bird* by Wyoming-born abstractionist (Paul) Jackson Pollock, 29; *Women with Bulldog* by French painter Francis Picabia, 62; *Divers Against Yellow Background* by Fernand Léger. Robert Delaunay dies at Montpelier, France, October 25 at age 56.

Theater *Arsenic and Old Lace* by U.S. playwright Joseph Kesselring, 39, 1/10 at New York's Fulton Theater with Josephine Hull, Jean Adair, Boris Karloff, 1,437 perfs.; *Native Son* by Richard Wright and Paul Green 3/24 at New York's St. James Theater, with Canada Lee, 114 perfs. (*see* 1940 novel); *Watch on the Rhine* by Lillian Hellman 4/1 at New York's Music Box Theater, with Paul Lukas, Mady Christians, Anne Blyth, George Coulouris, 378 perfs.; *Mother Courage, A Chronicle of the Thirty Years' War (Mutter Courage und ihre Kinder)* by Bertolt Brecht 4/19 at Zurich's Schauspielhaus; *Blithe Spirit* by Noël Coward 7/2 at London's Piccadilly Theatre, 1,997 perfs.; *Candle in the Wind* by Maxwell Anderson 10/22 at New York's Shubert Theater, with Helen Hayes, Lotte Lenya, 95 perfs.; *Junior Miss* by Jerome Chodorov and Joseph Fields 11/18 at New York's Lyceum Theater, with Patricia Peardon, 16, as Judy Graves, Lenore Lonergan as Fuffy Adams, 710 perfs.; *Angel Street* by Patrick Hamilton 12/5 at New York's Golden Theater, with Judith Evelyn, Vincent Price, 1,295 perfs.

Films Orson Welles's *Citizen Kane* with Welles, Joseph Cotten, Everett Sloane, Agnes Moorehead, Evelyn Keyes, George Coulouris opens May 1 at New York's Palace Theater. Publisher William Randolph Hearst has tried to block showing of Welles's allegory on the theme of idealism corrupted by power (the film is not reviewed or advertised in any Hearst newspaper).

Other films Alexander Hall's *Here Comes Mr. Jordan* with Robert Montgomery, Evelyn Keyes; John Ford's *How Green Was My Valley* with Roddy McDowell, Donald Crisp, Sara Allgood, Walter Pidgeon; Gabriel Pascal's *Major Barbara* with Wendy Hiller, Rex Harrison; John Huston's *The Maltese Falcon* with Humphrey Bogart, Sydney Greenstreet, Peter Lorre, Mary Astor, Elisha Cook, Jr.; Alfred Hitchcock's *Rebecca* with Joan Fontaine, Laurence Oliver, George Sanders; Preston Sturges's *Sullivan's Travels* with Joel McCrae, Veronica Lake. Also: William Dieterle's *The Devil and Daniel Webster* with Edward Arnold, Walter Huston; Sam Wood's *The Devil and Miss Jones* with Jean Arthur, Robert Cummings; Michael Powell's *Forty-Ninth Parallel* with Anton Walbrook, Eric Portman, Leslie Howard; Charles Chaplin's *The Great Dictator* with Chaplin, Paulette Goddard, Jack Oakie; King Vidor's *H.M. Pulham, Esq.* with Hedy Lamarr, Robert Young, Ruth Hussey; Mitchell Leisen's *Hold Back the Dawn* with Charles Boyer, Olivia de Havilland; Sam Wood's *King's Row* with Ann Sheridan, Robert Cummings, Ronald Reagan; Preston Sturges's *The Lady Eve* with Barbara Stanwyck, Henry Fonda; William Wyler's *The Little Foxes* with Bette Davis; William Keighley's *The Man Who Came to Dinner* with Bette Davis, Ann Sheridan, Monty Woolley; Edward Cline's *Never Give a Sucker an Even Break* with W. C. Fields, Gloria Jean; Irving Rapper's *One Foot in Heaven* with Fredric March, Martha Scott; George Stevens's *Penny Serenade* with Cary Grant, Irene Dunne; Michael Curtiz's *The Sea Wolf* with Edward G. Robinson, John Garfield,

Ida Lupino; Howard Hawks's *Sergeant York* with Gary Cooper; Mark Sandrich's *Skylark* with Claudette Colbert, Ray Milland; John Cromwell's *So Ends Our Night* with Fredric March, Margaret Sullavan; Alfred Hitchcock's *Suspicion* with Cary Grant, Joan Fontaine; Garson Kanin's *Tom, Dick and Harry* with Ginger Rogers, George Murphy, Burgess Meredith; Marcel Pagnol's *The Well-Digger's Daughter* with Raimu, Fernandel, Josette Day, Charpin; George Waggner's *The Wolf Man* with Lon Chaney, Jr., Evelyn Ankers.

Hollywood musicals Ben Sharpsteen's *Dumbo* with Walt Disney animation, music by Frank Churchill, lyrics by Oliver Wallace.

West Coast musical *Jump for Joy* is an all-black revue with songs that include "I Got It Bad and That Ain't Good" by Duke Ellington, lyrics by Paul Francis Webster, Henry Newman, Irving Mills.

Broadway musicals *Lady in the Dark* 1/23 at the Alvin Theater with Gertrude Lawrence, Danny Kaye (*né* David Daniel Kominsky), music by Kurt Weill, lyrics by Ira Gershwin, songs that include "Jenny (the Saga of)," "My Ship," "Tchaikovsky," 162 perfs. (Kaye, 28, has married Sylvia Fine while working in the Catskill Mountain "borscht circuit" and she has helped him with her witty songs); *Best Foot Forward* 10/1 at the Ethel Barrymore Theater, with Rosemary Lane, Nancy Walker, June Allyson, songs by Hugh Martin and Ralph Blane that include "Buckle Down, Winsocki," 326 perfs.; *Let's Face It* 10/29 at the Imperial Theater with Danny Kaye, Eve Arden, Nanette Fabray, music and lyrics by Cole Porter, songs that include "You Irritate Me So," and Danny Kaye patter songs with lyrics by Sylvia Fine (above), 547 perfs.; *Sons O' Fun* 12/1 at the Winter Garden Theater with Olsen and Johnson, Carmen Miranda, Ella Logan, music by Sammy Fain and Will Irwin, lyrics by Jack Yellen and Irving Kahal, songs that include "Happy in Love," 742 perfs.

First performances *Symphonie Dances* for Orchestra by Sergei Rachmaninoff 1/4 at Philadelphia's Academy of Music; Concerto for Violoncello and Orchestra by Paul Hindemith 2/7 at Boston's Symphony Hall, with Gregor Piatigorsky as soloist; Concerto for Violin and Orchestra by Samuel Barber 2/7 at Philadelphia's Academy of Music, with Albert Spalding as soloist; Symphony No. 1 by U.S. composer Paul Creston (*né* Joseph Guttoveggio), 34, 2/22 at the Brooklyn Academy of Music; *Latin-American Symphonette* by Morton Gould 2/22 at Brooklyn; *Ballad of a Railroad Man* for Chorus and Orchestra by Roy Harris 2/22 at Brooklyn; *Sinfonia da Requiem* by Benjamin Britten 3/29 at New York's Carnegie Hall (Tokyo commissioned the work in 1939 for the 2,600th anniversary of the imperial dynasty in 1940 but rejected it, calling it too Christian); *A Symphony in D for the Dodgers* by U.S. composer Robert Russell Bennett, 46, 5/16 in a radio concert from New York; Piano Sonata by Aaron Copland 10/21 at Buenos Aires; Symphony No. 2 by Virgil Thomson 11/17 at Seattle; Symphony in E flat major by Paul Hindemith 11/21 at Minneapolis; *Scottish Ballad* for Two Pianos and Orchestra by Benjamin Britten 11/28 at Cincinnati; Symphony No. 1 by U.S. composer David Diamond, 26, 12/21 at New York's Carnegie Hall.

Luftwaffe night raiders destroy the Queen's Hall, London, with incendiary bombs May 10.

A BBC broadcast from London June 27 has urged the subjugated peoples of Europe to adopt the Morse code signal for the letter V (for Victory)—three dots and a dash—and to whistle the opening motif of Symphony No. 5 by Ludwig van Beethoven whenever Nazi soldiers are around. London's Westminster chimes have added the four notes on the hour.

Popular songs "Lili Marlene" by German composer Norbert Schultze (who also wrote "Bombs on England"), lyrics by World War I German soldier Hans Leip, now 45, who wrote them before going to the Russian front in 1916, combining the name of his girlfriend with that of a friend's girl), English lyrics by London publisher Jimmy Phillips and lyricist Tommy Connor, who adapt the song Allied troops have learned from prisoners of war and from German radio (German publishers will receive royalties beginning in 1958); "Blues in the Night" by Harold Arlen, lyrics by Johnny Mercer; "(There'll Be Blue Birds Over) The White Cliffs of Dover" by English songwriters Nat Burton and Walter Kent; "We Did It Before (and We Can Do It Again)" by Cliff Friend, lyrics by Charlie Tobias; "I Don't Want to Walk Without You" by London-born composer Jule Styne, 36, and Frank Loesser; "I Don't Want to Set the World on Fire" by Eddie Seiler, Sol Marcus, Bennie Benjamin, Eddie Durham; "Take the A Train" by Duke Ellington and Billy Strayhorn (*see* New York subway, 1933); "(I Like New York in June) How About You?" by Burton Lane, lyrics by Ralph Freed; "Elmer's Tune" by Chicago undertaker's assistant Elmer Albrecht, lyrics by Sammy Gallop and bandleader Dick Jurgens; "Racing with the Moon" by Johnny Watson, lyrics by Watson's wife Pauline Pope and bandleader Vaughn Monroe; "This Love of Mine" by Sol Parker and Henry Sanicola, lyrics by Frank Sinatra; "Why Don't You Do Right" by blues songwriter Joe McCoy; "Yes Indeed!" by Sy Oliver; "Deep in the Heart of Texas" by Don Swander, 36, and his wife June, 32 (who has never been in Texas); "Jersey Bounce" by Billy Plate, Tony Bradshaw, Edward Johnson, Robert B. Wright; "The Hut-Sut Song" by Leo V. Killion; "Cow-Cow Boogie" by Don Raye, Gene DePaul, and Benny Carter; "Chattanooga Choo-Choo" by Harry Warren, lyrics by Mack Gordon; "Why Don't We Do This More Often" by Allie Wrubell, lyrics by Charlie Newman; "Let's Get Away From It All" by Tommy Dorsey protégé Matt Dennis, lyrics by Tom Adair; "Boogie Woogie Bugle Boy" by Don Raye and Hughie Prince; "Anniversary Waltz" by Al Dubin and Dave Franklin.

Bobby Riggs wins in men's singles at Forest Hills, Sarah Palfrey Cooke, 18, in women's singles.

Stan Musial goes to bat for the St. Louis Cardinals beginning late in the season and has a .426 average for 12 games. Stanley Frank Musial, 22, will have a lifetime average of .331 in 22 seasons and a career total of 3,630 hits, a record second only to that of Ty Cobb (*see* 1905).

1941 ***(cont.)*** The New York Yankees win the World Series by defeating the Brooklyn Dodgers 4 games to 1.

New York gangster Abe "Kid Twist" Reles "falls" to his death November 12 from a sixth-story Coney Island hotel window in the presence of a six-man police bodyguard. Reles last year informed to the Brooklyn district attorney, his testimony attributed 130 hired killings between 1930 and 1940 to a national crime syndicate called Murder Incorporated, and his testimony will lead to the execution of mob figures who include Louis "Lepke" Buchalter (*see* Kefauver Committee hearings, 1950).

Mexico's wheat harvest falls by half as stem rust devastates crops in the nation's breadbasket—Queretaro, Guanajuato, Michoacan, and Jalisco. Mexico is obliged to spend $30 million (100 million pesos) each year to import corn and wheat, leaving less for desperately needed power generators, machinery, and chemicals from abroad.

The Rockefeller Foundation begins a program to improve Mexican agriculture at the urging of Secretary of Agriculture Henry A. Wallace. It sends University of Minnesota rust specialist Elvin C. Stakman, 56, Cornell soil expert Richard Bradford, 46, and Harvard plant geneticist Paul C. Mangelsdorf, 42, to Mexico, and they submit a blueprint for redeveloping the country's rural economy.

Britain receives her first U.S. Lend-Lease shipments of U.S. food April 16, just in time to avert a drastic food shortage. By December, 1 million tons of U.S. foodstuffs have arrived.

Russia has her most severe winter in 30 years, and Leningrad (above) has no heat, no electricity, and little food. The Luftwaffe destroys food stocks stored in wooden buildings, and besieged Leningraders eat cats, dogs, birds, jelly made of cosmetics, and soup made from boiled leather wallets. Those able to work receive a bread ration of half a pound per day, others get only two slices, and the bread is often made of cheap rye flour mixed with sawdust, cellulose residues, and cottonseed cake. Thousands die of starvation; before the siege is lifted in 1944, 20 to 40 percent of Leningrad's 3 million people will have died in 900 days of hunger and related disease (*see* Stalingrad, 1942).

Germany is the best-fed of Europe's combatant nations. A rationing plan provides Germans with at least 95 percent of the calories received in peacetime—2,000 calories per day.

Soviet citizens outside Leningrad receive an estimated 1,800 calories per day, but people in occupied Finland, Norway, Belgium, and the Netherlands receive fewer than 1,800, and those in the Baltic states, Poland, France, Italy, Greece (and the other Balkan countries) are lucky to get 1,500.

A National Nutritional Conference for Defense convened by President Roosevelt examines causes for the physical defects found in so many young men called up by the draft. Mayo Clinic nutrition specialist Russell M. Wilder heads a group of experts in a study of the eating habits of 2,000 representative U.S. families (*see* 1943).

Only 30 percent of U.S. white bread is enriched with vitamins and iron. South Carolina becomes the first state to require enrichment.

The average fat content of U.S. frankfurters is 19 percent, up from 18 in 1935 (*see* 1969).

Daniel Gerber's Fremont Canning Co. sells 1 million cans of baby food per week (*see* 1929; 1948).

Cheerios breakfast food, introduced by General Mills, is 2.2 percent sugar.

The Chemex coffee maker designed by Peter Schlumbohn is introduced by Chemex Corp. of New York.

Basic U.S. food prices by December are 61 percent above prewar prices.

New York's Le Pavillon opens October 15 at 51 East 55th Street under the direction of Henri Soule, 38, who ran the restaurant at the World's Fair's French Pavilion. The new eating place has a staff of 40 and charges $1.75 for a table d'hôte luncheon of hors d'oeuvre, plat du jour, dessert, and coffee.

The U.S. population reaches 132 million, Britain has 47 million, France 40, Germany with its annexed territories 110, the U.S.S.R. with its annexed territories 181, Japan 105 (up from 89 million in 1936), Brazil 41, India 389, China about 500.

1942 Japanese forces take Manila January 2, invade the Dutch East Indies January 10, but suffer their first major sea loss in late January at the Battle of Macassar Strait when U.S. and Dutch naval and air forces attack a Japanese convoy.

President Roosevelt calls in January for production in 1942 of 60,000 planes, 45,000 tanks, 20,000 anti-aircraft guns, and 6 million deadweight tons of merchant shipping. His $59 billion budget submitted January 7 has more than $52 billion earmarked for the war effort.

An Inter-American Conference assembles 21 representatives at Rio de Janeiro January 15 to coordinate Western Hemisphere defenses against aggression. Delegates adopt a unanimous resolution January 21 calling for severance of relations with the Axis powers, and the group sets up an Inter-American Defense Board in March.

The bazooka antitank weapon developed at the Frankfort Arsenal by Edward J. Uhl, 24, and Leslie A. Skinner, 40, is tested for the first time in May.

Napalm (napthenic acid and palmetate), developed by Harvard chemist Louis F. Fieser, 43, in response to a U.S. Army request, is a jellied gasoline incendiary that will increase the effective range of flamethrowers and slow down the rate of burning. Fieser produces a cheap derivative of petroleum and palm oils that will find wide use as a canister filler for bombs.

Singapore falls to the Japanese February 15. The big guns of the British naval base have fixed emplacements and point out to sea, making them useless against the Japanese who approach from the Malayan interior under Gen. Tomoyuki Yamashita, 36. He receives the

surrender of more than 130,000 British and Commonwealth troops.

Japanese forces wipe out an Allied squadron in the Battle of the Java Sea late in February and land in the Solomon Islands March 13, threatening the vital route to Australia.

U.S. troops on the Bataan peninsula in the Philippines surrender to Gen. Yamashita April 9. Most of the 36,000 men are killed on the "death march" to internment camps or in the camps.

Tokyo is raided April 18 by 16 B-25 bombers from the U.S. carrier *Hornet* under the command of Major "Jimmy" Doolittle, now 45, who also bombs Yokohama and other cities before landing in China. Three U.S. flyers die in crash landings; the Japanese capture 11 others and execute three as "war criminals."

Japanese forces cut the Burma Road and capture Lashio April 29.

China's Chiang Kai-shek gets U.S. support in his resistance to Japanese aggression. U.S. cargo planes deliver supplies from India over "the hump" (Himalayas); Chiang's chief of staff Gen. Joseph Warren "Vinegar Joe" Stilwell, 59, begins the 478-mile Burma-India Ledo Road to link up with the old Burma Road.

Mandalay falls to the Japanese May 2 as they complete their conquest of Burma.

The U.S. garrison at Corregidor in the Philippines surrenders May 6 after 300 air attacks. Gen. Jonathan Mayhew Wainwright III, 59, and his 10,000 men are taken prisoner. The Japanese sustain heavy losses in the Battle of the Coral Sea, May 6, but sink the U.S. carrier *Lexington*.

The RAF raids Cologne on the night of May 30 in the first 1,000-bomber attack on German industrial targets.

U.S. carrier-based planes stop the Japanese at the Battle of Midway in early June.

Japanese forces occupy Attu and Kiska in the Aleutians June 7.

Eight German saboteurs are captured by the FBI June 28, two weeks after being landed by a U-boat on Long Island and Florida beaches.

Tobruk in Libya falls June 21 to German field marshal Erwin Rommel, 50, who takes 25,000 British prisoners. Axis forces sweep east to El Alamein, 70 miles from Alexandria, before being checked by the forces of Gen. Bernard L. Montgomery, 54.

Canadian commandos raid Dieppe on the French coast August 19, but German defenders beat off the attacks; 3,500 Canadians are lost.

The Battle of Stalingrad begins August 22. The Germans have captured Sevastopol in the Crimea July 2 after an 8-month siege and launch an offensive against the Volga River center for shipment of oil from the Caucasus, a city of 500,000 whose population will dwindle to 1,515 in the barbaric 5-month battle. The Russians will lose 750,000 troops, the Germans 400,000, the Romanians nearly 200,000, the Italians 130,000, the Hungarians 120,000, and Axis survivors will be sustained in many cases only by cannibalism before 22 German divisions, reduced to 80,000 men, surrender early next February.

The German battleships *Scharnhorst, Gneisenau,* and *Prinz Eugen* have escaped from the occupied French harbor of Brest to the Baltic in February but avoid Allied naval vessels and aircraft.

U-boat activity off the U.S. Atlantic Coast takes a heavy toll of merchant ships bound for British and Russian ports with war matériel, foodstuffs, and men. Washington orders a dim-out extending 15 miles from the coast April 28 to make it harder for the U-boats to sight their quarry at night.

The Women's Auxiliary Army Corps (WAAC) established by act of Congress May 14 is headed by *Houston Post* editor Oveta Culp Hobby, 37.

The Office of Strategic Services (OSS) created by executive order June 13 is headed by World War I hero William Joseph "Wild Bill" Donovan, 59, whose military intelligence group will be the basis of the CIA (*see* 1947).

The WAVES (Women Accepted for Voluntary Emergency Service) authorized by act of Congress July 30 to support the Navy is headed by Wellesley College president Mildred H. McAfee, 42.

Marshal Pétain has reinstated Pierre Laval as French premier under pressure from German occupation forces April 14. He appoints Laval as his successor November 17 and empowers him with making laws and issuing decrees.

A British-U.S. force of 400,000 commanded by Gen. Dwight D. Eisenhower lands at Casablanca, Oran, and Algiers November 7 and 8 in an armada of 500 transports convoyed by 350 naval vessels. Vichy French garrisons at Casablanca, Oran, and Algiers are overpowered after brief fighting, the Vichy government's Admiral Jean-François Darlan, 61, arranges an armistice and is assassinated December 24.

French crews scuttle most of the fleet at Toulon November 27 to keep it from falling into German hands.

"I have not become the King's First Minister in order to preside over the liquidation of the British Empire," says Prime Minister Churchill November 10 in a speech at the Mansion House. Immediate independence of India has been demanded by Mahatma Gandhi and other leaders who have been arrested but later released (*see* 1947).

German troops occupy unoccupied France November 11.

The first surface-to-surface guided missile is launched December 24 at Peenemunde by German rocket engineer Wernher von Braun, 30, who tests his buzz bomb (*see* 1944).

1942 ***(cont.)*** Czech patriots with help from British agents assassinate deputy Gestapo chief and Reich "protector" of Czechoslovakia Reinhard Heydrich, 38, May 31. President Beneš in exile has given orders for the hit. The death of *der Henker* (the Hangman) brings swift and terrible retribution.

The Germans burn the Czech village of Lidice in Bohemia June 6 after executing every male in reprisal for the assassination of Reinhard Heydrich (above). Only one man reportedly escapes; the female population is abused.

Some 30,000 Parisian Jews are rounded up by 2,000 police officers July 16. The Germans bus them out of the city to Nazi concentration camps; only 30 will survive.

Some 7,000 French Jews including infants and children are locked into a stadium for 8 days without water or toilet facilities.

Hitler and his Gestapo chief Heinrich Himmler begin a methodical annihilation of all European Jews within their reach. Some 8,000 Greek Jews from Salonika are transferred to concentration camps in the mountains of Macedonia; another 45,000 will be deported in the next 3 years, leaving scarcely 500 in the city (*see* Poland, 1943).

Executive Order 9066 issued by President Roosevelt February 19 calls for internment of some 110,000 Japanese-Americans living in coastal Pacific areas. California's attorney general Earl Warren, 50, is running for governor and has urged the measure. No similar moves are taken against Japanese-Americans in the Hawaiian Islands or the U.S. East and Midwest, nor against alien Germans or Italians, but native-born U.S. citizens of Japanese ancestry are rounded up along with Japanese aliens and interned in remote areas of Arizona, Arkansas, inland California, Colorado, Idaho, Utah, and Wyoming, where they live in tar-paper barracks heated with coal stoves and guarded by military police armed with rifles. The interned Japanese-Americans (two-thirds are U.S. citizens) lose an estimated $400 million in property of which Washington will repay $38.5 million (*see* 1959).

The Congress of Racial Equality (CORE), founded by University of Chicago students, will use passive resistance tactics pioneered by Mahatma Gandhi in India and employ sit-in techniques to end racial segregation and discrimination. James Leonard Farmer, 22, is national chairman (*see* 1961).

The Beveridge Plan advanced by English economist William Henry Beveridge, 63, is a comprehensive report embodying proposals for postwar "cradle-to-grave" social security in Britain. Master of Oxford's University College, Beveridge is chairman of Britain's Inter-Departmental Committee on Social Insurance and Allied Services (*see* National Health Service, 1946).

An Emergency Price Control Act voted by Congress January 30 gives the Office of Price Administration (OPA) power to control prices (*see* 1941; 1946).

A General Maximum Price Regulation Act voted by Congress April 28 freezes 60 percent of U.S. food items at store-by-store March price levels.

Congress appropriates $26.5 billion for the Navy February 2, bringing total war costs since June 1940 to $116 billion. Another $42.8 billion is appropriated for the Army June 30, and by July 1 the nation is spending $150 million per day.

The Revenue Act of 1942 is the largest in U.S. history. Designed to produce $9 billion, it contains provision for a 5 percent victory tax on all incomes above $624 for the duration of the war.

President Roosevelt creates an Office of Economic Stabilization and asks Supreme Court Justice James F. Byrnes, 63, to resign and become its director. An order issued by Byrnes October 27 limits salaries to $25,000 per year.

More than 3.6 million American men remain unemployed, but the number has fallen from 9.5 million men and women in 1939. The ranks of the unemployed dwindle rapidly as war plants, shipyards, oil fields, and recruiting offices clamor for manpower.

The U.S. Mint issues zinc-coated steel pennies under an Emergency Coinage Act designed to release copper for war production (silver nickels are issued to free nickel for the war effort). By the time the last wartime coins are issued in 1945, they will be made from salvaged shell casings.

The last new U.S. automobile that will be produced until 1945 rolls off the Ford assembly line February 10 as auto plants turn to producing tanks, Jeeps, aircraft, and other war matériel on cost-plus contracts under the direction of the War Production Board set up by executive order January 16 with former Sears, Roebuck executive Donald M. Nelson, 49, as WPB director.

A tire-rationing plan has been issued in January by the OPA (above), but experts warn that stronger measures are needed. "Unless corrective measures are taken

U.S. factories poured out a torrent of war matériel to help Allied troops defeat the Axis powers.

immediately [to conserve rubber], this country will face both a military and a civilian collapse," say Wall Street sage Bernard Baruch, now 72, Harvard University president James Bryant Conant, 49, and Chicago University physicist Arthur H. Compton.

Far Eastern sources now cut off by the Japanese have supplied all but 3 percent of U.S. crude rubber needs and the nation's stockpile of rubber is only 660,000 tons. Civilian consumption has been running at the rate of between 600,000 and 700,000 tons per year, and U.S. synthetic rubber plants produced only 12,000 tons last year (*see* 1940).

Mexico agrees in September to sell its entire production of guayule for U.S. rubber production until the end of 1946.

Nationwide gasoline rationing is ordered in September, chiefly to reduce rubber consumption. Rationing has begun in May on the East Coast, where U-boat sinkings have reduced tanker shipments, but more than 200 congressmen have requested and received X stickers entitling them to unlimited supplies of gasoline. All U.S. motorists are assigned A, B, or C stickers as of December 1. Those with A stickers are allowed 4 gallons per week, which will later be reduced to 3 gallons, but nearly half of all motorists obtain B or C stickers, which entitle them to supplementary rations because their driving is essential to the war effort, or public health, or for similar reasons. Truckers receive T stickers permitting them unlimited amounts of gasoline or diesel fuel, but pleasure driving is banned and a 35-mile-per-hour speed limit established on highways.

New York's East River Drive opens May 26 with ceremonies led by Mayor La Guardia. The roadbed between 23rd and 34th streets is filled with bricks and rubble left from air raids on London and donated by the British, but the new drive that cuts Manhattan off from the East River is poorly drained and will flood in heavy rains.

The Alcan International Highway opens officially November 30. Some 30,000 U.S. and Canadian troops and civilians have built the 1,500-mile road in 8 months to link Alaska with Alberta.

The French Line passenger ship S.S. *Normandie* that went into service in 1935 has been gutted by fire and has capsized in February at her New York pier while being converted for troop transport service. Renamed the U.S.S. *Lafayette* January 1 after U.S. seizure, she will be righted, towed away, and scrapped.

The Liberty ship *Robert E. Peary* launched November 9 by Kaiser Shipyards on the Pacific Coast is the first of some 1,460 Liberty-class cargo vessels and other ships that Kaiser will build during the war. She is delivered November 12—7½ days after her keel was laid (*see* 1941).

Henry Kaiser has built the first steel mill on the Pacific Coast to produce steel for his shipyards, and he has built a magnesium plant as well. He manufactures aircraft, Jeeps, and other war matériel in plants he has acquired for the purpose (*see* aluminum, 1945; Kaiser Foundation Hospital, below).

The IL-2 Stormovik dive bomber designed by Soviet aircraft designer Sergei Vladimirovich Ilyushin, 48, goes into volume production for use against the U.S.S.R.'s German invaders. Some 36,000 of the heavily armored planes will be produced, and it will serve as the Soviet equivalent of the U.S. DC-3 introduced in 1936.

The entire world supply of penicillin is barely enough to cure one serious case of meningitis (*see* 1943; Florey, 1940).

French medical researcher André Loubatiére pioneers oral drugs for diabetics with his finding that sulfa drugs are closely related chemically to para-amino benzoic acid (PABA) and produce a lowering of blood sugar levels (*see* Domagk, 1935).

The Kaiser Foundation Health Plan, a pioneer health maintenance organization, has its beginnings in a 54-bed hospital at Oakland, Calif., dedicated by U.S. industrialist Henry Kaiser (above). The Foundation will often be called Kaiser-Permanente after the creek in the Santa Cruz Mountains where Kaiser built his first cement plant.

The Office of War Information (OWI) created by President Roosevelt June 13 is headed by CBS commentator and novelist Elmer Holmes Davis, 52.

Yank begins publication June 17 under the direction of Col. Egbert White, 48, who puts out the weekly armed forces magazine with help from *Chattanooga Times* general manager Adolph Shelby Ochs, *Saturday Evening Post* editor Robert Martin Fuoss, and *Liberty* magazine art editor Alfred Strasser. White worked as a sergeant on *Stars and Stripes* in 1918 and has risen through the ranks.

Negro Digest begins monthly publication at Chicago. Local publisher-insurance executive John Harold Johnson, 24, has worked his way up from office boy at the Supreme Liberty Life Insurance Co. while studying at the University of Chicago and Northwestern. The first issue has sales of 3,000, but circulation will soar to 100,000 when Eleanor Roosevelt writes a piece for its feature series "If I Were a Negro" (*see Ebony,* 1945).

Nonfiction *Victory Through Airpower* by Alexander de Seversky (*see* Republic Aviation, 1939); *Admiral of the Ocean Sea: A Life of Christopher Columbus* by Harvard history professor Samuel Eliot Morison, 55, who has made several voyages following the route of the Great Navigator. Morison is appointed historian of naval operations with the rank of lieutenant commander and will write a 15-volume *History of U.S. Naval Operations in World War II; The Spanish Labyrinth* by émigré English writer Gerald Brenan, 48, who was caught at Malaga when the Spanish Civil War erupted; "The Myth of Sisyphus," a philosophical essay by French man of letters Albert Camus, 29, who has joined the Resistance against Nazi occupation forces; *Generation of Vipers* by U.S. author Philip Gordon Wylie, 40, introduces the term "Momism" to describe emasculating American matriarchy.

Fiction *Flight to Arras* (*Pilote de la guerre*) by Antoine de Saint-Exupéry; *The Stranger* (*L'Etranger*) by Albert

1942 ***(cont.)*** Camus; *The Family of Pascal Duarte (La familia di Pascal Duarte)* by Spanish novelist Camilo José Cela, 26; *Go Down, Moses* by William Faulkner; *The Company She Keeps* by U.S. novelist Mary McCarthy, 30; *The Robber Bridegroom* (novella) by Eudora Welty; *Pied Piper* by English novelist Nevil Shute (Nevil Shute Norway), 43; *To Be a Pilgrim* by Joyce Cary; *The Moon Is Down* by John Steinbeck; *The Just and the Unjust* by James Gould Cozzens; *Breakfast With the Nikolides* by Rumer Godden; *Put Out More Flags* and *Work Suspended* by Evelyn Waugh; "The Catbird Seat" (story) by James Thurber in the November 14 *New Yorker*.

Poetry *And Suddenly It's Night* (*Ed e subito sera*) by Italian poet Salvatore Quasimodo, 41; *The Man Without a Way (Mennen a tan vag)* by Swedish poet Eric Lindegren, 32; *Blood for a Stranger* by U.S. poet Randall Jarrell, 28.

Painting *Patience* by Georges Braque; *L'Oiseau bleu* by Pierre Bonnard; *Male and Female* by Jackson Pollock. Walter Sickert dies at Bath January 23 at age 81; Grant Wood dies at Iowa City February 12 at age 49.

Theater *The Apollo of Bellac* (*L'Apollon de Bellac*) by Jean Genet 6/16 at Rio de Janeiro; *The Eve of St. Mark* by Maxwell Anderson 10/7 at New York's Cort Theater, with Aline MacMahon, William Prince, Matt Crowley, Martin Ritt, 291 perfs.; *Without Love* by Philip Barry 11/10 at New York's St. James Theater, with Katharine Hepburn, Elliot Nugent, Audrey Christie, 113 perfs.; *The Skin of Our Teeth* by Thornton Wilder 11/18 at New York's Plymouth Theater, with E. G. Marshall, Florence Eldredge, Fredric March, Tallulah Bankhead, Montgomery Clift, 359 perfs.

Films David Hand's *Bambi* with Walt Disney animation; Michael Curtiz's *Casablanca* with Humphrey Bogart, Ingrid Bergman; Lucas Demare's *The Gaucho War* (*La Guerra Gaucha*); Noël Coward and David Lean's *In Which We Serve* with Coward, John Mills, Celia Johnson, Michael Wilding (score by Coward); Orson Welles's *The Magnificent Ambersons* with Joseph Cotten, Tim Holt, Anne Baxter, Dolores Costello, Agnes Moorehead; George Stevens's *The Talk of the Town* with Jean Arthur, Ronald Colman, Cary Grant. Also: Raoul Walsh's *Gentleman Jim* with Errol Flynn, Alexis Smith; Stuart Heisler's *The Glass Key* with Brian Donlevy, Veronica Lake, Alan Ladd; Robert Stevenson's *Joan of Paris* with Michelle Morgan, Paul Henreid; Billy Wilder's *The Major and the Minor* with Ginger Rogers, Ray Milland; Elliot Nugent's *The Male Animal* with Henry Fonda, Olivia de Havilland; William Wyler's *Mrs. Miniver* with Greer Garson, Walter Pidgeon; Irving Rapper's *Now, Voyager* with Bette Davis, Paul Henreid, Claude Rains; Luchino Visconti's *Ossessione* with Massimo Girotti, Clara Calamia; Michael Powell and Emeric Pressburger's *One of Our Aircraft Is Missing* with Godfrey Tearle, Eric Portman; Preston Sturges's *The Palm Beach Story* with Claudette Colbert, Joel McCrae, Rudy Vallee; Mervyn LeRoy's *Random Harvest* with Ronald Colman, Greer Garson; Ernst Lubitsch's *To Be or Not To Be* with Jack Benny, Carole Lombard (who has died in a plane crash January 16 at age 32); George Stevens's *Woman of the Year* with Katharine Hepburn, Spencer Tracy.

John Barrymore dies at Hollywood May 29 at age 63.

Hollywood musicals Michael Curtiz's *Yankee Doodle Dandy* with James Cagney as George M. Cohan (who dies at New York November 5 at age 64); Busby Berkeley's *For Me and My Gal* with Judy Garland, Gene Kelly; David Butler's *The Road to Morocco* with Bing Crosby, Bob Hope, Dorothy Lamour, music by Jimmy Van Heusen, lyrics by Johnny Burke, songs that include "Constantly," "Moonlight Becomes You," and the title song; George Marshall's *Star-Spangled Rhythm* with Bing Crosby, Bob Hope, Dorothy Lamour, Ray Milland, Veronica Lake, Susan Hayward, Alan Ladd, Paulette Goddard, Cecil B. DeMille, songs that include "That Old Black Magic" by Harold Arlen, music by Johnny Mercer; William A. Seiter's *You Were Never Lovelier* with Fred Astaire, Rita Hayworth, music by Jerome Kern, lyrics by Johnny Mercer, songs that include "Dearly Beloved," "I'm Old Fashioned," and the title song; Victor Schertzinger's *The Fleet's In* with William Holden, Dorothy Lamour, Betty Hutton, Eddie Bracken, music by Schertzinger, lyrics by Johnny Mercer, songs that include "Arthur Murray Taught Me Dancing in a Hurry" and "Tangerine"; Irving Cummings's *Springtime in the Rockies* with Betty Grable, Carmen Miranda, Cesar Romero, Jackie Gleason, Charlotte Greenwood, songs that include "I Had the Craziest Dream" by Harry James, lyrics by Mack Gordon; Mark Sandrich's *Holiday Inn* with Bing Crosby, Fred Astaire, music and lyrics by Irving Berlin, songs that include "White Christmas," a holiday number that will break all marks for record sales.

First performances Diversions on a Theme for Piano for Left Hand and Orchestra by Benjamin Britten 1/16 at Philadelphia's Academy of Music (*see* Wittgenstein, 1917); Imaginary Landscape No. 3 by U.S. composer John Milton Cage, Jr., 29, 3/1 at Chicago (scored for electric oscillator, buzzers of variable frequency, Balinese gongs, generator whine, coil, tin cans, marimba); Symphony No. 7 *(Leningrad)* by Dmitri Shostakovich (who was airlifted out of the beleaguered city last year) 3/1 at Moscow; Second Essay for Orchestra by Samuel Barber 4/16 at New York's Carnegie Hall; *Lincoln Portrait* (Symphonic Poem) by Aaron Copland, *Mark Twain* by Jerome Kern, and *The Mayor La Guardia Waltzes* by Virgil Thomson 5/14 at Cincinnati; *Bachianos Brasileiras* No. 9 for Orchestra by Heitor Villa-Lobos 6/6 at New York; *Choros* No. 6, No. 9, and No. 11 by Villa-Lobos 7/5 at Rio de Janeiro.

Ballet *Romeo and Juliet* 4/6 at New York's Metropolitan Opera House, with Jerome Robbins, Hugh Laing, Antony Tudor, Alicia Markova, Sono Osato, music by the late Frederick Delius, choreography by Tudor; *Pillar of Fire* 4/8 at New York's Metropolitan Opera House, with Hugh Laing, Antony Tudor, Nora Kaye (as Hagar), music by Arnold Schoenberg, choreography by Tudor; *Rodeo* 10/16 at New York's Metropolitan Opera House, with Agnes George de Mille, 37, music by

Aaron Copland, choreography by de Mille; *Metamorphoses* 11/25 at New York's City Center, with Tanaquil LeClerq, Todd Bolender, music by Paul Hindemith, choreography by George Balanchine; *Gayeneh* 12/9 at Molotov, with music (including a "Saber Dance") by Armenian Soviet composer Aram Khatchaturian, 39.

Broadway musicals *Star and Garter* 1/24 at the Music Box Theater, with Bobby Clark, Gypsy Rose Lee, Georgia Sothern in an extravaganza mounted by showman Mike Todd, songs that include Harold Arlen's "Blues in the Night," 609 perfs.; *By Jupiter* 6/2 at the Shubert Theater, with Ray Bolger, music by Richard Rodgers, lyrics by Lorenz Hart, 427 perfs.; *This Is the Army* 7/24 at the Broadway Theater, with an all-soldier cast for the benefit of the Army Emergency Relief Fund, music, book, and lyrics by Irving Berlin, songs that include "I Left My Heart at the Stage-Door Canteen," the title song, and songs from Berlin's 1918 musical *Yip-Yip-Yaphank,* 113 perfs.

Popular songs "Warsaw Concerto" by English composer Richard Addinsell, 38; "When the Lights Go on Again" by Eddie Seiler, Sol Marcus, Bennie Benjemen; "Praise the Lord and Pass the Ammunition" by Frank Loesser; "Don't Get Around Much Anymore" by Duke Ellington, lyrics by Bob Russell; "A String of Pearls" by Jerry Gray, lyrics by Eddie De Lange; "In the Blue of Evening" by Alfred A. D'Artega, lyrics by Tom Adair; "My Devotion" by Ric Holman and Johnny Napton; "There Will Never Be Another You" by Harry Warren, lyrics by Mack Gordon (for the film *Iceland);* "I've Heard That Song Before" by Jule Styne, lyrics by Sammy Cahn (for the film *Youth on Parade);* "Serenade in Blue" by Harry Warren, lyrics by Mack Gordon; "Who Wouldn't Love You?" by Carl Fischer, lyrics by Bill Carey; "Jingle Jangle Jingle" by Joseph J. Lelley, lyrics by Frank Loesser; "One Dozen Roses" by bandleader Dick Jergens and Walter Donovan, lyrics by Roger Lewis and Country Washburn.

The first "golden record" is presented February 10 to Glenn Miller, whose 1941 hit "Chattanooga Choo Choo" has been sprayed with gold by RCA-Victor in recognition of the record's having sold more than 1 million copies. Miller disbands his orchestra late in the year, will join the Army Air Force as a major, organize a large orchestra, broadcast to troops in every theater of the war, but disappear in mid-September 1944 on a flight from England to Paris.

Frederick R. "Ted" Schroeder, 20, wins in men's singles at Forest Hills, Pauline Betz, 22, in women's singles.

The St. Louis Cardinals win the World Series by defeating the New York Yankees 4 games to 1.

The Supreme Court rules Nevada divorces valid throughout the United States.

New York's City Council outlaws pinball machines in a law that takes effect in January. Mayor La Guardia has pushed through the law, saying schoolboys are stealing nickels and dimes from their mothers' pocketbooks to feed the machines that are feeding the underworld. The mayor smashes machines with a sledgehammer for news photographers; most of the confiscated machines are dumped in the ocean.

Boston's Cocoanut Grove nightclub becomes a roaring holocaust Saturday night, November 28. The club is packed with nearly 1,000 patrons of whom 492 die and 270 are injured.

Camp David outside Thurmont, Md., has its beginnings in the Shangri-La retreat built for President Roosevelt on a 134-acre site cleared 3 years ago in Catoctin Mountain Park, a 90-minute drive from the White House. When Gen. Eisenhower becomes president in 1953 he will rename the hideaway after a grandson; later presidents will use Camp David to varying degrees, sometimes for summit conferences.

Americans cultivate "Victory Gardens" in backyards and communal plots as vegetables become scarce, especially in California, where two-thirds of the vegetable crop has been grown by Japanese-Americans (*see* above). Forty percent of all U.S. vegetables are produced in nearly 20 million Victory Gardens but the number will fall as interest flags.

Britons "Dig for Victory" and raise vegetables in backyard gardens.

Soviet geneticist Nikolai Vavilov dies of malnutrition in the prison where he has been confined since 1940. Vavilov's anti-intellectual rival T. D. Lysenko advises Josef Stalin to switch from traditional grain crops to millet, which requires less moisture.

Florida becomes the leading U.S. producer of oranges, passing California.

Famine kills some 1.6 million Bengalese as fungus disease ruins the rice crop near Bombay.

Oxfam is founded to fight world famine by Oxford University classical scholar Gilbert Murray, 76.

Millions of Europeans live in semistarvation as German troops cut off areas in the Ukraine and North Caucasus that have produced half of Soviet wheat and pork production. Food supplies fall to starvation levels in German-occupied Greece, Poland, and parts of Yugoslavia.

U.S. sugar rationing begins in May after consumers have created scarcities by hoarding 100-pound bags and commercial users have filled their warehouses. One-sixth of U.S. sugar supplies have come from the Philippines, which are in Japanese hands, U.S. householders are asked by ration boards to state how much sugar they have stockpiled, ration stamps are deducted to compensate, the weekly ration averages 8 ounces per person but will rise to 12 ounces.

Hoarding of coffee leads to coffee rationing, which begins in November.

Forty-two percent of U.S. white bread is enriched with B vitamins and iron, up from 30 percent last year. Louisiana and South Carolina enact laws requiring that corn staples be enriched.

British flour extraction rates rise to 85 percent to stretch wheat supplies. The higher extraction rate raises

1942 *(cont.)* levels of vitamins and minerals in British bread (which remains unrationed; *see* 1917).

U.S. troops stationed in Britain are forbidden to drink local milk, which is not pasteurized. Many diseases that affect Britons are ascribed to unpasteurized milk *(see* 1900).

K rations, packed for U.S. troops by Chicago's Wrigley Co., contain "defense" biscuits and compressed Graham biscuits, canned meat or substitute, three tablets of sugar, four cigarettes, and a stick of chewing gum in each combat ration (*see* Wrigley, 1893). The breakfast ration includes also a fruit bar and soluble coffee *(see* below), the dinner ration flavored and plain dextrose tablets and a packet of lemon-juice powder, the supper ration bouillon powder and a bar of concentrated chocolate called "Ration D."

Instant Maxwell House coffee has its beginnings in the soluble coffee for K rations (above) that has been developed for the armed forces by General Foods at its Hoboken, N.J., Maxwell House coffee factory. The coffee will be introduced to the public after the war as Instant Maxwell House *(see* Nescafé, 1938).

U.S. soft drink companies are allowed enough sugar to meet quotas of 50 to 80 percent of the production attained in the base year 1941 but with no limit on sales to the armed forces.

Kellogg's Raisin Bran, introduced by Kellogg Co., is 10.6 percent sugar *(see* 1928; 1950).

Milk deliveries in most U.S. cities are reduced to alternate days; in some cities horse-drawn milk wagons reappear.

1943 The tide of war turns against the Axis in North Africa, the Pacific, Italy, and on the Russian front.

Tripoli falls to the British Eighth Army January 23. Axis forces in Tunisia retreat with the British in hot pursuit.

The Casablanca Conference attended by President Roosevelt, Prime Minister Churchill, Gen. Giraud, and Gen. de Gaulle ends January 27 with the appointment of Gen. Eisenhower as commander of unified forces in North Africa.

Berlin rushes reinforcements to Tunisia, the Germans take Kasserine Pass February 22, but U.S. troops retake the pass 4 days later. Tunis falls to the British May 7, Bizerte falls to the Americans the same day, and Axis resistance ends in North Africa.

British paratroopers and U.S. airborne troops invade Sicily July 9 and 10. More than 500 U.S. bombers raid Rome July 19.

Benito Mussolini and his cabinet resign under pressure July 25. Marshal Pietro Badoglio, 72, takes over.

Allied armies take Messina August 17, cross the Straits of Messina, and invade Southern Italy as representatives of the new Badoglio regime (above) sign an armistice with Allied officers at Algiers.

Italy has surrendered unconditionally, the Allied high command announces September 8, but German forces in Italy resist the Allied advance.

The U.S. Fifth Army lands at Salerno September 9 and sustains heavy losses. The Americans take Naples October 1 as German forces seize Rome and other major Italian cities.

Anzac (Australian-New Zealand-Canadian) and U.S. forces take the southeastern tip of New Guinea from the Japanese January 22, assuring the safety of Australia from Japanese invasion.

Allied forces take Guadalcanal in the Solomon Islands February 8 after heavy fighting.

The Battle of the Bismarck Sea in early March ends with U.S. Liberator and Flying Fortress bombers sinking an estimated 21 Japanese transports bound for New Guinea with 15,000 troops.

Allied forces take the Japanese air base at Munda in the Solomons August 5 and destroy more than 300 Japanese planes 2 weeks later in attacks on the Wewak airfield in New Guinea.

U.S. intelligence intercepts and decodes Japanese messages about an inspection trip by Admiral Yamamoto out of Rabaul, P-38s shoot down Yamamoto's plane August 18, and Japan loses the man who opposed the Axis alliance yet triumphed at Pearl Harbor.

U.S. troops that landed at Sitka in the Aleutians in May retake Kiska August 15.

Heavy bombing of industrial centers in Germany and occupied France has begun on a continuous basis in January.

Soviet troops relieve Leningrad's 17-month siege in February, but the Germans will blockade the narrow corridor to the city more than 1,200 times in the next year and starvation will continue.

The Germans lose 500,000 men in 3 months of heavy winter fighting as the Russians retake Kharkov and other key cities, but they mount a spring offensive and recapture Kharkov March 15.

The Kremlin quietly dissolves the Comintern in May in consideration of the wartime alliance between the Communists and Western allies (*see* 1935).

The Battle of Kursk that begins July 5 involves 6,000 German and Russian tanks and 4,000 planes. It ends after a week of heavy fighting in a victory for the Soviet Fifth Army, but while the Germans have lost 70,000 men, 2,000 tanks, 1,392 planes, and 5,000 vehicles the Russian losses are at least comparable.

Soviet forces retake Kharkov August 23 with help from increased Soviet industrial output and with U.S. war matériel including steel, industrial machinery, planes, and motor vehicles supplied via Archangel, Vladivostok, and the Persian Gulf.

Soviet forces retake Smolensk September 25 and Kiev on the Dnieper River November 6.

The Moscow Conference in late October establishes a European advisory commission on terms of German surrender, separation of Austria from Germany, and destruction of Italy's Fascist regime.

The United Nations has its beginnings in a brief congressional resolution drafted by freshman congressman J. William Fulbright, 38, (D. Ark.) (*see* agriculture, below; Fulbright awards, 1946).

The United Nations Relief and Rehabilitation Administration (UNRRA) is established November 9 by an agreement signed at Washington.

The Teheran Conference in late November discusses an Allied military landing in France. President Roosevelt, Prime Minister Churchill, and Premier Stalin agree to set up an advisory commission to study European problems.

Allied forces have landed on Bougainville in the Solomons November 1 following Allied bombing of Rabaul in New Britain since October 12.

U.S. troops take Tarawa and Makin in the Gilbert Islands in late November. Allied forces land on Cape Gloucester, New Britain, December 26.

France's Pas de Calais area is attacked December 24 by 3,000 Allied planes that include 1,300 from the U.S. Eighth Air Force.

The Battle of the Warsaw Ghetto that begins Passover Eve, April 18, ends 6 weeks later with 5,000 German troops killed and wounded, but 5,000 Jews have been killed defending themselves against German tanks and artillery. Some 500,000 Jews had been locked into an area that formerly accommodated half that number, and while thousands have escaped to join the Polish resistance, some 20,000 are deported to death camps such as Auschwitz (Oswiecim), Birkenau, Belzec, Chmelno, Maidenek, and Sobibor. Jews at Bialystock, Tarnow, and other Polish cities offer resistance, but few will survive the genocidal Nazi Holocaust.

The plan for the "extermination of the Jews" is well advanced, says SS *führer* Heinrich Himmler October 4 in a speech to his *Gruppenführer* (lieutenant generals), despite pleas to spare this or that "exceptional" Jew. The Nazi elite force will not be deflected from their objective, says Himmler, and while he says "we will never speak of it" in public, the destruction of the Jews will remain forever "an unwritten and never-to-be-written page of glory."

The SS in Denmark begins rounding up Danish Jews in October after having left them untouched since 1940, but the Danes help most of their Jewish compatriots escape to safety in Sweden.

President Roosevelt orders Secretary of the Interior Harold L. Ickes to take over strike-threatened soft coal mines. United Mine Workers boss John L. Lewis defies threats to send in troops, saying, "They can't dig coal with bayonets."

A West Virginia state law requiring schoolchildren to salute the flag or face expulsion is ruled invalid by the Supreme Court June 14 in the case of *West Virginia Board of Education v. Bernette.*

Federal troops quell a Detroit race riot June 22 after 34 have died in 2 days of disturbances that have involved thousands.

Harlem has a race riot August 1.

The $109 billion budget submitted to Congress by President Roosevelt January 11 has $100 billion earmarked for the war effort.

The president issues a "hold-the-line" anti-inflation order in April. The wholesale price index published monthly by the Labor Department's Bureau of Statistics since 1913 stands at 103.1; by August 1945 it will have risen only to 105.8, thanks to price controls.

A "Pay-As-You-Go" Current Tax Payment Act voted by Congress June 9 follows a plan proposed by R. H. Macy chairman Beardsley Ruml, 49. The act provides for income taxes on wages and salaries to be withheld by employers from paychecks (*see* Britain, 1944).

Ford Motor Company president Edsel Ford has died May 26 at age 49 and bequeathed 90 percent of his stock in the company to the Ford Foundation he set up in 1936. He has owned more than 40 percent of the stock and his will employs a tax avoidance scheme created by Wall Street investment banker Sidney Weinberg, 52, of Goldman, Sachs (*see* 1947).

The Nuffield Foundation for medical, scientific, and social research is created by British auto maker William Richard Morris, 66, Viscount Nuffield, who has made a fortune with his Morris Motors, Ltd.

A U.S. federal rent control law takes effect in November. The controls will expire in 1950 but will be continued by New York State and some other states.

The "Big Inch" pipeline to carry Texas oil to the eastern seaboard is completed, and the ban on nonessential driving is revoked in September following a dramatic decline in highway traffic deaths.

Chicago's first subway—5 miles long—is dedicated October 16. Rapid transit helps conserve rubber.

Penicillin is applied for the first time to the treatment of chronic diseases (*see* 1942; 1945).

Rutgers University microbiology professor Selman Abraham Waksman, 43, finds the actinomycete *Streptomyces giisens* in the soil of a field and the throat of a child and coins the word "antibiotic" (see streptomycin, 1944).

The medical establishment recognizes the "Pap" test for detecting cervical cancer after 15 years of work by Greek-American physician George Nicholas Papanicolaou, 60, who developed the vaginal smear test at Cornell in 1928 to diagnose the cancer that has been the leading cause of death among U.S. women. It is based on exfoliative cytology (microscopic study of cells shed by bodily organs), and within 20 years will have reduced cancer of the cervix to number three as a cause of death among U.S. women.

The Infant and Child in the Culture of Today by Yale University psychoclinician Arnold L. Gesell, 63, and his assistant Frances Ilg says the infant should not be sub-

1943 ***(cont.)*** jected to totalitarian rule but rather given autonomy (*see* Spock, 1946).

Swiss chemist Albert Hofmann of Sandoz A.G. discovers by accident the hallucinogenic properties of LSD (lysergic acid diethylamide) when he is "seized by a peculiar sensation of vertigo and restlessness. . . . With my eyes closed, fantastic pictures of extraordinary plasticity and intensive color seemed to surge toward me. After two hours this state gradually wore off." Hofmann has swallowed a derivative of ergot, he has spectacular hallucinations when he ingests a quarter milligram of the compound, and his findings will spur research in psychopharmacology (*see* ergot, 857 A.D.; Gaddum, 1953).

Zenith Radio introduces a $40 electric hearing aid. Eugene McDonald has lost the hearing in one ear following an automobile accident and found that the only available hearing aids sell for $150 to $200 (*see* 1935; 1952).

American Broadcasting Co. (ABC) is created by Life Savers millionaire Edward Noble (*see* 1920). The Federal Communications Commission has ordered NBC to divest itself of one of its two radio networks. Noble writes a check for $8 million to acquire the 16-year-old Blue Network, turns it into ABC, and sets out to rival NBC and CBS.

Cartoonist Bill Mauldin is shipped overseas and joins the Mediterranean staff of the U.S. Army newspaper *Stars and Snipes.* William Henry Mauldin, 21, will cover campaigns in Italy, France, and Germany, and the miseries of his grimy G.I. characters Willie and Joe will reflect those of countless enlisted men enduring the hardships of war.

Nonfiction *Christianity and Democracy* by French philosopher Jacques Maritain, 61, who became a Roman Catholic in 1906 and champions the cause of liberal Catholicism; *Being and Nothingness (L'Etre et le Néant)* by Jean Paul Sartre, who was captured at the fall of France in June 1940, escaped in April 1941, and has been engaged in the resistance against occupation forces.

Fiction *Arrival and Departure* by Arthur Koestler; *The Man Without Qualities* (*Der Mann ohne Eigenschaften)* by the late Robert Musil, who fled Germany after the 1938 Anschluss, was refused refuge by the United States, and died last year in Switzerland at age 62, leaving his novel unfinished; *Dead Fires (Fogo Morto)* by Brazilian novelist José Lins do Rego; *The Fountainhead* by Russian-American novelist Ayn Rand, 38; *The Big Rock Candy Mountain* by Wallace Stegner; *Perelandra* by C. S. Lewis; *The Apostle* by Sholem Asch.

Juvenile *The Little Prince* (*Le Petit Prince*) by Antoine de Saint-Exupéry who has joined the Free French forces in North Africa; *Johnny Tremain: A Novel for Young and Old* by U.S. author Esther Forbes, 52.

Poetry *Four Quartets* by T. S. Eliot; *Western Star* by Stephen Vincent Benét who dies at New York March 13 at age 44.

Bill Mauldin's sardonic cartoons deglamorized war and found humor at the expense of army brass.

Painting *Broadway Boogie-Woogie* by Piet Mondrian; *Odysseus and Calypso* by Max Beckmann; *Welders* by Ben Shahn; *The She-Wolf* by Jackson Pollock; *Queen of Hearts* by Willem de Kooning; *The Thanksgiving Turkey* by Grandma Moses. Chaim Soutine dies August 9 at age 49 while undergoing surgery in a Paris hospital; Marsden Hartley dies at Ellsworth, Me., September 2 at age 66.

Theater *The Patriots* by Sidney Kingsley 1/29 at New York's National Theater, with Madge Evans, Francis Reid, 157 perfs.; *The Good Woman of Setzuan (Der gute Mensch von Sezuan*) by Bertolt Brecht 2/4 at Zurich's Schauspielhaus; *Harriet* by F. Ryerson and Cohn Claues 3/3 at New York's Henry Miller Theater, with Helen Hayes, 377 perfs.; *Kiss and Tell* by F. Hugh Herbert 3/17 at New York's Biltmore Theater, with Joan Caulfield as Corliss Archer, Robert White as Dexter Perkins, 103 perfs.; *Tomorrow the World* by U.S. playwrights James Gow and Arnaud d'Usseau 4/14 at New York's Ethel Barrymore Theater, with Ralph Bellamy, Shirley Booth, Skippy Homeier, 500 perfs.; *Present Laughter* by Noël Coward 4/29 at London's Haymarket Theatre, with Coward, Joyce Carey, 38 perfs. (plus 528 beginning 4/16/47); *This Happy Breed* by Nöel Coward 4/30 at London's Haymarket Theatre, with Coward, Joyce Carey, 38 perfs. (alternating with *Present Laughter*, above); *Galileo* (*Leben des Galilei*) by Bertolt Brecht 8/9 at Zurich's Schauspielhaus; *The Voice of the Turtle* by John Van Druten 12/8 at New York's Morosco Theater, with Margaret Sullavan,

Audrey Christie, Elliott Nugent, 1,557 perfs.; *While the Sun Shines* by Terence Rattigan 12/24 at London's Globe Theatre, 1,154 perfs.

New York Times theatrical caricaturist Al Hirschfeld, 40, introduces the name of his newborn daughter Nina into his drawings and begins a puzzle game that will continue for more than 35 years. Hirschfeld's drawings have appeared on the drama pages of the *Times* for 20 years, and his circus sketch includes a poster with the legend "Nina the Wonder Child."

Films René Clair's *Forever and a Day* with some 80 British stars; Michael Powell and Emeric Pressburger's *The Life and Death of Colonel Blimp* with Roger Livesey, Deborah Kerr; William Wellman's *The Ox-Bow Incident* with Henry Fonda, Dana Andrews; Henry King's *The Song of Bernadette* with Jennifer Jones. Also: Edmund Goulding's *The Constant Nymph* with Charles Boyer, Joan Fontaine, Alexis Smith; Henri-Georges Clouzot's *Le Corbeau* with Pierre Fresnay, Ginette Leclerc; Carl Theodor Dreyer's *Day of Wrath* with Thorkild Roose, Lisbeth Movin; Billy Wilder's *Five Graves to Cairo* with Franchot Tone, Anne Baxter, Akim Tamiroff, Erich von Stroheim; Sam Wood's *For Whom the Bell Tolls* with Gary Cooper, Ingrid Bergman, Katina Paxinou; Lewis Seiler's *Guadalcanal Diary* with Preston Foster, Lloyd Nolan; Ernst Lubitsch's *Heaven Can Wait* with Gene Tierney, Don Ameche; Clarence Brown's *The Human Comedy* with Mickey Rooney, Frank Morgan; Sergei Eisenstein's *Ivan the Terrible (I)* with Nikolai Cherkasov, score by Serge Prokofiev; Michael Curtiz's *Mission to Moscow* with Walter Huston, Ann Harding; Richard Wallace's *A Night to Remember* with Loretta Young, Brian Aherne; Zoltan Korda's *Sahara* with Humphrey Bogart, Rex Ingram, J. Carrol Naish; Akira Kurosawa's *Sanshiro Sugata*; Alfred Hitchcock's *Shadow of a Doubt* with Teresa Wright, Joseph Cotten; Herman Shumlin's *Watch on the Rhine* with Bette Davis, Paul Lukas.

Hollywood musicals Edward H. Griffith's *The Sky's the Limit* with Fred Astaire, Joan Leslie, Robert Benchley, songs that include "My Shining Hour" and "One for My Baby" by Harold Arlen, lyrics by Johnny Mercer; Edward Sutherland's *Dixie* with Bing Crosby, Dorothy Lamour, songs that include "Sunday, Monday or Always" by Jimmy Van Heusen, lyrics by Johnny Mercer.

First performances *Freedom Morning* Symphonic Poem for Orchestra, Tenor, and Chorus by Marc Blitzstein 4/28 at London, with Richard Hayes as soloist and a chorus of 200 black U.S. aviation engineers; Symphony No. 8 by Dmitri Shostakovich 11/4 at the Moscow Conservatory; Symphony No. 4 by Howard Hanson 12/3 at Boston's Symphony Hall.

Ballet *Mother Goose Suite* in October at New York's Central High School of the Needle Trades Auditorium, with music by the late Maurice Ravel.

Broadway musicals *Something for the Boys* 1/7 at the Alvin Theater, with Ethel Merman, book by Harold and Dorothy Fields, music and lyrics by Cole Porter, songs that include "Hey, Good Lookin'" 422 perfs.; *Oklahoma!* 3/31 at the St. James Theater, with Alfred Drake, Celeste Holm, Howard da Silva, a book based on the 1931 Broadway play *Green Grow the Lilacs*, music by Richard Rodgers, lyrics by Oscar Hammerstein II, songs that include "Oh, What a Beautiful Morning," "People Will Say We're in Love," "Kansas City," "I Cain't Say No," "Pore Jud," "The Surrey with the Fringe on Top," and the title song (lyrics by Otto Harbach), 2,212 perfs.; *The Ziegfeld Follies* 4/1 at the Winter Garden Theater, with Milton Berle, music by Ray Henderson with music by Harold Rome interpolated, lyrics by Jack Yellen and others, 553 perfs.; *One Touch of Venus* 10/7 at the Imperial Theater, with Mary Martin, Kenny Baker, music by Kurt Weill, libretto by S. J. Perelman and Ogden Nash, lyrics by Nash, songs that include "Speak Low," "West Wind," 567 perfs.; *Carmen Jones* 12/2 at the Broadway Theater, with Muriel Smith, drummer Cosy Cole, music by Georges Bizet, book and lyrics by Oscar Hammerstein II, 503 perfs.

Popular songs "You'd Be So Nice to Come Home To" by Cole Porter (for the film *Something to Shout About*); "Do Nothin' Till You Hear from Me" by Duke Ellington, lyrics by Bob Russell; "Comin' in on a Wing and a Prayer" by Jimmy McHugh, lyrics by Harold Adamson; "Mairzy Dotes" by Milton Drake, Al Hoffman, Jerry Livingston; "How Many Hearts Have You Broken?" by Al Kaufman, lyrics by Marty Symes; "San Fernando Valley" by Gordon Jenkins (title song for film); "They're Either Too Young or Too Old" by Arthur Schwartz, lyrics by Frank Loesser (for the film *Thank Your Lucky Stars*); "No Love, No Nothin'," by Harry Warren, lyrics by Leo Robin (for the film *The Gang's All Here*); "I Couldn't Sleep a Wink Last Night" by Jimmy McHugh, lyrics by Harold Adamson (for the film *Higher and Higher*); "Let's Get Lost" and " 'Murder,' He Says" by Jimmy McHugh, lyrics by Frank Loesser (for the film *Happy Go Lucky*); "Tico-Tico" by Brazilian composer Zequinha Abreu, lyrics by Aloysio Oliveira; "Besame Mucho" by Mexican songwriter Consuelo Velazquez.

Count Fleet wins the Kentucky Derby and goes on to win U.S. racing's Triple Crown but injures his leg coming down the stretch at Belmont Park. He is retired to stud.

U.S. Navy Lieut. Joseph R. Hunt, 24, wins in men's singles at Forest Hills, Pauline Betz in women's singles.

The New York Yankees win the World Series by defeating the St. Louis Cardinals 4 games to 1.

Americans are told to "use it up, wear it out, make it do or do without." Shoes are rationed beginning February 7, but while Britons are allowed only one pair per year, the U.S. ration is three pairs, only slightly below the 3.43 bought on average in 1941.

Sneakers are impossible to find; schoolboys have to buy shoes with reclaimed rubber soles that leave black marks on gymnasium floors.

Rubber, metal, paper, silk, and nylon are all collected for recycling, and housewives who wash and flatten tins

1943 ***(cont.)*** for recycling save kitchen fats to be exchanged for red ration points at the butcher's (*see* below).

The Pentagon, completed January 15 at Arlington, Va., is the world's largest office building and provides space for the vast and growing number of U.S. war administrators. The $83 million, five-story, five-sided building has 6.5 million square feet of floor space, 17 miles of corridors, and 7,748 windows.

The Jefferson Memorial is dedicated at Washington, D.C., on land reclaimed from the Potomac. The monument contains a 19-foot bronze statue of the nation's third president.

Rent controls (above) prevent landlords from raising rents by more than a small percentage each year and will be a factor in discouraging new housing construction in New York.

The Mexican volcano Paricutin emerges in a Michoacan cornfield in February and begins eruptions that will continue for 7 years. The new volcano buries some nearby populated areas under lava and ashes, threatens a wide area, and will grow to a height of 820 feet.

A United Nations Conference on Food and Agriculture at Hot Springs, Ark., from May 19 to June 3 provides for a UN Food and Agricultural Organization (FAO; *see* 1945).

DDT is introduced to fight insect pests that destroy U.S. crops (*see* Muller, 1939).

Mexican wheat production is spurred by research undertaken with support from the Rockefeller Foundation. Mexican agricultural output will increase by 300 percent in the next 25 years, while the country's population increases by 70 percent (*see* 1941; Borlaug, 1944).

Rinderpest kills most of Burma's livestock. Since animals provide most of the country's power for irrigation and cultivation, the Burmese rice crop declines, and little is left for export to Bengal (*see* 1944).

Japan has her worst rice crop in 50 years. The daily 1,500-calorie subsistence level cannot always be met.

U.S. meat rationing begins March 29, but the ration is 28 ounces per week, and meat production rises by 50 percent. Meat consumption actually rises to 128.9 pounds per capita on an annual basis as the wartime economic boom puts more money into working people's pockets and as meat prices are rolled back to September 1942 levels. G.I.'s are served 4.5 pounds of meat per week (navy men get 7 pounds), and creamed chipped beef is a staple of service chow lines as Washington takes 60 percent of prime and choice beef and 80 percent of utility grade beef.

An estimated 20 percent of U.S. beef goes into black market channels, bacon virtually disappears from stores, western cattle rustlers kill and dress beef in mobile slaughterhouses and sell the meat to packing houses, and wholesalers force butchers to buy hearts, kidneys, lungs, and tripe in order to get good cuts of meat.

Butchers upgrade meat, selling low grades at ceiling prices and at the ration point levels of top grade.

Less than one-fourth of Americans have "good" diets according to a nutrition study undertaken in 1941.

Texas and Alabama adopt bread-enrichment laws. Seventy-five percent of U.S. white bread is enriched with iron and some B vitamins, up from 30 percent in 1941.

Japan acts to prevent beriberi, which is disabling many civilians. The Tokyo government distributes no white rice and orders citizens to eat *haigamai* (brown rice) or *hichibuzuki mai* (70 percent polished rice), but many, if not most, Japanese laboriously hull their *hichibuzuki mai* to obtain the white, glutinous, nutritionally deficient rice they prefer.

U.S. butter consumption falls to 11 pounds per capita, down from 17 pounds in the 1930s, as Americans forgo some of their 4 ounces per week ration in order to save red stamps for meat. Butter is often unavailable, since most butterfat is employed to make cheese for Lend-Lease aid.

U.S. margarine is subject to a 10 percent federal tax if artificially colored (25¢ per pound if colored by the consumer or used uncolored). Millions of householders use vegetable dye to color their white margarine yellow, but as more consumers turn to margarine, Eleanor Roosevelt campaigns for a repeal of the margarine tax (*see* 1950).

Cheese is rationed at the rate of four pounds per week per capita but requires red stamps that may also be used for meat. Americans ate cheese at a rate of 2.5 pounds per week before the war, but home economists estimate that the country's most affluent one-third ate cheese at the rate of five pounds per week, while the poor ate scarcely one pound.

Sale of sliced bread is banned by Agriculture Secretary Claude Wickard in a move to hold down prices.

Flour, fish, and canned goods join the list of rationed foods, but coffee is derationed in July. Americans eat far better than do citizens of the other belligerent nations (Britons enjoy only about two-thirds of what Americans are allowed under rationing programs).

U.S. distillers produce alcohol for synthetic rubber production (*see* 1933). Liquor is generally scarce.

A Chinese Act signed by President Roosevelt December 17 repeals the Chinese Exclusion Acts of 1882 and 1902. It makes Chinese residents of the United States eligible for naturalization and permits immigration of an annual quota of 105 Chinese (*see* 1965).

U.S. chemist Russell Marker, 41, pioneers development of oral contraceptives with the discovery of a cheap source of the hormone progesterone in the barbasco plant that grows wild in Mexico. Using an extract from the root of this member of the Dioscorea family, Marker works in a Mexico City pottery shop, produces 2,000 grams of progesterone worth $80 per gram in just 2 months, and will join a small Mexican drug firm to form Syntex, S.A. Another group of chemists will buy him out but Syntex, S.A. will become a major supplier of raw materials for oral contraceptives (*see* Pincus, 1955; Enovid 10, 1960).

1944 Allied troops storm ashore 25 miles from Rome on Anzio beach January 22 after the start of a new assault on the Gustav Line to the south. The Germans have just withdrawn their patrols in the Anzio area to reinforce the Gustav Line, but Gen. J. P. Lucas, commanding the U.S. VI Corps, misses his chance to advance on Rome. He has his men dig foxholes for 2 days to consolidate their beachhead and the Germans rush in reinforcements, firing from the Alban Hills to pin down Allied troops. Allied bombers February 15 destroy the historic Benedictine abbey at Monte Cassino, 1702 feet above the main road from Naples to Rome. Gen. Frido von Senger und Etterlin has not used the abbey as an observation post but employs its ruins as a fortress. Thousands of Allied troops die at Monte Cassino before Polish forces take it in May.

D-Day, June 6, sees 176,000 Allied troops landing at Omaha Beach, Utah Beach, and other Normandy beaches under the supreme command of Gen. Eisenhower whose forces take Cherbourg June 6.

An intrepid German spy at Istanbul has stolen allied plans for Operation Overlord; they are in Nazi hands, but the German high command has been led to believe that the Normandy landings are a diversion to draw Hitler's panzer divisions south to Normandy while 30 divisions strike at Calais. Gen. Rommel tries to persuade Gen. von Runstedt and *der führer* that the Normandy landing is the real thing, Allied intelligence uses its knowledge of the secret German code to keep abreast of the split in German thinking, the flow of false information is redoubled, and Rommel does not receive the support that would have driven the invaders back into the sea.

D-Day on the Normandy beachheads cost many Allied lives, but the successful landings opened the road to Paris.

Soviet troops have recaptured Novgorod January 20, relieved Leningrad completely a week later, reached the border of prewar Poland in February, taken Krivoi Rog in the Ukraine February 22 after trapping 10 German divisions near Cherkassy, crossed the Dniester March 19, retaken Odessa April 10, and retaken Sevastopol May 9.

From 60,000 to 100,000 Germans are trapped at Sevastopol and surrender, but the Russians take no prisoners. Soviet troops round up 300,000 to 500,000 Crimean Tatars who welcomed the Germans and send them into exile in Central Asia as they clear the Ukraine of German invaders.

Minsk falls to the Russians July 3; 100,000 Germans are captured.

The RAF has bombed Berlin January 20, dropping 2,300 tons of bombs in an action that provoked protests February 9 in the House of Lords. U.S. high-level bombers have begun daylight attacks March 6, using the Norden bombsight first developed in 1927 and still a U.S. secret.

Buzz bombs with warheads of nearly 1 ton each begin falling on Britain September 8 as the Germans retaliate (*see* below).

Allied forces in the Pacific take Kwajalein Island February 6, attack the main Japanese Central Pacific base at Truk in mid-February, land on Eniwetok in the Marshalls February 17, and at Hollandia in New Guinea April 22. B-29 bombers attack Japan's home island of Kyushu June 15.

The Battle of the Philippine Sea June 19 ends with a loss of 402 Japanese planes, while only 27 U.S. planes are downed.

The U.S. Army Air Corps has nearly 80,000 aircraft by July and employs more than 2.4 million men.

Saipan in the Marianas falls to U.S. troops July 9 after Lieut. Gen. Yoshitsugo Saito and 31,000 troops have fought to the death in a 23-day battle that has cost 3,500 U.S. lives. Hundreds of Japanese civilians throw themselves and their children off cliffs at the island's northern tip. Japan's 473 carrier-based planes are outmatched by 956 U.S. planes and 3 of Japan's 9 carriers are sunk. Losing more than 400 planes and nearly that many flyers effectively ends her viability as an air power. New B-29 bombers can reach Tokyo from Saipan, and the island's conquest marks a turning point in the war. The Tokyo government falls July 18.

Iceland attains the status of an independent republic July 17. A sovereign state since 1918 but a dependency of Denmark, the country discovered by the Vikings in 874 has been occupied since May 1940 by British and then by U.S. forces to prevent seizure by Germany, but withdrawal of U.S. troops will not begin until the end of 1959.

Adolf Hitler survives an assassination attempt July 20 and continues to direct the *Wehrmacht* as it falls back before advancing Allied armies. Some of Hitler's generals have plotted against him, a bomb explodes, but the *führer* is unhurt.

1944 ***(cont.)*** Soviet troops cross the Curzon line in Poland July 23, the Kremlin recognizes the Lublin Committee of Polish Liberation in Moscow July 26 as the governing authority of a liberated Poland, Warsaw rises against the Germans August 1 on orders from Gen. Bor-Komorowski (Tadeusz Komorowski), 49, of the Polish underground, who responds to an appeal by Moscow Radio, but Soviet troops are unable to come to the support of Bor-Komorowski, the Germans crush the uprising and inflict heavy losses, and Bor-Komorowski and a few survivors are forced to surrender after 2 months of ferocious fighting.

Allied forces cross the Loire August 11, the U.S. Seventh Army lands in southern France August 15 and begins moving up the Rhine Valley, the Third Army of Gen. George S. Patton, Jr., reaches the Seine August 19, Allied forces exterminate the German Seventh Army in the Falaise Gap from August 13 to August 20, French troops retake Marseilles August 23 and liberate Paris August 25 after more than 4 years of German occupation. Gen. de Gaulle enters Paris August 25; the French provisional government moves there from Algiers 5 days later.

Romania surrenders to Soviet forces August 24, the Red Army enters Bucharest August 30, Moscow declares war on Bulgaria September 5, Soviet columns enter Sofia September 16, and Belgrade falls to Soviet and Yugoslav partisan forces October 20.

Antwerp falls to the Allies September 4, Brussels is liberated September 5, but British airborne troops landed September 17 at Eindhoven and Arnhem are unable to outflank the German Westwall defenses and sustain heavy losses before the survivors are withdrawn.

Buzz bombs (above) take an increasing toll in Britain despite efforts by Allied bombers to destroy launching sites and knock out the German factories producing the bombs. Designed by German rocket expert Wernher von Braun, the liquid-fueled missiles are called V-2 (*Vergeltungswaffe Zwei*, revenge weapon two) bombs; more than 1,000 will land in Britain, killing more than 2,700, maiming and injuring another 6,500.

British and Greek Partisan forces retake Athens October 13, and the Allies retake Aachen in the Netherlands October 21.

Guam in the Pacific has been retaken in August by Allied forces, which take Pelelieu in the Palaus in October and begin carrier-based raids on Formosa (Taiwan), where 300 Japanese planes are destroyed October 12.

Allied forces invade the Philippines October 20.

President Roosevelt wins reelection to a fourth term with 53 percent of the popular vote and 432 electoral votes. His Republican opponent is New York governor Thomas E. Dewey, 42, who receives 46 percent of the popular vote but fails to carry his home state and winds up with only 99 electoral votes. Roosevelt has dumped his vice president Henry A. Wallace because he is considered too liberal for southern voters, the president has been discouraged from naming WPB boss Donald Nelson as his running mate because of Nelson's womanizing, and so the vice president-elect is Sen. Harry S. Truman, 60, (Mo.) who has gained national recognition by chairing a special Senate "watchdog" committee to investigate possible profiteering on national defense projects.

The German battleship *Tirpitz* is sunk November 12.

A U.S. task force commanded by Admiral Daniel V. Gallery, 43, of the carrier *Guadalcanal* has seized the German submarine *U-505* June 4 off French West Africa in the first capture at sea of any foreign warship since 1815, obtaining valuable military secrets that included a German radio code and a new type of torpedo, but the coup is kept secret so that the Germans will believe the submarine has sunk.

Superfortress raids on Tokyo from Saipan begin November 24.

Strasbourg falls to the Allies November 24, but the Battle of the Bulge that begins December 16 takes a heavy toll of U.S. troops as the Germans launch an offensive in the Ardennes Forest of Belgium that will continue into January of next year.

"Nuts," says Gen. Anthony McAuliffe, 46, December 22 to German demands that he surrender his 101st Airborne Division at Bastogne, which he has held against overwhelming odds; he is relieved December 26 by units of Gen. Patton's Third Army headed by Col. Wendell Blanchard, 41, whose tanks have come 150 miles in 19 hours.

Soviet forces surround Budapest December 27.

Nazi troops gun down 335 Italians in the Fosse Ardeatine caves beside the Appian Way outside Rome March 24 in reprisal for the killing of 35 German soldiers; by partisans. Of the dead, 255 are civilians, 68 soldiers; the status of 12 will remain a mystery, but all are victims of Herbert Kappler, 37, the SS colonel who has ordered the massacre.

Nazi authorities jail Hungarian bishop Jozsef Mindszenty, 52, for prohibiting Roman Catholics to say a Mass and sing a government-ordered Te Deum "in thanksgiving for the successful liberation of the town [Budapest] from the Jews" (*see* 1948).

Amsterdam Jew Otto Frank and his family are betrayed to the Gestapo August 4 after more than 2 years in hiding and deported with eight others in the last convoy of cattle trucks to the extermination camp at Auschwitz. Frank's daughter Anne, 15, has her head shaved at Auschwitz (the Reich uses women's hair for packing round U-boat pipe joints and for other purposes), but is not gassed. She will be shipped to the Bergen-Belsen concentration camp and die there in March of next year, probably of typhus, but three notebooks left behind at 263 Prinsengracht in Amsterdam will be found to contain her diary chronicling the period in which she and her family hid from the Gestapo (*see* Broadway play, 1955).

An American cannot be denied the right to vote because of color, the Supreme Court decides April 3.

The case of *Smith v. Allwright* involves a Texas primary election.

An American Dilemma by Swedish sociologist Gunnar Myrdal, 44, explores the history of U.S. black-white relationships and examines the moral and psychological dilemmas posed by years of professing equality while practicing inequality.

Britain institutes a Pay-As-You-Earn (PAYE) income tax program beginning April 6 (*see* Ruml plan, 1943).

A United Nations Monetary and Financial Conference at Bretton Woods, N.H., in July establishes the International Bank for Reconstruction (World Bank) and International Monetary Fund as the Western powers look to postwar economic problems. The monetary system formulated at Bretton Woods pledges every participating country to keep its currency within a percentage point or two of an agreed dollar value, and the IMF is designed to provide moral suasion and credit to keep the system alive. The system will remain in effect until March 1973, when stability rules will be generally disregarded following a 1971 U.S. move unlinking the dollar from any gold value.

The $70 billion budget submitted to Congress January 10 is sharply reduced from last year's budget, but President Roosevelt observes in November that the war is costing the United States $250 million per day, and he opens a sixth War Loan Drive for $14 billion.

Theory of Games and Economic Behavior by Hungarian-American mathematician John (Janos) von Neumann, 40, and German-American economist Oskar Morgenstern, 42, presents a theory that can be applied to any situation in business, diplomacy, or military strategy in which a decision maker confronts a problem in which he is not in complete control of all the variables affecting the outcome. Von Neumann's 25-page paper "Toward a Theory of Games" appeared in 1928, Morgenstern approached the concept the same year in his first book on business forecasting, and the two met in 1939.

Road to Serfdom by Viennese-American economist Friedrich August von Hayek, 45, attacks Keynesian economic theories.

Textron, Inc., is founded by U.S. parachute maker Royal Little, 48, who who will turn it into the first business "conglomerate" in 1955. A nephew of rayon pioneer Arthur D. Little, Royal dropped out of Harvard after 1 year to serve as an officer in World War I, apprenticed himself without pay to learn silk manufacturing with Cheney Brothers (*see* 1838), built up his Atlantic Rayon Corp. through mergers, took over his New Hampshire competitor Suncook Mills last year, and now makes plans to create a giant that will dominate the textile industry (*see* 1948).

Liquid natural gas (LNG) bursts out of two East Ohio Gas Co. storage tanks at Cleveland October 20, creating a firestorm that kills 131 people; 2 million gallons of LNG flow down the streets and into the sewers, exploding at the slightest spark, gutting 29 acres of houses and stores, and creating anxiety about the safety of LNG.

Carbon monoxide fumes kill more than 100 Italians and possibly twice that many March 2 in a railway tunnel 2 miles from Balvano. Most of the victims are black-market operators who have been obtaining chocolate, cigarettes, and other scarcities (often from Allied troops) and trading them for eggs, meat, and olive oil from country people for sale in Naples.

A Federal Highway Act passed by Congress November 29 establishes a new U.S. National System of Interstate Highways. The arterial network of 40,000 miles is planned to reach 42 state capitals and to serve 182 of the 199 U.S. cities with populations above 50,000 (*see* 1955; 1956).

The first automatic, general-purpose digital computer is completed at Harvard University, where it has been built under the aegis of Engineering School mathematics professor Howard Hathaway Aiken, 44, with a $5 million grant from IBM (*see* Watson, 1924; Bush, 1930). The Harvard-IBM Mark I Automatic Sequence Controlled Calculator has 760,000 parts and 500 miles of wire, requires 4 seconds to perform a simple multiplication, 11 seconds for a simple division, and is subject to frequent breakdown (*see* ENIAC, 1946).

Quinine is synthesized by Harvard chemist Robert Burns Woodward and Columbia chemist William von Eggers Doering, both 26 (*see* 1820).

Dutch physician Willem J. Kolff, 33, devises the first kidney machine to filter noxious wastes out of the blood of patients with kidney disease. He uses the clinical apparatus in secret to save the lives of Dutch partisans in German-occupied Holland. A surgical operation is required to insert the machine's large tubes into an artery and vein, and the machine is used only for brief emergency treatment of waste-product buildup (*see* Scribner, 1960).

A typhus epidemic threatens Naples but widespread spraying of DDT kills the lice responsible (*see* Muller, 1939).

The first operation to save a "blue baby" is performed November 9 at Baltimore's Johns Hopkins Children's Hospital by surgeon Alfred Blalock, 45, who proceeds on the premise advanced by his colleague Helen Taussig, 46, that anoxemia can be cured by bypassing the pulmonary artery. Prevented by a congenital pulmonary artery defect from getting enough blood into their lungs, many infants have been born blue, grown progressively weaker, and either died or been doomed to chronic invalidism. The surgical technique developed by Blalock and Taussig will permit blue babies to live normal lives (and have normal color).

Selman Waksman introduces the antibiotic streptomycin (*see* 1943).

The G.I. Bill of Rights (Servicemen's Readjustment Act) voted by Congress June 22 will finance college educations for millions of U.S. war veterans.

The University of California opens a Santa Barbara campus (*see* 1919; 1965).

1944 *(cont.)* *Al-Akhbar* begins publication at Cairo. Twin publishers Aly and Mustapha Amin, 30, begin the weekly *Akhbar el Yom* in addition to their daily newspaper.

Parisien Liberé begins publication at Paris under the direction of advertising man Emilien Amauray, 33, who has started clandestine Resistance papers while serving as Vichy government director of propaganda favoring large families. The de Gaulle government has expropriated the daily *Le Petit Parisien* on charges of collaborating with the enemy and placed Amauray in charge of the 81-year-old paper; he has renamed it and will make it a commercial success with 22 regional editions and a daily circulation of more than 800,000.

Le Monde begins publication at Paris December 19 to succeed the prewar paper *Le Temps*, which had become a mouthpiece for the French steel trust, big private banks, and the foreign office. Directeur Hubert Beuve-Mery has Gen. de Gaulle's blessings to make *Le Monde* an independent paper.

Seventeen magazine begins publication in September. Publishing heir Walter H. Annenberg, 36, has started the periodical for young girls *(see* 1939; *TV Guide*, 1953).

Fiction *Winter Tales* by Isak Dinesen; *The Ballad and the Source* by Rosamond Lehmann; *The Golden Fleece* by Robert Graves; *There Were No Windows* by Norah Hoult; *The Dwarf (Dvargen)* by Swedish novelist Pär Lagerkvist, 53, *Dangling Man* by Canadian-American novelist Saul Bellow, 29; *A Bell for Adano* by *Time* magazine journalist-novelist John Richard Hersey, 30; *A Walk in the Sun* by U.S. war novelist Harry Peter M'Nab Brown, 27; *Strange Fruit* by U.S. novelist Lillian Smith, 47, is about a lynching. *The Lost Weekend* by U.S. novelist Charles Reginald Jackson, 41, is about alcoholism; *The Razor's Edge* by W. Somerset Maugham.

Painting *Three Studies for Figures at the Base of a Crucifix* by English painter Francis Bacon, 34, whose asthma has kept him out of military service; *Death's Head* by Pablo Picasso; *How My Mother's Embroidered Apron Unfolds in My Life* by Arshile Gorky; *Pelvis III* by Georgia O'Keeffe; *Factory Workers* by Harlem painter Romare Bearden, 31; *Mother and Child* by New York painter Milton Avery, 51; *The Broken Column* by Frida Kahlo; *The King Playing with the Queen* by Max Ernst. Edvard Munch dies at his estate outside Oslo January 23 at age 80; Piet Mondrian dies at New York February 1 at age 71; sculptor Aristide Maillol dies in an auto accident near Banyuls-sur-Mer October 5 at age 82; Wassily Kandinsky dies at Paris December 17 at age 78.

Kodacolor, introduced by Eastman Kodak, is a color negative film that makes it possible to take color snapshots with low-priced cameras *(see* Kodachrome, 1936; Ektachrome, 1946).

Theater *Decision* by U.S. playwright Edward Chodorov, 39, 2/2 at New York's Belasco Theater, with Georgia Burke is a melodrama about fascism in America and U.S. race relations, 160 perfs.; *Antigone* by Jean Anouilh 2/4 at the Théâtre de l'Atelier in occupied Paris; *The Searching Wind* by Lillian Hellman 4/12 at New York's Fulton Theater, with Cornelia Otis Skinner, Dudley Digges, Montgomery Clift, 318 perfs.; *No Exit (Huis Clos)* by Jean-Paul Sartre 5/27 at the Théâtre du Vieux-Colombier in occupied Paris; *Anna Lucasta* by U.S. playwright Philip Yordan, 30, 8/30 at New York's Mansfield Theater, with Hilda Simons, Canada Lee, John Tate, 957 perfs.; *I Remember Mama* by John Van Druten (who has adapted Kathryn Farber's book *Mama's Bank Account)* 10/19 at New York's Music Box Theater, with Mady Christians, Frances Heflin, Joan Tetzel, Marlon Brando, Oscar Homolka, 714 perfs.; *Harvey* by U.S. playwright Mary Coyle Chase, 37, 11/1 at New York's 48th Street Theater, with Frank Fay, 1,775 perfs.; *Dear Ruth* by U.S. playwright Norman Krasna, 35, 12/13 at New York's Henry Miller Theater, 683 perfs.; *O Mistress Mine* (*Love in Idleness)* by Terence Rattigan 12/20 at London's Lyric Theatre, with Alfred Lunt and Lynn Fontanne.

Films Marcel Carne's *Children of Paradise* (*Les Enfants du Paradis)* with Jean-Louis Barrault, Arletty; Billy Wilder's *Double Indemnity* with Barbara Stanwyck, Fred MacMurray; Preston Sturges's *Hail the Conquering Hero* with Eddie Bracken, Ella Raines, William Demarest; Otto Preminger's *Laura* with Clifton Webb, Gene Tierney, Dana Andrews; Preston Sturges's *The Miracle of Morgan's Creek* with Eddie Bracken, Betty Hutton, William Demarest; Clarence Brown's *National Velvet* with Mickey Rooney, Elizabeth Taylor, 12. Also: Frank Capra's *Arsenic and Old Lace* with Cary Grant, Priscilla Lane, Josephine Hull, Raymond Massey, Peter Lorre; Alfred Hitchcock's *Lifeboat* with Tallulah Bankhead, William Bendix, John Hodiak; John Cromwell's *Since You Went Away* with Claudette Colbert, Jennifer Jones, Joseph Cotten; Robert Siodmak's *The Suspect* with Charles Laughton, Ella Raines; David Lean's *This Happy Breed* with Robert Newton, Celia Johnson, John Mills; Leslie Fenton's *Tomorrow the World* with Fredric March, Betty Field, Skippy Homeier, Agnes Moorehead; Alf Sjoberg's *Torment* with Mai Zetterling, Stig Jarrel; Carol Reed's *The Way Ahead* with David Niven, Stanley Holloway, Peter Ustinov, Trevor Howard; Henry King's *Wilson* with Alexander Knox, Charles Coburn, Geraldine Fitzgerald; Fritz Lang's *The Woman in the Window* with Joan Bennett, Edward G. Robinson.

Hollywood musicals Vincente Minnelli's *Meet Me in St. Louis* with Judy Garland, songs by Hugh Martin and Ralph Blane that include "The Trolley Song," "The Boy Next Door," the title song, and "Have Yourself a Merry Little Christmas"; Leo McCarey's *Going My Way* with Bing Crosby, Barry Fitzgerald, songs that include "Would You Like to Swing on a Star?" and "Too-ra-Loo-ra-Loo-ra"; Mark Sandrich's *Here Come the Waves* with Bing Crosby, music and lyrics by Harold Arlen and Johnny Mercer, songs that include "Ac-cent-tchu-ate the Positive"; Charles Vidor's *Cover Girl* with Rita Hayworth, Gene Kelly, music by Jerome Kern, lyrics by Ira Gershwin, songs that include "Long Ago (and Far Away)"; Robert Siodmak's *Christmas Holiday* with Gene Kelly, Deanna Durbin, songs by Frank Loesser that include "Spring Will Be a Little Late This Year."

First performances *Symphonic Metamorphosis of Themes by Carl Maria von Weber* by Paul Hindemith 1/20 at New York's Carnegie Hall; *Ode to Napoleon* by Arnold Schoenberg 1/23 at Carnegie Hall (the ode is based on a poem by Lord Byron); *Jeremiah* by New York Philharmonic assistant conductor Leonard Bernstein, 25, 1/28 in a concert by the new Pittsburgh Symphony; Concerto for Piano and Orchestra by Schoenberg 2/6 in an NBC Symphony radio concert from New York; Symphony No. 2 (dedicated to the Army Air Forces) by Cpl. Samuel Barber 3/3 at Boston's Symphony Hall; Symphony No. 2 by Walter Piston 3/5 at Washington, D.C.; *Theme with Variations According to the Four Temperaments* by Hindemith 9/3 at Boston's New England Mutual Hall; *Capricorn Concerto* by Barber 10/8 at New York's Town Hall; Theme and Variations in G minor for Orchestra by Schoenberg 10/20 at Boston's Symphony Hall; *Fugue for a Victory Tune* by Piston 10/21 at New York's Carnegie Hall; Concerto for Orchestra by Béla Bartòk 12/1 at Boston's Symphony Hall.

Ballet *Fancy Free* 4/18 at New York's Metropolitan Opera House with Jerome Robbins, John Kriza, and Hugh Laing as the Three Sailors, music by Leonard Bernstein, choreography by Robbins; *Appalachian Spring* 10/30 at the Library of Congress in Washington, with Martha Graham, music by Aaron Copland; *Herodiade* 10/30 at the Library of Congress, with Graham, music by Paul Hindemith, text from a poem by Stéphane Mallarmé.

Broadway musicals *Mexican Hayride* 1/28 at the Winter Garden Theater, with Bobby Clark, June Havoc, book by Herbert and Dorothy Fields, music and lyrics by Cole Porter, songs that include "Count Your Blessings," 481 perfs.; *Hats Off to Ice* 6/22 at the Center Theater, with ice skaters performing to music and lyrics by John Fortis and James Littlefield, 889 perfs.; *The Song of Norway* 8/21 at the Imperial Theater, with music based on works by Edvard Grieg, book and lyrics by Robert Work and George "Chet" Forest, songs that include "Strange Music," 860 perfs.; *Bloomer Girl* 10/5 at the Shubert Theater, with Celeste Holm as Evalina, music by Harold Arlen, lyrics by E. Y. Harburg, songs that include "Evalina," "It Was Good Enough for Grandma," 653 perfs.; *On the Town* 12/28 at the Adelphi Theater, with Sono Osato, Betty Comden, Adolph Green, Nancy Walker, music by Leonard Bernstein, dances derived from the ballet *Fancy Free*, 563 perfs.

Popular songs "I'll Walk Alone" by Jule Styne, lyrics by Sammy Cahn (for the film *Three Cheers for the Boys*); "I'm Making Believe" by James V. Monaco, lyrics by Mack Gordon (for the film *Sweet and Low Down*); "Don't Fence Me In" by Cole Porter (for the film *Hollywood Canteen*); "Jealous Heart" by Nashville country music singer Jenny Lou Carson; "That's What I Like About the South" by Andy Razaf; "Moonlight in Vermont" by Karl Senssdorf, lyrics by John Blackburn; "Straighten Up and Fly Right" by Irving Mills, lyrics by singer Nat "King" Cole (*né* Nathaniel Adams Coles), 25; "Linda" by Ann Ronell; "Sentimental Journey" by Ben Homer, Bud Green, and Les Brown; "Candy" by Mack Davis, Joan Whitney, and Alex Kramer; "Twilight Time" by Buck Ram, Morty Nevins, and Artie Dunn; "It Could Happen to You" by Jimmy Van Heusen, lyrics by Johnny Mercer (for the film *And the Angels Sing*); "Nancy (with the Laughing Face)" by Jimmy Van Heusen and comedian Phil Silvers; "Rum and Coca-Cola" by Jeri Sullivan and Paul Baron, lyrics by comedian Morey Amsterdam, 29; "You're Nobody Till Somebody Loves You" by Russ Morgan, Larry Stock, and James Cavanaugh; "You Always Hurt the One You Love" by Robert Allan and Doris Fisher; "I Should Care" by Paul Westerly, Sammy Cahn, and Alex Stordahl (for the film *Thrill of a Romance*); "I've Got a Lovely Bunch of Cocoanuts" by English songwriter Fred Heatherton.

Sgt. Frank Parker, 28, U.S. Army, wins in men's singles at Forest Hills, Pauline Betz in women's singles.

The St. Louis Cardinals win the World Series by defeating the St. Louis Browns 4 games to 2 (both teams are comprised mainly of over-aged athletes or those with physical defects; most ballplayers are in the armed forces).

The G.I. Bill of Rights (above) will provide 4 percent home loans to veterans with no down payment required. The law will subsidize a postwar building boom and encourage a mass exodus from U.S. cities that will deplete major metropolitan areas of middle-income taxpayers (see Levittown, 1947).

A large-scale migration of Americans from rural to urban areas begins. The shift will create major problems in cities.

Congress establishes Big Bend National Park in a bend of the Rio Grande River of Texas. It occupies 708,221 acres of mountain-land and desert.

Britain cuts her food imports to half their prewar levels. Domestic wheat production has increased by 90 percent, potato by 87 percent, vegetable by 45, sugar beet by 19 despite the manpower shortage.

A "Green Revolution" moves forward outside Mexico City as former E. I. du Pont plant pathologist Norman Borlaug, 30, joins the Rockefeller Foundation effort to improve Mexican agricultural production (*see* 1943; 1949; Salmon, 1945).

The average yield per acre in U.S. corn fields reaches 33.2 bushels, up from 22.8 in 1933 (*see* 1968).

U.S. soybean acreage reaches 12 million as new uses for the beans are found in livestock feed, sausage filler, breakfast foods, enamel, solvent, printing ink, plastics, insecticides, steel-hardening, and beer (*see* McMillen, 1934).

U.S. farm acreage will decline by 7.3 percent in the next 20 years, dropping by 1.3 million acres per year. Some 27 million acres of non-croplands will be con-

1944 ***(cont.)*** verted to farm use (mostly in Florida, California, Washington, Montana, and Texas), but 53 million acres of croplands will go out of production and be used for home and factory sites, highways, and the like.

Bengal's rice crop fails again as in 1942. Millions starve, some because they refuse to accept wheat flour (*see* Burma, 1943).

Kentucky and Mississippi enact bread enrichment laws.

Bread, flour, oatmeal, potatoes, fish, fresh vegetables, and fruit other than oranges remain unrationed in Britain. Prices are controlled so that the average householder has about half the food budget available for unrationed foods after buying rationed foods plus foods whose distribution is controlled or allocated on a "points" basis.

U.S. meat consumption rises to 140 pounds per capita, up from 128.9 last year, as national income rises to $181 billion, up from $72.5 billion in 1939. A large proportion of the meat is sold through black-market channels organized by criminal elements.

Ohio State University tests prepackaged foods in cooperation with companies looking beyond the war to explore ways to reduce losses in food distribution. By 1961, 88 percent of U.S. supermarkets will be prepackaging all or some of their produce, and 90 percent will be prepackaging all fish, smoked meats, and table-ready meats in transparent film (*see* polyethylene, 1935).

Only 10 U.S. grocers have completely self-service meat markets. The figure will increase to 5,600 by 1951, 11,500 by 1956, and 24,100 by 1960, when 35 percent of all meat sold at retail will be from self-service cases.

Chiquita Banana is introduced by the United Fruit Company in a move to make bananas a brand-name item rather than a generic commodity (*see* Sunkist, 1919). The bananas are advertised on radio with a tune composed by Len MacKenzie and performed by Ray Bloch's orchestra with vocalist Patti Clayton singing Garth Montgomery's lyrics: "I'm Chiquita Banana/ And I've come to say,/ Bananas have to ripen in a certain way:/ When they are fleck'd with brown and have a golden hue/ Bananas taste the best and are the best for you. / You can put them in a salad/ You can put them in a pie-aye/ Any way you want to eat them/ It's impossible to beat them/ But bananas love the climate of the very, very tropical equator/ So you should never put bananas in the refrigerator" (copyright 1945 Maxwell-Wirges Publications, Inc.) (*see* 1932). (The skins of refrigerated bananas turn brown but the fruit keeps longer.)

1945 World War II ends in Europe May 8 and in the Pacific August 14, but only after major new military and naval activity (*see* below) and after the deaths of three of the world's five leading heads of state.

Massive operations in both theaters mark the closing months of the war. U.S. forces invade the Philippines in force January 9 under the command of Gen. MacArthur who enters Manila February 4 and completes recovery of the city within 3 weeks.

1,000 U.S. bombers raid Berlin February 3 and drop tons of explosives, British Lancasters raid Dresden February 13 with phosphorus and high explosive bombs, creating a firestorm that kills an estimated 135,000.

The U.S. 9th Armored Division seizes the Ludendorff Bridge at Remagen, last Rhine crossing left standing by the retreating *Wehrmacht*. The 1st Armored Division crosses the bridge and occupies Cologne; by March 10 the Allies control the west bank of the Rhine from Nijmegen in the Netherlands to Koblenz.

The U.S. 9th Armored Division reaches the Rhine opposite Düsseldorf by March 16 and joins with the 1st Armored Division to take more than 325,000 German prisoners in the Battle of the Ruhr.

More than 100 U.S. B-24 bombers raid Tokyo the night of March 9, rain incendiaries on the city, and start fires that kill more than 124,000.

U.S. troops invade Mindanao in the Philippines beginning March 10.

U.S. forces take Iwo Jima March 16 after 35 days of heavy fighting have cost nearly 6,000 U.S. and 19,000 Japanese lives on the 8-square-mile volcanic island.

The League of Arab States organized March 22 unites Egypt, Iraq, Jordan, Lebanon, Saudi Arabia, Syria, and Yemen in coordinating efforts toward achievement of complete independence and mutual cooperation. By 1974 the League will include 20 Arab states including Algeria, Tunisia, Morocco, Lebanon, Kuwait, Qatar, Bahrain, Oman, Somalia, Sudan, and Mauritania.

Jewish terrorists in Palestine dynamite railroads and scuttle British police launches used to apprehend illegal immigrants. Polish-born militant Menahem Begin, 32, leads the Irgun Zvai Leumi that spearheads opposition to British control (*see* 1941; 1946).

President Roosevelt dies April 12 of a cerebral hemorrhage at Warm Springs, Ga., 2 months after a Yalta Conference with Prime Minister Churchill and Premier Stalin in which the three powers have pledged "unity of purpose and of action." Dead at age 63, FDR had been preparing for a conference at San Francisco where the charter of a United Nations organization is to be drafted.

FDR (*see* above) is succeeded by his vice president Harry S. Truman.

Nuremberg falls to the U.S. 7th Army April 21 as Russian forces reach the suburbs of Berlin.

Russian and U.S. patrols meet April 25 on the Elbe just south of Berlin and celebrate far into the night with whiskey, vodka, and accordion music. Adolf Hitler's hope of a collapse in the alliance between capitalists and Communists is shattered, and Heinrich Himmler begins surrender negotiations.

Benito Mussolini and 12 of his former Cabinet officers are executed April 28 at Lake Como. German armies in Italy surrender unconditionally April 29.

Adolf Hitler commits suicide April 30 at age 56 in his bunker at the Reich Chancellery at Berlin as Soviet troops converge on the city. Joseph Goebbels also commits suicide.

Hitler aide Martin Bormann, 45, is seen May 2 in a tank crossing Berlin's Weidenammer Bridge. Bormann will not be positively identified hereafter but rumors will persist that he is alive and living in Argentina or Chile.

Germany surrenders unconditionally May 7. President Truman proclaims May 8 V-E (Victory in Europe) Day.

U.S. forces regain Okinawa in the Ryukyus from the Japanese June 21 after nearly 3 months of fighting. It has taken half a million Americans to subdue 110,000 Japanese; The U.S. death toll is 49,151.

Washington announces reconquest of the Philippines July 5.

Full air war against Japan's home islands of Honshu and Kyushu begins July 10 from bases in the Marianas and Okinawa.

Carrier-based air attacks on Tokyo begin July 17 after U.S. planes have dropped leaflets threatening Japan with aerial destruction unless she surrenders unconditionally.

An atomic bomb is tested successfully in the desert near Alamogordo, N.M., July 18 (*see* below).

The U.S.S. *Indianapolis* is torpedoed July 30 just after midnight and sinks 12 minutes later in the Indian Ocean 3 days after having delivered essential components of the atomic bomb to Tinian in the Marianas. The heavy cruiser's main communications center has been destroyed, she has sent no S.O.S., no survivors are spotted until August 2, and by that time most of her 1,996 officers and men have drowned, been killed by sharks, or in some cases been murdered by crazed brothers in arms. Only 316 survive, and although Capt. Charles Butler McVay, III, the ship's commander, is court-martialed in December and found guilty of having failed to run a zigzag course in a war zone, the charges will be dropped in February of next year, and McVay will be promoted to rear admiral after he retires in 1949.

Japanese air ace Saburo Sakai downs a U.S. B-29 over Japan in early August to end a career in which he has blasted 64 Allied aircraft out of the skies from the cockpit of his Mitsubishi Zero fighter. Sakai lost one eye in combat, has flown half-blind, and has only recently accepted a commission as commander after years of refusal.

The Potsdam Conference ends August 2 after 16 days of deliberations by President Truman, Generalissimo Stalin, and Prime Minister Churchill who was replaced July 28 (Clement R. Attlee's Labour party has defeated Churchill's Conservatives in the general elections). The world leaders agree that Germany is to be disarmed and demilitarized, her National Socialist institutions dissolved, her leaders tried as war criminals, and democratic ideals encouraged with restoration of local self-government and freedom of speech, press, and religion subject to military security requirements. Manufacture of war matériel is to be prohibited and strict controls placed on production of metals, chemicals, and machinery essential to war.

Tokyo rejects the Potsdam Declaration calling upon Japan to surrender. U.S. planes drop leaflets over Hiroshima August 4 warning, "Your city will be obliterated unless your Government surrenders."

The atomic bomb (code name "Little Boy") dropped on Hiroshima August 6 is a 10-foot-long, 9,000-pound device with the explosive power of 20,000 tons of TNT. The U.S. Army Air Force Boeing B-29 bomber *Enola Gay* piloted by Paul Tibbets, Jr., has flown from the Pacific atoll of Tinian, the bombardier drops "Little Boy" from an altitude of 32,000 feet. The bomb explodes 660 yards above the city and flattens 4 square miles of Hiroshima, 100,000 Japanese die outright, close to another 100,000 will die from burns and radiation sickness, hundreds will be disfigured.

Soviet Russia declares war on Japan August 8 and sends troops into Manchuria.

The atomic bomb (code name "Big Boy") dropped on Nagasaki in Kyushu August 9 by the B-29 bomber *Bock's Car* is a plutonium-core bomb with the same explosive power as "Little Boy." It kills 75,000 Japanese outright and many of the 75,000 survivors will die from burns and radiation sickness. Leaflets dropped earlier had threatened "a rain of ruin the like of which has never been seen on earth."

Japan sues for peace August 10. President Truman proclaims V-J (Victory over Japan) Day August 14.

World War II ends after nearly 6 traumatic years in which an estimated 54.8 million have died, most of them civilians. Many millions more are left crippled, blind, mutilated, homeless, orphaned, and impoverished. Europe has 10 million displaced persons. The war has directly involved 57 nations, but the Soviet Union, Germany, China, and Japan have borne the lion's share of casualties. Counting only military personnel, the U.S.S.R. has lost 7.5 million, Germany nearly 2.9 million, China 2.2, Japan 1.5, Britain nearly 398,000, Italy 300,000, the United States more than 290,000, France nearly 211,000 including colonial troops (but not including 36,877 prisoners of war who died in captivity), Canada 39,139, Australia 29,395, New Zealand 12,262, South Africa 8,681, India 36,092, the rest of the British Empire 30,776.

U.S. forces land in Japan August 28. Gen. MacArthur is named supreme commander of Allied occupation forces.

Formal Japanese surrender terms signed September 2 aboard the U.S.S. *Missouri* in Tokyo Bay provide for Soviet control of Outer Mongolia, Chinese sovereignty over Inner Mongolia, Manchuria, Taiwan, and Hainan, U.S. and Soviet occupation of Korea until a democratic government can be established. Japan cedes the Kuril Islands and southern part of Sakhalin to the U.S.S.R. Emperor Hirohito is to remain Japanese head of state, but the Japanese high command and military establish-

1945 *(cont.)* ment is to be disbanded, and a U.S. army of occupation is to rule Japan.

Japanese forces in China, estimated to number 1 million, capitulate formally September 9 at Nanjing. Their leaders sign surrender papers with representatives of Chiang Kai-shek.

The British at Singapore accept formal surrender of Japanese forces in Southeast Asia (estimated to number 585,000) September 12. The British reoccupy Hong Kong.

Former French premier Pierre Laval is executed for collaborating with the enemy. Marshal Pétain has been sentenced to death August 15 for treason but his sentence has been commuted to life imprisonment (now 89, he will live to July 1951). French Communists and Socialists who have led the Resistance prevail in elections for the constituent assembly October 21. Gen. de Gaulle outmaneuvers them, is elected by the assembly as president of the provisional government November 16, forms a cabinet November 21, but will resign early next year in the face of continuing Leftist opposition (see 1958).

Brazil's President Vargas resigns October 25 after nearly 15 years of dictatorship as pressures mount for a more liberal government. Chief Justice José Linhares becomes president pro tem; voters elect Gen. Enrico Dutra president December 2 (*see* 1950).

Soviet troops liberate Auschwitz January 26 but find fewer than 3,000 prisoners at the Polish death camp where more than 1 million have died in SS gas chambers. The SS has removed the rest to camps inside Germany. Gen. Patton liberates Buchenwald April 11 after 60,000 have died there, Dachau is liberated April 24, but the Allies are able to rescue only 500,000, many of whom will die of the effects of hunger and disease.

The Nazi genocide has killed an estimated 14 million "racial inferiors" including Poles, Slavs, gypsies, and close to 6 million Jews. One third of the world's Jews have died in Nazi death camps in the last 6 years, including those killed with cyanide gas at the Auschwitz-Birkenau extermination camp in southern Poland (*see* above).

The World Zionist Congress demands admission of 1 million Jews to Palestine in a statement issued August 13 (*see* terrorists above).

French women gain the right to vote for the first time under terms of a new enfranchisement bill.

U.S. private Eddie Slovik has died January 31 before a firing squad, the first American to be executed for desertion since the Civil War. A Detroit plumber before the war, Slovik was drafted last year after having been rejected earlier, his commanding officer in basic training recognized his "total inability" and tried to obtain his discharge or at least a transfer to a noncombat unit, it took Slovik 6 weeks to locate his unit after being shipped overseas, he claimed to have deserted and said he would desert again, he signed a "confession" calculated to keep him out of combat, and Gen. Eisenhower approved his execution.

The Banque de France is nationalized.

Britain introduces family allowances.

A U.S. tariff act empowers the President to encourage reciprocal trade by reducing tariffs to 25 percent of original schedule rates (*see* 1934; GATT, 1947).

A revenue act passed by Congress November 8 provides for nearly $6 billion in tax reductions.

The U.S. Gross National Product (GNP) for the year is $211 billion, double the booming 1929 figure, even though strikes have closed down Ford, General Motors, and other plants. Strikers shut down bituminous coal mines through much of the fall.

The International Bank for Reconstruction and Development comes into being December 27; 21 countries have subscribed nearly $7.2 billion, nearly $3.2 billion of it from the United States. The bank's first loans will be made to France in May 1947 at interest rates of 3 and 3.5 percent (*see* Bretton Woods, 1944).

A U.S. Court of Appeals finds that Aluminum Co. of America (Alcoa) held a 90 percent monopoly in U.S. aluminum ingot production before the war, but the monopoly enjoyed by the Mellons for more than half a century will soon be ended (*see* 1891). Reynolds Metals produced ingots with government encouragement during the war (*see* 1916), Henry Kaiser creates Kaiser Aluminum to take over some government-built aluminum plants, and several U.S. copper companies will soon enter the industry.

World copper reserves are estimated at 100 million tons. The world will use roughly that much copper in the next 30 years, but the reserve will then be 300 million tons as a result of new discoveries.

Fuji Steel is founded by Japanese industrialist Shigeo Nagano, who will merge his company with Yawata Steel in 1950 to make Fuji a major world steel producer (*see* Yawata, 1896; Nippon Steel, 1950). Japan's steel industry has suffered relatively little damage in the war.

The top 8.5 percent of Americans hold 20.9 percent of U.S. personal wealth, down from 32.4 percent of personal wealth in 1929.

The atomic bomb tested successfully July 18 at Alamogordo, N. Mex., and then dropped on Hiroshima and Nagasaki (above) raises the possibility of a new energy source (as well as of a nuclear holocaust that would end civilization).

Physicist Igor Kurchatov achieves the first Soviet atomic chain reaction (*see* 1949).

U.S. gasoline and fuel oil rationing ends August 19.

A B-25 light bomber flies into New York's Empire State Building July 28, tearing a hole between the 78th and 79th floors, killing all three men aboard plus 10 early Sunday morning churchgoers.

Former U.S. fighter pilot Robert W. Prescott, 32, founds Flying Tiger Line, taking the nickname of Gen. Claire Chennault's American Volunteer Group with which he flew from late 1941 until July 1942 when the group was disbanded after many sorties against the

Nuclear fission raised fearsome questions: Would it end dependence on fossil fuels or would it destroy the world?

Japanese. Starting with four planes and 16 employees, Prescott will make Flying Tiger Line the world's largest air cargo firm (*see* Federal Express, 1977).

U.S. automobile companies convert to passenger car production. A redesigned CJ-2A Universal Jeep goes into production in September at the Willys-Overland plant in Toledo, Ohio (*see* 1940; Brazil, 1965).

Henry Ford steps down from the presidency of Ford Motor Company at age 82 as production of civilian passenger cars resumes. He yields control of the shattered organization to his grandson Henry Ford II, 28, son of Edsel Ford's widow Eleanor Clay Ford (she and her mother had threatened to sell their stock in the company if the founder did not relinquish control). Young Henry immediately dismisses Harry Bennett, who has ruled the Ford empire through sycophants by fear and terror. He then raids other companies to create a new management team: Bendix Aviation president Ernest R. Breech, 48, will join Ford next year as executive vice president (*see* 1947).

Penicillin is introduced on a commercial basis following work by a large Anglo-American research team.

Streptomycin is introduced commercially.

New Orleans surgeon Alton Ochsner, 49, observes "a distinct parallelism between the incidence of cancer of the lung and the sale of cigarettes" in an address at Duke University (which has been endowed with a fortune derived from tobacco sales; *see* 1924). "The increase is due to the increased incidence of smoking and . . . smoking is a factor because of the chronic irritation it produces."

Ballpoint pens go on sale October 29 at New York's Gimbel Bros. Chicago promotor Milton Reynolds, 53, has seen the Ladislao Biro pen while visiting Buenos Aires on business in June (*see* 1938), developed a pen different enough to get around existing patents, gone into production October 6, and is producing 70 pens per day. Gimbels quickly sells out at $12.50 each with the initial promise that the pens will write underwater. Some banks suggest that ballpoint pen signatures may not be legal, but the new pens do not leak at high altitudes and will make it practical to handwrite multicopy business and government forms using carbon paper.

Meet the Press debuts 10/5 on a New York radio station with *American Mercury* magazine publisher Lawrence E. Spivak, 44, moderating a "spontaneous, unrehearsed weekly news conference of the air." Journalists interview prominent news figures on the show which will go on NBC television beginning in November 1947 and be moderated by Spivak until November 1975.

Some 5,000 U.S. homes have television sets—bulky receivers with tiny screens that pick up what little programming is available from the handful of stations in operation. But the world is on the edge of a communications revolution that will see television sets in nearly every home of every developed country (*see* 1939; 1948).

Ebony magazine begins publication in November; the black-oriented U.S. picture monthly sells out its initial press run of 25,000 copies. John H. Johnson of Johnson Publishing introduced *Negro Digest* in 1942 and will introduce *Tan* in 1950 and *Jet* in 1951 as he increases his holdings in Supreme Liberty Life Insurance Co.

Nonfiction *The Yogi and the Commissar* by Arthur Koestler; *Black Boy* (autobiography) by Richard Wright.

Fiction *The Animal Farm: A Fairy Story* by George Orwell whose allegorical fable contains the line, "All animals are created equal, but some animals are more equal than others"; *The Age of Reason* (*L'age de raison*) and *The Reprieve* (*Le Sursis*) by Jean-Paul Sartre, first two novels in a trilogy; *Christ Stopped at Eboli (Christo si e fermato ad eboli)* by Italian novelist Carlo Levi, 43, who reveals the social and human problems of Lucania where he was confined in the mid-1930s for antifascist activities; *The Serpent* (*Ormen*) by Swedish novelist Stig Dagerman, 22; *Krane's Café (Kranes Konditori)* by Cora Sandel; *Grape-harvest (Vindima)* by Portuguese novelist-poet-physician Miguel Torga (Adolfo Coelho Da Rocha), 38; *The Crack-Up*, an incomplete novel by the late F. Scott Fitzgerald whose widow Zelda will die in 1948 when a fire destroys the Asheville, N.C., mental hospital where she is a patient; *Prater Violet* by Christopher Isherwood; *Brideshead Revisited* by Evelyn Waugh; *That Hideous Strength* by C.S. Lewis; *Loving* by Henry Green; *A Fearful Joy* by Joyce Cary; *Cannery Row* by John Steinbeck; *Forever Amber* by U.S. novelist Kathleen Winsor, 29.

Juvenile *Stuart Little* by E. B. White of the *New Yorker* magazine whose protagonist is a mouse; *The Famous Invasion of Sicily by the Bears (La Famosa Invasione degli Ursi in Sicilia)* by Dino Buzzati.

Painting *If This Be Not I* by Canadian-American painter Philip Guston, 32; *The Unattainable* and *The*

1945 *(cont.)* *Diary of a Seducer* by Arshile Gorky; *The Temptation of St. Anthony* by Max Ernst; *For Internal Use Only* by Stuart Davis; *Great Tenochtitlan* and *The Market in Tiangucio* by Diego Rivera for Mexico City's Palacio Nacional. German graphic artist Käthe Kollwitz dies outside Dresden at Moritzburg Castle April 22 at age 77.

Sculpture *Red Pyramid* (mobile) by Alexander Calder; *Family Group* by Henry Moore.

Theater *The Hasty Heart* by U.S. playwright John Patrick, 39, 1/3 at New York's Hudson Theater, with John Lund, Richard Basehart, 207 perfs.; *The House of Bernarda Alba* (*La casa de Bernarda Alba*) by the late Federico García Lorca 3/6 at the Teatro Avenida, Buenos Aires; *The Glass Menagerie* by Mississippi-born playwright Tennessee (Thomas Lanier) Williams, 34, 3/31 at New York's Plymouth Theater, with Laurette Taylor, Eddie Dowling, Julie Haydon, Anthony Rose, 561 perfs.; *State of the Union* by Howard Lindsay and Russel Crouse 11/14 at New York's Hudson Theater, with Myron McCormack, Ralph Bellamy, 765 perfs.; *Dream Girl* by Elmer Rice 12/14 at New York's Coronet Theater, with Wendell Corey, Betty Field (Mrs. Rice), 348 perfs.; *The Madwoman of Chaillot (La Folle de Chaillot)* by the late Jean Giraudoux 12/19 at Paris; *Home of the Brave* by U.S. playwright Arthur Laurents, 27, 12/27 at New York's Belasco Theater, is about anti-Semitism in the U.S. Army, 69 perfs.

Films René Clair's *And Then There Were None* with Barry Fitzgerald, Walter Huston, Louis Hayward, Judith Anderson, June Duprez; Cavalcanti, Basil Dearden, Robert Hamer, and Charles Crichton's *Dead of Night* with Mervyn Johns, Roland Culver, Michael Redgrave, Sally Ann Howes, Googie Withers; Laurence Olivier's *Henry V* with Olivier; Billy Wilder's *The Lost Weekend* with Ray Milland, Jane Wyman; Jean Renoir's *The Southerner* with Zachary Scott, Betty Field; John Ford's *They Were Expendable* with Robert Montgomery, Donna Reed; Elia Kazan's *A Tree Grows in Brooklyn* with Dorothy McGuire, Joan Blondell; Anthony Asquith's *The Way to the Stars* with John Mills, Michael Redgrave, Trevor Howard, Felix Aylmer. Also: Henry King's *A Bell for Adano* with Gene Tierney, John Hodiak, William Bendix; David Lean's *Blithe Spirit* with Rex Harrison, Constance Cummings, Margaret Rutherford; Robert Wise's *The Body Snatcher* with Henry Daniell, Boris Karloff, Bela Lugosi; Vincente Minnelli's *The Clock* with Judy Garland, Robert Walker; Henry Hathaway's *The House on 92nd Street* with William Eythe, Lloyd Nolan, Signe Hasso; Serge Eisenstein's *Ivan the Terrible (II)*; Akira Kurosawa's *The Man Who Tread the Tiger's Tail*; Michael Curtiz's *Mildred Pierce* with Joan Crawford, Jack Carson, Zachary Scott; Raoul Walsh's *Objective, Burma!* with Erroll Flynn; Roy Rowland's *Our Vines Have Tender Grapes* with Edward G. Robinson, Margaret O'Brien; Compton Bennett's *The Seventh Veil* with James Mason, Ann Todd; Alfred Hitchcock's *Spellbound* with Ingrid Bergman, Gregory Peck.

Hollywood musicals Walter Lang's *State Fair* with Jeanne Crain, Dana Andrews, music by Richard Rodgers, lyrics by Oscar Hammerstein II, songs that include "It Might as Well Be Spring," "It's a Grand Night for Singing."

Stage musicals *Up in Central Park* 1/27 at New York's Century Theater, with music by Sigmund Romberg, book by Herbert and Dorothy Fields, lyrics by Dorothy Fields, 504 perfs.; *Carousel* 4/19 at New York's Majestic Theater, with John Raitt as Billy Bigelow (Liliom), Jan Clayton as Julie Jordan, music by Richard Rodgers, lyrics by Oscar Hammerstein II, book based on the 1909 Ferenc Molnár play *Liliom*, songs that include "If I Loved You," "June Is Bustin' Out All Over," "Soliloquy," "You'll Never Walk Alone," 890 perfs.; *Perchance to Dream* 4/21 at the London Hippodrome, with Ivor Novello, Muriel Barron, Margaret Rutherford, book, music, and lyrics by Novello, 1,022 perfs.; *Sigh No More* (revue) 8/22 at London's Piccadilly Theatre, with Joyce Grenfell, Madge Elliott, Cyril Ritchard, music and lyrics by Noël Coward, 213 perfs.; *Billion Dollar Baby* 12/21 at New York's Alvin Theater, with Joan McCracken, Mitzi Green, Bill Talbert, music by Morton Gould, lyrics by Betty Comden and Adolph Green, 219 perfs.

Opera *Peter Grimes* 6/7 at the Sadler's Wells Theatre, London, with music by Benjamin Britten, libretto from the 1810 poem "The Borough" by English poet George Crabbe.

First performances Symphony No. 5 in B flat major by Serge Prokofiev 1/13 at Moscow while an artillery salute outside the concert hall celebrates news of a great Soviet victory on the Vistula; Symphony No. 9 by Dmitri Shostakovich 11/3 at Leningrad; Prelude for Orchestra and Mixed Choir by Arnold Schoenberg 11/18 at Los Angeles; String Quartet No. 2 in C by Benjamin Britten 11/21 at London.

Austrian composer Anton von Webern dies at Mittersill, near Salzburg, September 15 at age 61 after being shot by a trigger-happy U.S. military policeman outside his son-in-law's villa.

Ballet *Undertow* 4/10 at New York's Metropolitan Opera House, with Hugh Laing, Alicia Alonso, music by William Schuman, and choreography by Antony Tudor.

Popular songs "It's Been a Long, Long Time" by Jule Styne, lyrics by Sammy Cahn; "Till the End of Time" by Buddy Kay and Ted Mossman who have adapted Chopin's Polonaise in A flat; "For Sentimental Reasons" by William Best, lyrics by Deke Watson; "I'm Beginning to See the Light" by Don George, Johnny Hodges, Duke Ellington, and Harry James; "Les Trois Cloches" ("While the Angelus Was Ringing") by French songwriter Jean Villard; "La Mer" by French singer-songwriter Charles Trenet, 32; "Laura" by David Raksin, lyrics by Johnny Mercer (title song for last year's film); "Let It Snow! Let It Snow! Let It Snow!" by Jule Styne, lyrics by Sammy Cahn.

U.S. gospel singer Mahalia Jackson, 34, records "Move on Up a Little Higher" and scores a huge success. Seven other hymns recorded by Jackson will have sales of more than a million copies each, including "I Believe," "I Can Put My Trust in Jesus," and "He's Got the Whole World in His Hands."

Frank Parker wins in men's singles at Forest Hills, Sarah Cooke in women's singles.

Pitcher Warren Spahn, 24, returns from the European war to resume a career with the Boston Braves that was interrupted after four games in 1942.

The Detroit Tigers win the World Series by defeating the Chicago Cubs 4 games to 3.

Aerosol spray insecticides begin a revolution in packaging. The commercial "bug bombs" employ a Freon-12 propellant gas developed by two U.S. Department of Agriculture researchers in 1942 and have been used during the war to protect troops from malaria-carrying mosquitoes; the new spray cans are heavy and costly to produce but will soon be replaced by lightweight tin and aluminum cans employing lower pressure propellants, leading to widespread use of aerosol spray products (*see* Freon-12 refrigerant gas, 1931; RediWip, 1947; Ablanalp, 1953).

Conrad Hilton opens the Conrad Hilton Hotel at Chicago, buying control of (and renaming) the Stevens Hotel built in 1927 and occupied by the Army during the war.

Hilton also takes over Chicago's 20-year-old Palmer House with backing from local building-supply magnate Henry Crown, 49 (*see* 1919).

The Food and Agriculture Organization (FAO) of the United Nations establishes headquarters at Rome with English agriculturist John Boyd Orr as director.

U.S. Department of Agriculture agronomist Samuel C. Salmon, 60, discovers the semidwarf wheat variety Norin 10 whose 2-foot stems respond quickly to water and fertilizer and do not fall over (lodge) from the weight of their grain heads. Salmon is a cereal improvement expert helping Japanese reconstruction as a member of Gen. MacArthur's occupation force. He finds Norin 10 on a visit to the Morioka experimental station in northern Honshu, returns home with some seeds which he grows in quarantine. Norin 10 proves to be extremely susceptible to leaf stripe and powdery mildew, but Salmon's colleague Orville Vogel crosses the Japanese wheat with resistant strains of U.S. wheat. Grains developed from Norin 10 will increase wheat harvests in India and Pakistan by more than 60 percent.

The U.S. soybean crop reaches 193 million bushels, up from 78 million in 1940.

The University strawberry developed by the University of California agriculture station will make strawberries a long-season widely grown commodity.

Japan mobilizes schoolchildren to gather more than a million metric tons of acorns for use in flour making to supplement scarce wheat and rice stocks.

CARE (Cooperative for American Remittances to Europe) is founded as a private relief organization to help deal with the misery widespread on the Continent. Ready-assembled U.S. food packages of good quality are shipped to the European families and friends of people in America for $10 each, including transportation, with delivery guaranteed at a time when packages sent through normal channels are highly susceptible to loss, theft, or damage.

Italians only reluctantly accept pea-soup powder sent to relieve hunger because it is so unfamiliar, but powdered eggs gain universal acceptance.

U.S. food rationing on all items except sugar ends November 23, but food remains scarce in most of the world. Black markets exist throughout Europe.

A fluoridation program at Grand Rapids, Mich. is the first attempt to fluoridate community water supplies to reduce the incidence of dental caries in children. Newburgh, N.Y., follows suit, but the moves arouse widespread political opposition (*see* Britain, 1955).

Alabama, Georgia, Hawaii, Indiana, Maine, New Hampshire, New York, North Carolina, North Dakota, South Dakota, Washington, West Virginia, and Wyoming enact bread-enrichment laws.

Some 85 percent of U.S. bread is commercially baked, up from 66 percent in 1939.

U.S. food processors pioneer frozen orange juice, using knowledge gained in wartime production of powdered orange juice. They develop an easily frozen sludge of concentrated juice that can be reconstituted to taste far more like fresh-squeezed juice than does ordinary canned juice (*see* 1946).

Welch Grape Juice Co. is acquired by wine maker J. M. Kaplan who has built his upstate New York winery into National Grape Corp. (*see* 1933; Welch, 1897). Kaplan has encouraged the six small farmers' cooperatives that supply him with grapes to form a new, stronger cooperative (*see* 1956).

Tupperware Corp. is founded by former E. I. DuPont chemist Earl W. Tupper, who has designed plastic refrigerator bowls and cannisters with a patented seal that prevents leaking, as he will demonstrate at Tupperware Home Parties.

Coca-Cola Company registers the name "Coke" as a trademark (*see* 1942; 1955).

Omaha's C. A. Swanson & Sons begins to develop a line of canned and frozen chicken and turkey products under the Swanson label using experience gained in World War II when nearly all of the firm's poultry was shipped to the armed forces. Called the Jerpe Commission Co. until now, the 45-year-old company has been

1945 ***(cont.)*** owned since 1928 by Carl A. Swanson (*see* "TV Dinners," 1954).

World War II has killed an estimated 25 million in the Soviet Union, 7.8 in China, and more than 6 in Poland which has lost more than 22 percent of her people.

1946 "From Stettin in the Baltic to Trieste in the Adriatic, an iron curtain has descended across the Continent," says former prime minister Winston Churchill March 5 in an address at Westminster College in Fulton, Mo. President Truman is in the audience that hears the British statesman declare that Moscow's totalitarian dominance has produced a decline of confidence in "the haggard world."

An Austrian Republic with frontiers as of 1937 has gained recognition from the Western powers January 7, Albania has proclaimed herself a People's Republic January 11, Yugoslavia has adopted a new Soviet model constitution January 31, Hungary has proclaimed herself a republic February 1.

Italy's Victor Emmanuel III abdicates May 9 at age 76 after a 46-year reign, his son Umberto, 41, proclaims himself king, a referendum June 2 rejects the monarchy 12.7 million to 10.7 million, Umberto II joins his family at Lisbon June 3, and the nation becomes a republic. Italy loses the Dodecanese Islands to Greece and parts of her northern territory to France June 27.

Bulgaria ousts her monarchy and becomes a People's Republic September 15 following a referendum. Bulgarian Communist Georgi Dimitrov returns from Moscow November 21 to become premier (*see* Reichstag fire, 1933).

Greece's George II returns to Athens September 28 following a plebiscite favoring retention of the monarchy.

The United Nations General Assembly opens its first session January 10 at London with former Belgian premier Paul Henri Spaak, 46, as president. The UN Security Council meets for the first time January 17 at London, Norwegian Socialist Trygve Lie, 49, is elected Secretary-General February 1, the Security Council meets March 25 at New York's Hunter College, the League of Nations assembly dissolves itself April 18, the UN General Assembly meets at New York October 23, selects New York as permanent UN site December 5, and 9 days later accepts a gift of $8.5 million from John D. Rockefeller, Jr., toward purchase of property on the East River for permanent headquarters.

British and French forces begin evacuating Lebanon March 10. The country gains full independence after 26 years of French colonial rule.

Moscow agrees to withdraw troops from Iran April 25 on the promise of reforms in Azerbaijan.

Transjordanian independence gains British recognition March 22. Transjordan proclaims herself a kingdom May 25, Emir Abdullah will rule until 1951 as King Abdullah (*see* 1949) but Moscow vetoes Jordanian admission to the United Nations.

The United Nations struggled to find peaceful solutions to the problems that divided a war-weary world.

Romania's wartime premier Ion Antonescu is sentenced to death May 17.

The Nuremberg Tribunal returns verdicts September 30. Twelve leading Nazis are sentenced to death including Joachim von Ribbentrop and Hermann Goering (who commits suicide by taking poison). Rudolf Hess and Walter Funk are sentenced to life imprisonment, Franz von Papen and Hjalmar Schacht acquitted.

Paris recognizes Vietnamese independence within the French Union March 6, but hostilities with native Communists continue as Ho Chi Minh tries to drive out the French and unite Indo-China. The French retaliate by proclaiming an autonomous Republic of Cochin China at Saigon June 1; they bombard Haiphong November 23 and kill 6,000, beginning a long struggle.

China's civil war resumes April 14 after a truce of more than 3 months negotiated by Gen. George C. Marshall, the Nationalist government returns from Chungking to Nanjing May 1, another truce stops hostilities for 6 weeks beginning May 12, the Guomindang reelects Chiang Kai-shek president October 10, a Sino-American treaty of friendship, commerce, and navigation is signed November 4, a national assembly meets November 15 without Communist representation, and the Chinese adopt a new constitution December 25, but Mao Zedong has 1 million men in uniform plus 2 million guerrillas and has broadcast a radio message in August ordering all-out war against the Nationalists.

Siam's Rama VIII is found dead of a bullet wound June 9 at age 21 after an 11-year reign. His brother Phumiphon Aduldet will reign as Rama IX beginning May 5, 1950, after returning from school in Switzerland. The Siamese accept a UN verdict October 13 that they return to Indo-China the provinces acquired in 1941 as allies of Japan.

The Philippines gain independence from the United States July 4 under provisions of the McDuffie-Tydings Act of 1934. The Philippine Republic headed by President Manuel A. Roxas tries to subdue the Communist-led Huks (Hukbalahaps) peasant party that has appropriated lands in central Luzon.

The Irgun Tzvai Leumi led by Menachen Begin bombs Jerusalem's King David Hotel July 22 killing 91 in a protest against British rule in Palestine (*see* 1945; 1948).

U.S. military authorities conduct the first Bikini Atoll atomic bomb tests in the Pacific in July.

Mathematician John von Neumann and German-born physicist Klaus Emil Julius Fuchs, 35, file application for a U.S. patent on a nuclear fusion (hydrogen) bomb (*see* von Neumann, 1944; Eniwetok, 1952; Fuchs, 1950).

A new Japanese constitution, drafted by young lawyers on Gen. MacArthur's staff and promulgated in October, vests sovereignty not with the emperor but with the people of Japan, which gets democracy after centuries of absolutism. Article 9 abrogates war (*see* 1889; women's rights, below).

The Cheribon Agreement initialed by Dutch and Indonesian diplomats November 15 follows conclusion of a truce in the fighting between Indonesians and Dutch and British forces, the Dutch agree to recognize the Indonesian Republic (Java, Sumatra, Madura) and to the establishment of the United States of Indonesia which will include also Borneo, the Celebes, the Sunda Islands, and the Moluccas to be joined in equal partnership with the Netherlands under the Dutch Crown. The last British troops leave November 29 (*see* 1949).

Sarawak is ceded to the British Crown by the rajah Sir Charles Brooke (*see* 1888).

Juan Perón wins election as president of Argentina February 24 at age 50 and begins a rule that will continue until 1955.

Mexico elects her first civilian president July 7. Miguel Alemán Valdés, 44, will have closer U.S. ties than his predecessors. He appoints a cabinet of economic experts and begins a program of industrialization, electrification, irrigation, and improvement of transportation.

Japanese women gain the right to vote as U.S. occupation authorities make sweeping changes in Japanese society; 89 women stand for election to the Diet, 34 win seats in the April 10 elections.

Italian women gain the vote for the first time.

France's "Loi Marthe Richard" closes Paris brothels. Assemblywoman Richard, 58, has campaigned against the enslavement of women in houses of prostitution. She became the sixth woman in the world to fly an airplane 34 years ago, served as a spy in World War I and stole the German plan for U-boat attacks on U.S. troopship convoys from the bedroom of naval attaché Hans von Krohn in Spain. Her legislation does not outlaw prostitution, but bordellos are no longer licensed, medical examinations are dropped, and demimondaines have to solicit in the streets.

A Polish pogrom at Kielce July 14 kills Jews who have survived the Holocaust.

Bombay removes social disabilities of the city's untouchables, but prejudice against the *harijans* continues (*see* 1947; Gandhi, 1932).

Congress establishes an Indian Claims Commission to settle once and for all the disputes that have been smoldering for nearly a century over lands taken by the white man.

$ Housing shortages and pent-up demand for consumer goods lead to runaway prices and rising wage demands in much of the world.

Strikes idle some 4.6 million U.S. workers during the year with a loss of 116 million man-days, the worst stoppage since 1919: 260,000 electrical workers strike General Electric, Westinghouse, and General Motors January 15; 263,000 meat packers strike January 16; 750,000 steel workers strike January 20.

Parliament repeals the British Trades Disputes Act of 1927 February 13, legalizes certain kinds of strikes, and lifts most restrictions on labor unions' political activities.

The British Government takes over the Bank of England February 14.

A wage-pattern formula announced by President Truman February 14 entitles U.S. labor to 33 percent wage increases that equal the rise in the cost of living since January 1941.

An employment act passed by Congress February 20 declares that maximum employment must be the government's policy goal.

President Truman reestablishes the Office of Economic Stabilization February 21 to control inflation. He names former advertising executive Chester Bowles, 45, to head the office.

U.S. troops seize the nation's railroads May 17 in the face of imminent strikes. A brief strike ends when President Truman recommends wage boosts.

Parliament nationalizes the British coal industry in May with a law that is to take effect January 1, 1947.

U.S. troops seize American soft coal mines May 22 to end a strike begun April 1. John L. Lewis, who has brought his United Mine Workers back into the AF of L late in January (*see* 1936), leads his membership out on strike November 21 in defiance of a federal injunction, he is fined $3.5 million December 4, the workers return to the pits December 7.

The Office of Price Administration (OPA) expires June 29 when the President vetoes a Compromise Price Control act; Congress revives the OPA July 25 with a new Price Control act.

Canada institutes family allowances (*see* Britain, 1945). Sales of children's shoes jump from 762,000 pairs per month to 1,180,000 as parents receive financial aid.

1946 ***(cont.)*** Hungary suffers the worst inflation in world history as the gold pengo of 1931 falls in value to 130 trillion paper pengos in June. The government prints 100-trillion pengo notes.

Wall Street's Dow Jones Industrial Average reaches a post-1929 high of 212.50, but falls to 163.12.

High meat prices bring demands that President Truman resign, but prices soften when the government withdraws from the market in the summer and prices break when the OPA (*see* above) removes price ceilings in the fall.

Meat price controls end October 15, and President Truman issues an executive order November 9 lifting all wage and price controls except those on rents, sugar, and rice.

The U.S. inflation that will continue for decades begins December 14 as President Truman removes curbs on housing priorities and prices by executive order. President Roosevelt's inflation control order of April 1943 kept prices from climbing more than 29 percent from 1939 to 1945 as compared with a 63 percent jump in the 1914 to 1918 period, but Truman is more worried about a possible postwar depression than about inflation.

The McMahon Act passed by Congress August 1 establishes a civilian U.S. Atomic Energy Commission. Former TVA director David E. Lilienthal is sworn in November 1 as AEC chairman.

Researchers at the California Institute of Technology's Jet Propulsion Laboratories find that a liquid polysulfide polymer is the ideal propellant for rockets: it not only works as a fuel but also bonds directly to the walls of rocket shells, eliminating the need for heavy mechanical supports.

Soviet petroleum production begins a major expansion that will make the U.S.S.R. the world's largest oil producer within 30 years. Engineers exploit deposits between the Volga River and Ural Mountains where prospectors first found oil in the 1930s.

Kuwait Oil Co. is founded on a 50–50 basis by Gulf and Anglo-Iranian which will make Kuwait the Middle East's largest oil-producing nation (*see* 1938). The company will produce 1 billion barrels in the next 8 years and another billion in the following 3 (*see* Getty, 1949).

Dutch windmills dwindle to 1,400, down from 9,000 in the 1870s. The number will fall to 991 by 1960 and 950 by 1973 despite government efforts to preserve the polder mills that have kept the Netherlands from being flooded during the war.

London's Heathrow Airport opens on a formal basis May 21 (*see* 1929). The airport has been used since January for flights to Buenos Aires.

British transatlantic passenger air service to North America starts July 1 as BOAC begins flying Lockheed Constellations between London and New York. Scheduled flying time is 19 hours, 45 minutes.

Pan American Airways inaugurates the great circle route to Tokyo.

SAS (Scandinavian Airlines System) is created by a consolidation of Sweden's AB Aerotransport, Denmark's DDL, and Norway's DNL to operate on transatlantic routes.

Air India is created July 29 by a reorganization of Tata Air Lines which flies passengers at low rates to all the largest cities of India. Weekly Constellation service from Bombay to London via Cairo and Geneva will begin June 8, 1948, Air India will become a state corporation in 1953, will be serving Nairobi, Hong Kong, and Tokyo by 1955, and by 1960 will be serving Djakarta, Darwin, Sydney, Tashkent, Moscow, Prague, Frankfurt, and New York.

Alitalia (Aerolinee Italiane Internazionali) is created September 15 with British European Airways holding 30 percent of the shares in the new company.

IATA (International Air Transport Association) is founded. The trade association has 63 member airlines within a year and begins to set policies related to traffic, rates, and fares.

The DC-6 introduced by Douglas Aircraft can carry 70 passengers at 300 miles per hour with cargo, mail, and luggage (*see* 1938). Douglas has been the leading U.S. aircraft producer during the war but is only slightly ahead of Lockheed in sales (*see* DC-7, 1953).

Kaiser-Frazer automobiles are introduced in a venture that will mark the last major bid to challenge the established powers of the U.S. automobile industry. Henry J. Kaiser of Kaiser Aluminum has started the new company with former Willys-Overland president Joseph Washington Frazer, 54 (*see* 1950).

Bermuda permits motorcars to operate on the island after a half-century ban on vehicles other than bicycles, horse-drawn carriages and wagons, and motortrucks for special purposes.

Vespa motor scooters are introduced in Italy to provide cheap if noisy transportation. Aircraft maker Enrico Piaggio, whose works at Pontadera were destroyed by Allied bombers, has had his chief engineer Corradino d'Ascanio design the scooter to give him a product with which to resume production, and he calls it by the Italian name for "wasp."

ENIAC (electronic numerical integrator and computer) is the world's first automatic electronic digital computer and makes the Mark I electromechanical computer of 1944 obsolete. A massive device that employs 18,000 radio tubes and large amounts of electric power, operates on the principle that vacuum tubes can be turned on and off thousands of times faster than mechanical relays, and can make some 4,500 additions per second; although its tubes are not reliable and it functions only for brief periods, ENIAC begins a revolution in industrial technology. University of Pennsylvania electrical engineers John Presper Eckert, 27, and John William Mauchly, 39, have developed ENIAC following Mauchly's visit with University of Iowa scientist John Vincent Atanasoff, now 41, who has invented basic concepts of computer technology (*see* transistor, 1948; Univac, 1951).

Ford Motor Company engineer Delmar S. Harder, 54, coins the word "automation" for the system he has devised to manufacture automobile engines. His completely automatic process produces a new engine every 14 minutes and is the first complete self-regulating system applied to manufacturing (*see* assembly line, 1913).

A National Health Service Bill enacted by Parliament makes medical services free to all Britons. The effect of the new measure will be to lower dramatically Britain's infant mortality rate and maternal death rate to levels below those in the United States, and reduce death rates from bronchitis, influenza, pneumonia, tuberculosis, and some other diseases to levels below U.S. rates.

U.S. chemist Vincent Du Vigneaud, 45, synthesizes penicillin.

The Common Sense Book of Baby and Child Care by New York psychiatrist Benjamin Spock, 43, will be used to raise a generation of children along permissive lines of behavioral standards. It will be retitled *Baby and Child Care* and become an all-time best seller.

The Revised Version of the New Testament, published after 17 years of work, revises the *American Standard Version* of 1901. Forty-four Protestant denominations have appointed a committee of 22 biblical scholars headed by Yale University Divinity School dean Luther A. Weigle, 65. Its *Revised Standard Version of the Old Testament* will appear in 1952 (both bibles are based on the King James version of 1611).

Colleges and universities throughout the world struggle to cope with swollen enrollments after years in which many have gone bankrupt or come close. U.S. college enrollments reach an all-time high of more than 2 million as returning veterans crowd classrooms with help from the G.I. Bill of Rights (*see* 1944).

Fulbright awards for an international exchange of students and professors are initiated August 1 as President Truman signs into law a bill introduced by Sen. J. William Fulbright who was a Rhodes scholar in 1925 (*see* 1902). In the next 30 years, 120,000 Fulbright fellowships will be awarded; a master of Oxford's Pembroke College will tell Fulbright he is "responsible for the largest and most significant movement of scholars across the face of the earth since the fall of Constantinople in 1453," and Fulbright alumni will include heads of state by the dozen.

The BBC Third Programme begins in September to give British radio listeners more variety.

Xerography wins support from the Rochester, N.Y., firm Haloid Co. whose research director John H. Dessauer has seen an article on "electrophotography" by Chester Carlson in a July 1944 issue of *Radio News* (*see* 1938). Dessauer and his boss, Joseph C. Wilson, 36, travel to Columbus, Ohio, and see experiments conducted by the Batelle Memorial Institute, Haloid invests $10,000 to acquire production rights, and within 6 years the firm will raise more than $3.5 million to develop what will be called the Xerox copier (*see* 1950).

Die Welt (The World) begins publication at Hamburg under supervision of British occupation authorities. German publisher Axel C. Springer will acquire the daily in 1953.

Der Spiegel (*The Mirror*) begins publication at Hamburg. Two young British officers hire editor Rudolf Augstein, 23, who models the weekly on *Time* and *Newsweek* and will make it a powerful voice in publishing (*see* 1962).

Scientific American magazine is acquired after 101 years of publication by former LIFE magazine science editor Gerard Piel, 31, who will be joined by LIFE staffmen Dennis Flanagan, 27, and Donald H. Miller, Jr., in broadening the appeal of the monthly as science grows in its impact on the lives and careers of more Americans.

Family Circle magazine begins monthly publication in September after 14 years as a weekly supermarket giveaway. The cover price is 5¢, the September issue has 96 pages, some chains drop the magazine, others accept it as a profit-making item.

Nonfiction *The Chrysanthemum and the Sword: Patterns of Japanese Culture* by Ruth Benedict who has never been to Japan but whose book will nevertheless be a classic in its field; *Hiroshima* by John Hersey whose documentary account of last year's nuclear attack on the Japanese city occupies the entire August 31 issue of the *New Yorker* magazine and then appears in book form; *Journal, 1939–42* by André Gide; *The Art of Plain Talk* by Viennese-American author Rudolph Flesch, 35.

Fiction *Zorba the Greek (Vois Kai Politeia Tou Alexi Zorba)* by Nikos Kazantzakis; *The President (El Señor Presidente)* by Guatemalan novelist Miguel Asturias, 47; *The Berlin Stories* by Christopher Isherwood; *Delta Wedding* by Eudora Welty; *The Member of the Wedding* by Carson McCullers; *Memoirs of Hecate County* (stories) by Edmund Wilson; *Ladders to Fire* by U.S. novelist Anaïs Nin, 32; *Willowaw* by U.S. novelist Gore Vidal, 21; *All the King's Men* by Robert Penn Warren; *The Big Clock* by New York poet-novelist Kenneth Fearing, 44; *Back* by Henry Green; *Miracle de la rose* by playwright Jean Gênet; "The Idiot" ("Hakucid") by Japanese short-story writer Ango Sakaguchi, 40, who during the war came out against nationalism in "My Personal View of Japanese Culture" ("Nippon Bunka Shikan"); "The Courtesy Call" by Osamu Dazai, who has been the victim of a wartime air raid, has said that he was deceived by the emperor, and expresses a nihilistic feeling of despair (he will commit suicide in 1948 by throwing himself into a Tokyo reservoir); *The Foxes of Harrow* by U.S. novelist Frank Garvin Yerby, 36; *This Side of Innocence* by Taylor Caldwell; *B. F.'s Daughter* by John P. Marquand; *Mr. Blandings Builds His Dream House* by *Fortune* editor Eric Hodgins, 46; *I the Jury* by U.S. comic-strip writer Frank Morrison "Mickey" Spillane, 28 (critics scorn his raunchy Mike

1946 *(cont.)* Hammer detective novel emphasizing sex, sadism, and violence).

Poetry *Lord Weary's Castle* by U.S. poet Robert Lowell, 29, who has served a term in prison for his pacifism in World War II; *The Double Image* by English poet Denise Levertov, 23; *Deaths and Entrances* by Dylan Thomas; *Paterson* by William Carlos Williams.

Painting *Faun Playing the Pipe* and *Françoise with a Yellow Necklace* by Pablo Picasso whose mistress Françoise Gilot will give birth next year to his son Claude; *Composition with Branch* by Fernand Léger; *Eyes in the Heat* by Jackson Pollock. Joseph Stella dies at New York November 5 at age 69.

Pablo Picasso founds a pottery at Vallauris, France.

Ektachrome, introduced by Eastman Kodak, is the first color film a photographer can process himself. Much faster than the Kodachrome introduced in 1935 and 1936, the new film is available initially only in transparency sheet form, but Eastman will introduce it in roll film sizes next year and increase its speed several times *(see* Kodacolor, 1944; Tri-X, 1954).

Photographer Alfred Stieglitz dies at New York July 13 at age 82 survived by Georgia O'Keeffe, now 58.

Theater *Those Ghosts! (Questi fantasmi!)* by Italian playwright Edouardo De Filippo, 46, 1/12 at Rome's Teatro Eliseo; *Born Yesterday* by U.S. playwright-screenwriter-director Garson Kanin, 34, 2/4 at New York's Lyceum Theater, with Judy Holliday, Paul Douglas, Gary Merrill, 1,642 perfs.; *Santa Cruz* by Swiss playwright Max Frisch, 34, 3/7 at Zurich's Schauspielhaus; *Beaumarchais, or the Birth of Figaro* by Friedrich Wolf 3/8 at Berlin's Deutsches Theater; *A Phoenix Too Frequent* by English playwright Christopher Fry, 38, 4/25 at London's Mercury Theatre; *The Winslow Boy* by Terence Rattigan 9/6 at London's Lyric Theatre, 215 perfs.; *The Iceman Cometh* by Eugene O'Neill 10/9 at New York's Martin Beck Theater, with James Barton, Carl Benton Reid, Dudley Digges, 136 perfs.; *The Chinese Wall (Die chinesische Mauer)* by Max Frisch 10/10 at Zurich's Schauspielhaus; *Filomenca Marchurano by* Edouardo De Filippo 11/7 at Naples; *The Respectful Prostitute (La putain respecteuse)* by Jean-Paul Sartre 11/8 at the Théâtre Antoine, Paris; *Another Part of the Forest* by Lillian Hellman 11/20 at New York's Fulton Theater, with Patricia Neal, Mildred Dunnock, 182 perfs.

Films Jean Cocteau's *Beauty and the Beast* with Jean Marais; William Wyler's *The Best Years of Our Lives* with Fredric March, Myrna Loy, Teresa Wright, Dana Andrews; Howard Hawks's *The Big Sleep* with Humphrey Bogart, Lauren Bacall, Martha Vickers; David Lean's *Brief Encounter* with Trevor Howard, Celia Johnson; David Lean's *Great Expectations* with John Mills, Valerie Hobson; Sidney Gilliat's *Green for Danger* with Alistair Sim; Frank Capra's *It's a Wonderful Life* with James Stewart, Lionel Barrymore, Thomas Mitchell, Donna Reed; Robert Siodmak's *The Killers* with Burt Lancaster, Ava Gardner; John Ford's *My Darling Clementine* with Henry Fonda, Victor Mature, Linda Darnell; Carol Reed's *Odd Man Out* with James Mason; Roberto Rosselini's *Open City* with Anna Magnani, Aldo Fabrizi; Tay Garnett's *The Postman Always Rings Twice* with Lana Turner, John Garfield; Vittorio De Sica's *Shoeshine* with Rinaldo Smerdoni, Franco Interlenghi; Michael Powell and Emeric Pressburger's *Stairway to Heaven* with David Niven, Kim Hunter. Also: John Cromwell's *Anna and the King of Siam* with Irene Dunne, Rex Harrison; Basil Dearden's *The Captive Heart* with Michael Redgrave; Ernst Lubitsch's *Cluny Brown* with Charles Boyer, Jennifer Jones; Irving Rapper's *Deception* with Bette Davis, Claude Rains, Paul Henreid; Claude Autant-Lara's *Devil in the Flesh* with Gerard Philipe, Micheline Presle; Jean Negulesco's *Humoresque* with Joan Crawford, John Garfield; Zoltan Korda's *The Macomber Affair* with Gregory Peck, Joan Bennett; Alfred Hitchcock's *Notorious* with Cary Grant, Ingrid Bergman, Claude Rains; Edmund Goulding's *The Razor's Edge* with Tyrone Power, Gene Tierney; Wilfred Jackson's (animated) and Harve Foster's *Song of the South* with Bobby Driscoll, James Baskett, Ruth Warrick; Robert Siodmak's *The Spiral Staircase* with Dorothy McGuire, George Brent, Ethel Barrymore; Jean Delannoy's *La Symphonie Pastorale* with Pierre Blanchar, Michele Morgan; Jean Negulesco's *Three Strangers* with Sydney Greenstreet, Geraldine Fitzgerald, Peter Lorre; Kenji Mizoguchi's *Utamaro and His Five Women*; Clarence Brown's *The Yearling* with Gregory Peck, Jane Wyman, Claude Jarman, Jr.

Hollywood musicals Stuart Heisler's *Blue Skies* with Bing Crosby, Fred Astaire, Joan Caulfield, music and lyrics by Irving Berlin, songs that include "A Couple of Song and Dance Men;" Vincente Minnelli's *Ziegfeld Follies* with William Powell, Judy Garland, Fred Astaire, Gene Kelly; George Sidney's *The Harvey Girls* with Judy Garland, music by Harry Warren, lyrics by Johnny Mercer, songs that include "The Atchison, Topeka and the Santa Fe" *(see* Fred Harvey, 1876); Alfred E. Green's *The Jolson Story* with Larry Parks and with the voice of Al Jolson, now 60, singing "April Showers," "Mammy," "Swanee," and other songs he made famous; H. Bruce Humberstone's *Three Little Girls in Blue* with June Haver, Vivian Blaine, Celeste Holm, songs that include "You Make Me Feel So Good" and "On the Boardwalk at Atlantic City" by Joseph Myron, lyrics by Mack Gordon.

Stage musicals *Lute Song* 2/6 at New York's Plymouth Theater, with Mary Martin, Yul Brynner, music by Raymond Scott, lyrics by Bernard Harrigan, songs that include "Mountain High, Valley Low," 142 perfs.; *St. Louis Woman* 3/30 at New York's Martin Beck Theater, with Pearl Bailey heading an all-black cast, music by Harold Arlen, lyrics by Johnny Mercer, songs that include "Come Rain or Come Shine," 113 perfs.; *Call Me Mister* 4/18 at New York's National Theater, with Betty Garrett, music and lyrics by Harold Rome, book by Arnold Auerbach and Arnold Horwitt, songs that include "South America, Take It Away," 734 perfs.; *Annie Get Your Gun* 5/16 at New York's Imperial Theater, with Ethel Merman, Ray Middleton, book based on the life of markswoman Phoebe "Annie Oakley" Mozee of Buffalo Bill's Wild West Show (*see* 1883), music and lyrics by Irving Berlin, songs that include "Anything You Can Do," "Doin' What Comes Naturally," "The Girl That I Marry," "You Can't Get a Man with a Gun," "I Got the Sun in the Morning," "There's No Business Like Show Business," 1,147 perfs.; *Pacific 1860* 12/19 at London's Theatre Royal in Drury Lane, with Mary Martin, music and lyrics by Noël Coward, 129 perfs.

Opera *The Medium* 5/8 at Columbia University's Brander Matthew Theater, with music by Italian-American composer Gian-Carlo Menotti, 34; *The Rape of Lucretia* 7/12 at Glyndebourne, with music by Benjamin Britten.

Ballet *The Serpent Heart* 5/10 at Columbia University's McMillin Theater, with Martha Graham, music by Samuel Barber, choreography by Graham; *The Four Temperaments* 11/20 at New York's Central High School of the Needle Trades, with Gisella Caccialanza, Tanaquil LeClerq, Todd Bolender, music by Paul Hindemith, choreography by George Balanchine.

First performances Symphony in Three Movements by Igor Stravinsky 1/24 at New York's Carnegie Hall; *Metamorphosen* by Richard Strauss 1/25 at Zurich; Concerto for Clarinet and Orchestra by Darius Milhaud 1/30 at the Marine Barracks, Washington, D.C.; Concerto No. 3 for Piano and Orchestra by the late Béla Bartòk 2/8 at Philadelphia's Academy of Music. Bartòk died in poverty at New York last September at age 64; Concerto for Oboe and Chamber Orchestra by Strauss 2/26 at Zurich; Scherzo a la Russe by Stra-vinsky 3/22 at San Francisco; Sonatina No. 2 for 16 Wind Instruments by Strauss 3/25 at Winterthur, Switzerland; *Ebony Concerto* by Stravinsky 3/25 at New York's Carnegie Hall, with Woody Herman as clarinet soloist; Concerto for Violoncello and Orchestra by Samuel Barber 4/5 at Boston's Symphony Hall; Symphony No. 3 by Charles Ives 4/5 at New York; *The Bells* symphonic suite by Milhaud 4/26 at Chicago; *The Unanswered Question (A Comic Landscape)* by Ives (who wrote it in 1908) 5/11 at Columbia University; *For Those We Love* requiem by Paul Hindemith 5/14 at New York's City Center, with Robert Shaw's Collegiate Chorale, text from the Walt Whitman poem "When Lilacs Last in the Dooryard Bloom'd"; Concerto No. 3 for Piano and Orchestra by Milhaud 5/26 at Prague; Symphony for Strings by William Schuman (above) 7/12 at London; Symphonie No. 3 *(Liturgique) by* Arthur Honegger 8/17 at Zurich's Tonhalle; String Quartet No. 2 by Ives (who wrote it between 1907 and 1913) 10/7 at New York's Town Hall; Quartet No. 2 for Strings by Ernest Bloch 10/9 at London's Wigmore Hall; Symphony No. 3 by Aaron Copland 10/18 at Boston's Symphony Hall; Concerto for Cello in E major by Aram Khatchaturian 10/30 at Moscow; Concerto No. 2 for Violoncello and Orchestra by Milhaud 11/28 at New York's Carnegie Hall; Ricercari for Piano and Orchestra by U.S. composer Norman Dello Joio, 33, 12/19 at New York's Carnegie Hall; Symphony No. 2 by Milhaud 12/20 at Boston's Symphony Hall; *Minstrel Show by* Morton Gould 12/21 at Indianapolis.

The first vinylite phonograph record appears in October. RCA-Victor issues a new recording of the 1895 Richard Strauss work *Till Eulenspiegels Lustige Streiche,* but vinylite will not displace shellac until the perfection of long-playing records (see 1948).

Popular songs "La Vie en Rose" by Italian-born Paris composer Louiguy (Louis Guglielmi), 30, lyrics by French chanteuse Edith Piaf (Edith Givanna Gassion), 31; "(Get Your Kicks on) Route 66" by Bob Troup; "To Each His Own" by Raymond B. Evans, 31, lyrics by Jay Livingston, 31 (title song for film); "Put the Blame on Mame" by Allan Robert and Doris Fischer (for the film *Gilda*); "Tenderly" by Walter Gross, lyrics by Jack Lawrence; "You Call Everybody Darling" by Sam Martin, Ben Trace, Clem Walls; "Ole Buttermilk Sky" by Hoagy Carmichael and Jack Brooks (for the film *Canyon Passage);* "Stella by Starlight" by Victor Young, lyrics by Ned Washington (for the film *The Uninvited*); "Seems Like Old Times" by Carmen Lombardo and John Jacob Loeb; "The Christmas Song (Chestnuts Roasting on an Open Fire)" by U.S. songwriter-vocalist Mel Tormé, 21, and Robert Wells.

World chess champion Aleksandr Alekhine dies March 24 at age 54 on the eve of a title match against Soviet chess master Mikhail Botvinnik, 35, who has beaten five other seeded World Chess Federation players including Polish-American master Samuel Reshevsky, 35, who held the U.S. title from 1936 to 1944. The world title will remain vacant until 1948.

King Ranch 3-year-old Assault wins U.S. racing's Triple Crown.

Yvon Petra, 30 (Fr), wins in men's singles at Wimbledon as tournament play resumes after a 6-year lapse; Pauline Betz wins in women's singles. Jack Kramer, 24, wins in men's singles at Forest Hills, Betz in women's singles.

The St. Louis Cardinals win the World Series by defeating the Boston Red Sox 4 games to 3.

An automatic Pinspotter displayed in prototype at a national bowling tournament in Buffalo, N.Y., by American Machine and Foundry Co. (AMF) begins a revolution in bowling. Devised by Pearl River, N.Y., inventor Fred Schmidt, the patented vacuum Pinspotter has been rejected by Brunswick Balke-Collender, which will scramble to catch up with AMF as bowling becomes the leading U.S. participation sport.

The skimpy two-piece bikini swimsuit designed by French couturier Louis Reard is modeled (by a stripper) at a Paris fashion show July 5, 4 days after the first U.S. atomic bomb test (above) and creates a sensation. Banned at Biarritz and other resorts, it will not be seen on U.S. beaches until the early 1960s.

Tide, introduced by Procter & Gamble, is the first detergent strong enough for washing clothes as well as dishes (*see* Persil, 1907). By the end of 1949 one out of every four U.S. washing machines will be using Tide.

1946 *(cont.)* The new Westinghouse Laundromat is a front-loading machine that requires a low-sudsing soap or detergent.

British-made nylon hosiery go on sale in December. The stockings are the first nylon consumer goods made in Britain (*see* 1940).

Timex watches are introduced by Norwegian-American entrepreneur Joakim Lehmkuhl, 51, who in 1940 was sent to New York by his government in exile to take charge of the Norwegian Shipping Center and in 1943 was elected president of Waterbury Watch Co., famous for its Ingersoll watch (*see* 1892). Lehmkuhl made it a wartime producer of timing mechanisms for bomb and artillery shell fuses and now moves it back into watch production, dropping the Ingersoll name because it is associated with the $1 price. While $1 will no longer buy a watch, Lehmkuhl's mass-produced Timex at $6.95 and up will soon account for 40 percent of U.S. wristwatch sales as Waterbury Watch becomes U.S. Time Corp. (*see* Polaroid, 1950).

Seiko watches are introduced in Japan by K. Hattori & Co. Founder Kintaro Hattori began in 1881 selling foreign-made watches to high government officials. He made his own clocks starting in 1892.

Estée Lauder makes her first sale to Saks Fifth Avenue. The New York beautician, born Josephine Esther Mentzer, has joined with her husband Joseph, 38, to make and market cosmetics that will grow to outsell those of Elizabeth Arden, Helena Rubinstein, or Revlon.

War-torn Europe and Japan rebuild their cities out of the rubble left by the worst destruction thus far in the history of warfare. Some European cities such as Warsaw are reconstructed in part from photographs to resemble their prewar appearance, Tokyo is rebuilt as it was after the 1923 earthquake.

The Flamingo Hotel goes up at Las Vegas, beginning the transformation of the Nevada city into a resort of grandiose hotel-casinos (*see* legalized gambling, 1931). "Murder Incorporated" veteran Benjamin "Buggsy" Siegel has built the Flamingo with backing from syndicate boss Meyer Lansky, 42. (Siegel will be killed in June 1947 by Chicago and New York mobsters.)

An earthquake in the Aleutian Trench April 1 shifts the sea floor and creates a seismic wave (tsunami). It plucks the Scotch Cap lighthouse from a rock 45 feet above the sea on Unimak, reaches Hawaii 5 hours later, kills 173, and injures 163.

The worst red tide in U.S. history begins in November on Florida's west coast, killing fish and shellfish by the millions and spurring efforts by the Fish and Wildlife Service to seek ways to the control *G. breve* (*see* 1916; 1932).

A Japanese land reform act drafted by U.S. occupation authorities disposesses absentee landlords and sharply limits the amount of cultivable land an individual may own. Tenant farmers decline from nearly one-half the total to one-tenth.

Blizzards destroy wheat crops throughout Europe. Freak storms sweep over much of Britain at harvest time, wrecking crops in Suffolk after a wet summer has ruined hay crops in the north, in the mountains of Wales, and on the southern moors.

A world wheat shortage forces Britain to ration bread which was never rationed during the war. Restrictions on most other staple foods ensue.

A National School Lunch Act is signed into law June 4 as food prices soar (*see* Britain, 1906).

Court rulings invalidate vitamin D irradiation patents granted to Harry Steenbock in 1927. Steenbock's finding was a "discovery" rather than a patentable invention, say the courts. Even a farmer who lets his alfalfa be exposed to the ultraviolet rays of the sun would technically be infringing on the patents which have produced $9 million in royalties to the Wisconsin Alumni Foundation and funded research programs and professors' salaries.

U.S. bread consumption drops as wheat and flour are exported to starving Europe. Domestic shortages lead to a compulsory long-extraction program, and demand for flour whiteners jumps as Americans are put off by bread that is not as white as usual.

Self-rising corn meal is marketed for the first time.

An electric rice saucepan introduced by Konoskuke Matsushita, now 55, is the first low-priced Japanese electrical cooking appliance; it revolutionizes Japanese cooking (*see* 1918).

U.S. frozen orange juice sales reach 4.8 million 6-ounce cans, but most consumers find the product unacceptable. A "cutback" technique developed by Louis Gardner MacDowell over-concentrates the juice and then adds fresh juice to the concentrate. C. D. Atkins and E. L. Moore help him refine the process to produce a concentrate whose sweetness, flavor, and acidity are more uniform after dilution than fresh-squeezed orange juice. The new frozen concentrate will find wide acceptance and boom demand for oranges, which are now in such oversupply that Florida growers have been cutting down trees and replacing them with avocado trees (*see* 1945; 1947).

U.S. birthrates soar as returning servicemen begin families and add to existing ones. The nation has 3,411,000 births, up from 2,858,000 last year.

The United States has 140 million people, Britain 46 million, France 40, West Germany 48, East Germany 18, Spain 21, Brazil 45, Mexico 22, the U.S.S.R. 194, Japan 73, Korea 24, China 555, India 311.

Some 2.5 million Sudeten Germans are transferred to Germany and 415,000 Karelian Finns to Finland.

Palestine has 650,000 Jews, many of whom have immigrated illegally to find sanctuary from European persecution, and 1.05 million Arabs (*see* 1948).

1947 Arabs and Jews reject a final British proposal for division of Palestine into Arab and Jewish zones administered as a trusteeship February 7. Britain refers the question to the United Nations, whose general assembly votes November 29 for partition, with Jerusalem to be under a UN trusteeship. The Jews approve the plan, the Arabs reject it. The Arab League announces December 17 that it will use force to resist partition, and raids begin against Jewish communities in the Holy Land (*see* Israel, 1948).

The Truman Doctrine outlined in a presidential message to Congress March 12 announces plans to aid Greece and Turkey and proposes economic (and military) aid to countries threatened by Communist takeover.

Greece's George II dies April 1 at age 56 after a second reign of 12 years. He is succeeded by his brother, 45, who will reign until 1964 as Paul I.

Denmark's Christian X dies April 20 at age 76 after a 35-year reign and is succeeded by his son, 48, who will reign until 1972 as Frederick IX.

Vincent Auriol has become president of France January 16, Gen. de Gaulle has assumed control of the nationwide R.P.F. (Rassemblement du Peuple Français) party April 14, it emerges as the strongest group in the October municipal elections (the Communists are second), many peasants refuse to deliver their grain after a poor harvest, strikes in November affect nearly 2 million workers, and a new cabinet takes office November 23 under Robert Schuman, 61.

Hungarian Communists backed by the Red Army seize power in a coup d'état May 30 while Prime Minister Ferenc Nagy, 43, is on holiday in Switzerland. Hungary becomes a Soviet satellite (*see* 1956).

An article signed "X" in the July issue of *Foreign Affairs* magazine proposes a policy of "containment" toward the Soviet Union. The author is the new head of the State Department's policy planning staff George Frost Kennan, 43, who helped set up the U.S. embassy at Moscow in 1943. Secretary of State Marshall and his successors will adopt the essentials of Kennan's policy.

The Central Intelligence Agency (CIA) authorized by Congress in a National Security Act July 26 in response to an order by President Truman works to counter activities by Moscow, which is pouring money into Western Europe and attempting through local Communist parties to establish governments (*see* 1949). The president has acted on the advice of Secretary of State Marshall and James V. Forrestal, the new secretary of defense (below).

India gains independence from Great Britain August 15 and becomes a dominion following endorsement of a plan to partition the subcontinent by the Muslim League and the All-India Congress. Jawaharlal Nehru, 58, becomes first prime minister of Hindu India.

Britain sets up Pakistan as an independent Muslim state bordering India to the west and east with her capital at Karachi and names Mohammed Ali Jinnah, 71, of the Muslim League as governor general. A dispute over control of Kashmir is referred to the United Nations December 30 after millions have died in bloody riots following partition (*see* 1948).

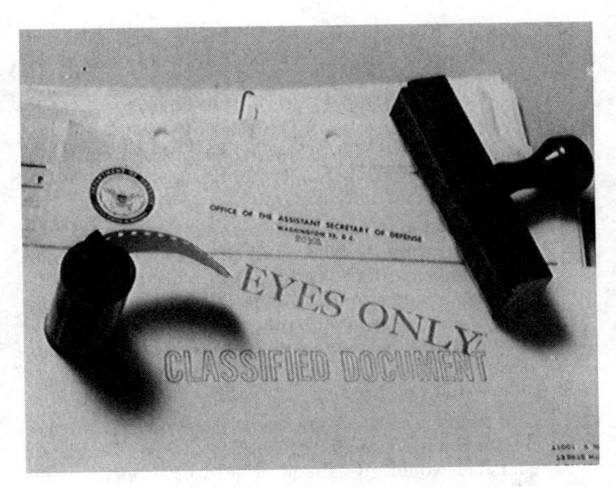

The CIA was created to counter Soviet espionage activities, but it would sometimes exceed its authority.

Romania's Michael abdicates under Communist pressure December 30 after a 7-year reign. The Romanian monarchy begun by Carol I in 1866 is ended and Romania becomes a Communist state.

India outlaws "untouchability" in the age-old caste system, but discrimination against the people Mahatma Gandhi calls *harijans*—who represent 15 percent of the population—will continue for decades (*see* 1932; 1946).

The first U.S. major league black baseball player signs with the Brooklyn Dodgers (below).

The Taft-Hartley Act passed over President Truman's veto June 23 restricts organized labor's power to strike, outlaws the closed shop (which requires that employers hire only union members), prohibits use of union funds for political purposes, introduces an 80-day "cooling off" period before a strike or lockout can begin, and empowers the government to obtain injunctions where strikes "will imperil the national health or safety" if allowed to occur or continue.

Congress forbids the CIA (above) to have "police, subpoena, law enforcement powers or internal security functions" inside the United States lest the agency conflict with the FBI and jeopardize the privacy of American citizens, but beginning in the 1950s and especially after 1967 CIA operatives will undertake domestic surveillance activities, including break-ins, surreptitious mail inspections, and telephone wiretaps, and will compile intelligence files on at least 10,000 Americans in violation of the National Security Act.

A "Hollywood Black List" of alleged Communist sympathizers compiled at a conference of studio executives meeting at New York's Waldorf-Astoria Hotel names an estimated 300 writers, directors, actors, and others known or suspected to have Communist party affiliations or of having invoked the Fifth Amendment against self-incrimination when questioned by the

1947 ***(cont.)*** House Committee to Investigate Un-American Activities. The "Hollywood Ten" who refuse to tell the committee whether or not they have been Communists are Alvah Bessie, Herbert Biberman, Lester Cole, Edward Dmytryk, Ring Lardner, Jr., John Howard Lawson, Albert Maltz, Samuel Ornitz, Adrian Scott, and screenwriter Dalton Trumbo, 39, who has in fact been a party member since 1943.

The film industry blacklists the "Hollywood Ten" (above) November 25 and all draw short prison sentences for refusing to testify. Dalton Trumbo's screen credits include the 1943 film *Tender Comrade* with Ginger Rogers whose mother has tearfully testified before the Dies Committee that her daughter had to utter the "Communist line" in the film "Share and share alike—that's democracy" (*see* 1950).

$ The Marshall Plan proposed at Harvard commencement exercises June 5 by Secretary of State George C. Marshall would give financial aid to European countries "willing to assist in the task of recovery."

The General Agreement on Tariffs and Trade (GATT) signed by the major world powers will lead to a significant lowering of tariff barriers, end some tariff discrimination, and help revitalize world trade.

President Truman asks Congress November 17 to appropriate $597 million for immediate aid to France, Italy, and Austria. Congress authorizes $540 million December 23 for interim aid to France, Italy, Austria, and China and in the next 40 months will authorize $12.5 billion in Marshall Plan aid to restore the economic health of free Europe (and halt the spread of Communism).

U.S. occupation authorities in Japan break up the *zaibatsu* that has controlled the nation's industry, but the old family companies that formed the consortium will remain strong factors in the Japanese economy (they include Mitsubishi, Mitsui, Sumitomo, and others).

Henry Ford dies April 7 at age 83; his will leaves 90 percent of his Ford Motor Company stock to the Ford Foundation (*see* Edsel Ford, 1943). The federal tax paid by Ford's heirs amounts to only $21 million on a taxable estate of $70 million since the bulk of the $625 million fortune has been placed tax free in the Ford Foundation which becomes the richest philanthropic organization in the history of the world. Ford's heirs are left in control of Ford Motor Company, and the Ford Foundation is provided with more funds for useful projects in education and other fields (*see* 1950).

U.S. coal mines return to private ownership June 30 after operation by the federal government since May 22 of last year. UMW workers threaten a new strike but receive a wage boost of 44.375¢ per hour July 7.

The United Mine Workers withdraws from the AF of L in December as it did in 1936.

Britain nationalizes electricity production under a new Electricity Authority that takes over 550 private electric companies.

Britain's first atomic pile comes into operation in August at Harwell.

Britain nationalizes transportation. Parliament gives a new Transportation Company responsibility for organizing and rehabilitating the nation's bankrupt and antiquated railroads and reorganizing 3,000 trucking firms (*see* 1948; 1953).

The first tubeless automobile tires, introduced by B. F. Goodrich, seal themselves when punctured.

The round-the-world speed record set by Howard Hughes in 1938 falls August 10 when U.S. pilot William P. Odum arrives at Chicago's Douglas Airport after a flight of 73 hours, 5 minutes, 11 seconds—more than 18 hours faster than the 1938 record.

A U.S. Bell X-1 rocket plane piloted by U.S. Air Force captain Chuck Yeager, 24, reaches Mach 1.06 (750 miles per hour) October 14 and breaks the sound barrier broken up to now only by planes diving earthward with help from gravity.

The "Spruce Goose" built and flown by Howard Hughes taxis across Long Beach Harbor in California November 2 and takes off for a one-mile flight at a maximum altitude of 70 feet. Largest aircraft ever built, the 140-ton eight-engine seaplane, made largely of birch, has a wingspan of 320 feet. Hughes has built it as a prototype troop transport, but the Pentagon rejects his design and the huge plane goes into storage, never to fly again.

Avions Marcel Dassault Breguet Aviation is founded by French aviation pioneer Marcel-Ferdinand (Bloch) Dassault of 1916 variable pitch propeller fame who survived 3 years' imprisonment at Buchenwald and has converted to Roman Catholicism, adopting the pseudonym used by one of his brothers in the French resistance ("d'assault" means literally "on the attack"). Now 55, Dassault will produce Mirage fighter planes and make his company France's leading aeronautical firm.

The finding that sexual reproduction occurs in bacteria will open up a whole new world of study and lay the groundwork for future work in bacterial genetics. Columbia University graduate student Joshua Lederberg, 22, and Yale geneticist Edward Lawrie Tatum, 37, make the discovery.

British physicist Patrick Maynard Stuart Blackett, 50, at the University of Manchester advances the theory that "all massive rotating bodies are magnetic." He has worked on cosmic rays and especially on the electrical particles known as "mesons."

British archaeologist Francis Steele reconstructs the 18th century B.C. Hammurabi law code from excavations made at Nippur before 1900.

The discovery by U.S. chemist Willard Frank Libby, 38, that all organic materials contain carbon-14 atoms which decay at a measurable rate begins development of an "atomic clock" that will determine geological age and clear up many mysteries of archaeology and anthropology (see 1940). Libby's finding will make it possible for the first time to date organic archaeological remains within narrow limits of time.

A Bedouin boy exploring a cave at Qumran, northwest of Palestine's Dead Sea, discovers an earthenware jar containing scrolls of parchment containing all but two small parts of the Old Testament Book of Isaiah. Written in the 1st century B.C. by Jews of the obscure, ascetic Essene sect which was later wiped out by the Romans, the parchments have been wrapped in yards of cloth and covered with pitch. Sold piecemeal by the boy who found them, they will greatly expand knowledge of ancient Judaism, and will be followed by several more finds of biblical manuscripts in the area.

Oklahoma faith healer Oral Roberts, 29, claims to pray for God's help in healing the ill and deformed who are then healed by God himself. Evangelist Roberts will start the healing ministry Healing Waters next year and appear frequently on radio and television; his Pentecostal Holiness Church will grow to have 2 million members by 1960.

Sony Corp. has its beginnings in the Tokyo Telecommunications Co. (Tokyo Tsushin Kogyo) started by Japanese electrical engineer Masaru Ibuka, with backing from sake brewing heir Akio Morita, 26. Ibuka has worked on infrared detection devices and a telephone scrambler for the military, taken over a gutted and boarded-up Tokyo department store, and started a factory to produce shortwave converters for radio sets that will enable listeners to receive news from abroad. Morita has read about Ibuka's device and joins him in founding the company that will be renamed Sony Corp. in 1958 (*see* tape recorder, 1950).

Publisher Marshall Field III acquires the *Chicago Sun* and merges it with the *Times* to create the *Sun-Times* (*see* 1959).

"Steve Canyon" makes its debut on newspaper cartoon pages. After 13 years of drawing "Terry and the Pirates," Milton Caniff leaves the "Terry" strip to be carried on by others.

Nonfiction *Gamesmanship* by BBC writer Stephen Potter, 47; *The Proper Bostonians* by former *Harvard Crimson* editor Cleveland Amory, 29.

Fiction *The Plague (La Peste)* by Albert Camus; *The Twins of Nuremberg (Die Zwillingen von Nurnberg)* by German novelist Hermann Kesten, 47; *Nekya (Nekyia)* by German novelist Hans Erich Nossack, 46; *The Woman of Rome (La Romana)* by Alberto Moravia; *The Path of the Nest of Spiders (Il sentiero de nidi di ragno)* by Italian novelist Italo Calvino, 24; *Under the Volcano* by English novelist Malcolm Lowry, 38; *Tea with Mrs. Goodman* by English novelist Philip Toynbee, 31; *Querelle of Brest (Querelle de Brest)* by Jean Genet; *Hetty Dorval by* Canadian novelist Ethel Wilson, 57; *The Middle of the Journey* by Columbia University professor Lionel Trilling, 42; *Euridice* by José Lins do Rego; *Thousand Cranes* by Yasunari Kawabata; *The Setting Sun (Shayo)* by Osamu Dazai examines an aristocratic family which has been forced at the end of the war to cope with a new life devoid of its former wealth and prestige. Japan's new poor will be called Setting Sun people (*shayo-zoku*); *Tales of the South Pacific* by U.S. writer James (Albert) Michener, 40, who served in the Pacific during the war; *The Victim* by Saul Bellow; *Gentleman's Agreement* by U.S. novelist Laura Zametkin Hobson, 47, examines the covert anti-Semitic practices institutionalized in U.S. society; *The Wayward Bus* by John Steinbeck.

Poetry *The Age of Anxiety* by W. H. Auden; "Day After Day" ("*Giorno dopo Giorno*") by Salvatore Quasimodo.

Painting *Ulysses with His Sirens* by Pablo Picasso; *Young English Girl* by Henri Matisse; *M. Plume, Portrait of Henri Michaux* by French artist Jean Dubuffet; *Das Matterhorn* by Oskar Kokoschka; *The Blind Leading the Blind* (wood) and *Quarantania* (painted wood on wooden base) by French painter Louise Bourgeois, 36; *Y* by U.S. painter Clyfford Still, 41; *Full Fathom Five* by Jackson Pollock; *Betrothal II*, *Dark Green Painting*, *The Plan and the Song*, and *Agony* by Arshile Gorky. Pierre Bonnard dies at Le Cannet north of Cannes January 23 at age 79.

Sculpture *Three Standing Figures* by Henry Moore; *Pointing Man* by Alberto Giacometti.

The Polaroid Land Camera patented by Edwin H. Land of 1932 polaroid lens fame develops its own films within its body and produces a sepia print in 60 seconds (*see* 1948).

ASA ratings developed by the American Standards Association standardize U.S. film speeds.

Theater *All My Sons* by U.S. playwright Arthur Miller, 32, 1/29 at New York's Coronet Theater, with Ed Begley, 328 perfs.; *The Prophet's Diamond (Il diamante del profeta)* by Italian playwright Carlo Terron, 33, 4/2 at Rome's Teatro Valle; *Command Decision* by U.S. playwright William Wister Haines, 39, 10/1 at New York's Fulton Theater, with James Whitmore, Paul Kelly, 408 perfs.; *Medea* by Robinson Jeffers 10/20 at New York's National Theater, with Judith Anderson, John Gielgud, Florence Reed, 214 perfs.; *Ring Around the Moon (L'invitation au château)* by Jean Anouilh 11/4 at the Théâtre de l'Atelier, Paris; *A Streetcar Named Desire* by Tennessee Williams 12/3 at New York's Ethel Barrymore Theater, with Marlon Brando, 23, as Stanley Kowalski, Jessica Tandy as Blanche DuBois, Kim Hunter, Karl Malden, 855 perfs.

Tony Awards established by the American Theatre Wing honor outstanding Broadway plays, directors, performers, scenic designers, costumers, etc. The name Tony honors Antoinette Perry who headed the Theatre Wing during World War II; the awards rival the Oscars given since 1928 by the Motion Picture Academy of Arts and Sciences.

Films Michael Powell and Emeric Pressburger's *Black Narcissus* with Deborah Kerr, Flora Robson, Sabu, Jean Simmons; Robert Rossen's *Body and Soul* with John Garfield, Lilli Palmer; Elia Kazan's *Boomerang* with Dana Andrews, Jane Wyatt, Lee J. Cobb; Michael Powell and Emeric Pressburger's *I Know Where I'm Going* with Wendy Hiller, Roger Livesey, Finlay Currie; Michael Curtiz's *Life With Father* with Irene Dunne, William Powell; George Seaton's *Miracle on 34th Street* with Edmund Gwenn, Maureen O'Hara, John Payne, Natalie Wood; Luchino Visconti's *La Terra Trema*. Also: Jules Dassin's *Brute Force* with

1947 ***(cont.)*** Burt Lancaster, Hume Cronyn; George Cukor's *A Double Life* with Ronald Colman, Signe Hasso; H. C. Potter's *The Farmer's Daughter* with Loretta Young, Joseph Cotten; Joseph L. Mankiewicz's *The Ghost and Mrs. Muir* with Gene Tierney, Rex Harrison; Charles Chaplin's *Monsieur Verdoux* with Chaplin, Martha Raye; Edmund Goulding's *Nightmare Alley* with Tyrone Power, Joan Blondell; Jacques Tourneur's *Out of the Past* with Robert Mitchum, Kirk Douglas, Jane Greer; Roberto Rosselini's *Paisan* with Harriet White; Albert Lewis's *The Private Affairs of Bel Ami* with George Sanders, Angela Lansbury; Robert Montgomery's *Ride the Pink Horse* with Montgomery, Wanda Hendrix; Irving Pichel's *They Won't Believe Me* with Susan Hayward, Robert Young; Irving Rapper's *The Voice of the Turtle* with Ronald Reagan, Eleanor Parker.

The Edinburgh Festival of the Arts has its first season. The annual summer event will attract performing artists to Scotland from all over the world.

Opera *The Telephone* 2/18 at New York, with music by Gian Carlo Menotti together with a revised version of his last year's opera *The Medium*; *The Trial of Lucullus* 4/18 at the University of California, Berkeley, with music by Roger Sessions, libretto from the 1939 Bertolt Brecht radio play; *The Mother of Us All* 5/7 at Columbia University's Brander Matthews Hall with Dorothy Dow as Susan B. Anthony, music by Virgil Thomson, libretto by the late Gertrude Stein who died at Paris last July at age 72; *Les Mamelles de Tiresias* 6/3 at the Opéra-Comique, Paris, with music by Francis Poulenc, libretto from a surrealist play by Guillaume Apollinaire; *Albert Heering* 6/20 at Glyndebourne, with music by Benjamin Britten.

Ballet *Night Journey* 5/3 at Cambridge, Mass., with Martha Graham, music by William Schuman; *The Seasons* 5/17 at New York, with music by John Cage.

First performances Symphony No. 2 by Roger Sessions 1/9 at San Francisco; Symphony No. 4 by Arthur Honegger 1/21 at Basel; *Symphonia Serena* by Paul Hindemith 2/1 at Dallas; Concerto for Violin and Orchestra in D minor by George Antheil 2/9 at Dallas; Symphony No. 3 by Morton Gould 2/16 at New York's Carnegie Hall; *Bachianas Brasilieras No.* 3 for Piano and Orchestra by Heitor Villa-Lobos 2/19 in a CBS Orchestra broadcast; Concerto for Piano and Orchestra by Hindemith 2/27 at Cleveland's Severance Hall, with Jesus Maria Sanroma as soloist; Suite for Harmonica and Orchestra by Darius Milhaud 5/28 at Paris with Larry Adler as soloist; *Bachianas Brasilieras No. 8* by Villa-Lobos 8/6 at Rome; Symphony No. 6 in E flat minor by Serge Prokofiev 10/10 at Leningrad; Symphony No. 3 *(Hymnus Ambrosianus)* by Milhaud 10/30 at Paris.

Broadway musicals *Finian's Rainbow* 1/10 at the 46th Street Theater, with Ella Logan, David Wayne, music by Burton Lane, lyrics by E. Y. Harburg, songs that include "If This Isn't Love," "How Are Things in Glocca Mora," "Old Devil Moon," "Look to the Rainbow," "When I'm Not Near the Girl I Love," "That Great Come-and-Get-It-Day," 725 perfs. The musical breaks new ground with social commentary on subjects ranging from the population explosion to the maldistribution of wealth; *Brigadoon* 3/13 at the Ziegfeld Theater, with James Mitchell, David Brooks, music by Frederick Loewe, book and lyrics by Alan Jay Lerner, 28, songs that include "The Heather on the Hill," "Come to Me, Bend to Me," "Almost Like Being in Love," "There But for You Go I," 581 perfs.; *High Button Shoes* 10/9 at the Century Theater, with Phil Silvers, Nanette Fabray, Helen Gallagher, music by Jule Styne, lyrics by Sammy Cahn, songs that include "Papa, Won't You Dance with Me," 727 perfs.; *Allegro* 10/10 at the Majestic Theater, with John Battles, John Conte, music by Richard Rodgers, lyrics by Oscar Hammerstein II, songs that include "A Fellow Needs a Girl," "The Gentleman Is a Dope," 315 perfs.; *Angel in the Wings* 12/11 at the Coronet Theater, with Paul and Grace Hartman, Elaine Stritch, music by Bob Hilliard, lyrics by Carl Sigman, songs that include "Civilization (Bongo, Bongo, Bongo)," 197 perfs.

Popular songs "Autumn Leaves" ("Les Feuilles Mort") by French composer Joseph Kosma, lyrics by Jacques Prevest (English lyrics by Johnny Mercer); "Ballerina" by Carl Sigman, lyrics by Bob Russell; "Golden Earrings" by Ray Evans and Victor Young, lyrics by Jay Livingston (title song for the film); "The Back Pay Polka" and "For You, For Me, For Evermore" by the late George Gershwin, lyrics by Ira Gershwin (for the film *The Shocking Miss Pilgrim*); "Green Dolphin Street" by Bronislaw Kaper, lyrics by Ned Washington (title song for film); "Feudin' and Fightin'" by Burton Lane and Al Dubin; "I'll Dance at Your Wedding" by Ben Oakland, lyrics by Herb Magidson; "Ivy" by Hoagy Carmichael; "Open the Door, Richard" by Jack McKea and Dan Howell, lyrics by Dusty Fletcher and John Mason; "Too Fat Polka" by Ross MacLean and Arthur Richardson; "Woody Woodpecker" by George Tibbles and Ramey ldriss.

Jack Kramer wins in men's singles at Wimbledon, Margaret Osborne, 29 (U.S.), in women's singles; Kramer wins in men's singles at Forest Hills, Althea Louise Brough, 24, in women's singles.

Jackie Robinson signs with the Brooklyn Dodgers and starts the season at first base, the first black baseball player in the major leagues. John Roosevelt Robinson, 28, will continue through the 1956 season and have a lifetime batting average of .311.

The New York Yankees win the World Series by defeating the Brooklyn Dodgers 4 games to 3.

English cricket attracts 2 million paying customers. The figure will drop to 500,000 by 1970 as the game loses its attraction for many Britons.

The death of Manolete August 28 plunges Spain into mourning. The multimillionaire matador Manuel Rodriguez has been fatally gored by a Miura bull in the small 8,268-seat ring at Linares and dies at age 30 after a career that has made him a legend.

New York's Collyer brothers make headlines March 21 when Homer Collyer, 71, is found dead of malnutrition in a cluttered brownstone at 5th Avenue and 128th Street. The rat-gnawed body of his brother Langley, 61, is found dead in the house April 8 after searchers have removed 120 tons of rubbish including bicycles, sleds, most of a Model T Ford, a car generator and radiator, the top of a horse-drawn carriage, kerosene stoves, umbrellas, 10 clocks, 14 grand pianos, an organ, a trombone and cornet, three bugles, five violins, 15,000 medical books, thousands of other books, mountains of yellowed newspapers dating to 1918, etc.

The "New Look" designed by Paris couturier Christian Dior, 42, late last year lowers skirt lengths to 12 inches from the floor, pads brassieres, unpads shoulders, adds hats, makes present wardrobes obsolete, and wins quick support from fashion magazines and the $3 billion U.S. garment industry. U.S. women resist the new fashion briefly, but then succumb and slavishly adopt not only long, full peg-top skirts, V-necks, curving waists, sloping shoulders, and frothy blouses, but also clogs, espadrilles, spike-heeled "naked sandals," and fezzes.

UFOs (Unidentified Flying Objects) make headlines. Boise, Idaho, businessman Kenneth Arnold, 32, claims to have seen nine shiny, pulsating objects flying over the Cascade Mountains at speeds of up to 1,700 miles per hour while flying his two-seat plane from Chehalis to Yakima June 24. "They seemed to be alive in the center, to have the ability to change their density," he says, the Civil Aeronautics Administration expresses doubts that "anything would be flying that fast," other UFO sightings are reported, some 15 million Americans will claim to have seen UFOs in the next 25 years, more than half of all Americans will say they believe in the existence of such objects which many will say are manned by creatures from other planets, but professional airline pilots will have more mundane explanations.

Ajax cleanser, introduced by Colgate-Palmolive-Peet, is a silica-sand product that is more likely to scratch than feldspar cleansers such as Bon Ami and Dutch cleanser but requires less elbow grease and soon outsells its rivals (*see* Comet, 1956).

New York's postwar building boom begins with a 21-story office building at 445 Park Avenue erected by Tishman Brothers.

New York's 33-story Esso building, designed by Carson and Lundin, opens in Rockefeller Center.

Levittown goes up on Long Island to help satisfy the booming demand for housing. Builder Abraham Levitt and his sons employ wartime experience gained by building houses for the Navy. Their nearly identical houses are built on concrete slabs with no basements, crews assemble the houses using precut materials. They erect the mass-produced houses round village greens with shops, a playground, and a community swimming pool. The modest prices include major appliances, and by 1951 there will be more than 17,000 Levittown houses (*see* 1951).

The San Juan Hilton put up by Conrad Hilton begins a Puerto Rico building boom.

The Eames side chair designed by Charles Eames, 40, is introduced by Herman Miller Furniture of Zeeland, Mich. The revolutionary chair made of contour-molded plywood on a frame of aluminum tubing won top prize in a 1940 Museum of Modern Art Organic Design competition, but the aluminum has been replaced by chrome-plated steel.

Everglades National Park, established by act of Congress, is a 1.4-million-acre reserve of subtropical Florida wilderness embracing open prairies, mangrove forests, freshwater and saltwater areas, and abundant wildlife.

Norwegian anthropologist-explorer Thor Heyerdahl, 32, crosses 4,300 miles of open Pacific in 101 days on a balsa raft he calls *Kon-Tiki*, finding no trace of man in the "crystal clear" water.

Pittsburgh begins a program of huge proportions to clean up and modernize (*see* 1816). The city's air is so thick with smoke that street lights must often be turned on all day even in fair weather (*see* Donora smog, 1948).

Iceland's Mount Hekla volcano erupts as it has 13 times since 1104, causing widespread destruction.

New York City and its environs are crippled December 26 by a 25.8-inch snowfall, the worst since 1888 when 20.9 inches fell. The blizzard lasts 16 hours, but while it stops suburban trains the storm is not accompanied by the savage winds of 1888 and the drifts are not so high.

British agriculture sustains £20 million in losses March 16 as a gale blows across England's Fens knocking down trees and creating floods that destroy potatoes, root crops, and poultry flocks. Homes and farm buildings are ruined and floods make planting impossible.

Snowstorms isolate hundreds of British farms, killing more than a quarter of the sheep and lamb flock. Two-thirds of the valuable hill wool is lost and 30,000 head of cattle die or have to be shot.

The rat poison Warfarin discovered by University of Wisconsin biochemist Karl Paul Link, 46, is introduced to fight the rodent which consumes a large portion of the world's grain production each year. The anticoagulant is based on the chemical coumarin which occurs in spoiled sweet clover and causes animals to bleed to death by interfering with vitamin K which enables the liver to produce the blood-clotting chemical prothrombin.

Reports begin to come in of fly and mosquito strains that have developed resistance to DDT and the British-developed benzene hexachloride.

U.S. agricultural chemical production reaches nearly 2 billion pounds, up from 100 million in 1934.

A rice-farming boom begins in the lower Mississippi valley.

Widespread food shortages continue in the wake of World War II and crop failures exacerbate the situa-

1947 *(cont.)* tion. President Truman urges meatless and eggless days October 5 to conserve grain for hungry Europe, and a Friendship Train leaves Los Angeles November 8 for a cross-country tour to collect food for European relief.

The Soviet Union continues food rationing until December but exports grain despite widespread hunger at home just as czarist Russia did in 1911.

Philippine health authorities begin the "Bataan experiment" in an effort to solve the problem of beriberi (*see* thiamine, 1936, 1937). The study will show that beriberi incidence is reduced by nearly 90 percent in an area where people are given rice fortified with thiamine, niacin, and iron, while a control population has no reduction.

U.S. sugar rationing ends June 11.

Almond Joy is introduced to augment the popular Mounds bar. Peter Paul has survived the war by importing coconuts via small schooners from Honduras, El Salvador, Nicaragua, and other Caribbean countries (*see* 1932).

The 5¢ Hershey bar contains 1 1/8 ounces of chocolate (but *see* 1968).

Reddi-Wip, introduced by Reddi-Wip, Inc., is the first major U.S. aerosol food product. Founder Marcus Lipsky advertises aerated "real" whipped cream in pressurized cans (*see* 1945; Ablanalp, 1953).

U.S. frozen orange juice concentrate sales reach 7 million cans, up from 4.8 million last year, but a glut of fresh oranges that has dropped prices from $4 to 50¢ per box and an oversupply of canned single-strength juice brings the fledgling concentrate industry to the brink of ruin.

Minute Maid Corp. has its beginnings in Vacuum Foods Co., headed by John M. Fox, 34, whose pioneer orange concentrate producing firm has lost $371,000 in the past year. With more than $500,000 tied up in retail packages bearing the Snow Crop label, Fox goes door to door at Hingham, Mass., handing out free cans of concentrate and the names of local grocery stores where the product can be purchased; when stocks are cleaned out he is convinced that he has a desirable product. Fox obtains a $50,000 loan from William A. Coolidge, whose Cambridge, Mass., firm initiated the development of citrus concentrates for the U.S. Army in 1942, and Coolidge gets venture capitalist John Hay (Jock) Whitney interested in Vacuum Foods (*see* 1949).

Monosodium glutamate (MSG) is marketed for the first time under the Accent label (*see* 1908; 1934).

The antioxidant BHA (butylated hydroxyanisole) is introduced commercially in the United States to retard spoilage in foods.

The first commercial microwave oven is introduced by the Raytheon Co. of Waltham, Mass., whose Percy LeBaron Spencer, now 53, discovered in 1942 that microwaves used for signal transmission would cook food (they agitated molecules of a chocolate bar in his pocket, melting it). Raytheon's $3,000 Radarange restaurant oven employs an electronic tube (magnetron) developed in 1940 by John Randall and J. A. H. Boot of Birmingham University for British radar. It cooks quickly but the results are unappetizing (*see* Amana, 1967).

Seagram's 7 Crown becomes the world's largest selling brand of whiskey (*see* 1934). By 1971 Seagram will be selling 7.5 million cases of 7 Crown per year, half a fifth for every American man, woman, and child.

1948 Burma gains independence January 4 after more than 60 years of British colonial rule.

A Hindu extremist assassinates Mahatma Gandhi, now 78, January 30. Many Hindus resent Gandhi's agreement to last year's partition of India and Pakistan.

Communists take over Czechoslovakia in a coup d'état February 25; President Beneš resigns June 7 and dies September 3 at age 65.

The State of Israel is proclaimed May 14 as the British mandate over Palestine expires. Britain has withdrawn her forces under pressure from the terrorist Irgun, "Stern Gang," and Haganah, most Arabs have fled the country, Polish-born British biochemist and Zionist leader Chaim Weizmann, now 73, takes office as provisional president, Polish-Palestinian David Ben-Gurion, 61, who has been chairman of the Jewish Agency for Palestine since 1935, is prime minister and minister of defense. The new state occupies four-fifths of Palestine and has an Arab Palestinian population of 200,000 after 500,000 have left. Israel opens her doors to the world's Jews.

President Truman recognizes the new Jewish homeland against the advice of Secretary of State Marshall (who says the president is acting for reasons of domestic polities).

Israel gained independence, offered refuge to the world's oppressed Jews, and fought off Arab invaders.

Transjordan's Arab Legion enters Jerusalem, Egypt joins the attack on Israel May 15 and bombs the temporary capital at Tel Aviv, but Israeli defenders throw back the invaders. Jewish terrorists assassinate Swedish diplomat and UN mediator Count Folke Bernadotte September 21 in the Israeli-held quarter of Jerusalem.

Yugoslavia is expelled June 28 from the Communist Information Bureau (Cominform) for alleged doctrinal errors and hostility to Moscow. The Yugoslav Communist party supports Marshal Tito and purges itself of all Cominform supporters in a clear break with Stalin.

Soviet occupation forces in Germany set up a blockade July 24 to cut off rail and highway traffic between West Germany and Berlin. An airlift begins July 25 with U.S. and British aircraft flying in food and supplies for the more than 2 million people of West Berlin, the airlift is carrying 4,500 tons per day by September, and it will continue until September of next year.

The phrase "cold war" is coined by U.S. presidential adviser Bernard Baruch, now 78.

A new Selective Service act passed by Congress June 24 provides for the registration of all U.S. men between 18 and 25 with men between 19 and 25 to be inducted for 21 months' service. Registration begins August 30 to initiate a program that will continue for 25 years.

British Commonwealth nation citizens gain the status of British subjects in the British Citizenship Act passed by Parliament July 30.

Alger Hiss, 43, supplied Soviet agents with classified U.S. documents while working in the State Department in the 1930s, says *Time* magazine senior editor Whittaker Chambers, 47, in testimony before the House Un-American Activities Committee August 3. Hiss is president of the Carnegie Endowment for International Peace; he sues Chambers for slander. Chambers will produce a microfilm he has kept hidden in a pumpkin at his Westminster, Md., farm. A federal jury at New York indicts Hiss for perjury December 15, and his first trial will end in a hung jury (but *see* 1950).

Rep. Richard Milhous Nixon, 35, (R. Calif.) pushes a congressional investigation of the Hiss affair (above). Nixon won election in 1946 by suggesting that his opponent H. Jerry Voorhis had Communist support.

The Republic of Korea is proclaimed at Seoul August 15 with Synghman Rhee, 73, as president, but the Korean People's Democratic Republic proclaimed September 9 in North Korea challenges the Rhee regime and claims dominion over the entire country (*see* 1949).

Indian troops sent into Hyderabad by Jawaharlal Nehru September 13 enter the city of Hyderabad 4 days later. The Nizam Sir Osman Ali Khan Bahadur, 62, has ruled the 82,000-square-mile state since 1907 and refused to join India or grant parliamentary government; he has appealed to the United Nations but bows to force and is permitted to retain his title, palaces, and private property (which includes a harem of 42 concubines). His annual income of $50 million is reduced to an allowance of $900,000.

Wilhelmina of the Netherlands abdicates September 4 at age 68 after a 58-year reign. Her daughter, 39, will reign as Juliana until 1980.

President Truman wins reelection with 303 electoral votes to 189 for his Republican opponent Thomas E. Dewey of New York who receives 45 percent of the popular vote to Truman's 49.5 percent. States' Rights Democratic (Dixiecrat) candidate (James) Strom Thurmond, 45, of South Carolina has received 39 electoral votes (but less than 3 percent of the popular vote) and other minority party candidates including Henry Wallace (Progressive) and Norman Thomas (Socialist), splinter the Democratic vote, but Truman has campaigned vigorously to confound poll-takers and embarrass the *Chicago Tribune,* which has hit the streets with a front-page headline proclaiming Dewey the winner.

Japan's wartime prime minister Hedeki Tojo and six others are convicted of war crimes after a 2½-year trial and hanged by U.S. occupation authorities December 23.

An apartheid platform favoring separation of the races in South Africa wins election May 26 for a Nationalist Afrikaner bloc which defeats a United and Labour party coalition headed by Jan Christiaan Smuts, now 78 (*see* 1949).

The NAACP convention at Kansas City decides to give up the fight for equalization of separate black facilities and push instead for integration, especially of schools (*see* 1917; 1954).

Executive Order 9981, signed by President Truman July 26 in response to a civil disobedience campaign organized by A. Philip Randolph, ends racial segregation in the U.S. armed forces.

The Supreme Court rules May 3 that the government may not enforce private acts of discrimination (*Shelley v. Kramer).* Restrictive covenants in deeds have prohibited sales of houses to minorities, but such covenants are not legally enforceable, the Court rules.

Hungarian police arrest Jozsef Cardinal Mindszenty December 26 for his anti-Communist statements (*see* 1944). Cardinal Mindszenty will have a show trial next year and be sentenced to death, but the sentence will be commuted to life imprisonment (*see* 1956).

A Universal Declaration of Human Rights adopted by the United Nations General Assembly at Paris December 10 at 3 A.M. is based on the U.S. Bill of Rights, the Magna Carta, and the French Declaration of the Rights of Man. Eleanor Roosevelt has fought for the Declaration (authored primarily by René Cassin, 61) and wins a standing ovation.

A Foreign Assistance Act passed by Congress April 3 implements the 1947 Marshall Plan. The act authorizes spending $5.3 billion in the first year for economic aid to 16 European countries, creating the European Recovery Program (ERP) and Economic Cooperation Administration (ECA). Studebaker boss Paul G. Hoff-

1948 ***(cont.)*** man is appointed Economic Cooperation Administrator of the ERP.

U.S. production, employment, and national income reach new highs but renewed strikes bring a third round of inflationary wage boosts.

An Income Tax Reduction Act has become law April 2 over President Truman's veto and given further impetus to inflation. Speaking at Spokane, Wash., June 9, Truman calls the 80th Congress "the worst we've ever had."

Some 360,000 U.S. soft coal workers strike from mid-March to mid-April demanding $100 per month in retirement benefits at age 62.

President Truman orders the Army to operate the railroads beginning May 10 to prevent a nationwide rail strike. The railroads will be operated under U.S. Army control until 1952.

General Motors grants an 11¢-per-hour wage increase to the United Automobile Workers. The contract signed May 25 by GM's Charles E. Wilson and the UAW's Walter P. Reuther includes a cost-of-living clause based on the Consumer Price Index of the Bureau of Labor Statistics (*see* 1913; 1949).

The U.S. cost-of-living index reaches a record high in August (173 against 1935–1939 average) but the 80th Congress resists President Truman's repeated appeals for anti-inflationary legislation.

West German economist Ludwig Erhard, 51, reforms the occupied nation's currency. Germans line up June 20 to receive 40 new Deutschemarks, printed in the United States (business enterprises receive an additional 60 per employee). Erhard announces an end to most rationing June 21, letting market forces govern.

French gold reserves fall to 478 tons, down from 5,000 in 1932, and the war-weary nation adopts protectionist measures (*see* Schuman Plan, 1950; Monnet, 1953).

Textron becomes the world's first business conglomerate as Royal Little diversifies by acquiring Cleveland Pneumatic Tool, a leading producer of aircraft landing struts (*see* 1944). Finding textile industry too cyclical, Little proposes the idea of a conglomerate that will employ management skills to improve the performance of firms in completely unrelated fields and he will spin off some unprofitable textile subsidiaries late in 1952 (*see* Textron American, 1955).

A U.S. energy crisis in mid-January brings urgent requests for voluntary reductions in the use of gasoline, fuel oil, and natural gas. Long a net exporter of oil, the United States has become a net importer through the development of low-cost petroleum sources in Venezuela and the Middle East.

Foreign crude oil, mostly from Venezuela, will capture 18 percent of the U.S. market in the next 10 years (*see* 1958).

Arabian-American Oil (Aramco) sells a 30 percent interest to Standard Oil of New Jersey and a 10 percent interest to Standard Oil of New York in order to broaden marketing of its Saudi-Arabian oil (*see* 1933). SoCal and Texas Company each retain a 30 percent interest (*see* 1973).

U.S. railroads shift from coal-fired steam locomotives to diesel-electric locomotives in a move that will eliminate a major market for coal and that will lead to a 36 percent decline in coal mine output in the next 25 years and an increase in demand for oil (*see* 1924).

British railroads are nationalized January 1.

The Army operates U.S. railroads beginning May 10 on presidential orders (above).

The Canadian Pacific adds an airline to its railway and shipping line. The CP acquires routes that will link Canada's major cities to much of the world.

New York City transit fares advance to 10¢ March 30, having remained at 5¢ since the subway opened its first line in 1904 (Fifth Avenue buses have charged 10¢ for years, and fares in most other U.S. cities have been 10¢ for some time). Irish-American Transport Workers Union president Michael J. Quill, 42, has made a deal with Mayor O'Dwyer's labor relations assistant Theodore W. Kheel, 33, to support the 10¢ fare in return for a generous labor contract (*see* 1953).

The first jet aircraft to fly the Atlantic arrive in Labrador July 12. Six RAF de Havilland Vampires complete the crossing from Britain (*see* 1952; 1958).

The Vickers Viscount flown July 16 is the first British turboprop airliner.

The Soviet Air Force adopts the IL-28 jet bomber designed by S. V. Ilyushin (see 1942). The new aircraft will be the backbone of Soviet airpower for 15 years.

President Truman dedicates New York's Idlewild International Airport July 31. The world's largest commercial airport relieves some of the pressure from the city's 9-year-old La Guardia but is much farther from midtown Manhattan (*see* 1963).

The Land-Rover introduced at the Amsterdam Motor Show April 30 by England's Rover Co. of Solihull, Warwickshire, is a Jeep-like vehicle designed for civilians (*see* Universal Jeep, 1945).

The first automobile air conditioner goes on the market. The crude affair is designed to be installed under the dashboard; within 25 years most new cars will have factory-installed air conditioning.

A new 1949 Ford four-door sedan sells in the fall for $1,236 F.O.B. Detroit.

The Honda motorcycle is introduced by Japanese entrepreneur Soichiro Honda, 38, whose company will be the world's leading motorbike producer by 1958. The first Honda motorcar will appear in 1963.

Michelin Cie. introduces the world's first radial tires.

The Liberian Maritime Law adopted by the Liberian Legislature in December will make ships flying the Liberian "flag of convenience" the largest merchant fleet in the world. The "flag of convenience" system began in 1940 to permit the United States to send aid to

Britain while preserving technical neutrality, and it has been exploited by former Secretary of State Edward R. Stettinius, Jr., to permit major shipping (and oil) companies to operate more cheaply and profitably, avoiding use of U.S. crews and shipyards that would be required by U.S. registry, freeing themselves from certain inspection requirements, and taking advantage of Liberian tax laws to conceal income.

Cybernetics by MIT mathematician Norbert Wiener, 53, summarizes results of studies in communication and information control. Derived from the Greek word for *steersman*, Wiener's title will come into common use to cover the whole field of information control by automated machines such as computers while Wiener works over the next 16 years to elaborate on the possibilities of cybernetics and to warn of its dangers (*see* 1950).

The transistor announced by Bell Telephone Laboratories will permit miniaturization of electronic devices such as computers, radios, and television sets and lead to the development of guided missiles. The tiny but rugged three-electrode solid-state transistor developed by physicists William Shockley, 38, John Bardeen, 40, and Walter H. Brattain, 46, will replace the glass vacuum tube pioneered by Bell Labs physicist H. D. Arnold in 1912 (*see* Texas Instruments, 1954; mesa, 1954).

U.S. physicists Richard Phillips Feynman and Julian Schwinger, both 30, develop a quantum theory of electrodynamics far more powerful than the 1926 Dirac theory or 1927 Heisenberg theory.

Moscow purges Soviet scientific committees following T. D. Lysenko's denunciation of hostile geneticists.

The United Nations establishes a World Health Organization (WHO) with headquarters at Geneva. The United States accepts membership June 14.

Britain's National Health Insurance Program goes into operation July 5. The program operates under legislation enacted by Parliament in 1946 and provides Britons with free medical treatment.

The adrenal hormone cortisone synthesized by Mayo Foundation biochemist Philip Showalter Hench, 52, and his colleague Edward C. Kendall of 1915 thyroxine fame brings new relief to arthritis victims. Hench administers a dose of his compound E September 21 to a woman of 20 incapacitated by arthritis, she is active again 4 days later, and Hench then treats 13 other arthritics with similar dramatic results. By early next year he and Kendall will show motion pictures of patients treated with synthetic cortisone who are able to run after being bedridden for years with arthritis (*see* ACTH, 1949; prednisone, 1955).

The antibiotics aureomycin and chloromycin are developed.

Dramamine is developed by Johns Hopkins physicians who discover accidentally that an anti-allergy drug relieves motion sickness. A clinic patient being treated for hives (urticaria) with beta-diaminoethyl-bendohydryl-ether-8-chlorothyeophyllinate reports that taking 50 milligrams by mouth before boarding a streetcar prevents the motion sickness she has usually experienced. Control groups on the U.S. troop transport *General Ballou* test the drug while en route to Bremerhaven for occupation duty.

Sexual Behavior in the Human Male by Indiana University zoologist Alfred Charles Kinsey, 54, indicates that many sex acts thought heretofore to be perversions are so common as to be considered almost normal. Kinsey has conducted interviews with some 18,500 men and women throughout America with funds provided by the National Research Council and the Rockefeller Foundation. The "Kinsey Report" arouses great controversy.

Brandeis University is founded at Waltham, Mass.

Congress funds Voice of America radio broadcasts to foreign countries January 27 as part of an overseas information program established by the Mundt Act (*see* Radio Free Europe, 1949).

One million U.S. homes have television sets, up from 5,000 in 1945, and while the sets are still large cabinets with small screens, the programming begins to improve and more TV stations go into operation (*see* Ed Sullivan, Bill Boyd, below).

U.S. News and World Report begins publication to compete with *Time* and *Newsweek*. Editor-columnist David Lawrence, 59, has edited *U.S. News* since 1933 and merges it with the *World Report* he founded 2 years ago.

"Pogo" makes its debut in the *New York Star*, successor to the Marshall Field tabloid PM. Cartoonist Walt (Walter Crawford) Kelly, 35, is a former Walt Disney animator whose Okefenokee Swamp opossum will move to the *New York Post* upon the demise of the *Star* next year, and be in 450 papers worldwide by the late 1960s. Commenting on the ecological crisis, Pogo will say, "We have met the enemy and he is us."

Nonfiction *The Seven Storey Mountain* by French-American Trappist monk Thomas Merton, 33, creates a sensation with its autobiographical revelations.

Fiction *Intruder in the Dust* by William Faulkner; *The Young Lions* by U.S. novelist Irwin Shaw, 35; *The Naked and the Dead* by U.S. infantry veteran Norman Mailer, 25, whose novel excoriates the hypocrisy of war; *Guard of Honor* by James Gould Cozzens whose war novel about a Florida military air base brings attention to his neglected earlier novels; *Remembrance Rock* by Carl Sandburg; *Confessions of a Mask (Kamen no kokuhaku)* by Japanese novelist Yukio Mishima (Kimitake Hiraoka), 23, who gains quick celebrity and some notoriety for his semi-autobiographical novel of a homosexual with strong sadistic impulses; *Other Voices, Other Rooms* by U.S. novelist Truman Capote, 23; *The World Is a Wedding* (stories) by Delmore Schwartz; *Circus in the Attic* (stories) by Robert Penn Warren; *The Atom Station (Atom stooin)* by Icelandic novelist Halldor Laxness, 46; *The Aunt's Story* by Aus-

1948 *(cont.)* tralian novelist Patrick White, 36; *The Corner That Held Them* by English novelist Sylvia Townsend Warner, 55; *Concluding* by Henry *Green; Portrait of a Man Unknown (Portrait d'un inconnu)* by Nathalie Sarraute; *The City and the Pillar* by Gore Vidal; *The Heart of the Matter* by Graham Greene; *Cry the Beloved Country* by South African novelist Alan Stewart Paton, 45, principal of the Diepkloof Reformatory since 1935, who indicts the apartheid society of his native land; *The Stain in the Snow (La Neige était sale)* by Georges Simenon; "The Lottery" by U.S. writer Shirley Jackson, 29, in the June 26 *New Yorker*.

Poetry *The Pisan Cantos* (72 to 84) by Ezra Pound who wrote them during his imprisonment at the end of World War II. Now 63, Pound has been arrested by U.S. military authorities in Italy and indicted for treason in connection with profascist anti-Semitic propaganda broadcasts he made for the Italians. Judged mentally unfit to face trial, he will be confined to St. Elizabeth's Hospital at Washington, D.C., until 1958.

Abstract expressionist painting is pioneered by Jackson Pollock, now 36, whose *Composition No. 1 (tachisma)* combines splashes and splotches of multihued paints on canvas to help launch a new school of "action painting" that will radically alter the direction of American art. Pollock says his splashes are controlled by personal moods and unconscious forces.

Christina's World by U.S. painter Andrew Nelson Wyeth, 31, captures the youthful vigor and anguish of Cushing, Me., cripple Christina Olsen, 55, in a work that will be widely reproduced. Wyeth's late father, the illustrator N. C. Wyeth, was killed 3 years ago with a grandson when a train struck his automobile. The younger Wyeth studied under N. C., illustrated the Brandywine edition of *The Merry Adventures of Robin Hood* at age 12, and has been working largely in tempera since 1946.

Other paintings *Natura Morta* by Italian painter Giorgio Morandi, 58; *Woman* by Willem de Kooning; *Painting* by Jackson Pollock (above); *Elegy to the Spanish Republic* by U.S. abstractionist Robert Motherwell, 33. Arshile Gorky takes his life at New York July 3 at age 43.

The Nikon camera introduced by Nippon Kogaku K.K. of Japan is a 35-mm rangefinder camera designed to compete with the German Leica that has been famous since 1925. Organized by the Mitsubishi group in 1917, Nippon Kogaku has been making lenses under the Nikkor trade name since 1932 but has never before made cameras (*see* 1959).

The Polaroid Land Camera goes on sale for the first time November 26 at Boston's Jordan Marsh department store (*see* 1947; 1950).

Theater *The Antigone of Sophocles (Die Antigone des Sophokles)* by Bertolt Brecht 2/15 at the Stadttheater in Chur, Switzerland; *Mr. Roberts* by U.S. playwrights Thomas Heggen, 26, and Joshua Logan, 39, 2/18 at New York's Alvin Theater, with Henry Fonda, David Wayne, Robert Keith, 1,157 perfs.; *The Lady's Not for Burning* by Christopher Fry 3/11 at the London Art Theatre; *The Caucasian Chalk Circle (Der kaukasische Kreidekreis)* by Bertolt Brecht (based on the Chinese drama *The Circle of Chalk)* 5/4 at Carleton College, Northfield, Minn. (in English); *Mr. Puntila and His Hired Man (Herr Puntila und sein Knecht Matti)* by Bertolt Brecht 6/5 at Zurich's Schauspielhaus; *Summer and Smoke* by Tennessee Williams 10/6 at New York's Music Box Theater, with Anne Jackson, Margaret Phillips, 100 perfs.; *State of Siege (L'état de siège)* by Albert Camus 10/27 at the Théâtre Marigny, Paris; *The Cry of the Peacock* (*Ardele, ou La marguerite*) by Jean Anouilh 11/4 at the Théâtre de l'Atelier, Paris; *Playbill* (*The Browning Version* and *Harlequinade*) by Terence Rattigan 11/8 at London's Phoenix Theatre; *Anne of the Thousand Days* by Maxwell Anderson 12/8 at New York's Shubert Theater, with Joyce Redman as Anne Boleyn, Rex Harrison as Henry VIII, 286 perfs.

Television *The Ed Sullivan Show* (initially *The Toast of the Town*) in June with *New York Daily News* columnist Edward Vincent Sullivan, 45, as master of ceremonies. CBS program development manager Worthington Miner, 47, has hired Sullivan to emcee the new TV variety show that introduces comedians Dean Martin and Jerry Lewis in its first program and will continue on Sunday nights until 1971; *Hopalong Cassidy* 11/28 on NBC TV with actor Bill Boyd, 53, in the title role is television's first Western series.

Films Vittorio de Sica's *The Bicycle Thief* with Lamberto Maggiorani; Laurence Olivier's *Hamlet* with Olivier; Joseph Mankiewicz's *A Letter to Three Wives* with Jeanne Crain, Linda Darnell, Ann Sothern; David Lean's *Oliver Twist* with Alec Guinness, Robert Newton, John Howard Davies; Howard Hawks's *Red River* with John Wayne, Montgomery Clift; Michael Powell and Emeric Pressburger's *The Red Shoes* with Anton Walbrook, Marius Goering, Moira Shearer; Fred Zinneman's *The Search* with Montgomery Clift; Alexander McKendrick's *Tight Little Island* with Basil Radford, Joan Greenwood; John Huston's *The Treasure of the Sierra Madre* with Humphrey Bogart, Walter Huston, Tim Holt; Preston Sturges's *Unfaithfully Yours* with Rex Harrison, Linda Darnell. Also: Sam Wood's *Command Decision* with Clark Gable, Walter Pidgeon, Van Johnson; Billy Wilder's *A Foreign Affair* with Jean Arthur, Marlene Dietrich; George Stevens's *I Remember Mama* with Irene Dunne, Barbara Bel Geddes; Jean Negulesco's *Johnny Belinda* with Jane Wyman, Lew Ayres; John Huston's *Key Largo* with Humphrey Bogart, Edward G. Robinson, Lauren Bacall; Robert Flaherty's documentary *Louisiana Story*; Henry Cornelius's *Passport to Pimlico* with Stanley Holloway, Margaret Rutherford; Walter Lang's *Sitting Pretty* with Robert Young, Maureen O'Hara; Anatole Litvak's *The Snake Pit* with Olivia de Havilland.

Hollywood musicals Vincente Minnelli's *The Pirate* with Judy Garland, Gene Kelly, music and lyrics by Cole Porter, songs that include "Be a Clown"; John Berry's *Casbah* with Yvonne de Carlo, Tony Martin,

Peter Lorre, music by Harold Arlen, lyrics by Leo Robin, songs that include "For Every Man There's a Woman."

Broadway musicals *Look Ma I'm Dancing* 1/29 at the Adelphi Theater, with Nancy Walker, Harold Lang, music and lyrics by Hugh Martin, songs that include "Shauny O'Shay," 188 perfs.; *Where's Charley* 10/11 at the St. James Theater, with Ray Bolger in an adaptation of the 1892 English comedy *Charley's Aunt*, music and lyrics by Frank Loesser, songs that include "Once in Love with Amy," "My Darling, My Darling," "The New Ashmolean Marching Society and Students' Conservatory Band," 792 perfs.; *As the Girls Go* 11/13 at the Winter Garden Theater, with Bobby Clark, Irene Rich, music by Jimmy McHugh, lyrics by Harold Adamson, songs that include "It Takes a Woman to Make a Man," 420 perfs.; *Lend an Ear* 12/16 at the National Theater, with music and lyrics by Charles Gaynor, songs that include "When Someone You Love Loves You," 460 perfs.; *Kiss Me Kate* 12/30 at the New Century Theater, with Alfred Drake, Patricia Morrison, Lisa Kirk, Harold Lang in an adaptation of the 1596 Shakespeare comedy *The Taming of the Shrew*, music and lyrics by Cole Porter, songs that include "We Open in Venice," "I've Come to Wive It Wealthily," "I Hate Men," "Another Opening, Another Show," "Why Can't You Behave," "Were Thine that Special Face," "Too Darn Hot," "Where Is the Life that Late I Led," "Always True to You in My Fashion," "So in Love," "Tom, Dick or Harry," "Wunderbar," "Brush up Your Shakespeare," 1,077 perfs.

Soviet composers Aram Khatchaturian, Serge Prokoviev, and Dmitri Shostakovich are rebuked February 11 when the Central Committee of the Communist Party calls their work an expression of "bourgeois decadence." The scolding comes on the heels of a denunciation by Soviet cultural czar Andrei Zhdanof who has accused the composers of "failing in their duties to the Soviet people."

The Aldeburgh Festival founded by composer Benjamin Britten has its first season 100 miles northeast of London.

Opera *The Beggar's Opera* 5/24 at England's Cambridge Arts Theatre, with music by Benjamin Britten; *Magdalena* 7/28 at Los Angeles, with music by Heitor Villa-Lobos.

Cantatas *Knoxville—Summer of 1915* by Samuel Barber 4/19 at Boston's Symphony Hall, with text by James Agee; *Saint Nicolas* by Benjamin Britten 6/5 at the Aldeburgh Festival (above) and in July at the 100th anniversary of St. Nicolas College, Lancing, Sussex; *A Survivor of Warsaw* by Arnold Schoenberg 11/4 at Albuquerque, N. M.

Ballet *Fall River Legend* 4/22 at New York's Metropolitan Opera House, with Alicia Alonso, John Kriza, music by Morton Gould, choreography by Agnes de Mille; *Orpheus* 4/28 at the New York City Center, with Maria Tallchief as Eurydice, Tanaquil LeClercq as the leader of the Bacchantes, music by Igor Stravinsky, choreography by George Ballanchine; *Cinderella* 12/27 at London's Royal Opera House in Covent Garden, with Moira Shearer, music by Serge Prokofiev, choreography by Frederick Ashton.

First performances Symphony No. 3 in E major by Walter Piston 1/9 at Boston's Symphony Hall; Symphony No. 4 by David Diamond 1/23 at Boston's Symphony Hall; *Mandu-Carará* (Symphonic Poem) by Heitor Villa-Lobos 1/23 at New York; *Fantasy for Trombone and Orchestra by* Paul Creston 2/12 at Los Angeles; *The Seine at Night* (Symphonic Poem) by Virgil Thomson 2/24 at Kansas City; Concerto No. 1 for Violin and Orchestra by Diamond 2/29 at Vancouver; Symphony No. 6 by Ralph Vaughan Williams, now 75, 4/21 at London; Symphony No. 4 *(1848)* by Darius Milhaud 5/20 at Paris; Symphony No. 2 by Roger Sessions 6/9 at Amsterdam; *Toccata* by Piston 10/14 at Bridgeport, Conn.; *Sinfonietta* by Francis Poulenc 10/24 in a BBC Orchestra concert from London; *Mass* by Igor Stravinsky 10/27 at Milan; Concerto No. 2 for Violin and Orchestra by Milhaud 11/7 at Paris; *Wheat Field at Noon* by Thomson 12/7 at Louisville; Concerto for Piano by Howard Hanson 12/31 at Boston's Symphony Hall; Symphony No. 5 by George Antheil 12/31 at Philadelphia's Academy of Music.

The long-playing 12-inch vinyl plastic phonograph record demonstrated at New York June 18 by CBS engineer Peter Goldmark turns at a rate of 33⅓ revolutions per minute instead of the usual 78, has 250 "Microgrooves" to the inch, plays 45 minutes of music, and begins a revolution in the record industry *(see* 1946; Goldmark's color television, 1940). Goldmark also unveils a lightweight pickup arm and a silent turntable for his LP records (*see* 1949).

California guitar and amplifier maker C. (Clarence) Leo Fender, 40, goes into mass production with an electric guitar he calls the Broadcaster (it will be renamed the Telecaster in 1950). Its solid body reduces feedback because it does not resonate inside and will help define the rock 'n' roll sound soon to dominate the pop music scene (*see* 1954).

Popular songs "Tennessee Waltz" by Redd Stewart and Pee Wee King; "On a Slow Boat to China" by Frank Loesser; "Manana—Is Soon Enough for Me" by Peggy Lee and Dave Barbour; "Nature Boy" by Eden Ahbez; "It's a Most Unusual Day" by Jimmy McHugh, lyrics by Harold Adamson (for the film *A Date with Judy);* "Candy Kisses" by George Morgan; "'A'- You're Adorable" by Buddy Kaye, Fred Wise, and Sidney Lippman; "Red Roses for a Blue Lady" by Sid Tepper and Roy Beaumont; "Baby, It's Cold Outside" by Frank Loesser; "You're Breaking My Heart" by Pat Genaro and Sunny Skylar; "Buttons and Bows" by Jay Livingston and Ray Evans (for the film *Paleface*); "Enjoy Yourself—It's Later than You Think" by Carl Sigman, lyrics by Herb Magidson; "Pigalle" by French songwriter Georges Ulmer; "Sleigh Ride" by Leroy Anderson, 39, lyrics by Mitchell Parish; "I'll Be Home for Christmas" by Kim Gannon, Walter Kent, Buck Ram.

1948 ***(cont.)*** Robert Falkenburg, 22 (U.S.), wins in men's singles at Wimbledon, Louise Brough in women's singles; Richard A. "Pancho" Gonzalez, 20, wins in men's singles at Forest Hills, Margaret Osborne duPont in women's singles.

The Olympic games are held for the first time since 1936. A record 6,005 contestants from 59 nations compete at London with U.S. athletes taking the lion's share of medals.

Soviet chess master Mikhail Botvinnik wins the world title he will hold until 1957, regain the following year, and then hold until 1960.

Citation with Eddie Arcaro up wins U.S. racing's Triple Crown. No horse will win it again until 1973.

The Cleveland Indians win the World Series, defeating the Boston Braves 4 games to 2 with pitching from Bob Feller and Satchel Paige, now 42 (*see* 1934).

Scrabble is copyrighted by Newtown, Conn., entrepreneur James Brunot who has taken over the "crossword game" played with wooden tiles on a board from his friend Alfred M. Butts, who invented the game in 1931 and called it Criss-Cross. Brunot renames the game and begins manufacturing it, Scrabble will be produced beginning in 1952 by the Bay Shore, Long Island, firm Selchow & Righter, Brunot will receive a royalty of 5 percent on the retail price until 1971 when Selchow & Righter will acquire the game outright, Scrabble will be produced in six languages, and sales will rival those of the Parker Bros. game Monopoly introduced in 1935.

Dial soap, introduced by Chicago's Armour and Co., is the world's first deodorant soap. It employs the bacteria-killing chemical hexachlorophene discovered in World War II.

A killer smog on the Monongahela River south of Pittsburgh affects 43 percent of the 12,000 residents of Donora, Pa. Sulfur from burning coal at the local steel mill combines with dampness in the air to produce sulfuric acid that causes headache, nausea, vomiting, and irritation of the eyes, nose, and throat; 22 die within a few days, and U.S. Steel closes down its Donora mill.

Our Plundered Planet by U.S. zoologist Fairfield Osborne, 51, expresses concern about the growing use of DDT (*see* 1943; Carson, 1962).

Maine's Atlantic Sea-Run Salmon Commission begins a program to remove old dams that bar fish from migrating upstream to spawning grounds (*see* Columbia River, 1938). The Commission starts to restock the Aroostook, Dennys, Machias, East Machias, Narraguagus, Penobscot, Pleasant, Sheepscot, Union, and other rivers (*see* Anadromous Fish Act, 1965).

Lysenkoites opposed to hybridization take over the U.S.S.R.'s last remaining centers of pure biological research (*see* Lysenko, 1940). The centers were organized at the Lenin Academy of Agricultural Science and at the Institute of Genetics by the late Nikolai Vavilov (*see* 1954).

U.S. corn is now 75 percent hybrid (*see* Golden Cross Bantam, 1933).

Vitamin B_{12} (cyanocobalamin) is isolated from animal livers as a red crystalline substance, making it possible to contain pernicious anemia with monthly injections (*see* Castle, 1929). The vitamin will be synthesized in 1955.

All 13 vitamins considered essential to human health have now been isolated and some synthesized.

The first full-sized British supermarket is opened January 12 at Manor Park, London, by the London Co-Operative Society (*see* 1951).

U.S. sales of frozen orange juice concentrate leap to more than 12 million 6-ounce cans (*see* 1947; Minute Maid, 1949).

"V-8" Cocktail Vegetable juice, introduced by Campbell Soup Co., is a mixture of tomato, carrot, celery, beet, parsley, lettuce, watercress, and spinach juices (*see* Swanson, 1955).

Gerber Products Co. sells 2 million cans and jars of baby food per week (*see* 1941; MSG, 1951).

U.S. sales of oregano will increase by 5,200 percent in the next 8 years as demand is boosted by the growing popularity of pizza pies and other Italian specialties discovered by servicemen in Europe (*see* 1953).

The McDonald hamburger stand that opened in 1940 at Pasadena, Calif., is turned into a self-service restaurant by the McDonald brothers who begin to franchise the name McDonald to other fast food entrepreneurs. The McDonalds install infrared heat lamps to keep French fries warm (*see* 1954).

The Baskin-Robbins ice cream chain begins its growth as California entrepreneur Burton "Butch" Baskin merges his small chain with one begun by the Snowbird Ice Cream stand at Adams and Palmer streets in Glendale opened December 7, 1945, by Irvine Robbins, now 30. Baskin-Robbins will soon be selling ice cream in more than 100 flavors including Blueberries 'N Cream, Pralines 'N Cream, Huckleberry Finn, Baseball Nut, New England Maple, Chocolate Divinity, and Nuts to You, and by 1976 they will have 1,600 franchised outlets in the United States, Canada, Japan, and Europe.

A Displaced Persons act voted by Congress June 25 authorizes 400,000 homeless people to settle in the United States, but the D.P.s are charged to the quotas of their countries under the Johnson-Reed Law of 1924 (*see* McCarran-Walter Act, 1952).

Birth-control pioneer Margaret Sanger founds the International Planned Parenthood Federation (*see* 1927; Pincus, 1951).

A "Eugenic Protection" law enacted in occupied Japan authorizes abortion on demand. Japan's population is nearly 80 million, up from 64.5 million in 1930.

1949 The Western powers pledge cooperation as Moscow breaks the U.S. nuclear monopoly and Communists take over China.

Dean Gooderham Acheson, 55, replaces George C. Marshall as U.S. secretary of state January 7, Andrei Vyshinsky replaces V. M. Molotov as Soviet foreign minister March 4.

President Truman proposes four major courses of U.S. action in his inaugural address January 20. Point Four calls for "a bold new program for making the benefits of our scientific advances and industrial progress available for the improvement and growth of underprivileged areas."

Newfoundland joins the Dominion of Canada March 31 as the tenth province.

The North Atlantic Treaty Organization (NATO) created by a treaty signed at Washington April 4 joins the United States, Canada, Iceland, Britain, France, Denmark, Norway, Belgium, the Netherlands, Luxembourg, Italy, and Portugal in a pledge of mutual assistance against aggression within the North Atlantic area and of cooperation in military training, strategic planning, and arms production.

The North Atlantic Treaty Organization (above) gives approval to President Truman's Point Four program of technical assistance to aid world peace (above).

The International Court of Justice of the United Nations hands down its first decision April 9. The court holds Albania responsible for incidents that occurred in the Corfu Channel in 1946 and awards damages to Britain.

The Republic of Eire is formally proclaimed April 18 at Dublin. London recognizes Irish independence May 17 but reaffirms the position of Northern Ireland within the United Kingdom.

The Council of Europe statute signed May 5 at London establishes a Committee of Ministers and a Consultative Assembly with Council headquarters at Strasbourg. The statute is signed by Belgium, Denmark, France, Britain, Ireland (Eire), Italy, Luxembourg, the Netherlands, Norway, and Sweden who will be joined by Greece, Iceland, and Turkey.

Soviet authorities officially lift the Berlin blockade May 12, but the Berlin airlift continues until September 30 when it ends after having completed 277,264 flights (*see* 1948; Laker Airways, 1966).

The German Federal Republic (West Germany) is established May 23 with headquarters at Bonn.

The German Democratic Republic (East Germany) is established October 7 under Soviet control.

The Hashemite Kingdom of Jordan is created June 2 by a renaming of the 3-year-old Kingdom of Transjordan ruled by Abdullah (*see* 1951).

China's Chiang Kai-shek resigns his presidency January 21 as Nationalist armies suffer reverses at the hands of the Communists who have taken Tientsin January 15. He loses whole divisions by desertion to the victorious Communists and begins removing his Nationalist forces to Formosa (Taiwan). U.S. aid to Nationalist China ends August 5.

The People's Republic of China is proclaimed October 1 at Beijing with Mao Zedong as chairman of the central people's administrative council, Zhou En-lai as premier and foreign minister. Removal of Nationalist forces from the mainland to Taiwan is completed by December 8.

Paris recognizes Vietnamese independence within the French Union March 8. The former emperor of Annam Bao Dai heads the state which includes Cochin China, but France retains the right to maintain military bases (*see* 1946; 1950).

Siam renames herself Thailand May 11.

Civil war looms in Korea, reports a UN Commission September 2 (*see* 1950).

Indonesia elects Achmed Sukarno, 48, president December 16, the Dutch grant independence December 27 after 4 years of hostilities, the United States of Indonesia is established with its capital at Jakarta (Batavia), and President Sukarno begins an administration that will continue until 1968 (*see* 1959).

The U.S. Defense Department is created August 10 by a retitling of the War Department under terms of the National Security Act of 1947. The first secretary of defense James V. Forrestal resigned in March with symptoms of nervous exhaustion and depression, entered Bethesda Naval Hospital, and jumped from a window there May 22, dying at age 57.

"We have evidence that in recent weeks an atomic explosion occurred in the U.S.S.R.," President Truman announces September 23. Developers of the Soviet bomb include chiefly Andrei Dmitriyevich Sakharov, 28, who won a doctorate at age 26, and 1925 German Nobel laureate Gustav Hertz, 62, whose uncle Heinrich Hertz pioneered the wireless in 1887.

China became a Communist people's republic. Mao Zedong's entry into Shanghai ended the power of Chiang Kai-shek.

1949 ***(cont.)*** President Truman pledges continued U.S. support of the United Nations October 24 in ceremonies dedicating the new UN site at New York.

Washington imposes stringent controls November 8 on export of all strategic commodities to prevent reshipment to Soviet bloc countries.

The Geneva Conventions adopted April 12 revise the conventions of 1864, 1907, and 1929. They provide for "free passage of all consignments of essential foodstuffs, clothing, and tonics intended for children under 15, expectant mothers, and maternity cases" in event of war but do not specifically outlaw sieges, blockades, or "resource denial" operations and do not address conflicts that are partly internal and partly international (*see* Vietnam, 1962).

A South African apartheid program takes effect in June under terms of the South African Citizenship Act which suspends automatic granting of citizenship to Commonwealth immigrants after 5 years and bans marriages between Europeans (meaning whites) and non-Europeans (meaning blacks or "coloreds") (*see* 1948; 1951; 1960).

"Tokyo Rose" goes on trial for treason in July at San Francisco. Los Angeles-born UCLA graduate Iva Toguri D'Aquino, 34, is one of at least a dozen Tokyo radio announcers who were called "Tokyo Rose" by English-speaking listeners in the Pacific during the war. She did the work under pressure from Japanese secret police after being caught in Tokyo at the outbreak of the war, refused to renounce her U.S. citizenship as did two other women announcers (who thus could not be charged with treason), and married a Portuguese national in 1945. Government officials threaten and intimidate defense witnesses, the prosecution bribes a witness to give false testimony and tries to bribe an AP reporter to lie on the witness stand, the trial lasts 13 weeks and costs $750,000, the judge's instructions make it impossible for the jury to acquit, D'Aquino is found innocent of eight alleged overt acts of treason but guilty on one count of trying to undermine U.S. morale. Sentenced to 6 years in prison, she will serve 6½.

New York dancer Paul Draper, 39, and harmonica player Larry Adler sue a Greenwich, Conn., woman for libel after Red-baiting columnist Westbrook Pegler has picked up her accusations that they have Communist sympathies (*see* Trumbo, 1947). The trial will end in a hung jury, but Draper and Adler (who have been making $100,000 per year) are unable to obtain further bookings (*see Red Channels*, 1950).

Hungarian officials arrest foreign minister Laszlo Rajk June 16 on conspiracy charges and begin a wholesale purge of Hungarian Communists accused of deviating from pro-Soviet policy.

Poland has a purge in November of prominent members of the United Workers' party central committee accused of having "Titoist" sympathies (*see* Tito, 1948).

A Communist bloc Council for Mutual Economic Assistance created January 18 at Moscow tries to further economic cooperation between the Soviet Union and her satellites. Poland joins a week later.

Congress raises the salary of the president to $100,000 per year January 19 with a tax-free allowance of $50,000 for expenses.

Congress creates the General Services Administration to serve as an independent federal agency with responsibilities for managing government property and records economically and efficiently. The GSA will construct and operate buildings, distribute supplies, stockpile critical materials, and dispose of government surpluses.

United Automobile Workers at General Motors plants accept a slight wage cut as a business recession produces a decline in the cost of living (*see* 1948).

U.S. unemployment reaches 5.9 percent, up from 3.8 percent last year, and Wall Street's Dow Jones Industrial Average falls to 161 at midyear, down from 193 a year earlier, but while most consumer prices drop, housing and health-care costs increase.

The average American steel worker has $3,000 per year to spend after taxes, the average social worker $3,500, a high school teacher $4,700, a car salesman $8,000, a dentist $10,000. Typical prices include a new Cadillac for $5,000, a gallon of gasoline 25¢, a man's gabardine suit $50, a 10-inch table TV set $250, a pack of cigarettes 21¢, a pound of pork 57¢, a pound of lamb chops $1.15, a bottle of Coca-Cola 5¢, a quart of milk 21¢, a loaf of bread 15¢, a dozen eggs 80¢.

A Foreign Assistance bill enacted by Congress April 19 authorizes U.S. aid amounting to $5.43 billion for the European Recovery Program (*see* 1947).

The Far Eastern Commission terminates Japanese reparation payments May 12 to spur Japan's economic recovery.

The U.S. Export-Import Bank extends a $20 million loan to Yugoslavia September 8 to help the Tito government resist Soviet domination (*see* 1948).

Britain devalues the pound September 18 from $4.03 to $2.80. Most European nations follow the British move and devalue their currencies.

Nationalization of Britain's iron and steel industries takes effect November 24.

British dock workers strike in June and close ports.

Australian coal workers strike June 27 and stay out until mid-August when emergency legislation authorizes the government to send in troops to work the mines.

U.S. coal workers strike September 19, President Truman invokes the Taft-Hartley Act September 30, the miners refuse to obey the injunction, and 20 UMW leaders are cited for contempt. They win acquittal and the strike is settled.

A new Minimum Wage act passed by Congress October 26 amends the Fair Labor Standards Act of 1938 and raises the minimum hourly wage from 40¢ to 75¢ (*see* 1955).

U.S. steel workers strike October 1 and do not return until November 11. The 500,000 workers win company-paid pensions but no wage boosts.

Seven major oil companies control 90 percent of world petroleum reserves outside the United States and the Soviet bloc territories: Standard Oil of New Jersey, Standard Oil of California, Gulf Oil, the Texas Company, Socony-Vacuum, Royal Dutch-Shell, and Anglo-Iranian control 88 percent of crude oil production, 77 percent of refining capacity, 85 percent of cracking capacity, two-thirds of all privately owned tankers, and every important pipeline.

Jean Paul Getty, 56, obtains a concession for his Pacific Western Co. to drill for petroleum in the Neutral Zone established in 1924 between Saudi Arabia and Kuwait. The oil magnate made his first million by age 23 and has been battling for years to gain control of the Tide Water Associated Oil Co. owned largely by Standard Oil of New Jersey (above). The desert in his new concession will turn out to cover one of the world's largest petroleum reserves, and the oil will make Getty the world's richest man (*see* 1951).

Hughes Tool Co. supplies more than 75 percent of all bits used in oil drilling anywhere in the free world. Hughes leases the drills in the United States to keep dulled bits from being re-tipped and sold at lower prices (*see* 1908). Mounted on three cones, the powerful Hughes rock bit can drill into rock formations at speeds of up to 180 feet per hour.

An Air Force XB-47 jet bomber sets a new U.S. transcontinental speed record February 8, crossing the country in 3 hours, 46 minutes at an average speed of 607 miles per hour (*see* Doolittle, 1922; Glenn, 1957).

Air Force pilots flying the B-50 Superfortress *Lucky Lady II* complete the first nonstop round-the-world flight March 2. They have refueled in midair and arrive at Fort Worth, Tex., after a 23,452-mile flight of 94 hours, 1 minute (see 1947).

Chicago's O'Hare Airport is named in honor of the late Lieut. Commander Edward H. "Butch" O'Hare who earned a Congressional Medal of Honor in 1942 for having shot down five Japanese bombers and crippled a sixth, but died in 1943 at age 29. Originally called Orchard Place, O'Hare will surpass Chicago's Midway Airport by 1961 to become the world's busiest air travel facility; by 1972 it will be handling 2,000 flights per day and a decade later will have to expand once again.

A Japanese train wreck August 17 kills three persons. Someone has removed a length of rail, and 20 workers are arrested on charges of sabotage; 14 are Communist party members. The government was about to announce a mass layoff of workers and uses the incident to suppress the railway workers' union. It is widely believed that U.S. occupation authorities are involved. Japan's Supreme Court will exonerate the convicted men but not until 1963.

The *Chicago Zephyr* begins service between Chicago and San Francisco via the Burlington, the Rio Grande, and the Western Pacific Railroads with Vista Dome coaches for sightseeing (*see* 1934).

General Motors, Standard Oil of California, Firestone Tire, and other companies are convicted of criminal conspiracy to replace electric transit lines with gasoline or diesel buses (*see* 1932; 1936; 1938). GM has replaced more than 100 electric transit systems in 45 cities with GM buses and will continue this program despite the court action (the company is fined $5,000, its treasurer $1) (*see* 1955).

U.S. auto production reaches 5.1 million and catches up after 20 years with the 1929 record.

Germany's Volkswagen begins commercial production and is introduced into the United States, but only two of the odd-looking "beetles" are sold in America (*see* 1938; 1953; 1955).

Saab-Scania AB is founded in Sweden to compete with the 22-year-old Volvo group.

Reynolds Metals acquires bauxite deposits in Jamaica, British West Indies, as a source of raw material for aluminum production (*see* 1945). The Reynolds move will be followed by Alcoa, Kaiser Aluminum, Canada's Aluminium Co., and others who will buy 225,000 acres of Jamaica land rich in aluminum ore.

Merck Laboratory researchers synthesize from liver bile the adrenocorticotropic hormone ACTH produced by the human pituitary gland to stimulate the adrenal glands (*see* cortisone, 1948). Upjohn researchers will use microbes to produce ACTH more cheaply.

Indian physician Rustom Jal Vakil reports in the prestigious *British Heart Journal* that he has lowered blood pressure effectively using powdered root from the tropical plant *Rauwolfia serpentina* (*see* 1952).

Pfizer biochemists develop the antibiotic oxytetracycline; Pfizer will be awarded a patent on the drug in 1955, but American Cyanamid's Lederle Laboratories will introduce tetracycline under the trade name Achromycin in 1953 and will maintain a large share of the market.

Parke, Davis introduces the antibiotic chloramphenicol under the trade name Chloromycetin; it is hailed as the first major breakthrough against typhoid fever.

Selman Waksman isolates the antibiotic neomycin (*see* streptomycin, 1943).

North Carolina evangelist Billy Graham, 31, gains prominence with a tent crusade at Los Angeles that leads to the conversions of an Olympic track star, a noted gambler, and a notorious underworld syndicate figure. Graham has just become first vice president of Youth for Christ International.

Radio Free Europe begins beaming world news to listeners behind the Iron Curtain from an operations base at Munich (*see* Voice of America, 1948). Gen. Lucius D. Clay has helped start the organization and is chairman of the board, he asks for private contributions to help fund Radio Free Europe, but nearly all its support comes from the CIA.

1949 ***(cont.)*** The Department of Justice charges American Telephone & Telegraph with having monopolized the telephone instrument market in violation of the Sherman Act of 1890. AT&T may be a "natural monopoly," says Justice, but the manufacture of telephones is by no means a natural monopoly. It asks the company to break its Western Electric division into three separate companies so as to permit competition in telephone instrument production and installation.

The Intertype Fotosetter Photographic Line Composing Machine installed by Intertype Corp. of Brooklyn, N.Y., at the Rochester, N.Y., plant of Stecher-Traung Lithograph is the first typesetting machine that dispenses with metal type (*see* Mergenthaler, 1884; Photon process, 1953).

Nonfiction *The Second Sex* (*Le Deuxième Sexe*) by French philosopher-novelist Simone de Beauvoir, 41, will be the bible of feminists (*see* 1953); *The Need for Roots (L'Enracinement)* by the late French writer Simone Weil who died in 1943 at age 34 of voluntary starvation in England after refusing to eat more than her compatriots were receiving from Nazi occupation authorities; *The Mediterranean* and *The Mediterranean World in the Age of Philip II* by French historian Fernand Braudel, 47.

Fiction *Conjugal Love* (*L'Amore Coniugale e Altri Racconti)* by Alberto Moravia whose prostitutes are his most sympathetic characters; *The House on the Hill (La Casa in Collina)* by Italian novelist Cesare Pavese, 41; *Iron in the Soul* (*The Troubled Sleep* or *La Mort dans l'ame*) by Jean-Paul Sartre; *The Oasis* by Mary McCarthy; *The House of Incest* by Anaïs Nin; *The Sheltering Sky* by émigré U.S. novelist Paul Bowles, 39; *The Man with the Golden Arm* by Chicago novelist Nelson Algren, 40, is about drug addiction; *The Beginning and the End* by Egyptian novelist Naguib Mahfouz, 38; *Barabas* by Pär Lagerkvist; *Mary* by Sholem Asch; *Point of No Return* by John P. Marquand; "No Consultation Today" ("Honjitsu Kyushin") by Japanese short story writer Masuji Ibuse, 51; *The Third Man* by Graham Greene; *Nineteen Eighty-Four* by George Orwell who will die of tuberculosis at age 46 in January 1950.

Nineteen Eighty-Four by George Orwell (above) is a chilling projection of a totalitarian state whose authorities exercise mind control ("Big Brother Is Watching You"). Orwell introduces the inverted graffiti "Ignorance Is Strength" and "Freedom Is Slavery" along with such portmanteau words as "newspeak," "bellyfeel," and "doublethink." "Who controls the past controls the future; who controls the present controls the past," says one of Orwell's characters (compare Santayana, 1905).

Poetry *The Arrivistes* by U.S. poet Louis Simpson, 26.

Painting *The Kitchen* by Pablo Picasso; *Number 2* by Jackson Pollock; *Onement III* by U.S. abstractionist Barnett Newman, 44. José Clemente Orozco dies at Mexico City September 7 at age 65; James Ensor dies at Ostend November 19 at age 89.

Theater *Death of a Salesman* by Arthur Miller 2/10 at New York's Morosco Theater, with Lee J. Cobb as salesman Willie Loman, Mildred Dunnock, Arthur Kennedy, Cameron Mitchell, 742 perfs.; *Detective Story* by Sidney Kingsley 3/23 at New York's Hudson Theater, with Ralph Bellamy, Les Tremayne, Alexander Scourby, Maureen Stapleton, Joseph Wiseman, 581 perfs.; *Romulus the Great (Romulus der Grosse)* by Swiss playwright Friedrich Dürrenmatt, 28, 4/25 at Basel's Stadttheater; *The Just Assassins* (*Les Justes*) by Albert Camus 12/15 at the Théâtre Hebertot, Paris.

Television drama *The Goldbergs* 1/17 on CBS is the first TV situation comedy. Derived from the radio show first aired in 1929, it stars Molly Berg, will continue until late June 1951, and will be followed by dozens of "sitcoms"; *Amos 'n' Andy* becomes a weekly TV program with black actors (*see* 1928; 1965).

The National Academy of Television Arts and Sciences confers its first Emmy Awards.

Films George Cukor's *Adam's Rib* with Spencer Tracy, Katharine Hepburn; Robert Rossen's *All the King's Men* with Broderick Crawford, Mercedes McCambridge; William Wyler's *The Heiress* with Olivia de Havilland, Ralph Richardson, Montgomery Clift; Ken Annakin, Arthur Crabtree, Harold French, and Ralph Smart's *Quartet* with Basil Radford, Mai Zetterling, Ian Fleming, Dirk Bogarde, Naunton Wayne; Akira Kurosawa's *Stray Dog* with Toshiro Mifune; Carol Reed's *The Third Man* with Orson Welles, Joseph Cotten, Trevor Howard, Valli; Henry King's *Twelve O'Clock High* with Gregory Peck, Gary Merrill, Dean Jagger. Also: Carol Reed's *The Fallen Idol* with Ralph Richardson, Michele Morgan, Bobby Henrey; Joseph H. Lewis's *Gun Crazy* with Peggy Cummins, John Dall; Lloyd Bacon's *It Happens Every Spring* with Ray Milland, Jean Peters, Paul Douglas; Jacques Tati's *Jour de Fête* with Tati; Vincente Minnelli's *Madame Bovary* with Jennifer Jones, James Mason, Van Heflin, Louis Jourdan; Michael Powell and Emeric Pressburger's *The Small Back Room* with David Farrar, Jack Hawkins; Sam Wood's *The Stratton Story* with James Stewart, June Allyson; Nicholas Ray's *They Live By Night* with Farley Granger, Cathy O'Donnell; Raoul Walsh's *White Heat* with James Cagney, Virginia Mayo.

Hollywood musical Gene Kelly and Stanley Donen's *On the Town* with Kelly, Frank Sinatra, music by Leonard Bernstein, lyrics by Betty Comden and Adolph Green.

Broadway musicals *South Pacific* 4/7 at the Majestic Theater, with Ezio Pinza, Mary Martin, Myron McCormick, Juanita Hall, Betta St. John, William Tabbert, music by Richard Rodgers, lyrics by Oscar Hammerstein II, book based on James Michener's tales of U.S. military personnel in World War II posts, songs that include "Some Enchanted Evening," "Younger Than Springtime," "I'm in Love with a Wonderful Guy," "This Nearly Was Mine," "I'm Gonna Wash that Man Right Out of My Hair," "A Cockeyed Optimist," "Honey Bun," "You've Got to Be Carefully Taught," "Dites-moi Pourquoi," "There Is Nothing Like a

Dame," "Happy Talk," "Bali Ha'i," 1,925 perfs.; *Miss Liberty* 6/15 at the Imperial Theater, with Philip Borneuf, music and lyrics by Irving Berlin, songs that include "Let's Take an Old Fashioned Walk," 308 perfs.; *Lost in the Stars* 10/30 at the Music Box Theater, with Todd Duncan, music by Kurt Weill, lyrics by Maxwell Anderson, book based on Alan Paton's 1948 novel *Cry the Beloved Country,* 273 perfs.; *Gentlemen Prefer Blondes* 12/8 at the Ziegfeld Theater, with Carol Channing, music by Jule Styne, lyrics by Leo Robin, book based on the 1925 novel, songs that include "A Little Girl from Little Rock," "Diamonds Are a Girl's Best Friend," "Bye Bye Baby," 740 perfs.

Opera *Let's Make an Opera* 6/14 at Aldeburgh's Jubilee Hall, with music by Benjamin Britten, libretto by Eric Crozier.

Ballet *Beauty and the Beast* 12/20 at the Sadler's Wells Theatre, London, with music by the late Maurice Ravel, choreography by South African-born dancer-choreographer John Cranko, 22.

First performances Spring Symphony for Soprano, Alto, and Tenor Soli, Mixed Chorus, Boys' Choir, and Orchestra by Benjamin Britten 7/9 at Amsterdam; *Concerto Symphonique* for Piano and Orchestra by Ernest Bloch 9/3 at the Edinburgh Festival; Phantasy for Violin and Piano Accompaniment by Arnold Schoenberg 9/13 at Los Angeles; Concerto for Organ, Brasses, and Woodwinds by Paul Hindemith 11/14 at Boston's Symphony Hall.

CBS introduces improved long-playing vinyl plastic phonograph records (see RCA, 1946; Goldmark, 1948). RCA introduces small 45 rpm LPs that require large spindles.

Stereo components (amplifiers, turntables, speakers) enjoy a sales boom.

The first LP record catalog is published in October by Cambridge, Mass., record shop proprietor William Schwann whose 26-page listing of 674 entries from 11 companies will grow in 25 years to list some 50,000 LPs in a book of more than 250 pages.

Popular songs "Melodie d'Amour" by French songwriter Henri Salvador; "Bonaparte's Retreat" by Peewee King; "Dear Hearts and Gentle People" by Sammy Fain, lyrics by Bob Hilliard; "I Don't Care If the Sun Don't Shine" by Mark David; "Scarlet Ribbons (for Her Hair)" by Evelyn Danzig, lyrics by Jack Segal; "Mona Lisa" by Jay Livingston and Ray Evans; "Huckle-Buck" by Andy Gibson, lyrics by Roy Alfred; "Daddy's Little Girl" by Bobby Burke and Horace Gerlah; "The Harry Lime Theme" by Anton Karas (for the film *The Third Man);* "Rudolph the Red-Nosed Reindeer" by U.S. songwriter Johnny Marks, 40, who has adapted a verse written in 1939 by his brother-in-law, Robert May, for a Montgomery Ward promotional comic book.

U.S. prizefighter Ezzard Charles, 27, gains the world heavyweight championship June 22 by winning a 15-round decision over Joe Walcott at Chicago following the retirement of Joe Louis March 1.

Ted Schroeder wins in men's singles at Wimbledon, Louise Brough in women's singles; Pancho Gonzalez wins in men's singles at Forest Hills, Mrs. duPont in women's singles.

The New York Yankees win the World Series by defeating the Brooklyn Dodgers 4 games to 1.

Silly Putty is introduced by New Haven, Conn., advertising man Peter C. L. Hodgson, 37, who has discovered a substance developed by General Electric researchers looking for a viable synthetic rubber. The useless silicone substance can be molded like soft clay, stretched like taffy, bounced like a rubber ball, and can pick up printed matter when pressed down on newsprint and transfer it, but the stuff has no market until Hodgson borrows $147 to buy a batch from GE, hires a Yale student to separate it into 1-oz. globs, packages it in clear compact plastic cases at $1 each, advertises it in a catalogue of toys he is preparing for a local store, and finds that Silly Putty is an immediate success.

Clothes rationing ends in Britain.

Congress passes a federal Housing Act July 15 to fund slum-clearance and low-rent public housing projects. Sen. Robert Taft (R. Ohio) sponsored the legislation.

Title I of the new Housing Act (above) encourages municipalities to acquire and resell substandard areas at prices below cost for private redevelopment.

A two-bedroom U.S. house sells typically for $10,000 while a five-bedroom New York apartment rents for $110 per month. A four-bedroom duplex cooperative apartment in the East 60's near Park Avenue with two-story living room and wood-burning fireplace in its 16x21-foot library sells for $8,250 with annual maintenance of $2,970, an eight-room co-op on Fifth Avenue in the 70's with three bedrooms, 30x17-foot living room, and a view of Central Park sells for $7,434 with annual maintenance of $3,591.

New York architect Philip Johnson puts up a glass house at New Canaan, Conn., to serve as his residence.

Houston's Shamrock Hotel opens March 17 (St. Patrick's Day). Wildcat oil millionaire Glenn McCarthy has built the place.

The Rust cotton picker of 1927 goes into mass production at Allis-Chalmers Corp. in Milwaukee and Ben Pearson, Inc., in Pine Bluff, Ark.

The Brannan Plan advanced by U.S. Secretary of Agriculture Charles F. Brannan, 45, proposes to increase U.S. food production without taking the profit out of farming. The plan employs a formula used by sugar beet and wool producers and would pay farmers directly the difference between the market price of a commodity and the price needed to give the farmer a fair profit; it gives the consumer the benefit of lower prices produced by greater supply, but provides no payment for the 2 percent of farm operators who earn more than $20,000 per year and sell 25 percent of U.S. farm products. Brannan Plan supporters compare it to the minimum wage law of 1938 and Social Security law of 1935, but opponents call it socialism and Congress votes it down. The Price Parity Act passed October 31

1949 ***(cont.)*** supports wheat, corn, cotton, rice, and peanut prices at 90 percent of 1910–1914 levels through 1950, 80 to 90 percent through 1951, 75 to 90 percent on a sliding scale thereafter.

Latin America becomes a net grain importer after years as a net exporter, but some places begin to feel the effects of a "green revolution" (*see* Borlaug, 1944).

Famine ravages the new People's Republic of China (above). The nation's cereal grain production falls to 110 million tons, down from 150 million before the war when the population was smaller (*see* 1952).

Meat, dairy products, and sugar remain in short supply in Britain and sales are restricted.

The first edible vegetable-protein fiber made from spun soy isolate is introduced (*see* 1937). Chemical engineer Robert Boyer files for patents on a process for de-hulling soybeans, turning them into flakes, milling the flakes into a flour that is more than 50 percent protein, and further processing the flour to make it 90 percent protein (*see* 1957).

General Mills and Pillsbury introduce prepared cake mixes.

Sara Lee Cheese Cake is introduced by Chicago baker Charles Lubin, 44, whose refrigerated product will make his Kitchens of Sara Lee (named after his daughter) one of the world's largest bakeries.

Frozen orange juice concentrate sales continue to soar (*see* 1948). Minute Maid Orange juice is promoted by crooner Bing Crosby who acquires 20,000 shares of stock in Vacuum Foods Co. on the advice of John Hay Whitney in exchange for a daily 15-minute radio show that pushes concentrate sales. Vacuum Foods renames itself Minute Maid Corp. (*see* 1947; 1953).

Pepsi-Cola reduces its sugar content and standardizes its recipe. Former Coca-Cola executive Alfred N. Steele, 48, replaces Walter Mack as head of Pepsi and sells the drink as "the light refreshment" (*see* 1934; 1939). Pepsi bottlers add their own sugar while Coca-Cola supplies bottlers with syrup containing sugar.

1950

A Sino-Soviet treaty of friendship, alliance, and mutual assistance signed February 14 names Japan and the United States as common enemies and pledges joint action against "Japanese imperialism."

Communist North Korean forces invade the Republic of South Korea June 25, beginning a 3-year Korean War that will involve 16 nations against the Communists. UN Secretary-General Trygve Lie urges UN members to support South Korea June 27 and President Truman that day orders U.S. air and sea forces to "give the Korean government troops cover and support" (below). Seoul falls to the North Koreans June 28.

Gen. MacArthur visits Korea June 29 and resolves not only to drive the Communists back but also to unite Korea. He takes command of UN forces July 9.

UN forces land at Inchon north of the 38th parallel September 14, retake Seoul September 26 but fail to trap the North Korean army. The UN General Assembly sanctions a move across the 38th parallel October 9, ROK troops occupy the North Korean capital of Pyongyang October 19, U.S. troops reach the Yalu River November 21 and expect to be home by Christmas.

Chinese forces have crossed the Yalu; some 300,000 Chinese and North Korean forces attack UN lines in sub-zero weather November 26, inflicting (and sustaining) heavy losses. U.S. troops retreat in wild disorder, 15,000 marines are trapped (3,000 die, 7,000 are wounded), and the U.S. 8th Army abandons Pyongyang December 8 (little is left of the city). U.S. Gen. Matthew Bunker Ridgway, 55, takes command of defeated UN forces December 25 but Chinese forces cross the 38th Parallel December 28 and UN forces retreat to the 37th as thousands of refugees stream south (*see* 1951).

Alger Hiss has been found guilty January 25 of committing perjury when he denied the allegations made by Whittaker Chambers in 1948. He is sentenced to 5 years' imprisonment and will begin serving time in March 1951 after exhausting all appeals.

German-born physicist Klaus Fuchs is found guilty March 1 of giving British atomic secrets to Soviet agents. The British hired Fuchs to do nuclear research in 1941 knowing he was a Communist, he was a member of the British-American team that developed the atomic bomb beginning in 1943, his work at Los Alamos, N.Mex., made him privy to the bomb's design, construction, components, and detonating devices, and he will serve 10 years in prison. His U.S. accomplice Harry Gold gets 30 years.

Marshal K. E. Voroshilov announces Soviet possession of an atomic bomb March 8 (*see* 1949).

President Truman has advised the Atomic Energy Commission January 31 to proceed with development of the hydrogen bomb (*see* 1946, 1952).

The king of the Belgians Leopold III returns July 22 after 6 years in exile (he was a prisoner of war from 1940 until 1945), Socialist demonstrations against him break out July 23 at Brussels, he abdicates August 1 after a 16-year reign and is succeeded by his son Baudouin, 19.

Chiang Kai-shek has resumed the presidency of Nationalist China following British recognition of the People's Republic (*see* 1949).

The United States recognizes Vietnam, supplies arms to Saigon, sends a military mission to advise the Vietnamese on how to use the arms, and signs a military assistance pact with France, Cambodia, Laos, and Vietnam (*see* 1949; 1951).

Chinese Communist forces invade Tibet October 21 following rejection of Beijing's offer to permit Tibet regional autonomy if she will join the Communist system.

Brazil reelects Getulio Vargas president October 3 amidst economic difficulties, notably rising inflation, that have persisted since 1945 (*see* 1954).

Sweden's Gustavus V dies October 29 at age 92 after a 43-year reign. His son of 66 will rule until 1973 as Gustav VI Adolf.

President Truman escapes an assassination attempt November 1. White House guards outside the Blair-Lee House, occupied by the Trumans during a White House remodeling, shoot two Puerto Ricans, one fatally, after the Puerto Ricans have killed one guard and wounded another (*see* 1954).

French forces withdraw from the northern frontier of Indochina November 3.

Libya prepares for the independence that will take effect in 1952 in accordance with last year's vote by the UN General Assembly. Mohammed Idris el-Senussi is proclaimed king December 3 and will rule as Idris I until 1969.

The UN votes December 2 to unite Ethiopia with Eritrea, which has been administered by British authorities.

South Africa refuses to place South-West Africa (Namibia) under UN trusteeship.

Washington and Madrid resume diplomatic relations at year's end.

Johannesburg has riots January 29 as blacks begin to protest the apartheid program that went into effect last year (*see* 1951).

North and South Koreans commit atrocities, as do UN forces, and subject prisoners of war to unspeakable cruelty. Both sides execute suspected collaborators and dissidents without trial.

Sen. Joseph McCarthy, 41, (R. Wis.) addresses a Republican women's club at Wheeling, W. Va., and claims to have a list of a great many "known Communists" employed by the State Department. He starts a Communist "witch-hunt" that will continue for the next 4 years; the Senate Foreign Relations Committee denies his charges July 20; McCarthy claims congressional privilege to protect himself from retaliation for his campaign of character assassination (*see* 1951).

The McCarran Act (Control of Communists Act) passed by Congress September 20 over President Truman's veto calls for severe restrictions against suspected Communists, especially in sensitive positions and during emergencies.

Red Channels makes sweeping accusations of Communist subversion in the American entertainment industry (*see* Hollywood Black List, 1947; Adler and Draper, 1949). Written by former U.S. Naval Intelligence officer Vincent Hartnett but published anonymously, the paperback book will lead to hearings before the Dies Committee, scores of actors, choreographers, playwrights, musicians, producers, directors, and screenwriters will be defamed and given no opportunity to defend themselves, dozens will be barred from employment on suspicion of using the films, stage, radio, and television as vehicles for Communist propaganda. Abe Burrows, Edward Dmytryk, Elia Kazan, Clifford Odets, Jerome Robbins, and others will survive

"I Have Here In My Hand . . ." Herblock attacked Sen. McCarthy for his reckless witch-hunt against "Communists."

by collaborating with the blacklisters, others will serve prison terms or at best have their careers crippled for years by the *Red Channels* accusations. They will include Bertolt Brecht, Alvah Bessie, Charles Chaplin, Norman Corwin, José Ferrer, John Garfield, Jack Gilford, Lee Grant, Dashiell Hammett, Canada Lee, Ring Lardner, Jr., Arthur Miller, Zero Mostel, Larry Parks, Doré Schary, and Dalton Trumbo (who has moved to Mexico after a year in prison and is grinding out screenplays under a pseudonym) (*see* Lillian Hellman, 1952).

The Schuman Plan proposed May 9 to French foreign minister Robert Schuman by statesman Jean Monnet, 71, calls for a pooling of Western Europe's coal and steel resources. The Plan will evolve into the European Economic Community (*see* 1948; 1953).

Nippon Steel Corp., founded at Tokyo, combines Yawata Iron & Steel with Fuji Steel to create an enterprise with the world's largest crude steel-making capacity (*see* Fuji, 1945).

President Truman requests a $10 billion war fund plus internal economic controls July 10. The Defense Production Act passed by Congress September 8 establishes a system of priorities for materials, provides for wage and price stabilization, and curbs installment buying.

The Revenue Act passed by Congress September 23 increases income and corporation taxes.

The Celler-Kefauver Amendment to the Clayton Anti-Trust Act of 1914 "puts teeth" into the Clayton Act by curbing mergers of U.S. business firms. Written by Rep. Emmanuel Celler, 62, (D. N.Y.) and Sen. Estes Kefauver, 47, (D. Tenn.), the new law stops companies from buying up stock in other companies, but large corporations will find other ways to achieve mergers and the number of mergers in years to come will dwarf this year's 219 (*see* 1960).

1950 ***(cont.)*** General Motors announces 1949 profits of nearly $636.5 million, a new high for any U.S. corporation. GM signs a 5-year contract May 23 granting UAW employees pensions and wage boosts.

AF of L membership reaches roughly 8 million, CIO membership 6 million.

The U.S. Gross National Product (GNP) reaches $284 billion, up from $99 billion in 1940 and $103 billion in 1929. Government spending accounts for 21 percent of the total, up from 18 percent in 1940, 10 percent in 1929 (*see* 1960).

The U.S. Consumer Price Index for all goods and services will rise by 10 percent in the next decade.

The Federal Reserve Board estimates December 25 that four out of 10 U.S. families are worth at least $5,000 and 1 in 10 has assets of at least $25,000.

The Ford Foundation distributes $24 million in grants a year after coming into the bulk of its 1943 and 1947 bequests (*see* 1954).

London dock workers strike from April 19 to May 1.

Federal troops seize U.S. railroads August 25 on orders from President Truman to avert a scheduled strike (*see* 1948).

Le Mistral goes into service for the French National Railways, which has welded and polished seams between track lengths to eliminate the clickety-clack of the wheels. The new electric luxury passenger train between Paris and Nice has a barbershop and other amenities not found on U.S. trains and covers the 676-mile route in 9 hours, 8 minutes.

Stockholm's T-line subway opens with modern stations, inaugurating a system that will grow to have 60 miles of track with 94 stations.

U.S. railroad trackage for passenger traffic will decline in the next 22 years from 150,000 to 68,000, and electric railway trackage will decline from 9,600 miles to 790 as bus lines replace rail transit and more and more Americans drive automobiles (*see* General Motors, 1949).

Mitsubishi Heavy Industries, founded at Tokyo, will be a major shipbuilder and producer of rolling stock, aircraft, and automobiles. The automobile division will become independent in 1970 under the name Mitsubishi Motors.

U.S. auto production reaches 6.7 million, up from 5.1 million last year. Used car sales are above 13 million.

Auto registrations show one passenger car for every 3.75 Americans, up from one for every 5.5 in 1930 (*see* 1960). Americans own some 40 million cars, up from 32.6 in 1941, and the figure will more than double in the next 25 years until every U.S. family has an average of 1.4 automobiles.

The Henry J introduced by Henry J. Kaiser and his son Edgar has a 4-cylinder engine and gets 25 miles per gallon (*see* 1946). A 6-cylinder engine will be added, but the car costs nearly $2,000, Americans who want compacts prefer Volkswagens, and while about 127,000 Henry Js will be sold the make will be discontinued in 1954. The larger Kaiser make will expire in 1955.

The United States has 1.68 million miles of surfaced road, up from 1.34 million in 1940, and only 1.31 million miles of dirt road, down from 1.65 (*see* 1960).

A second Tacoma-Narrows bridge opens to traffic across Washington's Puget Sound. The new suspension bridge is a 2,799-foot span.

New York's Brooklyn-Battery Tunnel opens May 25 and carries nearly twice the traffic anticipated; construction was interrupted by the war.

New York's Port Authority Bus Terminal opens on Eighth Avenue between 40th and 41st Streets. The facility will handle upwards of 7,000 buses and 200,000 passengers per week and be the world's busiest (it will be enlarged in 1989).

E. I. du Pont introduces Orlon. It began developing the wool-like polymerized acrylonitrile fiber in 1941 under the direction of William Hale Church of 1926 waterproof cellophane fame (*see* Dacron, Terylene, 1941).

Tranquilizer drugs (ataraxics) that eliminate anxiety and excitement without making users too drowsy are developed by Wallace Laboratories of Carter Products, Inc., and by Wyeth Laboratories. Chemists B. J. Ludwig and E. C. Piech at Wallace in New Brunswick, N.J., have synthesized meprobamate (methyl+propyl+carbamate) to produce the new tranquilizers (*see* Miltown, Equanil, 1954).

A poll of U.S. physicians reveals that penicillin is prescribed for roughly 60 percent of all patients.

Blue Cross programs cover 37 million Americans, up from 6 million in 1940, but most Americans have no health insurance (*see* 1969; Britain, 1948; Medicare, 1965).

The U.S. ranks tenth among all countries in terms of life expectancy for men, but its ranking will drop sharply in the next decade and a half.

Pope Pius XII proclaims the dogma of the bodily assumption of the Virgin Mary.

The National Council of the Churches of Christ in the United States is created by 25 Protestant and four Eastern Orthodox church groups with 32 million members.

Haloid Co. of Rochester, N.Y. produces the first Xerox copying machine (*see* Carlson, 1938; Haloid, 1946; model 914, 1960).

The first Japanese tape recorder, produced by Tokyo Tsushin Kogyo (Sony), weighs nearly 40 pounds, uses tape made from rice paper, and sells for nearly $500 (*see* 1947). Masaru Ibuka has seen an American tape recorder and tried to improve on it. Although his machine has few buyers at first, some 50,000 orders come in when Akio Morita tours Japanese schools and gives demonstrations that persuade teachers to buy the recorder as a teaching device (*see* 1940; 1952).

U.S. television set sales begin a rapid rise (*see* 1948). By June more than 100 TV stations operate in 38 states, and while the U.S. census for the year shows 5 million homes with sets, the figure is belied by sales figures that show 8 million sets in use (45 million U.S. homes have radios).

"Peanuts" by *St. Paul Pioneer Press* cartoonist Charles Monroe Schulz, 27, begins appearing in eight newspapers. Syndicated by the United Press, Schulz's comic-strip characters Charlie Brown ("good grief"), Lucy, Snoopy (a dog), Linus, Schroeder, and Pig Pen will make "Peanuts" popular with readers of more than 900 newspapers by 1970 and be the basis of books, television shows, and countless manufactured items.

Nonfiction *The Lonely Crowd* by University of Chicago social sciences professor David Riesman, Jr., 41, and sociologists Reuel Denney and Nathan Glazer attracts wide attention with its analysis of inner-directed, outer-directed, and other-directed character types. Twentieth-century Americans are more likely to be other-directed, say the authors, working to get ahead within a group and adjusting to the needs of others, so where earlier, inner-directed, Americans pioneered new production efficiency, the emphasis now is on efficient administration (*see* Whyte, 1956); *The Human Use of Human Beings* by Norbert Wiener warns against abuse of the new technology; *Worlds in Collision* by Russian-American physician Immanuel Velikovsky, 55, draws on mythology, archaeology, astronomy, and other disciplines to evolve a theory that the Earth barely avoided colliding with several other celestial bodies sometime about 3,200 B.C.

Fiction "For Esme—With Love and Squalor" (story) by U.S. writer J.D. (Jerome David) Salinger, 31, in the *New Yorker* April 8; *The Short Life* (*La vida breve*) by Uruguayan novelist Juan Carlos Onetti, 41; *The Moon and the Bonfire (La luna e ilfalo)* by Cesare Pavese; *Anton Wachter* (eight novels) by Simon Vestdijk; *The Wall* by John Hersey deals with the 1943 Warsaw ghetto uprising against the Nazis; *A Town Like Alice* by Nevil Shute; *The Grass Is Singing* by Persian-born Rhodesian novelist Doris Lessing, 31; *Some Tame Gazelle* by English novelist Barbara Pym, 37; *The Preacher and the Slave* by Wallace Stegner deals with the late labor organizer Joe Hill; *Across the River and into the Trees* by Ernest Hemingway; *Nothing* by Henry Green.

Television changed people's lives more than radio ever had done. Not all of TV's contributions were beneficial.

Poetry *Canto General* by Pablo Neruda who lived through the Civil War in Spain from 1935 to 1939, returned to Chile in 1945 after working for Spanish Republican refugees at Paris, and has been publishing his epic hymn to virgin America since 1947.

Juvenile *The Thirteen Clocks* by James Thurber; *Yertle the Turtle* by Dr. Seuss.

Painting *Chief* by New York "action" painter Franz Kline, 40; *The Constructors* by Fernand Léger; *Excavation* by Willem de Kooning; *Lavender Mist* by Jackson Pollock; *Tundra* by Barnett Newman; *Zulma* by Henri Matisse. Max Beckmann dies at New York December 27 at age 66.

Sculpture *Seven Figures and a Head* by Alberto Giacometti; *The Goat* by Pablo Picasso; *Blackburn—Song of an Irish Blacksmith* by U.S. sculptor David Smith, 44.

Polaroid replaces its sepia print of 1947 with a black-and-white print, but the new print fades and a crisis develops until company chemists come up with a new film. The company introduces a new "electric eye" shutter that will automatically select shutter speeds between 1/10th and 1/1,000th of a second for the camera's fixed F/5.4 lens. Polaroid contracts with U.S. Time Corp. to produce Polaroid Land Cameras (*see* Timex, 1946; high-speed film, 1960).

Theater *The Member of the Wedding* by Carson McCullers 1/5 at New York's Empire Theater, with Ethel Waters, Julie Harris, and Brandon de Wilde is based on the 1946 McCullers novel, 501 perfs.; *Venus Observed* by Christopher Fry 1/18 at the St. James's Theatre, London; *The Cocktail Party* by T. S. Eliot 1/21 at New York's Henry Miller Theater, with Alec Guinness, Kathleen Nesbitt, Irene Worth, 409 perfs.; *Come Back, Little Sheba* by U.S. playwright William Motte Inge, 37, 2/15 at New York's Booth Theater, with Shirley Booth, Sidney Blackmer, Joan Loring, 191 perfs.; *Judith (Giuditta)* by Carlo Terron 5/2 at Milan's Teatro Nuovo; *The Bald Prima Donna (La Cantatrice chauve)* by French playwright Eugène Ionesco, 38, 5/11 at the Théâtre des Noctambules, Paris (Ionesco uses meaningless platitudes to convey the sterility of modern life); *Affairs of State* by French playwright Louis Verneuil 9/25 at New York's Royale Theater, with Reginald Owen, Celeste Holm, Sheppard Strudwick, 610 perfs.; *The Rehearsal* (*La repetition, ou L'amour Pani*) by Jean Anouilh 10/25 at the Théâtre Marigny, Paris; *Trial of the Innocent (Processo agli innocenti)* by Carlo Terron 11/7 at Milan's Teatro Oden; *The Country Girl* by Clifford Odets 11/10 at New York's Lyceum Theater, with Paul Kelly, Uta Hagen, 235 perfs.; *Bell, Book, and Candle* by John Van Druten 11/14 at New

1950 ***(cont.)*** York's Ethel Barrymore Theater, with Lili Palmer, Rex Harrison, 233 perfs.

Films Joseph L. Mankiewicz's *All About Eve* with Bette Davis, Anne Baxter, Celeste Holm; Michael Gordon's *Cyrano de Bergerac* with José Ferrer; Vincente Minnelli's *The Father of the Bride* with Spencer Tracy, Elizabeth Taylor, Joan Bennett; John Boulting's *Seven Days to Noon* with Barry Jones, Olive Sloane; Billy Wilder's *Sunset Boulevard* with Gloria Swanson, William Holden, Erich von Stroheim. Also: John Huston's *The Asphalt Jungle* with Sterling Hayden, Louis Calhern, Jean Hagen; George Cukor's *Born Yesterday* with Judy Holliday, William Holden, Broderick Crawford; Michael Curtiz's *The Breaking Point* with John Garfield, Patricia Neal; David Lean's *Breaking the Sound Barrier* with Ralph Richardson, Ann Todd, Nigel Patrick; Robert Bresson's *Diary of a Country Priest* with Claude Laydu; Henry Koster's *Harvey* with James Stewart, Josephine Hull; Robert Hamer's *Kind Hearts and Coronets* with Alec Guinness; Fred Zimmerman's *The Men* with Marlon Brando; Elia Kazan's *Panic in the Streets* with Richard Widmark, Paul Douglas, Barbara Bel Geddes; Anthony Pelissier's *The Rocking Horse Winner* with Valerie Hobson, John Howard Davies, John Mills; Max Ophuls's *La Ronde* with Anton Walbrook, Simone Simon, Simone Signoret, Serge Reggiani, Danielle Darrieux, Jean-Louis Barrault; Jean Negulesco's *Three Came Home* with Claudette Colbert, Sessue Hayakawa; Anthony Mann's *Winchester 73* with James Stewart, Shelley Winters; Anthony Asquith's *The Winslow Boy* with Robert Donat, Margaret Leighton.

Television shows *What's My Line* 2/2 on CBS stations, with emcee John Charles Daley, Jr., 35, and panelists: actress Arlene Francis (*née* Kazanjian), 41; columnist Dorothy Kilgallen, 36; and anthologist Louis Untermeyer, 64, who try to guess the occupations of guests. Publisher Bennet A. Cerf, 51, will replace Untermeyer in 1952 and the show will continue until 9/3/67; *Your Show of Shows* debuts on NBC stations with comedians Sid Caesar, Imogene Coca, Carl Reiner, and Howard Morris. Developed by NBC executive Sylvester "Pat" Weaver, Jr., 41, it will continue with 160 weekly programs until 1954.

Hollywood musicals Wilfred Jackson, Hamilton Luske, and Clyde Geronomi's *Cinderella* with Walt Disney animation, music by Al Hoffman and Jerry Livingston, lyrics by Mack David, songs that include "A Dream Is a Wish Your Heart Makes," "Bibbidi Bobbidi Boo."

Opera *The Consul* 3/15 at New York, with music by Gian Carlo Menotti; *The Triumph of St. Joan* 5/9 at Sarah Lawrence College, Bronxville, N.Y., with music by Norman Dello Joio; *Bolivar* 5/12 at the Paris Opéra, with music by Darius Milhaud.

Ballet *Judith* 1/4 at Louisville, with Martha Graham, music by William Schuman; *The Age of Anxiety* 2/26 at the New York City Center, with Jerome Robbins, music by Leonard Bernstein.

First performances Concerto for Piano and Orchestra by Francis Poulenc 1/6 at Boston's Symphony Hall, with Poulenc as soloist; Concerto for Piano and Orchestra by Paul Creston 1/11 at Washington, D.C.; *Timon of Athens* symphonic portrait by David Diamond 2/1 at Louisville; Concerto for Violin and Orchestra by William Schuman 2/10 at Boston's Symphony Hall, with Isaac Stern as soloist; Sinfonietta in E major by Paul Hindemith 3/1 at Louisville; Concerto No. 4 for Piano and Orchestra by Darius Milhaud 3/3 at Boston's Symphony Hall; Concerto for Cello by Virgil Thomson 3/21 at Philadelphia's Academy of Music; *Variations and Fugue on a Theme by Purcell* by Benjamin Britten 5/2 at Boston's Symphony Hall; Concerto for French Horn and Orchestra by Hindemith 6/8 at Baden-Baden; *Fantasia (Quasi Variazione)* on the "Old 104th" Psalm Tune for Piano, Orchestra, Organ, and Chorus by Ralph Vaughan Williams 9/16 at Gloucester's Three Choirs Festival; Symphony No. 3 by Creston 10/27 at Worcester, Mass.; Symphony No. 3 by Diamond 10/30 at Boston's Symphony Hall; Concerto for Clarinet, String Orchestra, Harp, and Piano by Aaron Copland 11/6 in an NBC Symphony Orchestra broadcast, with Benny Goodman as soloist; *Concerto Grosso* by Williams 11/18 at London; Short Symphony by U.S. composer Howard Swanson, 41, 11/23 at New York's Carnegie Hall; Concerto for Clarinet and Orchestra by Hindemith 12/11 at Philadelphia's Academy of Music, with Benny Goodman as soloist.

Cantata *On Guard for Peace* by Serge Prokofiev 12/19 at Moscow.

Washington Post critic Paul Hume writes a disparaging review of a song recital by Margaret Truman and receives a letter in longhand from the President on White House stationery dated December 6: "Mr. Hume: I have just read your lousy review of Margaret's concert. I've come to the conclusion that you are an eight-ulcer man on four-ulcer pay. . . Some day I hope to meet you. When that happens, you'll need a new nose, a lot of beefsteak for black eyes, and perhaps a supporter below."

Broadway musicals *Call Me Madam* 10/12 at the Imperial Theater, with Ethel Merman as U.S. Ambassador to Luxembourg Perle Mesta, music and lyrics by Irving Berlin, songs that include "Hostess with the Mostes' on the Ball," "It's a Lovely Day Today," "The Ocarina," "The Best Thing for You," 644 perfs.; *Guys and Dolls* 11/24 at the 46th Street Theater, with Robert Alda, Vivian Blaine, Sam Levene, Isabel Bigley, music and lyrics by Frank Loesser, songs that include "Luck Be a Lady," "Fugue for Tinhorns," "I've Never Been in Love Before," "A Bushel and a Peck," "Adelaide's Lament," "Sue Me," "If I Were a Bell," 1,200 perfs.; *Out of This World* 12/21 at the New Century Theater, with William Eythe, Charlotte Greenwood, Peggy Rea, music and lyrics by Cole Porter, songs that include "From This Moment On" and "Use Your Imagination," 157 perfs.

The Mambo is introduced from Cuba to U.S. dance floors.

Popular songs "Dearie" by Bob Hilliard and Dave Mann; "Hoop-Dee-Doo" by Frank Loesser; "If I Knew

You Were Comin' I'd 'Ave Baked a Cake" by Al Hoffman, Bob Merrill, Clem Watts; "It's So Nice to Have a Man Around the House" by Harold Spina, lyrics by Jack Elliott; "Music! Music! Music!" by Stephen Weiss and Bernie Baum; "My Heart Cries for You" by Percy Faith and Carl Sigman; "Rag Mop" by Johnnie Lee Wills and Deacon Anderson; "Sam's Song" by Lew Quadling, lyrics by Jack Elliott; "Sentimental Me" by Jim Morehead and Jimmy Cassin; "Sunshine Cake" by Jimmy Van Heusen, lyrics by Jimmy Burke (for the film *Riding High);* "I Don't Care If the Sun Don't Shine" by Mark David; "Cherry Pink and Apple Blossom White" by Louiguy, lyrics by Jacques Larme (English lyrics by Mack David); "Silver Bells" by Jay Livingston and Ray Evans (for the film *The Lemon Drop Kid).*

New York's Birdland opens on Broadway with alto saxophonist Charles Christopher "Charlie" Parker, 30, who moved to Harlem from his native Kansas City in 1939 and has become known as "Yardbird" or simply "Bird" through his "bebop" progressive jazz style. Parker will be a leading exponent of bebop until his death March 12, 1955, after making his last appearance at Birdland.

J. Edward "Budge" Patty, 26 (U.S.), wins in men's singles at Wimbledon, Louise Brough in women's singles; Arthur Larsen, 25, wins in men's singles at Forest Hills, Mrs. Du Pont in women's singles.

Florence Chadwick, 31, swims the English Channel August 20 and beats the record set by Gertrude Ederle in 1926. The San Diego stenographer crosses from France to England in 13 hours, 20 minutes (*see* 1951).

The New York Yankees win the World Series, defeating the Philadelphia Phillies 4 games to 0. The Phillies have won the National League pennant on the last day of the regular season beating the Brooklyn Dodgers with a 10th inning three-run homer by Dick Sisler.

Uruguay wins the fourth World Cup football (soccer) championship, edging Brazil 5 to 4 in the first competition since 1938. 200,000 spectators fill the three tiers of Rio's unfinished Maracana Stadium for the opening games and the finals; Germany is excluded.

Cotton's share of the U.S. textile market falls to 65 percent, down from 80 percent in 1940, and man-made fibers—mostly rayons and acetates—increase their share to more than 20 percent *(see* 1960).

Miss Clairol, introduced by Clairol Co., takes half the application time needed by other hair colorings *(see* 1931).

The Brink's robbery at Boston January 17 breaks all previous records for losses to armed robbers. Seven men wearing Navy pea jackets, chauffeurs' caps, and crepe-soled shoes enter Boston headquarters of the 91-year-old Brink's Express Co. shortly after 7 o'clock in the evening, don Halloween masks, and force five Brink's employees to turn over $1,218,211 in cash plus $1,557,000 in money orders. J. Edgar Hoover calls the job a Communist conspiracy, the FBI will spend $129 million trying to apprehend the perpetrators, it will arrest 10 men only days before the statute of limitations expires, two will die before coming to trial, eight will be sentenced, but only $50,000 of the take will ever be recovered.

Sen. Kefauver begins hearings May 26 at Miami in an investigation of U.S. criminal activities. His committee studies testimony by the late Abe "Kid Twist" Reles *(see* 1941) and when it moves its hearings to Washington, N.J., it hears gambler-racketeer William "Willie" Moretti testify that he never heard the word "Mafia" in his life.

The FBI issues its first "Ten Most Wanted Criminals" list in a publicity move. Several criminals are arrested after being identified by citizens who saw their pictures in newspapers, magazines, or post offices, and in the next 24 years some 300 criminals on the "most wanted" list will be located.

The Federal Bureau of Investigation by U.S. critic Max Lowenthal condemns FBI activities. The Dies Committee subpoenas Lowenthal and copies of the book are soon hard to find at bookstores.

Otis Elevator installs the first passenger elevators with self-opening doors in the Atlantic Refining building at Dallas. Self-service elevators will force thousands of operators to seek other means of employment.

New York's United Nations Secretariat building is completed by Wallace K. Harrison and consultants to provide offices for the UN's 3,400 employees on land overlooking the East River.

Twelve new office buildings are completed in New York with more than 4 million square feet of floor space.

U.S. suburban land values will rise in the next decade by 100 percent to 3,760 percent as middle-class whites desert the cities *(see* Levittown, 1947).

Smokey the Bear becomes a symbol for forest fire prevention. Found clinging to a charred tree in New Mexico's Lincoln National Forest, the badly burned little orphan black bear cub is flown by rangers to Santa Fe, nursed back to health at the home of a game warden, and shipped to the National Zoo at Washington, D.C. Posters appear showing Smokey with the message, "Only You Can Prevent Forest Fires." Smokey will be officially retired as the Forest Service symbol in May 1975 and will die in late 1976.

Park District National Park, designated December 28, is the first British national park.

World fisheries' production regains its prewar level of 20 million tons per year (*see* 1960).

Oyster production from the Connecticut coast south to New Jersey reaches 3.3 million bushels (see 1966).

Only 82 Atlantic salmon are landed on the Maine coast, less than a thousand pounds as compared with 150,000 in 1889. Dams and pollution have reduced spawning (see Danish fishermen, 1964).

"The industrialization of China should be based on the vast market of rural China," says party leader Liu Shao-Chi, who promulgates a new agrarian reform law, main-

1950 *(cont.)* taining that industrialization will be impossible without such reform. But while Mao Zedong says China's future lies in mobilizing the peasantry, Liu follows Marxist orthodoxy, insisting that large farms depend on chemical fertilizers, electrification, tractors, and other farm machinery, agronomists, and engineers. Most of the nation's budget goes to industrialization, but China's agricultural output nevertheless begins a rapid rise *(see* 1949; 1952).

U.S. farm prices will rise 28 percent in the next year as the Korean War (above) fuels inflation and boosts food prices.

The average U.S. farm worker produces enough food and fiber for 15.5 people, up from 7 in 1900 *(see* 1963).

The percentage of the U.S. available work force employed on the land falls to 11.6 percent, down from 25 percent in 1933, and 20 percent of U.S. farmers will quit the land in the next 12 years *(see* 1960).

Insects damage $4 billion worth of U.S. crops, consuming more food than is grown in New England, New York, New Jersey, and Pennsylvania combined.

The boll weevil, which has destroyed at least $200 million worth of U.S. cotton each year since 1909, destroys $750 million worth.

More than 75 percent of U.S. farms are electrified, up from 33 percent in 1940.

Sucaryl, introduced by Abbott Laboratories, is a cyclamate-based artificial sweetener *(see* Sveda, 1937; No-Cal, 1952).

Sugar Pops breakfast food is introduced by Kellogg *(see* 1942; 1952).

U.S. foods use 19 synthetic colors; the number will decline to 11 by 1967.

Green Giant Co. is adopted as the name of the Minnesota Valley Canning Co., whose Green Giant symbol has helped make it the largest U.S. canner of peas and corn *(see* 1936; beans, 1958).

A federal tax of 10¢/lb. on U.S. margarine is removed as are all federal restrictions on coloring margarine yellow. Butter is in such short supply that retail prices often top $1/lb. and U.S. consumers turn increasingly to margarine, whose average retail price is 33¢/lb. versus an average of 73¢ for butter.

General Foods introduces Minute Rice.

The Diners Club is founded to give credit card privileges at a group of 27 New York area restaurants. Local lawyer Frank X. McNamara starts with 200 card holders who pay an annual membership fee for the card that will be accepted within a few years at hotels, motels, car rental agencies, airline ticket counters, and retail shops as well as at restaurants in most of the United States and in many foreign countries *(see* 1965).

The population of the world reaches 2.52 billion.

The U.S. population has doubled since 1900. While the urban population will increase by only 1.5 percent in the decade ahead, the suburban population will increase by 44 percent as Americans flee to the suburbs, partly to escape the decaying social milieu of the inner cities.

1951 Seoul falls again to Communist forces January 4 and—80 percent destroyed—is retaken by UN forces March 14 as superior UN air power knocks Russian-manned MIGs out of the skies and uses bombs, rockets, strafing, and napalm to destroy the enemy. North Korea's supply lines are not cut and she builds underground factories, dormitories, and schools to carry on the war.

Gen. MacArthur offers to discuss a truce, Beijing rejects his bid March 29, MacArthur makes public his call for air attacks on Chinese cities, and President Truman relieves him of his command April 11 (Gen. Ridgway takes over). Now 71, MacArthur returns to America, is hailed as a hero and urged to run for president, but retires to private life after a farewell address to a joint session of Congress in which he quotes a World War I British Army song: "Old soldiers never die; they simply fade away."

Japan regains her autonomy May 3 (Constitution Day) after nearly 6 years of Allied military government *(see* below).

Soviet UN representative Jacob A. Malik proposes a Korean cease-fire June 23, Gen. Ridgway broadcasts an offer to negotiate, talks open July 8 at Kaesong but break off August 23. Heartbreak Ridge falls to UN forces September 23 after 37 days of bitter trench warfare to gain the strategic heights north of Yanggu, UN forces grow to 500,000 and outnumber the Communists, new peace talks begin October 25 at Panmunjom, but a 30-day "trial" armistice ends December 27.

Korean hostilities lowered the temperature of the cold war to chilly new levels, raising fears of an atomic war.

A U.S.-Philippine mutual defense pact signed August 30 is the first of several security pacts among anti-Communist powers in the Far East.

A U.S.-Japanese mutual security pact signed September 8 permits U.S. troops to remain in Japan indefinitely to assist UN operations in the Far East; no other nation is to have Japanese bases without U.S. consent.

French forces under Gen. Jean De Lattre de Tassigny, 62, repulse Communist attacks on Hanoi in French Indochina; the Communists revert to guerrilla tactics (*see* 1953).

A federal judge at New York finds Ethel Rosenberg, 35, her husband Julius, 34, and their friend Morton Sobell guilty March 30 of having sold atomic secrets to Soviet agents. Mrs. Rosenberg's brother David Greenglass worked at the Los Alamos nuclear research station in New Mexico; the couple is sentenced to death April 5 (*see* 1953; Fuchs, Gold, 1950).

U.S. nuclear scientists headed by Edward Teller set off the world's first thermonuclear reaction May 8 in a test at the mid-Pacific atoll Eniwetok.

British diplomats Guy Burgess and Donald Maclean flee to Moscow in late June amidst suspicions that they transmitted classified information to the Communists. British double agent H. A. R. "Kim" Philby has told Burgess and Maclean that their cover was blown and British intelligence agents were closing in (*see* 1963).

President Truman directs federal agencies August 10 to support a program that would disperse industrial plants to prevent total destruction in the event of a nuclear war.

Truman demands more stringent classification of security information by government agencies September 25.

The U.S. Army explodes an atomic device over the Nevada desert November 1 in the first war maneuver involving troops and nuclear weapons.

Iran's senate and majlis name Mohammed Mossadegh premier April 29. The leader of the extremist National Front will hold power for nearly 15 months (*see* oil industry, industrialization below).

Jordan's King Abdullah is assassinated July 20 at Jerusalem, his heir Emir Talal is proclaimed king by the national assembly September 5, but the new king will be declared unfit to rule within the next year (*see* 1949; 1952).

The UN Security Council protests Egyptian restrictions that prevent Israel-bound ships from using the Suez Canal (the U.S.S.R., the People's Republic of China, and India abstain). Egypt abrogates her 1936 alliance with Britain October 27 and abrogates also the condominium agreement of 1899 covering the Sudan.

Britain's Labour government has fallen in the general elections October 26 after 6 years in power. Winston Churchill becomes prime minister once again at age 77, but most of the social programs innovated by the Labour government remain in place under the new Conservative government that takes office October 27.

The Twenty-second Amendment ratified February 27 has limited U.S. presidential terms to two.

Sen. McCarthy calls George C. Marshall, now 70, a Communist agent, but the junior senator from Wisconsin has been unable to prove, or even show any real evidence, that anyone in the State Department is guilty of any subversive activity (*see* 1950). Sen. Millard E. Tydings, 61, (D. Md.) has attacked McCarthy for perpetrating "a fraud and a hoax," but he continues his campaign of accusations (*see* 1954).

South Africa's Department of the Interior color-classifies residents and issues cards to prove that they are white, black, or colored as the government moves to enforce its apartheid policies (*see* 1949).

Florida NAACP state secretary Harry T. Moore and his wife Harriet are killed Christmas night in a bombing of their house at Mims. Moore has protested the failure of Florida's governor Millard F. Caldwell to "take effective action" following the lynching of Jesse James Payne in 1946, his killing is thought to be associated with that protest 3 years ago, but no arrests will ever be made.

President Truman submits the largest peacetime budget in history January 15 with much of the $71.6 billion earmarked to meet the Korean emergency (above).

The Federal Reserve Board raises stock-purchase margin requirements January 16 from 50 to 75 percent in a move to discourage credit expansion.

The Wage Stabilization Board freezes wages and salaries January 26, the second wage-salary freeze in less than 9 years.

The Office of Price Stabilization orders "margin" profit ceilings on more than 200,000 consumer items February 27.

State "fair trade" price-fixing laws are not binding on retailers, the Supreme Court rules May 21. Price wars begin immediately and department stores are mobbed.

Congress extends the Defense Production Act July 31 but with reduced anti-inflation powers.

U.S. income tax receipts reach a record high of $56.1 billion.

The U.S. Atomic Energy Commission established in 1946 builds the first power-producing nuclear fission (atomic) reactor (*see* 1958; Calder Hall, 1956; Jersey Central Power and Light, 1963).

Iran's extremist new National Front government (above) nationalizes the oil industry, abrogating a 1933 concession treaty signed with London. The decree of nationalization takes effect May 2 retroactive to March 20, Britain appeals to the International Court of Justice May 26, the court rules against Iran July 5, President Truman appeals for a compromise and sends W. Averell Harriman to urge a settlement, Iranian forces occupy the Abadan oil fields September 27, Anglo-Iranian Oil completes evacuation of its personnel from

1951 ***(cont.)*** Abadan October 4, and Iran agrees December 10 to resubmit the issue to the Court of Justice; the dispute will continue until August 1954.

Jean Paul Getty gains control of Skelly Oil and Tidewater Oil (now called Mission Corp.) (*see* 1949; Getty Oil, 1956).

A Pennsylvania Railroad derailment at Woodbridge, N.J., February 6 kills 84 and injures 330 as a temporary track proves inadequate for the train's speed.

The United States has 8.62 million trucks, up from 4.85 million in 1941 (*see* 1961).

Chrysler installs power steering in 10,000 Crown Imperial sedans and convertibles. The mechanism patented by Francis Wright Davis in 1926 was used in U.S. Army armored vehicles during World War II, the patents have expired, and General Motors will begin mass production of power steering next year using improved Davis designs.

New York's First Avenue becomes one-way northbound and Second Avenue becomes one-way southbound beginning June 4 as U.S. cities try to speed up traffic flow in streets choked with trucks and passenger cars (*see* parking, 1955; Seventh and Eighth Avenues, 1954).

The New Jersey Turnpike opens November 5 to speed traffic between New York and Philadelphia. The 118-mile toll road will receive 2 billion tolls in its first 25 years.

Remington Rand introduces the Univac computer on a commercial basis for use by business firms and scientists. The company has acquired the ENIAC electronic computer of 1946 and renamed it (*see* IBM, 1953).

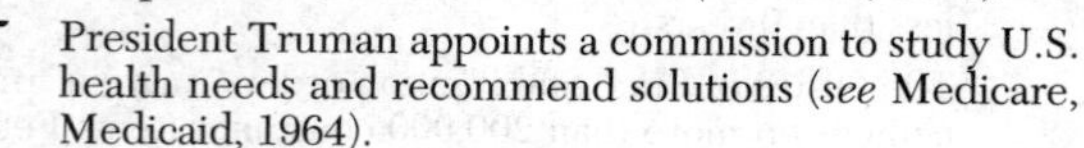

President Truman appoints a commission to study U.S. health needs and recommend solutions (*see* Medicare, Medicaid, 1964).

Ergotism breaks out August 17 in the south of France where people in the town of Pont-Saint-Esprit near Avignon have eaten bread containing illegal amounts of rye flour. Three hundred people are affected, 31 go mad, and four die in the largest epidemic since the Russian outbreak of 1926 (*see* LSD, 1943). French wage levels only now regain their Depression levels of 1938, and the ergotism episode is in part a consequence of heavy dietary reliance on cheap rye bread (*see* Monet, 1953).

Numbers in Color (Nombres en Couleur) by Belgian educator Emile-Georges Cuisenaire, 60, explains the author's method of teaching children to count by having them associate numbers and colors and thus learn the basics of addition and subtraction. Colored wooden Cuisenaire rods range in length from 1 to 10 centimeters and will be used throughout the world.

"Dennis the Menace" appears February 11 in 16 newspapers. U.S. cartoonist Henry King "Hank" Ketcham, 30, has been inspired by a neighbor's comment on his 5-year-old son. The new cartoon strip meets with almost instant success and will eventually be syndicated to some 1,600 papers worldwide.

Juan Perón shuts down *La Prensa* April 13, reacting to criticism of his Argentine dictatorship.

CBS broadcasts color television programs on a commercial basis beginning June 25, but conventional sets cannot pick up the signals in black and white (*see* RCA, 1954).

A new coaxial cable carries the first transcontinental U.S. television broadcast September 4.

See It Now with Edward R. Murrow debuts live coast to coast November 18 on CBS stations via coaxial cable (above).

Bell Telephone gives transistors their first commercial application October 10, using them in a trunk dialing apparatus and initiating long-distance "direct dial" telephone service (*see* transistor, 1948).

The volume of U.S. telephone calls is so great that without dial systems it would require nearly half the nation's adult population to operate switchboards (*see* 1919).

Local U.S. telephone call rates begin jumping from 5¢ to 10¢.

Nonfiction *God and Man at Yale—The Superstition of Academic Freedom* by 1950 graduate William Frank Buckley, 25.

Fiction *The Rebel (L'Homme Révolté)* by Albert Camus, who refutes communism because it leads to totalitarianism; *Adam, Where Art Thou? (Wo warst du Adam?)* by German novelist Heinrich Böll, 34, who was wounded in the war and gives a vivid account of the Nazi collapse; *The Conformist (Il Conformiste)* by Alberto Moravia; *The Catcher in the Rye* by J. D. Salinger whose hero Holden Caulfield seeks the pure and the good and eschews the "phonies"; *From Here to Eternity* by U.S. novelist James Jones, 30, who has served in the regular army and received encouragement from the late Maxwell Perkins at Scribner's; *The Caine Mutiny* by U.S. novelist Herman Wouk, 36; *Lie Down in Darkness* by U.S. novelist William Styron, 26; *Mr. Beluncle* by English novelist V. S. (Victor Sawden) Pritchett, 51; *The End of the Affair* by Graham Greene; *The Loved and the Lost* by Morley Callaghan; *The Troubled Air* by Irwin Shaw deals with the Communist witch-hunt in America; *A Question of Upbringing* by Anthony Powell begins a 12-volume *roman à fleuve; The Ballad of the Sad Café* (stories) by Carson McCullers; *The Grass Harp* (stories) by Truman Capote.

Poetry *The Mills of the Kavanaughs* by Robert Lowell.

Painting *Massacre in Korea* by Pablo Picasso; *Black and White Painting* by Jackson Pollock; *#3—1951* by Clyfford Still; *First Row Orchestra* by Edward Hopper; Diego Rivera completes murals for Mexico City's waterworks. John F. Sloan dies at Hanover, N.H., September 7 at age 80.

Sculpture *Baboon and Young* by Pablo Picasso; *Hudson River Landscape* by David Smith.

Theater *Second Threshold* by the late Philip Barry 1/2 at New York's Morosco Theater, with Clive Brook,

Betsy von Furstenberg, 126 perfs.; *The Rose Tattoo* by Tennessee Williams 2/3 at New York's Martin Beck Theater, with Maureen Stapleton, Don Murray, 306 perfs.; *Count Oderland (Graf Oderland)* by Max Frisch 2/10 at Zurich's Schauspielhaus; *The Lesson (La leçon)* by Eugène Ionesco 2/20 at the Théâtre de Poche, Paris; *The Autumn Garden* by Lillian Hellman 3/7 at New York's Coronet Theater, with Florence Eldredge, Kent Smith, Jane Wyatt, 101 perfs.; *A Sleep of Prisoners* by Christopher Fry 4/3 at Oxford's University Church; *Don Juan's Wife (La moglie di don Giovanni)* by Edouardo De Filippo 5/5 at Turin; *The Love of Four Colonels* by English playwright Peter Ustinov, 29, 5/23 at Wyndham's Theatre, London; *The Four Poster* by Jan de Hartog 10/24 at New York's Ethel Barrymore Theater, with Jessica Tandy, Hume Cronyn, 632 perfs.; *I Am a Camera* by John Van Druten 11/28 at New York's Empire Theater, with Julie Harris as Sally Bowles, William Prince as Christopher Isherwood, 262 perfs.

Television *Search for Tomorrow* 9/19 on CBS (daytime); *Love of Life* 9/9 on CBS (daytime); *I Love Lucy* 10/15 on CBS, with Lucille Ball (to 5/6/1957); *Kukla, Fran & Ollie* becomes an NBC Network show after having appeared on local Chicago television. Fran Allison will play opposite Burr Tillstrom's puppets for 6 years on NBC and ABC-TV.

Films John Huston's *The African Queen* with Katharine Hepburn, Humphrey Bogart, Robert Morley; Alessandro Blasetti's *Fabiola* with Michele Morgan, Henry Vidal; John Boulting's *The Magic Box* with Robert Donat and Maria Schell in a biography of motion picture pioneer William Friese-Greene (who actually contributed little to the development of movies); Akira Kurosawa's *Rashomon* with Toshiro Mifume; Jean Renoir's *The River* with Patricia Walters; Alfred Hitchcock's *Strangers on a Train* with Robert Walker, Farley Granger; Elia Kazan's *A Streetcar Named Desire* with Marlon Brando, Vivien Leigh. Also: Budd Boetticher's *The Bullfighter and the Lady* with Robert Stack, Joy Page, Gilbert Roland, Katy Jurado; Anthony Asquith's *The Browning Version* with Michael Redgrave, Jean Kent, Nigel Patrick; Zoltan Korda's *Cry the Beloved Country* with Canada Lee, Sidney Poitier; Laslo Benedick's *Death of a Salesman* with Fredric March, Mildred Dunnock; Rene Clement's *Forbidden Games (Jeux Interdits)* with Brigitte Fossey; Charles Crichton's *The Lavender Hill Mob* with Alec Guinness, Stanley Holloway; Alexander Mackendrick's *The Man in the White Suit* with Alec Guinness, Joan Greenwood; Mitchell Leisen's *The Mating Season* with Gene Tierney, John Lund, Thelma Ritter; Vittorio de Sica's *Miracle in Milan* with Francesco Golisano; Christian Nyby's *The Thing (From Another World)* with Kenneth Tobey; Gordon Parry's *Tom Brown's Schooldays* with John Howard Davies, Robert Newton, Hermione Baddeley.

Hollywood musicals Vincente Minnelli's *An American in Paris* with Gene Kelly, Leslie Caron, Oscar Levant, music by the late George Gershwin; Clyde Geronimi, Hamilton Luske, and Wilfred Jackson's *Alice in Wonderland* with Walt Disney animation, voices of Ed Wynn, Richard Haydn, Sterling Holloway, Jerry Colonna, and others, music by Sammy Fain, lyrics by Bob Hilliard, songs that include "I'm Late."

Broadway musicals *The King and I* 3/29 at the St. James Theater, with Gertrude Lawrence, Yul Brynner, music by Richard Rodgers, lyrics by Oscar Hammerstein II, book based on Margaret London's novel *Anna and the King of Siam* based in turn on diaries kept by the late Anna Lewin-Owen (*see* Phra Chom Klao Mongkut, 1851), songs that include "Getting to Know You," "Hello, Young Lovers," "We Kiss in a Shadow," "Shall We Dance," "Whistle a Happy Tune," 1,246 perfs.; *A Tree Grows in Brooklyn* 4/19 at the Alvin Theater, with Shirley Booth, Johnny Johnston, music by Arthur Schwartz, lyrics by Dorothy Fields, songs that include "Make the Man Love Me," "Look Who's Dancing," "I'll Buy You a Star," 267 perfs.; *Flahooley* 5/14 at the Broadway Theater, with Bil Baird, Cora Baird, Yma Sumac, music by Sammy Fain, lyrics by E. Y. Harburg, songs that include "Here's to Your Illusions," 40 perfs.; *Top Banana* 11/1 at the Winter Garden Theater, with Phil Silvers, music and lyrics by Johnny Mercer, 350 perfs.; *Paint Your Wagon* 11/12 at the Shubert Theater, with Eddie Dowling, Ann Crowley, music by Frederick Loewe, lyrics by Alan Jay Lerner, songs that include "They Call the Wind Maria," "I Talk to the Trees," "I Still See Elisa," 289 perfs.

Opera *The Pilgrim's Progress* 5/26 at London's Covent Garden, with music by Ralph Vaughan Williams; *The Rake's Progress* 9/11 at the Venice Festival, with music by Igor Stravinsky; *Ivan IV (Ivan the Terrible)* 10/12 at the Grand Théâtre de Bordeaux, with music by the late Georges Bizet; *Billy Budd* 12/1 at London's Covent Garden, with music by Benjamin Britten, libretto from Herman Melville's last complete work (Melville finished *Billy Budd, Foretopman* in 1891 but it was not published until 1924); *Amahl and the Night Visitors* 12/24 on the NBC Television Opera Theater, with music by Gian Carlo Menotti who has written the first opera for television.

Ballet *The Miraculous Mandarin* 9/6 at the New York City Center, with Melissa Hayden, Hugh Laing, music by the late Béla Bartòk, choreography by Todd Bolender; *The Pied Piper* 12/4 at the New York City Center, with Tanaquil LeClerq, Jerome Robbins, music by Aaron Copland, choreography by Robbins.

Cantata *The Sons of Light* by Ralph Vaughan Williams 5/6 at London, with a chorus of 1,000 schoolchildren.

First performances Symphony No. 2 by Charles Ives (who wrote it between 1897 and 1901) 2/23 at New York's Carnegie Hall (Ives declines Leonard Bernstein's invitation to attend, but he does hear part of the performance on his housemaid's radio—he does not own one himself—and he comes out of her room dancing a jig); *Suite Archaïque* by Arthur Honegger 2/28 at Louisville; Symphony No. 5 *(Di tre re)* by Honegger 3/9 at Boston's Symphony Hall; Symphony No. 4 by Walter Piston 3/30 at Minneapolis; *Monopartita* by Honegger 6/12 at Zurich; *Stabat Mater* by Francis Poulenc 6/13 at Strasbourg.

1951 ***(cont.)*** Popular songs "In the Cool, Cool, Cool of the Evening" by Hoagy Carmichael, lyrics by Johnny Mercer (for the film *Here Comes the Groom*); "Kisses Sweeter Than Wine" by Huddie "Lead Belly" Ledbetter, lyrics by The Weavers (Pete Seeger, 32; Lee Hays, Fred Hellerman, Ronni Gilbert); "Unforgettable" by Irving Gordon; "I'm in Love Again" by Cole Porter; "Mockin' Bird Hill" by Vaughn Horton; "Too Young" by Sid Lippman, lyrics by Sylvia Dee; "If" by Tolchard Evans, lyrics by Robert Hargreaves and Stanley J. Domerell; "It's All in the Game" by former U.S. Vice President Charles Gates Dawes, lyrics by Carl Sigman; "Mixed Emotions" by Stuart F. Loucheim; "Domino" by Louis Ferrani, lyrics by Jacques Plante (English lyrics by Don Raye); "Be My Love" by Nicholas Brodszky, lyrics by Sammy Cahn (for the film *The Toast of New Orleans);* "Be My Life's Companion" by Milton De Lugg and Bob Hilliard; "Come on-a My House" by Ross Bagdasarian and William Saroyan; "Cold, Cold Heart" by Hank Williams; "It's Beginning to Look a Lot Like Christmas" by Meredith Willson.

"Sugar Ray" Robinson (*né* Walker Smith), 30, wins the world middleweight boxing title February 4 at Chicago by beating Jake LaMotta (who gave Robinson his only defeat in 123 professional bouts 8 years ago); Sugar Ray loses the title July 10 to Randy Turpin, regains it September 12, and relinquishes it to try for the light-heavyweight crown.

"Jersey Joe" Walcott (*né* Arnold Raymond Cream), 37, wins the world heavyweight title July 18, knocking out Ezzard Charles in the seventh round of a championship bout at Pittsburgh.

Richard Savitt, 24 (U.S.), wins in men's singles at Wimbledon, Doris Hart, 26 (U.S.), in women's singles; Frank Sedgman, 23 (Australia), wins in men's singles at Forest Hills, Maureen Connolly, 16, in women's singles.

A college basketball scandal at New York rocks the sports world. The courts will find several college players guilty of taking payoffs from gamblers to throw games, some will serve jail terms, all will lose their chances to play professional basketball.

Willie Mays joins the New York Giants lineup in center field. Willie Howard Mays, Jr., 20, will play his first full season in 1954, win the National League's batting championship with a .345 average, continue with the Giants when they move from the Polo Grounds to San Francisco in 1957, and remain a stalwart of the team until 1972.

Mickey Mantle joins the New York Yankees at center field, playing alongside Joe DiMaggio, who handles left field (*see* 1936). Mickey Charles Mantle, 19, plays only 96 games and sets a league record of 111 strikeouts but shows he can hit the ball out of the park; he will be a Yankee star for 17 years beginning next season and his batting average in his 10 peak years will always be above .300.

New York Giants third baseman Robert Brown "Bobby" Thompson, 27, hits a three-run homer in the last half of the ninth inning of the third playoff game of the National League pennant race to beat the Brooklyn Dodgers 5 to 4 at the Polo Grounds. The Giants win the pennant after having come from 13½ games behind to tie the Dodgers and force the playoff.

The New York Yankees win the World Series by defeating the New York Giants 4 games to 2.

Florence Chadwick becomes the first woman to cross the English Channel from England to France (*see* 1950). She makes the crossing from Dover September 11 in 16 hours, 22 minutes, and will break that record 10/12, 1955, when she does it in 13 hours, 33 minutes.

The first Pan-American Games are held at Buenos Aires. Only Western Hemisphere athletes may compete.

The world's first sky-diving championship is held in Yugoslavia. Parachute jumpers compete for prizes.

U.S. imports of the Chemise Lacoste begin. Izod Cie. has contracted with René Lacoste to use the "alligator" symbol associated with him since he helped France win the Davis Cup in 1927. The long-tailed all-cotton tennis shirt will become a U.S. status symbol.

Victor Gruen Associates, founded by Austrian-American architect Victor Gruen, 48, will have a major impact on U.S. building design and city planning (*see* Fort Worth Plan, 1952). Gruen has designed shopping centers, store fronts, and some of the structures at the 1939 New York World's Fair.

Pittsburgh's Alcoa building is completed by Harrison & Abramovitz with 30 stories clad in aluminum.

Pittsburgh's U.S. Steel and Mellon building is completed by Harrison & Abramovitz and rises 41 stories.

A second Levittown goes up in Bucks County, Pa., with houses for 17,000 families plus schools and churches to be built on land donated by the developers (*see* 1947). A third Levittown will have 12,000 homes in New Jersey by 1965 with houses of three basic designs in a variety of color schemes commingled on each street. The three- to four-bedroom houses sell for between $11,500 and $14,500 each including major appliances, and the low prices encourage thousands of Americans to leave the city and move into suburban developments that sprawl across the landscape as scores of builders adopt the methods pioneered by Levitt.

The worst floods in U.S. history inundate Kansas and Missouri; 41 die between July 2 and 19, 200,000 are left homeless, property damage reaches $1 billion.

Britain produces enough food to supply half her domestic needs, up from 42 percent in 1914 when the population was 35 million versus 50 million today.

British farms have 300,000 tractors, up from 55,000 in 1949.

A 2-year drought begins in Australia that will kill off millions of head of livestock and will force rationing of butter.

Australian sheep-raisers introduce the virus disease myxomatosis that is endemic in South America in an effort to kill off the rabbits which are consuming enough grass to feed 40 million sheep (*see* Austin,

1859). The introduction of hawks, weasels, snakes, and other predators has not appreciably reduced the rabbits' numbers, hundreds of thousands of miles of costly rabbit-proof fencing has not stopped them, but rabbits infected with the virus are released, the disease is communicated by mosquitoes and other insects, and in some areas 99 percent of the rabbits are eliminated before natural resistance to myxomatosis becomes dominant (*see* France, 1952).

U.S. Department of Agriculture entomologist Edward F. Knipling, 42, finds a new way to control insect populations biologically rather than chemically; he develops an irradiation method for sterilizing male insects and releasing them to mate with females who will die without reproducing (*see* 1954).

U.S. food prices climb steeply in the Korean War inflation. The nation suffers a beef shortage as affluent consumers increase their demand.

A British beef shortage leads to the consumption of 53,000 horses for food.

Congress votes June 15 to loan India $190 million to buy U.S. grain (*see* P.L. 480, 1954).

The United States agrees to supply Yugoslavia with $38 million in food aid.

Gerber Products starts using MSG (monosodium glutamate) in its baby foods to make them taste better to mothers (*see* MSG, 1934, 1969).

Tropicana Products is founded at Bradentown, Fla., to produce chilled, pasteurized grapefruit and orange juice. Italian-American entrepreneur Anthony T. Rossi, 51, established a business in the late 1940s to pack chilled fruit sections in glass jars and will develop ways to extract juice by varying degrees depending on the quality of the fruit (*see* 1955).

Britain's first supermarket chain begins operations in September. Premier Supermarkets opens its first store in London's Earls Court (*see* 1948).

The U.S. population reaches 153 million, the U.S.S.R. has 172 million, China 583 million, India 357, Pakistan 76, Japan 85, Indonesia 78, Britain 50, West Germany 50, Italy 47, France 42, Brazil 52.

Infant mortality in India is 116 per thousand, down from 232 in 1900. In the United States it is 29, down from 122 in 1900, and in Britain it is 30.

New York (with suburban areas) has 12.3 million people, London 8.4 million, Paris 6.4, Tokyo 6.3, Shanghai 6.2, Chicago 5, Buenos Aires 4.8, Calcutta 4.6, Moscow 4.2, Los Angeles 4, Berlin 4.4, Leningrad 3.2, Mexico City 3.1, Bombay 2.9, Beijing 2.8.

Margaret Sanger urges development of an oral contraceptive for humans. She visits Gregory Pincus at the Worcester Foundation and asks him to follow up on his experiments on fertility in animals (*see* 1948; 1954).

1952 Britain's George VI dies of lung cancer February 6 at age 56 after a reign of more than 15 years (the king was a heavy smoker). His elder daughter, 25, flies home from a visit to East Africa and ascends the throne as Elizabeth II to begin a long reign in which the British Empire will decline from 40 nations to no more than 12 with the British monarch having an effective voice in only one.

Canada's first Canadian governor-general has taken office January 24. Vincent Massey, 64, will serve until 1959 (a scion of the Massey-Harris farm equipment family, his brother Raymond is a prominent actor).

A Cuban military coup March 10 overthrows President Prio Socarras and replaces him with Gen. Fulgencio Batista y Zaldivar, who ruled as dictator from 1933 to 1940 and will rule as dictator again until 1959.

Puerto Rico adopts a new Constitution July 25 and becomes the first U.S. commonwealth; residents obtain all rights of U.S. citizenship except voting in federal elections and need not pay federal income taxes (*see* 1898; 1954).

Beijing accuses U.S. forces in Korea of using germ warfare. U.S. Air Force planes bomb North Korean hydroelectric plants June 23.

Egypt's dissolute king Farouk I abdicates July 26 after a 16-year reign and 3 days after a coup by Gen. Mohammed Naguib, who forms a government September 7. Egypt and Sudan sign an agreement October 13 over use of water from the Nile, and the Egyptian Constitution of 1923 is abolished December 10 (*see* 1954).

Jordan's schizophrenic king Talal is deposed August 11 after a brief reign. His playboy son Hussein, now 16, will ascend the throne next year and reign for more than 38 years despite at least 10 attempts on his life and countless conspiracies to depose him.

Ethiopia takes over Eritrea from the British September 11 (*see* 1950).

Moscow ousts U.S. Ambassador George F. Kennan October 3. Kennan has commented on the isolation of Western diplomats in Moscow and the Kremlin demands his recall.

President Truman has relieved Gen. Eisenhower of his post as Supreme Allied Commander at Ike's request in April and named Gen. Matthew Ridgway to succeed him. Gen. Mark Clark has replaced Ridgway in the Far East, and the Republican party nominates Eisenhower to run for president. Democrats nominate Gov. Adlai E. Stevenson, 52, of Illinois.

Critics accuse Gen. Eisenhower's running mate, Sen. Richard Nixon, of taking a "slush fund" of $18,000 from California businessmen. Nixon appears on television September 23 and says, "I come before you tonight as a candidate for the vice presidency and as a man whose honesty and integrity have been questioned. . ."; he denies that any of the money in question went for his personal use and adds, "I did get something, a gift after the nomination. It was a little cocker spaniel dog, black and white, spotted. Our little girl Tricia, 6, named it 'Checkers.' The kids, like all kids, love the dog. Regardless of what they say about it, we are going to keep it." The speech produces more than 1 million favorable letters and telegrams. Eisenhower wins 55 percent of the

1952 ***(cont.)*** popular vote and 442 electoral votes, Stevenson only 89 electoral votes with 44 percent of the popular vote.

Britain tests an atomic (nuclear fission) bomb developed by her physicists October 2 over the Monte Bello Islands near Australia. She joins the United States and the U.S.S.R. as a nuclear power.

The U.S. Atomic Energy Commission tests a nuclear-fusion (hydrogen) device November 1 (*see* von Neumann, Fuchs, 1946; Truman, 1950). The test at the Eniwetok proving grounds in the Pacific is the first full-scale thermonuclear explosion in history (*see* 1951; Bikini, 1954).

A Mau Mau insurrection against Kenya's white settlers begins October 20. Kikuyu tribal leader Jomo Kenyatta (Kamau wa Ngengi), 63, has organized the Mau Mau ("Hidden Ones"). London declares a state of emergency and dispatches a cruiser and battalion of troops (*see* 1953).

Israel's president Chaim Weizmann dies November 7 at age 77. The Knesset replaces him December 8 with Itzhak Ben-Zvi.

Former Czech Communist party secretary Rudolf Slansky and former Czech foreign minister Vladimir Clementis go on trial for treason November 27.

President-elect Eisenhower visits Korea December 2 in fulfillment of a campaign promise. The UN adopts an Indian proposal for a Korean armistice December 3 but Beijing rejects it December 15.

South Africa's Supreme Court invalidates apartheid racial legislation March 20, Prime Minister Daniel F. Malan introduces legislation April 22 to make the parliament the highest court, the bill passes, demonstrations against "unjust" racial laws begin June 26, police arrest demonstrators through July, and leaders of the movement say deliberate violations of the law will continue until all the jails in the country are filled.

Malcolm X joins U.S. Black Muslim leader Elijah Mohammed following his release from prison after having served 6 years on a conviction of armed robbery. The disciple, now 27, has educated himself in prison, read the teachings of Elijah Mohammed, adopted the new faith, changed his name from Malcolm Little, and will become the sect's first "national minister" in 1963 (*see* 1931; 1965).

Playwright Lillian Hellman defies the Dies Committee May 22, testifying that she is not a "Red" but will not say whether or not she was 3 or 4 years ago because such testimony "would hurt innocent people in order to save myself. . . . I cannot and will not cut my conscience to fit this year's fashions." But the Communist "witch-hunt" continues, ruining many careers (*see* 1950; 1951).

President Truman orders federal seizure of U.S. steel mills April 8 to avert a nationwide strike, the steel companies take legal action to block government operation of the mills, the Supreme Court rules June 2 that the executive order was illegal, and 600,000 CIO steel workers walk out June 23 to begin a 53-day strike.

The U.S. government returns the nation's railroads to private control May 23 after 21 months of operation by federal troops (*see* 1950).

U.S. income tax receipts reach a record $69.6 billion but the nation has a record peacetime deficit of $9.3 billion as the Korean "police action" continues to drain U.S. resources.

Congress amends the Social Security Act of 1935 to increase benefits to the elderly by 12.5 percent and to permit pension recipients to earn as much as $75 per month without loss of benefits.

Australian cattleman-prospector Langley George Hancock, 43, discovers a mountain of solid iron ore in the Hammersley Range when his light plane is forced off course and he sees rust-colored outcroppings. Hancock will keep the location secret for 10 years until a change in state mining laws permits him to stake his claim, but he gets in touch with the worldwide mining concern Rio Tinto which will make an agreement with Kaiser Aluminum and exploit the ore deposit that will bring Hancock an estimated $250 million, making him the richest man in Australia (*see* 1964).

The United States Lines passenger ship S.S. *United States* leaves New York July 3 on her first transatlantic voyage and sets a new record. Built with immense 240,000-horsepower steam turbines that can push her at 50 miles per hour and convertible to a troopship that can transport 14,000 men, the $79 million 53,000-ton vessel is 990 feet in length overall, can carry 1,750 passengers, and makes the crossing of 2,949 nautical miles in 3 days, 10 hours, 40 minutes, averaging 35.59 knots per hour (more than 40 mph).

The Volga-Don Ship Canal opens between the Volga and Don Rivers in the Soviet Union. The 62-mile waterway provides a link for the Baltic-Black Sea route.

Europe's Arlberg-Orient Express is rerouted via Salonica following last year's closing of the Turkish-Bulgarian border (*see* 1932; 1962).

London's last tram runs July 6 as motorbuses replace streetcars.

British Motors is created by a merger of Austin and Morris (*see* Austin, 1905).

Volkseigener Betrieb Sachsenring in East Germany begins production of the Trabant, a small, plastic-bodied car for which demand will far exceed supply.

Jet aircraft passenger service is inaugurated by a British De Havilland Comet which jets from London to Johannesburg, covering the 6,724-mile distance in less than 24 hours (*see* 1948; 1958).

Archaeologist Michael Ventris deciphers Myceanean texts of 1450 B.C., doing for the inscriptions of the Minoan civilization what Champollion's work in 1822 and earlier did for the inscriptions of ancient Egypt.

King Solomon's Ring by Austrian animal behaviorist Konrad Lorenz, 45, shows that Darwinian natural selection processes operate not only in shaping physical form but also in determining behavioral characteristics.

A poliomyelitis epidemic strikes more than 50,000 Americans, 3,300 die, thousands are left crippled.

U.S. microbiologist Jonas Edward Salk, 38, tests a vaccine against polio, administering a hypodermic solution based on killed viruses. It employs a method of virus culture developed in the laboratory of Children's Hospital, Boston, by bacteriologists John Franklin Enders, 55, with Frederick Chapman Robbins, 36, and Thomas Huckle Weller, 37 (they will all share in a 1954 Nobel prize). Enders helped Harvard bacteriology professor Hans Zinsser develop the first anti-typhus vaccine in 1930 and will apply the Enders-Robbins-Weller culture method to developing a measles vaccine (*see* 1962; Salk, 1954).

Reserpine is isolated by Swiss chemists employed by Ciba under the direction of Emil Schlitter; they produce pure crystals of the active ingredient in *Rauwolfia serpentina* (*see* Vakil, 1949). Boston heart specialist Robert Wallace Wilkins, 45, observes that reserpine not only reduces high blood pressure but also reduces anxiety (*see* LSD and serotinin, 1953).

Isoniazid, introduced to fight tuberculosis, is Hoffman-La Roche's brand of isonicotinic acid hydrozide. Research chemist H. Herbert Fox, 37, has employed work in organic chemistry to develop the drug, which will be found to produce tumors in test mice but not in humans. So effective will it be as a prophylactic against TB that it will remain in use despite some potentially risky side effects.

Chlorpromazine is developed to intensify the action of standard sedatives. French chemists at Rhone-Poulenc have found new potential in an antihistamine synthesized in the late 1940s (*see* Thorazine, 1954).

Sonotone Corp. of Elmsford, N.Y. introduces the first transistorized hearing aids December 29 (*see* 1935; transistor, 1948).

The first pocket-size transistor radios are introduced under the name Sony by Japanese tape-recorder maker Masaru Ibuka who returns from a visit to America with a license from AT&T's Western Electric division (*see* 1950; Shockley, 1948). Their engineers have told him that 95 of 100 transistors coming off production lines are defective and must be discarded, and that to make a profit it would be necessary to charge $6 for a transistor alone when vacuum-tube radios sell for as little as $6.95 complete, but Ibuka has improved the technology of transistor production, cut the reject rate from 95 percent to 2 percent, and reduced the cost from $6 to a fraction of one dollar (*see* 1959).

The Today Show debuts 1/14 on NBC television stations with Chicago radio personality Dave Garroway, 38, serving as master of ceremonies for the 2-hour morning news and interview show developed by Pat Weaver. Working out of a street-level studio with a window facing on New York's Rockefeller Center street traffic, Garroway is soon joined by chimpanzee J. Fred Muggs whose presence encourages children to turn on the TV set.

Most major countries of the world have TV stations and receiving sets, but while the U.S. TV system employs 525 lines to the inch and produces a sharper image than the one produced by the 450-line BBC system, European countries have adopted a 625-line system that produces far better resolution. France and Monaco will institute 819-line systems, but Britain, the United States, Japan, and Latin America will be locked into their coarser image systems, thus paying the penalty of leadership (*see* Japan, 1953; BBC-2, 1964).

U.S. inventors John Multin and Wayne Johnson demonstrate the first videotape November 11 at Beverly Hills, Calif. (*see* 1956).

Some 17 million U.S. homes have TV sets by year's end, up from 5 to 8 million in 1950 (*see* 1958).

The *National Enquirer* founded in 1926 by a former Hearst advertising executive is taken over by Generoso Pope, Jr., 25, whose father made a fortune in sand and gravel and has published the New York Italian-language newspaper *Il Progresso*. The *Enquirer* has become little more than a tout sheet, but Pope will emphasize crime, gore, miracle cures, gossip, and sex to build circulation that will pass 4 million per week by 1975 and will multiply his $75,000 investment hundreds of times over.

Mad magazine has its beginnings in the *Mad* comic book that appears in May with 32 pages in full color satirizing other comic books and even making fun of itself. Founder-editor Harvey Kurtzman will develop a freckled, big-eared, cow-licked idiot boy character, comedian Ernie Kovaks will name the boy Melvin Cowznofski, and he will become infamous as Alfred E. Neuman ("What—Me Worry?"). Publisher William Gaines, 32, will sell his E. C. (Entertaining Comics) Publications to Warner Communications.

Nonfiction *Witness* by Whittaker Chambers; *The Irony of American History* by U.S. theologian Reinhold Niebuhr, 60; *The Power of Positive Thinking* by New York clergyman Norman Vincent Peale, 54, of the Marble Collegiate Church.

Fiction *Men of Good Will* (*Les Hommes de bonne volonté*) by Jules Romains; *The Works of Love* by U.S. novelist Wright Morris, 32; *Wise Blood* by U.S. novelist Flannery O'Connor, 27; *The Groves of Academe* by Mary McCarthy; *Martha Quest* by Doris Lessing; *Excellent Women* by Barbara Pym; *A Buyer's Market* by Anthony Powell; *Doting* by Henry Green; *East of Eden* by John Steinbeck; *The Natural* by New York novelist Bernard Malamud, 38; *The Long March* by William Styron; *The Old Man and the Sea* (novella) by Ernest Hemingway; *Player Piano* by U.S. novelist Kurt Vonnegut, Jr., 29.

Juvenile *Charlotte's Web* by E. B. White.

Poetry *A Mask for Janus* by U.S. poet W. S. Merwin, 25; *Rod of Incantation* by English poet Francis King, 29; "Do not go gentle into that good night" by Dylan Thomas whose failing eyesight has inspired him to write, "Rage, rage against the dying of the light."

1952 ***(cont.)*** Painting *Red Painting* by New York minimalist Ad (Adolf F.) Reinhardt, 38; *Woman and Bicycle* by Willem de Kooning; *Number Three* by Jackson Pollock; *Adam* by Barnett Newman; *Mountains and Sea* by U.S. painter Helen Frankenthaler, 25.

The Asahiflex 1, made by Asahi Optical Co., is Japan's first 35 mm. single lens reflex camera (*see* 1936; 1954).

Theater *The Waltz of the Toreadors (La valse des toréadors)* by Jean Anouilh 1/10 at the Comédies de Champs-Elysées, Paris; *The Shrike* by U.S. playwright Joseph Kramm 1/15 at New York's Cort Theater, with José Ferrer, Judith Evelyn, 161 perfs.; *Jane* by S. N. Behrman 2/1 at New York's Coronet Theater with Edna Best, Basil Rathbone, 100 perfs.; *The Deep Blue Sea* by Terence Rattigan 3/5 at London's Duchess Theatre; *The Marriage of Mr. Mississippi (Die Ehe des Herrn Mississippi)* by Friedrich Dürrenmatt 3/26 at Munich's Kammerspiele; *The Chairs (Les chaises)* by Eugène Ionesco 4/22 at the Théâtre Lanery, Paris; *Dial 'M' for Murder* by English playwright Frederick Knott 6/5 at London's Westminster Theatre, with John Williamson; *Quadrille* by Noël Coward 9/12 at London's Phoenix Theatre, with Alfred Lunt, Joyce Carey; *The Time of the Cuckoo* by Arthur Laurents 10/15 at New York's Empire Theater, with Shirley Booth, Dino DiLuca, 263 perfs.; *No Peace for the Ancient Faun (Non ce pace per Lantico fauno)* by Carlo Terron 11/7 at Milan's Teatro di Via Mazzoni; *The Mousetrap* by Agatha Christie 11/25 at London's 453-seat Ambassadors Theatre, with Richard Attenborough and Sheila Sim in a melodrama that will move March 25, 1974, to the larger St. Martin's Theatre and still be playing when Dame Agatha dies early in 1976; *The Seven-Year Itch* by U.S. playwright George Axelrod, 30, 11/20 at New York's Fulton Theater, with Tom Ewell, Neva Patterson, 1,141 perfs.

Films Fred Zinneman's *High Noon* with Gary Cooper, Grace Kelly; John Ford's *The Quiet Man* with John Wayne, Maureen O'Hara, Barry Fitzgerald; Vittorio De Sica's *Umberto D* with Carlo Battisti; Elia Kazan's *Viva Zapata!* with Marlon Brando. Also: Vincente Minnelli's *The Bad and the Beautiful* with Kirk Douglas, Lana Turner, Dick Powell; Jacques Becker's *Casque d'Or* with Simone Signoret; Daniel Mann's *Come Back, Little Sheba* with Burt Lancaster, Shirley Booth; Robert Siodmak's *The Crimson Pirate* with Burt Lancaster, Nick Cravat; Yasujiro Ozu's *The Flavor of Green Tea Over Rice;* Irving Reis's *The Four Poster* with Rex Harrison, Lili Palmer; Jean Renoir's *The Golden Coach* with Anna Magnani; Cecil B. De Mille's *The Greatest Show on Earth* with Betty Hutton, Charlton Heston, Cornel Wilde, James Stewart; Akira Kurosawa's *Ikuru;* Kenji Mizoguchi's *The Life of O-Haru;* John Huston's *Moulin Rouge* with José Ferrer as Henri de Toulouse-Lautrec; Luis Buñuel's *Los Olvidados (The Young and the Damned)* with Alfonso Mejia; Orson Welles's *Othello* with Welles, Michael MacLiammoir, Suzanne Cloutier; John Huston's *The Red Badge of Courage* with Audie Murphy, Bill Mauldin; George Sidney's *Scaramouche* with Stewart Granger, Eleanor Parker; Stuart Heisler's *The Star* with Bette Davis, Sterling Hayden, Natalie Wood; Ingmar Bergman's *Summer With Monika* with Harriet Andersson, Lars Ekborg; Henri-Georges Clouzot's *The Wages of Fear* with Yves Montand.

Hollywood musicals Gene Kelly and Stanley Donen's *Singin' in the Rain* with Kelly, Debbie Reynolds, Donald O'Connor, Jean Hagen, Cyd Charisse in a spoof on early talking pictures with songs that include "My Lucky Star," "Broadway Melody," "Good Morning," and the title song; Charles Vidor's *Hans Christian Andersen* with Danny Kaye, music and lyrics by Frank Loesser, songs that include "Anywhere I Wander," "Ugly Duckling," "Inchworm," "Thumbelina."

Broadway musicals *New Faces* 5/16 at the Royale Theater, with Eartha Kitt, Alice Ghostley, songs that include "Love Is a Simple Thing" by Jane Carroll, lyrics by Arthur Siegal, "Monotonous" by Arthur Siegal, lyrics by Jane Carroll, Nancy Graham, 365 perfs.; *Wish You Were Here* 6/25 at the Imperial Theater, with Sheila Bond, Jack Cassidy, music and lyrics by Harold Rome, songs that include "Where Did the Night Go?" and the title song, 598 perfs.

Opera *Trouble in Tahiti* 6/12 at Brandeis University, Waltham, Mass., with music and libretto by Leonard Bernstein.

First performances Symphony, *Die Harmonie der Welt* by Paul Hindemith 1/24 at Basel; Symphony No. 4 by Paul Creston 1/30 at Washington, D.C.; *Symphonic Concertante* in E minor by Serge Prokofiev 2/18 at Moscow, with Mstislav Rostropovich as soloist; *The Meeting of the Volga and the Don Rivers* by Serge Prokofiev 2/22 on Moscow Radio; Three Spiritual Folk Songs by the late Anton von Webern 3/16 at Columbia University's McMillin Theater; *Walt Whitman* symphonic essay by Paul Creston 3/28 at Cincinnati; Symphony No. 4 by Morton Gould 4/13 at West Point, N.Y.; *Water Music* by John Cage 5/2 at New York; Romance in D flat for Harmonica, String Orchestra, and Piano by Ralph Vaughan Williams 5/3 at New York, with Larry Adler as soloist; *West Point Suite* by Darius Milhaud 5/30 at West Point, N.Y.; Symphony No. 7 in C sharp minor by Prokofiev 10/11 at Moscow; Concerto for Tap Dancer and Orchestra by Morton Gould 11/16 at Rochester, N.Y.; Symphony No. 7 by Roy Harris 11/20 at Chicago; Concerto for Violin by Gian Carlo Menotti 12/5 at Philadelphia's Academy of Music; *Sea Piece with Birds* by Virgil Thomson 12/10 at Dallas.

Television *The American Bandstand* debuts in January on ABC network stations. Host Dick Clark, 22, will continue to emcee the show for more than 25 years.

Dave Brubeck and his San Francisco quartet pioneer "modern" or "progressive" jazz. David Warren Brubeck, 31, makes his New York debut and begins a 15-year career of innovation that will give jazz music a new complexity of rhythm and counterpoint.

Popular songs "Lullaby of Birdland" by George Shearing, lyrics by B. Y. Forster; "Takes Two to Tango" by Al Hoffman and Dick Manning; "Do Not Forsake Me" by

Dmitri Tiomkin, lyrics by Ned Washington (for the film *High Noon*); "Don't Let the Stars Get in Your Eyes" by Slim Willet; "Your Cheatin' Heart" and "Jambalaya—in the Bayou" by Hank Williams; "Wheel of Fortune" by Bennie Benjamin and George Weiss; "Blue Tango" by Leroy Anderson; "Pittsburgh, Pennsylvania" by U.S. songwriter Bob Merrill (*né* Henry Lavin), 30; "Delicado" by Waldyr Azevedo, lyrics by Jack Laurence; "Count Your Blessings (Instead of Sheep)" by Irving Berlin (for the 1954 film *White Christmas*); "I Saw Momma Kissing Santa Claus" by Tommy Conner.

Eddie Arcaro wins his fifth Kentucky Derby, riding Hill Gail.

Frank Sedgman wins in men's singles at Wimbledon and Forest Hills, Maureen Connolly in women's singles.

The Olympic Games at Helsinki attract 5,867 contestants from 69 nations. Bob Mathias, 21 (U.S.), triumphs in the decathlon event; U.S. athletes win the most medals.

Rocky Marciano wins the world heavyweight title that he will defend six times before retiring undefeated in 1956. Brockton, Mass., fighter Rocco Francis Marchegiano, 28, knocks out Jersey Joe Walcott September 23 in the 13th round of a title bout at Philadelphia.

The New York Yankees win the World Series by defeating the Brooklyn Dodgers 4 games to 2.

"Never wear a white shirt before sundown," says an advertisement for Hathaway shirts in the October 18 *New Yorker*. Four out of five shirts sold in America are white but the ratio will fall to two out of five in the next 15 years, and all-cotton shirts will give way to blends of cotton and synthetic fibers.

Gleem toothpaste, introduced by Procter & Gamble, begins a major P&G effort to penetrate the dentifrice business (*see* Crest, 1955).

New York's Lever House opens April 29 on Park Avenue between 53rd and 54th Streets. Designed by Gordon Bunshaft of Skidmore, Owings & Merrill, the 24-story glass-walled building takes up far less of its site than the law permits, critics hail its aesthetics (although some find it incongruous in a street of conventional buildings), a traveling gondola suspended from its roof enables window cleaners to wash its 1,404 panes of heat-resistant blue-green glass, but the windows are sealed, the centrally air-conditioned building makes profligate use of energy, and it will be a model for energy-wasting architectural extravangazas that will be constructed throughout much of the world.

The Fort Worth Plan presented by Victor Gruen Associates is the first U.S. proposal for reclaiming a downtown city area by banning automobiles from the central business district and turning it into a pedestrian shopping mall. The plan will not be implemented (*see* Gruen, 1951; Northland, 1954).

The first Holiday Inn opens on U.S. Highway 70 at Memphis where local builder Kemmons Wilson puts up a motel with a swimming pool, air conditioning and a television set in every room, free ice, free baby cribs, free kennels and dog food for family pets, and no charge for children under 12 who share their parents' accommodations. Wilson took his wife and five young children on a trip to Washington, D.C., last year and decided that the motel business was "the greatest untouched industry in America." He will go into partnership next year with builder Wallace E. Johnson, now 51, to found the motel chain.

A 4-day London smog in December raises the city's death toll to three times its normal level. Ninety percent of the 4,703 dead are over age 45 and the Ministry of Health blames the contaminated air on oxides of sulfur and other irritants from coal smoke. One pollution expert estimates that in a bad fog year such as this a heaped teaspoonful of black powder enters the lungs of each Londoner (*see* 1955).

Rabbits infected with myxomatosis are released in France and the disease brings a sharp decrease in the French wild rabbit population (*see* 1951; 1953).

A giant Soviet tree-planting program undertaken by T. D. Lysenko is a fiasco but Lysenkoism continues to dominate Soviet agriculture (*see* 1948; 1954).

The Declaration of Santiago extends the maritime and fisheries jurisdictions of Chile, Ecuador, and Peru to 200 miles from their coastlines, but the United States maintains her 3-mile limit and most other nations limit their jurisdictions to 12 miles (*see* 1966; Jefferson, 1793).

China's cereal grain production rises to 163 million tons, up from 110 million in 1949, but Asia's rice crop falls below the level of prewar harvests while populations have increased by roughly 10 percent from prewar levels (*see* 1956).

Drought reduces U.S. grain harvests. Farmers in Alabama, Georgia, Kentucky, Maine, Massachusetts, South Carolina, and Tennessee are declared eligible for disaster loans August 1, but while the drought will last for 5 years and be even more intense than the one in the 1930s, its effects will be tempered by tree stands planted in the Depression years and by contour plowing and new tilling methods which combine to conserve available moisture and prevent creation of a new Dust Bowl (*see* 1953).

Nearly half of U.S. farms have tractors, up from one quarter in 1940.

War-inflated food prices are so high that Gen. Eisenhower uses them as a campaign issue in his presidential campaign (above), but farm prices will fall 22 percent in the next 2 years and food prices will fall as a consequence.

Kellogg's Sugar Frosted Flakes, 29 percent sugar, are introduced by Kellogg Co. (*see* 1950; 1953).

No-Cal Ginger Ale, introduced by Kirsch Beverages of Brooklyn, N.Y., uses cyclamates in place of sugar, is the first palatable sugar-free soft drink, and begins a revolution in the beverage industry (see cyclamates, 1950).

1952 ***(cont.)*** Russian-American Hyman Kirsch, now 75, started a soft-drink business in 1904 with a 14x30-foot store in Brooklyn's Williamsburg section. Physicians at the Kingsbrook Medical Center he founded asked Kirsch in 1949 to develop a sugar-free, salt-free soft drink for obese, diabetic, and hypertensive patients (*see* Diet-Rite Cola, 1962).

Pream powdered instant cream for coffee is introduced by M. and R. Dietetic Laboratories, Inc., of Columbus, Ohio (*see* Coffee-Mate, 1961).

New York State repeals its law against selling yellow-colored margarine; other dairy states continue to prohibit its sale (*see* 1943; 1967).

John D. Rockefeller, Jr., founds the Population Council "to stimulate, encourage, promote, conduct, and support significant activities in the broad field of population."

A contraceptive tablet developed by Chicago's G. D. Searle laboratories is made of phosphated hesperiden (*see* 1960; Pincus, Hoagland, 1954).

The McCarran-Walter Immigration and Nationality Act, passed over President Truman's veto, becomes law June 27. It removes the ban on Asian and African immigrants to the United States and permits spouses and minor children to enter as nonquota immigrants, but it extends the national origins quota system set up by the Johnson-Reed Act of 1924. Critics point out that while Britain uses only half her quota of 65,000, Italy has a 20-year waiting list for her quota of 5,500 (emerging nations of Africa and Asia have token quotas of 100 each) (*see* 1965).

1953 Josef Stalin dies March 5 at age 73. Having ruled the Soviet Union since 1928, he is succeeded as chairman of the Council of Ministers by World War II aircraft and tank production chief Georgi Maximilianovich Malenkov, 51, who will head the U.S.S.R. until 1958. Soviet Minister of Internal Affairs Lavrenti Pavlovich Beria, 54, is dismissed July 10 and shot as a traitor December 23; Nikita Sergeevich Khrushchev, 58, is named first secretary of the Communist party.

Sick and disabled Korean War prisoners are exchanged following Stalin's funeral but hostilities soon escalate (below).

Swedish diplomat Dag Hammarskjöld, 48, is elected Secretary-General of the UN March 31 and will hold the post until his death in 1961.

U.S. planes bomb North Korean dams in May, flooding rice fields; President Synghman Rhee releases North Korean prisoners of war June 18 in a move to stall peace talks, 60,000 Chinese attack July 13 and 45,000 UN troops counterattack 2 days later, but an armistice signed July 27 at Panmunjom near the 38th parallel ends a 3-year conflict that has left both sides in ruins. North Korean and Chinese casualties (dead and wounded): 1,540,000; UN casualties: 344,227 including 25,604 U.S. dead, 103,492 wounded. Some 2 million Korean civilians, North and South, have been killed, thousands left homeless.

Yugoslavia has adopted a new constitution January 12 and elected Marshal Tito two days later as first president of a Yugoslav Republic (*see* 1948).

Ethel Rosenberg and her husband Julius are executed June 19 for transmitting U.S. atomic secrets to Soviet agents (*see* 1951); a new series of U.S. atomic tests begins in the Nevada desert.

The U.S.S.R. explodes a hydrogen device August 12 and elevates physicist Andrei Sakharov to full membership in the Soviet Academy of Sciences.

U.S. atomic scientist J. Robert Oppenheimer is charged in December with having Communist sympathies and with possible treason. Now 49, Oppenheimer led the Manhattan Project that built the 1945 atomic bomb but has opposed building a hydrogen bomb (*see* 1954).

Iran's Shah Mohammed Reza Pahlevi regains power August 19 through a coup engineered and financed by the CIA to prevent a Soviet takeover (*see* 1951; 1978).

Saudi Arabia's ibn Saud dies November 9 at age 73. He founded the kingdom in 1926, saw it renamed Saudi Arabia in his honor 6 years later, and is succeeded by his son who will reign ineptly until 1964 as Saud ibn Abdel Aziz.

Norodom Sihanouk seizes all government buildings in a bid for complete independence after French colonial officers sign protocols May 9 giving Cambodia "full sovereignty" in military, judicial, and economic matters. The French start building a huge entrenched camp at Dienbienphu in an effort to retain Vietnam and Cochin China; their 250,000-man army faces a Communist army only half as large (*see* 1951; 1954).

A 10-year campaign for independence of Kenya from Britain begins in East Africa. Jomo Kenyatta and five other Kikuyu are convicted April 8 of masterminding the terrorist effort to drive the whites from Kenya, Kenya's Supreme Court quashes the conviction July 15, and the East African Court of Appeals sustains the ruling September 22, but Jomo Kenyatta will be banished until 1961 (*see* 1963).

French voters elect René Coty president December 23 to succeed Vincent Auriol.

The South African Parliament votes February 24 to give Prime Minister Daniel F. Malan dictatorial powers to oppose black and Indian movements against the nation's "unjust" apartheid laws (*see* 1952; Sharpeville, 1960).

East Berlin workers rise up June 17 to protest low wages and bad conditions. Soviet tanks mow down the strikers with machine gun fire, ending illusions that communism is a workers' paradise.

The U.S. Department of Justice tells Charlie Chaplin that he cannot re-enter the United States until he can satisfy the Immigration Office that he is not a dangerous and unwholesome character. Now 64 and still a British subject, the actor-producer joins the long list of motion picture people blacklisted because of alleged Communist sympathies and "subversive, un-American" opinions.

The new Eisenhower administration gives orders in early February that all U.S. federal agencies are to curtail new requests for personnel and construction and recommend ways that the Truman budget may be cut.

Federal tax reductions should be postponed until the budget is balanced, says President Eisenhower.

Per capita state taxes have increased from $29.50 to $68.04 since 1943, the Census Bureau reveals.

Wall Street's Dow Jones Industrial Average begins the year at 293.79—87.38 short of the 1929 high—but falls back to 255.49 by year's end (*see* 1954).

The American Stock Exchange is created by a renaming of the New York Curb Exchange whose members have not traded from curbs since 1921.

Some 63.4 million Americans are gainfully employed by September; unemployment falls to its lowest point since the close of World War II.

France must industrialize if she is to surmount poverty, says General Planning Commission chief Jean Monnet, whose Monnet Plan will lead to increased industrialization (*see* 1950; 1959).

Gen. Omar Bradley tells outgoing President Truman that a criminal investigation into the "international oil cartel" threatens national security; Truman drops his attack on Standard Oil of New Jersey, Gulf, the Texas Company, Socony-Mobil, Standard Oil of California, and their foreign colleagues Anglo-Iranian and Royal Dutch-Shell; the Justice Department drops its grand jury probe in April and files a civil complaint, accusing the companies of a conspiracy to monopolize the industry (*see* 1960).

Congress cancels President Truman's executive order of January 18 creating a U.S. Navy petroleum reserve out of offshore oil lands. The Tidelands Oil Act passed May 22 gives coastal states title to the submerged lands. The Supreme Court will uphold the act within a year.

Petroleo Brasileiro (Petrobrás) is established under a bill signed in October by Brazil's President Vargas creating a government monopoly in oil exploration, development, refining, and transport.

A giant uranium deposit found in Ontario's Algoma Basin will make Canada a leading world supplier of the ore used to produce fuel for nuclear energy. Latvian-born Toronto mining stock speculator Joseph Herman Hirshhorn, 54, has organized a search by some 80 field men using Geiger counters and stakes 1,400 claims covering 56,000 acres. He obtained rights 3 years ago to mine uranium on 470 square miles of territory round Lake Athabasca and established Rix Athabasca, the first Canadian uranium mine to operate on private risk capital. Hirshhorn will share his Algoma Basin find with the Rothschild-backed Rio Tinto Co. of London, a 90-year-old firm that is now selling its Spanish copper, sulfur, and iron pyrite properties and looking for new ventures; he will acquire a 55 percent interest in Rio Tinto of Canada (*see* iron ore, 1964).

President Eisenhower proposes an Atoms for Peace program December 8 in a speech to the UN General Assembly. The United States, he says, can help other nations harness the power of nuclear energy for peaceful purposes that will divert emphasis from exploitation of nuclear fission for military applications; the new Soviet regime (above) agrees to exploratory talks.

The Vickers Viscount powered by Rolls-Royce turboprop Dart engines goes into service for British Overseas Airways (BOAC). Nearly 440 of the new passenger planes will be sold, including 82 to U.S. airlines, and Rolls-Royce will make more than 6,000 Dart engines.

The DC-7 propeller plane introduced by Douglas Aircraft sells for well over $1.5 million as compared with less than $100,000 for the DC-3 of 1936. Douglas has been slow to move into production of commercial jets and will not build its DC-8 jet until forced by pressure from Boeing, whose 707 will give it industry leadership (*see* 1956; Comet, 1952).

Japan Airlines (JAL) is organized to give the Japanese a national overseas airline.

Britain de-nationalizes road transport in May, just before the coronation of Elizabeth II (*see* 1947).

German Volkswagens are sold in Britain for the first time (*see* 1949).

New York City transit fares rise to 15¢ July 25, having remained 10¢ since 1948. Subway ridership will decline by 33 percent in the next 23 years, and on Eighth Avenue line trains such as the A Train it will decline by 42 percent as fares continue to rise and frequency of trains decline (*see* 1966).

The last of New York's double-decker Fifth Avenue buses go out of service (*see* 1936). Rising labor costs have doomed buses that require fare collectors in addition to drivers.

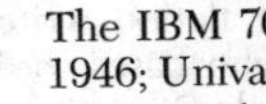

The IBM 701 is the first IBM computer (*see* ENIAC, 1946; Univac, 1951). The scientific electronic computer competes in the market with Remington Rand's scientific and business computer (*see* 1955).

A new catalytic process for producing polyethylene plastic invented by German chemist Karl Ziegler uses atmospheric pressure instead of the 30,000 pounds per square inch pressure required by the I.C.I. process of 1935. Ziegler is director of the Max Planck Institute for Coal Research at Mülheim in West Germany, his process ushers in a new era of low-cost plastics, and foreign companies besiege the Institute with applications for rights to use the Ziegler process.

A plastic valve mechanism for aerosol cans developed by U.S. inventor Robert H. Abplanalp, 30, sharply lowers production costs. The Abplanalp valve will lead to the marketing of countless consumer products propelled by freon gas from low-cost containers (*see* 1945; ozone question, 1958).

"We wish to suggest a structure for the salt of deoxyribose nucleic acid (DNA)," begins a one-page article in the April 25 *Nature*. U.S. genetic researcher James Dewey Watson, 25, and English geneticist Francis H. C. Crick, 37, of Cambridge University write, "This structure has novel features that are of considerable

1953 *(cont.)* biological interest" and in the following (May 30) issue they develop some of the implications of their model which has the basic structure of a double helix and shows how the genetic material in animal and human cells can duplicate itself. Watson and Crick show that chromosomes consist of long helical strands of the substance DNA, research studies conducted throughout the world confirm their experiments, and breaking the genetic code that determines the inheritance of all physical characteristics opens new possibilities for preventing inherited disorders (*see* 1926; Ochos, 1955).

A new U.S. Department of Health, Education and Welfare established by Congress April 11 incorporates most of the functions of the Federal Security Agency. Former WAC commander Oveta Culp Hobby is named first secretary of HEW.

Cryosurgery, pioneered by U.S. physician Henry Swan II, 40, lowers body temperature to slow circulation and permit dry-heart surgery.

LSD is a potent "antagonist" of the chemical substance serotinin that has just been found to exist in the human brain (*see* Hofmann, 1943). English biochemist John Gaddum shows that while serotinin makes muscles contract, LSD blocks its ability to stimulate the contraction of certain muscles (*see* 1955).

"Cancer by the Carton," published in the *Reader's Digest*, warns against smoking.

Tobacco-tar condensates can induce cancer on the skin of mice, report Ernest L. Wynder, 31, and Evarts Graham of the Sloan-Kettering Institute of Cancer Research at New York.

The Church of Scientology is founded at Washington, D.C., by former U.S. science fiction writer L. (Lafayette) Ronald Hubbard, 42, who bases the new church on his best-selling 1950 book *Dianetics—The Modern Science of Mental Health.* Man is essentially a free and immortal spirit who can achieve his true nature only by freeing himself of emotional encumbrances of the past through counseling ("auditing"), says Hubbard, his "applied religious philosophy" will make him a fortune, but British authorities in 1968 will refuse entry to Scientology students and teachers on the ground that Scientology is "socially harmful" and that its "authoritarian principles and practices are a potential menace to the well-being of those so deluded as to become followers."

Japanese television broadcasting begins February 1 as NHK airs its first TV programs. A pioneer radio firm that was reorganized on a nonprofit basis in 1950, NHK will grow to have two radio networks and two TV networks whose stations will outnumber those of commercial networks.

The U.S. television industry has revenues of $538 million thanks in part to heavy support by cigarette advertisers. Radio advertising revenues amount to $451 million and take their first downturn since the Depression of the 1930s (*see* 1968).

TV Guide begins publication April 3 with pocket-size weekly program listings and has a circulation of 1.5 million by year's end. Publisher Walter H. Annenberg has merged his 9-year-old *Seventeen* magazine with his father's *Philadelphia Inquirer* and racing publications to create Triangle Publications (*see* 1922; 1936). He has bought out small TV-list publishers in New York, Philadelphia, Washington, Baltimore, and Chicago and created 10 regional editions.

L'Express begins publication at Paris under the direction of *Elle* editor Françoise Giroud, 37, and her economist lover Jean-Jacques Servan-Schreiber, 30. In 10 years they will turn the journal of leftist opinion into a highly successful newsmagazine modeled on *Time* and *Newsweek*.

Playboy magazine begins publication in December with a nude calendar photograph of Marilyn Monroe. Chicago publisher Hugh Hefner, 27, has started his frankly sexist magazine with an initial investment of $10,000. He features nude photography and ribald cartoons (*see Penthouse*, 1969).

Canadian newspaper proprietor Roy Thompson, 59, moves to Britain and begins an empire that will include hundreds of newspapers and magazines worldwide and other enterprises.

The Higonnet-Moyroud photographic type-composing machine introduced by Photon, Inc. is far more advanced than the 1949 Fotosetter. Produced under license from the Graphic Arts Research Foundation, it is operated from a standard typewriter keyboard at full electric typewriter speed and delivers film negatives instead of metal type.

Nonfiction *The Second Sex* in an English translation of the 1949 book by Simone de Beauvoir; *A Stillness at Appomattox* by U.S. historian (Charles) Bruce Catton, 54, is the third in a trilogy of Civil War histories.

Fiction *Go Tell It on the Mountain* by U.S. émigré novelist James Baldwin, 29, who has gone to Paris to escape U.S. prejudice against blacks and homosexuals; *Invisible Man* by U.S. novelist Ralph Waldo Ellison, 39, is about black identity; *For Esme—With Love and Squalor* (nine stories) by J. D. Salinger, whose story "Teddy" appears in the *New Yorker* January 31; *Acquainted with the Night* (*Undsagte kein einziges Wort*) by Heinrich Böll who points for the first time to the moral vacuum that underlies the "economic miracle" of postwar Western Germany; *Chapel Road* (*De Kapellekensbaan*) by Flemish novelist Louis Paul Boon, 41; *The Erasers* (*Les Gommes*) by French novelist Alain Robbe-Grillet, 31, who rejects conventional character and plot; *The Adventures of Augie March* by Saul Bellow; *The Garden to the Sea* by Philip Toynbee; *Someone Like You* (stories) by British writer Roald Dahl, 37; *Too Late the Phalarope* by Alan Paton; *Martereau* by Nathalie Sarraute; *Fahrenheit 451* by U.S. novelist Ray Bradbury, 33; *Casino Royale* by London actor-novelist Ian (Lancaster) Fleming, 45, whose hero James Bond is a Secret Service agent allowed to kill in line of duty (007); *Battle Cry* by U.S. Marine Corps veteran Leon

Uris, 29; *Junkie* by U.S. novelist William Lee (William Seward Burroughs), 39, a grandson of the adding machine inventor. He has been a heroin addict since 1944 and will remain one until 1957.

Poetry *The Waking* by U.S. poet Theodore Roethke, 45; *Poems, 1940–1953* by U.S. poet Karl Shapiro, 40; *Poems* by English poet Elizabeth Jennings, 27.

Painting *Study after Velàzquez: Pope Innocent X* by Francis Bacon; *Blue Poles* by Jackson Pollock; *Washington Crossing the Delaware* by U.S. "beatnik" painter Larry Rivers (*né* Yizroch Lotza Grossberg), 29; *Apples* by Georges Braque. Raoul Dufy dies at Forcalquier, France, March 23 at age 75; John Marin dies at Addison, Me., October 1 at age 82; Francis Picabia dies at Paris November 30 at age 74.

Sculpture *King and Queen* by Henry Moore.

Theater *Waiting for Godot (En attendant Godot)* by Anglo-Irish playwright Samuel Beckett, 46, 1/5 at the Théâtre de Babylone, Paris (written in French, it will open at London in August 1955); *The Crucible* by Arthur Miller 1/22 at New York's Martin Beck Theater, with Arthur Kennedy, Jennie Egan, Walter Hampden in an account of the 1692 Salem witch trials intended as a parallel to the persecution of alleged Communist sympathizers in the United States, 197 perfs.; *Picnic* by William Inge 2/19 at New York's Music Box Theater, with Ralph Meeker, Janice Rule, Paul Newman, Kim Stanley, Eileen Heckart, 477 perfs.; *Camino Real* by Tennessee Williams 3/19 at New York's Martin Beck Theater, with Eli Wallach, Frank Silvera, Jo Van Fleet, Martin Balsam, Barbara Baxley, Hurd Hatfield, Michael Griggs, 60 perfs.; *Medea (Medée)* by Jean Anouilh 3/26 at the Théâtre de l'Atelier, Paris; *Under Milk Wood* by Dylan Thomas 5/3 at Harvard's Fogg Museum; *Don Juan, or The Love of Geometry (Don Juan oder Die Liebe zur Geometrie)* by Max Frisch 5/5 at Berlin's Schillertheater; *Tea and Sympathy* by Robert Anderson 9/30 at New York's Ethel Barrymore Theater, with Deborah Kerr, John Kerr, 712 perfs.; *The Teahouse of the August Moon* by John Patrick 10/15 at New York's Martin Beck Theater, with John Forsythe, David Wayne, 1,027 perfs.; *The Sleeping Prince* by Terence Rattigan 11/5 at London's Phoenix Theatre, with Laurence Olivier, Vivien Leigh; *The Angel Comes to Babylon* by Friedrich Dürrenmatt 12/22 at Munich's Kammerspiele.

Films Henry Koster's *The Robe* with Richard Burton, Jean Simmons, and Victor Mature is the first film to be produced in CinemaScope. Designed to counter the inroads that television is making on movie theater receipts, CinemaScope employs screens much wider than those used for conventional films and has a stereophonic soundtrack; Arch Oboler's *Bwana Devil* with Robert Stack is the first three-dimensional film and represents another effort to compete with television (audiences need Polaroid viewers to see the 3-D film).

More notable films Fred Zinneman's *From Here to Eternity* with Burt Lancaster, Montgomery Clift, Deborah Kerr, Frank Sinatra; Federico Fellini's *I Vitelloni* with Alberto Sordi, Franco Interlenghi, Franco Fabrizi; Charles Walters's *Lili* with Leslie Caron, Mel Ferrer; Kefiji Mizoguchi's *Princess Yang Kwei Fei* with Machiko Kyo, Masayuki Mori; Ingmar Bergman's *Sawdust and Tinsel* with Harriet Andersson, Ake Gronberg; George Stevens's *Shane* with Alan Ladd, Jean Arthur, Brandon De Wilde; Billy Wilder's *Stalag 17* with William Holden, Don Taylor, Otto Preminger; Yasujiro Ozu's *Tokyo Story* with Chishu Ryu, Chieko Higashiyama. Also: Charles Frend's *The Cruel Sea* with Jack Hawkins; Sidney Gilliat's *The Story of Gilbert and Sullivan* with Robert Morley, Maurice Evans, Martyn Green; Arne Sucksdorff's *The Great Adventure* with Anders Norberg; Joseph L. Mankiewicz's *Julius Caesar* with Marlon Brando, James Mason, John Gielgud; Philip Leacock's *The Little Kidnappers* with Jon Whitely, Vincent Winter; Jacques Tati's *Mr. Hulot's Holiday (Les Vacances de M. Hulot)* with Tati; John Ford's *Mogambo* with Clark Gable, Ava Gardner; Anthony Mann's *The Naked Spur* with James Stewart, Janet Leigh, Ralph Meeker, Robert Ryan; William Wyler's *Roman Holiday* with Audrey Hepburn, Gregory Peck; Herbert Biberman's *Salt of the Earth* with Juan Chacon, Rosoura Revueltas, Will Geer; Ken Annakin's *The Sword and the Rose* with Richard Todd, Glynis Johns as Mary Tudor; Kenji Mizoguchi's *Ugetsu* with Machiko Kyo, Masayuki Mori.

Hollywood musical Vincente Minnelli's *The Band Wagon* with Fred Astaire, Cyd Charisse, Jack Buchanan, Nanette Fabray, Oscar Levant, music by Arthur Schwartz, lyrics by Howard Dietz, songs that include "That's Entertainment," "Triplets," "By Myself," "Shine on Your Shoes."

Broadway musicals *Hazel Flagg* 2/11 at the Mark Hellinger Theater, with Helen Gallagher, music by Jule Styne, lyrics by Bob Hilliard, songs that include "Every Street's a Boulevard in Old New York," 190 perfs.; *Wonderful Town* 2/25 at the Winter Garden Theater, with Rosalind Russell, music by Leonard Bernstein, book based on the 1940 stage play *My Sister Eileen*, lyrics by Betty Comden and Adolph Green, songs that include "Ohio," "A Quiet Girl," "Conga!" 559 perfs.; *Can Can* 5/7 at the Shubert Theater, with Gwen Verdon, book by Abe Burrows, music and lyrics by Cole Porter, songs that include "I Love Paris," "C'est Magnifique," "It's All Right with Me," 892 perfs.; *Me and Juliet* 5/28 at the Majestic Theater, with Isabel Bigley, Joan McCracken, music by Richard Rodgers, lyrics by Oscar Hammerstein II, songs that include "No Other Love," 358 perfs.; *Comedy in Music* 10/2 at the John Golden Theater, with Danish pianist Victor Borge, 44, in a one-man show, 849 perfs.; *Kismet* 12/3 at the Ziegfeld Theater, with Alfred Drake, Richard Kiley, music based on the works of Aleksandr Borodin, lyrics by Robert Wright and George "Chet" Forest (Forrest Chichester, Jr., 38), songs that include "Baubles, Bangles and Beads," "Stranger in Paradise," "This Is My Beloved," 538 perfs.

First performances *Suite Hebraique* by Ernest Bloch 1/1 at Chicago; Symphony No. 7 *(Sinfonia Antarctica)*

1953 ***(cont.)*** by Ralph Vaughan Williams 1/14 at Manchester; *Sinfonia breve* and Concerto No. 2 in G minor for Strings by Bloch 4/11 in a BBC Symphony concert broadcast; *Te Deum* and *Orb and Sceptre* march by Sir William Walton 6/2 at London for the coronation of Elizabeth II; Symphony No. 5 by Darius Milhaud 10/16 in a radio concert broadcast from Turin; *Atomic Bomb* symphonic fantasy by Japanese composer Masao Oki, 52, 11/6 at Tokyo; Symphony No. 10 by Dmitri Shostakovich 12/17 at Leningrad.

Cantata *Une Cantata de Nöel* by Arthur Honegger 12/12 at Basel.

Opera *The Harpies* 5/25 at New York, with music and lyrics by Marc Blitzstein (who wrote it in 1931); *Gloriana* 6/8 at London's Covent Garden, with music by Benjamin Britten, libretto by William Plomer.

Ballet *Fanfare* 6/2 at London's Covent Garden, with music by Benjamin Britten, choreography by Jerome Robbins.

Popular songs "I'm Walking Behind You" by English songwriter Billy Reid; "I Believe" by Erwin Drake, Irvin Graham, Jimmy Shirl, and Al Stillman; "Ricochet" by Larry Coleman, Joe Darion, and Norman Gimbel; "Rags to Riches" by Richard Adler and Jerry Ross; "Ruby" by Heinz Roemheld, lyrics by Mitchell Parish (for the film *Ruby Gentry*); "Oh! My Pa-pa" by Paul Burkhard (for the Swiss film *Fireworks*), English lyrics by John Turner and Geoffrey Parsons; "Eternally" ("The Terry Theme") by Charles Chaplin (for his film *Limelight*), lyrics by Geoffrey Parsons; "You, You, You" by Lotar Olias, lyrics by Robert Mellin; "Hi-Lili, Hi-Lo" by Bronislau Kapes, lyrics by Helen Deutsch (for the film *Lili*); "That's Amore" by Harry Warren, lyrics by Jack Brooks (for the film *The Caddy*); "That Doggie in the Window" by Bob Merrill.

Dark Star wins the Kentucky Derby to give Native Dancer his only defeat in 22 races.

New Zealand climber Edmund Hillary, 33, and his Sherpa guide Tenzing Norkay, 39, reach the summit of Mount Everest May 29 in the first successful ascent of the world's highest (29,000 feet) peak.

Elias Victor "Vic" Seixas, Jr., 29 (U.S.), wins in men's singles at Wimbledon, Maureen Connolly in women's singles. Marion Anthony "Tony" Trabert, 22, wins in men's tennis at Forest Hills, Connolly in women's singles (Connolly wins the "grand slam," taking the Australian, French, English, and U.S. women's singles championships).

Ben Hogan wins the U.S. Open and Masters golf tournaments, breaks the Masters record by five strokes, and wins his first British Open.

The Boston Braves become the Milwaukee Braves as U.S. major league baseball teams begin a geographical realignment spurred in part by the fact that the airplane has reduced travel time (*see* Aaron, 1954; Atlanta, 1966).

The New York Yankees win the World Series by defeating the Brooklyn Dodgers 4 games to 3 and gain the championship for an unprecedented fifth consecutive time.

L&M cigarettes, introduced by Liggett & Myers and advertised as "just what the doctor ordered," have "Alpha-Cellulose" filter tips (*see* medicine, above).

Tareyton filter-tipped cigarettes are introduced by American Tobacco, but U.S. per-capita cigarette consumption will decline 8.8 percent in the next 2 years.

Winter storms on the North Sea wreak havoc; Dutch dikes burst February 1, drowning 1,800, leaving 100,000 homeless, and flooding large areas.

French rabbits infected with myxomatosis are released in Britain. The disease will wipe out most of the country's wild rabbits (*see* 1952).

A Nevada nuclear test May 19 yields fallout that is deposited 100 miles to the east, exposing St. George, Utah, residents to radiation levels of 6,000 rems—higher than ever before measured in a populated area. Childhood leukemia, thyroid cancer, other cancers, and birth defects will show marked increases in Utah.

Nikita Khrushchev defies warnings by agronomists that rainfall in the Kazakhstan Soviet Republic is undependable. He orders virgin lands in Kazakhstan to be plowed and planted with grain that will increase Soviet food production (*see* 1963).

Drought worsens in the U.S. Midwest; sections of 13 states are declared disaster areas (*see* 1952). Southern Missouri has an even dryer season than in 1936, pastures are dead to the grass roots, and farmers dump cattle on the market at sacrifice prices.

University of Minnesota physiologist Ancel Keys, 49, points out a correlation between coronary heart disease and diets high in animal fats and observes that the incidence of heart disease tends to be higher in communities where animal fats comprise a large part of people's diets (*see* Anichkov, 1913).

The U.S. Food and Drug Administration (FDA) is placed in the new Department of Health, Education and Welfare (above).

Some 93 Japanese will be fatally poisoned in the next 7 years as a result of eating fish and shellfish contaminated with mercury wastes discharged into rivers at Minamata Bay in Kyushu and at Niigata in Honshu. Many more will go blind, lose the use of limbs, or suffer brain damage from mercury poisoning.

U.S. dairy scientist David D. Peebles develops a process that converts nonfat dry milk into crystals that retain all the protein, mineral, and vitamin content of fresh skim milk.

U.S. meat packers begin moving out of Chicago to plants closer to western feedlots (*see* 1865; 1960).

The first underground freezer storage room is opened at Kansas City by local mine operator Leonard Strauss who sees potential for food storage in the city's mined-out limestone quarries. Strauss employs the constant 45 to 54 degree temperature of the caves to reduce cooling costs, and his room will grow in the next 14 years into the world's largest refrigerated warehouse, handling 8 million pounds per day.

U.S. pizzerias grow to number at least 15,000, and there are at least 100,000 stores where ready-made refrigerated or frozen pizzas may be purchased (*see* oregano, 1948). The Americanized pizza pies are thin, round layers of bread dough laden with tomato paste, mozzarella cheese, olive oil, pepper, and oregano with optional extras that may include anchovies, eggs, mushrooms, onions, or sausages, the pizzas are baked on bricks or in metal ovens, and a few pizzerias offer the Sicilian pizza (baked in a pan and slightly thicker and more breadlike).

Sugar Smacks breakfast food, introduced by Kellogg, is 56 percent sugar (*see* 1952; 1955).

Candy rationing ends in Britain.

Schweppes Tonic Water is bottled for the first time in the United States and sold under the name Schweppes Quinine Water to comply with an FDA order based on the allegation that "Tonic Water" implies therapeutic virtues.

White Rose Redi-Tea, introduced by New York's Seeman Brothers, is the world's first instant iced tea.

Ninety-eight-cent frozen "TV Dinners"—turkey, cornbread dressing, gravy, peas topped with butter, whipped sweet potatoes flavored with orange juice and butter—on 3-section aluminum trays are introduced in December by C. A. Swanson & Sons of Omaha which 2 years ago began selling frozen pot pies (*see* 1945; Campbell, 1955).

1954 France appeals for U.S. aid to relieve her troops surrounded at Dienbienphu in Indochina (*see* 1953). President Eisenhower declares that defeat of Communist aggression in Southeast Asia is vitally important to the United States but declines to deploy U.S. airpower to relieve the siege.

President Eisenhower outlines the "domino theory" first voiced by newspaper columnist Joseph W. Alsop, Jr., 43: "You have a row of dominoes set up. You knock over the first one, and what will happen to the last is that it will go over very quickly." But Eisenhower resists "hawkish" suggestions by Vice President Nixon, Secretary of State John Foster Dulles, and Joint Chiefs of Staff Admiral Arthur W. Radford; Ike refuses to send U.S. forces to Vietnam.

A nuclear-fusion (hydrogen) "device" exploded by the U.S. Atomic Energy Commission March 1 at Bikini Atoll in the South Pacific is hundreds of times more powerful than the atom bomb (*see* 1956; Eniwetok, 1952).

A Geneva Conference of world powers meeting from April 26 to July 21 divides Vietnam at the 17th parallel into North Vietnam and South Vietnam. Ho Chi Minh, now 64, takes power as president of the Communist "Democratic Republic" of North Vietnam after 9 years of leading guerilla forces.

South Vietnam gains "complete independence" in "free association" with France under terms of a treaty signed June 4 by the French premier Joseph Laniel and South Vietnam's Prince Bau Loa.

The French government falls June 12; Premier Laniel is replaced by Radical-Socialist party leader Pierre Mendès-France, 47, who favors greater self-rule by Tunisia and Morocco and French withdrawal from Indo-China.

South Vietnam's Prince Bau Loa is replaced June 14 by Ngo Dinh Diem, 53 (*see* 1953; 1955).

Laos gains full sovereignty December 29 by agreement with Paris.

"Free Puerto Rico!" cries Lolita Lebron, 28, March 1 from the gallery of the House of Representatives and she fires the first of eight shots that injure five congressmen. The divorcée and her three confederates have pulled out concealed weapons to stage their demonstration; each will serve at least 23 years in prison, rejecting parole offers lest he or she appear repentant or acceptive of U.S. law.

Guatemala dissents from a resolution to bar communism from the Western Hemisphere approved in March by the Tenth Inter-American Conference meeting at Caracas.

Guatemala has charged January 29 that Nicaragua is planning to invade her with support from several other Latin American states and tacit U.S. consent. The U.S. State Department reports a major shipment of Communist-made arms to Guatemala March 17; her foreign minister declares May 21 that a U.S. boycott has left his country defenseless, forcing it to buy arms from Communist sources.

Paraguay's President Felipe Molás López dies at Asunciòn March 2 at age 54. Voters in a one-party election name the army's commander-in-chief Gen. Alfredo Stroessner, 41, president to serve until February 1958, but Stroessner will rig elections to remain in power until 1989, heading a military dictatorship that will give refuge to Nazi war criminals.

Brazil's President Vargas, now 71, resigns under pressure from the army August 24 after a government scandal and commits suicide. He is succeeded by his vice president João Cafe Filho (*see* 1955).

Egypt's military government replaces Gen. Mohammed Naguib Bey as premier. Col. Gamal Abdel Nasser, 36, takes over April 17 and signs a treaty with Britain October 19 providing for British withdrawal from the Suez Canal Zone within 20 months. Britain is to give up her rights to the base at Suez, but the Egyptians agree to keep it in combat readiness and permit British entry in the event of an attack on Turkey or any Arab state by an outside power (*see* 1956).

An Algerian revolt against France breaks out October 31 under the leadership of the Front de Libération (FLN) (*see* 1957).

Moscow rejects German unification in a meeting at Berlin of British, French, U.S., and Soviet foreign ministers.

1954 ***(cont.)*** Ottawa and Washington agree to build a "DEW" line (Distant Early Warning Line) of radar stations across northern Canada to alert authorities to the approach of aircraft or missiles over the Arctic.

Sen. McCarthy conducts hearings from April 22 to June 17 as chairman of the Senate Permanent Investigations Subcommittee and charges that a Communist spy ring is operating at the U.S. Army Signal Corps installation at Fort Monmouth, N. J. (*see* 1951). The hearings are televised across the country, McCarthy accuses the secretary of the Army of concealing evidence, and the secretary retains Boston lawyer Joseph Nye Welch, 63, to represent him. When McCarthy makes a vicious charge against one of his assistants, Welch says, "Until this moment, Senator, I think I never really gauged your cruelty or recklessness. . . . Have you no sense of decency, sir, at long last? Have you left no sense of decency?"

The Senate votes 67 to 22 December 2 to condemn Sen. McCarthy for misconduct.

The Atomic Energy Commission revokes J. Robert Oppenheimer's security clearance, President Eisenhower follows suit, the AEC appoints a board to investigate, it clears Oppenheimer of disloyalty charges but votes 2 to 1 that Oppenheimer's imprudent associations have made him a security risk and upholds revocation of the physicist's security clearance (*see* 1953). (Oppenheimer will receive the AEC's Fermi Award in 1963 after the anti-Communist frenzy stirred up by Sen. McCarthy has abated.)

The Supreme Court rules 9 to 0 May 17 in *Brown v. Board of Education* that racial segregation in public schools is unconstitutional. The court overturns the "separate but equal" doctrine laid down in the 1896 case *Plessy v. Ferguson,* Chief Justice Earl Warren orders the states to proceed "with all deliberate speed" to integrate educational facilities, but Topeka, Kan., requests time to formulate a plan for racial balance. It was at Topeka that the father of Linda Brown, now 11, brought the 1951

"Separate but equal" was accepted in the South until the Supreme Court ruled segregation unconstitutional.

action that has led to the court's landmark decision, and Topeka will repeat its request for more time for the next 20 years (*see* Little Rock, 1957, 1958).

Only 154 Americans have incomes of $1 million or more, down from 513 in 1929.

A U.S. taxpayer with an income of $100,000 may now pay more than $67,000 in taxes, up from $16,000 or less in 1929, and the individual tax exemption is $600, down from $1,500 in 1929. Income taxes will be reduced slightly next year, but the average tax rate, including surtax, is just above 20 percent and taxpayers in top brackets pay 87 percent.

The Ford Foundation gives away $67.8 million (*see* 1950). Some $34 million goes to education, $18 million to international programs, the rest to programs in economics, public affairs, and the behavioral sciences (*see* 1956).

Wall Street's Dow Jones Industrial Average finally passes its 381.17 high of 1929 November 23 and closes the year above 404, up from 280.

Detroit's Northland, opened in march with 100 stores, is the world's largest shopping center and a monument to the American exodus from the inner city. The $30-million project designed by Victor Gruen Associates includes a J. L. Hudson branch that is the largest department store erected since the 1920s (*see* Gruen, 1951; Southdale, 1957).

The *Nautilus* launched January 21 at Groton, Conn., is the first nuclear-powered submarine. Built at the insistence of Rear Admiral Hyman George Rickover, 54, it has been preceded by the Soviet ice-breaker *Lenin*, a nuclear-powered surface vessel (see 1958).

The world's first nuclear power station begins producing electricity for Soviet industry and agriculture June 27. The station 55 miles from Moscow at Obninsk has an effective capacity of 5,000 kilowatts.

Participation of U.S. private industry in the production of nuclear power gains congressional approval August 30 in amendments to the McMahon Atomic Energy Act of 1946.

The Supreme Court upholds Federal Power Commission price regulation of U.S. natural gas piped interstate June 7 in a 5-to-3 decision handed down in *Phillips Petroleum v. U.S.* The Court rules that Phillips is a natural gas company within the meaning of the Natural Gas Act of 1938, the FPC freezes the well-head price of natural gas, 106 companies say the commission has gone beyond the court decision and that the price freeze is illegal and unworkable, Sen. Lyndon Baines Johnson, 45 (D. Tex.), says he will urge congressional action to solve problems raised by the court decision, the Supreme Court refuses to reverse its ruling in the Phillips case, the FPC chairman denies that his commission seeks to socialize the industry, saying it will set rates high enough to provide profits necessary for risking capital to explore and develop new natural gas reserves. The industry will work for decades to have wellhead prices deregulated, meanwhile promoting wasteful uses of natural gas including its use as a boiler fuel to generate electricity.

President Eisenhower signs the Dixon-Yates contract November 11 permitting private companies to supply energy to the Tennessee Valley Authority, but the propriety of the contract comes under attack and the president will cancel it.

The UN votes unanimously December 4 to approve a November 15 U.S. move to donate more than 200 pounds of fissionable material, implementing President Eisenhower's 1953 Atoms for Peace plan.

A solar battery developed by Bell Laboratories makes it possible to convert sunlight directly to electric power, but solar power is far too costly to compete with other forms.

U.S. gasoline prices average 29¢ per gallon in October, up from 21¢ in 1944.

British Petroleum (BP) is created by a renaming of Anglo-Iranian, founded in 1908 as Anglo-Persian.

The New York State Thruway opens to traffic June 24. Toll revenue from the 559-mile road between New York and Buffalo is projected to reach $68 million by 1975 and its vehicular traffic to reach 108 million trips by that time, but revenues will exceed the projections by 1963 and traffic by 1965.

New York makes Seventh Avenue one-way southbound and Eighth Avenue one-way northbound (*see* 1951). Traffic flow is speeded by 25 to 40 percent, but the Fifth Avenue Coach Co. complains that it has lost customers on its Seventh and Eighth Avenue routes and the Transport Workers Union opposes further one-way avenues (*see* Broadway, 1956).

American Motors Corp. (AMC) is created by a merger of Hudson Motor Car and Nash-Kelvinator (*see* 1922; Rambler, 1959).

Studebaker-Packard Corp. is created by a merger (*see* 1930; 1964).

The Mercedes 300 SL introduced by Daimler-Benz employs the first fuel-injection system for automobiles. Benz engineer Rudolf Uhlenhaut has adapted a system developed by the firm for aircraft in 1935.

The IL-14 introduced in Soviet Russia replaces the IL-12 that has been the Soviet DC-3 (*see* 1948; 1962).

The first practical silicon transistors, introduced by Texas Instruments, are much cheaper than germanium transistors and will sharply increase use of solid-state electronic components, booming sales for Texas Instruments, whose president John Erik Jonsson, 53, has brought the price of germanium transistors down from $16 to $2.50 (*see* Shockley, 1948; Sony, 1952).

Electronic computers are applied for the first time to business uses (*see* Univac, 1950; IBM, 1955).

Wang Laboratories is founded at Lowell, Mass., by Chinese-American computer engineer An Wang, 34, to make small business calculators. Wang has done pioneering work on magnetic core memory, the basis for all modern computer technology (*see* word processing, 1976).

A small Detroit steel mill installs the first U.S. oxygen steel-making furnace. Perfected in 1950 by a tiny Austrian company, the furnace is the first major technological breakthrough at the ingot level for steelmaking since the 19th century, West German and Japanese steel makers are using it to modernize their production facilities, but McLouth Steel has less than 1 percent of U.S. steel ingot capacity. Jones & Laughlin will not adopt the oxygen furnace until 1957, Bethlehem and United States Steel not until 1964.

Pittsburgh school children receive the first mass poliomyelitis immunization shots February 23 (*see* 1955; Salk, 1952).

Minneapolis physician C. (Clarence) Walton Lillehei, 35, advances open-heart surgery with pumps to keep the blood circulating (*see* 1893).

Montreal physicians Heinz Lehmann and G. E. Hanrahan report success in treating psychotic patients with the drug chlorpromazine, which they have obtained from Rhone-Poulenc in France (*see* 1952). Smith-Kline & French researchers have found chlorpromazine useful in preventing vomiting and the firm will introduce it under the brand name Thorazine. It will effect a large-scale reduction in the hospitalization of mental patients by dramatically increasing recovery from schizophrenia.

Miltown, introduced by Wallace Laboratories, and Equanil, introduced by Wyeth Laboratories, are tranquilizers (*see* 1950; Librium, 1960).

President Eisenhower proposes a "reinsurance" plan that will encourage private initiative to strengthen U.S. health services. A survey of medical costs reveals that 16 percent of U.S. families go into debt each year to pay for treatment.

Harvard physicians perform the world's first successful kidney transplant December 23 at Peter Brent Brigham Hospital. Their patient will live for 7 years with a kidney from his identical twin brother.

Dramatic evidence emerges that smoking is hazardous to health. U.S. epidemiologists E. (Edward) Cuyler Hammond, 42, and Daniel Horn, 38, report on a 2-year study of 187,783 U.S. men aged 50 to 69 of whom 11,870 have died. Of the deceased, 7,316 were smokers; 11,870 is 2,265 more statistically than the number would have been if none of the men had smoked. More than half the deaths have been from coronary heart disease, 14 percent from lung cancer, another 14 percent from all other forms of cancer (including carcinomas of the esophagus and larynx). Dr. Horn gives up smoking (*see* Surgeon General's Report, 1964).

The Unification Church is founded by Korean evangelist Sun Myung Moon, 34, who claims to have received a vision on a hillside in Korea at age 16 that gave him the "key to righteousness and restoration of the Kingdom of Heaven on Earth." Moon will develop a theology based on platitudes, rabid anti-Communism, and pop history supported by bits of numerology, astrology,

1954 *(cont.)* and scientology (*see* Hubbard, 1953). Moon's supporters will claim that the Bible is written in code and that only Moon has broken the code.

President Eisenhower modifies the U.S. pledge of allegiance June 14, ordering that the phrase "one nation, under God" replace "one nation, indivisible" (*see* 1892).

The U.S. Air Force Academy, created by act of Congress, will move into permanent quarters north of Colorado Springs in 1958. The first class of 306 cadets will be sworn in July 11 of next year at Lowry Air Force Base, Denver.

RCA introduces the first U.S. color television sets but color reception is unreliable at best. No other company will enter the color-TV market on a sustained basis until 1959, when the courts will settle patent suits brought against RCA by Zenith and others (*see* Zenith, 1963).

U.S. black and white TV sets with 19-inch screens retail at $187.

The world has 94 million telephones, 52 million of them in the United States.

The *Washington Post* buys out the *Washington Times-Herald*, its only morning competition, March 17 for $8.5 million (*see* 1933; *Newsweek*, 1961).

Sports Illustrated begins publication August 16. The new Time-Life weekly will lose $26 million before it becomes profitable in 1964.

Nonfiction *McCarthy and His Enemies* by William F. Buckley, Jr., recognizes the Wisconsin senator's excesses but supports his campaign against U.S. Communists.

Fiction *The Thaw (Ottepel)* by Soviet novelist Ilya Grigoryevich Erenburg, 63, supplies the name for a new trend in Soviet literature and refers for the first time to the iniquities of the Stalin regime; *Felix Krull (Die Bekenntnisse des Hochstaplers Felix Krull)* by Thomas Mann; *The Unguarded House (Haus ohne Hater)* by Heinrich Böll; *The Sound of the Mountain* by Yasunari Kawabata; *People of the City* by Nigerian (Ibo) novelist Cyprian Ekwenski, 33; *The Radiance of the King (Le Regard du Roi)* by West African novelist Camara Laye, 26; *Lucky Jim* by English novelist Kingsley Amis, 32; *Lord of the Flies* by English novelist William Golding, 43; *The Mandarins* by Simone de Beauvoir; *Bonjour Tristesse* by French novelist Françoise Sagan, 19; *A Proper Marriage* by Doris Lessing; *Under the Net* by English novelist Iris Murdoch, 35; *Live and Let Die* by Ian Fleming; *The Blackboard Jungle* by New York novelist Evan Hunter, 28; *Tunnel of Love* by U.S. comic novelist Peter De Vries, 44; *The Ponder Heart* by Eudora Welty; *The Bride of Innisfallen* (stories) by Welty; *Pictures from an Institution* by poet Randall Jarrell.

Poetry *Collected Poems* by Wallace Stevens; *The Desert Music* by William Carlos Williams.

Painting *Les Vagabonds* by Jean Dubuffet; *Colonial Cubism* by Stuart Davis; *Collection* and *Charlene* by U.S. painter Robert Rauschenberg, 28; *Painting* by Philip Guston; *White Light* by Jackson Pollock; *Acrobat and Horse* by Fernand Léger. Reginald Marsh dies of a heart attack outside Bennington, Vt., July 3 at age 56; Frida Kahlo dies of cancer at Coyoacan, Mexico, July 13 at age 44 (her coffin, draped in the Soviet flag, lies in state in the rotunda of the National Institute of Fine Arts, guarded by notables including former President Cárdenas); André Derain dies at Chambourcy, France, September 10 at age 74; Henri Matisse dies at Nice November 3 at age 84.

Sculpture *The Caliph of Baghdad* by Joseph Cornell; The Iwo Jima Memorial Monument dedicated at Washington, D.C., honors the U.S. Marine dead of all wars. Based on a World War II photograph taken in March 1945 by Joe Rosenthal, the statue by Felix de Weldon has six bronze figures, each about 32 feet high.

Tri-X film, introduced by Eastman Kodak, is a high-speed (200 ASA) black and white roll film that permits photography in dim light (it can be rated at more than 1,000 ASA and pushed in the darkroom). Eastman also introduces Tri-X motion picture film and Royal Pan sheet film.

The Asahiflex II, introduced by Asashi Optical, revolutionizes single-lens reflex cameras with a quick-return mirror system that permits a photographer to view the subject immediately before the picture is taken without making a separate adjustment to clear the viewfinder (*see* 1952). A pentaprism viewing system in 1957 will further the concept of eye-level viewfinding and be the basis of the brand name Pentax in America.

Theater *The Caine Mutiny Court-Martial* by Herman Wouk 1/2 at New York's Plymouth Theater, with Lloyd Nolan as Lieut. Commander Philip Francis Queeg, Henry Fonda, John Hodiak, 415 perfs.; *The Reluctant Debutante* by English playwright William Heatherton, 30, 8/24 at London's Cambridge Theatre, 752 perfs.; *Separate Tables* (*Table by the Window* and *Table Number Seven*) by Terence Rattigan 9/22 at the St. James's Theatre, London, with Eric Portman, 726 perfs.; *The Tender Trap* by U.S. playwrights Robert Paul Smith and Max Shulman 10/13 at New York's Longacre Theater, with Robert Preston, 101 perfs.; *Ladies of the Corridor* by Dorothy Parker and Arnaud d'Usseau 10/21 at New York's Longacre Theater, with Betty Field, June Walker, Edna Best, Vera Allen, Walter Matthau, Sheppard Strudwick, 45 perfs.; *The Rainmaker* by N. Richard Nash 10/28 at New York's Cort Theater, with Geraldine Page, Richard Coogan, 125 perfs.; *The Bad Seed* by Maxwell Anderson (from the novel by William March) 12/8 at New York's 46th Street Theater, with Patty McCormack, Nancy Kelly, Eileen Heckart, 332 perfs.

Films Edward Dmytryk's *The Caine Mutiny* with Humphrey Bogart, José Ferrer, Van Johnson; Robert Wise's *The Day the Earth Stood Still* with Michael Rennie, Patricia Neal; Elia Kazan's *On the Waterfront* with Marlon Brando, Eva Marie Saint, Rod Steiger, Karl Malden; Alfred Hitchcock's *Rear Window* with James Stewart, Grace Kelly, Wendell Corey; Kenji Mizoguchi's *Sansho the Bailiff* with Kinuyo Tanaka, Kisho Hanayagi, Kyoko Kagawa; Akira Kurosawa's *The Seven Samurai* with Toshiro Mifume; Richard Fleischer's

20,000 Leagues Under the Sea with Kirk Douglas, James Mason. Also: John Sturges's *Bad Day at Black Rock* with Spencer Tracy, Robert Ryan, Ernest Borgnine; Mark Robson's *The Bridges at Toko-Ri* with William Holden, Grace Kelly; Edward Dmytryk's *Broken Lance* with Spencer Tracy, Robert Wagner, Jean Peters; Kenji Mizoguchi's *Chikamatsu Monogatori* (*The Crucified Lovers*) with Kazuo Hasegawa, Kyoko Kagawa; Henri-Georges Clouzot's *Diabolique* with Clouzot's wife Vera, Simone Signoret; Ralph Thomas's *Doctor in the House* with Dirk Bogarde; Teinosuke Kinugasa's *Gate of Hell* with Machiko Kyo, Kazuo Hasegawa; Henry Cornelius's *Genevieve* with Kay Kendall, Kenneth More; William A. Wellman's *The High and the Mighty* with John Wayne, Claire Trevor, Lorraine Day, Robert Stack; Marcel Pagnol's *Letters From My Windmill* with Rellys; Satyajit Ray's *Pather Panchali* with Karuna and Subir Banerji; Billy Wilder's *Sabrina* with Humphrey Bogart, Audrey Hepburn, William Holden; Lewis Allen's *Suddenly* with Frank Sinatra, Sterling Hayden; Gordon Douglas's *Them!* with James Whitmore, Edmund Gwenn; Keisuke Kinoshita's *24 Eyes* with Hideko Takamine.

Television play *The Bachelor Party* by Paddy Chayefsky.

Opera *The Threepenny Opera* 3/10 at New York's off-Broadway Theatre de Lys, with Kurt Weill's widow Lotte Lenya who played Jenny in the original 1928 Berlin production. Marc Blitzstein has written the new English language version; *The Turn of the Screw* 9/14 at Venice's Teatro la Fenice, with music by Benjamin Britten, libretto from the Henry James story; *Troilus and Cressida* 12/3 at London's Covent Garden, with music by William Walton, libretto from the Chaucer poetry of the late 14th century; *The Saint of Bleeker Street* 12/27 at New York's Broadway Theater, with music by Gian Carlo Menotti.

Oratorios *Moses and Aaron* by the late Arnold Schoenberg 3/12 at Hamburg's Musikhalle; *Prayers of Kierkegaard* by Samuel Barber 12/3 at Boston's Symphony Hall.

Ballet *The Legend of the Stone Flower* 2/12 at Moscow's Bolshoi Theater, with music by the late Serge Prokofiev.

Cantata *Hodie* by Ralph Vaughan Williams 9/8 at Worcester Cathedral.

First performances De Profundis for Mixed Chorus a capella by the late Arnold Schoenberg 1/29 at Cologne; *Washington's Birthday* by Charles Ives 2/21 in a CBS radio broadcast of chamber music (Ives dies at New York May 19 at age 79); Concerto No. 4 for Piano and Orchestra by Darius Milhaud 4/30 at Haifa, Israel; *The Odyssey of a Race (Odisseia de uma raca)* by Heitor Villa-Lobos 5/30 at Haifa; String Quartet No. 2 by Roger Sessions 5/31 at Haifa; Concerto in F minor for Bass Tuba and Orchestra by Ralph Vaughan Williams, now 82, 6/13 at London; Quartet No. 4 for Strings by Ernest Bloch 7/28 at Lenox, Mass.; *In Memoriam Dylan Thomas*, Dirge, Canons, and Song by Igor Stravinsky 9/20 at Los Angeles.

Stage musicals *The Boy Friend* 1/14 at Wyndham's Theatre, London, with music and lyrics by English songwriter Sandy (Alexander Galbraith) Wilson, 29, songs that include "I Could Be Happy with You," 2,084 perfs.; *The Pajama Game* 5/13 at New York's St. James Theater, with John Raitt, Janis Paige, music and lyrics by Richard Adler and Jerry Ross, book by Richard Bissell, songs that include "Hey, There," "There Once Was a Man," "7½ Cents," "Hernando's Hideaway," 1,063 perfs.; *Peter Pan* 10/20 at New York's Winter Garden Theater, with Mary Martin, Cyril Ritchard, Margalo Gilmore, music by Mark Charlap with additional music by Jule Styne, lyrics by Carolyn Leigh, 27, additional lyrics by Betty Comden and Adolph Green, songs that include "I'm Flying," "Never Never Land," "Tender Shepherd," "I Won't Grow Up," "I've Gotta Crow," 152 perf.; *Fanny* 11/4 at New York's Majestic Theater, with Ezio Pinza, William Talbert, book by S. N. Behrman and Joshua Logan based on sketches by Marcel Pagnol, music and lyrics by Harold Rome, 888 perfs.; *House of Flowers* 12/30 at the Alvin Theater, with Pearl Bailey, Diahann Carroll, Juanita Hall, music by Harold Arlen, lyrics by Arlen and Truman Capote, 165 perfs.

Hollywood musicals Stanley Donen's *Seven Brides For Seven Brothers* with Howard Keel, Jane Powell, music and lyrics by Johnny Mercer and Gene Paul, choreography by Michael Kidd; George Cukor's *A Star Is Born* with Judy Garland, James Mason, music by Harold Arlen, lyrics by Ira Gershwin.

Popular songs "Three Coins in the Fountain" by Jule Styne, lyrics by Sammy Cahn (title song for Jean Negulesco's film); "Mister Sandman" by Pat Ballard; "If I Give My Heart to You" by Jimmie Greene, Al Jacobs, and Jimmy Brewster; "The Naughty Lady of Shady Lane" by Sid Tepper and Roy C. Bennett; "Little Things Mean a Lot" by Edith Lindeman and Carl Stutz; "Cross Over the Bridge" by Bennie Benjamin and George Weiss; "Sh-Boom" by James Keyes, Claude Feaster, Carl Feaster, Floyd F. McRae, and James Edwards; "Shake, Rattle and Roll" by Charles Calhoun.

Elvis Presley makes his first commercial recording at age 19 and achieves some success with "That's All Right, Mama" and "Blue Moon of Kentucky." Guitarist-vocalist Elvis Aaron Presley will sign an agreement next year with RCA-Victor which will promote "Elvis the Pelvis" with TV appearances, Presley will begin a motion picture career in 1956 and be a major factor in the rock 'n' roll music style.

The Stratocaster electric guitar introduced by Leo Fender has three magnetic pickups for a range of tones, a vibrato tailpiece, and other features that will make it a favorite of Buddy Holly, Jimi Hendrix, Eric Clapton, and other rock 'n' roll pioneers (*see* 1948).

The ChaChaCha based on the classic Cuban Danzon is introduced to U.S. dance floors where it will be a standard Latin dance step.

Quebec City's first Winter Carnival opens in February. Local businessmen have organized the 10-day event to attract visitors with street dances, costume balls,

1954 ***(cont.)*** ice-sculpture championships, a canoe race across the ice-clogged St. Lawrence River, speed skating, barrel-jumping, skiing, snow-shoeing contests, and other attractions.

English track star Dr. Roger Bannister, 25, runs the mile in 3 minutes, 59.4 seconds May 6 at Oxford, breaking the 4-minute barrier.

Jaroslav Drobny, 22 (Czech), wins in men's singles at Wimbledon, Maureen Connolly in women's singles; Vic Seixas wins in men's singles at Forest Hills, Doris Hart in women's singles.

The St. Louis Browns become the Baltimore Orioles.

Hank Aaron joins the Milwaukee Braves to begin a baseball career that will span 2 decades. Henry Louis Aaron, 20, will surpass Babe Ruth's lifetime home-run record (*see* 1974).

Sportswriter Grantland Rice dies of a stroke at New York July 13 at age 73. Obituaries quote his immortal verse, "When the Great Scorer comes/ To mark against your name,/ He'll write not 'won' or 'lost,'/ But how you played the game."

The New York Giants win the World Series by defeating the Cleveland Indians 4 games to 0.

West Germany wins the World Cup in football (soccer), upsetting a heavily favored Hungarian team 3–2 at Bern's Wanddorf Stadium.

Veterans Day becomes a national holiday to replace Armistice Day, a legal holiday celebrated each November 11 since 1928 under terms of a 1926 resolution of Congress to commemorate the end of World War I (*see* Uniform Monday Holiday Act, 1968).

Geodesics, Inc., and Synergetics, Inc., are established in a reorganization of Dymaxion Dwelling Machines, founded in 1944 to produce seven-room circular aluminum Dymaxion houses and geodesic domes designed by Buckminster Fuller (*see* 1927).

The Fontainebleau Hotel opens at Miami Beach, Fla., with 900 rooms. Russian-American architect Morris Lapidus, 51, has designed the $13 million 14-story resort hotel.

Clouds of radioactive coral dust rain down on the Japanese fishing boat *Lucky Tiger* following the U.S. explosion of a hydrogen device at Bikini Atoll (above); the dust falls subsequently on small Micronesian Islands downwind. People exposed to the fallout will be observed in the next 2 decades to suffer from thyroid gland nodules and cancerous tumors, loss of thyroid function in some cases, and leukemia as a result of exposure to radioactive iodine 131 isotopes.

An International Convention for the Prevention of Pollution of the Sea with Oil addresses problems of water contamination.

Lake Erie produces 75 million pounds of commercial fish and remains relatively unpolluted.

Congress acts to relieve world hunger and put a prop under U.S. farm prices. The Agricultural Trade and Development Act (Public Law 480) signed into law by President Eisenhower July 10 empowers the Department of Agriculture to buy surplus wheat, butter, cheese, dry skim milk, vegetable oils, and other surplus foods for donation abroad, barter, or to sell for native currency. P.L. 480 protects U.S. farmers from world market collapse.

Congress passes a Watershed and Flood Prevention Act to protect farmers from water erosion.

The Department of Agriculture fires its attaché at Tokyo because he does not meet "security requirements." Russian-American agricultural economist Wolf Isaac Ladejinsky, 55, had charge of Japanese land reform under Gen. MacArthur, his only interest in revolution is in a green one, the government will admit its error, and Ladejinsky will work on land reform in South Vietnam at the urging of President Eisenhower.

A breakthrough in wheat genetics achieved by University of Missouri plant geneticist Ernest Robert Sears, 44, does for wheat genetics what Russia's Dmitri Mendeléev did for chemistry in 1870. Sears shows that specific chromosomes in wheat can be substituted to achieve desired changes, and he raises hopes for new hybrid wheat strains that will raise yields and will increase resistance to disease and drought.

Nikita Khrushchev orders large-scale planting of new hybrid grain varieties in the Soviet Union, overriding Lysenkoite objections (*see* 1952).

A test program using sterilized males wipes out screwworm flies on the Caribbean island of Curaçao in 4 months (*see* Knipling, 1951).

The FDA approves use of the antioxidant BHA (butyrated hydroxyanisole) in U.S. foods (*see* 1947). The additive will be widely employed to avoid rancidity and the breakdown of fatty acids and ascorbic acid (vitamin C) in many food products.

Trix breakfast food, introduced by General Mills, is 46.6 percent sugar (*see* 1941; 1958).

McDonald's begins the proliferation that will make it the world's largest food-service company. Milkshake-machine salesman Raymond A. Kroc, 52, receives such a large order from the California hamburger chain that he flies out to investigate, persuades the McDonald brothers to sell him franchise rights, and soon has hamburger stands with golden arches opening everywhere (*see* 1948; 1961).

U.S. births will remain above 4 million per year for the next decade.

An oral contraceptive pill developed by Gregory G. Pincus, Hudson Hoagland, and Min-Cheh Chang at the Worcester Foundation employs the hormone norethisterone (*see* 1955; Sanger, 1951).

1955 Moscow recognizes the independence of West Germany January 15 and Soviet premier Georgi M. Malenkov resigns February 8. Marshal Nikolai Aleksandrovich Bulganin, 59, his successor, reaffirms Sino-Soviet ties and appoints Marshal Georgi Konstantinovich Zhukov, now 59, minister of defense.

France's Mendès-France government falls February 5 over the question of North-African insurgency. Edgar Faure forms a new cabinet February 23.

Britain's prime minister Winston S. Churchill resigns April 5 at age 81 and is succeeded by former secretary of state for foreign affairs Sir Anthony Eden, 57.

The presidium of the Supreme Soviet cancels treaties of friendship with Britain and France May 7, West Germany is admitted to NATO May 9, and eight European Communist powers sign the Warsaw Pact May 14.

Austria formally regains her sovereignty July 27 by the Treaty of Vienna; occupation troops withdraw from the country.

South Vietnam has civil war beginning at Saigon April 28. Supporters of Premier Ngo Dinh Diem defeat supporters of Bao Dai and force Binh Xuyen rebels out of the capital May 2 after 5 days of heavy fighting. Diem asks France to move her troops to the northern frontier for action against the Communist Vietminh or to withdraw completely; Paris agrees May 20 to deploy troops against the Vietminh. Beijing promises North Vietnam $338 million in economic aid July 7 in an agreement with President Ho Chi Minh. South Vietnam's chief of state Bao Dai dismisses Premier Diem October 18, Diem refuses to resign, a referendum October 23 gives him an overwhelming vote over Bao Dai, and he proclaims a South Vietnamese Republic October 26 with himself as president (*see* 1954; 1960).

U.S. occupation of Japan ends.

Perónist power in Argentina ends September 19 as a military junta stages a coup that overthrows dictator Juan Perón after 9 years in which Perón has created Latin America's first and only labor movement, a force of considerable economic and political power. Perón has nationalized Argentina's British-owned railroads and established a "Third World" bloc in international affairs but has lost popularity since his suppression of *La Prensa* in 1951 and the death of his wife Eva to cancer the following year. He goes into exile, first in Paraguay, then in Nicaragua (*see* 1973).

Cook County Democratic party chairman Richard J. Daley, 53, wins the Chicago mayoralty race and begins a 21-year career as mayor of the second largest U.S. city. A machine politician in the old tradition, Daley will use patronage to control the Illinois state vote and obtain tax breaks and zoning-law favors for real estate interests and others that support him.

Black political leader Lamar D. Smith, 63, is shot to death in front of the Lincoln County Courthouse at Brookhaven, Miss., after seeking to qualify blacks to vote. More than 20 people witness the shooting, including several blacks, but nobody admits to having seen anything and no witnesses testify against the three white men charged with the murder.

Black minister George W. Lee is killed gangland style at Belzoni, Miss., after a week of terror during which whites have vandalized blacks' property. The blacks have refused to send their children to racially segregated schools, the whites have retaliated by refusing credit to blacks at local stores, and Lee had campaigned for black voting rights.

The Supreme Court orders desegregation of public golf courses, parks, swimming pools, and playgrounds November 7 in a decision based on the 1954 ruling in *Brown v. Board of Education.*

The Interstate Commerce Commission acts November 25 to ban racial segregation on interstate buses and trains and in terminal waiting rooms. Montgomery, Ala., seamstress Rosa Parks, 42, refuses to give up her seat on a downtown bus to a white man December 1 (*see* King, 1956).

The U.S. federal minimum wage rises from 75¢ per hour to $1 August 12 by act of Congress (*see* 1961).

A contract signed with American Can and Continental Can August 13 wins the United Steel Workers the first 52-week guaranteed annual wage in any major U.S. industry.

The AF of L and CIO merge December 2 in a pact engineered by Walter Reuther and George Meany who at age 61 assumes the presidency of the combined AFL-CIO that he will hold for more than 23 years. Some 12 million U.S. workers are unionized, a number that will increase to 20 million in the next 20 years.

A staff report issued by the U.S. Senate Committee on Banking and Currency observes that "less than one percent of all American families owned over four-fifths of all publicly held stocks owned by individuals."

H & R Block is founded by Kansas City tax accountants Henry Wohlman Bloch, 32, and his brother Richard, 29, who use the firm name Block to avoid mispronunciation. Specializing in low-cost mass-produced income tax returns, the firm will open franchised offices in every major U.S. city. By 1977 it will be preparing 10 percent of all U.S. tax returns (25 percent in some cities).

Textron American is incorporated as the first business conglomerate. Royal Little has acquired the giant 56-year-old American Woolen Co. but will now move Textron out of textiles which account for more than 80 percent of sales (*see* 1948). By the end of next year Little will have acquired firms making test equipment, heavy machinery, chain saws, plastics, electromechanical equipment, cement, aluminum, plywood, and veneers, setting a course that other conglomerates will follow.

U.S shopping centers increase in number to 1,800, proliferating rapidly as Americans move out of the central cities to live in the sprawling suburbs (*see* Country Club Plaza, 1922; Northland, 1954).

U.S. motorcar sales reach an unprecedented 7,169,908 of which fewer than 52,000 are imported, but sales of small foreign cars led by Volkswagen will now begin a rapid rise (*see* 1949; 1959; 1968).

General Motors introduces the first Chevrolet V8, but half of all new cars have six-cylinder engines which are superior for city driving, only marginally less powerful in accelerating from 0 to 30 miles per hour, and far

1955 *(cont.)* more efficient in fuel consumption (*see* Ford V8, 1932). By 1976 nearly 70 percent of all U.S. motorcars will have V8 engines, only 21.8 percent sixes, 8.5 percent four-cylinder engines.

The Ford Thunderbird is introduced in the form of a two-seat sportscar.

The Citroën DS-19 introduced at the October Automobile Show in Paris is the first new model in 2 decades. The $2,000 car has been designed by G. Berton for S.A. André Citroën and some 1.4 million of the cars will be sold at steadily rising prices before Citroën is absorbed in 1974 by Peugeot.

Yamaha Motor Co. is founded in Japan under the aegis of Yamaha Musical Instrument Co. It will be a leading maker of motorcycles.

New York's Long Island Expressway opens and is inadequate for the volume of traffic.

The Tappan Zee Bridge opens to traffic at Tarrytown, N.Y. The cantilever span across the Hudson is 3,793 feet in length.

President Eisenhower submits a 10-year $101-billion highway program to Congress (*see* 1956).

The President's Advisory Committee on Transportation recommends reduction of federal controls to stimulate a return to competitive conditions in transportation but in intracity transit free enterprise has worked to restrict competition. Gasoline and diesel buses have replaced some 88 percent of U.S. electric streetcar systems (*see* 1949; General Motors, 1956).

New York installs "Walk/Don't Walk" signals at busy intersections April 19 and institutes alternate side of the street parking regulations to facilitate street-cleaning. All major cities seek solutions to the problem of traffic choked by private automobiles.

New York's Third Avenue "El" makes its last run May 12 and is razed after 77 years of operation (*see* Sixth Avenue "El," 1940).

Rome's first subway opens to link the city's central railroad station to a convention center and government building complex south of the city. The subway was planned by the late Benito Mussolini (*see* 1980).

The mesa introduced by Bell Telephone Laboratories is a new kind of transistor (*see* 1948).

The IBM 752 computer shipped in February is the first IBM business computer; by August IBM has reduced Univac's lead to 30 installations versus 24 (*see* 1951). By August 1956 IBM will have 76 to Univac's 46 with orders for 193 machines versus 65 for Univac; while there are few intrinsic differences between the computers, IBM employs superior salesmanship and will be far ahead by late 1956.

Spanish-born New York University biochemist Severo Ochoa, 49, announces the synthesis of ribonucleic acid (RNA), a basic constituent of all living tissues. It is a giant step toward the creation of life in the laboratory out of inert materials (*see* Watson, Crick, 1953; Kornberg, 1957).

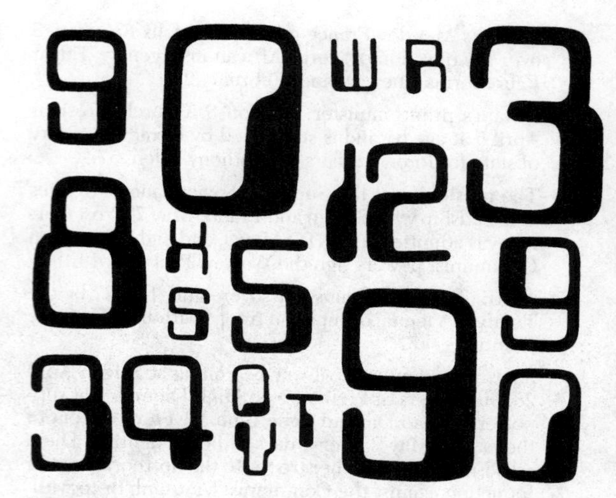
Computer technology used electronic data processing to solve problems previously insoluble and to end work tedium.

Prednisone is introduced for treating arthritis and other conditions. The new steroid drug has far fewer side effects than cortisone (*see* 1948; 1949).

More than 100 cases of paralytic poliomyelitis develop among Americans injected with certain lots of Salk vaccine produced by Cutter Laboratories, which has failed to inactivate the polio virus through a manufacturing error. Yet the Salk vaccine is judged highly effective (*see* 1954).

U.S. National Heart Institute researchers find that reserpine depletes the brain's supply of the chemical substance serotinin (*see* 1952; LSD, 1953).

The *National Review* begins publication. William F. Buckley, Jr., edits and publishes the biweekly journal of conservative political opinion.

The *Village Voice* begins publication at New York in October. Daniel Wolf and Edward Fancher have started the 12-page 5¢ weekly with $15,000, its initial circulation is 2,500, but Wolf and Fancher will increase circulation and sell the *Voice* in 15 years for $3 million.

The *Reader's Digest* abandons its 33-year-old position against running ads but will not accept cigarette advertising.

"Ann Landers Says" is introduced in the *Chicago Sun-Times* by journalist Esther Pauline Friedman Lederer, 37, whose confidential column will be syndicated to hundreds of newspapers as will that of her twin sister Pauline who will write the "Dear Abby" column under the name Abigail Van Buren. Lederer has seen her husband Jules found Budget Rent-a-Car and prosper but will divorce him in 1975 after 36 years of marriage.

Nonfiction *The Guinness Book of World Records* by London sportswriter Ross McWhirter, 30, and his twin brother Norris whose book will have sales of more than 24 million copies in 14 languages in the next 20 years;

Auntie Mame by *Foreign Affairs* promotion manager Patrick Dennis (Edward E. Tanner III), 34, whose story of a rich young orphan and his eccentric aunt has been rejected by 10 publishers.

Fiction *Lolita* by Vladimir Nabokov whose tale of a middle-aged European's love for a passionate 12-year-old American "nymphet" is a satire that reverses the usual contrast between American innocence and European worldliness; *The Voyeur* by Alain Robbe-Grillet; *The Inheritors* by William Golding; *The Acceptance World* by Anthony Powell; *The Recognition* by U.S. novelist William Gaddis, 33; *The Ginger Man* by Brooklyn-born Irish novelist J. P. (James Patrick) Donleavy, 29; *The Quiet American* by Graham Greene; *Andersonville* by MacKinlay Kantor; *A Good Man Is Hard to Find* (stories) by Flannery O'Connor; *A Charmed Life* by Mary McCarthy; *Lord of the Rings* by J. R. R. Tolkien; *Witness for the Prosecution* by Agatha Christie.

Juvenile *Eloise* by U.S. entertainer Kay Thompson, 43, whose heroine lives at the Plaza Hotel.

Poetry "Howl" by U.S. beatnik poet Allen Ginsberg, 29, who reads it at Berkeley, Calif., and wins acclaim in the youth underground; *Journey to Love* by William Carlos Williams; *The Less Deceived* by English poet Philip Larkin, 33; *Good News of Death* by Louis Simpson.

Painting *Flag* and *Target with Four Faces* (painted assemblage) by U.S. painter Jasper Johns, 25; *Bed* and *Rebus* by Robert Rauschenberg; *Scent* by Jackson Pollock; *Portrait* (box construction) by Joseph Cornell; *Double Portrait of Birdie* by Larry Rivers, who has painted two views of his mother-in-law in the nude; *One and Others* (painted wood) by Louise Bourgeois; monochromatic surfaces by French painter Yves (*né* Raymond) Klein, 27; *The Women of Algiers* by Pablo Picasso; *Italian Square* by Giorgio de Chirico. Fernand Léger dies at Gif-sur-Yvette, Seine-et-Oise August 17 at age 74; Maurice Utrillo dies at Dax in Landes November 5 at age 71.

Theater *Bus Stop* by William Inge 3/2 at New York's Music Box Theater, with Kim Stanley, Albert Salmi, 478 perfs.; *Cat on a Hot Tin Roof* by Tennessee Williams 3/24 at New York's Morosco Theater, with Barbara Bel Geddes, Ben Gazzara, Mildred Dunnock, folk singer Burl Ives as "Big Daddy," 694 perfs.; *Inherit the Wind* by U.S. playwrights Jerome Lawrence, 39, and Robert E. Lee, 37, 4/21 at New York's National Theater, with Paul Muni as Clarence Darrow, Ed Begley as William Jennings Bryan is based on the 1925 Scopes "monkey trial," 806 perfs.; *A View from the Bridge* by Arthur Miller 9/29 at New York's Coronet Theater, with J. Carroll Naish, Van Heflin, Eileen Heckart, Jack Warden, 149 perfs.; *The Diary of Anne Frank* by Frances Goodrich and Albert Hackett 10/5 at New York's Cort Theater, with Joseph Schildkraut, Susan Strasberg, 717 perfs.; *No Time for Sergeants* by Ira Levin 10/20 at New York's Alvin Theater, with Andy Griffith, Don Knotts, 796 perfs.; *A Hatful of Rain* by U.S. playwright Michael Gazzo, 32, 11/9 at New York's Lyceum Theater, with Ben Gazzara, Shelley Winters, Frank Silvera, Anthony Franciosa is about drug addiction, 398 perfs.

Films Elia Kazan's *East of Eden* with Julie Harris, James Dean, Raymond Massey, Burl Ives; John Ford and Mervyn LeRoy's *Mister Roberts* with Henry Fonda, James Cagney, Jack Lemmon, William Powell; Leslie Norman's *The Night My Number Came Up* with Michael Redgrave, Sheila Sim, Alexander Knox, Denholm Elliott; Carl Dreyer's *Ordet* with Henrik Malberg, Emil Hass, Christiansen Preben, Lendorff Rye; Nicholas Ray's *Rebel Without a Cause* with Natalie Wood, Sal Mineo, James Dean (who is killed September 30 in an automobile accident at age 24); Ingmar Bergman's *Smiles of a Summer Night* with Ulla Jacobsen, Harriet Andersson, Eva Dahlbeck, Jarl Kulle. Also: Frank Launder's *The Belles of St. Trinians* with Alastair Sim, Joyce Grenfell, Hermione Baddeley, Beryl Reid; Richard Brooks's *Blackboard Jungle* with Glenn Ford, Ann Francis; Daniel Mann's *I'll Cry Tomorrow* with Susan Hayward, Richard Conte; Robert Aldrich's *Kiss Me Deadly* with Ralph Meeker; Hamilton Luske, Clyde Geronimi, and Wilfred Jackson's *The Lady and the Tramp* with Walt Disney animation, voices of Peggy Lee, Stan Freberg, et al.; Delbert Mann's *Marty* with Ernest Borgnine in an adaptation of the 1953 Paddy Chayefsky television script; Joshua Logan's *Picnic* with William Holden, Kim Novak, Rosalind Russell; Daniel Mann's *The Rose Tattoo* with Anna Magnani, Burt Lancaster; Brian Desmond Hurst's *Simba* with Dirk Bogarde, Virginia McKenna; David Lean's *Summertime* with Katharine Hepburn, Rossano Brazzi in a film version of the Arthur Laurents play *Time of the Cuckoo*.

Disneyland opens July 17 at Anaheim 25 miles south of Los Angeles. Mickey Mouse originator Walt Disney has mortgaged his life insurance, stock holdings, house, and furniture to acquire an orange grove, cut it down, and finance construction of the Jungle River Ride, Mark Twain paddle-wheeler ride, and other attractions of Adventureland, Frontierland, Fantasyland, and Tomorrowland for the $17 million 244-acre amusement park. He will assemble a 27,500-acre tract near Orlando, Fla., to build a Disneyland East that will be part of a vacation resort, but chain-smoker Disney will die of lung cancer at age 65 in mid-December 1965 and never see Walt Disney World.

Television *Captain Kangaroo* 10/3 on CBS network stations, with actor Robert James Keeshan, 28, hosting the morning show for preschoolers; *The Mickey Mouse Club* in October on ABC network stations, with "Mousketeers" Annette Funicello, Cubby O'Brien, Karen Pendleton, Cheryl Holdridge, and host Jimmie Dodd who has composed the march "M-I-C-K-E-Y M-O-U-S-E" and will keep the afternoon show going until 1958.

Hollywood musicals Fred Zinneman's *Oklahoma!* with Gordon MacRae, Shirley Jones, Charlotte Greenwood; Richard Quine's *My Sister Eileen* with Betty Garrett, Janet Leigh, Jack Lemmon, Bob Fosse.

1955 *(cont.)* Broadway musicals *Silk Stockings* 2/24 at the Imperial Theater, with Hildegarde Neff as Ninotchka, George Tobias, Julie Newmar, Don Ameche, David Opatashu, music and lyrics by Cole Porter, songs that include "All of You," 478 perfs.; *Damn Yankees* 5/5 at the 46th Street Theater, with Gwen Verdon, music and lyrics by Richard Adler and Jerry Ross, songs that include "Whatever Lola Wants," "You've Got to Have Heart," "Two Lost Souls," 1,019 perfs.; *Pipe Dream* 11/3 at the Sam S. Shubert Theater, with Helen Traubel, music by Richard Rodgers, lyrics by Oscar Hammerstein II, based on the John Steinbeck novel *Sweet Thursday*, songs that include "All at Once You Love Her."

The Vienna State Opera House reopens after having been nearly completely destroyed by wartime bombing and gunfire.

First performances Symphony No. 8 by Heitor Villa-Lobos 1/14 at Philadelphia's Academy of Music; Symphony No. 5 *(Sinfonia Sacra)* by Howard Hanson 2/18 at Philadelphia's Academy of Music; Symphony No. 6 by Walter Piston 11/25 at Boston's Symphony Hall.

The Merengue introduced to U.S. dance floors by New York teacher Albert Butler, 61, and his wife Josephine is an adaptation of a Dominican one-step combining the rumba and *paso doble*.

Popular songs "Rock Around the Clock" by William "Bill" Haley; "Maybellene" by U.S. guitarist-blues singer Charles Edward Anderson "Chuck" Berry, 29, Russ Frato, and Alan Freed. Berry scores an immediate hit that he will follow with "Roll Over, Beethoven," "Rock 'n' Roll Music," "Memphis," and other triumphs; "Something's Gotta Give" by Johnny Mercer (for the film *Daddy Long Legs*); "Memories Are Made of This" by Terry Gilkyson, Richard Debny, and Frank Miller; "Love Is a Many-Splendored Thing" by Sammy Fain, lyrics by Paul Francis Webster (title song for film); "Cry Me a River" by Arthur Hamilton; "Love and Marriage" by James Van Heusen, lyrics by Sammy Cahn (for a TV production of Thornton Wilder's *Our Town*); "Dance with Me, Henry" by Etta James; "Domani" by Italian composer Ulpio Minucci, lyrics by Tony Velona.

Swaps beats the favorite Nashua to win the Kentucky Derby but Nashua comes back to defeat Swaps in a match race and will be retired to stud next year, the first racehorse to be syndicated for more than $1 million.

Tony Trabert wins in men's singles at Wimbledon and Forest Hills, Louise Brough in women's singles at Wimbledon, Doris Hart at Forest Hills.

The Philadelphia Athletics become the Kansas City Athletics as baseball broadens its geographical base.

The Brooklyn Dodgers win their first World Series, defeating the New York Yankees 4 games to 3.

U.S. cigarette consumption resumes its rise after a 2-year drop as the industry increases its advertising, especially on network television. R. J. Reynolds promotes its filter-tipped Winstons, American Tobacco its cork-tipped king-size Herbert Tareytons with activated charcoal filter, P. Lorillard its Old Gold filters, Philip Morris its Benson & Hedges and Marlboro filters, Brown & Williamson its new filter-tipped Kools and Raleighs.

Le Corbusier completes Notre Dame du Haut at Ronchamp, France.

The United States has 30,000 motels or motor courts, up from 10,000 in 1935.

The London Clean Air Act passed in October bans the burning of untreated coal to prevent a recurrence of the killer smog of 1952 and save an estimated $4.5 million per year to repair the corrosion damage and clean up the 75,000 tons of soot that rained down on the city. The new law will permit 80 percent more sunshine to reach London, birds will return, grass grow, churches and other buildings be cleaned and returned to their original white stone appearance for the first time in centuries as workers remove soot packed up to 9 inches deep.

Mao Zedong proposes rapid collectivization of Chinese agriculture (*see* 1950). Other leaders fear a repetition of the Soviet Union's experience in the late 1920s and early '30s but Mao prevails and peasants who resist are liquidated.

Dwarf indica rice, introduced into Taiwan (Formosa), is higher yielding than most varieties but requires lots of fertilizer and insecticides.

Hurricane Janet hits Grenada in the British West Indies, destroying 75 percent of the island's nutmeg trees (*see* 1858). Grenada's trees have long supplied 40 percent of world nutmeg and mace production, the only other major source is the tiny island of Siame north of the Celebes whose shipments are delayed by troubles in Indonesia, and since it takes 2 months for the nutmeg to reach Singapore and another 2 to reach New York, the price of nutmeg rises from 35¢/lb. to over $2.

The National Farmers Organization (NFO) is founded by farmers from northwest Missouri and southwest Iowa disgruntled by livestock prices lower than production costs. It will engage in collective bargaining in livestock, grain, and milk nationwide.

A USDA survey finds that one-tenth of U.S. families live on nutritionally "poor" diets, a healthy improvement over 1943 findings.

Britain's first fluoridation of community drinking water begins November 17 in Anglesey (*see* 1945).

Crest toothpaste, introduced by Procter & Gamble, contains stannous fluoride (*see* Gleem, 1952). It will become the top-selling U.S. dentifrice, encouraging other makers to introduce fluoride toothpastes.

Campbell's Soup Co. enters the frozen foods business by acquiring control of C. A. Swanson & Sons, originator of the "TV Dinner" (*see* 1954). Campbell's will expand the Swanson line of 11 prepared frozen items to 65 in the next 17 years.

Special K breakfast food, introduced by Kellogg, is 4.4 percent sugar (*see* 1953; 1958).

Coca-Cola Co. uses the name "Coke" officially for the first time (*see* 1945; Minute Maid, 1960; Tab, 1963).

Tropicana's Anthony Rossi builds a port at Cape Canaveral, Fla. (*see* 1951). Unable to expand his fleet of trucks for shipping chilled orange juice to northern markets, Rossi buys an 8,000-ton ship with stainless steel tanks, builds a bottling plant at Whitestone, Queens, and begins sending juice north by ship, a practice he will later abandon in favor of rail and truck shipment.

Colonel Sanders' Kentucky Fried Chicken has its beginnings as Corbin, Ky., restaurateur Harland Sanders, 65, travels the country in an old car loaded with pots and pans and a "secret blend of herbs and spices" looking for prospective licensees. The roadside restaurant he ran for 25 years has been ruined by a new highway that diverted traffic from his location, his only income is a $105 monthly Social Security check, but Sanders will franchise hundreds of "finger-lickin' good" fast food operators in the next 9 years and receive 3 percent of their gross sales in return for use of his name, spices, milk and egg dip, gravy mix, and paper supplies (*see* 1964).

The U.S. Census Bureau reveals that the American population increased by 2.8 million last year, the largest 1-year advance on record.

The Soviet Union resumes legalized abortion on demand subject to certain safeguards but discourages abortion and birth control (*see* 1936).

The Planned Parenthood Federation of America holds conferences on abortion.

The Fifth International Conference on Planned Parenthood convenes at Tokyo and hears Gregory Pincus speak about inhibition of ovulation in women who have taken progesterone or norethynodrel, active ingredients in newly developed contraceptive pills that will soon be spoken of collectively as The Pill (*see* 1954). The new oral contraceptive has a failure rate of only one pregnancy per thousand women per year (*see* Puerto Rico study, 1957).

1956 Soviet authorities crack down on Poles and Hungarians who revolt following Nikita Khrushchev's denunciation of the late Josef Stalin and his policies February 14 at the 20th Communist Party Conference (below).

Marshal Zhukov orders occupation of Hungary after a long debate in the Politburo where a majority has initially opposed military action. Soviet ordnance contains only intermediate ballistic missiles, they can reach certain strategic areas of southern Europe only if launched from bases in Hungary, and Zhukov insists on having those bases. Moscow responds also to pressure from its allies at Beijing who urge suppression of the Hungarian revolution lest world communism disintegrate.

The Polaris missile developed at the Woods Hole, Mass., Oceanographic Institute can be launched underwater from submarines and carry a nuclear warhead.

The U.S. Atomic Energy Commission explodes the world's first airborne hydrogen bomb May 21 as it conducts a new series of nuclear tests in the Pacific (*see* 1954; Britain, 1957).

Soviet troops crushed uprisings in Poland and Hungary, a reminder that satellite nations were puppets of Moscow.

Sudan is proclaimed an independent democratic republic January 1 and joins the Arab League January 19, increasing membership in the league to nine.

Jordan and Israel accept UN truce proposals January 24, the Declaration of Washington issued February 1 by President Eisenhower and Prime Minister Eden reaffirms joint Anglo-American policy in the Mideast, Moscow states February 12 that sending U.S. or British troops to the Mideast would violate the UN Charter, a military alliance signed April 21 at Jedda joins Egypt, Saudi Arabia, and Yemen, UN Secretary General Dag Hammarskjöld arranges a cease-fire between Israel and Jordan that takes effect April 29, cease-fires with Lebanon and Syria take effect May 1.

Egypt announces June 4 that she will not renew the Suez Canal Company's concession after it expires in 1968, the last British troops leave the Canal base June 13, President Nasser seizes the Canal July 26 under a decree outlawing the company, British and French nationals leave Egypt August 2, British families are airlifted out of the Canal Zone August 9, President Nasser boycotts a London conference on the Suez August 16 and rejects 18-nation proposals on the Suez September 10.

Britain and France submit the Suez dispute to the UN Security Council September 23, Israeli troops invade the Sinai Peninsula October 29, Israel accepts an October 30 Anglo-French ultimatum calling for a cease-fire and withdrawal of troops 10 miles from the Suez Canal, but Egypt does not accept, French and British planes bomb Egyptian airfields October 31, Washington sends aid to Israel, Jordan refuses to permit RAF planes to use Jordanian bases for operations against Egypt, Gaza falls to British forces November 2, the UN General Assembly adopts a Canadian resolution November 4 to send an international force to the Mideast, but

1956 ***(cont.)*** Britain and France abstain. British paratroops land at Port Said November 5 to recover control of the canal that Prime Minister Eden says is vital if his country is not to starve, Moscow threatens to use rockets if Britain and France do not accept a cease-fire in Egypt, the cease-fire begins November 7, but Britain says she will evacuate her troops only upon the arrival of a UN force. The UN force arrives November 15, the last Anglo-French forces leave Port Said December 22, and the UN fleet begins clearing the canal of scuttled ships December 27.

Pakistan becomes an Islamic republic February 29 but remains within the British Commonwealth.

Tunisia gains independence March 20; Habib Bourguiba accepts the premiership April 10 at the invitation of the bey of Tunis *(see* 1957).

Morocco gains independence from France March 1 and from Spain April 7 (*see* 1957).

Britain grants Gold Coast independence September 18 (*see* Ghana, 1957).

President Eisenhower wins reelection against a second challenge from Democratic hopeful Adlai Stevenson despite the president's illness and warnings that in the event of his death he would be succeeded by his controversial vice president Richard Nixon. Ike polls 57 percent of the popular vote and wins 457 electoral votes, Stevenson gets only 42 percent of the popular vote and 73 electoral votes, but the Democrats retain control of Congress.

"History is on our side. We will bury you!" says Nikita Khrushchev to Western ambassadors November 17 at a Kremlin reception.

Polish workers riot at Poznan June 28 to protest social and economic conditions under the Communist regime. More than 100 demonstrators are killed as the militia moves in to suppress the riots, trials of the rioters end abruptly October 10, several Polish Communists demand removal of Soviet officers from the Polish Army October 16, but Polish and Soviet frontier troops exchange fire as former Soviet officer Marshal Konstantin K. Rokossovsky orders his Polish troops to take positions near Warsaw.

Former Polish Workers' party secretary general Wladyslaw Gomulka has been freed and rehabilitated after 5 years in prison and becomes first secretary of Poland's Communist party October 21.

Hungarian university students meeting October 22 put together a list of 16 demands while expressing solidarity with the Polish rebels (above). They post the petition throughout Budapest during the night on trees and walls and at tram stops, and the students make plans to present their demands to the minister of internal affairs who is also the head of the secret police.

Students mob Budapest's Bem Square beginning at 4:30 in the afternoon of October 23 as the ministry of internal affairs retracts its denial of permission to hold a rally, truckloads of workers arrive, and the crowd moves toward Parliament Square 2 miles away as its numbers swell to more than 100,000 men and women demanding democratic government, the return of former premier Imre Nagy to power, withdrawal of Soviet troops, and the release of Jozsef Cardinal Mindszenty, who has been held in solitary confinement since the end of 1948.

Imre Nagy is restored as premier October 24 to appease the Hungarian demonstrators, the Stalinist head of the Communist party Ernö Gerö is replaced October 25 by János Kádár, but the revolt begins to spread across Hungary.

Soviet troop withdrawal from Poland begins October 25, but several Soviet divisions from East Germany have entered Poland a few days earlier and remain. Marshal Rokossovsky plots to overthrow Party Secretary Gomulka, Polish militia using tear gas suppress attacks on Soviet Army installations, the Soviet military coup against Gomulka is aborted by betrayal of his plans, Rokossovsky returns to Moscow October 28, and the Polish prelate Stefan Cardinal Wyszinski is released from custody October 29.

Soviet forces withdraw from Budapest October 30, Cardinal Mindszenty is released from prison the same day, Premier Nagy goes on the radio to promise Hungarians free elections and a prompt end to one-party dictatorship, Nagy denounces the Warsaw Pact of 1954 November 2, and 16 Soviet divisions move in 2 days later with 2,000 tanks to crush the Hungarian defiance. Premier Nagy is replaced by János Kádár (above) who will continue as premier until 1965, and Cardinal Mindszenty takes refuge in the U.S. embassy at Budapest, where he will occupy the top floor for 15 years, supported by funds from U.S. bishops.

The United Nations General Assembly condemns Soviet interference in Hungary November 4 and calls for an investigation, the last rebel stronghold on Csepel Island in the Danube falls to Soviet forces November 14, and the Russians seize Imre Nagy November 22 as he leaves the Yugoslav embassy at Budapest.

The Eisenhower administration offers asylum to Hungarian "freedom fighters" November 29 but makes no other effort to intervene and takes no action until after the president's reelection (above).

The University of Alabama has expelled its first black student Autherine Lucy March 1 in defiance of a federal court order reversing her suspension by the university's president and trustees.

Southern congressmen issue a manifesto March 11 pledging to use "all lawful means" to upset the Supreme Court's 1954 desegregation ruling.

A Supreme Court ruling April 23 outlaws racial segregation in intrastate public transportation.

Rev. Martin Luther King, Jr., 27, organizes a boycott of Montgomery, Ala., public transportation. The black clergyman leads a protest against racial discrimination.

South Africa's Nationalist government has revealed plans January 13 to remove 60,000 mixed-blood "coloreds" from the Cape Province voting roll; on August

25 the government orders more than 100,000 nonwhites in Johannesburg to leave their homes within a year to make room for whites (*see* 1953; 1961).

A U.S. Supreme Court ruling March 26 upholds a federal law passed in 1954 to compel witnesses to testify in cases involving national security. The court rules in *Ullmann v. United States* that the witnesses will have immunity from prosecution equivalent to the protection guaranteed by the Fifth Amendment.

The Supreme Court excludes states from punishing persons for sedition. It rules April 3 in *Pennsylvania v. Nelson* that the federal government has "occupied the field" under the Smith Act of 1940.

Chinese who resist communization continue to be liquidated as they have been since 1949 (and especially since 1952) and will be for another 4 years. By 1960 some 26.3 million Chinese will have been killed according to some estimates, the largest massacre in world history.

Hungarians stage a general strike in December to protest the János Kádár regime (above). Some 150,000 people leave the country, including many of the best minds.

The Ford Foundation sells 20 percent of its Ford Motor stock to Wall Street investors, permitting the public to buy into Ford for the first time and bringing in some $643 million to the Foundation, which announces that it will give away more than $500 million in the next 18 months, $150 million more than in the 20 years since 1936 (*see* 1950; 1954).

Wall Street's Dow Jones Industrial Average peaks at 521.05.

IBM signs a consent decree agreeing to sell its tabulating and computer machines as well as leasing them, thus ending a legal skirmish with the Department of Justice (*see* 1955; 1969).

England's Calder Hall initiates full-scale use of nuclear fuel to produce electricity August 20 when the facility's first turbine goes into operation. It feeds power into the grid of the Central Electricity Authority beginning October 17. Designed to generate some 90,000 kilowatts of power, it also manufactures the artificial nuclear fuel plutonium for military purposes (*see* 1954; United States, 1958).

Parliament passes a Finance Act giving the British government direct responsibility for long-term financing of the electric utilities industry, which has been nationalized since 1948.

Uranium ore bodies found north of Saskatchewan's Lake Athabasca will make the province the world's leading uranium producer (*see* 1953).

Canada pledges to help India develop nuclear energy for peaceful purposes in an agreement signed by Prime Minister Louis St. Laurent with Prime Minister Jawaharlal Nehru. India will complete nuclear power reactors north of Bombay at Trapur in 1969 but peaceful application of nuclear energy for India will not come until later (*see* 1974).

Libya's first oil well comes into production on a concession granted by Idris I to the Libyan American Oil Co. Rights to the oil go to U.S. explorer-archaeologist Wendell Phillips, 35, who has persuaded the king that an independent company will work harder to develop Libya's petroleum resources than will any of the seven world oil giants.

Getty Oil, created by a reorganization of Jean Paul Getty's Pacific Co., is the world's largest personally owned oil producer and distributor (*see* 1951). Getty's Saudi-Arabian wells have been producing since 1953; Getty Oil will coordinate his interests in other areas.

Britain imposes petrol (gasoline) rationing December 17 in the Suez crisis (above).

The Federal Aid Highway Act passed by Congress June 29 authorizes construction of a 42,500-mile network of roads to link major U.S. urban centers, with 90 percent of the $33.5 billion cost to be borne by the federal government (*see* 1944). The new legislation indirectly subsidizes trucking firms, intercity bus lines, motor vehicle producers, and oil companies, while railroads remain unsubsidized and will depreciate and abandon their unprofitable passenger services. The system is to be completed by 1972, and while that target date will prove beyond reach, some 38,000 miles of the Interstate Highway System will be open by 1976.

U.S. traffic accidents kill 6.28 people for every 100 million miles traveled. The rate will drop to 3.57 by 1974 and on interstate highways (above) it will be 1.55.

The world's first containership port opens at Elizabeth, N.J., where the Port of New York Authority has paid $3.5 million to buy 40 acres for the new facility. The first container ship is a converted tanker with simple trailer vans on her decks, but containerships will revolutionize cargo handling by reducing the need for longshoremen, and Elizabeth will remain the world's leading containerport.

The Italian passenger liner S.S. *Andrea Doria* collides in a heavy fog 60 miles off Nantucket Island, Mass., July 24 with the Swedish liner S.S. *Stockholm.* The 29,500-gross-ton Italian ship has used improper radar procedure, she has turned to her left at the last moment rather than to her right, her engine room is missing a watertight door, she has diminished her stability by failing to ballast her empty fuel tanks, and she goes down in 225 feet of water the next morning with a loss of 52 lives.

General Motors announces 1955 earnings of more than $1 billion after taxes and is indicted for using anticompetitive actions to obtain an 85 percent monopoly on the U.S. market for new buses (*see* 1949; 1955). The Justice Department charges GM with inducing municipal transit systems to adopt specifications which only GM buses could meet, entering into long-range requirement contracts with bus company operators, and installing a GM officer and director as board chairman and leading stockholder of GM's chief competitor in bus-making (*see* 1965).

1956 ***(cont.)*** New York makes Broadway one-way southbound below 47th Street (*see* 1954; 1957).

Boeing and Douglas battle for leadership in the commercial jet aircraft industry (*see* 1953). American Airlines, Air France, Air India, and eight other airlines order Boeing 707s; United, Trans Canada, SAS, and nine other airlines order Douglas DC-8s, but Boeing orders will exceed Douglas orders by three to one within a year (*see* 1958).

An Aeroflot Tupolev-104 airliner takes off from Moscow for Irkutsk September 15 on the world's first scheduled passenger jet flight.

The first successful videotape recorder (VTR) is demonstrated in February in an Ampex Corp. laboratory at Redwood City, Calif. (*see* 1952). Most TV shows quickly move to videotape, and U.S. companies race to develop home VCRs (videocassette recorders) (*see* 1965).

Collier's ceases publication at year's end (*see* 1895).

Nonfiction *The Organization Man* by *Fortune* editor William Hollingsworth Whyte, Jr., 39, argues that a new collective ethic has arisen from the bureaucratization of society and is replacing the old Protestant individualist code. "Belongingness" rather than personal fulfillment has become the ultimate need of the individual (compare Riesman, 1950); *The Art of Loving* by Erich Fromm; *Profiles in Courage* by Sen. John Fitzgerald Kennedy, 39 (D. Mass.) (actually ghostwritten, mostly by historian Alan Nevins, 66, and lawyer Theodore C. Sorenson, 28), examines men prominent in American history.

Fiction *The Floating Opera* by U.S. novelist John Barth, 26, whose nihilist hero decides after a comic analysis not to commit suicide; *The Last Hurrah* by Boston novelist Edwin O'Connor, 38, whose fictional Frank Skeffington is based on the politician James Michael Curley, now 81, whose autobiography *I'd Do It Again* will appear next year; *The Fall (La Chute)* by Albert Camus; *A Certain Smile* (*Un Certain Sourire*) by Françoise Sagan; *The Towers of Trebizond* by Rose Macaulay who produces her most successful novel at age 75; *Thin Ice* by Compton Mackenzie, now 73; *Anglo-Saxon Attitudes* by English novelist-playwright Angus Wilson, 43; *A Tangled Web* by Irish novelist C. Day Lewis, now 52, who gains great success with his detective novel; *The Butterfly of Dinard (La farfalla di Dinard)* (prose sketches) by poet Eugenio Montale; *The Last of the Wine* by English novelist Mary Renault (Mary Challens), 51; *Pitcher Martin* by William Golding; *Sybil* by Pär Lagerkvist; *Diamonds Are Forever* by Ian Fleming; *Peyton Place* by U.S. novelist Grace Metalious (*née* Repentigny), 31.

Poetry "Zima junction" ("Stantsiya Zima") by Russian poet Yevgeny Yevtushenko, 23, who presents problems that tormented him after Stalin's death in 1953 when the poem was written and reopens long forbidden themes; *The False and the True Green* by Salvatore Quasimodo; "Chaka" by Senegalese poet Leopold Sedar Senghor, 50; *Platero and I* (*Platero y yo*) by Juan Ramon Jimenez; *Collected Poems* by English poet Kathleen Raine, 48.

Painting *Just what is it that makes today's homes so different, so appealing?* (photo-collage) by English pop artist Richard Hamilton, 34; *Pins* (tar and oil) by Italian artist Piero Manzoni, 23; *Easter Monday* by Willem de Kooning. Lyonel Feininger dies at New York January 13 at age 84; Jackson Pollock is killed in an auto accident on Long Island, N.Y., August 17 at age 44.

Theater *The Visit* (*Der Besuch der alten Dame*) by Friedrich Dürrenmatt 1/29 at Zurich's Schauspielhaus; *Romanoff and Juliet* by Peter Ustinov 4/2 at the Manchester Opera House, England; *The Chalk Garden* by Enid Bagnold 4/4 at London's Theatre Royal, 658 perfs.; *Look Back in Anger* by English playwright John Osborne, 27, 5/8 at London's Royal Court Theatre, with Kenneth Haigh, Mary Ure, Alan Bates expresses the frustration of radical youth in a strident attack on the middle class; *The Quare Fellow* by Irish playwright Brendan Behan, 33, 7/24 at London's Comedy Theatre, with Maxwell Shaw; *A Long Day's Journey into Night* by the late Eugene O'Neill 11/7 at New York's Helen Hayes Theater, with Fredric March, Florence Eldredge, Jason Robards, Jr., Bradford Dilman, Katharine Ross in an autobiographical play about alcoholism and drug addiction by the playwright who died late in 1953 at age 65, 390 perfs.; *The Matchmaker* by Thornton Wilder 12/5 at New York's Royale Theatre, with Ruth Gordon, Arthur Hill in a play that will be the basis of the 1964 musical *Hello, Dolly!*, 486 perfs.

Television *As the World Turns* 4/2 on CBS; *Edge of Night* 4/2 on ABC (both daytime).

Films Anatole Litvak's *Anastasia* with Ingrid Bergman, Yul Brynner, Helen Hayes, Akim Tamiroff; Norman Panama and Melvin Frank's *The Court Jester* with Danny Kaye, Glynis Johns, Basil Rathbone; William Wyler's *Friendly Persuasion* with Gary Cooper, Dorothy McGuire, Marjorie Main; George Stevens's *Giant* with Rock Hudson, Elizabeth Taylor, the late James Dean; Vincente Minnelli's *Lust for Life* with Kirk Douglas as Vincent Van Gogh, Anthony Quinn as Gauguin; Jules Dassin's *Rififi* with Jean Servais, Carl Mohner; John Ford's *The Searchers* with John Wayne, Jeffrey Hunter; Ingmar Bergman's *The Seventh Seal* with Max von Sydow; Federico Fellini's *La Strada* with Anthony Quinn, Giuletta Massina, Richard Basehart. Also: Satyajit Ray's *Aparajito* with Smaran Ghosal; Mario Monicelli's *Big Deal on Madonna Street* with Vittorio Gassman, Marcello Mastroianni; Joshua Logan's *Bus Stop* with Marilyn Monroe, Eileen Heckart, Don Murray; Helmut Kautner's *The Devil's General* with Curt Jurgens; Kon Ichikawa's *Harp of Burma* with Rentaro Mikuni, Shoji Yasui; Don Siegel's *Invasion of the Body Snatchers* with Kevin McCarthy, Dana Wynter; Stanley Kubrick's *The Killing* with Sterling Hayden; Alexander Mackendrick's *The Ladykillers* with Alec Guinness, Peter Sellers; Nunnally Johnson's *The Man in the Grey Flannel Suit* with Gregory Peck, Jennifer Jones, Frederic March;

Albert Lamorisse's *The Red Balloon* with Lamorisse's son Pascal; Laurence Olivier's *Richard III* with Olivier, John Gielgud, Ralph Richardson, Claire Bloom; Robert Wise's *Somebody up There Likes Me* with Paul Newman, Pier Angeli; Kenji Mizoguchi's *Street of Shame* with Machiko Kyo; Alexander Mackendrick's *Sweet Smell of Success* with Burt Lancaster, Tony Curtis; Daniel Mann's *The Teahouse of the August Moon* with Marlon Brando, Glenn Ford, Michiko Kyo; Frank Launder's *Wee Geordie* with Bill Travers, Alastair Sim. Plus Terry Morse and Inoshiro Honda's *Godzilla, King of the Monsters* with Raymond Burr, Takashi Shimura.

Stage musicals *My Fair Lady* 3/15 at New York's Mark Hellinger Theater, with Julie Andrews, 20, as Eliza Doolittle, Rex Harrison as Professor Higgins, book based on the 1914 Bernard Shaw play *Pygmalion*, music by Frederick Loewe, lyrics by Alan Jay Lerner, songs that include "Why Can't the English Teach Their Children How to Speak," "The Rain in Spain," "The Street Where You Live," "I'm Getting Married in the Morning," "I've Grown Accustomed to Her Face," "I Could Have Danced All Night," 2,717 perfs.; *Mr. Wonderful* 3/22 at New York's Broadway Theater, with Sammy Davis, Jr. and Sr., Chita Rivera, music by Jerry Bock, 27, lyrics by Larry Holofcener and George Weiss, songs that include "Too Close for Comfort," 388 perfs.; *South Sea Bubble* 4/25 at London's Lyric Theatre, with Vivien Leigh, Alan Webb, Joyce Carey, music and lyrics by Noël Coward, 276 perfs.; *The Most Happy Fella* 5/3 at New York's Imperial Theater, with Robert Weede, Art Lund, Jo Sullivan, book based on the 1924 Sidney Howard play *They Knew What They Wanted*, music and lyrics by Frank Loesser, songs that include "Big D" and "Standing on the Corner," 676 perfs.; *Li'l Abner* 11/10 at the St. James Theater, with Peter Palmer as Abner Yokum, Edith Adams as Daisy Mae, Charlotte Rae as Mammy Yokum, Stubby Kaye, Julie Newmar, Tina Louise, music by Gene de Paul, lyrics by John Meyer, songs that include "Jubilation T. Cornpone," 693 perfs.; *Bells Are Ringing* 11/29 at the Sam S. Shubert Theater, with Judy Holliday, Jean Stapleton, music by Jule Styne, lyrics by Betty Comden and Adolph Green, songs that include "Just in Time," "The Party's Over," 924 perfs.

Opera *The Ballad of Baby Doe* 7/7 at Colorado's Central City Opera House with music by U.S. composer Douglas Stuart Moore, 61, libretto by John Latouche based on the career of silver magnate Horace Tabor and his widow (*see* 1880; 1893).

First performances Proclamation for Trumpet and Orchestra by Ernest Bloch 2/3 at Houston; *Modern Psalm* for Speaker, Mixed Chorus, and Orchestra by the late Arnold Schoenberg 5/29 at Cologne; *Johannesburg Festival* Overture by William Walton 9/25 at Johannesburg; *Robert Browning* Overture by the late Charles Ives 10/14 at New York's Carnegie Hall.

Popular songs "Love Me Tender" by Elvis Presley and Vera Walson; "Hound Dog" by Jerry Leiber and Mike Stoller; "Heartbreak Hotel" by Elvis Presley, Mae Boren Axton, Tommy Durden; "I Walk the Line" by U.S. singer-composer John R. "Johnny" Cash, 24, who last year made his first record "Hey, Porter" and followed it with "Folsom Prison Blues;" "Since I Met You, Baby" and "I Almost Lost My Mind" by U.S. composer Ivory Joe Hunter, 45, whose music and rhythm-and-blues numbers will include "My Wish Came True" and "I Need You So" for Elvis Presley, Pat Boone, and Sonny James; "Around the World in 80 Days" by Victor Jones, lyrics by Harold Adamson (title song for film); "Que Será, Será (What Will Be, Will Be)" by Jay Livingston and Ray Evans.

Lewis A. "Lew" Hoad, 21 (Australia), wins in men's singles at Wimbledon, Shirley Fry, 29 (U.S.), in women's singles; Kenneth R. "Ken" Rosewall, 21 (Australia), wins in men's singles at Forest Hills, Fry in women's singles.

The Olympic games at Melbourne attract 3,539 contestants from 67 nations. Soviet athletes win the most medals.

Brazilian soccer player Edson Arantes "Pele" do Nascimento, 15, signs with the Santos team to begin an 18-year career of 1,253 games in which he will score 1,216 goals to become the world-famous athletic figure known in Brazil as Perola Negra (Black Pearl).

The New York Yankees win the World Series by defeating the Brooklyn Dodgers 4 games to 3 after losing last year's Series to the Dodgers by the same one-game margin. Don Larsen of the Yankees pitches the first "perfect" game in Series history, allowing no hits in the fifth game (*see* Cy Young, 1904).

U.S. prizefighter Floyd Patterson, 21, wins the world heavyweight title November 30 by knocking out Archie Moore, 42, in the fifth round of a championship match at Chicago.

"Does she or doesn't she?" ask advertisements for "Miss Clairol" hair coloring that show children with mothers to convey the idea that hair-coloring is not just for women of dubious virtue (*see* 1950).

Comet, a silica-sand cleanser introduced by Procter & Gamble to compete with Colgate's Ajax (*see* 1947), will soon be the top seller.

Salem cigarettes, introduced by R. J. Reynolds, will overtake Brown & Williamson's Kools and be the leading mentholated brand within 18 months.

Britain's Clean Air Act begins a systematic ban on burning of soft coal and other smoky fuels (*see* 1955).

Georgia's Buford Dam, completed on the Chattahoochee River, creates a reservoir and makes Lake Lanier a recreation area with 540 miles of shoreline.

Virgin Islands National Park, established by Congress on 15,150 acres donated by Laurence Rockefeller, embraces most of the island of St. John.

Rock paintings found 900 miles southeast of Algiers establish that the Sahara was once fertile land. French archaeologist Henri Lhote, 53, dates the paintings in the Tibesti Mountains to 3,500 B.C.

1956 ***(cont.)*** Some 100 million Chinese peasant families are forced into large collective farms called Agricultural Producers' Cooperatives.

Poland ends efforts to collectivize agriculture after a decade marked by little success. Wladyslaw Gomulka (above) reverses collectivization policies.

President Eisenhower proposes a soil-bank plan to ease the problem of declining U.S. farm income in a special January 9 message to Congress. Congress votes May 28 to authorize a program to pay farmers for withdrawing land from production (*see* AAA, 1938).

India receives a $300 million food loan August 29 to buy surplus U.S. farm products that are depressing domestic American prices (*see* P.L. 480, 1954).

Banana production shifts to Ecuador, which was not a major exporter before the war but rarely has high winds, provides good growing conditions, and will become the world's leading exporter within a decade, accounting for half the bananas shipped to the United States.

La Leche League International, Inc., is founded with headquarters at Franklin Park, Ill., to "help mothers who wish to breast-feed their infants" and encourage breast feeding.

Welch's Grape Juice has sales of $40 million for the year. National Grape Cooperative Association acquires the company from Jacob Kaplan; its 4,500 member grape-growers become the first farmers in history to own an international business (*see* 1945).

U.S. canners ship 700 million cases of food, up from 400 million in 1940.

Busch Bavarian beer, introduced by Anheuser-Busch, augments its 80-year-old Budweiser and 60-year-old Michelob brands.

1957 Britain's discredited prime minister Sir Anthony Eden resigns January 9. (Maurice) Harold Macmillan, 62, of the publishing family heads a new Cabinet and will be PM until 1963.

Premier Bulganin of the U.S.S.R. and Premier Zhou En-lai of the People's Republic announce January 18 that they will support the Mideast against Western aggression.

President Eisenhower begins a second term by extending the Truman Doctrine of 1947 to the Mideast. The president enunciates the Eisenhower Doctrine for protecting the Mideast from Communist aggression, sends aid to Jordan, and asks Congress to authorize him to send U.S. armed forces to the Mideast if necessary.

Israel rejects a February 2 UN resolution calling upon her to withdraw from Egypt's Gaza Strip and other occupied Egyptian territory unless she receives more UN assurance that her own territory will be protected. Washington indicates that economic sanctions will be applied against Israel if the UN requires such action but assures Tel Aviv that it will support "free passage" through the Gulf of Aqaba.

Israel announces withdrawal of her troops from the Gaza Strip and the Gulf of Aqaba area March 1 on the "assumption" that the UN Emergency Force will administer the Gaza Strip and that navigation in the Gulf of Aqaba will continue.

Prime Minister Macmillan and President Eisenhower meet in the Bermuda Conference from March 21 to 24, reestablishing the relationship strained by last year's Suez crisis. The United States agrees to make certain guided missiles available to Britain.

West German nuclear physicists announce April 12 that they will not cooperate in producing or testing nuclear weapons, Japan sends Moscow a note April 20 protesting Soviet nuclear tests, and the British explode their first thermonuclear bomb in the megaton range May 15 at Christmas Island in the Pacific.

The National Committee for a Sane Nuclear Policy (SANE) is started by U.S. founders who include psychoanalyst Erich Fromm.

Ghana becomes the first African state south of the Sahara to attain independence. The newly independent Gold Coast unites with the UN trust territory of British Togoland March 6, Ghana takes that name from the Sudanic empire that flourished between the 4th and 10th centuries, and Kwame N. Nkrumah begins a 15-year rule.

The Aga Khan dies July 11 at age 79 after 73 years as spiritual and temporal leader of the Ismailis whose numbers have grown to some 20 million. The immensely rich Aga Sultan Sir Mohammed Shah is succeeded by his grandson Shah Karim, 21, a Harvard student who becomes Aga Khan IV.

Algerian FLN terrorists disrupt France but President René Coty restates the government's refusal to grant independence (*see* 1954). The French Parlement votes Premier Bourges-Mannoury special powers July 18 to suppress the FLN (*see* de Gaulle, 1958).

Tunisia deposes the bey of Tunis July 25, proclaims herself a republic, and elects Habib Bourguiba president (*see* 1956). President Bourguiba asks for U.S. aid September 12 following border clashes with French and Algerian troops, Washington and London agree November 14 to supply small arms, and Tunisia announces November 18 that she has rejected a Soviet arms offer.

Morocco's Sherifian Empire becomes the Kingdom of Morocco August 11, 5 months after signing a treaty of friendship and alliance with Tunisia (above). The sultan Sidi Mohammed II becomes King Mohammed V and will reign until his death in 1961.

Syria ousts three U.S. embassy officials August 13 on charges of plotting to overthrow President Shukri al-Kuwatly. Washington asks Syrian embassy officials to leave, President Eisenhower charges Moscow with trying to take over Syria, and he reaffirms the Eisenhower Doctrine September 21. The Syrian president meets at Cairo with President Nasser, he meets in Syria September 25 with the premier of Iraq and the king of Saudi Arabia, and the meetings bolster Arab solidarity.

Norway's Haakon VII dies September 21 at age 85 after a 52-year reign and is succeeded by his son, 54, who will reign until 1991 as Olaf V. Norway's Labor party is returned to power October 7.

Moscow has complained to Istanbul September 11 of Turkish troop concentrations on the Syrian border, Nikita Khrushchev writes October 12 to British and European labor and socialist parties urging them to try to stop U.S. and Turkish aggression in the Mideast, Syria declares a state of emergency October 16, and U.S. secretary of state John F. Dulles warns Moscow against an attack on Turkey.

Moscow relieves Marshal Zhukov of his duties October 26 and exiles Vyacheslav Molotov to Siberia for conspiring against Nikita Krushchev.

Ku Klux Klansmen accuse Alabama grocery-chain truck driver Willie Edwards, 25, of having made remarks to a white woman and force him at pistol point January 23 to jump to his death from the Tyler Goodwin Bridge into the Alabama River. His body is found downriver in Lowndes County in late April. Since January 23 was his first day on the truck route, it will be demonstrated that Edwards was not the man sought by the Klan.

Little Rock, Ark., has a school integration struggle as officials try to implement the *Brown v. Board of Education* decision by the Supreme Court in 1954. Jeering crowds led by Arkansas governor Orval Faubus prevent nine black children from entering Central High School in September; President Eisenhower sends in federal troops to quell the disorder (*see* 1958).

The first U.S. civil rights bill since Civil War reconstruction days, passed by Congress September 9, establishes a Civil Rights Commission and provides federal safeguards for voting rights. Many Southerners opposed the bill (Sen. Strom Thurmond of South Carolina spoke against it continuously for 24 hours and 18 minutes August 29, a new filibuster record) (*see* 1964).

Britain's *Report of the Committee on Homosexual Offences and Prostitution*, issued in September, recommends an end to punitive laws against homosexuality "between consenting adults in private." Reading University chancellor Sir John Wolfenden, 51, heads the committee; the Wolfenden Report also calls for an end to the requirement that "annoyance" be established before a constable may book a prostitute for soliciting a customer above the age of 21, but the recommendation on homosexuality is what creates a controversy. The Church of England supports the recommendations (*see* Street Offences Act, 1959).

A Great Leap Forward launched by Mao Zedong in the People's Republic of China puts more than half a billion peasants into 24,000 "people's communes." The people are guaranteed food, clothing, shelter, and child care, but deprived of all private property.

Sputnik I, launched by the Soviet Union October 4, is the world's first manmade Earth satellite. The 184-pound sphere orbits the Earth once every 90 minutes in an elliptical orbit and is followed in November by Sputnik II, which weighs more than 1,000 pounds and carries a live dog, but *New York Times* correspondent Harrison Salisbury in Moscow hears a Russian say, "Better to learn to feed your people at home before starting to explore the moon."

Moscow and Paris have signed a trade pact February 11.

The Treaty of Rome signed March 25 establishes a European Economic Community (the Common Market). Belgium, France, West Germany, Italy, Luxembourg, and the Netherlands remove mutual tariff barriers to promote the economy of Europe and make it a viable competitor with Britain and the United States.

Britain relaxes restrictions on trade with the People's Republic of China May 30, British tourists traveling to dollar areas receive an allowance increase June 4, and Britain decontrols most rents June 6 despite opposition from the Labour party.

A University of Wisconsin study shows that 20 percent of Americans still live below the "poverty line" (*see* 1959).

Sen. Estes Kefauver investigates the effect on consumers of increasing mergers by U.S. auto and steel makers, bread bakers, and pharmaceutical firms.

E. I. du Pont's 23 percent ownership of General Motors stock creates conditions that violate the antitrust laws, the Supreme Court rules June 3. Du Pont is ordered to divest itself of its GM stock (*see* 1915; 1929).

Union Carbide is created by a shortening of the name Union Carbide and Carbon (*see* 1920). The company has some 400 U.S. and Canadian plants plus affiliated companies.

Fried. Krupp agrees to supply the Soviet Union with a chemical plant and a synthetic fiber complex. The German colossus has diversified from steel making.

Southdale at Minneapolis is completed by Victor Gruen whose firm will develop auto-free environmental plans for Rochester and other U.S. cities. Southdale

The Common Market eliminated trade barriers that had put European companies at a commercial disadvantage.

1957 ***(cont.)*** is a covered, heated, air-conditioned shopping center with 75 shops (*see* Northland, 1954).

A treaty signed at Rome March 25 creates the European Atomic Energy Community (Euratom) (*see* 1958).

The Wankel rotary engine produced at Lindau on Lake Constance by German engineer Fritz Wankel, 55, is the first new internal combustion engine since the Benz and Daimler engines of the 1880s and the 1895 diesel engine. Wankel replaces the up-down movement of oblong pistons in cylinders with a triangular disk that rotates inside a round cylinder. The slightly curved sides of the disk leave some space for the moving and expanding gases on at least two of its three sides, and the rotating disk opens and closes intake and exhaust valves automatically.

The Wankel engine (above) has only two moving parts and while it consumes slightly more fuel than conventional engines, it is smaller, weighs 25 percent less, works more smoothly, and costs less to build in mass production. The German NSU will be the first automobile to employ the new engine, but Toyo Kogyo in Japan will use a Wankel engine to power its Mazda beginning in 1968 as foreign auto makers rush to obtain licenses to use the rotary engine that will be employed also in pumps and other nonautomotive applications.

Armand Hammer comes out of retirement at age 60 to head the failing Occidental Petroleum Co. in which he has invested $60,000 "for fun" after being told by his tax counsel to find a money-losing tax shelter. Occidental sold last year at 18¢ a share on the Los Angeles Stock Exchange, its net worth was $34,000, but Hammer will build its revenues from $1 million per year to more than $1 billion, and he will make it the 11th largest U.S. oil producer and the 20th largest U.S. firm of any kind (*see* Hammer, 1921).

Three U.S. Air Force jets complete a nonstop round-the-world flight January 18 having averaged more than 500 miles per hour.

U.S. Marine Corps pilot Major John Herschel Glenn, 36, sets a new transcontinental speed record of 3 hours, 20 minutes, 8.4 seconds across the United States (*see* 1949; earth orbits, 1962).

Diesel power on U.S. railroads eclipses steam power. The last Atchison, Topeka and Santa Fe steam locomotive is retired from service August 27 after 88 years of Santa Fe steam operation (*see* 1941).

New York's last trolley car is retired from service across the Queensboro Bridge as motorbuses replace streetcars in most major U.S. cities (*see* General Motors, 1956).

New York makes its Avenue of the Americas (Sixth Avenue) one-way northbound despite objections from the Fifth Avenue Coach Co. and the Transport Workers Union (*see* 1956; 1960).

Michigan's Mackinac Straits Bridge opens at Mackinaw City. The $100 million 3,691-foot span is the world's longest suspension bridge.

The Edsel, introduced by Ford Motor Company to compete with General Motors' Oldsmobile, will be a colossal failure. Ford has invested $250 million to develop a new line of cars; its failure will raise doubts about the effectiveness of market research.

Detroit labor leader Jimmy (James Riddle) Hoffa, 44, gains control of the International Brotherhood of Teamsters which is ousted from the AFL-CIO on charges of corruption. Hoffa organized his fellow Kroger grocery chain warehousemen when he was 17, won a brief strike, obtained an AF of L charter for his local the following year, and will make the Teamsters one of the country's most powerful labor unions (*see* 1964).

Control Data Corp. is organized to produce computers designed by Minneapolis engineer-mathematician Seymour R. Cray, 32, whose Model 1604 will be one of the first transistorized computers and whose Model 6600 will be the first commercially successful "super" computer (*see* transistor, 1948). The 6600 will be a mainstay of atomic energy laboratories and other scientific research labs.

Some 1,000 electronic computers are shipped to U.S. and foreign customers, up from 20 in 1954 (*see* 1960).

An electric watch introduced by Hamilton Watch Co. uses a conventional balance wheel powered by a tiny battery (*see* Accutron, 1960).

Synthetic DNA, produced by Stanford University biochemist Arthur Kornberg, 39, and his associates, is identical in chemical and physical structure with the actual genetic material but biologically inert (*see* 1961; 1966; Watson, Crick, 1953; RNA, 1955).

Congress funds a National Cancer Institute to seek cures for the disease that is second only to heart disease as a cause of death in the United States.

The antibiotic rifampia that will be used successfully to cure "incurable" tuberculosis is developed in Italy.

Eli Lilly introduces Darvon (propoxyphene) as an alternative to the opiate codeine. Chemically related to the drug methadone discovered in 1943, the new pain killer will be prescribed by physicians as a nonaddictive alternative to the opiate codeine, will be one of the most widely prescribed drugs although it costs 10 times as much as aspirin, is no more effective for most uses, and can be addictive.

A Senate investigation led by Estes Kefauver (above) finds the pharmaceutical industry peculiarly vulnerable to monopolistic control. World War II injuries have spurred development of miracle drugs which were mostly unknown 20 years ago, and Kefauver says the industry "is charging all the traffic will bear in selling its new drugs," making profits of roughly 20 percent of total investment, double the average for other manufacturers.

Prednisone has a captive market of arthritis sufferers, says Sen. Kefauver (above), and the 2-year-old drug fetches 30¢ per tablet while small, independent laboratories sell it in wholesale lots for as little as 2¢. Large pharmaceutical firms point out that their development costs are enormous (*see* Kennedy, 1962).

A high-speed (350,000 rpm) dental drill devised by Washington, D.C., dentist John V. Borden reduces patient time (and pain) but has problems. Refined, it will come into common use by 1971.

The S.I. Newhouse newspaper chain acquires Condé Nast Publications (*see Vogue*, 1909; Newhouse, 1912, 1967).

Nonfiction *Memoirs of a Catholic Girlhood* by Mary McCarthy; *Where Did You Go? Out. What Did You Do? Nothing* by U.S. author Robert Paul Smith.

Fiction *Doktor Zhivago* by Boris Pasternak; *Arturo's Island (L'Isola di Arturo)* by Italian novelist Elsa Morante, 39, wife of Alberto Moravia; *Pnin* by Vladimir Nabokov; *Jealousy (La jalousie)* by Alain Robbe-Grillet; *The Wind (Le Vent)* by French novelist Claude Simon, 44; *On the Beach* by Nevil Shute who foresees the last days of a world destroyed by nuclear holocaust; *A Choice of Enemies* by Canadian novelist Mordecai Richler, 26; *The Assistant* by Bernard Malamud; *That Awful Mess on Via Merulana (Que Pasticciaccio Brutto de Via Merulana)* by Italian novelist Carlo Emilio Gadda, 64; *The Fountain Overflows* by Rebecca West; *At Lady Molly's* by Anthony Powell; *Gimpel the Fool* (stories) by Polish-American Yiddish novelist Isaac Bashevis Singer, 53; *O Pays, Mon Beau Peuple* by Senegalese novelist Sembene Ousmane, 34; *The Wapshot Chronicle* by New England novelist John Cheever, 45; *From Russia with Love* by Ian Fleming; *On the Road* by U.S. novelist Jack (Jean-Louis) Kerouac, 35, who wrote it in 3 weeks in 1951; *Atlas Shrugged* by Ayn Rand.

Poetry *Opus Posthumous* by Wallace Stevens who died in early August 1955 at age 75; "Sunstone" ("Piedra del Sol") by Mexican poet Octavio Paz, 43; *Collected Poems* by the late Scottish poet Norman Cameron who died in 1953 at age 48.

Juvenile *How the Grinch Stole Christmas* by Dr. Seuss; *The Cat in the Hat* by Dr. Seuss begins a series of "Beginner Books," "Bright and Early Books," and "Big Beginner Books." His new book will be translated into many foreign languages and will have sales of between 8 and 9 million in the next 20 years; *Little Bear* by Danish-American author Else Holmelund Minarik will be translated into 17 foreign languages including Urdu and Bengali, will have U.S. sales of 1 million in the next 20 years, and begins a series of "I Can Read Books."

Painting *Betelgeuse* by Hungarian-American painter Victor Vasarely, 49; *Berkeley #26* by U.S. painter Richard Diebenkorn, 35; *Black on Black* by Ad Reinhardt, who mixes subtle tones of olive, violet, and other deep colors in his blacks (he promulgates rules for what the artist should avoid, including texture, brushwork, drawing, forms, design, colors, light, space, time, size and scale, movement, and objects and symbols); *Achromes* (gesso and kaolin on canvas) by Piero Manzoni; *Painting with Red Letter 'S'* by Robert Rauschenberg; *New York, N.Y.* by U.S. painter Ellsworth Kelly, 34; *1975D No. 1* by Clyfford Still; *Annette* by Alberto Giacometti; *Mujeres peinandose* and *Sandias* by Diego Rivera who dies at Mexico City November 25 at age 70.

Sculptor Constantin Brancusi dies at Paris March 16 at age 81.

Theater *The Potting Shed* by novelist Graham Greene 1/29 at New York's Bijou Theater, with Sybil Thorndike, Leueen MacGrath, Carol Lynley, Frank Conroy, 157 perfs.; *Orpheus Descending* by Tennessee Williams 3/21 at New York's Martin Beck Theater, with Maureen Stapleton, Cliff Robertson, 68 perfs.; *Endgame* by Samuel Beckett 4/3 at London's Royal Court Theatre; *The Entertainer* by John Osborne 4/10 at London's Royal Court Theatre, with music by John Addison; *A Moon for the Misbegotten* by the late Eugene O'Neill 5/2 at New York's Bijou Theater, with Wendy Hiller, Franchot Tone, Cyril Cusack, 68 perfs.; *The Rope Dancers* by Morton Wishengrad 11/20 at New York's Cort Theater, with Siobhan McKenna, Art Carney, Theodore Bikel, 189 perfs.; *Look Homeward, Angel* 11/28 at New York's Ethel Barrymore Theater, with Jo Van Fleet, Anthony Perkins, Arthur Hill is based on the 1929 Thomas Wolfe novel, 564 perfs.; *The Dark at the Top of the Stairs* by William Inge 12/5 at New York's Music Box Theater, with Pat Hingle, Teresa Wright, Eileen Heckart, 468 perfs.

Films David Lean's *Bridge on the River Kwai* with William Holden, Alec Guinness, Jack Hawkins, Sessue Hayakawa; Martin Ritt's *Edge of the City* with John Cassavetes, Sidney Poitier, Jack Warden, Ruby Dee; Federico Fellini's *Nights of Cabiria* with Giulietta Massina; Stanley Kubrick's *Paths Of Glory* with Kirk Douglas, Ralph Meeker; Akira Kurosawa's *Throne of Blood* with Toshiro Mifune; Sidney Lumet's *12 Angry Men* with Henry Fonda, Lee J. Cobb; Ingmar Bergman's *Wild Strawberries* with Victor Sjostrom, Ingrid Thulin, Bibi Andersson; Billy Wilder's *Witness for the Prosecution* with Marlene Dietrich, Tyrone Power, Charles Laughton, Elsa Lanchester. Also: Mikhail Kalatozov's *The Cranes Are Flying* with Tatyana Samoilova, Alexei Batalov; Walter Lang's *Desk Set* with Spencer Tracy, Katharine Hepburn; Elia Kazan's *A Face in the Crowd* with Andy Griffith, Patricia Neal, Lee Remick, Walter Matthau; Helmut Kautner's *The Last Bridge* with Maria Schell; Joseph Pevney's *Man of a Thousand Faces* with James Cagney, Dorothy Malone, Jane Greer; Mark Robson's *Peyton Place* with Lana Turner, Hope Lange, Arthur Kennedy; Joshua Logan's *Sayonara* with Marlon Brando, Ricardo Montalban, Miiko Taka, Miyoshi Umeki; Nunnally Johnson's *The Three Faces of Eve* with Joanne Woodward, David Wayne, Lee J. Cobb.

Hollywood musicals Billy Wilder's *Love in the Afternoon* with Gary Cooper, Audrey Hepburn, Maurice Chevalier, songs that include "Fascination" by F. D. Marchetti and Maurice de Feraudy and "C'est Si Bon" by Henri Betti and André Hornez; George Cukor's *Les Girls* with Gene Kelly, Mitzi Gaynor, Taina Elg; Stanley Donen's *Funny Face* with Fred Astaire, Audrey Hepburn, Kay Thompson, model Suzy Parker; George

1957 ***(cont.)*** Abbott and Stanley Donen's *The Pajama Game* with Doris Day, John Raitt, Carol Haney.

Broadway musicals *The Ziegfeld Follies* 3/11 at the Winter Garden Theater, with Beatrice Lillie, Billie de Wolfe, Jane Morgan in the 24th and final edition of the *Follies* 50 years after its first opening. Music by Sammy Fain, Jack Lawrence, Michael Myers, and others, lyrics by Howard Dietz, Carolyn Leigh, and others, 123 perfs.; *New Girl in Town* 5/14 at the 46th Street Theater, with Gwen Verdon, Thelma Ritter in a musical version of the 1921 Eugene O'Neill play *Anna Christie*, music and lyrics by Bob Merrill, songs that include "Sunshine Girl," 431 perfs.; *West Side Story* 9/26 at the Winter Garden Theater, with a book based on the 1595 Shakespeare tragedy *Romeo and Juliet*, music by Leonard Bernstein, lyrics by Stephen Sondheim, 27; songs that include "Tonight," "Maria," 732 perfs.; *Jamaica* 10/31 at the Imperial Theater with Lena Horne, Ricardo Montalban, Ossie Davis, music by Harold Arlen, lyrics by E. Y. Harburg, 558 perfs.; *The Music Man* 12/19 at the Majestic Theater, with Robert Pres-ton, music and lyrics by Meredith Willson, songs that include "Seventy-Six Trombones," "Gary, Indiana," "Till There Was You," 1,375 perfs.

Opera *Dialogues des Carmelites* 1/26 at Milan's Teatro alia Scala, with music by Francis Poulenc, libretto from the novel and play by the late Georges Bernanos; *Die Harmonie der Welt* 8/11 at Munich, with music by Paul Hindemith (*see* Kepler, 1619).

Ballet *The Prince and the Pagodas* 1/1 at London, with music by Benjamin Britten; *Agon* 6/17 at Los Angeles, with music by Igor Stravinsky, now 75.

First performances *Declaration* symphonic variations by Morton Gould 1/20 at Washington's Constitution Hall; Concerto for Cello by William Walton 1/25 at Boston's Symphony Hall; Jekyll and Hyde Variations by Gould 2/2 at New York's Carnegie Hall; Symphony No. 2 by the late Charles Ives (who wrote it between 1897 and 1901) 2/22 at New York's Carnegie Hall; *Lydian Ode* by Paul Creston 2/24 at Wichita; Symphony No. 6 by David Diamond 3/8 at Boston's Symphony Hall; Symphony No. 5 by Arthur Honegger 3/9 at Boston's Symphony Hall; Symphony No. 4 by Walter Piston 3/30 at Minneapolis; Symphony No. 10 by Heitor Villa-Lobos 4/4 at Paris; *Song of Democracy* by Howard Hanson 4/9 at Washington, D.C.; Concerto No. 2 for Piano and Orchestra by Dmitri Shostakovich 5/10 at Moscow, with Maxim Shostakovich as soloist; *Monopartita* by Honegger 6/12 at Zurich; *Stabat Mater* by Francis Poulenc 6/13 at Strasbourg; Polyphonie X for 17 Solo Instruments by French composer Pierre Boulez, 26, 10/6 at the Donaueschippu Festival of Contemporary Music; Toccata for Orchestra by Creston 10/18 at Cleveland's Severance Hall; *Apocalypse* (Symphonic Poem) by Gian Carlo Menotti 10/19 at Pittsburgh; Symphony No. 11 *(1905)* by Shostakovich 10/30 at Moscow; *Erosion, or The Origin of the Amazon River* (Symphonic Poem) by Villa-Lobos 11/7 at Louisville; Symphony No. 3 by Roger Sessions 12/26 at Boston's Symphony Hall.

Popular songs "All the Way" by Jimmy Van Heusen, lyrics by Sammy Cahn (for the film *The Joker Is Wild* with Frank Sinatra); "April Love" by Sammy Fain, lyrics by Paul Francis Webster (title song for film); "Bye Bye, Love" by Felice and Boudleaux Bryant; "Young Love" by Carole Joyner and Ric Cartey; "Tammy" by Jay Livingston and Ray Evans (for the film *Tammy and the Bachelor);* "School Day" by Chuck Berry; "A White Sport Coat and a Pink Carnation" by Marty Robbins; "Jailhouse Rock" by Jerry Leiber and Mike Stoller; "Old Cape Cod" by Claire Hothrock, Milt Yakers, and Allan Jeffrey.

Motown Corp. is founded by entrepreneur Berry Gordy, Jr., 30, who invests $700 to start a recording company whose "Motown Sound" will figure large in popular music for more than 2 decades.

Middleweight boxing champion Sugar Ray Robinson, now 36, loses the title January 2 to Gene Fullmer, regains it from Fullmer May 2, and loses it again September 23 to Carmen Basilio.

Lew Hoad wins in men's singles at Wimbledon, Althea Gibson, 29 (U.S.), in women's singles (the first black American to be invited); Malcolm James "Mal" Anderson, 22, at Forest Hills, Gibson in women's singles.

The Milwaukee Braves win the World Series by defeating the New York Yankees 4 games to 3.

The Frisbee is introduced by the Wham-O Manufacturing Co. of San Gabriel, Calif., whose public relations man has found Yale students tossing metal pie tins produced since 1871 by Frisbee Pie Co. of Bridgeport, Conn. The plastic Frisbee launches a new sport whose first world champion will be crowned in 1968.

A .32 caliber bullet grazes New York mobster Frank Costello in the head May 2 in the lobby of the Majestic apartment house where he lives. His would-be assassin Vincent (the Chin) Gigante escapes, Costello rushes to the emergency room of Roosevelt Hospital, he is not badly hurt, but police search his pockets and find a slip showing more than $1 million in winnings from his gambling casinos.

Mafia lieutenant Frank (Don Cheech) Scalice is shot four times in the head June 17 while shopping for peaches at a Bronx fruit stand. Scalice is a henchman of Murder, Inc., boss Albert Anastasia.

Albert Anastasia is killed October 25 in the barbershop of New York's Park-Sheraton Hotel by two gunmen reputedly in the pay of Vito Genovese.

Apalachin, N.Y., makes headlines when police find 58 Mafia members from all over the United States gathered November 14 at the home of Joseph Barbara. Included are Vito Genovese, Joseph Bonanno, Carlo Gambino, Joseph Profaci, Jerry Catena, and Mike Miranda.

George "Bugs" Moran dies of lung cancer while serving a 10-year prison term for bank robbery.

Florida's $17 million Americana Hotel at Bal Harbour, designed by Morris Lapidus, is completed by hotel-men Laurence Tisch, 33, and his brother Robert

Preston, 30, who started in 1946 with $124,000 in capital, have operated two hotels in Atlantic City plus one in New York, and will create an empire of hotels, movie theaters, tobacco, and broadcasting (*see* Loew's, 1959).

Los Angeles adopts a revised building code that permits construction of high-rise buildings. The code reflects earthquake-stress engineering technology.

China's Great Leap Forward (above) virtually eliminates houseflies, mosquitoes, rats, and bedbugs over wide areas in a prodigious outburst of human energy.

The world's longest river dam is completed on India's Mahanadi River. The 83,523-foot Hirakud Dam stretches for nearly 16 miles.

African bees imported into Brazil escape from a breeding experiment and begin heading north at a rate of about 200 miles per year. Stings of aggressive "killer" bees will kill hundreds of people in the next 33 years.

Chinese cereal grain production rises to 200 million tons after having increased in the past 7 years at an annual rate of 8 percent while the rest of the world has never exceeded a rate of 3 percent.

Huge irrigation projects add 100 million acres of irrigated cropland to China's agricultural resource and give the nation 60 times as much irrigated cropland as Europe. The Chinese work to bring under control the Huanghe (Yellow) River that has for centuries overrun its banks 2 years out of 3 to create floods that have produced havoc, famine, and desolation.

U.S. per capita margarine consumption overtakes butter consumption for the first time. The average American uses 8.6 pounds of margarine per year versus 8.3 pounds of butter (*see* 1960).

A new process mixes high-protein soy flour with an alkaline liquid to create a spinning solution which is fed under pressure into spinning machines (*see* 1949). Food company engineers will develop improvements on this spinning technique, others will adapt extrusion methods from the plastics industry to develop new meat analogs from soy protein isolate (*see* Bac*Os, 1966).

The United States has a record 4.3 million births.

Gregory Pincus and Boston gynecologist John Rock, 67, begin an intensive trial of birth-control pills to prevent unwanted births in Puerto Rico (*see* 1955; Enovid 10, 1960).

Seventy-one world cities have populations of more than 1 million, up from 10 in 1914.

1958

Nikita Krushchev replaces Nikolai Bulganin as chairman of the Soviet Council of Ministers March 27, he visits Beijing in August, and Bulganin is dismissed from the Communist Party Presidium September 6.

A United Arab Republic has been proclaimed February 1 in a union of Egypt and the Sudan with Syria under the leadership of Egypt's President Nasser.

The Arab Federation created February 14 joins Iraq and Jordan under Iraq's Faisal II, but a coup engineered by Gen. Abdul Karim al-Kassem assassinates Faisal July 14 along with his infant son and his premier, Gen. Karim proclaims a republic, Jordan's Hussein succeeds Faisal as head of state, but he dissolves the Arab Federation August 1.

Lebanon is torn by riots allegedly provoked by the United Arab Republic (above), UN observers move in to guard against illegal movement of troops or arms into the country, U.S. troops from the 6th Fleet land near Beirut beginning July 15 as President Eisenhower vows to protect U.S. lives and property and to defend Lebanese sovereignty and independence, a new government comes to power in Lebanon, and U.S. troops are withdrawn October 25.

France names Gen. de Gaulle premier May 31 after an Algerian crisis has threatened civil war. Agents of the Algerian rebels commit acts of violence within France as the open revolt that began in 1954 continues under the direction of French military leaders who have seized control of Algeria.

Paris withdraws French troops from all of Tunisia except for Bizerte June 17 after months of conflict in which French planes have bombed the Tunisian village of Sakiet-Sidi-Youssef, killing 79. Tunisia joins the Arab League October 1.

France approves a Fifth Republic by a vote of more than four to one in a popular referendum September 28, the Gaullist Union gains control of the French assembly in the November elections, Gen. de Gaulle is named president of the republic for a 7-year term to begin January 8 of next year, and France gives her overseas possessions 6 months to decide whether to become departments of the republic or to become autonomous members of a French Community (*see* 1959).

Guinea becomes an independent republic October 2 with Sékou Touré as president following a popular referendum in which the people have rejected membership in the French Community (above), the only French colony in Africa to take such action.

The Malagasy Republic (Madagascar) becomes an autonomous state within the French Community (above) October 14, Senegal gains autonomy November 25, Gabon, the new Republic of the Congo (Middle Congo), Mauritania, and Mali (French Sudan) November 28, the new Central African Republic (Ubangi Shari) December 1, Dahomey and Ivory Coast December 4.

The Federation of the West Indies has come into being January 3 with 3 million people scattered over 77,000 square miles that embrace the British colonies Antigua, Barbados, Dominica, Grenada, Jamaica, Montserrat, St. Kitts-Nevis-Anguilla, St. Lucia, St. Vincent, Tobago, and Trinidad.

A symbol for total nuclear disarmament introduced in an Easter march at Aldermaston, England, by English philosopher Bertrand Russell, now 86, will become a universal peace symbol, but Washington rejects a plan for a denuclearized zone in Central Europe in April and Britain follows suit in May.

Peace symbols mushroomed amidst rattling of atomic swords and anxieties about a war that could destroy mankind.

President Eisenhower proposes mutual inspection to enforce an atomic test ban April 8.

Ottawa and Washington establish the North American Air Defense Command May 12.

Hungary's Communist regime executes former premier Imre Nagy in mid-June after a secret trial (*see* 1956).

An Anglo-U.S. agreement to cooperate in the development of nuclear weapons is signed July 3.

The Soviet Union resumes nuclear testing September 30 after having suspended such tests in June, but a voluntary moratorium on nuclear weapons testing takes effect November 4 with London, Moscow, and Washington in accord (*see* France, 1960).

The John Birch Society is founded by U.S. right-wing candy manufacturer Robert Welch, 58, at Belmont, Mass., who names his political group to honor a U.S. Army officer who was killed in China by Communists 10 days after the end of World War II while working as an intelligence operative for the Office of Strategic Services (OSS). President Eisenhower is a conscious agent of the international Communist conspiracy, says Robert Welch. The John Birch Society will publish the monthly magazine *American Opinion*, support more than 400 American Opinion bookstores, and grow to have more than 60,000 members.

Moroccan women gain the right to choose their own husbands. Rabat restricts polygamy.

Britain's First Offenders Act prohibits magistrates from imprisoning any adult first offender if there is a more appropriate way of dealing with the offense. Youth crimes have increased sharply since 1955.

The U.S. Civil Rights Commission swears in six members January 3 and begins operations.

An employee of the Bethel Baptist Church at Birmingham, Ala., finds a dynamite bomb beside the church June 29 and moves it to an open area where it explodes without injuring anyone. Pastor Fred L. Shuttlesworth is a civil rights leader and authorities will arrest a Ku Klux Klan leader, now 34, in 1977 on charges of trying to blow up the black church.

The Supreme Court meeting in special term rules unanimously September 29 that schools at Little Rock, Ark., must integrate according to schedule (*see* 1957). The Court's decision in *Cooper v. Aaron* implements its ruling in the 1954 case of *Brown v. Board of Education*.

U.S. unemployment reaches a postwar high of more than 5.1 million, and the Department of Labor reports that a record 3.1 million Americans are receiving unemployment insurance benefits. Economic recession grips the nation with nearly one-third of major industrial centers classified as having "substantial" unemployment.

The upper 1 percent of Americans enjoys nearly 9 percent of the nation's total disposable income, down from 19 percent of disposable income in 1929. Sixty-four percent of households have incomes above $4,000 per year, up from 40 percent in 1929, but white families average twice as much as nonwhite families.

The median U.S. family income is $5,087, up from $3,187 in 1948 (half of all families have incomes below the median, half above), but prices have climbed along with incomes.

A house that cost $47,409 in 1948 sells for $59,558, a family size Chevrolet that sold for $1,255 sells for $2,081, a gallon of gasoline has climbed from 25.9¢ to 30.4¢, a pair of blue jeans that sold for $3.45 sells for $3.75, a pair of men's shoes that was $9.95 is now $11.95, a daily newspaper that cost 3¢ now costs 5¢, a year's tuition at Harvard that cost $455 costs $1,250, a hospital room that cost $13.09 per day costs $28.17, a pound of round steak that cost 90.5¢ costs $1.04, a Nathan's hot dog that cost 20¢ costs 25¢, a ticket to a Broadway musical that cost $6.00 costs $8.05.

Some prices have come down: a ranch mink coat that cost $4,200 in 1948 costs $4,000, a roundtrip flight to London from New York that cost $630 costs $453.60, a phone call from New York to Topeka, Kan., that cost $1.90 costs $1.80 (day rate), a pound of chicken that cost 61.2¢ has come down to 46.5¢.

The Affluent Society by Harvard economics professor John Kenneth Galbraith, 49, decries the overemphasis on consumer goods in the U.S. economy and the use of advertising to create artificial demand for such goods. More of the nation's wealth should be allocated to public purposes, says Galbraith.

Visa has its beginnings in the BankAmericard introduced by California's Bank of America. It costs the consumer nothing and relieves subscribing merchants of credit worries in return for a small percentage of each retail sale charged to the card. Card users must pay 18 percent interest on unpaid balances and 12 percent on cash advances obtained by presenting the card at teller windows (*see* 1966).

The American Express Card, introduced by American Express Co., requires an annual fee from "members" who use the card to charge air fares, auto rentals, hotel and motel rooms, restaurant meals, and other expenses. American Express sets out to overtake Diners Club, which started the travel and entertainment card business in 1950 (*see* travelers cheques, 1891).

"The torrent of foreign oil robs Texas of her oil market" and costs the state $1 million per day, says the chairman of the Texas Railroad Commission which controls production in the state. The Commission reduces Texas oil wells to 8 producing days per month (*see* 1959).

Indonesia's President Sukarno seizes Royal Dutch-Shell concessions February 3 in an anti-Dutch campaign as the Netherlands retains Western New Guinea. Dutch refugees pour out of the country.

The first U.S. atomic power station for peaceful purposes is dedicated May 26.

The European Atomic Energy Community (Euratom), consisting of Belgium, France, Italy, Luxembourg, the Netherlands, and West Germany, receives a U.S. pledge November 8 to supply uranium for 20 years and extend a loan of $135 million.

The Boeing 707 goes into service to challenge the British-built Comet for leadership in the aircraft industry. The first U.S.-built commercial jet, the 707 beats out Douglas, which will not have its DC-8 in service until next year, and Consolidated-Vultee, whose Convair 880 will not start flying until 1960. Lockheed has stuck with its Electra turboprop propeller jet and loses out in the race as pure jets begin to make other planes obsolete (*see* McDonnell-Douglas, 1967).

Gen. Curtis Emerson Lemay, 51, U.S. Air Force, flies a converted KC-135 7,100 miles non-stop from Tokyo to Andrews Field September 12 in 12 hours, 28 minutes—a new record.

Two de Havilland Comet IVs complete the first commercial transatlantic jet flights October 4, landing in well under six hours.

Pan Am and BOAC inaugurate transatlantic jet service in October (*see* London-Johannesburg, 1952). By the end of next year, airlines will be carrying 63 percent of all cross-Atlantic traffic—more than 1.5 million air passengers versus 881,894 sea passengers, but European companies will continue to launch new passenger liners.

The first domestic U.S. 707 flight takes off December 10. National Airlines has rented two of the big Boeing jets from Pan Am for the winter season in a bid to take business away from Eastern.

The first voyage under the North Pole is completed by the 4-year-old U.S. atomic submarine *Nautilus.*

London gets its first parking meters.

The M-14 .30-caliber rifle adopted by the U.S. military is a fully automatic version of the M-1 Garand rifle used in World War II and in Korea.

The cost of 100,000 computerized multiplication computations falls to 26¢, down from $1.26 in 1952, says IBM (*see* 1964).

Synanon opens in a rented storefront at Ocean Park, Calif., where former alcoholic Charles "Chuck" Dederich, 44, starts weekly discussion groups with friends from Alcoholics Anonymous in an effort to rehabilitate alcoholics and drug addicts (*see* AA, 1935). Dederich and his colleagues evolve a new form of therapy that will revolutionize rehabilitation programs and lead to the establishment of Synanon communities in several parts of the United States, Puerto Rico, and England as drug abuse increases (*see* Phoenix House, 1967).

Acupuncture is used for the first time as an anaesthetic after nearly 5,000 years of use in medical therapy in China (*see* Shen Nung, 2700 B.C.).

A U.S. National Defense Education Act authorizes federal spending of $480 million over 4 years for "strengthening science, mathematics, and foreign language instruction." Title III of the act gives a boost to sales of textbooks and tape recorders.

The first U.S. Earth satellite is launched.

U.S. first class postal rates climb to 4¢ per ounce August 1, up from the 3¢ level that has held since July 1932 (*see* 1963).

U.S. television sets number an estimated 41 million, up from 5 to 8 million in 1950.

U.S. daily newspapers number 357, down from 2,600 in 1909, as more Americans rely on radio and television for the news.

UPI (United Press International) is created by a merger of the 51-year-old United Press with the 52-year-old Hearst International News Service.

Nonfiction *Parkinson's Law* by Englishman Cyril Northcote Parkinson, 49, satirizes the growth of bureaucracy: "Work expands so as to fill the time available for its completion"; "Expenditure rises to meet income"; "Expansion means complexity, and complexity, decay. Or: the more complex, the sooner dead"; *Only in America* by North Carolina newspaper editor Harry Golden, 56.

Fiction *The Ugly American* by U.S. novelists Eugene Burdick and William J. Lederer who are attacked by the State Department for their revelations of arrogance, insensitivity, and stupidity on the part of U.S. diplomats in the developing countries; *Things Fall Apart* by Nigerian novelist Chinua Achebe, 28; *The City and the Dogs* (*La ciudad y los perros*) by Peruvian novelist Mario Vargas Llosa, 22; *Memed, My Hawk* (*Ince Memed*) by Turkish novelist Yasar Kemal, 36; *The Leopard (Il gatto pardo)* by the late Sicilian novelist Prince Giuseppe Tomasi di Lampedusa who died last year at age 61; *Leech (Kjop ikke Dondi)* by Cora Sandel; *The Grass (L'Herbe)* by Claude Simon; *Moderato Cantabile* by French novelist Marguerite Duras, 44; *The Middle Age of Mrs. Eliot* by Angus Wilson; *Spinster* by New Zealand novelist Sylvia Ashton-Warner, 53; *Saturday Night and Sunday Morning* by English novelist Alan Sillitoe, 30, who draws on his experiences as a

1958 ***(cont.)*** Nottingham factory worker; *A Ripple from the Storm* by Doris Lessing; *I'm Not Stiller* by Max Frisch; *The Ordeal of Gilbert Pinfold* by Evelyn Waugh; *The Dharma Bums* by Jack Kerouac; *Some Came Running* by James Jones; *Breakfast at Tiffany's* (stories) by Truman Capote; *Exodus* by Leon Uris; *Dr. No* by Ian Fleming.

Poetry *Selected Poems 1928–1958* by U.S. poet Stanley Kunitz, 53; *Words for the Wind* by Theodore Roethke.

Juvenile *The Cat in the Hat Comes Back* by Dr. Seuss.

Robert Rauschenberg pioneers "Pop Art" with a semi-abstract hole into which he inserts four Coca-Cola bottles.

Painting *Three Flags* by Jasper Johns; *Four Darks on Red* by U.S. abstractionist Mark Rothko, 54; *Peace* by Pablo Picasso whose hand extending a bouquet will be widely reproduced; *Homage to Malkevich* by Victor Vasarely. Georges Rouault dies at Paris February 13 at age 86; Maurice de Vlaminck dies at Rueil-la-Gadeliére outside Paris October 11 at age 82.

Theater *The Garden District* (*Suddenly Last Summer* and *Something Unspoken)* by Tennessee Williams 1/7 at New York's York Theater, with Anne Meacham; *Two for the Seesaw* by U.S. playwright William Gibson, 42, 1/16 at New York's Booth Theater, with Henry Fonda, Anne Bancroft as Gittel Mosca, 750 perfs.; *Sunrise at Campobello* by U.S. playwright (and former M-G-M production chief Dore Schary, 52, 1/30 at New York's Cort Theater, with Ralph Bellamy as Franklin D. Roosevelt, Mary Fickett as Eleanor, 556 perfs.; *Epitaph for George Dillon* by John Osborne and Anthony Creighton 2/11 at London's Royal Court Theatre; *Biedermann and the Firebugs (Biedermann und die Brandstifter)* by Max Frisch 3/29 at Zurich's Schauspielhaus; *The Killer (Tueur sans gages)* by Eugène Ionesco 4/4 at Darmstadt's Landestheater; *Variations on a Theme* by Terence Rattigan 5/8 at London's Globe Theatre; *The Birthday Party* by English playwright Harold Pinter, 27, 5/19 at London's Lyric Theatre, with Beatrix Lehmann; *Five Finger Exercise* by English playwright Peter Shaffer, 32, 7/17 at London's Comedy Theatre; *A Touch of the Poet* by the late Eugene O'Neill 10/2 at New York's Helen Hayes Theater with Tom Clancy, Eric Portman, Helen Hayes, Kim Stanley, Betty Field, 284 perfs.; *The Hostage* by Brendan Behan 10/14 at London's Theatre Royal, Stratford; *The World of Suzie Wong* by Paul Osborn 10/14 at New York's Broadhurst Theater, with France Nuyen, William Shatner, 508 perfs.; *The Pleasure of His Company* by Cornelia Otis Skinner and Samuel Taylor 10/22 at New York's Longacre Theater, with Skinner, Cyril Ritchard, 474 perfs.; *Krapp's Last Tape* by Samuel Beckett 10/28 at London's Royal Court Theatre; *The Resistible Rise of Arturo Ui (Der aufhaltsame Aufstieg des Arturo Ui)* by Bertolt Brecht 11/10 at Stuttgart; *The Disenchanted* by Budd Schulberg and Harvey Breit (based on Schulberg's novel) 12/3 at New York's Coronet Theater, with Jason Robards, Jr. (and Sr.), Rosemary Harris, Salome Jens, George Grizzard, 189 perfs.; *The Cold Wind and the Warm* by S. N. Behrman 12/8 at New York's Morosco Theater, with Eli Wallach, Maureen Stapleton, Vincent Gardenia, Morris Carnovsky, Suzanne Pleshette, 120 perfs.; *JB* by Archibald MacLeish 12/11 at New York's ANTA Theater, with Raymond Massey, Christopher Plummer, Pat Hingle, 364 perfs.

Films Stanley Kramer's *The Defiant Ones* with Tony Curtis, Sidney Poitier; Jacques Tati's *Mon Oncle* with Tati; Roy Baker's *A Night to Remember* with Kenneth More; Jack Clayton's *Room at the Top* with Laurence Harvey, Simone Signoret; Delbert Mann's *Separate Tables* with Rita Hayworth, Deborah Kerr, David Niven, Wendy Hiller, Burt Lancaster; Orson Welles's *Touch of Evil* with Welles, Charlton Heston, Janet Leigh; Alfred Hitchcock's *Vertigo* with James Stewart, Kim Novak, Barbara Bel Geddes. Also: Andrzej Wajda's *Ashes and Diamonds* with Zbigniew Cybulski; Richard Brooks's *Cat on a Hot Tin Roof* with Elizabeth Taylor, Paul Newman, Burl Ives; Jacques Tourneur's *Curse of the Demon* with Dana Andrews; Akira Kurosawa's *The Hidden Fortress* with Toshiro Mifune; Ronald Neame's *The Horse's Mouth* with Alec Guinness; Robert Wise's *I Want to Live!* with Susan Hayward; John Guillermin's *I Was Monty's Double* with M. E. Clifton-James, John Mills; John Ford's *The Last Hurrah* with Spencer Tracy; Tony Richardson's *Look Back in Anger* with Richard Burton, Claire Bloom, Edith Evans; Satyajit Ray's *The Music Room;* Douglas Sirk's *The Tarnished Angels* with Rock Hudson, Dorothy Malone, Robert Stack; George Pal's *tom thumb* with Russ Tamblyn, June Thornton; Edward Dmytryk's *The Young Lions* with Marlon Brando, Montgomery Clift, Dean Martin, Hope Lange.

Hollywood musical Vincente Minnelli's *Gigi* with Leslie Caron, Maurice Chevalier, Louis Jourdan, music by Frederick Loewe, lyrics by Alan Jay Lerner, songs that include "Thank Heaven for Little Girls," "The Night They Invented Champagne," and the title song.

Stage musicals *Irma la Douce* 7/17 at London's Lyric Theatre with Elizabeth Seal, Keith Mitchell, Clive Revill, music by French composer Marguerite Monmot, book and lyrics by Alexandre Breffet, Julian More, Monty Norman, David Heath, 1,512 perfs.; *Flower Drum Song* 12/1 at New York's St. James Theater, with Pat Suzuki, Juanita Hall, Myosuki Umeki, Larry Blyden, music and lyrics by Richard Rodgers and Oscar Hammerstein II, 601 perfs.

Opera *Vanessa* 1/15 at New York's Metropolitan Opera with music by Samuel Barber, libretto by Gian Carlo Menotti; *Noye's Fludde (Noah's Flood)* 6/18 at Orford Church, Suffolk, with music by Benjamin Britten, libretto from the 14th century *Chester Miracle Play.*

First performances Concerto No. 2 for Piano and Orchestra by Dmitri Shostakovich 5/10 at Moscow; *Nocturne* song cycle for tenor and small orchestra by Benjamin Britten 10/16 at Leeds.

Popular songs "Diana" by Canadian rock singer-composer Paul Anka, 15, who begins a meteoric rise to stardom (he will have made his first $1 million by age 17); "Satin Doll" by Duke Ellington, Billy Strayhorn, and Johnny Mercer; "Volare" ("Nel Blu, Dipinto di Blu") by Italian composer Dominico Modugno, English lyrics by Mitchell Parish; "Arrivederci Roma" by Carl Sigman, Renato Rascel, S. Giovanni, and Jack Fishman; "Splish Splash" by Bobby Darin and Jean Murray; "Everybody Loves a Lover" by U.S. composer Robert Allen, lyrics by Richard Adler; "Sugartime" by Charlie Phillips and Odis Echols; "Twilight Time" by Morty Lewis, lyrics by Buck Ram; "Catch a Falling Star" by Paul Vance and Lee Pockriss; "The Chipmunk Song (Christmas Don't Be Late)" by Ross Bagdasarian; "Jingle Bell Rock" by U.S. songwriters Joe Beal and Jim Boothe.

The first Grammy Award given by the National Academy of Recording Arts and Sciences goes to the Italian song "Volare" (above), but the Burbank, Calif., organization comes under fire from much of the recording industry for favoring older, more conservative, white, middle-of-the-road artists over youth-oriented performers.

Sugar Ray Robinson wins the middleweight boxing title for a record fifth time March 25 at Chicago Stadium in a 15-round decision over Carmen Basilio.

Ashley John Cooper, 21 (Australia), wins in men's singles at Wimbledon and Forest Hills, Althea Gibson in women's singles.

Arnold Palmer wins the Masters title at Augusta, Ga., plus two other golf championships and earns an impressive $42,000. Now 28, Palmer won the U.S. amateur title 4 years ago, turned professional soon after, and will be the most popular golfer since Bobby Jones.

The U.S. ocean yacht *Columbia* designed by Olin J. Stephens, 50, fends off the English challenger *Sceptre* in the first America's Cup competition since 1937. The 12-meter yachts are much smaller than the prewar "J"-class boats that nobody can now afford to build.

The New York Giants become the San Francisco Giants and play their first season at Candlestick Park.

The Brooklyn Dodgers become the Los Angeles Dodgers and play their first season at the Coliseum.

The New York Yankees win the World Series by defeating the Milwaukee Braves 4 games to 3.

Brazil wins her first World Cup in football (soccer), defeating Sweden 5 to 2 at Stockholm. Pele scores two of the Brazilian goals (*see* 1956).

U.S. intercollegiate football rules change to give teams the option of trying for a two-point conversion after touchdown by running the ball or passing it from the 3-yard line. A kick between the goal posts from the 2-yard line is still worth only one point.

Americans buy 100 million Hula Hoops, introduced by Wham-O Mfg., but the fad is short-lived.

Nebraska police and National Guardsmen mount roadblocks in late January and organize posses to hunt down a "mad dog" killer who has shot Marion Bartlett of Lincoln and his wife Velda through the head, clubbed their younger daughter to death, and run off with Carol Fugate, 14, Mrs. Bartlett's daughter by a previous marriage. Charles Starkweather, 19, has just lost his job as helper on a garbage truck, he proceeds to kill seven more people in the next three days before Wyoming police catch him January 30, and he will die in the electric chair June 25 of next year.

The Havana Hilton opens March 27. The $21 million hotel will be the Havana Libre beginning next year.

Brasília's Presidential Palace is completed by Oskar Niemeyer.

Milan's Pirelli building is completed by Pier Luigi Nervi and G. Ponti.

Rome's Palazzetto dello Sport is completed by Pier Luigi Nervi.

New York's Polo Grounds and Brooklyn's Ebbets Field become available for new housing sites as the baseball teams move to California (above).

U.S. scientists begin testing Earth's radiation shield of ozone to discover what effects if any have been caused by atmospheric testing of nuclear weapons and by the growing number of high-altitude flights by military and commercial jet aircraft. Ozone is an unstable form of oxygen containing three atoms instead of the usual two and serves as a barrier at altitudes of 50,000 to 135,000 feet against undue exposure to the sun's ultraviolet rays, which can cause human skin cancer. The ozone layers will be found to increase in the next 13 years, but will then begin to shrink, possibly as a result of being broken up by chlorine gas released from freon gas by ultraviolet rays in the atmosphere at altitudes of 12 to 15 miles (*see* freon 12 refrigerant, 1931; plastic aerosol valve, 1953). Nearly a million tons of freon will be released into the atmosphere each year by the 1970s, mostly from aerosol cans, and environmentalists will suggest the possibility that the Earth's ozone layer is being depleted.

U.S. Atomic Energy Commission physicist Willard Libby says the United States is the "hottest place in the world" in terms of radioactivity largely as a result of fallout from recent Soviet and British nuclear tests in the atmosphere (*see* Libby, 1947).

Iceland extends her fishery limits to 12 miles offshore. The move will produce conflicts with British fishing vessels (*see* 1961).

Whalers kill 6,908 blue whales, the largest creatures ever to inhabit the Earth. When hunting ceases in 1965, only one blue whale will be found, and it will be estimated that fewer than 1,000 remain in the seas.

China's wheat crop reaches 40 million tons, 2 million more than the U.S. crop, and cereal grain production jumps 35 percent above last year's levels despite a poor rice crop, but total food production falls far short of estimates. The dearth of food encourages peasants to neglect the grain crops of the collectives and raise vegetables and livestock which they can sell privately, if illegally (in some communes half the land is privately cultivated) (*see* 1957; 1959).

United Fruit Co. agrees February 4 to establish a competitor in the banana industry. The agreement settles a

1958 ***(cont.)*** 4-year antitrust suit brought by the U.S. government (*see* 1944).

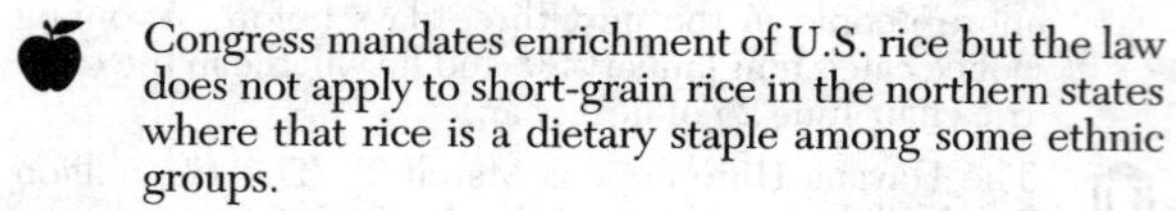

Congress mandates enrichment of U.S. rice but the law does not apply to short-grain rice in the northern states where that rice is a dietary staple among some ethnic groups.

A food additives amendment to the Food, Drug and Cosmetic Act of 1938 passed by Congress permits no food additives other than those used widely for many years and "generally recognized as safe" (GRAS list) unless the FDA agrees after a thorough review of test data that the new additive is safe at the intended level of use.

The Delaney "cancer clause" inserted in the new amendment (above) by a Brooklyn, N.Y., congressman states that if *any* amount of any additive can be shown to produce cancer when ingested by humans or test animals, no amount of that additive may be used in foods for human consumption (*see* cyclamates, 1969).

Sweet 'n Low sugarless sweetener, introduced by Cumberland Packing Co. of Brooklyn, N.Y., uses saccharin in place of sugar.

Cocoa Puffs breakfast food, introduced by General Mills, is 43 percent sugar (*see* 1954; 1959).

Cocoa Krispies breakfast food, introduced by Kellogg, is 45.9 percent sugar (*see* 1955; 1959).

Green Giant canned beans are introduced (*see* 1950; 1961).

Pizza Hut opens at Kansas City to begin a franchise chain that will grow into the largest group of U.S. pizzerias

China's birth control program of 1954 ends as China begins a Great Leap Forward (above; *see* 1962).

English Roman Catholic economist Colin Clark deplores birth control. He writes in the magazine *Nature*, "When we look at the British in the seventeenth and eighteenth centuries, at the Greeks in the sixth century B.C., the Dutch in the seventeenth century, and the Japanese in the nineteenth century, we must conclude that the pressure of population upon limited agricultural resources provides a painful but ultimately beneficial stimulus, provoking unenterprising agrarian communities into greater efforts in the fields of industry, commerce, political leadership, colonization, science, and [sometimes. . .] the arts."

Abortion in the United States by Great Neck, Long Island, physician's wife Mary Steichen Calderone, 53, reports on conferences of the Planned Parenthood Federation but is almost totally ignored (*see* 1955; Colorado, 1967).

1959 Cuban dictator Fulgencio Batista resigns January 1 after nearly 7 years in power and flees to Dominica and thence to Miami as rebel leader Fidel Castro, 32, captures Santiago a day after taking the provincial capital of Santa Clara. Castro roars into Havana January 3 and assumes office as premier February 16 after a 2-year rebellion. He arrives at Washington on an unofficial visit April 15 and 2 days later calls his revolution "humanistic," not Communist.

Alaska and Hawaii are admitted to the Union as the 49th and 50th states.

Tibet's Dalai Lama escapes to India March 31 and receives asylum after a rising against the Chinese garrison at Lhasa. The new Chinese puppet government seals the border to India, which refuses to recognize the Dalai Lama as head of a "separate" Tibetan government functioning in India. Chinese and Indian forces have border clashes.

Mao Zedong steps down as China's chief of state April 27 in favor of Liu Shao-chi but remains chairman of the Communist party.

Moscow urges Japan to end her agreement permitting U.S. bases on Japanese soil. The May 4 note from the Kremlin invites Tokyo to accept a Soviet guarantee of permanent neutrality.

Singapore becomes self-governing June 3 after 13 years as a British crown colony. Lee Kuan Yew is prime minister of the new republic and will rule wisely but strictly until 1990 (*see* 1963).

Indonesia's President Sukarno dissolves the constituent assembly July 5 and moves his nation toward a new "Guided Democracy" regime that will grow progressively more authoritarian as Sukarno's Communist Party (PKI) gains more power.

The Federal State of Mali is created January 17 by a merger of the African republics Senegal and French Sudan a week after Michel Debre is installed as premier of France's Fifth Republic.

Cuba si! Yanqui no! Fidel Castro replaced a dictatorship with a Marxist regime, the first in the Western Hemisphere.

Premiers of the 12 autonomous republics in Africa's French Community confer at Paris in early February with President de Gaulle (*see* 1958; 1960).

Rwanda is torn in November by a great uprising of Hutu (Bahutu) tribesmen against the minority Tutsi (Batutsi) aristocracy. Belgium governs the nation under a UN mandate. Rwanda's monarchy is abolished and thousands of Tutsi refugees flee across the borders into the Belgian Congo (where anti-European riots at Leopoldville killed 71 in early January), Uganda, and Tanganyika.

Ceylon's prime minister Sirimavo Bandaranaike dies September 26 of wounds inflicted by a Buddhist monk 3 months after conclusion of a trade pact with Beijing.

Switzerland's electorate votes February 1 to reject a proposed constitutional amendment that would permit women to vote in national elections and run for national office.

Japanese-Americans who renounced their citizenship in 1942 when placed in concentration camps under Executive Order 9066 regain full citizenship May 20 but receive no compensation for the losses they incurred as a result of their relocation (*see* 1989).

The Street Offences Act passed by the British Parliament July 16 makes it an offense "for a common prostitute to loiter or solicit in a secret or public place for the purpose of prostitution," and says "A constable may arrest without warrant anyone he finds in a street or public place and suspects, with reasonable cause, to be committing an offence under this section." English and Welsh prostitutes will circumvent the law by posting their telephone numbers at news kiosks or buying dogs and "walking" them as an excuse for being in the streets (*see* 1869; 1957).

Atlanta integrates its buses January 21 but the governor of Georgia asks citizens to continue "voluntary" segregation. Buses in most other Southern states remain segregated, with blacks obliged to sit in the back, and segregation continues at airport and railroad terminals, bus depots, and recreational facilities (*see* "Freedom Riders," 1961).

The UN General Assembly votes November 10 to condemn racial discrimination anywhere in the world, including South Africa.

Cologne has anti-Semitic riots December 24.

President Eisenhower speaks out February 18 against continued emergency aid to the unemployed. He says the nation's economy is on a "curve of rising prosperity."

A University of Michigan study shows that 10 percent of U.S. families live on the "poverty line" and 20 percent live below it (*see* 1957).

One American with an income of $20 million pays no federal taxes, five with incomes of more than $5 million each pay no federal taxes, one with an annual income of nearly $2 million has paid no federal taxes since 1949 (*see* 1960).

Vice President Nixon opens an American exhibit at Moscow July 25 and engages in a public debate with Premier Khrushchev. The dialogue takes place in a model kitchen and Nixon boasts of America's material progress and abundance of consumer goods (*see* Galbraith, 1958).

France demonstrates the industrial strides she has taken under the Monnet Plan of 1953 by producing twice as much steel as in 1929. The nation uses four times as much electricity as in 1929 and manufactures nearly five times as many cars and trucks.

A nationwide U.S. steel strike begins July 15.

The Landrum-Griffin Act passed by Congress September 14 marks the first major change in U.S. labor law since the Taft-Hartley Act of 1947. The new law requires labor unions to file annual financial reports with the secretary of labor but includes a "bill of rights" for labor.

Washington seeks a Taft-Hartley Act injunction to halt the steel strike (above) October 19. The United Steel Workers challenge the validity of the injunction provision but the Supreme Court upholds the entire act November 7.

International Telephone and Telegraph (ITT) names former Raytheon executive vice president Harold Sydney Geneen, 49, president (see 1925). Geneen will turn ITT into a giant conglomerate (*see* Textron, 1955).

Reynolds Metals acquires British Aluminium, the U.K.'s largest aluminum producer (*see* 1945).

President Eisenhower imposes mandatory quotas on U.S. oil imports in response to pleas from domestic producers, saying the move is based on reasons of national security. The quotas effectively subsidize domestic producers of crude oil and force U.S. consumers to pay higher prices for gasoline and home heating oil.

The St. Lawrence & Great Lakes Waterway dedicated by Queen Elizabeth June 26 gives ocean-going vessels access to Great Lakes ports as far west as Duluth, some 2,342 miles from the Atlantic. The Seaway can accommodate vessels of up to 25-foot 9-inch draught—80 percent of the world's saltwater fleet—but will prove costly to maintain.

The S.S. *Rotterdam*, launched by Holland-American Line, is a 38,645-ton luxury liner.

The first U.S. nuclear-powered merchant ship is launched July 21 at Camden, N.J.

English engineer Christopher Sydney Cockerell, 46, demonstrates his SRN-1 hovercraft with a crossing of the English Channel on a cushion of air. He has been experimenting since 1953 with ways to reduce the friction round the hulls of boats. Within 6 years the industry will be well established as hovercraft ferries and patrol boats come into use.

The Bristol Britannia wins approval for use in commercial aviation. Powered by Rolls-Royce engines, the 133-passenger plane is the first large turboprop aircraft.

Japanese auto makers produce 79,000 cars, up from just 110 in 1947. Most are Nissans and Toyotas and the number will increase to 3.2 million by 1970.

1959 ***(cont.)*** U.S. Volkswagen sales reach 120,000, up from 29,000 in 1955, while total sales of foreign-made U.S. cars such as British Fords, the General Motors Opel, and Chrysler's Simca top 120,000.

The Ford Falcon is Detroit's first major response to the competition from Volkswagen and other foreign compact cars. Chevrolet tools up to market a rear-engined Corvair compact (*see* Mustang, 1964).

The Rambler, introduced by American Motors, is a 14-foot 10-inch compact designed to compete with Volkswagen and the new Ford Falcon (above; *see* 1954).

The average U.S. automobile wholesales at $1,880, up from $1,300 in 1949.

The microchip invented by Texas Instruments engineer Jack Kilby and Fairchild Semiconductor engineer Robert N. Noyce, 32, will open the way to electronic wristwatches and a host of miniature products. Working independently, Kilby has encased an integrated circuit in a single silicon wafer, Noyce has found a way to join the circuits by printing, thus eliminating thousands of man-hours and sharply lowering the size, weight, and cost of electronic components. Noyce will found Intel Corp. which will give the microchip memory and logic functions to produce the microprocessor that will, in turn, make possible the personal computer.

The RCA 501 computer introduced by Radio Corp. of America is the world's first fully transistorized computer (*see* 1948).

Skull fragments with crude stone tools found in Tanganyika's Olduvai Gorge by British anthropologist-paleontologist Louis S. B. Leakey, 56, suggest that Australopithecene man-ape lived at least 1.78 million years ago. Potassium-argon dating of volcanic ash surrounding the fossils indicates their age.

Russian archaeologist Tatiana Proskouriakov finds a pattern of dates indicating a list of important events in the lives of certain Mayan individuals in the Yucatán. The find enables scholars to decipher the periods in which certain rulers reigned and thus establish dynasties, but scholars will argue about whether the Mayan glyphs represent ideas, words, or both.

The first complete amino acid sequence of the enzyme ribonuclease sheds some light on the chemical composition of the enzyme that helps break down ribonucleic acid (RNA), one of the two fundamental chemicals that determine normal and abnormal growth in living things. Rockefeller University scientists William H. Stein, 47, Stanford Moore, 45, and others work up the sequence (*see* Crick and Watson's DNA research, 1953, 1966).

West German clinics observe 12 cases of the birth defect phocomelia. Infants are born with flipper-like stubs in place of one or more normal limbs. Not one case has been seen in the past 5 years and the large number sounds alarms (*see* 1960).

Saudi Arabia's King Faisal permits education for girls despite protests from some religious groups.

The first transistorized TV set is introduced by Sony Corp. which last year changed its name from Tokyo Tsushin Kogyo (*see* 1952). The portable black-and-white receiver will be followed next year by the first transistorized color set, employing a picture tube developed years ago by Ernest O. Lawrence of 1932 cyclotron fame who played a leading role in developing the atomic bombs that destroyed Hiroshima and Nagasaki.

India gets her first television; villagers travel for hundreds of miles to visit six community TV centers at New Delhi where mounted police are required to keep the crowds away from the receivers. TV stations will be built at Bombay, Amritsar in Punjab, Srinigar in Kashmir, Madras in South India, and Calcutta in East India by 1966, but locally made receiving sets will cost $425 and the average per capita income is below $80 per year.

Color TV is introduced in Cuba.

South Africa decides not to introduce any television. The decision will stand for 16 years.

Chicago becomes a two-publisher city as Marshall Field's *Sun-Times* buys the *Daily News* from publisher John S. Knight and the *Tribune* buys the *Herald-American* from Hearst.

France legalizes press censorship, adopting a new article in the Constitution. Newspapers that disagree with the Gaullist government will be seized and closed down; papers will be impounded on frequent occasion until March 1965.

A federal district court at New York July 21 lifts a U.S. Post Office ban on distributing the 1928 D. H. Lawrence novel *Lady Chatterley's Lover* despite protests that the book uses such words as *fuck* and *cunt* and is explicit in its descriptions of the sex act. Grove Press has distributed an unexpurgated version of the book, Postmaster General Arthur E. Summerfield has banned it from the mails, and Judge Frederick van Pelt Bryan, 55, rules in Grove's favor. His 30-page decision in *Roth v. the United States* says not only that the book is not obscene but also that the Postmaster General is neither qualified nor authorized to judge the obscenity of material to be sent through the mails; he is empowered only to halt delivery of matter already judged obscene.

Nonfiction *The Elements of Style* by E. B. White who has adapted a manual by the late William Strunk, Jr., who was White's teacher at Cornell in 1919 and died in 1946.

Fiction *The Poorhouse Fair* by U.S. novelist John Hoyer Updike, 27; *Goodbye, Columbus* (stories) by U.S. author Philip Milton Roth, 26; *Advertisements for Myself* by Norman Mailer; *Henderson the Rain King* by Saul Bellow; *The Naked Lunch* by William Burroughs; *Mrs. Bridge* by U.S. novelist Evan S. Connell, 25; *The Apprenticeship of Duddy Kravitz* by Mordecai Richler; *Malcolm* by U.S. novelist James Purdy, 36; *The Mansion* by William Faulkner; *Memento Mori* by Scottish novelist Muriel Spark, 41; *Billiards at Half-Past Nine* by Heinrich Böll; *The Impossible Proof (Unmögliche Beweisaufnahme)* by Hans Erich Nossack; *Homo Faber* by Max Frisch; *Zazie* by Raymond Queneau; *The Planetarium (Le Planetarium)* by Nathalie Sarraute; *Aimez-Vous Brahms?* by Françoise Sagan; "The Loneliness of the Long Distance Runner" (story) by Alan Sillitoe; *Dear and Glorious Physician* by Taylor Caldwell; *Advise and Consent* by U.S. novelist Alan Drury, 41; *Hawaii* by James Michener; *Goldfinger* by Ian Fleming.

Poetry *A Dream of Governors* by Louis Simpson; *Heart's Needle* by U.S. poet W. D. Snodgrass, 33.

Painting *Numbers in Color* by Jasper Johns; *Jill* by U.S. painter Frank Stella, 23; *Virginia Site* by U.S. painter Kenneth Noland, 35; *Zinc Yellow* by Franz Kline.

Sculpture *White on White* by U.S. sculptor Louise Nevelson, 59.

New York's Guggenheim Museum opens on Fifth Avenue between 88th and 89th streets to house the collection of the late copper magnate Solomon R. Guggenheim whose mentor Hilla Rebay induced him to buy dozens of canvases by the late abstractionist Wassily Kandinsky (*see* Wright, below).

The Nikon F 35-mm. single-lens reflex camera is introduced by Nippon Kogaku K.K. whose rangefinder cameras enabled *Life* magazine photographers to provide sharp coverage of the Korean conflict (*see* 1948).

Theater *A Taste of Honey* by English playwright Shelagh Delaney, 20, 2/10 at London's Wyndham Theatre, 617 perfs.; *A Majority of One* by Hollywood screenwriter Leonard Spigelgass, 50, 2/16 at New York's Shubert Theater, with Gertrude Berg, Cedric Hardwicke, 558 perfs.; *Sweet Bird of Youth* by Tennessee Williams 3/10 at New York's Martin Beck Theater, with Paul Newman, Geraldine Page, 378 perfs.; *Raisin in the Sun* by U.S. playwright Lorraine Hansberry, 29, 3/11 at New York's Ethel Barrymore Theater, with Sidney Poitier, Ruby Dee, Diana Sands, Claudia McNeil, 530 perfs. (Hansberry has taken her title from a Langston Hughes poem containing the line, "What happens to a dream deferred?/Does it dry up/Like a raisin in the sun?"); *The Connection* by U.S. playwright Jack Gelber, 27, 7/5 at New York's off-Broadway Living Theater, with Leonard Hicks, Ira Lewis is about drug addiction, 722 perfs.; *The Zoo Story* by U.S. playwright Edward Franklin Albee, 31, 9/28 at Berlin's Schillertheater Wehrstatt; *The Miracle Worker* by William Gibson 10/19 at New York's Playhouse Theater, with Patty Duke as Helen Keller, Anne Bancroft as Annie Sullivan in a stage version of the 1957 *Playhouse 90* television drama, 702 perfs.; *Serjeant Musgrave's Dance* by English playwright John Arden, 29, 10/22 at London's Royal Court Theatre; *The Blacks (Les nigres)* by Jean Genet 10/28 at the Théâtre de Lutèce, Paris; *The Tenth Man* by Paddy Chayefsky 11/5 at New York's Booth Theater, 623 perfs.; *The Andersonville Trial* by U.S. playwright Saul Levitt, 46, 12/29 at Henry Miller's Theater, New York, with George C. Scott, Jr., Albert Dekker.

Hotelmen Laurence and Robert Preston Tisch acquire the Loew's theater chain from M-G-M; they will rename their parent company Loew's, Inc.

Films Otto Preminger's *Anatomy of a Murder* with James Stewart, Lee Remick; William Wyler's *Ben Hur* with Charlton Heston, Jack Hawkins, Stephen Boyd; François Truffaut's *The 400 Blows* with Jean-Pierre Leaud; Alfred Hitchcock's *North by Northwest* with Cary Grant, Eva Marie Saint, James Mason; Stanley Kramer's *On the Beach* with Gregory Peck, Ava Gardner, Fred Astaire; Alexei Batalov's *The Overcoat* with Roland Bykov; Billy Wilder's *Some Like It Hot* with Jack Lemmon, Tony Curtis, Marilyn Monroe. Also: Marcel Camus's *Black Orpheus* with Breno Mello, Marpessa Dawn; Jean-Luc Godard's *Breathless* with Jean-Paul Belmondo, Jean Seberg; Frank Launder's *The Bridal Path* with Bill Travers, Bernadette O'Farrell; Richard Fleischer's *Compulsion* with Orson Welles, Diane Varsi, Dean Stockwell, Bradford Dilman; Claude Chabrol's *The Cousins* with Jean-Claude Brialy, Gerard Blain; Guy Hamilton's *The Devil's Disciple* with Burt Lancaster, Kirk Douglas, Laurence Olivier; George Stevens's *The Diary of Anne Frank* with Millie Perkins, Joseph Schildkraut, Shelley Winters; Kon Ichikawa's *Fires on the Plain* with Eiji Funakoshi, Mantaro Ushio; Yasujiro Ozu's *Floating Weeds* with Ganjiro Nakamura, Machiko Kyo; John and Ray Boulting's *I'm All Right, Jack* with Peter Sellers satirizes Britain's worsening labor-management relations; Jack Arnold's *The Mouse That Roared* with Peter Sellers, Jean Seberg; Fred Zinneman's *The Nun's Story* with Audrey Hepburn, Peter Finch, Edith Evans; Blake Edwards's *Operation Petticoat* with Cary Grant, Tony Curtis; Michael Gordon's *Pillow Talk* with Doris Day, Rock Hudson; Howard Hawks's *Rio Bravo* with John Wayne, Dean Martin; Michael Anderson's *Shake Hands with the Devil* with James Cagney; Joseph L. Mankiewicz's *Suddenly Last Summer* with Elizabeth Taylor, Katharine Hepburn, Montgomery Clift; Satyajit Ray's *The World of Apu* with Soumitra Chatterjee.

Rock 'n' roll barnstormer Buddy Holly, 22, flies into a snowstorm February 2 after a concert at Clear Lake, Iowa, and dies in a crash. In his 18-month career he has written songs that include "Peggy Sue," "True Love Ways," and "That'll Be the Day."

Stage musicals *Once Upon a Mattress* 5/11 at New York's Phoenix Theater, with Carol Burnett, Jack Gilford, music by Mary Rodgers, lyrics by Marshall Barer, 458 perfs.; *Gypsy* 5/21 at New York's Broadway Theater, with Ethel Merman, Jack Klugman, book by Arthur Laurents based on the life of stripteaser Gypsy Rose Lee, music by Jule Styne, lyrics by Stephen Sondheim, songs that include "Small World," "Together Wherever We Go," "Everything's Coming Up Roses," 702 perfs.; *Lock Up Your Daughters* 5/28 at London's new Mermaid Theatre, with music by Laurie Johnson, lyrics by Lionel Bart (*né* Lionel Begleiter), 28, 328 perfs.; *At the Drop of a Hat* 10/8 at New York's John Golden Theater, with Englishmen Michael Flanders and Donald Swan, music and lyrics by Flanders and

1959 (cont.) Swan, 217 perfs.; *Take Me Along* 10/22 at New York's Shubert Theater, with Jackie Gleason, Walter Pidgeon, Robert Morse, music and lyrics by Bob Merrill, 448 perfs.; *The Sound of Music* 11/16 at New York's Lunt-Fontanne Theater, with Mary Martin, Theodore Bikel, Kurt Kaszner, music by Richard Rodgers, lyrics by Oscar Hammerstein II, songs that include "Climb Every Mountain," "Do-Re-Mi," "Edelweiss," "My Favorite Things," and the title song, 1,443 perfs.; *Fiorello!* 11/23 at New York's Broadhurst Theater, with Tom Bosley as the late Mayor La Guardia, Howard da Sylva, Ellen Hanley, music by Jerry Bock, lyrics by Sheldon Harnick, songs that include "Politics and Poker," 796 perfs.

First performances *Pittsburgh* Symphony by Paul Hindemith 1/30 at Pittsburgh; Missa Brevis in D for treble voices and organ by Benjamin Britten 7/22 at Aldeburgh; Concerto in E flat for Violoncello and Orchestra No. 1 by Dmitri Shostakovich 10/4 at Leningrad, with Mstislav Rostropovich as soloist.

Popular songs "The Children's Marching Song" by M. Arnold (for the film *The Inn of the Sixth Happiness*); "High Hopes" by James Van Heusen, lyrics by Sammy Cahn (for the film *A Hole in the Head*); "Lonely Boy" and "Put Your Head On My Shoulder" by Paul Anka; "Breaking Up is Hard to Do" by Neil Sedaka and Howard Greenfield; "Waterloo" by John Loudermilk and Marijohn Wilkin.

Swedish prizefighter Ingemar Johansson, 26, wins the world heavyweight crown June 26 by knocking out Floyd Patterson in the third round of a title match at New York.

Alejandro "Alex" Olmedo, 23 (Peru), wins in men's singles at Wimbledon, Maria Bueno, 19 (Brazil), in women's singles; Neale Andrew Fraser, 25 (Australia), wins in men's singles at Forest Hills, Bueno in women's singles.

The Los Angeles Dodgers win the World Series by defeating the Chicago White Sox 4 games to 2 after the White Sox have taken their first pennant since the "Black Sox" scandal of 1919.

The Barbie doll introduced by California entrepreneur Ruth Handler, 42, and her artist husband Eliot is allegedly based on dolls handed out to patrons of a West Berlin brothel. The busty doll with her endless wardrobe enriches the Handlers' toy firm Mattel, Inc.

Pantyhose—waist-high nylon hose requiring no garters, garter-belts, or corsets—are introduced by Glen Raven Mills of Altamahaw, N.C. (*see* 1967)

Spring filter-tipped cigarettes are introduced by P. Lorillard.

A housing bill passed by Congress September 23 authorizes expenditure of $1 billion over a 2-year period with $650 million earmarked for slum clearance.

New York's Guggenheim Museum (above) has been designed by Frank Lloyd Wright who dies April 9 at age 89. The building has a dramatic circular inner ramp employed by Wright earlier on a smaller scale.

New York's Seagram building is completed opposite the 41-year-old Racquet and Tennis Club on Park Avenue. Designed by Mies van de Rohe and Philip Johnson, the bronze-clad, 40-story tower rises from a plaza landscaped with fountains and trees.

China suffers catastrophic crop failures (*see* 1958, 1960).

A corn farmer must have at least 1,000 acres to be a viable producer, says the U.S. Department of Agriculture, but 1 million U.S. farms contain fewer than 50 acres. The country has 3.7 million farms, down from more than 6 million in 1940 when 2.2 million farms had fewer than 50 acres, and 136,000 farms have 1,000 acres or more, up from 100,000 in 1940.

The Food and Drug Administration seizes 0.25 percent of the U.S. cranberry crop and orders cranberries from Washington and Oregon off the market. Residues of the weed-killer aminotriazole have contaminated a tiny fraction of the crop, but the headlines alarm consumers and all cranberry-sauce sales drop sharply.

Congress approves a 2-year extension of a program for foreign disposal of surplus U.S. farm commodities. Some $250 million worth of surplus food is to go to needy Americans through food stamps (*see* 1939; 1964).

Supermarkets account for 69 percent of all U.S. food-store sales even though they represent only 11 percent of food stores.

Minute Maid proves to the Florida Citrus Commission that frost-damaged oranges can be used successfully in concentrates (*see* 1949). After several years of wild fluctuations in profit and loss, the company uses earth-moving machines to build 10-foot walls round 7,000 acres of marsh to reclaim savannah land that has been under water 9 months of the year. It plants citrus trees from whose fruit it will make frozen juice concentrates and develops programs that permit growers to participate in the retail prices of concentrates, thus assuring a constant supply of fruit. Coca Cola Company will acquire Minute Maid next year.

Frosty O's sugar-coated breakfast food is introduced by General Mills (*see* 1958; 1961).

The National Cranberry Association changes its name to Ocean Spray, expands production and distribution of Cranberry Juice Cocktail, and introduces Dietetic Cranberries to overcome consumer resistance to cranberry sauce (above; 1930). The association, which harvested its first million-barrel crop in 1953, adopts new promotion ideas that will boost consumption in years to come (*see* Cranapple, 1965).

Hawaii (above) has a native population of 12,000, down from 37,000 in 1860.

Congress authorizes admission of some 57,000 additional immigrants without regard to quotas established by the 1952 McCarran-Walter Act. Preference is given to resident aliens and to relatives of U.S. citizens (*see* 1965).

1960 Soviet ground-to-air missiles at Sverdlovsk down a U.S. supersonic U-2 spy plane flying at 60,000 feet May 1. The Russians capture CIA agent pilot Francis Gary Powers, 30, with his electronic sensing equipment, Washington admits to having sent aerial reconnaisance flights over Soviet territory, and Premier Khrushchev cancels a Paris summit meeting with President Eisenhower.

Fidel Castro has signed an agreement at Havana February 13 with Soviet first deputy premier Anastas I. Mikoyan providing $100 million in Soviet credit to Cuba and the Soviet purchase of 5 million tons of Cuban sugar, Castro threatens June 23 to seize all American-owned property and business interests to counter U.S. "economic aggression," Eisenhower cuts Cuba's sugar quota by 95 percent July 6 and declares that the United States will never permit a regime "dominated by international Communism" to exist in the Western Hemisphere. Khrushchev threatens July 9 at Moscow to use Soviet rockets to protect Cuba from U.S. military intervention.

The Monroe Doctrine of 1823 has died a "natural death," says Premier Khrushchev at a news conference July 12. Washington reaffirms the Doctrine July 14 and accuses Khrushchev of trying to set up a "Bolshevik doctrine" for worldwide Communist expansion.

Havana nationalizes all banks and large commercial and industrial enterprises October 14, Washington imposes an embargo October 19 on all exports to Cuba except medical supplies and most foodstuffs, Washington sends a note to the Organization of American States October 28 charging that Cuba has received substantial arms shipments from the Soviet bloc (*see* 1961).

France has exploded her first atomic bomb February 13 over the Sahara Desert in southwestern Algeria, joining the "atomic club" of the United States, the U.S.S.R., and Britain. She begins a series of atmospheric nuclear tests that will continue for some years in Africa and the Pacific.

French and Belgian colonies in Africa gain independence: French Cameroun January 1, Togo April 27, the Malagasy Republic (Madagascar) June 26, the Independent Congo Republic June 30, Somalia (French and Italian Somaliland) July 1, Ghana July 1, Dahomey August 1, Upper Volta in August, Ivory Coast August 7, Chad August 11, the Central African Republic August 13, Gabon August 17, Mali August 20, Niger September 3, Senegal September 5, Nigeria October 11, the Islamic Republic of Mauritania November 28.

The new Independent Congo Republic (above) headed by President Joseph Kasavubu and Premier Patrice Lumumba quickly dissolves into chaos. Separatist Moise Tshombe proclaims an independent Katanga, separatist Albert Kalonji proclaims an independent Kasai, Congolese troops mutiny, Premier Lumumba appeals to the UN for aid, the UN demands withdrawal of Belgian forces, the Security Council votes to send in UN troops, Secretary-General Dag Hammarskjöld leads UN forces into Katanga, President Kasavubu and Premier Lumumba dismiss each other, and the Congo army commander Col. Joseph D. Mobutu takes over (*see* 1961).

Cyprus gains independence August 16 after 88 years of British colonial rule. The new Cypriot republic elects Archbishop Makarios president, but Greek and Turkish interests will vie for control of the Mediterranean island (*see* 1964).

The United Nations establishes the International Development Association (IDA) with headquarters at Washington, D.C.

South Korean President Synghman Rhee wins election to a fourth term March 15 (he ran unopposed), police fire on demonstrators at Seoul protesting the "rigged" elections April 19, 127 are reported killed, Rhee resigns April 27, new elections are held July 29 (*see* 1961).

Japan's premier asks President Eisenhower June 16 to cancel a scheduled visit following 3 weeks of anti-American protest demonstrations by leftist groups. U.S. and Japanese diplomats have signed a treaty of mutual security and cooperation January 19 at Washington, the Japanese Diet approves the treaty June 19, and it takes effect June 23.

Japanese Socialist party leader Inajiro Asanuma, 61, is assassinated on a public stage at Tokyo October 12 by a 17-year-old right-wing extremist with a foot-long sword. Asanuma has supported the U.S.-Japanese mutual-defense treaty.

John F. Kennedy, now 43, wins election to the presidency by defeating Vice President Nixon, but the first Roman Catholic president-elect squeaks in with a margin of only 113,057 votes out of more than 69 million cast; 502,773 go to minority party candidates including U.S. Sen. Harry F. Byrd, Jr., 45 (D. Va.), who captures 15 electoral votes. Nixon gets 219 to Kennedy's 303.

South Vietnam's President Ngo Dinh Diem regains power November 12 following a coup by the paratroop brigade at Saigon. Dissident groups, collectively called the Vietcong (Vietnamese Communists), meet secretly December 20 and organize the National Front for the Liberation of South Vietnam (*see* 1962).

South Africa's Sharpeville massacre March 21 draws a sharp U.S. State Department protest. Some 20,000 blacks have besieged a police station in a Johannesburg suburb to demonstrate against a law requiring that all blacks carry papers. The police open fire, killing 56, wounding 162 (of whom 16 die). The pass law is suspended March 26 but violence continues. Whites arm themselves, 30,000 blacks march on Cape Town March 30 to demand the release of their leaders.

U.S. blacks begin "sit-in" demonstrations February 1 at Greensboro, N.C., lunch counters. The city desegregates eating places July 25.

Southern senators begin a filibuster February 29 to block civil rights legislation. Sessions continue round the clock until March 5 to set a filibuster record of 82 hours, 3 minutes, but Congress passes a Civil Rights Act May 6 authorizing federal referees where patterns

Indochina's rice fields became battlefields in a surrogate war between Communists and non-Communists.

of discrimination against black voters exist. The Department of Justice brings its first sweeping civil rights suit September 13, charging a plot to obstruct black voting in Tennessee.

A state may not change the boundaries of a city to exclude black voters, the Supreme Court rules November 14 in the Tuskegee, Ala., gerrymander case *Gomillion v. Lightfoot* (*see* 1812).

The U.S. Gross National Product (GNP) is $503 billion, up from $284 billion in 1950; government spending accounts for 27 percent of the total, up from 21 percent in 1950.

President Eisenhower acts to curb a rising deficit in the nation's balance of payments and a drain on U.S. gold reserves. He orders a reduction of government spending abroad November 16 and the Defense Department 9 days later takes steps to limit sharply the number of dependents accompanying servicemen stationed abroad.

President Eisenhower warns against the "military-industrial-complex" that works to maintain high levels of spending for the defense establishment, but President-elect Kennedy (above) has campaigned on a promise to close a phantom "missile gap" and will push through measures that increase military spending.

More than 25 million Americans declare incomes of $5,000 or more, 5.3 million declare incomes of more than $10,000, more than 500,000 declare more than $25,000 (up from 42,500 in 1939). Thirty-two million taxpayers pay 90 percent of all income taxes collected with more than 25 percent of the population paying some income tax.

Britain's richest 10 percent holds 83 percent of the nation's wealth, down from 88 percent in 1939.

West Germany's industrial production reaches 176 percent of Germany's 1936 level as an "economic miracle" progresses in Bremen, Cologne, Düsseldorf, Frankfurt, Hamburg, Leipzig, and Stuttgart.

U.S. corporate mergers total 844, up from 219 in 1950 when Congress passed the Celler-Kefauver amendment to strengthen the Clayton Anti-Trust Act of 1914 (*see* 1966).

The first privately financed nuclear power plant opens at Dresden, Ill., just south of Chicago. Within 14 years Commonwealth Edison will be producing more than 30 percent of its power from nuclear reactors but most utility companies will continue to use fossil fuels (*see* Consolidated Edison, 1962).

The Yankee Atomic Plant starts up August 22 at Rowe, Mass.

Coal supplies 45 percent of U.S. energy needs, a figure that will decline in the next 15 years to less than 18 percent as power plants and homeowners switch to oil and natural gas.

Twenty-nine U.S. oil companies go on trial in a U.S. district court at Tulsa, Okla., February 1 on charges of conspiring to raise and fix crude oil and gasoline prices. Federal judge Royce H. Savage rules for the defendants February 13, saying, "The evidence does not rise above the level of suspicion."

The Organization of Petroleum Exporting Countries (OPEC) meets for the first time September 14 at Baghdad and forces a retraction of the decrease in oil prices by Standard Oil of New Jersey, which has unilaterally rolled back prices by 4¢ to 14¢ per barrel. The five charter OPEC members include Saudi Arabia, Iran, Iraq, Kuwait, and Qatar; the organization will grow to include Abu Dhabi, Algeria, Gabon, Libya, Nigeria, Indonesia, Ecuador, and Venezuela.

Grand juries indict General Electric, Westinghouse, Allis-Chalmers, and 26 other U.S. producers of heavy electrical equipment for involvement in the largest conspiracy in restraint of trade in the 70-year history of the Sherman Act. The court will levy fines totaling more than $1.92 million and hand down seven jail sentences plus 24 suspended sentences.

The Erie Lackawanna Railroad is created October 15 by a merger of the 99-year-old Erie and slightly younger Delaware, Lackawanna and Western.

U.S. auto registration figures show one passenger car for every three Americans, a number that will increase in the next decade; 15 percent of families have more than one car, up from 7 percent in 1950.

There are 74 million motorcars on U.S. roads, up from 32.6 million in 1941.

The United States has 2.17 million miles of surfaced road by year's end, up from 1.68 million in 1950, and 951,100 miles of dirt road, down from 1.31 million.

New York makes Third Avenue one-way northbound July 17 and Lexington Avenue one-way southbound, but merchants, the Fifth Avenue Coach Co. and the Transport Workers Union fight conversion of Fifth and Madison Avenues to one-way traffic (*see* 1957; 1966).

Harbor Freeway opens at Los Angeles, giving access to the west side of town.

Britain's Daimler Motor merges into Jaguar.

The Datsun motorcar, introduced in the United States by Japan's Nissan Motors, is underpowered, hard to start and stop, but will rank sixth among imported U.S. cars by 1966, third by 1970 (*see* 1912; 1983).

Some 2,000 electronic computers are delivered to U.S. business offices, universities, laboratories, and other buyers. The figure will more than double in the next 4 years and debate will rage as to whether computers wipe out jobs or create new ones (*see* Luddites, 1811–1813).

The laser (light amplification by stimulated emission of radiation) perfected by research physicist Theodore Maiman at the Hughes Laboratory in Malibu, Calif., can cut metal quickly and will find wide use in delicate welding (including retinal eye operations). U.S. inventor R. Gordon Gould, 40, coined the term *laser* 3 years ago at Columbia University and had his notebook showing the basic laser concept notarized, but physicists Charles H. Townes and Arthur Schawlow applied for a patent first and were first to publish their findings in scientific journals. Gould has tried to interest defense officials in his potential "death ray," but he was involved in some left-wing political activities in the early 1940s, the Defense Department has classified his patent application secret, denied him security clearance, and confiscated his notebooks. Lasers will find uses in determining distances (e.g., to missile targets), relaying communications, projecting three-dimensional holograph pictures, surgery, exploring nuclear applications, and computer printing.

The Bulova Accutron tuning-fork watch introduced October 25 is the world's first electronic wristwatch. Its tuning fork vibrates 360 times per second—144 times as fast as the balance oscillators in a conventional handwound, automatic, or electric watch (*see* 1957; quartz, 1967).

Librium receives FDA approval February 24. Polish-American research chemist Leo H. Sternbach, 52, and his associates at Roche Laboratories have developed the anti-anxiety drug by treating a quinazoline with methylamine. Librium sales will far surpass those of the meprobamates (*see* Miltown, 1954; Valium, 1963).

Indonesia begins a 10-year campaign against malaria, which kills an estimated 120,000 per year and accounts for between 10 and 15 percent of all infant mortalities. Equipment supplied by the World Health Organization, largely with U.S. funds, will drain mosquito-infested swamps, workers will go house-to-house spraying with DDT, and the country will adopt a program to detect and treat victims; by 1964 Java will be free of malaria except for a few isolated areas on the southern coast (*see* 1969).

Seattle physician Belding Scribner improves the kidney machine devised by Dutch physician Willem J. Kolff in 1944. He develops a way to put tubes into a large artery and a fat vein so that they can be left in place for months or even years to facilitate treatment with artificial kidneys of patients with permanently impaired organs.

Echo I, launched by the United States August 12, is the world's first communications satellite (*see* 1962).

Television debates give Sen. Kennedy the edge that enables him to defeat Vice President Nixon (above) more by his appearance and manner than anything he says (radio listeners believe Nixon won the debates). TV debates will not be used in a presidential election again until 1976.

The Xerox 914 copier begins a revolution in paperwork reproduction (*see* 1950). The first production line Xerox copier is called the 914 because it makes copies of up to 9 by 14 inches on ordinary paper.

The Imperimerie National at Paris introduces typesetting by computer (*see* 1953; 1970).

The Pentel introduced by Tokyo's Stationery Co. in August is the world's first felt-tip pen.

U.S. paperback book sales reach an annual rate of more than 300 million.

Nonfiction *Growing Up Absurd* by U.S. novelist-poet-psychologist-philosopher Paul Goodman, 49, is a study of youth and delinquency that uses materials from literature, psychology, and political theory.

Fiction *The Trial Begins (Sud Idyot)* by Russian novelist Andrei Sinyavsky, 25; *To Kill a Mockingbird* by Alabama novelist Harper Lee, 34; *Set This House on Fire* by William Styron; *Rabbit, Run* by John Updike who captures the tedium of smalltown life; *The Sotweed Factor* by John Barth (*see* 1708); *A Separate Peace* by U.S. novelist John Knowles, 23; *The Violent Bear It Away* by Flannery O'Connor; *This Sporting Life* by English novelist David Storey, 27; *The Country Girls* by Irish novelist Edna O'Brien, 28; *Casanova's Chinese Restaurant* by Anthony Powell; *Kiss Kiss* (stories) by Roald Dahl; *The Nephew* by James Purdy; *The Magician of Lublin* by Isaac Bashevis Singer; *The Human Season* by U.S. novelist Edward Lewis Wallant, 34; *All or Nothing* by John Cowper Powys, now 88; *The Flanders Road (La Route des Flandres)* by Claude Simon; *The Key (Kagi)* by Junichiro Tanizaki.

Poetry *What a Kingdom It Was* by U. S. poet Galway Kinnell, 33; *To Bedlam and Half Way Back* by U.S. poet Anne Sexton, 32.

Painting *Door to the River* by Willem de Kooning; *Painted Bronze* (ale cans) by Jasper Johns; *Campbell's Soup Can (Tomato and Rice)* by U.S. pop artist Andy Warhol, 32; *The Postcard* by René Magritte.

Sculpture *Bathtub* by German artist Joseph Beuys, 39, who has covered the metal tub in which he was bathed as a child and covered it with sticking plaster and gauze soaked in fat (a Luftwaffe pilot in 1943, Beuys was shot down over the Crimea, saved by Tatars who wrapped him in fat to save his life); *Walking Man* by Albert Giacometti.

Polaroid introduces a new high-speed film with an ASA rating of 3,000—10 times faster than the average speed

1960 *(cont.)* of previous Polaroid films (*see* 1950; color film, 1962).

Theater *Toys in the Attic* by Lillian Hellman 2/25 at New York's Hudson Theater, with Maureen Stapleton, Jason Robards, Jr., 556 perfs.; *The Room* and *Dumb Waiter* by Harold Pinter 3/8 at London's Royal Court Theatre; *The Best Man* by Gore Vidal 3/31 at New York's Morosco Theater, with Melvyn Douglas, Lee Tracy, Frank Lovejoy, 520 perfs.; *The Balcony (Le balcon)* by Jean Genet 5/18 at the Théâtre du Gymnase, Paris; *A Man for All Seasons* by English playwright Robert Bolt, 36, 9/1 at London's Globe Theatre; *Period of Adjustment* by Tennessee Williams 11/10 at New York's Helen Hayes Theater, with James Daley, Barbara Baxley, 132 perfs.; *All the Way Home* by Tad Mosel 11/30 at New York's Belasco Theater, with Coleen Dewhurst, Arthur Hill, Lillian Gish in a play based on the late James Agee's only novel, 334 perfs.

Films Billy Wilder's *The Apartment* with Jack Lemmon, Shirley MacLaine, Fred MacMurray; Alfred Hitchcock's *Psycho* with Janet Leigh, Anthony Perkins; Fred Zinneman's *The Sundowners* with Deborah Kerr, Robert Mitchum, Peter Ustinov; Ronald Neame's *Tunes of Glory* with Alec Guinness, John Mills. Also: Michelangelo Antonioni's *L'Avventura* with Monica Vitti; Federico Fellini's *La Dolce Vita* with Marcello Mastroianni, Anita Ekberg, Anouk Aimee; Richard Brooks's *Elmer Gantry* with Burt Lancaster, Jean Simmons; Tony Richardson's *The Entertainer* with Laurence Olivier; Roberto Rossellini's *General Della Rovere* with Vittorio De Sica; Alain Resnais's *Hiroshima Mon Amour* with Emmanuelle Riva, Eiji Okada; Roger Corman's *House of Usher* with Vincent Price; Stanley Kramer's *Inherit the Wind* with Spencer Tracy, Fredric March, Gene Kelly; Basil Dearden's *The League of Gentlemen* with Jack Hawkins, Nigel Patrick, Bryan Forbes, Richard Attenborough; Roger Corman's *The Little Shop of Horrors* with Jonathan Haze, Jackie Joseph; John Sturges's *The Magnificent Seven* with Yul Brynner, Steve McQueen, Eli Wallach, Horst Buchholz, James Coburn, Charles Bronson; John and Faith Hubley's *Moonbird* (animated short); Jules Dassin's *Never on Sunday* with Melina Mercouri; Luchino Visconti's *Rocco and His Brothers* with Alain Delon, Renato Salvatori, Claudia Cardinale; Karel Reisz's *Saturday Night and Sunday Morning* with Albert Finney; Jack Cardiff's *Sons and Lovers* with Dean Stockwell, Trevor Howard, Wendy Hiller; Stanley Kubrick's *Spartacus* with Kirk Douglas, Laurence Olivier; Ken Annakin's *The Swiss Family Robinson* with John Mills, Dorothy McGuire, Sessue Hayakawa.

Opera *A Midsummer Night's Dream* 6/11 at Aldeburgh's Jubilee Hall, with music by Benjamin Britten, libretto from the 1595 Shakespeare comedy.

Cantata *Carmen Baseliense, Cantata Academics* by Benjamin Britten 7/1 at Basel University, for the university's 500th anniversary.

Oratorio *The Manger (El Pesebrio)* by Pablo Casals, now nearly 84, 12/17 at Acapulco.

First performances Symphony No. 4 by Roger Sessions 1/2 at Minneapolis; *Lincoln, The Great Commoner* by the late Charles Ives (who wrote it in 1912) 2/10 at New York's Carnegie Hall; *Music for Amplified Toy Pianos* by John Cage 2/25 at Wesleyan University, Middletown, Conn.; Sinfonietta in E major by Paul Hindemith 3/1 at Louisville; Symphony No. 9 by Darius Milhaud 3/29 at Fort Lauderdale, Fla.; *San Francisco Suite* by Ferde Grofé 4/23 at San Francisco; *Missa pro Defunctis* by Virgil Thomson 5/14 at Pottstown, N.Y.: Symphony No. 2 by Sir William Walton 9/2 at Edinburgh; *Tocata Festive* by Samuel Barber 9/30 at Philadelphia's Academy of Music; Concerto No. 2 for Violin and Orchestra by Walter Piston 10/14 at Pittsburgh; Symphony No. 7 by William Schuman 10/21 at Boston's Symphony Hall; Concerto No. 2 for Violin and Orchestra by Paul Creston 11/12 at Los Angeles.

Stage musicals *Fings Ain't Wot They Used t'Be* 2/4 at London's Garrick Theatre, with music and lyrics by Lionel Bart, 886 perfs.; *Bye Bye Birdie* 4/14 at New York's Martin Beck Theater, with Dick Van Dyke, Chita Rivera, Kay Medford, music and lyrics by Charles Strouse and Lee Adams, songs that include "Put on a Happy Face," 607 perfs.; *The Fantasticks* 5/3 at New York's 153-seat off-Broadway Sullivan Street Playhouse, with music by Harvey Schmidt, lyrics by Tom Jones, songs that include "Try to Remember," 10,000+ perfs.; *Oliver* 6/30 at London's New Theatre, with Keith Hamshere as Oliver, music and lyrics by Lionel Bart, book based on the Charles Dickens novel of 1837, songs that include "Food, Glorious Food," "Consider Yourself," "You've Got to Pick a Pocket or Two," "As Long as He Needs Me," "I'd Do Anything," "Who Will Buy," "Reviewing the Situation," 2,618 perfs.; *The Unsinkable Molly Brown* 11/3 at New York's Winter Garden Theater, with Tammy Grimes, music and lyrics by Meredith Willson, 532 perfs.; *Camelot* 12/3 at New York's Majestic Theater, with Richard Burton, Julie Andrews, Robert Goulet, music by Frederick Loewe, lyrics by Alan Jay Lerner, songs that include "If Ever I Would Leave You," the title song, 873 perfs. (the new musical is based on Arthurian legend, President-elect Kennedy attends a performance, and his administration will be identified by some with the romance of Camelot); *Wildcat* 12/11 at New York's Alvin Theater, with Lucille Ball, Keith Andes, music by Cy Coleman, lyrics by Carolyn Leigh, songs that include "Hey, Look Me Over," 171 perfs.; *Do Re Mi* 12/26 at New York's St. James Theater, with Phil Silvers, Nancy Walker, music by Jule Styne, lyrics by Betty Comden and Adolph Green, book by Garson Kanin, 400 perfs.

Popular songs "The Twist" by Hank Ballard is recorded by Ernest "Chubby Checker" Evans, 19, and launches an international dance craze. Checker performs at the small Peppermint Lounge bar off New York's Times Square, he moves on to the Copacabana night club, and discothèques where patrons dance to phonograph records blossom to cash in on the new teenage dance sensation; "Never on Sunday (The Children of Piraeus)" by Greek songwriter Manos Hadjidakis, 35 (title song for film); "The Second Time Around" by

Sammy Cahn, lyrics by Jimmy Van Heusen; "Itsy Bitsy Teenie Weenie Yellow Polka Dot Bikini" by Lee Pockriss and Paul J. Vance; "Only the Lonely" by Texas songwriter Roy Orbison, 24, with Joe Melson; "Cathy's Clown" by Don Everly and his brother Phil; "Green Fields" by Terry Gilkyson, Richard Dehr, and Frank Miller.

World chess champion Mikhail Botvinnik loses the title for the second time May 10, having held it from 1948 to 1957 and again from 1958 until now. The new champion is Soviet chess master Mikhail Tal, 23, who will hold the title for just over a year.

Floyd Patterson regains the world heavyweight boxing championship June 20 by knocking out Sweden's Ingemar Johansson in the fifth round of a title bout at New York.

Neale Fraser wins in men's singles at Wimbledon and Forest Hills, Maria Bueno in women's singles at Wimbledon, Darlene Hard, 24, at Forest Hills.

The summer Olympic Games at Rome attract 5,396 contestants from 84 countries. Soviet athletes repeat their triumph of 1956.

The Pittsburgh Pirates win the World Series by defeating the New York Yankees 4 games to 3.

Cotton's share of the U.S. textile market falls to 65 percent, down from 68 percent in 1950, while manmade fabrics increase their share to 28 percent with polyesters commanding 11 percent of the market (*see* 1970).

Brooklyn, N.Y., Mafia leaders Albert and Lawrence Gallo revolt from the Joseph Profaci organization and force Profaci to share more income with them.

Convicted kidnapper-rapist Caryl Chessman dies in the gas chamber at San Quentin May 2 at age 38 after nine stays of execution since he was sent to Death Row July 3, 1948.

Brasília becomes Brazil's federal capital by order of President Kubitshek. The new city in the uplands occupies a site that 3 years ago had only three non-Indian inhabitants, planner Lucio Costa has laid it out, Oscar Niemeyer has designed its buildings, and in 6 years it will have a population of 200,000 including 90,000 civil servants.

Boston's Mayor John F. Collins invites U.S. urban planning expert Edward J. Logue, 38, to create and administer an ambitious program for the onetime "hub of the universe." Logue has masterminded a pioneering urban renewal program in New Haven, Conn., and in the next few years will transform shabby Scollay Square into Government Center, will revitalize Roxbury, and will modernize other parts of Boston, using somewhat ruthless measures that employ the powers of eminent domain to override local zoning and building codes, relocating residents without giving them any voice in the planning.

An earthquake near Concepciòn, Chile, May 22 creates seismic waves (*tsunamis*) that shatter every coastal town between the 36th and 44th parallels. Traveling 442 miles per hour, the waves reach Hilo, Hawaii, after midnight and move on to Japan. The death toll is roughly 1,000.

Large Soviet fishing fleets move south from Newfoundland's Grand Banks using equipment far superior to that employed by U.S. fishermen to pursue the herring which is plentiful off the U.S. Atlantic Coast. Followed by well-equipped Canadian and eastern European fleets, the Soviet trawlers and purse seiners will reduce herring populations by 90 percent, virtually wiping out the haddock that has been the lifeblood of Boston fishermen.

The world fisheries catch reaches 40 million tons, up from half that amount in 1950.

Fish in the Mississippi begin to die by the millions as pollution lowers oxygen levels in the water.

China's grain production falls below 1952 levels as the Great Leap Forward program of 1957 reduces harvests and produces starvation in the country that now has 100 million more mouths to feed than in 1952, but strict rationing avoids the famine tolls of pre-Mao times (*see* 1961; 1963).

Grain worth $6 billion piles up in U.S. government-owned storage facilities, congressmen file vigorous complaints about storage costs, but the reserves will drop sharply in the decade ahead as U.S. grain relieves world hunger (*see* India, 1966).

Ten percent of the U.S. work force is on the farm, down from 18 percent in 1940, 11.6 percent in 1950.

U.S. corn yields per acre are up 75 percent over 1940, wheat yields are up 63 percent, livestock productivity is up 45 percent, milk production per cow, 30 percent, egg production per hen, 65 percent.

It takes 8 to 10 weeks and just 7 pounds of feed to produce a meaty broiler chicken in the United States, down from 12 to 15 weeks and 12 pounds of feed for a scrawnier (but tastier) broiler in 1940.

FDA researcher Frances Kelsey keeps thalidomide off the U.S. market by delaying approval of a Cincinnati firm's application to market the tranquilizer under the brand name Kevadon. Produced by a West German pharmaceutical house and widely used in Britain and West Germany for sleeplessness, nervous tension, asthma, and relief of nausea in early pregnancy, thalidomide is said to have no side effects, but Kelsey notes that it is not effective in making animals sleepy, observes that some British patients have complained of numbness in feet and fingers after using it, and points out that phocomelia is becoming endemic in West Germany where 83 such birth defects appear at the clinics that saw 12 last year (*see* 1961).

Aluminum cans, used commercially for the first time for food and beverages, will come to be the single largest use of aluminum. They are not biodegradable (tin-plated steel cans rust in time) and present environmental problems of litter; 95 percent of U.S. soft drinks and 50 percent of beer is sold in returnable bottles typically used 40 to 50 times each (*see* 1962).

1960 *(cont.)* U.S. margarine consumption reaches 9.4 pounds per capita, up from 8.6 pounds in 1957; butter consumption falls to 7.5 pounds, down from 8.3 (*see* 1963).

Annual U.S. beef consumption reaches 99 pounds per capita.

Iowa Beef Processors (IBP) is founded at Denison, Iowa, by former Sioux City cattle buyers who include Currier J. Holman, 46. IBP will grow from a single slaughterhouse to become the world's largest meat packer (larger than Swift, Armour, Wilson, Morrell, and Cudahy combined) but Holman will be convicted of having ties with a distributor who bribed supermarket executives and butchers' union officials.

Chicago's last packing house closes as packers shift activities to the West (*see* 1953).

Dole Corp. is created by a renaming of the 58-year-old Hawaiian Pineapple Co. started by James Dole (*see* 1922; Castle & Cooke, 1964).

Howard Johnson has 607 independently owned restaurants in his franchise, making it the third largest U.S. food distributor, surpassed only by the U.S. Army and Navy (*see* 1937). Now 63, Johnson heads a family-owned enterprise that operates 296 restaurants.

Domino's Pizza opens at Detroit. Local entrepreneur Thomas S. Monaghan, 23, has borrowed $500 to buy a pizza parlor that will grow to have thousands of franchised outlets.

Enovid 10, introduced in August by G. D. Searle, is the first commercially available oral contraceptive. Searle biochemist Byron Riegel, 53, has headed the group that developed it (*see* Puerto Rican test, 1957). Fifty women at Birmingham, England, cooperate with Searle to make the first British test of an oral contraceptive (*see* 1961).

The Pill (above) sells at 55¢ each and costs a woman $11 per month. Condoms continue to account for $150 million of the $200 million U.S. contraceptive business with diaphragms and spermicidal creams and gels accounting for most of the rest.

More than 34 percent of U.S. women over age 14 and 31 percent of all married women are in the labor force, up from 25 percent of women over age 14 in 1940. Only 10 percent of working women are farm hands or servants, down from 50 percent in 1900.

141 world cities record populations of 1 million or more, up from 16 in 1900. Tokyo has 9.6 million, New York and London 7.7 million each, Shanghai 6.2, Moscow 5, Mexico City 4.8, Buenos Aires 4.5, Bombay 4.1, São Paulo more than 4 million, up from 250,000 in 1900, 1 million in 1929.

The world population tops 3 billion, up from 2 billion in 1930.

Forty-eight percent of U.S. blacks live outside the 11 states of the old Confederacy, up from 30 percent in 1940.

U.S. population density approaches 50 per square mile, up from 10.6 in 1860; nearly 70 percent of the population is urban.

1961 Fidel Castro demands that the U.S. embassy staff at Havana be reduced to 11, Washington severs relations January 3, two major Cuban opposition groups set up a revolutionary council at New York in late March with former Cuban premier José Miro Cardona as president, he urges all Cubans to revolt against Castro, and Cuban Foreign Minister Raul Roa charges at the UN April 15 that U.S. and Latin American forces are preparing to invade Cuba.

The Bay of Pigs invasion ends in disaster and embarrassment for the United States. Some 1,600 Cuban exiles trained by the CIA in Florida, Louisiana, and Guatemala land near the Bahia de los Cochinos April 17 in an inept and ill-supported effort to overthrow the Castro government, Castro's forces repel the invaders with heavy loss of life.

Premier Khrushchev demands April 18 that the invasion of Cuba be halted, he promises to aid Cuba, President Kennedy replies that the United States will not permit outside military intervention, and Kennedy asserts April 20 that the United States will take steps to halt Communist expansion if U.S. security is threatened, but survivors of the Bay of Pigs withdraw.

Premier Castro announces May 17 that Cuba will exchange prisoners taken at the Bay of Pigs (above) for 500 U.S. bulldozers, negotiations break down June 30, and Castro declares himself a Marxist-Leninist December 2. He announces formation of a united party to bring communism to Cuba.

Dominican dictator Rafael Leonidas Trujillo, now 70, is killed May 30 at Ciudad Trujillo by an eight-man assassination team that has caught the generalissimo unguarded after 30 years in which he has controlled the country directly or indirectly. Ciudad Trujillo is renamed Santo Domingo.

President Kennedy and Premier Khrushchev confer at Vienna June 3–4 and issue a joint communiqué reaffirming support for the neutrality of Laos and stating that they have discussed disarmament, a nuclear test ban, and the German question.

A Soviet memorandum issued June 4 urges demilitarization of Berlin. Premier Khrushchev declares June 15 that Moscow will conclude a treaty by year's end and "rebuff" any Western move to enforce rights of access to West Berlin. London, Paris, and Washington reject Soviet proposals to make Berlin a free city. President Kennedy delivers a national address July 25 proposing a 217,000-man increase in the armed forces and a $3.4 billion boost in defense spending to meet the "worldwide Soviet threat."

East German authorities close the border between East and West Berlin August 13, and the Berlin Wall erected from August 15 to August 17 stops movement from East Berlin to West Berlin. The East German parliament has built the Wall in response to a communiqué from Warsaw Pact nations appealing for a halt in the mass exodus of East Berliners to the West.

Some 1,500 troops enter Berlin in mid-August to reinforce the Western garrison, East Berlin authorities

The Berlin Wall exemplified cold war hostilities between East and West. Stalin's ghost still hung over Europe.

block entry of U.S. civilians October 26 demanding that they show identity papers, U.S. and Soviet tanks move briefly to the Friedrichstrasse crossing point but are withdrawn October 28.

China's Premier Zhou En-lai walks out of a Soviet party congress at Moscow October 23 heralding a break in Sino-Soviet relations.

A thermonuclear device tested in the Novaya Zemlya area of the U.S.S.R. October 30 creates a shock wave that circles the earth in 36 hours, 27 minutes. Experts estimate the power of the device at upwards of 57 megatons (57 million tons of trinitrotoluene—TNT) and two more shock waves follow the first.

Premier Khrushchev declares November 7 that he is willing to postpone settlement of the Berlin issue.

The African charter of Casablanca proclaimed January 7 by the heads of state of Ghana, Guinea, Mali, Morocco, and the United Arab Republic meeting at Casablanca announces that a NATO-like organization of African states will be established.

Congolese ex-premier Patrice Lumumba is killed January 17, the Katanga government announces his death February 13 saying that hostile tribesmen murdered Lumumba, but Moscow charges UN Secretary-General Hammarskjöld with having been an "accomplice" in the murder (*see* below). The UN Security Council demands an inquiry into Lumumba's death and urges use of force to prevent civil war in the Congo.

Rwanda (Ruanda-Urundi in the former Belgian Congo) proclaims herself a republic January 28 and Gregoire Kayibanda is proclaimed president October 26 following new elections in which the Hutu party has gained victory (*see* 1960).

The Peace Corps of Young Americans for overseas service is created March 1 by President Kennedy who has said in his inaugural address, "Ask not what your country can do for you—ask what you can do for your country." Peace Corps volunteers will work to improve education, agriculture, and living standards in Latin America, Asia, and Africa.

Algiers is seized April 21 by insurgent French troops led by Gen. Maurice Challe, loyal French troops crush the insurrection and reoccupy the city April 26, the rebel leaders are tried and convicted July 11, Gen. Salan and others are sentenced in absentia to death as Algerian independence talks proceed (*see* 1958; 1962).

Sierra Leone gains independence April 27. The monarchy will become a republic in 1971.

Tanganyika gains full internal self-government May 1 with Julius Nyerere as premier.

The United States of the Congo is founded May 12 with her capital at Leopoldville as hostilities continue in the Congo.

Angola begins an insurrection against the Portuguese; hostilities will continue until Lisbon offers independence in 1974.

South Africa severs her ties with the British Commonwealth May 31 and becomes a republic with Charles R. Swart as president. Ghana refuses to recognize the new republic.

UN Secretary-General Dag Hammarskjöld (above) is killed September 18 at age 56 when his plane crashes in the Congo en route to a meeting with the governor of Katanga Moise Tshombe whose army is fighting UN forces attempting to disarm Katanga troops. Burmese diplomat U Thant, 52, has served as acting secretary general and ordered UN commanders in the Congo to take all necessary action to restore the UN position at the Katanga capital of Elizabethville. He is unanimously elected to succeed Hammarskjöld November 3 and will hold the post until 1972. A cease-fire takes effect December 21.

A South Korean military junta overthrows the democratic government May 16. The "anti-Communist" junta decrees absolute military dictatorship June 6 and Gen. Chung Hee Park, 44, takes over July 3, beginning a repressive regime that will continue even after his assassination in 1979.

Syrian troops revolt September 28. The revolutionary command sets up a civilian government and proclaims independence from the United Arab Republic September 29.

The Twenty-Third Amendment ratified March 29 provides for congressional representation of Washington, D.C., which is now largely black.

The UN General Assembly condemns South Africa's apartheid policies April 13.

U.S. "Freedom riders" begin a civil rights demonstration at Birmingham, Ala., May 4 in a move organized by the biracial Congress of Racial Equality (CORE) to overturn segregation practices in the Deep South. A white mob attacks the freedom riders May 20 at Montgomery, Ala., U.S. marshals sent in by Attorney Gen-

1961 ***(cont.)*** eral Robert F. Kennedy restore order and maintain safe passage for travelers in interstate commerce, but similar demonstrations occur in Louisiana and other states (*see* Supreme Court, 1962).

The first manned space ship circles the earth April 12 in 89.1 minutes at an altitude of 187.7 miles. Soviet astronaut Yuri Alekseyevich Gagarin, 27, makes the orbit in the space vehicle Vostok I, and astronaut (or cosmonaut) Gherman Titov, orbits the earth 17 times less than 4 months later (*see* Glenn, 1962).

Navy Commander Alan B. Shepard, Jr., makes the first U.S. manned space expedition May 5.

An Alliance for Progress formed at Punta del Este, Uruguay, and announced by President Kennedy in March will spur economic and social development of Latin-American countries which cooperate with the United States. Agrarian reform and tax reform are major objectives of the new *Alianza* created in August by an agreement signed with 19 Latin-American countries which are promised $10 billion in U.S. aid over a 10-year period.

A Latin-American free trade association comes into force June 2.

U.S. ambassador to the United Nations Adlai Stevenson urges that the developed countries of the world each contribute 1 percent of their Gross National Product to the development of the emerging nations.

South Africa (above) adopts the decimal system; the rand becomes legal tender February 14.

Britain's Chancellor of the Exchequer Selwyn Lloyd announces an austerity program July 17 to improve the nation's trade deficit. His emergency budget imposes a hold on wages and raises the bank rate from 5 percent to 7 percent.

Rocket ships soared from science fiction into reality. A U.S.-Soviet space race propelled space-age development.

A new Fair Labor Standards Act signed into law by President Kennedy May 5 takes effect in September, raising the minimum wage to $1.15 per hour.

New York's First National City Bank offers fixed-term certificates of deposit paying a higher rate of interest than is permitted on savings accounts. Citibank offers the CDs in denominations as low as $500 and other banks soon follow suit as depositors seek ways to keep inflation from eroding their savings.

The U.S. Gross National Product reaches $521 billion, up 60 percent since World War II as measured in constant dollars.

The S.S. *France* launched by the Compagnie Générale Transatlantique is the longest and last of the great transatlantic passenger liners. Measuring 1,035.2 feet in overall length, the $81.3 million 66,348-ton French Line ship will make her maiden voyage from Le Havre to New York beginning February 3 of next year.

The S.S. *Canberra,* launched for the Cunard Line, is a 44,807-ton passenger liner 818.5 feet in length.

The first London minicabs, introduced March 6 by Carline of Wimbledon, begin to replace the city's large, comfortable Austin taxicabs.

The Department of Justice indicts General Motors for having forced most U.S. railroads to buy locomotives made by GM's Electro-Motive Division or risk having GM cars shipped by other carriers. The criminal proceedings will be dropped in December 1964.

The United States has 11.7 million trucks, up from 8.62 million in 1951.

Britain's minister of health Enoch Powell increases British Health Service charges February 1.

Acetaminophen tablets gain FDA approval in July as an alternative to aspirin. The analgesic introduced under the brand name Tylenol by McNeill Laboratories division of Johnson & Johnson is far less likely to cause gastric irritation, reduces fever, is effective against headaches and toothaches, but does not have acetylsalicylic acid's other properties (e.g., reducing muscular and arthritic pain).

Phocomelia deforms 302 newborn infants in West Germany, a Hamburg physician notes that mothers of several of the infants have taken the tranquilizer drug thalidomide, and the German Ministry of Health issues a warning to physicians (*see* 1960; 1962).

U.S. television programming is a "vast wasteland," says FCC chairman Newton N. Minow, 35, in a May address to the National Association of Broadcasters convention at Washington, D.C. "I invite you to sit down in front of your television set when your station goes on the air, and stay there. You will see a vast wasteland—a procession of game shows, violence, audience participation shows, formula comedies about totally unbelievable families. . . blood and thunder. . . mayhem, violence, sadism, murder. . . private eyes, more violence, and cartoons. . . and, endlessly, commercials—many screaming, cajoling, and offending."

The IBM Selectric typewriter designed by Eliot Fette Noyes, 51, is introduced by International Business Machines (*see* computer, 1955). The Selectric has a moving "golf ball" cluster of interchangeable type, will be linked in 1964 to a magnetic tape recorder that permits automated, individually addressed original copies of any letter, and by 1975 will account for an estimated 70 to 80 percent of the electric typewriter market (*see* word processor, 1974).

Newsweek magazine is purchased from the Vincent Astor estate by the *Washington Post*, whose president Philip L. Graham, 46, is a son-in-law of the late Eugene Meyer (*see* 1937).

Nonfiction *Nobody Knows My Name* (essays) by James Baldwin who returned to the United States from France in 1957 but will live much of his life abroad; *A Study of History* (12th and final volume) by English historian Arnold Toynbee, 72, whose first volume appeared in 1934.

Fiction *Cat and Mouse (Katze und Maus)* by German novelist Günter Grass, 33; *A House for Mr. Biswas* by Trinidadian novelist V. S. (Vidiadhur Surajprasad) Naipaul, 29; *Catch-22* by U.S. novelist Joseph Heller, 38, who has taken 7 years to write his work about a World War II bomber group that is a metaphor for U.S. society: "There was only one catch and that was Catch-22, which specified that a concern for one's own safety in the face of dangers that were real and immediate was the process of a rational mind. Orr was crazy and he could be grounded. All he had to do was ask: and as soon as he did he would no longer be crazy and would have to fly more missions. Or be crazy to fly more missions and sane if he didn't, but if he was sane he had to fly them. If he flew them he was crazy, didn't have to; but if he didn't want to he was sane and had to . . ."; *The Fox in the Attic* by Richard Hughes; *A New Life* by Bernard Malamud; *A Shooting Star* by Wallace Stegner; *Report to Greco* by the late Nikos Kazantzakis who died in 1957; *The Pawnbroker* by Edward Lewis Wallant; *The Prime of Miss Jean Brodie* by Muriel Spark; *A Burnt Out Case* by Graham Greene; *Riders in the Chariot* by Patrick White; *The Moviegoer* by U.S. writer Walker Percy, 45; *The Lime Twig* by U.S. novelist John Hawkes, 36; *The Spinoza of Market Street* (stories) by Isaac Bashevis Singer; *Mrs. Golightly* (stories) by Ethel Wilson; *Seduction of the Minotaur* by Anaïs Nin; *The Shipyard (El astillero)* by Juan Carlos Onetti; *A Passion in Rome* by Morley Callahan; *Franny and Zoey* by J. D. Salinger; *Thunderball* by Ian Fleming.

Poetry Soviet authorities denounce Yevgeny Yevtushenko for a poem about the Nazi massacre of Ukrainians at Babi Yar in 1941. "No monument stands over Babi Yar./A drop sheer as a crude gravestone./I am afraid," the poem begins, and critics interpret the verse as a protest against Soviet anti-Semitism.

Painting *Still Life with Lamp Light* by Pablo Picasso; *Blue II* by Joan Miró; *New Madrid* by Frank Stella; *Delta Nu* by U.S. painter Morris Louis, 49; *The Italians* by U.S. painter Cy Twombly, 22; *Switchsky's Syntax* by Stuart Davis. Grandma Moses dies at Hoosick Falls, N.Y., December 13 at age 101.

Sculpture *Magic Base* (any person or thing is to be considered art as long as it rests on this base), *Base of the World*, bread rolls covered with kaolin, and *Line 1,000 Meters Long* by Piero Manzoni; *Box With Sound of Its Own Making* by U.S. sculptor Robert Morris, 30.

Kodachrome II color film introduced by Eastman Kodak is 2½ times faster than the Kodachrome introduced in 1936 (see Ektachrome, 1946; Instamatic, 1963).

Theater *The American Dream* by Edward Albee 1/24 at New York's off-Broadway York Theater, 370 perfs.; *Mary, Mary* by U.S. playwright Jean Kerr, 37, 3/8 at New York's Helen Hayes Theater, with Barbara Bel Geddes, Barry Nelson, 1,572 perfs.; *Luther* by John Osborne 6/26 at Nottingham's Theatre Royal; *Happy Days* by Samuel Beckett 9/17 at New York's off-Broadway Cherry Lane Theater; *Purlie Victorious* by U.S. playwright Ossie Davis, 44, 9/28 at New York's Cort Theater, with Davis as Purlie Victorious Judson, Ruby Dee, Godfrey Cambridge, Alan Alda, 261 perfs.; *Gideon* by Paddy Chayefsky 11/9 at New York's Plymouth Theater, with Fredric March, George Segal, 236 perfs.; *Andorra* by Max Frisch 11/2 at Zurich's Schauspielhaus; *The Detour (Der Abstecher)* by German playwright Martin Walser, 34, 11/28 at Munich's Werkramm Theater der Kammerspiele; *The Night of the Iguana* by Tennessee Williams 12/28 at New York's Royale Theater, with Patrick O'Neal, Bette Davis, Margaret Leighton, Alan Webb, 316 perfs.

Films Robert Rossen's *The Hustler* with Jackie Gleason (as "Minnesota Fats"), Paul Newman, Piper Laurie, George C. Scott; Stanley Kramer's *Judgment at Nuremberg* with Spencer Tracy, Burt Lancaster, Richard Widmark, Maximilian Schell, Montgomery Clift; François Truffaut's *Jules and Jim* with Jeanne Moreau, Oskar Werner, Henri Serre; Billy Wilder's *One, Two, Three* with James Cagney, Arlene Francis, Horst Buchholz; Daniel Petrie's *Raisin in the Sun* with Sidney Poitier, Claudia McNeil, Ruby Dee, Diana Sands, Louis Gossett; Vittorio De Sica's *Two Women* with Sophia Loren, Eleanora Brown, Jean-Paul Belmondo. Also: Blake Edwards's *Breakfast at Tiffany's* with Audrey Hepburn, George Peppard, Martin Balsam, Patricia Neal, Mickey Rooney; Shirley Clark's *The Connection* with William Redfield; J. Lee Thompson's *The Guns of Navarone* with Gregory Peck, David Niven, Anthony Quinn; Jack Clayton's *The Innocents* with Deborah Kerr, Michael Redgrave, Martin Stephens, Pamela Franklin in a film version of the 1898 Henry James short story "The Turn of the Screw"; Delbert Mann's *Lover Come Back* with Rock Hudson, Doris Day, Edie Adams; John Huston's *The Misfits* with Marilyn Monroe, Montgomery Clift, Clark Gable (who dies 11/16 at age 59); Roger Corman's *The Pit and the Pendulum* with Vincent Price; Peter Glenville's *Summer and Smoke* with Geraldine Fitzgerald, Laurence Harvey; Tony Richardson's *A Taste of Honey* with Rita Tushingham, Robert Stephens; Basil Dearden's *Victim* with Dirk Bogarde,

1961 ***(cont.)*** Sylvia Sims; Luis Buñuel's *Viridiana* with Silvia Pinal, Fernando Rey; Bryan Forbes's *Whistle Down the Wind* with Hayley Mills, Alan Bates.

Hollywood musicals Robert Wise's *West Side Story* with Natalie Wood, Rita Moreno, Richard Beymer, choreography by Jerome Robbins.

Stage musicals *Carnival* 4/13 at New York's Imperial Theater, with Anna Maria Alberghetti, Kaye Ballard, Jerry Orbach, music and lyrics by Bob Merrill, 719 perfs.; *Beyond the Fringe* (revue) 5/10 at London's Fortune Theatre, with former Oxbridge students Alan Bennett, Peter Look, Jonathan Miller, and Dudley Moore, all in their mid-20s; *Stop the World I Want to Get Off* 7/23 at the Queen's Theatre, London, with Anthony Newley, 29, music and lyrics by Leslie Bricusse, 30, and Newley, songs that include "Gonna Build Me a Mountain," "What Kind of Fool Am I?" 485 perfs.; *Milk and Honey* 10/10 at New York's Martin Beck Theater, with Robert Weede, Mimi Benzell, Molly Picon, music and lyrics by Jerry Herman, 543 perfs.; *How to Succeed in Business Without Really Trying* 10/14 at New York's 46th Street Theater, with Robert Morse, Rudy Vallée, book by Shepherd Mead, music and lyrics by Frank Loesser, songs that include "I Believe in You," "Brotherhood of Man," 1,417 perfs.; *Subways Are for Sleeping* 12/27 at New York's St. James Theater, with Sydney Chaplin, Carol Lawrence, Orson Bean, music by Jule Styne, lyrics by Betty Comden and Adolph Green, songs that include "Comes Once in a Lifetime," 205 perfs.

Oratorio *Jacob's Ladder (Die Jakobsleiter)* by the late Arnold Schoenberg 6/17 at Vienna.

First performances Gloria by Francis Poulenc 1/20 at Boston's Symphony Hall; Symphony No. 7 by Walter Piston 2/10 at Philadelphia's Academy of Music; Symphony No. 12 *(The Year 1917)* by Dmitri Shostakovich 10/9 at Leningrad.

Popular songs "Moon River" by Henry Mancini, now 37, lyrics by Johnny Mercer (for the film *Breakfast at Tiffany's*); "It Was a Very Good Year" by Ervin Drake; "Running Scared," "Crying," and "I'm Hurting" by Roy Orbison.

Bob Dylan begins singing at Greenwich Village, New York, coffeehouses, is discovered by Columbia Records vice-president John Hammond, and releases his first album. Now 20, the Minnesotan has changed his name from Robert Zimmerman in honor of the late Welsh poet Dylan Thomas and will provide civil rights demonstrators and student protest movements of the 1960s with their anthems "Blowin' in the Wind" and "The Times, They Are A-Changin'."

The Supremes sign a contract with Berry Gordy's 4-year-old Motown Corp. and cut their first records. Detroit singers Diana Ross, Mary Wilson, and Florence Ballard are still in their teens but as The Supremes they will make eight gold records in less than 2 years and have seven top records.

U.S. saxophonist John William Coltrane gains his first wide acclaim at age 35 with his John Coltrane Quartet.

Mikhail Botvinnik regains the world chess title May 12 by defeating Mikhail Tal and winning the championship for an unprecedented third time at age 49 *(see* 1960; 1963).

Rodney George "Rod" Laver, 22 (Australia), wins in men's singles at Wimbledon, Angela Mortimer, 29, (Brit), in women's singles; Roy Emerson, 24 (Australia), wins in men's singles at Forest Hills, Darlene Hard in women's singles.

The Minnesota Twins and Los Angeles Angels play their first season.

Roger Maris of the New York Yankees breaks Babe Ruth's home run record of 60 October 1 at Yankee Stadium, but purists maintain that the ball Ruth hit was heavier and Ruth set his record in a 154-game season while the October 1 game is the 162nd (and last) Yankee game for 1961.

The New York Yankees win the World Series by defeating the Cincinnati Reds 4 games to 1.

U.S. cigarette makers spend $115 million on television advertising, up from $40 million in 1957 (*see* Surgeon General's Report, 1964; Banzhaf, 1966).

The Death and Life of Great American Cities by U.S. social critic Jane Jacobs, 45, observes that cities were safer and more pleasant when they consisted of neighborhood communities where people lived in relatively low-priced buildings, knew their neighbors, and lived in the streets and on their doorsteps rather than in the depersonalized environment characteristic of modern cities. Critic Lewis Mumford, 65, notes that congested 18th-century cities were hardly safer or healthier.

New York adopts a new zoning resolution that permits buildings to contain a maximum of 12 times as much floor space as the area of the original site with special bonuses for enlightened land use (*see* 1916: the Equitable Building of 1915 contained nearly 30 times as much floor space as was contained in its land site). Chief effect of the new resolution will be to encourage architects to set off their buildings with open plazas.

New York's Chase Manhattan Bank building, completed by Skidmore, Owings & Merrill in lower Manhattan, is a 60-story glass and aluminum tower that gives the financial district its first open plaza.

Century City goes up on a 180-acre site in west Los Angeles formerly owned by Twentieth Century Fox studios. Aluminum Corp. of America (Alcoa) pursues a grandiose scheme for a city within a city that will provide homes for 12,000 with office and retail space for 20,000, a scheme projected by New York real estate operator William Zeckendorf, 56.

Haleakala National Park is established by act of Congress in the new state of Hawaii. Its 26,403 acres include the dormant 10,023-foot Haleakala Volcano.

Hawaii's legislature enacts the first statewide land-use zoning plan, setting aside some land for agriculture, some for conservation, and some for urban development.

A study by the California Wildland Research Center reveals that only 17 million acres of U.S. wilderness remain, down from 55 million in 1926. Virgin land has been disappearing for 35 years at the rate of a million acres per year; the largest single unit of wilderness remaining embraces 2 million acres (*see* 1964).

A conference aimed at preserving Africa's wildlife convenes in Tanganyika in September.

Britain and Iceland settle a fisheries dispute but controversy over fishing rights in the North Atlantic will continue with "cod wars" between the two.

China suffers further crop failures (*see* 1960; 1963).

Widespread mechanical harvesting of processing tomatoes for use in canning, ketchup, paste, sauces, tomato juice, and tomato soup begins in California as farmers plant the tough-skinned VF 145-B7879 variety developed by researchers at the Davis campus of the University of California. Rising labor costs have threatened the industry in California which produces much of the world's processing tomatoes. By 1975 California will be producing more than 7 million tons of processing tomatoes per year, up from 1.3 million in 1954 (table tomatoes will continue to be hand-picked, partly because machine-harvested tomatoes must be in rows separated by wide spaces to permit passage of machines).

Frito-Lay, Inc., is created by a merger of Atlanta's H. W. Lay Co. and the Frito Co. of Dallas (*see* 1932; 1939; PepsiCo, 1965).

Sprite, introduced by Coca-Cola Co., is a lemon-lime drink that competes with 7-Up (*see* 1933; Tab, 1963).

Coffee-Mate nondairy creamer is introduced by Carnation Co. (see Pream, 1952). The powder is made of corn syrup solids, vegetable fat, sodium caseinate, and various additives (*see* "contented cows," 1906).

Total breakfast food, introduced by General Mills, will be promoted for its nutritional qualities (*see* Lucky Charms, 1964).

Green Giant enters the frozen foods business with frozen June peas, niblet corn, green beans, and baby lima beans frozen in a pouch with butter sauce (*see* 1958).

Canned pet foods are among the three top-selling categories in U.S. grocery stores. Americans feed an estimated 25 million pet dogs, 20 million pet cats.

McDonald's hamburger stands begin a vast proliferation as Ray Kroc buys out the McDonald brothers and acquires all rights to the McDonald name for $14 million, including interest costs (*see* 1954). Kroc has established more than 200 stands and will build McDonald's into a worldwide chain.

Conovid pills, made by G.D. Searle with 5 mg. of hormone, are the first oral contraceptives sold in Britain.

A birth control clinic opens at New Haven, Conn., but is forced to close after 9 days. The order will not be rescinded for nearly 4 years (*see* New York, 1929; Supreme Court, 1965).

The Lippes Loop intrauterine device for contraception developed by Buffalo, N.Y., physician Jack Lippes will be implanted within 8 years in an estimated 8 million women worldwide.

China's population reaches an estimated 650 million, India has 445 million, the U.S.S.R. 215, the United States 185 (up from 141 at the close of World War II), Indonesia 100, Pakistan 97, Japan 95, Brazil 73, West Germany 58, the U.K. 54.

New York and its suburban environs have a population of 14.2 million, Tokyo 10 million, London 8.3, Paris 7.7, Shanghai more than 7, Buenos Aires 6.9, Los Angeles 6.6, Chicago more than 6, Moscow 5.2, Mexico City 5, Calcutta 4.5, Bombay 4.2, Beijing 4.2, Philadelphia 3.7.

1962 The Cuban missile crisis in October produces a tense nuclear confrontation between Washington and Moscow. U.S. aerial surveillance has discovered Soviet offensive missile and bomber bases in Cuba, President Kennedy orders an air and sea "quarantine" of Cuba to prevent shipment of arms to Fidel Castro, Attorney General Kennedy meets with Soviet ambassador Dobrynin at his embassy and learns that Moscow will withdraw the missiles with their atomic warheads from Cuba if U.S. nuclear missiles are withdrawn from Turkey, President Kennedy rejects the deal when Premier Khrushchev offers it publicly, but U.S. general Lauris J. Norstad (who opposes quick removal of the 15 obsolescent Jupiter rockets from Turkey) is unexpectedly retired from his post as NATO commander, Khrushchev agrees to dismantle the Cuban missile sites and remove them, the Cuban blockade ends, and the U.S. missiles in Turkey are quietly removed.

Soviet authorities have released U.S. espionage pilot Gary Powers February 10 in exchange for Soviet espionage agent Col. Rudolf Abel who has been in U.S. hands since 1957 (*see* Powers, 1960).

South Vietnamese forces launch "Operation Sunrise" March 22 to eliminate Vietcong guerrillas. U.S. money, arms, and field observers help the South Vietnamese. The International Control Commission on Indo-China, composed of Canadian, Indian, and Polish representatives, reports June 2 that North Vietnam is supplying the Vietcong rebels in violation of the 1954 Geneva agreement on Vietnam, but Polish representatives do not sign the report.

Civil war threatens in the Independent Congo Republic following President Joseph Kasavubu's repudiation of the Fundamental Law of May 1960. Fighting breaks out at Stanleyville January 13, UN Secretary-General U Thant orders UN troops in the Congo to stop the fighting, the Katanga assembly agrees February 15 to reunite Katanga with the Congo, but new fighting breaks out in October between Katanga troops and the forces of the central government (*see* 1965).

An organization of African states is established in early February by leaders of 20 nations meeting at Lagos, Nigeria.

Ghana, Guinea, Mali, Morocco, and the United Arab Republic boycott the Lagos meeting; their ministers

1962 ***(cont.)*** meet at Casablanca in April and agree to set up an African common market (below).

Ghana's Kwame N. Nkrumah declares a general amnesty for refugees May 5 and orders the release of many political prisoners.

Burundi in Central Africa gains full independence from Belgium July 1 with a monarchy dominated by the Tutsi (Batusi and Watusi) minority. Attempts to overthrow the government by Hutu tribesmen will fail, and Tutsi from Burundi will attack Rwanda until 1967.

Algeria's rebel provisional government receives no invitation to Lagos (above) and violence continues there; the head of the illegal Secret Army Organization (OAS) has issued a manifesto calling for mobilization of Algerians against a cease-fire with France.

Paris proclaims Algerian independence July 3 after a national referendum in which the Algerians have voted for independence by nearly 6 million to 16,534. Leaders of the provisional government have a falling out that leads to civil war, but in late September the national assembly asks Mohammed Ben Bella to form a cabinet and he will be Algeria's president until 1965.

Eritrea becomes an integral part of Ethiopia after 10 years of union on a federal basis but conflict will persist between Muslim Eritreans and Christian Ethiopians.

Jamaica, B.W.I., gains full independence August 6.

Trinidad and Tobago gains independence August 31.

Western Samoa has gained independence January 1 after more than 47 years of New Zealand colonial rule.

Uganda becomes an independent state within the British Commonwealth October 9. The country adopts a federal form of government to overcome the reluctance of Buganda's king Sir Edward Mutesa II to abandon the privileged position of his tribe and his country, but conflict will continue between President Mutesa and Milton Obote (*see* 1966).

Tanganyika becomes a republic December 9 with Julius K. Nyerere as president (*see* 1961; 1964).

Israeli authorities hang Adolf Eichmann May 31 in punishment for his concentration-camp activities during World War II.

U.S. manufacturing firms with federal contracts of $50,000 or more are ordered January 29 to report the number of blacks on their payrolls.

The U.S. Department of Labor establishes minimum wages for migratory Mexican workers March 9.

The National Farm Workers Association (NFWA), founded by community leader César Estrado Chavez, 32, represents stoop-labor in California's Coachella, Imperial, and San Joaquin valleys (*see* 1965).

The Department of Justice orders court action to halt racial segregation in hospitals built with federal funds May 8.

Southern School News reports that at least some token school integration has taken place in response to the Supreme Court decision of 1954 in all Southern states except Alabama, Mississippi, and South Carolina but that practically no action has been taken in Northern school districts.

Racial tension grips the South as black student James Howard Meredith, 29, attempts to enter the University of Mississippi which has been ordered to admit the Air Force veteran by a federal appellate court whose order has been upheld by the Supreme Court. Rioting breaks out on the "Old Miss" campus September 30 despite a televised appeal by President Kennedy for peaceful desegregation, U.S. marshals accompany Meredith, federal troops stand guard beginning October 1, and Meredith begins 10 months living on campus under constant federal guard (*see* 1966).

A New Orleans federal appeals court orders the Justice Department to bring criminal proceedings against Mississippi's governor Ross Barnett who has interfered with the court's order to admit black students to "Old Miss" (above).

The Supreme Court reverses the convictions of six Freedom Riders who were convicted in Louisiana after trying to desegregate a bus terminal.

City College of New York psychology professor Kenneth Bancroft Clark, 48, founds Harlem Youth Opportunities Unlimited (HARYOU) to combat unemployment among youth in the New York ghetto and supplement local teaching facilities.

Marine Corps pilot John Glenn makes the first U.S. Earth orbits February 7; launched into space in the Mercury capsule Friendship 7, he completes three orbits, covering 81,000 miles at an altitude of 160 miles (*see* 1961; 1969; Glenn, 1957).

President Kennedy embargoes nearly all trade with Cuba February 3 (*see* missile crisis, above; cigars, below).

President Kennedy reduces tariff duties on some 1,000 items March 7 to increase foreign trade.

United States Steel raises prices $6 per ton April 10, President Kennedy reacts angrily, two firms do not follow Big Steel's lead, and White House pressure forces a price rollback April 14.

African ministers meeting at Casablanca in April (above) agree not only to establish a common market but also to set up an African payments union and an African development bank.

Wall Street's Dow Jones Industrial Average falls to 535.76, down from 734.91 late last year.

Consolidated Edison announces plans to build a million-kilowatt nuclear generating station at Ravenswood, Queens, in the heart of New York City. The Atomic Energy Commission rejects the proposal but approves construction of a plant 28 miles north of the city at Indian Point on the Hudson (*see* 1963; Chicago, 1960).

Europe's Arlberg-Orient Express goes out of service May 27 after nearly 79 years of operation between Paris and Istanbul and the Simplon-Orient Express ends service as well. Both have been victims of the airplane that has cut travel time between Paris and Istanbul to 2 hours.

The IL-62 jet designed by S. V. Ilyushin goes into service for Aeroflot (*see* 1954; 1968).

Paris and London sign an agreement November 29 for joint design and development of a supersonic passenger plane (*see* 1969).

John Foster Dulles International Airport opens 24 miles from downtown Washington to supplement Baltimore's Friendship and Washington's National airports. Dulles is the first civil airport specifically designed for jets; its mobile lounges, which adjust to the height of any aircraft door, spare passengers the long walk to loading gates.

The Lear jet, introduced by aviation pioneer William P. Lear, will be the leading make of private jet aircraft within 5 years.

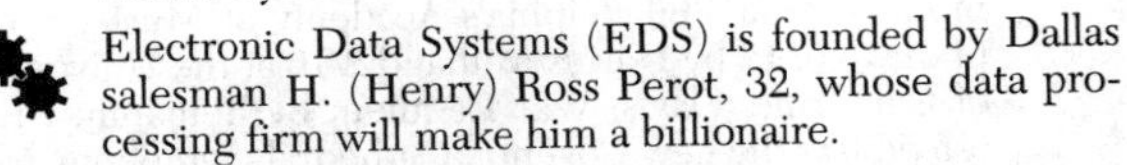

Electronic Data Systems (EDS) is founded by Dallas salesman H. (Henry) Ross Perot, 32, whose data processing firm will make him a billionaire.

John F. Enders, now 65, produces the first successful measles vaccine, but it will not be introduced until 1966 (*see* Salk, 1952).

The Esalen Institute encounter-group mental therapy center, founded at Big Sur, Calif., by entrepreneurs Michael Murphy, 32, and Richard Price, will attract such psychologists and psychotherapists as Fritz Perls and Ida Rolf.

Congress creates COMSAT (Communications Satellite Corp.) to handle space communications on a profit-making basis under government supervision (*see* Echo 1, 1960; Early Bird, 1965).

Telstar I is launched the night of July 10. The new satellite is used to transmit the first live transatlantic telecasts between the United States and Britain.

Ninety percent of U.S. households have at least one TV set; 13 percent have more than one.

ABC begins color telecasts for 3½ hours per week beginning in September, 68 percent of NBC prime evening time, programming is in color, but CBS confines itself to black and white after having transmitted in color earlier. All three networks will be transmitting entirely in color by 1967.

Der Spiegel publishes an exposé of West Germany's Defense Minister Franz Josef Strauss, revealing the weakness of German armed forces despite expenditure of large sums of money. Staff members are arrested in late October on orders from Strauss, Chancellor Adenauer accepts Strauss's resignation under pressure from public opinion, but *Der Spiegel* publisher Rudolf Augstein will not be released for 4 months (*see* 1946).

Former Vice President Nixon holds a press conference in November after losing a bid to defeat California governor Edmund G. "Pat" Brown by nearly 300,000 votes: "As I leave you I want you to know—just think how much you're going to be losing: You won't have Nixon to kick around any more, because, gentlemen, this is my last press conference." He complains that the press has been against him ever since his attacks on Alger Hiss in the late 1940s.

Nonfiction *The Rich Nations and the Poor Nations* by English economist Barbara Mary Ward, 48; *The Other America: Poverty in the United States* by former St. Louis welfare worker Michael Harrington, 34, says the nation has a huge "underclass" of employed people living below the poverty line and neglected by society; *Travels with Charley in Search of America* by John Steinbeck; *The Gutenberg Galaxy* by Canadian critic Marshall McLuhan, 51, says, "The medium is the message": since spoken words are richer in meaning ("hotter") than written words TV can save mankind by turning the world into a global village.

Fiction *One Day in the Life of Ivan Denisovich (Odin den v zhizni Ivana Denisovicha)* by Soviet novelist Aleksandr Isayevich Solzhenitsyn, 44, is based on years spent in Siberian forced-labor camps; *One Flew Over the Cuckoo's Nest* by U.S. novelist Ken Kesey, 27, is a satire on the dehumanization of Western society; *King Rat* by English-American novelist James Clavell, 37; *The Reivers* by William Faulkner, who dies at his home in Oxford, Miss., July 6 at age 64; *Letting Go* by Philip Roth; *Ship of Fools* by Katherine Anne Porter; *The Kindly Ones* by Anthony Powell; *The Lonely Girl* by Edna O'Brien; *Aura* by Mexican novelist Carlos Fuentes, 33; *A Clockwork Orange* by English novelist Anthony Burgess (John Anthony Burgess Wilson), 45; *Morte d'Urban* by U.S. novelist J. F. (John Farl) Powers, 45; *Another Country* by James Baldwin; *The Golden Notebook* by Doris Lessing; *Wolf Willow* by Wallace Stegner; *The Garden of the Finzi Continis (Il giardino dei Finzi Contini)* by Italian novelist Giorgio Bassani, 46; *Pale Fire* by Vladimir Nabokov; *Reinhart in Love* by U.S. novelist Thomas (Louis) Berger, 38.

Juvenile *James and the Giant Peach* by Roald Dahl.

Poetry *The Jacob's Ladder* by Denise Levertov; *Pictures from Bruegel* by William Carlos Williams; *All My Pretty Ones* by Anne Sexton; *A Sad Heart at the Supermarket* by Randall Jarrell; *Silence in the Snowy Fields* by U.S. poet Robert Bly, 36.

Painting *100 Cans* (of Campbell's Beef Noodle Soup) and *Green Coca-Cola Bottles* by Andy Warhol; *Blam! and Head, Red and Yellow* by U.S. "pop" artist Roy Lichtenstein, 39, who has employed comic-strip techniques; *Silver Skies* by U.S. painter James Rosenquist, 28; *Actual Size* by U.S. painter Edward Joseph Ruscha, 24; *Fool's House* and *Diver* by Jasper Johns; *Ace* by Robert Rauschenberg; *White-Dash Blue* by Ellsworth Kelly; *Movements with Squares* by English artist Bridget Riley, 31; *Cantabile* by Kenneth Noland; *Stripes* by Morris Louis who dies of lung cancer at Washington, D.C., September 7 at age 50. Franz Kline has died in New York May 13 at age 51. Yves Klein dies at Paris June 6 at age 34.

Polaroid Corp. introduces color film invented by Edwin H. Land (*see* 1947). The high-speed film produces

1962 ***(cont.)*** color prints in 60 seconds (Polaroid's black-and-white film produces prints in 10 seconds).

Theater *The Physicist (Die Physiker)* by Friedrich Dürrenmatt 2/21 at Zurich's Schauspielhaus; *Oh, Dad, Poor Dad, Mama's Hung You in the Closet and We're Feeling So Sad* by U.S. playwright Arthur Kopit, 24, 2/26 at New York's off-Broadway Phoenix Theater, 454 perfs.; *The Knack* by English playwright Ann Jellicoe, 34, 3/27 at London's Royal Court Theatre; *A Thousand Clowns* by U.S. playwright-cartoonist Herb Gardner, 27, 4/5 at New York's Eugene O'Neill Theater, with Jason Robards, Jr., Sandy Dennis, 428 perfs.; *Chips with Everything* by English playwright Arnold Wesker, 29, 4/27 at London's Royal Court Theatre, with Frank Finlay, John Kelland; *The Rabbit Race (Esche und Angora)* by Martin Walser 9/23 at Berlin's Schillertheater; *Who's Afraid of Virginia Woolf?* by Edward Albee 10/13 at New York's Billy Rose Theater, with Arthur Hill as George, Uta Hagen as Martha, 664 perfs.; *Lord Pengo* by S. N. Behrman 11/19 at New York's Royale Theater, with Charles Boyer, Agnes Moorehead, Henry Daniell, 175 perfs.; *Never Too Late* by U.S. playwright Sumner Arthur Long, 41, 11/27 at New York's Playhouse, with Paul Ford, Maureen O'Sullivan, Orson Bean, 1,007 perfs.; *Exit the King (Le roi se meurt)* by Eugène Ionesco 12/5 at the Théâtre de L'Alliance, Paris.

Films Roman Polanski's *Knife in the Water* with Leon Niemczyk, Jolanta Umecka, Zygmunt Malanowicz; David Lean's *Lawrence of Arabia* with Peter O'Toole, Omar Sharif, Alec Guinness; John Ford's *The Man Who Shot Liberty Valance* with James Stewart, John Wayne; Sam Peckinpah's *Ride the High Country* with Randolph Scott, Joel McCrea; François Truffaut's *Shoot the Piano Player* with Charles Aznavour; Akira Kurosawa's *Yojimbo* with Toshiro Mifune. Also: Peter Ustinov's *Billy Budd* with Robert Ryan, Ustinov, Melvyn Douglas; Hiroshi Imagaki's *Chushingura* with Koshiro Matsumoto, Yuzo Kayama, Toshiro Mifune; George Seaton's *The Counterfeit Traitor* with William Holden, Lilli Palmer; Blake Edwards's *Days of Wine and Roses* with Jack Lemmon, Lee Remick; Pietro Germi's *Divorce, Italian Style* with Marcello Mastroianni; Luis Buñuel's *The Exterminating Angel* with Silvia Pinal, Enrique Rambal; John Huston's *Freud* with Montgomery Clift; Tony Richardson's *The Loneliness of the Long Distance Runner* with Michael Redgrave, Tom Courtenay; John Frankenheimer's *The Manchurian Candidate* with Laurence Harvey, Frank Sinatra, Janet Leigh, Angela Lansbury; Arthur Penn's *The Miracle Worker* with Anne Bancroft, Patty Duke; Kon Ichikawa's *The Outcast* with Raizo Ichikawa; Richard Brooks's *Sweet Bird of Youth* with Paul Newman, Geraldine Page; Robert Mulligan's *To Kill a Mockingbird* with Gregory Peck; Robert Aldrich's *What Ever Happened to Baby Jane?* with Bette Davis, Joan Crawford; Ingmar Bergman's *Winter Light* with Ingrid Thulin, Gunnar Bjornstrand, Max von Sydow.

Marilyn Monroe takes an overdose of sleeping pills and dies at her Hollywood home August 5 at age 36.

Hollywood musicals Morton Da Costa's *The Music Man* with Robert Preston, Shirley Jones, music and lyrics by Meredith Willson, songs that include "Trouble."

Broadway musicals *No Strings* 3/15 at the 54th Street Theater, with Diahann Carroll, Richard Kiley, music and lyrics by Richard Rodgers, songs that include "The Sweetest Sounds," 580 perfs.; *A Funny Thing Happened on the Way to the Forum* 5/8 at the Alvin Theater, with Zero Mostel, Jack Gilford, book by Burt Shevelove and Larry Gelbart, music and lyrics by Stephen Sondheim, songs that include "Comedy Tonight," 964 perfs.; *Little Me* 11/17 at the Lunt-Fontanne Theater, with Sid Caesar, Joey Faye, Virginia Martin, book by Neil Simon, music by Cy Coleman, lyrics by Carolyn Leigh, songs that include "Real Live Girl," 257 perfs.

First performances Symphony No. 8 by Roy Harris 1/17 at San Francisco; Symphony No. 7 by David Diamond 1/26 at Philadelphia's Academy of Music; Symphony No. 12 by Darius Milhaud 2/16 at the University of California, Davis; War Requiem by Benjamin Britten 5/30 at the new Coventry Cathedral; Symphony No. 8 by Lincoln Center president William Schuman 10/4 at New York's new Philharmonic Hall (below); Symphony No. 13 by Dmitri Shostakovich 12/18 at Moscow, with voices singing words by poet Yevgeny Yevtushenko that deal harshly with anti-Semitism. Premier Khrushchev suppresses the work.

New York's Philharmonic Hall opens September 23 as part of the new Lincoln Center for the Performing Arts nearing completion on Columbus Avenue between 62nd and 66th streets on a site acquired with federal aid and built with contributions of more than $165 million. The 2,863-seat concert hall designed by Max Abramovitz becomes the new home of the New York Philharmonic after 69 years at Carnegie Hall, it will be renamed Avery Fisher Hall in 1974 (*see* Fisher, 1937, 1969), but its acoustics are so poor that it will undergo repeated guttings and redesigns until 1976 (*see* Opera House, 1966).

Opera *Katerina Ismailova* 12/20 at Moscow, with music by Dmitri Shostakovich who has revised his politically unacceptable 1934 opera *Lady Macbeth of the Mzensk District*.

"We Shall Overcome" is copyrighted by Nashville, Tenn., schoolteacher Guy Carawan who has added rhythmic punch to a song that will be the hymn of the U.S. civil rights movement, sung in protest demonstrations throughout the world. The late Zilphia Horton, co-director of the Highlander Folk Center at Monteagle, Tenn., learned it from striking black tobacco workers at Charleston, S.C., in the mid-1940s and taught it to folk singer Peter Seeger in 1947 and to Frank Hamilton (who added a verse).

Popular songs "Those Lazy, Hazy, Crazy Days of Summer" by Hoagy Carmichael, lyrics by Charles Tobias; "Days of Wine and Roses" by Henry Mancini, lyrics by Johnny Mercer (title song for film about alcoholism); "Dream Baby" and "Leah" by Roy Orbison; "Ramblin' Rose" by Noel and Joe Sherman; "Roses Are Red, My

Love" by Al Byron and Hugh Evans; "The Wanderer" by Ernest Maresca; "I Left My Heart in San Francisco" by George Cory, lyrics by Douglas Cross; "The Lonely Bull" by California trumpet player-vocalist-composer Herb Alpert, 27; "Surfin' Safari" by the Beach Boys Brian Wilson, 20, Dennis Wilson, 17, Mike Love, 21, Al Jardine, 19, and Carl Wilson, 25.

Rod Laver wins the "grand slam" in tennis (Australia, France, Britain, and the United States); Mrs. Karen Hantze Susman, 19 (U.S.), wins in women's singles at Wimbledon, Margaret Smith, 19 (Australia), at Forest Hills.

Ohio golfer Jack William Nicklaus, 22, wins the U.S. Open by defeating Arnold Palmer in a playoff.

The New York Mets and the Houston Colt 45s (later Astros) play their first seasons. The Mets replace the New York Giants and Brooklyn Dodgers, who moved to California in 1958.

The New York Yankees win the World Series by defeating the San Francisco Giants 4 games to 3.

Australia's first bid for the America's Cup in ocean yacht racing fails as the U.S. defender Weatherly defeats *Gretel* 4 races to 1.

Sonny Liston wins the world heavyweight boxing title September 25. Now 28, he knocks out Floyd Patterson in the first round of a championship bout at Chicago.

Brazil retains the World Cup in football (soccer) by defeating Czechoslovakia 3 to 1 at Santiago, Chile, and wins without Pele who has torn a thigh muscle in an earlier game with the Czechs.

Cigar smokers are the chief U.S. victims of President Kennedy's embargo on trade with Cuba (above). U.S. cigar sales exceed 6 billion per year with 95 percent of Cuban cigars rolled and wrapped in U.S. plants, but without Cuban tobacco cigar sales will fall to 5.3 billion per year by 1976 despite population growth.

Philip Morris introduces "Marlboro Country" to advertise its top filter-tip cigarette against R. J. Reynolds's Winston brand. The cowboy theme will make Marlboro the leading brand worldwide.

The first Wal-Mart store opens July 2 at Rogers, Arkansas. Retail merchant Sam Moore Walton, 44, has run a Ben Franklin store with his brother James at Bentonville; he proposed a chain of discount stores in small towns; Ben Franklin dismissed the idea and Walton has gone into business for himself. His chain will grow to have more than 1,500 stores by 1990. Sales will pass those of Sears, Roebuck by 1991.

K Mart discount stores are opened by the 63-year-old S.S. Kresge Co., whose five-and-ten-cent stores are losing money. By 1977 Kresge will have sales second only to those of Sears but Wal-Mart (above) will pass it in the 1980s.

Montreal's Place Ville Marie gets its first high-rise buildings with underground galleries built according to a plan put forward by New York developer William Zeckendorf.

Silent Spring by U.S. biologist Rachel Louise Carson, 55, warns of dangers to wildlife in the indiscriminate use of persistent pesticides such as DDT (*see* 1943; Osborne, 1948).

Alaskan Eskimos are found to have high concentrations of Cesium-137 in their bodies as a result of eating caribou meat, their staple food. The caribou have grazed on lichens and have absorbed fallout dust from atmospheric nuclear tests.

Petrified Forest National Park in Arizona is established by act of Congress. The 94,189-acre park embraces Indian ruins, petrified wood, and colorful desert land.

Maine's Baxter State Park is designated as such to honor former governor Percival P. Baxter who has spent 31 years acquiring piecemeal a 202,000-acre tract which he presents to the state of Maine. The park encompasses scenic forest land surrounding 5,273-foot Mt. Katahdin, Maine's highest peak.

Britain has a cold spell that freezes outdoor house pipes, interferes with plumbing, and creates general misery (*see* 1963).

Florida loses nearly 8 million citrus trees as temperatures fall to 24° F. and remain below freezing.

An International Rice Research Institute (IRRI) is established in the Philippines at Los Baños with support from both the Ford and Rockefeller foundations and the Philippine government (*see* IR5, IR8, 1964).

The average Asian eats more than 300 pounds of rice per year as compared with 6 pounds in the West and many rice-consuming nations including the Philippines do not produce enough for domestic needs. IRRI (above) will advance the "Green Revolution" but while per capita food production in the developing countries has increased in the past decade at an annual rate of 0.7 percent it will increase at an annual rate of only 0.2 percent in the decade ahead.

Britain and France remove thalidomide from the market (*see* 1960; 1961). The *Washington Post* credits FDA researcher Frances Kelsey with having prevented thalidomide birth defects in thousands of U.S. infants (July 15 editorial), she receives praise in the Senate for her "courage and devotion to the public interest," President Kennedy gives her a medal for distinguished service in a White House ceremony, but U.S. physicians have received thousands of sample packages of Kevadon and have given a few to expectant mothers.

The thalidomide affair (above) spurs Congress to pass the Kefauver legislation giving the FDA extensive new authority to regulate the introduction of new drugs. The new law also contains provisions to prevent drug firms from overcharging and from confusing buyers of prescription drugs.

President Kennedy has requested a law that will police the pharmaceutical industry in a message that asked also for a broad program of consumer protection, a "Consumer Bill of Rights" that would include a "Truth in Lending" law and a "Truth in Packaging" law (*see* 1968; 1966).

1962 ***(cont.)*** Diet-Rite Cola, introduced by Royal Crown Cola, is the first sugar-free soft drink to be sold nationwide to the general public. The cyclamate-sweetened cola will soon have powerful competitors (*see* No-Cal, 1952; Tab, 1963).

Tab-opening aluminum end cans for soft drinks and beer developed with backing from Aluminum Corp. of America make their first appearance (*see* 1960). Pittsburgh's Iron City Beer in tab-opening cans is test-marketed in Virginia, many brewers question whether the cans are worth their extra cost, consumers cut their fingers opening the cans, but designers improve the tabs. More than 90 percent of beer cans will be self-opening by 1970. Discarded tabs will be a hazard for bare feet until designs are further improved.

Frozen, dehydrated, and canned potatoes account for 25 percent of U.S. potato consumption.

China resumes the birth control campaign suspended in her Great Leap Forward Program of 1958, but a proposal by the World Health Organization that it offer family-planning advice brings rebukes by 30 nations (but *see* 1965).

Pope John XXIII convenes Vatican II October 11, less than a century after Vatican I was convened in 1869. The world's population has tripled since 1869 to 3.1 billion.

1963 The assassination of President John F. Kennedy November 22 ends a 34-month administration that has initiated domestic social programs which Kennedy's successor will carry out. The president slumps in the seat of his open Lincoln Continental as his motorcade approaches the Dallas World Trade Center for a scheduled speech, the rifle bullets that killed him are thought to have come from the sixth floor of the Texas School Book Depository in Elm Street, Lyndon B. Johnson is sworn in as president on the plane to Washington.

Johnson Takes Nation's Helm, Pages 4 and 5

The Dallas Morning News

John F. Kennedy Life History, Pages 16 and 17

KENNEDY SLAIN ON DALLAS STREET

Receives Oath on Aircraft

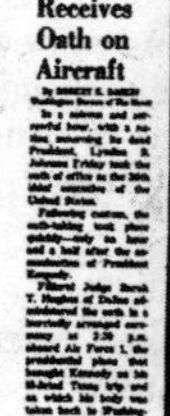

Pro-Communist Charged With Act

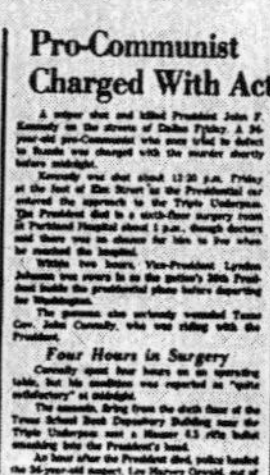

Four Hours in Surgery

An assassin's bullet ended the brief presidency of John F. Kennedy. Conspiracy theories gained wide credence.

President Kennedy's alleged assassin Lee Harvey Oswald, 24, is a Marine Corps veteran who has spent some time in the U.S.S.R., married the daughter of a KGB colonel, and handed out literature for the Fair Play for Cuba Committee at New Orleans. Dallas police arrest him 80 minutes after the assassination on charges of having killed patrolman J. D. Tippit. Dallas nightclub operator Jack Ruby, 52, shoots Oswald to death November 24 in the basement of the Dallas city jail as police are moving the suspect to safer quarters, and millions watch the shooting on television (*see* Warren Commission, 1964).

"Ich bin ein Berliner," President Kennedy has said June 26 in a speech at West Berlin during a 4-day visit to West Germany. Kennedy has pledged support for efforts to defend West Berlin from Communist encroachment, reunify Germany, and work toward European unity.

West German chancellor Konrad Adenauer resigns October 16 at age 87 and is succeeded by Ludwig Erhard, now 66.

Britain's Profumo-Christine Keeler scandal makes world headlines. The war minister Lord John Dennis Profumo is charged with having been intimate with call girl Christine Keeler, 21, to whom he was introduced naked in a swimming pool by Viscount Astor's osteopath friend Stephen Ward who has his own cottage at Astor's Cliveden estate. Profumo is married to actress Valerie Hobson, but the real scandal arises from the fact that Keeler is sleeping with Soviet naval attaché Evgeny "Honeybear" Ivanov, a known spy attached to the Russian embassy who has asked Keeler to find out from Profumo when nuclear warheads will be delivered to West Germany. Profumo confesses to the affair and resigns in early June, Keeler draws a 9-month sentence in December for perjury and conspiracy to obstruct justice.

British journalist H. A. R. Philby disappears from Beirut and the Soviet Union gives him asylum July 30.

Israeli premier David Ben Gurion has resigned June 16 at age 76 and been succeeded by former finance minister Levi Eshkol. Israeli and Syrian forces clash August 20 along the demilitarized zone north of the Sea of Galilee but UN truce observers persuade both sides to accept a cease-fire August 25.

The Federation of Malaysia formally established September 16 joins Malaya, Singapore, Sarawak, and North Borneo. Tunku Abdul Rahman, 60, is prime minister.

Britain's prime minister Harold Macmillan, now 69, resigns for reasons of health October 18. Queen Elizabeth asks the earl of Home, Sir Alexander Douglas Home, 60, to form a new cabinet, he resigns his peerages, calls himself Sir Alec Douglas-Home, and is the first prime minister to have a seat neither in the House of Lords nor the Commons.

The South Vietnam government of Ngo Dinh Diem falls the night of November 1 in a coup engineered by Gen. Duong Van Minh and other anti-Communist officers. Diem has been president since the republic was proclaimed in 1955 but his opponents kill him along with his security chief as civil war continues in Vietnam.

Kenya gains independence December 12 after 43 years as a British crown colony. Kikuyu leader Jomo Kenyatta is president of the new republic.

"I have a dream," says Martin Luther King, Jr., in a ceremony January 1 at the Lincoln Memorial in Washington to commemorate the centennial of the Emancipation Proclamation: "I have a dream that one day, on the red hills of Georgia, sons of former slaves and the sons of former slaveowners will be able to sit down together at the table of brotherhood."

The Supreme Court orders that indigent defendants in all U.S. criminal cases be provided by the states with the services of defense attorneys. The ruling comes March 18 in the case of *Gideon v. Wainwright*.

The Supreme Court rules May 20 that convictions by lower courts in cases of sit-ins to protest discriminatory practices by retail establishments were unconstitutional.

Congress votes June 10 to guarantee women equal pay for equal work, but the law will prove difficult to enforce.

President Kennedy federalizes Alabama's National Guard June 11 and orders Gov. George C. Wallace to allow two black students to be enrolled at the University of Alabama.

NAACP leader Medgar Evers, 37, is murdered June 12 in the doorway of his home at Jackson, Miss., right after a presidential broadcast on civil rights.

President Kennedy asks Congress June 19 to enact far-reaching civil rights legislation that will include provisions to bar discrimination in the use of privately owned public facilities.

More than 200,000 black and white Americans conduct a "march on Washington" August 28 to demonstrate support for civil rights.

Three Alabama cities desegregate public schools after President Kennedy has again federalized Alabama's National Guard in the face of Gov. Wallace's continued opposition.

Four black Alabama schoolchildren are killed and 19 people injured September 15 when a bomb explodes at Birmingham's 16th Street Baptist Church while 200 are attending Sunday services. The deaths of Denise McNair, 11, Carole Robertson, 14, Addie Mae Collins, 14, and Cynthia Wesley, 14, provoke racial riots, police dogs are used to attack civil rights demonstrators, and two black schoolboys—Virgil Wade, 13, and Johnnie Robinson, 16—are killed later the same day.

The European Economic Community vetoes British entry January 14 after President de Gaulle raises objections.

U.S. factory workers average more than $100 per week for the first time in history but unemployment reaches 6.1 percent by February.

More than 50 million U.S. taxpayers support the Treasury, up from 4 million in 1939. Most income taxes are paid by payroll withholdings (*see* 1943).

New Hampshire conducts the first state lottery—a sweepstakes ostensibly to raise money for state education. The state ranks last in terms of state aid to education, the lottery will not raise this ranking nor will it contribute more than 3 percent of state education aid in any year, but other states will follow New Hampshire's example as politicians seek substitutes for income taxes, postponing fundamental tax reforms and adequate funding for social services.

The U.S. federal budget is nearly $100 billion, up from $9 billion in 1939. Almost half the total goes for military appropriations as U.S. involvement in Southeast Asia begins to escalate.

An inflation begins in Indonesia that will depreciate the currency 688-fold in the next 7 years (*see* 1965).

U.S. federal employees number 2.5 million, up from 1.13 million in 1940.

Wall Street's Dow Jones Industrial Average plummets 21 points in 30 minutes at news of the Kennedy assassination (above), but recovers quickly.

A nuclear reactor installed by Jersey Central Power and Light is the first commercial reactor and the first nuclear power plant large enough to compete with coal and oil fuel. Jersey Central's 640-million-kilowatt Oyster Creek plant will be followed by much larger installations, 41 nuclear plants will be ordered in the next decade, 56 will account for 8 percent of total U.S. energy production by 1975, but the rising price of uranium, anxieties about nuclear disasters, and rising construction costs will delay the growth of nuclear power and demand for electric power will fall far short of forecasts.

Con Edison opens a nuclear power station on the Hudson River at Indian Point, N.Y. (*see* 1962; fish kill, below).

Enel, founded in February, consolidates 1,200 Italian electric plants into a single national agency under terms of a nationalization law.

The nuclear-powered submarine U.S.S. *Thresher* sinks April 10 in 8,400 feet of water 220 miles off Cape Cod killing all 129 men aboard in the worst submarine disaster of all time. The ship has put to sea with no sure way of blowing water out of her ballast tanks in an emergency at low depths; the Portsmouth Naval Shipyard has failed to test silver-brazed joints with sound waves.

Two-thirds of the world's automobiles are in the United States (which has 6 percent of the world's population).

Trolley car service ends at Baltimore where electric trolleys were first introduced just 88 years and 84 days earlier. Buses replace the trolleys November 4.

New York renames Idlewild Airport John F. Kennedy (JFK) Airport.

Houston surgeon Michael Ellis De Bakey, 54, uses an artificial heart for the first time to take over the functions of blood circulation during heart surgery.

1963 *(cont.)* Valium, introduced in December, is five to 10 times more potent than Librium as a muscle-relaxant and anti-convulsant, has no more unwanted side effects, costs more, and will soon be even more widely prescribed than Librium (*see* 1960). Roche Laboratories chemist Leo Sternbach synthesized the benzodiazepine anti-anxiety drug in 1959 and has been awaiting FDA approval.

Pope John XXIII dies June 3 at age 81 after a 5-year papacy and is succeeded by Cardinal Giovanni Battista Montini, 66, who becomes Pope Paul VI.

Reading the Lord's Prayer or verses from the Bible in U.S. public schools is unconstitutional, the Supreme Court rules June 17 in *School District of Abington Township v. Schempp.* The decision caps a campaign by atheist Madalyn Murray O'Hair, 44.

U.S. first class postal rates go to 5¢ per ounce January 7 (*see* 1958). U.S. Postmaster General J. Edward Day orders that 5-digit ZIP codes be used by July 2 to speed sorting of mail (*see* 1968).

A "hot line" emergency communications link between Washington and Moscow goes into service August 30 under an agreement signed June 20 to reduce the risk of accidental war. The line will operate via satellite beginning early in 1978.

AT&T introduces transistorized electronic Touch-Tone telephones November 18 in two Pennsylvania communities on an optional basis at extra cost. They have far more capabilities than electro-mechanical dial phones introduced in 1919.

New York City newspapers resume publication April 1 after a 114-day strike but Hearst's financially weakened *New York Mirror* ceases publication October 15 after 41 years in which it has built up a circulation surpassed in the United States only by that of the *New York Daily News* (*see* 1966).

Zenith Radio and Sylvania begin producing their own color picture tubes for television sets after having bought the tubes for years from RCA. Both companies pay royalties to RCA, which controls basic patents, but Zenith will grow to surpass RCA as the leading U.S. producer of color TV sets.

Nonfiction *The Feminine Mystique* by U.S. feminist Betty Goldstein Friedan, 42, argues that women as a class suffer various forms of discrimination but are victimized especially by a system of delusions and false values that encourages them to find personal fulfillment through their husbands and children (*see* NOW, 1966); *The Fire Next Time* (essays) by James Baldwin.

Fiction *Dog Years* (*Hundejahre*) by Günter Grass; *Acquainted With Grief* (*La cognizione del dolore*) by Carlo Emilio Gadda; *The Clown* by Heinrich Böll; *Hopscotch* by Argentinian novelist Julio Cortázar, 49; *Honey for the Bears* by Anthony Burgess; *Cat's Cradle* by Kurt Vonnegut, Jr.; *Big Sur* by Jack Kerouac; *The Second Stone* by U.S. novelist-critic Leslie Fiedler, 46; *Raise High the Roofbeam, Carpenters* and *Seymour—An Introduction* by J. D. Salinger; *The Tenants of Moonbloom* by the late Edward Lewis Wallant, who died last year at age 36; *The Bell Jar* by the late U.S. poet Sylvia Plath, who takes her own life February 11 at age 31 (her autobiographical novel is an account of manic depression); *The Icicle* (*Fantasticheskiye Povesti*) (stories) by Andrei Sinyavsky; *The Spy Who Came in from the Cold* by English novelist John Le Carré (David Cornwell), 31; *V* by U.S. novelist Thomas Pynchon, 26; *The Centaur* by John Updike; *The Group* by Mary McCarthy; *The Collector* by English novelist John Fowles, 45; *The Sailor Who Fell from Grace with the Sea* by Yukio Mishima.

Juvenile *Where the Wild Things Are* by New York illustrator-writer Maurice Sendak, 35; *The Wolves of Willoughby Chase* by Joan Aiken, daughter of poet-novelist Conrad Aiken.

Poetry *At the End of the Open Road* by Louis Simpson.

The *New York Review of Books* begins publication on a bi-weekly basis September 26.

Painting *Homage to the Square "Curious"* by German-American painter Josef Albers, now 75, who was a master at Weimar's Bauhaus for 10 years until 1933 and has been a major force on U.S. painters and architects ever since; *Second Marriage* by English painter David Hockney, 26; *Jackie* (*The Week That Was*) (silkscreen) by Andy Warhol; *Nomad* by U.S. painter James Rosenquist, 30; *Map* by Jasper Johns; *Red, Blue, Green* by Ellsworth Kelly; *Whaam!* and *Hopeless* by Roy Lichtenstein; *Man and Child* by Francis Bacon. Georges Braque dies at Paris August 31 at age 81.

Sculpture *Fountain, Wheels, Litanies,* and *Swift Night Ruler* by Robert Morris. Piero Manzoni dies in his Milan studio February 6 at age 29.

Eastman Kodak introduces Instamatic cameras that can be loaded with film cartridges. Eastman also introduces Kodachrome-X, Ektachrome-X, and Kodacolor-X films with twice the speed of their predecessors (*see* 1936; 1944; 1946; Kodachrome II, 1961).

Theater *The Milk Train Doesn't Stop Here Anymore* by Tennessee Williams 1/16 at New York's Morosco Theater, with Hermione Baddeley, Mildred Dunnock, 69 perfs.; *The Deputy* (*Der Stellvertrecar*) by German playwright Rolf Hochhuth, 31, 2/23 at Berlin's Theater am Kurfurstendamm; *Barefoot in the Park* by Neil Simon 10/23 at New York's Biltmore Theater, with Elizabeth Ashley, Robert Redford, Mildred Natwick, Kurt Kasznar, 1,502 perfs.; *Mr. Krott, Larger Than Life* (*Über lebengross Herr Krott*) by Martin Walser 11/30 at Stuttgart's Staatstheater.

Films Elia Kazan's *America, America* with Stathis Giallelis, Frank Wolff, Elene Karam, Lou Antonio; Federico Fellini's *8½* with Marcello Mastroianni; John Sturges's *The Great Escape* with Steve McQueen, Richard Attenborough, James Garner, James Coburn, Donald Pleasence; Martin Ritt's *Hud* with Patricia Neal, Paul Newman, Melvyn Douglas, Brandon de Wilde; Luchino Visconti's *The Leopard* with Burt Lancaster, Claudia Cardinale, Alain Delon; Tony Richardson's *Tom Jones* with Albert Finney. Also: Alfred Hitchcock's *The Birds* with Rod Taylor, Jessica Tandy, Tippi Hedren, Suzanne Pleshette; Stanley Donen's *Charade* with Cary Grant, Audrey Hepburn, Walter Matthau; Louis Malle's *The Fire Within* with Maurice Ronet; Robert Wise's *The Haunting* with Julie Harris, Claire Bloom; George Pollock's *Murder at the Gallop* with Margaret Rutherford (as Miss Marple), Robert Morley, Flora Robson; Ingmar Bergman's *The Silence* with Ingrid Thulin, Gunnel Lindblom; Lindsay Anderson's *This Sporting Life* with Richard Harris; John and Faith Hubley's *The Hole* (animated short).

Berlin's Philharmonic Hall opens in Tiergarten Park with 2,200 seats.

Munich's reconstructed National Theater with its Staatsoper opens November 23.

Opera *The Long Christmas Dinner* 3/13 at New York's Juilliard School of Music with music by Paul Hindemith, libretto from a play by Thornton Wilder.

Cantatas *Cantata Misericordia* by Benjamin Britten in September at Geneva for the hundredth anniversary of the Red Cross; *Song of Human Rights* by Howard Hanson 12/10 at Washington, D.C., with a text that includes excerpts from President Kennedy's inaugural address.

Stage musicals *Half a Sixpence* 3/21 at London's Cambridge Theatre, with Tommy Steele, music and lyrics by David Hendon, book by Douglas Cross based on the 1905 H. G. Wells novel *Kipps*, 677 perfs.; *She Loves Me* 4/23 at New York's Eugene O'Neill Theater, with Barbara Cook, Jack Cassidy, music by Jerry Bock, lyrics by Sheldon Harnick, 39, songs that include "Days Gone By," 301 perfs.

The Beatles score their first big success with a recording of "I Want to Hold Your Hand." The Liverpool rock group includes songwriters John Lennon, 22, and Paul McCartney, 20, who are supported by George Harrison, 20, and drummer Ringo Starr (*né* Richard Starkey) 24, who has replaced the original Beatles drummer Peter Best. Their long hair will be widely imitated and their music will raise rock to new heights of artistry and popular appeal.

Popular songs "Fly Me to the Moon" by New York songwriter Bart Howard, 47; "Blame It on the Bossa Nova" by U.S. songwriters Barry Mann and Cynthia Weil; "In Dreams," "Falling," and "Blue Bayou" by Roy Orbison.

World chess champion Mikhail Botvinnik is defeated at Moscow May 20 by Soviet chess master Tigran Petrosian, 33, who will hold the title until 1969.

Charles "Chuck" McKinley, 22 (U.S.), wins in men's singles at Wimbledon, Margaret Smith in women's singles; Rafael Osuna, 24 (Mexico), wins in men's singles at Forest Hills, Maria Bueno in women's singles.

The Los Angeles Dodgers win the World Series by defeating the New York Yankees 4 games to 0.

A British train robbery August 8 nets $7 million, mostly in banknotes. A team of robbers stops a train near Cheddington, 36 miles northwest of London.

Bayonne, N.J., salad oil king Anthony De Angelis, 47, is indicted for fraud. His Allied Crude Vegetable Oil & Refining Corp. has rigged its tanks, using seawater in place of salad oil which served as collateral for warehouse receipts. The deficiency runs to 827,000 tons of oil valued at $175 million. De Angelis goes bankrupt and some investors are ruined.

Pecos, Tex., financier Billie Sol Estes, 43, is convicted of selling finance companies $24 million worth of mortgages on nonexistent fertilizer tanks. He will serve 6 years of a 15-year sentence for mail fraud and conspiracy to defraud.

New York's Pan Am building, completed above Grand Central Terminal, is the world's largest commercial office building. Designed by Emery Roth & Sons with help from architects Pietro Belluschi and Walter Gropius, the 59-story structure has 2.4 million square feet of office space, but its pre-cast concrete curtain wall raises protests because it blocks the vista of Park Avenue.

Houston's Tenneco building is completed by Skidmore, Owings & Merrill.

U.S. motels have a 70 percent occupancy rate versus 64 percent for conventional hotels. Motel rooms top the million mark, up from 600,000 in 1958.

Tokyo's first high-rise building goes up following removal of a government ban on structures more than 31 meters tall (*see* earthquake, 1923; Los Angeles, 1957).

Thermal pollution kills thousands of striped bass at the new Con Edison nuclear generating plant on the Hudson (above) before a mesh screen is erected to fence out the anadromous fish, letting them swim upriver to spawn.

Soviet crops fail both in Kazakhstan and in the Ukraine (*see* 1953). Crops fail in the People's Republic of China and both China and the U.S.S.R. are forced to seek grain from the west.

Canada makes a major sale of wheat to the P.R.C. in August and follows it with large sales to the U.S.S.R.

"Comrade Khrushchev has performed a miracle," Muscovites joke: "He has sown wheat in Kazakhstan and harvested it in Canada."

Khrushchev declares the Kazakhstan virgin lands experiment a failure and seeks nearly 2 million tons of U.S. wheat. President Kennedy approves the sale, Richard Nixon and most other Republicans oppose it, but it goes through at year's end with support from President Johnson.

Typhoons destroy nearly half of Japan's standing crops. The resulting food shortages boost Japanese demand for imported foodstuffs.

The U.S. corn crop tops 5 billion bushels for the first time, up from 3 billion in 1906.

The number of U.S. farmers falls to 13.7 million—7.1 percent of the population.

1963 ***(cont.)*** Weight Watchers is founded by Queens, N.Y., housewife Jean Neditch, 39, who has reduced her own weight from 213 pounds to 142 with help from a high-protein diet developed by Norman Jolliffe, 62, of the New York City Department of Health. Using a form of group therapy, completely proscribing some foods while permitting others without restriction, Weight Watchers will grow into a worldwide operation with average weekly attendance of half a million.

Average U.S. per capita meat consumption reaches 170.6 pounds, but veal consumption drops to 5 pounds, down from 9.7 in 1949. Chicken consumption is 37.8 pounds, up from 23.5 in 1945 when chicken was more costly than beef.

Average U.S. per capita butter consumption falls to 6.7 pounds, down from 18.3 in 1934 and 7.5 in 1960; margarine 9.3 pounds, up from 8.6 in 1957; cheese 9.4 pounds, up from 7 in 1949.

Tab, introduced by Coca-Cola, is a cyclamate-sweetened cola drink that competes with Royal Crown's Diet-Rite (*see* 1962; Diet Pepsi, 1965).

1964 The Tonkin Gulf Resolution approved by Congress August 7 authorizes President Johnson to "take all necessary measures to repel any armed attack against forces of the United States and to prevent further aggression." Three North Vietnamese PT boats have allegedly fired torpedoes August 2 at a U.S. destroyer in the international waters of Tonkin Gulf 30 miles off the coast of North Vietnam following 6 months of covert U.S. naval operations, President Johnson has ordered retaliatory action after a second such alleged attack, U.S. aircraft have bombed North Vietnamese bases August 5, and the resolution has been approved 88 to 2 in the Senate and 416 to 0 in the House of Representatives.

Brazil's President Goulart has been overthrown April 1 and fled to Uruguay after a military coup that followed a presidential order distributing federal lands to landless peasants, doubling the minimum wage, and expropriating lands adjacent to federal highways. A U.S. naval force assembles in the Caribbean and heads south as President Johnson prepares to intervene to prevent a leftist takeover.

An anti-Communist purge follows the military coup in Brazil (above), U.S. ambassador to Brazil Lincoln Gordon requests the recall of U.S. naval forces before they reach Brazilian waters, and Brazil's congress elects army chief of staff Gen. Humberto Castelo Branco April 11 to serve out the balance of President Goulart's term.

Greece's Paul I has died March 6 after a 17-year reign. His son of 23 will reign as Constantine II until 1967 but George Papandreou has become premier, fighting has broken out between Greeks and Turks in Cyprus, Archbishop Makarios abrogates a 1960 treaty April 4, heavy fighting follows in northwestern Cyprus, Athens rejects direct talks with Ankara June 11, Turkish planes attack Greek Cypriot positions August 8, the UN orders a cease-fire August 9, Greece withdraws her units from NATO August 17, and the UN extends its mandate for force in Cyprus December 18 (*see* 1967).

Malta gains independence September 21 after 140 years of British colonial rule. She will become a republic in December 1974.

Jawaharlal Nehru dies suddenly May 27 at age 74 after nearly 17 years as prime minister of India. The "Lion of Kashmir" Sheik Mohammed Abdullah has been released from prison April 8 after 6 years' confinement and has denounced Indian policy toward Kashmir, which is claimed by Pakistan, but New Delhi's policy continues under Nehru's successor Lal Bahadur Shastri, 60 (*see* 1966).

Malawi gains independence July 6 after 73 years of British colonial rule. Formerly Nyasaland, she has broken her ties with Rhodesia.

Rhodesian finance minister Ian D. Smith, 45, has succeeded white-supremacy extremist Winston J. Field as prime minister April 13 following dissolution of the Federation of Rhodesia and Nyasaland.

Zambia is created October 24 out of Northern Rhodesia and Barotseland with Kenneth Kaunda as president of the new independent state. The British South African Co. has released mineral rights in the area on the promise of compensation from London and from the new Zambian government.

The United Republic of Tanzania takes that name October 29 with Julius K. Nyerere as president. African nationalists in Zanzibar overthrew the predominantly Arab government January 12 and set up a People's Republic with Abeid A. Karume as president and three Communist-trained cabinet members, Zanzibar merged with Tanganyika April 26 at the suggestion of President Nyerere, and the new republic is Communist-oriented.

A Soviet coup d'état October 13 strips Nikita Khrushchev, now 70, of all power. Leonid Ilych Brezhnev, 57, becomes party leader, Aleksei Nikolaevich Kosygin, 60, premier October 14.

The Warren Commission Report issued September 27 finds that Lee Harvey Oswald alone was responsible for last year's assassination of President Kennedy and that no conspiracy was involved. The report by the presidential commission headed by Chief Justice Earl Warren meets with skepticism in many quarters as various lawyers and publicists come out with books that seek to discredit the Warren Commission.

Britain's Labour party wins the general elections in October, Prime Minister Douglas-Home resigns, and (James) Harold Wilson, 48, begins a ministry that will continue until 1970.

Canada adopts the Maple Leaf flag October 22. Queen Elizabeth will make it official early next year.

Saudi Arabia's Saud ibn Abdel Aziz is deposed after a 12-year reign. His brother Faisal is proclaimed king November 2 and will reign until 1975.

President Johnson wins reelection with the largest popular vote plurality in U.S. history, receiving 61 percent of the popular vote and 486 electoral votes in a landslide victory over the conservative Sen. Barry M. Goldwater, 55, of Arizona. Goldwater has narrowly defeated the more liberal New York governor Nelson A. Rockefeller in the California primary and has been chosen to lead the Republican ticket, he has attacked Democrats for their interference in the life of the individual and their "big government" competition with private industry, but his suggestions of escalating the Vietnam war has raised wide fears and he receives only 38 percent of the popular vote, 52 electoral votes (only Arizona and 5 Deep South states go Republican).

President Johnson announces a vast increase in U.S. aid to South Vietnam December 11 "to restrain the mounting infiltration of men and equipment by the Hanoi regime in support of the Vietcong"; a military coup December 19 overthrows South Vietnam's High National Council.

South Africa expands her apartheid racial segregation laws May 6 with a Bantu Laws amendment bill that empowers her minister of Bantu administration to declare "prescribed" areas in which the number of Bantus (Africans) to be employed can be specified. Eight native leaders including Nelson Mandela are sentenced to life imprisonment June 12 for sabotage and subversion and in July the police make massive arrests under the General Laws Amendment Act which allows them to hold suspects for up to 6 months without reporting their arrests.

The Twenty-Fourth Amendment that takes effect February 24 makes U.S. poll taxes unconstitutional.

A 75-day filibuster by southern senators opposed to a civil rights bill ends June 10 when the Senate invokes cloture. The bill wins approval June 19 by a vote of 73 to 27 and goes back to the House; President Johnson signs it into law July 2.

Racial discrimination by a labor union constitutes an unfair labor practice, the National Labor Relations Board rules by a 3 to 2 vote July 2.

Philadelphia, Mississippi, makes headlines in early August with the discovery of the bodies of three civil rights workers killed by white supremacists. James E. Chaney, 21; Michael H. Schwerner, 24; and Andrew Goodman, 20, have been missing since June 21.

Harlem has a race riot July 18; Philadelphia has race riots beginning August 28.

Atlanta restaurateur Lester G. Maddox, 48, closes his Pickrick Restaurant rather than submit to federal government orders that he serve blacks as well as whites. His opposition to integration will propel Maddox into the governorship of Georgia in 1967, and when he is unable to succeed himself he will continue as lieutenant governor (Maddox has passed out pickax handles on the street in front of his restaurant to partisans who will cudgel any blacks who try to enter).

A Free Speech Movement begins among University of California students at Berkeley under the leadership of Bettina Aptheker, Mario Savio, and Jack Weinberg, the movement climaxes in the occupation of Sproul Hall, the occupation is ended December 3 by police acting on orders from Governor Edmund G. "Pat" Brown, 59, more than 800 people are hauled out of Sproul Hall and some 732 sit-in demonstrators are arrested in the largest U.S. mass-arrest thus far in history. Suppression of the California demonstration begins a long period of U.S. campus unrest that will develop into an antiwar movement.

President Johnson calls for "total victory" in a "national war on poverty" March 16. He signs an Economic Opportunity Act August 20 and appoints R. Sargent Shriver, 48, to head the new Office of Economic Opportunity (OEO). Brother-in-law of the late President Kennedy, Shriver will coordinate such agencies as the Job Corps, the Neighborhood Youth Corps, Volunteers in Service to America (VISTA), community action programs, and a Head Start program designed to help preschool children achieve higher levels of health, nutrition, and preparedness for school.

Negotiations begin at Geneva November 16 to reduce world trade tariffs in the so-called first Kennedy Round of discussions (*see* 1963).

The largest iron-ore contract in world history is signed to supply Japan's major steel firms with 65.5 million tons of ore in the next 16 years from Hamersley Range deposits discovered by Lang Hancock in Australia in 1952. The deposits have been acquired by Australia's Conzinc Rio Tinto owned jointly by Kaiser Industries of the United States with backing from French financier Baron Guy de Rothschild, Joseph Hirschhorn's Rio Tinto Mining Co. of Canada, and Consolidated Zinc. Nearly 6 million tons of the ore will have been shipped to Japan by the end of 1967, and the iron will permit a major expansion of the Japanese steel industry that will help make Japan an industrial superpower.

U.S. sales through retail vending machines total $3.5 billion as Americans drop 83 million coins into the machines every 24 hours.

The U.S. Federal Power Commission relaxes its rule that natural gas pipeline companies must have 12 years' reserve before accepting new customers. The FPC acts in response to huge surpluses that have resulted from newly discovered sources and its action produces widescale promotion by utility companies of natural gas for heating and appliances.

U.S. gasoline prices in October are 30.3¢ per gallon, up from 29¢ in 1954.

Jimmy Hoffa achieves his goal of bringing all U.S. truckers under a single Teamsters Union contract and raises fears that he may paralyze the country with a nationwide strike (*see* 1957). Hoffa has made the Teamsters the most powerful union in America and heightened his power potential by pursuing efforts to unite railway, airline, canal boat, and merchant marine workers with his Teamsters. Attorney General Kennedy has made Hoffa his chief target since 1962, a Senate investigating committee has found him hard to

1964 *(cont.)* question, but he is found guilty of having tampered with the jury in a 1962 trial.

U.S. lawyer Ralph Nader, 30, submits a report on what the government should do about auto safety. Nader has come to Washington to work as a consultant to the Department of Labor (*see* 1965).

Studebaker-Packard breaks with the majors, becomes the first U.S. maker to offer seat belts as standard equipment, but quits the auto industry by year's end (*see* 1954).

The Mustang, introduced by Ford Motor Company, is a sporty compact that is essentially a Falcon with different exterior sheet metal (*see* 1959).

Washington's Capitol Beltway opens April 11 with 66 miles of six-lane highway designed to handle 49,000 vehicles per day through suburban Maryland and Virginia.

The Chesapeake Bay Bridge-Tunnel opens April 15 to carry traffic over and under 17.6 miles of ocean in 23 minutes, saving motorists a 2-hour ferry crossing and closing the last water gap on the coastal highway between Canada and Florida.

Scotland's Forth Road Bridge—3,276 feet long—opens to traffic at Queensferry September 4.

New York's Verrazano Narrows Bridge opens across the harbor November 21 to link Brooklyn with Staten Island (*see* Brooklyn Bridge, 1883; George Washington Bridge, 1931). Designed by Swiss-American engineer Othmar A. Ammans, now 85, the 4,260-foot bridge is the world's largest single span suspension bridge.

The Volga-Baltic Ship Canal opens to connect the Caspian Sea with Leningrad on the Baltic 224 miles away.

Japan's Tokaido Shin Kansu (New Tokaido Line) begins service October 1 with "bullet" trains that average 102 miles per hour and cut the 309-mile trip between Tokyo and Osaka from 390 minutes to 190 minutes. The new train will bring a sharp decrease in air travel between the two cities, a second line will open in March of 1972 between Osaka and Okayama, and the line will be extended 3 years later to Hakata on the island of Kyushu 668 miles from Tokyo.

The Surgeon General's Report issued January 11 by Dr. Luther L. Terry, 52, links cigarette smoking to lung cancer and other diseases (*see* 1954). The lung cancer rate among U.S. men will increase in the next decade from 30 per 100,000 to 50 and will more than double among women to 10 as young people increase cigarette smoking despite warnings (*see* FTC, 1968).

The Medicare Act signed by President Johnson July 30 at Independence, Mo., sets up the first government-operated health insurance program for Americans age 65 and over. Harry Truman, now 84, urged such coverage in 1949, the American Medical Association has opposed the amendment to the Social Security Act of 1935. Funded by payroll deductions, federal subsidies, and (initially) $3 per month in individual premiums, Medicare covers 20 million seniors.

Methadone helps rehabilitate heroin addicts (*see* Bayer, 1898). Rockefeller University researcher Vincent P. Dole, 51, and psychiatrist Marie Nyswander use the synthetic opiate invented by German chemists during World War II, some 1,000 addicts will be enrolled in methadone maintenance programs by 1968, nearly 10,000 will be on methadone by 1970, but heroin addiction will remain a source of crime in U.S. cities where the need to support their costly habit will drive addicts to burglary, robbery, and prostitution.

The Roman Catholic liturgy in the United States changes November 29 to include use of English in some prayers; the entire mass will be in English by Easter 1970 but some priests will defy the Vatican and stick to Latin.

U.S. parochial-school enrollment reaches an all-time high of 5.6 million pupils. The figure will fall to below 3.5 million within 10 years.

The U.S. Supreme Court rules unanimously March 9 that a public official cannot recover libel damages for criticism of his public performance without proving deliberate malice. The decision in *New York Times v. Sullivan* bolsters press freedom in America.

BBC-2 begins broadcasting April 30. The second British Broadcasting Corporation television channel transmits on 625 lines to provide high-definition pictures for sets equipped to pick up 625-line signals.

Radio Manx begins broadcasting November 23 to give Britain her first commercial television.

Nonfiction *Understanding Media* by Marshall McLuhan; *Anti-Intellectualism in American Life* by Columbia University history professor Richard Hofstadter, 48; *One-Dimensional Man* by German-American Brandeis University philosophy professor Herbert Marcuse, 66, tries to explain the repressive nature of American society and its potential for totalitarianism (many students will use the Marcuse book to justify abandoning democratic processes in order to achieve radical goals); *Games People Play* by Canadian-American psychiatrist Eric Berne (*né* Eric Lennard Bernstein), 54, who pioneers non-Freudian "transactional analysis."

Fiction *Herzog* by Saul Bellow; *Last Exit to Brooklyn* by U.S. novelist Hubert Selby, 38 (a London court convicts Selby of obscenity, but he wins reversal on appeal); *Arrow of God* by Chinua Achebe; *Once a Great Notion* by Ken Kesey; *Nothing Like the Sun* by Anthony Burgess; *The Snow Ball* by Irish novelist Brigid Brophy, 33; *The Spire* by William Golding; *The Keepers of the House* by U.S. novelist Shirley Ann Grau, 35; *The Children at the Gate* by the late Edward Lewis Wallant; *Little Big Man* by Thomas Berger; *Julian* by Gore Vidal; *The Wapshot Scandal* by John Cheever; *Corridors of Power* by C. P. Snow; *Girls in Their Married Bliss* by Edna O'Brien; *The Valley of Bones* by Anthony Powell; *The Ravishing of Lol Stein* by Marguerite Duras; *The Palace (La Palace) by* Claude Simon; *The Makepiece Experiment* (*Lyubimov*) by Andrei Sinyavsky.

Juvenile *Harriet the Spy* by U.S. author Louise Fitzhugh; *Charlie and the Chocolate Factory* by Roald Dahl.

Poetry *For the Union Dead* by Robert Lowell; *The Bourgeois Poet* by Karl Shapiro; *The Whitsun Wedding* by Philip Larkin; *Rediscoveries* by Elizabeth Jennings.

Painting *Retroactive II* by Robert Rauschenberg; *The Man in the Bowler Hat*, *The Great War*, and *The Sin of Man* by René Magritte; *According to What* by Jasper Johns; *Fez* by Frank Stella; *Rising Moon* by Hans Hofmann; *Brillo Boxes* and *Shot Orange Marilyn* by Andy Warhol. Stuart Davis dies in New York June 24 at age 69.

Sculpture *Classic Figure* by Jean Arp, now 77; *Cubi* by David Smith; *Homage to the 6,000,000* by Louise Nevelson.

Theater *The Persecution and Assassination of Marat as Performed by the Inmates of the Asylum of Charenton under the Direction of the Marquis de Sade* (*Die Verfolgung und Ermordung Jean-Paul Marats, dargestelt durch die Schauspielgruppe des Hospizes a Charenton unter Auleiting der Herrn de Sade*) by émigré German playwright-novelist-painter Peter Weiss, 48; *After the Fall* by Arthur Miller 1/23 at New York's ANTA Theater-Washington Square, with Jason Robards, Jr., Barbara Loden, David Wayne, Hal Holbrook, Salome Jens, Ruth Attaway, Faye Dunaway, Zohra Lampert, Ralph Meeker, 208 perfs.; *Any Wednesday* by U.S. playwright Muriel Resnik 2/18 at New York's Music Box Theater, with Sandy Dennis, Gene Hackman, Rosemary Murphy, 982 perfs.; *But for Whom Charlie* by S. N. Behrman 3/12 at New York's ANTA Theater-Washington Square, with Salome Jens, Jason Robards, Jr., Barbara Loden, Ralph Meeker, David Wayne, 39 perfs.; *Benito Cereno* by Robert Lowell 4/1 at New York's American Place Theater in St. Clements Church; *Blues for Mr. Charlie* by James Baldwin 4/23 at New York's ANTA Theater, with Pat Hingle, Rip Torn, Diana Sands, 148 perfs.; *The Entertaining Mr. Sloan* by English playwright Joe Orton, 31, 5/6 at London's New Arts Theatre; *The Subject Was Roses* by U.S. playwright Frank D. Gilroy, 39, 5/25 at New York's Royale Theater, with Martin Sheen, Jack Albertson, Irene Dailey, 832 perfs.; *Inadmissable Evidence* by John Osborne 9/9 at London's Royal Court Theatre, with Nicol Williamson; *The Black Swan* (*Der Schwarze Schwan*) by Martin Walser 10/14 at Stuttgart's Staatstheater; *The Sign in Sidney Brustein's Window* by Lorraine Hansberry 10/15 at New York's Longacre Theater, 101 perfs. (Hansberry will die of cancer early next year at age 36); *Slow Dance on the Killing Ground* by U.S. playwright William Hanley, 33, 11/3 at New York's Plymouth Theater, 88 perfs.; *Luv* by U.S. playwright Murray Schisgal, 37, 11/11 at New York's Booth Theater, with Alan Arkin, Eli Wallach, Anne Jackson, 901 perfs.; *The Royal Hunt of the Sun* by Peter Shaffer 2/8 at London's Old Vic Theatre, with Christopher Plummer as Francisco Pizarro; *The Plebeians Rehearse the Uprising* (*Die Plebem proben den Aufstrand*) by Günter Grass 9/20 at Berlin deals with the June 17, 1953 revolt in East Berlin; *Hunger and Thirst* (*La soif et la faim*) by Eugène Ionesco 12/30 at Düsseldorf's Schauspielhaus; *Tiny Alice* by Edward Albee 12/30 at New York's Billy Rose Theater, with John Gielgud, Irene Worth, 167 perfs.

New York State Theater opens in New York's Lincoln Center April 23 with 2,729 seats. Philip Johnson has designed the house.

The Dorothy Chandler Pavilion with 3,250 seats gives Los Angeles the start of a new cultural center. It will be augmented in the next 5 years by the 2,100-seat Ahmanson Theater and the 750-seat Mark Taper Forum.

Films Peter Glenville's *Becket* with Richard Burton, Peter O'Toole, John Gielgud; Stanley Kubrick's *Dr. Strangelove, or How I Learned to Stop Worrying and Love the Bomb* with Peter Sellers, George C. Scott, Sterling Hayden, Keenan Wynn; Robert Stevenson's *Mary Poppins* with Julie Andrews, Dick Van Dyke; Bryan Forbes's *Seance on a Wet Afternoon* with Kim Stanley, Richard Attenborough; Blake Edwards's *A Shot in the Dark* with Peter Sellers, Elke Sommer; Jules Dassin's *Topkapi* with Melina Mercouri, Robert Morley, Peter Ustinov; Jacques Cousteau's documentary *World Without Sun*; Vittorio De Sica's *Yesterday, Today, and Tomorrow* with Sophia Loren, Marcello Mastroianni. Also: Arthur Hiller's *The Americanization of Emily* with James Garner, Julie Andrews; Philippe de Broca's *Cartouche* with Jean-Paul Belmondo, Claudia Cardinale; Jean-Luc Godard's *Contempt* with Brigitte Bardot, Fritz Lang, Michel Piecoli; Sidney Lumet's *Fail-Safe* with Henry Fonda, Walter Matthau; Anthony Mann's *The Fall of the Roman Empire* with Sophia Loren, Stephen Boyd, James Mason; Guy Hamilton's *Goldfinger* with Sean Connery, Gert Frobe, Harold Sakata; Joseph Losey's *King and Country* with Dirk Bogarde, Tom Courtenay; Clive Donner's *Nothing but the Best* with Alan Bates, Denholm Elliott; Larry Peerce's *One Potato, Two Potato* with Barbara Barrie, Bernie Hamilton; Pietro Germi's *Seduced and Abandoned* with Stefania Sandrelli, Saro Urzi; Jacques Demy's *Umbrellas of Cherbourg* with Catherine Deneuve; Hiroshi Teshigahara's *Woman in the Dunes* with Eiji Okada, Kyoko Kishida; George Roy Hill's *The World of Henry Orient* with Peter Sellers, Tippy Walker, Angela Lansbury; Michael Cocoyannis's *Zorba the Greek* with Anthony Quinn, Alan Bates.

Hollywood musicals George Cukor's *My Fair Lady* with Audrey Hepburn, Rex Harrison; Richard Lester's *A Hard Day's Night* with the Beatles; Charles Walters's *The Unsinkable Molly Brown* with Debbie Reynolds.

Broadway musicals *Hello, Dolly!* 1/16 at the St. James Theater with Carol Channing, music and lyrics by Jerry Herman, 2,844 perfs.; *Funny Girl* 3/26 at the Winter Garden Theater, with Barbra Streisand as the late Fanny Brice, music and lyrics by Jule Styne and Bob Merrill, songs that include "People," "Don't Rain on My Parade," 1,348 perfs.; *Fiddler on the Roof* 9/22 at the Imperial Theater, with Zero Mostel, music by Jerry Bock, lyrics by Sheldon Harnick, songs that include "If

1964 ***(cont.)*** I Were a Rich Man," "Tradition," "Matchmaker, Matchmaker," "Sunrise, Sunset," 3,242 perfs.

Opera *Montezuma* 4/19 in West Berlin, with music by Roger Sessions.

First performances Symphony No. 5 by Roger Sessions 2/7 at the Philadelphia Academy of Music; *Curlew River* (church parable) by Benjamin Britten in July at Aldeburgh, with text by William Plower inspired by a 15th century Japanese Nō drama.

Popular songs "I Feel Fine," "Love Me, Do," "Please, Please Me," and "She Loves You" by John Lennon and Paul McCartney of The Beatles; "I Get Around" by Brian Wilson of the Beach Boys; "My Kind of Town" by Jimmy Van Heusen, lyrics by Sammy Cahn (for the film *Robin and the Seven Hoods*); "It Ain't Me, Babe" and "Mr. Tambourine Man" by Bob Dylan; "The Girl From Ipanema" by Brazilian composer Antonio Carlos Jobim, lyrics by Vinicius de Moraes, English lyrics by Norman Gimbel; "Oh, Pretty Woman" and "It's Over" by Roy Orbison and William Dees; "Look of Love" by Jeff Barry and Ellie Greenwich; "King of the Road" by Roger Miller; "I Will Wait for You" by French composer Michel Legrand, lyrics by Jacques Demy (English lyrics by Norman Gimbel) (for the film *The Umbrellas of Cherbourg*); "Pass Me By" by Cy Coleman, lyrics by Carolyn Leigh (for the film *Father Goose*).

Cassius Marcellus Clay, 22, knocks out world heavyweight champion Sonny Liston February 25 at Miami to win the title he will hold until 1967.

Kelso wins the Washington, D.C., International at age 7 to cap a 5-year racing career in which the big gelding has won 39 times in 83 starts to earn an all-time record of nearly $2 million.

A football stadium riot at Lima, Peru, May 24 ends with 300 dead after a soccer match between Peru and Argentina.

Roy Emerson wins in men's singles at Wimbledon and Forest Hills, Maria Bueno in women's singles.

The Olympic Games at Tokyo attract 5,541 contestants from 94 nations. Soviet athletes collect 41 gold medals, U.S. athletes, 37.

The U.S. America's Cup defender *Constellation* defeats Britain's *Sovereign* 4 to 0 in the sixteenth British effort to regain the cup.

The St. Louis Cardinals win the World Series by defeating the New York Yankees 4 games to 3.

The topless bathing suit introduced by California designer Rudi Gernreich will lead to a general abandonment of brassieres by young women.

A San Francisco bar introduces topless dancers June 19. The dancers will be bottomless beginning September 3, 1969.

Dynel is introduced by Union Carbide, whose new synthetic fiber will be used in textiles, fake furs, and hairpieces.

G.I. Joe, a doll for boys, is introduced by Hasbro, a major U.S. toy maker.

U.S. cigarette consumption reaches 524 billion—more than 4,300 smokes for every American over age 18.

The tobacco industry halts advertising in college newspapers, magazines, sports programs, and on college radio stations, in response to public pressure (*see* Banzhaf, 1966).

Carlton cigarettes are introduced by the American Tobacco Company but low-tar cigarettes will not begin to gain popularity until 1970.

"The Boston Strangler" makes his final attack in January. He has broken into a dozen apartments since mid-1962 to rob, violate, and kill women (not always by strangling them). Cambridge police take Albert Henry DeSalvo, 32, into custody November 4 but do not say he is the "strangler." DeSalvo was sentenced to prison on robbery counts in 1961, was paroled after 11 months, claims to have raped at least 1,000 women, and will boast to inmates at Bridgewater State Hospital that he has killed 13 women.

The Kitty Genovese case raises alarms about America's growing isolation, callousness, and inhumanity (*see* Jacobs book on cities, 1961). An attacker stalks Queens, N.Y., bar manager Catherine Genovese early in the morning of March 13; 38 of her Kew Gardens neighbors hear her wild calls for help; nobody interferes for fear of "getting involved," the neighbors watch from windows while Genovese is stabbed to death; nobody phones the police until half an hour later.

Boston's Prudential Tower, completed for the Newark-based Prudential Insurance Co., is twice the height of the John Hancock building completed in 1949. The 52-story Prudential Tower spurs John Hancock management to proceed with plans for an even taller structure (*see* 1968).

The Washington Cathedral in the nation's capital gets a new tower that makes it visible from 30 miles away (*see* 1907). Soaring 301 feet above the ground, the Gothic tower has spires that make it taller than the Washington Monument of 1888.

Chicago's Marina City apartment houses and offices are completed on the Chicago River. The 60-story round twin towers designed by Bertrand Goldberg are of concrete construction with weight loads carried chiefly by cylindrical cores, the pie-shaped rooms extend into rings of semicircular balconies, and the first 18 floors are taken up by parking space.

The Mauna Kea Beach Hotel opens on a tract leased from the 117-year-old Parker Ranch at the foot of 13,796-foot Mauna Kea on the island of Hawaii. The $15 million, 154-room resort hotel has been financed by Laurance Rockefeller.

More than $1 billion in U.S. federal aid for housing and urban renewal is authorized through September 30 of next year by a bill signed into law by President Johnson September 2 (*see* HUD, 1965).

The worst earthquake felt anywhere in the world since 1960 rocks Alaska March 28. The quake measures 8.4 on the Richter scale (which will be revised in 1977 to give the quake a rating of 9.2), it creates a seismic "tidal" wave *(tsunami)* in the southwest part of the state, and the 220-foot high wave is the largest such wave ever recorded.

High waters in the Adriatic Sea in November cause the Venetian Lagoon to rise 6 feet above normal levels. Disastrous floods wash out the foundations of many buildings, ruin ground-floor frescoes, cause some structures to collapse, and hasten the deterioration of the once powerful city of Venice.

Congress passes a Wilderness Act to protect the country's last remaining wild lands from development (*see* California Wildland Research Center, 1961). Wilderness Society executive director Howard Zahniser, 58, has exhorted the congressmen to do something. Their act provides for immediate protection of 9.1 million acres classified as wilderness by the U.S. Forest Service; it establishes a procedure to review every 10 years some 5.4 million acres in the national forests that are classed as "primitive" and gives Congress sole power to designate a national wilderness or declassify an established one.

Canyonlands National Park, created by act of Congress, embraces 257,640 acres of Utah mesas and rock spires.

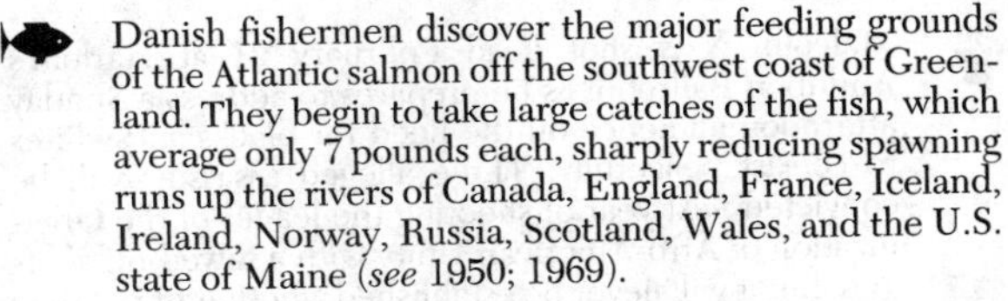

Danish fishermen discover the major feeding grounds of the Atlantic salmon off the southwest coast of Greenland. They begin to take large catches of the fish, which average only 7 pounds each, sharply reducing spawning runs up the rivers of Canada, England, France, Iceland, Ireland, Norway, Russia, Scotland, Wales, and the U.S. state of Maine (*see* 1950; 1969).

High-yielding dwarf strains of *indica* rice are introduced on experimental basis under the names IR5 and IR8 by the International Rice Research Institute at Los Baños in the Philippines. The new "miracle" rice for tropical cultivation has been developed by crossing ordinary indica rice with Japan's high-yielding *japonica* variety (*see* 1962; 1965).

The U.S. food stamp program conducted at Rochester, N.Y., from 1939 to 1943 is reactivated on a broad scale by the U.S. Department of Agriculture to help feed needy Americans (*see* 1959; 1967).

Awake is introduced by General Foods which promotes the synthetic orange juice with a budget of $5 million—more than is spent to promote pure orange juice, whether frozen, chilled, or fresh.

Pop-Tarts toaster pastries are introduced by Kellogg.

Lucky Charms breakfast food, introduced by General Mills, is 50.4 percent sugar (*see* 1961; 1966).

Dole Corp. is acquired by Honolulu's 113-year-old Castle & Cooke which for the first time markets fresh pineapple under the Dole Royal Hawaiian label, using jet planes to rush the fruit to East Coast U.S. markets, although most still goes by ship to West Coast ports.

Colonel Sanders has more than 600 licensees offering his "finger-lickin' good" Kentucky Fried Chicken (*see* 1955). Now 74, Sanders sells the franchise business for $2 million plus a salary of $40,000 per year for life to act as a goodwill ambassador.

The Time Has Come by Boston Catholic physician John Rock, now 74, rejects the Church's position against artificial contraceptive methods. Rock helped develop the progesterone contraceptive pill.

A dozen U.S. states have tax-supported birth control programs, most of them in the South where Catholic influence is weak and where whites try to hold down black birth rates. There are 450 public birth-control clinics in the nation (*see* 1965).

1965

U.S. bombers pound North Vietnamese targets February 7 and 8, retaliating against a National Liberation Front (NLF) attack on U.S. ground forces in South Vietnam. Washington announces a general policy of bombing North Vietnam February 11 and President Johnson says, "The people of South Vietnam have chosen to resist [North Vietnamese aggression]. At their request the United States has taken its place beside them in this struggle."

Some 160 U.S. planes bomb North Vietnam March 2, and 3,500 U.S. Marines land at Da Nang March 8 to 9 in the first deployment of U.S. combat troops in Vietnam. A bomb explodes in the U.S. embassy at Saigon March 30.

North Vietnamese MIG fighter planes shoot down U.S. jets April 4, a student demonstration in Washington, D.C., April 17 protests the U.S. bombing of North Vietnam, but U.S. planes raid North Vietnam in force April 23.

Australia decides April 29 to send troops to aid South Vietnam.

A "teach-in" broadcast May 15 to more than 100 U.S. colleges opposes the war in Vietnam, but Congress authorizes use of ground troops in direct combat June 8 if the South Vietnamese army so requests. The first full-scale combat offensive by U.S. troops begins June 28.

Some 125,000 U.S. troops are in Vietnam by July 28 and President Johnson announces a doubling of draft calls. While he asks the UN to help negotiate a peace, U.S. troops engage in their first major battle as an independent force in mid-August and destroy a Viet Cong stronghold near Van Tuong August 19.

The Indonesian army defeats a Communist takeover attempt in late September. Indonesia has withdrawn from the United Nations January 2 (the first nation to do so). Jakarta has seized U.S. oil company properties and those of Goodyear Tire & Rubber March 19 and seizes all remaining foreign-owned properties April 24. Communists kidnap the army chief of staff and five generals at the end of September, but other generals escape capture and thwart the attempted coup. A general massacre of Communists begins October 8, estimates of the dead range up to 400,000, but many of those killed are ethnic Chinese who dominate the

President Johnson escalated U.S. involvement in the war between South Vietnam and Communist North Vietnam.

Indonesian economy and are slain because they are landlords or creditors.

"The Vietcong are going to collapse within weeks," says President Johnson's National Security Adviser Walt Whitman Rostow, 48, "Not months, but weeks."

Antiwar rallies October 15 attract crowds in four U.S. cities. Poet Allen Ginsberg at Berkeley, Calif., introduces the term "flower power" to describe a strategy of friendly cooperation. Oakland police have blocked the Berkeley peace marchers from entering the city and the Hell's Angels motorcycle gang has attacked the marchers, calling them "un-American."

Look magazine reveals in November that Washington rejected secret peace talks with North Vietnam arranged in September of last year by UN Secretary General U Thant.

The AFL-CIO pledges "unstinting support" for the Vietnamese war effort December 15.

U.S. Marines have landed in the Dominican Republic April 28 to protect American citizens and prevent an allegedly imminent Communist takeover of the Santo Domingo government. An Inter-American Peace Force from the Organization of American States (OAS) takes over peacekeeping operations beginning May 23, but 20,000 U.S. Marines remain for several months.

Romania's Premier Gheorghe Gheorghiu-Dej has died in March after nearly 13 years and been succeeded as head of state by Nicolae Ceausescu, 47, who will rule despotically until late 1989.

The Gambia has gained independence February 18 after nearly 122 years of British colonial rule. The kingdom will become a republic in 1970.

The Maldives in the Indian Ocean gain independence July 26 after 78 years of British colonial rule.

Rhodesia's Prime Minister Ian Smith visits London in early October and demands immediate independence, Britain refuses unless the Salisbury government first agrees to expand representation of native Africans in the government with a view to eventual majority rule, Ian Smith declares Rhodesian independence unilaterally November 11 while reaffirming loyalty to the queen, the British governor at Salisbury declares Smith and his government deposed, London calls the declaration illegal and treasonable and proclaims economic sanctions against Rhodesia.

The UN Security Council calls on all nations November 12 to withhold recognition of the new Rhodesian regime and refuse it aid.

Guinea severs diplomatic relations with France November 15 after discovering a plot to assassinate Sékou Touré and overthrow his regime.

The Independent Congo Republic has a bloodless coup November 24 to 25, 6 weeks after President Kasavubu has dismissed Moise Tshombe as premier. Gen. Joseph Mobutu deposes President Kasavubu and makes himself president and proceeds to rule by decree (*see* 1962; 1966).

The Organization of African Unity threatens to break relations with Britain December 5 unless London applies force to suppress Ian Smith's rebellion (above) by December 15.

Malcolm X is shot dead February 21 at Harlem's Audubon Ballroom as he prepares to address a Sunday afternoon audience on the need for blacks and whites to coexist peacefully. Three alleged assassins will be convicted next year of shooting the leader of the Organization of Afro-American Unity with a sawed-off shotgun, but it will never be established whether or not they were members of the Black Muslim sect with which Malcolm X broke last year.

Selma, Ala., is the focus of civil rights demonstrations throughout February and March. Martin Luther King, Jr., and 770 others are arrested February 1 at Selma during demonstrations against state regulations relating to voter registration, black marchers leave Selma March 7 for the state capital at Montgomery after 2,000 prospective voters have been arrested in registration lines or in demonstrations, the marchers are attacked by 200 Alabama state police using tear gas, whips, and night sticks. Gov. Wallace refuses police protection for a second march. President Johnson sends in 3,000 federalized National Guardsmen and military police, the marchers leave Selma March 21 and arrive at Montgomery March 25. White civil rights leader Viola Liuzzo, 38, is overtaken after a high-speed chase and shot dead March 25 by four Ku Klux Klansmen outside Montgomery while driving a car with a black passenger. 25,000 attend the rally at Montgomery that day.

Chicago police arrest 526 anti-segregation demonstrators from June 11 to 15 after the rehiring of a school superintendent. The demonstrations continue nevertheless and Martin Luther King, Jr., leads a march of 20,000 to City Hall July 26.

The Voting Rights Act becomes law August 10 (*see* 1964); federal examiners begin registering black voters in Alabama, Louisiana, and Mississippi.

The Watts section of Los Angeles has violent race riots beginning August 12 as upwards of 10,000 blacks burn and loot an area of 500 square blocks and destroy an estimated $40 million worth of property. Some 15,000 police and National Guardsmen are called in, 34 persons killed (28 of them blacks), nearly 4,000 arrested, and more than 200 business establishments totally destroyed.

North Carolina judge James B. McMillan orders busing of schoolchildren to achieve racial desegregation as required by the 1954 Supreme Court decision. His order for crosstown busing in the Charlotte-Mecklenburg County school system starts a pattern that will be followed in much of the country, but although the Supreme Court will uphold McMillan's ruling in 1971 the use of busing creates a storm of controversy.

Amos 'n Andy is withdrawn from syndication following protests against its stereotyped images of blacks. Started as a radio show in 1928, *Amos 'n Andy* has been a leading television program since 1949 with blacks playing the roles originally created by whites.

President Johnson outlines programs for a "Great Society" that will eliminate poverty in America in his State of the Union message January 4; he signs a $1.4 billion program of federal-state economic aid to Appalachia into law March 9, but U.S. military involvement in Southeast Asia escalates (above), draining the U.S. economy.

New York City's welfare roll grows to 480,000. The number of welfare recipients will be 1.2 million by 1975 and the city's welfare agency will account for more than a quarter of the city's $12 billion budget with half the welfare aid reimbursed by the federal government.

The United States Mint switches to "clad" coins in a move to conserve silver by eliminating much of that metal in dollars, half-dollars, quarters, and dimes.

The United States contributes roughly one percent of its Gross National Product (GNP) to foreign aid as proposed by Adlai Stevenson in 1961, but the percentage will decline sharply (*see* 1967).

Britain freezes wages, salaries, and prices in an effort to check inflation and improve the nation's worsening trade deficit (*see* 1961; 1967).

The Diners Club of America founded in 1950 reaches a membership of 1.3 million, American Express has 1.2 million after just 7 years, and major U.S. banks prepare to issue credit cards of their own (*see* BankAmericard, 1958; 1966; Master Charge, 1966). Not only restaurants but also hotels, motels, airlines, travel agencies, car rental agencies, and many retail stores now honor the credit cards.

The worst power failure in history blacks out most of seven states and Ontario November 9, affecting 30 million people in an 80,000-square-mile area. New York's power fails at 5:27 in the afternoon, Brooklyn regains it at 2 the next morning, Queens at 4:20 in the morning, Manhattan at 6:58. Telephone companies maintain service and New York has a record 62 million phone calls in one day, nearly double the city's weekly average. Con Edison assures the public that no blackouts will recur (but *see* 1977).

The S.S. *Michelangelo* and the S.S. *Raffaello* go into service for the Italian line. The 45,911-ton sister ships are each 904 feet in length overall.

The Zeeland Bridge opened by Queen Juliana of the Netherlands is the longest bridge in Europe; it connects North Beveland with Schouwen-Duiveland more than a mile away and cuts the Rotterdam-Flushing run from 90 miles to 70.

Germany's Emmerich Bridge opens to span the Rhine. It is 1,644 feet in length.

The Soviet Antonov AN-22 makes a flight carrying 720 passengers.

Brazil produces her 1 millionth automobile. Willys-Overland do Brasil, S.A., is Brazil's largest; more than 50 percent owned by Brazilians, it also makes trucks and Jeeps (*see* 1945).

A court order requires General Motors to make its bus patents freely available to other bus makers and to sell parts, engines, and designs to other busmakers in settlement of the suit brought by the Justice Department in 1956, but GM's grip on the bus market will remain virtually undiminished.

U.S. production of soft-top convertibles peaks at 507,000 for the model year 1965. The figure will drop to 28,000 by 1974 as air-pollution, vandalism in the form of top-slashing, higher driving speeds, body rattles, and factory-installed air-conditioning discourage sales of convertibles.

Unsafe at Any Speed by consumer advocate Ralph Nader notes that more than 51,000 Americans are killed each year by automobiles, a figure the Department of Commerce had projected for 1975. Nader has quit his job with the Department of Labor to crusade for consumer protection and he writes, "For more than half a century, the automobile has brought death, injury and the most inestimable sorrow and deprivation to millions of people."

The U.S. death rate falls to 943.2 per 100,000, down from 1,719 in 1900.

Congress establishes a National Clearinghouse for Smoking and Health and orders that cigarette packages be labeled "Caution: cigarette smoking may be hazardous to your health."

The $2 million appropriated by Congress for the National Clearinghouse (above) is less than 1 percent of the amount spent on TV advertising by the tobacco industry, and the legislation frees the industry from Federal Trade Commission regulations for the next 4 years. The major tobacco companies continue to diversify by acquiring food companies.

Britain bans cigarette advertising from commercial television beginning August 1.

1965 ***(cont.)*** The International Society for Krishna Consciousness is founded at New York by Calcutta chemist and Sanskrit scholar A. C. Bhaktivedanta, 59, whose followers call him Swami Prabhupada and regard him as the successor to an unbroken chain of Hindu spiritual teachers dating back 5,000 years. The swami has arrived in New York with $50 in rupees and a pair of cymbals, determined to spread the teachings of Lord Krishna, a supreme deity in Hindu mythology. He sits on a sidewalk in the East Village, begins the "Hare Krishna" chant that will soon become familiar throughout the world, offers young people a relatively ascetic life of devotion and proselytizing, an alternative to conventional society and drugs, and is soon holding religious classes in an empty storefront on Second Avenue, attracting youths who shave their heads, wear saffron-colored dhatis, and chant, "Hare Krishna . . ."

U.S. university enrollments swell as young Americans take advantage of draft deferrals for college students to escape the expanding war in Vietnam, but campuses are tense with unrest.

The University of California opens new Irvine and Santa Cruz campuses (*see* 1944).

Congress votes $1.3 billion in federal aid to elementary and secondary schools.

Britain's Universities of Kent and Warwick are founded.

University College is founded at Britain's Cambridge University.

"Early Bird" is put into orbit by the 3-year-old Communications Satellite Corporation (COMSAT) and relays telephone messages and television programs between Europe and the United States. The world's first commercial satellite, it begins a global network of space communications.

Sony introduces Betamax, a small home "videocorder"; by the late 1970s Japanese companies will own the VCR industry (*see* 1956).

New York's first all-news radio programming begins April 19 on Westinghouse Broadcasting's WINS. Other stations across the country will follow suit as radio becomes more specialized.

Nonfiction *The Autobiography of Malcolm X* by the late Afro-American Unity leader (above) and writer Alex Haley, 43. A cook in the U.S. Coast Guard from 1939 to 1959, Haley begins research in the National Archives at Washington to learn the history of his own family (see *Roots*, 1976); *Manchild in the Promised Land* by U.S. reform school veteran Claude Brown; *The Psychedelic Reader* by Harvard psychology professor Timothy Leary, 44, who has experimented with marijuana and other behavior-changing drugs. Leary advises readers to "turn on, tune in, drop out."

Fiction *Everything That Rises Must Converge* by the late Flannery O'Connor who died last year at age 39 of a rare disease; *The Painted Bird* by Polish-American novelist Jerzy Kosinski, 32; *The House of Assignation* (*La Maison de rendezvous*) by Alain Robbe-Grillet; *The Last of the Pleasure Gardens* by Francis King; *Landlocked* by Doris Lessing; *God Bless You, Mr. Rosewater* by Kurt Vonnegut, Jr.; *The Magus* by John Fowles.

Poetry *The Old Glory* (three verse dramas based on stories by Herman Melville and Nathaniel Hawthorne) by Robert Lowell; *The Lost World* by Randall Jarrell, who walks in front of an oncoming car at Chapel Hill, N.C., October 14 and dies at age 51.

Painting *Every Building on the Sunset Strip* by Ed Ruscha; *Campbell's Tomato Soup Can* and *'65 Liz* by Andy Warhol; "Op" art that creates optical illusions by using color, form, and perspective in bizarre ways becomes fashionable; *Self-portrait* by Pablo Picasso. Milton Avery dies at New York January 3 at age 71.

Sculpture *How to Explain Pictures to a Dead Hare* by Joseph Beuys (performance); *Untitled* (perforated steel) by U.S. sculptor Donald Judd, 37. David Smith dies at Albany, N.Y., May 24 at age 59 following an automobile accident.

The Federal Aid to the Arts Act signed by President Johnson September 30 establishes a National Endowment for the Arts and the Humanities, funded by an initial 3-year appropriation of $63 million.

Theater *Loot* by Joe Orton 2/2 at Brighton, England; *The Odd Couple* by Neil Simon 3/10 at New York's Plymouth Theater, with Walter Matthau as Oscar Madison, Art Carney as Felix Ungar, 964 perfs.; *The Amen Corner* by James Baldwin 4/15 at New York's Ethel Barrymore Theater, with Bea Richards, Juanita Hall, Frank Silvera, 84 perfs.; *The Homecoming* by Harold Pinter 6/3 at London's Aldwych Theatre, with Vivien Merchant, Paul Rogers, Ian Holm, Pinter; *Generation* by U.S. playwright William Goodhart 10/6 at New York's Morosco Theater, with Henry Fonda, 299 perfs.; *Hogan's Goat* by U.S. playwright William Alfred, 43, 11/11 at New York's American Place Theater in St. Clements Church, with Ralph Waite, Cliff Gorman, Faye Dunaway, 607 perfs.; *Cactus Flower* by Abe Burrows 12/8 at New York's Royale Theater, with Lauren Bacall, Barry Nelson, Brenda Vaccaro, 1,234 perfs.

New York's Vivian Beaumont Theater opens in Lincoln Center October 21. Designed by Eero Saarinen, it has 1,140 seats.

Films Stuart Burge's *Othello* with Laurence Olivier, Frank Finlay, Maggie Smith, Joyce Redman; Sidney Lumet's *The Pawnbroker* with Rod Steiger; Roman Polanski's *Repulsion* with Catherine Deneuve; Jan Kadar's *The Shop on Main Street* with Elmar Klos, Josef Kroner, Ida Kaminska. Also: John Schlesinger's *Darling* with Julie Christie, Dirk Bogarde, Laurence Harvey; Luis Buñuel's *Diary of a Chambermaid* with Jeanne Moreau; Sidney Lumet's *The Hill* with Sean Connery, Michael Redgrave, Ossie Davis; Sidney J. Furie's *The Ipcress File* with Michael Caine; Tony Richardson's *The Loved One* with Robert Morse, Jonathan Winters; Miloš Forman's *Loves of a Blonde* with Hana Brejchova, Josef Sebanek; Fred Coe's *A Thousand Clowns* with Jason Robards, Jr., Barbara Harris, Barry Gordon, Martin Balsam, Gene Saks.

Stage musicals *Baker Street* 2/16 at New York's Broadway Theater, with Fritz Weaver as Sherlock Holmes, music and lyrics by Marian Grudeff and Raymond Jessel, 313 perfs.; *The Roar of the Greasepaint—The Smell of the Crowd* 5/16 at New York's Shubert Theater, with Anthony Newley, Cyril Ritchard, Sally Smith, book, music, and lyrics by Leslie Bricusse and Newley, 231 perfs.; *On a Clear Day You Can See Forever* 10/17 at New York's Mark Hellinger Theater, with Barbara Harris, music by Burton Lane, lyrics by Alan Jay Lerner, songs that include the title song, 272 perfs.; *Man of La Mancha* 11/22 at New York's ANTA-Washington Square Theater, with Richard Kiley, Joan Diener, Irving Jacobson, Robert Rounseville, book based on the 1615 Cervantes novel *Don Quijote de la Mancha*, music by Mitch Leigh, 37, lyrics by Joe Darian, 48, songs that include "The Impossible Dream," "Dulcinea," and the title song, 2,329 perfs.; *Charlie Girl* 12/15 at London's Adelphi Theatre, with Christine Holmes, Anna Neagle, music and lyrics by David Heneker and John Taylor, 2,202 perfs.

First performances Symphony No. 8 by Walter Piston 3/5 at Boston's Symphony Hall; *Gemini Variations on an Epigram of Kodaly* for Flute, Violin, and Piano Duet by Benjamin Britten in June at Aldeburgh's Jubilee Hall; *From the Steeples and Mountains* by the late Charles Ives (who wrote it in 1901) 7/30 at New York's Philharmonic Hall; *The Golden Brown and the Green Apple* by Duke Ellington 7/30 at New York's Philharmonic Hall; *Voices for Today* by Britten 10/24 (20th anniversary of the United Nations) at New York, London, and Paris.

The Rolling Stones gain huge success with a recording of "[I Can't Get No] Satisfaction." The 2-year-old English rock group has taken its name from the Muddy Waters song "Rolling Stone Blues," leader Michael Philip "Mick" Jagger, 22, is backed by drummer Charlie Watts and bass guitarist Bill Wyman, and the group invokes a moody, frankly sexual tone derivative of Chuck Berry that makes the Beatles sound like choirboys by comparison.

The Grateful Dead has its beginnings in a San Francisco acid-rock group started at 710 Ashbury Street by local electric guitarist Jerry Garcia, 24, with drummer Mickey Hart (whose father will vanish with the group's profits), Ron "Pigpen" McKernan, and others, Garcia has dropped acid with Ken Kesey, he has taken the name Grateful Dead from an Oxford dictionary notation on the burial of Egyptian pharaohs, his admirers call him Captain Trips, but McKernan will die of alcohol and drugs and Garcia will receive a year's probation in New Jersey in 1973 for possession of LSD, marijuana, and cocaine.

Big Brother and the Holding Company is founded by San Francisco rock musicians Peter Albin, 20, Sam Andrew, 23, James Martin Gurley, 23, and Brooklyn-born drummer David Getz, 26, who will soon add singer Janis Joplin, 22, to their group.

Popular songs "Yesterday" and "Michelle" by John Lennon and Paul McCartney; "Sounds of Silence" by U.S. songwriters Paul Simon, 24, and Art Garfunkel, 25; "I Got You Babe" by U.S. singer-songwriter Sonny Bono, 30, who has married singer Cher Sirkisian, 21; "What the World Needs Now Is Love" by Burt Bacharach, lyrics by Hal David; "It's Not Unusual" by Gordon Mills; "Like a Rolling Stone" by Bob Dylan; "The Shadow of Your Smile" by Johnny Mandel, lyrics by Paul Francis Webster (for the film *The Sandpiper*); "Lara's Theme (Somewhere My Love)" by Maurice Jarre, lyrics by Paul Francis Webster (for the film *Dr. Zhivago*); "If I Ruled the World" by Cyril Ornadel, lyrics by Leslie Bricusse (for the Broadway musical *Pickwick*); "She (He) Touched Me" by Milton Schafer, lyrics by Ira Levin (for the Broadway musical *Drat! The Cat!*); "Goodnight" by Roy Orbison.

A British Court convicts 10 professional soccer players of fixing matches.

Roy Emerson wins in men's singles at Wimbledon, Margaret Smith in women's singles; Manuel Santana, 27 (Spain), wins in men's singles at Forest Hills, Smith in women's singles.

The Los Angeles Dodgers win the World Series by defeating the Minnesota Twins 4 games to 3.

The miniskirt appears in December in "swinging" London. Designed by Mary Quant, the provocative new skirt comes to 6 inches above the knee, it will help Quant's 10-year-old Bazaar on King's Road, Chelsea, become an international conglomerate of fashion, cosmetics, fabrics, bed linens, children's books, dolls, and wine (Mary Quant Limited), and the skirt eventually will dwindle to micro-mini lengths.

A U.S. Department of Housing and Urban Development (HUD) is inaugurated September 9.

St. Louis's Gateway Arch, completed by Eero Saarinen for the city's waterfront, commemorates the Louisiana Purchase of 1803.

New York's CBS building is completed by Eero Saarinen & Associates on the Avenue of the Americas (Sixth Avenue) at 52nd Street. The building's 38 floors are mainly supported from a central core.

Congress appropriates funds to remove U.S. highway billboards at the urging of "Lady Bird" Johnson, wife of the president. All billboards on sections of interstate and primary highways not zoned "commercial or industrial" are to be razed by July 1, 1970, and states are to pass conforming laws and prepare laws. The "Lady Bird Bill" will not begin to take effect until 1970, when the Senate will vote unanimously to apply $100 million of the $5.5 billion per year highway trust fund to compensate billboard companies for removal of their signs.

Hurricane Betsy roars across Florida September 8 with winds of up to 145 miles per hour and moves into Louisiana and Mississippi, killing 23 in 15 days.

A major drought in the northeastern United States forces New York City to turn off air-conditioning in sealed skyscrapers in order to conserve water. City fountains are turned off, lawn-watering is forbidden, and signs appear reading, "Save water: shower with a friend."

1965 *(cont.)* Congress enacts a 5-year, $25-million anadromous fish program to conserve alewives, river sturgeon, salmon, shad, and striped bass which swim upriver to spawn (*see* Maine, 1948).

Soviet Russia suffers another crop failure as it did in 1963 and is forced to pay gold for wheat from Australia and Canada. Moscow is discouraged from buying U.S. wheat by a presidential requirement that half of all shipments be made in U.S. vessels (at high cost) and by a longshoremen's threat not to load any grain for shipment to Communist Russia.

Dwarf Indica rice with higher per-acre yields is introduced in India, the Philippines, and other Asian nations (*see* Taiwan, 1955).

U.S. farms fall in number to 3.5 million, down from 6.84 million in 1935, 3.7 million in 1959.

Pakistan and much of India suffer widespread starvation as monsoon rains fail and crops wither in a drought of unprecedented proportions.

Half of all Americans enjoy "good" diets, up from less than one quarter in 1943, according to a new study.

Cranapple fruit juice is introduced by Ocean Spray Cranberries (*see* 1959).

Diet Pepsi is introduced by Pepsi-Cola Co. (*see* Tab, 1963). Pepsi president Donald M. Kendall, 44, engineers a merger with Frito-Lay, Inc., to create Pepsico, Inc. (*see* 1950; 1961).

Apple Jacks breakfast food, introduced by Kellogg, is 55 percent sugar (*see* 1964; 1966).

Home-delivered milk accounts for 25 percent of U.S. milk sales, down from more than 50 percent before World War II. The figure will be 15 percent by 1975.

The World Health Organization finds that family planning advice is welcome where such advice was protested in 1962.

Oxfam decides to support family planning projects as well as the programs for famine relief and food production it has backed since its founding in 1942.

Connecticut's 1879 law prohibiting sale of birth control devices is unconstitutional, the Supreme Court rules 7 to 2 June 7 in *Griswold v. Connecticut*. The case involved a New Haven clinic run by leaders of the state's Planned Parenthood League.

The United States has 700 public birth control clinics with 33 states giving, or about to give, tax support to birth control (*see* 1964).

U.S. live births total some 300,000 less than in the peak year of 1957 and in some months are at a rate lower than in 1939.

A new U.S. immigration act signed into law by President Johnson October 3 at the Statue of Liberty in New York Harbor abolishes the national origins quota system of 1952. The new law permits entry by any alien who meets qualifications of education and skill provided such entry will not jeopardize the job of an American.

The new U.S. immigration act (above) imposes an overall limit of 120,000 visas per year for Western Hemisphere countries and 170,000 per year for the rest of the world, but immediate relatives of U.S. citizens may enter without regard to these limits.

The population of the world reaches 3.3 billion.

1966 An encyclical issued by Pope Paul VI January 1 during a 37-day truce in Vietnam asks for an end to hostilities in Southeast Asia, Japanese prime minister Eisaku Sato announces an international peace mission January 25, the Senate Foreign Relations Committee's chairman J. W. Fulbright challenges the legality of U.S. military intervention January 28, Sen. Fulbright questions Secretary of State Dean Rusk, but U.S. bombing of North Vietnam begins by the end of the month.

International Days of Protest in many world cities criticize U.S. policy in Vietnam.

India's prime minister Lal Bahadur Shastri dies of a heart attack January 11 at age 61. Mrs. Indira Nehru Gandhi, 48, daughter of the late Jawaharlal Nehru, is elected to succeed Shastri January 19.

President de Gaulle proposes a "Europeanized Europe" free of U.S. and Soviet domination. He announces March 11 that France will withdraw her troops from NATO and requests that NATO remove all its bases and headquarters from French soil by April 1 of next year (*see* 1949). He sends his foreign minister to visit Eastern European capitals and visits the U.S.S.R. himself from June 20 to July 1, on which date Supreme Headquarters Allied Powers Europe (SHAPE) moves from Paris to Casteau, outside Brussels.

Romania's premier Nicolae Ceausescu proposes dissolution of both NATO and the Warsaw Pact alliance in a meeting of Warsaw Pact powers at Bucharest July 4 to 6. Ceausescu also asks that all nations withdraw their troops from the soil of all other nations.

Ghana's army and police officers stage a coup February 24 and oust President Nkrumah who is away on a visit to Beijing. Given refuge and named co-president by Guinea's president Sékou Touré, Nkrumah threatens military action to regain the power he held for 15 years.

Uganda's prime minister Milton Obote assumes full powers February 22 and deposes Sir Edward Mutesa II from the presidency March 2 (*see* 1962; 1971).

President Mobutu of the Democratic Republic of Congo takes over all legislative powers from Parliament in March and renames the nation's cities July 1. Leopoldville becomes Kinshasa, Stanleyville Kisingani, Elisabethville Lubumbashi (*see* Zaire, 1971).

South Africa's prime minister Henrik F. Verwoerd is assassinated September 6 after 8 years in power. Balthazar Johannes Vorster, 50, succeeds him a week later and will continue the Verwoerd policies of apartheid and support for the white regime in Rhodesia.

The UN General Assembly terminates South Africa's mandate in South-West Africa (Namibia), but South Africa calls the action illegal, ignores it, and refuses a

UN administrative commission entry into the mandate territory.

Botswana becomes an independent republic within the British Commonwealth September 30 and elects Sir Seretse Khama first president of the bleak territory known heretofore as British Bechuanaland.

Lesotho becomes an independent kingdom within the British Commonwealth October 4. Formerly British Basutoland, the new state ruled by King Moshoeshoe runs into difficulties when the king tries to establish more effective control. He is imprisoned, and released only after promising to abide by the constitution.

Rhodesia's president Ian Smith meets with Britain's prime minister Wilson on a warship off Gibraltar in early December, they make a tentative agreement that Rhodesia will have majority rule within 10 or 15 years, the Salisbury government rejects the agreement December 5, London appeals December 6 for UN sanctions against the Smith government, the UN Security Council imposes mandatory sanctions, but South Africa and Portugal refuse to participate.

Guyana becomes an independent state within the British Commonwealth May 26. Black leader Forbes Burnham becomes first prime minister of the South American country once called British Guiana.

The Dominican Republic elects moderate Joaquin Balaguer president over Juan Bosch after more than a year of occupation by an inter-American peacekeeping force. Balaguer has U.S. support and embarks on a program of economic and social reform as the OAS withdraws its force in October.

Jordan suspends relations with the Palestine Liberation Organization in July but is unable to stop PLO raids across the Jordan into Israel. Israeli tanks and aircraft attack the Jordanian village of Sammu November 13 in reprisal for the PLO raids and the undeclared war brings agitation by Palestinians for the overthrow of Jordan's Hussein, who has Saudi and U.S. support (*see* Six-Day war, 1967).

President Johnson visits New Zealand, Australia, the Philippines, South Vietnam, Thailand, Malaysia, and South Korea from October 19 to November 2, leaders of the allied nations pledge support for the war in Vietnam in a conference at Manila October 24 to 25, targets around Hanoi are bombed intensively in early December, and by year's end 389,000 U.S. troops are in South Vietnam.

The People's Republic of China fires her first nuclear bomb from a guided missile October 27.

Mao Zedong's wife Chiang Chin gets her first political job December 5. Now 52, she is made cultural consultant to the General Political Department of the Chinese Army as Mao acts to end the insolence of Red Guard youths, but Mme. Mao will become increasingly sympathetic toward the youths.

The U.S. Department of Commerce orders economic sanctions against Ian Smith's Rhodesia March 18, prohibiting export of anything that may be useful (*see* 1965).

University of Wisconsin students protest draft deferment examinations. They occupy administration buildings May 16 to begin a sit-in demonstration (*see* 1965; Columbia, 1968).

The National Organization for Women (NOW) is founded to help U.S. women gain equal rights. Founder and president of the new civil rights organization is Betty Friedan.

Secretary of Health, Education and Welfare John W. Gardner, 53, orders that federal funds be withheld from 12 Alabama, Louisiana, and Mississippi school districts that are in violation of 1964 Civil Rights Act guidelines; the May 13 order ignores Northern school districts that have done nothing to desegregate schools.

The University of Mississippi's first black graduate James Meredith is shot from ambush and wounded June 6 while walking from Memphis to Jackson, Miss., in a voting rights demonstration (*see* 1962). The march continues until June 26 when 15,000 demonstrators rally before the state capitol.

U.S. police officers must warn anyone taken into custody that he or she has a right to counsel and to remain silent and the courts must provide lawyers for those too poor to pay, the Supreme Court rules June 13 in the case of *Miranda v. Arizona*. The 5-to-4 decision handed down by Chief Justice Warren declares that the privilege against self-incrimination under the Fifth Amendment rules out confessions by persons in police custody unless careful steps are taken to protect the rights of suspects. "No statement obtained [in the atmosphere of the police station] can truly be the product of [the defendant's] free choice," says the majority decision, but Justice Harlan denounces the decision as "dangerous experimentation" at a time of a "high crime rate that is a matter of growing concern" (*see* 1971).

A month of racial riots and looting begins June 23 at Cleveland and blacks in Chicago's West Side riot for three nights in mid-July.

Atlanta has race riots in its black Summerhill section after having been the first city in the South to integrate its public schools without disturbance.

Gov. Wallace signs a bill September 2 forbidding Alabama's public schools to comply with the Office of Education's desegregation guidelines.

The U.S. Senate votes 49 to 37 September 21 to prohibit voluntary prayers in U.S. public schools (*see* Supreme Court decision, 1963).

Massachusetts voters elect the first black U.S. senator since Reconstruction. They send Edward W. Brooke, 47, to Washington to join Edward M. Kennedy.

The United States has 2,377 corporate mergers, up from 844 in 1960 (*see* 1967).

California's Bank of America creates BankAmerica Service Corp. to license other banks to issue the BankAmericard and participate in the system (*see*

1966 ***(cont.)*** 1958). By year's end there are 2 million BankAmericard holders and 64,000 merchant outlets (*see* 1968; Master Charge, below).

Mastercard has its beginnings in the Master Charge credit card introduced by New York's Marine Midland Bank at the urging of Buffalo banker Karl H. Hinke, 60, to compete with BankAmericard. Other banks will be licensed to issue the card which, like BankAmericard, will be available to bank customers without charge and will be accepted by hotels, restaurants, auto rental agencies, and airlines as well as by retail merchants.

The Tennessee Valley Authority orders construction of a 1-million-kilowatt nuclear power plant at Decatur, Ala., in the heart of the coal country. By late 1972 there will be 30 nuclear plants in the United States with 51 more under construction and 72 on order.

Pan Am orders 25 Boeing 747 jumbo jets, setting a lead that other carriers will have to follow. Depending on seat configuration, the new planes will carry from 342 to 490 passengers, numbers that will tax the capacities of existing terminal facilities. Few airlines will be able to keep their 747 jets filled and many, including Pan Am, will suffer financial reverses as a result of adopting the unprofitable jumbo jets (*see* 1970).

Laker Airways is founded by English aviation executive Freddie Laker, 44, who made his first fortune ferrying cargo to Berlin during the 1948-1949 Berlin Blockade. Laker has bought 12 obsolete bombers with a $100,000 loan from a friend, converted them to cargo carriers, and used his profits to help start Britain's largest independent airline (British United), but he has quarreled with BU's chairman and left to start his own charter airline, carrying passengers but no cargo.

FIAT chief Vittorio Valletta, 82, signs an agreement to build a $1-billion automobile plant in the Soviet Union. It is to be constructed at Togliattigrad, named for Italian Communist leader Palmiro Togliatti (*see* 1967).

New York makes Fifth Avenue one-way southbound and Madison Avenue one-way northbound beginning January 14 to ease congestion caused by a transit strike (*see* Third and Lexington Avenues, 1960); Tiffany & Co. president Walter Hoving, 68, has led opposition to the one-way traffic scheme and says Fifth Avenue is now a "superhighway."

New York City transit fares rise to 20¢ July 5; they have been 15¢ since 1953 (*see* 1970).

Complete decipherment of the genetic code is announced by MIT biochemist Har Gobind Khorana, 44, who synthesized vitamin A in 1959. He has built on the work of National Institutes of Health biochemist Marshall Warren Nirenberg, 39, and Cornell University biochemist Robert William Holley, 44 (*see* Watson and Crick, 1953).

The U.S. infant mortality rate falls to 24.3 per thousand live births, down from 29 in 1951 but the British rate falls to 20, down from 30, and the Swedish rate to 15. Inadequate health care delivery systems and poor nutrition among expectant mothers in low income groups are held responsible for the relatively poor U.S. showing.

The *Times* of London changes its format after 178 years to run news on its front page rather than classified advertisements.

The *New York Herald-Tribune, Journal-American,* and *World-Telegram & Sun* announce a merger March 21 after 2 months of fighting a Newspaper Guild charge of unfair labor practices by all three. The *Herald-Tribune* is to continue daily morning publication, the *World-Journal* to appear afternoons, and the *World-Journal & Tribune* to be a combined Sunday paper. The Guild strikes April 23 and all three papers cease publication. The *World-Journal & Tribune* appears for the first time September 12 and quickly sells out its 930,000 press run, but the Department of Justice forces the new company to offer its syndicated 19 columns and features to other papers (*see* 1967).

The Supreme Court rules that material with redeeming social value is uncensorable. The court modifies its anti-obscenity stand of 1957, but still defines obscene material as any matter in which "to the average person applying contemporary standards, the dominant theme taken as a whole appeals to a prurient interest" (*see* 1967).

Nonfiction *Quotations of Chairman Mao* is published at Beijing; *On Aggression* by Konrad Lorenz; *Division Street America* by Chicago journalist Louis "Studs" Terkel, 54.

Fiction *In Cold Blood* by Truman Capote; *The Fixer* by Bernard Malamud; *The Comedians* by Graham Greene; *Giles Goat-Boy* by John Barth; *The Crying of Lot 49* by Thomas Pynchon; *Black Light* by Galway Kinnell; *The Solid Mandela* by Patrick White; *The Jewel in the Crown* by English novelist Paul (Mark) Scott, 46; *The Birds Fall Down* by Rebecca West; *The Soldier's Art* by Anthony Powell; *The Green House* by Mario Vargas Llosa; *The Last Picture Show* by U.S. novelist Larry McMurtry, 30; *Omensetter's Luck* by U.S. novelist William Gass, 42; *Wide Sargasso Sea* by Jean Rhys; *Paradise* by Cuban novelist-poet José Lezama Lima, 56; *Silence* by Japanese novelist Shusaku Endo, 43; *The Mask of Apollo* by Mary Renault; *Tai-Pei* by James Clavell; *The Valley of the Dolls* by U.S. novelist Jacqueline Susann, 45.

Painting *Who's Afraid of Red, Yellow, Blue* by Barnett Newman; *Yellow and Red Brushstrokes* by Roy Lichtenstein; *Flowers* (silkscreen) by Andy Warhol; *Hatos II* by Victor Vasarely; *Le triomph de la musique* by Marc Chagall for New York's new Metropolitan Opera House (below). Alberto Giacometti dies at Chur, Switzerland, January 11 at age 64; Hans Hofmann dies at New York February 17 at age 85; Maxfield Parrish dies in New Hampshire March 30 at age 95; Jean Arp dies at Basel June 17 at age 78.

Sculpture Platform made up of two rectilinear boxes on the floor (fiberglas) by U.S. sculptor Bruce Nauman, 25.

New York's Whitney Museum of American Art moves

September 2 into a new building by Marcel Breuer on upper Madison Avenue (*see* 1931).

Theater *The Lion in Winter* by U.S. playwright James Goldman, 38, 3/3 at New York's Ambassador Theater, with Robert Preston as Henry II, Rosemary Harris as Eleanor of Aquitaine, 92 perfs.; *A Delicate Balance* by Edward Albee 9/22 at New York's Martin Beck Theater, with Jessica Tandy, Hume Cronyn, 132 perfs.; *America Hurrah* by Belgian-American playwright Jean-Claude van Itallie, 31, 11/7 at New York's off-Broadway Pocket Theater.

U.S. nightclub comedian Lenny Bruce (*né* Leonard Alfred Schneider) is found dead of a drug overdose in his Hollywood, Calif., house August 3 at age 39. A long-time heroin addict, Bruce was imprisoned for obscenity in 1961, a Los Angeles jury found him guilty of possessing narcotics in 1963, but he had become a cult hero among alienated U.S. youths.

Television drama *Star Trek* 9/8 on NBC stations with William Shatner as Captain James T. Kirk, Leonard Nimoy as Mr. Spock in a science-fiction series that will continue for 78 episodes.

Films Andrei Tarkovsky's *Andrei Rublev* with Anatol Solonitzine as the 15th century icon painter; Marco Bellochio's *Fist in the Pocket* with Lou Castel, Paola Pitagora; Bruce Herschensohn's documentary *John F. Kennedy: Years of Lightning, Day of Drums*; Fred Zinneman's *A Man for All Seasons* with Paul Scofield, Robert Shaw. Also: Lewis Gilbert's *Alfie* with Michael Caine, Shelley Winters; Michelangelo Antonioni's *Blow-Up* with Vanessa Redgrave, David Hemmings; James Hill's *Born Free* with Virginia McKenna, Bill Travers; Richard Fleischer's *Fantastic Voyage* with Stephen Boyd, Raquel Welch; Silvio Narizzano's *Georgy Girl* with Lynn Redgrave, James Mason, Alan Bates, Charlotte Rampling; Pier Paolo Passolini's *The Gospel According to St. Matthew* with Enrique Irazoqui; Alain Resnais's *La Guerre Est Finie* with Yves Montand, Ingrid Thulin; Claude Lelouch's *A Man and a Woman* with Anouk Aimée, Jean-Louis Trintignant; Karel Reisz's *Morgan!* with Vanessa Redgrave; John Sturges's *The Satan Bug* with George Maharis, Richard Basehart, Anne Francis; John Frankenheimer's *Seconds* with Rock Hudson, Salome Jens; Mike Nichols's *Who's Afraid of Virginia Woolf?* with Richard Burton, Elizabeth Taylor; Bryan Forbes's *The Wrong Box* with Ralph Richardson, John Mills, Peter Cook, Dudley Moore, Peter Sellers, Michael Caine; Francis Ford Coppola's *You're a Big Boy Now* with Peter Kastner, Elizabeth Hartman.

New York's Metropolitan Opera House opens September 16 to replace the 83-year-old Met that will be razed next year. Designed by Abby Rockefeller's brother-in-law Wallace K. Harrison, the new $45.7 million Met has 2,729 seats and is the largest building in Lincoln Center, but its premiere of the opera *Antony and Cleopatra* with Leontyne Price, music by Samuel Barber, is a disaster.

Broadway musicals *Sweet Charity* 1/29 at the Palace Theater, with Gwen Verdon, book by Neil Simon based on the 1957 Federico Fellini film *The Nights of Cabiria,* music by Cy Coleman, lyrics by Dorothy Fields, songs that include "Big Spender," "If My Friends Could See Me Now," 608 perfs.; *It's A Bird! It's A Plane! It's Superman!* 3/29 at the Alvin Theater, with Bob Holliday, music by Charles Strouse, lyrics by Lee Adams, songs that include "You've Got Possibilities," "We Don't Matter at All," 75 perfs.; *Mame* 5/24 at the Winter Garden Theater, with Angela Lansbury, music and lyrics by Jerry Herman, songs that include "If He Walks into My Life," "Open a New Window," "We Need a Little Christmas," 1,508 perfs.; *The Apple Tree* 10/18 at the Shubert Theater, with Barbara Harris, Larry Blyden, Alan Alda, music by Jerry Bock, lyrics by Sheldon Harnick, 463 perfs.; *Cabaret* 11/20 at the Broadhurst Theater, with Jill Haworth as Sally Bowles, Jack Gilford, Lotte Lenya, Joel Grey, book based on Christopher Isherwood's *Berlin Stories,* music by John Kander, 39, lyrics by Fred Ebb, 31, songs that include "Wilkommen," "Money, Money," and the title song, 1,165 perfs.; *I Do! I Do!* 12/5 at the 46th Street Theater, with Mary Martin, Robert Preston, music by Harvey Schmidt, lyrics by Tom Jones, songs that include "My Cup Runneth Over," 584 perfs.

First performances Concerto for Violoncello and Orchestra No. 2 by Dmitri Shostakovich 9/25 at Moscow with Mstislav Rostropovich as soloist; Symphony No. 6 by Roger Sessions 11/19 at Newark, N.J.

Popular songs "Alice's Restaurant" by U.S. folk singer Arlo Guthrie, 23, whose father Woody will die next year at age 55; "Yellow Submarine," "Nowhere Man," and "Eleanor Rigby" by John Lennon and Paul McCartney of The Beatles (who give their last public concert August 29 in San Francisco's Candlestick Park); "Sunshine Superman" and "Mellow Yellow" by Scottish rock singer Donovan Leitch; "If I Were a Carpenter" by Tim Hardin; "Scarborough Fair—Canticle" by Paul Simon and Art Garfunkel; "Monday, Monday" and "California Dreamin" by John E. A. Phillips, 30, of The Mamas and the Papas (Phillips, Cass Elliott, 22, Dennis Doherty, 24, Holly Michelle Gilham, 22); *Jefferson Airplane Takes Off* (album) by the Jefferson Airplane (Grace Slick, 22, guitarist-banjoist Paul Kantner, 24, bass guitarist Jack Casady, 22, guitarist Jorma Kaukonen, 25, drummer Spencer Dryden, 23), a rock group that has been playing at San Francisco's Fillmore Auditorium; "Summer in the City" by John B. Sebastian, Mark Sebastian, Joe Butler; "These Boots Are Made for Walking" by U.S. singer Lee Hazlewood; "Guantanamero" by Pete Seeger and Hector Angulo, lyrics from a poem by the 19th century Cuban patriot José Martí; "Winchester Cathedral" by English songwriter Geoffrey Stephen; "What Now My Love" ("El Manthout") by French singer-composer Gilbert Becaud, lyrics by P. DeLanoe, English lyrics by Carl Sigman; "Alfie" by Burt Bacharach, lyrics by Hal David (title song for film); "Georgy Girl" by Tom Springfield, lyrics by Jim Dale (title song for film); "Born Free" by John Barry, lyrics by Don Black (title song for film); "Strangers in the Night" by Bert Kaempfort, lyrics by Eddie Snyder (for the film *A Man Could Get Killed*);

1966 *(cont.)* "Happy Together" by Alan Lee Gordon and Garry Bonner; "Good Vibrations" by Brian Wilson of the Beach Boys.

The electric guitar gains prominence in England where U.S. rock musician Jimi Hendrix, 23, begins to exploit the full potential of the relatively new instrument. Hendrix uses imagination, virtuosity, invention, and sexual pantomime in his stage appearances.

Manuel Santana wins in men's singles at Wimbledon, Mrs. Billie Jean King (*née* Moffitt), 22 (U.S)., in women's singles; Fred Stolle, 27 (Australia), wins in men's singles at Forest Hills, Maria Bueno in women's singles.

Houston's $31-million Astrodome opens to provide the Astros and Oilers with a 44,500-seat air-conditioned stadium. Louisiana voters approve a constitutional amendment that would permit construction of a New Orleans stadium estimated to cost $35 million, but by the time the 72,675-seat New Orleans Superdome opens in 1975 it will have cost upwards of $175 million.

The Milwaukee Braves become the Atlanta Braves but retain Hank Aaron (*see* 1954; Brewers, 1970).

The Baltimore Orioles win the World Series by defeating the Los Angeles Dodgers 4 games to 0.

England wins her first World Cup in football (soccer) by defeating West Germany 4 to 2 at Wembley Stadium in the first victory for a home team since World Cup play began in 1930.

Pampers disposable diaper pads are successfully test marketed in Sacramento by Procter & Gamble whose 6¢ pad begins a revolution in baby diapering.

Washington, D.C., law clerk John Banzhaf III, 26, of the U.S. District Court writes a letter to WCBS-TV, New York, citing commercials that present smoking as "socially acceptable and desirable, manly, and a necessary part of a rich full life." He requests free time roughly equal to the time spent promoting "the virtues and values of smoking" to "present contrasting views on the issue of the benefits and advisability of smoking" (*see* 1967).

Some 13 million Americans will give up smoking in the next 4 years. The percentage of male smokers will drop from 52 percent to 42, of female smokers from 34 percent to 31.

Congress approves a plan to send millions of pounds of tobacco to famine-stricken India under terms of the 1954 law P.L. 480 at the recommendation of Harold S. Cooley (D. N.C.). Chairman of the House Agriculture Committee, Cooley says the tobacco will ease the tension of starving people and will enable them to eat and assimilate their food better (his remark will be stricken from the *Congressional Record*).

Eight student nurses in a Chicago dormitory die July 13 at the hands of Richard F. Speck, 24, who has served time in Texas for theft, forgery, and parole violations that included threatening a woman with a knife. A Peoria jury will find Speck guilty on all eight counts of murder next year and recommend execution. A psychiatrist who examined Speck for 100 hours will tell newsmen that brain damage in conjunction with drugs and alcohol had left Speck irresponsible for his acts. The Supreme Court will overrule the death sentence in 1971, and in 1972 a judge will impose eight sentences of 50 to 150 years each.

New York's Pennsylvania Station of 1910 comes down to make way for a graceless new rail terminal designed by Charles Luckman Associates plus a 29-story office building, a new 20,000-seat Madison Square Garden sports arena to replace the Garden of 1925, a 1,000-seat Felt Forum, a 500-seat movie theater, and a 48-lane bowling alley.

Chicago's Civic Center is completed by Jacques Brownson of C. F. Murphy Associates.

Atlanta's Hyatt Regency Hotel opens in Peachtree Center with a dramatic 22-story atrium of rough poured concrete filled with hanging plants and fountains in which lighted glass-walled elevators fly up and down. Local architect John Portman, 42, has designed the 23-story 1,000-room hotel, which sets a new pattern of hotel construction.

The 1924 U.S. Oil Pollution Act is emasculated by a revision requiring that government prosecutors prove gross and willful negligence.

California legislators respond to complaints about smog by imposing limitations on the amounts of carbon monoxide and hydrocarbons that may be emitted from automobile exhausts. The new standards are to take effect on 1969 model cars and to raise prices by no more than $45 per vehicle.

The first rare and endangered species list, issued by the U.S. Department of the Interior, contains 78 species. The number will increase in 3 years to 89.

Guadelupe Mountains National Park is authorized by act of Congress but not opened to the public. It embraces 77,518 acres of Permian limestone mountain rising from the Texas desert.

A Welsh landslide October 21 at Landsfford Aberfan kills 116 children and 28 adults as water-soaked coal-mine wastes plow into the Pantglas Junior School and 16 homes. Slurry from the Merthyr Vale mine has been "tipped" onto Aberfan's hillsides since 1870 and a subterranean spring under coal tip No. 7 has soaked the waste rock, precipitating the slide.

Washington yields to pressure from commercial fishermen and extends jurisdiction over territorial waters to 12 miles (*see* Declaration of Santiago, 1952). The United States has been the last major nation to maintain the three-mile limit.

The Vatican rescinds the rule forbidding U.S. Catholics to eat meat on Friday, but fish sales drop only briefly.

The Hudson River shad catch falls to 116,000 pounds; the Bureau of Commercial Fisheries in the Department of the Interior stops keeping records of Hudson

River shad which are in any case generally tainted with oil and other pollutants.

Kazakhstan in the Soviet Union harvests a record 25.5 million tons of grain (*see* 1963).

The Rockefeller Foundation and Mexican government establish the International Maize and Wheat Improvement Center (CIMMYT) under the direction of Norman Borlaug (*see* 1944).

India suffers her worst famine in more than 20 years and imports more than 8 million tons of U.S. wheat and other foodstuffs, but rice-eaters in Karala state riot in protest against eating wheat flour (*see* 1944).

The world food crisis reaches new intensity as total food production falls 2 percent below last year's output. Food production in Africa, Latin America, and the Far East falls well below prewar levels.

Drought begins in the Sahel—the 2,600-mile semi-desert strip south of Africa's Sahara Desert. The Sahel embraces Mauritania, Senegal, Mali, Upper Volta, Niger, Chad, northern Nigeria, Cameroon, and parts of Ethiopia, the drought will kill thousands of head of cattle by 1975, and it will bring famine.

Congress passes a Fair Packaging and Labeling Act (*see* Kennedy, 1962). The new "Truth in Packaging" law calls for clear labeling of the net weight of every package, bans phony "cents off" labels and phony "economy size" packages, and imposes controls over the confusing proliferation of package sizes, but food will continue to be sold in packages that make it hard for supermarket customers to know how much they are paying per pound.

Cola drink bottlers and canners receive FDA dispensation not to list caffeine as an ingredient. Cola drinks generally have 4 milligrams of caffeine per fluid ounce, coffee 12 to 16.

Food prices are higher in poor neighborhoods of U.S. cities than in better neighborhoods according to a study. Ghetto food merchants charge more to compensate for "shrinkage."

Consumers boycott supermarkets at Denver and other cities, protesting high prices. The National Commission of Food Marketing concludes that food store profits are generally higher than for comparable industries and that in 20 years the grocery chains' returns on investment have averaged 12.5 percent and have never been lower than for other industries.

Bac•Os, introduced by General Mills, are bits of soy protein isolate flavored to taste like bacon (*see* 1957; Boyer, 1949).

U.S. per-capita consumption of processed potatoes reaches 44.2 pounds per year, up from 6.3 pounds in 1950.

New York assemblymen introduce a bill calling for reform of New York State's 19th century abortion law, responding to an appeal by Manhattan borough president Percy Sutton who has seen the costs of illegal abortions in lives and maimings in the New York ghettos (*see* 1970; Colorado, 1967).

Prospects of global famine spurred efforts to increase food production and curb the population explosion.

The Food and Drug Administration studies "the Pill" and reports "no adequate scientific data at this time proving these compounds unsafe for human use." But most U.S. women are reluctant to use "the Pill" (*see* 1960; 1970).

Japan's birth rate falls to 14 per thousand, down from 34 per thousand in the late 1940s (*see* 1948).

China's birth rate is estimated at between 38 and 43 per thousand. Mao Zedong's Great Proletarian Cultural Revolution has interrupted the nation's widespread birth control campaign.

1967

U.S. popular sentiment turns increasingly against the war in Vietnam as more troops are shipped overseas and as casualties mount. Martin Luther King, Jr., speaks out against the war in February, 5,000 scientists petition for a bombing halt, University of Wisconsin students push Dow Chemical recruiters off the campus to protest Dow's production of napalm, a Women's Strike for Peace demonstrates outside the Pentagon, and Senator Robert F. Kennedy proposes that bombing of North Vietnam be halted so that troop withdrawal may be negotiated.

President Ho Chi Minh responds March 15 to President Johnson's proposal for direct U.S.-North Vietnam peace talks by demanding that bombing be halted and U.S. troops withdrawn from South Vietnam before the start of any talks.

U.S. officials announce March 22 that Thailand has given permission to use Thai bases for B-52 bombers formerly based on Guam.

The U.S. government is "the greatest purveyor of violence in the world," says Martin Luther King, Jr., April

1967 *(cont.)* 4. He encourages draft evasion and proposes a merger between the anti-war and civil rights movements.

Anti-war demonstrations April 15 at New York and San Francisco bring out upwards of 100,000 at New York, 50,000 at San Francisco.

Greece has a right-wing coup April 21 as colonels led by George Papadopoulos and Brig. Gen. Styliano Patakos begin a 7-year military dictatorship. They arrest leftist leaders including George Papandreou and his son Andreas, releasing the elder Papandreou October 7. Constantine II fails to overthrow the junta and restore Greece's democratic institutions, the king and his family flee to Rome December 14. A Christmas amnesty frees Andreas Papandreou, who calls on the world's democracies to help overthrow Premier Papadopoulos.

A Six-Day Arab-Israeli War begins June 5 following months of conflict between Israel and Syria in which Israeli tanks have crossed into Syria and Israeli Mirage fighters have shot down six Syrian MIG-21 fighters. Israeli jets and armor abort an Arab invasion of Israel, Egyptian and Syrian air forces, whose equipment has been supplied largely by Moscow, are wiped out, the Israelis take Arab Jerusalem June 7 and incorporate it with the rest of the city June 27, guaranteeing freedom of access to the Holy Places for people of all faiths. The UN asks July 4 that the action be rescinded, Moscow severs diplomatic relations with Tel Aviv June 10, Tel Aviv rejects the UN request July 14 and retains the strategic Golan Heights in Syria and the West Bank of the Jordan, which along with Arab Jerusalem contains half the population of Jordan and half her economic resources.

President Nasser is persuaded not to resign, begins a purge of the Egyptian army and air force, and receives Soviet president Podgorny, who promises military and economic assistance to help rebuild Egyptian power, but Egypt's Suez Canal is closed (below).

Resolution 242, approved unanimously by the UN Security Council November 22, calls for "withdrawal of Israeli armed forces from territories occupied," an end to belligerency, and recognition that every state in the area has a "right to live in peace within secure and recognized boundaries" (but *see* 1970).

"Vive le Quebec libre," says President de Gaulle July 25 on a state visit to Canada, which is celebrating her centennial as a dominion with a great world exposition (Expo 67) at Montreal. De Gaulle openly promises French support for an independent Quebec, is rebuked by Premier Pearson, cancels a projected visit to Ottawa, and returns home, but his remarks encourage separatist leader René Levesque in the province of Quebec.

The council of the Organization of African Unity (OAU) meeting from February 27 to March 4 has urged use of force to end South Africa's mandate over South-West Africa (Namibia) and topple Rhodesia's Ian Smith regime. OAU ministers meet in September at Kinshasa and demand the departure of white mercenaries from the Congo.

The Republic of Biafra is proclaimed May 30 by Nigerian general Odumegwu Ojukwu who leads the Ibo tribespeople out of the 13-year-old Nigerian Federation. The Lagos government calls the secession a rebellion, Lagos is supported by all other African states and buys arms from Britain, the Ibos buy arms and supplies from France, Nigerian troops take the Biafran capital Enugu October 4, but hostilities will continue until 1970 between the Ibos and the Muslim Hansa-Fulani conservatives to the north and the Yorubas to the west.

Former Congo premier Moise Tshombe is sentenced to death March 13 by a military court which has convicted him in absentia of inciting to rebellion, Tshombe's plane is hijacked June 30 over the Mediterranean, he is flown to Algeria and held captive, pro-Tshombe European mercenaries revolt in July at Kisangani, the rebels are driven across the Rwanda border in early November with support from U.S. transport planes.

Rwanda and Burundi have effected a reconciliation March 20; disarmed Tutsi refugees return to Rwanda after nearly 4 years of attacks on the nation from Burundi.

Che Guevara is killed October 9 at age 39 by Bolivian troops. Born in Argentina, Ernesto Guevara helped Fidel and Raul Castro oust Cuba's Batista regime in the late 1950s but resigned his Cuban cabinet post 2 years ago to lead Bolivian guerrilla revolutionists.

U.S. bombers pound targets around Hanoi, trying to break the Ho Chi Minh supply route that maintains North Vietnam's guerrillas to the south. North Vietnamese forces are vastly outnumbered by more than 1 million South Vietnamese, U.S., and allied troops but some 50,000 Chinese work to repair damage done by the U.S. air attacks.

Protests against the Vietnam war and the draft continue in the United States. Among the 260 demonstrators arrested December 5 at New York are physician Benjamin Spock and poet Allen Ginsberg.

The Saigon government threatens December 26 to pursue Communist troops into Cambodia if that country is used as a base for infiltration into South Vietnam. Beijing replies 3 days later by promising Cambodia Chinese support if U.S. operations are extended there.

The People's Republic of China has exploded her first hydrogen bomb June 17, increasing Soviet fears of a nuclear confrontation with Beijing.

The Twenty-Fifth Amendment ratified February 10 has clarified questions of presidential succession.

The CIA initiates Operation Chaos, exceeding its statutory authority in response to a presidential request that the agency unearth any ties between U.S. anti-war groups and foreign interests. The operation will index 300,000 names, keep 13,000 subject files, and intercept large numbers of letters and cables to compile information on the domestic activities of U.S. citizens.

Heavyweight champion Muhammad Ali (*né* Cassius Clay) is denied conscientious objector status, refuses induction into the U.S. Army, and is arrested April 28.

Race riots rock 127 U.S. cities, killing at least 77 and injuring at least 4,000. Atlanta, Boston, Buffalo, Cincinnati, and Tampa have riots in June, riots in July disrupt Birmingham, Chicago, Detroit, New York, Milwaukee, Minneapolis, New Britain, Conn., Plainfield, N.J., and Rochester, N.Y., but the worst are in Newark and Detroit.

The Newark riots follow the beating of a black man after a traffic arrest July 12, they continue for 5 days over an area of 10 square miles, police and National Guardsmen shoot indiscriminately at blacks, 24 of the 26 killed are blacks, more than 1,500 are injured, 1,397 arrested, property damage amounts to some $10 million.

The Detroit riots July 23 to 30 follow a police raid on an after-hours club at which they make 73 arrests, enraging the black community. Whites as well as blacks loot 1,700 stores, rioters set 1,142 fires that leave 5,000 homeless, 36 of the 43 killed are black, more than 2,000 are injured, police arrest 5,000, and the disturbances end only after Detroit summons federal troops—the first use of federal troops to quell a civil disturbance since 1942.

A Black Power Conference at Newark July 23 adopts anti-white, anti-Christian, and anti-draft resolutions.

"Burn this town down," cries black militant H. "Rap" Brown (Hubert Gerold Brown) of the Student National Coordinating Committee (SNCC) July 25 at Cambridge, Md. Police arrest him for inciting to riot.

A black ghetto section in Washington, D.C., has an episode of arson and rock-throwing August 1.

Martin Luther King, Jr., calls August 15 for a campaign of massive disobedience to bring pressure on Washington to meet black demands (*see* 1968).

SNCC militant leader Stokely Carmichael urges blacks August 17 to arm for "total revolution." Martin Luther King, Jr., (above) has rejected Carmichael's "Black Power" separatist movement.

Dr. Martin Luther King, Jr., gave the U.S. civil rights movement effective leadership that produced results.

President Johnson names Thurgood Marshall, 59, to the Supreme Court seat resigned by Justice Thomas C. Clark following appointment of Clark's son Ramsey, 39, to the post of U.S. Attorney General. Marshall represented the plaintiff in the 1954 *Brown v. Board of Education* case. The first black Supreme Court justice, he is sworn in October 2.

Anti-war demonstrators march on the Pentagon October 21. Police arrest 647 of the 50,000 to 150,000 involved, and similar demonstrations occur at Chicago, Philadelphia, Los Angeles, and Oakland (where police arrest 125 including singer Joan Baez at the Oakland Draft Induction Center).

College students arrested in anti-war demonstrations will lose their draft deferments, Selective Service director Lewis B. Hershey announces November 7.

Urban Coalitions are organized in 48 U.S. metropolitan areas late in the year following an appeal by former Health, Education and Welfare Secretary John W. Gardner. He has resigned to become head of the National Urban Coalition that will mobilize the private sector to join in social-action projects with representatives of the cities' dispossessed minorities.

"If we don't give the Negro of this country some measure of hope," says Mobil Oil vice president Christian A. Herter, Jr., "the people at the extreme left of the Negro community will ultimately gain control and we won't have a chance ever to do this job again." Herter has helped found the New York Urban Coalition.

French workers, squeezed by inflation, strike France's largest and most profitable shipyard. The strikers at Saint-Azaire in Brittany win support from other workers and from students (*see* 1968).

$ The United Auto Workers union with 1.6 million members quits the AFL-CIO April 22 charging a lack of democratic leadership and organizing effort.

The United States contributes one-seventh of 1 percent of its Gross National Product (GNP) to foreign aid, down from a full 1 percent in 1965 (*see* 1961).

Master Charge card holders number 5.7 million and charge $312 million worth of purchases (*see* 1966). By 1976 there will be 40 million Master Charge card holders and they will run up bills of $13.5 billion.

The United States has 2,975 corporate mergers, up from 2,377 last year.

Britain devalues the pound November 18 from $2.80 to $2.40 in an effort to check inflation and improve the nation's trade deficit (*see* 1965).

A Census Bureau report in December shows that 41 percent of nonwhite families in the United States make less than $3,300 per year versus 12 percent of white families, that 7.3 percent of nonwhites are unemployed versus 3.4 percent of whites, and that 29 percent of blacks live in substandard housing versus 8 percent of whites.

1967 *(cont.)* U.S. wage rates will rise by 92 percent in the next 10 years, buying power by only 8 percent.

Mining of the Athabasca tar sands begins in northern Alberta. The field contains some 300 billion barrels of recoverable petroleum, but development of the oil extraction project will be slow and costly.

Atlantic Richfield is created by a merger.

Soviet engineers complete the world's largest hydroelectric power project on the Yenisei River in Siberia. The Krasnoyarsk Dam will begin generating 6,096 megawatts next year (the Grand Coulee Dam completed in 1941 generates 2,161).

Egypt's Suez Canal is closed by scuttled ships and by mines in the Six-Day War (above), depriving the nation of some $250 million per year in revenues. Roughly 70 percent of the world's tankers have been able to use the Suez fully laden and all but 1 percent of tankers have been able to go through it in ballast, but shipbuilders in the next 7 years will concentrate on building supertankers and by the mid-1970s only 35 percent of the world tanker fleet will be able to go through the Suez fully laden (*see* 1968).

McDonnell-Douglas Corp. is created April 28 in a takeover of Douglas Aircraft by the 39-year-old McDonnell Aircraft Corp., which produces military aircraft. Douglas lost $27.6 million last year on sales of more than $1 billion and its working capital has shrunk to $34 million from $187 million in 1958 when the Boeing 707 was introduced.

Montreal gets a new subway system with rubber-tired, air-conditioned cars.

U.S. mass transit rides fall to 8 billion, down from 23 billion in 1945, as prosperous Americans rely at an ever-growing rate on private cars to reach suburban homes and shopping centers.

New York State's Adirondack Northway opens from Albany north to the Canadian border.

The Pennsylvania Railroad sends a $1 million experimental train down its tracks May 24 at 156 miles per hour in a public test but the New York Central discontinues its crack *Twentieth Century Limited* December 2 after 65 years on the Chicago run. Both railroads are in deep financial trouble and have received authorization to merge (*see* 1968).

Canada's 1,097-foot steel arch bridge across the St. Lawrence opens at Trois Rivieres, Quebec.

Venezuela's 2,336-foot Angostura Bridge opens at Ciudad Bolivar.

Czechoslovakia's 1,082-foot steel arch Zdakov Bridge opens to span the Vitava River.

FIAT auto production surpasses that of Volkswagen. The great Italian industrial colossus has been managed for the past year by Giovanni "Gianni" Agnelli, 46, grandson and namesake of FIAT's founder (*see* 1899).

Sweden switches from driving on the left side of the road to driving on the right, but British and Japanese drivers continue to drive on the left.

A new Chevrolet sells for less than $2,500 in the United States.

The electronic quartz wristwatch announced in December by the Swiss Horological Electronic Center has a tiny rod of quartz crystal that vibrates 8,192 times per second when activated by a battery (*see* microchip, 1959; Accutron, 1960). An integrated circuit counts the oscillations and every 1/256th of a second sends power to the micromotor which then drives gears that move the watch hands. 31 Swiss firms pooled $7 million in 1962 to develop the watch, which retails at $550 and up.

New York's Phoenix House is started on the fifth floor of a West 85th Street tenement by five detoxified former drug addicts who are all former patients of the Morris Bernstein Institute, one of the few city hospitals that admit addicts. The rehabilitation center for addicts is joined in July by psychiatrist Mitchell S. Rosenthal who has come from California full of enthusiasm for the 9-year-old Syanon program. By 1970 Phoenix House will have returned 150 graduates to society and established seminar rap session techniques that other rehabilitation centers will follow.

Smoking-withdrawal clinics proliferate across the country but Americans buy 572.6 billion cigarettes—210 packs per adult (*see* 1966; 1969).

Parkinson's disease victims get relief with cryogenic surgery developed by U.S. physician Irving S. Cooper, and with L-dopa therapy developed by Greek-American neurologist George C. Cotzias, 49, whose drug counters the deficiency of dopamine in the brain and will be introduced into medical practice in 1970 (*see* 1817). Asked by the World Health Organization to help investigate chronic manganese poisoning among miners in Chile, Cotzias has found their neurological symptoms—rigid facial expression, clenched hands, speech and balance difficulties—similar to those seen in Parkinson's disease. He succeeds in treating patients with L-dopa where others have failed because he uses much larger doses for longer periods of time.

Cape Town, South Africa, surgeon Christiaan Barnard, 45, performs the world's first heart transplant December 3; his patient lives for 18 days. The first U.S. heart transplant is performed 4 days later by New York surgeon Adrian Kantrowitz, 49, whose patient survives for only a few hours. Surgeons will attempt some 260 operations on more than 250 patients in the next 7 years, some heart recipients will live for more than 6 years with hearts obtained from donors killed in accidents, but only 20 percent of the recipients will survive for more than 1 year.

The *New York World-Journal & Tribune* closes May 5 after less than 8 months during which the company has had 18 work stoppages (management says the unions have forced it to employ 500 more people than necessary). The combined circulation has been 700,000 daily, 900,000 Sundays, and although the *New York Times* and the *Washington Post* will continue the Paris edition of the *Herald-Tribune*, the end of the *World-Journal &*

Tribune leaves New York with only three regular dailies—the *Times, News,* and *Post.*

S. I. Newhouse buys the *Cleveland Plain-Dealer* for a record $54.3 million.

Rolling Stone magazine begins publication in November. San Francisco entrepreneur Jann Wenner, 21, has started the rock and roll record publication with an initial investment of $7,500, most of it borrowed, his first edition sells 6,000 copies, and circulation will grow to 400,000 in 8 years as *Rolling Stone's* volunteer staff grows to a paid staff of 80 and as the magazine becomes a journal of the counterculture.

Britain's first color TV broadcasting begins July 1 as BBC-2 transmits 7 hours of programming, most of it coverage of lawn tennis from Wimbledon.

The Public Broadcasting Act signed into law by President Johnson November 7 creates a Corporation for Public Broadcasting to broaden the scope of noncommercial radio and TV beyond its educational role. Federal grants (plus funds from foundations, business, and private contributions) will within 3 years rival NBC, CBS, and ABC with National Public Radio and Public Broadcasting Service (TV) networks.

The Federal Communications Commission notifies CBS that its programs dealing with the effects of smoking on health are not sufficient to offset the influence of the 5 to 10 minutes of cigarette commercials the network's New York television station is broadcasting each day. The FCC acts in response to a formal complaint against WCBS-TV filed by John Banzhaf (*see* 1966). While the FCC does not agree with Banzhaf's request for equal time, it does order that all radio and TV cigarette commercials carry a notice of possible danger in cigarette smoking, and it asks WCBS-TV to provide free each week "a significant amount of time for the other viewpoint. . . . This requirement will not preclude or curtail presentation by stations of cigarette advertising which they choose to carry. . . . We hold that the fairness doctrine is applicable to such advertisements" (*see* 1969).

Congress creates a Commission on Obscenity and Pornography. It will conclude that pornography does not contribute to crime or sexual deviation and will recommend repeal of all federal, state, and local laws that "interfere with the right of adults who wish to do so to read, obtain, or view explicit sexual materials," but President Nixon will call the commission "morally bankrupt" and say that "so long as I am in the White House, there will be no relaxation of the national effort to control and eliminate smut from our national life" (*see* 1969).

Nonfiction *Children of Crisis* by Harvard psychiatrist-humanist Robert Coles, 37, who begins a decade in which he will write 25 books and nearly 400 articles observing socio-ethical problems affecting U.S. life, destroying one-dimensional stereotypes about blacks, white Southerners, police officers, farm workers, and American youth; *Division Street* by Studs Terkel; *Our Crowd* by former New York advertising copywriter Stephen Birmingham traces the histories of some prominent U.S. Jewish families.

Fiction *One Hundred Years of Solitidude (Cien anos de soledad)* by Colombian novelist Gabriel García Márquez, 39; *The Confessions of Nat Turner* by William Styron (*see* 1831); *The Chosen* by U.S. novelist Chaim Potok, 38; *Up Above the World* by Paul Bowles; *Snow White* by U.S. novelist Donald Barthelme, 36; *A Grain of Wheat* by Kenyan novelist Ngugi wa Thiong'o, 29; *Washington, D.C.* by Gore Vidal; *Rosemary's Baby* by Ira Levin; *Histoire* by Claude Simon; *Ausgefragt* by Günter Grass; *The Eighth Day* by Thornton Wilder, now 70; *Topaz* by Leon Uris.

Poetry *The Light Around the Body* by Robert Bly; *The Lice* by W. S. Merwin.

Painting *Three Folk Musicians* (collage) by Romare Bearden. Edward Hopper dies at New York May 15 at age 84; René Magritte dies at Brussels August 15 at age 68; Ad Reinhardt dies at New York August 30 at age 53.

Sculpture *Broken Obelisk* by Barnett Newman; *Homage to Bernini* by Louise Bourgeois; *Henry Moore Bound to Fail* by Bruce Nauman; *Borne au logos VII* (cast urethane) by Jean Dubuffet.

Theater *MacBird* by U.S. playwright Barbara Garson 2/22 at New York's off-Broadway Village Gate Theater, with Stacy Keach, 386 perfs.; *Fortune and Men's Eyes* by U.S. playwright John Herbert 2/23 at New York's off-Broadway Actor's Playhouse, 382 perfs.; *A Day in the Death of Joe Egg* by English playwright Peter Nichols, 39, at London, 148 perfs.; *You Know I Can't Hear You When the Water's Running* by Robert Anderson 3/13 at New York's Ambassador Theater, with George Grizzard, Eileen Heckart, Martin Balsam, 755 perfs.; *Rosencrantz and Guildenstern Are Dead* by Czech-born British playwright Tom Stoppard, 29, 4/11 at London's Old Vic Theatre; *Little Murders* by U.S. cartoonist-playwright Jules Feiffer, 38, 4/25 at New York's Broadhurst Theater, with Elliott Gould, Heywood Hale Broun, Barbara Cook; *Soldaten* by Rolf Hochhuth 10/7 at West Berlin's Freie Volksbuhre; *Borstal Boy* by Brendan Behan 10/10 at Dublin's Abbey Theatre.

Television drama *The Forsyte Saga* with Eric Porter as Soames Forsyte, Kenneth More as his cousin, Susan Hampshire as his daughter Fleur in a serialized dramatization of the John Galsworthy novels on BBC and U.S. Public Broadcasting stations.

Films Luis Buñuel's *Belle de Jour* with Catherine Deneuve; Arthur Penn's *Bonnie and Clyde* with Warren Beatty, Faye Dunaway; Mike Nichols's *The Graduate* with Dustin Hoffman, Katharine Ross, Ann Bancroft; Richard Brooks's semidocumentary *In Cold Blood* with Robert Blake, Scott Wilson; Norman Jewison's *In the Heat of the Night* with Rod Steiger, Sidney Poitier; Peter Brooks's *Marat/Sade* with the Royal Shakespeare Company; Theodore J. Flicker's *The President's Analyst* with James Coburn, Godfrey Cambridge; Sergei Bondarchuk's *War and Peace* with Ludmila Savelyeva, Vyacheslav Tihonov. Also: Joseph

1967 ***(cont.)*** Losey's *Accident* with Dirk Bogarde; Ralph Nelson's *Charly* with Cliff Robertson, Claire Bloom; Stuart Rosenberg's *Cool Hand Luke* with Paul Newman, Jo Van Fleet, George Kennedy; Robert Aldrich's *The Dirty Dozen* with John Cassavetes, Lee Marvin, Ernest Borgnine, Donald Sutherland, Jim Brown; D.A. Pennebaker's documentary *Don't Look Back* with Bob Dylan, Joan Baez, Donovan, Alan Ginsberg; John Schlesinger's *Far from the Madding Crowd* with Julie Christie, Peter Finch, Alan Bates, Terence Stamp; Gene Kelly's *A Guide for the Married Man* with Walter Matthau, Inger Stevens; Michael Winner's *I'll Never Forget What's 'is Name* with Orson Welles, Oliver Reed; Ingmar Bergman's *Persona* with Bibi Andersson, Liv Ullmann; John Boorman's *Point Blank* with Lee Marvin, Angie Dickinson; Luchino Visconti's *The Stranger* with Marcello Mastroianni; Franco Zeffirelli's *The Taming of the Shrew* with Elizabeth Taylor, Richard Burton; Fred Wiseman's documentary *The Titicut Follies;* James Clavell's *To Sir with Love* with Sidney Poitier, Judy Gleeson; Buzz Kulik's *Warning Shot* with David Janssen.

Broadway and off-Broadway musicals *You're a Good Man, Charlie Brown* 3/7 at the Theater 80 St. Marks, with Gary Burghoff, Reva Rose, book based on Charles Schultz's 17-year-old comic strip *Peanuts,* music and lyrics by Clark Gesner, 1,579 perfs.; *Hair* 10/29 at the off-Broadway Public Theater, with Gerome Ragni as Bezar, music by Galt MacDermot, lyrics by Ragni and James Rado, songs that include "Hare Krishna" and "Aquarius," 94 perfs. (plus 1,742 beginning 4/29/68 at the Biltmore Theater with some nudity).

The Monterey Pop Festival held at Monterey, Calif., is the first large rock gathering. Participating rock stars include the Grateful Dead and Janis Joplin's Big Brother and the Holding Company.

Popular songs "Ode to Billy Joe" by U.S. country singer-songwriter Bobbie Gentry, 23; "Up, Up and Away" by Oklahoma songwriter Jimmy Webb, 21; "Gentle on My Mind" by John Hartford; "Can't Take My Eyes Off of You" by Bob Crewe and Bob Gandio; "Somebody to Love" by the Jefferson Airplane makes the San Francisco rock group the first to gain wide acclaim; "Ruby Tuesday" by Mick Jagger and Keith Richard of The Rolling Stones; "All You Need Is Love" and "Penny Lane" by John Lennon and Paul McCartney of The Beatles; *Sergeant Pepper's Lonely Hearts Club Band* (album) by The Beatles reflects the growing drug culture in the song "Lucy in the Sky with Diamonds"; *Magical Mystery Tour* (album) by The Beatles is similarly drug-oriented (positive references to drugs or drug use occur in 16 of the top 40 U.S. phonograph records); "Light My Fire" by Robert Krieger, James Morrison, Raymond Manzarek, and John Densmore of The Doors; "Release Me" by Eddie Miller and W. S. Stevenson; "Don't Sleep in the Subway" by U.S. songwriter Jackie Trent; "Respect" by Otis Redding who dies December 10 at age 28 along with four younger members of his troupe when his plane crashes into an icy Wisconsin lake; *Surrealistic Pillow* (album) by the Jefferson Airplane; *Between the Buttons* (album) by The Rolling Stones; *Happy Jack* (album) by The Who (Peter Townshend, 22; John Entwhistle, 21; Roger Daltrey, 22; and Keith Moon, 20), a new English rock group whose members smash up their instruments on stage; "New York Mining Disaster" by the Bee Gees—English rock composer Barry Gibb, 20, and his brothers Robin and Maurice, 17 (unidentical twins); "This Is My Song" by Charlie Chaplin, now 78 (for his film *A Countess* from Hong Kong); "A Man and a Woman" by French composer Francis Lai, lyrics by Jerry Keck (title song for film); "The Beat Goes On" by Sonny Bono.

The World Boxing Association strips Muhammad Ali (above) of his world heavyweight title (*see* 1968).

The Green Bay Packers of the National Football League defeat the Kansas City Chiefs of the American League January 15 at Los Angeles in Super Bowl I. The event will grow to have even more TV viewers than baseball's World Series.

John Newcombe, 23 (Australia), wins in men's singles at Wimbledon and Forest Hills, Mrs. King in women's singles.

Mickey Mantle of the New York Yankees hits his five-hundredth home run in league competition.

The St. Louis Cardinals win the World Series by defeating the Boston Red Sox 4 games to 3.

Australia's *Dame Pattie* loses 4 to 0 to the U.S. defender *Intrepid* in the America's Cup yacht races.

British yachtsman Sir Francis Chichester, 65, sails his 53-foot ketch *Gipsy Moth* into Plymouth harbor May 28 to complete a 28,500-mile, 220-day one-man voyage round the world.

U.S. and British pantyhose sales climb as women adopt the miniskirt (*see* 1959, 1969; Quant, 1965).

Montreal's Bonaventure complex is completed beside Place Ville Marie by Boston-born planner Vincent Ponte, 48, who has pushed for a multilayer plan to avoid congestion by having cars on one level, pedestrians on a level below (where they are protected from the elements), trains and trucks on a lower level still (*see* 1963).

Habitat is completed for Montreal's Expo 67 by Israeli-born architect Moshe Safde, 28, whose idea of an apartment house runs counter to the regimentation of modern apartment blocks. Safde creates what he calls a "sense of house" with private outdoor space for each family in an arrangement that represents a radical departure from prevailing apartment house design.

Chicago's John Hancock Center goes up. Rising 1,107 feet, the skyscraper will be second in height only to New York's Empire State Building of 1931 (*see* 1969).

The *Torrey Canyon* wreck at Easter in the English Channel off the coast of Cornwall in southern England creates the biggest oil spill thus far. The 118,000-ton tanker strikes a submerged reef March 18 and spills 860,000 barrels of crude oil, taking a heavy toll of marine life, especially of sea fowl, but 80 to 90 percent of the damage to beaches and coves is cleaned up in 6 weeks and by the end of the season all trace of the spill is gone.

Congress passes an Endangered Species Act; the Department of the Interior classifies various U.S. fish and game species based on estimated population.

A Fisherman's Protective Act passed by Congress appropriates funds to reimburse U.S. commercial fishermen for fines, lost time, and fish spoilage incurred through apprehension by foreign powers in waters the United States considers international (*see* Declaration of Santiago, 1952).

Mao Zedong brings Chinese agriculture into the Great Proletarian Cultural Revolution, but the country is wracked by virtual civil war. Peasants abandon collective grain fields to cultivate private gardens and sell on the black market. China continues to import from 4 to 7 million tons of wheat, mostly from Australia and Canada, as she has done since 1962.

European fruit and vegetable surpluses are destroyed to maintain prices.

U.S. farm prices fall as huge crops are harvested and Agriculture Secretary Orville H. Freeman comes under attack from farmers. He had urged heavy planting in case India and Pakistan should need more U.S. grain.

Nearly 10,000 U.S. farmers receive more than $20,000 each in subsidies, more than 6,500 receive more than $25,000 each, 15 receive between $500,000 and $1 million each, five receive more than $1 million each, one receives more than $4 million.

Iran, South Africa, and Turkey have record grain crops but hunger remains widespread in India.

More than 6 million tons of U.S. wheat are shipped to India under terms of the P.L. 480 of 1954. Reserves of wheat stored in U.S. grain elevators fall sharply as more than one-fourth of the export crop goes to India.

Widespread hunger and malnutrition in the Mississippi Delta is reported to a Senate subcommittee by a medical team that has surveyed the region.

The Department of Agriculture returns $200 million of unused food aid funds to the Treasury.

Some 2.7 million Americans receive food stamp assistance as of Thanksgiving (*see* 1964; 1969).

A U.S. Federal Meat Inspection Act takes effect after a campaign by consumer advocate Ralph Nader has persuaded Congress to strengthen the law of 1906.

Annual U.S. beef consumption reaches 105.6 pounds per capita, up from 99 pounds in 1960, and the average American eats 71 pounds of other red meat.

Wisconsin permits sale of yellow margarine, becoming the last state to repeal laws against it, but like some other states it continues to impose special taxes on margarine at the behest of dairy interests (*see* New York, 1952).

U.S. bread sells at 22–25¢ per 1-lb. loaf.

Imitation milk appears in Arizona supermarkets. Made from corn-syrup solids, vegetable fat, sugar, salt, artificial thickeners, colors, and flavors, sodium caseinate, and water, the "milk" will be sold under such brand names as Moo, Farmer's Daughter, and Country Cousin.

Del Monte Corp. is created by a reorganization of California Packing Corp. (*see* 1917). Del Monte has become an integrated international company that controls vast acreages of U.S. farmland and directly or indirectly employs thousands of farm workers.

The first compact microwave oven for U.S. home use is introduced by Amana Refrigeration, a subsidiary of Raytheon with facilities at Amana, Iowa, which applies its consumer marketing experience to Raytheon's microwave technology. Engineer Keishi Ogura of Japan Radio developed an improved electron tube 3 years ago, making possible a compact microwave oven that retails at $495 (*see* 1947; Amana, 1855).

Japanese microwave oven production reaches 50,000 up from 15,000 last year. Many Japanese households move directly from hibachi grills to microwave ovens.

Japan's birth rate climbs to 19.3 per 1,000 after its fall to 14 last year in the Year of the Fiery Horse.

France legalizes birth control, but diaphragms, intrauterine devices, and spermicidal creams and jellies remain generally unavailable.

Israel's population tops 3.5 million with some 2.4 million Jews and more than 1 million Arabs as the Six-Day War (above) expands her territory (*see* 1948).

The U.S. population passes 200 million November 20, having doubled in just 50 years. Demographers predict a population of 300 million by the year 2000 if present growth rates continue.

1968 A great Tet offensive begins January 30 as Vietcong and North Vietnamese forces attack some 30 South Vietnamese cities including Hue and Saigon in an effort to topple the regime of generals Nguyen Cao Ky and Nguyen Van Thieu who are supported by the United States. The Communists besiege Khesanh, a base that commands a major road junction and infiltration route, and while South Vietnamese forces recapture the ancient palace grounds at Hue February 24 and the troops manning Khesanh are successfully evacuated June 27, the power of North Vietnam has impressed the world.

The Tet offensive (above) comes one week after the seizure by North Korea of the spy ship U.S.S. *Pueblo* whose 83-man crew sustains one casualty in the capture of the ship off the port of Wonsan by a torpedo boat and subchaser. North Korea had broadcast warnings that she would not tolerate spy ships, but *Pueblo* commander Lloyd M. Bucher will deny that his ship came within 12 miles of the Korean coast and will be released with his surviving men after 11 months of captivity. A U.S. intelligence plane is shot down April 15 some 90 miles off the Korean coast.

The United States loses her 10,000th plane over Vietnam January 5.

My Lai village in South Vietnam is the scene of a massacre March 16. U.S. troops of C Company, First Bat-

1968 ***(cont.)*** talion, Twentieth Infantry, Eleventh Brigade, Americal Division enter the village, gather hundreds of men, women, and children into groups, and "waste" them with automatic weapons fire. The soldiers will later say that they acted on orders from Lieut. William L. Calley, Jr., but news of the My Lai massacre will be suppressed for 20 months.

Opposition to the Vietnam War enables Sen. Eugene McCarthy, 52 (D. Wis.), to make a strong showing in the New Hampshire primary; his success persuades President Johnson to announce March 31 that he will not be a candidate for reelection.

President Johnson announces cessation of U.S. air and naval bombardment north of the 20th parallel in Vietnam March 31, but Gen. William C. Westmoreland's statement that "the enemy has been defeated at every turn" has a hollow ring.

Nine Roman Catholic priests enter Selective Service offices at Catonsville, Md., in May, dump hundreds of 1-A classification records into trash baskets, take the records outside, burn them, and await arrest for their "symbolic act" protesting the Vietnam War. Jesuit priest-poet Daniel Berrigan, 47, his brother Philip, 45, and their fellow-priests will be sentenced to prison terms of from 2 to 3½ years but Daniel Berrigan will avoid apprehension after his conviction and exhaustion of appeals; he will continue speaking out against the war until rearrested in August 1970.

Sen. Robert F. Kennedy (D. N.Y.) makes a bid for the presidency after seeing the success of Eugene McCarthy in New Hampshire (above). Now 42, he captures 174 delegate votes (winning the Indiana, Nebraska, and California primaries, losing in Oregon), but is assassinated June 5 in a Los Angeles hotel kitchen pantry after leaving a victory celebration. Jordanian-American Sirhan Bishara Sirhan, 24, is seized as he empties his .22 caliber, 8-shot Ivor Johnson pistol, wounding five others in the Ambassador Hotel pantry.

Cambodia's Prince Norodom Sihanouk warns American ships to stay out of his part of the Mekong River. His forces seize a U.S. patrol boat July 17.

Czech Communist party secretary Alexander Dubcek declines invitations to attend conferences at Warsaw or Moscow, he receives support from presidents Tito and Ceausescu, but the "Prague Spring" that has seen a relaxation of oppression ends in August. Some 200,000 Soviet and satellite troops invade August 20 on orders from Moscow; Romanian troops do not participate.

Popular demonstrations in Czechoslovakia raise a threat of revolution like the one in Hungary 12 years ago. Moscow increases the army of occupation to 650,000 and summons Czech leaders to the Kremlin. They return to Prague August 27 and announce the annulment of several important reforms. Soviet foreign minister Kuznetsov arrives at Prague, Party Secretary Dubcek bans political clubs September 6, Czech authorities introduce a censorship system September 13, and the foreign minister who had presented the Czech case at the UN resigns under pressure September 19.

Australia's prime minister Harold Holt has drowned January 10 while swimming; his successor is Liberal party leader John Gorton.

Nauru in the South Pacific gains independence January 31. Once a German colony, the island with its rich phosphate deposits has been a British mandate and a United Nations Trust territory.

Mauritius in the Indian Ocean gains independence March 12 after more than 153 years of British colonial rule.

Swaziland gains independence September 6 after 66 years of British rule.

Manila lays claim to Saba on the island of Borneo and passes a law September 18 incorporating the territory into the Philippine Republic after talks at Bangkok with the Malaysian government have broken down.

Communist guerrilla activity in northern Malaysia has resumed in June; Kuala Lampur breaks relations with Manila.

Equatorial Guinea in West Africa gains independence October 12 after 124 years of Spanish colonial rule.

The African continent that was completely white-controlled 10 years ago is controlled south of the Sahara by black regimes in every country except Angola, Mozambique, Rhodesia, South Africa, South-West Africa (Namibia), and Equatorial Guinea. Within 6 years only Rhodesia, South Africa, and South-West Africa will remain white-controlled.

Bloody police confrontations mark the Democratic party convention at Chicago in August, with demonstrators protesting U.S. military involvement in Southeast Asia and many domestic policies. Some 10,000 militants protest the rising death toll in Vietnam with a "Festival of Life" in Grant and Lincoln parks that includes rock concerts, marijuana smoking, public lovemaking, beach nude-ins, and draft card burnings.

Self-styled revolutionists who include Abbot H. "Abbie" Hoffman, 30, and Jerry C. Rubin, 29, of the "Yippies" (Youth International Party), Rennie Davis, 27, and Thomas E. Hayden, 27, of the Students for a Democratic Society (SDS), Black Panther leader Bobby Seale, 31, and civil rights advocate David Dellinger, 52, have called for a mobilization of 500,000 at Chicago. Their forces are far outnumbered by 16,000 Chicago police officers, 4,000 state police officers, and 4,000 National Guardsmen armed with tear gas grenades, night sticks, and firearms who act on orders from Chicago's Mayor Richard Daley and brutally crack heads to prevent demonstrators from remaining overnight in city parks.

Democrats at Chicago nominate Vice President Hubert Horatio Humphrey, 57, to succeed President Johnson.

Former Vice President Nixon wins the Republican nomination on the first ballot, having actively campaigned for Republican candidates in 1966 and regained favor. He claims to have a "secret plan" for ending the war in Vietnam and wins election by the narrowest margin since his own defeat by John F. Kennedy in 1960—43.4 percent of the popular vote to Humphrey's 43 percent (but 302 electoral votes to Humphrey's 191). Former Alabama governor George Wallace carries five Southern states with a combined electoral vote of 45; his wife (and successor) Lurleen has died in May and he has run as the candidate of the American Independent Party.

President Johnson has announced complete cessation of U.S. aerial, artillery, and naval bombardment of Vietnam north of the 20th parallel October 31 in a move to further the peace talks at Paris and help Vice President Humphrey's chances for the presidency, but the talks produce no results.

The gap between rich nations and underdeveloped nations must be narrowed or "men and women will be impelled to revolt," warns India's prime minister Indira Gandhi in New Delhi.

The Kerner Report February 29 says, "Our nation is moving toward two societies, one black, one white, separate and unequal." The National Advisory Commission on Civil Disorders headed by Gov. Otto Kerner, Jr., 59, of Illinois charges white society with condoning the black ghetto it has created.

Martin Luther King, Jr., is shot dead April 4 as he steps out on the balcony of his Memphis motel room. The civil rights leader has for 6 years been under surveillance by the FBI whose director J. Edgar Hoover has used wiretaps, electronic bugs, and paid informants to gain information on King's private life and circulated it in an effort to discredit him. Still only 39, King has been picked off with one shot fired from a sniper's 30.06 Remington rifle. Fingerprints indicate that the assassin was ex-convict James Earl Ray, 39, who escaped last year from Mississippi State Penitentiary. He is indicted for murder and arrested June 8 by Scotland Yard detectives at a London airport. Extradited to stand trial, he will plead guilty and be sentenced next year to 99 years in prison, but doubts will remain as to whether he acted alone.

Race riots erupt at Baltimore, Boston, Chicago, Detroit, Kansas City, Newark, Washington, D.C., and scores of other cities following the King assassination (above). Chicago's Mayor Daley gives police "shoot to kill" orders to put down the rioting, 46 deaths result across the country, 55,000 federal troops and National Guardsmen are called out, 21,270 arrests made.

A March 3 memo from J. Edgar Hoover has spelled out FBI goals in a "Counter-Intelligence Program" against "Black Nationalist Hate-Groups:" "1. Prevent the coalition of militant black nationalist groups . . . [which] might be the first step toward a real 'Mau Mau' in America, the beginning of a true black revolution. 2. Prevent the rise of a 'messiah' who could unify and electrify the militant black nationalist movement. . . King could be a very real contender for this position should he abandon his supposed 'obedience' to 'white, liberal doctrines' (nonviolence) and embrace black nationalism. . . ."

A U.S. Civil Rights bill signed into law by President Johnson April 11 stresses open housing.

A Poor People's March on Washington gets under way May 3 (below).

Militant Columbia University students shut down the school to protest building a gymnasium in an area needed for low-cost housing. The protesters include members of the radical Students for a Democratic Society (SDS) and black activists H. Rap Brown and Stokely Carmichael. Gym construction is halted April 25, students take over five buildings in a week-long sit-in, police storm the buildings April 30 after wanton destruction of property and make 628 arrests; classes are formally suspended May 5.

French universities close down in May after widespread street fighting that began with violent student demonstrations at the University of Nanterre and spread quickly to the Sorbonne and other schools. Strikes in various industries produce a crisis, President de Gaulle asks for restoration of order in a radio appeal broadcast May 24, French-born German student Daniel Cohn-Bendit ("Danny the Red"), 23, is exiled May 24 as a threat to public order, Parlement is dissolved May 30, but the Communists and other radical parties lose seats in the June elections and the Gaullist party wins a clear majority.

Mexico City riot police and troops open fire on student demonstrators October 2 in Tlatelolco Square. Officials report 40 dead, others count 700 bodies.

San Francisco State College students strike November 6 demanding open admission and a Third World Studies Department. The college is closed November 19 following daily confrontations between students and police. Semanticist S. I. Hayakawa is named president of the college, reopens it December 2, then closes it early for the Christmas vacation to avoid having high school students join the protest. The strike will continue for 5 months.

National Turn in Your Draft Card Day November 14 features burning of draft cards and war protest rallies at many campuses as the U.S. Vietnam death toll approaches 30,000 and U.S. troop strength in Vietnam reach its peak of 550,000.

Japan's Gross National Product (GNP) climbs 12 percent to $140 billion and passes that of West Germany by $10 billion to make Japan the free world's second strongest economic power after the United States (whose GNP is $860 billion), but per capita income is $921, little more than that of Ireland.

U.S.-owned multinationals have been the chief beneficiaries of the European Common Market, says French economist Jean-Jacques Servan-Schreiber in his book *The American Challenge*.

Economic inflation continues to mount in much of the world. The United States and six West European nations agree March 18 to supply no more gold to private buyers.

Britain's first 5- and 10-pence coins are issued April 23 as the nation shifts to the decimal system.

The United States has 4,462 corporate business mergers, up from 2,975 last year (*see* 1969).

1968 ***(cont.)*** President Johnson signs a bill June 28 adding a 10 percent surcharge to income taxes and reducing government spending but the war in Indochina is costing the United States millions of dollars per day.

The Pentagon announces August 1 that it will buy steel only from firms that have not raised prices, but wholesale steel prices beginning at the end of this year will climb in 15 of the next 18 months as American steel makers negotiate agreements with European and Japanese mills to reduce their shipments to the United States and U.S. industry raises prices in anticipation of possible price controls.

Congress enacts a Consumer Credit Protection Law. The "Truth in Lending" Act requires banks and other lending institutions to disclose clearly the true annual rate of interest and other financing costs on most types of loans.

BankAmericard holders number 14 million by year's end, up from 2 million at the end of 1966, and 316,000 U.S. merchants accept the card, up from 64,000 at the end of 1966. By 1974 there will be more than 26 million BankAmericard holders and at least 33 million Master Charge card holders, bank credit cards will be used to finance $13 billion worth of business per year in the United States (roughly 2.6 percent of all retail sales), 970,000 U.S. merchants will honor BankAmericard and more thousands will honor Master Charge, thousands of merchants in other countries will honor the cards, and approximately 3.8 percent of U.S. consumer debt will be in loans outstanding on bank credit cards, much of it at interest rates as high as 18 percent per year (*see* VISA, 1975).

First Philadelphia Bank installs a one-way cash dispenser imported from Britain; Canton, Ohio, safe manufacturer Deibold & Co. begins making automatic teller machines; New York's Chemical Bank will put in a cash dispenser next year; and Atlanta's Citizens & Southern Bank will install a two-way electronic teller (*see* 1973).

U.S. natural gas consumption begins to exceed new gas discoveries and reserves for U.S. interstate pipelines begin falling (*see* 1964). The industry blames artificially low prices established by the Federal Power Commission for having discouraged exploration for new reserves at a time when FPC actions have increased demand for natural gas. Some independent economists agree.

Oil is discovered on Alaska's North Slope, the new reserve proves to be the largest north of the Mexican border, and petroleum companies join forces to establish the Alyeska Pipeline Service Co. to bring the oil south to the ice-free port of Valdez, which has been rebuilt since the earthquake of 1964 and is favored as a port for shipment by tanker to world markets, but a debate begins as to the environmental impact of the pipeline (*see* 1969).

Tenneco Corp. acquires Kern County Land Co., whose management has rejected offers by Armand Hammer's Occidental Petroleum (*see* 1890; 1957). Tenneco is a giant U.S. complex of natural gas, packing, and chemical interests with more than $3 billion in assets.

A Delta Airlines jet carrying 169 passengers is hijacked February 21 over southern Florida, the hijacker holds a gun on the pilot and forces him to land in Cuba, similar episodes follow, and the hijackings will lead to airport searches of passengers and luggage.

Pan Am and Aeroflot begin direct service between New York and the Soviet Union July 15. Aeroflot uses the four-jet Il–62 designed by Sergei V. Ilyushin, now 73.

The Tu-144 demonstrated by Aeroflot December 31 is the first supersonic airliner.

The S.S. *QE 2* launched by the Cunard line replaces the 83,673-ton *Queen Elizabeth* launched in 1940. The new 66,850-ton passenger liner is 963 feet in length overall, carries nearly 3,000 people including crew, and has four swimming pools, 13 decks, 24 elevators, and a 531-seat theater.

A 326,000-ton supertanker goes into service on charter to Gulf Oil. She will be followed by five sister ships (*see* 1972).

Penn Central is created February 14 by a merger of the Pennsylvania Railroad with the New York Central. Both are in financial trouble as a result of competition from trucks and automobiles that use publicly financed highways and airlines that use publicly financed airports. The new $5 billion corporation will be bankrupt within 2 years.

U.S. automobile production reaches 8.8 million; truck and bus production approaches 2 million.

Volkswagen captures 57 percent of the U.S. import market, selling 569,292 vehicles in the United States, up from 120,000 in 1959. Seventy percent of VWs sold are Beetle models priced at less than $1,800 and VWs outsell many U.S. makes including Pontiac, Chevrolet Chevelle, Ford Fairlane, Plymouth Fury, Buick, Ford Mustang, Oldsmobile, and Chrysler.

West Germany produces 2.5 million cars and nearly 600,000 trucks; Japan 2.1 million cars, 2 million trucks; Britain 1.7 million cars, 400,000 trucks and buses; France 1.8 million cars, 243,000 trucks; Italy 1.5 million cars, 115,000 trucks.

British-Leyland is created by a merger of British Motor Holdings and Leyland Motors. The new automotive giant produces Austin, Jaguar, MG, Rover, and Triumph automobiles, employing 160,000, but in 6 years will be in such financial straits that only a huge government-guaranteed loan will save it from bankruptcy.

The German NSU Ro 80 introduced in Britain is the first Wankel-engine automobile (*see* Wankel, 1957). Toyo Kogyo in Japan uses the Wankel engine for its new Mazda cars.

Christiaan Barnard performs the most successful heart transplant to date January 2, giving retired Cape Town dentist Philip Blaiberg, 58, a heart that will continue to function for 19 months (*see* 1990).

Houston cardiovascular surgeon Denton Arthur Cooley, 47, performs the first successful U.S. heart transplant May 2, removing the damaged heart of a 47-year-old male patient and replacing it with the heart of a 15-year-old female who has died of a brain injury. Cooley performs four similar transplants within 4 weeks (two patients die subsequently of other causes).

The United States has only 22,231 reported cases of measles, down from 400,000 in 1962, as a result of the Enders vaccine (*see* 1962).

The U.S. first class postal rate climbs to 6¢ January 7 (*see* 1963; 1971).

World television set ownership nears 200 million with 78 million sets in the United States, 25 million in the Soviet Union, 20.5 million in Japan, 19 in Britain, 13.5 in West Germany, 10 in France.

The U.S. television industry has advertising revenues of $2 billion, roughly twice the total of radio advertising revenues. Both industries derive heavy support from cigarette advertising (*see* 1969).

Television *60 Minutes* debuts 9/14 on CBS TV with journalists Dan Rather, Harry Reasoner, Morley Safer, and Mike Wallace.

The 911 emergency telephone number instituted in New York to summon emergency police, fire, or ambulance assistance is the first such system in the United States (*see* Britain, 1937). By 1977 some 600 U.S. localities with a total population of 38 million will have 911 systems.

Nonfiction *Soul on Ice* by Black Panther leader Eldridge Cleaver who flees to Cuba in November to avoid going to prison for parole violations. Cleaver begins a 7-year exile; *The Armies of the Night* by Norman Mailer; *Slouching Towards Bethlehem* (essays) by U.S. journalist-novelist Joan Didion, 33; *The Whole Earth Catalog* by Menlo Park, Calif., Truck Store operator Stewart Brand, 29.

Fiction *First Circle* (*V kruge pyervom*) and *Cancer Ward* (*Rakovy korpus*) by Aleksandr Solzhenitsyn; *Enderby* by Anthony Burgess; *The Military Philosophers* by Anthony Powell; *The Day of the Scorpions* by Paul Scott; *Expensive People* by U.S. novelist Joyce Carol Oates, 30; *My Michael* by Israeli novelist Amos Oz, 29; *The Electric Kool-Aid Acid Test* by U.S. novelist Tom Wolfe (Thomas Kennerly Wolfe, Jr.), 37; *Welcome to the Monkey House* (stories) by Kurt Vonnegut, Jr.; *Couples* by John Updike; *Myra Breckenridge* by Gore Vidal.

Poetry *White Haired Lover* by Karl Shapiro.

Painting *Untitled* by Frank Stella; *The Beatles* (photo montage) by Richard Hamilton. Marcel Duchamp dies at Neuilly outside Paris October 1 at age 81.

Sculpture *Orchestra of Rags* by Italian artist Michelangelo Pistoletto, 35, of the "arte primavera" movement possessed of a minimalist aesthetic but using found materials (earth, cloth, used industrial objects) in an unaltered state where U.S. minimalists use rigorous industrial fabrications; *Accession III* (fiberglas and plastic tubing) by German sculptor Eva Hesse, 32; *Earthwork* by Robert Morris.

Theater *The Prime of Miss Jean Brodie* by Jay Allen (who has adapted the novel by Muriel Spark) 1/16 at New York's Helen Hayes Theater, with Zoë Caldwell, 378 perfs.; *I Never Sang for my Father* by Robert Anderson 1/25 at New York's Longacre Theater, with Hal Holbrook, Lillian Gish, Matt Crowley, Teresa Wright, 124 perfs.; *The Price* by Arthur Miller 2/7 at New York's Morosco Theater, with Pat Hingle, Kate Reid, Arthur Kennedy, 429 perfs.; *Hadrian VII* by English playwright Peter Luke, 48, (who has adapted the 1904 novel) 4/18 at London's Mermaid Theatre, with Alec McCowen; *The Boys in the Band* by Mart Crowley 4/18 at New York's off-Broadway Theater Four, with Cliff Gorman, 1,000 perfs.; *Indians* by Arthur Kopit 7/4 at London's Aldwych Theatre, with Barrie Ingham as "Buffalo Bill" Cody; *The Great White Hope* by U.S. playwright Howard Sackler, 39, 10/3 at New York's Alvin Theater, with James Earl Jones, Jane Alexander, 276 perfs.; *Forty Carats* by Jay Allen (who has adapted a French play) 12/26 at New York's Morosco Theater, with Julie Harris, Glenda Farrell, 780 perfs.

Britain's Theatres Act effectively abolishes the Lord Chamberlain's powers of censorship (*see* 1737).

Television *Hawaii Five-O* 9/26 on CBS with Jack Lord, High Dheigh (to 4/5/1980).

Films Anthony Harvey's *The Lion in Winter* with Peter O'Toole, Katharine Hepburn; Richard Lester's *Petulia* with Julie Christie, George C. Scott, Richard Chamberlain; Jacques Tati's *Playtime* with Tati; Roman Polanski's *Rosemary's Baby* with Mia Farrow, John Cassavetes; Ingmar Bergman's *Shame* with Liv Ullmann, Max von Sydow; François Truffaut's *Stolen Kisses* with Jean-Pierre Leaud; Stanley Kubrick's *2001: A Space Odyssey* with Keir Dullea. Also: François Truffaut's *The Bride Wore Black* with Jeanne Moreau, Claude Rich, Jean-Claude Brialy; Peter Yates's *Bullitt* with Steve McQueen, Robert Vaughn, Jacqueline Bisset; Robert Altman's *Countdown* with Robert Duvall, James Caan; John Cassavetes's *Faces* with John Marley, Gena Rowlands; Donald Siegel's *Madigan* with Richard Widmark, Henry Fonda, Harry Giardino, James Whitmore; George A. Romero's *Night of the Living Dead* with Duane Jones, Judith O'Dea; Jack Smight's *No Way to Treat a Lady* with Rod Steiger, George Segal, Lee Remick; Franklin J. Shaffner's *Planet of the Apes* with Charlton Heston, Roddy McDowell, Kim Hunter; Mel Brooks's *The Producers* with Zero Mostel, Gene Wilder; Paul Newman's *Rachel, Rachel* with Joanne Woodward; Franco Zeffirelli's *Romeo and Juliet* with Leonard Whiting, Olivia Hussey; Joseph Losey's *Secret Ceremony* with Elizabeth Taylor, Mia Farrow, Robert Mitchum; Frank and Eleanor Perry's *The Swimmer* with Burt Lancaster, Janice Rule, Kim Hunter; Peter Bogdanovich's *Targets* with Boris Karloff; Jean-Luc Godard's *Weekend* with

1968 *(cont.)* Mireille Dare; Tom Gries's *Will Penny* with Charlton Heston.

Film musicals George Dunning's *Yellow Submarine* with animated drawings that show the Beatles trying to save Pepperland from the Blue Meanies, music and lyrics by John Lennon and Paul McCartney; Carol Reed's *Oliver* with Ron Moody, Mark Lester, Oliver Reed, Hugh Griffith; William Wyler's *Funny Girl* with Barbra Streisand as Fanny Brice.

Stage musicals *Your Own Thing* 1/18 at New York's off-Broadway Orpheum Theater, with music and lyrics by Hal Hester and Danny Appolinar, 933 perfs.; *Canterbury Tales* 3/21 at London's Phoenix Theatre, with music by Richard Hill and John Hawkins, lyrics by Noell Coghill, 2,082 perfs.; *Zorba* 11/17 at New York's Imperial Theater, with Herschel Bernardi, music by John Kander, lyrics by Fred Ebb, 224 perfs.; *Promises, Promises* 12/1 at New York's Shubert Theater, with Jerry Orbach, Jill O'Hara, music and lyrics by Burt Bacharach and Hal David, songs that include "Wherever You Are," "I'll Never Fall in Love Again," 1,281 perfs.

First performances Symphony No. 6 by Howard Hanson 2/29 at New York's Philharmonic Hall; Symphony No. 8 by Roger Sessions 5/2 at New York's Philharmonic Hall; *The Prodigal Son* (church parable) by Benjamin Britten 6/10 at Orford, Suffolk.

U.S. guitar sales reach $130 million, up from $35 million in 1960.

Popular songs "Both Sides Now" by Canadian-born singer-guitarist-songwriter Joni Mitchell (*née* Roberta Joan Anderson), 25; "Mrs. Robinson" by Paul Simon (for the film *The Graduate*); "Galveston" by Jimmy Webb; "Spinning Wheel" by David Clayton Thomas; "Over You" by Jerry Filler; "Do Your Own Thing" by Jerry Leiber and Mike Stoller; "Hello, I Love You (Won't You Tell Me Your Name") by Robert Krieger, John Densmore, James Morrison, and Raymond Manzarek of The Doors; "John Wesley Hardin" by Bob Dylan who becomes a country singer after coming close to death in a motorcycle accident; "Hey, Jude" and "Lady Madonna" by John Lennon and Paul McCartney; "Jumpin' Jack Flash" by Mick Jagger and Keith Richard; *Led Zeppelin No. 1* (album) by Led Zeppelin, a new English rock group (Jimmy Page, 23, John Paul Jones, 22, John Bonham, 21, and Robert Plant, 21); *Papas and the Mamas—Mamas and the Papas* (album); *Cheap Thrills* (album) by Big Brother and The Holding Company; "Hurdy Gurdy Man" by Donovan Leitch; "Little Green Apples" by Bobby Russell; "The Windmills of Your Mind" by Michel Legrand, lyrics by Alan and Marilyn Bergman (for the film *The Thomas Crown Affair*); "Gotta Get a Message to You" and "I Started a Joke" by the Bee Gees; "Those Were the Days" by Gene Reskin.

Green Bay beats Oakland 33 to 14 at Miami January 14 in Super Bowl II.

Joe Frazier wins the world heavyweight boxing crown March 24 at age 24 by knocking out Buster Mathis in the eleventh round of a title bout at New York nearly 10 months after the World Boxing Association took the title away from Muhammad Ali for refusing to accept induction into the U.S. Army (*see* Cassius Clay, 1964). Jimmy Ellis, 28, wins an eight-man title tournament staged by the WBA, defeating Jerry Quarry, 22, in a 15-round decision April 27 at Oakland, but Frazier will gain undisputed right to the title early in 1970.

Rod Laver wins in men's singles at Wimbledon, Mrs. King in women's singles; Arthur Ashe, 25, wins in men's singles at Forest Hills (the first black to do so), Margaret Smith Court in women's singles. Ashe wins the first U.S. Open men's singles, (Sarah) Virginia Wade, 22 (Brit.), the women's singles.

The Olympic Games at Mexico City attract 6,082 contestants from 109 countries. U.S. athletes win the most gold medals.

The Detroit Tigers win the World Series by defeating the St. Louis Cardinals 4 games to 3.

Congress enacts a Uniform Monday Holiday Law to give Americans 3-day holidays. Scheduled to take effect beginning in 1971, the new law follows centuries-old European laws and orders that Washington's Birthday, Memorial Day, Columbus Day, and Veterans Day be observed on Mondays regardless of what day February 22, May 30, October 30, or November 11 may be.

The Jacuzzi Whirlpool bath introduced in June at California's Orange County Fair by the 53-year-old farm-pump maker Jacuzzi Bros. begins a fad only slightly related to hydrotherapy. Jacuzzi's $700 "Roman Bath" will spawn a host of imitators.

Ralph Lauren is founded by New York fashion designer Ralph Lauren (*né* Lifshitz), 29, whose "old money" and "old West" Polo brand looks will make him a fortune.

U.S. cigarette sales decline slightly to 571.1 billion as adults smoke an average of 205 packs, down from 210 last year. FTC studies show that while filter-tips dominate the market, most cigarettes yield more tars and nicotine in their smoke because tobacco companies are using tobacco leaf higher in tars and nicotine.

A Tokyo bank van robbery December 10 nets the equivalent of $1 million for a motorcycle gang that intercepts an armored truck carrying cash. The perpetrators will not be caught but Japan's crime rate is low by U.S. and European standards and is declining thanks in part to family social structure, low availability of narcotics, effective gun control laws, and a police force that apprehends 96 percent of suspects in murder cases and 92 percent of suspects in other criminal assaults. Japan has as many murders per year as New York City (which has one-tenth the population of Japan).

Construction begins in Boston's Copley Square of a 60-story John Hancock building. Designed by Henry Cobb of I. M. Pei & Partners, the Hancock Building will run into problems of falling glass panels and will not open until 1974.

North Cascades National Park is established by act of Congress which sets aside 505,000 acres of Washington State glaciers, jagged peaks, and mountain lakes.

Redwood National Park, established by act of Congress, embraces 58,000 acres along 40 miles of California's Pacific Coast containing forests with virgin groves that include a 369.2-foot redwood—the world's tallest tree.

Enzyme detergents introduced by Procter & Gamble, Lever Brothers, and Colgate-Palmolive create problems in U.S. water and sewage systems.

The U.S. Coast Guard reports that 714 major oil spills have occurred during the year, up from 371 in 1966.

The U.S. Department of the Interior estimates that 15 million fish are killed by pollution each year, two-thirds of them commercial varieties.

Fifty-eight percent of U.S. fish is imported, up from 41.4 percent in 1966.

The average American eats 11 pounds of fish per year, the highest since the mid-1950s.

Improved IR-8 rice strains from the IRRI in the Philippines produce record yields in Asia (*see* 1962; 1964). But the "miracle" rice requires more fertilizer and water than do such traditional strains as Bengawan, Intam, Peta, and Sigadio, IR-8 has little innate resistance to a virus carried by green leaf hoppers, and Filipinos do not like the cooking and eating qualities of the sticky new rice milled from IR-8.

India's wheat production is 50 percent above last year's level as a result of intensive aid by Ford Foundation workers who have introduced new Pitic 62 and Penjamo 62 wheat strains from Mexico.

Desert locusts devastate crops in Saudi Arabia and other countries along the Red Sea in the first major locust plague since 1944.

Some 6,000 Utah sheep die when VX nerve gas from the U.S. Army's Dugway Proving Ground blows across the range.

Tenneco (above) will be a major U.S. producer of lettuce and other field crops.

The average U.S. farm acre can produce 70 bushels of corn, up from 25 in 1916, and some farmers get 200 bushels (*see* 1973).

U.S. crop acreage produces yields 80 percent above those in 1920; the output per breeding animal has roughly doubled since then.

U.S. farms have 5 million tractors, 900,000 grain combines, 780,000 hay balers, 660,000 corn pickers and shellers. Major crops are all harvested by machine.

Farm labor represents only 7 percent of the U.S. work force, down from 10 percent in 1960, although another 32 percent of the work force is engaged in supplying the farmer or handling his produce.

The average U.S. farm subsidy is nearly $1,000, up from $175 in 1960, as the number of U.S. farms continues to drop off substantially. Many farmers receive far in excess of the average subsidy (*see* 1967).

A nationwide boycott of table grapes organized by Cesar Chavez of the United Farm Workers Organizing Committee gains support from much of the public (*see* 1962). Chavez dramatizes "La Causa" ("The Cause") with long fasts.

Thousands die of starvation in Biafra, whose Ibo tribespeople have seceded from Nigeria.

"Hunger in America," telecast by CBS May 21, documents conditions of deficiency diseases in the world's most affluent nation.

A Poor People's March on Washington focuses its protest on U.S. hunger conditions. Originally planned by Martin Luther King, Jr. (above), the demonstration proceeds from May 3 to June 23 under the leadership of Ralph D. Abernathy, Jr.

The Citizens Board of Inquiry into Hunger and Malnutrition in the United States observes that federal food-aid programs reach only 18 percent of the nation's poor.

The Department of Agriculture liberalizes its food stamp program (*see* 1967). It expands from two to 42 the number of counties where federal authorities will handle the program, which in some areas is resisted by local authorities (*see* 1969).

India's food minister Chidambara Subramaniam relates malnutrition to brain damage: "On the basis of studies in my own state of Madras, where I was Minister of Education, it has been estimated that between 35 and 40 percent of the children of India have suffered permanent brain damage by the time they reach school age because of protein deficiency. This means that we are, in effect, producing subhuman beings at the rate of 35 million per year. By the time they reach school age they are unable to concentrate sufficiently to absorb and retain knowledge."

Malnutrition in some parts of the United States is as grim as any he has seen in India or any other country, reports U.S. nutrition investigator Arnold E. Schaefer, 51, who at one school has found vitamin A deficiencies worse than those in children who have gone blind from keratomalacia. These Head Start Program children could go blind at any time, "five minutes from now or a year from now," says Schaefer.

The average U.S. diet provides an estimated 3,200 calories per day with 98 grams of protein but millions of U.S. diets do not approach the average.

Hershey Foods discontinues the nickel Hershey bar sold since 1894. It has shrunk the bar from 1 1/8 ounces in 1947 down to 3/4 oz. and Hershey announces that further size reductions are impractical. The 10¢ Hershey bar weighs 1.5 ounces.

Pope Paul VI condemns artificial measures for birth control in his encyclical *Humanae Vitae*. Laypersons and even some priests attack the condemnation.

Britain legalizes abortion April 27 as the Abortion Act that received Royal Assent late last year overturns an

1968 ***(cont.)*** 1861 law that made abortion a crime under all circumstances. The new law makes all abortion legal if two registered physicians find that "continuance of pregnancy would involve risks to the life or injury to the physical or mental health of the pregnant woman or the future well-being of herself or of the child or her children." Opponents of the new law predict that 400,000 abortions will be performed each year (but *see* 1969).

A British immigration act excludes thousands of Asians in Kenya from residency in Britain even though they have been offered and have accepted British nationality.

1969 U.S. B-52s secretly attack Communist bases in Cambodia in March as more than 600,000 U.S. and allied troops escalate the war in Vietnam.

President Nixon meets with President Thieu at Midway June 8 and announces the start of U.S. troop withdrawal and a new "Vietnamization" policy that will help the Indochinese nation deal with her own problems.

North Vietnamese leader Ho Chi Minh dies September 3 at age 79 after 15 years as president.

Premier Kosygin meets at Beijing Airport with Zhou En-Lai en route home from the funeral of Ho Chi Minh at Hanoi and the two Communist leaders discuss the Sino-Soviet border clashes that have occurred earlier in the year in East and Central Asia. Formal border conferences begin at Beijing October 19 amid signs of deepening divisions in the Communist world.

Sen. Edward M. Kennedy, 36 (D. Mass.), has defied tradition by challenging Russell Long of Louisiana after only 6 years' tenure and defeated the 20-year veteran January 4 to become assistant majority leader. But Kennedy's credibility as a statesman is blighted in July when the body of a former campaign worker for the late Robert Kennedy is retrieved on the morning of July 19 from a 1967 Oldsmobile sedan that plunged into Poucha Pond on Chappaquiddick Island off Martha's Vineyard some hours earlier and is upside down in 8 feet of water. Mary Jo Kopechne, 28, left a party at midnight of July 18 with Sen. Kennedy; he drove off Dike Bridge, allegedly tried to save the young woman, but has failed unaccountably to report the accident to police for 10 hours and thus raised questions as to his judgment.

"I will say confidently that looking ahead just three years the war will be over," says President Nixon October 12. "It will be over on a lasting basis that will promote lasting peace in the Pacific."

The New Mobilization Committee to End the War in Vietnam demands a moratorium on the war October 15 and masses hundreds of thousands of demonstrators in Washington, D.C., November 15. Police surround the White House with D.C. Transit buses parked bumper to bumper to protect the executive mansion where President Nixon watches a football game on television. The police use tear gas to disperse demonstrators gathered in front of the Justice Department, and the antiwar left charges the FBI and the CIA with spying on them and with breaking into their offices to gather information.

Chicago judge Julius Hoffman, 74, hears testimony in the case of *U.S. v. Dellinger and others* that opened September 24 (*see* 1968). Attorney William Moses Kunstler, 50, defends David Dellinger, Rennie Davis, Tom Hayden, Abbie Hoffman, Black Panther leader Bobby Seale, Jerry Rubin, university lecturer Lee Weiner, 32, and chemistry professor John R. Froins. Justice Hoffman has Seale gagged and manacled to his chair in the courtroom and sentences him in November to 4 years in prison for contempt of court.

Chicago police raid an apartment before dawn December 4 pursuant to a court order. The 14 officers are armed with a Thompson submachine gun, five shotguns, a .357 pistol, and 19 or 20 .38 caliber pistols, Black Panther leader Fred Hampton or Mark Clark fires one shot at the police, the police fire at least 82 shots, killing both Hampton and Clark, and recover illegal weapons.

U.S. radicals who have broken with the Students for a Democratic Society and call themselves the Weathermen change their name to the Weather Underground. The original name came from the Bob Dylan song "Subterranean Homesick Blues" which contains the words, "You don't need a weatherman to know which way the wind blows." The Weather Underground will plant bombs to protest the continuing war in Vietnam.

Israel's Premier Levi Eshkol has died in February and been succeeded by Russian-born Milwaukee-raised Palestine pioneer Golda Mabovitch Meir, 70, who will hold office for 5 embattled years.

Somalia's president Abdi Rashid Ali Shermarke, 49, is assassinated October 15 while visiting the drought-stricken northeast. Premier Mohammed Ibrahim Egal rushes home from Palm Springs, Calif., where he has been visiting actor William Holden, but Gen. Mohamed Siad Barre dissolves the legislature, arrests government leaders, sets himself up as dictator of a renamed Somali Democratic Republic, and asks the Peace Corps to leave (*see* 1960; 1974).

Sudan's Council for the Revolution president Gen. Gaafar Mohamed Nimeiri takes power as prime minister in October (*see* 1971).

New York's Stonewall Inn riot launches a "gay rights" movement as homosexuals protest a June 27th police raid on a Greenwich Village dance club and bar on Christopher Street.

Man walks on the moon for the first time July 21 as U.S. astronaut Neil A. Armstrong steps out of the lunar module from Apollo 11 and is joined by his companion Edwin E. "Buzz" Aldrin, Jr., 39. Armstrong is watched on television by much of the world but millions of Americans believe the moon walk was staged in a studio to divert attention from the Vietnam War (above); many millions more demand that money and technology be applied to more socially productive purposes.

Italy's "economic miracle" postwar recovery grinds to a stop after a boom that pulled the country out of a recession early in the decade. Splintered until now by fights between Communists and non-Communists, Italy's trade union movement achieves some unity and employs strikes and violence to win major pay raises but productivity does not increase, agriculture declines, Italy is forced to increase imports to meet the demand fueled by rising living standards, and a slide toward bankruptcy begins.

The U.S. Department of Justice files an antitrust suit against IBM—the last official act under the outgoing Johnson administration (*see* 1956). Largest antitrust action ever taken, the case will not come to trial until 1975 by which time International Business Machines will be the world's largest company in terms of the value of its stock (which will be worth more than the combined value of all the stock of all companies listed on the American Stock Exchange).

A booming U.S. economy employs a record number of workers, unemployment falls to its lowest level in 15 years, the prime interest rate is 7 percent, the dollar is strong in world money markets, and Wall Street's Dow Jones Industrial Average rises above 1,000 for the first time in history.

Washington announces December 19 that it will relax restrictions on U.S. trade with Beijing.

Phillips Petroleum drillers discover a giant oil field off the coast of Norway. The North Sea basin will prove to be the largest reservoir of oil outside the Middle East and will be found to lie 60 percent in British waters, 40 percent in Norwegian waters.

Exploitation of the Alaskan North Slope petroleum reserve remains the subject of hot debate as storage tanks are built at Prudhoe Bay on the Beaufort Sea. Delays in building pipelines to carry the oil up over the Brooks Range and south to Prince Charles Sound will result in a better designed pipeline with fewer potential danger spots for leakage.

Congress votes December 22 to reduce the oil depletion allowance granted to U.S. petroleum producers since 1926 as a tax incentive to encourage exploration for new reserves, lowering it from 27.5 percent to 22 percent despite President Nixon's campaign promise last year to maintain the old rate. The new tax allowance, like the old one, is based on the selling price of oil rather than on the capital invested to establish the well being depleted.

The United States Lines retires its passenger ship S.S. *United States* after 17 years as competition from transatlantic air carriers and foreign flag liners makes U.S. passenger vessels unprofitable.

The Concorde supersonic jet makes its first flight March 2 from Toulouse and its first supersonic flight October 1 (*see* 1962; 1970).

A Venezuelan DC-9 crashes after takeoff from Maracaibo March 16 killing all 84 aboard plus 71 on the ground; a United Arab Ilyushin-18 crashes at Aswan Airport March 20 killing 87; a Mexican Boeing 727 flies into a mountain near Monterrey June 4 killing 79; an Allegheny Airlines DC-9 collides with a student plane at Shelbyville, Ind., September 9 killing 83; a Nigerian

Richmond Times-Dispatch

Man Walks on Moon

Astronauts Begin Exploration Early

Luna Lower, May Overfly Eagle Site

Today's Events On Apollo 11

People Around Globe Acclaim Feat, Rejoice in Crew's Safety

Full-Scale Air War Waged Over Suez

"One small step for a man, one giant leap for mankind" was what astronaut Neil Armstrong meant to say.

VC-10 crashes near Iju November 20 killing 87; an Olympic Airways DC-6B crashes in a storm near Athens December 8 killing 93.

The Trans-Australian Railway marks completion of a track standardization program with a golden spike ceremony November 29 at Broken Hill in New South Wales. The road includes a 29-mile stretch of continuously welded track—the longest such stretch of track in the world (*see* Indian-Pacific Express, 1970).

The Ford Maverick introduced April 17 is a compact car designed to compete with Volkswagen and other foreign makes. General Motors prepares to introduce its Chevrolet Camaro compact.

The average U.S. automobile wholesales at $2,280, up from $1,880 in 1959.

Blue Cross health insurance programs cover some 68 million Americans, up from 37 million in 1950, and Blue Shield surgical insurance covers roughly 60 million. Hospital insurance policies issued by private, for-profit companies (often as part of a package that includes life insurance) cover another 100 million, and some 7 million are covered by nonprofit programs other than Blue Cross; third-party groups pay more than 90 percent of most U.S. city hospital bills.

The cost of medical care in the United States escalates and a crisis in health care delivery looms in large part because patients can in many cases receive insurance benefits only if hospitalized, because they often are hospitalized unnecessarily by sympathetic physicians, because Blue Cross pays hospitals on a cost-plus basis without scrutinizing costs too carefully, because physicians order countless tests to protect themselves from malpractice suits, because hospital administrators install costly equipment and facilities that are under-

1969 ***(cont.)*** utilized, and because hospital workers receive higher wages.

President Nixon bans production of chemical and biological warfare agents in November. Included are bacteria that produce anthrax, tick-borne encephalitis, bubonic plague, psittacosis (parrot fever), Q-fever, brucellosis, tularemia, Rocky Mountain spotted fever, and botulism along with the chemical agents mustard gas, phosgene, and the VX nerve gas that killed 6,000 sheep last year, but riot control agents such as tear gas remain in production.

The National Association of Broadcasters announces a plan July 8 to phase out cigarette advertising on radio and television over a 3-year period beginning January 1, 1970 (*see* 1967).

Sesame Street debuts in November on U.S. Public Service television stations and starts to revolutionize children's attitudes toward learning and adults' attitudes about what children are capable of learning. Designed by the Children's Television Workshop and funded by the Ford Foundation, Carnegie Corp., and U.S. Office of Education, *Sesame Street* teaches preschool children letters and numbers with the same techniques used in commercial television programs such as the 14-year-old *Captain Kangaroo Show. Sesame Street* introduces characters such as Oscar the Grouch, Big Bird, the Cookie Monster, Ernie, and Grover.

U.S. newspapers and television commentators assail President Nixon's position on Vietnam, his appointment of Clement F. Haynsworth, Jr., to the Supreme Court, and his November 3 appeal to "the great Silent Majority of my fellow Americans." White House speech writer Patrick Buchanan writes an address for Vice President Agnew to deliver at Des Moines, President Nixon goes over the speech, and Vice President Agnew stands up November 13 to attack the "dozen anchormen, commentators, and executive producers [who] . . . decide what 40 to 50 million Americans will learn of the day's events in the nation and in the world . . . read the same newspapers . . . draw their political and social views from the same sources . . . talk constantly to one another, thereby providing artificial reinforcement to their shared viewpoint." President Nixon begins a campaign of intimidation against the press and electronic media.

"Doonesbury" by U.S. cartoonist Garry Trudeau, 20, comments on U.S. military involvement in Vietnam. Trudeau began the comic strip about one Michael J. Doonesbury under the title "Bull Tales" in the *Yale Daily News* last year, he syndicates it to 25 newspapers through the newly formed Universal Press Syndicate, and nearly 300 newspapers will pick it up.

The Supreme Court declares a Georgia antipornography law unconstitutional April 7, saying, "If the First Amendment means anything it means that a state has no business telling a man, sitting alone in his own house, what books he may read or what films he may watch."

Penthouse magazine begins publication at New York in September. U.S. publisher Robert Guccione, 38, started the magazine at London in March 1965. He does not use an airbrush to eliminate pubic hair from nude photographs, and he challenges Hugh Hefner's *Playboy* whose newsstand sales he will overtake by 1975 (*see* 1953).

The Saturday Evening Post ceases publication February 8 after 148 years.

Nonfiction *The Peter Principle* by Canadian-born University of Southern California education professor Laurence J. Peter, 50, and Raymond Hull unmasks pretensions to power among middle-management bureaucrats: "In a hierarchy every employee tends to rise to his level of incompetence"; *The Emerging Republican Majority* by U.S. political pundit Kevin Phillips, 29, whose "Southern strategy" is credited with helping Nixon win last year's presidential election.

Fiction *Ada* by Vladimir Nabokov; *The Monster and Margarita* by the late Soviet novelist Mikhail Bulgakov, who died in 1940 at age 48; *The Four-Gated City* by Doris Lessing; *them* by Joyce Carol Oates; *Spring Snow* by Yukio Mishima; *Portnoy's Complaint* by Philip Roth; *Slaughterhouse Five, or The Children's Crusade* by Kurt Vonnegut, Jr.; *Bullet Park* by John Cheever; *The Godfather* by U.S. novelist Mario Puzo, 48, who writes about the Mafia; *Between Life and Death (Entre la vie et la mort)* by Nathalie Sarraute; *La vita e gioco* by Alberto Moravia; *The Andromeda Strain* by Harvard Medical School student Michael Crichton, 26.

Painting *Orange Yellow Orange* by Mark Rothko; *Montauk* by Willem de Kooning; *City Limits* by Philip Guston, who breaks with abstract expressionism and adopts an allegorical, mock-childlike style. Ben Shahn dies at New York March 14 at age 70.

Sculpture *One-Ton Prop* (a house of cards) by U.S. sculptor Richard Serra, 29; *Mirror Displacement: Cayuga Salt Mine Project* by U.S. sculptor Robert Smithson, 31; *Continuous Project Altered Daily* by Robert Morris; *Cumul I* (marble) by Louise Bourgeois.

Theater *To Be Young Gifted and Black* by the late Lorraine Hansberry (adapted by Robert Nemiroff) 1/2 at New York's off-Broadway Cherry Lane Theater, with Barbara Baxley, Cicely Tyson, 380 perfs.; *Ceremonies in Dark Old Men* by U.S. playwright Lonnie Elder, 37, 2/4 at New York's St. Marks Playhouse; *What the Butler Saw* by the late Joe Orton 3/5 at the Queen's Theatre, London, with Ralph Richardson (Orton was murdered last summer by his roommate who then committed suicide); *No Place to Be Somebody* by U.S. playwright Charles Gordone, 41, 5/4 at New York's off-Broadway Public Theater; *Butterflies Are Free* by U.S. playwright Leonard Gersh 10/21 at New York's Booth Theater, with Keir Dullea, Blythe Danner, Eileen Heckart, 1,128 perfs.; *Last of the Red Hot Lovers* by Neil Simon 12/28 at New York's Eugene O'Neill Theater, with James Coco, Linda Lavin, 706 perfs.

Films George Roy Hill's *Butch Cassidy and the Sundance Kid* with Paul Newman, Robert Redford, Katharine Ross, Cloris Leachman, Jeff Corey; Lindsay

Anderson's *If* with Malcolm McDowall, David Wood; Haskell Wexler's *Medium Cool* with Robert Forster, footage of last year's Democratic convention in Chicago; John Schlesinger's *Midnight Cowboy* with Dustin Hoffman, Jon Voight; Sam Peckinpah's *The Wild Bunch* with William Holden, Robert Ryan, Ernest Borgnine. Also: Masahiro Shinoda's *Double Suicide* with Kachienon Nakamura, Shira Iwashita; Dennis Hopper's *Easy Rider* with Hopper, Peter Fonda, Jack Nicholson dramatizes the alienation between the hippie generation and established rural American social values; Frank Perry's *Last Summer* with Richard Thomas, Barbara Hershey, Cathy Burns; Karel Reisz's *The Loves of Isadora* with Vanessa Redgrave as Isadora Duncan, Jason Robards, James Fox; Sergio Leone's *Once Upon a Time in the West* with Charles Bronson, Henry Fonda, Claudia Cardinale, Jason Robards; Peter R. Hunt's *On Her Majesty's Secret Service* with George Lazenby, Diana Rigg; Ingmar Bergman's *The Passion of Anna* with Liv Ullmann, Bibi Andersson, Max von Sydow; Ronald Neame's *The Prime of Miss Jean Brodie* with Maggie Smith; Mark Rydell's *The Reivers* with Steve McQueen, Rupert Crosse; Burt Kennedy's *Support Your Local Sheriff* with James Garner, Joan Hackett; Sidney Pollack's *They Shoot Horses, Don't They?* with Jane Fonda, Michael Sarrazin, Susannah York, Red Buttons, Gig Young.

Television *Marcus Welby M.D.* 9/23 on ABC with Robert Young (to 5/11/1976); *Monty Python's Flying Circus* 10/5 on BBC with Oxford and Cambridge-educated comedians John Cleese, Graham Chapman, Eric Idle, Michael Palin, and others in sophisticated, surrealistic skits (to late 1974).

Broadway and off-Broadway musicals *Dear World* 2/6 at the Mark Hellinger Theater, with Angela Lansbury, Milo O'Shea in an adaptation of the 1945 Jean Giraudoux play *The Madwoman of Chaillot*, music and lyrics by Jerry Herman, 132 perfs.; *1776* 3/16 at the 46th Street Theater, with book by Peter Stone, music and lyrics by Sherman Edwards, 1,217 perfs.; *Oh! Calcutta!* 6/17 at the off-Broadway Eden Theater, a revue devised by English critic Kenneth Peacock Tynan, 42, with contributions by Samuel Beckett, Jules Feiffer, Dan Greenberg, John Lennon, and others, music and lyrics by the Open Door, scenes that include frontal nudity and simulated sex acts, 704 perfs. (plus 610 more at the Belasco Theater beginning in late February, 1971); *Coco* 12/18 at the Mark Hellinger Theater, with Katharine Hepburn as the late Coco Chanel, music by André Previn, book and lyrics by Alan Jay Lerner, 332 perfs.

Avery Fisher sells his 32-year-old hi-fi business for $31 million.

First performances Symphony No. 14 by Dmitri Shostakovich in September at Leningrad.

The Woodstock Music and Art Fair in the Catskill Mountains at Bethel, N.Y., draws 300,000 youths from all over America for 4 days in August to hear Jimi Hendrix, Joan Baez, Ritchie Havens, The Jefferson Airplane, The Who, The Grateful Dead, Carlos Santana, and other rock stars. Despite traffic jams, thunderstorms, and shortages of food, water, and medical facilities the gathering is orderly with a sense of loving and sharing, but thousands in the audience are stoned or tripping on marijuana ("grass," "pot," "maryjane"), hashish ("hash"), lysergic acid diethylamide (LSD), barbiturates ("downs"), amphetamines ("uppers"), mescaline, cocaine, and other drugs.

The Altamont Music Festival outside San Francisco December 6 draws more than 300,000 to a free Rolling Stones concert. Members of the Hell's Angels motorcycle gang, hired to provide security, administer several beatings and stab a boy to death when he tries to reach the stage.

Popular songs "Hot Fun in the Summertime" by Sylvester Stewart; "Get Back" by John Lennon and Paul McCartney; "Honky Tonk Women" by Mick Jagger and Keith Richard; "Lay Lady Lay" by Bob Dylan; "It's Your Thing" by Rudolph, Ronald, and O'Kelley Isley; "Games People Play" by Joe South; "Oh Happy Days" by Edwin R. Hawkins; "Come Saturday Morning" by Dory Previn, lyrics by Fred Kurlin (for the film *The Sterile Cuckoo);* "Raindrops Keep Falling on My Head" by Burt Bacharach (for the film *Butch Cassidy and the Sundance Kid*); "Okie from Muskogee" by U.S. composer Roy Edward Burris with California-born songwriter Merle Haggard, 32.

The New York Jets beat the Baltimore Colts 16 to 7 January 12 at Miami to win Super Bowl III.

Rod Laver, now 31, wins the "grand slam" in tennis for a second time; Mrs. Adrianne Shirley Hardon "Anne" Jones, 30 (Brit.), wins in women's singles at Wimbledon, Mrs. Court at Forest Hills.

Baseball's two major leagues split into eastern and western divisions with two new expansion teams each. The National League adds the Montreal Expos and the San Diego Padres, the American League adds the Kansas City Royals and the Seattle Pilots.

The New York Mets win their first World Series, defeating the Baltimore Orioles 4 games to 1.

World chess champion Tigran Petrosian is defeated by Soviet grandmaster Boris Spassky, 32, who will hold the title until 1972.

U.S. pantyhose production reaches 624 million pair, up from 200 million last year, as American women switch from nylon hosiery.

The New Jersey Chapter of the American Civil Liberties Union files suit in federal court charging that marijuana should not be classed with heroin and other dangerous drugs. Marijuana, says the ACLU, is harmless both to the user and to society.

Heroin sales to New York schoolchildren have jumped as a result of the federal government's Operation Intercept program to restrict the flow of marijuana from Mexico, says an expert testifying before a joint legislative committee at Washington. The price of marijuana, he says, has climbed so high that heroin sells at a competitive price.

1969 ***(cont.)*** The Tate-LaBianca murders make headlines in August. Screen actress Sharon Tate Polanski, 26, is murdered at her Bel-Air home in Benedict Canyon early in the morning of August 10 along with coffee heiress Abigail Folger, 25, her common-law husband Wojiciech "Voytek" Frykowski, 32, Hollywood hair stylist Jay Sebring, 35, and delivery boy Steven Earl Parent. Supermarket chain president Leno LaBianca, 44, and his wife Rosemary, 38, are murdered later in the day at Los Angeles. Police say the murders are unrelated, but a jury late next year will find ex-convict Charles M. Manson, 32, and his hippie cult family guilty of all seven murders. (Manson has allegedly mesmerized his followers with drugs, sex, and religion.)

The National Commission on Urban Growth recommends development of 10 new U.S. communities of 100,000 each to accommodate an expected 20 million or more in additional population but no government action is taken.

Boston's new city hall is completed by Kallman, McKinnell, and Knowles.

Chicago's John Hancock Center with 100 floors is completed by Skidmore, Owings & Merrill.

An offshore oil well blowout in California's Santa Barbara Channel covers a 30-mile stretch of shoreline with 235,000 gallons of crude oil beginning January 28. The offshore drilling platforms of Union Oil Co. kill fish and wildlife, and while the blowout is contained after 11 days, the platforms will continue to leak oil for years.

Thor Heyerdahl of 1947 *Kon Tiki* fame crosses much of the Atlantic in a papyrus raft he calls *Ra* and observes pollution not seen in the Pacific. Empty bottles, cans, plastic containers, nylon, and—most conspicuously—clots of solidified black petroleum.

Hurricane Camille hits the Mississippi Gulf Coast August 17 with winds 190 miles per hour. Strongest hurricane since 1935, the storm strikes even harder in Virginia and West Virginia, leaves 248 dead, 200,000 homeless, and property damage of $1.5 billion.

California mud slides, produced by heavy rains, destroy or damage 10,000 homes; 100 die.

Europe's Rhine River has a massive fish kill in June 2 years after the disappearance of two 50-pound canisters containing the insecticide Thiodan. The canisters evidently fell overboard in transit from Frankfurt to the Netherlands and experts theorize that the poison has finally leaked into the water to cause the death of millions of fish.

The Northeast Atlantic Fisheries Commission urges a ban on salmon fishing outside national fishing boundaries as the salmon catch declines in rivers of Canada, Britain, Europe, and the United States.

The average U.S. farm worker produces enough food and fiber for 47 people, up from 40 last year, as agricultural productivity continues to climb.

The average Wisconsin dairy cow yields 10 quarts of milk per day, up from 6 in 1940, as a result of breed improvements.

A White House Conference on Food, Nutrition and Health opens in December under the chairmanship of Harvard's French-American nutritionist Jean Mayer. Funding for food stamp programs will increase in the next 5 years from $400 million to more than $3 billion.

Twenty-one million U.S. children participate in the National School Lunch Program. About 3.8 million receive lunch free or at substantially reduced prices and the figure will soon rise to 8 million.

Japanese lathe operator Takako Nakamura, 28, throws herself off a speeding train after discovering that she has been poisoned by inhaling cadmium fumes while working for the Toho Zinc Co. Prime Minister Eisaku Sato announces tearfully that he is determined to secure passage of strong antipollution laws.

Arizona orders a 1-year moratorium on use of DDT after milk in the state proves to have high levels of the pesticide.

Mother's milk contains four times the amount of DDT permitted in cows' milk, says Sierra Club executive vice president David Brower in testimony before the House Merchant Marine and Fisheries Committee. "Some wit suggested that if [mother's milk] were packaged in some other container we wouldn't allow it across state lines," says Brower.

Coho salmon in Michigan lakes and streams prove to have DDT concentrations of 20 parts per million. The Food and Drug Administration seizes 28,000 pounds of fish and Michigan restricts spraying of crops with DDT.

The FDA (above) notes that 90 percent of fish sold in the United States contain less than 1 part per million of DDT. Some critics demand a zero level but commercial fishermen in some areas demand a higher level and even critics cannot demonstrate any real evidence that DDT is harmful to humans.

U.S. frankfurters have an average fat content of 33 percent, up from 19 percent in 1941, and some franks are more than half fat.

Major U.S. baby-food makers halt use of monosodium glutamate (MSG) after tests show that mice fed large amounts of the flavor enhancer suffer brain damage.

The Food and Drug Administration removes cyclamates from its GRAS (generally recognized as safe) list in October and reveals plans to remove cyclamate-sweetened products from stores. It cites bladder cancers in test rats fed excessive amounts of cyclamates, but serious doubts will be raised as to the validity of the tests (*see* Delaney clause, 1958).

Paris tears down its Les Halles market and moves it 9 miles south of the city to Rungis near Orly Airport. Les Halles had been "the belly of France" (in novelist Emile Zola's phrase) since 1137.

Cyclamates (above) are employed in U.S. foods to the tune of 20 million pounds per year as compared with 20 billion pounds of sugar: 70 percent of the cyclamates are used in sugar-free soft drinks.

Kaboom breakfast food, introduced by General Mills, is 43.8 percent sugar.

Frosted Mini-Wheats breakfast food, introduced by Kellogg, is 28 percent sugar.

Miller beer is acquired June 12 by Philip Morris, which pays W. R. Grace $150 million for Grace's 53 percent share and will buy the remainder next year for $97 million. By 1978, Miller will be second only to Anheuser-Busch.

Forty-three percent of U.S. women over age 16 and 41 percent of all married women are in the labor force, up from 34 percent of women over age 14 and 31 percent of married women in 1960, but only 4 percent are farm hands or domestic servants.

Britain has 28,859 abortions in the 10 months ending February 25 and exceed 1,000 per week in England and Wales by late July. The National Health Service pays for 60 percent of the abortions (*see* 1968).

California's supreme court rules in September that the state's anti-abortion law is unconstitutional. It infringes on a woman's right to decide whether to risk childbirth and bear children, says the court.

1970 Paris peace talks to end the Vietnam war continue for a second year without progress but Washington reduces U.S. troop strength in Vietnam below 400,000 in response to mounting public pressure as casualties rise.

North Vietnamese troops and tanks seize a key Laotian stronghold in the Plaine des Jarres in mid-February.

U.S. military activity in Laos clearly "violates the spirit" of congressional measures aimed at barring use of American ground forces there, says Sen. Charles McC. Mathias (D. Md.) February 25. Mathias cites a report that the CIA has hired hundreds of former Green Beret troops to serve in Laos.

U.S. authorities at Saigon charge five U.S. Marines with murdering 11 South Vietnamese women and 5 children while on patrol south of Danang February 19.

Capt. Ernest L. Medina and five other soldiers are charged with premeditated murder and rape at the South Vietnamese village of My Lai (Songmy) in 1968 and with maiming a suspect during questioning.

Gen. Samuel W. Koster resigns as superintendent of the U.S. Military Academy at West Point following accusations that he and 13 other officers suppressed information. Gen. Koster commanded the Americal Division whose First Battalion C Company was involved in the 1968 My Lai massacre of 47 civilians. A secret army investigation has reportedly found that the number of victims dwindled as information moved up the chain of command but that U.S. troops did indeed commit acts of murder, rape, sodomy, and maiming against "non-combatants."

"The Senate must not remain silent now while the President uses the armed forces of the United States to fight an undeclared and undisclosed war in Laos," says Sen. Fulbright March 11. He proposes a resolution challenging Nixon's authority to commit U.S. forces to combat in or over Laos.

Cambodians stage peaceful protest demonstrations at Phnom Penh over the presence of North Vietnamese forces in the country. Prince Norodom Sihanouk, who has held secret discussions with the Vietcong and North Vietnamese, is overthrown March 18 while away on a visit to Moscow. The premier and defense minister Lon Nol seizes power and begins a reign of terror against Cambodia's 400,000 Vietnamese residents, appealing for U.S. aid to stop the North Vietnamese from taking over.

A massive South Vietnamese move into Cambodia begins April 29 with support from U.S. planes and advisers.

President Nixon makes a television address April 30 saying he has ordered U.S. combat troops into part of Cambodia to destroy North Vietnamese "headquarters" and "sanctuaries." The action is intended to save U.S. lives, he insists, and is essential to his plan for "Vietnamizing" the war.

College campus radicals who oppose his policies in Vietnam are "bums," says President Nixon May 1.

Kent State University students in Ohio rally at noon May 4 to protest the widening of the war in Southeast Asia. National Guardsmen open fire on the 1,000 students and four fall dead, including two young women; eight others are wounded.

U.S. colleges close down in anti-war demonstrations and some will remain closed for the balance of the spring term as students coordinate plans for strikes and demonstrations. Secretary of the Interior Walter J. Hickel sends a letter to President Nixon warning that the administration is contributing to anarchy and revolt by turning its back on American youth and that further attacks on the motives of young people by Vice President Agnew will solidify hostility and make communication impossible.

New York construction workers break up an anti-war rally in the Wall Street area May 8, force City Hall officials to raise the American flag to full staff (it had been lowered to half staff in memory of the Kent State dead), and invade Pace College. But President Nixon holds his first press conference in 3 months and announces that U.S. troops will be out of Cambodia by mid-June.

An anti-war rally May 9 brings 75,000 to 100,000 peaceful demonstrators to Washington, D.C. President Nixon is unable to sleep and drives to the Lincoln Memorial before dawn to talk for an hour with students protesting the war.

Portuguese dictator Antonio Salazar dies July 27 at age 81, not knowing that he has been replaced as premier by Marcello Caetano after nearly 40 years in power.

Nigeria's civil war ends January 12 with the capitulation of Biafran chief of staff Brig. Gen. Philip Effiong after more than 30 months of conflict in which some 2 million people have died (*see* 1967). Effiong has assumed

Vehement opposition to continued U.S. troop deployment in Southeast Asia inspired demonstrations in many cities.

leadership following the flight of Gen. Odumegwu Ojukwu to the Ivory Coast.

Libyan military leader Col. Muammar Qadaffi, 27, assumes power as premier January 16, 4½ months after seizing control of the country. French defense minister Michel Debre announces 5 days later that France will provide Libya with 100 military aircraft—twice the number originally announced—following Libya's promise to end her support of rebels in neighboring Chad.

Israeli jets raid Cairo suburbs in January; commandos strike within 37 miles of Cairo January 16, destroying power and telephone pylons on the main road between Cairo and Port Suez.

Representatives of five Arab nations meet at Cairo and vow to continue fighting to recover territory occupied by Israel since the 1967 war. They blame the United States for Israel's refusal to give up the territory, allude to profitable U.S. oil investments, and warn that the Arabs will not permit their "resources and wealth" to be exploited to help Israel.

Egypt's President Nasser accepts a U.S. peace formula for the Middle East July 24, Jordan announces her acceptance 2 days later, Syria makes a show of rejecting the formula, Israel announces her acceptance July 31 as Palestinians meet 25,000 strong at Amman and cheer a guerrilla leader's call for rejection of the formula and "liberation" of all Palestine.

Arab and Israeli forces clash on three fronts August 2 as diplomats in world capitals work to end hostilities, a cease-fire goes into effect August 7, guerrilla spokesmen at Amman say they will work to undermine the 90-day truce, and while the cease-fire remains intact along the Suez Canal Israeli jets attack guerrilla bases in Lebanon August 9, Israeli forces fight infiltrators from Syria in the Golan Heights, intelligence reports installation of new Soviet antiaircraft missiles on the Egyptian side of the Suez Canal 4 hours after the cease-fire went into effect, Israeli planes bomb and strafe Jordanian army posts that are said to make guerrilla raids possible, and concerted efforts begin at the UN in New York to settle the disputes in the Middle East.

Jordan has civil war from September 15 to 26. King Hussein escapes an assassination attempt, his Bedouin troops eject Palestine Liberation Organization forces with considerable bloodshed, and the PLO moves to Lebanon (*see* 1975). Syrians invade in Soviet-built tanks but withdraw after threats of U.S. and Israeli intervention.

Egypt's President Nasser dies September 28 at age 52. His friend Anwar el-Sadat, 51, is elected president October 14 by an overwhelming vote.

A Syrian military coup November 13 replaces the civilian government with a rightist regime. Defense Minister Lieut. Gen. Hafez el-Assad, 40, takes over as premier November 19, beginning a repressive dictatorship.

The British general elections January 18 have given the Conservative party a 30-seat majority in the House of Commons. Prime Minister Harold Wilson is turned out after more than 5 years of Labour Party government, and a new cabinet takes office with Conservative Edward Richard George Heath, 53, as prime minister.

An explosion in New York's Greenwich Village March 6 completely obliterates a townhouse at 18 West 12th Street allegedly used by members of the Weather Underground to produce bombs. Three people are killed.

University of Wisconsin students protesting the university's participation in government war research blow up a campus laboratory August 24 killing a research graduate student, injuring four others, and destroying a $1.5-million computer.

Tonga gains independence June 4 after 70 years as a British protectorate.

Fiji gains independence October 10 after 96 years of British colonial rule.

Chilean president Salvador Allende Gossens takes office November 3, restates his campaign promise to nationalize much of Chile's economy, and extends recognition to Cuba's Castro government. Allende, 62, is the first Marxist to be elected head of a government in the Western Hemisphere by a democratic majority; the CIA has tried to block his election (*see* 1973).

Charles de Gaulle dies November 9 at age 79 at Colombey-les-Deux-Eglises, 18 months after resigning as president.

Japanese novelist Yukio Mishima harangues 1,000 troops November 25 on the "disgrace" of having lost the Pacific war in 1945, urges them to support him and his private army in a coup d'état, arouses no interest, and dies by his own hand in a ceremonial act of *seppuku* at age 45.

Poland has riots beginning December 14 at Gdansk (formerly Danzig) one week after signing a treaty with the German Federal Republic that gives provisional recognition to the Oder-Neisse line as Poland's western frontier. Warsaw has given assent to the repatriation of Germans living east of the Oder-Neisse line. The riots, arising from food shortages and higher prices of food and other commodities, spread to other cities, police and troops put them down with heavy loss of life, Wladyslaw Gomulka and other members of the Polish politburo resign under pressure December 20, and Edward Gierek, 57, the party chief in Upper Silesia, succeeds to Gomulka's offices.

The U.S. Commission on Civil Rights calls a recent presidential policy statement on school integration inadequate, overcautious, and possibly the signal for a major retreat.

Yale president Kingman Brewster, Jr., expresses doubt April 24 that black revolutionaries can get "a fair trial anywhere in the United States." A special coroner's jury at Chicago has ruled January 21 that last year's killing of Fred Hampton and another Black Panther in a predawn police raid was "justifiable."

A student dormitory at Jackson State College in Mississippi is riddled with police bullets May 14, killing a student and a local high school senior, both blacks.

George Wallace urges Southern governors to defy federal integration orders. Addressing a noisy Birmingham rally February 8, the former Alabama governor says he will run for the presidency again in 1972 "if Nixon doesn't do something about the mess our schools are in."

Northern liberals should drop their "monumental hypocrisy" and concede that de facto segregation exists in the North, says Sen. Abraham A. Ribicoff, 59 (D. Conn.), February 9.

Congress approves education appropriation bills containing amendments designed to halt busing of children to achieve racial balance (*see* 1971).

A group of civil rights leaders attacks the policy of "benign neglect" toward blacks advocated by presidential adviser Daniel Patrick Moynihan, 43. They call the policy "a calculated, aggressive, and systematic" effort by the administration to "wipe out" gains made by the civil rights movement.

The U.S. Gross National Product (GNP) reaches $977 billion, up from $503 billion in 1960. Government spending accounts for 32 percent, up from 27 percent in 1960.

Wall Street's Dow Jones Industrial Average bottoms out at 631 and jumps 32.04 points May 27 to close at 663.20—the largest one-day advance ever recorded. Daily volume on the New York Stock Exchange averages 11.6 million shares, up from 2.6 million in 1955, 3 million in 1960. Brokerage houses struggle to automate their back rooms to keep up with mounting paperwork.

Gold prices in the world market at London fall below the official U.S. price of $35 per ounce.

A U.S. Office of Management and Budget, proposed by President Nixon, is created by Congress in May. The president appoints Secretary of Labor Charles Schultz first OMB director.

The Department of Labor reports June 5 that 5 percent of the U.S. work force is unemployed, the highest rate since 1965. Hardest hit by the industrial slowdown are skilled workers in aircraft, aerospace, weapons, and auto making.

President Nixon goes on television June 17 to ask that business and labor end inflation by voluntarily resisting wage and profit increases. The president says he will not impose direct wage and price controls but he creates a new national commission and asks it to suggest ways for increasing output per worker.

Some 25.5 million Americans live below the poverty line—$3,908 per year for a family of four—and another 10.2 million live only slightly above the line. Nearly half the 35.7 million total are in the South.

United Automobile Workers strike General Motors plants November 2, beginning a 67-day walkout.

A rupture in Syria's Tapline May 3 stops the flow of Saudi Arabian crude oil that has been coming through at the rate of 500,000 barrels per day. A bulldozer has cut the pipeline by accident, says Damascus, but the Syrians refuse to allow Tapline technicians into the country to repair the break.

Col. Qadaffi orders cutbacks in Libyan oil production to conserve petroleum and push up prices.

OPEC nation delegates meeting at Caracas, Venezuela, in December agree to raise the posted prices of Persian Gulf oil and increase taxes on the oil.

Boeing 747 jumbo jets go into transatlantic service for Pan Am beginning January 21 (*see* 1966).

Two armed men hijack a Pan Am Boeing 747 September 6 en route from Amsterdam to New York, reroute the plane to Beirut, take dynamite aboard, fly on to Cairo, evacuate all passengers via emergency exits, and blow up the aircraft 2 minutes later.

An armed man and woman commandeer an Israeli El Al flight September 6 en route from Tel Aviv to London, but security guards on the plane mortally wound the man and passengers subdue the woman. Jailed at London, she turns out to be Leila Khaled, 24, a former student at Beirut's American University who took part in a hijacking last year.

Palestinian militants hijack a TWA 707 and a Swissair DC-8 September 6 and force them to land outside Amman, Jordan. Militants hijack a BOAC VC-10 a few days later and force it to land on the same strip; they blow up all three planes after removing the passengers and hold the passengers hostage for several weeks until British, West German, Swiss, and Israeli authorities release Leila Khaled (*see* above) and other Arabs.

A Dominican DC-9 crashes into the sea February 15 on takeoff from Santo Domingo killing 102; a British charter jet crashes near Barcelona July 3 killing 112; an Air

1970 ***(cont.)*** Canada DC-8 crashes near Toronto July 5 killing 108; a Peruvian turbojet crashes after takeoff from Cuzco August 9 killing 101, including some on the ground.

The Concorde supersonic jet exceeds twice the speed of sound for the first time November 4.

A survey reports that rail travel is 2.5 times as safe as air travel, 1.5 times as safe as bus travel, 23 times as safe as automobile travel.

Burlington Northern, Inc., is created in March by a merger of the Great Northern, Northern Pacific, and Chicago, Burlington & Quincy railroads with the Spokane, Portland, and Seattle. With 25,000 miles of track, BN is the longest railway system in the free world; its 6-mile-long Flathead Tunnel opens in November.

The Rail Passenger Service Act passed by Congress October 14 creates the National Rail Passenger Corp. to improve U.S. rail travel (*see* Amtrak, 1971).

The Indian-Pacific Express between Sydney and Perth begins service in March on the Trans-Australian Railway (*see* 1969). The twice-a-week transcontinental train runs 2,460 miles through forest, desert, mountains, and wheat lands and is so popular that space must be booked well in advance.

New York City's transit fare goes to 30¢ January 4; it has been 20¢ since 1966 (*see* 1972).

Mexico City's Metro subway line opens with 537 bright orange cars carrying 1 million passengers per day on a 22.5-mile route with 47 stations, 10 of them above ground. Built with the help of a substantial French loan, the $400 million subway employs rubber-tired, French-built trains, the standard fare is 8¢, but while the handsome new subway eases congestion, some 40 percent of the city's commuters continue to use private vehicles, going home at noon for lunch and a siesta, returning to work in midafternoon, and creating enormous traffic (and pollution) problems.

A restriction enzyme discovered by U.S. biochemist Hamilton O. Smith, 39, always breaks certain DNA molecules at the same place; U.S. biologist Daniel Nathans, now 41, will find next year that it can break up the DNA of a tumor virus, which will lead to a complete genetic mapping of the virus.

U.S. microbiologist David Baltimore, 32, demonstrates the existence of "reverse transcriptase," a viral enzyme that reverses the normal DNA-to-RNA process.

U.S. public schools are for the most part "grim," "joyless," and "oppressive," says a study commissioned by the Carnegie Corporation, and they fail to educate children adequately.

No U.S. scholar specializing in Vietnamese studies has a tenured professorship, no scholar is devoting most of his time to studying current affairs in North Vietnam, and fewer than 30 Americans are studying the Vietnamese language, a survey reveals.

U.S. colleges close down in anti-war protests (*see* above). Ivy League schools begin to go co-ed.

British undergraduates demonstrate against keeping files on students' political activities.

Paris students protest the banning of a Maoist splinter group and riot in the Latin Quarter to protest prison sentences handed down against two Maoist student leaders.

Some 152,000 U.S. postal workers strike 671 locations in March, the Army is sent in to sort the mail, and the Postal Reorganization Act, signed into law August 12, converts the Post Office Department into the patronage-free U.S. Postal Service, which loses $6.3 billion this year on 85 billion pieces of mail (*see* 1971).

Cleveland's Harris Corp. introduces the first electronic editing terminal for newspapers.

Nonfiction *The Making of a Counter Culture* by California State University professor Theodore Roszak, 37; *The Greening of America* by Yale professor Charles Reich, 34; *Up the Organization* by former Avis Rent-a-Car president Robert Townsend, 50; *Future Shock* by U.S. sociologist Alvin Toffler, 41; *Hard Times* by Studs Terkel; *Sexual Politics* by U.S. feminist Kate (Katherine Murray) Millett, 35.

Fiction *Time and Again* by U.S. science fiction novelist Jack (Walter Braden) Finney, 59; *Fifth Business* by Canadian novelist Robertson Davies, 57; *A Guest of Honor* by South African novelist Nadine Gordimer, 47; *Mr. Sammler's Planet* by Saul Bellow; *Vital Parts* by Thomas Berger; *Losing Battles* by Eudora Welty; *Play It as It Lays* by Joan Didion; *Deliverance* by U.S. poet-novelist James Dickey, 47; *Islands in the Stream* by the late Ernest Hemingway who took his own life in July at age 61; *The Blood Oranges* by John Hawkes; *Bech: A Book* by John Updike; *Runaway Horses* by Yukio Mishima (above); *Travels with My Aunt* by Graham Greene; *The French Lieutenant's Woman* by John Fowles; *Love Story* by Yale classics professor Erich Segal, 33; *Jonathan Livingston Seagull* by former U.S. Navy jet pilot Richard Bach, 33.

Juvenile *The Trumpet of the Swan* by E. B. White.

Poetry *Eye-Beaters, Blood, Victory, Madness, Buckhead and Mercy* by James Dickey.

Painting *Andy Warhol* by Alice Neel; *Patchwork Quilt* (collage) by Romare Bearden. Mark Rothko dies by his own hand February 25 at New York in a fit of depression at age 66; Barnett Newman dies at New York July 3 at age 65.

Sculpture *Spiral Jetty* by Robert Smithson; *Untitled* (rope piece) by Eva Hesse who dies of a brain tumor at New York May 29 at 34.

Theater *Sleuth* by English playwright Anthony Shaffer, 43 (Peter's twin brother), 2/12 at St. Martin's Theatre, London, with Anthony Quayle, Keith Baxter; *Child's Play* by New York playwright Robert Marasco, 32, 2/17 at New York's Royale Theater, with Pat Hingle, Fritz Weaver, Michael McGuire, 342 perfs.; *The Effect of Gamma Rays on Man-in-the-Moon Marigolds* by U.S. playwright Paul Zindel, 33, 4/7 at New York's off-

Broadway Mercer O'Casey Theater, 819 perfs.; *Home* by David Storey 6/17 at London's Royal Court Theatre, with John Gielgud, Ralph Richardson; *The Philanthropist* by English playwright Christopher Hampton, 24, 8/3 at London's Royal Court Theatre, with Alec McCowen; *How the Other Half Loves* by English playwright Alan Ayckbourne, 31; 8/5 at London's Lyric Theatre, with Robert Morley; *The Gingerbread Lady* by Neil Simon 12/13 at New York's Plymouth Theater, with Maureen Stapleton, 193 perfs.

Television *The Mary Tyler Moore Show* 9/19 on NBC with Moore, 32, Ed Asner, 40, Ted Knight, 46 (to 9/3/1977).

Films Bob Rafelson's *Five Easy Pieces* with Jack Nicholson, Karen Black, Susan Anspach, Billy Green Bush; Bernardo Bertolucci's *The Garden of the Finzi-Continis* with Dominique Sanda; Arthur Penn's *Little Big Man* with Dustin Hoffman, Faye Dunaway; Robert Altman's *M*A*S*H* with Elliott Gould, Donald Sutherland, Sally Kellerman; Luis Buñuel's *The Milky Way* with Paul Frankens, Laret Terzieff; Franklin Shaffner's *Patton* with George C. Scott, Karl Malden; Marcel Ophuls's documentary *The Sorrow and the Pity*; Laurence Olivier's *Three Sisters* with Olivier, John Sichel, Joan Plowright, Alan Bates. Also: François Truffaut's *Bed and Board* with Jean-Pierre Leaud, Claude Jade; William Friedkin's *The Boys in the Band* with Kenneth Nelson, Peter White; Aram Avakian's *End of the Road* with Stacy Keach, James Earl Jones; Gilbert Cates's *I Never Sang for My Father* with Melvyn Douglas, Gene Hackman; Eric van Zuylen's *In for Treatment* with Marja Kok; Hal Ashby's *The Landlord* with Beau Bridges, Pearl Bailey; Alan Cooke's *The Mind of Mrs. Soames* with Terrence Stamp; Billy Wilder's *The Private Life of Sherlock Holmes* with Robert Stephens, Colin Blakely; René Clement's *Rider on the Rain* with Charles Bronson, Marlene Jobert; Otto Preminger's *Tell Me That You Love Me, Junie Moon* with Liza Minnelli, Ken Howard, Robert Moore; Joseph L. Mankiewicz's *There Was a Crooked Man . . .* with Kirk Douglas, Henry Fonda; Claude Chabrol's *This Man Must Die* with Jean Yanne, Michael Duchaussoy, Caroline Cellier; Luis Buñuel's *Tristana* with Catherine Deneuve, Fernando Rey; Joseph Strick's *Tropic of Cancer* with Rip Torn, James Callahan, Ellen Burstyn; Ken Russell's *Women in Love* with Alan Bates, Oliver Reed, Glenda Jackson, Eleanor Brown.

Film musicals Michael Wadleigh's *Woodstock* with Joan Baez, Richie Havens, Crosby, Stills and Nash, The Jefferson Airplane, Joe Cocker, Sly and the Family Stone, The Who, and others (a documentary account of last year's bash at Bethel, N.Y.); David Maysles, Albert Maysles, and Charlotte Swerin's *Gimme Shelter* with The Rolling Stones (a documentary account of last year's Altamont rock concert); Gene Kelly's *Hello, Dolly* with Barbra Streisand.

Broadway and off-Broadway musicals *The Last Sweet Days of Isaac* 1/26 at the Eastside Playhouse with Austin Pendleton, music by Nancy Ford, lyrics by Gretchen Cryer, 465 perfs.; *Purlie* 3/15 at the Broadway Theater with Melba Moore, music by Gary Geld, lyrics by Peter Udell, songs that include "I Got Love," 688 perfs.; *Applause* 3/30 at the Palace with Lauren Bacall, music by Charles Strouse, lyrics by Lee Adams, book by Betty Comden and Adolph Green, 896 perfs.; *Company* 4/26 at the Alvin Theater with Elaine Stritch, music and lyrics by Stephen Sondheim, 706 perfs.; *The Me Nobody Knows* 5/18 at the Orpheum Theater, with music by Gary William Friedman, lyrics by Will Holt, 587 perfs.; *The Rothschilds* 10/19 at the Lunt-Fontanne Theater, with music by Jerry Bock, lyrics by Sheldon Harnick, 507 perfs.; *Two by Two* 11/10 at the Imperial Theater, with Danny Kaye, music by Richard Rodgers, lyrics by Martin Charnin, 352 perfs.

First performances *The Yale-Princeton Football Game* orchestral suite by the late Charles Ives (completed by Gunther Schuller) 11/29 at New York's Carnegie Hall.

Popular songs "Bridge over Troubled Water" by Simon and Garfunkel (who will break up their partnership next year); "Your Song" by English singer-pianist-songwriter Elton John (Reginald Kenneth Dwight), 23; "I'll Never Fall in Love Again" by Bobbie Gentry; *Deja Vu* (album) by Crosby, Stills, Nash & Young (Canadian artist Neil Young, 24, has joined the group); *Self Portrait* (album) by Bob Dylan; *Cosmos Factory* (album) by John Fogerty of Creedence Clearwater Revival; *Blood, Sweat and Tears III* (album) by Blood, Sweat and Tears; *Led Zeppelin III* (album) by Led Zeppelin; *Let It Be* (album) by John Lennon and Paul McCartney of The Beatles; *Abraxas* (album) by Santana; *Black Sabbath* and *Paranoid* (albums) by Birmingham (England) rock group Black Sabbath (John Osbourne, Tony Iommi, Geezer Butler, and Bill Ward). Jimi Hendrix dazzles audiences at a rock festival on the Isle of Wight in August but dies of drugs or alcohol in his London apartment in mid-September at age 27; Janis Joplin dies of a drug overdose at Hollywood, Calif., October 3 at age 27.

Kansas City beats Minnesota 23 to 7 at New Orleans January 11 in Super Bowl IV.

Joe Frazier regains the world heavyweight boxing title February 16 by knocking out Jimmy Ellis in the fifth round of a championship bout at New York.

John Newcombe wins in men's singles at Wimbledon, Mrs. Court in women's singles; Ken Rosewall wins in men's singles at Forest Hills, Mrs. Court in women's singles.

Golfer Tony Jacklin becomes the first Briton to win the U.S. Open in 50 years.

The Australian ocean yacht *Gretel II* loses her bid for the America's Cup. The U.S. defender *Intrepid* wins 4 races to the *Gretel's* 1.

Brazil wins the World Cup football (soccer) championship by defeating Italy 4 to 1 at Mexico City.

The first New York Marathon September 23 attracts 126 starters who run around Central Park 4 times. Far Rockaway fireman Gary Muhrcke, 30, wins the event, which will grow to attract more than 25,000 runners

1970 ***(cont.)*** who start on Staten Island and finish in Central Park.

The Seattle Pilots become the Milwaukee Brewers and Milwaukee becomes a major league city once again (*see* 1966).

The Baltimore Orioles win the World Series by defeating the Cincinnati Reds 4 games to 1.

U.S. women balk at a new midi-skirt decreed by fashion arbiters. Unsold garments are returned to manufacturers, women wear their skirts as long or short as they like, and slaves to fashion fade from the scene.

Man-made fabrics raise their share of the U.S. textile market to 56 percent, up from 28 percent in 1960, with polyesters enjoying a 41 percent share of the market and cotton only 40 percent, down from 65 percent in 1960. E. I. Du Pont's patent on polyester has run out, other companies have entered the market, and some big chemical companies have helped mills that use polyester-cotton blends with massive consumer advertising to proclaim the virtues of durable press fabrics.

Adult Americans give up cigarettes in growing numbers but smoking among teenagers increases: 36.3 percent of Americans aged 21 and over smoke cigarettes, down from 42.5 percent in 1964, 42.3 percent of adult males smoke cigarettes, down from 52.5 percent in 1964, 30.5 percent of adult females smoke cigarettes, down from 31.5 percent.

The Racketeer Influenced and Corrupt Organizations (RICO) Act signed by President Nixon October 15 will be used in the 1980s to prosecute both Mafia kingpins and white-collar criminals, notably Wall Street traders using privileged information.

U.S. conservationists win a battle to prevent construction of a giant international jetport near Florida's Everglades.

President Nixon signs an executive order February 4 calling for elimination of all air and water pollution caused by federal agencies. He authorizes expenditure of $359 million to carry out the order with a 3-year deadline to meet state pollution standards.

The Clean Air Act signed by President Nixon in December is the toughest such measure yet, even after the compromises with auto makers. They are given 6 years to develop engines that are 90 percent emission free.

U.S. Army engineers sink an obsolete Liberty ship carrying 12,540 canisters of nerve gas in 16,000 feet of water off the Bahamas. Legal action has been taken to prevent the action but initial tests show that no gas has escaped from the ship (critics say the canisters will eventually rust).

Earth Day April 21 sees the first mass demonstrations against pollution and other desecrations of the planet's ecology. Environmentalists block off streets and employ other means to raise U.S. awareness of threats to the environment of spaceship Earth (in the phrase of the late Adlai Stevenson).

The most destructive earthquake in the history of the Western Hemisphere rocks Peru May 31 from an epicenter 15 miles west of Chimbote. Part of the west face of Mt. Huascaran, Peru's highest mountain, is jarred loose, an estimated 50 million cubic yards of ice and rock tumble down on the town of Yungay at 200 miles per hour, burying the town 20 feet deep, upwards of 15,000 are killed, and Peru's total death toll reaches 70,000 with 50,000 injured and 186,000 buildings destroyed—30 percent of all structures in the region.

Los Angeles buildings sway August 12 as a series of sharp rolling earthquakes make California shudder as far south as San Diego, breaking windows, blocking some highways, and increasing anxieties (*see* 1971).

Hurricane Celia strikes Corpus Christi, Tex., August 3 with winds of up to 145 miles per hour; 90 percent of downtown Corpus is damaged or destroyed.

Southern California has its worst brush fires in history. Thousands in San Diego County are driven from their homes, Los Angeles suburbs are threatened, and the fires in late September strike Sequoia National Forest north of Bakersfield.

A cyclone devastates East Pakistan November 13. Great waves engulf the coast and sweep over islands in the Bay of Bengal, 300,000 are feared dead, and the death toll mounts in the weeks following as poor transportation slows distribution of food to survivors (*see* 1991).

The Environmental Protection Agency created by Congress December 2 will be the largest U.S. regulatory agency within 5 years with 9,000 employees and a budget of $2 million per day.

World cotton production tops 50 million 50-pound bales, up from 21 million in 1920, but U.S. planters account for only 10 million, down from 13 in 1920.

The U.S. corn harvest falls off as a result of a new strain of the fungus *Helminthosporum maydis*. Since most U.S. corn is now genetically similar in its lack of resistance to the blight, much of the crop is lost, raising meat and poultry prices.

High food prices in Poland have triggered the riot of shipyard and factory workers (*see* above) that brings down the government.

The National Research Council warns expectant mothers in the United States not to restrict weight gain too severely. A Committee on Maternal Nutrition of the NRC says that at least half of infant mortalities are preventable through use of prenatal care, adequately trained personnel at delivery, correction of dietary deficiencies, proper hygiene, and health education and it says that a weight gain of 20 to 25 pounds in pregnancy is permissible.

U.S. breakfast foods come under fire July 23 from former Nixon administration hunger consultant Robert Burnett Choate, Jr., 46, who testifies before a Senate subcommittee that 40 of the top 60 dry cereals have little nutritional content. "The worst cereals are huckstered to children" on television, says Choate.

High levels of mercury are found in livers of Alaskan fur seals in the Pribilof Islands. Liver tissue samples show a mercury content of 58 parts per million and iron supplement pills made from freeze-dried seal liver are withdrawn from the market.

The FDA orders the recall of all lots of canned tunafish in which mercury levels above 0.5 parts per million have been discovered. The December recall order is said to affect nearly one quarter of all canned tuna in U.S. markets but by spring of next year it will turn out that only 3 percent of the canned tuna pack (nearly 200,000 cases) exceeds the 0.5 ppm FDA guideline.

A U.S. Occupational Safety and Health Act signed by President Nixon December 29 compromises differences between labor and management and establishes an office that will work to minimize hazards in industry.

The Poison Prevention Packaging Act passed by Congress December 30 requires that manufacturers of drugs, sulfuric acid, turpentine, and other potentially dangerous products put safety tops on their containers so that children will not be able to open them. The act will take effect in 1972 and fatalities from aspirin and other potential poisons will drop.

Birth control pills may produce blood clots, warns the Food and Drug Administration. The FDA sends letters to more than 300,000 physicians urging them to pay close attention to the risks involved in taking "the Pill" and to inform their patients.

The most liberal abortion law in the United States goes into effect July 1 in New York State. At least 147 women undergo abortions, more than 200 register at municipal hospitals for abortions bringing the application total to more than 1,200, but "Right to Life" groups continue to protest the new law.

The Italian Senate votes October 9 to legalize divorce for the first time.

The world's population reaches 3.63 billion with 1.13 billion in South Asia, 929.9 million in East Asia (including Japan), 462.1 million in Europe, 344.4 million in Africa, 283.3 million in Latin America, 227.6 million in North America, 19.4 million in Oceania.

The U.S.S.R. has only 9 urban centers with populations of 1 million or more versus 35 such urban centers in the United States. Only 8.5 percent of the Soviet population is in these large urban centers versus 41.5 percent of the U.S. population.

Soviet census figures show that Muslim populations in the Uzbek S.S.R., the Tadzhik S.S.R., and other Muslim republics have increased by roughly 50 percent since the 1959 census but that the number of Great Russians in these republics has declined. Ethnic Russians remain the largest single national group within the U.S.S.R., they are roughly three times as numerous as the Ukrainians, but they will become a minority by 1976.

Population in the Soviet Union reaches 242.6 million, in the People's Republic of China 760 million, in India 550 million, in the United States 205 million. The United States has 85 people per square mile, the P.R.C. 305, India 655, Japan 1,083.

Note: In terms of using natural and irreplaceable resources and of contaminating air and water with chemical waste, one American is comparable to at least 25 Chinese or Indians. In that light the U.S. population is by some measures effectively more than 4 billion.

1971 Uganda ousts President Obote January 25 while he is returning from a conference at Singapore. A military government headed by Major Gen. Idi Amin, 44, seizes power and begins liquidating political opponents.

South Vietnamese troops invade Laos February 13 with U.S. air and artillery support to root out North Vietnamese supply depots but are driven out within 6 weeks after taking heavy casualties. A military tribunal convicts Lt. William Calley March 31 of having killed 20 civilians at My Lai in 1968 but he is freed by President Nixon April 6 pending further review of his guilt. Capt. Ernest Medina is acquitted September 22. U.S. troop withdrawals from South Vietnam reduce the total to about 200,000, down from 534,000 in mid-1969. Australia and New Zealand withdraw their forces by fall.

Haiti's President Duvalier dies April 22 after a dictatorship of more than 13 years during which "Papa Doc's" Ton Ton Macoute have terrorized the populace. Duvalier's son Jean-Claude, 19, is sworn in as "president for life" and continues the repression.

East German Communist party leader Walter Ulbricht resigns in May at age 78 after having headed the party for 25 years. He is succeeded as first secretary by Erich Honecker, 58, whose corrupt regime will continue until 1989.

East Pakistan has widespread riots and strikes following the announcement March 1 of a delay in convening the new national assembly elected last year. The Awani League whose strength is concentrated in Bengal has won 167 of the 313 seats to be filled, the League's leader protests the delay, a strike cripples the key port of Chittagong, Bengali extremists murder many non-Bengalis, Pakistan's president Yahya Khan flies to East Pakistan to negotiate with the Awami League's leader Sheik Mujib, he denounces Sheik Mujib as a traitor and orders an invasion of Bengal March 25, the East Bengal Regiment defects to the sheik's cause, Bengal separatists blow up major bridges, railroads, and communications lines, the separatists are subdued, but guerrillas resist the Pakistani army of occupation.

India concludes a 20-year friendship pact with the Soviet Union in August to deter any Pakistani attack, Indian troops invade Bengal December 4, Pakistani forces in Bengal surrender December 16 (*see* Bangladesh, 1972).

Chinese defense minister Lin Pao, 65, leads an abortive coup against Chairman Mao Zedong with support from members of the armed forces. Officially designated to succeed Mao, Lin is a veteran of the Long March that began in 1934 and has been a member of the Chinese Politburo since 1955, but he dies September 13 in a

1971 *(cont.)* Mongolian plane crash while fleeing the retribution of Chairman Mao.

The People's Republic of China is formally seated as a UN member November 15.

Sudan elects Prime Minister Nimeiri as her first president (*see* 1969;1985).

The Republic of Zaire is created October 27. President Mobutu Sese Seko (formerly Joseph D. Mobutu) renames the Democratic Republic of Congo and has the Congo River renamed the Zaire (*see* 1966).

Wilmington, N.C., has racial violence beginning in late January with incidents of arson, dynamitings, and shootings. Black activists headed by United Church of Christ field director for racial justice Ben Chavis, 22, barricade themselves inside Mike's Grocery February 6, saying the police do not protect them from roving whites, a policeman shoots a black teenager that night and unknown assailants kill an armed middle-aged white man near the grocery store. A court early next year will convict Chavis and nine others of firebombing the store, and the ten will draw sentences of from 29 to 34 years despite international clamor for their release (*see* 1980).

The Supreme Court weakens its 1966 *Miranda* decision in a 5-to-4 decision handed down February 24. It rules that although a suspect's statement is inadmissible as evidence if the arresting police officers have not advised the suspect of his or her rights, the prosecution may still use the statement to contradict the defendant's testimony in a trial.

Conscientious objectors who seek draft exemption must show that they are opposed to all wars, not just to the Vietnam War, the Supreme Court rules March 8 in an 8 to 1 decision.

So-called "objective" criteria for hiring employees are actually discriminatory, the Supreme Court rules in a case involving a written examination for employment. The criteria are illegal, the court rules, since they result in a relative disadvantage to minorities without "compelling business interest," and future rulings will extend the concept to recruitment practices, job placement, transfers, and promotions. The 8 to 0 decision handed down March 8 in *Griggs v. Duke Power Co.* is the first major interpretation of the job bias provisions (Title VII) of the Civil Rights Act of 1964.

The Supreme Court upholds busing of schoolchildren to achieve racial balance where segregation has official sanction and where school authorities have offered no acceptable alternative to busing. The court hands down the unanimous decision April 20 in the case of *Swann v. Charlotte-Mecklenburg Board of Education* (*see* 1965).

Federal authorities impose a strict busing plan on the Austin, Tex., school system in May, a federal judge rejects the plan in late July, Alabama's Gov. Wallace says the Nixon administration has done more than any preceding administration to desegregate public schools, and President Nixon publicly repudiates the Austin plan August 3 and orders that busing be limited

Busing black children to white schools and vice versa helped achieve racial balance but raised cries of protest.

"to the minimum required by law" (the busing ordered in Charlotte, N.C., is actually less than that used to preserve segregation).

The Twenty-Sixth Amendment ratified July 1 lowers the U.S. voting age from 21 to 18.

Attica Correctional Facility at Attica, N.Y., is the scene of the bloodiest one-day encounter between Americans since the Indian massacres of the late nineteenth century. Inmates of the overcrowded prison discover inexplicable differences in sentences and parole decisions that appear to have a racial bias, outraged prisoners take over cell blocks beginning September 9 and kill several trusty guards, state police move in September 13 by order of Gov. Rockefeller, 39 inmates are killed and more than 80 wounded in the 15 minutes it takes for the police to retake the prison; the total death toll is 43.

A "New Economic Policy" announced by President Nixon August 15 imposes a 90-day freeze on U.S. wages and prices, temporarily suspends conversion of dollars into gold, and asks Congress to impose a 10 percent import surcharge in an effort to strengthen the dollar as the Vietnam War increases inflationary pressures.

West Germany, Austria, Switzerland, and the Netherlands have either revalued their currencies upward or floated them against the fixed value of the dollar, but while the European move paves the way for more flexible exchange rates the U.S. action unlinking the dollar from any gold value begins the breakdown of the international monetary system established by the Bretton Woods Conference of 1944.

Wall Street responds to President Nixon's economic message (above) with enthusiasm. The Dow Jones Industrial Average makes a record one-day leap of 32.93 points on a record volume of 31.7 million shares, but the AFL-CIO says it has "absolutely no faith in the ability of President Nixon to successfully manage the economy of this nation" and refuses to cooperate in the president's wage freeze.

Britain's Parliament votes October 28 to join the European Economic Community (*see* 1972).

U.S. imports top exports by $2.05 billion—the first trade deficit since 1888. President Nixon kills the 10 percent surcharge on imports December 20 and raises the official gold price, thus devaluing the dollar by 8.57 percent.

The average U.S. taxpayer gives the government $400 for defense, $125 to fight the war in Indochina, $40 to build highways, $30 to explore outer space, $315 for health activities ($7 for medical research).

J. C. Penney dies at New York February 12 at age 95 leaving an empire of 1,660 stores with annual sales of more than $4 billion, the largest non-food retail enterprise after Sears, Roebuck (*see* 1927).

Egypt's Aswan High Dam (Saad-el-Aali), completed in January, generates 2,100 megawatts of power. Moscow has financed the project.

The *North Central Power Study* produced by a group of U.S. utility companies with help from the Department of the Interior proposes strip-mining of coal in the northern Great Plains and constructing huge mine-mouth generating plants (21 in Montana alone), with a web of transmission lines to serve the Midwest and Pacific Coast, but environmentalists fight plans to exploit the Fort Union Coal Formation that contains an estimated 1.3 trillion tons of soft coal, most of it low-sulfur.

Rolls-Royce, Ltd., declares bankruptcy February 4, having lost huge sums developing a jet engine for a new Lockheed plane; the British government bails out Rolls-Royce to maintain national prestige and avoid loss of jobs.

The U.S. Senate votes 51 to 46 to stop all further federal funding of SST (supersonic transport) development. The March 24 vote comes 1 week after a similar vote in the House, Boeing lays off 62,000 workers at Seattle, environmentalists hail the Senate action, which leaves Britain and France free to develop their Concorde SST without U.S. competition.

Amtrak (The National Railroad Passenger Corp.) takes over virtually all U.S. passenger railroad traffic May 1 in a federally funded effort to halt the decline in rail passenger service. The takeover effectively ends service to many cities.

The Brotherhood of Locomotive Engineers agrees May 13 to let U.S. railroads scrap the age-old "divisional rule" that has called 100 miles a day's work and required freight trains to stop every 100 miles to change crews. The rule has enabled diesel locomotive engineers to earn a day's pay in 150 minutes.

The human growth hormone somatropin produced by the pituitary glands is synthesized by Chinese-American biochemist-endocrinologist Choh Hao Li at the University of California in January. The synthetic hormone has roughly 10 percent of the growth-producing properties of the natural hormone but it has great potential for treating pituitary dwarf children who have each required hormone extract obtained from the pituitary glands of 650 cadavers.

Researchers at the Anderson Tumor Institute in Texas isolate the cold-sore herpes virus from the lymph cell cancer known as Burkitts lymphoma. One form of herpes virus has been associated with the earliest stage of cervical cancer in women (*see* 1960).

Soft contact lenses win FDA approval March 18. Invented in 1962 by Czech technician Otto Wichterle, the $300 lenses are more comfortable than hard lenses, introduced in 1939, but are intended to last only a year.

Cigarette smoking has become a cause of death comparable to the great epidemic diseases typhoid and cholera in the nineteenth century, says Britain's Royal College of Physicians in a sharply worded report.

U.S. physicians receive an FDA warning in November against administering DES (diethylstilbestrol) or related hormones to pregnant women. Massachusetts General Hospital researchers have concluded that daughters of women who took the synthetic hormone to prevent miscarriage have an increased risk of clear-cell adenocarcinoma of the vagina.

The National Cancer Act signed into law by President Nixon December 24 authorizes appropriations of $1.5 billion per year by the National Cancer Institute to combat America's second leading cause of death; the massive effort will yield few practical results in the next decade.

All Things Considered debuts May 3 on U.S. Public Radio with in-depth news analyses and features.

The Electric Company debuts in October on U.S. Public Broadcasting Service TV stations. Like *Sesame Street*, it is the product of Joan Ganz Cooney's Children's Television Workshop.

The "Pentagon Papers" excerpted in the *New York Times* beginning June 13 give details of U.S. involvement in Vietnam from the end of World War II to 1968. The highly classified 3,000-page study was written at the order of former Secretary of Defense Robert S. McNamara, the *Times* has received a copy from former Defense Department official Daniel Ellsberg, 39. Attorney General John Mitchell asks the *Times* to cease further publication on grounds that the information revealed will cause "irreparable injury to the defense interests of the United States," the *Times* refuses, a federal judge orders it to stop, pending a hearing, a federal appeals court stops the *Washington Post* from publishing the material, Ellsberg admits he gave the study to the press and surrenders to federal authorities at Boston, a federal grand jury at Los Angeles indicts him on charges of stealing the secret 47-volume study showing that federal officials have consistently lied to the American people about Vietnam, Ellsberg and co-worker Anthony Russo are indicted December 29 for espionage and conspiracy.

1971 *(cont.)* A British postal strike halts mail deliveries for 47 days.

U.S. first-class postal rates rise to 8¢ per ounce May 19 (*see* 1968; 1974). The U.S. Postal Service replaces the Post Office Department July 1 and optimistic speeches hail the semi-independent corporation.

Look magazine ceases publication October 19 after 34 years; other magazines shrink their formats to save postage in accordance with new postal regulations.

Nonfiction *Bury My Heart at Wounded Knee, An Indian History of the American West* by Illinois historian Dee Brown.

Fiction *Rabbit Redux* by John Updike; *Maurice* by the late E. M. Forster; *The Tenants* by Bernard Malamud; *The Book of Daniel* by Sarah Lawrence College teacher E. L. (Edgar Lawrence) Doctorow, 39, is based on the 1951 Rosenberg trial; *St. Urbain's Horseman* by Mordecai Richler; *Love in the Ruins* by Walker Percy; *Grendel* by U.S. novelist John Gardner, 38; *The Wanderers* by South African novelist Es'kia (Ezekiel) Mphahlele, 51; *Birds of America* by Mary McCarthy; *Group Portrait with Lady* by Heinrich Böll; *The Decay of the Angel* and *The Temple at Dawn* by the late Yukio Mishima; *Winds of War* by Herman Wouk; *Books Do Furnish a Room* by Anthony Powell; *The Scarlatti Inheritance* by U.S. novelist Robert Ludlum, 44.

Juvenile *The Slightly Irregular Fire Engine* by Donald Barthelme.

Painting *Amityville* by Willem de Kooning; *Shoot* (artist shot in left arm by friend) by U.S. artist Chris Burden, 25.

Theater *The House of Blue Leaves* by U.S. playwright John Guare, 33, 2/10 at New York's off-Broadway Truck and Warehouse Theater, 337 perfs.; *And Miss Reardon Drinks a Little* by Paul Zindel 2/25 at New York's Morosco Theater, with Estelle Parsons, Nancy Marchand, Julie Harris, 108 perfs.; *Butley* by English playwright Simon Gray, 35, 7/5 at London's Criterion Theatre, with Alan Bates; *Where Has Tommy Flowers Gone?* by U.S. playwright Terrence McNally, 31, 10/7 at New York's off-Broadway Eastside Playhouse, 78 perfs.; *The Changing Room* by David Storey, 11/9 at London's Royal Court Theatre; *The Prisoner of Second Avenue* by Neil Simon 11/11 at New York's Eugene O'Neill Theater, with Peter Falk, Lee Grant, Vincent Gardenia, 780 perfs.

Television *Masterpiece Theater* (initially, *The First Churchills*) 1/10 on U.S. Public Television with English-American journalist Alistair Cooke, 62, as host; *All in the Family* 1/12 on CBS with Carroll O'Connor as Archie Bunker in a series devised by Norman Lear, 48. Co-starring Jean Stapleton, Sally Struthers, and Rob Reiner, the show violates sacrosanct taboos against ethnic and bathroom humor; *Columbo* 9/15 on NBC with Peter Falk (to 9/1/78).

Films Woody Allen's *Bananas* with Allen, Louise Lasser; William Friedkin's *The French Connection* with Gene Hackman, Fernando Rey; Peter Bogdanovich's *The Last Picture Show* with Timothy Bottoms, Jeff Bridges, Ben Johnson, Cloris Leachman, Ellen Burstyn, Cybill Shepherd, Eileen Brennan. Also: Eric Rohmer's *Claire's Knee* with Jean-Claude Brialy, Aurora Cornu; Stanley Kubrick's *A Clockwork Orange* with Malcolm McDowell, Patrick Magee; Federico Fellini's *The Clowns;* Bernardo Bertolucci's *The Conformist* with Jean-Louis Trintignant, Stefania Sandrelli, Dominique Sanda; Luchino Visconti's *Death in Venice* with Dirk Bogarde; Arthur Hiller's *The Hospital* with George C. Scott, Diana Rigg; Peter Brook's *King Lear* with Paul Scofield, Irene Worth; Alan J. Pakula's *Klute* with Jane Fonda as a call girl, Donald Sutherland; Jack Lemmon's *Kotch* with Walter Matthau; Roman Polanski's *Macbeth* with Jon Finch, Francesca Annis; Robert Altman's *McCabe and Mrs. Miller* with Warren Beatty, Julie Christie; Louis Malle's *Murmur of the Heart* with Lea Massari; Paul Bogart's *Skin Game* with James Garner, Louis Gossett; Miloš Forman's *Taking Off* with Lynn Carlin, Buck Henry, Linnea Heacock; Monte Hellman's *Two-Lane Blacktop* with James Taylor, Warren Oates; Barbara Loden's *Wanda* with Loden, Michael Higgins; Blake Edwards's *Wild Rovers* with William Holden, Ryan O'Neal, Karl Malden, Lynn Carlin.

Hollywood musical Norman Jewison's *Fiddler on the Roof* with Chaim Topol as Tevye, violin on sound track by Isaac Stern.

Broadway and off-Broadway musicals *Follies* 4/14 at the Winter Garden Theater, with Gene Nelson, Yvonne De Carlo, Alexis Smith, Dorothy Collins, music and lyrics by Stephen Sondheim, songs that include "Broadway Baby," 521 perfs.; *Godspell* 5/17 at the Cherry Lane Theater, with music and lyrics by Stephen Schwartz, songs that include "Day by Day," 2,605 perfs. counting Promenade and Broadhurst Theaters; *Jesus Christ Superstar* 10/10 at the Mark Hellinger Theater, with Jeff Fenholt, music by Andrew Lloyd Webber, 23, lyrics by Tim Rice, book by Tom O'Horgan, 711 perfs.; *Ain't Supposed to Die a Natural Death* 10/20 at the Ethel Barrymore Theater, with music and lyrics by Melvin Van Peebles, 325 perfs.; *Two Gentlemen of Verona* 12/1 at the St. James Theater (after 14 performances at the off-Broadway Public Theater), with Raul Julia, Clifton Davis, Diana Davila, music by Galt MacDermot, lyrics by John Guare, 627 perfs.

Opera *Owen Wingrave* 5/16 at Aldwych, with music by Benjamin Britten, libretto from an 1892 pacifist short story by Henry James.

Washington's John F. Kennedy Center for the Performing Arts opens September 8 with a performance of Leonard Bernstein's new Mass. When complete, the center will seat 6,100 in three halls.

Popular songs *Tapestry* (album) by U.S. singer-songwriter Carole King, 29, includes her songs "It's Too Late" and "You've Got a Friend"; "Friends with You," "Take Me Home, Country Roads," and "Poems, Prayers, and Promises" by John Denver; *Blue* (album) by Joni Mitchell; "If Not for You" by Australian vocalist Olivia Newton-John, 21; *Sticky Fingers* (album) by The

Rolling Stones; *Imagine* (album) by John Lennon; *Runes* (album) by Led Zeppelin; *Who's Next* (album) by The Who; "Shaft" theme by Isaac Hayes (for Gordon Parks's film); *Tumbleweed Connection* (album) by Elton John; "Lonely Days" by The Bee Gees; The Doors' lead singer Jim Morrison dies of a probable drug overdose at Paris July 3 at age 27.

Baltimore beats Dallas 16 to 3 January 17 at Miami in Super Bowl V.

New York's Off-Track Betting Corp. (OTB) is created April 8 as legislation takes effect legalizing off-track betting in an effort to take business away from organized crime and produce revenues for the city and state. But the "numbers game" based on smaller bets will continue to flourish, fattening the profits of underworld syndicates.

John Newcombe wins in men's singles at Wimbledon, Evonne Goolagong, 21 (Australia), in women's singles; Stanley Roger "Stan" Smith, 24 (U.S.), wins in men's singles at Forest Hills, Mrs. King in women's singles.

The Pittsburgh Pirates win the World Series by defeating the Baltimore Orioles 4 games to 3.

Washington, D.C., loses its major league baseball team at year's end. The Senators become the Texas Rangers.

U.S. cigarette sales reach 547.2 billion despite the new ban on radio and television advertising.

New York hoodlum Joseph Gallo recruits blacks to replace the depleted ranks of the Mafia. Gallo has rebelled from the leadership of Joseph Colombo, who helps organize an Italian-American Unity Day parade June 28 at New York's Columbus Circle, is shot three times by a black gunman and left crippled; an unidentified gunman immediately kills Colombo's assailant, but there is disagreement as to whether the assailant was in the pay of Joey Gallo, reputed Mafia leader Carlo Gambino, or someone else.

Chicago's Standard Oil of Indiana building is completed. The 1,107-foot structure is taller than the 4-year-old John Hancock Center.

An earthquake measuring 6.6 on the Richter scale rocks Los Angeles February 10, killing 51, injuring 880.

The U.S. Geological Survey reports that cadmium levels in drinking water are "dangerous" at Birmingham and Huntsville and above acceptable limits at Hot Springs, Little Rock, East St. Louis, Shreveport, Cape Girardeau, St. Joseph, Wilkes-Barre, Scranton, and Pottsville.

The FDA advises Americans to stop eating swordfish. The advisory issued in early May follows a study of more than 853 swordfish samples of which 811 had an average mercury content of 1 part per million and 8 percent had levels above 1.5 parts per million. Swordfish caught between 1878 and 1909 turn out to have about the same mercury content, and few people can afford to eat enough swordfish to make the mercury content a real danger (average per capita consumption is 2 ounces per year).

The Chicago Union Stock Yards that opened Christmas Day 1865 close July 30 as meat packers continue to move their slaughterhouses closer to their sources of supply. Operations at the Stock Yards peaked in 1924 and have been declining ever since.

Annual U.S. beef consumption reaches 113 pounds per capita, up from 85.1 pounds in 1960. Consumption will peak at 128.5 pounds in 1976 as Americans eat more than 50 billion hamburgers, paying more than $25 billion for beef in various forms.

The hormone LHRH isolated by Polish-American biochemist Andrew Schally, 44, is essential to human ovulation.

1972 President Nixon arrives at Beijing February 20 and confers with Chairman Mao Zedong and Premier Zhou En-lai, ending the U.S. hostility toward the People's Republic of China that has persisted since 1949. The president has 6 days earlier ordered that U.S. trade with Beijing be on the same basis as trade with Moscow and Soviet-bloc nations.

Beijing's UN ambassador lays claim March 3 to Hong Kong and Macao and reasserts claims to the uninhabited Senkalu Islands claimed also by Japan; a P.R.C. UN delegate charges Japan with expansionism.

Premier Zhou En-lai meets with North Vietnam's premier Pham Van Dong to assure him that China has made no secret commitments to President Nixon with regard to Indochina, where hostilities continue.

U.S. planes bomb Haiphong and Hanoi April 16. The B-52 raids are the first on the big North Vietnamese cities since 1968 and heated debate on the resumption of bombing begins in the U.S. Senate April 19 as North Vietnamese MIG fighters attack U.S. destroyers, which are shelling coastal positions.

The provincial capital Quang Tri falls to the North Vietnamese May 1, 80 U.S. advisers are evacuated, South Vietnam's Third Division flees to the South, Saigon relieves the commander of the division who later claims to have resigned, more than 150,000 flee the imperial capital of Hue as deserters loot the city, engage in gunfights among themselves, and set fire to the marketplace, President Thieu visits Hue and gives military police authority to shoot looters and arsonists, Communist forces surround Kontum and Pleiku, Saigon fails in efforts to reopen the supply route between the two beleaguered cities, and U.S. planes mine the approaches to Haiphong while intensifying raids on Communist transport lines.

Chinese, Soviet, North Vietnamese, and Mongolian representatives meet at Beijing May 19 to speed aid to North Vietnam following the U.S. action in mining Haiphong and other North Vietnamese ports.

President Nixon arrives at Moscow May 22 and confers with Party Secretary Leonid Brezhnev in the first visit of a U.S. president to the Soviet Union since 1945. Nixon also visits Teheran and Warsaw.

Okinawa is returned to Japan after 27 years of U.S. occupation.

Détente eased East-West political tensions, facilitating increased trade between Communist and capitalist nations.

Philippine president Ferdinand Marcos declares martial law in response to an alleged "Communist rebellion" and assumes near-dictatorial powers.

Ceylon becomes a republic and changes her name to Sri Lanka May 22.

Bangladesh (formerly East Pakistan) proclaims herself a sovereign state with Sheik Mujibur Rahman as prime minister (*see* 1971).

Denmark's Frederick IX dies January 14 at age 72 after a 25-year reign. He is succeeded by his daughter, 31, who will reign as Margrethe II.

Britain imposes direct rule over Northern Ireland March 30 after years of violence between Catholics and Protestants: 467 Northern Irish are killed in the course of the year (*see* Bloody Sunday, below).

Gunmen hired by Palestinian guerrillas shoot up Lod Airport near Tel Aviv May 30 killing 24 and wounding 76. Two of the gunmen are Japanese, two are killed by security guards, an Israeli court convicts Kozo Okamoto and sentences him July 17 to life imprisonment.

A U.S. Federal Election Campaign Act signed by President Nixon February 7 limits campaign spending in the media to 10¢ per person of voting age in the candidate's constituency and requires that all campaign contributions be reported. Both parties but especially the Republicans receive millions in contributions before the new law takes effect April 7.

A "confidential" memorandum released to newspapers February 29 by columnist Jack Anderson links a Justice Department settlement favoring ITT in pending antitrust suits to an ITT commitment to supply funds for the Republican National Convention to be held at San Diego.

Gov. Wallace of Alabama campaigns for the Democratic presidential nomination but is shot May 15 by would-be assassin Arthur H. Bremer, 22, at Laurel, Md., while addressing a crowd before a forthcoming primary election. Wallace will be a paraplegic for life.

The Watergate affair that will grow into the greatest constitutional crisis in U.S. history has its beginnings at 2 in the morning June 17 when District of Columbia police alerted by security guard Frank Wills, 24, arrest five men inside Democratic party national headquarters in Washington's new Watergate apartment complex. Bernard L. Barker, 55, James W. McCord, 42, Eugenio R. Martinez, 48, Frank A. Sturgis (Fiorini), 47, and Virgilio R. Gonzalez, 46, are seized with cameras and electronic surveillance equipment, President Nixon's campaign manager John Mitchell states June 18 that they were not "operating either on our behalf or with our consent," but Nixon's office confirms June 19 that Barker met earlier in June with CIA official E. Howard Hunt, 53, who until March 29 had been acting as consultant to presidential counsel Charles W. Colson, 38. Nixon tells Colson June 20 that he is involved in a "dangerous job." Hunt has earlier directed CIA activities against Cuban prime minister Fidel Castro, and three of the men arrested are Cubans, but the motive for their break-in and the source of their support remains a mystery.

Sen. George S. McGovern, 49 (S.D.), receives the Democratic nomination for the presidency at Miami Beach (Republican operatives have aborted the campaign of Sen. Edmund S. Muskie, 58 (Me.), who had led Nixon in opinion polls). McGovern selects Sen. Thomas F. Eagleton, 42 (Mo.), as his running mate but switches to former Peace Corps director R. Sargent Shriver, Jr., 56, when Eagleton turns out to have been treated for manic depression. "Peace is at hand" in Vietnam, says Henry Kissinger on the eve of election, and President Nixon wins reelection despite gossip about the Watergate break-in (above). McGovern carries only Massachusetts with its 17 electoral votes, Nixon receives 47 million votes, 521 electoral votes, to 29 million for McGovern in the most one-sided presidential election since 1936.

"Bloody Sunday" January 30 in Northern Ireland sees 13 Roman Catholics shot dead by British troops at Londonderry where riots have followed a civil rights march conducted in defiance of a government ban. The Irish Republican Army calls a general strike January 31 to protest the shootings, and an estimated 25,000 demonstrators rally in protest at Dublin February 2, destroying the British Embassy by fire (*see* direct rule, above).

The death penalty as now administered in the United States constitutes "cruel and unusual punishment" and is therefore unconstitutional, the Supreme Court rules June 29 in a 5 to 4 decision (but *see* 1976).

President Nixon signs a bill January 5 authorizing a $5.5 billion 6-year program to develop a space-shuttle craft that will lift off as a rocket and return to earth as an airplane.

The unmanned U.S. spacecraft Pioneer 10 lifts off from Cape Kennedy March 2 on a 639-day, 620 million-mile journey past the planet Jupiter.

The Soviet space craft Venus 8 makes a soft landing on the planet Venus in March.

U.S. Apollo 16 astronauts Charles M. Duke, Jr., Thomas K. Mattingly, and John W. Young blast off April 16 from Cape Kennedy on a flight of nearly 255 hours. Young and Duke spend a record 71 hours, 2 minutes on the surface of the moon beginning April 20 and return with 214 pounds of lunar soil and rock.

The European Economic Community (Common Market) created in 1957 by the Treaty of Rome accepts Britain, Ireland, Denmark, and Norway to membership with the Treaty of Brussels signed January 22. The House of Commons acts July 14 to permit British participation and the EEC expands to embrace 20 nations with 257 million people in the world's most powerful trading block, but Belgian premier Paul Henri Spaak who helped to found the EEC dies July 31 at age 73 and Norway votes to remain outside the EEC.

British coal miners walk out in January for the first time in nearly 50 years, demanding an 11 percent raise in their basic wage which remains below $50 per week. The strike continues for 7 weeks before the 280,000 miners accept a settlement.

London area rail workers strike in April in an effort to bring their wages to an average of $78 per week, the strike spreads a week later to stop rail traffic throughout the country, the workers agree to a 14-day cooling off period after 4 days but wildcat strikes continue to slow service.

Britain's richest 10 percent holds 51 percent of the nation's wealth, down from 83 percent in 1960. After-tax income commanded by the richest 10 percent falls to 23.6 percent, down from 34.6 percent in 1939. The richest 1 percent still holds one-quarter of the nation's wealth despite high inheritance taxes.

The strike of London dock workers and others (*see* above) idles 500 to 600 ships, preventing export goods from leaving the country.

U.S. wages, prices, and profits remain controlled by Phase II economic measures.

Chile's president Salvador Allende Gossens continues to nationalize his country's large industrial firms.

The Star of Sierra Leone discovered February 14 weighs 969.8 carats, making it the third largest gem-quality diamond ever found.

The United States enters the international money market July 19, selling German marks at decreasing prices in the first move since August of last year to shore up the dollar.

Wall Street's Dow Jones Industrial Average closes at 1003.16 November 14, up 6.00 to close above the 1,000 mark for the first time in history; it climbs to 1036.27 December 11 but falls sharply December 18 at news of a breakdown in Vietnam peace talks.

An international uranium cartel organized in the spring has Canadian, Australian, and South African sponsorship with tacit support from France. Members will try to eliminate outside competition, and the price of raw uranium ("yellowcake") will jump from $6 a pound to $41 in the next few years, but rising oil prices imposed by the Organization of Petroleum Exporting Countries (OPEC), not any actions by the cartel, will be largely responsible for the sudden boom in demand for nuclear fuels and the climb in uranium prices.

Mexican prospectors strike oil outside Villahermosa in May. The new Chiapas-Tabasco field will prove the largest in the Western Hemisphere. Mexico's proven reserves will total 16.8 billion barrels by 1978, and the potential will be estimated at 120 billion.

Baghdad nationalizes Iraq Petroleum's Kirkuk field June 1 (*see* 1927). The field produces 1.1 million barrels of crude oil per day, it is the first key field to be seized by an Arab Persian Gulf nation, Iraq Petroleum threatens action against anyone who buys the oil, and by year's end Iraq is producing only 660,000 barrels per day (Baghdad leaves Basrah Petroleum untouched), down from 1.9 million at the beginning of the year (*see* 1973).

Nearly 30 percent of U.S. petroleum is imported, up from 20 percent in 1967.

Japan's San-Yo Shin Kansen (New San-Yo Line) railroad opens in April to link Osaka with Okayama 103 miles to the south. Built to even higher standards than the 8-year-old Tokaido line, the new line can handle speeds of 155 miles per hour and is the first stage of a new expansion that will extend the wide-track railroad to Kyushu.

Amtrak increases Metroliner service between New York and Washington to 14 daily round trips, up from 6 in 1969. (Eastern Airlines operates 16 daily shuttle flights between the two cities.)

New York City transit fares rise January 5 to 35¢, up from 30¢ (*see* 1970; 1975).

The San Francisco Bay Area Rapid Transit System (BART) that goes into service September 11 is the first new U.S. regional transit system in more than 50 years, but mechanical problems plague the sleek aluminum-bodied trains from the start. The automated system has been completed 3 years behind schedule and has cost $1.6 billion instead of the projected $120 million. Its 71 miles of track link San Francisco, Oakland, and Berkeley with subway and elevated lines that extend to a score of smaller cities to the south and east, but some trains stall beneath the bay and accidents occur.

New Jersey voters reject a $650 million transportation bond issue in November. Environmentalists have objected that too much of the money was earmarked for highways, too little for mass transit.

The Globtik Tokyo launched in Japan is a $56 million 476,025-ton supertanker that dwarfs the 366,813-ton *Nisseki Maru* launched last year, raising the possibility of a "megaton tanker" with a capacity of a million tons.

1972 ***(cont.)*** Indian entrepreneur Ravi Tikkoo, 40, founded Globtik Tankers with $2,500 in 1967 after 3 years of working at London as a middleman between ship owners and bankers.

British Airways (formerly BOAC) orders five Concorde supersonic jets July 18; Air France orders four *(see* 1970).

A human skull found in northern Kenya by Richard Leakey and Glynn Isaac allegedly dates the first humans to 2.5 million B.C. and opens a new controversy on the age of man. Leakey's father Louis S. B. Leakey dies October 1 at age 69 *(see* 1959).

A new Surgeon General's Report on smoking issued January 10 warns that nonsmokers exposed to cigarette smoke may suffer health hazards *(see* 1964). Later evidence will show that carbon monoxide and other toxins in "sidestream" smoke actually present greater perils to nonsmokers than to smokers, bringing new pressure to protect airline passengers, office workers, and others from the minority whose cigarette smoke threatens their health and comfort.

The papal encyclical *Pacem in Terris* issued April 9 appeals to all Christians of good will, not just Roman Catholics, but the Greek Orthodox archbishop rejects the bid for unity, saying his followers will never accept papal infallibility.

Federal Express is founded at Memphis by local millionaire Vietnam veteran Frederick W. Smith, 27, whose father built the Greyhound bus system in the South. Smith has raised $72 million in venture capital—the largest such capital assemblage yet—to start an overnight delivery service with its own aircraft (14 French-built Falcon jets) and fleet of trucks. On its first night of operation next year it will deliver 16 packages. By 1981, when it adds letter delivery, the company will be handling 100,000 parcels and letters nightly, a number that will grow to 1 million by 1989.

Washington Post reporters Bob Woodward, 29, and Carl Bernstein, 28, begin to crack open the Watergate affair (above), working 12 to 18 hours per day 7 days a week. On August 1 they report a financial link between the Watergate break-in and the Committee for the Re-Election of the President (CREEP); on September 16 they report that campaign finance chairman Maurice H. Stans, 64, and CREEP aides control a "secret fund;" on September 17 they report withdrawals from the fund by CREEP executive Jeb Stuart Magruder, 37, and his aide Herbert L. Porter, on September 29 they report that former Attorney General John N. Mitchell actually controls the fund, and their story October 10 begins, "F.B.I. agents have established that the Watergate bugging incident stemmed from a massive campaign of political spying and sabotage conducted on behalf of President Nixon's reelection and directed by officials of the White House and the Committee for the Re-Election of the President." White House spokesmen denounce the *Post* stories as "shabby journalism," "mud-slinging," "unfounded and unsubstantiated allegations," and "a political effort by the *Washington Post*, well conceived and coordinated, to discredit this administration and individuals in it." The Federal Communications Commission receives three license challenges in the next 3 months against Florida TV stations owned by the *Post*.

Most other newspapers, magazines, and television networks give short shrift to the Watergate break-in story, dismissing it as a "caper." The White House continues the intimidation of the press begun in 1969, but President Nixon has the support of 753 U.S. dailies in his bid for reelection; only 56 endorse McGovern.

Ms. magazine begins publication in July with former *Look* editor Patricia Carbine as publisher, feminist Gloria Steinem, 38, as editor.

Life magazine suspends weekly publication December 29 after 36 years; more magazines shrink their formats to conform with new postal regulations.

Nonfiction *The Foxfire Book* by Georgia schoolteacher Eliot Wigginston, 28, who has enlisted the help of his students at the Raburn Gap-Nachoche School to record with camera and tape recorder the traditions, crafts, and folklore of Appalachia; *The Limits to Growth, A Report for the Club of Rome's Project on the Predicament of Mankind* by a team of MIT scientists using computer techniques developed by MIT systems engineer Jay Forrester.

Fiction *August 1914* by Aleksandr Solzhenitsyn; *Invisible Cities* by Italo Calvino; *Manticore* by Robertson Davies; *The Sunlight Dialogues* by John Gardner; *Mumbo Jumbo* by U.S. novelist Ishmael Reed, 34; *The Optimist's Daughter* by Eudora Welty; *The Needle's Eye* by English novelist Margaret Drabble, 33; *The Chant of Jimmie Blacksmith* by Australian novelist Thomas Keneally, 37; *Watership Down* by English novelist Richard George Adams; *The Friends of Eddie Coyle* by Boston journalist-novelist George V. Higgins, 33; *The Terminal Man* by Michael Crichton; *The Exorcist* by U.S. novelist William P. Blatty, 44; *The Day of the Jackal* by English novelist Frederick Forsyth, 34.

Painting *Mao* (silkscreen and paint on canvas) by Andy Warhol.

Sculpture *Seedbed* by Italian artist Vito Acconci, 32. Joseph Cornell dies at his Flushing, Queens, home December 29 at age 69.

The Polaroid SX-70 system unveiled in April produces a color print that develops outside the camera while the photographer watches.

Theater *Jumpers* by Tom Stoppard 2/2 at London's National Theatre (Old Vic), with Michael Hordern, Diana Rigg; *Moonchildren* by U.S. playwright Michael Weller, 29, 2/21 at New York's Royale Theater, with Kevin Conway, 16 perfs.; *Sticks and Bones* by David Rabe 3/1 at New York's John Golden Theater (after 121 perfs. at the off-Broadway Public Theater), with Tom Aldridge, Elizabeth Wilson, 366 perfs. (total); *Small Craft Warnings* by Tennessee Williams 4/2 at New York's off-Broadway Truck and Warehouse Theater, 200 perfs.; *That Championship Season* by U.S. playwright Jason Miller, 33, 9/14 at New York's Booth Theater (after 144 perfs. at the off-Broadway Public Theater), with Michael McGuire, Walter McGinn, Richard Dysart, 844 perfs. (total); *6 Rooms Riv Vu* by U.S. playwright Bob Randall (Stanley B. Goldstein), 35, 10/17 at New York's Helen Hayes Theater, with Jerry Orbach, Jane Alexander, 247 perfs.; *The River Niger* by U.S. playwright Joseph A. Walker, 37, 12/5 at New York's St. Marks Playhouse (the play will move uptown next spring; 400 perfs. total); *The Sunshine Boys* by Neil Simon 12/20 at New York's Broadhurst Theater, with Jack Albertson, Sam Levene, Lewis J. Stadlen, 538 perfs.

Television *The Waltons* 9/13 on CBS with Richard Thomas, Ralph Waite, Michael Learned, Will Geer, and Ellen Corby (to 8/13/1981); *M*A*S*H* 9/17 on NBC stations with Alan Alda, Wayne Rogers, Loretta Swit, Gary Burghoff, Larry Linville, and McLean Stevenson (to 2/28/1983, with Mike Farrell replacing Rogers, Harry Morgan replacing Stevenson, David Ogden Stiers replacing Linville).

Films John Boorman's *Deliverance* with Jon Voight, Burt Reynolds, Ned Beatty, author James Dickey; Luis Buñuel's *The Discreet Charm of the Bourgeoisie* with Fernando Rey, Delphine Seyrig, Stephane Audran, Bulle Ogler; Peter Medak's *The Ruling Class* with Peter O'Toole, Alastair Sim; Joseph L. Mankiewicz's *Sleuth* with Laurence Olivier, Michael Caine; Martin Ritt's *Sounder* with Cicely Tyson, Paul Winfield, Kevin Hooks. Also: Billy Wilder's *Avanti!* with Jack Lemmon; Peter Medak's *A Day in the Death of Joe Egg* with Alan Bates, Janet Suzman; Alfred Hitchcock's *Frenzy* with Jon Finch, Barry Foster, Barbara Leigh-Hunt; Francis Ford Coppola's *The Godfather* with Marlon Brando, Al Pacino, James Caan, Diane Keaton; Cliff Robertson's *J. W. Coop* with Robertson, Geraldine Page; Bob Rafelson's *The King of Marvin Gardens* with Jack Nicholson, Bruce Dern, Ellen Burstyn; Jacques Demy's *The Pied Piper* with Donovan, Donald Pleasence; Herbert Ross's *Play It Again, Sam* with Woody Allen, Diane Keaton; Sam Peckinpah's *Straw Dogs* with Dustin Hoffman; Joseph Anthony's *Tomorrow* with Robert Duvall, Olga Bellin; François Truffaut's *Two English Girls* with Jean-Pierre Leaud; Robert Aldrich's *Ulzana's Raid* with Burt Lancaster.

Hollywood musical Bob Fosse's *Cabaret* with Liza Minnelli, Michael York, Joel Grey, Helmut Griem.

Broadway musicals *Sugar* 4/9 at the Majestic Theater, with Robert Morse, Cyril Ritchard, music by Jule Styne, lyrics by Bob Merrill, book from the 1959 film *Some Like It Hot*, 505 perfs.; *Don't Bother Me, I Can't Cope* 4/19 at the Playhouse, with Micki Grant, music and lyrics based on ballads, calypso songs, Gospel music, 1,065 perfs.; *Grease* 6/7 at the Broadhurst Theater (after 128 perfs. at the off-Broadway Martin Eden Theater), with music and lyrics by Jim Jacobs, book by Warren Casey, songs that include "Look at Me, I'm Sandra Dee," "We Go Together," "Alone at a Drive-In Movie," "Shakin' at the High School Hop," 3,388 perfs.; *Pippin* 10/23 at the Imperial Theater, with John Rubinstein, Ben Vereen, music and lyrics by Stephen Schwartz, 1,944 perfs.

Popular songs "American Pie" by U.S. singer-songwriter-guitarist Don McLean, 26; "The First Time Ever I Saw Your Face" by Ewan MacColl; *You Don't Mess Around with Jim* (album) by U.S. singer-songwriter-guitarist Jim Croce, 30; "Operator" by Jim Croce; *Honky Chateau* (album) by Elton John.

Dallas beats Miami 24 to 3 at Miami January 16 in Super Bowl VI.

Stan Smith wins in men's singles at Wimbledon, Mrs. King in women's singles; Ilie Nastase, 26 (Romania), wins in men's singles at Forest Hills, Mrs. King in women's singles.

Bobby Fischer becomes the first American to win the world chess title. He defeats Soviet grandmaster Boris Spassky 12.5 games to 8.5 at Reykjavik, Iceland, winning a record purse of $250,000; the event focuses unprecedented world attention on chess.

The Olympic Games at Sapporo, Japan, and Munich attract a record 8,512 athletes from a record 121 countries. Soviet athletes win the most gold medals, U.S. swimmer Mark Spitz, 22, wins a record seven gold medals, but the games at Munich are marred by the murder of 11 Israeli athletes at the hands of Palestine Liberation Organization terrorists organized by Mohammed Daoud Mohammed Auda, 35.

The Oakland Athletics win the World Series by defeating the Cincinnati Reds 4 games to 2.

Nike Inc. is founded under the name Blue Ribbon Sports by Portland (Ore.) entrepreneur Philip H. Knight, 33, and his former Oregon University track coach William J. Bowerman, who have been importing Japanese-made running shoes since 1964. Bowerman will develop a "waffle sole" in 1975 and add padding, they will have their shoes made in Korea and Taiwan, and by 1990 Nike will be the world's largest sneaker maker, overtaking West Germany's Adidas and making Knight a billionaire.

Reputed Mafia leader Joseph "Crazy Joe" Gallo is gunned down April 7 during a birthday party at Umberto's Clam House in New York's "Little Italy." Six other men with alleged gangland connections meet with violent deaths in 11 days.

Reputed Mafia leader Thomas "Tommy Ryan" Eboli is found dead July 10 on a Brooklyn, N.Y. sidewalk. He has been shot five times in the head.

New York's World Trade Center opens its first offices in one of two 110-story towers designed by Troy, Mich., architect Minoru Yamasaki, 58. Soaring 100 feet higher than the 1,250-foot Empire State Building of 1931, the Trade Center will remain the world's tallest building only briefly.

San Francisco's pyramid-shaped Transamerica Corp. building is completed by William L. Pereira Associates. The 48-story tower dominates the city's skyline.

1972 ***(cont.)*** Congress passes a Water Pollution Control Act, setting a 1977 deadline for the installation of the "best practicable" pollution control equipment for treating fluid waste discharges and sets a 1983 deadline for the installation of the "best available" equipment.

Congress appropriates $18 billion for assistance grants to help municipalities build sewage treatment facilities but the massive correction action has the handicap of statutory ambiguities that must be resolved in the courts.

An earthquake rocks Iran April 10 leveling 45 villages within a 250-mile radius and killing 5,000.

Tropical storm Agnes strikes the eastern United States from June 10 to June 20, creating what the U.S. Army Corps of Engineers calls the worst natural disaster in U.S. history. Parts of Florida, Virginia, Maryland, Pennsylvania, and New York are declared disaster areas as rivers overflow their banks, crippling transportation, destroying crops, buildings, bridges, and roads, isolating communities, forcing thousands to flee their homes, and killing 134.

The Susquehanna River breaks through its dikes June 23, flooding the Wyoming Valley and creating havoc in the Wilkes-Barre area.

A Nicaragua earthquake December 22 shatters Managua, killing thousands. The government cuts off food supplies to the city December 25 in an effort to force survivors to leave before decaying bodies beneath the rubble can produce an epidemic.

The worst drought since 1963 withers Soviet and Chinese grain crops, forcing Moscow to buy American grain or reduce livestock herds, thus risking the political consequences of reduced meat supplies (*see* Poland, 1970). U.S. Secretary of Agriculture Earl L. Butz has visited the Crimea in April and has suggested to the Soviet minister of agriculture Vladimir V. Matskevich that the Ukraine's rich black soil be irrigated but the Soviet budget has allocated few rubles for agricultural progress.

Soviet grain buyers arrive in the United States in late June, find U.S. wheat for July delivery selling at $1.40 per bushel, and place orders. Continental Grain receives assurance from an assistant U.S. secretary of agriculture that the USDA will pay export subsidies to maintain the export price of U.S. wheat at $1.63 per bushel ($60 per metric ton), Moscow's buyers begin signing contracts, first with Continental, then with Cargill and others. In 6 weeks they buy a million tons of soybeans, several million tons of corn, and 20 million tons of wheat—more than half of it U.S. wheat, one quarter of the entire U.S. wheat crop.

Black sigatoka fungus is found in Honduran banana fields and begins to spread throughout Central America, attacking leaves and preventing photosynthesis. By 1979 it will be in Africa, threatening famine.

A Massachusetts law prohibiting sale or dispensing of contraceptives to single persons is unconstitutional, the Supreme Court rules March 22.

The National Center for Health Statistics reports May 23 that the U.S. birth rate has fallen to 15.8 per 1,000—the lowest since the survey began in 1917.

1973 A ceasefire in Vietnam January 28 ends direct involvement of U.S. ground troops in Indochinese hostilities. America's combat death toll has reached 45,958, the last U.S. troops leave South Vietnam March 29, but U.S. bombing of Cambodia continues as prisoners of war are repatriated and President Nixon vetoes a Senate measure that would halt the bombing (*see* 1975).

Former CIA employee James W. McCord, Jr., 48, implicates Republican party officials in last year's Watergate break-in and pleads guilty with four other defendants before Justice John W. Sirica, 70. President Nixon announces "major developments" in the Watergate case April 17, his aides H. R. Haldeman, 46, and John Erlichman, 48, resign under pressure, former White House legal aid John W. Dean III, 34, implicates other Nixon intimates in testimony before a Senate investigating committee chaired by Sen. Samuel J. Ervin, Jr., 76 (D. N.C.).

President Nixon names Harvard law professor Archibald Cox, 60, special Watergate prosecutor (but *see* below).

Vice President Agnew resigns under pressure October 10, pleading no contest to charges of income tax evasion in connection with money received during his tenure as governor of Maryland. President Nixon names House minority leader Gerald R. Ford, 60, vice president and Ford takes office December 3.

Special Watergate prosecutor Cox is discharged the night of October 20 when he insists that the president turn over tape recordings of conversations with his aides relevant to the Watergate break-in. Attorney General Elliot L. Richardson, 53, who has succeeded John Mitchell in that post resigns in protest, President Nixon names Leon Jaworski, 68, to succeed Cox as Watergate prosecutor. The White House releases tapes of the president's conversations in response to a subpoena, but some of the key tapes contain gaps and the White House claims that some missing tapes do not exist.

The Bahamas gain full independence July 10 after 256 years as a British crown colony.

Argentina has a rash of assassinations and kidnappings early in the year; Perònist Hector J. Campora wins election to the presidency in March, he takes office May 25, and former dictator Juan Perón, now 77, returns June 20 after nearly 18 years in exile.

Chile has a wave of strikes that shut down shops and transportation in nine provinces, a violent coup September 11 overthrows Chile's Marxist president Salvador Allende Gossens who reportedly takes his own life. A military junta names Gen. Augusto Pinochet Ugarte, 57, president; he breaks off relations with Cuba, vows to "exterminate Marxism" from Chile, and begins a repressive 16-year rule.

Spain's dictator Francisco Franco, now 80, names Admiral Luis Carrero Blanco, 70, premier but remains chief of state. Assassins kill Premier Carrero Blanco.

Sweden's Gustavus VI dies September 15 at age 90 after a 23-year reign. He is succeeded by his grandson of 23 who will reign as Carl Gustavus XVI, not of the Swedes, Goths, and Wends but simply as king of Sweden.

The Yom Kippur War that begins in the Middle East October 6 on the Jewish Holy Day of Atonement is the fourth and fiercest Arab-Israeli war since 1948. Hostilities have persisted since January when Israeli fighter planes shot down 13 Syrian MIG-21 jets, both sides accuse the enemy of having begun the new fighting which erupts along the 103-mile-long Suez Canal and on the Golan Heights, UN observers report that Egyptian forces crossed the canal at five points and that Syrian forces attacked at two points on the Golan Heights, Israeli troops push the Syrians back to the 1967 cease-fire line by October 10 despite the arrival of Iraqi troops to support the Syrians, the Israelis push to within 18 miles of Damascus October 12, and Jordan's best troops arrive October 13 to help defend Damascus. Egyptian troops meanwhile force the Israelis to give up the Bar Lev defense line on the East Bank of the Suez Canal, Egyptian SAM-6 missiles stymie Israel's air attacks, Soviet planes airlift equipment to help Arab forces on both fronts, Moscow announces October 15 that it will "assist in every way" the Arab effort to regain the territory taken by Israel in 1967.

Washington announces that it has begun supplying military equipment to Israel to counter the Soviet airlift of arms to the Arabs, Israel sends a task force across the Suez Canal to attack Egyptian tanks, missile sites, and artillery on the West Bank, heavy tank battles begin in the Sinai October 17, Moscow tries to persuade Egypt and Syria to resolve the Middle East conflict through diplomacy, four Arab foreign ministers meet with President Nixon at Washington and urge U.S. mediation of the Arab-Israeli dispute, Soviet premier Aleksei Kosygin meets at Cairo with President Sadat, U.S. Secretary of State Henry Kissinger confers at Moscow October 20 with Soviet party leader Leonid A. Brezhnev, a resolution sponsored jointly by the United States and the U.S.S.R. calling for a cease-fire in place receives a 14–0 vote of approval in the UN Security Council early in the morning of October 22, heavy fighting resumes 12 hours after the cease-fire takes effect, Israel and Egypt agree to a new cease-fire October 24, U.S. forces are put on "precautionary alert" October 25, Secretary Kissinger justifies the action by asserting that "ambiguous" signs from Moscow have suggested possible Soviet intervention in the Middle East, the crisis passes, the cease-fire holds, but Israel has lost 4,100 men killed or wounded in 18 days of fighting, Egypt has lost 7,500, Syria, 7,300.

A congressional War Powers Resolution passed over President Nixon's veto November 7 limits a president's authority to commit troops in a foreign conflict without congressional approval, affirming Article I, Section 8, Paragraph 11 of the Constitution. Nixon has said the resolution would impose unconstitutional and dangerous restrictions on presidential power and "seriously undermine this nation's ability to act decisively and convincingly in times of international crisis." Future presidents will ignore the measure.

Britain joins the European Economic Community after a decade of controversy (*see* 1972).

The oil shock (above) and soaring grain prices precipitate a world monetary crisis and then a worldwide economic recession, the worst since the Great Depression of the 1930s.

Speculative selling of U.S. dollars on foreign exchanges forces the second devaluation of the dollar in 14 months. Secretary of the Treasury George P. Schultz announces February 12 that the dollar is to be devalued by 10 percent against nearly all other major world currencies in a move to make U.S. goods more competitive in foreign trade. The devaluation brings the price of the wheat bought by Moscow last year down to $1.48 per bushel, making it a bigger bargain than ever for the U.S.S.R. (which has thus far paid only $330 to $400 million), and fast-rising gold prices in the London market enable the Russians to sell their bullion at much more favorable rates than they had expected, but the devaluation in the dollar produces consternation in Japan and boosts the price of imported goods (and imported oil) to U.S. consumers and businessmen.

The Chicago Board Options Exchange opens April 26 in a former Board of Trade lunchroom with 282 members dealing in options to buy (calls) 16 stocks on the New York Stock Exchange. By year's end, the CBOE is trading in options on 50 NYSE stocks and it will add sell options (puts) in 1977. Options trading will increase volatility on the New York exchanges.

NCR (formerly National Cash Register) begins manufacturing automatic teller machines as more banks install the devices to offer 24-hour service (*see* 1968).

An energy crisis grips the world; an oil embargo by Arab nations in the fall exacerbates the problem.

Saudi Arabia acquires a 25 percent interest in the 40-year-old Aramco petroleum colossus January 1 and moves to enlarge its equity in the company to 51 percent by 1982 (*see* 1948).

Iraqi Oil Minister Sadoon Hammadi negotiates a settlement with Iraq Petroleum (see 1972). The company recognizes Baghdad's nationalization of the fields and agrees to pay $345 million in back taxes and help boost Iraqi oil output to 3 million barrels per day by 1975; Baghdad agrees to give the company 15 million tons of oil in compensation. The Compagnie Française des Pétroles agrees to buy 23.75 percent of Iraq's annual production.

President Nixon acts May 1 to end the oil import quota system imposed in 1959, urges additional tax credits to subsidize the oil industry's efforts to find new sources, and asks Congress to end federal regulation of natural

1973 *(cont.)* gas prices to provide incentive for exploration (*see* 1954). He urges states to encourage use of coal, even if it means delaying air pollution controls, and asks Americans to conserve energy.

"America faces a serious energy problem," President Nixon tells Congress June 29: "While we have 6 percent of the world's population, we consume one-third of the world's energy output. The supply of domestic energy resources available to us is not keeping pace with our ever-growing demand." U.S. demand for oil runs at about 17 million barrels per day, domestic output is little more than 11 million barrels per day, Venezuela and Canada supply most of the imported oil, and at present rates of increase demand will reach 24 million barrels per day by 1980 with domestic sources supplying only half that much.

Critics charge that there is no real shortage of petroleum but that major oil companies are engaged in a conspiracy to force up prices and drive out competition.

Reza Shah Pahlevi takes over all properties of the multinational Iranian Consortium including the huge Abadan Refinery, guaranteeing the participating companies a 20-year supply of oil in proportion to their interests in the consortium (British Petroleum has the lion's share); the Majlis ratifies the shah's action August 3.

Saudi Arabia's Faisal announces September 4 that his country will not increase oil production so long as U.S. policy favors Israel at the expense of the Arab countries (above). Saudi crude oil production is 6.5 million barrels per day (the United States produces 9.5 million, Iran 5 million, Kuwait 3 million, Libya 2.2 million, Abu Dhabi 1.2 million, Iraq 1.1 million, Qatar 550,000, Oman 300,000.

Other Arab nations cut back oil production in part for political reasons, in part to conserve resources, in part to force up prices in a sellers' market. The Arab oil-producing nations have enough money in reserve to shut down production completely for 18 months while maintaining normal imports of food and other necessities but Arab oil cutbacks force Europe and Japan to reduce airline service, cut down heating in offices, limit driving, and in other ways conserve dwindling supplies of fuel.

Moscow agrees to abide by the terms of the Universal Copyright Convention and to cease publishing pirated editions of Western works.

Nonfiction *The Best and the Brightest* by U.S. journalist David Halberstam, 39, probes U.S. involvement in Vietnam; *Fire in the Lake—The Vietnamese and the Americans in Vietnam* by U.S. journalist Frances Fitzgerald.

Fiction *Gravity's Rainbow* by Thomas Pynchon; *The Castle of Crossed Destinies* by Italo Calvino; *Temporary Kings* by Anthony Powell; *The Black Prince* by Iris Murdoch; *Fear of Flying* by U.S. novelist Erica Jong, 31; *Ninety-Two in the Shade* by U.S. novelist Thomas McGuane, 34; *Burr* by Gore Vidal; *Breakfast of Champions* by Kurt Vonnegut, Jr.

Poetry *The Bow and the Lyre* and *Alternating Current* by Octavio Paz.

Pablo Picasso dies at Mongins, France, April 8 at age 91 leaving works whose value is estimated at more than $1 billion. Active almost to the end, Picasso is hailed as the greatest artist of the twentieth century.

Sculpture *Amarillo Ramp* by Robert Smithson who dies in the crash of a small plane outside Amarillo, Texas, July 20; Jacques Lipschitz has died at Capri May 26 at age 81.

Theater *Finishing Touches* by Jean Kerr 2/8 at New York's Plymouth Theater, with Barbara Bel Geddes, Robert Lansing, 164 perfs.; *The Hot l Baltimore* by U.S. playwright Lanford Wilson, 35, 3/22 at New York's Circle in the Square Theater; *Absurd Person Singular* by Alan Ayckbourne, 7/4 at London's Criterion Theatre, with Richard Biers, Sheila Hancock; *Equus* by Peter Shaffer 7/26 at London's National Theater (Old Vic), with Alec McCowen; *Cromwell* by David Storey 8/15 at London's Royal Court Theatre, with Albert Finney; *When You Comin' Back, Red Ryder* by U.S. playwright Mark Medoff, 33, 11/4 at New York's Circle Repertory Theater, with Kevin Quinn, Robin Goodman, Addison Powell, James Kierman; *The Good Doctor* by Neil Simon 11/27 at New York's Eugene O'Neill Theater, with Christopher Plummer, Marsha Mason, 208 perfs.

Films Martin Scorsese's *Mean Streets* with Robert De Niro, Harvey Keitel; Lindsay Anderson's *O Lucky Man!* with Malcolm McDowell, Rachel Roberts; Peter Bogdanovich's *Paper Moon* with Ryan O'Neal, Tatum O'Neal, Madeline Kahn; Ingmar Bergman's *Scenes from a Marriage* with Liv Ullmann, Erland Josephson, Bibi Andersson. Also: George Lucas's *American Graffiti* with Richard Dreyfuss, Candy Clark; John Hancock's *Bang the Drum Slowly* with Michael Moriarty, Robert De Niro; Don Siegel's *Charley Varrick* with Walter Matthau; François Truffaut's *Day for Night* with Jacqueline Bisset, Jean-Pierre Aumont, Truffaut; Fred Zinneman's *The Day of the Jackal* with Edward Fox, Alan Badel; Robert Aldrich's *Emperor of the North Pole* with Lee Marvin, Ernest Borgnine, Keith Carradine; William Friedkin's *The Exorcist* with Ellen Burstyn, Max von Sydow, Linda Blair; Peter Yates's *The Friends of Eddie Coyle* with Robert Mitchum, Peter Boyle; Waris Hussein's *Henry VIII and His Six Wives* with Keith Mitchell, Donald Pleasence, Charlotte Rampling; Alan Bridges's *The Hireling* with Robert Shaw, Sarah Miles; John Frankenheimer's *The Iceman Cometh* with Lee Marvin, Fredric March, Robert Ryan, Jeff Bridges; Hal Ashby's *The Last Detail* with Jack Nicholson, Otis Young, Randy Quaid; Bernardo Bertolucci's *Last Tango in Paris* with Marlon Brando, Maria Schneider; Jan Troell's *The New Land* with Max von Sydow, Liv Ullmann; James Bridges's *The Paper Chase* with Timothy Bottoms, Lindsay Wagner, John Houseman; Victor Erice's *Spirit of the Beehive* with Fernando Ferman Gomez, Teresa Gimpera; George Roy Hill's *The Sting* with Paul Newman, Robert Redford; Robin

Hardy's *The Wicker Man* with Edward Woodward, Christopher Lee, Britt Ekland; Douglas N. Schwarz's *Your Three Minutes Are Up* with Beau Bridges, Ron Leibman, Janet Margolin.

Broadway and off-Broadway musicals *El Grande de Coca-Cola* 2/13 at the Mercer Arts Center, with Ron House, Diz White, music and lyrics by the cast, 1,114 perfs.; *A Little Night Music* 2/25 at the Shubert Theater, with Len Cariou, Hermione Gingold, Glynis Johns, music and lyrics by Stephen Sondheim, book from the 1955 Ingmar Bergman film *Smiles of a Summer Night*, songs that include "Send in the Clowns," 600 perfs.; *Seesaw* 3/18 at the Uris Theater, with Michele Lee, music by Cy Coleman, lyrics by Dorothy Fields, book from the 1958 William Gibson play *Two for the Seesaw*, 296 perfs.; *Raisin* 10/18 at the 46th Street Theater, with Joe Morton, Ernestine Jackson, music by Judd Woldon, lyrics by Robert Britten, book from the 1959 Lorraine Hansberry play *Raisin in the Sun*, 847 perfs.

Popular songs "I Shot the Sheriff" by Jamaican reggae composer-performer Robert Nesta "Bob" Marley, 28; "Tie a Yellow Ribbon Round the Ole Oak Tree" by Irwin Levine and L. Russell Brown; "Killing Me Softly with His Song" by Charles Fox, lyrics by Norman Gimbel; *Goodbye Yellow Brick Road* and *Don't Shoot Me I'm Only the Piano Player* (albums) by Elton John; *Goat's Head Soup* (album) by The Rolling Stones; *The Material World* (album) and "Give Me Love (Give Me Peace)" by former Beatles singer-instrumentalist George Harrison; *Life and Times* (album) and "Bad, Bad Leroy Brown" by Jim Croce; "You Are the Sunshine of My Life" and "Superstition" by U.S. songwriter Stevie Wonder (Stevland Morris).

Miami beats Washington 14 to 7 at Los Angeles January 14 in Super Bowl VII.

George Foreman, 24 (U.S.), gains the world heavyweight boxing championship January 22 by knocking out Joe Frazier in the second round of a title bout at Kingston, Jamaica.

Jan Kodes, 27 (Czech), wins in men's singles at Wimbledon, Mrs. King in women's singles; John Newcombe wins in men's singles at Forest Hills, Mrs. Court in women's singles.

A tennis match promoted as the "battle of the sexes" ends in defeat for former Wimbledon champion Bobby Riggs, now 55, who loses 6–4, 6–3, 6–3 to Mrs. King September 20 at the Houston Astrodome.

Secretariat wins U.S. horse racing's Triple Crown.

U.S. baseball's American League permits teams to field a tenth player—a "designated hitter" to bat in place of the pitcher.

The Oakland A's win the World Series by defeating the New York Mets 4 games to 3.

U.S. textile mills produce 482 million square yards of cotton denim, up from 437 million in 1971. The figure will soar to 820 million by 1976 as demand booms for blue jeans and denim jackets.

The U.S. median sales price of an existing single-family house reaches $28,900, up from $20,000 in 1968. By 1976 the price will be $38,100.

Chicago's Sears Tower opens with 3.6 million square feet of rentable space. Designed by Skidmore, Owings & Merrill, the $200 million 110-story structure rises 1,455 feet into the sky.

U.S. farmers plant 322 million acres of wheat, up from 293 million acres last year, in response to higher prices. Wheat production hits a record 1.71 billion bushels—nearly 50 million metric tons.

President Nixon announces a temporary embargo June 27 on exports of soybeans and cottonseeds, shocking Japan, Korea, and other traditional customers for U.S. oilseeds. At least 92 percent of the soybeans Japan uses for *tofu* (bean curd), soy sauce, and cooking oil comes from the United States, soybean prices jump by 40 percent in less than a week in Japan, the White House lifts the embargo after 5 days in response to State Department pressure, the Department of Commerce approves shipment of all orders received prior to June 13 but announces that special licenses will be required for all subsequent orders and says contracted amounts will be cut in half (by 40 percent for soybean oil, cake, and meal).

Foreign buyers redouble their purchases of U.S. grain lest further controls be applied, more than 30 million tons of U.S. wheat are sold for export by the end of July, U.S. farmers hold back their crops as buyers bid up prices, other farmers and ranchers cull flocks and herds as poultry raising and cattle production become unprofitable.

U.S. corn yields average 96.9 bushels per acre, up from 70 in 1968.

The average U.S. farm worker produces enough food and fiber for 50 people, up from 31 in 1963.

Farm labor represents 5 percent of the U.S. work force, down from 7 percent in 1968.

Nutrition labeling regulations promulgated by the Food and Drug Administration standardize the type of information to be presented on U.S. food packages.

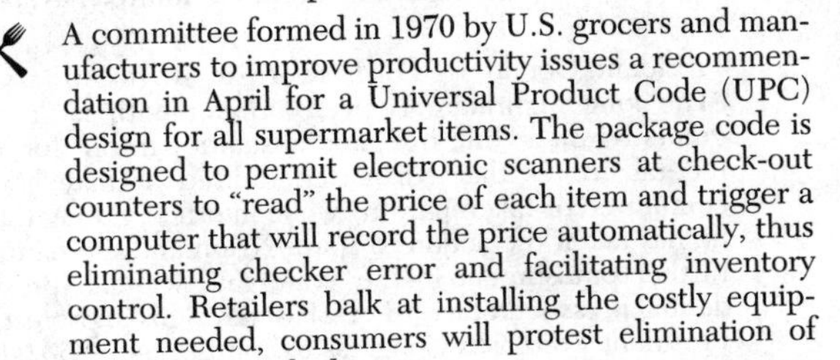

A committee formed in 1970 by U.S. grocers and manufacturers to improve productivity issues a recommendation in April for a Universal Product Code (UPC) design for all supermarket items. The package code is designed to permit electronic scanners at check-out counters to "read" the price of each item and trigger a computer that will record the price automatically, thus eliminating checker error and facilitating inventory control. Retailers balk at installing the costly equipment needed, consumers will protest elimination of individually marked prices.

Food prices soar in the United States, Japan, and Europe in the wake of last year's Soviet wheat and soybean purchases which have forced up the price of feed grains and consequently of meat, poultry, eggs, and dairy products as well as of baked goods. U.S. consumer groups organize boycotts to protest rising prices but

The Universal Product Code and electronic scanners shortened check-out lines at U.S. supermarkets.

prices will continue to rise even without the excuse of a "Russian wheat deal."

President Nixon orders a freeze on all retail food prices June 13, saying he will "not let foreign sales price meat and eggs off the American table."

Vodka outsells whiskey for the first time in the United States.

Abortion should be a decision between a woman and her physician, the Supreme Court rules January 22 in *Roe v. Wade*. The court's 7-to-2 decision upholds a woman's right to privacy in opting for abortion during the first 6 months of pregnancy. "Right-to-life" groups work to undermine the ruling.

1974 President Nixon resigns in disgrace August 9—the first U.S. chief of state ever to quit office. The Supreme Court has ruled 8 to 0 July 24 that Nixon must turn over 64 White House tape recordings to a special prosecutor. The House Judiciary Committee has voted July 30 to adopt three articles of impeachment charging Nixon with obstruction of justice, failure to uphold laws, and refusal to produce material that the committee has subpoenaed.

President Gerald R. Ford, sworn in August 9, says, "The long nightmare is over." One month later he grants Nixon a "full, free, and absolute pardon" for all federal crimes that Nixon "committed or may have committed or taken part in" while in office, noting that he has taken the action to spare Nixon and the nation further punishment in the Watergate scandal (presidential press secretary J. F. terHorst resigns in protest). President Ford asks Congress to appropriate $850,000 to facilitate Nixon's transition to private life; Congress trims the grant to $200,000.

The Election Reform Act passed by Congress 355 to 48 just hours before President Nixon's resignation (above) limits to $1,000 the amount that any individual may contribute to a candidate for federal office, limits to $20 million what any presidential candidate may spend on a bid for election or reelection, provides or a $1 tax check-off on individual federal income tax returns to provide federal funding of presidential elections, and contains other provisions to minimize the impact of large company campaign contributions and thus prevent the kinds of abuses that characterized the Watergate scandal. The legislation crowns efforts by Common Cause, a private citizens' group founded in 1970, but political action committees (PACs) will raise large amounts of money for candidates, thereby blunting the effects of the new law.

Britain's Conservatives have lost the general elections February 28 in the midst of a coal miners' strike that has forced the nation to go on a fuel-conserving 3-day work week. Former Prime Minister Harold Wilson has called for nationalization of North Sea oil, his Labour party has won 301 seats to 296 for the Conservatives, Edward Heath resigns, and a second Wilson ministry begins March 5, the first in 45 years to lack a majority in the House of Commons. Labour retains power in the general elections October 10, winning 319 seats to 276 for the Conservatives, but inflation (19.1 percent for the year) and economic decline continue to plague Britain.

France's President Pompidou dies of cancer April 2 at age 62. Former finance minister Valéry Giscard d'Estaing, 48, is elected May 19 to continue the Gaullist regime that has ruled since 1958, narrowly defeating the leftist François Mitterand.

President Nixon resigned to escape impeachment proceedings after trying to cover up the Watergate break-in scandal.

Portugal has a coup d'état April 25. A leftist military junta headed by General Antonio Sebastião Riveiro de Spinola takes power from Premier Marcelo Caetano, 68, who has been dictator since Premier Antonio de Oliveiro Salazar suffered a stroke in 1968. The new regime resumes diplomatic relations with Moscow for the first time since 1917, raising fears in the NATO alliance that Communists in the new Lisbon government will leak secrets.

West German terrorists Bernd Andreas Baader, 31, and Ulrike Meinhof, 40, of the notorious Bader-Meinhof gang begin their third hunger strike September 13 at Düsseldorf's Schwalmstadt Prison and Cologne's Ossendorf Prison; 40 other prisoners join the strike. Captured in June 1972 after a series of bank robberies and bombings, the leftist Red Army Faction (*Rote Armee Faction*, or RAF) members have led one of several gangs that are disrupting West Germany. RAF militant Holger Meins, 33, dies November 9 after 5 weeks of forced feeding. Four youths carrying flowers call on the president of West Berlin's court of appeals November 10, Judge Gunter von Drenkmann opens his front door, and he is shot dead (*see* 1975).

Israel and Egypt have signed a disengagement agreement January 18 after negotiations by U.S. Secretary of State Henry Kissinger (*see* 1973). Israel withdraws from the west bank of the Suez Canal, Egypt reoccupies the east bank, and a UN buffer zone is created between the two. Israel agrees in June to withdraw from Syria and from part of the Golan Heights (*see* Lebanon, 1975).

Cypriot patriot Gen. George Grivas has died January 28 at age 75. Greek Cypriot troops overthrow the Makarios government July 15, Athens denies any link to the uprising, Archbishop Makarios arrives at New York and charges the Greek military regime with complicity in the coup. Turkish forces invade Cyprus July 20, vowing to restore Makarios and defend the island's ethnic minority. Greece mobilizes troops, Moscow puts 40,000 men on alert, the UN Security Council calls for a halt in hostilities, but heavy fighting continues.

Greece's military junta resigns July 23, former Premier Constantine Caramanlis returns from exile to head the first civilian government at Athens since 1967, he announces in mid-August that Greece will not go to war to stop the Turkish invasion but will not negotiate under pressure. U.S. Ambassador Rodger P. Davies is shot dead August 19 during a Greek Cypriot demonstration outside the U.S. embassy at Nicosia; President Ford vetoes a congressional bill that would cut off military aid to Turkey. President Makarios returns to Nicosia December 7, Turkish forces occupy 45 percent of the island, and tensions continue.

Angola, Mozambique, and Portuguese Guinea redouble efforts to regain independence following the change at Lisbon (above). Portuguese Guinea becomes Guinea-Bissau September 10.

Somalia's Marxist government signs a treaty of friendship and cooperation with Moscow, the first black African nation to do so, and Somalia becomes a Soviet satellite (*see* 1969; 1991).

Ethiopia's army seizes Addis Ababa in late June after months of riots and mutinies to protest government handling of a famine that has killed 100,000 peasants, has inundated the cities with refugees from drought-stricken areas, and produced food shortages and inflation. Emperor Haile Selassie, now 82, is deposed September 12 after a 44-year reign interrupted by the Italians from 1936 to 1941, and the new Soviet-dominated regime announces December 20 that Ethiopia will become a socialist state directed by one political council.

The U.S. Army grants a parole to Lieut. William L. Calley, Jr., who has been serving a 10-year term for his part in the My Lai massacre of 1968 in South Vietnam.

Grenada gains independence February 7 after more than 200 years of British rule. Prime Minister Eric M. Gairy curbs civil liberties to reduce violence on the Caribbean island (*see* 1979).

Argentine dictator Juan Peròn dies July 1 at age 78. His vice president and third wife Maria Estela (Isabel) Martinez de Peròn, 43, becomes the hemisphere's first woman chief of state (*see* 1976).

New Delhi announces May 18 that India has conducted a successful test of a 10-to 15-kiloton atomic device, joining the United States, the U.S.S.R., Britain, France, and China in the world nuclear club. Ottawa protests the underground Indian test and suspends Canadian aid to India's atomic energy program (*see* 1956). Paris and Washington agree to supply Iran with nuclear reactors, but the Indian test dramatizes the need to halt the proliferation of fissionable materials and nuclear weapons technology.

A car carrying Pakistan President Zulfikar Ali Bhutto's vigorous parliamentary critic Ahmad Raza Kasuri is ambushed November 11 at Lahore. Kasuri is unhurt but his father is killed. The perpetrators turn out to be members of the Bhutto government's security agency.

Japan's Prime Minister Kakuei Tanaka, 56, resigns November 26 in the face of financial scandals (*see* 1976). Tanaka's Liberal-Democratic party has barely survived the July 7 elections, it is feared that the party will break up if either of the two leading candidates is chosen to succeed Tanaka, and the Diet names Takeo Miki prime minister December 28.

Federal judge W. Arthur Gerrity rules June 21 that the Boston School Committee has deliberately segregated schools by race; he adopts a plan calling for exchange of students in black Roxbury and white South Boston, but buses carrying blacks to South Boston September 12 encounter white mobs shouting, "Nigger, go home!" Violence ensues, and in October Governor Francis W. Sargent calls out the National Guard to prevent a race war.

British coal miners strike February 10 with Communist encouragement after a months-long slowdown has forced the nation to adopt a 3-day work week (above). The workers return to the pits March 6 after winning a

1974 *(cont.)* 35 percent wage increase that gives underground workers $122 per week, up from $85, and surface workers $73, up from $58.

Economic recession deepens in the world following last year's hike in oil prices by major petroleum producers. Inflation, meanwhile, raises prices in most of the free world. Double-digit inflation is worst in Israel, India, Brazil, and Japan.

The U.S. Consumer Price Index rises by 12.2 percent following last year's 8.8 percent increase. Increases averaged less than 2.4 percent in the 25 years from 1948 to 1972, and the CPI actually declined in 1949 and 1954, but it will go up another 7 percent next year, 4.8 percent in 1976, and 6.8 percent in 1977.

Dreyfus Liquid Assets, launched in February, is the first money-market fund advertised to the public. It permits small investors to enjoy the same high rates of interest heretofore available only to individuals and institutions rich enough to buy financial instruments in denominations of $100,000 and more. Investors can write checks on their balances. Money-market funds will number more than 100 by 1980.

The Employee Retirement Income Security Act signed by President Ford September 2 provides for federal regulation and insurance of private pension benefits with broader protection for workers. It establishes tax-deductible Individual Retirement Accounts (IRAs) whose assets will total $26 billion by 1981, $400 billion by 1989.

New York's Franklin National Bank is declared insolvent September 8, the biggest bank failure in U.S. history. Many larger banks receive no interest on outstanding loans. Italian financier Michel Sindona, 53, is a director of the bank's parent company and will be convicted in March 1980 of looting the twentieth largest U.S. bank of $45 million, thus causing its collapse.

Britain's National Coal Board boosts coal prices 28 percent in September. Coal workers reject proposals for incentives to boost productivity.

Wall Street's Dow Jones Industrial Average bottoms out December 9 at 570.01, down from 1003.16 late in 1972.

World oil prices escalate in the wake of last year's action by the Organization of Petroleum Exporting Countries. OPEC crude oil prices reach $11.25 per barrel by year's end, up from $2.50 at the beginning of 1973. Efforts increase to find new sources of energy as higher coal prices bring a boom to depressed coal-mining areas.

Nuclear fuel facility laboratory technician Karen Gay Silkwood, 28, dies in an automobile crash near Oklahoma City November 13 on her way to meet with a *New York Times* reporter and a union official. She has planned to document her allegations that Kerr-McGee Nuclear Corp. has falsified quality control reports on fuel rods and that 40 pounds of highly dangerous plutonium are missing from the Kerr-McGee plant near Crescent, Okla. Investigators find high levels of radiation in Silkwood's apartment.

Solar power remains far too costly and inefficient to compete with power from electric utilities. Comsat engineer Joseph Lindmayer has devised a silicon photovoltaic cell 50 percent more efficient than anything previously made and last year founded Solarex, Inc., believing that solar cells are ripe for further development.

A nationwide 55-mile-per-hour highway speed limit act signed by President Nixon January 2 makes federal aid for highways conditional on state enforcement of the new speed limit. Nixon had requested a 50 mph limit, and although motorists in western states routinely exceed the speed limit and jeopardize federal highway funds, the new law will conserve 3.4 billion gallons of fuel per year. Highway fatalities fall to 45,196, down from 54,052 last year, but will climb to 50,700 by 1979 as motorists flout the law.

The Airbus A300B assembled at Toulouse, France, begins to challenge Boeing for the world jet aircraft market. Airbus Industrie is a consortium of government-owned British and French aircraft makers with some private German companies and a 4 percent Spanish participation.

A Turkish DC-10 crashes outside Paris March 3, killing 346. A Pan Am 707 has crashed January 31 while landing at Pago Pago, Samoa, killing 97 out of 101 aboard; a TWA 727 crashes outside Upperville, Va., December 1 killing 92; a chartered Dutch DC-8 carrying Indonesian Muslims home from Mecca crashes at Colombo, Sri Lanka, December 4 killing 191.

Charles de Gaulle Airport opens 12 miles north of Paris at Roissy to relieve pressure on Le Bourget and Orly.

Prague's first subway opens. A 3-mile east-west line will open in the summer of 1978 to supplement the initial 4.5-mile north-south system.

Kyoto's Minato Ohashi Bridge is completed—the world's third longest cantilever bridge (933 feet).

Japan's Kuronoseto Bridge, completed at Nagashima in Kyushu, is the world's fourth longest continuous truss bridge (932 feet).

The National Academy of Sciences urges a temporary worldwide ban on certain types of genetic manipulation, especially experiments involving the bacterium *Escherechia coli* found in the human digestive tract, lest scientists create a virulent organism more deadly and more resistant than any found in nature (*see* 1976).

The computed axial tomography (CAT) scanner developed in England by EMI, Ltd. (formerly Electrical Musical Instruments, Ltd.) with money from sales of Beatles records gains wide use not only for diagnosing brain damage but also for whole-body scanning. The device assembles thousands of X-ray images into a single, remarkably detailed picture of the body's interior. It revolutionizes diagnostic medicine but costs upwards of $500,000 (*see* NMR, 1982).

The "Heimlich maneuver" described in the June issue of *Emergency Medicine* by Cincinnati surgeon Henry Jay Heimlich, 54, will save thousands of people from choking to death on food: Place heel of hand on victim's abdomen slightly above the navel and well below the ribs; use other hand to press with sharp upward thrusts until obstructing food pops out. Heimlich says so-called "café coronaries" are easily distinguished from heart attacks—the victim cannot speak, turns blue, and collapses. Mouth-to-mouth resuscitation will only push the food farther down, oxygen deprivation may cause permanent brain damage before an ambulance can get the victim to a hospital, hence the need for prompt use of the Heimlich maneuver.

U.S. insurance companies raise rates on malpractice policies, forcing up physicians' fees and hospital rates. More companies will hike rates next year and some will stop writing malpractice policies as the unique U.S. tort system boosts health care costs.

South Carolina evangelist Jim Bakker, 34, founds the PTL (Praise the Lord) television ministry that will become a multimillion-dollar religious empire (*see* 1987).

Word processors with cathode-ray tube displays and speedy printers begin to replace typewriters as the economic recession (above) encourages business managers to automate offices. The IBM Selectric typewriter introduced in 1961 was given a magnetic tape and turned into a primitive word processor in 1964, but IBM's machine has cost $10,000, vs. $600 for a regular office typewriter, and has actually cut productivity because secretaries sat back and watched it type. Vydek is first to introduce a text-editing computer with a CRT screen and printer. By 1976 impact printers with bidirectional "daisy wheels" will be printing documents at 30 to 55 characters per second, vs. 15 for the typing ball on a power typewriter, while secretaries key in material for other documents.

Japanese newspapers ignore the story of Prime Minister Tanaka's corruption (above) until foreign publications circulating in Japan pick up facts uncovered by a Japanese monthly magazine. Tanaka, whose conviction of accepting bribes from coal-mining interests early in his career was later overturned, has been a notorious wheeler-dealer and has steadily increased his wealth through construction and real estate operations while in office, yet none of the major dailies has investigated his personal finances nor his innovative use of political contributions.

The U.S. first class postal rate rises March 2 to 10¢ per ounce; higher rates for lower-class mail forces magazines to downsize (*see* 1971, 1975).

Knight-Ridder Newspapers is created as publisher John S. Knight, 79, merges his *Akron Beacon Journal* and 15 other papers with those of the 82-year-old Ridder chain (*see* 1903).

People magazine begins publication March 4 at New York. Time, Inc. has started the new 35¢ weekly in a bid to recoup circulation lost when *Life* ceased weekly publication late in 1972.

High Times magazine begins publication at New York in late summer with advertising from E-Z Wider and JOB rolling papers. "Yippie" Tom Forcade has started the magazine with $20,000 allegedly derived from drug smuggling, his stories refer to marijuana as a "medical wonder drug," he ridicules the Drug Enforcement Administration, and he will shoot himself to death in November 1978 following the fatal crash of his best friend's plane on a smuggling mission.

A U.S. Freedom of Information Act passed by Congress November 21 over President Ford's veto assures broader public access to public information.

Nonfiction *The Gulag Archipelago* by Aleksandr Solzhenitsyn, who is arrested by the KGB and expelled from the Soviet Union (he will take up residence in Vermont); *The Ultra Secret* by former British intelligence officer Frederick W. Winterbotham, 76, reveals World War II secrets of cracking the German code; *All the President's Men* by Bob Woodward and Carl Bernstein; *The Ascent of Man* by Polish-born English mathematician-humanist Jacob Bronowski who dies August 22 at age 66 in a Long Island, N.Y., automobile accident; *The Lives of a Cell* by New York biologist Lewis Thomas, 60; *The Joy of Sex* by English gerontology expert Alex Comfort, 54; *Flying* by Kate Millett; *Working* by Studs Terkel.

Fiction *The Conservationist* by Nadine Gordimer; *Ebony Tower* by John Fowles; *The Eye of the Storm* by Patrick White; *The Sacred and Profane Love* by Iris Murdoch; *The Honorary Consul* by Graham Greene; *Napoleon Symphony: A Novel in Four Movements* by Anthony Burgess; *Zen and the Art of Motorcycle Maintenance* by U.S. novelist Robert M. Pirsig, 46; *Something Happened* by Joseph Heller; *Centennial* by James Michener; *Theophilus North* by Thornton Wilder, now 77; *Look at the Harlequins* by Vladimir Nabokov, now 75; *Jaws* by U.S. novelist Peter Benchley, 33; *The Dog Soldiers* by U.S. novelist Robert Stone, 37; *The Dogs of War* by Frederick Forsyth; *Tinker, Tailor, Soldier, Spy* by John Le Carré; *Celestial Navigation* by U.S. novelist Anne Tyler, 32; *The War Between the Tates* by U.S. novelist Alison Lurie, 45; *Cogan's Trade* by George V. Higgins; *52 Pick-Up* by Detroit novelist Elmore Leonard, 51.

Poetry *Conjunctions and Disjunctions* by Octavio Paz; *The Dolphin* by Robert Lowell; *High Window* by Philip Larkin; *The Death Notebooks* by Anne Sexton who dies at Weston, Mass., October 4 at age 45 in an apparent suicide.

Painting *Corpse and Mirror* (oil, encaustic, collage on canvas) by Jasper Johns.

Sculpture *The Destruction of the Father* (mixed media) by Louise Bourgeois; *Coyote: I Like America and America Likes Me* by Joseph Beuys who makes his first visit to the United States and instead of mounting a conventional exhibit fences himself in for a week with a

1974 ***(cont.)*** live coyote at his dealer's gallery; *Trans-Fixed* (artist crucified on Volkswagen "beetle") by Chris Burden.

Theater *Short Eyes* by New York playwright Miguel Piñero, 27, 1/3 at New York's Riverside Church Theater, 102 perfs.; *Knuckle* by English playwright David Hare, 26, 3/4 at London's Comedy Theatre, with Edward Fox, Kate Nelligan; *Thieves* by Herb Gardner 4/7 at New York's Broadhurst Theater, with Dick Van Patten, Marlo Thomas, Irwin Corey, 312 perfs.; *Bad Habits* by Terrence McNally 5/5 at New York's Booth Theater (after 96 perfs. at the Astor Place Theater), with Cynthia Harris, Doris Rafelo, Emory Bass, J. Frank Luca, 273 perfs.; *Travesties* by Tom Stoppard 6/10 at London's Aldwych Theatre, with John Wood, Tom Bell as James Joyce, Frank Windsor as Lenin, Maria Aitken, 39 perfs. in repertory; *All over Town* by Murray Schisgal 12/29 at New York's Booth Theater, with Barnard Hughes, Cleavon Little, 233 perfs.

Television *The Rockford Files* 9/13 on NBC with James Garner (to 1/10/80).

Films Harold Pinter's *Butley* with Alan Bates, Jessica Tandy; Roman Polanski's *Chinatown* with Jack Nicholson, Faye Dunaway, John Huston; Francis Ford Coppola's *The Conversation* with Gene Hackman; Rainer Werner Fassbinder's *Fox and His Friends* with Fassbinder; Francis Ford Coppola's *The Godfather II* with Al Pacino, Robert De Niro; Louis Malle's *Lacombe, Lucien* with Pierre Blaise; Claude Sautet's *Vincent, François, Paul and the Others* with Yves Montand, Michel Piccoli, Gerard Depardieu, Stephane Audran. Also: Martin Scorsese's *Alice Doesn't Live Here Anymore* with Ellen Burstyn, Kris Kristofferson; Federico Fellini's *Amarcord* with Bruno Zanin; Ted Kotchoff's *The Apprenticeship of Duddy Kravitz* with Richard Dreyfuss; Joe Camp's *Benji* with Peter Breck; Mel Brooks's *Blazing Saddles* with Gene Wilder, Brooks, Cleavon Little, Madeline Kahn; Akira Kurosawa's *Dersu Uzala* with Maxim Munzuki, Yuri Solomin; Peter Davis's Vietnam war documentary *Hearts and Minds*; Bob Fosse's *Lenny* with Dustin Hoffman as the late Lenny Bruce; Robert Aldrich's *The Longest Yard* with Burt Reynolds, Eddie Albert; Alan J. Pakula's *The Parallax View* with Warren Beatty, Paula Prentiss; Maximilian Schell's *The Pedestrian* with Gustav Rudolf Sellner, Peter Hall, Schell; Luis Buñuel's *The Phantom of Liberty* with Jean-Claude Brialy, Monica Vitti; Alain Resnais's *Stavisky* with Jean-Paul Belmondo; Steven Spielberg's *Sugarland Express* with Goldie Hawn, William Atherton; Joseph Sargent's *The Taking of Pelham One, Two, Three* with Walter Matthau, Robert Shaw, Martin Balsam; Robert Altman's *Thieves Like Us* with Keith Carradine, Shelley Duvall; Mel Brooks's *Young Frankenstein* with Gene Wilder.

Hollywood musical Jack Haley, Jr.'s *That's Entertainment* with Fred Astaire, Bing Crosby, Gene Kelly, Peter Lawford, Liza Minnelli, Donald O'Connor, et. al. in scenes from old musicals.

Film opera Ingmar Bergman's *The Magic Flute* with Ulric Cold, Irma Urrila, music by Mozart.

New York musicals *Let My People Come* 1/8 at the Village Gate, with music and lyrics by Earl Wilson, Jr., 31, songs that include "Dirty Words," "Give It to Me," and "Whatever Turns You On;" the sexually-oriented show will move uptown to the Morosco Theater 7/27/1976, 1,167 perfs.; *Lorelei* 1/27 at the Palace Theater, with Carol Channing in a revised version of the 1949 musical *Gentlemen Prefer Blondes*, music by Jule Styne, lyrics by Betty Comden and Adolph Green, 321 perfs.; *The Magic Show* 5/28 at the Cort Theater, with Doug Henning, music and lyrics by Stephen Schwartz, 1,859 perfs.

Popular songs "The Way We Were" by U.S. composer Marvin Hamlisch, 30, lyrics by Alan and Marilyn Bergman (title song for film); *Band on the Run* (album) and "Jet" by former Beatles singer-composer Paul McCartney for his back-up group Wings (McCartney, wife Linda, 32, and Denny Laine, 30); "Feel Like Makin' Love" by U.S. songwriter Eugene McDaniels; "Behind Closed Doors" by U.S. songwriter Kenny O'Dell (Kenneth Gist, Jr.); "I Honestly Love You" by Australian composer Peter Allen, lyrics by U.S. songwriter Jeff Barry, 35; "(You're) Havin' My Baby" by Paul Anka; "Rhinestone Cowboy" by U.S. songwriter Larry Weiss, 33.

Miami beats Minnesota 24 to 7 at Houston January 13 in Super Bowl VIII.

James Scott "Jimmy" Connors, 21 (U.S.), wins in men's singles at Wimbledon and Forest Hills, Christine Marie "Chris" Evert, 19 (U.S.), in women's singles at Wimbledon, Mrs. King at Forest Hills.

The U.S. yacht *Courageous* retains the America's Cup, defeating Australian challenger *Southern Cross* 4 to 0.

West Germany wins the World Cup in football (soccer), defeating the Dutch 2 to 1 at Munich.

Hank Aaron breaks Babe Ruth's career home run record April 8 at Atlanta, hitting his 715th off a pitch by Al Downing of the Los Angeles Dodgers.

The Oakland Athletics win the World Series, defeating the Los Angeles Dodgers 4 games to 1.

Oakland pitcher Jim "Catfish" Hunter, 28, is declared a free agent December 16 when a three-man arbitration panel rules that Athletics owner Charles O. Finley, 56, has not honored his contract to pay Hunter $100,000 (*see* 1975).

Muhammad Ali regains the world heavyweight boxing crown October 20 by knocking out George Foreman in the eighth round at Kinshasa, Zaire. Ali will hold the title until February 1978.

"Streaking" becomes a popular U.S. fad as male and female college students dash naked between dormitories.

San Francisco black militants calling themselves the Symbionese Liberation Army kidnap publishing heiress Patricia Hearst, 19, February 5 and demand $2 million in ransom. They then demand $230 million in free food for the poor of California. A San Francisco bank robbery April 15 nets $10,960 and an automatic camera at the bank records Hearst holding a submachine gun for the robbers (*see* 1975).

"Zebra" killings terrorize San Francisco for 179 days as black fanatics armed with .32 Baretta handguns and other weapons shoot whites at random in the streets.

Former United Mine Workers president W. A. "Tony" Boyle is convicted April 9 of having ordered the New Year's Eve 1969 killing of his rival Joseph A. "Jake" Yablonski who was found slain with his wife and daughter at their Clarksdale, Pa., home early in 1970. Boyle will win acquittal on appeal but another court will find him guilty again early in 1978.

A tornado hits Xenia, Ohio, April 3, cutting a quarter-mile swath through the Dayton suburb, killing 34, and causing $500 million property damage.

A cyclone destroys Darwin, Australia, December 25, forcing virtual abandonment of the city.

An earthquake in northern Pakistan December 28 leaves more than 5,000 dead over an area of 1,000 square miles.

The Union of Banana Exporting Countries, formed by Central and South American nations, pushes for higher fruit prices to offset climbing fuel costs. Honduras imposes a 50¢ tax on each 40 lb. box of bananas in April; the tax is halved to 25¢ within 5 months after $2.5 million is deposited to the Swiss bank account of former economic minister Abraham Bennaton Ramos (*see* Black suicide, 1975).

Hawaiian pineapples account for 33 percent of the world crop, down from 72 percent in 1950, as the industry shifts to the Philippines and Thailand.

Bangladesh has famine; hundreds of thousands die.

A World Food Conference at Rome in November hears a U.S. refusal to make commitments for specific increases in food aid to needy countries. Impact on U.S. consumer prices in the face of tight supplies is a major reason, but there is also an emphasis on the importance of self-help in the needy countries, including material advances in population control.

London's Covent Garden fruit, vegetable, and flower market moves after more than 300 years in the heart of town to a new site at Nine Elms in south London (*see* 1552).

Average U.S. food prices: white bread 34.5¢ 1 lb. loaf (up from 27.6¢ last year); sugar 32.3¢ lb., up from 15.1¢; rice 44¢ lb., up from 26¢; potatoes 24.9¢ lb., up from 20.5¢; coffee $1.28 lb., up from $1.04.

Hong Kong rescinds a 6-year-old policy of accepting illegal immigrants from China. Thousands of Vietnamese and other refugees have poured into the crowded city.

1975 North Vietnamese forces close in on Saigon, President Thieu resigns April 21 and flees the country, his successor resigns a week later, and Gen. Duong van Minh surrenders the city April 30 as U.S. helicopters complete evacuation of 1,373 Americans and 5,595 Vietnamese, ending the war that has cost 1.3 million Vietnamese and 56,000 U.S. lives, $141 billion in U.S. aid. Congress appropriates $405 million to resettle 130,000 refugees in America.

Nationalist China's Chiang Kai-shek has died April 5 at age 87 after 26 years as president of the Republic (Taiwan). His son Chiang Chingkuo continues as premier and will assume the presidency in 1978.

Cambodia's Lon Nol government has fallen April 16, ending a 5-year war with the Communist Khmer Rouge. The new regime, headed by French-educated revolutionist Pol Pot and others of peasant origin, takes Phnom Penh April 17 and seizes the U.S. merchant ship *Mayaguez* May 13. U.S. Navy and Marine units move in for a rescue operation May 15, sustain 38 casualties, but recover ship and crew.

The Comoros in the Indian Ocean proclaim independence July 6 after 89 years of French colonial rule. One island, Mayotte, has a Christian majority and remains French.

Laos becomes a "people's democratic republic" December 2 as the Pathet Lao leader Prince Souphanovong ends the monarchy, makes his half brother Souvanna Phouma, premier since 1962, an "adviser," and forces King Sisavang Vatthana to abdicate after nearly 25 years of U.S. efforts to block a takeover by the indigenous Communists of the Oregon-sized nation of 3.4 million.

Iran and Iraq have agreed at Algiers March 6 to fix their border where it was set in 1947. Iran takes over half the Shatt Al-Arab estuary formed by the confluence of the Tigris and Euphrates Rivers about 100 miles from

Communist forces triumphed in Vietnam after decades of conflict, first with the French, then with the Americans.

1975 ***(cont.)*** the Persian Gulf. Iraq has conceded the territory in return for Iran's promise to stop supporting Kurdish rebels in Kurdistan, but Iraqi nationalists vow to regain full control of the 40-mile long waterway (*see* 1980).

Saudi Arabia's King Faisal, 69, is assassinated March 25 by his U.S.-educated nephew Prince Faisal Musad Abdel Aziza, 27, who is beheaded within 3 months. Arrested by Colorado authorities in 1969 for selling drugs, the prince was released after confessing. Faisal is succeeded after an 11-year reign by his brother Khalid, who will rule until 1982, continuing Faisal's relatively moderate policies within the Organization of Petroleum Exporting Countries (OPEC).

The Helsinki Accord adopted August 1 by 36 nations formalizes détente between East and West. The declaration issued by the Conference on Security and Cooperation in Europe emphasizes the inviolability of frontiers, full support for the United Nations, and mutual respect for "sovereign equality and individuality. . . ." The participants renounce "threat or use of force" and subversion in settling international disputes.

The United Nations General Assembly votes November 10 to approve an Arab-inspired resolution defining Zionism as "a form of racism and racial discrimination." U.S. Ambassador Daniel Patrick Moynihan says the United States will never "acquiesce in this infamous act," but the resolution carries 72 to 35, with 32 abstentions.

Lebanon is plunged into civil war after 5 years of confrontations, often bloody, between Lebanese and Palestinian refugees (*see* 1970). The Palestine Liberation Organization has used refugee camps as bases for guerrilla attacks on Israel, and there have been reprisals. Christians take up arms against leftist "Islamo-Progressivists," violence escalates as both sides begin using artillery, PLO leader Yasir Arafat, 46, pledges that his people will not involve themselves in Lebanon's affairs, but President Frangieh accuses the PLO in mid-December of violating its agreements and bringing on the civil war which is wrecking Beirut. Palestinian guerrillas more radical than the PLO invade an OPEC conference at Vienna December 21, kill three, and seize 81 hostages, including 11 OPEC ministers (*see* 1976).

West Germany's Baader-Meinhof gang has ended its hunger strike February 5 and goes on trial at Stuttgart in May (*see* 1974). Baader has lost 46 pounds, Meinhof 28, and court physicians confirm that prison conditions have endangered the lives of the defendants (*see* 1976).

Nixon cronies John Mitchell, H. R. Haldeman, John Erlichman, and Robert Mardian have drawn prison sentences of up to 8 years each February 21 for their participation in covering up White House involvement in the 1972 Watergate break-in.

A Rockefeller Commission report reveals excesses committed by the CIA. President Ford dismisses Secretary of Defense James R. Schlesinger, 46, and CIA Director William E. Colby, 62, November 2. Ford has escaped two assassination attempts in September, one by a 26-year-old member of the Charles Manson "family" (*see* crime, 1969), one by a 45-year-old former FBI informant, both women.

Cape Verde in the South Atlantic west of Senegal gains independence July 5 after 480 years of Portuguese colonial rule. Aristides Pereira is president.

Mozambique gains independence June 24 after 470 years of Portuguese rule. A Marxist government takes power with Maoist Samora Moises Machel as president.

The People's Republic of Angola is founded November 11 as the last Portuguese African colony gains independence, but Cuban troops with Soviet weapons and advisers are required to resist non-Marxist rivals for the power of President Aghostino Neto (*see* 1979).

Dahomey changes her name to Benin November 30 to commemorate an African kingdom that flourished in the seventeenth century.

Morocco invades Spanish Sahara after an International Court of Justice ruling in November that the people of Western Sahara have a right to self-determination and independence. Mauritania occupies the southern part, Spain withdraws, Algeria protests.

Spain's Generalissimo Francisco Franco dies November 20 at age 82 after a 36-year dictatorship. Prince Juan Carlos Alfonso Victor Maria de Borbon, 39, is proclaimed king to succeed his late grandfather Alfonso XIII (*see* 1931).

Suriname (Dutch Guiana) becomes an independent republic November 25 after 160 years of Dutch rule.

Papua New Guinea has attained independence September 16 after 74 years of Australian trusteeship.

Indonesian troops invade East Timor December 6 with the tacit approval of President Ford. A short-wave radio reports that troops are gunning down women and children in the streets. Exiles escaping from the mountainous island 400 miles off the Australian coast will appeal to the United Nations, but the Indonesians will prevent UN observers from visiting guerrillas who continue resistance in the jungles. Half the country's 600,000 people will have been killed or starved to death by 1980.

New Delhi has massive demonstrations against the Indira Gandhi government March 6 as at least 100,000 people march through the city. India's high court rules June 11 that Gandhi used corrupt practices to gain election to Parliament in 1971, that her election was invalid, and that she must resign. Gandhi vows to remain in office and has more than 750 political opponents arrested. Anti-government violence breaks out at New Delhi June 30, Gandhi announces steps to lower prices, reduce peasants' debts, and achieve fairer distribution of land in an appeal for political support, but she suppresses dissent and imposes strict press censorship.

Cambodia's new Khmer Rouge government (above) begins a wholesale slaughter of intellectuals, political enemies, dissidents, and peasants guilty of "mistakes."

Military vans equipped with loudspeakers urge doctors, technicians, and other professionals in Phnom Penh to turn out for "reconstruction;" all who respond to the April 18 appeal are murdered, as are all who refuse to leave their homes. Evacuees are settled in rural "communes," families are separated, marriage is abolished, and everyone over age 10 is put to work in the fields. Phnom Penh's Lycée Tuol Svay Prey is turned into the Tuol Sleng extermination camp in December.

The American Civil Liberties Union wins a $12 million damage suit January 16 on behalf of 1,200 clients whose rights were violated in 1971 when they were arrested during anti-war demonstrations at Washington, D.C.

The Supreme Court rules 5 to 4 January 22 in *Goss v. Lopez* that unless their presence poses a physical threat, students may not be temporarily suspended from school for misconduct without some attention to due process.

Congress votes July 28 to extend the Voting Rights Act of 1965 and broaden it to include Spanish-speaking Americans and other "language minorities."

The Age Discrimination in Employment Act passed by Congress November 28 strengthens an earlier law against age discrimination in the workplace.

The first U.S.-Soviet space linkup takes place July 18. Astronauts Thomas P. Stafford, Donald K. Slayton, and Van D. Brand exchange visits 140 miles above Earth with cosmonauts Aleksei A. Leonov and Valery N. Kubasov whose Soyuz spacecraft lands safely in the Soviet Union July 21. The Apollo astronauts splash down in the Pacific 3 days later, ending the Apollo missions.

Economic recovery begins in much of the world despite higher oil prices and continued high unemployment. Japan announces that her Gross National Product dropped by .08 percent in fiscal 1974, the first decline in "real" terms since the end of World War II. When the inflation rate subsides, the Japanese government announces a program of spending for public works, aid to small and medium sized business, and a lowering of the bank rate to catapult the nation out of recession, but a new 10 percent increase in OPEC prices dampens optimism.

British unemployment reaches 1.25 million in August. Prime Minister Wilson says November 5 that government aid for industrial development and rejuvenation must take precedence over social welfare programs such as nationalized health and subsidized housing.

U.S. investors fail to respond to the first opportunity they have had to buy gold since 1933. Gold prices fall when the expected U.S. "gold rush" does not materialize, dropping to below $140 per ounce by October, down from $174.50 in the London market January 2.

Wall Street's fixed commission rate ends May 1 by order of the Securities and Exchange Commission. Institutional investors (bank trust departments, insurance companies, and the like) begin negotiating lower rates, some will enjoy rates up to 90 percent lower than those paid at fixed rates, many brokers and dealers will be forced out of business in the next two years or will survive only by merging, and the change will encourage more individual investors to buy mutual funds. The reduction in the cost of doing business will make more money available for investment purposes. Brokerage houses will cut back on services to compensate for the loss in revenue.

Hourly wages for production workers average $6.22 in the United States, up from $3.15 in 1965; $7.12 in Sweden, up from $1.86; $6.46 in Belgium, up from $1.32; $6.19 in West Germany, up from $1.41; $4.57 in France, up from $1.19; $4.52 in Italy, up from $1.10; $3.20 in Britain, up from $1.13; $3.10 in Japan, up from 48¢.

The VISA credit card replaces the BankAmericard introduced in 1958 (it has been called VISA in some countries for some years).

New York City narrowly avoids defaulting on its bonds. "FORD TO CITY: DROP DEAD," headlines the *Daily News* October 30. Legislation passed by the House of Representatives 275 to 130 December 15 and signed by President Ford December 18 includes a special federal loan of nearly $2 billion for the city.

Discovery of oil and gas in the Pine View Field of Summit County, Utah, near the Wyoming border early in the year focuses attention on the Overthrust Belt, a 2,300-mile-long geological fold stretching from Montana south through Arizona. It will prove to be a major new energy source with an estimated 9.73 trillion cubic feet of natural gas (and possibly more than 20 trillion) plus at least 1 billion barrels of crude oil—30 percent of the crude oil reserves in Alaska's Prudhoe Bay.

North Sea oil pumping begins June 11 but provides for only a negligible part of Britain's oil needs. Oil supplies 42 percent of British energy, coal 30 percent, natural gas 17 percent, nuclear and hydroelectric power 11 percent.

The Energy Policy and Conservation Act of 1976 signed by President Ford December 22 sets gasoline mileage standards for automobiles and establishes strategic petroleum reserves. The reserves are to hold 1 billion barrels of oil underground in Louisiana and Texas caverns, but by the end of 1980 only 100 million barrels will be in storage, less than 2½ weeks' supply of imported oil.

The Great Uhuru Railway completed by Chinese engineers links Kapiri Mposhi, Zambia, with Dar es Salaam, Tanzania, 1,162 miles away, and collects $18 million in revenue from 530,000 tons of freight in its first 6 months. Soviet and Western engineers have refused to touch the project, but 45,000 black Africans and 15,000 People's Republic of China technicians have driven the railroad through some of Africa's roughest, most remote country, building 300 bridges, 23 tunnels, 147 stations.

The Chicago Rock Island & Pacific founded in 1852 files for bankruptcy March 18 as trucks and barges continue to take business away from U.S. railroads.

1975 ***(cont.)*** Egypt's Suez Canal reopens June 5, just 8 years after the outbreak of the 6-day Arab-Israeli War.

A U.S. Air Force Galaxy C-58 crashes April 4 after taking off from Saigon (most of the 172 killed are Vietnamese children); an Eastern Airlines Boeing 747 from New Orleans crashes June 24 on arrival at New York's Kennedy Airport, killing 113; a chartered Boeing 707 carrying Moroccan workers home from a vacation in France crashes August 3 on a mountain outside Agadir, killing 188; a Czech airliner crashes while landing at Damascus August 20, killing 126 of 128 aboard.

New York City transit fares rise from 35¢ to 50¢ September 1 (*see* 1972, 1980).

The Metric Conversion Act signed by President Ford moves the United States toward the metric system (*see* 1801). Compliance is voluntary, and few Americans take the law seriously, but the Pentagon and U.S. industries selling or manufacturing abroad have already adopted the metric system.

Microsoft is founded at Seattle by computer whiz William Henry Gates III, 19, and his friend Paul Gardner Allen, 22. Gates, who wrote his first computer program at 13 and scored a perfect 800 on his math S.A.T., has dropped out of Harvard to start what will be the biggest seller of computer software and will make Gates a billionaire before he is 30.

The first monoclonal antibodies—laboratory-made versions of the antibodies produced by the body to fend off viruses and other "foreign" substances—open a new era in diagnostic and therapeutic medicine.

Karen Ann Quinlan goes into coma April 15 after drinking alcohol mixed with small doses of Librium and Valium. The 21-year-old Landing, N.J., woman will be kept alive in a respirator for more than 8 years, even after her respirator is turned off June 10, 1976, following a court battle. She will be fed by a nasal tube and given antibiotics to ward off infections despite the fact that there is no hope of recovery, and her much-publicized case will raise continuing arguments about the right to die.

Lyme disease is identified in Lyme, Conn. Transmitted primarily by the bite of a tick found on white-tailed deer, white-footed field mice, and other animals, the bacterial infection can lead to serious neurological, cardiac, or arthritic complications. It will spread quickly throughout most of the Northeast and Middle Atlantic states, Wisconsin, Minnesota, and the West Coast.

Mexican authorities begin spraying illegal marijuana fields with the broad-leaf weed herbicide paraquat in a $35-million U.S.-funded program to wipe out the narcotic plant whose use is increasing among young Americans. Harvesters will market the weed after it has been sprayed, the Center for Disease Control at Atlanta will issue warnings that cannabis tainted with paraquat may cause irreversible lung damage, no case of paraquat poisoning will ever be confirmed, demonstrations by pot smokers at Washington will force the government to cut off support for the program in October 1979.

The U.S. first-class postal rate goes to 13¢ per ounce December 31 (*see* 1974, 1978).

Nonfiction *Roll, Jordan, Roll: The World the Slaves Made* by University of Rochester historian Eugene D. Genovese, 45; *Passage to Arrarat* by London writer Michael J. Arlen, 44; *The Great War and Modern Memory* by Rutgers English professor Paul Fussell, 51; *The Great Railway Bazaar* by U.S. writer Paul Theroux, 34.

Fiction *Humboldt's Gift* by Saul Bellow; *Ragtime* by E. L. Doctorow; *JR* by William Gaddis; *Hearing Secret Harmonies* by Anthony Powell; *A Division of Spoils* by Paul Scott completes his Raj Quartet; *Terms of Endearment* by Larry McMurtry; *Autumn of the Patriarch* by Gabriel García Márquez; *Conversation in the Cathedral* by Mario Vargas Llosa; *Guerillas* by V. S. Naipaul; *The Cockatoo* by Patrick White; *Shōgun* by James Clavell; *Dry Tortuga* by U.S. novelist Peter Matthiessen, 48.

Poetry *Self-Portrait in a Convex Mirror* by U.S. poet John Ashbery, 41.

Painting *Weeping Women, The Barber's Tree*, and five other encaustic cross-hatched works by Jasper Johns; *Whose Name Was Writ in Water* by Willem de Kooning; *Cubist Still Life with Lemons*, a "commercialized" reinterpretation of Picasso by Roy Lichtenstein; *Morro Da Viuva II, Montenegro*, and other brightly colored aluminum reliefs by Frank Stella; *The Magnet* and *Blue Light* by Philip Guston; stage sets for a production of the 1951 Stravinsky opera *The Rake's Progress* at Glyndebourne by David Hockney. Thomas Hart Benton dies at Kansas City January 19 at age 85.

Theater *The Ritz* by Terrence McNally 1/20 at New York's Longacre Theater, with Rita Moreno, Jerry Stiller, 406 perfs.; *Seascape* by Edward Albee 1/26 at New York's Shubert Theater, with Deborah Kerr, Barry Nelson, Frank Langella, 63 perfs.; *Same Time, Next Year* by Canadian playwright Bernard Slade, 44, 3/13 at New York's Brooks Atkinson Theater, with Ellen Burstyn, Charles Grodin, 1,444 perfs.; *No Man's Land* by Harold Pinter 4/23 at London's Old Vic Theatre, with Sir John Gielgud, Sir Ralph Richardson, Michael Feast, Terence Rigby; *The Taking of Miss Janie* by U.S playwright Ed Bullins, 39, 5/4 at New York's Mitzi E. Newhouse Theater, with Hilary Jean Beane, Diane Oyama Dixon, 42 perfs.; *Absent Friends* by Alan Ayckbourn 7/23 at London's Garrick Theatre, with Richard Brier, Peter Bowles; *Otherwise Engaged* by English playwright Simon Gray, 38, 7/30 at the Queen's Theatre, London, with Alan Bates, Jacqueline Pearce, Julian Glover.

Television *The Jeffersons* 1/18 on CBS with Sherman Hemsley; *NBC's Saturday Night Live* 10/11 with Chevy Chase, John Belushi, Gilda Radner, Bill Murray, and Dan Aykroyd; *One Day at a Time* 12/16 on CBS with Chuck McCann, Bonnie Franklin.

Films Richard Brooks's *Bite the Bullet* with Gene Hackman, Candice Bergen, James Coburn; Werner Herzog's *Every Man for Himself and God Against All* with Bruno S. as Kaspar Hauser; Steven Spielberg's *Jaws* with Roy Scheider, Robert Shaw, Richard Dreyfuss; Robert Altman's *Nashville* with Keith Carradine, Lily Tomlin, Shelley Duvall; Miloš Forman's *One Flew over the Cuckoo's Nest* with Jack Nicholson, Louise Fletcher; Lina Wertmuller's *Swept Away . . .* with Giancarlo Giannini, Mariangela Melato. Also: Claude Lelouch's *And Now My Love* with Marthe Keller, André Dussolier; Stanley Kubrick's *Barry Lyndon* with Ryan O'Neal, Marisa Berenson; Claude Lelouch's *Cat and Mouse* with Michele Morgan, Serge Reggiani; John Schlesinger's *The Day of the Locust* with Donald Sutherland, Karen Black, Burgess Meredith, William Atherton; Sidney Lumet's *Dog Day Afternoon* with Al Pacino; Joan Micklin Silver's *Hester Street* with Steven Keats, Carol Kane; Jan Kadar's *Lies My Father Told Me* with Yossi Yadin; Volker Schlondorff's *The Lost Honor of Katharina Blum* with Margarethe von Trotta; John Huston's *The Man Who Would Be King* with Sean Connery, Michael Caine, Christopher Plummer; Arthur Penn's *Night Moves* with Gene Hackman, Jennifer Warren; Melvin Frank's *The Prisoner of Second Avenue* with Jack Lemmon, Anne Bancroft; Frank Perry's *Rancho Deluxe* with Jeff Bridges, Sam Waterston, Elizabeth Ashley; Michael Richie's *Smile* with Bruce Dern, Barbara Feldon, Michael Kidd; Satsuo Yamomoto's *Solar Eclipse* (*Kinkanshoku*).

Broadway musicals *The Wiz* 1/5 at the Majestic Theater, with Geoffrey Holder, Stephanie Mills, music and lyrics by Charlie Smalls, songs that include "Ease on Down the Road," book based on L. Frank Baum's *The Wizard of Oz*, 1,666 perfs.; *Shenandoah* 1/7 at the Alvin Theater, with music by Gary Geld, lyrics by Peter Udell, 557 perfs.; *Chicago* 6/3 at the Forty-Sixth Street Theater, with Chita Rivera, Gwen Verdon, Jerry Orbach, music by John Kander, lyrics by Fred Ebb, 922 perfs.; *A Chorus Line* 10/19 at the Shubert Theater (after 101 perfs. at the off-Broadway Public Theater plus previews), with music by Marvin Hamlisch, lyrics by Edward Kleban, 36, 6,137 perfs.

Discothèques enjoy a resurgence unequaled since the early 1960s in major U.S. and European cities. Highly formula-conforming disco records by Van McCoy, Donna Summer, The Bee Gees, and others will dominate popular music until 1979.

Popular songs "Feelings" by U.S. singer/songwriter Morris Albert; "The Hustle" by U.S. songwriter Van McCoy, 35; *Frampton* (album) by English songwriter-singer-guitarist Peter Frampton, 25; "Love Will Keep Us Together" by early 1960s pop star Neil Sedaka, now 36, lyrics by Howard Greenfield; "At Seventeen" by U.S. folk singer-songwriter Janis Ian, 24; *Blood on the Tracks* (album) and "Tangled Up in Blue" by Bob Dylan; *On the Border* (album) and "Lyin' Eyes" by Los Angeles rock guitarist Glenn Frey, 27, and drummer Don Henley, 28, of The Eagles (Frey, Henley, guitarist Don Felder, 28, and bass guitarist Randy Meisner, 29) who popularize the California Sound—pop music with country and western influence; *Born to Run* (album) by U.S. singer-composer Bruce Springsteen, 26, whose energetic anthems to his blue collar background will make him one of the decade's most influential rock stylists; "Jive Talkin'" by The Bee Gees; "I'm Not in Love" by Graham Gouldman, 29, and Eric Stewart, 30, of the British pop group 10cc (Gouldman, Stewart, Lol Creme, 28, and Kevin Godley, 29).

Pittsburgh beats Minnesota 16 to 6 at New Orleans January 12 in Super Bowl IX.

Arthur Ashe, 32 (U.S.), wins in men's singles at Wimbledon, Mrs. King in women's singles; Manuel Orantes, 26 (Spain), wins in men's singles at Forest Hills, Chris Evert in women's singles.

U.S. tennis racquet sales peak at 9.2 million units; they will fall to 5 million by 1978.

New Zealand miler John Walker, 23, runs the mile in 3 minutes, 49.4 seconds August 12 at Göteborg, Sweden, beating Roger Bannister's 1954 mark by 10 seconds.

Muhammad Ali retains his world heavyweight boxing title September 30 by knocking out former champion Joe Frazier in the 14th round at Manila.

The Cincinnati Reds win the World Series by defeating the Boston Red Sox 4 games to 3.

The arbitration panel that made Jim "Catfish" Hunter a free agent last year rules December 23 that Montreal Expos pitcher Dave McNally, 33, and Los Angeles Dodger pitcher Andy Messersmith, 30, are no longer bound by their contracts and are free to negotiate with other teams. The ruling strikes at baseball's reserve clause which has allowed a team owner to renew contracts unilaterally on a yearly basis and has prevented a player who rejected the contract from playing for any other team (*see* 1976). Hunter signs a 5-year contract with the Yankees at $200,000 per year plus $2.5 million in bonuses, insurance, and lawyer fees.

Bobby Fischer relinquishes his world chess title at year's end (*see* 1972). Soviet grand master Anatoly Karpov, 23, succeeds as world champion.

Aim toothpaste, introduced by Lever Brothers, is a fluoridated gel that supplements Lever's Pepsodent and the Close-Up brand introduced in 1969. Aim and Close-Up will have 17.3 percent of the U.S. market by 1981, but competition from Procter & Gamble will prevent Lever from making any profit on the products.

Teamster boss Jimmy Hoffa disappears July 30 and is believed to have been murdered by underworld figures (*see* 1964).

FBI agents at San Francisco apprehend Patricia Hearst September 18 along with remnants of the Symbionese Liberation Army (*see* 1974). She is held on bank robbery charges.

London model Norman Scott, 35, walks his Great Dane bitch Rink along a Somerset road the night of October 24, an airline pilot who has driven him to the scene shoots the dog and tries to shoot Scott, and the gun jams. Scott charges that Liberal party leader Jeremy Thorpe, 46, has paid the pilot £5,000 to kill him and thus end rumors of an old homosexual relationship that are endangering Thorpe's political career (*see* 1979).

1975 ***(cont.)*** The Toxic Substances Control Act signed by President Ford October 12 requires phasing out all production and sale of PCBs (polychlorinated biphenyls) within 3 years. The measure imposes strict technology requirements on the chemical industry and will oblige smaller firms to quit production. PCBs have been linked to cancer and birth defects, and Congress has voted 319 to 45 August 23 to adopt the tough new environmental law.

The Indoor Air Act passed by the Minnesota state legislature requires all public places to have smoke-free areas, but enforcement will prove difficult. Arizona banned smoking in public places in May 1973, declaring smoking a public nuisance and a health hazard. It extended the ban last year but tobacco industry lobbyists work to prevent passage of such laws and to defeat popular referenda on smoking in public places, warning voters against "Big Brother" (*see* Orwell novel, 1949).

India's wheat crop is 24.1 million metric tons, up from 12.3 in 1965.

A U.S. agreement to sell the Soviet Union 6 to 8 million tons of wheat and corn per year is announced October 20, 10 days after President Ford has lifted an embargo on grain sales to Poland. Moscow agrees to buy a minimum of 6 million tons per year through 1981 (*see* 1980 embargo).

United Fruit Co. president Eli M. Black throws himself through the window of his office in New York's Pan Am Building February 3 and plummets 44 floors to his death. United Fruit has lost its leadership in the banana business to Castle & Cook, earnings have declined, and company executives who have not quit have been fomenting an insurrection against Black. Improper payments by United Fruit to foreign governments will come to light to help explain Black's suicide.

U.S. soft drinks edge past coffee in popularity and will pass milk next year.

Miller Lite Beer is introduced by Philip Morris, which acquired the brewery in 1970. The low-calorie beer will lift Miller from 7th place to 2nd, just behind Anheuser-Busch.

Japan's population reaches nearly 112 million October 1, up from 56 million in 1920. China has at least 843 million, India 615, the U.S.S.R. 255, the United States 213.

Cambodia's population of 7 to 9 million will decline to 4.8 million by 1980 as the Pol Pot regime (above) starves and brutalizes the people. Under normal circumstances, the population would have increased by about 15 percent.

1976 China's Premier Zhou Enlai dies of cancer January 8 at age 77 and is succeeded by former minister of public security Hua Guofeng, 54. Dogmatic Maoists revile Zhou and take away wreaths honoring him, Beijing has a riot April 5 as 100,000 people fill the main square of Tienanmen to join or watch a public protest against those who neglect the memory of Premier Zhou.

Chairman Mao Zedong dies of Parkinson's disease September 10 at age 82. His widow Jiang Qing is imprisoned early in October along with other "radicals" in her "Gang of Four" (Shanghai boss Zhang Chunqiao, 59; propagandist Yao Wenyuan, 45, and agitator Wang Hongwen, 41) for having undermined the party, government, and economy.

Cambodia's Prince Sihanouk resigns April 2 and Khmer Rouge leader Pol Pot becomes prime minister (*see* 1975).

A UN Security Council resolution calls for an independent Palestinian state and for total Israeli withdrawal from Arab territories occupied since 1967. The United States vetoes the resolution January 27; another U.S. veto February 25 blocks a resolution deploring Israeli policies in Jerusalem and in occupied Arab lands.

Civil war continues in Lebanon and threatens to involve Libya, Jordan, Egypt, Saudi Arabia, Israel, and the world powers. Libya's Muammar Qaddafi supplies the Palestine Liberation Organization with arms and money. Syria's President Hafez el-Hassad intervenes April 9 under an Arab League mandate to oppose the PLO's Yasir Arafat. U.S. Ambassador Francis E. Meloy, Jr., 59, is assassinated at Beirut June 17 and Washington advises all Americans to leave Lebanon. Lebanon's 19-month civil war ends in mid-November with the Syrian army in control; 35,000 have been killed, the once-beautiful city of Beirut is in ruins, and unrest continues.

Pro-Palestinian terrorists hijack an Air France plane and force its pilot to land at Uganda's Entebbe Airport. Airborne Israeli commandos storm the plane at Entebbe July 4, free 104 hostages, kill several Ugandan soldiers, and escape with only one casualty.

Israel signs an accord with Egypt October 10 agreeing to withdraw from 1,900 square miles of Sinai territory within 5 months; agitation continues over Israeli occupation of other Egyptian, Jordanian, and Syrian lands.

Mozambique closes her borders with Rhodesia March 3, saying that a state of war exists. Rhodesian Prime Minister Ian D. Smith orders scores of raids on guerrilla camps in Mozambique, U.S. Secretary of State Henry Kissinger says at Lusaka, Zambia, April 27 that Rhodesia must achieve black majority rule within 2 years, he announces a stiffening of U.S. economic and political sanctions against Rhodesia, and Ian Smith agrees September 24 to accept a plan for black rule based on a temporary biracial government with majority rule by 1978.

Spain has withdrawn from Spanish Sahara in February. Morocco annexes two-thirds of the phosphate-rich country April 14, Mauritania has annexed the other one-third, but a guerrilla group has proclaimed the region independent February 27 and has obtained Algerian support.

Transkei gains nominal independence from South Africa October 26. The Organization of African States and a UN Special Committee call the new state a sham since South Africa has declared 1.3 million Xhosa tribespeople Transkei citizens and deprived them of South African citizenship.

Argentina has a bloodless coup March 24 after months of terrorist attacks, robberies, and murders. Public officials and legislators have left office, scoffing openly at Juan Peròn's widow and her ministers. A military junta arrests Isabel Martinez de Peròn and declares martial law.

Mexican voters elect former financial secretary José Lopez Portillo, 56, president July 4. Lame-duck President Luis Echeverria Alvarez, whose 6-year term does not end until December 1, works to stabilize the nation's economy and to bolster his populist image, thus securing a power base among Mexico's peons although many hate him as the man who ordered the October 1968 massacre.

Britain's Prime Minister Harold Wilson resigns March 16. Foreign Secretary (Leonard) James Callaghan, 64, heads a new Labour ministry beginning April 5.

Violence continues in Ulster as it has since 1969. The British ambassador Christopher T. E. Ewart Biggs and his aide are killed July 22 by a land mine beneath their car.

Ulrike Meinhof hangs herself in her cell at Stuttgart's Stammheim Prison May 9 at age 42 after nearly 44 months in prison as West Germany's Baader-Meinhof gang trial continues (*see* 1975). Meinhof's burial in West Berlin a week later attracts a crowd of 4,000, many of them sympathizers wearing masks or with their faces painted white (*see* 1977).

Bernhard of the Netherlands resigns all his military and political posts August 26 after accusations that he was involved in a Lockheed bribery scandal.

The Seychelles in the Indian Ocean gain independence June 29 after 166 years of British colonial rule.

Japanese authorities arrest former prime minister Kakuei Tanaka July 26 on charges that he accepted a $1.6 million bribe from Lockheed Aircraft Corp. A U.S. Senate subcommittee on multinational corporations disclosed February 4 that Lockheed paid $7 million to a Japanese rightist with both political influence and underworld connections. Tanaka is indicted August 16 along with former officers of the Marubeni Trading Corp. Japan's Liberal-Democratic party loses its parliamentary majority in the December elections.

Swedish voters end 44 years of Socialist rule September 19. A conservative coalition headed by sheep-raiser Thorbjorn Falldin, 50, rallies the public against the Socialists' nuclear power program and against a plan that would give labor unions control of all business within 20 years.

Former Chilean ambassador to the United States Orlando Letelier, 44, dies at Washington, D.C., September 21 along with a colleague in a car bomb blast. Letelier has been an outspoken critic of Chile's President Pinochet, who overthrew President Allende in 1973, and of the Chilean secret police.

Jimmy Carter wins the U.S. presidential election with 297 electoral votes to 240 for President Ford. Former Georgia governor James Earl Carter, Jr., 52, has campaigned with a "born again" Christian religious fervor, has capitalized on the widespread distrust of Republicans in the wake of the Watergate affair, has won widespread black support, receives 51 percent of the popular vote to Ford's 48 percent, and will be the first Southern president since Zachary Taylor.

Chicago's Mayor Richard Daley dies December 20 at age 84 after more than two decades of dominating Illinois politics.

China's new leadership releases more than 100,000 political prisoners.

India's Supreme Court upholds the right of Prime Minister Gandhi's government to imprison political opponents without legal hearings. The court hands down a 34 to l ruling April 29 amidst charges that political prisoners are being tortured.

South African authorities suppress the worst racial violence in the nation's history after 5 days of rioting that began June 16. Soweto township 10 miles southwest of Johannesburg has more than a million inhabitants, the most populous urban concentration in southern Africa. Some 10,000 Soweto students, protesting public school instruction in Afrikaans, go on a rampage of looting, burning, and stone-throwing until police open fire. Riot squads restore order in 11 black townships after 176 people (including 2 whites) have been killed, 1,139 (including 22 police officers) injured. The language requirement is revoked July 6.

Blacks and other minorities are entitled to retroactive job security, the U.S. Supreme Court rules March 24.

Capital punishment does not constitute "cruel and unusual punishment," the U.S. Supreme Court rules in a 7-to-2 decision handed down July 2. The court held in 1972 that the death penalty was unconstitutional, Justice Thurgood Marshall citing evidence that the death penalty did not deter crime. Congress and most states have drafted new death penalty laws for murder-

South Africa's bloody Soweto riots brought decades of repression to a head, beginning the end of apartheid.

1976 ***(cont.)*** ers, and the court upholds such laws in Georgia, Florida, and Texas (*see* Gilmore, 1977).

Mexico devalues the peso August 31, allowing it to float after being pegged for 22 years at 12.5 to the U.S. dollar. By week's end, the peso has fallen to more than 20 per dollar. Thousands of U.S. and other foreign investors, attracted by interest rates of up to 12 percent, have bought more than $1 billion worth of fixed-interest peso bonds and sustain losses. Mexico had a $3.7 billion trade deficit in 1975, her foreign debt is $18 billion, her inflation rate is 45 percent. Devaluation quadruples prices of goods imported from the United States and inflation will be down to a 20 percent rate by mid-1977.

Mexican labor unions call off a nationwide strike set for September 24 and accept a 23 percent emergency wage increase to compensate for inflationary increases caused by devaluation (above). They warn that they will make new wage demands unless prices are controlled; 18 international banks meet at New York October 1 to arrange an $800 million Eurodollar loan to help Mexico's economic development.

Wall Street has a record 44.5-million share day February 20 and the Dow Jones Industrial Average peaks at 1,014.79 September 21, short of the 1,051.70 high reached January 11, 1973, but a high that will not be seen again for 6 years. The Dow Jones Average gains nearly 18 percent in the year, most of the rise coming in a big January spurt in which volume averaged 40 million shares per day. Daily volume averages 21 million.

Federal Trade Commission figures show that the 451 largest U.S. companies control 70 percent of the nation's manufacturing assets, up from about 50 percent in 1960, and earn 72 percent of the profits, up from 59 percent.

Kohlberg Kravis Roberts, an investment banking firm that will concentrate on mergers and acquisitions, is created in New York by bankers Jerome Kohlberg, Jr., 51, Henry Kravis, 32, and Kravis's cousin George Roberts, 32.

Drexel Burnham Lambert, an investment banking firm that will soon concentrate on high-yield "junk" bonds to finance corporate takeovers, is founded in New York through a merger.

Four new U.S. nuclear power reactors begin commercial production of electricity.

The Energy Policy and Conservation Act signed by President Ford December 22 sets gasoline mileage standards for automobiles, establishes strategic petroleum reserves, and authorizes the president to develop contingency plans for energy emergencies.

Regularly scheduled Air France and British Airways commercial supersonic Concorde flights begin January 21 between Paris and Rio and London and Bahrain, respectively. Both airlines begin Concorde service to Washington, D.C., May 24.

Pan Am begins nonstop New York-Tokyo service via Boeing 747 April 26.

A chartered 707 hits a mountain near Agadir, Morocco, August 3 killing 188; a British Airways Trident collides with a Yugoslav DC-9 in mid-air near Zagreb September 10 killing 176 in history's worst mid-air collision; a bomb explodes aboard a Cuban DC-8 near Barbados October 6 and the plane crashes, killing 73; an Indian Caravelle jet crashes after takeoff from Bombay October 12 killing 95; a Bolivian 707 cargo jet crashes October 13 killing the crew of 3 plus 97, mostly children, on the ground at Santa Cruz, Bolivia; an Egyptair 707 explodes and crashes at Bangkok December 25 killing 81, including many on the ground; an Aeroflot TU-104 crashes at Moscow's Sheremetyevo airport December 28 killing 72.

A $6.4-billion Railroad Revitalization and Reform Act signed by President Ford February 5 is followed March 30 by a $2.14-billion measure designed to improve rail service on the Boston-New York-Washington, D.C. line.

Washington, D.C., inaugurates its $5 billion Metro subway March 27 with a 4.6-mile line that will grow to have 75 miles by next year, reaching the city's Maryland and Virginia suburbs. Fares vary according to distance traveled.

Conrail (Consolidated Rail Corp.) begins operations May 1 with 88,000 freight workers as the federal government attempts to maintain service on lines served by the now bankrupt Penn Central, Ann Arbor, Boston & Maine, Central of New Jersey, Erie Lackawanna, Lehigh and Hudson Valley, and Reading. Railroads carry less than 37 percent of U.S. freight. Until 1981, Conrail will carry New York area commuters in addition to freight.

Brussels opens its first subway September 20.

A Cadillac convertible rolls off the assembly line at Detroit April 21, the last production model U.S. convertible for more than 5 years (*see* 1965).

Peugeot completes acquisition of Citroën, the French auto maker previously controlled by Michelin, to create P.S.A. Peugeot-Citroën (*see* 1895; 1955). The Peugeot family controls more than half the stock of the company which will soon rival FIAT as Europe's largest auto maker.

Libya buys a 10 percent interest in FIAT for $415 million.

Cincinnati Milacron enters the industrial robot business, challenging the pioneer firm Condec Corp. to produce robots that can be programmed to weld, drill, paint, and perform other assembly line functions (*see* 1887). U.S. companies will abandon the field to the Japanese in the late 1980s.

Apple Computer is founded in a California garage to produce personal computers. Founders Stephen G. Wozniak, 26, and Steven Jobs, 21, are college dropouts who have raised $1,500 and spent 6 months designing the crude prototype for Apple I (*see* 1977).

Guidelines issued June 23 by the National Institutes of Health end a voluntary moratorium on recombinant

DNA research in effect since 1974 but ban certain forms of genetic experimentation and tightly regulating others lest they create virulent new organisms which may take a deadly toll. The NIH wants to prevent transfer of drug-resistant properties to dangerous bacteria and release of synthetic genetic material. The Cambridge, Mass., City Council votes July 7 to place a 3-month moratorium on recombinant DNA research at Harvard, whose officers have approved plans for a genetics research laboratory to explore possible cures for cancer, production of rare hormones, and self-fertilizing plants that would increase world food production (George Wald has led opposition to the research).

M.I.T. researcher Har Gobind Khorana reports August 28 that his team has successfully constructed a bacterial gene, complete with regulatory mechanisms, and has implanted it in a living cell where it has functioned normally. They have produced a tyrosine transfer RNA gene from *Escherichia coli* bacterium nucleotides (the four basic chemicals of the genetic code), a breakthrough in genetic engineering (*see* 1966).

The International Council of Scientific Unions announces October 14 that it has formed a nongovernmental Committee of Genetic Experimentation to produce uniform regulations for recombinant DNA research.

Genentech is founded by University of California, San Francisco, biochemist Herbert W. Boyer, 40, and venture capitalist Robert Swanson, 28 (an M.I.T.-trained chemist), whose first commercial gene-splicing product is somatostatin, a brain protein.

A "swine flu" epidemic threatens the United States, warns President Ford on the advice of medical authorities. Washington mounts a $135-million inoculation program, 48 million Americans receive influenza shots, but the warning turns out to be a false alarm. Only six cases of swine flu are recorded, 535 of those inoculated develop the rare paralytic affliction Guillain-Barre Syndrome, hundreds bring suit against the government.

"Legionnaire's disease" kills 29 members of the American Legion who have gathered at Philadelphia's 72-year-old Bellevue Stratford Hotel for the Legion's annual meeting. The disease will prove to be a rare form of pneumonia.

Asia is free of smallpox for the first time in history, the World Health Organization reports November 13 (*see* 1977).

The FDA approves use of Inderal, a beta-blocker effective in treating high blood pressure.

Bavarian epileptic student Anneliese Michel, 23, dies of starvation July 1 after months of ritual by "exorcists" Wilhelm Rens, 65, Ernst Alt, 38, and the woman's parents Josef Michel, 59, and Anna, 45. Bishop Josef Stangl approved the exorcism to rid Anneliese of "demons" at the recommendation of Jesuit priest Adolf Rodewiyk of Frankfurt and appointed Renz and Alt to carry out the ritual.

Chinese universities reopen following the death of Mao Zedong (above). Most closed at the start of the Cultural Revolution in 1966. Those that remained open admitted students on political rather than academic grounds, scholars were sent to work in the countryside, and library acquisitions were halted. Academic degrees, abolished as an invidious status distinction, will not be reintroduced until 1980, and only 1 percent of college-age men and women will be able to attend college as compared to nearly 50 percent in America.

New York's multi-college City University of New York (CUNY) levies tuitions for the first time (*see* 1849). Under pressure from blacks, Hispanics, and community groups, CUNY adopted an open-admissions policy in 1970 and its enrollment has swelled from 174,000 to 268,000, mostly black and Hispanic. Its faculty has grown from 7,800 to 12,800. Budget cuts will reduce enrollment to 172,000 by 1980 and the faculty will be cut by some 3,000.

Word processors made by Wang Laboratories begin to revolutionize offices with work stations that share central computers (*see* 1974; Wang, 1951).

Fax (facsimile transmission) machines gain ground as second-generation technology cuts transmission time from 6 minutes per page to 3. The devices translate a printed page or graphics into electronic signals, transmit them over telephone lines, and print out signals received from other fax machines thousands of miles (or one block) away. Government offices, law enforcement agencies, news agencies, publishers, and banks are the major users. Prices fall for machines, but quality remains poor (*see* 1982).

U.S. citizens' band radio sales reach 11.3 million units with a retail value of over $3 billion.

The MacNeil/Lehrer Report debuts on U.S. Public Television stations. Canadian-American journalist Robert MacNeil, 44, and Texas-born journalist James Lehrer, 41, launched the Robert MacNeil Report last year with in-depth examination of the day's news events.

Atlanta's WTBS superstation is founded by local sportsman-entrepreneur Robert Edward "Ted" Turner III, 37.

Arizona Republic reporter Don Bolles, 47, dies June 13—11 days after a bomb ripped through his car in a Phoenix hotel parking lot. Bolles had been investigating fraudulent land deals and an underworld group associated with Arizona dog racing.

Mexico City newspaper editor Julio Schere Garcia and managing editor Hero Rodriguez are ousted July 8 in an apparent government effort to silence *Excelsior*, the nation's only outspoken central newspaper.

The *New York Post* is acquired in November by Australian publisher Rupert Murdoch, 45, whose late father Sir Keith Murdoch headed the *Melbourne Herald*. The younger Murdoch has built up a worldwide empire of 83 newspapers and 11 magazines with strong emphasis on scandals, sex, crime, and sports. His

1976 *(cont.)* lurid tabloid weekly *The Star* competes with *The National Enquirer* (*see* 1952) and is his most lucrative property.

Nonfiction *Blind Ambition* by former White House counsel John W. Dean 3rd; *The Russians* by *New York Times* reporter Hedrick Smith; *The Twilight of Capitalism* by Michael Harrington; *The Hite Report: A Nationwide Study of Female Sexuality* by U.S. cultural historian Shere D. Hite, 33.

Fiction *Roots* by Alex Haley; *Meridian* by U.S. novelist Alice Walker, 32; *Flight to Canada* by Ishmael Reed; *Henry and Cato* by Iris Murdoch; *The Kiss of the Spider Woman* by Argentinian novelist Manuel Puig, 44; *Les Flamboyants* by French novelist Patrick Grainville, 29; *Lady Oracle* by Canadian novelist Margaret Atwood, 37; *Woman on the Edge of Time* by U.S. novelist Marge Piercy, 40; *October Light* by John Gardner; *The Boys from Brazil* by Ira Levin; *The Takeover* by Muriel Spark; *Searching for Caleb* by Anne Tyler; *1876* by Gore Vidal.

Poetry *The Afterlife* by U.S. poet Larry Levis, 39.

Painting *Skull* (silkscreen) by Andy Warhol; *Portrait of Andy* (Warhol) by Andrew Wyeth; *Corpse and Mirror* by Jasper Johns; *Beginner* by U.S. painter Elizabeth Murray, 36; *Our Family in 1976* by U.S. painter Alfred Leslie, 49. Josef Albers dies at New Haven March 24 at age 88; Max Ernst dies at Paris April 1 at age 85; Man Ray dies at Paris November 18 at age 86.

Sculpture *Running Fence* (nylon, steel poles and cable, 24.5 miles long, 18 feet high) by Bulgarian-American artist Christo (Javacheff), 40, who has sold original drawings and collages of his plan in order to raise the $2 million needed for labor, materials, a 450-page environmental impact report, and legal fees in connection with 17 public hearings in California's Marin and Sonoma counties plus three superior court sessions; *Batcolumn* (grey-painted Corten steel, 100 feet tall) by Swedish-American sculptor Claes Oldenburg, 47, for Washington's new Social Security Administration building. Alexander Calder dies at New York November 11 at age 78.

Theater *American Buffalo* by U.S. playwright David Mamet, 28, 1/26 at New York's off-Broadway St. Clement's Theater, 135 perfs.; *Knock Knock* by Jules Feiffer 2/24 at New York's Biltmore Theater, with Lynn Redgrave, John Heffernan, 38 perfs.; *Streamers* by David Rabe 4/2 at New York's off-Broadway Mitzi E. Newhouse Theater, 478 perfs.; *Dirty Linen* and *New-Found Land* by Tom Stoppard 4/5 at London's Almost Free Theatre; *The Runner Stumbles* by U.S. playwright Milan Stitt, 34, 5/18 at New York's off-Broadway Little Theater, 191 perfs.; *California Suite* by Neil Simon 6/10 at New York's Eugene O'Neill Theater, with Tammy Grimes, George Grizzard, 445 perfs.; *For Colored Girls Who Have Considered Suicide/ When the Rainbow Is Enuf* by U.S. playwright Ntozake Shange (*née* Paulette Williams), 27, 9/15 at New York's Booth Theater (after 120 perfs. at the Public Theater), with Trazana Beverly, Laurie Carlos, Rise Collins, Aku Kadogo, June League, Paul Moss, and Shange, 867 perfs. (total); *A Texas Trilogy* (*Lu Ann Hampton Laverty Oberlander*, *The Oldest Living Graduate*, and *The Last Meeting of the Knights of the White Magnolia*) by U.S. playwright Preston Jones, 40, 9/22 at New York's Broadhurst Theater, with Fred Gwynne, Henderson Forsythe, Diane Ladd, 22 perfs. in repertory.

Films Alan J. Pakula's *All the President's Men* with Robert Redford, Dustin Hoffman, Jason Robards; Martin Ritt's *The Front* with Woody Allen, Zero Mostel (about blacklisting of alleged Communists in the 1950s); Marcel Ophuls's documentary *The Memory of Justice* about the Nuremberg trials after World War II; Sidney Lumet's *Network* with Faye Dunaway, Peter Finch, William Holden, Robert Duvall; Lina Wertmuller's *Seven Beauties* with Giancarlo Giannini; François Truffaut's *Small Change* with Geory Desmouceaux, Philippe Goldman. Also: Hal Ashby's *Bound for Glory* with Keith Carradine as Woody Guthrie; Bruce Beresford's *Don's Party* with John Hargreaves; Alain Tanner's *Jonah Who Will Be 25 in the Year 2000* with Myriam Boyer, Jean-Luc Bideau; Wim Wenders's *Kings of the Road* with Rudiger Vogler; Elia Kazan's *The Last Tycoon* with Robert De Niro; Gordon Parks's *Leadbelly* with Roger E. Mosley as the legendary blues singer, convict, and 12-string guitar player Huddie Ledbetter; Nicholas Roeg's *The Man Who Fell to Earth* with David Bowie, Rip Torn; John G. Avildsen's *Rocky* with Sylvester Stallone; Don Siegel's *The Shootist* with John Wayne, Lauren Bacall; Martin Scorsese's *Taxi Driver* with Robert De Niro, Cybill Shepherd, Harvey Keitel; Roman Polanski's *The Tenant* with Polanski, Isabelle Adjani.

Broadway musicals *Pacific Overtures* 1/11 at the Winter Garden Theater, with music and lyrics by Stephen Sondheim, 193 perfs.; *Bubblin' Brown Sugar* 3/2 at the ANTA Theater, with music and lyrics by Danny Holgate, Enice Kemp, and Lillian Lopez plus old songs by Duke Ellington, Noble Sissle, Eubie Blake, Andy Razaf, Fats Waller, and others, 766 perfs.; *Your Arms Too Short to Box with God* 12/22 at the Lyceum Theater, with Vinette Carroll, music and lyrics by Alex Bradford, additional music and lyrics by Micki Grant, 427 perfs.

Popular songs "I Write the Songs" by former Beach Boys arranger Bruce Johnston; "All by Myself" by U.S. singer-songwriter Eric Carmen, 27; "Love Hangover" by U.S. songwriters Pam Sawyer, 38, and Marilyn McLeod, 34; "Disco Lady" by H. Scales, L. Vance, and Don Davis; *Frampton Comes Alive* (album) and "Show Me the Way" by Peter Frampton who gains phenomenal popularity but whose youthful following will largely outgrow him within 3 years; *Fleetwood Mac* (album) by rock group Fleetwood Mac (drummer Mick Fleetwood, 34; bass guitarist John McVie, 31; singer Stevie Nicks, 28; and guitarist Lindsey Buckingham, 29).

Punk rock gains favor among England's working class youth with its highly amplified, politically rebellious style. U.S. bands performing at New York's CBGB's and Max's Kansas City rock clubs will develop this new wave rock with its back-to-basics approach.

Opera *Einstein on the Beach* 11/21 at New York's Metropolitan Opera House (after 29 European performances) with electronic music by U.S. composer Philip Glass, 49, staging by Robert Wilson.

Pittsburgh beats Dallas 21 to 17 at Miami January 18 in Super Bowl X.

Björn Borg, 20 (Swed), wins in men's singles at Wimbledon, Chris Evert in women's singles; Jimmy Connors wins in men's singles at Forest Hills, Evert in women's singles.

Soviet athletes win 142 medals at the Olympic Games held at Innsbruck and Montreal, East Germans 109, Americans 104.

A U.S. District court judge at Kansas City upholds last year's arbitration panel ruling on free agents February 4, a three-judge panel of the U.S. Court of Appeals at St. Louis upholds the federal judge's ruling March 9, and the first draft of free agents is held November 4 at New York's Plaza Hotel.

The Cincinnati Reds win the World Series, shutting out the New York Yankees.

Merit low-tar cigarettes are introduced by Philip Morris with a $40 million advertising budget and capture 1.5 percent of the market.

New York's .44-caliber killer claims his first victim early in the morning of July 29. A man who will prove to be David Berkowitz, 24, of Yonkers approaches an Oldsmobile double-parked in front of a Bronx apartment house and opens fire with a .44-caliber Bulldog revolver on medical technician Donna Lauria, 18, who has just returned from a Manhattan discothèque with her friend Jody Valenti, 19, a nurse; Berkowitz begins a 12-month career in which he will terrorize the city, killing five women and one man, leaving seven wounded (*see* 1977).

The Los Angeles World Trade Center goes up on a four-acre complex at 333 South Flower Street.

The Los Angeles Bonaventure Hotel designed by John Portman opens at Figueroa and 5th Streets with five bronze glass towers.

Detroit's 1,500-room Detroit Plaza hotel opens in the city's $337 million Renaissance Center.

Boston's Quincy Market reopens August 26, just 150 years after Josiah Quincy opened it, as Baltimore developer James Rouse, 62, restores the historic heart of the city. In addition to more than 25 suburban shopping malls, Rouse has built America's only "new city" (Columbia, Md.). He will reopen a second Quincy building next year and a third in 1978 while working on a mall for downtown Philadelphia and one for New York's South Street Seaport.

The worst earthquake in modern history shatters 20 square miles around Tangshan, China, July 28 leaving an estimated 655,000 dead. A Guatemala quake February 4 has taken nearly 23,000 lives, Indonesian quakes and landslides June 26 have left 3,000 dead and 3,000 missing, Mindanao in the Philippines loses up to 8,000 in an earthquake and tidal wave August 17 (the disaster temporarily quells a rebellion by the Muslim majority), and a quake in eastern Turkey November 24 kills 4,000.

An explosion at the Icmesa chemical factory in Seveso, north of Milan, July 10 releases a poisonous cloud of dioxin. Italian authorities evacuate 739 residents, but not until 13 days later. Thousands of rabbits, chickens, and dogs die after eating dioxin-coated vegetables, children develop skin rashes, the evacuees are shunned, nobody except decontamination experts will return to Seveso for 16 months and fears will linger that exposure to the dioxin may lead to liver disease, nervous disorders, stillbirths, birth defects, even cancer.

The U.S. Environmental Protection Agency has ordered U.S. Steel to close its Gary, Ind., coke plants January 1 because emissions violated clean air standards.

A California drought forces compulsory water rationing February 1 in Marin County north of San Francisco. Water prices go up 40 percent March 1 (*see* 1977).

The U.S. Department of Transportation issues an order November 18 requiring airlines to muffle 1,000 noisy plane engines that exceed noise limits established after they were built.

The Liberian tanker *Argo Merchant* runs aground December 16 off Nantucket, spilling heavy crude oil into the Atlantic. The vessel's bow splits in half the next day, spilling more, 180,000 barrels are lost, but strong winds move the slick away from resort areas.

President Ford signs legislation April 13 to extend U.S. jurisdiction over fishing rights to 200 miles offshore, effective March 1, 1977, and to ban fishing of 14 species unless they are shown to be in surpluses beyond the capacity of the U.S. fishing fleet. Japan's foreign ministry says that it regrets the unilateral U.S. action. Iceland has broken diplomatic relations with Britain over cod-fishing rights. Moscow imposes a 200-mile limit on foreign fishermen December 10.

Perrier water is introduced in U.S. markets (*see* 1863). Although consumers prefer cheaper brands like Canada Dry in blind taste tests, Perrier sales will reach $177 million in a decade as fitness-minded Americans switch from alcoholic beverages.

The Hyde Amendment to the Health, Education and Welfare appropriations bill clears Congress by a 256–114 vote September 16 and bars use of federal funds for abortions "except where the life of the mother would be endangered if the fetus were brought to term." From 250,000 to 300,000 U.S. women received Medicaid-funded abortions in 1975, critics call the new legislation discriminatory and unconstitutional, supporters—including Rep. Henry Hyde, 52 (R. Ill.)—contend that the government should not use tax revenues to fund operations which a substantial percentage of Americans consider immoral (*see* 1980).

1977 China expels her "Gang of Four" from the Communist party and restores purged leader Deng Xiaoping, 73, to power July 2. The first Communist Party Congress

1977 ***(cont.)*** since Mao's death last year elects pragmatic new Chinese leadership August 20.

Prime Minister Indira Gandhi frees most of India's political prisoners, but voters repudiate her repressive 18-month "emergency" rule that has tried to stifle political opposition in the world's largest democracy; former Prime Minister Morarji R. Desai, 81, returns to power and promises to restore morality to government.

Prime Minister Zulfikar Ali Bhutto of Pakistan loses power July 3 as his army chief of staff ousts the civilian government and imposes martial law, dissolving the national and state assemblies and banning all political parties. Authorities arrest Bhutto September 3 on charges of having conspired to murder his parliamentary critic Ahmad Raza Kasuri in 1974.

A Nuclear Non-Proliferation Treaty signed September 21 seeks to curb the spread of nuclear materials. The fifteen signatories include the United States and the U.S.S.R.

President Carter acts in January to pardon virtually all Vietnam era draft evaders and permit those living abroad to return without threat of persecution, but deserters are not pardoned. A spokesman for the National Council for Universal and Unconditional Amnesty says the action discriminates against deserters, who "are primarily and disproportionately from the poor and minority groups," whereas resisters "are essentially white, middle-class, and well-educated."

Two Panama Canal treaties signed by President Carter September 7 with Panama's head of state Brig. Gen. Omar Torrijos Herrera after 13 years of negotiations provide for a phasing out of U.S. control. Panamanians approve the treaties 2 to 1 in an October 24 plebiscite. U.S. opponents include former California governor Ronald Reagan who appropriates California state senator S.I. Hayakawa's remark about the Canal Zone, "Its ours. We stole it fair and square" (*see* 1978).

President Carter's friend Bert Lance, director of the Office of Management and Budget, comes under attack for questionable debts and financial commitments, the Controller of the Currency finds "unsafe and unsound" banking practices but no criminal behavior, a second Senate committee conducts hearings, and Carter tearfully accepts Lance's resignation September 21.

West German terrorists murder the attorney general in charge of the Baader-Meinhof gang prosecution April 7 along with his driver and bodyguard (*see* 1976). Andreas Baader and two accomplices are convicted 3 weeks later and sentenced to life terms for murder, complicity in 34 attempted murders, and forming a criminal association. "So-called political motives" are no excuse for terrorism, says the judge, and no reason for clemency. The head of the Dresdner Bank, Jurgen Ponto, is murdered at Frankfurt July 30 by his granddaughter, 26, an RAF member (*see* 1974). Five terrorists at Cologne on September 5 seize the head of the German Industries Federation, Hanns Martin Schleyer; they kill his driver and three bodyguards, demand a ransom, and demand the release of Baader and ten other RAF members. Further outrages ensue, and Baader is found shot dead in his cell October 18 at Stuttgart's Stammheim Prison; his girlfriend Gudrun Ensslin, 37, is found hanging from a bar in her cell window; other RAF members are killed or wounded. Schleyer is "executed" October 18, West Germany mobilizes 30,000 police officers and restricts civil liberties.

Ethiopia's president Gen. Teferi Benti, 55, and 10 others are killed February in a gunfight at a council meeting in Addis Ababa. Lieut. Col. Mengistu Haile Mariam is named head of state February 11; he ejects U.S. officials April 23 and brings in Cuban advisers, Somali forces threaten Harrar, and Moscow announces in October that it will cease military aid to Somalia, backing Ethiopia instead.

Djibouti (French Somaliland) gains independence June 27.

The Central African Republic becomes the Central African Empire December 4 as President Jean-Bedel Bokassa has himself crowned emperor in a $20-million ceremony funded largely by France. Bokassa I seized power at the end of 1963, replacing the elected government of his cousin David Dacko (*see* 1979).

Charter 77 draws worldwide attention in January as Czech authorities begin to crack down on its signatories—dissidents dedicated to safeguarding human and civil rights "in our country and the world." The authorities have violated last year's Helsinki Agreement, the U.S. State Department charges January 26. A U.S. statement January 28 warns Moscow not to attempt to silence Nobel scientist Andre Sakharov, who has accused the KGB of planting a bomb that killed several people in a Moscow subway car as an excuse to intensify repression of Soviet dissidents. President Carter meets at the White House March 1 with exiled dissident Vladimir Bukovsky.

Soviet authorities arrest Jewish human rights activist Anatoly Shcharansky, 28, March 15 after an open letter by a former political prisoner, published in *Izvestia,* has accused Shcharansky of working with alleged CIA agents. A U.S. protest brings charges by the U.S.S.R. and other countries that Washington is intruding in the internal affairs of foreign nations.

A South African magistrate rules December 2 that security police were blameless in the death of black leader Steve Biko, who was arrested in a crackdown on dissidents early in the year, kept naked for 19 days, shackled in handcuffs and leg irons for 50 consecutive hours, and driven 700 miles by Land Rover without medical attention before dying in a military hospital.

The U.S. Supreme Court rules 6 to 3 January 25 in *Oregon v. Mathiason* that a criminal subject who enters a police station voluntarily and is not under arrest may be interrogated without being informed of his legal rights as required in the *Miranda* decision of 1966.

The Supreme Court upholds racial quotas used in reapportioning legislative districts to comply with the Voting Rights Act of 1963. It hands down the 7-to-1 decision March 1 in *United Jewish Organizations of Williamsburg, N.Y. v. Carey*.

Kohlberg Kravis Roberts pioneers the leveraged buy-out April 7, using mostly bank loans to buy a small maker of truck suspensions (*see* 1976). KKR will employ high-yield "junk" bonds to finance future LBOs.

Nearly half of all Chinese urban workers receive 5 to 10 percent wage increases late in the year. Deputy Premier Deng Xiaoping (*see* above) begins de-communizing the nation with capitalist programs that increase productivity.

The Foreign Corrupt Practices Act passed by Congress December 7 provides for a fine of up to $1 million for any U.S. corporation found to have paid a bribe to a foreign government, political party official, or political candidate. A corporate official or employee is subject to imprisonment for up to 5 years and a $10,000 fine if convicted of involvement in such a bribe (*see* Lockheed scandal, 1976).

President Carter proposes a national energy program April 21 as U.S. imports of foreign oil continue to rise despite higher prices. Calling the situation "the moral equivalent of war," Carter urges major conservation efforts coupled with waste reduction and higher fuel prices to discourage consumption, but millions of Americans insist that the "energy crisis" has been fabricated by large oil companies to obtain price increases. Gasoline prices average less than 70¢ per gallon in most areas.

A power failure even worse than that of 1965 blacks out New York July 13 and continues for 25 hours during a heat wave. Looters in ghetto areas break into shops; business losses from theft and property damage come to nearly $150 million. Con Edison will be found guilty of negligence.

Oil from Alaska's Prudhoe Bay fields on the Arctic Ocean arrives July 28 at the ice-free port of Valdez on Prince William Sound, the first oil to come through the $9.7-billion, 789-mile pipeline (*see* 1969). Some 5.5-billion barrels will flow to U.S. refineries in the next 10 years, less than half the recoverable oil in the largest source by far of domestic petroleum.

The U.S. Department of Energy created by an act of Congress signed into law August 4 broadens federal control over all forms of energy. Former CIA director James Schlesinger is the first Secretary of Energy and his DOE will have 20,000 employees and an annual budget of $11 billion; critics will say the DOE only worsens the U.S. energy dilemma.

A new Japanese atomic energy plant opens September 1 and 7 new U.S. nuclear power reactors begin commercial operation.

An Australian train carrying 600 Sydney-bound commuters derails January 18; a bridge collapses and two wooden cars are crushed, killing 82, injuring 89.

A Chicago rush-hour elevated train, struck from behind February 4, plunges into a busy Loop intersection, killing 11 passengers and pedestrians, injuring 189.

The Orient Express that began service in 1883 makes its last trip into Istanbul from Paris May 22. Most travelers prefer to cover the 1,900 miles in 3 hours by air rather than take 60 hours by rail.

Canada creates Via Rail to take over the failing passenger services of Canadian Pacific and the government-owned Canadian National. Heavily subsidized, Via Rail will never show a profit or even break even (*see* 1989; Amtrak, 1971).

West Virginia's New River Gorge bridge is completed at Fayetteville; 1,700 feet long, it is the world's longest steel arch bridge.

A KLM Boeing 747 pilot misreads tower control instructions at Santa Cruz de Tenerife in the Canary Islands March 27, his plane collides on takeoff with a fog-shrouded Pan American 747 still on the ground, and the accident kills all 249 aboard the KLM jet plus 333 of the 394 aboard the Pan Am jet.

Laker Airways Skytrain service between London and New York begins in September with low fares that buck the International Air Transport Association cartel's fixed ticket price policy (which is based on operating profitably at 45 to 60 percent of seat capacity). Skytrain passengers line up for seats and most Laker flights are full (*see* 1966). Service on Laker Airways will continue for only a few years before other carriers force it out of business.

U.S. semiconductor makers Motorola, Advanced Micro Devices, Fairchild, Intel, and National Semiconductor form the Semiconductor Industry Association to lobby against government-subsidized Japanese efforts to dominate the industry.

The Apple II personal computer introduced by Stephen Wozniak and Steven Jobs requires that users employ their TV sets as screens and store data on audiocassettes, but it is an advance over Apple I and retails at only $1,298 (*see* 1976).

U.S. scientists report May 23 that they have produced insulin from bacteria in the laboratory, using recombinant DNA techniques of genetic engineering to change the bacteria (*see* 1937; 1976).

The first MRI (Magnetic Resonance Imaging) scanner is tested July 2 by Brooklyn, N.Y. medical researcher Raymond V. Damadian, 39, whose diagnostic tool will be widely used to detect cancer and other abnormalities without exposing patients to X-ray radiation. The scanner is based on the phenomenon that nuclei of some atoms line up in the presence of an electromagnetic field (*see* science, 1937). The FDA will approve commercial sale of MRI scanners in 1984.

Tagamet (cimetidine), a revolutionary new ulcer treatment drug developed in Britain, wins FDA approval August 23 (it has been available in Britain, Canada, and Mexico). SmithKline's drug blocks action of the body chemical histamine that stimulates secretion of gastric acids, the chief cause of ulcers (*see* 1746).

A cholera epidemic in the Arab states is reported September 10.

1977 *(cont.)* The world's last known natural case of smallpox is reported October 26 in Somalia. When no further cases are reported after 2 years, the disease that once killed an estimated 500,000 per year will be considered eradicated.

U.S. scientists identify a previously unknown bacterium that caused the "Legionnaire's disease" first reported last year.

Lung cancer deaths among U.S. women (14.9 per 100,000, up from 1.5 per 100,000 in 1930) pass colorectal cancer deaths (14.3 per 100,000) and begin to approach breast cancer deaths.

Moscow moves to discourage smoking with a ban on lighting up in the dining areas of all restaurants. Other cities prohibit smoking in government offices, shops, cinemas, sports arenas, hotel lobbies, even on beaches, but heavy smokers defy the ban. Party chief Leonid I. Brezhnev finally kicks his heavy smoking habit. Lung cancer, heart disease, and respiratory ailments have risen steadily in the U.S.S.R.

Buenos Aires newspaper publisher Jacobo Timerman of *La Opinion* is arrested April 15 on unspecified charges as Argentina's ruling junta cracks down on critics. Timerman is tortured and held incommunicado for several weeks, a military tribunal orders his release in October, the junta refuses, the Supreme Court will issue the same order in September 1979, and the junta will then strip Timerman of his citizenship and deport him.

South Africans see television for the first time May 10 as test transmissions begin from the state-backed South Africa Broadcast Co. The Pretoria government has yielded to public pressure after years of banning TV on grounds that it was morally corrupting. Half the broadcasts are in English, half in Afrikaans, but no more than 10,000 sets have been sold since sales began early in the year.

Nonfiction *A Rumor of War* by Vietnam veteran Philip Caputo, 36, who began the work as a novel; *Samuel Johnson* by Harvard scholar W. Jackson Bate, 59; *The Path Between the Seas: The Creation of the Panama Canal, 1870–1914* by U.S. historian David McCullough, 44; *The Dragons of Eden* by Cornell astronomer Carl Sagan, 42; *The Age of Uncertainty* by John Kenneth Galbraith; *The Complete Book of Running* by U.S. writer James S. Fixx, 45, who capitalizes on the new U.S. passion for jogging.

Fiction *The Flounder* (*Der Butt*) by Günter Grass; *Lancelot* by Walker Percy; *The Professor of Desire* by Philip Roth; *The Ice Age* by Margaret Drabble; *Daniel Martin* by John Fowles; *The Thorn Birds* by Australian novelist Colleen McCullogh, 39; *A Season in Purgatory* by Thomas Keneally; *Song of Solomon* by U.S. novelist Toni Morrison, 46; *The Silmarillion* by J. R. R. Tolkien; *Elbow Room* by U.S. novelist James Alan McPherson, 34; *Ackroyd* by Jules Feiffer; *The Honourable Schoolboy* by John Le Carré; *Blood Ties* by U.S. novelist Mary Lee Settle, 34; *Earthly Possessions* by Anne Tyler; *A Book of Common Prayer* by Joan Didion.

Painting *Looking at Pictures on a Screen* and *My Parents* by David Hockney, who employs a deadpan, subtly satiric style; *The Swimmer* by U.S. painter Robert Moskowitz, 42; *Nude in Profile* by French painter Balthus (Count Balthasaiklossowski de Rola), 69; Untitled Film Stills (black-and-white photographs) by U.S. artist Cindy Sherman, 23; *Self-Portrait* by U.S. photo-realist Chuck Close, 67, who has painted a photograph taken with a wide aperture and with the focus on the eyes; *Ocean Park* by Richard Diebenkorn continues a series begun in 1967. William Gropper dies at Manhasset, N.Y., January 6 at age 79.

Paris hails the opening of its Pompidou Center (Centre National d'Art et de Culture Georges-Pompidou, commonly called Beauborg for the neighborhood it graces). English architect Richard Rogers and Italian architect Renzo Piano have designed the combination art museum and performing arts center.

Theater *The Shadow Box* by U.S. playwright Michael Cristofer (*né* Procaccino), 32, 3/31 at New York's Morosco Theater, with Laurence Luckinbill as a terminal cancer patient, 315 perfs.; *The Basic Training of Pavlo Hummel* by David Rabe 4/24 at New York's Longacre Theater, with Al Pacino, 117 perfs.; *Gemini* by U.S. playwright Albert Innaurato, 28, 5/21 at New York's Little Theater, with Danny Aiello, 1,789 perfs.; *Da* by Irish playwright Hugh Leonard (John Keyes Byrne), 49, 7/18 at the King's Head Theatre, London, with Eamon Kelly, Tony Doyle, Mike McCabe; *The Gin Game* by U.S. playwright D. L. Coburn, 36, 10/6 at New York's John Golden Theater, with Hume Cronyn, Jessica Tandy, 518 perfs.; *The Night of the Tribades* (*Tribadernes natt, or Lesbian's Night*) by Swedish novelist-journalist-playwright Per Olov Enquist 10/12 at New York's Helen Hayes Theater, with Max von Sydow as August von Strindberg, Bibi Andersson, Eileen Atkins, Werner Klemperer, 12 perfs.; *A Life in the Theater* by David Mamet 10/20 at New York's off-Broadway Theater de Lys, with Ellis Rabb, Peter Evans, 288 perfs.; *Dracula* adapted by U.S. playwrights Hamilton Deane and John L. Balderston 10/20 at New York's Martin Beck Theater, with Frank Langella, 925 perfs.; *The Elephant Man* by U.S. playwright Bernard Pomerance, 36, 11/17 at London's Hampstead Theater, with David Schofield as John Merrick, the horribly misshapen victim of Proteus Syndrome (*not* neurofibromatosis) who was a patient at London Hospital from 1886 until his death in 1890; *Chapter Two* by Neil Simon 12/4 at New York's Imperial Theater, with Cliff Gorman, Anita Gillette, Judd Hirsch, Ann Wedgeworth, 857 perfs.; *Cold Storage* by U.S. playwright Ronald Ribman, 45, 12/29 at New York's Lyceum Theater, with Martin Balsam, Len Cariou, 227 perfs (including 47 at the off-Broadway American Place Theater).

Television *Roots* 1/27–30 on ABC with LeVar Burton, Cicely Tyson, Maya Angelou, Ben Vereen; *The Love Boat* 9/24 on ABC with Gavin McLeod.

Films Woody Allen's *Annie Hall* with Allen, Diane Keaton, Shelley Duvall; Steven Spielberg's *Close*

Encounters of the Third Kind with Richard Dreyfuss, François Truffaut; Fred Zinneman's *Julia* with Jane Fonda, Vanessa Redgrave, Jason Robards; Andrzej Wajda's *Man of Marble* with Krystyna Jancia. Also: Joan Micklin Silver's *Between the Lines* with John Heard, Lindsay Crouse; Pierre Schoendoerffer's *Le Crabe Tambour* with Jean Rochefort, Claude Rich, Jacques Dufilho, Jacques Perrin; Donald Cammell's *Demon Seed* with Julie Christie, Fritz Weaver; Kenji Mizoguchi's *A Geisha* with Michiyo Kogure; Herbert Ross's *The Goodbye Girl* with Richard Dreyfuss, Marsha Mason; Barbara Koppel's documentary *Harlan County, U.S.A.*; Robert Benton's *The Late Show* with Art Carney, Lily Tomlin; Bo Widerberg's *Man on the Roof* with Gustav Lindstedt; Bernardo Bertolucci's *1900* with Robert De Niro, Gerard Depardieu, Donald Sutherland, Burt Lancaster, Dominique Sanda, Stefania Sandrelli, Sterling Hayden; Kidlat Tahimik's *The Perfumed Nightmare* with Tahimik, Dolores Santamaria; George Lucas's *Star Wars* with Peter Cushing; Robert Altman's *Three Women* with Sissy Spacek, Shelley Duvall, Janice Rule; Ettore Scola's *We All Loved Each Other So Much* with Vittorio Gassman, Nino Manfredi, Stefania Sandrelli.

Charlie Chaplin dies at his Swiss estate December 25 at age 88.

Hollywood musicals John Badham's *Saturday Night Fever* with John Travolta, Karen Lynn Gorley, music and lyrics by The Bee Gees, songs that include "Stayin' Alive," "How Deep Is Your Love," and the title song; Martin Scorsese's documentary *The Last Waltz* about The Band's farewell concert at Thanksgiving, 1976.

Broadway musicals *Side by Side by Sondheim* 4/18 at the Music Box Theater, with music and lyrics by Stephen Sondheim, 384 perfs; *Annie* 4/21 at the Alvin Theater, with Andrea McArdle as the cartoon character "Little Orphan Annie" (*see* 1924), music by Charles Strouse, lyrics by Martin Charnin, 2,377 perfs.

Popular songs "You Light Up My Life" by U.S. composer-film director Joe Brooks, 37 (title song for his film); *Rumours* (album) by Fleetwood Mac becomes the largest-selling pop album thus far, indicating a dramatic increase in record sales that will continue until 1979; "Tonight's the Night" by English rock singer-songwriter-guitarist Rod Stewart, now 32; *Hotel California* (album) by The Eagles; "All Alone" by U.S. singer-songwriter Boz Scaggs (William Royce), 33; "Fly Like an Eagle" by U.S. rock singer-composer Steve Miller, 34; *Songs in the Key of Life* (album) by Stevie Wonder.

Rock music pioneer Elvis Presley dies at his Memphis mansion August 16 at age 42 after a game of racquetball; it is commonly accepted that Presley's addiction to prescribed barbiturates contributed to his early death. Bing Crosby collapses on a golf course at Madrid October 14 and dies at age 73.

Oakland beats Minnesota 32 to 14 at Pasadena January 9 in Super Bowl XI.

Björn Borg wins in men's singles at Wimbledon, Virginia Wade in women's singles; Guillermo Villas, 25, (Arg) wins in U.S. Open men's singles, Chris Evert in women's singles.

The U.S. yacht *Courageous* retains the America's Cup by defeating her Australian challenger *Australia* 4 to 0.

Seattle Slew wins U.S. horse racing's Triple Crown.

The New York Yankees win the World Series, defeating the Los Angeles Dodgers 4 games to 2.

U.S. blue jean sales top 500 million pairs, up from 150 million in 1957 and just over 200 million in 1967. Levi Strauss & Co. remains the largest producer, but higher priced designer jeans increase their share of the market and counterfeit labels proliferate.

Convicted murderer Gary Gilmore, 36, goes before a firing squad at his own request January 17 in Utah after a 10-year moratorium on capital punishment in America. His execution will be followed by 150 more in the next 15 years.

New York's .44-caliber killer continues his murders (*see* 1976). He kills Wall street clerk Christine Freund, 25, January 29 as she sits with her boyfriend in his car in Ridgewood, Queens. He shoots Virginia Voskerichain, 19, in the face at point-blank range March 8 less than 100 yards from the Freund shooting. He shoots Bronx student Valentine Suriani, 18, and her boyfriend Alexander Esau, 20, of Manhattan in a parked car April 17 a few blocks from last year's Lauria murder. He shoots Bronx student Judy Placido, 17, and her boyfriend Sal Lupo, 20, of Brooklyn June 26 in a parked car outside a Bayside, Queens, discothèque but both survive. Flatbush, Brooklyn, woman Stacy Moscowitz, 20, and her Bensonhurst boyfriend Robert Violante, 20, are shot July 31; she dies after extensive brain surgery, he is blinded. Psychotic Yonkers postal worker David Berkowitz, 24, is arrested August 10 and claims he has acted on orders from the dog of his neighbor Sam Carr, 64, who does not know him.

Canada stops granting licenses to carry handguns for protection of property and requires licenses for rifles and shotguns as well as for handguns.

Boston's Federal Reserve Bank building, completed by Cambridge, Mass., architects Hugh Stubbins & Associates, is an aluminum-sheathed 33-story tower.

New York's 60-story, aluminum-clad Citicorp Center, by Hugh Stubbins & Associates, is completed on Lexington and Third avenues between 53rd and 54th streets.

New York's 51-story Olympic Tower apartment and office building, designed by Skidmore, Owings & Merrill, is completed at 645 Fifth Avenue, just north of St. Patrick's Cathedral.

President Carter visits Charlotte Street in New York's rubble-strewn South Bronx October 5 and promises to revive urban renewal after 8 years of Republican neglect, but a $500-million program for the South Bronx, unveiled in April of next year, will take years to implement. Meanwile, the section will remain a disaster area of arson-gutted buildings.

1977 *(cont.)*  Major U.S. rivers freeze over in January and February as some cities record all-time low temperatures (Cincinnati -25° F., Miami Beach +25° F.) in the coldest winter on record.

A Romanian earthquake March 5 shatters most of downtown Bucharest, destroying more than 20,000 houses, killing at least 1,541, injuring more than 11,000.

A North Sea oil-well blowout in late April creates a 20-mile slick.

The Surface Mining and Reclamation Act passed by Congress July 21 requires that companies restore stripped coal lands to approximate original contours. A company must obtain a permit before it can strip-mine coal from farmland and must demonstrate its technological capability to restore the land to productivity.

A cyclone and flood from the Bay of Bengal November 19 leaves 7,000 to 10,000 dead in India's Andhra Pradesh state.

The Environmental Protection Agency has authority to establish industry-wide standards for discharging pollutants into waterways, the Supreme Court rules.

Congress amends the Clean Air Act of 1970 with new restrictions on air pollution but allows California to set even stricter limits in an effort to reduce that state's mounting smog problem. Auto makers will install catalytic converters that reduce tailpipe emissions by 90 percent to meet the California law (*see* 1989).

California has its worst drought year in history (*see* 1976).

Congress moves in December to ban U.S. manufacture of nearly all aerosol products containing fluorocarbons which pose a threat to the atmospheric ozone layer shielding the earth from much of the sun's ultraviolet radiation. Without that ozone protection, scientists say, there would be a sharp increase in skin cancers (*see* 1957). The new law exempts many products and other countries do not follow the U.S. move.

U.S. and Soviet control of the sea is extended to 200 miles offshore March 1, matching the limit set by Chile, Ecuador, and Peru in their 1952 Declaration of Santiago (*see* 1966). Japan does not recognize Soviet claims to the waters surrounding Soviet-occupied islands claimed by the Japanese.

A Canadian study linking saccharin intake with bladder cancer in rats leads the U.S. Food and Drug Administration to announce March 9 that it will ban use of saccharin in foods, soft drinks, chewing gum, and toothpaste (*see* 1907; cyclamate ban, 1969). Congress votes to delay the ban for 18 months, the British medical journal *Lancet* raises doubts that saccharin causes bladder cancer in humans, the FDA proposes label warnings and store signs pointing out the danger. Future studies will show that saccharin is at worst a weak animal carcinogen but may tend to enhance cancer-causing properties of other chemicals, especially to heavy smokers.

The U.S. Department of Agriculture begins efforts to reduce the amount of nitrites used by meat processors to color bacon following reports that crisply fried bacon contains nitrosamines—powerful carcinogens formed when the nitrites combine with amines produced naturally in the body. Some critics insist that no nitrites at all should be permitted, others point out that most human nitrite and nitrate intake is from leafy green vegetables. The meat industry says nitrates are needed to prevent development of botulinum toxins but will agree to use smaller amounts.

Nothing in the Constitution requires that states use Medicaid money to fund elective abortions, nor does any federal law, says the Supreme Court in a 6 to 3 ruling handed down June 20. The senate votes 56 to 42 June 29 to bar the funding of elective abortions except in cases of rape, incest, or medical necessity. The House votes 238 to 162 August 2 to bar such funding except where childbirth would endanger the life of the mother. Right-to-life groups have mobilized to defeat politicians supporting legalized abortion; critics object that abortion, while legal, is becoming a privilege for the rich. President Carter concedes that discriminating against the poor in abortion matters is unfair but echoes the late President Kennedy in saying, "There are many things in life that are not fair." Rosaura Jiminez, 27, dies in pain October 3 at the hands of an illegal McAllen, Texas, abortionist, leaving a 5-year-old daughter. The cutback in federal Medicaid funds for abortion drove her to seek out the local woman, and hers is the first recorded death by illegal abortion since the cutback.

India's birth control efforts will collapse in the wake of Indira Gandhi's defeat (above) when it is revealed that 500 unmarried women were forcibly sterilized during Gandhi's "emergency" and that 1,500 men died as a result of improper vasectomies. The scandal will make the very words "family planning" taboo, the Ministry of Health and Family Planning will become the Ministry of Health and Family Welfare, and while surgical sterilization will remain the surest and most prevalent form of birth control, voluntary sterilization will decline to 1.8 million per year, down from more than a million per month in 1975. Millions of women will become pregnant who would not have done had government programs not been cut back.

The U.S. State Department proposes December 3 that 10,000 Vietnamese "boat people" be admitted on an emergency basis. Since the fall of Vietnam 2 years ago, the United States has approved admission of 165,000 Indochinese refugees.

1978  Iran's holy city of Qom has religious riots January 7—fifteenth anniversary of the shah's land reform and women's emancipation decrees, both despised by the nation's 180,000 Muslim preachers. The shah has the chief of the Savak secret police arrested June 9 on charges of corruption and torturing prisoners. A packed movie theater at Abadan burns August 20 with a loss of 377 lives; opponents of the shah charge that Savak agents set the fire, the government blames Islamic Marxists. The prime minister resigns August 27 and the shah, hoping to appease his opponents, closes gambling casinos and dismisses high-ranking members of the Bahai sect, including his personal physician. Martial law is imposed in the capital and 11 other cities after 100,000 march in an anti-shah demonstration at Teheran. Troops open fire in Teheran's Jaleh Square, killing 121 demonstrators, wounding 200 others. The cabinet resigns November 6 and Iran's first military government since 1953 comes to power. Moscow cautions Washington against military intervention, Washington warns Moscow to stay out. A Teheran mob trying to storm the U.S. embassy December 24 is driven away by Marine guards using tear gas. The shah asks a leader of the opposition National Front to form a new civilian government December 29 and Shahpur Bakhtiar, 62, assumes power (*see* 1979).

Afghanistan has a bloody coup April 27 as pro-Soviet leftists oust President Mohammed Daud, a former prime minister who in 1973 ended the monarchy of Mohammed Zahir Shah, abolished all royal titles including his own, and proclaimed a republic. Mur Muhammad Taraki, 60, succeeds Daud as president and concludes a 20-year economic and military treaty with Moscow (*see* 1979).

The president of North Yemen (The Yemen Arab Republic) is killed by a bomb June 24 as he receives the credentials of the new ambassador from South Yemen (The People's Democratic Republic of Yemen), whose presidential council head is ousted and executed June 26. A new president of South Yemen is elected by the Supreme Council December 27, he reverses moves toward reconciliation with North Yemen, acquiesces to a continuing Soviet military buildup in his country, but will embarrass Moscow and resign abruptly in 1980 (*see* 1979).

Pakistan's President Fazel Elahi Chaudhry leaves office September 16 after a 5-year term; the chief martial law administrator General Mohammed Zia ul-Haq declares himself president.

Syrians have begun fighting with Lebanon's Christian militia in February, an Al Fatah guerrilla assault on the Haifa-Tel Aviv road March 11 kills 30 Israeli civilians, Israel invades Lebanon March 14, pushes back Palestinians who have been harrassing Israelis, establishes a "security zone," and begins a phased pullback April 11 as a UN peacekeeping force occupies a buffer zone between the border and the Litani River. Yasir Arafat of the PLO agrees May 24 to keep his forces out of the UN buffer zone, and Israelis complete their withdawal June 13, turning over posts to the Christian militia who declare an independent enclave but permit UN forces to occupy some sites.

The Camp David accord reached September 17 after 13 days of negotiating by Egypt's President Sadat, Israel's Prime Minister Begin, and President Carter provides what all parties call a "framework for peace" in the Middle East. President Carter has summoned Sadat and Begin to Camp David and worked intensively to keep the two parties talking (*see* 1977; 1979).

A Soviet fighter jet has attacked a Korean Air Lines commercial jet bound from Paris to Seoul April 20

The Camp David Accord ended 2 decades of Israeli-Egyptian hostility but other Arab states remained adamant.

when defective navigational equipment caused the KAL 707 to stray over Russian territory. Two passengers were killed and 10 injured by the Soviet missile; the plane crash-landed on a frozen lake near Murmansk.

A Sino-Japanese treaty of friendship signed August 12 brings Soviet charges that the pact is hostile to Moscow. Beijing and Washington then announce that they will reopen full diplomatic relations January 1, 1979.

The Solomon Islands gain independence July 7 after 85 years of British rule.

Washington recognizes the People's Republic of China December 15, announcing that it will sever diplomatic ties with Taiwan as of January 1, 1979, despite objections that this will abrogate a 1954 treaty of mutual security and will mean deserting an old friend.

The Panama Canal treaties approved by the U.S. Senate 68 to 32 March 16 and April 18 provide for Panamanian operation of the canal beginning December 31, 1999. Neutrality of the canal is guaranteed. The action is intended to begin a new era in U.S.-Latin American relations.

Nicaraguan leftist guerrillas seize the National Palace at Managua August 22 and hold hundreds hostage for two days in a bid to oust dictator Anastasia Somoza. The Sandinista Liberal Front takes its name from guerrilla Gen. César Augusto Sandino (*see* 1934). Sandinista rebels free all but eight hostages August 24 in exchange for the release of political prisoners held by the Somoza regime, $71,000 in cash, and safe passage to Panama, where they seek asylum (*see* 1979).

Guatemala elects Gen. Romeo Lucas Garcia president. Washington cuts off military aid as Lucas Garcia begins a brutal and corrupt 4-year reign.

Dominica in the Caribbean gains independence November 3 after 153 years of British colonial rule.

1978 *(cont.)* Italian Red Brigade terrorists kidnap former premier Aldo Moro, 61, March 15 after killing Moro's five bodyguards. The government rejects demands to release imprisoned terrorists, the Red Brigades announce that they have held a "people's trial" and have found Moro guilty. His bullet-riddled body is found in a parked car in Rome May 9, and the election of socialist Sandro Martini, 81, to the presidency July 8 assures continued cooperation between Italy's Communists and the Roman Catholic Christian Democrats.

Kenya's president Jomo Kenyatta (*né* Kaman Ngengi) dies August 22 at age 86 after 15 years in office and is succeeded by Daniel Arap Mori, 53.

South Africa approves a UN plan to set up an independent government in Namibia (South-West Africa). Pretoria agrees December 22 to let a peacekeeping force of 1,500 South Africans and 7,500 UN troops police the 318,261-square mile area (estimated population: 1 million) and to hold UN-supervised elections in 1979. Differences will develop over UN proposals for monitoring guerrillas of the South-West Africa People's Organization (Swapo) during a cease-fire and over South African demands that the internal parties have a formal part in negotiations (*see* 1974).

Tuvalu (Ellice Islands) in the western Pacific gains independence September 30 after 86 years as a British protectorate.

Vietnamese troops invade Cambodia (Kampuchea) December 25, using Soviet-supplied arms to drive out Pol Pot's regime (*see* 1975; 1979).

Algeria's President Houari Boumedienne (*né* Mohammed Ben Braham Boukharuba) dies December 27 at age 46 after 13 years in office. He is succeeded by Chadi Benjedid, 49.

President Carter signs legislation April 6 raising the mandatory retirement age for most U.S. workers to 70.

The Bakke decision handed down by the Supreme Court June 28 upholds a reverse-discrimination ruling and thus jeopardizes affirmative action programs designed to help minority students gain admittance to U.S. colleges and graduate schools. National Aeronautics and Space Administration engineer and Marine Corps Vietnam veteran Allen Paul Bakke, now 38, applied in 1973 to the University of California Medical School at Davis and was twice rejected, not because of age or lack of any qualification but because a special-admissions minority program had reduced the number of places open to white applicants (16 of the 100 seats were reserved for minority applicants, and the total number of applicants was 3,737). The Court rules 5 to 4 that the special-admissions program at Davis violates Title VI of the Civil Rights Act. A different majority of five justices condones some consideration of race in the admissions process, so while the Bakke decision clearly forbids racial quotas and non-competitive evaluations of minority candidates, it does not bar schools from considering race and ethnic origin in their efforts to obtain diversity among student bodies.

U.S. Ambassador to the United Nations Andrew Young compares Soviet dissidents to U.S. civil rights activists and says July 12 that there are "political prisoners in both countries." The White House disavows Young's views and President Carter rebukes him.

A Moscow court sentences Soviet dissident Anatoly Shcharansky to 13 years of prison and hard labor in a ruling handed down July 14. Shcharansky has upheld the Final Act of the Helsinki Agreement of 1975, "the right to know and act upon one's rights." In the courtroom he has cried out, "Next year in Jerusalem." World opinion has failed to sway the judges.

Unemployment rises throughout the world with a U.S. rate of 6 percent, up from 4.9 in 1973 (twice as many blacks are jobless as in 1968). Britain's rate is 6.1 percent, up from 2.9; France's 5.5, up from 2.7; West Germany's 3.4, up from 0.8.

California voters approve a $7-billion (57 percent) cut in property taxes (Proposition 13), and the support by 65 percent of the electorate June 6 sparks similar tax revolts in other states.

U.S. inflation pressures force President Carter to act. He announces a program of voluntary wage-price guidelines October 24, resisting demands that he impose mandatory controls and raising fears that inflation will worsen. Wall Street's Dow Jones Industrial Average nevertheless leaps a record 35.4 points November 1.

President Carter signs an $18.7-billion tax-cut bill November 6 despite feelings that it favors middle- and upper-income taxpayers.

Tokyo's Nikkei Dow Jones average closes above 6,000 for the first time December 1. It reached 5,000 for the first time in December 1972, then plunged in the wake of the 1973 oil crisis and did not reach 5,000 again until January 1978.

Cleveland defaults on $14 million in debts December 16 and another $1.5 million in notes held by the city treasury are extended.

U.S. money market funds, paying interest rates far higher than those obtainable at banks, credit unions, or savings and loan institutions, have assets of $10 billion by year's end. They will have $100 billion by March 1981, all of it invested in large-denomination, high-yielding financial instruments not previously available as investment vehicles except to large institutions (*see* 1974).

The Nuclear Exports Control Act (Nuclear Non-Proliferation Act) signed by President Carter March 10 imposes strict new controls on export of U.S. nuclear technology to prevent the spread of atomic weapons. A receiving nation may not further enrich reactor-grade U.S. uranium to the 90 percent-plus concentration needed for a bomb. India will not agree to such a limitation and the Nuclear Regulatory Commission acts April 20 to bar the sale of 7.6 tons of enriched uranium for two nuclear projects in India. President Carter approves the sale April 27.

Five new U.S. nuclear reactors begin generating electricity on a commercial basis.

Nuclear fusion enthusiasts claim a breakthrough July 4 when the Princeton Large Torus test reactor, fueled by deuterium gas, reaches a temperature of 60 million degrees F. and holds it for one-twentieth of a second. Scientists strive to attain a temperature that will match the sun's—100 million° F.—and sustain it for pulses of a full second or more, making it commercially feasible to produce electricity from heavy isotopes of hydrogen, extractable in abundance from seawater.

The Nuclear Regulatory Commission suspends construction of a Seabrook, N.H., atomic energy facility but rules August 10 that construction may resume despite opposition from the "Clamshell Alliance" of environmentalists who vow to continue nonviolent civil disobedience protests. West Coast opponents of nuclear energy demonstrate against the Trojan nuclear plant near Rainier, Ore., and the Diablo Canyon plant near San Luis Obispo, Calif.

President Carter signs a National Energy Act November 4 containing a controversial regulation deregulating natural gas but forbidding electric utilities to burn natural gas. Within a year there will be such a glut of natural gas that the Department of Energy will be urging utilities to use gas.

Ministers of 13 OPEC nations meet in Abu Dhabi and agree December 17 to raise oil prices by 14.5 percent in four stages by the end of 1979, from $12.70 per barrel to $14.54 by October 1, 1979. Libya and Iraq have demanded increases of 20 percent or more but agree to the lower hike, thus ending an 18-month price freeze. The OPEC accord affects about half of all world crude oil production.

Petroleos Mexicanos (Pemex) announces December 20 that it will gradually raise its export price of crude oil 10.7 percent to $14.50 per barrel by the end of 1979.

Germany's last Volkswagen "Beetle" sedan rolls off the assembly line January 19 at Emden. The Wolfsburg plant switched over to producing Rabbits and other new models in July 1974 and the Emden plant's cars are the last of 19.2 million Beetles made since 1949 (*see* 1934), a number that surpasses the production record of 15,007 million Ford Model Ts set in the 1920s. Volkswagen plants in Brazil, Mexico, Nigeria, and South Africa continue to turn out Porsche-designed Beetles.

A 1973 Ford Pinto explodes in flames August 10 after being struck from behind by a van near Goshen, Ind. Three young women are fatally burned. Criminal charges will be brought against Ford Motor Company and a Winimac, Ind., jury will decide for acquittal in March 1980 after Ford shows it did everything it could to recall the Pinto beginning in June 1978 following a government investigation of complaints about the car.

Tokyo's new International Airport opens in May at Narita, 41 miles outside the city, after violent protests by farmers and students that have delayed the opening 2 months.

An Air India Boeing 747 takes off from Bombay January 1 en route to Dubai in the Middle East; it crashes into the sea, killing 213. A Pacific Southwest Boeing 727 collides with a Cessna while landing at San Diego's Lindbergh Field September 25, killing all 135 aboard plus 2 in the Cessna and 4 on the ground. An Icelandic Airlines DC-8 carrying Indonesian Muslims home from Mecca crashes on landing at Colombo, Sri Lanka, November 15, killing 183.

The Airline Deregulation Act signed by President Carter October 24 provides for a phasing out of federal regulation of the U.S. industry (*see* CAA, 1934). Civil Aeronautics Authority control of routes will end in 1982, pricing power in 1983, and the Board will expire January 1, 1985. Airlines begin dropping shorter, less profitable routes and competing for business on longer routes and routes with high volume, cutting fares to attract passengers despite higher fuel costs.

Cigarette smoking is "slow motion suicide," says HEW Secretary Joseph A. Califano, Jr., January 11. A onetime heavy smoker who has quit at the urging of his children, Califano announces a new government campaign to discourage children and teenagers from smoking and to help 53 million U.S. smokers quit the habit. He does not challenge the $80 million Department of Agriculture program to support tobacco prices. The American Tobacco Institute calls the Califano plan an intrusion on civil liberties. President Carter visits North Carolina August 4 and undercuts Califano by pledging government support of efforts to make cigarettes "even safer than they are."

The first recombinant DNA product—human insulin—is produced at the City of Hope Medical Center, Duarte, California, in a joint effort with Genentech, Inc. The U.S. Food and Drug Administration will approve its use in 1982.

Pope Paul VI dies August 6 at age 80 after a 15-year pontificate; his 65-year-old successor John Paul I (Albino Luciani) dies September 28 after 34 days; and Polish Cardinal Karol Wojtyla, 58, is elected pope October 16, becoming John Paul II—the first non-Italian pontiff since 1523.

Jonestown, Guyana, is the scene of the most sickening religious mayhem since John of Leyden's excesses in 1535. U.S. Congressman Leo J. Ryan of California arrives at the remote settlement November 17 to investigate complaints from constituents about treatment of relatives in the commune established by former San Francisco clergyman Jim Jones, 47. Accompanied by 17 staff members and several newsmen, Ryan meets with Jones, who denies any mistreatment and keeps most commune members away from Ryan. Some 20 members of the People's Temple tell Ryan they want to leave and accompany him to Port Kaituma airport November 18, but as the group starts to board two small planes cult member Larry Layton pulls a handgun and fires, wounding two people before fleeing. Three other men arrive, wound Ryan and four others, then shoot each in the head. After a planeload of survivors takes off for Georgetown, Jones has 911 of his

1978 *(cont.)* followers drink Kool-Aid laced with cyanide. Those who refuse are shot or forcibly injected with cyanide; Jones shoots himself in the head or is murdered; only 32 escape.

Congress entitles all U.S. college students to federally insured, subsidized loans regardless of family income.

Beijing adopts the Pinyin (transcription) system for spelling most Chinese names in Roman letters, replacing the old Wade-Giles system in an effort to render the sounds of Mandarin Chinese more accurately. Peking becomes Beijing, Tibet Xisang, etc., but most Western publishers will retain a few Wade-Giles spellings. Some Pinyin letters have unique sounds (e.g., x = sh, as in she; q = ch, as in cheek; zh = j, as in jump).

Nicaraguan publisher Pedro Joaquin Chamorro is gunned down in the streets of Managua January 10. His newspaper *La Prensa* has bitterly opposed dictator Anastasio Somoza (*see* above), and his death gains support for Sandinista rebels as members of the Chamorro family continue publication.

Self magazine begins publication at New York in January as Condé Nast expands.

Working Woman magazine begins publication at New York in March.

The U.S. first-class postal rate goes to 15¢ per ounce May 29 (*see* 1975, 1981).

"Garfield" by U.S. cartoonist Jim Davis begins syndication to newspapers. Garfield is a cat.

New York newspapers resume publication November 6 after an 88-day strike by 10 unions representing some 11,000 workers. Automation was the underlying issue.

The *Times* of London and the *Sunday Times* suspend publication November 30 after failure to reach agreement with the printing unions over use of new computer typesetting equipment, a reduction in jobs, and changes in the procedures for settling disputes to eliminate slowdowns and wildcat strikes that have cost the paper nearly £4 million in lost revenues since January 1. Publication will not resume for nearly 11 months.

Ninety-eight percent of U.S. households have television sets, up from 9 percent in 1950, 83.2 percent in 1958; 78 percent have color TV, up from 3.1 percent in 1964, 24.2 percent in 1968.

A new U.S. copyright law takes effect January 1, superseding the 1909 law and extending protection for up to 75 years from the date of publication (28 years plus a renewal term of 47 years, or for the author's lifetime plus 50 years).

Fiction *Falconer* by John Cheever; *The Coup* by John Updike; *The World According to Garp* by U.S. novelist John Irving, 36; *An Imaginary Life* by Australian novelist David Malouf, 44; *The Sea, The Sea* by Iris Murdoch; *War and Remembrance* by Herman Wouk; *Chesapeake* by James Michener; *The Girl with a Squint* by Georges Simenon, now 75; *Scruples* by Beverly Hills novelist Judith Tarcher Krantz, 50.

Soviet authorities permit a Moscow show of avant garde paintings to open March 7 but only after a score of works have been removed for ideological reasons. Painters represented include Vitaly Limtsky, Vladislav Pomotorov, and Nikolai N. Smirnov; but Vladimir N. Petrov-Gladki, Nikolei N. Rumyantsev, and others are excluded.

Painting *Children Meeting* by Elizabeth Murray; *Self-Portrait* by Andy Warhol. Norman Rockwell dies at Stockbridge, Mass., November 13 at age 84; Giorgio de Chirico dies at Rome November 20 at age 90.

Sculptor Christo covers 3 miles of winding footpaths in Kansas City's Loose Park with nylon October 3 in a $100,000 project.

Theater *Deathtrap* by Ira Levin 2/26 at New York's Music Box Theater, with John Wood, Marian Seldes, Victor Garber, 1,793 perfs.; *Plenty* by David Hare 4/14 at London's National Theatre (Lyttleton); *Tribute* by Bernard Slade 6/1 at New York's Brooks Atkinson Theater with Jack Lemmon, 212 perfs.; *Night and Day* by Tom Stoppard 11/8 at London's Phoenix Theatre, with Diana Rigg, Peter Machin, William Marlowe, Ohu Jacobs, David Langton; *Betrayal* by Harold Pinter 11/15 at London's National Theatre (Lyttleton), with Michael Gambon, Daniel Massey, Penelope Wilton; *Buried Child* by U.S. playwright Sam Shepard (*né* Samuel Shepard Rogers), 35, 12/5 at New York's off-Broadway Theater de Lys, with Richard Hamilton, Mary McDonnell, Tom Noonan, Jacqueline Brooks, 152 perfs.

Television *Dallas* 4/2 on CBS with Larry Hagman (as J.R. Ewing), Barbara Bel Geddes, Victoria Principal (to 5/3/91); *Diff'rent Strokes* 11/3 on NBC with Gary Coleman, Dody Goodman, Conrad Bain.

Films Michael Cimino's *The Deer Hunter* with Robert De Niro, Meryl Streep, Christopher Walken, John Savage; Permanno Olmi's *The Tree of the Wooden Clogs* with Luigi Ornaghi; Paul Mazursky's *An Unmarried Woman* with Jill Clayburgh, Alan Bates. Also: Paul Schrader's *Blue Collar* with Richard Pryor; Franco Brusati's *Bread and Chocolate* with Nino Manfredi, Anna Karina; Fred Schepisi's *The Chant of Jimmie Blacksmith* with Tommy Lewis; Hal Ashby's *Coming Home* with Jane Fonda, Jon Voight; Terrence Malick's *Days of Heaven* with Richard Gere, Brooke Adams, Linda Manz; Bertrand Blier's *Get out Your Handkerchiefs* with Gerard Depardieu, Carol Laure, Riton, Patrick Deware; Ted Post's *Go Tell the Spartans* with Burt Lancaster, Craig Wasson; Howard Zieff's *House Calls* with Walter Matthau, Glenda Jackson, Art Carney; Woody Allen's *Interiors* with Kristin Griffith, Mary Beth Hurt, Diane Keaton, Richard Jordan; Reinhard Hauff's *Knife in the Head* with Bruno Ganz; Alan Parker's *Midnight Express* with Brad Davis; Geoff Steven's *Skin Deep* with Deryn Cooper, Ken Blackburn; Paul Verhoeven's *Soldier of Orange* with Rutger Hauer; Richard Donner's *Superman* with Christopher Reeve as Clark Kent, and Leonard Nimoy.

Hollywood musicals Randal Kleiser's *Grease*, with John Travolta, songs that include the title song by Barry Gibb; Steve Rash's *The Buddy Holly Story* with Gary Busey as Holly; Bruno Barreto's *Dona Flor and Her Two Husbands* with Sonia Braga, Jose Wilkes.

Stage musicals *On the 20th Century* 2/19 at New York's St. James Theater, with Imogene Coca, Kevin Kline, John Cullum, Madeline Kahn, music by Cy Coleman, lyrics by Betty Comden and Adolph Green, 453 perfs.; *Dancin'* 3/27 at New York's Broadhurst Theater, with 18 dancers, choreography by Bob Fosse, music by 25 composers from J. S. Bach to Neil Diamond, 1,774 perfs.; *Ain't Misbehavin'* 5/9 at New York's Longacre Theater, with Ken Page, Amelia McQueen, André De Shields, Charlotte Woodward, music and lyrics mostly by the late Thomas Wright "Fats" Waller who died in 1943, songs that include "Honeysuckle Rose," "Mean to Me," "The Joint Is Jumpin'," and the title song, 1,604 perfs.; *Runaways* 5/13 at New York's Plymouth Theater (after 76 perfs. at the Public/Cabaret Theater), with Elizabeth Swados, music and lyrics by Swados, 274 perfs.; *The Best Little Whorehouse* in Texas 6/19 at New York's 46th Street Theater, with Carlin Glynn, Henderson Forsythe, Delores Hall, music and lyrics by Carol Hall, choreography by Tommy Tune, book based on a *Playboy* magazine story by journalist Larry L. King about the closing of the "Chicken Ranch" at La Grange, Texas, 1,584 perfs.; *Evita* 6/21 at London's Prince Edward Theatre, with Elaine Paige as Eva Peròn, music by Andrew Lloyd Webber, lyrics by Tim Rice, songs that include "Don't Cry For Me, Argentina;" *The Act* 10/9 at New York's Majestic Theater, with Liza Minnelli, music by John Kander, lyrics by Fred Ebb, 233 perfs.; *Ballroom* 12/14 at New York's Majestic Theater, with Dorothy Loudon, Vincent Gardenia, music by Gilly Goldenburg, lyrics by Alan and Marilyn Bergman, 116 perfs.

Popular songs "Just the Way You Are" by New York composer-piano bar performer Billy Joel, 29; "Miss You" by Mick Jagger and Keith Richards of The Rolling Stones; "Shadow Dancing" and "Love Is Thicker Than Water" by Andy Gibb; "Last Dance" by U.S. songwriter Paul Jabara; "Here You Come Again" by Barry Mann and Cynthia Weill; "Three Times a Lady" by U.S. songwriter Lionel Ritchie and the soul group The Commodores (Ritchie, Walter Orange, 31; Thomas McClary, 28; Ronald La Pread, 28; William King, 29; and Milan Williams, 29); "Feels So Good" by U.S. songwriter-jazz flugelhornist Chuck Mangione; "Peg" by U.S. songwriters Walter Becker and Donald Fagen of the rock group Steely Dan.

Dallas beats Denver 27 to 10 at New Orleans January 15 in Super Bowl XII.

Muhammad Ali loses his heavyweight boxing title February 15 to U.S. Olympic champion Leon Spinks, 25, in a 15-round decision at Las Vegas, the World Boxing Council withdraws recognition of Spinks March 18 and awards its title to Ken Norton, 32, who loses it June 9 to Larry Holmes, 28, in a 15-round decision at Las Vegas. Ali, now 36, regains the World Boxing Association title September 15 by winning an easy 15-round decision over Spinks in the Superdome at New Orleans.

Björn Borg wins in men's singles at Wimbledon, Martina Navritilova, 21 (Czech), in women's singles; Jimmy Connors wins in men's singles at the new USTA stadium in Flushing Meadow, New York, Chris Evert in women's singles.

Affirmed wins U.S. racing's Triple Crown under the whip of jockey Steve Cauthen, 17.

The New York Yankees beat the Boston Red Sox 5 to 4 in a one-game playoff for the American League eastern division title and go on to win the World Series, defeating the Los Angeles Dodgers 4 games to 2.

Argentina wins the World Cup football (soccer) championship, defeating the Dutch 3 to 1 in overtime at Buenos Aires.

Americans buy 13 million pairs of running shoes and 42 million pairs of "look-alike" jogger-type sneakers.

Atlantic City, N.J., hails the opening May 26 of the first legal U.S. gambling casino outside Nevada (*see* 1931). The state's voters approved gambling last year, Resorts International opens the casino in a Boardwalk hotel, and the take for the first 6 days is $2.6 million.

Huggies disposable diapers, introduced by Kimberly-Clark, have an hourglass shape and elastic fit that challenge Procter & Gamble's Pampers (*see* 1966, 1980).

United Brands pleads guilty July 19 to charges that it conspired to pay a $2.5 million bribe to a prominent Honduran official for a reduction in the export tax on bananas and a 20-year extension of certain favorable terms (*see* Black suicide, 1975).

An Illinois prison riot July 22 at the Pontiac Correctional Center 100 miles south of Chicago ends with three guards dead and a fourth with 31 knife wounds. The 107-year-old facility has 2,000 prisoners in space designed for 1,400 with 300 guards doing the work of 400. The 5-hour rampage causes $4 million in damage before state police quell it with tear gas. Prison officials say they're surprised the riot did not occur earlier.

San Francisco supervisor Dan White resigns November 10, saying he can no longer afford the $9,600-per-year job. Mayor George Moscone calls a news conference November 27 to name White's successor. White changes his mind and meets privately with the mayor minutes before the scheduled conference. Moscone is found shot to death and shots then ring out in the office of Supervisor Harvey Milk, an avowed homosexual who has been at odds with White over his law-and-order positions. White turns himself in to police 35 minutes later and is charged with killing Milk and Mayor Moscone.

A precision robbery in the morning of December 11 at the Lufthansa cargo terminal of New York's JFK Airport nets more than $5 million in unmarked currency and $850,000 in jewelry for six or seven bandits.

1978 ***(cont.)*** The U.S. tanker *Amoco Cadiz* runs aground off the northwestern coast of France March 16 and breaks up, spilling 1.62 million barrels of crude oil into the sea.

A tornado April 16 in India's Orissa state kills 600. Floods in northern India kill 1,200 from May to September.

Love Canal east of Niagara Falls, N.Y., makes headlines in August as scores of residents are evacuated from houses built over an abandoned excavation site used from 1947 to 1953 to dump toxic chemical waste. A high incidence of birth defects and illnesses have been reported in the neighborhood, and there will be further evacuations in the next few years.

An earthquake in northeastern Iran September 16 kills 25,000.

Michigan and Maine voters approve a ban on non-deposit, no-return bottles—a victory for environmentalists. The Glass Packaging Institute and other lobby groups work to prevent passage of "bottle bills" against non-returnables. Oregon, Vermont, Iowa, and Connecticut have banned such bottles. Industry groups claim that litter-recycling laws are more effective than outright bans.

Coca-Cola signs an exclusive agreement to bottle its drinks in the People's Republic of China. Pepsi-Cola has exclusive rights in the U.S.S.R.

7-Up is acquired by Philip Morris for $520 million. It will break up the company in 1986 and sell the parts after taking a big loss.

Italy votes May 18 to legalize abortion in the first 90 days of pregnancy. The Vatican continues to call abortion homicide.

The world's first "test tube baby," Louise Brown, is born July 25 at London's Oldham Hospital where consultant gynecologist Patric Steptoe and physiologist Robert Edwards have fertilized an egg from Mrs. Lesley Brown's womb with sperm from her husband and have re-implanted the fertilized egg in the mother's womb. Surgeons had been unable to remove a blockage in the Fallopian tubes leading to her uterus.

"Boat people" pour out of Vietnam in search of asylum. Canada, France, West Germany, Taiwan, and the United States take thousands of refugees. Malaysia reverses policy December 4 to permit entry of Vietnamese; hundreds have drowned in boat capsizings after being refused entry.

Population pressures contribute to tensions in the Middle East (above). Egypt has 39.8 million people, up from 20.5 in 1950; Israel 3.7, up from 1.3; Jordan 2.9, up from 1.3; Lebanon 2.9, up from 1.4; Saudi Arabia 7.9, up from 3.4; Syria 8.1, up from 3.5; Iraq 12.5, up from 5.2; Iran 38.2, up from 17.4. Few of these countries have much arable land, most depend on food imports, and present growth rates indicate that populations will double in most by 2005.

1979 Iran's Mohammed Reza Shah Pahlevi appoints Prime Minister Shahpur Bakhtiar to head a regency and flees January 16 to Egypt after nearly 38 years in power. The Shiite Muslim leader Ayatollah Ruholla Khomeini, 78, flies into Teheran February 1 after 15 years in exile, his supporters clash with government troops and rout the elite Imperial Guard February 11, Bakhtiar resigns, and turmoil continues throughout the year with thousands killed in rioting and mass executions (*see* 1978; Shiites, 680). Khomeini accuses the "satanic" United States and her "agents" of fomenting disunity and sends troops to crush a rebellion of Kurdish guerrillas seeking autonomy. The shah moves on to Morocco, the Bahamas, and Cuernavaca, Mexico. Ill with cancer, he is permitted entry to the United States October 22 at the insistence of former Secretary of State Henry Kissinger and Chase Manhattan president David Rockefeller despite warnings from the U.S. ambassador in Teheran. The shah is admitted to New York Hospital for removal of his gall bladder and Iranian terrorists seize the U.S. embassy at Teheran November 4, taking 66 hostages and demanding extradition of the shah. The "student" terrorists release five women and eight black male hostages November 19 to 20. The shah departs for Panama via Texas December 16 (*see* 1980).

North Yemen and South Yemen have a border war beginning February 24, President Carter sends $390 million worth of arms with military advisers to North Yemen, he dispatches a naval force to the Arabian Sea, and nearly 3,000 Cuban and Soviet troops arrive in South Yemen.

A terrorist bomb explodes in a Jerusalem marketplace January 18, Israeli forces retaliate the next day with their heaviest strike into Lebanon since March 1978, they kill 40 Palestinians, a truce halts shelling across the border January 24, but Palestinian guerrillas attack an Israeli settlement in early May and 400 Israeli troops cross into Lebanon in pursuit May 9 with tanks and armored cars.

Iran's fundamentalists toppled the shah's regime, established a theocracy, and execrated the U.S. "Satan."

The peace treaty signed by Egypt's President Sadat and Israel's Prime Minister Begin at Washington March 26 ends a state of war that has existed for nearly 31 years. Both leaders credit President Carter, whose negotiations at Camp David last September have continued with visits to Cairo and Jerusalem. Other Arabs denounce the treaty with demonstrations, strikes, and bombings; some countries break relations with Cairo and impose an economic boycott of Egypt.

Prime Minister Callaghan loses a vote of confidence by one vote in Britain's House of Commons March 28, the first time such a vote has defeated a government since 1924. Conservative leader Margaret Hilda Roberts Thatcher, 53, becomes the nation's first woman prime minister when her party regains power May 3, winning the general elections by the largest majority any party has received since 1966. Mrs. Thatcher has promised to cut income taxes, scale down social services, and reduce the role of the state in daily life.

A bomb planted by Irish Republican Army terrorists explodes August 27 on the fishing boat of Lord Mountbatten off the coast of County Sligo, killing the 79-year-old cousin of Elizabeth II with his grandson and a 15-year-old passenger. Four others are seriously hurt (one dies the next day), and an IRA ambush 35 miles south of Belfast kills 18 British soldiers. A leading Conservative MP has been killed by the IRA outside the House of Commons March 30 and the violence continues.

A bomb of a different sort explodes in Parliament November 15 when art historian Sir Anthony Blunt, 72, is revealed to have been a Soviet spy. Blunt confessed to treason in 1964 and was given immunity from prosecution and permitted to remain curator to the queen. He is stripped of his knighthood.

Washington has broken ties with Taiwan January 1 and established diplomatic relations with Beijing (*see* 1978).

Cambodia's capital city Phnom Penh falls to Vietnamese forces January 7 as does the country's only seaport, Kompong Som. Moscow congratulates the Cambodian rebels on their "remarkable victory," but Romania breaks with her Warsaw Pact allies to denounce Vietnam for intervening in Cambodia (Kampuchea), calling it a "heavy blow for the prestige of socialism" and a threat to peace. Prince Sihanouk says 14 Vietnamese divisions have invaded his country with Soviet backing and entreats the UN Security Council to get Vietnam out of Cambodia. Chinese troops invade Vietnam in March.

Beijing advises Moscow April 3 that China will not renew her 1950 treaty of friendship, due to expire in 1980. Moscow replies April 4 that the decision was taken "contrary to the will and interests of the Chinese people."

South Korean opposition leader King Young Sam is ousted from the National Assembly October 4, all 70 members resign in protest a few days later, and President Park Chung Hee, now 62, is assassinated October 26 by the director of his Central Intelligence Agency. Park's wife was killed in an attempt on her husband's life in August 1974. Premier Choi Kyu Hah becomes president and begins releasing imprisoned dissidents.

Pakistan's former prime minister Zulfikar Ali Bhutto, 51, is hanged April 4 at Rawalpindi for plotting the murder of a political opponent in 1974. President Carter, Pope John Paul II, and other world leaders have appealed for clemency to no avail. Demonstrations throughout Pakistan protest the execution of Ali Bhutto, who was sentenced by the Lahore High Court in mid-March 1978.

Afghan Muslim extremists abduct U.S. Ambassador Adolph Dubs, 58, at Kabul February 14, local police storm the hotel in which he is being held, and Dubs is killed along with several of his abductors in an exchange of gunfire. Soviet agents try to oust President Taraki's rival Prime Minister Hafizullah Amin in mid-September, but Taraki himself is killed in a shootout.

Soviet troops invade Afghanistan December 24, allegedly at the invitation of the new president Amin, who is convicted December 27 of "crimes against the state" and executed along with members of his family by a "revolutionary tribunal." He is replaced by Soviet puppet Babrak Karmal. U.S.-supplied Afghan guerrillas will resist Soviet occupation forces for more than 9 years, forcing them to retreat into fortified cities (*see* Olympic boycott, grain embargo, 1980).

U.S. Senate leaders block a Strategic Arms Limitation Treaty signed at Vienna June 18 by Jimmy Carter and Leonid Brezhnev after nearly 5 years of negotiations. The United States has 2,283 missiles and bombers as of June 18, the Soviet Union 2,504. President Carter has approved construction of a new generation of smaller aircraft carriers January 8 and scrapped plans for another 90,000-ton supercarrier. He has approved development June 7 of a $30 billion MX missile plan that would deploy large missiles located in any of 8,800 underground shelters connected by miles of track under the desert of Utah, Nevada, or another western state. The MX will survive a "first strike" by Moscow, say its supporters, provide counterforce capability, and for the first time give the United States "first strike" ability to eliminate large numbers of Soviet land-based missiles.

Rhodesian whites vote January 30 to ratify a new constitution enfranchising all blacks, establishing a black majority in the Senate and Assembly, and renaming the country Zimbabwe Rhodesia. Delegates from 30 Commonwealth countries meeting at Lusaka, Zambia, August 5 approve a new proposal to end the 6-year-old civil war in Zimbabwe Rhodesia (*see* 1980).

Ugandan exiles and Tanzanian troops occupy Kampala April 10 and force Uganda's President Idi Amin Dada into exile. His 8 years of bloody oppression have left his country $250 million in debt with only $200,000 in foreign exchange left in the central bank. Although Uganda is 85 percent Christian, Amin has given preference to his own Muslim Kakwa tribe and other Sudanic

1979 *(cont.)* tribes. Nilotic Acholi, Langi, Baganda, and other tribesmen (Uganda has at least 32 distinct tribes) vie for supremacy (*see* 1980).

South Africa's President John Vorster, 63, resigns June 4 after the release of a report accusing him of covering up irregularities in government spending on secret propaganda efforts in the United States. Vorster was elected president after leaving the prime ministry last September. He is succeeded by Marais Viljoen.

Ghana has a coup June 4; rebels install flight lieutenant Jerry John Rawlings, 32, as head of state and execute Gen. Frederick Akuffo and two other former chief executives. An admirer of Libya's Muammar Qadaffi, Rawlings turns over the government to an elected president in July (but *see* 1981).

Equatorial Guinea has a bloodless coup August 3 that ends the brutal 11-year rule of President Masie Nguema Biyogo Negue Ndong, 57, who has allied himself with Moscow but kept foreigners out of the 10,000-square-mile former Spanish colony. As president for life, Masie Nguema has put tens of thousands to death, scared more than half the country's 300,000 people into fleeing abroad, and shattered the nation's economy while spending lavishly on palaces and other symbols of power. His nephew Theodore Nguema Mbasago heads a new ruling junta. A civilian-military tribunal convicts Masie Nguema of genocide, treason, systematic violation of human rights, and embezzlement of public funds. He and six aides are executed by firing squad.

Angola's President Agostinho Neto, 56, dies at Moscow September 10 following surgery for pancreatic cancer; he is succeeded by his planning minister José Eduardo dos Santos.

The Central African Emperor Bokassa I is overthrown in a bloodless coup September 20 (*see* 1977). Accused of bankrupting his country and joining in a massacre of schoolchildren, Bokassa is rebuffed by France but given asylum by the Ivory Coast. Former president David Dacko regains power, revives the Central African Republic, and pledges a restoration of democracy.

St. Kitts and Nevis in the Caribbean gains full independence February 22 after more than 175 years of British colonial rule. St. Lucia gains full independence the same day. St. Vincent and the Grenadines gains full independence October 26 after nearly 196 years of British rule.

Grenada has a leftist coup March 13 as the New Jewel Movement headed by Maurice Bishop takes power while Prime Minister Gairy is in New York, accusing Gairy of fiscal responsibility. Bishop is a protégé of Cuba's Fidel Castro (*see* 1974, 1983).

Nicaraguan dictator Anastasio Somoza resigns his presidency July 17 after a 7-week civil war and takes refuge at Miami with 45 aides, ending the 46-year Somoza family dynasty (*see* 1978). Sandinista rebels enter Managua July 20. The new five-man junta expropriates the vast business empire of the Somozas and their supporters (*see* 1980).

El Salvador's President Carlos Humberto Romero, who came to power after disputed elections in February 1977, is deposed in a military coup October 15 after months of violence following the killing of 23 leftist demonstrators May 8. A new military junta assumes power, begins to redistribute lands to the peasants, and makes other reforms after 47 years of military dictatorship. Violence continues (*see* 1980).

The Gilbert Islands (Kiribati) gain independence July 12 after 64 years of British colonial rule.

Iranian women march in Teheran to protest the Ayatollah Khomeini's revocation of the 1975 Family Protection Law, his abolition of coeducational schools, and government pressure to wear the *chador*, a heavy veil that obscures the face.

Soviet authorities release five prominent dissidents and fly them to the United States April 27. Aleksandr Ginzburg, Mark Dymshitz, Eduard S. Kuzentsov, Valentyn Moroz, and the Rev. George Gins have spent years at various labor camps and prisons in punishment for their human rights activities. Moscow permits a record 51,320 Jews to emigrate, up from 28,864 last year and 16,737 in 1977, but will crack down on such emigration in 1980.

Brazil's Economic Development Council recommends a new salary program that will adjust the lowest salaries in the private sector every 6 months to 10 percent more than the rate of inflation, while adjusting higher salaries at a rate much lower than that of price increases. In 15 years of military rule, the share of the national wealth enjoyed by the lowest-paid workers has steadily declined.

A reverse discrimination suit ends June 27 in a rebuff for Kaiser Aluminum and Chemical lab technician Brian F. Weber, 32, who has sued his union and employer over a job-training program that gave preference to blacks. Steel workers Local 5702 and the company announced in 1974 that half the openings in a training program for higher-paying craft jobs would be reserved for blacks regardless of seniority. The Supreme Court's 5 to 2 decision in *United Steel Workers v. Weber* is regarded as the definitive statement on affirmative action programs.

Ku Klux Klansmen at Greensboro, N.C., fire on a group of self-styled Communists of the Workers Viewpoint Organization November 3, killing 3 and wounding 12 in an episode that is really gang warfare although the world press relates it to the civil rights movement of the 1960s.

Miami insurance salesman Arthur McDuffie, 33, borrows a cousin's motorcycle after midnight December 17, three white Dade County police officers stop him after a chase through red lights and stop signs at speeds of up to 100 miles per hour, other officers arrive at the scene and, without provocation, yank off McDuffie's helmet, handcuff him behind his back, and proceed to crack his skull with their night sticks and heavy metal flashlights. McDuffie is killed, several officers are dismissed amidst charges that their attack was in part racially motivated, some are indicted for manslaughter and for making it appear that McDuffie was killed in an accident (*see* 1980).

The U.S. Supreme Court rules unanimously January 16 in the case of *Thor Power Tool Co. v. the Commissioner of Internal Revenue* that the valuation of warehouse stock may not be reduced for tax purposes unless it is disposed of or sold at reduced prices (*see* book publishers, below).

A pact signed July 7 grants China most-favored-nation status, allowing her the lower tariff rates available to most other U.S. trading partners. U.S. exports to China will reach $1.7 billion next year, imports from China will reach $600 million, and total trade by 1990 will be nearly $4 billion.

Gold prices top $300 per ounce for the first time in history July 18 and $400 per ounce September 27 as world financial markets react to inflation worries.

New York banker Paul A. Volcker, 52, is appointed chairman of the Federal Reserve Board by President Carter in August and on October 6 he announces a 1 percent increase in the discount interest rate that Federal Reserve banks charge member institutions. The move to halt inflation sends stock prices sharply lower and begins a short recession.

Double- and even triple-digit inflation plagues much of the world. U.S. prices increase 13.3 percent for the year, largest jump in 33 years, and the Federal Reserve Board's move in October to tighten the money supply sparks a jump in loan rates that will continue for 6 months. Banks raise their prime loan rate to 14.5 percent October 9, Wall Street's Dow Jones Industrial Average falls 26.48 points that day, and the New York Stock Exchange has a record 81.6 million share day October 10 as small investors panic. The U.S. Gross National Product has risen by more than a third in constant dollars since 1969 and unemployment has averaged less than 6 percent (it topped 9 percent in only one calendar year, versus a peak of 25 percent in the 1930s when the GNP rose by only 4 percent and when stock prices declined by only 31 percent as compared to 42 percent in the 1970s).

An accident at Unit II of the Three Mile Island nuclear generating station near Harrisburg, Pa., March 28 raises alarms that the year-old reactor may explode and release radioactive cesium. The overheated reactor shuts down automatically but Metropolitan Edison Company operators, misled by ambiguous indicators, think water pressure is building and shut down pumps still operating; the reactor heats up further and tons of water that have poured out through the stuck-open valve overflow into an auxiliary building through a valve that should have been shut. The reactor core does not melt down despite exposure and damage. Some 144,000 people, mostly pregnant women and small children, are evacuated from the Middletown area. Little radiation is released.

Nine new U.S. nuclear reactors begin commercial production; France, Japan, Soviet Russia, and other nations continue to expand nuclear energy capabilities and reduce dependence on petroleum, but the malfunction at Three Mile Island (*see* above) discourages new atomic energy facilities. U.S. utility companies cancel 11 reactor orders and will cancel more next year, installation of other reactors is indefinitely delayed. Inflation, high interest rates, and declining demand for electricity limit construction of new reactors as much as does public opinion.

Iran's new government announces February 17 that exports of oil will resume March 5 at a price about 30 percent higher than that set by the OPEC nations in December 1978. Iranian oil production averages only 3.4 million barrels per day for the year, down from 5.4 million last year. Since the 900,000 barrels of Iranian crude imported daily by the United States last year supplied 6 percent of U.S. consumption, the drop in Iranian imports creates genuine fuel shortages in many states.

President Carter announces April 5 that he is acting under new legal authority to decontrol domestic oil prices, allowing U.S. producers to compete with the new OPEC base price of $14.54 per barrel. Price controls on 80 percent of marginally productive U.S. oil wells come off June 1. Carter proposes a windfall profits tax to recover unearned profits gained by oil companies and an Energy Security Fund to help low-income families pay higher fuel bills, fund construction of more public transportation facilities, and promote research and development of alternative energy sources.

U.S. motorists line up at filling stations from spring through summer and are often unable to obtain more than a few gallons at a time.

OPEC representatives raise prices 50 percent between January 1 and June 28 to about $20 per barrel, an economic summit meeting of the seven major industrial democracies June 29 at Tokyo agrees to set limits on imports, and U.S. imports for the year fall to 7.9 million barrels, down from a record 8.6 million in 1977, as prices for fuel oil and gasoline rise. At the 55th OPEC meeting in mid-December at Caracas, most OPEC members raise crude oil prices to $24 per barrel; Iran and Libya raise prices to well above $28, and crude oil in the spot market brings as much as $40.

Panama takes over operation of the Panama Canal October 1 under terms of last year's treaty establishing a 20-year transition from U.S. control.

Two Soviet planes collide in mid-air August 15 over the Ukraine, killing 150; a Western Airlines DC-10 crashes on landing at Mexico City October 15, killing 73; a Pakistan International Airlines 707 carrying pilgrims from Mecca crashes November 26 at Jidda, Saudi Arabia, killing 156; an Air New Zealand DC-10 carrying sightseers over Antarctica crashes into Mt. Erebus November 28, killing 257.

Motorcycle accidents claim 4,893 lives in the United States, up from 3,312 in 1976 when Congress struck down a strong federal regulation requiring helmets. Many states have repealed or weakened their helmet laws.

1979 *(cont.)* Atlanta's metropolitan rapid transit system (MARTA) begins service June 30 with a 6.7-mile high-speed commuter line linking the downtown business district with eastern suburbs. The 54-mile system is still at least 10 years short of completion and will cost about $2.7 billion more than the $1.3 billion projected in 1971.

Renault saves American Motors from bankruptcy by purchasing a controlling interest and arranging to have AM dealers market Renaults (*see* 1987).

The Chrysler Loan Guarantee Bill passed by Congress December 27 saves the 54-year-old auto maker, America's seventeenth largest company, from bankruptcy. The measure guarantees $1.2 billion in loans to Chrysler.

Medicare-funded kidney dialysis treatment costs $851 million for 46,000 U.S. patients, nearly 20 percent of them bedridden more or less permanently and about half unable to work (*see* 1960). The escalating costs of hemodialysis raise questions about how much the nation can afford without slighting other health needs.

The Moral Majority is founded by Lynchburg, Va., evangelist Jerry Falwell, 46, of the Thomas Road Baptist Church whose weekly *Old Time Gospel Hour* airs on more than 300 U.S. television stations and 64 foreign stations. The new political action group will register millions of new voters by November 1980 in an effort to block the Equal Rights Amendment, impede reform of the criminal code, disrupt the White House Conference on the Family, and fight abortion liberalization. It will continue until 1989.

U.S. spending on education reaches $151.5 billion, up from $8.3 billion in 1950, with $126.5 billion coming from government sources. But student proficiency scores will fall sharply in the next decade as the nation falls behind countries that spend much less.

A Beijing court sentences Chinese journalist Wei Jingsheng, 29, to 15 years in prison October 16, ending the brief "Beijing Spring" in which political dissent was tolerated. Wei edited the underground journal *Explorations* which published his essays, including "The Fifth Modernization: Democracy." These were also posted on Democracy Wall where they attracted worldwide attention by calling for the establishment of political democracy as the necessary basic for economic and social reform.

The British Post Office inaugurates a Prestel system giving subscribers access to 160,000 pages, or television screenfuls, of information. Using telephones, computers, and TV sets, subscribers can obtain news and information such as rail and air schedules and stock and commodity quotations, make purchases, buy airline tickets, reserve hotel rooms, and book theater seats by remote control. The government has spent $30 million to develop the system. A Prestel set, which can also receive ordinary TV programming, costs at least $2,000 U.S. and the subscriber is billed for the time the set is used.

Morning Edition debuts November 1 on U.S. National Public Radio with host Bob Edwards.

U.S. book publishers destroy or remainder stocks of scholarly and scientific works because the *Thor Power Tool* decision (*see* above) prevents them from depreciating the value of unsold books, although the case actually involved spare parts inventories.

Nonfiction *Munich: The Price of Peace* by historian Telford Taylor, 71, who was chief U.S. counsel at the 1946 Nuremberg trials; *Civilization and Capitalism: 15th-18th Centuries* by Fernand Braudel, now 77; *Japan as Number One* by Harvard sociologist Ezra F. Vogel, 48; *The Gnostic Gospels* by Barnard College religious historian Elaine Hiesey Pagels, 36; *The Medusa and the Snail* by Lewis Thomas; *The Executioner's Song* by Norman Mailer; *The Old Patagonian Express* by Paul Theroux.

Fiction *The Book of Laughter and Forgetting* by Czech novelist Milan Kundera, 50; *November* by exiled East German poet-singer-novelist Rolf Schneider, 48; *A Bend in the River* by V. S. Naipaul; *Chirundu* by Es'kia Mphahlele; *Sophie's Choice* by William Styron; *The Ghost Writer* by Philip Roth; *Good as Gold* by Joseph Heller; *Territorial Rights* by Muriel Spark; *Burger's Daughter* by Nadine Gordimer; *The Year of the French* by U.S. novelist Thomas Flanagan, 63, deals with the Irish-French military effort against the British in 1798; *Life Before Man* by Margaret Atwood; *Sleepless Nights* by U.S. novelist Elizabeth Hardwick, 63, whose ex-husband, Robert Lowell, died in September 1977 at age 60; *The Confederates* by Thomas Keneally; *The White Album* by Joan Didion; *Smiley's People* by John Le Carré; *So Long, See You Tomorrow* by former *New Yorker* writer-editor William Maxwell, 71.

Juvenile *The Garden of Abdul Gasazi* by Providence, R.I., art teacher Chris Van Allsburg, 30; *A Christmas Card* by Paul Theroux.

Painting *Paintsplats* (on a wall by Philip Johnson)—*Model* and *Crusoe Umbrella* (mural) by Claes Oldenburg; *Elephant* by U.S. minimalist Milton Resnick, 62.

Sculpture *The People Machine* by Vito Acconci.

Theater *Wings* by Arthur Kopit 1/28 at New York's Lyceum Theater, with Constance Cummings, 113 perfs.; *On Golden Pond* by U.S. playwright Ernest Thompson, 29, 2/28 at New York's New Apollo Theater, with Tom Aldredge, Frances Sternhagen, 126 perfs.; *Cloud Nine* by English playwright Caryl Churchill, 40, 3/29 at London's Royal Court Theatre, with Anthony Sher, Jim Hooper, Carol Hayman; *Whose Life Is It Anyway?* by English playwright Brian Clark, 46, 4/17 at London's Trafalgar Theatre, with Tom Conti, Jean Marsh; *Bent* by U.S. playwright Martin Sherman, 37, 5/3 at London's Royal Court Theatre, with Ian McKellen, Tom Bell is about Nazi persecution of homosexuals, 240 perfs.; *Knockout* by U.S. playwright Louis LaRusso II, 43, 5/6 at New York's Helen Hayes Theater, with Danny Aiello, 154 perfs.; *Loose Ends* by Michael Weller 6/7 at New York's Circle in the Square Theater, with Kevin Kline, Roxanne Hart, 270 perfs.; *A Life* by Hugh Leonard 10/4 at Dublin's Abbey Theatre, with Cyril Cusack, Philip O'Flynn; *One Mo'*

Time by U.S. actor-playwright Vernel Bagners 10/22 at New York's Village Gate Downstairs, with Bagners, 1,372 perfs.; *Romantic Comedy* by Bernard Slade 11/8 at New York's Ethel Barrymore Theater, with Anthony Perkins, Mia Farrow, Carole Cook, 396 perfs.; *Amadeus* by Peter Shaffer 11/2 at London's Olivier Theatre, with Paul Scofield as the jealous mediocrity Antonio Salieri, Simon Callow as W. A. Mozart.

Television *Knots Landing* 12/27 on CBS with Donna Mills, Ted Shackelford (to 5/12/93).

Radio *A Prairie Home Companion* in February on National Public Radio with Minnesota Lutheran Garrison Keillor, 30 (to February 1987).

Films Ira Wohl's documentary *Best Boy* with Philly Wohl, Zero Mostel; James Bridges's *The China Syndrome* with Jane Fonda, Jack Lemmon; Luchino Visconti's *The Innocent* with Giancarlo Giannini, Laura Antonelli, Jennifer O'Neill; Robert Benton's *Kramer vs. Kramer* with Dustin Hoffman, Meryl Streep; Martin Ritt's *Norma Rae* with Sally Field as textile union organizer Crystal Lee; Volker Schlondorff's *The Tin Drum* with David Bennent. Also: Robert M. Young's *Alambrista!* (*The Illegal*) with Domingo Ambriz as a Mexican farm worker in California; Francis Ford Coppola's *Apocalypse Now* with Martin Sheen, Robert Duvall, Marlon Brando; Bruce Beresford's *Breaker Morant* with Edward Woodward, Jack Thompson; Peter Yates's *Breaking Away* with Paul Dooley; Istvan Szabo's *Confidence* with Ildiko Bansagi, Peter Andorai; Peter Lilienthal's *David* with Mario Fischel; George A. Romero's *Dawn of the Dead* with David Emge, Tom Savini; Paul Verhoeven's *The Fourth Man* with Jeroene Krabbe, Renée Soutendijk; Martin Brest's *Going in Style* with George Burns, Art Carney, Lee Strasberg; Richard Pearce's *Heartland* with Conchata Ferrell, Rip Torn; George Miller's *Mad Max* with Mel Gibson; Woody Allen's *Manhattan* with Allen, Mariel Hemingway, Diane Keaton, Marshall Brickman; Rainer Werner Fassbinder's *The Marriage of Maria Braun* with Hanna Schygulla; Gilian Armstrong's *My Brilliant Career* with Judy Davis; Ted Kotcheff's *North Dallas Forty* with Nick Nolte, Mac Davis; John Hanson and Rob Nilsson's *Northern Lights* with Robert Behling; Roman Polanski's *Tess* with Nastassia Kinski, Peter Firth, Leigh Lawson; Rainer Werner Fassbinder's *Third Generation* with Eddie Constantine, Hanna Schygulla; Philip Kaufman's *The Wanderers* with Ken Wahl, John Friedrich; John Huston's *Wise Blood* with Brad Dourif, Ned Beatty, Amy Wright, Daniel Shor.

John Wayne dies of lung and stomach cancer at Los Angeles, June 11, at age 72.

Film musicals Bob Fosse's *All That Jazz* with Roy Scheider, Jessica Lange, Ben Vereen; Franc Roddam's *Quadrophenia* with Phil Daniels, Mark Wingett, Sting.

Stage musicals *They're Playing Our Song* 2/11 at New York's Imperial Theater, with Lucie Arnaz, Robert Klein, book by Neil Simon, music by Marvin Hamlisch, lyrics by Carole Bayer Sager, 1,082 perfs.; *Sweeney Todd* 3/1 at New York's Uris Theater, with Len Cariou as the "demon barber of Fleet Street" (*see* 1847 play), Angela Lansbury, music and lyrics by Stephen Sondheim, songs that include "Pretty Women," 558 perfs.; *A Day in Hollywood & A Night in the Ukraine* 3/28 at London's Mayfair Theatre, with Paddie O'Neal, John Bay as Groucho Marx, Franz Lazarus as Chico, music by Lazarus, lyrics by Dick Vosburgh, 588 perfs.; *Scrambled Feet* (revue) 6/11 at New York's Village Gate, with Jeffrey Haddon, 31, John Driver, 32, Evalyn Barron, Roger Neil, skits and songs by Haddon and Driver; *Sugar Babies* 10/8 at New York's Mark Hellinger Theater, with Ann Miller, Mickey Rooney, songs by Jimmy McHugh, additional lyrics by Arthur Malvin, 1,208 perfs.; *Joseph and the Amazing Technicolor Dreamcoat* 11/1 at London's Westminster Theatre, with Paul Jones, music by Andrew Lloyd Webber, lyrics by Tim Rice.

Popular songs "Do Ya Think I'm Sexy?" by Rod Stewart; "What a Fool Believes" by Kenny Loggins and Michael McDonald; "Bad Girls" by U.S. singer-songwriter Donna Summer (Donna Gaines), 30, who first gained fame as a disco singer in Munich; "Heart of Glass" by Deborah Harry, 29, and Chris Stein of the New York new wave rock group Blondie (Harry, guitarist Stein, drummer Clem Burke, bass guitarist Nigel Harrison, guitarist Frank Infante, keyboardist Jimmy Destri); "Too Much Heaven" by The Bee Gees; *52nd Street* (album) by Billy Joel; *Armed Forces* (album) by rebellious British singer-composer Elvis Costello (Declan McManus), 24; *C'est Chic* (album) by U.S. songwriters Nile Rodgers and Bernard Edwards of the manufactured disco group Chic whose songs "Le Freak" and "We Are Family" gain wide popularity.

Eleven concertgoers are crushed to death December 3 in a stampede for seats to a concert by The Who at Cincinnati's Riverfront Coliseum, casting a pall over the $2-billion-per-year rock concert business.

The Walkman cassette player introduced by Sony Corp. is a $200 pocket stereo with two pairs of earphones, making it possible to hear high-fidelity sound in any location without disturbing one's neighbors. It is the brainchild of Sony chairman Akio Morita, now 58, and will soon have an FM radio version.

Pittsburgh beats Dallas 35 to 31 at Miami January 21 in Super Bowl XIII.

Björn Borg wins in men's singles at Wimbledon, Martina Navratilova in women's singles; John McEnroe, 20, wins in men's singles at Forest Hills, Tracy Ann Austin, 16, in women's singles.

Spectacular Bid wins the Kentucky Derby and Belmont Stakes, Coastal the Preakness.

Britain's annual Fastnet yacht race in the English Channel and Irish Sea ends in disaster August 14 as gale-force winds and mountainous waves strike more than 300 boats, drowning 15 sailors.

The Pittsburgh Pirates win the World Series, defeating the Baltimore Orioles 4 games to 3.

1979 ***(cont.)*** British Liberal party leader Jeremy Thorpe goes on trial May 8 at London's Old Bailey on charges of hiring an assassin to kill his alleged ex-lover Norman Scott (*see* 1975). Thorpe wins acquittal after 6 weeks of testimony and deliberation; his career is ruined.

Mafia boss Carmine Galante, 69, has lunch at a Brooklyn, N.Y., café July 12 and is gunned down by hitmen of an underworld rival. Head of a 200-member "family" formerly led by Joseph Bonnano, Sr., Galante had hoped to succeed the late Carlo Gambino as the "boss of bosses;" police say he had been marked for execution for more than a year.

Atlanta youth Edward H. Smith, 14, vanishes July 20; his body, shot in the chest, is found July 28. Alfred J. Evans, 13, vanishes July 25 and his body is also found July 28. Milton Harvey, 14, vanishes September 4; his skeletal remains are found November 8 and his mother presses for a broad investigation of the killings which will continue for almost 2 years as more than 20 local black children, aged 7 to 15, almost all boys, die at the hands of an unknown killer or killers (*see* 1980).

A New York State law effective September 1 modifies the severe penalties (mandatory life sentences for anyone convicted of selling any amount of heroin, cocaine, morphine, or other "hard" drug) of the law put through by former governor Nelson Rockefeller in 1973. Hundreds of persons are serving longer terms for selling or possessing small amounts of drugs than for rape or robbery, the old law has not helped to control the drug traffic, and state courts are backlogged with cases of people standing trial rather than working out plea bargains.

The United States has 21,456 reported murders, half involving handguns, 13 percent involving rifles or shotguns. Canada has 207 shooting deaths, down from 292 such homicides in 1975. France has 1,645 homicides, just over half involving firearms. West Germany has 69 crimes involving murder or robbery with a firearm, Japan has 171. London police fire guns only 8 times.

Chicago's 40-story Xerox Centre is completed by German-American architect Helmut Jahn, 40, of C. F. Murphy Associates. Jahn has departed from the rectangular steel and glass office "boxes" that have dominated commercial construction since the end of World War II when the ideas of the late Ludwig Mies van der Rohe took over the world of architecture.

Boston's John F. Kennedy Library designed by I. M. Pei and Partners opens in Dorchester. Chinese-American architect Ieoh Ming Pei, 63, works on a new east wing to Washington's National Gallery.

An earthquake in eastern Iran January 16 kills an estimated 1,000.

Accidental release of dry anthrax spores in early April at the Microbiology and Virology Institute, a Soviet biological warfare facility in Sverdlovsk, contaminates an area with a radius of at least 3 kilometers and by some estimates kills several hundred persons. Tight censorship is imposed, but hundreds, perhaps thousands, of residents and military personnel reportedly die after inhaling the spores and contracting pulmonary anthrax. Soviet authorities say only that illegal sales of anthrax-contaminated meat have caused a public health problem. Critics charge Moscow with violating the 1975 Helsinki ban on developing biological weapons.

A Mexican offshore oil-well blowout June 3 in the Gulf of Mexico contaminates Gulf fisheries and beaches with millions of gallons of oil in the largest spill ever recorded. After 3 months of uncontrolled spillage, the oil has traveled 600 miles from the hole drilled by Pemex explorers, shaking confidence in the ability of so-called blowout "preventers" and backup systems to protect the environment, straining U.S.-Mexican relations as the State Department conducts delicate negotiations over U.S. purchases of Mexican oil and natural gas. Ixtoc 1 continues to run out of control, defying efforts by the most experienced well cappers. An estimated 3.5 million barrels will have spilled into the sea by the time the flow is stopped in late March of next year.

The tankers *Atlantic Express* and *Aegean Captain* collide off Trinidad and Tobago July 19, spilling an estimated 1.02 million barrels of crude oil into the sea. The *Burmah Agate* burns after a collision in Galveston Bay, Texas, November 1, spilling 62,000 barrels.

Java's Mount Metrap erupts without warning and kills 149 in a village 6,000 feet up the slopes (*see* 1006).

Paris police drop breathalizer tests for drivers after a public outcry and protests by restaurant owners that the spot checks are hurting business. Wine is sold legally at highway service stations and some 5 million of France's 53.5 million people are heavy drinkers, 2 million alcoholics. Per-capita consumption of pure alcohol is almost 17 quarts per year, the highest in the world, and alcoholism costs the nation some $4.5 billion per year in medical and social service costs.

1980 India's former prime minister Indira Gandhi regains power January 6 in an election victory engineered by her son Sanjay, 33, only 33 months after a humiliating defeat. Called ruthless and autocratic for pushing slum clearance projects that left thousands homeless and family planning programs that included forced sterilizations, Sanjay has been convicted on one of more than a dozen criminal charges that he reaped huge profits from a state project to produce small, cheap automobiles, none of which ever came off the assembly line. Sanjay and a flight instructor die June 23 in a plane crash while doing illegal aerial acrobatics.

The shah of Iran leaves Panama for Cairo March 23 at the invitation of President Sadat, his enlarged spleen and part of his liver are removed by surgeons March 28, and the cancer-riddled shah dies July 26 at age 60, ending the Pahlevi dynasty that ruled Iran from 1921 until last year.

A U.S. attempt to rescue the 53 hostages held at Teheran since November ends in disaster April 24 and the Ayatollah Khomeini threatens to kill the hostages if another "silly maneuver" is tried. Six U.S. C-130 transport planes from a base in southern Egypt have landed

90 commandos in the desert 300 miles southeast of Teheran; mechanical problems and a sandstorm knock out three of the operation's eight helicopters, the mission is aborted, eight men are killed as a fourth helicopter collides on the ground with a C-130, and the survivors beat a hasty retreat. The hostages remain in custody at year's end as negotiations proceed for their release, which will not come until January 20.

Iraqi planes hit 10 Iranian airfields September 22 after months of border skirmishes, troops cross into Iran September 23 and besiege the huge oil refinery at Abadan, beginning an 8-year war over the Shatt Al-Arab estuary (*see* 1975).

Turkish strikes, terrorism, inflation, and rising unemployment bring the country to the verge of anarchy until a military government establishes some order in September.

Zimbabwe (formerly Rhodesia) gains independence April 17. A new government headed by Marxist Robert Mugabe, 56, takes power after years of civil war; he appeals to Zimbabwe's 7 million blacks for fair treatment of the new nation's 230,000 whites.

Liberia's President William R. Tolbert, Jr., is ousted in a military coup April 12 and has been executed along with 27 other high officials. A 17-member People's Redemptive Council suspends the Constitution April 25 and assumes all executive power with General (formerly Master Sergeant) Samuel K. Doe, 28, as president.

Senegal's first president Léopold Sédar Senghor, now 74, steps down after 20 years in power. He will be succeeded beginning next year by Abdou Diouf.

Uganda returns to constitutional government after elections in December, the first in 18 years, put former president Milton Obote back in power.

Juliana of the Netherlands abdicates on her seventy-first birthday April 30 and is succeeded after a 32-year reign by her daughter Beatrix, 42. The new queen, more pretentious than her mother, is married to a former officer in Hitler's SS. Demonstrations rock Amsterdam, Rotterdam, and Utrecht.

Yugoslavia's President Josip Broz Tito dies May 4 at age 87 after 35 years in power. His death leaves a power vacuum that raises fears of a break-up of Yugoslavia into her oldtime states (Croatia, Montenegro, Serbia, Slovenia, Bosnia, and Herzogovina).

Japan's Prime Minster Masayoshi Ohira dies June 12. Zenko Suzuki, 66, becomes prime minister after elections held June 22.

Vanuatu (formerly the New Hebrides) in the South Pacific gains independence July 30 after 93 years of joint British and French colonial rule.

A draft registration measure signed by President Carter June 27 requires that some 4 million U.S. men aged 19 and 20 register. Congress has excluded women, despite a request by Carter that they be included.

U.S. voters turn Carter out of office and elect former California governor Ronald Wilson Reagan, 69, an ex-Hollywood actor who campaigns with slick television commercials and wins 489 electoral votes to Carter's 49, with 51 percent of the popular vote (43.2 million) vs. 42.5 percent (34.9 million) for Carter, 6.5 percent (5.6 million) for independent candidate John Anderson, an Illinois Republican congressman. Leading liberal Democrats lose as the Republicans gain control of the Senate for the first time since the 1950s.

Peru's President Fernando Belaunde Terry, now 67, has been reelected May 18 after 12 years of military rule and heads one of the few remaining civilian governments in Latin America. Strikes, economic problems, and terrorist insurgency continue as in so much of the world.

Former Nicaraguan dictator Anastasio Somoza is assassinated at Asunción, Paraguay, September 17 at age 54. Gunmen firing a bazooka and machine guns hit Somoza's Mercedes, killing also his driver and a financial adviser.

Polish shipyard workers at Gdansk quit August 14 to protest a new rise in meat prices. The strike at the Lenin Shipyard spreads as some 350,000 workers demand the right to strike and to form self-governing unions independent of Communist party control. Other demands include wage raises, release of political prisoners, a curb on censorship, and meat rationing. Led by electrician Lech Walesa, 37, the strikers do not stage street demonstrations as in 1970, when they gave authorities an excuse to use force and kill at least 55. Party leader Edward Gierek agrees to the demands September 1, he releases dissidents who have been arrested, and Poland's Solidarity, with 10 million members, becomes the first independent labor union in a Soviet bloc country. Gierek is replaced by Stanislaw Kania, 53, as Moscow masses 55 divisions on Poland's frontiers, fearing the deviation from orthodox Marxist-Leninist philosophy will spread to other Soviet satellites and even to Soviet Russia.

Moscow has exiled physicist and human rights leader Andrei Sakharov in January to the remote city of Gorki

Polish shipyard workers defied Communist authorities to gain at least temporary recognition for their union.

1980 *(cont.)* and has increased suppression of dissent (*see* 1977). Only 21,147 Jews emigrate by way of Vienna, the chief exit route, down from a peak of 51,320 permitted to leave last year.

El Salvador's leading human rights activist Archbishop Oscar Arnulfo Romero, 62, is murdered by a sniper March 24 while saying mass in a hospital chapel.

A bomb explosion outside a Paris synagogue October 4 leaves four dead and ten seriously injured, raising fears that anti-Semitism is reviving in France. Parisians of all faiths walk in a great procession demonstrating support and sympathy, President Valery Giscard d'Estaing and his ministers ban neo-Nazi meetings and vow to dissolve "racist organizations" and increase police protection, but the terrorists who planted the bomb remain unknown and at large.

Miami has riots beginning May 17 after blacks hear that an all-white six-man Tampa jury has acquitted four white ex-policemen accused of beating Arthur McDuffie to death last year and making it look like an accident. The riots leave 14 people beaten or shot to death in a 40- by 60-block area—a ghetto of 233,000 where it is generally believed that a black cannot get a fair trial and a white man can "get away with anything."

The Supreme Court rules 6 to 3 July 2 in *Fullilove v. Klutznick* that Congress has authority to redress past racial discrimination through the use of quotas in government contract awards. The court upholds the constitutionality of a provision in a $4-billion 1977 emergency public works program requiring that 10 percent of the contracts be awarded to "minority business enterprises."

A U.S. Circuit Court of Appeals panel rules December 4 that the prosecution's key witness perjured himself in the 1972 trial of the "Wilmington 10" in North Carolina (*see* 1971). Amnesty International called Ben Chavis and the others "political prisoners," the Department of Justice in 1978 called for a reversal of the convictions, 55 members of Congress signed a "friend of the court" brief, but the civil rights workers served up to 4 years before the last was paroled.

The U.S. spacecraft *Voyager I* explores Saturn November 12, making new discoveries about the planet's 14 moons and more than 1,000 rings. She is on a 3-year journey of 1.3 billion miles.

Brazilian miners discover the Serra Palada mine in January when a tree falls over in a rainstorm, baring rocks of gold. Prospector José Maria da Silva, 34, arrives in April, borrows money for food, finds 22 pounds of gold within 2 weeks, and on 1 day in September extracts 700 pounds of gold worth $4.75 million. The government bars all but Brazilians from the area.

Gold peaks at $875 per ounce in January amidst predictions of $2,000 per ounce; it falls to $600 by year's end.

The Banking Deregulation Act signed by President Carter March 31 establishes a universal system of banking reserves and effects reforms supposedly designed to favor consumers. The measure phases out ceilings on interest paid to small depositors; it authorizes payment of interest on checking accounts and on similar accounts at thrift institutions.

U.S. banks raise their prime rate (the rate on loans granted to favored customers) to 20 percent April 2 as the Federal Reserve tightens money. The rate falls to 12 percent by October but peaks at 21.5 percent in mid-December.

Double-digit inflation continues in the United States with prices rising 12.4 percent by year's end as compared to 13.3 percent last year, fueling opposition to President Carter. Some countries have triple-digit inflation. A U.S. recession in the second quarter cuts real output by 9.9 percent; the economy is on the rise again by fall.

President-elect Reagan (above) has campaigned on what his running mate George Bush called "voodoo economics" (based on supply-side ideology) during the primary elections, but Bush drops his opposition at the convention.

British unemployment rises above 2 million for the first time since 1935 (when the work force was one-third smaller) as recession depresses the economies of many countries. By year's end, unemployment reaches nearly 2.5 million, up from 800,000 early in 1975, and industrial production falls 5 percent as the government's monetarist policies try to stem a new burst of inflation, which again climbs above 20 percent, double the rate when Thatcher took office.

West Germany has a currency deficit of $14.2 billion, up from $5.4 billion last year (there was a surplus of nearly $9 billion in 1978) as energy costs climb, interest rates rise, and consumer spending eases. Imports grow more costly as the mark falls 15 percent in relation to the U.S. dollar.

Poland's Western debts soar to $23 billion and industrial production falls 1.3 percent as a result of labor unrest (above) and shortages of fuel, raw materials, and parts. Average monthly wages rise 20 percent to about $207 (at official exchange rates); personal income, adjusted for inflation, rises only 1 percent.

U.S. personal bankruptcies jump to 367,000, up from 209,500 last year. A new federal bankruptcy law that went into effect October 1, 1979, enables individuals to protect much more of their property against seizure by creditors.

Some 36 million Americans receive monthly Social Security checks, 26 million Medicare benefits, 22 million Medicaid benefits, 18 million food stamps, 15 million veterans' benefits, 11 million Aid to Families with Dependent Children funds; millions of students receive federal scholarship aid (*see* education, below); 27 million children benefit from school lunch programs (*see* 1946), and most of these categories overlap. Ronald Reagan (above) promises to reduce the size of government.

Japan's oil imports cost $39.5 billion for the year ending March 31, producing a record trade deficit of $14.4 billion as compared to a $13.4 billion surplus last year.

The Crude Oil Windfall Profit Tax Act signed by President Carter April 2 complements his decontrol of domestic U.S. oil prices last year. The new law actually applies to price increases above the controlled levels prior to June 1, 1979, not to profits. Oil producers complain, but they will receive a projected $1 trillion in extra revenues and will still have an additional $221 billion after all federal, state, and local taxes. Domestic oil averages about $22 per barrel, up from $8.57 in 1977, and the industry drills an estimated 59,475 wells, up from a low of 27,300 in 1971 (the record, set in 1956, is 58,160). But proven domestic U.S. petroleum reserves fall below 27 billion barrels, down from 39 billion in 1970.

OPEC ministers meeting at Algiers in June set a price ceiling of $32 per barrel on crude oil, with some grades to sell for as high as $37. Saudi Arabia pumps 9.5 million barrels per day (up from 8.5 billion before the fall of Iran's shah) and accounts for one-third of all OPEC production, but cannot persuade other oil exporters to hold the price at $28. Some Western analysts say market demand and supply factors would force prices even higher without OPEC.

U.S. gasoline prices average $1.20 to $1.23 per gallon for most of the year, up from 66.1¢ in 1978 but still less than half the price in most countries.

Rome's Metropolitan "A" line, the city's first really serviceable subway, begins operations February 16 between Cinecetta, in the populous suburbs, and Ottaviano, near the Vatican in the city's northwest. By autumn, 6-coach trains running at 3-minute intervals cover the distance in 25 minutes. Rome's first subway, planned by Mussolini, opened in 1955 but did not serve much of the city until urban expansion made its location more central. Plagued by vandalism and violence, this "B" line remains unpopular.

A New York transit strike ends April 11 after 35,000 workers have stopped the city's 6,400 subway cars and 4,550 buses for 11 days, forcing 5.4 million people to walk, bike, or find other means of transportation. Fares rise from 50¢ to 60¢; the archaic, crime-ridden, graffiti-defaced subway remains dirty and dangerous.

Pittsburgh raises transit fares November 2 from 60¢ to 75¢.

The Motor Carrier Act signed by President Carter July 1 curbs federal controls over interstate trucking.

The Staggers Rail Act signed October 14 gives railroads more flexibility in setting rates and more authority to enter into long-term contracts with freight shippers, lets railroads drop unprofitable routes more easily, and deregulates in other ways (*see* airline deregulation, 1978).

Soviet engineers push forward a new $10 billion Siberian railway that will give access to areas of coal, oil, and mineral development. Extending nearly 2,000 miles from Lake Baikal to the Amur River, the new road lies north of the Trans-Siberian Railway, completed in 1906, and is scheduled to open in 1983.

A Polish train wreck August 19 kills 62 and injures 50 as a freight train misses a stop signal, heads down the wrong track, and collides with a crowded passenger train before dawn. The passengers were returning to Lodz from a holiday on the Baltic.

First class rail service between London and Brussels/Paris ends October 31. British Rail took over the old blue and gold sleeping cars from French Wagons Lits in 1977 but air travel has largely replaced rail travel.

CSX Corp., created in November by a merger of the Chessie System and Seaboard Coast Line Industries, becomes the largest U.S. railroad company, with 27,000 miles of tracks. The Chessie System embraces the Chesapeake & Ohio, Baltimore & Ohio, and Western Maryland roads; Seaboard Coast Line embraces the old Seaboard Airline, Atlantic Coast Line, Louisville & Nashville, and other roads.

The Burlington Northern becomes a 30,000-mile system November 21. Merger with the St. Louis-San Francisco Railway (Frisco Line) gives it track from Florida to the Northwest.

U.S. sales of foreign cars fall 15.2 percent from 1979 levels, domestic car sales fall 20 percent despite the introduction of fuel-efficient models. Japanese automobile production (11 million cars and trucks) rises 10 percent and for the first time overtakes U.S. production (7.8 million cars and trucks), which declines by 30 percent (*see* 1981). Imported cars and trucks, 78 percent of them Japanese, capture nearly one-fourth of the U.S. market.

Chrysler loses nearly $1.8 billion, Ford $1.5 billion, the largest losses ever sustained by any U.S. corporations (Chrysler's 1979 loss of $1.1 billion was the old record). Even General Motors loses $763 million as U.S. automakers pay the costs of retooling for downsized cars to achieve fuel economy and meet Japanese competition. Front-wheel drive Chrysler K cars, Ford Escorts, and Chevrolet Chevettes sell briskly despite high sticker prices (if often with the help of cash rebates).

British Leyland loses $1.21 billion, up from $372.4 million last year, and reduces its work force from more than 155,000 to 130,000. August sees Japanese imports outsell BL cars in Britain for the first time ever. The state-owned BL Ltd. introduces the subcompact front-wheel drive Austin Mini Metro October 8 and says it will average 46.1 miles per gallon in town, 63.7 mpg at 56 mph, 50 mpg at 75 mph. BL works on the Triumph Acclaim, developed with Honda, to be introduced in the fall of 1981.

The Rolls-Royce Silver Spirit unveiled at the Paris Auto Show in October is the first new 4-door Rolls in 15 years. Price: $119,500.

The Philippine passenger ferry M.V. *Don Juan* collides April 22 with the government oil tanker M.T. *Talcloban City* while the ferry captain is drinking beer and playing mahjongg. The luxury ship has a legal capacity of 810 passengers and is carrying 1,349; 113 bodies are recovered, another 200 persons are missing and presumed dead.

1980 *(cont.)* Switzerland's Goschenen-Airolo Tunnel opens September 25 beneath the St. Gotthard Pass after 11 years of construction. More than 300 Italian, Swiss, Spanish, Turkish, West German, and Austrian miners have built the $420-million tunnel (19 were killed), and the 10-mile project—longest road tunnel in the world—cuts 4 hours off a truck trip from West Germany or the Netherlands to Italy. Critics say the new tunnel will encourage private cars at the expense of public transport and have negative environmental effects.

Atlanta's $750-million Hartsfield International Airport opens September 21 with 138 passenger gates (versus 94 at Chicago's O'Hare) and with a station for the city's fledgling rapid transit system MARTA, whose tracks are scheduled to reach the airport by 1985.

Pan Am acquires 50-year-old National Airlines, gaining its first domestic routes.

Major U.S. airlines lose an estimated $200 million (operating profits were $215 million in 1979, $1.4 billion in 1978) as passenger miles flown fall 5 percent and fuel prices climb from 57¢ per gallon to $1. Many companies cut routes (*see* deregulation, 1978).

An Iran Air Boeing 727 crashes into mountains near Laskgarak January 21, killing all 128 aboard; a LOT Polish Airlines Ilyushin 62 crashes at Warsaw March 14, killing 87, including 22 boxers and officials of a U.S. amateur boxing team; a chartered Boeing 727 crashes at Santa Cruz de Tenerife in the Canary Islands April 25, killing 138 British vacationers and a crew of 8; a Saudi-Arabian Lockheed 1011 Tristar jet catches fire after takeoff from Riyadh August 9, her pilot turns back, but the fire kills 301 people (a Muslim pilgrim's butane gas stove is blamed).

The Supreme Court rules 5 to 4 June 16 in *Diamond v. Chakrabarty* that a man-made life form—specifically, a genetically engineered bacterium capable of breaking down multiple components in crude oil—may be patented.

A rabies vaccine approved by the U.S. Food and Drug Administration June 9 requires only five shots in the arm instead of 23 in the abdomen (*see* Pasteur, 1885). Developed by Philadelphia's Wistar Institute and produced in France, the new vaccine is made from viruses grown in human cell cultures rather than in duck eggs.

The Church of England adopts the *Alternative Service Book* in November, replacing *The Book of Common Prayer* used since 1549. "Our Father in heaven, hallowed be your name . . . ," says Elizabeth II, instead of "Our Father, which art in heaven, hallowed be thy name . . ."

A new U.S. Department of Education begins operations May 4. Health, Education and Welfare becomes Health and Human Services pursuant to a law enacted last year in fulfillment of a Carter campaign pledge to teachers. Critics say the new department, with 17,000 full-time employees and a $17-billion annual budget, will curtail local control of schools.

Only 29,019 U.S. high school students score above 650 on the verbal part of the Scholastic Aptitude Test (SAT; *see* 1900), down from 53,794 in 1972. Only 73,386 score above 650 on the mathematics part, down from 93,868. Educators suggest that young people are reading less because of TV and changing social values, that students are taking fewer difficult courses, that college admissions standards and course requirements have fallen. A general decline in SAT scores began in 1964.

More than a third of 11 million U.S. college students receive some federal financial aid, up from 14 percent in 1970, as tuition and other costs rise to average $3,500 per year at state universities, more than $7,500 at private ones (tuition, room, and board at Harvard come to $9,170, at Yale $9,110). Federal aid to college students reaches $4.5 billion, up from $600 million in 1970, and the higher education bill signed by President Carter October 3 lets parents borrow up to $3,000 per year per student at 9 percent interest in addition to the $2,500 per year that a student can borrow on his/her own (*see* 1978; 1990).

The French Post Office earmarks $100 million to develop the telématique—a new system linking telephones with central computers to replace telephone directories and make all sorts of information readily accessible to subscribers (*see* Britain's Prestel, 1979). The 2-year development effort is part of a crash program that has increased telephone subscribers to 15 million, up from only 6 million in 1974. The post office raises the price of paper phone directories by 500 percent to encourage use of telématique terminals consisting of typewriter keyboards attached to television screens. The terminals are free; asking for information costs money each time.

Cable News Network (CNN) goes on the air June 1 with a speech by owner Ted Turner.

Moscow resumes selective jamming of Western radio broadcasts August 20 in an apparent effort to lessen the impact of reports of Polish industrial unrest (above). The first Soviet jamming since 1973 violates a provision of the 1975 Helsinki accords.

The U.S. Supreme Court has ruled July 2 in *Richmond Newspapers v. Virginia* that the press and the public have a right to attend criminal trials. An act of Congress signed into law by President Carter October 14 forbids unannounced searches of newsrooms except in narrowly defined circumstances.

Discover magazine begins publication at Chicago in October as Time-Life launches a new science monthly.

Nonfiction *The Zero-Sum Society* by MIT economist Lester Thurow, 42; *Cosmos* by Carl Sagan.

Fiction *The Name of the Rose* by University of Bologna semiotics professor Umberto Eco, 48; *Man in the Holocene* by Max Frisch; *Earthly Powers* by Anthony Burgess; *Rites of Passage* by William Golding; *The Transit of Venus* by Australian-American novelist Shirley Hazzard, 49; *Nuns and Soldiers* by Iris Murdoch; *Morgan's Passing* by Anne Tyler; *Neighbors* by Thomas Berger; *The Second Coming* by Walker Percy; *Loon Lake* by E. L. Doctorow; *Covenant* by James Michener.

Poetry *The Morning of the Poem* by U.S. poet James Marcus Schuyler, 56.

Painting *To the Unknown Painter* by U.S. painter Anselm Keifer, 35. Oscar Kokoschka dies at Montreux, Switzerland, February 22 at age 93; Clyfford Still dies at Baltimore June 23 at age 75, 5 months after a Metropolitan Museum show of his abstract paintings—the largest one-man exhibition by a living artist in Met history.

Theater *Piaf!* by English playwright Pam Gems 1/15 at Wyndham's Theatre, London, with Jane Lapotaire as France's legendary "little sparrow" Edith Piaf (*see* 1946) who died 10/11/63 at age 47 (as Jean Cocteau was dying at age 74); *The Lady from Dubuque* by Edward Albee 1/30 at New York's Morosco Theater, 12 perfs.; *Talley's Folly* by Lanford Wilson 2/10 at New York's Brooks Atkinson Theater, with Judd Hirsch, Trish Hawkins, 279 perfs.; *Children of a Lesser God* by Mark Medoff 3/30 at New York's Longacre Theater, with John Rubinstein as the speech therapist for hearing-impaired students, Phyllis Freilich (totally deaf since birth) as the student he marries, 854 perfs.; *The Dresser* by English playwright Ronald Harwood (*né* Horwitz), 45, 4/30 at the Queen's Theatre, London, with Tom Courtenay, Freddie Jones; *Home* by U.S. actor-playwright Samm-Art Williams, 33, 5/7 at New York's Cort Theater, with Charles Brown, L. Scott Caldwell, Michelle Shay, 279 perfs.; *The Life and Adventures of Nicholas Nickleby* as adapted from the 1838 Dickens novel by English playwright David Edgar, 32, 6/6 at London's Aldwych Theatre, with Roger Rees, 35, as Nicholas, David Threlfell, and 40 other members of the Royal Shakespeare Company (in two parts, the play runs 8½ hours); *Educating Rita* by Willy Russell 8/19 at London's Piccadilly Theatre (after a run at the Warehouse Theatre), with Mark Kingston, Julie Walters; *The Fifth of July* by Lanford Wilson 11/3 at New York's New Apollo Theater, with Christopher Reeve, Swoosie Kurtz, 511 perfs.; *True West* by Sam Shepard 12/23 at New York's Public Theater, with Peter Boyle, Tommy Lee Jones, 52 perfs.

Television *Shōgun* 9/15–19 on NBC with Richard Chamberlain, Toshiro Mifune.

Films Louis Malle's *Atlantic City* with Burt Lancaster, Susan Sarandon; Rainer Werner Fassbinder's 15½-hour-long *Berlin Alexanderplatz* with Gunter Lamprecht; Irvin Kershner's *The Empire Strikes Back* with Mark Hamill, Harrison Ford, Carrie Fisher; Akira Kurosawa's *Kagemusha* (*The Phantom Samurai*) with Tatsuya Nakadai; Robert Redford's *Ordinary People* with Donald Sutherland, Mary Tyler Moore, Timothy Hutton; Martin Scorsese's *Raging Bull* with Robert De Niro as prizefighter Jake La Motta; Richard Rush's *The Stunt Man* with Peter O'Toole, Steve Railsback. Also: Nicolas Roeg's *Bad Timing: A Sensual Obsession* with Art Garfunkel, Theresa Russell; Samuel Fuller's *The Big Red One* with Lee Marvin; Michael Apted's *Coal Miner's Daughter* with Sissy Spacek as country singer Loretta Lynn; Krzysztof Zanussi's *Contract* with Leslie Caron, Maja Komorowska; Brian De Palma's *Dressed to Kill* with Michael Caine, Angie Dickinson; Francesco Rossi's *Eboli* with Gian Maria Volante; David Lynch's *The Elephant Man* with Anthony Hopkins, John Hurt; Ingmar Bergman's *From the Life of the Marionettes* with Robert Atzorn, Christine Buchegger, Martin Benath; Micheline Lanctot's *The Handyman* with Jocelyn Berube; John Mackenzie's *The Long Good Friday* with Bob Hoskins; Maurice Pialat's *Loulou* with Isabelle Huppert, Gerard Depardieu; Robert Sickinger's *Love in a Taxi* with Diane Sommerfield, James H. Jacobs; Andrzej Wajda's *Man of Iron* with Wajda, Jerzy Radziwilowicz; Jonathan Demme's *Melvin and Howard* with Jason Robards as Howard Hughes, Paul Le Mat; Alain Resnais's *Mon Oncle d'Amerique* with Gerard Depardieu, Nicole Garcia; William Peter Blatty's *The Ninth Configuration* with Stacy Keach; Nikita Mikhalkov's *Oblomov* with Oleg Tabakov; Daniel Petrie's *Resurrection* with Eva LeGallienne, Ellen Burstyn, Sam Shepard; Stephen Wallace's *Stir* with Bryan Brown.

Anthem "Oh, Canada" is officially adopted as the Canadian national anthem June 27 by the House of Commons (*see* 1880).

Broadway musicals *Barnum* 4/20 at the St. James Theater, with Jim Dale as P. T. Barnum, music by Cy Coleman, lyrics by Michael Stewart, 854 perfs.; *42nd Street* 8/25 at the Winter Garden Theater, with Jerry Orbach, Wanda Richert, Tammy Grimes, music and lyrics from the 1933 Hollywood musical, choreography by Gower Champion (who dies opening night at age 61 of a rare blood cancer), 3,486 perfs.

San Francisco's Louise M. Davis Symphony Hall opens September 16 with 3,000 seats across the street from the 48-year-old War Memorial Opera House.

Popular songs *Christopher Cross* (album) by San Antonio songwriter Christopher Cross, 29, who makes a hit with his song "Sailing"; *Gaucho* (album) by Steely Dan (Walter Becker and Donald Fagen); *Second Edition* (album) by Public Image Ltd. (John Lydon, guitarist Keith Levene, bassist Jah Wobble, drummer Martin Atkins); *Unknown Pleasures* (album) by English rock group Joy Division whose lead singer Ian Curtis hangs himself in May; *Illusions* (album) by rock group Arthur Blythe; *Get Happy* (album) by Elvis Costello and the Attractions; *Remain in Light* (album) by the Talking Heads; *Fourth World Vol. 1/Possible Musics* (album) by Brian Eno and trumpeter Jon Hasell; *Doc at the Radar Station* (album) by Captain Beefheart (Don Van Vliet) and the Magic Band; *Crawfish Fiesta* (album) by New Orleans rhythm and blues group Professor Longhair; *Seconds of Pleasure* (album) by Rockpile (Dave Edmunds, Billy Bremner, Nick Lowe, Terry Williams); *Double Fantasy* (album) by John Lennon and Yoko Ono (who sees her husband shot to death; *see* crime, below).

1980 ***(cont.)*** Pittsburgh beats Los Angeles 31 to 19 at Pasadena January 20 in Super Bowl XIV.

The U.S. Olympic ice hockey team scores a 4 to 3 victory over a heavily favored Soviet team at Lake Placid February 22. President Carter has banned U.S. participation in the summer Olympics at Moscow to demonstrate disapproval of continued Soviet military activity in Afghanistan; 81 countries send athletes to the Moscow games, 57 follow the U.S. example.

Björn Borg defeats John McEnroe to win his fifth consecutive Wimbledon singles title, Evonne Goolagong Cawley, 28 (Australia), wins in women's singles; McEnroe defeats Borg to win in U.S. men's singles, Chris Evert-Lloyd wins in U.S. women's singles.

The Philadelphia Phillies win their first World Series, defeating the Kansas City Royals 4 games to 2.

Pennsylvania-born pacer Niatross wins the Triple Crown (Cane Pace, Little Brown Jug, and Messenger Stakes) after winning 19 consecutive races. Undefeated (13 for 13) as a 2-year-old, Niatross eclipses the harness racing records of Hambletonian (*see* 1876) and Dan Patch (*see* 1902) with a world record pacing mile of 1 minute 49.02 seconds.

The U.S. yacht *Freedom* retains the America's Cup, defeating *Australia* 4 races to 1.

Luvs disposable diapers, introduced by Procter & Gamble, are "super-absorbent" and have elastic gathers. Huggies, introduced by Kimberly-Clark 2 years ago, have cut into sales of P&G's Pampers.

The U.S. Supreme Court rules 6 to 3 in October against hearing an appeal from an Illinois Supreme Court decision that a woman with a sleep-in boyfriend has violated the state's antifornication statute and has thus forfeited her right to custody of her three children. An estimated 1.1 million Americans live together out of wedlock and an estimated 25 percent or more of these households include children.

Rollerblade, Inc. is founded near Minneapolis by Canadian hockey player Scott Olsen, 20, who has bought a Chicago company's patent for an in-line roller skate and perfected the design with his brother Brennan, 16, using a blade of polyurethane wheels and a molded ski-type boot.

U.S. cigarette sales rise slightly to 614.5 billion with low-tar brands accounting for nearly 49 percent, up from 16.7 percent in 1976, despite growing evidence that the "safer" cigarettes offer only limited health benefits and may even pose new hazards because of flavor-boosting additives. Smoking has dropped 28 percent since 1970 among men 20 and older, 13 percent among adult women, 20 percent among teenage boys. Smoking among teenage girls has increased by 51 percent since 1968.

The New Mexico State Penitentiary outside Santa Fe has a 36-hour riot beginning February 2. Designed to hold 850 men, the facility has more than 1,300, many of whom are obliged to sleep on the floor in a nightmare of blaring radios, harassment, and sexual attacks. Prison authorities have used intimidation and favors to encourage a system of informants—"snitches" who are prime targets when inmates seize control, invade the pharmacy, shoot drugs, and vent their rage. The riot ends with 33 inmates dead, some of drug overdoses, some brutally murdered, and more than 100 seriously wounded by fellow prisoners. Property damages total more than $25 million.

Murders of black Atlanta children resume in March (*see* 1979). By year's end two girls and 10 more boys have been found dead and another boy is missing. Police have no clues; the city's black community is terrorized (*see* 1982).

"Scarsdale Diet" creator Dr. Herman Tarnower, 69, is shot fatally March 10 in his Harrison house by Madeira School headmistress Jean Struven Harris, 56, who claims it was an accident.

Former U.S. congressman Allard Lowenstein, 51, is shot fatally March 14 in his New York office by a deranged former associate in the civil rights movement.

Washington, D.C., physician-author Michael Halberstam, 48, is shot fatally December 5 by a burglar whom he has surprised in his house.

John Lennon of Beatles fame is shot fatally December 8 outside the Dakota apartment building in New York. Lennon's death at age 40 increases demands for gun control laws, but president-elect Reagan says more effective laws would not have prevented Lennon's shooting by a deranged fan. U.S. handgun killings average 29 per day, and 55 million "Saturday night specials" and other handguns are believed to be in circulation, easily obtainable through 175,000 federally licensed dealers (often just by showing a driver's license, which need not be checked for authenticity in many states) or through illicit channels.

A U.S. Coast Guard cutter opens fire October 9 on a fishing boat off Miami after the 50-foot boat, loaded with 12.5 tons of marijuana, failed to comply with a request to board and has ignored warning shots. Marijuana from Jamaica and Latin America continues to pour into the United States as smugglers evade law enforcement officers. The gunners aboard the cutter *Point Francis* have been the first to aim directly at their prey in peacetime, they disable the boat and tow her into Miami, but the incident does little to discourage the flourishing traffic that has made Miami banks the nation's largest users of $100 bills and other U.S. currency.

Crystal Cathedral, completed at Garden Grove, Calif., seats 2,890. The spectacular $18-million structure designed by Philip Johnson and John Burgee has no air conditioning but rises 128 feet high with a mirrored glass exterior and open panes that permit convection currents to keep the interior comfortable.

U.S. housing starts fall below 1.3 million, down 33 percent from last year, as prices rise and high interest rates put mortgages out of reach for many prospective buyers. Government programs assist 44.1 percent of the housing starts, but more than 15 percent of builders are forced to quit the industry.

New York's Grand Hyatt Hotel opens September 25 with 1,400 rooms. Designed by Gruzen & Partners on the skeleton of the 1919 Commodore Hotel, the 30-story hotel has a mirrored glass façade and an atrium 4 stories high. Queens developer Donald Trump, 34, has bought the Commodore from the bankrupt Penn Central on condition that the city give him a huge tax break.

A fire at the 7-year-old MGM Grand Hotel at Las Vegas November 21 traps 3,500 guests and kills 84 people. Helicopters lift more than 1,000 from the roof and carry them to safety. The hotel has no smoke alarms.

Mount St. Helens in Washington State erupts May 18, spewing forth volcanic ash that kills scores of people, wreaks havoc on farmlands, fouls automobile and truck engines, and blocks the Columbia River with 51 million cubic yards of sand, dirt, and rocks.

Brazil offers incentives to colonists who will settle in the nation's northwest state of Rondonia, hoping that tens of thousands will respond. More than a million landless farmers and urban slum dwellers will travel up BR364 in the next decade, the World Bank will finance paving of the highway, the settlers will cut and burn the rain forest, Indians in the region will die from epidemics of flu and measles, the soil will not support agriculture, and the disastrous scheme will bring threats of global warming (*see* 1988).

Hurricane David hits Dominica in the Caribbean August 30 with winds well over 100 miles per hour, killing at least 22 and leaving 60,000 homeless. David batters the Dominican Republic August 31 and kills at least 600, leaving 150,000 homeless and causing more than $1 billion in losses.

An earthquake in southern Italy November 23 kills several thousand people and leaves tens of thousands without food, shelter, or medical care.

Mexico unilaterally abrogates all fishing treaties with the United States December 29 after 15 meetings since 1977 have failed to settle disputes over tuna fishing rights.

President Carter orders a partial embargo of U.S. grain sales to the U.S.S.R. January 4 in response to the Soviet invasion of Afghanistan. Argentina ships grain to Russia, which has her second bad harvest in a row. Normal customers for Argentine grain receive shipments from other countries, and critics say the embargo of 17 million metric tons of grain hurts U.S. farmers more than it hurts Moscow. Grain shipments guaranteed by the 1975 agreement are not affected.

Drought reaches into western Iowa, most of Illinois, all of Indiana, and much of Ohio, shortening some crops, especially peanuts. Conditions in the Corn Belt and Cotton Belt rival those of 1934, 1936, and 1953, and endless weeks of high temperatures and cloudless skies exhaust subsurface moisture in much of the Southeast, Southwest, and plains states. The Dakotas and Minnesota have their second dry year in a row. Winter wheat, harvested before the drought, is up 16 percent to a record 1.9 billion bushels, but feed grains are badly hurt, with the corn crop dropping 14 percent to 6.6 billion bushels.

U.S. corn yields average 90 bushels per acre, up from 43 in 1960, 25 in 1940, although Nebraska loses 75 percent of its corn crop. Wheat yields average 30.5 bushels per acre, up from 20.5 in 1960, 13 in 1940. Sorghum yields average 53 bushels, up from 24 in 1960, 12.5 in 1940. Soybean yields average 28 bushels, up from 22 in 1960, 16.2 in 1940. High yields and large reserves (1.7 billion bushels of corn, 901 million bushels of wheat) from past years keep prices down, but prices for corn and soybeans rise above the level before the embargo (above) and wheat nears the pre-embargo price of $3.80 per bushel by August. Farm income falls sharply from the 1979 peak of $33 billion as interest costs (above), higher diesel fuel prices, and other factors squeeze the capital-intensive U.S. farm economy.

U.S. farmland averages $609 per acre for the year, up from $525 last year. Acreage prices climb 20 percent or more in 13 states despite falling farm incomes.

The Supreme Court rules unanimously June 16 in *Bryant v. Yellen* that farmers in California's Imperial Valley may continue to receive Colorado River water regardless of the size of their farms and despite a 1926 law limiting to 160 acres the size of farms eligible for water from federal irrigation projects.

European Economic Community countries ban the use of hormones in cattle feed October 1 in response to a French boycott on veal that Britain and Belgium have echoed. The ban goes beyond a U.S. ban on using DES early in the year. Its effect will be to make meat costlier while making veal taste better.

Poland imposes meat rationing in December for the first time since World War II, partly to meet a demand by striking workers (above). Meat shortages stem from the need to export meat for Western currency used to pay off Poland's huge foreign debt (above).

World sugar prices rise to 24¢ per lb. by midyear, up 60 percent from 1979. Coca-Cola Co. is the world's largest user of sugar and leads in a switch to high fructose corn sweeteners, about 25 percent cheaper than sugar. Many foods containing sugar, notably Jell-O, simply rise in price.

Rum outsells vodka in the United States and outsells whiskey for the first time since the early nineteenth century (*see* vodka, 1973). All whiskeys except Canadian brands have declined in the 1970s while vodka, rum, and tequila have gained in popularity.

Some 10,800 Cubans rush onto the grounds of Havana's Peruvian Embassy in mid-April seeking asylum; Washington agrees to admit more immigrants; beginning April 21, 125,262 Cubans leave the country until Fidel Castro closes down the port of Mariel September 26, leaving an estimated 375,000 would-be emigrants still in Cuba.

1980 ***(cont.)*** The U.S. Supreme Court upholds the Hyde Amendment of 1976 in a 5 to 4 ruling handed down June 30 to the delight of anti-abortion groups. Abortions paid for by Medicaid funds fell from 295,000 to only 2,100 for fiscal 1978, 3,900 for fiscal 1979, but some states continue to pay for abortions (average cost: $200) entirely out of their own funds and without federal reimbursement (*see* 1989).

1981 Iran releases all U.S. hostages January 20 (they are flown to Algiers following 444 days in captivity) after U.S. negotiators agree to unblock certain Iranian funds and Iran agrees to repay U.S. bank loans. Iran's President Abolhassan Bani-Sadr is removed from office June 22 and flees to France. Ayatollah Mohammed Beheshti, chief justice and head of the Islamic Republican Party, is killed along with four key government ministers in a bombing attack at Teheran June 28. Iran's President Ali Rajai, Prime Minister Hojatolislam Javad Bahonar, and Col. Houshang Dagsgerdi die in a bombing attack at Teheran August 30; a grenade kills Ayatollah Assadolah Madani, an aide to the Ayatollah Khomeini, at Teheran September 11.

President Reagan and three others are wounded by pistol bullets March 31 at Washington, D.C., in an assassination attempt with no evident political motive. Reagan's press secretary James Brady suffers permanent brain damage. Handguns remain readily obtainable in most of the United States.

Pope John Paul II is wounded at Rome May 13 in an assassination attempt in St. Paul's Square. The attacker is a Bulgarian-trained Turk; KGB complicity is widely suspected.

Italy's cabinet resigns May 26 after revelations linking 953 Cabinet officers, legislators, judges, and bankers to a secret Masonic organization.

Army officers in Bangladesh fail in an attempted coup May 30 but kill President Ziaur Rahman, the army chief of staff who took power late in 1976.

Six Afghan guerrilla groups continue strong opposition to Soviet occupation forces, which have suffered at least 10,000 casualties since the invasion that began in December 1979.

Lebanese Christian militiamen aided by Israeli forces shoot down two Syrian helicopters, the Syrians move surface-to-air missiles into the Bekaa valley east of Beirut, Israel threatens to knock out the missiles, and President Reagan summons former Under Secretary of State Philip C. Habib, 61, out of retirement to help negotiate a truce.

Israeli jets destroy Iraq's Osirak nuclear reactor June 7 in a preemptive strike to prevent production of plutonium.

Israeli and PLO forces clash through June and July with several weeks of heavy fighting, shellfire falls on Israeli settlements, and Israeli jets strike targets in Beirut and southern Lebanon before a cease-fire is negotiated with Habib's help July 24. Israel charges in August that Palestinians are moving artillery and ammunition within Lebanon's UN zone (*see* 1978), and Israel annexes the Golan Heights December 14, raising strong protests.

Egypt's President Anwar el-Sadat cracks down on dissidents, has 1,600 arrested in a single night in September, and is assassinated by Islamic extremists at Cairo October 6 while watching a parade of troops. Dead at age 62, he is succeeded by his vice president Hosni Mubarak, 53, who was on the reviewing stand with Sadat; he has several hundred fanatics arrested and 24 tried for murder (five will be executed). Mubarak immediately affirms Egypt's commitment to Sadat's peace treaty with Israel but makes friendly overtures to other Arab states and initiates release of political prisoners.

Northern Ireland's former parliamentary speaker Sir Norman Stronge and his son have been assassinated by terrorists January 21—4 days after an attempt at Belfast on northern Irish nationalist Bernadette Devlin McAlisky and her husband. Ten hunger strikers die in a Belfast prison protest from May to July. Bobby Sands, 27, who dies May 5 after 65 days without food, had recently been elected to Parliament despite the fact that he is serving a 14-year sentence for firearms possession.

French voters elect socialist François Maurice Mitterand, 64, president in June balloting.

Panamanian strongman Brig. Gen. Omar Torrijos Herrera is killed with six others in a plane crash July 31 at age 62. Col. Manuel Antonio Noriega, 48, a longtime CIA informant, emerges as the most powerful figure in Panama (*see* 1988).

Belize (formerly British Honduras) becomes a fully independent commonwealth September 21.

Antigua and Barbuda in the Caribbean gain full independence November 1.

Ghana has a coup December 31. Jerry J. Rawlings seizes power, accusing President Hilla Limann, 47, of taking the country "down the road to economic ruin";

Egypt's Anwar Sadat paid with his life for his 1978 rapprochement with Israel.

he institutes an austerity program to reduce budget deficits (*see* 1979).

President Ferdinand Marcos of the Philippines ends 8 years of martial law January 17 and wins election to a second 6-year term June 16, but Marcos has effectively ended democracy in the country and stifled opposition, using anti-Communism to mask a policy of suppression (*see* 1983).

Polish general Wojciech Jaruzelski, 57, becomes prime minister in February, the fourth in a year, as social unrest continues (*see* 1980). Jaruzelski orders an army-police crackdown in September on lawlessness and anti-Soviet activity, he succeeds Stanislaw Kania as First Secretary of the Communist Party October 18, and he imposes martial law December 13 to squelch strikes. Solidarity is outlawed, martial law will continue for 19 months, and opposition leaders, including Lech Walesa, will either go underground or serve time in prison.

President Reagan appoints Arizona appellate court judge Sandra Day O'Connor, 51, to the Supreme Court July 7. She is confirmed by the Senate September 21 and becomes the first woman justice on the high court. Her appointment begins a tilt of the court toward right-wing statism.

President Reagan signals a tough new policy toward organized labor August 6, by dismissing air traffic controllers who have defied his return-to-work order. Patco (Professional Air Traffic Controllers Organization) has struck August 3, demanding a 4-day week and a $10,000-a-year raise. Only 2,000 Patco members remain on the job. Patco is decertified as the bargaining body for air traffic controllers in October and files for bankruptcy in November.

The U.S. prime-interest rate reaches 21.5 percent, highest since the Civil War, as double-digit inflation and high unemployment plague the economy.

President Reagan signs a bill August 13 mandating the deepest tax and budget cuts in U.S. history. It follows "supply-side" economic theories that reject Keynesian ideas popular since the 1930s. Supply-siders, led by California economist Arthur Laffer, 40, claim that reducing taxes will encourage business and the rich to invest in taxable activities rather than parking income in nonproductive tax shelters and will thus help the overall economy. While "Reagonomics" will be credited with producing the longest peacetime boom in history, it will also lead to neglect of cities, deterioration of infrastructure, and massive deficits financed by foreign borrowing.

President Reagan addresses the nation on television September 29 appealing for fiscal austerity and asking for an additional $13 billion in spending cuts for fiscal 1982. He astonishes supply-siders by requesting $3 billion in tax increases.

Standard Oil Co. (of Ohio) has record earnings of $1.95 billion and pays $1.77 billion for Kennecott, the biggest U.S. copper producer.

E. I. Du Pont acquires Conoco (formerly Continental Oil Co.) for $6.8 billion September 1 (*see* 1929). Conoco directors have acted June 17 to approve "golden parachutes"—special pay or bonuses—for nine company officers should they quit or be fired after a takeover.

Chicago transit fares rise January 1 from 60¢ to 80¢, Dallas fares from 60¢ to 65¢, Miami fares from 60¢ to 75¢, Milwaukee fares from 50¢ to 65¢, Salt Lake City fares from 40¢ to 50¢. Washington, D.C., fares rise January 4 from 55¢ to 60¢. Toledo fares rise February 1 from 35¢ to 50¢. New York fares rise July 3 from 60¢ to 75¢.

Conrail labor unions and management agree to forego $290 million in wages and benefits per year to make the heavily subsidized system profitable and head off a Reagan administration threat to break up Conrail and sell it.

France's TGV train begins service to Lyons September 22. Powered by electricity and capable of 236-m.p.h. speeds, it is Europe's first super-high-speed passenger line and will reach Marseilles by 1983.

Britain's 4,626-foot Humber Bridge, completed at Hull, is the world's longest suspension bridge.

Sales of U.S.-made autos fall to 6.2 million, lowest level in 20 years (*see* 1980). Japanese makers placate U.S. makers by voluntarily limiting exports to America to 1.68 million units.

U.S. air service drops 25 percent following the Patco strike (*see* above) as new air traffic controllers are trained to supplement the 2,000 remaining Patco workers and 2,500 non-union workers and military personnel.

People Express begins flying from Newark to Boston with three planes and 250 employees as deregulation of U.S. airlines takes effect. Founder Calvin Burr, 40, has bought used planes from airlines with excess capacity and taken over an empty terminal building with low rent. Within 4 years the discount-fare airline will be serving 30 cities with more than 60 planes and 4,000 employees, but it will rarely show a profit and will be merged into Texas Air early in 1987.

IBM introduces its first personal computer August 12 and soon has 75 percent of the market. Its PC uses a Microsoft disk-operating system (MS-DOS) and competitors quickly develop lower-priced "clones."

AIDS (Acquired Immune Deficiency Syndrome) begins taking a worldwide toll that will be compared to that of the Black Death in the fourteenth century. San Francisco and New York physicians report that a few dozen previously healthy homosexual men have died of Kaposi's sarcoma, a form of cancer endemic in Africa but rare in the rest of the world. The men have suffered abnormalities of the immune system. New York doctors realize they have seen a number of similar cases in the past few years, all unexplained. More cases appear each month. Drug addicts, mostly black and Hispanic, in New York, Newark, and other northeastern cities begin

1981 *(cont.)* dying of a previously rare pneumonia and other diseases brought on by a collapse of the body's disease-fighting ability. Invariably fatal, AIDS will be diagnosed in more than 32,000 Americans by early 1987, and nearly 60 percent will have died. Spread by exchange of bodily fluids containing a retrovirus, AIDS finds its victims almost exclusively among homosexual males, drug addicts using contaminated needles, and people given prophylactic medical shots with such needles. AIDS will be epidemic in parts of Africa, with as many female victims as male, and within a decade will be killing people throughout Asia, Europe, and the Americas.

CBS newsman Walter Cronkite, 64, goes off the air March 6 after 19 years as top U.S. "anchorman." Dan Rather, 49, succeeds him.

British Telecom, created in October under legislation separating the government telephone company from the Post Office, works to complete the world's longest high-speed optical fiber link, connecting Birmingham and London.

The U.S. first-class postal rate goes to 18¢ per ounce March 22 and to 20¢ per ounce November 1 (*see* 1978, 1985).

Fiction *Midnight's Children* by Bombay-born British novelist Salman Rushdie, 34; *Gorky Park* by U.S. novelist Martin Cruz Smith, 38; *Rabbit Is Rich* by John Updike; *July's People* by Nadine Gordimer; *Funeral Games* by Mary Renault; *The Hotel New Hampshire* by John Irving; *The Mosquito Coast* by Paul Theroux; *Noble House* by James Clavell; *The Thirty Years Peace* by West German writer Peter O. Chotjewitz.

Juvenile *Jumanji* by Chris Van Allsburg.

Painting *Paramount* (painting with metal brackets) by U.S. artist Robert Ryman, 51; *Red Mill* (pastel on paper) by Robert Moskowitz; *Self-Portrait* by Alice Neel, now 81, who shows herself in her armchair wearing only her eyeglasses; *Artist with Painting and Model* (collage) by Romare Bearden. The Fun Gallery opens on 10th Street in New York's East Village, the first commercial venture in the hybrid of punk rock and visual art.

Human immunodeficiency virus appeared out of nowhere to plague much of the world with a deadly AIDS epidemic.

Sculpture *Tilted Arc* (steel) by Richard Serra is installed in New York's Federal Plaza (commissioned by the General Services Administration; *see* 1989); *Untitled* (plywood installation) by Donald Judd.

Theater *Steaming* by English playwright Nell Dunn, 45, 7/1 at London's Theatre Royal, Stratford East; *Key Exchange* by U.S. playwright Kevin Wade, 27, 7/14 at New York's off-Broadway Orpheum Theater, with Brooke Adams, Mark Blum, Ben Masters; *Quartermaine's Terms* by Simon Gray 7/30 at the Queen's Theatre, London, with Edward Fox; *A Talent for Murder* by Jerome Chodorov and Norman Panama 10/1 at New York's Biltmore Theater, with Claudette Colbert, Jean-Pierre Aumont, 77 perfs.; *Sister Mary Ignatius Explains It All for You* by U.S. playwright Christopher Durang 10/16 at New York's Playwrights Horizons Theater, with Polly Draper, Elizabeth Franz, 947 perfs.; *Torch Song Trilogy* by U.S. playwright Harvey (Forbes) Fierstein, 27, 10/16 at New York's off-off-Broadway Richard Allen Center, with Fierstein as the drag queen Arnold, 117 perfs.; *Crimes of the Heart* by U.S. playwright Beth Henley, 29, 11/4 at New York's John Golden Theater, with Mia Dillon, Holly Hunter, Lizbeth Mackay, and Peter MacNicol, 535 perfs.; *A Soldier's Play* by U.S. playwright Charles Fuller, 42, 11/5 at New York's off-Broadway Theater Four, 468 perfs.; *Mass Appeal* by U.S. playwright Bill C. Davis, 30, 11/12 at New York's Booth Theater, with Milo O'Shea, Michael O'Keefe, 212 perfs.; *The West Side Waltz* by Ernest Thompson 11/10 at New York's Ethel Barrymore Theater, with Katharine Hepburn, Dorothy Loudon, 126 perfs.; *Grownups* by Jules Feiffer 12/10 at New York's Lyceum Theater, with Frances Steegmuller, Harold Gould, Bob Dishy, 13 perfs.

Films Volker Schlöndorff's *Circle of Deceit* with Bruno Ganz, Hanna Schygulla; Hector Babenco's *Pixote* with Fernando Ramos da Silva, Marilia Pera; Steven Spielberg's *Raiders of the Lost Ark* with Harrison Ford as archaeologist Indiana Jones. Also: Jean-Claude Tramont's *All Night Long* with Gene Hackman, Barbra Streisand; Steve Gordon's *Arthur* with John Gielgud, Dudley Moore, Liza Minnelli; Wolfgang Peterson's *Das Boot* with Jurgen Prochnow; Hugh Hudson's *Chariots of Fire* with Ben Cross as 1924 Olympic runner Harold Abrahams, Ian Charleson as runner Eric Liddell, Lindsay Anderson; Paul Kagan's *The Chosen* with Maximilian Schell, Rod Steiger, Robby Benson, Barry Miller; Bertrand Tavernier's *Coup de Torchon* with Philippe Noiret, Isabelle Huppert; Shohei Imamura's *Eijanaka* with Shigeru Izumiya; Karel Reisz's *The French Lieutenant's Woman* with Meryl Streep, Jeremy Irons; Peter Weir's *Gallipoli* with Mark Lee, Mel Gibson; Moshe Mizrahi's *I Sent a Letter to My Love* with Simone Signoret, Jean Rochefort; George

Miller's *Mad Max II* with Mel Gibson; Warren Beatty's *Reds* with Beatty as John Reed, Diane Keaton as Louise Bryant, Maureen Stapleton as Emma Goldman, Jack Nicholson as Eugene O'Neill; Kenji Misumi's *Shogun Assassin* with Robert Houston, Tomisaburo Wakiyama; John Badham's *Whose Life Is It, Anyway?* with Richard Dreyfuss, John Cassavetes.

Television *Dynasty* 1/12 on ABC with John Forsythe, Linda Evans; *Hill Street Blues* 1/15 on NBC with Daniel J. Travanti (as Capt. Furillo) and others in dramas, based on a police precinct house, conceived by Steve Bochco; *Miami Vice* 9/16 on NBC with Don Johnson; *Cagney and Lacey* 10/8 on ABC with Loretta Swit, Tyne Daley.

Stage musicals *Sophisticated Ladies* 2/1 at New York's Lunt-Fontanne Theater, with Phyllis Hyman, P. J. Benjamin, music by the late Duke Ellington, 767 perfs.; *Marry Me a Little* 3/12 at New York's Actor's Playhouse, with Craig Lucas, songs by Stephen Sondheim that include "Can That Boy Foxtrot!," "Happily Ever After," and "There Won't Be Trumpets," 96 perfs.; *Woman of the Year* 3/29 at New York's Palace Theater, with Lauren Bacall, Harry Guardino, music and lyrics by John Kander and Fred Ebb, book by Peter Stone based on a screenplay by Ring Lardner, Jr., and Michael Kanin, 770 perfs.; *March of the Falsettos* 4/19 at New York's Playwrights Horizons Theater, with Michael Rupert, Stephen Bogardus, Alison Fraser, music and lyrics by U.S. composer-playwright William Finn, 29, 170 perfs.; *Cats* 5/11 at London's New London Theatre, with music by Andrew Lloyd Webber, lyrics by the late T. S. Eliot, songs that include "Memory"; *Dreamgirls* 12/21 at New York's Imperial Theater, with Jennifer Holliday, music by Henry Krieger, book and lyrics by Tom Eyen, 40, 1,522 perfs.

Popular songs "Bette Davis Eyes" by U.S. songwriters Donna Weiss and Jackie De Shannon; "The Tide Is High" by U.S. songwriter John Holt; "Arthur's Theme (Best That You Can Do)" by Christopher Cross; "Not a Day Goes By" by Stephen Sondheim (for his short-lived musical *Merrily We Roll Along*); "Boy from New York City" by the rock group Manhattan Transfer.

MTV (Music Television) goes out to cable TV subscribers beginning August 1 with rock 'n' roll numbers.

Oakland beats Philadelphia 27 to 10 at New Orleans January 25 in Super Bowl XV.

John McEnroe wins the British and U.S. men's singles titles, Chris Evert-Lloyd the women's singles title at Wimbledon, Tracy Austin the U.S. women's title.

The Los Angeles Dodgers win the World Series 4 games to 2 over the New York Yankees after a season interrupted by a 7-week players' strike.

Britain's Prince Charles, 32, marries Lady Diana Spencer, 20, at St. Paul's Cathedral, London, July 29.

Dallas hails a new city hall designed by I. M. Pei.

Two walkways at Kansas City's new Hyatt Regency Hotel collapse July 18, killing 113, injuring 186.

The U.S. Department of Agriculture responds to cuts in the school lunch program by announcing in September that ketchup can be counted as a vegetable. The public outcry forces President Reagan to restore funds for school lunches.

Aspartame gains FDA approval for tabletop use October 22. U.S. chemist James M. Schlatter discovered in 1955 while trying to develop an anti-ulcer drug that a mixture of the amino acids aspartic acid and phenylalinine had a sweet taste. Two teaspoons of sugar contain 32 calories; aspartame provides the same sweetening with 4 calories. Marketed by G. D. Searle under the brand name NutraSweet, the artificial sweetener is far costlier than saccharin but does not have saccharin's bitter aftertaste. Other countries have permitted the sale of aspartame, which is also made by Ajinimoto in Japan (*see* 1983).

Nabisco Brands is created by a merger of Nabisco (National Biscuit Co.) and Standard Brands (*see* RJR-Nabisco, 1985).

Kellogg introduces Nutri-Grain wheat cereal.

The world's population reaches 4.5 billion, up from 2.5 billion in 1950, with at least 957 million in the People's Republic of China, where female infanticide increases. China's severe population-control policy limits families to one child each; without any social security programs it becomes vital that the child be a male who can support his parents in their old age. India has an estimated 664 million, the U.S.S.R. 266 million, Indonesia 152, Japan 117, Bangladesh 88, Vietnam 52, Brazil 122, Mexico 72, Canada 24, Nigeria 77, Egypt 42, Britain 56, France 54, West Germany 61.4, Italy 57, Poland 35, Spain 38.

Spain legalizes divorce.

The United States has 228 million people, up from 203.2 in 1970. Blacks number 26.5 million, up from 22.6 in 1970; Hispanics, many of them illegal immigrants, number 14.6 million.

1982

Terrorists at Paris assassinate an Israeli diplomat April 3, Israel's Prime Minister Menachem Begin warns of PLO guerrilla activities and arms buildups, and Israel hits PLO strongholds in Lebanon April 21—the first Israeli strike since last year's cease-fire. The PLO has allegedly staged 130 guerrilla attacks inside Israel during the cease-fire, and Israeli planes raid PLO bases south of Beirut May 9. PLO forces respond with artillery fire across the border.

Israeli forces have completed withdrawal from the Sinai April 25 under terms of the 1978 Camp David accord.

Iranian forces recover the port city of Khurramshahr May 24, taking 30,000 Iraqi prisoners in the ongoing war. Syria has reportedly supplied Iran with Soviet-built weapons.

Israel's ambassador to Britain is critically wounded at London June 3 by terrorists more extreme than the PLO.

Israel invades Lebanon June 6, captures medieval Beaufort Castle June 7, downs dozens of Soviet-built

1982 ***(cont.)*** Syrian MIGs (weaponry made in Russia proves itself no match for Israel's U.S.- and French-made arms and aircraft) destroy Syrian surface-to-air missiles in the Bekaa Valley, and reach the outskirts of Beirut June 10. Israeli jets bomb civilian West Beirut July 27, killing 120 and injuring 232.

U.S. ships land 800 Marines at Beirut August 25 to evacuate 8,000 PLO guerrillas after heavy fighting has brought mediation by U.S. envoy Philip Habib. Lebanon's president-elect Bashir Gemayel is killed in a bomb explosion September 14 at the headquarters of his Christian Falangist party (he is later succeeded by his brother Amin Gemayel, 40).

Israeli forces move into West Beirut September 16, too late to prevent a massacre of Palestinians by Christian Falangists. The bloodshed brings fresh demands for Prime Minister Begin's resignation.

Twelve hundred U.S. Marines land in Lebanon September 29 as part of an international peace-keeping force and take up positions round Beirut International Airport.

Saudi Arabia's Khalid ibn Abdel Aziz Al Saud has died at Tali June 13 at age 68. The new king is Abdel's diabetic brother Fahd, 60, who has called Israel's Menachem Begin a "fanatic Zionist."

Israeli forces invaded Lebanon to root out the PLO terrorists who were harassing the Jewish state.

A Vietnam War memorial dedicated at Washington November 13 displays the names of all 57,692 killed or missing U.S. soldiers, sailors, marines and airmen etched into black granite. (The monument was designed last year by Yale architecture student Maya Ying Lin, now 22).

President Reagan's budget address February 6 has called for much higher military appropriations and less spending on social programs. Congress votes 346 to 68 to increase military spending by 6 percent after inflation (Reagan had asked a 13 percent boost) over fiscal 1982. But the Boland Amendment to the defense appropriations bill, approved unanimously by Congress December 8, bans use of defense funds to support CIA efforts to overthrow Nicaragua's Sandinista government. Congressmen Edward P. Boland (D. Mass.) and Tom Hakins (D. Iowa) have introduced the measure.

Argentine forces invade Britain's Falkland (Malvina) Islands April 2 and seize South Georgia Island April 3, Britain imposes a blockade April 12, British commandos invade South Georgia April 25, Argentina's only cruiser is sunk by a British submarine May 2 with a loss of more than 320 lives, the *QE2* is used to bring troops to the South Atlantic, Washington expresses support for its NATO ally, British troops return in force to the Falklands May 14, fierce fighting brings heavy casualties to both sides in the next few weeks, and Argentine forces surrender June 14. Argentina has lost more than 1,000 men including those who went down with the cruiser *General Belgrano*, Britain 243. Argentina has lost 74 planes and 7 helicopters, says Britain; Britain has lost 48 planes, says Argentina.

Guatemalan dictator Romeo Lucas Garcia is overthrown in a coup by a three-man military junta March 23 and charged by Amnesty International with responsibility for at least 5,000 political murders. Gen. José Efrain Rios Montt, 55, assumes dictatorial power in June.

Panama formally assumes responsibility for policing the Canal Zone April 1 under terms of the 1977 treaty with the United States.

Canada gets her own Constitution April 17. Britain's Elizabeth II signs the Constitution Act at Ottawa, replacing the North America Act of 1867.

A rally against the nuclear arms race brings 800,000 demonstrators into New York's Central Park June 12.

Mexican Secretary of Planning and Budget (and Deputy Director General of Pemex) Miguel de la Madrid Hurtado, 47, is elected president July 4 as falling oil prices force peso devaluations and push Mexico to the edge of financial crisis (below).

West German voters unseat Helmut Schmidt October 1 and make Helmut Kohl chancellor of the Bonn republic. Schmidt has opposed nuclear freeze proposals.

Spanish voters elect a Socialist government October 29 for the first time since 1937. Felipe Gonzalez becomes prime minister and there is dancing in the streets of Madrid.

Leonid I. Brezhnev dies November 10 at age 75 after 17 years as Soviet party secretary. He is succeeded in that position by former KGB head Yuri V. Andropov, 68, who will rule for only 15 months before succumbing to a chronic kidney ailment.

The White House announces January 8 that it approves giving tax-exempt status to South Carolina's Bob Jones University and other schools alleged to practice racial discrimination. President Reagan has reversed an 11-year policy but softens his stance January 12 in response to a storm of protest.

An Equal Rights Amendment to the U.S. Constitution comes within three states of being ratified but the deadline for ratification passes June 30.

Washington announces June 11 that heavy tariffs will be imposed on some steel imports in order to help struggling U.S. steel makers whose foreign competition receives government subsidies.

U.S. Steel acquires Marathon Oil for $3 billion.

An Oklahoma City bank goes into receivership July 5. Its receiver, the Federal Deposit Insurance Corp., begins paying off $271 million in insured deposits as examiners uncover a morass of bad loans to wildcat drillers and others made by Penn Square, based in a shopping center. Penn Square's collapse hurts "upstream" banks, chiefly Continental Illinois of Chicago (*see* 1984).

A U.S. tax reform measure approved by Congress August 19 cuts back on the tax reductions enacted last year, raises taxes on cigarettes, tightens loopholes that have permitted rich people to avoid taxes. Critics call it the end of the "supply-side" economic experiment.

Mexican Treasury Secretary Jesus Silva Herzog tells foreign bankers August 20 that Mexico cannot repay her $60 billion foreign debt, the first Third World country to default. Brazil, Argentina, and more than 20 other countries follow suit. U.S. and other foreign banks grant delays in interest payments and make new loans to tide the debtor nations over.

Recession continues throughout most of the world, international trade declines, unemployment in the United States reaches 10.8 percent in November—the highest since 1940—and the number of Americans living below the poverty line is the highest in 17 years.

Paul Volcker announces October 2 that the war against inflation has gone too far and that he is abandoning his experiment with monetarism. An 18-month recession ends in November.

The Garth-St Germain Depository Institutions Act signed into law by President Reagan October 15 removes "artificial" regulatory restraints on federally insured savings and loan companies. S&Ls have been hurt by soaring interest rates. Allowed to pay depositors higher rates of interest, freed to make more aggressive loans and investments, and largely unregulated because Reagan budget cuts have reduced inspection personnel, S&L officers will in many cases make reckless deals, accruing losses that will cost U.S. taxpayers at least $150 billion to "resolve" failed thrift institutions plus at least $350 billion in interest by 2020 (*see* 1988).

Wall Street stages a rally on the strength of lower inflation and lower interest rates. The Dow Jones Industrial Average sets a record high of 1072.55 December 27, up from 776.92 August 12, after trading a record 147.1 million shares October 7.

A Boeing 737 crashes into Washington's 14th Street bridge January 13 after takeoff from National Airport; only five of the 79 aboard are rescued and 4 others are killed on the bridge. A Japan Air Lines DC-8 from Fukuoka plunges into Tokyo Bay February 9, killing 24 of its 174 passengers and crew; neurotic pilot Seiji Katagiri has thrown two of the plane's four engines into reverse as it approached Haneda Airport despite efforts by the flight engineer to restrain him. A Chinese jetliner from Canton crashes near Guilin in April killing all 112 aboard; a Brazilian VASP Airlines jet crashes near Fortaleza June 8, killing all 137 aboard; a Pan Am 727 takes off in a rainstorm from Moisant Airport, New Orleans, July 9 and crashes, killing all 145 aboard plus 4 on the ground.

Braniff International Airways files for bankruptcy May 14 after 52 years in business, victim of over-expansion and weak economy; the first major U.S. airline to fold.

The Boeing 767 makes its commercial debut September 8 on a United Airlines flight from Chicago to Denver. The plane can seat 211 as compared with 147 on the 707 introduced in 1958, 145 on the 727 introduced in 1964, and 452 on the 747 introduced in 1970. There are 1,760 727s in service, 530 747s, 600 707s. Overcapacity plagues the world's financially troubled airlines; few order 767s.

Union Pacific receives ICC approval September 13 to merge with Missouri Pacific and Western Pacific to create a 22,800-mile system.

Honda starts making cars at Marysville, Ohio, in November as Japanese autos take 22.6 percent of U.S. sales, up from 3.7 percent in 1970, despite voluntary quotas on imports.

MRI (Magnetic Resonance Imaging) machines introduced in Britain cost 50 percent more than CAT-scanning devices (*see* 1974) but give physicians a superior new diagnostic tool, permitting them to monitor blood flowing through an artery, to see the reaction of a malignant tumor to medication, etc. (*see* 1977).

A new Surgeon General's Report issued in March by U.S. Surgeon General C. Everett Koop, 65, calls cigarette smoking the chief preventable cause of death. Lung cancer kills about 111,000 Americans, up from 18,313 in 1950. More than 30 million Americans have quit smoking since 1964, but smoking-related health care costs the nation $13 billion, while lost production and wages costs another $25 billion. More U.S. women than men smoke for the second year in a row.

1982 *(cont.)* A Chicago assassin laces bottles of Tylenol capsules with cyanide, seven die in late September, and Tylenol maker Johnson & Johnson promptly recalls the product October 5, destroying 31 million Tylenol capsules on store shelves and in home medicine chests. Reintroduced in triple-sealed safety packages, Tylenol will regain 95 percent of its top market share in 3 months.

An artificial heart is implanted for the first time December 2 at the Utah Medical Center in Salt Lake City; patient Barney B. Clark, 61, will live until March 23, 1983.

American Telephone & Telegraph agrees January 8 to be broken up in settlement of an anti-trust suit filed in 1974. AT&T will retain its long-distance lines, its Western Electric manufacturing arm, and Bell Laboratories, spinning off its 22 regional and local companies (*see* 1983).

The *Philadelphia Bulletin* ceases publication January 29.

USA Today begins publication September 15. Keeping stories brief and making wide use of color, Gannett's paper is the only national daily except for the *Christian Science Monitor* and *Wall Street Journal* (although the *New York Times* prints editions in several cities and has worldwide distribution).

"Electronic mail" via fax machines gains popularity as third-generation Japanese technology cuts transmission time to 20 seconds per page, down from 6 minutes with first-generation machines, and thus reduces telephone charges from $4 per page to less than $1 (*see* 1976). The new machines are cheaper and compatible with earlier models. By year's end, there are 350,000 U.S. fax installations, up from 69,000 in 1975, and the first directory of users will be out next year.

The U.S. space shuttle *Columbia* deploys two communications satellites and lands at Edwards Air Force Base, California, November 16 after a successful inaugural mission.

Fiction *The Color Purple* by Alice Walker; *The Safety Net* by Heinrich Böll; *A Moving Target* by William Golding; *Berry Patches* (his first novel) by Soviet poet Yevgeny A. Yevtushenko; *Chronicle of a Death Foretold* by Gabriel García-Márquez (who wins the Nobel prize for literature); *Aunt Julia and the Scriptwriter* by Mario Vargas Llosa; *The Dean's December* by Saul Bellow; *Dinner at the Homesick Restaurant* by Anne Tyler; *Bech Is Back* by John Updike; *A Bloodsmoor Romance* by Joyce Carol Oates; *The Portage to San Cristobal of A.H.* by University of Geneva literature professor George Steiner, 52; *Space* by James Michener.

Painting *Keyhole* by Elizabeth Murray; *Monuments to the Stag* by Joseph Beuys.

Theater *Pump Boys and Dinettes* by U.S. playwrights John Foley, Mark Hardwick, Debra Monk, Cass Morgan, John Schwind, and Tim Warner 2/4 at New York's off-Broadway Princess Theater, 573 perfs.; *The Dining Room* by U.S. playwright A.R. (Albert Ramsdell) Gurney, 42, 2/11 at New York's Playwrights Horizons Theater, with John Shea, 607 perfs.; *Noises Off* by English playwright Michael Frayn, 48, 2/23 at London's Lyric Theatre, Hammersmith, with Paul Eddington, Patricia Routledge; *Agnes of God* by U.S. playwright John Pielmeier, 33, 3/30 at New York's Music Box Theater, with Elizabeth Ashley, Geraldine Page, Amanda Plummer, 599 perfs.; *Good* by the late English playwright C.P. Taylor (who died 12/9/81) 4/20 at London's Aldwych Theatre (after opening 9/2/81 at The Warehouse), with Alan Howard as a physician in Nazi Germany; *"Master Harold" . . . and the Boys* by South African playwright Athol Fugard, 49, 5/4 at New York's Lyceum Theater with Zakes Mokae, Lonny Price, 344 perfs.; *Top Girls* by Caryl Churchill 9/1 at London's Royal Court Theatre, with Gwen Taylor, Deborah Findlay, Carol Hayman; *True West* by Sam Shepard 10/17 at New York's Cherry Lane Theater, with John Malkovich, 762 perfs.; *Foxfire* by U.S. playwrights Susan Cooper and Hume Cronyn 11/10 at New York's Ethel Barrymore Theater, with Cronyn, Jessica Tandy, 213 perfs.; *The Real Thing* by Tom Stoppard 11/16 at London's Strand Theatre, with Felicity Kendal, Roger Rees; *Extremities* by U.S. playwright William Mastrosimone 12/22 at New York's West Side Arts Theater, with Susan Sarandon, James Russo, 317 perfs.; *Whodunnit* by Anthony Shaffer 12/30 at New York's Biltmore Theater, with George Hearn, Barbara Baxley, 157 perfs.

Films Andrzej Wajda's *Danton* with Gerard Depardieu; Steven Spielberg's *E.T., The Extra-Terrestrial* with Dee Wallace, Henry Thomas; Werner Herzog's *Fitzcarraldo* with Klaus Kinski, Claudia Cardinale; Jerzy Skolimowski's *Moonlighting* with Jeremy Irons; Sydney Pollack's *Tootsie* with Dustin Hoffman, Jessica Lange; George Roy Hill's *The World According to Garp* with Robin Williams, Mary Beth Hurt, Glenn Close. Also: Eric Rohmer's *Le Beau Mariage* with Beatrice Romand; Jean-Jacques Beineix's *Diva* with Frederic Andrei, Wilhelmina Wiggins Fernandez; Paul Bartel's *Eating Raoul* with Bartel, Mary Woronov; Walter Hill's *48 HRS* with Nick Nolte, Eddie Murphy; Richard Attenborough's *Gandhi* with Ben Kingsley; Rainer Maria Fassbinder's *Lola* with Barbara Sukowa, Armin Mueller-Stahl; George Miller's *The Man from Snowy River* with Kirk Douglas, Tom Burlinson; Margarethe von Trotta's *Marianne and Juliane* with Jutta Lampe, Barbara Sukowa; Constantin Costa-Gravas's *Missing* with Jack Lemmon, Sissy Spacek; Anne Claire Poirier's *Over Forty* with Roger Blay, Monique Mercure; Robert Towne's *Personal Best* with Mariel Hemingway, Scott Glenn, Patrice Donnelly; Ed Stabile's *Plainsong* with Jessica Nelson, Teresanne Joseph, Lyn Traverse; Tobe Hooper's *Poltergeist* with Craig T. Nelson, JoBeth Williams; Susan Seidelman's *Smithereens* with Susan Berman; Sidney Lumet's *The Verdict* with Paul Newman, Charlotte Rampling.

Screen star Romy Schneider is found dead May 29 at age 43; Henry Fonda dies August 12 at age 77; Ingrid Bergman dies of cancer August 29 at age 67; Princess Grace of Monaco (*née* Grace Kelly) dies of a brain hemorrhage September 14 at age 52 following an automobile accident.

Television *Cheers* 9/29 on NBC with Ted Danson, Shelley Long, Rhea Perlman (to 5/12/93).

Film opera Franco Zefirelli's *La Traviata* with Teresa Stratas, Placido Domingo, Cornell MacNeil.

Film musicals George T. Nierenberg's documentary *Say Amen, Somebody* with gospel singers Willie Mae Ford Smith, Thomas A. Dorsey; David Leivick, Frederick A. Ritzenberg, and James Cleveland's concert film *Gospel* with the Southern California Community Choir, Walter Hawkins and the Hawkins Family, Mighty Clouds of Joy, Shirley Caesar, Twinkie Clark and the Clark Sisters; Jim Brown's documentary *The Weavers: Wasn't That a Time!* with Lee Hays, Pete Seeger; *Victor/Victoria* with Julie Andrews, James Garner, Robert Preston, music by Henry Mancini, lyrics by Leslie Bricusse.

Stage musicals *Forbidden Broadway* (revue) 1/15 at Palsson's Theater, New York, with Gerard Alessandrini, music and lyrics by Alessandrini; *Nine* 5/9 at the 46th Street Theater with Raul Julia, music and lyrics by Maury Yeston, book by Arthur Kopit based on the 1963 Fellini film 8½, 739 perfs.; *Windy City* 7/20 at London's Victoria Palace Theatre, with Dennis Waterman as Hildy Johnson, Anton Rodgers as Walter Burns, music by English composer Tony Macaulay, 38, book and lyrics by U.S. writer Dick Vosburgh who has adapted *The Front Page* of 1928; *Little Shop of Horrors* 7/27 at New York's off-Broadway Orpheum Theater, with Ellen Green, music by Alan Menken, books and lyrics by Howard Ashman, 2,209 perfs.

Popular songs "Up Where We Belong" by Jack Nitzsche and Buffy Sainte-Marie, lyrics by Will Hennings (for the film *An Officer and a Gentleman*); *Toto IV* (album) by the veteran rock group Toto; "Always on My Mind" by Johnny Christopher, Mark James, and Wayne Carson; "Ebony and Ivory" by Paul McCartney and Stevie Wonder; *American Fool* (album) by John Cougar; *The Nylon Curtain* (album) by Billy Joel; *Tug of War* (album) by Paul McCartney with his memorial to John Lennon "Here Today;" *The Nightfly* (album) by Donald Fagen.

San Francisco beats Cincinnati 26 to 21 at Pontiac, Mich., January 24 in Super Bowl XVI.

Jimmy Connors wins in men's singles at Wimbledon, Martina Navratilova in women's singles; Connors wins his fourth U.S. Open singles title, Chris Evert-Lloyd wins in women's singles.

Italy wins the World Cup (soccer) championship, defeating West Germany 3 to 1 at Madrid's 90,089-seat Santiago Bernabeu Stadium.

The St. Louis Cardinals win the World Series, defeating the Milwaukee Brewers 4 games to 3.

Reebok aerobic shoes gain on Nike running shoes (*see* 1972). Boston camping-equipment distributor Paul B. Fireman, now 38, obtained the U.S. license 3 years ago for British Reebok running shoes, started Reebok International with British financing, and introduces the $45 Freestyle glove-leather aerobic-dance shoe in fashion colors. Reebok will overtake Nike in sales for several years before Nike regains the lead.

Atlanta photographer Wayne B. Williams, 23, is found guilty February 27 of having killed two boys and is sentenced to two consecutive life terms (*see* 1980). The Public Safety Commissioner says the conviction "clears" 23 of 30 killings.

More than 25 million Americans smoke marijuana, spending $24 billion on the controlled substance.

President Reagan announces a war on drugs October 14 (*see* 1983).

Houston's 75-story Texas Commerce Tower, designed by I. M. Pei, is completed.

Beijing's Fragrant Hills Hotel, designed by I. M. Pei, opens just outside town.

The Soviet Union draws down grain reserves and increases imports to maintain livestock herds as crops fail for the fourth consecutive year.

China announces the results of a census October 27: her population has grown to 1.008 billion.

1983 President Reagan proposes a "Strategic Defense Initiative" to protect America and her allies with a high-tech shield against nuclear missiles. His March 23 speech envisions flocks of satellites that will shoot down incoming missiles with lasers, but projected costs of what Sen. Kennedy calls a "Star Wars" program are staggering, and since it will have to be virtually 100 percent effective, few scientists believe it is feasible at any cost. Moscow views SDI as a move to escalate the arms war into outer space.

The Soviet Union is "the focus of evil in the modern world," Reagan has told an evangelical group at Orlando, Fla., March 8. He calls the U.S.S.R. "an evil empire."

Europeans turn out by the thousands April 1 in a "Green" movement to protest the presence of U.S. nuclear weapons on the Continent.

Former Philippine senator Benigno S. Aquino, Jr., 50, returns from exile to Manila August 21 and is shot dead upon arrival by an unknown gunman who is himself immediately shot dead. The last national leader still held in detention in 1980, Aquino was permitted to leave the country that year for open-heart surgery in the United States. He formed an anti-Marcos coalition in January 1982 and worked from abroad to restore democracy to the Philippines. Despite warnings that ailing President Ferdinand Marcos, his wife, Imelda, or their political allies (or opponents) would kill him, Aquino had decided it was time to organize opposition to Marcos at home (*see* 1986).

A Korean Air Lines Boeing 747 carrying 269 passengers and crew bound from New York to Seoul violates Soviet air space near Sakhalin Island in the North Pacific and is shot down at 3:30 in the morning of September 1 by an air-to-air missile fired from a Soviet fighter jet. All aboard are killed. Moscow makes no apology, insists the commercial jetliner was on a U.S. spy mission, and criticizes the United States for casting aspersions on the

1983 *(cont.)* "peace-loving" Soviets. Washington reveals that a U.S. reconnaissance plane was in the vicinity earlier. U.S.-Japanese efforts to recover the 747's "black box" recording all its moves are unsuccessful. Arms control talks resume at Geneva despite growing tensions between the world's superpowers.

Terrorists in Lebanon blow up the U.S. Embassy at Beirut April 18, killing 63 people. Two U.S. marines are killed and 13 wounded August 29 when mortar shells and rockets land in an airport compound during clashes between Lebanese Army and Shiite Muslim and Druse forces. Israeli forces withdraw from Lebanon's central mountains September 4 as rival Christian and Druse militia intensify their battles.

Israel's ailing Prime Minister Menachem Begin resigns September 15 and is succeeded as leader of the Herut party by Foreign Minister Yitzhak Shamir, 67, who, like Begin, was born in Poland and fought in the underground against Palestine's British authorities in the 1930s and '40s.

U.S. Marines in Lebanon come under increasing attack in September and October. A terrorist drives a truck packed with explosives into a building full of sleeping Marines and sailors October 23 while another bomb-laden truck slams into a French paratroop barracks; the U.S. death toll is 241, the French toll 58. A suicide truck bomber blows up an Israeli military installation November 4 killing 60, including 28 Israelis.

A time-bomb planted by North Korean terrorists explodes October 9 at the Martyrs' Mausoleum in Rangoon, killing 19 and wounding 49. The dead include 16 South Koreans, among them four cabinet ministers and the ambassador to Burma. South Korea's President Chun Doo Hwan is en route to the mausoleum for a wreath-laying ceremony when the bomb goes off and is saved by a hitch in his schedule that has delayed him. The mausoleum commemorates the assassination July 19, 1947, of seven members of Burma's Executive Council, including its head U Aung San, 33, during the transition between internal autonomy and full Burmese independence from Britain.

A Soviet fighter plane shot down a Korean Air Lines 747 passenger plane suspected of spying on military installations.

Former prime minister Kakuei Tanaka is convicted in Tokyo District Court October 12 of having accepted a $2.2 million bribe from Lockheed Corp. to use his influence to persuade All Nippon Airways to use Lockheed Tristar jets (*see* 1976). Fined the amount of the bribe and sentenced to 4 years in prison, Tanaka files an appeal as opposition parties threaten to boycott Diet proceedings unless he resigns his seat.

Beijing begins in October to purge China's Communist party of leftist extremists who remain from the Mao Zedong era. The move is an effort to ensure that party members adhere to the more pragmatic policies of Deng Xiaoping.

Grenada has a coup October 12 as Deputy Prime Minister Bernard Coard, a Marxist hardliner, overthrows Prime Minister Maurice Bishop, 39. When his supporters engineer a prison break a week later, Bishop is assassinated along with most of his cabinet. Growing ties between the little Caribbean island (population: 110,000) and Havana have worried Washington, as has construction of a 10,000-foot runway which could be used as a way-station for shipping Soviet and Cuban arms to Central America. U.S. Marines and Army Rangers land on Grenada October 25 (2 days after the disaster to Marines in Beirut) with 300 military personnel from Antigua, Barbados, Dominica, Jamaica, St. Kitts-Nevis, St. Lucia, and St. Vincent supporting the Americans, who soon number more than 3,000. President Reagan justifies the action on grounds that political thugs had taken over Grenada, that U.S. medical students on the island were in danger, that Cuba was bent on making Grenada a new bastion for Communism in the Caribbean and Central America (where the United States has been supporting anti-Sandinista forces in Nicaragua). Longtime U.S. friends condemn the action.

Argentina returns to civilian rule in December after an 8-year military regime and disastrous war with Britain have plunged the nation into deep financial straits. Political moderate Raul Ricardo Alfonsín, 56, has won election to the presidency.

Nigeria's 5-year-old democratic experiment ends December 31 in a military coup. Gen. Mohammed Buhari, 41, deposes President Alhaji Shehu Shagan, 58, and says his armed forces have saved the nation from "total collapse."

February riots protesting immigration of Muslim refugees from Bangladesh have left 600 dead in India's state of Assam. Student agitators have tried to stop state elections.

A congressional committee reports February 24 that the World War II imprisonment of Japanese-Americans was a "grave injustice" prompted by "racial prejudice, war hysteria, and failure of political leadership" (*see* 1942; 1989).

The U.S. Supreme Court rules 8 to 1 in May that it is unconstitutional to grant tax exemptions to private schools that practice discrimination.

The U.S. Supreme Court rules June 28 that a sentence of life imprisonment with no chance of parole is unconstitutional.

Legislation signed by President Reagan November 2 makes the third Monday in January a national holiday beginning in 1986 to celebrate Martin Luther King's birthday (*see* 1968). Reagan had originally opposed the holiday, which will be ignored by the financial markets and most business firms.

Social Security legislation signed by President Reagan April 20 delays the 1983 cost-of-living increase for 6 months, boosts payroll deductions beginning in 1984, gradually raises the minimum retirement age to 67 by 2027, requires that new federal employees join the system, and mandates that some benefits of higher-income retirees be subject to federal incomes taxes. The reforms are designed to assure the system's solvency for the next 75 years.

Drexel Burnham Lambert executive Michael R. Milken, 37, suggests in November that high-yield "junk" bonds be used to facilitate both friendly and hostile takeovers, and for buy-outs of companies "going private" (buying up publicly owned stock). Assets of a target company are pledged to repay the principle of the junk bonds, which yield 13 to 30 percent and are bought by many insurance and savings and loan companies.

Economic recovery in the United States sends Wall Street's Dow Jones Industrial Average to new heights (it closes the year at 1258.64, up from 1046.54 at the end of 1982). Inflation remains low, unemployment begins to drop, but the Census Bureau reports that 35.3 million Americans live in poverty—the highest rate in 19 years.

Washington State's Public Power Supply System (WPPS) defaults on $2.25 billion in municipal bonds, the largest governmental failure in U.S. history.

Pennzoil buys 590,000 shares of Getty Oil stock on the open market in December. On December 27 it launches a tender offer to raise its holdings to 20 percent (*see* 1984).

An Avianca Boeing 747 crashes November 27 near Madrid's Barajas Airport, killing 183.

Santa Fe Southern Pacific is created December 23 by a merger of two leading U.S. railroads amidst predictions that the country will soon have just two large coast-to-coast rail carriers.

Ibuprofen receives nonprescription drug status in Britain; over-the-counter products containing the drug begin gaining on aspirin and acetaminophen products even though they are far more likely to produce ulcers. British researcher Stewart S. Adams of Boots Co. developed the non-steroidal anti-inflammatory drug, used heretofore only in high-dose prescription drugs.

Crack—crystallized cocaine that can be smoked to produce a short but intense high—is developed by drug traffickers, probably Dominicans, in the Bahamas and soon appears in West Coast U.S. cities. The low-priced, highly addictive drug (cocaine hydrochloride + baking soda + a "comeback" filler boiled with water) opens a mass market for cocaine among adolescents and young adults, increasing crime rates, devastating families and communities, multiplying health emergencies and the incidence of syphilis and AIDS as users engage in indiscriminate sex.

"Just Say No" is the slogan for a new program to combat drug use unveiled by First Lady Nancy Reagan in October. Critics say she has picked up the issue to counter her negative image as a woman of wealth.

AT&T issues new shares January 1 to its 3 million stockholders pursuant to the consent decree of January 1982; holders receive stock in seven new phone companies, with Ma Bell keeping only its long-distance service plus its Bell Laboratories research facilities and its Western Electric manufacturing facilities.

Chicago motorists begin in December to talk in their cars on cellular telephones available at $3,000 plus $150 per month for service. The Federal Communications Commission has authorized a test of a Motorola phone system using low-power transmitters scattered across the city plus a computer system to pass along calls to moving vehicles. Mobile phones have existed since the 1940s but with only one transmitter per city the service was unsatisfactory until now.

Nonfiction *Modern Times* by English journalist Paul Johnson, 53; *Vietnam: A History* by U.S. journalist Stanley Karnow, 59.

Fiction *Ironweed* by former Albany, N.Y., journalist William Kennedy, 53; *The Life and Times of Michael K.* by South African novelist J. M. (John Michael) Coetzee, 43; *Meditations in Green* by U.S. novelist Stephen Wright, 37; *The Anatomy Lesson* by Philip Roth; *Stanley and the Women* by Kingsley Amis; *At the Bottom of the River* by Antigua-born U.S. novelist Jamaica Kincaid (*née* Elaine Potter Richardson), 34.

Poetry *The Changing Light* by U.S. poet James Merrill, 49.

Tokyo's Metropolitan Teien Art Museum opens October 8 in a renovated 1932 Art Deco mansion designed by French architect Henri Rapin for a member of the imperial family.

Painting *Racing Thoughts* by Jasper Johns; *Elements IV* by U.S. painter Brice Marden, 44. Joan Miró dies at Palma, Majorca, December 25 at age 90.

Theater *Painting Churches* by U.S. playwright Tina Howe, 45, 2/8 at New York's off-Broadway South Street Theater, with Marian Seldes, Donald Moffatt, Frances Conroy, 206 perfs.; *Fen* by Caryl Churchill 3/9 at

1983 *(cont.)* London's Almeida Theatre, with Jenny Stoller, Bernard Strother; *Brighton Beach Memoirs* by Neil Simon 3/27 at New York's Alvin (later Neil Simon) Theater, with Matthew Broderick, 1,530 perfs. (moved to 46th Street Theater 2/26/85); *Run for Your Wives* by English playwright Ray (Raymond George Alfred) Cooney, 50, 3/30 at London's Shaftesbury Theatre, with Richard Briers, Bernard Cribbins; *'night, Mother* by U.S. playwright Marsha Norman, 35, 3/31 at New York's Golden Theater, with Kathy Bates, Anne Pitoniak, 380 perfs.; *Glengarry Glen Ross* by David Mamet 9/21 at London's Cottlesloe Theatre, with Jack Shepherd; *Pack of Lies* by English playwright Hugh Williams 10/26 at London's Lyric Theatre, with Mary Miller, Frank Windsor; *Fool for Love* by Sam Shepard 11/30 at New York's off-Broadway Douglas Fairbanks Theater, with Ed Harris, 1,000 perfs.; *Isn't It Romantic* by U.S. playwright Wendy Wasserstein, 33, 12/15 at New York's Playwrights Horizons Theater, with Lisa Banes, Betty Comden, Chip Zuin, 233 perfs.

Films Ingmar Bergman's *Fanny and Alexander* with Pernilla Allwin, Bertil Guve; Godfrey Reggio's *Koyaanisqatsi* with a score by Philip Glass; Gregory Nava's *El Norte* with Aide Silvia Gutierrez, David Villalpando; James L. Brooks's *Terms of Endearment* with Shirley MacLaine, Debra Winger, Jack Nicholson. Also: Robert Bresson's *L'Argent* with Christian Patey, Sylvie van den Essen; Ann Hui's *Boat People* with Lam Chi-Cheung; Peter Yates's *The Dresser* with Albert Finney, Tom Courtenay; Mary Dore, Sam Sills, and Noel Buckner's Spanish Civil War documentary *The Good Fight;* Philip Borsos's *The Grey Fox* with Richard Farnsworth; Alain Tanner's *In the White City* with Bruno Ganz, Teresa Madruga; Martin Scorsese's *The King of Comedy* with Robert De Niro, Jerry Lewis; Alain Resnais's *Life Is a Bed of Roses* with Vittorio Gassman, Ruggero Raimondi, Geraldine Chaplin; Bill Forsyth's *Local Hero* with Peter Riegert, Burt Lancaster; Paolo and Vittorio Taviani's *The Night of the Shooting Stars* with Omero Antonutti; Richard Marquand's *Return of the Jedi* with Mark Hamill, Harrison Ford, Carrie Fisher; Euzhan Paloys's *Sugar Cane Alley* with Garry Ladenat; Lynne Litman's *Testament* with Jane Alexander; John Korty and Charles Swenson's *Twice upon a Time* (animated); Roger Spottiswoode's *Under Fire* with Gene Hackman, Nick Nolte, Joanna Cassidy.

Hollywood musicals Adrian Lyne's *Flashdance* with Jennifer Beals, music by Irene Cara; Barbra Streisand's *Yentl* with Streisand, Mandy Patinkin, Amy Irving, music by Michel Legrand, lyrics by Alan and Marilyn Bergman.

Broadway and off-Broadway musicals *Merlin* 2/13 at the Mark Hellinger Theater, with Doug Henning, Chito Rivera, music by Elmer Bernstein, book and lyrics by Michael Levinson and William Link, 199 perfs.; *Mama, I Want to Sing* 3/25 at the 667-seat off-Broadway Hecksher Theater, with Tisha Campbell, 14, gospel music by Rudolph V. Hawkins, book and lyrics by Vy Higginsen and Ken Wydro, 2,213 perfs.; *My One and Only* 5/1 at the St. James Theater, with Tommy Tune, Twiggy (Leslie Hornby), dancer Charles "Honi" Coles, music from old Gershwin musicals, 762 perfs.; *La Cage aux Folles* 8/21 at the Palace Theater, with George Hearn, Gene Barry, music and lyrics by Jerry Herman, 1,761 perfs.; *Baby* 12/4 at New York's Ethel Barrymore Theater, with music by David Shire, lyrics by Richard Maltby, Jr., 241 perfs.; *The Tap Dance Kid* 12/21 at the Broadhurst Theater, with Hinton Battle, 699 perfs.

Popular songs *Thriller* (album) by Michael Jackson; "All Night Long" by Lionel Ritchie; "Beat It" by Michael Jackson; "Every Breath You Take" by Sting (Gordon Sumner); "Karma Chameleon" by British songwriters George O'Dowd, John Moss, Michael Craig, Roy Hay, and Phil Picket; "Let's Dance" by David Bowie (David Jones); "That's Livin' Alright," by British songwriters David McKay, Ken Ashby; "Say Say Say" by Paul McCartney and Michael Jackson; *Can't Slow Down* (album) by Lionel Ritchie; *An Innocent Man* (album) by Billy Joel.

Washington beats Miami 27 to 17 at Pasadena January 30 in Rose Bowl XVII.

John McEnroe wins in men's singles at Wimbledon, Martina Navratilova in women's singles; Jimmy Connors wins in U.S. Open men's singles, Navratilova in women's singles.

Australia II defeats *Liberty* 4 races to 3 and takes away the America's Cup held by the United States since 1851, ending the longest winning streak in sports history.

The Baltimore Orioles win the World Series, defeating the Philadelphia Phillies 4 games to 1.

Cabbage Patch dolls marketed by Coleco Industries become black market items as stores sell out to eager parents and grandparents. The craze will reach its peak in 1985, with buyers paying nearly $600 million at retail.

Three masked gunmen steal $39 million in gold at London's Heathrow Airport November 26.

New York's American Telephone and Telegraph building, designed by Philip Johnson, is completed at Madison Avenue and 56th Street.

New York's IBM building, designed by U.S. architect Edward Larrabee Barnes, is completed at Madison Avenue and 57th Street.

The Norton house is completed at Venice, Calif., by U.S. architect Frank Gehry, 54.

Environmental Protection Agency scandals make headlines. Rita M. Lavelle, who heads the EPA's Superfund program to clean up toxic waste, is fired by President Reagan February 7 (she will serve 4 months in prison for lying to Congress); EPA admnistrator Anne McGill Burford resigns under fire March 9; and Secretary of the Interior James G. Watt resigns October 9 after fighting to open federal lands to private exploitation, including oil drilling. Watt has offended with a lighthearted comment that his coal advisory commission was a well-balanced mix: "I have a black, a woman, two

Jews, and a cripple." President Reagan shocks environmentalists by naming his longtime rancher friend William P. Clark, 51, to replace Watt.

The tanker *Castillo de Bellver* catches fire off Cape Town, South Africa, August 6, and spills 250,000 tons of crude oil into the sea.

An El Niño (a body of warm water) off the coasts of Ecuador and Peru upsets weather patterns worldwide by acting as a catalyst for prevailing conditions.

An earthquake in eastern Turkey October 30 levels more than 35 villages, killing more than 1,300.

Soviet fisherman stop hauling nets on the Aral Sea, which 10 years ago supplied 10 to 15 percent of the nation's freshwater catch. The 800-mile Kara Kum Canal, completed in the 1970s as part of an ill-conceived plan to turn the surrounding deserts into cotton- and rice-growing farmland, has siphoned water from the inland sea's two source rivers, creating flood-control problems in some areas while lowering the Aral Sea's water level by 40 feet, shrinking it by one third, doubling its salinity, killing most aquatic life, and creating an ecological disaster as winds blow dust and salt from the sea bottom, contaminated with pesticide residues, on surrounding fields, poisoning water supplies and even mothers' milk.

Soviet harvests are good for the first time since 1978.

The worst drought since 1936 combines with the largest acreage diversion in history to reduce the U.S. corn harvest by some 2 billion bushels.

A payment-in-kind (PIK) program rewards U.S. farmers for not planting. This and other farm support programs cost taxpayers an unprecedented $21.5 billion and encourage marginally efficient farmers to continue.

A new U.S.-Soviet grain agreement formally signed at Moscow August 25 pledges the U.S.S.R. to buy at least 9 million metric tons of U.S. grain per year for the next 5 years and pledges the United States to supply up to 12 million tons. The figures are 50 percent higher than in the 1975 agreement and there is no escape clause, as in 1975, suspending shipments in the event of shortages that would raise domestic food prices.

U.S. soft drink makers begin using NutraSweet, initially in combination with saccharin, to sweeten diet beverages (*see* 1981).

The Supreme Court rules June 15 that many local abortion restrictions are unconstitutional.

1984 Soviet leader Yuri V. Andropov dies February 9 at age 69. He is succeeded as general secretary of the Communist Party Central Committee by Politburo member Konstantin U. Chernenko, 72.

Beirut terrorist gunmen kill American University president Malcolm H. Kerr January 18 and vow to rid Lebanon of Westerners. U.S., French, and Italian peacekeeping forces leave Lebanon in the spring but Syria refuses to withdraw her troops from the Bekaa Valley. Israeli forces occupy southern Lebanon.

Israeli parliamentary elections July 23 end with the Labor Alignment party, headed by Shimon Peres, winning 44 seats in the Knesset. The Likud party, headed by Prime Minister Yitzhak Shamir, wins 41 seats. Since neither party has a 61-seat majority, the Knesset votes September 14 to have a coalition government in which Peres will serve as PM for 25 months followed by a 25-month term for Shamir.

Iran-Iraq hostilities spread to the Persian Gulf. Iraq uses French Exocet air-to-surface missiles against tankers loading at Iran's Kharg Island, Iran strikes back at tankers loading oil from Saudi Arabia and smaller Arab oil states. Foreign military analysts estimate that more than 100,000 Iranians and 50,000 Iraqis have been killed. Teheran refuses to make peace unless Iraq withdraws to the prewar border, pays war damages, and punishes her leaders.

Mozambique and South Africa end hostilities with the Accord of Nkomati, signed March 16. It is the first agreement between white South Africa and any black nation.

Upper Volta becomes Burkina Fasso ("land of upright men") August 3 as the African nation's military government moves to exorcise its colonial past.

Sikh extremists occupy the Golden Temple at Amritsar; India's Prime Minister Indira Gandhi sends in troops June 5 to 6 and 600 to 1,200 are killed in a bloody takeover of the temple. The Sikhs, who control prosperous Punjab state, have pressed for independence, as have some other Indian states, and Mrs. Gandhi is determined to keep the nation united by whatever means. She is assassinated by two Sikh members of her personal guard October 31 at age 66. Some 1,000 are killed in anti-Sikh riots. Gandhi is succeeded by her son Rajiv, 40, who has had little political experience but who wins election as prime minister in his own right at year's end.

Speedboats from a CIA mother ship mine Nicaraguan harbors in a move to block import of Cuban and Soviet arms allegedly destined for rebels in El Salvador. A

Iran repelled her Iraqi invaders, sustaining terrible losses but refusing peace with an Iraq headed by Saddam Hussein.

1984 *(cont.)* Soviet freighter and other foreign ships are damaged in a clear violation of the 1982 Boland Amendment. The World Court denounces the U.S. action, the White House denies World Court jurisdiction in the matter, and the Senate votes 84 to 12 April 10 to cut off funds for any further mining of the harbors.

Salvadoran junta leader José Napoleon Duarte is elected president in May, defeating ultra-rightist candidate Roberto D'Aubuisson who has been linked to death squads. Duarte succeeds Alvaro Alfredo Magana, visits Washington in July, and persuades Congress to provide increased economic and military aid.

Canada's Liberal Prime Minister Pierre Elliott Trudeau resigns June 30 after 16 years in office (interrupted for 9 months in 1979–80). Corporate lawyer Brian Mulroney, 45, leads the Progressive Conservatives in an election sweep September 4 as his party takes 211 of the 282 seats in the House of Commons.

President Reagan, now 73, is reelected with 525 electoral votes to 13 for former vice-president Walter Mondale, 56, who has run with Queens, N.Y., congresswoman Geraldine Ferraro, 48, but carries only the District of Columbia and his home state of Minnesota. Reagan wins 59 percent of the popular vote.

An agreement signed at Beijing December 19 by British Prime Minister Thatcher and Chinese Premier Zhao Ziyang provides for transfer of Hong Kong to Chinese sovereignty in 1997. Hong Kong is to retain its capitalist way of life until 1947.

The U.S. Commission on Civil Rights votes January 17 to discontinue numerical quotas for promotion of black workers and executives. "Such racial preferences merely constitute another form of discrimination," the commission says in a sharp reversal of its previous position. President Reagan has appointed its members and they reflect his views.

Polish security police abduct pro-Solidarity priest Jerzy Popieluszko, 37, and murder him October 19. His body is discovered in a reservoir on the Vistula River and public reaction forces the government to hold a public trial of the perpetrators. Four security officials will be convicted in February of next year.

Chilean police raid the Santiago slum district La Victoria November 15, round up 32,000 suspected leftists, and hold them in a soccer stadium for questioning about recent demonstrations against the absolutist rule of Gen. Pinochet.

Britain's National Union of Mine Workers, headed by Arthur Scargill, bans overtime work in January as pit closings loom. Ian MacGregor, an American who heads the National Coal Board which manages the industry, insists on the government's right to shut down unprofitable pits, he closes a Yorkshire colliery in early March, and 55,000 miners strike. Since 1977, workers who have mined more have earned more, a bonus plan favoring those in coal-rich Nottinghamshire but hard on those in north Derbyshire, Kent, Scotland, south Wales, and Yorkshire, where many miners join the strike. Violence spreads and other unions begin to support the strikers. Some 3 to 4 million are idle by year's end although many miners drift back to work.

U.S. economic growth rises at a 6.8 percent rate, highest since 1951, while the Soviet economy, with grain harvests below target, grows by only 2.6 percent, lowest since World War II. The inflation rate, 3.7 percent, is the lowest since 1967. But U.S. budget and trade deficits rise to record levels, and the Department of Housing and Urban Development reports May 1 that 250,000 to 350,000 Americans are homeless.

Continental Illinois Bank escapes failure May 10 through a record $4.5 billion in federal loan guarantees after the biggest run on a U.S. bank since the Great Depression has threatened a chain reaction of bank closings. Oil and gas drilling ventures, condominium developments, and Latin American projects financed by the bank have turned sour. The Federal Deposit Insurance Co. takes 80 percent of the bank's stock in late July and buys $4.5 billion in bad loans for $3.5 billion. Public confidence in banks declines.

Economic reforms announced by Beijing October 20 extend to urban areas the incentive system granted to farmers in 1979: wages and bonuses will be linked to job performance, prices will not be held to "irrational" levels, each enterprise will be permitted more independence to plan production and marketing. A degree of capitalism in Chinese rural areas has boosted production sharply and raised living standards.

France gets her first deliveries of Soviet natural gas January 1.

Texaco's board acts January 5 to authorize management to negotiate for the acquisition of Getty Oil for as much as $125 a share (*see* Pennzoil 1983).

Standard Oil of California acquires Gulf Oil in a $13.3-billion cash merger and changes its name to Chevron as world oil prices decline.

New York's transit fares rise January 1 from 60¢ to 75¢ (*see* 1980, 1986).

The *New York Times* publishes its "Shipping/Mails" column for the last time April 15. Passenger volume in the port has dwindled to 400,000, down from 900,000 in 1960, and most of the arrivals and departures are cruise ships. Jet planes have almost completely supplanted transatlantic passenger liners.

Britain's Thatcher government sells Jaguar Motors to private investors (*see* Ford, 1989).

The Macintosh, introduced by Apple January 24, is a "user-friendly" personal computer with superior graphics capabilities.

Bell Laboratories announces December 20 that it has perfected a one-megabit random access memory chip able to store on a tiny sliver of silicon four times as much information as anything now available.

American Telephone & Telegraph Co. divests itself January 1 of its 22 Bell operating companies pursuant to a federal court order. AT&T retains its Western Electric division and remains in the long-distance telephone and computer business. Regional holding companies—Ameritech, Bell Atlantic, BellSouth, Nynex, Pacific Telesis, Southwestern Bell, and USWest—take over 22 Bell units and thrive as local telephone rates go up across the country and service deteriorates.

Prime Minister Thatcher forbids union membership at Britain's General Communications Headquarters and offers £1,000 for each union card turned in. All but 150 GCHQ employees accept the offer.

British Telecom shares go on sale as Britain moves to "privatize" telephone service.

Japan moves to end government ownership of telephone service.

The U.S. Supreme Court rules January 17 that home videotape recording does not infringe copyrights. Sony Corp., makers of Betamax, and other VCR makers hail the 5 to 4 decision, Hollywood film makers bewail it.

Nonfiction *Son of the Morning Star: Custer and the Little Big Horn* by Evan S. Connell, now 60; *Weapons and Hope* by U.S. writer Freeman Dyson, 61.

Fiction *The Unbearable Lightness of Being* by Milan Kundera; *L'Amant* by Marguerite Duras, now 70; *Down from the Hill* by Alan Sillitoe; *Foreign Affairs* by Alison Lurie; *Parachutes and Kisses* by Erica Jong; *Love Medicine* by U.S. novelist Louise Erdrich, 30; *God Knows* by Joseph Heller; *Fragments* by U.S. novelist Jack Fuller; *Him with His Foot in His Mouth and Other Stories* by Saul Bellow; *Lincoln* by Gore Vidal; *The War of the End of the World* by Mario Vargas Llosa.

Painting *Polestar* by Andy Warhol, French-American painter Jean-Michel Basquiat, 24, and Italian-American painter Francesco Clemente, 32. Alice Neel dies of cancer at New York October 13 at age 84.

Sculpture *Bad Dream House No. 1* by Vito Acconci; *Monument with Standing Beast* by Jean Dubuffet for Chicago's State of Illinois Building.

The Dallas Museum of Art opens in January, replacing the 81-year-old Museum of Fine Arts and beginning a 60-acre Arts District. Edward Larrabee Barnes has designed the giant new limestone building.

New York's Museum of Modern Art opens a new (west) wing that more than doubles its gallery space (*see* 1939).

U.S. landscape photographer Ansel Adams dies at Carmel, Calif., April 23 at age 82.

Theater *Benefactors* by Michael Frayn 4/4 at London's Vaudeville Theatre, with Polly Adams, Clive Francis; *The Miss Firecracker Contest* by Beth Henley 5/27 at New York's off-Broadway Manhattan Theater Club, with Holly Hunter, Mark Linn-Baker; *The War at Home* by U.S. playwright James Duff, 29, 6/13 at London's Hampstead Theatre, with David Threlfell, Frances Sternhagen; *Hurlyburly* by David Rabe 8/7 at New York's Ethel Barrymore Theater, with William Hurt, Harvey Keitel, Christopher Walken, Jerry Stiller, Sigourney Weaver, 343 perfs.; *Balm in Gilead* by Lanford Wilson 9/6 at New York's off-Broadway Minetta Lane Theater, with Steven Bauer, Glenn Headley, Laurie Metcalf, 143 perfs.; *Ma Rainey's Black Bottom* by U.S. playwright August Wilson, 38, 10/11 at New York's Cort Theater, with Theresa Merritt, Charles S. Dutton, 225 perfs.; *The Foreigner* by U.S. actor-playwright Larry Shue, 38, 11/1 at New York's Astor Place Theater, with Shue, Anthony Heald, 686 perfs.

Films Miloš Forman's *Amadeus* with Tom Hulce, Murray Abraham; Martin Brest's *Beverly Hills Cop* with Eddie Murphy; Alan Parker's *Birdy* with Matthew Modine, Nicolas Cage; Alan Rudolph's *Choose Me* with Genevieve Bujold, Keith Carradine, Lesley Ann Warren; Roland Joffe's *The Killing Fields* with Sam Waterston, Dr. Haing S. Ngor, Athol Fugard; Paul Mazursky's *Moscow on the Hudson* with Robin Williams, Maria Conchita Alonso; Robert Altman's *Secret Honor* with Philip Baker Hall as Richard Nixon; Norman Jewison's *A Soldier's Story* with Howard E. Rollins, Jr., Adolph Caesar; Martin Bell's documentary *Streetwise* about teenage vagrants; Bertrand Tavernier's *A Sunday in the Country* with Louis Ducreux; James Cameron's *The Terminator* with Arnold Schwarzenegger; Robert Epstein's documentary *The Times of Harvey Milk* narrated by Harvey Fierstein; John Huston's *Under the Volcano* with Albert Finney, Jacqueline Bisset; John Hanson's *Wildrose* with Lisa Eichhorn.

Television *The Cosby Show* debuts 9/20 on NBC with comedian Bill Cosby, 47, (to 4/30/92).

Stage musicals *The Rink* 2/9 at New York's Martin Beck Theater, with Liza Minnelli, Chita Rivera, music by John Kander, lyrics by Fred Ebb, book by Terrence McNally, 204 perfs.; *Starlight Express* 3/27 at London's Apollo Theatre, with performers on roller skates imitating locomotives, music by Andrew Lloyd Weber, lyrics by Richard Stilgoe; *Sunday in the Park with George* 5/2 at New York's Booth Theater with Mandy Patinkin as painter Georges Seurat, music and lyrics by Stephen Sondheim, 540 perfs.; *The Hired Man* 10/31 at London's Astoria Theatre, with music by English composer Howard Goodall, 29, book and lyrics by Melvyn Bragg, 45.

Hollywood musicals Jonathan Demme's *Stop Making Sense* with the Talking Heads.

Popular songs "What's Love Got To Do With It" by U.S. songwriters Graham Lyle and Terry Britten; "Can't Slow Down" by Lionel Ritchie; *Born in the U.S.A.* (album) Bruce Springsteen; *Purple Rain* (album) by Minnesota funk rock singer Prince (Rogers Nelson), 26; "Like a Virgin" by pop singer Madonna (Madonna Louise Veronica Ciccone), 26; "Against All Odds (Take a Look at Me Now)" by British songwriter Phil Collins; "Careless Whisper" by British songwriters George Michael and Andrew Ridgley; "Do They Know It's Christmas" by British songwriters Bob Geldof and Midge Ure.

1984 ***(cont.)*** Los Angeles beats Washington 38 to 9 at Tampa January 22 in Super Bowl XVIII.

John McEnroe wins both the U.S. and British men's singles titles, Martina Navratilova the women's titles.

The Olympic Games at Los Angeles attract a record 7,800 from 140 nations despite a boycott by 14 Soviet bloc countries. U.S. athletes win 174 gold medals, West German athletes 59.

The Detroit Tigers win the World Series, defeating the San Diego Padres 4 games to 1.

Running enthusiast James Fixx dies of a heart attack July 20 at age 52 while jogging at Hardwick, Vt. An autopsy shows he has suffered previous myocardial infarctions.

Trivial Pursuit revives the board-game industry. Developed by a Canadian entrepreneur, the game is introduced in U.S. stores and has sales of $777 million.

Japan has a candy scare as extortionists announce that confectionery in retail outlets has been poisoned.

Four gunmen get away with $21.8 million in Rome March 24.

Three men with revolvers seize two Merrill Lynch Canada couriers December 21 at Montreal and escape with $51.3 million in securities.

Unemployed security guard Oliver Huberty, 41, of San Ysidro, Calif., walks into the local McDonald's July 18 with a semiautomatic rifle, shotgun, and pistol, begins firing at anything that moves, and kills 20, wounds 16, before police sharpshooters kill him.

A New York "subway vigilante" shoots four black youths December 22, climbs off the train, and disappears into a tunnel after telling a motorman that the teenagers had tried to rob him. Crime has been rampant in the subways and public support rallies at first behind the unknown gunman, especially when his victims all turn out to have criminal records. Engineer Bernhard Hugo Goetz, 37, will turn himself in to New Hampshire police early in January and confess to the shootings, which have left one youth paralyzed from the waist down. A Manhattan grand jury will indict Goetz only on charges of illegal weapon possession (he will be convicted and serve 8 months). A second grand jury will indict Goetz for attempted murder but he will be acquitted.

The average price of a new U.S. single-family house tops $101,000 in May, crossing the six-figure mark for the first time.

Mexico City loses some 300 homes November 19 as 50,000 barrels of gas explode at a depot of the state oil company, Pemex, killing at least 500.

A Union Carbide pesticide plant operated entirely by Indians at Bhopal, India, leaks the lethal gas methyl isocyanate December 3, killing more than 2,000 outright and injuring 200,000. The death toll will rise to 3,500. India's Supreme Court in February 1989 will order Union Carbide to pay $470 million in damages.

U.S. agriculture remains in distress as world markets shrink, partly because the dollar is so high. Costly federal farm programs come under fire.

International Harvester avoids bankruptcy by selling its farm-equipment division to Tenneco. It confines itself to making trucks and will change its name in 1986 to Navistar.

Famine kills 300,000 Ethiopians as drought worsens in sub-Sahara Africa. Civil war and poor roads prevent aid from reaching the hungry. Some 800,000 will die before foreign grain comes to the rescue next year.

U.S. funding of international birth control programs is halted by order of the Reagan administration.

1985 Mikhail Sergeyevich Gorbachev, 54, becomes general secretary of the Soviet Communist Party after Konstantin Chernenko dies of emphysema March 11 at age 73. Moscow and Washington have reached a compromise agreement January 8 to resume negotiations toward limiting and reducing nuclear weapons and preventing an arms race in space. Gorbachev, an agricultural specialist with no experience in foreign affairs, meets with President Reagan November 21; the two agree to speed up arms control talks and renew cultural relations.

Albania's dictator Enver Hoxha dies April 11 at age 78 after 41 years in power. He is succeeded as Communist party chief by Ramiz Alia, 59, who has been president since 1982.

Terrorist attacks by Arab, French guerrilla, Islamic, and Palestinian groups kill 107, wound more than 428 in

Mikhail Gorbachev struggled to overcome Marxist ideology's resistance to capitalist production efficiency.

Europe and the Mediterranean. Bombs explode in Madrid, Paris, Athens, Frankfurt, a U.S. air base near Frankfurt, and Rome; grenades are thrown in Rome; a TWA jetliner, hijacked June 14 between Athens and Rome, is diverted to Beirut, where passengers are held hostage for 17 days; hijackers seize the cruise ship *Achille Lauro* in the Mediterranean October 7, killing a U.S. passenger; an Egyptair jetliner, hijacked November 23 between Athens and Cairo, is forced to land in Malta, 2 passengers are killed, 58 people are killed when Egyptian commandos storm the plane; gunmen attack Rome and Vienna airports December 27 and 20 people are killed, including 4 terrorists. Libya aided the attackers, says President Reagan.

Iraqi jets armed with French Exocet missiles bomb Iran's strategic Kharg Island oil terminal August 17 in the ongoing Persian Gulf war.

France's government totters in a scandal over the sinking of a French anti-nuclear ship off New Zealand September 22. New Zealand has refused entry to a U.S. warship February 4 after Washington refused to say whether the ship carried nuclear arms.

Italy's government falls October 16 after a political crisis triggered by the *Achille Lauro* hijacking (above).

Vietnamese forces in Cambodia drive the Khmer Rouge from the last of their bases in mid-February.

Uruguay returns to civilian rule March 1 after 12 years of military dictatorship that has taken thousands of political prisoners in an attempt to demoralize the people. President Julio Maria Sanguinetti, 49, heads the new government after unemployment has risen to 30 percent, the inflation rate to 66 percent, and foreign debts to $5 billion.

Brazil returns to civilian rule March 15 after 21 years of military dictatorship. Opposition candidate Tancredo Neves has won overwhelmingly in the January election but dies of complications following intestinal surgery before he can take office. His running mate José Sarney, 54, former governor of Maranhão State, becomes president.

Nicaragua's President Daniel Ortega Saavedra offers peace initiatives in February but President Reagan says March 1 that the Nicaragua contras are "the moral equal of our Founding Fathers." Ortega compares Reagan to Hitler, U.S. critics say Reagan is obsessed with Nicaragua, and Congress votes July 18 to prevent Reagan from supplying the contras with anything but "non-lethal" aid.

Peru has her first constitutional transfer of power since 1945. President Belaunde Terry ends his 5-year term, Social Democrat Alan Garcia Pérez, 36, is elected to succeed him, and Garcia takes office July 28. Much of the country remains under military control as Maoist guerrillas of the Sendero Luminosa (Shining Path) ravage the countryside.

Guyana's president Forbes Burnham dies August 6 following throat surgery after 21 years in power. He is succeeded by Prime Minister Desmond Hoyte.

Uganda's president Obote flees into exile July 27 following a military coup. General Tito Okello is installed as president but will serve only 7 months (*see* 1986).

Nigeria's president Buhari is overthrown August 27 in a bloodless coup. Maj. Gen. Ibrahim Babangida proclaims himself president (*see* 1983).

Sudan has a military coup April 6 while President Nimeiri is abroad. He has survived three previous coup attempts in his 16-year rule but his country is ravaged by civil war, tribal conflicts, and famine. His defense minister Gen. Abdel Rahman Siwar el-Dahab seizes power.

Tanzania's president Julius Nyerere resigns in November after 21 years in power. His vice president, Ali Hassan Myinyi, succeeds him.

Ethiopian Jews have sought refuge in neigboring Sudan (*see* above) where 2,000 of them perish in January.

President Reagan visits West Germany in May to mark the fortieth anniversary of the liberation of the Buchenwald death camp April 13 but offends many by visiting Bitburg Cemetery, thus honoring the graves of 49 Waffen SS officers.

Philadelphia police try to dislodge members of MOVE, an organization of armed blacks. They firebomb a house from the air May 13 and the fire spreads to adjacent houses, killing 11 and leaving 200 homeless.

South Africa declares a state of emergency July 20, giving police and the army almost absolute power in black townships. Police may make arrests without warrants and hold people indefinitely without trial, but interracial marriages are legalized and some movie theaters are opened to patrons of all races.

President Reagan calls upon Congress February 4 to make major budget reductions. His State of the Union message 2 days later emphasizes tax reform and economic growth.

The collapse of world oil prices (*see* below) puts pressure on banks and savings institutions in Texas, Oklahoma, and other energy sector states. Many will fail, affecting money-center banks, real estate owners, and taxpayers.

Home State Savings & Loan in Ohio fails in March, panicky depositors make runs on other S&Ls not federally insured. Governor Richard Celeste closes 70 other Ohio thrifts, and federal regulators take over a Beverly Hills, Calif., S&L. S&Ls have been obliged to pay 15 percent interest to keep depositors, their income from mortgages has fallen far short of needs, and worse trouble lies ahead (see Garth-St Germain Act, 1982).

Mikhail Gorbachev calls for sweeping economic changes in June, indirectly criticizing his predecessors.

The United States has become a debtor nation for the first time since 1914, the Department of Commerce announces September 16. After years of deficits in the balance of payments, the nation has relied on foreign buying of U.S. Treasury bonds and notes instead of on taxation.

1985 *(cont.)* The Gramm-Rudman-Hollings Act signed by President Reagan December 12 mandates congressional spending limits in an effort to eliminate the federal deficit by 1991. Gramm-Rudman will cut budget projections but will be ineffective in cutting actual spending. Its sponsors are congressmen Phil Gramm (R. Tex.), and Warren B. Rudman (R. N.H.) and Sen. Ernest F. Hollings (D. S.C.).

U.S. corporate mergers and acquisitions increase, with 24 involving more than $1 billion each. High-yield "junk" bonds are used to finance many takeovers.

A pricing discount system announced by Saudi Arabian oil minister Ahmad Zaki Yamani, 45, September 13 at the annual energy seminar at Oxford University begins a price war that will glut oil markets and slash world oil prices by 60 percent in the next 6 months. With more than a quarter of the earth's oil reserves, Saudi Arabia aims to encourage worldwide use of oil while forcing British, Norwegian, Canadian, U.S., Mexican, Venezuelan, Nigerian, Egyptian, and Algerian competitors to shut their higher-cost wells.

A Texas jury November 19 orders Texaco to pay a record $10.53 billion to Pennzoil for interfering with Pennzoil's agreement to acquire Getty Oil in 1984.

U.S. federal tax credits for installing solar heating in homes, instituted in the Carter administration, end December 31 under Reagan budget cuts.

An Ethiopian train bound for Addis Ababa jumps the rails and plunges into a ravine January 13, killing 392 and injuring 370 in Africa's worst train accident, the world's third worst ever.

An Air India Boeing 747 crashes off the Irish coast June 23, killing 329; a Delta L-1011 tries to land in a violent thunderstorm at Dallas-Fort Worth August 2 and crashes, killing 140 of the 161 aboard; a Japan Air Lines Boeing 747 crashes into a mountain on a domestic flight August 12, killing 520 in the worst single plane accident ever; a chartered Arrow Air DC-8 carrying members of the U.S. 101st Airborne Division crashes on takeoff at Gander, Newfoundland, December 12, killing 256.

The scanning tunneling microscope developed by IBM researchers at Zurich makes it possible to obtain atomic resolution pictures of surfaces of materials. Work by Gerd Binnig, 38, and Heidrich Rohrer, 52, to image individual atoms will have wide applications in scientific research.

Karen Ann Quinlan, comatose since 1976, dies of pneumonia June 11 at age 31 after a court permits removal of her respirator.

Movie actor Rock Hudson (*né* Roy Scherer, Jr.), collapses at the Paris Ritz July 21 and dies of AIDS at Beverly Hills October 2 at age 59, shocking Americans into heightened awareness of the disease that is killing tens of thousands of men each year.

Procter & Gamble acquires Richardson-Vicks for $1.24 billion after an attempted hostile takeover by Unilever. This and other acquisitions will make P&G a major over-the-counter pharmaceutical company.

U.S. authorities arrest Indian guru Bhagwan Shree Rajneesh, 53, in Oregon October 28. Rich followers have joined the guru's ashram and given him funds to buy a fleet of Rolls-Royces.

The U.S. Supreme Court rules 5 to 4 July 1 that public school teachers may not teach in parochial schools.

A federal court in New York clears *Time* magazine January 24 of libel charges brought by Israeli leader Ariel Sharon. The magazine did not deliberately lie, the court rules.

Gen. William Westmoreland settles a libel suit brought against CBS February 18.

The U.S. first-class postal rate goes to 22¢ February 17 (*see* 1981, 1988).

Newhouse Publications acquires the *New Yorker* magazine March 8 for $142 million.

New York Newsday begins publication as the Long Island tabloid establishes a Manhattan presence (*see* 1940).

Capital Cities Communications buys American Broadcasting Co. for $3.5 billion in March.

General Electric acquires RCA and its National Broadcasting Co. for $6.3 billion in December.

Nonfiction *Common Ground: A Turbulent Decade in the Lives of Three American Families* by U.S. journalist J. Anthony Lukas, 52; *The Good War* by Studs Terkel.

Fiction *The Good Apprentice* by Iris Murdoch; *Annie John* by Jamaica Kincaid; *The Accidental Tourist* by Anne Tyler; *Zuckerman Bound* by Philip Roth; *Lonesome Dove* by Larry McMurtry; *What's Bred in the Bone* by Robertson Davies; *Lake Woebegone Days* by Garrison Keillor; *Glitz* by Elmore Leonard.

Juvenile *The Polar Express* by Chris Van Allsburg.

Poetry *Elegies* by Scottish poet Douglas Dunn, 43.

Painting *Van Heusen* (Ronald Reagan, silkscreen) by Andy Warhol; *Ocean Park No. 139* by Richard Diebenkorn; *Interior Perspective* (*Discordant Harmony*) by Canadian-American painter Dorothea Rockburne; *Grotesques* (photographic series) by Cindy Sherman; *Capri—Batterie* by Joseph Beuys. Marc Chagall dies March 28 at St. Paul de Venne, France, at age 97; Jean Dubuffet dies of emphysema May 12 at Paris at age 83.

Theater *Biloxi Blues* by Neil Simon 3/28 at New York's Neil Simon Theater, 524 perfs.; *As Is* by U.S. playwright William M. Hoffman, 46, 5/1 at New York's Lyceum Theater, with Jonathan Hadery, Jonathan Hogan is about homosexuals and AIDS, 285 perfs.; *I'm Not Rappaport* by Herb Gardner 11/19 at New York's Booth Theater, with Judd Hirsch, Cleavon Little, 1,071 perfs.; *A Lie of the Mind* by Sam Shepard 12/5 at New York's Promenade Theater, 185 perfs.

Films Luis Puenzo's *The Official Story* with Norma Aleandro, Hector Alterio about Argentina's "disappeared;" Claude Lanzman's Holocaust documentary *Shoah*. Also: Norman Jewison's *Agnes of God* with Jane Fonda, Anne Bancroft; Ron Howard's *Cocoon* with

Don Ameche, Hume Cronyn, Jack Gilford, Gwen Verdon, Maureen Stapleton, Jessica Tandy, Steve Guttenberg; Steven Spielberg's *The Color Purple* with Danny Glover, Whoopi Goldberg, Adolph Caesar; Elem Klimov's *Come and See* with Alexei Kravchenko; Mike Newell's *Dance with a Stranger* with Miranda Richardson; David Drury's *Defense of the Realm* with Gabriel Byrne, Greta Scacchi, Denholm Elliott; Marion Hansel's *Dust* with Jane Birkin, Trevor Howard; Nicolas Roeg's *Insignificance* with Gary Busey, Tony Curtis; Masahiro Shinoda's *MacArthur's Children* with Masako Natsume, Shima Iwashata; Stephen Frears's *My Beautiful Laundrette* with Saeed Jaffray, Gordon Warnecke, Daniel Day Lewis; Sidney Pollack's *Out of Africa* with Meryl Streep, Robert Redford; John Huston's *Prizzi's Honor* with Jack Nicholson, Kathleen Turner, Anjelica Huston; Woody Allen's *Purple Rose of Cairo* with Allen, Mia Farrow; Akira Kurosawa's *Ran* with Tatsuya Nakadai in an adaptation of *King Lear*; Giles Foster's *Silas Marner* with Ben Kingsley; Christopher Cain's *The Stone Boy* with Robert Duvall, Glenn Close; Agnes Varda's *Vagabond* with Sandrine Bonnaire.

Stage musicals *Big River* 4/25 at New York's Eugene O'Neill Theater with music and lyrics by Roger Miller, 1,005 perfs.; *Les Misérables* 10/8 at London's Palace Theatre, with Colin Wilkinson as Jean Valjean, music by Claude-Michel Schonberg, lyrics by Alain Boubil and Herbert Kretzmer; *Black and Blue* 11/25 at Paris's Chatelet Théatre, with Ruth Brown, Linda Hopkins, and a cast of 41 performing 21 vaudeville songs from the 1920s, '30s, and '40s; *The Mystery of Edwin Drood* 12/2 at New York's Imperial Theater, with George Rose, Karen Morrow, Howard McGillin, music and lyrics by Rupert Holmes, 39, songs that include "Perfect Strangers," "Don't Quit While You're Ahead," "Moonfall," book based on the unfinished Dickens novel, 608 perfs.

Live Aid, a marathon rock concert at Philadelphia July 13, raises $70 million for starving Africans. Organized by Irish singer Bob Geldof, the concert features such stars as Joan Baez, Phil Collins, Bob Dylan, Mick Jagger, Madonna, Paul McCartney, Sting, Tina Turner, and U2.

Popular music *We Are the World* (album) by Michael Jackson and Lionel Ritchie; *No Jacket Required* (album) by Phil Collins; *Sun City* (album) by Artists United Against Apartheid; *Escenas* (album) by Ruben Blades y Seis del Solar; *Biograph* by Bob Dylan; *Who's Zoomin' Who?* by Aretha Franklin; *In Square Circle* by Stevie Wonder; *Centerfield* by John Fogerty; "We All Stood Together" by Paul McCartney; "Edge of Darkness" by British songwriters Eric Clapton and Michael Kamen.

Compact discs and CD players are introduced with superior sound qualities. Music lovers hail the improvement and shift to the new technology.

San Francisco beats Miami 38 to 16 at Palo Alto January 20 in Super Bowl XIX.

British football (soccer) fans riot March 4 at Chelsea and March 13 at Luton, Prime Minister Thatcher appoints a ministerial group to address the problem of hooliganism, a White Paper issued May 16 proposes measures to curb violence at sporting events, but a Liverpool fan club charges into Italian supporters of Turin's Juventus Club before the European Cup Finals at Brussels May 29, 38 are killed (most of them in the collapse of a brick wall in the 60,000-seat Heysel Stadium. Juventus defeats Liverpool 1 to 0 and Belgium bans play by British teams on her soil until such time as Britain shall "put her house in order."

British miler Steve Cram sets a new record July 27 at Oslo, running the distance in 3:46:31 (*see* Bannister, 1954).

Boris Becker, 17 (W. Ger.), wins in men's singles at Wimbledon, Martina Navratilova in women's singles; Ivan Lendl, 25 (Czech.), wins in U.S. men's singles, Hana Mandlikova, 23 (Czech.), in women's singles.

Cincinnati Reds first baseman Pete Rose passes Ty Cobb's record of 4,191 career hits September 11.

The Kansas City Royals win the World Series, defeating the St. Louis Cardinals 4 games to 3.

Former Olympic medal winner Michael Spinks, 29, wins the world heavyweight boxing title September 21 at Las Vegas in a 15-round decision over Larry Holmes, 35.

World chess master Anatoly Karpov loses his title November 9 at Moscow's Tchaikovsky Concert Hall to dissident Soviet chess master Gary Kasparov, 22.

Boston-born Los Angeles merchant Robert Y. Greenberg, 45, sees that less than 20 percent of athletic shoes are bought for athletic use, closes his L.A. Gear apparel store and concentrates on importing Korean-made fashion sneakers to sell under the L.A. Gear name. His sales for the year are $11 million, will more than triple to $36 million next year as he designs fashion hightops for girls, will hit $224 by 1988, and by 1989 will be third in the business behind Nike and Reebok (*see* 1972; 1982).

Merrill Lynch in New York receives a letter May 25 alleging insider trading in its Caracas office. Typed in broken English, the one-page letter sparks an internal inquiry that leads to a bank in Nassau and thence to Dennis B. Levine, 32, a managing director of acquisitions for Drexel Burnham Lambert in New York (*see* 1986).

New York Mafia boss Paul Castellano, 71, and his bodyguard Thomas Bibotti are shot dead in the street December 16 while Castellano is on trial with nine others for auto-theft conspiracy. John Gotti, 45, is suspected of having them rubbed out so he can seize control of the Gambino crime family.

British scientists report in March that a giant "hole" in the earth's ozone layer is opening each spring over Antarctica (*see* 1958).

A Mexico City area earthquake September 19 measuring 7.8 on the Richter scale kills more than 5,000.

A Colombian volcano erupts November 14, leaving 25,000 dead and missing.

1985 ***(cont.)*** A farm bill signed by President Reagan December 23 provides for payments estimated to cost $52 billion over 3 years but favors large producers as smaller growers continue to go under.

R. J. Reynolds acquires Nabisco Brands for $4.9 billion and becomes RJR Nabisco (*see* 1981; 1988). It sells Canada Dry to Britain's Cadbury Schweppes and sells Kentucky Fried Chicken to PepsiCo for $850 million, making PepsiCo the largest U.S. restaurant concern.

Philip Morris acquires General Foods for $5.7 billion and becomes the largest U.S. consumer products company.

Coca-Cola announces in April that it is replacing its famous 99-year-old formula with a sweeter Coca-Cola designed for younger tastes. Protests from longtime Coke drinkers force the company to reintroduce its traditional beverage under the name Classic.

U.S. drug maker A. H. Robins announces April 2 that it has set aside $615 million to settle claims brought by users of its contraceptive device the Dalkon shield.

1986 Corazon C. Aquino, 53, assumes the presidency of the Philippines February 26 after winning election amidst charges of ballot tampering by Ferdinand Marcos. Widow of slain opposition leader Benigno Aquino (*see* 1983), "Cory" receives support from key military leaders and Marcos is flown to Guam after U.S. pressure has been applied to make him leave Manila. He is given sanctuary in Hawaii (where he will die late in 1989) after a 20-year rule that has bled the country of perhaps $5 billion.

Haiti's President Jean-Claude Duvalier, 34, has resigned February 6 and been given "temporary" sanctuary in France after 15 years of repressive rule in which he looted the nation's treasury. Haitians rejoice, but Duvalier cronies retain power.

Swedish Prime Minister Olof Palme, 59, is assassinated at Stockholm February 28 after leaving a movie theater with his wife Lisbet. Carl Gustav Christer Pettersson will be arrested in December 1988, charged with the crime, and convicted, but a court will overturn his conviction in October 1989.

U.S. warplanes from Britain bomb Libya's Muammar Qadaffi's headquarters at Tripoli April 15 in an 11-minute strike that leaves 15 civilians dead. President Reagan has ordered the attack in retaliation for the terrorist bombing of a West Berlin discothèque that killed a U.S. soldier and a Turkish woman and wounded 230. An American F-111 with two airmen is lost in the attack on Libya, and three hostages are killed in Lebanon in reprisal for the U.S. action.

Former U.S. Navy analyst Jonathan Pollard, 31, pleads guilty June 4 to supplying classified information to Israel.

Congress votes June 26 to approve $100 million in aid to the military adventurers trying to overthrow Nicaragua's Sandinista regime (*see* 1987).

Iran and Iraq continue their bloody war, with Iran receiving covert aid in the form of U.S. arms and aircraft replacement parts. Israel is the chief source but a Beirut magazine reveals in November that the United States has sent spare parts and ammunition to Iran in hopes that "moderates" there would help obtain the release of U.S. hostages. Further investigation will show that other arms sales were made to Iran with the profits going to fund Contra forces in Nicaragua. Marine Lieut. Col. Oliver L. North, 43, and National Security Council adviser Vice Admiral John F. Poindexter, 58, resign their positions and refuse to answer congressional investigators' questions about their activities in the affair (*see* 1987).

Terrorists continue to take their toll. A bomb aboard a TWA plane over Athens kills four Americans April 2; a West Berlin discothèque explosion April 5 kills two, injures 230 (*see* Libya attack, above); guards at London's Heathrow Airport avert a tragedy April 17 when they arrest a British woman with explosives in her luggage, planted there by her Jordanian fiancé in an effort to blow up a Tel Aviv-bound El Al flight; Arabic-speaking gunmen seize a Pan Am jet at Karachi September 5, kill 15 passengers, wound 127; two Arabs fire submachine guns into worshipers at an Istanbul synagogue September 6, killing 21; a bomb at a Parisian department store September 17 kills five after four earlier explosions in September have killed three, injured 170.

French voters vote March 15 to elect Paris Mayor Jacques Chirac, 53, to head a Conservative Parliament and share power with President Jacques Miterrand, whose Socialist party has ruled since 1981.

Uganda's government has fallen January 29 after seizure of Kampala by an anti-Obote group that has been excluded from the new regime. Yoweri Museveni is declared president but fighting continues in the northern provinces.

Nuclear space weapons are a stumbling block to disarmament as President Reagan and Soviet Party Secretary Gorbachev hold summit meetings at Reykjavik, Iceland, in October. They reach conditional agreements to ban medium-range missiles, fail to agree on strategic forces, and wind up in icy disharmony when Reagan rejects Soviet demands that he restrict development of his Strategic Defense Initiative ("Star Wars").

Soviet authorities free political prisoners Anatoly Scharansky, Yuri F. Orlov, and two others February 11 and permit them to leave the country in exchange for prisoners held by the West. Mikhail Gorbachev phones Andrei D. Sakharov in December to let him know that he may return to Moscow from exile in Gorky.

Chinese university students demand democratic freedoms guaranteed by the nation's constitution but denied by her leaders. Tens of thousands engage in demonstrations that risk their future careers, but traditional values preclude rooting of democratic principles in the People's Republic (*see* 1989).

South African police fire on crowds of demonstrators March 6, killing 30.

Black civil rights leader Desmond Tutu, 54, is elected Anglican Archbishop of South Africa April 14.

More than 1.5 million South African blacks strike May 1 to protest apartheid in the nation's largest job action ever.

South Africa declares a second state of emergency June 12 (*see* 1985). It covers the whole nation and bans "subversive" press reports. Millions of blacks strike June 16 to commemorate the 1976 Soweto uprisings and protest apartheid. Pass laws are abolished, permitting blacks to move to cities, and a new law entitles residents of so-called tribal homelands to South African citizenship.

U.S. Senators vote 78 to 21 October 3 to override President Reagan's veto of economic sanctions against South Africa. Pretoria takes further measures to quell violence as General Motors and IBM join the Western companies divesting themselves of South African subsidiaries.

U.S. women professionals outnumber men for the first time but average substantially less in pay than their male counterparts.

All seven astronauts aboard the U.S. space shuttle *Challenger* perish January 28 as their craft explodes 73 seconds after liftoff from Florida's Cape Canaveral. The unmanned spacecraft *Voyager 2* launched in 1977 has come close to the planet Neptune January 24 and made new discoveries of moons and rings, but the *Challenger* tragedy is a setback for space exploration. NASA (National Space and Aeronautics Administration) has dismissed warnings by engineers at Morton-Thiokol, makers of its booster rockets, not to launch in very cold weather.

Spain and Portugal join the European Common Market January 1.

The U.S. national debt tops $2 trillion, up from $1 trillion in 1981. The nation's trade deficit worsens despite a weakening dollar, setting a record of over $170 billion. The budget deficit worsens.

Congress restructures the federal income tax system, consolidating 15 brackets into 2 (15 percent and 28 percent), eliminating many tax breaks for the rich, removing lowest bracket earners from the tax rolls, but hiking taxes on businesses, which will in many cases pass along their higher tax costs in higher prices. President Reagan signs the legislation October 22.

The Dow Jones Industrial Average soars past 1900, up from 1546.

Mergers and acquisitions continue as Wells Fargo Bank acquires Crocker National for $1.08 billion, Burroughs Corp. acquires Sperry for $4.8 billion (and becomes Unisys), Unilever acquires Cheseborough-Pond's for $3.1 billion, etc.

U.S. Steel becomes USX July 9.

Singer stops making sewing machines. It announces plans to spin off its sewing operations to a separate firm and concentrate on aerospace.

May Stores acquires Associated Dry Goods for $2.7 billion. Included is Lord & Taylor.

Campeau Corp., a Canadian firm headed by self-made real estate developer Robert Campeau, 63, buys Allied Stores for $3.6 billion, most of it borrowed by issuing high-yield "junk" bonds. Included are Brooks Brothers (which Campeau will sell to Britain's Marks & Spencer), Bonwit Teller, and Boston's Jordan Marsh (*see* Federated, 1988).

R.H. Macy chairman Edward Finkelstein, 61, takes his department store chain private in a $3.7 billion leveraged buyout—the largest management-led buyout ever. Macy's will borrow another $1 billion to acquire more stores, including Federated's Bullocks, Bullocks-Wilshire, and I. Magnin divisions (*see* 1988).

Superconductivity makes news in January as Swiss physicist K. Alex Müller, 58, and German physicist J. Georg Bednorz, 35, of IBM's Zurich Research Laboratory discover zero resistivity in a ceramic material (copper oxide mixed with barium and lanthanum) that permits superconductivity at -397° F.—a much higher temperature than was ever before possible. Their breakthrough, which promptly wins them a Nobel award, opens potentials for more energy-efficient motors, computers, and the like as other scientists, such as University of Houston physicist Paul C. W. Chu will find ways to achieve superconductivity at -283° F.

Nuclear energy receives a setback April 26 when the Soviet Union's Chernobyl power plant near Kiev in the Ukraine explodes, sending clouds of radioactive fallout across much of Europe. More than 30 fire fighters and plant workers die in the first weeks after the accident; predictions of future cancer deaths due to radioactive exposure range from 6,500 to 45,000. Vast tracts of Soviet land will remain uninhabitable and unarable for thousands of years.

World oil prices collapse, bottoming out at $7.20 per barrel in July.

Saudi Arabia's King Fahd removes Sheik Ahmed Zaki Yamani from his oil ministry post October 28 after 24

The Chernobyl accident in the Ukraine confirmed fears of many that nuclear energy was inherently hazardous.

1986 *(cont.)* years as the most powerful figure in OPEC. Yamani has been discounting Saudi Arabian oil.

New York transit fares rise January 1 from 75¢ to $1 (*see* 1984; 1990).

Vancouver's Annacis Bridge is completed; 1,525 feet long, it is the world's longest cable-stayed bridge.

Texas Air acquires Eastern Airlines for $676 million February 24 and becomes the largest U.S. airline. It acquires People Express in September.

Romania completes the Danube-Black Sea Canal from Cernavoda to Constanza. President Ceausescu has used political prisoners and army conscripts to dig the 40-mile waterway that saves ships 150 miles.

The U.S. Food and Drug Administration approves the first genetically engineered vaccine. It will be used to immunize against hepatitis B.

A U.S. law effective in August makes it illegal for hospitals to turn out patients who can no longer afford to pay, but nearly one third of Americans have inadequate health insurance or none at all, and Medicaid payments are so low that many physicians avoid taking Medicaid patients.

Columbia Broadcasting (CBS) is acquired in a leveraged buyout by investor Laurence A. Tisch and CBS founder William S. Paley.

The *Independent* begins publication October 7 at London. Former *Telegraph* editor Andreas Whittam Smith heads the politically neutral daily.

Nonfiction *The Reckoning* by David Halberstam compares Ford Motor Company with Nissan; *Cities and the Wealth of Nations* by Jane Jacobs.

Fiction *Roger's Version* by John Updike; *The Handmaid's Tale* by Margaret Atwood; *The Real Life of Alejandro Mayta* by Mario Vargas Llosa; *The Old Devils* by Kingsley Amis; *A Perfect Spy* by John Le Carré; *The Sportswriter* by U.S. novelist Richard Ford, 42; *A Summons to Memphis* by U.S. novelist Peter Taylor, 70.

Painting *Mural with Blue Brushstroke* by Roy Lichtenstein (for the five-story lobby of New York's Equitable Life Assurance Center). Georgia O'Keeffe dies at Santa Fe, New Mexico, March 6 at age 98.

Sculpture *Nature Study* (bronze) by Louise Bourgeois; *Untitled '88* by Donald Judd. Joseph Beuys dies at Düsseldorf January 23 at age 64; Henry Moore dies at Hadham, England, August 31 at age 88.

Theater *Lend Me a Tenor* by U.S. playwright Ken Ludwig, 35, 3/6 at London's Globe Theatre, with Denis Lawson, John Barron, Ronald Holgate; *A Woman in Mind* by Alan Ayckbourn 9/3 at London's Vaudeville Theatre, with Julia McKenzie, Martha Jarvis; *Breaking the Code* by English playwright Hugh Whitemore, 50, (who has adapted Andrew Hodges's book *Alan Turing: The Enigma of Intelligence*) 10/21 at London's Haymarket Theater, with Derek Jacobi as the man who cracked the Enigma code in World War II, laying the groundwork for computer theory and artificial intelligence, was then hounded by the law for his homosexuality, and committed suicide at age 42; *Coastal Disturbances* by Tina Howe 11/19 at New York's off-Broadway Second Stage Theater, with Annette Bening, Timothy Daly, Rosemary Murphy, Addison Powell, 350 perfs.; *Broadway Bound* by Neil Simon 12/4 at New York's Broadhurst Theater, with Jonathan Silverman, Linda Lavin, Phyllis Newman, Jason Alexander, John Randolph, Philip Sterling, 756 perfs.

Television *L.A. Law* 9/15 on NBC with Harry Hamlin, Jill Eikenberry, Richard Dysart.

Films Ross McElwee's *Sherman's March* with McElwee. Also: James Cameron's *Aliens* with Sigourney Weaver; Fons Rademaker's *The Assault* with Derek de Lint, Marc van Uchelen; Eugene Corr's *Desert Bloom* with Jon Voight, JoBeth Williams, Annabeth Gish; Woody Allen's *Hannah and Her Sisters* with Allen, Michael Caine; Claude Berri's *Jean de Florette* with Yves Montand, Gerard Depardieu; Peter Weir's *The Mosquito Coast* with Harrison Ford; Oliver Stone's *Platoon* with Tom Berenger, Willem Dafoe, Charlie Sheen; Margarethe von Trotta's *Rosa Luxemburg* with Barbara Sukowa; Fielder Cook's *Seize the Day* with Robin Williams, Joseph Wiseman, Jerry Stiller; Erich Rohmer's *Summer* with Maria Riviere; Juzo Itami's *Tampopo* with Ken Watanabe, Tsutomu Yamakazi; Lizzie Borden's *Working Girls* with Louise Smith, Ellen McElduff, Amanda Goodwin.

Cary Grant dies at Davenport, Iowa, November 30 at age 80.

Film musicals Frank Oz's *Little Shop of Horrors* with Rick Moranis, Ellen Greene, Vincent Gardenia; Bertrand Tavernier's *Round Midnight* with François Cluzot, bebop tenor saxophonist Dexter Gordon, music by Herbie Hancock.

Stage musicals *Phantom of the Opera* 10/9 at Her Majesty's Theatre, London, with Michael Crawford, Sarah Brightman, music by Andrew Lloyd Weber, lyrics by Charles Hart, book based on the 1911 Gaston Leroux novel.

Popular songs "That's What Friends Are For" by Burt Bacharach and Carol Bayer Sager; *Graceland* (album) by Paul Simon; *Dancing on the Ceiling* (album) by Lionel Ritchie; *Whitney Houston* (album) by Houston; *Bruce Springsteen & The E Street Band Live/1975–85* (album) by Springsteen; "Nikita" by Elton John and Bernie Taupin; "We Don't Need Another Hero" by British songwriters Graham Lyle and Terry Britten.

Chicago beats New England 46 to 10 at New Orleans January 26 in Super Bowl XX.

Texas-born jockey Willie Shoemaker wins the Kentucky Derby at age 54 riding a 17-to-1 shot (Ferdinand) to crown a 37-year racing career.

University of Maryland basketball star Len Bias, 22, is selected by the Boston Celtics June 17 but dies of a cocaine overdose 2 days later.

Boris Becker wins in men's singles at Wimbledon, Martina Navratilova in women's singles; Ivan Lendl wins in U.S. men's singles, Navratilova in women's singles.

Argentina wins the 13th World Cup football (soccer) championship, defeating West Germany 3 to 2 at Mexico City.

The New York Mets win the World Series, beating the Boston Red Sox 4 games to 3.

Mike Tyson, 20, wins the world heavyweight boxing title November 23, knocking out Trevor Berbick, 29, in the second round of a bout at Las Vegas.

Nintendo video games debut in America and wow the youngsters with sophisticated graphics and entries like "The Legend of Zelda," in which the hero, Link, must rescue Zelda. Founded in 1898 to manufacture playing cards, Nintendo has U.S. sales of $300 million as kids demand the $100 players and $35–$40 cartridges. Sales will hit $830 million next year and top $3.4 billion by 1990.

Grossinger's resort in the Catskills is demolished after 72 years of serving a New York area social group that has become increasingly assimilated into the American mainstream.

Insider trading scandals rock Wall Street (*see* 1985). The Securities and Exchange Commission accuses Dennis B. Levine May 12 of making $12.6 million by trading on non-public information; he cuts a deal, pleading guilty to felony charges and agreeing to cooperate with government investigators (he will serve only 18 months of a 2-year sentence). Arbitrageur Ivan F. Boesky, 49, pleads guilty November 14 to being tipped off about forthcoming merger bids and then buying huge blocks of stock (*see* 1987).

Nearly 75 metric tons of cocaine come into the United States, up from 19 in 1976, and prices drop from $125 per gram to $75. Colombia's Cali and Medellín cartels operate drug rings that net billions of dollars per week—money that must be "laundered" through accommodating banks (*see* 1989).

Congress passes new anti-drug legislation October 17 as use of crack cocaine spreads across the nation.

New York's four-block Jacob K. Javits Convention Center opens in the West 30s, replacing the 25-year-old Coliseum at Columbus Circle.

The new U.S. tax law (*see* above) will lead to depressed prices in the housing market by eliminating tax shelters, although interest on mortgages and home-improvement loans remains tax deductible.

Toxic gas kills 1,524 in Cameroon August 26 after an underwater volcanic explosion. Entire villages are reported devoid of life.

A fire at a warehouse of the Sandoz pharmaceutical firm at Basel November 1 results in 1,000 tons of toxic chemicals being discharged into the Rhine, killing millions of fish and contaminating water supplies. European environmentalists demonstrate in protest as the chemicals make their way out to the North Sea.

The U.S. Department of Agriculture approves release of the first genetically altered virus and the first outdoor test of genetically altered plants. The virus is used to fight a form of swine herpes, the plants are high-yield tobacco plants.

More than 60,000 U.S. farms are sold or foreclosed as depression continues in the rural West and Midwest. Foreign grain exporters undersell U.S. exporters despite the falling dollar, and the $14.21 billion U.S. trade deficit in May includes the first agricultural deficit in 20 years.

The steroid abortifacient drug RU486 (mifepristone) developed in 1980 by French endocrinologist Etienne-Emile Baulieu, 60, wins approval in September for testing in France and the People's Republic of China. Roussel-Uclaf withdraws the drug in October but promptly resumes sale on orders from the French government.

Congress enacts sweeping revisions in the U.S. immigration law. The Simpson-Mazzoli Act signed into law by President Reagan November 7 permits millions of illegal immigrants to remain in the country legally and imposes criminal sanctions on employers who hire undocumented workers; illegal aliens continue to cross into the country from Mexico.

1987 Soviet Party Secretary Mikhail Gorbachev demands reforms January 27 and on June 25 announces plans for a new direction in economic policy. Moscow's vast central-planning system is braking the economy rather than stimulating it, he says. Beginning January 1, 1988, Soviet factories should have the chance to exercise local initiative and assume risks of failure. Moscow Communist chief, Boris Nikolayevich Yeltsin, 56, has gained popularity by sacking corrupt officials but is ousted November 10 after complaining of the slow pace of *perestroika* (reform); he is given a senior position in construction November 18 after criticism of his dismissal (*see* 1989).

Gorbachev arrives at Washington December 7 for a three-day summit conference on arms reduction. He and President Reagan sign the first treaty to reduce the size of nuclear arsenals, agreeing to dismantle all Soviet and U.S. medium- and shorter-range missiles, with extensive weapons inspection on both sides. Western experts have inspected a heretofore top-secret radar site at Krasnoyarsk September 5 to see if it violates the 1972 Anti-Ballistic Missile Treaty as Reagan officials have said.

China's Communist party has expelled dissidents January 14.

Syrian troops occupy West Beirut February 22, ending 3 years of anarchy during which terrorists, mainly pro-Iranian Shiites, have kidnapped dozens of foreigners, but Lebanese prime minister Rashid Karami, 55, is assassinated June 1.

Kuwait asks for U.S. naval protection of her tankers against Iranian attacks in the Persian Gulf as the Iraq-Iran war continues. President Reagan complies, knowing that refusal will result in Kuwait asking Moscow and a spasmodic "tanker war" begins in the Gulf. A Soviet vessel is attacked for the first time May 8, the U.S. frigate *Stark* is hit by Iraqi missiles May 17 with a loss

1987 *(cont.)* of 37 men, Iraqi president Saddam Hussein apologizes.

The Meech Lake Accord signed by Canada's provincial prime ministers outside Ottawa in April recognizes Quebec as a "distinct society" and makes special concessions to that province, but the accord requires ratification by all ten provincial legislatures by June 23, 1990 (*see* 1990).

Britain's Prime Minister Thatcher is reelected to a third term June 11.

India and Sri Lanka sign a treaty July 29 designed to end the ethnic violence that has persisted for 4 years, but a bomb explosion at Colombo November 9 leaves 32 dead, more than 75 injured. Tamil and Indian forces clash in the north, extreme Sinhala nationalists commit acts of violence in the south.

Recriminations over the "Iran-Contra" deal (selling weapons to Iran and using the funds to supply contra forces in Nicaragua) embroil U.S. Cabinet officers (*see* 1986).

The Tower report submitted January 29 by a Senate investigating committee charges that members of the administration deceived Congress and each other. Former CIA director William J. Casey dies May 6 at age 74 leaving many touchy political questions unanswered and perhaps unanswerable. Oliver North testifies before a congressional committee in July that his secret operations had approval from higher-ups, John J. Poindexter testifies that he authorized use of profits from Iran arms sale to support the contras, Secretary of State George P. Shultz testifies that he was repeatedly deceived, Secretary of Defense Caspar W. Weinberger testifies to official intrigue and deception, and President Reagan says August 12 that U.S. policy in the affair went astray. The final congressional committee report November 18 after 3 months of hearings charges Reagan with failing to obey the constitutional requirement that the President execute the laws. It documents distribution of nearly $48 million from arms sales to the contras and says the President bears "ultimate responsibility" for the wrongdoing of his aides.

Five Central American nations sign an accord August 7 at Guatemala City agreeing to cooperate with Costa Rican president Oscar Arias Sánchez in finding ways to resolve conflicts in the region. Arias, whose country has not had an army since 1948, is awarded the Nobel Peace Prize October 13 but hostilities continue between Nicaragua's Sandinista government and U.S.-supported contras.

South Korea's military dictator Chun Doo Wha appoints Roh Toe Woo, 55, as his successor in June, students stage violent protests, Roh—and, later, Chun—agree to direct elections, opposition forces split, and Roh wins election December 16 with 36.6 percent of the popular vote.

Tibetan demonstrators stone police to protest Chinese rule and the October 1 riots leave 6 dead, scores seriously injured. Beijing accuses the Dalai Lama of stirring up anti-Chinese feeling by "criminal" actions. The White House expresses support for Beijing October 6, but the U.S. Senate votes 98 to 0 to condemn the Chinese crackdown.

Oliver North admitted arming Nicaraguan contras with proceeds of arms sales to Iran. Some thought him a hero.

Tunisia's prime minister Gen. Zine al-Abidine Ben Ali declares President Bourguiba, now 84, senile and unfit to continue his 30-year rule. Ben Ali takes over (legally, under the constitution) and promises democratic reforms.

Zulu prince Mangosuthu Buthelezi, 58, launches a bloody civil war against South Africa's African National Congress. Less militant against apartheid than the still-imprisoned Nelson Mandela, Buthelezi has opposed economic sanctions against Pretoria, favors capitalism, and has enlisted 1.7 million members, mostly Zulus, in Inkatha, which he founded in 1975 (*see* Mandela, 1990).

Niger's dictator Seyni Kountche dies November 10 after a 13-year rule; he is succeeded by his chief of staff Col. Ali Saibou.

President Reagan and Party Leader Gorbachev sign a treaty at the White House December 8 agreeing to eliminate medium-range intermediate nuclear weapons in both superpower arsenals.

A truck driven by an Israeli kills four people in the occupied Gaza Strip December 8. Rock-throwing young Palestinians begin violent protests and the uprising, or *intifada*, spreads throughout the occupied territories. Civil disobedience will cost more than 300 Arab lives in the next 12 months in a struggle to oust Israeli occupation forces and establish a separate Palestinian state (*see* 1988).

Czech Communist Party leader Gustav Husak resigns December 17 after 18 years in power. He is replaced by economic specialist Miloš Jakes, 65, who is urged by Soviet leader Gorbachev to promote Soviet-style *perestroika* (restructuring) and "democratization" (*see* 1989).

South Africa acts April 11 to ban protests aimed at winning release for detainees.

Black miners in South Africa return to the pits August 30 after a strike that has gained them nothing.

The U.S. Supreme Court rules May 4 that Rotary clubs must admit women.

A French court rules July 4 that Klaus Barbie, 73, was guilty of war crimes and sentences him to life imprisonment. Gestapo chief in Lyons from 1941 to 1945, Barbie was arrested in Bolivia 4 years ago.

Homosexuals demonstrate at Washington October 11 to demand an end to discrimination and more federal funds for the fight against AIDS. Some 600 are arrested trying to enter the Supreme Court to protest a sodomy decision.

Wappingers Falls, N.Y., schoolgirl Tawana Brawley, 15, is found November 28 half naked with "KKK" and "Nigger" smeared with dog feces on her body. She claims six white men, one of them wearing a police badge, kidnapped her November 24 and repeatedly raped her. Self-serving lawyers take up her case, which inflames the black media, but a grand jury investigation will show that she left home for 4 days and staged her condition to avoid violent punishment from her stepfather.

Romanian tractor factory workers at Brasov lead a protest in November, complaining of a decline in living standards after months of energy rationing. President Ceausescu's *Securitate* (secret police) arrest hundreds (*see* 1989).

Brazil announces in February that it is suspending interest payments on loans from foreign banks. The action by the Third World's largest debtor nation signals a worsening debt crisis.

Burma's Ne Win government decrees in March that 75-, 35-, and 25-kyat notes are now worthless and will be replaced by 90- and 45-kyat bills. Few Burmese keep money in banks; the voiding of the three highest-denomination banknotes wipes out many people's savings.

The U.S. trade deficit hits a record $16.5 billion in July. President Reagan has imposed a 100 percent retaliatory tariff on certain Japanese imports April 17.

Citicorp, Manufacturers Hanover Trust, and Bankers Trust report major losses July 21, blaming nonpayment of foreign debts.

Congress acts under the Graham-Rudman Act in late September to enforce budget-balancing by imposing automatic spending cuts. President Reagan, adamantly opposing any tax increases, signs an overall spending measure December 22 designed to hold down the federal deficit.

Wall Street's Dow Jones average peaks at 2722.42 August 25, falls to 2346 the third week of October (after rising a record 75.23 points September 22), then plunges 508 points—22.6 percent—on October 19 to 1738.74, a drop even sharper than that of October 1929, as New York Stock Exchange volume exceeds 604 million shares. Pundits blame computerized trading programs, the U.S. trade and budget deficits. By August 24, 1989, the Dow will have topped its 1987 high.

President Reagan and Canada's Prime Minister Mulroney sign a free-trade agreement October 3. Canada's Liberal and New Democratic parties have opposed the pact.

The AFL-CIO executive council votes October 24 to permit the Brotherhood of Teamsters, expelled in 1957 for reasons of ethics, to rejoin.

A powerful superconductor made of ceramic and capable of operating at relatively low temperatures is announced March 17 but while its potential for cheap power is enormous experts say it will be years, perhaps decades, before that potential can be realized.

Texaco files for bankruptcy April 12 despite assets of $34.9 billion, the largest bankruptcy filing in history. The company agrees December 18 to pay Pennzoil $3 billion after a 4-year legal battle over Texaco's acquisition of Getty Oil.

A fire in London's King's Cross underground station creates a panic in which 32 are killed, about 50 injured.

President Reagan vetoes a highway appropriations bill; Congress overrides his veto March 27.

Chrysler acquires American Motors for 1.2 billion (*see* 1979).

A Polish LOT airliner crashes near Warsaw May 9, killing 183; a Northwest Airlines McDonnell Douglas MD-30 crashes on a heavily-traveled Detroit boulevard August 16, killing 153; a South African Airways Boeing 747 goes into the sea south of Mauritius November 26, killing 160; a bomb planted by North Korean terrorists blows up a Korean Air Boeing 747 November 29 and the plane goes into the sea off Burma, killing all 115 aboard.

The English Channel ferry *Herald of Free Enterprise* leaves Zeebrugge, Belgium, March 6 with its bow loading doors open, and capsizes a few hundred yards out, killing 192; a Filipino passenger ferry collides with a tanker off Mindoro Island December 20 and at least 1,500 are drowned.

U.S. Surgeon General C. Everett Koop tells a House subcommitee February 10 that condom commercials should be permitted on TV to help stop the AIDS epidemic.

AZT wins FDA approval March 20. Made by Burroughs Wellcome and used for treating AIDS, the drug costs more than $10,000 per year per patient. It does not cure the disease but extends lives of victims and relieves some symptoms, although for some it has terrible side effects.

1987 *(cont.)* Lovastatin (Mevacor), a cholesterol-lowering drug developed by Merck, is approved by the FDA in September.

U.S. spending on health care rises to $500 billion, up 9.8 percent over 1986 (the 1987 inflation rate is 4.4 percent).

PTL minister Jim Bakker resigns March 19 after revelations that he cheated on his wife, Tammy Faye, in 1980 with church secretary Jessica Hahn and used ministry money to buy Hahn's silence (*see* 1974). Rev. Jerry Falwell takes over the PTL Ministry (*see* 1989).

Clashes between Shiites and other Muslims at Mecca August 1 leave at least 400 dead as Saudi police open fire on Shiites visiting the city on their annual *hadj* (pilgrimage) demonstrate against Sunnis, who back Iraq in the ongoing war with Iran.

A $2 million libel judgment against the *Washington Post* is reversed on appeal March 13.

Nicaragua's anti-Sandinista newspaper *La Prensa* resumes publication October 1 after a 451-day shutdown and the Roman Catholic radio station, off the air by government order since January 1985, resumes broadcasting following a five-nation peace accord.

Nonfiction *The Closing of the American Mind* by University of Chicago political science professor Allan Bloom, 56; *The Making of the Atomic Bomb* by U.S. author Richard Rhodes, 50; *A History of the Jews* by Paul Johnson.

Fiction *The Counterlife* by Philip Roth; *Moon Tiger* by English novelist Penelope Lively, 54; *Beloved* by Toni Morrison; *Presumed Innocent* by U.S. lawyer-novelist Scott Frederic Turow, 38; *Paco's Story* by U.S. novelist Larry Heinemann, 43; *The Tenants of Time* by Thomas Flanagan; *The Sacred Night* (*La Nuit Sacrée*) by Moroccan novelist Tahar Ben Jelloun, 42; *The Bonfire of the Vanities* by Tom Wolfe; *Empire* by Gore Vidal; *Touch* and *Bandits* by Elmore Leonard.

Painting *Diptych* by Brice Marden; *The Hunger Artist* by Elizabeth Murray; *Untitled* by Willem de Kooning; *Constant* by Robert Ryman. Andy Warhol dies after gall-bladder surgery at New York Hospital February 23 at age 58.

Theater *Fences* by August Wilson 3/26 at New York's 46th Street Theater, with James Earl Jones, Mary Alice, 526 perfs.; *Driving Miss Daisy* by U.S. playwright Alfred Uhry, 51, 4/15 at New York's Playwrights Horizons Theater, with Dana Ivey, Morgan Freeman, Ray Gill, 80 perfs.; *Burn This* by Lanford Wilson 10/15 at New York's Plymouth Theater, with John Malkovich, 437 perfs.; *Lettice and Lovage* by Peter Shaffer 10/27 at London's Globe Theatre, with Maggie Smith, Margaret Tyzack.

Films Gabriel Axe's *Babette's Feast* with Stéphane Audran; Norman Jewison's *Moonstruck* with Cher, Nicolas Cage, Olympia Dukakis, Danny Aiello. Also: John Huston's *The Dead* with Anjelica Huston; Peter Greenaway's *Drowning by Numbers* with Joan Plowright, Bernard Hill; David Jones's *84 Charing Cross Road* with Anne Bancroft, Anthony Hopkins; John Boorman's *Hope and Glory* with Sarah Miles, David Hayman; Bernardo Bertolucci's *The Last Emperor* with Peter O'Toole, Joan Chen, John Lone; John Sayles's *Matewan* with Chris Cooper; Lasse Hallestrōm's *My Life as a Dog* with Anton Glanzelius; Joel Coen's *Raising Arizona* with Nicolas Cage, Holly Hunter; Brian De Palma's *The Untouchables* with Robert De Niro, Kevin Costner, Sean Connery.

Stage musicals *Sarafina!* in June at Johannesburg's Market Theater, with a cast of 23, music and lyrics by Mbongeni Ngema, 35; *Into the Woods* 11/5 at New York's Martin Beck Theater with Bernadette Peters, Joanna Gleason, music by Stephen Sondheim, lyrics by James Lapine, 764 perfs.

Popular songs "Somewhere out There" by James Horner, Barry Mann, and Cynthia Weil; *Bad* (album) by Michael Jackson; *Tunnel of Love* (album) by Bruce Springsteen; *Whitney* (album) by Whitney Houston; *A Momentary Lapse of Reason* (album) by Pink Floyd; *Tango in the Night* (album) by Fleetwood Mac; *The Joshua Tree* (album) by the Irish rock group U2.

The New York Giants beat Denver 39 to 20 at Pasadena January 25 in Super Bowl XXI.

Mike Tyson wins the World Boxing Association heavyweight title from James Smith March 7 in a 12-round decision at Las Vegas (*see* 1986, 1990).

Pat Cash, 22 (Australia), wins in men's singles at Wimbledon, Martina Navratilova in women's singles; Ivan Lendl wins in U.S. men's singles, Navratilova in women's singles.

Stars and Stripes regains the America's Cup, defeating Australia's *Kookaburra III* 4 to 0 (*see* 1983).

The Minnesota Twins win the World Series, beating St. Louis Cardinals 4 games to 3.

Gary Kasparov retains his chess title December 19 against challenger Anatoly Karpov at Seville. The match ends in a tie, with both players sharing the purse.

Three Wall Street traders are charged February 12 with illegal "insider" trading that has given them millions in profits (*see* 1986). Ivan F. Boesky is sentenced December 18 to 3 years in light security prison for conspiring to falsify stock trading records connected with insider deals. He is also fined $100 million, half of which is tax-deductible, and is spared a heavier sentence because he has named others who have profited from insider trading.

Former U.S. Treasury Secretary Robert B. Anderson, 77, has been sentenced for tax evasion June 25. The sentence: 1 month in prison, 5 months of house arrest, and 5 years' probation.

A Sicilian Mafia trial ends December 16 with 338 of the 452 defendants sentenced to prison, 19 of them for life. Their criminal empire has been built primarily on heroin trafficking in the United States.

Colombia angers U.S. officials December 31 by releasing Medéllin cartel leader Jorge Luis Ochoa, wanted for wholesale cocaine trafficking (*see* 1989).

Nations meeting at Montreal agree September 16 on measures to protect the environment, notably a gradual ban on chlorofluorocarbons that deplete the earth's ozone layer and increase incidence of skin cancer. Researchers report December 31 that the ozone shield declined sharply from 1979 through 1986.

Brazilian landowners burn 80,000 square miles of Amazon rain forest in 79 days (July 15 through October 2), heightening environmentalist fears that loss of oxygen from the forest will create a "greenhouse effect," increasing global temperatures and raising sea levels. Tax incentives encourage turning jungle into ranch land (*see* 1989).

A Colombian avalanche September 28 kills at least 120.

An earthquake registering 6.1 on the Richter scale strikes Los Angeles October 1, killing 6, injuring 100.

Britain's worst storm in memory strikes just after midnight October 16, knocking down thousands of trees and causing other damage.

Gene-altered bacteria to aid agriculture are tested April 24 despite alarms by some that scientists have unloosed a monster.

U.S. microwave oven sales reach a record 12.6 million (*see* 1967). Sears, Roebuck's Kenmore is the largest-selling brand, followed by Sharp and General Electric. Food companies rush to develop microwavable food products.

Kellogg introduces Just Right breakfast food with raisins, nuts, and dates.

1988

Moscow agrees April 14 to withdraw Soviet forces from Afghanistan (the first group leaves May 17), promises to have all 115,000 out by mid-February 1989, and agrees to restore a nonaligned Afghan state. Occupation of the country since December 1979 has cost at least 15,000 Soviet and more than 1 million Afghan lives. *Mujahedeen* resistance fighters, covertly supplied by the CIA, step up efforts to oust the puppet regime at Kabul.

President Reagan visits Moscow from May 29 to June 1 but accomplishes little besides antagonizing Party Leader Gorbachev with appeals for increased civil and religious liberties.

The Communist Central Committee has voted May 26 to limit the terms of officials. Delegates to a Communist conference at Moscow July 1 endorse Gorbachev's proposals, including partial transfer of power from the party to democratically elected legislatures, and approve inauguration of the position of President. Gorbachev is named President October 1 and addresses the United Nations in New York December 7, promising unilateral reduction of Soviet troops, missiles, and munitions on the western frontiers of the Warsaw Bloc.

Romania's President Nicolae Ceausescu announces in March that he will undertake a program to demolish 8,000 of the country's 13,000 villages and resettle residents, including ethnic Hungarians, in urban housing complexes (*see* 1989).

Hungary's Communist party ousts János Kádár May 22 after nearly 32 years of power. Károly Gròsz takes over as Prime Minister.

Iraqi forces recover Fao April 18 after a two-day battle that has cost 10,000 Iraqi lives (at least 53,000 Iraqis and possibly 120,000 Iranians have died fighting for the city). German engineers have modified Iraq's Soviet-built Scud missiles to extend their range from 190 miles to 375 (the Hussein) and even 560 (the Abbas) but with payloads of 1,102 and 661 pounds of explosives, respectively (the standard Scud carries about 2,100 pounds) and with an accuracy margin of only 1,100 to 3,300 yards (*see* 1991).

Iran regains most of the territory lost earlier as the combatants reach a stalemate. President Khomeini has insisted that Iraq's President Saddam Hussein must step down and Iraq pay war damages before there can be peace, but he agrees July 20 to a cease-fire following some Iranian military setbacks. Direct negotiations begin after nearly 8 years of hostilities and Iran accepts an Iraqi truce plan August 8, agreeing to a ceasefire followed by direct talks to end the conflict that has cost 105,000 Iraqi lives (Iran has lost at least 1 million) and left the country with $85 billion in war debt.

The skipper of the U.S. warship *Vincennes* in the Persian Gulf has mistaken an Iran Air A300 Airbus for an attacking plane July 3 and shot it down. Washington, embarrassed, will offer reparations to survivors of the 290 dead next year.

Israel begins deporting nationalists seized in January rioting as Israeli police and troops kill hundreds of rock-throwing young Arab demonstrators in a continuation of the violence that roils the country. Prime Minister Yitzhak Shamir's Likud party promises a hard line against the Arabs and wins the November elections.

PLO senior official Abu Jihad (Khalil al-Wazir), believed to have directed the uprising (*see* above), is assassinated April 16 along with two bodyguards and a driver at Tunisia. Israel reportedly ordered the killing.

Jordan's King Hussein announces July 31 that he is ceding the West Bank to the PLO and abandoning the area ruled by his family from 1948 to 1967. He questions the effectiveness of U.S. peace efforts in the Middle East.

Taiwan's President Chiang Ching-kuo, son of the late Chiang Kai-shek, dies January 13 at age 77 and is succeeded by Lee Teng-hui, 65, the first native-born Taiwanese to hold the position. U.S. educated, Lee is an expert in agricultural planning.

French voters reelect President Mitterand to a 7-year term July 8, rejecting his right-wing challenger Jacques Chirac.

Panamanian dictator Gen. Manuel Antonio Noriega is indicted by federal grand juries in Tampa and Miami February 5 on charges of accepting millions of dollars in bribes from drug traffickers, but when President Delvalle tries to oust Noriega he is dismissed by the National Assembly. Noriega's opponents stage a general strike in March, the government closes the banks,

1988 *(cont.)* U.S. sanctions are imposed, and civil disorders follow as workers are unpaid and the government seizes flour mills and Canal docks (*see* 1989).

Mexico's ruling PRI party succeeds in having its candidate Carlos Salinas de Gortari elected president but the margin is the narrowest ever; opposition parties challenge the result.

Angola, South Africa, and Cuba agree August 8 to an immediate truce in Angola and neighboring Namibia after mediation by U.S. diplomats in the long conflict.

Pakistan's Gen. Mohammad Zia ul-Haq, 64, deposes Prime Minister Mohammed Junejo in May and dissolves the National Assembly, saying it has not moved swiftly enough to establish Islamic law or address ethnic conflicts. Zia is killed August 18 when his plane explodes in midair (U.S. Ambassador Arnold I. Raphel, 45, is also killed). Benazir Bhutto, 35, whose father, Prime Minister Zulfikar Ali Bhutto, was executed by Zia in 1979, is elected prime minister in December and becomes the first woman to head a Muslim state.

Vice President George Herbert Walker Bush, 64, wins the U.S. presidential election with 53 percent of the popular vote to 46 percent for Massachusetts Governor Michael Dukakis, 54, who takes 10 states. Republican Bush is the first sitting vice president to win election since 1836.

Iraqi forces use poison gas March 16 against Kurdish civilians in the town of Halabja, killing at least 4,000 men, women, and children (some estimates say 12,000). German companies have built facilities in Iraq to produce the gas, which is also used to kill an estimated 10,000 Iranian soldiers.

Rangoon police club a student to death in March and let 41 suffocate in a crammed van. Aung Sang Suu Kyi, 44, daughter of the Burmese hero who was assassinated in 1947, returns from exile in England in April. A political protest in front of Convocation Hall on the Rangoon University campus June 21 leads to a march by 1,000 students in downtown Rangoon. Police crack down, killing as many as 300. Gen. Ne Win closes all schools, he resigns July 23, and is succeeded as head of state by Gen. Sein Lwin, who has directed the major blood baths since 1962. Burmese by the tens of thousands demonstrate in Rangoon streets against Sein Lwin, thousands are killed, but monks control Mandalay and Ne Win, who retains actual power, has Sein Lwin replaced August 19 by his attorney general U Maung Maung. Prisoners are released, freedom of the press is permitted for the first time since 1962, but Gen. Saw Maung, who is prime minister and foreign minister, has his troops open fire on demonstrators September 19, orders strikers back to work, and kills hundreds in a renewal of savagery directed by Ne Win.

South African blacks strike June 6 in a three-day walkout to protest a crackdown on anti-apartheid groups. Employers and unions agree to negotiate on a bill to curb labor's powers.

Polish workers strike for 3 weeks in August demanding return of the outlawed Solidarity organization and political and economic reforms. They go back to work September 3.

President Reagan and Canadian Prime Minister Mulroney sign a trade agreement January 2 that eliminates tariffs and lowers other trade barriers. Canada's House of Commons approves the accord August 31, ending a century of economic nationalism.

President Reagan signs a trade bill in August giving him broad powers to retaliate against countries found to be engaged in unfair trade practices. A protectionist trade bill to limit textile imports passes the House 248 to 150 and the Senate 59 to 36 but the President vetoes the measure September 28.

The U.S. Supreme Court rules 5 to 4 April 20 that Congress may tax interest on municipal bonds issued by states and local governments.

U.S. unemployment falls in April to 5.4 percent, lowest since 1974.

Median weekly U.S. earnings: lawyer $914, pharmacist $718, engineer $717, physician $716, college teacher $676, computer programmer $588, high school teacher $521, registered nurse $516, accountant $501, editor, reporter, $494, actor, director, $488, writer, artist, entertainer, athlete, $483, mechanic $424, truck driver (heavy) $387, carpenter $365, bus driver $335, laborer $308, secretary $299, truck driver (light) $298, machine operator $284, janitor $258, hotel clerk $214, cashier $192 (source: Bureau of Labor Statistics).

U.S. S&Ls lose $13.44 billion.

President-elect George Bush (above) has campaigned on a pledge ("Read my lips") to impose no new taxes and to tax capital gains at a lower rate.

Campeau Corp. outbids R.H. Macy and pays $6.6 billion to acquire Federated Department Stores. It sells off Brooks Brothers and Bonwit Teller to reduce its staggering $11 billion debt but its losses outpace its profits (*see* 1986, 1989).

A Soviet-built Ilyushin Il-18 plows into a Chinese hillside January 19 killing 108; an Avianca Boeing 727 crashes after takeoff on a domestic flight March 17 killing 136, including two soccer teams; a gaping hole opens in the fuselage of a 19-year-old Aloha Airlines Boeing 737 April 28, flight attendant C. B. Lansing is swept to her death, but the plane lands safely at Maui Airport and the airline industry institutes new maintenance procedures; a Pan Am 747 explodes in midair over Lockerbie, Scotland, December 21, killing all 259 aboard plus 11 on the ground (a bomb planted by a Mideastern terrorist in Frankfurt is blamed).

Italy inaugurates 155-mile-per-hour rail service on the Direttisima between Rome and Florence.

British Rail introduces the Electra locomotive on its London-Leeds route, increasing speed to 140 miles per hour.

U.S. health-care spending reaches $51,926 per capita as costs run out of control, accounting for 11.1 percent of gross national product. Sweden spends 9.1 percent, Canada and France 8.5, the Netherlands 8.3, West

Germany 8.1, Austria and Switzerland 8, Ireland 7.9, Finland and Iceland 7.5, Belgium 7.1, Luxembourg and New Zealand 6.9, Australia and Norway 6.8, Italy and Japan 6.7, Britain 6.2, Denmark 6.1, Spain 6, Portugal 5.6, Greece 3.9, Turkey 3.6. Every industrial nation except the United States and South Africa has a national health-care program.

Baton Rouge, La., television evangelist Jimmy Swaggart, 52, visits Nicaragua's President Daniel Ortega February 12, confesses sin February 21, and is removed from his pulpit by the Assemblies of God after revelations that he has had sex with a prostitute. Swaggart has lost 69 percent of his viewers and 72 percent of the enrollment at his bible college. He is defrocked April 8 and ordered to stay off TV for a year, but returns in 3 months.

The U.S. Supreme Court has ruled unanimously February 24 that *Hustler* magazine's criticism of evangelist Jerry Falwell was within the rules protecting attacks on public figures.

The U.S. first-class postal rate goes to 25¢ per ounce April 3 (*see* 1985; 1991).

Rupert Murdoch agrees August 7 to pay Walter H. Annenberg $3 billion for Triangle Publications (*TV Guide*, *Daily Racing Form*, *Seventeen*).

Turner Network Television (TNT) is founded by Ted Turner who has purchased the M-G-M library of old films.

Britain's Thatcher government imposes a "Sinn Fein ban" in October on BBC and commercial radio and TV stations, forbidding voices of Irish political activists to be heard on the air.

Nonfiction *Battle Cry of Freedom* by Princeton historian James McPherson, 51; *Parting the Waters: America in the King Years* by U.S. writer Taylor Branch, 41; *A Bright Shining Lie: John Paul Vann and America in Vietnam* by U.S. journalist Neil Sheehan, 51; *Day of Reckoning: The Consequences of American Economic Policy Under Reagan and After* by Harvard economist Benjamin Friedman.

Fiction *The Satanic Verses* by Salman Rushdie incenses Muslim readers with its alleged "blasphemies" (*see* 1989); *Paris Trout* by U.S. novelist Pete (Peter Whittemore) Dexter, 45; *The Middleman and Other Stories* by Indian novelist Bharati Mukherjee, 46; *Foucault's Pendulum* by Umberto Eco; *Love in the Time of Cholera* by Gabriel Garcia Marquez; *Breathing Lessons* by Anne Tyler; *. . . And Members of the Club* by Ohio novelist Helen Hooven Santmyer, 88; *Freaky Deaky* by Elmore Leonard.

Painting *Diagrammed Couplet No. 1* by Brice Marden. Romare Bearden dies at New York March 12 at age 75; Jean-Michel Basquiat dies of a drug overdose in New York August 12 at age 27.

Sculptor Louise Nevelson dies at New York April 17 at age 88.

Theater *The Piano Lesson* by August Wilson 1/9 at Boston's Huntington Theater, with Carl Gordon, Rock Dutton, Starletta DuPois; *M. Butterfly* by Hong Kong-born U.S. playwright David Henry Hwang, 30, 3/20 at New York's Eugene O'Neill Theater with John Lithgow, B. D. Wong, John Getz, 777 perfs.; *The Heidi Chronicles* by Wendy Wasserstein 4/15 at New York's off-Broadway Playwrights Horizons Theater, with John Allen, Peter Friedman, Boyd Gaines; *Speed-the-Plow* by David Mamet 5/2 at New York's Royale Theater with David Rasche, Bob Balaban, Felicity Huffman, 278 perfs.; *Our Country's Good* by U.S.-British playwright Timberlake (Lael Louisiana) Wertenbaker (based on *The Playmaker* by Thomas Keneally) 9/1 at London's Royal Court Theatre, with Nick Dunning, Ron Cook, Linda Bassett, Lesley Sharp; *Rumors* by Neil Simon 11/17 at New York's Broadhurst Theater, with Joyce Van Patten, André Gregory, Ken Howard, 531 perfs.

Films Wim Wenders's *Wings of Desire* with Bruno Ganz, Otto Sander. Also: Woody Allen's *Another Woman* with Gena Rowlands; Louis Malle's *Au Revoir Les Enfants* with Gaspard Manesse, Raphael Fejito; Penny Marshall's *Big* with Tom Hanks; Fred Schepisi's *A Cry in the Dark* with Meryl Streep, Sam Neill; Stephen Frears's *Dangerous Liaisons* with Glenn Close, John Malkovich, Michelle Pfeiffer; Charles Sturridge's *A Handful of Dust* with James Wilbym, Kristin Scott Thomas; Istvan Szabo's *Hanussen* with Klaus Maria Brandauer, Erland Josephson; Mike Leigh's *High Hopes* with Philip Davis, Ruth Sheen; Marcel Ophuls's documentary *Hotel Terminus: The Life and Times of Klaus Barbie*; Martin Scorsese's *The Last Temptation of Christ* with Willem Dafoe; Martin Brest's *Midnight Run* with Robert De Niro, Charles Grodin; Alan Parker's *Mississippi Burning* with Gene Hackman; Bille August's *Pelle the Conqueror* with Pelle Hvenegaard, Max von Sydow; Barry Levinson's *Rain Man* with Dustin Hoffman, Tom Cruise; Sidney Lumet's *Running on Empty* with Christine Lahti, Judd Hirsch; Marina Goldovskaya's documentary *Solovki Power* about a Soviet gulag (prison camp) in the White Sea's Solovetsky archipelago. Juzo Itami's *A Taxing Woman* with Nobuko Miyamoto, Tsutomou Yamazaki; Errol Morris's documentary *The Thin Blue Line*; Antony Thomas's documentary *Thy Kingdom Come . . . Thy Will Be Done* about "born-again" Christianity; Francis Ford Coppola's *Tucker: The Man and His Dream* with Jeff Bridges; Philip Kaufman's *The Unbearable Lightness of Being* with Daniel Day Lewis, Lena Olin, Juliette Binoche; Robert Zemeckis's *Who Framed Roger Rabbit* with Bob Hoskins interacting with animated characters; Pedro Almodóvar's *Women on the Verge of a Nervous Breakdown* with Carmen Marua.

Film musicals Dennis Potter's *The Singing Detective* (made for BBC television). Clint Eastwood's *Bird* with Forrest Whitaker as Charlie Parker.

Popular songs "Don't Worry, Be Happy" by Bobby McFerrin; *Rattle and Hum* (album) by U2; "So Emotional" by Whitney Houston; "Sweet Child O' Mine" by Guns N' Roses; *Roll with It* (album) by English rocker Steve Winwood; *Appetite for*

1988 *(cont.)* *Destruction* (album) by Guns N' Roses; *Faith* (album) by George Michael; *Talk Is Cheap* (album) by Keith Richards of The Rolling Stones.

Washington beats Denver 42 to 10 at San Diego January 31 in Super Bowl XXII.

Stefan Edberg, 22 (Sweden), wins in men's singles at Wimbledon, Steffi Graf, 19 (W. Ger.), wins tennis's first "grand slam" since Margaret Court of England did it in 1970. Mats Wilander, 23, becomes the first Swede to win the U.S. singles title.

Stars and Stripes retains the America's Cup, defeating *New Zealand* 2 to 0 off San Diego, but New Zealand protests. A New York State Supreme Court judge will rule in March 1989 that the San Diego Yacht Club's use of a catamaran was unfair and that San Diego must forfeit yachting's most prestigious trophy to the giant mono-hulled *New Zealand*; a New York appeals court will reverse the decision 6 months later.

The Los Angeles Dodgers win the World Series, defeating the Oakland Athletics 4 games to 1.

Soviet athletes win 132 medals in the Olympic Games at Seoul, East German athletes 102, U.S. athletes 94. Canadian runner Ben Johnson wins the 100-meter dash, setting a 9.79-second record, but is stripped of his gold medal September 27 for using performance-enhancing anabolic steroids.

The SEC accuses Drexel Burnham Lambert September 7 of having a secret agreement to defraud clients by trading on inside information. The firm, whose high-yield "junk" bond expert Michael Milken in Beverly Hills has been instrumental in effecting scores of mergers and acquisitions, pleads guilty in December to 6 felony charges and agrees to pay a record $650 in fines. Drexel says it will withhold $200 million due Milken for 1988 under terms of his employment contract.

Brazilian floods and mud slides kill 117 February 7.

Global warming threatens mankind, NASA climatologist James E. Hansen, 47, tells the Senate Committee on Energy and Natural Resources June 23. Increased atmospheric levels of carbon dioxide and other heat-trapping "greenhouse" gases are probably to blame, he says.

Chinese floods in early August kill thousands along the eastern coast, leave hundreds of thousands homeless.

Bangladesh has floods in early September that cover much of the country, the worst in 70 years, with 1,000 dead and millions homeless. Donor nations rush aid, and the government asks that international experts work on flood control projects.

An earthquake registering 6.9 on the Richter scale rocks Soviet Armenia December 7, killing at least 25,000.

Brazilian rubber tapper Francisco "Chico" Mendes Filho is shot dead December 22 at his home in Xapuri, raising a worldwide storm of protest against ranchers who are clearing the western Amazonian rain forest. Mendes has rallied families to stand up against the chain saws and bulldozers (*see* 1989).

Drought reduces North American crops. The United States is obliged for the first time in history to import grain for domestic needs.

Leaders of the seven largest industrial democracies meet for three days at Toronto in June but rebuff President Reagan's demand that high priority be given to ending government farm subsidies by the year 2000. U.S. farm subsidies are $26 billion, up from $3 billion in 1981.

Philip Morris buys Kraft for $13.1 billion.

Kohlberg Kravis Roberts agrees in October after a bidding contest to pay $24.9 billion for RJR-Nabisco, the largest leveraged buyout ever.

Kellogg introduces Common Sense Oat Bran cereal.

Canada's Supreme Court rules January 28 that a law restricting abortion is unconstitutional.

The Reagan administration acts January 29 to bar most family planning clinics from providing abortion assistance if they receive federal funds.

The FDA gives approval May 23 to cervical cap contraceptives long available in Europe.

A federal jury finds G. D. Searle guilty in a case that involves testing and marketing the Copper-7 intrauterine contraceptive device. The jury awards plaintiffs $8.7 million.

France and China act September 24 to authorize use under medical supervision of the steroid drug RU486 (Mifepristone) which induces abortion in the first months of pregnancy. Hoechst-Roussel, U.S. subsidiary of the West German maker Roussel-Uclaf, does not apply for FDA approval lest pro-life groups boycott the company's other products (*see* 1990).

Illegal U.S. immigrants flood agency offices prior to May 4, expiration date for the amnesty program set up under the 1986 Immigration Control and Reform Act.

1989 Soviet citizens gain rights and other eastern Europeans overthrow despots in spontaneous uprisings after Beijing cracks down on dissidents with a bloody massacre (*see* below).

Japan's Emperor Hirohito dies January 7 at age 87 after a 62-year reign. The Showa emperor is succeeded on the "Chrysanthemum Throne" by his son Akihito, 55, who will reign as the Heisei emperor.

Japan's dominant political party loses at the polls in July following the resignation of a prime minister in one scandal and the tainting of his successor in a scandal involving a geisha.

Cuban troops begin pulling out of Angola January 10 pursuant to a December agreement.

Soviet troops complete their withdrawal from Afghanistan in February.

Soviet voters elect opposition candidates in March to the Presidium of the Supreme Soviet, a newly reconstituted parliament. Boris Yeltsin wins a landslide victory and warns that Mikhail Gorbachev is gaining too much power. Yeltsin visits America in September and says that if Gorbachev does not show more progress within a year he will face revolt, as demonstrated by widespread strikes (*see* below) that have already crippled production in some areas.

Poland ends 40 years of strict Communist rule August 18. Party candidates have been roundly defeated in June parliamentary elections. A new cabinet headed by *Tygodnik Solidarnosc* editor Tadeusz Mazowiecki, 62, takes over with support from Lech Walesa and Roman Catholic Primate Jozef Cardinal Glemp, but Communists retain the interior and defense ministries. Walesa visits the United States in November and receives a hero's welcome.

Lithuania, Latvia, and Estonia demand autonomy; Moscow admits to secret protocols in the 1939 Stalin-Molotov pact under which the U.S.S.R. was to annex the then-independent Baltic republics. Hundreds of thousands of Latvians, Lithuanians, and Estonians join hands August 23 in a human chain stretching across the three republics. Lithuania acts December 7 to change her constitution, ending the guarantee of Communist party domination (*see* 1990).

Armenians, Azerbaijanis, Georgians, and Ukrainians agitate for autonomy as ethnic divisions threaten to dismember the Soviet Union.

Hungary permits thousands of East German "holiday visitors" to cross her frontier into Austria (and thence to West Germany) in September despite a 1969 treaty in which she agreed to prevent any such exodus. The German Democratic Republic allows East German visitors in Czechoslovakia to leave for West Germany in October.

East Germany (the GDR) celebrates her fortieth anniversary in October with a 2-day visit by Mikhail Gorbachev but arrests demonstrators after Gorbachev leaves October 7. More than 170,000 East Germans emigrate to the West. President Erich Honecker, now 77, issues live ammunition to his forces at Leipzig but is forced to resign October 18. He is succeeded by his security chief and protégé Egon Krenz, 52, a hardline Stalinist like Honecker who makes some conciliatory moves but says there will be no sharing of power with pro-democracy groups. GDR authorities permit citizens to exit without visas November 9, joyful East Germans by the millions seize the opportunity to visit the West, and demolition begins of the Berlin Wall erected in 1961. Revelations of corruption force Krenz to resign in early December (Honecker has lived lavishly and his labor minister has kept a 5,000-acre estate on the Baltic with a large staff of servants and groundskeepers). Berlin's Brandenburg Gate is opened December 22 and the city reunites (*see* 1990).

Hungary's ruling Communist party renames itself the Socialist party and the country's parliament adopts a quasi-democratic constitution October 18. The country proclaims itself a democratic republic October 23 and plans multi-party elections.

Soviet Foreign Minister Eduard A. Shevardnadze tells the legislature October 23 that the country's invasion of Afghanistan was illegal and that a radar station near Krasnoyarsk in Siberia was a violation of the Anti-Ballistic Missile Agreement with the United States. Moscow agreed in September to dismantle the facility. President Bush meets with President Gorbachev off Malta in early December.

West Germany's ultra-leftist Red Army Faction kills Deutsche Bank chief executive Alfred Herrhausen, 59, at Bad Homburg December 1. A bomb blows up his armored, chauffeur-driven Mercedes-Benz 500SE.

Bulgaria's president and party leader Todor I. Zhivkov, now 78, resigns November 10 after 35 years in power. He is replaced as party secretary by Foreign Minister Petar T. Mladenov, 53, who says "there is no alternative to restructuring" the nation's economy and tightly-controlled political apparatus. A pro-democracy rally at Sofia December 10 brings out 50,000 people demanding that the constitution be changed to eliminate Communist monopoly on power (*see* Czechoslovakia, Romania, below).

Vietnamese troops leave Cambodia (the last ones exit September 26) after nearly 11 years of occupation. Civil war ensues as the Khmer Rouge tries to regain control.

Iran's Ayatollah Khomeini dies June 4 at age 86 (89 by some accounts). Hashemi Rafsanjani takes office as president.

Paraguay's dictator Alfredo Stroessner, now 76, is overthrown in a bloody coup February 2 to 3 after a brutal 35-year "presidency." Gen. Andres Rodriguez, 64, makes himself president, wins office in the first multi-candidate election since 1958, and promises to relinquish power to an elected civilian president in 1993. Rodriguez has made himself one of the richest men in Latin America on a $400-per-month salary but denies that he has trafficked in drugs.

Argentine voters elect Perónist leader Carlos Saul Menem, 58, president May 14, the first peaceful transfer of power in the country since 1927. Menem vows to privatize about 25 industries nationalized in the Perón years, reform the tax system, and stop Argentina's hyperinflation (below).

El Salvador's 10-year-old civil war sees its most severe rebel attack since 1981 as the Farabundo Martin National Liberation Front mounts a "final offensive" November 11 but fails in its attempt to kill President Alfredo Cristiani or Vice President Francisco Merino. More than 70,000 have died and thousands been maimed since 1979, and the murder November 16 by government forces of six Jesuit priests brings demands in Washington that Congress stop supporting the Christiani regime at a cost of nearly $1 million per day.

Panama's voters oust strongman Manuel Antonio Noriega in free elections May 7, Noriega ignores the election results and retains power, he quells an attempted military coup October 3. The Bush administration comes under fire for not giving the rebels more support. Noriega's National Assembly declares war

1989 ***(cont.)*** December 15 and formalizes Noriega's position as head of state, an unarmed U.S. army officer is killed December 16, a Navy officer and his wife are harrassed, and airborne U.S. troops invade Panama December 20, offering $1 million reward for information leading to Noriega's arrest. As many as 4,000 Panamanian civilians are killed (Washington says 202), 23 U.S. servicemen. Other Latin countries express outrage at the U.S. invasion; the UN General Assembly denounces it as a "flagrant violation of international law." Noriega eludes capture, turns himself in to Vatican authorities in Panama City December 24, and receives political asylum for 10 days before surrendering to U.S. authorities for trial at Miami on drug charges.

Chile's brutal 16-year Pinochet regime nears its end December 14 as voters elect coalition candidate Patricio Aylwin, 71, president in a return to democratic tradition. General Pinochet remains military chief of staff.

Brazil holds her first democratic elections in 29 years. Obscure state governor Fernando Collor de Mello, 40, a former model, wins the presidency December 17 by inveighing against the nation's *maharadjahs*—overpaid, underworked civil servants.

Lebanon's Muslim and Christian factions reach an accord in October at Taif, Saudi Arabia. Christian Maronite René Moawad, 64, is elected president November 5 at a special session of Parliament, but President Moawad is assassinated with 23 others in a Beirut bombing November 22. Moawad is replaced by Catholic Maronite Elias Hrawi, 64, but Christian army commander Gen. Michel Aoun considers himself the legitimate president and begins an 11-month rebellion.

India's Congress (I) party loses power in December elections. Rajiv Gandhi is replaced as prime minister by his former minister of finance and defense Vishwanath Pratap Singh, 58, who has launched a crusade against corruption.

China's Deng Xiaoping, now 85, resigns his last political post November 9 and pledges not to meddle in politics. He has been succeeded as chairman of the party's military commission in June by Jiang Zemin, 63.

Chinese Politburo member Hu Yaobang dies April 15 at age 73, university students gather in Beijing's Tiananmen Square ostensibly to mourn Hu's death (he was forced to resign as General Secretary in January 1987 by hard-liners for not cracking down on student unrest) but actually to demand more democracy and demonstrate against the abuses of corrupt government officials. They remain in the square night and day for weeks. Students in at least six other cities demand political reform and the resignation of Premier Li Peng. Troops sent to clear Tiananmen Square are won over by the students until June 6, when Deng Xiaoping sends in young Mongolian soldiers who fire into the crowd with AK-47 assault rifles, killing hundreds if not thousands. Leaders of the democracy movement are executed despite appeals from Western powers for leniency. Congress imposes sanctions against Beijing but President Bush secretly sends an emissary in July to meet with China's political leaders; he acts in December to veto a bill that would extend the visas of some 40,000 Chinese students in the United States and waives some congressional sanctions.

Soviet coal miners strike in the Ukraine's Donets Basin and other Soviet coalfields in the biggest Soviet industrial walkout since the 1920s. Moscow quickly accedes to worker demands, and while independent miners take steps in September to seize control of the official coal workers' union, ousting some officials, the All-Union Council of Trade Unions, dominated by the Communist party, retains control. Laws are passed forbidding strikes in key industries, but some coal workers strike again in late October, saying Moscow has failed to make good on its promises.

Czech authorities crush a demonstration October 28 and arrest leading dissidents, including playwright and Charter 77 founder Vaclav Havel who have led chants of "Freedom!" and "We want democracy!" Mikhail Gorbachev urges Czechoslovakia to respond to the need for change, officials announce November 14 that Czechs will be permitted free travel to the West, but Prague police beat student demonstrators November 17. Huge demonstrations follow in Prague's Wenceslas Square and in other major cities demanding the resignation of Communist Party General Secretary Miloš Jakes, now 67. Former party leader Alexander Dubcek, now nearly 68, speaks out for the first time since the suppression of the "Prague Spring" in 1968. Jakes is replaced November 24, but Czechs, unappeased, demand more rights. President Gustav Husak resigns December 10, a new cabinet with a Communist minority is installed in what some call a "velvet revolution," the parliament votes December 19 to move toward Western-style democracy, and on December 29 it elects playwright and longtime dissident Vaclav Havel, 53, president, making Dubcek chairman of parliament.

Beijing students demonstrated in Tiananmen Square until Chinese authorities finally crushed their movement.

Romania's president Nicolae Ceausescu, now 71, has praised Beijing's action in Tiananmen Square (above) and continues to suppress dissent at home, maintaining a Stalinist hard line even though his country suffers the worst food and fuel shortages in Eastern Europe. Demonstrators in Timisoara surround a church to prevent Ceausescu's *Securitate* (secret police) from arresting a clergyman who has supported rights of ethnic Hungarians in Transylvania, the *Securitate* shoots down protestors by the thousands beginning the night of December 16, and more are shot in Bucharest demonstrations December 21. Army personnel quickly go over to the side of the demonstrators and Ceausescu is ousted after 24 years in power. His *Securitate*, which outnumbers the military 2 to 1, battles with army units, but Ceausescu and his wife Elena are captured December 22 and executed by a firing squad December 25 after a military court has convicted them of "genocide" and plundering more than $1 billion from the state. Onetime official Ion Iliescu, 59, heads a new, provisional government.

Soviet scientist, congressman, and civil rights champion Andrei D. Sakharov dies December 14 at age 68, hours after warning fellow deputies that the U.S.S.R. is headed for catastrophe.

South Africa's President Pieter W. Botha, 73, suffers a mild stroke January 18 and resigns August 15. He is succeeded in September by F. (Frederick) W. de Klerk, 53, who permits anti-apartheid marches and pledges to make the government more representative. President de Klerk releases some leading political prisoners in mid-October and meets with members of the outlawed African National Congress (*see* Mandela, 1990).

A 6-to-3 Supreme Court ruling January 23 invalidates a Richmond, Virginia, program requiring that 30 percent of the city's public works funds be set aside for minority-owned construction firms. The court calls it reverse discrimination and says such programs can be justified only if they serve the "compelling state interests" of redressing "identified discrimination" (*City of Richmond v. J.A. Croson*).

The first elected black governor since Reconstruction wins office in Virginia November 7 and New York City elects its first black mayor. Former lieutenant governor L. Douglas Wilder, 58, is elected governor, former Manhattan borough president David N. Dinkins, 63, mayor.

The Internment Compensation Act signed by President Bush in November awards $20,000 to each Japanese-American surviving victim of President Roosevelt's infamous February 1942 Executive Order 9066.

Poland suffers hyperinflation as prices escalate by 600 percent. The zloty is devalued at least 12 times, Parliament acts October 16 to compensate workers and farmers for rising prices but critics contend that this will only put more pressure on prices.

East Germany (above) has the Eastern bloc's strongest economy, but corruption is rife and her gross national product is less than one-fourth that of West Germany.

Yugoslavia suffers hyperinflation as prices increase at the rate of 10,000 percent annually.

Argentina suffers hyperinflation, with prices doubling and even tripling from month to month. The nation has less than $100 million in foreign reserves in April and $60 billion in foreign debts, her nationalized oil, telephone, and transportation companies are losing millions, only 31,000 people pay taxes, and her currency has depreciated to 6.5 billion to the dollar, down from about 1,750 in 1980. President Menem brings businessmen and free-market economists into his government, drastically lowers the inflation rate, cracks down on tax evaders, and builds up foreign reserves to $2 billion.

U.S. banks write off billions of dollars in uncollectible Latin loans.

The Financial Institutions Rescue, Recovery and Enforcement Act signed by President Bush August 9 "bails out" the nation's federally insured savings and loan associations at taxpayers' expense, but it does not address crucial issues of deposit insurance, will fail in its purpose, and will inadvertently jeopardize commercial banks. The government sells many S&Ls to private investors and banks at bargain prices.

Wall Street's Dow Jones Industrial Average drops 190.58 points Friday, October 13, as confidence weakens in "junk" bond financing of mergers and acquisitions, but prices rebound the following week and the Dow closes the year at 2753, up 584.63.

Abraham & Straus opens on New York's Greeley Square in mid-September, replacing Gimbels (*see* 1893, 1910) but Campeau Corporation, its owner, battles creditors and puts the 17-store Bloomingdale's chain up for sale.

The British conglomerate B.A.T. Industries announces in October that it will sell its U.S. retailing operations, which include 46 Saks Fifth Avenue, 24 Marshall Field's, 17 Breuner's, and 23 Ivey's stores.

A chartered Boeing 707 crashes February 8 in the Azores killing 137 Italian tourists and 7 U.S. crew members en route from Bergamo, Italy, to the Dominican Republic and Jamaica; a cargo door rips away from a United Airlines Boeing 747 out of Honolulu February 24 and 9 people aboard are sucked out; a United Airlines DC-10 crashes into a cornfield near Sioux City, Iowa, July 19, killing 112 of the 296 aboard; a terrorist bomb blows up a French DC-10 en route from Chad to Paris September 19 and it crashes in Niger, killing all 171 aboard.

New York real estate developer Donald Trump acquires the Eastern Airlines shuttle and renames it the Trump Shuttle.

The Baikal-Amur Mainline, a new trans-Siberia railroad, opens in September (5 years behind schedule) 200 miles north of the original line opened in 1904. Crossing 5 mountain ranges and 17 rivers with 3,000 bridge and 4 major tunnels (the long Svero-Muisky tunnel at the western end remains unfinished and is bypassed), the costly rail link was intended to serve

1989 *(cont.)* industries to be established on its route, but development of the mineral-rich region has been delayed.

France's state-owned railway begins service September 20 on the T.G.V. (Train à Grande Vitesse) Atlantique, carrying passengers from Paris to Le Mans at speeds up to 186 miles per hour (T.G.V. trains between Paris and Lyons have run at 168 m.p.h. since 1983). Service to Tours in the Loire Valley and points south will begin next year.

Canada announces October 4 that passenger rail service will be cut in mid-January from 405 trains per week to 191 as the nation tries to reduce a national deficit that is 50 percent higher per capita than its southern neighbor's. Prime Minister Mulroney says the heavily subsidized service in some cases costs the government more than $400 every time a passenger boards a Via Rail train (*see* 1977).

Ford Motor Company acquires Jaguar Motors for $2.5 billion.

General Motors aquires half of Sweden's Saab.

Bristol-Meyers merges with Squibb to create a pharmaceutical giant second in size only to Merck.

Burning the American flag in public to protest goverment policies is a right protected by the First Amendment, the U.S. Supreme Court rules in a 5 to 4 decision handed down June 21. President Bush asks for a constitutional amendment to prohibit flag burning. Some patriots support the court's decision, others favor a law or amendment to countermand it. Congress passes a law in October but the Senate rejects a constitutional amendment.

The *Los Angeles Herald Examiner* ceases publication November 2 after 86 years.

Iran's Ayatollah Khomeini (above) has offered a $3 million reward for the death of author Salman Rushdie (*see* Fiction, 1988). British authorities protect Rushdie and break relations with Iran.

Nonfiction *From Beirut to Jerusalem* by New York Times reporter Thomas L. Friedman, 36; *Citizens: A Chronicle of the French Revolution* by Harvard history professor Simon Schama, 44.

Fiction *The Joy Luck Club* by U.S. novelist Amy Tan, 37; *The Remains of the Day* by Japanese-English novelist Kazuo Ishiguro, 34; *Spartina* by U.S. novelist John Casey, 50; *Jasmine* by Bharati Mukherjee; *Oldest Living Confederate Widow Tells All* by U.S. novelist Allan Gurganus, 42; *Billy Bathgate* by E. L. Doctorow; *Killshot* by Elmore Leonard.

Historical Portraits by Cindy Sherman is a photographic series. Salvador Dali dies January 23 in Spain at age 84; Hans Hartung dies at Antibes, France, December 7 at age 85.

Sculptor Richard Serra's 1981 work *Tilted Arc* is removed from New York's Federal Plaza March 15 amidst much controversy.

The Louvre Museum in Paris opens a new metal-and-glass pyramid entrance by I. M. Pei March 30.

Washington's Corcoran Gallery announces June 12 that it has canceled an exhibition of work by the late U.S. photographer Robert Mapplethorpe who died last year of AIDS. A few homoerotic pictures are included in the show and Sen. Jesse Helms (R. N.C.) introduces legislation that would bar the National Endowment for the Arts from funding "obscene" work (pieces by U.S. artist André Serrano, 39, have also aroused his ire). Congress votes in September to establish a panel that will evaluate standards for judging if art is obscene.

Theater *Another Time* by Ronald Harwood 9/25 at Wyndham's Theatre, London, with Albert Finney; *Love Letters* by A. R. Gurney 10/31 at New York's Edison Theater with a rotating cast (Jason Robards, Colleen Dewhurst, Swoosie Kurtz, Richard Thomas); *Shadowlands* by English playwright William Nicholson 10/23 at the Queen's Theatre, London, with Nigel Hawthorne as C. S. Lewis, Jane Lapotaire as U.S. poet Joy Davidman Gresham; *A Few Good Men* by U.S. playwright Aaron Sorkin, 28, 11/15 at New York's Music Box Theater, with Tom Hulce, Roxanne Hart; *My Children, My Africa* by Athol Fugard 12/18 at New York's off-off-Broadway Perry Street Theater, with John Kani, Lisa Fugard, Courtney B. Vance.

Films Edward Zwick's *Glory* with Denzel Washington, Morgan Freeman, Matthew Broderick as Civil War colonel Robert Gould Shaw. Also: Lawrence Kasdan's *The Accidental Tourist* with William Hurt, Kathleen Turner; Terry Gilliam's *The Adventures of Baron Munchhausen* with John Neville, Eric Idle, Oliver Reed; Oliver Stone's *Born on the Fourth of July* with Tom Cruise as Vietnam War veteran Ron Kovik; Giuseppe Tornatore's *Cinema Paradiso* with Philippe Noiret; Woody Allen's *Crimes and Misdemeanors* with Allen, Martin Landau, Mia Farrow, Anjelica Huston; Spike Lee's *Do the Right Thing* with Lee, Danny Aiello, Ossie Davis; Bruce Beresford's *Driving Miss Daisy* with Jessica Tandy, Morgan Freeman; Gus Van Sant's *Drugstore Cowboy* with Matt Dillon, Kelly Lynch; Paul Mazursky's *Enemies, A Love Story* with Anjelica Huston, Ron Silver; Kenneth Branagh's *Henry V* with Branagh; Patrice Leconte's *Monsieur Hire* with Michel Blanc, Sandrine Bonnaire; Jim Sheridan's *My Left Foot* with Daniel Day Lewis; Ron Howard's *Parenthood* with Steve Martin, Mary Steenburgen; Jon Amiel's *Queen of Hearts* with Anita Zagaria, Joseph Long; Hiroshi Teshigahara's *Rikyu* with Rentaro Mikuni, Tsutomo Yamazaki; Michael Moore's *Roger and Me* with Moore; Harold Becker's *Sea of Love* with Al Pacino, Ellen Barkin; Nigel Noble's *Voices of Sarafina!* with Miram Makeba.

Film musicals Walt Disney's *The Little Mermaid* with music and lyrics by Howard Ashman and Alan Menken; Terence Davies's *Distant Voices, Still Lives* with Dean Williams, Pete Postelthwaite, songs from the 1940s and '50s.

Stage musicals *Miss Saigon* 9/20 at London's Theatre Royal Drury Lane, with Jonathan Pryce, Lea Salonga, book and music by Alain Boubil and Claude-Michel Schönberg, lyrics by Richard Maltby, Jr.; *Buddy* 10/12 at London's Victoria Palace Theatre, with Paul Hipp as

Buddy Holly, songs by the late rock 'n' roll pioneer; *Dangerous Games* 10/19 at New York's Nederlander Theater, with tango music by Astor Piazzolla, lyrics by William Finn; *Grand Hotel* 11/12 at New York's Martin Beck Theater, with Karen Akers, Gerrit de Beer, David Carroll, music and lyrics by Robert Wright and George Forest; *City of Angels* 12/7 at New York's Virginia Theater, with James Naughton, Gregg Edelman, music by Cy Coleman, lyrics by David Zippel, 35.

Dallas's $110 million Morton H. Meyerson Symphony Hall, designed by I. M. Pei, opens September 8 with a concert featuring pianist Van Cliburn. Contributor H. Ross Perot has stipulated that the hall be named for the former president of Electronic Data Systems.

Popular songs *Don't Be Cruel* (album) by Bobby Brown; *Hangin' Tough* (album) by New Kids on the Block (a quintet from Boston); *Forever Your Girl* (album) by Paula Abdul; *Appetite for Destruction* and *G N' R Lies* (albums) by Guns N' Roses; *The Raw and the Cooked* (album) by Fine Young Cannibals; *The Traveling Wilburys* (album) by The Traveling Wilburys; *Hysteria* by Def Leppard; *Girl You Know It's True* (album) by Milli Vanilli; "Look Away" by Chicago; "My Prerogative" by Bobby Brown; "Every Rose Has Its Thorn" by Poison; "Straight Up" and "Fool Hearted" by Paula Abdul.

San Francisco beats Cincinnati 20 to 16 at Miami January 22 in Super Bowl XXIII.

Boris Becker and Steffi Graf win the British and U.S. Open singles titles.

Cincinnati Reds player-manager Pete Rose is suspended for life from baseball when evidence is produced that he has wagered on baseball games.

The Oakland Athletics win the World Series, defeating the San Francisco Giants 4 games to 0 after an earthquake (below) causes an 11-day postponement of Game 3.

Former Drexel Burnham Lambert "junk" bond guru Michael Milken, now 42, his brother Lowell, 40, and a 31-year-old former colleague are indicted in March on 98 counts of conspiracy, stock manipulation, racketeering, and securities fraud. The government estimates that Milken's salary and bonuses for 1984–1986 came to $554 million, and that he was paid $550 million in 1987. Prominent business leaders rise to Milken's defense (*see* 1990).

A 29-year-old investment banker jogging in New York's Central Park is raped and left for dead April 19 by a band of black and Hispanic youths. Her attackers are caught and indicted.

Four black New York youths visiting a Bensonhurst (Brooklyn) neighborhood August 23 to look at a used car are attacked by seven white youths; Yusef K. Hawkins, 16, is shot dead in a racial incident that plays a role in the mayoral election contest.

New York hotel operator Leona Helmsley, 69, is convicted August 30 on 33 counts of income tax evasion and massive tax fraud; she is sentenced to 4 years in prison and fined $7.1 million. A former housekeeper has testified that Mrs. Helmsley told her, "Only the little people pay taxes."

Former PTL minister Jim Bakker, now 49, is convicted October 5 on 24 counts of fraud and conspiracy after revelations that he has bilked followers out of $158 million and diverted nearly $4 million to supply himself with mansions, an air-conditioned doghouse, and fleets of Mercedes and Rolls-Royce automobiles. Bakker is sentenced to 45 years in prison and a $500,000 fine.

Cuba's Fidel Castro conducts a show trial of military hero Gen. Arnaldo Ochoa and others on charges of corruption and drug trafficking. Ochoa is executed July 13. The Castro government has reportedly been involved for years in the drug trade.

Colombian presidential candidate Luis Carlos Galan is assassinated August 18, President Virgilio Barco Vargas declares war on the Medellín and Cali drug cartels that have been responsible for murdering scores of judges, government officials, and newspaper editors. Washington sends military hardware, much of it inappropriate.

President Bush names former Secretary of Education William J. Bennett Director of National Drug Control Policy and uses his first presidential TV address September 5 to announce a war on drugs.

Colombian police make nearly 500 arrests, nine suspects are extradited to the United States, and $250 million worth of property is confiscated, along with drugs and weapons. Drug "kingpins" strike back with 265 bombings that kill 187 civilians and officials, they cause more than $500 million in property damage, but billionaire drug trafficker José Gonazalo Rodriguez Gacha, 42, is shot down by Colombian police December 15 along with his son of 17 and 15 bodyguards in a rural area near Cartagena. Drug trafficker Pablo Escobar Gaviria, 39, remains at large, as do three brothers of Medellín's Ochoa family (*see* 1991). Escobar and Rodriguez Gacha have been suspected of ordering the November bombing of an Avianca plane that exploded in mid-air, killing all 107 aboard.

New York's 47-story Morgan Building, designed by Kevin Roche, is completed at 60 Wall Street.

Hong Kong's 70-story Bank of China Building, designed by I. M. Pei and the tallest structure in the crown colony, is completed to house Beijing's main overseas branch.

Warsaw's 41-story Marriott Hotel is completed with 520 guest rooms, office space, a shopping mall, and a swimming pool.

The Ivory Coast's air-conditioned Basilica of Our Lady of Peace is completed in Yamoussoukro, 135 miles from Abidjan on the coast, for President Houphouët Boigny, now 83, whose nation of 10 million has about 1 million Christians. Designed by architect Pierre Fakhoury and modeled after St. Peter's in Rome, it is the tallest church in Christendom (its dome rises 525 feet in the air), and its 7.4-acre granite and marble plaza can accommodate 300,000 pilgrims.

1989 ***(cont.)*** The tanker *Exxon Valdez* takes on 1.26 million barrels of crude oil at Valdez, southern terminus of the Alaska Pipeline, and runs aground March 24 on Bligh Reef, releasing 240,000 barrels of oil into Prince William Sound, an area rich in otters, whales, porpoises, seafowl, and fish. Exxon sacks the skipper for drinking on the job after the worst U.S. tanker spill thus far.

An explosion and fire aboard the Iranian tanker *Khark 5* off Morocco December 19 cause leakage of a quarter of the ship's cargo—20 million gallons of crude oil—in the Atlantic.

Africa's elephant population falls to 625,000 or less, down from 1.2 million in 1981, as illegal trade in ivory flourishes despite restrictions.

Brazil responds to world environmentalist opinion and suspends (but does not end) tax incentives that have favored land clearance in the Amazon jungle (*see* 1987). A low-interest World Bank loan of $8 million plus $8 million appropriated by the Brazilian congress, provides funds to hire forestry agents, rent helicopters, and buy trucks, but a helicopter sent to investigate an illegal forest fire in Pará state is fired upon, and hired gunmen kill a forestry agent. Landowners who burn trees without permits are fined some $10 million by agents of the new Institute of Environment and Renewable Natural Resources. August rains put an early end to burning but chain saws continue to fell trees, legally and illegally. Less than 88 percent of Brazil's forests are gone by year's end, down from 99 percent in 1975.

Arab militants burn an estimated 250,000 trees in Israel's Carmel National Park September 20, blackening 2,000 acres in the heart of the 20,000-acre natural pine and oak forest.

Hurricane Hugo in mid-September slams into the Virgin Islands, Puerto Rico, and the Carolinas, killing more than 70 all told, leaving hundreds of thousands homeless, and wreaking havoc on trees and buildings. Cost: over $4 billion.

New York, New Jersey, and the eight New England states adopt air pollution standards matching those in California (*see* 1977). California announces a phasing in of draconian measures that will outlaw gas-powered lawnmowers and outdoor barbecues and require many vehicles to run on alternate fuels (*see* 1990). Some European countries begin to follow the U.S. example.

San Francisco trembles October 17 in North America's most destructive earthquake since 1906. Measuring 7.1 on the Richter scale (less than 1/10th the 8.3 of 1906), the 15-second tremor at 5:04 in the afternoon buckles highways and the Bay Bridge, kills an estimated 90 people (most of them crushed in cars when the upper level of the Nimitz Highway collapses), and causes at least $6 billion in property damage.

Moscow offers Soviet wheat growers hard currency incentives if they exceed certain production goal but continues to buy millions of tons of U.S. and other foreign grain.

The U.S. Supreme Court returns the issue of abortion to the political arena, ruling July 3 in *Webster v. Reproductive Health Services* that states can limit access to abortion. Right-to-life groups hail the 5 to 4 decision, pro-choice groups anticipate further erosion of the 1973 *Roe v. Wade* decision, the Florida legislature votes October 12 to reject measures proposed by the governor to restrict abortion, but Pennsylvania adopts strict new laws on abortion. Congress votes in October to weaken the Hyde Amendment of 1976 by authorizing Medicare payment for abortions in victims of rape and incest; President Bush vetoes the measure and Congress fails to override.

Political candidates supported by right-to-life advocates all lose in the November elections.

Romania permits abortion for the first time in 24 years following the death of Nicolae Ceausescu (*see* above).

1990 Germany reunites and the U.S.S.R. crumbles as Iraqi aggression threatens to ignite a Mideast conflagration.

Iraqi forces invade Kuwait August 2 after Kuwait refuses demands by President Saddam Hussein that she pay compensation for allegedly drilling oil on Iraqi territory, cede disputed land, reduce oil output, and raise prices. The Bush administration has told Saddam Hussein that it would not take sides, but Washington, Moscow, Tokyo, London, Teheran, and Beijing unite in denouncing his move and the United Nations Security Council votes 13 to 0 August 6 to impose economic sanctions (Yemen and Cuba abstain). Iraq masses troops on the border of Saudi Arabia, Riyadh agrees to receive U.S. ground and air forces. President Bush says Iraq's aggression "will not stand" and dispatches forces to Saudi Arabia August 7, risking his presidency; Iraq annexes Kuwait August 8 and proceeds to loot the country; Egypt, Syria, Morocco, and nine other Arab states vote August 10 to oppose Iraq with military force; Saddam Hussein calls for a "holy war" against Westerners and Zionists, gaining wide popular support among Arabs; he holds more than 10,000 foreigners hostage beginning August 18 but permits women and children to leave August 29 and releases all the others by early December as the standoff continues. Bush ups the ante November 8 (2 days after the elections), committing far more U.S. forces to "Operation Desert Shield," but popular opposition grows to launching any offensive action.

Kuwait has rebuffed Iraqi demands that it forgive $15 billion in loans extended during the Iraq-Iran war (*see* oil, below). Saddam Hussein has received assurances from a U.S. diplomat in July that Washington has no treaty obligation to defend Kuwait.

The United Nations Security Council votes November 29 to authorize members to use all necessary force to expel Iraqi forces from Kuwait if they remain there after January 15, the first such resolution since the Korean conflict in 1950. President Bush reverses his position November 30 and agrees to talks with Saddam Hussein and his foreign minister.

Kuwait's billionaire emir Sheik Jaber al-Ahmed al-Sabah, 64, has narrowly escaped capture and fled to

Saudi Arabia; he addresses the United Nations General Assembly September 27, urging it to stand by the sanctions it has imposed. His relatives have acted swiftly to keep Kuwaiti funds abroad out of Saddam Hussein's hands.

The Republic of Yemen created May 23 has united the People's Democratic Republic of Yemen in the south with the Yemen Arab Republic in the north after more than 400 years of separation. President Ali Abdullah Saleh, 48, casts his lot with Iraq (above), denouncing Western sanctions and military threats, but then bars Iraqi ships from unloading.

Lebanon's 15-year-old civil war ends in the fall with the surrender of Christian forces led by Gen. Michel Aoun (the allied embargo against Iraq, above, has cut off his supply of arms) After some weeks of murders to settle old scores, Syria begins to withdraw her militia and Beirut's barricades come down.

Pakistan's President Ghulam Ishaq Khan dismisses Prime Minister Benazir Bhutto August 6, dissolves the National Assembly, and declares a state of emergency, saying that the Bhutto government is corrupt and inefficient.

Soviet leaders have agreed February 7 to surrender the Communist Party's 72-year monopoly on power. The party's governing Central Committee ends a stormy 3-day meeting with a strong endorsement of President Gorbachev's proposal for political pluralism. Gorbachev critic Boris Yeltsin is elected president of the Russian Republic in May; he quits the party in July, followed by the mayors of Moscow and Leningrad. Gorbachev asks for special powers November 17 as the Soviet economy collapses, he is granted the powers despite fears of a new dictatorship, the liberal minister of the interior is replaced by a KGB officer, and Foreign Minister Eduard A. Shevardnadze announces his resignation December 20, warning the Congress of the People's Deputies against "reactionaries." The Parliament shrugs off Shevardnadze's warning and votes December 25 to give Gorbachev almost dictatorial powers, including powers over the 15 republics.

Lithuania's Parliament has voted March 11 to secede from the U.S.S.R. President Gorbachev denounces the move, Soviet tanks move into Vilnius, Gorbachev says there will be no Tiananmen Square, but his troops take over buildings and in April he cuts off oil and other supplies in an effort to force Lithuania to rescind her declaration of independence. President Bush comes under pressure for backing Gorbachev instead of Lithuania. Lithuanian youths resist conscription into the Soviet army, as do youths in Armenia, Georgia, and other rebellious Soviet republics.

East Germans vote March 18 in the first free elections since 1932, approving a parliament that restores the borders of Saxony, Thuringia, Saxony-Anhalt, Mecklenburg, and Brandenburg, paving the way for reunification.

Germany reunites October 3 after 43 years of separation. A 3-day meeting at Ottawa has ended February 13 with an accord by Soviet, British, French, and U.S. foreign ministers on a framework for negotiating reunification. Foreign Ministers Shevardnadze and Baker agree to reduce Soviet and U.S. strength in Central Europe to 195,000 troops each, while permitting an additional 30,000 U.S. troops to be stationed in England, Portugal, Spain, Greece, and Turkey. President Gorbachev has met with President Kohl in July and agreed to permit membership of a unified Germany in NATO.

The Conventional Forces in Europe (CFE) treaty signed by 22 world leaders at Paris November 19 ends the "era of confrontation and division" that has followed World War II. NATO and Warsaw Pact countries agree to reduce weapons (Moscow will scrap 19,000 tanks, NATO 2,000), no one country may have more than one-third the total number of arms in a single category. But Yugoslavia verges on civil war between its component states and tensions persist elsewhere on the Continent (e.g., Moldavia, Romania, Hungary, Catalonia).

British Prime Minister Margaret Thatcher is forced out after 11½ years in office, the longest ministry of the century. She is replaced November 27 by her Chancellor of the Exchequer and hand-picked successor John Major, 47, son of a circus acrobat and youngest prime minister of this century.

Poland has her first free elections since before World War II. Voters give Solidarity leader Lech Walesa 40 percent of the presidential vote November 25 and repudiate Prime Minster Mazowiecki, who has eschewed demagogic promises, receives only 18.1 percent of the vote, and promptly resigns. Emigré entrepreneur Stanislaw Tyminski, 42, who returned to Poland in September after 21 years in Canada and Peru, wins 23.1 percent of the vote but loses to Walesa in a run-off December 9.

Romanian voters have elected Ion Iliescu president in June; the government of former Communists brings in armed miners to put down street demonstrations against the regime in Bucharest.

Bulgaria's Prime Minister Andrei Lukanov resigns November 29 after 9 months in office following 2 weeks of anti-Communist demonstrations by striking workers to protest the nation's economic disarray.

Nicaraguan voters defeat Sandinista leader Daniel Ortega Saavedra's bid for re-election February 25 and elect coalition leader Violetta Barrios de Chamorro, 60, to the presidency. Widow of *La Prensa* editor Antonio de Chamorro, who opposed the Samoza regime and was shot to death in 1978, the president-elect has no political experience.

Canada's Meech Lake Accord of 1987 expires June 23 as two provincial legislatures refuse to ratify it.

Haiti holds her first free election since 1957. Voters go to the polls December 18 and choose as president radical priest Jean-Bertrand Aristide, 37, who has denounced the country's military as brutal and corrupt.

Namibia has gained independence March 21 after a period of German colonial rule followed by 74 years of

1990 ***(cont.)*** South African rule. Former South-West Africa People's Organization (SWAPO) leader Sam Nujoma is president of the new republic, which begins life as a democracy.

Liberia's President Doe is killed September 9 after 10 years of U.S.-subsidized misrule as rival invading forces battle for control. Prince Yealu Johnson declares himself head of state pending elections but is opposed by Charles Taylor, 42, as a peacekeeping force sent into the country in August tries to restore order amidst tribal warfare. Backed by Libya's Muammar Qadafi, Taylor was a high-level official in Doe's administration who fled to the United States after being charged with embezzling up to $1 million in state funds. He escaped from a Massachusetts jail in 1985 while awaiting extradition and has gained popular support in 8 months since landing in northeastern Liberia with about 100 Libyan-trained troops. An estimated 400,000 Liberians have fled the country.

Singapore's prime minister Lee K. Yew, now 67, resigns November 26 after 31 years of strict rule that have seen the former colonial outpost transformed into a major metropolis. His hand-picked successor Goh Chok Tong, 49, takes over but will continue to take orders from Lee.

South African resistance leader Nelson Mandela, now 71, is released from Victor Verster Prison outside Cape Town February 11 after more than 27 years of incarceration on charges of high treason. President F. W. de Klerk asks Mandela to help negotiate a political settlement between whites and blacks. Mandela travels to Canada and the United States in June and addresses both the United Nations General Assembly and Congress. The African National Congress agrees August 7 to stop infiltrating trained guerrillas and weapons into South Africa, which agrees to begin a phased release of political prisoners and grant amnesty to some 20,000 ANC exiles, paving the way for negotiating a new constitution based on sharing of power with blacks. But supporters of Zulu prince Mangosuthu Buthelesi attack ANC strongholds in murderous acts of Zulu versus Xhosa tribal violence that disrupt the country.

Repeal of South Africa's Separate Amenities Act in October ends any legal basis for segregating community swimming pools, libraries, and other public places.

The U.S. Supreme Court rules 5 to 4 June 25 that a state may sustain the life of a comatose patient in the absence of "clear and convincing evidence" that the patient would have wanted treatment stopped (*see* Quinlan, 1985).

The Americans with Disabilities Act signed by President Bush July 26 bans discrimination in employment, public accommodations, transportation, and telecommunications against the nation's 43 million disabled persons. The law provides new protection for workers with AIDS.

Salvadoran leftist guerrillas and government negotiators agree July 26 to let a United Nations team verify and publicize human rights abuses in the event of a cease-fire. Some 40,000 civilians have allegedly been killed by the army and government death squads but killings have dropped dramatically in the past 6 years; no officer has ever been convicted of any rights abuse.

South Africa released black nationalist Nelson Mandela after 21 years of imprisonment and scrapped most apartheid laws.

The Supreme Soviet ends decades of religious repression September 26, forbidding government interference in religious activities and giving citizens the right to study religion in homes and private schools.

Israeli security forces open fire on stone-throwing Palestinians October 8, killing 17, wounding more than 140. The Palestinians have attacked worshipers at Jerusalem's Wailing Wall on the Temple Mount close to the Dome of the Rock and Al Aqsa mosques. Israel says the violence was inspired by Iraq's Saddam Hussein; the United Nations Security Council votes unanimously October 13 to support a U.S.-sponsored resolution condemning Israel's excessive use of force.

Polish free-market rules instituted January 1 create a glut of food and consumer goods but at prices few can afford. Warsaw University economist Leszek Balcerowicz, 43, and Harvard economics professor Jeffrey Sachs, 34, have been the leading architect of the "shock therapy," which Polish voters reject in the November elections and some other Eastern bloc nations adopt late in the year as they struggle to change from decades of state-run economies with artificial prices and wages.

The Soviet Parliament approves a property law March 6, voting 350 to 3 to give private citizens the right to own the means of production—or at least small factories and other business enterprises—for the first time since the early 1920s.

President Gorbachev comes under increasing attack as Soviet citizens try to cope with shortages. The Parliament gives him virtually free rein September 25 to decontrol the economy but Gorbachev moves cautiously and prices escalate.

East German Ostmarks have become convertible to West German marks July 1 as the prosperous Federal

Republic finances economic union with the ailing Democratic Republic. Most GDR enterprises fail by year's end, producing massive unemployment.

President Bush has conceded June 26 that "tax revenue increases" are needed along with spending cuts to reduce a projected $160-billion budget deficit—19 months after winning election on a pledge of "no new taxes." Congress rejects a budget reconciliation measure October 5, federal employees are furloughed briefly for lack of money, and a compromise tax bill is not signed until November. Some Republican losses in the polls are blamed on Bush's flip-flop and he then repeats his "no new taxes" pledge despite growing evidence that the "supply-side" economic experiment of the 1980s served only to pile up a massive national debt.

Wall Street's Dow Jones Industrial Average closes at 2999.75 July 16 and 17, falls to 2365.10 in October, and ends the year at 2633.66, down 4.3 percent since January 1. Tokyo's Nikkei Dow Jones Average loses 39 percent, having been down nearly 50 percent in late September.

America's record 8-year economic boom ends in July as the country goes into recession. Britain and France also slump, Germany and Japan remain economically robust.

The Federal Reserve Board acts September 20 to authorize J. P. Morgan & Co. to underwrite stocks, the first time a bank has had that power since the 1933 Glass-Steagall Act.

GATT (General Agreement on Tariffs and Trade) talks at Brussels collapse December 11 over the issue of farm subsidies. Farm products account for 14 percent of world trade.

Oil prices soar following Iraq's seizure of Kuwait (above). Iraq is second only to Saudi Arabia in oil reserves, and with Kuwait controls nearly 20 percent of world oil.

Iraq has blamed Kuwait for a fall in oil prices, Kuwait has rebuffed demands that she reduce pumping from a disputed oil field and discuss sharing the field with Baghdad.

Iraq's pipelines through Turkey and Saudi Arabia are shut down and the world boycotts Iraqi oil.

New York City transit fares rise January 1 to $1.15 but the subways are now graffiti-free and 94 percent of the 6,200 cars are air-conditioned.

An Indian Airlines A320 Airbus crashes February 14 at Bangalore killing 91 of the 146 aboard; a Chinese Boeing 737 is hijacked October 2 and crashes on landing at Canton into another 737, killing 120; an Aeroflot Ilyushin-62 crashes at Yakutsk November 21, killing all 176 aboard.

Smoking is banned on virtually all U.S. domestic flights February 25 by act of Congress.

The Saturn car introduced by General Motors in October challenges Japanese auto makers, who have taken one third of the U.S. market while GM's share has fallen from 45 percent to 35. Made in Spring Hill, Tenn., the three plastic-bodied Saturn models have no GM identification.

Muslim pilgrims visiting Mecca July 2 jam a pedestrian tunnel, the ventilating system fails, and 1,426 are trampled to death or die of suffocation in the ensuing stampede.

Tuition at Harvard, Yale, Stanford, and other top U.S. colleges tops $14,000 per year, total expenses exceed $20,000, but 80 percent of undergraduates attend public universities, where tuition averages less than $2,000 per year, another 16 percent go to private colleges where tuition is below $10,000, scarcely 4 percent pay more, and up to two-thirds of these receive scholarships, subsidized loans, or both.

Entertainment Weekly begins publication February 12. Time Warner spends $150 million to launch its first new magazine since *People* in 1974.

Nonfiction *The Politics of Rich and Poor* by Kevin Phillips; *The Japan That Can Say No* by Japanese pundit Shintaro Ishihara, 57, with Sony founder Akio Morita.

Fiction *Middle Passage* by U.S. novelist Charles Johnson, 42; *Solomon Gursky Was Here* by Mordecai Richler; *Vineland* by Thomas Pynchon; *Rabbit at Rest* by John Updike; *My Son's Story* by Nadine Gordimer; *The Burden of Proof* by Scott Turow; *Lucy* by Jamaica Kincaid; *Buffalo Girls* by Larry McMurtry.

Juvenile *Haroun and the Sea of Stories* by Salman Rushdie, who remains in hiding in Britain under an Iranian death threat.

Poetry *Omeros* by St. Lucia-born poet Derek Walcott, 60.

Theater *Racing Demon* by David Hare 2/8 at London's Cottesloe Theatre, with Michael Bryant, Richard Pasco; *Man of the Moment* by Alan Ayckbourn 2/14 at London's Globe Theatre, with Michael Gambon, Peter Bowles; *Dancing at Lughnasa* by Brian Friel 4/24 at Dublin's Abbey Theater, with Gerard McSorky, Frances Tometty; *Prelude to a Kiss* by New York playwright Craig Lucas, 38, 5/1 at New York's Helen Hayes Theater, with Barnard Hughes, Mary-Lou Parker, Timothy Hutton is about AIDS, 440 perfs.; *Six Degrees of Separation* by John Guare 6/14 at New York's Mitzi E. Newhouse Theater, with James McDaniel, Stockard Channing, John Cunningham.

Films Stephen Frears's *The Grifters* with Anjelica Huston, John Cusack, Annette Bening; Fred Schepisi's *The Russia House* with Sean Connery, Michelle Pfeiffer. Also: Barbara Kopple's documentary *American Dream* about meat packers; Jean-Paul Rappeneau's *Cyrano de Bergerac* with Gerard Depardieu; Kevin Costner's *Dances with Wolves* with Costner; Jerry Zucker's *Ghost* with Patrick Swayze, Demi Moore, Whoopi Goldberg; Martin Scorsese's *Goodfellas* with Robert De Niro, Paul Sorvino, Joe Pesci; Franco Zeffirelli's *Hamlet* with Mel Gibson, Glenn Close, Alan Bates, Paul Scofield; Philip Kaufman's *Henry & June*

1990 ***(cont.)*** with Fred Ward, Maria de Medeiros; John McNaughton's *Henry: Portrait of a Serial Killer* with Michael Rooker as Henry Lee Lucas; Xavier Koller's documentary *Journey of Hope* about Kurds seeking emigration to Switzerland; Whit Stillman's *Metropolitan* with Edward Clements, Carolyn Farina; Joel Coen's *Miller's Crossing* with Gabriel Byrne, Albert Finney, John Turturro; James Ivory's *Mr. and Mrs. Bridge* with Paul Newman, Joanne Woodward; Bob Rafelson's *Mountains of the Moon* with Patrick Bergin as explorer Richard Burton, Iain Glen as John Speke; Michael Verhoeven's *The Nasty Girl* with Lena Stolze. Barbet Schroeder's *Reversal of Fortune* with Jeremy Irons, Glenn Close, Ron Silver.

Stage musicals *Once on This Island* 10/18 at the Booth Theater, with La Chanze, Jerry Dixon, music by Stephen Flaherty, book and lyrics by Lynn Ahrens, choreography by Graciela Daniele; *Five Guys Named Moe* (revue) 12/14 at London's Lyric Theatre, with U.S. actor Clarke Peters, 38, songs ("There Ain't Nobody Here But Us Chickens," "Messy Bessy,"etc.), written or popularized by jazzman Louis Jordan (1909–1975).

Popular songs *Time's Up* (album) by the rock group Living Colour; *Goo* (album) by the rock group Sonic Youth; *World Clique* (album) by the group Deee-Lite; *Listen Without Prejudice*, Vol. 1 (album) by George Michael; "Sooner or Later (I Always Get My Man)" by Stephen Sondheim (for the film *Dick Tracy*).

San Francisco beats Denver 55 to 10 at New Orleans January 28 in Super Bowl XXIV.

Journeyman U.S. boxer James "Buster" Douglas, 29, wins the world heavyweight crown February 10 at Tokyo, knocking out Mike Tyson in the 10th round. Douglas loses the title October 25 at Las Vegas to Evander Holyfield, 28.

Stefan Edberg wins in men's singles at Wimbledon, Martina Navratilova wins her ninth women's singles title (a record). Pete Sampras, 19 (U.S.), wins in U.S. Open men's singles, Gabriela Sabatini, 20 (Arg), in women's singles.

The Cincinnati Reds win the World Series, defeating the Oakland A's 4 games to 0.

West Germany wins in World Cup football (soccer), defeating Argentina 1 to 0 at Rome.

The Gillette Sensor razor, introduced in January after 10 years and $200 million of development, is the first significant mechanical advance in razor design in years. Laser-welded twin blades, mounted on springs, hug the face more closely than conventional blades ever did.

An illegal social club in the Bronx is set afire March 25, killing 87 in the worst New York conflagration since the Triangle Shirtwaist Factory fire of March 25, 1911. A Cuban immigrant is charged with 87 counts of murder.

"Junk bond" king Michael Milken pleads guilty to insider trading and is sentenced November 21 to 10 years in prison (*see* 1989). The sentence will be reduced to 3 years.

U.S. prisons have 1.3 million inmates (51 percent nonwhite), twice as many as in 1980 and more than in the Soviet Union or any other country, yet crime rates remain undiminished. It costs more per year to maintain a prisoner than to send him to Harvard but construction of jail cells continues at the expense of education, health care, and other budget items.

Washington's National Cathedral (Cathedral Church of St. Peter and St. Paul) is completed after 80 years of construction. The Gothic structure designed by Philip Hubert Frohman rises above Mount St. Albans.

Tokyo's 48-story twin-tower City Hall is completed by architect Kenzo Tange in the Shinjuku section.

Hurricane-force winds batter Britain and the Continent January 25, February 3, and February 26, uproot trees, overturn trucks, blow off roofs, kill more than 140, and cause about $1 billion in damage.

Earthquakes in northern Iran June 21 and 24 kill an estimated 50,000, injure 200,000, and leave 500,000 homeless.

The U.S. Fish and Wildlife Service acts June 22 to put the Northern spotted owl on the Endangered Species List but delays implementation of rules that would stop logging on federal land in the owl's Pacific Northwest habitat, areas that timber companies have been stripping of old-growth Douglas firs, redwoods, spruce, and hemlock—the oldest trees on earth. Woodsmen complain that a halt in tree-cutting will cost nearly 30,000 jobs in the next decade; environmentalists counter that destroying the last great American forests would end the jobs anyway. President Bush acts June 26 to delay any halt in logging on the federal lands.

President Bush breaks a long deadlock with Congress June 26 and makes sharp reductions in offshore acreage available for oil and gas drilling until at least the year 2000.

Iraq's Saddam Hussein annexed Kuwait, refused to yield, and brought the United Nations down on his head.

The Clean Air Act signed into law by President Bush November 15 phases in new tailpipe emission standards, requires special gasoline pump nozzles to reduce smog-related fumes in nearly 60 areas, requires auto makers to begin producing cars that will run on alternative fuels by 1995 and to install gauges that will alert drivers to problems in pollution-control equipment, mandates cleaner-burning, reformulated gasoline that will cut emissions of hydrocarbons and toxic pollutants, requires utilities to cut nitrogen-oxide emissions, etc.

U.S. and Soviet leaders oppose demands by environmental ministers meeting at Geneva in November that all nations burn less oil to avert global warning. The United States accounts for 24 percent of world carbon dioxide emissions.

Maine fishermen trap 28 million pounds of lobster, topping the 1889 record of 24 million pounds.

Bumper wheat crops in America, China, and the U.S.S.R. force prices down from $3.72 per bushel last year to $2.20.

A bumper Soviet potato crop rots in the fields amidst economic wrangles (above) and political charges and counter-charges. Moscow and Leningrad stores run out of bread and state food stocks fall so low as to raise a threat of famine.

Economic Community ministers agree November 7 to reduce farm subsidies and other barriers to agricultural trade, but collapse of the GATT talks in December (above) threatens to reduce the subsidy cuts by nearly half.

The Supreme Court rules 5 to 4 June 25 in *Hodgson v. Minnesota* that a state may require a pregnant minor to inform both her parents before having an abortion.

Roussel Uclaf expands marketing of its abortifacient drug RU-486 to Britain but political opposition blocks moves to market it in China, the U.S.S.R., Scandinavia, and the United States (*see* 1988).

The Norplant contraceptive approved by the FDA December 10 is the first really new birth-control measure since the Pill of the mid-1960s. Devised by Rockefeller Foundation researcher Sheldon J. Segal, 64, and already used in 14 other countries, it is surgically implanted under the skin of a woman's arm and slowly releases the female hormone progestin over a 5-year period.

Cities worldwide grow to unwieldy size. The Tokyo-Yokohama metropolitan area has 27 million people, Mexico City 23, São Paulo 18, Seoul 16, New York 14, Istanbul, Bombay, and Calcutta 12 each, Buenos Aires 11.5, Rio 11, Moscow and Los Angeles about 10 each, Cairo 9.8, Teheran 9.3, London 9, Paris 8.7.

1991 Soviet troops in Vilnius seize the local television station by force January 13, killing 15, wounding hundreds, and signaling a harsh new attitude by Moscow toward the republics. Boris N. Yeltsin wins easy election June 13 as president of the Russian Republic in the first democratic elections ever held in Russia. Leningraders vote to rename their city St. Petersburg.

A coup attempt by Communist hard liners August 19 ends August 23 after President Yeltsin in Moscow calls for a general strike to resist the coup. Some tank commanders support Yeltsin, and the coup leaders flee. Soviet President Gorbachev returns from brief detention in his Crimean summer home August 23 and suspends the Communist party August 24, ending 74 years of Communist rule.

President Gorbachev persuades the all-Soviet Congress to surrender power September 5. It has lost authority over the 15 constituent republics, the nation verges on political and economic collapse, and Gorbachev works with leaders of the republics to restore order, draft a new constitution, and create a new, non-Communist political order. He recognizes the independence of the Baltic republics September 6 as other republics gain autonomy. Gorbachev resigns December 25; Yeltsin becomes president.

Norway's Olaf V dies January 17 at age 87 after a 33-year reign. His son of 53 assumes the throne as Harald V.

Germany's Bundestag votes June 20 to move the nation's capital from Bonn to Berlin despite the high costs involved. Berlin is closer to East Germany, supporters of the move argue.

Croatia and Slovenia declare independence from Yugoslavia June 25 but no world power recognizes them. Serbo-Croat battles erupt, Belgrade sends in troops, the European Community tries to mediate.

U.S. and allied missiles and planes bomb targets in Iraq and Kuwait beginning at 3 o'clock in the morning of January 17. Congress has voted January 12 to approve legislation permitting President Bush to make war on Iraq if it does not withdraw from Kuwait by January 15 in accordance with UN resolutions (*see* 1990). Pilots fly more than 1,000 missions per day in the first weeks of the Persian Gulf War, dropping thousands of pounds of TNT with computer-guided accuracy in history's heaviest bombing, and meeting little resistance. U.S. and allied casualties are minimal.

Anti-war demonstrations ("No blood for oil") have been staged in American cities since fall and increase beginning the night of January 16 when news of the outbreak of hostilities reaches America. European cities also have peace demonstrations, but polls show Americans almost solidly united behind President Bush, who has

Communism died a natural death as the Union of Soviet Socialist Republics began to dissolve.

1991 ***(cont.)*** the highest approval rating since that enjoyed by Franklin Roosevelt in December 1941.

Turkey's parliament votes January 17 to let U.S. and allied planes use Turkish air bases for attacks on Iraq, but the war is costing the country billions in lost revenue and most Turks side with Iraq.

Iraqi missiles strike Tel Aviv and Haifa beginning January 18, causing little damage. Israel has refrained from taking any pre-emptive strike against Iraqi missile sites and does not retaliate lest it destroy the allied coalition. Washington sends in Patriot surface-to-air missiles manned by U.S. servicemen when these prove effective in destroying airborne Iraqi Scud missiles over Saudi Arabia.

Operation Desert Storm begins February 24 and ends in 100 hours with Iraqi forces defeated. President Bush has spurned the advice of Gen. Colin Powell, chairman of the Joint Chiefs of Staff, to give economic sanctions more time to work. U.S. Gen. H. Norman Schwarzkopf, 56, has planned the combined air and ground attack, sending 270,000 U.S., British, and French troops in a sweep around the Iraqi's western flank while air attacks sever the main highway route from Baghdad to Basra. More than 100,000 Iraqi troops surrender, at least 100,000 Iraqis are killed, but President Saddam Hussein remains in power. When Shiite and Kurdish forces rebel against his regime in some cities, he sends in helicopter gunships in an effort to suppress the rebels.

Israeli and Arab delegates confer together at Madrid October 30 and 31 under pressure from Washington and Moscow to begin settling their differences. Bilateral talks follow.

Thailand's military has staged a coup February 23, ousting Prime Minister Chatichai Choonhaven after 15 years in office and replacing him with Anand Panyarachun, 58, as interim prime minister until a new constitution is written and elections held.

Former Indian prime minister Rajiv Gandhi, now 46, is killed May 21 in a bombing southwest of Madras while campaigning for re-election. His death ends the dynasty begun by his grandfather Pandit Nehru in 1947 and leaves the country in turmoil.

Jiang Qing, widow of the late Mao Zedong, reportedly takes her own life in early June at age 77. Other members of her "Gang of Four"—Wang Hongwen, 56; Zhang Chunqiao, 74; and Yao Wenyuan, 60—remain in detention.

Somalia's president Mohammed Siad Barre flees his capital, Mogadishu, January 26 as rebels of the United Somali Congress end his 21-year dictatorship. Ali Mahdi Mohammed, the rebel leader, is sworn in as interim president January 29.

Haiti's President Aristide takes office in February, tries to control the military, but is ousted September 30 in a bloody coup led by Brig. Gen. Raoul Cedras, 43. Other nations cut off all but humanitarian aid to the new regime.

Los Angeles police brutalize an unarmed black motorist, March 3. The incident involving Rodney G. King, 25, is videotaped, the officers are indicted, but critics demand the dismissal of Police Chief Daryl F. Gates.

The Supreme Court rules 5 to 4 March 26 that using as evidence a confession extracted from a defendant by third-degree or other coercive methods can be "harmless error." A 6-to-3 decision April 16 places severe limitations on a state prisoner's right to raise constitutional claims in federal habeas corpus proceedings. The Court rules 5 to 4 June 17 that prison conditions such as overcrowding, poor sanitation, and exposure to violence do not violate the 8th Amendment's prohibition of cruel and unusual punishment unless prisoners can show that administrators have acted with "deliberate indifference" to human needs. And the Court rules 6 to 3 June 20 that elections for judges are covered by the federal Voting Rights Act, a decision that will put more black judges on benches in Southern states and counties.

Sexual harassment in the U.S. workplace is the focus of public attention in mid-October as University of Oklahoma Law School professor Anita Hill, 35, charges that Supreme Court nominee Clarence Thomas, 45, made indecent remarks to her while head of the Equal Opportunity Commission 8 years ago and, earlier, in the Department of Education. Both are black, but Thomas claims he is being "lynched" as an "uppity" black; the Senate votes 52 to 48 to confirm his appointment to succeed Thurgood Marshall.

European nations send aid to relieve distress among Kurds seeking refuge in remote areas from the killing squads of Iraq's Saddam Hussein (above). President Bush comes under attack for encouraging revolt and then not supporting it, he balks at interfering in Iraq's internal affairs, but finally approves sending troops and supplies to help. Turkey accepts some refugees, Iran accepts far more, but both have problems with their own Kurdish minorities.

South Africa's Parliament votes June 17 to repeal the nation's 41-year-old Population Registration Act, which has classified every newborn infant according to race (whether one was white, black, Asian, or of mixed parentage has determined whether one could vote or own land, where one could work, eat, or enjoy recreation, whom one could marry, and the like). But schools remain for the most part racially segregated, disparities continue between blacks and whites in government benefits such as old-age pensions, and many U.S. congressmen oppose any lifting of economic sanctions as premature (*see* 1986). President Bush lifts sanctions July 10. South African government aid to the Inkatha Freedom party soon comes to light.

Soviet citizens panic January 29 when the evening news reports that savings accounts have been frozen and 50- and 100-ruble banknotes will be withdrawn from circulation. Many have their savings in such notes, but the government decrees that large bills may be exchanged for their equivalent only up to a maximum of 1,000 rubles (about one month's salary), or 200 in the case of pensioners. The move is designed to halt inflation and quash black-market currency traders.

Inflation moderates in the United States but rents are 35 percent above the 1982–1984 average, electricity 28 percent higher, medical care 56 percent higher, food 34 percent higher, entertainment 28 percent higher, footwear 24 percent higher, apparel nearly 9 percent higher. Public transit is virtually unchanged on average, gasoline slightly lower, fuel oil nearly 7 percent lower.

Wall Street's Dow Jones Industrial Average closes above 3,000 for the first time April 17 in anticipation of an early recovery from the 9-month recession but has trouble holding there amidst forecasts that the recession is not about to end. Executives and professionals feel the pinch as well as blue-collar workers.

Iraq sets fire to two Kuwaiti oil refineries January 22 and fires an oil field near Kuwait's border with Saudi Arabia. By war's end, 732 Kuwaiti wells are ablaze, raising doubts about the future of the vital oilfield.

Oil prices drop sharply early in the year as the threat of an Iraqi military success fades. In the United States, No. 2 heating fuel is 90¢ gal. in January, down from $1.29 in December; gasoline $1.43 gal., down from $1.53 (in most other countries it is $2.50 to $4.50 gal.).

Japan's worst nuclear-power accident February 9 raises new anxieties about the safety of the nation's 37 atomic-energy plants. A leak of radioactive water contaminates the water in the steam generator at the Mihama plant of the Kansai Electric Power Co. in Fukui prefecture, about 220 miles west of Tokyo. A gauge registers an alarming increase in radioactivity in a cooling chamber, but operators delay taking action, thinking it may be a false reading. An emergency system shuts down the plant and injects water into the reactor's core to prevent a meltdown of nuclear fuel, the radioactivity released is about 8 percent of the plant's annual emissions.

Eastern Airlines ceases operations January 18 after 62 years of operation following a 22-month strike by machinists. PanAm ceases operations December 4.

A Lauda Air Boeing 767 leaves Bangkok for Vienna May 26, explodes 16 minutes later, and crashes in the jungle, killing all 223 aboard.

New York's 78-year-old Grand Central becomes strictly a commuter terminal April 7 as Amtrak reroutes its remaining long-distance trains (to Albany, Montreal, Toronto, Schenectady, Niagara Falls, Buffalo, and Chicago). They now leave from Penn Station and move up the West Side to the Spuyten Duyvil Bridge.

Cholera strikes Peru January 23, beginning the first epidemic in a century. Chile works to block spread of the disease but it reaches Brazil in March and by May has killed more than 700 Peruvians, making 100,000 ill.

Thousands of Iraqis fall ill from contaminated water and food in the wake of the Gulf War (above), which has knocked out water systems in Baghdad and other cities.

The FDA acts October 9 to approve DDI, a Bristol-Myers AIDS drug that is much cheaper than AZT and can be taken by patients unable to tolerate AZT.

L.A. Lakers superstar Earvin "Magic" Johnson, 32, stuns the world November 7 by announcing that he has the AIDS virus and is retiring from basketball to devote his efforts to ending the spread of the disease that is killing 100 Americans every day. The Bush administration has been mostly silent about AIDS lest it appear to condone casual sex.

The U.S. first-class postal rate goes to 29¢ February 3 (*see* 1988) but remains lower than in most countries.

The Supreme Court rules June 21 that states and localities may prohibit nude dancing without violating First Amendment rights. Laws may require that dancers wear at least G-strings and pasties, the Court rules in a 5-to-4 decision. Critics fear that the ruling may lead to further censorship.

Czech-born London publisher Robert Maxwell (*né* Ludvik Hoch), 68, buys the strike-bound *New York Daily News* for $40 million in March but dies of a heart attack aboard his yacht November 5 and falls into the Atlantic off the Canary Islands.

Nonfiction *The Promised Land: The Great Black Migration And How It Changed America* by U.S. journalist Nicholas Lemann; *The Work of Nations* by Harvard economist Robert B. Reich.

Fiction *Immortality* by Milan Kundera; *St. Maybe* by Anne Tyler; *Brotherly Love* by Pete Dexter.

Rufino Tamayo dies at Mexico City June 24 at age 91; Robert Motherwell dies on Cape Cod July 17 at age 76.

Sculpture *Adjustable Well Bras* by Vito Acconci; *Cleavage* (marble) by Louise Bourgeois; *Medusa Head* by Chris Burden.

Theater *Lost in Yonkers* by Neil Simon 2/21 at New York's Richard Rodgers Theater, with Irene Worth, Mercedes Ruehl, Kevin Spacey; *Silly Cow* by English playwright Ben Elton 2/27 at London's Haymarket Theatre, with Patrick Barlow, Victoria Carling, Kevin Allen; *The Substance of Fire* by U.S. playwright Jon Robin Baitz, 3/17 at New York's Playwrights Horizon Theater, with Sarah Jessica Parker, Patrick Breen, Ron Rifkin; *Don't Dress for Dinner* by French playwright Marc Camoletti 3/26 at London's Apollo Theatre, with John Quayle, Simon Cadell, Su Pollard.

Films Warren Beatty's *Bugsy* with Beatty, Annette Bening; Agnieszka Holland's *Europa Europa* with Marco Hofschneider; James Lapine's *Impromptu* with Judy Davis as George Sand, Hugh Grant as Frédéric Chopin; Neil Jordan's *The Miracle* with Niall Byrne, Lorraine Pilkington; Jennie Livingston's documentary *Paris Is Burning* about homosexual black and Latino men; Yves Simoneau's *Perfectly Normal* with Michael Riley, Robbie Coltraine; Jonathan Demme's *The Silence of the Lambs* with Anthony Hopkins, Jodie

1991 ***(cont.)*** Foster; Ridley Scott's *Thelma and Louise* with Geena Davis, Susan Sarandon; Anthony Minghella's *Truly, Madly, Deeply* with Juliet Stevenson, Alan Rickman.

Stage musicals *The Will Rogers Follies* 3/31 at New York's Palace Theater, with Keith Carradine, music by Cy Coleman, lyrics by Betty Comden and Adolph Green, choreography by Tommy Tune; *Matador* 4/16 at the Queen's Theatre, London, with Nicky Henson, John Barrowman, Stephanie Powers, music by Michael Leander, lyrics by Edward Seago; *The Secret Garden* 4/25 at New York's St. James Theater, with Daisy Eagan, Mandy Patinkin, music by Lucy Simon, book and lyrics by Marsha Norman based on the 1903 Burnett story.

Popular songs *Use Your Illusions* (1 and 2) by Guns N' Roses.

The New York Giants beat Buffalo 21 to 19 at Tampa January 27 in Super Bowl XXV.

Michael Stich, 22 (Ger.), wins in men's singles at Wimbledon, Steffi Graf in women's singles; Stefan Edberg wins in U.S. men's singles, Monica Seles, 17 (Yugo.), in women's singles.

The Minneapolis Twins win the World Series by defeating the Atlanta Braves 4 games to 2.

Colombian cocaine kingpin Pablo Escobar Gavira turns himself in to authorities June 19 on the promise of President César Gaviria Trujillo that he will not be extradited to the United States for trial despite evidence that he and his cohorts have killed hundreds in a 2-year orgy of bombings and assassinations. The multibillionaire is held in a luxurious prison at his hometown of Engivado, 10 miles from Médelin; the Cali cartel takes over and cocaine exports continue.

The U.S. Supreme Court rules 5 to 4 June 20 that police may board long-distance buses without a warrant and ask passengers for permission to search their luggage for drugs.

Switzerland abolishes secret numbered bank accounts in July except in cases that involve ongoing litigation (*see* 1934). The numbered accounts have been used to hide drug profits and the loot of dictators worldwide; the Swiss end the practice in response to appeals by law-enforcement officials.

The Bank of Commerce and Credit International (BCCI) is indicted at New York July 29 on criminal charges of fraud, theft, and money-laundering. New York State seeks extradition of BCCI's founder Aga Hassan Abedi, a former Pakistani banker, and Swaleh Naqvi, the bank's former chief operating officer.

Iraq pumps Kuwaiti crude oil into the Persian Gulf beginning January 24, causing pollution far worse than the 1979 blowout in the Mexican Gulf. The vast slick threatens to foul Saudi Arabian water desalination plants before being blown out to sea, taking a heavy toll on marine life.

Fallout from blazing Kuwait oil wells (*see* above) contaminates air, water, and ground surface throughout the Persian Gulf are, creating severe health hazards to humans as well as to plant and animal life.

A cyclone out of the Bay of Bengal strikes Bangladesh April 30, killing more than 138,000, flooding croplands, and destroying 80 percent of livestock.

Mount Pinatubo on Luzon in the Philippines erupts in beginning of June, caking fields, roads, and vehicles with talc-like gray ash, closing airports, and forcing evacuation of 20,000 Americans from Clark Air Base and Subic Bay Naval Station.

California has its fifth straight year of drought. Farmers control 83 percent of the water delivered by vast federal, state, and City of Los Angeles dams, aqueducts, and reservoirs; they pay as little as $2.50 per acre foot to irrigate their crops while cutbacks are imposed on angry residents of the booming cities who resent use of precious Sierra Nevada and Colorado river water to grow crops such as rice.

Average U.S. food prices in January: white bread 70.5¢ 1 lb. loaf, French bread $1.28, eggs (grade A, large) $1.10 doz., milk $1.38 ½-gal., chicken 89¢ lb., ground beef $1.65 lb. (all figures higher in the Northeast, lower in the South).

The world's population reaches 5.5 billion, up from 3.63 billion in 1970.

The U.S. population tops 250 million. China has 1.15 billion, India 850 million, the Soviet Union 293, Poland 39, Japan 125, Bangladesh 116, South Korea 44, North Korea 24, Taiwan 20, Vietnam 68, Indonesia 185.5, Germany (united) 77, France 57, Britain 58, Spain 39.5, Brazil 150, Mexico 88, Canada 26.5, Nigeria 117, Egypt 56, Turkey 56, Iran 53.5, Iraq 17, Israel 5 , Saudi Arabia 15.3.

The U.S. Supreme Court rules 5 to 4 May 23 in *Rust v. Sullivan* that Congress did not violate the Constitution in 1970 when it barred employees of federally financed family-planning clinics from providing information about abortion. President Bush vetoes new legislation allowing abortion to be discussed.

The Senate Foreign Relations Committee votes 18 to 0 June 12 to remove almost all of the 250,000 names on a secret State Department list of aliens whose ideological views have been grounds for barring admission to the United States even for a visit under the McCarran-Walter Immigration Act of 1952. The Bush Administration agrees to the move almost immediately.

1992 Egyptian Deputy Prime Minister for Foreign Affairs Boutros Boutros-Ghali, 69, takes office January 1 as secretary-general of the United Nations, participates January 31 in the first Security Council Summit, and is invited to analyze the UN's capacity for preventive diplomacy, peace-making, and peace-keeping, and to recommend ways to strengthen that capacity in Africa, Asia, Central America, and Europe. Within 2 years the number of peacekeepers will have grown from 11,500 to 72,000, straining the UN's financial resources.

Voters in Bosnia-Herzegovina opt for independence from Yugoslavia February 29, provoking fresh hostilities in the Balkans as Europe tries to deal with the end of the cold-war bipolarity on which political relationships have been based for nearly 50 years. Serbia and Montenegro form a new Yugoslavia April 17. Croatia, Slovenia, and Bosnia-Herzegovina gain UN membership May 22; Washington that day revokes landing rights for Yugoslav national airline planes and orders expulsion of Yugoslav military attachés to punish Serbia's President Slobodan Milosevic, 50, who has sent troops into Bosnia. They besiege Sarajevo for most of the year as both sides commit bloody atrocities while other European countries dither about how to end the fighting. Despite widespread condemnation in the West, Milosevic wins reelection December 21.

British voters reelect the Conservatives April 9 despite an economic recession that is worse than America's.

Czechoslovkia's President Vaclav Havel resigns July 17 following June elections in which the voters have decided to end the 74-year federation and create two independent republics, a break to become official January 1, 1993. Prime Minister Vladimir Meciar will head the Slovak republic, Vaclav Klaus the Czech republic. Former Czech president Alexander Dubcek dies at Prague November 7 at age 70.

Peru's President Alberto Fujimori suspends the Constitution April 5 and assumes dictatorial powers as he struggles to fight corruption and the Maoist Sendera Luminosa guerrillas. Washington suspends aid to Peru.

Lima police capture Shining Path leader Abimael Guzmán Reynoso September 12, but some of his cohorts remain at large and continue to disrupt the country.

Brazil's Chamber of Deputies impeaches President Fernando Collor de Mello September 29 on charges of having accepted millions of dollars in illegal payments. Relieved of his powers pending trial by the Senate, he resigns December 29 and his vice president, Itamar Augusto Cautiero Franco, 62, takes over as recession and near-hyperinflation continue to wrack the country.

The Cuban Democracy Act signed by President Bush October 23 tightens the 30-year-old embargo against trade with the Communist-controlled Caribbean nation. Congress has passed the law under pressure from Florida's politically powerful Cuban community. The UN General Assembly votes 59 to 3 November 23 to rebuke the United States, whose new law covers foreign subsidiaries of U.S. companies.

Afghan rebels seize Kabul in April, having forced out the communist president Mohammad Najibullah, but disputes arise over how fundamentalist the new Muslim regime should be.

Cambodian statesman Prince Norodom Sihanouk heads a Supreme National Council under terms of a UN agreement signed at Paris calling for the Council to hold power until free elections can be held next year, but Sihanouk surprises the nation by siding with the Vietnamese-backed government against the Khmer Rouge as the UN deploys troops to keep order and clear away land mines.

Presidents Bush and Yeltsin agree June 16 to drastic cuts in their respective nuclear arsenals, scrapping key land-based missiles and reducing long-range warheads. A second Strategic Arms Reduction Treaty (START 2) announced at Geneva December 29 calls for mutual reductions of nuclear warheads to 1960s levels by 2003.

Algeria's President Chadli Benjedid resigns January 11 following Islamic fundamentalists' victory at the polls, runoff elections are canceled January 12, former National Liberation Front dissident Mohammed Boudiaf, 73, returns from 27 years of exile and is sworn in as president January 16, a court dissolves the Islamic Salvation Front March 4, and President Boudiaf is assassinated June 29 at Annaba, leaving the nation in turmoil.

Israeli helicopter gunships fire on a motorcade in southern Lebanon February 16, killing Hisballah (Party of God) leader Sheik Abbas-al-Musawi (the Shiite organization has engaged in terrorist activities, including kidnappings, against Westerners since the early 1980s). Israel's Labor party defeats the Likud party in elections June 23. Yitzhak Rabin becomes prime minister after campaigning on a willingness to exchange land for peace with Israel's Arab neighbors. PLO leader Yasir Arafat has lost his Soviet patron and his rich Gulf State patrons have cut off funding since his support of Saddam Hussein in last year's Gulf War. Official Mideast peace negotiations falter in December as Israel deports 415 Palestinians following the murder of an Israeli policeman by militants, bringing condemnation from the United Nations; Lebanon refuses to accept the deportees; Middle East history professor Yair Hirschfeld breaks Israeli law by meeting in a London hotel with a Palestine Liberation Organization member (Ahmed Kriah, head of the PLO's economics department) (*see* 1993).

Washington bans Iraqi flights south of the 32nd parallel in August to protect Shiite Muslims from air attacks, Iraqi jets breach the no-flight zone December 27, and a U.S. F-16 shoots one down.

Thai troops fire on pro-democracy demonstrators in May, killing at least 52 and perhaps as many as 200. More than 400 people disappear, and rumors abound that their bodies have been fed to crocodiles. A coalition led by pro-democracy forces wins election in September and installs a new government, headed by Prime Minister Chuan Leekpai, which works to suppress drug trafficking and child prostitution.

Philippine voters elect Gen. Fidel V. Ramos, 64, to succeed Corazon Aquino as president; he takes office June 30 in the first peaceful change of Filipino government since November 1965.

Japanese Liberal Democratic Party vice president Shin Kanemaru, 77, resigns August 28 after admitting that he accepted nearly $4 million in illegal donations from Sagawa Kyubin, a trucking company that sought exemptions from rules and approval of new routes, and distributed the money to lawmakers; Japan's most powerful politician, Kanemaru is convicted only of a misdemeanor and fined $1,700—less than the highest penalty

1992 *(cont.)* for overnight parking in Tokyo. Accusations then surface that he employed gangsters to help install Noboru Takeshita as prime minister in 1987 (*see* 1993).

Angola's Marxist president José Eduardo dos Santos retains office after UN-sponsored elections in September but U.S.- and South African-sponsored UNITA rebel leader Jonas Savimbi rejects the election results in October and resumes the civil war that has devastated the country since 1975.

U.S. voters elect Arkansas governor William Jefferson Blythe "Bill" Clinton, 46, to the presidency, rejecting George Bush's reelection bid as economic recession shows few signs of abating. Bush wins 18 states with 168 electoral votes to Clinton's 370 while taking 38 million popular votes to Clinton's 44 million (Dallas billionaire Ross Perot, who entered the race October 1, gets 19 million).

President Bush grants pardons December 24 to former Secretary of Defense Caspar W. Weinberger, 75, and five other Reagan administration officials who have been indicted (and in some cases convicted) of lying to Congress in connection with the Iran-Contra affair of the mid-1980s. Independent Counsel Lawrence E. Walsh, 80, condemns the pardons, which fuel doubts about Bush's own involvement in trading arms for hostages in the 1980s and using proceeds of arms sales to arm Nicaragua's contras.

South Africa's whites vote 2 to 1 March 18 to give President de Klerk a mandate to end white-minority rule. A massacre in the black township of Boipatong June 17 ends with more than 40 residents shot or hacked to death. Nelson Mandela charges police involvement, the African National Congress pulls out of talks on majority rule for the nation, and de Klerk later admits police participation in township violence.

A California jury acquits Los Angeles policemen of charges brought in connection with last year's beating of Rodney King. The April 29 verdict triggers the worst violence and looting in U.S. urban history; more than 50 are killed, 2,000 injured, 50 square miles of south-central L.A. devastated; property damage exceeds $1 billion, far worse than in the 1965 riots (from which the Watts section has never recovered).

Sexual harassment in the Japanese workplace draws censure April 16 as a district court rules in favor of an unmarried woman in Fukuoka who claimed that a male employee at the small publishing company for which she worked had spread rumors that she was promiscuous. It is the first such case ever filed in Japan.

The Pentagon issues a sharp criticism September 24 of a Navy inquiry into the Tailhook sexual harassment scandal, charging that high-ranking officers tried to suppress the facts about last year's Tailhook convention of naval aviators at Las Vegas at which at least 26 women, including 14 officers, were made to run the "gantlet," a practice begun in 1986. Women passing down a particular hallway were encircled by drunken male officers and sexually molested to varying degrees. Three rear admirals are forced to resign.

Serbian forces in Bosnia open concentration camps and impose "ethnic cleansing" measures to rid the country of Muslims and other opponents. Serbians rape and impregnate thousands of Muslim women as a matter of policy.

German neo-Nazis, "skinheads," and other xenophobes attack resident gypsies and Turkish working-class families in an upsurge of violence that includes anti-Semitic outbursts (even though the nation's Jewish population has fallen to 40,000). Hundreds of thousands of Germans march to protest the bigotry.

Militant Hindus demolish a four-century-old Muslim mosque in northern India December 6, triggering waves of shootings, stabbings, and arson that leave more than 1,000 dead (*see* 1993).

Lloyd's of London announces in June that over-all losses for 1989 were the worst in the insurance market's history (*see* 1993).

Britain cuts its benchmark interest rate one point to 9 percent September 22, withdrawing from the European monetary system and abandoning efforts to stabilize the pound sterling in a move to reinvigorate the flagging economy. British interest rates fall to their lowest level since 1988 and are lower than German rates for the first time in a decade.

Prime Minister Major announces October 18 that 10 coal mines will be closed at a cost of 7,000 jobs, bowing to mass popular protest and Conservative objections to closing all 31 of Britain's uneconomic mines.

The top 20 percent of Britons commands 42 percent of the nation's total income, the bottom fifth 8 percent—the biggest gap since 1949, but the top 20 percent of Americans commands 47 percent of total income, the bottom fifth only 3.9 percent.

Queen Elizabeth agrees in late November to start paying taxes on her private income and to pay $1.3 million of the royal family's expenses, breaking with tradition, but British taxpayers will bear the estimated $90 million cost of repairs to Windsor Castle, which has been damaged by fire.

The U.S. national debt tops $3 trillion, up from $735 billion in January 1981.

Russia and other former Soviet nations struggle with inflation and unemployment as they try to follow Poland's more successful example of moving from a state economy to a market economy.

Japanese companies lay off workers and cut salaries as recession deepens in Japan and Germany even while it eases in America.

Mall of America opens in August at Bloomington, Minn., after 7 years of construction. The $500 million shopping mall, largest in the world, will soon have 400 stores (including Bloomingdale's, Macy's, Nordstrom, Sears, The Limited, Victoria's Secret, and Benetton) encircling an amusement park (Knott's Camp Snoopy), attracting shoppers from all over the world (many come from Japan on package tours), and taking in $2 million per day.

Lawrence Berkeley Laboratory astrophysicist George Smoot, 47, announces April 23 that microwave receivers mounted by his team on satellites have located the sites where huge clusters of galaxies began to form about 300,000 years after the universe began. The discovery advances the controversial theory that the universe began with a "Big Bang" and then ballooned in a fraction of a second from a size smaller than an atom to one much larger than astronomers can see with the most powerful telescopes.

France bans smoking in public places beginning November 1 but French smokers generally defy the ban.

U.S. healthcare spending tops 14 percent of national economic output, up from 9.6 percent in 1981, 12.2 percent in 1990 (when Canada, France, Germany, and Sweden spent 8 to 9 percent, Japan 6.5 percent, and Britain 6.1). An estimated 37 million Americans have no health insurance, and the number grows as layoffs surge. President-elect Clinton says health care is the major crisis facing America.

Nonfiction *Two Nations, Black and White* by Queens College professor Andrew Hacker, 62; *Head to Head* by Lester Thurow; *Reinventing Government* by U.S. writers David Osborne and Ted Gaebler.

Fiction *All the Pretty Horses* by U.S. novelist Cormac McCarthy, 59; *A Thousand Acres* by U.S. novelist Jane Smiley, 43; *Outerbridge Reach* by U.S. novelist Robert (Anthony) Stone, 55.

Painter Francis Bacon dies of a heart attack at Madrid April 28 at age 82.

Theater *The Kentucky Cycle* by Tennessee-born playwright Robert Schenkkan, 39, 1/6 at the Mark Taper Forum, Los Angeles, with a cast of 20 in nine interconnected one-act plays set in the years 1775 to 1975; *Conversations with My Father* by Herb Gardner 3/29 at New York's Royale Theater, with Judd Hirsch; *Two Trains Running* by August Wilson 4/13 at New York's Walter Kerr Theater, with Larry Fishburne, Roscoe Lee Brown; *The Sisters Rosensweig* by Wendy Wasserstein 10/22 at New York's Mitze E. Newhouse Theater, with Jane Alexander, Madeline Kahn, Frances McDormand; *Angels in America: A Gay Fantasia on National Themes* by U.S. playwright Tony Kushner, 34, 11/8 at the Mark Taper Forum, Los Angeles with Ron Liebman, Stephen Spinella, Cynthia Mace, and Ellen McLaughlin (the play about AIDS is in two parts, *Millennium Approaches* and *Perestroika*).

Films Neil Jordan's *The Crying Game* with Stephen Rea, Jaye Davidson, Forest Whittaker; Mike Newell's *Enchanted April* with Josie Lawrence, Miranda Richardson, Joan Plowright, Polly Walker; James Foley's *Glengarry Glen Ross* with Al Pacino, Jack Lemmon, Alec Baldwin, Ed Harris; Stephen Frears's *Hero* with Dustin Hoffman, Geena Davis; James Ivory's *Howards End* with Emma Thompson, Vanessa Redgrave, Anthony Hopkins; Woody Allen's *Husbands and Wives* with Allen, Blythe Danner, Judy Davis, Mia Farrow; Michael Mann's *The Last of the Mohicans* with Daniel Day-Lewis; Spike Lee's *Malcolm X* with Denzel Washington; Robert Altman's *The Player* with Tim Robbins; Martin Brest's *Scent of a Woman* with Al Pacino; Clint Eastwood's *Unforgiven* with Eastwood, Gene Hackman, Morgan Freeman; Ron Shelton's *White Men Can't Jump* with Wesley Snipes, Woody Harrelson.

Marlene Dietrich dies at Paris May 6 at age 90.

Hollywood musicals Ron Clements and John Musker's *Aladdin*, with Disney animation, Robin Williams's voice (the Genie), music by Alan Menken, lyrics by Tim Rice, songs that include "Friend Like Me," and "Whole New World."

Broadway musicals *Crazy for You* 2/19 at the Shubert Theater, with Harry Groener, Jodi Benson, old Gershwin songs, book by Ken Ludwig; *Jelly's Last Jam* 4/26 at the Virginia Theater, with Tonya Pinkins, Savion Glover, Gregory Hines as the late "Jelly Roll" Morton, music by Morton, lyrics by Susan Birkenhead; *Falsettos* 4/29 at the John Golden Theater, with Michael Rupert, Stephen Bogardus, Chip Zien, music and lyrics by William Finn, book by Finn and James Lapine.

Popular songs "Tears in Heaven" by English rock star Eric Clapton, now 47, and Will Jennings; *Unplugged* (album) by Clapton; "Layla" by Clapton and Jim Gordon; "End of the Road" by the rock group Boyz II Men (for the film *Boomerang*); *Ten* (album) by Seattle alternative-metal band Pearl Jam (with San Diego singer Eddie Vedder, 26); *Ropin' the Wind* (album) and *No Fences* (album) by U.S. songwriter-performer Garth Brooks; *Dangerous* (album) by Michael Jackson.

Washington beats Buffalo 37 to 24 at Minneapolis January 26 to win Super Bowl XXVI.

Andre Agassi, 22, (U.S.) wins in men's singles at Wimbledon, Steffi Graf in women's singles; Stefan Edberg wins the U.S. Open men's singles title, Monica Seles the women's.

The Olympic Games at Barcelona attract more than 14,000 athletes from 172 nations. Contestants from the Unified Team from former Soviet nations win 45 gold medals, U.S. athletes 37.

The Toronto Blue Jays win the World Series, defeating the Atlanta Braves 4 games to 2.

Evander Holyfield loses his heavyweight boxing title to Riddick Bowe, 25, November 13 at Las Vegas.

Prime Minister Major announces December 9 that Prince Charles and Princess Diana have separated after 11 years of marriage but will not divorce.

Japan moves in March to crack down on the *yakuza*—gangsters who have for decades engaged more or less openly in extortion, prostitution, gambling, and drug dealing.

Former Panamanian strongman Manuel Noriega is convicted by a Miami jury April 9 of having assisted Colombia's cocaine cartel.

New York Mafia boss John Gotti is convicted of murder April 2 and sentenced June 23 to life imprisonment without parole. The Gambino crime family tries to find a new, less conspicuous boss.

1992 ***(cont.)*** Italian crime fighter Judge Giovanni Falcone, 53, is murdered May 23 by a remote-controlled bomb that tears up a highway and kills Falcone's wife and three others. Some 40,000 protesters march in Palermo June 27, demanding an end to *omertà*—the Mafia vow of silence that has kept criminals from being identified. Rome sends 7,000 troops into Sicily to suppress the Mafia, but chief prosecutor Paolo Borsellino is murdered by a remote-controlled bomb July 19, and investigator Giovanni Lizzio is gunned down July 28, allegedly on orders from Mafia boss Salvatore "Totó" Riina, 61 (*see* politics, 1993).

Robbers get into an armored car warehouse in Brooklyn's Greenpoint section December 27 and take $8.3 million.

Hong Kong's 78-story Central Plaza, completed at year's end, is Asia's tallest building.

Earthquakes in eastern Turkey March 13 and 15 kill 4,000, but a much stronger (7.4) one June 28 in California's Yucca Valley kills only one.

The Chicago River breaks into the city's 60-mile-long freight tunnels April 13, 250 million gallons of water rise into sub-basements of downtown buildings, the flood produces power blackouts, and it shuts down subways, the Board of Trade, the Mercantile Exchange, and many businesses.

An Earth Summit at Rio de Janeiro ends June 14 with agreements to increase aid to third-world countries in their efforts to clean up the global environment.

Hurricane Andrew strikes the Bahamas August 22, killing 4, and hits south of Miami August 24, killing 15, leaving 250,000 homeless, and causing $20 billion in damage before blowing into Louisiana; Hurricane Iniki flattens the Hawaiian island of Kauai September 11, killing two, injuring 98.

A Nor'easter strikes New Jersey, Pennsylvania, New York, and New England December 11 with hurricane-force winds, killing at least 15, destroying 12,000 homes, and causing widespread beach erosion.

Famine kills more than 300,000 in Somalia as the nation falls into anarchy and armed thugs prevent world food aid from relieving starvation. Scenes of starving people appear on world TV screens, the UN Security Council moves December 3 to approve U.S.-led military intervention, and the first of some 28,000 troops arrive in Somalia December 9 (*see* politics, 1993).

Civil war in Sudan starves hundreds of thousands, similar conditions prevail in Angola and Mozambique, but television does not show them.

The U.S. Coast Guard begins returning Haitian refugees to Haiti May 24, saying the detention camp at the Navy base in Guantánamo, Cuba, cannot accommodate any more. Thousands have left Haiti to escape political oppression and economic collapse.

The U.S. Supreme Court reaffirms its 1973 *Roe v. Wade* decision on abortion but its 5-to-4 ruling June 29 in *Planned Parenthood v. Casey* supports a Pennsylvania law limiting a woman's right to abortion, which in any case is unavailable in 83 percent of America. The Court December 7 lets stand a Mississippi law requiring a 24-hour waiting period that effectively bars many poorer women from obtaining legal abortions.

The injectable contraceptive hormone Depo-Provera approved by the FDA November 4 has been available for years in many other countries. Developed by Upjohn in 1957, it inhibits ovulation.

1993 Israel's Supreme Court rules January 28 that the deportation of 415 Palestinians to Lebanon in mid-December was legitimate, but Prime Minister Yitzhak Rabin, yielding to UN and U.S. pressure, announces February 1 that 100 of the deportees may return to their homes immediately and the rest within the year. Rockets fired by pro-Iranian Hezbollah (Party of God) guerrillas in southern Lebanon kill eight Israeli soldiers in early July, and Israeli planes and artillery retaliate beginning July 25 in the biggest effort since 1982, forcing hundreds of thousands of Lebanese to flee northward. But secret negotiations between Israeli and Palestine Liberation Organization officials have been going on since January under Norwegian auspices outside Oslo, and Israel's Cabinet announces September 9 that it has agreed unanimously to recognize the PLO, grant limited self-rule to 770,000 Palestinians in the Gaza Strip plus 1 million more in the West Bank, beginning with the oasis of Jericho, and to withdraw its occupation forces from those areas in 6 months. The PLO renounces terrorism and recognizes Israel's right to live in peace and security. Both sides are more fearful of right-wing fundamentalists than of each other. Yasir Arafat, now 64, and Prime Minister Rabin come to Washington, D.C., September 13 to witness the signing of an agreement, but extremists on both sides oppose the deal.

A bomb explosion at New York's World Trade Center February 26 kills six, and starts a fire that sends black smoke through the 110-story twin towers, injuring hun-

Israel and the Palestine Liberation Organization began a rapprochement after decades of hostility.

dreds and forcing 100,000 to evacuate the premises. Mohammed A. Salameh, 25, an illegal Jordanian immigrant, is arrested in Jersey City March 4 and proves to be a follower of self-exiled Islamic fundamentalist Sheik Omar Abdel Rahman, 55, who is wanted by Egypt for inciting anti-government riots in 1989. FBI agents make further arrests, and in June seize Arab terrorists accused of plotting to blow up the United Nations headquarters and New York's Holland and Lincoln tunnels. U.S. authorities arrest Rahman and imprison him 72 miles northwest of New York on suspicion of complicity in the World Trade Center bombing. Egyptian authorities request extradition of the blind, diabetic cleric; his Islamic supporters threaten retaliation if he is extradited.

Tomahawk cruise missiles from U.S. Navy ships in the Persian Gulf and Red Sea hit Iraqi intelligence headquarters at Baghdad June 26 following revelations of an Iraqi-engineered plot to assassinate former President George Bush on a visit to Kuwait in mid-April.

The Czech and Slovak republics have become separate officially January 1 after 74 years as Czechoslovakia. Prague is capital of the Czech Republic, Bratislava of the Slovak.

Serbian aggression in Bosnia continues with violence on both sides despite UN efforts to halt the killings and provide humanitarian relief. U.S. troops join UN peacekeeping forces in Macedonia in July to help prevent further spread of the conflict.

The People's Congress votes 623 to 252 March 11 to impose sharp curbs on the power of Russia's President Yeltsin. Communists and other hard-line opponents of Yeltsin's sweeping economic changes have gained control of the Supreme Soviet (Parliament) but Yeltsin shows his defiance March 20 by assuming virtually

Serbian troops besieged Sarajevo, killing Bosnians and recalling the "Balkan horrors" of the 1870s.

unlimited power and calling for an April 25 plebiscite. He wins approval in the plebiscite despite continued economic depression and growing social problems connected with inflation, joblessness, and crime. Yeltsin orders disbanding of the Supreme Soviet September 21, the Supreme Soviet tries to depose him, he receives support from world leaders and orders elections to be held in December; when his opponents defy him in an armed uprising he uses tanks and troops to crush them, more than 60 are killed in street fighting October 3, tanks shell the Parliament building October 4, setting it afire, and about 30 opposition leaders are arrested on orders from Yeltsin. Voters approve a new Constitution increasing presidential powers in the December 12 election, but the strength at the polls of right-wing opponents to economic reform, led by neo-fascist lawyer Vladimir Volfovich Zhirinovsky, 47, raises fears that ultranationalists may gain power.

Poland's Democratic Left Alliance and Polish Peasants Party win in the national elections September 19 as Poles vote to restore former Communists, stalling progress toward a full market economy.

Italy's continuing political scandal forces the resignations of key figures in industry and government amidst revelations of corruption, kickbacks, alleged vote-rigging, and assertions that former Prime Minister Giulio Andreotti and others in the Christian Democratic party had Mafia connections (*see* crime, 1992 and below). An April 18-19 referendum shows that 82.7 percent of Italians want electoral reforms that would eliminate small parties and bring more accountability.

Baudoin I, fifth king of the Belgians, dies of a heart attack July 31 at age 62 while vacationing in Spain. He has no direct heir and is succeeded after a 42-year reign by his brother, 59, who will reign as Albert II.

Japanese police arrest former Liberal Democratic Party vice president Shin Kanemaru, now 78, March 6 on tax-evasion charges in the nation's biggest political scandal since the arrest of former Prime Minister Kakuei Tanaka in 1976 (*see* 1992). Raids on the power broker's home and offices turn up hundreds of pounds of gold bars plus roughly $50 million in cash and securities. Prime Minister Kiichi Myazawa's LDP (Liberal Democratic Party) is defeated in the June 18 elections and loses power for the first time in 38 years; conservative populist Morihiro Hosokawa, 55, is elected August 6 to head a new coalition government.

Hindu nationalists and Muslims wrangle in India: 13 bombs go off in Bombay's financial district March 11, killing 232 and injuring more than 1,400; police blame "Tiger" (Ibrahim Abdul Razaq Memon), a reputed drug dealer who has fled to Dubai, but many believe Muslims have planted the bombs to retaliate for violence against them in January. Enlightened Hindus who act to protect Muslims are themselves targets of violence, which is encouraged by the 40,000 activists of the Maharashtra state party Shiv Sena. Prime Minister P. V. Narasimha Rao works to revive the nation's economy but does not provide forceful leadership against religious bigotry.

1993 ***(cont.)*** A suicidal Tamil militant straps explosives to his body and kills Sri Lanka's president Ranasinghe Premadas, 68, along with 23 others May 1 at Colombo.

Pakistan's Parliament reelects former prime minister Benazir Bhutto October 19 after her party wins the national elections.

Cambodia has free elections in May under the aegis of the United Nations Transitional Authority (Untac); 90 percent of eligible voters come to the polls despite threats from Khmer Rouge militants and a new government takes power September 24 with Norodom Sihanouk, now 70, as king. His son, Prince Norodom Ranariddh, is first prime minister, and Hun Sen, who led the government installed by Vietnam, second prime minister.

Eritrea gains independence as a new nation May 24.

Somali guerrillas ambush and kill 24 Pakistani UN troops June 5 as last year's humanitarian aid mission becomes a political disaster. More killings follow, the UN offers a $25,000 reward for the capture of fugitive strongman Gen. Mohammed Farah Aidid, 57, and President Clinton sends 400 Army Rangers from Fort Benning, Ga., to Mogadishu beginning August 26 to protect the area from looters, snipers, and Gen. Aidid's militia. His men shoot down a U.S. helicopter September 25, killing three G.I.s., U.S. gunships kill at least 80 Somalis October 2 when they fire into a crowd, Somali militiamen kill at least 12 Americans and wound 78, President Clinton sends in more troops to augment the 4,500-man force that supports 28,500 UN peacekeepers, but he then withdraws the Rangers and vows to pull out of Somalia by March 31, 1994.

Nigerians oust Gen. Ibrahim Babangida, now 52, in free elections June 12 but Gen. Babangida annuls the election results, refuses to step down, closes a radio station, and suppresses several newspapers, including one owned by Social Democratic party leader Moshood Abiola, the apparent election winner. Gen. Babangida turns over power August 26 to an interim government headed by Harvard-educated businessman Ernest Shonekan, 57; Abiola, who left the country August 3, vows to form a new government, and the threat of civil war impels tens of thousands to seek refuge in their tribal homelands.

Burundi's first elected president and the first president from the Hutu tribe is assassinated October 21 along with some of his Cabinet as Tutsi troops storm the presidential palace in a coup attempt. Ethnic clashes follow the death of President Melchior Ndadaye, 40, anarchy reigns, and some 800,000 people flee to neighboring Rwanda, Tanzania, and Zaire, where thousands die of disease and malnutrition in refugee camps while thousands more die in fighting between the Hutu and Tutsi.

Angola's civil war kills 1,000 people per day by autumn as Jonas Savimbi's forces prevent government forces from supplying most of the country with food and medical supplies (*see* 1992). Land mines kill and maim tens of thousands.

The Ivory Coast's first president, Félix Houphouët-Boigny, dies December 7 at age 88 (or 95 by some accounts) after a 33-year administration in which his nation's per-capita income has increased from $90 to $900.

Canada's Progressive Conservative party votes June 13 to make Defense Minister Kim Campbell, 46, the nation's first woman prime minister, succeeding Brian Mulroney, now 53, who has announced his resignation February 24 after nearly 9 years in office. Canadian voters oust the Conservative party in elections October 25 as recession continues; Liberal leader Jean Chrétien, 59, becomes prime minister.

Opponents of Haiti's deposed president Jena-Bertrand Aristide murder scores of his supporters (*see* 1991), the UN imposes an embargo, Gen. Cédras agrees in July at Governor's Island in New York Harbor to step down October 30 and let President Aristide resume power in exchange for amnesty and $35 million, but Cédras's thugs and local police block the troopship U.S.S. *Harlan County* from landing U.S. and Canadian technicians at Port au Prince, Haiti's Justice Minister is murdered, and a new UN embargo begins October 19. Gen. Cédras continues to defy efforts to restore President Aristide despite dwindling fuel supplies.

A federal district judge in Los Angeles rules January 28 that the Pentagon's 11-year-old ban on homosexuals is unconstitutional and permanently enjoins the military from discharging or denying enlistment to gay men or lesbians "in the absence of sexual conduct which interferes with the military mission." President Clinton, who has campaigned on a promise to reverse the ban, announces January 29 that such discharges are suspended and recruits will no longer be asked questions about their sexual orientation, but he yields to pressure and authorizes the Pentagon to continue its ban for 6 months pending an executive order, to be drafted by the Defense Department, that would lift the ban. One of the largest civil rights demonstrations ever held at Washington brings out close to 1 million gays and lesbians April 25.

South African Communist party leader Chris Hani, 50, is murdered April 10 at Johannesburg after a career of opposition to apartheid that has made him a hero to blacks. Police arrest Januzu Jakub Waluz, 40, a Polish-born South African with links to the Afrikaner Resistance Movement, a militant white nationalist group; Clive Derby-Lewis, a leader of the pro-apartheid Conservative Party, is later arrested, and the killing of Hani brings new waves of violence in the racially troubled nation.

Neo-Nazi German right-wing extremists attack foreign workers and their families, notably Turks at Mölln and Solingen but also asylum-seekers from Poland, Yugoslavia, and elsewhere, in a surge of xenophobic violence that brings decent Germans to the streets in protest marches. Germany has nearly 1.7 million Turks, "skinheads" were held responsible for more than 2,280 racial attacks last year, and Ankara demands something more than apologies.

Machete-wielding Brazilian gold miners massacre as many as 75 Yanomani Indians, including women and children, in late August, producing international protests.

The European Community permits free movement of goods across national borders beginning January 1, but passport controls for travelers remain in effect (*see* population, below).

Germany's Bonn government announces August 11 that it will begin cutting unemployment and welfare benefits. About 9 percent of the work force is unemployed, and unemployment benefits are cut by 3 percentage points to 55 or 53 percent of the recipient's last pay, depending on whether he or she has dependents. Costs of rebuilding east Germany, which exceed $60 billion per year, have put severe financial pressure on the reunited nation.

Russia borders on hyperinflation as she tries to adjust to the new economic order imposed by President Yeltsin.

IBM announces in January that it lost $4.6 billion last year, the largest operating loss of any company in history. Competitors have taken away the popular personal computer market, leaving Big Blue dominant only in larger, costlier computers that are in less demand.

Sears Roebuck announces January 25 that it will discontinue its 97-year-old general merchandise catalogue and close 113 "unprofitable" stores, moves that will eliminate 50,000 jobs. Sears catalogues, which once offered everything from groceries to houses to tombstones, went out to 14 million Americans in 1992 and catalogue sales totaled $3.3 billion, but the operation has been losing money.

Lloyd's of London announces June 22 that over-all losses for 1990 were £1,910,000,000, the worst in the market's history (*see* 1992). It has lost nearly £5.5 billion in the past three years, far more than it has earned since 1955, and investors (called "names") in two-fifths of the syndicates are in many cases forced to declare bankruptcy.

The Revenue Reconciliation Act signed into law by President Clinton August 10 seeks to reduce by nearly $500 billion the growth in the federal deficit (which has ballooned since 1981) in the fiscal years 1994 through 1998, less than half of it through modest tax increases.

The North American Free Trade Agreement (NAFTA) signed into law by President Clinton December 8 provides for a phasing out of all tariffs and other trade barriers over a 14-year period. Negotiated by the Bush administration, NAFTA has been approved by Congress 234 to 200 November 17 despite fierce opposition from labor unions, Ross Perot (who has preyed upon fears that U.S. jobs will be lost), protectionists, and some environmentalists, but President Clinton has pulled out all stops to win support for the agreement, mostly from Republicans.

The Miami-bound Amtrak *Sunset Limited* from Los Angeles hurtles off a 12-foot-high trestle September 22, plunges into Alabama's Big Bayou Canot north of Mobile at 70 miles per hour, and catches fire, killing 47, including many of the train's 189 sleeping passengers and several of its 17 crew members. A barge loaded with cement and coal had broken loose from a tow and collided with the trestle, weakening it just before the accident.

Secondhand smoke causes lung cancer that kills an estimated 3,000 nonsmokers per year, the Environmental Protection Agency announces January 7 (EPA administrator William K. Reilly says 434,000 Americans die from diseases caused, or aggravated, by smoking, including 140,000 lung cancer victims). The EPA statement, based on a 4-year report that the tobacco industry challenges as inconclusive, calls second-hand smoke a Class A carcinogen and blames indirect smoking for increasing the severity of symptoms in 200,000 to 1 million child asthmatics and causing 150,000 to 300,000 cases of respiratory infections (e.g., bronchitis, pneumonia), in infants under 18 months of age. First Lady Hillary Rodham Clinton bans smoking in the White House February 1 but little or no opprobrium attaches to smoking in most parts of the world.

President Clinton lashes out at the pharmaceutical drug industry February 12, charging drug makers with pursuing "profits at the expense of our children." Comparing prices of the same drugs in America and abroad, he says, "Our prices are shocking. The pharmaceutical industry is spending $1 billion more each year on advertising and lobbying than it does on developing new and better drugs. Meanwhile, its profits are rising at four times the rate of the average *Fortune* 500 company."

A paper by Austrian researchers published in a February issue of the *New England Journal of Medicine* supports the theory advanced 9 years ago by Australian-American medical researcher Barry J. Marshall, 41, relating *Helicobacter pylori* infection to gastritis and recurrent duodenal ulcer. Antibiotics can wipe out the infection, cure ulcers, and generally prevent their recurrence, say the Austrians. Dr. Marshall says antibiotic treatment can cure ulcers at a total cost of $650 whereas it costs $1,200 per year for a standard regimen of Zantac, the world's leading prescription drug.

The World Health Organization declares a tuberculosis global emergency in April and says November 15 that TB threatens to kill 30 million people in the next 10 years and could become incurable if efforts are not increased to control the disease. Populations in developing countries are most at risk, and drug-resistant strains are multiplying in those countries, but even in the United States some 15 million people are infected, says the WHO.

The Branch Davidian fundamentalist religious cult led by fanatic David Koresh (Vernon Howell), 33, stockpiles an arsenal outside Waco, Texas, and comes under scrutiny of the Treasury Department's Alcohol, Tobacco and Firearms Unit after reports of child abuse. Four ATF officers are killed February 28, Koresh is wounded and at least two of his followers killed, a 51-day standoff ensues between the Branch Davidians and law-enforcement officers, the FBI moves in with tear gas April 19, and cult members set fire to their compound, killing more than 80, including 24 children.

1993 ***(cont.)*** AT&T announces August 16 that it will pay $12.6 billion to acquire McCaw Cellular Communications, the nation's largest cellular telephone company. The move raises possibilities that as wireless phones replace wired lines, AT&T will be able to bypass the regional telephone companies, thus avoiding their fees, which amount to billions of dollars per year.

Pocket-size telephones become commonplace in U.S. cities. Major companies work to structure multibillion-dollar megadeals with a view toward creating a huge information superhighway offering on-demand video, telephone calls on cable, and TV programming on phone lines.

The *New York Times* acquires the *Boston Globe* October 1 for $1.1 billion.

Nonfiction *Spider's Web: The Secret History of How the White House Illegally Armed Iraq* by U.S. journalist Alan Friedman; *A Different Person* (memoir) by poet James Merrill.

Fiction *The Night Manager* by John Le Carré; *Operation Shylock: A Confession* by Philip Roth; *Streets of Laredo* by Larry McMurtry.

Painter Richard Diebenkorn dies of respiratory failure at Berkeley, Calif., March 30 at age 70.

Sculpture *Intersection II* (hot-rolled steel) by Richard Serra.

Theater *Redwood Curtain* by Lanford Wilson 3/30 at New York's Brooks Atkinson Theater, with Jeff Daniels, Debra Monk, Sung Yun Cho; *The Madness of George III* by Alan Bennett 4/1 at London's Lyttleton National Theatre, with Nigel Hawthorne; *Arcadia* by Tom Stoppard 4/13 at London's Lyttleton National Theatre, with Felicity Kendal, Harriet Walter, Rufus Sewell, Bill Nighy; *Playboy of the West Indies* by Trinidadian playwright Mustapha Matura 5/9 at New York's Mitzi E. Newhouse Theater; *A Perfect Ganesh* by Terrence McNally 6/27 at New York's Manhattan Theater Club, with Frances Sternhagen, Zoë Caldwell.

Joyce Carey dies at London February 28 at age 94; Helen Hayes at Nyack, N.Y., March 17 at age 92.

President Clinton names actress Jane Alexander, now 53, to head the National Endowment for the Arts.

Films Steven Spielberg's *Schindler's List* with Liam Neeson, Ben Kingsley, Ralph Fiennes, Embeth Davidtz. Also: Martin Scorsese's *The Age of Innocence* with Daniel Day-Lewis, Michelle Pfeiffer, Winona Ryder; Les Blair's *Bad Behavior* with Stephen Rea, Sinead Cusack; Claude Sautet's *Un Coeur en Hiver* with Daniel Auteuil, Emmanuelle Héart, André Dussolier; Andrew Davis's *The Fugitive* with Harrison Ford, Tommy Lee Jones; Wolfgang Petersen's *In the Line of Fire* with Clint Eastwood; Jim Sheridan's *In the Name of the Father* with Daniel Day-Lewis, Neil Postelthwaite; Agnes Varda's *Jacquot* with Jacques Demy; Wayne Wang's *The Joy Luck Club* with Tsai Chin, Tamlyn Tomita; Steven Spielberg's *Jurassic Park* with fake dinosaurs; Steven Soderberg's *King of the Hill* with Jesse Bradford, Jeroen Krabbe, Lisa Eichhorn; Alfonso Arau's *Like Water for Chocolate* with Regina Torne and Lumi Cavazos; Robert Rodriguez's *El Mariachi* with Carlos Gallardo; Kenneth Branagh's *Much Ado About Nothing* with Emma Thompson, Branagh, Denzel Washington, Michael Keaton; Mike Leigh's *Naked* with David Thewlis; Clint Eastwood's *A Perfect Life* with Kevin Costner; Jane Campion's *The Piano* with Holly Hunter, Harvey Keitel; James Ivory's *The Remains of the Day* with Anthony Hopkins, Emma Thompson; Robert Altman's *Short Cuts* with Bruce Davison, Peter Gallagher, Jack Lemmon, Jarrett Lennon, Frances McDormand, Matthew Modine, Annie Ross, Lori Singer, Lily Tomlin; Fred Schepisi's *Six Degrees of Separation* with Stockard Channing, Donald Sutherland, Will Smith; Ross McElwee's *Time Indefinite* with McElwee, Marilyn Levine; Ang Lee's *The Wedding Banquet* with Winston Chao, May Chin; Brian Gibson's *What's Love Got to Do with It?* with Angela Basset as singer Tina Turner, Laurence Fishburne as Ike Turner.

Audrey Hepburn dies of colon cancer at age 63 January 20 at Tolocherraz, Switzerland; Lillian Gish at New York February 28 at age 99; Ruby Keeler of cancer at Palm Springs, Calif., February 28 at age 82.

Hollywood musical Agnieszka Holland's *The Secret Garden* with Kate Maberly, Heydon Prowse, Maggie Smith, music by Zbigniew Preisner.

Stage musicals *Tommy* 4/23 at New York's St. James Theater, with Michael Cerveris, Buddy Smith, Paul Kandel, music by Peter Townshend, who has recycled songs from a 1969 rock opera by The Who; *Kiss of the Spider Woman* 5/3 at New York's Broadhurst Theater, with Brent Carver, Anthony Crivello, Chita Rivera, music and lyrics by John Kander and Fred Ebb; *Sunset Boulevard* 7/12 at London's Adelphi Theatre, with Kevin Anderson, Patti LuPone, Betty Schaefer, music by Sir Andrew Lloyd Weber, book and lyrics by Christopher Hampton and Don Black.

Popular songs *Vs.* (album) by Pearl Jam; *In Utero* (album) by the Seattle trio Nirvana (singer-guitarist Kurt Cobain, bassist Krist Novoselic, drummer Dave Grohl); *River of Dreams* (album) by Billy Joel.

Dallas beats Buffalo 52 to 17 at Pasadena January 31 in Super Bowl XXVII.

Britain's Grand National Handicap Steeplechase at Aintree April 3 is cancelled after two false starts. About 30 animal-rights demonstrators have rushed onto the course just before the first start, an official waving a red flag to abort the second start is taken for a demonstrator, and the general confusion convinces many that the nation's aristocratic class is inept.

Pete Sampras wins in men's singles at Wimbledon and Flushing Meadow, Steffi Graf in women's singles.

The Florida Marlins (Miami) and Colorado Rockies (Denver), new National League extension teams, play their first seasons.

The Toronto Blue Jays win the World Series, beating the Philadelphia Phillies 4 games to 2.

Evander Holyfield regains his heavyweight boxing title November 6, winning a 12-round decision over Riddick Bowe at Las Vegas.

Robbers get into a Brinks warehouse at Rochester, N.Y., January 5 and take $7.4 million. More than $2 million will be recovered from a New York City apartment after F.B.I. agents arrest a retired Rochester police officer, a priest, and an Irish Republican Army arms smuggler.

Former Lincoln Savings & Loan head Charles H. Keating, Jr., 69, is convicted January 6 on 73 counts of securities and wire fraud, racketeering, and conspiracy (he is sentenced to 12 years' imprisonment and a fine of $125 million). His son, Charles H. 3rd, 37, is found guilty on 64 similar counts. Their Phoenix S&L was seized by federal regulators in 1989.

Italian carabinieri arrest Mafia boss Salvatore "Totó" Riina in downtown Palermo January 15 (*see* 1992; politics, above).

Liverpool toddler James Bulgur, 2, disappears from a local shopping center February 10, his disfigured body is found 2 days later, two 10-year-old boys are charged with his murder (they are convicted in November), and the incident raises alarms of moral and social decline in a nation struggling with economic recession.

Russia and Poland experience a frightening upsurge in crime, with organized mobs committing acts of violence. Many associate the crime wave with the revival of capitalism.

The "Brady Bill" signed by President Clinton November 30 requires a 5-day waiting period for handgun purchases (it is named for former President Reagan's press secretary James Brady who was so severely wounded by a would-be presidential assassin in early 1981 that he was left a paraplegic). The U.S. Senate votes November 17 to ban the sale and manufacture of most assault rifles, this despite opposition from the National Rifle Association, which insists that widespread ownership of such weapons does not contribute to the nation's soaring homicide rate. Demands for federal licensing of handguns, steep taxes on ammunition, etc. rise after a gunman massacres passengers aboard a Long Island Rail Road commuter train December 7, killing five and wounding 18. Police arrest Jamaican-American Colin Ferguson, 35, and charge him with the murders.

Billionaire Colombian drug king Pablo Escobar, now 44, is shot dead by police at a Medellín shopping center December 2, 16 months after escaping from a maximum-security prison (*see* 1991). The Cali cartel, far more sophisticated than the Medellín cartel, continues to pour cocaine into the United States and Europe.

Las Vegas hails the opening of three new casino-hotels: the $375-million pyramid-shaped Luxor, the $475-million Treasure Island, and the $1 billion MGM Grand—world's largest hotel.

The worst storm to hit the eastern United States in 17 years roars up from Cuba to Canada March 13 and 14, dropping temperatures to 2° in Birmingham, Ala., and covering that area with 13 inches of snow, depositing up to 50 inches elsewhere, killing at least 240, and causing hundreds of millions in property damage.

The Mississippi and Missouri rivers rise along with their tributaries beginning in April and flood their banks in July and early August, breaching levees, halting barge traffic, killing 50, inundating vast areas of eight Midwestern states, and causing an estimated $12 billion in crop and property damage in the worst such flooding ever seen. The southeastern United States has a withering drought.

Monsoon rains in northern India, Bangladesh, and Nepal swell rivers, cause massive landslides, wash away railroads, kill more than 2,000, leave tens of thousands homeless, and destroy millions of acres of wheat and rice crops in Punjab, Haryana, and elsewhere.

A pre-dawn earthquake measuring 6.4 on the Richter scale rocks central India September 29, killing an estimated 20,000.

Russian officials halt pollock fishing June 15 in their 200-mile-wide territorial zone in the Sea of Okhotsk, but while Japan agrees to join in a 3-year moratorium on fishing in a 35- by 300-mile area of converging boundaries, Poland, South Korea, and China question whether pollock stocks are in peril. They announce in late July that they will reduce their fishing activities in the area by 25 percent.

Japan has her worst rice harvest in decades and announces that she will import foreign rice to make up for the shortfall.

The U.S. Food and Drug Administration gives approval November 5 to bovine somatropin (BST), a genetically engineered synthetic hormone intended to increase the amount of milk produced by dairy cows. Overproduction of milk has depressed prices, forcing many small family dairy farmers out of business, and agricultural economists say widespread use of BST will mean a further reduction in the number of cows and dairy farms.

The Flavr Savr tomato awaiting FDA approval is the first genetically altered food. Created by U.S. molecular biologists using recombinant DNA, the new tomato encounters opposition from fear-mongers who encourage restaurateurs to insist on "natural" tomatoes.

More than 10 percent of Americans—26.6 million people—rely on food stamps in January to help them get enough to eat. It is the highest number since the program began in 1964 and an indication that while the nation may be emerging from recession the economy is not generating new jobs.

U.S. immigration authorities try to stem the influx of illegal aliens, including potential terrorists (*see* World Trade Center bombing, above). Chinese smuggling gangs cram would-be émigrés, most of them from Fujian province, into freighter holds and try to get them into the United States; ten people die when 200 jump into the water from a grounded freighter off Long Island June 6.

The new Czech Republic (above) has a population of about 10 million, Slovakia about 5 million. The 12 European Economic Community countries have knocked down barriers to trade in goods as of January 1 (above) but continue to resist immigration of Eastern Europeans, Turks, Algerians, Moroccans, Pakistanis, and other foreigners lest they compete for jobs and force reductions in wage scales. France's foreign population (6 percent) is no greater than in 1931, but Paris sets a goal of "zero immigration" in June. Germany stops hearing new pleas for asylum as of July 1.

President Lech Walesa signs legislation in February that makes Poland one of Europe's most restrictive nations with regard to abortion. Abortions may be performed only in cases of rape and incest, when the mother's health is seriously endangered, or where tests have revealed severe fetal defects.

Pensacola, Fla., physician David Gunn, 47, is shot three times in the back during a demonstration outside his Women's Medical Services Clinic March 10 and dies 2 hours later while in surgery at a local hospital. Other abortion clinics in Florida and Texas have been burned by arsonists or sprayed with noxious chemicals, but Gunn's murder is the first of its kind.

Index

Song and story titles are in quotation marks. Italics indicate U.S. Supreme Court decision, name of ship or train, magazine or newspaper title, or title of (a) rock music album, (b) ballet, (f) film, (h) Hollywood musical, (j) juvenile fiction, (m) stage musical, (n) novel, (o) opera, (p) play, (r) radio program/series, (s) story or stories, or (t) television program/series. Space limitations prevent indexing of every name mentioned in text.